chemical
and process
technology
encyclopedia

other reference works of interest

American Society of Mechanical Engineers   ASME HANDBOOKS:
    Engineering Tables   Metals Engineering—Processes
    Metals Engineering—Design   Metals Properties
Baumeister and Marks   STANDARD HANDBOOK FOR MECHANICAL ENGINEERS
Beeman   INDUSTRIAL POWER SYSTEMS HANDBOOK
Brady   MATERIALS HANDBOOK
Burington and May   HANDBOOK OF PROBABILITY AND STATISTICS WITH TABLES
Callender   TIME-SAVER STANDARDS FOR ARCHITECTURAL DESIGN DATA
Carrier Air Conditioning Company   HANDBOOK OF AIR CONDITIONING SYSTEM DESIGN
Carroll   INDUSTRIAL INSTRUMENT SERVICING HANDBOOK
Considine   ENCYCLOPEDIA OF INSTRUMENTATION AND CONTROL
Considine   PROCESS INSTRUMENTS AND CONTROLS HANDBOOK
Considine and Ross   HANDBOOK OF APPLIED INSTRUMENTATION
Crocker and King   PIPING HANDBOOK
De Chiara   TIME-SAVER STANDARDS FOR BUILDING TYPES
Dudley   GEAR HANDBOOK
Emerick   HANDBOOK OF MECHANICAL SPECIFICATIONS FOR BUILDINGS AND PLANTS
Emerick   HEATING HANDBOOK
Emerick   TROUBLESHOOTERS' HANDBOOK FOR MECHANICAL SYSTEMS
Factory Mutual Engineering Division   HANDBOOK OF INDUSTRIAL LOSS PREVENTION
Fink and Carroll   STANDARD HANDBOOK FOR ELECTRICAL ENGINEERS
Flügge   HANDBOOK OF ENGINEERING MECHANICS
Gartmann   DE LAVAL ENGINEERING HANDBOOK
Harris   HANDBOOK OF NOISE CONTROL
Harris and Crede   SHOCK AND VIBRATION HANDBOOK
Heyel   THE FOREMAN'S HANDBOOK
Hicks   STANDARD HANDBOOK OF ENGINEERING CALCULATIONS
Kallen   HANDBOOK OF INSTRUMENTATION AND CONTROLS
King and Brater   HANDBOOK OF HYDRAULICS
Klerer and Korn   DIGITAL COMPUTER USER'S HANDBOOK
Korn and Korn   MATHEMATICAL HANDBOOK FOR SCIENTISTS AND ENGINEERS
LeGrand   THE NEW AMERICAN MACHINISTS' HANDBOOK
Lund   INDUSTRIAL POLLUTION CONTROL HANDBOOK
Machol   SYSTEM ENGINEERING HANDBOOK
Manas   NATIONAL PLUMBING CODE HANDBOOK
Mantell   ENGINEERING MATERIALS HANDBOOK
Maynard   INDUSTRIAL ENGINEERING HANDBOOK
Merritt   BUILDING CONSTRUCTION HANDBOOK
Morrow   MAINTENANCE ENGINEERING HANDBOOK
Perry   CHEMICAL ENGINEERS' HANDBOOK
Perry   ENGINEERING MANUAL
Rossnagel   HANDBOOK OF RIGGING
Rothbart   MECHANICAL DESIGN AND SYSTEMS HANDBOOK
Shand   GLASS ENGINEERING HANDBOOK
Society of Manufacturing Engineers:
    Die Design Handbook   Manufacturing Planning and
    Handbook of Fixture Design   Estimating Handbook
                      Tool Engineers Handbook
Staniar   PLANT ENGINEERING HANDBOOK
Streeter   HANDBOOK OF FLUID DYNAMICS
Truxal   CONTROL ENGINEERS' HANDBOOK

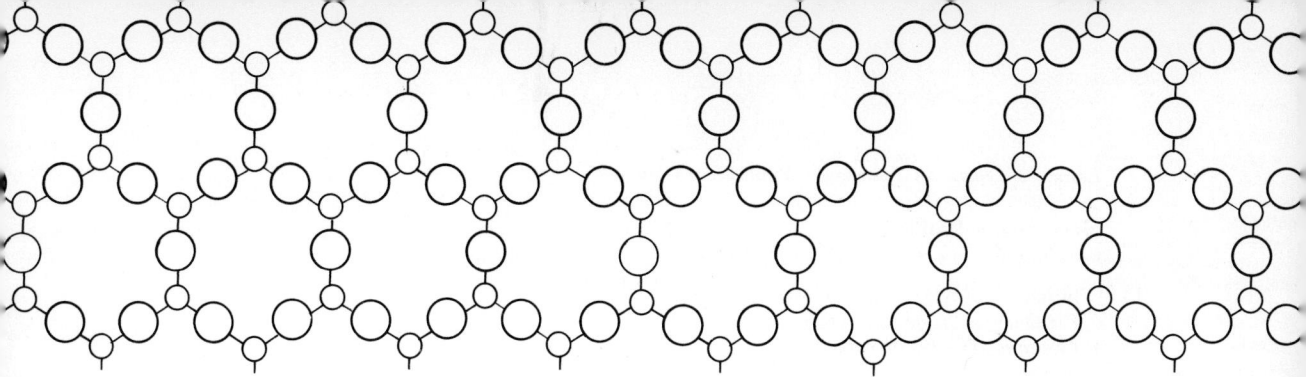

chemical
and process
technology
encyclopedia

editor-in-chief

Douglas M. Considine

Consulting Engineer, Los Angeles, California

McGRAW-HILL BOOK COMPANY

New York   St. Louis   San Francisco   London
Düsseldorf   Johannesburg   Kuala Lumpur
São Paulo   Singapore   Toronto
Montreal   Mexico   Sydney
Panama   New Delhi

Library of Congress Cataloging in Publication Data

Main entry under title:
Chemical and process technology encyclopedia.

   Includes bibliographical references.
   1. Chemistry, Technical—Dictionaries.
I. Considine, Douglas Maxwell, ed.
TP9.C49   660'.03   73-12913
ISBN 0-07-012423-X

1234567890VHVH7654

*The editors for this book were Harold B. Crawford and Lila M.
Gardner, the designer was Naomi Auerbach, and its production
was supervised by George Oechsner. It was set in Baskerville
by York Graphic Services, Inc.*

It was printed and bound by Von Hoffmann Press, Inc.

contents

contributors

SHIRO AKABORI, D.Sc. *Consultant, Ajinomoto Co., Inc., and formerly President of Osaka University. President of the Protein Research Foundation, Osaka, Japan. Member, Japan Academy of Sciences.* (PEPTIDES AND PROTEINS)

TURNER ALFREY, Jr., Ph.D. *Physical Research Laboratory, The Dow Chemical Company, Midland, Mich.* (POLYMERIZATION)

HERBERT G. ARLT, Jr., Ph.D. *Manager of Research, Turpentine and Tall Oil Chemicals, Arizona Chemical Company, Stamford, Conn. Member, American Chemical Society, American Oil Chemists' Society, Technical Association of the Pulp and Paper Industry.* (TERPENES)

RALPH S. ARMSTRONG, B.S.Chem. *Section Supervisor, Emulsion Research Laboratory, The Sherwin-Williams Co., Cleveland, Ohio. Member, American Society for Testing and Materials.* (PAINT)

JOHN K. BACKUS, Ph.D. *Research Manager, Mobay Chemical Company, Pittsburgh, Pa. Fellow, American Institute of Chemists. Member, American Chemical Society, New York Academy of Sciences.* (URETHANES)

ROBERT S. BARKER, Ph.D. *Research Associate, Catalyst Development Corporation, Little Ferry, N.J. Member, American Association for the Advancement of Science, American Chemical Society.* (CATALYSTS)

ROBERT Q. BARR, B.S.E. *Associate Director, Technical Information, Climax Molybdenum Company, New York. Member, American Institute of Metallurgical Engineers, American Society for Metals, National Association of Corrosion Engineers.* (MOLYBDENUM; TECHNETIUM)

W. F. BEACH *Union Carbide Corporation, New York.* (PARYLENE POLYMERS)

RICHARD C. BENNETT, B.S. *Manager, Crystallizer Department, Swenson Division of Whiting Corporation, Harvey, Ill. Member, American Chemical Society, American Institute of Chemical Engineers. Registered Professional Engineer (Illinois).* (CRYSTALLIZATION)

JOHN G. BENSON, B.A. *Analytical Chemist, Allied Chemical Corporation, Industrial Chemicals Division, Solvay, N.Y. Member, American Chemical Society.* (POTASSIUM)

R. E. BERGSTROM, M.B.A. *Manager, Swenson Division of Whiting Corporation, Harvey, Ill. Member, American Institute of Chemical Engineers, Technical Association of the Pulp and Paper Industry.* (EVAPORATION)

L. BERTRAND, B.S.Ch.E. *Engineering Service Division, Engineering Department, E. I. du Pont de Nemours & Company, Inc., Wilmington, Del. Member, American Institute of Chemical Engineers, American Society of Mechanical Engineers.* (DISTILLATION)

CHARLES G. BIGELOW, Jr., B.S.M.E. *Vice President–Research, Selas Corporation of America, Dresher, Pa. Member, American Society of Mechanical Engineers, Industrial Research Institute. Registered Professional Engineer (Pennsylvania).* (FUELS)

EDWARD C. BINGHAM, B.S., M.B.A. *Formerly Technical Advisor, Farmers Chemical Association, Inc., Chattanooga, Tenn. Member, Air Pollution Control Association, American Institute of Chemical Engineers, Water Pollution Control Federation.* (NITRIC ACID)

MELVIN BLUM, B.S., M.S. (Chemistry), M.S. (Metallurgy) *General Manager, Atomergic Chemetals Co., Division of Gallard-Schlesinger Chemical Mfg. Corp., Carle Place, N.Y. President, Fluoride Research Foundation. Member, American Association for Crystal Growth, American Ceramic Society, American Chemical Society, American Institute of Physics, American Nuclear Society, American Society for Metals, American Vacuum Society, International Union for Vacuum Science, National Science Foundation, Society of Applied Spectroscopy.* (HAFNIUM)

G. BOUISSIÈRES, D.Sc. *Professor, University of Paris, Institute of Nuclear Physics, Orsay, France.* (ACTINIUM; FRANCIUM; POLONIUM; RADIUM)

FRANK A. BOWER, Ph.D. *Head of Aerosol and Development Division, Freon Products Laboratory, E. I. du Pont de Nemours & Company, Inc., Wilmington, Del.* (CHLOROFLUOROCARBONS)

JAMES L. BREWBAKER, Ph.D. *Research Chemist, The Dow Chemical Company, Midland, Mich. Member, American Chemical Society, Scientific Research Society of America.* (MONOCHLORO-STYRENE)

ROBERT M. BROWN, B.S.Ch.E. *Senior Project Manager, Scientific Design Company, Inc., New York. Member, American Institute of Chemical Engineers.* (POLYVINYL CHLORIDE)

N. W. BROWNE, M.A. *Senior Engineer, Davy Powergas Ltd, London, England. Member, Institute of Chemical Engineers, Institution of Gas Engineers.* (ISOPROPYL ALCOHOL)

WILLIAM A. BRUINSMA, B.S.M.E. *Senior Mechanical Engineer, Fluor Corporation, Los Angeles, Calif.* (DESALINATION)

WALTER M. BRUNER, Ph.D. *Consultant, E. I. du Pont de Nemours & Company, Inc., Wilmington, Del. Member, American Chemical Society, Society of Plastics Engineers.* (ACETALS)

J. WESLEY BURGESS, B.S.Civ.Eng. *Manager of Projects, Fluor Utah, Inc., San Mateo, Calif. Member, American Institute of Mining, Metallurgical and Petroleum Engineers, Missouri Society of Professional Engineers, National Society of Professional Engineers. Registered Professional Engineer (Texas).* (BENEFICIATION, ORE; COPPER PRODUCTION)

JOSEPH G. CANNON, B.S.Ch.E. *Manager, Modern Metals, Molybdenum Corporation of America, White Plains, N.Y. Member, American Institute of Chemical Engineers, American Institute of Metallurgical Engineers, Institute of Electrical and Electronics Engineers.* (DYSPROSIUM; ERBIUM; EUROPIUM; GADOLINIUM; HOLMIUM; LANTHANUM; LUTETIUM; NEODYMIUM; PRASEODYMIUM; PROMETHIUM; RARE-EARTH ELEMENTS AND METALS; SAMARIUM; TERBIUM; THULIUM; YTTRIUM)

S. C. CARAPELLA, Jr., Ph.D. *Central Research Laboratories, American Smelting and Refining Company, South Plainfield, N.J. Member, American Institute of Metallurgical Engineers, American Society for Metals, Institute of Metals (Britain), Metal Science Club of New York, Scientific Research Society of America.* (ARSENIC; BISMUTH; INDIUM; THALLIUM; TELLURIUM)

V. B. CARLSON, B.S. (Food Technology) *Project Manager, Cherry-Burrell Corporation, Cedar Rapids, Iowa. Member, American Association of Contamination Control, American Dairy Science Association, International Association of Milk and Food Sanitarians, Mayonnaise and Salad Dressings Institute, National Canners Association.* (PASTEURIZATION)

LEO J. CARROLL, B.S.M.E. *Associate Manager, Water and Waste Industries, Fischer & Porter Company, Warminster, Pa. Member, American Water Works Association, Instrument Society of America, Water Pollution Control Federation.* (CHLORINATION; FLUORIDATION)

D. D. CARSWELL *Product Manager, The Polymer Corporation, Reading, Pa.* (NYLONS)

LORING COES, Jr., B.S., M.S. (Chemistry) *Consultant, Research and Development, Grinding Wheel Division, Norton Company, Worcester, Mass. Fellow, American Institute of Chemists. Member, Worcester (Mass.) Engineering Society.* (ABRASIVES)

ALVIN S. COHAN, B.S. (Metallurgy) *deceased, formerly Editorial Director, Scientific Design Company, Inc., New York. Member, American Institute of Chemical Engineers, American Institute of Mining, Metallurgical and Petroleum Engineers, American Society for Metals.* (ETHANOLAMINES; ETHYLENE OXIDE; PROPYLENE OXIDE)

DAVID COOPER, B.Ch.E. *Engineering Department, Maurice A. Knight Company, Akron, Ohio.* (ABSORPTION)

ANTHONY T. COSCIA, Ph.D. *Manager, Organic and Polymer Research Department, Industrial Chemicals Division, American Cyanamid Company, Stamford, Conn. Member, American Chemical Society.* (ACRYLAMIDE POLYMERS)

ROBERT A. DAANE, Ph.D. *Vice President, Research, Beloit Corporation, Paper Machinery Division, Beloit, Wis. Member, American Society of Mechanical Engineers, Technical Association of the Pulp and Paper Industry.* (PAPERMAKING AND FINISHING)

J. L. S. DALEY, B.S. (Physics) *deceased, formerly Development Associate, Battery Engineering Department, Union Carbide Corporation, Cleveland, Ohio.* (BATTERIES)

R. L. DANIEL, B.A., M.S. (Chemistry) *Manager of Solvents Laboratories, Texas Division, The Dow Chemical Company, Freeport, Tex. Member, American Chemical Society.* (CARBON TETRACHLORIDE)

S. C. DESAI, B.Sc. (Metallurgical Engineering) *Assistant Chief Engineer, Davy Ashmore International Ltd, Stockton-on-Tees, Teesside, England.* (IRON; IRON AND STEEL; IRON ORES; IRON- AND STEELMAKING)

HAROLD D. DeSHON *Technical Service Representative, Inorganic Chemicals Department, The Dow Chemical Company, Midland, Mich. Member, National Institute of Drycleaning.* (PERCHLOROETHYLENE)

RONALD L. DICKENSON, B.S., M.S.Ch.E. *Process Engineer, Shell Development Company, Houston, Tex. Member, American Institute of Chemical Engineers.* (ETHYLENE ALKYLATION; ETHYLENE GLYCOL)

JOSEF DIETL, Dr. Eng., Dipl. Physics *Member of Research Staff, Wacker Chemie, GmbH, Munich, West Germany.* (BORON)

EARL D. DIETZ, Ph.D. *Director, Corporate Research Laboratories, Owens-Illinois, Inc., Toledo, Ohio. United States representative on Executive Committee, International Commission on Glass. Member, American Ceramic Society, American Geophysical Union, National Institute of Ceramic Engineers, Scientific Research Society of America, Society Glass Technology (England).* (GLASS)

WOLFGANG DIETZ, Dipl. Physics *Vice President, Wacker Chemie, GmbH, Munich, West Germany. Member, Society of German Physicists.* (BORON)

D. E. DODD, B.S.Ch.E. *Engineering Department, Shell Development Company, Houston, Tex.* (ETHYL ALCOHOL)

BERNARD A. DOMBROW, Ph.D. *Senior Research Associate, Nopco Chemical Division, Diamond Shamrock Chemical Company, Morristown, N.J. Member, American Association for the Advancement of Science, American Chemical Society, New York Academy of Science, Society of Plastics Engineers, Society of the Plastics Industry.* [ANIONIC SURFACTANTS (SULFURBEARING)]

FRANK J. DONNELLY, B.S.M.E. *Consultant, Fluor Corporation, Los Angeles, Calif. Fellow, American Society of Mechanical Engineers. Member, Water and Power Committee, Los Angeles Chamber of Commerce. Member, Air and Water Resources Committee and Fuel and Utilities Committee, California Manufacturers Association. Registered Professional Engineer (California, New York).* (DESALINATION)

THOMAS P. DOWLING, B.S.C.E. *Manager, Explosives Technical Division, Trojan–U.S. Powder, Division of Commercial Solvents Corporation, Allentown, Pa. Member, American Institute of Mining, Metallurgical and Petroleum Engineers, American Ordnance Association, National Society of Professional Engineers, Pennsylvania Society of Professional Engineers, Professional Engineers* *in Industry, Society of American Military Engineers. Registered Professional Engineer (Pennsylvania, Virginia).* (EXPLOSIVES; PYROTECHNICS)

HANS DRESSLER, Ph.D. *Senior Group Manager, New Chemical Process Group, Koppers Company, Inc., Monroeville, Pa. Member, American Chemical Society, Chemical Society (London), New York Academy of Sciences.* (PYRIDINE AND DERIVATIVES)

HENRY E. DUCKHAM, Jr., B.S.Ch.E. *Process Manager in Cryogenics, The M. W. Kellogg Company, a Division of Pullman Incorporated, Houston, Tex. Member, American Institute of Chemical Engineers. Registered Professional Engineer (New York).* (CRYOGENIC PROCESSES; HELIUM; NATURAL GAS)

LLOYD DUWELIUS, B.S.Ch.E. *Senior Engineering Scientist, B. F. Goodrich Chemical Company, Cleveland, Ohio. Registered Professional Engineer (Commonwealth of Kentucky).* (VINYL CHLORIDE MONOMER)

JAMES P. DUX, Ph.D. *Section Leader, Synthetic Staple and Industrial Yarn Section, Fibers Research and Development, American Viscose Division, FMC Corporation, Marcus Hook, Pa. Member, American Chemical Society, Fiber Society.* (ACETATE FIBERS; POLYESTER FIBERS)

R. B. ELLESTAD, Ph.D. *Senior Scientist, Research Department, Lithium Corporation of America, a subsidiary of Gulf Resources & Chemical Corporation, Bessemer City, N.C.* (LITHIUM)

WAYNE P. ELLIS, B.S. *Research Director, Foster Division, Amchem Products, Inc., Ambler, Pa. Member, American Chemical Society, American Society of Heating, Refrigerating and Air-Conditioning Engineers, American Society for Testing and Materials, Franklin Institute, International Institute of Refrigeration.* (INSULATION, THERMAL)

FRANK EMLEY, B.S. (Metallurgy) *Director of Research, Pfizer Metals & Composite Products, Pfizer Inc., Wallingford, Conn. Member, American Institute of Metallurgical Engineers, American Plastic Manufacturers Institute, American Society for Metals, American Society for Testing and Materials, Institute of Metals.* (BARIUM; CALCIUM; STRONTIUM)

HARRY A. FENNICK, B.S.Ch.E. *Administrative Assistant to the Vice President of Refining, Pennzoil Company, Oil City, Pa.* (LUBRICATING OILS)

HELMUT FENNINGER, Dr. Philos. *Member of Research Staff, Wacker Chemie, GmbH, Munich, West Germany. Member, German Society for Air and Space Travel, Association of Austrian Chemists.* (BORON)

ROBERT A. FIEDLER, B.Ch.E. *Development Engineer, Dorr-Oliver Incorporated, Stamford, Conn. Member, American Institute of Chemical Engineers, Institute of Food Technology.* (MEMBRANE FILTERS)

FREDERIC C. FLINDT, B.A. *Senior Presentation Specialist, Textiles Division, Monsanto Company, Decatur, Ala. Member, American Association for the Advancement of Science, American Chemical Society.* (ACRYLIC FIBERS)

ROBERT C. FREY, B.S. *Manager, Dryer Division, The C. M. Kemp Manufacturing Company, Glen Burnie, Md. Member, American Society for Heating, Refrigeration and Air-Conditioning Engineers, American Society of Mechanical Engineers.* (DRYING, GASES)

CARL W. FULLER, B.S. *Manager of Technical Service, Mapico Iron Oxide Operations, Cities Service Company Incorporated, Trenton, N.J. Member, American Ceramic Society, American Society for Testing and Materials, Dry Color Manufacturing Association, Federation of Societies for Paint Technology, Institute of Electrical and Electronics Engineers, National Paint Varnish and Lacquer Association, New York Pigment Club.* (IRON OXIDES, SYNTHETIC)

NEOPHYTOS GANIARIS, B.S., M.S.Ch.E. *Manager, Process Systems Division, Struthers Scientific and International Corporation, New York. Member, American Chemical Society, American Institute of Chemical Engineers.* (FREEZE-DRYING)

JOHN A. GARMAN, Ph.D. *Director, Research and Development, Great Lakes Chemical Corporation, West Lafayette, Ind. Member, American Chemical Society.* (BROMINE)

B. A. GEISERT, B.S.Ch.E. *The Dow Chemical Company, Midland, Mich. Member, Society of Plastics Engineers.* (POLYETHYLENE)

H. E. GIBBS, B.Sc. (Metallurgy) *Senior Metallurgical Engineer, Davy Powergas Ltd, London, England. Member, Institute of Mining and Metallurgy. Associate, Institute of Mining.* (ZINC AND ZINC/LEAD SMELTING)

CLARENCE L. GRANT, Ph.D.   *Research Professor, University of New Hampshire, Durham, N.H. Fellow, American Institute of Chemists. Member and Past President, Society for Applied Spectroscopy.* (PLASMA, CHEMICAL PROCESSING WITH)

J. DENNIS GRIFFIN, Ph.D.   *deceased, formerly Manager of Styrene Molding Polymers Research and Development, The Dow Chemical Company, Midland, Mich. Member, American Chemical Society, American Institute of Chemical Engineers.* (POLYSTYRENES)

WILLIAM H. GROSS, B.S. (Metallurgy)   *formerly Technical Editor, Corporate Communications, The Dow Chemical Company, Midland, Mich. Member, American Society for Metals.* (MAGNESIUM)

KARL A. GSCHNEIDNER, Jr., Ph.D.   *Professor of Metallurgy and Senior Metallurgist, Iowa State University; Group Leader in Ames Laboratory, U.S. Atomic Energy Commission, and Director, Rare-Earth Information Center, Ames, Iowa. Member, American Association for the Advancement of Science, American Chemical Society, American Crystallography Association, American Nuclear Society, American Society for Metals, and The Metallurgical Society of AIME (American Institute of Mining, Metallurgical and Petroleum Engineers).* (CERIUM; YTTERBIUM)

WILBUR S. HALL, B.Ch.E.   *Research Chemist, Amchem Products, Inc., Ambler, Pa. Member, American Chemical Society, Federation of Societies for Paint Technology, National Association of Corrosion Engineers.* (CONVERSION COATINGS)

G. M. HAMPSON, B.S.Ch. Eng., A.M.I. Chem. Eng.   *formerly Senior Process Design Engineer, Davy Powergas Ltd., London, England. Associate Member, Institute of Chemical Engineers. (Professional Engineer.)* (SYNTHESIS GAS)

A. E. HANDLOS, B.S. Chem., M.S.Ch.E.   *Chemical Engineer, Shell Development Company, Houston, Tex. Member, American Chemical Society.* (BUTADIENE; ISOPRENE)

E. S. HILL, Ph.D.   *Process Engineering Department, Shell Development Company, Houston, Tex. Fellow, American Association for the Advancement of Science. Member, American Institute of Chemical Engineers. Registered Professional Engineer (California).* (ABSORPTION, ACIDIC GASES)

DAVID B. HOISINGTON, S.B., M.S. (Electrical Engineering)   *Professor of Electronics, Naval Postgraduate School, Monterey, Calif. Consulting Engineer, Monterey, Calif. Author of "Nucleonics Fundamentals," McGraw-Hill, New York, 1959. Senior Member, Institute of Electrical and Electronics Engineers.* (ATOMIC STRUCTURE)

MELVIN E. HOOVER, B.Ch.E.   *Section Manager, Polymer Converting, Marbon Division, Borg-Warner Corporation, Washington, W.Va. Member, American Chemical Society, American Institute of Chemical Engineers.* (ACRYLONITRILE-BUTADIENE-STYRENE RESINS)

HERBERT S. HOPKINS, B.Ch.E.   *Senior Technical Service Engineer, Chemicals Division, Olin Corporation, Stamford, Conn. Member, Association for Computing Machinery, Chlorine Institute, Compressed Gas Association, Technical Association of the Pulp and Paper Industry.* (CAUSTIC POTASH; CAUSTIC SODA; CHLORINE)

ERNEST W. HORVICK, B.S.Ch.E.   *Director of Technical Services, Zinc Institute, Inc., New York. Member, American Chemical Society, American Society for Metals, American Society for Testing and Materials, National Association of Corrosion Engineers, Society for Die Casting Engineers.* (ZINC)

JOHN S. HOUSTON, B.S.Chem., B.S.Met.Eng.   *Specialty Products Manager, Amoco Chemicals Division, Standard Oil of Indiana, Naperville, Ill. Member, American Electroplater's Society, American Society for Metals, Society of Plastics Engineers. Registered Professional Engineer (Illinois).* (POLYPROPYLENE)

CLIFFORD H. HULLINGER, B.S., M.S.   *Director of Research, American Maize-Products Company, Hammond, Ind. Member, American Chemical Society, Association of Cereal Chemists, Institute of Food Technologists.* (STARCH)

J. B. JONES, B.S.Ch.E.   *Engineering Service Division, Engineering Department, E. I. du Pont de Nemours & Company, Inc., Wilmington, Del. Member, American Institute of Chemical Engineers.* (DISTILLATION)

MARJORY L. JOSEPH, Ph.D.   *Chairman, Department of Home Economics, California State University, Northridge, Northridge, Calif. Fellow, American Institute of Chemists. Member, American Association of Textile Chemists and Colorists, American Home Economics Association, American*

Society for Testing and Materials, Association of Dyers and Colourists, California Home Economics Association. [DYESTUFFS AND DYEING (TEXTILES)]

W. L. KAISER, B.S.Ch.E.   *Process Engineering-Licensing Department, Shell Development Company, Houston, Tex. Member, American Institute of Chemical Engineers, Association for Computing Machinery. Registered Professional Engineer (New York).* (SULFUR DIOXIDE REMOVAL)

C. W. KAMIENSKI, Ph.D.   *Director, Chemical Division, Research Department, Lithium Corporation of America, a subsidiary of Gulf Resources & Chemical Corporation, Bessemer City, N.C.* (LITHIUM)

THOMAS J. KEHOE, B.S.   *Manager, Application Engineering, Process Instruments and Controls Division, Beckman Instruments, Inc., Fullerton, Calif. Past President, Instrument Society of America. Member, American Chemical Society, American Institute of Chemical Engineers, American Institute of Chemists.* [pH (HYDROGEN-ION CONCENTRATION)]

MYRON KIN, M.S. (Pulp and Paper Engineering)   *Facilities Coordinator, Dow Corning Corporation, Midland, Mich.* (SILICONES)

MALCOLM H. KNAPP, B.S.Chem.   *Group Leader, Synthetic Lubricants Laboratory, Intermediates Division, Tenneco Chemicals, Inc., Piscataway, N.J. Member, American Society of Lubrication Engineers.* (LUBRICANTS, SYNTHETIC)

JEROME KOHL, B.S.Ch.E.   *Nuclear Engineering Extension Specialist, Department of Nuclear Engineering, North Carolina State University, Raleigh, N.C. Member, American Institute of Chemical Engineers, American Nuclear Society. Registered Professional Engineer (California).* (RADIOISOTOPES)

E. G. KOMINEK, B.S., M.B.A.   *Manager, Industrial Water and Waste, Eimco PMD Division, Envirotech Corporation, Salt Lake City, Utah. Member, American Chemical Society, American Institute of Chemical Engineers, American Petroleum Institute, Association of Iron and Steel Engineers, National Society of Professional Engineers, Technical Association of the Pulp and Paper Industry, Water Pollution Control Federation. Registered Professional Engineer (Arizona, Illinois, Ohio).* (WASTE TREATMENT)

RICHARD E. KOWAL   *U.S. Industrial Chemicals Co., a division of National Distillers and Chemical Corporation, New York.* (ETHYLENE–VINYL ACETATE COPOLYMERS)

JOSEPH C. KOZIAR, B.A.   *Research Chemist, Chemetals Division, Diamond Shamrock Chemical Co., Baltimore, Md.* (MANGANESE)

KARL L. KRUMEL, Ph.D.   *Project Leader in the Cellulose Projects Group, Designed Polymers Research Laboratory, The Dow Chemical Company, Midland, Mich. Member, American Chemical Society.* (ETHYL CELLULOSE)

CALVIN J. KUHRE, B.S., M.E. (Chemical Engineering)   *Process Engineer, Shell Development Company, Houston, Tex. Member, American Chemical Society, American Institute of Chemical Engineers. Registered Professional Engineer (California).* (GASIFICATION)

WILLIAM D. LANG, B.S., M.S.,   *formerly Manager, Chemical Preparations Research Department, Pigments and Chemicals Divisions, NL Industries, Inc., Hightstown, N.J. Member, American Chemical Society.* (LEAD)

GEORGE G. LAUER, Ph.D.   *Consultant, Koppers Company, Inc., Monroeville, Pa. Member, American Association for the Advancement of Science, American Chemical Society.* (COAL TAR AND DERIVATIVES)

WALTER WM. LAWRENCE, Jr., Ph.D.   *Research Chemist, Ethyl Corporation, Baton Rouge, La. Member, American Chemical Society, Catalysis Society, Combustion Institute.* (CHLORINE ORGANICS)

IRWIN A. LICHTMAN, Ph.D.   *Manager, Physical Chemistry Laboratory, Nopco Chemical Division, Diamond Shamrock Chemical Company, Morristown, N.J. Member, American Chemical Society, American Institute of Chemists.* (DEFOAMING AGENTS)

JOHN LOMARTIRE, Sc.B.Chem.   *Director, Apparel and Industrial Technology, Textiles Division, Monsanto Company, New York. Member, American Chemical Society.* (ACRYLIC FIBERS)

ALLAN J. LUNDEEN, Ph.D.   *Supervising Research Scientist, Research and Development Department, Continental Oil Company, Ponca City, Okla. Member, American Chemical Society, New York Academy of Sciences.* (ALCOHOLS; ALCOHOL PROCESS, ALFOL)

WILLIAM A. LUTZ, M.S.Ch.E.   *Consulting Chemical Engineer. President, Weston Process Design Corporation, Louisville, Ky. Member, American Chemical Society. Registered Professional Engineer (Connecticut).* (PHOSPHORIC ACID)

ROBERT J. MacDONALD, B.A., B.S.   *President, Greenback Industries, Inc., Greenback, Tenn. Formerly Director of Metallurgy and Research, Handy & Harman, Fairfield, Conn. Member, American Institute of Metallurgical Engineers, American Society for Metals, Metal Powders Industry Federation.* (GOLD; SILVER)

C. E. MacKINNON, S.B. (Electrochemical Engineering)   *Technical Director, United Salt Corporation, Houston, Tex. Member, American Association for the Advancement of Science, American Chemical Society, American Institute of Chemical Engineers, American Institute of Chemists, American Society for Testing and Materials, Electrochemical Society, Institute of Food Technologists, National Association of Corrosion Engineers, New York Academy of Sciences.* (SALT, NaCl)

JOHN J. MADDEN   *Program Manager, Thermosetting Resins and Compounds, Union Carbide Corporation, New York.* (EPOXY RESINS)

RICHARD L. MARCELL, B.S.   *Manager of Process Development, Scientific Design Company, Inc., New York. Member, American Institute of Chemical Engineers, National Society of Professional Engineers. Licensed Professional Engineer (New York).* [CYCLOHEXANOL/CYCLOHEXANONE (KA OIL)]

JOHN A. MARTUCCI, B.S. (Chemistry; Industrial Management)   *Manager, Chemistry, Nuclear Laboratories, Nuclear Power Department, Combustion Engineering, Inc., Windsor, Conn. Fellow, American Institute of Chemists. Member, American Nuclear Society, American Society for Testing and Materials, National Association of Corrosion Engineers.* (NUCLEAR POWER PLANTS)

JOHN W. McBROOM, B.S.Chem.   *Group Leader, Plastics Industry Laboratory, Intermediates Division, Tenneco Chemicals, Inc., Piscataway, N.J.* [STABILIZERS (PVC)]

JAMES A. McMASTER, B.W.E.   *formerly Manager Commercial Research, Titanium Metals Corporation of America, West Caldwell, N.J. Member, American Society of Mechanical Engineers, American Welding Society.* (TITANIUM)

L. R. MEADOWS, B.A., M.A. (Chemistry)   *Research Supervisor, National Zinc Company, Inc., Bartlesville, Okla. Member, American Institute of Mining, Metallurgical and Petroleum Engineers.* (CADMIUM)

JOHN R. MEIDINGER, B.S.M.E.   *Director, Research and Development, Wemco Division, Envirotech Corporation, Sacramento, Calif.* (FLOTATION)

ROBERT MERIMS, B.S.Ch.E.   *Assistant Vice President, Scientific Design Company, Inc., New York. Member, American Chemical Society, American Institute of Chemical Engineers.* (FUMARIC ACID; MALEIC ANHYDRIDE)

T. W. MERRILL, B.S.   *formerly Manager Metallurgical Applications Research, Foote Mineral Company, Exton, Pa.* (VANADIUM)

WAYNE R. MERRIMAN, B.S.Chem.   *Assistant Director, Marketing, Great Lakes Chemical Corporation, West Lafayette, Ind. Member, American Chemical Society, Chemical Marketing Research Association, Commercial Development Association.* (BROMINE)

GLEN E. MEYER, Ph.D.   *Manager, Emulsion Polymer Department, Research Division, The Goodyear Tire & Rubber Company, Akron, Ohio. Member, American Chemical Society.* (ELASTOMERS)

HARRY V. MILES, B.Ch.E.   *Manager of Technical Information Systems, Dorr-Oliver Incorporated, Stamford, Conn. Member, American Chemical Society, American Institute of Chemical Engineers, Technical Association of the Pulp and Paper Industry. Registered Professional Engineer (Connecticut).* (FILTRATION)

NORBERT L. MILLER, M.S. (Dairy Science)   *Laboratory Manager, Cherry-Burrell Corporation, Cedar Rapids, Iowa. Member, Institute of Food Technologists.* (HOMOGENIZATION)

HOWARD A. MILLS, Jr., B.S.Ch.E.   *Process Development, The Procter & Gamble Company, Cincinnati, Ohio. Member, American Institute of Chemical Engineers, Society of American Oil Chemists.* [ALCOHOLS, FATTY (VIA HYDROGENATION)]

ELBERT J. MINARCIK, B.Ch.E.   *Coordinator Metal Division Services, NL Industries, Inc., Hightstown, N.J. Member, American Institute of Mining, Metallurgical and Petroleum Engineers, American Society for Testing and Materials.* (LEAD)

JOHN H. MINSKER, B.S.Ch.E. *Development Specialist, The Dow Chemical Company, Midland, Mich.* (VINYL ESTER RESINS)

L. DOW MOORE, B.S.Ch.E. *Supervisor of Product Development, Fiber Glass Division, PPG Industries, Inc., Pittsburgh, Pa. Member, Society of Plastics Engineers, Society of the Plastics Industry.* (FIBER GLASS)

FRANCIS CARROLL MORAN, B.S.Met.Eng. *Process Engineer, Fluor Utah, Inc., San Mateo, Calif. Member, American Institute of Mining, Metallurgical and Petroleum Engineers.* (BENEFICIATION, ORE; COPPER PRODUCTION)

MARGUERITE K. MORAN, B.S., M.S., M.L.S. *Manager, Technical Information Center, M&T Chemicals Inc., subsidiary of American Can Company, Rahway, N.J. Member, American Chemical Society, American Society for Information Science, American Society for Testing and Materials, Special Libraries Association.* (GRIGNARD REAGENTS; MAGNESIUM ORGANICS; TIN)

OSCAR P. MULLER, Ph.D. *Technical Manager and Assistant to the General Manager, Paint Division, NL Industries, Inc., New York. Member, American Chemical Society, National Paint, Varnish & Lacquer Association.* (TITANIUM DIOXIDE)

PAUL A. MUNTER, Ph.D. *Associate Manager, Research and Development, Pennwalt Corporation, King of Prussia, Pa. Member, American Association for the Advancement of Science, American Chemical Society, American Microchemical Society, Coblentz Society, Society for Applied Spectroscopy.* (HYDROGEN FLUORIDE)

WILLIAM J. MURRAY, B.S.Ch.E., M.B.A. *Eimco Processing Machinery Division, Envirotech Corporation, Salt Lake City, Utah. Member, American Institute of Chemical Engineers, National Society of Professional Engineers.* (EXPRESSION)

ROBERT W. NEWKIRK, B.S.M.E., A.E. *Manager, Mechanical Engineering, Fluor Corporation, Los Angeles, Calif. Member, American Society of Mechanical Engineers. Registered Professional Engineer (California).* (DESALINATION)

EDWIN B. NYQUIST, Ph.D. *Development Specialist, Organic Chemicals Department, Reactive Organic Intermediates, The Dow Chemical Company, Midland, Mich. Member, American Chemical Society, Federation of Societies for Paint Technology, Midland Coatings Society, Scientific Research Society of America.* (HYDROXYALKYL ACRYLATES AND METHACRYLATES)

HARUOMI OEDA, D.Sc. *deceased, formerly Advisor to the President, Ajinomoto Co., Inc., Tokyo, Japan.* (AMINO ACIDS)

L. L. PALM, M.S. *Product Manager, Sedimentation Equipment, Eimco PMD Division, Envirotech Corporation, Salt Lake City, Utah. Member, American Institute of Mining Engineers, National Society of Professional Engineers. Registered Professional Engineer (California).* (LIQUID-SOLIDS SEPARATIONS; SEDIMENTATION)

A. C. PATSAVAS, B.S.Ch.E. *Manager, Dryer Department, Swenson Division of Whiting Corporation, Harvey, Ill.* (SPRAY-DRYING)

E. PELITTI, Ph.D. *Vice President, Bechtel France, Paris, France.* (PHOSPHORIC ACID; PROCESS PLANTS, PACKAGED; UREA)

MARY E. PISKLAK, B.S. *Technical Coordinator, Specialty Gas Department, Industrial Gas Division, Air Products and Chemicals, Inc., Allentown, Pa. Member, American Chemical Society.* (APATITE; FLUORINE; FLUORITE; HYDROGEN SULFIDE)

CHRISTOPHER J. PRATT *Manager, Commercial Development, Hoechst-Uhde Corporation, Englewood Cliffs, N.J.* (FERTILIZER; PHOSPHORUS; SULFUR)

DUANE B. PRIDDY, Ph.D. *Research Chemist, The Dow Chemical Company, Midland, Mich.* (CARBOXYLIC ACIDS)

WILLIAM V. RATHBONE, Jr., B.S. *Associate Technical Director, Plastics and Chemicals Department, Marbon Division, Borg-Warner Corporation, Washington, W.Va. Member, Society of Plastics Engineers, Society of the Plastics Industry.* (ACRYLONITRILE-BUTADIENE-STYRENE RESINS)

GERALD REED, Ph.D. *Vice President, Research and Development, Universal Foods Corporation, Milwaukee, Wis.* (YEASTS)

ROBERT M. REED, Ph.D. *Consulting Chemical Engineer, C & I/Girdler Incorporated, Louisville, Ky. Fellow, American Institute of Chemists. Member, American Chemical Society, American*

Institute of Chemical Engineers. Registered Professional Engineer (California, Georgia, Indiana, Kentucky, Maine, Wyoming). (HYDROGEN; PROCESS PLANTS, PACKAGED; UREA)

A. CURTIS REENTS, M.S.Ch.E. *President, Techni-Chem, Inc., Cherry Valley, Ill. Member, American Institute of Chemical Engineers. Member and former Chairman, Division of Water and Waste Chemistry, American Chemical Society. Member, American Water Works Association.* (CHELATING AGENTS; ION-EXCHANGE RESINS; WATER TREATMENT)

WILLIAM S. REVEAL, Ph.D. *Process Engineer, Shell Development Company, Houston, Tex. Member, American Institute of Chemical Engineers.* (ETHYLENE ALKYLATION)

STEPHEN D. ROACH *formerly Product Manager, Centrifuges, Eimco Processing Machinery Division, Envirotech Corporation, Salt Lake City, Utah.* (CENTRIFUGING)

MICHAEL ROBIN, B.S., M.S. *Manager, Chemical Products Division Laboratory, Ashland Chemical Company Division, Ashland Oil and Refining Company, Fords, N.J. Fellow, American Association for the Advancement of Science, American Institute of Chemists. Member, American Chemical Society, Society of Plastics Engineers.* (ANTIOXIDANTS)

P. G. ROBINSON (Chemical Engineer) *Senior Process Design Engineer, Davy Powergas Ltd, London.* (METHYL ALCOHOL)

THOMAS H. ROGERS, Jr., B.S.Ch.E. *Research Manager of Rubber and Plastics Applications, The Goodyear Tire & Rubber Company, Akron, Ohio. Fellow, Institute of the Rubber Industry (London). Chairman, Committee on Flexible Cellular Materials, American Society for Testing and Materials. Member, American Chemical Society, American Institute of Chemical Engineers. Registered Professional Engineer (Ohio).* (RUBBER, NATURAL)

ZOLTAN R. ROSZTOCZY, Ph.D. *Manager, Safety Analysis, Nuclear Power Department, Combustion Engineering, Inc., Windsor, Conn. Member, American Nuclear Society.* (NUCLEAR POWER PLANTS)

S. JOHN SANSONETTI, B.S. (Education), B.S. (Chemistry and Metallurgy) *Director, Research Technology and Applied Science, Metallurgical Research Division, Reynolds Metals Company, Richmond, Va. Fellow, The American Society for Metals. Member, American Chemical Society, American Institute of Metallurgical Engineers, American Optical Society, Electrochemical Society, Faraday Society, National Association of Corrosion Engineers.* (ALUMINUM; BAUXITE; CRYOLITE; GALLIUM)

KAZUO SATAKE, D.Sc. *Consultant, Teijin Limited, Tokyo, Japan. Formerly Professor at Tokyo Metropolitan University. Member, Editorial Board,* The Journal of Biochemistry, *and The Japanese Biochemical Society.* (PEPTIDES AND PROTEINS)

JOSEPH W. SCHAPPEL, B.S.Ch.E. *Section Leader, Research and Development, American Viscose Division, FMC Corporation, Marcus Hook, Pa. Member, American Association of Textile Chemists and Colorists, American Association of Textile Technologists, American Chemical Society, Fiber Society.* (RAYON)

JOHN R. SCHEMEL, B.S. (Metallurgical Engineering) *Chief Metallurgist, Amax Specialty Metals Division, American Metal Climax, Inc., Akron, N.Y. Member, American Society for Metals, American Society for Testing and Materials, Institute of Metallurgical Engineers, National Association of Corrosion Engineers.* (ZIRCONIUM)

WOLF-RUEDIGER SCHILLER, Dr. Philos. Nat., Dipl. Min. *Member of Research Staff, Wacker Chemie, GmbH, Munich, West Germany.* (BORON)

HERBERT M. SCHROEDER, B.S. *Manager of New Product Development, Spencer Kellogg Division of Textron Inc., Buffalo, N.Y. Member, American Chemical Society, American Oil Chemists Society, Federation of Societies for Paint Technology.* (VEGETABLE OILS)

MORTIMER SCHUSSLER, B.S.Met.Eng. *Senior Scientist, Fansteel Inc., North Chicago, Ill. Member, American Institute of Mining, Metallurgical and Petroleum Engineers, American Society for Metals, American Society for Testing and Materials.* (TUNGSTEN)

FRANK C. SCIAVOLINO, Ph.D. *Research Chemist, Medical Research Laboratories, Pfizer Inc., Groton, Conn. Postdoctoral Research Fellow, University of Illinois. Member, American Chemical Society, New York Academy of Sciences.* (ANTIBIOTICS)

CHESTER S. SHEPPARD, Ph.D. *Supervisor of Research, Lucidol Division, Pennwalt Corporation,*

Buffalo, N.Y. Fellow, American Institute of Chemists. Member, American Chemical Society, The Chemical Society (London). (PEROXIDES, ORGANIC)

EDWARD SHERMAN, Ph.D. *Section Leader, Chemical Research Department, Chemicals Division, The Quaker Oats Company, Barrington, Ill. Member, American Association for the Advancement of Science, American Chemical Society.* (FURAN GROUP)

LIONEL E. SIMMONS *President, Simmons Refining Company, Chicago, Ill.* (IRIDIUM; OSMIUM; PALLADIUM; PLATINUM METALS AND PLATINUM; RHODIUM; RUTHENIUM)

D. G. SLEEMAN (Chemical Engineer) *Senior Process Design Engineer, Davy Powergas Ltd, London, England. Associate Member, Institute of Chemical Engineers. (Professional Engineer).* (FORMALDEHYDE)

GEORGE S. SPEIDEL, III, B.Ch.E. *Bar Soap and Household Cleaning Products Division, The Procter & Gamble Company, Cincinnati, Ohio.* (SOAPS)

CHARLES STEENBERGEN, B.S., M.S. *Technical Representative, Penreco Inc., Butler, Pa. Member, American Chemical Society, American Pharmaceutical Association, Society of Cosmetic Chemists.* (MINERAL OILS)

MELVIN J. STERBA, B.S.Ch.E. *Assistant to the Vice President, Engineering and Development, UOP Process Division, Universal Oil Products Company, Des Plaines, Ill. Member, American Chemical Society, American Institute of Chemical Engineers.* (ALKYLATION; CRACKING; FLUID CATALYTIC CRACKING; HYDROCRACKING; HYDROTREATING; ISOMERIZATION; PETROCHEMICAL COMPLEX; PETROLEUM PROCESSING; REFORMING; THERMAL CRACKING)

KENNETH R. STERRETT, B.S.Ch.E. *Sprout, Waldron & Company, Inc., Muncy, Pa. Member, American Chemical Society, American Institute of Chemical Engineers.* (AGGLOMERATION; MIXING AND BLENDING, SOLIDS/SOLIDS; SIZE REDUCTION)

HANS STRIEBEL, Ph.D. *Consultant, Wacker Chemie, GmbH, Munich, West Germany.* (SILICON)

PAUL J. STUBER, B.Sc., M.Sc.Ch.Eng. *Engineering Specialist, Monsanto Enviro-Chem Systems Inc., Chicago, Ill. Member, American Institute of Chemical Engineers. Registered Professional Engineer (Illinois, Missouri).* (SULFURIC ACID)

JOHN M. SUMMERFIELD, B.S.Ch.E., M.B.A. *Chemical Research Manager, United States Gypsum Company, Des Plaines, Ill. Fellow, American Institute of Chemists. Member, American Chemical Society, American Institute of Chemical Engineers.* (GYPSUM; LIME; LIMESTONE)

KENNETH S. SUPRENANT, B.S.Chem. *Solvents Development Specialist, The Dow Chemical Company, Midland, Mich.* (TRICHLOROETHYLENE; 1,1,1-TRICHLOROETHANE)

HENRY F. SZEPAN, B.S.M.E. *Manager, New Products Application Group, Improved Machinery Inc. (subsidiary of Ingersoll-Rand Company), Nashua, N.H. Member, American Society of Mechanical Engineers, National Society of Professional Engineers, Paper Industry Management Association, Technical Association of the Pulp and Paper Industry. Registered Professional Engineer (New Hampshire).* [PULP (WOOD) PRODUCTION AND PROCESSING]

DUNBAR G. TERRY, Ph.D. *Assistant Manager, New Products Application Group, Improved Machinery Inc. (subsidiary of Ingersoll-Rand Company), Nashua, N.H. Member, Technical Association of the Pulp and Paper Industry.* [PULP (WOOD) PRODUCTION AND PROCESSING]

DAVID P. THORNTON, Jr., B.S.Ch.E. *Manager of Technical Publications, Universal Oil Products Company, Des Plaines, Ill. Member, Association of Petroleum Writers, Society for Technical Communication.* (PETROLEUM)

MERLE L. THORPE, A.B., M.S. *Executive Vice President, Ionarc Smelters Ltd., and President, Humphreys Corporation, Bow, N.H. Member, American Institute of Aeronautics and Astronautics, American Society of Mechanical Engineers, Electrochemical Society, Instrument Society of America.* (PLASMA, CHEMICAL PROCESSING WITH)

J. RICHARD TONRY, B.S.Ch.E. *Manager, Quality Control, Alpha Portland Cement Company, Easton, Pa. Member, American Ceramic Society, American Institute of Mining, Metallurgical and Petroleum Engineers, American Petroleum Institute, American Society for Testing and Materials.* (CEMENT)

LOUIS J. TROSTEL, Jr., Ph.D. *Research Associate, Research and Development, Refractories, Industrial Ceramics Division, Norton Company, Worcester, Mass. Fellow, American Ceramic Society. Member, American Society for Testing and Materials.* (CERAMICS)

GAILEN T. VANDEL, B.S.Min.Eng. *Consulting Engineer, Fluor Utah, Inc., San Mateo, Calif. Member, American Institute of Mining, Metallurgical and Petroleum Engineers. Registered Professional Engineer (California, Oregon).* (BENEFICIATION, ORE; COPPER PRODUCTION)

L. W. VERMEULEN *Chemical Engineer, Rio Algom Mines Limited, Toronto, Canada.* (URANIUM)

RONALD C. VICKERY, Ph.D. *Research Consultant, Director, Hudson Laboratories, Hudson, Fla. Fellow, American Association for the Advancement of Science. Member, American Ceramic Society, American Chemical Society, American Society for Testing and Materials, New York Academy of Sciences.* (SCANDIUM)

RICHARD VILLALOBOS, B.S.Chem. *Principal Application Engineer, Beckman Instruments, Inc., Fullerton, Calif. Member, American Association for the Advancement of Science, American Chemical Society, Instrument Society of America.* (CHROMATOGRAPHY)

PETER A. WALDHEIM, B.S., M.S.Ch.E. *Process Engineer, The M. W. Kellogg Company, a Division of Pullman Incorporated, Houston, Tex. Member, American Institute of Chemical Engineers, Research Society of America.* (AMMONIA)

RICHARD D. WALKER, B.A. (Mathematics and Physics), B.S.Ch.E. *Packaged Soap and Detergent Division, The Procter & Gamble Company, Cincinnati, Ohio. Member, American Institute of Chemical Engineers.* (SURFACTANTS)

JAMES E. WALLACE, B.Ch.E., M.Ch.E. *Process Engineer, Organic Chemicals Processing, The M. W. Kellogg Company, a Division of Pullman Incorporated, Houston, Tex. Member, American Institute of Chemical Engineers, Research Society of America. Registered Professional Engineer (New Jersey).* (ETHYLENE)

R. K. WALTON *Group Leader, Research and Development Department, Plastics Division, Union Carbide Corporation, Bound Brook, N.J.* (POLYSULFONES)

HELMUT H. WELDES, Ph.D. *Director, Marketing, PQ International Inc., subsidiary of Philadelphia Quartz Company, Philadelphia, Pa. Member, American Association of Cost Engineers, American Chemical Society, American Foundrymen's Society, American Institute of Chemical Engineers, American Society for Testing and Materials, Gesellschaft Deutscher Chemiker, National Association of Corrosion Engineers, Society of Chemical Industry, Society of Glass Technology.* [SILICATES (SOLUBLE)]

J. Y. WELSH, Ph.D. *Director of Research, Chemetals Division, Diamond Shamrock Chemical Company, Baltimore, Md.* (MANGANESE)

MICHAEL B. WHITEMAN, Ph.D. *formerly General Manager, Cleveland Refractory Metals, Division of Chase Brass & Copper Co., Inc., subsidiary of Kennecott Copper Corporation, Solon, Ohio.* (RHENIUM)

R. N. WILKINSON, B.E. (Chemical Engineering), M.S. *Packaged Soap and Detergent Division, The Procter & Gamble Company, Cincinnati, Ohio.* (DETERGENTS)

PHILLIP H. WILKS, Ph.D. *Supervisor, Materials and Process Development, Ionarc Smelters Ltd., Bow, N.H. Member, American Chemical Society. Registered Professional Engineer (Ohio).* (PLASMA, CHEMICAL PROCESSING WITH)

E. WILLIAMS, B.Sc., A.I.M. *Cobalt Information Centre, London. Member, Institute of Metallurgists, Institute of Metals, Iron and Steel Institute.* (COBALT)

RICHARD B. YERMAN, B.S.Chem. *Applications Chemist, The Dow Chemical Company, Midland, Mich. Member, Midland Coatings Society.* (EPICHLOROHYDRIN)

TOICHI YOSHIDA, D.Sc. *Advisor, Ajinomoto Co., Inc., Central Research Laboratories, Kawasaki, Japan.* (AMINO ACIDS)

J. YU, B.S., M.S.Ch.E. *formerly Process Engineer, Shell Development Company; Houston, Tex. Member, American Institute of Chemical Engineers.* (ETHYLENE GLYCOL ETHERS)

preface

This Encyclopedia brings together in a formal reference work the many facets of inorganic, organic, and physical chemistry—and of chemical, metallurgical, and process engineering—which, when viewed as an integrated body of scientific knowledge, comprise *chemical and process technology*. In a marked departure from the traditional editorial format, this book combines the detailed coverage of a handbook with the convenience and scope of an alphabetized encyclopedia.

It not only embraces the traditional spheres of interest in industrial chemistry and chemical technology as reflected by the petroleum, petrochemical, chemical, paper, textile, and other long-established process industries, but also more recent applications of an advancing and broadening chemical technology, including, as examples, the materials and processes now required by the electronics, optics, and aerospace industries. The materials sciences are given much attention.

The Encyclopedia is designed to serve as a focal information center for the traffic of knowledge from the specialist to the generalist and, in fact, from specialist to specialist. The content and style are directed to scientific, engineering, and technical business people and industrialists at the professional and management levels; regular and frequent use by educators and librarians is also a major objective.

Because of the wide range of chemical and process technology, there are few if any professionals who have a working familiarity with all the branches. When the professional seeks information on a subject with which he may not have been in touch for a long period, if at all, he seeks rapid, concise, extremely clear summaries that will provide initial orientation and lead him to further, more detailed information. Thus the reader will find strong editorial emphasis in this book on definitions of terms, clarification of nomenclature, and classification of subject matter, leading him directly and quickly to more detailed shelf literature and other information sources.

This Encyclopedia aims to provide within one volume not only the attendant convenience and economy of a single source as contrasted with multivolume works which rarely provide conclusive detail, but also a very large portion of highly select information which an interested reader may seek in his initial concern with a given topic. In many instances, because of the manner in which the information content was selected, the reader will find all the answers he is seeking within this one volume. Where further information is required, extensive reference lists, carefully culled for value, are included. Exhaustive bibliographies containing references of marginal value are not included. In most instances, references have been selected which, in themselves, contain long lists of references, so that the reader can assemble his own information bank rapidly, using this book as a starting base.

This Encyclopedia is *not* a compilation of generalities, but rather is packed with detail information, carefully selected on a priority basis by each author-specialist. Following a terse overview of a topic in most cases, the author presents details which, in his expert opinion, are the most important and timely. In outlining the content of each editorial entry, the staff and authors attempted to visualize the information needs of representative readers and, accordingly, to adjust both tone and depth of the entry. This *customized editorial approach* adjusts the technical approach and content of each topic to reflect inherent complexity and relative importance and frequency of occurrence in the information spectrum.

Classified Index

The structure of the Classified Index is shown in the table on page 1188. Each entry of this Encyclopedia is listed *once* and *only once* in the Classified Index. Cross references are used extensively throughout the book, and these references, coupled with the detailed Subject Index at the end of the volume, provide a webbing to hold all topics together even though all topics are entered into the book alphabetically. Admittedly, some of the decisions involved in placing topics under the major categories of the table were arbitrary. This situation could have been avoided only by listing some entries under several categories. Such multiple listings would have resulted in a much longer and less convenient topical classification.

Some of the major cross-reference entries in the volume are included in the Classified Index, but again, they never are duplicated. These cross-reference headings are shown in italic.

Seldom can a classified index render full justice to the content of a volume, and therefore extensive use of the Subject Index is encouraged.

DOUGLAS M. CONSIDINE
Editor-in-Chief

acknowledgments

The preparation of a book that is concerned with such a diverse body of scientific information as chemical and process technology is a rather massive task. While care and thoughtfulness are prerequisites for the generation of a useful, long-term reference, a relatively short time span for preparation and production also is an important directive. Thus, the broad scope, the complexities and interrelationships of the science, and the emphasis on timing dictate that an effort of this kind must be one of extensive cooperation among hundreds of experienced authors and advisors. The pleasant interchanges enjoyed by the editor with these hundreds of interested individuals and institutions, in the United States and throughout the world, prove once again how people will cooperate, how they will take time from otherwise busy schedules to further a cause that is dedicated to the advancement of their professional pursuits. Thus, this general acknowledgment of assistance and of appreciation for that help rendered by numerous individuals who in some way sponsored and encouraged this work.

The important roles of the over 170 contributors who signed the majority of entries in this volume are obvious. There is a list of contributors, their affiliations, and brief biographies included in the book. But, the roles of the contributors should not be glossed over quickly simply because they are self-evident. The tasks did not involve simply writing down a few facts and opinions, but rather in numerous instances also involved searching out new information and, in particular, of organizing information in the highly condensed, yet comprehensive format required here. The contributors were very helpful in making decisions regarding those topics which should be self-contained alphabetical entries—and those associated and interrelated subjects that would give the required visibility through cross referencing and indexing. Examples of this coordination include:

Agricultural chemicals, fertilizers, phosphorus, and *sulfur* by Christopher J. Pratt, Hoechst-Uhde Corporation, Englewood Cliffs, N.J.

Alkaline earths by Frank Emley, Pfizer Metals & Composite Products, Pfizer Inc., Wallingford, Conn.

Aluminum chemicals and the metallurgy of aluminum by S. John Sansonetti, Reynolds Metals Company, Richmond, Va.

Amino acids by Dr. Haruomi Oeda, Ajinomoto Co., Inc., Tokyo.

Chlorine organics by Dr. Walter Wm. Lawrence, Jr., Ethyl Corporation, Baton Rouge, La.

Detergents, fatty alcohols, soaps, and surfactants by Howard A. Mills, Jr., George S. Speidel, III, Richard D. Walker, and R. N. Wilkinson, Procter & Gamble Company, Cincinnati, Ohio.

Heavy industrial chemicals by Herbert S. Hopkins, Olin Corporation, Stamford, Conn.

Industrial solvents by R. L. Daniel, The Dow Chemical Company, Freeport, Tex., and Harold D. DeShon and Kenneth S. Suprenant, The Dow Chemical Company, Midland, Mich.

Iron and iron and steel processing by S. C. Desai, Davy Ashmore International Ltd, Stockton-on-Tees, Teesside, England.

Peptides and proteins by Dr. Shiro Akabori, President of the Protein Research Foundation, Osaka, Japan, and Dr. Kazuo Satake, Teijin Limited, Tokyo.

Petroleum and petrochemical processing by Melvin J. Sterba and D. P. Thornton, Jr., Universal Oil Products Company, Des Plaines, Ill.

Plasma processing by Merle L. Thorpe and Dr. Phillip H. Wilks, Ionarc Smelters Ltd., Bow, N.H.

Radioactive elements by Dr. G. Bouissières, Institute of Nuclear Physics, University of Paris, Orsay, France.

Smelter by-product metal elements by Dr. S. C. Carapella, Jr., American Smelting and Refining Company, South Plainfield, N.J.

Uncommon nonferrous elements by Technical staff of Kawecki Berylco Industries, Inc., Reading, Pa.

Water and aqueous processes by A. Curtis Reents, Techni-Chem, Inc., Cherry Valley, Ill.

Special acknowledgment is extended to Joseph G. Cannon, Molybdenum Corporation of America, White Plains, N.Y., who, with the very able cooperation of Dr. Karl A. Gschneidner, Jr., Professor of Metallurgy at Iowa State University and a staff executive of the Ames Laboratory of the United States Atomic Energy Commission, pioneered the preparation and release of new information in the area of the *rare-earth elements and metals*. New data also were contributed by Dr. Karl J. Strnat, Professor, University of Dayton, Dayton, Ohio.

The editors are deeply indebted to Ernest W. Horvick, Zinc Institute, Inc., New York, for his assistance in the editorial coordination of the major nonferrous metals.

In numerous instances, a number of editorial entries emanated from the same institution, requiring special coordinating efforts among contributors. Individuals who worked behind the scenes in this manner included:

Herman W. Andre, Great Lakes Chemical Corporation, West Lafayette, Ind.

Dr. Gerhard Barth-Wehrenalp, Pennwalt Corporation, Philadelphia, Pa.

Gene Boyo, Olin Corporation, New York.

Ronald T. Cambio, E. G. Foster, and James A. Sykes, Jr., Shell Development Company, Houston, Tex.

E. J. Canty, Norton Company, Worcester, Mass.

Alvin S. Cohan, Scientific Design Company, Inc., New York.

James R. Corbin, Marbon Division, Borg-Warner Corporation, Washington, W.Va.

Dr. D. A. Dahlstrom, Envirotech Corporation, Salt Lake City, Utah.

Dr. Wolfgang Dietz, Wacker Chemie, GmbH, Munich, West Germany.

A. B. Dumain, P. Ingman, Thomas Sharp, and F. R. Smith, Davy Powergas Ltd, London.

Jack D. Eadie, James C. Hansen, Robert McKellar, and Gordon Sears, The Dow Chemical Company, Midland, Mich.

Jack Gardner, Fluor Utah, Inc., San Mateo, Calif.

James E. King, The Procter & Gamble Company, Cincinnati, Ohio.

M. A. Knight, Jr., Maurice A. Knight Company, Akron, Ohio.

A. R. Koenig, Swenson Division, Whiting Corporation, Harvey, Ill.

Dr. John J. Miskel, Jr., Nopco Chemical Division, Diamond Shamrock Chemical Company, Morristown, N.J.

Dr. I. Mockrin, Kawecki Berylco Industries, Inc., Boyertown, Pa.

J. B. Orr, Davy Ashmore International Ltd, Stockton-on-Tees, Teesside, England.

W. J. Pedrick, Jr., Air Products and Chemicals, Inc., Allentown, Pa.

Dr. E. Pelitti, Bechtel France, Paris.

Dr. Marvin Rosen, Tenneco Chemicals, Inc., Piscataway, N.J.

D. H. Stormont, Fluor Corporation, Los Angeles, Calif.

Ray Waters, The M. W. Kellogg Company, a division of Pullman Incorporated, Houston, Tex.

Dr. Toichi Yoshida, Ajinomoto Co., Inc., Kawasaki, Japan.

The editors extend special appreciation to Dr. Ronald C. Vickery, Research Consultant and Director, Hudson Laboratories, Hudson, Fla., who provided critical reviews of several of the editorial topics in the area of physical chemistry; and to Dr. W. H. Thomas, Ethyl Corporation, Baton Rouge, La., who provided much of the information on sodium metal and its applications.

Particular appreciation is extended to Glenn D. Considine, Mountain View, Calif., who prepared the renditions of many of the flowsheets and other illustrations used in this volume.

Finally, the editors are indebted to the late Dr. Sidney D. Kirkpatrick, whose achievements both in the technical and humane aspects of chemical technology and chemical engineering would require numerous pages to list, for his input during the early planning stages of this volume and for his inspiring counsel prior to his passing early in 1973.

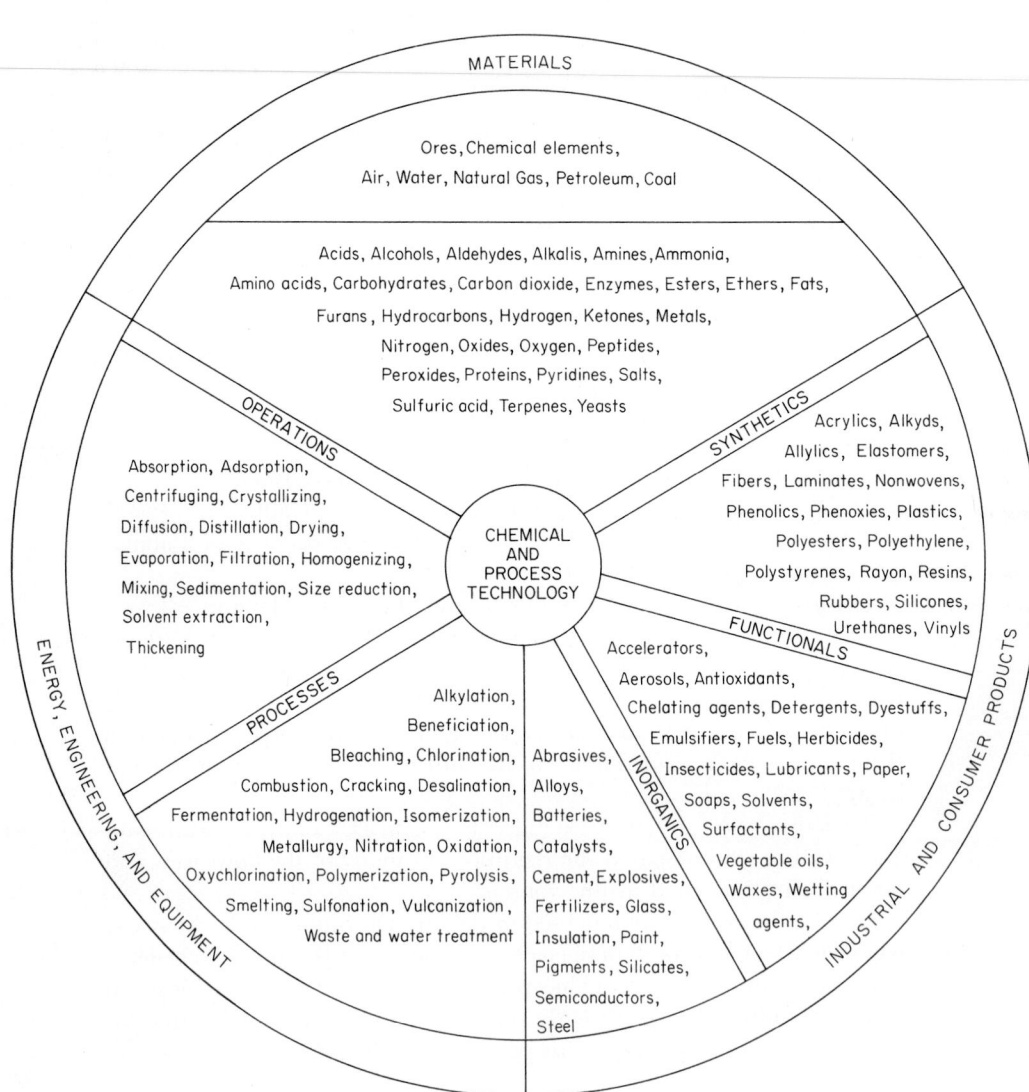

MATERIALS

Ores, Chemical elements,
Air, Water, Natural Gas, Petroleum, Coal

Acids, Alcohols, Aldehydes, Alkalis, Amines, Ammonia,
Amino acids, Carbohydrates, Carbon dioxide, Enzymes, Esters, Ethers, Fats,
Furans, Hydrocarbons, Hydrogen, Ketones, Metals,
Nitrogen, Oxides, Oxygen, Peptides,
Peroxides, Proteins, Pyridines, Salts,
Sulfuric acid, Terpenes, Yeasts

OPERATIONS

SYNTHETICS

Absorption, Adsorption,
Centrifuging, Crystallizing,
Diffusion, Distillation, Drying,
Evaporation, Filtration, Homogenizing,
Mixing, Sedimentation, Size reduction,
Solvent extraction,
Thickening

CHEMICAL
AND
PROCESS
TECHNOLOGY

Acrylics, Alkyds,
Allylics, Elastomers,
Fibers, Laminates, Nonwovens,
Phenolics, Phenoxies, Plastics,
Polyesters, Polyethylene,
Polystyrenes, Rayon, Resins,
Rubbers, Silicones,
Urethanes, Vinyls

FUNCTIONALS

Accelerators,
Aerosols, Antioxidants,
Chelating agents, Detergents, Dyestuffs,
Emulsifiers, Fuels, Herbicides,
Insecticides, Lubricants, Paper,
Soaps, Solvents,
Surfactants,
Vegetable oils,
Waxes, Wetting
agents,

PROCESSES

Alkylation,
Beneficiation,
Bleaching, Chlorination,
Combustion, Cracking, Desalination,
Fermentation, Hydrogenation, Isomerization,
Metallurgy, Nitration, Oxidation,
Oxychlorination, Polymerization, Pyrolysis,
Smelting, Sulfonation, Vulcanization,
Waste and water treatment

INORGANICS

Abrasives,
Alloys,
Batteries,
Catalysts,
Cement, Explosives,
Fertilizers, Glass,
Insulation, Paint,
Pigments, Silicates,
Semiconductors,
Steel

ENERGY, ENGINEERING, AND EQUIPMENT

INDUSTRIAL AND CONSUMER PRODUCTS

graphic representation of contents

This Encyclopedia is divided into approximately three equal portions, as indicated by the accompanying pie chart, or contents wheel: (1) materials, (2) energy, engineering, and equipment, and (3) industrial and consumer products that derive from the application of chemical and process technology.

Only a relatively few examples from the total content are included in this chart, hopefully, a sufficient number to give the reader some appreciation for both the scope and depth of the volume.

**chemical
and process
technology
encyclopedia**

A

Abrasives Abrasives are commercially useful hard substances utilized in the form of granules or powder for shaping, cleaning, or polishing surfaces. The substances are characterized by high hardness, high melting point, and chemical inertness. See Table A-1. In addition to grinding and polishing, the general properties of abrasives make them well suited for other applications, including refractories, wear-resistant surfaces, and in chemically inert shapes for catalyst supports and tower packings.

Natural Abrasives: These materials are recovered from natural deposits and may be used after sorting, crushing, and sizing operations are performed.

Quartz, as silica sand, sandstone, or flint, is the earliest known abrasive and is widely used because of low cost. Current consumption (United States)* is 1.8 million short tons/yr, principally in loose form, for grinding glass and stone and in the pressure blasting of metals. In bonded form, its main use is in sandpaper for finishing wood. Because of relatively poor abrasive qualities and rising labor costs, quartz is being replaced

*All consumption figures in this description are for the United States.

slowly in most uses by silicon carbide and aluminum oxide.

Emery occurs naturally as an impure granular corundum. The principal impurities are magnetite and spinel. Consumption of emery is about 8,000 short tons/yr and is declining. Major uses are in nonskid floors and in coated abrasives.

Corundum, formerly used extensively in grinding wheels, now is used almost wholly in loose form in grinding optical glass. Consumption is about 8,000 short tons/yr.

Garnet occurs as a high-quality abrasive grade, used chiefly in coated abrasive products for wood finishing, and as an alluvial garnet having much poorer abrasive quality, used in pressure-blasting operations. Consumption of garnet is about 15,000 short tons/yr.

Tripoli occurs as a fine silica and is used in various polishing operations and as a powder or buffing compound on soft metals, wood, and glass. Tripoli is also known as *rotten stone.* Consumption is about 70,000 short tons/yr.

Diatomaceous earth, also known as *kieselguhr, diatomite,* and *tripolite,* consists of the siliceous skeletons of small marine organisms called *diatoms.* Its abrasive uses are similar to those of tripoli. The

1

TABLE A-1. *Key Properties of Major Abrasives*

Abrasive	Composition	Knoop Hardness*	Mp, °C	Sp gr
Quartz	SiO_2	820	1700	2.65
Emery	$Al_2O_3(i)$	2,000	1900	4.00
Corundum	$Al_2O_3(i)$	2,000	2050	3.95
Garnet		1,360	1200(d)	4.25
Tripoli	SiO_2 (98%)	820	1700	2.50
Diatomite	SiO_2 (89%)	820	1700	2.50
Pumice	SiO_2 (70%) + oxides	. . . . .	. . . . . .	2.50
Diamond	C′	6,500	1000(d)	3.51
Fused alumina†	Al_2O_3 (93–97%)	2,000	2050	3.95
Silicon carbide‡	SiC	2,450	2400(u)	3.20
Cubic boron§ nitride . . .	BN	4,700	2000(d)	

*Knoop hardness = diamond indentation. This is similar to the Brinnel hardness test, but suitable for hard, brittle material.
†Such as Alundum, Aloxite, and Lionite.
‡Such as Carborundum and Crystolon.
§Known as Borazon.
i = impure form; d = decomposes; u = uncertain.

material is used mainly in nonabrasive applications such as absorbents, filter aids, and catalysts.

Pumice, a volcanic glass, is used mostly in loose form for polishing operations on painted surfaces, metals, and lithograph stones. Consumption of pumice is approximately 14,000 short tons/yr and declining.

Diamond, the hardest of all abrasives, is found in a variety of forms: (1) gem diamonds; (2) *industrials,* which are nearly perfect crystals, but have a color making them unsuitable for gems but well suited for drill bits, truing diamonds, styli, and turning tools; and (3) bort, which is comprised of all diamond that is not classed as gem or industrials. Industrial diamond comprises all bort plus industrials. The principal use for bort is in crushed form as a loose or bonded abrasive. Some finely polycrystalline diamond, which is opaque and often colored, finds use in drill bits. Diamond consumption is approximately 22 million carats/yr.

Synthetic Abrasives: These materials are made commercially from nonabrasive materials. Since 1900, several types of synthetic materials have been introduced for improving on the abrasive quality and uniformity of available natural abrasives.

Fused aluminum oxide was introduced in 1901 by C. B. Jacobs and now is made in various forms, all of which originate from bauxite. See also

Bauxite. Bauxite for abrasive manufacture must be of the highest quality, high in aluminum oxide content and low in silica and titania. A typical analysis of calcined material would be Al_2O_3, 80%; SiO_2, 8%; Fe_2O_3, 8%; and TiO_2, 4%. The production of alumina is described under **Bauxite.** A 10-ton pig of aluminum oxide abrasive is shown in Fig. A-1. The consumption of fused alumina is about 350,000 short tons/yr.

Alumina abrasives also are made from bauxite without fusion. In this process, the bauxite is milled to a submicron size, pressed, granulated, sized, and fired to produce a strong, dense abrasive grain. This abrasive is characterized by its toughness and ability to produce hard, slow-wearing grinding wheels which are particularly suited for steel-mill applications. This type is referred to as *sintered alumina.*

Alloy types of fused-alumina abrasives also have been commercialized. The most successful is a eutectoid structure resulting from the cofusion of alumina with 20–40% zirconium oxide. These abrasives are very tough and well suited for severe application.

Silicon carbide is a synthetic abrasive second in importance only to fused alumina. E. G. Acheson first synthesized silicon carbide in a furnace of his own devising and a design which still supplies the entire demand. Acheson named the material *carborundum.* The material is formed from coke and sand by the reaction $SiO_2 + 3C \rightarrow SiC + 2CO$.

Fig. A-1. Ten-ton pigs of aluminum oxide abrasive are set aside to cool at the Chippawa, Ontario, Canada, plant of Norton Company. Temperatures exceeding 2038°C (3700°F) are required to melt impurities from the mineral bauxite that is used as a raw material.

In practice, the raw mix is packed around a horizontal graphite rod resistor. Power through the resistor heats the mix to 2000–2200°C to accomplish the foregoing reaction. On cooling, the abrasive-grade material in the central zones is separated from the outer zones of the partially reacted mix. The material is cleaned, crushed, and sized for use. The outer zones then are recycled. The product is green to black, depending upon the purity of the starting materials. Green silicon carbide is made from all-new mix, while the common gray or black variety is made from runs containing recycled material. Consumption of silicon carbide is about 180,000 short tons/yr.

Synthetic diamond of abrasive quality competes pricewise with natural diamond of bort grade. The synthesis is accomplished by a process discovered in 1955 and commercialized in 1960.* Graphite and a metal catalyst are heated to 1500–2000°C under a pressure of 65–90 kilobars. Metals which function as catalysts form carbides melting within the operating range. Nickel is commonly employed. By manipulation of the

*General Electric Co.

temperature and pressure, the crystal size and habit can be varied to suit abrasive needs.

Boron nitride in the cubic variety also is made synthetically for abrasive purposes. This is a high-pressure synthesis with the conditions for processing at a temperature of 2000°C and a pressure of about 80 kilobars. The starting material is hexagonal boron nitride. Lithium or lithium nitride may be used as catalyst. The abrasive is somewhat softer than diamond, but is more resistant to oxidation. The material shows great promise in the grinding of high-speed steels.

Other abrasives of minor importance, prepared synthetically for polishing, include iron oxide (as rouge or crocus), zirconia, tin oxide, and cerium oxide. The synthetic processes involve the precipitation and calcination of the carbonates and hydroxides.

Metallic Abrasives: Steel in the form of shot, cracked shot, or wool is used extensively in abrasive applications. The shot is used in blasting operations for cleaning metal surfaces and for imparting strength by peening. Wool is used mainly in polishing operations. A small amount of granular tungsten carbide is used in coated products, primarily on metal backings.

Preparation and Sizing of Loose Abrasives: Crude abrasives produced in electric-furnace operations first are reduced to large lumps by sledges or "skull crackers." Poorly crystallized and incompletely reacted material is rejected. The remaining lumps then are fed to a jaw crusher and reduced to a size suitable as feed for roll or impact crushers. Impact crushing tends to produce a higher yield of the coarse, blocky material, while the product from the roll tends to a more splintery habit. Crushing usually is continued until the entire product will pass an 8-mesh screen. The material then is separated on vibrating screens into many sizes, only a few of which are listed in Table A-2. Sizes smaller than 320 mesh are prepared by hydraulic classification.

Use of Abrasives: There are three major configurations in which abrasives are applied: (1) loose, (2) bonded in a single layer on a backing of cloth or paper, and (3) bonded in solid shapes, such as wheels, segments, sticks, and blocks.

In *loose form,* important uses include: (1) *Blasting,* which is performed primarily to remove and clean rust, scale, and paint from metal surfaces. Quartz sand and, to a lesser extent, garnet have been used primarily for this purpose. Because of increasing labor costs and the consequent desirability of improving abrasive efficiency, the trend is toward

TABLE A-2. *Average Particle Size of Abrasive Grain Used in Grinding Wheels*

Grit Size	Inches	Micrometers
8	0.1817	4,620
10	0.1366	3,460
12	0.1003	2,550
14	0.0830	2,100
16	0.0655	1,660
24	0.0408	1,035
30	0.0365	930
36	0.0280	710
46	0.0200	508
60	0.0160	406
100	0.0068	173
180	0.0034	86
280	0.00175	44
400	0.00090	23
600	0.00033	8

the use of synthetic abrasives, particularly fused alumina, for this purpose. (2) Rubbing on the workpiece by means of a rotating or reciprocating tool, as in *honing* or *lapping*. (3) Agitating the workpiece in a bed of loose abrasive, as in *tumbling* or *barrel finishing*.

Single-layer or *coated abrasive products* in the form of sheets, disks, and belts are used extensively for smoothing of metal, wood, and plastic surfaces. Quartz, emery, and garnet, previously used in coated abrasive products, have largely been replaced by fused alumina and silicon carbide. Alumina is used extensively for the finishing of metal, wood, and plastics, while the use of silicon carbide is restricted mostly to painted surfaces, glass, and ceramics.

The use of abrasives in *bonded shapes,* such as wheels, segments, sticks, and blocks, is divided into two main categories: (1) in steel mills for the removal of scale and defects from slabs or billets prior to rolling, and (2) in foundries for the removal of excess metal from rough castings. Other important categories of rough grinding are cutting-off operations and tool sharpening. Precision grinding is performed by the controlled approach of the grinding wheel to the workpiece with the object of generating geometrically accurate surfaces. This is the largest single category of abrasive use.

Manufacture of Abrasive Products: Coated abrasive products are made by coating one side of a backing material, such as cloth or paper, with an adhesive which is usually hide glue or an "A"-stage phe-

nolic resin. The abrasive then is deposited on the tacky surface by dusting or electrostatically, the latter being preferred because the abrasive is oriented on the backing so as to produce a sharper cutting surface. The sheet then is partially dried or cured to fix the abrasive. Finally, the sheet is coated with sizing, to provide a stronger binding, and given a final cure.

Solid-bonded abrasive products, such as grinding wheels, usually are separated into two categories, depending upon whether the bonding material is organic or inorganic. Organic products may be bonded with shellac, rubber, or synthetic resins, especially of the phenol-formaldehyde type. The abrasive is mixed with a resin solvent such as furfural. Powdered "B"-stage phenolic resin, mixed with appropriate fillers, is added to the moist abrasive, and the mix is completed. The granular mixture is spread in a mold and pressed to a predetermined weight/volume, or controlled porosity. The molded product then is cured at a temperature usually between 160 and 200°C. In this process the proper hardness or *grinding grade* for the intended application is achieved. A typical medium-grade wheel might contain, in percent by volume: abrasive, 54; bond, 26; pores or voids, 20. The harder grades are less porous. The molded wheel is trued for geometrical accuracy, tested for speed strength, and packaged. In speed testing, the wheel is run at 50% greater speed than the recommended operating speed, the latter usually being 9,500 surface ft/min, but may be as high as 16,000 surface ft/min, or even higher.

Vitrified bond wheels are made by mixing the abrasive with a temporary binder, such as dextrine and water, and powdered bonding material comprising clays, feldspars, or frits, followed by pressing to a controlled-volume structure and then firing at 800–1200°C to develop the bond, which may be glassy or porcelaneous in nature. Operating speeds for vitrified wheels usually are 6,500 surface ft/min, but may be as high as 12,000 surface ft/min in special products.

Nature of Grinding: The grinding process appears to be a mixture of chemical and physical processes. Chips are cut from the metal, for example, by the sharp abrasive points which are heated in the process of the melting temperature of the metal. At this temperature, chemical reactions occur which involve the abrasive, the metal, and the surrounding atmosphere. These reactions cause dulling of the abrasive points, necessitating wearing of the wheel structure to expose new points. The reactions also have a beneficial effect in preventing the rewelding of the chips to the base metal and their adhesion to the wheel structure. When the latter occurs, it is termed *loading*. The detrimental reactions involving the abrasive are minimized by the control of its purity. The beneficial reactions can be augmented by the incorporation of chemical aids, either into the wheel structure or into a fluid applied to the point of contact. Substances commonly used for this purpose are organic and inorganic sulfides and chlorides.

Polishing appears to be a quasi-chemical process in which the metal (or other material) is removed in particles approaching molecular size.

REFERENCE

Coes, Loring, Jr.: "Abrasives," Springer-Verlag, Bonn, 1971.

—Loring Coes, Jr., *Norton Company, Worcester, Mass.*

Absorption In chemical engineering terms, absorption may be defined as the transfer of a soluble component (solute) of a gas mixture to a liquid (solvent). The reverse operation—the transfer of a solute from the solvent to the gas mixture—is called *stripping* or *desorption*.

Absorption may be (1) purely physical, a simple application of the solubility of the solute in the solvent; or (2) a chemical reaction may occur between the solute and the solvent, or a reagent may be added to the solvent, thus increasing the solvent's ability to absorb solute. If the reagent is nonvolatile, the solute must be absorbed physically before reaction with the reagent can occur.

In many instances, reagents can be selected such that the reactions are reversible and the reagent can be regenerated and recycled. An example of absorption with chemical reaction is the absorption of CO_2 from a flue gas with aqueous NaOH solutions. Na_2CO_3 is formed. This reaction is irreversible. Continued absorption of the CO_2 with the Na_2CO_3 solution, however, results in the formation of $NaHCO_3$. The $NaHCO_3$ solution when heated will decompose to CO_2, H_2O, and Na_2CO_3. Thus the Na_2CO_3 solution can be recycled.

Nearly all absorption operations are accompanied by heat transfer, such as the absorption of HCl in water or the absorption of H_2O from wet gas in strong H_2SO_4. This ability to transfer heat simultaneously with absorption is put to good use in water-cooling towers where water is cooled by desorbing water (solute) from water (solvent) into air. The heat required to evaporate the water (solute) is supplied by the sensible heat of the water (solvent), thus cooling the water.

Major Equipment Elements: For gas absorption they may be (1) a packed tower filled with a packing material; (2) a spray tower in which the absorbing liquid is sprayed into an otherwise empty tower; (3) a tray tower containing bubble caps, sieve trays, or valve trays; (4) a falling-film absorber or wetted-wall column; and (5) stirred vessels. Towers are the most commonly used configuration, and packed towers are the most widely used of the towers.

General Design Factors: To design an absorption plant, it is necessary to know (1) gas rate, (2) gas composition, (3) gas temperature, (4) operating pressure and pressure drop permissible across the absorber, (5) minimum acceptable degree of recovery of the solutes and the specific solvent to be used, (6) desired concentration of solute in solvent, and (7) whether or not solvent recovery is required. Further, the designer is required to determine (1) the best solvent, (2) vessel diameter, (3) height and type of packing or number of trays, (4) optimum liquid rate through absorber and solvent recovery system, (5) temperatures of streams entering and leaving the absorber and quantity of heat to be removed to account for heat of solution and other heat effects, (6) operating pressure for the stripper and absorber, (7) mechanical design of the absorption and stripping towers, including flow distributors and packing supports, and (8) materials of construction.

Packed Towers. Widely used, these units require a lower capital investment than tray towers and provide turbulence in the gas phase, resulting in good mass-transfer efficiency for gas-phase-controlling systems. With the advent of plastic packing materials, the inherent disadvantages of weight of packed columns compared with tray columns was overcome. In comparison with a tray column, the packed column is relatively simple, as illustrated in Fig. A-2.

A typical column consists of a cylindrical shell containing a support plate and redistributor plates to support the packing. A liquid distributor is set above the packing and designed to provide effective irrigation of the packing. Many types of packing are commercially available, each having specific advantages for liquid-gas contacting, from the aspects of cost, surface availability, liquid-surface regeneration, pressure drop, weight, and corrosion resistance. Typical packings are shown in Fig. A-3.

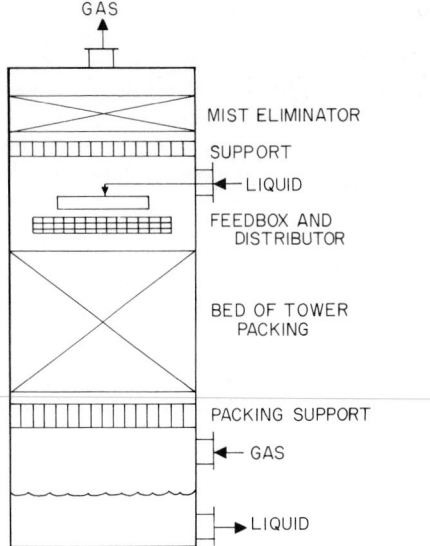

GAS

MIST ELIMINATOR

SUPPORT

LIQUID

FEEDBOX AND DISTRIBUTOR

BED OF TOWER PACKING

PACKING SUPPORT

GAS

LIQUID

Fig. A-2. Typical packed tower. (*Maurice A. Knight Company.*)

Design Factors: The *diameter* of the packed tower is a function of the liquid and gas rates, physical properties of the liquid and gas, size and geometry of the packing material, and the method of installation of the packing material. Sherwood, Ship-

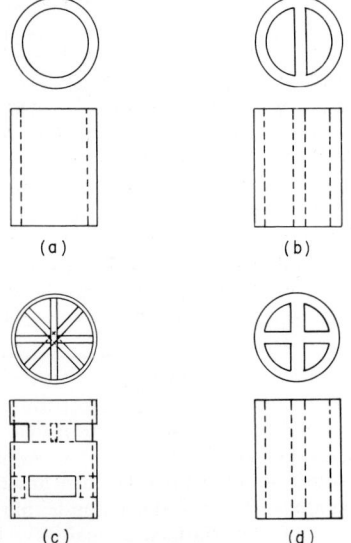

Fig. A-3. Representative tower packings. (a) Ceramic Raschig ring, (b) ceramic Lessing ring, (c) plastic Flexiring (*Koch Engineering Co.*), (d) ceramic cross-partition ring.

ley, and Holloway[1] have developed correlations for both dumped and stacked packings from which can be calculated the *limiting* or *flooding* gas velocity for a given liquid, based on the physical properties of the liquid and gas and the size and geometry of the packing. In this plot, the abscissa is

$$\frac{L}{G}\sqrt{\frac{\rho_g}{\rho_L}}$$

and the ordinate is

$$\frac{V^2}{g}\frac{a}{\varepsilon^3}\frac{\rho_g}{\rho_L}\mu^{0.2}$$

where L = liquid rate
   G = gas rate, both in terms of lb/(hr)(ft^2) of tower cross section
   ρ_g = gas density
   ρ_L = liquid density, both in terms of lb/ft^3
   μ = viscosity of liquid, cP
   g = acceleration due to gravity, 32 ft/(s)(s)
   V = gas velocity at flooding, ft/s
   a/ε^3 = packing factor

The factor a/ε^3 is defined as the surface area of the packing per cubic foot of packing divided by the cube of the free space. The lower the value of this factor, the higher is the limiting gas velocity allowable.

Leva et al.[2] have modified the foregoing correlation by replacing the ordinate with the group:

$$\frac{V^2}{g}\frac{a}{\varepsilon^3}\frac{\psi}{\rho_L\rho_g}\mu^{0.2}$$

where ψ is a liquid-density correction factor and is defined as the ratio of the density of water to the density of the liquid. Lerner,[3] from the published data, has developed a mathematical equation relating the limiting gas rate to the liquid rate. The general equation is

$$G_F = \phi k(1 - mL^{0.57})\left(\frac{\sigma}{\sigma_w}\right)^n$$

where σ = surface tension of liquid, dyne-cm
   σ_w = surface tension of water, dyne-cm
   k, m = constants related to size and shape of packing
   n = constant = 0.58 at low-liquid rates and 1.0 for high-liquid rates

The values of k and m for various packings are given in Table A-3. Most packed towers will be designed to operate at a maximum of 70% of the

TABLE A-3. *Constants for Flooding-Velocity Equation*

$$G_0 = k\,(1 - mL_0^{0.57})$$

Packing Type and Size, in.	Nominal Range	Experimental L_0 Range	k	$m \times 10^3$
Ceramic Raschig rings:				
$\frac{3}{4}$ in.	Overall	2,500–13,000	1,566	3.06
	Low	2,500– 7,500	1,683	3.36
1 in..	Overall	2,500–18,300	1,595	2.78
	Low	2,500–10,000	1,870	3.46
	High	10,000–18,300	1,420	2.67
$1\frac{1}{2}$ in.	Low	2,500–13,000	2,118	2.58
2 in..	Low	7,500–17,000	2,280	2.21
Ceramic Berl saddles:				
$\frac{3}{4}$ in.	Overall	2,500–18,000	1,645	2.74
	Low	2,500–13,000	1,825	3.00
	High	13,000–18,000	1,405	2.56
1 in..	Overall	2,500–18,000	2,360	2.70
	Low	2,500–15,000	2,340	2.66
$1\frac{1}{2}$ in.	Low	5,000–17,800	2,740	2.13

limiting gas velocity or less, depending on the maximum pressure drop allowable across the tower.

Intrinsic pressure drop data for packing materials can be found in the manufacturer's literature or in Perry et al.[4] Leva et al.[2] have plotted lines of constant pressure drop on the flooding-correlation diagram of Sherwood et al.[1] Lerner[3] has developed mathematical equations for calculating the intrinsic pressure drop for various sizes and shapes of packing by dividing the liquid and gas rates into four zones, as follows:

ZONE 1. Preload and precritical
ZONE 2. Preload and supercritical
ZONE 3. Load to flood and precritical
ZONE 4. Load to flood and supercritical

The generalized pressure-drop equation is

$$\frac{\Delta P}{N} = a(10^{b(L/\rho_L)}) \left(\frac{G}{\phi}\right)^n + K\left(\frac{L}{\rho_L}\right)^m$$

where $\Delta P/N$ = pressure drop, in. H_2O/ft of packing

a = a constant depending upon packing size and shape and zone

b = a constant depending upon packing size and shape and zone

K = a constant depending upon packing size and shape = zero in Zones 1 and 3

n = 1.8 for Zones 1 and 2, 3.0 for Zones 3 and 4, except $\frac{3}{4}$-in. ceramic Berl saddles, where $n = 2.75$

m = a constant depending upon Zones 2 or 4 and packing size and shape

Values of these constants are given in Table A-4.

The *height of packing* required can be obtained from:

1. A point-to-point calculation, which is tedious and time-consuming.

2. Graphically determining the transfer units and multiplying the *number of transfer units* (NTU) by the *height of a transfer unit* (HTU). Height of packing = (HTU)(NTU). NTU can be graphically determined using the methods of White[5] or Baker[6]. HTU is either directly available in the literature for various solute-solvent systems and varying gas and liquid rates, or where HTUs are not available, they may be calculated or estimated from the *mass-transfer coefficient* $k_G aP$. When $k_G aP$ is nearly proportional to the first power of the gas mass velocity G_M, then

$$\text{HTU} = \frac{G_M}{k_G aP(1 - y)}$$

where G_M = moles/(hr)/(ft^2) of inert gas

$k_G a$ = moles/(hr)/(ft^3)(atm) of solute transferred

P = total pressure, atm

$(1 - y)$ = mole fraction of inert gas in bulk of gas

TABLE A-4. *Values of Constants Used for Pressure-Drop Calculations in Design of Columns Using Ceramic Packing*

Packing	L/ρ_L (Critical)	K		m		Zone 1		Zone 2		Zone 3		Zone 4	
		Zone 2	Zone 4	Zone 2	Zone 4	$a \times 10^{-6}$	$b \times 10^3$	$a \times 10^6$	$b \times 10^3$	$a \times 10^{10}$	$b \times 10^3$	$a \times 10^{10}$	$b \times 10^3$
Raschig rings:													
¾ in.	136.5	5.15×10^{-6}	3.68×10^{-3}	1.90	0.75	2.64	4.83	2.18	5.12	10.1	6.47	8.64	6.38
1 in.	144.5	1.70×10^{-6}	3.63×10^{-4}	2.24	1.35	1.29	6.60	2.57	3.67	6.95	6.42	9.11	4.77
1½ in.	209	1.38×10^{-11}	4.55×10^{-5}	3.72	1.50	1.45	2.60	2.83	2.035	4.40	4.10	5.77	2.98
2 in.	240	8.27×10^{-5}	7.02×10^{-5}	5.3	1.43	1.0	3.05	1.265	2.36	2.55	3.91	4.33	2.51
Berl saddles:													
¾ in.	160	5.77×10^{-9}	1.283×10^{-5}	3.0	1.90	2.08	3.25	1.379	4.68	27.1	5.607	101.8	2.20
1 in.	176.5	4.98×10^{-12}	1.425×10^{-5}	4.3	1.67	1.36	3.34	2.28	2.06	2.83	4.51	5.8	3.42
1½ in..	285	*	*	*	*	0.790	2.13	*	*	1.612	2.977	*	*
2 in.	300	*	*	*	*	0.428	3.57	*	*	0.744	3.46	*	*

*Data not available.

The design of the packing support is critical. It makes no sense to design a tower with a certain type and size of packing and then select a packing support of less (percent) open area than the packing void space. Flooding may well occur at the packing support. Thus the support plate should have an open area equal to or greater than the packing, if possible. Common types of packing supports are shown in Fig. A-4.

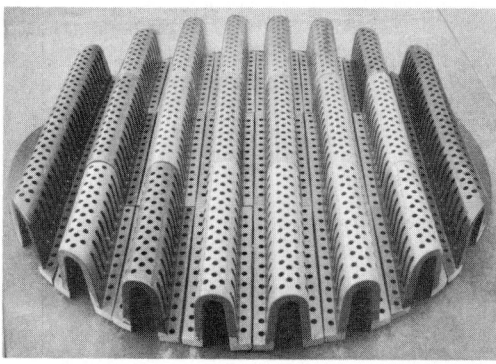

(a)

(b)

Fig. A-4. Representative packing supports. (a) Ceramic multi-beam support, (b) metal bar-type support. Glass bar-type packing supports and other configurations also are obtainable. (*Maurice A. Knight Company.*)

The efficiency of the packing for mass transfer depends upon the effective continually wetted area of the packing. The extent to which the effective continually wetted area approaches the total available surface area depends upon the effective distribution of the liquid and gas streams. Three factors effect distribution: (1) plumbness of the tower when erected, (2) initial liquid-distribution device, and (3) length of liquid flow. As the length of liquid travel increases, the liquid tends to flow toward the walls; thus redistribution is necessary if the packed height exceeds 15 ft.

Distributors may be (1) spray nozzles, (2) multiple weir units, and (3) multiple nozzles. Redistributors may be wall wipers or weir-type distributors with packing supports. Common types of liquid distributors are shown in Fig. A-5.

Tray Columns. Plate columns used for liquid-gas contacting, such as absorption, are of two classifications: (1) cross-flow plate and (2) counterflow plate. The cross-flow tray (Fig. A-6a) requires a liquid downcomer and generally is used more than the counterflow tray (Fig. A-6b) because of trans-

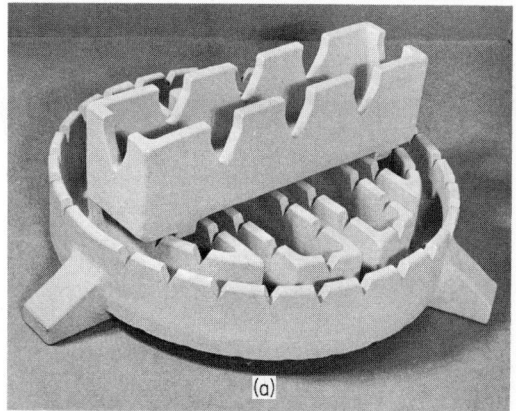

(a)

(b)

Fig. A-5. Typical liquid distributors used in packed columns. (a) Pan-type V-weir distributor with feedbox made of chemical stoneware, (b) trough-type V-weir distributor with feedboxes of metal construction.

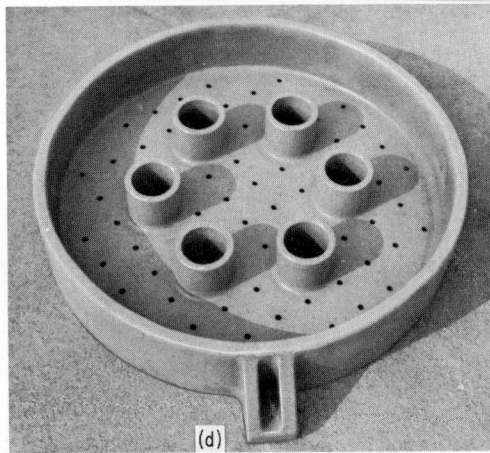

Fig. A-5. (cont.) Typical liquid distributors used in packed columns. (c) Weir flow-tube distributor made of chemical stoneware, (d) pan-type orifice distributor made of chemical stoneware. (*Maurice A. Knight Company.*)

fer-efficiency advantages and greater operating range. The flow pattern of the liquid on a cross-flow tray may be controlled by variation in placement of downcomers in order to increase stability

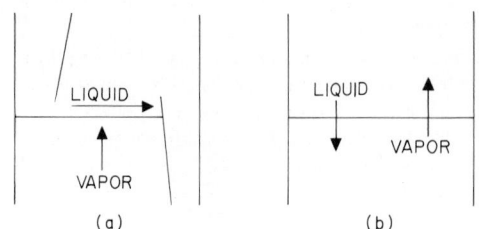

Fig. A-6. Tray arrangements. (a) Cross-flow tray, (b) counterflow tray.

of operation and to improve the mass-transfer efficiency. The internals used in plate towers for the purpose of gas dispersion are bubble caps, siever perforations, or modifications of these basic configurations.

The physical seal dispersion system consists of the bubble cap and its modification. The bubble-cap seal consists of a riser which acts as a liquid seal and through which the vapor rises. The vapor proceeds through a reversal path and is then dispersed via slots in the cap. The bubble cap is shown in Fig. A-7a, and its basic modifica-

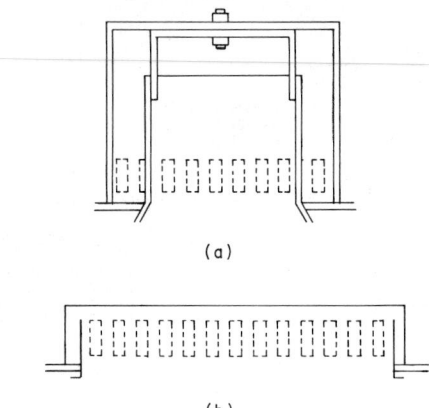

Fig. A-7. (a) Circular or bell bubble cap. (b) Tunnel cap. (*Maurice A. Knight Company.*)

tion, the tunnel cap, is shown in Fig. A-7b. The vapor area of the slots varies between 8 and 15% of the tray area.

The tray construction wherein the liquid is maintained on the tray surface by the kinetic energy of the vapor is called the *sieve tray.* The openings may take the form of circles or slits formed by mechanical punching of the tray (Fig. A-8).

Much research has gone into the design of more efficient trays, and numerous proprietary designs

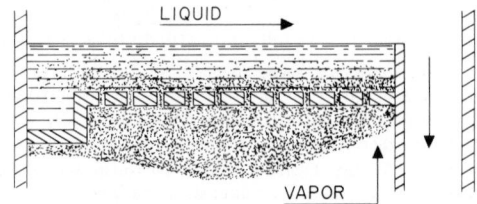

Fig. A-8. Action of sieve tray.

have been developed, all of which cannot be described here. Valve-type trays provide an interesting example. See Figs. A-9–A-11.

Column-operating Problems: Column instability may result from flooding or priming. A rapid decrease in column efficiency may result from excessive entrainment. Thus any one of three factors may establish the limiting capacity of a cross-flow-tray column, and these may not be completely interdependent. *Flooding* occurs when the pressure drop through a tray exceeds the liquid head available in the downcomer. In such cases, the vapor will tend to short-circuit the normal path and rise through the downcomer.

Fig. A-11. Tray constructed with rows of perforation clusters alternating with rows of valves (see Fig. A-10), which provides the economy of a sieve tray but minimizes loss of efficiency when flow rates differ widely from design specifications. (*Flexitray, Koch Engineering Co.*)

Priming occurs when the foam on the tray reaches the tray above. This condition causes an additional resistance to the passage of gas through the vapor-dispersion section on the tray above. As a result, *excessive vapor-pressure drop* occurs and effective entrainment increases, resulting in a limiting column capacity due to either a rapid decrease in efficiency or flooding.

Behavior of Various Systems and Packings. In view of the incomplete development of the theory of mass-transfer behavior, much of the estimation of rate of transfer is based on comparison with systems already evaluated. Comparison generally is based on equivalent geometry and flow characteristics, variations related primarily to the physical properties of the systems. See Table A-5.

Packed Towers versus Plate Towers. The relative advantages and disadvantages of packed and plate towers may be summarized: (1) packed towers may be advantageous for vacuum operations because the pressure drop through a packed tower can be less than for a plate tower; (2) packed towers may be preferred in the case of liquids that foam; (3) liquid hold-up generally is less in a packed tower; (4) plate towers may be preferred when there are deposits of solid material that must be removed periodically (a plate tower can be fitted with manholes, and the plates may be spaced far enough apart to facilitate cleaning); (5) the total weight of a plate tower usually is less than for a packed tower designed for the same duty (the limited crushing strength of packing materials may make it impossible for one packing support plate to bear the weight of a tall column

Fig. A-9. Close-up of round movable cap shown mounted on portion of perforated deck. Many caps are mounted on a deck, as shown in Fig. A-11. Caps operate like check valves and have a limited lift, which is accomplished by integral guide legs and lift stops. The valves are fabricated in different metal gages and normally are installed in alternating rows of light and heavy valves, parallel to the outlet weir to provide good vapor distribution over a wide range of vapor flow rates. At low vapor flow rates, the lighter valves are lifted to an open position. As the vapor flow increases, the lighter-weight and then the heavier-weight valves open progressively wider to their full-open positions. (*Flexitray, Koch Engineering Co.*)

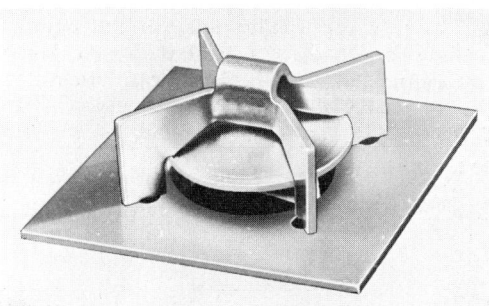

Fig. A-10. Close-up of round movable cap mounted on portion of perforated deck. Cap, approximately 2 in. diam, operates like that shown in Fig. A-9, but construction is not fully integral in that a special, four-legged hold-down fixture is used to attach the assembly to the tray deck. This is a more rugged configuration, designed for more corrosive, dirtier service. (*Flexitray, Koch Engineering Co.*)

TABLE A-5. *Mass-Transfer Behavior of Representative Absorption Systems*

System	Influences
Ammonia-air-water	Liquid and gas phases contributing; chemical reaction contributing
Air-water	Gas phase controlling
Sulfur dioxide–air–water	Liquid and gas phases contributing
Carbon dioxide–air–water	Liquid phase controlling
Carbon dioxide–air–mono-ethanolamine	Absorption accompanied by chemical reaction
Carbon dioxide–air–carbonate	Absorption accompanied by chemical reaction
Ethanol-air-water	Liquid and gas phases contributing
SiF_4–air–water	Gas phase controlling; chemical reaction contributing

of packing); (6) plate towers may be more suitable when the operation is carried out intermittently at temperatures either higher or lower than atmospheric temperature (alternate expansion and contraction of the shell in such cases may crush the packing); (7) cooling coils are installed readily on plates, making plate towers more desirable when heat of solution requires internal cooling; (8) plate columns may be preferable for operations that require a large number of transfer units or theoretical plates (packed towers are subject to channeling of the vapors or liquid streams and thus are limited in the amount of material that can be transferred); (9) higher liquid rates usually can be handled in plate towers, provided that the distance the liquid must travel to cross each plate is not greater than a few feet; (10) construction of packed towers usually is simpler and cheaper when corrosive substances must be handled.

Spray Towers. The simplest spray absorber consists of an empty tower into which liquid is sprayed at the top and gas is introduced at the bottom. Such a unit has the advantages of very low pressure drop and inexpensive construction. Because of mixing of the gas within the chamber (promoted by entrainment of gas by the sprays) and entrainment of fine spray droplets, spray absorbers are unsuitable where true countercurrent action is needed to obtain a large number of transfer units. Although not so effective, a spray chamber closely approaches the performance of a packed tower for very short heights (both for liquid- and gas-phase controlled systems). Thus, where only a few transfer units are needed, spray absorbers may be more economical, particularly when low pressure drop is important.

Cyclone Scrubbers: High gas rates cannot be obtained in simple spray absorbers owing to entrainment. This can be markedly reduced by resorting to cyclone spray devices in which the droplets are thrown to the wall by centrifugal force (owing to their tangential entry) before they are swept out the top of the chamber by the gas stream. Such devices give a cross-flow type of contact, however, and so are limited to operations requiring no more than one theoretical stage.

Stirred Absorber. Rates of absorption in stirred vessels are most commonly needed for application to chemical reaction in which a gas is first absorbed and then reacts with a component in the solution. The feature of long liquid residence time cannot be so easily provided in tower-type absorbers. In general, an agitated absorber also has a slight advantage in size and power consumption over a packed tower, but it is limited to low gas throughputs (under 10 ft/min).

Wetted-Wall Columns. These devices find use particularly when heat release is high and heat-transfer surface must be provided adjacent to the liquid. A primary example of this application is the HCl cooler-absorber, usually consisting of multiple impervious graphite tubes arranged in a bundle. These units also are referred to as *falling-film absorbers.*

REFERENCES
1. Sherwood, Shipley, and Holloway: *Ind. Eng. Chem.,* vol. 42, p. 768, 1938.
2. Leva et al.: *Ind. Eng. Chem.,* vol. 46, p. 1225, 1954.
3. Lerner: A New Look at Pressure Drop and Flow Behavior in Packed Towers, Maurice A. Knight Company Bull. TP-107, Akron, Ohio, 1966.
4. Perry and Chilton: "Chemical Engineers' Handbook," 5th ed., sec. 14, McGraw-Hill, New York, 1973.
5. White, G. E.: *Trans. AIChE,* vol. 36, p. 359, 1940.
6. Baker: *Ind. Eng. Chem.,* vol. 27, p. 977, 1935.

—David Cooper, *Maurice A. Knight Company, Akron, Ohio*

Absorption, Acidic Gases The treatment of gases to remove acidic components, CO_2, H_2S, and other sulfur compounds, has acquired increasing importance. Large amounts of H_2, which require removal of acid gases, are generated to produce NH_3 and to desulfurize liquid fuels. The demand for natural gas has grown faster than the rate of discovery of new gas reservoirs. This has brought about treating of natural gases to remove acid gases up to over one-half the volume of the gas produced from the wells in some fields. This has resulted in the development of several widely used processes to remove the acid gases. One* of these processes (Fig. A-12) was developed to remove large amounts of H_2S and associated quantities of CO_2, COS, mercaptans, and other sulfur compounds from sour gases. The process is used to treat natural gas, H_2, ethane, and propane in the vapor phase.

The process illustrated uses a conventional absorption and regeneration cycle. The sour gas is contacted countercurrently with lean solvent at essentially ambient temperature. The rich solution is flashed at an intermediate pressure to allow

*SULFINOL^{SM} (process) is a service mark of the Shell Oil Company.

absorbed hydrocarbons or H_2 to escape. The solvent is regenerated by heating in a stripper at a pressure slightly above atmospheric.

There are numerous variations in the details of the process flow diagram, depending upon the particular requirements of a project. For example, where substantial amounts of H_2S are present, the flash from the solvent flash vessel cannot be vented, and it is treated in a fuel-gas contactor. The treated flash gas is suitable for plant fuel or recompression to the sweet-gas line. Where the acid gas produced is CO_2, the purity can be improved by heating the rich solvent in the solvent flash vessel to drive off coabsorbed gases. Where the process is used to treat the regeneration gases from mole-sieve, sour-gas treating units, the process can be designed to absorb the H_2S, COS, and mercaptans evolved.

The process uses a solvent that combines the physical solvent-absorption capacity of sulfolane for acid gases with the chemical absorption ca-

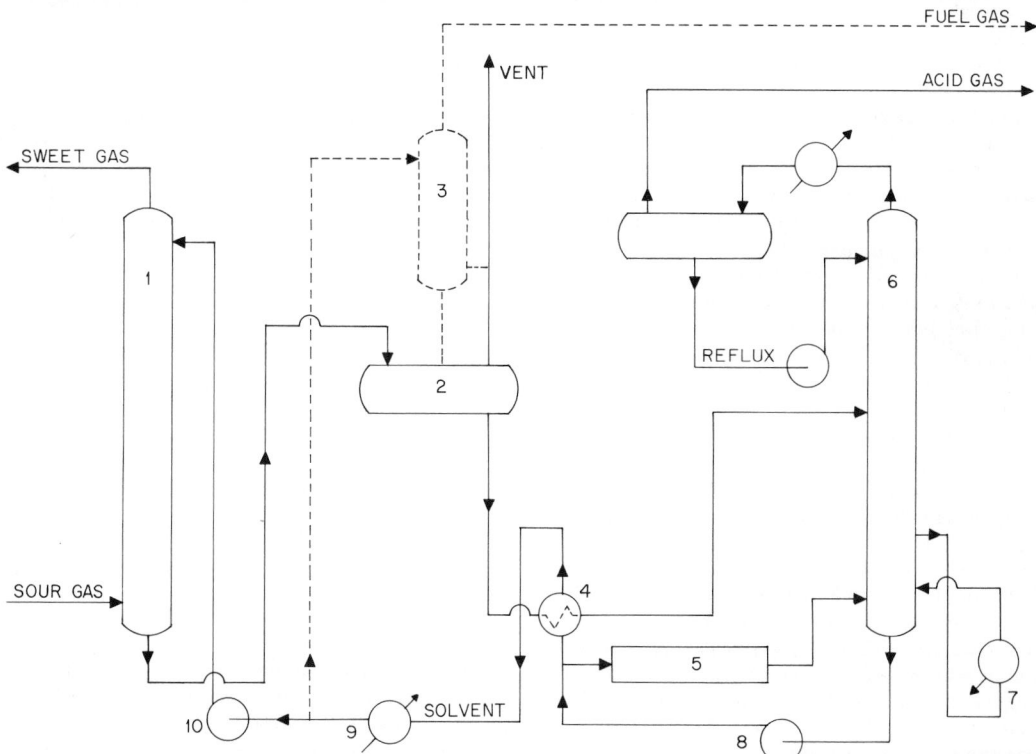

Fig. A-12. Process for treatment of gases to remove acidic components. (1) Contactor, (2) solvent flash vessel, (3) fuel-gas contactor, (4) lean-rich exchanger, (5) reclaimer, (6) stripper, (7) reboiler, (8) solvent booster pump, (9) solvent cooler, (10) high-pressure solvent pump. (SULFINOL^{SM} is a service mark of the Shell Oil Company)

pacity of diisopropanolamine. This combination of absorption capabilities offers advantages for both absorption and regeneration. The diisopropanolamine combines with sour components in an acid-base reaction, essentially nonsensitive to pressure, and sulfolane adds physical solubility which is proportional to pressure. The net result is a solvent having good absorption for sour components at low to medium partial pressures and very high absorption of these components at high partial pressures. A substantial portion of the absorbed gases is released from the solution on pressure reduction, requiring less heat for stripping.

The process solvent is nonfoaming in that it can be operated with high contactor loadings on the solution gas from crude oil without foaming problems, even when some crude oil enters the contactor. A few pipeline corrosion inhibitors have caused foaming in SULFINOL[SM] contactors, and a satisfactory antifoam agent has been used until the type of corrosion inhibitor could be changed to one of the many that have no effect on SULFINOL[SM] solvent.

The process is noncorrosive to carbon steel in applications where some H_2S is present. Where no H_2S is present, a limited amount of stainless steel is required. A corrosion inhibitor is not used in the SULFINOL[SM] process. The results of measurements on corrosion in SULFINOL[SM] systems were published by MacNab and Treseder.[1]

The heat requirements for regenerating the solvent are low, with the steam requirement per pound of acid gas removed somewhat less than half the quantity used for monoethanolamine solutions used for acid-gas treating.

Auxiliary facilities used in the process are a small side-stream filter to remove solids from the solvent and a small reclaimer. The stable degradation products have boiling points higher than sulfolane and diisopropanolamine so that the solvent components can be recovered in a reclaimer.

Losses of solvent components are distributed between vapor, chemical, and mechanical. Sulfolane does not degrade, and typical losses of sulfolane in operating plants have been substantially less than 1 lb/million scf of sour gas treated, some plants adding little or no sulfolane during the first year of operation. CO_2 reacts to form diisopropanolamine-oxazolidone in quantities varying from less than 5 to 20 lb/million scf of CO_2; therefore losses vary considerably between plants.

Applications: The use of a multicomponent solvent permits adjustment of the composition for the optimum treating of a wide variety of sour gases, such as removal of CO_2 from H_2 produced by steam-methane reforming and removal of CO_2, H_2S, and COS from H_2 produced by gasification of high-sulfur fuels.

The process is used in natural-gas applications to treat sour gases where the H_2S content has varied from a few ppm to 53 mole % and the CO_2 content has varied from 1.1 to 28%. The SULFINOL[SM] process should be considered for the following natural-gas applications:

1. Where there are no hydrocarbon liquids produced with the sour gas. Also, for special situations where hydrocarbon liquids are produced and foaming could be a problem with other processes. Generally, the acid-gas partial pressure should be above 50 psi.

2. Where the removal of mercaptans, COS, and CS_2 is an important consideration at acid-gas partial pressures as low as 5 psi. The total pressure should be above 500 psig, except in special cases, such as treating propane in the vapor phase where the COS can be reduced to a few parts per million.

3. Where low utility requirements for regeneration are necessary.

Natural-gas pipeline specifications that can be readily attained for applications of SULFINOL[SM] solvent are (1) H_2S content, below 0.25 grain/100 scf; (2) CO_2 content, below 0.3 mole %; (3) mercaptan content, below 0.2 grain/100 scf; and (4) total sulfur, below 1.0 grain/100 scf.

With minor modifications to the normal plant design, the H_2S can be reduced to less than 0.05 grain/100 scf and the CO_2 can be lowered to less than 100 ppm.

Some natural-gas pipeline specifications provide for CO_2 levels up to 2 or 3 mole % with up to 1.0 grain of H_2S/100 scf. This is attained with SULFINOL[SM] solvent at high ratios of CO_2 to H_2S in the feed gas.

The sulfolane component of SULFINOL[SM] solvent has an affinity for heavy hydrocarbons, especially aromatics, in the feed gas. Where substantial amounts of heavy hydrocarbons or aromatics would enter the acid gas and a sulfur plant would be used, it is necessary to use a separation process (usually chilling) on the feed gas to reduce the condensible hydrocarbons to an acceptable level. [The alternative is to use charcoal to remove the aromatics from the acid-gas feed to the sulfur plant.

Some typical applications for the SULFINOL[SM] process for natural-gas treating are shown

TABLE A-6. *Typical Circulation and Reboiler Steam Rates of SULFINOL^SM Units for Natural-Gas Applications*

Item	Gas "A"	Gas "B"	Gas "C"	Gas "D"
Contactor pressure, psia.	1,000	1,000	1,000	1,000
Feed-gas temperature, °F (43°C)	110	110	110	110
Feed-gas flow (dry), million scf/day	100	100	100	100
Feed-gas composition, mole %:				
H_2S	0.65	20.1	51.5	0.10
CO_2	8.73	2.0	3.5	18.00
N_2	2.37	1.4	8.6	0.70
C_1	87.90	71.5	25.8	80.94
C_2	0.35	2.0	5.8	0.17
C_3	. . .	1.7	3.2	0.05
C_4	. . .	1.1	1.6	0.04
C_{5+}	. . .	0.2		
Total	100.00	100.00	100.00	100.00
COS, grains/100 scf	3.0	7.3	8.4	
RSH, grains/100 scf	2.1	1.5	3.1	
H_2S/CO_2 ratio	0.0744	10.05	14.71	0.0056
Sweet-gas H_2S content, grains/100 scf	<0.25	<0.25	<0.25	<0.25
Sweet-gas CO_2 content, mole %	<1.0	<1.0	<1.0	2.0*
Sweet-gas total sulfur, grains/100 scf	<1.0	<1.0	<1.0	<1.0
Solution circulation rate, gal/min at 110°F	1,483	1,748	2,366	2,167
Solution net pickup, scf acid gas/gal	4.39	8.78	16.14	5.24
Reboiler steam rate, lb/hr	69,740	111,440	156,010	99,600
Reboiler steam rate, lb/gal solution	0.78	1.06	1.10	0.77

SOURCE: B. G. Goar, *Oil and Gas J.*, June 30, 1969.
*The SULFINOL^SM system for gas "D" was designed to leave 2.0 mole % CO_2 in the sweet gas, while reducing the H_2S content to 0.25 grain/100 scf.

in Table A-6 by Goar[2] for four different gases. These illustrations show the wide range of acid-gas solubility in the SULFINOL^SM solvent and the low utility requirements. The reboiler steam rates depend upon the amount of heat recovered in a lean-rich solvent heat exchanger, if one is used. Generally, no lean-rich exchangers are needed for gases "B" and "C" because the steam generated in the sulfur plant is adequate for re-generating the SULFINOL^SM solvent without any process heat recovery.

The removal of COS and acid gases from synthesis gas generated by partial oxidation of heavy fuel oil is reported by Klein.[3] The sour synthesis gas contained 0.3–1.2% H_2S and 100–500 ppm COS in addition to CO_2. The removal of the COS and sour-acid-gas components from typical crude synthesis gases is shown in Table A-7.

TABLE A-7. *Typical Operating Data of SULFINOL^SM Units for Synthesis-Gas Applications*

Gas throughput, million scf/day	13.71
H_2S content, vol %	0.36
CO_2, vol %.	5.20
COS content, ppm	125
Absorption pressure, psia	588
Absorption temperature, °F	104
Number of trays	30
Purified gas, ppm:	
H_2S content	0.5
COS content.	0.3
CO_2 content	25
Steam consumption at 51.5 psia, lb/lb acid gas removed	1.9
Current consumption, kWh/lb acid gas removed	0.015

REFERENCES

1. MacNab, A. J., and R. S. Treseder: Materials Experience in the Sulfinol Gas-treating Process, *Materials Protection and Performance*, vol. 10, pp. 21–26, January 1971.
2. Goar, B. G.: *Oil and Gas J.*, June 30, 1969, pp. 117–120.
3. Klein, J. P.: Developments in Sulfinol and Adip Processes Increase Uses, *Oil and Gas Int.*, vol. 10, no. 9, pp. 109–112, September 1970.

—Earl S. Hill, *Shell Development Company, Houston, Tex.*

Absorption, Radiation   (See **Dysprosium; Erbium; Gadolinium; Hafnium;** and **Radioisotopes**.)

Absorption Spectrum   (See **Atomic structure**.)

Accelerators   (See **Rubber, natural;** and **Tellurium**.)

Acceptors   (See **Catalysts**.)

Acetaldehyde   (See **Aldehydes;** and **Ethyl alcohol**.)

Acetals   Acetals are thermoplastic resins, available as homopolymers or copolymers, made mainly from formaldehyde or a formaldehyde derivative. The properties of these resins are summarized in Table A-8.

Manufacture: Polymerization of formaldehyde can proceed with a variety of ionic initiators, which include tertiary amines and quaternary ammonium salts. Molecular weight is controlled by chain transfer, as illustrated by the following equations:

$$-CH_2OCH_2O^- + H_2O \rightarrow$$
$$-CH_2OCH_2OH + OH^- \quad (1)$$

(Growing chain reacts with water,
releasing hydroxyl ion.)

$$OH^- + CH_2O \rightarrow HOCH_2O^- \quad (2)$$

(New chain
is initiated.)

$$HOCH_2O^- + nCH_2O \rightarrow$$
$$HOCH_2O(CH_2O)_{n-1}CH_2O^- \quad (3)$$

End capping of the high-molecular-weight polyoxymethylene glycol gives the stable commercial acetal resin.

Either formaldehyde or the cyclic trimer trioxane, $CH_2OCH_2OCH_2O$, may be used as starting material to form polymers that are similar in properties. However, with trioxane the polymerization catalyst normally preferred is boron trifluoride or other Lewis acids. Polymerization in the presence of ethylene oxide or 1,3-dioxolane comonomer results in ethyleneoxy, $-OCH_2CH_2-$, groups in the polymer chain, and the product is an acetal copolymer.

Chemical Resistance: Acetals resist common organic compounds, including most esters, aldehydes, alcohols, ethers, glycols, hydrocarbons, soaps, and ketones, unless exposure is prolonged at elevated temperatures, but resistance to oxidizing agents and strong acids is limited. Copolymers and some homopolymers resist weak bases. Generally, acetals are not used where resistance to radiation, weathering and burning, or exposure to strong acids and bases is required.

Fabrication and Uses: Products are fabricated by standard injection molding, extruding, or machining. Acetals are widely used for mechanical parts, bearings, cams, gears, springs, sprockets, electrical parts, hardware, and housings.

—Walter M. Bruner, *E. I. du Pont de Nemours & Co., Wilmington, Del.*

Acetamide   (See **Amides and imides**.)

Acetanilide   (See **Amides and imides;** and **Amines**.)

TABLE A-8. *Properties of Acetal Resins*

	Homopolymer	Copolymer
Tensile strength, psi, at 23°C (73°F)	10,000	8,800
Elongation, % (at break), at 23°C	25	60
Impact strength, ft-lb/in., at 23°C (unnotched) . .	20.5	20.0
Modulus (tensile) of elasticity, psi, at 23°C	520,000	410,000
Flexural yield strength, psi, at 23°C	14,100	13,000
Flexural endurance limit, psi, at 10^7 Hz	4,400	3,300
Heat-distortion temperature, 264-psi load.	124°C (255°F)	110°C (230°F)
Melting point, crystalline	175°C (347°F)	165°C (329°F)
Water absorption, % (24 hr)	0.25	0.25
Hardness, Rockwell M	94	80
Specific gravity (density)	1.42	1.41
Dielectric constant, 50% RH at 23°C	3.7	3.7–3.8
Dissipation factor, 10^6 Hz, at 23°C	0.0048	0.006

NOTE: Acetals have the highest fatigue endurance of commercial thermoplastics.

Acetate Fibers The term acetate fiber refers to all textile fibers manufactured using cellulose acetate as the basic polymer. There are two general types of acetate fibers: those made from the fully acetylated cellulose, called *triacetate* fibers, and those made from partially hydrolyzed cellulose triacetate, called *secondary acetate*, or simply acetate fiber. The latter is the most common type. Acetate fibers are a high-production commodity (United States, 1969, 498 million lb). Continuous-filament, fine-denier textile yarns represent about 90% of this quantity. The remainder is in the form of staple, or tow for cigarette filters.

Chemistry of Manufacture: The basic raw material for the manufacture of cellulose acetate is highly purified cellulose. Cellulose is a polymer of high molecular weight whose basic repeating unit is cellobiose:

Cellobiose, in turn, is made up of two anhydroglucose units connected by a β-1,4-glucoside linkage. Each anhydroglucose unit is a trihydric alcohol containing two secondary alcohols in the 2 and 3 positions, and one primary alcohol in the 6 position.

Cellulose acetate is prepared by esterifying these three hydroxyl groups. One of the most important properties of the product is the degree of substitution (DS), which is the average number of hydroxyl groups esterified per anhydroglucose unit. The maximum DS, of course, is 3.0, which corresponds to cellulose triacetate. In practice, the commercial product has a DS of 2.90–2.95. Secondary cellulose acetate has a DS of 2.35–2.40.

Cellulose and acetic acid do not react directly to any appreciable extent. To prepare the ester the cellulose is reacted with acetic anhydride, in the presence of a catalyst, usually sulfuric acid, and a solvent, usually glacial acetic acid. The overall reaction to produce the triacetate may be written

$$C_6H_7O_2(OH)_3 + (CH_3CO)_2O \rightarrow$$
$$C_6H_7O_2(OCOCH_3)_3 + 3CH_3COOH$$

where $C_6H_7O_2(OH)_3$ stands for one anhydroglucose unit in the cellulose molecule.

Secondary acetate is produced from the solution of triacetate, by adding water in the form of dilute acetic acid, to hydrolyze off about one out of every five acetate groups from the ester. The reaction is then stopped and the product precipitated by addition of more water. The reason for arriving at the secondary acetate in this roundabout manner is to ensure a random distribution of acetyl groups and hydroxyl groups on the resulting product. This is necessary to enhance the solubility in acetone, which is the solvent used for spinning. In the manufacture of triacetate the product is simply precipitated without hydrolysis.

Manufacturing Process: Although at least one firm has installed a continuous operation,* most manufacturers use a batch process. Until World War II, the pulp used was made from cotton linters, but currently most acetate is made from wood pulp. The producers of wood pulp now refine their product well, and this source is not subject to the price instability that characterized the linters-pulp product. In general, the wood pulp will be a high-alpha type ($>95\%$) which has been dried carefully. Overdrying leads to hard spots which are difficultly accessible to the acetylating agent, resulting in insoluble gel particles in the spinning dope.

The anhydride used in making cellulose acetate is produced by the reaction of ketene with acetic acid. Ketene in turn is made by the catalytic pyrolysis of either acetic acid or acetone. Since production of cellulose acetate is the major end use for acetic anhydride, most acetate producers in the United States manufacture their own anhydride. This also enables them to recycle the acetic acid produced in the acetylation reaction.

The process begins with the shredding and "fluffing" of the pulp, which is received in sheet or roll form. The acetylation reaction is mainly diffusion-controlled, and this shredding step ensures rapid penetration of the pulp by the reactants. The equipment used must not compact or squeeze the pulp fibers, or unacetylated fibers and gel particles will result. The shredded pulp is next treated with glacial acetic acid, which may contain part or all of the sulfuric acid used as catalyst. The purpose of this "pretreatment," or "activation," step is to swell the cellulose fibers and thus increase the accessibility to the acetylating agent. The ratio of acetic acid to cellulose is 1:1 to 3:1 in this step, and the time of treatment may vary from $\frac{1}{2}$ hr to several hours, depending

*Rhone-Poulenc Textile Cie.

on the temperature. This important pretreatment step may take place in a separate vessel or in the same vessel used for acetylation. Agitation is often employed to ensure the rapid and homogeneous penetration of the pulp by the acetic and sulfuric acids. If sulfuric acid is present, it will begin to react with cellulose and lower the initially high molecular weight. This degradation reaction continues throughout the acetylation and hydrolysis and is necessary in order to bring the molecular weight down so that a reasonable viscosity is obtained in the spinning solution.

The activated pulp next is treated with the acetylating mixture, which consists of acetic anhydride and acetic acid in about a 2:3 ratio and containing the remainder of the sulfuric acid catalyst (5–20%, based on the cellulose). Because the reaction is very exothermic and the temperature must be kept low ($<50°C$), the acetylating mixture is usually cooled to about $0°C$ before addition, and vigorous agitation and external cooling are required of the reaction vessel. The resulting mixture is about 10–14% cellulose and contains about 5% more anhydride than is required by the stoichiometry (including whatever water is initially present in the pulp).

The reaction is allowed to proceed under careful temperature control 5–10 hr, or until the pulp mass has completely dissolved in the acetylating mixture. At this point the cellulose has become completely acetylated and is in the form of triacetate. If triacetate is the desired product, the excess anhydride is destroyed with aqueous acetic acid, and the product precipitated, washed, and dried.

To make secondary acetate, the excess anhydride is carefully reacted with water in the form of aqueous acetic acid, keeping close control of temperature. When all the anhydride has reacted, excess water is added to the extent of 10–30% of the mixture, and the acetate allowed to hydrolyze for a period of time, until a DS of about 2.35–2.40 is reached. The time of hydrolysis depends strongly on temperature, being 24–60 hr at room temperature and 4–8 hr if the solution is heated to 70–80°C, which is the modern practice. During this time, the sulfo esters of the cellulose are also hydrolyzed, and sodium or magnesium acetate may be introduced with the hydrolysis water to neutralize the acid produced.

After hydrolysis, the acetate is precipitated by addition of more water. There are many variations in the methods used to add the water and effect precipitation. The object is to obtain the acetate as a porous, open "flake" rather than a powder or pellet-type material. This is to expose the maximum amount of surface to the subsequent washing step which removes the excess acetic acid. Washing is in countercurrent steps, and the wash liquor is distilled to recover acetic acid. The flake finally is dried in a conventional commercial dryer and conveyed to flake storage bins.

Fiber Spinning: Cellulose acetate is spun into fibers from an acetone solution, using the technique known as *dry spinning*. In this process, the solution of the polymer is forced through small holes in a metal plate into a long cabinet containing warm air which evaporates the solvent. The solidified filaments are oiled and wound up on a bobbin at the bottom of the spinning tube.

The process begins with the blending of several batches of flake to ensure a uniformity of chemical composition, particularly acetyl content, which is quite important for uniform dyeing properties of the spun yarn. The blended flake is mixed and dissolved in an acetone-water solution in large horizontal or vertical mixers. The mixing may be done as either a batch or continuous operation. The composition of the final solution is about 25% acetate, 72% acetone, and 3% water. The water is present as a "cosolvent" with the acetone; i.e., its presence leads to a marked reduction in the viscosity of the spinning solution, which is of the order of 900–1,000 P when solution is complete. During this mixing step TiO_2 to the extent of 1 or 2% may be added to the "dope" to act as a delustering agent. The major portion of acetate yarns spun today is the "dull," or delustered, type. Colored pigments may also be added to the dope at this stage to make "spun-dyed" yarns.

The viscous solution next is carefully filtered several times through large plate and frame presses containing wood pulp and cotton or rayon fabric as filter media. Since the holes in the spinnerets used in spinning acetate are quite small (0.0015–0.0025 in. diam), careful filtration is necessary to ensure continuity of spinning. After filtration the dope is transported to storage tanks to enable the air bubbles present in the liquid to spontaneously rise to the top of the dope and disperse.

The dope next is pumped to the headers of the spinning machines. Each machine has a hundred or more spinning positions where a single yarn of acetate fiber is being produced. The dope is circulated to a metering pump for accurate control of yarn denier, through a small filter and a heat exchanger which raises the temperature to 50–

70°C. This lowers the viscosity of the dope, and hence the pressure across the spinneret. A small circular filter is usually placed just behind the face of the spinneret to further remove any gel particles or dirt before the dope contacts the small holes.

The spinnerets, or *jets,* are made of stainless steel and contain 14–100 or more holes, depending on the yarn denier being spun. The holes are arranged in a circular pattern, are 0.0015–0.0025 in. in diameter, and are usually slightly shorter than their diameter.

After leaving the jet the nascent filaments traverse the vertical spinning tube or chamber, which may be 10–20 ft long and 6–12 in. in diameter and which contains hot air moving either cocurrent with or countercurrent to the path of the filaments. The hot air evaporates most of the acetone and thus solidifies the liquid stream into an acetate filament. The acetone-laden air is removed from the tube and passes to absorbers filled with activated charcoal which remove the acetone. The charcoal absorbers are steam-stripped, and the acetone recovered by distillation. Recovery is very efficient, 99% overall efficiency being common in the industry.

Leaving the spinning tube, the yarn is passed over an oiling device which coats the yarn with finish. This finish has, in general, two functions: to lubricate the yarn, and thus reduce friction in subsequent textile processing, and to prevent or diminish static electrification of the somewhat hydrophobic acetate yarns. Finish pickup from the oiler is about 2–3% by weight of the yarn. After finishing, the yarn passes to a rotating *godet,* or wheel, which supplies the force required to pull the yarn from the jet through the tube, and then to the take-up device which winds the yarn on a bobbin or tube.

Spinning speeds for acetate yarn range from 500 to 750 m/min. The major limitations on spinning speed is the speed with which the acetone can be removed from the yarn and keeping the acetone content of the emerging air below the lower explosive limit. Thus the speed depends on the denier of the yarn, air velocity and temperature, dope temperature and composition, and spinning-tube length.

Cellulose triacetate yarns are spun in essentially the same manner as secondary acetate, except that triacetate requires a different solvent. The solvent used is generally a 90:10 mixture of methylene chloride and methanol. Other than this, and minor variations in temperatures and yarn speeds, the fiber is spun substantially the same as secondary acetate.

The yarn, after spinning, is usually processed further before shipment to the customer. Some of it may be rewound on cardboard cones or tubes for use as filling or weft yarns in the shuttles of textile looms, while the remainder may be rewound on large metal spools, called *beams,* each of which has thousands of individual yarns wound side by side. The latter are used directly to form the warp used in a loom, or they may be used as the feed yarn for tricot knitting machines. The empty beams are shipped back to the yarn manufacturer.

Staple and tow are spun the same as continuous filament, but are not taken up on bobbins. Instead, a large number of yarns are collected in a bundle, or tow, which is then mechanically crimped and cut into short lengths (1.5–2.0 in.). Acetate and triacetate staple are shipped in bales, usually 400 lb.

Acetate-Yarn Properties: Commercial acetate and triacetate yarns are produced in a range of 45–900 deniers, the largest portion being in the range 55–150 (denier is a measure of the fineness of a yarn and is the weight in grams of 9,000 m of yarn). The number of filaments in these yarns runs from 11 to 240. The denier per filament ranges from 3 to 4, apparently because this range is the most efficient denier for economical operation of the dry-spinning process. The yarns are produced as "bright" yarns, "delustered" yarns containing 1–2% TiO_2, and a small quantity of "solution-dyed," or pigmented, yarns. Most of the filaments are produced with a cross section which is approximately circular, containing deep crenellations caused by the evaporation of solvent through the rigid skin which forms when the filaments are first contacted with warm air in the spinning tube. However, shaped cross sections are also available which serve to increase the bulk of fabrics without increasing the weight. For example, if the holes in the spinneret are triangular instead of circular, a Y-shaped cross section can be spun.

Thermal Properties: The two most important thermal properties of a textile fiber are its glass-transition temperature and its melting point. The former determines the temperature above which crystallization, and hence *heat setting,* can be induced, and the latter determines the maximum safe temperature to which the fabric can be exposed, as for example in ironing. Secondary acetate, because of the random substitution of the hydroxyl groups, shows very little crystallinity and cannot be induced to crystallize or heat-set. Triacetate, in contrast, is a stereospecific molecule and is capable of crystallizing, and will do so if heated above approximately 205°C for a short time. Herein lies the major difference between the two fibers. Triacetate fabrics, heated at 240°C for 30 s, in a state of tension, will tend to retain this flat geometry at all lower temperatures, and thus will resist wrinkling. Pleats and creases deliberately introduced into the fabric at this temperature will be retained during normal usage. This factor leads to the use of triacetate in the "easy-care" fabrics. This type of heat setting cannot be done with secondary acetate because of its inability to crystallize. On the negative side, heat-set triacetate fabrics tend to be somewhat stiffer, and hence show less drapability and a harsher hand than secondary acetates.

Although the melting point of triacetate fiber is about 300°C, it begins to soften at about 200°C, as does secondary acetate. For this reason, the safe ironing temperature of acetate usually is considered to be about 175°C, with heat-set triacetate fibers able to withstand ironing temperatures somewhat higher.

Physical Properties: The physical properties of a textile fiber depend on the molecular properties of the basic polymer and the degree of orientation and crystallinity of the fiber. The acetate molecule is relatively stiff and inflexible, which is responsible for the high softening temperature, as compared with molecules such as polyethylene. Key physical properties of acetate and triacetate fibers are shown in Table A-9. Tensile properties, as shown, are rather poor when compared with such fibers as nylon and polyester.

Chemical Properties: Acetate and triacetate fibers, being organic esters, are susceptible to hydrolysis in strong alkaline solutions. Weakly basic or acid solutions have little effect on these fibers. Strong mineral acids, in addition to causing hydrolysis, also attack the basic cellulose backbone, causing severe degradation. Both types of fiber are strongly resistant to chlorine bleach in moderate concentrations and are not affected by normal dry-cleaning solvents.

Dyeing: Both fibers normally are dyed with the class of dyestuffs known as *disperse dyes*. These materials are pigments which are dispersed in water before being applied to the fabric. In contact with the fabric, the dyes dissolve preferentially in the acetate yarns. Triacetate dyes at a significantly lower rate than secondary acetate, but bright colors and high depth of dyeing are possible on both types of fiber.

—James P. Dux, *American Viscose Division, FMC Corporation, Marcus Hook, Pa.*

Acetate Plastics   (See **Cellulose ester plastics, organic.**)

Acetic Acid   Pure acetic acid (glacial acetic acid) is a very hygroscopic, colorless, crystalline solid which melts at 16.60°C and boils at 118.1°C and has a density of 1.051 at 20°C. The material has a sour taste and a sharp, penetrating acidlike (strong-vinegar) odor. The compound can cause painful burns in contact with the skin. Chemically, acetic acid is a weak monobasic acid. The three hydrogen atoms linked directly to the carbon atom are not replaceable by metals. See also **Carboxylic acids.** Acetic acid is miscible in all proportions with H_2O, alcohol, and ether. Acetic acid dissolves sulfur and phosphorus and numerous organic compounds. In the chemical industry, acetic acid plays a major role as a solvent

$$
\begin{array}{ccc}
\text{H} & \text{O} \\
| & \| \\
\text{H}-\text{C}-\text{C}-\text{O}-\text{H} \\
| \\
\text{H}
\end{array}
$$

Reagent-grade glacial acetic acid assays at 99.7% CH_3COOH minimum with the following maximum allowable impurities: chloride, 0.0001%; heavy metals, such as Pb, 0.00005%; Fe, 0.00002%; reducing substances, such as SO_2, 0.015%; sulfate, 0.001%; and a residue after evaporation, 0.001%. U.S.P. grade glacial acetic acid contains 99.5% CH_3COOH.

Commercial aqueous acetic acid solutions are available in several concentrations and purities,

TABLE A-9. *Selected Physical Properties of Acetate Yarn*

Property	Secondary Acetate	Triacetate
Tenacity, g/denier:		
Conditioned	1.2–1.4	1.2–1.4
Wet	0.8–1.0	0.8–1.0
Breaking elongation, %		
Conditioned	25–45	25–45
Wet	35–50	35–50
Density, g/cm³	1.32	1.30
Percent moisture regain, 65% RH at 22°C	6–8	2–4

including 28, 36, 56, 70, 80, 85, and 90% CH₃COOH. Laboratory reagent aqueous acetic acid usually contains approximately 36% CH₃COOH.

Acetic acid occurs naturally in many fruit juices and in the stems and woody parts of plants. Improperly prepared or preserved alcoholic beverages often contain acetic acid as a degradation product.

Acetic acid is the principal ingredient of vinegar, which contains 2–10% or more of CH₃COOH. Often additional ingredients are incorporated in cooking and salad vinegars to impart desired flavors. Cider vinegar contains traces of malic acid. At one time small amounts of H_2SO_4 were added to vinegar as a preservative, but this practice no longer continues, the H_2SO_4 now being regarded as an adulterant.

In the production of naturally derived vinegars, several alcohol-bearing starting materials may be used. In the United States, apple cider predominates. Wine and spirit vinegars are most common in Europe. Malt vinegar is used extensively in Great Britain.

Reactions: Acetates are the salts of acetic acid, and all are soluble in H_2O, with the exception of Hg(ous) and Ag acetates. Many acetates are soluble in alcohol. Some basic acetates (Fe^{3+} and Al^{3+}) are insoluble in H_2O. Strong acids convert acetates into acetic acid. Acetic acid is used widely in the preparation of esters, such as ethyl acetate, $C_2H_3O_2 \cdot C_2H_5$, which can be prepared from ethyl alcohol and acetic acid. With phosphorus trichloride, PCl_3, acetic acid forms acetyl chloride, $CH_3 \cdot CO \cdot Cl$, a compound widely used to transfer the acetyl group $CH_3CO—$ in various syntheses. Acetic acid reacts with Cl_2 and Br_2 to form the corresponding mono-, di-, or trisubstitution products, as bromacetic acid or trichloracetic acid. In turn, hydroxy- and amino-aldehydic-dibasic acids can be made from the aforementioned substitution products. Acetamide is formed when ammonium acetate is distilled.

The ability of acetic acid as a solvent, plus the

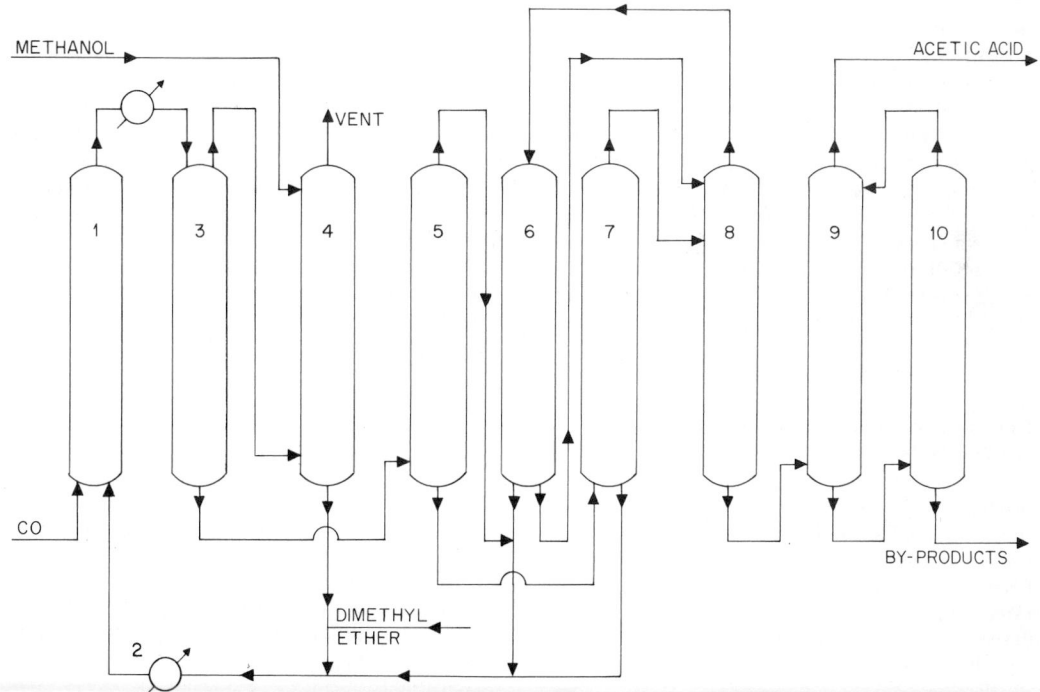

Fig. A-13. Production of acetic acid from methanol and carbon monoxide. Highly schematic representation of materials flow in a type of process designed by Badische Anilin- & Soda-Fabrik. (1) Reactor, (2) preheater, (3) purifier, (4) wash column, (5) degasser, (6) separator, (7) catalyzer separator, (8) dehydrator, (9) purifier, (10) finisher.

great reactive versatility and large number of derivatives only briefly described here, attests to the importance of this high-tonnage acid for hundreds of uses in the chemical industry. Applications include the production of synthetic fibers and resins, pharmaceuticals, flavorants, bleaches, preservatives, photographic chemicals, etching compounds, and numerous intermediates. See also **Petrochemical complex.**

Production: Earlier high-tonnage production processes included (1) the action of *Bacterium aceti* and air (oxidation-fermentation) on ethyl alcohol to produce dilute solutions of acetic acid and (2) the destructive distillation of wood. Aside from vinegar production, acetic acid of higher concentration for industrial purposes now is produced synthetically. The wood-distillation route has been phased out over a period of years. Two synthetic processes include the oxidation of acetaldehyde and the direct synthesis from methanol and carbon monoxide. A process of the latter type* is shown in Fig. A-13.

The charge stock is composed of methanol and CO, although some of the methanol may be replaced by dimethyl ether. The reaction takes place at about 650 atm in the liquid phase (water present) and at about 250°C. A dissolved cobaltous iodide is used as catalyst. About 530,000 kcal is released per ton of acetic acid made. This heat is absorbed by the cold feed and gas, and a small deficit of heat is made up by preheating the feed to a temperature of 40–80°C. Unreacted gas and the crude acid produced are taken from the top of the reactor, whereupon they are cooled and then reduced in pressure down to 10 atm. The vent gas is washed with the feed methanol so that methyl iodide in the unreacted gas may be recovered. The methanol is returned to feed, while the washed gas goes to the fuel supply. The reaction takes place in the following manner: $CH_3OH + CO \rightarrow CH_3COOH$; and $CH_3OCH_3 + H_2O + 2CO \rightarrow 2CH_3COOH$.

In a so-called *working-up* section, the by-products which have a higher boiling point than the acetic acid are separated from the acid. After degassing, the crude acid is subject to a first distillation where low-boiling components are separated. This is followed by a catalyst-separation operation. The crude acid, free of catalyst, is dehydrated and purified by means of azeotropic distillation. Acetic acid produced is claimed to be over 99.8% pure. The by-product mixture is

*Badische Anilin- & Soda-Fabrik AG, W. Reppe.

free of acetic acid. Two parts of propionic acid and two parts of compounds with boiling points higher than propionic acid are produced for each 100 parts of acetic acid made.

Acetic Anhydride (See **Acetate fibers;** and **Cellulose ester plastics, organic.**)

Acetone (See **Isopropyl alcohol; Ketones;** and **Petrochemical complex.**)

Acetonitrile (See **Acrylonitrile;** and **Isoprene.**)

Acetophenone (See **Ketones.**)

Acetylation (See **Acetate fibers;** and **Cellulose ester plastics, organic.**)

Acetylene Acetylene is a colorless gas of the formula $CH:CH$ and is the simplest of the acetylene series of hydrocarbons. The gas contains 92.3% carbon by weight and sometimes is referred to as *gaseous carbon.* Pure acetylene has a sweet odor, but is noxious when small amounts of impurities, such as H_2S, are present. The gas burns brightly in air, accounting for its use as an illuminant prior to electric lighting. Acetylene still is used in some forms of portable lanterns, wherein the gas is generated in situ from water and CaC_2. $CaC_2 + 2H_2O \rightarrow C_2H_2 + Ca(OH)_2$. Also identified as ethyne or ethine, acetylene has a formula weight of 26.04 and a specific gravity of 0.906 (air), liquefies at 1°C (48 atm), and boils at −84°C (760 mm). The gas generally is considered nontoxic and is relatively soluble in water, alcohol, and acetone (100 cm^3 in 100 g H_2O at 18°C; 600 cm^3 in 100 g C_2H_5OH at 18°C). Acetylene forms explosive mixtures with air, the maximum explosive effect occurring at 7.7% acetylene. The gas is marketed in transportable form in cylinders, where the gas is dissolved in acetone. One volume of acetone will dissolve 25 volumes of C_2H_2 at atmospheric pressure, or 250 volumes at 10 atm.

Although the relative importance of various organic chemicals vacillates because of technical and economic factors, the importance of acetylene as a basic raw material has diminished in recent years. The gas still is used as a building block for certain organics and, of course, is used for oxyacetylene welding. Extensive uses of C_2H_2 in synthesis have included halogen derivatives, acetaldehyde, acrylonitrile, and vinyl chloride. Much acrylonitrile is now prepared from propylene.

Acetylene is expensive compared with ethylene and propylene.

The early commercial process for manufacturing C_2H_2 was from CaC_2, as previously described. A large percentage of C_2H_2 now comes from hydrocarbon feedstocks, primarily natural gas.

Acetylene may be made commercially from the pyrolysis of naphtha. In a two-stage cracking process, the first stage generates an extremely hot energy carrier by virtually complete combustion of fuel and oxygen. In a second stage, vaporized naphtha is heated to the cracking temperature by intimate mixing with the gaseous heat carrier and immediately quenched to a lower level. The acetylene-ethylene ratio can be altered by varying the naphtha feed rate. Thermodynamically and economically, a 40:60 acetylene-ethylene ratio appears most satisfactory. Quenched cracked gases from the reactor contain some heavy hydrocarbons that are removed by fractionation. Successive absorption steps then extract pure C_2H_2 from the product gas. Potassium methyl taurinate solution has been used as the CO_2 absorbent. Cracked gas also contains C_3 and heavier hydrocarbons. These are mainly unsaturated compounds, such as higher acetylenes and dienes, quite susceptible to polymerization. Because of their reactivity and dangerous properties at high concentrations, they are not isolated, but are withdrawn by a naphtha scrubbing unit that operates at a low temperature. The C_2 hydrocarbons, along with CH_4, CO, and H_2, go into the overhead, while C_3 and heavier unsaturates are dissolved by incoming naphtha and fed back to the cracking reactors. From the naphtha scrubbers, cracked gas enters the absorber, where acetylene is extracted by means of acetone solution. Acetylene of over 99.9% purity, suitable for vinyl acetate manufacture, can be produced in this manner.[1]

A submerged-flame process also can be used for making C_2H_2 from crude oil. Crude oil is gasified by means of a flame supported by oxygen below the oil's surface. Reactions between hydrocarbons and O_2 occur in the submerged-flame burner. Once electrically ignited, the burner projects a flame into which O_2 and oil are fed. Combustion and cracking of the oil occurs at the boundaries of the flame. The resulting gases are quenched by surrounding oil. Hot cracked gases leave the top of the reactor at about 480°F (249°C) and 120 psi and pass through a mechanical scrubber, where light gasoline removes entrained soot and aerosols. The gasoline is separated in a subsequent cyclone. Part of it is passed through a cooler and reused in the scrubber, while the remainder is recycled to the reactor. Still containing some gasoline, the gases flow into a spray cooler that cools them while condensing the remaining gasoline. The latter is returned to the cyclone. Water used for spraying flows into a vessel, where it is separated from entrained gasoline. After cooling, cracked gas is separated in suitable combinations of gas scrubbers or by means of partial condensation at low temperatures. Degree of separation depends on the purity required of the acetylene and ethylene. Composition of the cracked gas (6.3% acetylene, 6.7% ethylene by volume) is virtually independent of the nature of the crude-oil feedstock. The gas contains traces[2] of carbon oxysulfide and H_2S.

REFERENCES
1. Kamptner, H. K., W. R. Krause, and H. P. Schilken: Acetylene from Naphtha Pyrolysis, *Chem. Eng.*, Feb. 28, 1966, p. 80.
2. Baur, K. G.: Acetylene from Crude Oil Makes Debut in Italy, *Chem. Eng.*, Feb. 10, 1969, p. 82.

Acetylenes   (See **Hydrocarbons.**)

Acid   In a very general way, an acid may be defined as a substance that yields hydrogen ions. Acids generally are quite active chemically (corrosive) and sour to the taste, and are electrolytes. Water also furnishes hydrogen ions, but in equal amounts with hydroxyl ions; hence pure water is a neutral substance rather than an acid or a base. The ionization of acids (and bases) is described under **pH (hydrogen-ion concentration).** Acidification is the operation of causing hydrogen ions to be in excess, usually accomplished by adding an acid to a neutral or alkaline material. Neutralization is the operation of creating a balance between the hydrogen and hydroxyl ions present. Thus an acid may be neutralized by adding a suitable basic substance in proper amounts. The result is a salt and water.

Inorganic acids, such as hydrochloric acid, nitric acid, and sulfuric acid, are high-tonnage industrial chemicals, each of which is described elsewhere in this volume. Consult the Index. In terms of traditional qualitative inorganic chemical analysis, the acid radicals include the arsenate, bromide, carbonate, chlorate, chloride, iodide, nitrate, oxalate, phosphate, sulfate, and sulfite radicals.

There are several classes of organic substances

that are termed acids, including (1) the *carboxylic acids* (many of which also are high-tonnage chemicals) which contain the COOH group found in *aliphatic acids*, such as the fatty acids (acetic acid, etc.), in carbocyclic or *aromatic acids*, such as benzoic acid, and in *heterocyclic acids*, such as pyromucic or furoic acids (see **Carboxylic acids**); (2) the *amino acids* (see **Amino acids**); and (3) the *nucleic acids* (see **Peptides and proteins**).

Acid Copper Lead   (See **Lead.**)

Acid Dyes   [See **Dyestuffs and dyeing, (textile).**]

Acoustic Collectors, Gas-Solids   (See **Separation operations.**)

Acoustic Materials   (See **Insulation, thermal;** and **Lead.**)

Acrolein   (See **Aldehydes;** and **Chlorine organics.**)

Acrylamide Polymers   During the 1940s, work in Germany and the United States led to the development of the commercially important monomer acrylonitrile. In the years 1950–1960, the amide of acrylonitrile, acrylamide or propenoic acid amide, was made available on a commercial scale, and intensive work was initiated on its conversion into polymers and copolymers. Many applications were first reported for these polymers, while use for the monomer as such was essentially nonexistent. In the last decade, use of both homopolymers and copolymers of acrylamide has shown significant growth, and these products have become commercially important.

Acrylamide is derived from acrylonitrile by a hydration reaction using acid or base catalysis. Commercial processes for its preparation employ H_2SO_4, and the acrylamide-sulfate salt initially formed may be isolated. Neutralization of aqueous solutions of this product with bases such as NH_3, NaOH, and lime yields acrylamide.

$$CH_2{=}CHCN + H_2O \xrightarrow{H_2SO_4}$$

$$CH_2{=}CHCONH_2 \cdot H_2SO_4 \xrightarrow{2NH_3}$$

$$CH_2{=}CHCONH_2 + (NH_4)_2SO_4$$

The acrylamide may be isolated from aqueous solution by concentration and crystallization.

A number of catalytic processes for the direct hydration of acrylonitrile over solid surfaces of metals, metallic salts, and oxides have been reported in the last five years.

Acrylamide is a white crystalline solid with a melting point of 84.5°C. It can be distilled or sublimed, although extreme care must be taken to assure that impurities are absent because polymerization, especially when initiated by free radicals, can proceed rapidly and often violently. The physical and chemical properties of acrylamide have been thoroughly covered in the chemical literature[1] and in commercial technical bulletins.[2]

Acrylamide Polymerization: Under controlled conditions, acrylamide may be polymerized by a large number of techniques. These include use of various forms of radiation, photopolymerization, and ultrasonic waves. Standard free-radical initiation of acrylamide in aqueous solution is most frequently used, and the catalysts employed include azo catalysts and many inorganic redox couples. Polymerization of acrylamide may also be effected by ionizing radiations in the solid state. In contrast to free-radical initiation, a number of strong base catalysts, e.g., alcoholates, effect formation of a poly-β-alanine (nylon 3) rather than a vinyl polymer.

The kinetics of free-radical polymerization have been investigated by a number of workers. The rate constant for propagation has been reported to be 1.8×10^4 l/mole-s, and that for termination 1.45×10^7 l/mole-s, indicating an exceptionally high overall rate of polymerization $R_p/R_t^{1/2}$. Radical chain transfer to monomer and to backbone is generally very slow, particularly at moderate temperatures (below 60°C). A consequence of this is that polyacrylamide may be "grown" to extremely high molecular size, and molecular weights above 10 million are frequently obtained. The heat of reaction of acrylamide polymer formation by free-radical mechanism is 19.5 kcal/mole.

Acrylamide copolymerizes readily with a large number of polar vinyl monomers. It generally exhibits high Q and e values (the latter positive) in the Alfrey-Price scheme, indicating significant resonance-stabilization factors and an electropositive double bond by virtue of a strong inductive effect into the amide group. Reactivity ratios and also chain-transfer constants with a variety of organic compounds in acrylamide polymerization have been tabulated in recent literature.

Polyacrylamide: The vinyl polymer of acrylamide is a white solid with a high glass-transition temperature of 165°C and a softening temperature above 200°C. It is very soluble in water, but insoluble in most organic solvents. Homopolymers of acrylamide exhibit a polyelectrolyte effect, perhaps due to the presence of small amounts of acrylic acid often present in the monomer, or to hydrolysis, which can occur during polymer formation. Nevertheless, the polymers show an exceptionally high salt tolerance. Compatibility with a number of other ionic water-soluble polymers is also common, in contrast to the general rules of polymer incompatibility which apply to nonionic organosoluble polymers.

Chemistry of Polyacrylamides: Polyacrylamides undergo all the typical reactions of simple aliphatic amides. These reactions are conveniently run in aqueous medium. One of the most commercially important is hydrolysis, which leads to a carboxylated species useful in a number of applications. The hydrolysis is conveniently accomplished in alkaline medium, and the neighboring group effects which can occur during such a reaction are unique to the polymeric structure.

Treatment of polyacrylamide with strong acids results in imide formation; in concentrated solutions intermolecular imidization leads to crosslinking.

Reactions with aldehydes are also widely employed. Formaldehyde is most often used, providing methylol derivatives. These have a number of uses besides being intermediates for formation of cationic derivatives (by reaction with secondary amines) and anionic polymers (by further condensation with sodium bisulfite). Polyacrylamides may also be condensed with glyoxal to give thermosetting species which have become articles of commerce.

Halogenation of polyacrylamide has been reported but is less important. Chlorination or bromination under alkaline conditions leads to a Hofmann rearrangement, and a polyvinyl amine is formed. Most often hydrolysis is concurrent, and a polyampholyte is obtained. Other reactions reported are hydrogenolysis and cross-linking via free-radical mechanisms, or by heavy-metal ions such as chromium salts.

Industrial Uses of Polyacrylamide: Polyacrylamides almost always are employed in aqueous solution; fabrication of the bulk polymers into fibers, films, sheets, or molded products has not achieved commercial importance. Solutions of very high molecular weight polymers tend to decrease in viscosity on aging; various explanations have been advanced,[3] including disentanglement,

physical (shear) degradation, and chemical degradation.

The most important applications are in flocculation, wherein suspended matter is either to be removed or concentrated in aqueous system. Two mechanisms of action have been reported during flocculation: a charge neutralization, and a bridging effect between separate particles of substrate which results in coagulation.

The more important uses of polyacrylamides are given in Table A-10.

Paper Industry Applications: Polyacrylamides have been applied to various types of cellulose fibers in the process of papermaking. On-machine application results in a web which, when formed and dried into a paper sheet, has improved internal strength. The polymer usually is synthesized with a positive or negative charge so that adsorption to cellulose fibers during beater addition is efficient. Alum frequently is used to promote attachment of the anionic derivatives to the negative cellulose. Polyvinylamides with thermosetting characteristics, by virtue of appended aldehyde functions, provide a paper which has wet as well as dry strength.

Application in the paper industry also includes a number of examples of flocculation. The result of bridging of the fibers themselves is a "freeing" of the furnish and improved drainage results. Often clays and other pigments are flocculated, and their loss into the circulating white-water system is significantly reduced. The polymers also may be used as effective aids in save-all flotation.

Water and Waste Treatment: In recent years, polymers of acrylamide and anionic and cationic copolymers have found uses in water and waste treatment, especially in the control of pollution. The polymers may be used as primary flocculants or as coagulant aids. They also are used effectively as sewage-sludge dewatering agents. Combinations of hydrolyzed polymers with cationic bridging agents, such as alum or lower-molecular-weight polyamines, sometimes are used.

Mining Chemicals: Two general types of operation include thickening of mineral concentrates and the flocculation of tailings or slimes for pollution prevention or water reuse. Optimum levels of polymer must be applied, ranging from 0.01 to 0.1 lb/ton of suspended solids. Overuse of ionic derivatives of polyacrylamides will result in charge reversal and at least partial dispersion.

Oil Production: In addition to use as flocculants, polyacrylamides are used in oil production as flooding agents for secondary-oil recovery, as drill-

TABLE A-10. *Uses of Polyacrylamides*

Application	Molecular Features, Size, and Charge	Function
Dry-strength additives for paper . .	Copolymers containing charged comonomer, moderate molecular weight	Internal hydrogen bonding with cellulose
Wet-strength additives for paper . .	Aldehyde-modified copolymers, ionic, low to moderate molecular weight	Thermosetting through aldehyde functionality
Retention and drainage aids for fibers	Anionic or cationic, high-molecular-weight copolymers	Flocculation and bridging of particles to fibers
Mineral flocculants	High-molecular-weight homopolymers, cationic and anionic copolymers	Segregation of ore values and settling of tailings and fines
Influent-plant-water coagulant aids (various industries)	High-molecular-weight anionic copolymers and homopolymers	Settling of suspended matter
Waste-water-treatment agents . . .	Anionic or cationic high-molecular-weight copolymers	Flocculation of waste particles
Sewage-sludge-dewatering agents .	Cationic, high-molecular-weight copolymers	To increase sludge solids
Flooding agents in oil recovery . . .	Anionic copolymers, high molecular weight	To decrease oil saturation
Drilling-mud additives	Anionic copolymers of moderate molecular weight	To control fluid loss in drilling
Friction reducers	Very high molecular weight homopolymers and anionic copolymers	To reduce power requirements in fracturing

ing-mud additives, and as friction reducers in hydraulic fracturing. As drilling-mud additives, the polyacrylamides help to reduce fluid loss.

Other Uses: Acrylamide is widely used in polymers in which it is a minor but important component, as in emulsion polymer technology. The latices may be used for formulation into coating compositions, including baking enamels, and as pigment binders and textile-treating agents. The acrylamide, which often is used at levels of 1 to 3 mole %, offers improved adhesion, stability, and oil and solvent resistance and may act as a crosslinking site for additional upgrading.

Homopolymers of acrylamide find lesser use as intermediates for dispersants and thickeners and act as protective colloids. As engineering chemicals, polyacrylamides cross-linked in situ with agents, such as methylenebisacrylamide, have been used effectively in grouting operations to prevent water loss. Polyacrylamide which has been cross-linked with varying amounts of methylenebisacrylamide also is used as a substrate in the analytical techniques of gel-permeation chromatography and gel electrophoresis. A wide range of miscellaneous uses has been cataloged by Thomas.[1]

REFERENCES

1. Thomas, W.M.: Acrylamide Polymers, "Encyclopedia of Polymer Science and Technology," vol. 1, p. 177, Interscience, New York, 1964.
2. Chemistry of Acrylamide, Technical Bulletin, American Cyanamid Co., Wayne, N.J., 1969.
3. Shyluk, W. P., and F. S. Stow: *J. Appl. Polymer Sci.*, vol. 13, p. 1023, 1969.

—Anthony T. Coscia, *Industrial Chemicals Division, American Cyanamid Company, Stamford, Conn.*

Acrylates   (See **Hydroxyalkyl acrylates and methacrylates.**)

Acrylic Fibers   An acrylic fiber is defined by the U.S. Federal Trade Commission* as one "in which the fiber-forming substance is any long-chain synthetic polymer composed of at least 85 percent by weight of acrylonitrile units ($-CH_2-C-$)."

$$CN$$

When a fiber is "composed of less than 85 percent but at least 35 percent by weight of acrylo-

Rules and Regulations under the Textile Fiber Products Identification Act, effective Mar. 3, 1960, p. 4, U.S. Federal Trade Commission, Washington, D.C., 1960.

nitrile units, it is properly known as a modacrylic fiber."

Acrylic fibers are known for their aesthetic properties: a soft warm hand, reminiscent of fine wool, and supple drape. They can be dyed to clear bright colors or muted somber hues with excellent fastness properties. Fabrics made from acrylic fibers have excellent dimensional stability and warmth. Acrylic fibers excel in many properties, particularly in their resistance to weathering, but it is their exceptionally good *balance* of properties that has led to a steady broadening of their utility.

Acrylonitrile polymers have been known in the German patent literature since the late 1920s, but because of their instability at the melting point and the lack of suitable solvents, their conversion into synthetic fibers was not possible by either melt-spinning or solution-spinning routes. In 1938, H. Rein of I. G. Farbenindustrie described fibers obtained from polymer dissolved in aqueous solutions of quaternary ammonium compounds, such as benzylpyridinium chloride,* or of metal salts, such as lithium bromide, sodium thiocyanate, or aluminum perchlorate.† Du Pont, after studying the suitability of many solvents,‡ selected dimethyl formamide and started developing "fiber A" in the early 1940s.

The first commercial acrylic fiber was produced by Du Pont in 1949.§ Since then, acrylic fiber production has experienced a very rapid growth.¶ World production of acrylic fibers (1972) was approximately 2.8 billion lb, including the modacrylic fibers. At present, acrylics account for more than 80% of the total, but modacrylics are increasing in importance as special properties, such as decreased flammability, are built into the fibers. Acrylic fibers are used extensively in carpets, circular-knit apparel, pile and fleece fabrics, sweaters, and half hose.

Chemical Composition: Acrylonitrile, the chief raw material for acrylic fibers, is a colorless liquid which is easily handled in most standard unit processes. See also **Acrylonitrile.**

*United States Patent 2,117,210, May 10, 1938, H. Rein (to I. G. Farbenindustrie).

†United States Patent 2,140,921, Dec. 20, 1938, H. Rein (to I. G. Farbenindustrie).

‡R. C. Houtz, *Textile Res. J.,* vol. 20, pp. 786 and 801, 1950.

§Under the name of Orlon.

¶Monsanto started producing Acrilan fibers in 1952; Dow introduced Zefran fibers in 1958; and American Cyanamid introduced Creslan fibers in 1959.

Polymerization of acrylonitrile proceeds by the successive addition of acrylonitrile monomer units, resulting in large linear molecules:

$$n(CH_2=CHCN) \rightarrow \left(\begin{array}{c} -CH_2CH- \\ | \\ CN \end{array} \right)_n$$

where n may range from 500 to 5,000. For commercial fibers, n is normally in the range 600–2,000, equivalent to molecular weights of 32,000–110,000. Commercially, polymerization is carried out in the presence of redox catalyst systems.

Fibers of 100 percent polyacrylonitrile (homopolymers) are rarely used commercially because of difficulty in dyeing, but do have industrial application due to their excellent chemical and weathering resistance. Consequently, acrylonitrile is usually copolymerized with at least one other monomer. The resulting more open structure aids dye penetration of the fiber. Alternatively, other constituents (other polymers or copolymers) may be incorporated by physical blending with polyacrylonitrile at a point before spinning. Dye sites for dyestuffs may be introduced either by additives or by the generation of residual end groups by means of the redox catalyst system used.

Commercially important comonomers include methyl acrylate, methyl methacrylate, vinyl acetate, and vinyl benzene methylmethacrylate. Terpolymers and tetrapolymers based on these and other comonomers, plus other compounds, particularly those containing dye-receptive groups, also are commercially important. See also **Hydroxyalkyl acrylates and methacrylates;** and **Polymerization.**

Manufacture: The properties of the fibers themselves are largely dependent upon several factors: (1) the inherent structure of polyacrylonitrile, (2) the kind and amount of modifier used, (3) the spinning method, and (4) the degree of stretching during fiber formation. The last two factors affect the molecular orientation and crystallinity of the fiber to a large degree.

A flow diagram (Fig. A-14) shows the main steps in the manufacture of acrylic fibers: (1) polymerizing the acrylonitrile, (2) dissolving the polymer, (3) spinning, and (4) aftertreating. Acrylic fibers are produced by wet- or dry-spinning methods. Melt spinning is not possible because of polymer degradation at temperatures below or at the melting point. Both wet and dry spinning require that the polymer be dissolved in a suitable solvent to form a viscous solution that is forced through a spinnerette. The spinnerette

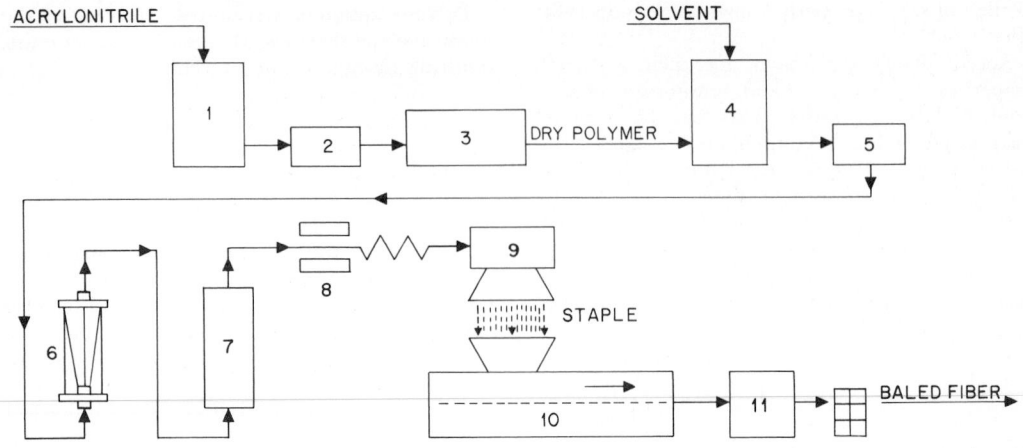

Fig. A-14. Schematic representation of materials flow in a representative acrylic fiber production process. (1) Polymerization, (2) filtration, (3) drying, (4) dissolving, (5) filtration, (6) spinning, (7) fiber processing, (8) crimping, (9) cutting, (10) drying, (11) fiber baling.

is a metal plate or dish perforated with holes of approximately 0.003 in. diam.

Upon extrusion of the fiber from the spinnerette hole, solvent is removed from the plastic mass, thereby regenerating the acrylic polymer in filamentary form. If solvent is removed by hot gases, the system is called dry spinning. If the solvent is leached out by another liquid, the process is called wet spinning. In either case, the extruded fibers must be drawn (extended) to impart satisfactory strength and elongation, and then crimped to make processing on textile staple equipment possible.

Acrylic fibers usually are manufactured in the form of staple and tow. Staple length varies depending upon the type of yarn spinning system used to convert the fiber into spun yarn. In the United States, lengths range from $1\frac{1}{8}$ in. for the cotton system to 6 in. for the worsted system. A small amount of continuous-filament yarn is produced in Europe and Japan.

Fiber fineness, reported by denier (weight in grams of 9,000 m of fiber) varies commercially from the apparel range (1–6) up to the carpet range (8–20). Microfibers (less than 0.75 denier per filament) also have been produced.

Properties: In general, all acrylics are resistant to ordinary chemicals, but are degraded readily by hot, concentrated alkalies. The acrylics, as a class, are very resistant to light, insects, and to microbiological attack. Moth and carpet beetle larvae, for example, have virtually no effect on acrylic fibers. The fibers are not weakened by

mold or mildew and generally are considered nonallergenic.

Important physical properties of acrylic fibers include (1) *specific gravity,* 1.16–1.18; (2) *tensile strength,* 32,000–54,000 psi; (3) *breaking tenacity,* dry, 2.0–3.6 g/denier, wet, 1.6–2.9 g/denier; (4) *breaking elongation,* dry, 20–50%, wet, 26–60%; (5) *elastic recovery,* 99% after 2% stretch and 89% after 5% stretch; and (6) *water absorbency,* 1–2.5% (70°F; 21°C; 65% RH) and 2–5% (95% RH).

The homopolymers of acrylonitrile have an unusual degree of chemical resistance, but are difficult to dye. The homopolymers also have higher strength (about 4 g/denier). For these reasons, the homopolymers are quite useful in certain industrial applications. The use of comonomers, which provide the needed dye sites, tends to interfere with orientation during stretching and thereby lowers the strength of the fine textile deniers to about 3 g/denier and the heavier carpet deniers to about 2 g/denier.

The cross section of dry-spun fiber usually is dog-bone in shape (Fig. A-15). In wet spinning, coagulation results in a round or bean-shaped cross section (Fig. A-16). Modified cross sections are obtainable by either dry or wet spinning. Particles of TiO_2, used as a delustrant, can be observed in many acrylic fibers, although the fiber also is used in bright form without a delustrant.

Dyeing: The earlier acrylic fibers, particularly those containing greater than 95% acrylonitrile, posed a rather difficult dyeing problem, but fiber modification and the development of new dye-

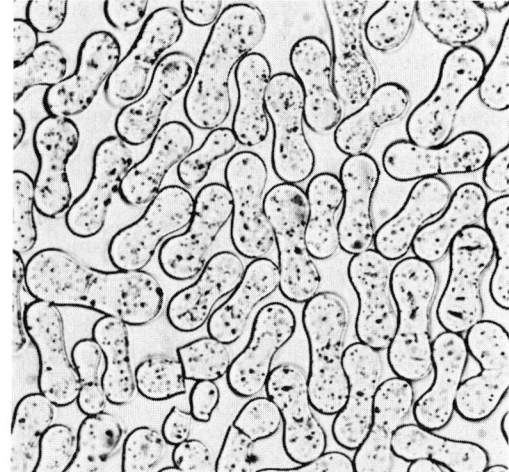

Fig. A-15. Dry-spun, semidull acrylic fibers possess the characteristic dog-bone shape and clearly show the particles used as a delustrant. (*Dralon acrylic fiber made by Farbenfabriken Bayer AG.*)

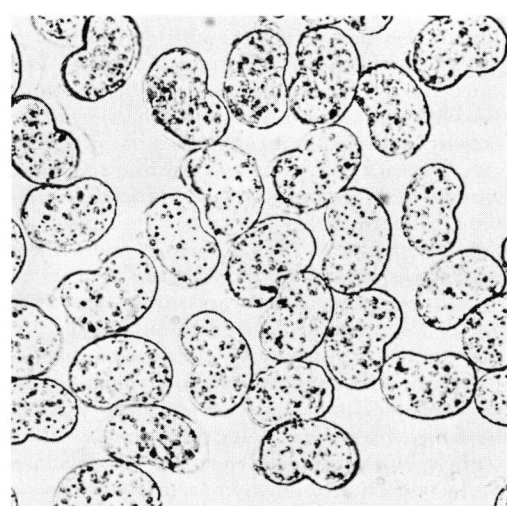

Fig. A-16. Bean-shaped cross section that is typical of wet-spun acrylic fibers. Round examples also are quite common. (*Five-denier, type* S-16 *Acrilan acrylic fiber made by Monsanto Company.*)

stuffs now permit fully satisfactory dyeing for virtually every end use at reasonable costs. Acid and neutral premetalized and chrome dyes applied by the cuprous-ion method were widely used at first. Early *cationic dyes* (formerly called *basic dyes*) gave generally inadequate lightfastness, but certain selected varieties now resist fading. Fastness to various cleaning methods, alkaline and acid perspiration, ozone, and other similar consumer requirements also can be controlled through dyestuff selection. Disperse dyes for acrylics are being made to higher fastness standards and are fully adequate in light to medium shades for many end uses.

Acrylics are stock-, yarn-, or piece-dyed. Most textile products take the piece or yarn route, whereas most carpets take the stock or yarn routes. Acrylics also may be dyed under pressure in stock, yarn, or piece form for shorter dye cycles and better leveling.

Acrylic fibers which will accept acid dyes and resist cationic dyes are made by most producers to permit cross-dyeing. In cross-dyeing, cationic and acid dyes are combined in a single bath (set at the proper pH with suitable auxiliaries) along with fabric specially prepared from regular and acid-dyeable acrylic fibers in heather or pattern styles. In addition to two-color combinations, color-and-white effects are also quite practical in both textile and carpet products. Several suppliers make producer-colored acrylic fibers in dope-pigmented or dyed varieties. Dope-pigmented types combine colorfastness to light with superior resistance to degradation by sunlight.

Fiber Modifications: Certain properties can be built into or excluded from basic forms of the fiber. Some properties that are manipulated include staple length, residual shrinkage, elongation to break, strength, hand, chemical resistance, and several other factors important to a particular end use. Staple length varies to suit the yarn spinning system chosen, variable cuts being produced for special uses, such as blankets, carpets, pile, and fleece.

High-Bulk Fibers: Tow is made in two-stretch types for direct tow-to-sliver conversion which produces yarns of unusual softness, bulk, and cover, by what is known as a *high-bulking process.* It is well known that for every high polymer there is a narrow temperature region at which the polymer characteristically changes from a more or less highly resilient state to a viscoelastic state. If acrylic fiber is heated to 140°F (60°C) or above, depending upon the individual fiber, stretched 10 to 20%, and then cooled under tension, the fiber is put into a metastable state. In this form, the fiber is called "hibulk" and is stable at ordinary temperatures and humidities for a long time. This modified fiber may be processed either alone or, preferably, in blends with regular acrylic or other fibers. When the hibulk fiber in the blended yarn is relaxed in a high-temperature dyebath or by steaming, it shrinks to its original length and thus shortens the entire yarn. This causes the non-shrinking fibers to pucker. The unusual softness, lightness, pleasing appearance, and high cover thus obtained are important to certain styles of garments.

Bicomponent Acrylic Fibers: Introduced in 1959,* these fibers are increasing in commercial importance. In one common type of bicomponent acrylic, a homopolymer and a copolymer are brought together as separate entities, the two sides being joined together along the entire length of the fiber. When such a fiber shrinks, it develops a helical twist because the two sides shrink by a

*By Du Pont.

Fig. A-17. A bicomponent acrylic carpet fiber after crimp development. Note how the helix direction reverses. Reversals of this type add to carpet loft, bulk, and resilience. The mechanism of crimp development is similar to that which causes curvature in a bimetallic thermostat. Due to large recovery forces, helical bicomponent crimp is more energetic in resisting and recovering from deformation than the planar crimp that is mechanically induced in monocomponent fibers. (*Monsanto Company.*)

different amount. Skin-core, biconstituent, and other types of two-element fibers also are known. When the two elements belong to different fiber classes, the product must be designated a biconstituent fiber, according to a Federal Trade Commission regulation. Bicomponent fibers have improved bulk, cover, and resilience, generally because of built-in, permanent crimp (Fig. A-17). The newer bicomponent acrylic fibers, produced by either wet or dry spinning, have cross sections ranging from bean to popcorn to worm shapes.

Flame-retardant Fibers: Although the flammability of acrylics has been considered acceptable, being generally classified with the cellulosics, new evolving user needs, led by government standards set for carpets and children's sleepwear, have led to the development of flame-retardant acrylics and acryliclike modacrylics. These newer forms are now in general use.

—John Lomartire, *Monsanto Company, New York,*
and Frederic C. Flindt, *Monsanto Company,*
Decatur, Ala.

Acrylic Paints   (See **Paints.**)

Acrylic Plastics   Acrylics are produced from methyl methacrylate monomers. See also **Hydroxyalkyl acrylates and methacrylates.** Major product groupings include cast sheet, molding powder, and high-impact molding powder. Cast acrylic sheet is stable, formable, transparent, and strong. It is used for internally lighted outdoor signs, architectural panels, aircraft glazing, skylights, and product prototypes and models. The regular acrylic molding powders are used in the mass production of intricate shapes, including automotive lights, lenses for lighting fixtures, and dials and control panels for automobiles and appliances. Although less transparent than regular acrylics, high-impact acrylic molding powders have unusual toughness, and applications include outboard-motor shrouds, housings and containers, nameplates, toys, business-machine components, and blow-molded bottles. Acrylic resins are widely used in coatings. See also **Paints.**

Typical properties of the acrylic plastics are given in Table A-11.

Acrylonitrile   Sometimes referred to as *vinyl cyanide* or *propene nitrile*, acrylonitrile, $CH_2 : CHCN$, is an important intermediate in the manufacture of acrylic fibers, acrylonitrile-based plastics, nitrile rubbers, insecticides, and other products of organic synthesis. Acrylonitrile is a liquid with a

TABLE A-11. *Typical Properties of Representative Acrylic Plastics*

		Molded Parts		
	Cast Sheet	Standard	High-impact	High-temperature
Tensile strength, psi	10,500	10,500	5,600	10,000
Elongation, %	5	6	38	3
Modulus tension, psi	450,000	430,000	225,000	470,000
Compressive strength, psi	18,000	17,000	7,300	18,000
Flexural strength, psi	16,000	16,000	8,500	13,000
Impact strength:				
Charpy unnotched, ft-lb, $\frac{1}{2}$- $\times$ 1-in. section	7.0	7.5	35	6.2
Izod notched, ft-lb/in. of notch	0.4	0.4	2.3	0.3
Hardness, Rockwell, $\frac{1}{4}$-in. ball, 100-kg load	M93	M92	M20	M100
Specific gravity	1.19	1.19	1.11	1.16
Water absorption, % (24-hr immersion) . . .	0.2	0.3	0.3	0.2
Heat-distortion temperature:				
66 psi, °F	225	214	194	239
°C	107	101	90	115
264 psi, °F	205	198	169	221
°C	96	92	76	105
Refractive index (ASTM D 542)	1.49	1.49		
Transmittance, 0.125 in., %	93	93	25	95
Dielectric constant, 60 Hz	3.7	3.7	3.5	

Commercial names for some acrylic plastics include Gafite, Lucite, Plexiglas, Vernonite, and Zerlon.

boiling point of 78°C. Production of acrylonitrile (United States) is well in excess of $\frac{1}{2}$ million tons/yr. See also **Acrylic fibers; Acrylonitrile-butadiene-styrene resins; Elastomers;** and **Petrochemical complex.**

A number of processes have appeared since the early 1960s, most of which are based upon the *ammonoxidation* (sometimes called *oxyamination*) of propylene with NH_3 and air in the manner of

$$CH_3CH\!:\!CH_2 + NH_3 + 1\frac{1}{2}O_2 \rightarrow$$
Propylene

$$CH_2\!:\!CHCN + 3H_2O$$
Acrylonitrile

One process uses a supported catalytic system containing P, Mo, and Bi and carries out the reaction in a fluidized-bed reactor. Another process involved the one-stage oxidation of propylene to acrolein over a catalytic system of Sn and Mo oxides which later was modified to yield acrylonitrile from propylene, NH_3, and O_2 over a catalyst containing Sn and Sb, or Mo and Co. In the SNAM (Italy) process, high-purity acrylonitrile (fiber grade) is made from unrefined propylene feedstocks, using a catalyst in the form of vanadium-bismuth molybdate. The reaction takes

place at 750–1000°F (400–538°C) and atmospheric pressure with NH_3/propylene feed ratios of about 1:1 and O_2/propylene around 1.5:1. NH_3, propylene, and air are initially mixed with steam, and this gaseous mixture is preheated and fed to the reactor, where the heat developed is used for steam generation. The two main byproducts are acetonitrile ($CH_3 \cdot CN$) and HCN, with the formation of small amounts of acrolein, acetone, and acetaldehyde. After heat exchange with the incoming feed stream, acrylonitrile and other reaction products are sent to a neutralizer where unreacted NH_3 is absorbed by a countercurrent flow of aqueous H_2SO_4 and discharged as $(NH_4)_2SO_4$ solution from the bottom. The overhead stream passes to a second absorber that eliminates the stream of inerts, such as CO_2, CO, and N_2. The liquid phase that collects at the bottom of this tower is preheated and sent to a stripping column. A portion of the absorption water flowing to the bottom of the stripper is partially recycled, while the overhead product stream, after condensation, feeds a column where HCN is separated. An extractive distillation step follows wherein water is used as an extraction solvent to remove acetonitrile from the process stream. The overhead from this column (acrylo-nitrile, light products, and water) is condensed and fed to a dehydration column and then to a set of purifiers for separation of pure acrylonitrile from light and heavy boilers, such as acetone, acetaldehyde, and some nitrile polymers. Bottoms from the extractive distillation column contain the water solvent and acetonitrile. The latter is sepa-rated and recovered in a two-column system. It is estimated that the raw-material requirements per metric ton of polymerization-grade acrylo-nitrile are 1.15 tons of 100% propylene and 0.50 ton of 100% NH_3. The by-products recovery reaches 110 lb of 99% pure acetonitrile and 110 lb of 98% pure HCN per metric ton of acrylo-nitrile produced. In the process* shown in Fig. A-18, propylene, anhydrous fertilizer grade NH_3, and air are introduced into a fluid-bed catalytic reactor operating at 5–30 psig and 750–950°F (400–510°C). The reactor effluent is scrubbed in a countercurrent absorber, and the organic mate-rials are recovered from the absorber water by distillation. HCN, H_2O, light ends, and high-boiling impurities are removed from the crude acrylonitrile by fractionation to produce specifica-

*A process of The Standard Oil Company (Ohio), Cleveland, Ohio—the Sohio acrylonitrile process. Plants of this type with capacities up to 250 million lb/yr have been constructed.

tion acrylonitrile product. With the once-through fluid-bed reactor, no separation or recycling of unreacted raw materials is necessary. The original catalyst was based principally on Mo and Bi. A new, proprietary catalyst is claimed to produce relatively more acrylonitrile and less by-product acetonitrile. Yields in excess of 0.80 lb acrylo-nitrile/lb propylene feed are achieved. About 0.15–0.20 lb of by-product HCN can be recovered per pound of acrylonitrile.

Acrylonitrile-Butadiene Rubber   (See **Elastomers.**)

Acrylonitrile-Butadiene-Styrene Resins   These thermoplastic resins, commonly referred to as *ABS plastics,* are manufactured by grafting styrene and acrylonitrile onto a diene-rubber backbone. The diene rubber used normally is PBR (polybuta-diene rubber), SBR (styrene-butadiene rubber), or NBR (acrylonitrile-butadiene rubber). See also **Elastomers.** Polybutadiene most often is the pre-ferred substrate because of its low glass-transition temperature, which is just above −80°C. The diene usually is a high-gel or cross-linked latex made by a "hot"-emulsion polymerization formu-lation, although stereospecific diene rubber made by solution polymerization is preferred for ABS resin made by suspension or mass-polymerization processes.

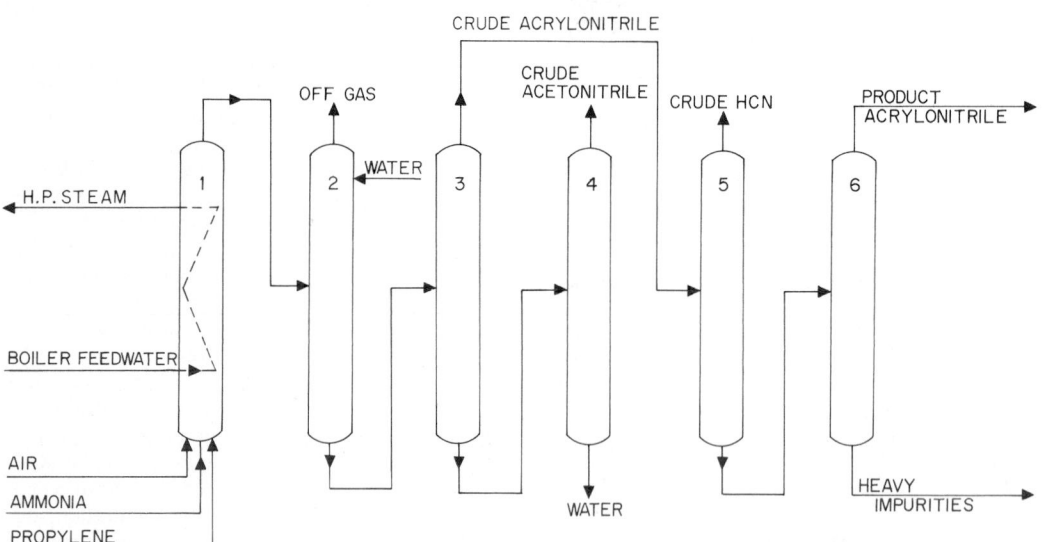

Fig. A-18. Acrylonitrile production process. (1) Reactor, (2) absorber, (3) acrylonitrile recovery column, (4) acetonitrile recovery column, (5) lights column, (6) product column. [*The Standard Oil Company (Ohio).*]

ABS resins offer a good balance of impact resistance, tensile strength, hardness, and elastic modulus properties over a range of temperatures -40 to $107°C$ (-40 to $225°F$). The high gloss, chemical resistance, and nonstaining characteristics of ABS further enhance its appeal. ABS exhibits good dimensional stability with excellent creep resistance and low water absorption or volume change at varying humidities. The material is available in the form of custom color-matched compounded pellets, or as a granular resin for compounding or alloying with other plastics, such as polyvinyl chloride (PVC). The specific gravity of ABS pellets ranges from 1.03 to 1.13, depending upon the pigmentation. Bulk densities of the ABS blending resins used in modifying PVC range from 15.0 to 21.0 lb/ft^3, with the particular size below 100 mesh.

ABS plastics can be processed into finished parts by almost any of the standard thermoplastic converting processes. Typical parts produced by injection molding include housings for radio, TV, appliances, automobile parts, plumbing fixtures and fittings, telephone sets, and refrigerator parts. Extrusion of ABS into sheet, pipe, and profile is common. Sheet up to 12 ft wide can be extruded easily. Parts with large surface areas and deep draws can be thermoformed from sheet stock. Vehicle bodies, lawn-mower housings, refrigerator liners, pipe and conduit, snowmobile shrouds, and camper bodies are typical applications involving extrusion and subsequent thermoforming. All the basic thermoforming techniques are applicable to ABS, including vacuum forming, drape forming, vacuum-plug forming, plug and air assist, and vacuum snapback.

ABS parts can be fabricated by blow molding, cold stamping of thin sheet, calendering, machining, and compression molding. Many methods of postdecoration also are applicable, including painting, vacuum metalizing or chrome plating, embossing, laminating, polishing, silk screening, hot stamping, and offset printing.

Due to the good compatibility of ABS with other plastics, these resins are used as impact modifiers and as processing additives with many other polymers. ABS resins make efficient modifiers of flexible and rigid PVC compounds, such as flexible automotive-instrument panel covers and clear, rigid, blow-molded PVC bottles. With many degrees of freedom, the structure and property relationship for an amorphous terpolymer, such as ABS, can be tailored to meet numerous specific needs. Constituents can be altered, as, for example, the substitution of α-methyl styrene for styrene to increase heat-distortion temperatures or methacrylonitrile for acrylonitrile to improve barrier properties to gases like CO_2 in carbonated-beverage containers.

Over 75 grades of ABS now are available. Self-extinguishing, electroplating, antistatic-expandable, glass-reinforced, high-heat, cold-forming, and low-gloss sheet grades are available, as well as alloys of ABS/PVC, ABS/PC (polycarbonate), and ABS/PU (polyurethane). The medium-high-heat products are used in automotive components requiring as much as $17°C$ ($30°F$) higher temperature resistance than the general-purpose ABS at $88°C$ ($190°F$). The high-impact ABS is used in the manufacture of pipe and fittings, underground conduit, and drainage pipe, as well as sheet products used in large vacuum-formed recreational vehicles.

Emulsion Polymerization: ABS is manufactured in three polymerization steps, using peroxy-type catalysts such as potassium persulfate to initiate the reaction through the formation of free-radical species (Fig. A-19). In the first polymerization step, polybutadiene rubber is made by feeding butadiene, water, an emulsifier, and catalyst into a 5,000-gal glass-lined reactor. The reaction takes place with the evolution of 580 Btu/lb of butadiene, which is reacted to about 80% conversion over a period of some 50 hr. Residual butadiene monomer is then steam-stripped from the polybutadiene-rubber latex, compressed, purified, and recycled for further use.

ABS Grafting Reaction: In the second polymerization step, the polybutadiene rubber is further polymerized in the presence of styrene and acrylonitrile monomers. Polymerization is accomplished in closed, low-pressure reactors under a N_2 atmosphere. The monomers are grafted onto the rubber backbone through the residual unsaturation remaining from the first polymerization step. SAN (styrene-acrylonitrile) resin is manufactured separately by emulsion, suspension, or mass polymerization in a third step by free-radical techniques. Typical emulsion polymerization of SAN involves charging monomers, emulsifier, catalyst, and water into 3,500-gal stainless-steel reactors operated at $75°C$ ($167°F$) and 5 psig for 6–7 hr. By blending the ABS graft phase with the SAN resin and incorporating various antioxidants, lubricants, stabilizers, and pigments, a wide range of ABS grades are made. The blended ABS latex is combined with water and a chemical flocculating agent under strong agitation in a stainless-steel tank under atmospheric pressure at temperatures of 60–100°C (140–212°F). The flocculated

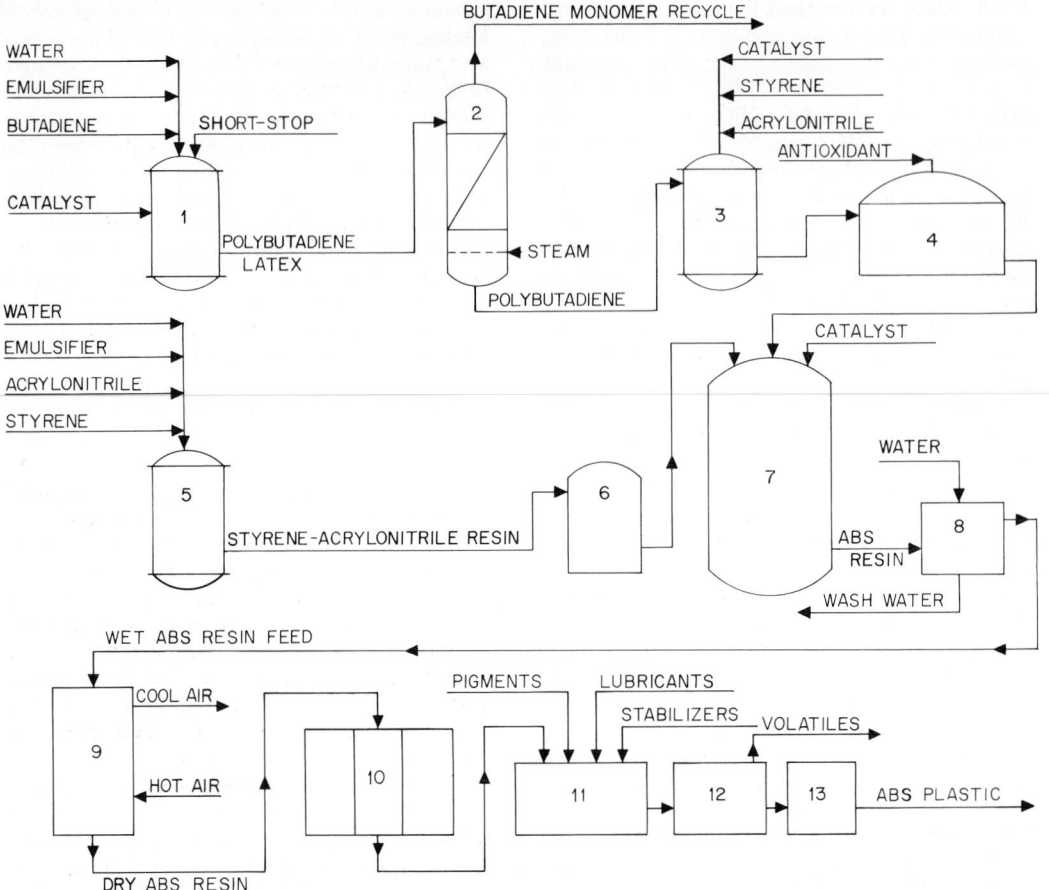

Fig. A-19. Schematic representation of materials flow in the manufacture of acrylonitrile-butadiene-styrene (ABS) resins. (1) Poly-butadiene polymerization reactor, (2) steam stripper, (3) ABS-graft-phase polymerization reactor, (4) ABS-graft-phase storage and blending (catalyst also added), (5) styrene-acrylonitrile (SAN) polymerization reactor, (6) SAN storage and blending, (7) latex-flocculation tank, (8) filter, (9) fluid-bed dryer, (10) ABS-resin storage, (11) ABS-resin blender, (12) devolatilizing extruder, (13) ABS-resin pelletizer. Reactor 1 operates at 66°C (150°F) and 150 psig. Reactor 5 operates at 75°C (167°F) and 5 psig. Lubricants are added (11) to improve processing. Stabilizers are added for color stability, and pigments for color matching.

latex becomes a slurry of fine resin particles in water and is passed to rotary-drum filters, where water is separated from the resin. Washing is done with fresh water on the filter. The cake is discharged at 40–55% water content to the dryer. Drying to 1.0% final moisture content is accomplished in a standard fluid-bed dryer, operating at 121–132°C (250–270°F) inlet and 49–60°C (120–140°F) outlet air temperatures.

The resin is stored as an intermediate in epoxy-lined silos of 25,000–250,000 lb capacity. The finished ABS pellet is produced by blending

pigments and resin in stainless-steel double-cone blenders and discharging to single-screw compounding extruders. The compounding extruders discharge melted strands at 232–288°C (450–550°F) into water at 27–38°C (80–100°F), where the strands cool and are subsequently granulated into $\frac{3}{32}$-in. lengths. See also **Acrylonitrile;** and **Butadiene.**

—William V. Rathbone, Jr., and Melvin E. Hoover, *Marbon Division, Borg-Warner Corporation, Washington, W.Va.*

ACTH, Adrenocorticotropic Hormone (See **Peptides and proteins**.)

Actinium (Gk. *aktis*, ray.) **Ac** = 227* (at. wt.); 89 (at. no.). In 1899, one year after the discovery of polonium and radium by P. Curie and M. Curie, A. Debierne showed the existence of a new radioactive substance in the residue after treatment of pitchblende. The method of investigation used was that elaborated by the Curies in pioneering the basis of radiochemistry. The only observable property of the newly researched substance being its radioactivity, chemical separations were effected on the uranium ore, and the radioactivity of each portion was measured. In this way, Debierne succeeded in concentrating the newly found radioactive element with the cerium earths. He named the new element actinium after the Greek word for ray. Independently, in 1902, F. Giesel observed the presence of a radioactive material in the rare-earth extracts of pitchblende which gave birth to an emanation. For this reason, Giesel named the material "emanium." In Paris in 1904, Debierne and Giesel compared their preparations and established the identical behavior of the two radioactive materials. At the time, it was not possible to determine their atomic weights. It was not until ten years later, when Fajans and Soddy had established the law of radioactive displacement, that Ac definitely could be classed in the periodic system as a higher homologue of lanthanum.

Isotopes of Actinium: At present, 24 isotopes of Ac have been identified. Their mass numbers range from 207 to 230. All are radioactive. The isotope discovered by Debierne was ^{227}Ac. Its half-life, 21.7 years, is the longest. ^{227}Ac results from the decay of ^{235}U (AcU—actino-uranium) and is present in natural uranium to the extent of 0.715%. The proportion of Ac/U in uranium ores is about 2.10^{-10} at radioactive equilibrium (Table A-12). Abundance of Ac in the earth's crust is estimated at 8.10^{-14}%.

The existence in nature of a second isotope, ^{228}Ac, was established in 1908 by O. Hahn. This is a product of the decay of thorium, and for this reason is also called *meso*-thorium 2. It has a half-life of 6.13 hr. The proportion of $MsTh_2$/Th in thorium ores is only 5.10^{-14}. The other isotopes were discovered synthetically by bombardment of

*Mass number of the isotope of longest known half-life.

thorium targets mainly with diverse particles. Among them, ^{225}Ac offers a particular interest because of its half-life of 10 days, which is the longest of the artificial isotopes. This isotope is present in nature as a member of the neptunium family, but only in infinitesimal quantities difficult to detect.

^{227}Ac is the only isotope that is reasonably stable and obtainable on a macroscopic scale. It can be extracted from the uranium ores (0.2 mg/ton uranium). This, however, is a laborious method of preparation because, during the treatment, Ac accompanies the rare earths which are relatively abundant and to which Ac is very similar chemically. Enrichment of the mixtures of rare earths containing Ac was for a long time accomplished by fractional crystallization or precipitation of appropriate compounds. However, the application of these methods has not permitted the complete separation of actinium from lanthanum. See also **Lanthanum.** The most concentrated preparation obtained, starting from ore residues, was a sample of approximately 0.1 mg lanthanum oxide containing 0.007 mg actinium. In modern practice, the Ac-La separation can be accomplished much more easily by ion-exchange methods, using a cationic resin and elution, notably with a solution of ammonium citrate or ammonium-α-hydroxyisobutyrate.

Nevertheless, to avoid the difficulties associated with the treatment of ores, ^{227}Ac is now obtained on the gram scale by transmutation of radium by neutron irradiation in the core of a nuclear reactor. Actinium is formed by the following process:

$$^{226}\text{Ra }(n,\ \gamma)^{227}\text{Ra} \xrightarrow{\beta^-}\ ^{227}\text{Ac}$$

The cross section for the capture of thermal neutrons by radium is 23 barns (23×10^{-24} cm^2). The irradiation should not be too prolonged because the accumulation of Ac is limited by the reaction ($\sigma = 500$ barns)

$$^{227}\text{Ac }(n,\ \gamma)^{228}\text{Ac (MsTh}_2) \xrightarrow{\beta^-}\ ^{228}\text{Th (RdTh)}$$

An irradiation of 25 g of radium carbonate, $RaCO_3$, for 13 days at a flux of 2.6×10^{14} ncm^{-2} s^{-1} yields approximately 108 mg ^{227}Ac (8 Ci) and 13 mg ^{228}Th (11 Ci). F. Hagemann succeeded in producing 1 mg Ac by this method and in isolating for the first time, in 1947, a pure compound of the element.

A vast program in Belgium, developed by the

TABLE A-12. *The Actinium Series*

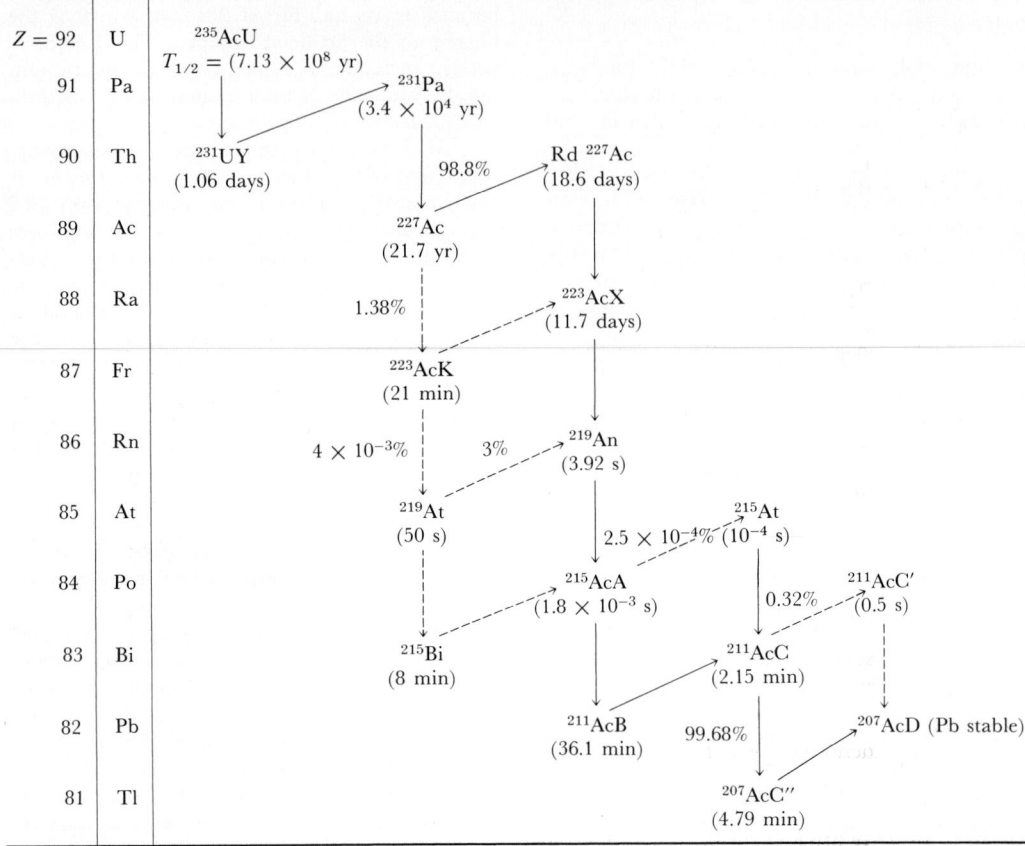

Centre d'Étude de l'Énergie Nucléaire Belge, Union Minière, brought about the production of more than 10 g Ac in 1970–1971. The treatment of the irradiated products is technologically very difficult because of their high radioactivity, and also because of the evolution of radon, a gas which results from the disintegration of radium. The chemical procedures adopted in Belgium to bring about the separation of ^{226}Ra, ^{227}Ac, and ^{228}Th, at this level of activity, is essentially the precipitation of radium nitrate from concentrated HNO_3, followed by the elimination of Th, which is adsorbed on a mineral ion exchanger, zirconium phosphate, which can stand high levels of radiation without decomposition.

The two other isotopes that present an interest for the chemical study of actinium, ^{228}Ac and ^{225}Ac, are, respectively, prepared by separation from *meso*-thorium 1 (^{228}Ra) sources, and by the irradiation of thorium with high-energy protons.

Metallic Actinium: Too electropositive to be obtained by electrolysis of aqueous solutions, metallic Ac has been prepared on the milligram scale by reducing AcF_3 in vacuo with lithium vapor at a temperature between 1000 and 1275°C. Similarly, $AcCl_3$ can be reduced with potassium vapor at 350°C. Metallic Ac is a white, silvery metal that emits a faint light of a blue tint, visible in the dark, due to its radioactivity. The metal crystallizes in a face-centered cubic lattice with a melting point of 1050 ± 50°C. The boiling point is estimated by extrapolation at 3300°C.

Metallic Ac easily forms an amalgam, either (1) by the action of a lithium amalgam on a citrate solution of the element at a pH between 1.7 and 6.8 or (2) by electrolysis on a mercury cathode.

This amalgamation involves an intermediate step which is the reduction of trivalent Ac to an unstable bivalent state. The half-wave potential for the process $Ac^{++} + 2e^- + Hg \rightarrow Ac^0(Hg)$ is -1.96 V versus the normal hydrogen electrode. The normal potentials of the Ac^{3+}/Ac^{++} and the Ac^{3+}/Ac^0 couples are -1.6 and -2.6 V, respectively.

Electronic Structure: This may be described as $Rn + 6d^1 7s^2$. As an element of group IIIA, actinium shows chemical properties that are very similar to those of its lower homologue, lanthanum, a trivalent element. The ionic radius of Ac^{3+} is 1.11 Å, somewhat greater than that of La^{3+}, which is 1.04 Å. See also **Chemical elements.**

Chemical Behavior: Actinium behaves ordinarily as an element even more basic than lanthanum, which is the most basic element of the lanthanide series. As a result, if one fractionates mixtures of rare earths containing Ac by fractional precipitation of the hydroxides, Ac concentrates in the mother liquor. Conversely, by fractional crystallization of the double nitrates of lanthanides and ammonium, Ac is separated with La in the least soluble fraction. However, in the fractional crystallization of the double nitrates with magnesium or manganese, Ac has an abnormal and still unexplained behavior, in that it does not follow lanthanum but concentrates itself between neodymium and samarium. As could be expected from its ionic radius. Ac is eluted from a cationic-on exchanger after lanthanum.

The mineral salts of Ac are difficultly extracted from their aqueous solution by an organic solvent. Therefore it is generally extracted in the form of a chelate with trifluoroacetone (TTA) or diethylhexylphosphoric acid (DEHPA). The extraction in the form of $Ac(TTA)_3$ by benzene is practically quantitative at a pH of 5–6. This indicates a slightly weaker complex formation than for lanthanum, which can be extracted at somewhat higher acidities.

The Ac salts that are insoluble in water correspond with those of lanthanum, namely, the hydroxide, fluoride, fluosilicate, carbonate, oxalate, phosphate, double sulfate of potassium, and so on. With the exception of the sulfide, which is black, all other Ac compounds are white and form colorless solutions. Since the ionic radii of actinium and lanthanum are similar, the crystalline compounds are isomorphic.

Besides its marked resemblance to lanthanum, actinium presents a striking chemical analogy to the group of trivalent transuranium elements curium ($Z = 96$) to lawrencium ($Z = 103$). It is this analogy that led G. T. Seaborg to formulate the actinide theory, according to which Ac begins a new series of rare earths characterized by the filling of the $5f$ inner electron shell, as the filling of the $4f$ electron shell characterizes the lanthanide elements. This classification has brought about controversy because the chemical properties of the first elements of the actinide series are considerably different from those of actinium. One observes, notably, a multiplicity of valences which is without equivalent in the case of the lanthanides.

Measurements: The determination of ^{227}Ac is carried out by radioactive methods more sensitive than spectral analysis. However, this isotope emits a beta radiation of very weak energy ($E_{max} = 46$ keV) and an alpha radiation ($E = 4.94$ MeV) of very low intensity (1.38 α/100 disintegrations). Thus Ac concentration generally is determined by measuring the radiation of its daughters AcK(^{223}Fr), Ac(^{219}Rn), AcA(^{215}Po) or of the active deposit. See Table A-12.

Measurement of the penetrating beta and gamma radiation of the active deposit has the advantage of permitting measurement of preparations kept in a sealed tube, and is successful provided the material is pure and the Ac is in radioactive equilibrium with its daughters. Equilibrium is attained in approximately five months. If the equilibrium is established, one can alternatively measure the 7.36-MeV alpha radiation, characteristic of AcA, on thin-layer samples. On the other hand, measurement of the beta radiation from AcK, with a 21-min half-life, which can be easily separated, permits a rapid measurement of Ac in the presence of other radioelements. The radioactive equilibrium is in this case established in approximately three hours.

The selective counting of the 4.94-MeV alpha particles emitted by ^{227}Ac can be applied only to the measurement of preparations of rather high specific activity, after radiochemical purification. To avoid these analytical difficulties, the isotopes ^{228}Ac (α, β, γ emitter) and ^{225}Ac (an α emitter) are currently used instead of ^{227}Ac in chemical studies of Ac at tracer-level concentrations.

Manipulations required in the preparation of Ac require great precautions. Its rather long half-life, the accumulation of daughters (which are α, β, and γ), and an unfavorable metabolism (selective fixation with long retention in the skeleton) give the element a high radiotoxicity. The maximum permitted dose in the human body is 0.01 μCi (1.4 × 10⁻¹⁰ g).

Uses: The use of ²²⁷Ac as a source of heat in space satellite vehicles has been advocated. The heat resulting from the absorption of the radiation emitted by 1 g Ac, in equilibrium with its daughters, is 12,500 cal/hr. This application of Ac in the fabrication of thermionic generators may lead to a reevaluation of radium, the starting material for the synthesis of Ac.

REFERENCES

Bagnall, K. W.: "Chemistry of the Rare Radioelements," pp. 150–165, Academic, New York, 1957.

Bouissières, G.: "Actinium dans le nouveau traité de chimie minérale de P. Pascal," pp. 7 and 1413–1446, Masson, Paris, 1960.

Centre d'Étude de l'Énergie Nucléaire Belge, Union Minière: Programme Actinium, Annual Report, 1969, Brussels.

Hageman, F.: The Chemistry of Actinium, in G. T. Seaborg and J. J. Katz (eds.), "The Actinide Elements," National Nuclear Energy Series, IV–14A, p. 14, McGraw-Hill, New York, 1954.

Katz, J. J. and G. T. Seaborg: "The Chemistry of the Actinide Elements," pp. 5–15, Methuen, London, 1957.

Kirby, H. W.: "The Analytical Chemistry of Actinium," Mound Laboratory, Miamisburg, Ohio, 1966.

—G. Bouissières, *The University of Paris, France*

Activated Aluminas (See **Adsorption;** and **Bauxite.**)

Activated Carbon (See **Adsorption;** and **Carbon.**)

Activated Silica Sols (See **Waste treatment.**)

Activated Sludge Process (See **Waste treatment.**)

Activation, Water (See **Nuclear power plants.**)

Activation Energy The quantity of energy that must be furnished to start a reaction: (1) energy per molecule for a chemical reaction and (2) energy per atom for a nuclear reaction. See also **Molecule.**

Acyls (See **Radicals.**)

Addition Polymerization (See **Polymerization.**)

Additives (See **Antioxidants; Paint;** and **Salt, NaCl.**)

Adhesives An adhesive is a material for use in bonding two or more materials to provide some form of geometrical continuity. Adhesives are of three basic types: *structural* adhesives where the bond is as strong if not stronger than the materials of the parts; *holding* adhesives where the adhesive functions to keep materials in their proper position, as tiles on a floor or wall where the forces acting upon the adhesive bond generally are relatively weak; and *calking or sealing* adhesives where the adhesive is used to fill in voids and cracks, functioning both in an adhesive capacity and as an extra material.

Adhesive bonds are *chemical,* in which the adhesive ingredients react chemically with the material being bonded and the strength is obtained through intermolecular attraction, and *mechanical,* in which the material attaches in a physical sense through the penetration into the pores and interstices of the materials being bonded. Structural adhesives almost always are the chemical type.

Adhesives are used widely for metal-to-metal, metal-to-nonmetal, and nonmetal-to-nonmetal bonding and in recent years have been adapted to high-production assembly operations, displacing conventional methods of fastening (rivets, bolts, welding, soldering, clips) in numerous instances. The use of high-strength adhesives received its impetus from the aircraft and automotive industries, where such properties as greater strength with lighter weight, smoother surfaces, and greater fatigue resistance were sought.

High-strength adhesives derive their unusually strong and rugged properties in many instances as the result of alloying two or more adhesive resins of the types (1) thermosetting adhesives, such as phenol-formaldehyde and the epoxies, (2) the elastomeric adhesives, such as neoprene or nitrile rubbers, and (3) the thermoplastic adhesives, such as polyvinyl formal and polyvinyl butyral.

Phenol-formaldehyde adhesives are used in high tonnage in making plywood, other wood products, and paper products. Although important, epoxy adhesives are used in somewhat lesser quantities. Both of these high adhesives form strong bonds, even in metal-to-metal applications, but such bonds often are brittle, and the materials in their unalloyed form are not considered true high-strength adhesives. Elastomeric adhesives of neoprene or nitrile rubbers are important contact cements forming medium-strength, high-impact-resistant bonds in most uses. Deformation of the bond in metal-to-metal use is considerable under deadweight loadings. Bond deformation decreases significantly when the elastomeric adhesives are alloyed with the more brittle phenolic adhesives.

Polyvinyl butyral is of major importance as a sandwich adhesive for automotive safety glass. Phenolic modifications of polyvinyl formal, acetal, and butyral are high-strength adhesives, although bond strengths at elevated temperatures do not quite compare with those of other high-strength adhesives. High-strength adhesives are highly formulated products and may contain several different types of adhesive resins and many other components, such as solvents, pigments, fillers, extenders, and catalysts.

Numerous nonferrous metal-bonding applications include (1) magnesium metal sandwiches with plastic-foam cores for air-transportable communications-equipment housings; (2) aluminum sandwiches with cores of foam plastics, glass, high-pressure laminates, and paper or aluminum honeycomb for communications equipment, aircraft, and missiles; (3) copper foil for high-pressure laminate sandwiches for electronics printed circuitry; and (4) assemblies of castings, special alloys, tungsten carbides, and numerous other materials otherwise difficult to join.

Ferrous-to-ferrous metal bonding includes bonding stainless steel for helicopter rotor blades, special hard steels for joining tool bits to holders, and steel-to-steel for magnet motors, motor laminations, and fire doors. The automotive industry depends upon high-strength adhesives for bonding ferrous metals to dissimilar materials. In replacement brakes, friction materials are adhesive-bonded to metal brake shoes. Automatic transmissions and clutches have adhesive-bonded bands and facings of friction materials. In the process industries, metal tanks can be plastic-lined to resist corrosion wherein adhesives provide the bond between steel tank frame and plastic lining. Ceramic parts and coatings may be adhesive-joined to steel equipment to provide resistance to wear and corrosion. Abrasive fabrics can be bonded to metal backing plates.

Although most of the major progress in adhesives development has paralleled the development of new resinous materials, sodium silicate adhesives continue to be used in high tonnage by the paper and board industries, for corrugated paperboard, insulating boards, and other products. See also **Silicates (soluble).**

Also included in the category of adhesives are pastes of various types made from starches or dextrines and sometimes mixed with gums, resins, or glue to add strength, and containing antioxidants. These are the least costly adhesives and continue to be used in certain paper-converting operations and for household applications. Tapioca paste is used in the manufacture of low-cost plywoods. Glues, usually water solutions of animal gelatin, still find use in packing and shipping operations. Latex cements are solvent solutions of rubber latex and are excellent for bonding paper, leather, and certain fabrics, but are subject to rather rapid disintegration unless cured. In the shoe industry, cellulose acetate cements are used for bonding soles to shoes. A pyroxylin cement may be a simple solution of nitrocellulose in various solvents, or it may contain additional resins and plasticizers.

Several of the resins used in the manufacture of adhesives are described elsewhere in this volume. Consult the Subject Index. See also **Waxes.**

REFERENCES

Bikerman, J. J.: "The Science of Adhesive Joints," Academic, New York, 1961.

Bodnar, M. J. (ed.): "Symposium on Adhesives for Structural Applications," Wiley, New York, 1962.

Guttmann, Werner H.: "Concise Guide to Structural Adhesives," Van Nostrand, New York, 1961.

Simonds, Herbert R. (ed.): "Concise Guide to Plastics," 2d ed., Van Nostrand, New York, 1963.

Skeist, Irving (ed.): "Handbook of Adhesives," Van Nostrand, New York, 1962.

Thompson, M. S.: "Gum Plastics," Van Nostrand, New York, 1958.

Adiabatic Process A process in which there are changes in matter but, theoretically, without the transfer of heat is termed an adiabatic process. The adiabatic compression of a gas is an example. When a gas is compressed adiabatically, the temperature rises, causing the gas to expand. Thus the compressing force, and therefore the work required to effect a given change in volume, are higher by adiabatic compression than is the case by isothermal compression (where heat would be removed).

Adipic Acid (See **Carboxylic acids;** and **Nylons.**)

Admiralty Brass (See **Copper.**)

Adsorption Adsorption sometimes is defined as the tendency of a solid substance to condense and retain on its surface a layer of a gaseous or liquid substance. The degree of adsorption that takes place depends upon (1) the chemical composition of the adsorbing material; (2) the condition of the surface of the adsorbing material, including area exposed and other geometric and physical characteristics; (3) the nature of the material being adsorbed; (4) the temperature; and (5) in gaseous systems, the pressure of the gas being adsorbed.

The phenomenon of adsorption is expressed by the Freundlich equation

$$\left(\frac{x}{m}\right)^n = kc$$

where

x = weight of adsorbed material
m = weight of adsorbing material
c = concentration in equilibrium with adsorbed material
k and n = constants to be determined experimentally for each temperature

Adsorption utilizes natural or synthetic materials of microcrystalline structure. Selective combination of solid and solute occurs on the pore surfaces throughout this structure. Surface areas up to 100 cm^2/cm^3 are encountered, corresponding to an effective cylindrical pore diameter of less than 200 Å, or 2×10^{-6} cm. At higher temperatures (usually above 200°C), adsorption may occur through a true reaction or chemical bonding, in which case the process is termed *chemisorption*.

Separations usually are carried out under conditions where the attractive forces are weaker (such as van der Waals forces) and less specific than those of chemical bonds; here the combining effect is identified as *physical adsorption*.

Adsorption usually is analogous to a condensation of gas molecules, or to crystallization from a liquid. Its selective action is most pronounced in a monomolecular layer next to the solid surface, but at times selectivity may persist to a height of three or four molecules. Adsorption capacity of a solid for a solute tends to increase with the fluid-phase concentration of the solute. Adsorbents in large-scale installations include activated carbon, silica gel, activated alumina, fuller's earth, and numerous clays. Several adsorbing operations are described in this volume. Consult the Subject Index and, in particular, see **Aluminum; Catalysts; Chromatography; Drying, gases; Ion-exchange resins;** and **Silicates (soluble).** Physical properties of representative adsorbents used in the process industries are given in Table A-13.

Gas-Solids Systems: The simplest form of gas-solids contactor is a static bed in which particulate solids are held in fixed positions, one particle resting upon another with no relative motion among particles. Gas is made to flow over, impinge upon, or flow through the voids among the particles. Contacting is confined to the interface between the gas and solid phases. Channeling, a difficult condition to correct, may result from nonuniformities in bed packing or density and pose a serious operating problem. Fixed beds almost always require cyclic operations because of the difficulty of adding or removing solids during operation. However, as contrasted with moving beds, the solids in a static bed do not have to move throughout a portion of the system, and hence there are fewer losses, as from abrasion. In a

TABLE A-13. *Physical Properties of Representative Adsorbents*

	Internal Porosity, %	External Void Fraction, %	Bulk Dry Density, lb/ft³	Average Pore Diam, Å	Surface Area, m²/g	Adsorptive Capacity, g/g dry solid
Active alumina	25	49	50	34	250	0.14[a]
CaCl₂-impregnated alumina	30–47	~42	~51	136:99	90:190	
Activated bauxite	35	40	~53	~50	. . . .	0.04–0.2[b]
Fuller's earth	~54	40	30–40	. . . .	130–250	
Silica gel	~70	30–40	40–48	25–50	~320	1.0[c]
Shell-base carbon	~50	~37	27–32	20	800–1,100	45[d]
Wood-base carbon	55–75	~40	10–35	20–40	625–1,400	6–9[e]
Coal-base carbon	65–75	45–70	20–30	20–38	500–1,200	~0.40[f]
Petroleum-base carbon . .	70–85	26–34	28–34	18–22	800–1,100	0.6–0.7[g]
Anhydrous CaSO₄	38	~45	60	. . . .	. . . .	0.1[h]

[a] Water at 60% RH.
[b] Water; test condition not specified.
[c] Water at 100% RH.
[d] Accelerated chloropicrin test.
[e] Phenol value.
[f] Benzene at 20°C and 7.5 mm partial pressure.
[g] Carbon tetrachloride; test condition not specified.
[h] Water; test condition not specified.

moving-bed adsorption system, there is relative movement among the solid particles, and this exposes new surfaces for gas contacting, prevents short-circuiting, and reduces the formation of stagnant pockets. Such moving-bed systems are used not only for adsorption, but also for effecting chemical reactions, providing for heat transfer, mixing of solids with gases, drying solids and gases, heat treatment, and other operations involving contact between the gas and solid phases.

Sorption Operations: Three major types of solid-fluid sorption equipment are in use: (1) batch units, (2) semicontinuous units involving fixed beds, and (3) fully continuous units which provide for countercurrent or cocurrent movement of sorbent and fluid.

Batch Operations: It is often advantageous to carry out sorbent-liquid contact in batch equipment, but such methods are infrequently used for the treatment of gases. In purifying the products of organic chemical syntheses, decolorizing carbons and clays frequently are used as *contact adsorbents;* i.e., they are stirred directly into a liquid-phase mixture or solution and subsequently

separated by filtration. Contact filtration of lubricating oils also is used to remove colored and carbon-forming materials from lubricant stocks, as well as the traces of products formed in H_2SO_4 treatment. In some cases, either vacuum distillates or residuum fractions from crude petroleum may require only acid and clay treatment in order to meet product specifications. Elsewhere, solvent extraction will be the principal means of treatment, followed by refining with acid and clay. See also **Lubricating oils.** A contact-filter plant used for long-residuum or cylinder stock having a flash point of 230–240°C and a Saybolt Universal viscosity of 80–85 s at 100°C is shown in Fig. A-20. The residuum is first treated with H_2SO_4 at the rate of 40–45 lb of 66 °Bé acid/bbl oil at about 60°C. After this treatment, the sludge is settled at the bottom of the agitator and drawn off. The oil, with a small amount of added clay, is filtered to remove emulsified acid. It is then mixed with about 0.5 lb clay/gal in mixing agitators and pumped to a pipe still, where it is brought to about 240°C. It is held at this temperature for several minutes and then cooled to

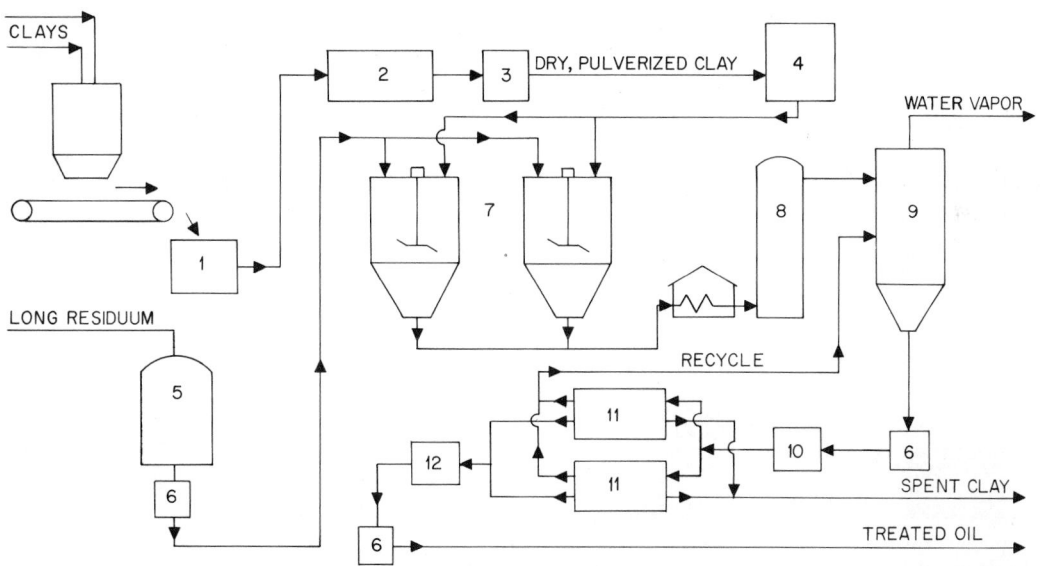

Fig. A-20. Highly schematic representation of materials flow in contact filtration process for treatment of lube-oil residuum with clay.
(1) Hammer mill, (2) pulverizer, (3) blower—all used to pretreat and condition the raw clays. (4) Clay storage, (5) long-residuum storage, (6) reciprocating pumps, (7) mixing agitators, where residuum and clays are mixed. Effluent goes from (7) to pipe still (8), where oil and vapors exit at approximately 232°C (450°F) and enter vapor separator (9). Bottoms from the vaporizer pass through precooler (10), where the temperature is lowered from a range of 220–230°C (425–450°F) to 150°C (300°F) before entering filters (11). After filtration, the product is cooled further by cooling box (12) equipped with continuous pipe-cooling units, to approximately 66°C (150°F) before going to storage. Depending on the overall filtration operations, a portion of the product may be returned as recycle to (9).

150°C and filtered. Particle size of the clay usually is 80–100 mesh. Diatomaceous earth may be used as a filter precoat.

Mixing and filtering can be accomplished in the same vessel by using large-particle sorbent material (50 mesh and coarser) through which a liquid can drain quite readily. The solid sorbent is re-used a number of times before it is regenerated or replaced. With the vessel filled with a charge of process liquor, gentle agitation usually is obtained by sparging the slurry with air. Draining of the liquor is hastened by use of a reverse gas flow, which serves to expel the interstitial solution from the settled beds, as indicated in Fig. A-21.

Where the process liquor is a slurry rather than a clear liquid, batch contact with sorbent materials generally is preferred to percolation-type contact. The latter involves less agitation and is more likely to give a progressive accumulation of slurry particles in the body of sorbent. If the slurry particles are larger than those of the sorbent, the sorbent may be incorporated in the slurry. After sufficient contact, the process liquor may be centrifuged or filtered. Subsequently, the

sorbent is recovered by hydraulic classification or by wet screening. This method is used for pectin recovery from grapefruit peel.

Fixed Beds: The most frequently used method of fluid-solid contact for sorption operations is in columnar units, with the solid particles closely packed in a relatively fixed arrangement. Adsorbent particle sizes in such equipment usually lie in a narrow range, but may average from as large as 4 mesh to as small as 250 mesh. Pressure drop often is taken as a determining factor in particle-size specification. In simple fixed-bed operations, the solute undergoing adsorption is removed continuously from the carrier fluid and accumulated in the solid phase, as shown in Fig. A-22. Such transfer proceeds until the concentration on the solid reaches a value corresponding to equilibrium with the concentration in the feed stream. At this point, the fluid just leaving that solid layer reaches the feed concentration. However, until the last layer of the sorbent is nearly saturated, the column effluent remains practically free of the solute. This change in effluent concentration with time is known as the *breakthrough curve*, or *concen-*

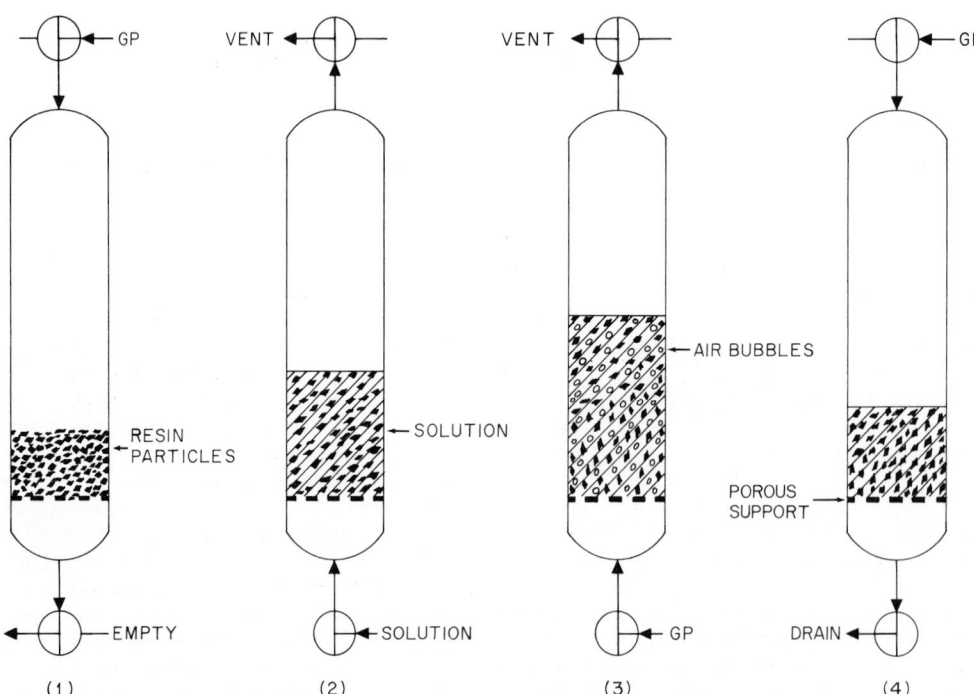

Fig. A-21. Sequence of operations in a batch mixer-settler. (1) Empty, (2) filling, (3) equilibrating, (4) draining. GP = gas pressure.

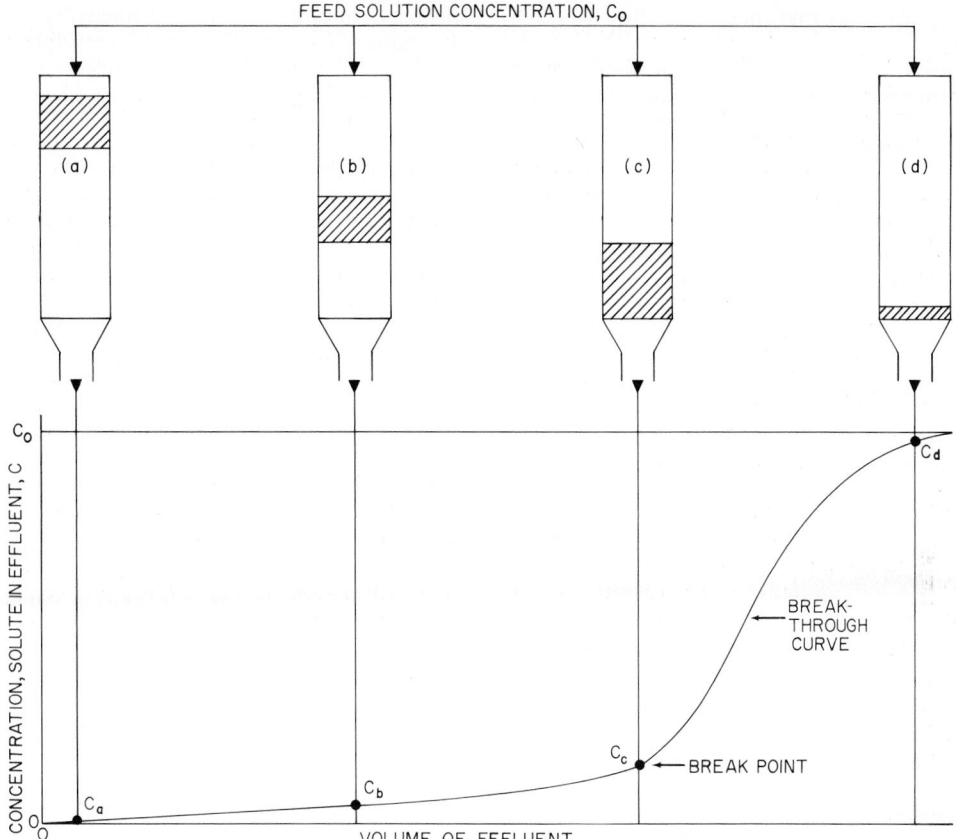

Fig. A-22. Passage of adsorption wave through a fixed bed. Adsorption zone is shown by crosshatching. (*Adapted from a concept by Treybal, "Mass Transfer Operations," McGraw-Hill, New York, 1955.*)

tration history. Mathematical models frequently are used to make determinations in advance of equipment design. See Perry and Chilton, "Chemical Engineers' Handbook," 5th ed., pp. 16–26, McGraw-Hill, New York, 1973.

Continuous Systems: The primary problems to be overcome in continuous countercurrent sorption operations include (1) mechanical complexity of the equipment, (2) gradual attrition of the solid sorbent, (3) limitations in particle-size range to avoid either classification or excessive pressure drop, (4) channeling (nonuniform flow) of either fluid or solid, and (5) contamination between functional sections of the equipment due to granular and porous structure of the solid.

The general equipment configurations for sorption operations follow along the lines of fluidized beds which have been designed predominantly for reactions, both catalytic and noncatalytic. See also **Fluidized-bed operations.**

Moving beds have been used in connection with ethylene recovery from methane and other gases of lower molecular weight. The adsorbent, usually activated carbon, moves downward in a tower through a cooler, a rectifying section, a reflux section, and a heated stripping section, into a discharge mechanism and a sealing leg. A gas lift then returns the adsorbent to the top of the tower, where its flow reverses in an impactless separator and from which it passes through a storage hopper back into the column. The stripping section is fed with steam, and the mixture of steam and desorbed bottom product is disengaged from the bed at a tray just below the reflux section. A side stream of steam-free bottom product is introduced in the reflux section to displace any of the overhead components that may originally have been adsorbed.

Above the feed plate, the process gases undergo countercurrent contact with the main bed of adsorbent. The major part of this stream is disengaged below the cooler, while a minor part of it serves to dehydrate the stripped carbon and is then added to the lift-gas circulation stream. Part of the latter, in turn, is bled off with carbon dust at the top of the unit. External reactivation at higher temperature can be carried out continuously on a small part of the solid inventory. A process of this type sometimes is referred to as *hypersorption*.

An adsorption tower for drying gases of about 10 ft diam and about 70 ft in height has been used. The adsorption section contains a bubble-cap entrance tray and seven perforated trays, five of which are designated as adsorption plates and two as heat-transfer plates. The latter cool the incoming hot-silica-gel adsorbent, which is at a temperature of about 120°C. The gel is 6–12 mesh when new and moves downward at a rate of 80 lb/min. The superficial gas velocity is 4.5 ft/s. The regeneration section consists of four perforated plates and one bubble-cap entrance plate. The plates have a gel loading of 6 lb/ft². However, only 50–60% of the quantity of gas used in the adsorber section is required. In this system, part of the hot gas to be dried is first used as the regenerating medium, and a condensing cooler is provided between the regenerator- and adsorber-plate sections. The remainder of hot gas is bypassed around the regenerator plates and goes directly into the cooler. See also **Drying, gases.**

Most other continuous adsorption systems involve ion-exchange resins. See also **Ion-exchange resins.**

Adsorption, of course, enters into various dyeing processes. For example, when colloidal hydroxides, such as $Al(OH)_3$, are precipitated out of solutions of acidic dyes (containing —OH or —COOH groups), the dye tends to adhere to the precipitate. The result is termed a *lake*. See also **Dyestuffs and dyeing.**

Molecular Sieves: Alumosilicates which have undergone heating to remove water of hydration possess high porosity and fall into a classification of substances known as molecular sieves. The pores (lattice vacancies) are of uniform size and essentially are of molecular dimensions. These materials adsorb small molecules only, are selective on molecular shape, and have a particular affinity for unsaturated and polar molecules. The lattice vacancy may have a diameter of 4–5 Å, depending upon whether the alumosilicate is of the Na or Ca form. Molecular sieves are used mainly in gas treatment, but also are effective for drying organic liquids.

Particulate-gel dialysis is another type of separation which is based upon molecular size, carried out with a liquid substrate. The process involves granules of a permeable and highly solvated material, such as is used in membrane dialysis, e.g., cellulose. The granule admits crystalloid solutes, either ionic or molecular, up to about 10–15 Å diam, but completely rejects colloidal material larger than this limiting size. Regeneration with solute-free liquid enables the crystalloid to quit the particle and be recovered in the effluent, leaving the solid ready for reuse.

REFERENCES
American Institute of Chemical Engineers: "Adsorption from Liquids," Symposium, Philadelphia, 1969.
Bikerman, J. J.: "Physical Surfaces," Academic, New York, 1970.
Clark, Alfred: "The Theory of Adsorption and Catalysts," Academic, New York, 1970.
Fowkes, Frederick M. (ed.): "Hydrophobic Surfaces," Academic, New York, 1969.

Aeration (See **Waste treatment** and **Water treatment**)

Aerosols (See **Air pollution; Chlorofluorocarbons;** and **Paint.**)

Agar-Agar (See **Gums.**)

Agglomeration Agglomeration, simply stated, is size enlargement or upgrading of otherwise finer particles. The agglomerate may take many forms, depending on the process used. The forms may be spheroids, pills or tablets, pillow-shaped, cylindrical, irregular extrusions, or merely loosely bound aggregates or clusters. See also **Crystallization.**

There are many reasons for agglomerating. There are the obvious advantages from the standpoints of storage and packaging due to the increase in bulk density and reduction of objectionable dust. The process provides an advantage of improving the free-flow characteristics of otherwise difficult-to-handle materials, and it increases the control over many processing variables. The conversion of solid-waste materials into a more useful, easier-to-handle form often is accomplished by agglomeration.

A summary of the advantages of agglomeration would include (1) increase in bulk density, (2) reduction of storage-space requirements, (3) improvement in bulk-handling characteristics, (4) prevention of segregation of mixed and combined ingredients, (5) improvement in heat-transfer properties, (6) reduction of drying and cooling costs, (7) better control over solubility, (8) reduction of objectionable dusting, (9) reduction of material loss and minimization of pollution, (10) reduction in labor due to more efficient handling, and (11) conversion of waste materials into useful form.

Classification of Agglomeration Operations: Basically, all agglomeration falls into one of four categories: compaction, extrusion, agitation, and fusion. *Compaction* may be defined as the compression of a loose powder either between two opposing surfaces or moving members or within a die or cavity. *Extrusion* is a variation of the compacting force, the material being forced through a predetermined orifice by some mechanism, such as roll, screw, or piston. *Agitation* may be either mechanical, such as a rotating drum or disk, or it may be accomplished in an airstream. The agglomerate is formed by the collision and adherence of particles, usually in the presence of a binder or wetting agent. *Fusion* is the formation of larger particles through partial fusion of the material itself, with the addition of a binder. Fusion sometimes is accompanied by pressure or even chemical reaction.

Most agglomeration operations require the use of a binder or some other additive. The exceptions would be materials with a sufficiently low melting point so that semifusion takes place and provides the necessary binding or adhesion. Additives may be either liquid or powder and, in certain instances, may be necessary to give a lubricating property to facilitate removal from the die or cavity.

The more common types of agglomerating equipment include:

Type	Principle used
Tableting	Compaction by plunger
Flaking rolls . . .	Roll compaction
Pellet mill	Ring-roll extrusion
Screw extruder . .	Extrusion through plate
Briquetting	Pocketed-roll press
Rotary drum . . .	Balling by agitation
Rotary disk	Balling by agitation
Sintering	Partial fusion
Spraying	Wetting and cohesion in turbulence

Basis for Selecting Equipment: Factors that must be considered in the effective selection of agglomerating equipment include:

1. *Feedstock:* (*a*) Identification and chemical composition; (*b*) particle-size distribution; (*c*) bulk density; (*d*) mechanical characteristics, including moisture or other liquid content, hardness or abrasiveness, fibrous nature, adhesive or sticky nature, flow characteristics, temperature sensitivity or melting point; and (*e*) physical characteristics, including corrosiveness, volatility, explosiveness, flammability, hygroscopic nature, and toxicity.

2. *Capacity Required:* (*a*) Rate (lb/hr) and (*b*) specification of rate, based on feedstock or finished and graded agglomerates.

3. *Finished Product:* (*a*) Size required (such as diameter and mesh); (*b*) shape required (such as spheres or cylinders); and (*c*) indication of whether screening or classification will be used.

4. *Special Considerations:* (*a*) Can fusion be tolerated? (*b*) May additives be used? Specifying kind, maximum percentage and, if used, whether additive must be removed by drying after agglomeration. (*c*) Can discoloration be tolerated? (*d*) Is surface hardening (case hardening) of product permissible? (*e*) Is solubility of product a factor?

5. *Equipment Specifications:* (*a*) Materials of construction (carbon steel, stainless steel, other alloys, surface finish or polish) and special sanitation codes or requirements; and (*b*) electrical characteristics [voltage; cycles (Hz)], enclosures (TEFC, explosion-proof), starting equipment (NEMA standard).

Tableting Machines: A tableting machine compresses loose material by the action of two opposing plungers operating within a cavity. Tablets may vary in size from $\frac{1}{8}$ to 4 in. diam. The products of tableting are precise and uniform, an advantage for pharmaceutical products as well as industrial catalyst and metallic powders.

Flaking generally is accomplished by large-diameter rolls, usually 18 in. and larger. The surface of the rolls is generally smooth, and the compacted material quite often is removed from the surface by a knife or doctor blade. The flakes are irregular in size and frequently require post-crushing or granulation. This process most frequently is used for certain clay-type minerals, but also has application with certain types of plastics, such as phenolic molding compound.

Pelleting is an example of the extrusion process. As shown in Fig. A-23, a pellet mill operates on the ring-roll principle, with a cylindrical ring or die having carefully machined holes spaced uniformly and drilled radially. The force of extrusion is obtained by the rollers acting on the inner face of the die, the friction of the material in the die holes supplying the necessary resistance for agglomeration.

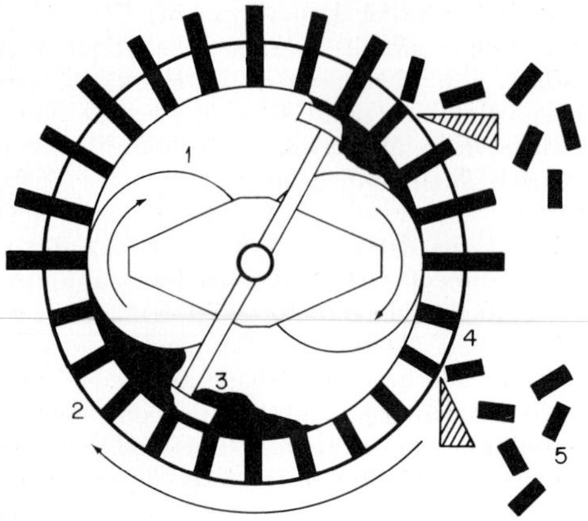

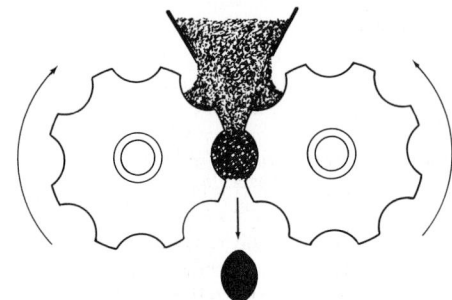

Fig. A-23. **Pellet mill.** Revolving die cover (not shown) with welded flights helps to direct the feed into the die chamber. An actual mill will have many more slots than shown schematically here. The extruding or pelletizing of the feed is accomplished by squeezing the feed between the rotating die and the friction-driven rolls. (1) Rollers, (2) die, (3) spreader flight, (4) knives, (5) pellets.

glomeration. The pellet mill may operate either by applying power to the die and rotating it about free-turning rollers, or the die may be fixed and the roller assembly rotated within the die. As the material is extruded from the die, cutoff knives at the surface control the length of the pellets. See also **Iron ores.**

Screw Extruder: These machines may be either single- or double-screw and either single- or multiple-stage, depending on the function and use. Special shapes and profiles of extrusions may be obtained by the design of the die or orifice plate. The material is forced through the orifice by the pressure developed by a screw or auger, as shown in Fig. A-24. A variation of this may be a solid ram operated by mechanical, hydraulic, or pneumatic force. The unit generally has a variable-speed motor and frequently heating and cooling jackets. Mixing may be accomplished in the same unit, but usually auxiliary mixing equipment is required.

Briquetting Machine: This is an example of the compaction method where the material is compressed into pillow-shaped, tear-drop-shaped, or even spherical-shaped blocks between two oppos-

ing rollers with machined pockets (Fig. A-25). Roll diameters generally are large (24–36 in.) and require pressures up to 300 tons/in.^{2}. Capacities are large, ranging up to 25–50 tons/hr. Many materials may be briquetted without the benefit

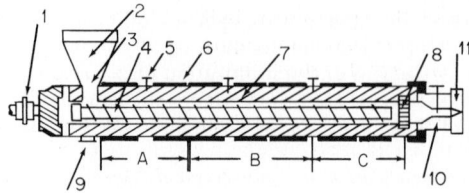

Fig. A-24. **Various components of a single-screw extruder used in agglomeration operations.** (1) Gear-reducer flexible coupling, (2) hopper, (3) feed throat, (4) screw, (5) thermocouple(s), (6) heater bands, (7) heated barrel, (8) breaker plate, (9) port-to-hopper cooling jacket, (10) adapter, (11) die; (A) rear heat zone, (B) center heat zone, (C) front heat zone.

of a binder or lubricant. Additives, however, generally increase the strength of the briquette and often reduce the coefficient of friction between the briquette and the rolls. See also **Iron ores.**

Fig. A-25. **Operating principle of briquetting machine.**

Rotating Drum: This, perhaps, is the simplest and most fundamental method of agglomeration. Aggregates are formed by the collision and adherence of the particles in the presence of a liquid binder or wetting agent to produce, in essence, a simple "snowballing" effect. The longer the operation continues, the larger the spheroids become. The binder or wetting agent influences the strength and hardness of the particles. Generally, this operation is accompanied by a screening process, with the unpelletized material or fines recirculated to the drum.

Rotating Pan or Disk: These devices are variations of the rotating drum. The most common version is the inclined pan, which operates at an inclination from the horizontal of 30–65°. Here the larger agglomerated particles move to the surface of the particle bed at the lower side of the pan and are discharged over the rim. The pan itself acts as a classifier, and the fine particles remain in the pan for further agglomeration. The size of the spheroid or ball is a function of the recycle size and throughput.

Sintering: In the sintering process, which is primarily used for ores and minerals as well as powdered metals, heated air is passed through a loose bed of finely ground material. The particles are bound together through partial fusion with or without a binder. The process may be carried to the extreme in a rotary kiln, as in the formation of a cement clinker. Often the process is accompanied by the volatilization of impurities and the removal of objectionable moisture. See also **Iron ores.**

Spray-Type Agglomeration: Several principles are combined to make this a unique process. Loosely bound clusters or aggregates are formed by the collision and coherence of the fine particles and a liquid binder in a turbulent stream. The mixing vessel consists of a vertical tank, around the lower periphery of which are mounted spray nozzles for the introduction of the liquid. See Fig. A-26. A suction fan draws air through the bottom of the tank and creates an updraft within the mixing vessel. This air passes through the mixing zone and eventually leaves the top of the vessel through the duct. The intensity of this draft can be controlled by a regulating valve. The solid materials are preblended and conveyed by negative pressure into a cyclone separator in the top of the mixing chamber. Materials spiral downward through the mixing chamber, where they meet the updraft and are held in suspension near the portion of the vessel where the liquids are injected. Liquids are introduced in a fine mist, and the individual droplets gather the solid particles until the resulting agglomerate overcomes the force of the updraft, and then falls to the bottom of the vessel as finished product.

Balers: Compacters, such as baling machines and hydraulic presses, widely used throughout industry for general waste handling, are not particularly germane to the process industries and are not described here.

REFERENCES

Agglomeration, *Chem. Eng.*, Dec. 4, 1967.
Farrell, R. J.: Extrusion Equipment: Types, Functions and Applications, *Proc. Amer. Assoc. Cereal Chemists 12th Ann. Symp.*, February 1971.
Pellet Mill Makers Woo CPI, *Chem. Week*, Feb. 13, 1960.

—Kenneth R. Sterrett, *Sprout, Waldron & Company, Inc., Muncy, Pa.*

Aggregation   (See **Agglomeration.**)

Agitators   (See **Absorption; Mixing, fluids; Mixing and blending, solids/solids;** and **Waste treatment.**)

Agricultural Chemicals   (See **Chlorine organics; Fertilizers; Gypsum; Limestone;** and **Sulfur.**)

Air   Air is a major industrial medium and chemical raw material. The average composition of dry air at sea level, disregarding unusual buildup of certain pollutive constituents in active areas (highly populated, industrialized, intensively agricultural) is shown in Table A-14. See also **Nitrogen;** and **Oxygen.** The amount of water vapor in the air varies seasonally and geographically and is a factor of large importance where air in stoichiometric quantities is required for reaction processes, or where water vapor must be removed in air-conditioning and compressed-air systems. The water content of air for varying conditions of temperature and pressure is shown in Fig. A-27.

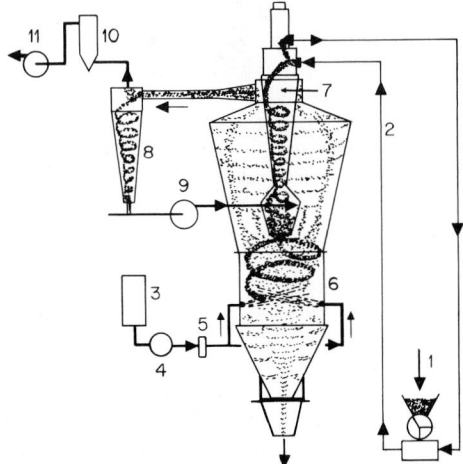

Fig. A-26. Highly schematic representation of a spray-type agglomerator. (1) Raw-material feed, (2) pneumatic dry-materials loading system, (3) liquid-holding tank, (4) metering pump, (5) liquid filter, (6) spray nozzles, (7) cyclone separator, (8) recovery separations, (9) fines-reintroduction system, (10) bag filter, (11) updraft suction fan. (*Spraymix, Sprout, Waldron & Company, Inc.*)

TABLE A-14. *Composition of Dry Air*

Constituent	Percent by Weight	Percent by Volume
Oxygen, O_2	23.15	20.98
Ozone, O_3	1.7×10^{-6}	0.00005
Carbon dioxide, CO_2 .	0.05	0.03
Nitrogen, N_2	75.54	78.09
Hydrogen, H_2	0.000004	0.00005
Argon, Ar	1.26	0.90
Neon, Ne	0.0012	0.0018
Helium, He	0.00007	0.0005
Krypton, Kr	0.0003	0.0001
Xenon, Xe	5.6×10^{-5}	0.000008

TABLE A-15. *Water Content of Saturated Air*

Temperature		Pounds in 1 lb, Kilograms in 1 kg Water
°F	°C	
40	4.44	0.00520
45	7.22	0.00632
50	10.0	0.00765
55	12.8	0.00920
60	15.6	0.01105
65	18.3	0.01322
70	21.1	0.01578
75	23.9	0.01877
80	26.7	0.02226
85	29.4	0.02634
90	32.2	0.03108
95	35.0	0.03662
100	37.8	0.04305
105	40.6	0.05052

The water content of saturated air at various temperatures is shown in Table A-15. See also **Air pollution; Drying, gases;** and **Oxygen.**

Air Filters   (See **Separation operations.**)

Air Pollution   Major air pollutants from industrial sources are summarized in Table A-16. Most of these pollutants are described under various topical headings throughout this volume. Consult the Subject Index.

Nitrogen Oxide Emissions: Calculations for the year 1968 indicate that 16×10^6 tons of nitrogen oxides (NO_x), calculated as NO_2, were emitted to the atmosphere in the United States.* Of this total, it is estimated that 60% of the emissions were produced by stationary sources, the remaining 40% deriving from vehicles of all types. Of the 60% total (or 9.6×10^6 tons of NO_x), 37.5% of the emissions were derived from the electric power generating industry; 29.2% from industrial boilers; 20.8% from internal-combustion engines of the stationary type, as used in gas plants and pipeline gas transmission; 10.4% from small commercial and industrial combustion facilities; and 2.1% from noncombustion sources. The important observation is that 98% of NO_x emissions coming from stationary sources are products of fossil-fuel combustion processes, with only 2% of the total derived from chemical processes. It is estimated

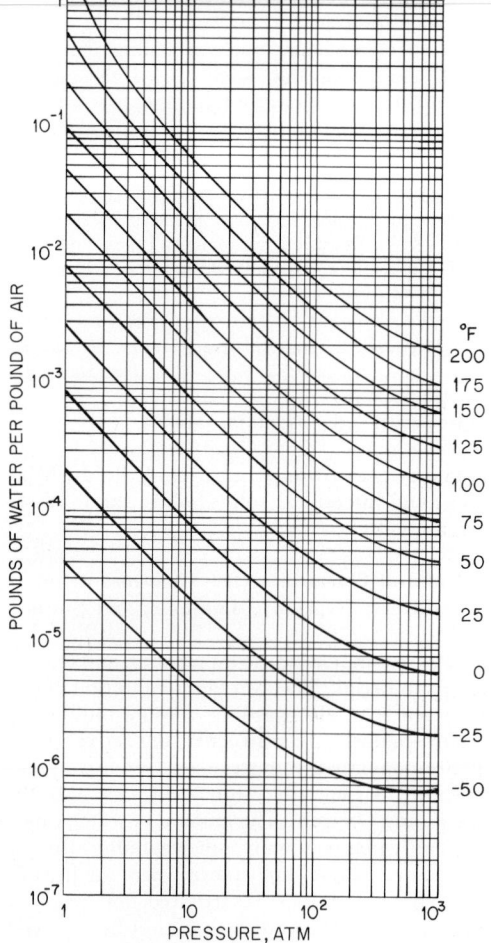

Fig. A-27. Water content of air. (*Based upon original data developed by Landsbaum, Dadds, and Stutzman,* Ind. Eng. Chem., *vol. 47, no. 1, January 1955.*)

*From a report prepared under contract by Esso Research and Engineering Company for the National Air Pollution Control Administration (United States).

TABLE A-16. *Representative Major Air Pollutants and Sources*

Pollutant	Source
Beryllium dust	Ore preparation, metal-working.
Carbon monoxide	Fuel-burning operations—vehicles, power plants, incinerators, dump burning
Carbonaceous smoke, soot, and flyash	Fuel-burning operations, waste recovery, field burning, smudge pots
Chlorinated hydrocarbons	Crop dusting and spraying, chemical plants
Fluorides	Chemical and metallurgical plants
Hydrocarbons	Fuel-burning operations, paint spraying, solvent cleaning, printing, chemical and metallurgical plants
Metal fumes, such as arsenic, lead, and zinc	Metallurgical plants, crop dusting and spraying
Nitrogen oxides (NO_x)	Fuel-burning operations; chemical plants, particularly HNO_3 plants
Organic phosphates	Crop dusting and spraying
Organic vapors	Fuel-burning operations, chemical plants, paint spraying, waste recovery
Particulates (mineral and organic)	Milling, crushing, screening, grinding, and demolition operations; quarries and cement plants.
Radioactive fallout, such as ^{14}C, ^{137}Ce, and ^{90}Sr	Nuclear-device testing
Sulfur oxides	Chemical and metallurgical plants, fuel-burning operations
Uranium dust	Ore preparation

that a large portion of this small segment of emissions comes from HNO_3 plants. It is further estimated that emanations from HNO_3 plants essentially create highly localized problems, whereas combustion processes are found wherever industry is found.

Current fuel consumption in the United States for stationary combustion installations is divided approximately as follows: 40% coal, 40% gas, 15% oil, and 5% refuse and minor fuels.

NO_x results from all fossil-fuel combustion processes where air is used as the oxidant. O_2 from the air and N_2 combine at combustion flame temperatures to form nitric oxide, NO: $N_2 + O_2 \rightleftharpoons 2NO$. The rate at which NO is formed and decomposed depends largely upon temperature. For the majority of stationary combustion processes, there is too short a residence time for the full oxidation of NO to NO_2, an estimated average of only 5–10% of this reaction occurring. It is important to observe, then, that although NO_x emissions generally are given as "equivalent NO_2," the predominant NO_x in combustion gases is NO.

A number of factors affect the generation of NO_x pollutants. Those factors which tend to decrease NO_x emissions include (1) a decrease in excess air for combustion, (2) a decrease in preheat temperature, (3) a decrease in the heat-release rate, (4) an increase in the heat-removal rate, (5) an increase in back-mixing, and (6) a decrease in fuel nitrogen content. With the exception of very large installations, coal appears to generate more NO_x than oil; and oil generates more NO_x than gas.

Although low excess air firing reduces NO_x emission, the air level can be no lower than that required to fully burn the fuel without creating other pollutants, such as smoke and CO, not to mention the limitations imposed by combustion efficiency and equipment design. As each of the foregoing factors is reviewed, it becomes obvious that numerous trade-offs are involved in establishing the most effective combustion-process conditions. In addition to combustion efficiency and economy, factors which always must be given full consideration where energy costs are a significant percentage of production costs, there is the risk of solving one pollution problem while creating other pollution problems.

Current findings indicate that effective steps toward reducing the overall emission of NO_x from stationary combustion sources are: (1) employ low excess air firing; (2) provide for two-stage combustion; (3) utilize flue-gas recirculation; (4) use water injection. And of course use combinations of these techniques. These objectives, reduced to terms of hardware, mean changes in the configuration, location, and spacing of burners, and the kinds of firing and combustion techniques used.

The advantages of low excess air firing for gas and oil combustion in reducing NO_x have been documented well. Further research is required pertaining to commercial coal-fired equipment.

Two-stage combustion may be defined as firing all fuel with below-stoichiometric amounts of primary air in a first stage of combustion, followed by injecting air in a second stage, whereupon burnout of the fuel is completed. There is removal of heat between the two stages. The formation of NO in the first stage is limited because the available O_2 for combination with N_2 is limited. The removal of heat between stages kinetically limits the formation of NO when excess air is added to the second stage. Reports indicate that a 90% reduction in NO_x emission can be achieved in this manner. By recirculating flue gas, both the peak flame temperature and O_2 content are lowered. Injecting low-temperature steam or water also provides a diluting effect. Although probably of limited value for electric utility boilers (because thermal efficiency is lowered), the water-injection technique may be one of the better ways to reduce NO_x emissions in connection with internal-combustion engines of the stationary type.

Tangential firing (the furnace proper is used as the burner) appears to reduce NO_x emissions by as much as 60% as compared with highly turbulent cyclone burners in coal-fired plants, attributed to lower-peak flame temperatures because the flame front is spread out. The objective of lower flame temperatures also leads to consideration of fluidized-bed combustion. It is possible, of course, that chemically bound nitrogen in the fuel may be oxidized to an extent exceeding that from the fixation of N_2 in the atmosphere occurring in current combustion systems.

Treatment of Flue Gas for Removal of NO_x: The NO_x pollutant problem should be approached from the prior standpoint of reducing emissions to the greatest practical extent, and also from the standpoint of removing as much NO_x, once produced, from flue gases to the greatest practical extent. It is estimated that a conventional gas-fired plant of 750-mW capacity will produce approximately 5 tons NO_x/hr, which will be contained in a total volume of flue gas of approximately 70×10^6 scf/hr. Thus, although the NO_x concentration is relatively low, the amount of gas to be treated is extremely large. Further, the problem is complicated by the fact that other objectionable components also must be removed from the flue gas, including SO_2, plus the presence of components which in themselves may not be targets for removal (moisture, CO_2, and O_2) but which may interfere with NO_x removal.

Processes which may be used for NO_x removal include (1) catalytic decomposition, (2) catalytic reduction, (3) adsorption using solids, (4) absorption using liquids, and (5) various physical separations. It is highly desirable, of course, to design a system that will remove both sulfur oxides (SO_x) and NO_x.

Possibly the most attractive method is the catalytic decomposition of NO (thermodynamics are favorable), but unfortunately no catalyst is currently available that will provide sufficient activity within reasonable temperature ranges. For the higher temperatures encountered in fluidized-bed combustion, some alkali or alkaline-earth materials hold promise as decomposition catalysts. Extensive research for a suitable catalyst has been conducted, particularly in connection with auto emissions.

Ammonia has been established as being capable of selectively reducing NO in an atmosphere containing O_2, but practiced in only a limited way in HNO_3 plants. It is attractive because it also may be useful in controlling SO_x emissions. Additional selective reductants include H_2 and H_2S, but for all these approaches suitable catalysts remain to be developed.

A number of common adsorbents[1] have been considered for NO_x removal from flue gas, including alumina, silica gel, molecular sieves, and ion-exchange resins. Each material was to some extent capable of converting NO to NO_2 and then adsorbing the NO_2, but with low efficiencies at the levels of NO concentration in flue gas. Of materials studied, aqueous absorption systems (sulfuric acid or alkaline solutions) appear to have the best potential, especially with recycling of NO_2 back into the flue gas. Sorbent attrition, however, remains a serious problem. Potentially useful alkaline scrubbing systems employ lime-water and magnesium hydroxide. The latter offers easier regeneration of the oxide.

Nitric Acid Plants: The tail gas or off gas emanating from a conventional HNO_3 plant contains approximately 3% O_2, 0.6% H_2O, 0.3% NO plus NO_2, and 96.1% N_2. The reddish plume from HNO_3 plant stacks, of course, is the telltale mark of NO_2. Where the tail gas is treated catalytically with Pt, Pd, or Rh, the reactions are described[2] as $CH_4 + 4NO_2 \rightarrow 4NO + CO_2 + 2H_2O$ (by converting NO_2 to NO, the red telltale is eliminated); $CH_4 + 2O_2 \rightarrow CO_2 + 2H_2O$ (combustion); and $CH_4 + 4NO \rightarrow CO_2 + 2H_2O + 2N_2$ (abatement). The first reaction can be effected simply by adding fuel beyond stoichiometric requirements. Decolorization, while effective from a public relations standpoint, still leaves NO_x as a pollutant and additionally creates another, lesser pollutant, CO_2. Before the complete abatement reaction (the third reaction) can occur, fuel is required well in excess of stoichiometric requirements and the second reaction must be fully completed. It is therefore apparent that achieving complete abatement is many times more difficult than simple decolorization of the red plume.

The use of NH_3 in a selective way with a Pt catalyst[2] is an alternative, this process proceeding along the lines $8NH_3 + 6NO_2 \rightarrow 7N_2 + 12H_2O$; $4NH_3 + 6NO \rightarrow 5N_2 + 6H_2O$; and followed by $NH_3 + 3O_2 \rightarrow 2N_2 + 6H_2O$. NH_3 is required in slight excess over stoichiometric requirements. The Pt catalyst is more costly than Pd (used in nonselective processes). With these cost considerations, the process may be feasible only if low-cost NH_3 is immediately available from other plant operations.

Fuels considered for both decolorizing and abatement include natural gas, methane, synthesis gas, and naphtha. It has been found that catalysts which perform well for decolorization may not be satisfactory for abatement reactions. To date, the most satisfactory catalyst configurations are spherical and honeycomb. The process variables recognized as most important in abatement reactions are tail-gas composition, the inlet and outlet temperatures of the reactor, and stability of the catalyst.

In addition to treatment of NO_x emissions from HNO_3 plants, reduction of the amount of NO_x produced by way of changing the process design and operating conditions must be considered and may, in the long term, provide the most effective solution. A logical target for reducing NO_x emissions is the absorption tower, where additional capacity could be provided. In considering this approach, two conclusions have been drawn:[3] for economic reasons, NO_x emissions should be maintained within the range 1,500–2,000 ppm; and the costs to increase absorber size to reduce NO_x below 500 ppm become excessive. See also **Nitric acid.**

Desulfurization: Efforts to greatly reduce the emission of SO_x, particularly SO_2, to the atmosphere follow patterns similar to those mentioned in connection with NO_x, namely, a three-pronged approach: (1) exerting better control and selectivity over the root source of the contaminants, essentially the sulfur content of fuels; (2) redesigning processes to reduce the amounts of SO_x produced; and (3) developing processes to partially or fully convert pollutants to less objectionable forms. The removal of SO_2 from flue gases is described under **Sulfur dioxide removal.** The removal of hydrogen sulfide and other sulfur compounds, as well as acidic components and CO_2, is described under **Absorption, acidic gases.**

Limestone is a candidate for control of SO_2 pollution from industrial combustion processes. This is a dry process in which pulverized limestone is injected into the boiler, where it calcines and reacts in the gas phase to absorb SO_x. In another process, dry limestone injection is combined with wet scrubbing; the calcined limestone is removed by a scrubber and becomes the reactant that removes SO_2 from flue gases. Efficiency is a major consideration in the limestone processes. About 50% of SO_x can be removed with twice the theoretical amount of limestone required, which is about as much extra dust loading as a boiler can tolerate. The effects of different limestone-dolomite materials are under intensive investigation.

An alkalized alumina process uses an absorbent solid, $Na_2OAl_2O_3$, to remove SO_2 from effluent gas. After absorption, the solid can be regenerated via contact with reducing gases at high temperatures, and the resulting H_2S can be converted to elemental sulfur. In another process, caustic soda is used to absorb SO_2 from flue gas. The resulting sodium bisulfite solution is stripped to release SO_2, which goes to a H_2SO_4 plant. The sodium sulfate solution left over after stripping is sent to an electrolytic cell, where use of a special membrane enables the cell to produce caustic soda, sodium acid sulfate, dilute H_2SO_4, O_2, and H_2. See also **Sulfuric acid.**

Solvent Pollutants: Exclusive of motor exhausts, the discharge of organic solvents into the atmosphere in industrialized and populated areas represents a significant pollution problem. In a major United States city, for example, industrial and commercial sources of organic solvent fumes will average 300–600 tons/day. Major sources of these pollutants are various painting and coating and dry-cleaning operations. Without legal restrictions, unfortunately, some operators have found it most economical to allow vapor-laden air to escape to the atmosphere rather than to attempt to recover the solvents used. Of course, as an operation increases in size, the economic justification for solvent recovery is strengthened.

Major methods used for removing organic vapors from air are (1) carbon adsorption, (2) air incineration, and (3) liquid scrubbing. The costs of all these operations rise as the concentration of vapors in the air treated goes down. Carbon-adsorption systems usually comprise two adsorbers, one on stream while the other is being regenerated with steam. After the steam and stripped solvent are condensed, the solvent is recovered by

TABLE A-17. *Characteristics of Separation Equipment for Aerosols and Dusts*

Separation Equipment	Optimum* Size Particle, μm	Optimum Concentration, grains/ft³	Temperature Limitations °F	Temperature Limitations °C	Pressure Drop†	Efficiency	Space Requirements‡	Collected Pollutant	Remarks
Mechanical collectors:									
Settling chamber	>50	>5	700	371	<0.1	<50	L	Dry dust	Good as precleaner
Cyclone	5–25	>1	700	371	1–5	50–90	M	Dry dust	Low initial cost
Dynamic precipitator	>10	>1	700	371	Fan	<80	M	Dry dust	
Impingement separator	>10	>1	700	371	<4	<80	S	Dry dust	
Bag filter	<1	>0.1	500	260	<4	<99	L	Dry dust	Bags sensitive to humidity, filter velocity, and temperature
Wet collectors:									
Spray tower	25	>1	40–700	4–371	0.5	<80	L	Liquid	Waste treatment required; visible plume possible; corrosion; high-temperature operation possible
Cyclonic	>5	>1	40–700	4–371	<2	<80	L	Liquid	
Impingement	>5	>1	40–700	4–371	<2	<80	L	Liquid	
Venturi	<1	>0.1	40–700	4–371	1–60	<99	S	Liquid	
Electrostatic precipitator	<1	>0.1	850	454	<1	95–99	L	Dry or wet dust	Sensitive to varying conditions and particle properties

source: G. J. Celenza: Designing Air Pollution Control Systems, *Chem. Eng. Prog.*, vol. 66, no. II, p. 32, November 1970.

*Minimum particle size (collected at approximately 90% efficiency under usual operation conditions).

†In. H₂O, gage.

‡S = small; M = moderate; L = large.

distillation or decantation. If the solvent concentration is reduced below 200 ppm, steam requirements rise to about 30 lb steam/lb solvent recovered.

Incineration systems often are used where the solvent-laden air contains a significant amount of preheat. In thermal incineration systems, a temperature in the range 650–760°C (1200–1400°F) must be reached; and in catalytic systems, the effective temperature range is 425–650°C (800–1200°F), which means that a significant amount of fuel is required, particularly without preheat.

Water scrubbers can be used to remove water-soluble organic vapors from effluent air, but very large quantities of water normally are required, particularly if the organic vapors are in low concentration. This, then, can tend to convert an air-pollution problem into a water-pollution problem. Absorption oils are used for vapors which are not soluble in water. Such systems are complicated by the fact that the absorbent oil itself has a tendency to escape to the atmosphere and create a pollution problem. This is true even of mineral oils that have relatively high boiling points. Oil absorption systems generally are impractical unless solvent concentration is in excess of 1% of effluent air.

A combination of carbon-adsorbent–air incineration methods appears to have good potential from the standpoint of overall economics, particularly where solvents of relatively high value and in relatively high volumes are involved.

Much can be accomplished toward reducing solvent pollution through the proper selection of solvents and in the design of the equipment which uses the solvents, striking at the root of the problem, namely, lowering the amount of solvent pollutants created. See **Perchloroethylene; 1,1,1-Trichloroethane;** and **Trichloroethylene.**

Aerosols and Dusts: Aerosols generally are considered to be very tiny spherical droplets of a liquid that may be as small as 0.01 μm in diameter. These small liquid particles and the larger liquid particles, including mists and sprays, along with dusts, permit numerous physical separating and isolating means that generally do not apply to gases and vapors as previously described. A number of rather conventional chemical engineering unit operations, equipments, and processes are applicable in handling these kinds of pollutants. Frequently, there is an economic as well as sociological incentive for producers to separate out valuable by-products that once were pollutants

discharged to the atmosphere. For example, in the processing of numerous ores, many years ago a number of the lesser elements, for which very useful purposes now have been found, simply were lost in stack discharges. Following earlier work of Sir Oliver Lodge, Frederick Gardner Cottrell applied the rudimentary principles of electrical precipitation to the collection of H_2SO_4 mists as early as 1906. Cottrell's first major installations were in connection with Cu smelters. Prior to installation of precipitation and other recovery techniques, it is interesting to note that the Cu smelter located at Anaconda, Mont. (ca. 1910), in a day of normal operation, emitted 3,200 tons of SO_2, 200 tons of SO_3, 30 tons of arsenic trioxide, 3 tons of zinc, and over 2 tons each of Cu, Pb, and antimony trioxide.[4] Years after another smelter had been dismantled, it is reported that a large area centering around the former installation was stripped and sent to concentrators, whereupon over $1 million in Cu and other metals was recovered.

Gas-solids separation techniques are described briefly under **Separation operations.**

A comparison of characteristics and relative

TABLE A-18. *Factors in Physical Characterization of Particulates*

I. Particle size
   A. Significance (discriminating property used)
   B. Size distribution or uniformity
      1. Bases
         a. Count or number
         b. Surface
         c. Mass or volume
         d. Conversion between bases (uniform shape assumed)
      2. Representation of distribution data
         a. Graphical
            (1) Frequency
            (2) Cumulative
         b. Analytical
   C. Representation of average sizes
      1. Median size
      2. Mean size
      3. Conversion between averages
   D. Special problems
      1. Mixture of phases
      2. Flocs and agglomerates
II. Particle shape
III. Particle density and structure
IV. Pore structure and distribution

SOURCE: C. E. Lapple, Particle-Size Analysis and Analyzers, *Chem. Eng.,* May 20, 1968. Used with permission of McGraw-Hill, Inc.

effectiveness of various types of physical separation methods for aerosols and dusts is given in Table A-17.

Air-Pollution Instrumentation: The basic needs for instruments and controls as these relate to air-pollution problems are two: (1) the devices required by pollution-control researchers and engineers, to determine degrees and sources of pollution and to enforce pollution regulations; and (2) the systems required by the organizations that generate the pollutant materials, principally manufacturers of various types, but also including various modes of transportation that depend upon combustion processes for their motive power. An identifiable body of science specifically slanted toward pollution instrumentation has been slow in forming. Consequently, the instruments and controls used largely are adaptations of apparatus that has been developed for other reasons. Pollution engineers have become quite expert in adapting conventional instruments to their specific needs.

Sources of atmospheric contaminants provide clues to the kinds of instrumentation required. Three factors are critical in source testing and analysis for air-pollution control: (1) measuring gas properties, including suspended particulate materials, in situ, in the atmosphere, ducts, chimneys, stacks, and so on; (2) withdrawing representative samples; and (3) accurate analyses.

TABLE A-19. *Particulate-Analysis Techniques*

Size-discriminating Property		Type of Measurement Technique
Type	Character	
Geometric.	Physical barrier	Sieving Ultrafiltration
Mechanical or dynamic (in fluids)	Inertia	Impaction on surface Pressure pulse (sonic)
	Terminal settling velocity	Elutriation Sedimentation Decantation Photosedimentation
	Diffusion	Particle displacement Particle deposition
Optical	Imaging	Light microscopy Ultramicroscopy Electron microscopy
	Transmission	Spectrum
	Scattering	Single-particle count Macroscopic
	Diffraction	Light X-ray Laser
Electrical	Resistance	Current alteration
	Capacitance	Potential pulse
	Charge	Tribocharging Induction charging Corona charging
Magnetic	Magnetic particles	Particle migration, pulse
Thermal.	Particle deposition	Particle migration
Physicochemical	Condensation	Growth of nuclei

SOURCE: C. E. Lapple, Particle-Size Analysis and Analyzers, *Chem. Eng.*, May 20, 1968. Used with permission of McGraw-Hill, Inc.

To date, considerably more automation has been used in the measurement and control of air pollution by the manufacturers (sources of pollutants) than by pollution enforcers. The latter depend heavily upon laboratory techniques rather than continuous analyses. The mobile laboratory, often well instrumented, has added much to the pollution engineer's effectiveness.

Estimation of pollutant emission rates is extremely important in studying air-pollution problems. These rates are determined in two basic ways: (1) by making a materials balance across an entire plant and by reference to shelf information which indicates the weight of contaminants that result from fuel burning, for example; and (2) by direct measurement. Because of the volumes involved, generally it is impractical to "meter" total pollutants. Hence accurate sampling techniques must be developed. Accurate estimate of duct velocities, for example, becomes very important. In some cases, traverse sampling can be used where ports are located to provide a representative cross section of total flow. This introduces the interesting concept of *isokinetic rate*, which can be defined as a sample rate that precisely matches the average velocity of the mainstream.

The measurement and analysis of particulates has progressed rather rapidly because of the emphasis on air-pollution control. The general properties of particulates of interest to the pollution engineer are listed by Lapple[5] and summarized in Table A-18. A short summary of particulate analysis techniques is given in Table A-19.

REFERENCES

1. Bartok, W., A. R. Crawford, and A. Skopp: Control of NO_x Emissions from Stationary Sources, *Chem. Eng. Prog.*, vol. 67, no. 2, pp. 64–72, February 1971.
2. Adlhart, O. J., S. G. Hindin, and R. E. Kenson: Processing Nitric Acid Tail Gas, *Chem. Eng. Prog.*, vol. 67, no. 2, pp. 73–78, February 1971.
3. Newman, D. J.: Nitric Acid Plant Pollutants, *Chem. Eng. Prog.*, vol. 67, no. 2, pp. 79–84, February 1971.
4. Cameron, Frank: "Cottrell: Samaritan of Science," Doubleday, Garden City, N.Y., 1952.
5. Lapple, C. E.: Particle-Size Analysis and Analyzers, *Chem. Eng.*, May 20, 1968, pp. 149–156.
6. Lee, Robert E., Jr.: The Size of Suspended Particulate Matter in Air, *Science*, vol. 178, no. 4061, pp. 567–575, November 10, 1972.
See also references listed under **Chemical engineering.**

Air-Type Mixers (see **Mixing and blending, solids/solids.**)

Alabaster (See **Gypsum.**)

Alanine (See **Amino acids;** and **Peptides and proteins.**)

Alberger Process (See **Salt, NaCl.**)

Albumin (See **Peptides and proteins.**)

Alcohols Alcohols may be regarded as hydrocarbon derivatives in which the hydroxyl group, OH, replaces hydrogen on a saturated carbon. Alcohols are classified as primary, secondary, or tertiary according to the number of hydrogens on the carbon with the hydroxyl substituent. Alcohols also may be regarded as alkyl derivatives of water. Thus alcohols with a small hydrocarbon group tend to be more like water in properties than a hydrocarbon of the same carbon number. Alcohols with a large hydrocarbon group are found to have physical properties similar to a hydrocarbon of the same carbon structure. Some comparisons are shown in Table A-20.

$R' = H$, alkyl, aryl
$R =$ alkyl, aryl

In addition to the basic classification as primary, secondary, or tertiary, alcohols may be grouped according to other structural features. Aromatic alcohols contain an aryl group attached to the carbon having the hydroxyl function; aliphatic alcohols contain only aliphatic groups. The prefix *iso-* usually indicates branching of the carbon chain.

Many alcohols are known by common and IUPAC names. A third system, which is widely used but is unofficial, names alcohols as derivatives of methanol or carbinol. The isomeric butanols and names under the three systems are listed in Table A-21.

Alcohols with two hydroxyl groups are called *dihydric alcohols* or *glycols*. Ethylene glycol, $HOCH_2CH_2OH$, trimethylene glycol, $HOCH_2 \cdot CH_2CH_2CH_2OH$, and 1,4-butanediol are examples of industrially important glycols. See also **Ethylene glycol.** Glycerol, $HOCH_2CHOHCH_2OH$, has three hydroxyl

TABLE A-20. *Comparison of Physical Properties of Alcohols and Hydrocarbons*

Alcohol	Hydrocarbon	Formula	Properties
Methanol	. . . . . .	CH_3OH	Liquid; bp 65°C; sol. H_2O
	Methane	$CH_3 \cdot H$	Gas; insol. H_2O
Ethanol	. . . . .	CH_3CH_2OH	Liquid; bp 78.5°C; sol. H_2O
	Ethane	$CH_3CH_2 \cdot H$	Gas; insol. H_2O
Tetradecanol	. . . .	$CH_3(CH_2)_{12}CH_2OH$	bp 263.2°C; insol. H_2O
	Tetradecane	$CH_3(CH_2)_{12}CH_2 \cdot H$	bp 253.5°C; insol. H_2O

TABLE A-21. *Nomenclature of the Butyl Alcohols*

Formula	IUPAC	Common	Carbinol
$CH_3CH_2CH_2CH_2OH$ . . .	1-Butanol	*n*-Butyl alcohol	*n*-Propylcarbinol
$(CH_3)_2CHCH_2OH$	2-Methyl-1-propanol	Isobutyl alcohol	Isopropylcarbinol
$CH_3CH_2CHOHCH_3$ . . .	2-Butanol	*sec*-Butyl alcohol	Methylethylcarbinol
$(CH_3)_3COH$	2-Methyl-2-propanol	*t*-Butyl alcohol	Trimethylcarbinol

groups per molecule and is a trihydric alcohol. Physical properties of alcohols containing more than one hydroxyl group can be estimated considering the number of carbons for each hydroxyl group, as was done for simple alcohols (Table A-20).

Reactions: At first glance, alcohols appear to undergo a very large number of reactions. However, a closer look reveals that these reactions may be grouped into a few general types. Reactions of alcohols may involve the O—H, but not the C—O, bond. Ester or ether formation is an example of this type of reaction. Primary alcohols generally are easier to convert to these derivatives than secondary alcohols.

$$ROH + CH_3COOH \rightleftharpoons$$
$$CH_3COOR + H_2O \quad (1)$$

$$ROH + SO_3 \rightarrow ROSO_3H \quad (2)$$

$$ROH + nCH_2\!\!-\!\!CH_2 \rightarrow$$
$$RO(CH_2CH_2O)_nH \quad (3)$$

Tertiary alcohols do not give satisfactory yields of esters by the direct reaction.

Primary, secondary, or tertiary alcohols can be converted to salt—another reaction which involves only the O—H bond.

$$ROH + K \rightarrow RO^-K^+ + \tfrac{1}{2}H_2 \quad (4)$$

A reaction which involves the C—O bond is the conversion of an alcohol to an alkyl halide.[1]

$$RCH_2OH + HCl \xrightarrow[170°C]{ZnCl_2} RCH_2Cl + H_2O \quad (5)$$

This is a substitution reaction in which hydroxyl has been replaced by halogen. Primary alcohols undergo substitution in solution by displacement on carbon (Sn_2 mechanism).[2] Tertiary alcohols undergo substitution reactions according to Eq. (5), but usually under somewhat different conditions than are required for primary alcohols.[3] Tertiary alcohols react by an ionization process (Sn_1 mechanism). Secondary alcohols undergo substitution by both Sn_1 and Sn_2 mechanisms.

Many industrially important substitution reactions of alcohols are conducted in the vapor phase over a catalyst. Only primary alcohols give satisfactory yields of product under these conditions.[4,5]

$$CH_3OH + H_2S \xrightarrow{K_2WO_4} CH_3SH + H_2O \quad (6)$$
$$RCH_2OH + (CH_3)_2NH \xrightarrow{Al_2O_3}$$
$$RCH_2N(CH_3)_2 + H_2O \quad (7)$$

A reaction of alcohols which involves the O—H and adjacent C—H bonds is dehydrogenation. This is a reversible reaction which is important in alcohol manufacture as part of the oxo synthesis.

$$R_2CHOH \underset{cat.}{\rightleftharpoons} R_2C\!\!=\!\!O + H_2$$
$$R = H, \text{ alkyl, aryl} \quad (8)$$

Manufacture: Many lower alcohols (amyl and lower) are prepared by hydrogenation of carbon monoxide (methanol), olefin hydration (ethanol, isopropanol, *sec-* and *t-*butanol), hydrolysis of alkyl chlorides, direct oxidation, or the oxo process.

$$C{=}O + 2H_2 \rightarrow CH_3OH \qquad (9)$$

$$CH_3CH{=}CH_2 + H_2O \xrightarrow{H^+} CH_3CHOHCH_3 \quad (10)$$

$$C_5H_{11}Cl + H_2O \rightarrow \underset{\text{(mixture of isomers)}}{C_5H_{11}OH} \quad (11)$$

Most higher alcohols (hexanol or higher) and primary alcohols of three carbons or more are synthesized by one of four general processes or derived from a structurally related natural product.

1. *The Oxo Process:* An olefin may be hydroformylated to a mixture of aldehydes.[6] The aldehydes are readily converted to alcohols by hydrogenation [reverse of Eq. (8)]. Many olefins from ethylene to dodecenes are used in the oxo reaction. The oxo alcohols are typically a mixture of linear and methyl branched primary alcohols. These reactions are discussed under **Oxo process.**

$$RCH{=}CH_2 + CO + H_2 \rightarrow$$
$$RCH_2CH_2CHO + RCH(CHO)CH_3 \quad (12)$$

2. *Aldol Condensation:* Aldehydes also may be dimerized by an aldol condensation reaction to give a branched unsaturated aldehyde. The latter may be converted to a branched alcohol by hydrogenation.

$$2CH_3CH_2CH_2CHO \rightarrow$$
$$CH_3CH_2CH_2CH{=}\underset{\overset{|}{C_2H_5}}{C}CHO \xrightarrow{H_2}$$
$$CH_3(CH_2)_3\underset{\overset{|}{C_2H_5}}{C}HCH_2OH \quad (13)$$

Alcohols from an aldol reaction may be linear if acetaldehyde is a reactant, but usually aldol alcohols are branched primary alcohols. An aldol condensation sometimes is done with an oxo reaction. The combined process is called the *Aldox process.*[7]

3. *Oxidation of Hydrocarbons:* Oxidation of hydrocarbons by air generally results in a mixture of oxygenated compounds and is not a useful synthesis of alcohols except under special circumstances. Cyclohexanol may be prepared by air oxidation of cyclohexane since only one isomer can result.

The yield of alcohol from normal paraffin oxidation may be improved to a commercially useful level by oxidizing in the presence of boric acid.[8]

$$3RH + \tfrac{3}{2}O_2 + H_3BO_3 \rightarrow (RO)_3B + 3H_2O$$
$$(RO)_3B + 3H_2O \rightarrow H_3BO_3 + 3ROH \quad (14)$$

A borate ester is formed which is more stable to further oxidation than the free alcohol. This is easily hydrolyzed to recover the alcohol. These alcohols, which are predominantly secondary, are used in surfactant manufacture. See also **Surfactants.**

4. *Synthesis from Alkylaluminums:* Fundamental work on organoaluminum chemistry by Karl Ziegler and coworkers at the Max-Planck Institute provided the basis for a commercial synthesis of even-carbon-numbered straight-chain primary alcohols.[9] These alcohols are identical with products derived from naturally occurring fats. See also **Alcohol process, ALFOL.**

$$(C_2H_5)_3Al \xrightarrow{CH_2{=}CH_2}$$
$$[CH_3(CH_2CH_2)_nCH_2]_3Al \xrightarrow[(2)\ H_2O]{(1)\ O_2}$$
$$3CH_3(CH_2CH_2)_nCH_2OH + Al(OH)_3 \quad (15)$$

5. *Synthesis from Natural Products:* Many higher alcohols are prepared by reduction of the corresponding acids, which are components of animal or vegetable fats. These alcohols are straight-chain, even-carbon-numbered compounds. Tallow and coconut oil are two major raw materials for higher-alcohol manufacture. See also **Alcohols, fatty (via hydrogenation).** Alcohols from these sources are being augmented by the corresponding ethylene-derived materials as described under the ALFOL process.

A few higher alcohols occur in nature as esters. Spermaceti, a waxy solid obtained from sperm whale oil, is largely cetyl palmitate. Cetyl alcohol is obtained by hydrolysis.

$$CH_3(CH_2)_{14}COO(CH_2)_{15}CH_3 \xrightarrow{H_2O}$$
$$CH_3(CH_2)_{14}COOH + CH_3(CH_2)_{14}CH_2OH \quad (16)$$

Pyrolytic decomposition of sodium ricinoleate results in 2-octanol and sebacic acid.

$$CH_3(CH_2)_5CHCH=CH(CH_2)_7COONa \xrightarrow[\text{(2) neutralization}]{\text{(1) NaOH, O}_2}$$
$$\underset{\displaystyle OH}{|}$$

$$CH_3(CH_2)_5CHCH_3 + (CH_2)_8 \overset{\displaystyle COOH}{\underset{\displaystyle COOH}{\big\langle}} \qquad (17)$$
$$\underset{\displaystyle OH}{|}$$

Most higher alcohols are converted to derivatives and used as plasticizers, lubricants, and detergents. A partial list of the higher alcohols, selected because they are manufactured in large volume, the major process or natural source, and the most important derivatives are shown in Table A-22.

Fermentation Processes: Fermentation remains an important process for ethyl alcohol manufacture although synthetic routes represent most of the industrial alcohol production in the United States. The ethyl alcohol present in most alcoholic beverages is the result of fermentation processes. See also **Enzymes; Ethyl alcohol;** and **Yeasts.**

Properties of a number of representative alcohols are given in Table A-23. See also **Alcohol process, ALFOL; Cyclohexane/cyclohexanol; Detergents; Ethyl alcohol; Ethylene glycol; Isopropyl alcohol;** and **Methyl alcohol.**

TABLE A-22. *Representative Higher Alcohols*

Alcohol	Process and/or Raw Material	Major Derivatives
1-Octanol	Trialkylaluminum, coconut oil	Phthalate, adipate
2-Ethylhexanol	Aldol condensation of oxo aldehyde	Phthalate, adipate
"Isooctanol"	Oxo	Phthalate
C_7–C_{11} mixed, primary alcohols	Oxo	Phthalate
1-Decanol	Trialkylaluminum, coconut oil	Phthalate
"Isodecanol"	Oxo	Phthalate
C_{11}–C_{15} mixed, secondary alcohols . . .	Paraffin oxidation	Ethoxylate
1-Dodecanol	Trialkylaluminum, coconut oil	Ethoxylate, sulfate, acrylates
"Isotridecanol"	Oxo	Phthalate, ethoxylate
1-Tetradecanol	Trialkylaluminum, coconut oil	Ethoxylate, sulfate, acrylates
1-Hexadecanol	Trialkylaluminum, coconut oil	Ethoxylate, sulfate, acrylates
1-Octadecanol	Trialkylaluminum, coconut oil	Ethoxylate, sulfate, acrylates

TABLE A-23. *Properties of Representative Alcohols*

Compound*	Formula	Formula Weight	Sp gr	Mp, °C	Bp, °C
Allyl	$CH_2{:}CH{\cdot}CH_2OH$	58.08	0.854	−129	96.6
Amyl (n-)	$CH_3(CH_2)_3CH_2OH$	88.15	0.817	−78.5	137.9
Amyl (s-, n-)	$C_2H_5CH_2CH(OH)CH_3$	88.15	0.810	. . . .	119.5
Amyl (pri-, iso-)	$(CH_3)_2CHCH_2CH_2OH$	88.15	0.813	−117.2	132.0
Amyl (s-, iso-)	$(CH_3)_2CHCH(OH)CH_3$	88.15	0.819	. . . .	113.4
Amyl (t-)	$(CH_3)_2C(OH)C_2H_5$	88.15	0.809	−11.9	102
Amyl, active	$C_2H_5CH(CH_3)CH_2OH$	88.15	0.816	. . . .	128
Arabitol	$CH_2OH(CHOH)_3CH_2OH$	152.14	. . . .	103	
Benzhydrol	$(C_6H_5)_2CHOH$	184.24	. . . .	68	299
Benzyl	$C_6H_5CH_2OH$	108.13	1.043	−15.3	204.7
Borneol	$C_{10}H_{17}OH$	154.25	1.011	209	213
Butyl (n-)	$C_2H_5CH_2CH_2OH$	74.12	0.810	−79.9	117
Butyl (s-)	$C_2H_5CH(OH)CH_3$	74.12	0.808	−114.7	99.5
Butyl (iso-)	$(CH_3)_2CHCH_2OH$	74.12	0.805	−108	107–108
Butyl (t-)	$(CH_3)_3COH$	74.12	0.779	25.5	82.9

*With exception of those compounds that end with *ol,* the term *alcohol* should follow to fully designate the compound. Examples: allyl alcohol, amyl alcohol, butyl alcohol, and so on.

d, decomposes; m, meta; n, normal; pri-, primary; s, secondary; t, tertiary.

TABLE A-23. *Properties of Representative Alcohols* (*Continued*)

Compound*	Formula	Formula Weight	Sp gr	Mp, °C	Bp, °C
Ceryl	$C_{25}H_{51}CH_2OH$	382.72	. . . .	80	
Cetyl	$CH_3(CH_2)_{14}CH_2OH$	242.43	0.818	49–50	189.5
Cholesterol (*n*-)	$C_{27}H_{45}OH$	386.66	1.067	148	<360
Cholesterol (iso-)	$C_{27}H_{45}OH$	386.66	. . . .	138	
Cinnamyl	$C_6H_5CH:CHCH_2OH$	134.18	1.040	33	254
Crotonyl	$CH_3CH:CHCH_2OH$	72.11	. . . .	. . . .	118
Cycloheptanol	$CH_2(CH_2)_5CHOH$	114.19	0.957	. . . .	185
Cyclohexanol	$CH_2(CH_2)_4CHOH$	100.16	0.962	24	162
Decyl (*n*-)	$CH_3(CH_2)_8CH_2OH$	158.28	0.830	7	232.9
Diacetone	$(CH_3)_2C(OH)\cdot CH_2COCH_3$	116.16	0.931	−47	167.9
Diglycerol	$[(HO)_2C_3H_5]_2O$	166.17	. . . .	. . . .	220–230
Dulcitol	$CH_2OH(CHOH)_4CH_2OH$	182.17	1.466	189	290–295
Eicosyl	$C_{19}H_{39}\cdot CH_2OH$	299.57	. . . .	68	
Ergosterol	$C_{27}H_{41}OH$	382.63	. . . .	160	
Erythritol	$CH_2OH(CHOH)_2CH_2OH$	122.12	1.451	126	329–331
Ethyl	CH_3CH_2OH	46.07	0.789	−112	78.4
Fenchyl	$C_{10}H_{17}OH$	154.24	0.935	35	201
Fluroenoyl	$(C_6H_4)_2CHOH$	182.22	. . . .	116	156
Furfuryl	$C_4H_3O\cdot CH_2OH$	98.10	1.129	−14.6	169.5
Geraniol	$C_{10}H_{17}OH$	154.25	0.883	7–15	229
Glycerol	$CH_2OH\cdot CHOH\cdot CH_2OH$	92.09	1.260	17.9	290
Glycol (ethylene)	$CH_2OH\cdot CH_2OH$	62.07	1.113	−15.6	197.4
Heptyl (*n*-)	$CH_3(CH_2)_5CH_2OH$	116.20	0.824	34.6	175
Hexamethylene glycol . .	$HO(CH_2)_6OH$	118.17	. . . .	42	250
Hexyl (*n*-)	$CH_3(CH_2)_4CH_2OH$	102.17	0.820	−51.6	157.2
Lauryl	$CH_3(CH_2)_{10}CH_2OH$	186.33	0.831	24	255–259
Mannitol	$CH_2OH(CHOH)_4CH_2OH$	182.17	1.489	166	290–295
Methyl	CH_3OH	32.04	0.729	−97.8	64.7
Methylphenyl carbinol .	$(CH_3)(C_6H_5)CHOH$	122.17	1.003	. . . .	205
Myricyl	$C_{31}H_{63}OH(?)$	452.82	0.777	88	
Myristyl	$CH_3(CH_2)_{12}CH_2OH$	214.38	0.824	38	167
Nitrobenzyl (*m*-)	$NO_2\cdot C_6H_4CH_2OH$	153.13	. . . .	27	175–180
Nonyl(*n*-)	$C_8H_{17}CH_2OH$	144.26	0.828	−5	215
Octadecyl	$C_{17}H_{35}CH_2OH$	270.50	. . . .	59	
Octyl (*n*-)	$CH_3(CH_2)_6CH_2OH$	130.22	0.827	−16	194–195
Octyl (*s-, n*-)	$CH_3(CH_2)_5CH(OH)CH_3$	130.22	0.822	38.6	179–180
1,4-Pentadiol	$HOCH_2(CH_2)_3CH_2OH$	104.15	0.994	. . . .	239.4
Phenylethyl	$C_6H_5CH_2CH_2OH$	122.16	1.023	. . . .	219–221
Phenylpropyl (*n*)	$C_6H_5(CH_2)_3OH$	136.19	1.008	<−18	235–237
Phytol	$C_{20}H_{37}CH_2OH$	308.55	0.850	. . . .	145
Pinacol	$[(CH_3)_2C\cdot OH]_2$	118.17	0.967	43	171–172
Propargyl	$CH_3:C\cdot CH_2OH$	58.08	0.922	−17	115
Propyl (*n*-)	$CH_3CH_2CH_2OH$	60.09	0.804	−127	97–98
Propyl (iso-)	$(CH_3)_2CHOH$	60.09	0.789	−85.8	82.5
Propylene glycol	$CH_3CH(OH)CH_2OH$	76.09	1.040	. . . .	188–189
Sorbitol	$[CH_2OH(CHOH)_2]_2$	182.17	1.47	110–112	
Terpineol	$C_{10}H_{17}OH$	154.25	0.933	35	220
Tetrahydrofurfuryl	$C_4H_4O\cdot CH_2OH$	102.13	1.050	. . . .	177.8
Triethylene glycol	$(\cdot CH_2OCH_2CH_2OH)_2$	150.17	1.125	−5	290
Trimethyl carbinol	$(CH_3)_3COH$	74.12	. . . .	25.5	83
Trimethylene glycol . . .	$HOCH_2CH_2CH_2OH$	76.09	1.060	. . . .	214
Triphenyl carbinol	$(C_6H_5)_3COH$	260.32	1.188	162	<360
Vanillic	$CH_3O(OH)C_6H_3CH_2OH$	154.16	. . . .	115	d
Vinyl	$CH_2:CHOH$	44.06	. . . .	. . . .	
Vinyl (poly-)	$(CH_2:CHOH)_x$	$(44.06)_x$	1.3	220d	

*With exception of those compounds that end with *ol,* the term *alcohol* should follow to fully designate the compound. Examples: allyl alcohol, amyl alcohol, butyl alcohol, and so on.

d, decomposes; *m,* meta; *n,* normal; pri-, primary; *s,* secondary; *t,* tertiary.

REFERENCES

1. Kingsley, H. A., Jr., and H. Bliss: *Ind. Eng. Chem.,* vol. 44, p. 2479, 1952.
2. March, J.: "Advanced Organic Chemistry," p. 344, McGraw-Hill, New York, 1968.
3. Hine, J.: "Physical Organic Chemistry," 2d ed., pp. 123–185, McGraw-Hill, New York, 1962.
4. Folkins, H. O., and E. L. Miller: *Ind. Eng. Chem., Process Des. Develop.,* vol. 4, p. 271, 1962.
5. Samykal, K.: United States Patent 2,043,965, 1936.
6. Hatch, L. F.: "Higher OXO Alcohols," Wiley, New York, 1957.
7. Jaros, S. E., and C. Roming, Jr.: United States Patent 3,119,876, 1964.
8. Bashkirov, A. N., and V. V. Kamzolken: *Proc. 5th World Pet. Congr.,* sec. IV, pp. 175–83, 1959.
9. Ziegler, K.: In H. Zeiss (ed.), "Organometallic Chemistry," ACS Monograph 147, Van Nostrand, New York, 1960.

—Allan J. Lundeen, *Continental Oil Company, Ponca City, Okla.*

Alcohols, Fatty (via Hydrogenation)   Fatty alcohols long have been a major raw material for synthetic detergents. Long-chain unbranched primary aliphatic alcohols, RH_2COH, are called fatty alcohols. The primary source of these alcohols is natural fats and oils. The majority of natural fats and oils consist of fatty triglycerides, i.e., glycerine esterified with 3 moles of a fatty acid. The proportions of the various chain lengths are fairly uniform in fats and oils of common origin. Any fatty triglyceride can be used as a raw material for fatty alcohol manufacture. The most common starting materials are coconut oil and tallow. Alcohols with 12–18 carbon atoms find extensive use in detergents, solvents, and petroleum additives.

Two basic processes of fatty alcohol making via reduction have been used. The earlier process was sodium reduction of either fats and oils or sodium reduction of fatty acid esters. More recently, the high-pressure catalytic hydrogenation of fats, fatty acids, and fatty acid esters has eclipsed fatty alcohols from sodium reduction.

One early raw-material source for hydrogenation which has been practical commercially is the hydrogenolysis of triglyceride.

The glycerine so made is subject to further reduction to propane diols, and this loss in yield of a valuable coproduct results in economic losses that have eliminated direct hydrogenation of triglyceride.

The yield of by-product glycerine can be maxi-

Hydrogenolysis of triglyceride

mized by separation prior to hydrogenation. Triglyceride hydrolysis or interesterification with a simple alcohol (such as methanol) have both been practiced.

The early large-scale hydrogenation systems hydrogenated fatty acid. The unit involved the catalytic hydrogenation of fatty acids under high pressure (up to 5,000 psig) and temperature (up to 600°F, 316°C).

$$\underset{\text{Fatty acid}}{\overset{O}{\underset{\|}{RCOH}}} + 2H_2 \rightarrow \underset{\text{Fatty alcohol}}{\overset{H}{\underset{H}{RCOH}}} + H_2O \qquad (1)$$

Soon after the construction of fatty acid hydrogenation plants, it was found that the methyl esters of fatty acids could be hydrogenated more advantageously than the fatty acids themselves. The main advantages of ester hydrogenation over acid hydrogenation are lower pressure, lower temperature, lower catalyst usage, and lower-cost materials of construction. Today the major source of fatty alcohol is the hydrogenation of methyl esters, and more detailed process description follows. See Fig. A-28.

Methyl Ester Making: Methyl esters are prepared by reacting methanol and coconut or tallow triglyceride in the presence of catalytic amounts of sodium.

Catalyst Making: The catalyst used for high-pressure hydrogenation is an Adkins type catalyst. Catalyst making consists in reacting copper nitrate and chromic oxide with ammonia, followed by vacuum filtration, water washing, and finally roasting. In the resultant complex mixture, copper oxide is considered the true catalyst, while the copper chromite, very stable at high temperatures, acts as a stabilizer of the copper oxide. The stability of this catalyst is probably due to a molecular dispersion of the copper oxide throughout the copper chromite.

$$
\begin{array}{c}
\underset{\displaystyle R-\overset{O}{\overset{\|}{C}}O-\overset{\displaystyle H}{\underset{\displaystyle |}{C}H}}{} \\[2pt]
\underset{\displaystyle R-\overset{O}{\overset{\|}{C}}O-\underset{\displaystyle |}{C}H}{} \;+\; 3CH_3OH \;\xrightarrow{\;Na\;} \\[2pt]
R-\overset{O}{\overset{\|}{C}}O-\overset{\displaystyle H}{\underset{\displaystyle |}{C}H}
\end{array}
$$

Triglyceride   Methanol
(refined coconut oil)

$$
3R\overset{O}{\overset{\|}{C}}OCH_3 + C_3H_8O_3 \quad (2)
$$

Methyl esters   Glycerine

Hydrogenation: The primary reaction is the reduction of methyl esters to fatty alcohol and methanol:

$$
R\overset{O}{\overset{\|}{C}}OCH_3 + 2H_2 \xrightarrow{\;cat.\;} R\overset{\displaystyle H}{\underset{\displaystyle H}{C}}OH + CH_3OH \quad (3)
$$

Methyl ester   Fatty alcohol   Methanol

Hydrogenation is accomplished at about 3,000 psig at 550–600°F (290–315°C) in a continuous reactor. A series of three or four vertical reactors is used to provide minimum back-mixing. The catalyst powder, slurried in fatty alcohol, the catalyst slurry, and the methyl esters are introduced into the bottom of the first reactor. Hydrogen is compressed to operating pressure; after compression it is heated to operating temperature and introduced into the bottom of the first reactor through a distributing pipe. The reaction takes place as the materials rise in the successive reactors. Approximately 30 moles of hydrogen is fed per mole of ester. The hydrogen serves not only as the reducing agent, but also as a principal source of heat and agitation.

The outlet stream from the last reactor contains fatty alcohol, methanol, hydrogen, catalyst, and a small level of unreacted ester. In the first separator the liquid portion containing the bulk of the fatty alcohol and catalyst drops out, while the vapors, including hydrogen, pass overhead. The liquid portion, called *crude alcohol,* is cooled and depressurized. It is then centrifuged to concentrate about 90% of the catalyst for recycle to high pressure. The remainder of the catalyst, which is in the effluent from the centrifuge, is removed by filtration.

The vapors from the crude separator are cooled and sent to a second separator, where the methanol and the remainder of the fatty alcohol are removed. The gaseous hydrogen is recycled to high pressure, joining the incoming hydrogen stream. The methanol or overheads stream is filtered.

The crude and overheads stream are combined, polish-filtered, and sent to a stripper, where the methanol is removed. This methanol is added to another recycle methanol stream, dried, and recycled to ester making. The crude alcohol is then fractionated as necessary to provide products with the proper chain lengths. Also during fractionation, any unreacted esters are removed as fatty alcohol–fatty acid esters in the still bottoms.

Two side reactions of importance occur during hydrogenation:

$$
R\overset{\displaystyle H}{\underset{\displaystyle H}{C}}OH + H_2 \rightarrow R\overset{\displaystyle H}{\underset{\displaystyle H}{C}}H + H_2O \quad (4)
$$

$$
R\overset{O}{\overset{\|}{C}}OCH_3 + R'OH \rightarrow R\overset{O}{\overset{\|}{C}}OR' + CH_3OH \quad (5)
$$

The reduction of fatty alcohol to hydrocarbon not only is a loss in yield, but also introduces a product impurity. The formation of a fatty-fatty ester, separable during distillation, requires recycle to avoid loss of yield. The process is controlled mainly by changing reaction temperature, catalyst level, and/or catalyst usage. Increasing temperature increases by-product hydrocarbon formation more than it increases completeness. Lowering rate (increasing reaction time) increases both hydrocarbon formation and completeness of ester reduction, except that, given lower throughputs, the temperature can be lowered to maintain product quality. Increasing catalyst level increases completeness without increasing hydrocarbon formation. The system is run at the highest rate, lowest temperature, and lowest catalyst usage commensurate with yield and alcohol capacity.

REFERENCES

Adkins, Homer: "Reactions of Hydrogen," University of Wisconsin Press, Madison, 1937.

Groggins, P. H.: "Unit Processes in Organic Synthesis," 5th ed., McGraw-Hill, New York, 1958.

Monick, John A.: "Alcohols, Their Chemistry, Properties, and Manufacture," Van Nostrand-Reinhold, New York, 1968.

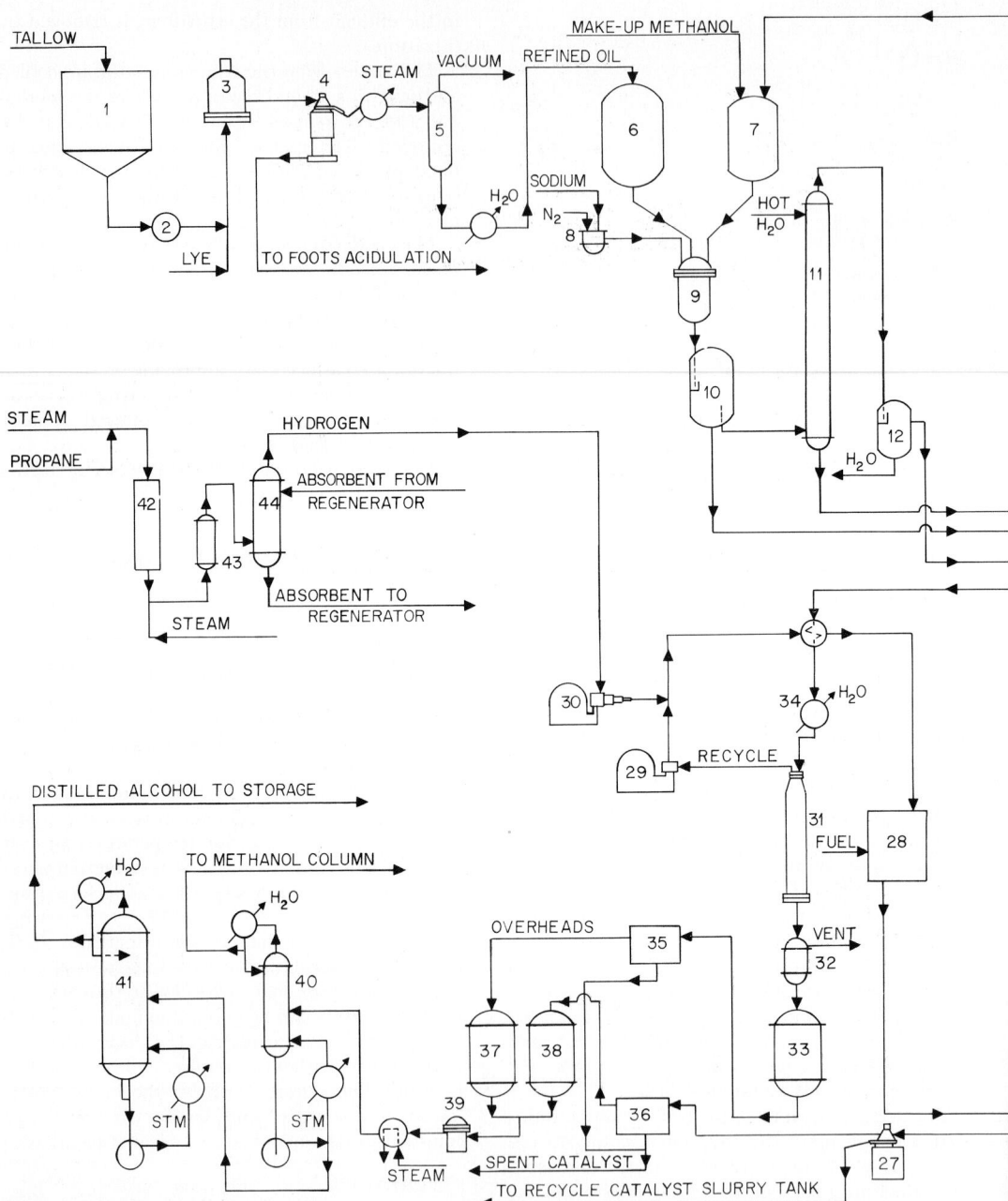

Fig. A-28. Schematic representation of materials flow in the production of fatty alcohols. The process is divided into several major areas: (a) refining—operations 1–5; (b) methyl ester production—operations 6–12; (c) methanol and glycerine recovery—operations 13–18; (d) ester distillation—operations 20 and 21; (e) hydrogenation—operations 23–39; (f) alcohol distillation—operations 40 and 41; (g) hydrogen production—operations 42–44.

Major equipments include (1) feed tank, (2) heater, (3) lye-oil mixer, (4) refining centrifuge followed by steam **heater,** (5) vacuum dryer, (6) dry oil storage, (7) dry methanol storage, (8) sodium melt tank, (9) reactor, (10) settling tank, (11) methyl ester washing column, (12) settling tank, (13) glycerine column feed tank, (14) glycerine column, (15) acidulation vessel, (16) methanol-drying-column feed tank, (17) methanol drying column, (18) settling tank, (19)

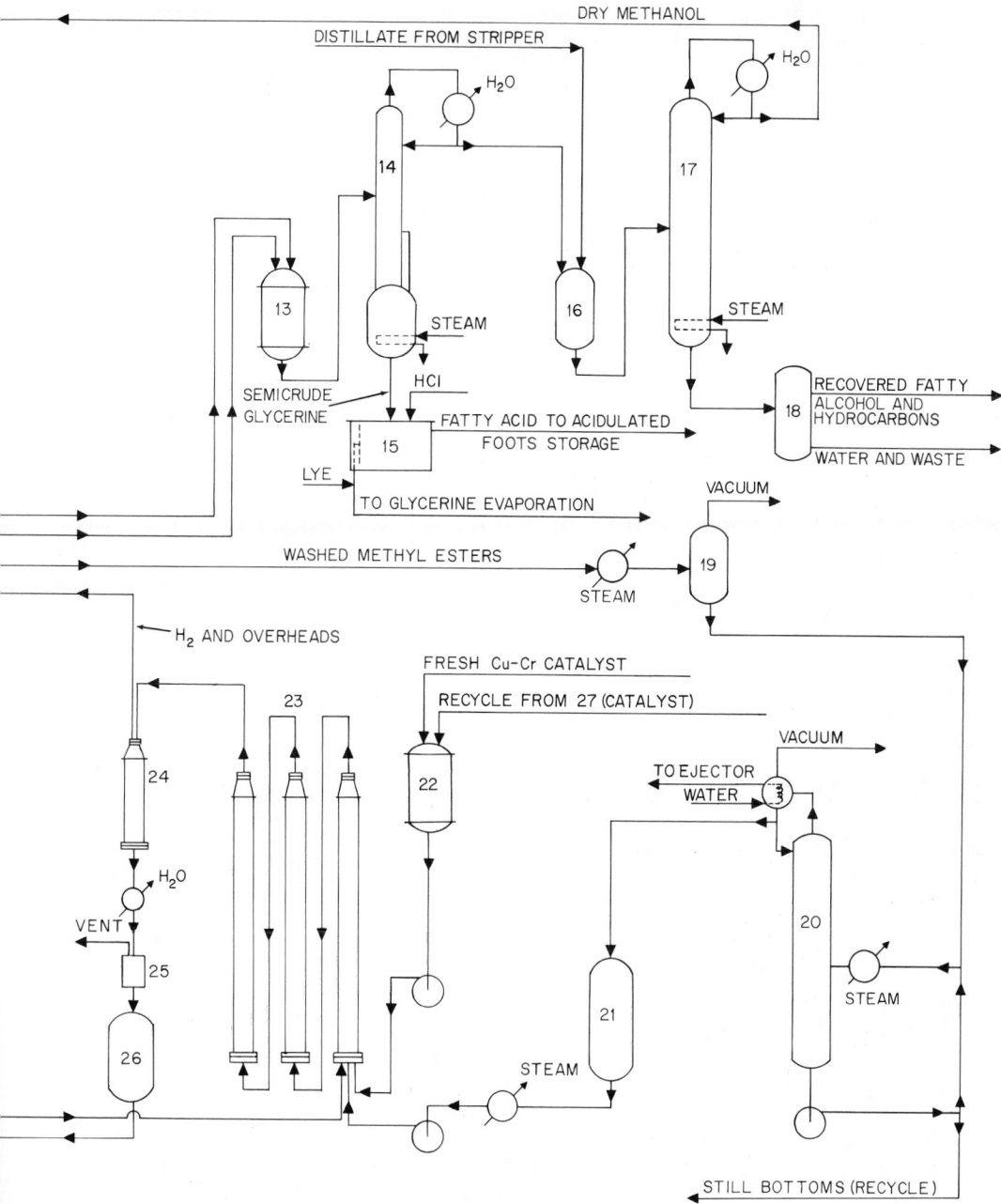

vacuum dryer, (20) ester still, (21) methyl ester feed tank, (22) catalyst slurry tank, (23) hydrogenation reactors (3,000 psi, 550°F), (24) hydrogenation underflow separator, (25) underflow blowdown, (26) underflow centrifuge feed, (27) catalyst centrifuge, (28) hydrogen heater, (29) hydrogen recirculation compressor, (30) hydrogen compressor (makeup hydrogen), (31) overheads separator, (32) overheads blowdown, (33) overheads filter tank, (34) hydrogen cooler, (35) overheads filter, (36) underflow filter, (37) overheads still feed, (38) underflow still feed, (39) polish filter, (40) methanol stripper column, (41) distillation column, (42) reformer furnace, (43) carbon dioxide converter, (44) carbon dioxide absorber. Bottoms from distillation column (41) go to rehydrogenation.

Thomas, J. M. and W. J. Thomas: "Introduction to the Principles of Heterogenous Catalysis," Academic Press, London, New York, 1967.

—Howard A. Mills, Jr., *The Procter & Gamble Company, Cincinnati, Ohio*

Alcohol Process, ALFOL   Fundamental research in organoaluminum chemistry by Karl Ziegler and his associates (Max Planck Institute) provided the chemical basis for a process to synthesize even-carbon-numbered straight-chain alcohols.[*,1,2]

Chemistry of Process: This alcohol process begins with the synthesis of triethylaluminum from aluminum, hydrogen, and ethylene. This is done in two stages, with recycle of two-thirds of the triethylaluminum produced.

$$2Al(C_2H_5)_3 + Al + \tfrac{3}{2}H_2 \rightarrow 3Al(C_2H_5)_2H \quad (1)$$

$$3Al(C_2H_5)_2H + 3CH_2{=}CH_2 \rightarrow 3Al(C_2H_5)_3 \quad (2)$$

The hydrogenation reaction of Eq. (1) is catalyzed by Ti and Zr.[3] This reaction probably does not occur with ultrapure aluminum.

The next step is the addition of ethylene to triethylaluminum to form higher alkylaluminum compounds. This chain-lengthening reaction is called the *growth reaction*. The average degree of polymerization is indicated by m.

$$Al(C_2H_5) + 3mC_2H_4 \rightarrow Al[CH_2(C_2H_4)_mCH_3]_3 \quad (3)$$

Thus the m value of the growth product determines the distribution of products from the process, as shown in Fig. A-29. The curves represent a Poisson distribution.[4]

Addition of ethylene to triethylaluminum is first order in monomeric triethylaluminum and ethylene.[5] Since triethylaluminum is largely dimeric in the liquid phases, kinetics of the growth reaction have the form of Eq. (5).

$$[Al(C_2H_5)_3]_2 \rightleftharpoons$$

$$2Al(C_2H_5)_3 \xrightarrow{\;C_2H_4\;} \text{"growth product"} \quad (4)$$

$$\frac{d(C_2H_4)}{dT} = K[Al(C_2H_5)_3]^{1/2}C_2H_4 \quad (5)$$

*Shortly after Ziegler's initial results were reported, work began in the Ponca City, Okla., laboratories of Continental Oil Company to develop the basic chemistry to a point where commercialization would be possible. The first plant (Lake Charles, La.) was completed in 1962.

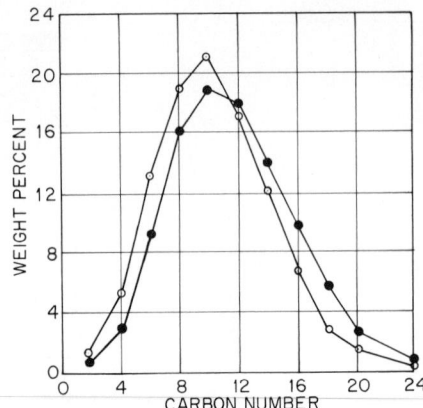

Fig. A-29. Distribution of ALFOL alcohols produced by process shown in Fig. A-30. Where points are represented by solid dots, $m = 4.0$; where represented by circles, $m = 3.5$.

A small amount of 1-olefin is a by-product in the growth reaction. This is a consequence of the facile thermal reaction of trialkylaluminum compounds to form dialkylaluminum hydrides and terminal olefins.[4]

$$(RCH_2CH_2)_3Al \rightleftharpoons (RCH_2CH_2)_2AlH + RCH{:}CH_2 \quad (6)$$

In the presence of excess ethylene, the reverse reaction is insignificant because ethylene reacts with the dialkylaluminum hydride much more rapidly than the terminal olefin.

Growth product is oxidized with air to form an aluminum alkoxide. This step is simple mechanically, but is the most complex chemical step in the process. Dry air is passed into the growth product with vigorous agitation of the mixture. The reaction begins spontaneously and continues as long as air is supplied, or until the reaction is complete.

Oxidized growth product is heated under vacuum to remove solvent and by-products. The resulting stripped oxidized growth product can be hydrolyzed with dilute H_2SO_4 (as in the Lake Charles, La., plant) or with water [as in a later German plant (Condea)] to produce aluminum sulfate and high-purity alumina, respectively. The mixture of alcohols obtained from hydrolysis then is fractionated.

Plant Installation:[6] A simplified flow diagram of the Lake Charles plant is shown in Fig. A-30. Annual capacity is 150 million lb.[7] Aluminum

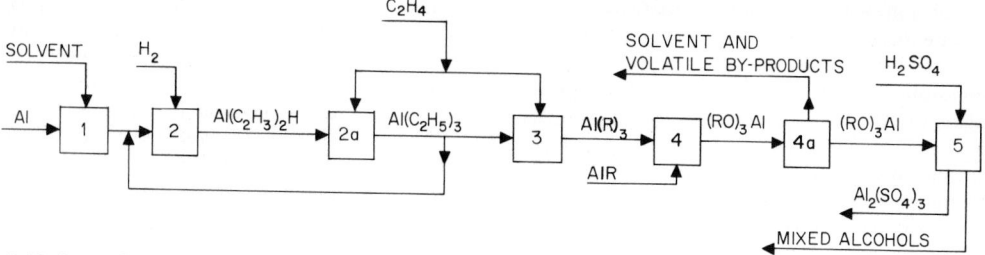

Fig. A-30. Process for manufacturing ALFOL alcohols. (*Continental Oil Company*)

powder containing a catalytic amount[3] of Ti is activated (1) by milling with solvent and triethylaluminum to remove surface oxide. The hydrogenation reaction (2) is followed by reaction with ethylene under mild conditions (2a) to form triethylaluminum.

Triethylaluminum is reacted with ethylene under higher temperature and pressure in a continuous reactor (2) to form growth product.[8] The growth product with solvent and by-product olefins is oxidized (4) and stripped under vacuum (4a) to give "stripped oxidized growth product."[9] Stripped oxidized growth product is hydrolyzed (5) to give a mixture of alcohols. This mixture is separated into blends or pure components by distillation.

The ALFOL alcohols and a few of the major-volume derivatives are listed in Table A-24. Most of the alcohols are furnished as blends of two or more components.

REFERENCES

1. Ziegler, K.: In H. Zeiss (ed.), "Organometallic Chemistry," ACS Monograph 147, pp. 194–269, Van Nostrand, New York, 1960.
2. Köster, R., and P. Binger: *Adv. Inorg. Chem. Radio Chem.*, vol. 7, p. 263, 1965. Comprehensive review of organoaluminum chemistry.
3. Radd, F. G., and W. W. Woods: United States Patent 3,104,252, 1963.
4. Ziegler, K., H. G. Gellert, K. Zosel, E. Holzkamp, J. Schneider, M. Söll, and W. R. Kroll: *Ann.*, vol. 628, p. 121, 1960.
5. Allen, P. E., G. R. Jones, and J. C. Robb: *Trans. Faraday Soc.*, vol. 63, p. 1936, 1963.
6. Lobo, P. A., D. C. Coldiron, L. N. Vernon, and A. T. Ashton: *Chem. Eng. Prog.*, vol. 58, p. 85, 1962.
7. *Oil Paint Drug Rep.*, vol. 198, no. 3, July 20, 1970.
8. Lobo, P. A.: United States Patent 2,971,969, 1961.
9. Foster, C. V., and J. A. Acciarri: United States Patent 3,104,251, 1963.

—Allan J. Lundeen, *Continental Oil Company, Ponca City, Okla.*

TABLE A-24. *Representative Products of ALFOL Alcohol Process*

Generic Name	ALFOL Designation	Major Derivatives and (Uses)
1-Butanol	4	Phthalate (solvent, chemical intermediate)
1-Hexanol	6	Trimellitate (chemical intermediate)
1-Octanol	8	(Mercaptan and amine synthesis)
	610*	Phthalate, adipate
	810*	Phthalate, adipate
1-Dodecanol	12	Sulfate, ethoxylate (mercaptan and amine synthesis)
1-Tetradecanol	14	Sulfate (amine synthesis)
	1214†	Sulfate
	1412†	Ethoxylate
1-Hexadecanol	16	Sulfate, ethoxylate (amine synthesis)
	1216*	Sulfate
1-Octadecanol	18	Sulfate, ethoxylate (amine synthesis)
	1618*	Ethoxylate
	1620*	Acrylate, sulfate
	1218*	Acrylate, ethoxylate

*Mixtures of ALFOL 6, 8, and 10 alcohols (610); ALFOL 8 and 10 alcohols (810); and so on.
†Mixtures of ALFOL 12 and 14 alcohols. The major component is first listed.

Alcohol Sulfates   [See **Anionic surfactants (sulfur-bearing).**]

Alcoholate   An alcoholate results when the hydrogen of the hydroxyl group of an alcohol is replaced by a metal, commonly a metal that forms a strong base, such as sodium or potassium. Sodium ethylate, C_2H_5ONa, also known as sodium ethoxide, is an example. Another example is CH_3OK, potassium methylate or potassium methoxide.

Alcoholysis   (See **Vegetable oils.**)

Aldehyde Carboxylic Acids   (See **Carboxylic acids.**)

Aldehydes   The term aldehyde provides a clue to the derivation of these compounds—*alcohol dehydro*genation. This signifies that removal of two hydrogens from an alcohol will yield an aldehyde, as

$$CH_3 \cdot C\boxed{H_2}OH \rightarrow CH_3 \cdot CHO$$

Ethyl alcohol   Formaldehyde

$$C_2H_5 \cdot C\boxed{H_2}OH \rightarrow C_2H_5 \cdot CHO$$

Propyl alcohol   Propaldehyde

The general formula for the homologous series of aldehydes is $C_nH_{2n}O$. Structurally, an aldehyde is R*CHO*, the *CHO* being the *chemical signature* of the aldehyde. The aldehydes contain a carbonyl group ($C{=}O$), as do the ketones, accounting for certain behavioral similarities with the ketones. Just as the ethers represent a first stage of oxidation of the alcohols to acids, the aldehydes, like the ketones, represent a mid-stage of oxidation. Consequently, an aldehyde may result from the oxidation of an alcohol, or the reduction of an acid.

Nomenclature: The common or trivial names of the aldehydes generally are derived after the fatty acids which the aldehyde yields upon oxidation. Thus the oxidation of *form*aldehyde yields *form*ic acid; the oxidation of *acet*aldehyde yields *acet*ic acid; and so on. In another system, the aldehyde may be named after the alcohol from which it may be derived, so that formaldehyde may be referred to as *meth*aldehyde (from *meth*yl alcohol), or acetaldehyde may be referred to as *eth*aldehyde (from *eth*yl alcohol). Further, the aldehyde may be named after the hydrocarbon from which it theoretically is derived and by adding the ter-

mination -*al*. Thus propaldehyde may be referred to as *propanal* (from *propane*); butaldehyde may be referred to as *butanal* (from *butane*). The use of the termination -*al* here is similar to the use of -*ol* for the naming of the alcohols by this same system.

Profile: The specific properties of a representative sampling of aldehydes are given in Table A-25. The following observations generally apply to most aldehydes:

1. Except for formaldehyde, which is a gas, the aldehydes up to about C_{11} are mobile, neutral, volatile liquids. The higher carbon-bearing compounds are solids.

2. They exude an unpleasant, irritating odor.

3. The lower-carbon aldehydes are readily soluble in water. The solubility rapidly decreases with rise in formula weight.

4. The higher-carbon aldehydes are practically insoluble in water but are miscible with alcohol and ether.

Chemical Reactivity: Although essentially of theoretical importance, it is interesting to note that the aldehydes, like the ketones, are unsaturated because of the double bond between the C and O of the carbonyl group. It should be emphasized that the hydrogen atom that is directly attached to the carbonyl group is not easily displaced. The aldehydes are readily acted on by reducing agents to form primary alcohols. Catalytic reduction is used widely for the commercial production of certain compounds. The aldehydes combine readily with alcohols, with elimination of water, to form *acetals*. Aldehydes undergo a variety of condensation reactions. See **aldol condensation** under **Alcohols.** Aldehydes combine easily with hydrogen cyanide to form *cyanohydrins*. The lower-carbon aldehydes form crystalline additive compounds with solutions of sodium bisulfite, a valuable property in the purification of aldehydes. Aldehydes combine with hydroxylamine to yield *aldoximes*. With hydrazine, aldehydes react to form *hydrazones*. The aldehydes react with semicarbazine to form *semicarbazones*. Upon oxidation, the aldehydes are converted into fatty acids that contain the same number of carbons as found in the originating aldehyde.

Acetals:

$$CH_3CHO + CH_3CH_2OH \rightarrow$$

Acetaldehyde   Ethyl alcohol

$$CH_3 \cdot CH(OC_2H_5)_2 + H_2O$$

Acetaldehyde diethyl
acetal or acetal

Cyanohydrins:

$$CH_3CHO + HCN \rightarrow CH_3 \cdot CH(OH) \cdot CN$$

Acetaldehyde   Ethylidene cyanohydrin or
                      acetaldehyde cyanohydrin

Aldoximes:

$$CH_3 \cdot CHO + H_2N \cdot OH \rightarrow$$

Acetaldehyde   Hydroxylamine

$$CH_3 \cdot CH{:}N \cdot OH + H_2O$$

Acetaldoxime

Hydrazones:

$$CH_3 \cdot CHO + H_2N \cdot NHC_6H_5 \rightarrow$$

Acetaldehyde   Phenylhydrazine

$$CH_3 \cdot CH{:}N \cdot NH \cdot C_6H_5 + H_2O$$

Acetaldehyde hydrazone
or aldehyde phenylhydrazone

Semicarbazones:

$$CH_3 \cdot CHO + NH_2 \cdot CO \cdot NH \cdot NH_2 \rightarrow$$

Acetaldehyde   Semicarbazide

$$CH_3CH{:}N \cdot NH \cdot CO \cdot NH_2 + H_2O$$

Acetaldehyde semicarbazone

The cyanohydrin reaction is important because cyanohydrins readily hydrolyze to form hydroxy acids. Ethylidene cyanohydrin, for example, yields lactic acid; propionaldehyde cyanohydrin yields α-hydroxybutyric acid.

The chemical reactivity of formaldehyde is described under **Formaldehyde**.

Acetaldehyde reacts with acids at relatively low temperature to form metaldehyde, $(C_2H_4O)_x$; with alcohol and dry hydrogen chloride to form acetal, $CH_3 \cdot CH(OC_2H_5)_2$; with ammonio–silver nitrate solution as a reductant to form metallic silver; with anhydrous ammonia to yield aldehyde-ammonia, $CH_3 \cdot CHOH \cdot NH_2$; with chlorine to form chloral (trichloroacetaldehyde), $CCl_3 \cdot CHO$; with formaldehyde and slaked lime to form pentaerythritol, $C(CH_2 \cdot OH)_4$; with hydrocyanic acid to form acetaldehyde cyanohydrin; with hydrogen sulfide to yield thioacetaldehyde, $CH_3 \cdot CHS$; with hydroxylamine hydrochloride to form acetaldoxime, $CH_3CH{:}NOH$; with phenylhydrazine to form acetaldehyde phenylhydrazone, $CH_3 \cdot CH{:}N \cdot NH \cdot C_6H_5$; and with semicarbazide to form acetaldehyde semicarbazone, $CH_3CH{:}N \cdot NH \cdot CO \cdot NH_2$. Evaporation of an aqueous solution of acetaldehyde forms paraldehyde, a cyclic trimer. Magnesium methyl iodide in anhydrous ether, followed by reaction with water, when mixed with acetaldehyde, forms isopropyl alcohol, $(CH_3)_2CHOH$. Phosphorus pentachloride when mixed with acetaldehyde yields ethylidene chloride, $CH_3 \cdot CHCl_2$.

Acetaldehyde may be produced by: (1) *Direct oxidation of ethylene*, using air or oxygen, depending

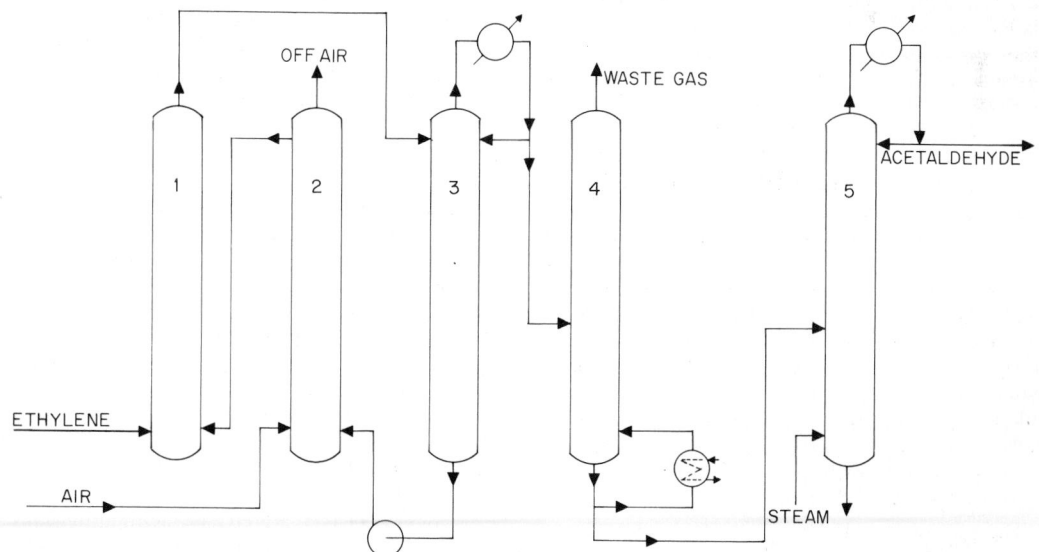

Fig. A-31. Production of acetaldehyde. Highly schematic representation of materials flow in a type of process designed by Hoechst-Uhde Corp. (1) Reactor, (2) oxidizer, (3) crude acetaldehyde still, (4) degasser, (5) final still.

TABLE A-25. *Properties of Representative Aldehydes*

Aldehyde	Formula	Formula Weight	Sp gr	Mp, °C	Bp, °C
Acetaldehyde	CH_3CHO	44.05	0.783	-123.5	20.2
Acrylic aldehyde	$CH_2:CH \cdot CHO$	56.06	0.841	-87.7	52.5
Aldo acetaldol	$CH_3CH(OH)CH_2CO_2H$	88.10	1.103	. . . .	83
Anisaldehyde (p-)	$CH_3OC_6H_4CHO$	136.14	1.123	2.5	247–248
Benzaldehyde	C_6H_5CHO	106.12	1.046	-26	179
Benzil	$C_6H_5COCOC_6H_5$	210.22	1.23	95	348d
Butyraldehyde (n-)	$CH_3CH_2CH_2CHO$	72.10	0.817	-99	75.7
Butyraldehyde (iso-)	$(CH_3)_2CHCHO$	72.10	0.794	-65.9	64
Capric aldehyde	$CH_3(CH_2)_8CHO$	156.26	. . . .	. . . .	208
Caproic aldehyde	$CH_3(CH_2)_4CHO$	100.15	. . . .	. . . .	130
Caprylic aldehyde	$CH_3(CH_2)_6CHO$	128.18	. . . .	. . . .	168
Chloral	$CCl_3 \cdot CHO$	147.40	1.505	-57	97.6
Chlorobenzaldehyde (o-)	ClC_6H_4CHO	140.57	1.29	11	208
Chlorobenzaldehyde (m-)	ClC_6H_4CHO	140.57	1.25	17–18	213–214
Chlorobenzaldehyde (p-)	ClC_6H_4CHO	140.57	1.2	47.8	213
Cinnamic aldehyde	$C_6H_5CH:CHCHO$	132.15	1.10	-7.5	252
Citral (α-)	$C_9H_{15}CHO$	152.23	0.890	. . . .	229
Citronellal (d-)	$C_9H_{17} \cdot CHO$	154.24	0.855	. . . .	204–208
Crotonic aldehyde (α-)	$CH_3CH:CHCHO$	70.09	0.853	-69	102.2
Formaldehyde	$HCHO$	30.03	0.815	-92	-21
Formaldehyde (m-)	$(CH_2O)_3$	90.08	1.17	64	114.5
Formaldehyde (p-)	$(CH_2O)_x \cdot xH_2O$	$(30.03)_x$	. . . .	150–160	s
Furfuraldehyde	$C_4H_3O \cdot CHO$	96.09	1.16	-36.5	161.7
Glyceric aldehyde	$CH_2OHCHOHCHO$	90.04	. . . .	138	
Glycol aldehyde	CH_2OHCHO	60.03	. . . .	97	
Glyoxal	$CHOCHO$	58.01	1.14	15	50
Heptoic aldehyde	$CH_3(CH_2)_5CHO$	114.18	0.850	-42	155
Hydroxybenzaldehyde (p-) . . .	$HO \cdot C_6H_4CHO$	122.12	1.129	116–117	s
Lauric aldehyde	$CH_3(CH_2)_{10}CHO$	184.30	. . . .	45	185
Nitrobenzaldehyde (m-)	$NO_2 \cdot C_6H_4CHO$	151.12	. . . .	58	164
Paraldehyde	$(C_2H_4O)_3$	132.16	0.994	105–112	124.4
Phenylacetaldehyde	$C_6H_5CH_2CHO$	120.14	1.025	. . . .	193–194
Propanal	CH_3CH_2CHO	58.08	0.807	-81	49.5
Propargylic aldehyde	$CH:CCHO$	54.03	. . . .	. . . .	61
Salicylaldehyde	$HO \cdot C_6H_4CHO$	122.12	1.153	-7	196.5
Valeric aldehyde (n-)	$C_2H_5CH_2CH_2CHO$	86.13	0.819	-92	103.4
Valeric aldehyde (iso-)	$(CH_3)_2CHCH_2CHO$	86.13	0.803	-51	92.5
Vanillin	$CH_3O(OH)C_6H_3CHO$	152.14	1.056	81.2	285

d, dextrorotatory; *m*, meta; *n*, normal; *o*, ortho; *p*, para; *s*, sublimes.

upon local oxygen availability and cost and the purity of the ethylene supply. A catalytic solution of copper chloride containing small quantities of palladium chloride is used. The palladium chloride is reduced to elemental Pd and HCl and is reoxidized by the cupric chloride. During the regeneration of the catalyst, the cuprous chloride is reoxidized. See Fig. A-31.

$$C_2H_4 + 2CuCl_2 + H_2O \rightarrow$$
$$CH_3CHO + 2HCl + 2CuCl$$

$$2CuCl + 2HCl + \tfrac{1}{2}O_2 \xrightarrow{PdCl_2} 2CuCl_2 + H_2O$$

(2) *Oxidation of ethyl alcohol* with sodium dichromate. (3) *Dry distillation of calcium acetate* with calcium formate.

Benzaldehyde reacts with acetaldehyde in a slightly alkaline solution to form cinnamic aldehyde, $C_6H_5CH:CHCHO$; with nitric acid to form metanitrobenzaldehyde, $C_6H_4CHO(NO_2)$; with phenylhydrazine to form benzaldehyde phenylhydrazone, $C_6H_5CH:NNHC_6H_5$; with anhydrous sodium acetate and acetic anhydride to yield sodium cinnamate, C_6H_5COONa; with phosphorus pentachloride to form benzylidene chloride,

$C_6H_5CHCl_2$; with sodium cyanide in alcohol solution to form benzoin, $C_6H_5 \cdot CHOHCOC_6H_5$; with sodium hydrogen sulfite to form benzaldehyde sodium bisulfite, $C_6H_5CHOHSO_3Na$; with sodium hydroxide to yield benzyl alcohol and sodium benzoate; and with sulfuric acid to form metabenzaldehyde sulfonic acid, $C_6H_4CHO \cdot (SO_3H)$. Like acetaldehyde, benzaldehyde reduces an ammonio–silver nitrate solution to form metallic silver.

Benzaldehyde may be produced from (1) benzal chloride and calcium hydroxide with heat, (2) calcium benzoate and formate with heat, and (3) boiling dilute acid with glucoside amygdalin of bitter almonds.

Acrolein may be produced by (1) oxidation of allyl alcohol, (2) distillation of fats, and (3) heating glycerol with anhydrous magnesium sulfate.

Synthesis of Alcohols: Because of the growing demand for some of the alcohols (above the butyls), the oxo process for catalytically converting olefins into aldehydes that contain one more carbon than the olefin feedstock is becoming commonplace. The aldehydes serve as an intermediate in a total process that also involves aldol condensation and hydrogenation. See also **Oxo process; Peroxides, organic.**

Amino Acids: Aldehydes also serve as a starting source for certain amino acids. See also **Amino acids.**

Aldol Condensation (See **Alcohols; Aldehydes;** and **Polymerization.**)

Aldoximes (See **Aldehydes.**)

Alfin Copolymer (See **Elastomers.**)

Algin (See **Gums.**)

Aliphatic Compounds A compound that can be considered as a derivative of methane, CH_4, may be termed an aliphatic compound.

The majority of aliphatic compounds are comprised of open carbon chains (straight and branched), saturated or unsaturated. These structures are to be contrasted with the ring structures that characterize the *aromatics* or benzenoids. See also **Hydrocarbons.**

Typical open-chain aliphatic compound
Propionic acid

Typical aromatic compound
Benzoic acid

The distinction pertaining to chain versus ring structure, however, is not precise. There are some exceptions. The aliphatic compound succinimide is of a closed-chain or ring construction, but succinimide definitely is aliphatic because of its relationship to the aliphatic compound succinic acid, into which succinimide can easily be converted.

Ring-type aliphatic compound
Succinimide

Chain-type aliphatic compound
Succinic acid

The term aliphatic (or fatty) derives from the Greek (ἄλειφαρ = oil) and was used originally in connection with the higher acids of the $C_nH_{2n}O_2$ series, such as stearic acid, $C_{18}H_{36}O_2$.

Alkalies [See **Base; Caustic potash; Caustic soda; Lime; Limestone; pH (hydrogen-ion concentration);** and **Soda ash.**]

Alkali Metals (See **Chemical elements.**)

Alkaline Earths (See **Chemical elements.**)

Alkaloids Sometimes referred to as *vegetable alkaloids,* these complex compounds are organic bases, usually crystalline and colorless. In plants, alkaloids usually are found as the crystalline salts of carboxylic acids, such as malic, oxalic, citric, succinic, quinic (cinchona alkaloids), and meconic acid (opium alkaloids). The compounds are best known for their marked physiological activity, each alkaloid producing a specific pharmacological effect. A truly scientific classification of the alkaloids remains to be fully developed, and they most often are classified by their physiological activity. Morphine and related compounds provide relief of pain; curare alkaloids paralyze voluntary muscular activity, and hence sometimes are used as an adjunct in anesthesia; ergot alkaloids are used clinically to induce motility of the uterus in the last stages of pregnancy; belladonna alkaloids prevent normal response of smooth muscle to nervous impulses and are used in the control of excess activity of the gastrointestinal tract and to paralyze the accommodation muscle of the eye in ophthalmic practice; and hyoscine or scopolamine (sometimes referred to as truth serum) causes loss of part of the normal inhibition control.

As will be noted from the following very partial list of alkaloids, most of these compounds are combinations of carbon, hydrogen, oxygen, and nitrogen, with relatively high molecular weights.

Aconine, $C_{25}H_{41}NO_9$
Allantoin, $C_4H_6N_4O_3$
Atropine, $C_{17}H_{23}NO_3$
Belladonnine, $C_{17}H_{21}NO_2$
Caffeine (theine), $C_8H_{10}N_4O_2 \cdot H_2O$
Cocaine, $C_{17}H_{21}NO_4$
Codeine, $C_{18}H_{21}NO_3 \cdot H_2O$
Morphine, $C_{17}H_{19}NO_3 \cdot H_2O$
Papaverine, $C_{20}H_{21}NO_4$
Strychnine, $C_{21}H_{22}N_2O_2$
Thyroxine, $C_{15}H_{11}O_4I_4N$ (also classified as a hormone)
Tropine, $C_8H_{15}NO$

From the standpoint of chemical technology, the principal interest is the synthesis of these materials. For example, derivatives of pyridine include *meperidine hydrochloride* (1-methyl-4-carbethoxy-4-phenyl piperidine), Demerol, an important narcotic and analgesic; *nicotinic acid* (pyridine-3-carboxylic acid), *niacin*, the amide of which is a member of the vitamin B group; *nikethamide* (*N,N*-diethylnicotinamide), a respiratory and heart stimulant; *pipadrol* (α,α-diphenyl-2-piperidinemethanol), a stimulant for the central nervous system; and *piperocaine hydrochloride* [*d*,1-(2-methylpiperidino) propyl benzoate hydrochloride], a local anesthetic. See also **Pyridine and derivatives.**

The useful function of alkaloids as present in several higher seed-bearing plants has not been fully explained, but it is now suspected that they may play an allelopathic role, i.e., the poisoning and suppression of growth of other plant species. In addition to alkaloids, other suspected allelopathic substances include phenolic acids, flavonoids, terpenoid substances, steroids, and organic cyanides.

Alkyd Paints (See **Paint.**)

Alkyd Resins Thermosetting hydroxycarboxylic resins, produced by the esterification of a polybasic acid with a polyhydric alcohol, are also referred to as alkyd resins. A widely used member of this series is prepared from phthalic anhydride and glycerol. Phthalic acid imparts hardness and stability. Alkyd resins made with maleic acid have a higher melting point. A less brittle and softer resin is obtained with azelaic acid. Adipic acid and other long-chain dibasic acids yield resins of considerable toughness and stability. Pentaerythritol or glycol may be used instead of glycerol. The rather wide range of starting ingredients and fillers that can be used provide alkyd resins with a broad scope of properties and applications. Advantages of alkyd resins include high adhesion to metals, ease of coloration, transparency, toughness, heat and chemical resistance, and good dielectric strength. The alkyds are easily blended, and consequently are used widely in paints and finishes. See also **Paints.** The alkyds may be reacted with fatty acids, oils, and other resins, such as melamine and urea, to provide them with the compatibility required by drying oils. High-impact alkyd compounds, which have a high glass content, find applications in switchgear, electrical terminals, relay housings, and electronic encapsulations. Mineral-filled grades are applicable to automotive ignition parts, small switch housings, and electronic insulation. The properties of alkyd resins are summarized in Table A-26 (opposite).

Alkylaluminums (See **Alcohols;** and **Alcohol process, ALFOL.**)

Alkylanilines (See **Amines.**)

Alkylation For the production of a motor-gasoline blending component, alkylation means the chemical combination of *isobutane* with any one or a combination of propylene, butylenes, and amylenes to form a mixture of highly branched paraffin that has a high antiknock rating and good stability. Alkylation reactions are conducted at or slightly below normal cooling-water temperatures and at pressures sufficiently high to keep the feed and reaction mixture in a liquid phase. Either sulfuric acid or hydrofluoric acid is used as a catalyst.

Most all alkylation practiced by the petroleum refining industry employs blends of butylenes and propylenes as the olefin portion of the feedstock. Only a minor amount of amylene is alkylated. The resulting alkylate consists of a mixture of

TABLE A-26. *Properties of Alkyd Resins*

Specific gravity: 1.20–2.30, depending upon type and amount of filler used.

Bonding: After degreasing or sandblasting, molded alkyd parts can be bonded with epoxy bonding resins. Common cements and glues are not effective.

Moisture absorption: Low at room temperature, an advantage for electronic parts. Resins with electrical-grade fillers may absorb as low as 0.05% moisture.

Colorability: Usual colors produced are white, blue, green, medium tan, red, and gray. A strong pigment, such as titanium dioxide, makes opaque light shades possible.

Types of molding compounds:

Fibrous, compounded with long glass fibers (~0.5 in.) to impart strength
Rope (extruded)
Granular, compounded with short ($1/16$-in.) glass, cellulose, or asbestos fibers

	Impact strength, ft-lb/in. of notch	Tensile strength, psi
Fibrous	8–10	6,000–7,000
Rope	1–4	4,000–9,000
Granular	0.3–0.5	3,000–5,000

	Dielectric constant	Continuous max. service temp.
Mineral filler	5.5–6.0	135°C, 275°F
Mineral and cellulose filler	5.8–6.5	121°C, 250°F
Glass filler	5.2–6.0	149°C, 300°F

Design: Molded parts can be held to dimensional tolerances of ±0.001 in./in. Postmold shrinkage usually is small. Sectional thicknesses less than 0.040 in. are not recommended.

---

isoparaffins, ranging from pentanes to decanes and higher, regardless of which olefin is used as a reactant. The simple addition of isobutane to an olefin does not explain the formation of this wide range of compounds found analytically in alkylates. The overall reaction mechanism is exceedingly complex.

Properties of Alkylates: Motor fuel alkylates can be typified as having unleaded octane ratings* in the low and middle 90s, a low sensitivity (Motor and Research Methods octane numbers† nearly equal), and an excellent octane number response to the addition of lead alkyls. These properties characterize alkylates as being a high-quality blending stock for the refiner's gasoline pools and especially for premium grade gasolines. Even though motor fuel alkylates are a mixture of paraffinic isomers having a wide range of molecular weights, they do fall within the boiling range of commercial gasolines. The composition by carbon number of a particular depentanized alkylate, made from a mixed propylene-butylene feed in a commercial unit using hydrofluoric acid as a catalyst, is shown in Table A-27 to give an indication of the range in molecular weights. Although the alkylate is typically predominant in C_8 paraffins, the C_9 and heavier portion ranged up to dodecanes, which were present in the small amount of 0.1%.

*The octane rating of a motor fuel is defined in terms of its knocking characteristics relative to those of blends of isooctane (2,2,4-trimethylpentane) and *n*-heptane. Arbitrarily, an octane number of zero has been assigned to *n*-heptane, and a rating of 100 to isooctane. The octane number of an unknown fuel is numerically equal to the volume percent of isooctane in a blend with *n*-heptane which has the same knocking tendency as the unknown fuel when both the unknown and the reference blend are run in a standard single-cylinder engine operated at specified conditions.

†Motor Method octane numbers are measured at more severe engine conditions and are numerically lower than those determined by the milder Research Method. The difference between the two numbers is termed "sensitivity."

TABLE A-27. *Composition and Carbon Number of a Particular Depentanized Alkylate**

Carbon No.	Composition, vol %
C_6	2.7
C_7	26.8
C_8	63.8
C_9 (and heavier)	6.7

*Made from a mixed propylene-butylene feed in a commercial unit using hydrofluoric acid as a catalyst.

TABLE A-28. *Composition of C_8 Fraction in an Alkylate**

C_8 Isomer	Vol %
2,2,4-Trimethylpentane	58.6
2,3,4-Trimethylpentane	19.3
2,3,3-Trimethylpentane	9.3
2,2,3-Trimethylpentane	0.8
2,4-Dimethylhexane	3.9
2,2- and 2,5-Dimethylhexane	2.3
3,4-Dimethylhexane	0.5
3,3-Dimethylhexane 4-Methylheptane 2-Methylheptane	0.8
3-Methylheptane 2,3-Dimethylhexane	4.5

*Same sample as described in Table A-27.

Also of interest is the isomeric composition of the C_8 paraffins in this alkylate, as given in Table A-28. Well over one-half of the C_8 fraction in the sample was found to be 2,2,4-*trimethylpentane,* which by definition has a numerical value of 100 on the octane number scale. The highly branched trimethylpentanes comprised nearly 90% of the C_8 fraction, and no *n*-octane was found.

Octane numbers and other properties of a typical hydrofluoric acid (HF) alkylate are given in Table A-29. On an unleaded basis, the Motor octane number is only slightly less than the Research octane number. But with 3 ml of tetraethyl lead per gallon, the Motor octane number exceeds the Research octane number. This relationship is typical of alkylates, which, by this characteristic, are noted to have a low octane number sensitivity,

TABLE A-29. *Properties of a Typical Hydrofluoric Acid Alkylate*

Gravity, °API	71.5	
ASTM distillation	°F	°C
Initial boiling point	116	(46.7)
10% point	163	(72.6)
30% point	210	(98.9)
50% point	220	(104.4)
70% point	226	(107.8)
90% point	246	(118.9)
End point	380	(193.3)
Octane ratings:		
Research Method, unleaded	95.0	
Research Method, with 3 ml tetraethyl lead/gal	106.3	
Motor Method, unleaded	92.8	
Motor Method, with 3 ml tetraethyl lead/gal	107.4	

as distinct from gasolines that contain olefins and/or aromatics whose Motor octane numbers can be several numbers lower than their Research octane ratings. Octane quality of alkylates can vary over a small range of up to four or five numbers, depending on the relative amounts of the several olefins usually present in the feedstock as well as on the catalyst condition and on the plant design and operating conditions used to produce the alkylate.

Alkylate-Production Capacity: Since its introduction during the early 1940s to produce a high-quality aviation-gasoline blending component, alkylation capacity has increased steadily. In 1970 (United States), just short of 750,000 bbl/day of alkylation capacity was installed, representing about 5.9% of the crude being refined. About 71% of the alkylation plants use sulfuric acid as a catalyst; the remainder use hydrofluoric acid as a catalyst.*

Feedstocks for Alkylation: Catalytic cracking units are the major source of olefinic feeds for alkylation in the refining industry. A typical C_3/C_4 feedstock produced by catalytic cracking will have the composition indicated in Table A-30. Alkylates produced from butylenes will have octane numbers about three units higher than those made from propylene. From this standpoint, 2-butylene is superior to 1-butylene. C_3/C_4 fractions from catalytic cracking are considered to be excellent alkylation feedstocks because the butylenes amount to nearly 60% of the total olefin content. The isobutane in the C_3/C_4 fraction from cracking, however, is insufficient in amount to alkylate the olefins. Additional quantities must be supplied from catalytic reforming,

*Total alkylation capacity in the free world amounted to about 890,000 bbl/day in 1970.

TABLE A-30. *Composition of Typical C_3/C_4 Feedstock Produced by Catalytic Cracking**

Component	Vol %
Propane	12.7
Propylene	23.6
Isobutane	25.0
n-Butane	6.9
Isobutylene	8.8
1-Butylene	6.9
2-Butylene	16.1

*Composition of feedstock will vary to some extent, depending upon stock being cracked, the catalyst, and operating conditions.

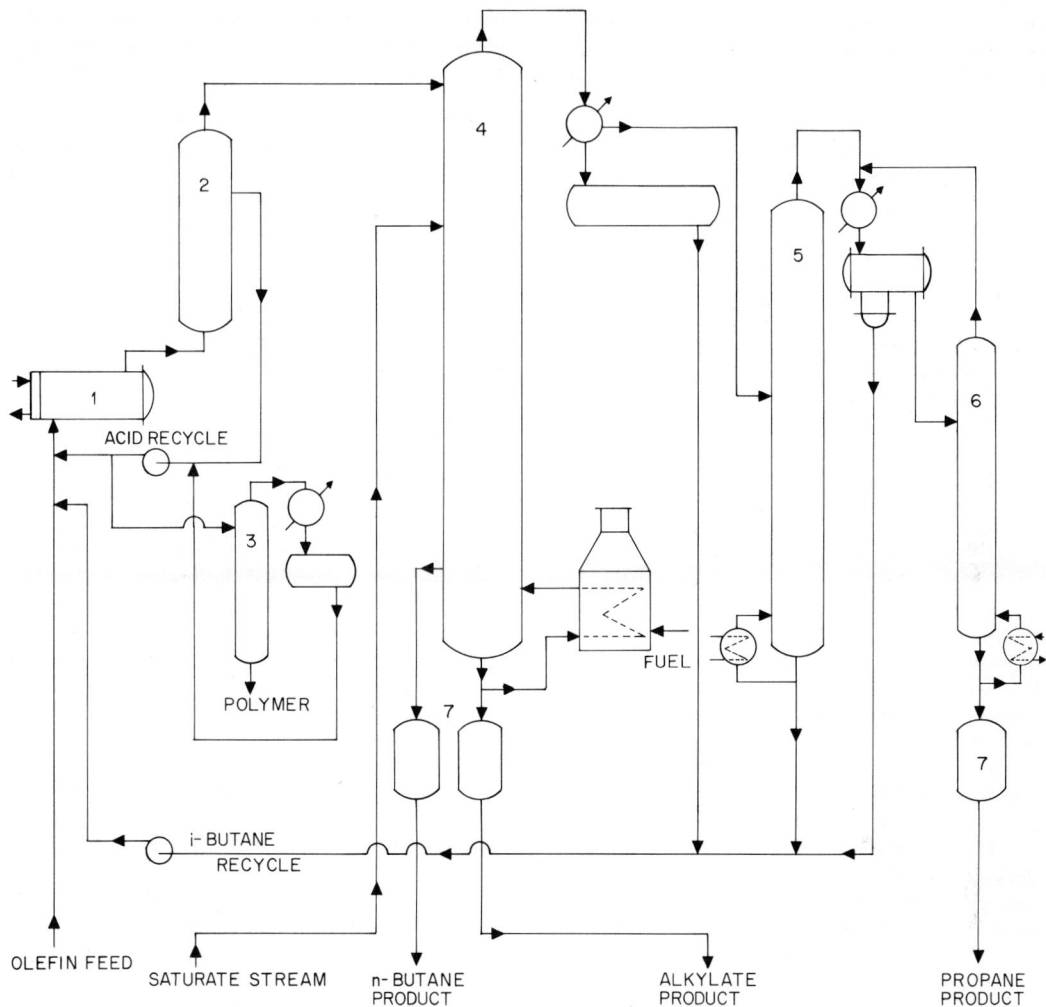

Fig. A-32. Hydrofluoric acid alkylation unit. (1) Reactor, (2) settler, (3) regenerator, (4) isostripper, (5) depropanizer, (6) HF stripper, (7) KOH treaters. (*Universal Oil Products Company.*)

hydrocracking, or from natural-gas sources. See also **Fluid catalytic cracking; Hydrocracking; Isomerization;** and **Reforming.** Under normal HF alkylation conditions, about 1.15 *and* 1.33 volumes of isobutane are required for reaction with 1 volume of C_4 and C_3 olefins, respectively. The resulting yield of alkylate will be 1.75 and 1.77 volumes per volume of the respective olefins. These factors for the isobutane requirement and alkylate yield vary, depending upon operating conditions employed, and perhaps reflect a more meaningful practical description of the overall

reaction than statements of chemical mechanisms, which are complex in accounting for the broad spectrum of hydrocarbons found in alkylates. For instance, to account for the appearance of slightly greater amounts of propane and *n*-butane in the total alkylation-unit products than are brought in with the feed, obviously the principle of hydrogen transfer is involved as a part of the overall reaction mechanism.

HF Alkylation Process: One of several possible configurations of an HF alkylation process is illustrated in Fig. A-32. The principal components of this unit are (1) the reactor, in which intimate contact is brought about between the mixture of C_3 and C_4 olefins, the isobutane, and the HF catalyst and in which the heat of reaction is removed; (2) the acid system, which includes a settler to separate the acid and hydrocarbon phases, provision to recycle acid to the reactor, an acid regenerator, and a stripper to remove acid from the net propane product; and (3) the hydrocarbon fractionation system, which provides a stream rich in isobutane for recycle to the reactor and which yields n-butane and propane as separate products.

Before introduction to the alkylation unit, the feedstocks are treated, primarily to remove sulfur compounds and water. The olefin-containing stream, the HF acid recycle, and a recycle stream that contains a controlled excess of isobutane over that necessary to react with the olefins are fed in various proprietary arrangements to the reactor vessel. In the reactor, the hydrocarbons and acid are intimately contacted to form an emulsion, and herein the reactions occur to produce the alkylate. A substantial heat of reaction is removed in one of several ways. In HF alkylation, cooling water in a heat-exchange-tube bundle within the reactor is commonly used, and thereby controls the emulsion temperature.

The emulsion effluent from the reactor passes to a settler in which the hydrocarbon phase rises and is routed to the fractionation system. The acid phase is recycled to the reactor to maintain the desired ratio of acid to hydrocarbons. A small portion of the recycled acid is diverted to a regenerator in which relatively pure HF is distilled from a minor amount of heavy organic compounds and water. In many units, the acid-regeneration system is operated on an intermittent schedule.

Hydrocarbons leaving the settler are fed to a fractionator that separates an overhead stream rich in isobutane. This stream is recycled to the reactor system to supply isobutane considerably in excess of that required to react with the feed olefins. If the olefin-bearing feed system is deficient in isobutane, an outside saturate stream is brought to the alkylation unit, generally as a blend of isobutane and normal butane. This stream is fed to the isostripper column as shown in Fig. A-32. Normal butane that comes to the alkylation unit with the olefin feed or in the saturate stream is rejected from the system by the isostripper column as a side cut near the bottom. In some cases, this n-butane is processed in an isomerization unit from which the effluent mixture of n- and isobutane is brought back to the isostripper as the saturate stream. See also **Isomerization.** The alkylate is produced as a bottom stream by the isostripper column.

When the olefin feed contains propylene and propane, a portion of the isostripper overhead is routed to a depropanizer which separates propane as an overhead stream. The bottoms stream contains butanes and joins the isobutane recycle stream as shown in Fig. A-32. Because the propane stream will contain a small amount of dissolved HF, an HF stripper is required to recover the acid so that it may be returned to the reactor section with the isobutane recycle. Propane product is removed from the unit as a bottoms stream from the HF stripper.

HF alkylation processes differ from one another primarily in the reactor-section design: in the manner in which acid and hydrocarbons are brought together; in the mechanics of producing an emulsion; in the design, configuration, internals, and arrangement of the reactor and settler; and in the technique for removing exothermic heats of reaction. Some units separate isobutane recycle of relatively high purity in a center-fed, refluxed fractionator, while other units employ a top-fed, unrefluxed isostripper to provide a recycle of somewhat lower isobutane purity, but also at a lower capital and operating cost.

Sulfuric Acid Alkylation: The predominant difference between HF and H_2SO_4 alkylation processes, of course, is the type of acid catalyst used. Other differences include the manner of producing the emulsion, with provision for an extended interfacial surface for the reaction; the way in which the reactor section is designed to accept the hydrocarbon feed and acid; and the geometrical relationship between the reactor and the settler. Important differences result from the technique of removing the heat of reaction and of controlling the reactor temperature by means of evaporative refrigeration. In earlier designs, external refrigerant and a tubular exchanger were used. More modern, proprietary designs involve a refrigerated cascade reactor in one instance, and in another case, a portion of the reactor effluent is vaporized by pressure reduction to provide cooling for the reactor.

Mixed Olefinic Feeds: In the alkylation of mixed olefinic feeds, both H_2SO_4 and HF processes generally employ fractional distillation to separate propane and butane as separate product streams. Unlike H_2SO_4, HF is slightly soluble in hydrocarbons. Thus, in the HF process, provision is made, as shown by Fig. A-32, to strip and recover HF from the hydrocarbon product for return to the reactor section.

Aromatic Alkylation: Although not intended for motor fuels, the refining industry makes a number of petrochemical intermediates by the use of aromatic alkylation with olefins. These applications include (1) the alkylation of benzenes with ethylene to produce ethylbenzene, using special catalysts* (see also **Ethylene alkylation**); (2) benzene alkylation with propylene, using a catalyst,† to make cumene; and (3) detergent alkylate production by reacting benzene with tetramer (a selected cut from a wide-boiling-range polymer), or with *n*-olefins of appropriate molecular weights for use in biodegradable-detergent manufacture. See also **Detergents;** and **Petrochemical complex.**

Process Variables: Conditions considered to be important either to H_2SO_4 or HF alkylation include (1) the isobutane-to-olefin ratio in the reactor feed, (2) the acid-to-hydrocarbon ratio in the reaction zone, (3) the reactor temperature, and (4) acid characteristics. Pressure has a negligible effect on the process, but is maintained sufficiently high to keep the reaction mixture in the liquid phase at its prevailing temperature.

H_2SO_4 alkylation proceeds at temperatures of 4–10°C (39–50°F), a range which yields the best economical combination of butylene alkylate quality and acid consumption. Above 21°C (70°F), alkylate quality diminishes and acid consumption increases.

HF alkylation is conducted at temperatures in the range 24–38°C (75–100°F). Although alkylate qualities are better in the lower regions, there is a tendency toward the formation of alkyl fluorides as the temperature is lowered.

Because of the differences in the manner of maintaining catalyst acidity within the reaction system, the consumption of HF amounts to 0.1–0.3 lb/bbl of product, while H_2SO_4 consumption is in the order of 6–20 lb/bbl of product.

—Melvin J. Sterba, *UOP Process Division, Universal Oil Products Company, Des Plaines, Ill.*

Alkylperoxymetalloids (See **Peroxides, organic.**)

Alkyls (See **Radicals.**)

Allanite (See **Cerium.**)

UOP solid phosphoric acid (SPA) and UOP Alkar catalyst.

†*UOP solid phosphoric acid (SPA).*

Allelopathic Substances (See **Alkaloids.**)

Allophanate (See **Urethanes.**)

Allotropes (See **Carbon; Chemical elements; Iron and Steel; Selenium;** and **Sulfur.**)

Alloys (Consult **Subject Index.**)

Allylic Resins Thermosetting allylic resins include diallyl phthalate, diallyl isophthalate, diallyl maleate, and diallyl chlorendate. These resins shrink little during and after molding and have good electrical properties. Self-extinguishing and highly flame-resistant molding compounds are obtainable. Monomeric allylics serve as nonvolatile cross-linking agents in polyester compounds. The mechanical and electrical properties of polyesters are upgraded by the use of allylic monomers. The allylics find application where the combination of excellent electrical properties, good temperature, and dimensional stability is required. Dimensional precision is aided by the excellent flow characteristics and ease of molding at low temperatures. Preimpregnated glass cloth (known as "prepregs") are made with the allylic prepolymer. Prepregs have the advantage of long storage life prior to molding. They are used widely to make lightweight, intricate, short-run parts. Diallyl phthalate prepolymer is also used in decorative laminates. The properties of diallyl phthalate and diallyl isophthalate prepolymers are summarized in Table A-31.

TABLE A-31. *Properties of Allylic Resins*

	Diallyl Phthalate	Diallyl Isophthalate
Impact strength, ft-lb/in. of notch . . .	0.2–0.3	0.2–0.3
Tensile strength, psi . .	. . . . . .	4,300
Deflection temperature, °C, at 264 psi	155	238
Specific gravity at 25°C	1.27	1.264
Dielectric constant . . .	3.9	3.4
Flexural strength, psi .	7,000–9,000	7,400–8,300
Moisture absorption, %, 24 hr, at 25°C	0–0.2	0.1

Alpha Brass (See **Copper.**)

Alpha Particle An alpha particle is a positively charged subatomic particle having two protons and two neutrons. The nucleus of the helium atom contains two protons and two neutrons. An alpha particle is emitted from certain radioactive elements or isotopes, has high ionizing power but little penetrating ability, and can damage living tissue. An alpha ray comprises a stream of alpha particles, high-velocity particles originating in particle accelerators or in radioactive atoms.

Alpha decay is the radioactive transformation that occurs when an alpha particle is emitted by a nuclide. The decay product is a new nuclide having a mass number four units smaller and an atomic number two units smaller than the original nuclide. The disintegration energy of an alpha disintegration is equal to the sum of the kinetic energy of the alpha particles and the kinetic energy of recoil of the product atom. A radionuclide that undergoes transformation by alpha-particle emission is termed an *alpha emitter*. See also **Radioactive isotopes.**

Alphyls   (See **Radicals.**)

Altaite   (See **Tellurium.**)

Aluminas   (See **Bauxite;** and **Ceramics.**)

Aluminum   (L. *alumen*, alum; F. Brit., G., *aluminium*.) **Al** = 26.98 (at. wt.); 13 (at. no.). Aluminum is a malleable and ductile, light, silvery metal. Al is found in group IIIb in the periodic table along with boron, gallium, indium, and thallium. There are several isotopes of Al, ranging from ^{23}Al with a half-life of only 0.13 s to ^{27}Al, which is 100% stable. Pure Al has a melting point of $660 \pm 1°C$ ($1220 \pm 1.8°F$) and a boiling point of $2452 \pm 15°C$ ($4445 \pm 27°F$). Al is an excellent reducing agent and, because of its great affinity for oxygen, is often used to reduce metals from their oxides. Al is amphoteric. Aluminum hydroxide acts primarily as a base, but also ionizes as a weak acid. Aluminum halides are important catalysts in certain organic reactions. Al forms hydrated double salts of aluminum sulfate and alkali sulfates (alums) that are important commercial chemicals. Although Al is an active metal, an invisible, transparent, tightly adhering oxide film provides stability and corrosion resistance. Commercially pure Al (99.0% Al minimum) is characterized by ductility, good weldability, electrical conductivity, and excellent corrosion

resistance, and commercial applications include food containers, decorative packaging, electrical conductor wire and cable, foil, utensils, powder, and general sheet applications where low mechanical properties are adequate and where good formability is required. Al forms a series of excellent alloys, with a wide range of properties and applications, with such metals as copper, manganese, magnesium, silicon, and zinc.

The important physical properties of Al are summarized in Table A-32. The properties of Al alloys and chemicals are summarized later in this description.

Discovery and Early Production: While Al was predicted as the metal contained in aluminum oxide (alumina) by Lavoisier of France in 1782, Al was first produced in metallic form by H. C. Oersted, a Dane, in 1825. Oersted produced an impure form of Al by heating potassium amalgam with anhydrous aluminum chloride and then distilling away the mercury. Friedrich Wöhler of Germany repeated Oersted's method and produced an aluminum powder in 1827.

The first of two interesting coincidences in the history of Al occurred in 1854. Both the French scientist Henri Sainte-Claire Deville and Robert Wilhelm Bunsen, professor of chemistry at Heidelberg University, working separately, discovered how to isolate Al by using sodium instead of potassium in the amalgam that was used by Oersted and Wöhler.

At the Paris Exposition of 1855, some of the first metal produced by Deville was exhibited next to the crown jewels. Napoleon III was interested in the metal for his army and commissioned Deville to improve his process and to reduce the cost. By 1855, plants were operating in France. The price declined from $115 to $17 per pound in 1859, and although still too expensive for military applications, the metal saw uses as jewelry, table service, and novelties. By 1885, the price of aluminum dropped to $8 per pound.

Electrolytic Process: In 1886, the second interesting coincidence occurred in aluminum research. Charles Martin Hall, while a student at Oberlin College, Oberlin, Ohio, became interested in aluminum and began experimenting to find a better way to produce it. Hall continued his work after graduation and in 1886 discovered that metallic Al could be produced by dissolving alumina, Al_2O_3, in molten cryolite, $3NaF \cdot AlF_3$, at 960°C (1760°F) in a carbon-lined box and then passing an electric current through the bath. Hall applied for a U.S. patent, which was granted in 1889.

TABLE A-32. *Properties of Aluminum*

Atomic radius . . . . 1.48×10^{-8} cm
Atomic volume . . . . 10.0 cm/g atom
Boiling point . . . . $2452 \pm 15°C$; $4445 \pm 27°F$
Covalent radius . . . . 1.18 Å
 Ionic radius . . . . 0.50 Å
Crystal structure . . . . Face-centered cube (fcc)
 Lattice edge . . . . 4.0495×10^{-8} at 25°C (77°F)
Density, liquid . . . . 2.368 g/cm³; 0.0856 lb/in.³ at 660°C (1220°F)
Density, solid . . . . 2.6989 g/cm³; 0.0975 lb/in.³ at 20°C (68°F)
Electrical conductivity, mass . . . . 212.6% annealed copper at 20°C (68°F)
Electrical conductivity, volume . . . . 64.94% annealed copper at 20°C (68°F)
Electrical resistivity . . . . 2.6548 $\mu\Omega$/cm; 14.7 Ω/million ft at 0°C (32°F)†
 2.65 $\mu\Omega$/cm; 16.0 Ω/million ft at 20°C (68°F)†

Temperature coefficient of electrical
 resistance . . . . 0.00429 at 20°C (68°F)
Electrochemical equivalent . . . . 0.3354 g/Ah
Electrode potential (standard hydrogen scale) . . . . -1.66 V at 25°C (77°F)
Electron arrangement in atom . . . . (2) (8) 3
Electronic structure . . . . $(1s^2, 2s^2, 2p^6, 3s^2, 3p^1)$
Emissivity . . . . 3% at 9.3 μm
Heat of combustion . . . . 399,000 cal/g mole; 720,000 Btu/lb mole
Heat capacity . . . . 5.82 cal/(mole)(°C) at 25°C
Ionization potential . . . . 5.98 V, 18.8 V, 28.5V
Isotopes*

	Half-life	Type of decay	Particle energy, eV	Mode of formation‡
^{23}Al	0.13 s	. . . .	. . . . .	. . . . . .
^{24}Al	2.1 s	$\beta+^{26}$	~8.5	^{24}Mg (p, n)
		γ	1.38–7.1	. . . . . .
^{25}Al	7.6 s	$\beta+$	3.2	^{25}Mg (p, n)
^{26m}Al	6.6 s	$\beta+$	3.2	^{25}Mg (d, n)
				^{26}Mg $(d, 2n)$
				^{25}Mg (d, n)
^{26}Al	8×10^5 yr	Electron capture	. . . . .	^{26}Mg (p, n)
		β	1.16	^{27}Al $(n, 2n)$
		γ	1.83–1.14	
^{27}Al	100% stable	. . . .	. . . . .	. . . . . .
^{28}Al	2.27 min	$\beta-$	2.86	^{27}Al (n, γ)
		γ	1.78–1.27	^{27}Al (d, p)
				^{31}P (n, a)
				Al $(\alpha, 2p)$
^{29}Al	6.56 min	$\beta-$	2.5, 1.4	Mg (α, p)
		γ	1.28, 2.43	

Latent heat of fusion . . . . 94.6 cal/g; 167 Btu/lb
Latent heat of vaporization (est.) . . . . 2000 cal/g; 3500 Btu/lb
Magnetic susceptibility . . . . 0.6276×10^{-6} g⁻¹
Mechanical properties 99.99% purity at 20°C (68°F)
 Brinell hardness 12–16
 Percent in 2 in. elongation 65
 0.2% offset yield strength . . . . 3,200 psi
 Tensile strength . . . . 7,000 psi
Melting point . . . . $660 \pm 1°C$ ($1220 \pm 1.8°F$)
 Contraction in volume:
 From liquid to solid at melting point 6.7%
 From liquid at melting point to solid
 at 20°C (68°F) 11.9%
 From solid at melting point to solid
 at 20°C (68°F) 5.6%

*SOURCE: J. E. Lewis, National Academy of Science, Nuclear Science Series, 3032, p. 3.
†Ohms per mil thickness per foot length.
‡p = proton; n = neutron; α = alpha particle; γ = gamma; d = deuteron.

TABLE A-32. *Properties of Aluminum* *(Continued)*

Modulus of elasticity	7,240 kg/mm^2; 10.3 × 10^6 lb/in.^{2}
Modulus of rigidity	2,710 kg/mm^2; 3.85 × 10^6 lb/in.^{2}
Poisson's ratio	0.33
Reflectivity for heat	85–95%
Reflectivity for white light (bright polished metal) . . .	85–90%
Reflectivity for light from tungsten filament	90%
2500 Å	85%
10,000 Å	95%
Specific heat	0.214 cal/g at 100°C (212°F)
Spectral lines	3961.53 Å; 3944.03 Å; 3092.71 Å; 3092.84 Å; 3082.16 Å
Surface tension	900 dynes/cm at 700°C (1292°F)
Thermal conductivity	0.59 (cal)(cm^2)/(s)(cm)(°C) at 25°C
	142.7 (Btu)(ft^2)/(hr)(ft)(°F) at 77°F
Thermal expansion, linear coefficient of	22.5 × 10^{-6} cm/(cm)(°C)
	12.5 × 10^{-6} in./(in.)(°F)
Average coefficient of linear expansion	23.6 × 10^{-6} cm/(cm)(°C) 20–100°C
	13.1 × 10^{-6} in./(in.)(°F) 68–212°F
Thermal-neutron-absorption cross section	0.215 barns (10^{-24} cm^2)
Valence	2, 3
Viscosity	0.01275 poise at 700°C (1292°F)

Paul Heroult, a young man in Gentilly, France, conceived the same process in the same year (1886). It was indeed a remarkable coincidence that Heroult should discover the same process at almost exactly the same time. But the coincidence does not end there. Both men were born in 1863, and thus each man was 23 years old when he made his discovery, and both men died in the same year (1914).

Neither man knew of the other's work until Heroult applied for a patent in the United States. While Hall obtained U.S. patents, Heroult secured patent rights in France and some other European countries.

Modern Production Plants: The commercial electrolytic plants of today are based on the process that was simultaneously discovered by Hall and Heroult. The cryolite used to dissolve the alumina in the process is a snow-white, translucent, sodium-aluminum-fluoride mineral compound found only in Greenland. A method for producing synthetic cryolite has been developed. See also **Cryolite.** Held at about 980°C (1796°F), the molten cryolite dissolves up to about 20% of alumina readily. See also **Bauxite.**

Modern electrolytic cells (Figs. A-33 and A-34) that hold the molten cryolite are large steel boxes lined with carbon which serves as the cathode (negative pole) in the electric circuit. Carbon anodes (positive pole) are immersed in the fused-cryolite bath. A large electric current is passed through the molten bath between these two sets of electrodes. The current breaks down the dissolved alumina into aluminum and oxygen. The molten Al collects at the bottom of the cell and is siphoned off every few days as a sufficient amount accumulates. The oxygen combines with the carbon at the anodes and passes off as carbon dioxide gas.

The Al that collects at the bottom of the cell is molten because the cells are operated at a temperature (see above) well in excess of the melting point of Al. The intense heat is developed by the very large amount of electricity consumed by the cells. Alumina is added intermittently to the fused bath to make up for its decomposition. Carbon anodes are replaced as the oxygen consumes them. Cryolite is added as required to make up for "drag-out" or volatilization losses. Direct current is supplied continuously so that there is no interruption of the electrolysis. The numerous chemical and electrochemical reactions that take place during electrolysis still are not completely understood.

Modern reduction plants may contain up to eight lines of electrolytic cells—"pot lines" of the type illustrated in Fig. A-35. The tapping of molten Al from a Söderberg cell by means of a vacuum ladle is shown in Fig. A-36. Each pot line may contain 100 or more cells of 50,000–150,000 A each, operating at only 4.0–5.5 V. The cells are electrically connected in series and are supplied with electric current from numerous types of electric generation stations, including hydraulic and

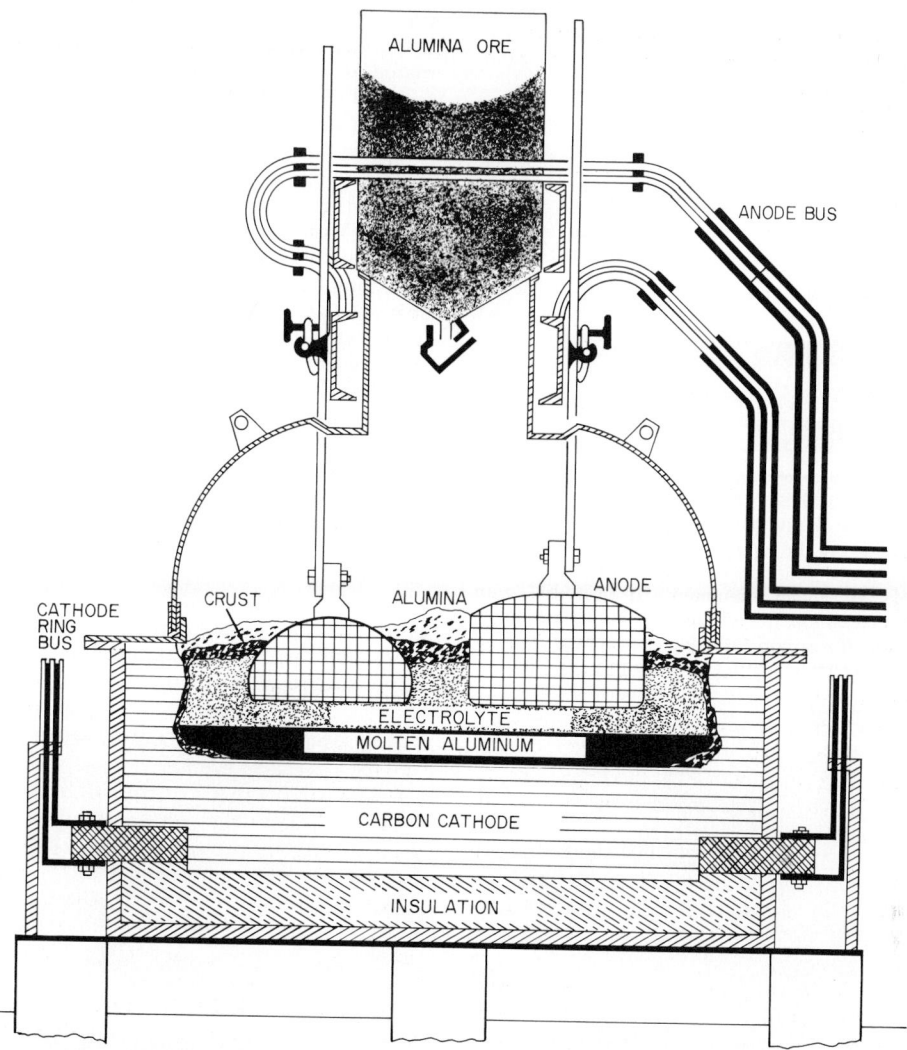

ALUMINA ORE

ANODE BUS

CRUST ALUMINA ANODE

CATHODE
RING
BUS

ELECTROLYTE

MOLTEN ALUMINUM

CARBON CATHODE

INSULATION

Fig. A-33. Prebake-type aluminum-reduction cell showing main parts of a typical Niagara pot. The anodes are separate blocks of carbon that have been prebaked in a separate operation. Each block has a lead to the main bus bar.

atomic power plants. Each pound of Al produced from the modern cells requires 6–8 kWh power, 0.45–0.6 lb carbon, 1.9 lb alumina, and 0.1 lb cryolite.

A new process that could reduce by 30% the amount of electricity required for the electrolytic reduction of aluminum was announced in 1973. The new system eliminates the need for fluoride and operates at lower temperatures in a *completely* enclosed cell. The process combines alumina (refined from bauxite) with chlorine in a chemical reactor to form aluminum chloride. The chloride is electrolyzed in a closed cell, releasing molten aluminum and chlorine. The gaseous chlorine is recycled to the chemical reactor. A 15,000-ton/yr plant is scheduled for production in the United States by 1975.

The lowest price of Al (in the history of the industry) was $0.15 per pound in the 1941–1948 period. In subsequent years, the price has fluctuated from $0.18 to $0.28 per pound. Statistics for 1970 indicate that there were 120 primary reduc-

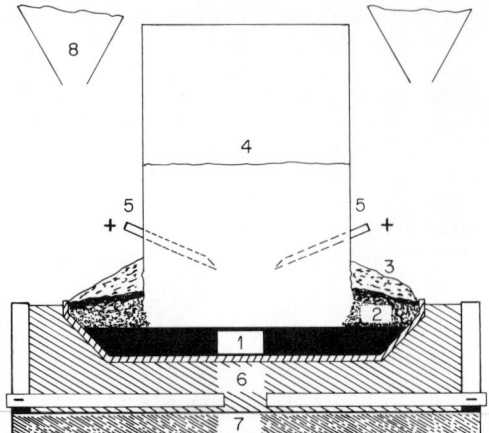

Fig. A-34. Main parts of a typical Söderberg pot. The anode is a single large carbon block which moves down at a controlled rate during electrolysis. New carbon-anode paste is added to the top of the mold so that the anode is being formed continuously and baked out while in position. (1) Molten aluminum, (2) electrolyte with crust formed on top, (3) alumina, (4) carbon anode, (5) anode pins, (6) carbon cathode with cathode bars located directly underneath, (7) insulation, (8) alumina-ore bins.

tion plants in 31 nations around the world. Total world aluminum production for 1970 was estimated at just over 17 billion lb with projections for 1984 of 65 billion lb. In 1970, approximately 49.2% of world Al production took place in North America, 37.1% in Europe, and 9.3% in Asia. The eight leading producers in 1970 were the United States, the U.S.S.R., Canada, Japan, Norway, France, West Germany, and Italy.

Because of its wide range of properties, making it adaptable for many applications, and because of its abundance (third most abundant element), Al will continue to be a metal of growing major importance. See also **Chemical elements.**

Thermal Reduction Process: There has been a continuing search for processes to produce Al where lower capital investments are required and where a greater concentration of electric energy can be brought into the refining process. Present large electrolytic cells are basically low-level energy cells, the largest of which may approach 650 kW.

Electric arc furnaces, on the other hand, make possible a concentration of high energy in a small space, and therefore this approach has been heavily researched. The thermal reduction processes

Fig. A-35. A Söderberg type "pot line" for aluminum production. Heavy-aluminum bus conductors bring current to the pot at the ends of the cell. There are two alumina bins per cell. Automatic jacks raise and lower the Soderberg anode. Fume-control system that links all cells is shown along top of view. (*Reynolds Metals Company.*)

Fig. A-36. Tapping molten aluminum from bottom of Söderberg cell by means of a vacuum ladle. Note crane scale for weight control. (*Reynolds Metals Company.*)

which have been brought to commercial practicability are based upon feeding bauxite, clays such as kaolin or kyanite or partially refined bauxite mixed with carbon or charcoal in briquetted form, into an electric dc or ac arc furnace. Because of the impurities in the raw clays and ores, only impure aluminum alloys are produced, containing 65–70% aluminum, 25–30% silicon, and 1% or more of iron, as well as other impurities, such as titanium, carbides of aluminum and iron, and the nitrides of these same metals.

Distillation Process: The impure output from the thermal reduction furnaces can be refined further to obtain pure Al. Some of these processes include the distillation of relatively pure Al at high temperatures from zinc, magnesium, or mercury baths in which the impure alloy has been dissolved. Aluminum dissolves in these metals and leaves the impurities as residues in solid form, which are removed by filtration. The mercury process is receiving attention because it provides a possibility of producing 99.99+ purity aluminum.

Gas-Reaction Purification: A process based upon reactions of gas phases at high temperature has received considerable attention in recent years. The process is based upon reacting aluminum trichloride gas with molten Al at 1000°C (1832°F) to produce aluminum monochloride, AlCl, gas. The aluminum chloride gas is then cooled outside of the reacting chamber by bringing the gas vapor into contact with molten Al at a lower temperature. The same process also can be used with aluminum fluoride, AlF_3, gas instead of $AlCl_3$ gas.

The feed materials for these processes can be scrap aluminum, aluminum produced by the thermal reduction process; aluminum carbide, Al_4C_3; or aluminum nitride, AlN.

Aluminum carbide is produced in an electric arc furnace by reacting carbon and alumina at approximately 2400°C (4352°F). The alloy is allowed to cool slowly after removal from the furnace so that the carbide crystallizes in a network within which pure Al separates and freezes. The Al within this network of the carbide crystals can be leached out with molten chloride, or fed directly into the gas processes just described. The carbide residue can be recycled in the arc furnace.

Aluminum nitride is produced in an electric furnace by bringing carbon and nitrogen and alumina into reaction at 1750–1800°C (3180–3270°F). The aluminum nitride then is dissociated in a vacuum furnace at 1800°C (3270°F), from which the Al vapor is condensed on cool surfaces. The nitrogen is removed as a gas and recycled.

The principal problems associated with all but the electrolytic processes are concerned with materials of construction because of the very high temperatures involved. Several thermal reduction processes producing impure Al are operating commercially in Europe.

Superpurity Aluminum: In the electrolytic process, Al purity as high as 99.94% can be produced provided that the alumina used is refined to high purity. Pure synthetic cryolite is required, and pure carbon must be used in the anodes of the cell. For normal commercial production, these restrictions are not practical to overcome. Therefore the normal purity from electrolytic cells ranges from 99.35 to 99.9% Al.

To satisfy a demand for superpurity Al, the metal from the conventional electrolytic cells can be superrefined in a Hoopes type cell (Fig. A-37) which was developed in 1924 and is based upon a three-layer principle. The impure Al is alloyed with 33% copper to give the eutectic composition which serves as the anode in the cell. A fused-salt layer, containing 60% barium chloride, $BaCl_2$, and 40% chiolite, $AlF_3 \cdot 1.5NaF$, with a melting point of 720°C (1328°F), floats on the aluminum-copper alloy.

The density of each layer of the cell must be carefully maintained for the cell to operate effectively. This cell differs from the normal electrolytic cell in that only the base of the cell is made of carbon blocks. The walls of the cell are nonconducting magnesia brick. The top layer of the

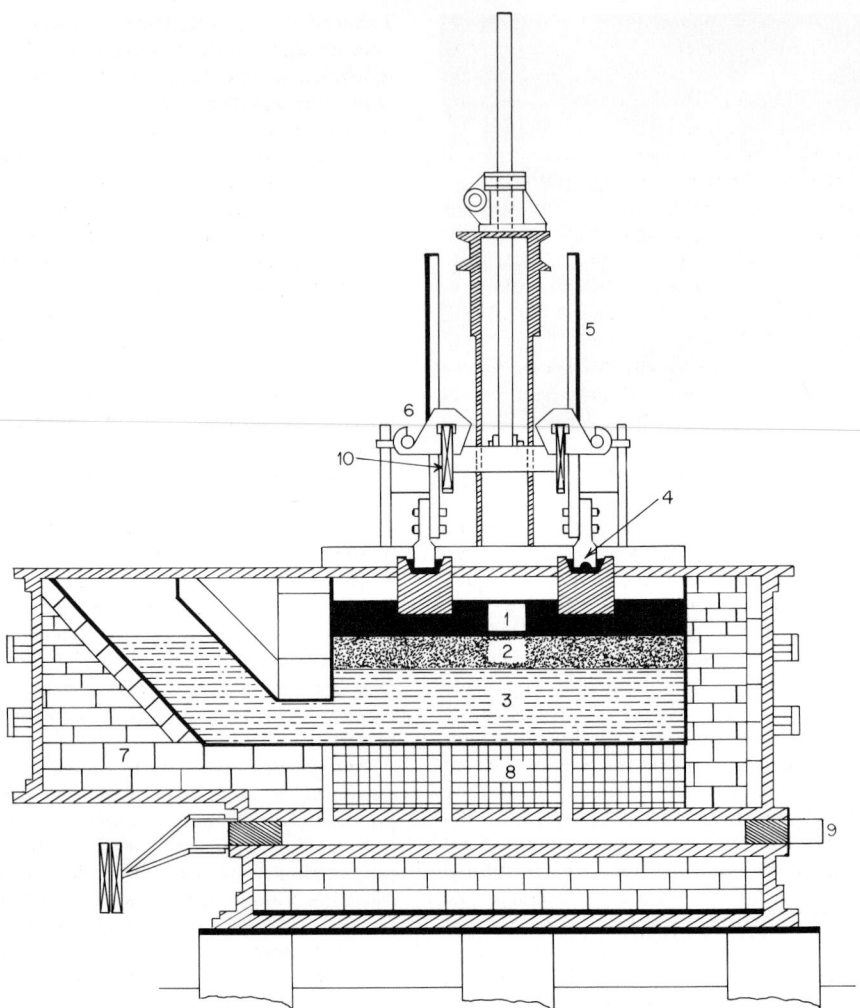

Fig. A-37. Hoopes type cell for producing superpurity aluminum. Superrefined metal accumulates in top layer (1). Middle layer (2) is fused salt (60% $BaCl_2$, 38.5% AlF_3, 1.5% NaF) with a melting point of 720°C (1328°F). Bottom layer (3) is impure aluminum alloyed with 33% copper. Cell components include (4) graphite cathode and stub, (5) electrode bar(s), (6) electrode bar clamps, (7) anode (block lining), (8) carbon-anode block, (9) anode bar, (10) cathode bus.

three-layer cell is superrefined Al which floats on the fused-salt layer. Graphite electrodes for the cathode are immersed in this third layer. These refining cells are operated at 750°C (1380°F) at 6 V. Cells with current consumptions as high as 20,000 A are in commercial production.

Hoopes cells operate with an efficiency of 98% and can produce metal with a purity of 99.995% Al. The cells are operated on a continuous basis and are provided with an inlet to feed more impure Al into the anode layer. Periodically, some

of the anode metal which builds up with impurities is removed and replaced with fresh eutectic alloy. To maintain the high purity of the refined metal, the molten product usually is cast into graphite equipment to avoid contamination by iron, since iron is readily soluble in molten Al.

Zone Refinement: Extreme-purity Al (99.9996+) can be produced by zone-melting refinement of Al that has been produced by the Hoopes cell. In the zone-melting process, a rod of impure metal is melted in a localized area with a narrow induction coil. The coil is moved along the rod slowly, thus transferring the melted zone from one end of the rod to the other. With each pass, always in the same direction, the impurities move with the liquid zone, resulting in concentration of impurities at one end of the rod. The remainder of the rod increases, incrementally, in purity.

Aluminum Alloys. Pure Al possesses many desirable characteristics: light weight, pleasing appearance, good malleability, high electrical and thermal conductivity, and excellent resistance to corrosion. In addition, Al is nonmagnetic and nonsparking and has superb reflectivity for light of all wavelengths, as well as radiant heat. Aluminum can be strengthened and hardened by (1) addition of other metals to form alloys, (2) heat treatment of some types of alloys, and (3) strain hardening by cold working. Over 200 alloys have been introduced. Some Al alloys were developed for cast products, such as sand castings, permanent mold castings, die castings, and centrifugal castings; other Al alloys were developed for wrought products such as sheet, plate, wire, rod, bar, tube, pipe, forgings, and other shapes.

Alloying Elements: Silicon, iron, copper, manganese, magnesium, chromium, nickel, zirconium, and zinc are the most common alloying elements used in commercially produced Al alloys. A number of elements are used in trace quantities to improve alloy properties and characteristics. Some high-strength alloys, capable of developing 100,000 psi tensile strength, may contain as many as six elements in addition to those present in controlled amounts as impurities.

Undesirable Elements: Alkaline metals, such as sodium, potassium, and cesium, are almost completely insoluble and undesirable in Al alloys. Hydrogen is appreciably soluble in molten Al and can be generated in reactions of molten Al with moisture, which results in hydrogen and aluminum oxide. Hydrogen is partially rejected and partially coalesced upon freezing and results in gross porosity or voids in the solidified metal. Thus hydrogen is deliberately removed by special fluxing treatments before the metal is cast into solid forms.

Metallurgical Treatments: Aluminum alloys become harder upon cold working and become softer when annealed at temperatures up to 400°C (750°F). Metallurgists take advantage of these properties in optimizing the characteristics of each alloy.

Classification of Aluminum Alloys: Because Al alloys are classified in an orderly manner, con-

sumers are able to find the correct and most economical alloy for each application. Primary producers make the following fabricated products: ingot, forgings, sheet and plate, powder, foil, cast bar and rod, bare wire and cable, covered and insulated wire and cable, drawn tubing, welding tubing, extrusions, and castings. The major Al-alloy systems are described in Table A-33.

Aluminum Particles and Powders: These products are prepared by atomizing molten Al metal or by hammering and ball-milling Al foil. Aluminum powders are used for paints, fillers in plastics, and inks, to form aluminum halides and alloys. Aluminum powders have a high ratio of surface to volume, and therefore ignite easily and burn with intense heat. Because of this characteristic, Al powders are used in pyrotechnics, flares, incendiaries, explosives, and in solid fuels for rocket propulsion.

Aluminum-alloy particles and powders also can be pressed into solid shapes and diffusion-bonded to form finished parts, such as cams, gears, fittings, bearings, and decorative parts. Aluminum-alloy acicular particles can be continuously rolled into sheet for the same applications found for conventionally produced sheet from rolled ingots. Particles also can be compacted and extruded into shapes by conventional processes.

Aluminides: These materials are high-melting-point phases in metallic systems with Al. One system of potential high value is the nickel-aluminum system where Al is present as 14–34% by weight in Ni_3Al and NiAl. Powder-metallurgy techniques for pressing and diffusion bonding and sintering are used. The 17.5% Al alloyed with nickel exhibits the best strength properties— 80,000 psi at 20°C and 50,000 psi at 815°C. This system can be rolled at 1315°C. Aluminides are present in certain permanent magnets, such as CoAl, FeAl, and NiAl.

Joining Methods. Aluminum can be used to produce any type of metal fasteners, including rivets, screws, nuts, bolts, clips, and studs. As with other applications, the proper Al alloy should be selected carefully for the job to be done.

Welding: Eight welding processes are used with Al. The most common methods are TIG (inert-gas tungsten arc), MIG (inert-gas-shielded metal arc), electron-beam welding, friction welding, and resistance spot welding. Special alloy filler wires are used for the metal-arc processes. No filler wires are required by the other methods. Cold welding of thin foils and sheets and of wire can be accomplished by applying ultrasonic energy or

TABLE A-33. *Major Aluminum-Alloy Systems**

IXXX Series Unalloyed, non-heat-treatable	Commercially pure aluminum (99.0% min.). This series is characterized by purity, ductility, good weldability, electrical conductivity, and excellent corrosion resistance. Commercial applications include food containers, decorative packaging, electrical-conductor wire and cable, foil, utensils, powder, and general sheet applications where low mechanical properties are adequate (8,000–24,000 psi tensile strength, 11,000–22,000 psi yield strength) and good formability is required. Superpure (99.999% aluminum) is used primarily for research purposes in chemical reactions, for special capacitors in electronics, for superconducting magnets at cryogenic temperatures, in optics for vapor-plated mirrors, and in physics for fiber research.
2XXX Series Aluminum-copper, heat-treatable	The mechanical properties of this series match or exceed those of mild steel. These alloys (2014, 2124, 2024, 2117, and 2219) are used in structural applications for aircraft and aerospace hardware because of their strength (43,000–68,000 psi tensile strength; 24,000–50,000 psi yield strength), toughness, formability, and weldability (2219). Most alloys of this series have excellent machining characteristics and therefore are used for fasteners. Although corrosion resistance is good, protection is required in aggressive solutions and environments. Pure-aluminum claddings are applied to sheet and plate products to provide cathodic protection.
3XXX Series Aluminum-manganese, non-heat-treatable	Alloy 3003 is one of the most widely used alloys in industry. It has higher strength than the 1XXX series and has good formability, weldability, and toughness. Alloy 3003 is used widely for cryogenic applications, utensils, building sheet, and chemical equipment. Alloy 3004 has wide applications for cans, containers, and building sheet. For this series, the mechanical properties are tensile strength, 16,000–41,000 psi; yield strength, 6,000–36,000 psi.
4XXX Series Aluminum-silicon, non-heat-treatable	Silicon alloys are used as welding wires and brazing sheet because of their lower melting points and good fluidity. With additions of magnesium and copper, members of this series are excellent casting alloys. Mechanical properties are tensile strength, 19,000–55,000 psi; yield strength, 8,000–46,000 psi.
5XXX Series Aluminum-magnesium, non-heat-treatable	As a group, these alloys have excellent weldability, strength, and toughness. Alloys 5083, 5086, and 5456 have wide applications for cryogenic tanks, chemical equipment, fence wire, boat and ship hulls, armored vehicles, and railroad cars. Mechanical properties are tensile strength, 18,000–63,000 psi; yield strength, 8,000–59,000 psi.
6XXX Series Aluminum-magnesium-silicon, heat-treatable	This series provides medium-strength alloys with good formability, weldability, and extrudability and excellent corrosion resistance. Alloy 6061, the oldest of this group, is highly versatile and has been used for pipe and pipe fittings, chemical equipment, cryogenic equipment, truck bodies, and furniture. Alloy 6063 has had wide architectural applications, including storefronts, curtain walls for skyscrapers, irrigation pipe, light poles, and automotive decorative extrusions. Mechanical properties are tensile strength, 13,000–45,000 psi; yield strength, 7,000–40,000 psi.
7XXX Series Aluminum-zinc, heat-treatable	Aircraft and aerospace are the principal uses of these alloys, especially 7075, 7079, 7178, and the newer alloys 7001 and 7475. This series is the source of the highest-strength aluminum alloys for sheet, plate, forgings, rod, bar, tubes, and extrusions. Aircraft alloys are not considered readily weldable. These particular alloys have only fair corrosion resistance. Therefore they usually are protected by painting or anodizing. Mechanical properties are tensile strength, 33,000–100,000 psi; yield strength, 15,000–90,000 psi. In recent years, copper-free Al-Zn-Mg alloys (7004, 7005, 7039, 7106) have been introduced as sheet, plate, and extrusions. These alloys are relatively insensitive to rate of quenching from heat-treatment temperatures and thus provide relatively uniform properties even for heavy sections. These copper-free alloys are weldable, and the welded areas approach the strength of the parent metal. Applications include cryogenic tanks and equipment, military vehicles, railroad cars, bridges, highway guard rails, and numerous other structural applications. Mechanical properties are tensile strength, 28,000–60,000 psi; yield strength, 12,000–50,000 psi.

*Aluminum Association alloy designation system.

by direct, localized high pressure to clean metal and causing a 30% or greater reduction in cross-section thickness.

Brazing: A joining method employing a low-melting alloy usually containing 7–13% silicon and operated at temperatures above 425°C (800°F). The low-melting alloy is usually clad on a core alloy with a higher melting point, or the brazing alloy is placed in the joint to be made. Vacuum-brazing methods, which eliminate fluxes and air pollutants, have recently been introduced for brazing Al. The older methods were carried out either by immersion in molten-salt baths or in furnace brazing of flux-coated components. Since fluxes are very corrosive, they must be removed immediately after brazing to prevent corrosion of the parts in service. With the vacuum-brazing system, no postcleaning is required.

Soldering: Refers to joining at temperatures below 425°C (800°F). Aluminum-alloy fillers are not used because their melting points are too high. Most solders for Al are alloys of zinc, tin, cadmium, or lead. Soldered joints are usually less corrosion-resistant than other methods of joining. Protective measures are recommended.

Adhesive Bonding: Refers to nonmetallic joints formed by solidification of organic materials such as resins by vaporization or by thermal treatments, usually below 200°C (392°F). The metal surfaces must be cleaned and the curing temperatures must not be exceeded, or the quality of the bond will deteriorate. Adhesive-bonding technology has developed rapidly in the past decade and is used today to assemble aircraft sections of Al and other materials because it provides superior resistance to corrosion and fatigue characteristics and permits the retention of full strength of the materials assembled.

Chemical Properties of Aluminum. Aluminum is an excellent reducing agent, $Al(s) \rightarrow Al^{+3} + 3e^-$, with an oxidation-reduction potential of $+1.66$ V. Because of its great affinity for oxygen, $2Al(s) + \frac{3}{2}O_2(g) \rightarrow Al_2O_3(s) + 399$ kcal, it is often used to reduce metals from their oxides. Thermite reaction using Al powder mixed with iron oxide can be used to produce iron in weldments. The Goldschmidt process uses aluminum to reduce manganese and chromium from their oxides.

Aluminum is amphoteric. It is attacked by solutions of strong acids and alkalies, releasing hydrogen and forming soluble aluminates, except for the phosphate $AlPO_4$. Aluminum hydroxide, $Al(OH)_3$, acts primarily as a base, but it also

ionizes as a weak acid. AlN, Al_2S_3, Al_4C_3, and AlH_3 are readily hydrolized and release NH_3, H_2S, CH_4, and H_2, respectively. Aluminas and corundum are described under **Bauxite.**

Halogens: Chlorine, bromine, and iodine in gaseous or solution form react with Al to form halides which are water-soluble. Fluorine forms both soluble and insoluble salts with Al. The halides are important catalysts in organic reactions such as isomerization of *n*-butane to isobutane. $AlBr_3$ is used for bromination of propylene. $AlCl_3$ is used in Friedel-Crafts reactions to convert toluene to xylene and benzene, to prepare alkyls, alkylates, naphthalene, anthracene, *o*-benzoic acid, diphenylmethane, ethylbenzenes, propylene, butylene, and many others.

Alums: These are hydrated double salts of aluminum sulfate and alkali sulfates. Potassium alum is represented as $K_2SO_4 \cdot Al_2(SO_4)_3 \cdot 24H_2O$. In addition to potassium, sodium, lithium, cesium, ammonia, silver, rubidium, thallium; and numerous organic radicals can enter the formation. The aluminum alums are astringent and slightly acid. They have wide application, including baking powders, color dyeing, water purification, papermaking, and additions to plaster, to mention a few.

See Table A-34 for listing of properties of representative Al compounds.

Organometallic Compounds: Trialkyls of aluminum can be made by the action of alpha olefins on AlH_3:

$$AlH_3 + 3CH_2 = CH \cdot R \xrightarrow[248°F]{120°C} (CH_2CH_2R)_3Al$$

Magnesium-aluminum alloys react with alkyl halides in the presence of ethers to give trialkyls, $R_3Al \cdot OEt_2$. See also **Alcohol process, ALFOL.**

Trimethylaluminum, a dimer, is a solid with a melting point of 15°C (59°F); however, triethylaluminum is a liquid. In the presence of oxygen, trialkyls are spontaneously inflammable. Alkyls react vigorously with water:

$$Et_3Al + 3H_2O \rightarrow Al(OH)_3 + 3EtH$$

Triphenylaluminum, $(C_6H_5)_3Al$, is a product of reacting aluminum and diphenyl mercury, $(C_6H_5)_2Hg$.

Aluminum organometallic alkyls and aryls behave as Lewis acids and form compounds with electron-donating substances. These compounds are used as organometallic bases for forming polymers of the Al-N type.

Aluminum Palmitate: $Al(C_{16}H_{31}O_{32})_3 \cdot H_2O$ is one of the important metallic soaps. It is soluble

TABLE A-34. *Properties of Representative Aluminum Compounds*

Compound	Formula	Formula Weight	Sp gr, g/cm^3 (25°C)	Mp, °C	Bp, °C
Aluminum ammonium chloride . .	$AlCl_3 \cdot NH_4Cl$	186.84			
Aluminum ammonium sulfate . . .	$Al(NH_4)(SO_4)_2$	237.13	2.04		
	$Al(NH_4)(SO_4)_2 \cdot 12H_2O$	453.32	1.64	93.5	120 ($-10H_2O$)
Aluminum bromide	$AlBr_3$	266.72	3.01	97.5	263
Aluminum butoxide	$Al(OC_4H_9)_3$	246.18	1.03	102	285
Aluminum carbide	Al_4C_3	143.88	2.36	1400	
Aluminum chloride	$AlCl_3$	133.34	2.44	190	(s)
Aluminum ethoxide	$Al(OC_2H_5)_3$	162.09	1.14	134	205
Aluminum fluoride	AlF_3	83.97	2.88	. . . . . .	1260(s)
Aluminum monohydroxide	$Al_2O_3 \cdot H_2O$	119.96	3.41	360	
Aluminum dihydroxide	$Al_2O_3 \cdot 2H_2O$	137.97			
Aluminum trihydroxide	$Al(OH)_3$	77.99	2.42	300 ($-2H_2O$)	
Aluminum iodide	AlI_3	407.73	3.98	191	360
Aluminum nitrate	$Al(NO_3)_3 \cdot 9H_2O$	375.13	. . .	73	135(d)
Aluminum nitride	AlN	40.98	3.05	2200	(s)
Aluminum oxide	Al_2O_3	101.94	3.5	2200	
Aluminum oxide (corundum) . . .	Al_2O_3	101.94	4.0	2050	2250
Aluminum orthophosphate	$AlPO_4$	121.99	2.57	>1500	
Aluminum phenolate	$Al(OC_6H_5)_3$	306.09	1.23	265	
Aluminum propoxide	$Al(OC_3H_7)_3$	204.13	1.06	105	248
Aluminum silicate	$Al_6O_4Si_2O_4$	345.95	1.87	99	
Aluminum sulfate	$Al_2(SO_4)_3 \cdot 9H_2O$	504.26	1.71		
	$Al_2(SO_4)_3$	342.12	2.71	770	
Aluminum sulfide	Al_2S_3	150.12	2.02	1100	1550(s)

d, decomposes; s, sublimes.

in oils, alkalies, and benzol, but not soluble in water, and is used to waterproof fabrics, paper, and leather and is also used in lubricating oils. Other similar organics include aluminum resinate, aluminum oleate, and aluminum stearate. Aluminum stearate is also used in similar applications as palmitate, and is also used in cosmetic ointments.

Corrosion Resistance. The invisible, transparent, tightly adhering oxide film (0.2–0.6 μin. thick) on the surface of Al provides exceptional protection. Although Al is an active metal, this oxide film provides stability and corrosion resistance in air, rain environments, and most waters. If this oxide film is disrupted, the film begins to reform immediately in most environments.

Pure Al and Al alloys used for architectural and building products resist the weathering action of most rural, urban, industrial, and marine atmospheres. The metal develops a uniform, graylike patina in most cases. In some environments, slight pitting may begin, but these become self-stopping.

In aggressive environments, the corrosion of Al, like that of other common structural metals, is primarily caused by electrochemical reactions. An electric current flows between various cathode and anode sites in the metal or between the components—different alloys or different metals in the structure. If the electropotential differences between the anode (+) and cathode (−) sites are large and a good electrolyte is present, galvanic current will flow, resulting in pitting action at the anode sites.

Alloying constituents have a direct bearing on the amount and distribution of corrosion. Alloying elements such as copper, iron, and nickel form cathode sites, while alloying elements of zinc and magnesium form more anodic sites. Joining aluminum to copper, nickel, iron, lead, . . ., and alloys of these metals results in a high galvanic current flowing because of the large potential difference between Al and these metals. Special attention must be given to avoid contact of these metals with Al and its alloys.

High-strength alloys of the 2000, 5000, and 7000 series, if not properly fabricated, can exhibit aggressive types of corrosion, including exfoliation, stress corrosion cracking, and intergranular corrosion, in addition to pitting. Corrosion prevention begins with proper design and selection of the Al alloys for the application. Protective measures should be instituted by qualified corrosion engineers who are familiar with methods available, such as cathodic protection, inhibitors and passivators, and protective coatings.

Finishes: All the usual finishes—paints, enamels, lacquers, porcelain enamels, and platings—perform very well when applied to Al. Mechanical finishes, such as abrasive blasting, hammering, scratch brushing, and polishing, are used for decorative effects. Chemical conversion coatings and electrochemical finishes not only enhance appearance but also increase resistance to wear and corrosion. Electroformed oxides (anodized surfaces) provide hard, abrasion-resistant, and corrosion-resistant surfaces. Anodizing also provides a wide variety of integral colors, as well as dyed colors, for interior decorative and for exterior architectural applications. See also **Conversion coatings.**

Aluminum and Health. Since Al is present in approximately 8% of the earth's crust, much food contains traces of Al. Absolutely no ill effects on life have been established for this element. Aluminum hydroxide is prescribed for the treatment of stomach ulcers. It has been established that Al does not accelerate the destruction of vitamins during cooking. Vitamin C in milk pasteurized in Al loses no more concentration than in milk processed in glass containers. By contrast, in copper containers, vitamin C is totally destroyed. Clinical tests with animals and human beings show that inhaling aluminum dust into the lungs reduces the irritation caused by silicosis. Aluminum foil is used to cover severe burns to expedite the healing process.

REFERENCES

Aluminum Association: *Aluminum Standards and Data* (*1971*); *Aluminum Statistical Review* (*1969*), New York.

American Society for Metals: Metals Handbooks, 8th ed., vols. 1–5, 1961–1970, Metals Park, Ohio.

Goddard, H. P.: "Corrosion of Light Metals," Corrosion Monograph Series of The Electrochemical Society, Wiley, New York, 1967.

Van Horn, K. R. (ed.): "Aluminum," vols. 1–3, American Society for Metals, Metals Park, Ohio, 1967.

Wernick and Pinner: "Surface Treatment of Aluminum and Its Alloys," 3d ed., Robert Draper, Teddington, England, 1954.

—S. John Sansonetti, *Reynolds Metals Company, Richmond, Va.*

Aluminum Bronze (See **Copper.**)

Alums (See **Potassium; Salt;** and **Waste treatment.**)

Alunite (See **Potassium.**)

Amalgam An alloy containing mercury. See **Mercury;** and **Potassium.**

Amatols (See **Explosives.**)

Americium (See **Chemical elements.**)

Amethyst (See **Bauxite.**)

Amides and Imides Amides may be considered as derivatives of ammonia in which one or more of the hydrogens of the ammonia molecule are replaced by one or more acyl radicals, such as acetyl, $-CO \cdot CH_3$, or benzoyl, $-COC_6H_5$. (Amides are to be contrasted with the structurally similar amines in which the hydrogens are replaced by alkyl or aryl groups.) See **Radicals.**

Primary amides contain one acyl radical and the *amido* group ($-NH_2$); *secondary* amides contain two acyl radicals and the *imido* group ($=NH$); and *tertiary* amides contain three acyl radicals and the N radical. Parenthetically, the amino and amido groups are exactly the same, the latter term preferred when discussing amides. Similarly, the imino and imido groups are identical. There are no compounds formed among the amides that are comparable with the quaternary ammonium compounds as in the case of the amines.

Ammonia

Primary amide
Acetamide

Secondary amide
Diacetamide

Tertiary amide
Triacetamide

As contrasted with the amines which are strongly basic, the amides are very weak bases, and although the amides form salts with strong acids, the salts are quite unstable.

Amides also may be considered to be derivatives of corresponding acids by the introduction of the amido group to replace the hydroxyl radical of the carboxylic group, COOH. In the case of monobasic acids:

$$CH_3COOH ---\!\!\rightarrow CH_3CONH_2$$
Acetic acid   Acetamide

$$HCOOH ---\!\!\rightarrow HCONH_2$$
Formic acid   Formamide

The commercially important compound urea, or carbamide, may be considered to be the normal amide of the theoretical carbonic acid, $O:C(OH)_2$ namely, NH_2CONH_2. See also **Urea.** This is an example of a dibasic acid in which the amido groups have replaced the hydrogens in both hydroxyl groups. In a similar fashion, the dibasic acid, malic acid, $OHCO \cdot CH_2CH(OH) \cdot COOH$, provides the amide, malamide, $NH_2CO \cdot CH_2CH(OH) \cdot CONH_2$.

Among the aromatic amides (sometimes called arylamides), benzamide corresponds with acetamide (aliphatic).

$$N\!\!<\!\!\begin{array}{l} COC_6H_5 \\ H \\ H \end{array} \qquad N\!\!<\!\!\begin{array}{l} COC_6H_5 \\ COC_6H_5 \\ H \end{array} \qquad N\!\!<\!\!\begin{array}{l} COC_6H_5 \\ COC_6H_5 \\ COC_6H_5 \end{array}$$

Primary amide   Secondary amide   Tertiary amide
Benzamide   Dibenzamide   Tribenzamide

Benzamide is related to benzoic acid just as acetamide is related to acetic acid, namely, by the substitution of the hydroxyl radical of the carboxylic group by the amido group:

$$C_6H_5COOH \qquad C_6H_5CONH_2$$
Benzoic acid   Benzamide

The sequence of compound formation in the case of the dihydroxy carboxylic acids, including the formation of *imides,* is demonstrated by oxalic,

TABLE A-35. *Properties of Representative Amides and Imides*

Compound	Formula	Formula Weight	Sp gr	Mp, °C	Bp, °C
Acetamide	CH_2CONH_2	59.05	1.159	81	222
Acetanilide	$C_6H_5NHCOCH_3$	135.08	1.211	114.2	303.8
m-Acetotoluide	$CH_3C_6H_4NHCOCH_3$	149.09	. . . . .	65.5	303
o-Acetotoluide	$CH_3C_6H_4NHCOCH_3$	149.09	1.168	110	296
p-Acetotoluide	$CH_3C_6H_4NHCOCH_3$	149.09	. . . . .	152	307
Acrylamide	$CH_2:CH \cdot CONH_2$	71.08	. . . . .	84	
Benzamide	$C_6H_5CONH_2$	121.06	1.341	130	290
Benzanilide	$C_6H_5CONHC_6H_5$	197.09	1.321	161	
Benzene sulfanilide	$C_6H_5SO_2NHC_6H_5$	235.15	. . . . .	110	
Benzylacetamide	$C_6H_5CH_2NHCOCH_3$	149.09	. . . . .	61	300
Butamide	See **Succinimide**				
Carbamide	See **Urea**				
Diacetamide	$(C_2H_3O)_2NH$	101.06	. . . . .	78	223.5
Diacetanilide	$C_6H_5N(COCH_3)_2$	177.09	. . . . .	37	142
Dibenzamide	$(C_6H_5CO)_2NH$	225.24	. . . . .	148	
Dimethylbenzamide	$C_6H_5CON(CH_3)_2$	149.20	. . . . .	41	272
Diphenylacetamide	$(C_6H_5)_2NCOCH_3$	211.11	. . . . .	103	
Ethylacetamide	$CH_3CONHC_2H_5$	87.08	0.942		205
Ethylbenzamide	$C_6H_5CONC_2H_5$	148.19	. . . . .	70	299
Formamide	$HCONH_2$	45.03	1.139	2	193
Malonamide	$CH_2(CONH_2)_2$	102.06		170	
Methylacetamide	$CH_3CONHCH_3$	73.10		28	206
Methylacetanilide	$C_6H_5N(CH_3)(C_2H_3O)$	149.09		102	254.7
Oxamide	$CONH_2CONH_2$	88.05	1.667	419	
Oximide	$NHCOCO$	71.02	. . . . .		
o-Phthalimide	$C_6H_4(CO)_2NH$	147.05	. . . . .	238	
Propionamide	$CH_3CH_2CONH_2$	73.06	1.042	79	213
Propionanilide	$C_6H_5NHCOCH_2CH_3$	149.09		104	
Succinamide	$NH_2COCH_2CH_2CONH_2$	116.08	. . . . .	243	
Succinimide	$(CH_2CO)_2NH$	99.05	1.412	124	288
Thiacetamide	CH_3CSNH_2	75.11	. . . . .	108.5	
Tribenzamide	$(C_6H_5CO)_3N$	329.34	. . . . .	207	
Urea	NH_2CONH_2	60.05	1.335	132.7	

TABLE A-36. *Some Reactions of the Amides*

React With:	To Yield:
Sodium hydroxide, NaOH . .	NH_3 plus the corresponding carboxylic acid
Phosphorus pentoxide, P_2O_5	Cyanide by loss of water
Nitrous acid, HNO_2	N_2 plus the corresponding carboxylic acid
Hypobromite, BrO, in NaOH	Amines of one carbon less than in the amide reactant

The properties of representative amides and imides are given in Table A-35. Some reactions of the amides are given in Table A-36, and processes of preparation are given in Table A-37.

The general process for introducing the amido group is related to amine chemistry. See also **Amines.**

Thiamides are derived from amides by replacement of oxygen with sulfur:

$$NH_2 \cdot CO \cdot CH_3 \qquad NH_2 \cdot CS \cdot CH_3$$
Acetamide   Thiacetamide

$$C_6H_5 \cdot NH \cdot CO \cdot CH_3 \qquad C_6H_5 \cdot NH \cdot CS \cdot CH_3$$
Acetanilide   Thiacetanilide

TABLE A-37. *Some Processes for Making the Amides*

Formamide . . . . . Distillation of ammonium formate.

Acetamide . . . . . Treatment of ethyl acetate with a concentrated aqueous solution of NH_4OH:

$$CH_3 \cdot CO \cdot OC_2H_5 + NH_3 \rightarrow CH_3 \cdot CO \cdot NH_2 + C_2H_5OH$$

Distillation of ammonium acetate in a stream of dry NH_3:

$$CH_3 \cdot CO \cdot ONH_4 \rightarrow CH_3 \cdot CO \cdot NH_2 + H_2O$$

Benzamide . . . . . Treatment of benzoyl chloride with an excess of dry ammonium carbonate:

$$C_6H_5 \cdot COCl + 2(NH_4)HCO_3 \rightarrow C_6H_5 \cdot CO \cdot NH_2 + 2CO_2 + 2H_2O + NH_4Cl$$

Treatment of ethyl benzoate with concentrated NH_3:

$$C_6H_5 \cdot COOC_2H_5 + NH_3 \rightarrow C_6H_5 \cdot CO \cdot NH_2 + C_2H_5OH$$

Oxamide . . . . . . Treatment of dimethyl oxalate or diethyl oxalate with concentrated NH_3:

$$C_2O_4(C_2H_5)_2 + 2NH_3 \rightarrow C_2O_2(NH_2)_2 + 2C_2H_5OH$$

Succinamide . . . . Treatment of ethyl succinate with concentrated NH_3.

Phthalimide . . . . Heating a mixture of phthalic anhydride and dry ammonium carbonate in the ratio of 5 to 6 parts, respectively. In the synthesis of indigo from naphthalene, phthalimide is an intermediate product.

succinic, and phthalic acids:

Oxalic acid   Oxamide   Oxamic acid   Oximide

Succinic acid   Succinamide

Succinamic acid   Succinimide

Phthalic acid   Phthalimide

Sulfonamides are derived from sulfonic acids:

$$C_6H_5 \cdot SO_2 \cdot OH \qquad C_6H_5 \cdot SO_2 \cdot NH_2$$
Benzene-sulfonic acid   Benzene-sulfonamide

The sulfonamides came into prominence in 1935 when a German scientist, Domagk, observed the clinical value of *prontosil*. This is a red compound that is derived from azo dyes. The effective part of the prontosil molecule was demonstrated to be paraaminobenzene sulfonamide (named sulfanilamide). The use of this drug in treating hemolytic streptococcal and staphylococcal infections was quickly followed by a series of related sulfa drugs.

Amination (See **Amines**; and **Catalysts**.)

Amines *Aliphatic amines* are nitrogen bases of the alkyl radicals, such as methyl, —CH_3, and ethyl,

—C_2H_5, and sometimes are referred to as *ammonia bases.* Amines may be considered as derivatives of ammonia in which one or more of the hydrogens in the ammonia molecule are replaced by one or more alkyl radicals. (Amines are to be contrasted with the structurally similar amides in which the hydrogens are replaced by an acyl group.) See **Radicals.**

Primary amines contain one alkyl radical and the *amino* group (—NH_2); *secondary* amines contain two alkyl radicals and the *imino* group (=NH); and *tertiary* amines contain three alkyl radicals and the N radical. *Quaternary compounds* contain four alkyl radicals and are considered as derivatives of ammonium hydroxide.

Ammonia   Primary amine   Secondary amine
            Methylamine   Dimethylamine

Tertiary amine   Quaternary
Trimethylamine   ammonium
                  compound
                  Tetramethyl
                  ammonium iodide

Mixed amines contain different alkyl groups, such as

$$NH(CH_3)(C_3H_7) \qquad N(CH_3)(C_2H_5)(C_3H_7)$$
Methylpropylamine   Methylethylpropylamine

The term mixed also applies to amines which may contain an alkyl group and an aryl (aromatic) group, such as phenyl, —C_6H_5, and naphthyl, —$C_{10}H_7$.

Amines that contain small alkyl groups closely resemble ammonia and are in fact more strongly basic than ammonia. They have a pronounced ammoniacal odor. Amines range from combustible gases (the lowest members) through liquids to solids, such as tricetylamine, $(C_{16}H_{33})_3N$, which has a high boiling point and is insoluble in water. Solubility in water, as well as volatility, decreases as the molecular weight of the compounds increases. Most amines are lighter than water. Like ammonia, the amines form salts. Ethylamine hydrochloride, $C_2H_5 \cdot NH_3Cl$ or $C_2H_5 \cdot NH_2 \cdot HCl$, is an example.

Amines form a homologous series, $C_nH_{2n+3}N$. They exhibit isomerism. For example, there are four compounds with the formula C_3H_9N, namely, isopropylamine, methylethylamine, propylamine, and trimethylamine. For a summary of the key properties of the principal amines, see Table A-38. Specific amines also are discussed elsewhere in this volume. Consult the Subject Index.

Aromatic amines in their simplest form are represented by aniline (aminobenzene or phenylamine). Aniline is derived by replacing one of the hydrogens of benzene with an amino group. Thus the formula for aniline is $C_6H_5 \cdot NH_2$. Aniline, used in the manufacture of dyes, medical products, and in the preparation of numerous benzene derivatives, yields *acetanilide* when treated with acetic anhydride or acetyl chloride and *thiocarbanilide* when carbon disulfide vapor is passed into aniline. By displacement of the hydrogen atoms of the amino group in aniline with phenyl radicals, *diphenylamine*, $(C_6H_5)_2NH$, and *triphenylamine*, $(C_6H_5)_3N$, may be produced. The amino group also may be united with a carbon of the side chain to produce compounds, such as *benzylamine*, $C_6H_5 \cdot CH_2 \cdot NH$.

Aminotoluenes, or *toluidines*, $C_6H_4(CH_3) \cdot NH_2$, are derived from o-, m-, and p-nitrotoluenes. Treatment of toluidines with nitrous acid yields diazonium salts.

See also **Ion-exchange resins.**

Alkylanilines result when amino compounds are treated with alkyl halides to displace one or both of the hydrogen atoms of the amino group. These compounds include methylaniline, $C_6H_5 \cdot NH \cdot CH_3$; ethylaniline, $C_6H_5 \cdot NH \cdot C_2H_5$; and dimethylaniline and diethylaniline.

Hydroxylamines are derived from hydroxylamine, $NH_2 \cdot OH$, in much the same manner that amines are derived from ammonia. Depending upon the structure of the source substance, hydroxylamine, the compound will be either

$$NH_2OCH_3 \qquad or \qquad CH_3 \cdot NH \cdot OH$$
α-Methylhydroxylamine   β-Methylhydroxylamine

Hydrazines resemble the amines, but differ in that hydrazines contain two atoms of nitrogen, as

$$CH_3 \cdot NH \cdot NH_2 \qquad C_2H_5 \cdot NH \cdot NH_2$$
Methyl hydrazine   Ethyl hydrazine

$$(C_2H_5)N \cdot NH_2$$
Diethyl hydrazine

Parenthetically, it is interesting to note that the formation of *phosphines, arsines,* and *stibines* resem-

TABLE A-38. *Properties of Representative Amines*

Amine	Formula	Formula Weight	Sp gr	Mp	Bp
Primary Amines					
Allylamine	$CH_2:CHCH_2NH_2$	57.06	0.761	. . . .	253.2
Aminobenzene	See **Aniline**				
Amylamine (iso)	$C_5H_{11}NH_2$	87.11	. . . .	. . . .	95
Amylamine (n)	$C_5H_{11}NH_2$	87.11	0.761	−55	103
Aniline	$C_6H_5NH_2$	93.06	1.022	−6.2	184.4
Benzidine	$(4')H_2NC_6H_4C_6H_4NH_2(4)$	184.11	. . . .	218	
Benzylamine	$C_6H_5CH_2NH_2$	107.08	0.980	. . . .	184
Butylamine (iso)	$C_4H_9NH_2$	74.08	. . . .	−85	68
Butylamine (n)	$C_4H_9HN_2$	73.09	0.740	−30	78
Cadaverine	See **Pentamethylenediamine**				
1,2-Diaminopropane	$CH_3CHNH_2CH_2NH_2$	74.13	. . . .	. . . .	119
Dimethylenediamine	$CH_2NH_2 \cdot CH_2NH_2$	60.12	. . . .	8	116
Ethanolamine	$NH_2CH_2CH_2OH$	61.06	1.018	10.5	172
Ethylamine	$C_2H_5NH_2$	45.06	0.689	−81	17
Ethylenediamine	See **Dimethylenediamine**				
Hexamethylenediamine	$(CH_2)_6 \cdot (NH_2)_2$	116.14	. . . .	42	204
Hexamethylenetetramine	$(CH_2)_6N_4$	140.13	. . . .	263	
Mesidine	See **trimethylaniline**				
Methylamine	CH_3NH_2	31.05	0.699	−93	−7
α-Naphthylamine	$C_{10}H_7NH_2(1)$	143.08	1.131	50	301
β-Naphthylamine	$C_{10}H_7NH_2(2)$	143.08	1.061	111	306
Pentamethylenediamine	$(CH_2)_5(NH_2)_2$	102.13	0.885	9	179
p-Phenylaniline	$C_6H_5 \cdot C_6H_4 \cdot NH_2$	169.24	. . . .	51	302
m-Phenylenediamine	$C_6H_4(NH_2)_2(1,3)$	108.08	. . . .	63	285
o-Phenylenediamine	$C_6H_4(NH_2)_2(1,2)$	108.08	. . . .	103	257
p-Phenylenediamine	$C_6H_4(NH_2)_2(1,4)$	108.08	. . . .	140	267
α-phenylethylamine	$C_6H_5 \cdot CHNH_2 \cdot CH_3$	121.09	0.940	. . . .	187
β-phenylethylamine	$C_6H_5 \cdot CH_2CH_2NH_2$	121.09	0.958	. . . .	198
Propylamine (iso)	$C_3H_7NH_2$	59.08	0.694	−101	33
Propylamine (n)	$C_3H_7NH_2$	59.08	0.719	−83	50
o-Tolidine	$[(NH_3)_2(CH_3)C_6H_3]_2$	212.14	. . . .	129	
m-Toluidine	$(3)CH_3C_6H_4NH_2(1)$	107.08	0.984	−31	203
o-Toluidine	$(2)CH_3C_6H_4NH_2(1)$	107.08	0.998	−16	200
p-Toluidine	$(4)CH_3C_6H_4NH_2(1)$	107.08	1.046	44	200
1,2,3-Triaminobenzene	$C_6H_3(NH_2)_3$	123.09	. . . .	103	336
1,2,4-Triaminobenzene	$C_6H_2(NH_2)_3$	123.09	. . . .	100	340
2,4,6-Trimethylaniline	$(CH_3)_3C_6H_2NH_2(1)$	135.23	. . . .	. . . .	229
Tetramethylenediamine	$(CH_2)_4 \cdot (NH_2)_2$	88.11	. . . .	27	158
Trimethylenediamine	$(CH_2)_3 \cdot (NH_2)_2$	74.15	. . . .	. . . .	135
Urotropine	See **Hexamethylenetetramine**				
Vinylamine	$CH_2:CHNH_2$	43.05	0.832	. . . .	56
Secondary Amines					
Allylaniline	$C_6H_5NHCH_2CH:CH_2$	133.09	0.982	. . . .	209
o-Aminodiphenylamine	$(2)NH_2C_6H_4NHC_6H_5$	184.11	. . . .	79	
p-Aminodiphenylamine	$(4)NH_2C_6H_4NHC_6H_5$	184.11	. . . .	66	
Benzylidene aniline	See **Phenylbenzylaniline**				
Butylaniline (iso)	$C_6H_5NHC_4H_9$	149.13	. . . .	. . . .	231
Butylaniline (n)	$C_6H_5NHC_4H_9$	149.13	. . . .	. . . .	240.9
Dibenzylamine	$(C_6H_5CH_2)_2NH$	197.13	1.026	−26	300
Dibutylamine (iso)	$(C_4H_9)_2NH$	129.16	. . . .	. . . .	139
Dibutylamine (n)	$(C_4H_9)_2NH$	129.16	0.767	. . . .	161

TABLE A-38. *Properties of Representative Amines* *(Continued)*

Amine	Formula	Formula Weight	Sp gr	Mp	Bp
Diethanolamine	$NH(CH_2CH_2OH)_2$	105.09	. . . .	28	270
Diethylamine	$(C_2H_5)_2NH$	73.09	0.711	−39	55
Dimethylamine	$(CH_3)_2NH$	45.06	0.680	−96	7
Diphenylamine	$(C_6H_5)_2NH$	169.09	1.159	53	302
Dipropylamine (iso)	$(C_3H_7)_2NH$	101.13	. . . .	. . . .	83
Dipropylamine (n)	$(C_3H_7)_2NH$	101.13	0.738	−40	110
Ethylaniline	$C_6H_5NHC_2H_5$	121.09	0.963	−63.5	204.7
Methylaniline	$C_6H_5NHCH_3$	107.08	0.986	−57	195.7
Methylethylamine	$(CH_3)(C_2H_5)NH$	59.13	. . . .	. . . .	34
Phenylbenzylamine	$C_6H_5NHCH_2C_6H_5$	183.11	1.038	37	300
Propylaniline (n)	$C_6H_5NHC_3H_7$	135.11	0.949	. . . .	222

Tertiary Amines

Amine	Formula	Formula Weight	Sp gr	Mp	Bp
Dibenzylaniline	$C_6H_5N(CH_2C_6H_5)_2$	273.16	. . . .	70	
Dibutylaniline (n)	$C_6H_5N(C_4H_9)_2$	205.19	0.907	. . . .	262.8
Dimethylaniline	$C_6H_5N(CH_3)_2$	121.09	0.956	1.7	193.5
Dimethylethylamine	$(CH_3)_2NC_2H_5$	73.16	. . . .	. . . .	37
Diphenylbenzylamine	$(C_6H_5)_2NCH_2C_6H_5$	259.37	. . . .	86	
Dipropylaniline (iso)	$C_6H_5N(C_3H_7)_2$	177.16			
Dipropylaniline (n)	$C_6H_5N(C_3H_7)_2$	177.16	0.910	. . . .	241
Ethyldiphenylamine	$C_2H_5N(C_6H_5)_2$	197.13	. . . .	. . . .	297
Methyldiethylamine	$CH_3N(C_2H_5)_2$	87.11	. . . .	. . . .	65
Methyldiphenylamine	$CH_3N(C_6H_5)_2$	183.11	1.047	−7.6	293.4
Methylethylaniline	$(CH_3)(C_2H_5)NC_6H_5$	135.11	. . . .	. . . .	201
Tribenzylamine	$(C_6H_5CH_2)_3N$	287.17	0.991	92	385
Tributylamine (iso)	$(C_4H_9)_3N$	185.22	. . . .	. . . .	190
Tributylamine (n)	$(C_4H_9)_3N$	185.22	0.778	. . . .	214
Triethanolamine	$N(CH_2CH_2OH)_3$	149.13	1.126	21.2	277
Triethylamine	$(C_2H_5)_3N$	101.13	0.728	−114.8	89.5
Trimethylamine	$(CH_3)_3N$	59.08	0.662	−124	3.5
Tripropylamine (iso)	$(C_3H_7)_3N$	143.17			
Tripropylamine (n)	$(C_3H_7)_3N$	143.17	0.757	−93.5	156

Quaternary Ammonium Derivatives

Amine	Formula	Formula Weight	Sp gr	Mp	Bp
Tetraethylammonium bromide	$(C_2H_5)_4NBr$	210.08	1.397		
Tetraethylammonium hydroxide	$(C_2H_5)_2NOH$	147.17	. . . .	190	
Tetramethylammonium hydroxide	$(CH_3)_4NOH \cdot 5H_2O$	181.19	. . . .	63	
Trimethylphenylammonium hydroxide	$C_6H_5 \cdot N(CH_3)_3 \cdot OH$	153.25			

bles the amines in their formation and structure. Primary, secondary, and tertiary phosphines, for example, stem from phosphuretted hydrogen, PH_3, in which alkyl radicals replace the hydrogens, just as in amines the alkyl radicals replace the hydrogens of ammonia. Triethyl phosphine, $P(C_2H_5)_3$, is an example.

Some reactions of the amines are given in Table A-39, and processes for making the amines are given in Table A-40.

Amination: The general process for introducing the amino group, $-NH_2$, into organic compounds is termed *amination*. Processes for the high-tonnage production of amines and related compounds are described elsewhere. See also **Amino acids; Ethanolamines; Hydrazine;** and **Melamine.**

TABLE A-39. *Some Reactions of the Amines*

React With:	To Yield:
Primary Amines	

Concentrated HNO_3 . . . . Nitramines
   Example: $C_2H_5NH_2$ (ethylamine) . . . . $C_2H_5NHNO_2$ (ethylnitramine)
Chloroform plus a base . . . . Isocyanides
Benzene–sulfonyl chloride, $C_6H_5SO_2Cl$ . . . . Substituted benzene sulfonamides
   Example: $C_2H_5NH_2$ (ethylamine) . . . . $C_6H_5SO_2NHC_2H_5$ (*N*-ethylbenzene sulfonamide)
Acetylchloride or benzoyl chloride. . . . . A substituted amide
   Example: $C_2H_5NH_2$ (ethylamine) . . . . $C_2H_5NHOCCH_3$ (*N*-ethylacetamide)
Nitrous acid, HNO_2, plus alkyl amine . . . . N_2 plus alcohol
HNO_2 plus aryl amine (warm). . . . . N_2 plus phenol
HNO_2 plus aryl amine (cold) . . . . Diazonium compounds

| Secondary Amines | |

Benzene–sulfonyl chloride. . . . . Benzene sulfonamides
   Example: $(C_2H_5)_2NH$ (diethylamine) . . . . $(C_6H_5 \cdot SO_2N(C_2H_5)_2$ (*N, N*-diethylbenzene-
                              sulfonamide)
Acetyl chloride or benzyl chloride . . . . Substituted amide
   Example: $(C_2H_5)_2NH$ (diethylamine) plus acetyl . . . . . $(C_2H_5)_2NOCCH_3$ (*N, N*-diethylacetamide)
   chloride
Nitrous acid (HNO_2) . . . . Nitrosoamines

| Tertiary Amines | |

Alkyl haloids . . . . Quaternary ammonium haloids

TABLE A-40. *Some Processes for Making the Amines*

Primary amines:
   Reduction of nitro compounds (*example:* aniline
     from nitrobenzene)
   Reduction of cyanides, hydroxylamines, hydra-
     zones, nitroso compounds, and oximes
   Hydrolysis (alkaline–NaOH) of isocyanates or
     isocyanides
   Alcohols (vapor) plus NH_3 in presence of catalyst
   Natural decomposition of amino acids
Secondary amines:
   Alkylation of primary amines
   Natural decomposition of amino acids
Tertiary amines:
   Heating quaternary ammonium hydroxides, also
     yielding alcohol (lower members) or olefin plus
     water (higher members)
   Alkylation of secondary amines
   Natural decomposition of amino acids

Amino Acids This description is confined to those amino acids that occur as constituents of natural proteins. Each amino acid has a carboxyl group as well as an amino group in its α position. There are two natural acids which have an *imino* group instead of an amino group. Those amino acids which have an asymmetric carbon (indicated by an asterisk in the diagram) can exist as two optical isomers. All amino acids of protein constituents have been proved to have the same configuration, although the absolute configuration was not determined until suitable x-ray–diffraction analytical methods were developed (after 1950). The L-form shown in the diagram is the natural-occurring form and coincides with the earlier assumptions of E. Fischer.

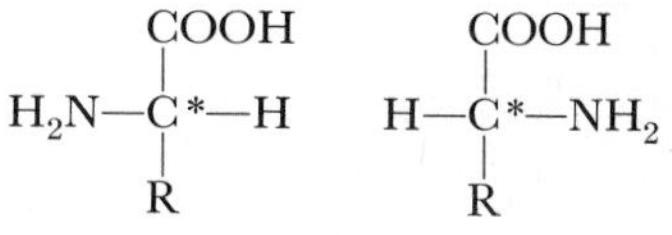

In this description, amino acids are discussed in the following order: (1) a broad classification, (2) physical properties, (3) chemical properties, (4) industrial production, (5) role of amino acids in nutrition, and (5) a summary of the progress that has been made by the amino acid industry.

Classification. Twenty-two amino acids are used as examples here to show how they fit into broad classifications, i.e., into eight groups, according to the structure of the R-group. The formulas are given in Table A-41, where the amino acids, along with their properties, are listed alphabetically for convenient reference. There are several amino acids composing protein other than these, but they are omitted because they are less common. Among those omitted are thyroxine (in thyroid protein), di-iodotyrosine (in sponge, coral protein), and hydroxylysine (a small amount in gelatin), which sometimes are cited as protein-bound amino acids.

Neutral amino acids	Acidic amino acids
Aliphatic	Aspartic acid
Glycine	Glutamic acid
Alanine	Basic amino acids
Valine	Histidine
Leucine	Lysine
Isoleucine	Arginine
Hydroxy-	Imino acids
Serine	Proline
Threonine	Hydroxyproline
Sulfure-containing	
Cysteine	
Cystine	
Methionine	
Amide	
Asparagine	
Glutamine	
Aromatic	
Phenylalanine	
Tyrosine	
Tryptophan	

Physical Properties

Dipolar Structure: Generally, amino acids exist as dipolar ions, $RCH(NH_3^+)COO^-$, in a neutral state, where both amino and carboxyl groups are ionized. Although a nonpolar form, $RCH(NH_2)COOH$, may be taken into account, the dipolar form is greatly predominant for the usual monoamino monocarboxylic acid, and is estimated as 10^5–10^6 times greater than the nonpolar form. Nevertheless, the nonpolar form often is used as the structural formula. The remarkable difference between amino acids and usual organic compounds is due to this dipolar structure. Amino acids do not show any defined melting point, and they decompose thermally at a remarkably high temperature (200–300°C). Amino acids have low vapor pressure and are soluble in water, but are practically insoluble in organic solvents. The characteristics of amino acid derivatives that only slightly assume a dipolar structure are much closer to ordinary organic compounds.

Thus their amides show a defined melting point at lower temperatures; their esters can be distilled under reduced pressure; and they are soluble in some organic solvents.

The ionic states of a simple α-amino acid may be described as

$$RCH(NH_3^+)COOH \underset{+H^+}{\overset{-H^+(K_1)}{\rightleftharpoons}}$$

Cationic form (acidic)

$$RCH(NH_3^+)COO^- \underset{+H^+}{\overset{-H^+(K_2)}{\rightleftharpoons}} RCH(NH_2)COO^-$$

Dipolar form (neutral)   Anionic form (basic)

Corresponding to the change of the ionic state, dissociation constants are

$$K_1(COOH) = \frac{[H^+][RCH(NH_3^+)COO^-]}{[RCH(NH_3^+)COOH]}$$

$$K_2(NH_3^+) = \frac{[H^+][RCH(NH_2)COO^-]}{[RCH(NH_3^+)COO^-]}$$

As $pK = -\log K$, glycine has the values of $pK_1 = 2.34$ and $pK_2 = 9.60$ (aqueous solution at 25°C), and the homologous amino acids show similar values. The pH at which acidic ionization balances basic ionization is called the *isoelectric point* (pH_I):

$$[RCH(NH_3^+)COOH] = [RCH(NH_2)COO^-]$$

From these formulas, the pH_I is

$$pH_I = \tfrac{1}{2}(pK_1 + pK_2)$$

The pH_I value is about 6.0 for homologues of glycine and alanine—smaller for the acidic and larger for the basic amino acids. In the usual pH region, amino acids show minimum solubility at their isoelectric points. This characteristic is utilized as the optimum pH for the crystallization of each amino acid. The ionic form of lysine with no formal charge is $NH_3^+(CH_2)_5(NH_2)COO^-$, and that of glutamic acid is $HOOC(CH_2)_2CH(NH_3^+)COO^-$.

Specific Optical Rotation: $[\alpha]_D^{20}$ in Table A-41 indicates the specific optical rotation measured at 20°C with the sodium D line. The value expresses the rotation (in degrees) of an amino acid per unit path length (in 10 cm) and concentration (in 100 g/100 ml of solution).

Inasmuch as the specific optical rotation varies to some extent, depending upon the measuring concentration, the percentage concentration of the amino acid should also be cited. For ordinary organic compounds, the melting point may be used as a criterion of purity. However, because

TABLE A-41. *Properties of Natural Amino Acids*

Name and Formula	Abbreviation	Isoelectric Point	Solubility, g in 100 g H_2O at 20°C	Specific Optical Rotation			Discoverers and Dates
				C	Solution	Angle	
Alanine $CH_3-CH-COOH$ 　　　NH_2	Ala	6.0	16	10	6N HCl	+14.7°	Weyl, 1888; Schutzenberger, 1879
Arginine $H_2N-C-NH(CH_2)_3-CH-COOH$ 　　NH　　　　　NH_2	Arg	11.2	15*	8	6N HCl	+27.2°	Hedin, 1895
Asparagine $H_2NOC-CH_2-CH-COOH$ 　　　　　　NH_2	Asn	5.4	2	10	3N HCl	+35.0°	Damodaran, 1932
Aspartic acid $HOOC-CH_2-CH-COOH$ 　　　　　NH_2	Asp	2.8	0.4	10	2N HCl	+26.2°	Ritthausen, 1868
Citrulline $H_2NC-NH-(CH_2)_3CH-COOH$ 　O　　　　　NH_2	Cit	5.9	12	8	6N HCl	+25.8°	Koga, 1914; Odake, 1914; Wada, 1930
Cysteine $HS-CH_2-CH-COOH$ 　　　　NH_2	Cys	5.1	8†	8	1N HCl	+8.1°	
Cystine $(-SCH_2-CH-COOH)_2$ 　　　　NH_2	Cys Cys	4.6	0.01	2	1N HCl	−223°	Mörner, 1899; Emden, 1899
Dihydroxyphenylalanine OH $HO-\bigcirc-CH_2-CH-COOH$ 　　　　　　NH_2	Dopa	5.5	0.5	2	1N HCl	−11.9°	Torquati, 1913; Guggenheim, 1913
Glutamic acid $HOOC-CH_2-CH_2-CH-COOH$ 　　　　　　　NH_2	Glu	3.2	0.7	10	2N HCl	+32.0°	Ritthausen, 1866
Glutamine $H_2NOC-CH_2-CH_2-CH-COOH$ 　　　　　　　NH_2	Gln	5.7	4	4	H_2O	+6.5°	Damodaran, Jaaback, and Chibnall, 1932
Glycine NH_2-CH_2-COOH	Gly	6.0	23		‡	‡	Braconnot, 1820
Histidine H N $\diagdown-CH_2-CH-COOH$ N　　　　NH_2	His	7.6	4*	11	6N HCl	+12.3°	Kossel, 1896; Hedin, 1896

*Anhydrous base.　†At 25°C.

‡Glycine has no optical rotatory power because it has no asymmetric carbon atom.

TABLE A-41. *Properties of Natural Amino Acids*   *(Continued)*

Name and Formula	Abbreviation	Isoelectric Point	Solubility, g in 100 g H_2O at 20°C	Specific Optical Rotation			Discoverers and Dates
				C	Solution	Angle	
Hydroxyproline HO—⟨pyrrolidine⟩—COOH	Hyp	5.8	35	4	H_2O	−75.1°	Fischer, 1902
Isoleucine C_2H_5—C(H)(CH_3)—CH(NH_2)—COOH	Ile	6.0	4	4	6N HCl	+40.7°	Ehrlich, 1904
Leucine $(CH_3)_2CH$—CH_2—CH(NH_2)—COOH	Leu	6.0	2	4	6N HCl	+15.2°	Braconnot, 1820
Lysine H_2N—$(CH_2)_4$CH(NH_2)—COOH	Lys	9.7	70§	8	6N HCl‡	+21.2°	Drechsel, 1889
Methionine CH_3—S—CH_2—CH_2—CH(NH_2)—COOH	Met	5.7	5	8	6N HCl	+24.2°	Mueller, 1922
Ornithine CH_2—CH_2—CH_2—CH(NH_2)—COOH, NH_2	Orn	9.7	55§	4	6N HCl	+23.5°	Riesser, 1906 (from arginine)
Phenylalanine ⟨C_6H_5⟩—CH_2—CH(NH_2)—COOH	Phe	5.5	3	2	H_2O	−34.3°	Schulze and Barbieri, 1881
Proline ⟨pyrrolidine⟩—COOH	Pro	6.3	155	4	H_2O	−84.8°	Fischer, 1901
Serine HO—CH_2—CH(NH_2)—COOH	Ser	5.7	38	10	2N HCl	+15.1°	Cramer, 1865
Threonine CH_3—CH(OH)—CH(NH_2)—COOH	Thr	6.2	9	6	H_2O	−28.5°	Schryver and Buston, 1925; Gortner and Hoffmann, 1925
Tyrosine HO—⟨C_6H_4⟩—CH_2—CH(NH_2)—COOH	Tyr	5.7	0.04	5	1N HCl	−12.0°	Bopp, 1849

§Anhydrous HCl salt.

TABLE A-41. *Properties of Natural Amino Acids* (*Continued*)

Name and Formula	Abbreviation	Isoelectric Point	Solubility, g in 100 g H_2O at 20°C	Specific Optical Rotation			Discoverers and Dates
				C	Solution	Angle	
Tryptophan 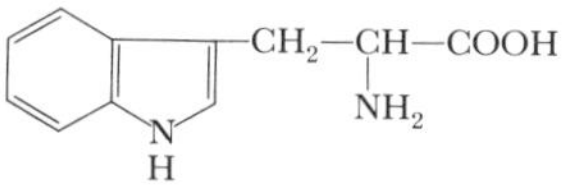 CH_2—CH—COOH with NH$_2$ (indole ring, N—H)	Trp	5.9 ·	1	1	H_2O	−31.5°	Hopkins and Cole, 1902
Valine $(CH_3)_2CH$—CH—COOH with NH_2	Val	6.0	9	8	6N HCl	+28.0°	Fischer, 1901

amino acids show no defined melting point, specific rotation is used for this purpose. All the natural amino acids belong to the L-series, irrespective of showing a plus or minus sign of specific rotation.

Specific rotation also depends upon the pH, and therefore the concentration of HCl used should also be cited. For the amino acids with $[\alpha] > 0$, the specific rotation increases with the strength of the acidity, and so it is measured preferably in 6N HCl. For some amino acids, where the solubility is too small to obtain an accurate measurement, a more dilute HCl solution is used. For the amino acids with $[\alpha] < 0$, the specific rotation reduces in magnitude with the strength of the acidity, and so an aqueous solution generally is used. In the case of cystine and dopa, 1N HCl is used because of their exceptionally low solubility in a neutral solution. Dopa is the abbreviation for *d*ihydroxyphenylalanine.

Gas-chromatographic analytical techniques have been developed in connection with several volatile derivatives of the amino acids. These derivatives are esters (such as amino acid *n*-butyl esters or N,O-trimethyl silyl esters), amides (such as N,O-trifluor acetates), and aldehydes formed by the reaction of ninhydrin on amino acids.

Chemical Properties. Because amino acids are amphoteric, they have certain characteristics of both organic bases and organic acids.

Salt Formation: Amino acids, being amines, form stable salts, such as hydrochlorides or aromatic sulfonic acid salts, which are used as selective precipitants of certain amino acids. On the other hand, amino acids, being also organic acids, form complex salts with heavy metals, of which

the less soluble salt is utilized for amino acid separation.

Esterification: Amino acids form esters by heating with the equivalent amount of strong acid (usually HCl) in absolute alcohol. Esters are obtained as hydrochlorides. The concentrated aqueous solution is cooled, neutralized with NaOH, and salted out with K_2CO_3 to gain the isolated ester which is soluble in organic solvents.

Amide Formation: Amino acid esters are condensed with NH_3 or amines to form acid amides:

$$RCH—CONH_2$$
$$|$$
$$NH_2$$

Amino acid ester + NH_3

$$\left(\begin{array}{c} RCH—CONH—R' \\ | \\ NH_2 \end{array} \right)$$

Amino acid ester + amine (R'—NH_2)

Acyl Compounds: Amino acids in alkaline solution react with acid chlorides or acid anhydrides to form acyl compounds:

$$RCH—COONa$$
$$|$$
$$NHCOR'$$

Deamination (Van Slyke Reaction): α-Amino acids react with excess nitrous acid to form α-hydroxyl acids on a quantitative basis and simultaneously generate nitrogen gas:

$$RCH—COOH + HNO_2 \rightarrow$$
$$|$$
$$NH_2$$

$$RCH—COOH + N_2 + H_2O$$
$$|$$
$$OH$$

Inasmuch as this reaction on the α-amino group is completed within 5 min at room temperature, a measurement of the volume of generated N_2 can be used for amino acid determination.

The reaction of the ω-amino takes longer, and an imino acid does not generate N_2. Thus the Van Slyke reaction cannot be used for both determinations. Because NH_3 also produces N_2 gas gradually by reacting with dilute HNO_2, NH_3 in an amino acid solution should be excluded prior to the determination.

Decarboxylation: Amino acids form amines when heated with inert solvents, such as kerosene:

$$RCH—COOH \rightarrow RCH_2 + CO_2$$
$$\underset{NH_2}{|} \qquad\qquad \underset{NH_2}{|}$$

Glycerol is not suitable for this reaction because of its dehydration reaction. Decarboxylative enzymes are known, which react specifically with amino acids having free polar groups at the ω position. The examples are cadaverine from lysine, histamine from histidine, γ-amino acid from glutamic acid, and tyramine from tyrosine.

Oxidation: α-Amino acids decompose easily with oxidizing agents to form the corresponding fatty acid with the carbon number less by one.

$$RCH—COOH \xrightarrow{O} RCHO \xrightarrow{O}$$
$$\underset{NH_2}{|}$$

$$RCOOH + NH_3 + CO_2$$

Ninhydrin Reaction: Neutral solutions of an amino acid react with ninhydrin (triketohydrindene hydrate),

CO—H_2O, by heat-

ing to cause oxidative decarboxylation.

$$RCH—COOH \rightarrow RCHO + NH_3 + CO_2 + H_2O$$
$$\underset{NH_2}{|} \qquad \text{Aldehyde}$$

The central carbonyl of this triketone is reduced to an alcohol,

CHOH. This latter alcohol reacts further with the NH_3 formed from the amino acid and causes a red-purple color. Inasmuch as this reaction proceeds quantitatively, the optical density of the color may be used for amino acid determination.

Imino acids, such as proline and hydroxyproline, react with ninhydrin to develop a yellow color, the density of which may be used for imino acid determination. An autoanalyzer using this technique for determining amino acid mixtures currently is widely used. The mixture is separated into each amino acid with the aid of ion-exchange resins, followed by determination using the ninhydrin reaction.

Maillard Reaction: The amino group in amino acids has a characteristic of forming condensation products with aldehydes, which are regarded as the cause of the browning reaction when an amino acid and a sugar coexist. As a characteristic flavor is evolved as well as the color, this reaction sometimes is useful in food preparations.

Ion-Exchange Resin: Since amino acids are amphoteric, they behave as an acid as well as a base according to the pH of the solution. Thus all amino acids dissolved in water can be adsorbed either on a strong-acid cation-exchange resin or a strong-base anion-exchange resin. The affinity to an ion-exchange resin depends on the kind of amino acids and the pH of the solution. This characteristic is utilized in the separation of each amino acid from an amino acid mixture with the aid of an ion-exchange-resin column.

The ion-exchange resin for the separation of a specific amino acid from a protein hydrolysate has been utilized on an industrial scale. This method often is used for the purpose of purification of an amino acid produced by a fermentation process. For example, the industrial separation of basic amino acids, such as histidine, arginine, and lysine, may occur as follows. Only basic amino acids are adsorbed on the Na-form of a strong-acid cation exchanger from a protein hydrolysate. Acidic or neutral amino acids pass through the column without adsorption. The adsorbate then is eluted with an aqueous caustic soda. The eluate is passed through the H-form of a weak-acid cationic exchanger. Arginine and lysine are adsorbed, while histidine passes through. Absorbed arginine and lysine are eluted with an aqueous ammonia solution and passed through a strong-base anion exchanger, by which only lysine is adsorbed and arginine is passed through. Lysine can be isolated by elution with hydrochloric acid.

There are several kinds of instruments (referred to as "amino acid analyzers") for this type of determination. As an example of this technique, an amino acid mixture is adsorbed by a strong-acid cationic-resin column; the pH value of the buffered eluant gradually is raised from acidic to neutral, eluting each component of amino acid separately. Ninhydrin solution continuously added to the eluate develops color, the absorbance of which is recorded. Each amino acid is identified from the retention volume, and the quantity is determined from the peak area. See also **Chromatography.**

Industrial Production. There are three principal ways to manufacture commercially valuable amino acids in substantial quantities: (1) by extraction from natural proteins, (2) by fermentation, and (3) by chemical synthesis.

Extraction Methods: Although most high-production amino acids, such as monosodium glutamate (MSG), now are produced by other methods,* there are four amino acids, as described next, which still are produced by the extraction method because no bacterial strains have been discovered to produce them in high yields by fermentation, or because no appreciable market has been found to stimulate consideration of the chemical synthesis route.

L-Leucine, easily extractable in quantity from any kind of vegetable protein hydrolyzates, may continue to be produced by this method. L-Cystine generally is extracted from the human-hair hydrolyzate. L-Histidine is available from blood of beasts in quantities, but it may be produced by fermentation in the near future, since some suitable artificial mutants of bacteria have already been discovered. The sole commercial source of L-hydroxyproline is now gelatin.

Almost all the 22 kinds of amino acids previously listed and included in Table A-41 are manufactured commercially.† Three natural amino acids, which usually are not contained in proteins but which are effective in medicine, also are listed in Table A-41. These are citrulline, ornithine, and dihydroxyphenylalanine.

Fermentation Methods: There are many kinds of microorganisms which are able to synthesize amino acids necessary to support their life from a simple carbon source and an inorganic nitrogen source, such as ammonium or nitrate salts, or nitrogen gas. With the remarkable progress carried out recently in the field of biochemistry and microbiology, a large amount of information on the biosynthetic pathways of amino acids in these microorganisms and also on the controlling mechanisms working in these pathways has been

accumulated. The production of amino acids through microbiological processes has been accomplished by making practical use of this accumulated information.

In 1956, Japanese microbiologists succeeded in developing the first industrial production of L-glutamic acid through a microbiological process. Since then, the development and improvement of microbiological processes have continued. Currently, almost all common amino acids can be produced by *amino acid fermentation* at a low cost on an industrial scale. For example, monosodium glutamate was produced by the fermentation route at a rate of 180,000 metric tons in 1969, accounting for 90% of world production. See Fig. A-38.

The amino acid fermentation process is relatively simple. A physiologically active isomer of amino acid, L-amino acid, can be obtained exclu-

*Some MSG production by extraction may remain in connection with utilizing glutamine contained in sugar beets.

†As, for example, by Ajinomoto Co., Inc., Kawasaki, Japan.

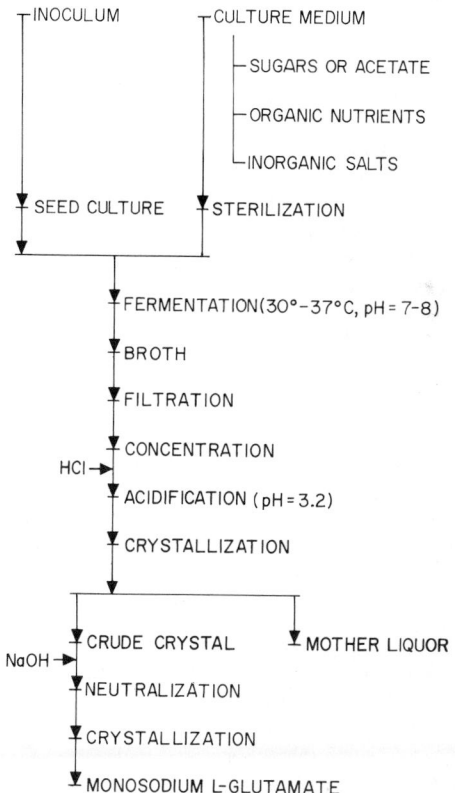

Fig. A-38. Preparation of monosodium L-glutamate by fermentation.

sively and directly. This is an advantage over other methods of amino acid manufacture.

The amino acid fermentation processes are classified as follows:

A. Methods involving simply the cultivation of a microbial strain
1. Isolated from a natural source, or a wild microbial strain, or
2. Improved by some genetic methods, or an artificial mutant
B. Method involving the conversion of a certain precursor to the corresponding amino acid with microbiological enzymes

Method with a Wild Microbial Strain: Through a series of microbiological studies, it has been found that some microbial strains isolated from natural sources possess excellent abilities to excrete and accumulate a large amount of a particular amino acid in the cultural broth under specifically controlled conditions.

The production of glutamic acid previously mentioned is a typical case. When a certain bacterial strain is cultured aerobically for about 24–48 hr in a chemically defined medium containing carbon sources, such as sugar or acetate, and nitrogen sources, such as ammonium salt, more than 50 wt % of the carbon sources is converted to glutamate. S. Kinoshita and his coworkers discovered that such a high yield of glutamate can be attained only when the growth of the used bacterium is controlled to a certain level by the limited supply of biotin, a vitamin required by the glutamate-producing bacterium. An excess amount of biotin in the medium can be controlled by the addition of certain antibiotics or detergents. This phenomenon has been explained in terms of the change of cellular permeability of glutamate. The limited supply of biotin or the addition of antibiotic or detergent increases the cellular permeability of glutamate. The glutamate thus formed inside the cells is excreted easily through the cell membrane and accumulated in the cultural broth.

Other amino acids produced by this method on a commercial scale are listed in Table A-42.

Method with an Artificial Mutant: Soon after the glutamate-producing bacterium was isolated and the concept of amino acid fermentation first materialized, microbiologists studied ways to improve the ability of microorganisms to accumulate amino acids by the use of *genetic techniques*. From biochemical observations, it is known that the accumulation of amino acids takes place (1) if a metabolic flow of amino acids in the cells is blocked at a certain step by a genetic method and/or (2) if some biochemically controlled mechanisms (such as feedback inhibition and repression, regulating the biosynthesis of amino acids in the cells) are modified or destroyed. In fact, some excellent strains have been found among auxotrophic and/or drug-resistant mutants obtained so that the above conditions are fulfilled in these mutants.

Some amino acids produced in this way also are listed in Table A-42.

Conversion of a Precursor to Amino Acid: Several amino acids are manufactured from their direct precursors by the use of microbially produced enzymes. Bacterial L-aspartate β-carboxylase is used for the production of L-alanine from L-aspartic acid. Bacterial aspartase responsible for the amination of fumaric acid is employed to produce L-aspartic acid. Tyrosine is produced through a process in which the condensation of phenol and serine is catalyzed by bacterial β-tyrosinase.

Some amino acids are produced by means of a slightly modified method of this type. The conversion of a certain precursor to the aimed amino acid takes a longer and rather complicated pathway than the action of a single enzyme. In the case of the fermentative production of isoleucine, α-aminobutyric acid is added to the culture. Similarly, anthranilic acid is used for the production of tryptophan.

The amination of α-keto acids to form their corresponding amino acids with transaminase or dehydrogenase will be utilized in the future.

As previously outlined, all protein-constituting amino acids except sulfur-containing amino acids can be produced by the amino acid fermentation process. The microbiological process for the production of sulfur-containing amino acids is now under investigation.

Isolation of Amino Acids from the Fermentation Broth: Amino acid accumulated in the fermentation broth should be isolated as pure crystals for the market. There are several physicochemical methods available for this purpose, but only a few are used for the industrial production of amino acids.

1. *Ion-Exchange Resin:* The chromatographic separation of an amino acid mixture using ion-exchange resins is one of the most common purification methods. Usual fermentation broths contain contaminating amino acids as well as bacterial cells. Isolation of the relevant amino acid from other components is easily done by

TABLE A-42. *Characteristics of Important Natural Amino Acids in Descending Order of Production Tonnage*

Name	Characteristics	Present Mode of Manufacture	World Annual Production, tons
Essential			
DL-Methionine	First limiting amino acid for soybean	Synthesis from acrolein and mercaptan	10^4
L-Lysine·HCl	First limiting amino acid for all cereals	Fermentation (AM)	10^3
L-Threonine	Second limiting amino acid for rice	Fermentation (AM)	10
L-Tryptophan	Second limiting amino acid for corn	Synthesis from acrylonitrile and resolution	10
L-Phenylalanine	Rich in plant protein	Synthesis from phenyl-acetaldehyde and resolution	10
L-Valine		Fermentation (AM)	10
L-Leucine		Extraction from protein	10
L-Isoleucine	Deficient in some cases	Fermentation (WS)	10
Quasi-essential			
L-Arginine·HCl	Essential to human infants	Synthesis from *L*-ornithine Fermentation (AM)	10^2
L-Histidine·HCl		Extraction from protein	10
L-Tyrosine	Limited substitute for phenylalanine	Enzymation of phenol and serine	10
L-Cysteine L-Cystine	Limited substitute for methionine	Extraction from human hair	10
Nonessential			
L-Glutamic acid	MSG, taste enhancer	Fermentation (WS) Synthesis from acrylonitrile and resolution	10^5
Glycine	Sweetener	Synthesis from formaldehyde	10^3
DL-Alanine		Synthesis from acetaldehyde	10^2
L-Aspartic acid	Hygienic drug	Enzymation of fumaric acid	10^2
L-Glutamine	Anti–gastroduodenal ulcer drug	Fermentation (WS)	10^2
L-Serine	Rich in raw silk	Synthesis from glycolonitrile and resolution	<10
L-Proline	Rich in gelatin	Fermentation (AM)	<10
L-Hydroxyproline		Extraction from gelatin	<10
L-Asparagine	Neurotropic metabolic regulator	Synthesis from L-aspartic acid	<10
L-Alanine	Rich in degummed white silk	Enzymation of L-aspartic acid	<10
L-Dihydroxy-phenylalanine	Specific drug for Parkinson's disease	Synthesis from piperonal, vanillin, or acrylonitrile and resolution	10^2
L-Citrulline	Ammonia detoxicant	Fermentation (AM)	<10
L-Ornithine		Fermentation (AM)	<10

AM = artificial mutant; WS = wild strain.

flowing the broth through an ion-exchange resin column. This method also is used for the recovery of an amino acid from the mother liquor of the amino acid crystal.

2. *Precipitation:* Compounds which give insoluble salts with amino acids are selected for the separation and purification of amino acids. Picrate has been used for the purification of proline, which is difficult to purify because of its high solubility.

3. *Crystallization:* An amino acid is least soluble when the pH of the solution is adjusted to the isoelectric point. If the solubility is sufficiently low at this point (as in the case of glutamic acid), adjustment of the pH of fermentation broths serves to separate a pure amino acid in crystalline form. The polymorphism and polytypism of the crystals often are observed. The way of controlling supersaturation (agitation, temperature, existence of minute amounts of impurities, seed crystals, hysteresis of the solution, and other factors) is important in obtaining the required purity of the separated crystals. In the case of glutamic acid, the L-form crystal is unstable, and thus it is necessary to observe rigorously the subtle optimal conditions needed for the production of good pure crystals. Concentration and cooling are used most frequently in the final stage of production of pure amino acid crystals. See also **Crystallization.**

Some physiological testing is done for the production of medical preparations. As test animals, cats are chosen for checking blood-pressure depression and rabbits for the pyrogen test.

Chemical Synthesis. There are many laboratory methods for amino acid synthesis, but only a few appear to offer industrial potential. The synthetically produced amino acids listed in Table A-42 are made mostly from aldehydes. Glycine and DL-alanine are produced by the Strecker synthesis, starting from formaldehyde and acetaldehyde, respectively. β-Cyanopropionaldehyde, the raw material for glutamic acid, is produced by hydroformylation of acrylonitrile. This aldehyde contains isomeric α-cyanopropionaldehyde, but the ratio of α to β is smaller than 1:10. Following are two ways for the synthetic production of amino acids from aldehydes. The hydantoin process has an advantage over the Strecker synthesis in giving higher yields generally. In the case of glutamic acid, the conversion of glutamic acid (in solution) from β-cyanopropionaldehyde by the hydantoin process is nearly quantitative.

Strecker Synthesis: Aldehydes react with hydrogen cyanide and excess ammonia to give aminonitriles, which in turn are converted into α-amino acids on hydrolysis.

$$RCH \xrightarrow{HCN} RCH-CN \xrightarrow{NH_3}$$
$$\underset{O}{\Vert} \qquad \underset{OH}{\vert}$$

$$RCH-CN \xrightarrow{NaOH} RCH-COONa$$
$$\underset{NH_2}{\vert} \qquad \underset{NH_2}{\vert}$$

Hydantoin Process: Aldehydes react with sodium cyanide and ammonium carbonate to give hydantoins. It will give α-amino acids on hydrolysis with alkali.

$$RCHO \xrightarrow{NaCN, (NH_4)_2CO_3} RCH-CO \xrightarrow{NaOH}$$
$$\underset{NH \quad NH}{\vert \qquad \vert}$$
$$\underset{CO}{\diagdown \quad \diagup}$$

$$RCH-COONa + (NH_4)HCO_3$$
$$\underset{NH_2}{\vert}$$

In this case, excesses of carbon dioxide and ammonia are easily removed from the reaction mixture by treating with a little amount of alkali under high temperature and pressure.

Optical Resolution: α-Amino acids produced by chemical synthesis are obtained in racemic forms (DL-forms). D-Form of glutamic acid has no flavor-enhancing properties, and so it should be resolved into optically active forms so far as MSG is concerned. In the case of DL-alanine, both forms have sweet taste; so the resolution is not necessary for usage as a sweetener. DL-Methionine also need not be resolved, because the D-form is equally effective with L-form for human and animal nutrition. Three industrial methods for the resolution of optical isomers are as follows:

1. *Preferential Inoculation Method:* The supersaturation of a racemic acid solution caused by lowering the temperature, for example, gives the preferential growth of the optical active seed crystals and leaves the antipode dissolved in the solution. After the crystals have grown to a certain extent, they are filtered off and the seed crystals of antipode are added to grow them in the filtrate. Thus the racemic mixtures successively give both optical isomers as crystals. In industrial MSG production, seed crystals of the D- and L-forms will grow separately in each section of a resolution tank, which is divided into two parts by a suitable net. By vigorous agitation, the supersaturation of the D- and L-forms is equalized without mixing of

antipode crystals to enable simultaneous growth of crystals of both optically active forms. As a necessary condition for the performance of the inoculation method, the solubility of the racemic mixture should be larger than that of the optically active form.

2. *Diastereoisomer Method:* For example, synthesized DL-lysine (base) gives equimolar amounts of salts; D-lysine L-glutamate and L-lysine L-glutamate, by the addition of L-glutamic acid.* These two are not optical antipodes and have different solubilities. As L-lysine L-glutamate has a smaller solubility, in this case, this crystallizes first. L-Lysine is easily obtained from L-lysine L-glutamate. Mother liquor is concentrated to give crystals of D-lysine L-glutamate.

3. *Acylase Method:* N-Acyl compound made from DL-amino acid is hydrolyzed using an acylase (enzyme) to give only L-amino acid, whereas D-form remains as an acyl compound (unchanged).

The foregoing three methods have merit in not causing loss of both forms on resolving, but for industrial application, the inoculation method† has the advantage that it needs no chemical reagent for the resolution.

Racemization: Hydrolysis of proteins with alkali at about 100°C gives completely racemized amino acids. Isolated amino acids cannot be racemized so easily, and heating to a higher temperature is required for complete racemization. An alkaline condition is not always necessary. When a neutral aqueous solution is heated to 200°C, complete racemization is achieved with no appreciable decomposition. In the case of glutamic acid, it is dehydrated to pyroglutamic acid at a lower temperature than that required by racemization, and so hydrolysis using acid or alkali at a temperature a little higher than 100°C is necessary after racemization.

$$\underset{\text{D-Glutamic acid}}{\overset{\displaystyle \text{H}_2\text{C}-\text{CH}_2}{\underset{\displaystyle \text{H}_2\text{N}\quad\text{COOH}}{\text{HOOC}\quad\text{CH}}}} \xrightarrow[\text{at 200°C}]{-\text{H}_2\text{O}}$$

$$\underset{\text{DL-Pyroglutamic acid}}{\overset{\displaystyle \text{H}_2\text{C}-\text{CH}_2}{\underset{\displaystyle \text{NH}}{\text{OC}\quad\text{CH}-\text{COOH}}}} \xrightarrow[\text{acid or alkali}]{+\text{H}_2\text{O}} \underset{\text{acid}}{\text{DL-glutamic}}$$

*Data from a U.S. patent.
†Resolution of glutamic acid in Ajinomoto Co., Inc., production is performed by this method.

Thus formed, DL-glutamic acid again may be resolved into the L- and D-forms, the latter of which may be racemized again. As a result, all DL-forms can be converted into L-forms.

Glutamic acid synthesis and resolution‡ are shown in Fig. A-39. The synthetic route to methionine is shown in Fig. A-40. All methionine is produced by synthesis because a fermentation method has not been discovered. Inasmuch as D-methionine is equally effective in the animal body, the racemic form is marketed.

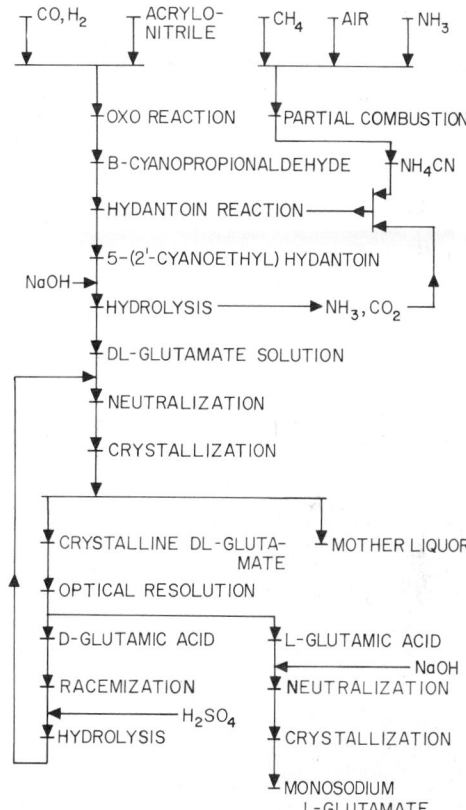

Fig. A-39. Preparation of monosodium L-glutamate by chemical synthesis.

The commercial synthetic production of lysine from caprolactam is being investigated, but has thus far not been realized. As shown in Table A-42, lysine occupies the second position in world production of amino acids. All this production (mostly by Japanese firms) is carried out by fer-

‡Ajinomoto Co., Inc.

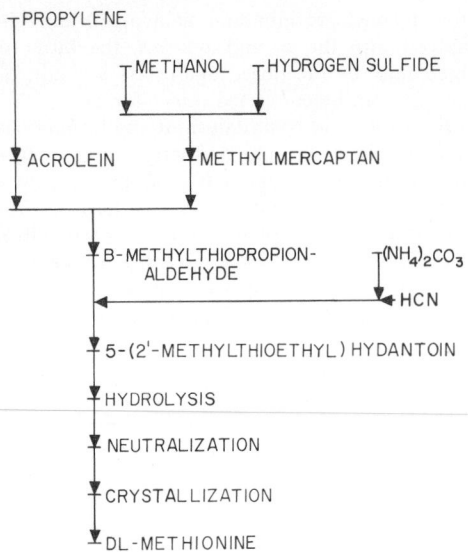

Fig. A-40. Preparation of DL-methionine by chemical synthesis.

The amount of lactic acid is directly proportional to the concentration of the amino acid to be assayed and is accurately determined by the direct titration with standard alkali.

Amino Acid Requirements and Pattern: It has long been known that proteins are required in the diets of animals, and some proteins are more effective than others in supporting life. In reality, the need is not for proteins as such, but rather, for the amino acids present in the proteins and released by hydrolysis.

In a report by FAO/WHO (Protein Requirements, 1965), the following conclusion was drawn. "To determine the quality of a protein, two factors have to be distinguished, namely, the proportion of essential to non-essential amino acids and, secondly, the relative amounts of the essential amino acids. It was concluded that the best patterns of essential amino acids for meeting human requirements was that found in whole egg protein or human milk, and comparisons of pro-

mentation. See flow sheet, Fig. A-41. The fermentation method permits easy scale-up to meet increasing demands.

Role in Nutrition

Microbiological Assay: Plants can synthesize all common amino acids from simple carbon and nitrogen sources and a few kinds of minerals. In contrast, unicellular microorganisms show a remarkable diversity in their nutritional needs. Whereas certain organisms can grow on relatively simple media composed of glucose, ammonia, and inorganic salts, other organisms are more complicated in that they require a variety of amino acids, vitamins, and minerals. For example, *Leuconostoc mesenteroides* P-60 requires 17 amino acids for growth. This fact has led directly to the use of this microorganism for the quantitative determination of amino acids. Amino acids shown in Table A-43, except cysteine and hydroxyproline, can be determined by the microbiological-assay method.

In routine practice, a base medium is prepared with the known essential nutrients required for optimal growth of the above organism or other lactobacilli, but without the amino acid to be assayed. Graded doses of the sample to be determined are added to the basal medium, which in turn are sterilized, inoculated into the appropriate assay organism, and incubated at 37°C. During the course of metabolism they produce lactic acid.

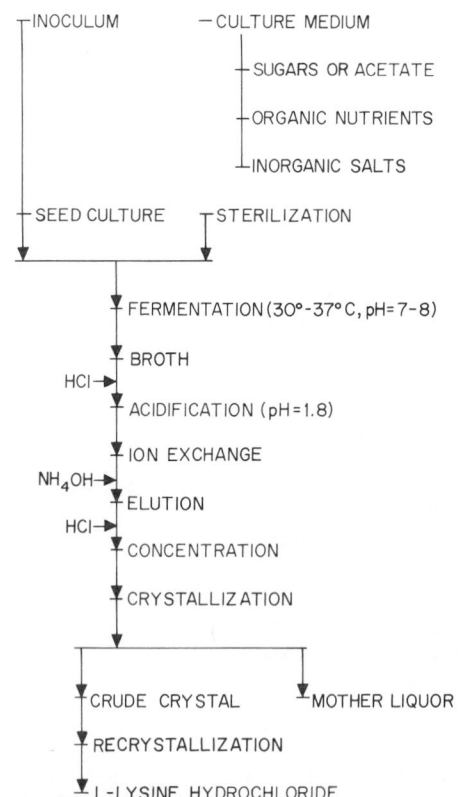

Fig. A-41. Preparation of L-lysine hydrochloride by fermentation.

TABLE A-43. *Selected Essential Amino Acid Patterns A/E Ratio, Milligrams per Gram of Total Essential Amino Acids*

	Hen's Egg (Whole)	Human Milk	Cow's Milk
Total "aromatic" amino acids	195	226	197
Phenylalanine	(114)	(114)	(97)
Tyrosine	(81)	(112)	(100)
Leucine	172	184	196
Valine	141	147	137
Isoleucine	129	132	127
Lysine	125	128	155
Total "S"	107	87	65
Cystine	(46)	(43)	(17)
Methionine	(61)	(44)	(48)
Threonine.	99	99	91
Tryptophan	31	34	28

SOURCE: Cited from Protein Requirements, Report of a joint FAO/WHO expert group, WHO Technical Report Series, No. 301, and FAO Nutrition Meeting Report Series, No. 37, World Health Organization, Geneva, 1965.

tein quality should be made by reference to the essential amino acid patterns of either of these two proteins."

In Table A-43, the ratio of each essential amino acid to the total sum is given for hen's egg and human and cow's milk. In the first two proteins nearly the same total amount of nonessential amino acids are contained. Differences of A/E* ratio in these proteins (Table A-43) are probably not significant, but cow's milk is somewhat inferior, as each A/E value shows larger deviation from the standard. Cow's casein, when fed to children, is not completely absorbed in their digestive organs. Hen's egg and human milk can be fed to persons of all ages in experiments, and show the same effect.

In the human body, tyrosine and cysteine can be formed from phenylalanine and methionine, respectively, but the reverse transformations do not occur. Total "aromatic" amino acids and sulfur-containing amino acids can be substituted by phenylalanine and methionine, respectively. But in reality, tyrosine and cystine are contained in these proteins, as shown in Table A-43. The requirement for these two essential amino acids can be reduced to the figures shown in the table.

*A/E ratio equals ten times percentage of single essential amino acid to the total essential amino acids contained.

Human infants have an ability to synthesize arginine and histidine in their bodies, but the speed of the synthesis is very slow compared with their requirements. Since they can be regarded as essential to growing infants, tyrosine and cystine, as well as arginine and histidine, occasionally are grouped as "quasi-essential amino acids." See Table A-41.

In plant proteins, A/E ratios markedly differ from the above-mentioned standard proteins. An essential amino acid which shows the largest deficiency is named the first limiting amino acid. An amino acid showing the next largest deficiency is named the second limiting one. If plant proteins are fortified by adding these limiting amino acids, their quality can be largely improved.

As to the other species of mammals, histidine is also essential for the dog, rat, mouse, and chick; arginine for the chick; and glutamic acid and proline for the young chick.

Fortification of Human Foods: As shown in Table A-43, several essential amino acids have been proved to be the limiting factor of nutrition in plant proteins. In advanced countries, the ratio of vegetable proteins to animal proteins in foods is 1.4:1. In underdeveloped countries, the ratio is 3.5:1, which means that people in underdeveloped countries depend upon vegetable proteins.

Among vegetable staple foods, wheat easily can be fortified. It is used as flour all over the world. L-Lysine hydrochloride (0.2%) is added to the flour. Wheat bread fortified with lysine already is used in several areas in the world. In Japan it is now supplied as a school ration.

In rice the fortification situation is somewhat complicated. Before cooking, rice must be washed (polished) with water, and in some countries, the cooking water is allowed to boil over or is discarded. Such significant loss of fortified amino acids must be considered. L-Lysine hydrochloride (0.2%) and L-threonine (0.1%) are shaped like rice grain with other nutrients and enveloped in a film. The added materials must hold the initial shape and never dissolve out during boiling, and be easily freed from their coating in the digestive organs. Rice-fortification tests now are going on in a certain area of Thailand by the U.S. Agency for International Development, with the cooperation of Ajinomoto Co., Inc., which is supplying the fortified rice.

Feed Supplementation: Essential amino acids are arranged in Table A-42 in the order of their production tonnage. Each of the four amino acids at the beginning of the table are all limiting factors of various vegetable proteins. Chick feed usually is supplemented with fish meals, but where the latter is in limited supply, soybean meals are substituted.

The demand for DL-methionine, limiting amino acid in soybean meals, is now increasing. When seed meals, such as corn and sorghum, are used as feeds for chickens or pigs, L-lysine hydrochloride must be added for fortification. Lysine production is increasing upward to the level of methionine.

Progress in the Amino Acid Industry. Although the chemistry of amino acids, as shown by the listing of discoveries in Table A-41, goes back at least as early as 1820, when Braconnot discovered glycine and leucine from protein in hydrolysate, the beginnings of a true amino acid industry had to await the development of a need and market for large quantities of a commercially produced amino acid. In this regard, an important event occurred in 1908, when K. Ikeda of the Imperial University of Tokyo invented a taste enhancer, the original monosodium glutamate (MSG).*

During and after World War II new methods of amino acid determination, such as the isotope dilution method, microbiological assay, and a little later paper partition chromatography had been developed. These new analytical methods gave clues leading to the discovery of new amino acids, especially in the field of microorganisms. Without the limitation of being components of proteins and being α-amino acids, the number of new natural amino acids discovered up to 1967 had reached 250, of which structure determination and identification with a synthetic sample already had been completed.

This worldwide progress of chemistry had some influence upon the amino acid industry. For example, D-phenylglycine (an unnatural amino acid) is now synthesized and resolved into D-form on a commercial scale. It is used as a raw material of semisynthesized penicillin.

---

*The new product was called AJI-NO-MOTO, for which industrial production was started by the Ajinomoto Co., Inc., in 1910. Extracted from vegetable protein exclusively through 1959, when production was 13,600 metric tons/yr, the fermentative process was introduced in 1960, and the synthetic process in 1962, both of which have replaced vegetable-protein sources, and now account for an annual production of about 74,000 metric tons in 1970. (The figures represent the production of Ajinomoto's domestic and overseas factories.)

See also **Enzymes; Proteins and peptides;** and **Yeasts.**

REFERENCES
Albanese, A. A. (ed.): "Protein and Protein Nutrition," Academic, New York, 1959.
Greenstein, J. P., and M. Winitz: "Chemistry of the Amino Acids," 3 vols., Wiley, New York, 1961.
Meister, A.: "Biochemistry of the Amino Acids," 2d ed., 2 vols., Academic, New York, 1965.

—Haruomi Oeda, *Ajinomoto Co., Inc., Kawasaki, Japan*

Amino Carboxylic Acids   (See **Carboxylic acids.**)

Amino Resins   Produced by an addition reaction between formaldehyde and such compounds as melamine, urea, aniline, ethylene urea, sulfonamide, and dicyandiamide, these thermosetting resins have a rather long history of use as molding compounds, wood adhesives, laminating coatings, wet-strength paper resins, and textile-treating resins. The most widely used amino resins are the urea and melamine compounds.

The production of urea-formaldehyde resins proceeds in steps. In the first stage, urea, NH_2CONH_2, formalin (40% solution of paraformaldehyde), some thiourea, NH_2CSNH_2, and ammonia are mixed and permitted to stand for approximately two hours, during which time a transparent liquid (estimated to be dimethylolurea, $HOCH_2NHCONHCH_2OH$) is formed. Without further reacting, this solution is used as an impregnant for wood, paper, fabric, and similar materials and as an adhesive. When the transparent fluid is heated further with the addition of an accelerator acid catalyst, dimethylene urea, $CH_2:NCON:CH_2$, is formed. Further heating transforms the gelatinous material into a harder resinous mass.

To prepare melamine resins, dicyanodiamide, $H_2NC(:NH)—NHC:N$, is heated with formaldehyde under pressure. See also **Melamine.**

The basic color of both urea and melamine resins is water white (transparent). Translucent and opaque colors are easily obtained by the addition of pigments and opacifying agents. Light transmission is reduced by the addition of cellulose filler. Particularly where color is unimportant, asbestos, wood flour, macerated fabric, and glass fiber may be added to the melamine resins for numerous industrial applications. Likewise, the addition of wood flour to the urea resins produces a low-cost industrial raw material.

TABLE A-44. *Properties of Amino Resin Compounds*

	Urea		Melamine	
	Type of Filler			
Property	Alpha Cellulose	Wood Flour	Alpha Cellulose	Glass Fiber
Specific gravity....................	1.5	1.5	1.5	1.94–2.0
Density, g/in.^{3}....................	24.6	24.6	24.6	31.8–32.8
Hardness, Rockwell E..............	94–97	95	110	
Molding shrinkage, in./in.	0.006–0.009	0.006–0.014	0.008–0.009	0.002–0.004
Deflection temperature, 264 psi	130°C, 266°F	132°C, 270°F	183°C, 361°F	204°C, 400°F
Heat resistance (continuous)	77°C, 170°F	77°C, 170°F	99°C, 210°F	149°C, 300°F
Water absorption, %, 24 hr at 23°C (73.4°F) .	0.04–0.8	0.7	0.3–0.5	0.09–0.3
Color range.....................	Unlimited	Brown/black	Unlimited	Natural gray

Among the advantages of amino resins are: (1) easily fabricated by economical molding methods; (2) hardness; (3) rigidity; (4) abrasion resistance; (5) high resistance to deformation under load; (6) no tendency to become brittle at subzero (Fahrenheit) temperatures; (7) self-extinguishing characteristics; (8) resistance to attack by common organic solvents, oils and greases, and weak alkalies and acids; (9) no tendency to impart tastes and odors to foods; and (10) good electrical insulation characteristics. In their resistance to heat and boiling water, acids, and alkalies, melamines are superior to the ureas. The amino resins do exhibit fairly high mold shrinkage and also shrink upon aging. When cycled frequently between dry and wet conditions, urea moldings tend to develop cracks. Both color and strength are affected adversely when the amino resins are subjected to prolonged elevated temperatures. The urea resins are not suited to outdoor exposure. The melamines are more rugged in this respect, but may discolor. Amino resins are fabricated mainly by transfer and compression molding. There is limited use of injection molding and extrusion.

The properties of some urea and melamine amino compounds are summarized in Table A-44.

Aminoglycosides (See **Antibiotics**.)

Ammonia Known since ancient times, ammonia, NH_3, has been commercially important for well over one hundred years and has become the second largest chemical in tonnage and the first chemical in value of production. The first practical plant of any magnitude was built in 1913. Worldwide production of NH_3 (1972) is estimated at 48 million short tons/yr, the United States

accounting for approximately 14 million short tons/yr. Ammonia production is growing at a rate of 10–12%/yr.

The major consuming areas for NH_3 are summarized in Table A-45. The properties of the main commercial grades of NH_3 are given in Table A-46.

Properties: At standard temperature and pressure, NH_3 is a colorless gas with a penetrating, pungent-sharp odor in small concentrations which, in heavy concentrations, produces a smothering sensation when inhaled. The formula weight is 17.03, melting point is $-77.7°C$, boiling point is $-33.35°C$, and specific gravity is 0.817 at $-79°C$ and 0.617 at 15°C. Ammonia is very soluble in water, a saturated solution containing approximately 45% NH_3 by weight at the freezing temperature of the solution and about 30% by weight at standard conditions. NH_3 dissolved in water forms a strongly alkaline solution of ammonium hydroxide, NH_4OH. The univalent radical NH_3^+ behaves in many respects like K^+ and Na^+ in vigorously reacting with acids to form salts. NH_3 is an excellent nonaqueous electrolytic solvent, its ionizing power approaching that of water. Ammonia burns with a greenish-yellow flame.

Ammonia derives its name from *sal ammoniac,* NH_4Cl, the latter material having been produced at the Temple of Jupiter Ammon (Libya) by distilling camel dung. During the Middle Ages, NH_3 was referred to as the spirits of hartshorn because it was produced by heating the hoofs and horns of oxen. The composition of ammonia first was established by Claude Louis Berthollet (France, ca. 1777). The first significant commercial source of NH_3 (during the 1880s) was its

TABLE A-45. *Major Areas of Ammonia Consumption**

Consuming Areas	Percentage of Consumption			
	Fertilizer	Fibers and Plastics Intermediates	Nonfertilizer	Total
Ammonia—direct application	26.6	. . .	. . .	26.6
Ammonium nitrate (AN)	17.0	. . .	4.0	21.0
Nitric acid for non-AN uses	1.1	1.6	4.2	6.9
Urea.	10.1	0.8	1.6	12.5
Ammonium phosphate	11.2	. . .	. . .	11.2
Ammonium sulfate	5.9	. . .	. . .	5.9
Nitrogen solutions and mixed fertilizers.	4.4	. . .	. . .	4.4
Acrylonitrile	. . .	2.2	. . .	2.2
Hexamethylenediamine	. . .	0.7	. . .	0.7
Amides and nitriles	. . .	. . .	0.6	0.6
Caprolactam	. . .	0.2	. . .	0.2
Losses (transportation, handling, and storage).	. . .	. . .	. . .	3.0
Miscellaneous other uses	. . .	. . .	4.8	4.8
Percentage of market	76.3	5.5	15.2	100.0

*Based on practice in the United States.

TABLE A-46. *Typical Anhydrous Ammonia Specifications*

Commercial or fertilizer grade:
Ammonia content, % by wt . .	99.5 min.
Water, % by wt	0.5 max., 0.2 min.
Oil, ppm	5.0 max.

Refrigeration grade:
Ammonia content, % by wt . .	99.98 min.
Water, % by wt	0.015 max.
Oil, ppm	3.0 max.
Noncondensable gas, ml/g . . .	0.2 max.

Metallurgical grade:
Ammonia content, % by wt . .	99.99 min.
Water, ppm	33.00 max.
Noncondensable gas, ml/g . . .	10.00 max.
Oil, ppm	2.00 max.
Dew point	$-79°F (-61.6°C)$

production as a by-product in the making of manufactured gas through the destructive distillation of coal. See also **Coal tar and derivatives.**

Nitrogen *fixation* is a term assigned to the process of converting N_2 in the air to nitrogen compounds. Although some bacteria in soil are capable of this process, N_2 as an ingredient of fertilizer is required for soils that are depleted by crop production. The production of synthetic NH_3 is the most important industrial nitrogen-fixation process. See also **Fertilizers.**

Synthesis: The first breakthrough in the large-scale synthesis of ammonia resulted from the work of Fritz Haber (Germany, 1913), who found that ammonia could be produced by the direct combination of two elements, nitrogen and hydrogen, $(N_2 + 3H_2 \rightleftharpoons 2NH_3)$ in the presence of a catalyst (iron oxide with small quantities of cerium and chromium) at a relatively high temperature (550°C) and under a pressure of about 200 atm, representing difficult processing conditions for that era. Largely because of the urgent requirements for ammonia in the manufacture of explosives during World War I, the process was adapted for industrial-quantity production by Karl Bosch, who received one-half of the 1931 Nobel prize for chemistry in recognition of these achievements. Thereafter, many improved ammonia-synthesis systems, based on the Haber-Bosch process, were commercialized, using various operating conditions and synthesis-loop designs.

The principal features of an NH_3 synthesis process system are the converter design, operating conditions, method of product recovery, and type of recirculation equipment. Most current systems operate at or above the pressure used in the original Haber-Bosch process. Converter designs have either a single continuous catalyst bed, which may or may not have heat-exchange cooling for controlling reaction heat, or several catalyst beds with provision for removing or controlling reaction heat between the beds.

Claude Process: The original Claude process was one of the first systems to use a high operating pressure (1,000 atm), achieving 40% conversion without recycling. This system used multiple converters in a series-parallel arrangement. The present * Claude process operates at 340–650 atm, using a single converter with continuous catalyst-charged tubes externally cooled by heat exchange. Approximate H_2 conversion is 30–34 mole %. The pressure is increased gradually to compensate for catalyst aging and loss in activity. Product recovery is by simple condensation in a water-cooled condenser. Unreacted gas is recycled by compressor.

Casale Process: This is another high-pressure conversion system, using synthesis pressures of 450–600 atm, which also permits conversions in the 30 mole % range. As in the Claude process, the high pressure allows NH_3 to be recovered from the converter effluent by water cooling. The Casale converter uses a single catalyst bed with internal heat-exchange surfaces. Reaction rate and temperature rise across the catalyst is controlled by retaining 2–3 mole % NH_3 in the converter feed. Recycling of unreacted gases with some NH_3 is done by an ejector system. This eliminates the need for a mechanical recycle compressor, but requires high feed-gas pressures to supply the energy required for the ejector.

Low-Pressure Processes: Several systems† use low synthesis pressures with H_2 conversion below 30 mole % and product recovery by water and refrigeration.

Synthesis-Gas-Production Processes: These processes were improved and developed as a result of changes in feedstock availability and economics. Before World War II, most NH_3 plants obtained H_2 by reacting coal or coke and steam in the water-gas process. A small number of plants used water electrolysis or coke-oven by-product H_2. The subsequent low-cost availability of natural gas brought about steam-hydrocarbon reforming as the major source of H_2 for the NH_3 synthesis gas.

Partial oxidation processes to produce H_2 from natural gas and liquid hydrocarbons were also developed after World War II and accounted for 15% of the synthetic NH_3 capacity by 1962. The steam-hydrocarbon reforming process‡ was developed in 1930. In this process, methane was mixed with an excess of steam at atmospheric pressure, and the mixture reformed inside Ni-catalyst-filled alloy furnace tubes. The heat of reaction was supplied by externally heating the catalyst-filled tubes to about 1600°F (871°C). Since the late 1950s, improvements in tubular-reforming technology and metallurgy have brought about the utilization of high-pressure (>350 psig) reforming, which cut synthesis-gas-compression costs and increased heat recovery. The first pressure reformer§ was built in 1953. In addition, the higher pressures allowed improvements in the efficiency of synthesis-gas-purification systems. High-pressure steam-reforming technology also has been extended to cover heavier hydrocarbon gases, including propane, butane, reformer gases, and streams containing a high amount of olefins. In 1962, a process‖ for reforming straight-run liquid distillates (naphthas) was commercialized. This process is based on the use of an alkali oxide–modified Ni catalyst¶ which permits reforming of desulfurized naphthas at low ($\sim$3.5:1) steam carbon ratios, without significant carbon deposition on the catalyst.

Noncatalytic partial oxidation processes designed to produce H_2 from a wide range of hydrocarbon liquids, including heavy fuel oils, crudes, naphthas, coal tar, and pulverized bituminous

*Developed by Grande Paroisse and L'Air Liquide.

†Fauser-Montecatini, Tennessee Valley Authority, Chemico, Kellogg, Lummus, Imperial Chemical Industries, Topsoe, Société Belge de l'Azote, Oesterreichische Stickstoffwerke, AG (OSW), and Friedrich Uhde. Uhde also offers a high-pressure system. All the aforementioned systems use mechanical recycle compressors, except Fauser-Montecatini, which sometimes uses a fresh-feed-gas injector to circulate the recycle gas. The Fauser-Montecatini converter is a multibed type with pressurized water-cooling coils to remove the reaction heat. Converter ammonia concentrations are approximately 1.5 mole % in and 20 mole % out. The TVA converter consists of a calandria-type feed-effluent interchanger and a single catalyst bed with countercurrent cooling tubes. Chemico's converter is similar to TVA's, except that cooling is done by using cocurrent cooling tubes. Kellogg uses a multiple-bed-type converter with provisions to inject cold gas as quench between the layers. Feed is heated by a feed-effluent interchanger in the converter. Lummus also uses a TVA type converter, the effluent containing 14–18 mole % ammonia. ICI uses a quench-type converter with a patented distributor which operates submerged in the catalyst. ICI also offers a converter design with a tube-cooled arrangement. For its 300–350-atm system, Haldor Topsoe uses a tube-cooled converter of the TVA type. OSW developed a converter design based on cooling partly converted gases by indirect heat exchange between catalyst beds and injection of cold feed gas at the catalyst inlet. Uhde uses a quench-type converter.

‡Originally developed by Standard Oil Company of New Jersey.

§Built by M. W. Kellogg for Shell Chemical Corp. (Ventura, Calif.).

‖M. W. Kellogg and Imperial Chemical Industries.

¶Developed by M. W. Kellogg.

coal, were commercialized in 1954* and 1956.†
In both these processes, the hydrocarbon feed is
oxidized and reformed in a refractory-lined pressure vessel. The required O_2 usually is supplied
by an air separation plant from which N_2 also can
be used as feed for the synthesis gas. The main
differences between the two processes are in the
reactor design, feeding method, burner design, and
heat recovery. The partial oxidation processes and
the steam-naphtha reforming process are favored
in areas with short supplies of natural gas.

The source of nitrogen for the synthesis gas has
always been air, either supplied directly from a
liquid-air separation plant or by burning a small
amount of the hydrogen in the synthesis gas. The
need for air separation plants or burning of hydrogen has been eliminated in modern ammonia
plants by use of secondary reforming, where
methane is burned with sufficient air to produce
a 3:1 mole ratio hydrogen-nitrogen synthesis gas.

Most ammonia plants built since the early
1960s are in the 600–1,500 short tons/day range
and are based on new integrated designs that have
cut the cost of ammonia manufacture in half.
Today's plants, in fact, have reached a peak in
overall efficiency by combining all the separate
units (e.g., synthesis-gas preparation, purification,
and ammonia synthesis) in one single design.
High-pressure reforming has reduced the synthesis-gas compression load and allows the recovery of heat at high enough levels to generate the
steam required to drive the compressors. This
compactness in design has also led to increased
plant size at reduced investment and operating
cost.

Use of Multistage Centrifugal Compressors: The
largest single factor contributing to the improved
economics of ammonia plants is in the application
of multistage centrifugal compressors, which have
replaced the reciprocating compressors traditionally used in the synthesis feed and recycle
service. A single centrifugal compressor can do
the job of several banks of reciprocating compressors, thus reducing equipment cost, floor space,
supporting foundations, and maintenance.

The use of multistage centrifugal compressors
was made possible by redesigning the synthesis
loop to operate at low pressures (150–240 atm)

and by increasing plant capacity to above the
compressor's minimum-flow restriction. (Most
synthesis loops using reciprocating compressors
had been operating at intermediate pressures of
300–350 atm.) Centrifugal compressors capable of
developing pressures up to 340 atm are already
being offered and used in some large-capacity
(1,000 short tons/day) plants, where the loss in
efficiency associated with higher pressures is offset
by the increased compressor and driver efficiencies
associated with higher flow rates.

An operating NH_3 plant using the aforementioned improvements is illustrated schematically
in Fig. A-42.‡ This plant has a capacity of 1,000
short tons/day and uses natural gas as feedstock.
The plant can be divided into the following
integrated-process sections: synthesis-gas preparation; synthesis-gas purification, and compression
and ammonia synthesis.

Synthesis-Gas Preparation: The desulfurized
natural gas is fed to the primary reformer, where
it is reacted with steam in Ni-catalyst-filled tubes
to produce a major percentage of the H_2 required.
The principal reactions taking place are

$$CH_4 + H_2O \rightleftharpoons CO + 3H_2$$
$$\Delta H_{298} = +49.3 \text{ kcal/mole} \quad (1)§$$
$$CO + H_2O \rightleftharpoons CO_2 + H_2$$
$$\Delta H_{298} = -9.8 \text{ kcal/mole} \quad (2)§$$

Reaction (1) is the principal reforming reaction,
and reaction (2) is the water-gas shift reaction.
The net reactions are highly endothermic. The
partially reformed gas leaves the primary reformer
containing approximately 10% methane, on a
mole dry-gas basis, at 400–500 psig and up to
1500°F (816°C). The required heat of reaction
is supplied by natural-gas-fired arch burners,
which are designed to also burn purge and flash
gases from the synthesis section. Waste heat from
the primary-reformer flue is recovered by generating high-pressure superheated steam, which
along with waste-heat-process boilers and an appended auxiliary boiler assure a steam system that
is always in balance, while providing high-
pressure steam to compressor turbine drivers and

---

‡Designed by The M. W. Kellogg Company (Houston, Tex.), for which Kellogg received the 1967 Kirkpatrick Chemical Engineering Achievement Award.

§Heats of reaction at 298K (25°C), 1 atm pressure,
gaseous substances in ideal state.

*Texaco partial oxidation process.
†Shell gasification process.

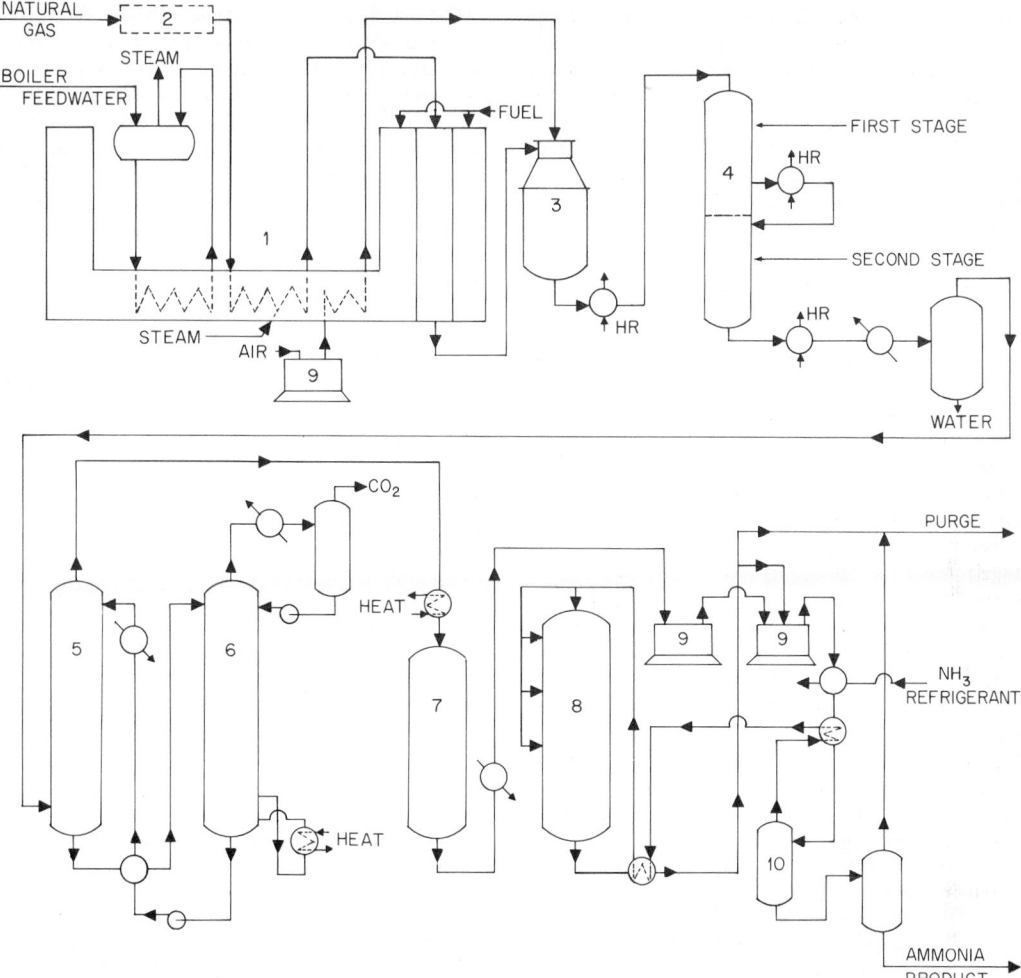

Fig. A-42. Ammonia production process. (1) Primary reformer, (2) desulfurization, (3) secondary reformer, (4) CO shift converter (in two stages), (5) CO$_2$ absorber, (6) CO$_2$ stripper, (7) methanator, (8) NH$_3$ converter, (9) compressor, (10) separator. HR = heat recovery. (*The M. W. Kellogg Company, a division of Pullman Incorporated.*)

low-pressure steam to pump drivers. Further waste heat is recovered by preheating the natural-gas–steam feed mixture, steam-air for secondary reforming, and fuel. Economies are also achieved by using a common convection section, furnace draft fan, and stack.

The primary reforming step is followed by conversion of the residual methane to hydrogen and carbon oxides over a bed of high-temperature chrome and nickel catalysts in the secondary re-

former. Sufficient preheated air, supplied by a steam-driven centrifugal compressor, is introduced in the combustion-chamber portion of the refractory-lined secondary reformer to produce a raw synthesis gas with a 3:1 H/N mole ratio. The secondary reforming step reduces fuel-gas input and overall reforming cost by shifting part of the required hydrocarbon conversion from the high-cost primary reformer to the lower-cost secondary reformer. It also permits an increase in the residual methane level at the primary effluent, which results in lower operating temperatures, reduced steam requirements, and milder tube-metal conditions.

Process waste-heat boilers then cool the reformed gas and steam to 700°F (371°C) while generating high-pressure steam. The cooled gas-steam mixture enters a two-stage shift converter. The purpose of shift conversion is to remove CO from the gas and produce an equivalent amount of H_2 by the reaction $CO + H_2O \rightleftharpoons CO_2 + H_2$. Since the reaction rate in the shift converter is favored by high temperatures but equilibrium is favored by low temperatures, two conversion stages, each with a different catalyst, provide the optimum conditions for maximum CO shift. The high-temperature shift is performed in the first stage, using an iron catalyst. The low-temperature shift in the second stage uses a copper catalyst. Interstage heat recovery is provided to remove the heat of reaction generated in the first stage. Heat is recovered by additional high-pressure steam generation and warming of the methanator feed. Gas from the shift converter is the raw synthesis gas, which, after purification, becomes the feed to the NH_3 synthesis section.

Purification of Synthesis Gas: This involves the removal of carbon oxides to prevent poisoning of the NH_3 synthesis catalyst. An absorption process is used to remove the bulk of the CO_2, followed by methanation of the residual carbon oxides in the methanator.

The absorption process consists in countercurrent-contacting the raw synthesis gas with a liquid absorbent solution in a packed or trayed column. The original Bosch process used water as an absorbent to remove CO_2 from the raw synthesis gas, but low efficiency and excessive hydrogen losses have rendered this process obsolete. Modern ammonia plants use a variety of CO_2-removal processes with effective absorbent solutions. The principal absorbent solutions currently in use are hot carbonates and ethanolamines. Other solutions used include methanol, acetone, liquid N_2, glycols, and other organic solvents. Commercialized CO_2-removal processes* are characterized by the type of absorbent solution or solvent, solution activators, and corrosion inhibitors used.

*Processes using hot potassium carbonate solutions include Catacarb, Benfield, Carsol, and Giamarco-Vetrocoke. Shell's SULFINOL[SM] process is based on the

The partially purified synthesis gas leaves the CO_2 absorber containing approximately 0.1% CO_2 and 0.5% CO. This gas is preheated at the methanator inlet by heat exchange with the synthesis-gas-compressor interface cooler and the primary-shift-converter effluent and reacted over a Ni oxide catalyst bed in the methanator. The methanation reactions shown here are highly exothermic and are favored by low temperatures and high pressures:

$$CO + 3H_2 \rightleftharpoons CH_4 + H_2O$$
$$CO_2 + 4H_2 \rightleftharpoons CH_4 + 2H_2O$$

The activity of the catalyst is favored by high temperature and pressure; therefore methanator feed is preheated to reduce catalyst volume and equipment size. The synthesis gas leaves the methanator containing less than 10 ppm of carbon oxides. The methanator effluent is cooled by heat exchange with boiler feedwater and cooling water. Water condensed from the process stream by cooling is separated in and discharged from the synthesis-gas-compressor suction drum.

Compression and Synthesis: The purified synthesis gas, containing H_2 and N_2 in a 3:1 mole ratio and with an inert gas (methane and argon) content of about 1.3 mole %, is delivered to the suction of the synthesis-gas compressor. The synthesis gas is compressed in a steam-turbine-driven two-case centrifugal compressor, to which the recycle is added as a side stream for compression with the feed in the final compressor wheel. Interstage cooling is provided by heat exchange with methanator feed, cooling water, and NH_3 refrigerant.

Anhydrous ammonia is catalytically synthesized in a loop system in which unconverted reactables are separated from both the reacted and the inert components leaving the converter and recycled for

use of an aqueous sulfolane (tetrahydrothiophene dioxide)-alkanolamine solution. BASF offers a CO_2 removal process using a triethylene glycol and diethyl ether solution. Lurgi's Rectisol process uses cold methanol as absorbent; their Purisol process employs *N*-methyl-2-pyrrolidone (NMP or M-Pyrol), a true physical solvent. The Fluor solvent process employs an anhydrous organic compound, propylene carbonate, as a physical solvent. The Fluor Econamine process employs an aqueous solution of an alkanolamine trade-named Diglycolamine (DGA). Ethanolamines most commonly used for CO_2 removal are monoethanolamine (MEA) and diethanolamine (DEA).

eventual conversion to ammonia. The arrangement of flows in this loop system is such that the recovery of ammonia is made after the recycle mixture is combined with the fresh feed and both are compressed as a mixture. The purpose of this arrangement is to permit scrubbing of converter feed with ammonia in the separator, thus removing oxides (CO, CO_2, H_2O) poisonous to the synthesis catalyst. The total compressor discharge is cooled with cooling water and NH_3 refrigerant to achieve partial condensation of NH_3. The partially condensed mixture enters a separator, where the NH_3 condensate exists as product and the disengaged vapor becomes the feed to the synthesis converter.

The synthesis converter has both a promoted iron oxide–catalyst–filled section or basket, and a heat-exchanger section housed in a high-pressure shell. The catalyst section is a cylindrical insulated shell (basket) coaxial to the converter shell and separated from it by an open annulus. Several beds of catalyst of varying heights are optimized to give minimum catalyst volume consistent with suitable temperatures and long catalyst life. Since the synthesis reaction is highly exothermic (13 kcal evolved per mole of ammonia formed), interbed injection of cool feed gas is provided as a means of controlling temperatures and concentrations of the feed to the beds. Volumes of beds are graduated according to the temperature rises occurring across them. With a small top bed, the temperature rise is limited and the quench becomes effective. Succeeding beds are larger, since NH_3 concentrations are higher and the temperature rises are consequently smaller.

Feed from the separator is heated by heat exchange with the compressor discharge stream and the converter effluent. It then enters the bottom of the converter and flows upward between the pressure shell and basket. The feed receives preheat from heat conducted, and radiated to the converter shell from the catalyst basket. The feed is finally preheated to the reaction initiation temperature by the feed-effluent interchanger at the top of the converter. For temperature control to the top bed, a portion of the feed gas may be introduced directly to the converter, where it bypasses the exchanger and is mixed with the preheated feed. The gas flows downward through the beds, optimum temperature control for maximum yields being obtained by interbed injection of a portion of cold feed gas. Reaction takes place at approximately 800–900°F (427–482°C). The

effluent then flows through a central riser to the converter interchanger and, after heating the feed, leaves the converter. Heat exchange with boiler feedwater and synthesis feed gas provides further cooling of the effluent gas. A portion of this gas is withdrawn from the loop to purge the inerts brought in with the fresh feed; the remainder is sent to the compressor for recycle in the loop.

Ammonia liquid, separated from the loop in the separator and from the recycle purge, contains dissolved synthesis gas, which is released when the combined stream is flashed into the letdown drum. The flashed gas is then separated in the letdown drum and combined with the vapors from the purge separator to form a stream of waste fuel gases.

Liquid ammonia in the letdown drum still contains some dissolved gases which must be disengaged. This is effected in the ammonia-refrigeration cycle (not shown), to which this stream is added directly.

Materials of Construction: The production of synthesis gas and ammonia synthesis is carried out at conditions which require the use of special materials in most sections of an ammonia plant. The high temperatures and pressures now used in the primary reformers require the use of centrifugally cast tubes of 25 chrome–20 nickel alloy HK 40. The high temperatures in the secondary reformer require the vessel to be lined with a high-alumina castable or gunned-type refractory. All equipment handling water condensate saturated with CO_2 must be of, or lined with, stainless steel (18–8), aluminum, or phenolic-epoxy coatings, because this condensate is extremely corrosive to carbon steels. CO_2-removal systems use solution inhibitors and alloys to prevent corrosion attack. Copper, brass, bronze, and aluminum bearing alloys cannot be used for equipment in contact with solutions of hot carbonate, ammonia, and alkalies. In addition, copper and copper-base alloys are avoided where possible for parts exposed to the atmosphere.

The ammonia-synthesis section operates at conditions under which steel materials are subject to attack by hydrogen. Hydrogen at certain partial pressures and temperatures can cause embrittlement and loss of ductility by diffusion into the steel microstructure, or cracks and blisters due to high-pressure methane pockets formed by reaction of hydrogen with carbides in the steel. Carbide stabilizers such as chromium and molybdenum are used in varying percentages to prevent hydrogen attack. Required stabilizer compositions in steel as a function of temperature and hydrogen partial pressure are determined by the Nelson chart. Formation of brittle layers of iron nitride due to nitrogen and/or ammonia attack on some steels has been experienced in some areas of ammonia plants, but this has not been of real concern. Water used for generation of high-pressure steam is required to be of exceptionally high purity to prevent erosion and corrosion of steam-turbine drive blades.

See also **Gasification; Hydrogen;** and **Synthesis gas.**

REFERENCES

Allen, J. B.: Ammonia Manufacture, *Chem. Proc. Eng.,* September 1965.

Axelrod, L. C., and T. E. O'Hare: Production of Synthetic Ammonia, chap. 5 in V. Sauchelli (ed.), "Fertilizer Nitrogen: Its Chemistry and Technology," Van Nostrand, New York, 1964.

Quartulli, O. J., et al.: Best Pressure for Ammonia Plants, *Hydrocarbon Proc.*, vol. 47, p. 11, November 1968.

—Peter A. Waldheim, *The M. W. Kellogg Company, a Division of Pullman Incorporated, Houston, Tex.*

Ammonium Carbamate   (See **Urea**.)

Ammonium Nitrate   (See **Explosives;** and **Fertilizers**.)

Ammonium Phosphates   (See **Fertilizers**.)

Ammonium Picrate   (See **Explosives**.)

Ammonium Sulfate   (See **Caprolactam;** and **Fertilizers**.)

Ammoxidation   (See **Ammonia;** and **Catalysts**.)

Ammunition   (See **Explosives**.)

Ampholytic Surfactants   (See **Detergents**.)

Amyl Alcohols   (See **Alcohols**.)

Amylenes   (See **Alkylation;** and **Isoprene**.)

Amylose   (See **Starch**.)

Anaerobic Processes   (See **Waste treatment**.)

Analyzers, Process   Numerous chemical processes depend upon accurate and reliable instruments for determining chemical composition directly or for measuring various physical parameters that are indicative of composition. While space is not available in this volume to describe the many types of instruments used, some of which are continuous *on-stream* analyzers, the instrumental methods are classified in terms of their interactions between matter and energy. All physical and chemical analysis techniques, whether instrumental or not, are fundamentally based upon these relationships. This follows directly from the fact that all known matter is made up of a complex but systematic arrangement of particles having mass and electric charge. Energy states, which are characteristic of the composition of any particular substance, can be most readily inferred by observing the consequences of interaction between the substance and an external source of energy. This external energy source may be in any of the following basic groups: (1) electromagnetic radiation, (2) chemical affinity or reactivity, (3) electric or magnetic fields, and (4) thermal or mechanical energy.

The foregoing groups differ fundamentally in their modes of interaction with matter. Moreover, the types of information which these interactions afford may vary considerably in specificity or uniqueness, a situation which sometimes can be controlled by combining techniques. Many properties can be measured or inferred by more than one type of interaction, as can be readily observed by inspection of Table A-47.

Electromagnetic Radiation: Interaction of electromagnetic radiation with matter affords information of a most basic kind, owing to the fact that photons of electromagnetic radiation are emitted or absorbed whenever changes occur in the quantitized energy states occupied by the electrons associated with atoms and molecules. For example, x-rays, which consist of photons or electromagnetic wave packets having relatively high energy, penetrate deeply into the electron orbits in an atom and provide, upon absorption, the large amount of energy that is required to excite one of the innermost electrons. The pattern of x-ray excitation or absorption is related to the identity of the atoms whose orbital electrons are excited, making the x-ray technique useful for determining the presence of atoms and elements in dense samples. However, because of their great penetrating power, x-rays are not adapted to the excitation or observation of low-energy states corresponding to outer-shell or valence electrons, or of interatomic bonds involving vibration or rotation. In such an application, the use of x-rays

TABLE A-47. *Generalized Relationships between Matter and Energy that Can Be Measured to Ascertain Chemical Composition*

Phenomenon to Be Measured	Interaction with Electromagnetic Radiation	Interaction with Other Chemicals	Reaction to Electric and Magnetic Fields	Interaction with Thermal or Mechanical Energy
Definition	Measurement of the quantity and quality of electromagnetic radiation emitted, reflected, transmitted, or diffracted by the sample.	Measurement of the results of reaction with other chemicals in terms of amount of sample or reactant consumed, product formed, or thermal energy liberated or determination of equilibrium attained.	Measurement of the current, voltage, or flux changes produced in energized electric and magnetic circuits containing the sample.	Measurement of the results of applying thermal or mechanical energy to a system in terms of energy transmission, work done, or changes in physical state.
Relation of measurements to chemical variables	Electromagnetic radiation varies in energy with radiation frequency, that of the highest frequency or shortest wavelength having the highest energy and penetration into matter. Radiation of the shortest wavelengths (gamma rays) interacts with atomic nuclei, x-rays with the inner-shell electrons, visible and ultraviolet with valence electrons and strong interatomic bonds, while infrared radiation and microwaves interact with the weaker interatomic bonds and with molecular vibrations and rotation. Most of these interactions are structurally related and completely unique. They may be used to detect and measure the elemental or molecular composition of gas-liquid and solid substances within the limitations of the available equipment.	The selectivity inherent in the chemical affinity of one element or compound for another, together with their known stoichiometric and thermodynamic behavior, permits positive identification and analysis under many circumstances. In a somewhat opposite sense, the apparent dissociation of substances at equilibrium in chemical solution gives rise to electrically measurable valence potentials, called oxidation-reduction potentials, whose magnitude is indicative of the concentration and composition of the substance. While individually all the above effects are unique for each element or compound, many are readily masked by the presence of more reactive substances, and so they can be applied only to systems of known composition limits.	The production of net electric charges on atoms or molecules by bombardment with ionizing particles or radiation or by electrolysis or dissociation in solution or the induction of dipoles by strong fields establishes measurable relationships between these ionized or polarized substances and electric and magnetic energy. Ionized gases and vapors can be accelerated by applying electric fields, focused or deflected in magnetic fields, and collected and measured as an electric current in mass spectroscopy. Ions in solution can be transported, and deposited if desired, under the influence of various applied potentials for coulometric or polarographic analysis and for electrical-conductivity measurements. Inherent and induced magnetic properties give rise to specialized techniques, such as oxygen analysis based on its paramagnetic properties and nuclear magnetic resonance, which is exceedingly precise and selective for the determination of the compounds of many elements.	The thermodynamic relationship involving the physical state and thermal-energy content of any substance permits analysis and identification of mixtures of solids, liquids, and gases to be based upon the determination of freezing or boiling points and upon the quantitative measurement of physically separated fractions. Useful information often can be derived from thermal-conductivity and viscosity measurements, involving the transmission of thermal and mechanical energy, respectively.

TABLE A-47. *Generalized Relationships between Matter and Energy that Can Be Measured to Ascertain Chemical Composition* (*Continued*)

Phenomenon to Be Measured	Interaction with Electromagnetic Radiation	Interaction with Other Chemicals	Reaction to Electric and Magnetic Fields	Interaction with Thermal or Mechanical Energy

Specific technique and types of information they provide	1. Emitted radiation
	  a. Thermally excited
	    (1) Optical-emission spectrochemical analysis
	    (2) Flame photometry
	  b. Electromagnetically excited
	    (1) Fluorescence
	    (2) Raman spectrophotometry
	    (3) Induced radioactivity
	    (4) X-ray fluorescence
	2. Transmission and reflection measurements
	  a. X-ray analysis
	  b. Ultraviolet spectrophotometry
	  c. Conventional photometry
	  d. Colorimetry
	  e. Light scattering
	  f. Optical rotation (polarimetry)
	  g. Refractive index
	  h. Infrared spectrophotometry
	  i. Microwave spectroscopy
	  j. Gamma-ray spectroscopy
	  k. Nuclear quadrupole moment
		1. Consumption of sample or reactant
		  a. Orsat analyzers
		  b. Automatic titrators
		2. Measurement of reaction products
		  a. Impregnated-paper tape devices
		  b. Continuous chemical reactions
		3. Thermal-energy liberation
		  a. Combustion types
		  b. Quantitative exothermic reaction of unknown with reactant
		4. Equilibrium solution potentials (oxidation-reduction)
		  a. Redox potentiometry
		  b. pH (hydrogen-ion concentration)
		  c. Metal-ion equilibria
			1. Mass spectroscopy
			2. Electrochemical analysis
			  a. Controlled-potential electrolysis
			  b. Polarography
			  c. Coulometry
			  d. Amperometry
			  e. "Dead-stop" method
			3. Electrical properties
			  a. Electrical conductivity
			  b. Dielectric constant and loss factor
			  c. Oscillometry
			  d. Gaseous conduction
			4. Magnetic properties
			  a. Paramagnetism
			  b. Nuclear magnetic resonance
			  c. Electron paramagnetic resonance
				1. Effects of thermal energy
				  a. Thermal conductivity
				  b. Melting- and boiling-point determinations
				  c. Ice point (crystallization)
				  d. Dew point
				  e. Vapor pressure
				  f. Fractionation
				  g. Chromatography
				  h. Thermal expansion
				2. Effects of mechanical energy or forces
				  a. Viscosity
				  b. Sound velocity
				  c. Density

would be analogous to attempting to stop a windmill with a rifle bullet. On the other hand, electromagnetic energy at longer wavelengths, in the infrared region, is made up of photons having relatively low energy, corresponding to the energy transformations involved in the vibration of atoms in a molecule due to the stretching or twisting of the interatomic bonds. Using the windmill model to represent a molecule, infrared radiation would correspond more to a breeze, which can indeed have a profound effect upon the windmill operation.

Chemical Affinity or Reactivity: In a very real sense, chemical reactions contain an inherent ability for the recognition of certain substances because, by mixing, the two parties to the reaction can be brought into rather intimate contact so that the valence potential or activity coefficient, or what might be called the "potential driving force toward reaction," can come into play. This situation can be characterized by a high degree of specificity and permits composition determinations in liquids, slurries, and the like, where the information that could be afforded by electromagnetic radiation absorption would be somewhat less significant.

Electric or Magnetic Fields: This is a powerful method for the determination of chemical composition when it is possible to rely upon some inherent or conferred electrical or magnetic distinction between the sought-after components. It is employed by incorporating the sample in a suitable electric or magnetic circuit so that the distinguishing feature can be sorted out and measured. The mass spectrometer, which sorts out the constituent ions in a sample, according to their mass and conferred charge, in a combination of electric and magnetic fields, can produce a complete, although empirical, chemical analysis of gas or vapor samples. Techniques also have been devised for the analysis of both liquids and solids. A more commonly encountered system for the determination of ions in solution is that of electrical-conductivity measurement. In this case, however, no distinction is afforded between different ions.

Thermal or Mechanical Energy: This technique involves interactions of a gross nature compared with the other three techniques described. For example, the distinguishing ability of some gas molecules to become highly excited in vibration, twisting, and rotation enables them to conduct larger amounts of heat away from heated bodies with which they collide than other gas molecules.

The gross cooling effect upon the heated body can be used to determine the quantity of the particular molecule present in a mixture. The simple and widely used thermal-conductivity analyzers that depend upon this principle are indeed limited in their specificity or ability to recognize just one molecule, but where a gas of high thermal conductivity, such as H_2, occurs in a gas of lower thermal conductivity, such as N_2, the method is an excellent choice. Another example of gross energy transfer is the measurement of viscosity of a substance by doing work upon it with such devices as rotating disks or paddles or dropping a weight through it. Here again the measurement affords an insight regarding the actual intermolecular forces that must be overcome, whether the measurement is made for determining concentration, degree of polymerization, or composition.

A basic framework for viewing the foregoing four types of interaction of energy with matter is given in greater detail in Table A-47. See also **Chromatography**; and **pH (hydrogen-ion concentration)**.

REFERENCES
Considine, Douglas M. (ed.): "Encyclopedia of Instrumentation and Control," McGraw-Hill, New York, 1971.
Considine, Douglas M. (ed.): "Instruments and Controls Handbook," 2d ed., McGraw-Hill, New York, 1974.
Considine, Douglas M., and S. D. Ross (eds.): "Handbook of Applied Instrumentation," McGraw-Hill, New York, 1963.

Anatase   (See **Titanium dioxide.**)

Anergic Process   An anergic process is one in which $W = 0$, where W is defined as any form of energy other than heat which is in transition between the system and the surroundings as a result of a difference in mechanical, electrical, and chemical forces. As used here, W does not include energy associated with a transfer of mass across the boundaries of the system. W is *positive* (by convention) when the system *loses* work energy, and W is *negative* when the system *receives* work energy.

Anglesite   (See **Lead.**)

Anhydrous Ethyl Alcohol   (See **Ethyl alcohol.**)

Anhydrous Silica   (See **Ceramics.**)

Anidex Fibers   (See **Fibers.**)

Aniline [See **Amines;** and **Cyclohexanol/Cyclo-hexanone (KA oil).**]

Animal Fats [See **Alcohols; Alcohols, fatty (via hydrogenation); Soaps;** and **Vegetable oils.**]

Anion-Exchange Resins (See **Ion-exchange resins;** and **Water treatment.**)

Anionic Surfactants (Sulfur-bearing) Two different sulfur-containing solutizing groups are being employed to produce anionic surfactants. The first nonsoap surfactant, Turkey red oil (introduced commercially about 1875), was a sulfated castor oil. Its active group was an H_2SO_4 ester

salt, $-O-\overset{\overset{\textstyle ONa}{|}}{\underset{\underset{\textstyle O}{\|}}{S}}=O$. The other hydrophilic group

was the sulfonate, $-\overset{\overset{\textstyle O}{\|}}{\underset{\underset{\textstyle O}{\|}}{S}}-ONa$, which was first in-

troduced by the Germans during World War I in the form of an alkyl naphthalene sulfonate. The most striking difference between the resulting two types of surfactants is their hydrolytic stabilities.

Because sulfated compounds are H_2SO_4 esters, they are subject to hydrolysis under both acid and alkaline conditions. In particular, acidic pH's are so effective that the quantitative analytical method for the organic-sulfur-group content is based upon this factor. Sulfonates, on the other hand, having a carbon-to-sulfur link, are much more resistant to hydrolysis. A somewhat similar situation holds for thermal stability. Sulfates tend to split off the inorganic group, leaving an olefin, while again the sulfonate is much more stable. Judging by the Krafft points,* the sulfate group is more hydrophilic than the sulfonate group.[7] The balance of hydrophilic and hydrophobic groups determines the water solubility and application potential as well.

The commercially most significant general types of surfactants, described here, are (1) *sulfated compounds,* including (*a*) sulfated fats and oils and (*b*) alcohol sulfates in general; and (2) *sulfonated compounds,* including (*a*) alkyl aryl sulfonates, (*b*) sulfosuccinates, (*c*) sulfoethyl esters and amides, and (*d*) aliphatic sulfonates.

The individual general types will be described in the foregoing order, centering upon model

*Krafft point is the temperature at which a 1% solution of the surfactant becomes clear upon heating.

compounds, with generalized preparations and variants. More detailed preparations and specific variations may be found in Refs. 1–3, which contain extensive bibliographies.

Sulfated Compounds

Sulfated Fats and Oils:[3] Sulfuric acid can react in two ways with fatty oils to form water-miscible products. The first reaction is the addition of the mineral acid across the double bond of the unsaturated fatty acid moiety, forming an H_2SO_4 ester

$$CH_3(CH_2)_7\overset{\overset{\textstyle H}{|}}{C}=\overset{\overset{\textstyle H}{|}}{C}-(CH_2)_7C\underset{\displaystyle OR}{\overset{\displaystyle O}{\diagup}} + H_2SO_4 \rightarrow$$

$$CH_3(CH_2)_7\overset{\overset{\textstyle H}{|}}{\underset{\underset{\textstyle H}{|}}{C}}-\overset{\overset{\textstyle H}{|}}{\underset{\underset{\textstyle OSO_3H}{|}}{C}}-(CH_2)_7C\underset{\displaystyle OR}{\overset{\displaystyle O}{\diagup}} \quad (1)$$

where R = $\frac{1}{3}$ glyceryl oleate molecule.

The general procedure for this preparation is to add H_2SO_4 dropwise to a well-chilled sample of oil, using good stirring, and holding the temperature of the reaction mass preferably below 20°C (with fats, lowest possible liquid temperature). The mineral acid blends completely clear with the oil, bodying it up. After this acid is added completely, the reaction mass is freed of the unreacted H_2SO_4 by washing with salt solutions. The salted-out surfactant is neutralized somewhat beyond the point required to neutralize all the sulfuric esters and mineral acid in order to obtain a clear product. The final product is a mixture of sodio salts of H_2SO_4 esters, sodio carboxylates, unreacted fatty oils, and salt in an aqueous medium.

The second reaction of H_2SO_4 is the esterification of a hydroxy fatty oil, such as castor oil, to form the H_2SO_4 ester (Turkey red oil, a textile dye leveler).

$$CH_3(CH_2)_5\overset{\overset{\textstyle H}{|}}{\underset{\underset{\textstyle OH}{|}}{C}}-CH_2-\overset{\overset{\textstyle H}{|}}{C}=\overset{\overset{\textstyle H}{|}}{C}-(CH_2)_7C\underset{\displaystyle OR}{\overset{\displaystyle O}{\diagup}}$$

$$+ H_2SO_4 \rightarrow$$

$$CH_3(CH_2)_5\overset{\overset{\textstyle H}{|}}{\underset{\underset{\textstyle OSO_3H}{|}}{C}}-CH_2\overset{\overset{\textstyle H}{|}}{C}=\overset{\overset{\textstyle H}{|}}{C}-(CH_2)_7C\underset{\displaystyle OR}{\overset{\displaystyle O}{\diagup}} \quad (2)$$

The preparation method is similar to the foregoing except that, since the reaction is an esterification, the mineral acid is left in contact with the oil for several hours, and frequently at higher temperatures, up to approximately 35°C. The reaction mass is handled similarly with reference to washing, neutralizing, and other details.

It should be recognized that these sulfated products, which often are misnamed sulfonated oils, are in no sense single components (even in a most crude form), and in many cases their applications depend upon this so-called state of impurity. The reactions portrayed by Eqs. (1) and (2) do not proceed to high conversion and, under the best of conditions, much unreacted fatty oil remains.

In the case of castor oil, the hydroxy group reacts in preference to the double bond. This occurs despite the fact that the addition to the double bond is almost instantaneous, while the esterification reaction is far slower. Purification of some degree has been found possible by means of partition between organic solvents and the water contained in the sulfated oil.

The fatty oils that are used as raw materials for sulfated fats and oils are, besides the castor oils, those with iodine values from about 45 to about 130. This comprises the gamut of fatty oils from lard (or tallow) through olive oil up to corn oil (or cottonseed oil). Higher unsaturated oils tend to polymerize, forming undesirable by-products, and hence are not generally used. However, sulfated fish oils are employed in the leather industry as raw oil carriers in fat liquors. The most general use for the sulfated oils is as emulsifiers for clear self-emulsifiable oils. Monobasic esters of oleic acid, such as butyl oleate, are sulfated also, producing wetting and rewetting agents especially useful in the textile industry.

Recently, linear alpha olefins have become commercial products. It has been found possible to sulfate them in a manner similar to that for glycerol oleate or oleic acid. However, most of the studies based upon these chemicals have been directed toward sulfonation rather than sulfation.

Alcohol Sulfates in General: This group of sulfates, which has the model formula RCH_2—O—SO_3Na, can be produced with a high degree of conversion. The RCH_2 portion of the molecule may be obtained from a large collection of primary alcohols, such as aliphatic alcohols (linear or branched), ethoxylated alcohols (primary and secondary), ethoxylated alkyl phenols, ethoxylated amides, amines and monoglycerides, and other materials.

Secondary alcohols and propoxylates also have been suggested as substrates. Chlorosulfonic acid

(maximum reaction temperature about 20°C) and sulfamic acid (range of reaction temperature 100–140°C) are the usual sulfating agents. The use of H_2SO_4 or oleum tends to lower the conversion to ester, besides causing undesirable side reactions.

$$ROH + ClSO_3H \rightarrow ROSO_3H + HCl\uparrow \quad (3)$$
$$ROH + NH_2SO_3H \rightarrow ROSO_3NH_4 \quad (4)$$

For reaction (3), provision is made to remove the by-product HCl gas. The reaction mass is neutralized by adding the mass to a chilled solution of alkali with efficient stirring.

Reaction (4) produces an ammonium salt; conversion by fixed alkali is employed to produce either the Na or K salts, releasing gaseous NH_3. Ethoxylated fatty alcohols are a favorite hydroxy substrate. It has been found that the addition of a few moles of ethylene oxide to a fatty alcohol before sulfation produces a more water-soluble compound with enhanced foaming ability. Not only is the quantity of foam increased, but more important, the stability of the foam is improved. Thus, in this group of surfactants, there are at least three parameters to control surface-active properties, namely a hydrophobic chain, an ethylene oxide chain, and a sulfuric acid ester group. Additional variations are possible due to amino and amido groups. The use of monoglycerides points to other types that may be employed.

Sulfonated compounds

Alkyl Aryl Sulfonates: The original surfactant based upon sulfonic acids came from this class of compounds. This product was principally isopropyl naphthalene sulfonate contaminated with some diisopropyl and polyisopropyl derivatives. It was made by a one-batch method wherein isopropyl alcohol, naphthalene, and a large excess of H_2SO_4 were caused to react by heating, followed by neutralization. The final dried product contained much Na_2SO_4 because of the large excess of H_2SO_4 used in the reaction. This product was used, as is, as a wetting agent and general surfactant.

Later, processes were developed to produce sulfonates with varying quantities of salt impurities. Many sulfonic acids have been found to be insoluble to an appreciable extent in H_2SO_4 in a comparatively narrow concentration range (50–80%). The simplest method of taking advantage of this factor is to add small quantities of water to the acidic reaction mass, causing a two-layer separation. The amount of water necessary to create a salting-out condition depends on the structure of the hydrocarbon (alkyl aryl) that is sulfonated. The lower mineral acid layer is discarded; the upper layer finished as before. A liming process is used in conjunction with the water method to obtain higher purity or where the latter method fails to produce a separation due to high water solubility of the sulfonate in question.

With the higher molecular weight alkyl-substituted aryl sulfonate, the usual technique is a two-kettle process. First the aryl hydrocarbon is alkylated, and then this product, purified, is sulfonated. Until recently, the most popular detergent was a dodecyl benzene sulfonate. The parent hydrocarbon was produced by alkylating benzene with the tetramer of propylene by means of the Friedel-Crafts reaction. However, the biodegradability of this detergent is poor because the alkyl chain is highly branched. Now linear alkyl sulfonate (LAS) has replaced this. The new alkyl group is derived from alpha linear olefin; hence the biodegradability of the detergent is far greater.

R is CH_3—$(CH_2)_n$—CH_2, and n may equal 8 or 10.

There are surfactants in this class based upon other aryl hydrocarbons, such as diphenyl and diphenyl ether. The lower molecular weight sulfonates, based upon toluene, xylene, cumene, and methyl naphthalene, are hydrotropes.

Mahogany Soaps: These also may be considered to be in this class. In the refining of petroleum oil fractions, a process involving the addition of fuming H_2SO_4 is used to reduce the unsaturation present therein. An oil-soluble sulfonate fraction (termed *mahogany soaps*), as well as an oil-insoluble but water-soluble sulfonate fraction (termed *green soaps*) are extractable from the neutralized acid sludge reclaimed from the process.

These sulfonates are highly complex materials.[6] Mahogany soaps, which find many applications as emulsifiers, contain, in general, one aromatic ring which has been sulfonated and is also highly substituted. On the other hand, the oil-insoluble type is of lower molecular weight with monosulfonation, and where there are higher-molecular-weight-substituted aromatics, they are found to be disulfonated. Reference 6 contains a series of possible parent structures.

Sulfosuccinates: One of the better wetting agents is sodium dioctyl sulfosuccinate. This product is synthesized by esterifying maleic anhydride to form first dioctyl maleate.

Usually, 2-ethyl hexanol is employed as the octyl alcohol. This unsaturated ester is converted to the sulfosuccinate by its reaction with a strong aqueous solution of sodium bisulfite.

R = 2-ethylhexyl.

Many derivatives of maleic acid are converted to those of sulfosuccinic acid by means of reactions similar to Eq. (8) in order to form useful surfactants. Where maleic anhydride is reacted so as merely to open up the anhydride group, forming half ester or half amide, the disodium sulfite salt is employed to neutralize the free carboxylic group, liberating the acid sulfite for addition to the double bond.

Sulfoethyl Esters and Amides: Hydroxyethyl sulfonic acid (isethionic acid, $HO-CH_2CH_2SO_3H$) and *N*-alkyl amino ethyl sulfonic acids (taurines, $RNH-CH_2CH_2SO_3H$) have been used as the bases for two series of surfactants, by forming the esters and amides, respectively, with fatty acids, preferably the acid chlorides. The most important commercial member of this group is the oleic acid derivative of methyl taurine.

$$C_{17}H_{33}-C\overset{\displaystyle O}{\underset{\displaystyle Cl}{}} \ +$$

$$CH_3NHCH_2CH_2SO_3Na + NaOH \rightarrow$$

$$C_{17}H_{33}\overset{\displaystyle O}{C}-\underset{\displaystyle CH_3}{N}-CH_2CH_2SO_3Na + NaCl \quad (9)$$

The general method of reacting the fatty-acid chloride with the sodium salt of methyl taurine in the presence of alkali may also be used for the hydroxy derivative. The product of Eq. (9) is used as a general surfactant, being good for wetting and dispersing applications in the textile industry in particular. Other fatty acids, such as coconut fatty acids, palmitic acid, and tall oils also have been employed.

Aliphatic Sulfonates: The classical method of placing a sulfonate group on an aliphatic chain is to react an aliphatic halide with sodium sulfite.

$$RBr + Na_2SO_3 \rightarrow RSO_3Na + NaBr \quad (10)$$

A more feasible method commercially is based upon the Reed reaction. In the presence of actinic light, paraffins react with a mixture of SO_2 and Cl_2 to produce an aliphatic sulfonyl chloride.

$$RH + SO_2 + Cl_2 \rightarrow RSO_2Cl + HCl \quad (11)$$

These acid chlorides are hydrolyzed to the salts of the sulfonic acids. The ultimate uses are dependent upon the structure of the initial paraffins. The surfactants may be purified either before or after the hydrolysis stage.

With the advent of linear alpha olefins, another basis for biodegradable aliphatic sulfonates becomes commercially possible. The addition of SO_3 to the terminal double bond is somewhat complex.[4]

$$\underset{O\rule{1cm}{0.4pt}SO_2}{RCH-CH_2-CH_2} \ + \ \underset{O_2S\rule{0.6cm}{0.4pt}O}{RCH_2-\underset{O}{CH}-\underset{SO_2}{CH_2}} \xrightarrow{\text{hydr.}}$$

$$\underset{OH}{RCH-CH_2-CH_2SO_3H} \ +$$

$$RC\overset{H}{=}\overset{H}{C}-CH_2SO_3H \cdots \text{etc.} \quad (12)$$

Two main SO_3 addition products are formed, a sultone and a carbyl sulfate. Upon hydrolysis, these produce a group of hydroxyl sulfonic and unsaturated sulfonic acids. A study of the mechanism and products of this reaction may be found in Ref. 4. Unsaturated fatty acids, such as oleic acid, may be sulfonated in a similar manner. Saturated fatty acids have been found to sulfonate differently; SO_3 produces alpha sulfo acids.[5]

$$RCH_2C\overset{\displaystyle O}{\underset{\displaystyle OH}{}} + SO_3 \rightarrow RC\overset{H}{\underset{SO_3H}{}}-\overset{\displaystyle O}{C}-OH$$

It has been suggested that esters and amides of the sulfo acids may have promise as surfactants.

REFERENCES

1. Schwartz, A. M., and J. W. Perry: "Surface Active Agents," vol. 1, Wiley-Interscience, New York, 1949; Schwartz, A. M., J. W. Perry, and J. Berch: "Surface Active Agents," vol. 2, Wiley-Interscience, New York, 1958.
2. Sisley, J. P.: "Encyclopedia of Surface Active Agents," Chemical Publishing Company, New York, vol. 1, 1952, vol. 2, 1964. Index des Huiles Sulfonées et Détergents, vol. 3, Editions Modernes Teintex, Paris, 1961.
3. Dombrow, B.: Sulfated Fats and Oils, chap. 8 in W. M. Linfield (ed.), "Anionic Surfactants," Marcel Dekker, New York, 1973.
4. Puschel, F.: *Tenside* 4, pp. 286–292, 1967. Rubinfield, J., and H. D. Cross: *Soap Chem. Spec.*, vol. 43, no. 3, p. 104, 1967.
5. Weil, J. K., R. G. Bistline, Jr., and A. J. Stirton: "Organic Syntheses," vol. 4, p. 862, Wiley, New York, 1963.
6. Brown, A. B., and J. C. Knoblock (eds.): Symposium on Composition of Petroleum Oils, *ASTM Spec. Tech. Pub.* vol. 224, pp. 213–229, 1958.
7. Sperling, R.: *Ind. Eng. Chem.*, vol. 40, pp. 890–897, 1948.
8. Gotte, E., Fette, and Seifen: *Anstrichm.*, vol. 71, no. 3, pp. 219–223, 1969.

—Bernard A. Dombrow, *Nopco Chemical Division, Diamond Shamrock Chemical Company, Morristown, N.J.*

Annealing   (See **Copper; Glass; Iron and steel;** and **Ironmaking and steelmaking.**)

Anodizing   (See **Conversion coatings.**)

Anthracene   Anthracene, $C_{14}H_{10}$, sometimes called *paranaphthalene* or *anthracin green oil*, is a colorless solid of monoclinic crystalline form. When crystallized from benzene, colorless, lustrous plates are formed which exhibit a blue fluorescence. Physical properties include formula weight = 178.22; melting point = 217–218°C; boiling point = 340–342°C; specific gravity = 1.25 at 27°C (referred to water at 4°C); insoluble in water; sparingly soluble in alcohol (1.5 parts in 100 parts at 20°C). Like benzene and naphthalene, anthracene is a closed-chain aromatic compound. For diagram, see **Carbon compounds.**

Anthracene is present in the creosote fraction of coal tar in the form of "anthracene oil" in sufficient concentration to recover a crude material by crystallization. Recrystallization yields a material suitable for oxidation to anthraquinone by reaction with HNO_3 or chromates. Anthraquinone and other anthracene derivatives are important in dye chemistry. Anthraquinone also is produced by a reaction of phthalic anhydride with benzene and a Friedel-Crafts catalyst. Other derivatives of anthracene include anthraquinone-β-monosulfonic acid, $C_6H_4 \cdot (CO)_2 \cdot C_6H_3 \cdot SO_3H$; alizarin, $C_6H_4 \cdot (CO)_2 \cdot C_6H_2(OH)_2$ (1,2-dihydroxyanthraquinone); phenanthrene, $C_{14}H_{10}$; phenanthraquinone, $C_{14}H_8O_2$; and diphenic acid, $C_{14}H_{10}O_4$.

Anthracene is transformed by sunlight to paraanthracene, $(C_{14}H_{10})_2$.

See also **Coal tar and derivatives.**

Anthracite Coal   (See **Carbon.**)

Anthraquinone   (See **Coal tar and derivatives;** and **Hydrogen peroxide.**)

Antibiotics   Antibiotics are chemical substances derived from microorganisms which have the capacity of inhibiting the growth of, and even killing, other microorganisms.

Although the use of antibiotics as chemotherapeutic agents is relatively recent when compared with some other areas of medicine, the existence of such substances has been recognized for many decades. The first scientific demonstration of microbial antagonism was made by Pasteur and Joubert in 1877 when they observed that certain common bacteria inhibited the growth of anthrax bacilli. This basic phenomenon by which one microorganism destroys the life of another to preserve its own was entitled *antibiosis* by Vuillemin

in 1889. In the decades that followed, the therapeutic efficacy of antibiotics to control infectious disease was eventually demonstrated, after which the pursuit of microbial antagonists rapidly became an organized applied science.

Pyocyanase was the first microbially derived antibiotic product to be used in treating bacterial infections in man. Although it had only limited clinical use, it is interesting historically because it demonstrated as early as 1906 the principle of selective toxicity, i.e., specificity of action against the invading pathogen and a correlative lack of toxic action in the host. Following the decline in use of pyocyanase, it was almost a quarter of a century before interest in anti-infective agents from microbial sources was renewed.

In 1929, the British bacteriologist Alexander Fleming published his observations on the inhibition of a staphylococcus culture by growing colonies of *Penicillium notatum*. This report went largely unpursued for a decade, after which Florey and Chain reinvestigated Fleming's work and, in 1941, demonstrated the clinical usefulness of penicillin. In 1939, Dubos, by careful, well-planned studies, obtained the antibiotic tyrothricin from the soil organism *Bacillus brevis*. Although tyrothricin has had only limited use, the work of Dubos on the chemical, biological, and physical properties of this antibiotic contributed immeasurably toward forcing a realization of the potentialities of antibiotic substances. Similarly, Waksman undertook a systematic search for antimicrobial substances in a group of soil-inhabiting microbes known as *Streptomyces* and announced the discovery of streptomycin in 1944.

The foregoing discoveries stimulated worldwide interest, and the discovery of useful new antibiotics during the period 1939–1959 was prolific. During this period, every useful class of antibacterial antibiotic now known was recognized. Many specific drugs which presently occupy major places in therapeutic practice were discovered directly in microbial fermentations. In contrast, since 1959, most major discoveries have been by chemical modification of existing antibiotics.

Classification: The antibiotics comprise an unusually diverse group of substances, differing not only in chemical structure but also in their mode of action, antibacterial spectra, origin, and other points. In spite of the variety of chemical structures involved, most antibiotics appear to arise from a limited number of biogenetic themes, and may be divided into three major groups according to their potential derivation from amino acids, sugars, and acetate or propionate units. A classification of some useful antibiotics based on possible biogenetic origins and chemical structure is given in Table A-48.

In general, antibiotics exert their toxic action on bacteria by impairing one of the following processes: synthesis of the bacterial cell wall; synthesis of intracellular protein; synthesis of deoxyribonucleic acid; function of the cytoplasmic membrane. Grouping antibiotics by the latter criteria bears little resemblance to their classification according to biogenetic origin and chemical structure. Thus, while cycloserine and chloramphenicol are biogenetically derivable from a single amino acid, cycloserine inhibits cell-wall synthesis, while chloramphenicol acts on protein synthesis. In contrast, the mechanisms of action of the macrolide antibiotics and lincomycin are very similar, but their chemical structures are markedly different.

Antibiotics are used to combat a large number of pathogens that range in complexity from viruses to protozoa. Bacteria constitute the largest single class of microorganisms which are susceptible to antibiotics. While this description is concerned principally with antibacterial antibiotics, there are antifungal, antiprotozoal, anthelmintic, antiviral, antineoplastic, and animal-growth-promoting antibiotics. The microbial origin of antibiotics serves as yet another means of classification. For example, approximately 2,000 antibiotics have been discovered since 1940, of which 68% were isolated from the family Strepto-

mycetaceae, 12% from Bacillaceae, and 20% from various other fungi. These in turn may be further subclassified into various genera, species, and varieties of microbes. Thus antibiotics may be classified on the basis of many different criteria, each of which represents an integral part of antibiotics technology.

Major Clinically Useful Classes. In general, there are currently four major groups of clinically useful antibiotics: (1) the β-lactams, which include penicillins and cephalosporins; (2) the tetracyclines; (3) the macrolides; and (4) the aminoglycosides.

Penicillins: The penicillins are chemically characterized by a four-membered lactam ring fused to a thiazolidine ring and are differentiated by the side-chain substituent R (Table A-49) attached to this bicyclic nucleus. Penicillins are most frequently named by attaching the chemical name of the R-substituent as a prefix to the word penicillin. Thus, in the case where R is $C_6H_5CH_2-$, the compound is called *benzylpenicillin.*

Seven naturally occurring penicillins have been discovered in the fermentation broths of *Penicillium* and *Cephalosporium* cultures. Of these, only benzylpenicillin, also called *penicillin G,* or simple *penicillin,* has achieved wide application as an anti-infective drug because of its superior in vivo activity and amenability to large-scale manufacture. In spite of its wide use, benzylpenicillin has a number of limitations, including acid instability, allergenicity, and susceptibility to enzymatic inactivation by penicillinases.

In 1947, it was discovered that addition of

TABLE A-48. *Classification of Antibiotics by Biogenetic Origins and Chemical Structure*

Amino Acid Units	Acetate/Propionate	Sugar Units
Amino acid congeners	Fused-ring systems	Aminoglycoside
D-Cycloserine	Tetracyclines	Streptomycin
Chloramphenicol	Oxytetracycline	Kanamycins
β-Lactams	Chlortetracycline	Gentamicins
Penicillins	Steroidal antibiotics	
Cephalosporin C	Fusidic acid	
Polypeptides	Griseofulvin	
Bacitracins	Antibacterial macrolides	
Polymixins	Erythromycin	
Viomycin	Oleandomycin	
Capreomycin	Leucomycins	
	Spiramycins	
	Polyene macrolides	
	Nystatin	
	Amphotericins	
	Ansa-macrolides	
	Rifamycins	

TABLE A-49. *Structures of Some Typical Penicillins*

Generic Name	R	X
Benzylpenicillin	—CH₂— (phenyl)	C=O
Phenoxymethylpenicillin . . .	—O—CH₂— (phenyl)	C=O
6-Aminopenicillanic acid . . .	. . . .	H
Methicillin	(dimethoxyphenyl)	C=O
Cloxacillin	(chlorophenyl isoxazole-CH₃)	C=O
Ampicillin	—CH— NH₂ (phenyl)	C=O
Carbenicillin	—CH— CO₂H (phenyl)	C=O

phenylacetic acid to penicillin fermentation media increased the yield of benzylpenicillin at the expense of other less desirable natural penicillins. Following this observation, a new generation of biosynthetic penicillins was prepared by addition of monosubstituted acetic acid derivatives to penicillin fermentations. The most important of these biosynthetic derivatives is penicillin V, obtained by adding phenoxyacetic acid to penicillin growth media. This penicillin is relatively stable in dilute acid, is not destroyed by the acidic contents of the stomach, and consequently is effective by oral administration.

Although this biosynthetic approach created many new penicillins, it was severely limited in the type of side chain (R) that could be introduced. Only derivatives with an unsubstituted methylene adjacent to the amide carbonyl (X) could be generated.

The next major breakthrough in penicillin research came in 1959 with the isolation of the penicillin nucleus 6-aminopenicillanic acid (6-

APA) from fermentation mixtures to which no side-chain precursor had been added. Although chemical synthesis of 6-APA and its utility for the preparation of new penicillins by acylation were announced by Sheehan in 1958, the fermentation method provided the first practical means of obtaining large quantities of 6-APA. Chemical acylation of 6-APA allowed introduction of almost unlimited varieties of side chains and gave rise to a third generation of penicillins called *semisynthetic penicillins*. The nature of the acyl side chain has been found to have a profound effect on the properties of the penicillins, influencing such therapeutically important properties as acid stability, oral absorption, serum protein binding, penicillinase resistance, and gram-negative activity.

One of the major developments resulting from the availability of 6-APA has been the creation of semisynthetic penicillins that resist destruction by penicillinases. The empirical finding that triphenylmethylpenicillin was resistant to penicillinase led to screening of other penicillins with sterically hindered side chains, partly because the presence of a bulky group near the β-lactam ring resulted in reduced affinity of these substances for the enzyme. Methicillin and cloxacillin are compounds with such side chains which have proved clinically useful.

The single most important advance in penicillin research since the availability of 6-APA was the development of broad-spectrum penicillins. Until recently the penicillins were used primarily for treating gram-positive infections. It is of interest that two substances which have helped to extend the range of the penicillins to gram-negative bacteria have an ionizable group in their side chains. Ampicillin, D-(−)-α-aminobenzylpenicillin, has proved useful for the treatment of infections caused by a variety of gram-negative organisms. Carbenicillin, α-carboxybenzylpenicillin, was found to have considerable activity against *Pseudomonas* and *Proteus* species, two bacteria which have been resistant to attack by a wide range of other antibiotics.

Cephalosporins: These substances constitute the second major class of β-lactam antibiotics and are chemically characterized by a β-lactam fused to a dihydrothiazine ring. In contrast to the penicillins, where the side chain of the antibiotic varies, depending on precursors present in the fermentation mixture, fermentation-derived cephalosporins contain the same side chain. (See Table A-50.) Cephalosporin C, the parent antibiotic of this class, is clinically useless. For example, it is

about 0.1% as active as benzylpenicillin against staphylococci. However, cephalosporin C did exhibit certain interesting properties which provoked further investigation. It was more stable toward acid than penicillin; it was unaffected by penicillinases; it exhibited appreciable activity against some gram-negative bacteria; and it appeared to have no cross-allergenicity with the penicillins. Consequently, in the late 1950s many laboratories were investigating both chemical and microbiological methods for removing the aminoadipoyl side chain of cephalosporin C to obtain the cephalosporin nucleus 7-aminocephalosporanic acid (7-ACA).

A practical chemical process for accomplishing this transformation was announced in 1962. Like 6-APA, 7-ACA is readily acylated, and a large number of semisynthetic cephalosoporins have been prepared. Cephalothin was the first clinically useful broad-spectrum cephalosporin to emerge from synthetic studies. Cephaloridine soon followed cephalothin and is 2–8 times more active than the latter against gram-positive organisms. In 1970, cephaloglycin became commercially available as the first orally effective broad-spectrum cephalosporin. Most recently, cephalexin was announced as a metabolically more stable analog of cephaloglycin. Although cephalexin is intrinsically less active than cephaloglycin, its excellent oral-absorption and low-serum-binding characteristics compensate to a large extent for its low in vitro antibacterial activity.

Tetracyclines: These substances comprise a family of broad-spectrum antibiotics possessing a common perhydronaphthacene skeleton. They have a wider range of antimicrobial activity than any other clinically useful class of antibiotics. (See Table A-51.) They are active against many species of gram-positive and gram-negative bacteria, spirochetes, rickettsiae, and some of the larger viruses. Chlortetracycline, the first member of this class to be isolated, was discovered in 1948 among the metabolites of *Streptomyces aureofaciens;* oxytetracycline was isolated two years later from a *S. rimosus* fermentation. Both antibiotics quickly found wide medical use, not only because they were effective orally, but because they were useful against a much wider spectrum of bacteria than penicillin G.

Chemical studies on chlortetracycline and oxytetracycline, which provided a basis for structure assignment, in general led to products with diminished or no antibacterial activity. In 1953, the

TABLE A-50. *Structures of Some Typical Cephalosporins*

Generic Name	R	X	Y
Cephalosporin C	H_2N -(CH$_2$)$_3$- with CO$_2$H	C=O	CH_3CO_2-
7-Aminocephalosporanic acid	. . . .	H	CH_3CO_2-
Cephalothin	thiophene-CH$_2$-	C=O	CH_3CO_2-
Cephaloridine	thiophene-CH$_2$-	C=O	pyridinium N$\oplus$-
Cephaloglycin	phenyl-CH(NH$_2$)-	C=O	CH_3-CO_2-
Cephalexin	phenyl-CH(NH$_2$)-	C=O	H

TABLE A-51. *Structures of Some Typical Tetracyclines*

Generic Name	R$_1$	R$_2$	R$_3$	R$_4$	R$_5$
Tetracycline	H	OH	CH$_3$	H	H
Chlortetracycline	Cl	OH	CH$_3$	H	H
Oxytetracycline	H	OH	CH$_3$	OH	H
Demethylchlortetracycline	Cl	OH	H	H	H
Methacycline	H	=CH$_2$		OH	H
Doxycycline	H	H	CH$_3$	OH	H
Rolitetracycline	H	OH	CH$_3$	H	pyrrolidine N-CH$_2$-

first scientific reports appeared describing an active tetracycline prepared by chemical modification of a fermentation product. This was tetracycline, the parent member of this family of antibiotics, prepared by catalytic hydrogenolysis of chlortetracycline. It was more stable and better tolerated than its fermentation-produced progenitor, and almost completely displaced chlortetracycline from medical practice. Interestingly, tetracycline was later found in fermentation broths of a mutant strain of *S. aureofaciens* and may be manufactured by this method also.

Following the discovery of tetracycline, useful new drugs from chemical modification of tetracycline antibiotics were slow in coming, for the complexity and chemical lability of the tetracyclines did not render them amenable to facile systematic studies of the relationships between chemical structure and biological properties. Unlike the β-lactam antibiotics where structural modifications were being sought mainly to improve their antibacterial spectra and potency, superior semisynthetic tetracyclines were obtained, in general, with improved pharmacokinetic properties, i.e., through such factors as rate of oral absorption, degree of serum protein binding, rate of urinary excretion, and biological half-life.

The most effective approach to the discovery of superior tetracycline antibiotics has come from studies yielding tetracyclines modified at the C-6 position. Thus demethylchlortetracycline and methacycline are somewhat superior to tetracycline in terms of a longer serum half-life, and more recently doxycycline has been shown to exhibit near-ideal pharmacokinetics.

Macrolides: These substances comprise a family of antibiotics chemically characterized by a macrocyclic lactone to which one or more sugars are attached. The compounds are often divided into various subgroupings, but the group discussed in this section can be termed the *antibacterial* macrolides. They are distinguished chemically by having, in addition to the large lactone, various ketonic and hydroxyl functions and glycosidically bound deoxy sugars. A second grouping of commercial importance is known as the *polyene* macrolides, chemically characterized by extended conjugated double-bond systems. The polyenes are devoid of antibacterial activity but are potent antifungal agents.

A number of the antibacterial macrolides have been found to be clinically useful chemotherapeutic substances, falling generally under the title of *medium-spectrum antibiotics.* This term is taken to mean that these substances are effective against most gram-positive bacteria and have a degree of activity against certain gram-negative organisms such as *Haemophilus, Brucella,* and *Neisseria* species. The antibacterial macrolides also appear to inhibit certain pleuropneumonia-like organisms.

Erythromycin and oleandomycin are the major clinically useful drugs in this macrolide class. Although both antibiotics are commercially available as the parent entities, semisynthetic derivatives of these macrolides have proved to be clinically superior to their natural congeners. Like the tetracyclines, synthetic transformations in the macrolide series have not significantly altered their antibacterial spectra, but have improved their pharmacodynamic properties. For example, while the triacetyl derivative of oleandomycin exhibits somewhat less antimicrobial activity in vitro than does the parent compound, greater antibacterial activity occurs in the blood as a result of more complete enteric absorption.

Likewise, the propionate ester of erythromycin lauryl sulfate, erythromycin estolate, offers greater acid stability than the unesterified parent substance, so that time of use relative to meals is no longer an important factor. Although the estolate appears in the blood somewhat more slowly, the peak serum levels reached are higher and persist longer than other forms of the drug. The macrolides offer certain advantages over other types of antibiotic therapy.

For example, the profound changes in the intestinal flora usually associated with prolonged use of broad-spectrum antibiotics is rarely a problem. There is generally a low incidence of toxic manifestation and allergic reactions to the antibacterial macrolides.

Aminoglycosides: These substances comprise a class of potent broad-spectrum antibiotics which are chemically characterized by basic carbohydrate moieties glycosidically bound to a cyclitol unit. In general, the aminoglycosides are effective against most gram-positive and gram-negative bacteria, as well as *Mycobacterium tuberculosis.* Because of their highly ionic nature, the aminoglycosides are not absorbed from the gastrointestinal tract and must be administered parenterally. In a small percentage of patients, prolonged use of this class of antibiotics can adversely affect the eighth cranial nerve, causing some impairment of hearing and balance.

The discovery of streptomycin in 1944 drew immediate interest because it was the least toxic of the broad-spectrum antibiotics known at that time. Indeed, streptomycin was used to treat many gram-negative microbial infections, but because of the ease with which organisms developed resistance to it during treatment, many of these applications were abandoned when the tetra-cyclines became available. Today, streptomycin is used primarily in the treatment of tuberculosis.

Unlike other major classes of clinically useful antibiotics where chemical modifications have contributed heavily to the discovery of useful new analogs, nature has thus far been the dominant

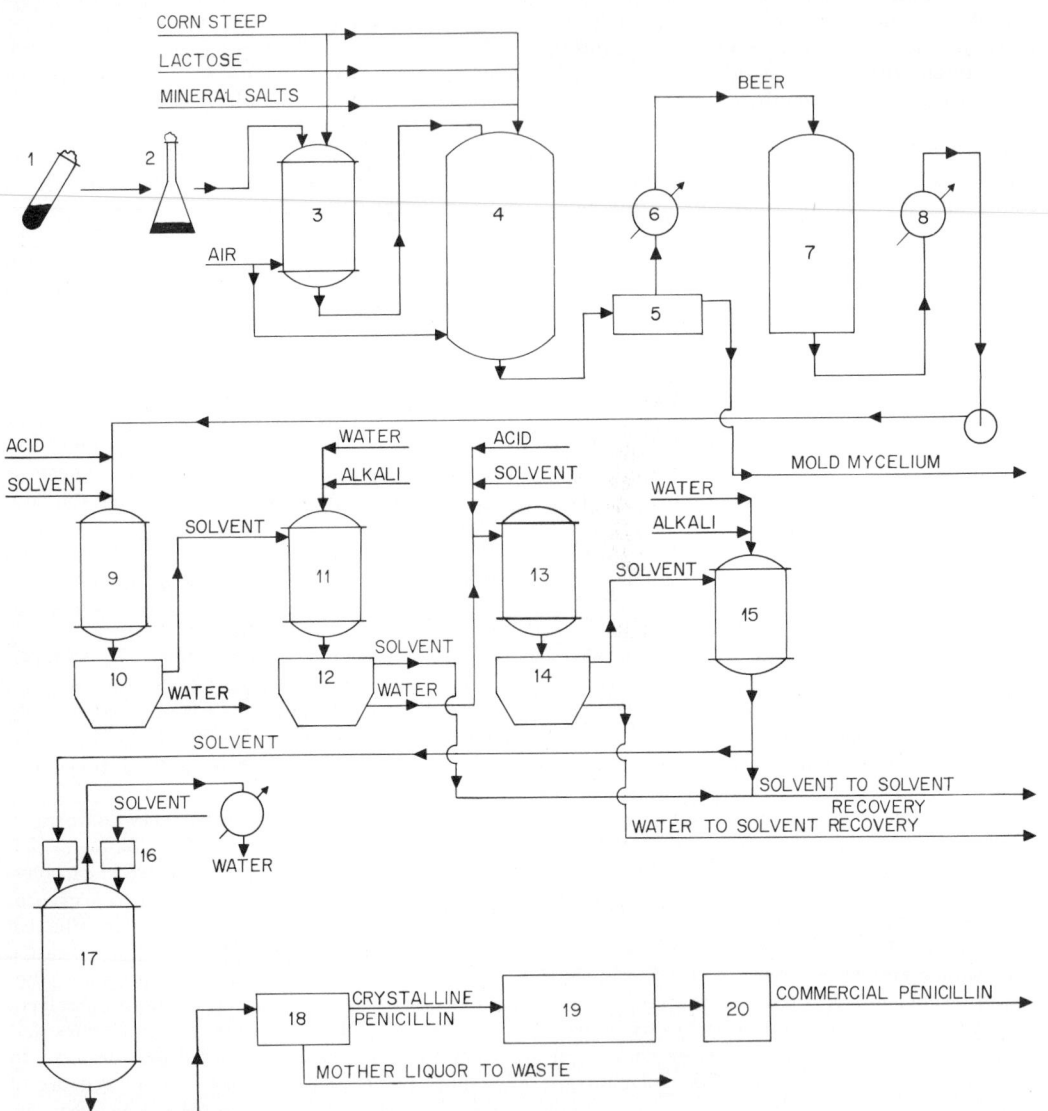

Fig. A-43. Schematic representation of materials flow in a commercial penicillin manufacturing process. (1) Agar slant culture, (2) bran spore culture, (3) seed tank, (4) fermentor, (5) filter, (6) brine cooler, (7) storage tank, (8) brine cooler, (9) mixing tank, (10)–(15) separation operations, (16) bacteriological filters, (17) crystallizer, (18) filter, (19) dryer, (20) finishing operations.

creator of superior aminoglycosides. All commercially available aminoglycoside antibiotics are manufactured by fermentation. Following the discovery of streptomycin, the next clinically significant discovery in the aminoglycoside area came in 1957 with the isolation of kanamycin. This antibiotic was more potent than streptomycin against most gram-positive and gram-negative bacteria and somewhat less toxic, thus resulting in a more favorable therapeutic index. The discovery of kanamycin at a time when penicillin-resistant staphylococcal infections were becoming a significant clinical problem resulted in its becoming a useful new drug; its lack of cross-resistance with other antibiotics made it useful not only for staphylococcal infections, but also against gram-negative infections refractory to other chemotherapeutic agents.

A breakthrough in the aminoglycoside area came in 1963 with the discovery of gentamicin. This complex of three closely related congeners was isolated from a *Micromonospora* fermentation and is the most potent aminoglycoside currently in clinical use. In particular, gentamicin is highly effective in the treatment of infections caused by *Pseudomonas* and *Proteus* organisms, two genera of bacteria mentioned briefly in the β-lactam section as being resistant to most known antibiotics.

Manufacture: With few exceptions, most antibiotics and starting materials for semisynthetic antibiotics are manufactured by fermentation. Fermentation can be regarded as a unit process in which the fermentation equipment, producing microorganisms, fermentation procedure itself, and extraction, purification, and crystallization processes must all be considered. Commercial fermentors are carbon-steel or stainless-steel enclosed tanks which may have a capacity of 50,000 gal or more and a height/diameter ratio in the order of 2:1 to 3:1. These tanks are built and instrumented in such a manner that maximum control over the growth of the microorganism may be maintained. Such factors as aeration, agitation, temperature, and hydrogen-ion concentration must be constantly monitored.

The antibiotic-producing microorganism is grown in submerged culture in a fermentation medium which contains various carbon, nitrogen, and trace-metal sources, required by the organism for its nutrition. The organism is grown under conditions of pure culture; that is, other microorganisms are excluded from the fermentation, since they compete for nutrients and may produce enzymes which are detrimental to the desired antibiotic, thus reducing the yield of product. When the fermentation has reached peak potency in 4–6 days, the antibiotic may be recovered by an extraction technique, such as distribution into a water-immiscible solvent, ion-exchange chromatography, or precipitation. Following extraction, purification and crystallization are carried out by procedures compatible with the physicochemical properties of the particular antibiotic being produced. A commercial process for the manufacture of penicillin is illustrated schematically in Fig. A-43.

See also **Amino acids; Ion-exchange resins;** and **Peptides and proteins.**

REFERENCES

Evans, R. M.: "The Chemistry of the Antibiotics Used in Medicine," Pergamon, New York, 1965.

Florey, H. W., E. Chain, N. G. Heatley, M. A. Jennings, A. G. Sanders, E. P. Abraham, and M. E. Florey: "Antibiotics," vols. 1 and 2, Oxford, London, 1949.

Goldberg, H. S.: "Antibiotics: Their Chemistry and Non-medical Uses," Van Nostrand, New York, 1959.

Perlman, D.: In A. Burger (ed.), "Medicinal Chemistry," 3d ed., part I, chap. 17, Wiley-Interscience, New York, 1970.

—F. C. Sciavolino, *Pfizer Inc., Groton, Conn.*

Antiknock Compounds (See **Alkylation; Bromine; Isomerization; Lead; Manganese; Petroleum; Petroleum processing;** and **Reforming.**)

Antimatter A form of matter in which protons, electrons, and other particles have charges opposite those with which they are normally associated.

Antineutron: A neutron that has the same mass as a neutron, with no electric charge, but that has the ability to annihilate another neutron.

Antiproton: An elementary particle that differs from a proton only in that its charge is negative and has the ability to annihilate a proton. The particle sometimes is termed a *negative proton.*

Antimonial Lead (See **Antimony;** and **Lead.**)

Antimony (L. *antimonium,* L. *stibium.*) **Sb** = 121.75 (at. wt.); 51 (at. no.). Antimony is a lustrous, silvery, blue-white metal, extremely brittle, of a flaky or scalelike crystalline texture, and easy to pulverize. Sb crystals are rhombohedral. Sb is found in group VB (along with As and Bi), period 5, of the periodic table. There are two natural isotopes of Sb (121 and 123) and ten radioactive isotopes (116–120, 122, and 124–127). See also **Chemical elements.** Sb exists in several allotropic forms. Valence = 3 and 5.

The specific gravity of Sb = 6.684 (25°C); melting point = 630.5°C; boiling point = 2624°C (760 mm Hg). Pure Sb metal has a hardness of 55 on the Brinell scale and 3.0–3.3 on the Mohs' scale. The latent heat of fusion = 38.3 cal/g. Among the more common metals, Sb is the poorest electrical conductor (4.5 on a scale where Cu = 100). The electrical resistivity of Sb = 41.7×10^{-6} Ω-cm at 20°C with a temperature coefficient of 0.0036 at 20°C. Ionization potential = 8.64 eV. Sb also is a very poor thermal conductor (5 on a scale where Cu = 100). Sb has the unusual property (like Bi, Ga, and H_2O) of expanding upon solidifying from the molten state; this advantage is utilized in type metal, yielding castings of exceptionally sharp configurations.

Antimony was described by Johann Thölden (1450), who published under the pseudonym of Basil Valentine. It is generally believed that the element was known prior to Valentine's disclosures, and he did not claim discovery.

Antimony cannot be rolled, forged, drawn, or extruded. When cast, Sb is brittle, showing low yield strength, below that of Sn and Pb. However, Sb does impart useful characteristics when alloyed with other metals, contributing to hardness, lower melting point, and less shrinkage upon freezing.

Chemical Properties: Antimony metal is insoluble in H_2O and does not react with dry air at room temperature, forming only a very slight tarnish. The metal will oxidize slowly with moist air at room temperatures. If heated, Sb will burn to form the white oxide Sb_2O_3. Molten Sb will attack most metals. Graphite is one of the most suitable containers for molten Sb.

Antimony will dissolve in aqua regia or hot concentrated H_2SO_4 to form SO_2 and $Sb_2(SO_4)_3$. Antimony is insoluble in HCl, but is converted by HNO_3 to the oxides Sb_2O_3(-ous) or Sb_2O_5(-ic), depending upon acid concentration. The most effective solvent is HNO_3, followed by HCl, containing only a small quantity of HNO_3. A dilute solution of HNO_3 and HF also is an effective solvent for Sb and most Sb alloys. Chlorine reacts with Sb to form $SbCl_3$ (butter of antimony) or $SbCl_5$, depending upon Cl_2 concentration, temperature, and time. Antimony reacts with NaOH to form sodium antimonite, $Na \cdot SbO_2 \cdot 3H_2O$. An antimonite also is formed with Ag. Stibine, SbH_3, a colorless, odorless gas (similar to arsine, but even more poisonous) may be formed by reducing Sb to the metallic state (by the action of Zn and HCl or H_2SO_4), with attendant generation of H_2. $SbCl_3 + 3Zn + 3HCl \rightarrow 3ZnCl_2 + SbH_3$. Upon

burning, SbH_3 produces Sb metal (a test for Sb). As shown by Table A-52, Sb forms a number of oxides, sulfides, and salts.

Potassium antimonyl tartrate (tartar emetic), $K(SbO)(C_4H_4O_6) \cdot \frac{1}{2}H_2O$, used in medicine and as a mordant in dyeing leather and textiles, is a white crystalline compound formed by reacting potassium hydrogen tartrate with antimony trioxide. Numerous other organic compounds containing Sb have been prepared. See Table A-52.

Sources: Stibnite, impure Sb_2S_3, is the principal Sb ore, containing approximately 45–60% Sb. After concentration, the material is supplied as matte, containing 92% Sb. Stibnite occurs mainly in China, but also in Mexico, Japan, Bolivia, and the United States (western states, including Alaska). The stibnite is reduced to Sb metal and then separated from the fused Fe_2S. The ore also is roasted to produce Sb_2O_3, which then is reduced by fusing with carbon and Na_2CO_3. Other ores include kermesite (red antimony, $2Sb_2S_3 \cdot Sb_2O_3$), valentinite (white antimony, Sb_2O_3), senarmontite (Sb_2O_3), and cervantite ($Sb_2O_3 \cdot Sb_2O_2$). Substantial amounts of Sb are recovered during the smelting of Pb.

Uses: Numerous alloys contain Sb. Antimonial lead, or *hard lead*, is ordinary Pb which contains up to 12% Sb. The Sb reduces the melting point of Pb and hardens the resulting alloy. Below a temperature of 140°C (284°F), 6% antimonial Pb shows better abrasion resistance than chemical lead. Antimonial leads are age-hardenable. The melting point of the Pb/Sb eutectic (12.5% Sb by weight) = 251°C. Antimony also is used in Sn/Pb solders up to an Sb content of 1%. Type metals are Pb-base alloys containing 3–9% Sn and 3–19% Sb. Lead-base diecasting alloys contain Sb: ASTM No. 4, 14–16% Sb; ASTM No. 5, 9.25–10.75% Sb; bearing alloy, 15% Sb, CT metal, 12.5% Sb; and Sn-free alloy, 10% Sb. Standard Babbitt (bearing) metals contain Sb: SAE 10, 4–5% Sb; SAE 11, 6–7.5% Sb; SAE 12, 7–8.5% Sb; SAE 13, 9.25–10.25% Sb; SAE 14, 14–16% Sb; and SAE 15, 14.5–15.5% Sb. Antimony is contained in a large number of low-melting alloys of Pb and Sn and, in some cases, with Bi, Cd, In, and Tl present. Utensils sometimes are spun from Britannia metal which contains 5% Sb, 93% Sn, and 2% Cu. Pewter contains up to 7% Sb, with the major content as Sn, and up to 20% Pb and 4% Cu.

Antimony sulfides are used in the manufacture of fireworks, as pigments and colors, and in rubber as a vulcanizing and coloring agent. *Antimony*

TABLE A-52. *Properties of Selected Antimony Compounds*

Compound and Formula	Formula Weight	Sp gr	Mp, °C	Bp, °C	Solubility
Antimony trichloride, $SbCl_3$	228.13	3.14	73.4	220.2	Very soluble in H_2O
Antimony trioxide, Sb_2O_3 (valentinite)	291.52	5.67	656	1570	Very slightly soluble in H_2O
Antimony trioxide, Sb_2O_3 (senarmontite)	291.52	5.2	652		
Antimony tetraoxide, Sb_2O_4	307.52	3.9	930	. . .	Insoluble in H_2O
Antimony pentaoxide, Sb_2O_5	323.52	3.8	380	930	Insoluble in H_2O
Antimony trisulfide, Sb_2S_3	339.70	4.6	550	. . .	Insoluble in cold H_2O; decomposes in hot H_2O
Antimony pentasulfide, Sb_2S_5	403.82	4.1	135	. . .	Insoluble in H_2O
Antimonyl potassium tartrate, $(SbO)KC_4H_4O_6 \cdot \frac{1}{2}H_2O$	333.94	2.6	100		Slightly soluble in cold H_2O; soluble in hot H_2O
Antimony sulfate, $(SbO)_2SO_4$	371.58	4.9	. . .	. . .	Decomposes in H_2O
Antimony bromide, $SbBr_3$	361.51	4.1	97	280	Decomposes in H_2O
Antimonic acid, $o\text{-}H_3SbO_4$	188.78	6.6	100d	. . .	Slightly soluble in H_2O
Antimonous acid, $o\text{-}H_3SbO_3$	172.78	. . .	d		Insoluble in H_2O
Antimony oxychloride(-ous), $SbOCl$ . . .	173.22	. . .	170d	. . .	Insoluble in cold H_2O; decomposes in hot H_2O
Antimony selenide, Sb_2Se_3	481.12	. . .	611	. . . . . .	Very slightly soluble in H_2O
Triethyl antimony, $Sb(C_2H_5)_3$	208.88	1.3	−29	159	Insoluble in H_2O; soluble in alcohol and ether
Trimethyl antimony, $Sb(CH_3)_3$	166.83	1.5	. . .	81	Slightly soluble in H_2O
Antimony lactate, $Sb(C_3H_5O_3)_3$	388.88	. . .	. . .	. . .	Slightly soluble in H_2O
Antimony tartrate, $Sb_2(C_4H_4O_6)_3 \cdot 6H_2O$	795.71	. . .	. . .	. . .	Slightly soluble in H_2O

d = decomposes.

trisulfide (also known as *antimony vermilion*) is used in paints and in the preparation of red lakes through precipitation of various dyestuffs, such as madder, cochineal, and aniline dyes when used with inorganic salts, such as Sn and Al. The compound also is used for wall finishes and in printing inks. Butter of antimony, $SbCl_3$, is used for bronzing steel, in medicine, and as mordant in dyeing. Mixtures of Sb oxides and sulfides are used as yellow pigments in glass and porcelain production.

Antioxidants Oxygen and oxidation can produce damage in organic materials during manufacture, processing, storage, and in service, with adverse effects upon appearance and properties. The speed of such deterioration depends on a number of factors, which include composition of the material and the conditions of exposure to oxygen, heat, and light. These adverse effects may be greatly retarded in most organic (hydrocarbon) materials by the use of antioxidants. See Table A-53.

Antioxidants may be defined as substances that oppose oxidation, or inhibit or retard reactions promoted by oxygen or peroxides, and therefore its effects. They are substances that retard atmospheric oxidation or the degradative effects of oxidation and extend the useful life of the substrate when they are added in very low concentrations on the order of fractions of 1%. Composition of the substrate, processing conditions, impurities, and use of the product must be considered in choosing the best antioxidant system.

Desirable features of antioxidants include: effective at low concentrations; compatible with the substrate; stable; comparatively nonvolatile and nonextractable under the conditions of use; nontoxic; easy and safe to handle; low in cost; and must not impart any undesirable features to the substrate (e.g., color and odor).

It would appear that, since oxidative degrada-

TABLE A-53. *Major Antioxidants*

Structural Type	Example	Source	Properties	Major Applications
Hindered Phenolic—Nonstaining; nondiscoloring:				
Alkylated phenol	2,6-di-*t*-Butyl-*p*-cresol	1	Crystalline solid	Direct food additive; fuels, oils, rubber, plastics
Alkylidene bisphenol	2,2'-Methylene-bis-(6-*t*-butyl-*p*-cresol)	2	Crystalline solid	Rubber, plastics, oils
	2,2'-Butylidene-bis-(6-*t*-butyl-*m*-cresol)	3	Crystalline solid	Rubber, plastics, oils
Thiobisphenol	2,2'-Thiobis-(6-*t*-butyl-*p*-cresol)	4	Crystalline solid	Rubber, plastics, oils
	4,4'-Thiobis-(6-*t*-butyl-*m*-cresol)	5	Crystalline solid	Rubber, plastics, oils
Multiphenolic	1,1,3-Tris-(2-methyl-4-hydroxy-5-*t*-butylphenyl) butane	6	Crystalline solid	Plastics
	Tetrakis (methylene(3,5-di-*t*-butyl-4-hydroxy cinnamate)) methane	7	Crystalline solid	Plastics
Amine—Staining; discoloring:				
Diphenylamine	Octyl diphenylamine	8	Waxy solid	Rubber
Phenylenediamine	*N*-Isopropyl-*N'*-phenyl-*p*-phenylenediamine	9	Liquid	Rubber
Alkylenediamine	*N,N'*-Diphenylethylenediamine	10	Solid	Rubber
Naphthylamines	Phenyl-*β*-naphthylamine	11	Solid	Rubber
Propionic esters:				
Dialkyl-thio-dipropionate . .	Dilaurylthiodipropionate	12	Solid	Synergist in plastics, rubber
Organophosphites	(Tris)nonyl phenyl phosphite	13	Liquid	Rubber, plastics
	Hindered phenol phosphite	14	Solid	Rubber, hot melt, plastics

SOURCES:
1. Ashland CAO-1, CAO-3; Hercules Dalpac; Koppers DBPC, Impruvol; Shell Ionol.
2. Ashland CAO-5, CAO-14; Cyanamid 2246.
3. Monsanto Santowhite Powder.
4. Ashland CAO-4 and CAO-6.
5. Monsanto Santonox.
6. ICI Topanol CA.
7. Geigy Irganox 1010.
8. Cyanamid Cyanox 8, Uniroyal Octamine.
9. ICI Nonox CN, Uniroyal Flexzone 3-C, Monsanto Santoflex IP.
10. ICI Nonox DED.
11. Du Pont PBNA, ICI Nonox D, Goodrich.
12. Cyanamid Plastonox DLTDP, Du Pont Milban, ICI DLTDP.
13. Uniroyal Polygard.
14. Ashland CAO-35.

tion occurs in a variety of hydrocarbon materials which are dissimilar in appearance and have entirely different applications and also different properties, and the degrading effect results in a loss of different desirable properties, the oxidation mechanisms would be different. However, the mechanism for oxidative degradation, in our present state of knowledge, is considered to be the same for all hydrocarbon materials, whether it be for oils used in food, gasoline for powering a car, rubber for gloves, or polyethylene for food wrap. They degrade by the same free-radical mechanism.

Mechanisms of Hydrocarbon Oxidation and Inhibition: The mechanisms of oxidation and antioxidant action are still the subject of investigation. Oxidation has been described as a free-radical, chain-type reaction. At processing temperatures and more slowly at ambient temperatures, hydrocarbon free radicals $(R \cdot)$ are formed. These react with oxygen to form peroxy radicals $(ROO \cdot)$, which can abstract a hydrogen atom from the hydrocarbon to form a hydroperoxide $(ROOH)$ and another hydrocarbon free radical.

The cycle repeats itself with the addition of oxygen to the new free radical. The unstable hydroperoxides left along the hydrocarbon molecule are the major source for degradation. Under the influence of heat, light, and certain metals, they decompose to form carbonyl groups. When this happens, the hydrocarbon molecule breaks and splits off another hydrocarbon free radical. Ultimately, this type of degradation can lead to rancidity and color development in oils, fats, waxes, and materials containing them; loss of nutrients in and palatability of foods and feeds; decomposition and gum formation in gasolines; sludging in lubricants (petroleum products); loss of strength; stiffening and cracking in rubber; and degradation of plastics as evidenced by color formation, odor formation, stiffening, softening, cracking, and loss of many desirable properties.

The antioxidants used act either to tie up the peroxy radicals so that they are incapable of propagating the reaction chain or to decompose the hydroperoxides in such a manner that carbonyl groups and additional free radicals are not formed. The former, which are called *chain-breaking antioxidants, free-radical scavengers,* or *inhibitors,* are usually hindered phenols or amines. The latter, called *peroxide decomposers,* are generally sulfur compounds (sulfides and thiodipropionates) or phosphorous compounds (organophosphites).

A general outline of the reactions involved in oxidation and inhibition is given on next page.

Synergism: By using two or more different types of antioxidants or additives, the resistance to oxidation or deterioration of an organic material may be improved to a greater extent than would be predicted on the basis of strict additivity. The two additives are then said to show a *synergistic* effect toward one another. The converse of synergism is *antagonism.*

Probably the most generally effective mixtures of antioxidants are those in which one compound functions as a decomposer of peroxides (sulfides, thiodipropionates) and the other as an inhibitor of free radicals (hindered phenols, amines). Although the latter retards the formation of reaction chains, some hydroperoxide is nevertheless formed. If this hydroperoxide then reacts with a decomposer of peroxides, instead of decomposing into free radicals, the two antioxidants act together to complement each other. Moreover, the peroxide decomposer may itself be subject to oxidation by peroxy radicals, and its efficiency will therefore be increased in the presence of an inhibitor of free radicals. In the case of phenol-sulfide mixtures, the sulfide (peroxide decomposer) also continuously regenerates the phenol (radical scavenger) to accentuate the synergistic nature of the mixture. Metal chelators or deactivators (e.g., citric and phosphoric acids) of prooxidant metals (iron, copper, nickel, tin), ultraviolet-light absorbers (carbon black, substituted benzophenones, benzotriazoles, and salicylates), and antiozonants (substituted phenylenediamines) also develop synergistic effects with antioxidants.

Government Regulations: The use of antioxidants in foods, pharmaceuticals, and animal feeds (direct food additives), as well as their use in food-contact surfaces (indirect additives), is closely regulated in the United States as well as in many other countries. The Food and Drug Administration (21 Code of Federal Regulations 121) regulates all applications involving nonanimal products; the Department of Agriculture regulates meat and meat products. Antioxidants are approved only after extensive extraction, toxicological, and feeding studies. Very few antioxidants have this type of approval and, as such, at only maximum levels of 0.005% addition to the food or 0.02% of the fat or essential oil content. The antioxidant that has found the greatest use in food and food-contact applications is butylated hydroxy toluene (BHT). It is generally used by itself in foods, or in combination with butylated hy-

Representative Reactions Involved in Oxidation and Inhibition

Initiation	$RH \xrightarrow[\text{heat, light, metal}]{\text{activation}} R\cdot + H\cdot$ Free radical
Propagation with O_2	$R\cdot + O_2 \longrightarrow ROO\cdot$ Peroxy radical $ROO\cdot + RH \longrightarrow ROOH + R\cdot$ Hydroperoxide
Chain branching	Where double bonds, $ROO\cdot + C{=}C \longrightarrow ROOC-\overset{\cdot}{C}$ $ROOH \xrightarrow[\text{prooxidant metals}]{\text{heat, light}} RO\cdot + HO\cdot$ $RO\cdot + RH \longrightarrow ROH + R\cdot$ $HO\cdot + RH \longrightarrow HOH + R\cdot$
Termination (small amount nonradical products)	$ROO\cdot + ROO\cdot$ $ROO\cdot + R\cdot$ $\rangle$ inert products $R\cdot + R\cdot$
Inhibition (AH is *antioxidant*)	$\left.\begin{array}{l} ROO\cdot \\ RO\cdot \\ R\cdot \\ HO\cdot \end{array}\right\} \xrightarrow{AH} \left\{\begin{array}{l} ROOH \\ ROH \\ RH \\ HOH \end{array}\right. + A\cdot$ (stable)
Peroxide decomposition	$ROOH + S \longrightarrow$ nonradical products where R = hydrocarbon

droxy anisole (BHA), propyl gallate, citric, or phosphoric acids, to obtain a synergistic effect. In food-contact surfaces BHT is generally used by itself or in combination with thiodipropionates and/or organophosphites to obtain a synergistic effect also.

Foods: Oxidative rancidity and color deterioration is a problem of great importance in fats and fatty foods. Rancidity not only interposes undesirable flavors and odors, but exerts harmful nutritional and physiological effects. Oxidative deterioration leads to the destruction of fat-soluble vitamins and essential fatty acids, as well as to direct toxicological effects of various types. The use of both natural and synthetic antioxidants has made important contributions to better nutrition and to conservation of foods by improving the nutrient values, palatability, and shelf life of the food. The antioxidant may be added directly to the food, or may be incorporated into the packaging material, which will serve to protect the packaging material and the food contacting it from

deterioration, as well as permitting the antioxidant to migrate into the food and thus protect it. BHT and BHA have "carry-through" antioxidant effectiveness, while propyl gallate does not. The maximum antioxidant level generally permitted is 0.02% of the fat content.

Rubber: The stabilization of rubber is complicated by the variation in raw material. Numerous synthetic rubbers are prepared from unsaturated hydrocarbons and require antioxidants for proper functioning and long life. Natural rubber has a degree of oxidation resistance from natural antioxidants, but the major effects of these are lost during processing, so that synthetic antioxidants are added to maintain desirable properties in the finished product.

The major causes responsible for the degradation of both natural and synthetic rubber are shelf aging, oxidation aided by metal catalysts, the effects of heat, light, ozone, and flex or atmospheric cracking.

The antioxidants most widely used in rubber

are of the amine and hindered-phenol type. The amines are considered to be staining and discoloring and are generally used in carbon-filled (black) rubbers. The hindered phenolics are nonstaining and nondiscoloring and are used in white or light-colored rubber products, either by themselves or in synergistic combination with organophosphites. The latter are also used by themselves as rubber stabilizers (gel retarders). The amine-type antioxidants also exhibit antiozonant properties. Antioxidant-use levels vary from about 0.5 to 2 PHR (parts per hundred of rubber).

Plastics: The antioxidant requirements for plastics vary with the structure of the polymer, those containing unsaturation being less resistant to oxidation than saturated polymers. Plastics such as fluorocarbons, polyacrylates, phenolformaldehyde, polyesters, and polyamides require relatively little stabilization against oxidation, while vinyls, high-impact polystyrene, ABS, polyolefins (polyethylene, polypropylene), and polyacetals need antioxidants to maintain their useful properties during manufacture, processing, and use. The primary antioxidants used for the stabilization of plastics are of the hindered-phenolic type. Amines have limited use because of their staining and discoloring properties. Under ordinary manufacturing, processing and use conditions hindered phenolics, such as BHT, and alkylidene bisphenols are used by themselves or in combination with thiodipropionates and/or organophosphites to obtain an enhanced effect, depending on the degree of stabilization desired. In black-filled polyolefins, thiobisphenols are used, since they develop a synergistic effect with channel black.

Under more severe conditions and in polymers more susceptible to degradation, greater concentrations of antioxidants would be required; or those with greater staying power and higher molecular weight (multiphenolic), or a multifunctional group containing antioxidants, would be used. Antioxidant use levels vary from about 0.01 to 1.0%.

See also **Elastomers;** and **Rubber, natural.**

REFERENCES
Ingold, K. U.: Inhibition of the Autoxidation of Organic Substances in the Liquid Phase, *Chem. Rev.,* vol. 61, pp. 563–589, 1961.
Lundberg, W. O.: "Autoxidation and Antioxidants," vols. 1 and 2, Wiley-Interscience, New York, 1962.
Scott, G.: "Atmospheric Oxidation and Antioxidants," Elsevier, New York, 1965.
—Michael Robin, *Ashland Chemical Company Division, Ashland Oil and Refining Company, Fords, N.J.*

Antiozonants (See **Rubber, natural.**)

Apatite Apatite is a mineral having the general formula $Ca_5(F,Cl)(PO_4)_3$ and belonging to the hexagonal crystal system, tripyramidal class. The main varieties are fluorapatite, $Ca_5F(PO_4)_3$, and chlorapatite, $Ca_5Cl(PO_4)_3$. Compounds of intermediate composition containing both chlorine and fluorine are known. The most common form is fluorapatite. The mineral usually occurs as prismatic crystals of varying lengths. It exhibits imperfect cleavage. The specific gravity ranges from 3.15 to 3.20. On the Mohs' scale of hardness (1–10), it has a relative value of 5. The mineral has a vitreous (glasslike) to subresinous luster and is transparent to translucent. The color is generally green or brown, but blue, violet, and colorless varieties have been noted. Apatite is characterized by its crystal structure, color, and hardness.

The mineral is widely distributed. It occurs as an accessory constituent in igneous, sedimentary, and metamorphic rocks. The world's largest deposits are found in the U.S.S.R. near the town of Kirov on the Kola Peninsula. Other sites of deposits are Norway, Sweden, and Canada. In the United States, apatite is found in California, Connecticut, Maine, Massachusetts, New Hampshire, New Jersey, and New York. The mineral has been used extensively as a fertilizer. The transparent varieties have limited use as gemstones.

—Mary E. Pisklak, *Industrial Gas Division, Air Products and Chemicals, Inc., Allentown, Pa.*

Aqua Regia (See **Nitric acid;** and **Platinum metals and platinum.**)

Aragonite (See **Limestone.**)

Arc Discharge Processing (See **Plasma, chemical processing with.**)

Arginine (See **Amino acids;** and **Peptides and proteins.**)

Argon (See **Chemical elements.**)

Aroma Chemicals Usually associated with perfumes and colognes, aroma chemicals are incorporated in numerous products—cosmetics, toilet preparations, soaps, foods (along with flavorants), scented papers, tobacco, and many types of household products, particularly those packaged in aerosol form. Although most aromas are designed with a pleasing effect in mind, other functions can be served, as, for example, the case of adding odors to otherwise odorless fuel gases to provide warning of gas leaks. Aromas can be selected to set moods, associations, and reactions, in addition to the strictly aesthetically pleasing.

Originally, the volatile oils and other scent-producing substances used in perfumes were derived from natural substances exclusively. Such compounds were used for these purposes long before their composition, often complex, was identified. However, once determined, their composition can be duplicated in many instances via organic synthesis.

TABLE A-54. *Representative Aroma Chemicals*

Scent	Compound
Almond, bitter	Benzaldehyde
Banana	Benzyl propionate
Berry	Methylnaphthyl ether
Carnation , ...	Isoeugenol
Clove	Eugenol
Clover	Amyl salicylate
Crab apple	Geranyl
Eucalyptus	Cineole
Gardenia	Methylphenyl acetate
Grape	Methyl anthranilate
Hay	Coumarin
Heliotrope	Heliotropin
Hyacinth	Phenylacetic aldehyde
Jasmin	Benzyl acetate, *p*-cresyl phenylacetate
Lavender	Linalyl alcohol
Lemon	Citral
Lilac	Terpineol, hydroxycitronellal
Lily of the valley	Linalool
Magnolia	Nerol
May blossom	Anisic aldehyde
Narcissus	Methylnaphthyl ketone
Orange blossom	β-Naphthol ethyl ether
Peach	Phenethyl salicylate
Pine	Bornyl isovalerate
Raspberry	Methylionone
Rose	Phenethyl alcohol, phenyl ether, geraniol ester, citronellol esters
Spice	Cinnamaldehyde
Strawberry	Ethylmethylphenyl glycidate
Sweet pea	Benzylidene acetone
Vanilla	Vanillin, ethyl protocatechnic aldehyde, bourbonal
Violet	Ionone
Wintergreen	Ethyl or methyl salicylate
Woodsy	Cedrol, cedryl acetate, isobutyl cinnamate

The chemical composition of a number of compounds, along with their characteristic scent, is given in Table A-54. See also **Terpenes**.

Aromatic Compounds   A compound that contains a closed-chain or ring (nucleus), as contrasted with an open-chain structure (the aliphatics), may be termed an aromatic compound. Benzene and benzene derivatives are classical examples of the aromatics. The term *benzenoid* sometimes is used with reference to aromatic compounds.

Numerous aromatic compounds incorporate open chains (termed *side chains*) which attach to one of the atoms in the nucleus or ring. An example is

Phenylbutylene dibromide

All ring-type compounds are not aromatics, as in the instances of the cycloparaffins and cyclo-olefins, which, as in the case of other aliphatics, are considered to be derivatives of methane. See also **Aliphatic compounds;** and **Hydrocarbons.**

A majority of aromatic compounds are unsaturated because of the presence of double bonds ·between carbon atoms. The benzene ring, for example, contains three pairs of single-bonded and three pairs of double-bonded carbon atoms. Some aromatics contain several adjacent, interlocked rings, as found in naphthalene and anthracene. Because of the presence of several carbons in the ring, with several options for the formation of substitution products, and because of the presence in numerous instances of more than one ring, the naming of some aromatics can be complex. See also **Carbon compounds.**

The term *aromatic* derives from the distinctive odor of several benzene derivatives. Because the carbon content percentagewise is larger in most aromatic compounds than aliphatic compounds, a majority of aromatics are crystalline solids under standard conditions. Aromatics form halogen-substitution products readily, as do the aliphatics. However, the two classes of compounds differ with respect to the actions of nitric acid and sulfuric acid. Most aromatics yield nitro derivatives when exposed to concentrated nitric acid, and sulfonic acids when treated with concentrated sulfuric acid. These types of derivatives are rare among the aliphatics.

Numerous aromatic compounds are described throughout this volume. See, particularly, **Petrochemical complex.**

Arrhenius Equation (See **Catalysts.**)

Arsenic (L. *arsenicum,* yellow orpiment.) **As** = 74.9216 (at. wt.), 33 (at. no.). Arsenic is a member of group VA of the periodic table, which includes the elements nitrogen, phosphorus, antimony, and bismuth. It exhibits one stable isotope with a mass number of 75.

Elemental As has been reported to exist in several allotropic forms. Those most commonly encountered are the stable α-form, which is metallic and has a rhombohedral crystal structure; the β-form, which is vitreous; and the yellow form, which is metastable.

Arsenic has a steel-gray appearance in the crystalline form and is quite brittle. It is one of the few metals which on heating under normal conditions sublimes rather than melts. Table A-55 lists some of its selected physical properties. See also **Chemical elements.**

TABLE A-55. *Selected Physical Properties of Arsenic*

Density, g/cm^3	5.72
Atomic volume, cm^3/g	13.09
Melting point, 28 atm, °C	817
Boiling point, °C	618(s)
Specific heat, cal/(g)(°C)	0.082
Coefficient of thermal expansion, in./(in.)(°C)	4.7×10^{-6}
Electrical resistivity, Ω-cm	33×10^{-6}

s = sublimes.

Discovery: Arsenic in the form of the sulfide mineral had been recognized as early as 400 B.C. Aristotle referred to realgar, the red form of arsenic sulfide, As_2S_2, as sandarache. Albert Magmus in 1250 was reported to have prepared elemental arsenic; however, it was not until the seventeenth century that sufficient documentation on the preparation of the element was presented first by J. Schroder, and later by N. Lémery. By the eighteenth century the characteristics of arsenic were well enough known to classify it as a semimetal.

Occurrence: Arsenic is found widely disseminated in small amounts throughout nature. The crustal abundance has been reported to be 2.5–5 ppm. Native As is occasionally found; however,

it is most commonly encountered in the minerals arsenopyrite, $FeAsS$; loellingite, $FeAs_2$; realgar, As_2S_2; orpiment, As_2S_3, and enargite, $3Cu_2S \cdot As_2S_5$. Over 150 minerals of As have been identified. In addition, As replaces sulfur to some extent, and is found associated with many sulfide minerals, such as iron pyrite, galena, and chalcopyrite. When the As content in these minerals exceeds a few hundred parts per million, microscopic inclusions of As minerals may often be observed. Most of the As of industry is obtained as a by-product from the treatment of copper and, to a lesser extent, from lead, cobalt, and gold and silver ores.

Recovery: When rich concentrations of arsenopyrite ore are encountered, metallic As may be recovered by direct smelting of the ore at 650–700°C in the absence of air. Commercially, the metal is obtained by the reduction of arsenic trioxide with carbon.

The major demand for As has been in the form of arsenic trioxide. It is recovered from the treatment of flue dusts resulting from the smelting of copper and lead concentrates. The initial flue dust obtained contains about 30% arsenic trioxide. This material is mixed with pyrite or galena and roasted to volatilize the As, leaving behind a clinkered residue suitable for further processing. The arsenic oxide vapor formed is condensed in a series of kitchens. The resulting product, which is a powder and off-white in color, referred to as *crude arsenic,* is sold commercially to a minimum purity of 95% As_2O_3. A refined *white arsenic* with a minimum purity of 99% As_2O_3 is produced by resubliming the crude arsenic.

Chemical Properties: In some of its chemical behavior, As resembles phosphorus, antimony, and bismuth in that they all exhibit principal valence states of 3^-, 0, 3^+, and 5^+. The 3^+ oxidation state is the most common for arsenic compounds.

Elemental metallic arsenic is relatively stable in dry air; however, on exposure to moist air it will oxidize, initially developing a golden-bronze tarnish, which on further exposure changes to a black coating. Elemental As does not melt on heating under normal atmospheric conditions, but sublimes and oxidizes readily to *arsenic sesquioxide,* As_4O_6, which more commonly is referred to as *arsenic trioxide.* In the metallic state As is not readily attacked by water, alkaline solutions, or nonoxidizing acids. It will react with nitric acid forming orthoarsenic acid, H_3AsO_4. In the presence of an oxidant, it will be attacked by hydrochloric acid.

Some arsenic compounds of interest are:

HYDROGEN ARSENIDE. Arsine, H_3As, may be formed by the hydrolysis of AlAs. It may also be formed from As compounds by either electrolytic reduction or chemical reduction with zinc or magnesium. It can be thermally decomposed by heating above 250°C. It is an extremely toxic gas, and proper precautions must be taken when handling. There is no direct reaction between hydrogen and arsenic to form arsine.

ARSENIC TRIOXIDE, As_2O_3-As_4O_6, may be formed by the ignition of arsenic vapor in air or oxygen. There are two crystal modifications of the oxide, cubic and monoclinic.

ARSENIC PENTOXIDE, As_2O_5, may be produced by the reaction of As_2O_3 with nitric acid to form orthoarsenic acid, H_3AsO_4, which is crystallized from solution as a monohydrate. Upon heating, it decomposes to As_2O_5.

ORTHOARSENIC ACID, H_3AsO_4, is prepared by dissolving As_2O_3 in nitric acid. Pyroarsenic acid, $H_4As_2O_7$, and metaarsenic acid, $HAsO_3$, may be obtained by heating orthoarsenic acid.

SODIUM ARSENITE, Na_3AsO_3, and sodium meta arsenite, $NaAsO_2$, may be formed by the reaction of As_2O_3 with sodium hydroxide.

SODIUM ARSENATE, Na_3AsO_4, may be obtained by the reaction of sodium hydroxide with orthoarsenic acid.

TRICALCIUM ARSENATE, $Ca_3(AsO_4)_2$, may be prepared from the reaction of calcium hydroxide with orthoarsenic acid.

LEAD ARSENATE, $Pb_3(AsO_4)_2$, may be prepared by the reaction of a soluble lead salt with sodium arsenate.

ARSENIC HALIDES of type AsX_3 are formed by the direct combination of the elements. AsF_5 is the only stable 5 + halogen compound, and can be formed by reacting AsF_3 with antimony fluoride and bromine. AsI_2 is formed by the direct combination of the elements. Some properties of the halides follow:

Compound	Appearance	Mp, °C	Bp, °C
$AsBr_3$	Yellowish solid	32.8	221
$AsCl_3$	Oily liquid	−18	130
AsF_3	Oily liquid	−8.5	63
AsF_5	Colorless gas	−80	−53
AsI_2	Red solid	136(d)	
AsI_3	Red solid	146	403

d = decomposes.

Arsenic will react directly with sulfur to form the compounds As_2S_3, As_2S_2, and As_2S_5 and mixtures of various proportions. As_2S_3 is found in nature as the yellow mineral orpiment, and As_2S_2 as the red mineral realgar.

Because of its electronegative position in the periodic table, arsenic will combine with many of the metallic elements to form compounds. Two compounds which have interesting semiconductor properties are indium arsenide, InAs, and gallium arsenide, GaAs. See also **Gallium.**

Organic compounds of As are quite numerous. The largest group of compounds is the arsonic acids of the type $RAsO(OH)_2$, in which R may be aliphatic, aromatic, or heterocyclic, and their salts. Some organic arsenic compounds of commercial interest are disodium methylarsonate, ammonium arsonate, arsanilic acid (paraaminophenylarsonic acid), and cacodylic acid (dimethylarsonic acid).

Economy and Uses: There are no current statistics published on the world production of arsenic trioxide. In 1968 the world production was estimated to be 56,520 tons. In early 1971, the American Smelting and Refining Company was the only domestic producer of arsenic trioxide. The trioxide is sold in a crude form 95% As_2O_3, and in refined form, white arsenic, 99% As_2O_3. Commercial metallic arsenic is imported primarily from Sweden. High-purity As with a purity of 99.999 + % is produced domestically.

Metallic As, because of its metalloid characteristics, finds limited use. In metallurgical applications it is used primarily as an additive metal. In quantities of $\frac{1}{2}$-2% it is used in the manufacture of lead shot to improve sphericity. It is added to lead-base bearing alloys to improve their physical and elevated temperature properties. A small amount of As is added to lead-base battery gridmetal and cable-sheathing alloys to improve hardness. Small additions of arsenic (0.02–0.05%) to brass minimize or prevent dezincification.

The major use of As is in the form of the oxide, which is used to prepare various chemical compounds. Many of these compounds are used for herbicide and pesticide control. Some typical compounds for these applications are noted. Calcium arsenate is used in the control of boll weevils and crabgrass. Lead arsenate is used in the control of fruit pests. Sodium arsenite is used to kill weeds, to control growth of underbrush, to control leaf rot on potatoes due to fungicide, to debark trees, to control aquatic weeds, and to protect sheep from ticks and disease carriers. Sodium arsenate is an active ingredient of Wolman salts, which are used to preserve wood. Arsenic acid and cacodylic acid (dimethylarsonic acid) are used in the defoliation of cotton crops and as soil sterilants.

Disodium methylarsonate and ammonium methanearsonate are used for crabgrass control and also as selective herbicides. Arsenilic acid is a useful feed additive for swine and poultry feed.

Refined white arsenic trioxide is used quite extensively as a decolorizing and refining agent in the manufacture of glass.

Arsenic trisulfide is used in the manufacture of infrared lenses, and arsenic sulfide is an ingredient of fireworks.

High-purity As in the form of indium arsenide, gallium arsenide, and gallium arsenide phosphide is used in semiconductor applications.

When formulated with various combinations of the elements sulfur, thallium, selenium, iodine, germanium, and tellurium, arsenic will form a series of low-melting-point glasses. High-purity arsenic hydride and arsenic trichloride are also used in the manufacture of epitaxial gallium arsenide.

Hygienic Considerations: Metallic arsenic and arsenic trisulfide may be handled, but in general skin contact with arsenical compounds should be avoided. In those operations where fumes and dust are present, proper ventilation should be provided and respirators should be worn. Extreme caution should be exercised in handling arsine.

—S. C. Carapella, Jr., *American Smelting and Refining Company, South Plainfield, N.J.*

Arsenical Copper   (See **Copper.**)

Arsines   (See **Amines;** and **Arsenic.**)

Aryls   (See **Radicals.**)

Aseptic Packaging   (See **Pasteurization.**)

Asparagine   (See **Amino acids.**)

Aspartic Acid   (See **Amino acids;** and **Peptides and Proteins.**)

Asphalt   (See **Coal tar and derivatives;** and **Petroleum.**)

Astatine   (See **Chemical elements.**)

Aston Process   (See **Iron and steel.**)

Atactic   (See **Polymerization.**)

Atomic Diameter   (See **Chemical elements.**)

Atomic Energy   (See **Atomic structure; Nuclear power plants; Radioactive isotopes;** and **Uranium.**)

Atomic Explosives   (See **Explosives.**)

Atomic Fusion   (See **Fusion, nuclear;** and **Nuclear power plants.**)

Atomic Number   The atomic number Z of any element equals the number of protons in the nucleus and also equals the number of orbital electrons. For example, helium ($Z = 2$) has two protons in the nucleus and two orbital electrons. The atomic numbers for all elements are given under **Chemical elements.** See also **Atomic structure.**

Atomic Power   (See **Nuclear power plants.**)

Atomic Structure   The fact that many atomic masses are very nearly whole numbers hints that all atoms may be built of the same smaller particles. As early as 1815, William Prout proposed the theory that all atoms were combinations of hydrogen atoms. This assumption was based on atomic masses known at that time, all of which were very nearly integral numbers. Prout's theory was close to the present model of the atom, but was soon discarded as additional atomic masses were determined. Note, for example, the atomic mass of chlorine, which is very nearly 35.5. See **isotopes** under **Chemical elements;** and **Radioisotopes.**

For the present we can consider atoms to be constructed from combinations of three elementary particles: electrons, protons, and neutrons. Protons and neutrons have approximately the same mass, about the same as the hydrogen atom, but they differ in electric charge. The neutron is electrically neutral, but the proton carries one positive unit of electric charge, 1.60210×10^{-19} C. The mass of the electron is only $1/1{,}836.10$ as great as that of the proton, but it carries a negative unit of electric charge equal in magnitude to the positive charge on the proton. Experience indicates that this basic quantity of electric charge, represented by the symbol e, cannot be divided. The quantity of electricity carried by any body must therefore be some multiple of e, positive or negative, depending on whether the body contains an excess of protons or of electrons. Electrons, protons, and neutrons spin about their own axes, each with an angular momentum of $\frac{1}{2}\hbar$, where $\hbar = 1.0544 \times 10^{-34}$ J·s. Numerically, $\hbar$ is $1/2\pi$ times Planck's universal radiation constant h. The important properties of the elementary particles are listed in Table A-56.

TABLE A-56. *Properties of Elementary Particles*

Particle	Mass, amu	Charge, C	Spin
Neutron . . .	1.0086654	0	$\frac{1}{2}\hbar$
Proton	1.0072766	1.60210×10^{-19}	$\frac{1}{2}\hbar$
Electron . . .	0.000548597	-1.60210×10^{-19}	$\frac{1}{2}\hbar$

In its normal state an atom contains equal numbers of electrons and protons, and is therefore electrically neutral. The protons and neutrons are located in the central core or nucleus of the atom, while the electrons rotate in orbits about the nucleus. The nuclear particles, often called *nucleons,* are closely packed; thus the nucleus is relatively small and very dense. The electron orbits are at relatively large distances from the nucleus, making the total volume of the atom about 10^{14} times the nuclear volume. Atomic diameters range from about 0.6×10^{-8} cm for helium to about 5.4×10^{-8} cm for cesium. Nuclear diameters range from about 3×10^{-13} cm for hydrogen to 1.9×10^{-12} cm for uranium. Since nearly all the mass of the atom is in the relatively small nucleus, its density is extremely high, about 10^{11} kg/cm^3! If all the empty spaces in the electron orbits of atoms making up the earth were eliminated, its diameter would be less than 600 m! The pressures in certain heavy stars are so high that their atoms are partially collapsed. For this reason Sirius B has a density of 50 kg/cm^3, and some stars are believed to have a density greater than 1,000 kg/cm^3 (18 tons/in.^{3})!

The simplest atom is the hydrogen atom, at. no. 1, consisting of but one proton in the nucleus and one orbital electron. Its atomic mass is approximately 1 atomic mass unit (amu). Figure A-44 is a representation of the hydrogen atom. The lone electron moves about the nucleus in an orbit we can assume to be circular. It is held in place by the force of attraction which exists between electric charges of opposite sign. Note that the orbit and the particles are *not* drawn to the same scale. If drawn to scale, the nuclear proton and the electron would have a diameter of only about 2×10^{-5} cm in the diagram.

Each successive element in the periodic table (see **Periodic law**) *has an additional proton plus neutrons in the nucleus and an additional orbital electron.* The atomic number of any element equals the number of protons in the nucleus and also the number of orbital electrons. Helium, at. no. 2, therefore has

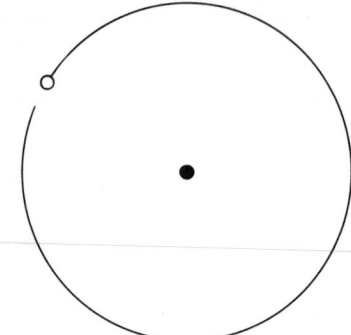

Fig. A-44. Highly simplified concept of a hydrogen atom.

two protons in the nucleus and two orbital electrons. Since its atomic mass is approximately 4, it follows that there are also two neutrons in the nucleus. The circular orbits of the two electrons are approximately equidistant from the nucleus (Fig. A-45), but because of the greater force of attraction of the double nuclear charge, they rotate at a smaller radius than the lone electron in hydrogen.

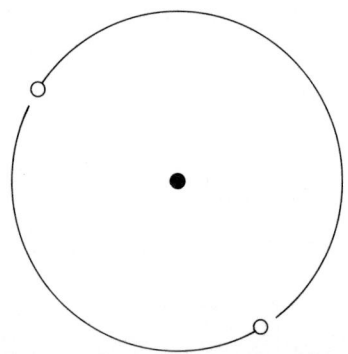

Fig. A-45. Highly simplified concept of a helium atom.

Lithium, at. no. 3, has an atomic mass of approximately 7. Four neutrons therefore normally accompany the three protons in the nucleus. (The number of neutrons in the nuclei of atoms of an element may vary from one isotope to another.) As with helium, two electrons rotate in circular orbits, but closer to the nucleus because of the greater attraction of the triple charge. The third electron is found to rotate in a complicated orbit at a greater average distance from the nucleus than the first two. This orbit, shown in Fig. A-46, cuts inside the circular orbits during each revolution and never closes on itself.

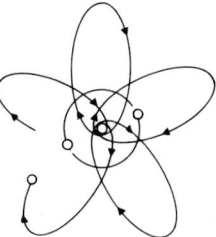

Fig. A-46. Highly simplified concept of a lithium atom.

According to a model consistent with the quantum theory, the electrons in any atom occupy orbits or states in one or more of several shells. Each shell is at a greater average distance from the nucleus. Closest to the nucleus is the K shell, which contains two electron orbits. The L shell contains two subshells with two and six orbits, respectively, or eight states in all. The successive shells and subshells and their states are M, $2 + 6 + 10 = 18$ states; N, $2 + 6 + 10 + 14 = 32$ states; O, $2 + 6 + 10 + 14 + 18 = 50$ states; P, $2 + 6 + 10 + 14 + 18 + 22 = 72$ states; and Q, $2 + 6 + 10 + 14 + 18 + 22 + 26 = 98$ states. No atom has enough electrons to occupy all these states. Electrons normally occupy first those states with lowest energy. Although states in succeeding shells have greater *average* energy, there is considerable overlap of energy levels between the M, N, O, P, and Q shells. The first eight states in each shell having more than eight states have lower energy than any states in succeeding shells. These eight states are therefore filled in unexcited atoms before electrons occupy any of the states in the succeeding shell. Additional electrons may be added to a shell after those states of lower energy level have been occupied in succeeding shells.

In heavy atoms the K-shell electrons are held relatively close to the nucleus by the strong electrostatic field of the nuclear charge. This is one factor preventing the existence of extremely heavy atoms, for the innermost electron orbits would lie partly inside the nucleus.

The manner in which the shells are filled and the correlation between the shell structure and chemical properties of the elements is clearly indicated by a comparison of the periodic table (see **Periodic law**) with Table A-57. The first period in the periodic table corresponds to filling the K shell, the second period to filling the L shell, and the third period to filling the first eight states in the M shell. Note from the periodic table that each element in group IA has one electron in the outermost occupied shell; group IIA, two electrons; group IIIB, three electrons; and so on. Since elements in each group have similar chemical properties, it follows that *the chemical properties of an element are determined primarily by the number of electrons in the outermost occupied shell.* Each element in group VIIIB except helium has eight electrons in the outermost occupied shell. In helium the outermost shell is complete with two electrons. All elements in this column are inert gases; that is to say, they are gases that do not readily react chemically to form compounds.

There are 18 elements in the fourth period. The 10 additional elements, numbers 21–30, are due to the completion of the M shell with 10 additional electrons. Each of these 10 elements has one or two electrons in the outer (N) shell. Iron, cobalt, and nickel, elements 26–28, are grouped together because of their very similar properties. Period 5 is similar to period 4, elements 39–48 corresponding to the addition of 10 more electrons to the N shell.

Period 6 includes 32 elements. The additional 14 elements are due to the completion of the N shell during the period. The 14 elements 58–71 are the first group of the rare earths. They are grouped together in the table because of their very similar chemical properties. Period 7 is similar to period 6. The rare earths in period 7 include thorium, uranium, and the artificially produced transuranium elements.

Chemical compounds may be formed by means of a sharing of the outer or valence electrons of the atoms. Since eight electrons in the outermost shell is a very stable configuration, many compounds share valence electrons in such a way that the total number of outer electrons is eight. Potassium chloride, KCl, is an example. The one N electron of potassium and the seven M electrons of chlorine make a total of eight electrons shared by the two atoms. In water molecules, H_2O, the six L electrons of oxygen and the K electron from each hydrogen atom are shared.

Ionization of Atoms: The electrons in an atom normally occupy the states of lowest energy level.

TABLE A-57. *Electronic Shell Structure of the Elements*

Period	Element		Shell Electrons						
		K_2	L_8	M_{18}	N_{32}	O_{50}	P_{72}	Q_{98}
1	1 H	1						
	2 He	2						
2	3 Li	2	1					
	4 Be	2	2					
	5 B	2	3					
	6 C	2	4					
	7 N	2	5					
	8 O	2	6					
	9 F	2	7					
	10 Ne	2	8					
3	11 Na	2	8	1				
	12 Mg	2	8	2				
	13 Al	2	8	3				
	14 Si	2	8	4				
	15 P	2	8	5				
	16 S	2	8	6				
	17 Cl	2	8	7				
	18 Ar	2	8	8				
4	19 K	2	8	8	1			
	20 Ca	2	8	8	2			
	21 Sc	2	8	9	2			
	22 Ti	2	8	10	2			
	23 V	2	8	11	2			
	24 Cr	2	8	13	1			
	25 Mn	2	8	13	2			
	26 Fe	2	8	14	2			
	27 Co	2	8	15	2			
	28 Ni	2	8	16	2			
	29 Cu	2	8	18	1			
	30 Zn	2	8	18	2			
	31 Ga	2	8	18	3			
	32 Ge	2	8	18	4			
	33 As	2	8	18	5			
	34 Se	2	8	18	6			
	35 Br	2	8	18	7			
	36 Kr	2	8	18	8			
5	37 Rb	2	8	18	8	1		
	38 Sr	2	8	18	8	2		
	39 Y	2	8	18	9	2		
	40 Zr	2	8	18	10	2		
	41 Nb	2	8	18	12	1		
	42 Mo	2	8	18	13	1		
	43 Tc	2	8	18	13	2		
	44 Ru	2	8	18	15	1		
	45 Rh	2	8	18	16	1		
	46 Pd	2	8	18	18	0		
	47 Ag	2	8	18	18	1		
	48 Cd	2	8	18	18	2		
	49 In	2	8	18	18	3		
	50 Sn	2	8	18	18	4		
	51 Sb	2	8	18	18	5		
	52 Te	2	8	18	18	6		

TABLE A-57. *Electronic Shell Structure of the Elements* (*Continued*)

Period	Element	Shell Electrons						
		K_2	L_8	M_{18}	N_{32}	O_{50}	P_{72}	Q_{98}
5	53 I	2	8	18	18	7		
	54 Xe	2	8	18	18	8		
6	55 Cs	2	8	18	18	8	1	
	56 Ba	2	8	18	18	8	2	
	57 La	2	8	18	18	9	2	
	58 Ce	2	8	18	20	8	2	
	59 Pr	2	8	18	21	8	2	
	60 Nd	2	8	18	22	8	2	
	61 Pm	2	8	18	23	8	2	
	62 Sm	2	8	18	24	8	2	
	63 Eu	2	8	18	25	8	2	
	64 Gd	2	8	18	25	9	2	
	65 Tb	2	8	18	27	8	2	
	66 Dy	2	8	18	28	8	2	
	67 Ho	2	8	18	29	8	2	
	68 Er	2	8	18	30	8	2	
	69 Tm	2	8	18	31	8	2	
	70 Yb	2	8	18	32	8	2	
	71 Lu	2	8	18	32	9	2	
	72 Hf	2	8	18	32	10	2	
	73 Ta	2	8	18	32	11	2	
	74 W	2	8	18	32	12	2	
	75 Re	2	8	18	32	13	2	
	76 Os	2	8	18	32	14	2	
	77 Ir	2	8	18	32	15	2	
	78 Pt	2	8	18	32	17	1	
	79 Au	2	8	18	32	18	1	
	80 Hg	2	8	18	32	18	2	
	81 Tl	2	8	18	32	18	3	
	82 Pb	2	8	18	32	18	4	
	83 Bi	2	8	18	32	18	5	
	84 Po	2	8	18	32	18	6	
	85 At	2	8	18	32	18	7	
	86 Rn	2	8	18	32	18	8	
7	87 Fr	2	8	18	32	18	8	1
	88 Ra	2	8	18	32	18	8	2
	89 Ac	2	8	18	32	18	9	2
	90 Th	2	8	18	32	18	10	2
	91 Pa	2	8	18	32	20	9	2
	92 U	2	8	18	32	21	9	2
	93 Np	2	8	18	32	22	9	2
	94 Pu	2	8	18	32	23	9	2
	95 Am	2	8	18	32	24	9	2
	96 Cm	2	8	18	32	25	9	2
	97 Bk	2	8	18	32	26	9	2
	98 Cf	2	8	18	32	27	9	2
	99 Es	2	8	18	32	29	8	2
	100 Fm	2	8	18	32	30	8	2

SOURCE: W. E. Forsythe, "Smithsonian Physical Tables," 9th ed., pp. 622–623, Smithsonian Institution, Washington, D.C., 1954; and current literature.

It is possible, however, to move one or more electrons to states of higher energy level by supplying an amount of energy equal to the difference in the energy levels of the states involved. The atom is then said to be in an *excited* state. If sufficient energy is supplied, one or more electrons may be completely removed from the atom, in which case it is said to be *ionized*. It is convenient to express electron energies in *electron volts* (eV), where an electron volt is the amount of work required to move one electronic unit of charge through a potential difference of one volt. Since the charge on an electron is 1.60210×10^{-19} C, it follows that 1 eV $= 1.60210 \times 10^{-19}$ J. The ionization energies of several elements are listed in Table A-58. In general, the energies are lowest for the alkali metals, highest for the inert gases.

TABLE A-58. *Ionization Energies of Certain Elements* (*Atomic Form*)

Element	Ionization Energy, eV
Cesium	3.894
Potassium	4.341
Radon	10.748
Xenon	12.130
Hydrogen	13.598
Oxygen	13.618
Nitrogen	14.534
Argon	15.759
Neon	21.564
Helium	24.587

SOURCE: National Bureau of Standards, Washington, NSRDS-NBS 34, 1970.

The excitation or ionization energy may be supplied in any one of several ways. A strong electric field may supply the necessary energy. Such a field may be set up either by a high voltage applied between electrodes or by an ion passing through or near the atom. The energy may also be supplied thermally, or by electromagnetic radiation such as light, ultraviolet, or x- or gamma rays, or most commonly, by atomic collision. When excited electrons fall back to their normal states, the excess energy is liberated in the form of a pulse of electromagnetic radiation, or photon. Each element is characterized by a unique radiation pattern, or *spectrum,* when its atoms are excited. Systematic study of atomic spectra is largely responsible for a detailed knowledge of the electron states.

The Spectrum of Hydrogen: If a beam of light is passed through a prism or diffraction grating, the colors contained in the light are separated into a spectrum. If all colors are present, the spectrum is a *continuous spectrum.* If only a limited number of colors are present, each characterized by a certain wavelength or narrowband of wavelengths, the spectrum is called a *line spectrum.* If an atomic gas is made luminous by heating it to a high temperature or by passing an electric current through it, and it is viewed through a spectroscope, a line spectrum is seen. Each kind of gas is characterized by a different spectrum. The visible spectrum of atomic hydrogen is shown in Fig. A-47a. Other atomic spectra are more complicated (Fig. A-47b), and the molecular spectra are still more complicated (Fig. A-47c). The wavelengths of lines in atomic and molecular spectra may be measured precisely.

When a continuous band of radiation is passed through a gas, certain wavelengths are absorbed by the gas. The resulting light with these lines removed is called an *absorption spectrum.* Lines in the absorption spectrum of the gas are found to have the same wavelengths as certain of the lines in the emission spectrum. The spectrum of the sun is seen in this way. The glowing mass of the sun emits a continuous spectrum, certain lines of which are absorbed in the sun's atmosphere. The element helium was recognized in the solar spectrum before its discovery on the earth, and some 30 other elements have been identified in the sun's atmosphere.

In 1885, Johann J. Balmer succeeded in writing a simple formula for the wavelengths of the visible lines of the hydrogen spectrum. Balmer's equation may be written in the form

$$\frac{1}{\lambda} = R\left(\frac{1}{2^2} - \frac{1}{n'^2}\right) \tag{1}$$

where n' may have any integral value greater than 2, and R is the Rydberg constant, 1.09678×10^7 m^{-1}. Later, other series of lines were discovered for hydrogen, all of which are given by the formula

$$\frac{1}{\lambda} = R\left(\frac{1}{n^2} - \frac{1}{n'^2}\right) \tag{2}$$

For each value of n $(1, 2, 3, \ldots)$ a series of lines is given by substituting all integral values of n' greater than n in the equation. The Balmer, or visible, series is given by $n = 2$. The Lyman series of lines in the ultraviolet region is given by $n = 1$. Series of lines in the infrared region are given by $n = 3$ (Paschen series), $n = 4$ (Brackett series),

$n = 5$ (Pfund series), and $n = 6$ (Humphrey series).

Of all the elements, hydrogen is the only one with a spectrum that can be represented by one simple formula, but W. Ritz was able to show in 1908 that the wavelengths of the many lines in any spectrum can be expressed in terms of the differences of a relatively few numbers characteristic of the particular element or compound.

Bohr's Model of the Hydrogen Atom: By bombarding thin metal foils with the alpha particles emitted by certain radioactive substances and noting that most of them passed through the foil with little or no deflection, Ernest Rutherford had concluded in 1911 that the atom consists of a very small dense nucleus surrounded by electrons rotating in orbits at a relatively large distance from the nucleus. One aspect of this model was puzzling to physicists of the time. According to classical electromagnetic theory, an electric charge rotating in an orbit must continuously radiate energy at a frequency equal to the frequency of rotation. An electron would therefore radiate energy at a steadily increasing frequency until it fell into the nucleus. It was known, of course, that atoms radiate only when excited, and then only certain discrete frequencies rather than the continuous band demanded by the classical theory. In 1913, Niels Bohr proposed his model of the hydrogen atom which gave an explanation of the then known phenomena. Bohr assumed that the electrons could occupy only certain discrete states around the nucleus, and that instead of radiating continuously, energy is emitted in the form of a photon when an electron falls from one state to another of lower energy level. Bohr calculated the states and found that he could get a formula similar to Eq. (2) for the characteristic wavelengths, provided the angular momentum of the allowed states is given by

$$mvr = \frac{nh}{2\pi} \tag{3}$$

where $n = 1, 2, 3, \ldots$ is the number of the orbit. He was able to derive the numerical value of the Rydberg constant of Eq. (1) in terms of the mass and charge of the electron, Planck's constant, the velocity of light, and the permittivity of free space, all of which could be measured independently of the hydrogen atom or its spectrum. The calculated and measured values agreed within experimental accuracy.

Figure A-48 shows the first five Bohr orbital states of the hydrogen atom. Bohr calculated the

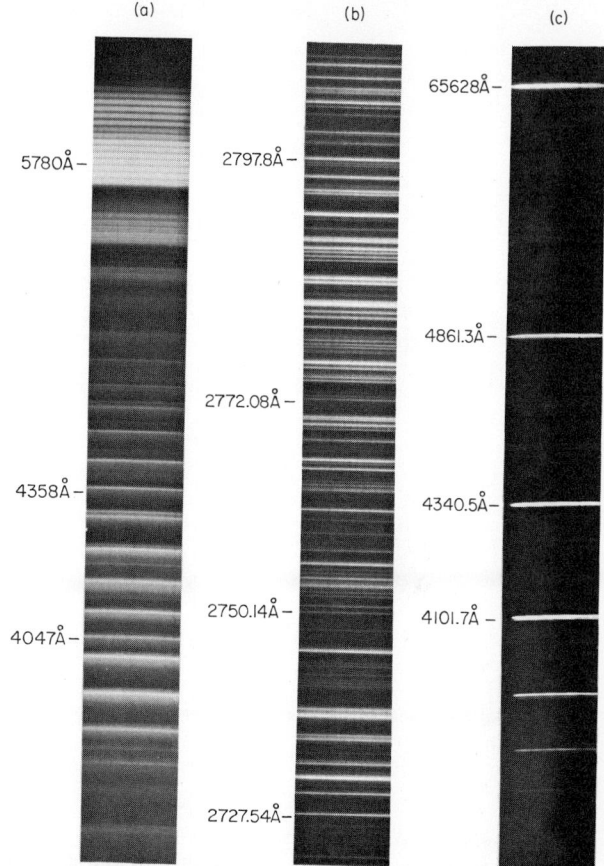

Fig. A-47. Atomic and molecular spectra. (a) A portion of the visible spectrum of molecular chlorine, (b) a portion of the ultraviolet spectrum of iron, (c) the Balmer (visible) series for hydrogen. (*Courtesy of Sidney H. Kalmbach.*)

diameter d_0 of the first orbit, or ground, state to be 1.06×10^{-10} m, a value which agrees reasonably well with currently accepted values of the diameter of the hydrogen atom. Diameters of the other orbits are given by $d = n^2 d_0$. The energies of electrons in the orbits are given by

$$E_n = -\frac{Rch}{n^2} \tag{4}$$

where the energy of an electron outside the atom ($n = \infty$) is zero, and the negative sign shows that electrons in orbits have less energy. When an excited electron falls to a state closer to the ground state, energy equal to the difference between the energy levels of the two states is emitted as a

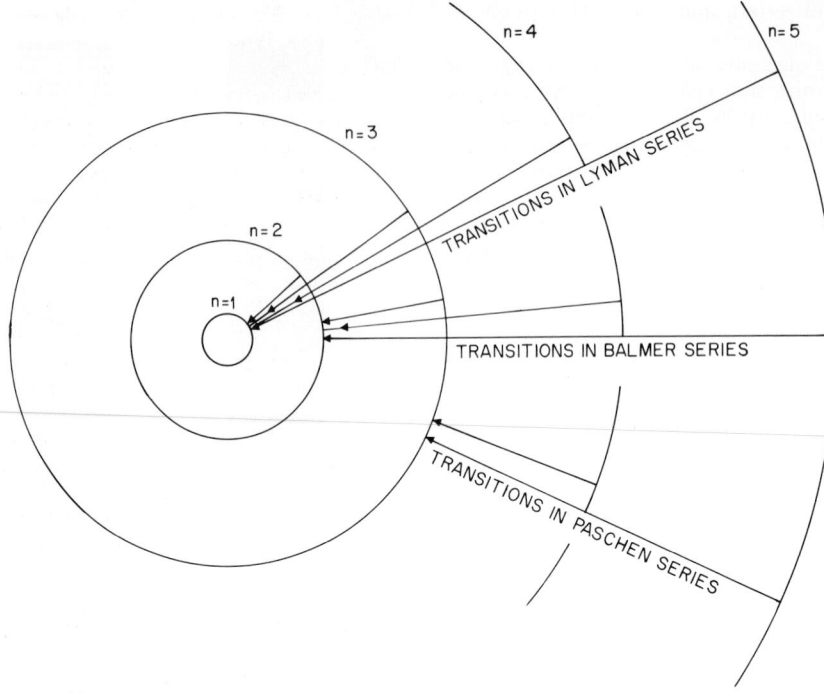

Fig. A-48. Inner orbital states of hydrogen atoms. Only a portion of the fourth and fifth states is shown. The lower transitions in three of the spectral series are shown.

photon. As shown by Figs. A-48 and A-49, an excited electron may fall directly to the ground state or may fall in two or more steps. Electrons falling directly to the ground state ($n = 1$) emit photons in the Lyman (ultraviolet) series; electrons falling to the second state emit photons in the Balmer series; and so on.

The absorption spectrum is formed when photons of particular wavelengths are absorbed, the energy being used to raise electrons to a higher energy level. Note that hydrogen atoms in the ground state cannot absorb visible light directly. Ultraviolet photons must first raise the electrons from $n = 1$ to $n = 2$. An absorption spectrum therefore usually does not include prominently all lines seen in the emission spectrum.

The spectra of the heavier atoms are much more complicated than the hydrogen spectrum. The large numbers of electrons in these atoms is one factor leading to this greater complexity. Because of the greater nuclear charge, the energies associated with the inner orbits are much greater than for hydrogen. Spectral lines of heavy atoms therefore extend through the ultraviolet and x-ray regions.

Extension of the Bohr Model—Wave Mechanics: The study of atomic and molecular spectra continued during the years following the development of Bohr's model of the atom. New features of spectra were discovered that could not be explained by the Bohr model.

In 1924 Louis de Broglie postulated that moving particles have properties characteristic of electromagnetic waves and that the wavelength associated with the moving particle is given by

$$\lambda = \frac{h}{mv} \tag{5}$$

where h = Planck's constant. Electrons passing through crystalline solids are diffracted much as are the energetic electromagnetic waves called x-rays, confirming the wave nature of electrons. If mv is eliminated between Eqs. (3) and (5), Bohr's condition for hydrogen states may be written

$$2\pi r = n\lambda \tag{6}$$

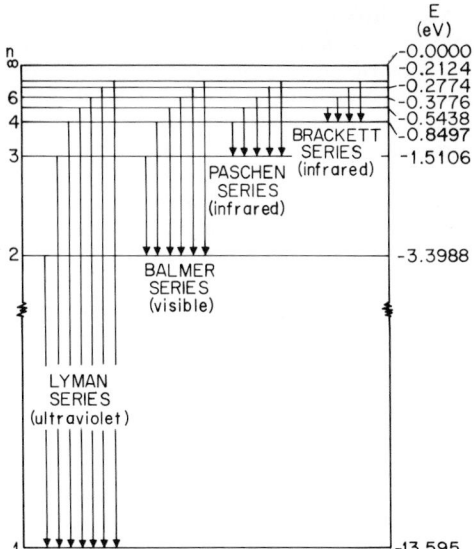

Fig. A-49. Energy levels in the hydrogen atom. $E_n = -13.595/n$ eV. The Lyman series is ultraviolet; Balmer series, visible; Paschen and Brackett series, infrared.

In words, this equation states that *the circumference of each of the hydrogen orbits is equal to an integral number of wavelengths of the electron.* The situation is somewhat analogous to the vibrations of a string fixed at both ends: there must be an integral number of wavelengths so that the wave can "fit" in the circular orbit.

As a result of de Broglie's postulate and mathematical extensions worked out by Werner Heisenberg, Erwin Schrödinger, and others, the *wave-,* or *quantum-mechanics,* model of the atom was developed. This theory is able to explain all phenomena associated with the electronic states that were known then or have since been discovered. In this theory, the electrons are viewed as waves that fill the regions corresponding to Bohr's orbits. This explains why there is electromagnetic radiation from atoms only when electrons jump from one state to another. As long as electrons remain in given states, the charge density throughout the region of the orbits remains constant. There is no change in the electric charge at any point, as there would be if the electrons were small charged spheres rotating in orbits, and so there is no radiation.

Although the orbits have lost much of the physical significance assigned them by Bohr, it is convenient to retain the term. Each state may be defined in terms of a set of four *quantum numbers: n, l, m,* and *s.* The *principal quantum number n* defines the shell in which the electron is located. Thus, for the K shell, $n = 1$; for L, $n = 2$, for M, $n = 3$; etc. The *azimuthal quantum number l* determines the orbital angular momentum of the electron and, together with n, the eccentricity of the orbit. For most nearly circular orbits, $l = n - 1$. The smaller the value of l, the greater the eccentricity. In magnitude, l can take on only positive integral values less than n. Thus l must be zero in the K shell; may be either 0 or 1 in the L shell; 0, 1, or 2 in the M shell; etc. The *orientation quantum number m* defines the orientation of the magnetic field of the orbit. In magnitude, m may have any integral value from $-l$ to $+l$. Thus, if $l = 2$, m may be -2, -1, 0, $+1$, or $+2$. The *spin quantum number s* defines the direction of the electron spin about its own axis. The two possible spin momenta are $+\frac{1}{2}h$ and $-\frac{1}{2}h$, corresponding to spin in opposite directions. Numerically, h equals Planck's constant divided by 2π.

The number of possible states in any atom is limited by *Pauli's exclusion principle,* which states that *no two electrons in an atom may have identical sets of quantum numbers.* This quantum mechanical law imposes a limit on the number of states in each shell of an atom. In the K shell $n = 1$. Since states l and m are both zero when $n = 1$, the only possible combinations of quantum numbers n, l, m, and s are 1, 0, 0, $+\frac{1}{2}$, and 1, 0, 0, $-\frac{1}{2}$. The K shell can then contain at the most two electrons. In the L shell $n = 2$, l may be 0 or 1. When $l = 1$, m may be -1, 0, or 1. Again s may be either $+\frac{1}{2}$ or $-\frac{1}{2}$. There are eight possible combinations of these numbers; therefore the L shell may contain up to eight electrons.

The quantum numbers of the states in the K, L, and M shells are given in Table A-59. The subshell groupings in each shell are also indicated. The successive subshells each have four additional electrons, and each shell has one more subshell than is contained in the previous one. Each group of subshells (s, p, d, f) has a characteristic set of spectral lines from which the subshells receive their designation. Note that the stable configuration of eight electrons in the outer orbit of the inert gases consists of completed s and p subshells. See also **Atomic number; Atomic weight; Ceramics; Chemical elements; Nuclear power plants;** and **Periodic law.**

[*Note:* The information in this description was extracted with minor changes from David B. Hoisington, "Nucleonic Fundamentals," chap.

TABLE A-59. *Quantum Numbers of K, L, and M Shell Electrons*

Shell	Subshell	Quantum Numbers			
		n	l	m	s
K	1s	1	0	0	$+\frac{1}{2}$
		1	0	0	$-\frac{1}{2}$
L	2s	2	0	0	$+\frac{1}{2}$
		2	0	0	$-\frac{1}{2}$
	2p	2	1	-1	$+\frac{1}{2}$
		2	1	-1	$-\frac{1}{2}$
		2	1	0	$+\frac{1}{2}$
		2	1	0	$-\frac{1}{2}$
		2	1	1	$+\frac{1}{2}$
		2	1	1	$-\frac{1}{2}$
M	3s	3	0	0	$+\frac{1}{2}$
		3	0	0	$-\frac{1}{2}$
	3p	3	1	-1	$+\frac{1}{2}$
		3	1	-1	$-\frac{1}{2}$
		3	1	0	$+\frac{1}{2}$
		3	1	0	$-\frac{1}{2}$
		3	1	1	$+\frac{1}{2}$
		3	1	1	$-\frac{1}{2}$
	3d	3	2	-2	$+\frac{1}{2}$
		3	2	-2	$-\frac{1}{2}$
		3	2	-1	$+\frac{1}{2}$
		3	2	-1	$-\frac{1}{2}$
		3	2	0	$+\frac{1}{2}$
		3	2	0	$-\frac{1}{2}$
		3	2	1	$+\frac{1}{2}$
		3	2	1	$-\frac{1}{2}$
		3	2	2	$+\frac{1}{2}$
		3	2	2	$-\frac{1}{2}$

2, McGraw-Hill Book Company, New York, 1959 (with permission of the publisher).]
—David B. Hoisington, *Naval Postgraduate School, Monterey, Calif.*

Atomic Weight The atomic weight (or *atomic mass*) of an element is a *relative number* (for which one specific element is used as a standard) rather than an absolute number. The practical importance of atomic-weight data pertains to stoichiometric calculations where, with prior knowledge of the equations of reaction, the weight proportions of both reactants and products of the reaction can be predicted. See **Chemical elements** for a tabulation of the atomic weights of all known chemical elements. *Formula weights* (simply the sum of the weights of the elemental constituents) for hundreds of chemical compounds are given in numerous tables throughout this volume.

Avogadro's law (1811) states that *equal volumes of ideal gases at the same temperature and pressure contain the same number of molecules.* This law provided the basis for comparing the weights of atoms and molecules. If equal volumes of hydrogen, helium, nitrogen, oxygen, fluorine, and neon gases, as examples, are compared, their relative masses are found to be very close to 2, 4, 28, 32, 38, and 20, respectively. If adjusted for the fact that He and Ne are monatomic gases whereas H_2, N_2, O_2, and F_2 are diatomic, the relative masses of these atoms then become close to 1, 4, 14, 16, 19, and 20. For elements that cannot be converted readily into gaseous form, such as uranium which has a boiling point in excess of $3800°C$, a simple compound often can be found for use in such determinations. For example, uranium hexafluoride, UF_6, will sublime at $56°C$. Since the information on fluorine is known accurately, the relative mass of ^{238}U can be inferred from the determined weight of the compound (352) by subtracting $6 \times 19(F)$.

Since the International Atomic Weight tabulation of 1961, carbon has been used as the reference element—with a relative mass of 12. Earlier tabulations had been based upon oxygen with a relative mass of 16.0000. Over the years, there

TABLE A-60. *Comparison of Atomic Weight Tables of 1923 and 1966 (Abridged)*

Element	1923 Values O = 16.000	1966 Values C = 12
Aluminum	26.97	26.9815
Barium	137.37	137.34
Calcium	40.07	40.08
Carbon	12.000	12.01115
Copper	63.57	63.546
Fluorine	19.000	18.9984
Gold	197.2	196.967
Helium	4.00	4.0026
Iodine	126.932	126.9044
Iron	55.84	55.847
Lead	207.20	207.19
Mercury	200.61	200.59
Nickel	58.69	58.71
Nitrogen	14.008	14.0067
Oxygen	16.000	15.9994
Phosphorus	31.027	30.9738
Potassium	39.096	39.102
Silver	107.880	107.868
Sodium	22.997	22.9898
Sulfur	32.064	32.064
Tin	118.70	118.69
Tungsten	184.00	183.85
Zinc	65.38	65.37

have been rather slight variations in atomic-weight figures because of the effects of new and better information, resulting from improved procedures and, particularly, from increased information pertaining to isotopes and their relative abundance. Changing from oxygen to carbon also created certain minor alterations. A comparison of the Atomic Weight Table for 1923 (oxygen = 16.000) with the Atomic Weight Table for 1966 (carbon = 12) indicates that the alterations have been relatively minor. See Table A-60.

Actually, atomic-weight figures in most instances are average figures because most elements are composed of two or more stable isotopes, each with a different atomic mass. Isotopes are described under **Chemical elements.**

In calculations, the term *gram atom* or *gram atomic mass* may be used. A gram atom simply is an amount of the element, in grams, equal to the atomic weight. Thus a gram atom of barium, for example, will weigh 137.34 g.

Attrition   (See **Size reduction.**)

Attrition Mills   (See **Mixing and blending, solids/solids.**)

Austenite   (See **Iron and steel.**)

Autoxidation   (See **Antioxidants;** and **Peroxides, organic.**)

Autunite   (See **Uranium.**)

Avogadro's Number   Even though Amedeo Avogadro (ca. 1811) had no experimental means available to determine the figure, he predicted that the number of atoms in a gram atom and the number of molecules in a mole for all elements and compounds should be the same. Avogadro's law, which states that equal volumes of ideal gases at the same temperature and pressure contain the same number of molecules, inferred the foregoing conclusion. As the result of several different determinations, the number (known as Avogadro's number) N_0 is 6.0247×10^{23} atoms per gram atom and 6.0247×10^{23} molecules per mole.

Azeotropic Distillation   (See **Distillation.**)

Azines   (See **Ketones.**)

Azlon   (See **Fibers.**)

B

Babbitt (See **Antimony;** and **Tin.**)

Backscattering (See **Radioisotopes.**)

Bacterial Cultures (See **Amino acids;** and **Peptides and proteins.**)

Bacterial Sterilization (See **Pasteurization.**)

Bacterial Waste Treatment (See **Waste treatment.**)

Bacterium Aceti (See **Acetic acid;** and **Carboxylic acids.**)

Bag Filters (See **Separation operations.**)

Bakelite (See **Phenolic resins;** and **Resins.**)

Baker's Yeast (See **Yeasts.**)

Ball Mill (See **Paint;** and **Size reduction.**)

Banbury Mixer (See **Mixing, pastes.**)

150

Barium (Gk. *barys,* heavy.) **Ba** $= 137.34$ (at. wt.); 56 (at. no.). Barium is a silvery-white metal of the alkaline-earth group (IIA) of the periodic table. Chemically, it is similar to its neighbors calcium and strontium, having a valence of $2+$ and being very reactive and forming compounds having high free energy of formation. The metal is relatively soft and ductile and may be readily worked.

Barium was first produced as a mercury amalgam by Sir Humphry Davy in 1808. A relatively pure form of the metal was produced by Guntz in 1905 by distillation of the amalgam.

Barium occurs chiefly in nature as the sulfate baryta or barytes, and as the carbonate witherite. These ores are mined chiefly for production of Ba compounds since very little metal is used commercially as such. The metal *is produced by thermal reduction of the oxide with aluminum metal in a high-temperature, high-vacuum, closed-retort process similar to that used in the commercial production of calcium.

*The metal is produced commercially in the United States by Pfizer Inc., and in Canada by Dominion Magnesium, Ltd.

The basic process for this reduction is described in the literature and depends upon the reaction

$$4BaO + 2Al = BaOAl_2O_3 + 3Ba$$

in which the gaseous Ba is removed from the reaction by condensation.

The purity of the reduced metal is a function of the starting oxide purity and the process conditions.

As with calcium, Ba is sufficiently reactive to render it difficult to produce and to handle in a relatively pure state. For this reason it is difficult to measure many of its physical properties with accuracy. Table B-1 is a compilation of the most reliable values from the literature.

TABLE B-1. *Selected Physical Properties of Barium*

Chemical valence	2+
Isotopes (stable)	130, 132, 134, 135, 136, 137, 138
Electron configuration.	2-8-18-18-8-2
Density, g/cm^3, at 20°C	3.5
Melting point, °C	725 ± 5
Boiling point, °C	1637
Specific heat, cal/g, at 20°C . . .	0.068
Heat of fusion, cal/g	13.3
Heat of vaporization, cal/g, at 25°C	304
Thermal expansion coefficient, in./(in.)(°C), 0–100°C	18 × 10^{-6}
Electrical resistivity, 10^{-6} Ω-cm, at 20°C	33
Crystal form	Body-centered cubic
Lattice constant, Å, at 20°C . . .	5.025
Closest approach of atoms, Å . .	4.348
Electron work function, eV	2.5

There is little commercial demand for Ba metal. Numerous patents and other literature describe its use, usually in alloy form, as a getter in vacuum equipment to tie up trace gases to very low partial pressures. It is used also as an electron emitter usually in a compound form. Combined with calcium, silicon, and iron, Ba is used in treating molten metals.

Barium compounds are used widely in industry for ceramics, in hard magnet ferrites, glass, lubricants, and pigments, as a purifying agent in chemical reactions, and as heat-treating media (molten-salt baths).

Barium and all its compounds are highly toxic to human beings. Barium sulfate, however, is so insoluble that it can be ingested into the body without harm, and in medicine it is used as an x-ray opaque medium for diagnostic purposes.

REFERENCES
American Society for Metals: "Metals Handbook," 8th ed., vol. 1, Metals Park, Ohio, 1961.
Barium Bibliography, FMC Corp., New York, 1961.
Hampel, C. A. (ed.): "Rare Metals Handbook," 2d ed., Van Nostrand, New York, 1961.
Rashid, and Kayer: *J. Less Common Metals,* vol. 24, 1971.

—Frank Emley, *Pfizer Metals & Composite Products, Pfizer Inc., Wallingford, Conn.*

Base   In a very general way, a base may be defined as a substance that yields hydroxyl, OH$^-$, ions. Bases generally are active chemically (corrosive), and bitter to the taste and are electrolytes. Water also furnishes hydroxyl ions, but in equal amounts with hydrogen ions; hence pure water is a neutral substance rather than a base or an acid. The ionization of bases (and acids) is described under **pH (hydrogen ion concentration).** Alkalization (not to be confused with alkylation) is the operation of causing hydroxyl ions to be in excess, usually accomplished by adding a base to a neutral or acid material. Neutralization is the operation of creating a balance between the hydroxyl and hydrogen ions present. Thus a base may be neutralized by adding a suitable acidic substance in proper amounts.

The common inorganic bases or alkalies are termed *hydroxides,* such as potassium hydroxide and sodium hydroxide; they are high-tonnage industrial chemicals, described elsewhere in this volume. Consult the Subject Index. Quaternary ammonium derivatives, such as tetraethylammonium hydroxide, $N(C_2H_5)_4 \cdot OH$, are examples of organic hydroxides. See **Amines.**

Basic Oxygen Steelmaking   (See **Ironmaking and steelmaking;** and **Oxygen.**)

Bastanite   (See **Cerium; Lanthanum; Neodymium; Praseodymium; Rare-earth elements and metals;** and **Samarium.**)

Batteries   A battery consists of two or more series- or parallel-connected galvanic cells. A *primary* galvanic cell converts chemical energy directly into electric energy and consists of two electrodes of dissimilar material isolated from one another electronically, in a common ionically conductive electrolyte. The electrolyte may be solid or liquid, but usually is an aqueous salt solution. If the cell is of a *secondary,* or rechargeable, variety, input

TABLE B-2. *Principal Types of Batteries and Characteristics*

Electrochemical System	Negative Electrode	Positive Electrode	Electrolyte	Type	Overall Equations of Reaction
Zinc–manganese dioxide (usually called Leclanché or carbon–zinc)	Zinc	Manganese dioxide	Aqueous solution of ammonium chloride, zinc chloride	Primary	$2MnO_2 + 2NH_4Cl + Zn \rightarrow$ $ZnCl_2 \cdot 2NH_3 + H_2O + Mn_2O_3$
Zinc–alkaline–manganese dioxide	Zinc	Manganese dioxide	Aqueous solution of potassium hydroxide	Primary	$2Zn + 2KOH + 3MnO_2 \rightarrow$ $2ZnO + 2KOH + Mn_3O_4$
				Rechargeable	$Zn + KOH + 2MnO_2 \rightleftarrows$ $ZnO + Mn_2O_3 + KOH$
Zinc–mercuric oxide	Zinc	Mercuric oxide	Aqueous solution of potassium hydroxide	Primary	$Zn + HgO + KOH \rightarrow ZnO + Hg + KOH$
Zinc–silver oxide	Zinc	Monovalent silver oxide	Aqueous solution of potassium hydroxide or sodium hydroxide	Primary	$Zn + Ag_2O + KOH \rightarrow ZnO + 2Ag + KOH$
		Divalent silver oxide	Aqueous solution of potassium hydroxide	Rechargeable	$Zn + Ag + KOH \leftrightharpoons ZnO + Ag + KOH$
Lead–lead dioxide (usually called lead-acid)	Lead	Lead dioxide	Aqueous solution of sulfuric acid	Rechargeable	$2Pb + 2PbO_2 + 2H_2SO_4 + H_2O \rightleftarrows$ $PbSO_4 + 2PbO + 3H_2O$
Nickel–cadmium	Cadmium	Nickelic hydroxide	Aqueous solution of potassium hydroxide	Rechargeable	$Cd + 2NiOOH + KOH + 2H_2O \rightleftarrows$ $Cd(OH)_2 + 2Ni(OH)_2 + KOH$
Nickel–iron	Iron	Nickelic hydroxide	Aqueous solution of potassium hydroxide	Rechargeable	$Fe + 2NiOOH + KOH + 2H_2O \rightleftarrows$ $Fe(OH)_2 + 2Ni(OH)_2 + KOH$
Magnesium–manganese dioxide	Magnesium	Manganese dioxide	Aqueous solution of magnesium perchlorate	Primary	$2Mg + 2MnO_2 + 4H_2O + Mg(ClO_4) \rightarrow$ $Mn_2O_3 \cdot H_2O + 2Mg(OH)_2 + Mg(ClO_4) + H_2$
Silver–cadmium	Cadmium	Divalent silver oxide	Aqueous solution of potassium hydroxide	Rechargeable	$Cd + AgO + KOH + H_2O \rightleftarrows$ $Cd(OH)_2 + Ag + KOH$
Zinc–air (oxygen)	Zinc	Oxygen	Aqueous solution of potassium hydroxide	Primary	$2Zn + O_2 + 4KOH + 2H_2O \rightarrow 2K_2Zn(OH)_4$

Note: Wh = watt hours

Nominal Voltage per Cell	Typical Commerical Service Capacities	Input if Rechargeable	Energy Density (Commercial)		Features	Limitations
			Wh/lb	Wh/in.³		
1.5	Several hundred mAh to 30 Ah		5–40	1–3	Low cost; variety of shapes and sizes; excellent shelf life	Efficiency decreases at high current drains; poor low-temperature performance
1.5	Several hundred mAh to 23 Ah		20–40	2–3	High efficiency under moderate continuous drain conditions; good low-temperature performance; low impedance; long shelf life	Expensive for low drains
		Approximately 100% of energy withdrawn	10	1.0–1.2		Rechargeable-limited cycle life; voltage-limited taper current charging
1.35	16 mAh–14 Ah		10–50	4–8	High service capacity/ volume ratio; flat voltage discharge characteristic; good high-temperature performance; good storage life	Poor low-temperature performance on some types
1.5	38–190 mAh		30–60	4–8	Moderately flat voltage discharge characteristic; good storage life	
1.8/1.5 (two-step)	Vented, 5 Ah to several thousand Ah	Minimum of 110% of energy withdrawn	40–70	2–8	High-energy density	Expensive; short storage life after activation; short cycle life; two-step discharge curve
2	Vented, 1–10,000 Ah	Minimum of 110% of energy withdrawn	Sealed, 10–15	0.8–1.1	Spill-resistant	Limited low-temperature performance; vented cells require servicing
			Vented, 7–12	0.5–2	Low cost	
1.25	Sealed, 20 mAh–100 Ah	Sealed, minimum of 140% of energy withdrawn	Sealed, 12–17	Sealed, 1–1.5	Excellent cycle life; flat voltage discharge characteristic; good high- and low-temperature performance; high resistance to shock and vibration; can be stored indefinitely in any charge state	High initial cost; only fair charge retention
	Vented—few Ah to over 500 Ah	Vented, minimum of 125–150% of energy withdrawn	Vented, 12–20	Vented, 1–1.5		
1.2	Vented, several hundred Ah	Minimum of 125% of energy withdrawn	8–14	0.9	Excellent cycle life; rugged	Poor charge retention; usually restricted to low-rate use; heavy gassing on overcharge
1.8	Sealed, up to 20 Ah		15–50	1–3	Excellent high-temperature storage life; good in moderate-drain continuous uses	Poor low-temperature performance; delay after circuit is closed before battery operates; not good for intermittent use
1.4	Sealed, up to 300 Ah	Minimum of 110% of energy withdrawn	22–34	1.8–2.5	Good energy/weight ratio; good charge retention; long wet-stand life	Expensive; poor low-temperature performance; two-step discharge curve
1.25	Vented, ½–2,000 Ah		80–100	3.2	Flat voltage discharge characteristic	Restricted to low-rate use

electric energy can be converted to chemical energy and thus stored. Generally, it is impractical to attempt to recharge a cell which is intended for primary use.

For any cell system to be commercially attractive, the electrode-electrolyte combination must be such that it will deliver a reasonable quantity of electric energy at useful voltage and current. The voltage exhibited by a particular system is the algebraic sum of the observed individual electrode emf values, compared with a standard, in the chosen electrolyte. As will be observed from Table B-2, the range of nominal voltages available per cell is not wide, all major cells being in the 1–2 V/cell span. The quantity of energy available per unit cell volume is almost solely a function of the electrochemical equivalents and densities of the active electrode materials selected. As the table shows, there is a much wider range among the major cells in this regard.

Aside from the effects of cell size, current is limited throughout discharge by cell-element configuration, activities of the electrode materials, reaction-product solubilities, and both ionic and electronic conductivities of all components.

The two electrodes of the galvanic cell are termed *negative* and *positive* here to avoid confusion which can result from use of the terms anode and cathode, the latter often being used loosely in the industry. The negative electrode of the primary (or charged secondary) cell is metallic and is oxidized (increased in valence) during discharge, giving up electrons to the external circuit. The positive electrode initially is an oxygen donor, usually a metal oxide, which is reduced as it receives electrons from the external circuit. Charge transfer from one electrode to the other within the cell is via the ions of the electrolyte salt. In certain cells, the electrolyte serves only as a charge carrier. In other systems, the electrolyte enters further into reactions and actually changes in composition during use. The overall equations of reaction for major primary and rechargeable cells are given in Table B-2. Plates for lead-acid cells are described under **Lead.**

Major points of difference between primary and secondary cells include (1) features of electrode design (including choice of active materials used) which assist in maintaining secondary-electrode integrity through extended cycling of charging and discharging; (2) use of auxiliary structural materials which are especially oxidation-resistant and durable in secondary cells; and (3) in the most refined, hermetically sealed secondary cells, inclu-

sion of materials in the system to prevent excessive gas-pressure buildup during any phase of use. Features designed to enhance rechargeability generally add less by proportion to consumer cost than they add to overall usefulness.

Cells designed for complete portability range in utility from the commonplace, inexpensive Leclanché primary cell to the hermetically sealed nickel-cadmium cell, which will serve very efficiently through hundreds of full-capacity cycles. Testing of portable batteries is shown in Fig. B-1.

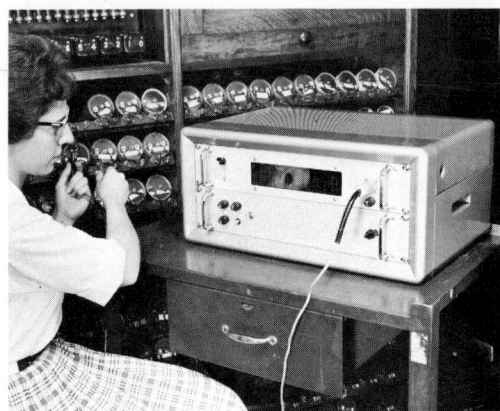

Fig. B-1. Testing electrical characteristics of portable batteries. (*Union Carbide Corp.*)

Cells are marketed which deliver as little as a few milliampere-hours at 1.2 V (0.05 oz) to as much as several thousand ampere-hours at 2 V and weigh upward of a ton. Most cells larger than about 25 Ah capacity require correct orientation during use, and the secondaries must be serviced (water addition) occasionally. Probably the largest battery in common usage is that for submarine propulsion, which weighs several hundred tons and will deliver more than 5 million Wh.

See also **Fuel cells;** and **Solar cells.**

—J. L. S. Daley, *Union Carbide Corporation, Cleveland, Ohio*

Bauxite Bauxite, the predominant aluminum ore used by the aluminum industry, was discovered by a French chemist, P. Berthier, in 1821 near Les Baux in southern France. Ores containing high aluminum oxide concentration are usually termed bauxite, a carry-over from the name of this area, where it was first discovered.

Aluminum-containing compounds (aluminum-silicate) have been used by man as far back as 5300 B.C. in making pottery in northern Iraq. "Alums" were used by Egyptians and Babylonians as early as 2000 B.C. in medicines, dyes, and chemical processes. See also **Aluminum.**

Unlike most metals, aluminum-ore supplies are available the world over. Aluminum ores have been found in large deposits on all continents except Antarctica. Two principal types of ores, gibbsite, $Al_2O_3 \cdot 3H_2O$, and boehmite, $Al_2O_3 \cdot H_2O$, are of primary interest in today's commercial processes because they contain 40–60% Al_2O_3. If these two minerals become depleted in the far future, clays rich in alumina (30–40%) are available in great abundance all over the world. Processing methods for clays, though slightly more costly than for bauxite, are already developed for commercial use.

The gibbsite-type bauxite deposits are to be found in tropical and subtropical areas such as Jamaica, Haiti, Surinam, Brazil, Guinea, Ghana, Zaire, and Australia and are also found in southern United States. The largest boehmite deposits are to be found in southern Europe, Russia, Turkey, and China.

The color of the ores varies, depending on the content or iron oxide and other impurities. Ores high in iron (Jamaica, Haiti) are dark red or brown. Lower iron content gives ores that are light red, light brown, yellow, or pink.

Other bauxite ores can be white, gray, or cream-colored. The ores being used commercially today have a wide range of chemistries. Acceptable compositions include the following range: Al_2O_3, 35–60%; SiO_2, 1–15%; Fe_2O_3, 5–40%; TiO_2, 1–4%; H_2O, 10–35%; others, 0–2.0%.

Formation and Mining: Geologists believe that the alumina in bauxites and clays had its origin in the igneous siliceous rocks in the earth's crust—basalts, dolerites, micaschists, syenites, feldspar, etc. In past geologic ages, the weathering processes which produced soil were also responsible, under certain environmental conditions, for leaching out silica, limestones, and other soluble compounds, leaving behind or redepositing alumina in hydrated form (bauxites), along with the contaminants present or as clay deposits containing substantial quantities of alumina.

Bauxite may be dense and hard or soft and powdery. Since most bauxites lie near the surface, open-pit mining methods are generally used. After the overburden is removed with large dragline shovels, the ore is dug with power shovels and

transported by truck, aerial cars, or railroad cars to a central point or port. If the ore is to be transported by ocean vessels great distances, it is usually crushed and dried to avoid hauling water and to improve handling the ore with automatic equipment.

Bayer Process: An Austrian chemist, Karl Josef Bayer (1888), evolved the process used almost universally to extract alumina from the ores. The sequential processing steps required to extract pure Al_2O_3 in dry form from the impurities are shown in Fig. B-2.

The principal ingredients in the process are caustic soda, NaOH or Na_2CO_3, lime, steam, water, and heat. Bauxite is reacted under pressure with hot caustic to dissolve the $Al_2O_3 \cdot xH_2O$ alumina as sodium aluminate. Filtration removes the insolubles, which are discarded, and the solution is cooled and agitated with a small addition of aluminum hydrate to expedite the precipitation of crystalline hydrate. This is then filtered, washed, and kiln-dried at 1100°C (2000°F) to remove the combined water to form Al_2O_3. Other processes based on variations of the Bayer process or on acid extraction have been developed but are not competitive with the Bayer process.

Alumina: The purity of Al_2O_3 entering the electrolytic process determines the purity of the aluminum produced. Commercial grades of Al_2O_3 contain 99.0–99.5% Al_2O_3 plus such impurities as H_2O (0.10%), SiO_2 (0.005–0.01%), Fe_2O_3 (0.005–0.03%), TiO_2 (0.005%), ZnO (0.01–03%), and traces of V_2O_3, Na_2O, P_2O_5, NiO, CuO, Cr_2O_3, K_2O, CaO, and others. More refined, hence more expensive, aluminas are produced for electrolytic cells where 99.9%+ aluminum is desired. The refined aluminas, however, find greater use in the chemical and ceramic industries.

Alumina products are among the most versatile in modern industry. The properties of alumina and hydrated aluminas can be varied almost endlessly—from the hardness of ruby and sapphire to a softness equal to that of talc, from a weight of over 200 lb/ft^3 to about 5 lb/ft^3, from insolubility and inertness to ready solubility and great activity.

For this reason aluminas are used in virtually every important industry. To name just a few, aluminas are used to purify water, to make glass, to produce steel alloys, for waterproofing textiles, for coating ceramics, and for making abrasives, refractory materials, cosmetics, and electronic equipment.

Hydrated Aluminas: These are crystalline α-

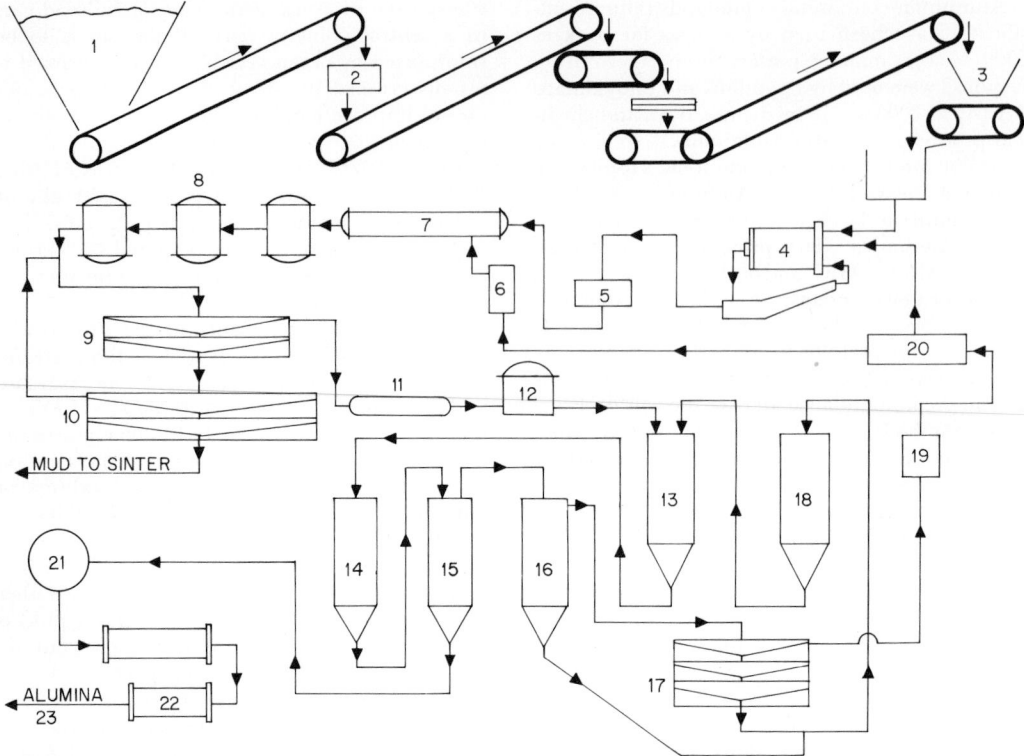

Fig. B-2. Bayer process for producing alumina. (1) Bauxite feed, (2) hammer mill, (3) bauxite weighing, (4) rod mill, (5) slurry tank, (6) heater, (7) digester, (8) pressure relief tanks, (9) mud settler, (10) mud washer, (11) filter, (12) cooling tank, (13) precipitator, (14) relay tank, (15) primary classifier, (16) secondary classifier, (17) tray thickener, (18) seed storage tank, (19) evaporator, (20) strong liquor storage (test tanks), (21) continuous rotary filter, (22) rotary kiln and cooler, (23) to alumina storage.

alumina trihydrates, represented by the formula $\alpha\text{-Al}_2\text{O}_3 \cdot 3\text{H}_2\text{O}$ or $\alpha\text{-Al(OH)}_3$. They are dry, free-flowing, snow-white crystalline powders available in a range of particle sizes.

Since hydrated alumina reacts with strong acids and alkalies, it is extensively used by the chemical industry for the production of important aluminum salts, such as aluminum sulfate, sodium aluminate, basic aluminum sulfate, aluminum chloride, and aluminum phosphate. These compounds are used in the production of the water-treating agent zeolite, napalm, wet-strength papers, paper whiteners, antiperspirants, carbonizing agents for wool, pharmaceuticals, catalysts, titanium-pigment production, precipitation of pectins from fruit wastes, soaps, and other industrially important compounds.

Hydrated aluminas are used in ceramic glazes and glass to add sparkle as well as resistance to chemical attack. Hydrates absorb pigments; hence are used to color ceramic objects.

Hydrates are used for mold coatings, as a reinforcement for pigment in rubber, as a pigment extender in paints, as a color mordant and a filler in vinyl plastics, and in foam rubber, adhesive tapes, synthetic-resin varnishes, and cosmetics. Hydrates can also be used as mild abrasives in polishes and buffing compounds.

Alumina ($\alpha\text{-Al}_2\text{O}_3$): The Bayer process produces α-aluminum trihydrate ($\alpha\text{-Al}_2\text{O}_3 \cdot 3\text{H}_2\text{O}$), also known as gibbsite. It is marketed as dried hydrate, as mentioned above, or as calcined alumina, $\alpha\text{-Al}_2\text{O}_3$, made by heating the gibbsite to over $1000°C$. As dehydration takes place, the gibbsite converts to several transition aluminas with partially disordered structures which can be identified by x-ray techniques. As the temperature is raised, the structures become more ordered, until final transformation produces α-Al_2O_3 (also known as corundum). The equation shows the progress of change with dehydration to final calcination as α-Al_2O_3:

$$\alpha\text{-}Al_2O_3 \cdot 3H_2O \xrightarrow[200\text{-}400°C]{\text{heat in air}}$$

$$\chi\text{-}Al_2O_3 \cdot xH_2O \xrightarrow{900\text{-}1000°C}$$

Chialumina,
cubic structure

$$k\text{-}Al_2O_3 \cdot xH_2O \xrightarrow{1100\text{-}1200°C} \alpha\text{-}Al_2O_3$$

Kappa alumina,
orthorhombic
structure

Alpha alumina,
hexagonal
rhombohedral
structure,
stable form

α-Alumina can also be produced by dehydrating other aluminum hydroxides, including

α-Alumina trihydrate (gibbsite):

$$\alpha\text{-}Al_2O_3 \cdot 3H_2O \xrightarrow[1100\text{-}1200°C]{\text{heat in air}} \alpha\text{-}Al_2O_3$$

β-Alumina trihydrate (bayerite):

$$\beta\text{-}Al_2O_3 \cdot 3H_2O \xrightarrow[1100\text{-}1200°C]{\text{heat in air}} \alpha\text{-}Al_2O_3$$

α-Alumina monohydrate (boehmite):

$$\alpha\text{-}Al_2O_3 \cdot H_2O \xrightarrow[1100\text{-}1200°C]{\text{heat in air}} \alpha\text{-}Al_2O_3$$

β-Alumina monohydrate (diaspore):

$$\beta\text{-}Al_2O_3 \cdot H_2O \xrightarrow[1100\text{-}1200°C]{\text{heat in air}} \alpha\text{-}Al_2O_3$$

α-Alumina, α-Al_2O_3, is dense, hard, and resistant to chemical attack. Ruby, sapphire, amethyst, and emerald are α-Al_2O_3 with traces of specific impurities. Artificial forms of jewelry are prepared by fusing finely powdered alumina with traces of coloring oxide (Cr_2O_3 for ruby) in an oxygen-hydrogen flame.

Activated Aluminas: Essentially highly porous aluminum oxide (Fig. B-3). Activated alumina, γ-Al_2O_3, is produced by heating the hydrate to a temperature sufficient to drive off most of the combined water. The heating must be carefully controlled to produce a uniform product with maximum surface area. Aluminas are available both as granules and as fine powders; hence they provide a large surface area and absorptive capacity per unit volume. Alumina is inert chemically to most gases and vapors of commercial importance and is nontoxic. After being saturated with moisture, activated aluminas can be reactivated many times by controlled heating.

Commercial applications include drying gases (15% of dry weight of alumina), including air, nitrogen, hydrogen, oxygen, carbon dioxide, chlorine, sulfur dioxide, ethylene, butane, Freon, natural gas, and others. See also **Drying, gases.** It is also used for dehydrating liquids such as pyridene, ethyl acetate, gasoline, benzol, toluol, alcohol, carbon tetrachloride, and vegetable and animal oils and is used as a filtration medium for lubricants and transformer oils.

The affinity of activated alumina for fluorine makes it an effective defluoridation agent for potable water. In petroleum refining, it is used in the defluoridation of the alkylates yielded from hydrofluoric acid alkylation. See also **Alkylation.**

Catalysts: Stability, chemical inertness, and large surface area make activated aluminas excellent catalysts for such reactions as reforming, dehydrogenation, desulfurization, and the cracking of petroleum products.

Ceramics: Calcined aluminas are prepared from hydrated aluminas by firing in kilns at temperatures in excess of 1100°C (2000°F). Calcined aluminas are ground to granules and fine powders. Applications include abrasive and fast-cutting grinding wheels; polishing and buffing compounds; laboratory ware; additive to glass for improvement in luster, thermal shock, and chemical resistance in dinnerware; and can be fused and cast into bricks of special shapes for linings for tanks or high-temperature furnaces.

A large volume of ceramic-grade aluminas are used for spark plugs, electrical substrates, protective armor for military personnel, and in areas where wear resistance is important.

Corundum: An aluminum oxide α-Al_2O_3 with a hexagonal crystal structure produced by fusion of alumina or bauxite in an electric arc furnace at above 2200°C (4000°F), cast into molds, and slowly cooled. After cooling, the product is crushed and sold for abrasives and refractories. The specific gravity of corundum is 3.95, and its hardness is 2000 on the Knoop scale.

REFERENCES
Gerard, Gary, and P. T. Stroup (eds.): "Extractive Metallurgy of Aluminum," vol 1, "Alumina," Interscience, New York, 1963.
Patterson, Sam H.: Bauxite Reserves and Potential Alu-

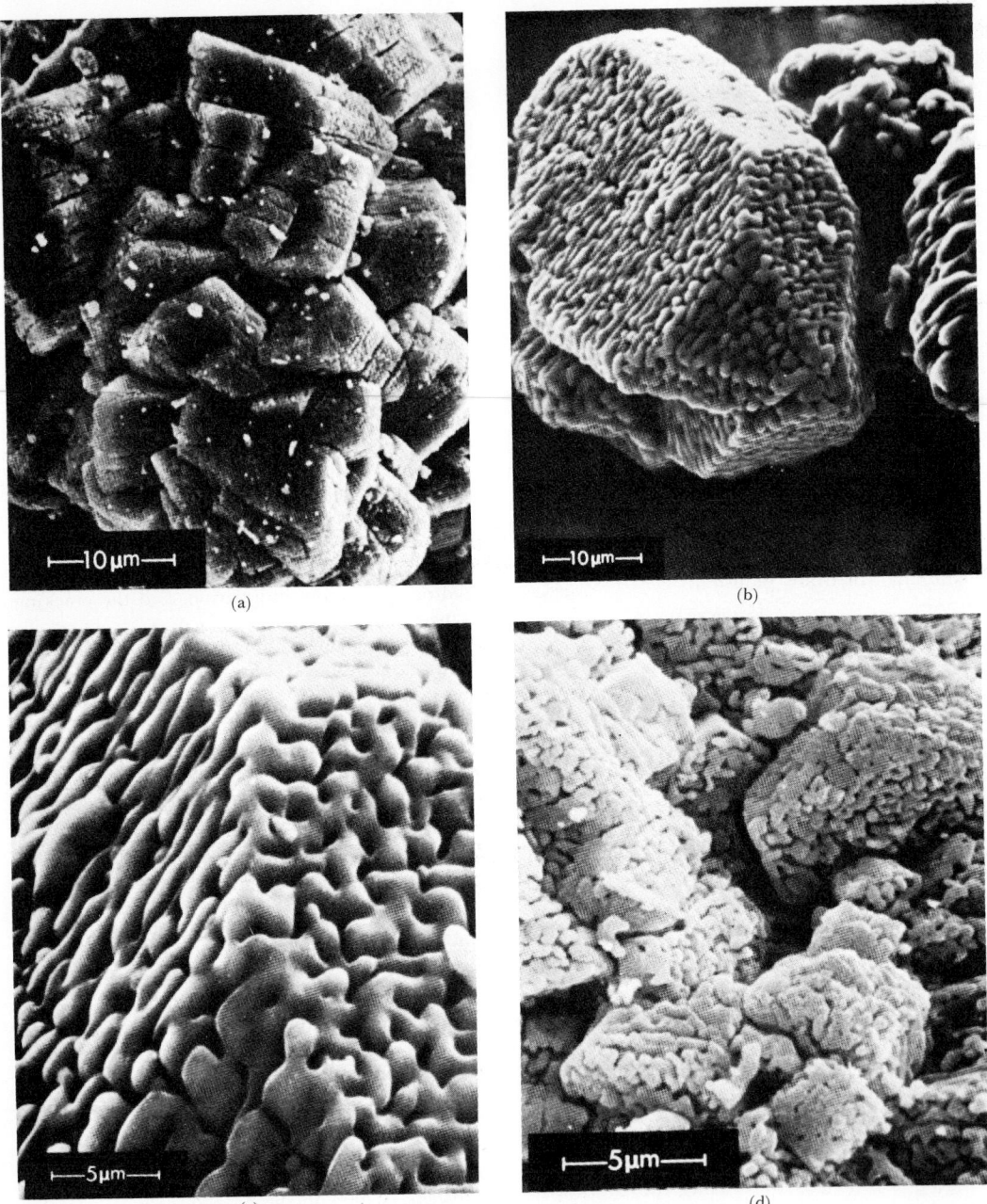

Fig. B-3. Texture modifications of gibbsite, α-Al$_2$O$_3$·3H$_2$O, during calcination as seen with a scanning electron microscope. (a) As dehydration proceeds, some of the fissures develop into major partings (800–1250°C, 1472–2282°F) and the cell wall between the pores becomes rounded into small corundum crystallites. (b) and (c) The general relationship of the skeleton of the original gibbsite now rounded into small corundum crystallites. (d) Crystallites viewed end on down the "[0001]" axis. The random nature of the tubular pores is well illustrated. This type of structure is called *coral texture*. (*Photographs furnished by The British Aluminum Co., Research Division.*)

minum Resources of the World, *U.S. G. S., Bull.* 1228, 1967.

Sherwin, R. S.: Extractive Metallurgy of Aluminum, *J. Metals,* April 1950.

Smith, Bracewell: "Bauxite, Alumina and Aluminum," Overseas Geological Surveys, Mineral Resources Division, H. M. Stationery Office, London, 1962.

Tieman, T. D.: Extraction of Alumina from Haiti and Jamaica Bauxites, *J. Metals,* May 1951.

U.S. Bureau of Mines, Washington, reports:

Peters, F. A., P. W. Johnson, and R. C. Kirby: Method of Producing Alumina from Clay: An Evaluation of the Sulfurous Acid Caustic Purification-Acid Caustic Purification Process, RI-5997, 1962; Methods for Producing Alumina from Clay: An Evaluation of Five Hydrochloric Acid Processes, RI-6133, 1962; Methods for Producing Alumina from Clay: An Evaluation of Three Sulfuric Acid Processes, RI-6229, 1963.

Johnson, P. W., F. A. Peters, and R. C. Kirby: Methods for Producing Alumina from Clay: An Evaluation of Nitric Acid Processes, RI-6431, 1964; A Cost Estimate for the Bayer Process for Producing Alumina, RI-6730, 1966.

—S. John Sansonetti, *Reynolds Metals Company, Richmond, Va.*

Bayer Process (See **Bauxite.**)

Bearing Metals (See **Antimony;** and **Tin.**)

Beckmann Rearrangement (See **Caprolactam.**)

Becquerelite (See **Uranium.**)

Beer (See **Pasteurization.**)

Beeswax (See **Waxes.**)

Belt Filter (See **Filtration.**)

Beneficiation, Ore Beneficiation of an ore requires the separation of the bulk material into one or more valuable components, generally called *concentrates,* and a waste component (*gangue*). Concentration is accomplished by processes which take advantage of the differences in various properties of the mineral components, including appearance, specific gravity, preferential wetting, and magnetic and electrical properties.

Hand sorting, i.e., utilizing the difference in the appearance of an easily recognizable, high-grade product and the mass of broken ore, is the oldest method of concentrating and is still practiced in a few instances, such as hand picking of diamonds.

Gravity Concentration: One of the earlier recognized differential properties of minerals was the relative specific gravities of these materials, as compared with gangue materials and also other minerals. For example, waste rock composed of silica or carbonates may have a specific gravity of 2.7, whereas the specific gravity of sphalerite, ZnS, is 3.9–4.1 and of galena, PbS, 7.4–7.6.

Jigs: One of the earliest gravity-concentrating methods, still in use for coal cleaning and separation of some heavy metals, is known as *jigging.* A jig is constructed in a series of compartments which are, in effect, boxes with a screen at the bottom. A bed of crushed ore is supported by the screen. A pulsating flow of water up through the screen stratifies the bed, with the heaviest material in the bottom layer and the lightest material in the top layer. A submerged weir on the side of the machine removes the heavy product as a concentrate; the lighter tailing product cascades over the end of the jig.

Concentrating Tables: The first concentrating table was built by A. F. Wilfley in 1895 and rapidly became one of the most important pieces of equipment in the metallurgical industry. These machines consist of a flat deck, or table, covered with linoleum or sheet rubber and with a series of small parallel wood strips fastened to its surface. Tables generally range in size from 4 x 8 ft to 6 x 16 ft. The longitudinal, reciprocating motion imparted to the table is such that the forward motion is at a lower velocity than the return. This action causes an object to move across the table in a series of short movements. The plane of the table may be adjusted to tilt longitudinally down from the feed end and to tilt down across the table. The general magnitude of these slopes is 0–1½ in. longitudinally and ⅜–⅝ in. transversely.

A slurry of crushed ore in water, at a pulp density of 25–30% solids, is fed onto the table through a distribution box on the high end of the table. The motion of the table causes the water and ore particles to move diagonally across the table. The heavy particles settle into the riffles formed by the wood strips and are moved across the table to the other end. The lightweight particles ride on top of the concentrate and are washed across the riffles into a side-tailing launder.

Tables ase designated as *sand tables,* treating about minus-8-mesh material, and *slime tables,* treating about minus-80-mesh material. The riffles cover from two-thirds to the full length of the table deck. Riffles are lower and wider on a slime table and terminate in a diagonal pattern which leaves the far upper corner of the deck unriffled.

Tables work best with clean minerals, properly sized and with a reasonable spread between the specific gravities of the two components. Applications include gold, tin, chromite, coal, tungsten, oil-treated phosphate rock, graphite, and various nonmetallics. Concentration by tabling also finds application for low-tonnage, short-life mineral deposits wiich cannot support a large capital expenditure.

Humphrey's Spiral: This is a flowing-film gravity concentrator, consisting of a trough (basic construction cast iron) spiraling around a vertical axis. Feed is a slurry of 20–40% solids which flows down the trough. Heavy mineral particles settle to the bottom of the trough, while lighter particles ride up the outer side of the spiral. Discharge openings in the bottom of the trough draw off the heavy concentrates or middlings; the lighter fraction rides the side to discharge at the bottom end. Favorable sizing for the operation is no larger than 10 mesh and no smaller than 200 mesh. The capacity of one spiral, about 2 ft outside diameter, generally is 1–2 tons/hr. When processing large quantities of material, as is usually the case, a large number of spirals are required for an installation. Beach sands, dredged mineral deposits, hematitic iron ores, etc., are concentrated with spirals.

Sink-Float Concentration: This process is based on the premise that a slurry of finely ground solids will behave as a heavy liquid. Materials used in preparation of the heavy media, the density of which can be made as high as 3.3, are many, with barite, galena, magnetite, or ferrosilicon commonly used.

Ore to be concentrated by this process is crushed and charged to a vessel filled with a heavy medium having a specific gravity between that of the two ore components to be separated. The lightweight component floats to the surface and overflows to a drainage screen, then to a washing screen. Liquid from the drainage screen returns to the vessel. Liquid from the washing screen (diluted medium) is densified by magnetic separation or flotation before being returned to the vessel.

The heavy component sinks to the bottom and is lifted by means of an airlift, bucket-elevator classifier spiral to a drainage screen and then to a washing screen. The heavy-medium slurry removed by the drainage screen returns to the vessel. Liquid from the washing screen (diluted medium) joins the similar discharge from the tailings washing screen for densification and return to the vessel.

Froth Flotation: This process represents one of the great discoveries in the history of the metallurgical industry. Probably no other process has contributed so much to the recovery of minerals from their ores. It is now the most generally used concentration process, expressed both as tonnage and as number of ores treated. This process is described in detail under **Flotation.**

Magnetic Concentration: Magnetic pulleys, drums, or other configurations are used for separating a limited number of materials which are influenced by a magnetic field. The technique is used extensively in large-scale plants for the recovery of magnetite from iron ores.

Electrostatic Concentration: For susceptible minerals, it is possible to make a separation based on the attraction or repulsion of charged particles. If one or more components of a granular mixture receive a surface charge before entering an electrostatic field, the grains of that material will be attracted to one electrode and repelled from the other. Selection of the proper locations of the receivers therefore completes the separation of the falling stream of particles. Success of the process depends upon dry particle surfaces. The black sand deposits containing garnet, rutile, zircon, and monazite usually are treated by electrostatic methods.

Other major operations involved in beneficiation processes include clarifying, thickening, and filtration, described separately in this volume. Several examples of specific ore-beneficiation processes are included. Consult the Subject Index. The beneficiation of ferrous ores tends to become quite specialized. See **Iron ores;** and **Nickel.**

REFERENCES

Newton, Joseph: "Extractive Metallurgy," Wiley, New York, 1959.

Taggart, Arthur F.: "Handbook of Mineral Dressing," Wiley, New York, 1945.

Wraith, William, Jr., and T. G. Fulmor: Anaconda's Butte Concentrator, *Mining Eng.,* May 1964.

—J. Wesley Burgess, Francis Carroll Moran, and Gailen T. Vandel, *Fluor Utah, Inc., San Mateo, Calif.*

Benzal (See **Radicals.**)

Benzaldehyde (See **Aldehydes.**)

Benzamide (See **Amides and imides.**)

Benzene Although benzene, C_6H_6, is a very high tonnage chemical and of tremendous industrial importance, particularly as a starting ingredient of many reactions, it is also, as the simplest hydrocarbon of the aromatic group, of much theoretical interest. The current comprehension of the structure and chemistry of aromatic compounds stems from very early investigations of benzene. Although benzene is isomeric with the aliphatic compound dipropargyl, $CH:C \cdot CH_2 \cdot CH_2 \cdot C:CH$, the chemical properties of benzene are very different. See also **Carbon compounds; and Isomerism.**

Kekulé (1865) concluded that the six carbons in benzene form a closed, symmetrical chain or nucleus and that each carbon atom is united with one and only one hydrogen atom. Over the years, several other theoretical concepts of the benzene structure were proposed, including those of Claus (diagonal formula), Ladenburg (prism formula), and Armstrong-Baeyer (centric formula).

Benzene is the first member of a homologous series of compounds, C_nH_{2n-6}. Methylbenzene or toluene, $C_6H_5 \cdot CH_3$, is the only homologue with the formula C_7H_8. The next highest member of the homologous series, C_8H_{10}, exists in four isomeric forms: ethylbenzene, $C_6H_5 \cdot C_2H_5$, and ortho-, meta-, and paradimethylbenzene, $C_6H_4(CH_3)_2$. Eight isomerides of the formula C_9H_{12} are possible. The number of possible isomerides of course increases rapidly as the carbon count increases.

Properties: Sometimes referred to as *benzol, phenyl hydride,* or *cyclohexatriene,* benzene is a colorless, highly flammable liquid that burns with a smoky flame. Some of the physical properties are formula weight = 78.11; melting point = 5.5°C; boiling point = 80.1°C; specific gravity = 0.879 at 20°C (referred to water at 4°C); practically insoluble in water (0.07 parts in 100 parts at 22°C); and fully miscible with alcohol, ether, and many other organic liquids.

Industrially pure benzene has a specific gravity of 0.875–0.876 and a distillation range of 78.1–82.1°C.

Production: Many of the C_nH_{2n-6} series of hydrocarbons occur in coal gas and coal tar, from which they can be extracted and which represented the major early sources for these materials. Today, almost 90% of the benzene produced comes from petroleum sources. Separation of benzene from coal tar is difficult because of the presence of scores of isomerides, with close boiling points. In one process for making benzene of high purity (99.94% or higher) from coke-oven light oil (the

cut boiling between 60 and 150°C), the light oil and a stream of hydrogen are heated to reaction temperature and passed through fixed-bed reactors that contain a catalyst (undisclosed), whereupon the nonaromatics present are converted to light hydrocarbon gases, and any sulfur compounds present are converted to H_2S. Some dealkylation of the higher aromatics present also produces benzene in addition to that contained in the feedstock. From the reactor, the vapors are cooled and passed to a stabilizer tower where dissolved H_2S and light hydrocarbons (boiling below benzene) are removed. The bottoms from the stabilizer, containing benzene, toluene, and xylene, are clay-treated, followed by a series of fractionations to produce benzene, toluene, and xylene, as well as additional higher-boiling hydrocarbons. If a portion of the hydrocarbons in the product fuel gas is reformed, no external hydrogen is required.

For the synthetic preparation of benzene, the dealkylation of toluene accounts for the most growth in recent years.

1. In one noncatalytic process, a hydrogen-rich gas is mixed with liquid toluene feed and preheated prior to charging the reactor. Toluene reacts with the hydrogen to form benzene and methane. The reaction is exothermic, and the operating conditions are approximately 500–1,000 psig and 595–760°C. The process operates to about 98% of theoretical yield of benzene. Toluene is recycled.

2. In a catalytic dealkylation process, toluene or C_8 aromatics (alkylbenzenes) are fed to a reactor, along with a hydrogen-containing gas. The hydrogen source is not critical and may be manufactured hydrogen or off-gas from a reforming or other refining unit. See Fig. B-4. Effluent from the reactor, after cooling, is charged to a separator, from which hydrogen is removed and recycled to the reactor. Liquid phase from the separator is stripped of hydrocarbons (boiling lower than benzene) in a stabilizing column. One further fractionating step yields product-benzene overhead. The bottoms from this tower are recycled to the reactor for dealkylation. Yields of 98% of the theoretical are claimed.

3. In another process, mixtures of aromatics and nonaromatic hydrocarbons constitute the charge. In a first step, aromatics are continuously extracted from the feed by using an aqueous solution of *N*-methylpyrrolidone. A multistage countercurrent extractor tower is used. This operation is carried out at very modest temperatures and

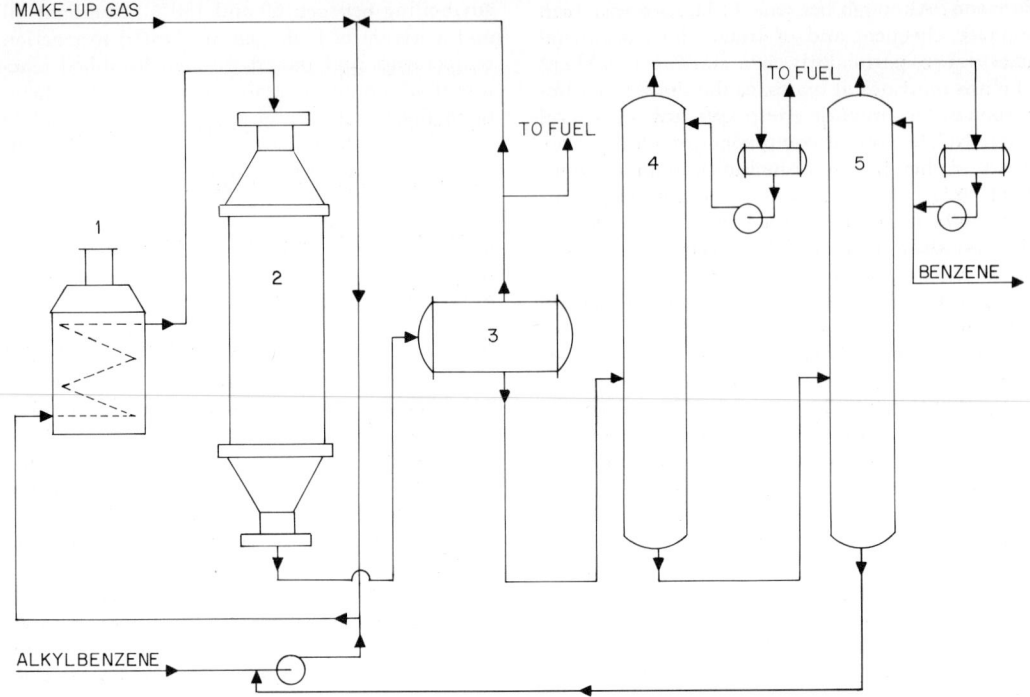

Fig. B-4. Catalytic dealkylation of toluene to produce benzene. (1) Heater, (2) reactor, (3) separator, (4) stabilizer, (5) benzene column. (*Universal Oil Products Co.*)

pressures. The rich aromatic extract phase then proceeds to a stripper, where pentane and a part of the benzene are removed overhead and recycled to the extractor. The bottoms from the stripper are free of nonaromatics and enter a second stripper for further separation. The distillate from the second stripper contains pure aromatics and water. The bottoms contain aromatics-free solvent, which is returned to the extractor. One or more further fractions yield benzene, toluene, and xylenes of desired specification. A typical feedstock range may contain an aromatics mixture of the following proportions: benzene, 26–60%; toluene, 14–22%; xylenes plus ethylbenzene, 15–5%. A somewhat similar process uses dimethyl sulfoxide as a solvent.

Consumption: Currently, the production of styrene is the major use of benzene, followed by the production of cyclohexane. Until relatively recently, phenol production represented the second largest need for benzene. Of course, benzene and cyclohexane are intimately related economically because cyclohexane can be manufactured by reacting benzene with hydrogen. See also **Petrochemical complex.**

United States production of benzene is approximately 0.4 million tons/yr.

As becomes obvious from a cursory look at the homologues and derivatives of benzene, the use of benzene extends over hundreds of important products, several of which are described elsewhere in this volume.

Homologues and derivatives of benzene include toluene, $C_6H_5CH_3$; the four xylenes, C_8H_{10}—ortho-, meta-, and paraxylene and ethylbenzene; mesitylene, $C_6H_3(CH_3)_3$ (1-,3-,5-, or symmetrical trimethylbenzene); pseudo-cumene (1,2,4-trimethylbenzene); cumene, $C_6H_5 \cdot CH(CH_3)_2$; and cymene, $C_6H_4(CH_3) \cdot C_3H_7$ (paramethylisopropylbenzene). The aforementioned compounds contain only one benzene nucleus. Other important benzene-related compounds containing two benzene nuclei include diphenyl, $C_6H_5 \cdot C_6H_5$, and diphenylmethane, $C_6H_5 \cdot CH_2 \cdot C_6H_5$. Triphenylmethane, $(C_6H_5)_3CH$, contains three benzene nuclei.

Important halogen derivatives of benzene include chlorobenzene, C_6H_5Cl (phenyl chloride); bromobenzene, C_6H_5Br; benzyl chloride, $C_6H_5 \cdot CH_2Cl$; benzal chloride, $C_6H_5 \cdot CHCl_2$

(benzylidene dichloride); and benzotrichloride, $C_6H_5 \cdot CCl_3$ (phenylchloroform). See also **Chlorine organics.**

Nitro derivatives of benzene include nitrobenzene, $C_6H_5 \cdot NO_2$, and metadinitrobenzene, $C_6H_4(NO_2)_2$. Important amino compounds derived from benzene include aminobenzene, $C_6H_5 \cdot NH_2$ (aniline); diaminobenzene, $C_6H_4(NH_2)_2$; and triaminobenzene, $C_6H_3(NH_2)_3$. Phenol, $C_6H_5 \cdot OH$, is hydroxybenzene. Resorcinol, catechol, and quinol, $C_6H_4(OH)_2$, may be considered to be dihydroxybenzenes, and pyrogallol and phloroglucinol, $C_6H_3(OH)_3$, trihydroxybenzenes. Benzene-related alcohols, aldehydes, and ketones include benzyl alcohol, $C_6H_5 \cdot CH_2 \cdot OH$ (phenylcarbinol); benzaldehyde, $C_6H_5 \cdot CHO$; benzoin, $C_6H_5 \cdot CO \cdot CH(OH) \cdot C_6H_5$; salicylaldehyde, $C_6H_4(OH) \cdot CHO$ (o-hydroxybenzaldehyde); anisaldehyde, $C_6H_4(OCH_3) \cdot CHO$ (p-methoxybenzaldehyde); acetophenone, $C_6H_5 \cdot CO \cdot CH_3$ (phenylmethyl ketone); benzophenone, $C_6H_5 \cdot CO \cdot C_6H_5$ (diphenyl ketone or benzoyl benzene); and quinone, $C_6H_4O_2$.

Benzene-related acids and salts include benzoic acid, $C_6H_5 \cdot COOH$; ethyl benzoate, $C_6H_5 \cdot COOC_2H_5$; benzoyl chloride, $C_6H_5 \cdot COCl$; benzoic anhydride, $(C_6H_5 \cdot CO)_2O$; benzamide, $C_6H_5 \cdot CO \cdot NH_2$; benzonitrile, $C_6H_5 \cdot CN$ (phenyl cyanide); anthranilic acid, $C_6H_4(NH_2) \cdot COOH$; phthalic acid, $C_6H_4(COOH)_2$; phthalic anhydride, $C_6H_4(CO)_2O$; phthalimide, $C_6H_4(CO)_2NH$; isophthalic acid, $C_6H_4(COOH)_2$; terephthalic acid, $C_6H_4(COOH)_2$; benzenehexacarboxylic acid, $C_6(COOH)_6$; phenylacetic acid, $C_6H_5 \cdot CH_2 \cdot COOH$; cinnamic acid, $C_6H_5 \cdot CH \colon CH \cdot COOH$ (β-phenylacrylic acid); salicylic acid, $C_6H_4(OH) \cdot COOH$ (o-hydroxybenzoic acid); methyl salicylate, $C_6H_4(OH) \cdot COOCH_3$; phenyl salicylate, $C_6H_4(OH) \cdot COOC_6H_5$; acetylsalicylic acid, $C_6H_4(OAc) \cdot COOH$; anisic acid, $C_6H_4(OCH_3) \cdot COOH$ (p-methoxybenzoic acid); gallic acid, $C_6H_2(OH)_3 \cdot COOH[3OH = 3,4,5]$ (pyrogallol carboxylic acid); and mandelic acid, $C_6H_5 \cdot CH(OH) \cdot COOH$ (phenylglycollic acid).

Benzenoids   (See **Hydrocarbons.**)

Benzofuran   (See **Furan group.**)

Benzophenone   (See **Ketones.**)

Benzoyl   (See **Radicals.**)

Benzyl   (See **Radicals.**)

Benzyl Bromide   (See **Bromine.**)

Benzyl Chloride   (See **Chlorine organics.**)

Benzylpenicillin   (See **Antibiotics.**)

Berkelium   (See **Chemical elements.**)

Berl Saddle   (See **Absorption.**)

Beryllium (L. *beryllus.*) **Be** $= 9.015043$ (at. wt.);* 4 (at. no.). Beryllium appears in group IIA of the periodic table along with Mg, Ca, Sr, Ba, and Ra. All natural beryllium consists of the ^9Be isotope. ^6Be, ^7Be, ^8Be, and ^{10}Be can be made artificially; some occur in reactor operations. Beryllium is a grayish metal which can be polished to a bright metallic luster. The crystal structure is close-packed hexagonal. This α-form transforms to a body-centered cubic structure at a temperature very close to the melting point. Beryllium is the only light metal with a high melting point. Important physical properties are summarized in Table B-3.

TABLE B-3. *Some Physical Properties of Beryllium*

Melting point, °C	1277–1284
Boiling point, °C	2770–2970
Density, g/cm³, at 20°C	1.848
Thermal conductivity, cgs units, 0–100°C	0.35–0.40
Mean specific heat, (cal)(g)/°C, 0–100°C	0.45–0.49
Resistivity, $\mu\Omega/cm$, at 20°C	4–6
Coefficient of resistivity, $\times 10^3$, 0–100°C	6.0
Coefficient of expansion, $\times 10^6$, 0–100°C	11.6–12
Heat of fusion, cal/g	260

SOURCES: Include Metals Reference Book, 4th ed. (C. J. Smithells) and ASM Metals Handbook, 8th ed. Authorities do not agree in connection with all the physical properties of Be. The boiling point has not been firmly established.

Beryllium was discovered by Vauquelin (1797) during the course of work with the mineral beryl (emerald). The metal did not become available in commercial forms until the 1940s.

Chemical Properties: Beryllium, with a valence of 2 and as a member of the alkaline group of metals, is very reactive, due to its high free energy of reaction with oxygen. At ordinary tempera-

*On the physical scale. The figure is 9.013 ± 0.0004 on the chemical scale.

tures, highly polished surfaces retain their brilliance for years, due to a nonporous protective oxide film (approx. 100 Å thick) which forms upon exposure to air. However, oxidation becomes quite noticeable at 700–800°C and above.

Nitrogen attacks Be above 900°C, and CO_2 interacts with Be to form a corrosive product. H_2SO_4 and HCl attack Be readily. The reactions with HNO_3, H_3PO_4, and glacial CH_3COOH are much slower. Fluorides, including HF, dissolve Be, and the element reacts vigorously with all halide acids, strong hydroxides, and molten alkalies. Beryllium is resistant to attack by pure H_2O, but corrodes severely in H_2O containing only 0.5 ppm of halide ion. Corrosion in O_2-free Na is slight. Molten Be attacks most of the oxides.

The characteristics of selected Be compounds are summarized in Table B-4.

Production: Beryllium is available as powder, beads, electrolytic flakes, ingot, strip, sheet, plate, foil, wire, rod, bar, tube, castings, and fabricated parts. In recovering the metal from the usual ores (beryl and bertrandite), the sulfate process and the fluoride process produce beryllium oxide or hydroxide. This step is followed either by a thermal reduction or an electrolytic process. Thermal reduction involves the magnesium reduction of a halide, generally beryllium fluoride. The electrolytic process requires conversion to the chloride for the fused-salt bath and produces Be of higher purity.

Beryllium Alloys: Most important among the Be alloys is beryllium copper. Production starts with the manufacture (from an oxide) of a beryllium-copper master alloy containing about 4% Be. Other alloys are made from this master material. The beryllium-copper alloy that contains about 2% Be and 0.25% Co is among the most important. In the annealed condition, this alloy is ductile and can be readily formed. It attains strengths of nearly 300,000 psi when heat-treated. The unique value of beryllium copper lies in its yield strength and spring properties, in combination with good electrical conductivity. These alloys are used in the marine, automotive, electronic, aerospace, instrument, and electrical industries for springs, precision-instrument parts, high-integrity microtolerance circuit interconnections, antennas in space vehicles, high-strength undersea housings, and nonsparking tools.

Beryllium-nickel alloys, which contain about 2% Be, are ductile and strong. They possess good strength at elevated temperatures when heat-treated. Because of the high elastic and endurance properties, wear resistance, and resistance to oxidation, these alloys are used for heat-resistant springs, instruments, bellows, retainer clips, and guides.

In addition to the use of Be in the 5000 and 7000 series of Al alloys (also contain Mg), where Be adds strength and forms an oxide on the surface of the liquid metal to prevent oxidation and loss of Mg, Be also is used widely for structural members in missiles, where it provides crucial weight savings. Beryllium is used to reinforce or stiffen Al and Ti alloy matrices in various struc-

TABLE B-4. *Properties of Selected Beryllium Compounds*

Compound	Formula	Formula Weight	Sp gr	Mp, °C	Bp, °C
Beryllium aluminum silicate (beryl)	$3BeO \cdot Al_2O_36SiO_2$	537.36	2.66	1410 ± 100	
Beryllium bromide.	$BeBr_2$	168.85	3.465	490 ± 10	520
Beryllium chloride	$BeCl_2$	79.93	1.9	400 ± 10	520
Beryllium fluoride (basic)	$2BeO \cdot 5BeF_2$	285.14	2.01		
Beryllium oxide (bromellite)	BeO	25.02	3.025	2570	3900
Beryllium silicate (phenacite)	Be_2SiO_4	110.10	3.0		
Beryllium silicate (bertrandite).	$2Be_2(SiO_4)H_2O$	238.22	2.6		
Beryllium sulfate.	$BeSO_4$	105.08	2.44	540d	
	$BeSO_4 \cdot 4H_2O$	177.14	1.71	100	250
Beryllium sulfide.	BeS	41.08	2.36	271	1470
Beryllium acetate	$Be(C_2H_3O_2)_2$	127.07	. . .	300d	
Beryllium acetylacetonate.	$Be(C_5H_7O_2)_2$	207.13	1.17	108	270
Beryllium propionate (basic)	$BeO \cdot 3Be(C_3H_5O_2)_2$	490.31	. . .	120	
Diethyl beryllium	$Be(C_2H_5)_2$	67.10	. . .	12	110

d = decomposes.

tural shapes and composites. Honeycomb and truss-core panels have been developed to withstand very high temperatures.

Miscellaneous Uses: Included among other applications for Be are its use as a moderator, reflector, and cladding material in nuclear reactors, and as a constituent of very high thrust fuels, where Be is second only to H_2 in its high heat of combustion per unit of fuel (17.2 kcal/g). An early application for Be was in connection with windows in x-ray tubes, due to its high permeability to x-rays (because of low atomic weight). Beryllium also has been used as a source of neutrons in nuclear reactors. Beryllium is attractive as a heat sink or capacitor because of its high specific heat and thermal conductivity combined with lightness. The good dimensional stability, high elastic modulus, low density, and low coefficient of expansion make Be a well-suited material for base supports for optical systems in aerospace use.

Oxides of Beryllium: Only one oxide is known, BeO (beryllia), which is a refractory material. Two varieties exist, α-BeO and β-BeO. Beryllia is of great interest in nuclear-energy applications because of its low-neutron-absorption cross section, very high melting point, and very low vapor pressure at high temperatures, coupled with its high moderating capacity. Beryllia also has high thermal conductivity, high electrical resistivity, excellent dielectric properties, good thermal-shock resistance, high mechanical strength, and good chemical resistance. Beryllia ceramics therefore are used as heat sinks and electrical insulators (substrate for microcircuits and semiconductor components) and as rocket nozzles, crucibles, thermocouple wells, and insulators, and its transparency to microwaves, plus other properties, makes it an ideal material for radomes. The theoretical density of beryllia is given as 3.008 g/cm^3 (Ref. 1) and as 2.69 g/cm^3 (Ref. 2). The melting point is given as $2450° \pm 20°C$ and the boiling point as 3900–$4260°C$ (Ref. 3); thermal conductivity at $0°C$ is given as 5.5 cal/(mole)/($°C$) (Ref. 4); specific heat at $200°C$, 0.35 cal/(g)($°C$), and at $900°C$, 0.50 cal/(g)($°C$) (Ref. 5); modulus of elasticity E, 49×10^{-6} psi at $200°C$ and 41×10^{-6} psi at $900°C$ (Ref. 5); and thermal conductivity K, 0.35 (cal)(deg)/(cm)(s) at $200°C$ and 0.06 (cal)(deg)/(cm)(s) at $900°C$ (Ref. 5).

Beryllia is extracted from the mineral beryl ($3BeO \cdot Al_2O_3 \cdot 6SiO_2$), one of few Be materials of commercial importance, via a sulfate process and a fluoride process, both of which yield technical grades of BeO, sufficiently pure for the production

of commercial Be alloys, such as beryllium copper and beryllium aluminum. $Be(OH)_2$ and $BeSO_4 \cdot 4H_2O$, each obtainable from the two processes, serve as starting materials for making BeO. High-purity grades of BeO are necessary for reactor and other applications and are prepared by several processes (Ref. 11). The extent of impurities in high-grade BeO made by the sulfate process include $<0.002\%$ Fe, $<0.003\%$ Al, 0.020% Si, 0.005% Na, 0.002% Mg, $<0.0001\%$ Li, $<0.001\%$ Cu, $<0.001\%$ Ni, and $<0.0001\%$ B.

Beryllium Hydroxide: There are at least two forms, β- (a gelatinous material, produced by adding the stoichiometric quantity of alkali to a cold beryllium salt solution) and α- (a granular form, obtained by making the precipitation from a hot solution or by boiling a beryllate solution). Methods are described in Refs. 6–8. Commercial practice has favored precipitation of the β-form because alkali consumption is only half as great as in the case when the α-hydroxide is produced.

Halides of Beryllium: Known halides of Be include BeF_2, $BeCl_2$, $BeCl_2 \cdot 4H_2O$, $BeBr_2$, and BeI_2. The halides generally are obtained by reacting the metal, hydroxide or oxide, with the halogen under heat and pressure. Methods are described in Ref. 9. In no case can wet methods be used because beryllium halides hydrolyze easily to form the corresponding acid. The conversion of BeO to $BeCl_2$ has been fully described in the literature on the Kroll process for beryllium, including Refs. 9 and 14–17.

BeF_2, important commercially because it serves as the source of Be (reduction by Mg), cannot be readily prepared by wet reactions. BeF_2 can be prepared by the direct action of gaseous hydrogen fluoride on BeO at $220°C$ (Ref. 18), but it is generally obtained commercially by Lebeau's method, in which ammonium fluoroberyllate, $(NH_4)_2BeF_4$, is thermally decomposed (Ref. 19).

Other Beryllium Compounds: Some other known Be compounds, not included in Table B-4, are the nitride Be_3N_2; oxo salts—the carbonate $BeCO_3 \cdot 4H_2O$, the nitrate $Be(NO_3)_2 \cdot 4H_2O$, and the phosphate $Be_3(PO_4)_2 \cdot 4H_2O$; a number of salts of organic acids, including the oxalate, acetate, formate, and stearate; and several double salts, such as diammonoberyllium chloride, a double chloride of Pd and Be, and a double chloride of Pt and Be. Organo-Be compounds include dimethylberyllium, $(CH_3)_2Be$, diethylberyllium, diisopropylberyllium, and diphenylberyllium. Beryllium also is known to form alcoholates and phenolates.

Safe Handling of Beryllium: Extensive experience indicates that Be can be processed safely if certain work codes are observed. Most important are the safe-breathing concentrations (in-plant and out-plant) established by the Atomic Energy Commission (Ref. 20). The major hazard of the industry arises from the inhalation of Be and certain Be compounds. This can cause irritation of the upper respiratory tract and affect the lungs. The terms "beryllium poisoning," "berylliosis," "beryllium disease," and "beryllium pneumonitis" generally refer to conditions produced by inhalation. Minor hazards include skin reactions caused by contact with soluble Be compounds. See References.

REFERENCES

1. Ryshkewitch, E.: "Oxide Ceramics: Physical Chemistry and Technology," Academic, New York, 1960.
2. Smith, D. K., et al.: *J. Nucl. Materials,* vol. 14, pp. 237–238, 1964.
3. Budnikov, P. S., et al.: *Z. Priklad. Khim.,* vol. 33, p. 1901, 1960.
4. Glassner, A.: "Thermochemical Properties of the Oxides, Fluorides, and Chlorides to 2500°K," Argonne National Laboratory, Chicago.
5. Smith, R. S., and J. P. Howe: "Beryllium Oxide," p. 407, North-Holland Publishing Co., Amsterdam, 1964.
6. Fricke, R., and H. Humme: *Z. Anorg. Chem.,* vol. 178, p. 400, 1929.
7. Havestat, L., and R. Fricke: *Z. Anorg. Chem.,* vol. 188, p. 357, 1930.
8. Fricke, R., and H. Severn: *Z. Anorg. Chem.,* vol. 205, p. 287, 1950.
9. Darwin, G. E., and J. H. Buddery: "Beryllium," Butterworth, London, 1960.
10. The Institute of Metals: "The Metallurgy of Beryllium," Chapman & Hall, London, 1963.
11. Hausner, H. H. (ed.): "Beryllium: Its Metallurgy and Properties," University of California Press, Berkeley, Calif., 1965.
12. Pinto, N. P., and J. Greenspan: "Beryllium," vol. 6, pp. 320–372, in Bruce Gonser (ed.), "Modern Materials: Advances in Development and Applications," Academic, New York, 1968.
13. Proceedings of the Beryllium Conference, vol. 1, A Report of the National Materials Advisory Board, National Academy of Sciences–National Academy of Engineers, Pub. NMAB-272, Washington, D.C., July 1970.
14. Block, F. E., et al.: "Advances in Extractive Metallurgy," pp. 551–571, The Institute of Mining and Metallurgy, London, 1968.
15. Kalischer, P. R.: *Brush Beryllium Co., Prog. Rep.* BBC-6, Cleveland.
16. May, J. T., and C. L. Hoatson: U.S. Bureau of Mines, Washington, Pub. 6037, 1962.
17. White, D. W., and J. E. Burke: "The Metal Beryllium," pp. 102–114, American Society for Metals, Metals Park, Ohio, 1955.
18. Hyde, K. R., et al.: *J. Inorg, Nucl. Chem.,* vol. 6, p. 14, 1958.
19. Lebeau, P.: *C.R.,* vol. 126, p. 1418, 1898.
20. Eisenbud, M.: "The Metal Beryllium," p. 620, American Society for Metals, Metals Park, Ohio, 1955.
21. Browning, E.: "Toxicity of Industrial Metals," pp. 55–71, Butterworth, London, 1961.
22. Ferreira, L. E.: "Recommended Practices for Safe Handling of Beryllium Oxide Ceramics," ASTM, *Spec. Tech. Pub.* 300, Philadelphia, 1961.
23. Cholak, J., et al: Toxicity of Beryllium, *U.S. Dept. Commerce Rep.* ASD-TR-62-7-665, Washington, 1962.
24. Strokinger, H. E. (ed.): "Beryllium: Its Industrial Hygiene Aspects," Academic, New York, 1966.

—Technical Staff, Kawecki Berylco Industries, Inc., *Reading, Pa.*

Beryllium Copper   (See **Copper.**)

Bessemer Process   (See **Ironmaking and steelmaking.**)

Beta Brass   (See **Copper.**)

Beta Particle   A beta particle is a positive electron (*positron*) or negative electron (sometimes called *negatron*) that is emitted from a nucleus during beta decay. A beta ray is composed of a stream of beta particles, high-velocity electrons (usually negative) originating in particle accelerators or in radioactive atoms.

Beta decay is the radioactive transformation of a nuclide in which the atomic number increases or decreases by 1 and the mass number remains unchanged. The atomic number increases when a negative beta particle (negatron) is emitted, and decreases when a positive beta particle (positron) is emitted, or when an electron is captured. The beta disintegration energy for negatron emission is equal to the sum of the kinetic energies of the beta particle, the neutrino, and the recoil atom. For positron emission, the energy equivalence of two electron rest masses must be added. For electron capture, the disintegration energy is equal to the sum of the kinetic energy of the neutrino and the electronic excitation energy of the product atom. A radionuclide that disintegrates by beta-particle emission is termed a *beta emitter.* See also **Radioactive isotopes.**

Betterton-Kroll Process   (See **Bismuth.**)

Bicomponent Fibers (See **Acrylic fibers.**)

Binary Systems (See **Distillation.**)

Biochemistry Biochemistry is concerned with the chemistry of life. Biochemical engineering is concerned with designing and operating equipments and processes in which advantage may be taken of biochemical principles and phenomena for the production of raw, intermediate, and final chemical products and, very importantly, in the production of substances (commonly referred to as *biologicals*) that are used by medical doctors, veterinarians, and botanists, for research and in the treatment and prevention of diseases and malfunctions of the organisms of man, animals, and plants. Numerous aspects of biochemistry and biochemical engineering are described in this volume, notably under such topics as **Amino acids; Antibiotics; Enzymes; Peptides and proteins, Photosynthesis; Polymerization;** and **Yeasts.** Consult the Subject Index.

Biodegradability [See **Anionic surfactants (sulfur-bearing); and Detergents.**]

Biological Waste Treatment (See **Waste treatment.**)

Biopolymers (See **Amino acids; Peptides and proteins; and Polymerization.**)

Biosynthesis (See **Amino acids; Peptides and proteins; and Polymerization.**)

Biotite (See **Potassium.**)

Bismuth (*G. weisse Masse,* white mass.) **Bi** = 208.980 (at. wt.); 83 (at. no.). Bi has one stable isotope with a mass number of 209.

Bismuth is located in group VA of the periodic table, which includes N, P, As, and Sb. In the elemental form Bi exhibits a rhombohedral crystal structure.

Bismuth is a silvery-white crystalline metal which exhibits some limited ductility. It is one of the few metals which undergoes an increase in volume (3.32%) upon solidification. It is the most diamagnetic of all the metals, with a mass susceptibility of -1.35×10^{-6}. Some of its more common physical properties are listed in Table B-5. See also **Chemical elements.**

Discovery: Bismuth was used by the early Greeks

TABLE B-5. *Selected Physical Properties of Bismuth*

Density, g/cm³, at 20°C	9.8
Atomic volume, cm³/g atom	21.3
Melting point, °C	271.3
Boiling point, °C	1560
Specific heat, cal/(g)(°C), at 20°C	0.0294
Thermal conductivity, (cal)(cm²)/ (cm)/(°C)(s), at 20°C	0.020
Latent heat of fusion, cal/g	12.5
Coefficient of thermal expansion, in./(in.)(°C)	13.3×10^{-6}
Electrical resistivity, Ω-cm, at 0°C	106.8×10^{-6}

and Romans, but it was not recognized as a distinct element until the Middle Ages, when Basil Valentine (pseudonym of Johann Thölden) in the fifteenth century observed some of its peculiarities and referred to it as bismuth, which was later Latinized to bisemutum. It was not until 1753 that Bi was characteristically defined as a new element by C. Geoffroy and T. Bergman.

Occurrence: Bismuth is ranked 65th in the order of terrestrial abundance and occurs at a level of about 0.2 ppm.

Bismuth is chalcophyllic in nature. Some of the common minerals of Bi are bismuthinite, Bi_2S_3, bismite, Bi_2O_3, bismuthite, $Bi_2O_2 \cdot CO_3$, and tetradymite, Bi_2Te_2S. Over 75 minerals containing Bi have been encountered in nature. Native Bi is not uncommon. From an economic consideration, there are no significant primary ores of Bi, and for this reason the major source of Bi is as a by-product derived from the treatment of lead, copper, and molybdenum ores. In lead mineralization, Bi substitutes readily for the lead in galena, PbS. In copper ores the Bi may be present as a complex sulfide mineral.

Recovery: At one time Bi was recovered from enriched bismuth sulfide and bismuth carbonate ores by direct smelting. Most of these deposits were limited and have been exploited, and as a result Bi is now recovered mainly as a by-product metal from the treatment of lead, copper, and molybdenum ores.

In the treatment of lead ores, the Bi accompanies the lead during processing. The Bi is recovered from the lead by either a kettle-refining operation or an electrolytic process. Most of the kettle refining is based on the Betterton-Kroll process in which softened and desilverized lead is treated with magnesium and calcium to form insoluble high-melting compounds such as Ca_3Bi_2 and Mg_3Bi_2, forming a dross that is skimmed from the bath. Lead is further liquated from the dross to enrich it, and it is then treated with lead chloride or chlorine to remove the calcium and magnesium. The resulting lead-bismuth alloy is desilverized by the Parkes process and chlorinated to remove the lead. The deleaded Bi is given a final caustic treatment, resulting in a metal with a purity in excess of 99.99%.

Most of the electrolytic refining of lead is done using the Betts process, which employs an electrolyte of lead fluosilicate and free fluosilicic acid. The Bi which is present in the impure lead anodes is collected as slimes, which also contain silver and gold. These slimes are melted down in a reverberatory furnace to fume off the arsenic and antimony. The metal is transferred to a cupelling furnace in which a litharge slag containing the Bi is formed, along with doré metal containing the silver and gold. The litharge slag is reduced with coke, forming a crude lead-bismuth alloy ($\sim$25% Bi). This metal is desilverized and then cast into anodes for electrolytic refining. The enriched Bi slimes which result are melted down and chlorinated to remove the lead. A caustic treatment follows which removes the residual chlorides. The purity of the Bi is in excess of 99.99%.

Chemical Properties: In its chemical behavior Bi is similar to arsenic and antimony. Its properties are the most metallic of the group. It exhibits valences of 3^-, 0, 3^+, and 5^+. The 3^+ compounds of Bi are largely covalent in the solid state. In the elemental form Bi is relatively stable and does not oxidize readily when exposed to the atmosphere for prolonged periods of time. Bismuth will oxidize at its melting point and form a protective oxide layer. Metallic Bi vapor oxidizes readily to form Bi_2O_3.

Bismuth is insoluble in cold sulfuric acid and is slightly attacked by concentrated sulfuric acid. It is attacked by HNO_3, aqua regia, but not by cold HCl. Hot concentrated sulfuric and hydrochloric acid are reported to attack Bi. The attack of Bi by HNO_3 is accelerated with the addition of 1–5% Na_2CrO_4. Elemental Bi is relatively inert to alkalies.

Some typical compounds of Bi are listed as follows:

HYDROGEN BISMUTHIDE (bismuthine) may be formed by the reaction of Mg_2Bi_3 by hydrolysis or on reaction with dilute HCl. Because of its lack of stability, its properties have not been studied.

BISMUTH SESQUIOXIDE, Bi_4O_6, normally referred to as bismuth trioxide, may be formed by the direct oxidation of Bi with air. The pentoxide will not form under ordinary conditions of oxidation. The trioxide may also be produced by the decomposition of basic bismuth nitrate. It is yellow in color, with a melting point of 817°C and boiling point of 1900°C.

BISMUTH PENTOXIDE, Bi_2O_5, may be formed by the oxidation of an alkaline suspension of the trioxide, with a strong oxidizing agent such as peroxide and hypochlorite. The fusion of Bi_2O_3 with potassium hydroxide and potassium chlorate in air will form Bi_2O_5. It is reddish brown in appearance. Upon heating it will decompose to Bi_2O_3 at about 360°C.

BISMUTH TETROXIDE, Bi_2O_4, has been reported; however, it probably is a mixture of Bi_2O_3 and Bi_2O_5.

BISMUTH MONOXIDE, BiO, is reported to be prepared by heating basic bismuth oxylate. It oxidizes readily to Bi_2O_3.

BISMUTH HYDROXIDE, $Bi(OH)_3$, or oxide hydrate, BiO(OH), may be obtained by the dissolution of bismuth in nitric acid and precipitating it in the cold with sodium hydroxide.

The halogen compounds of Bi normally can be prepared by the direct combination of the elements. Compounds of the type BiX_3, in addition to BiCl, $BiCl_2$, $BiBr_2$, $BiCl_4$, and BiF_5, have been reported. The various compounds of BiX_3 are listed as follows:

Compound	Appearance	Mp, °C	Bp, °C
BiF_3	Gray crystalline	727	
$BiCl_3$	White crystalline	232	447
$BiBr_3$	Yellow crystalline	218	453
BiI_3	Reddish crystalline	408	$\sim$500

When halides of bismuth are hydrolyzed with water, oxyhalides are often formed.

Bismuth nitrate, usually obtained as pentahydrate, $Bi(NO_3)_3 \cdot 5H_2O$, is prepared by dissolving elemental Bi in concentrated HNO_3 and crystallizing it from solution. The pentahydrate crystals on dilution with water will form an oxysalt.

Basic bismuth nitrate or subnitrate, $Bi(OH)_2NO_3$, or $xBi_2O_3 \cdot yN_2O_5 \cdot zH_2O$, is prepared by the hydrolysis of bismuth nitrate from a given HNO_3 range maintained by using $NaHCO_3$. The hydrolysis is carried out in the temperature range 30–70°C.

Basic bismuth carbonate or subcarbonate, $Bi(OH)_2CO_3$ or $(BiO)_2CO_3$, may be produced under controlled conditions by the treatment of a suspension of bismuth subnitrate with Na_2CO_3.

Bismuth sulfate, $Bi_2(SO_4)_3$, may be obtained by dissolving Bi_2O_3 in H_2SO_4 and evaporating the acid solution. The sulfate will convert to basic sulfate, $(BiO)_2SO_4$, upon hydrolysis.

Bismuth sulfide, Bi_2S_3, is formed by the reaction of H_2S with a Bi salt. It can also be formed from a direct reaction of the elements.

A number of Bi organic compounds have been prepared. Some of the more familiar organic compounds of Bi are the tartrates and camphorates.

Bismuth is electronegative in nature and will react with many metals to intermetallic compounds.

Economy and Uses: The total free-world production of refined Bi in 1968 was approximately 8 million lb, of which about one-third was consumed by the United States. The production of Bi is related in part to the production of lead, and will be affected by the uncertainty of future applications of tetraethyl lead in gasoline. Up until the summer of 1971 the price of commercial quantities of Bi offered in purities of 99.99% was $6 per lb, at which time it fell to $4.75 per lb. Quantity price for 99.999 + % Bi is $6.50 per lb.

The major uses of Bi are in metallurgical, pharmaceutical, cosmetic, catalytic, and electronic applications.

In metallurgical applications, it is used in the formulation of low-melting-point fusible alloys and as an additive to aluminum, steel, and cast iron.

Fusible alloys which contain about 50% Bi and different combinations of lead, tin, cadmium, and indium are used in a variety of applications, which include anchoring, joining and sealing, heat-protective devices, and short-life dyes. Alloys with zero liquid-to-solid volume changes can also be formulated for special applications.

As a metallurgical additive, it is added to both steel and aluminum in quantities of about 0.2%, along with similar amounts of lead to improve machinability. In the manufacture of malleable cast iron, it is used in quantities of 0.02% to stabilize carbides on solidification, especially in heavy cross-section castings.

In pharmaceutical applications, it is used in proprietary formulations as a remedy for some types of gastrointestinal disturbances. At one time its chemotherapeutic properties were useful in the treatment of syphilis.

The pearlescent nature of bismuth oxychloride has found wide application as an ingredient in cosmetics. It imparts a frosty look to lipstick, nail polish, and eye shadows.

As a catalyst, bismuth phosphomolybdate earlier was an important development in the production of acrylonitrile for use in plastic fibers and paints.

In the field of electronics, Bi-Sn and Bi-Cd alloys have been used as counterelectrode alloys in the manufacture of selenium rectifiers. The compounds bismuth telluride, Bi_2Te_3, and bismuth selenide, Bi_2Se_3, exhibit thermoelectric properties. Modifications of these compounds are used as solid-state devices in some commercial and military applications for power-generation and refrigeration purposes.

Hygienic Considerations: Under normal conditions, Bi metal may be handled with impunity. It is one of the least toxic of the heavy metals.

—S. C. Carapella, Jr., *American Smelting and Refining Company, South Plainfield, N.J.*

Bisphenol A (See **Epichlorohydrin; Epoxy resins; Phenoxy resins;** and **Polycarbonates.**)

Bituminous Coal (See **Carbon;** and **Coal tar and derivatives.**)

Biuret (See **Urea;** and **Urethanes.**)

Black Liquor [See **Pulp (wood) production and processing.**]

Black Powder (See **Explosives.**)

Blast Furnace (See **Coal tar and derivatives; Energy systems for processes;** and **Ironmaking and steelmaking.**)

Blasting Agents (See **Explosives.**)

Bleaching [See **Chlorine; Detergents; Peroxides, organic;** and **Pulp (wood) production and processing.**]

Bleaching Powder (See **Chlorine.**)

Blending (See **Homogenization; Mixing, fluids; Mixing, pastes;** and **Mixing and blending, solids/solids.**)

Block Copolymers (See **Polymerization.**)

Blood Plasma (See **Peptides and proteins.**)

Blooming Mill (See **Ironmaking and steelmaking**.)

Blow Molding (See **Plastics**.)

Blowing (See **Glass**.)

Blowing Agent (See **Mixing, fluids**; and **Urethanes**.)

Bohr's Atomic Model (See **Atomic structure**.)

Boilers (See **Air pollution; Energy systems for processes; Fuels; Sulfur dioxide removal**; and **Water treatment**.)

Boiling Points (See **Chemical elements**.)

Bonding (See **Atomic structure; Ceramics**; and **Molecule**.)

Bonding Agents [See **Silicates (soluble)**.]

Bornite (See **Copper production**.)

Boron (Arabian *būraq*, white.) **B** = 10.82 (at. wt.); 5 (at. no.). The elemental form of boron is a black-brown powder, or black or red very hard crystals which display only slight reactivity at room temperature. The electrical qualities of boron designate it as a semiconductor. The elemental form does not occur naturally.

Boron is a member of IIIA, period 2, of the periodic table. The melting point of elemental B is approximately 2200°C, boiling point = approximately 4200 K, and density 2.35 g/cm³ in the amorphous form. Other important properties of B are summarized in Tables B-6 and B-7. Also see **Chemical elements**.

Around 1807, Davy[1] accomplished the first production of elemental B in amorphous form by means of electrolysis of boric acid. At approximately the same time, Gay-Lussac and Thénard[2] obtained a similar product by reducing boric acid with potassium. It was not until 1892 that Moissan[3] achieved B with a purity of over 90% by reducing B_2O_3 with Mg. Moissan's work on the chemical behavior of the then so-called "Moissan boron" revealed that the products of earlier works could, at best, have been compounds of B.

Weintraub[4] melted B powder, produced in accordance with Moissan's method, thereby vaporizing the impurities and obtaining lumps with a purity of approximately 98%. Weintraub also tested the reduction of BCl_3 with H_2 in the electric arc and mentioned the decomposition of gas on the surface of hot B rods. Warth[5] developed the latter method further. Earlier, Stock[6] obtained small quantities of highly pure B by thermally decomposing boranes. Van Arkel[7] also used the method of pyrolysis for boron tribromide to deposit B on very thin metal wires.

The very pure base substances for the gas-phase decomposition facilitated the production of sufficiently pure B for examining the complex structure and physical characteristics of the element. Since 1945, enriched ^{10}B also has been available in larger quantities.

Commercial Production and Purification. Boron occurs naturally mainly in the form of boric acid (more or less aqueous borates) and in rich natural deposits of boric silicates.[8]

Reduction Using Metals: In an exothermic reaction, the abundant B_2O_3 is reduced by Li, Na, K, Mg, Be, Ca, or Al, the most efficient of which is Mg. Since the time of Moissan, Mg has been used for the technical production[9] of B. This method produces a brown powder with at least a 90–95% B purity. Boric oxide, MgO, and magnesium boride also are present in varying proportions. At a temperature of approximately 2000°C, under vacuum, a 99.5% B is obtained,[10] with traces of Mg and O_2. Sodium will reduce KBF_4 and also BF_3. The use of Na results in a boron of about 90% purity.

Reduction with Compounds: In the electric arc, B_2O_3 and also BCl_3 have been reduced with H_2. Calcium carbide and tungsten carbide also have been used for the reduction of B_2O_3. BF_3 has been transformed with CaH_2. These reactions did not prove economically significant.

Reduction of Gaseous Compounds with Hydrogen: Boron deposition will occur on metallic filaments or bars heated to a surface temperature of 800–1600°C in an atmosphere of boron halide in hydrogen.[11-13] The deposit ranges from amorphous (by x-ray diffraction) to the various crystalline modifications, depending upon temperature and pressure during deposition. The purity of B extracted by this process is high because the starting materials can be made quite pure,[14] although the deposition rate is low. A variation of this method is used to produce high-strength B filaments.[15]

Thermal Decomposition of Boron Compounds: Well suited to this process are the very poisonous boranes, which, in combination with O_2 or H_2O, also are very strongly reactive. The boron halides, some borides, boron sulfide, boron phosphide, and alkali metal borohydrides (at a higher temperature) can be thermally decomposed.

To produce the hyperpure polycrystalline B which is used in semiconductor technology,[16] B_2H_6 is decomposed on the hot surface of a B rod under reduced pressure. By zone-refining such rods in an H_2 plasma, generated by RF currents, monocrystals of β-rhombohedral B are obtained.[17] These methods are suitable for obtaining small quantities of highly pure materials, in single-crystal form.

Electrochemical Reduction of Boron Compounds: In this method, smeltings of metallic borates or metallic fluoroborates are decomposed between electrodes of metal. Similarly, boron oxide alkali metaloxide–alkali chloride compounds are decomposed between carbon electrodes. It is assumed that the first separated metal will react in smelting, thereby forming B, which will be found near the cathode in the form of thinly dispersed powder.[18] By this method, the enrichment of isotope [10]B (important in reactor technology) was achieved.

Elemental B also can be deposited on the cathode by dissolving an anode of boron carbide. A compound of NaCl–KCl and KBF_4 is used as the electrolyte.[19] Electrochemical methods are well suited for the technical production of B powder with a purity of approximately 99.8%, and often are preferred to the direct reduction with metals.

Purification: Boron deposited from the gas phase on hot surfaces contains only a small portion of impurities inasmuch as very pure raw materials are decomposed. Attempts have been made to remove the pure deposit from the substrate by breaking, sorting, and etching. If the impurities are dissolved, other methods of purification are required.

Chemical Methods: Raw B, produced in accordance with Moissan, is treated with acids, leaving a residue of at least 90% B and about 8% Mg. The residue is heated with approximately equal parts of KBF_4, KF, and HF under argon or helium for approximately 30 min at 1000°C. This is followed by another leaching with acid. The resultant powder (95–97% purity) is melted in vacuum and thereby refined to B with a purity of over 99%.[20] Gaseous HCl at 550°C and BF_3 gas or HF also can be used individually, as well as meltings of KF or of fluoroborates.

The high-temperature treatment in vacuum, just described, can be used at 1300–2000°C for a distinct depletion of the various impurities. Al, Cu, Mg, and Sn evaporate almost completely; Fe, Si, and Ti evaporate partially.[21]

Through sublimation of B, a product of 99.95% purity can be achieved.[22]

Float zoning is the preferred method of purification, especially for final purification and simultaneous molding. Boron nitride is used as the crucible material,[23] if crucible-free float zoning is not preferred.[24,25] A B rod, produced by pyrolysis, or a rod of B powder condensed with boric acid as agent, is zone-refined several times if necessary. The main impurity of the resultant 99.99% B is carbon in a concentration of less than 30 ppm. This carbon is difficult to remove even by repeated float zoning.

Structure

Bondings: Because of its structure, crystalline B has a unique role among the chemical elements. The B atom, having only three valence electrons, is able to form only three classic bondings between atoms with three equal neighbors. However, no stable octet of electrons results. Therefore there is no possibility of constructing a B lattice by normal covalent bondings. Conversely, the atoms in the lattice cannot hold together by displacement of all bondings because the bond strength of the valence electrons is too high. Bondings by displacements would make a metallic B, *but crystalline B is not a metal.*

The structure of B is explained by the so-called *delta bonding,* where a common pair of electrons bonds not only two, but also more (usually three) B atoms. Accordingly, the single B atom is not the most important block, but instead, a very stable unit contains 12 B atoms. This unit is an almost regular icosahedron. Each of the 12 corners of this polyhedron marks a fivefold axis. Because these points have a fivefold symmetry, each atom is surrounded by five other atoms at about the same distance. Since there are only three valence electrons available for the five bindings of each atom, there are no bondings belonging to one another in an icosahedron. In an icosahedron, the 12 B atoms would own 36 valence electrons. But only 26 are used for the icosahedron (B—B, 1.73–1.79Å). The remaining 10 are available for outside bondings.[26]

Modifications: A three-dimensional crystalline lattice cannot be built—without holes—from quasi-fivefold structural elements like the B icosahedron. Even by dense packing of the icosahedron, the steric capacity is not used up and shows large spaces which can be filled with individual atoms—B or other elements.

These individual atoms, in turn, influence the arrangement of the B icosahedra, and thereby the final structure of the lattice. Therefore it cannot be determined at once which of the lattice types of boron thus far discovered are "genuine" and represent pure B lattices.[27] The existence of four modifications (Table B-6) the structures of which are known seems to be assured. All four modifications have demonstrated that the B icosahedron is the most important building block. This applies also to the glasslike amorphous B which in short-range order also consists of B icosahedra. This material has not yet yielded a stereoperiodic structure.

While the lattices of the α-rhombohedral and the II-tetragonal boron are formed only from these icosahedra, the I-tetragonal and β-rhombohedral boron (containing icosahedra combined together in a complex way) need, in addition, some individual B atoms.

Transformations: At present, it is not possible to name well-defined stability zones for the various modifications of B. Although a series of transformations has been observed, none of these were reversible.[18] The reaction conditions under which the individual B lattices form also do not indicate stability limits. Apart from the crystallization out of the melt, these reactions occur in temperature zones (approximately 800–1600°C) in which kinetic factors are of much greater importance than theoretical lattice energies.

Comparatively recent test results point out that some lattice types, thus far accepted as "genuine" B lattices, are induced only by foreign atoms, or stabilized beyond their normal existence range; i.e., in reality they represent B compounds.[33]

Only so far as rhombohedral modifications are concerned can reliable observations be made. For instance, only the β-rhombohedral B crystallizes from the B melt. However, below 1000°C, only α-rhombohedral B forms as the crystalline B phase. This latter substance[34] is thermally unstable and changes irreversibly into β-rhombohedral boron above 1200°C.

To date, it has not been decided, either through experiments or thermodynamic considerations, whether the α-rhombohedral B and the tetragonal modifications, in comparison with the β-rhombohedral B, display their own thermodynamic stabilization zones or represent only monotropic forms.

Chemical Properties. The reactivity of B depends strongly upon the surface area, structure, and degree of purity.

Reactions with Nonmetals: Boron reacts vigorously with F_2, especially in fine dispersion. The reactions of coarse crystalline B with halogens of a higher periodic number are progressively slower, yet crystalline B transforms quickly[35] with I_2 at 900°C.

Finely dispersed B oxidizes in atmospheric O_2 at 20°C and causes ignition at 800°C. Coarse crystals are guarded from further oxidation by the formation of a continuous layer of boric oxide.

Sulfur causes the creation of B_2S_3 at 600°C. Selenium reacts at still higher temperatures. With melted Te (mp = 452°C), no reaction is observed.

BN is formed at approximately 1000°C. Boron phosphide forms at approximately the same temperature. Formation of the carbide commences above 1000°C. Reactions with Si require even higher temperatures.

Reactions with Metals: Although no compound formation is described with alkali metals, B starts to react with Be at approximately 1100°C and with Hg at 800°C. The reactions with alkali-earth metals occur more readily at lower temperatures. Aluminum boride is formed from the elements, commencing at 800°C. Boron and the transition elements frequently form compounds.[36]

Reactions to Oxides, Acids, and Bases: At a red heat, a reaction occurs between B and H_2O. At 1050°C, boron monoxide is formed from B and B_2O_3. H_2O_2 and concentrated HNO_3 oxidize B into B_2O_3. Above 1050°C, B reduces SiO_2; above 1200°C, B reduces CO; and at a bright-red heat,

TABLE B-6. *Modifications of Elemental Boron*

Modification	Steric Group	a_0	c_0	α	Unit Cell	Building Unit
α-Rhombohedral	R $\bar{3}$ m	5.057	. . .	58.06	12	B_{12}
β-Rhombohedral	R $\bar{3}$ m	10.14	. . .	65.28	105	$B_{84} + 2B_{10} + 1B$
I-Tetragonal	$P4_2/nnm$	8.75	5.06	. . .	50	$4B_{12} + 2B$
II-Tetragonal	$P4_1$ 22	10.12	14.14	. . .	192	$16B_{12}$ (?)

CO_2 also is reduced. Under these conditions, N_2O or NO_2 creates B_2O_3 and BN. At temperatures above 800°C, P_2O_5, As_2O_3, As_2O_5, and SO_2 are reduced. Thinly dispersed B partially reduces some metallic oxides at 20°C.

Boron is very resistant to nonoxidizing acids. Especially under heat, B is rapidly attacked by oxidizing acids. Boron reacts similarly with aqueous solutions of oxidizing compounds or its meltings.

Aqueous solutions of alkali remain ineffective (likewise alkali meltings) below 500°C.

Physical Properties. Until relatively recently, most of the physical properties of B could not be established on an exact basis because of the difficulty of producing large specimens of pure B containing only one particular modification. Most of the measurements were conducted with specimens of unknown structure and frequently of doubtful purity. For several years, β-rhombohedral B, in large crystals of high purity, was the only available modification for study. Thus this form of B is the most studied modification insofar as its physical properties are concerned. Unless especially noted, the physical data given in Table B-7 deal with β-rhombohedral B.

Although the II-tetragonal form and the α-form are red, the other modifications of B appear almost black and with a high metallic luster.

Among the elements, only the diamond surpasses B in hardness. At a melting point of approximately 2200°C, B is considered a refractory material.

As early as 1909, it was perceived that B belongs to the group of semiconductor elements.[4] Although single crystals have become available since that date,[17] the examined semiconductor properties are based almost exclusively on analyses of polycrystalline materials. It has been mentioned that β-rhombohedral B, in its purest form, still contains impurities at a level[16] which, in other semiconductors, would be considered high doping. For instance, the carbon concentration for the purest B crystals is on the order of 10^{18} cm^{-3}.

The unusual optical and electrical phenomena displayed by the semiconductor β-rhombohedral

TABLE B-7. *Some Important Properties of Boron*

Atomic properties:
  Atomic number 5
  Atomic weight 10.82
Isotopes, %:
  10 . . . . .19.8
  11 . . . . .80.2
Cross section (thermal neutrons):
  ^{10}B 3,840
  ^{11}B 0.005
Atomic radius, Å 0.86
Mechanical properties:
Compressibility, at 30°C:
  Linear . . . .1.8×10^{-7}
  Volume . . . .$<5.5 \times 10^{-7}$
Density, g/cm³:
  Amorphous 2.35
  α-Rhombohedral 2.46
  β-Rhombohedral 2.35
  ^{10}B 1.73
Hardness:
  Mohs' hardness 9.3
  Vickers' hardness (100-g load), kg/mm²:
    At 25°C 2,800
    500°C . . . .~2,000
    900°C . . . .~400
  Knoop hardness (100-g load), kg/mm² 2,410
Tensile strength (amorphous), psi . . . .$2.3–3.5 \times 10^5$
Young's modulus (amorphous), psi . . . . .64×10^6

TABLE B-7. *Some Important Properties of Boron*   (*Continued*)

Thermal properties:
  Melting point, °C . . . . ~2200
  Heat of melting, kcal/g atom 5.3
  Boiling point, K . . . . ~4200
  Heat of sublimation, kcal/g atom:
    At 25°C 137
    4200 K 129
  Vapor pressure, atm:
    At 2115 K . . . . 3.5×10^{-7}
    2413 K . . . . 156×10^{-7}
  Debye temperature, K 1219
  Thermal conductivity:
    Polycrystalline, cal/(cm)(s)(deg) 0.003
    Amorphous, cal/(cm)(s)(deg) 0.0077
  Thermal coefficient of expansion, deg^{-1}, at 20–750°C . . . . . . 8.3×10^{-6}
  Specific heat:
    Amorphous, cal/(g)(deg), at 25°C 0.263
    β-Rhombohedral, cal/(g)(deg), at 25°C 0.245
  Heat of transition, kcal/g atom:
    Amorphous → β-rhombohedral 0.4
Optical properties:
  Refractive index:
    At 0.25/μm 2.7
    0.5/μm 3.3
    1/μm 3.0
  Reflection coefficient, 1–20/μm 0.24–0.22
  Infrared absorption, 0.9–4.5/μm . . . . Rather transparent
  Absorption coefficient, cm^{-1}, at 3/μ 7
  Absorption edge, μm:
    α-Rhombohedral 0.6
    β-Rhombohedral 0.8
Electrical properties:
  Band gap, eV:
    α-Rhombohedral 2.0
    β-Rhombohedral 1.56
  Hole concentration, cm^{-3}, at 500 K . . . . 10^{16}
  Hole mobility (hopping process), cm^2/(V)(s) . . . . <0.1
  Specific resistance, Ω-cm, at 300 K . . . . ~10^6
  Temperature coefficient of resistance, deg^{-1}, at 300 K . . . . −0.079
  Thermoelectric power, μV/deg:
    At 25°C 700
    200°C 600
  Photoconductivity, μm, at 300 K . . . . Peaks at 0.35, 0.8, and 5.5
  DC dielectric constant 14

B in present available purity arise from local energy levels or traps for the electrons and holes in the crystal.[37,75,76] These levels presumably result from foreign atoms and from structural defects on the atomic scale in the crystal.

The activation energy of the traps for the electrons is much higher than for the holes. Therefore, up to very high temperatures, the electrons do not noticeably contribute to the transport of charges. At lower temperatures, hopping processes between the trap levels of holes represent the prevailing mechanism of conductivity. At higher temperatures, the conductivity in the valence band is predominant.

Applications. Presently, elemental B remains a material for which there is not a large market. Comparatively recently, B filaments gained significance as a reinforcing fiber for plastics and light metals.[66]

In the field of metallurgy, B is used to degas copper[67] and to reduce other nonferrous metals.[68] A better resistance against corrosion as well as a greater hardness at high temperatures is reached if steel is alloyed with 0.0005–0.004% B.[69]

In the form of thin surface layers, B provides protection because of its hardness and resistance to abrasion and corrosion.[70] Because of the high effective cross section of ^{10}B for thermal neutrons $[^{10}B(n,\alpha)Li^7]$, natural and enriched B is applied for standard test bars and shieldings in nuclear reactors.[71] Boron-layered thermocouples are used as neutron flux counters, thereby utilizing the heating of the layered welding area by absorbed neutrons.[72]

A number of propositions have been made for the use of B in the electrical field, but to date there are only a few examples. Its use as a thermistor[63] or as a component of a complex semiconductor alloy[73] for ignition electrodes in rectifiers and vacuum gages is worthy of mention. Because of its current-voltage characteristic, B can be used as a switch.[74] For various purposes, B also is used as a dopant (*p* type) for *p-n* junctions in silicon. The main obstacle affecting the greater use of B as a semiconductor is the high-lattice-defect concentration in the B crystals now available.

See also **Boron compounds.**

REFERENCES

1. Davy, H.: *Phil. Trans.,* vol. 98, p. 333, 1808, and vol. 99, p. 32, 1809.
2. Gay-Lussac, J. L., and L. J. Thénard: *Rech. phys.-chim. (Paris),* I, p. 278, 1811; *Ann. Chim. Phys.,* I, vol. 68, p. 169, 1808.
3. Moissan, H.: Ct. R., vol. 114, p. 319, 1892.
4. Weintraub, E.: *Trans. Am. Electrochem. Soc.,* vol. 16, p. 165, 1909.
5. Warth, A. H.: *Maryland Acad. Sci. Bull.,* vol. 3, no. 3, p. 8, 1923.
6. Stock, A.: "Hydrides of Boron and Silicon," Cornell University Press, Ithaca, N.Y., 1933.
7. Van Arkel, A. E.: *Reine Metalle,* (Berlin), 1939; *Metallw.,* vol. 13, 405/8, p. 511, 1934.
8. Rankma, K., and T. G. Samaha: "Geochemistry," Chicago, 1950.
9. Smatko, J. S.: *FIAT Rep.* 738 (PB-34726), Feb. 21, 1946.
10. Markovskii, L. Y. A.: *Proc. Int. Symp. on Boron,* p. 95, Warsaw, 1968.
11. Laubengayer, A. W., D. T. Hurd, A. E. Newkirk, and J. L. Hoard: *J. Am. Chem. Soc.,* vol. 65, p. 1924, 1943.
12. Ellis, R. C.: In "Boron: Synthesis, Structure, and Properties," p. 42, Plenum, New York, 1960.
13. Bean, K. E., and W. E. Medcalf: In ibid., p. 48.
14. Armington, A. F., J. T. Buford, and R. J. Starks: In G. K. Gaulé (ed.), "Boron," vol. 2, p. 21, Plenum, New York, 1965.
15. Lasday, A. H., and C. P. Talley: *SAMPE 10th Natl. Symp.,* vol. 10, D-1, San Diego, Calif., November 1966.
16. Hinz, Ingeborg, and H. Wirth: In G. K. Gaulé (ed.), "Boron," vol. 2, p. 9, Plenum, New York, 1965.
17. Borchert, W., W. Dietz, and H. Kölker: *Z. angew. Phys.,* vol. 29, p. 277, 1970.
18. Newkirk, A. E.: In R. Adams (ed.), "Metallo-Boron Compounds and Boranes," p. 233, Wiley, New York, 1964.
19. Stern, D. R.: In "Boron: Synthesis, Structure, and Properties," p. 27, Plenum, New York, 1960.
20. Yannacakis, J., and N. P. Nies: In ibid., p. 39.
21. Newkirk, A. E.: In R. Adams (ed.), "Metallo-Boron Compounds and Boranes," p. 253, Wiley, New York, 1964.
22. Searcy, A. W., and C. E. Meyers: *J. Phys. Chem.,* vol. 61, p. 957, 1957.
23. Horn, F. H.: In "Boron: Synthesis, Structure, and Properties," p. 70, Plenum, New York, 1960.
24. Starks, R. J., and W. E. Medcalf: In ibid., p. 59.
25. Greiner, E. S.: In ibid., p. 105.
26. Longet-Higgins, M. C., and M. de V. Roberts: *Proc. Roy. Soc.,* ser. A, vol. 230, p. 110, 1955.
27. Hoard, J. L., and A. E. Newkirk: *J. Am. Chem. Soc.,* vol. 82, p. 70, 1960.
28. Decker, B. F., and J. S. Kasper: *Acta Cryst.,* vol. 12, p. 503, 1959.
29. Hughes, R. E., C. H. L. Kennard, D. B. Sullenger, H. A. Weakliem, D. E. Sands, and J. L. Hoard: *J. Am. Chem. Soc.,* vol. 85, p. 361, 1963.
30. Hoard, J. L., R. E. Hughes, and D. E. Sands: *J. Am. Chem. Soc.,* vol. 80, p. 4507, 1958.
31. Talley, C. P., S. LaPlaca, and B. Post: *Acta Cryst.,* vol. 13, p. 271, 1960.
32. Amberger, E., and K. Ploog: *J. Less Common Metals,* vol. 23, p. 21, 1971.
33. Ploog, K., and E. Amberger: *J. Less Common Metals,* vol. 23, p. 33, 1971.
34. Hoard, J. L., and R. E. Hughes: In E. L. Muetterties (ed.), "The Chemistry of Boron and Its Compounds," p. 25, Wiley, New York, 1967.
35. Newkirk, A. E.: In R. Adams (ed.), "Metallo-Boron Compounds and Boranes," p. 276, Wiley, New York, 1960.
36. Samsonov, G. V. (ed.): "Boron: Its Compounds and Alloys," p. 278, Publishing House of the Academy of Sciences of the Ukrainian SSR, Kiev, 1960.
37. Werheit, H.: Festkörperprobleme X, pp. 189–226, Pergamon/Vieweg, Braunschweig, 1960.
38. Bridgman, P. W.: *Proc. Am. Acad. Arts Sci.,* vol. 64, p. 51, 1929.
39. Talley, C. P., L. E. Line, Jr., and Q. D. Overman, Jr.: In "Boron: Synthesis, Structure, and Properties," p. 94, Plenum, New York, 1960.
40 McCarty, L. V., and D. R. Carpenter: *J. Electrochem. Soc.,* vol. 107, p. 38, 1960.
41. Hoard, J. L., D. B. Sullenger, C. H. L. Kennard, and R. E. Hughes: *J. Solid State Chem.,* vol. 1, p. 268, 1970.

42. King, L. D. P.: *Proc. U.N. Int. Conf. Peaceful Uses Atomic Energy,* vol. 2, p. 372, 1956.
43. Seybolt, A. U.: *Trans. Am. Soc. Metals,* vol. 52, p. 971, 1960.
44. Cline, C. F.: *J. Electrochem. Soc.,* vol. 106, p. 322, 1959.
45. Talley, C. P.: *J. Appl. Phys.,* vol. 30, p. 1114, 1959.
46. Talley, C. P.: *J. Appl. Phys.,* vol. 32, p. 1787, 1961.
47. Stull, D. R., and G. C. Sinke: Thermodynamic Properties of the Elements, *J. Am. Chem. Soc.,* 1956, pp. 12–13, 58–59.
48. Paule, R. C., and J. L. Margrave: *J. Phys. Chem.,* vol. 67, p. 1368, 1963.
49. Searcy, A. W., and C. E. Meyers: *J. Phys. Chem.,* vol. 61, p. 957, 1957.
50. Johnston, H. L., H. N. Hersh, and E. C. Kerr: *J. Am. Chem. Soc.,* vol. 73, p. 1112, 1951.
51. Talley, C. P.: *J. Phys. Chem.,* vol. 63, p. 311, 1959.
52. *Proc. Conf. on Boron,* Fort Monmouth, N.J., 1959, Plenum, New York, 1960.
53. Dupuy, E., and L. Hackspill: *C. R.,* vol 197, p. 229, 1933.
54. Rossini, F. D., D. D. Wagman, W. H. Evans, S. Levine, and J. Jaffee: *Natl. Bur. Standards Circ.* 500, p. 313, 1952.
55. Kierzek-Pecold, E., J. Kolodziejczak, and J. Pracka: *Phys. Stat. Sol.,* vol. 22, K147, 1967.
56. Spitzer, W. G., and W. Kaiser: *Phys. Rev. Letters,* vol. 1, pp. 230, 382, 1958.
57. Jaumann, J., and H. Werheit: *Phys. Stat. Sol.,* vol. 33, p. 587, 1969.
58. Horn, F. H.: *J. Appl. Phys.,* vol. 30, p. 1611, 1959.
59. Medcalf, W. E., K. E. Bean, and R. J. Starks: *Met. Soc. Conf.,* vol. 12, p. 381, 1962.
60. Dzhamagidze, S. Z., Y. A. Mal'tsev, and R. R. Shavangiradze: *Sov. Phys. Semicond.,* vol. 2, p. 320, 1968.
61. Neft, W., and K. Seiler: In G. K. Gaulé (ed.), "Boron," vol. 2, p. 89, Plenum, New York, 1965.
62. Shaw, W. C., D. E. Hudson, and G. C. Danielson: *Phys. Rev.,* vol. 107, p. 419, 1957.
63. Dietz, W., and H. Herrmann: International Symposium on Boron, *Electron Tech.,* vol. 3, no. 1, p. 2, P. A. Sci., Warsaw, p. 195, 1970.
64. Borchert, W., W. Dietz, and H. Herrmann: *Z. Angew. Phys.,* vol. 19, p. 485, 1965.
65. Werheit, H.: *Phys. Stat. Sol.,* vol. 39, p. 109, 1970.
66. United Aircraft Corp; Hamilton Standard Div.: *Composite Mater. Dept. Tech. Data Sheet,* 1970.
67. British Thompson-Houston Co., Ltd.: British Patent 193,844, Oct. 1, 1921.
68. Stipp, H. E.: *U.S. Bur. Mines Bull.* 585, p. 141, 1960.
69. Grange, R. A.: Boron in Iron and Steel, pp. 1–57 in "Boron, Calcium, Columbium, and Zirconium in Iron and Steel," Wiley, New York, 1957.
70. Powell, C. F., J. E. Campbell, and B. W. Gonser: "Vapor Plating," pp. 103–119, Wiley, New York, 1955.
71. Anderson, W. K.: *Nucl. Sci. Eng.,* vol. 4, p. 357, 1958.
72. Jaques, T. A. J., H. A. Ballinger, and F. Wade: *Proc. Inst. Elec. Eng.* (*London*), vol. 100, p. 110, 1952.
73. Cooper, H. S.: Boron, in C. A. Hampel, (ed.), "Rare Metals Handbook," Reinhold, New York, 1961.
74. Dietz, W., and H. Helmberger: In G. K. Gaulé, (ed.), "Boron," vol. 2, p. 301, Plenum, New York, 1965.
75. Gaulé, G. K., J. T. Breslin, and R. R. Patty: In ibid., p. 169.
76. Prudenziati, M., G. Majni, and A. Alberig: *Quaranta J. Phys. Chem. Solids,* Pergamon, London, 1971.

—Josef Dietl, Wolfgang Dietz, Helmut Fenninger, and Wolf-R. Schiller, *Wacker Chemie, GmbH, Munich, West Germany*

Boron Compounds   Alfred Stock, the renowned German chemist, wrote in 1901, "It was evident that boron, the close neighbor of carbon in the periodic system, might be expected to form a much greater variety of interesting compounds than merely boric acid and the borates, which were almost the only ones known at that time." During the thirty years of research which followed, Stock synthesized nearly all the important *boranes* (compounds of hydrogen and boron). In addition to the familiar applications of borax and boric acid, boron compounds find application in high-energy rocket fuels and jet-engine and automotive fuels, in heat-resistant glasses, ceramics, and lubricants.

Stock considered the possibility of a series of boron hydrides because of the close proximity of boron to carbon in the periodic table. After a number of years of difficult experimentation with these compounds, Stock found that boron hydrides fall into two series, with the general formulas of B_nH_{n+4} and B_nH_{n+6}. During his lifetime, Stock discovered eight of the common boranes. Neoborane, B_9H_{15}, the ninth compound, was not firmly identified until 1958.

Potentially, boranes are better than liquid H_2 in terms of efficient rocket fuels. Despite their lower heats of combustion (significantly less than liquid H_2), the boranes yield 50% greater energy per unit mass than most conventional liquid and solid fuels. In addition to favorable application in solid propellants, such as derivatives of decaborane, $B_{10}H_{14}$, the organoboranes have been used for automotive and jet fuels. When added to leaded gasoline, for example, a fuel is produced with a lower octane requirement. Further, the resistance to premature ignition is greater.

Borohydride chemistry was further pioneered by Schlesinger and Burg (University of Chicago) in the late 1940s. In borohydrides, BH_4 units are attached to metals. Sodium borohydride, $NaBH_4$, is used as a reducing agent in producing certain synthetics. Uranium borohydride, $U(BH_4)_4$, and the methyl derivative, $U(CH_3BH_3)_4$, were tried during the Manhattan Project to separate the isotopes of uranium. Because of instability, these compounds were discarded in favor of the only other more volatile uranium compound, uranium hexafluoride, UF_6.

Boron carbide, B_4C, has found use as a neutron-absorbing material in nuclear reactors. ^{10}B has been used in research pertaining to the treatment of brain tumors. Borax, for example, when given to a patient intravenously, will cause boron to collect in a higher concentration in the tumor than in normal tissue.

Of interest is the similarity (chemical analog) of *borazine*, $B_3N_3H_6$, with benzene, C_6H_6.

Borazine   Benzene

Other interesting examples of this type are bibora-zinyl, $B_6N_6H_{10}$, analogous to biphenyl, $C_{12}H_{10}$, and borazanaphthalene, $B_5N_5H_8$, analogous to naphthalene, $C_{10}H_8$. Because of boron's tendency to link with itself, it may be the foundation of inorganic polymeric materials. For example, boron-phosphorus polymers, prepared by reacting diborane with phosphone derivatives, have been found to have high resistance to heat, but poor mechanical strength.

Borazine yields boron nitride at a very high temperature. Boron nitride takes two forms: (1) hexagonal boron nitride with a graphiticlike layer structure with excellent heat resistance and lubricating properties, and (2) *borazon*, which is a cubic crystalline material produced under great pressure (85,000 atm and greater) and high temperature (1800°C and greater). The transformation is similar to that of converting graphite into diamond. Tests have shown borazon to be harder than diamond and unaffected by temperatures below 1500°C.

Sources: The earliest sources of boron compounds were borax and boric acid. Ores of more recent importance include *ulexite*, $NaCaB_5O_9 \cdot 8H_2O$, *kernite*, $Na_2B_4O_7 \cdot 4H_2O$, and *colemanite*, $Ca_2B_6O_{11} \cdot 5H_2O$—all found in the United States—and *boracite*, $Mg_7Cl_2B_{16}O_{30}$, which occurs in Germany.

BORIC ACID, H_3BO_3, formerly sometimes called *boracic acid*, is a tonnage material, the predominant use of which (except as a raw material for boron and boron compounds) is in medical and pharmaceutical compounds, such as eye- and mouthwashes and nasal sprays. Solutions of boric acid are nonirritating and slightly astringent, with antiseptic qualities. A saturated solution of H_3BO_3 contains approximately 2% of the compound at 0°C, increasing to 39% at 100°C. H_3BO_3 also is soluble in alcohol. The compound no longer is used as a food (primarily meats) preservative because harmful effects have been observed. Boric acid, as well as borax, has been used in fire-retardant chemicals, such as *Minalith* (diammonium phosphate, ammonium sulfate, sodium tetraborate, and boric acid) and *Pyresote* (zinc chloride, ammonium sulfate, boric acid, and sodium bichromate). Boric acid also is used in the tanning industry for deliming skins by forming calcium borates soluble in water, as a flux in soldering and brazing, and in glass and pottery mixes. Commercially, boric acid has the composition $B_2O_3 \cdot 3H_2O$, derived by adding HCl or H_2SO_4 to a solution of borax, followed by crystallization. The resultant white crystalline powder has a specific gravity of 1.435 at 15°C and decomposes at 185°C.

BORAX (sodium tetraborate), $Na_2B_4O_7 \cdot 10H_2O$, is a very high tonnage material with many uses. Uses include (1) flux for welding and soldering wherein borax dissolves oxide coatings, leaving a clean metal surface; (2) filler in paints and paper; (3) cleaning compounds of many types; (4) ingredient of glass and ceramics, particularly heat-resistant glass where as much as 40 lb of borax may be used per 100 lb of finished glass; (5) source of elemental boron and other boron compounds; (6) constituent of fertilizers; and (7) as a corrosion inhibitor in antifreeze compounds. Like boric acid, borax also has fire-retardant properties and uses.

Borax is a white or colorless crystalline material. As found in nature (mainly western United States), borax has a specific gravity of 1.75, a hardness of 2–2.5, and a melting point of 75°C. When used in water softeners, detergents, and disinfectant products, the borax often is in water solution. An aqueous solution of borax may be described as being mildly alkaline and antiseptic.

BORON TRICHLORIDE, BCl_3, finds use primarily as a catalyst and for controlling magnesium fires. The compound is a gas above $12.5°C$, but also is used as the dihydrate (a fuming liquid).

BORON TRIFLUORIDE, BF_3, is a gas used in the polymerization of epoxy resins. Usually, the solid form (boron ethylamine), $BF_3 \cdot C_2H_5NH_2$, is used to release BF_3 at elevated temperatures.

BORON TRIBROMIDE, BBr_3, is highly reactive and used to form boron hydrides as sources of hydrogen.

TRIMETHYL BOROXINE, $(CH_3O)_3B \cdot B_2O_3$, a liquid, is used for extinguishing metal fires. Upon heating, the compound decomposes, causing a molten coating of boric acid which smothers the fire.

HYDRAZINE DIBORIDE, $BH_3 \cdot NH_2 \cdot NH_2 \cdot BH_3$, has been used as a source of H_2 in rocket fuels. The compound is a white crystalline free-flowing powder.

BORON CARBIDE, B_6C or B_4C, is a black crystalline powder and very hard. It is used as an abrasive and also pressed into wear-resistant products, such as drawing dies and gages. It also is used in heat-resistant parts, such as nozzles. Other uses have included shielding against neutrons (in combination with Al), as a grinding and lapping material, and as a deoxidizing agent for casting copper. The melting point is very high, but above $980°C$ the material reacts with O_2.

BORON NITRIDE, BN, is a light, fluffy white powder used as a lubricant for high-pressure bearings. It also is used for compacting into mechanical and electrical parts.

See also **Boron.**

Borosilicate Glass   (See **Glass.**)

Box Press   (See **Expression.**)

Boyle's Law   (See **Ideal gas.**)

Brackish Water   (See **Desalination.**)

Brannerite   (See **Uranium.**)

Brass   (See **Copper; Tin;** and **Zinc.**)

Braunite   (See **Manganese.**)

Brazing   (See **Aluminum; Copper;** and **Silver.**)

Breeder Reactor   (See **Nuclear power plants.**)

Bremsstrahlung   (See **Radioisotope.**)

Brewer's Yeast   (See **Yeasts.**)

Brighteners   (See **Detergents.**)

Brines   (See **Bromine; Caustic soda; Chlorine; Magnesium;** and **Salt, NaCl.**)

Briquetting   (See **Agglomeration;** and **Iron ores.**)

Britannia Metal   (See **Antimony;** and **Tin.**)

Bromic Acid   (See **Bromine.**)

Bromine   (Gk. *bromos,* bad smell.) **Br** = 79.916 (at. wt.); 35 (at. no.). Bromine, the only non-metallic element which is liquid at ambient temperatures, was discovered (1826) by the young French scientist Antoine-Jérôme Balard, who found Br occurring in seawater bitterns. Balard's chlorination procedure for the recovery of Br was the basis of all production until the bromide-rich Strassfurt, Germany, salt deposits were discovered in 1858. Bromine occurs widely in nature as the bromide ion, but is a relatively rare element, accounting for only about 1.6 ppm of the earth's crust. The principal commercial sources today include seawater and bitterns, salt lakes, deep brines, and certain salt deposits.

Bromine exists as two stable isotopes, ^{79}Br and ^{81}Br, in nearly equal proportions, giving the average atomic weight of 79.916. The element is in group VIIB, period 3, of the periodic table and has physical properties between two other halogens, chlorine and iodine. Some of the important physical properties of Br are given in Table B-8.

Like other halogens, bromine has seven valence electrons. All the halogens, of course, exhibit high electronegativity. In reactions, Br exhibits valences of 1^- and 5^-.

Bromine is a reddish-brown, fuming, corrosive liquid and a moderately strong oxidant (oxidation potential of -1.087 V, compared with hydrogen). It readily oxidizes numerous metallic and nonmetallic elements to form salts or covalently bonded compounds. In its reactions with organic compounds, it is less reactive than chlorine, but

TABLE B-8. *Some Physical Properties of Bromine*

Freezing point, °C	−7.3
Boiling point, °C	58.8
Vapor pressure, mm Hg:	
At − 7.3	44
0	66
10	109
20	173
30	264
40	392
50	564
58.8	760
60	793
78.8	1,520
Density, g/cm³:	
At 0°C	3.19
20	3.12
25	3.10
30	3.08
Viscosity, cP, at 20°C	0.99
Heat capacity (liquid), cal/g, at 25°C	0.113
Heat of fusion, cal/g, at −7.3°C	15.8
Heat of vaporization, cal/g, at 58.8°C	44.8
Critical temperature, °C	311
Critical pressure, atm	102

more reactive than iodine, and forms compounds principally by addition reactions to unsaturated linkages or by replacement reactions, usually of hydrogen. In the latter case, hydrogen bromide is a by-product and must normally be utilized, most often by recycle to Br manufacture, for purposes of economy and to avoid problems of waste disposal. Because of its greater size and mass, Br reactions proceed more slowly and with less vigor than those of chlorine; steric hindrance is frequently encountered, so that the Br analogs of several chlorine compounds are synthesized with difficulty or not at all.

The reactivity of Br with organic matter is especially important because it represents serious hazards to human beings. Exposure of skin to elemental Br will cause serious chemical burns which are very slow to heal. The recommended emergency treatment following skin contact is with a saturated thiosulfate solution or with a lime slurry. Bromine vapor is also hazardous, and it is reported that exposure to concentrations of 500–1,000 ppm (by volume) even for a short time is fatal. Exposure to 40–60 ppm for 30 min to 1 hr is said to be dangerous to life. The maximum allowable concentration for extended exposure is less than 1 ppm, but even at this level Br can still be detected by its odor, thus providing warning of its presence. For the accidental inhalation of Br fumes, the recommended treatment is with ammonia vapors. Without treatment, serious inflammation of the lungs will occur, frequently resulting in pneumonia.

The handling, storage, and shipping of Br presents special problems because of its toxicity hazards and corrosivity. When dry (less than 30 ppm water), Br can be stored and shipped in Monel, nickel, lead-lined, glass-lined, and certain plastic vessels. The presence of larger concentrations of water necessitates the use of glass or ceramic for exposed surfaces. Bromine is shipped in glass bottles, Monel drums, and lead-, nickel-, or Monel-clad tank cars and tank trucks. Transfer is usually accomplished by pressuring with dry nitrogen. Special precautions are required to prevent the pickup of moisture.

Manufacture: All commercial processes for the manufacture of Br begin with a solution of bromide ion. Four basic steps are involved: oxidation of bromide to bromine, separation of bromine vapor from the brine, condensation of the vapor (or absorption), and purification of the bromine. Concentrations of bromide ion in the various economically important sources are 65–70 ppm in seawater; 1,300–2,100 ppm in Michigan brines; 3,800–5,000 ppm in Arkansas brines; and over 5,000 ppm in the Dead Sea.

In the United States, virtually all the Br is recovered from naturally occurring brines in Arkansas and Michigan. The more concentrated brines in Arkansas (4,000 ppm and higher) now represent well over half the total U.S. production and have completely supplanted seawater as a commercial source. The recovery process employed for these brines, and generally used for brines containing more than about 1,000 ppm bromide, is the so-called *steaming-out process.*

A typical steaming-out plant for Br is depicted schematically in Fig. B-5. Fresh brine, which may be preheated through one or more heat exchangers (1 and 2) is introduced to the top of a tower (3), at the bottom of which chlorine and steam are injected. The bromine-exhausted acidic (pH ≈ 1) brine from the bottom of the tower is neutralized (4), and the hot neutralized brine may be used to heat fresh brine through one of the heat exchangers before being disposed of. A mixture of Br, chlorine, and water vapors from the top of the tower is condensed (5), and the condensate separated into crude Br and aqueous phases in a decanter (6). The aqueous phase is returned to the tower since it contains significant quantities of

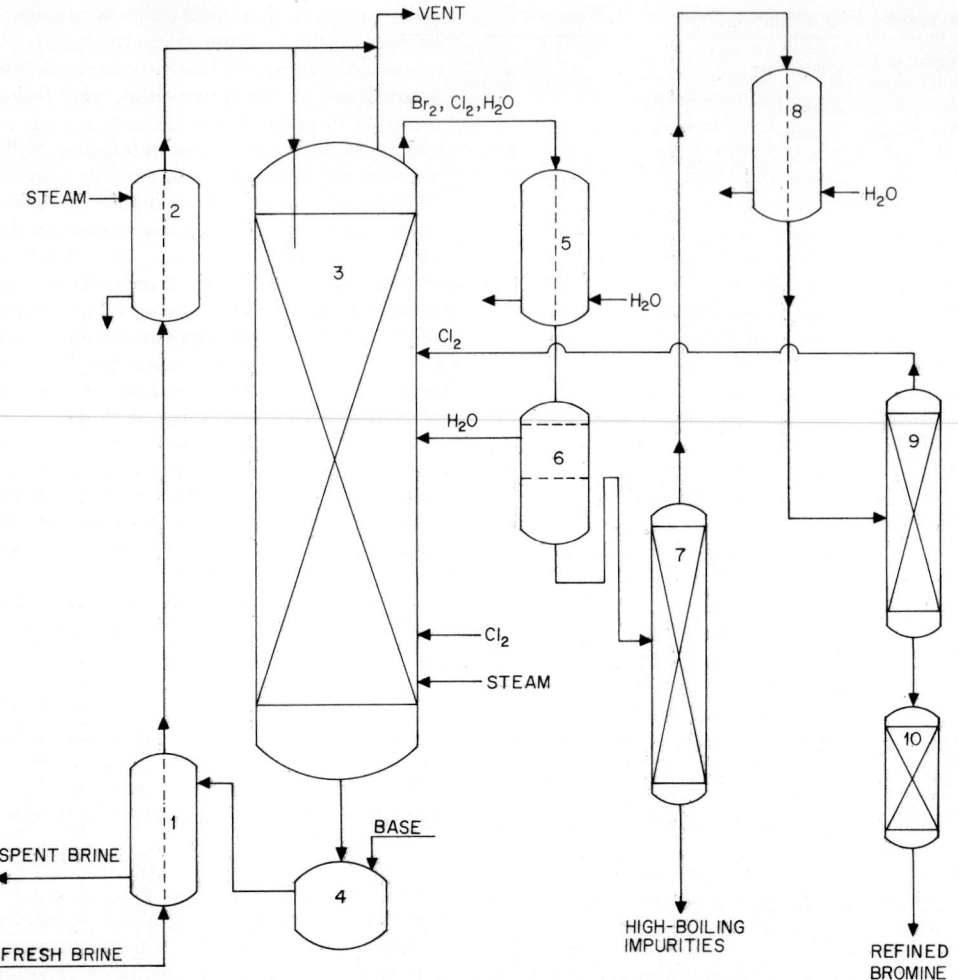

Fig. B-5. Bromine production using steaming-out process. (1) Heat exchanger, (2) heat exchanger, (3) steaming-out tower, (4) neutralizer, (5) condenser, (6) decanter, (7) distillation column, (8) condenser, (9) stripping column, (10) drying column.

chlorine and Br. The crude Br is fed to a distillation column (7), where it is taken overhead and separated from high-boiling impurities, mostly halogenated hydrocarbons. The vapors from this column are condensed (8). The Br is fed to the stripping column (9), where remaining traces of chlorine are distilled away for return to the tower, and refined Br is removed from the bottom of the column, dried from last traces of moisture (10), and packaged.

In theory, 1 lb chlorine will release 2.25 lb Br; in practice, considerably less than 2.25 lb Br/lb

chlorine is produced. The remaining chlorine is used in oxidation of other components of the brine, particularly organics and hydrogen sulfide, and in some processes to acidify the brine. Recoveries of Br of 95% and above, based on the bromide content of the fresh brine, are achieved.

When brines of low Br concentration, less than about 1,000 ppm, or seawater are used, the steaming-out process is uneconomic, and air is used to sweep out the Br vapors following chlorination. The bromine-laden air may be treated in various ways; for example, it may be scrubbed through sodium carbonate solution, where the Br is converted to a mixture of $NaBr$ and $NaBrO_3$, or the Br may be reduced with SO_2 to HBr. In the first case, acidification of the bromide-bromate solution releases Br, and in the second, the HBr is oxidized with additional chlorine to free the Br. In each case, Br refining is similar to that used in the steaming-out process.

Except for its use as a chemical intermediate, the major uses for elemental Br are as a sanitizing or disinfecting agent and as a bleaching agent. Bromine derivatives are used in antiknock additives for motor fuel, dyes, pharmaceuticals, photography, agricultural pesticides, and flame retardants.

The impetus for the development of commercial processes for Br recovery was supplied by the use of ethylene dibromide in combination with lead alkyls as an antiknock agent for motor fuels. Continuing research confirms that ethylene dibromide is the method of choice for minimizing lead deposits in engines. This use consumed 65–70% of all Br produced in 1970.

Agricultural uses are in second place in Br consumption, with demand concentrated in nematocides, fungicides, and selective herbicides. Dye and pharmaceutical manufacture remain as important but minor uses for Br. Photography is a steadily growing use for high-purity sodium bromide and certain organic Br compounds. Flame retardants represent the fastest-growing application for Br derivatives, with particular emphasis on needs in plastics and fibers.

Bromine Inorganics. The inorganic compounds of Br are, with but few exceptions, salts of the bromine acids—hydrobromic, hypobromous, and bromic acids. Compounds with P, S, B, Si, and N are among the principal exceptions. See Table B-9.

Hydrogen Bromide and Hydrobromic Acid: Hydrogen bromide, HBr, can be prepared by the direct combination of hydrogen and Br at high temperatures, for example, by combustion of hydrogen in a slight excess of Br vapors. Passage of the effluent gases through activated carbon removes the last traces of unreacted Br. Alternatively, the HBr can be prepared by the action of H_2SO_4 on a stable inorganic bromide, for example, NaBr, or as a by-product of the substitution reaction of Br with a hydrocarbon. The HBr so prepared can be absorbed in water to form hydrobromic acids of concentrations up to about 70%; however, HBr readily escapes from such concentrated solutions, and the highest-strength acid normally encountered is approximately 48%, the concentration of the constant boiling mixture of the components.

Hydrobromic acid can also be produced by reaction of Br in aqueous media with sulfur, SO_2, or H_2S. In each case the sulfur compound is oxidized ultimately to sulfuric acid. Hydrobromic acid is recovered from the resultant solution by distillation of the constant-boiling mixture.

TABLE B-9. *Characteristics of Major Inorganic Bromine Compounds*

Compound	Formula	Formula Weight	Form	Density or Sp gr	Mp, °C	Bp, °C
Hydrogen bromide	HBr	80.92	Colorless gas	3.5^0 g/l	−88.5	−67.0
Lithium bromide	LiBr	86.85	White crystals	3.464^{25}	547	1265
Sodium bromide	NaBr	102.90	White crystals	3.203_4^{25}	755	1390
Potassium bromide	KBr	119.01	White crystals	2.75^{25}	730	1435
Magnesium bromide	$MgBr_2$	184.13	White crystals	3.72_4^{25}	700	
Calcium bromide	$CaBr_2$	199.90	White crystals	3.353^{25}	730d	806
Strontium bromide	$SrBr_2$	247.44	White crystals	4.216^{24}	643	d
Zinc bromide	$ZnBr_2$	225.19	White crystals	4.201^{25}	394	650
Aluminum bromide	$AlBr_3$	266.71	White crystals	2.64^{10}	97.5	263^{747}
Ferric bromide	$FeBr_3$	295.57	Red-brown crystals	. . .	d	
Sulfur monobromide	S_2Br_2	223.95	Red liquid	263	−40	$54^{0.2}$
Boron tribromide	BBr_3	250.54	Colorless liquid	$2.643_4^{18.4}$	−46	91.3
Phosphorus tribromide	PBr_3	270.70	Colorless liquid	2.852^{15}	−40	172.9
Silicon tetrabromide	$SiBr_4$	347.72	Colorless liquid	2.772^{25}	5.4	154
Sodium bromate	$NaBrO_3$	159.90	White crystals	$3.34^{17.5}$	3.81d	
Potassium bromate	$KBrO_3$	167.01	White crystals	$3.27^{17.5}$	370d	

d = decomposes.

The principal uses of HBr and hydrobromic acids are in the production of some inorganic bromides and in the manufacture of certain alkyl bromides either by the replacement of alcoholic hydroxyl groups or by addition to olefins.

Sodium Bromide: Sodium bromide, NaBr, commercially the most important of the bromide salts, is usually manufactured by neutralization of HBr with sodium hydroxide, bicarbonate, or carbonate. Its major use is in the photographic industry, where especially high purity grades are employed in the preparation of silver bromide emulsions. It also finds use in combination with hypochlorites in bleaching systems, especially for cellulosics.

Lithium Bromide: Lithium bromide, LiBr, is also prepared by neutralization of the hydroxide or carbonate with HBr. It is an efficient desiccant and finds use as a humectant and in the industrial drying of air.

Zinc Bromide: Zinc bromide, $ZnBr_2$, is manufactured either by direct reduction of Br with Zn or by reaction of $ZnCO_3$ or oxide with HBr. It is extremely soluble in water and is commonly marketed as an 80% solution in water. It finds use as a rayon-finishing agent, as an absorbent in humidity control, as a catalyst, and as a gamma-radiation shield in viewing windows of nuclear-reactor installations.

Aluminum Bromide: Aluminum bromide, $AlBr_3$, is prepared by the direct reduction of Br with Al in an intensely exothermic reaction and finds use as a catalyst with activity similar to that found for $AlCl_3$.

Other Bromides: The bromides of calcium, $CaBr_2$, strontium, $SrBr_2$, barium, $BaBr_2$, and magnesium, $MgBr_2$, have minor commercial importance and are made usually by reaction of HBr with the corresponding carbonates or hydroxides. All but barium have pharmaceutical applications; it is used principally to make bromides from other inorganic sulfates. $CaBr_2$ has found some use in photography.

Phosphorus tribromide, PBr_3, is manufactured by the reduction of Br with phosphorus and finds use as a catalyst and as an intermediate in the preparation of phosphite esters. Silicon tetrabromide, $SiBr_4$, and boron tribromide, BBr_3, can be produced by reaction of the corresponding fluorides with $AlBr_3$. Both are useful as intermediates.

Hypobromous Acid and Hypobromites

Hypobromous Acid: Hypobromous acid, HOBr, exists only in aqueous solution and results from hydrolysis of Br with water. It decomposes either with liberation of O_2 or with simultaneous oxidation-reduction, producing bromic acid. Hydrobromic acid is a by-product in either case. Hypobromous acid has utility as a germicide, and for this reason it finds some use in water treatment. It is sometimes used as an oxidizing or brominating agent for certain organics.

Hypobromites: Although aqueous solutions of Br under normal conditions contain only small concentrations of hypobromous acid, relatively high concentrations of hypobromites can be formed by addition of Br to cooled solutions of alkalies. The hypobromites are reasonably stable in solution but, on standing or when heated, disproportionate to mixtures of bromide and bromate. Many oxidation reactions of Br in aqueous alkaline media are the consequence of the intermediate formation of alkali hypobromites.

Bromic Acid and Bromates

Bromic Acid: Bromic acid, $HBrO_3$, like hypobromous acid, exists only in solution; it can be prepared by the electrolysis of aqueous Br solutions or of HBr. It is a very strong oxidizing agent. On heating it decomposes to liberate Br_2 and O_2.

Bromates: Bromates are stable under normal conditions, and can be prepared by electrolytic oxidation of the corresponding bromides or by reaction of Br with aqueous solutions or suspensions of the appropriate hydroxides or carbonates. In the latter case bromides are formed simultaneously, and can be removed by crystallization of the less soluble bromates. Both potassium bromate, $KBrO_3$, and sodium bromate, $NaBrO_3$, have been used as oxidizers in hair-wave preparations, and both are of importance as general oxidants and bromine-releasing agents. Potassium bromate has been used as a flour-treating agent and in shrinkproofing wool.

Bromine Organics.

Organic compounds containing bromine may be formed in a variety of ways, but usually they result from direct reaction with Br in *addition:*

$$H_2C{=}CH_2 + Br_2 \rightarrow CH_2BrCH_2Br$$

Ethylene   1,2-Dibromoethane (ethylene dibromide)

or in *substitution* reactions:

Benzene   Bromobenzene

or from reaction of hydrogen bromide in addition:

$$CH_3(CH_2)_5CH{=}CH_2 + HBr \rightarrow$$
1-Octene

$$CH_3(CH_2)_5CHBrCH_3$$
2-Bromooctane

or *replacement* reactions:

$$CH_3OH + HBr \rightarrow CH_3Br + H_2O$$
Methanol   Methyl
bromide

Both liquid- and vapor-phase reactions are common, the former frequently employing a solvent to moderate the reaction or to dissolve the product. Often catalysts are of extreme importance, particularly where more than one bromine atom is to be introduced. Metals such as iron or aluminum, metal bromides, phosphorus tribromide, iodine, and sulfur are among the commonly used catalysts. Light is also frequently an effective catalyst. The choice of catalyst is often responsible for the course of the reaction, as for example in the bromination of toluene, where light-catalyzed bromination results in side-chain substitution, and metallic-bromide catalysis promotes bromination of the benzene nucleus, e.g.,

Benzyl bromide

4-Bromotoluene

Similarly, the course of the hydrogen bromide addition to olefins, which normally proceeds as depicted above for 1-octene, can be reversed in the presence of peroxidic catalysts:

$$CH_3(CH_2)_5\ CH{=}CH_2 + HBr \xrightarrow{\text{peroxide}}$$

$$CH_3(CH_2)_5CH_2CH_2Br$$
1-Bromooctane

Commercially important organic bromine compounds (Table B-10) include:

Methyl Bromide: Methyl bromide, CH_3Br, is normally manufactured by the reaction of methanol with hydrogen bromide or hydrobromic acid; however, it can be prepared by the direct catalytic bromination of methane at elevated temperatures. Polybrominated methanes are undesirable by-products by the latter reaction procedure, however. Methyl bromide is a gas at normal temperatures. It has a high toxicity and has found application as a soil and space fumigant to control insects, fungi, and unwanted plant growth. It is also employed as a methylating agent in numerous organic syntheses.

1,2-Dibromoethane (Ethylene Dibromide): Ethylene dibromide, CH_2BrCH_2Br, is by a wide margin the organic Br compound of greatest commercial importance and is readily produced by the addition of Br to ethylene at moderate temperatures. It is also used, as is CH_3Br, as a soil and space fumigant, but its major use by far is in combination with lead alkyls as antiknock agents for gasoline. The ethylene dibromide reacts with lead oxide deposits formed in the engines to yield lead bromide, which is sufficiently volatile to be swept out of the engine with the exhaust gases. With the increasing ecological concern over atmospheric lead pollution, this use of ethylene dibromide is expected to decline in coming years.

Bromochloromethane (Methylene Chlorobromide): Bromochloromethane, CH_2ClBr, is prepared from dichloromethane by reaction with Br and Al or with hydrogen bromide and aluminum bromide. It is a low-boiling liquid of low toxicity and is an effective fire-extinguishing agent frequently employed in aircraft systems and in portable fire extinguishers.

1,1,2,2-Tetrabromoethane (Acetylene Tetrabromide): Acetylene tetrabromide, $Br_2CHCHBr_2$, is prepared by the addition of Br to acetylene, with or without catalyst. It is used as a gage fluid and in specific-gravity separations of solids, but its principal use is as part of a catalyst system in the oxidation of *p*-xylene to terephthalic acid.

Tris(2,3-dibromopropyl) Phosphate: Tris(2,3-dibromopropyl) phosphate, $(BrCH_2CHBrCH_2O)_3 \cdot PO$, may be manufactured by the addition of Br to triallyl phosphate or by the reaction of phosphorus oxychloride with 2,3-dibromopropanol. The product is a viscous liquid which finds numerous applications as an additive in providing flame resistance to a variety of polymer systems, particularly fibers and plastic foams.

Tetrabromobisphenol A: Tetrabromobisphenol A, $HOC_6H_2Br_2C(CH_3)_2C_6H_2Br_2OH$, is manufactured by the direct bromination of bisphenol A.

TABLE B-10. *Characteristics of Major Organic Bromine Compounds*

Compound	Structure	Formula Wt.	Form	Density or Sp gr	Mp, °C	Bp, °C
Methyl bromide	CH_3Br	94.94	Colorless gas	1.734_4^0	−93.7	3.6
Bromochloromethane	CH_2BrCl	129.38	Colorless liquid	1.923_4^{25}	−88.0	67.8
Bromotrifluoromethane	CF_3Br	148.94	Colorless gas	2.027^{-65}	−174.4	−57.9
Ethyl bromide	CH_3CH_2Br	108.97	Colorless liquid	1.449_4^{25}	−119.3	38.4
1,2-Dibromoethane (ethylene dibromide)	CH_2BrCH_2Br	187.86	Colorless liquid	2.179_4^{20}	9.9	131.4
1,1,2,2-Tetrabromoethane (acetylene tetrabromide)	$CHBr_2CHBr_2$	345.65	Colorless liquid	2.964_4^{20}	0.1	151^{44}
Vinyl bromide	$CH_2{=}CHBr$	106.95	Colorless liquid	1.57^0	−139.2	15.8
Bromobenzene	C_6H_5Br	157.01	Colorless liquid	1.495_4^{20}	−30.6	156.1
2,4,6-Tribromophenol	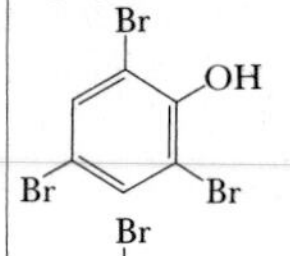	330.80	White crystals	2.55_4^{20}	96	162.0^{10}
Tetrabromophthalic anhydride	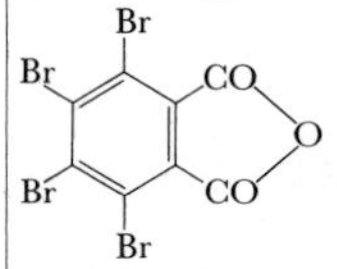	463.74	White to pale yellow crystals	. . .	280	. . .
Tetrabromobisphenol A	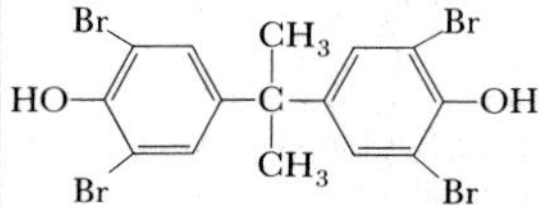	543.88	White crystals	. . .	181.2	316d
Tris (2,3-dibromopropyl) phosphate	$CH_2BrCHBrCH_2O$ $CH_2BrCHBrCH_2O$ ⟩$P \to O$ $CH_2BrCHBrCH_2O$	697.61	Colorless to light yellow liquid	2.25_{25}^{25}	. . .	d

d = decomposes.

It is widely used as a reactive flame retardant, actually incorporated in the polymer backbone, in epoxy resins, polycarbonates, and unsaturated polyesters.

Tetrabromophthalic Anhydride: Tetrabromophthalic anhydride, $Br_4C_6(CO)_2O$, is produced by the catalytic bromination of phthalic anhydride in fuming sulfuric acid. It is finding increasing application as a reactive flame retardant in the preparation of polyol systems for incorporation into polyurethane foams and is used also as a flame retardant in unsaturated polyesters.

REFERENCES

Faith, W. L., D. B. Keyes, and R. L. Clark: "Industrial Chemicals," 3d ed., Wiley, New York, 1965.

Jolles, Z. E. (ed.): "Bromine and Its Compounds," Academic, London, 1966.

Kreutzkamp, N., H. Meerwein, A. Roedig, and R. Stroh: "Methoden der organischen Chemie," 4th ed., vol. 5/4, Thieme, Stuttgart, 1960.

—Wayne R. Merriman and John A. Garman, *Great Lakes Chemical Corporation, West Lafayette, Ind.*

Bromobenzene (See **Bromine**)

Bronze (See **Copper;** and **Tin.**)

Brownian Movement (See **Colloid systems.**)

Bubble Behavior (See **Defoaming agents.**)

Bubble Cap (See **Absorption;** and **Distillation.**)

Builders (See **Detergents.**)

Burning (See **Fuels.**)

Busheling (See **Iron and steel.**)

Butadiene 1,3-Butadiene, C_4H_6, is a very reactive compound because of its conjugated double-bond structure. The bulk of butadiene production is used in producing polymers, such as SBR (styrene-butadiene rubber) and ABS (acrylonitrile-butadiene-styrene) plastics. The double-bond system provides a starting point for many organic syntheses, such as the Diels-Alder reaction. See also **Acrylonitrile-butadiene-styrene resins; Elastomers;** and **Rubber, natural.**

In 1970, the world production of butadiene amounted to approximately 5 billion lb/yr. The growth rate is estimated to be 5%/yr. The first large-scale effort to produce high-purity butadiene resulted from the World War II shortage of natural rubbers. During this period, the process* described here was developed to recover either butylenes for dehydrogenation to butadiene or for recovery of butadiene. The solvent used at that time was acetone. When acetonitrile, $CH_3 \cdot CN(ACN)$, became available as a by-product of acrylonitrile production, the butadiene recovery process was adapted to use acetonitrile because of its superior properties for this service. See also **Acrylonitrile.** Initially, butadiene was produced by dehydrogenation of butylenes. In recent years, naphtha cracking for ethylene and propylene production has produced a by-product C_4 stream, which is another important source of butadiene.

*Shell ACN process for butadiene recovery (Shell Development Company). First licensed in 1960, this process accounts for one-third of world production of butadiene.

TABLE B-11. *1,3-Butadiene Specifications and Properties*

Typical butadiene product specifications:	
1,3-Butadiene, percent (wt)	99.5
Peroxides as H_2O_2, ppm (wt)	5
Acetylenes as vinyl acetylene, ppm (wt)	50
Boiling range, °C	0.2
Butadiene dimer, percent (wt)	0.05
Nonvolatile residue, percent (wt)	0.02
Physical properties of 1,3-butadiene:	
Mole weight	54.09
Normal boiling point:	
°F	24.06
°C	−4.41
Specific gravity, 60°F/60°F	0.6272
Coefficient of expansion at 60°F, per °F	0.00113
Heat capacity (liquid at 1 atm), Btu/(lb)(°F)	0.5079
Heat of vaporization (1 atm), Btu/lb	174

Along with the growth of demand for butadiene, the product specifications have become more severe. A typical set of product specifications is shown in Table B-11. The physical properties of 1,3-butadiene also are given in Table B-11.

Butadiene-Recovery Process: A typical flow scheme for the recovery process* is given in Fig.

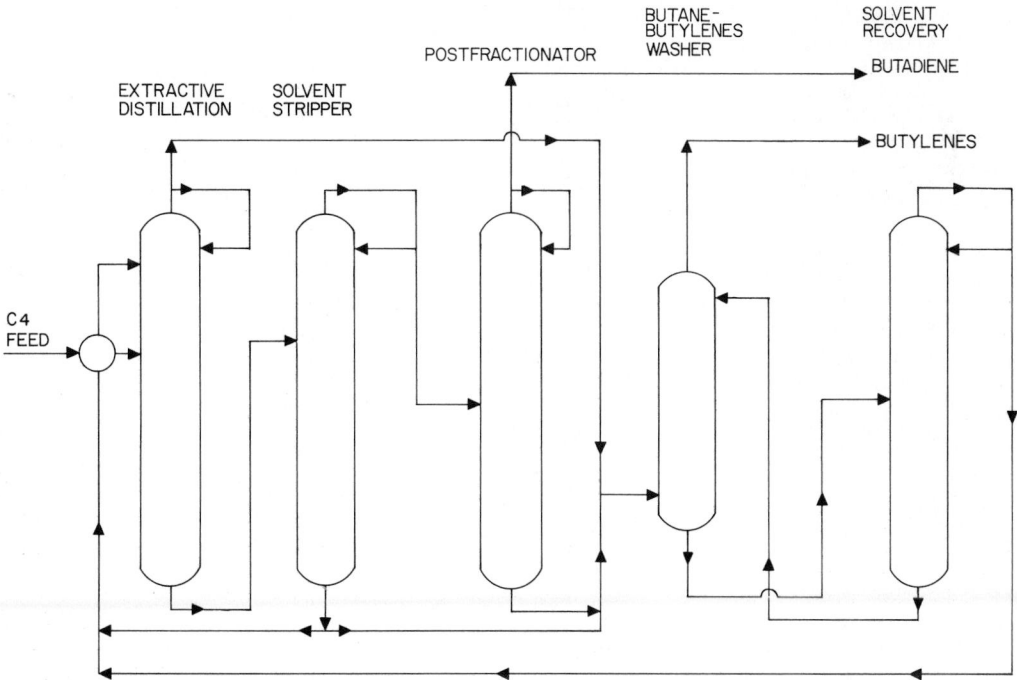

Fig. B-6. Process for butadiene recovery—the Shell ACN process. (*Shell Development Co.*)

B-6. In practice, each unit is designed to meet the requirements of a particular feedstock and product, as well as availability of utilities. The basis for the process is the change in relative volatility of C_4 hydrocarbons in the presence of acetonitrile solvent, which makes the separation easier. The C_4 mixture is fed to the extractive distillation column, where it is separated in a solvent environment into a by-product butane-butylenes (BB) stream overhead and a solvent-butadiene stream. The by-product butane-butylenes are washed to recover acetonitrile before leaving the process. Butadiene is stripped overhead from the fat solvent in the solvent stripper. The crude butadiene from the solvent stripper is fed to the postfractionator, where finished butadiene is recovered as the overhead product. A light ends column may be provided upstream of the postfractionator if required to meet a C_3 specification on the finished butadiene.

The characteristics and advantages of acetonitrile, as used for this and similar processes, are described under **Isoprene.**

Other butadiene-recovery processes are similar in their process flow sequence, but use other solvents in the extractive distillation. These solvents include *n*-methyl pyrrolidone (BASF), dimethyl formamide (Japan, GEON), furfural (Phillips), and dimethyl acetamide (Union Carbide). The furfural process was developed and used in the United States during the World War II period,

along with the cuprous ammonium acetate (Esso) liquid-liquid extraction process when butadiene production was based upon butylene dehydrogenation. The other aforementioned solvents have been applied to butadiene recovery more recently. See also **Petrochemical complex.**

—A. E. Handlos, *Shell Development Company, Houston, Tex.*

Butanes   (See **Hydrocarbons;** and **Isomerization.**)

Butanol   (See **Alcohols.**)

Butyl Rubber   (See **Elastomers.**)

Butylated Hydroxy Anisole (BHA)   (See **Antioxidants.**)

Butylated Hydroxy Toluene (BHT)   (See **Antioxidants.**)

Butylene   (See **Alkylation; Butadiene;** and **Isoprene.**)

Butyllithium   (See **Lithium.**)

Butyrate Plastics   (See **Cellulose ester plastics, organic.**)

Byers Process   (See **Iron and steel.**)

C

Cadmium (Gk. *kadmia,* earth.) **Cd** = 112.40 (at. wt.); 48 (at. no.). Cadmium is a member of group IIB, period 5, of the periodic table, along with Zn and Hg. The stable isotopes of Cd are 106, 108, 110, 111, 112, 113, 114, and 116. Radioactive isotopes are 104, 105, 107, 109, 115, and 117. The isotopic abundance of the element is given under **Chemical elements.**

Important physical properties of Cd include melting point = 320.9°C; boiling point = 767°C; specific gravity at 20°C = 8.65; and the crystalline structure is hexagonal.

Cadmium is a silver-white metal, moderately soft and extremely ductile. One of the minor nonferrous metals, ranking 57th in abundance in the earth's crust (0.15 ppm), it is never found alone in nature, but always associated with zinc. Greenockite, CdS, and otavite, $CdCO_3$, are the only known natural Cd minerals and are always found as minor constituents of sphalerite, ZnS, and smithsonite, $ZnCO_3$, respectively. The lack of a distinct deposit of Cd is probably the reason for its rather late discovery, 1817. In that year, F. Strohmeyer noticed a yellow tinge to an iron-free zinc carbonate produced at the Salzgitter Zinc Works, and correctly attributed the color to a yet unknown element. Dissolution of the sample and precipitation of the yellow sulfide in weak acid separated the unknown from the zinc. Reduction to the metal confirmed his theory and was the first known preparation of the element, which he named cadmium. Chemically, Cd resembles zinc, forming many analogous compounds, but unlike zinc, it is not amphoteric.

Production Processes: There are two major classes of processes for producing Cd, pyrohydrometallurgical and electrolytic. In all methods, zinc blende is roasted to eliminate sulfur and to produce a zinc oxide calcine.

As shown in Fig. C-1, in the *pyrohydrometallurgical process,* the calcined zinc oxide is mixed with coal, pelletized, and sintered. The primary purpose is to alter the physical condition of the calcine from a fine dust to a porous clinker. An additional benefit is the removal of volatile elements, such as Pb, As, and Cd. Chlorides are used in wetting

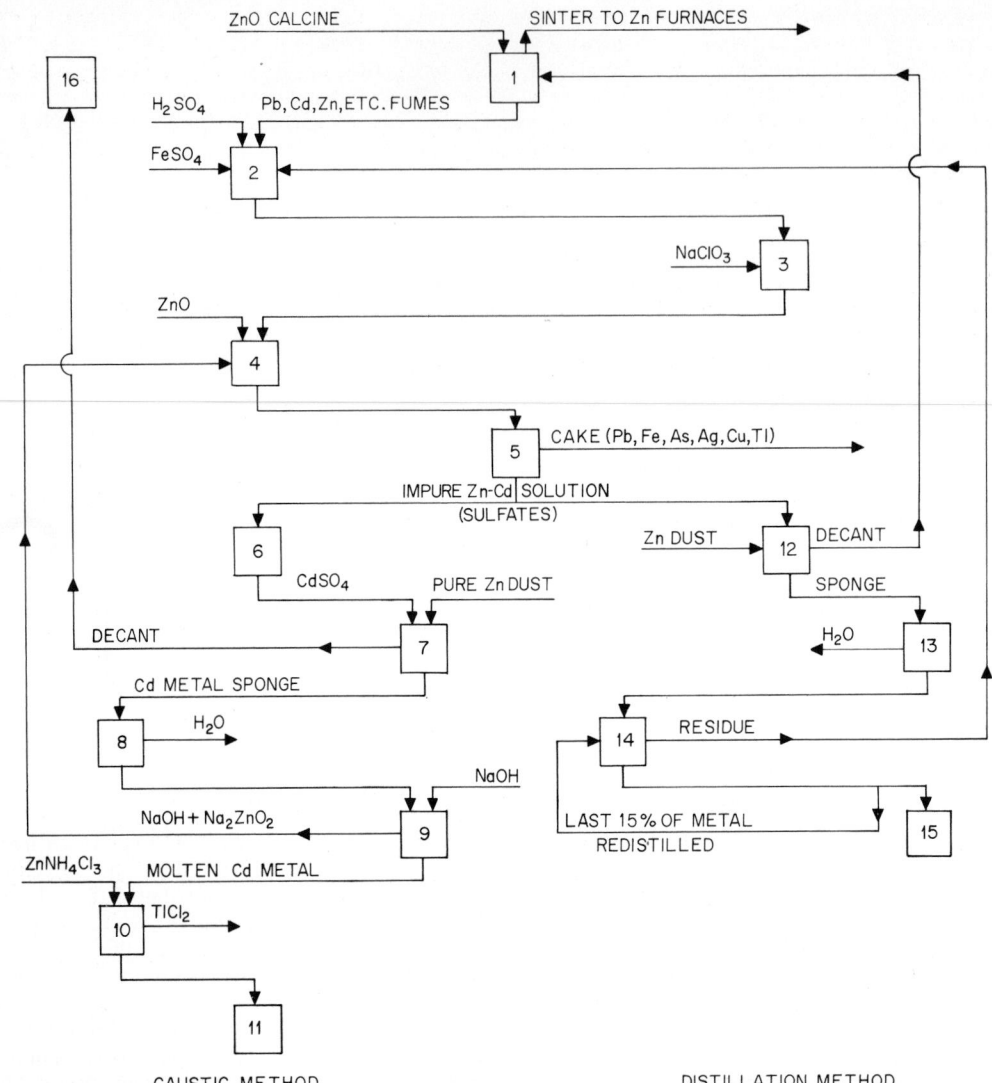

Fig. C-1. Pyrohydrometallurgical processes for producing cadmium. After preliminary operations, one of two methods may be used: (a) the caustic method or (b) the distillation method. (1) Sintering, (2) leaching, (3) oxidation, (4) neutralization to a pH of 5.0, (5) filtration, with the cake going to the lead smelter, (6) purification, (7) cadmium precipitation, (8) briquetting, (9) melting under caustic, (10) thallium elimination, (11) casting, (12) cadmium precipitation, (13) centrifuging, (14) furnace operations, (15) casting, (16) $ZnSO_4$ evaporator.

the calcine and eliminating 92–94% of the Cd, which is condensed and collected in an electrostatic precipitator. The fume, containing 5–25% Cd, is leached in H_2SO_4, to which iron sulfate has been added to control the As, and the slurry is oxidized, usually with $NaClO_3$. The slurry is neutralized to pH 5 with ZnO and filtered, and

the residue shipped to a Pb smelter. The remaining As must be less than 10 ppm to avoid formation of arsine gas, AsH_3. As shown in Fig. C-1, from this point in the process, there are two alternative routes.

In the *melting under caustic method,* last traces of heavy metals are removed, as in the electrolytic process. The purified solution is charged with high-purity zinc dust, decanted, and either returned to the sinter machine to wet the charge or treated to produce $ZnSO_4$ or $ZnCO_3$. The cadmium sponge is briquetted to remove excess water and melted under molten caustic to remove any zinc. The molten metal is treated with zinc ammonium chloride to remove thallium, and is then cast into shapes.

In the *distillation method,* regular zinc dust is used to make the sponge. The decant returned to the sinter machines for the recovery of Zn provides chlorides for wetting the charge. The sponge is washed, centrifuged to minimize water content, and charged to a retort. During heating and distillation, the charge must be kept in a reducing atmosphere. Because of Pb and Zn contamination, the last 10–15% of the distillate must be redistilled. The furnace residue is leached in H_2SO_4 and returned to the process. The furnace area must be well ventilated to maintain an acceptable level of toxic Cd fumes.

Equations covering the major steps in the foregoing processes are as follows.

Leaching:

$$CdO + H_2SO_4 \rightarrow \\ CdSO_4 + H_2O \quad \text{(Same for Pb)}$$

Oxidation:

$$3As_2O_3 + 2NaClO_3 \rightarrow 3As_2O_5 + 2NaCl$$
$$6FeSO_4 + NaClO_3 + 3H_2SO_4 \rightarrow \\ 3Fe_2(SO_4)_3 + NaCl + 3H_2O$$

Neutralization:

$$Fe_2(SO_4)_3 + As_2O_5 + 3ZnO + 8H_2O \rightarrow \\ 2FeAs(OH)_8 + 3ZnSO_4$$

Cadmium precipitation:

$$CdSO_4 + Zn \rightarrow Cd + ZnSO_4$$

Melting under caustic:

$$Zn + 2NaOH + \tfrac{1}{2}O_2 \rightarrow Na_2ZnO_2 + H_2O$$

In an *electrolytic plant,* the calcine is leached with H_2SO_4. To effect the electrolysis of Zn, the solution must be highly purified. One step of this purification is to charge Zn dust; this removes the Cd and other metals which are more electronegative than Zn. The resulting sponge is digested in H_2SO_4 and purified of all contaminants other than Zn. This step varies from plant to plant because of the different types and quantities of impurities encountered. High-purity Pb-free Zn

dust is added to the solution, precipitating a nearly pure Cd sponge, which is redigested in spent Cd electrolyte and deposited by electrolysis onto Al cathodes. After stripping, the Cd is melted and cast into the desired shapes. See Fig. C-2.

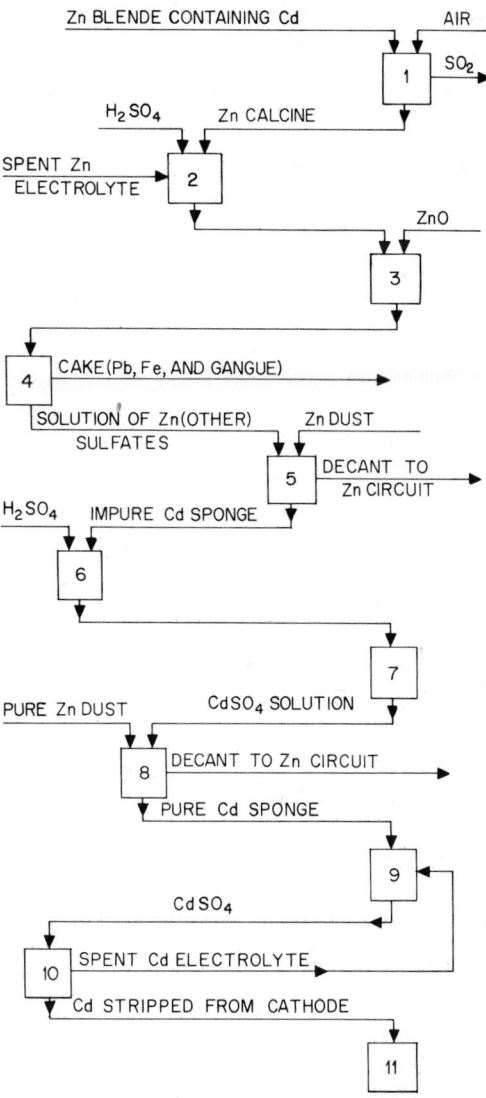

Fig. C-2. Electrolytic process for producing cadmium. (1) Roasting, (2) leaching, (3) neutralization to a pH of 5.2, (4) filtration, (5) first cadmium precipitation, (6) digestion, (7) purification, (8) second cadmium precipitation, (9) digestion, (10) electrolysis with a lead anode and aluminum cathode, (11) melting and casting.

Equations covering the major steps in the electrolytic process are as follows.

Roasting:

$$ZnS + 1\tfrac{1}{2}O_2 \rightarrow ZnO + SO_2 \qquad \text{(same for Cd)}$$

Leaching:

$$ZnO + H_2SO_4 \rightarrow ZnSO_4 + H_2O$$

Neutralization:

$$Fe_2(SO_4)_3 + 3ZnO + 3H_2O \rightarrow$$
$$2Fe(OH)_3 + 3ZnSO_4$$

Cadmium precipitation:

$$CdSO_4 + Zn \rightarrow Cd + ZnSO_4$$

Electrolysis:

$$CdSO_4 + H_2O \rightarrow Cd + H_2SO_4 + O_2$$

If Pb-Zn ores are treated in a blast furnace, the Cd is driven off with the Zn and recovered in rectifier columns when the Zn is refined by redistillation.

Metal Specifications: The ASTM specification for Cd metal (Designation B 440–66T) gives the following chemical requirements:

Cd	min. (by difference). . . .	.99.90% (99.95%)*
Zn	max.	0.035% (0.020%)*
Cu	max.	0.015%
Pb	max.	0.025%
Sn	max.	0.01%
Ag	max.	0.01%
Sb	max.	0.001%
As	max.	0.003%
Tl	max.	0.003%

*When specified by purchaser.

Cadmium metal is cast in several shapes to meet requirements of customers. A ball 2 in. in diameter is the most common and is used by many electroplaters. Production of cadmium compounds is either directly or indirectly from Cd metal. When the metal is to be digested to produce compounds, sponge is desired, but it can be used only when zinc and thallium will not cause problems with the final product. To produce an easily digested metal, the molten metal is poured through a drilled plate, forming many droplets, which fall into a water bath. Formed in this manner, the Cd metal has a tremendous surface area.

Alloys: Some of the more important Cd alloys and uses are given in Table C-1. Cadmium forms alloys with most common elements. The two notable exceptions are Al and Fe. There are no unusual problems in the production of Cd alloys other than the necessity to maintain adequate ventilation over the molten metal. During World War II, 35–40% of Cd production was consumed in alloys, but by 1970, this use had fallen to less than 5%, because of the high cost of Cd and the availability of less costly substitutes. See also **Copper.**

Other Uses of Cadmium Metal: The metal is used in the Ni-Cd battery, but because of high cost and limited supply, the use is confined to small-size batteries and special applications. See also **Batteries.** Cadmium metal has been used in nuclear reactors as a neutron absorber, but has been displaced by substitute materials. Electroplating of iron and steel remains the most important use of the metal, consuming half of the supply (1970). As recently as 1950, electroplating accounted for

TABLE C-1. *Principal Cadmium Alloys*

Alloying Elements	% Cadmium	Desired Property	Uses
Copper	1	Increase in tensile strength	Trolley lines
Bismuth	45	Eutectic 144°F, 62°C (low mp)	Fire protection
Gold	Varies	Green color	Jewelry
Lead-silver-copper	10–18	Low coefficient of friction	High-temperature and high-pressure bearings
Lead-silver-nickel	10–18	Low coefficient of friction	High-temperature and high-pressure bearings
Lead-bismuth-tin	10–12.5	55–70°F, 13–21°C mp	Low-pressure temperature solder
Lead-antimony-tin	0.25	Increase in fatigue limit of lead	Cable sheathing

90% of Cd consumption, but the increased demand by the pigment and plastic industries and the resulting high price level of Cd have sharply cut into this use. Cadmium plating is used widely for steel parts that are subject to continuous corrosive conditions of moisture and alkalies. In addition to high resistance to corrosion, the coating has the desirable characteristic of not peeling when scratched or otherwise damaged. Like Zn, the resistance to acid attack of Cd is poor.

Compounds: The most widely used Cd compound is CdS. This compound may be formed by the attack of sulfur on Cd metal, but commercial production is mostly by precipitation from solutions of Cd salts. The popularity of CdS for pigments is due to the wide range of color—lemon yellow through the oranges to a deep red—and to the intensity and stability of the colors. The variation is achieved by careful control of precipitating conditions—the acidity and temperature of the solution, the salt used (nitrate, chloride, sulfate, and so on), and the rate of addition of H_2S gas. CdS is insoluble in water, sublimes at 1350°C, and is dimorphous. The stable alpha hexagonal and the unstable beta cubic forms have specific gravities of 3.9 and 4.5, respectively. The cubic configuration is formed in normal H_2SO_4 and HNO_3 solutions. The color of the precipitate is not due to the crystalline forms because both show a wide range of color. The color is associated with the dispersion of the precipitate. Pigments account for 20–25% of the consumption of Cd.

Cadmium stearate is combined with barium stearate as a stabilizer in thermosetting plastics, and this use accounts for over 20% of the Cd consumption.

There is only one oxide of cadmium, namely, CdO, with a specific gravity of 8.2 and a subliming point of 1850°C. The compound is reddish brown in color and is formed by burning Cd in air. The oxide is insoluble in H_2O, absorbs CO_2 from air, and can be reduced to metallic Cd by carbon compounds at 550°C. Although early investigators had claimed the production of cadmium suboxide, Cd_2O, x-ray diffraction has demonstrated that the product was a mixture of $Cd \cdot CdO$. Claims also have been made for cadmium peroxides, but no stoichiometric compound has been produced, leading to the belief that these substances are wholly or partially dissociated.

Cadmiopone, $BaSO_4 + CdS$, produced by the addition of BaS to a $CdSO_4$ solution and containing about 30% Cd, varies from yellow to crimson and is used to color rubber goods and plastics.

Cadmium acetate and salicylate are used as external antiseptics. Cadmium arsenite is used as a semiconductor. Several additional Cd compounds, including cadmium nitrate, selenide, sulfoselenide, and tungstate, are used as pigments in paints, ceramics, and glass. The tungstate is used as a pigment in radioscopy. Cadmium borotungstate is used to form heavy media in mineral dressing. Cadmium halides are used in photography and pyrotechnics.

Cadmium is highly toxic to man. The metal at ambient temperatures presents no problem. Only when the metal is heated do hazards exist. At the melting point, 320.9°C, the vapor pressure is 0.1 mm Hg. At 400°C, the vapor pressure increases to 1.2 mm Hg. Maximum acceptable concentration of Cd is 0.1 mg/m³. Thus it is well to note that the saturated vapors from Cd at the two aforementioned temperatures are 660 and 8,000 mg/m³, respectively. When a worker is heating Cd metal or Cd-plated products, he should be protected with adequate ventilation and wear a respirator.

Economics: Even though the cost of cadmium has risen, causing users to seek substitutes in less costly materials, the production and demand have shown steady acceleration. For example, in 1926, the annual world production of Cd was 1.2 million lb; in 1950, 11.1 million lb; in 1962, 26.3 million lb; and in 1968, 31 million lb. Since cadmium, in essence, is a by-product of zinc production, the economy of Cd is closely related to that of Zn. The amount of Cd recovered per ton of Zn of course varies with both the raw materials and processes used. In the United States, approximately 10.4 lb of Cd is recovered per ton of Zn produced. The worldwide figure is 6.2 lb/ton.

REFERENCES

American Bureau of Labor Statistics: Year Book of the American Bureau of Metal Statistics, New York.

Chizhikov, D. M.: "Cadmium" (transl. by D. E. Hayler, Oxford, England), Pergamon, London, 1966.

Hempel, C. A.: "Rare Metals Handbook," Van Nostrand, New York, 1954.

Lange, N. A.: "Handbook of Chemistry," Handbook Publishing Co., Sandusky, Ohio, 1946.

Mellor, J. W.: "A Comprehensive Treatise on Inorganic and Theoretical Chemistry," vol. 4, Longmans, London, 1946.

Sneed, M. C., and R. C. Brasted: "Comprehensive Inorganic Chemistry," vol. 4, Van Nostrand, New York, 1955.

—L. R. Meadows, *National Zinc Company, Inc., Bartlesville, Okla.*

Cage Mill (See **Size reduction.**)

Cage Press (See **Expression.**)

Calandria (See **Evaporation.**)

Calaverite (See **Tellurium.**)

Calcining (See **Bauxite; Cement; Copper production;** and **Lime.**)

Calcite (See **Limestone;** and **Tungsten.**)

Calcium (L. *calx,* lime.) **Ca** = 40.08 (at. wt.); 20 (at. no.). Calcium is one of the alkaline-earth group of metals (group IIA in the periodic table), which consists of Ca, Ba, Sr, and Ra. The metal is the fifth most abundant element in the earth's crust. Calcium exists in the face-centered cubic form at room temperature, transforming to a body-centered cubic structure at 448°C and melting at 838°C.

Calcium is a silvery-white metal. Due to its high reactivity, it is relatively unstable in moist air, rapidly forming an oxidation coating. The metal can be stored readily in dry air (less than 30% RH) at room temperature. Calcium reacts spontaneously with water to displace hydrogen. When finely divided, it will ignite in air. Some important naturally occurring compounds include the carbonate (limestone) in various degrees of purity, the sulfate, and complex silicates.

Stable isotopes of Ca are 40, 42, 43, 44, 46, and 48, with 40 predominating. Calcium exhibits only one valence state, 2^+, in all its reactions. The metal is slightly less active than Ba and Sr in the same series. Calcium is a very ductile metal and can be formed readily by casting, extrusion, and rolling.

Selected physical properties of Ca are given in Table C-2. Calcium is especially useful for its high reactivity, notably as to the high heat of formation of some of its compounds. The low density and relatively low electrical resistivity make Ca one

TABLE C-2. *Selected Physical Properties of Calcium*

Electron configuration	2-8-8-2
Density, g/cc, at 20°C	1.55
Melting point, °C	838
Boiling point, °C	1440
Specific heat, cal/g, 0–100°C	0.149
Heat of fusion, cal/g	52
Heat of vaporization, cal/g	1000
Thermal expansion, in./(in.)(°C) 0–400°C	22.3×10^{-6}
Electrical resistivity, Ω-cm, at 0°C	3.91×10^{-6}
Thermal conductivity, (cal)(cm²)/(cm)(sec)(°C) at 20°C	0.3
Crystal form:	
Below 464°C	Face-centered cubic (fcc)
464°C to melting point	Body-centered cubic (bcc)
Lattice constant, Å, fcc	5.582
Closest approach of atoms, Å, fcc	3.946

Mechanical Properties	Annealed	Cold-worked
Tensile strength, psi	6,960	16,700
Yield strength, psi	1,990	12,300
Elongation, % in 1 in.	51	7
Modulus of elasticity, psi	$3.2–3.8 \times 10^6$	
Hardness, Rockwell B	16–18	

Heats of Formation of Calcium Compounds, kcal/mole

$CaBr_2$	−161.3
$CaCl_2$	−190.0
CaF_2	−290.3
CaH_2	−45.1
CaI_2	−127.8
Ca_3N_2	−103.2
CaO	−151.9
CaO_2	−157.5
Ca_3P_2	−120.5
CaS	−115.3

of the most efficient electrical conductors on a weight basis. See also **Chemical elements.**

Uses: The major uses of calcium stem from its active chemical nature. It is used in tonnage quantities in high-purity form as a reductant for the reactive metals, including zirconium, thorium, vanadium, and uranium. In the case of zirconium reduction, zirconium fluoride is reacted with calcium metal in a reactor. The high heat of reaction melts the zirconium, forming an ingot, which can be vacuum-remelted directly without further consolidation. The thorium and uranium oxides are reduced with an excess of calcium in reactors or trays under an atmosphere of argon. The resultant metal is leached with acetic acid to remove the lime.

An important use of calcium metal either pure or in alloy form is as a purifying reagent in steel, aluminum, and other metals. Its high affinity for oxygen, sulfur, and other deleterious impurities in molten metals results in better removal than is achieved by such common scavenger additions as manganese, silicon, and aluminum. Calcium is intermediate in cost as a molten-metal additive, and is used where it is important to achieve clean, inclusion-free metal or to change inclusion shape for better mechanical properties.

The literature of steel purification contains many references to calcium usage, with a great deal of disagreement as to results. This situation points up the need for careful control of operating conditions in using calcium. Care must be taken to achieve good reaction with the molten metal. Although tap-stream additions are effective, the most efficient means of using calcium or calcium alloys is the immersion technique. In this method the slag is removed and a fresh "cover of crushed lime is recommended over the melt to avoid reaction of CaS *'with FeO in the slag and oxygen in the air,'* thus allowing the sulfur to revert back into the metal bath." * See also **Ironmaking and steelmaking.**

Other metallurgical uses for calcium are as an efficient strengthening addition to lead and as an addition in a magnesium alloy used for etching. An alloy of 80% Ca–20% Mg is used to deoxidize magnesium castings.

Calcium is used as an electrode in certain battery applications. The metal is also converted to the hydride, which is used as a portable source of hydrogen gas (by reacting with water).

*Detailed procedural advice for the use of Ca in melt purification is available from the only U.S. supplier of Ca metal, Pfizer Inc., New York.

Metallic Ca is also used in the production of calcium pantothenate, an important B-complex vitamin.

Calcium is produced in the United States by Pfizer Inc., at Canaan, Conn., and in Canada by Dominion Magnesium, Ltd. The process used by both companies is the thermal reduction of lime with aluminum. Prior to World War II, the major production method was by electrolysis of fused calcium chloride, but this method has been largely discontinued.

The reactants, lime and aluminum powder, are briquetted and charged into high-temperature-alloy retorts which are maintained at a vacuum of less than 100 μm. The charge is heated to 1200°C, and the reaction takes place slowly to release Ca vapor, which is removed from the reaction by condensation, thus allowing the reaction to continue in the desired direction.

High-purity lime is required to produce high-purity Ca metal. The reduced Ca is contaminated with aluminum, which can be removed by a further vacuum-distillation step. This operation also reduces the level of other contaminants, such as manganese, as shown in Table C-3.

Commercial-grade Ca metal is available in several forms, including (1) full crowns (10-in.-diam × 20-in.-long crystalline deposits); (2) 6-mesh nodules; (3) turnings and lathe shavings; (4) ingots (6-lb notched pigs); (5) waffles (18 lb); and (6) billets (40 lb, 6 in. diam × 28 in. long). Ingots or waffles also are available with the composition 80% Ca and 20% Mg. Redistilled-grade Ca metal is available as $\frac{1}{8}$-in. nodules, broken crows, and 6-mesh nodules.

Calcium Compounds: Calcium silicide alloys, which are widely used in the metallurgical industry, are made by electric-arc-furnace reduction.

TABLE C-3. *Grades of Calcium Metal—Typical Chemical Analysis*

Element	Commercial Grade, %	Redistilled Grade, %
Mg	0.50	0.50
N	0.08	<0.02
Al	0.30	<0.001
Fe	0.008	<0.001
Mn	0.01	<0.002
Co	...	<0.0002
Li	...	<0.0001
Be	...	<0.0001
Cr	...	<0.0002
B	...	<0.0001
Ca and Mg	99.5	99.9

One reaction is $2SiO_2 + 2C + CaC_2 \rightarrow CaSi_2 + 4CO$ (-94 kcal).

The principal inorganic compounds of Ca are calcium carbonate (limestone and less pure forms), calcium magnesium carbonate (dolomite), calcium oxide, CaO (lime), and hydrated calcium oxide, $Ca(OH)_2$ (slaked lime), calcium nitrate, complex calcium silicates, calcium sulfate (gypsum), calcium phosphate, calcium chloride, calcium fluoride, calcium carbide, and many others. These important materials find extremely wide usage in both a natural and refined state and for many diverse applications. For example, limestone, as marble, is an important building material. The sedimentary deposits are used as an essential ingredient in the making of iron in the blast furnace, and agricultural usage is extensive, to name a few. See also **Gypsum; Lime;** and **Limestone.**

Precipitated calcium carbonate is widely used in pigments, paints, paper, and plastics. Flaked calcium chloride is used for melting snow and ice, in road-dust control, freezeproofing of coal and ores, refrigeration brines, and in winter-set concrete. Calcium hydroxide in the form of slaked lime is used in mortar. Calcium silicate is an ingredient of cement. See also **Cement.** Calcium fluoride in the form of fluorspar is used in metallurgical operations and chemical manufacture. Tricalcium phosphate is an ingredient of some fertilizers. See also **Fertilizers.** Calcium oxide is used in metallurgical slagging operations.

Calcium pantothenate is an important B-complex vitamin. Organocalcium compounds are used in lubricants, corrosion inhibitors, and detergents, and calcium salts of organic acids are used in chemical operations to produce ketones, such as acetone.

REFERENCES

American Society for Metals: "Metals Handbook," 8th ed., vol. 1, Metals Park, Ohio, 1961.

Hempel, C. A.: "Rare Metals Handbook," 2d ed., Van Nostrand, New York, 1961.

—Frank Emley, *Pfizer Metals & Composite Products, Pfizer Inc., Wallingford, Conn.*

Calcium Carbide   (See **Acetylene.**)

Calcium Carbonate   (See **Limestone.**)

Calcium Nitrate   (See **Fertilizers.**)

Calcium Oxide   (See **Lime.**)

Calcium Phosphate   (See **Fertilizers.**)

Calcium Silicates   (See **Cement;** and **Lime.**)

Calcium Sulfate   (See **Fertilizers;** and **Gypsum.**)

Calendering   (See **Papermaking and finishing.**)

Californium   (See **Chemical elements.**)

Calking   (See **Adhesives.**)

Camphene   (See **Terpenes.**)

Candelilla   (See **Waxes.**)

Candida Utilis   (See **Yeasts.**)

Capric Acid   (See **Vegetable oils.**)

Caprolactam   Caprolactam, $NH(CH_2)_5CO$, is a major ingredient in the manufacture of type 6 nylon. See also **Nylons.** Caprolactam is manufactured in several ways. One of the more recent processes involves a photochemical reaction. The heart of the process (PNC process*) is a photoreactor, where cyclohexane is converted into cyclohexanone oxime hydrochloride:

$$C_6H_{12} + NOCl \xrightarrow[\text{light}]{\text{HCl}} C_6H_{10}NOH \cdot 2HCl$$

The yield on cyclohexane is claimed to be greater than 86% by weight. The cyclohexanone oxime hydrochloride then is converted to ε-caprolactam in a Beckmann rearrangement:

$$C_6H_{10}NOH \cdot 2HCl \xrightarrow{H_2SO_4}$$

$$\underset{\underline{\hspace{1cm} NH \hspace{1cm}}}{CH_2-(CH_2)_4-C{=}O} + 2HCl$$

The nitrosyl chloride required for producing the oxime is obtained by the following series of reactions: (1) NH_3 is burned in air to produce NO_x: $2NH_3 + 3O_2 \rightarrow N_2O_3 + 3H_2O$; (2) nitrosyl sulfuric acid is obtained by reacting nitrogen trioxide with H_2SO_4: $2H_2SO_4 + N_2O_3 \rightarrow 2HNOSO_4 + H_2O$; (3) nitrosyl chloride then is produced by adding HCl to the nitrosyl sulfuric acid: $HNOSO_4 + HCl \rightarrow NOCl + H_2SO_4$.

To effect the photochemical reaction, high-pressure Hg lamps are used. Light of wavelengths

*Toyo Rayon Co., Ltd.

shorter than 3650 Å must be filtered out to avoid formation of tarry products. The factors that are vital to the photochemical reaction are (1) temperature, (2) distance between lamps and reactor, and (3) volume of the reactor.

The crude caprolactam solution is neutralized with NH_3, and the resulting mixture separates into an upper layer of crude caprolactam and a lower layer of aqueous $(NH_4)_2SO_4$. The PNC process is claimed to produce only about one-half as much $(NH_4)_2SO_4$ as from other processes.

In another very widely used process, benzene is hydrogenated to cyclohexane. See also **Cyclohexanol/Cyclohexanone (KA oil).** The latter is oxidized in the liquid phase with an air-O_2 mixture at about 160°C and 8–9 atm. Produced by the reaction are cyclohexanol, cyclohexanone, acids, CO_2, and esters. The efficiency of converting to cyclohexanol and cyclohexanone is claimed to be about 85%. After several purification steps and recycling, the cyclohexanone is reacted with hydroxylamine sulfate to produce cyclohexanoneoxime and an $(NH_4)_2SO_4$ byproduct. The hydroxylamine sulfate is made from NH_4NO_3 and SO_2. The aforementioned oxime is rearranged in H_2SO_4 (Beckmann) to produce caprolactam. Several purification steps are required to purify the crude caprolactam produced.

In still another process, the charge stock is toluene (nitration grade), air, H_2, anhydrous NH_3, and H_2SO_4. Toluene is oxidized to produce a 30% solution of benzoic acid plus intermediates and by-products. After fractionation and purification, the pure benzoic acid is hydrogenated (Pd catalyst) in stirred tank reactors (10 atm, 170°C). The cyclohexanecarboxylic acid produced is mixed with H_2SO_4 and reacted with nitrosylsulfuric acid to produce caprolactam. The nitrosylsulfuric acid is produced by absorption of nitrogen oxides, N_2O_3, in sulfuric acid: $N_2O_3 + H_2SO_4 + SO_3 \rightarrow 2NOHSO_4$. The acid solution is neutralized with NH_3 to produce $(NH_4)_2SO_4$ and a layer of crude caprolactam, which is further purified. The overall process reaction is

$$\langle\!\!\!\!\bigcirc\!\!\!\text{H}\rangle\!-\!\text{COOH} + \text{NOHSO}_4 \xrightarrow{SO_3}$$

$$\begin{array}{c} \text{CH}_2\text{---CH}_2\text{---CO} \\ | \qquad\qquad | \\ \text{CH}_2 \qquad\qquad | \\ | \qquad\qquad | \\ \text{CH}_3\text{---CH}_2\text{---NH} \end{array} + \text{CO}_2 + \text{H}_2\text{SO}_4$$

Caprylic Acid (See **Vegetable oils.**)

Carbanion (See **Polymerization.**)

Carbides (See **Aluminum; Iron and steel;** and **Tungsten.**)

Carbocyclic Compounds (See **Heterocyclic compounds.**)

Carbohydrates All carbohydrates contain carbon, hydrogen, and oxygen and are distinguished by the fact that the hydrogen and oxygen always are in the same proportion as in water, accounting for the general term *carbo-hydrate*. A general formula for all carbohydrates is $C_x(H_2O)_y$. However, the relationship can be misleading, because structurally the carbohydrates cannot be regarded as true hydrates of carbon, but rather the carbohydrates possess one of two basic structures, that of an aldehyde or that of a ketone.

A term synonymous with carbohydrate is *saccharide* (sometimes *saccharose*). When referring to saccharides, the basic molecular formula is considered to be $C_6H_{12}O_6$. Compounds with this general formula, such as glucose, mannose, and galactose, are known as monosaccharides because they contain one $C_6H_{12}O_6$. A disaccharide, as typified by sucrose, lactose, and maltose, has the general molecular formula $C_{12}H_{22}O_{11}$ and may be considered as containing two $C_6H_{12}O_6$ groupings that have been joined by one atom of oxygen with the elimination of one molecule of water. Similarly, the trisaccharides, such as raffinose, have the molecular formula $C_{18}H_{32}O_{16}$. Any larger molecules of the $C_x(H_2O)_y$ configuration are termed polysaccharides and include the starches, celluloses, dextrin, and glycogen.

Both the terms carbohydrate and saccharide are significant only by way of classifying these compounds, because neither term appears in whole or in part in any of the widely used names of these compounds. About the only point of nomenclature enjoyed in common by several of the saccharides is the termination -*ose*, as found, for example, in cellulose, dextrose, sucrose, and glucose.

Any saccharide having the structure of an aldehyde is termed an *aldose;* any saccharide with the structure of a ketone is termed a *ketose.* For those saccharides that contain 4–6 carbons, the number of carbons forms a nomenclature base, as a *tetrose,* $C_4H_8O_4$, a *pentose,* $C_5H_{10}O_5$, and a *hexose,* $C_6H_{12}O_6$. To be consistent with the relationship between a mono- and disaccharide, some author-

TABLE C-4. *Properties of Representative Carbohydrates*

Class and Examples	Formula Wt.	Mp, °C	Sp gr	Remarks
Monosaccharides:				
Tetroses, $C_4H_8O_4$	120.06			
Erythrose (an aldotetrose)	. . .	. . .	. . .	Obtained by oxidation of erythritol. Possibly a mixture of an aldose and a ketose.
Pentoses, $C_5H_{10}O_5$:				
Arabinose (α)	150.13	159.5	1.585	Occurs in bran and straw. Yields furfuraldehyde when distilled with HCl or dilute H_2SO_4.
Xylose ⎫ optically isomeric	150.13	153.5	1.535	
Ribose ⎬ aldopentoses	150.13			
Lyxose ⎭	150.13			
Hexoses, $C_6H_{12}O_6$ (aldohexoses):				
Glucose (dextrose, grape sugar)	180.16	(α) 146	1.544	Strong reducing agent. Precipitates Au and Ag from their salts. Forms glucosates with metal hydroxides. With *zymase*, produces C_2H_5OH and CO_2.
Galactose	180.16	(β) 150	1.562	
Mannose (*d-*) ⎤ optical isomerides	180.16	132	1.539	
Gulose (*d-*) ⎦ of glucose	180.16	168		
Idose.	180.16	156*		
Talose	180.16	188*		
Altrose	180.16	184*		
Allose	180.16	184*		
Hexoses, $C_6H_{12}O_6$ (ketohexoses):				
Fructose (levulose, fruit sugar)	180.16	95–105	1.669	Prepared from invert sugar and from inulin. Occurs in fruits and honey.
Sorbose	180.16	165	1.654	*d-*Sorbose obtained by oxidation of *d-*sorbitol.
Tagatose	180.16	124	. . .	Obtained by action of KOH on *d-*galactose.
Disaccharides, $C_{12}H_{22}O_{11}$:				
Sucrose (cane or beet sugar)	342.30	170–186d	1.588	With *invertase* produces fructose and glucose.
Lactose (milk sugar)	360.31†	202	1.525	With *lactose* produces glucose and galactose.
Maltose (malt sugar)	360.31†	d	1.540	With *maltose* produces glucose(s).
Cellobiose	. . .	. . .	. . .	With *maltose* or cellase produces glucose(s).
Trisaccharides, $C_{18}H_{32}O_{16}$:				
Raffinose	594.52‡	119	1.465	With *invertase* produces fructose and melibiose.
Polysaccharides, $(C_6H_{10}O_5)_x$:				
Starch	$(162.14)_x$	d	1.5	
Cellulose	$(162.14)_x$		1.3–1.4	
Dextrin.	$(162.14)_x$	178d	1.35	With *maltose* produces glucose.
Inulin.	. . .	. . .	. . .	With *invertase* produces fructose.
Glycogen (animal starch).	$(162.14)_x$	240	. . .	With *diastase* produces glucose and maltose. Occurs in liver, muscles, and white corpuscles.

(*d-*) = dextrorotary, d = decomposes.
*Of corresponding osazone.   †One molecule H_2O.   ‡Five molecules H_2O.

ities do not term a tetrose or a pentose a monosaccharide. By combining the *ald-* and *ket-* prefixes, certain compounds then may be called aldohexoses, such as glucose and galactose, or ketohexoses, such as fructose and sorbose.

The mono-, di-, and trisaccharides also are commonly termed *sugars*. A sugar generally is considered to possess the properties of a crystalline solid with a relatively low melting point (below 150°C), of being soluble in water, and of possessing a sweet taste. Thus the common names of several saccharides incorporate the term sugar, preceded by the common raw source of the substance, as glucose (grape sugar), sucrose (cane sugar or beet sugar), maltose (malt sugar), and lactose (milk sugar). The crosscurrents of the nomenclature employed for the carbohydrates will become clear upon inspection of Table C-4.

Importance of Carbohydrates: Several of the carbohydrates, such as sucrose, the starches, and the celluloses, represent extremely high production materials. Carbohydrates constitute a major source for several of the alcohols used in industrial and consumer products (particularly beverage alcohols derived from fermentation); they are a major component of the diet of man and livestock and, as such, are primary sources of heat and energy for mammals; and in addition to providing basic food values, several of the carbohydrates possess other properties (good taste) that make them attractive for incorporation into a wide variety of food products.

The starches are the basis for numerous products, including adhesives. See also **Starches.** Cellulose finds wide application in paper, wood products, and rayon. See also **Pulp (wood) production and processing;** and **Rayon.** The celluloses enter into a number of synthetic and semisynthetic materials. See also **Cellulose ester plastics, organic;** and **Ethyl cellulose.**

The carbohydrates, like so many of the chemical compounds derived from natural materials, are quite complex. Of particular interest are (1) the exceptionally large molecules that are represented by the polysaccharides, with 10 monosaccharide units in glycogen, 25 units in starch, and several hundred units in cellulose; (2) the dependence of the sugars and starches upon enzymes for their utilization in the metabolic process and for their conversion into other products in industrial processes—see also **Enzymes;** and (3) the manner in which higher saccharides can be synthesized from the simpler monosaccharides and the resulting large number of stereoisomeric forms that result. See also **Gums.**

Carbon (L. *carbo,* charcoal.) **C** = 12.01115 (at. wt.); 6 (at. no.). Carbon is one of the most remarkable of the chemical elements because of its versatility of behavior, reflected by the hundreds of thousands of compounds which contain carbon, and perhaps an equal or greater number of carbon-containing compounds as yet not isolated or synthesized. Carbon is exceeded only by hydrogen in the number of compounds which contain a given element.

As pointed out under **Carbon compounds,** this element is the anchor for all organic substances, and in the form of its oxide gases CO and CO_2 and salts notably the carbonates, cyanides, cyanates, sulfur compounds such as CS_2 and thiocyanates, and carbides, as a partial list—it plays an important role in inorganic materials as well. Unlike a number of chemical elements which essentially remain curiosities in their elemental state, the elemental forms of C are extremely utilitarian, as exemplified by diamond, graphite, and charcoal and coal (as an impure form of elemental C).

Elemental carbon exists in three allotropic forms:[*] (1) diamond and (2) graphite—both crystalline, and (3) amorphous forms as typified by C in charcoal, coal, and coke. Carbon occurs in group IVA of the periodic table, along with Si, Ge, Sn, and Pb, all these elements composing the carbon family of elements. Carbon is in period 2 of the table. There are two stable isotopes of C, 12 and 13. There are four unstable isotopes, 10, 11, 14, and 15. Although radioactive, ^{14}C occurs in minute quantities (10^{-10}% of the total C isotopic mixture) in naturally occurring forms of C. Because of accurate knowledge of the half-life of ^{14}C (5,570 years), the presence of this isotope in very old artifacts and documents makes a useful tool for archeological diagnosis.

Carbon (graphite) has a melting point of 3700 ± 100°C, a boiling point of 4830°C, and a density of 2.22 g/cm^3 at 20°C. Other important mechanical and physical properties of C are given in Table C-5.

[*]A so-called "white" allotropic form of C has been reported as produced at high temperature and low pressure during graphite sublimation. According to the reference cited here, under free-vaporization conditions above ~2550 K, the white C forms as small transparent crystals on the edges of the basal planes of graphite. The interplanar spacings of this material are reported to be identical with those of a C form noted in graphitic gneiss from the Ries Crater. A. Greenville Whittaker and P. L. Kintner, *Science,* vol. 165, pp. 589–591, Aug. 8, 1969.

TABLE C-5. *Some Properties of Elemental Carbon*

Graphite:

Thermal conductivity (cal)(cm²)/ (cm)(°C)(s), at 20°C	0.057
Electrical resistivity, Ω-cm, at 0°C . .	$1,375 \times 10^{-6}$
Modulus of elasticity in tension, 10^6 psi	0.7
Atomic volume, cm³/g atom	5.41
Crystal structure	Hexagonal
Lattice constant, kX units at 20°C:	
a	2.4564
c	6.6906
Color	Black
Closest approach of atoms	1.42
Ionization potential, eV	11.264
Density, g/cm³, at 20°C	2.22
Hardness, Mohs' scale	0.5–1

Amorphous:

Specific gravity, at 20°C	1.8–2.1
Color	Black

Diamond:

Crystal structure	Cubic
Index of refraction	2.417–2.4195
Specific gravity, at 20°C	3.51–3.521
Dielectric constant:	
10^4 Hz	16.5
10^8 Hz	5.5
Hardness, Mohs' scale	10
Hardness, Knoop scale	5,500–7,000

All forms of carbon are insoluble in water, acids, and alkalis.

Occurrence: The most common categories in which C occurs naturally include (1) *free carbon* as found in diamond, notably in South Africa; as graphite, in Ceylon; and along with impurities in coal, lignite, and peat, in scores of locations throughout the world; (2) *combined inorganic form,* as in the form of salts (notably carbonates) in rocks, minerals, and ores, such as limestone and dolomite, and with oxygen in the gases CO and CO_2;* and (3) *combined organic form,* principally as CH_4 (methane), the principal constituent of natural gas, and in a large series of CH_x (hydrocarbon) compounds in petroleum. These sources are described elsewhere in this volume. Consult the Subject Index.

The practical use of diamond in abrasives is described under **Abrasives.** The differences in the properties of diamond and graphite (Table C-5) are explained by their different crystalline configurations. In diamond, the normal single-bond

*A much more detailed description of these properties is contained in Donald I. Hamm, "Chemistry," pp. 793–794, Appleton-Century-Crofts, New York, 1965.

distances (C—C) apply; namely, they all are 1.54 Å. A diamond crystal may be described as a huge polymeric molecule, tightly packed with a density of 3.5 g/cm³ as compared with graphite (2.2 g/cm³), accounting for its extreme hardness. Surprisingly, the C—C bond distances are less in graphite than in diamond, being 1.42 Å. The findings of x-ray diffraction and other studies, however, point to a laminar construction of graphite with essentially two-dimensional molecules.† These structural differences are reflected in the physical qualities of diamond and graphite, including the electrical conductivity of graphite versus the nonconductivity of diamond. The lubricity of graphite also is attributed to its basic crystalline structure.

Certain previously assumed amorphous forms of C, including some of the carbon blacks, are now suspected to be microcrystalline, graphitic-type structures. It is interesting to note that when sublimed and condensed, all forms of C revert to graphite.

Carbon Black.‡ The term carbon black describes a rather wide spectrum of finely divided carbonaceous pigments which are derived from the pyrolysis of oils and hydrocarbon gases. Well over 90% of the carbon black produced goes into reinforcing and compounding agents for rubber, particularly for aircraft and motor-vehicle tires. It is estimated that the average life expectancy in normal usage of these tires is extended by a factor of 9 to 10 times as the result of the effective application of carbon blacks. Because of the deep and permanently black coloration of carbon black, the material also finds wide use in plastics, paper, paints, inks, and protective coatings. Adding only 1 to 2% carbon black to plastics often will minimize, if not fully overcome, the adverse effects of ultraviolet radiation (sunlight) on such materials.

Depending upon the method of manufacture, carbon black may be classified into five principal

†In resolving the spectral discrepancies from observations of the planet Mars, there is indication that the red coloration of Mars may be at least partially attributed to the presence of carbon suboxide, C_3O_2(bp = 7°C) in the Martian atmosphere. The linear molecules of C_3O_2 readily polymerize, whereupon they form several heavier molecules, with color gradations ranging from pale yellow to orange, reddish brown, violet, and nearly black. William T. Plummer and Robert K. Carson, *Science,* vol. 166, pp. 1141–1142, Nov. 28, 1969.

‡The assistance of the Columbian Division of Cities Service Company is gratefully acknowledged.

types: *lampblack, channel or impingement carbons, thermal blacks, acetylene black,* and *furnace carbon.* Most commercial carbon blacks are of the latter type.

The two properties of carbon black that are most important in commercial applications are particle size and high surface area. Particle sizes range from 100 to 5000 Å. Surface areas fall within a range from 6 to 1,100 m^2/g. Actual carbon content will vary from 83 to 99%. The arrangement of carbon particles may be described as comprising hexagonal nets of carbon atoms, *paracrystalline* in nature. When viewed with an electron microscope, the particles appear as rough spheres, although they usually are seen in clusters rather than as individual spheres. The aggregating or clustering behavior of carbon black is considered to result from both physical and chemical bonding effects. The manufacture of carbon black requires careful monitoring and control over surface area and particle size, and because of the wide range of control parameters, a very wide spectrum of carbon blacks can be created for varying requirements.

The human eye can distinguish 260 shades of blackness. The blackest carbon particle of the commercially available types will have a diameter of about 100 Å, whereas the grayest particle will have a diameter of approximately 5000 Å. The blackness of carbons often is referred to as *masstone,* those particles having the smaller diameters and greater surface areas exhibiting the highest masstone.

Methods of Manufacture: The production of *lampblack* dates back to antiquity. In the early production methods, coal-tar residues or petroleum substances were burned under very restricted combustion conditions, with the intent, of course, of producing large quantities of unoxidized carbon. Settling chambers, originally used for collection, were ultimately replaced by bag filters, cyclones, or electrical precipitators. Currently, most lampblack is made in oil furnaces.

In the production of *channel* or *impingement carbons,* natural gas or natural gas that may contain oil vapors is burned in literally hundreds of smaller burners, so that the small flames produced impinge upon a flat surface (channels) where the carbon deposits for periodic removal by scrapers. The burner house which contains all the apparatus incorporates dampers for regulating bottom and top drafting.

In the production of *thermal blacks,* natural gas, in the absence of air, is decomposed thermally. Cylindrically shaped furnaces (10–12 ft diam) are equipped with a brick open checkerwork, which is heated to the proper process temperature (1100–1650°C), after which natural gas is admitted to the furnace, where it decomposes into carbon and hydrogen. The C particles are removed from the checkerwork by a gas stream from which, after cooling, the C is collected in bag filters or cyclones. Two or more furnaces are used, so that one furnace may be on the heating cycle while another is on the producing, or "make," cycle. The reaction, of course, is endothermic. The hydrogen produced may be recycled back to the furnace as fuel. In the production of very fine thermal blacks, a part of the hydrogen may be used to dilute the natural-gas feed.

Acetylene black is produced from acetylene as the feed to high-temperature retorts which dissociate the acetylene into carbon and hydrogen. The reaction is exothermic, and thus controlled by holding the rate of acetylene feed at just the right point to obtain the desired dissociation temperature.

Furnace carbons range from very fine particles (as used in tire treads) to the coarser particles (tire carcasses). For production of fine grades, natural gas is burned with air in slight excess to support combustion. The liquid oil or other hydrocarbon feedstock to be decomposed into C is atomized and injected into the furnace at a point where the blast-flame gases are moving at a very high velocity. Upon such injection, the temperature is immediately increased to the point where the feed is instantly decomposed to carbon black. A direct water-quench spray is used to cool the products, after which the carbon black is collected. The process for making the coarser grades is similar, except that the furnace volume is larger, the residence time is longer, the oil/air ratio is higher, and the gas velocities within the furnace are lower.

Graphite: Over the years, graphite has been variously termed *plumbago, black lead,* and *Flanders stone.* The name stems from the Greek *graphite* (from *graphein,* to write), as proposed by A. G. Werner, a German mineralogist. The use of graphite in pencils (popularly termed lead pencils) has inaccurately associated graphite with lead in the minds of laymen for many years. Graphite occurs in high-grade lump form in Ceylon, a form often preferred for the manufacture of crucibles and vessels for chemical processing. The grade called No. 1 flake graphite contains a minimum of 90% graphitic carbon. Mexican graphite is an amorphous type, containing about 80% C. Some graphites from Malagasy contain up to 96% C.

In addition to its use in corrosion and high-temperature-resistant processing piping and vessels, graphite finds wide use in electrodes, as a pigment, and as a lubricant. Finely divided graphite particles will remain in solution for long periods (sometimes with alcohol present), providing an effective machine lubricant. Evaporation of the alcohol leaves a thin film of graphite on bearing surfaces. For the manufacture of pencils, graphite is mixed with clay, the clay contributing to the hardness of the "lead." Certain self-lubricative metals are prepared by mixing graphite with the oxides of Cu, Sn, and Pb, which, after sintering and powder-metallurgy operations, form alloys that will contain a significant volume of oil for long periods of service. Recrystallized molded graphite is used for rocket casings and other heat-resistant structures, permitting sustained temperatures up to 3100°C. A pyrolytic graphite with good electrical and thermal conductivity along the surface and a tensile strength up to 40,000 psi at temperatures of 2760°C and very impervious to liquids or gases is available in thin-sheet form (down to 0.001 in. thickness). By adding boron to this material, it is made suitable for atomic-radiation shielding.

Coal: Predominantly carbon (in terms of value), coal represents the remains of ancient vegetation that over past geologic ages has been subject to biochemical action, pressure, and heat. As is evident from Table C-6, coal is not a uniform material, but contains moisture and varying amounts of sulfur, hydrogen, and oxygen. Further, the physical and structural characteristics of coal range widely from one source to the next and also as the result of differences in processing, storing, and handling methods.

Coals, in approximate order of Btu value, range from (1) *lignite,* $H_2O = 43.4\%$, $C = 37.8\%$, 7400 Btu/lb; (2) *subbituminous,* $H_2O = 23.4\%$, $C = 42.4\%$, 9720 Btu/lb; (3) *low-rank bituminous,* $H_2O = 11.6\%$, $C = 47.0\%$, 12,880 Btu/lb; (4) *medium-rank bituminous,* $H_2O = 5.0\%$, $C = 54.2\%$, 13,880 Btu/lb; (5) *cannel coal,* $H_2O = 2.2\%$, $C = 63.0\%$, 13,250 Btu/lb; (6) *high-rank bituminous,* $H_2O = 3.2\%$, $C = 64.6\%$, 15,160 Btu/lb; (7) *low-rank semibituminous,* $H_2O = 3.0\%$, $C = 75.0\%$, 15,480 Btu/lb; (8) *high-rank semibituminous,* $H_2O = 5.0\%$, $C = 83.4\%$, 15,360 Btu/lb; (9) *semianthracite,* $H_2O = 6.0\%$, $C = 83.8\%$, 14,880 Btu/lb; (10) *anthracite,* $H_2O = 3.2\%$, $C = 95.6\%$, 14,440 Btu/lb.

Sulfur occurs in coal in three forms: (1) combined with iron as pyrite or marcasite, (2) organic sulfur, and (3) sulfate sulfur or sulfur combined with Fe or Ca together with O_2 as $FeSO_4$ or

TABLE C-6. *Typical Analyses of Representative Coals and Solid Fuels*

Type of Coal or Fuel	Source	Proximate Analysis			Ultimate Analysis					
		Moisture	Volatile Matter	Fixed Carbon	Ash	S	H	C	O	Btu/lb
Anthracite	Arkansas	2.1	9.8	78.8	9.3	1.7	3.6	80.3	3.6	13,700
	New Mexico	2.9	5.5	82.7	8.9	0.8	2.9	82.3	3.9	13,340
	Pennsylvania	4.1	3.5	81.7	10.7	0.5	2.2	81.6	4.4	12,590
	Pennsylvania	5.4	3.8	77.1	13.7	0.6	2.4	76.1	6.6	11,950
Subanthracite	Virginia	2.2	12.4	67.4	18.0	0.5	3.6	72.4	4.7	12,270
HV bituminous	Alabama	2.8	29.6	59.3	8.3	1.9	5.2	77.2	5.8	13,750
	Ohio	8.2	36.1	48.7	7.0	1.2	5.6	68.4	16.4	12,160
	Illinois	8.0	33.0	50.6	8.4	1.2	5.4	68.7	14.7	12,130
	Kentucky	7.2	39.8	48.8	4.2	2.6	5.8	71.5	14.3	12,950
	Indiana	12.4	36.6	42.3	8.7	2.3	5.7	63.4	18.6	11,420
MV bituminous	Virginia	3.1	21.8	67.9	7.2	1.0	5.0	80.1	5.2	14,030
LV bituminous	Maryland	3.2	18.2	70.4	8.2	1.0	4.5	79.0	5.7	13,870
	Oklahoma	2.6	16.5	72.2	8.7	1.0	4.3	80.1	4.2	13,800
Subbituminous.....	Colorado	19.6	30.5	45.9	4.0	0.3	6.0	58.8	29.6	10,130
	Wyoming	23.2	33.3	39.7	3.8	0.4	6.4	54.6	33.8	9,420
Lignite	North Dakota	34.8	28.2	30.8	6.2	0.7	6.7	42.4	43.3	7,210
	Texas	33.7	29.3	29.7	7.3	0.5	6.8	42.5	42.1	7,350
Petroleum coke	....	1.1	7.0	90.7	1.2	0.8	3.3	90.8	3.1	15,060
By-product coke	....	0.5	1.8	86.0	11.7	1.0	0.8	84.4	0.9	12,527

HV = high volatile, LV = low volatile, MV = medium volatile.

$CaSO_4$. Sulfur seldom occurs in the free state in coal. Normally, organic sulfur will constitute anywhere between 20 to 85% of the total sulfur content. Sulfate sulfur usually is found in weathered coal.

Peat, which may be described as partially carbonized vegetable matter, is believed to represent an early stage in the geological development of coal. Peat in its natural form is less than an ideal fuel because of its high moisture content (up to 80%). After drying, the moisture content will drop to about 71%, with a calorific value of about 5000 Btu/lb. Charred peat, which is peat that has been partially decomposed thermally, sometimes is marketed as a soil conditioner and fertilizer.

Activated Charcoal: Dense materials, such as hardwoods, coconut shells, and peach pits, are carbonized to produce a chemically pure amorphous carbon. Activated charcoal from coconut shells will adsorb up to 68% of its weight of CCl_4, which indicates its value as an adsorbent for various processes. The material was widely used for gas masks in connection with threats of gas warfare. Activated or filter carbons, used widely for purifying various chemicals, in solvent recovery systems, and for decolorizing liquids, usually are prepared from coal or petroleum. Activated charcoals also are obtained from wood distillation.

Carbon Arc   (See **Energy systems for processes; and Plasma, chemical processing with.**)

Carbon Blacks   (See **Carbon.**)

Carbon Compounds   Hundreds of thousands of carbon compounds are possible because of the versatility of the carbon atom. Although carbon has valences of 2^- and 4^-, the large majority of organic chemicals* comprise carbon atoms with a valence of 4^-. The versatility of the carbon atom stems from two main factors: (1) the ability of carbon atoms to bind one with another, sharing one, two, or three bonds between atoms; and (2) the availability of the four valence bonds in terms of (*a*) four single bonds, (*b*) two single bonds and one double bond, (*c*) one single bond and one triple bond, or (*d*) two double bonds. In effect, this provides the carbon atom with valences of 1^-,

---

*This immediate discussion is confined essentially to organic carbon compounds. Inorganic carbon compounds, including carbides, carbonates, cyanides, carbon dioxide, and carbon monoxide, are included with other descriptions. Consult the Subject Index.

2^-, and 3^- for combination with elements and radicals of like positive valence numbers. As shown in Table C-7, carbon combines readily with hydrogen, oxygen, nitrogen, and sulfur.

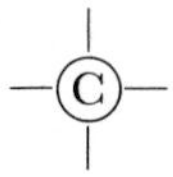

Four single bonds 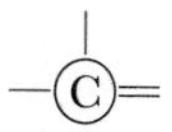   Two single bonds and one double bond

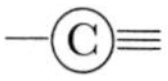

One single bond and one triple bond 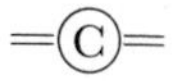   Two double bonds

Before considering compounds comprised of two or more carbon atoms, some insight into the behavior of the carbon atom can be gleaned from looking at a number of compounds in which a *single* carbon atom appears in various valence configurations. See Table C-8.

Chain Formation: Inasmuch as carbon atoms share bonds with each other, the formation of chains of atoms is a logical expectation. Where there are fewer than four atoms involved, as shown below, geometrically the chain must be of one continuous length, often termed a *straight chain*. Needless to say, the chain, in effect, need not be straight, but the latter term is used to identify those chains where there is no branching or forking.

With four or more atoms involved, branching occurs frequently, giving rise to a number of compounds (*isomers*) that contain the same type and number of atoms and hence have the same chemical formula and formula weight. Nevertheless, these compounds display differing chemical and physical properties, thus stressing the importance of structure. The two fundamental geometric arrangements possible with a linking of four atoms are shown below.

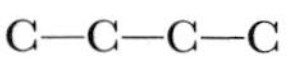

Straight chain 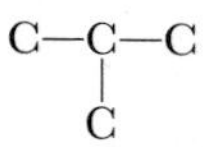   Branched chain

The effect of branching in carbon chains is demonstrated by the four butyl alcohols, shown on page 202. As the number of carbons in a chain increases, the probability for branching increases.

TABLE C-7. *Main Classes of Carbon Compounds in Terms of Other Elements Present*

C + O		C + H	
Oxides of carbon:		Hydrocarbons:	
Carbon dioxide		Acetylenes	Cycloparaffins
Carbon monoxide		Benzenoids	Olefins
Carbon suboxide		Cycloolefins	Paraffins

C + H + O		C + S $\pm$ H $\pm$ O $\pm$ N	
Acid anhydrides	Furans	Isothiocyanates	Thioacids
Acids	Ketones	Penthiophenes	Thioaldehydes
Alcohols	Lactides	Sulfinic acids	Thiocarbonic acid
Aldehydes	Lactones	Sulfinyl	Thiocyanates
Carbohydrates	Phenols	Sulfones	Thioethers
Esters	Quinones	Sulfonic acids	Thioketones
Ethers	Terpenes	Sulfonyls	Thiophenes
		Sulfoxides	Thiophenols
		Tertiary sulfoniums	

C + N $\pm$ H $\pm$ O			
Amides	Cyanates	Isocyanates	Proteins
Amines	Cyanides	Isocyanides	Purines
Amino acids	Diazo compounds	Nitrates	Pyridines
Amino guanidines	Fulminates	Nitrites	Pyrroles
Aminoazo compounds	Guanidines	Nitro compounds	Quaternary ammoniums
Anilides	Hydrazines	Nitrosamines	Semicarbazides
Azo compounds	Hydrazo compounds	Nitroso compounds	Semicarbazones
Azoxy compounds	Hydrazones	Osazones	Ureas
Carbamates	Hydroxyzao compounds	Oximes	Ureides
Cyanamides	Hydroxylamines	Polypeptides	

C $\pm$ H $\pm$ O $\pm$ N + Si or P or Metals		
Silicones (Si)	Fluorocarbons (F)	Organometallic compounds
Phosphoniums (P)	Halogen derivatives (Cl, Br, I, F)	(Al, As, B, Ca, Cd, Fe, Hg,
Grignard reagents (Mg)		K, Li, Mn, Na, Pb, Sb, Sn, Te, etc.)

Whereas there are only four isomeric butyl alcohols, there are eight isomeric amyl alcohols, which contain five carbon atoms.

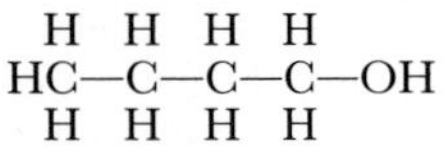

```
H  H  H  H
HC—C—C—C—OH
H  H  H  H
```

Normal butyl alcohol

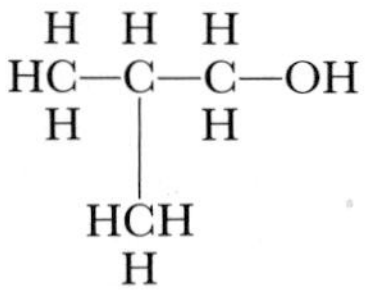

```
H  H  H
HC—C—C—OH
   H     H
      |
    HCH
      H
```

Isobutyl alcohol

```
H  H  H
HC—C—C—OH
   H  H
      |
    HCH
      H
```

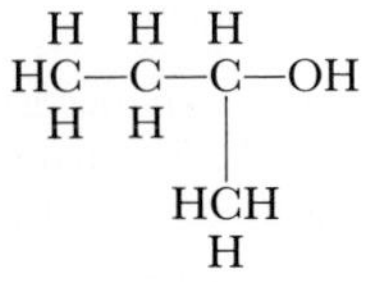

Secondary butyl alcohol

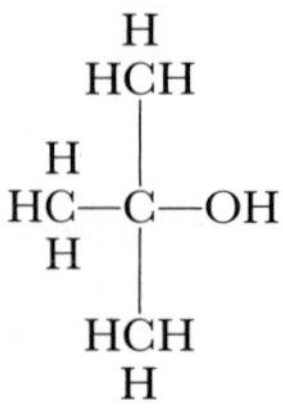

```
      H
    HCH
      |
H     |
HC—C—OH
   H  |
      |
    HCH
      H
```

Tertiary butyl alcohol

Ring Formation: The variety of organic compounds is increased tremendously by the ability of the carbon atom to form closed, ringlike formations as well as long-chain configurations. Just as it requires three points to make a plane, so it requires a minimum of three points (in this instance, carbon atoms) to form a ring. It appears that the majority of known organic compounds do not exceed six carbon atoms in the main framework of a given ring.

The possible bonding arrangements in a three-carbon ring are shown in Table C-9. Of these, only configuration *a*, with single bonds between framework carbon atoms, is represented by a stable compound, namely, *cyclopropane, cyclo-* designating that the compound is in the form of a ring.

The possible bonding arrangements in a four-carbon ring are shown in Table C-9. Of the several possible configurations, known, stable com-

TABLE C-8. *Principal Carbon Compounds Containing Only One Carbon Atom*

Four Single Bonds

Fill Valence Bonds with:	Resulting Compound Structure	Resulting Compound Name	Fill Valence Bonds with:	Resulting Compound Structure	Resulting Compound Name
4H	$H{-}C{-}H$ (H above, H below)	Methane	$4NO_2$	$NO_2{-}C{-}NO_2$ (NO_2 above, NO_2 below)	Tetranitromethane
$3H + Cl$	$H{-}C{-}Cl$ (H above, H below)	Methyl chloride (monochloromethane) *See Notes 1, 2*	$3H + PH_2$	$H{-}C{-}PH_2$ (H above, H below)	Methyl phosphine *See Note 6*
$2H + 2Cl$	$H{-}C{-}Cl$ (H above, Cl below)	Methylene chloride (methyl dichloride) (dichloromethane) *See Note 3*	$3H + HSO_3$	$H{-}C{-}HSO_3$ (H above, H below)	Methyl sulfonic acid *See Note 6*
$1H + 3Cl$	$Cl{-}C{-}Cl$ (H above, Cl below)	Chloroform (methyl trichloride) (trichloromethane)	$3H + SO_2Cl$	$H{-}C{-}SO_2Cl$ (H above, H below)	Methyl sulfonic chloride *See Note 7*
$4Cl$	$Cl{-}C{-}Cl$ (Cl above, Cl below)	Carbon tetrachloride (tetrachloromethane) *See Note 4*	$3H + SCN$	$H{-}C{-}SCN$ (H above, H below)	Methyl thiocyanate (methyl sulfocyanate) *See Note 6*
$3H + OH$	$H{-}C{-}OH$ (H above, H below)	Methyl alcohol (methanol) *See Note 5*	$3H + CN$	$H{-}C{-}CN$ (H above, H below)	Methyl cyanide
$3H + NH_2$	$H{-}C{-}NH_2$ (H above, H below)	Methyl amine (aminomethane) *See Note 6*	$3H + SiH_2Cl$	$H{-}C{-}SiH_2Cl$ (H above, H below)	Methyl chlorosilane

TABLE C-8. *Principal Carbon Compounds Containing Only One Carbon Atom* (Continued)

Fill Valence Bonds with:	Resulting Compound		Fill Valence Bonds with:	Resulting Compound	
	Structure	Name		Structure	Name
Four Single Bonds					
$3H + HSO_4$	H–C–HSO₄ (H top, H bottom)	Methyl hydrogen sulfate (methyl sulfuric acid) *See Note 6*	$3H + AsCl_2$	H–C–AsCl₂ (H top, H bottom)	Methyl arsine dichloride
$3H + NO_2$	H–C–NO₂ (H top, H bottom)	Nitromethane	$3H + AsCl_4$	H–C–AsCl₄ (H top, H bottom)	Methyl arsine tetrachloride
$2H + 2NO_2$	H–C–NO₂ (H top, NO₂ bottom)	Dinitromethane	$3H + AsO$	H–C–AsO (H top, H bottom)	Methyl arsine oxide
Two Single Bonds and One Double Bond					
$2H + O$	H–C=O (H)	Formaldehyde (methanal)	$NH_2 + NH + NHN_2O$	NH–N₂O, NH₂–C=NH	Nitroguanidine *See Note 8*
$H + OH + O$	OH–C=O (H)	Formic acid	$2Cl + O$	Cl, Cl–C=O	Carbonyl chloride (phosgene)
$H + O + ONa$	ONa, HC=O	Sodium formate *See Note 7*	$2OH + O$	OH, OH–C=O	Carbonic acid
$H + NH_2 + O$	NH₂, HC=O	Formamide (methanamide) *See Note 6*	$2NH_2 + O$	NH₂, NH₂–C=O	Urea (carbamide)
$NH + 2NH_2$	NH₂–C=NH (NH₂)	Iminocarbamide (guanidine)	$NH_2 + Cl + O$	NH₂–C=O, Cl	Carbamide chloride (carbamyl chloride)
$NH + NHNH_2 + NH_2$	NHNH₂, C=NH, NH₂	Amino guanidine *See Note 8*	$2NHNH_2 + O$	NHNH₂, NHNH₂–C=O	Carbazide

	Two Double Bonds			One Single Bond and One Triple Bond	
2O	O=C=O	Carbon dioxide *See Note 9*	H + N	H—C≡N	Hydrogen cyanide
2S	S=C=S	Carbon disulfide	NH₂ + N	NH₂—C≡N	Cyanamide (aminocyanide)
O + S	O=C=S	Carbon oxysulfide (carbonyl sulfide)	NCa + N	CaN—C≡N	Calcium cyanamid
HN + O	HN=C=O	Isocyanic acid	OH + N	HO—C≡N	Cyanic acid
			HS + N	HS—C≡N	Isothiocyanic acid
			KS + N	KS—C≡N	Potassium thiocyanate

NOTES:
1. Where the CH_3 configuration acts in concert (as a radical), it is referred to as the *methyl group*.
2. All halogens (Cl, Br, I, and F) form similar compounds.
3. Where the CH_2 configuration acts in concert (as a radical), it is referred to as the *methylene group*.
4. There are no stable mixed-halogen compounds, such as CCl_2Br_2 or $CBrI_3$, etc.
5. There are no known one-carbon (four single-valence) compounds that contain more than one hydroxyl, OH, radical, such as $CH_2(OH)_2$ or $CH(OH)_3$ or $C(OH)_4$.
6. There are no known one-carbon (four single-valence) compounds that contain additional elements or radicals of this type.
7. This is a derivative of the acid.
8. This is a derivative of guanidine.
9. Carbon monoxide is a rare example where carbon displays a valence of 2.

205

TABLE C-9. *Structure of Three- and Four-Carbon Rings*

Possible Bonding Arrangements in Three-Carbon Ring

(a) 6 bonds available   (b) 4 bonds available   (c) 2 bonds available   (d) 2 bonds available   (e) 0 bonds available

Cyclopropane, C_3H_6

Possible Bonding Arrangements in Four-Carbon Ring

(a) 8 bonds available   (b) 6 bonds available   (c) 4 bonds available   (d) 4 bonds available   (e) 4 bonds available

Cyclobutane, C_4H_8 Cyclobutene, C_4H_6

(f) 2 bonds available   (g) 2 bonds available   (h) 0 bonds available   (i) 0 bonds available

pounds are found in but a few of these configurations. Cyclobutane and cyclobutene are indicated in the table.

The possible bonding arrangements in a five-carbon ring are shown in Table C-10. The majority of compounds appear to utilize configurations *a–c*. Examples of all-carbon, nitrogen-carbon, oxygen-carbon, oxygen-nitrogen-carbon, sulfur-carbon, and sulfur-nitrogen-carbon rings, together with the numbering system for naming these compounds, also are given in Table C-10.

Many thousands of organic compounds are composed of six-carbon rings (Table C-11). Of course, all atoms of the framework need not be carbon, and frequently oxygen, nitrogen, and sulfur will be found in the main-ring framework. The all-carbon ring of cyclohexane features six single valences, while the all-carbon ring of benzene comprises three single and three double va-

lences. Gamma pyrone is an example of an oxygen-bearing ring, whereas pyridine is a nitrogen-bearing ring.

Joined-Ring Structures: There are many hundreds of examples of organic ring compounds in which the rings are joined. Representative compounds are illustrated in Table C-12. The familiar structure of the benzene ring, with its three double bonds and three single bonds, appears frequently in combined-ring formations. Naphthalene, for example, features two rings, each of which shares one of the double bonds of the benzene ring. This sharing removes four valences for bonding with hydrogen or other atoms and radicals and removes two carbon atoms. Thus the formula for naphthalene is not $2 \times C_6H_6$ (benzene), or $C_{12}H_{12}$, but rather the formula is $C_{10}H_8$.

The three-ring structure of anthracene is similar. In this case, four carbon atoms are shared

TABLE C-10. *Structure of Five-Carbon Rings*

Possible Bonding Arrangements in Five-Carbon Ring

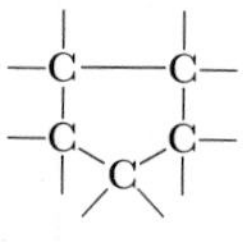

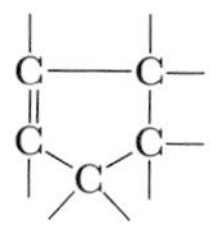

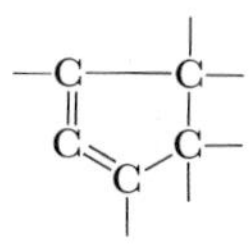

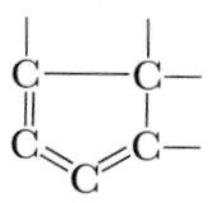

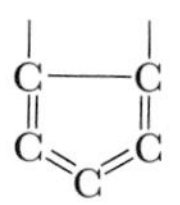

(*a*) 10 bonds available (*b*) 8 bonds available (*c*) 6 bonds available (*d*) 4 bonds available (*e*) 2 bonds available

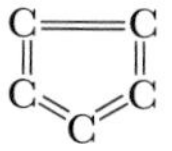

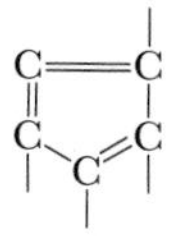

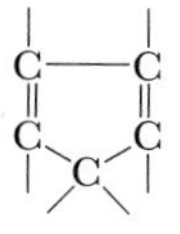

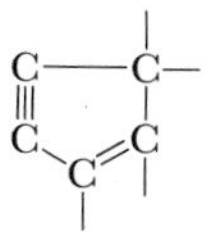

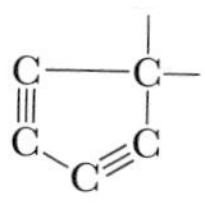

(*f*) 0 bonds available (*g*) 4 bonds available (*h*) 6 bonds available (*i*) 4 bonds available (*j*) 2 bonds available

Numbering System	All Carbon	Nitrogen-Carbon
4 3 5 1 2	Cyclopentane, C_5H_{10} Cyclopentene, C_5H_8	Pyrrole (imidol), C_4H_5N Pyrazole, $C_3H_4N_2$

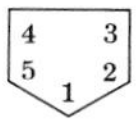

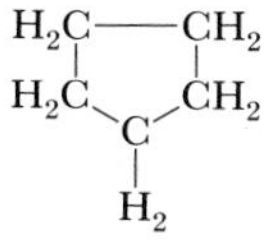

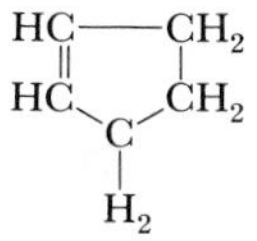

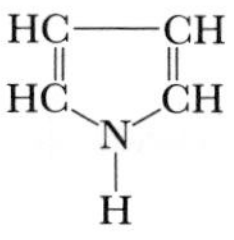

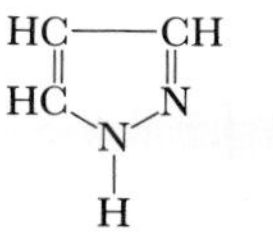

Oxygen-Carbon	Oxygen-Nitrogen-Carbon	Sulfur-Carbon	Sulfur-Nitrogen-Carbon
Furane, C_4H_4O	Oxazole, C_3H_3NO	Thiophene, C_4H_4S	Thiazole, C_3H_3NS

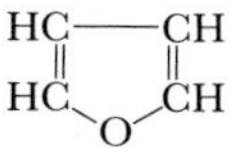

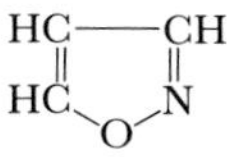

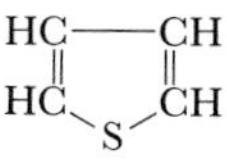

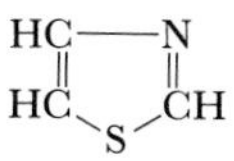

and the bondings available are reduced by eight. Thus the formula for anthracene is not $3 \times C_6H_6$, or $C_{18}H_{18}$, but rather $C_{14}H_{10}$.

In the instance of benzofurane and numerous other similarly structured compounds, the familiar six-atom ring is joined with a five-atom ring.

Both five- and six-atom rings can be joined in numerous ways. Several compounds of the format of carbazole appear in which various atoms and radicals, in addition to hydrogen, satisfy the outer valences. Numerous alkaloids, dyes, enzymes, and other complex compounds are structured along the lines of strychnine, as shown.

Bridged Rings: There are numerous examples among organic compounds of bridged rings in which there are interconnections between the framework atoms. As in this discussion of valences, the depiction of bridged-ring structures, and in fact all other structures, is symbolic short-hand that enables the reduction of a very complex topic into easily understood terms. It must be stressed, however, that these are symbolic representations and are a severe oversimplification of molecule formation.

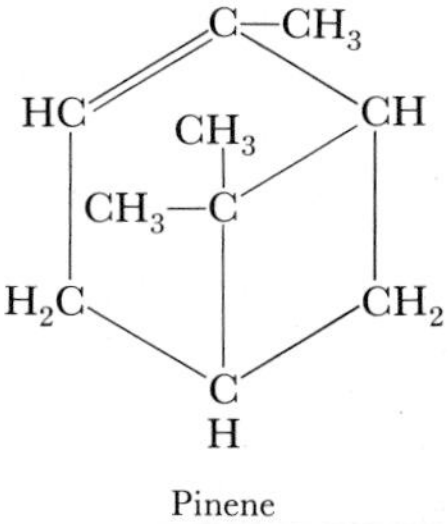

Pinene

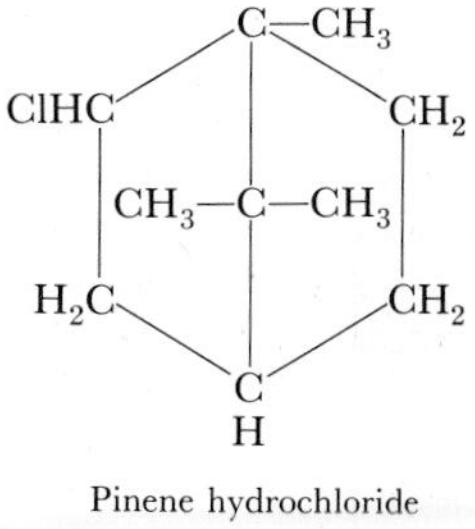

Pinene hydrochloride

Chain-Ring Compounds: To further extend the variety of carbon compounds, rings and chains

TABLE C-11. *Structure of Six-Carbon Rings*

Structure	Nomenclature	Representative Compounds			

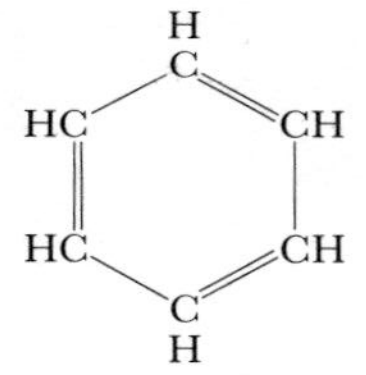

Benzene, C_6H_6

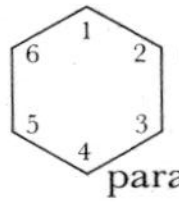

ortho (o) meta (m) para (p)

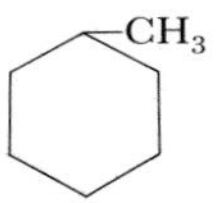

Representative Compounds:

Toluene (methylbenzene) (phenylmethane) C_7H_8

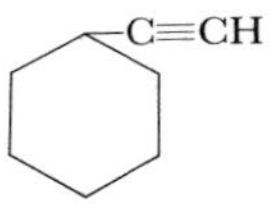

Phenylacetylene (ethinylbenzene) C_8H_6

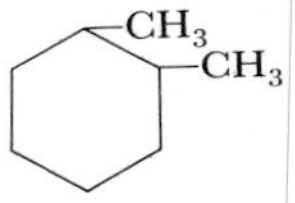

o-Xylene (o-dimethylbenzene) C_8H_{10}

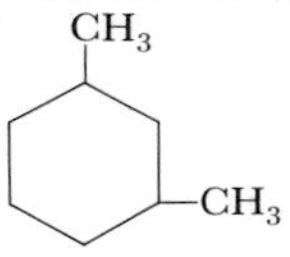

m-Xylene (m-dimethylbenzene) C_8H_{10}

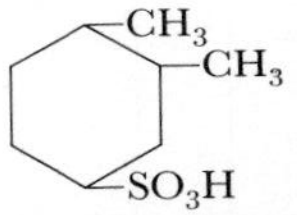

Xylene sulfonic acid (1,2-dimethylbenzene-4-sulfonic acid) $C_8H_{10}SO_3$

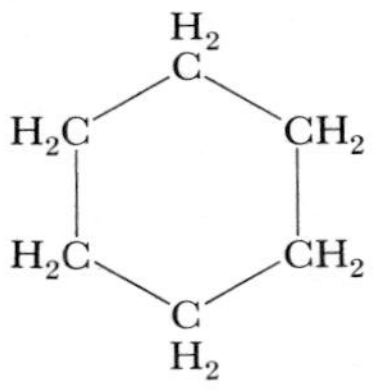

Cyclohexane, C_6H_{12}

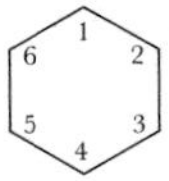

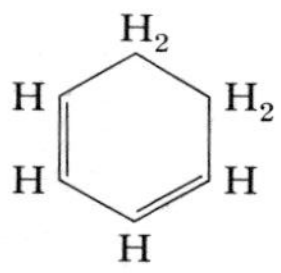

Cyclohexadiene (1,2) C_6H_8

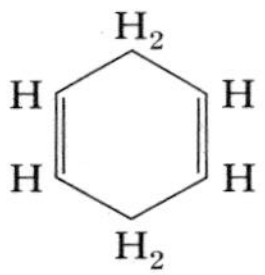

Cyclohexadiene (1,4) C_6H_8

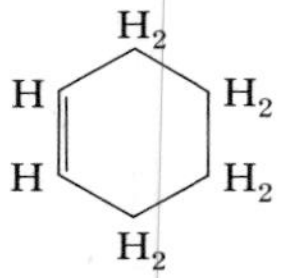

Cyclohexene C_6H_{10}

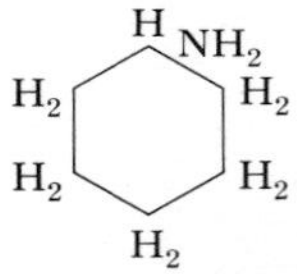

Cyclohexylamine (aminocyclohexane) $C_6H_{13}N$

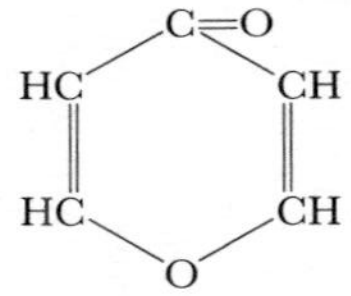

γ-Pyrone, $C_5H_4O_2$

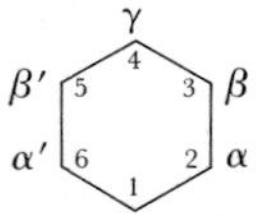

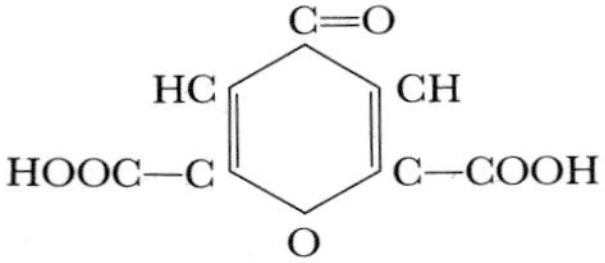

β γ β′ α α′

γ-Pyrone $\alpha\alpha'$ dicarboxylic acid $C_7H_4O_6$

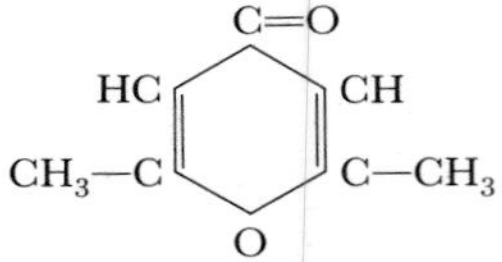

$\alpha\alpha'$-Dimethyl-γ-pyrone $C_7H_8O_2$

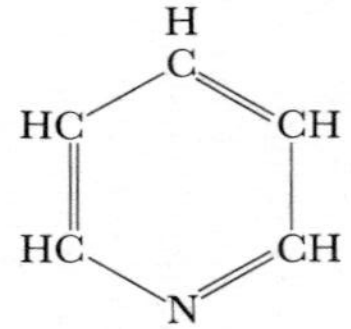

Pyridine, C_5H_5N

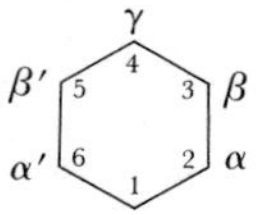

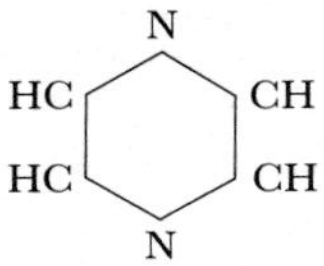

β γ β′ α α′

1,4-Pyrazine $C_4H_4N_2$

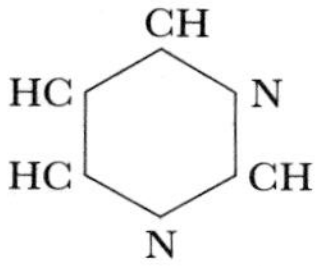

Pyrimidine (m-diazine) $C_4H_4N_2$

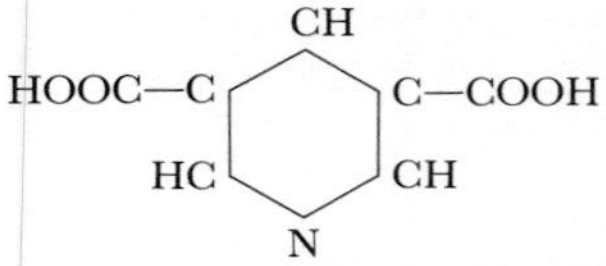

Pyridine dicarboxylic acid (3,5) (dinicotinic acid) $C_7H_5NO_4$

NOTE: Generally, the bonding and C and H atoms of the structural example at the left apply to the representative compounds unless otherwise indicated.

TABLE C-12. *Structure of Joined-Ring Compounds*

Structure	Nomenclature	Representative Compounds

Structure column:

Naphthalene
$C_{10}H_8$

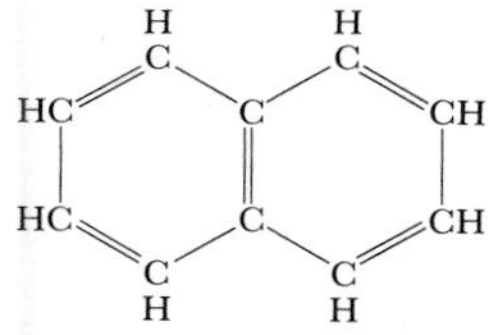

Anthracene
$C_{14}H_{10}$

Benzofurane
C_8H_6O

Carbazole
$C_{12}H_9N$

Nomenclature column:

(naphthalene numbering with α, β positions, 1–8, and α labels)

(anthracene numbering 1–10 with α, β, γ labels)

para- / ana- / meta- / ortho- (with positions 1–7, α, β)

(carbazole-type numbering 1–9)

Representative Compounds column:

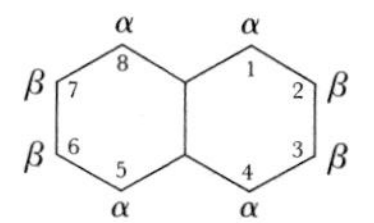

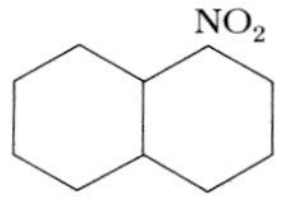

α-Nitronaphthalene
$C_{10}H_7NO_2$

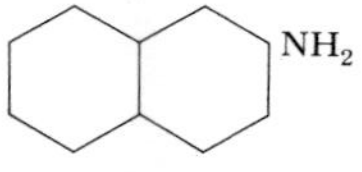

β-Naphthylamine
$C_{10}H_7\cdot NH_2$

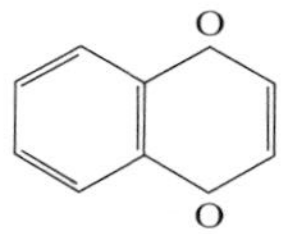

α-Naphthaquinone
$C_{10}H_6O_2$

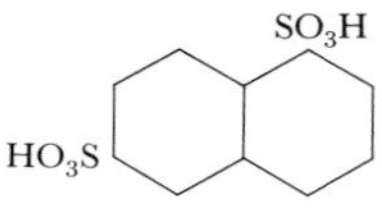

Naphthalene disulfonic acid (1,6)
$C_{10}H_6(SO_3H)_2$

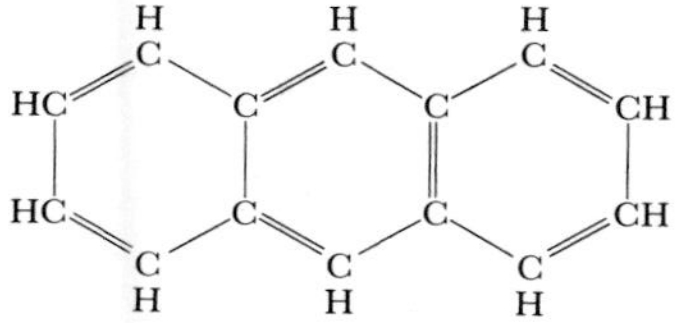

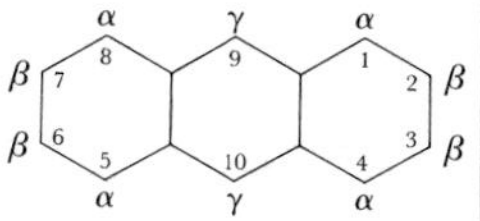

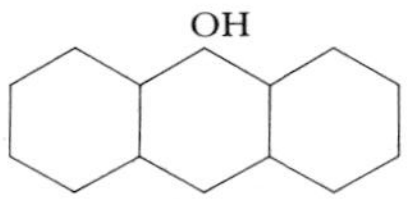

γ-Hydroxy anthracene
(9-hydroxyanthracene)
(9-anthol)

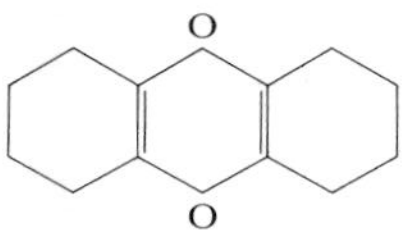

Anthraquinone
$C_{14}H_8O_2$

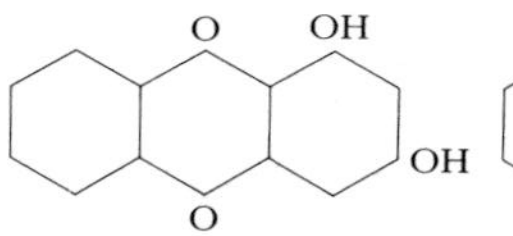

Alizarin
(1,2-dihydroxyanthraquinone)
$C_{14}H_8O_4$

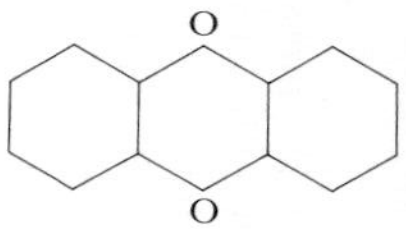

Xanthone
(dipheno-γ-pyrone)
$C_{13}H_8O_2$

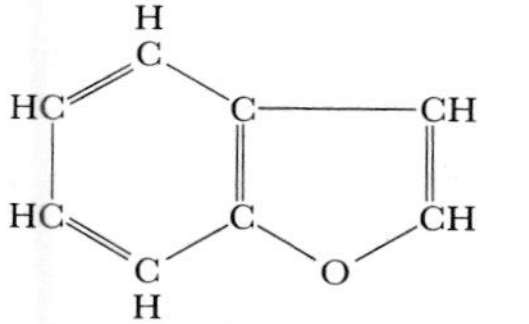

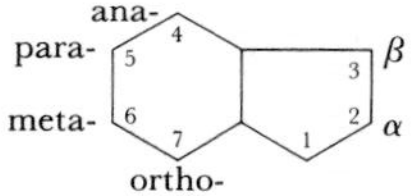

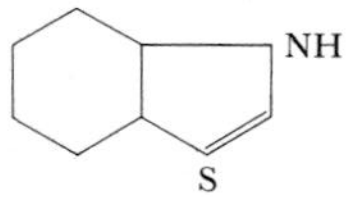

Benzthiazole
C_7H_6NS

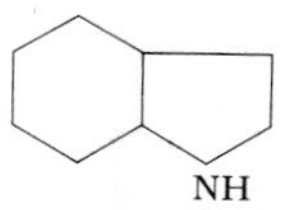

Indole
C_8H_7N

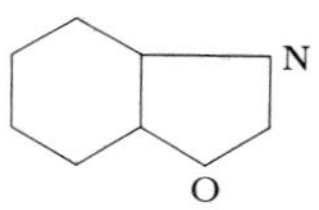

Benzoxazole
C_7H_5NO

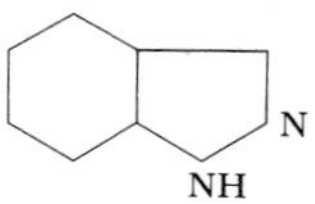

Isoindazole
$C_7H_6N_2$

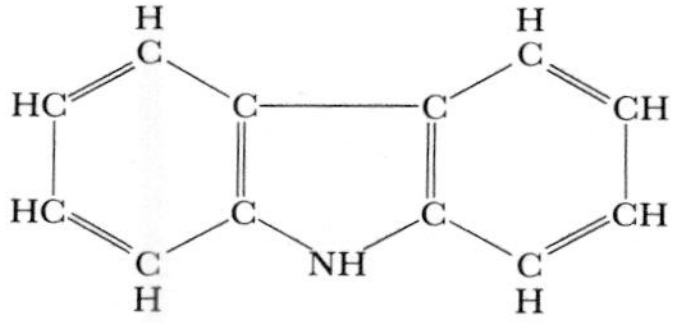

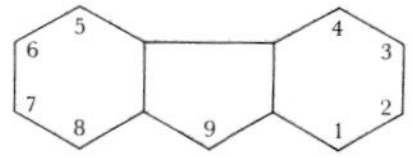

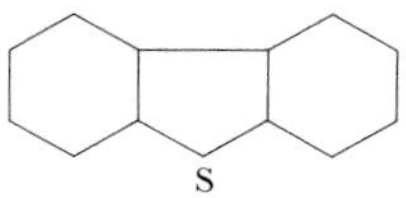

Diphenylene sulfide
$C_{12}H_8S$

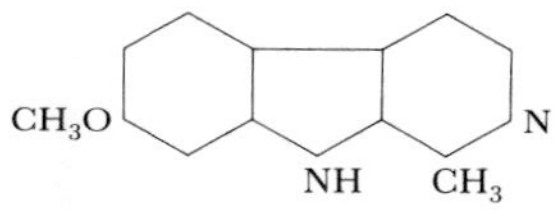

Harmine (an alkaloid)
$C_{13}H_{12}N_2O$

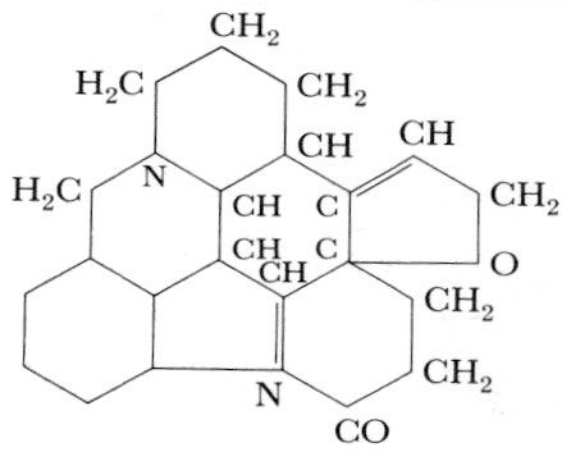

Strychnine
$C_{21}H_{22}O_2N_2$

NOTE: Generally, the bonding and C and H atoms of the structural example at the left apply to the representative compounds unless otherwise indicated.

often are joined. A very simple arrangement is shown by diphenyl. Nucleic acids are excellent examples of highly complex chain-ring structures. Only a portion of a nucleic acid is shown.

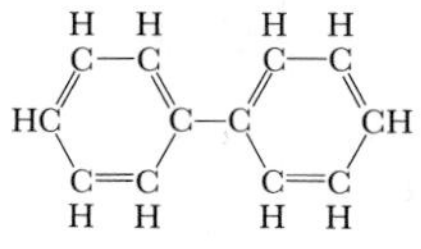

Diphenyl

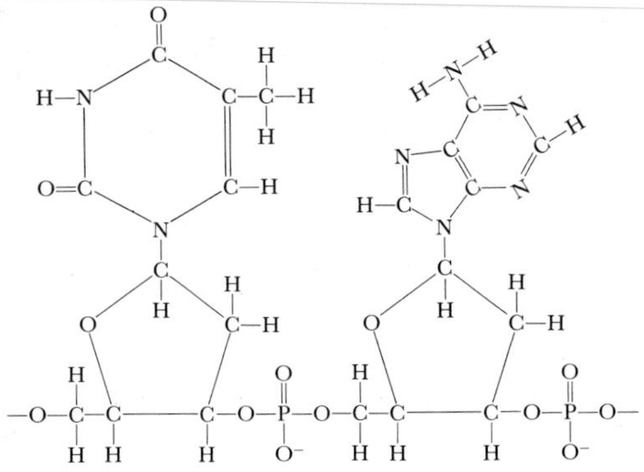

Portion of a nucleic acid

Properties of Carbon Compounds: Ranging from gases to liquids to solids, carbon compounds are so varied (containing from one to several thousand carbon-bearing molecules, as in the case of polymers) that generalization pertaining to their chemical behavior and physical properties is indeed difficult and, actually, not too meaningful. Table C-7 lists over 75 major classifications of carbon compounds, and each of these classes has many further subdivisions. A large percentage of the major classes of compounds are described elsewhere in this volume.

The behavior of the carbon atom, and in turn the behavior of molecules that contain carbon, present phenomena, such as isomerism, previously mentioned, for study and exploitation that seldom appear in the scope of inorganic chemicals.

Generally, the chemical activity of carbon compounds arises from two factors: (1) the activity of the noncarbon elements or radicals which attach themselves to the available carbon bonds,

and (2) the presence of specific linkages or combinations of linkages within the compounds. Again, these linkages usually involve elements and radicals other than carbon atoms exclusively. In many instances, the linked carbon atoms per se constitute the molecular framework to which more reactive elements and radicals are attached. This accounts, in part, for the numerous carbon-bearing radicals, such as methyl, CH_3, ethyl, C_2H_5, diphenyl, $C_6H_5 \cdot C_6H_5$, and so on, that proceed through vigorous chemical reactions untouched.

There are more reactions in which the total number of carbon atoms in an initial reactant remains the same or is increased than there are reactions where the number is reduced. There is a strong tendency of the carbon atoms to remain in chains or rings, rather than, for example, to be degraded to form carbon monoxide or carbon dioxide. A random sampling of organic reactions strengthens these observations:

$$CH_3COOH + C_2H_5OH \rightarrow$$

Acetic acid   Ethyl alcohol

$$CH_3 \cdot CO \cdot OC_2H_5 + H_2O$$

Ethyl acetate

$$C_2H_5NH_2 + CHCl_3 + 3KOH \rightarrow$$

Ethylamine   Chloroform

$$C_2H_5NC + 3KCl + 3H_2O$$

Ethylcarbylamine

$$C_2H_5 \cdot NO_2 + 6H \rightarrow C_2H_5 \cdot NH_2 + 2H_2O$$

Nitroethane   Ethylamine

$$CH_2 : CH \cdot CH_2 \cdot OH + 2O \rightarrow$$

Allyl alcohol

$$CH_2 : CH \cdot COOH + H_2O$$

Acrylic acid

$$C_6H_5COONa + NaOH \rightarrow C_6H_6 + Na_2CO_3$$

Sodium benzoate   Benzene

$$C_6H_6 + CH_3Cl \rightarrow C_6H_5CH_3 + HCl$$

Benzene   Methyl chloride Methylbenzene

$$(NO_2)(NH_2)C_6H_4 + 6H \rightarrow$$

Nitroaniline

$$(NH_2)(NH_2)C_6H_4 + 2H_2O$$

Phenylenediamine

$$C_6H_5 \cdot SO_2Cl + C_2H_5OH \rightarrow$$

Benzenesulfonyl   Ethyl alcohol
chloride

$$C_6H_5 \cdot SO_2 \cdot OC_2H_5 + HCl$$

Ethylbenzene sulfonate

$$C_6H_5 \cdot CHO + NH_2OH \rightarrow$$

Benzaldehyde   Hydroxylamine

$$C_6H_5CH : N \cdot OH + H_2O$$

Benzaldoxime

$$COCl_2 \quad + 2C_2H_5OH \rightarrow$$

Carbonyl chloride   Ethyl alcohol

$$CO(OC_2H_5)_2 + 2HCl$$

Ethyl carbonate

$$CHCl_3 + 3KOH + C_6H_5NH_2 \rightarrow$$

Chloroform   Aniline

$$C_6H_5NC + 3KCl + 3H_2O$$

Phenylcarbamine

$$CH_3Cl + KOH \rightarrow \quad CH_3OH + KCl$$

Methyl chloride   Methyl alcohol

$$2CH_3OH + H_2SO_4 \rightarrow (CH_3)_2SO_4 + 2H_2O$$

Methyl alcohol   Dimethyl sulfate

$$C_2H_5OH + HI \rightarrow \quad C_2H_5I + H_2O$$

Ethyl alcohol   Ethyl iodide

$$(CH_3)_2CO + PCl_5 \rightarrow \quad (CH_3)_2CCl_2 + POCl_3$$

Acetone   Dichloropropane

$$C_2H_5OH + 2O \rightarrow CH_3COOH + H_2O$$

Ethyl alcohol   Acetic acid

$$C_2H_5OH + 3O_2 \xrightarrow{\text{combustion}} 2CO_2 + 3H_2O$$

Ethyl alcohol

$$CH(OH) \cdot COOH$$
$$|$$
$$CH(OH) \cdot COOH \xrightarrow{\text{destructive distillation}}$$

Tartaric acid

$$CH_3 \cdot CO \cdot COOH + CO_2 + H_2O$$

Pyruvic or acetylformic acid

Examples of linkages that affect in large measure the chemical and physical properties of various classes of carbon compounds include:

1. The *oxygen linkage* of the ethers:

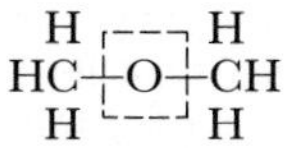

Dimethyl ether

2. The *carbonyl linkage* of the ketones:

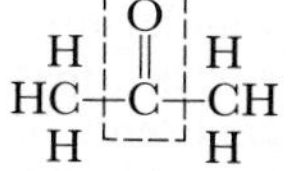

Acetone
(Dimethyl ketone)

3. The carbonyl linkage, acting in concert with a single hydrogen atom, to form the *aldehyde group:*

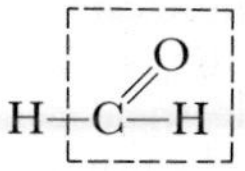

Formaldehyde

4. The carbonyl linkage, acting in concert with a hydroxyl radical, to form the *carboxyl group* of the acids:

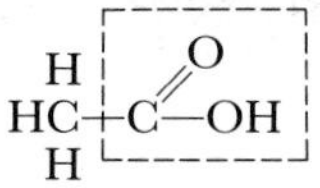

Acetic acid

5. The hydroxyl radical of the alcohols:

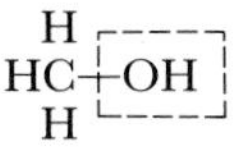

Methyl alcohol

The foregoing structures, as well as several others, are described further under each class of compounds, as will be found elsewhere in this volume. See also **Catalysts; Derivative; Hydrocarbons; Polymerization;** and **Radicals.**

Carbon Dioxide Carbon dioxide, CO_2, also known as *carbonic anhydride* or *carbonic acid gas,* is a colorless, odorless, and nontoxic gas at standard conditions. The physiological effects of CO_2 when breathed in mild or moderate concentrations serve in a measure as a warning of its presence. In concentrations much above 11 lb/1,000 ft^3, the physiological action may be too rapid in producing stupefaction or loss of consciousness to serve as adequate warning. Although nontoxic, the gas causes suffocation due to lack of ample O_2.

The formula weight of $CO_2 = 44.01$, density = 1.9769 g/l at 0°C and 760 mm Hg pressure. On the basis of air = 1.00, specific gravity = 1.53. The melting point of $CO_2 = -56.6$°C at 5.2 atm; subliming point = -78.5°C at 1 atm. Critical pressure = 73 atm; critical temperature = 31°C. CO_2 is quite soluble in H_2O (179.7 cm^3 CO_2 in 100 cm^3 H_2O at 0°C and 90.1 cm^3 CO_2 in 100 cm^3 H_2O at 20°C). CO_2 also is soluble in alcohol and is rapidly absorbed by most alkaline solutions. Some further properties of CO_2 are given in Table C-13.

Various salts and the carbonates are considered to be derived from the theoretical carbonic acid, classified by some authorities as the simplest dibasic organic acid. Various metal carbonates are described elsewhere in this volume. Consult the Subject Index; see also, particularly, **Limestone;** and **Sodium carbonate.**

Carbon dioxide, when not in excessive pollutive concentration in localized atmospheres, normally is present in the air to the extent of 0.05% by

TABLE C-13. *Some Properties of Carbon Dioxide*

Temperature		Pressure, psig	Vol. Vapor, ft³/lb	Density, lb/ft³	Heat Content, Btu/lb	
°F	°C				Liquid	Vapor
−69.9	−56.6	60.40*	1.157	74.6	−13.7	135.9
−60	−51.1	80.0	0.927	72.2	−10.1	136.6
−50	−45.6	103.5	0.749	71.0	− 4.7	137.2
−40	−40.0	131.1	0.611	69.6	0.0	137.8
−30	−34.4	163.1	0.503	68.2	4.5	138.2
−20	−28.9	200.2	0.417	66.8	9.1	138.5
−10	−23.3	242.6	0.347	65.2	13.9	138.7
0	−17.8	290.8	0.290	63.7	18.8	138.9
5†	−15.0†	319.7	0.267	62.8	21.3	138.8
10	−12.2	345.5	0.244	62.0	24.0	138.7
20	−6.67	407.1	0.205	60.1	29.4	138.3
30	−1.11	476.1	0.172	58.2	35.4	137.8
40	4.44	553.1	0.144	56.0	41.7	136.7
50	10.0	638.9	0.121	53.5	48.4	135.0
60	15.6	733.9	0.099	50.8	55.5	132.1
70	21.1	833.7	0.080	47.3	63.7	127.5
80	26.7	954.0	0.061	42.2	73.9	118.7
86†	30.0†	1,028.3	0.048	37.2	83.3	110.4
87.7‡	30.9‡	1,054.7‡	0.035	28.9	97.0	97.0

*Freezing point.
†Standard ton (refrigeration) temperatures.
‡Critical points.

weight. CO_2 is a normal product of combustion as described under **Fuels.** The atmospheric CO_2 balance is partially maintained by green plants, which, by a process of photosynthesis, use CO_2 as a raw material in constructing nutrition and tissue for the plants. There is concern that an increase in the content of CO_2 in the atmosphere may upset the energy balance of the earth as it relates to external radiation received from the sun.

In addition to fossil-fuel combustion processes, CO_2 is generated by various fermentation processes, by the oxidation of foods in the bodies of animals and plants, by the decay of organic matter, and by decomposing carbonates with acids or heat. Large quantities of CO_2 are generated during cement production. The usual recovery of CO_2 as a by-product from other operations and for marketing as CO_2 is by the route of dissolving CO_2 gas in absorbant sodium or potassium carbonate solutions, followed by the steam-heating of such solutions to free the CO_2, after which the gas is compressed, often into steel cylinders for the market.

The major industrial uses of CO_2 are (1) in carbonated beverages, (2) in the solid form as "Dry Ice," (3) in fire extinguishers, (4) as a refrigeration medium, and (5) as a raw material in the manufacture of various chemicals, including sodium bicarbonate, sodium carbonate, lead carbonate (white lead), and urea. CO_2 also is used for creating protective, inert atmospheres. Urea is prepared from CO_2 and NH_3. See also **Urea.**

It is estimated that 1,000 lb of Dry Ice will refrigerate a railroad car for the full trip between California and New York, as compared with 10,000 lb of water ice and salt for the same result. In connection with fire-extinguishing systems, CO_2 is heavier than air and can be used to blanket rooms, paint-spraying facilities, small tanks, and rotating machinery. In pipe trenches, it is far more effective than foam, which cannot be depended upon to flow into remote corners. CO_2 systems can be controlled manually or automatically with fixed piping leading to discharge points, from racks of pressure tanks connected to a piping manifold, or from trailer-mounted storage. Heat-actuated outlet closures similar to those on water sprays can be used. To minimize fire-control damage, electrical generating equipment usually is equipped with a CO_2 extinguishing system, fully automated with heat sensors.

Carbon Disulfide (See **Absorption, acidic gases; and Rayon.**)

Carbon Group (See **Chemical elements.**)

Carbon Monoxide Carbon monoxide, CO, also known as *carbonic oxide,* is a colorless, odorless, very toxic gas at standard conditions. Carbon monoxide has a great affinity for blood hemoglobin (300 times that of O_2), and thus stifles the ability of hemoglobin to carry O_2 throughout the body, causing death in excessive concentrations. Internal-combustion engines, the fumes from which may contain 7% CO or more, are the primary hazard of CO poisoning. Even in nonterminal CO-poisoning cases, deleterious effects of a prolonged nature may result.

CO is a product of imperfect combustion of fossil fuels and is of prime concern to air-pollution specialists. See also **Air pollution; and Fuels.**

The formula weight of CO = 28.01; density = 1.2504 g/l at 0°C and 760 mm Hg pressure. On the basis of air = 100, the specific gravity = 0.968. The melting point of CO = −207°C; boiling point = −192°C. Critical temperature = −139°C; critical pressure = 35 atm. CO is virtually insoluble in H_2O (0.0044 part CO in 100 parts H_2O at 0°C; 0.00018 part CO in 100 parts H_2O at 50°C). CO is soluble in alcohol and solutions of $CuCl_2$.

CO is an effective reducing agent and plays an effective role as reductant in various metal-smelting operations. CO reacts with alkali hydroxides to form formates: CO + NaOH → HCOONa. CO reacts with several metals to form carbonyls, such as $Fe(CO)_5$ and $Ni(CO)_4$. CO combines under sunlight directly with Cl_2 to form carbonyl chloride (phosgene), $COCl_2$. With HCl, CO reacts to form the unstable formyl chloride.

CO will burn to form CO_2 and is an important constituent of several fuels, such as blast-furnace gas, coal gas, producer gas, and water gas. CO is an important ingredient of several forms of synthesis gas: CO and H_2 for methanol synthesis and CO, H_2, and olefins for oxoalcohols. In some processes, CO is an intermediate as in the production of NH_3 synthesis gas. See also **Gasification; and Synthesis gas.**

Carbon Steel (See **Iron and steel.**)

Carbon Tetrachloride Carbon tetrachloride, CCl_4, is a heavy, colorless liquid at room temperature, with a characteristic nonirritating odor. It is non-flammable and noncombustible. Some of the physical properties of CCl_4 are given in Table C-14.

When dry, CCl_4 is noncorrosive to common metals except Al. CCl_4 hydrolyzes readily and therefore, when wet, is corrosive to Fe, Cu, Ni, and alloys containing these elements. Dry CCl_4 is thermally stable at temperatures below 400°C; between 900 and 1300°C, the equilibrium mixture contains significant quantities of perchloroethylene and chlorine, as shown by

$$2CCl_4 \leftrightharpoons C_2Cl_4 + 2Cl_2 \qquad (1)$$

When mixed with air and heated in the presence of Fe at 335°C, CCl_4 is partially oxidized to phosgene. Wet CCl_4 is hydrolyzed to phosgene and HCl when irradiated by ultraviolet light at room temperature or when heated to 250°C in the presence of a limited amount of water.

$$CCl_4 + H_2O \rightarrow COCl_2 + 2HCl \qquad (2)$$

When sufficient water is present, CO_2 and HCl are the reaction products at 250°C.

$$CCl_4 + 2H_2O \rightarrow CO_2 + 4HCl \qquad (3)$$

The most significant reaction of CCl_4 from a commercial standpoint is shown by

$$2CCl_4 + 3HF \xrightarrow{\text{cat.}}$$
$$CCl_2F_2 + CCl_3F + 3HCl \qquad (4)$$

Approximately 90% of the CCl_4 manufactured is used as a raw material for the production of the chlorofluorocarbons, as indicated by Eq. (4).

TABLE **C-14.** *Some Physical Properties of Carbon Tetrachloride*

Melting point, °C	−23
Boiling point, °C	76.75
Specific gravity, 25°C/25°C	1.588
Vapor density (air = 1)	5.32
Flash point	None
Explosive limits in air	None
Coefficient of cubical expansion, 0–40°C	0.00124
Critical temperature, °C	283.2
Critical pressure, psia	661
Solubility of CCl_4 in H_2O, g/100 g H_2O, at 25°C	0.08
Solubility of H_2O in CCl_4, g/100 g H_2O, at 25°C	0.013
Vapor pressure, mm, at 25°C	114.5
Threshold-limit value, ppm	10
Odor threshold, ppm	80

Carbon tetrachloride was first prepared by chlorination of chloroform in 1839 by Regnault. Liquid-phase chlorination of CS_2, developed by Müller and Dubois in 1893, was the basis for the first commercial process. Production on a large scale in the United States began about 1907. Early uses were as a metal-degreasing solvent, dry-cleaning fluid, fabric-spotting fluid, fire-extinguishing fluid, grain fumigant, and as a reaction medium. As the toxicity of CCl_4 was recognized, less toxic chlorinated hydrocarbons replaced CCl_4 in metal- and fabric-cleaning applications. Concurrently, demand for CCl_4 as a raw material for chlorofluorocarbon production developed in the 1940s, with the net result of continued growth for the product. In the United States, CCl_4 was banned in all consumer goods (1970). Current applications are chlorofluorocarbon manufacture, grain fumigation, and as a chlorination reaction medium.

The carbon disulfide process for making CCl_4 can be illustrated by

$$3C + 6S \rightarrow 3CS_2$$
$$2CS_2 + 6Cl_2 \rightarrow 2CCl_4 + 2S_2Cl_2$$
$$CS_2 + 2S_2Cl_2 \rightarrow CCl_4 + 6S$$

The reaction is carried out in a lead-lined reactor in a solution of CCl_4 at 30°C; iron filings are added as catalyst.

In the 1950s, chlorination of hydrocarbons, particularly methane, became the most important production route in the United States. Two basic reaction schemes are used.

1. Methane chlorination in the presence of excess methane:

$$CH_4 + Cl_2 \rightarrow$$
$$CH_3Cl + CH_2Cl_2 + CHCl_3$$
$$+ CCl_4 + HCl + \text{excess } CH_4$$

The reaction can be carried out in the liquid phase, that is, CCl_4 at moderate temperatures (30–40°C), using ultraviolet light to catalyze the reaction. Alternatively, the reaction can be carried out in the vapor phase at 450–500°C without a catalyst. Unreacted methane and partially chlorinated products are recycled to control the yield of CCl_4. Methylene chloride and chloroform are coproducts of this reaction scheme.

2. Methane chlorination in the presence of excess chlorine:

$$CH_4 + Cl_2 \rightarrow$$
$$CCl_4 + C_2Cl_4 + HCl + \text{excess } Cl_2$$

The reaction is carried out in the vapor phase at 550–650°C without a catalyst. The reactor temperature, and hence the product mix, are regulated by controlling the excess Cl_2 exiting the reaction, and the quantity of crude product recycled to the reactor. Perchloroethylene is a coproduct of this reaction scheme. Ethylene and propylene also have been substituted for methane in this process. This increases the yield of perchloroethylene and, in addition, results in by-product hexachloroethane, hexachlorobutadiene, and hexachlorobenzene formation.

Experimental exposure of laboratory animals to the vapors of CCl_4 have shown it to be highly toxic by inhalation at concentrations which are easily reached at ambient temperature. This has been confirmed by accidental exposure of human beings. The hazard can result from either single or repeated contacts.

—R. L. Daniel, *The Dow Chemical Company, Freeport, Tex.*

Carbonic Acid   (See **Water treatment.**)

Carbonium Ion   (See **Ethylene glycol; Polymerization;** and **Radicals.**)

Carbonization   (See **Coal tar and derivatives.**)

Carbonyl Linkage   (See **Aldehydes;** and **Ketones.**)

Carbonyl Process   (See **Nickel.**)

Carbonyl Sulfide, Removal of   (See **Absorption, acidic gases.**)

Carborundum   (See **Abrasives.**)

Carboxyl   (See **Radicals.**)

Carboxylic Acids   The carboxylic acids constitute one of the largest and most important groups of organic acids. The *chemical signature* and provider of the predominating characteristics of these acids is represented by the presence of one or more *carboxyl groups,* COOH, in the structure of all carboxylic acids. The general formula for a carboxylic acid is

$$R-C\!\!\begin{array}{c} O \\ \\ OH \end{array}$$

As will be noted from the classification of carboxylic acids which follows, these compounds range widely—from aliphatic to aromatic to heterocyclic structures—from containing one to as many as six COOH groups and, in the case of hydroxycarboxylic acids, to as many as six hydroxyl groups. A few of the carboxylic acids are gases; many are liquids; and a majority are solids under normal conditions of pressure and temperature. The carboxylic acids yield several derivatives, including esters, acid halides, anhydrides, and nitriles, and consequently play a large role in various organic syntheses, including those required for the manufacture of plastics, elastomers, and numerous other synthetic materials.

Several specific carboxylic acids are described elsewhere in this volume, including acetic acid, acetic anhydride, acrylic acid, adipic acid, benzoic acid, carbonic acid, citric acid, formic acid, fumaric acid, maleic acid, maleic anhydride, malonic acid, phthalic acid, phthalic anhydride, tannic acid, tartaric acid, and terephthalic acid. Consult the Subject Index. See also Table C-15, which gives the key properties of over 100 carboxylic acids.

Classification: Structurally, the carboxylic acids can be classified in several ways:

1. A carboxylic acid may be aliphatic, carbocyclic, or heterocyclic, depending upon the major *organic chemical family* of which it is a member, such as

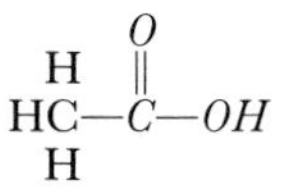

Aliphatic
Acetic acid

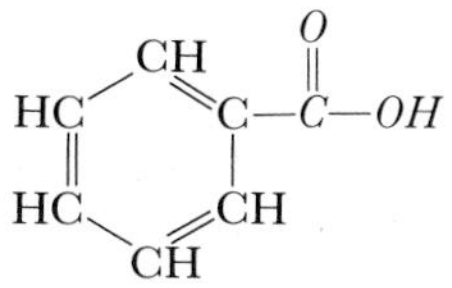

Carbocyclic or Aromatic
Benzoic acid

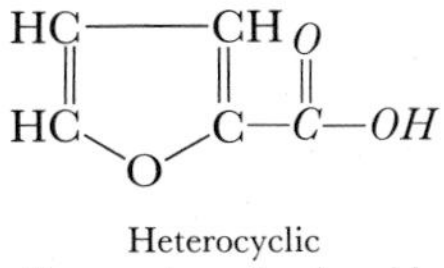

Heterocyclic
Pyromucic or furoic acid

2. Carboxylic acids may be classified by the *number of carboxyl, —COOH, groups* in the compound, giving rise to the terms:

a. Monocarboxylic, which signifies *one* carboxyl group. Fatty acids fall into this category, although there are other monocarboxylic acids. Examples include

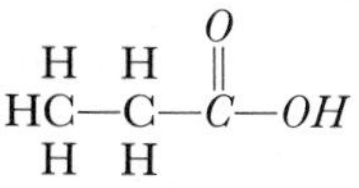

Propionic acid

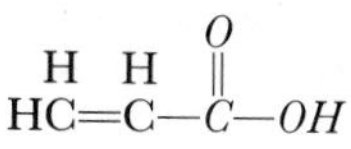

Acrylic acid

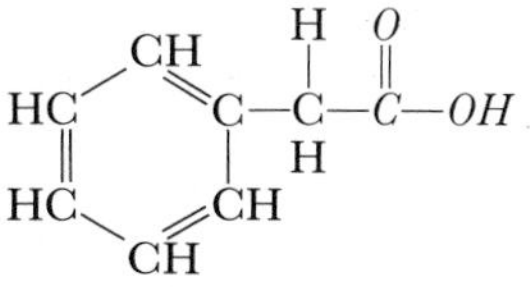

Phenylacetic acid

b. Dicarboxylic, which designates that there are *two* carboxyl groups in the compound. Examples are

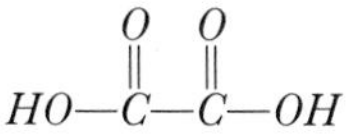

Oxalic acid

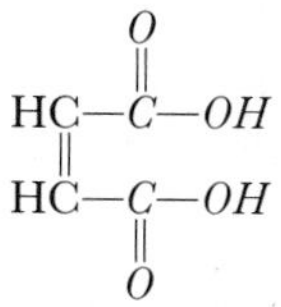

Maleic acid or
cis-ethylene dicarboxylic acid

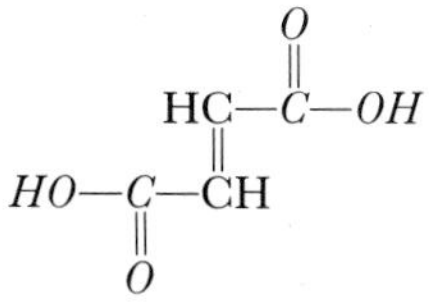

Fumaric acid or *trans*-ethylene dicarboxylic acid

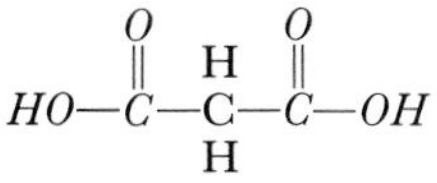

Malonic acid

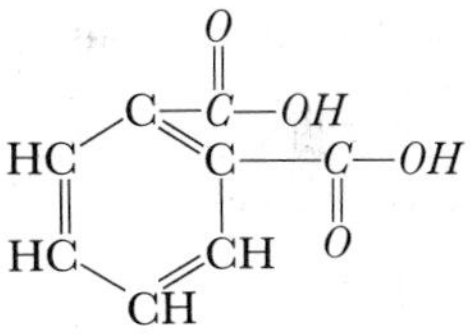

Phthalic acid or orthobenzene
dicarboxylic acid

c. Tricarboxylic, which indicates *three* carboxylic groups in the compound, as

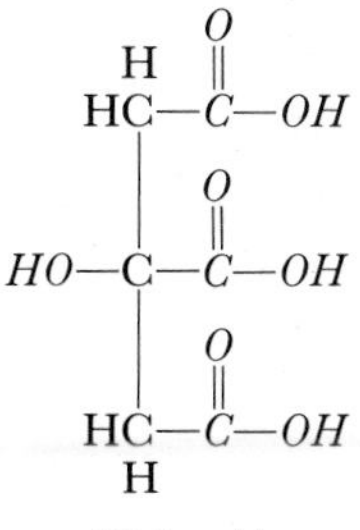

Citric acid

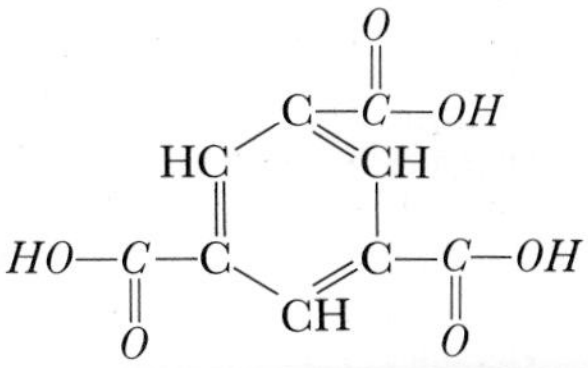

1,3,5-Benzenetricarboxylic acid
or trimesic acid

TABLE C-15. *Properties of Representative Carboxylic Acids*

Carboxylic Acid	Formula	Formula Wt.	Sp gr	Mp, °C	Bp, °C
Abietic	$C_{20}H_{35}O_2$	302.44	. . .	182	
Acetic	CH_3COOH	60.03	. . .	16.6	118.1
Acetoacetic	CH_3COCH_2COOH	102.06	. . .	. . .	100d
Aconitic	$C_3H_3(COOH)_3$	174.11	. . .	192d	
Acrylic	$CH_2:CHCOOH$	72.03	1.062	12.3	141.9
Adipic	$HOOC(CH_2)_4COOH$	146.14	1.360	151–153	265
Angelic	$CH_3CH:C(CH_3)COOH$	100.06	0.983	45	185
p-Anisic	$CH_3OC_6H_4COOH$	152.14	1.385	184.2	275–280
o-Anthranilic	$H_2NC_6H_4COOH$	137.13	. . .	144–145	s
Arabonic	$HOOC(CHOH)_3CH_2OH$	166.08	. . .	89	
Arachidic	$CH_3(CH_2)_{18}COOH$	312.52	. . .	77	328
Atropic	$C_6H_5CH(:CH)COOH$	148.15	. . .	106–107	267d
Behenic	$C_{21}H_{43}COOH$	340.34	. . .	84	306
1,2,3,4-Benzenetetracarboxylic	$C_6H_2(COOH)_4$	254.08	. . .	237d	
1,2,3,5-Benzenetetracarboxylic	$C_6H_2(COOH)_4$	254.08	. . .	238	
1,2,3-Benzenetricarboxylic	$C_6H_3(COOH)_3$	210.09	. . .	190	
1,2,4-Benzenetricarboxylic	$C_6H_3(COOH)_3$	210.09	. . .	216d	
1,3,5-Benzenetricarboxylic	$C_6H_3(COOH)_3$	210.09	. . .	350s	
Benzoic	C_6H_5COOH	122.05	1.266	121.7	249.2
Benzoylacetic	$C_6H_5COCH_2COOH$	164.06	. . .	104	
m-Benzoylbenzoic	$C_6H_5COC_6H_4COOH$	226.08	. . .	162	
o-Benzoylbenzoic	$C_6H_5COC_6H_4COOH$	226.08	. . .	127	s
p-Benzoylbenzoic	$C_6H_4COC_6H_4COOH$	226.08	. . .	194	
n-Butyric	$CH_3(CH_2)_2COOH$	88.06	0.959	−7.9	163.5
Isobutyric	$(CH_3)_2CHCOOH$	88.06	0.949	−47	154.4
Camphoric (d-)	$C_{10}H_{16}O$	152.23	0.999	178–179	209.1
Capric	$CH_3(CH_2)_8COOH$	172.26	0.889	31.5	268–270
Isocaproic	$(CH_3)_2CH(CH_2)_2COOH$	116.09	0.925	−35	207.7
n-Caproic	$CH_3(CH_2)_4COOH$	116.09	0.945	−1	205
n-Caprylic	$CH_3(CH_2)_6COOH$	144.21	0.910	16	237.5
Cerotic	$C_{25}H_{51}COOH$	396.41	0.836	82.5	d
Chloroacetic	$ClCH_2COOH$	94.50	1.370	61.2	189.5
o-Chlorobenzoic	ClC_6H_4COOH	156.57	1.544	141–142	
m-Chlorobenzoic	ClC_6H_4COOH	156.57	1.496	158	
p-Chlorobenzoic	ClC_6H_4COOH	156.57	1.541	242–243	s
α-Chloropropionic	$CH_3 \cdot CHCl \cdot COOH$	108.53	1.306	<−20	186
Cinnamic (cis-)	$C_6H_5CH:CHCOOH$	148.15	1.284	68	125
Citric	$C_3H_4(OH)(COOH)_3$	192.12	1.542	153	d
α-Crotonic	$CH_3CH:CHCOOH$	86.09	0.964	72	189
β-Crotonic (cis-)	$CH_3CH:CHCOOH$	86.09	1.031	15.5	170–171d
p-Cumic	$(CH_3)_2CH \cdot C_6H_4COOH$	164.20	1.162	116–117	s
Cyanoacetic	$CH_2(CN)COOH$	85.06	. . .	65.6	108
Dichloroacetic	$Cl_2CH \cdot COOH$	128.95	1.560	9.7	194.4
Diphenylacetic	$(C_6H_5)_2CHCOOH$	212.09	. . .	148	
Formic	$HCOOH$	46.02	1.226	8.4	100.5
Fumaric (trans-)	$HOOC \cdot CH:CH \cdot COOH$	116.07	1.635	286–287	290
Furoic	$C_4H_3O \cdot COOH$	112.03	. . .	133–134	230–232
Gallic (3,4,5)	$(HO)_3C_6H_2COOH \cdot H_2O$	188.06	1.694	220d	
Gluconic	$C_5H_6(OH)_5COOH$	196.09	. . .	. . .	s
Glutaric	$HOOC(CH_2)_3COOH$	132.06	1.429	97.5	200
Glyceric	$CH_2OHCHOHCOOH$	106.05			
Glycollic	$CH_2OH \cdot COOH$	76.03	. . .	79	d
Glyoxalic	$HCOCOOH$	74.02	. . .		
Heptoic	$CH_3(CH_2)_5COOH$	130.19	0.918	−10	221–222
Hippuric	$C_6H_5CONHCH_2COOH$	179.08	1.371	187–188	d
Hydracrylic	CH_2OHCH_2COOH	90.08	. . .	. . .	d
α-Hydroxybutyric	$CH_3CH_2CH(OH)COOH$	104.06	. . .	42.5	260
β-Hydroxybutyric	$CH_3CH(OH)CH_2COOH$	104.06	. . .	. . .	130
γ-Hydroxybutyric	$CH_2(OH)CH_2CH_2COOH$	104.06	. . .	−17	
Lactic	$CH_3CHOHCOOH$	90.05	1.248	18	122

TABLE C-15. *Properties of Representative Carboxylic Acids* (*Continued*)

Carboxylic Acid	Formula	Formula Wt.	Sp gr	Mp, °C	Bp, °C
Lauric	$CH_3(CH_2)_{10}COOH$	200.19	0.869	48	225
Levulinic	$CH_3CO(CH_2)_2COOH$	116.06	1.140	33.5	245–246
Lignoceric	$C_{23}H_{47}COOH$	368.37	. . .	81	
Linoleic	$C_{17}H_{31}COOH$	280.25	0.903	−9.5	229–230
Maleic	$HOOC \cdot CH : CH \cdot COOH$	116.03	1.609	130.5	135d
Malic (*dl*)	$HOOCCH_2CH(OH)COOH$	134.05	1.601	128–129	150d
Malic (*d or l*)	$HOOCCH_2CH(OH)COOH$	134.05	1.595	99–100	140d
Malonic	$HOOCCH_2COOH$	104.03	1.631	135.6	d
Mandelic (*dl*)	$C_6H_5CH(OH)COOH$	152.06	1.300	118.1	d
Margaric	$CH_3(CH_2)_{15}COOH$	270.44	0.853	60–61	227
Melissic	$C_{29}H_{59}COOH$	452.47	. . .	91	
Mellitic	$C_6(COOH)_6$	342.05	. . .	286	d
Mesoxalic	$HOOC \cdot CO \cdot COOH \cdot H_2O$	136.01	. . .	120	
α-Methylacrylic	$CH_2 : C(CH_3)COOH$	86.09	1.015	15–16	161–163
Mucic	$(\cdot CHOHCHOHCO_2H)_2$	210.14	. . .	206–214d	
Myristic	$CH_3(CH_2)_{12}COOH$	228.36	0.853	57–58	250.5
Nicotinic (3-)	C_5H_4NCOOH	123.11	. . .	235.2	s
Nicotinic (iso-)(4)	C_5H_4NCOOH	123.11	. . .	317	d
m-Nitrobenzoic	$NO_2 \cdot C_6H_4COOH$	167.12	1.494	140–141	
o-Nitrobenzoic	$NO_2 \cdot C_6H_4COOH$	167.12	1.575	147.5	
p-Nitrobenzoic	$NO_2 \cdot C_6H_4COOH$	167.12	1.550	240–242	s
Nondecyclic	$CH_3(CH_2)_{17}COOH$	298.30	. . .	66.5	299
Oleic	$CH_3(CH_2)_7CH : CH(CH_2)_7COOH$	282.27	0.895	14	286
Oxalic	$HOOCCOOH$ (anhydrous)	108.03	1.653	189	
Palmitic	$CH_3(CH_2)_{14}COOH$	256.25	0.853	64	380
Pelargonic	$CH_3(CH_2)_7COOH$	158.23	0.906	12.5	253–254
Pentadecyclic	$C_{14}H_{29}COOH$	242.39	. . .	52	257
Phenylacetic	$C_6H_5CH_2COOH$	136.14	1.081	76–77	265.5
Phthalic	$C_6H_4(COOH)_2$	166.13	1.593	208	d
Isophthalic	$C_6H_4(COOH)_2$	166.13	. . .	330	s
Pimelic	$HOOC(CH_2)_5COOH$	160.09	. . .	103	272
Propargylic	$CH_3 : CCOOH$	70.02	1.139	9	144d
Propionic	CH_3CH_2COOH	74.05	0.992	−22	141.1
Pyruvic	$CH_3COCOOH$	88.06	1.267	13.6	165
Ricinoleic	$C_{17}H_{32}(OH)COOH$	298.45	0.954	4–5	226–228
Saccharic (*d*)	$HOOC(CHOH)_4COOH$	210.08			
Isosaccharic	$HOOCCH(CHOH)_2CHCOOH$	192.06	. . .	185	d
Salicylic	HOC_6H_4COOH	138.05	1.443	159	s
Sebacic	$HOOC(CH_2)_8COOH$	202.14	. . .	−127	295
Stearic	$CH_3(CH_2)_{16}COOH$	284.28	0.847	69.3	383
Suberic	$HOOC(CH_2)_6COOH$	174.19	1.266	140–144	279
Succinic	$HOOC(CH_2)_2COOH$	118.05	1.564	185	235
Isosuccinic	$CH_3CH(COOH)_2$	118.05	1.455	135d	
Tartaric (*d or l*)	$(CHOHCO_2H)_2$	150.09	1.760	168–170	d
Tartaric (*meso-*)	$(CHOHCO_2H)_2$	150.09	1.737	159–160	
Tartaric (racemic)	$(CHOHCO_2H)_2H_2O$	168.10	1.697	205–206	
Tartronic	$CH(OH)(COOH)_2 \cdot \frac{1}{2}H_2O$	129.07	. . .	155.8d	s
Terephthalic	$C_6H_4(COOH)_2$	166.13	1.510	s	
o-Toluic	$CH_3 \cdot C_6H_4 \cdot COOH$	136.14	1.062	104–105	259
m-Toluic	$CH_3C_6H_4 \cdot COOH$	136.14	1.054	110–111	263
p-Toluic	$CH_3C_6H_4 \cdot COOH$	136.14	. . .	179–180	274–275
Tridecyclic	$CH_3(CH_2)_{11}COOH$	214.20	. . .	51	236
Triphenylacetic	$(C_6H_5)_3CCOOH$	288.12	. . .	265	
Undecyclic	$CH_3(CH_2)_9COOH$	186.17	. . .	29.3	228
n-Valeric	$CH_3(CH_2)_3COOH$	102.08	0.942	−59	187
Isovaleric	$(CH_3)_2CHCH_2COOH$	102.08	0.937	−37.6	176.7
Vinylacetic	$CH_2 : CH \cdot CH_2 \cdot COOH$	86.09	1.013	−39	163

d (in acid column) = dextrorotatory; d (in mp or bp column) = decomposes; *dl* = dextrolevorotatory; *l* = levorotatory; *m* = meta; *o* = ortho; *p* = para; *s* = sublimes.

d. Tetracarboxylic, which indicates *four* carboxylic groups in the compound, as

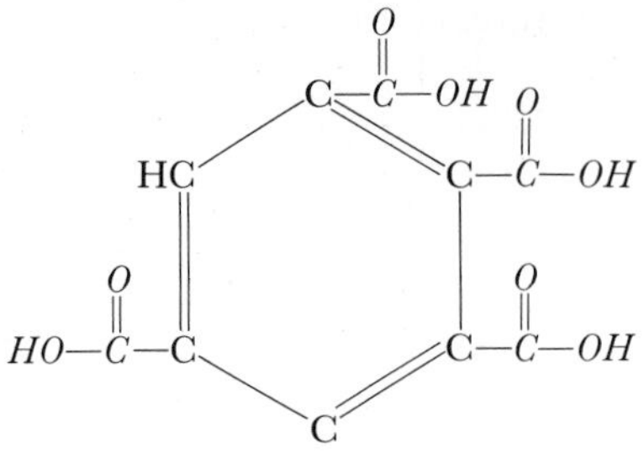

1,2,3,5-Benzenetetracarboxylic acid
or mellophanic acid

3. Carboxylic acids may be classified as to the number of separate *hydroxyl* groups, if any, contained in the compound.

a. No inclusion of the term *hydroxy* in the name of the acid signifies, of course, that the acid does *not* contain a hydroxyl group (other than the O—H in the carboxyl group). Of the acids mentioned thus far, only citric acid has the separate OH group. Hence citric acid is a hydroxytricarboxylic acid. As in the case of citric acid, the common or trivial name seldom reveals that the acid may or may not be a hydroxycarboxylic acid. It must be stressed that the term *hydroxy* should not be used alone because of possible confusion with the hydroxyamino acids. The term always should be combined, as hydroxycarboxylic.

b. Hydroxycarboxylic indicates the inclusion of *one* additional hydroxyl group in the compound. Examples include

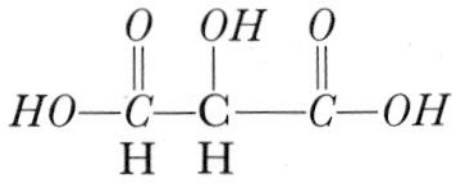

Tartronic acid
A hydroxydicarboxylic acid

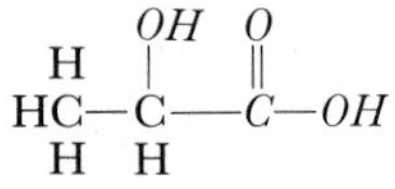

Lactic acid or
α-hydroxypropionic acid
A hydroxymonocarboxylic acid

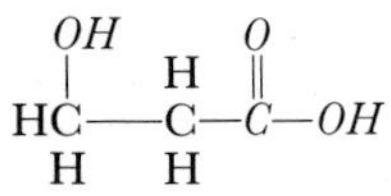

Hydracrylic acid or
β-hydroxypropionic acid
A hydroxymonocarboxylic acid

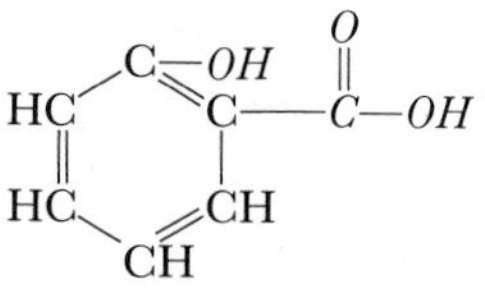

Salicylic acid
An aromatic hydroxymonocarboxylic acid

c. Dihydroxycarboxylic denotes the inclusion of *two* additional hydroxyl groups. Examples include

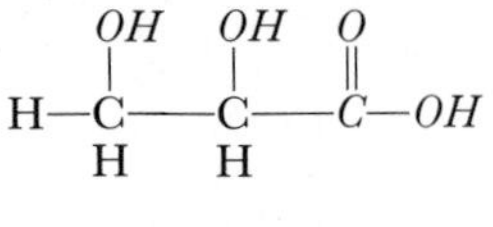

Glyceric acid

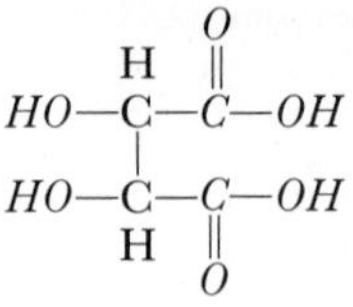

Tartaric acid

d. Trihydroxycarboxylic indicates inclusion of *three* additional hydroxyl groups. An example is

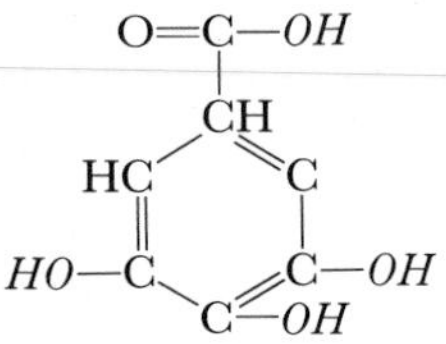

Gallic or pyrogallol carboxylic acid
or 3,4,5-trihydroxybenzoic acid
A trihydroxymonocarboxylic acid

4. Carboxylic acids may be classified in accordance with the *number of available hydrogens* for the formation of salts. These hydrogens may be available from the one or more carboxyl, —COOH, groups present and from the additional hydroxyl, OH, groups present, if any.

a. Monobasic carboxylic acids have only one available hydrogen for salt formation and, consequently, must always be a monocarboxylic acid, such as acetic acid, benzoic acid, propionic acid, acrylic acid, and phenylacetic acid, already shown.

b. Dibasic carboxylic acids have two available hydrogens. Both hydrogens may be available from the carboxyl groups, as in the case of dicarboxylic acids, such as oxalic acid, maleic acid, and fumaric acid. Or one hydrogen may be available from a carboxyl group and another hydrogen from a hydroxyl group. Hydroxycarboxylic acids, such as hydracrylic acid and salicylic acid already shown, are examples.

c. Polybasic carboxylic acids have three or more available hydrogens, from the total of carboxyl and hydroxyl groups present. Examples of *tribasic* carboxylic acids would include trimesic acid, tartronic acid, and glyceric acid. Examples of tetrabasic carboxylic acids would include citric acid, mellophanic acid, tartaric acid, and gallic acid.

5. Carboxylic acids may be classified by the *homologous series* of which they are a member, such as

a. Saturated monobasic fatty acids, $C_nH_{2n}O_2$

b. Unsaturated monobasic fatty acids, $C_nH_{2n-2}O_2$

c. Propiolic acid series, $C_nH_{2n-4}O_2$

d. Dicarboxylic acids, $C_nH_{2n}(COOH)_2$ $n = 0$ for oxalic acid

e. Hydroxymonocarboxylic acids, $C_nH_{2n} \cdot (OH) \cdot COOH$ $n = 0$ for carbonic acid

6. Carboxylic acids may be classified in terms of the *groups other than* the carboxyl and hydroxyl groups they may contain.

a. *Aldehydic* carboxylic acids contain the CHO group, as found in glyoxalic acid, $CHO \cdot COOH$, and in glycuronic acid, $CHO \cdot [CH(OH)]_4COOH$.

b. *Amino* carboxylic acids contain the NH_2 group, as found in glycine (aminoacetic acid), $CH_2(NH_2) \cdot COOH$, and in carbamic acid (aminoformic acid), NH_2COOH. Classically, the amino acids are usually considered as a separate, distinct class of compounds. See **Amino acids.**

c. *Ketonic* carboxylic acids contain the carbonyl (C—O) group. Monobasic ketonic acids include α-pyroracemic acid, $CH_3 \cdot CO \cdot COOH$, and β-acetoacetic acid, $CH_3 \cdot CO \cdot CH_2 \cdot COOH$. Monobasic aromatic ketonic acids include phenylglyoxylic (benzoylformic acid), $C_6H_5 \cdot CO \cdot COOH$, orthonitrobenzoylformic acid, $NO_2 \cdot C_6H_4 \cdot CO \cdot COOH$, and benzoylacetic acid, $C_6H_5 \cdot CO \cdot CH_2 \cdot COOH$. Dibasic ketonic carboxylic acids include mesoxalic acid, $CO(COOH)_2$, oxalacetic acid (butanone diacid), $COOH \cdot CH_2 \cdot CO \cdot COOH$, acetonedicarboxylic acid (pentanone diacid), $CO(CH_2 \cdot COOH)_2$, dihydroxytartaric acid, $COOH \cdot CO \cdot CO \cdot COOH$, and diacetoglutaric acid, $COOH \cdot CHAc \cdot CH_2 \cdot CHAc \cdot COOH$.

d. *Phenolic* carboxylic acids are derived from benzoic acid where the OH group is united with a carbon of the nucleus.

Nomenclature: Trivial or common names for these acids persist. In one system (IUPAC), the name of the parent hydrocarbon, such as propane, is adapted to the suffix *-oic,* followed by the term *acid.* Thus, *propanoic* acid. Benzoic acid is similarly constructed from the parent, benzene. In connection with the lower fatty acids, reference to acetic acid as a basic name is made. Thus *n*-butyric acid may be called ethylacetic acid, or propionic acid may be called methylacetic acid. Similar situations occur with other acids, such as *o*-toluic acid, which may be called *o*-methylben-

zoic acid, or salicylic acid, which may be called *o*-hydroxybenzoic acid. In some cases, the relevant hydrocarbon is followed by the term *carboxylic acid,* as the reference to acetic acid as methanecarboxylic acid; or acrylic acid as ethylenecarboxylic acid; or fumaric acid as *trans*-ethylene dicarboxylic acid. Some equivalent terms are given as follows:

Acetic acid	Methane carboxylic acid
Acrylic acid	Ethylenecarboxylic acid
Angelic acid	*Cis-α,β*-dimethylacrylic acid
Arachidic acid	Eicosanoic acid
n-Butyric acid	Ethylacetic acid
Capric acid	Decanoic acid
n-Caproic acid	*n*-Hexylic acid
n-Caprylic acid	Octanoic acid
Cinnamic acid	β-Phenylacrylic acid
Citric acid	Hydroxytricarboxylic acid
α-Crotonic acid	α-Butenic acid
β-Crotonic acid	Isocrotonic acid
Fumaric acid	*Trans*-ethylene dicarboxylic acid
Furoic acid	Pyromucic acid
Gluconic acid	Dextronic acid
Glyceric acid	2,3-Dihydroxypropionic acid
Glycollic acid	Hydroxyacetic acid, ethanolic acid
Hydracrylic acid	β-Hydroxypropionic acid, ethylene lactic acid
Lactic acid	α-Hydroxypropionic acid
Lauric acid	Dodecanoic acid
Levulinic acid	β-Acetopropionic acid
Linoleic acid	Linolic acid
Maleic acid	*Cis*-ethylenedicarboxylic acid
Malic acid	Hydroxysuccinic acid
Mellitic acid	Benzenehexacarboxylic acid
Palmitic acid	*n*-Hexadecyclic acid
Pelargonic acid	Nonanoic acid
Phenylacetic acid	α-Toluic acid
Isophthalic acid	*m*-Benzenedicarboxylic acid
Phthalic acid	*o*-Benzenedicarboxylic acid
Pimelic acid	Heptane diacid
Propargylic acid	Propiolic acid
Propionic acid	Methylacetic acid
Pyruvic acid	Pyroracemic acid
Saccharic acid	Tetrahydroxypentanedicarboxylic acid
Salicylic acid	*o*-Hydroxybenzoic acid
Sebacic acid	Decanedicarboxylic acid
Stearic acid	*n*-Octodecyclic acid
Suberic acid	Octanedicarboxylic acid
Isosuccinic acid	Methylmalonic acid
Tartaric acid	Racemic acid, dihydroxysuccinic acid
Tartronic acid	Hydroxymalonic acid
Terephthalic acid	*p*-Benzenedicarboxylic acid
o-Toluic acid	Methylbenzoic acid
Undecyclic acid	Decane-α-carbonic acid, undecanoic acid
Isovaleric acid	Isobutylformic acid

Profile: Although it is difficult to generalize a topic as varied and complex as represented by the carboxylic acids, some apparent order can be found by describing the major classical groupings:

1. *Fatty acids* generally are considered as resulting from the oxidation of saturated primary alcohols. They may be considered as the carboxy derivatives of the paraffins and members of the homologous series $C_nH_{2n}O_2$. The simplest and lowest member is formic acid (C_1), followed by acetic acid (C_2), propionic acid (C_3), butyric acid (C_4), valeric acid (C_5), to the long-chain compounds, such as palmitic acid (C_{16}), stearic acid (C_{18}), and melissic acid (C_{30}).

A profile of the fatty acids would include: (1) very stable and oxidized or converted to simpler compounds with difficulty (formic acid is an exception); (2) readily undergo double decomposition because of the carboxyl group; (3) although all monobasic, their acid qualities decrease as the formula weight increases; (4) react with alcohols to form esters and water; (5) convert to acid chlorides upon reaction with phosphorus pentachloride; and (6) yield halogen-substitution products. The lower-carbon members are relatively corrosive liquids with pungent odors and do not decompose when boiled. They are soluble in water, and aqueous solutions have decided acidic qualities. The intermediate fatty acids possess a rancid odor and are oily and only slightly soluble in water. The odor, solubility, and mobility of the fatty acids decrease as the carbon count rises. The boiling and melting points increase rather regularly with carbon count.

2. *Monohydroxy fatty acids* can be considered as the monohydroxy derivatives of the fatty acids and include hydroxyacetic acid (glycollic acid); β-hydroxypropionic acid (β-lactic acid). A profile of these acids would include: (1) decompose when volatilized; (2) syrupy liquids which readily give up water to form crystalline anhydrides; (3) soluble in water and generally in alcohol and ether. Glycollic acid occurs in unripe grapes. There are two isomeric lactic acids.

3. *Polyhydric monobasic acids* may be considered as resulting from the oxidation of the polyhydric alcohols. These compounds combine the properties of monobasic acids and polyhydric alcohols. Several of these acids may be formed from the oxidation of the sugars. Glyceric acid, a syrupy liquid, is an example of a dihydroxymonobasic acid and may be derived from the careful oxidation of glycerol. Arabonic and gluconic acids are named from the sugars arabinose and glucose, to which they are related, and are examples of tetrahydroxy and pentahydroxy acids, respectively.

4. *Aromatic acids* are similar to the fatty acids in many ways. Derived from benzene, these are mono-, di-, tri-, and up to hexabasic aromatic acids. A profile of the aromatic acids would include: (1) crystalline solids only slightly soluble in water, but usually soluble in alcohol and ether; and (2) the simpler aromatic acids may be sublimed or distilled without decomposition, but the more complex and, in particular, the phenolic and polycarboxylic aromatic acids yield carbon dioxide and a simpler compound when heated (salicylic acid breaks down into CO_2 and phenol). Many aromatic acids are found in natural resins, balsams, and animal organisms, usually in the form of nitrogen derivatives in the latter. Oxidation of primary alcohols, such as benzyl alcohol, yields saturated acids, such as benzoic acid.

Monobasic saturated aromatic acids include benzoic, hippuric, toluic acids (three in number), phenylacetic, phenylchloracetic, and dimethylbenzoic acid. Monobasic unsaturated acids include cinnamic, atropic, and phenylpropionic acids. Saturated phenolic acids include salicylic and gallic acids. Examples of alcohol acids include mandelic, amygdalic, and tropic acids. Coumaric acid is an example of an unsaturated monobasic phenolic acid.

Saturated dibasic aromatic acids are similar to those of the aliphatic oxalic acid series and include phthalic, isophthalic, and terephthalic acids. Trimesic and mellitic acids are examples of polybasic aromatic acids.

Chemical Reactivity: When a carboxylic acid is reacted with sodium hydroxide or other base, a corresponding salt (sodium acetate) is produced. Carboxylic acid salts also may be formed by heating the acids with a moist heavy-metal oxide or hydroxide, silver oxide often being used. Butanoic acid thus yields silver butanoate. Primary alcohols may be obtained by reducing the carboxylic acids with lithium aluminum hydride. For example, in this fashion, trimethylacetic acid will yield neopentyl alcohol, or *m*-chlorobenzoic acid will yield *m*-chlorobenzyl alcohol. Extensively used in the synthesis of amino acids, the Hell-Volhard-Zelinsky reaction involves replacing hydrogen atoms on a carbon atom adjacent to a carboxyl group with chlorine or bromine, with phosphorus present. For example, an α-bromo acid prepared in this manner then can be changed to an amino acid by using an excess of ammonia. The carboxylic acids react with the alcohols to produce corresponding esters. The lower fatty acids will react with ammoniacal silver hydroxide solutions to yield silver metal. Treatment of the acids with the halogens yields halogen-substitution products. Some of the reactions of a few important carboxylic acids are summarized in Table C-16.

TABLE C-16. *Some Reactions of Important Carboxylic Acids*

Reacts With:	To Yield:
Formic acid:	
Alcohols	Esters
Ammoniacal silver hydroxide	Silver metal, carbon dioxide, and water
Heat (in excess of 160°C	Carbon dioxide and hydrogen
Mercuric chloride solution	Mercurous chloride
Metallic hydroxides or carbonates	Salts
Potassium permanganate with H_2SO_4	Manganous salt and carbon dioxide
Sulfuric acid (concentrated) and heat	Carbon monoxide and water
Acetic acid:	
Alcohols	Esters
Halogens (chlorine) . . .	Halogen-substitution products

$$CH_3COOH + Cl_2 \rightarrow CH_2ClCOOH + HCl$$
$$CH_3COOH + 2Cl_2 \rightarrow CHCl_2COOH + 2HCl$$
$$CH_3COOH + 3Cl_2 \rightarrow CCl_3COOH + 3HCl$$

Lime plus catalyst . . .	Acetone and calcium carbonate
Phosphorus trichloride	Acetyl chloride

$$3CH_3COOH + PCl_3 \rightarrow 3CH_3COCl + P(OH)_3$$

Metallic hydroxides and carbonates	Salts
Benzoic acid:	
Ethyl alcohol and hydrogen chloride plus heat	Ethyl benzoate
Phosphorus pentachloride	Benzoyl chloride

Formation: The commercial production of several carboxylic acids is described in several separate entries in this book. Generally, the acids may be prepared:

1. By oxidizing the relevant alcohol. Acetic acid may be produced from ethanol and atmospheric oxygen in the presence of a catalyst.

2. By oxidizing the relevant aldehyde, as in the case of deriving acetic acid from acetaldehyde

with atmospheric oxygen in the presence of a catalyst.

3. From the bacterial fermentation of dilute alcohols, as in the case of acetic acid, where *Bacterium aceti* is used; or from the bacterial fermentation of other acids, as in the case of the production of butyric acid from lactic acid, where *lactic ferment* is used.

4. By using the haloform reaction, a methyl ketone may be reacted with sodium hypochlorite wherein chloroform is a by-product. For example, in this manner, *p*-methylacetophenone will yield *p*-methylbenzoic acid and chloroform.

5. Particularly useful for the synthesis of carboxylic acids is the carbonation of Grignard reagents. For example, *p*-bromochlorobenzene will yield *p*-chlorobenzoic acid in this manner.

6. By hydrolyzing nitriles.

7. By the malonic ester synthesis route.

Probably the most industrially important methods of preparing carboxylic acids are:

8. By oxidizing the relevant alkylaromatic, as in the case of deriving benzoic acid from toluene with atmospheric oxygen in the presence of a cobalt catalyst.

9. By treatment of an alkali metal phenolic with CO_2. For example, sodium phenoxide gives salicylic acid in nearly quantitative yield when treated with CO_2 at 150–180°C (302–356°F).

10. By the hydrocarboxylation of olefins with CO using nickel carbonyl catalyst. For example, in this manner, propylene will yield butyric acid.

See also **Ion exchange resins;** and **Peroxides, organic.**

—Duane B. Priddy, *The Dow Chemical Company, Midland, Mich.*

Carburizing (See **Iron and steel.**)

Carene (See **Terpenes.**)

Carnallite (See **Cesium; Fertilizers;** and **Potassium.**)

Carnauba (See **Waxes.**)

Carnotite (See **Uranium.**)

Carrollite (See **Cobalt.**)

Cartridge Brass (See **Copper.**)

Casale Process (See **Ammonia.**)

Casing, Polymer   (See **Plasma, chemical processing with.**)

Cassiterite   (See **Indium; Tin;** and **Tungsten.**)

Cast Iron   (See **Iron and steel;** and **Ironmaking and steelmaking.**)

Casting   (See **Epoxy resins; Glass; Ironmaking and steelmaking;** and **Plastics.**)

Castner Cell   (See **Chlorine.**)

Castor Oil   [See **Anionic surfactants (sulfur-bearing); Expression;** and **Vegetable oils.**]

Castor Wax   (See **Waxes.**)

Catalysts   Simply stated, a catalyst is a substance with whose aid the rate of one series of reactions among several is made to predominate, so that a desired product is produced, to the major exclusion of other products. The classical definition of a catalyst states that it is a substance which alters the rate of chemical reaction without appearing in the end product.

Catalysts are used in a wide spectrum of processes whose commercial applications are valued in billions of dollars.

Homogeneous and Heterogeneous Processes: There are two general types of catalytic processes, homogeneous and heterogeneous. In the homogeneous case, the chemical reaction is said to take place in a single phase, usually a liquid environment. In the heterogeneous case, the process takes place in a multiple phase, usually in a gaseous environment, and usually in the presence of a solid catalyst phase, which is either present as an undiluted material or on the surface of an inert substance (such as charcoal) called a *support*. The support allows the catalyst to have improved properties. For instance, a supported catalyst can be prepared with a higher active surface and greater physical strength than the original unsupported catalyst.

In many cases a catalyst cannot be used by itself, as for instance mercuric acetate in the reaction of acetylene and acetic acid in the vapor phase, to yield vinyl acetate, a component of many polymers and plastics. Mercuric acetate has poor physical characteristics, and in the pure state is quite costly. However, a solution containing mercuric acetate can be absorbed onto and into small pieces of activated carbon which can have a surface area of up to 1,200 m^2/g. The resultant supported catalyst now has mercuric acetate spread over 1,200 m^2/g with a great deal more activity than mercuric acetate powder, which normally has a small surface area of less than 1 m^2/g.

In one homogeneous process for making vinyl esters from vinyl acetate liquid and an organic acid, the catalyst mercuric acetate is dissolved in the mixture. In this case a maximum dispersion of mercury can be effected. One could also make a heterogeneous process out of the homogeneous one, by cycling the liquid components over the previously described mercury-on-carbon solid catalyst. However, this could probably result in the extraction of a large part of the mercury from the support after a period of time, creating a dual homogeneous-heterogeneous process.

In commercial processes liquid-phase reactions are carried out at lower temperatures than gas-phase reactions because of either the chemical nature of the reaction or the economics of the processes that are involved. In liquid-phase reactions carried out in open-to-atmosphere reactors or closed reactors subject to pressure (called *autoclaves*), the components of the reaction remain in contact with one another for times in the order of hours, so that the desired product can be formed, especially if the rate is slow. Where an edible product such as fruit jam is produced using an edible acid as catalyst, the liquid phase for the reaction is the only alternative.

Gas-phase reactions are of value where reaction rates of the desired intermediates are rapid, so that the desired product which is formed can be removed rapidly and in large quantities. It is possible to calculate by means of thermodynamics whether a new reaction is possible and save the time and effort of trying an impossible step.

Originally, many large-scale heterogeneous chemical processes were inorganic in nature: the manufacture of sulfuric acid, ammonia, and nitric acid. In the early 1900s methanol and phthalic anhydride were made by homogeneous processes: methanol from wood and phthalic anhydride from naphthalene. However, by the middle 1920s vapor-phase processes such as phthalic anhydride from naphthalene began to show commercial feasibility. And in the last twenty years the rise of catalysts in the chemical industries has grown exponentially, especially in those processes where raw materials are petroleum-derived.

For instance, in catalytic cracking operations, where hydrocarbons are changed from such substances as tar and asphalt to fuels such as light oils and gasoline, the consumption of catalysts (silica aluminas, processed clays, and zeolites) comes to over 100,000 tons/yr with a value of over $50 million. In isomerization reactions, where straight-chain paraffinic molecules are rearranged to branched-chain molecules in order to improve, in one example, octane numbers in gasoline, over $1\frac{1}{2}$ million tons of aluminum chloride, worth more than $100 million, is consumed annually.

In alkylation reactions, where an alkyl radical is introduced into a molecule by addition or substitution into an organic compound, in order to make such substances as gasoline, rubber antioxidants, dyes, flavors, and others, sulfuric acid worth $30 million/yr is consumed.

Further, in the treatment of organic materials with hydrogen to yield substances as diverse as gasoline and peanut butter, and in reactions such as polymerization, which is in essence the building of larger molecules (i.e., polyethylene) from smaller molecules (ethylene), various catalysts worth over $15 million are consumed.

Further reflection will show that catalysts are involved not only in our everyday chemical manufacture, but in our everyday biological existence, as for instance in our use of vitamins and minerals for proper physiological maintenance, and in our recognition and use of enzymes and hormones, without which no one could function.

Specifically, in some large-scale processes such as dehydrogenation, in which hydrogen atoms are removed with the aid of catalysts such as nickel, cobalt, platinum, palladium, and mixtures containing potassium, chromium, copper, aluminum, and others, solvents such as acetone (6 billion lb/yr) are made from isopropyl alcohol, and styrene from ethylbenzene (over 2 billion lb/yr). Styrene can be found in most of the polymers or plastics made today, especially in impact-resisting materials.

The process of oxidation, wherein air or oxygen is used to change a molecule to a more complex one having desirable properties, can be typified by the transformation of ethylene to ethylene oxide over a silver catalyst. Use of ethylene oxide in ethylene glycol antifreeze, polyurethane foams, plastics, polyesters, solvents, and chemical intermediates runs to over 1 billion lb/yr.

Examples of Catalytic Processes: The data in Table C-17, which represents a cross section of some major commercial catalytic processes, are presented as an example of the use of different catalysts, as well as the yields of products it is possible to obtain using these catalysts.

Initial Considerations in Selecting a Catalyst: How does one choose a catalyst so that screening programs do not become too expensive or time-consuming? In modern research laboratories, initial economic evaluation and literature and patent searches, as well as thermodynamic calculations for new reactions, tend to minimize time and effort in carrying out catalyst selection and evaluation.

If a reaction is reversible, thermodynamics can be used to calculate the equilibrium concentration of products, since this is the thermodynamic limit of conversion with or without a catalyst.

In general, a catalyst problem is either one of development of a new catalyst for a novel process or of a catalyst modification for an existing process. A catalyst cannot be considered by itself. It must be considered as part of an overall process, for when studying a chemical process, not only is the precise nature of a catalyst studied, but also the chemical-reaction mechanism, practical processing details, and production techniques.

Recent advances in physical chemistry, instrumental analysis, radioactive techniques, solid-state physics, and computer applications have helped catalytic work tremendously. However, published discovery of new processes and continued application of old catalysts to new reactions make it clear that catalytic chemistry is still an experimental art depending in large degree on the intuition, ingenuity, and perseverance of both chemist and chemical engineer.

Catalytic Action as a Rate Phenomenon—Use of Arrhenius Equation: A catalyst by definition can speed (or slow) a reaction but has no effect on equilibrium. It can modify the free-energy changes of the intermediate steps in a reaction. In essence, catalytic acceleration can be effected by the substitution of a sequence of steps, each having a low free energy of activation, in place of a single noncatalytic step involving a high free energy of activation. It is well to remember that the rate of a chemical reaction is proportional to a rate constant K, whose relationship to absolute temperature T, the gas constant R, and activation energy E is given by the Arrhenius equation, $K = Ae^{-E/RT}$. A is a constant depending upon whether the reaction is uni- or bimolecular. The ability of the catalyst to increase the velocity of the reaction is due to the lowering of the energy of activation E for the activated complex. Because of the exponential nature of the Arrhenius equation, a small decrease in E can bring about a marked increase in velocity. In practice it is often convenient to plot log K versus I/T, which usually gives a linear relation.

TABLE C-17. *Catalytic Processes**

Process & Product	Reactants	Catalyst	Yield
Amination:			
Amines	Alcohols + ammonia	$Al_2O_3(Co)$	90+
Ammoxidation:			
Acrylonitrile	Propylene + O_2 + NH_3	Bi-Mo-P	60–80
		CuO, SbSn	
Benzonitrile	Toluene + O_2 + NH_3	V_2O_5–Sb	90+
Phthalonitrile	*o*-Xylene + O_2 + NH_3	V_2O_5	90+
Chlorination:			
Chlorobenzene.	Benzene + Cl_2	Fe	70–75
Chloroacetic acid	Acetic acid + Cl_2	Red P	90
Benzyl chloride	Toluene + Cl_2	Ultraviolet light	95+
Hydration:			
Acetaldehyde	Acetylene + H_2O	Hg_2SO_4	95
Ethanol (alcohols).	Ethylene + H_2O	H_3PO_4, WO_3	95
Dehydration:			
Styrene	Methylethyl carbinol	TiO_2	80+
Ethylene (olefins)	Ethanol (alcohols)	Al_2O_3, ThO_2	90+
Acrylonitrile	Ethylene cyanohydrin	Al_2O_3	90+
Hydrogenation:			
Aniline	Nitrobenzene	Fe–HCl, Cu·SiO_2	90–95
Butanol.	Butyraldehyde	Co, Ni–SiO_2	98+
Cyclohexane	Benzene	Ni–Al_2O_3, PtO_2	96+
Ethylene	Acetylene	Fe, Ni, Cu, Pd–$BaSO_4$	99
Methanol.	Carbon monoxide	ZnO–CrO_3, Ni–Co	60
Dehydrogenation:			
Acetaldehyde	Ethanol + H_2	Cu, Ag, $FeMoO_4$	85–95
Benzene	Cyclohexane	Nu–Al_2O_3, Pt–Al_2O_3	95+
Butadiene	Butenes	Fe, Cr, K, $CaNiPO_4$	75–85+
Butene	Butane	Cr_2O_3–Al_2O_3	
Methyl ethyl ketone	*sec*-Butanol	ZnO–ZnCu	85–92
Styrene	Ethyl benzene	$ZnCrO_2$–FeMgO	86–92
		$CaNiPO_4$	
	Phenyl methyl carbinol	TiO_2	80+
Multiple reactions (see ammoxidation)			
Reductive amination (with hydrogen):			
Isopropylcyclohexylamine	Acetone + cyclohexylamine		
	Cyclohexanone + isopropylamine	Ni	90+
Reductive dehydration:			
Butane	Butanol + H_2	Ni–Al_2O_3	90+
Reforming:			
Aromatic.	Naphthenes + H_2	Mo–Al_2O_3	
		Pt–Al_2O_3–halides	
Desulfurization:			
Butane	Thiophene + H_2S + H_2	Co–Mo–Al_2O_3	
Oxidation:			
Acetaldehyde	Ethylene	$PdCl_2$–MgO–Cu	95+
Acetic acid.	Acetaldehyde	Mn^{++}	88–95
	Butane	Co, Bi	20–40
Acetic anhydride	Acetaldehyde	Cu, Co	70–75
Acetone.	Isopropanol	Cu, Ag, Zno	85–90
Adipic acid	Cyclohexanone	Cu–Mn, V–Cu	70–90
Benzoic acid	Toluene	Co^{++}	90
	Phenol	Cu	90

TABLE C-17. *Catalytic Processes* * (*Continued*)

Process & Product	Reactants	Catalyst	Yield
Benzaldehyde	Toluene	UO_2–MoO_3–Cu	30–50
Ethylene oxide.	Ethylene	Ag,AgO	70
Propylene oxide	Propylene	Mo, W, Ti, V	90
Phthalic anhydride	Naphthalene	V_2O_5–K_2SO_4	70–80
	o-Xylene		
Maleic anhydride	Benzene	Mo-V-P-Na	85
	Butene	V-P	60
Terephthalic acid	p-Xylene	Mn-Co	90+

SOURCE: Catalyst Development Corporation. Printed with permission.

*Discussed in detail in W. L. Faith, D. B. Keyes, and R. L. Clark, "Industrial Chemicals," Wiley, New York, 1965.

Requirements of a Catalyst: A catalyst must be active and selective and have chemical and physical stability.

Initial consideration must be given to selectivity, which is a measure of the degree the reaction is made to go in the direction that is desired. For instance, by dehydration (removal of water), ethanol can be transformed into either ethylene:

$$C_2H_5OH \rightarrow C_2H_4 + H_2O \qquad (1)$$

or into diethyl ether:

$$2C_2H_5OH \rightarrow C_2H_5OC_2H_5 + H_2O \qquad (2)$$

It is necessary to choose a proper catalyst: for step (1), it would be gamma alumina at 400°C; for step (2), sulfuric acid at 140°C. The catalyst and temperature determine the selectivity here.

It is also desirable to convert the reactant(s) as rapidly as possible. In other words, the catalyst must be active, for an active catalyst will have a desirably high production rate.

The concept of productivity is measured by the product of activity or conversion and selectivity per unit time and is written *Yt:* Conversion, $C \times$ selectivity, $S =$ yield, *Y.*

As previously mentioned, chemical and physical stability are necessary for a good catalyst. Valence states should be kept under control because changing valence states and migration of atoms in a crystal structure at elevated temperatures could change the physical structure, and thereby change the nature of a catalyst from one of high selectivity and activity to one less active and less selective.

It is important, too, that a catalyst have mechanical strength to withstand shipping stresses and transfer to catalytic reactors. Once in the reactor, a catalyst's activity and selectivity must be maintained during an economically useful time period. Catalysts can lose activity or selectivity for a number of reasons, some of which are reparable and some of which are not. In gas-phase reactions sudden surges of feed may overwhelm a catalyst by covering the active sites. The deposited material may be converted to coke because of high temperatures and inactivate the catalyst. If burning or removal of the coke can be done with air or other means without injury to the catalyst, it can be reused; otherwise a new batch is necessary. Sometimes, in cracking of petroleum or in hydrogenation reactions, a large amount of hydrogen is used with the petroleum feed. This prevents stripping of hydrogen from the feedstock molecules, and is done to prevent deposition of carbon, sulfur, or nitrogen tars on the catalyst surface. Further, overheating of catalyst can cause sintering, with loss of active crystal structure.

Physical Factors: In a heterogeneous catalytic reaction (by a solid surface) the following sequence of steps takes place:

1. Diffusion of reactants from the bulk or gas phase to the catalytic surface
2. Adsorption of reactants
3. Diffusion of adsorbed species to active sites
4. Electron transfer processes at active sites
5. Chemical interaction of neutral and charged species
6. Desorption of reaction products from active sites
7. Diffusion of products into bulk phase

Since catalyst activity is proportional to the ability of the catalyst to chemisorb reacting species, activity will normally increase as surface area increases. For very fast reactions (high-activity catalysts) only the external surface is normally involved, and the overall rate of reaction may become controlled by the rate of transfer of reagents to the catalyst surface. On the other hand, some reactions with porous catalysts may involve smaller reaction rates than rate of external diffusion; hence overall reaction rate will be noticeably affected by diffusion within catalyst pores.

Solid catalysts which contain networks of pores and create a large surface area will allow for diffusion of reacting molecules and for high catalytic activity. Selectivity, on the other hand, may or may not be affected adversely, depending on the nature of the reaction. As a rule, oxidation catalysts usually work best with large-diameter pores and low surface areas, whereas hydrogenation catalysts work better with high-surface-area catalysts.

As an example, the following common group of potential support materials are given in order of greatest porosity: mica, vermiculite, celite, kaolin, talc, pumice, sodium chloride, carborundum, sand, glass spheres.

Catalyst Preparation: The behavior of a solid catalyst depends not only on its chemical composition, but also on its method of preparation, both physical and chemical. Different methods of precipitation of reacting components, as for instance the mixing of cobalt salts and molybdenum salts to make cobalt molybdate, can lead to formation of high-surface-area gels or colloids, or conversely, crystals with a smaller surface area. Various temperatures of precipitation, as well as different rates of cooling, can lead to different-sized microproducts. Easily filtered agglomerated crystals can be created by high salt concentrations in a solution, whereas dispersed particles which may be difficult to filter can be caused by a too efficient washing technique.

The resultant powders, gels, or crystals are usually dried or activated in air or in a reducing atmosphere to form an active species of catalyst. Many vanadia catalysts are activated in air, while nickel-reduction catalysts are reduced in the presence of hydrogen.

Many commercial catalysts, after initial preparation, are formed into spheres or rods, by rolling, pelletization, or extrusion. The individual shapes are chosen either because of handling characteristics or by reason of optimum desirable behavior during the course of testing prior to commercial adaptation.

In pelletization, dry catalytic powders are mixed with binders and lubricants and compacted in a double-ended press. Inert and less costly ingredients may be added to the active powders to create separation of active sites as well as a less expensive product. On the other hand, the powders can be mixed with a suitable liquid to a doughlike consistency and extruded in spaghetti-like ribbons, which can be cut to desired lengths.

When coating a catalyst on a carrier, various techniques can be employed: (1) all ingredients can be charged to a heated rotating drum with a small amount of a suitable liquid, and rotated until the liquid evaporates, leaving the solid components on the surface of the support as well as in the internal structure; and (2) the catalyst can be impregnated by immersing a suitable carrier into a solution or colloidal suspension of the material. Other methods of deposition are spraying, electrodeposition, coprecipitation of carrier and catalyst, and vapor deposition.

Mechanisms in Catalyst Behavior: The concept of protonic acid (hydrogen-ion) and base (hydroxyl-ion) catalysis has been found applicable in discussions on heterogenous catalysis, wherein the mechanism is seen as the transfer of a proton from the catalyst to the reactant (acid catalysis) or from the reactant to the catalyst (base catalysis).

Intermediates may be of two types: *type I,* in which the reversible reaction that forms the intermediate is fast compared with further change into final products, so that the intermediate is always present in its equilibrium concentration, and *type II,* in which the intermediate is never present in appreciable concentration, and the velocity of the reaction is determined by the speed of formation of the complex.

A number of apparently different catalysts behave similarly because of their acidic nature, i.e., mineral acids, Friedel-Crafts catalysts, silica-alumina materials, and zeolite-cracking catalysts.

Because semiconductor theory has been used in attempting to explain catalytic action, a brief discussion of these concepts is presented.

Electronic Concepts: Solids, from an electronic point of view, can be divided into three groups: metals, semiconductors, and insulators. In semiconductors, which are usually oxides or sulfides, there are two kinds of carriers of electric current, electrons and holes. For example, in a covalently bonded carbon atom such as a diamond lattice, each carbon contributes four valence electrons to each bond and is tetrahedrally bonded also to four neighboring atoms. All the electrons are used to form the covalent bonds. Each carbon, therefore, is surrounded by eight electrons. In this situation no net flow of electrons through the solid is possible. If an electron is added through an interstitial position (from an outside source), it will be free to wander through the solid. It will move through the solid in an opposite direction from an applied electric field, and thus contribute to the electrical conductivity. Electrons which are thus not bound in the valence bonds and are free to move this way are called *conduction electrons*. If conduction electrons are produced in quantity, the insulator loses its insulation properties.

In the second case, the total number of electrons fails to match the number of available bonding sites; i.e., there are too few electrons, and hence there are gaps in the bonds. The missing bonding electron is described as a "hole," which is free to wander through the crystal. An electron in a bond adjacent to a hole can move into the empty position, leaving a vacancy behind it as it goes. As this process is repeated, with other electrons filling adjacent bonds, the net effect is for the hole to move through the crystal under the influence of an electric field, in a direction opposite to that of the conduction electron. The motion of the hole is opposite from that of the valence electrons. The hole thus behaves as a positively charged particle.

We note, then, that the conduction electron constitutes a local negative charge, and the hole a local positive charge, both of which are mobile.

These types can be written:

1. *n*-type semiconductors (or negatively charged dynamic electrons, where $n > p$)

2. *p*-type semiconductors (or positively charged dynamic holes, where $p > n$)

3. Intrinsic semiconductors (where $p = n$)

Positive and negative charges can be contributed to a semiconductor either by:

1. The addition of acceptor impurities (Ia), which contribute negative charges but whose atoms remain fixed or static in the lattice or by

2. Donor impurities (Id), which contribute a positive charge and whose atoms remain fixed in the crystal.

As a rule of thumb, Ia and Id can involve additions of impurities (guests) whose lattice radius can vary 10–15% from that of the host.

The impurity (doping agent) can be chosen on the basis of its lattice radius, electronegativity, ionization potential, and concentration.

In all solids, charge neutrality holds.

The ratio between *n* and *p* is a thermodynamic quantity which will determine the direction of flow of the electrons (from absorbed gases to solid, or vice versa). This ratio indicates the electrochemical potential and is referred to as *Fermi energy*.

Modifying Catalyst Behavior with Impurities

(*Dopants*): Since an optimum catalyst represents a balance between *n* and *p* relationships, both in support and catalyst, a standard relationship can first be attempted, and then modifications in this standard made in a systematic manner based on electron density characteristics.

It is possible that in some cases larger quantities of promoters or doping agents would be added than the usual 1–5%. In these cases the larger quantities can possibly give rise to a different form of crystalline structure than the standard (geometric modification), and hence the resultant catalyst could behave in a manner different than anticipated. Crystallographic techniques of analysis, as well as actual performance data, would show these variations.

As previously stated, it is possible to divide heterogeneous catalysts into groups based on their electronic properties. For example, metals (conductors) can be used in hydrogenation, dehydrogenation, and hydrogenolysis reactions; metal oxides or their sulfides (semiconductors) can be used in oxidation, reduction, dehydrogenation, or cyclization reactions; and salts or acid-site (insulators) catalysts can be used in cracking, dehydration, isomerization, polymerization, alkylation, dehalogenation, halogenation, and hydrogen transfer reactions.

A classical example of the effect of different catalysts on the same reagent is the reaction of ethyl alcohol, first over copper (a conductor) and then over alumina (insulator). In the first case either acetaldehyde or ethyl acetate plus hydrogen is obtained; in the latter case ethylene plus water or diethyl ether plus water is recovered.

Semiconductivity can also be produced by impurities as well as by nonstoichiometry, which provides a means of controlling this type of behavior. Examples are:

1. Solutions of small amounts of lithium oxide in nickel oxide, thereby substituting Li^+ ions for Ni^{++} in the lattice and producing controlled *p*-type semiconductivity by creating Ni^{3+}

2. Dissolving a small amount of alumina in zinc oxide so that electrons can be furnished for *n*-type semiconductivity

The general theory of catalysis by semiconductors has been discussed by F. F. Vol'kenshtein, *Adv. Catalysis,* vol. 12, pp. 189–264, 1960. Another excellent discussion is that in P. G. Ashmore, "Catalysis and Inhibition of Chemical Reactions," Butterworth, London, 1963, from which Table C-18 is drawn. See also Tables C-19 and C-20.

Catalytic processes are described in several

TABLE C-18. *Behavior of Nonstoichiometric Semiconductor Oxides*

	n Type	p Type
Oxides with interstitial ions	ZnO,CdO	UO_2
Oxides with lattice vacancies	TiO, ThO_2, CeO_2	Cu_2O, NiO, FeO
Effect of adding $(M^+)_2O$*	Decreases conductivity	Increases conductivity
Effect of adding $(M^{3+})_2O_3$*	Increases conductivity	Decreases conductivity
Effect of adsorption of O_2, N_2O	Decreases conductivity	Increases conductivity
Effect of adsorption of H_2, CO	Increases conductivity	Decreases conductivity
Mode of conduction	Electrons	Positive holes

* = cation.

SOURCE: P. G. Ashmore, "Catalysis and Inhibition of Chemical Reactions," Butterworth, London, 1963 (with permission of the publisher).

other places in this volume, including **Acetate fibers; Acrylonitrile; Air pollution; Alkylation; Cellulose ester plastics, organic; Chlorofluorocarbons; Elastomers; Fluid catalytic cracking; Formaldehyde; Hydrocracking; Hydrogen; Hydrogen fluoride; Hydrotreating; Ion-exchange resins; Iron oxides, synthetic; Isomerization; Lithium; Molybdenum; Nickel; Palladium; Plat-**

TABLE C-19. *Catalytic Properties of Some Individual Metal Oxides (on a Relative Basis)* in the Heterogeneous Catalytic Oxidation of Hydrocarbons*

Oxide	Absorption of Oxygen	Catalytic Activity for Oxidation of Hydrocarbons	Selectivity for Oxidation of Hydrocarbons
p type: Cu_2O NiO CoO	High	High	Generally low
Insulators: Al_2O_3 Ag_2O Cr_2O_3 CuO	Very low	Very low	
n type: Fe_2O_3 MoO_3 TiO_2 V_2O_5 ZnO	Low	Low	Generally high

*Properties shown can vary with type of hydrocarbon and with differences in temperature.

TABLE C-20. *Methods of Measurement of Some Catalyst Properties*

Property	Method of Measurement
Surface area	Physical absorption of gases such as argon, nitrogen, or krypton close to their boiling points
Porosity	Mercury penetration for large pores >500 Å; analysis of shape of adsorption or desorption isotherms in region of capillary condensation—for small pores <300 Å
Particle size and shape	Electron microscopy and x-ray diffraction
Phase identification	Optical microscopy; x-ray diffraction; differential thermal analysis
Valence of transition metal ions	Infrared absorption spectroscopy; magnetic susceptibility; electron spin resonance

inum; **Rare-earth elements and metals; Reforming; Rhenium; Rubber, natural; Sulfuric acid;** and **Tin.** Also consult the Subject Index.

REFERENCES

Anderson, Robert B. (ed.): "Experimental Methods in Catalytic Research," Academic, New York, 1968.

Ashmore, P. G.: "Catalysis and Inhibition of Chemical Reactions," Butterworth, London, 1963.

Balandin, A. A., et al.: "Catalysis and Chemical Kinetics," Academic, New York, 1964.

Bond, G. C.: "Catalysis by Metals," Academic, New York, 1962.

Frankenburg, W. G., V. I. Komarewsky, and E. K. Rideal (eds.): "Advances in Catalysis and Related Subjects" (15 vols., 1948–1964), Academic, New York.

Germain, J. E.: "Catalytic Conversion of Hydrocarbons," Academic, New York, 1969.

Jones, Mark M.: "Ligand Reactivity and Catalysis," Academic, New York, 1968.

Krylov, Oleg V.: "Catalysis by Nonmetals," Academic, New York, 1970.

Rylander, Paul N.: "Catalytic Hydrogenation over Platinum Metals," Academic, New York, 1967.

Thomas, Charles L.: "Catalytic Processes and Proven Catalysts," Academic, New York, 1970.

Topchiev, A. V., S. V. Zavgorodnii, and Y. M. Paushkin: "Boron Fluoride and Its Compounds as Catalysts in Organic Chemistry," Pergamon, Oxford, 1959.

Vol'kenshtein, F. F.: "The Electronic Theory of Catalysis on Semiconductors," Pergamon, Oxford, 1963.

Weissberger, Arnold (ed.): "Catalytic, Photochemical, and Electrolytic Reactions," 2d ed., Wiley, New York, 1956.

—Robert S. Barker, *Catalyst Development Corporation, Little Ferry, N.J.*

Catalytic Cracking   (See **Cracking; Fluid catalytic cracking;** and **Hydrocracking.**)

Cation-exchange Resins   (See **Ion-exchange resins;** and **Water treatment.**)

Cationic Dyes   [See **Dyestuffs and dyeing (textiles).**]

Cationic Surfactants   (See **Detergents;** and **Surfactants.**)

Caustic Potash   Caustic potash is the usual commercial name for potassium hydroxide, KOH, and its solutions. Caustic potash is produced by the electrolysis of KCl brines either in mercury cells or diaphragm cells in a manner analogous to caustic soda. See also **Caustic soda.** Production (United States) for 1970 is estimated at 188,000 short tons.

The principal use for caustic potash is in the manufacture of soft soaps and liquid detergents. Tetrapotassium pyrophosphate is produced from KOH and is used as a *building agent* in both light- and heavy-duty liquid detergents. The detergent use of caustic potash accounts for about 50% of the total production. Other uses are in textile dyeing, metal treating, and in the production of chemicals which require the greater solubility imparted by potassium as compared with sodium. Essential information for safe handling and use of caustic potash is given in Manufacturing Chemists Association (Washington, D.C.) *MCA Chem. Safety Data Sheet* SD-10. See also **Potassium.**
—Herbert S. Hopkins, *Olin Corporation, Stamford, Conn.*

Caustic Soda   Caustic soda is the usual commercial name for sodium hydroxide, NaOH, and its solutions. Selected physical constants of solid NaOH are given in Table C-21. Almost all caustic soda produced for commerce is a coproduct with chlorine (q.v.). Caustic soda is a major chemical raw material finding extensive use in chemical processing; manufacture of pulp and paper; production of soaps and detergents; processing of bauxite to produce alumina; rayon manufacture; textile processing; and other important segments of industry.

Large consumers of caustic soda usually procure and store it as a *solution.* Two strengths are commonly sold, namely, 50 and 73% NaOH concentration. Most growth in recent years has been in production of the 50% solution. The advantage of the 73% NaOH over the 50% NaOH solution

TABLE C-21. *Physical Constants of Solid Sodium Hydroxide*

Chemical formula	NaOH
Molecular weight	39.999
Freezing point (melting point):	
°F.	608
°C.	320
Latent heat of fusion:	
Btu/lb	76.5
gm-cal/gm	42.5
Heat capacity:	
Btu/(lb)(°F), at 77°F	0.480
kcal/(mole)(°C), at 25°C	19.2
Heat of formation:	
Btu/lb, at 77°F	−4590
kcal/mole, at 25°C	−101.99
Free energy of formation:	
Btu/lb, at 77°F	−4077
kcal/mole, at 25°C	−90.60
Specific gravity, at 70°F (21.1°C)	2.130

is the savings on freight costs (not having to ship another 23% of the product in the form of water). Since 73% caustic soda starts to freeze at about 62°C (143.6°F), it usually is diluted to some lower strength (frequently 50%) for storage. Large amounts of heat are generated in dilution, requiring heat exchange before the caustic soda can be put into storage tanks. Accordingly, most users prefer the 50% product for ease of handling when available. As new producing points have come into operation, freight savings have become a less important consideration. There is no uniform, industry-accepted nomenclature for grades. Strengths and commonly available shipping containers are listed in Table C-22.

Most producers of caustic soda sell two liquid grades: rayon and commercial grade. A variety of grades of dry or anhydrous caustic soda are produced, frequently tailored to a specific end use. Rayon-grade caustic soda was developed to provide a higher-quality, more uniform product to meet the needs of the rayon industry. Table C-23 shows typical specifications for several grades.

The expression Na_2O is frequently used to characterize caustic soda. Pure NaOH contains 77.47% Na_2O. However, when caustic soda first became an item of commerce, the purest material obtainable had 76% Na_2O. This became a standard for uniform billing and is used today for billing liquid forms of caustic soda. Dry forms are sold on a net-weight basis. The Na_2O content is frequently used to compare the alkalinity of different materials. For example, sodium carbon-

TABLE C-22. *Commercial Grades of Caustic Soda—Strengths and Shipping Containers*

	Form	Type of Containers	Contents	Equivalent Gallons of a 50% Solution	NaOH Content (Dry Basis)
Liquid......	50%	Tank cars	8,000; 10,000; 16,000 gal	8,000; 10,000; 16,000	51,000; 63,800; 102,000 lb
		Tank trucks	3,000; 3,500 gal	3,000; 3,500	19,100; 22,000 lb
		Barges	94,000–375,000 gal	94,000–375,000	300–1,200 tons
	73%	Tank cars	8,000; 10,000 gal	13,100; 16,400	82,400; 103,000 lb
		Tank trucks	3,000	4,900	30,900
Anhydrous ...	Solid	Steel drum	700 lb net	110	700 lb
	Flake	Steel drum	100; 400 lb net	15.7; 62.7	100; 400 lb
		Bags	50; 100 lb	7.9; 15.7	50; 100 lb
	Granular	Steel drum	400 lb net	62.7	400 lb
	Ground	Steel drum	400 lb net	62.7	400 lb

ate is sold as a material containing at least 58.0% Na$_2$O (theoretical, 58.48%).

Caustic soda is a heavy-tonnage chemical. Production in the United States in 1970 was 10,074,000 tons (liquid on 100% NaOH basis). Since 1950

TABLE C-23. *Specifications for Several Grades of Caustic Soda, Percent*

	50% Rayon	50% Commercial	73% Commercial	Flake
NaOH	50–52	49–52	72.0–74.0	98.4 min.
Na$_2$O	38.79–40.32	38.0–40.32	55.83–57.4	76.2 min.
Na$_2$CO$_3$	0.06	0.20	0.4	1.0 max.
NaCl	0.005	1.0	0.45	0.5
NaClO$_3$	0.0005	0.001	0.002	0.0001
Fe	0.0002	0.0004	0.001	0.003
Hg	0.00005	0.00005	0.00007	0.0001
Na$_2$SO$_4$	0.002	0.15	0.30	0.1
SiO$_2$	0.002	0.04	0.05	0.1
Al$_2$O$_3$	0.0006	0.004	0.007	0.001
CaO	0.002	0.006	0.006	0.005
MgO	0.0002	0.001	0.002	0.002
Mn	0.00004	0.0001	0.0001	0.0005
Ni	0.00003	0.0001	0.0002	0.0005
Cu	0.00003	0.0002	0.0002	0.0005

NOTES:

1. The above concentrations are percentages on an "as is" basis. Some producers report concentrations on the amount of NaOH as determined by assay.

2. These are typical specifications, not typical analyses. The actual concentration of impurities for any grade may be substantially lower than shown. Equally, some producers may not include all these constituents in their specifications. The reader is cautioned not to try to establish a purchasing specification from these data.

3. "Mercury-cell grade" is sometimes referred to in the literature as synonymous with "rayon grade." This is improper usage. Mercury-cell caustic soda is usually lower in impurities than diaphragm-cell caustic soda, and most rayon-grade caustic soda is produced by the mercury-cell process. Rayon-grade caustic soda can also be produced by purifying diaphragm-cell caustic soda.

4. Some of the elements shown here are mutually exclusive, depending on the process used.

its growth rate has been at about 7%/yr. This period has marked a decline in merchant production by the lime soda process.

The decade 1960–1969 saw a marked increase in demand for chlorine. Much of the demand for chlorine was to introduce a reactive position into a molecule. The chlorine atom would then subsequently be removed. Chlorinated insecticides also consumed large quantities. Production of vinyl chloride increased rapidly. As a result substantial quantities of caustic soda were available to move in the marketplace for whatever end use was available.

Aggressive marketing by caustic soda producers kept the supply picture in balance. They sought out end uses traditionally served by soda ash. By stressing ease of handling and by competitive pricing they were able to replace substantial tonnages of soda ash in the pulp and paper industry and even in the glass industry. All forecasts of the middle and late 1960s foresaw ever-mounting surpluses of caustic soda.

The situation rapidly reversed in 1970. The substitution of direct oxidation for chlorination in the manufacture of ethylene oxide freed large quantities of chlorine. More dramatic, however, was the fall from favor of chlorinated insecticides such as DDT. Very quickly the chloralkali industry switched from a caustic soda long position to a caustic soda short position. With reduced chlorine demand it was impossible for the industry to supply enough caustic soda. During periods when caustic soda was long, it was possible to store large quantities in large atmospheric-pressure tanks. Such tanks were available at many terminal locations. However, since chlorine requires pressure tanks, especially designed and installed, there was no way to readily expand the available storage. A number of producers were forced to cut back production.

Uses: Caustic soda is used widely in many segments of industry. Many of the identifiable uses, such as for bottle washing, consume only small fractions of the total produced, but still account for substantial tonnages (48,000 tons, or 0.5% of total produced 100% basis in 1967). While it is customary to show chemical processing as the single largest end use, many of the individual chemicals which compose the overall group require less than this.

Because caustic soda (as well as its coproduct chlorine) is a low-cost, high-tonnage commodity, there is a tendency for chloralkali plants to be located close to large identifiable users. The pulp and paper industry is a favorite for this purpose. In some cases the pulp mill itself produces its own chlorine and caustic, but merchant producers usually try to convince the pulp mill that they can achieve economies of scale which the pulp mill cannot. Bleached-kraft pulp mills are particularly attractive to producers because they consume chlorine and caustic soda in the approximate ratio in which they are produced. If the nature of the terrain permits, chlorine and caustic soda will be delivered by pipeline. If a pipeline is used, caustic will be delivered at about 20% to take advantage of the lowered freezing point (see phase diagram, Fig. C-3). The use of caustic soda in pulp and paper production is described under **Pulp (wood) production and processing.**

In textile processing, caustic soda is used for mercerization of cotton. When caustic soda is applied to cotton and then washed out while the cloth is under tension, the crystalline orientation of the fiber is changed. This results in a fabric having improved luster, greater strength, better dyeability, and in some cases greater dimensional stability. This use for caustic will decline but not be displaced as synthetic fibers take more of the cotton market.

In petroleum refining, caustic soda is used to remove sulfur compounds, such as H_2S and mercaptans. In food processing, caustic soda is used to peel fruits and vegetables, to process olives, and to refine vegetable oil. In alumina production, caustic soda is used to dissolve bauxite as a first step in the production of aluminum. In glass production, sodium hydroxide can be used to replace part of the soda ash as a source of Na_2O. The foregoing examples are not intended to be complete and do not cover very large uses of NaOH in the production of specific chemicals.

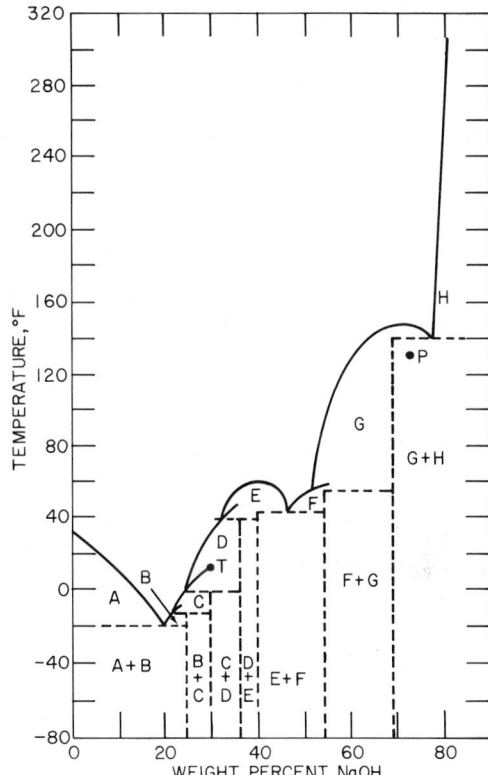

Key to Letters	Maximums		
	Temperature		
	°C	°F	Concentration, % Wt
A Ice	0	32	0.00
B NaOH·7H₂O	−23.3	−10	24.08
C NaOH·5H₂O	−12.2	10	30.75
D NaOH·4H₂O	7.78	46	35.70
E NaOH·3½H₂O	15.6	60	38.79
F NaOH·2H₂O	12.8	55	52.61
G NaOH·1H₂O	64.4	148	68.94
H NaOH	321	610	100.00
Transition points:			
A − B	−28	−18.4	19.09*
B − C	24.4	12.0	21.88
C − D	−18	−00.4	24.70
D − E	3.6	38.5	31.32
E − F	5.56	42.0	45.89*
F − G	12.1	54.0	50.50
G − H	60	140.0	74.76*

*Eutectic points.

Fig. C-3. Phase diagram for caustic soda solutions.

Manufacture: Almost all caustic soda manufactured today is produced by the electrolysis of NaCl brines.[1] Since chlorine is produced as a coproduct, the technology of caustic soda is intimately bound with that for Cl$_2$. See also **Chlorine.** About 65% of all NaOH produced is made by the diaphragm-cell process; the balance is largely produced by mercury cells.

The mercury-cell process requires solid salt—mined, solar, or from underground salt domes—for resaturation of the depleted process brine because essentially no water is purged in processing. The diaphragm process, however, produces a dilute caustic in a partially depleted brine. This product, commonly called *cell liquor,* is concentrated in evaporators which purge water from the system. See also **Evaporation.** Therefore, the source of salt may be a brine well, in addition to solid salt. The concentration of diaphragm-cell caustic soda precipitates a very pure grade of salt. A useful mode of operation has been to use mercury cells in conjunction with diaphragm cells, the salt from the diaphragm cells being used as feed to the mercury cells.

A necessary first step in the production of caustic soda is the purification of the brine. In the case of mercury-cell production, depleted brine from the cells, saturated with chlorine, is first acidified with hydrochloric acid to a pH of 2. This converts the hypochlorous acid and chlorate

to chlorine. Brine dechlorination is carried out in vacuum or air-blowing equipment of one or two stages. The recovered chlorine is returned to the chlorine-collection system. Caustic soda is then added to raise the pH to 9–10. Salt is added to saturate the brine. In some brine systems barium carbonate is added to precipitate the sulfate present. One patented system permits the brine to be saturated with calcium sulfate, and sulfate removal is not required.

For diaphragm cells, brine purification is necessary to prevent plugging of the diaphragm by precipitation of the metal hydroxides in the diaphragm pores. Also, impurities may be transmitted to the caustic soda. The pH is adjusted to about 10 to precipitate insoluble metal hydroxides. Soda ash is added to precipitate calcium impurities as calcium carbonate.

At the heart of the manufacture of caustic soda is the cell used. Mercury and diaphragm cells are quite different in construction and operation, and the processing of caustic soda from them differs. The mercury cell is described under **Chlorine.** Described herein are the diaphragm cell and the processing of caustic soda from both processes (Fig. C-4).

Most diaphragm cells in current use are similar

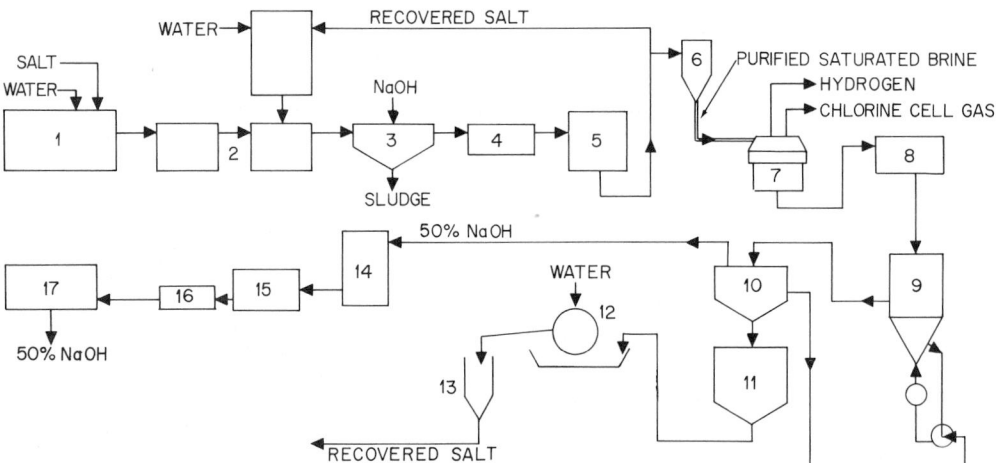

Fig. C-4. Simplified flowchart of brine treatment and production of 50% sodium hydroxide by diaphragm-cell method. Salt and water are mixed and stored as a solution in tank (1). The brine then goes to treatment tanks (2) to which fresh water and recovered salt are added. Sodium hydroxide is introduced into settling tank (3) from which sludge is removed as bottoms. The treated brine then is filtered (4) and passed to brine storage tank (5). Resaturated brine from (6) is passed to the diaphragm cell (7) where hydrogen and chlorine cell gas are produced, as well as cell liquor [latter goes to storage tank (8)]. The liquor is concentrated in a triple-effect evaporator (9). Salt is separated out in (10). Bottoms from (10) pass to slurry tank (11) and then to filters (12), where the recovered salt also is washed with water, collected in (13), and sent to storage. The 50% NaOH solution from (10) is cooled (14), further treated (15), and filtered (16) before going to storage (17).

to the Hooker type S cell. This cell consists of three parts. The bottom is a concrete base. Graphite anodes project vertically upward from the base. They are held in place in electrical contact with a copper connector by a lead casting. The lead is sealed with an asphaltic mastic. (Metal anodes have a different construction.) The cathode is a steel frame with flanged top and bottom. Inside the frame and welded to it are "fingers" which form an integral catholyte compartment. The fingers are a steel mesh and are so located that, with the cathode in place, they interleave the anodes. An asbestos diaphragm is applied to the fingers by immersing the cathode in an asbestos slurry and applying a vacuum. The cell is assembled by placing the cathode assembly on the base. The heavy concrete top is then placed on top of the cathode assembly and holds it in place.

Purified brine is fed to the cell through an opening in the top. The level in the cell adjusts itself so that the hydraulics of the cell are in balance with the withdrawal rate of cell liquor and the electrical input. As impurities deposit in the diaphragm, the diffusion rate through the diaphragm decreases and the level in the cell rises.

Chlorine formed at the anodes bubbles through the brine and is withdrawn through an outlet at the top of the cell. Cell liquor containing about 11.5% NaOH and 15% NaCl overflows through an adjustable overflow pipe which maintains the level in the cathode compartment. Such cells typically operate up to 55,000 A (up to 80,000 with metal anodes).

Cell liquor is concentrated in triple-effect evaporators. These today tend to be vertical-tube, forced circulation. As the concentration increases, salt is precipitated. Salt can be continuously removed by centrifugation as evaporation proceeds. When the product reaches 50%, it is allowed to cool and more salt is precipitated. Under the conditions normally prevailing, 50% caustic soda made by the diaphragm process contains 1% NaCl. This is suitable for most purposes.

If diaphragm-cell caustic is to be sold as rayon grade, it must be further processed. This can be done by an extraction with ammonia, which removes salt and chlorates approaching the level found in caustic soda produced by the mercury-cell process.

The precipitated salt is washed free of caustic soda. Some salt may be purged in order to control the buildup of sulfates. The balance of the salt is recycled to make fresh brine.

Diaphragm-cell caustic soda as produced is sometimes blue-colored due to iron complexes which may form. The addition of a small amount of chlorine will correct this. Finally, the caustic soda is filtered and put in storage ready for shipment.

In the mercury-cell process, caustic soda is formed outside the electrolyzer. The sodium amalgam from the primary cell (electrolyzer)—containing 0.2–0.4% Na in the mercury—flows into the decomposer (secondary cell). This is typically a steel tower containing graphite packing. Deionized water is fed to the bottom of the tower at such a rate that hydrogen and 50% NaOH is made directly from the reaction of the sodium in the amalgam with water. Hydrogen is collected from the top of the decomposer. It is passed through a demister and cooled. Entrained mercury is recovered in this step. Some hydrogen may be burned with chlorine to produce HCl for brine treatment. The balance may be used for further chemical synthesis (e.g., to make ammonia) or may be burned to furnish power.

The caustic soda from the decomposer is cooled from 120 to 50°C. It is then filtered to remove graphite particles and small traces of dispersed mercury. The filter is usually a porous-stone tube on which is deposited a filter aid. The filter aid is usually a two-part material, a precoat with either diatomaceous earth or cellulose and a final coat of activated carbon. The activated carbon reduces mercury to a level meeting the requirements of Food Chemicals Codex for food-grade sodium hydroxide solution: "Mercury, not more than 1 ppm (0.0001 percent) calculated on the basis of NaOH determined in the *Assay*."

Materials of Construction: Caustic soda normally is handled in carbon steel. Where prevention of iron pickup is important, nickel is used. Stainless steel shows little advantage over carbon steel insofar as iron pickup is concerned, and its use normally is not justified. Storage tanks normally are unlined, but if avoidance of iron pickup is important, the tanks may be lined with an epoxyphenolic or a neoprene latex coating. The pickup of iron increases greatly with increased temperature, and also with linear velocity across steel surfaces. Storage temperatures (in bare steel) should be kept below 65°C, but should be sufficiently high to prevent freezing. The start-to-freeze temperature for 50% NaOH is 12°C. Unnecessary recirculation of caustic soda through pipelines should be avoided.

Safety in Handling: Outer clothing and hats

should be made of cotton or of a suitable synthetic fabric. Caustic soda destroys wool and leather. Footwear and gloves should be made of rubber. The following equipment and clothing should be used when handling caustic soda: (1) chemical safety goggles or face shield, (2) wide-brimmed hat, (3) shirt with snug-fitting sleeves and collar which can be buttoned at the neck, (4) trousers with bottoms that extend over boot tops, (5) rubber boots with a safety toe if required by plant practice, and (6) rubber or polyethylene apron and gloves.

A respirator of a type approved by the U.S. Bureau of Mines for nuisance dusts and mists should be used whenever flake or ground caustic soda is handled. Respirator-hood units are available for use when the nature of the exposure warrants. Additional precautionary information is contained in Ref. 2.

REFERENCES

1. Sconce, James S.: "Chlorine: Its Manufacture, Properties and Uses," chaps. 5 and 6, ACS Monograph 154, Van Nostrand, New York, 1962.
2. Manufacturing Chemists Association: "Caustic Soda," *MCA Chem. Safety Data Sheet* SD-9, Washington, D.C., 1968.

—Herbert S. Hopkins, *Olin Corporation, Stamford, Conn.*

Causticizing [See **Pulp (wood) production and processing.**]

Cell, Living (See **Peptides and proteins.**)

Celluloid (See **Resins.**)

Cellulose The relative composition of cellulose may be represented by the formula $(C_6H_{10}O_5)_n$, although this is an oversimplification because in natural materials, such as woods, fibers, and films, the cellulose present usually is combined with other substances, such as fats and gums. Cellulose is confined almost exclusively to plants, and accounts for about 30% of all vegetable matter. Cellulose is the primary substance of which the walls of vegetable cells are constructed. The name stems from the Latin *cellula* (little cell). Relatively pure cellulose can be obtained easily from cotton fibers (90% cellulose) and flax fibers. Very small amounts of cellulose are found in insects, and none in animal tissues. Enzymes and digestive juices present in animal systems do not appear to attack cellulose; hence ingestion by man is relatively limited. By other biological processes (such as amoeboid protozoa present in the digestive tract), herbivora and insects digest and absorb some cellulose.

Structurally, cellulose may be considered a polysaccharide of glucose. See also **Carbohydrates.** Although insoluble in water, cellulosic structures normally are capable of soaking up relatively large quantities of water within their structure. Cellulose is insoluble in all ordinary solvents, but will dissolve in Schweitzer's reagent (an ammoniacal solution of cupric oxide). Mercerization is a process in which cellulose is treated with an alkali. When cellulose, treated in this fashion, is further exposed to CS_2, a cellulose xanthate solution results. These are reactions of which advantage is taken in the manufacture of rayon. See also **Rayon.** The action of acetic anhydride, in the presence of H_2SO_4, produces cellulose acetates, the basis for a whole line of synthetic materials. See also **Acetate fibers; Cellulose ester plastics, organic;** and **Ethyl cellulose.** Nitrocelluloses are produced by the action of HNO_3 and H_2SO_4 on cellulose, yielding compounds which are highly flammable and explosive. See also **Explosives.**

Cellulose is most important as a raw material on a heavy-tonnage basis in the production of pulp and paper. See also **Pulp (wood) production and processing.** Numerous other examples of the use of cellulose as a raw material for chemical and process technology are given throughout this volume. Consult the Subject Index.

Cellulose Ester Plastics, Organic The cellulosics, in contrast with most plastics, are not made from synthetic polymers. Rather, they are derivatives of the *natural polymer cellulose.* Cellulose, with its many hydroxyl groups, can react with organic reagents, such as acids, anhydrides, and acid chlorides, to form organic esters. Certain of these esters are the bases for *organic cellulose ester plastics.*

The first reported organic ester of cellulose was cellulose acetate, prepared by Schutzenberger (1865) by heating cotton and acetic anhydride to about 180°C in a sealed tube until the cotton dissolved. Franchimont (1879) accomplished this reaction at a lower temperature with the aid of H_2SO_4 as a catalyst. The product in both cases was very nearly the triester. Miles (1903) first described partially hydrolyzed (generally called *secondary*) cellulose acetate and distinguished it from the triacetate by its solubility in acetone. The solubility of secondary acetate in a low-cost and relatively nontoxic solvent such as acetone contributed greatly to the development and commercialization of this material.

Cellulose esters of the two-, three-, and four-carbon acids are prepared readily by the cellulose anhydride reaction. The acetate ester and the mixed acetate butyrate and acetate propionate esters are manufactured and used in large amounts. Esters of the higher acids require different synthesis techniques and tend to be prohibitively costly except as specialty products. An example is cellulose acetate phthalate, which is used as an enteric coating on pills.

Most commercial preparations of cellulose esters basically follow the earlier methods of Franchimont and Miles, namely, esterification with H_2SO_4 as a catalyst, followed by hydrolysis. The basic raw materials for preparing cellulose acetate and cellulose triacetate (the latter not used as a base for plastics, essentially because its softening temperature exceeds its decomposition temperature) are cellulose (highly purified cotton linters or wood pulp), acetic acid, acetic anhydride, and H_2SO_4. For the mixed ester cellulose acetate butyrate (CAB), butyric anhydride is added to the reactor; for the mixed ester cellulose acetate propionate (CAP), propionic anhydride is added.

Although we are concerned here with these esters as ingredients for plastic manufacture, it should be mentioned that they also find application in solution processes, as in the preparation of films, fibers, coatings, lacquers, and adhesives.

The properties of these esters are summarized in Table C-24. Generally, they are characterized by acyl content in weight percent and viscosity

in seconds. The viscosity measurement is obtained by timing the fall of a steel ball through a solution of the cellulose ester in accordance with ASTM Test Method D 1343. Low-viscosity esters generally are used in solution processes, whereas the high-viscosity esters are used for the production of plastics.

Manufacture: The appropriate ester is blended with plasticizer and other additives, such as stabilizers, ultraviolet inhibitors, dyes, and pigments, commonly in a sigma-blade mixer. The mixture is heated to its softening temperature and kneaded until it is homogeneous. Kneading is accomplished on hot milling rolls, in a compounding extruder, or in a Banbury mixer. The resulting molten mass of plastic is formed into small rods or strips, which then are cut into cylindrical or cubical pellets of about $\frac{1}{8}$ in. in size. Butyrate or propionate pellets may be ground into powder for use in rotational molding or powder-coating processes. Acetate currently is not used in powder form.

Properties: Depending upon plasticizer content, cellulose ester plastics range from soft, extremely tough materials to hard, strong, stiff compositions that still retain a considerable degree of toughness over a wide range of temperatures. These materials are basically transparent and virtually colorless, permitting their manufacture in almost any desired transparent, translucent, or opaque color. They are resistant to water and aqueous-salt solutions, but are attacked by aqueous solutions that are strongly acidic or basic. They resist several types of organic solvents, such as ethers and ali-

TABLE C-24. *Properties of Typical Commercial Cellulose Esters*

	Acyl Content, wt %			Hydroxyl Content, wt %	Viscosity Range*		Melting Range, °C	Typical Uses
	Acetyl	Butyryl	Propionyl		Seconds	Poises		
Cellulose acetate:								
Low acetyl, high viscosity . . .	39.4	. . .	. . .	3.7	22–38	83–144	235–255	Plastics
Medium acetyl, low viscosity. .	39.8	. . .	. . .	3.6	2–4	8–15	230–250	Lacquers
Medium acetyl, medium viscosity	39.8	. . .	. . .	3.4	8–13	30–49	230–250	Plastics
High acetyl	39.9	. . .	. . .	3.2	17–35	64–132	240–260	Plastics
Commercial triacetate	43.2	. . .	. . .	. . .	100–200	378–756	290–300	Film and fibers
Cellulose acetate butyrate:								
Low butyryl	29.5	17	. . .	1.3	22–30	81–115	230–240	Coatings
Low viscosity	13.5	37	. . .	1.5	0.3–0.5	1.15–2.05	155–165	Lacquers
High viscosity	13.5	37	. . .	2.0	17–28	64–105	195–205	Plastics
High butyryl.	5.0	49	. . .	0.9	4–6	15–23	165–175	Hot-melt coatings
Cellulose propionate:								
Low viscosity	2.5	. . .	45	2.8	0.3–0.5	1.15–2.05	200–210	Printing inks, coatings
High viscosity	2.5	. . .	45	2.1	17–28	72–95	200–210	Plastics

*ASTM D 817–65 (Formula·A) and D 1343–56.

phatic hydrocarbons, but they are dissolved or swollen by strongly polar liquid organic compounds, such as aromatic hydrocarbons, chlorinated hydrocarbons, ketones, and esters. The susceptibility of the plastics to attack decreases as the molecular weight of the attacking compound increases. The cellulose ester plastics are available in formulations that meet the requirements of the U.S. Food and Drug Administration for use in contact with food.

The concentration of plasticizer in a cellulose ester plastic determines its *flow temperature,* which in turn determines its *flow designation,* as defined by ASTM Test Method D 569. Flow designations range from various degrees of hardness through medium to various degrees of softness. At any given flow designation, the characteristics of the plastic will vary somewhat with the identity of the plasticizer used. Some plasticizers, for example, give very hard materials; some give very low water absorption; some permit unusual ease of processing. The flow temperature is used for quality control, and the corresponding flow designation is an important part of procurement specifications.

The three plastics described here are classified by the Underwriters' Laboratories as *slow-burning,* but acetate is available in self-extinguishing formulations and in formulations that burn at a rate about one-half that of general-purpose acetates.

Comparison of Acetate, Butyrate, and Propionate: Although these plastics resemble each other in many ways, there are significant differences among them. Acetate is the least costly of the three, but it also has the highest specific gravity. Thus some of the price advantage is negated by the need to use greater weight quantities. Butyrate and propionate generally are easier to process than acetate—a factor which also subtracts from the cost advantage of acetate. Acetate is available in very hard flows, and thus can be obtained with higher stiffness, hardness, and tensile strength than the other two plastics. Butyrate is available in the softest flows, and thus can be obtained in the toughest and easiest formulations to process. Butyrate and propionate generally are considered to be tougher than acetate, even though in some instances the measured impact strengths may be similar. Butyrate retains its toughness better than propionate at low temperatures. Butyrate and propionate use higher-boiling, less water-soluble plasticizers, which leads to better retention of plasticizer by butyrate and propionate when exposed to elevated temperatures or to the leaching action of water. This leads, in turn, to better

TABLE C-25. *Abridged Summary of Properties of Cellulosic Plastics*

	Acetate*	Butyrate*	Propionate*
Flexural strength at yield, psi	2,200–11,500	1,800–9,250	2,900–9,300
Hardness, Rockwell R scale:†			
Injection-molded specimens	7–122	17–112	47–113
Compression-molded specimens	27–116	59–113	53–117
Impact strength, Izod			
Injection-molded specimens:			
Ft-lb/in. of notch, at 23°C	1.0–9.0	2.2–11.0	1.4–10.7
Ft-lb/in. of notch, at −40°C	0.4–1.3	1.0–3.4	0.7–2.4
Compression-molded specimens:			
Ft-lb/in. of notch, at 23°C	0.5–4.1	0.8–6.3	0.9–6.4
Ft-lb/in. of notch, at −40°C	0.3–0.7	0.4–1.7	0.5–1.4
Stiffness in flexure, 10^5 psi	1.10–2.65	0.55–1.85	0.85–2.1
Tensile strength at fracture, psi:			
At 23°C	3,000–9,000	2,600–6,900	2,300–7,300
70°C	800–5,400	1,100–5,700	1,400–6,300
Tensile strength at yield, psi	2,200–7,400	1,400–6,200	1,700–6,800
Deflection temperature:			
At 264 psi fiber stress, °C	44–91	45–94	48–109
66 psi fiber stress, °C	53–98	54–108	64–121
Flow designation (as defined in ASTM Test Method D 569)	S2-H4	S2-H4	MH-H6

*Tenite.

†Rockwell hardness values shown for acetate of S flow and harder, for butyrate of MS flow and harder, and for propionate of H flow and harder.

permanence characteristics, i.e., smaller changes in dimensions and other properties with time.

Processing of Cellulose Ester Plastics: These materials are versatile and can be processed by almost any hot-processing technique used for thermoplastics. Injection molding and extrusion are the principal techniques used with all three plastics. Blow molding also is possible. Butyrate and propionate powder are used in fluidized-bed and electrostatic coating processes, as well as in the rotational molding process.

Any of these organic cellulose ester plastics can be cemented to itself, and propionate and butyrate generally can be cemented to each other, with appropriate solvents. Solvents most commonly used are mixtures of an active solvent, such as acetone or methyl ethyl ketone, a latent solvent, such as alcohol, and a diluent, such as benzene.

Representative properties of three resins are summarized in Table C-25. Because of the very large range of plasticizers and additives that may be used with these materials, coupled with the large range of processing methods, many properties can be obtained beyond those indicated.

—Technical Staff, *Plastics Division, Eastman Chemical Products, Inc., Kingsport, Tenn.*

Cellulose, Ethyl   (See **Ethyl cellulose.**)

Cement   Cement ranks with petroleum and steel as an indispensable material of an industrial economy. The mill value of cement shipments (United States) is well in excess of $1 billion/yr. Over 98% of this cement is portland cement, named by its inventor in 1824 because the color of the product resembled the limestone quarried on the Isle of Portland off the south coast of England.

Portland cement is a hydraulic cement, meaning that it will harden under water. It is a mixture of chemical compounds of Ca, Si, Al, Fe, and Mg, usually occurring together in the natural raw materials used in manufacturing. These materials are present in combinations as Ca silicates, Ca aluminates, calcium-alumina-ferrites, and lesser amounts of Na and K as sulfates, or as oxides in chemical combinations of other constituents. Generally, a cement will consist of about 75% Ca silicates, 5–10% Ca aluminates, 5% $CaSO_4$, 1% or less of oxides of Na and K, 2–4% MgO, and 5–10% calcium-alumina-iron compounds. In terms of chemical analysis, the materials would be reported as SiO_2, $Al_2O_3 \cdot Fe_2O_3$, CaO, MgO, SO_3, ignition loss, Na_2O, and K_2O. In terms of calculated potential composition, the materials would be tri-

calcium silicate, $3CaO \cdot SiO_2$, dicalcium silicate, $2CaO \cdot SiO_2$, tricalcium aluminate, $3CaO \cdot Al_2O_3$, tetracalcium-alumina-ferrite, $4CaO \cdot Al_2O_3Fe_2O_3$, and calcium sulfate, $CaSO_4$.

Major portland cement products include: *concrete,* a mixture of cement, aggregates consisting of sand and gravel or crushed rock, and water; *mortar,* a mixture of cement, sand, and water used for construction; also a masonry cement combined with sand and water for bonding masonry units, such as concrete blocks, bricks, and stones; *precast concrete,* concrete shapes, such as beams or floor slabs manufactured in one location and transported to an erection location for construction; and *cast concrete,* concrete placed in forms at the point of use in construction and allowed to harden, forming beams, floor slabs, and walls. There are many types of special cement products for specific applications, including high-speed and accelerated cements; various premixes, including mixes with aggregates, where the user need add only water; and special cements for use with particular materials, such as ceramic floors, tiles, and so on.

When water is added to portland cement, the water is taken into the crystal structure of the calcium silicates present, forming calcium hydroxide and a calcium silicate hydrate (sometimes referred to as tobermorite gel). The hydration process goes on for several days and, to a minor degree, for a much longer period. As the hydration proceeds, the concrete becomes stronger and harder.

Production Processes: There are two major processes for making cement: the *dry process* in which the raw materials are crushed, properly proportioned, ground, blended, and then fed to kiln to form clinker, after which gypsum is added to the clinker and this final mix ground to form portland cement; and the *wet process,* which differs from the former in that, after proportioning the raw materials, water is added before grinding and material is fed to the kiln in the form of a mixed slurry. A modern installation of the wet process is shown in Fig. C-5.

In this installation, with a capacity of 3 million bbl/yr, instrumentation, control, and automation are stressed. The plant has a closed-circuit television system for monitoring key operations, a sonic control system for materials grinding, and an x-ray fluorescent spectrographic instrument for raw materials and product analyses, and automation is so extensive that the entire basic cement-making system is operated by one man in the plant's central control room.

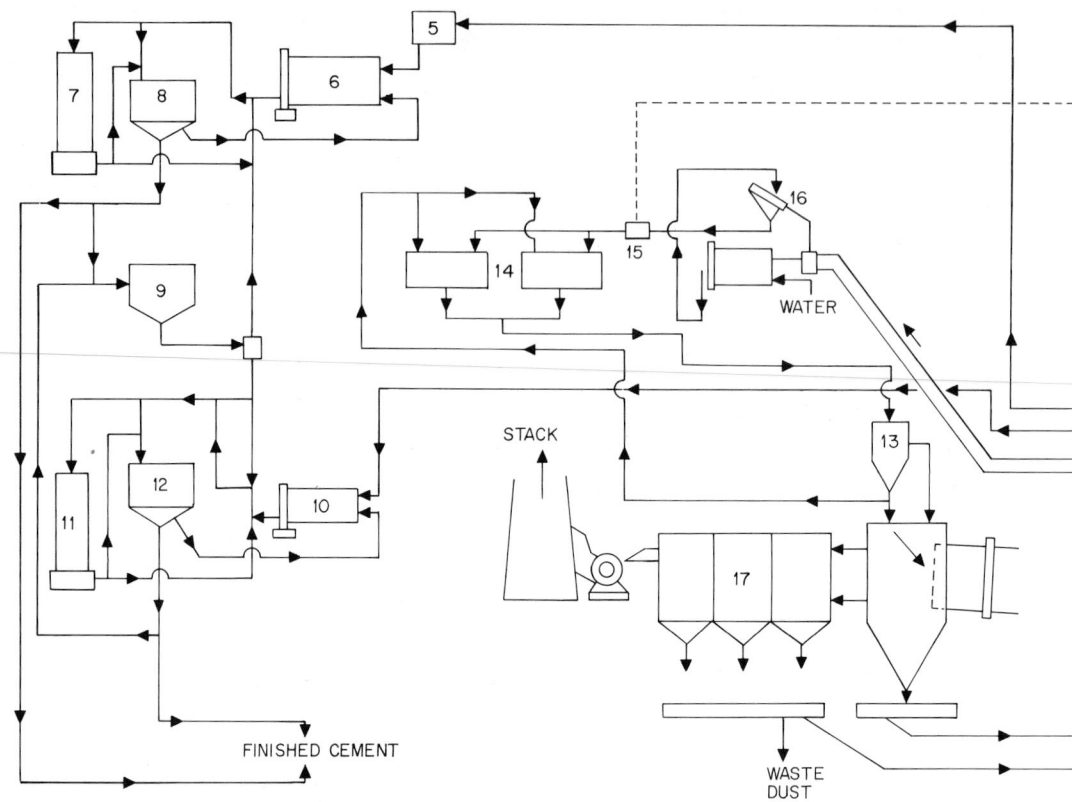

Fig. C-5. Schematic representation of materials flow in wet process for making cement. (1) Scalper, (2) conveyor scale, (3) mill-feed silos, (4) mill-weigh feeder, (5) crusher, (6) ball mill (coarse), (7) cement cooler, (8) air separator, (9) fringe bin, (10) ball mill (fine), (11) cement cooler, (12) air separator, (13) calibration tank, (14) slurry basins, (15) sampler, (16) screen, (17) electrostatic precipitator, (18) rotary kiln, (19) reclaimed-kiln-dust bin, (20) kiln-dust feed bin, (21) clinker-grate cooler, (22) fuel-oil storage.

Selective quarrying of raw materials for specific cement requirements is based upon chemical data. A television receiver, mounted on the quarry control panel, oversees the operations. Rocks sent to the crushing plant are reduced to particles smaller than 2 in. These are dropped onto a reversible shuttle conveyor that stockpiles the materials according to chemical composition in the raw-material storage located above a reclaiming tunnel.

From the stockpiles, vibrating feeders then withdraw specified amounts of raw materials and discharge them over a 1,230-ft-long belt conveyor that passes through the reclaiming tunnel. A television camera at the tunnel entrance observes the materials on their way to the next crushing stage. A second television camera supervises the screening operation, and a third camera observes the unloading of crushed material into the storage silos.

Raw materials, withdrawn from the silos, are conveyed to the raw-grinding mill, and water is added. A sonic system which "listens" to the sound of steel balls impacting inside the mill indicates the amount and fineness of the material being ground. Feedback from this system controls the total feed rate to the mill.

Raw-mill discharge (a slurry of about two-thirds solids content) is screened over a 50-mesh screen cloth and pumped to two slurry basins

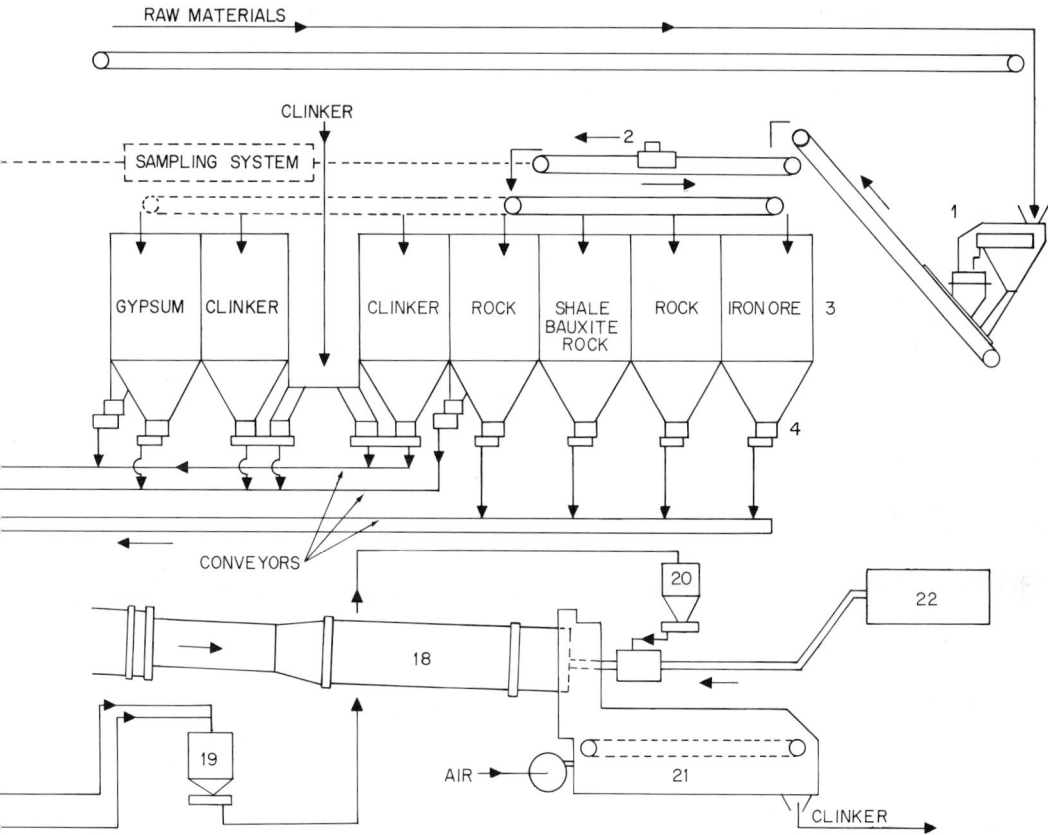

where it is thoroughly homogenized. Each basin (44 ft high × 85 ft diam) holds about a 3-day supply for the kiln.

The rotary kiln is 510 ft long and is fired by oil. The kiln's maximum daily output is 1,760 tons of clinker, equivalent to 9,700 bbl of cement. Operations are automatically controlled, and two television cameras, one at each end of the kiln, observe all material flow.

In its downward path through the kiln, the slurry passes first into a 91-ft-long drying zone (kept at 2000°F) and then into the hotter (above 2800°F) calcining and burning section. The drying zone is fitted with steel chains to improve fuel efficiency through better heat transfer. Feed rate of the oil is controlled by the gas temperature in the calcining zone. Thermocouples are placed at intervals inside the kiln to measure the temperature of solid particles and kiln gases. The critical

parameters are relayed to the plant's central control room.

Clinker and gypsum must be finely ground to produce the end product. Withdrawn from the mill-feed silos, the materials first are crushed and then fed to two ball mills at a rate that is controlled by sonic systems. The larger mill grinds cement for normal type 1 grade. For finer grinding, the smaller milling system is equipped with a special classifier-separator that uses two whirling fans to control fineness (the speed of the blades limits the size of particles that can pass through the fan). Tailings are recycled to the mill, and finished cement is pneumatically conveyed to storage or to a common fringe bin. The latter is used during a changeover from one grade of cement to another, to prevent the various grades from mixing.

The initial quarrying, primary and secondary

crushing, and screening of raw materials are not shown in Fig. C-5. See also **Gypsum.**

—J. Richard Tonry, Alpha Portland Cement Company, Easton, Pa.

Cement, Adhesive [See **Adhesives;** and **Silicates (soluble).**]

Cementite (See **Iron and steel.**)

Centrifuging A centrifuge is a machine designed to separate materials of different densities or for removing moisture from solids by the action of centrifugal force. The force acting on a particle within a centrifugal field is defined by Newton's fundamental force equation, $F = ma$. Acceleration acting on the particle, directed toward the center of rotation, is $a = rw^2$. Therefore the centrifugal force acting on the particle is $F = mrw^2$, or expressed as multiples of gravity,

$$F = 14.2 \times 10^{-6}DN^2$$

where m = mass of particle, g
   a = acceleration, cm/s^2
   r = radial distance of a particle in a centrifugal field from axis of rotation, cm
   w = angular velocity, rad/s
   D = inner diameter of centrifugal bowl, in.
   N = bowl speed, rpm
   A particle and a mixture introduced, confined,

and rotated within a circular enclosure accelerates as it moves from a neutral center toward the maximum diameter (inner periphery) of the enclosure. For example, the mixture, if introduced to the center of a 24-in.-diam solid-bowl centrifuge rotating at 1,500 rpm, will cause a particle to move at a speed of 32.2 fps at the center. At the maximum diameter, the particle will have a terminal velocity of 766.8 × 32.2 fps, or 24,690 fps. In the vernacular, one might say that the separation occurs at 766.8 × g. This, of course, is a simplistic explanation, because such factors as tensile stress of the bowl must be considered in applying theoretical potential to equipment design and practical centrifugal separation.

Sedimentation Centrifuges

Continuous Horizontal, Conveyor-Type Solid-Bowl Centrifuge: In this design (Fig. C-6), a continuous separation and discharge of a solid and liquid results when the solids have a heavier density than the liquor. Feed is continuously introduced through an axial tube. Solids settle to the bowl wall and are moved by the conveyor as the result of the differential rpm between the conveyor and bowl. The solids move up the sloping "beach" out of the liquid and are discharged at the smaller bowl diameter. A liquid level in the bowl is maintained by the adjustable ports. The liquid continuously overflows the desired setting at the larger bowl diameter.

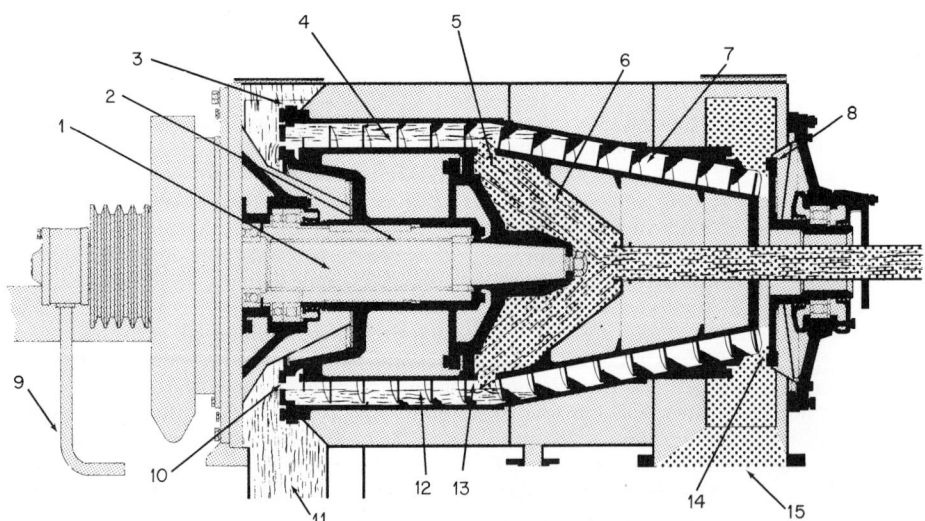

Fig. C-6. Continuous horizontal or vertical vibratory centrifuge. (1) Conveyor output eccentric shaft, (2) bowl output shaft, (3) effluent bowl head, (4) conveyor, (5) bowl, (6) feed compartment, (7) beach, (8) cake pump, (9) torque arm, (10) pool settling, (11) effluent, (12) pool volume, (13) feed ports, (14) cake ports, (15) cake solids. (*Envirotech Corp.*)

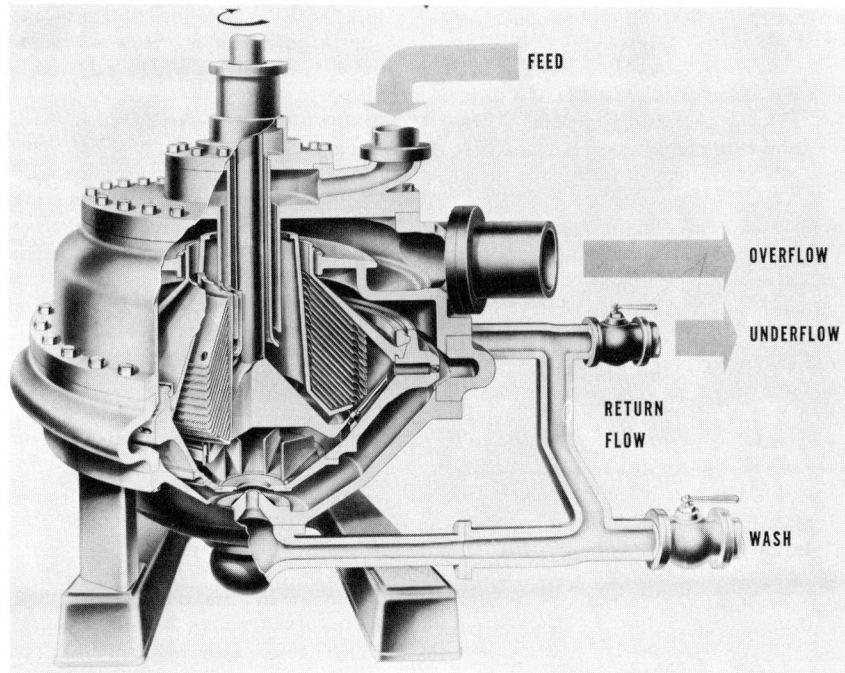

Fig. C-7. Peripheral-discharge-disk centrifuge. (*Dorr-Oliver Incorporated.*)

MAJOR APPLICATIONS: Chemical, food and beverage, plastics, and municipal industries.

SIZE RANGE: Bowl sizes 5.5–54 in. diam.

FORCE: Designed to operate at 770 to 6,000 × force of gravity. (Latter force times gravity simulates a 5-in. centrifuge.)

MAXIMUM OPERATING PRESSURE: 125 psig.

TEMPERATURE RANGE: −125 to 500°F(−87 to 260°C).

MATERIALS OF CONSTRUCTION: Carbon steel, 304–317 stainless steel, Monel, Hastelloy, titanium, which have at least 15–20% elongation. In abrasive-solids processing, the conveyor flights, feed ports, and cake ports are protected with hard-surfacing compounds, such as Stellite and carbide.

SOLID-PARTICLE SIZE RANGE: From a minimum of 1.0 μm (classification) to a maximum in excess of 100 μm.

CHARACTER OF CAKE: Examples: coarse NaCl and KCl result in cake solids 92–99% dry solids; industrial or municipal waste streams result in cake solids 18–35% by weight.

LIQUID-FLOW RATES: 5–500 gpm.

SOLIDS RATES: 0.25–75 tons/hr.

RELATIVE ADVANTAGES AND LIMITATIONS: Particularly designed for clarification (solids recovery), classification, thickening and solids drying. Initial costs are moderate.

Peripheral-Discharge-Disk Centrifuge: The continuous nozzle-disk centrifuge is used to clarify and thicken the solids from a low-solids-concentration waste stream (Fig. C-7). Feed is continuously introduced through a stationary tube. The solids settle to the bowl wall in a thickened cake concentration and are discharged through the peripheral nozzles. A recycle of thickened solids back to the centrifuge is a feature helpful to continuous processing. The liquid or effluent passes through the disks and is continuously discharged near the feed-slurry entrance point.

MAJOR APPLICATIONS: Food and beverage, petroleum, municipal, and chemical industries. Provides a continuous separation and discharge of a thickened solid and liquid when the solids have a heavier density than the liquor.

SIZE RANGE: Bowl sizes 9.0–42 in. diam.

FORCE: Designed to operate at 4,600–14,200 × force of gravity.

MAXIMUM OPERATING PRESSURE: 150 psig.

TEMPERATURE RANGE: −30 to 500°F(−34 to 260°C).

MATERIALS OF CONSTRUCTION: Usually 316 stainless steel. Also Monel and titanium. Abrasion usually is confined to nozzle area, where tungsten carbide is used to protect the metal surface.

SOLID-PARTICLE SIZE RANGE: 1.0 μm to 200 mesh. Large-size particles (for example, +100 mesh) will plug nozzles.

CHARACTER OF CAKE: A thickened slurry. In a municipal application, for example, concentration of cake may range from 6 to 10% by weight.

LIQUID-FLOW RATES: 10–400 gpm.

SOLIDS RATES: 0.1–9 tons/hr.

RELATIVE ADVANTAGES AND LIMITATIONS: Generally best suited to clarify and thicken the solids from a low-solids-concentration waste stream. Initial costs are moderately high.

Continuous Horizontal, Conveyor-Type Screen-Bowl Centrifuge: Feed slurry is introduced through a stationary pipe (Fig. C-8). The solids are sepa-

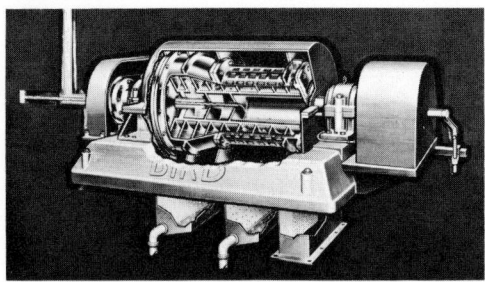

Fig. C-8. Continuous horizontal, conveyor-type screen-bowl centrifuge. (*Bird Machine Co.*)

rated from the liquid and settle at the bowl wall. The conveyor, while operating at a differential rpm to the bowl, transports the solids up the sloping "beach" out of the liquid, across the screen area, and discharges the solids at the smaller bowl diameter. The first-stage liquid continuously overflows the adjusted weir settings at the larger bowl diameter, and approximately 90% of the liquid is separated in this stage. The second-stage liquid separation takes place on the screens and results in a classification of fine solids and approximately 5% of the remaining liquid. The second-stage screen effluent is a separate stream that can be returned to the crystallizer as feed, or the fine

solids can be captured in a small-solid-bowl centrifuge as cake product. As a rule, some classification of fines is required in a screen-bowl application.

MAJOR APPLICATIONS: Chemical, food and beverage, and mineral industries. Provides a continuous separation, washing, and discharge of a solid and liquid when the solids have a heavier density than the liquor.

SIZE RANGES: Bowl sizes 5.5–54 in. diam.

FORCE: Designed to operate at 250–2,000 × force of gravity. (Latter force times gravity simulates a 5-in. centrifuge.)

MAXIMUM OPERATING PRESSURE: 125 psig.

TEMPERATURE RANGE: −125 to 500°F(−87 to 260°C).

MATERIALS OF CONSTRUCTION: Carbon steel, 304–317 stainless steel, Monel, Hastelloy, titanium, which have at least 15–20% elongation. In abrasive-solids processing, the conveyor flights, feed ports, and cake ports are protected with hard-surfacing compounds such as Stellite and carbide.

CHARACTER OF CAKE: Example: coarse NaCl and KCl result in cake solids 95–99% dry solids.

LIQUID-FLOW RATES: 5–500 gpm.

SOLIDS RATES: 0.25–75 tons/hr.

RELATIVE ADVANTAGES AND LIMITATIONS: Performs function of solid-bowl scroll-type centrifuge, but also is designed for continuous washing of crystalline solids and resulting drier cake solids. Initial costs are relatively high.

Filtering Centrifuges

Continuous Horizontal, Conveyor-Type Screen-Bowl Centrifuge: In this design (Fig. C-9), a continuous separation and discharge of a solid and liquid results when the solids have a heavier or lighter density than the mother liquor. Feed is continuously introduced at the small end of the screen basket through an axial tube. Solids settle on the screen basket and form a filter bed for the liquid to drain through. The liquid, following discharge from the basket, is continuously discharged from the machine, while the cake is moved by the conveyor across the remaining screen area to the discharge point at the large basket diameter. The conveyor operates at a differential rpm to the basket. The solids may be washed as they are conveyed across the screen surface. The horizontal filtering centrifuge shown in Fig. C-9 is selected to dewater and wash crystalline products. Classification of fine solids will discharge with the liquid, but generally these fines

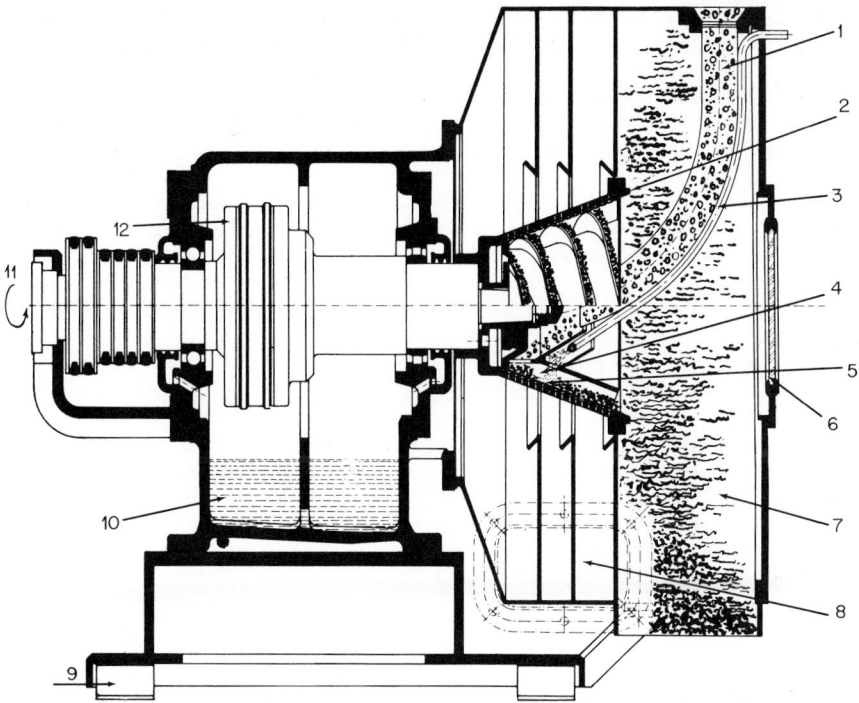

Fig. C-9. Continuous horizontal, conveyor-type screen-bowl filtering centrifuge. (1) Feed pipe, (2) perforated basket, (3) wash pipe, (4) conveyor spiral, (5) wash zone, (6) viewing fixture, (7) solids discharge, (8) effluent discharge, (9) rubber mounts, (10) oil sump, (11) conveyor-basket rotation, (12) cyclogear. (*Envirotech Corp.*)

are returned to the crystallizer as feed. Therefore some solids should be expected in the liquid discharge.

MAJOR APPLICATIONS: Chemical, food and beverage, plastics, and mineral industries. Provides a continuous separation, washing, and discharge of a solid and liquid when the solids have a heavier density than liquor.

SIZE RANGE: Bowl sizes 8.0–30 in. diam.

FORCE: Designed to operate at 250–2,000 × force of gravity. (Latter force times gravity simulates an 8-in. centrifuge.)

MAXIMUM OPERATING PRESSURE: 125 psig.

TEMPERATURE RANGE: -125 to $500°F(-87$ to $260°C)$.

MATERIALS OF CONSTRUCTION: Carbon steel, 304–317 stainless steel, Monel, Hastelloy, titanium, which have at least 15–20% elongation. In abrasive-solids processing, the conveyor flights, feed ports, and cake ports are protected with hard-surfacing compounds, such as Stellite and carbide.

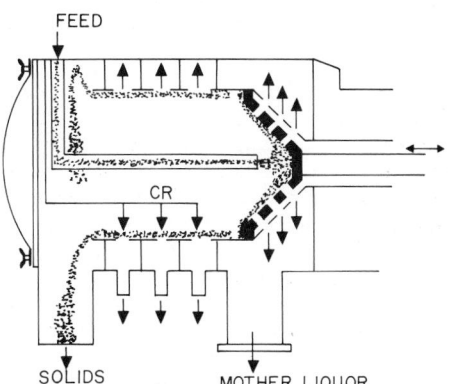

Fig. C-10. Continuous horizontal, pusher (reciprocating) filter centrifuge. CR = countercurrent rinses. (*Pennwalt Corporation.*)

SOLID-PARTICLE-SIZE RANGE: From a minimum of 150 mesh to a maximum in excess of 100 μm.

CHARACTER OF CAKE: Example: coarse NaCl and KCl result in cake solids 95–99% dry solids.

LIQUID-FLOW RATES: 5–300 gpm.

SOLIDS RATES: 0.25–50 tons/hr.

RELATIVE ADVANTAGES AND LIMITATIONS: Particularly designed for dewatering and washing of crystalline or fibrous solids. Prethickening of feed slurry is required (30–40% solids). Initial costs are moderate.

Continuous Horizontal, Pusher (Reciprocating) Filter Centrifuge: The continuous separation and discharge of solid and liquid takes place under the same conditions as the conveyor-type filtering centrifuge (Fig. C-10.) The feed is continuously introduced through a stationary pipe. The solids settle to the screen surface and form a filter bed through which the liquid passes. The liquid continues through the basket and is discharged from the machine, while the solids, having formed a uniform filter-bed thickness in the cylinder basket, are assisted to discharge by a pusher ram. The general maximum movement of cake solids on the screen basket by the pusher ram is 3 in.; thereby the solids end up pushing other solids until discharged. The limiting factor in pusher filtering centrifuges is the amount of solids the reciprocating pushing ram can move across the screen surface to discharge. Washing of cake solids, following the initial dewatering, is considered a desirable feature of this design.

MAJOR APPLICATIONS: Chemical, food and beverage, plastics, and mineral industries. Provides a continuous separation, washing, and discharge of a solid and liquid where the solids have a heavier density than the liquor.

SIZE RANGE: Bowl sizes 8.0–40 in. diam.

FORCE: Designed to operate at 250–1,100 × force of gravity. (Latter force times gravity simulates an 8.0-in. centrifuge.)

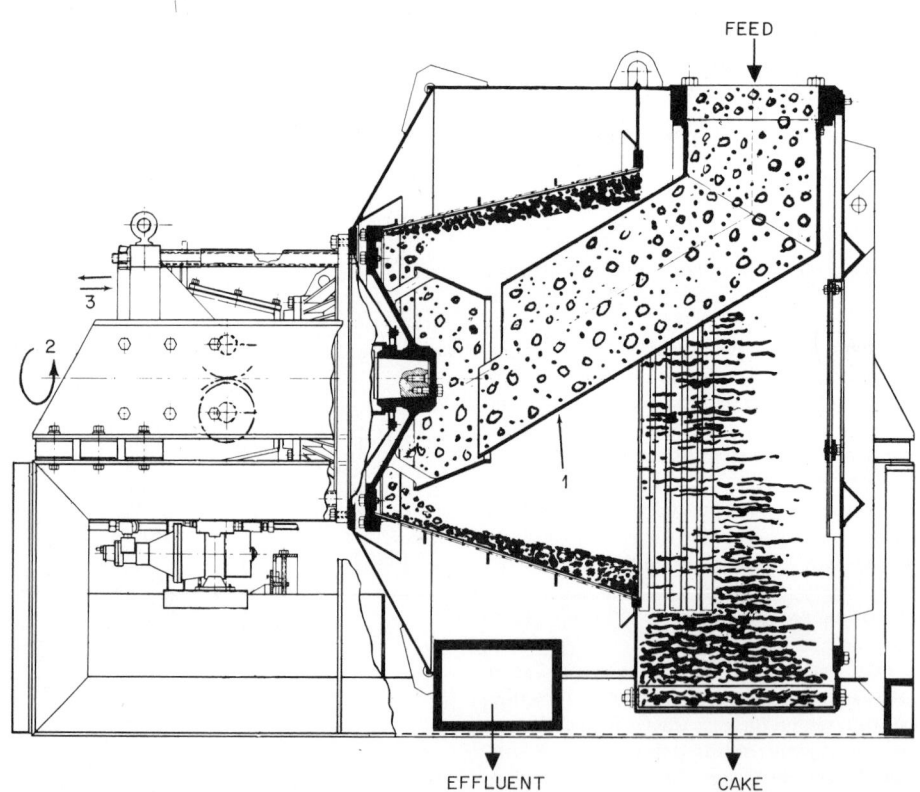

Fig. C-11. Continuous horizontal or vertical vibratory centrifuge. (1) Screen basket, (2) basket rotation, (3) vibration motion. (*Envirotech Corp.*)

MAXIMUM OPERATING PRESSURE: 125 psig.

TEMPERATURE RANGE: -125 to $500°F(-87$ to $260°C$).

MATERIALS OF CONSTRUCTION: Carbon steel, 304–317 stainless steel, Monel, Hastelloy, titanium, which have at least 15–20% elongation.

SOLID-PARTICLE-SIZE RANGE: From a minimum of 150 mesh to a maximum somewhat in excess of 100 μm.

CHARACTER OF CAKE: Example: coarse NaCl and KCl result in cake solids 92–99% dry solids.

LIQUID-FLOW RATES: 5–220 gpm.

SOLIDS RATES: 0.25–35 tons/hr.

RELATIVE ADVANTAGES AND LIMITATIONS: Designed for dewatering and washing crystalline solids. Prethickening of feed slurry is required (50–70% solids). Initial costs are moderately high.

Continuous Horizontal or Vertical Vibratory Centrifuge: In this design (Fig. C-11), feed is continuously introduced through a stationary pipe. The solids settle to the screen basket and form a uniform filter bed through which the liquid drains. The solids are moved along the conical screen basket by an amplitude vibration to discharge at the large diameter. The liquid, having drained through the solids and screen basket, is continuously discharged. Solids dryness and high tonnages exemplify the application of the vibratory centrifuge.

MAJOR APPLICATIONS: Chemical, mineral, and coal industries. Provides a continuous separation and discharge of a solid and liquid when the solids have a heavier density than the liquor.

SIZE RANGE: Bowl sizes 8.0–54 in. diam.

FORCE: Designed to operate at 75–150 $\times$ force of gravity. (Latter force times gravity simulates an 8-in. centrifuge.)

TEMPERATURE RANGE: 0–200°F (32–93°C).

MATERIALS OF CONSTRUCTION: Carbon steel, 304–317 stainless steel, Monel, Hastelloy, titanium, which have at least 15–20% elongation.

SOLID-PARTICLE-SIZE RANGE: From a minimum of $+28$ mesh to a maximum in excess of $\frac{1}{4}$ in.

CHARACTER OF CAKE: Coarse NaCl and coal result in cake solids 94–98% dry solids.

LIQUID-FLOW RATES: 10–100 gpm.

RELATIVE ADVANTAGES AND LIMITATIONS: Designed for dewatering of crystalline solids to a low product moisture. Feed solids are pre-

thickened on a screen to 80–90% solids. Initial costs are moderate.

—Stephen D. Roach, *EIMCO Processing Machinery Division, Envirotech Corporation, Salt Lake City, Utah*

Cephalosporins (See **Antibiotics.**)

Cephalothin (See **Antibiotics.**)

Ceramics The name ceramics stems from the ancient Greek word *keramos,* which meant "burnt stuff." This remains a good definition for this important field of science and technology, sometimes incorrectly thought of by the layman as restricted to pottery.

The ceramics industry (United States, 1971) is approximately a $10 billion industry, including (1) glass, $4.4 billion; (2) electronic and technical ceramics, $1.77 billion; (3) structural-clay products, $1.01 billion; (4) whitewares, $0.86 billion; (5) refractories, $0.81 billion; (6) porcelain-enamel coatings, $0.71 billion; and (7) abrasives, but not including organic and metal-bonded abrasives, $0.29 billion.

The scientific foundation for modern ceramics began with the chemical analyses of the clays, the raw material used in so many ceramics. These early analyses indicated that the clays were made up of varying proportions of silica, SiO_2, and alumina, Al_2O_3, combined with water, along with small amounts of other oxides such as iron oxide, which give red clays their color. Some of the early clay analyses by Herman A. Seger are shown in Table C-26.

Clays, which are the backbone of many ceramics, were studied in one of the outstanding early applications of the electron microscope to ceramics. The kaolinite particles, which average half a micrometer in size in their largest dimension, were found to exist as thin hexagonal plates or as stacks or books of these plates.

More recently, the scanning electron microscope has been developed and has found immediate applications in ceramic-microstructure study. In basic failure-mechanics studies, the fractured surfaces of the failed ceramics are well shown with much three-dimensional realism by this technique. Figure C-12 shows the fracture surface of a glassy bond post in a vitrified grinding wheel, revealing the complex nature of the fracture.

The internal structure of the crystalline phases that are the basic building blocks of many ceramic materials are shown by their x-ray–diffraction pat-

TABLE C-26. *Chemical Analysis of Several Clays, in Weight Percent*

Constituent	Ledez Kaolin	Kottikin Kaolin	Hoehr Stoneware Clay	Andenne, Belgium Fireclay	Puchan Common Brick Clay
SiO_2	49.16	49.91	70.12	49.64	69.25
Al_2O_3	36.73	35.99	21.43	34.78	15.56
Fe_2O_3	0.81	0.63	0.77	1.80	5.62
CaO	Trace	. . .	. . .	0.68	0.45
MgO	0.18	0.30	0.39	0.41	
$K_2O + Na_2O$	1.18	0.76	2.62	0.41	4.16
Water	12.41	12.34	4.92	12.00	4.97

SOURCE: A. V. Bleininger (ed.): "The Collected Writings of Hermann August Seger," vols. 1 and 2, Chemical Publishing Co., Easton, Pa., 1902.

Fig. C-12. Scanning electron photomicrographs. (Top) Fractured-glass bond post showing point of origin of fracture near center and glassy fractured patterns radiating toward the bottom. Top of fractured post shows evidence of crystallization (devitrification), magnification, ×690. (Bottom) Crystallized (devitrified) portion of glass bond post at higher magnification, ×7,100. (*Photographs by H. R. Baumgartner, Norton Co.*)

terns. The studies of these structures were pioneered by W. H. Bragg at the beginning of the present century.

The raw materials used for ceramics are both natural and synthetic, the synthetic man-made materials finding more application in special ceramics for the more extreme demands of high purity, extra strength, and crystal perfection.

The natural raw materials supply the bulk tonnage of all the ceramic constituents. The basis of most ceramic bodies are the clays. This is a family of hydrous aluminosilicates of fine particle size that develop plasticity when mixed with water. This aids forming the material into its desired shape. When dried, the fine clay particles help "glue" the shape together until final strength is developed during firing.

Other natural ceramic raw materials include talc, a hydrous magnesium silicate with a layer structure similar to the clay minerals, and the feldspars, which are anhydrous aluminosilicates containing K^+, Na^+, and Ca^{++}. These alkalies and alkali earths act to reduce the melting temperatures of the ceramic bodies, making them less expensive to produce.

Anhydrous silica in the form of the mineral quartz is a very important ceramic raw material as a major constituent in glass, glazes, enamels, some refractories, whitewares, and abrasives.

For refractories, special raw materials with high melting points are selected. These include special grades of more refractory clays called *fireclays*. Other materials are bauxite, a hydrated alumina; magnesite, a magnesium carbonate; dolomite, a calcium and magnesium hydrate; and chrome ore, primarily chrome oxide but containing considerable magnesia, alumina, and iron oxide.

Synthetic Materials: One of the synthetic ceramic materials used in large volume is silicon carbide, SiC. See also **Abrasives; and Silicon.** Numerous other carbides, nitrides, and borides are synthesized. The field of electronic ceramics, including ferrites, capacitors, transistors, diodes, and semiconductor devices depends upon the use of synthetic raw materials because of demands for extreme purity.

High-Temperature Chemistry: The study of ceramic processes is basically a study of high-temperature chemistry, much of it dealing with the solid-state reactions of diffusion, crystal growth, and sintering. There are also large areas dealing with liquid formation, solution, and precipitation with glasses. The reactions in the gas phases present are very important in crystal growing. The usual laws of thermodynamics apply, and indicate the directions of possible ceramic reactions. Some of these reactions, especially in the solid state and with highly viscous liquids, are extremely slow, so that ceramics are seldom at chemical equilibrium. With ceramics, because of their very strong bonding and their inert characteristics, the reactions must be carried out at quite high temperatures, well above those normally encountered by chemists.

Atomic Arrangements and Bonds: The properties of ceramics depend largely on how the atoms are arranged and the interatomic bonds they form. The most important bonding force in the crystalline phases in most ceramics is ionic bonding, the metallic atoms losing an outer electron to become positive ions and the nonmetallic atoms gaining an outer electron to form a negative ion. Ionic crystals are brittle and hard, melt at high temperature, and have low electrical conductivity at room temperature. Compounds of metals with oxygen ions that are largely ionic are MgO, Al_2O_3, and ZrO_2.

Another type of bonding force for ceramic crystalline materials is covalent bonding. In this case, a pair of electrons are shared by two atoms. Covalent crystals, such as diamond and silicon carbide, have high hardness, high melting points, and low electrical conductivities at low temperatures.

The basic building block for the silicate crystal structures is the silicon-oxygen tetrahedron with a silicon atom at the center and four oxygen atoms at the corners. The silicates are classed by the types of bonding existing between the tetrahedra in their crystal structures. In orthosilicates, the tetrahedra are independent of each other. These structures make good refractories because of their high melting points. This group includes the olivine minerals, garnets, zircon, kyanite, and

mullite. When the tetrahedra are joined at only one corner (oxygen atom), they form pyrosilicates, which are rare. In metasilicates, the tetrahedra share two corners to form a variety of ring or chain structures. Minerals of this type include the pyroxines, such as spondumene, and the amphiboles, such as asbestos. Sharing three corners, the tetrahedra form disilicates, which exist as sheets or planes, forming such minerals as mica. In the various forms of silica, such as quartz and cristobalite, all four tetrahedron corners are shared.

The important silicate structures such as feldspars and zeolites are essentially infinite three-dimensional silica frameworks. In feldspar, Al^{3+} ions replace some Si^{4+} ions to make a framework with a negative charge neutralized by large ions such as Na^+, Ca^{++}, and K^+ interstitial positions. Zeolites form similar but much more open alumina-silica frameworks. In their structures, the alkali or alkali-earth ions can be exchanged in aqueous solutions (water-softener applications). The channels in the zeolite structures are of the proper size to selectively trap out various molecules, leading to their use as molecular sieves.

The crystal structure of the most common clay mineral, kaolinite, consists of combinations of a layer of SiO_4 tetrahedra joined at the corners with a layer of alumina octahedra. With this type of structure, it can be readily visualized why kaolinite is found in thin platy hexagonal crystals whose thickness is usually only about one-tenth of its diameter.

The interesting application of ceramics as fuel-cell members is based on the lack of perfect crystalline structures in ceramic materials. One of the types of defects in these structures is called a *Frenkel disorder.* In this type of disorder one of the small metal ions leaves its normal site, becoming an interstitial ion and leaving a vacancy. This type of disordering movement occurs more readily at higher temperatures. Thus an increasing ionic current flow can be generated in a heated ceramic placed in an electric field, drawing the ions to one electrode, although the material may have had high resistivity at room temperature.

Most ceramic shapes do not consist of one single crystal but are composed of numerous crystals joined together to form polycrystalline structures. The characteristics of the grain boundaries between crystals can influence the strength, chemical stability, and electrical properties as much as do the crystalline structures within the individual grains.

The old saying, "A chain is only as strong as its weakest link," is appropriate. Fracture studies of pure aluminum oxide have shown that the fracture path at lower temperatures goes through the crystalline grains. However, at higher temperatures, the grain boundaries weaken more quickly, and the fracture now follows the lines of the grain boundaries around the grains.

Grain boundaries are weak points because of voids or pores which are located on the boundary of two or more grains, reducing the areas of contact and providing spaces through which gases may permeate, causing deleterious reactions. Also, the crystal orientation of one grain may not line up with that of the adjacent one, or entirely different grains may be joined, causing strains to be set up between the two.

Glass is a very important part of ceramics. The glass industry is the largest single element of the entire ceramic industry, and the glassy portions of many ceramic bodies are the bond that hold many ceramics together. See also **Glass.**

Porcelain enamels used to protect and decorate steel and aluminum metals are glasses specially designed to have high thermal expansions to match the base metal and to mature (become glassy) at temperatures low enough to prevent distortion of the metal sheet.

Glazes perform a similar function on ceramic substrates, again, as special glasses matching the thermal expansion of the base and maturing at the proper temperature.

Probably the majority of the ceramics produced are a mixture of crystalline grains and a glassy phase. The glass frequently acts as the bond between the crystalline grains. This is the basis of the vitrified-grinding-wheel business and much of the structural and whiteware branches of the ceramic industry.

The melting temperatures and the crystalline and amorphous phases formed in cooling from glassy melts are described for one, two, three, four, and some five oxide component systems in a large collection of phase diagrams published by the American Ceramic Society.[2,3] The basis for these studies stems from L. Willard Gibbs's original phase-rule work.

Composites: An increasingly important branch of special ceramics deals with composites. This field covers a variety of combinations, such as sapphire, Al_2O_3, whiskers in metals, metal-bonded carbides used in the machine-tool industry, and directionally solidified two-phase ceramic systems. In a system of the latter type, an oriented fibrous second phase is grown in a primary matrix phase to maximize in a selected direction various important characteristics, such as minimum long-term high-temperature creep. An example of use is in gas-turbine rotor blades.

REFERENCES

1. Bleininger, A. V. (ed.): "The Collected Writings of Hermann August Seger," vols. 1 and 2, Chemical Publishing Co., Easton, Pa., 1902.
2. Levin, E. M., C. R. Robbins, and H. F. McMurdie: "Phase Diagrams for Ceramists," The American Ceramic Society, Columbus, Ohio, 1964.
3. Levin, E. M., C. R. Robbins, and H. F. McMurdie: "Phase Diagrams for Ceramists," 1969 supplement, The American Ceramic Society, Columbus, Ohio, 1969.

—Louis J. Trostel, Jr., *Industrial Ceramics Division, Norton Company, Worcester, Mass.*

Cereals   (See **Amino acids.**)

Cerium* (From Ceres, an asteroid which had been sighted shortly before the discovery of "ceria," cerium oxide.) **Ce** = 140.12 (at. wt.); 58 (at. no.). In 1803, cerium was independently discovered by Klaproth and by Berzelius and Hisinger. In 1839, Mosander showed that this oxide actually consisted of at least two oxides, ceria and lanthana. The latter was later found to consist not only of lanthana, but also of praseodymia and neodymia. Metallic Ce, although rather impure since it contained other rare-earth metals, was first isolated by Mosander in 1825.

Cerium is in group III, period 6, of the periodic table of the elements and is the first of the elements in the "lanthanide" series, which is commonly shown as the first extra row of elements below the main body of the periodic table. Because of its unique position in the periodic table, Ce exhibits two valence states, 4+ with no $4f$ electrons and 3+ with one $4f$ electron. And as a result, metallic Ce has several unusual polymorphic phases, and the Ce^{4+}/Ce^{3+} couple has an oxidation-reduction potential of 1.28–1.87 V in various solutions.

Cerium melts at 798°C and boils at 3427°C. Its density at room temperature is 6.773 g/cm^3. Cerium is the most abundant element of the rare-earth group and ranks 28th in the abundances of the naturally occurring elements.

Cerium has a face-centered cubic structure at room temperature. This phase, which is called γ-Ce, upon cooling at atmospheric pressure, begins to transform to β-Ce, which has a hexagonal

*Work performed in the Ames Laboratory of the U.S. Atomic Energy Commission, *Contrib.* 3170.

structure. It is extremely difficult to obtain 100% conversion of γ-Ce to β-Ce, and thus, on continued cooling, the remaining γ-Ce transforms to another face-centered cubic structure, α-Ce, which is smaller than γ-Ce by 16% in volume. The γ-to-α transformation is believed to be accompanied by a transfer of the $4f$ electron to the valence level. α-Ce can be formed directly from γ-Ce at room temperature by applying a pressure of 7,500 atm or more. At liquid-helium temperature and very high pressures α-Ce transforms to α'-Ce, which is a superconductor. Since β-Ce orders magnetically, cerium is the only element in the periodic table which is both a superconductor and a magnetically ordering substance. Furthermore, Ce is also the only material known to have a solid-solid critical point, i.e., the termination of the phase boundary between two allotropic modifications. At high temperature, 70°C below cerium's melting point, γ-Ce transforms to a body-centered cubic form (δ-Ce).

Silvery-gray metallic Ce oxidizes very readily at room temperature, especially in moist air, to form CeO_2, which is easily recognized by its pale yellowish-green color. Below 300°C oxidation proceeds at a controlled rate, but above 300°C the rate of oxidation increases quite rapidly, and sometimes the Ce ignites to burn with a bright red glow. The pyrophoric and incendiary behaviors of Ce are quite evident by the shower of sparks emanating from the metal- or cerium-base alloys when filed, ground, or machined. The pyrophoric nature of iron-cerium alloys accounts for one of the most important uses of Ce, as lighter flint, and the incendiary nature of Ce for its use in ordnance as shell liners.

Cerium dissolves readily in dilute and concentrated mineral acids. A 1:1 mixture of 48% hydrofluoric acid and concentrated nitric acid does not attack Ce. Alkali solutions attack Ce. It reacts slowly at 25°C with the halogen gases to form the corresponding trihalide, but above 200°C it reacts very rapidly.

Cerium alloys with all the metals to the right of manganese and its cogeners, technetium and rhenium, in the periodic table to form intermetallic compounds. Magnesium, thorium, and the other rare-earth metals form extensive solid solutions with Ce. Immiscible liquids are formed with calcium, vanadium, chromium, manganese, niobium (columbium), and uranium.

Cerium and oxygen form a whole series of compounds between Ce_2O_3 and CeO_2, the number of which depends upon the temperature and oxygen pressure. Ce_2O_3 is easily dissolved in mineral acids, but CeO_2 is much more slowly dissolved. $CeCl_3$, $CeBr_3$, and CeI_3 are hygroscopic; usually exist as hydrated salts; and are soluble in water. CeF_3 is nonhygroscopic and insoluble in water. The only stable halide formed by tetravalent cerium is CeF_4. Cerium forms numerous compounds with the other nonmetallic elements. It reacts with hydrogen to form CeH_2 and CeH_3. The two cerium carbides Ce_2C_3 and CeC_2 react with water to form unsaturated hydrocarbons (mostly acetylene) and hydrogen. Boron, silicon, nitrogen, phosphorus, sulfur, selenium, and tellurium form at least one compound with cerium, most of which are high-melting. Cerium hydroxide is insoluble and formed by neutralizing a cerous or ceric solution with ammonium hydroxide. The cerous hydroxide is unstable and is oxidized by air to ceric hydroxide. Cerous sulfate is soluble in water, and its solubility decreases with increasing temperature. Ceric sulfate is quite soluble in water and is a strong oxidizing agent. Cerous carbonate is formed by precipitation from most aqueous solutions.

Cerium has 19 different isotopes, of which only four are naturally occurring, 136, 138, 140, and 142. Their respective natural abundances are 0.19, 0.25, 88.48, and 11.07%. Cerium has a low-thermal neutron-absorption cross section and a low toxicity rating.

Occurrence and Processing: The important minerals containing cerium are allanite, bastnasite, cerite, and monazite. Of these four minerals bastnasite and monazite are the most important sources of cerium. The major deposit of bastnasite, a rare-earth fluorocarbonate, is found in Southern California. Monazite, which is a phosphate containing thorium and the light lanthanides, is widely distributed over the world. Important deposits are found in Australia, Brazil, India, South Africa, and the United States (Florida, Idaho, and Montana).

Bastnasite is first concentrated by a flotation process to remove the barite, calcite, and quartz, and is then converted to the oxide by roasting. One of two methods is used to process monazite. In the acid process, monazite is converted to a thorium–rare-earth sulfate mixture by heating in sulfuric acid. By fractional precipitation with oxalic acid the light lanthanides are separated from thorium, yttrium, and heavy lanthanides. The alternative basic process involves digestion of the monazite in sodium hydroxide at high temperatures to form the insoluble rare-earth hydroxide in a trisodium phosphate solution. After filtering, the hydroxide is neutralized to dissolve the light lanthanides, and the thorium, yttrium, and heavy lanthanides remain an insoluble residue.

Cerium is separated chemically from the light rare earths by making use of its 3^+ and 4^+ valence states. The rare-earth mixture obtained as noted above is treated by an oxidizing agent to form the ceric ion, which is precipitated at a pH less than that required to precipitate the first trivalent rare earth, 6.3. Further purification is realized by repeating the process.

Cerium metal is prepared by metallothermic reduction of CeF_3 or $CeCl_3$ using a Ta or Mo crucible. After the reduction step, the slag is removed and the excess reducing agent (usually Ca) is removed by distillation. An alternative process is to reduce $CeCl_3$ or CeO_2 electrolytically at 800–1000°C. The chloride or oxide is soluble in a complex molten halide flux, and the Ce^{3+} is reduced to metal at a Mo cathode. The molten Ce is collected in a Mo crucible at the bottom of the electrolytic cell. The advantage of this method is its low cost and the capability of continuous operation.

Uses: The largest single use for CeO_2 in glass is for decolorizing soda-lime container glass. The CeO_2 imparts high brightness and reduces the amount of other complementary additives required for decolorizing.

CeO_2 is used to polish gemstones and glass, especially precision optical glasses. Cerium, because of its oxidizing power, is used in glass which is subjected to α-, γ-, and x-ray, light, and electron radiation. The cerium prevents the discoloration of glass due to the presence of Fe(II) by oxidizing Fe(II) as it is formed to Fe(III). This is important in color TV tubes, where electron radiation would discolor the glass and destroy the color quality. Cerium dioxide is used in capacitors, cathodes, ceramic coatings, phosphors, photochromic glasses,* refractory oxides, and semiconductors and as a catalyst and as an opacifier in porcelain enamels. CeO_2, because of its low nuclear cross section, is used as a diluent in oxide nuclear fuels.

In the metallic form Ce is used extensively as a mixed rare-earth alloy called *mischmetal* (50% Ce, 25% La, 18% Nd, 5% Pr, 2% other rare earths). The largest commercial uses of misch-

*Photochromic materials darken when light shines upon them, but when the light is removed they become more transparent.

metal are as an alloying agent to improve the malleability of ductile iron, as lighter flints (70% mischmetal–30% Fe alloy) and as shell linings in military projectiles. Minor uses include the use of mischmetal as alloying agents to improve the strength and creep resistance of Mg; strength of Al; oxidation resistance of Ni; and hardness of Cu, with little effect on the electrical conductivity of Cu. Cerium and mischmetal are being used as getters in vacuum tubes. $CeCo_5$ is one of the new rare-earth cobalt, RCo_5, permanent-magnet materials. These outstanding permanent magnets have properties which exceed those of the more common alnicos and ferrites by a wide margin. $CeCo_5$ permanent magnets are produced by casting and powder-metallurgy techniques.

Mixed rare-earth oxides and fluorides containing about 50% Ce are used as cores for carbon arcs, because the rare-earth mixture gives a tenfold increase in the intensity and improves the color balance. The mixed oxides are also used as catalysts for cracking petroleum and oxidation reactions and, as polishing materials, waterproofing agents, and fungicides.

—Karl A. Gschneidner, Jr., *Ames Laboratory, U.S. Atomic Energy Commission, Iowa State University, Ames, Iowa*

Cermets   (See **Chromium.**)

Cerussite   (See **Lead.**)

Cesium   (L. *coesius*, sky blue.) **Cs** = 132.913 (at. wt.); 55 (at. no.). Cesium is one of the alkali metals, a member of group IA, period 6, of the periodic table. It was discovered by Bunsen in Germany (1860), the first metal to be discovered by means of the spectroscope.

Cesium occurs in nature as the stable isotope ^{133}Cs. However, there are 15 radioactive isotopes

TABLE C-27. *Some Physical Properties of Cesium*

Melting point, °C	29.7
Boiling point, °C	700
Density, g/cm³, at 20°C	1.87
Mean specific heat, cal/(g)(°C), 0–100°C	0.056
Resistivity, Ω/cm, at 20°C	21×10^{-6}
Temperature coefficient of resistivity, 0–100°C	4.8×10^3
Coefficient of expansion, 0–100°C	97×10^6

Additional properties can be found in the Metals Reference Book. See Selected References.

TABLE C-28. *Some Cesium Compounds and Their Formation*

Oxygen compounds:
  Suboxides: Cs_3O, Cs_4O, Cs_7O
  Oxide: Cs_2O (orange-red); sp gr = 4.36; very soluble in cold H_2O; decomposes in hot H_2O
  Peroxide: Cs_2O_2 (yellow crystals); sp gr = 4.25; mp = 400°C; decomposes in H_2O
  Superoxides: Cs_2O_3 (chocolate-brown crystals); sp gr = 4.25; mp = 400°C; decomposes in H_2O
          Cs_2O_4 (yellow crystals); sp gr = 3.77; mp = 600°C; decomposes in H_2O
  Peroxide peroxyhydrate: $Cs_2O_2xH_2O$
  Ozonide: CsO_3
  With the exception of Cs_2O, all the foregoing compounds react with H_2O, forming CsOH and liberating O_2. Cs_2O forms only CsOH in contact with H_2O.
Halogen compounds:
  The halides of Cs are generally obtained by reacting CsOH or Cs_2CO_3 with the corresponding halogen acid (for example, $CsOH + HCl \rightarrow CsCl + H_2O$):
  CsBr (cubic); RI = 1.582; 1.6984; sp gr = 4.433; highly soluble in H_2O
  $CsBr_3$ (rhombic); soluble in H_2O
  CsCl (colorless, deliquescent, cubic); sp gr = 3.97; quite soluble in H_2O
  CsF (colorless, cubic); sp gr = 3.586; quite soluble in H_2O
  CsI (cubic); RI = 1.661; 1.7876; sp gr = 4.51; quite soluble in H_2O
  Cs also forms poly- and complex-halides (CsI_3, $CsBr_3$, CsI_4, CsI_9, CsI_2Br, $CsIFCl_3$, $CsCl_2I$, and others)
  CsI_3 can be obtained by dissolving CsI in sulfamic acid: $3CsI + 2NH_2SO_3H + H_2O \rightarrow CsI_3 + Cs_2SO_4 + (NH_4)_2SO_4$
  Because of its very high electropositivity, Cs readily combines with the very electronegative oxyhalions, giving compounds such as $CsClO_3$, $CsClO_4$, $CsBrO_3$, $CsIO_3$, and $CsIO_4$ (for example, $Cs_2CO_3 + 2HClO_4 \rightarrow 2CsClO_4 + H_2O + CO_2$).
Sulfur compounds:
  Sulfides: $Cs_2S \cdot 4H_2O$ (white, deliquescent); very soluble in H_2O
          $Cs_2S_2 \cdot H_2O$, Cs_2S_3, Cs_2S_5
  Sulfates: Cs_2SO_4 (white, rhombic needles); RI = 1.5644; sp gr = 4.243; quite soluble in H_2O
          $CsHSO_4$ (bisulfate or acid sulfate); (rhombic); sp gr = 3.352; soluble in H_2O
          $Cs_2S_2O_7$ (pyrosulfate)
Boron compounds:
  Cs generally exists in combination with B as borates: $CsBO_3 \cdot H_2O$ (perborate), $Cs_2B_4O_7 \cdot 5H_2O$ (tetraborate), $CsBO_2$ (metaborate)
Other compounds:
  $CsHCO_3$ (rhombic); soluble in H_2O
  Cs_2CO_3 (white, deliquescent); soluble in H_2O
  Cs_2CrO_4 (yellow, monoclinic); sp gr = 4.237; soluble in H_2O
  CsH (crystals); sp gr = 2.7; decomposes in H_2O
  CsOH (gray, deliquescent); sp gr = 3.675; quite soluble in H_2O
  $CsNO_3$ (white, hexagonal); RI = 1.55; sp gr = 3.687; moderately soluble in H_2O
  $CsNO_2$ (yellow crystals); very soluble in H_2O
  $CsMnO_4$ (violet, rhombic); sp gr = 3.597; practically insoluble in H_2O
Complex salts:
  Cs forms complex salts with other metals. These are frequently difficultly soluble compounds which may be used in the detection or estimation of the accompanying metal or of Cs itself. Examples are $CsAuCl_4$, $CsTiCl_6$, $CsBr_2 \cdot 2HgBr_2$, Cs_2PtCl_6

RI = refractive index

(125–132, 134–139) of Cs. See also **Chemical elements.** [137]Cs has a half-life of 33 years and is used as a source of gamma radiation, especially in radiography and therapy.

Cesium, is a silver-white, soft, ductile metal, and crystallizes in a body-centered cubic lattice. Principal physical properties are given in Table C-27.

Chemical Properties: Cesium oxidizes rapidly and will ignite in moist air. The metal reacts violently with water at ordinary temperatures with ignition of the H_2 evolved. The elemental form is rarely used. Small concentrations of oxide and nitride impart a bronze color to the element.

Cesium is the most strongly basic and electropositive metal known and shows strong ionization tendencies. Cesium dissolves in acids (forming salts) and in alcohols (forming alcoholates). Cesium is monovalent and is much like K and Rb in its behavior. Cesium hydroxide, CsOH, is a very powerful base. Cesium absorbs CO in the cold to form minute shining spheres of chamois-yellow color, insoluble in benzene, ether, or NH_3. Cesium monoxide, Cs_2O, is orange-red in color. A summary of Cs compounds is given in Table C-28.

Uses: The principal use for Cs is in photoelectric cells because of its high sensitivity to light. Cesium cells are suitable over a wide region of the spectrum and are used in motion picture, TV, radar, and instrumentation applications. Cesium also is suitable for use in luminescent tubes and screens.

Cesium also has been used as a catalyst for certain synthetic-resin reactions, for example, chloroprene. In this application, Cs is atomized and sprayed along with the resin. Interest has been shown in the use of Cs as a fuel in ion-propulsion engines of low thrust for spacecraft and as a thermal-transfer fluid in special applications.

Cesium also plays an important function in time measurement. The atomic second, officially defined in 1967 by the International Bureau of Weights and Measures, is equivalent to 9,192,631,770 oscillations of the atom of ^{133}Cs. The value expresses the ET (ephemeris time) second as closely as possible in terms of an atomic standard and was derived by a joint experiment of the U.S. Naval Observatory and Great Britain's National Physical Laboratory, using a dual-rate moon-position camera and a cesium-beam clock.

Alloys: Cesium forms a series of solid solutions with Rb. These alloys are used as *getters* to eliminate residual gases from vacuum tubes. The materials are difficult to handle because of their extreme reactivity in air. Alloys of Cs with Ca, Ba, or Sr are less reactive and more easily handled than pure metallic Cs. Ternary alloys of Cs and Al with Ba or Sr are used in photoelectric cells. The binary systems of Cs with Na (9%), K (25%), or Rb (14%) form eutectics. Alloys of Cs with Sb, Ag, Bi, and Au have photoelectric properties. As the Cs content increases, the electrical resistance of Sb-Cs increases sharply.

Production: Cesium generally is sold as its salts, especially chlorides, nitrates, or hydroxides. Cesium is obtained chiefly from carnallite. See also **Potassium.** An early process for producing Cs metal involved the reduction of a Cs salt, such as the chloride or hydroxide. Principally, Cs metal is obtained by electrolyzing the cyanide, which yields a pure metal.

SELECTED REFERENCES

Perelman, F. M.: "Rubidium and Caesium" (transl. by R. G. P. Townsdrown), Pergamon and Macmillan, New York, 1965.

Simons, Eric N.: "Uncommon Metals," Hart Publishing Co., New York, 1967.

Smithells, Colin J.: "Metals Reference Book," 4th ed., Plenum, New York, 1967.

—Technical Staff, Kawecki Berylco Industries, Inc., *Reading, Pa.*

Cetyl Alcohol (See **Alcohols.**)

Chain Compounds (See **Carbon compounds.**)

Chain Reaction (See **Nuclear power plants;** and **Polymerization.**)

Chalcocite (See **Copper production.**)

Chalcopyrite (See **Arsenic; Copper production; Indium;** and **Tellurium.**)

Chalk (See **Limestone.**)

Change-Can Mixers (See **Mixing, pastes.**)

Channel Carbon Black (See **Carbon.**)

Charcoal (See **Adsorption; Carbon;** and **Fuels.**)

Charles Law (See **Ideal gas.**)

Chelating Agents Chelating agents are organic chemicals which have the ability to combine with, or *chelate,* heavy and alkali metal cations in aqueous solution. A large family of these materials currently finds a variety of uses in industrial processes, water treatment, consumer and household products, and agriculture. These materials have the unique property of tightly binding the metal ion in a soluble complex compound. Chelating agents* as a class generally form chelates with metal ions which are stable over a relatively wide pH range. Generally, they are very soluble in water and insoluble in organic solvents. They are stable over a wide temperature range, but are attacked by strong oxidizing agents.

The word chelate is derived from the Greek word *chele,* which means "claw." Thus the word is descriptive of the very tight bond formed between the chelating agent and the metal ion.

*The terms "chelating agent" and "sequestering agent" essentially are synonymous. See also **Detergents.**

Two of the most commonly used chelating agents are the sodium salts of ethylenediaminetetracetic acid (EDTA) and nitrilotriacetic acid (NTA). The structure of the tetrasodium salt of EDTA is

$$\begin{array}{ccc}
\text{O} & & \text{O} \\
\| & & \| \\
\text{NaO—C—CH}_2 & & \text{CH}_2\text{—C—ONa} \\
& \diagdown \quad \diagup \quad \diagdown \quad \diagup & \\
& \text{N—CH}_2\text{—CH}_2\text{—N} & \\
& \diagup \quad \diagdown \quad \diagup \quad \diagdown & \\
\text{NaO—C—CH}_2 & & \text{CH}_2\text{—C—ONa} \\
\| & & \| \\
\text{O} & & \text{O}
\end{array}$$

When this compound reacts with the calcium ion, the resulting chelate has the structure

$$\begin{array}{ccc}
\text{O} & & \text{O} \\
\| & & \| \\
\text{NaO—C—CH}_2 & & \text{CH}_2\text{—C—ONa} \\
& \diagdown \qquad \diagup & \\
& \text{N—CH}_2\text{—CH}_2\text{—N} & \\
\text{CH}_2 & \quad \text{Ca} \quad & \text{CH}_2 \\
& \diagup \qquad \diagdown & \\
\text{O=C—O} & & \text{O—C=O}
\end{array}$$

This complex compound is generally sufficiently soluble in water to hold the calcium in solution. Other cations, such as magnesium, iron, manganese, and copper, are chelated and held in solution in the same manner.

During recent years, a considerable use for chelating agents has been created by research studies in the application of these materials in boiler-water treatment. When chelating agents are used in boiler-water treatment, the chelated metals remain in solution. Thus no sludges are formed. Heat transfer is more rapid due to clean metal surfaces. Chelating agents, such as EDTA and NTA, are used for scale prevention in hundreds of boilers operating in the pressure range 100–1,200 psig.

Chelating agents are more expensive than conventional phosphates used for boiler-water treatment, but the advantages of clean metal surfaces and no sludge formation outweigh increased chemical costs in many cases. The cost of chelate treatment can be minimized by careful handling of the condensate and by the use of new ion-exchange pretreatment processes on the boiler feedwater. Concentrations of metal cations in the feedwater can be reduced to fractional ppm values, which reduces the amount of chelating agents required.

When chelating agents are used in boiler-water treatment, care must be taken to ensure oxygen-free conditions in the boiler water. Also, overfeeding of the chelating agent can result in corrosion of the boiler system.

Current costs of chelating agents preclude their widespread use in water softening. Other methods, such as ion-exchange and chemical precipitation (lime and lime soda processes), remain the more economical. See also **Detergents.**

A partial list of *areas of use* of chelating agents includes:

Agriculture: As a carrier for micronutrients, such as Fe, Zn, Mn, and Cu, in the metabolism of both plants and animals.

Biological and food products: Traces of heavy metals are chelated to prevent adverse effects on the flavor, color, clarity, and stability of various products.

Cosmetics and toiletries: For maintaining stability, color, and clarity.

Metalworking and finishing: Effective in oxide control and removal, control of contaminating ions in plating solutions, immersion, and electroless plating.

Pulp and paper: To improve peroxide bleaching of pulp, in the removal of heavy-metal ions from pulp and paper, and in groundwood-pulp production.

Rubber and polymer production: Catalysts used for polymerization reactions, such as styrene-butadiene rubbers. Metal ions in polymer systems cause polymer breakdown, discoloration, and poor heat and light stability. Addition of a small amount of a chelating agent prevents such difficulties.

Textiles: In the cleaning, scouring, bleaching, and dyeing stages of textile manufacture to control undesirable effects of hardness and heavy-metal ions.

REFERENCES

Chaberek, Stanley, and Arthur E. Martell: "Organic Sequestering Agents: A Discussion of the Chemical Behavior and Applications of Metal Chelate Compounds in Aqueous Systems," Wiley, New York, 1959.

Dwyer, F. P., and D. P. Mellor (eds.): "Chelating Agents and Metal Chelates," Academic, New York, 1964.

Walton, Harold F.: Chelation, in "New Chemistry," Simon & Schuster, New York, 1957.

—A. Curtis Reents, *Techni-Chem. Inc., Cherry Valley, Ill.*

Chemical Compound   (See **Compound, chemical.**)

Chemical Elements   A chemical element is a substance that is made up of but one kind of atom. There are over 100 known atoms, and thus over 100 known chemical elements. Theoretical physicists do not all agree, but many believe that fission-stable nuclei should exist at atomic number 114. Predictions of half-lives of the superheavy elements vary widely. With the synthesis of element 105 (mid-1970),* the isolation and detection of additional elements enter into an area of extreme difficulty.

Of more than 100 known elements, only 90 are found in nature. The remaining elements are manufactured in nuclear reactors and particle accelerators.

A chemical element cannot be divided into simpler substances by *chemical* means; nor made by combinations of other substances. The characteristics of most of the known elements are summarized in Table C-29. The elements of large economic or scientific significance are discussed individually in this volume. Consult the Subject Index.

The chemical elements display a periodicity of properties when they are arranged in order of increasing atomic mass. This phenomenon is known as the *periodic law;* the resulting matrix arrangement is called the *periodic table.* See also **Periodic law.**

Abundance of the Chemical Elements: If one considers the cosmos—the stars and all components of the universe—it is estimated that hydrogen is the most abundant element. The second most abundant element is considered to be helium. In terms of the crust of the earth, however, oxygen is the most plentiful element, followed by silicon, aluminum, iron, calcium, sodium, potassium, magnesium, and titanium. It is estimated that these elements alone constitute over 99% of the earth's crust. In Table C-30, the terrestrial abundance of the elements is given in terms of grams per metric ton. A metric ton is equivalent to one million grams. It is interesting to compare these figures with the investigations of lunar soil.†

Isotopes: The atoms of any given stable element need not necessarily have the same exact mass. A variation in mass may arise from different numbers of neutrons in the nuclei of the atoms. For example, chlorine is composed of a mixture of two isotopes. One isotope has an atomic mass of approximately 35, with 17 protons and *18* neutrons in the nucleus. The other chlorine isotope has an atomic mass of approximately 37, with 17 protons and *20* neutrons in the nucleus. In this instance and in the case of isotopes of the other elements, the number of protons and orbital electrons (in the nonexcited, nonionized state) is the same for all isotopes and must be so to preserve the atom's unique chemical identity. It is only the number of neutrons that may vary, and these variations in turn cause slight differences in mass.

Ordinary hydrogen, for example, contains only a proton in the nucleus. There is no neutron. The heavy isotope of hydrogen, however, does contain a neutron, and hence is of greater mass. Only about 1 in 6,700 atoms of hydrogen will be found to be of the heavy variety. By concentrating these heavier isotopes of hydrogen, however, and then combining them with oxygen, heavy water (deuterium), which is approximately 11% heavier than ordinary water, is obtained.

A majority of the elements display a pattern of isotopes, as shown by Table C-31. Radioactive isotopes are not indicated in this tabulation.‡ Regardless of where an element may be found, the relative abundance of the isotopes usually is reasonably constant. Sulfur is one of the few exceptions, in that there is sufficient variation, depending upon the source of the sulfur, to cause a variation of its atomic mass by approximately $\pm 0.01\%$. For normal stoichiometric calculations, this small variation is of little practical significance. See also **Radioactive isotopes.**

---

*Researchers at the University of California, Berkeley's Lawrence Radiation Laboratory, claimed the synthesis of element 105. The heavy-ion linear accelerator at that location was used to bombard a target of ^{249}Cf with ^{15}N atoms stripped of their electrons. It is claimed that element $^{260}105$ was produced after the emission of four neutrons. The element is claimed to decay with a half-life of 1.6 ± 0.3 s to form ^{256}Lw and alpha particles. Soviet scientists at the Laboratory of Nuclear Reactions, Dubna, Moscow Region, U.S.S.R., claimed the "spontaneous fission of elements 103 and 105" a few months earlier.

†Abundance of elements in lunar rocks and soil returned by Apollo 11 averaged as follows: Si (as SiO_2), $19.3 \pm 0.9\%$; O, $39.9 \pm 1.1\%$; Ti (as TiO_2), $5.5 \pm 0.5\%$; Al (as Al_2O_3), $5.9 \pm 0.6\%$; Mg (as MgO), $4.6 \pm 0.6\%$; Fe (as Fe_2O_3 and FeO), $14.0 \pm 0.9\%$; Ca (as CaO), $7.4 \pm 0.3\%$; Na, $3,440 \pm 440$ ppm; Mn, $2,130 \pm 360$ ppm; Cu, 10 ± 2 ppm; and Cr, $2,180 \pm 360$ ppm. Very small amounts of P as P_2O_5 and of rare-earth elements also were detected. Organic compounds were reported as less than 1 ppm.

‡A nonradioactive isotope is termed a *stable isotope.*

TABLE C-29. *The Chemical Elements and Some of Their Characteristics*

| Name | | | Symbol | At. No. | At. Wt. | At. Diam.† | Valence | Date of Discovery |
English	French*	German*						
Actinium	. . .	. . .	Ac	89	[227]‡	. . .	3	1899
Alumin(i)um . .	Aluminium	Aluminium	Al	13	26.9815	2.82	3	1827
Americium . . .	Américium	. . .	Am	95	[243]‡	. . .	3	1945
Antimony	Antimoine	Antimon	Sb	51	121.75	3.228	3, 5	Early
Argon	. . .	. . .	Ar	18	39.948	3.82	0	1894
Arsenic	. . .	Arsen	As	33	74.9216	2.50	5, 3$^+$, 3$^-$	Early
Astatine	. . .	. . .	At	85	[210]‡	. . .	. . .	1940
Barium	. . .	Baryum	Ba	56	137.34	4.48	2	1808
Berkelium	. . .	. . .	Bk	97	[249]‡	. . .	4, 3	1950
Beryllium	Béryllium	. . .	Be	4	9.0122	2.25	2	1798
Bismuth	. . .	Wismut	Bi	83	208.980	3.64	3, 5	1753
Boron	Bore	Bor	B	5	10.811§	. . .	3	1808
Bromine	Brome	Brom	Br	35	79.909¶	2.26	1$^+$, 1$^-$, 5	1826
Cadmium	. . .	Kadmium	Cd	48	112.40	3.042	2	1817
Calcium	. . .	Kalzium	Ca	20	40.08	3.93	2	1808
Californium . . .	. . .	. . .	Cf	98	[251]‡	. . .	3	1950
Carbon	Carbone	Kohlenstoff	C	6	12.01115§	1.54	4$^+$, 4$^-$, 2	Early
Cerium	Cérium	Cer	Ce	58	140.12	3.64	3, 4	1803
Cesium	Césium	Caesium	Cs	55	132.905	5.40	1	1860
Chlorine	Chlore	Chlor	Cl	17	35.453¶	1.94	1$^+$, 1$^-$, 7, 5	1774
Chromium . . .	Chrome	Chrom	Cr	24	51.996¶	2.57	6, 3, 2	1797
Cobalt	. . .	Kobalt	Co	27	58.9332	2.50	3, 2	1735
Columbium	. . .	. . .	Cb	41	92.906	2.94	5, 3	1801
Copper	Cuivre	Kupfer	Cu	29	63.54	2.551	2, 1	Early
Curium	. . .	. . .	Cm	96	[247]‡	. . .	3	1944
Dysprosium . . .	. . .	. . .	Dy	66	162.50	3.54	3	1886
Einsteinium . . .	. . .	. . .	E	99	[254]‡	. . .	. . .	1955
Erbium	. . .	. . .	Er	68	167.26	3.50	3	1843
Europium	. . .	. . .	Eu	63	151.96	4.08	3, 2	1896
Fermium	. . .	. . .	Fm	100	[253]‡	. . .	. . .	1955
Fluorine	Fluor	Fluor	F	9	18.9984	1.36	1$^-$	1771
Francium	. . .	. . .	Fr	87	[223]‡	. . .	1	1939
Gadolinium . . .	. . .	. . .	Gd	64	157.25	3.59	3	1880
Gallium	. . .	. . .	Ga	31	69.72	2.7	3	1875
Germanium . . .	. . .	. . .	Ge	32	72.59	2.788	4	1886
Gold	Or	. . .	Au	79	196.967	2.878	3, 1	Early
Hafnium	. . .	. . .	Hf	72	178.49	3.17	4	1923
Helium	Hélium	. . .	He	2	4.0026	. . .	0	1895
Holmium	. . .	. . .	Ho	67	164.930	3.52	3	1879
Hydrogen	Hydrogène	Wasserstoff	H	1	1.00797§	3.0	1	1766
Indium	. . .	. . .	In	49	114.82	3.14	3	1863
Iodine	Iode	Jod	I	53	126.9044	2.7	1$^-$, 5, 7	1811
Iridium	. . .	. . .	Ir	77	192.2	2.709	3, 4, 6	1803
Iron	Fer	Eisen	Fe	26	55.847¶	2.52	3, 2	Early
Krypton	. . .	. . .	Kr	36	83.80	4.0	0	1898
Lanthanum . . .	Lanthane	Lanthan	La	57	138.91	3.741	3	1839
Lawrencium . .	. . .	. . .	Lw	103	257	. . .	. . .	1961
Lead	Plomb	Blei	Pb	82	207.19	3.49	2, 4	Early
Lithium	. . .	. . .	Li	3	6.939	3.13	1	1817
Lutetium	Lutétium	. . .	Lu	71	174.97	3.47	3	1907
Magnesium . . .	Magnésium	. . .	Mg	12	24.312	3.20	2	1755
Manganese . . .	Manganèse	Mangan	Mn	25	54.9380	2.5	7, 4, 2, 6, 3	1774
Mendelevium . .	. . .	. . .	Md	101	[256]‡	. . .	. . .	1955
Mercury	Mercure	Quecksilber	Hg	80	200.59	3.10	2, 1	Early
Molybdenum . .	Molybdène	Molybdän	Mo	42	95.94	2.80	6, 3, 5	1778
Neodymium . .	Néodyme	Neodym	Nd	60	144.24	3.63	3	1885

English	French*	German*	Symbol	At. No.	At. Wt.	At. Diam.†	Valence	Date of Discovery
Neon	Néon	. . .	Ne	10	20.183	3.20	0	1898
Neptunium . . .	. . .	. . .	Np	93	[237]‡	. . .	6, 5, 4, 3	1940
Nickel	. . .	. . .	Ni	28	58.71	2.49	2, 3	1751
Niobium (Nb) (see Colum- bium)								
Nitrogen	Nitrogène	Stickstoff	N	7	14.0067	1.06	3⁻, 5, 2	1772
Nobelium	. . .	. . .	No	102	[254]‡	. . .	. . .	1957
Osmium	. . .	. . .	Os	76	190.2	2.70	4, 6, 8	1803
Oxygen	Oxygène	Sauerstof	O	8	15.9994§	. . .	2⁻	1774
Palladium	. . .	. . .	Pd	46	106.4	2.745	2, 4	1803
Phosphorus . . .	Phosphore	Phosphor	P	15	30.9738	2.16	5, 3⁺, 3⁻	1669
Platinum	Platine	Platin	Pt	78	195.09	2.769	4, 2	1735
Plutonium . . .	. . .	. . .	Pu	94	[242]‡	. . .	6, 5, 4, 3	1940
Polonium	. . .	. . .	Po	84	[210]‡	. . .	2, 4	1898
Potassium	. . .	Kalium	K	19	39.102	4.76	1	1807
Praseodymium .	Praséodyme	Praseodym	Pr	59	140.907	3.65	3	1879
Promethium . .	Prométheum	. . .	Pm	61	[147]‡	. . .	3	1947
Protactinium . .	. . .	. . .	Pa	91	[231]‡	. . .	5	1917
Radium	. . .	. . .	Ra	88	[226]‡	. . .	2	1898
Radon	. . .	. . .	Rn	86	[222]‡	. . .	0	1900
Rhenium	. . .	. . .	Re	75	186.2	2.75	7, 4, 1⁻	1925
Rhodium	. . .	. . .	Rh	45	102.905	2.7	3, 4	1803
Rubidium	. . .	. . .	Rb	37	85.47	5.02	1	1861
Ruthenium . . .	. . .	. . .	Ru	44	101.07	2.67	3, 4, 6, 8	1844
Samarium	. . .	. . .	Sm	62	150.35	. . .	3	1879
Scandium	. . .	. . .	Sc	21	44.956	3.20	3	1879
Selenium	Sélénium	Selen	Se	34	78.96	2.32	6, 4, 2⁻	1817
Silicon	Silcium	Silicium	Si	14	28.086§	2.34	4	1823
Silver	Argent	Silber	Ag	47	107.870¶	2.883	1	Early
Sodium	. . .	Natrium	Na	11	22.9898	3.83	1	1807
Strontium	. . .	. . .	Sr	38	87.62	4.29	2	1790
Sulfur	Soufre	Schwefel	S	16	32.064§	2.12	6, 4, 2⁻	Early
Tantalum	Tantale	Tantal	Ta	73	180.948	2.94	5	1802
Technetium . . .	. . .	. . .	Tc	43	[99]‡	. . .	7	1937
Tellurium	Tellure	Tellur	Te	52	127.60	2.9	4, 6, 2⁻	1782
Terbium	. . .	. . .	Tb	65	158.924	3.54	3	1843
Thallium	. . .	. . .	Tl	81	204.37	3.42	1, 3	1861
Thorium	. . .	. . .	Th	90	232.038	3.6	4	1828
Thulium	. . .	. . .	Tm	69	168.934	3.48	3	1879
Tin	Étain	Zinn	Sn	50	118.69	3.164	4, 2	Early
Titanium	Titane	Titan	Ti	22	47.90	2.93	4, 3	1791
Tungsten	Tungstène	Wolfram	W	74	183.85	2.82	6	1781
Uranium	. . .	Uran	U	92	238.03	3.0	6, 5, 4, 3	1789
Vanadium . . .	. . .	. . .	V	23	50.942	2.71	5, 4, 2	1830
Xenon	Xénon	. . .	Xe	54	131.30	4.4	0	1898
Ytterbium . . .	. . .	. . .	Yb	70	173.04	3.87	3, 2	1878
Yttrium	. . .	. . .	Y	39	88.905	3.62	3	1794
Zinc	. . .	Zink	Zn	30	65.37	2.748	2	1746
Zirconium . . .	. . .	Zircon	Zr	40	91.22	3.19	4	1789

*Where there is no name in the column, the name is essentially the same as in English.

†Atomic diameters are expressed in 10^{-8} cm. Thus, for example, from this table it will be seen that the atomic diameter of zirconium is about 3.19×10^{-8} cm. It is interesting to note that the diameter of the largest atom is only about five times the diameter of the smallest.

‡Value in brackets indicates the mass number of the isotope of longest known half-life (or a better-known one, for Bk, Cf, Po, Pm, and Tc).

§Atomic weight varies because of natural variation in isotopic composition: B, ±0.003; C, ±0.00005; H, ±0.00001; Si, ±0.003.

¶Atomic weight is believed to have the following uncertainty: Br, ±0.002; Cl, ±0.001; Cr, ±0.001; Fe, ±0.003; Ag, ±0.003. For other elements, the last digit given for the atomic weight is believed correct and reliable to ±0.5.

TABLE C-30. *Terrestrial Abundance of the Chemical Elements*

Element	g/metric ton	Element	g/metric ton
O	466,000	Pr	5.53
Si	277,200	As	5
Al	81,300	Sc	5
Fe	50,000	Hf	4.5
Ca	36,300	Dy	4.47
Na	28,300	U	4
K	25,900	B	3
Mg	20,900	Tl	3
Ti	4,400	Yb	2.66
H	1,300	Er	2.47
P	1,180	Ta	2.1
Mn	1,000	Br	1.62
F	900	Ho	1.15
S	520	Eu	1.06
C	320	Sb	1
Cl	314	Tb	0.91
Rb	310	Lu	0.75
Sr	300	Hg	0.50
Ba	250	I	0.30
Zr	220	Tm	0.20
Cr	200	Bi	0.20
V	150	Cd	0.15
Zn	132	Ag	0.10
Ni	80	In	0.10
Cu	70	Se	0.09
W	69	Ar	0.04
Li	65	Pd	0.01
N	46.3	Au	0.005
Ce	46.1	Os	0.005
Sn	40	Pt	0.005
Y	28.1	Ru	0.004
Co	24	He	0.003
Nd	23.9	Te	0.002
Co	23	Re	0.001
La	18.3	Rh	0.001
Pb	16	Ir	0.001
Ga	15	Ne	7×10^{-5}
Mo	15	Ra	13×10^{-6}
Th	11.5	Kr	9.8×10^{-6}
Cs	7	Xe	1.2×10^{-6}
Ge	7	Pa	8×10^{-7}
Sm	6.47	Ac	3×10^{-10}
Gd	6.36	Po	3×10^{-10}
Be	6		

Aside from radioactive isotopes which can be produced artificially, the following elements do not display any isotopic behavior: helium, beryllium, fluorine, sodium, aluminum, phosphorus, scandium, manganese, cobalt, arsenic, yttrium, niobium (columbium), rhodium, iodine, praseodymium, terbium, holmium, thulium, gold, and bismuth. In the cases of the following elements, one isotope predominates in excess of 98%: hydrogen, carbon, nitrogen, oxygen, argon, vanadium, lanthanum, lutetium, and tantalum. Elements in which no one isotope is in excess of 80% of the total, and consequently where isotopic behavior is quite significant, include magnesium, chlorine, titanium, nickel, copper, zinc, gallium, germanium, selenium, bromine, krypton, rubidium, zirconium, molybdenum, ruthenium, palladium, silver, cadmium, tin, antimony, tellurium, xenon, barium, gadolinium, dysprosium, erbium, ytterbium, hafnium, tungsten, rhenium, osmium, iridium, platinum, mercury, tellurium, and lead.

As regards numbers of isotopes, tin leads with a total of 10. Xenon has 9 isotopes. Cadmium and tellurium each have 8 isotopes. Several elements have 7 isotopes: ytterbium, molybdenum, ruthenium, mercury, osmium, barium, neodymium, samarium, gadolinium, and dysprosium. Calcium, selenium, krypton, hafnium, palladium, platinum, and erbium each have 6 isotopes. Titanium, nickel, zinc, germanium, zirconium, and tungsten each have 5 isotopes, while chromium, iron, strontium, lead, and cerium have 4 isotopes.

Allotropes: Some of the chemical elements exist in two or more forms which display distinctly different physical properties, often accompanied by differences in chemical properties. Carbon, which occurs in the forms of diamond, graphite, lampblack, and charcoal, is an example of allotropic behavior. For any given element, the atoms must be distinct and alike. Therefore allotropy is attributed to differences in the manner in which the atoms are bound together. *Enantiomorphic allotropy* is the term used to designate that form of allotropy which is reversible and where transition from one allotropic form to the other occurs at a specific temperature. The α- and β-forms of sulfur exemplify this behavior. In *dynamic allotropy,* the transition still is reversible, but need not occur at a specific transition temperature. When transition is irreversible, the term *monotropic allotropy* applies. Examples of allotropic forms include:

ARSENIC: four forms (metallic, yellow, gray, and brown).

TIN: two forms (gray and white).

CARBON: three forms (diamond, graphite, and amorphous).

SELENIUM: four forms (amorphous, two crystalline monoclinic forms, red in color, and the stable, crystalline-gray metallic form).

THORIUM: two forms (crystalline and amorphous).

TABLE C-31. *Stable Isotopes and Isotopic Abundance of the Chemical Elements*

Element	Mass No. A	Isotopic Abundance, atomic %	Element	Mass No. A	Isotopic Abundance, atomic %	Element	Mass No. A	Isotopic Abundance, atomic %
H*	1	99.9851†	Cr	52	83.76	Co	93	100
	2	0.0149‡		53	9.55	Mo	92	15.86
He	4	~100		54	2.38		94	9.12
Li	6	7.5	Mn	55	100		95	15.70
	7	92.5	Fe	54	5.84		96	16.50
Be	9	100		56	91.68		97	9.45
B	10	18.7		57	2.17		98	23.75
	11	81.3		58	0.31		100	9.62
C	12	98.89	Co	59	100	Ru	96	5.5
	13	1.11	Ni	58	67.8		98	1.9
N	14	99.635		60	26.2		99	12.7
	15	0.365		61	1.25		100	12.7
O	16	99.759		62	3.66		101	17.0
	17	0.037		64	1.16		102	31.5
	18	0.204	Cu	63	69.1		104	18.7
F	19	100		65	30.9	Rh	103	100
Ne	20	90.92	Zn	64	48.89	Pd	102	0.96
	21	0.257		66	27.81		104	10.97
	22	8.82		67	4.11		105	22.23
Na	23	100		68	18.56		106	27.33
Mg	24	78.60		70	0.62		108	26.71
	25	10.11	Ga	69	60.2		110	11.81
	26	11.29		71	39.8	Ag	107	51.35
Al	27	100	Ge	70	20.55		109	48.65
Si	28	92.18		72	27.37	Cd	106	1.21
	29	4.71		73	7.67		108	0.88
	30	3.12		74	36.74		110	12.39
P	31	100		76	7.67		111	12.75
S	32	95.018	As	75	100		112	24.07
	33	0.750	Se	74	0.87		113	12.26
	34	4.215		76	9.02		114	28.86
	36	0.017		77	7.58		116	7.58
Cl	35	75.53		78	23.52	In	113	4.23
	37	24.47		80	49.82		115	95.77
Ar	36	0.34		82	9.19	Sn	112	0.95
	38	0.06	Br	79	50.54		114	0.65
	40	99.60		81	49.46		115	0.34
K	39	93.08	Kr	78	0.354		116	14.24
	40	0.0119		80	2.27		117	7.57
	41	6.91		82	11.56		118	24.01
Ca	40	96.96		83	11.55		119	8.58
	42	0.64		84	56.90		120	32.97
	43	0.145		86	17.37		122	4.71
	44	2.07	Rb	85	72.15		124	5.98
	46	0.0033		87	27.85	Sb	121	57.25
	48	0.185	Sr	84	0.55		123	42.75
Sc	45	100		86	9.87	Te	120	0.089
Ti	46	7.99		87	7.02		122	2.46
	47	7.32		88	82.56		123	0.87
	48	73.99	Y	89	100		124	4.61
	49	5.46	Zr	90	51.46		125	6.99
	50	5.25		91	11.23		126	18.71
V	50	0.24		92	17.11		128	31.79
	51	99.76		94	17.40		130	34.49
Cr	50	4.31		96	2.80	I	127	100

*Hydrogen has a third isotope (tritium) which has been prepared in nuclear reactors.
†Protium (average in ocean water).
‡Deuterium (varies).

TABLE C-31. *Stable Isotopes and Isotopic Abundance of the Chemical Elements* (*Continued*)

Element	Mass No. A	Isotopic Abundance, atomic %	Element	Mass No. A	Isotopic Abundance, atomic %	Element	Mass No. A	Isotopic Abundance, atomic %
Xe	124	0.096	Eu	153	52.23	Ta	180	0.012
	126	0.090	Gd	152	0.20		181	99.988
	128	1.919		154	2.15	W	180	0.135
	129	26.44		155	14.73		182	26.4
	130	4.08		156	20.47		183	14.4
	131	21.18		157	15.68		184	30.6
	132	26.89		158	24.87		186	28.4
	134	10.44		160	21.90	Re	185	37.07
	136	8.87	Tb	159	100		187	62.93
Cs	133	100	Dy	156	0.052	Os	184	0.018
Ba	130	0.101		158	0.090		186	1.59
	132	0.097		160	2.29		187	1.64
	134	2.42		161	18.88		188	13.3
	135	6.59		162	25.53		189	16.1
	136	7.81		163	24.97		190	26.4
	137	11.32		164	28.18		192	41.0
	138	71.66	Ho	165	100	Ir	191	38.5
La	138	0.089	Er	162	0.136		193	61.5
	139	99.911		164	1.56	Pt	190	0.012
Ce	136	0.193		166	33.4		192	0.78
	138	0.250		167	22.9		194	32.8
	140	88.48		168	27.1		195	33.7
	142	11.07		170	14.9		196	25.4
Pr	141	100	Tm	169	100		198	7.2
Nd	142	27.09	Yb	168	0.14	Au	197	100
	143	12.14		170	3.03	Hg	196	0.146
	144	23.83		171	14.31		198	10.02
	145	8.29		172	21.82		199	16.84
	146	17.26		173	16.13		200	23.13
	148	5.74		174	31.84		201	13.22
	150	5.63		176	12.73		202	29.80
Sm	144	3.16	Lu	175	97.4		204	6.85
	147	15.07		176	2.6	Tl	203	29.50
	148	11.27	Hf	174	0.19		205	70.50
	149	13.84		176	5.19	Pb	204	1.5
	150	7.47		177	18.47		206	23.6
	152	26.63		178	27.15		207	22.6
	154	22.53		179	13.75		208	52.3
Eu	151	47.77		180	35.22	Bi	209	100

BORON: two forms (crystalline and amorphous).

PHOSPHORUS: four forms (two white forms and violet and black). Red phosphorus is a mixture of the white and violet forms.

SULFUR: two forms, α-rhombic sulfur with a density of 2.07 and a melting point of 112.8°C, and β-monoclinic sulfur with a density of 1.96 and a melting point of 119°C. β-Monoclinic sulfur changes to α-rhombic sulfur below 96°C.

In the instances of some of the less frequently occurring elements, impure forms have been mistakenly identified as allotropic forms. A number of elements that once were considered to exist in both crystalline and amorphous forms were found to exist in only one form when perfectly pure.

Gaseous Elements: Several gaseous elements (at standard conditions of temperature and pressure) form molecules of two atoms each and are known as *diatomic gases*. These include hydrogen, H_2; nitrogen, N_2; oxygen O_2; and chlorine, Cl_2. The inert gases helium, neon, argon, krypton, and xenon are *monatomic gases,* and their symbols do not carry a subscript.

Melting and Boiling Points: The chemical elements are arranged in order of increasing melting point in Table C-32. It is interesting to note how relatively few gases and, in particular, how few liquids there are among the elements at standard temperatures and pressures.

Radioactive Elements: Elements that display radioactivity fall into two broad classes: (1) naturally occurring radioactive elements, and (2) artificially produced radioactive elements.

Naturally occurring radioactive elements are divided into three series:

1. The *uranium-radium* series, which begins with uranium 238 (^{238}U) with a half-life of 4.51×10^9

years and ends with lead 206 (^{206}Pb). The progression of radioactive decay (with type of radiation emitted shown above arrows) is

$$^{238}\text{U} \xrightarrow{\alpha} {}^{234}\text{Th} \xrightarrow{\beta} {}^{234}\text{Pa} \xrightarrow{\beta} {}^{234}\text{U} \xrightarrow{\alpha}$$

$$^{230}\text{Th} \xrightarrow{\alpha} {}^{226}\text{Ra} \xrightarrow{\alpha} {}^{222}\text{Rn} \xrightarrow{\alpha} {}^{218}\text{Po} \xrightarrow{\alpha \text{ and } \beta}$$

$$^{214}\text{Pb} \xrightarrow{\beta} {}^{218}\text{At} \xrightarrow{\alpha} {}^{214}\text{Bi} \xrightarrow{\beta \text{ and } \alpha} {}^{214}\text{Po} \xrightarrow{\alpha}$$

$$^{210}\text{Tl} \xrightarrow{\beta} {}^{210}\text{Pb} \xrightarrow{\beta} {}^{210}\text{Bi} \xrightarrow{\beta \text{ and } \alpha} {}^{210}\text{Po} \xrightarrow{\alpha}$$

$$^{206}\text{Tl} \xrightarrow{\beta} {}^{206}\text{Pb(stable)}$$

2. The *thorium series,* which starts with thorium 232 (^{232}Th) with a half-life of 1.39×10^{10} years and ends with lead 208 (^{208}Pb). The progression of the series is

TABLE C-32. *Melting Points and Boiling Points of Chemical Elements (Arranged in Order of Ascending Melting Point)*

Element	Mp, °C	Bp, °C	Element	Mp, °C	Bp, °C	Element	Mp, °C	Bp, °C
He	−269.7	−218.9	Zn	419.5	906	Dy	1407	2600
H	−259.2	−252.7	Te	449.5	989.8	Si	1410	2680
Ne	−248.6	−246	Sb	630.5	1380	Ni	1453	2730
F	−219.6	−188.2	Np	637	. . .	Ho	1461	2600
O	−218.8	−183	Pu	640	3235	Co	1495	2900
N	−210	−195.8	Mg	650	1107	Er	1497	2900
Ar	−189.4	−185.8	Al	660	2450	Y	1509	2927
Kr	−157.3	−152	Ra	700	1140	Fe	1536	3000
Xe	−111.9	−108.0	Ba	714	1640	Sc	1539	2730
Cl	−101.0	− 34.7	Sr	768	1380	Tm	1545	1727
Rn	− 71	− 61.8	Ce	795	3468	Pd	1552	3980
Hg	− 38.4	357	As	817*	616s	Lu	1652	3327
Br	− 7.2	58	Yb	824	1427	Ti	1668	3260
Cs	28.7	690	Eu	826	1439	Th	1750	3850
Ga	29.8	2237	Ca	838	1440	Pt	1769	4530
Rb	38.9	688	La	920	3470	Zr	1852	3580
P	44.2	280	Pr	935	3127	Cr	1875	2665
K	63.7	760	Ge	937.4	2830	V	1900	3450
Na	97.8	892	Ag	960.8	2210	Rh	1966	4500
Li	108.5	1330	Nd	1024	3027	B	2030	†
I	113.7	183	Pm	1027	2460	Tc	2200	2200‡
S	119.0	444.6	Ac	1050	3300	Hf	2222	5400
In	156.2	2000	Au	1063	2970	Nb	2415	3500
Se	217	685	Sm	1072	1900	Ir	2454	5300
Sn	231.9	2270	Cu	1083	2595	Ru	2500	4900
Po	254	. . .	U	1132	3818	Mo	2610	5560
Bi	271.3	1560	Pa	1230	2200	Os	2700	5500
At	302	. . .	Mn	1245	2150	Ta	2996	5425
Tl	303	1457	Be	1277	2770	Re	3180	5900
Cd	320.9	765	Gd	1312	3000	W	3410	5930
Pb	327.4	1725	Tb	1356	2800	C	3727	4830

*At 35.8 atm.
†Boron sublimes consideraly below its melting point.
‡±50°C.
s = sublimes.

$$^{232}Th \xrightarrow{\alpha} {}^{228}Ra \xrightarrow{\beta} {}^{228}Ac \xrightarrow{\beta} {}^{228}Th \xrightarrow{\alpha}$$

$$^{224}Ra \xrightarrow{\alpha} {}^{220}Rn \xrightarrow{\alpha} {}^{216}Po \xrightarrow{\alpha} {}^{212}Pb \xrightarrow{\beta \text{ and } \alpha}$$

$$^{216}At \xrightarrow{\alpha} {}^{212}Bi \xrightarrow{\beta \text{ and } \alpha} {}^{212}Po \xrightarrow{\alpha} {}^{208}Tl \xrightarrow{\beta}$$

$$^{208}Pb(stable)$$

3. The *actinium series,* which begins with uranium 235 (^{235}U) with a half-life of 7.13×10^8 years and ends with lead 207 (^{207}Pb). The progression of the series is

$$^{235}U \xrightarrow{\alpha} {}^{231}Th \xrightarrow{\beta} {}^{231}Pa \xrightarrow{\alpha} {}^{227}Ac \xrightarrow{\beta \text{ and } \alpha}$$

$$^{227}Th \xrightarrow{\alpha} {}^{223}Fr \xrightarrow{\beta} {}^{223}Ra \xrightarrow{\alpha} {}^{219}Rn \xrightarrow{\alpha}$$

$$^{215}Po \xrightarrow{\alpha \text{ and } \beta} {}^{211}Pb \xrightarrow{\beta} {}^{215}At \xrightarrow{\alpha}$$

$$^{211}Bi \xrightarrow{\beta \text{ and } \alpha} {}^{211}Po \xrightarrow{\alpha} {}^{207}Tl \xrightarrow{\beta} {}^{207}Pb(stable)$$

Taken altogether, the three series include 42 intermediates or radionuclides. Half-lives range from 3.04×10^{-7} s for ^{212}Po to 2.48×10^5 years for ^{234}U. In addition to the foregoing three radioactive series, there are 12 other naturally occurring long-lived radionuclides. Potassium, tungsten, and platinum are all slightly radioactive because of the presence of natural radioisotopes; and Bi, usually considered to be stable, actually may be an alpha emitter with a half-life of about 2×10^{17} years.

The production of artificially induced radioactive elements and artificially induced nuclear reactions commenced with Rutherford's work in 1919 when he discovered that alpha particles reacted with nitrogen atoms to yield protons and oxygen atoms. In 1933, Curie and Joliot discovered that when boron, magnesium, and aluminum were bombarded with alpha particles from polonium, these elements would emit neutrons, protons, and positrons. They also found that when the source of bombardment was removed, the protons and neutrons ceased to be emitted, but that the emission of positrons continued. The targets remained radioactive. Further, the radiation emitted fell off exponentially, as one would expect, from a naturally occurring radioactive element. A continuation of this investigation showed that nuclear reactions lead to the formation of radioactive isotopes.

Much additional work by numerous scientific investigators in this field ultimately led to the formulation of still another series, namely, the *neptunium series,* which commences with plutonium 241 (^{241}Pu) with a half-life of approximately ten years and ends with bismuth 209 (^{209}Bi). The progression of the series is

$$^{241}Pu \xrightarrow{\beta} {}^{241}Am \xrightarrow{\alpha} {}^{237}Np \xrightarrow{\alpha} {}^{233}Pa \xrightarrow{\beta}$$

$$^{233}U \xrightarrow{\alpha} {}^{229}Th \xrightarrow{\alpha} {}^{225}Ra \xrightarrow{\beta} {}^{225}Ac \xrightarrow{\alpha}$$

$$^{221}Fr \xrightarrow{\alpha} {}^{217}At \xrightarrow{\alpha} {}^{213}Bi \xrightarrow{\beta \text{ and } \alpha} {}^{213}Po \xrightarrow{\alpha}$$

$$^{209}Tl \xrightarrow{\beta} {}^{209}Pb \xrightarrow{\beta} {}^{209}Bi(stable)$$

Through continuing research in atomic physics and nucleonics, aided by particle accelerators and nuclear reactor research, essentially during the 1940s and 1950s, several new elements were confirmed.

Neptunium: With nuclide mass numbers 234–240 with half-lives ranging from 2.10 days to 2.20×10^6 years. Mass number 239 is produced by the absorption of resonance neutrons by atoms of uranium of mass number 238, followed by the emission of an electron from the nucleus. ^{237}Np is formed from ^{237}U, the latter created by the cyclotron bombardment of ^{238}U.

Plutonium: With nuclide mass numbers 236–243, with half-lives ranging from 4.98 hr to 3.8×10^5 years. Mass number 239 results from the emission of electrons from ^{239}Np.

Americium: With nuclide mass numbers 238–243, with half-lives ranging from 1.86 hr to 8×10^3 years. Mass number 241 results from the bombardment of plutonium with neutrons in a chain-reacting uranium-graphite structure. ^{241}Am is yielded by beta decay of the resulting ^{241}Pu.

Curium: With nuclide mass numbers 242–248, with half-lives ranging from 162.5 days to a period approaching, but not exceeding, 10^6 years. G. T. Seaborg, R. A. James, and A. Ghiorso, in 1944, noted the presence of ^{242}Cm in a product that resulted from the bombardment of ^{239}Pu with alpha particles of resonance energies. L. B. Werner and I. Perlman first isolated ^{242}Cm by the action of neutrons on ^{241}Am.

Promethium: With nuclide mass numbers 145–151, with half-lives ranging from 2.7 hr to approximately thirty years. Mass numbers 147 and 149 were identified among the fission products of uranium by J. A. Marinsky, L. E. Glendenin, and C. D. Coryell. The same isotopes were identified in the material resulting from the bombardment of stable neodymium with slow neutrons in the pile.

Berkelium: With nuclide mass numbers 246–250, with half-lives ranging from 3.1 hr to 7×10^3 years. S. G. Thompson, A. Ghiorso, and G. T.

Seaborg produced ^{243}Bk by the bombardment of ^{241}Am with 35-MeV alpha particles.

Californium: With nuclide mass numbers 248–254, with half-lives ranging from 18 days to approximately 700 years. S. G. Thompson, K. Street, Jr., A. Ghiroiso, and G. T. Seaborg discovered and produced ^{244}Cf by bombarding ^{242}Cm with 35-MeV alpha particles.

Einsteinium: With nuclide mass numbers 251–255, with half-lives ranging from 1.5 to approximately 300 days. Both einsteinium and fermium resulted from the decay of very heavy uranium isotopes formed by the action of the instantaneous neutron flux on the uranium in a device used for a thermonuclear explosion.

Fermium: With nuclide mass numbers 253–256, with half-lives ranging from 3.1 hr to 5 days.

Mendelevium: With nuclide mass number 256, although other nuclides have been reported. Half-life of mass number 256 is 1 hr. The element was produced by bombardment of einsteinium with a beam of 41-MeV alpha particles.

Nobelium: With nuclide mass number 253 and a half-life of approximately 11 min. An isotope of this element was first reported in 1957 by a team of scientists at the Nobel Institute for Physics in Stockholm. An isotope of nobelium was later identified by A. Ghiorso, T. Sikkeland, J. R. Walton, and G. T. Seaborg at the University of California Radiation Laboratory. The isotope resulted from the bombardment of an isotope of californium with ^{12}C ions accelerated in a heavy-ion linear accelerator at Berkeley.

Lawrencium: The discovery of this element was reported by A. Ghiorso, T. Sikkeland, A. E. Larsh, and R. M. Latimer at the Lawrence Radiation Laboratory of the University of California at Berkeley. Microgram quantities of californium were bombarded with boron nuclei having energies of about 70 MeV. The element is reported to have a half-life of about 8 s.

The properties of all the foregoing transuranium elements are similar. They form soluble chlorides, nitrates, sulfates, and perchlorates and acid-insoluble oxalates and trifluorides. They have electropositive trivalent ions.

Chemical Engineering Chemical engineering is primarily the *equipment* component of applied chemistry, particularly equipment for pilot and production plants as contrasted with apparatus for the chemical laboratory. Even the theoretical aspects of chemical engineering generally will be found to have an ultimate objective of diagnosing and improving the design and operation of equipment.

Chemical engineering is concerned with the fluid-bulk processing plants as contrasted with discrete-piece manufacturing. See **Process industries.** By no means limited to a few operations or to a few industries, such as the chemical and petroleum industries, chemical engineering plays an important part in the design and operation of all manufacturing processes that involve gases, liquids, and bulk solids. Such processes will be found in the following industries: pharmaceutical; explosives; fats and oils; fertilizers and agricultural chemicals; fibers and textile; foods and beverages; leather; lime and cement; metallurgical and metal products; mining and ore beneficiation; soap and detergent; stone, clay, glass, and ceramic; wood, pulp, and paper; and chemical and petroleum *industries.*

A relatively small percentage of the equipment designed and used by the chemical engineer is concerned with chemical reactions per se. Most of the equipment is essentially mechanical or mechanical-thermal in nature, and most of the operations performed are physical rather than chemical. A large percentage of the equipment is electrically powered, and a majority of the equipment is constructed of metals. Thus chemical engineering combines several disciplines, including inorganic, organic, and physical chemistry, with electrical, mechanical, metallurgical, and sometimes nuclear engineering.

The references at the end of this description are indicative of the technological content of chemical engineering. These references include publications available from the American Institute of Chemical Engineers.

Processes and Chemical Unit Operations. The term process normally connotes a series of operations, more appropriately expressed as a series of chemical unit operations. The design and operation of a processing plant essentially involves a serious exercise in the management of materials and energy. Unlike most other manufacturing situations where there is little or no energy generated within the process per se, the process industries do gain energy from some chemical reactions. On balance, however, these gains are relatively small, and frequently become more of a problem than a help because of the additional energy required for cooling. Because such a large portion of the process industries deals with materials in fluid form, energy recovery and recycling is relatively easy and, consequently, widely practiced. In some plants, a significant portion of the equipment investment may go into heat-recovery and heat-exchange equipment in an effort to keep energy costs to an absolute minimum. As shown in Fig. C-13, recycling is not confined to energy, but

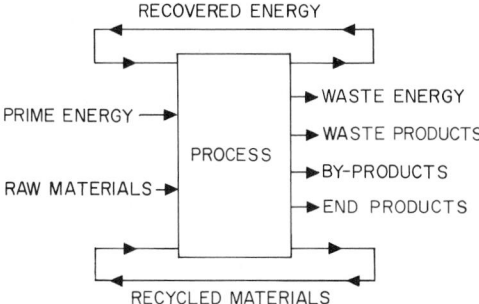

Fig. C-13. Highly generalized diagram of a process. A process converts raw materials into materials of different forms and properties, generally of higher economic value, for which there is a market. Seldom is there a fully complete (one-to-one) conversion of raw materials into desired end products. By-products and waste products are formed. A major process-design objective is to minimize resulting low-value by-products and to reduce waste products to an absolute minimum. Waste products not only represent a loss of material and applied energy, but they also require the expenditure of additional energy, and sometimes of materials to make them suitable for release to the environment. Wherever possible, energy is recovered via various heat-exchange media and recycled for further use in the process. Where waste heat must be removed, this usually is accomplished by way of evaporative cooling towers and ponds. Regulations are becoming increasingly rigid as regards dumping of heat into rivers and lakes, and even the ocean, because of the adverse effects of thermal pollution. To increase the yield of the most valuable products of a process, partially finished materials often are returned (as reflux to a distillation column, for example) to earlier steps of the process for further conversion, separation, and purification.

applies to materials as well. In many operations, by returning a portion of the product to the process, yields of purer products without the need for several process stages are possible. Much of the complexity of a processing plant is attributable to the extensive recycling of energy and materials.

As shown in Fig. C-14, a plant usually is comprised of several processing units. In the interest of conserving material-transportation costs, it is not infrequent to find two or more associated plants in adjacent locations where the by-products

of one plant become the feedstock of another. See **Petrochemical complex.**

As one analyzes the manufacture of products that involve gases, liquids, and bulk solids, many of the same operations appear again and again. Materials have to be heated, cooled, reacted, adsorbed, absorbed, dried, filtered, crushed, pumped, conveyed, mixed, separated, concentrated, diluted—whether the process relates to ore beneficiation, food processing, petroleum refining, textile dyeing, or chemical manufacture. These chemical unit operations, in essence, are the building blocks of a process. Sometimes, the term *unit process* is used also to refer to certain chemical reaction operations that appear with some frequency, such as esterification, halogenation, alkylation, and hydrogenation. As shown in Fig. C-15, processing-equipment design objectives fall into six fundamental categories; as follows.

Energy Systems: Equipment designed to maintain an energy balance within a process, adding or taking away energy, predominantly thermal and electric energy. See **Energy systems for processes.**

Separation Operations: Equipment designed to break a mixture into its components, as separating one gas from another; one liquid from another; a gas from a solid; a solid from a liquid; and so on. See **Separation operations.**

Dispersion Operations: Equipment designed to create mixtures and blends of gases with gases; liquids with liquids; solids with solids; gases with liquids; and so on. See **Dispersions.**

Size-Change Operations: Equipment required to increase or decrease the size (and usually shape) of solid materials, such as crushing and grinding coarse materials into powders or even colloidal particles; agglomerating and granulating to increase product size and to control product shape. See **Agglomeration; and Size reduction.**

Reaction Operations: Equipment designed to effect a chemical reaction between two or more materials, usually under very carefully controlled conditions. Many examples are found in this volume. See also **Plasma, chemical processing with.**

Handling Systems: Equipment designed to hold and to transport gases, liquids, and bulk solids throughout a process. The total handling responsibility ranges from receipt of raw materials to packaging and containerizing the final products for shipment and the handling of waste products.

Control and Economics. In the design and operation of equipment, there are three additional im-

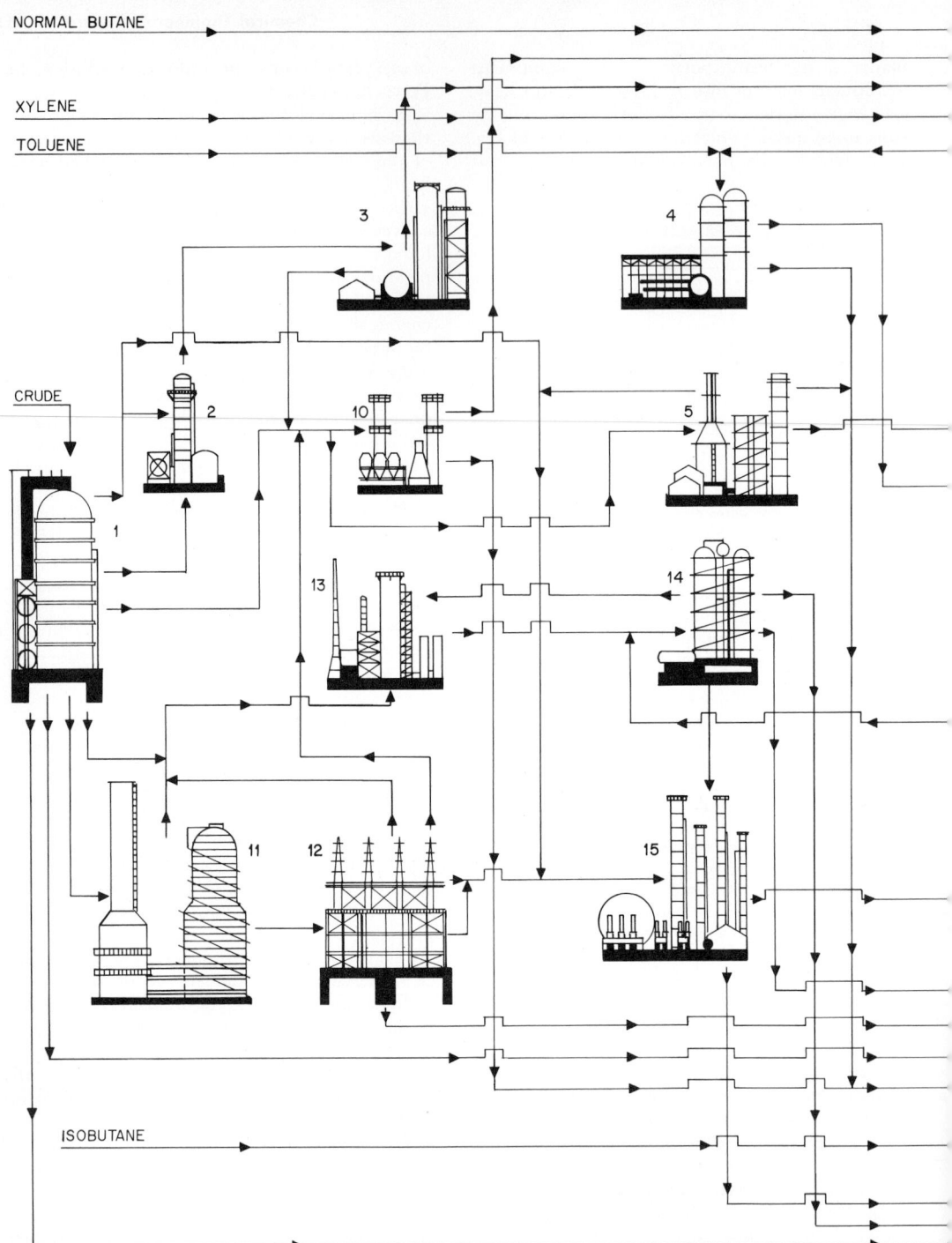

NORMAL BUTANE

XYLENE

TOLUENE

CRUDE

ISOBUTANE

Fig. C-14. A plant (in this example, a refinery) is comprised of several processes, each process, in turn, being made up of several chemical engineering unit operations and unit processes. The complex shown here is much abridged even for an average refinery. (1) Crude atmospheric tower, (2) debutanizer, (3) naphtha splitter, (4) hydrogen processing and purification, (5) reformer and preheater, (6) reformate fractionation, (7) extraction, (8) fractionation, (9) xylene fractionation, (10) reformer and preheater, (11) crude vacuum tower, (12) delayed coker, (13) fluid catalytic cracker, (14) fluid-catalytic-cracker fractionator, (15) gas recovery and fractionation, (16 and 17) hydrofluoric acid alkylation.

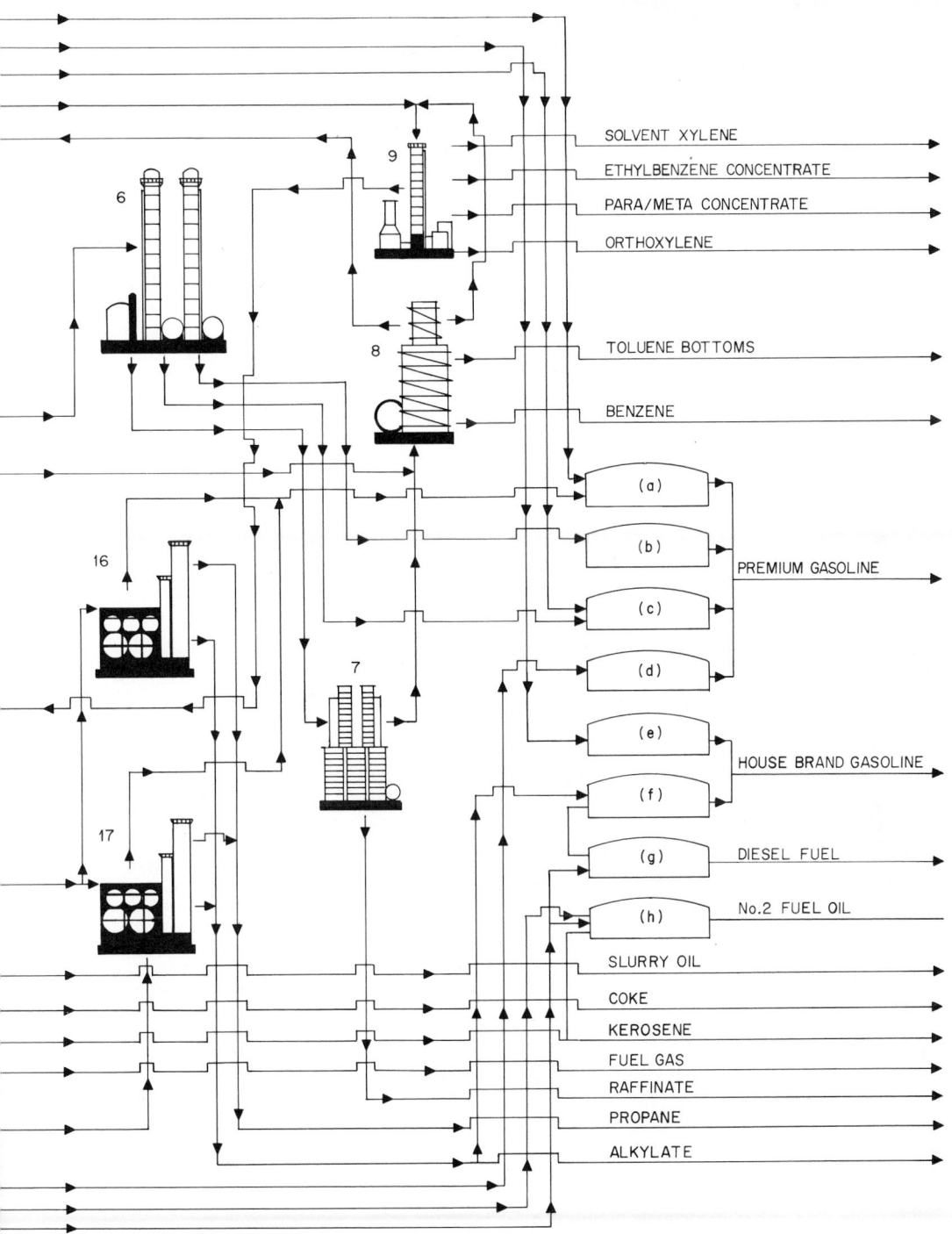

Finished products of the refinery are shown at the right. Of notable interest is the extensive dependence of the average refinery on blending various products to make up gasoline and other fuels to exact specification. For example, the different products contained in tanks (a) through (d) comprise the refinery's gasoline pool for making up premium gasoline. The contents of tanks (e) and (f) make up the pool for so-called house-brand gasoline. Even in a small refinery, of course, there will be many storage tanks for each symbolic tank shown on this diagram.

CHEMICAL ENGINEERING SYSTEMS

- ● ENERGY SYSTEMS

- ● REACTION OPERATIONS

- ● DISPERSION OPERATIONS

- ● SEPARATION OPERATIONS

- ● SIZE-CHANGE OPERATIONS

- ● HANDLING SYSTEMS

- ● MATERIALS OF CONSTRUCTION

- ● INSTRUMENTATION AND CONTROL

- ● ECONOMIC EVALUATION

- ● TOTAL SYSTEM PLANNING

Fig. C-15. Major categories of chemical engineering systems.

portant technological areas with which chemical engineering is concerned, as follows.

Instrumentation and Control: Determining and controlling those variable conditions of temperature, flow, pressure level, chemical composition, viscosity, moisture content, consistency, and so on, that vitally affect yield and safety. See **Analyzers, process.**

Materials of Construction: Selecting the most economically effective material from which to construct equipment that may have to withstand severe corrosion, abrasion, vibration, and high or low pressures and temperatures.

Economic Evaluation: Often involving numerous trade-offs that will affect profitability and competitive performance. Probably more important than all other considerations is the application of total-system thinking in the design of processes and plants. Economic evaluations tend to force system thinking.

Chemical Systems Engineering. Because of the fluid-bulk nature of processes, there is a natural inclination to think in terms of the total system rather than individual unit operations. Nevertheless, there are numerous examples where, because of a shortcoming in total-system thinking, serious errors in planning have resulted. Some of the problems of water, air, thermal, and noise pollution are the result of inadequate planning for the "tail end" of a process; or of not including sufficient funding for environmental correction costs in making economic trade-offs. In some

cases, too little attention is given to the packaging, warehousing, and shipping of materials. In other instances, inadequate flexibility is designed into a process to accommodate for changing market demands, or often, of equal importance, changing conditions that affect the cost and availability of raw materials, particularly materials of international commerce. Sometimes inadequate consideration is given to controllability of a process, resulting in the requirement for excessive instrumentation to compensate for basic design errors. Processes should be designed so that they can be tied into computer complexes and data communications networks without excessive alteration.

REFERENCES

Chemical Engineering, Comprehensive

American Institute of Chemical Engineers, New York: Principles of Information Storage and Retrieval, Manual ED-3, 1964.
Plant and Design Safety, Manual R-1, 1965.
Professionalism and the Individual, Manual R-5, 1966.
International Business, Manual T-26, 1967.
Loss Prevention, vol. 1, Manual T-30, 1967; vol. 2, Manual T-32, 1968; vol. 3, Manual T-34, 1969.
Aries, R. S., and R. D. Newton: "Chemical Engineering Cost Estimation," McGraw-Hill, New York, 1955.
Badger, W. L., and J. T. Banchero: "Introduction to Chemical Engineering," McGraw-Hill, New York, 1955.
Corcoran and Lacey: "Introduction to Chemical Engineering Problems," McGraw-Hill, New York, 1960.
Faith, W. L., Donald B. Keyes, and Ronald L. Clark: "Industrial Chemicals," Wiley, New York, 1957.
Henley, E. J., and H. Bieber: "Chemical Engineering Calculations: Mass and Energy," McGraw-Hill, New York, 1959.
Lapidus, L.: "Digital Computation for Chemical Engineers," McGraw-Hill, New York, 1962.
Lewis, W. K., Radasch and H. C. Lewis: "Industrial Stoichiometry," 2d ed., McGraw-Hill, New York, 1954.
Mickley, H. S., T. K. Sherwood, and C. E. Reed: "Applied Mathematics in Chemical Engineering," 2d ed., McGraw-Hill, New York, 1957.
Perry, R. H., and C. H. Chilton: "Chemical Engineers' Handbook," 5th ed., McGraw-Hill, New York, 1973.
Peters, M. S.: "Elementary Chemical Engineering," McGraw-Hill, New York, 1954.
Peters, M. S.: "Plant Design and Economics for Chemical Engineers," McGraw-Hill, New York, 1958.
Schoen, Herbert M. (ed.): "Interscience Library of Chemical Engineering and Processing," Wiley, New York, 1963.

Schweyer, H. E.: "Process Engineering Economics," McGraw-Hill, New York, 1955.

Shreve, R. N.: "The Chemical Process Industries," 2d ed., McGraw-Hill, New York, 1956.

Standen, Anthony (ed.): "Kirk-Othmer Encyclopedia of Chemical Technology," multivolume series (vol. 1, 1963), Wiley, New York.

Vilbrandt, F. C., and C. E. Dryden: "Chemical Engineering Plant Design," 4th ed., McGraw-Hill, New York, 1959.

Volk, W.: "Applied Statistics for Engineers," McGraw-Hill, New York, 1948.

Walker, W. H., Lewis, McAdams, and Gilliland: "Principles of Chemical Engineering," 3d ed., McGraw-Hill, New York, 1937.

Williams, E. T., and Johnson: "Stoichiometry for Chemical Engineers," McGraw-Hill, New York, 1958.

Electrical Engineering

Fink, D. G., and J. M. Carroll: "Standard Handbook for Electrical Engineers," McGraw-Hill, New York, 1968.

Fitzgerald, A. E., and D. E. Higginbotham: "Electrical and Electronic Engineering Fundamentals," McGraw-Hill, New York, 1964.

Lurch, E. Norman: "Fundamentals of Electronics," Wiley, New York, 1960.

Zeines, Ben: "Principles of Applied Electronics," Wiley, New York, 1963.

Heat Transfer, Energy Conversion, and Thermodynamics

American Institute of Chemical Engineers, New York: Heat Transfer Series, Manual S-5, 1953; Manual S-9, 1954; Manual S-17, 1955; Manual S-18, 1956; Manual S-29, 1959; Manual S-30, 1960; Manual S-32, 1961; Manual S-41, 1963; Manuals S-57 and S-59, 1965; Manual S-64, 1966; Manual S-82, 1968; Manual S-92, 1969.

Energy Conversion Systems, Manual S-75, 1967.

Heat Transfer with Phase Change, Manual S-79, 1967.

Advances in Cryogenic Heat Transfer, Manual S-87, 1968.

Benson, S. W.: "Foundations of Chemical Kinetics," McGraw-Hill, New York, 1960.

Branson, S. H.: "Applied Thermodynamics," Van Nostrand, New York, 1961.

Considine, Douglas M. (ed.): "Handbook of Energy Technology," McGraw-Hill, New York, publication scheduled 1975.

Dodge, B. F.: "Chemical Engineering Thermodynamics," McGraw-Hill, New York, 1944.

Ebaugh, Newton C.: "Engineering Thermodynamics," Van Nostrand, Princeton, N.J., 1952.

Eckert, E. R. G., and R. M. Drake: "Heat and Mass Transfer," 2d ed., McGraw-Hill, New York, 1959.

El-Saden, M. R.: "Engineering Thermodynamics," Van Nostrand, Princeton, N.J., 1966.

Fitts, D. D.: "Nonequilibrium Thermodynamics," McGraw-Hill, New York, 1962.

Geidt, Warren H.: "Principles of Engineering Heat Transfer," Van Nostrand, Princeton, N.J., 1957.

Griswold: "Fuels, Combustion, and Furnaces," McGraw-Hill, New York, 1946.

Grober, Erk, and Grigull: "Fundamentals of Heat Transfer," McGraw-Hill, New York, 1961.

Howerton, Murlin T.: "Engineering Thermodynamics," Van Nostrand, Princeton, N.J., 1962.

Hsu, S. T.: "Engineering Heat Transfer," Van Nostrand, Princeton, N.J., 1963.

Kays, W. M., and A. L. London: "Compact Heat Exchangers," 2d ed., McGraw-Hill, New York, 1964.

Kirkwood, J. G., and I. Oppenheim: "Chemical Thermodynamics," McGraw-Hill, New York, 1961.

Knudsen J. G., and D. L. Katz: "Fluid Dynamics and Heat Transfer," McGraw-Hill, New York, 1958.

Lewis, G. N., and Randall: "Thermodynamics," 2d ed., McGraw-Hill, New York, 1961.

Rose-Innes, A. C.: "Low Temperature Techniques," Van Nostrand, Princeton, N.J., 1964.

Scott, Russell B.: "Cryogenic Engineering," Van Nostrand, Princeton, N.J., 1959.

Spalding and Cole: "Engineering Thermodynamics," McGraw-Hill, New York, 1963.

Stoecker, W. F.: "Refrigeration and Air Conditioning," McGraw-Hill, New York, 1958.

Walas, S. M.: "Reaction Kinetics for Chemical Engineers," McGraw-Hill, New York, 1959.

Weber, Harold C., and Herman P. Meissner: "Thermodynamics for Chemical Engineers," Wiley, New York, 1957.

Inorganic Chemistry

Adams, Roy M.: "Boron, Metallo-Boron Compounds and Boranes," Wiley, New York, 1964.

Audrieth, F. L. (ed.): "Inorganic Syntheses," vol. 3, McGraw-Hill, New York, 1950.

Bailar, J. C., Jr. (ed.): "Inorganic Syntheses," vol. 4, McGraw-Hill, New York, 1953.

Basolo, Fred, and Ralph G. Pearson: "Mechanisms of Inorganic Reactions," Wiley, New York, 1958.

Booth, H. S. (ed.): "Inorganic Syntheses," vol. 1, McGraw-Hill, New York, 1939.

Cook, Gerhard A.: "Argon, Helium, and the Rare Gases," vols. 1 and 2, Wiley, New York, 1961.

Cotton, Albert F., and G. Wilkinson: "Advanced Inorganic Chemistry," Wiley, New York, 1962.

Emeleus, H. J., and J. S. Anderson: "Modern Aspects of Inorganic Chemistry," Van Nostrand, Princeton, N.J., 1960.

Fernelius, W. C. (ed.): "Inorganic Syntheses, "vol. 2, McGraw-Hill, New York, 1946.

Kleinberg, J. (ed.): "Inorganic Syntheses," vol. 7, McGraw-Hill, New York, 1963.

Moeller, T. (ed.): "Inorganic Syntheses," vol. 5, McGraw-Hill, New York, 1957.

Rochow, E. G. (ed.): "Inorganic Syntheses," vol. 6, McGraw-Hill, New York, 1960.

Sneed, M. Cannon (ed.): "Comprehensive Inorganic Chemistry," multivolume series (vol. 1, 1953), Van Nostrand, Princeton, N.J.

Instrumentation and Control

Ambrose D., and Barbara A. Ambrose: "Gas Chromatography," Van Nostrand, Princeton, N.J., 1962.

American Institute of Chemical Engineers, New York:
Advances in Computational and Mathematical Techniques in Chemical Engineering, Manual S-31, 1960.
Process Dynamics and Control, Manual S-36, 1962.
Computer Program Manuals C-6, C-7, C-9, C-10, C-20 on chemical processes, 1962–1964.
Experiences and Experiments with Process Dynamics, Manual S-36, 1964.
Optimization Techniques, Manual S-50, 1964.
Process Control and Applied Mathematics, Manual S-55, 1965.
Fundamentals of Process Analysis and Simulation, Manual ED-1, 1967.
Systems and Process Control, Manual T-27, 1967.

Bair, E. J.: "Introduction to Chemical Instrumentation," McGraw-Hill, New York, 1962.

Biemann, K.: "Mass Spectrometry: Organic Chemical Applications," McGraw-Hill, New York, 1962.

Buckley, Page S.: "Techniques of Process Control," Wiley, New York, 1964.

Campbell, Donald P.: "Process Dynamics: Dynamic Behavior of the Production Process," Wiley, New York, 1958.

Cassidy, Harold Gomes: "Adsorption and Chromatography," Wiley, New York, 1951.

Ceaglske, Norman H.: "Automatic Process Control for Chemical Engineers," Wiley, New York, 1956.

Considine, Douglas M. (ed.): "Encyclopedia of Instrumentation and Control," McGraw-Hill, New York, 1971.

Considine, Douglas M. (ed.): "Instruments and Controls Handbook," 2d ed., McGraw-Hill, New York, 1974.

Considine, Douglas M., and S. D. Ross (ed.): "Handbook of Applied Instrumentation," McGraw-Hill, New York, 1963.

Coughanowr, D. R., and L. B. Koppel: "Process Systems Analysis and Control," McGraw-Hill, New York, 1965.

Del Toro, V. and S. R. Parker: "Principles of Control System Engineering," McGraw-Hill, New York, 1960.

Goode, H. H., and Machol: "System Engineering," McGraw-Hill, New York, 1957.

Gordon, A. H., and J. E. Eastoe: "Practical Chromatographic Techniques," Van Nostrand, Princeton, N.J., 1964.

Grabbe, Eugene M., Simon Ramo, and Dean E. Wooldridge: "Handbook of Automation, Computation, and Control," vols. 1–3, Wiley, New York, 1958–1961.

Harriott, P.: "Process Control," McGraw-Hill, New York, 1964.

Ledley, R. S.: "Digital Computer and Control Engineering," McGraw-Hill, New York, 1960.

McDowell, C. A.: "Mass Spectrometry," McGraw-Hill, New York, 1963.

Pople, J. A., W. G. Schneider, and H. J. Bernstein: "High-Resolution Nuclear Magnetic Resonance," McGraw-Hill, New York, 1959.

Purnell, Howard: "Gas Chromatography," Wiley, New York, 1962.

Raven, F. H.: "Automatic Control Engineering," McGraw-Hill, New York, 1961.

Seifert, W. W., and C. W. Steeg: "Control Systems Engineering," McGraw-Hill, New York, 1960.

Smith, Ivor: "Chromatographic and Electrophoretic Techniques," Wiley, New York, 1960.

Truxal, J. G.: "Automatic Feedback Control System Synthesis," McGraw-Hill, New York, 1955.

Truxal, J. G.: "Control Engineer's Handbook," McGraw-Hill, New York, 1958.

Mechanical Engineering

American Institute of Chemical Engineers, New York:
Centrifugal Pumps (Newtonian Liquids), Testing Procedure E-5, 1959.
Safety in Air and Ammonia Plants, Manual T-4, 1960; Manuals T-1 and T-5, 1961; Manuals T-3 and T-8, 1964; Manual T-13, 1965; Manual T-31, 1968; Manual T-33, 1969.
Safety in Ammonium Nitrate Plants, Manual T-20, 1961.
Survey of Low Pressure Storage of Liquid Anhydrous Ammonia, Manual T-10, 1964.
Survey of Low Pressure Storage of Liquid Anhydrous Ammonia-Industry Comments, Manual T-14, 1965.
Heat Exchangers, 2d ed., Testing Procedure E-15, 1968.
Materials Engineering and Sciences Division (MESD) Biennial Conference, Manual P-3, 1970.

Baumeister, T., and L. S. Marks: "Standard Handbook for Mechanical Engineers," 7th ed., McGraw-Hill, New York, 1958.

Brownell, Lloyd E., and Edwin H. Young: "Process Equipment Design: Vessel Design," Wiley, New York, 1959.

Cambel A. B., and B. H. Jennings: "Gas Dynamics," McGraw-Hill, New York, 1958.

Cox, Glen N., and F. J. Germano: "Fluid Mechanics," Van Nostrand, Princeton, N.J., 1941.

Cox, Glen N., and William G. Plumtree: "Engineering Mechanics," Van Nostrand, Princeton, N.J., 1954.

Csanady, G. T.: "Theory of Turbomachines," McGraw-Hill, New York, 1964.

Eilon, Samuel: "Industrial Engineering Tables," Van Nostrand, Princeton, N.J., 1962.

Hagerty, William W., and Harold J. Plass: "Engineering Mechanics," Van Nostrand, Princeton, N.J., 1966.

Ham, C. W., E. J. Crane, and W. L. Rogers: "Mechanics of Machinery," 4th ed., McGraw-Hill, New York, 1958.

Harvey, John F.: "Pressure Vessel Design: Nuclear and Chemical Applications," Van Nostrand, Princeton, N.J., 1963.

Hesse, Herman C., and J. Henry Rushton: "Process Equipment Design," Van Nostrand, Princeton, N.J., 1945.

Hinze, J. O.: "Turbulence," McGraw-Hill, New York, 1959.

Johnson, W., and P. B. Mellor: "Plasticity for Mechanical Engineers," Van Nostrand, Princeton, N.J., 1962.

M. W. Kellogg Company: "Design of Piping Systems," Wiley, New York, 1964.

Lewis, Alexander D.: "Gas Power Dynamics," Van Nostrand, Princeton, N.J., 1962.

Morse, Frederick T.: "Power Plant Engineering," Van Nostrand, Princeton, N.J., 1953.

Phelan, R. M.: "Fundamentals of Mechanical Design," 2d ed., McGraw-Hill, New York, 1962.

Metallurgical Engineering

Abkowitz, Stanley, John J. Burke, and Ralph H. Hiltz, Jr.: "Titanium in Industry," Van Nostrand, Princeton, N.J., 1955.

American Institute of Chemical Engineers, New York: Mineral Engineering Techniques, Manual S-15, 1954. Liquid Metals Technology, Part I, Manual S-20, 1957. Recent Advances in Ferrous Metallurgy, Manual S-43, 1963.

Barrett, C., and T. B. Massalski: "Structure of Metals," 3d ed., McGraw-Hill, New York, 1966.

Birchenall: "Physical Metallurgy," McGraw-Hill, New York, 1959.

Clark, Donald S., and Wilbur R. Varney: "Physical Metallurgy for Engineers," Van Nostrand, Princeton, N.J., 1962.

Darken, L. S., and R. W. Gurry: "Physical Chemistry of Metals," McGraw-Hill, New York, 1953.

Howard-White, H. F.: "Nickel," Van Nostrand, Princeton, N.J., 1963.

Rhines, F. N.: "Phase Diagrams in Metallurgy: Their Development and Applications," McGraw-Hill, New York, 1956.

Seitz, F.: "The Modern Theory of Solids," McGraw-Hill, New York, 1940.

West, J. M.: "Electrodeposition and Corrosion Processes," Van Nostrand, Princeton, N.J., 1966.

Wise, Edmund M.: "Gold: Recovery, Properties and Applications," Van Nostrand, Princeton, N.J., 1964.

Nuclear Engineering

American Institute of Chemical Engineers, New York: Nuclear Engineering Manuals S-11 and S-13, 1954; Manual S-19, 1956; Manuals S-22, S-23, and S-27, 1959; Manual S-28, 1960; Manuals S-47, S-51, and S-53, 1964; Manuals S-56 and S-60, 1965; Manual S-65, 1966; Manual S-71, 1967; Manual S-68, 1968.

Ash, M.: "Nuclear Reactor Kinetics," McGraw-Hill, New York, 1965.

Benedict, M., and T. Pigford: "Nuclear Chemical Engineering," McGraw-Hill, New York, 1957.

Bonilla: "Nuclear Engineering," McGraw-Hill, New York, 1957.

Ellis, R. H.: "Nuclear Technology for Engineers," McGraw-Hill, New York, 1959.

Green, A. E. S.: "Nuclear Physics," McGraw-Hill, New York, 1955.

Hoisington, D. B.: "Nucleonics Fundamentals," McGraw-Hill, New York, 1959.

Organic Chemistry

Adams, Roger (ed.): "Organic Reactions," multivolume series (vol. 1, 1963), Wiley, New York.

Alfrey, Turner, Jr.: "Mechanical Behavior of High Polymers," Wiley, New York, 1948.

Alfrey, Turner, Jr., John J. Bohrer, and Herman Mark: "Copolymerization," Wiley, New York, 1952.

American Institute of Chemical Engineers, New York: Polymer Processing, Manual S-49, 1964. Hydrocarbons from Oil Shale, Oil Sands, and Coal, Manual S-54, 1965. Bioengineering and Food Processing, Manual S-69, 1966. Chemical Engineering in Medicine, Manual S-66, 1966. Bioengineering—Food, Manual S-86, 1968. Fossil Hydrocarbon and Mineral Processing, Manual S-85, 1968. Engineering of Unconventional Protein Production, Manual S-93, 1969.

Boekenoogen, H. A., W. O. Lundberg, and K. S. Markley (eds.): "Fats and Oils," multivolume series (commencing 1950), Wiley, New York.

Browning, B. L.: "The Chemistry of Wood," Wiley, New York, 1963.

Crook, E. M., and M. B. Donald (eds.): "Biotechnology and Bioengineering," multivolume series (commencing 1965), Wiley, New York.

Elderfield, Robert C. (ed.): "Heterocyclic Compounds," multivolume series (vol. 1, 1950), New York.

Eliel, E. L.: "Stereochemistry of Carbon Compounds," McGraw-Hill, New York, 1962.

Golding, Brage: "Polymers and Resins: Their Chemistry and Chemical Engineering," Van Nostrand, Princeton, N.J., 1959.

Hine, J.: "Physical Organic Chemistry," 2d ed., McGraw-Hill, New York, 1962.

Kaufman, Herbert C.: "Handbook of Organometallic Compounds," Van Nostrand, Princeton, N.J., 1961.

Kilner, E., and D. M. Samuel: "Applied Organic Chemistry," Wiley, New York, 1960.

Long, Cyril: "Biochemists' Handbook," Van Nostrand, Princeton, N.J., 1961.

Mark, Herman F., Norman G. Gaylord, and Norbert Bikales: "Encyclopedia of Polymer Science and Technology: Plastics, Resins, Rubbers, Fibers," Wiley, New York, 1964.

Maxwell, J. B.: "Data Book on Hydrocarbons," Van Nostrand, Princeton, N.J., 1950.

Meares, P.: "Polymers: Structure and Bulk Properties," Van Nostrand, Princeton, N.J., 1966.

Scheflan, Leopold, and Morris B. Jacobs: "The Handbook of Solvents," Van Nostrand, Princeton, N.J., 1953.

Schmidt, A. X., and C. A. Marlies: "Principles of High Polymer Theory and Practice," McGraw-Hill, New York, 1948.

Weissberger, Arnold: "The Chemistry of Heterocyclic Compounds," multivolume series, (vol. 1, 1953), Wiley, New York.

Physical Chemistry

American Institute of Chemical Engineers, New York:
Fluid Particle Technology, Manual S-62, 1966.
Phase Equilibria, Manual S-3, 1968.
Phase Equilbria and Gas Mixtures Properties, Manual S-88, 1968.

Daniels, Farrington, and Robert A. Alberty: "Physical Chemistry," Wiley, New York, 1961.

Evans, R. D.: "The Atomic Nucleus," McGraw-Hill, New York, 1955.

Henley, E., and W. Thirring: "Elementary Quantum Field Theory," McGraw-Hill, New York, 1962.

Jander, G.: "Chemistry in Nonaqueous Ionizing Solvents," multivolume series, Wiley, New York.

Kennard, E. H.: "Kinetic Theory of Gases," McGraw-Hill, New York, 1938.

Linke, William F.: "Solubilities," multivolume series (vol. 1, 1958), Van Nostrand, Princeton, N.J.

Present, R. D.: "Kinetic Theory of Gases," McGraw-Hill, New York, 1958.

Prigogine, I. (ed.): "Advances in Chemical Physics," multivolume series (vol. 1, 1958), Wiley, New York.

Slater, J. C.: "Introduction to Chemical Physics," McGraw-Hill, New York, 1939.

Slater, J. C.: "Quantum Theory of Atomic Structure," vols. 1 and 2, McGraw-Hill, New York, 1960.

Slater, J. C.: "Quantum Theory of Molecules and Solids," vol. 1, "Electronic Structure of Molecules," 1963; vol. 2, "Symmetry and Energy Bands in Crystals," 1965, McGraw-Hill, New York.

White, Harvey E.: "Introduction to Atomic and Nuclear Physics," Van Nostrand, Princeton, N.J., 1964.

Wilberg, Kenneth B.: "Physical Organic Chemistry," Wiley, New York, 1964.

Processes and Unit Operations

American Institute of Chemical Engineers, New York:
Reaction Kinetics in Chemical Engineering, Manual M-1, 1951.
Reaction Kinetics and Transfer Processes, Manual S-4, 1952.
Atomization and Spray Drying, Manual M-2, 1954.
Ion Exchange, Manual S-14, 1954.
Mass Transfer—Transport Properties, Manual S-16, 1955.
Bubble-tray Design Manual: Prediction of Fractionation Efficiency, Manual X-2, 1958.
Adsorption, Dialysis, and Ion Exchange, Manual S-24, 1959.
Reactor Kinetics and Unit Operations, Manual S-25, 1959.
The Manufacture of Nitric Acid by the Oxidation of Ammonia, Manual M-3, 1960.
Series on Equipment Testing Procedures:
Rotary Continuous Direct-Heat Dryers, E-4, 1960.
Standard Nomenclature for Mixing, E-6, 1959.
Mixing Equipment (Impeller Type), E-7, 1959.
Solids Mixing Equipment, E-8, 1961.
Evaporators, E-9, 1961.
Plate Distillation Columns, E-10, 1962.
Paste and Dough Mixing Equipment, E-11, 1964.
Packed Adsorption and Distillation Columns, E-12, 1965.
Batch Pressure Filters, E-13, 1967.
Rotary Positive Displacement Pumps, E-14, 1968.
Bubble Cap Tray Design, Computer Program Manual C-7, 1962.
Fluidization, Manual S-38, 1962.
Reaction Engineering, Manual T-9, 1964.
Distillation, Manual R-2, 1965.
Transport Phenomena, Manual S-58, 1965.
Fluidized Bed Technology, Manual S-67, 1966.
Vapor Deposition, Manual T-22, 1966.
Concepts of Polymer Processing, Manual ED-2, 1967.
Fundamental Research on Heat and Mass Transfer, Manual S-77, 1967.
Heat Transfer with Phase Change, Manual S-79, 1967.
High Pressure Technology, Manual S-76, 1967.
Physical Adsorption Processes and Principles, Manual S-74, 1967.
Recent Advances in Kinetics, Manual S-72, 1967.
Chemical and Food Applications of Radiation, Manual S-83, 1968.
Catalysis and Reactors, Manual M-6, 1969.
Developments in Physical Adsorption, Manual S-96, 1969.
Transport Phenomena, Manual ED-4, 1969.
Applications of Radioisotopes, Manual S-106, 1970.
Fluidization Fundamentals and Application, Manual S-105, 1970.
Fundamental Processes in Fluidized Beds, Manual S-101, 1970.
Hydrocarbon Production and Distribution, Manual S-103, 1970.
Proceedings Conference on Engineering Construction Contracts, Manual X-44, 1970.

Barron, Harry: "Modern Rubber Chemistry," Van Nostrand, Princeton, N.J., 1948.

Bell, H. S.: "American Petroleum Refining," Van Nostrand, Princeton, N.J., 1959.

Brown, George Granger: "Unit Operations," Wiley, New York, 1950.

DeNavarre, Maison G.: "The Chemistry and Manufacture of Cosmetics," Van Nostrand, Princeton, N.J., 1962.

Eckert, E. R. G., and R. M. Drake: "Heat and Mass Transfer," 2d ed., McGraw-Hill, New York, 1959.

Eckert, E. R. G., and J. H. Gross: "Introduction to Heat and Mass Transfer," McGraw-Hill, New York, 1963.

Gaudin, A. M.: "Principles of Mineral Dressing," McGraw-Hill, New York, 1939.

Groggins, P. H. (ed.): "Unit Processes in Organic Synthesis," 5th ed., McGraw-Hill, New York, 1958.

Hellferich, F. G.: "Ion Exchange," McGraw-Hill, New York, 1962.

Hoffman, E. J.: "Azeotropic and Extractive Distillation," Wiley, New York, 1964.

Leva, M.: "Fluidization," McGraw-Hill, New York, 1959.

Mantell, C. L.: "Adsorption," 2d ed., McGraw-Hill, New York, 1951.

Moilliet, John Lewis, Benjamin Collie, and William Black: "Surface Activity," Van Nostrand, Princeton, N.J., 1961.

Newton, Joseph: "Extractive Metallurgy," Wiley, New York, 1959.

Norman, W. S.: "Absorption, Distillation and Cooling Towers," Wiley, New York, 1962.

Rase, Howard F.: "Piping Design for Process Plants," Wiley, New York, 1963.

Rase, Howard F., and M. H. Barrow: "Project Engineering of Process Plants," Wiley, New York, 1957.

Reid, R. C., and T. K. Sherwood: "The Properties of Gases and Liquids," 2d ed., McGraw-Hill, New York, 1966.

Robinson, C. S., and E. R. Gilliland: "The Elements of Fractional Distillation," 4th ed, McGraw-Hill, New York, 1950.

Sherwood, T. K., and R. L. Pigford: "Absorption and Extraction," 2d ed., McGraw-Hill, New York, 1952.

Treybal, R. E.: "Liquid Extraction," 2d ed, McGraw-Hill, New York, 1963.

Zuiderweg, F. J.: "Laboratory Manual of Batch Distillation," Wiley, New York, 1957.

Water and Air Pollution and Abatement

American Institute of Chemical Engineers, New York:
Pollution and Environment Health, Manual S-35, 1961.
Pollution Control Engineering, Manual S-45, 1963.
Water Reuse, Manual S-78, 1967.
Water, Manual S-90, 1968; Manual S-97, 1969.
Industrial Process Design for Water Pollution Control, vol. 2, Manual W-2, 1970.
Research and Development Studies in Environmental Pollution and in Reactor Cooling Systems, Manual S-104, 1970.

Elkins, Hervey B.: "The Chemistry of Industrial Toxicology," Wiley, New York, 1959.

Faith, W. L.: "Air Pollution Control," Wiley, New York, 1959.

Jacobs, Morris B.: "The Chemical Analysis of Air Pollutants," Wiley, New York, 1960.

Linsley, R. K., and J. B. Franzini: "Water-Resources Engineering," McGraw-Hill, New York, 1964.

Lund, Herbert F. (ed.): "Industrial Pollution Control Handbook," McGraw-Hill, New York, 1971.

Sawyer, C. N.: "Chemistry for Sanitary Engineers," McGraw-Hill, New York, 1960.

Steel, E. W.: "Water Supply and Sewerage," McGraw-Hill, New York, 1960.

Chemical Lead (See **Lead.**)

Chemical Pulp [See **Pulp (wood) production and processing.**]

Chemical Recovery Process [See **Pulp (wood) production and processing.**]

Chemisorption (See **Adsorption.**)

Chillers (See **Energy systems for processes.**)

Chip Preparation [See **Pulp (wood) production and processing.**]

Chloral (See **Chlorine organics.**)

Chlorides (See **Hydrochloric acid; and Salt.**)

Chlorinated Ethanes (See **Chlorine organics; and 1,1,1-Trichloroethane.**)

Chlorinated Ethylenes (See **Chlorine organics; Perchloroethylene; and Trichloroethylene.**)

Chlorinated Methanes (See **Chlorine organics.**)

Chlorinated Rubber (See **Paint.**)

Chlorination Chlorination is a process wherein chlorine in liquid or gaseous form (or sometimes chlorine in combination with other materials) is added to a system. Chlorination is practiced extensively in industry primarily for bleaching in textile and pulp and paper manufacturing, in the manufacture and processing of organic chemicals, in metallurgical processes, and in bleach manufacturing. In addition to industrial applications, chlorination is the principal chemical treatment method for disinfection of municipal water supplies, treated waste water, and swimming pools.

Because of the extremely wide range of uses of chlorine, the needs of the user may vary from a few hundred pounds a season for a commercial swimming pool to several thousands of tons a year for a chemical manufacturer. To satisfy these various needs, liquid chlorine is packaged in pressurized 100- and 150-lb cylinders and 1-ton containers, and shipped in bulk in tank trailers, railroad tank cars, and tank barges.

Chlorine cylinders have a single valve and deliver gas when in an upright position and liquid when in an inverted position. Liquid withdrawal from cylinders is seldom practiced. Ton containers are stored and used in a horizontal position with the two valves on the end of the cylinder in a vertical line, gas being delivered from the upper valve and liquid from the lower valve. Tank cars, tank trucks, and barges are almost invariably emptied by discharging liquid.

Withdrawal of gas depends on vaporization of the liquid, which reduces the temperature and hence the chlorine vapor pressure. Thus gas withdrawal rates are limited to $1\frac{3}{4}$ lb/hr from a 150-lb cylinder and 15 lb/hr from a ton container. A number of cylinders or containers can be manifolded for increased capacity.

Continuous liquid discharge rates are about 200 lb/hr for cylinders, 400 lb/hr for ton containers, 7,000 lb/hr for a tank car discharging from one valve. Liquid is normally forced out of the container or tank by its own vapor pressure, but air padding up to 200 psi is often applied to tank cars for high discharge rates. When the chlorine is to be fed to the process as a gas, the discharged liquid is vaporized in a steam or electrically heated evaporator.

In industrial applications chlorine is applied to the process in both liquid and gaseous form. Liquid feed is more adaptable to batch processes where the rate of feed is not critical. Metering and control of liquid chlorine presents problems because of vapor flashing across any restriction in the line. Gaseous chlorine flow rate, on the other hand, is easily measured within accuracies of 2%, and the feed rate can be accurately controlled.

Basically, there are two methods for applying chlorine gas to a process. *Direct feed* involves metering of the gas and bubbling it under pressure into a liquid or introducing it through a series of jets into a stream of gas to be chlorinated. *Solution feed* involves metering of the gas under a vacuum, dissolving it in a stream of the chemical being chlorinated, or a carrier fluid, such as water, in an ejector or jet eductor. This is followed by injecting the solution into the reaction vessel or other point of application.

The feed rate of chlorine gas is controlled by (1) varying the pressure differential across a fixed orifice, (2) varying the orifice opening while maintaining a constant differential, or (3) a combination of both methods.

A solution-feed chlorinator of the type used in municipal water and waste-water-treatment plants is shown in Fig. C-16. The chlorine is metered under a vacuum created by the ejector, dissolved in water in the ejector, and discharged

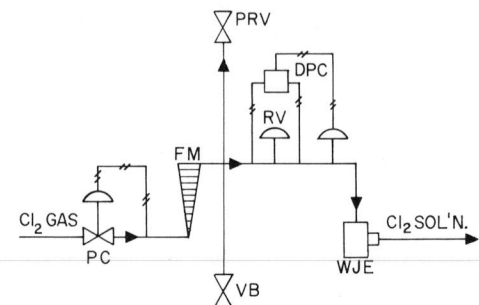

Fig. C-16. Chlorinator arrangement commonly used in municipal water and waste-water treatment plants. *PC,* metering pressure controller; *FM,* flowmeter; *PRV,* pressure-relief valve; *RV,* rate valve; *DPC,* differential pressure controller; *VB,* vacuum breaker; *WJE,* water jet ejector. (*Fischer & Porter Co.*)

to the application point as a high-strength solution. The chlorinator is operated under vacuum for safety reasons. In the event that a leak develops during operation, air will leak into the system rather than the highly toxic and corrosive chlorine leaking out. The chlorine gas supply is under pressure at P_0 of about 50–100 psi. This is dropped to a vacuum across the *metering pressure controller.* For automatic feeding of chlorine in proportion to the flow of the material being chlorinated, a transmitted flow signal can be applied either to one side of a diaphragm in the differential pressure controller or to the valve operator on an automatic rate valve.

A closed-loop chlorination system, commonly used in water and waste-water treatment, is shown in Fig. C-17. The transmitted flow signal and the control signal from the continuous residual chlorine analyzer, both are fed into the chlorinator, one to the differential pressure controller and one to the rate valve. Highly accurate control of chlorination thus is maintained even though both the flow and chlorine demand of the treated water are varying.

Chlorination processes are controlled by a variety of methods. Control of chlorine dosage in water treatment is determined by analysis of the chlorine residual in the water after a specified contact time. Disinfection, or other desired result

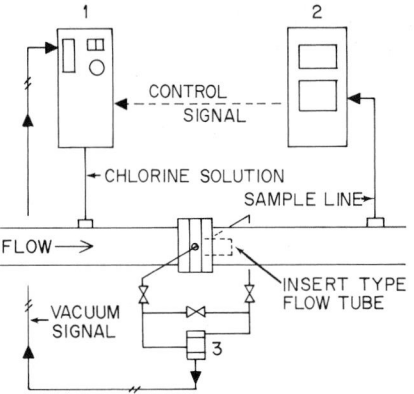

Fig. C-17. Closed-loop chlorination control system. (1) Chlorinator, (2) residual chlorine electronic control system, (3) differential pressure vacuum transmitter. Power supply required is 117 V, 50/60 Hz. (*Fischer & Porter Co.*)

TABLE C-33. *Physical Constants of Chlorine*

Molecular weight	70.906
Critical density, lb/ft^3	35.77
Critical pressure, psia	1,118.4
Critical temperature:	
°F	291.2
°C	144
Normal boiling point (liquefying point):	
°F	−29.29
°C	−34.05
Melting point (freezing point):	
°F	−149.76
°C	100.98
Density:	
Dry gas at 32°F and 14.696 psia, lb/ft^3 . .	0.2003
Liquid at 32°F (0°C), lb/ft^3	91.67

For other constants see Chlorine Manual, 1969, The Chlorine Institute, New York.

of chlorination of the particular water, is assumed to have been accomplished if a predetermined chlorine residual remains in the water after some established time interval, which may be between 10 min and several hours.

Residual analysis is also used for control of chlorination in some direct chlorine bleaching processes. In the manufacture of lime bleach in pulp mills the pH of the chlorinated lime is held at 11.2 or above to prevent overchlorination. In the preparation of sodium hypochlorite bleach, oxidation-reduction potential is often used for chlorination control by indicating the end point of the reaction between caustic and chlorine. In chlorinating hydrocarbons the temperature of the exothermic reaction is sometimes used as the control parameter for rate of chlorination.

Examples of industrial chlorination will be found under **Chlorine organics; Ethylene; Trichloroethylene;** and **Vinyl chloride monomer.**

> —Leo J. Carroll, *Fischer & Porter Company,*
> *Warminster, Pa.*

Chlorine (Gk. *chloros,* green.) **Cl** = 35.453 (at. wt.); 17 (at. no.). Chlorine occurs in group VIIA, period 3, of the periodic table and is a member of the halogen group of elements. Chlorine is a greenish-yellow gas at room temperature and atmospheric pressure. Selected physical properties of the element are given in Table C-33. Chlorine is the seventeenth most abundant element in the earth's crust. There are two stable isotopes of the element, 35 (75.53%) and 37 (24.47%), and six

unstable isotopes, 32–39. See also **Chemical elements.**

Chlorine was first produced as a relatively pure material in 1774 by Carl Wilhelm Scheele, who reacted MnO_2 with HCl. He named the pungent gas "dephlogisticated marine acid." For years, the leading chemists of those times could not agree whether the material was a compound (some strange and uncharacterizable acid) or an element. Finally, in 1810, Sir Humphry Davy established that the material was a chemical element and proposed its present name.

Chlorine is nonflammable, but supports combustion of some metals. It reacts violently with many organic materials, particularly in the presence of light or other catalytic agents. The element is toxic. Safe handling procedures are described below. When properly handled, chlorine is safe, and it is used in large tonnages to produce organic and inorganic chemicals, to bleach pulp for paper and rayon manufacture, to treat water and sewage, and for many other uses.

Almost all Cl_2 is made by electrolysis, principally of NaCl brine, accounting for 95% of the production. The coproduct is caustic soda. See also **Caustic soda.** Most of the remaining product is made by electrolysis of KCl brine, with caustic potash as a coproduct, or by electrolysis of fused NaCl, with metallic sodium a coproduct. Chlorine production in the United States has grown from a little over 2 million short tons in 1950 to nearly 10 million short tons in 1970.

Early Developments: As early as 1789, Cl_2 produced from MnO_2 and HCl was bubbled into potash to produce potassium hypochlorite, KClO, and used to bleach textiles. The commercial production of Cl_2 by electrolysis, discovered by Cruickshank in 1800 and described in principle by Faraday in 1834, had to await the development of adequate electric power generation. The world's first commercial production of Cl_2 was by the Griesheim Company (1890) in Germany. This company used KCl brine with KOH as a coproduct. The world's first commercial electrolytic unit to make Cl_2 and NaOH (from NaCl brine) was built in 1893 at Rumford Falls, Maine, using a diaphragm cell developed by Ernest A. Le Sueur. The installation was expanded fourfold in the following year.

The first mercury cell to operate commercially started up July 4, 1895, in Saltville, Va. This was the *Castner rocking cell,* named for its inventor, Hamilton Y. Castner, who was born in Brooklyn but developed his cell in England in order to produce caustic soda for the manufacture of aluminum. Before the cell could be commercialized, the aluminum process for which it was designed became obsolete. Thomas Mathieson founded the Mathieson Alkali Works (a predecessor of the Olin Corporation) and built at Saltville a 1 ton/day chlorine caustic plant based on the Castner cell. It was soon discovered that more power than anticipated was required, and the cells were moved to Niagara Falls to take advantage of the abundant, cheap power from a new hydroelectric plant. The new installation was designed for a production of approximately 30 tons/day. The cells were called "rocking cells" because a slow back-and-forth tilting motion was imparted to the cells to move the mercury from the electrolyzer to the decomposer and back again.

At about the time that Castner was developing his version of the rocking cell, Carl Kellner of Vienna made several important inventions in the field of mercury cells, among them the device of short-circuiting the soda cell. This principle is utilized in the amalgam decomposers of modern mercury-cell installations. In 1895 the Castner-Kellner Alkali Company was formed, and in 1897 this company came into full industrial-scale operation in Runcorn, Cheshire, England.

Shortly thereafter Kellner designed a mercury cell featuring a long, slightly inclined trough down which mercury and salt brine flowed by gravity. The denuded mercury was returned to the inlet by pump. This design, progressively improved, is the configuration of most modern mercury cells.[1]

During the 1960s consumption of Cl_2 grew at a rate of about 8.5%. This was faster than the natural growth of caustic soda, and accordingly producers of Cl_2 sought imaginative ways to market caustic soda in order to keep the coproducts in balance. At the same time they investigated alternative economic routes to Cl_2 which would not involve the production of caustic soda. One potential process, the *Deacon process,* is the oxidation of HCl with air or oxygen to produce Cl_2 and water:

$$4HCl + O_2 \rightleftharpoons 2Cl_2 + 2H_2O$$

The reaction rate is greater at higher temperatures, but high temperatures are unfavorable to the equilibrium point. The process has not been commercialized for the production of Cl_2 in the United States, although substantial development work has taken place. However, when a ready Cl_2 acceptor (such as ethylene) is present, the reaction proceeds smoothly to completion and permits the use of HCl for chlorination (oxychlorination). Substantial quantities of Cl_2 are produced in Europe by a modified Deacon process.

During 1970 the demand picture for Cl_2 changed. For the first time since Cl_2 began its spectacular growth pattern in the forties, the demand for caustic soda exceeded the demand for Cl_2. The available storage for Cl_2 is limited, and some producers of Cl_2 were forced to cut back production.

Uses: The principal use of Cl_2 is in the production of organic chemicals.[2] It is not possible to describe or discuss the many varied chemicals which use Cl_2 in their manufacture. The Cl_2 atom may or may not remain as part of the finished molecule. In many organic chlorinations one-half the Cl_2 molecule replaces a hydrogen atom, and by-product HCl is formed. The disposition of by-product HCl (formed from this or other reactions) has posed problems in the past. The development of technologies for recycling HCl has minimized this situation. For example, the production of vinyl chloride monomer (the single largest use of Cl_2) starts with ethylene. Ethylene is chlorinated to form ethylene dichloride and then dehydrochlorinated to form vinyl chloride. The HCl stripped is recycled and combined with additional ethylene in an oxychlorination step. This forms additional ethylene dichloride, and the process continues. An alternative route to vinyl chloride by the direct addition of HCl to acetylene is now largely abandoned, because a process starting with ethylene is cheaper if there are no problems with by-product HCl.

An important use for Cl_2 is water treatment.

Equipment for this purpose, usually called a *chlorinator,* has been highly developed. Typically, Cl_2, received in tank cars or in ton containers, is vaporized to Cl_2 gas. The Cl_2 gas is then absorbed in a circulating stream of water to form a solution

differences are in the configuration of the decomposer or denuder and in the technique for adjusting the anode gap. All designers of cells have successively improved their design to permit increased total current and current density. Some

TABLE C-34. *Chlorine Shipping Containers*

Type	Capacity	Typical Dimensions
Cylinders	100 lb	$8\frac{1}{2}$ in. diam $\times$ 53 in. long
	150 lb	$10\frac{1}{4}$ in. diam $\times$ 56 in. long
Ton containers	2,000 lb	30 in. diam $\times$ 80 in. long
Tank cars	16 tons	33 ft long $\times$ 11 ft high
	30 tons	35 ft long $\times$ 13 ft high
	55 tons	40 ft long $\times$ 15 ft high
	90 tons	46 ft long $\times$ 15 ft high
Barges	600–1,100 tons	

of about $\frac{1}{2}\%$ Cl_2 in water. Water operates an eductor which produces a negative pressure for the Cl_2 feed. The Cl_2 water solution is then mixed with the water supply. Chlorine is used at the rate of 3–4 ppm (weight per weight). At this application rate a "chlorine residual" of 0.3–0.5 ppm available chlorine can usually be produced. This ensures sanitary water suitable for domestic purposes. Chlorine is also used to treat industrial water supplies to prevent the growth of algae in cooling towers.

A summary of types of containers used for shipping Cl_2 is given in Table C-34.

Production Processes: Worldwide, slightly more than 50% of electrolytic Cl_2 production is by the mercury-cell process. In the United States, mercury cells account for about 30% of the production. Diaphragm cells and the associated process are described under **Caustic soda.** In both the diaphragm-cell and the mercury-cell processes, the treatment of Cl_2 after production is similar. Described herein is the mercury-cell process.

Concern has been expressed about the presence of Hg in plant effluents and in the products from mercury-cell plants. The amount of Hg in effluents has been substantially reduced, and many experts feel that the presence of Hg should not be a determining factor in the continued operation of such plants. Chlorine produced by the mercury-cell process contains typically less than 1 ppm (weight per weight) Hg and, accordingly, is suitable for all purposes for which Cl_2 normally is used.

Most modern mercury cells are similar in appearance and construction.[3] The most notable

have been more successful than others. Modern cells may operate above 300,000 A and at a current density at the cathode of over 10,000 A/m^2. Among the cells in operation in the United States are those by de Nora, Olin, Uhde, Solvay, and Krebs.

Production of Cl_2 by the mercury-cell process involves two cycles: the brine cycle and the mercury cycle (Fig. C-18). Brine is normally a sodium chloride brine. The brine is partially depleted of its sodium chloride in the electrolyzer and must be fortified using a source of anhydrous salt. The brine must be purified to ensure that harmful impurities from the salt and from the graphite anodes do not build up. For example, vanadium, molybdenum, and chromium cause hydrogen to be evolved at the mercury cathode in preference to sodium, and when hydrogen mixes with chlorine, an explosive mixture can be formed. The extent of the need for purification will depend upon the impurities in the salt and the operating conditions which have been established. The following partial brine treatment has been found suitable: Brine, recycled from the cell, is acidified and blown with air (or evacuated) to remove residual chlorine. The dechlorinated brine is then adjusted with dilute caustic soda to a pH of 9.0–10.5. It is resaturated with salt and settled. By using proper temperature of resaturation with excess sulfate ions present, calcium sulfate solubility is reduced and tends to be removed by precipitation and filtration. The final pH is adjusted to 2.5–4.0, at which pH the residual calcium sulfate does not cause excess hydrogen to be produced in the chlorine.[4]

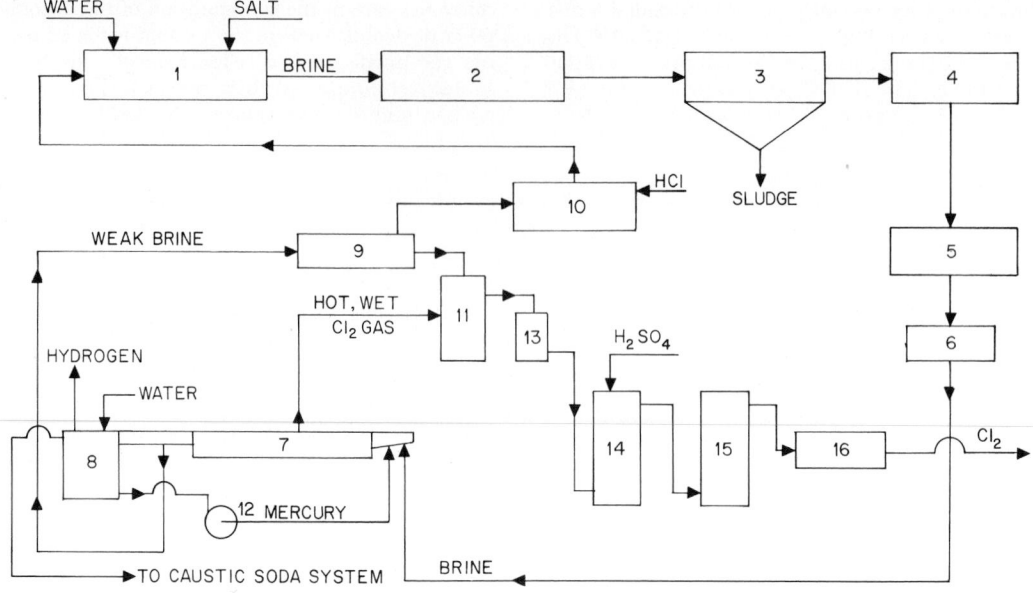

Fig. C-18. Simplified representation of materials flow in production of chlorine by mercury-cell process. (1) Salt dissolving, (2) brine treatment, (3) settling, (4) filtration, (5) brine storage, (6) heating, (7) electrolyzer, (8) decomposer, (9) dechlorination, (10) pH adjustment, (11) cooling, (12) mercury pump, (13) demisting, (14) drying, (15) scrubbing, (16) liquefaction.

The mercury cycle is part of the operation of the cell itself. Mercury flows by gravity in a thin layer along the bottom of the steel trough of the electrolyzer. Brine flows concurrently with it on top of the Hg. Metallic sodium formed at the mercury cathode immediately amalgamates with the Hg. The amalgam, containing up to 0.4% sodium, is removed at the end of the cell. It then goes to the decomposer (or denuder), where it is reacted with water. The decomposer acts as a short-circuited cell, the amalgam being the anode, graphite surfaces being the cathode, and caustic soda solution the electrolyte. In this reaction hydrogen is evolved and the sodium forms caustic soda. Caustic soda is normally produced at a concentration of 50%. The denuded Hg is then collected in a sump, where it is pumped back to begin its cycle over again.

Chlorine gas, saturated with water vapor and containing traces of organic impurities and hydrogen, collects in the cell chamber above the anodes. The gas goes from there to the drying and liquefaction part of the plant.

Where specific description or numbers are used in this article, they are based on the Olin E-510 cells as typical of most modern mercury cells.

Individual cells will not be described. That information is available elsewhere.[3] See Fig. C-19.

The electrolyzer is a long rectangular steel chamber with rubber-lined sides, top, and end boxes. It is about 4 ft wide × 40 ft long. It is supported on adjustable, insulated structural pedestals along the length. The bottom has a pitch of about 10 mm/m. The cell itself is 8–10 in. deep. The cell cover can be lifted from the cell by a crane to permit renewal of the anodes and cleaning of the cells. When in place, a gastight seal is made all around the edges with soft rubber gaskets and some clamping device.

The E-510 cell uses five anodes in a group supported from a bus running across the electrolyzer. The group is adjusted by jackscrews and sprockets at each side connected by a chain. The adjustment consists in lowering the anodes until there is an incipient short circuit. The anodes are then raised so that the clearance under dynamic, flowing conditions is 2 mm or more. The least practical separation between the anode and the mercury cathode is desirable in order to keep voltage drop per cell to a minimum while allowing for reasonable anode life (about 4.6 V with graphite anodes at 10 kA/m^3 cathode current density).

Fig. C-19. A block of 58 E-11 mercury cells at the chlorine-caustic plant of Olin Corporation at Augusta, Ga.

The Olin cell uses graphite electrodes 6 in. thick × 9 in. wide × 48 in. long. The anodes have two metal lead-ins screwed into the graphite block. They are protected from corrosion with porcelain sleeves. Flexible rubber seals and a flexible section in the anode bus permit a group of five anodes to be raised and lowered in a single operation. A recent development scans the voltage drop continuously over all the cells in a cell room and uses a computerized program to adjust the anode gap for optimum performance.

Most commonly at present, anodes are made of graphite. It is usual to machine grooves on the bottom of the electrodes and/or to provide holes in order to facilitate the escape of chlorine gas. As the anode operates it is consumed and its shape changes. The carbon which is consumed shows up largely as CO_2 (0.3–0.6%) in the Cl_2 gas.

Anode Developments: There has been much development work in recent years toward producing a metal anode. Desirable features of a metal anode would be reduced power consumption as a result of lowered overvoltage, dimensional stability, and easy Cl_2 release. Platinized titanium has these properties. However, platinum is easily amalgamated with mercury, and has not been found suitable for use in mercury cells. For mercury cells a coating of ruthenium oxide on titanium is suitable. An additional advantage of metal anodes would be the reduction of CO_2 and/or organic impurities in the finished Cl_2 which originates with graphite anodes. Metal anodes have achieved appreciable headway in the overall production of Cl_2. Most major producers of Cl_2 are known to be at least testing them, and a number of complete plants have been converted.

Postelectrolysis Processing: The first step in processing Cl_2 is to cool it. This can be done by direct contact in a packed tower or in a water-cooled titanium heat exchanger. In either case some finely divided salt gets through, and may cause problems in the drying step or with the compressors. A demister is often used to prevent as much brine mist getting through as possible. From the demister the Cl_2 goes to the dryers. These are usually two to four packed towers over which sulfuric acid is pumped to contact the Cl_2 countercurrently. Spent sulfuric acid is discharged at about 60%. Chlorine gas is dried so that it contains 50 parts per million moisture or less. Chlorine gas when dry is essentially noncorrosive to carbon steel at atmospheric temperatures.

Some producers scrub the dried gas by refluxing it with liquid Cl_2. This removes a large portion of the small amount of organic and inorganic impurities frequently found in Cl_2. The bottoms from this step (essentially a one-plate distillation) are neutralized with caustic soda and discharged.

The overhead gas is then compressed to the pressure at which it will be liquefied. Typically, the discharge pressure may be somewhere between 20 and 80 psig. For lower pressures substantial mechanical refrigeration is required. Dry air is usually added during liquefaction to replace the Cl_2 as it liquefies so as to maintain the gaseous mixture below 4% hydrogen. Noncondensable gases, principally air, hydrogen, and CO_2, are purged from the condenser system. The vent or sniff gas may be variously treated in order to remove the Cl_2. In some plants it is converted either to sodium or calcium hypochlorite, which may have a ready market. In other plants the sniff gas is scrubbed by an organic solvent or water and the Cl_2 stripped from the solvent or chilled water and returned to the process. Condensed liquefied Cl_2 is then transferred to storage tanks from which it may be loaded into transportation containers.

Transporting and Storing: Liquid Cl_2 is usually transferred by compressed air. Compressed air is dried in Cl_2-producing plants to a dew point of $-60°F$ or better. This is to prevent pickup of moisture in the transfer operation. Since pressures of 150 lb are not uncommon for transfer purposes, it is clear that tanks must be vented to permit the admission of fresh Cl_2. These are vented normally to the plant sniff system. In some cases where high transfer rates are required, submerged pumps designed for this service can be used. This is commonly the case when transferring from storage tanks to barges.

Chlorine storage tanks have traditionally been mounted on scales or load cells, and weight has been used as the primary indicator of whether the tank is full. Conventional float gages or differential-pressure measuring devices are felt not to be reliable in providing protection against overfilling. Carbon steel is the most usual material of construction for handling dry Cl_2, whether liquid or gas. Graphited bonded asbestos fiber is the most common gasket material. Two to four percent antimony lead is also used. Most flanges in use in Cl_2 service are tongue-and-groove, with the gasket fully retained inside the groove. There is increasing use of raised-face flanges, in which case spiral-wound gaskets using Monel and asbestos should be used.

Perhaps the most troublesome problem in the use and handling of Cl_2 is the admission of moisture into the system. This customarily occurs at the using point when making or breaking connections for hooking up a new container or for making repairs to the existing system. It is also a frequent problem in new installations, where failure to properly clean and dry the system may lead to unhappy results. New equipment, such as valves and unloading connections, must be carefully cleaned of all oil or grease because Cl_2 may react violently with them. The use of trichloroethylene is recommended for cleaning. Hydrocarbons or alcohols must not be used; any residual solvent may react with Cl_2. After being cleaned, the system should be steamed, and all condensate removed. While still hot, dried air, preferably of dew point $-40°F$ or better, should be used to dry the system. Particular attention should be given to any low point in the system, including shoulders behind screwed connections.

For many years Cl_2 producers, together with Cl_2 repackagers (nonproducers of Cl_2 who repackage Cl_2 from tank cars or barges into cylinders and ton containers), have worked through the Chlorine Institute to promote safety and the safe handling of Cl_2. Many of the regulations which apply specifically to Cl_2 and many suggestions for good practice have come out of deliberations of the Chlorine Institute. The Chlorine Institute continuously reviews through its committees and subcommittees recommendations on safe handling procedures as they apply to Cl_2. The Institute has published a series of pamphlets dealing with many aspects of Cl_2, its properties and handling. These are available at nominal cost from the Institute. In particular, the Institute has published a Chlorine Manual which describes different containers, handling, and unloading procedures and safety. This is an invaluable reference for anyone concerned with chlorine.[5]

Safety in Handling: Chlorine is handled in tremendous tonnages, and its safety record has been enviable. This is in spite of the fact that much Cl_2 finds its way to small users. Inevitably, however, in the course of any given year, there are emissions of Cl_2. Such emissions usually receive a great amount of publicity, largely on the basis that Cl_2 is a "war gas." The fact of the matter is that the use of Cl_2 never was very successful in achieving the purposes of a war gas, and aside from training purposes, it is not part of the chemical warfare arsenal of any country today. While Cl_2 is toxic and can cause death, there are many other industrial materials which are substantially more hazardous and can cause more damage.

Chlorine gas is primarily a respiratory irritant. It is so intensely irritating that low concentrations in the air are readily detectable by most persons. In higher concentrations the severely irritating effect of the gas makes it unlikely that any person will remain in a Cl_2-contaminated atmosphere unless he is unconscious or trapped.

Liquid chlorine will cause skin and eye burns upon contact. When exposed to normal atmospheric pressure and temperature, liquid Cl_2 vaporizes to Cl_2 gas. When a sufficient concentration of Cl_2 gas is present, the gas will irritate the mucous membranes, the respiratory system, and the skin. Large amounts cause irritation of eyes, coughing, and labored breathing. If the duration of exposure or the concentration of Cl_2 is excessive, general excitement of the person affected, accompanied by restlessness, throat irritation, sneezing, and copious salivation, results. The symptoms of exposure to high concentrations are retching and vomiting followed by difficult breathing. In extreme cases, difficulty in breathing may increase to the point where death can occur from suffocation. Liquid Cl_2 in contact with eyes or skin will cause local irritation and burns.

Chlorine produces no known systemic effect. All symptoms and signs result directly or indirectly from the local-irritation action.

The TLV (the permissible 8-hr time-weighted-average exposure limit) for Cl_2 is 1 part per million, or 3 mg/cm. Most literature references state that the least detectable odor by most people is 3.5 parts of Cl_2 per million parts of air. Recent data indicate that it may be substantially lower than this.

Suitable gas masks, particularly "self-contained breathing apparatuses," should be available wherever Cl_2 is handled. First-aid treatment involves removing the affected person from the exposed area and the administration of oxygen. A physician should be called. If breathing has apparently ceased, artificial respiration should begin immediately. The mouth-to-mouth method is preferred. Chlorine gas is heavier than air and tends to collect in low areas. Only trained persons ought to take emergency action to stop a Cl_2 leak. If a substantial leak which cannot be contained by normal procedures occurs, the only practical procedure is to evacuate people. They should move upwind from the source of contamination.

If Cl_2 containers are enveloped in fire, water should be used to cool the containers. Water should not be sprayed on a Cl_2 leak because water reacts with Cl_2 and forms hydrochloric acid, thus making the leak worse.

Hypochlorites: Hypochlorites, the reaction products of Cl with an alkali, are frequently used to provide the sanitizing and bleaching property of Cl_2 without the necessity of handling Cl_2 as a liquid or a gas. In fact, the term "liquid chlorine" is used in the swimming-pool trade to describe sodium hypochlorite solutions, and the term "dry chlorine" is part of the registered trademark of a proprietary calcium hypochlorite product containing 70% available Cl, also used in the swimming-pool trade. While these terms ought not to be used, their usage is well rooted in practice.

Sodium Hypochlorite: Commercial sodium hypochlorite, $NaClO$, is usually produced in one of two strengths. The familiar household liquid bleach is 5.25 wt % sodium hypochlorite. Commercial bleach, so-called 15% bleach, contains 150 g/l of available Cl, which is equivalent to 13.03 wt % sodium hypochlorite. Other strengths may be available in local areas for special purposes, such as 10% for swimming-pool sanitation.

Sodium hypochlorite is usually produced batchwise by diluting caustic soda to the necessary starting concentration (about 6.8% for 5.25% bleach and about 18.5% for 15% bleach). The caustic soda solution is allowed to cool either naturally or by the use of refrigeration. Chlorine is added through a sparger pipe until the required amount of Cl_2 has been added. The course of the reaction can be followed by titration. Experienced operators soon learn to know when the end point is approaching.

It is desirable to provide auxiliary agitation by air or other means, although this is not always the practice. High temperatures favor the production of chlorates, with consequent instability and loss of Cl_2. However, with adequate agitation, auxiliary cooling is seldom considered necessary today in the production of 5.25% bleach. This is not the case in producing 15% bleach. Most operators endeavor to keep the maximum temperature below some preset temperature, such as 100°F. Since there is a 0.6°F rise above the starting temperature for every gram per liter available Cl added, it is clear that auxiliary cooling is required to maintain this condition in making concentrated bleach.

Cooling may be achieved either by circulating bleach through an external heat exchanger, usually of titanium, or by prechilling the caustic soda to a suitable temperature by mechanical refrigeration prior to chlorination. Although the chlorination reaction is very rapid, some of the side reactions which result in the precipitation of heavy metals and other impurities are substantially slower. It is therefore desirable to allow the finished product to age prior to bottling. It is then either decanted or, preferably, filtered.

Calcium Hypochlorites: Bleaching powder, which is prepared by passing chlorine gas over slaked lime, was the first way that chlorine was made generally available commercially (the technique was patented in 1799). The product usually contained about 30% available chlorine. Although it was unstable and difficult to use, it was of enormous importance in bleaching of textiles, and later for sanitizing.

Bleaching powder has largely been supplanted in the United States by an improved calcium hypochlorite product containing about 70% available Cl (HTH, Percloron, Pittchlor).* It consists essentially of calcium hypochlorite dihydrate. Several commercial routes are available for its production. One such process chlorinates a slurry of lime and caustic soda. The slurry is cooled, and the precipitated crystals removed. The crystals are then mixed with calcium chloride and chlorinated lime. Upon warming, calcium hypochlorite dihydrate is precipitated, and sodium chloride remains in solution. The slurry is filtered, granulated, dried, and sized.

The largest use of the product is for swimming-pool sanitation, but substantial quantities are used for water purification, algae control, and other uses. Because of its relative stability, it is an ideal product as an emergency stand-by for Cl_2. It is particularly useful for sanitation at times of floods or other disasters.

Calcium hypochlorite, $Ca(ClO)_2$, (20–40 g/l available Cl) is frequently produced by pulp mills for pulp bleaching. The process may be batch or continuous. If batch, a slurry of lime of the right concentration is prepared, and Cl_2 injected until the product contains the desired amount of Cl_2. If continuous, a slurry of the correct concentration is pumped to a reactor (a holding vessel of perhaps 500–5,000 gal), and Cl_2 is bubbled in. Control of Cl_2 flow is by measurement of oxidation-reduction potential (ORP). The finished bleach liquor is allowed to overflow into storage.

REFERENCES

1. The Chlorine Institute: "Exceeding All Expectations: A Short History of Chlorine," New York, 1968.
2. Sconce, James S.: "Chlorine: Its Manufacture, Properties, and Uses," ACS Monograph 154, Van Nostrand, New York, 1962.
3. Somers, H. A.: The Chlor-Alkali Industry, *Chem. Eng. Prog.*, vol. 61, no. 3, March 1965.
4. Baker, J. E., and R. D. Burt: Startup and Operating Problems in Chlorine Plants, *Chem. Eng. Prog.*, vol. 63, no. 12, December 1967.
5. The Chlorine Institute: Chlorine Manual, 1969, New York (updated periodically).

—Herbert S. Hopkins, *Olin Corporation, Stamford, Conn.*

Chlorine Organics   Many hundreds of chlorine-containing carbon compounds have been cataloged; scores of these compounds are of commercial and industrial importance; and several represent very high tonnage production. As one scans the total realm of organic chemistry, few groups and families of organic materials are without some chlorine-bearing compounds.†

By far the greatest volume demand for chlorinated organics rests in the plastics markets, with polyvinyl chloride (PVC) the outstanding example. A series of thermoplastic polymers is based on copolymerization of vinyl chloride and vinyl acetate. The most common of these resins are the 85–95% vinyl chloride–vinyl acetate copolymers. Polyvinylidene chloride (Saran) is closely related to PVC. Penton, a polychloroether, is made by chlorination of pentaerythritol. The presence of Cl in a polymer makes it fire-resistant; thus these plastics will not support a flame once the ignition source is removed. Methylchlorosilanes are used as intermediates in the formation of silicone resins. Chlorparaffins, which are produced by random chlorination of mixed paraffin feeds (C_{10}—C_{30} straight-chain, saturated hydrocarbons), are used as plasticizers for PVC. Extremely chemically inert materials are derived from chlorofluorocarbons. For example, chlorotrifluoroethylene is polymerized to produce Kel-F.

The uses for chlorinated materials as chemical intermediates are very important and varied. Nearly the entire demand for ethyl chloride, for example, involves its use in making automotive antiknocks (tetraethyl lead and methyl ethyl leads). The major production of ethylene dichloride is consumed by the producers internally by dehydrochlorinating to vinyl chloride monomer. Methyl chloride is the intermediate for many materials, including mono- and disodium methyl arsenates, methoxychlor, silicones, butyl rubber, tetramethyl lead, and higher chlorinated methanes. Benzyl chloride is a reagent in the pharmaceutical industry for production of amphetamine,

*HTH is a registered trademark of the Olin Corporation; Pittchlor, PPG Industries; and Percloron, the Pennwalt Company.

†Because of the breadth of chlorine-bearing compounds, very few cross references are given in this description. Consult **Chlorine** in the Subject Index.

Demerol, phenobarbital, and many other medicinal compounds. Phosgene is reacted with diamines in the preparation of diisocyanates, which are then reacted with glycol to form polyurethanes (ethylene diamine is formed from ammonia and ethylene dichloride).

Other elastomers or rubbers are made which incorporate Cl in the product. Chloroprene (2-chlorobutadiene) is polymerized to elastomers which resemble natural rubber but are superior in some respects, such as oil resistance (neoprene). Protective rubber coatings are made from chlorosulfonated polyethylene (Hypalon).

Agricultural chemicals of widespread importance are found among the Cl organics. Aldrin, dieldrin, and endrin are chemically related hexachlorodimethanonaphthalenes; chlordan is an octachloromethanoindene; heptachlor is a heptachloromethanoindene.

The insecticide DDT (dichlorodiphenyltrichloroethane) and the herbicide 2,4-D (dichlorophenoxy acetic acid) were developed during World War II. DDT provided a cheap delousing agent, which saved countless lives through prevention of typhus fever epidemics.

Characteristics: Generally speaking, the presence of Cl in an organic molecule increases the density, viscosity, and chemical reactivity while decreasing the specific heat, solubility in water, and flammability. Many chloroorganics are very good as solvents for hydrophobic substances, such as fats, oils, and greases. Because of this, they are absorbed through the skin and lungs and should be handled carefully with gloves and in well-ventilated areas. When Cl organics burn, they have the potential of forming two toxic gases, phosgene, $COCl_2$, and hydrogen chloride, HCl, besides carbon monoxide. In compounds such as carbon tetrachloride and the higher chlorinated materials, there is some danger of forming phosgene in a fire or from high heat. For the materials which have a sufficient amount of hydrogen in the molecule, such as methyl chloride, vinyl chloride, and ethylene dichloride, hydrogen chloride will be formed in large amounts, with some possibility of forming small amounts of phosgene. This is an invaluable aid in driving personnel away from such a fire. Phosgene is more deadly but is not nearly as obnoxious.

Chlorine Derivatives of the Paraffins: Depending upon the conditions for reaction, Cl_2 will displace one, two, three, or more hydrogen atoms from the paraffins. Thus these substitution products are referred to as *mono* ($C_nH_{2n+1}Cl$), *di* ($C_nH_{2n}Cl_2$), *tri*

($C_nH_{2n-1}Cl_3$), and so on. Methyl chloride, CH_3Cl, ethyl chloride, C_2H_5Cl, and propyl chloride, C_3H_7Cl, are examples of *monochloro* derivatives of the paraffins—also referred to as *alkyl chlorides*. Methylene dichloride, CH_2Cl_2, and ethylene dichloride, $C_2H_4Cl_2$, are examples of *dichloro* derivatives. Chloroform, $CHCl_3$, and 1,1,1-trichloroethane, $C_2H_3Cl_3$, are *trichloro* derivatives, while carbon tetrachloride, CCl_4, is a *tetrachloro* derivative.

Olefins of the unsaturated ethylene series, C_nH_{2n}, of hydrocarbons combine with Cl_2 to form addition products or alkyl dichlorides, such as ethylene dichloride, propylene dichloride, and so on. They also combine with HCl or HOCl to form addition products such as ethyl chloride, propyl chloride, ethylene chlorohydrin, and propylene chlorohydrin. In the cases where the molecule has three or more carbons, various isomers are formed, but the Cl_2 will add preferentially to the carbon having the fewest hydrogens, when adding HCl, and the opposite for HOCl.

Substitution reactions of Cl_2 involve radical attack to remove a hydrogen, thus forming the hydrocarbon radical as an intermediate. Since a trisubstituted carbon, $-\overset{|}{\underset{|}{C}}-H$, gives the most stable intermediate, it reacts more readily than a secondary carbon, $-CH_2$, and that more readily than a methyl group. Once the first Cl is present in a hydrocarbon molecule, it reduces the activity of the hydrogens on the adjacent carbons. The second Cl radical will preferentially attack a hydrogen on the same carbon containing the first Cl in ethyl chloride, for example, yielding 1,1-dichloroethane. Specificity depends upon temperature: the higher the temperature, the more random the substitution. Thus, selection of the proper temperature is important to substitution reactions.

Chlorine-Substitution Products of Carbonyl Compounds: Chlorination of a ketone or aldehyde will occur most readily on the carbon adjacent to the carbonyl. This happens because the hydrogens on an α-methyl or methylene carbon are not very tightly bound, due to interaction with the carbonyl. Chloroacetone and chloral are produced by reaction of Cl_2 with acetone or acetaldehyde. Acrolein will react with dry hydrogen chloride at low temperatures to give β-chloropropionaldehyde.

Chlorine-Substitution Products of the Fatty Acids: Chlorination of carboxylic acids is much more difficult because the effect of the carbonyl group is offset by the electron donation effect from the hydroxyl group. This can be altered by reaction with the acid chloride or anhydride rather than the acid itself. Chlorination of an aliphatic acid is normally accomplished in the presence of a catalyst such as phosphorus trichloride. Monochloroacetic acid is an important industrial chemical. Chlorosuccinic acid is a dicarboxylic acid.

Chlorine-Substitution Products of the Ethers: Ethylene chlorohydrin reacts with sulfuric acid to form β,β'-dichloroethyl ether and is thus a by-product of the chlorohydrin process for the manufacture of ethylene glycol. The chlorines on this ether are quite inert. It has been used as a selective solvent in the refining of petroleum. It can be further chlorinated at 20–30°C to give α,β,β'-trichlorodiethyl ether, which can then be hydrolyzed to chloroacetaldehyde and ethylene chlorohydrin.

Chlorine Derivatives of the Benzenes: Chlorine can be substituted onto the benzene ring in the presence of a halogen carrier ($FeCl_3$, $AlCl_3$) to give chlorobenzene, which can then substitute additional Cl to give dichlorobenzenes. Although the second chlorine preferentially goes to the para position, ortho and meta isomers can be made. If the chlorination is carried out in the presence of ultraviolet light, addition rather than substitution occurs, yielding benzene hexachloride, a derivative of cyclohexane. The gamma isomer of BHC is the active ingredient in the insecticide lindane. Other functional groups may occur on the ring with Cl, such as chlorophenol, *p*-chlorobenzoic acid, and chlorobiphenyl.

Toluene can substitute Cl on the ring or on the methyl group. When the latter is desired, the reaction must be free of iron (glass- or lead-lined reactor); benzyl chloride is the first product, benzal chloride the second, and finally, benzotrichloride.

Chlorination. Chlorinations are carried out industrially by four general methods. The first uses molecular chlorine for substitution of hydrogens. These are chain reactions, which are initiated with ultraviolet lamps, catalysts, and thermal scission of the Cl_2 molecule. The second uses molecular Cl_2 for addition accross an unsaturated (double or triple) bond. The third utilizes hydrogen chloride, which also can add across an unsaturated bond or substitute an alcohol. The fourth is called oxychlorination, using hydrogen chloride, oxygen, and a catalyst. This latter is similar to producing chlorine, Cl_2, in situ, since the products are similar to those obtained from normal thermal chlorinations.

The methyl series of chlorinated hydrocarbons can be produced by direct reaction with Cl_2. Except for chloroform, methane is the most difficult to chlorinate; methyl chloride is chlorinated about twice as fast. Therefore, to increase the yield of methyl chloride, a large excess of methane is used. Approximately 10 volumes of methane to 1 of chlorine are reacted at about 450°C. Contact times vary from about 1 to 20 s. Since all four chloromethanes are formed in varying amounts, the desired product is favored by optimizing operating conditions. As the mole ratio of chlorine to hydrocarbon is increased, the mole fraction of the more chlorinated chlorohydrocarbon is increased. Below a ratio of 1.0, methyl chloride is favored; at 1.4, methylene chloride; at 2.6, chloroform; and above 3.4, carbon tetrachloride is the predominant product. Thus it is obvious that a plant chlorinating methane will be capable of producing methyl chloride, methylene chloride, chloroform, carbon tetrachloride, and hydrochloric acid.

The majority of the chlorinated methanes are produced by a modification of the simple methane chlorination process. This uses two stages such that the effluent from the first passes through aqueous zinc chloride and there contacts methanol at about 100°C. Thus the HCl produced in the first stage is used to displace the alcohol group and convert it to methyl chloride. Methyl chloride is then fed to a chlorination unit as described above.

Production of the chlorinated ethanes and ethylenes is similar to the chloromethane group but much more complex. This type of plant will normally produce most of the following: ethyl chloride, ethylene dichloride, vinyl chloride, vinylidene chloride, 1,1,1-trichloroethane, trichloroethylene, and perchloroethylene, $Cl_2C=CCl_2$. Other compounds, such as 1,1-dichloroethane, are also formed but are recycled as a feed to produce one of the other more desirable products.

The basic raw materials for producing chlorinated two-carbon products are ethylene and chlorine, although some processes use ethane.

Ethyl chloride is produced by hydrochlorination (HCl) of ethylene. The reaction is carried out at 125°F (about 52°C) and 125 psi in the presence of aluminum chloride catalyst, which is dissolved in the product ethyl chloride. Excess HCl is used to give excellent conversions and yields.

Ethyl chloride can also be produced by thermal chlorination of ethane. In this case ethane and chlorine are fed at 300–600°C into a fluidized bed of inert solids to maintain temperature control. Although conversions are kept low to reduce higher chlorinations, many products are formed. Since HCl is a product of this reaction, it must be recovered for use in an HCl-consuming process. Therefore chlorination of ethane is utilized only when the other products can be economically recycled.

Ethylene dichloride (EDC) is produced by reacting ethylene and chlorine in liquid EDC in the presence of $FeCl_3$. Low temperatures and pressures are suitable. Conversions and yields are very high.

Ethylene dichloride can also be produced by a process called *oxychlorination*. Ethylene, hydrogen chloride, and air are passed through a fluidized copper chloride–supported catalyst bed at about $250°C$. Molar ratios of air, HCl, and ethylene are about $2.2:1.7:1.0$ in the feed, and contact times are in the order of 4.4 s. The oxygen converts HCl to Cl_2, which then adds across the ethylene double bond. The unreacted HCl is scrubbed in caustic, and the product EDC collected and purified. Conversion of the HCl can be better than 95%, and EDC purity better than 98%.

The production of EDC from ethylene, HCl, and air requires a more expensive plant than from direct chlorination of ethylene with Cl_2. Also, the yields are somewhat lower. Thus it would be a more expensive process if by-product HCl were not cheaper than Cl_2. However, in an integrated chlorination plant, by-product HCl is generated in other sections. The overall economics of the entire plant are maximized by carefully balancing the utilization of Cl_2 and HCl.

Vinyl chloride can be produced by dehydrochlorination of EDC. One method is to pass pure EDC (1,2-dichloroethane) mixed with a fluidized catalyst containing a metal chloride, for example, $ZnCl_2$, in a heated reaction zone held at about $400°C$. Residence times are controlled to give about 60% conversion and about 99% purity of the vinyl chloride product.

Another reaction producing vinyl chloride is the thermal cracking of 1,1-dichloroethane. In this case the vapor is passed through the reactor at $400–450°C$ and about $1,500$ psig. Conversions are again maintained at about 60% to reduce undesirable by-product formation.

Vinylidene chloride is produced in the same reactor when 1,1,1-trichloroethane is fed or present in the previous feed.

1,1,1-Trichloroethane is produced in one process by first chlorinating ethane or ethyl chloride to 1,1-dichloroethane.

Another process is based on the reaction of vinyl chloride with HCl in the liquid product (1,1-dichloroethane) containing a catalyst, such as aluminum chloride. The 1,1-dichloroethane is then reacted with Cl_2 in a thermal chlorination to produce 1,1,1-trichloroethane.

Per- and trichloroethylene are produced by a series of chlorination-dehydrochlorination reactions using by-products from the various operations above. For example, Cl_2 can be added to 1,2-dichloroethylene to give 1,1,2,2-tetrachloroethane, which can then be thermally cracked to trichloroethylene and HCl. Pentachloroethane can be cracked in the presence of aqueous lime to yield perchloroethylene, $Cl_2C{=}CCl_2$.

Monochlorobenzene is produced by introducing Cl_2 into benzene containing $FeCl_3$ catalyst at about $80°C$. The product is removed to a continuous fractionation from which the unreacted benzene is returned to the chlorination reactor. By operating at low conversions, very little dichlorobenzenes are produced. The chlorobenzene is neutralized with caustic and redistilled.

Chloroprene is made by a vapor-phase chlorination of butadiene. The products from this reaction are mainly 3,4-dichloro-1-butene and cis- and trans-1,4-dichloro-2-butene. By heating this mixture in the presence of copper, isomerization occurs quite readily. By fractionating in the presence of copper, the desired 3,4-DCB is removed continuously (bp $123°C$) from the higher boiling cis or trans isomers (bp $155°C$). The 3,4-DCB is then dehydrochlorinated in the presence of aqueous alkali, and the chloroprene is fractionated in a purification step.

Chloroform: Chloroform is produced primarily from chlorination of methane. Its principal consumption (60%) is in production of fluorocarbon refrigerants and propellants. The successive replacement of Cl by fluorine is accomplished by treating with hydrogen fluoride in the presence of partially fluorinated antimony pentachloride catalyst. Refrigerant 22, $CHClF_2$, is used by the air-conditioning industry for small equipment, such as home air-conditioning units, and is also used as feed to produce tetrafluoroethylene, an intermediate for producing Teflon. Fluorocarbons are also used as propellants for certain aerosol applications. About 30% of chloroform production goes to making fluorocarbon plastics. Total demand for chloroform is 290 million lb/yr (1972).

Ethyl Chloride: Although ethyl chloride has several chemical, pharmaceutical, and solvent applications, there are only two important commercial demands. The ethyl cellulose industry consumes a fairly large amount, but 90% goes as an intermediate in the production of tetraethyl lead (TEL).

Ethyl chloride is reacted with sodium-lead alloy:

$$4PbNa + 4C_2H_5Cl \rightarrow$$
$$Pb(C_2H_5)_4 + 3Pb + 4NaCl$$

The TEL produced is then stripped out of the reaction mass with steam.

A source of cellulose, such as cotton linter, is digested in a dilute caustic solution to produce alkali cellulose. This is then treated with ethyl chloride to produce ethyl cellulose. One, two, or three ethyl ether stages can be made, yielding various grades of ethyl cellulose, which are used as synthetic gums and thickeners in the plastics and lacquer industries.

Ethyl chloride has been used to some extent as an alkylating agent on aromatics (benzene) in the presence of Friedel-Crafts catalysts ($AlCl_3$). In limited quantities it has use as a solvent, refrigerant, heat-transfer medium, in production of aerosols, and in inducing local and general anesthesia. Demand runs about 600 million lb/yr (1972).

Ethylene Dichloride: The principal importance of ethylene dichloride (1,2-dichloroethane) is in the manufacture of vinyl chloride. Second, it is used as a solvent intermediate. Of third importance is its use as a lead scavenger in TEL antiknock fluids. These mixtures normally contain EDC at about 30% of the weight of TEL, along with some ethylene dibromide (EDB). EDC is also used as an intermediate in the manufacture of ethylenediamine.

$$EDC + 4NH_3 \xrightarrow[120°C]{\text{Na oleate}}$$
$$H_2NCH_2CH_2NH_2 + 2NH_4Cl$$

It can also be used to produce succinic acid by reaction with sodium cyanide, followed by acid hydrolysis. Thiokol rubbers have been prepared by reaction of EDC with sodium tetrasulfide:

$$nEDC + nNa_2S_4 \rightarrow (CH_2CH_2—S_4)_n + 2nNaCl$$

Demand for EDC is about 7.5 billion lb/yr (1972).

Methyl Chloride: Methyl chloride is primarily used to produce silicone resins and rubbers. Excess methyl chloride is reacted with silicon metal at 300°C in the presence of a copper catalyst to give mono-, di- and trichloromethylsilanes. These are hydrolyzed with aqueous bicarbonate to give hydroxymethylsilanes, which are then polym-

erized to silicone resins and rubbers. About an equal volume goes to make tetramethyl lead, the balance to rubber manufacture. Demand is 440 million lb/yr (1972).

Methylene Chloride: Methylene chloride is practically nonflammable and is the least toxic of the chloromethanes. Principal demands, resulting from its outstanding solvent properties, include use as a paint remover and for solvent degreasing. About 13% is used for food and drug processing, 12% is used in the plastics industries, and 20% is used in aerosol formulations. Demand is 500 million lb/yr (1972).

Monochlorobenzene: About half of the current production of monochlorobenzene goes to formation of phenol, usually by hydrolysis with steam in the presence of a catalyst (SiO_2). Half of the rest is nitrated to nitrochlorobenzenes, which are used for the manufacture of azo and sulfur dyes, fungicides, preservatives, photochemicals, and pharmaceuticals. Nitrochlorobenzene is converted to nitro aniline by heating with ammonia. Production of DDT (by reaction of chlorobenzene with chloral in the presence of sulfuric acid) and other dye intermediates account for the rest. For example, aniline is produced from chlorobenzene by heating with ammonia in the presence of a copper oxide catalyst. Usage of chlorobenzene is between $\frac{1}{2}$ and 1 billion lb/yr.

Perchloroethylene: Perchloroethylene's main use is as a primary solvent for dry cleaning. Its fast solvent action for animal, mineral, vegetable, and body oils and for waxes, water-insoluble soil, and tars, coupled with gentle action on fabrics and dyes, makes it a valuable dry-cleaning agent. Furthermore, it is nonflammable, safe when handled in well-ventilated areas, and can be purified for reuse. See also **Perchloroethylene.**

Phosgene: Carbonyl chloride, $COCl_2$, or phosgene, is produced by direct reaction of carbon monoxide and chlorine. It is reacted with primary amines in the presence of basic materials, such as tertiary amines, to give isocyanates. Diisocyanates (from diamines) are reacted with dihydric alcohols to form polyurethanes. Isocyanates and polyurethanes represent about 70% of the consumption of phosgene. Agricultural chemicals (e.g., urea) take about 14%, leaving 16% for various other uses. Production is 840 million lb/yr (1972).

1,1,1-Trichloroethane: 1,1,1-Trichloroethane is used as a general solvent for vapor degreasing, cold cleaning, and as a solvent carrier for water- and stain-repellent compounds. Its boiling point is lower than most other highly chlorinated solvents; also, its specific heat and heat of vaporization are low. Demand is about 447 million lb/yr (1972). See also **Trichloroethane.**

Trichloroethylene: Trichloroethylene is the major solvent used in industrial degreasing and cleaning applications. It can be used on iron, aluminum, magnesium, copper, and various plating metals without harm. The relatively low boiling point and low heat of vaporization allow use of low temperatures for degreasing operations. It has a high specific heat, permitting its use as a heat-transfer medium. It is nonflammable and presents a relatively low toxicity hazard. It is the "heavy-duty" vapor degreaser. Demand is about 675 million lb/yr (1972). See also **Trichloroethylene.**

Vinyl Chloride: Vinyl chloride is polymerized in various ways and with various other monomers to make a large variety of useful resins. Polymerization technique includes bulk, emulsion, solution, and suspension polymerization. Vinyl chloride can be copolymerized with styrene, vinyl esters, olefins, dienes, acrylics, dicarboxylic acids, and other compounds. The copolymers with about 3 to 20% vinyl acetate are the most important. These resins are most often produced by a suspension technique, using acetone, butyl acetate, or ethylene dichloride. Demand is about 5 billion lb/yr (1972). See also **Vinyl chloride monomer;** and **Polyvinyl chloride.**

—Walter Wm. Lawrence, Jr., *Ethyl Corporation, Baton Rouge, La.*

Chloroacetone (See **Chlorine organics.**)

Chlorofluorocarbons For the purpose here, chlorofluorocarbons may be defined as methanes, ethanes, and ethylenes containing at least one fluorine atom per molecule. This definition includes all the commercially important chlorofluorocarbons, as well as compounds containing only fluorine and hydrogen. Certain members of the latter group have important commercial applications closely related to those of the chlorofluorocarbons. Perfluorocarbons have similar applications, but these materials fall into a special category and are not included here.

Nomenclature: Precise chemical nomenclature of the commercially important chlorofluorocarbons has proved to be cumbersome. To avoid this difficulty, a shorthand numbering system has been devised. The system consists of a four-digit number preceded by a generic term describing the product's application, for example, Refrigerant ABCD. In this system, D is the number of fluorine atoms in the molecule; C is 1 (one) plus the number of hydrogen atoms in the molecule; B is equal to the number of carbon atoms minus 1, and A equals the number of double bonds in the molecule. Whenever A and B equal zero, the digits are omitted from the number. Thus Refrigerant 12 is the shorthand name for dichlorodifluoromethane. The materials also are referred to by commonly known trade names.*

Uses: Most uses of saturated chlorofluorocarbons capitalize on the volatility, stability, and safety of this class of compounds. About 600 million lb is produced annually in the United States. Of this total, some 200 million lb is used as refrigerants, 350 million lb as aerosol propellants, and 50 million lb as solvents and other chemicals.

Refrigerant 11, Refrigerant 12, and Refrigerant 22 are applied to a variety of basic jobs. Refrigerant 11 is used in large centrifugal air-conditioning units in office buildings and industrial plants; Refrigerant 12 is chosen for household refrigerators and freezers, as well as for automobile air conditioners; Refrigerant 22 is extensively used in residential air conditioning, where high capacity and small unit size are important. A number of binary azeotropes are used in special situations. Components of some of the common binary azeotropes, with their numerical designations, are:

Refrigerant 500 . . . . R-12/difluoroethane
Refrigerant 501 . . . . R-12/R-22
Refrigerant 502 . . . . R-22/R-115
Refrigerant 503 . . . . R-13/R-23
Refrigerant 504 . . . . R-115/difluoromethane

The aerosol industry uses Propellant 12, Propellant 11, and Propellant 114 for popular aerosol dispensers. Propellant 12 provides pressure, while Propellant 11 and Propellant 114 are added to regulate the pressure of Propellant 12 to suit the needs of a particular product. A large variety of products, ranging from paints and insecticides through personal products such as perfumes and deodorants, are packed with fluorocarbon propellants.

In recent years, chlorofluorocarbons have been applied to specialized solvent-cleaning problems. Their selective solvent properties are advanta-

*Such as Freon, E. I. du Pont de Nemours & Company; Genetron, Allied Chemical Corp.; Ucon, Union Carbide Corp.; Isotron, Pennwalt Corp.; Kaiser, Kaiser Chemical Corp.; Racon, Racon, Inc.

geous in the electronics, aerospace, optical, and miniature-machine industries. Trichlorotrifluoroethane is especially suitable because of its lack of attack on paint, gaskets, and wire insulation. A variety of proprietary mixtures and azeotropes containing trichlorotrifluoroethane as the basic ingredient are used in special situations. Binary azeotropes containing methylene chloride, ethanol, or acetone provide variations from the solvent characteristics of trichlorotrifluoroethane proper.

Substantial quantities of Fluorocarbon 11 and Fluorocarbon 12 are used in plastic foams. Flexible polyurethane foams commonly are expanded with Fluorocarbon 11, while rigid foams frequently are prepared from polystyrene and Fluorocarbon 12.

Dichlorodifluoromethane has been adapted to a new method of food freezing in which food particles are frozen on contact with boiling chlorofluorocarbon. The resulting rapid heat transfer reduces the freezing time to seconds for most foods.

Fluoroolefins, such as chlorotrifluoroethylene, vinylidene fluoride, and vinyl fluoride are used extensively in the synthesis of high-performance lubricants, plastics, and elastomers.

Production: Chlorofluoromethanes normally are produced from carbon tetrachloride or chloroform and anhydrous hydrogen fluoride in the presence of a suitable catalyst, such as $SbCl_5$ or AlF_3.

$$CCl_4 + HF \rightarrow CCl_xF_y \qquad \text{where } x + y = 4$$
$$CHCl_3 + HF \rightarrow CHCl_xF_y \qquad \text{where } x + y = 3$$

Ethane derivatives normally are prepared from perchloroethylene, chlorine, and anhydrous hydrogen fluoride in the same manner.

$$CCl_2{=}CCl_2 + HF + Cl_2 \rightarrow$$
$$CClF_2{-}CCl_2F + CClF_2 + CClF_2$$
$$\text{Fluorocarbon 113} \qquad \text{Fluorocarbon 114}$$

Fluorocarbon 114 can be fluorinated further to give Fluorocarbon 115 and Fluorocarbon 116:

$$CClF_2 + CClF_2 \xrightarrow{\text{HF}}$$
$$\text{Fluorocarbon 114}$$

$$CClF_2{-}CF_3 \ + \ CF_3{-}CF_3$$
$$\text{Fluorocarbon 115} \qquad \text{Fluorocarbon 116}$$

Synthesis of the common chlorofluorocarbons generally is carried out in liquid or vapor phase at moderate temperatures and pressures. The most difficult fluorinations usually are conducted in the vapor phase at higher temperatures over appropriate catalysts.

Fluoroolefins are prepared by a variety of essentially unrelated reactions, such as

$$CClF_2{-}CCl_2F + Zn \rightarrow CClF{=}CF_2 + ZnCl_2$$
$$CH{\equiv}CH + HF \rightarrow CH_2{=}CHF + CH_3{-}CHF_2$$
$$CH_3{-}CClF_2 \rightarrow CH_2{=}CF_2 + HCl$$

Starting materials and by-products are separated by fractional distillation, and the product further purified by washing, followed by drying over a suitable desiccant. The several purification steps are performed carefully, with the result that the chlorofluorocarbons rank among the highest-purity organic materials marketed.

REFERENCES

American Society of Heating, Refrigerating, and Air Conditioning Engineers: Handbook of Fundamentals, New York, 1967.

Dossat, R.: "Principles of Refrigeration," Wiley, New York, 1961.

Eichler, O., et al.: "Pharmacology of Fluorides," Springer-Verlag, New York, 1966.

Hudlicky, M.: "Chemistry of Organic Fluorine Compounds," MacMillan, New York, 1962.

Sanders, P.: "Principles of Aerosol Technology," Van Nostrand, New York, 1970.

Sheppard, H.: "Organic Fluorine Chemistry," W. A. Benjamin, New York, 1969.

—*Frank A. Bower, E. I. du Pont de Nemours & Company, Inc., Wilmington, Del.*

Chloroform   (See **Chlorine organics.**)

Chloroprene   (See **Chlorine organics.**)

Chlorosilanes   (See **Chlorine organics; Silicon; and Silicones.**)

Chlorosulfonated Polyethylene   (See **Chlorine organics; and Elastomers.**)

Chlorosulfonic Acid   [See **Anionic surfactants (sulfur-bearing); and Surfactants.**]

Chromating   (See **Conversion coatings.**)

Chromatography   A procedure based on physical-adsorption principles for separating various components from a mixture of chemical substances. In its broad interpretation, chromatography is a combination of separation, identification, and quantitative measurement.

The procedure is broadly applied to mixtures of organic and inorganic materials and is particularly useful with mixtures of compounds whose chemical characteristics (composition and molecular structure) and whose physical properties (boiling point, density, etc.) are so nearly identical as to make other separation and analytical techniques difficult or impractical. Some forms of chromatography also offer the distinct advantage of requiring only very small samples—in terms of milligrams. Consequently, chromatography is used widely in the laboratory for organic chemical, biological, and medical studies; and in industry, notably in chemical, petroleum, and petrochemical plants, for quality and process control.

The recovery of pure compounds from mixtures in the laboratory often is as important as, or more important than, making quantitative determinations; hence, chromatography is widely used to separate and collect pure compounds for study by other methods.

For industrial applications, a chromatograph is a complex transducer which not only puts out a signal that identifies the types and amounts of given substances, but also must first separate out the target substances from a stream that may contain numerous other substances, some of which may be closely related physically or chemically. The transducer output can be used for off-line quality control, or on-line so that the chromatograph becomes part of the total control loop.

In 1906, a Russian botanist (Mikhail Tswett) used the chromatographic principle to separate plant pigments. He filled a vertical glass tube with an adsorbent, and as a sample of the pigments was washed through the tube with a solvent, a series of colored adsorption bands was produced—in essence, a graphic presentation of colors—thus the term *chromatography*. The term no longer is meaningful in this way, however, because chromatography now is used with colorless materials, and often with the operator no longer visually witnessing the separation process.

Basic Principle: Chromatography depends upon selective retardation and separation of substances by a bed of porous sorptive media as they are transported through the bed by a moving fluid. The degree of retardation, and hence the rate of migration of each substance, is determined by its relative affinity for the sorbent.

The sorbent bed often is referred to as the *stationary phase,* the moving fluid as the *moving phase.* The moving phase may be a liquid or a gas. The stationary phase may be a solid adsorbent which sorbs sample components on its surface, or it may be a liquid dispersed on a porous, inert solid support in which the sample components are soluble and are partitioned in equilibrium with the moving phase, hence the *solvent* * or *partition liquid.*

Chromatography may be classified according to a number of schemes:

1. *The form of technique:* (*a*) frontal analysis, (*b*) displacement, (*c*) development, and (*d*) elution.

2. *Some distinctive feature of the apparatus or method,* as shown schematically in Fig. C-20. Most commonly used for referring to a specific technique, e.g., paper, thin-layer, and gel-permeation chromatography.

3. *The nature of the moving and stationary phases,* as shown diagrammatically in Fig. C-21.

Frontal analysis and displacement chromatography are not discussed here because they are of limited value and use.

Development Chromatography: The sample is introduced onto a dry column or sorbent bed and washed through the bed with a solvent that is less strongly sorbed than the sample components. The solvent-washing process is continued until the solvent reaches a point just short of the exit or opposite end of the bed. The migration rate of each sample component is dependent upon its *partition coefficient* or distribution between sorbent and solvent. The most useful forms of development chromatography are paper and thin-layer chromatography. These have gained wide acceptance because of the ease and convenience of the method and the modest cost and availability of materials. The use of columns for development chromatography has declined because of difficulties in visualizing and documenting the developed column and in recovering the pure compounds from the sorbent bed.

Elution Chromatography: In this method the moving phase is passed through the sorbent bed until all sample components have been washed or eluted from the bed. In contrast to development chromatography, elution chromatography uses columns almost exclusively. In gas-elution systems and in most types of liquid-elution chromatography, the column can be reused many times and without interruption of flow of the moving phase. Components move through the column at rates depending upon the partition coefficient and are separated into bands which elute at characteristic times. A detector at the end of the column can generate an analog signal that

---

*The term *solvent* is used in both gas chromatography in reference to the stationary-liquid phase and liquid chromatography in reference to the moving-liquid phase.

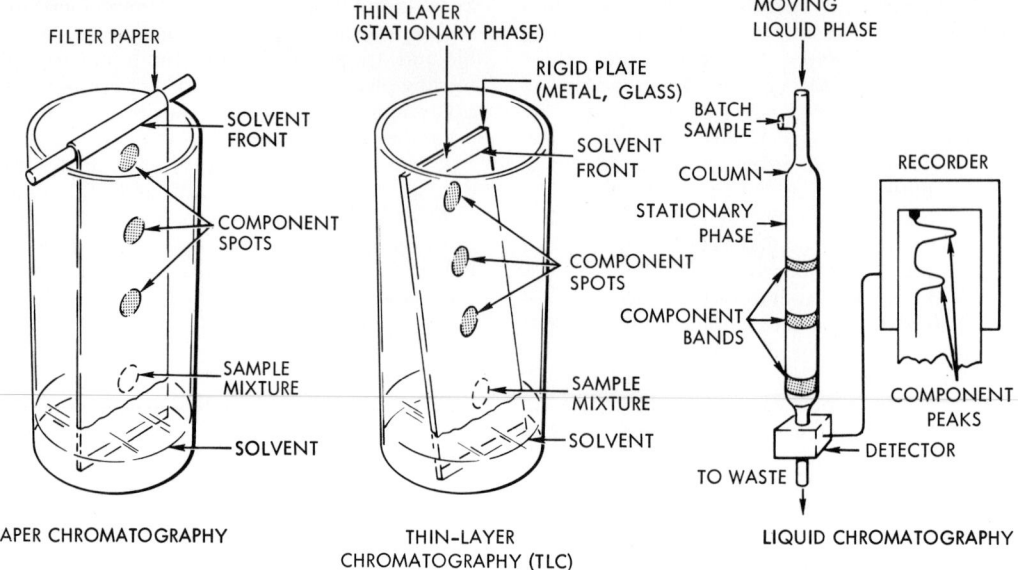

FILTER PAPER

SOLVENT FRONT

COMPONENT SPOTS

SAMPLE MIXTURE

SOLVENT

PAPER CHROMATOGRAPHY

THIN LAYER (STATIONARY PHASE)

RIGID PLATE (METAL, GLASS)

SOLVENT FRONT

COMPONENT SPOTS

SAMPLE MIXTURE

SOLVENT

THIN-LAYER CHROMATOGRAPHY (TLC)

MOVING LIQUID PHASE

BATCH SAMPLE

COLUMN

STATIONARY PHASE

COMPONENT BANDS

TO WASTE

DETECTOR

RECORDER

COMPONENT PEAKS

LIQUID CHROMATOGRAPHY

Fig. C-20. Basic chromatographic techniques.

is proportional to the concentration of the sample component in the moving phase. A time record of the detector signal is called a *chromatogram* and, as shown in Fig. C-22a contains the characteristic gaussian peaks.

Gas and liquid chromatography are the most popular forms of development chromatography. The theory, technique, and apparatus for both

methods have been highly developed. Completely automated versions (see Process-gas Chromatography,) have been developed and widely applied to the continuous, on-line analysis of chemical and petroleum-refining process streams.

Paper and Thin-Layer Chromatography: Both methods are similar in apparatus and technique. The sorbent bed is in the form of a thin sheet

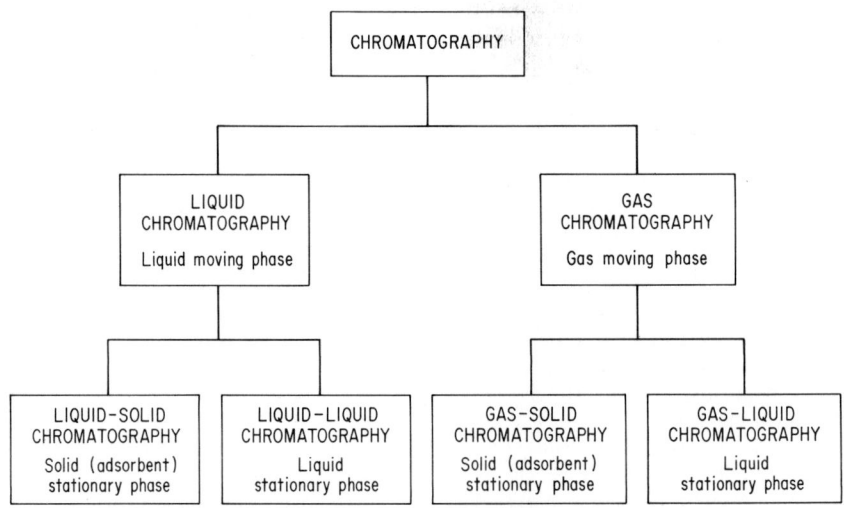

CHROMATOGRAPHY

LIQUID CHROMATOGRAPHY

Liquid moving phase

GAS CHROMATOGRAPHY

Gas moving phase

LIQUID–SOLID CHROMATOGRAPHY

Solid (adsorbent) stationary phase

LIQUID–LIQUID CHROMATOGRAPHY

Liquid stationary phase

GAS–SOLID CHROMATOGRAPHY

Solid (adsorbent) stationary phase

GAS–LIQUID CHROMATOGRAPHY

Liquid stationary phase

Fig. C-21. Chromatographic techniques classified according to the nature of moving and stationary phases.

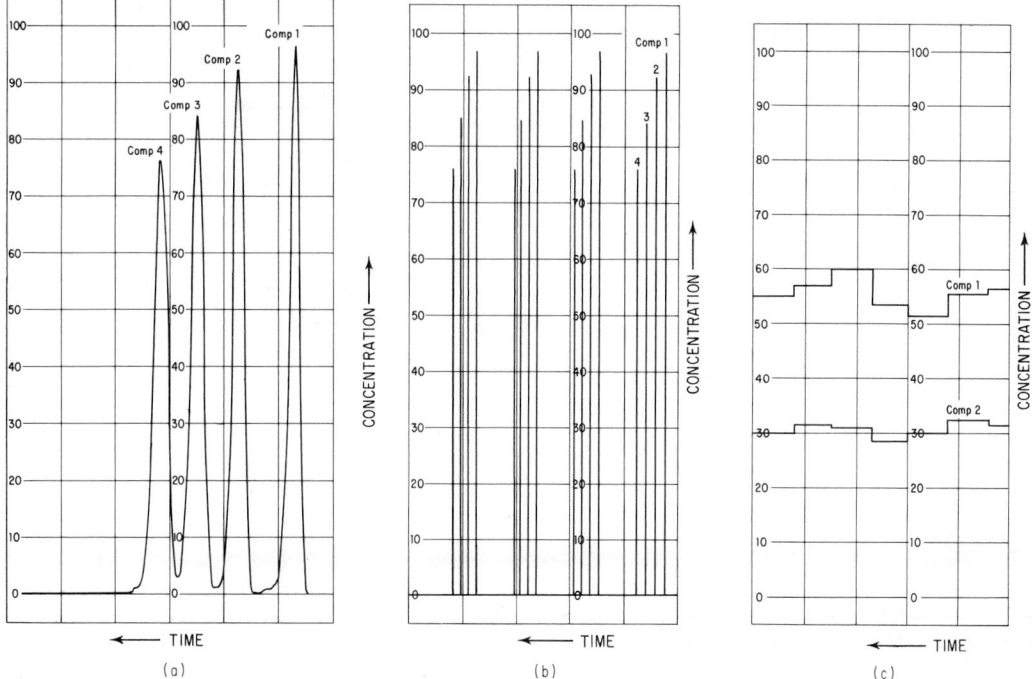

Fig. C-22. Typical chromatograph readouts. (a) Chromatogram, (b) bar graph, (c) trend.

of paper or a thin layer of a finely divided sorbent material deposited on a supporting metal, glass, or plastic plate. The sample is spotted near one end of the bed, which is then brought in contact with a source of solvent. As the solvent moves through the bed by capillary action, the sample components are washed through the bed at different rates and are separated into spots of pure compound, which can be recovered after the solvent is allowed to evaporate. Spots can be detected visually if the substances are colored, or with ultraviolet light if they are fluorescent. Components can be reacted to give colored or fluorescent derivatives by spraying with reagents. Conventional quantitative determinations can be made after recovery of the substance of interest with a solvent, or roughly estimated by comparison with spots produced with known quantities and concentrations. Recording densitometers are used for quantitative measurement in situ. The paper is drawn automatically past a slit in front of a photocell, and the transmitted or reflected light is recorded. Radioactive compounds are detected by contact exposure with x-ray film, which, after development, can be measured with a densitometer. Quantitative measurements in situ can be made with apparatus similar to a densitometer in which a radiation detector is used in place of the photocell.

Gas Chromatography: Almost any organic or inorganic compound that can be vaporized can be separated and analyzed with a gas chromatograph. As shown in Fig. C-23 the minimum requirements for a system include (1) a column which contains the *substrate* or stationary phase, (2) a supply of inert *carrier gas* (moving phase) which is continually passed through the column, (3) a means of *admitting* or *injecting* the sample into the carrier-gas stream, (4) a *detector* which senses the sample components as they elute, and (5) a *recorder.* The carrier gas may be any gas that does not react with the sample or adversely affect the detector. Helium, hydrogen, and argon are most commonly used.

Gas Chromatography in the Laboratory: A typical application is illustrated and described in Fig. C-24. Gas chromatography is so flexible that it has many end uses and apparatus configurations in the laboratory, some of which include:

1. *Programming of column temperature* to permit

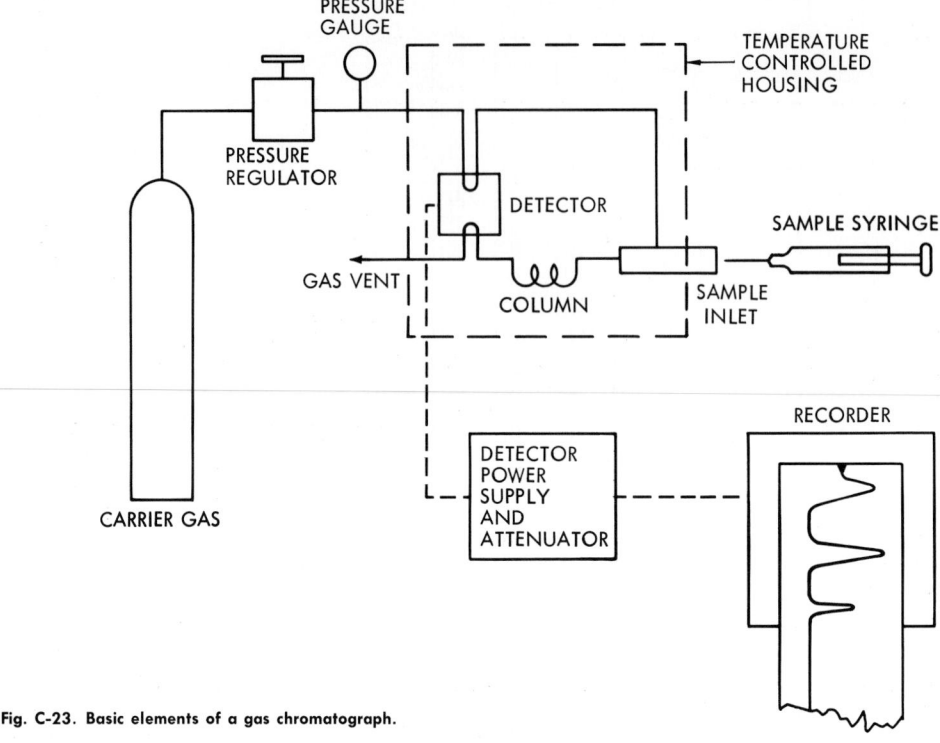

Fig. C-23. Basic elements of a gas chromatograph.

analysis of wide boiling-range fractions. A low starting temperature (sometimes subambient) affords separation of low-boiling components, whereas a later, higher column temperature reduces the elution time of high-boiling components and, in the overall, reduces analysis time. This technique is used widely for complex samples, such as essential oils and flavor and aroma concentrates and, in the petroleum industry, for simulated distillation analysis of crude oil and petroleum fractions.

2. *Multiple columns and column switching* to effect separations where a single column would be impractical and also to reduce analysis time.

3. *Fraction collecting or preparative chromatography* where separated compounds are collected for analysis by other methods.

4. *Pyrolysis chromatography* wherein the thermal degradation of high-molecular-weight materials, which yield characteristic fragment patterns, can be studied.

5. *Reaction chromatography* where compounds may be converted to other related compounds in

a reaction zone ahead of the chromatograph column.

6. *Subtraction chromatography* wherein a reaction zone in the system preferentially reacts with and removes compounds of a class of compounds, such as *n*-paraffins on a molecular sieve.

7. *Physicochemical measurements,* such as determination of activity coefficients, partition coefficients, *K* values, second virial coefficients, heat and entropy of solutions, adsorption, vaporization, and other physical and thermodynamic properties.

8. *Hyperpressure (supercritical) chromatography* in which a fluid above the critical point is used as the moving phase and which enhances the volatility of sample components by as much as 10^4. This permits the analysis of compounds of extremely high molecular weight (mol wt = 1,000) and of high boiling point (1000°C) at column temperatures below 300°C.

A large number of detectors have been used, many of which are highly specific and of limited application. The most widely used are the

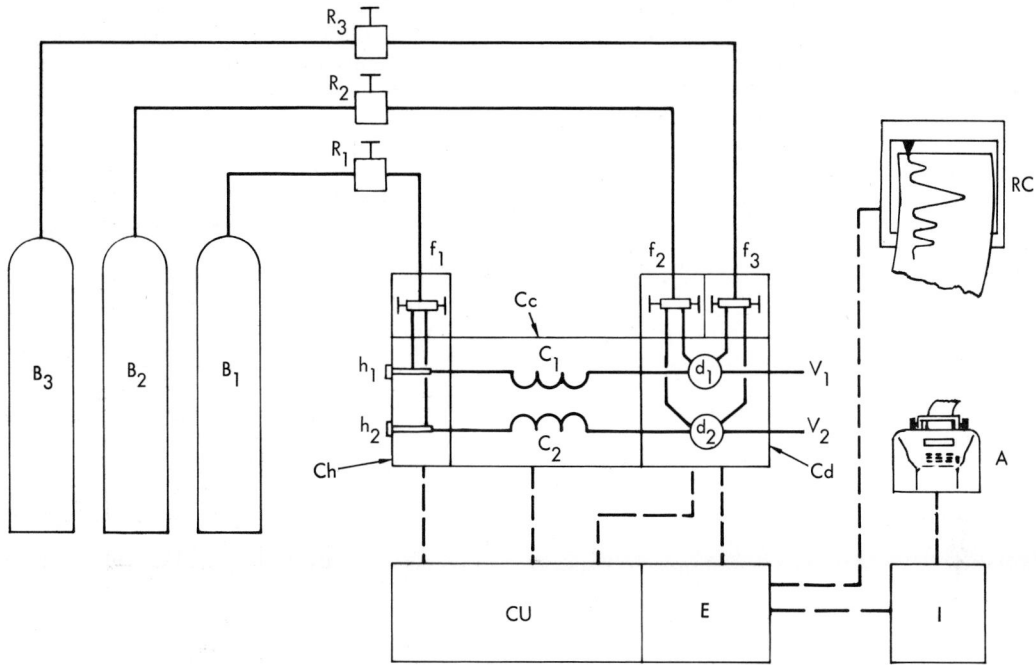

Fig. C-24. Basic elements of a gas chromatograph for research and laboratory purposes. Dual-column/dual-detector gas chromatograph shown is suitable for analysis of complex organic mixtures. Columns C_1 and C_2, with different characteristics, are enclosed in temperature-controlled column compartment Cc which may be isothermal and/or temperature-programmed during analysis for rapid elution of high-boiling compounds. Individual flash vaporization heaters h_1 and h_2 for injection and rapid vaporization of liquid samples are enclosed in temperature-controlled inlet compartment Ch. Carrier gas, supplied from high-pressure cylinder B_1, is maintained constant by pressure regulator R_1 and flow controller f_1. Sample components are detected by hydrogen-flame ionization detectors d_1 and d_2 in constant-temperature detector compartment Cd. Hydrogen fuel B_2 and combustion air B_3 are supplied to detectors through pressure regulators R_2 and R_3 and flow controllers f_2 and f_3. Combustion products are vented through V_1 and V_2. Control unit CU gives individual temperature control of sample inlets and detectors, and programs temperature of column compartment at rate and over range required. Electrometer amplifier E permits recording chromatogram on potentiometric recorder RC. Electronic integrator I detects peak maxima, integrates and digitizes peak areas, and prints out area and elution time of individual peaks on adding machine A.

thermal-conductivity, hydrogen-flame-ionization, argon-ionization, and electron-capture detectors.

Chromatograph Data Reduction: For a given substrate under given conditions, each compound has a characteristic retention time which can be used for tentative identification. However, two or more compounds may have the same time on a particular column. In such cases, the compound may be rerun on a different column with other characteristics to reduce ambiguity. Extensive compilations of individual compound-retention times on different substrates are available for reference. Positive identification can be made only by collecting the compound as it elutes and analyzing it by other means.

Quantitative analysis is based on the proportionality of detector response to the amount of component in the elution band. The most widely used measure of detector response is the area under the chromatogram peak. However, the peak height (amplitude of detector signal at peak maxima) also may be used. The area under the peak may be determined with a planimeter or from dimensions of the height and width of the peak. A ball-and-disk integrator, attached to the recorder, can be used to provide a permanent analog record of the area on the recorder chart.

Electronic integrators may sense the detector signal directly and provide digital printout of peak area and time of peak maxima. Special circuits detect peak maxima, the beginning and end of the peak, and minima between overlapping or incompletely resolved peaks, and compensate for zero drift and offset. This eliminates the need for operator attention other than for sample injection. Some systems may include an intermediate stage for converting the detector signal to a pulse-frequency-modulated signal and then recording this on magnetic tape for subsequent high-speed playback into an electronic integrator. This arrangement allows one integrator system to service several chromatographs.

An on-line computer may continuously monitor the output of several chromatographs and reduce all the collected data. The operator simply uses a console to identify the chromatograph and sample type, inject the sample, and signal the start of analysis. The output of each detector is sampled at 0.1- to 0.2-s intervals via high-speed multiplexer and analog-to-digital converter. The computer performs the integration of each peak by summation of individual amplitude measurements during elution of the peak. The computer identifies the component by comparing the time of peak maxima with a table of known values (stored in the computer memory), selects the applicable relative response factor, and calculates and prints out the composition of the sample. The program may include subroutines for resolving overlapping peaks and correcting for tailing, base-line drift, and other factors. A system may use a large computer for handling 20–40 chromatographs, or small, low-cost computers tied in with fewer instruments.

Process-Gas Chromatography: A system for continuous, repetitive, and fully automatic on-line analysis of process streams similar in all essential elements of the basic technique to the laboratory chromatograph, but different in design and appearance. Factors affecting design include (1) the need to comply with the National Electrical Code for operation in hazardous atmospheres, (2) the need to automate the procedure, and (3) the need for ready adaptability to closed-loop process control and communication with computers. Demand for maximum reliability and minimum maintenance has emphasized simplicity of hardware and methodology. Emphasis is placed on analyzing for a few rather than a number of components and on minimizing analysis time. These design targets have resulted in extensive use of multicolumn techniques for rapid separation of selected components, large portions of the sample being discarded. As shown in Fig. C-25, the major components of a process-gas chromatograph are the *analyzer,* the *programmer,* and one or more *recorders.* In some cases, a *sample conditioning* system and a *stream selector* are required.

Analyzer: This usually is located close to the process sample point in a housing that meets NEC Class I, Group D, Division 1. The controlled-temperature compartment will contain the column, sample- and column-switching valves, and detector.

Programmer: Commonly, this is located near or in the process-control room. The programmer sequences all analysis functions in the analyzer by means of a mechanical or electronic digital device. Peak-height measurement is most common, but electronic operational-amplifier integrators are also used, especially for summation of two or more components. The programmer may include a "peak picker" to convert the peak value of the component peak to a continuous analog signal for trend recording, for closed-loop control with conventional recorder-controllers, or for priority-interrupt transfer to a digital computer. The programmer also includes manual controls (sample and column-switching valves, attenuator, and zero adjustment) and a means for obtaining conventional chromatogram presentation for manual operation of the analyzer during start-up and maintenance periods.

Recorder: The simplest form of record is a bar graph, as shown in Fig. C-22b. The record consists of a series of bars, one for each component, of height proportional to the component concentration. An attenuator for each component is adjusted to give a full-scale reading equivalent to a convenient concentration value. Each component may have a different full-scale range. A large number of components in one or many streams may be recorded on the same instrument. Different streams may be identified by the height of a flat-top bar that precedes the series of bars for that stream.

Trend recording uses a conventional voltage, current, or pneumatic recorder or recorder-controller to display a continuous analog signal from the programmer. The reading is held constant at the last value recorded until a new value is recorded, causing a stepwise change in the record as the signal is updated (Fig. C-22c). A separate pen is required for each component trend so recorded.

Valves: Electrically or pneumatically operated valves are used for sample injection and column switching. Rotary, spool-and-O-ring, diaphragm, and sliding-plate valves with 4–12 ports are used.

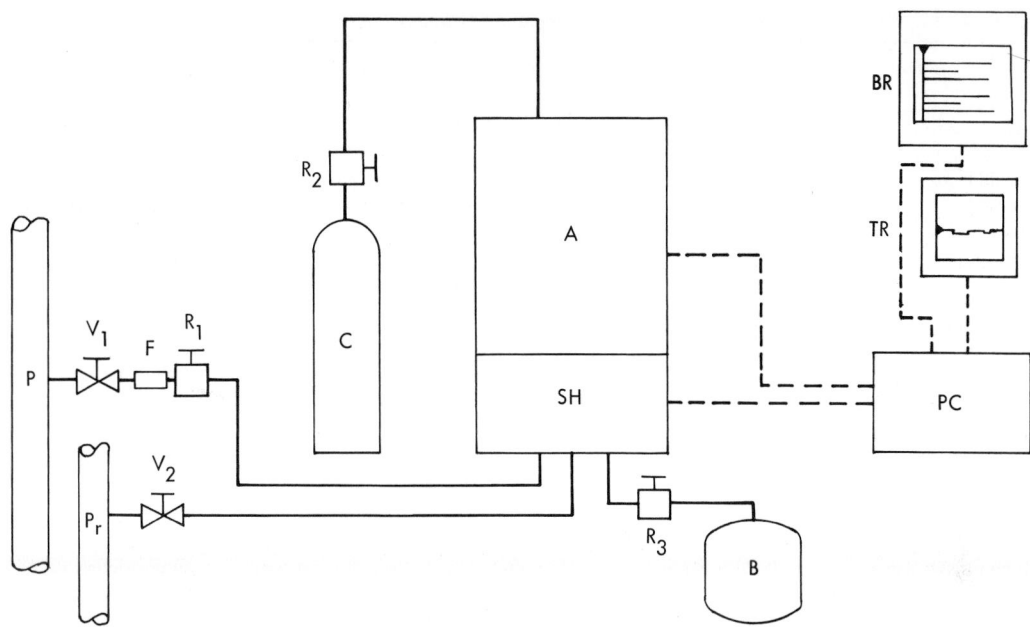

Fig. C-25. Basic elements of a process-gas chromatograph. Vapor sample is continuously withdrawn at a high rate from process line P, circulated through sample conditioner SH, and returned to lower-pressure point P_r, through shutoff valves V_1 and V_2. Particulate matter is removed by filter F, and pressure reduced to constant low level by regulator R_1. Sample conditioner contains flow control and other conditioning components and valve for switching to synthetic calibration blend B through pressure regulator R_3. Sample slipstream is circulated to sample valve in analyzer A, which also contains columns, detectors, and temperature-control system. Carrier gas C is controlled by regulator R_2. Programmer PC contains detector power supply, component attenuators, timer for controlling analyzer functions, and means for operating calibration blend valve in sample conditioner on demand. Programmer converts signal to form suitable for recording as bar graph on recorder BR or trend on trend recorder TR.

The sample valve meters a fixed volume of liquid or gas into the column with a repeatability of $\pm 0.25\%$. The sample size may vary from approximately 0.1 μl to 50 ml (with external sample loop).

Detectors: Thermal conductivity (for general use) and hydrogen-flame ionization (for trace organic analysis) are the most widely used detectors. Helium ionization (radioactive and photoionization) and thermal conductivity with amplification are used for trace inorganic and inert-gas analysis. Halogen detectors are used for applications that require sensitivity and specificity for halogen compounds (Table C-35).

Computer-controlled Chromatograph System: Large process chromatograph systems may use a dedicated low-cost computer to control all analyzer functions and perform all data reduction, thereby eliminating the individual programmers. The computer monitors each individual chromatograph detector, performs integration of peak areas, corrects for zero offset and drift, identifies components from elution times, applies response factors, and computes the composition of the sample by normalization of peak areas or by comparison with calibration standards for select components. Partial or complete stream analysis may be printed out on a teletypewriter, continuously or on demand, in accordance with program format. The computer may simultaneously provide analog trend output to conventional recorders for closed-loop control. The computer may also communicate analysis results to a larger supervisory computer (e.g., for material-balance calculations or for direct digital control). High- and low-level alarms for select components may also be set by the computer. The computer program may also include subroutines for detecting and alarming in case of malfunction or failure in chromatograph analyzers.

TABLE C-35. *Characteristics of Process-Gas Chromatography Detectors*

Detector	Compounds Detected	Best Full-Scale Sensitivity*	Advantages	Disadvantages
Thermal conductivity (low cost), two-element	All	0–1%	Low cost; rugged; reliable; low maintenance	Limited sensitivity
Thermal conductivity, four-element	All	0–0.1%	Moderate cost; rugged; reliable; high sensitivity	Requires excellent stability in balance of system
Thermal conductivity with amplifier	All	0–100 ppm	Rugged; reliable; superior sensitivity	Requires more sophisticated system, special components; higher cost
Gas-density balance	All	0–1%	Similar to thermal conductivity, but wider choice of carrier gas	Lower sensitivity; greater internal volume
Hydrogen-flame ionization	All organic except formic acid and formaldehyde	0–1 ppm	Ultrahigh sensitivity and rugged; wide choice of carrier gas	Cannot detect inorganic compounds and inert gases; requires more complex system
Helium-ionization detector (radioactive and photoionization)	All except neon in radioactive form	0.3 ppm	Ultrahigh sensitivity; universal	Requires ultrapure helium carrier and extreme sophistication; high cost

*Referred to concentration in process sample with nominal 1-cm^3 sample size; component peak approximately 0.25 min wide at base; He or H_2 carrier. Under special conditions sensitivity can be increased up to 50 times by use of large samples. See Process Gas Chromatography Detection Systems, *Chem. Eng. Prog.*, vol. 64, no. 4, p. 55, 1968.

Liquid Chromatography: This method is particularly useful for separation and analysis of high-molecular-weight compounds which are beyond the range of gas chromatography. It is generally classified according to type of stationary phase: (1) liquid-solid (adsorption), (2) liquid-liquid (partition), (3) ion exchange, and (4) gel-filtration (gel-permeation). Each type imposes specific requirements on apparatus design. The liquid is moved through the system by gravity or constant-flow pumps. Gradient devices make stepwise or continuous changes in the composition of moving phase during analysis (gradient elution). The differential refractometer probably is the most widely used detector; other detectors are based on visible, infrared, or ultraviolet photometry. Eluting compounds may be reacted with reagents to form colored substances for ease of detection. Other detectors used include a heat-of-adsorption detector and a modified hydrogen-flame-ionization detector.

See also **Amino acids; Ion-exchange resins;** and **Peptides and proteins.**

REFERENCES

Ettre, L. S., and A. Zlatkis: "The Practice of Gas Chromatography," Interscience-Wiley, New York, 1967.

Heftmann, E.: "Chromatography," 2d ed., Reinhold, New York, 1967.

Kobayashi, R., P. S. Chappelear, and H. A. Deans: Physico-chemical Measurements in Gas Chromatography, *Ind. Eng. Chem.*, vol. 59, no. 10, p. 63, 1967.

Purnell, J. H.: "Gas Chromatography," Wiley, New York, 1962.

—Richard Villalobos, *Beckman Instruments, Inc., Fullerton, Calif.*

Chromium (Gk. *chroma,* color.) **Cr** = 51.996 (at. wt.); 24 (at. no.). Chromium is a silvery-white crystalline (body-centered cubic) metal with a slight gray-blue tinge. Density = 7.19 g/cm^3 at 20°C; melting point = 1890° ± 10°C; boiling point = 2500°C; specific heat at 20°C = 0.11 cal/(g)(°C); heat of fusion = 75.6 cal/g or 136 Btu/lb; coefficient of linear expansion at 20°C = 6.2 μin./°C, or 3.4 μin./°F; thermal conductivity = 0.16 cal/cm^2/(cm)(°C)(s). Electrical resistivity = 13 $\mu\Omega$-cm at 28°C. Ionization potential = 6.764 eV. Valence is 2, 3, and 6. Modulus of elasticity in tension = 30 × 10^6 psi. Chromium has a hardness value of 9.0 on the Mohs' scale, exceeded by few materials, such as the carbides of boron, silicon, and tantalum.

Chromium is found in group VIB (along with Mo and W), period 4, of the periodic table. There are four stable isotopes of Cr (50 and 52–54) and four unstable isotopes (48, 49, 51, and 55). See also **Chemical elements.**

Chromium was discovered by Vauquelin (1797). The principal Cr ore is ferrous chromite, $Fe(CrO_2)_2$, which is found in Rhodesia and the Union of South Africa, the U.S.S.R., India, the Philippines, Japan, Turkey, Greece, Cuba, and the United States (California). In production of Cr, the chromite ore (1) is heated in an electric furnace with carbon to yield ferrochrome (used directly in alloys) or (2) is heated with Na_2CO_3 and $NaNO_3$ to form $NaCrO_4$, which then is extracted with H_2O. Chromium compounds usually are manufactured via the latter route.

Chemistry: HCl dissolves Cr to form a blue solution, $CrCl_2$, when air is excluded. If air is present, $CrCl_3$ is formed. Both Cl_2 and Br_2 attack Cr to form the relevant tervalent halide. HNO_3 in concentrated or dilute form tends to induce passivity in the metal. $Cr_2(SO_4)_3$ or $CrSO_4$ is formed when Cr is exposed to dilute H_2SO_4, depending upon concentration, time, and temperature. Chromic sulfate is used as a mordant in the dyeing of textiles, in tanning, in photographic fixing baths, and in ceramics. Water or moist air does not attack Cr at temperatures below 100°C, but Cr is oxidized to Cr_2O_3 above 200°C, particularly rapidly in the presence of NaOH.

Chromic oxide, Cr_2O_3 (green), may be formed by igniting various chromates and dichromates. The oxide is insoluble in H_2O, but is slowly soluble in acids, particularly HCl. Chromic hydroxide, $Cr(OH)_3$ (gray-green), precipitates out when an alkali is added to a chromic solution. The hydroxide is insoluble in H_2O, but soluble in acids. Chromous hydroxide, $Cr(OH)_2$ (brown), is made by adding NaOH to $CrCl_2$. Chromic sulfide, Cr_2S_3, is not precipitated out of solution because the compound is soluble in H_2O. Chromium forms two series of salts, chromates, M_2CrO_4 (yellow), and dichromates, $M_2Cr_2O_7$ (orange). Chromous compounds are strong reducing agents.

Chromic acid, H_2CrO_4 or $H_2Cr_2O_7$, is formed by the addition of concentrated H_2SO_4 to a dichromate solution to form chromic oxide, Cr_2O_3, and O_2, at a temperature of 190°C and acts as a powerful oxidizing agent. Cr_2O_3 is used as a mordant in dyeing silk and wool. Chromic oxide also is used in making yellow-green glass.

Fe(-ic), Mg, K, Na, and NH_4 chromates are soluble in H_2O; Cu(-ic) chromate is moderately soluble; Hg, Zn, Sr, and Ca chromates are moderately insoluble; and Ag, Pb, Bi, Co, Ni, and Ba chromates are insoluble. Lead chromate, $PbCrO_4$, is the pigment chrome yellow. Chromium acetate, bromide, chlorate, iodide, nitrate, and sulfate are soluble in H_2O; the oxalate is slightly soluble, and the arsenate, fluoride, phosphate, and silicate are insoluble.

Chromium Alloys: A major use of Cr is in alloys with numerous metals. Chromium is used in constructional steels primarily to increase hardness, improve hardenability, and promote the formation of carbides. Chromium steels are relatively stable at elevated temperatures and have exceptional wear resistance. Chromium enhances corrosion resistance and heat resistance, and it is the essential element in stainless and heat-resisting steels.

Stainless steels* comprise a group of ferrous alloys characterized by a high degree of resistance to chemical attack. This property, commonly referred to as *passivity,* occurs when Fe is alloyed with at least 11% Cr. Higher Cr and the addition of Ni and other elements further improve the corrosion resistance of stainless steels. The addition of this amount of Cr to low-carbon steels causes a marked and rather abrupt improvement in corrosion and heat resistance. A steel with 12% Cr will stain, but will not show progressive rusting in normal atmospheres. A steel with 18% Cr normally will not stain in conventional atmospheres, but may stain somewhat in heavy-industrial areas. Addition of 8% Ni makes an 18% Cr steel

*Excerpts from Stainless Steels, in Charles L. Mantell (ed.), "Engineering Materials Handbook," McGraw-Hill, New York, 1958.

stain-resistant in all but the worst atmospheres. Addition of Mo and other elements further enhances corrosion and heat resistance. The influence of Cr content on the resistance of low-carbon steel to boiling 65% HNO_3 is demonstrated by the fact that a steel containing 4.5% Cr will exhibit an average rate of corrosion of 12.9 in./month; with 8% Cr, this is reduced to 0.144 in./month; with 12% Cr, the rate is 0.010 in./month; with 18% Cr, the rate is 0.0025 in./month; and with 25% Cr, the rate is reduced to 0.00063 in./month—a rather dramatic reduction in corrodability. Although Fe-Cr alloys and their resistance to corrosion were known in England and France approximately 140 years ago, the phenomenon of passivity was not recognized until 1910, by Borchers and Monnartz in Germany. Following this discovery, commercial stainless steels were produced almost simultaneously in Germany, France, England, and the United States. Production now exceeds the million-ton mark annually. See also Table C-36.

Chromium also is alloyed with copper, vanadium, zirconium, and other metals to form many hundreds of Cr-bearing alloys.

Chromium Plating: A second major use for Cr is in electroplating. Although only a small percentage of the total tonnage of Cr used is consumed for electroplating, Cr is behind only copper, lead, and zinc in terms of special protective

TABLE C-36. *Numbering and Characteristics of Major Stainless Steels*

Austenitic stainless steels:

Type 302	Basic alloy for the group. 17–19% Cr; 6–8% Ni.
Type 302B	Si added to increase scaling resistance.
Type 303	Sulfur added for machinability.
Type 303Se	Selenium added for machinability.
Type 304L	Extra low carbon for improved weldability.
Type 304	Lower carbon content to improve weldability and inhibit carbide formation. 18–20% Cr; 8–12% Ni.
Type 305	Nickel increased to lower work-hardening. 17–19% Cr; 10–13% Ni.
Type 308	Cr and Ni increased for better corrosion and scaling resistance. 19–21% Cr; 10–12% Ni.
Type 309S	Lower carbon content for improved weldability.
Type 314	Si added for increased scaling resistance at high temperature. 23–26% Cr; 19–22% Ni; 2% Si max.
Type 316	Mo added to improve resistance to pitting corrosion and temperature strength. 16–18% Cr; 10–14% Ni; 2–3% Mo.
Type 316L	Extra low carbon for improved weldability.
Type 317	More Mo to further improve resistance to pitting corrosion. 18–20% Cr; 11–15% Ni; 3–4% Mo.
Type 321	Ti added to prevent Cr carbide precipitation. 17–19% Cr; 9–12% Ni; Ti = 5 × C content. C = 0.08% max.
Type 347	Columbium (niobium) or Ta added to prevent Cr carbide precipitation.
Type 348	Ta reduced for nuclear applications.

Martensitic stainless steels:

Type 410	Basic alloy for the group. 11.5–13.5% Cr; 0.5% Ni max.
Type 405	Al added to prevent weld-hardening.
Type 414	2% Ni added to improve corrosion resistance.
Type 416	Sulfur added to improve machinability.
Type 416Se	Selenium added to improve machinability.
Type 420	Carbon increased for higher hardness. 12–14% Cr; 0.5% Ni max.; C = over 0.15%.
Type 431	Cr increased to further improve corrosion resistance. 15–17% Cr; 1.25–2.50% Ni.
Type 440A	Carbon decreased slightly to improve toughness. 16–18% Cr; 0.5% Ni max.; C = 0.6–0.75%.
Type 440B	Carbon decreased slightly to improve toughness.
Type 440C	Carbon increased to increase hardness; Cr increased to make up loss in corrosion resistance. 16–18% Cr; 0.5% Ni max.; C = 0.95–1.2%.

Ferritic stainless steels:

Type 430	Basic alloy for the group. 14–18% Cr; 0.5% Ni max.
Type 430FSe	Se added to improve machinability.
Type 442	Cr increased to improve scaling and corrosion resistance. 18–23% Cr; 0.5% Ni max.
Type 446	Cr increased to improve scaling and corrosion resistance. 23–27% Cr; 0.5% Ni max.

coatings, including electroplating, applied to base metals. Chromium is plated for two major reasons—decorative and wear resistance. Chromium normally is plated over Ni, the nickel being 100 times thicker than the Cr. Automotive "bright work," plumbing fixtures, and electrical appliances, plated with Ni and Cr, are familiar examples of products that require attractive, durable finishes. Decorative Cr over Ni ranges 0.00001–0.00002 in. in thickness. In addition to plating, chromate coatings are formed on Zn, Cd, Al, Mg, and Cu by dipping these metals in solutions containing hexavalent Cr. Such coatings improve corrosion protection and are a good primer base for painting. Chromate conversion coatings are used on Zn die castings and hot-dipped or electrogalvanized parts. They are used extensively in Al parts for aircraft.

Cermets: Cermets are materials developed primarily for use at temperatures in the range 1600–2200°F (871–1204°C). The word "cermet" is a combination of the words ceramic and metal. Most cermets are produced using powder-metallurgy and related techniques. One form of cermet is chromium-bonded aluminum oxide.

Chromophores [See **Detergents;** and **Dyestuffs and dyeing (textiles).**]

Chromosomes (See **Peptides and proteins.**)

Chrysocolla (See **Copper production.**)

Chymotrypsinogen (See **Peptides and proteins.**)

Cinnabar (See **Mercury.**)

Cis Compounds (See **Isomerism.**)

Citric Acid (See **Carboxylic acids;** and **Fumaric acid.**)

Clarifiers (See **Sedimentation;** and **Waste treatment.**)

Classification (See **Screening.**)

Clathrates (See **Urea.**)

Claude Process (See **Ammonia.**)

Claus Process (See **Sulfur;** and **Sulfur-dioxide removal.**)

Clay (See **Bauxite; Ceramics;** and **Lime.**)

Cleaning Compounds [See **Anionic surfactants (sulfur-bearing); Detergents; Soaps;** and **Surfactants.**]

Cleaning Solvents (See **Air pollution; Perchloroethylene; 1,1,1-Trichloroethane;** and **Trichloroethylene.**)

Coagulation (See **Colloid systems; Elastomers; Rubber, natural; Waste treatment;** and **Water treatment.**)

Coal (See **Air pollution; Carbon;** and **Fuels.**)

Coal Tar and Derivatives Coal tar constitutes the major part of the liquid condensate obtained from the "dry" distillation or carbonization of coal (mostly bituminous) to coke. The three major products of this distillation are (1) metallurgical coke, (2) gas which is suitable as a fuel after appropriate chemical treatment, and (3) vaporized materials which leave the coke oven along with the gas and which are constituted principally of (*a*) ammonia and (*b*) coal tar. Generally, the ammonia is absorbed in water and neutralized with sulfuric acid, and the resulting ammonium sulfate, after evaporation and crystallization operations, is marketed as a by-product. A part of the gas, after it is freed of ammonia and tarry constituents, is returned to the coke oven as fuel for carbonization. If the coke oven is part of a steelmill operation, some of the gas may be burned, as, for example, in connection with open-hearth operations. Prior to the widespread use of natural gas as a domestic fuel, coke-oven gas was widely used for this market. The generalized yield of the foregoing products from a ton of coal is summarized in Table C-37, and it is of interest to note the relatively small amounts of chemicals recovered. However, considering the very high tonnage of coke produced, the resulting chemicals recov-

TABLE C-37. *Production and Yield of Coal Tar*

U.S. production of crude tar, million gal (1968)	750
Used as fuel	104
At least partially processed	644
From one ton of coal (average values):	
Coke, lb	1,500
Coke oven gas, scf	11,000
Tar, gal	10

ered are measured in terms of millions of gallons annually.

Economic Factors: In modern practice, metallurgical coke is the primary product of coal carbonization; all other products, including coal tar, are by-products. In fact, these installations often are termed *by-product coke ovens* to distinguish them from the earlier beehive type of coke ovens where no attempt whatever was made to recover gas as a fuel or tar as a source of chemicals and, consequently, represented a tremendous economic waste, as well as an intense source of air pollution.

Since metallurgical coke for use in blast furnaces is the prime product of coal carbonization, coal tar production is tied closely to the demand for metallurgical coke. Although steel production has increased progressively over the years, the demand for metallurgical coke has remained reasonably steady, for several reasons. Large improvements in blast-furnace efficiency have occurred. The amount of coke required to produce one ton of pig iron has dropped from 1,760 lb coke/ton pig iron (1955) to 1,260 lb/ton (1969), and in some modern blast furnaces, the rate is below 1,000 lb/ton. Further, coke has been partially replaced by lower-cost carbonaceous materials, such as petroleum oils, powder coal, and tar. Fundamental changes in the production of steel will reduce further the need for coke in the future.

A number of years ago, coal tar was the primary, if not the sole, source for hundreds of important organic chemicals and derivatives, notably the phenols, cresols, naphthalene, and anthracene, as well as other important coal tar end products, such as solvent naphtha and pitch. In recent years, synthetic processes for the production of phenol, the cresols, and most recently the xylenols have been developed, and thus, to a large extent, have pushed coal tar into the background as a source of materials for which it was at one time the major or only source. Nevertheless, as shown by Table C-38, the coal tar industry remains a high-tonnage industry.

TABLE C-38. *Production and Uses of Pitch*

U.S. pitch production, tons (1968)	2 million
Uses, tons (est.):	
Electrode binder	550,000–650,000
Roofing pitch	200,000–300,000
Graphite binder	150,000
Refractory brick	50,000–70,000
Target pitch	40,000
Coatings	40,000
Core pitch	15,000–25,000

Carbonization Processes: Present coking processes generally are of two types: (1) *high-temperature* (900–1200°C) carbonization for producing metallurgical coke and practically the only process practiced in the United States, and (2) *low-temperature* (500–750°C) carbonization, still practiced in some countries, notably Great Britain, where there is a market for "semicoke" as a smokeless home fuel. Low-temperature coking is discussed below.

Currently, in the United States, slot ovens with by-product recovery systems are used almost exclusively. These ovens are built in the shape of narrow chambers placed side by side with interspaced flues for heating. Usually, up to 90 chambers are placed together to form a battery. The chambers are charged individually with coal from the top. After carbonization is complete (14–20 hr), the chambers are discharged on one side by pushing with a ram from the opposite side. Each chamber is connected at the top to one or two collecting ducts, or *mains,* which carry the gas evolved and the distillate (tar) to suitable coolers and receivers. For data on tar production, see Table C-37.

Processing of Crude Tar: With reference to Fig. C-26, the crude tar, after being separated from ammonia and other gases, is subjected to an initial distillation (called *topping*) which separates the desired chemical constituents from the higher-boiling, more viscous tar constituents. In a typical case, the distillate from this operation (sometimes referred to as *chemical oil*) has an upper boiling point of about 250°C and contains (1) the phenols (*tar acids*); (2) naphthalene, which is the most prevalent single constituent of coal tar, 6–10%; (3) pyridine-type and primary bases (*tar bases*); and (4) neutral oils. The tar acids constitute about 1.5–3% of the coal tar.

Tar Acids: These materials are recovered by extraction of the chemical oil with aqueous alkali, usually caustic solution. The aqueous layer is separated from the dephenolized (*acid-free*) oil. The phenols then are recovered in crude form by acidification (*springing*) of the aqueous solution, usually by injecting CO_2, followed by gravity settling. The crude phenols then are fractionated to obtain phenol, cresols, and the higher-boiling phenols (mostly xylenols). See also **Phenol.**

Tar Bases: These materials are extracted from the dephenolized oil with aqueous solutions of mineral acids. This operation may be carried out on the entire neutral oil, or it may be done on the solvent naphtha fraction. In the latter case,

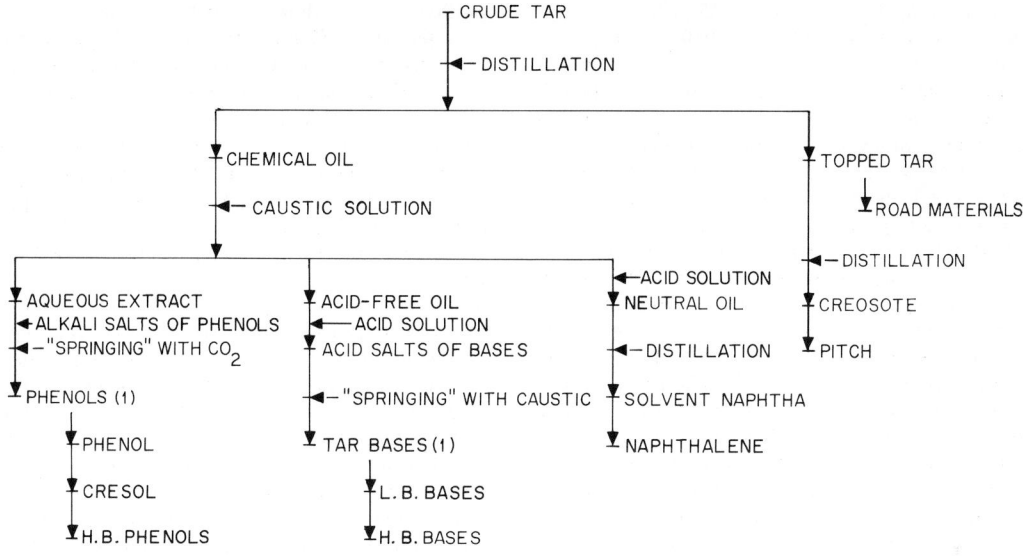

Fig. C-26. Bulk fractions from coal tar. (1) Plus salt solution. H.B. = high-boiling, L.B. = low-boiling.

only the lowest-boiling bases (picolines and lutidines) are recovered, and the higher-boiling bases (mostly quinoline and isoquinoline) can be recovered from postnaphthalene fractions or left in the residue for disposal. In European practice, the topping is carried out so that several fractions are obtained: *carbolic oil,* which yields the phenols and lower-boiling bases, and *naphthalene oil,* from which naphthalene is recovered by crystallization.

The tar bases form water-soluble salts with mineral acids which are separated from the oil. They are recovered from their salts by contacting with aqueous alkali (springing) and separating the crude bases from the salt solution. Uses for the bases vary with market demand. The lutidines constitute the major part of the lower-boiling bases. See also **Pyridine and derivatives.** Pyridine and 3-picoline are present in insufficient quantities to be of commercial significance. The higher-boiling bases are mainly quinoline and isoquinoline.

Solvent Naphtha: The lower-boiling fraction of the neutral oil is a very powerful solvent, particularly for coatings containing coal tar and pitch. The material also is a source of unsaturated compounds, such as indene and, in a lesser amount, coumarone and homologues of these compounds. Resins are formed in situ from these compounds when solvent naphtha is treated with Friedel-Crafts type catalysts. These resins are useful in

the manufacture of inexpensive floor tiles and coatings. The remaining solvent is recovered by distillation and used as a solvent.

Naphthalene: The most abundant single compound in coal tar, naphthalene finds a ready market principally for the production of phthalic anhydride. There is a great variety of processes to isolate naphthalene from the acid-free or neutral oils. Frequently, the naphthalene is first concentrated by distillation, and the enriched oil then is worked up by crystallization. This process is prevalent in Europe. It is also possible to isolate the naphthalene by careful fractionation. Depending on the purity desired, additional chemical treatment may be necessary. Naphthalene usually is traded with freezing point as a measure of purity (80.3°C for pure naphthalene). A good quality commonly used is "78°" naphthalene, which is about 96% pure. See also **Naphthalene.**

Topped Tar: With reference to Fig. C-26, it will be noted that topped tar is the residue remaining from the topping operation where the chemicals were separated as the distillate. The principal use of topped tar is in road materials. A number of standard grades (RT-1 to RT-12) are available, the grade depending on the "consistency" or viscosity of the tar. Road tar has excellent weather and skid resistance, but its use is limited by availability and price as compared with asphalt. This is borne out by the respective amounts used for road building (United States): 55 million gal tar versus 5 billion gal asphalt in 1969.

Creosote: Chemically, creosote is a mixture of a great number of compounds, almost exclusively of cyclic structure. Individual compounds present in creosote in concentrations of 2–4% are acenaphthene, fluorene, diphenylene oxide, anthracene, and carbazole. Only one compound, phenanthrene, is present in a larger concentration, 12–14%. For many years, chemists in many countries have tried to isolate individual compounds and to find profitable uses for them. None of these attempts has been successful, except for anthracene.

It has been the European practice to recover the creosote fraction in fairly narrow cuts. One of these is *anthracene oil,* in which anthracene is concentrated sufficiently to recover a crude material by crystallization. Recrystallization then will yield a material suitable for oxidation to anthraquinone by reaction with HNO_3 or with dichromate. The manufacture of anthraquinone dyestuffs, first carried out in Germany, was exclusively based on this "natural" anthracene. During World War I, when the United States was cut off from anthraquinone supplies, a synthetic process was developed based on the reaction of phthalic anhydride with benzene and a Friedel-Crafts catalyst to form o-benzoylbenzoic acid, which in turn is dehydrated to anthraquinone. This method had an economic advantage over anthracene oxidation, an advantage which it held for many years until relatively recently, when improvements in fractionation and oxidation techniques made using natural anthracene as a source of anthraquinone competitive with synthesis.

The principal use of creosote is for the preservation of wood. Railroad ties, poles, fence posts, marine pilings, and lumber for outdoor uses are impregnated with creosote in large cylindrical vessels. If properly treated, the life of the wood is greatly extended. The toxic action of creosote on soil and marine life has been and is being studied extensively. The action appears to be based on the combination of compounds present rather than on any one compound or group. Strict standards have been established for different grades of creosote in the United States by the American Wood-Preservers' Association (AWPA). Careful blending of various distillation fractions to meet these specifications is necessary. An important point with respect to creosote composition is the extent (depth) to which pitch is distilled. This is determined by the desired softening point of the pitch, the higher the softening point, the more high-boilers that have to be distilled off and thus are contained in the creosote. The pitch distillation sometimes is carried out so that several cuts of creosote are obtained to gain more freedom in the blending of the final product.

Materials that are competitive with creosote for wood-preservation purposes include various petroleum oils and pentachlorophenol. Pentachlorophenol is used in solutions of creosote or of petroleum oils. Blends of creosote with petroleum oils also are used for economic reasons.

Pitch. This is the residue from the processing of coal tar. Since pitch constitutes over 50% of the crude tar, its utilization has a major effect on the economics of tar processing. Coal tar contains an estimated 5,000–10,000 compounds, and it is reasonable to assume that about one-half of this number is contained in the pitch. Of the roughly 300 compounds identified in coal tar, about one-half must be in the pitch on the basis of their boiling points. It is probable that none of these compounds is present in pitch in concentrations of more than a fraction of 1%.

Coal tar pitch is a black, shiny material which is solid and brittle at low temperatures and liquid at high temperatures. Since it is composed of a great number of different compounds, many of which interact to form eutectic mixtures, it does not show a distinct melting or crystallizing point. Pitch usually is characterized by the *softening point,* which is determined by one of several standard methods (ring-and-ball, cube-in-air, cube-in-water, Kramer-Sarnow), each of which represents the temperature at which a given (very high) viscosity is reached. As mentioned before under *Creosote,* the depth of the distillation determines the softening point of the pitch residue.

Because of the importance of pitch in various industries, a great number of studies have been carried out to elucidate its composition. In many of these investigations, solvent fractionation was used to subdivide pitch into fractions by molecular weight, the higher-molecular-weight fractions requiring more powerful solvents. Since most of these fractions appeared as dark, resinlike substances, they were called *alpha resins, beta resins,* and *N-resins.* Although certain solvent fractions are still used today to characterize pitches, it is now realized that there are no real resins in pitch. Even the highest-molecular-weight constituents of pitch, with one exception, are substances of only moderate molecular weight, the highest consisting of up to six or seven condensed rings with molecular weights not higher than 350–400. It has been argued that, based on the glasslike, noncrystalline nature of pitch, it must contain some resinlike substances. However, the constituents isolated from pitch and identified were all crystalline, well-defined substances. It is generally believed today that the glasslike state of pitch is caused by association forces and by the mutual melting-point depression exerted by a large number of multiring compounds, many of which show eutectics when two or more are mixed.

Coal tar and pitch contain a minor but significant amount of an insoluble phase. This phase occurs in the form of small particles, about 1 μm in size, in high-temperature tar. Based on elemental composition, it appears that these insolubles are primarily polymerized ring systems with a certain amount of heterocyclic compounds. It is believed that this insoluble material is formed by secondary reactions in the hot zones of the coke oven and of the collecting main, probably by polymerization of low-molecular-weight, unsaturated compounds resulting from the primary cracking of the coal. The insoluble phase is called variously *quinoline insolubles,* C-1, *free carbon,* or *anthracene oil insolubles* (France), indicating sometimes the method of its quantitative determination. In the United States it is called *quinoline insolubles,* or QI. This insoluble phase should not be confused with the traces of coke breeze, mineral dust, and roof carbon from the coking chamber which sometimes can be found in pitch; it is a polymer of moderately high molecular weight composed of condensed-ring hydrocarbons and heterocyclic compounds. Unlike the trace impurities with which it is recovered, it has a very important function in many of the industrial uses of pitch.

Total U.S. production of coal tar pitch was nearly 2 million tons in 1968. The uses for pitch are in two general categories: (1) uses based on the *binder* properties of carbonized pitch (*pitch coke* or *binder coke*) and (2) uses based on the physical properties of pitch (Table C-38). The principal types of pitches and their uses are discussed here following the order of the two categories.

Carbon Pitch: Due to its composition—largely condensed aromatic rings—pitch forms a dense and strong coke, in good yields, when subjected to coking temperatures (950–1100°C). When mixed at elevated temperatures with an aggregate of coke, for instance, coke from the delayed coking of petroleum residues, the mix can be pressed into

suitable shapes which are "baked" to about 1100°C. The pitch forms a very strong coke (binder coke) which holds the shape together and imparts strength and good electrical conductivity to these shapes. This type of *carbon electrode* is used principally in the electrolytic reduction of alumina (*Hall process*), and its production constitutes the largest single use for pitch.

Pitches with softening points between 100 and 115°C are being used currently with a trend to somewhat higher-softening-point materials. See also **Aluminum.** Additional heat treatment of carbon electrodes, sometimes after impregnation with hot pitch for greater density, to temperatures of 2500–2800°C, results in the formation of graphite, a very dense and pure form of carbon. Graphite is used in a variety of shapes, ranging from carbon brushes for electric motors and generators to giant electrodes used in the production of steel and alloys. Graphite also is used in seals, bearings, acid-resistant construction materials, and in various nuclear and space applications.

Refractory Pitch: The inner lining of the basic oxygen furnace (BOF) consists of refractory brick, usually burned magnesite or dolomite, the pores of which are filled with pitch either by hot impregnation of the burned brick or by hot blending of the mineral with pitch followed by pressing. The brick is then placed as the inner lining into the BOF. By careful *firing,* the pitch in the brick is carbonized, and the pitch coke aids in retarding the penetration of molten slag and iron, thus prolonging the life of the brick lining.

Core Pitch: A very high melting pitch (150°C softening point) is used in the production of foundry cores. This pitch is pulverized and mixed with sand, and the mixture is wetted with water after addition of special clays and the cores are molded. After drying and baking they are ready for use. Upon pouring of hot metal into the forms, the pitch carbonizes and the coke provides strength to the core at the high temperatures.

The pitches previously described comprise the major types used where the pitch functions in the form of its coke. Following are the principal fields for pitch in which it is used because of its physical properties and resistance to weather, water, and chemicals.

Roofing Pitch: A substantial amount of pitch is used as covering membrane on flat roofs on industrial plants, large office and apartment buildings, parking garages, and similar structures. Pitches of 50–60°C softening point generally are used.

Fiber Pitch: A special pitch of about 70°C softening point is used to impregnate fiber pipe. This pipe is made from wastepaper only, typically with about ⅜-in. wall thickness with diameters ranging from 3 to 6 in. or larger. The fiber pipe is vacuum-impregnated with hot pitch, and the ends are machined after cooling for coupling. This type of pipe is used for drainage and as underground conduit for electrical cables where the smooth inner surface facilitates installation of the cables.

Target Pitch: Pitch with 125°C or higher softening point is used for the production of "clay pigeons," actually saucer-shaped targets for trap shooting. The targets are made by mixing hot pitch with pulverized minerals, usually some type of limestone, followed by molding. The high-melting pitch imparts brittleness to the targets so that they shatter easily when hit.

Pitches in Coatings: Substantial amounts of pitch are used in certain types of industrial coatings, in particular in hot-applied coatings ("enamels") for pipelines.

Low-Temperature Tar: When the carbonization of coal is carried out at lower temperatures, the resulting distillation product (tar) is of a more "primary" nature than that from high-temperature carbonization, mainly because the vapors emanating from the coal bed are not subjected to the high temperatures of the slot-type coke ovens which cause cracking and polymerization, thus giving rise to different, "secondary" products. The overall result of low-temperature carbonization is therefore a less aromatic nature of the tar, as shown by higher oxygen content, higher hydrogen and aliphatic content, and lower yield of a much less stable pitch. These general features are common to all low-temperature carbonization processes.

Such processes have been used mainly in countries where large amounts of subbituminous coals, lignite, or marginally coking coals are found. A major effort was made to utilize large amounts of lignite in the United States. The Bureau of Mines developed a fluidized-bed process to produce a dehydrated, pulverized fuel for power plants. This process is being used on a commercial scale. As a second stage, a process was developed to carbonize this fuel at about 550°C to produce a char with higher heating value and one which would provide additional income. However, all efforts to utilize this tar on a self-supporting basis were fruitless, and the second stage was not brought to commercial use.

Low-temperature carbonization took an entirely different course in England. A number of plants have been built to produce from bituminous coal a smokeless fuel suitable for burning in open fireplaces. The tar obtained has a different nature from that obtained from similar coal by high-temperature carbonization. It contains large amounts of oxygenated compounds, mainly mono- and dihydric phenols, for which profitable uses exist.

The neutral oil (creosote) is not as effective a wood preservative as high-temperature creosote. It is used for solvents and other bulk purposes. The yield of pitch is somewhat lower than in high-temperature carbonization, and its nature precludes its use in electrodes. Some of this material is used in road-binder blends where its easy miscibility with asphalt is an advantage.

REFERENCES

In spite of the importance of coal tar and derivatives, there is no comprehensive book which covers every aspect of the field. Three books contain excellent chapters on the subject:

Abraham, H.: "Asphalts and Allied Substances," 6th ed., vol. 2, Van Nostrand, New York, 1961.

Hoiberg, A. J. (ed.): "Bituminous Materials: Asphalt, Tars and Pitches," vol. 3, Wiley, New York, 1966.

Lowry, H. H. (ed.): "Chemistry of Coal Utilization," suppl. vol., Wiley, New York, 1963.

Additional references:

Franck, H. G.: Brennstoffchemie, vol. 34, p. 37, 1953. (Nature of tar oils.)

Franck, H. G.: Brennstoffchemie, vol. 36, p. 12, 1955. (Nature of coal tar pitch.)

Thomas, B. E. A.: *Gas World—Coking*, Apr. 2, 1960. (Electrode pitch.)

U.S. Bureau of Mines: Properties of Compounds in Coal-carbonization Products, *Bull.* 606, 1963.

—George G. Lauer, *Koppers Company, Inc., Monroeville, Pa.*

Coated Paper (See **Papermaking and finishing;** and **Waxes.**)

Coatings (See **Conversion coatings; Paint; Waxes;** and **Zinc.**)

Cobalt (G. *Kobold, goblin of mines.*) **Co** = 58.9332 (at. wt.); 27 (at. no.). Cobalt is a gray-white metal with a bluish tinge and exists in two allotropic forms. The close-packed hexagonal form ϵ (or α) is stable below 417°C, and the face-centered cubic form γ (or β) is stable from 417°C to the melting point. Table C-39 lists the currently accepted values of the various properties. The purity of the Co used for measuring these properties is impor-

TABLE C-39. *Properties of Cobalt*

Crystal Properties	Cph	Fcc
Atoms in cell.	2	4
Density, g/gm³ (RT)	8.85	8.80
Interatomic distances, Å	2.49–2.51	2.51
Lattice parameters (RT), Å . . .	2.5071	3.5441
c/a ratio.	1.628	

Thermal properties:
Transformation temperature, °C . . . . . 417 ± 7
Heat of fusion, cal/g 62
Melting point, °C. 1495
Boiling point (760 mm Hg), °C. 2802
Heat of vaporization, cal/g 1500
Specific heat, cal/(g)(°C):
   15–100°C 0.1056
   Molten. 0.141
Coefficient of thermal expansion,
  × 10⁻⁶/°C:
   Cph cobalt (RT) 12.5
   Fcc cobalt at transformation temp. . . 14.2
Thermal conductivity, cal/(s)(cm)(°C), at
  25°C. 0.215
Thermal neutron absorption, Bohr atom 34.8
Electrical and magnetic properties:
Electrical resistivity, Ω/cm, at 20°C . . . 5.8 × 10⁻⁶
Curie temperature, °C 1121
Saturation induction, 4πIₛ, G 18,700
Permeability:
   Initial, μc 68
   Maximum, μm. 245
Residual induction, G 4,900
Coercive force, Oe 8.9
Selected mechanical properties:
Young's modulus, psi . . . . 30.6 × 10⁶
Poisson's ratio 0.32
0.2% proof stress, psi 47,000
Ultimate tensile strength, psi 122,000
Elongation, % 20
Hardness (Diamond Pyramid) 215

RT = room temperature.
Cph = close-packed hexagonal α-form cobalt.
Fcc = face-centered cubic β-form cobalt.

tant and accounts for the scatter normally found in the published results. The mechanical properties are based on test data obtained with vacuum-melted, deoxidized metal and hot-rolled powder-metallurgical products.

The finely divided metal is pyrophoric, but in massive form Co is not attacked by air or water at temperatures below 300°C; above this temperature it is oxidized in air. It combines readily with the halogens to form the respective halides, and with most of the metalloids when heated or in the molten state. Cobalt does not combine directly with nitrogen but decomposes ammonia at elevated temperatures to form a nitride. Reaction with carbon monoxide above 225°C results in the formation of a carbide, Co_2C.

Metallic Co dissolves readily in dilute sulfuric, hydrochloric, and nitric acids, but only slowly in hydrofluoric acid, to form cobaltous salts. It is passivated by strong oxidizing agents such as the dichromates. A slow rate of attack is obtained with ammonium hydroxide, sodium hydroxide, and dilute acetic acid.

Cobalt is not very resistant to oxidation. A double-layered oxide is formed during exposure to air or oxygen at high temperatures. Between 300 and 900°C the scale consists of a thin layer of Co_3O_4 on the outside and a CoO layer next to the metal. Above 900°C, Co_3O_4 decomposes to CoO.

History: The German word *Kobold* became associated with material which was troublesome to refine and which did not necessarily contain the element. In the eighteenth century, G. Brandt, a Swedish scientist, isolated the metal, but for industrial purposes Co must be regarded as a twentieth-century metal. Cobalt minerals were known and used in Egyptian, Persian, and Chinese ceramic and glassware from about 2000 B.C. The rich blue color imparted to these materials is today derived from the metal oxides.

Occurrence: Although invariably associated in small amounts with other metal ores, such as copper, nickel, iron, or silver, Co is considered to make up about 0.001% of the igneous rocks of the earth's crust. At present, the sulfide, arsenide, and oxidized minerals form the principal economic source of Co, but it is anticipated that the extensive deposits of leached oxide and hydroxide ores found in subtropical and tropical areas (laterites) will become increasingly important.

The future economic potential of known large and valuable marine mineral deposits will depend on two factors: the availability of the appropriate engineering expertise and whether improved recovery efficiencies from the vast quantities of low-grade terrestrial deposits remain competitive.

Current production is restricted to a few countries, the most important being the Katanga area of the Republic of Zaire, Zambia, the United States, Canada, Germany, and Finland. The principal minerals and their occurrence are listed in Table C-40.

Mining and Recovery: The preliminary stages in the mining and recovery of Co are those normally applied for copper and nickel. Ore comminution

TABLE C-40. *Composition and Occurrence of Cobalt Minerals*

Mineral Name	Type	Composition	Occurrence
Cobaltite	Sulfide	$CoAsS$	Canada
Linnaeite	Sulfide	Co_3S_4	Katanga, United States, Finland
Carrollite	Sulfide	$CuCo_2S_4$	Zambia, Katanga, United States
Siegenite	Sulfide	$(Co,Ni)_3S_4$	Katanga
Safflorite	Arsenide	$CoAs_2$	Canada, Morocco
Skutterudite	Arsenide	$CoAs_3$	Canada, Morocco, United States
Smaltite	Arsenide	$CoAs_3$	Canada, Morocco, Germany
Asbolane	Oxidized	*	Canada, Zambia, Katanga
Heterogenite	Oxidized	$CoO \cdot OH$	Katanga, Zambia
Sphaerocobaltite	Oxidized	$CoCO_3$	United States, Katanga, Zambia
Erythrite	Oxidized	$Co_3(AsO_4)_2 8H_2O$	Canada, Morocco, Germany

*There is no available formula for this complex oxidized manganese-iron mineral which can contain up to 30% Co. A qualitative "formula" would appear to be $(Mn, Co, Fe, Cu)O \cdot OH$.

to an optimum size range (9–12 mm) precedes concentration. The ore is then either smelted to form a speiss or matte or, after further crushing and grinding, subjected to flotation prior to hydrometallurgical treatment. The subsequent extraction processes are varied and complex because the metallurgical properties of Co differ insufficiently from those of associated metals and also because the raw materials comprise the arsenide, sulfide, oxide, or a mixture of these. Pyrometallurgical techniques are not in themselves adequate, and although in almost all cases some preliminary heating is necessary, e.g., sulfate roasting to produce soluble salts ($Co_2CuS_4 + 6\frac{3}{4}O_2 \rightarrow 2CoSO_4 + \frac{1}{2}CuSO_4CuSO_4CuO + \frac{3}{2}SO_2$), various hydrometallurgical processes followed by electrolysis have been developed. Two of these processes are described briefly.

1. *The Katanga-Shituru (Luilu) Process—Hydrometallurgy and Electrolysis:* The Co derives from two sources, the closed copper hydrometallurgical cycle solution which contains 25–45 gal/l of Co and the direct leaching of the special oxide copper-cobalt ores. The trivalent oxide is only slightly soluble in sulfuric acid, and so it is reduced in the spent electrolyte by means of ferrous sulfate obtained from the leaching operation of sulfatized concentrates.

$$Co^{3+}(insol.) + Fe^{++} \rightarrow Co^{++}(sol.) + Fe^{3+}$$

The purification of the solutions require the removal of copper, phosphate, alumina, magnesium, manganese, zinc, iron, and nickel. Purification is based on the difference between the precipitation pH of iron, copper, and cobalt hydroxides and on the preferential sulfidation of nickel and zinc. The precipitated cobalt hydroxide becomes the feedstock of the electrolysis plant.

$$FeSO_4 + Ca(OH)_2 \rightarrow$$
$$Fe(OH)_2\downarrow + CaSO_4 \qquad pH\ 2.8–3.0$$
$$CuSO_4 + Ca(OH)_2 \rightarrow$$
$$Cu(OH)_2\downarrow + CaSO_4 \qquad pH\ 5.0–6.2$$
$$2NiSO_4 + 2NaHS \rightarrow$$
$$2NiS\downarrow + Na_2SO_4 + H_2SO_4$$
$$ZnSO_4 + H_2S \rightarrow ZnS\downarrow + H_2SO_4$$
$$CoSO_4 + Ca(OH)_2 \rightarrow$$
$$Co(OH)_2\downarrow + CaSO_4 \qquad pH\ 8.2$$

The cobalt hydrate is leached and the resultant electrolyte containing 50 gal/l of Co at 6.5 pH is rapidly circulated through the plant. The deposited Co has an average content of 99.85% Co. Main impurities are Ni (0.10%), Zn (20–40 ppm), C (40 ppm), Fe (70 ppm), H_2 (25 ppm). The hydrogen is finally reduced to 2 ppm by heat treatment at 1000°C for 1 hr.

The process adopted by the Nchanga Consolidated Copper Mines Limited in Zambia is broadly similar except that sulfide concentrates only are processed and an equally high-quality Co electrodeposit is obtained.

2. *Port Colborne Process:* The electrolytic method used in the extraction of nickel and copper from ore mined at Sudbury yields high-purity cobalt and cobalt oxide. The latter is in part reduced to the metal and marketed in the United States and in part exported to Clydach (Wales) for further refining.

At Port Colborne, Co is obtained from the electrolyte remaining after nickel recovery, iron being removed by precipitation. The solution is treated with chlorine and nickel carbonate, and

TABLE C-41. *Composition of Cobalt-Base Heat- and Creep-resistant Alloys*

Alloy	Cr	W	Mo	Ni	C	Co	Ta	Nb	Ti	B	Others
HS 21	27	. . .	5	3	0.25	Balance	. . .	. . .	. . .	0.01	
HS 31 (×40/45)	25	7.5	. . .	10	0.5/0.25	Balance					
WI 52	21	11	. . .	1	0.45	Balance	. . .	2			
HS 36	19	15	. . .	10	0.4	Balance	. . .	. . .	. . .	0.03	
HA 25	20	15	. . .	10	0.10	Balance					
G 32	19	. . .	2.0	12.5	0.28	Balance	. . .	1.4			
S 816	20	4	4	20	0.38	Balance	. . .	4	. . .	. . .	4Fe
MAR-M 302	21.5	10	. . .	. . .	0.85	Balance	9	. . .	0.20	0.005	
MAR-M 509	21.5	7.0	. . .	10	0.60	Balance	3.8	. . .	0.20	0.10	0.5Zr
FSX 414	30	7.0	. . .	10	0.25	Balance	. . .	. . .	. . .	0.10	
FSX 418	30	7.0	. . .	10	0.25	Balance	. . .	. . .	. . .	. . .	0.15Y
UMCo 50	27	. . .	. . .	. . .	0.1	Balance	. . .	. . .	. . .	. . .	20Fe
UMCo 51	27	. . .	. . .	. . .	0.3	Balance	. . .	2	. . .	. . .	20Fe

Co is precipitated in a cobalt-nickel slime. This is treated with sulfur dioxide to make the slime acid soluble and then dissolved in sulfuric acid. Any remaining copper or iron is removed from the solution, and cobalt hydroxide is precipitated with sodium hypochlorite. This is purified and roasted to the black oxide containing 70% Co, most of which is reduced to metal, cast into anodes, and electrolyzed.

Uses of Cobalt and Alloys: The metallurgical usage of Co represents about 75–80% of total world consumption; the remainder is taken up by chemical applications. Relatively limited use has been made of the pure metal, the most important being as the radioisotope cobalt 60 for teletherapy, radiation processing units, and gamma-ray radiography. Increasing availability of many wrought forms of pure Co will stimulate new applications, e.g., magnetic devices, the manufacture of hetero-

geneous welding rods for hard-facing, and electrodes for electrodeposition.

High-Temperature Materials: Improved efficiency of gas-turbine power units has created a demand for materials capable of reliable performance under the arduous conditions which prevail in the high-temperature zones of these engines. The development of a series of alloys based on Co (Table C-41) and Co containing nickel and iron base alloys has been an essential part of the progressive improvement in engine performance. Compositionally, simpler Co alloys, for example, UMCo 50 and 51, are increasingly used in furnace structures and components.

Magnetic Materials: As a consequence of cobalt's high magnetic properties, it features in a wide range of commercial magnet alloys. The more recent development of the Co–rare earth compound magnetic alloys offers, at a con-

TABLE C-42. *Composition and Hardness Properties of Some Cobalt Alloys*

Composition, %									Hardness, HV			
Co	Cr	W	C	Ni	Mo	Fe	Si	B	RT	300°C	600°C	900°C
50	33	13	2.5	. . .	. . .	1.0	. . .	. . .	606	508	406	140
53	31	14	1.0	. . .	. . .	. . .	. . .	. . .	560	450	340	130
66	26	5	1.0	. . .	. . .	2.5	. . .	. . .	457	345	235	110
63	30	. . .	0.2	. . .	6	. . .	. . .	. . .	320	225	200	100
59	29	9	1.8	. . .	. . .	. . .	. . .	. . .	546	390	362	153
45	19	13	1.0	13	. . .	. . .	3.0	2.5	730	600	330	45
50	19	8	1.0	13	. . .	. . .	2.5	1.5	490	415	310	85
52	19	9	1.0	13	. . .	. . .	2.5	1.5	525	460	290	60
42	19	15	1.5	13	. . .	. . .	3.0	3.0	950	740	420	30

RT = room temperature; HV = Vickers hardness number.

TABLE C-43. *Characteristics and Uses of Representative Cobalt Compounds*

Compound	Characteristics	Uses
Oxides......	Two main types, black and gray, available commercially. Black oxides (70–74% Co) are essentially Co_3O_4 with a few percent CoO. Cobaltous oxide, CoO, is the stable form above 900°C; cobaltic oxide, Co_3O_4 is stable below 900°C. In the gray oxide (75–78% Co), CoO is predominant constituent, although up to 40% Co_3O_4 can be present. A third oxide, Co_2O_3, is unstable and difficult to obtain in anhydrous state.	Ceramic, glass, and enamel industries. Also provide basis for manufacture of catalysts and drying agents.
Hydroxide ...	In commerical form, a pink powder prepared by precipitation with caustic soda from a solution of cobaltous salt. Insoluble in water and alkalies, but dissolves readily in most inorganic and organic acids. Exposure to air leads to formation of a brownish compound (probably hydrated cobaltic oxide) which dissolves with difficulty in acids.	Starting material for preparation of other cobalt compounds.
Acetate	Cobaltous acetate, $Co(CH_3COO)_2 \cdot 4H_2O$, contains 23.68% Co and forms monoclinic red, deliquescent crystals which are soluble in H_2O, dilute acids, alcohols, amyl acetate, fatty acids, drying oils, and resin. Prepared commercially by dissolving the hydroxide or carbonate in acetic acid.	Preparation of drying agents for inks, varnishes, and dressings for fabrics, such as canvas, heavy linen, sailcloth, etc. Also used in manufacture of catalysts and pigments and as a decolorizer in the ceramic industry.
Carbonate ...	Generally prepared by addition of an alkaline carbonate to a cobaltous salt soultion. The mauve precipitate is a basic carbonate with an intermediate formula, $xCoCO_3 \cdot yCo(OH)_2 \cdot zH_2O$. The material is insoluble in H_2O and alcohols, but dissolves easily in most inorganic and organic acids.	Preparation of other salts, e.g., chloride and nitrate. Decolorizing or pigment agent in the ceramic, pottery, and enameling industries. Also, dry-mixed into animal feeds as a mineral supplement.
Chloride.....	To date, the most important commercial Co halide. Prepared by dissolving Co metal, oxide, hydroxide, or carbonate in HCl. Evaporation yields the pink hexahydrate, and dehydration the blue anhydrous salt, which is very hygroscopic. Readily soluble in water and a number of organic solvents.	Salts can be used as humidity indicators and sympathetic inks. Electroplating, ceramic and glass industries, animal nutrition, and pharmaceuticals for the synthesis of vitamin B_{12}. Excellent catalyst for the oxidation in air of toxic waste solutions containing sulfites and some powerful antioxidants, such as cresols and phenols.
Nitrate	The hexahydrate, $Co(NO_3)_2 \cdot 6H_2O$, contains 20.26% Co and is a brown crystalline substance, deliquescent in moist air and efflorescent in dry air. It is prepared by dissolving Co metal, the oxide, hydroxide, or carbonate in dilute HNO_3 and concentrating the solution. The salt is very soluble in H_2O, alcohols, and acetone.	Preparation of catalysts for the petroleum industry.
Orthophosphate	A brilliant purple-red flocculent precipitate, $Co_3(PO_4)_2 \cdot 8H_2O$, containing 34.63% Co, is obtained upon reacting an alkaline phosphate with a solution of a Co salt. The material is insoluble in H_2O and alcohol, but dissolves in organic acids, particularly H_3PO_4.	Preparation of paints and ceramic pigments.

TABLE C-43. *Characteristics and Uses of Representative Cobalt Compounds (Continued)*

Compound	Characteristics	Uses
Oxalate	The compound, $CoC_2O_4 \cdot 2H_2O$, contains 32.23% Co and is marketed as a pink amorphous powder. Precipitation is by addition of oxalic acid or alkaline oxalate to a cobaltous salt solution. Insoluble in H_2O and slightly soluble in acids, but dissolves in NH_4OH and in some ammonium salt solutions.	Preparation of catalysts, and to some extent in the preparation of Co metal powder for powder-metallurgical applications.
Sulfate	This compound, $CoSO_2 \cdot 7H_2O$, contains 20.98% Co. The heptahydrate is a brownish-red efflorescent substance, whereas the monohydrate, obtained by heating above 70°C, is pink. More stable than the chloride or nitrate. Stores easily.	Extensive use in electroplating, ceramics, preparation of drying agents for paints and varnishes. Also used for pastureland top dressing.

servative estimate, a three- to fourfold increase in properties over the commercially available *alnicos.*

Hard-facing and Wear-resistant Alloys: These alloys are based on the cobalt-chromium-tungsten-carbon systems which saw first commercial application as a cutting-tool material during World War I. The ability of these better alloys to retain high hardness at elevated temperatures, coupled with outstanding abrasion resistance, resulted in a wider range of applications. This was promoted further by the ability to weld-deposit these alloys on cheaper material substrates, e.g., mild steel. The compositions and hardness properties of the most widely used grades are given in Table C-42.

Alloy Steels: The addition of Co in the range 6–12% to high-speed steel was one of the earliest metallurgical applications of the element, but these particular compositions have only recently come into wider usage. With regard to structural engineering steels, Co has little advantage over cheaper alloying elements, but it is increasingly featured in more specialized alloys, such as hot-working die steels, ultra-high-strength steels of the mar-aging type, and high-strength corrosion-resistant alloys.

Miscellaneous Alloys: An extensive range of Co-base and Co-containing alloys are utilized as low-expansion alloys, magnetostrictive alloys, dental and prosthetic alloys, spring materials for use up to 400°C, glass-to-metal seals, electrical-resistance alloys, and soft magnetic alloys.

Cobalt Compounds: In nearly all its compounds, Co exhibits a valence of 2^+ or 3^+, although in simple compounds the divalent form is the more stable. The cobaltous ion is quite basic, and therefore stable in aqueous solutions, whereas the cobaltic ion is such a powerful oxidizing agent that it cannot exist as such in aqueous media. Several important Co compounds are described in Table C-43.

Other important compounds include cobalt molybdate, used in the desulfurization, hydrogenation, and reforming of petroleum stock; cobalt carbonyls used in oxo synthesis; and the increasing usage of Co catalysts to provide starting material for the production of polyester. Complex inorganic salts, such as the silicate and aluminate, find extensive uses in the ceramic industry as glazes, and more recently the aluminate has found use as a nucleating agent in the investment casting of superalloys.

The trivalent ion of Co is one of the most prolific *complex* formers known. The important donor atoms in order of decreasing tendency to complex are nitrogen, carbon in the cyanides, oxygen, sulfur, and the halogens. By far the most numerous are the complexes of ammonia and amines.

A series of organic salts, such as cobalt resinate, oleate, linoleate, soyate, naphthenate, acetate, and tallate, are used extensively as drying agents for paints, inks, varnishes, and dressing for fabrics, as well as for catalysts.

—E. Williams, *Cobalt Information Centre, London, England*

Coconut Oil [See **Alcohols; Alcohols, fatty (via hydrogenation); Soaps;** and **Vegetable oils.**]

Code, Genetic (See **Peptides and proteins.**)

Coe-Clevenger Equation   (See **Liquid-solids separations; and Sedimentation.**)

Coffinite   (See **Uranium.**)

Coke   (See **Carbon; Coal tar and derivatives; Fuels; and Ironmaking and steelmaking.**)

Cold Working   (See **Iron and steel; and Soaps.**)

Collagens   (See **Peptides and proteins.**)

Collidines   (See **Pyridine and derivatives.**)

Colloid Systems   Although possessing a number of unique behavioral characteristics, colloids are defined on the basis of particle size, colloids occupying a mid-size range between that of the components of a true solution, molecular and ionic dispersoids (1 mμm and smaller) and coarse dispersions, such as emulsions and suspensions ($>$0.1 μm). The magnitude of particles in colloidal suspensions ranges between 0.1 and 1 mμm.

Colloidal dispersions also have the distinguishing characteristic of maintaining continuous random motion (sometimes referred to as *Brownian movement*). Where particles are submicroscopic, the system often is referred to as a *disperse system*. Unlike true solutions, colloidal materials exhibit negligible diffusion through parchment paper or animal membranes. Colloidal dispersions generally do not exhibit the usual phenomena shown by crystalloids or true solutions which are related to osmotic pressure, freezing-point depression, and vapor pressure.

Because colloidal particles possess an electric charge, they usually are subject to coagulation by application of an electric current or treatment with some electrolytes. The electric charges of colloids also form the basis for various instrumental and other techniques, including cataphoresis, dialysis, and electroendosmosis. At a neutral point (isoelectric point), the charge on colloidal particles is neutralized by ions of opposite charge in solution. At this point, the suspension has minimum stability, often causing precipitation. See also **Peptides and proteins.**

The properties of disperse systems essentially arise from the very large surface of the dispersed phase. For example, a 1-cm cube has a surface area of 6 cm^2; a cube 1 μm (0.001 mm) on edge has an area of 6,000 m^2, or about 1.5 acres. Thus surface effects of a material which may be negligi-

ble in macro dimension may become very significant when so magnified.

A *gel* is a colloidal disperse system in which is contained a dispersed component and a dispersion medium, both extending continuously throughout the system. Further, the system has equilibrium-elastic (time-independent) properties. The gel will support a static shear stress and not undergo permanent deformation. Thus, since they have a shear modulus of rigidity, gels are like solids, but in most other physical respects they behave like liquids. It is proposed that a gel has a three-dimensional network kept together by bonds or junction points which essentially have an unlimited lifetime. Junction points may be described as primary valence bonds, attractive forces of long range, or secondary valence bonds which maintain an association between parts of polymer chains or form submicroscopic crystalline regions. A gel may be defined as a flocculant and gelatinous precipitate. A jelly is a transparent elastic mass. Upon standing, a gel may shrink—a process known as *syneresis*.

Sometimes a colloidal system that resembles a liquid is termed a *sol,* whereas gel is applied to systems that resemble a solid, jellylike substance. Where water is one component, such systems may be termed *hydrosols* and *hydrogels*, respectively. The reversible transformation of sols into gels, and vice versa, are termed *solation* and *gelation.*

The term *lyophile* (*hydrophile* where water is the dispersion medium) is used where there is a distinct affinity between disperse phase and dispersion medium; and *lyophobe* (*hydrophobe*) where there is no such affinity. See also **Anionic surfactants (sulfur-bearing); Detergents;** and **Surfactants.**

As contrasted with true solutions, colloidal dispersions will clearly show the path of a beam of light when illuminated (known as *Tyndall phenomenon*).

A preponderance of living substances are present in the colloidal state.

Additive colloidal materials (protective colloids) may be incorporated in disperse systems to enhance stability. This is exemplified by a colloidal gold sol. If of red color, the addition of NaCl will change the color to blue because of resulting increase in particle size. Numerous substances (in order of decreasing effectiveness), including gelatin, casein in presence of NH$_3$, egg albumin, gum arabic, dextrin, wheat or potato starch, sodium stearate, sodium oleate, cane sugar, and urea, may be added to prevent this color change. This in turn can be used as a measure of the effectiveness of protective colloids, and many years ago, Zsigmondy proposed the "gold number," which may be defined as the weight (mg) of colloidal substance that is just sufficient to prevent the red-blue color change in 10 cc of gold sol (after addition of 1 cc of 10% NaCl). The gold number of gelatin is 0.005–0.01, whereas it is 8 for cane sugar and urea.

Some colloidal systems are *thixotropic;* i.e., they differ in their fluid behavior from pseudoplastic substances in that the flow rate increases with increasing duration of agitation as well as with increased shear stress. When agitation is stopped, internal shear stress exhibits hysteresis. Upon reagitation, generally less force is required to create a given flow than is required for the first agitation. Examples of thixotropic materials include silica gel, most paints, glue, molasses, lard, fruit-juice concentrates, and asphalts.

By rhythmic shaking or tapping of certain thixotropic suspensions, the suspensions will "set" or build up very rapidly. This type of nonnewtonian substance is said to be *rheopectic.* Bentonite sols and suspensions of gypsum in water are rheopectic. *Dilatant fluids* often are termed *inverted plastics,* or *inverted pseudoplastics.* Initial flow under a low shear stress is at a high rate; further increases in shear stress, however, result in lower flow rate. Some liquids may change from thixotropic to dilatant, or vice versa, as the temperature or concentration changes. Examples of dilatant materials include quicksand, peanut butter, and many candy compounds. For comparison, it should be recalled that a newtonian substance is a liquid or suspension which, when subjected to a shear stress, undergoes deformation wherein ratio of shear rate (flow) to shear stress (force) is constant.

These varying behavioral patterns of colloidal materials obviously become important considerations in the specifying of pumping and other handling and processing equipment.

An *emulsion* is a dispersion of minute drops of one liquid in another. Where an emulsion incorporates small amounts of the disperse phase, it resembles a lyophobe sol, including exhibition of Brownian movement and precipitation by electrolysis. Stable emulsions of two or more *pure* liquids require the addition of an *emulsifier* (emulsifying agent). For convenience, emulsions sometimes are classified as (1) oil-water types (oil or other immiscible material dispersed in water) and (2) water-oil types. The first type is the more common. Numerous emulsions are described throughout this volume. They are particularly important in mining processes and in food, drug, and cosmetic manufacture. See also **Detergents; Flotation; Homogenization; Soaps;** and **Surfactants.** Breaking up an emulsion is termed *deemulsification.*

Colloidal suspensions may be prepared by dispersion and condensation methods. The colloid mill in which a material is ground to the required degree of fineness is widely used and represents the dispersion method in which the relative surfaces of the system are increased. The reverse occurs in condensation; i.e., relative surfaces are decreased. Colloids may be prepared by nearly any chemical reaction involving precipitation, particularly where the concentration of electrolytes is low. Solution dilution also favors colloidal formations. Effectiveness of the procedure can be increased by adding protective colloidal materials and removing electrolytes through dialysis.

REFERENCES

Booth, F.: Recent Work on the Application of the Theory of the Ionic Double Layer to Colloidal Systems, in "Progress in Biophysics and Molecular Biology," vol. 3, Pergamon, London, 1953.

Fischer, Earl K.: "Colloidal Dispersions," Wiley, New York, 1950.

Jirgensons, B., and M. E. Straumanis: "A Short Textbook of Colloid Chemistry," Pergamon, London, 1962.

Marshall, C. Edmund: "The Colloid Chemistry of the Silicate Minerals," Academic, New York, 1949.

Mysels, Karol J.: "Introduction to Colloid Chemistry," Wiley, New York, 1959.

Shinoda, Kozo, et al.: "Colloidal Surfactants," Physical Chemistry Monograph Series, vol. 12, Academic, New York, 1963.

Van Olphen, H.: "An Introduction to Clay Colloid Chemistry," Wiley, New York, 1963.

Vold, Marjorie J., and Robert D. Vold: "Colloid Chemistry," Van Nostrand, New York, 1961.

Colloidal Solids Separation   (See **Waste treatment.**)

Colophony   (See **Resins.**)

Colors   (See **Antimony; Antioxidants; Bauxite; Cadmium; Glass; Iron oxides, synthetic;** and **Paint.**)

Colors, Glass   (See **Glass; Neodymium;** and **Praseodymium.**)

Colors, Textile   [See **Dyestuffs and dyeing (textiles).**]

Columbium (Columbia, United States.) **Cb** = 92.91 (at. wt.); 41 (at. no.). The historical, first established name for this element is columbium, and reference to this name is found widely dispersed throughout the literature. The element was discovered by C. Hatchett (1801) in Massachusetts while examining a mineral found in Connecticut. Hatchett named the element columbium. More recently, the element was renamed *niobium,* and in some official circles, the latter name has been formally accepted, with the symbol Nb. See also **Niobium.**

Columbium is a ductile, steel-gray metal with a natural isotope 93. There are several radioactive isotopes, 90–92 and 94–97. See also **Chemical elements** (under the symbol Nb). Columbium is found in group VB, period 5, of the periodic table. Columbium is a member of the refractory group of metals that also includes tungsten, tantalum, and molybdenum. Among these metals, the melting point of Cb is much lower than Ta and somewhat lower than Mo. The element has a slightly higher coefficient of thermal expansion, but only half the atomic weight and density of Ta. The crystal structure of Cb is body-centered cubic, with no allotropic transformations occurring. Important physical properties of Cb are summarized in Table C-44.

TABLE C-44. *Physical Properties of Columbium (Niobium)*

Crystallographic properties:
Crystal structure	Body-centered cubic; A_2
Lattice constant, Å	3.2941
Coordination number	8
Goldschmidt radius, CN.8, Å	1.426

Atomic and nuclear properties:
Density, at g/cm³, 20°C	8.66
Atomic volume, cm³/g atom	10.83
Cross section, thermal neutrons, barns per atom	1.1

Thermal properties:
Melting point, °C	2468 ± 10
Boiling point, °C	5127
Specific heat, cal/g, at 0°C	0.06430

Heat capacity, cal/mole:
At 0°C	6.012
At 25°C	5.95

Entropy, cal/mole,
At 25°C	8.73
Latent heat of fusion, cal/mole	6400
Latent heat of vaporization, kcal/g atom	166.5
Heat of combustion, cal/g	2379

Chemical Properties: Columbium is characterized by a high resistance to attack by corrosive aqueous solutions, although it is somewhat inferior to tantalum. Dilute mineral and organic acids and organic liquids do not affect Cb. However, the element is readily attacked by hot concentrated mineral acids, HF, and hot 5% NaOH. Oxidation in air commences at 230°C, and accelerated reaction begins at 338°C. H_2 and Cl_2 begin to attack Cb at about 200°C. The element resists attack by N_2 up to about 300°C. Fluorine attacks Cb at all temperatures. At high temperatures, Cb is rapidly attacked by all gases, even when the O_2 content is low. Columbium is very resistant to attack by molten metals, such as Li, Mg, K, and Na, with which reaction does not begin below 980°C. Important compounds of Cb are described below.

Production: The element is commercially available as sheet, foil, strip, wire, rod, and ingot. Most Cb is derived from minerals of the columbite-tantalite series. The columbite or tantalite is pulverized and fused with NaOH. This produces sodium tantalate and columbate and a mixture of Fe and Mn compounds which is washed with water to eliminate excess NaOH. Hot HCl dissolves the Fe and Mn, leaving a mixture of white acids of Ta and Cb which are extracted and separated by either of two methods. In an older process (*Marignac separation process*), advantage is taken of the fact that K_2CbOF_5 is more soluble than K_2TaF_7 in dilute HF solution. The other process involves liquid-liquid extraction and often is preferred. Probably the most widely used process utilizes the direct digestion of the ores in acid, followed by liquid-liquid extraction.

Alloys: Columbium has gained importance as an alloying element even though its use still may be considered to be in relatively early stages of development. Ferrocolumbium, a strong carbide-forming material, is added to 18-8 stainless steel to stabilize heat-affected areas during welding and subsequent intergranular corrosion. Heat-resistant steels containing Cb are used for gas-turbine rotors where temperatures up to 700°C must be withstood. Cb-base alloys are used for fast reactors, or as an alloying addition to U. Columbium is used in an alloy with 1% Zr for reactor use and is alloyed with Ni and Co to produce superalloys for military and other demanding applications.

Ti, Mo, and W additions increase the elevated-temperature hardness of Cb, while V or Zr additions have a potent strengthening effect at room temperature and up to 500°C, making alloys useful as high-temperature materials of construction. Columbium is particularly attractive because of its high melting point, density, good workability, and retention of tensile strength at high temperature. In the aerospace field, Cb is superior to most other metals on a strength-to-weight basis in the range 920–1200°C. The oxidation resistance of columbium-titanium-tungsten alloys is far superior to unalloyed Cb. In multicomponent Cb alloys, Zr and Hf appear to be more effective strengtheners than Mo or W, but have a deleterious effect on ductility.

Superconducting Material: At liquid-helium temperatures, Cb becomes a superconducting material. In the form of fine wire, the metal finds use in a superconducting cell which has advantages, both in terms of size and cost, over other electronic materials. Cb_3Sn is a fine superconductor material and becomes superconducting at a higher temperature than pure Cb, making it easier to apply. Nb-Ti alloys are used widely as superconductors, and there also is a potential application for Nb-Zr alloys in this area.

Columbium Compounds. The major classes of Cb compounds are presented as follows in alphabetical order.

Antimonides: Three are known, Cb_3Sb, Cb_5Sb_4, and $CbSb_2$.

Borides: Four are known, CbB_2, Cb_3B_4, CbB, and Cb_3B_2.

Carbides: These include CbC (cubic) and Cb_2C (hexagonal), both with very high melting points ~4000 K.

Columbates: When anhydrous Cb_2O_5 is fused with alkali-metal hydroxides or carbonates, columbates are formed. The specific formula varies with the ratio of base to acid (M_2O/Cb_2O_5), such ratios being variously reported as 1:1, 2:1; 5:1; 6:7, and so on. The best-known water-soluble potassium columbates are $K_8Cb_6O_{19} \cdot 16H_2O$ and $K_{14}Cb_{12}O_{37} \cdot xH_2O$. Other soluble columbates are known to exist, but compositions have not become firmly established. The following anhydrous columbates have been identified: $3K_2O$, Cb_2O_5; K_2O, Cb_2O_5; $2K_2O$, $3Cb_2O_5$; K_2O, $3Cb_2O_5$; $3K_2O$, $22Cb_2O_5$; $3Na_2O$, Cb_2O_5; Na_2O, Cb_2O_5; Na_2O, $4Cb_2O_5$; and Na_2O, $14Cb_2O_5$. Many other mixed oxide systems, which include Cb with metals other than alkali metals, also have been studied, and these include the series ACb_2O_{25} (in which A = P, As, or V).

Halogen Compounds: Include the *pentahalides:* $CbCl_5$ (yellow, deliquescent needles, sp gr 2.75,

decomposes in cold water); $CbBr_5$ (purple-red, decomposes in cold water); and CbF_5 (colorless, monoclinic, sp gr 3.29, decomposes in cold water). The *tetrahalides* include CbF_4 (black hygroscopic solid), $CbCl_4$ (violet-black crystals), $CbBr_4$ (crystals), and CbI_4 (plates and fine needles). The *trihalides* include $CbCl_3$ (black crystals), $CbBr_3$ (crystals), CbF_3 (crystals), and CbI_3 (crystals). *Oxyhalogen* compounds include $CbOCl_3$ (white needles, decomposes in cold water) and $CbOBr_3$ (yellow crystals, decomposes in cold water).

The *hydride* is CbH (gray plates, sp gr 6.0–6.6, soluble in HF). There are no simple *nitrates* of Cb known. The trinitrate is known: $CbCl_5 + 4N_2O_5 \rightarrow CbO(NO_3)_3 + 5NO_2Cl$. Two *nitrides* are known: CbN and Cb_2N. *Oxides*—three basic crystalline types of Cb_2O_5 are formed at various temperatures during calcination: the gamma form, prepared by calcining at 500°C, is white; the beta form is prepared at 1000°C and is yellow; and further heating results in the formation of the alpha form at 1100°C, which is white. Cb_2O_5 has a sp gr of 4.6. Other forms of Cb_2O_5 have been reported, but are not well documented. The monoxide CbO is black-cubic with sp gr 6.3–6.7 and is insoluble in water. The dioxide CbO_2 also is black and insoluble in water. *Suboxides* are formed at 300–500°C in O_2 at atmospheric pressure. Their approximate composition is Cb_6O, Cb_4O, and Cb_2O. Several *hydrous oxides* are known to exist, but compositions have not been firmly established. Three *oxide fluorides* are known: CbO_2F; Cb_3O_7F; and $Cb_5O_{12}F$.

Peroxo Compounds: When strongly alkaline solutions of soluble columbates are treated with a cold solution of H_2O_2, the corresponding peroxo salts may be obtained, such as K_3CbO_8. On treatment with dilute H_2SO_4, the peroxoortho salts give the corresponding peroxoortho acids, such as $HCbO_4$.

Phosphates: The metaphosphate, $CbOPO_4$, is known as well as the orthophosphate, $Cb_3(PO_4)_5$.

Phosphides: The mono- and diphosphides are known.

Selenium Compounds: Include Cb_5Se_4, Cb_3Se_4, $CbSe_4$, and $Cb_{1+x}Se_2$.

Silicides: Include Cb_4Si, α-Cb_5Si_3, β-Cb_5Si_3, $CbSi_2$, and $CbSi_3$.

Sulfate Complexes: The empirical formula for columbium sulfate is $Cb_2O(SO_4)_4$, $Cb_2O_5 \cdot 4SO_3$, or $(CbO_2)_2S_4O_{13}$. The existence of other columbium sulfates with the empirical formula $Cb_2O_5 \cdot SO_3$, $Cb_2O_5 \cdot 2SO_3$, or $Cb_2O_5 \cdot 3SO_3$, also have been demonstrated.

Double sulfates of Cb with $(NH_4)_2SO_4$ have

been described: $NH_4CbO(SO_4)_2$, $(NH_4)_6Cb_2O\text{-}(SO_4)_7$, and $(NH_4)_3Cb(SO_4)_4$.

Sulfides: The oxysulfide, Cb_2OS_3, is black and insoluble in water.

Tellurium Compounds: Include Cb_5Te_4, Cb_3Te_4, $CbTe_2$, and $CbTe_4$.

Organic Compounds: Columbium pentachloride reacts with alcohols, phenols, carboxylic acids, and other organic hydroxyl compounds to form Cb-O-R groups with the elimination of HCl: $CbCl_5 + 5HOC_{10}H_7 \rightarrow Cb(OC_{10}H_7)_5 + 5HCl$. Columbium acid oxalate, $Cb(HC_2O_4)_5$, is monoclinic and decomposes in cold water.

SELECTED REFERENCES

Simons, Eric N.: "Uncommon Metals," Hart Publishing Co., New York, 1967.
Sisco, Frank T., and Edward Epremian: "Columbium and Tantalum," Wiley, New York, 1963.

—Technical Staff, Kawecki-Berylco, Inc., *Reading, Pa.*

Columns   (See **Absorption**; and **Distillation**.)

Combustion   (See **Air pollution**; **Energy systems for processes**; **Fuels**; and **Sulfur-dioxide removal**.)

Compacting   (See **Agglomeration**.)

Composites   (See **Ceramics**; **Plastics**; **Polymerization**; and **Resins**.)

Compound, Chemical   A chemical compound is a pure material or substance that is composed of two or more of the chemical elements which are *chemically* combined in fixed and definite proportions.

Compression Molding   (See **Plastics**.)

Compressor Systems   (See **Cryogenic processes**; **Ethylene**; **Helium**; and **Natural gas**.)

Concrete   (See **Cement**.)

Condensation   (See **Colloid systems**; and **Polymerization**.)

Condensing   (See **Distillation**; **Energy systems for processes**; **Evaporation**; and **Separation operations**.)

Cone Crusher   (See **Size reduction**.)

Cone Mixer   (See **Mixing and blending, solids/solids**.)

Constant-boiling Solutions   (See **Hydrochloric acid**.)

Contact Adsorbents   (See **Adsorption**.)

Continuous Casting   (See **Ironmaking and steelmaking**.)

Controlled-Release Techniques   (See **Fertilizers**.)

Conversion Coatings   The industrial application of organic finishes to metals almost always requires the use of an intermediate conversion coating, particularly when the performance demands are high. Conversion coatings are formed chemically by causing the surface of the metal to be "converted" into a tightly adherent amorphous or crystalline coating, part or all of which consists of an oxidized form of the substrate metal. Conversion coatings can provide high corrosion resistance as well as strong affinity for organic coatings. They are also useful as lubricants for the drawing and forming of metals and sometimes are used for decorative purposes. The most important and widespread use of conversion coatings is on steel, zinc or galvanized steel, and aluminum alloys. The most widely used classes of conversion coatings for these metals are the phosphates and chromates. Depending on size, shape, volume of production, and other factors, the coating may be applied by spray, immersion, roll coat, or brush. A typical metal pretreatment sequence is (1) cleaning, (2) rinsing, (3) conversion coating, (4) rinsing, (5) final rinsing.

Phosphating of Steel: When a ferrous alloy is immersed in phosphoric acid, it initially forms a soluble phosphate. As the pH rises at the metal/solution interface, the phosphate becomes insoluble and crystallizes epitaxially on the substrate metal. The phosphate coating thus produced consists of a nonconductive layer of crystals that insulates the metal from any subsequently applied film and provides a topography with enhanced "tooth" for increased adhesion. The crystals insulate microanode and microcathode centers caused by stress or imperfections in the metal surface. This greatly reduces the severity of electrochemical corrosion.

Practically all phosphating processes involve patented proprietary solutions which produce superior coatings in shorter times and at lower temperatures than are obtainable with phosphoric acid alone. Treating times have been lowered from the earlier 1–2 hr to 1–5 min, while temperatures have been reduced from 98 to 20–30°C (208 to 68–86°F).

Iron phosphate coatings, weighing about 40–60 mg/ft^2, provide good paint adhesion but inferior heat and corrosion resistance. They are used when coatings performance is not very demanding. Iron phosphate coatings appear as a very thin blue or brown film to the naked eye.

Crystalline zinc phosphate coatings weigh about 200 mg/ft^2 when applied by spray, or as much as 3,000 mg/ft^2 when applied by immersion. They are a medium-gray color and are used when higher quality is mandatory. For even greater corrosion resistance and paint adhesion, *microcrystalline* zinc phosphate coatings are used. They are dark in color and give coating weights of about 150–200 mg/ft^2 when sprayed and up to 1,000 mg/ft^2 when applied by immersion. They are used as a paint base for products which are expected to last for years under varying environmental conditions.

Phosphating Chemistry: Iron and iron oxide react with phosphoric acid to form soluble primary iron phosphate, $Fe(H_2PO_4)_2$, liberating H_2 and H_2O, respectively. Due to the consumption of acid at the metal interface, there is a local rise in pH which causes insoluble secondary iron phosphate, $FeHPO_4$, to coat the metal. The iron in the coating is supplied by the substrate.

In zinc phosphating, a small amount of iron phosphate is formed initially, but the bath contains primary zinc phosphate, $Zn(H_2PO_4)_2$, which crystallizes on the metal surface as secondary and tertiary zinc phosphates, $ZnHPO_4$ and $Zn_3(PO_4)_2$, respectively, when the pH rises at the metal/solution interface. The most frequently used baths contain accelerators, preferably nitrates and nitrites, which oxidize the hydrogen formed by the pickling reactions. The fundamental zinc phosphating reactions occur in three steps, all in the same bath.

Pickling:

$$Fe^0 + 2H^+ \rightarrow Fe^{++} + H_2$$
$$3Fe^0 + 2NO_3^- + 8H^+ \rightarrow$$
$$3Fe^{++} + 2NO + 4H_2O$$

Coating:

$$3Fe^{++} + 2H_2PO_4^- \rightarrow 4H^+ + Fe_3(PO_4)_2$$
$$3Zn^{++} + 2H_2PO_4^- \rightarrow 4H^+ + Zn_3(PO_4)_2$$

Iron removal:

$$4Fe(H_2PO_4)_2 + O_2 \rightarrow$$
$$\underset{\text{Sludge}}{4FePO_4} + 4H_3PO_4 + 2H_2O$$
$$Fe(H_2PO_4)_2 + NaNO_2 \rightarrow$$
$$\underset{\text{Sludge}}{FePO_4} + NO + H_2O + NaH_2PO_4$$

In the coating reaction, each 3 moles of iron or zinc liberates 4 moles of hydrogen ion. However, in the pickling reaction, 8 moles of hydrogen ion is consumed. Thus the pH at the metal interface rises, and insoluble tertiary ferrous phosphate and zinc phosphate crystallize on the iron surface. The coating closest to the metal interface is largely iron phosphate, while that farther away is rich in zinc phosphate.

Iron buildup in the bath is objectionable, and in the iron-removal equations above it is seen that it can be removed by oxidation, slowly in air or more rapidly by peroxides or nitrite, as shown in the final equation. Note that the iron removed becomes fer*ric* phosphate, while iron in the coating is fer*rous* phosphate.

Accelerators speed phosphating reactions by reacting with hydrogen liberated at the metal. Were hydrogen not removed, it would form gas bubbles which would interfere with metal/solution contact. Strong oxidizer accelerators also serve to precipitate dissolved iron and, to a degree, act as metal cleaners by oxidizing residual organic soils. In spray application, mild accelerators are usually adequate for maintaining dissolved iron at safe levels because atomization of the solution permits it to absorb from air the oxygen needed for precipitation of iron.

Zinc phosphate coatings consist of varying ratios of hopeite, $Zn_3(PO_4)_2 \cdot 2H_2O$, and phosphophyllite, $Zn_2Fe(PO_4)_2 \cdot 4H_2O$, with hopeite usually predominant. Hopeite and phosphophyllite grow epitaxially on alpha-iron crystallites. Only a slight adaptation deformation is necessary for the lattice planes of both foreign phases compared with the alpha-iron lattice of the substrate. Good adhesive strength can be expected from such a bond.

The smaller the crystals, the better the adhesion: crystal to metal, crystal to crystal, and crystal to final finish. Also, the smaller the crystals, the tighter the packing, the denser the coating, the less total porosity area for corrosive reactions to take place.

Crystal size depends upon such factors as growth rate, agitation, and the effects of nucleating agents and foreign atoms in the crystal lattice. Spray application provides agitation which reduces crystal size. Accelerators increase the number of nucleation sites on the substrate and result in smaller crystals. The most refined technique for producing very small crystals is the introduction of foreign elements of different atomic radii into the crystal lattice. A calcium additive in a zinc phosphating solution produces scholzite, $CaZn_2(PO_4)_2 \cdot 2H_2O$, in which the crystals may average one-twentieth the size of the finest crystals produced by other methods. It is believed that the foreign elements cause uneven growth along one crystal face, creating stresses that either stunt growth at an early stage or rupture the crystal.

The following sequence is typical of a production spray phosphate system: (1) cleaning, 60 s, 71–77°C (160–170°F); (2) rinsing, 15–30 s, hot or cold; (3) phosphating, 60 s, 54–60°C (130–140°F); (4) rinsing, 15–30 s, cold; and (5) chromate-rinsing, 30–45 s, 27–60°C (80–140°F).

The final stage, an acid chromate rinse, is extremely important to the overall adhesion and corrosion resistance of the finish. One of its functions is to seal the pores in the phosphate coating. The effect can be demonstrated by placing a drop of chromic acid in the center of a phosphatized panel and subjecting it to corrosive exposure. Corrosion in the chromated area will be strongly inhibited.

Phosphating of Zinc: The chemistry in the phosphating of zinc alloys is similar, with the exception that iron does not play an important role in the coating or in the bath. The only cation involved is Zn. Zinc phosphate coatings are used widely in the treatment of galvanized steel for refrigerators, air conditioners, kitchen cabinets, and siding.

Pretreatment of Aluminum: Although Al protects itself against corrosion by forming a natural oxide, the protection is not complete. In the presence of moisture and electrolytes, Al alloys, particularly the high-copper alloys, corrode much more rapidly than pure Al.

Chemically produced oxides can be formed by treatment in 2–3% sodium carbonate containing 0.1% sodium dichromate for 10–20 min at 66°C (150°F), followed by "sealing" in 5% sodium dichromate at 82–88°C (180–190°F) for 10 min. Such coatings are softer, more porous, and not as effective as those produced by chromic acid anodizing.

Electrically produced anodic coatings from chromic, H_2SO_4, or oxalic acid electrolytes are more dense and less porous. Their corrosion resistance is improved by hot-water sealing, which is even more effective with the inclusion of dichromate.

Generally, the higher the alloy content of the base Al, the heavier the oxide present and the more difficult it is to remove prior to effective chemical pretreatment. A properly formulated deoxidizer will remove the oxide only to the desired degree, with minimum attack on the base metal. The baths usually consist of fairly large amounts (5–10%) of either HNO_3 or H_2SO_4, along with chromates and either free or complex fluorides. A deoxidizer will also remove the smut that forms on etching in strong alkali. Smut consists of the alkali-insoluble alloying elements and their oxides.

Amorphous phosphate coatings for Al were introduced in 1945. Their simplicity of operation, speed, and economy have resulted in wide commercial acceptance. They provide a continuous uniform green coating with excellent paint-bonding properties and underfilm corrosion protection. The coatings consist of varying ratios of chromic phosphate and hydrated aluminum oxide. The baths contain H_2F_2, which removes the natural oxide to permit contact of the coating-forming chemicals with the metal. The complexity of the reactions involved makes it difficult to present a simplified chemistry, but the results of many tests and analyses give the following coating composition: $xCrPO_4 \cdot yAl_2O_3 \cdot zH_2O$. The phosphate coatings vary from 10 to 300 mg/ft^2, depending on the end use. The lower coating weights are used for paint bonding; the higher range is used for decorative purposes.

Gold-colored conversion coatings are formed in baths containing hydrofluoric acid to remove the natural oxide, and chromic acid. The coating composition is chromic chromate plus varying amounts of hydrated aluminum oxide. Some baths also contain ferricyanide iron, which greatly accelerates the coating action and forms some chromic ferricyanide in the coating. This constitutes one of the most widely used conversion coatings on aluminum because of its high speed, excellent corrosion resistance, and high affinity for organic finishes. The hexavalent chromium content permits these coatings to withstand somewhat more severe corrosive environments than do the amorphous phosphate coatings. The baths have a pH of about 1.2–1.9 and can be applied by dip, brush, spray, or reverse roll coater.

Proprietary chromate rinses are frequently used

over conversion coatings on aluminum for increased corrosion resistance.

—Wilbur S. Hall, *Amchem Products, Inc.,*
Ambler, Pa.

Cooling Systems   (See **Energy systems for processes; Helium; Natural gas;** and **Nuclear power plants.**)

Copolymers   (See **Polymerization.**)

Copper   (L. *cuprum,* copper, and Gk. *kyprios,* Cyprus.) **Cu** = 63.54 (at. wt.); 29 (at. no.). Copper is a yellowish-red metal and is exceeded only by gold and silver in malleability. The metal gives a brilliant luster when polished and has a characteristic odor and taste. Copper is found in group IB (along with Ag and Au), period 4, of the periodic table. There are two natural isotopes of Cu (63 and 65) and seven radioactive isotopes (58–62, 64, and 66–67). See also **Chemical elements.** Pure copper melts at 1083°C and boils at 2595°C.

The specific gravity of Cu is 8.91, and it weighs 0.321 lb/in.³. Next to silver, Cu is the best conductor of electricity, with a conductivity approximately 97% that of Ag. The coefficient of expansion is 0.000017 per deg C.

Because of the versatility of Cu and its alloys, there are thousands of uses for this metal, some categories of which will be described below.

Chemical Properties. Although dry air has no effect upon Cu, moist air containing CO_2 forms an effective protecting coating of basic carbonate of a pale green, aesthetically pleasing color, sometimes referred to as *verdigris,* often apparent on buildings where Cu has been used for roofing or decor. The most effective solvent for Cu is dilute HNO_3, although it is even more readily dissolved by HNO_2. The reaction with HNO_3 is $3Cu + 8HNO_3 \rightarrow 3Cu(NO_3)_2 + 2NO + 4H_2O$. The effect of dilute H_2SO_4 on Cu is only slight, but hot, concentrated H_2SO_4 attacks Cu to produce SO_2, as $Cu + 2H_2SO_4 \rightarrow CuSO_4 + SO_2 + H_2O$. In this reaction, the Cu turns black, and this is best explained by the presence of three reactions in which Cu_2S and CuS are formed during the course of completing the total reaction. The halogen acids HF, HCl, and HBr have but a slight effect on Cu when dry. A hot solution of 15% HCl will dissolve the metal more readily than concentrated HCl to form Cu_2Cl_2 and H_2. The solubility of Cu in these reagents is markedly affected by the presence of impurities. At ordinary temperatures and on finely divided Cu, H_2S has little effect in the absence of air. The presence of air causes a vigorous reaction. A boiling solution of $Na_2S_2O_3$ readily changes Cu to Cu_2S. Cold acetic acid very slowly will dissolve Cu, with more rapid reaction from glacial acetic acid, particularly in the presence of H_2O_2.

Copper exists in several oxide forms: Cu_2O, CuO, Cu_2O_3, and CuO_2. CuO (black) results when Cu is heated in air or when various salts, such as the hydroxide, carbonate, sulfate, or nitrate, are ignited in air. Copper salts, except the sulfide, are soluble in NH_4OH. The cuprous salts generally are colorless and are soluble in HCl and cyanides; and are readily oxidized on exposure to moist air. The cupric salts, in crystalline form or in solution, are noted by their green or blue color. Their anhydrous forms are white.

The properties and uses of some of the more important Cu salts are summarized in Table C-45.

Unalloyed Copper. Of major importance to the applications of Cu metal are (1) high electrical conductivity, (2) large thermal conductivity, (3) corrosion resistance, (4) malleability, (5) formability, and (6) strength. For some uses, the nonmagnetic properties are desirable. The metal is easily finished by lacquering, painting, enameling, and overplating with other metals. Copper can be soldered, brazed, and welded.

The major use for unalloyed Cu is in the electric field, where its combination of electrical conductivity and corrosion resistance and strength gives it an overall economic advantage over most other conductors. Giving Cu a base of 100, the relative electrical conductivity is 106 for Ag, and then drops to 72 for Au, 62 for Al, 39 for Mg, 29 for Zn, 25 for Ni, 23 for Cd, 18 for Co, 17 for Fe, and 13–17 for steel.

Unalloyed Cu also is used as a structural material for roofing, sheathing, heat exchangers, and various large vessels used in processing chemicals, beverages, and foods. In some of these applications, thermal conductivity is very important. Again, using Cu as a base of 100, the relative thermal conductivity is 108 for Ag, 76 for Au, 56 for Al, 41 for Mg, 29 for Zn, 24 for Cd, 17 for Co, 17 for Fe, and 13–17 for steel.

Industrially, in the United States, the term copper signifies Cu containing less than 0.5% impurities or alloying elements. Copper-base alloys are those that contain Cu in no lesser amounts than 40%. Copper, of course, appears as a minor, but nevertheless important, ingredient of numerous other alloys, but these are not described here.

TABLE C-45. *Properties and Uses of Some Copper Compounds*

Compound and (Formula Weight)	Properties and Uses
Cupric acetate, $Cu(C_2H_3O_2)_2 \cdot H_2O$, (199.63), sp gr = 1.88, mp = 115°C, D = 240°C, RI = 1.545; 1.550	Dark brown powder. Slightly soluble in cold H_2O and alcohol; moderately soluble in hot H_2O and ether. Used as insecticide, fungicide, in pigments, and as a catalyst.
Cupric acetoarsenite (paris green), $(CuOAs_2O_3)_3 \cdot Cu(C_2H_3O_2)_2$, (1013.83)	Emerald-green powder. Very slightly soluble in cold H_2O. Soluble in alcohol and KCN. Used as a paint pigment, insecticide, and wood preservative.
Cupric acid orthoarsenite (Scheele's green), $CuHAsO_2$, (187.51)	Green powder. Insoluble in H_2O. Soluble in acids, alcohol, and NH_4OH.
Copper carbonate (basic), (malachite), $CuCO_3 \cdot Cu(OH)_2$, (221.17), sp gr = 3.9, D = 220°C, RI = 1.875	Dark green monoclinic crystals. Insoluble in cold H_2O; decomposes in hot H_2O. Soluble in KCN. Used as pigment. Also a Cu ore.
Cupric hydroxide, $Cu(OH)_2$, (97.59), sp gr = 3.37	Blue, gelatinous. Insoluble in cold H_2O; decomposes in hot H_2O. Soluble in alcohol, NH_4OH, and KCN. Sometimes used as pigment.
Cuprous cyanide, $Cu_2(CN)_2$, (179.16), sp gr = 2.9, mp = 475°C	White, monoclinic crystals. Insoluble in H_2O. Soluble in KCN, HCl, and NH_4OH. Used in Sandmeyer's reaction to synthesize benzenoid cyanides.
Cuprous iodide, Cu_2I_2, (380.98), sp gr = 5.63, mp = 605°C, bp = 1290°C, RI = 2.346	Cubic, white crystals. Practically insoluble in water and alcohol. Soluble in KI, KCN, and NH_4OH. Used in Sandmeyer's reaction to synthesize benzenoid chlorides.
Cupric oxide, CuO, (79.57), sp gr = 6.40, D = 1026°C	Black, cubic crystals. Insoluble in H_2O. Soluble in alcohol, KCN, and NH_4Cl. Used to color ceramics green and blue.
Cuprous oxide (cuprite), Cu_2O, (143.14), sp gr = 6.0, mp = 1235°C, bp = 1800°C, RI = 2.705	Red, cubic crystals. Insoluble in H_2O. Soluble in HCl, NH_4Cl, and NH_4OH. Used in rectifiers. Also a Cu ore.
Cupric sulfate (blue vitriol), $CuSO_4 \cdot 5H_2O$, (249.71), sp gr = 2.29, mp = 110°C, bp = 520°C, RI = 1.5368	Blue, triclinic crystals. Moderately soluble in cold H_2O; quite soluble in hot H_2O. Very slightly soluble in alcohol. Used in copper plating, dyestuffs, germicides, and for coppering steel.
Cupric chloride, $CuCl_2$, (134.48), sp gr = 3.054, mp = 498°C	Brown-yellow powder. Quite soluble in cold H_2O and alcohol. Very soluble in hot H_2O. Plays catalytic role in several organic syntheses, including production of vinyl chloride monomer.

sp gr = specific gravity; mp = melting point; D = decomposes; RI = refractive index; bp = boiling point.

Commercial copper is available in six general types. The cold- and hot-working properties of all types are excellent. The annealing temperature for all types ranges between 371 and 649°C. The hot-working temperature lies between 760 and 871°C. Machinability and polishing ratings are excellent for all types. The modulus of elasticity (tension) for all types is 17×10^6 psi. The melting point is essentially the same for all types (1083°C). The average coefficient of thermal expansion (range 68–570°F) is 9.8-millionths per deg F, except for arsenical Cu, for which it is 9.6. Hardness ranges from 40 to 55 on the Rockwell B scale.

Electrolytic Tough-Pitch Copper: [99.90% Cu (min.), 0.04% O_2 (nominal)] This metal has high electrical conductivity, high thermal conductivity, and excellent workability. It is excellent for soft soldering, good for silver-alloy brazing and metal arc welding, fair for carbon arc welding, and poor for resistance welding. Sp gr = 8.89–8.94; EC = 101% IACS; TC = 226; SH = 0.092; TS (sheet), hard = 50, soft = 32.* *Uses* include *archi-*

*Sp gr = specific gravity, g/cm³
   EC = electrical conductivity (International Annealed Copper Standard)
   TC = thermal conductivity, Btu (ft²)/(ft)(hr)(°F), at 68°F
   TS = tensile strength, 1,000 psi
   SH = specific heat, Btu/(lb)(°F), at 68°F

tectural (building fronts, downspouts, flashing, gutters, screening, and roofing); *automotive* (gaskets and radiators); *electrical* (bus bars, conductive-wire contacts, terminals, and switch parts); *hardware* (ball floats, cotter pins, nails and rivets, and soldering copper); *other* (die-pressed forgings, anodes, chemical process equipment, kettles, pans, printing rolls, rotation bands, and expansion plates).

Deoxidized Copper: [99.90% Cu (min.), 0.025% P (nominal)] This metal has higher forming and bending qualities than tough-pitch electrolytic copper and is preferred for coppersmithing and welding because of resistance to embrittlement at elevated temperatures. It is excellent for soft soldering and silver-alloy brazing, good for gas welding and gas-shielded arc welding, fair for carbon arc welding, and poor for metal arc and resistance welding. Sp gr = 8.94; EC = 80–90% IACS; TC = 185–205; SH = 0.092; TS (sheet), hard = 50, soft = 32–33. *Uses* include *industrial* (brewery vessels, condensers, evaporators, heat exchangers, dairy tubes, fractionating columns, kettles, pulp and paper lines, steam and water lines, and tanks); *transportation* (air, gasoline, oil, and hydraulic fluid lines and oil coolers); *miscellaneous* (die-pressed forgings, shell rotation bands, and gage lines).

Oxygen-free Copper: [99.92% Cu (min.), no residual deoxidants] This metal has high electrical conductivity and a good resistance to embrittlement by gases at high temperatures and is excellent for deep drawing, short-radius bending, and metal-to-glass sealing. It is excellent for soft soldering and good for silver-alloy brazing and gas-shielded arc welding, fair for carbon arc welding, and poor for gas and resistance welding. Sp gr = 8.89–8.94; EC = 101% IACS; TC = 226; SH = 0.092; TS (sheet), hard = 50, soft = 34. *Uses* include *electrical* (bus bars, conductors, electronic tubes, and waveguides operated at elevated temperatures in presence of reducing gases); *industrial* (gasoline supply lines, heater units, oil coolers, radiators, and refrigeration and watertubes).

Silver-bearing Copper: [99.90% Cu (min.), 8–25 oz Ag/ton] Compared with other coppers, this metal has an increased annealing (hence softening) temperature. It has high electrical and thermal conductivity and very good creep resistance. It is excellent for soft soldering and silver-alloy brazing and good for gas-shielded arc welding. It is fair for carbon arc welding and poor for gas and resistance welding. Sp gr = 8.91; EC = 100–101% IACS; TC = 223–227; TS (sheet), hard = 46–52, soft = 33. *Uses* include

electrical (commutator bars and heavy-duty motor windings where retention of strength at moderately elevated temperatures is desired); *miscellaneous* (die-pressed forgings and brazing solders).

Arsenical Copper: [99.68% Cu (nominal), 0.025% P (nominal), and 0.30% As (nominal)] This metal has increased resistance to corrosive media, lower electrical and thermal conductivities, and increased strength at elevated temperatures. It is excellent for soft soldering and silver-alloy brazing, good for carbon and metal arc welding, fair for gas welding and resistance welding. Sp gr = 8.94; EC = 45–80% IACS; TC = 110; TS (tube), hard = 45–60, soft = 35–37. *Uses* include heat-exchanger, boiler, condenser, and radiator tubes.

Free-cutting Copper: [99.4–99.5% Cu, 0.5–0.6% Te] This metal is similar to deoxidized copper, but with better machinability and high electrical and thermal conductivity. It is good for soft soldering and silver-alloy brazing, fair for carbon, metal, and gas-shielded arc welding, and poor for resistance welding. Sp gr = 8.94; EC = 90% IACS; TC = 205; TS (rod), hard = 45–48, soft = 32. *Uses* include *electrical* (connectors, motor and switch parts, and soldering coppers); *screw-machine parts* (particularly items requiring high conductivity, corrosion resistance, and copper color); *forgings;* and *welding-torch tips.*

Very High Copper Alloys. Alloys to which only a very small percentage of other ingredients is added to copper still are referred to as *coppers.* The elements, in addition to Ag, that are added include Cd, Cr, Be, Te, and Se.

Cadmium (0.6–1%) increases the softening temperature of Cu and also strengthens and toughens it, increasing its resistance to fatigue. The conductivity, when fully annealed, is about 95% IACS. Since cadmium copper is essentially free from O_2, the metal is not susceptible to gassing. Major uses include long-span overhead electric transmission lines and contact wires used in electrical transportation.

In *chromium copper* (about 0.5% Cr), both the thermal and electrical conductivities remain high and with some improvement in mechanical properties. Because the strength and hardness of chromium copper depend on heat treatment and not on cold working, the alloy can be used up to temperatures of about 454°C without the risk of softening.

Beryllium copper is of two types: (1) straight beryllium copper containing 2% Be and (2) an alloy containing only 0.4% Be, along with 2.6% Co, the latter added principally as a lower-cost substitute for Be. Beryllium copper is well suited for instrument springs, diaphragms, bellows, and bourdon tubes. The alloy allows the springs to be shaped in the soft condition, followed by hardening. Be-Cu tools are nonsparking and find use in hazardous locations. See also **Beryllium.**

Tellurium copper (up to 0.5% Te) is used widely where good machinability and electrical conductivity are important. Tellurium increases the softening temperature of work-hardened Cu. The alloy is well suited for electrical instruments, motor and switch parts, and electrical connectors where finish-dimension accuracy, smooth surfaces, and freedom from burrs are important. The alloy is used principally for screw-machine products and forgings.

Selenium copper combines high electrical conductivity with hot-working and free-machining properties. The alloy is excellent for copper-to-glass seals.

Brass. The various brasses essentially are alloys of Cu and Zn. As the percentages of Cu and Zn vary, there are accompanying, often pronounced, variations in properties. Alloy brasses contain additional metals, such as Mn, Sn, Si, Ni, Pb, and Al, but these additions seldom exceed 4%. An *alpha brass* contains 64% or more of Cu and 36% or less of Zn and is so named because a microscopic examination reveals only alpha crystals. *Beta brasses* contain 51–55% Cu and microscopically reveal only beta crystals. The beta brasses are very hard and strong, but are so brittle that they are of limited commercial value.

There are eight major categories of brasses (not including the leaded and alloy brasses). The cold-working fabrication properties of these brasses are excellent for all types except for Muntz metal, which is rated fair. As regards hot-working fabrication properties, Muntz metal is rated excellent; gilding brass, commercial bronze, red brass, and jewelry bronze are rated good; low brass and cartridge brass are rated fair; and yellow brass is rated poor. In terms of machinability, Muntz metal is rated 40;* red brass, jewelry bronze, low brass, and cartridge brass are rated 30; and gilding and commercial bronze are rated 20. For polishing, all are rated excellent.

All the eight categories of brasses are rated excellent for soft soldering and silver-alloy brazing, and all are rated good for gas welding. Jew-

elry bronze is rated fair-to-good, whereas all other types are rated only fair for carbon arc welding. All types are rated poor for metal arc welding. Gilding brass, commercial bronze, red brass, and low brass are rated good for gas-shielded arc welding, while the other types are rated fair. With the exception of cartridge brass, yellow brass, and Muntz metal, which are rated fair for resistance welding, all other types are rated poor.

Gilding Brass: [95% Cu, 5% Zn, 0.03% Pb (max.), and 0.05% Fe (max.)] Compared with copper, this alloy has higher tensile strength, equal ductility, and lower thermal properties. Mp = 1066°C; AT = 427–788°C; HWT = 760–871°C;† sp gr = 8.86; TS (sheet), hard = 50, soft = 34. *Uses* include *coinage* (coins, medals, and tokens); *munitions* (bullet jackets, firing-pin support shells, fuse caps, and primers); *novelties* (emblems, jewelry, and plaques); and a *base for gold plate* and a *base for vitreous enamel.*

Commercial Bronze: [90% Cu, 10% Zn, 0.05% Pb (max.), and 0.05% Fe (max.)] This metal has excellent cold-working properties and is very ductile. Mp = 1043°C; AT = 427–788°C; HWT = 760–871°C; sp gr = 8.80; TS (sheet), hard = 61, soft = 37. *Uses* include *architectural* (etching bronze, grillwork, screen cloth, and weather stripping); *cosmetics* (compacts and lipstick cases); *hardware* (escutcheons, kickplates, line clamps, marine hardware, rivets, and screws); *munitions* (primer caps and rotating bands); *miscellaneous* (costume jewelry, ornamental trim, screen wire, and as a vitreous-enamel base).

Red Brass: [85% Cu, 15% Zn, 0.06% Pb (max.), and 0.05% Fe (max.)] This metal has superior strength and ductility over copper and excellent corrosion resistance. Mp = 1027°C; AT = 427–732°C; HWT = 788–899°C; sp gr = 8.75; TS (sheet), hard = 70, soft = 40. *Uses* include *architectural* (etching parts, trim, and weather stripping); *electrical* (conduit, screw shells, and sockets); *hardware* (eyelets, fasteners, and fire extinguishers); *industrial* (condenser and heat-exchanger tubes, flexible hose, pickling crates, plumbing pipe, pump lines, and radiator cores); and *miscellaneous* (badges, compacts, costume jewelry, dials, etched articles, and lipstick cases).

Jewelry Bronze: [87.5% Cu, 12.5% Zn, 0.05% Pb (max.), and 0.10% Fe (max.)] Mp = 1035°C; AT = 427–760°C; HWT = 760–899°C; sp gr = 8.78; TS (sheet), hard = 66, soft = 39. *Uses*

---

*Where 100 is machinability rating for free-cutting brass.

†Mp = melting point; AT = annealing temperature; HWT = hot-working temperature.

include *architectural* (angles and channels); *hardware* (chains, eyelets, fasteners, and slide fasteners); *novelties* (compacts, costume jewelry, emblems, etched articles, lipstick cases, and plaques); and a *base for gold plate.*

Low Brass: [80% Cu, 20% Zn, 0.05% Pb (max.), and 0.05% Fe (max.)] This metal is quite similar to red brass in its properties. Mp = 999°C; AT = 427–704°C; HWT = 816–899°C; sp gr = 8.67; TS (sheet), hard = 74, soft = 44. *Uses* include *architectural* (ornamental metalwork, medallions, and spandrels); *electrical* (battery caps); *instruments* (bellows and musical instruments); and *hardware* (clock dials, flexible hose, pump lines, and tokens).

Cartridge Brass: [70% Cu, 30% Zn, 0.07% P (max.), and 0.05% Fe (max.)] This metal has the best combination of ductility and strength of any brass. It has excellent cold-working properties. Mp = 954°C; AT = 427–760°C; HWT = 732–843°C; sp gr = 8.53; TS (sheet), hard = 76, soft = 47. *Uses* include *automotive* (radiator cores and tanks, reflectors); *electrical* (bead chain, flashlight shells, lamp fixtures, socket shells, and screw shells); *hardware* (eyelets, fasteners, pins, rivets, springs, stampings, and tubes); and *munitions* (ammunition components). Admiralty brass contains 71% Cu, 28% Zn, and 1% Sn.

Yellow Brass: [65% Cu, 35% Zn, 0.15% Pb (max.), and 0.05% Fe (max.)] This metal has excellent cold-working properties combined with good corrosion resistance. Mp = 932°C; AT = 427–704°C; sp gr = 8.47; TS (sheet), hard = 74, soft = 47. *Uses* include *architectural* (grillwork); *automotive* (radiator cores and tanks, reflectors); *electrical* (flashlight shells, lamp fixtures, screw shells, and socket shells); *hardware* (bead chain, eyelets, fasteners, grommets, hinges, locks, push plates, and kick plates); *stencils; plumbing accessories;* and *wire* (pins, rivets, screws, and springs).

Muntz Metal: [60% Cu, 40% Zn, 0.30% Pb (max.), and 0.07% Fe (max.)] This metal has high strength combined with appreciably lowered ductility. Mp = 904°C; AT = 427–593°C; HWT = 621–788°C; sp gr = 8.39; TS (sheet), hard = 70, soft = 54. *Uses* include *architectural* (trimming); *hardware* (large nuts and bolts); *industrial* (brazing rod, condenser plates, condensers, evaporator and heat-exchanger tubes, valve stems, and hot forgings).

Leaded Brasses. Lead is frequently added to brass up to an amount of 4% to improve the machinability, but the Pb has practically no effect on the tensile strength or hardness of the metals. Lead does lower the ductility and shear strength, particularly of cold-worked materials.

Phosphor Bronzes. Tin is the principal alloying element in these metals, but they have received their name because phosphorus is added in small percentages as a deoxidizing agent in the casting of these alloys. The tensile strength of these alloys varies from moderate to very high, tending to decrease with the percentage of Sn added. The percentage of Sn will range from 1.25 to 10%. Phosphor bronzes have an extremely wide variety of uses. They exceed copper in duty where seawater and acid reagents are present.

Silicon Bronzes. Usually, these are of proprietary compositions. They range from 1.5 to 3.5% Si and 1.5% Zn or less. Sn, Mn, and Fe may be added in small amounts. They have become important construction materials because of their excellent strength, corrosion resistance, and ease of welding. Their corrosion resistance is equal to or better than Cu. They are susceptible to fire cracking as the Si content increases.

Aluminum Bronzes. These are high-copper alloys containing 4–10% Al. The metals are tough, moderately hard, and very ductile. Because of the presence of Al, the metals resist scaling or oxidation at high temperatures. This resistance increases with Al content. They also perform well in acids and alkalies. They are very good for seawater applications, particularly turbulent seawater.

Nickel Silvers. Nickel is added to Cu-Zn alloys, particularly to influence color. With a Ni content of around 18%, the alloy becomes silver white. Most mechanical properties and corrosion resistance are improved by the Ni. These alloys are widely used in operations requiring ductility in the cold condition, such as stamping, deep drawing, and spinning and for articles that may be plated. An alloy with 55% Cu, 18% Ni, and 28% Zn is widely used as a spring material because of its high tensile and fatigue properties. German silver contains 50% Cu, 30% Ni, and 20% Zn.

The Cupronickels. Whereas the nickel silvers may be classified as brasses, the cupronickels are basically copper-nickel alloys, with possibly minor additions of Mn, Fe, and Zn. These alloys do not work-harden rapidly and hence can be used for severe stamping, drawing, and spinning operations. They are widely used for condenser tubes and plates, heat exchangers, and a variety of process equipment.

Copper Production The discovery of copper dates from prehistoric times. Native Cu and Cu ores are said to have been mined for more than 5,000 years. The methods used for purifying the ores have improved with time, and new smelting processes are in various stages of development today.

During the early modern history of the metal, metallic Cu was commonly produced by direct smelting in blast or shaft furnaces. The feed for the furnace was high-grade copper oxide lump ores, or dead roasted sulfide ores which required a reducing agent. The latter normally was in the form of carbonaceous fuel, which was mixed with the charge.

Subsequent mining of the more plentiful copper sulfide ores, particularly low-grade ores which required concentration, produced a material totally unsuited for blast-furnace charge. Its fine particle size required that the flotation concentrates be sintered, usually on Dwight Lloyd downdraft machines, to produce an agglomerated material which could then be charged to the blast furnace. The product of such processing was impure "black copper" assaying 80–90% Cu.

The increasing rate of production of flotation concentrates led to the development of the reverberatory smelting furnace. Beginning about 1905, this development accelerated, and the *reverb* rapidly replaced the blast furnace in all but a few instances where special circumstances warranted its retention.

World Copper Sources: Most of the world's copper presently comes from large mining developments in the western United States, the west coast of South America (Chile and Peru), Central Africa (Zaire and Zambia), and the Soviet Union. The "porphyry coppers" are the most important sources of the world supply of this metal. Copper in these deposits normally occurs disseminated in granite-related intrusive rocks which have been loosely identified as "porphyries" and, to varying degrees, in the enclosing host rocks. The ore is low grade (usually less than 1% Cu) and is commonly associated with lesser amounts of Mo, Au, and Ag, or Ni, as in the case of the Canadian deposits at Sudbury, Ontario. Other minor metals may also be associated. Formerly important sources of Cu, such as the high-grade vein concentrations and native Cu deposits, have been largely mined out and now represent but a small percentage of total production and make up only a fraction of the long-range reserves.

A large number of copper-bearing minerals have been identified, but only a few are of real commercial importance. These have been divided into primary (sulfide) ores and secondary (oxidized) ores. The principal primary ores are chalcopyrite, $CuFeS_2$, containing about 35% Cu; bornite, $CuFeS_4$, containing about 63% Cu; chalcocite, Cu_2S, containing nearly 80% Cu; and enargite, $Cu_3As_5S_4$, with 48% Cu. The principal secondary or weathered ores consist chiefly of cuprite, $2CuCO_3Cu(OH)_2$, with 55% Cu, and chrysocolla, $CuSiO_3 \cdot 2H_2O$ with 36% Cu.

The porphyry-type deposits are a mixture of primary and secondary minerals. The secondary minerals are most abundant in the upper (weathered) portions of a given orebody and are represented in a lower, so-called *zone of secondary enrichment*. Primary minerals that once existed at higher elevations were weathered and oxidized, partly dissolved, conveyed downward by groundwater, and redeposited as secondary minerals. These secondary minerals tend to form an irregular capping above lower-grade but chemically undisturbed primary ores.

Fairly large reserves of Cu exist as low-value material rejected over the years onto mine dumps. As metallurgical and refining know-how improves, metal in such accumulations is being recovered. Nevertheless, as this source, together with the remaining ores from the more readily accessible weathered and enriched zones, becomes exhausted in the course of mining, and as more efficient recovery techniques are developed, the deeper primary ores acquire increasing importance as the world's major known reserve of Cu-producing minerals.

Modern Smelting: Modern Cu metallurgy is built around the mining of low-grade sulfide ores which are concentrated by flotation. The major part of these concentrates are smelted in reverberatory furnaces to produce a matte which is converted to blister copper for subsequent refining. See Fig. C-27.

Depending on the type of concentrates processed, they may or may not be roasted prior to smelting. Copper concentrates are roasted only to adjust the sulfur balance in the smelting operation to produce the desired grade of matte.

When the charge is fused in a reverberatory furnace, the sulfur preferentially combines with the Cu to form Cu_2S. By second preference, the remaining sulfur combines with iron to form FeS. Any additional sulfur combines with other metals present to form their sulfides. This mixture of sulfides is called *matte*.

Iron that does not enter the matte as FeS enters

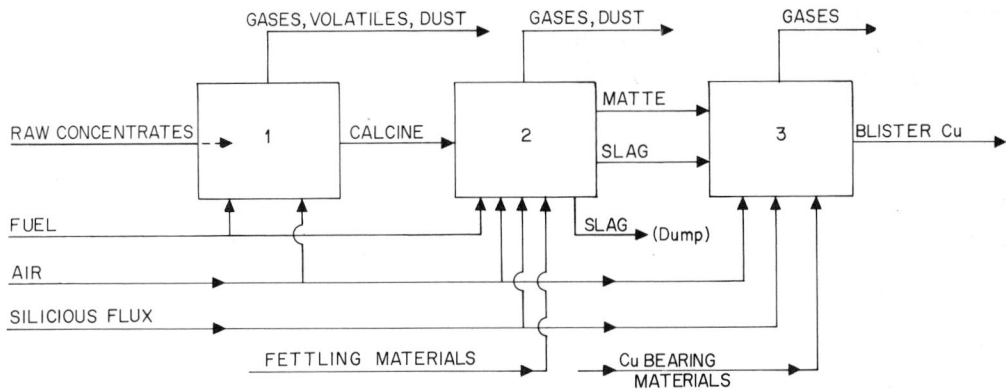

Fig. C-27. Schematic representation of materials flow in copper smelting. (1) Roaster, (2) reverberatory furnace, (3) converter. The raw concentrates contain 20–30% copper. Gases, volatiles, and dust from the roaster go to dust recovery and then to the acid plant or directly to stack. Gases and dust from the reverberatory furnace go to waste-heat boilers, dust recovery, and stack. Slag goes to the dump and contains 0.2–0.5% copper. The matte from the furnace contains 33–50% copper, and the slag transferred to the converter contains 2–5% copper. Blister copper produced by the converter is over 98% copper and goes to the refinery. The roaster sometimes is omitted where the concentrates are smelted raw. The percentages cited vary from one refinery and smelter to the next.

the slag as FeO to combine with the siliceous flux as $FeOSiO_2$. The slag is a complex fusible mass containing all the silicate-forming elements present. A fractional percent of Cu is lost to the reverb slag by inclusion. It is doubtful that any Cu exists in the slag as a silicate.

If the grade of matte is too high (grade referring to the Cu assay), Cu losses to the slag increase. If the grade of matte is too low, the relatively greater amount of FeS to be removed in the converter increases the converting costs. The optimum grade of matte is based on the economics of the individual plant and varies between plants.

Many Cu ores and concentrates contain relatively high amounts of sulfur and, if smelted directly, would produce a grade of matte lower than desired. It is under these circumstances that a roaster is added to the smelter flow sheet. Copper-concentrate roasters generally are multiple-hearth units. Since 1961, however, when the first fluid-bed roaster began operations at Copperhill, Tenn., some fluid-bed roasters have been installed.

The feed is only partially roasted in these units. The sulfur which is eliminated is carried off in the roaster exhaust gas as SO_2 plus a small fraction of SO_3. Roaster exhaust gases, after cleaning, may go to an acid plant or directly to the stack. The calcine product consists of a mixture of metallic oxides, sulfates, and sulfides plus gangue materials. The roasted concentrates, called *cal-*

cines—or raw concentrates if no roaster is in the process—are the primary feed to the reverberatory smelting furnace.

Reverberatory-Furnace Operations: Reverberatory furnaces were originally designed with sprung arches which limited their width to 25 or 30 ft. This construction has generally changed to the use of suspended arches, permitting somewhat wider structures. A general range of physical size of present-day reverbs is 18–38 ft wide × 70–125 ft long.

The furnace is usually charged along the sidewalls and is fired with burners located at one end. Fuel may be natural gas, fuel oil, or pulverized coal. Combustion air is preheated. Temperature of the exhaust gases is typically 1204–1316°C (2200–2400°F). Waste-heat boilers recover ±40% of the heat as steam; recuperators extract additional heat in preheating combustion air. After passing through electrostatic precipitators for dust removal, the reverb exhaust gases are discharged to the stack.

Reverb slag, the fluidity of which is maintained by the fluxes added to the charge, floats on top of the matte and is tapped off, usually continuously, into slag cars for discard. The matte, which contains all the precious metals in the charge, is tapped off into a ladle.

Refining Steps: The converting of molten matte, which is essentially a mixture of Cu_2S and FeS, depends on the fact that any air blown through it first oxidizes the FeS. Siliceous fluxes added to the charge slag the FeO, which may then be removed. Continuing the air blow, the sulfur is preferentially oxidized to SO_2, leaving metallic Cu (known as *blister Cu* at this stage). Blowing stops before the Cu is oxidized. The initial blow is exothermic and produces enough heat to maintain the liquidity of the charge during the balance of the cycle.

The Pierce-Smith converter is a refractory-lined cylindrical steel shell, mounted horizontally on trunnions so that it can be rotated. They range in size from 8 ft in diameter $\times$ 12 ft long to vessels as large as 13 ft $\times$ 30 ft.

Molten matte is poured into the mouth of the converter, which is then rotated to bring the mouth under an exhaust hood. Air is injected through a series of tuyeres located axially down the converter shell. After the iron is oxidized and slagged, the converter is rolled over to decant the slag. Then an additional ladle of molten matte is charged.

This charging and slagging is repeated until the converter holds 50–75 tons of Cu. The vessel is then skimmed clean of slag and blown to blister Cu. Converter exhaust gases usually pass through an electrostatic precipitator and then to the stack, but some smelters direct part or all of these gases to an acid plant. Converter slag contains too much Cu to be discarded, and is returned and charged molten to the reverb.

The blister Cu is transferred to a refining furnace where impurities are oxidized and slagged, and any oxides of Cu are reduced by polling with green timbers. Copper refined in this manner is cast into anodes by flowing the metal into molds on a casting machine, preparatory to electrolytic refining. Anodes generally assay in excess of 99% Cu.

The electrolytic cathodes are thin (about $\frac{1}{32}$ in.) Cu sheets produced by electrolytic deposition. The anodes and starting sheets are suspended in an acid copper sulfate solution electrolyte and, during the subsequent electrolysis, the Cu in the anode is removed and deposited on the starting sheet, which is the negative pole of the circuit.

Impurities in the anode drop to the bottom of the tank as a slime. This includes the precious-metal content, which is subject to subsequent treatment for recovery of the precious metals and other values. With a few exceptions, the current efficiency attained in electrolytic refining is in the range 90–95%; power consumption is approximately 0.1 kWh/lb Cu. Anodes generally weigh 500–700 lb, depending on individual plant practice. Anode life similarly is 26–30 days between plants. Cathodes are pulled at intervals of half the anode life.

A few plants operate the electrolytic refining cells as the *series process* instead of the *multiple process* described above. In the former there are no starting sheets, and only the first and last plate in the cell are directly connected to the electric circuit. The intervening plates, therefore, are bipolar, and the Cu is successively removed from one plate and deposited on the next until the last plate in the cell. The cathode metal is remelted and again fire-refined, as described earlier, before casting into commercial shapes.

Further Progress in Production Methods: A number of processes, aimed at displacing or improving the reverberatory smelting furnace, have been initiated. These processes are in various stages of development; some have had commercial application.

1. Outokumpu Oy developed a flash smelting technique involving injection of dried, fine-particle concentrates through a concentrate burner into a vertical reaction shaft. There the temperature rises high enough to smelt the particles before they fall into the settler section of the furnace. The reaction is largely autogenous, and the matte product can be a substantially higher grade than that produced by a reverb. Flash smelting has been in commercial operation in Finland since 1949. Several installations are operating in Japan, the first having begun in 1956.

2. Electric furnace smelting, requiring very low-cost power, has been practiced in Norway and Finland for many years.

3. Noranda Mines, Ltd., has developed a continuous smelting process. In a horizontal reactor, similar to a Pierce-Smith converter, concentrates and flux are fed into one end of the furnace. In the smelting process, matte and slag flows are controlled as they slowly move toward the tapping ports. FeS and Cu_2S are oxidized to permit continuous tapping of slag and metallic Cu.

4. International Nickel Co. has operated an autogenous flash smelting process for nickeliferous copper concentrates on a commercial basis since 1952.

5. The Worcra process, developed by Conzinc Riotinto of Australia, Ltd., is a continuous smelting process utilizing a pelletized feed and featuring a countercurrent flow between slag and matte or metal.

6. Use of oxygen-enriched air has led to some

success in smelting a portion of the concentrate feed directly in the converter. Development work in this area has been done primarily in Japan.

7. A number of hydrometallurgical processes for primary recovery of Cu are in various stages of development.

Mine and copper oxide dumps are leached with weak sulfuric acid, and cement Cu is precipitated from the resulting solutions by flowing over iron turnings or crushed cans. The cement copper at 80+% Cu is sent to a smelter for processing.

Several commercial installations in the United States have used a new technology in place of the cement-copper production noted above. In these plants, Cu is extracted from the leach liquors by solvent extraction. The electrolyte produced by stripping the pregnant organic fluid is advanced to an electrowinning section for production of electrolytic Cu.

> —J. Wesley Burgess, Francis Carroll Moran, and Gailen T. Vandel, *Fluor Utah, Inc.* *San Mateo, Calif.*

Copperas Reds   (See **Iron oxides, synthetic.**)

Copra   (See **Expression.**)

Corn   (See **Amino acids;** and **Starch.**)

Corn Oil   [See **Anionic surfactants (sulfur-bearing);** and **Vegetable oils.**]

Corona Chemistry   (See **Plasma, chemical processing with.**)

Corroding Lead   (See **Lead.**)

Corrosion   (See **Aluminum; Caustic soda; Chlorine; Conversion coatings; Iron and steel; Nuclear power plants; Paints; Titanium;** and **Zinc;** also consult the Subject Index.)

Corundum   (See **Abrasives;** and **Bauxite.**)

Cosmetic Materials   (See **Detergents; Soaps;** and **Waxes.**)

Cotton   (See **Acetate fibers; Cellulose ester plastics, organic;** and **Fibers.**)

Cottonseed Oil   [See **Anionic surfactants (sulfur-bearing); Expression;** and **Vegetable oils.**]

Coumarone-indene Resins   (See **Paint.**)

Covalent Bond   (See **Molecule.**)

Cracking   In petroleum technology, cracking denotes reactions in which a hydrocarbon molecule is fractured or broken into two or more smaller fragments. Other terms used to describe these reactions include cleavage, decomposition, fragmentation, pyrolysis, rupture, and scission. Chemical cleavage may occur at a carbon-hydrogen bond, a bond between carbon or hydrogen and an inorganic atom, such as sulfur or nitrogen, or at carbon-carbon bonds. Since the main objective of cracking is reduction in size of hydrocarbon molecules, the principal reaction involves the breaking of carbon-carbon bonds. There are three principal types of cracking: thermal cracking, catalytic cracking, and hydrocracking.

Thermal cracking for fuel production is performed by subjecting a feedstock to temperatures usually in excess of 455°C (850°F) and at pressures above atmospheric with the objective of converting a residual crude fraction or a heavy distillate into gasoline and light distillates; or reducing the viscosity of residual fractions; or for the production of coke. Since the widespread application of catalytic cracking processes, thermal cracking has been applied almost entirely to the processing of nondistillable crude residues which have experienced a diminishing market since World War II. The two basic thermal cracking processes now in use are (1) coking and (2) viscosity breaking (*visbreaking*), both of which are used to convert nondistillable residues into more valuable products. See also **Thermal cracking.**

Catalytic cracking is performed at temperatures in the region 455–540°C (850–1000°F), at pressures only slightly above atmospheric, and in the presence of a catalyst which, basically, is a specially prepared composite of silica and alumina. The process converts a distillate feedstock into gasoline as the primary product, with accompanying production of light hydrocarbons and distillate fuels that are lighter than the feedstock. As in the case of thermal cracking, the products will contain olefinic hydrocarbons. See also **Catalysts;** and **Fluid catalytic cracking.**

Hydrocracking processes operate at elevated pressures in the *presence of hydrogen* and catalyst at temperatures generally lower than 482°C (900°F). These processes produce gasoline and light distillates from feed distillates that are higher-boiling than the products. Hydrocracked products are *not* olefinic. Light gaseous hydrocarbons which are entirely paraffinic are produced by hydrocracking. See also **Hydrocracking.**

Trends: The application of the three principal types of cracking processes is summarized, in terms of the percentage of crude that each type of process has handled, in Table C-46.

TABLE C-46. *Use of Cracking Processes in the United States*

Type of Cracking Process	1930	1940	1950	1960	1970
	Capacity as Percent of Crude Capacity				
Thermal cracking . .	42	51	37	16	12
Catalytic cracking . .	. . .	3	23	36	37
Hydrocracking	. . .	. . .	. . .	. . .	5
Total	42	54	60	52	54
	Percent of Gasoline in National Motor Fuel Pool*				
Thermal cracking . .	34	45	32	10	5
Catalytic cracking . .	. . .	2	19	31	37
Hydrocracking	. . .	. . .	. . .	. . .	2
Total	34	47	51	41	44

*The national motor fuel pool (or *gasoline pool*) may be defined as the calculated average composition of all the gasoline produced by the nation's entire refining industry, if all of it was placed in one huge tank and mixed.

NOTES: The slowing of growth, as shown for the period 1960–1970, is the result of the growth of catalytic reforming. Cracking processes handled a smaller portion of the total crude refined, even though their daily throughput capacities continued to rise during the interval.

Thermal cracking first appeared during the early part of the twentieth century to increase the yield of gasoline above the amount of native gasoline in the crude. Many versions of the process were developed, and thermal cracking reached its peak of use in 1936 when about 55% of each average barrel of crude refined in the United States was processed by thermal cracking. At about that time catalytic cracking was introduced, and displaced thermal cracking to the extent that only 12% of the U.S. crude refined in 1970 was handled by thermal cracking methods.

Catalytic cracking first was introduced as a fixed-bed cyclic process. During World War II, continuous fluidized- and moving-bed processes were established on a commercial scale. The growth of catalytic cracking is dramatized by Table C-46. Octane numbers of catalytically

cracked gasolines are markedly higher than those of thermally cracked gasolines, which in turn are higher than those of the straight-run gasoline native in the crude. A characteristic of the catalytic cracking processes is that a carbonaceous deposit accumulates on the catalyst. It is regenerated continuously in the fluid- and moving-bed versions of the process by combustion of the deposit with air. This restores the performance of the catalyst.

Hydrocracking was introduced in the early 1960s, and by 1970 units representing several proprietary processes were handling about 5% of U.S. crude. Hydrocracking is conducted in an environment of sufficient hydrogen partial pressure to inhibit the formation of a catalyst deposit and thus can operate in a fixed-bed fashion for several months before regeneration is necessary. Although hydrocracking units may cost more to construct and operate than catalytic cracking units, the hydroprocess can handle heavier and "dirtier" feedstocks and also can be adapted to varying product ratios of gasoline to middle distillate. See also **Petroleum;** and **Petroleum processing.**

Because crude oil is converted to motor fuel to a much less extent outside the United States than is shown in Table C-46, the amount of cracking of all types in 1970 amounted to only 13% of the crude processed in the free world outside the United States and Canada.

—Melvin J. Sterba, *UOP Process Division, Universal Oil Products Company, Des Plaines, Ill.*

Crepe Rubber   (See **Rubber, natural.**)

Cresols   (See **Coal tar and derivatives.**)

Cresote   (See **Coal tar and derivatives.**)

Crookesite   (See **Thallium.**)

Cross-linking   (See **Peroxides, organic;** and **Polymerization.**)

Crude Oil   (See **Petroleum;** and **Petroleum processing.**)

Crushing   (See **Size reduction.**)

Crutching   (See **Soaps.**)

Cryogenic Processes   Cryogenic processes presently are defined as those processes which are concerned with temperatures below $-150°F$ (approximately $-100°C$). These processes are contrasted with conventional processes by (1) the use of materials of construction that do not become brittle at these low temperatures; (2) large and costly insulation systems designed to minimize heat ingress from the surroundings, and (3) specially designed vacuum-jacketed storage vessels to maintain the processed products in the liquid state at or near their atmospheric boiling points.

Cryogenic processing makes it possible to recover, liquefy, and store the so-called "fixed gases" economically. Without these processing techniques, frozen foods would not be possible, the exploration of space could not have been developed, the atomic physicists might well have been stymied in their quest for the knowledge of the nuclear particles which comprise the atom, cryosurgery for the care of Parkinson's disease would not have been possible, and many other scientific improvements which have and will continue to benefit mankind could not have been accomplished.

It may be stated that cryogenic processing began with the manufacture of ice in the late 1800s. As the advantages of low-temperature processing became evident, the topic attracted many researchers. These early investigations and experiments produced much of the fundamental data and equipment needed to obtain low temperatures. Carbon dioxide, air, and hydrogen, all were liquefied during the nineteenth century. It was not until 1908 that helium was liquefied. Since at that time there was little commercial use for these cryogens, they essentially remained research tools of the laboratory.

It was not until World War II that great emphasis was put on the manufacture of the cryogens, especially oxygen, which were used as a fuel by the Germans for their V-1 and V-2 rockets. Shortly after the war, the steel industry began using large quantities of oxygen in the manufacture of steel. This usage sparked a tremendous increase of technology of liquefying air to produce both oxygen and nitrogen. As a result of this process, a small constituent of the air, argon, was economically recovered and made available for industrial uses.

It was not until the advent of the space age that large quantities of liquid hydrogen and helium were made commercially available. Old cryogenic techniques developed by the early researchers, coupled with modifications and improvements of the necessary liquefaction equipment, were cli-

maxed when the first helium extraction and liquefaction plants were successfully put into operation. See also **Helium.**

Presently, a severe energy shortage has forced many countries into thinking of importing large quantities of liquefied natural gas. See also **Natural gas.** Since the main constituent of natural gas is methane, the normal boiling point of which is $-258.5°F$ (approximately $-160°C$), cryogenic processing techniques must be used to liquefy the gas. Liquefaction of the natural gas causes a volume reduction of almost $600:1$. Hence a smaller shipping volume is required to transport the required quantities of fuel. See also **Fuels.**

See also the Subject Index for reference to other cryogenic processes.

—Henry E. Duckham, Jr., *The M. W. Kellogg
Company, a division of Pullman Incorporated,
Houston, Tex.*

Cryolite Cryolite, $3NaF \cdot AlF_3$, is named from the Greek *krios* (frost) and *lithos* (stone). The discovery by Charles Martin Hall of the United States and Paul Heroult of France (both in 1886) that molten cryolite would dissolve alumina, Al_2O_3, making possible the electrolytic reduction of aluminum, brought special attention to this mineral, which had little use before that time. As the aluminum industry grew, the search for new natural deposits of cryolite was undertaken. Cryolite has been found in small deposits in Colorado (United States), Spain, and Russia. The most important deposit is at Ivigtut, Arsukfjord district of southwestern Greenland. The mineral deposit at this location is found embedded in granite that is covered with a metamorphic rock consisting essentially of granite (gneiss) in a foliated arrangement. Open-pit mining is used. Because of the severe, long winter, the mineral can be exported only during the summer months. Natural cryolite from this area is comprised of $F = 50\%$, $Na = 30\%$, $Al_2O_3 = 12\%$, $SiO_2 = 0.5\%$, the remainder being made up of minor impurities and moisture.

Synthetic Cryolite: The aluminum industry no longer depends on the natural mineral for normal operation of its cells, although natural cryolite sometimes is used to start new cells. Synthetic cryolite can be manufactured by a number of processes, depending upon the cost of the starting materials and their availability. Sodium aluminate from the Bayer process (see also **Bauxite**) can be reacted with hydrofluoric acid:

$$NaAlO_2 + 2NaOH + 6HF \rightarrow$$
$$3NaF \cdot AlF_3 + 4H_2O$$

or sodium carbonate can replace the sodium hydroxide in the reaction:

$$NaAlO_2 + Na_2CO_3 + 6HF \rightarrow$$
$$3NaF \cdot AlF_3 + CO_2 + 3H_2O$$

The hydrofluoric acid required can be produced from fluorspar:

$$CaF_2 + H_2SO_4 \rightarrow 2HF + CaSO_4$$

The principal properties of cryolite are given in Table C-47.

TABLE C-47. *Properties of Cryolite*

Molecular weight	209.94
Density, g/cm^3, at 25°C	2.97
Solid, at 1009°C.	2.62
Liquid, at 1009°C.	2.09
Solubility in H$_2$O, g/100 g, at 25°C . . .	0.04
pH.	6–7
Melting point, °C	1009
Surface tension, dynes/cm, at 1009°C. . .	125
Heat of vaporization, kcal/mole	54
Crystal structure	Monoclinic
Unit cells, Å:	
a	5.46
b	5.61
c	7.80
Composition, %:	
F	54.3
Al	12.9
Na	32.8
Hardness, Mohs' scale.	2.5
Soluble in H$_2$SO$_4$	
Refractive index	1.3389
Viscosity, p, at 1009°C	6.7
Heat of fusion, kcal/mole, at 1009°C . . .	26.7
Electrical conductivity, (liquid),	
Ω^{-1} cm^{-1}, at 1009°C	2.82
Special properties (x-ray): Atomic structure similar to garnet, cubic form at 570°C; colored by action of cathode rays	

Uses: Industrial uses for cryolite include imparting a milky hue to glass, as an ingredient in porcelain enamels, numerous metallurgical applications, as an ingredient of insecticides, and, of prime importance, in the electrolytic reduction cell by the aluminum industry. The following equation is believed to represent the action of cryolite on alumina:

$$2(3NaF \cdot AlF_3) + 2Al_2O_3 \xrightarrow[1040°C]{}$$
$$3NaAlO_2 + 3NaAlF_4$$

Ionized condition

In the electrolytic cell, electric current is primarily carried in the liquid by ionic transport (Na$^+$).

Health Hazards: Animals have survived oral dosages of several thousand milligrams per kilogram weight. Continual exposure leads to *fluorosis.* The product is toxic to insects.

—S. John Sansonetti, *Reynolds Metals Company, Richmond, Va.*

Crystallinity (See **Acetate fibers;** and **Polyester fibers.**)

Crystallization Crystallization* is a major chemical engineering unit operation, involving both inorganic and organic materials, that permits separation of pure materials from impure mixtures, usually at a comparatively low processing cost, and often yielding a product of uniform crystals that possess desirable end-product properties—high purity and good flow, handling, storage, and packaging characteristics. The primary industrial method is that of crystallizing a solute from solution. Additional approaches, more recently adapted to industrial processing, include crystallization from the melt (mainly for organic chemicals) and extractive crystallization (mainly for pharmaceuticals), the latter being an industrial adaptation of a common laboratory procedure. Research on *adduct*† and *clathrate*‡ formation is furthering the knowledge of phenomena that relate to crystallization and associated chemical engineering processes. Zone melting and single-crystal production were well established several years ago and became quite important in the production of various electronic industry materials. See also **Germanium;** and **Silicon.**

Crystalline State: Crystals are characterized by a complete regularity in the arrangement of the atoms or molecules of which the substance is constructed. The study of crystals involves (1) an examination of the external form and (2) the determination of the internal structure. The latter

---

*Portions of the introduction of this description are based on an excellent article by Dr. Donald Bruce Wilson (New Mexico State University), Crystallization, *Chem. Eng.,* Dec. 6, 1965.

†An adduct results from the formation of an addition group or compound by a chemical reaction that involves no change in valence—usually from the union of two binary molecules to form a more complex substance.

‡A clathrate is a substance which fixes gases and liquids as inclusion complexes (often up to 15% by weight), making it possible to handle the complex as a solid. The included ingredients subsequently may be released by melting or solvent action.

is more important from a chemical viewpoint, but a knowledge of the external form is of engineering importance since the crystal form governs ease of separation of the crystals from the mother liquor, and occasionally influences product marketability.

A general classification of crystalline material is by type, where the variation is in the kind and strength of the bond between the constituent atoms or ions, and in their electrical, magnetic, and mechanical properties. The five major types are (1) metallic, (2) ionic, (3) valence, (4) semiconductor, and (5) molecular.

Crystal symmetry and crystal systems are illustrated and described in Fig. C-28.

Crystallization Process: Crystallization as a separation process depends upon many system parameters. The solubility relationships between the components in the system are of prime importance. Because industrial emphasis has been on crystallization from solution, these solubility relationships have been classified as a solute dissolved in a suitable solvent. However, in the expanding scope of crystallization, consideration also must be given to the mutual solubility relationship, i.e., the solvent in the solute. The possibility of solid solutions separating from a solvent must be considered. In zone melting, the chemical engineer considers the preferential solubility of a solute between liquid and solid phases of a pure solvent.

Solubility: Solubility often is attributed to molecular forces although this is not fully correct, as shown by considering the mutual solubility of gases. Two gases mix in all proportions and have infinite mutual solubility. The mixing is not due to the interaction of the molecular forces, but rather to the motion of the gas molecules. However, as soon as the molecules are brought into close proximity, as in liquids and solids, the molecular forces have a decisive influence and either may increase or decrease the tendency toward mixing.

Since molecular motion is a function of temperature, the mutual solubility of a solid and a solvent will depend upon temperature. The equilibrium between a pure solid and its solution is illustrated and described in Fig. C-29. Solubility of the solid generally may be expected to increase with an increase of temperature, and the solubility of the solid may be expected to be greater the lower its melting point and the smaller its latent heat of melting. The effect of temperature rise on solubility is shown in Fig. C-30.

Another parameter of crystallization is the enhanced solubility of very fine particles. The larger crystals will grow rapidly at the expense of the smaller. This is evident from the degree of supersolubility existing in the solution. The supersolubility curve and its relation to the solubility curve is given in Fig. C-31.

In theory, a supersaturated solution or supercooled melt will not adjust spontaneously to the equilibrium condition unless the supersaturation or supercooling exceeds certain limits. This is a region of metastability. Within the metastable region, any appropriate seed will grow, but no new nuclei will form spontaneously. Beyond this limit lies the labile region, where the metastable condition is relieved automatically by the spontaneous generation of nuclei. In practice, industrial systems are always seeded and will reach equilibrium quite quickly.

Nucleation: Crystallization consists of two separate steps: nucleation and crystal growth. It is necessary for the solution or melt to be supersaturated for both phenomena, but the degree of supersaturation acts differently in each step. It is difficult to isolate the two mechanisms in real systems, but they are distinct.

Nucleation is the spontaneous or induced generation within a metastable phase of a more stable phase that is capable of growing. Spontaneous nucleation occurs when, during the cooling of a solution, the system reaches the labile region, as indicated in Fig. C-31. Specifically, the critical nucleation-determining embryo becomes a nucleus that differs from an equal number of ordinary molecules in solution by possessing sufficient excess surface energy to form a new phase. The energy needed to form a new phase within a homogeneous fluid is proportional to the product of the interfacial tension and the surface area of the new phase.

Adjustment of any metastable state (nucleation) consists of two steps: (1) surmounting the energy barrier that defines this condition and (2) subsequent passage to a state of lower energy and greater stability. Figure C-32 shows that either growth or dissolution will produce a state of lower energy. Quantitatively, the rate of spontaneous nucleation can be expressed by

$$R = Ve^{-\Delta G/kT}$$

where ΔG = free energy of activation
k = Boltzmann constant
T = absolute temperature
V = a frequency factor

If V is to be derived from theoretical considerations, it is necessary to postulate a mechanism for the nucleation: (1) simultaneous collisions, or

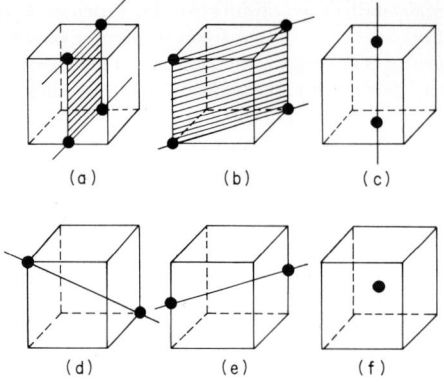

(a)   (b)   (c)

(d)   (e)   (f)

Elements of symmetry of a cube.

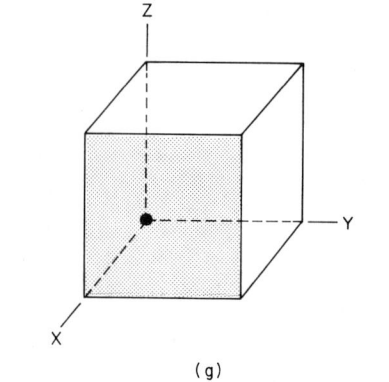

(g)

Shaded plane represents Miller index of 100 for the unit cube.

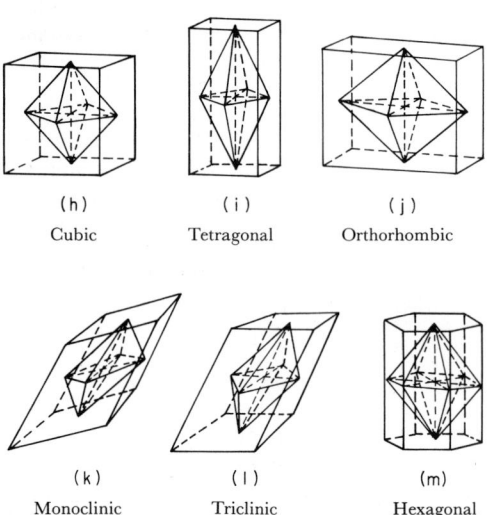

(h)   (i)   (j)

Cubic   Tetragonal   Orthorhombic

(k)   (l)   (m)

Monoclinic   Triclinic   Hexagonal

Fig. C-28. Crystal symmetry and crystal systems. The significance of the various elements of symmetry for the cube is indicated in diagrams (a)–(f). The elements of symmetry are (a) a rectangular plane of symmetry of which there are three (the others are at right angles to the one shown); (b) a diagonal plane of symmetry (six such planes pass diagonally through the cube); (c) one of the three axes of fourfold symmetry, at right angles to each other; (d) and (e) an axis of threefold symmetry, emerging from opposite edges; (f) center of symmetry at the mass center of the cube. A perfect cube has a total of 23 elements of symmetry.

The geometrical form of a crystal is described in terms of its crystallographic axes. These axes are three or four lines intersecting at a single point. The choice of lines is to some extent arbitrary, although a certain set will lead to a much better representation of a crystal than a second set of axes. A standard plane is selected to describe the faces of the crystal.

In order to represent any particular face in terms of the crystallographic axes, use is made of the intercept of that face on the axes. The method is called the *Miller indices*. Each crystal face, as shown in (g), is described by three integers that give the ratio of the intercepts of the unit plane to those of the given face. For a perfect cube, the crystallographic axes are chosen parallel to three edges at right angles, and meet at the center of the cube. Hence each face cuts only one axis, and the symbol for the cube form is 100.

Consideration of intercepts of other crystal faces with the crystallographic axes results in the *law of rational indices,* which states that there is an orderly arrangement of crystal units in space.

Geometric considerations, occurrences of different elements of symmetry passing through a single point, and the law of rational indices enable the establishment of 32 classes of crystal symmetry. These classes fall conveniently into six crystallographic systems that are based upon the angles between crystal faces [(h)–(m)].

(h) *Cubic* (regular or isometric). Three axes of equal length intersecting at mutually right angles to each other. (i) *Tetragonal*. Three axes intersecting at right angles. Two axes are equal in length, but the third is either longer or shorter than the other two. (j) *Orthorhombic* (rhombic, prismatic). Three axes of unequal length that intersect at right angles. (k) *Monoclinic*. Three axes of unequal length, two of which intersect at right angles, while the third axis is perpendicular to one and not perpendicular to the other. (l) *Triclinic* (anorthic, asymmetric). Three axes of unequal length, no two of which intersect at right angles. (m) *Hexagonal*. Three axes of equal length in the same plane and intersecting at 60° angles. A fourth axis is either longer or shorter and at right angles to the other three.

The types of crystal form are not related to the relative sizes of the crystal faces because the development of the faces is a characteristic of the specific material. The size and shape of the crystals of a given compound may vary

(Fig. C-28 legend continued on page 329.)

(Fig. C-28, continued from page 328.)

with the conditions under which crystallization occurs. Despite these differences, the angles between the faces remain constant. This relation is known as the *law of constancy of interfacial angles,* which states that the angles between specified external surfaces remain constant no matter how these faces develop.

The external shape or habit of a crystal of a given compound depends upon this relative development of the different faces. Crystal habit can be altered, and sometimes this can be convenient for further operations with crystals. The general procedure would be to add a second solute or impurity. The habit modification would occur in any of the following ways: (1) qualitatively the second component lowers the freezing point and decreases the supercooling, which in turn reduces the growth rate; (2) the second component forms an adsorbed layer of a particular face that impedes the otherwise normal crystallization; and (3) the second solute alters the viscosity and solvation effects, thus controlling the rate of crystal growth.

General nomenclature exists for describing crystal habit. The basic designation is the *regular cube.* Then there are *prismatic* or *columnar crystals,* which are further categorized as *tabular* or *lamellar.* A tabular crystal develops so that growth normal to the plane approximates the lengthwise growth of the face; a lamellar habit describes a crystal whose thickness is small compared with its length and breadth. This condition also is termed *platy* (or plates). Intermediate to tabular and lamellar crystals are those whose habit is described as a *blade* or a *lath.* Needles or acicular growth indicates the predominance of growth in one direction, and if extreme, the term *fibrous* is used. The foregoing terms apply to single crystals. In addition to the general terms describing habit, additional terms designate particular faces or combinations.

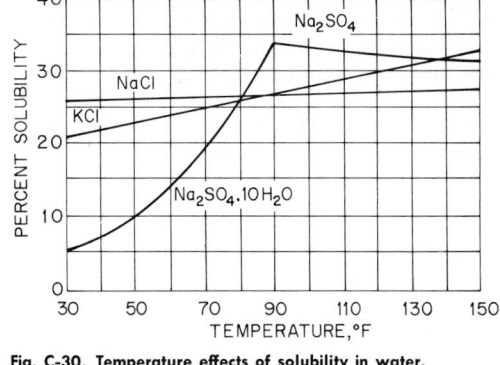

Fig. C-30. Temperature effects of solubility in water.

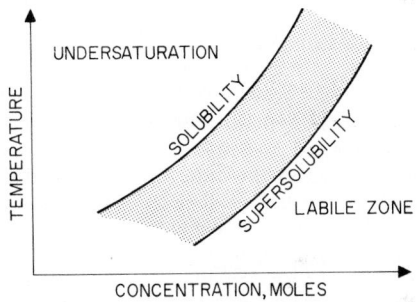

Fig. C-31. Effects of supersolubility on control of crystal growth.

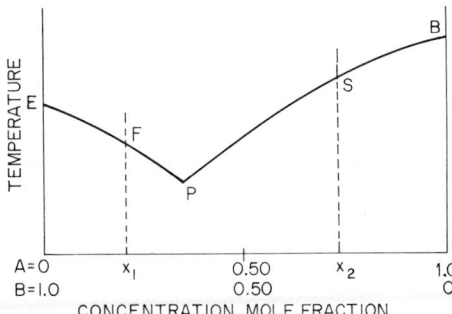

Fig. C-29. Mutual solubility of a solid-solvent system. A = solute, B = solvent.

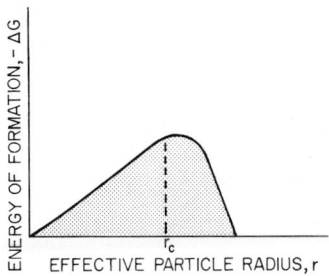

Fig. C-32. Production of a state of lower energy upon growing crystals or dissolving crystals.

(2) a stepwise series of collisions between embryos and single atoms, until a reasonable number of embryos grow to critical size and become nuclei. The latter mechanism usually is selected.

A second type of nucleation that can occur is *induced nucleation,* which may result from one or more of the following actions: (1) contact nucleation, such as may be produced by vigorous stirring or sharp collisions between crystals; (2) the catalytic effect of existing crystals, foreign nuclei, or container walls; (3) local variations in concentration that can be produced by uneven agitation, temperature gradients, or surface evaporation; (4) nuclei added to the solution in the form of small seed crystals that may be solute material or extraneous foreign material; and (5) nucleation induced by ultrasonic radiation.

Sonic irradiation provides a convenient method for rapidly multiplying growth centers and also produces a more uniform crystal product. Regardless of the mechanism used, the crystal growth for the given material proceeds in a similar manner for all nuclei.

Crystal Growth: An ideal crystal would have a perfectly regular arrangement throughout its entire structure. Real crystals deviate because of irregularities in growth. The resulting real crystal can be described as a mosaic structure composed of small blocks on the order of a few hundred to a few thousand lattice distances (10^{-4}–10^{-6} cm).

The mosaics are not in exact alignment, and their disarray constitutes the nonideality of the real crystal.

The two predominant mechanisms of growth are migration and spiral growth, both of which are illustrated in Fig. C-33. The adsorption theory (or migration theory of Kossel) states that a growth unit is not necessarily disposed to immediate incorporation in the growing base once it has arrived on location by diffusion. The reasoning is that the growth unit could diffuse away into the solution, or it may not have settled at a favorable point. Even when immediately at a preferred point of growth, the growth unit is assumed to be maneuvered or jockeyed into the final correct position on the growing lattice.

Kossel's model lists eight possible positions at which individual building units can be incorporated: (1) edge, (2) corner, (3) surface, (4) edge-adsorbed molecule, (5) corner-adsorbed molecule, (6) edge position or step, (7) surface-adsorbed molecule or center, and (8) surface hole. The preferred position is an edge where two or more faces are present.

The spiral-growth model (Frank's model) postulates that growth takes place at screw dislocations in the crystal. The crystal structure is assumed to be twisted, so that a ledge is always available for growth, thus making the formation of a two-dimensional nucleus on the crystal face

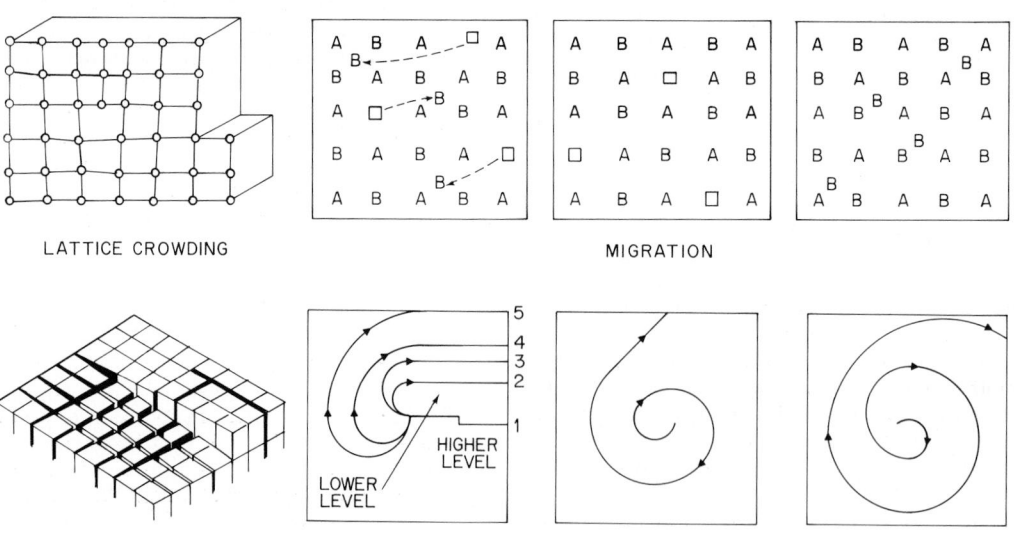

LATTICE CROWDING

MIGRATION

SCREW DISLOCATION

A SPIRAL GROWTH

Fig. C-33. Migration and spiral-growth mechanisms for crystal growth.

unnecessary. Dislocations in crystal structure are always present, as evidenced by the fact that actual crystals are only of the order of one-hundredth as strong as a perfect crystal should be.

Crystal Size and Purity: Large crystals are not necessarily purer than small crystals. Impurities may be entrapped in the growing crystal, actually incorporated into the lattice, or adsorbed on the surface. Mother liquor or suspended material may be included in the crystal, especially if the growth rate is high, and as the crystals become large.

If size is not important, short retention times will result in high production rates and small crystals. Conversely, coarse product is obtained at longer retention times, but at the expense of production rate. Even when the size and size distributions are not restrictive on the process, the filtration, washing, or centrifugation steps that usually follow crystallization may call for special crystal characteristics.

Crystal Aging: After the initial crystals have reached the size characteristics of the operating conditions for the system, they experience an aging process. Aging may comprise any of the following action: (1) *perfection of the primary particles* as a result of recrystallization in a liquid film around the particles (the primary particles usually are small and imperfect and are more so the higher the conditions of supersaturation under which they are formed); (2) *agglomeration* of the primary particles by sharing their liquid jackets (this process effectively cements together several primary particles, resulting in a decrease in surface, and at the same time, coprecipitated material is expelled with the exuded liquid film); and (3) *cementing together of the primary particles* in agglomerates as a result of the perfection process of recrystallization. Further factors that can cause agglomeration of the crystals into a granular mass are (1) unequally distributed stresses, (2) unequal size of crystals, (3) unstable or metastable forms, and (4) sheared or flawed crystal surfaces. In addition to the aging process, caking of the crystals can occur. Caking involves the formation of a film of saturated solution on the surface of the crystals either by adsorption or by migration of mother liquor that has not been removed in the separation process. This saturated solution concentrates at the point of contact between the crystals, and subsequent crystallization forms a solid bridge.

The tendency toward caking can be lessened by (1) conditioning of the crystal storage space (for water-soluble crystals, the humidity must be maintained below the critical humidity for deliquescence); (2) making larger crystals of uniform size, thus minimizing the number of points of contact; (3) modifying crystal habit to produce uniform symmetric crystals, thus reducing points of contact; (4) applying various coatings to absorb moisture preferentially, thus preventing intimate intergrowth of the bare crystals and (5) applying additions that cause future growth during storage to be of a very weak structure (dendritic) that will break readily.

Crystallization, Glass   (See **Glass.**)

Crystallizers   Since crystallization consists of the two steps of nucleation and growth described under **Crystallization** as well as conformance with the overall material and energy balance that any chemical reaction or operation must conform to, it is clear that any crystallizer will have at least these four boundary conditions. Although heat and energy balances have been in common use in crystallizer design for many years, it is only since 1962 that a comprehensive mathematical description of the combined effects of nucleation and growth have been available. Randolph and Larson in a pioneering work (*A.I.Ch.E. J.,* vol. 8, no. 5, November 1962) develops a comprehensive model for continuous crystallizers of the mixed suspension type which appear to be of practical and continuing use. For a comprehensive discussion of their work see Randolph and Larson, "Particulate Systems Analysis," Academic Press, 1972.

The underlying concept in this treatment of crystallizer analysis as developed by Randolph and Larson is the concept of a *population balance* which is an additional constraint on the system involving both nucleation and growth. A term called *the population density* may be defined as a function of crystal length *L,* such that

$$\lim_{\Delta L \to 0} \frac{\Delta N}{\Delta L} = \frac{dN}{dL} = n$$

The value *N* is the total number of particles up to size "*L*" per unit volume. The quantity *n* so defined is the population density and is measured in number of crystals per unit length per unit volume of crystallizer. Thus defined, *n* is a general property which may be obtained easily from a screen analysis.

In a mixed suspension crystallizer which is completely or uniformly mixed, wherein a product is removed under conditions such that it is representative of the suspension being circulated and in a system where growth occurs in accordance with McCabe's Delta L law, it has been shown by these authors that the population density is

$$n = n^o \, e^{(-L)/(Gt)}$$

By plotting log n versus the crystal length L, a straight line is obtained. From the slope of this line the growth rate G can be determined as long as the retention time t is known. The nucleation rate can be determined from the nuclei population density n^o, which is the intercept at the point where $L = 0$. The nucleation rate, thus defined, is

$$B = G \, n^o$$

From such information, changes in crystallizer operation and mechanical design can be evaluated in terms of the effect these changes produce on nucleation rate and growth rate. Using such information makes possible the design of equipment to produce predictable crystal size distributions.

Industrial Crystallization Equipment: The most common type of vacuum-crystallization equipment in industrial use today is the forced-circulation evaporative crystallizer shown in Fig. C-34. Vessels of this design have been built in sizes ranging from 18 in. to over 42 ft (diam). There is no inherent limit to the size of a vessel of this design. When vessels are of large size, it is common practice to use two or more circulating systems so that more readily obtainable pumps

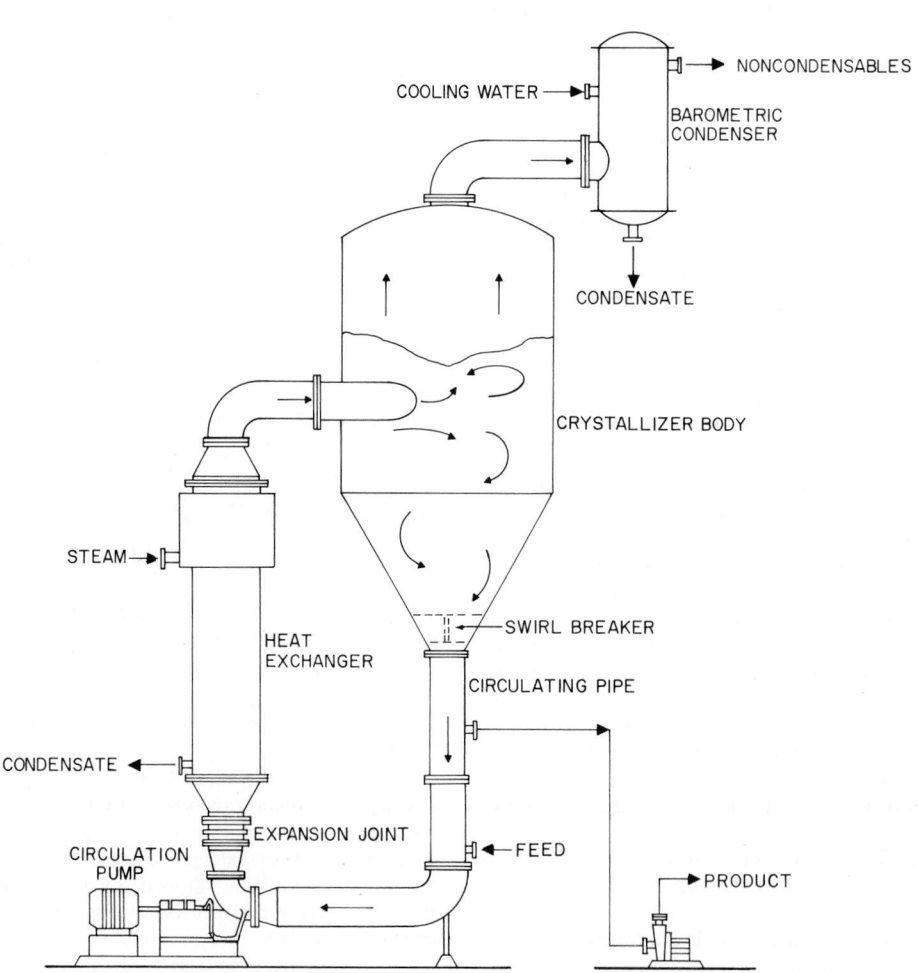

Fig. C-34. Forced-circulation evaporative crystallizer.

and heat-exchanger tubes of shorter length may be used. In this equipment, relatively low installed cost is obtained at the sacrifice of crystal-size-control flexibility.

This design is amenable to relatively larger evaporative loads than are most other types of crystallizing evaporators. The design normally is operated as a mixed-suspension–mixed-product removal crystallizer, although, in some cases, classification devices are incorporated into the machine to permit the removal of partially clarified mother liquor or to classify the product discharged. As normally operated, machines of this type are capable of long operating cycles and stable product size. Recent studies have indicated that the nucleation-rate equipment of this type is largely controlled by the type and speed of the circulation pump. Control of particle size in the product is achieved by regulating the retention time or the slurry density. With some systems (sodium carbonate and sodium chloride), the largest product size is not necessarily produced at the longest retention time and the highest slurry density.

Operation of the equipment can be understood best by visualizing the slurry being circulated in the vessel, withdrawn from the body through a swirl breaker, and mixed with incoming feed, as shown in Fig. C-34. This slurry is moved by the circulation pump through the heat exchanger, where it is subjected to a temperature rise of roughly 2–10°F (approx. 1–$5\frac{1}{2}$°C). The heated liquor is discharged tangentially into the body at a point sufficiently far beneath the surface so that the liquor entering tangentially is just at the boiling point for the liquid depth at which it is submerged. As the liquid rotates around the body and rises toward the surface, it starts to boil, and this boiling induces a secondary circulation which creates a spinning toroid of fluid within the body. Depending upon the location of tangent inlet with respect to the cone, this toroidal circulation can result in considerable secondary circulation and agitation within the body.

The purpose of the tangential inlet is to keep the solids within the body dispersed due to the agitation created by the secondary circulation and to distribute the boiling action across the vessel surface. In this way, the full cross-sectional area of the vapor head is utilized for vapor release. When properly designed, this type of vessel is capable of producing a smooth boiling action with relatively small amounts of wall salt while still maintaining a suitable suspension of product crystals within boiling zone and in the lower part of the vessel.

With inadequate or improperly directed circulation, this design can produce serious wall-salting problems and unduly high entrainment. Good examples of application of this equipment include the production of Na_2CO_3, Na_2SO_4, and NaCl. See also **Salt.** Operating cycles of this equipment between washouts to remove salt growth from the walls of the body or from the heat exchangers normally range from 30 to 90 days.

To achieve some control of the number of fine particles within the crystallizer body, and thereby increases the overall particle size, it is necessary to selectively remove the fine particles so that they can be destroyed by the action of heat or dilution. One design that achieves this objective, a draft-tube baffle crystallizer, is shown in Fig. C-35. In this design, a body of growing crystals is suspended by the circulation flowing up the draft tube from the propeller shown close to the bottom of the vessel. From the area surrounding this body of circulated slurry, a stream is removed at relatively low velocity so that gravitational settling will produce a separation between the product-size crystals and relatively fine crystals which are removed with the clarified mother liquor leaving by the circulating pipe.

In the case of the illustration, these fines are destroyed by the use of heat, and this heat is coincidentally used to produce the evaporation required for the material balance. Slurry leaving the vessel is removed by an elutriation leg in this example and pumped to succeeding slurry-handling equipment. Typically, the circulation through the external circulating loop would be relatively small, so that the temperature rise through the heater would be on the order of 10–25°F (approx. $5\frac{1}{2}$–14°C), thereby producing sufficient unsaturation with most inorganic salts to destroy fine crystals removed with the mother liquor.

Within the crystallizer body, the draft-tube circulation, which suspends the growing product crystals, would be relatively large and the temperature rise created by the addition of the hot liquor leaving the heater to the stream of circulated slurry would be on the order of $\frac{1}{2}$–2°F (approx. 0.3–1.1°C). Since the temperature rise in the draft tube is small, the boiling surface is relatively smooth and a very even vapor release is obtained. Wall salting with this type of equipment generally is very small, and typical operating cycles on $(NH_4)_2SO_4$ or KCl range from 3 to 10 weeks.

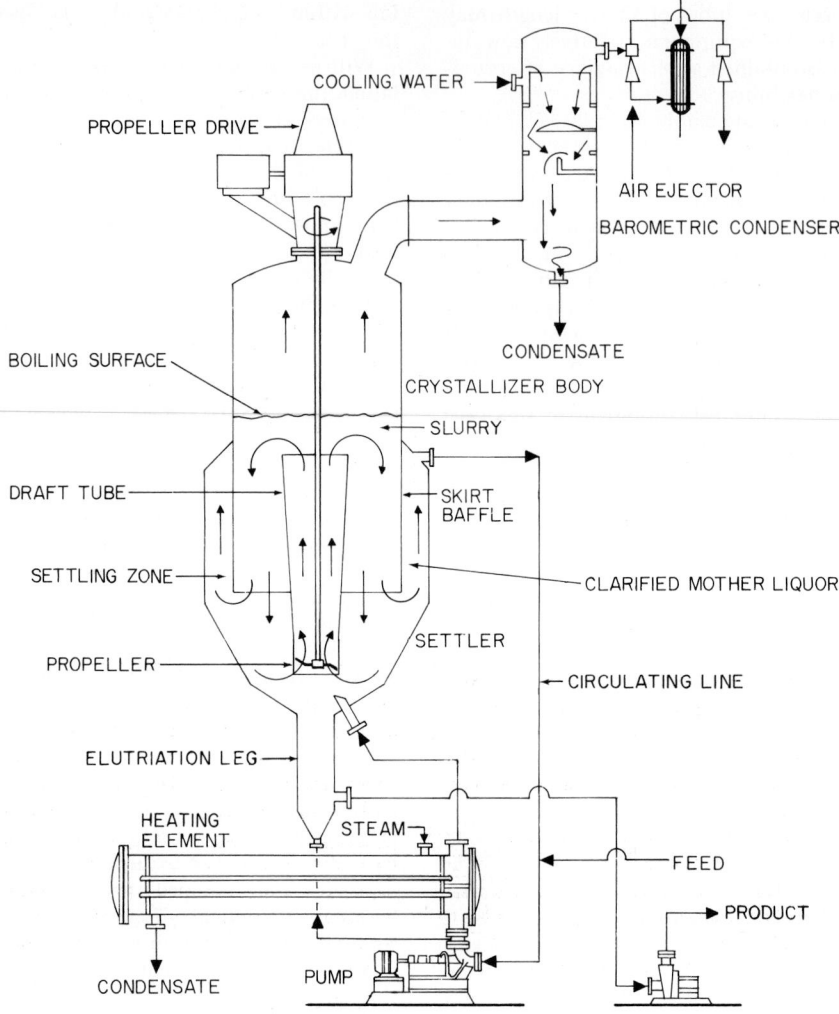

Fig. C-35. Draft-tube baffle crystallizer.

In the draft-tube baffle crystallizer, both the velocity of the liquor in the settling zone and the quantity of liquor removed by the circulating pump are important to ensure that the proper end-product size is achieved and that reasonable stability in particle size is obtained. Common applications for this equipment are crystallization of KCl, $(NH_4)_2SO_4$, $Na_2S_2O_3 \cdot 5H_2O$, and other relatively fast growing inorganic salts.

Some applications require surface cooling either because of the low temperature involved or because of the vapor-pressure characteristics of the salt system used. It is common in such systems (if small-scale) to use scraped-surface crystal-lizers* or some of the scraped-surface double-pipe exchangers. For larger applications, the type of equipment shown in Fig. C-36 can be used. In this system, a slurry is pumped through a tube-and-shell and heat exchanger and recirculated to a body which provides the retention time required for crystal growth. In the example shown, there is an internal skirt baffle which permits with-drawal of fine crystals so that product size can be controlled. In the heat-exchange element (called *surface cooler*), a refrigerant may be boiled by absorbing heat from the circulated slurry, or

*Such as the Swenson-Walker crystallizer.

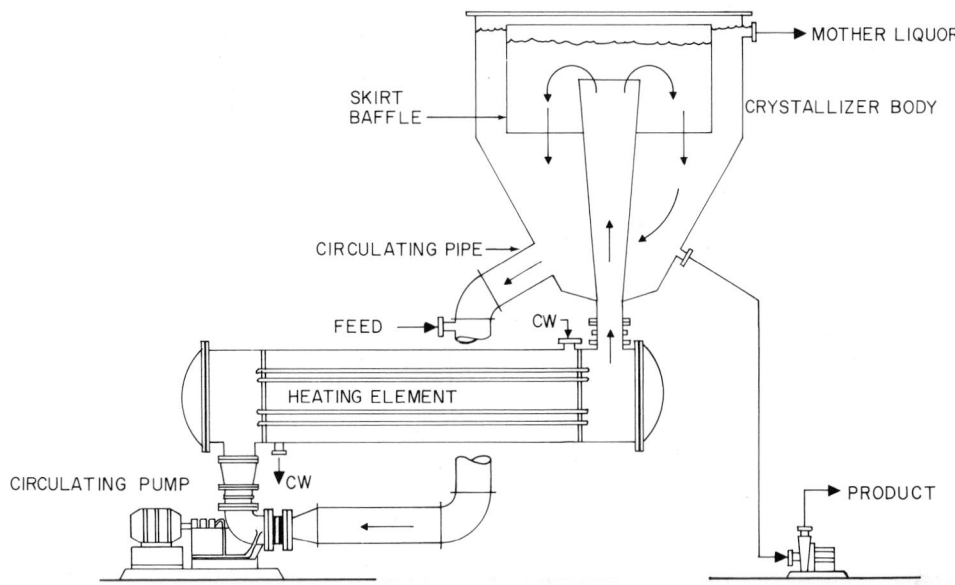

Fig. C-36. Crystallizer with provision for surface cooling. CW = cooling water.

cooling water or brine may be recirculated by means of a pump. The primary limitation of this equipment is that, with most salts, the allowable temperature difference between the cooling media and the slurry circulated through the tubes only can be on the order of 5–10°F (approx. 3–5½°C). This low ΔT, in combination with the low temperature, results in a relatively low heat-transfer coefficient, and therefore a large amount of heat-transfer surface is required.

Irrespective of design, ultimately all surface-cooled equipment must be "boiled out" to remove salt growth on the cooling surfaces. In well-designed equipment, this operating cycle may range from a few days to several weeks. These interruptions limit the application of this type of equipment.

A superior, although more costly, technique for low-temperature crystallization involves the direct mixing of refrigerant with the slurry being cooled (Fig. C-37). Here the refrigerant must be miscible with the slurry so that traces of refrigerant will not be lost in the slurry discharged from the system. It is possible with this technique to reach very low temperatures ($-100°F$, $-73.3°C$) and still preserve the advantages of transferring heat through a boiling-liquid surface rather than through the walls of a heat-exchanger tube. In other respects, this equipment operates like vacuum-crystallization equipment. The refriger-

ant vapor must be collected from the vapor outlet and compressed so that it can be condensed with ambient cooling media. The compressed refrigerant is then self-liquefied and remixed with the slurry in the body. Refrigerants used are commonly immiscible organics, such as Freon, CO_2, or propane. Considerable work has been done with this technique in the freezing of salt water to produce purified water (ice), and in this work, some of the longer-chain hydrocarbons have proved successful. See also **Desalination;** and **Salt.**

REFERENCES

Badger and Banchero: "Introduction to Chemical Engineering," McGraw-Hill, New York, 1955.

Buckley, H. E.: "Crystal Growth," Wiley, New York, 1951.

Buerger, Martin J.: "Crystal-Structure Analysis," Wiley, New York, 1960.

Bunn, Charles: "Crystals: Their Role in Nature and in Science," Academic, New York, 1964.

Chalmers, B. (ed.): "Progress in Materials Science," several volumes (1949–1963), Pergamon, New York.

Doremus, R. H., B. W. Roberts, and David Turnbull: "Growth and Perfection of Crystals," Wiley, New York, 1958.

Gray, G. W.: "Molecular Structure and the Properties of Liquid Crystals," Academic, New York, 1962.

Kröger, F. A.: "Chemistry of Imperfect Crystals," Wiley, New York, 1964.

Laurich, S. A., and H. Svanoe: Crystallizer Instru-

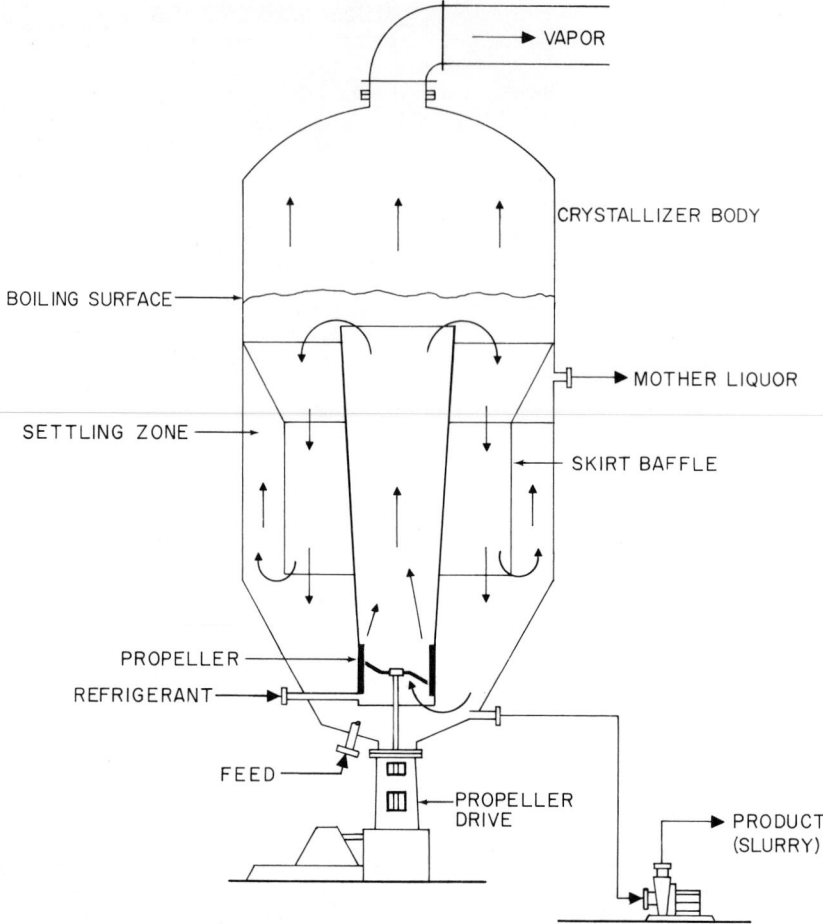

VAPOR

CRYSTALLIZER BODY

BOILING SURFACE

MOTHER LIQUOR

SETTLING ZONE

SKIRT BAFFLE

PROPELLER

REFRIGERANT

FEED

PROPELLER DRIVE

PRODUCT (SLURRY)

Fig. C-37. Low-temperature crystallizer using refrigerant.

mentation, in "Handbook of Applied Instrumentation," pp. 13–37–13–42, McGraw-Hill, New York, 1964.

Osburn, James O.: Crystallization, in "Chemical Engineers' Handbook," 5th ed., sec. 17–6, pp. 17–7–17–23, McGraw-Hill, New York, 1973.

Phillips, F. C.: "An Introduction to Crystallography," Wiley, New York, 1964.

Van Bueren, H. G.: "Imperfections in Crystals," Wiley, New York, 1960.

Van Hook, Andrew: "Crystallization: Theory and Practice," Van Nostrand, New York, 1961.

Wilson, Donald Bruce: Crystallization, *Chem. Eng.*, pp. 119–138, Dec. 6, 1965.

Wyckoff, Ralph W. G.: "Crystal Structures," vol. 1 (1963), vol. 2 (1964), Wiley, New York.

—R. C. Bennett, *Swenson Division of Whiting Corporation, Harvey, Ill.*

Cucairite   (See **Selenium.**)

Cultures, Bacterial   (See **Amino acids**; and **Antibiotics.**)

Cultures, Yeast   (See **Yeast.**)

Cumene   (See **Alkylation; Isomerism; Ketones; Petrochemical complex**; and **Phenol.**)

Cumene Hydroperoxide   (See **Peroxides, organic.**)

Cuprite   (See **Copper production.**)

Cupronickels   (See **Copper.**)

Curb Press   (See **Expression.**)

Curie   A curie is a unit of measurement to indicate the number of disintegrating radioactive atoms per unit time. One curie is defined as the quantity of any radioactive material having 3.7×10^{10} dps (disintegrations per second); the millicurie, 3.7×10^{7} dps; the microcurie, 3.7×10^{4} dps. See also **Radioactive isotopes.**

Curing Agents   (See **Epoxy resins; Paint; Peroxides, organic; Rubber, natural;** and **Tellurium.**)

Curium   (See **Chemical elements.**)

Cutting   (See **Size reduction.**)

Cyaniding   (See **Iron and steel;** and **Nitrogen.**)

Cyanogen   (See **Radicals.**)

Cyanohydrins   (See **Aldehydes;** and **Ketones.**)

Cyanuric Acid   (See **Urea.**)

Cycle Oil   (See **Fluid catalytic cracking; Hydrotreating;** and **Petroleum processing.**)

Cyclohexanol/Cyclohexanone (KA Oil)   Cyclohexanol (an alcohol) and cyclohexanone (a ketone) are the principal raw materials for the production of nylon 6 and nylon 66 fibers. These materials commonly are produced as a mixture, cyclohexanol/cyclohexanone (referred to as ketone-alcohol oil, or KA oil). While cyclohexanol and cyclohexanone once were derived mainly from phenol, most current production is based on the oxidation of cyclohexane. In the manufacture of nylon 66, the cyclohexanol/cyclohexanone mixture is converted to adipic acid and also can be used to make the other required ingredient, hexamethylene diamine. In the manufacture of nylon 6 or polycaprolactam, the KA oil is converted to

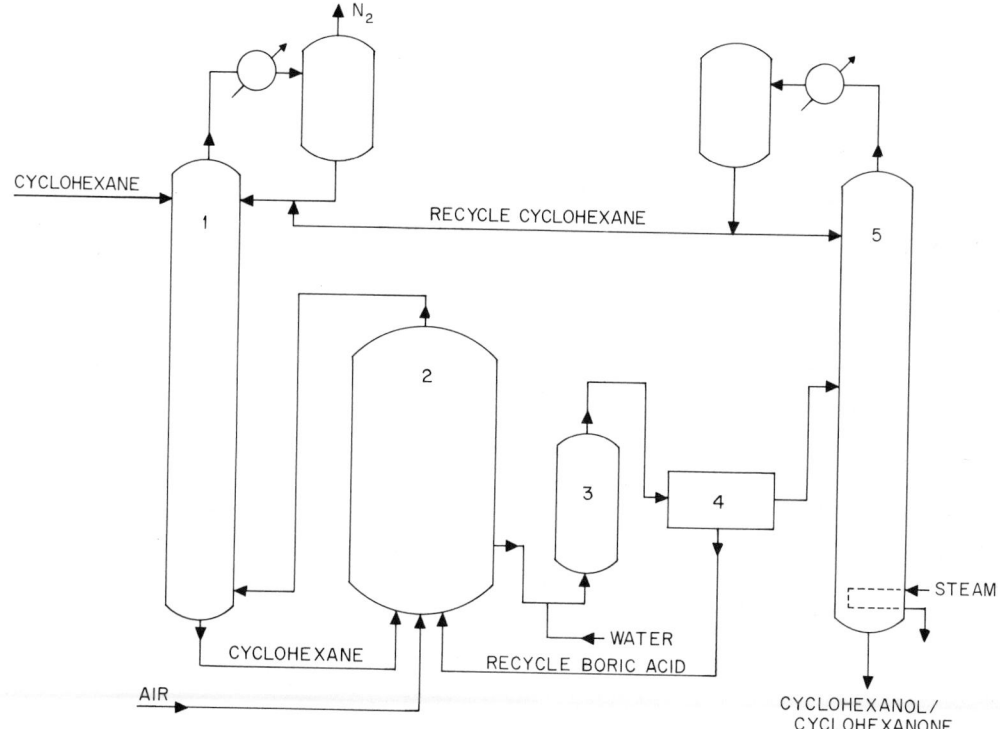

Fig. C-38. Cyclohexane oxidation to produce cyclohexanol/cyclohexanone. (1) Fractionator-refiner, (2) oxidizer, (3) hydrolyzer, (4) boric acid recovery, (5) cyclohexane column. (*Scientific Design Co.*)

caprolactam, the monomer and sole ingredient. See also **Caprolactam.**

In the oxidation of cyclohexane to cyclohexanol/cyclohexanone, the main object, other than low-production cost, is high purity of the KA oil. The higher the purity, the lower the conversion costs to adipic acid or caprolactam. Only about three processes for cyclohexane oxidation are used commercially. The process* described here is one of the major processes, accounting for nearly half of the world's nylon production. With this process, other derivatives, such as aniline, cyclohexylamine, pure cyclohexanol, and pure cyclohexanone also can be produced. See Fig. C-38.

The process can start by the direct oxidation of cyclohexane, which can in turn be produced by the hydrogenation of benzene. The process* can produce the alcohol and ketone in molar yields of 90–95% as compared with 65–70% in the older nonboron commercial processes. Thus cyclohexane can be commercially oxidized with essentially quantitative yield. A concomitant advantage of the higher yield is a lower production of deleterious by-products.

The oxidation of cyclohexane is most preferably carried out in the presence of metaboric acid or dehydration derivatives, although other special boron compounds can be employed. Molecular oxygen reacts with cyclohexane to form the cyclohexyl hydroperoxide, which on reaction with metaboric acid is believed to form a peroxyborate. This peroxyborate is thought to react subsequently to make cyclohexyl borate esters.†

*Scientific Design Co.
† Major patents covering the boron-assisted technology described have been issued to Halcon International, Inc.

One of the critical variables discovered was the effect of the partial pressure of water on the oxidation. During the course of the reaction, water is formed, and unless water partial pressure is maintained at a low level, this water has a deleterious effect on the selectivity of oxidation, with the result that less cyclohexanol/cyclohexanone is produced.

This is a process in which slurries and three-phase systems are handled, where inorganics are crystallized from streams containing organics and subsequently centrifuged and recycled. Furthermore, the kinetics of oxidation are affected by recycle of multiple streams to the oxidizing section. The process is obviously one whose scale-up requires very sophisticated engineering studies, and for full optimization, large-commercial-plant experience is essential.

—Richard L. Marcell, *Scientific Design Company, Inc., New York*

Cyclone Scrubber (See **Absorption.**)

Cyclones (See **Air pollution;** and **Separation operations.**)

Cyclonite (See **Explosives.**)

Cylinder Machine (See **Papermaking and finishing.**)

Cysteine (See **Amino acids;** and **Peptides and proteins.**)

Cystine (See **Amino acids.**)

Cytoplasma (See **Peptides and proteins.**)

D

Dairy Processes (See Homogenization; and Pasteurization.)

Danner Process (See Glass.)

Dative Bond (See Molecule.)

Deacon Process (See Chlorine.)

Dealkylation (See Petrochemical complex.)

Deamination (See Amino acids.)

Debarking [See Pulp (wood) production and processing.]

De Broglie Postulate (See Atomic structure.)

Decanting (See Liquid-solid separations; and Separation operations.)

Decarbonizing (See Thermal cracking.)

Decarboxylation (See Amino acids.)

Decolorizing (See Adsorption.)

Defiberization [See Pulp (wood) production and processing.]

Deflocculation [See Silicates (soluble).]

Defoaming Agents A defoaming agent is a formulation of surface-active materials used at low concentrations to prevent the formation of foam or to destroy foam which has formed.

Defoaming agents are used because the presence of foam severely limits the rate, if not the feasibility, of many production processes. The digestion of wood with caustic agents in the kraft process for papermaking results in a foamy pulp slurry; defoamers are required to wash the spent chemicals from the pulp without generating copious foam. Similarly, defoamers are required for the production of phosphoric acid from phosphate rock, the production of sugar from beets, the production and use of latex paints, and many fermentation processes.

A practical defoamer must satisfy a number of negative requirements in addition to the repression of foam. For instance, in the kraft process the defoamer should not cause pitch-deposition problems or interfere with the recycling of kraft chemicals. In the United States, beet sugar defoamers must be acceptable to the Food and Drug Administration.

Defoaming agents may operate via a number of mechanisms, but the most common ones appear to be those of entry and/or spreading. The defoamer must be insoluble in the foaming liquid for these mechanisms to function. Second, the surface tension of the defoamer, σ_D, must be as low as possible. The interfacial tension between defoamer-foamer, $\sigma_D \sigma_F$, should be low, but not so low that emulsification of the defoamer may occur. Third, the defoamer should be dispersible in the foaming liquid.

Robinson and Woods[1] showed thermodynamically that entry of the defoamer droplet into a bubble surface occurs when E, the entering coefficient, has a positive value (E is actually the negative of the decrease in surface free energy).

$$E = \sigma_F + \sigma_{DF} - \sigma_D$$

The entry in itself does not cause bubble rupture; it is only after the film has been thinned by drainage that a lens of defoamer droplet bridges the two surfaces of the bubble. At this point rupture will occur provided that the adhesion between defoamer and foaming liquid is poor; i.e., the value of σ_{DF} is relatively high.

A common type of defoamer consists of a dispersion in hydrocarbon oil of fine particles of silica coated with silicone; the silicone surface of the particles causes them to be hydrophobic. The defoaming action of such a formulation can be explained on the basis of the entry mechanism. Hydrophobic particles can act as an emulsifying agent where the defoamer oil constitutes the continuous phase and the foam constitutes the dispersed phase. It is interesting that excessively hydrophobic particles such as powdered Teflon do not function as well as the silicone-coated particles. An emulsifier particle must be wetted to some extent by the dispersed phase in order to function as an emulsifier.

The more efficient defoaming mechanism of spreading involves transport of underlying liquid[2] so that this liquid is replaced by a film of defoamer which does not support foam. A drop of oleic acid added to water spreads at a velocity of 30 mph;[3] the mechanical shock to a film by such a defoamer may be considerable. In addition to the foam-destroying aspect, spreading is also of value as a defoamer-dispersion method, particularly in viscous or poorly stirred systems.

A defoamer should be insoluble in the system to be defoamed, but even where solubilization of the defoamer occurs, there can be defoaming action to a lesser extent.[4]

A film surface freshly formed by extension of a film would have a surface tension greater than the equilibrium value; this represents a structurally weak area in the film surface. A defoamer may cause restoration of the equilibrium surface tension by accelerating the movement of surfactant micelles to the surface; however, the structural impairment remains and constitutes a vulnerable spot.

Foam may be destroyed by the replacement of molecules in the foam layer by other surfactant molecules. This is the case for the destruction of alkyl aryl sulfonate foam in the presence of calcium ion on addition of soap.[5] The calcium salt of the soap imparts brittleness to the film. Housewives observe this effect when a soap-embedded steel wool pad is used to clean dishes in the presence of foamy detergent solution.

Classification of Defoaming Agents: Defoaming agents may be categorized as solubilized surfactants, as dispersions of hard particles, and as dispersions of soft particles. The classifications more often than not overlap. In all cases a liquid nonaqueous vehicle is present, even where the defoamer is represented as a solid formulation.[6] Water may also be present, particularly in emulsified silicone formulations.

A common type of solubilized surfactant formulation is the fatty acid–fatty alcohol combination in hydrocarbon oil. The spreading rate of such formulations is very rapid, which is largely responsible for the popularity of this type. Either surfactant may be substituted by alkylene oxide adducts—mono-, di-, and triglycerides.[7] The fatty acid or a derivative thereof by itself in oil is widely used. Mixtures of esters, waxes, alkyl phosphates, fats, etc. have been used. Formulations of silicone and other polymers in oil are used, often in emulsion form.[8] The number of synthetic and natural materials used in defoamers is legion; however, they have in common a degree of surface activity.

Soft-particle formulations may consist of paraffinic waxes[9] or fatty amides among other components as the dispersed phase; a nonaqueous liquid serves as the vehicle. A fine particle size is generally desired, which is effected by grinding or chilling a hot solution rapidly. In addition to the particulate components, members of the solubilized surfactant class are generally present.

Hard-particle formulations most commonly consist of silica or a mineral coated with silicone which is dispersed in a vehicle. The optimal particle size[10] may be as small as 0.02 μm. A spreading agent, i.e., a surfactant, is usually present. Dispersants may be added to promote stability since the rate of settling in such formulations is greater than that for the soft-particle type.

Emulsifiers may be included in defoamer formulations to accelerate the dispersion of the defoamer throughout the foaming system. Such formulations are added to the foaming system neat or diluted with water. Additives may be present to lower the pour point, adjust the viscosity, raise the flash point, and alter other characteristics.

The evaluation of a defoamer for a given application requires samples of the foamy liquid tested under realistic conditions.

See also **Separation operations.**

REFERENCES
1. Robinson, J. V., and W. W. Woods: *J. Soc. Chem.,* vol. 67, pp. 361–365, 1948.
2. Schulman, J. H., and T. Teroell: *Trans. Faraday Soc.,* vol. 54, pp. 1337–1342, 1938.
3. Edser, E.: *Trans. Faraday Soc.,* vol. 17, p. 664, 1922.
4. Ross, S., and T. H. Bramfitt: *J. Phys. Chem.,* vol. 61, pp. 1261–1265, 1957.
5. Peper, H.: *J. Colloid Sci.,* vol. 13, pp. 199–207, 1958.
6. Canada Patent 867,321.
7. United States Patent 2,849,405.
8. United States Patent 3,340,193.
9. United States Patent 2,715,614.
10. United States Patent 3,076,768.

—*Irwin A. Lichtman, Nopco Chemical Division, Diamond Shamrock Chemical Company, Morristown, N.J.*

Degasification (See **Water treatment.**)

Degradation, Plastic [See **Stabilizers (PVC).**]

Degreasing (See **Perchloroethylene;** and **Trichloroethylene.**)

Degree of Substitution (See **Acetate fibers;** and **Cellulose ester plastics, organic.**)

Dehydration (See **Bauxite; Catalysts;** and **Ethylene.**)

Dehydrochlorination (See **Chlorine; Chlorine organics;** and **Trichloroethylene.**)

Dehydrogenation (See **Alcohols; Aldehydes; Catalysts; Isoprene; Ketones;** and **Petrochemical complex.**)

Deionization (See **Ion-exchange resins;** and **Water treatment.**)

Delayed Coking (See **Thermal cracking.**)

Delignification [See **Pulp (wood) production and processing.**]

Delustrants (See **Acrylic fibers.**)

Demolition Devices (See **Explosives.**)

Dense-Media Separation (See **Beneficiation, ore;** and **Separation operations.**)

Dental Materials (See **Gold; Mercury; Silver;** and **Waxes.**)

Deoxidized Copper (See **Copper.**)

Deoxidizing Agents (See **Copper; Ironmaking and steelmaking; Magnesium; Manganese;** and **Silicon.**)

Dephlegmation (See **Distillation.**)

Derivative In chemical terms, a derivative is a compound that results from the substitution of an element or radical in the structure of what might be termed a *parent compound.* For example, when a paraffin such as methane is halogenated, a series of halogen-substitution products may be formed. Depending upon conditions of the reaction, CH_4 plus Cl may yield the monosubstitution product CH_3Cl (methyl chloride), or the disubstitution product CH_2Cl_2 (methylene dichloride), or the trisubstitution product $CHCl_3$ (chloroform); or the tetrasubstitution product CCl_4 (carbon tetrachloride). In essence, these compounds are *derived* from methane and thus are methane derivatives as the result of chlorination. Normally, the term derivative applies to those compounds where the resulting compound is formed in one step, although a chain of steps may be involved in some cases, largely depending upon how easy it is to identify the "derivative" with the parent substance. Usually, where a chain of steps is involved, the intervening compounds are termed intermediates rather than derivatives. There are no clear-cut rules to follow in the use of these terms.

Desalination Salt can be removed from water by several different physical and chemical processes. The important separation processes by which

TABLE D-1. *Important Separation Processes for Removing Salt from Seawater*

Process	Constituent Removed from Saline Water	Phase to which Transported
Distillation	Water	Vapor
Reverse osmosis . . .	Water	Liquid
Freezing	Water	Solid
Hydrate process . . .	Water	Solid
Electrodialysis	Salt	Liquid

NOTE: Plants for desalting ocean water and brackish water are becoming economically competitive with the cost of transporting freshwater over long distances. Approximately 700 desalination plants (total capacity of about 250 million gal/day) had been constructed by the early 1970s. A plant in the Netherlands produces 7.6 million gal/day. Other larger plants are planned for European and American installation.

freshwater* can be made are listed in Table D-1.

Distillation. Distillation is the oldest and most common process for desalination. It is utilized for most of the existing total plant capacity. Numerous improvements have been made to the basic single-effect distillation system, resulting in the production of up to 21 lb freshwater for each pound of steam used. The multistage flash evapo-

*Water for human consumption should contain no more than 500 ppm dissolved solids, according to the U.S. Public Health Service. The purity of irrigation water varies with crop and soil conditions, but generally is about 1,200 ppm. Brackish water is generally classified as containing more than 1,000 ppm dissolved minerals, but less than seawater, which contains about 35,000 ppm.

ration process is the most popular distillation technique. The vertical-tube evaporator process and vapor compressor, as well as combinations of the three, show promise for future exploitation. Solar humidification is limited to areas with intense solar radiation.

Multistage Flash Evaporation Process: Seawater is treated with scale-control chemicals, deaerated, and combined with a stream of recycle brine, as shown in Fig. D-1. The saline solution then is pumped through tubes set lengthwise in the top half of horizontal vessels. Vertical baffles (stage dividers) divide each vessel into separate chambers, each constituting a stage of the process. The stages operate at progressively higher temperatures over a range of more than 150°F (84°C). Thus the saline water heats gradually as it passes in the tubes from stage to stage. The water does not boil because it is under pressure.

At the end of its first pass of a round trip through the evaporator, the seawater leaves the highest-temperature stage and enters a brine heater. There its temperature is further increased by steam supplied from either a conventional boiler or from a nuclear steam generator. After this heating, the seawater is introduced to the shell side of the highest temperature, or first stage.

When the brine enters the bottom of this spacious compartment, it suddenly boils (flashes) and releases steam, caused by a reduction of pressure on the brine. The steam rises and contacts the tubes, passing off heat to the seawater in them, and then condenses and drips into troughs beneath the tubes.

The second stage operates at slightly less pres-

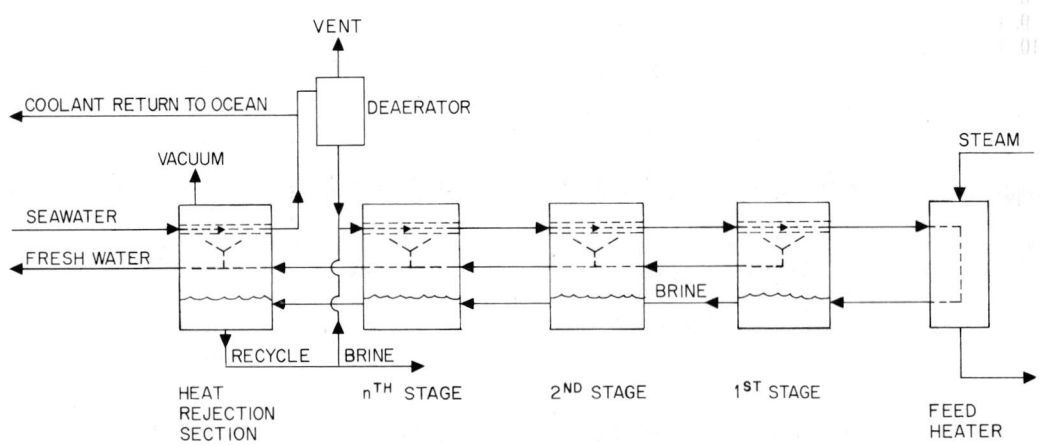

Fig. D-1. Multistage flash distillation with recycle for production of freshwater from seawater.

sure than the first. This causes the brine and the distilled water in the troughs to flow spontaneously through liquid pressure seals in the stage dividers and into the second chamber. The brine and product water again flashes, steam condenses on the tubes, and more freshwater is made.

The procedure repeats through the many stages of the system until all the available heat is removed from the brine. Concentrated seawater in excess of that needed for recycle is discharged to the ocean, and very pure freshwater is taken from the troughs as product.

Vertical-Tube Evaporator Process: As shown in Fig. D-2, the seawater is treated and then passed through several preheaters (not shown) and transferred into the top of a vertical-tube evaporator referred to as the *first effect*. Steam from a boiler passes on the outside of the tubes. Heat transferred to the seawater falling inside the tubes causes the steam to condense and the seawater to boil. The condensate collecting in the first effect is ordinarily returned as boiler feedwater. Steam from the boiling seawater is transferred to the second effect, where it is condensed on the outside of the vertical tubes. Some of the brine collecting at the bottom of the first effect is recycled to the top of the first effect. The remainder of the brine is transferred to the bottom of the second effect, where it flashes and releases steam because of a reduction in pressure.

Similarly, condensate from each effect, except

the first, flashes in the next effect and produces additional steam. Steam from the second effect is transferred to the next effect, where it heats the brine falling inside the tubes. The process is repeated through additional effects, where more condensate is produced. Each effect operates at a lower pressure than the preceding effect. Steam from the last effect is used to heat incoming seawater. Concentrated brine is discharged to the ocean, and condensate is removed as freshwater product. The vertical-tube evaporator may be combined with other processes.

Vapor-Compressor Distillation: Vapor compression can be integrated into either of the processes just described. An elementary diagram is shown in Fig. D-3. Seawater is pumped into a tubular heat exchanger within an evaporation chamber. There it is boiled. Steam from the boiling seawater is piped to a compressor, where its pressure and heat content are increased by compression. The compressed vapor then flows outside the tubes in the evaporator. There it is condensed to freshwater, giving up its latent heat to boil the incoming seawater. This method makes use of the fact that the saturation temperature of water increases with pressure.

Solar Distillation: Solar energy as a thermal energy source can be used effectively in certain areas of the world. A representative diagram of a simple solar still is shown in Fig. D-4. Vaporized seawater condenses on the surface of an air-

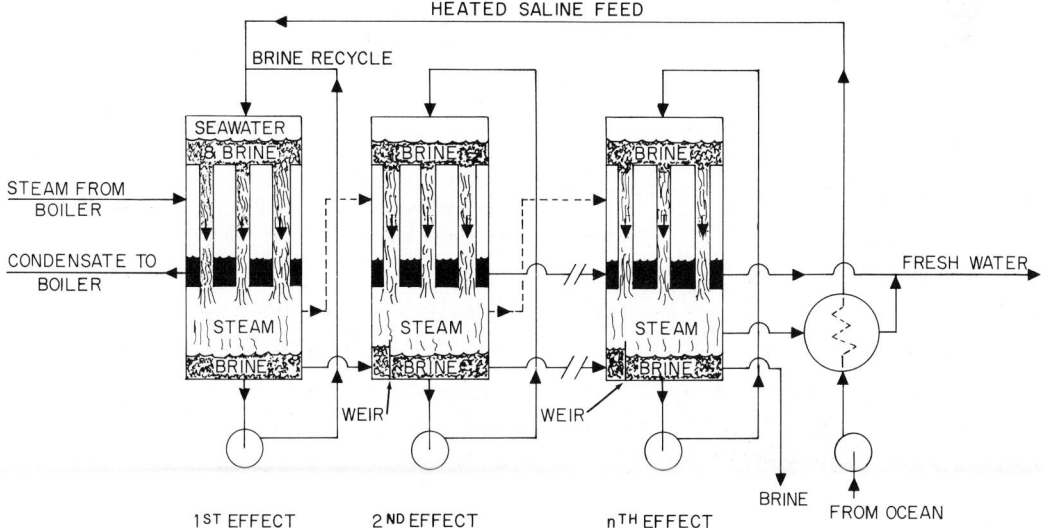

Fig. D-2. Vertical-tube evaporator used for producing freshwater from seawater.

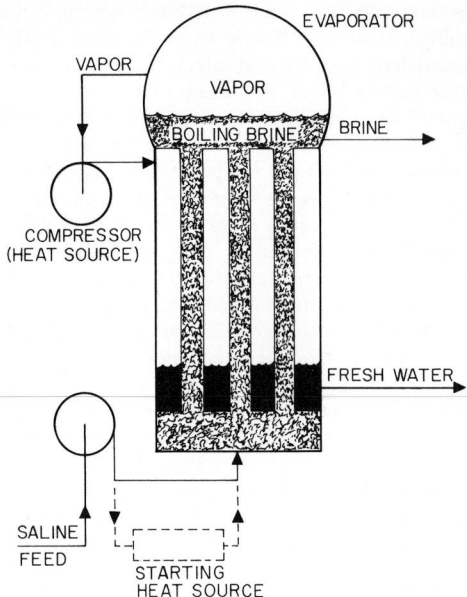

Fig. D-3. Elementary vapor-compression distillation.

supported plastic film, collects in troughs, and is removed as product water at selected points. These stills occupy large areas and operate at inefficient rates. They may be used to recover salt as well as freshwater.

Membrane Process. Membrane processes do not involve a change of phase and therefore should require less energy than distillation. These processes are becoming more popular for desalting brackish water since energy requirements are proportional to the quantity of the dissolved salts. In general, a membrane may be considered as a selective filter. See also **Membrane filters.**

Reverse Osmosis: When pure water and a saline solution are on opposite sides of a semipermeable

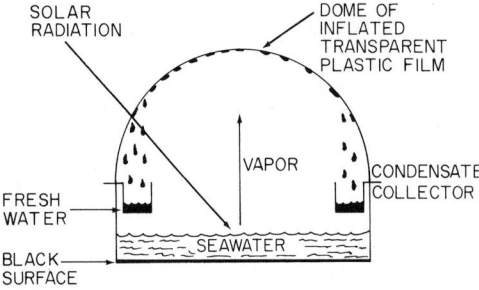

Fig. D-4. Simple solar still.

membrane, the pure water diffuses through the membrane and dilutes the saline water. This is called *osmosis,* and the effective driving force of any saline solution is called its *osmotic pressure.* By exerting pressure on the saline solution, the osmosis process can be reversed.

In the reverse osmosis process, pressure in excess of the osmotic pressure is applied to the saline solution. Freshwater permeates the membrane and collects on the opposite side, where it is drawn off as product. The pressure required to force the water through the membrane is a function of the salt content of the water. As an example, with brackish waters, it is normally in the range 300–800 psig, and it may go as high as 1,500 psig with seawater.

This process has long been recognized as one that had the potential of a high thermodynamic efficiency. The problem at hand for seawater involves the development of a suitable membrane with a flux rate adequate to reduce capital costs to an acceptable figure.

Electrodialysis: An electrodialysis cell consists of a sandwich of alternating cation- and anion-permeable membranes. Application of an electric current causes positively charged ions (such as sodium) to pass through cation-permeable membranes. Negatively charged ions (such as chloride) move in the opposite direction and pass through anion-permeable membranes. The water in the center chamber of each membrane sandwich thus is depleted of salt, while water passing through the adjacent chambers is enriched in salt.

The amount of electric current required and the resulting demineralization cost are proportional to the amount of salt to be removed. As a result, this process is favored for brackish waters up to about 2,500 ppm of total dissolved solids. Additional development on membranes and other equipment associated with the process is needed to make the electrodialysis process economically feasible for desalting of seawater.

See also **Separation operations;** and **Water treatment.**

Freezing Processes. When salt water is partially frozen, the solid component is pure ice. The liquid remaining is a saline solution of greater concentration than the original water. Theoretically, if the ice is separated from the brine, and then melted, it will yield pure water. However, in practice, some brine adheres to the surface of the ice crystals and is difficult to remove. Two types of process involving this principle have been investigated.

Direct Process: In this process, cold seawater is sprayed into a vacuum chamber to produce a suspension of ice crystals in brine. The low pressure flashes the seawater, and heat of vaporization removed from the water causes ice crystals to form. About one-half of the feed is frozen into ice. The resulting mixture of brine and ice is then transferred to the bottom of a separation column.

There the ice crystals float to the top, forming a bed of ice. The brine in the column is drawn off at the sides at selected points. The rising ice is washed by a small stream of product water before being taken by mechanical means to the melting tank. Vapor leaving the freezer is condensed in an auxiliary refrigeration system and is admitted to the top of the separation column.

Secondary Refrigerant Freezing Process: In this process, a hydrocarbon which is immiscible in water is the refrigerant. The incoming seawater is deaerated; then passed through brine and product coolers en route to the freezing chamber. In this chamber, a hydrocarbon, such as butane or isobutane, is flashed in direct contact with the seawater. The resulting ice crystals and brine are pumped to a separation column where the brine is removed and the ice crystals washed before being mechanically moved to the melter. The hydrocarbon vapor flashed from the freezer is compressed and recycled to the freezing chamber.

Hydrate Process: In this process, a low-molecular-weight hydrocarbon, such as propane, combines with water to form hydrate crystals. As with ice crystals, the hydrate crystals reject ionic constituents and accept only pure water in their lattices. The excess propane is vaporized to remove the heat of crystallization, then is recycled after having been compressed and condensed. The hydrate crystals are washed by product water and then decomposed to water and propane. Separation of these two is by decanting, or centrifuging. In many respects this process is similar to freezing.

REFERENCES

Much more information can be obtained from the U.S. Office of Saline Water, Washington.

Spiegler, K. S.: "Principles of Desalination," Academic, New York, 1966.

> —Robert W. Newkirk, Frank J. Donnelly, and William A. Bruinsma, *Fluor Corporation, Los Angeles, Calif.*

Desalting (See **Petroleum processing.**)

Desorption (See **Absorption;** and **Freeze-drying.**)

Desulfurization (See **Catalysts; Hydrotreating; Magnesium; Petroleum;** and **Petroleum processing.**)

Detergents For purposes of this discussion, detergents are defined as complete washing or cleansing products which contain among their ingredients an organic surface-active compound (surfactant) that possesses soil-removal properties. Frequently, the term detergent is used synonymously with *surfactant,* but common industry practice treats the surfactant as one component of a total detergent product, as is done here. Additionally, this discussion treats primarily only the so-called synthetic detergents, excluding those products in which soap is the sole or predominant surfactant. Soaps, the alkali salts of long-chain fatty acids, differ significantly in certain important performance properties from the synthetic surfactants and are described separately. See also **Soaps.**

The synthetic-detergent industry is one of the largest chemical process industries. The most recent estimates indicate an annual U.S. production of synthetic detergents of about 3 million tons, with an approximate annual value of $2 billion. The industry differs from many other chemical process industries, however, in that the bulk of its production is sold directly to individuals for household consumption, primarily as branded products, rather than to industrial or institutional users.

Detergent Ingredients and Their Functions. Detergent formulas vary greatly, depending upon the intended end-use application. Differentiation by surfactant type as well as selection of auxiliary ingredients is involved. Some of the ingredients most commonly used in detergent formulas and their functions are as follows.

Surfactants: By definition every detergent product contains one or more types of surfactants. Basically, every surfactant is an organic compound consisting of two parts: a hydrophobic portion, normally including a long hydrocarbon chain, and a hydrophilic portion, which renders the entire compound sufficiently soluble or dispersible in water or other polar solvent to serve its intended use. Together, these combined hydrophobic and hydrophilic moieties render the compound surface-active—able to concentrate at the interface between a surfactant solution and another phase, such as air, soil, and textile or other substrate to be cleaned.

Surfactants provide the detergent with the ability to penetrate and wet soiled surfaces, to

Fig. D-5. Pendant drop. The surface and interfacial tensions of surfactant solutions can be determined from the shape of the drop as seen on the ground-glass back of the camera. (*The Procter & Gamble Co.*)

displace, solubilize, or emulsify various soils, particularly oils and greases, and to disperse or suspend certain soils in solution to prevent their redeposition (Fig. D-5). In addition, the surfactants provide (to various degrees) whatever foaming or sudsing properties the detergent solution possesses, properties which are necessary for satisfactory use in certain industrial applications and highly desirable to many consumers in household-laundry, hand-dishwashing, and personal-care uses.

Surfactants in broad use may be classified into three general types: anionics, in which the hydrophilic portion of the molecule carries a negative charge; cationics, in which the charge of this portion is positive; and nonionics, which do not dissociate but commonly derive their hydrophilic portion from polyhydroxy or polyethoxy structures. Ampholytic and zwitterionic surfactants are also known but are not presently of significant commercial importance.

Of the three main types of surfactants the anionics are by far the most commercially important class, constituting, in particular, the major surfactant type represented in laundry and hand-dishwashing detergents. Among the anionics linear sodium alkyl benzene sulfonate (LAS), linear alkyl sulfates, and linear alkyl ethoxy sulfates are by far the most widely used compounds.

(The industry converted voluntarily to linear alkyl chains in the mid-1960s to obtain improved biodegradability relative to the branched-chain alkylates formerly used.)

$$CH_3(CH_2)_X$$

$$SO_3Na$$

$X = 9\text{--}15$ commonly

Linear alkyl benzene
sulfonate (LAS)

$$CH_3(CH_2)_XOSO_3Na$$

$X = 9\text{--}17$ commonly

Alkyl sulfate

$$CH_3(CH_2)_X(O{-}CH_2{-}CH_2)_YOSO_3Na$$

$X = 7\text{--}15$ commonly
$Y = 0\text{--}6$ commonly

Alkyl ethoxy sulfate

Nonionic surfactants are also used in substantial amounts in laundry detergents and in automatic-dishwashing detergents, both applications reflecting in particular their generally lower sudsing characteristics than the anionics. Commercially important examples of the nonionics include the alkyl ethoxylates, the ethoxylated alkyl phenols, the fatty acid ethanol amides, and complex polymers of ethylene oxide, propylene oxide, and alcohols.

$$CH_3(CH_2)_X(O{-}CH_2{-}CH_2)_YOH$$

$X = 9\text{--}15$ commonly
$Y = 4\text{--}20$ commonly

Alkyl ethoxylates

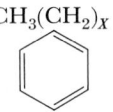

$$CH_3(CH_2)_X$$

$$(OCH_2CH_2)_YOH$$

$X = 7\text{--}13$ commonly
$Y = 3\text{--}12$ commonly

Ethoxylated alkyl phenols

$$CH_3(CH_2)_XCON(CH_2CH_2OH)_Y(H)_Z$$

$X = 9\text{--}17$ commonly
$Y = 1$ or 2
$Z = 1$ or 0

Fatty acid ethanol amide

Cationic surfactants are used in only limited tonnage for specialty detergent products, such as metal cleaners for electroplating, and more commonly in ancillary textile laundering products for their fabric-softening, antistatic, and germicidal properties. A typical cationic surfactant would be tallow trimethylammonium chloride.

$$CH_3(CH_2)_{13\text{-}17}N^+(CH_3)_3Cl^-$$

Tallow trimethylammonium chloride

In contrast with the soaps, virtually all the synthetic anionic and nonionic surfactants do not form readily visible insolubles in the presence of the Ca^{++} and Mg^{++} ions present in many water supplies—the familiar scum, curd, and "lime soap" of the laundry-soap era. This property, in combination with the discovery that the detergency performance of many synthetic surfactants could be greatly augmented by the addition of certain phosphate chelating agents, led to the "detergent revolution" in the laundry-product field.

Builders: Builder is the term used within the industry to designate materials in the synthetic detergent which chelate (sequester) or precipitate polyvalent metal ions present in the cleaning solution, particularly Ca^{++} and Mg^{++} ions which are present in substantial quantities in so-called hard-water supplies. Over 2 billion lb of builders is presently used in detergent products, particularly in textile-cleaning detergents, where they may constitute up to three-quarters of the total weight of the product. Laundry products containing significant quantities of builders are called *heavy-duty,* or *built,* detergents.

Builders perform several critical functions in present-day detergents:

1. They prevent polyvalent metal ions from combining with (many) surfactant(s) to form an adduct which is less effective than the unmodified surfactant in cleansing properties.

2. They prevent polyvalent metal ions from combining with various soils, such as lipid residues and clays, to form less dispersible residues which adhere tenaciously to the surface to be cleaned.

The above effects combine to provide greatly enhanced removal of many soils and stains by a formula with builder compared with a formula without.

3. They prevent the redeposition of soils removed from a surface back onto the surface through a dispersing action associated with chelating and charge-distribution effects.

4. They provide added and buffered alkalinity

to the wash solution, which is generally helpful to cleansing in most applications.

5. They provide enhanced removal and kill of microorganisms, a critical property in many applications involving sanitation of commercial facilities.

By far the most commonly used builders are the condensed polyphosphates, particularly pentasodium tripolyphosphate (STP) and, to a lesser extent, tetrasodium pyrophosphate. These polyphosphates chelate the polyvalent metal ions to form a soluble complex (under some conditions the pyrophosphate complexes are sparingly soluble). Precipitating builders, such as sodium carbonate, Na_2CO_3, are also used, but to a much less extent, since the precipitate formed can itself deposit on the surface to be cleaned unless special washing procedures are followed. Other chelating builders than phosphates, such as trisodium nitrilotriacetate (NTA), tetrasodium ethylenediamine tetraacetate (EDTA), and other polycarboxylates, have also been used in small to moderate quantities, but questions concerning their safety, efficacy, or economics have to date prevented major displacement of the phosphates.

STP

NTA

Tetrasodium pyrophosphate

EDTA

Bleaches: In many detergent applications the action of a supplementary bleaching agent is necessary or desirable. It is frequently preferable to incorporate the bleach directly into the detergent product for reasons of convenience or performance.

Two types of bleaching agents are in common use in detergent products: those based on a hypochlorite bleaching species and those based on a peroxygen species. The hypochlorite, or "chlorine," bleaches are considerably more powerful in their oxidizing action than the commonly used peroxygen bleaches under most U.S. usage conditions. Commercially important examples of hypochlorite species bleaches used in detergents include potassium dichloroisocyanurate (KDCC) and chlorinated trisodium phosphate (a physical combination of $NaOCl$, H_2O, and Na_3PO_4), while sodium perborate, $NaBO_3 \cdot 4H_2O$, is by far the most commonly used of the peroxygen bleaches. All current detergent products containing bleach are dry solids since aqueous bleach solutions have not been found to be sufficiently stable in the presence of other desired detergent ingredients.

Types of detergent products to which chlorine bleaches are frequently added include hard-surface cleaners, such as the scouring cleansers and the automatic-dishwashing detergents, and detergents for commercial sanitation, in which the disinfecting properties of the chlorine bleach are a critical performance attribute. In the cleaning of textiles, considerations of possible fiber and color damage prevent the universal use of a powerful oxidizing agent like hypochlorite. Consequently, a separate liquid solution of sodium hypochlorite is often used in conjunction with the detergent on those fabrics where its use is appropriate, while some textile detergents contain the milder sodium perborate bleach for more general application.

Corrosion Inhibitors: Early research indicated that unmodified alkaline detergents could be corrosive to certain hard surfaces such as aluminum, washing-machine porcelain, and the overglaze on fine china. It was quickly determined, however, that the addition of soluble silicates (silicates with varying ratios of SiO_2 to Na_2O are used, depending upon product processing and intended application) could essentially prevent this corrosive effect. Consequently, the soluble silicates are widely added in moderate amounts to alkaline detergents, and even contribute to detergency through their added alkalinity.

Sudsing Modifiers: In many applications the detergent surfactants which appear to provide optimum cleansing performance do not provide satisfactory product-sudsing characteristics from a functional or aesthetic standpoint. Accordingly, materials which depress or boost the sudsing of the basic surfactant system are often added in small amounts to such products. Examples of materials which can be used to boost the sudsing of common anionic surfactant systems are the mono- and diethanol amides of C_{10-16} fatty acids, while the long-chain, C_{16-22}, fatty acids themselves and certain nonionics, such as the ethoxylated fatty alcohols, are commonly used as suds depressors.

Fluorescent Whitening Agents (FWA): Fluorescent whitening agents (also called *fluorescers, brighteners, optical bleaches*) are organic chromophores which absorb incident light in the ultraviolet region and reemit part of the absorbed energy as visible light, generally in the blue region of the visible spectrum. In addition, the chromophore is modified with organic substituents to make it substantive to one or more textile substrates from a laundry-wash solution. Consequently, the brightness and whiteness of the fabrics on which FWA deposit is enhanced. In effect, an added portion of the incident light is reflected from the fabric.

Derivatives of sulfonated triazinylstilbenes are commonly used at low levels in laundry detergents as FWA for cellulosic fibers. Derivatives of other chromophores are used to provide wash brightening of synthetic fibers (although such brightening is now primarily accomplished by incorporating FWA in or on the synthetic fiber during its manufacture).

Enzymes: The most substantial advance in textile cleansing since the introduction of synthetic detergents has occurred in the last few years through the introduction of low levels of enzymes into laundry detergent and presoak products. Both proteolytic and amylolytic enzymes are used by the industry to attack and loosen from the textile substrate soils and stains with protein and carbohydrate substituents (such as body soils, many food stains, grass stains, blood, and many others). The enzymes act catalytically and relatively specifically, and thus may be used at low levels in products with excellent safety for fibers and textile colors (compared with a general chemical reactant such as hypochlorite bleach).

Presently, all detergent enzymes are derived from fermentation cultures of specific strains of the ubiquitous bacilli *B. subtilis* and *B. licheniformis*. (The enzymes themselves are, of course, nonliving proteins, biochemical products of the bacilli which are involved in their own metabolic processes.) It had been recognized since early in this century that enzymatic action could be a most useful mechanism in textile cleansing, but available enzymes were rendered ineffective by the elevated temperatures and alkalinity required for satisfactory general detergency. The discovery and production of the *B. subtilis* and *B. licheniformis* mutants and their metabolites which remained active under laundering conditions was the essential step required to make enzymatic action available in textile detergency.

Antiredeposition Agents: In textile laundering many soils once removed have a tendency to redeposit back onto the textile substrate during the remainder of the washing process. Certain agents, such as carboxymethyl cellulose and polyvinyl alcohol, have been found to be effective in preventing or minimizing this effect and are consequently added in very small amounts to many laundry detergents.

Other Materials: In addition to the basic performance ingredients previously discussed, other materials are commonly added to facilitate product manufacture (for example, hydrotropes such as xylene sulfonate) and to enhance product acceptability among consumers (colorants and perfumes). A few materials are formed or carried along into the finished product by the manufacturing process (such as sodium sulfate and water).

Types of Detergent Products. Synthetic detergents are manufactured to perform a wide variety of household and industrial cleansing operations. Within the commercially more important (in terms of tonnage and dollar volume of sales) household-product category, the following major functional areas may be defined.

Textile Cleansing: Household laundry detergents and presoak products represent the largest single category of detergent use, currently accounting for over two million tons per year in U.S. consumption. These products are used in conjunction with individually owned or self-service washing machines to launder most of the clothing, bed and bath linens, curtains, and other textiles used in the typical household. Prior to World War II laundry soaps (see **Soaps**) were almost exclusively used for this purpose, but the superior cleaning, absence of dulling deposits, and sudsing of the polyphosphate-built synthetic detergents have led to the almost complete displacement of the soap products from this market (less than 5% of current household laundry-product consumption).

Detergents in liquid, tablet, and granular-powder form are sold for household laundry use, the granules accounting for by far the major amount (90+%) of product sold. A household laundry detergent powder typically contains 10–20% surfactant, 30–60% builder, 5–10% sodium silicates, low levels of suds builders or suppressors, antiredeposition agents, fluorescent whitening agents, enzymes, peroxygen bleach, colorants, and perfume as required to meet product aesthetic and performance objectives, the balance being materials of processing (sodium sulfate, water). Powder bulk density varies among products, but most commonly is such as to provide a wash solution concentration of about 0.15% detergent at the product usage recommended. The presoak powders differ from the wash detergents in providing relatively higher wash-water concentrations of enzymes, builder, and peroxygen bleach (if used) but a relatively lower concentration of surfactant. The liquid laundry detergents include most of the types of materials present in the powders (the pyrophosphate form is required for extended stability in solution), along with hydrotropes, considerable water of processing, and a much higher bulk density.

Hard-Surface Cleaning: Products used for hard-surface cleaning around the home include the hand- and automatic-dishwashing products, the liquid and powder floor and wall cleaners, and the abrasive scouring cleansers.

Although the higher-sudsing laundry detergents may be used satisfactorily for hand dishwashing (and were so used extensively in the past), high-sudsing liquid products are predominantly used for this purpose. Relative to powders, the liquids offer superior ease and convenience of use, a product form into which surfactant types and levels better suited to the dishwashing task can be incorporated, and comparable economy of use since the builders required for superior laundry performance are not essential for hand-dishwashing use. Dishwashing liquids typically consist of 25–45% anionic or semipolar surfactants, the remainder consisting of solvents, hydrotropes, buffers, colorants, perfumes, and water of processing. Surfactant types used in this application include LAS, alkyl sulfates, alkyl polyethoxy sulfates, di- or monoethanol fatty acid amides, alkyl glyceryl ether sulfonates, and alkyl dimethylamine oxides. The alkyl portions of these surfactants commonly average to a coconut or C_{12} chain length, as opposed to laundry detergents, where slightly longer alkyl chains are used.

Automatic-dishwashing products are designed

solely for use in household mechanical dishwashers, machines in which a moving high-velocity water spray is used to clean the tableware and cooking utensils. Performance requirements for an automatic-dishwasher product differ substantially from those for laundry- and hand-dishwashing products and include (1) very low sudsing to prevent suds overflows with the high-velocity spray, (2) very complete rinsing to avoid residual deposits, (3) complete sequestration of calcium and magnesium ions in water supplies by the use of relatively large amounts of builder to avoid deposits of sparingly soluble calcium and magnesium salts, and (4) thorough removal of minute particles of food protein which can form nuclei for spot formation during drying. Consequently, automatic-dishwashing products typically contain a low level of a nonionic surfactant (commonly polyoxyethylene/polyoxypropylene condensates), a low level of a dry chlorine bleach (commonly KDCC or chlorinated trisodium phosphate), a high level of builder (STP and/or sodium carbonate), and moderate to high levels of auxiliary sources of alkalinity (sodium silicates and/or sodium carbonate). These products are all dry powders for reasons of machine design and product bleach stability during the period between manufacture and eventual use.

Abrasive cleaners are used to remove soils and stains from hard surfaces which are durable to the scouring action. Such surfaces include stainless steel and porcelain plumbing fixtures, metal and ceramic cooking utensils, and various stone, metal, and ceramic building surfaces. Typically, these products consist of a very high level of abrasive (commonly silica flour) with moderate to low levels of a dry chlorine bleach (KDCC or chlorinated trisodium phosphate) and low levels of surfactant (LAS) and builder (STP) for wetting action and improved stain removal.

Other hard-surface cleaners are formulated to clean larger surface areas which do not require or are less resistant to the action of an abrasive cleanser or from which the abrasive would be difficult to remove, such as floors, walls, woodwork, and large appliances. These products are sold in both liquid and powder form. The liquid products contain low levels of nonionic and anionic surfactants; moderate levels of a stable, highly soluble builder (commonly tetrapotassium pyrophosphate, $K_4P_2O_7$) for better stain removal; solvents; hydrotropes; and water of processing. The powdered products contain a very low level of surfactant (LAS), a moderate amount of builder (STP), and substantial amounts of sources of mild alkalinity (trisodium phosphate and mixtures of sodium carbonate and sodium bicarbonate) for improved soil removal. Low sudsing is a requirement for both types of products to aid in rinsing the surfaces after cleaning.

Personal-Care Products: Within the broad definition of synthetic detergents a variety of cleansing products are made for personal care. These include such products as cleansing bars, shampoos, bubble-bath products, cosmetic cleansers, and toothpastes. Formulations of these products vary widely, depending upon their intended use.

Although essentially pure soap products continue to dominate the cleansing-bar field, a few products contain synthetic surfactants in addition to soap to act as scum and curd dispersants. Synthetic surfactants used in this application include alkyl sulfates, alkyl glyceryl ether sulfonates, alkyl esters of sodium isethionate, and alkylamides of N-methyl tauride. Shampoos are commonly formulated in liquid, paste, and gel form and usually consist of high-sudsing anionic surfactant(s) (such as LAS and those listed immediately above for bars), along with specific ingredients for improved hair health or control (such as antidandruff agents and substantive collagen proteins). Bubble baths are provided in both liquid and powder form and commonly provide a high-sudsing surfactant(s). Cosmetic cleansers vary widely in formulation, depending upon intended use. Some provide mixtures of surfactants and heavy mineral oil (cold cream), while others provide an organic solvent (for mascara removal, for example). Toothpastes provide a low level of an anionic surfactant (several types used) and high levels of a moderate abrasive (insoluble pyrophosphates), along with other special ingredients such as anticaries agents.

Other Products: In addition to the products discussed above, many other products, based all or in part on synthetic detergents, are offered for specific household uses. These include, for example, the rug cleaners, automobile cleaners, scouring pads, and pet-care products.

Industrial and Institutional: In addition to the products offered for personal or household consumption, a wide variety of products based wholly or partially on synthetic detergents are also manufactured for industrial and institutional use. Included among these are products which are essentially direct analogies of the various household products and are used to perform similar functions by industry and institutions. In addition, however, there is a wide array of products which have no counterparts among the household detergents and which are manufactured specifically to meet special industrial requirements. Included among these would be, for example, detergents to scour raw yarns in the textile manufacturing process, detergents to clean metals prior to painting or electroplating, surgical preparation products, detergents to clean and disinfect poultry houses, and many more. Formulations of these products vary even more widely than do those of the household products, again reflecting the intended special use. Currently, the industrial and institutional products account for some 500 million lb of detergent industry production each year.

Manufacture of Detergents. The processes used for the manufacture of finished synthetic detergent products are relatively elementary in terms of the chemical engineering unit operations involved. They are quite complex, however, with respect to the scale of operations and the careful specification of operating conditions required to avoid unwanted interactions between ingredients. See Fig. D-6.

Powdered Detergents: Powdered detergents are commonly made using one of two processes: spray-drying, in which an essentially finished composition is blown in the spray tower, and simple mechanical mixing of dry ingredients, onto one or more of which is sprayed any liquid or waxy ingredient (such as surfactant). Of these two processes spray-drying is the more important, probably accounting for more than three-quarters of industry powder production (and the detergent producers represent by far the largest industrial users of the spray-drying process).

The spray-drying process for an anionic detergent commonly starts with the sulfation and/or sulfonation of anionic surfactant intermediates—fatty alcohol and/or linear alkyl benzene. This is accomplished through the introduction of oleum and the intermediates into a special acid-resistant alloy pump which acts as a simple mixer for the raw materials, as a reaction vessel, and as a pump for circulating the reaction mix. Heat of reaction is dissipated by introducing the surfactant intermediates into a much larger mass (perhaps 1:10) of recirculating externally cooled reaction product just prior to the sulfator. Key control variables to ensure completeness of reaction are materials temperature (high enough for pumpable viscosity and for rapid reaction, low enough to avoid charring the materials), the ratio of oleum

Fig. D-6. Detergent tower complex. Detergents are pumped to the top of the tower in liquid form and dried into crisp granules as the solution is broken up and dropped into a rising column of heated air. (*The Procter & Gamble Co.*)

to surfactant intermediate (excess acid required), mixing efficiency, and reaction time.

Following sulfation/sulfonation, the acid forms of the surfactants are neutralized with sodium hydroxide to form the sodium surfactant salts present in the final detergent product. Sodium sulfate is also formed from neutralization of the excess sulfuric acid and passes on into the finished product. As in the sulfation/sulfonation reaction, the heat of neutralization is removed by mixing the reactants into a larger externally cooled recirculating mass of neutralized material. Neutralized surfactant paste is passed through a cooler to a holding tank for introduction into the next stage of the process.

In the next stage the major detergent ingredients—surfactant paste, builder, and corrosion inhibitors—are mixed together into a thick paste in a large closed tank with a worm screw agitator, with the tank mounted on scales for accurate measurement of each ingredient. This mixer is called a *crutcher* in the industry, and the mixing process, *crutching*. Crutching remains commonly a batch operation. Consequently, the mixed materials pass from the crutcher into large tanks which can hold the output of two to three crutchers, so that the subsequent spray-drying operation can be operated continuously. Contents of the hold tank are also agitated to maintain uniform composition.

From the hold tank the mixed detergent paste is delivered by high-pressure pumps to the atomizing nozzles of a spray-drying tower under a pressure of 400–700 psi. Air may also be injected just prior to the nozzles to aid in puffing the droplets in the tower. Droplets exiting from the nozzles into the top of the tower are puffed into granules and dried by a current of hot [about 600°F (316°C) at entry] air flowing through the tower (countercurrently or concurrently, depending upon design). Spray-tower sizes vary widely within the industry. Some of the newer, larger towers are as large as 25 ft diam and 100 ft tall.

Granules exit from the bottom of the tower and are carried away by conveyor belts. The product may then be cooled by contact with ambient air, and then minor ingredients (such as perfume) which are too sensitive to pass through the spray-drying operation are admixed or sprayed on. The finished granules are then transferred by various handling systems to the finished-product packing lines.

Mechanically mixed powders are prepared through simple admixing of dry-detergent raw materials, commonly in large agitated drums. Any liquid or waxy components, such as the surfactant, are simply heated (if necessary) to obtain a liquid of proper viscosity and sprayed onto one or more of the dry ingredients as a carrier. One or more mixing operations with drying may be required to obtain the complete base formula, following which colorants or perfume may be sprayed on as the product passes toward final packing.

Liquid and Paste Detergents: Liquid detergents are commonly made through simple mixing of component raw materials in batch mixers of varying sizes and agitation designs. As in the spray-drying of powdered detergents, this mixing step may be preceded by sulfation of surfactant intermediates, or purchased finished surfactants may be used. Mix temperature, mixing time, agitation rate (shear) and configurations, and selection of appropriate hydrotropes are key process control factors.

Paste or gel detergents are either products with a high surfactant level which may naturally assume the paste form through the properties of the surfactant, or other more liquid formulas to which gelling agents or thickeners have been added. All detergent ingredients are commonly mixed at the time the paste is formed or, less frequently, worked into the paste by subsequent mixing equipment.

Further information on detergents will be found under **Alcohols, Alkylation, Ethanolamines, Ion-exchange resins, Petrochemical complex, Silicates (soluble),** and **Surfactants.**

REFERENCES

John W. McCutcheon, Inc.: Detergents and Emulsifiers, Annual Report, 1971, Morristown, N.J.

Schwartz, Anthony M., James W. Perry, and Julian Berch: "Surface Active Agents and Detergents," vol. 2, Interscience-Wiley, New York, 1958.

Shreve, R. Norris: "Chemical Process Industries," pp. 543–559, McGraw-Hill, New York, 1967.

—R. N. Wilkinson, *Packaged Soap and Detergent Division, The Procter & Gamble Company, Cincinnati, Ohio*

Detonation Testing   (See **Explosives.**)

Deuterium   Deuterium is the hydrogen isotope having mass number 2 (2.014740 amu). Deuterium is one form of heavy hydrogen, the other form being tritium. Deuterium is a stable isotope whereas tritium is not. The atomic abundance of deuterium in natural hydrogen (ocean-water reference) is 0.0149%. See also **Chemical elements.**

Deuteron   Deuteron is the nucleus of deuterium (heavy hydrogen). Also, a particle that contains one proton and one neutron.

Development Chromatography   (See **Chromatography.**)

Devitrification   (See **Ceramics.**)

Dewatering   (See **Expression;** and **Filtration.**)

Dextrorotatory Compounds   (See **Isomerism.**)

Diacetylenes   (See **Hydrocarbons.**)

Diacyl Peroxides   (See **Peroxides, organic.**)

Dialysis   (See **Adsorption; Desalination; Peptides and proteins;** and **Separation operations.**)

Diammonium Phosphate   (See **Fertilizers.**)

Diamond (See **Abrasives;** and **Carbon.**)

Diaphragm Cell (See **Caustic soda.**)

Diatomaceous Earth (See **Abrasives;** and **Adsorption.**)

Diazotization Diazotization is a reaction between a primary aromatic amine and nitrous acid to yield a diazo compound. In practice, an aqueous solution of $NaNO_2$ and a dilute mineral acid solution of the amine may be used. $ArNH_2 + NaNO_2 + 2HX \rightarrow ArN_2X + 2H_2O + NaX$. Few primary aromatic amines resist diazotization. The amino groups on benzene, naphthalene, and their substitutional derivatives may be diazotized, as well as heterocyclic amines, such as aminothiazoles and aminopyridines. Although used in other organic chemical syntheses, the coupling of diazo compounds with phenols, naphthols, and amines to form azo dyes is the most important commercial application of this reaction. See also **Dyestuffs and dyeing (textiles).**

Dibasic Acids (See **Carboxylic acids.**)

Dicarboxylic Acids (See **Carboxylic acids.**)

Dichlorodifluoromethane (See **Chlorofluoro-carbons.**)

Dichloroethyl Ether (See **Chlorine organics.**)

Diels-Alder Reaction (See **Butadiene;** and **Furan group.**)

Diene Monomers (See **Polymerization.**)

Diesel Fuel (See **Hydrocracking;** and **Petroleum processing.**)

Diethylene Glycol (See **Ethylene glycol.**)

Diffusional Operations Diffusion, the property of molecules and particles to migrate and intermingle as the result of random forces, which occurs in gases and solutions and, in a slower fashion, in solids, underlies several chemical engineering unit operations, notably absorption, adsorption, crystallization, dialysis, distillation, drying, electrodialysis, freeze-drying, gaseous diffusion, humidification, ion exchange, leaching, molecular distillation, solvent extraction, sublimation, and thermal diffusion.

When a homogeneous material—either gas, liquid, or solid—contains two or more components whose concentrations vary from point to point, there is a tendency for transfer of mass to take place in such a way as to cause the concentrations to become uniform. This phenomenon is associated with the thermal agitation of molecules. In a region where molecules of one kind are concentrated, there is a greater tendency for molecules of this kind to escape than to enter the region. The net rate of diffusion of material A, N_A, at a point in a stationary fluid, is found from experiment as well as from theory to be proportional to the concentration gradient at the point

$$N_A = -D_v \frac{\delta c}{\delta s}$$

where c = concentration, s = distance, and D_v = diffusivity. If c is expressed in g moles/cm^3, s in cm, and D_v in cm^2/s, then the units of N_A are g moles/(s)(cm^2). The rate of diffusion is rapid in gases and much slower in liquids.

Diffusion does not lead to conditions of constant concentration gradient unless a steady state is established. It is therefore often necessary to consider the change of concentration c with time t caused by diffusion as represented by the differential equation

$$\frac{\delta c}{\delta t} = D_v \frac{\delta^2 c}{\delta s^2}$$

The study of diffusion, along with viscosity and thermal conduction in gases, is a part of the well-developed kinetic theory of gases. The most advanced mathematical treatments predict that the diffusion coefficient should vary only slightly with composition, and this has been confirmed experimentally. The variation is greatest when the ratio of the masses of the molecules is large, but in no case has the maximum variation been found to exceed 13%.

The theory of diffusion in liquids is less well developed than for gases. Diffusion coefficients in liquids are typically about four orders of magnitude smaller than in gases, and are therefore more difficult to measure accurately. There are relatively few measurements of liquid diffusivities outside the temperature range 0–40°C. Diffusion coefficients of electrolytes can be predicted accurately at infinite dilution using the equation[1]

$$D_0 = 8.931 \times 10^{-10} T \frac{l_+^0 l_-^0}{\Lambda^0} \frac{z_+ + z_-}{z_+ z_-}$$

where D_0 = diffusivity of molecule, cm^2/s

l^0_+ = cationic conductance at infinite dilution, mhos/equivalent

l^0_- = anionic conductance at infinite dilution, mhos/equivalent

$\Lambda^0 = l^0_+ + l^0_-$ = electrolyte conductance at infinite dilution, mhos/equivalent

T = absolute temperature, K

z_+ = valence on cation (absolute, i.e., no sign)

z_- = valence on anion (absolute, i.e., no sign)

For determination of diffusivities at other than infinite dilution, the following relationship may be used:[2]

$$D_L = D_0 \left(1 + \frac{m\delta \ln \gamma_\pm}{\delta m} \right) \frac{1}{C_B \bar{V}_B} \frac{\mu_B}{\mu}$$

where m = molality

$\gamma_\pm$ = mean ionic activity coefficient based on molality[3]

C_B = g moles water/cm^3 solution

$\bar{V}_B$ = partial molal volume of water, cm^3/g mole

μ_B = viscosity of water

μ = viscosity of solution

In the case of diffusion with flow, if fluid motion is laminar, transfer of mass between adjacent layers of fluid takes place purely by molecular diffusion. If the velocity pattern of the flow is known, it is sometimes possible to calculate the overall rate of mass transfer into the moving fluid by the use of the basic equations of molecular diffusion. If the flow is turbulent, such calculations are generally difficult, since the laws that govern the transport of matter by turbulent mixing of small volumes of fluid are not well understood. Prediction of mass-transfer rates under such conditions frequently is based upon empirical methods.

REFERENCES
1. Nernst, *Zh. Physik. Chem.*, vol. 2, p. 613, 1888.
2. Gordon: *J. Chem. Phys.*, vol. 5, p. 522, 1937.
3. Glasstone, "Thermodynamics for Chemists," p. 402, Van Nostrand, New York, 1947.

Digesting, Pulp [See **Pulp (wood) production and processing.**]

Dilatant Fluids (See **Colloid systems.**)

Dimer (See **Polymerization.**)

Dimethyl Sulfoxide (See **Toluene.**)

Dimethyl Terephthalate (DMT) (See **Polyester fibers.**)

Diodes (See **Gallium; Germanium;** and **Silicon.**)

Diolefins (See **Hydrocarbons.**)

Dipentene (See **Terpenes.**)

Diphenylamine (See **Amines.**)

Dipolar Ions (See **Amino acids.**)

Dipropargyl (See **Isomerism.**)

Direct Dyes [See **Dyestuffs and dyeing (textiles).**]

Direct Reduction Process (See **Ironmaking and steelmaking.**)

Dishwashing Compounds (See **Detergents.**)

Disk Filter (See **Filtration.**)

Disk Mills (See **Mixing and blending, solids/solids;** and **Size reduction.**)

Disperse Dyes [See **Dyestuffs and dyeing (textiles).**]

Dispersing Agents [See **Anionic surfactants (sulfur-bearing); Colloid systems; Defoaming agents;** and **Surfactants.**]

Dispersions (See **Colloid systems; Mixing, fluids;** and **Paint.**)

Dissolved Air Flotation (See **Waste treatment.**)

Distillates (See **Fuels;** and **Petroleum processing.**)

Distillation The components of a liquid mixture can be separated by partially vaporizing the mixture and separately recovering the vapor and residue. As the operation proceeds, the lighter, more volatile components of the original mixture (*distilland*) concentrate in the vapor, while the less volatile components concentrate in the liquid residue (*bottoms*). Because of the similarity, it is well to emphasize that in distillation the vapors evolved are recovered by condensation, whereas in evaporation, the vapor (frequently water) is discarded, the solid residue being the prime target of the separation. The condensed vapors are termed the *distillate*.

The purification of materials by distillation is a prominent operation of numerous chemical processes. The problems of distillation control relate directly to the *relative volatility* of the materials being separated. For example, water and ethylene glycol are separated easily because of the high volatility of water relative to ethylene glycol. Similarly, water and methanol are easily separated. Closer-boiling mixtures, such as the isomers of xylene, are much more difficult to separate by distillation.

Basic Configurations: Flashing is a form of distillation in which the total vapor removed is in phase equilibrium with the residue liquid.

Rectification (or fractional distillation) involves the return of a portion of the condensed vapors (*reflux*) to countercurrently contact the rising vapors. This brings about an enrichment of the vapor in the more volatile components than otherwise would be accomplished with a single distillation and generally obviates the need for one or more redistillations to achieve the degree of separation desired. In rectifying columns, the feed usually is introduced at about the center or midlevel of the column or tower. The section above the feed entry is referred to as the *rectifying section;* that portion below is termed the *stripping section.* Where feed is added at the top of a column, the entire column is referred to as a *stripping column,* and with this arrangement, no reflux is used.

Dephlegmation is partial condensation and refers to the operation wherein a vapor stream is cooled to a desired temperature such that a portion of the less volatile components of the stream is removed from the vapor by condensation.

Steam distillation is a simple distillation where vaporization of the charge is achieved by blowing live steam directly through it. The practice has special value where it is desired to separate substances at a temperature lower than their normal boiling points because of heat sensitivity.

Some mixtures cannot be separated by conventional distillation methods because (1) they are of low volatility or (2) they contain a homogeneous azeotrope. An azeotropic mixture is one in which, over a certain composition range of two components, one component will be more volatile than the other and, after a point of convergence is exceeded, the volatility order of the two components will be reversed. These situations require *extractive,* or *azeotropic, distillation.*

Distillation may be *batch* or *continuous,* the latter comprising the majority of installations. In batch distillation, the charge material is boiled and vapors are removed continuously, condensed, and collected until such point is attained where there is the desired average composition. In continuous distillation, the rate of feed, removal and condensation of vapors, return of reflux, removal of bottoms, temperatures, and pressures, all are carefully controlled to maintain the column in an overall equilibrium condition. Such columns may operate for many months without downtime. Although much of the theory behind distillation-column design stems from studies of binary systems, many production distillation operations involve three or more components in the feedstock.

Distillation Column Control. * Whether a mixture is fundamentally easy or difficult to separate, the distillation equipment used generally is designed for a particular separation, with only modest margins of performance ability provided. Thus it is the responsibility of the distillation control system to safeguard the purity of distilled products. Although references are made here to the control of composition, temperature, pressure, and level, all these simply are instrumental methods for maintaining *purity of product.* Although some distillation control systems regulate productivity and effect economies, it is difficult to suggest many examples where composition of product is made subordinate to other control objectives. See **Desalination.**

The more elusive aspects of distillation control relate to control of product composition and its responses to disturbances, such as feed rate, feed composition, or feed enthalpy—factors that originate external to the distillation column. The McCabe-Thiele diagram originally was devised as an aid to the design of distillation columns. This device is helpful toward visualizing the effects of disturbances, several of which are described here.

Degrees of Freedom for Control: The concept of *key* component separation represents a means of treating a complex separation in terms of a simple binary equation. It is important to recognize the degrees of freedom of choice in specifying the variables to be controlled. The choice open is the natural consequence of the column heat and material balance and is analyzed here.

A typical distillation column is shown in Fig.

*Much of this description is extracted from L. Bertrand and J. B. Jones, Distillation-Column Control, in Douglas M. Considine (ed.), "Handbook of Applied Instrumentation," McGraw-Hill, New York, 1964.

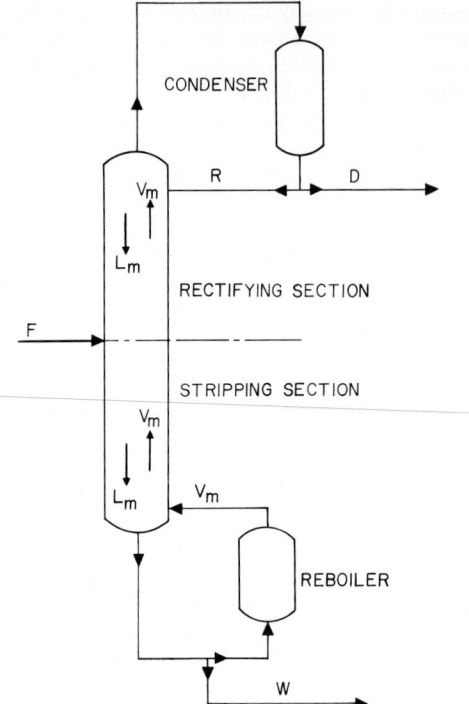

Fig. D-7. Typical distillation column. $F =$ feed rate, $W =$ bottom product, $D =$ distillate, $R =$ external reflux, $L_m =$ liquid rate in rectifying section, $V_m =$ vapor rate in stripping section, all in lb-moles/unit of time.

D-7. The overall material balance can be represented by

$$F = W + D \tag{1}$$
$$Fx_F = Wx_W + Dx_D \tag{2}$$

where $F =$ feed rate, lb-moles/unit of time
$W =$ bottom product, lb-moles/unit of time
$D =$ distillate, lb-moles/unit of time
$x_F =$ mole fraction of low boiler in the feed
$x_D =$ mole fraction of low boiler in the distillate
$x_W =$ mole fraction of low boiler in the bottom product

The assumption is made that the molar heat capacities and the latent heats of vaporization of all components are the same. Also, it is assumed that (1) heat losses from the column and (2) heats of mixing are both negligible. In consequence, the upward vapor flow and the downward liquid flow in both the rectifying and stripping sections are invariant within the sections. Also, the account-

ing for the column heat balance is independent of the compositions of the product streams.

Internal Material Balance: With these qualifications, the internal material balance and the heat balance can be accounted for by the following equations:

$$L_n = (1 + b)R \tag{3}$$
$$V_n = D + (1 + b)R \tag{4}$$
$$L_m = L_n + qF \tag{5}$$
$$V_m = L_m - W \tag{6}$$
$$x_W = f\left(\frac{L_m}{V_m}\right) \tag{7}$$
$$x_D = g\left(\frac{L_n}{V_n}\right) \tag{8}$$

where $V_n =$ vapor rate in rectifying section, lb-moles/unit of time
$V_m =$ vapor rate in stripping section, lb-moles/unit of time
$L_n =$ liquid rate in rectifying section, lb-moles/unit of time
$D =$ distillate rate, lb-moles/unit of time
$R =$ external reflux, lb-moles/unit of time
$b =$ a numerical factor, depending upon the reflux enthalpy or temperature (*Note:* $b > 0$ whenever the reflux temperature is below that at the top of the column.)
$q =$ a numerical factor, depending on the feed enthalpy whose value satisfies the following constraints:
$\quad q > 1 \qquad$ when the feed temperature is below that of the feed plate
$\quad q = 1 \qquad$ when the feed temperature and composition are identical with those of the feed plate
$\quad 1 > q > 0$ when the feed enters the column partially vaporized
$\quad q = 0 \qquad$ when the feed is completely vaporized and is at saturated temperature
$\quad q < 0 \qquad$ when the feed is superheated vapor

f and g = factors which account for functional relationships which are dependent upon column-design criteria, such as (1) number of plates in the column, (2) location of control plates, (3) location of feed plate, and (4) temperature and other criteria specified for the control plates

Fixed Controls: From the standpoint of controls, the foregoing factors are fixed. For example, the automatic controls are unable to shift the injection of feed up or down the column in response to departures from optimum column operation. Nor is the location of the control plates ordinarily regarded as a variable. Even the control-plate temperatures usually are considered fixed, although it is possible to regard these as variable and arrange feedback controls for their regulation.

The consequence of fixing all factors on which the functionals f and g depend is to fix the functionals in turn. Thus these items are excluded from the list of variables in the analysis which follows.

Free, Controllable, and Dependent Variables: Excluding f and g, Eqs. (1)–(8) include 13 variables. One possible classification of the 13 variables is given here.

FREE VARIABLES (usually not controllable): x_F, q, and b

CONTROLLABLE VARIABLES (whose values usually are controlled): F and x_D, or (alternatively) F and x_W

DEPENDENT VARIABLES: W, D, V_n, V_m, L_n, L_m, and x_W, or (alternatively), x_D

Thus 5 of the 13 variables, i.e., the free and controllable variables, are defined. This leaves 8 unknowns in 8 independent equations. There exists a unique solution for this set, and thereby the column performance is completely determined.

Possible Modes of Column Control: It is necessary to assign values to 5 of the 13 variables in order to reduce the system to 8 equations in 8 unknowns. Usually, some of the 5 variables represent conditions imposed by preceding process steps, such as feed composition, feed rate, and feed enthalpy. Usually, the overhead or the bottoms composition is specified. It is not permissible to overdefine the system; i.e., the sum of the *free* and *controllable* variables must equal 5. Conversely, the *dependent* variables must equal 8. Within this framework, there are a number of possible combinations that characterize different modes of column control. Several of these are shown in Table D-2.

Composition Measurement: Distillation is invariably concerned with composition control, of either the distillate or the bottoms product, or at times both. Because composition bears an unvarying relationship to boiling temperature (assuming fixed pressure), it is natural that product composition usually is controlled via temperature.

There are other possible methods of control, and they may assume more importance in the future. For example, product-stream composition could be measured continuously by other physical measurements, such as vapor-phase chromatography, ultraviolet photometry, or infrared photometry.

Differential-Vapor-Pressure Measurement: The

TABLE D-2. *Possible Modes of Distillation-Column Control*

Distillate and Feed	Distillate and Bottoms	Distillate and Boil-up	Distillate Bottoms and Boil-up	Bottoms and Boil-up	Bottoms Boil-up and Reflux	Feed-rate Distillate and Boil-up	Distillate Bottoms and Boil-up	Distillate Bottoms and Feed Enthalpy
Free Variables								
x_F, q, b	x_F, q, b	x_F, q, b	x_F, b	x_F, q, b	x_F, b	x_F, b	F, x_F, b	x_F, b
Specified Variables								
F, x_D	x_D, x_W	x_D, V_m	x_D, x_W, V_m	x_W, V_m	x_W, V_m, L_n	x_D, F, V_m	x_D, x_W, V_m	x_D, x_W, q
Dependent Variables								
$W, D, R,$ $V_m, L_m,$ V_n, L_n, x_W	$W, D, R,$ $V_m, L_m,$ V_n, L_n, F	$W, D, R,$ $L_m, V_n,$ L_n, x_W, F	$W, D, R,$ $L_m, V_n, L_n,$ F, q	$W, D, R,$ $L_m, V_n,$ $L_n, x_D,$ F	$W, D,$ $L_m, V_n,$ $x_D, F,$ q	$W, D, R,$ $L_m, V_n, L_n,$ q	$W, D, R,$ $L_m, V_n,$ L_n, q	$W, D, R,$ $L_m, V_m,$ V_n, L_n, F

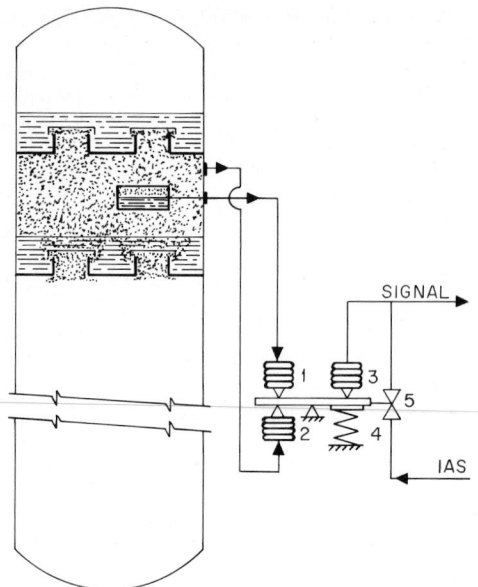

Fig. D-8. Differential-vapor-pressure type of temperature-measuring system used in distillation column control. (1) Bellows to sense differential pressure of sensor located in column between trays, (2) bellows to sense vapor pressure in column adjacent to the sensor, (3) balancing bellows, (4) restraining spring force, (5) pilot valve. *IAS* = instrument air supply.

tem shown affords a sensitive index of composition in the column as referred to the fixed composition in the bulb, which is independent of minor variations in absolute operating pressure.

Composition Control Systems Based on Temperature. The following systems are based on the use of temperature to indicate the composition of material. The techniques described, however, are equally applicable to control loops based on other composition-measurement methods. Regulation of product composition normally is accomplished by manipulating boil-up and/or reflux rates which change the relative rates of vapor and liquid flow throughout the column.

Distillate- and Bottoms-Composition Regulation: The control systems shown in Figs. D-9 to D-11

actual boiling temperature at the temperature-control point will change with changes in column pressure. This can be troublesome, particularly in low-pressure columns where minor variations in absolute pressure are substantial compared with the operating pressure.

A differential-vapor-pressure type of temperature-measuring system avoids the problem. Such a system is shown in Fig. D-8. To one of two opposing bellows is connected a thermometer bulb filled with liquid of the composition desired at that point in the column. When subjected to column temperature, this bulb develops an internal pressure dependent upon the temperature and the composition of liquid in the bulb. Simultaneously, the liquid in the column exerts a pressure dependent upon the temperature and the liquid in the column. Since the bulb and column contents are at the same temperature, the pressures exerted in the column and in the bulb are each a function of the liquid composition, i.e., that in the bulb and that in the column. Comparing the two pressures in the differential force-balance sys-

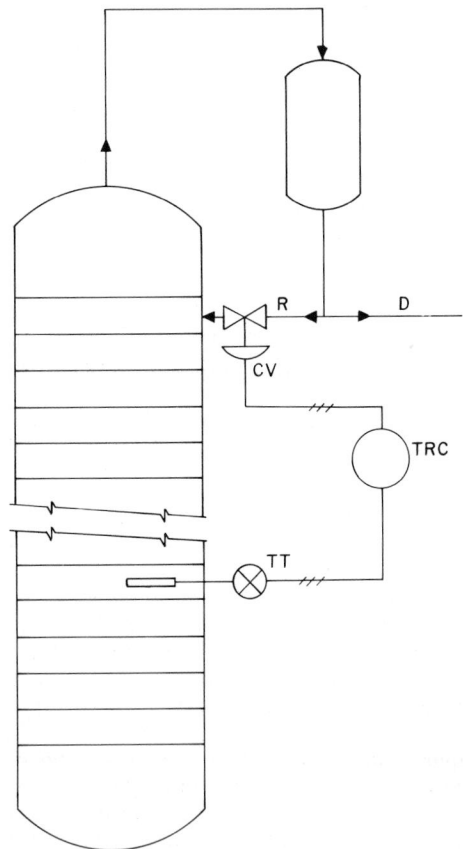

Fig. D-9. Distillate-composition control based upon temperature-sensing element located in column. R = reflux, D = distillate, TT = temperature transmitter, TRC = temperature recorder-controller, CV = control valve.

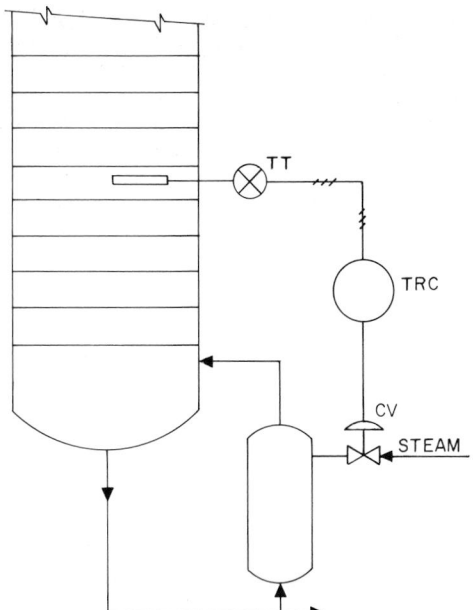

Fig. D-10. Bottoms-composition control by regulation of boil-up.
TT = temperature transmitter, TRC = temperature re-corder-controller (which indirectly controls composition), CV = control valve.

ature and purge the accumulated impurity (the so-called *pasteurization column*). Unless the accumulation is minor, however, it would be uneconomical to discard the purge without redistillation. An alternative is to allow the intermediate boiler to be discharged with either bottoms or distillate and provide an additional column for its separation if justified. The final choice rests, of course, on a comparison of the economics of the methods.

The foregoing description was based upon considerations of binary mixtures contaminated with intermediate-boiling impurities. It applies equally to the separation of key components from multi-component mixtures.

Location of Composition Temperature-sensing Element: Common to both distillate- and bottoms-temperature controllers is the fact that the pertinent temperature measurement is made, not on the product directly, but in the column a number of plates away. Usually, a product stream that is relatively pure is being separated from the binary mixture, so that boiling point or any other quality test of the product is often insensitive to

affect product composition only. Other aspects of distillation-column control are described later. In Fig. D-9, distillate composition, as measured by a temperature element in the column, is controlled by regulation of reflux. In Figs. D-10 and D-11, bottoms composition is controlled by regulation of either boil-up (Fig. D-10) or feed rate (Fig. D-11).

Relationship of Reflux and Boil-up: It would appear advantageous to maintain both reflux and boil-up on automatic control. If a simple binary separation is being controlled, this presents no problem, but rarely is this the case. If the feed contains an impurity having a volatility between those of the distillate and bottoms, automatic control on both reflux and boil-up presents problems. The reflux control operates to return the impurity down the column, while the boil-up control operates to send the impurity up the column. The result is that the intermediate boiler accumulates in mid-column and displaces the composition to such an extent that separation fails.

It is possible to devise controls, either continuous or intermittent, to sense mid-column temper-

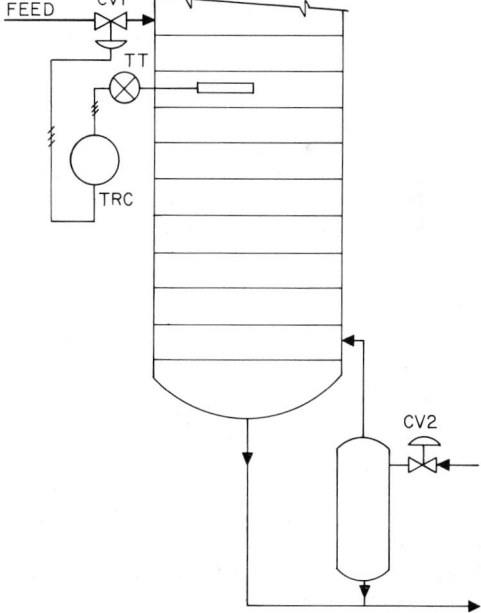

Fig. D-11. Bottoms-composition control by regulation of feed rate. TT = temperature transmitter, TRC = temperature recorder-controller, $CV1$ = feed control valve which acts in response to signal from TRC, $CV2$ = constant-steam-flow control valve.

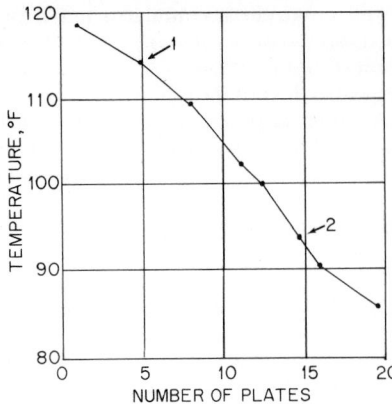

Fig. D-12. Typical temperature profile in a distillation column.
(1) Thermocouple for distillate control, (2) thermocouple for bottoms control.

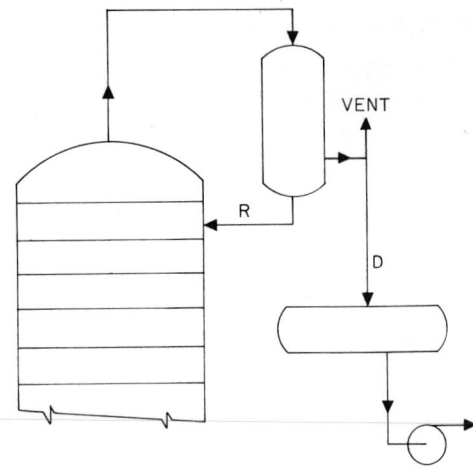

Fig. D-13. Distillation-column inventory-control system where column operates at atmospheric pressure, showing regulation of distillate. R = reflux, D = distillate.

changes in its concentrations. Location of the temperature-sensing element some plates away makes it possible to obtain a greater change in temperature for a fixed change in final composition. Fixing the composition at such a point in the column suffices to control the column-product composition within narrow limits, even with wide variations in other factors, such as vapor and liquid flows.

Selecting the optimum location can be done based on the column temperature profile, which the distillation engineer should provide. Figure D-12 shows a typical temperature profile in a distillation column. As indicated, the temperature-sensing element should be located where the temperature profile is steep, but not too far removed from the end of the column. The rectifying-section temperature measurement must be located above the feed plate; the stripping-section temperature measurement, below the feed plate.

Column-Inventory Control Systems. Distillation columns provide little or no surge capacity, so that it is necessary to remove the distillate and bottoms as fast as they accumulate. Controls for product drawoff should be no more complicated than absolutely necessary. In columns operating at atmospheric pressure, loop-sealed overflows may be all that is required. At positive pressures, level-controlled letdown valves are suitable. For vacuum operation, discharge pumps in combination with liquid-level-controlled valves are required.

Atmospheric Pressure: Figures D-13 and D-14 show systems for columns operating at atmospheric pressure with inventory control by loop-sealed

overflows on distillate and bottoms, respectively.

Positive Pressure: Figures D-15 and D-16 show systems for columns under positive pressure where liquid-level controllers are used.

Vacuum: A typical system for a column under vacuum is shown in Fig. D-17.

Low-Product Flow: Where the feed composition is very low or very high in low-boiler content (i.e., as x_F approaches 1.0 or 0), controlling the flow of the minor product stream can be difficult because of the low rates of flow involved. In such cases, a high-low level control system can be used, allowing the product to accumulate in a receiver

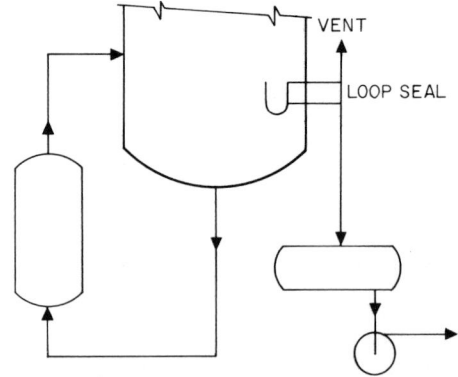

Fig. D-14. Distillation-column inventory-control system, showing loop-sealed overflow on bottoms. Column operates at atmospheric pressure.

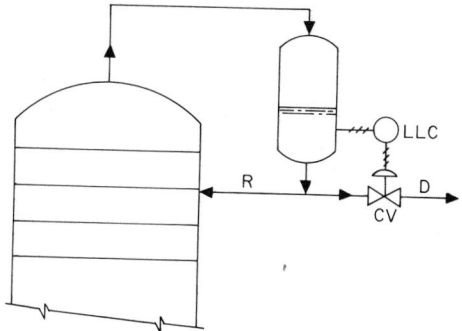

Fig. D-15. Distillation-column inventory-control system where column operates under positive pressure. R = reflux, D = distillate, LLC = liquid-level controller on condenser, CV = control valve.

until the upper limit of level is reached. Then the system opens a valve fully to discharge until the lower limit of level is reached. For pressure or vacuum operation, Fig. D-18 shows a typical control system employing a float switch with magnetic pickup to sense the high- and low-level points and to actuate electrically the discharge valve through a solenoid valve on the air-to-valve line. Note that, in the case of vacuum operation where a discharge pump is used, a sequence controller is employed to start the pump. Figure D-19 shows the use of an intermittent siphon to accomplish the high-low level control on a column operating at atmospheric pressure.

Pressure Control of Columns. Distillation systems invariably are designed for uniform-pressure operation, ranging from low vacuum through atmos-

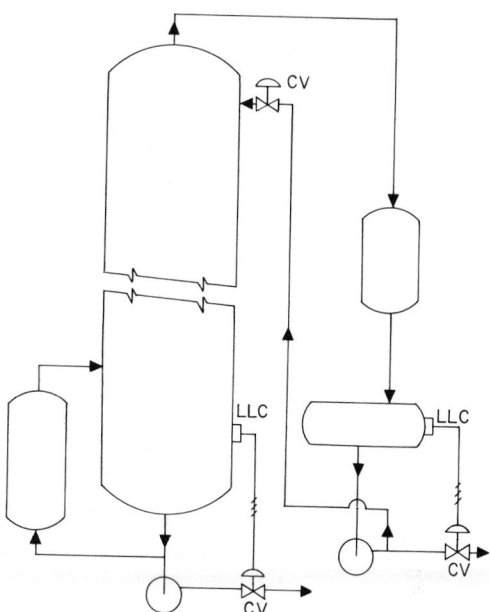

Fig. D-17. Distillation-column inventory-control system where column operates under vacuum. LLC = liquid-level controller, CV = control valve.

pheric to very high pressures. Atmospheric pressure usually presents no problem to the control-system designer. However, exposure to atmospheric contamination may not be permissible, in which case a modification of vacuum

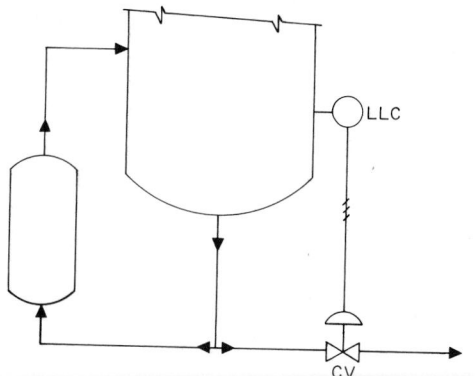

Fig. D-16. Distillation-column inventory-control system where column operates under positive pressure. LLC = liquid-level controller, CV = control valve.

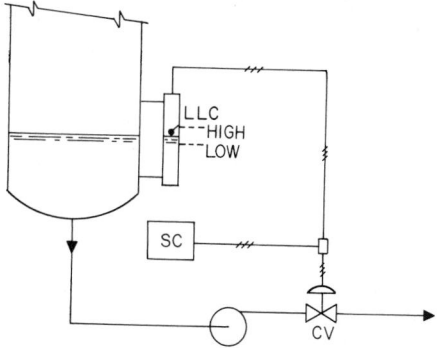

Fig. D-18. Distillation-column inventory-control system where control of flow of minor product stream is difficult. The system, for pressure or vacuum operation, uses a float switch with magnetic pickup to sense high- and low-level points. The pump is used in low-vacuum systems. LLC = liquid-level controller, CV = control valve, SC = sequence controller.

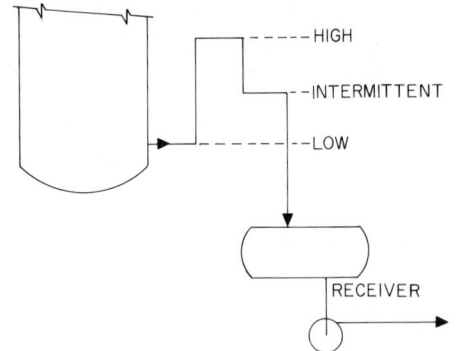

Fig. D-19. Distillation-column inventory-control system where control of flow of minor product stream is difficult. The system, for a column operating at atmospheric pressure, uses an intermittent siphon to effect high-low liquid-level control.

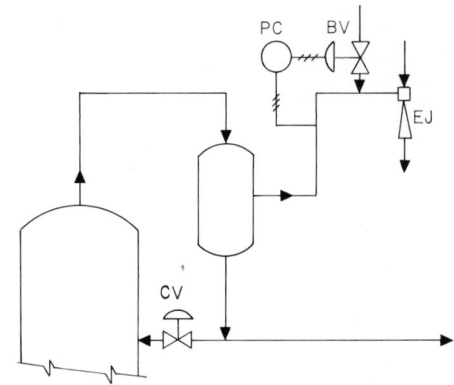

Fig. D-21. Vacuum control system for distillation column. Control is based upon the use of a vacuum-tempering valve ahead of the vacuum source. EJ = ejector, BV = bleed valve, CV = control valve, PC = pressure controller.

control can be used with a positive bleed-in of a suitable inert gas to keep the control active.

Figure D-20 shows a typical pressure control system for a distillation column. Here control is obtained by regulation of the amount of venting on the condenser. Figure D-21 shows a vacuum control system with control based on the use of a vacuum-tempering bleed valve ahead of the vacuum source.

Effect of Disturbances External to Column: Within the material-handling capacity of a given distillation column, i.e., between the excessive loadings at which flooding occurs and the very low loading

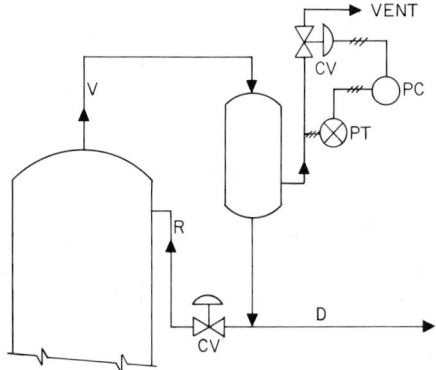

Fig. D-20. Pressure control system for a distillation column. Control is effected by regulation of venting from the condenser. V = vapor, R = reflux, D = distillate, PT = pressure transmitter, PC = pressure controller, CV = control valve.

at which the draindown occurs, the quality of separation depends to an important degree upon the column heat balance. The point at which the heat is applied also is of importance. For example, heat introduced at the base of the column as boil-up operates through a greater number of plates than does heat entering as feed enthalpy. Consequently, it is possible for separation to become poorer if boil-up heat supply is decreased by controls to compensate for an increase in feed enthalpy due to external conditions. External disturbances, therefore, are quite significant, to the extent that they upset column heat balance.

The composition controls are designed to mitigate the effects of such disturbances, but because of their customary location in the column, some number of plates from either the top or bottom of the column, they are deceived to a minor extent in assessing the purity of product and, in fact, to a major extent in assessing quantitative results of external disturbances.

The effects of several typical disturbances are analyzed here. A 15-plate toluene-benzene distillation is chosen as an illustration. Plate-by-plate calculations were made using a digital-computer program. Alternatively, the analysis can be made using the McCabe-Thiele graphical solution. The McCabe-Thiele diagrams are presented because they convey a clearer picture of the effects of disturbances than does a tabulation of column operating data.

McCabe-Thiele Diagram: Figure D-22 represents a typical McCabe-Thiele diagram of a binary distillation. Reference is made to the aforemen-

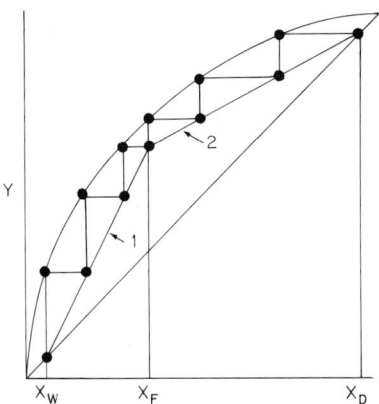

Fig. D-22. McCabe-Thiele diagram for a simple binary distilla-tion. X = mole fraction of low boiler in liquid, Y = mole fraction of low boiler in vapor. (1) Operating line (lower) with a slope of L_m/V_m, (2) upper operating line with a slope of L_n/V_n.

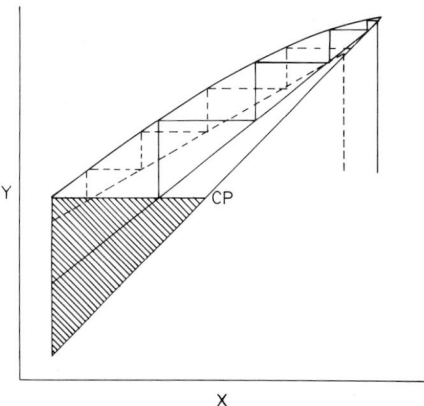

Fig. D-23. McCabe-Thiele diagram showing the effect of a reflux change under distillate-composition control. X = mole frac-tion of low boiler in liquid, Y = mole fraction of low boiler in vapor. CP = control plate.

tioned location of the composition-control tem-perature measurement in the column. The feed-back regulation of the composition on the control plate has the effect of maintaining the composi-tion at that point in the column at a preselected value. Once chosen, this composition is constant. Also fixed is the number of plates above and below the control plate and, of course, above and below the feed plate.

As indicated in Fig. D-22, the product composi-tion corresponds to the intersection of the oper-ating lines with the 45° diagonal, the upper oper-ating line for distillate composition, and the lower operating line for bottoms composition. The effect on the McCabe-Thiele diagram of external column heat-balance disturbances is to shift the location of the operating lines on the diagram.

The McCabe-Thiele diagram construction im-plicitly solves the column heat balance, so that this requirement, in analyzing the effect of heat-balance disturbances on composition control, is accounted for. Two other criteria must be met in the graphical solution of the effect of disturb-ances: (1) as previously noted, the composition on the control plate is held invariant by the com-position controls, and (2) the number of plates and the locations of the feed and control plates remain unchanged.

Effect of Reduced Reflux: Any response of a distillation column to external disturbances must entail shifts in the location of the column oper-ating lines; in fact the slopes will change. More-over, it is clear that the slope changes of the upper and lower operating lines are opposite in sign. The requirement that the control-plate composi-tion and the number of plates between it and the end of the column remain unchanged requires that the construction of the perturbed operating line be accomplished by displacing it about the control plate in such a way that the number of plates between it and the end of the column re-mains unchanged. This displacement of the oper-ating line is illustrated in Fig. D-23, which shows the construction of the upper-column diagram for the effect of reduced reflux.

The constraints fulfilled by the construction are: (1) the slope of the operating line declined, and (2) the graphical solution above the control plate corresponded to number of plates in the upper column above the control plate. Obviously, the decrease in slope of the upper-column line can only cause a drop in distillate purity. A similar analysis holds for the effect of altering the slope of the lower operating line where bottoms-composition control is in operation. Therefore, to estimate the effect of external disturbances on separation, it remains only to determine the slope change on the operating lines and construct the complete diagram, upper and lower, in such a way as to account for the total number of plates and their deployment.

Figures D-24 to D-26 represent the analysis of several typical external disturbances. Although the examples shown were precisely calculated by computer, a graphical analysis by means of the

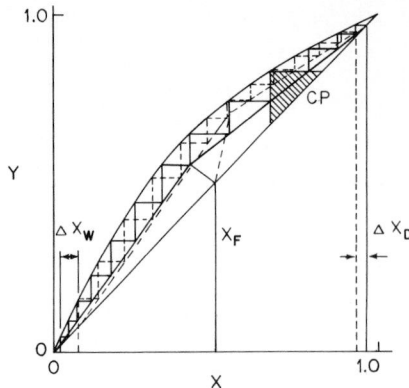

Fig. D-24. Effect of increased feed on a column operating with distillate-composition control of reflux and constant boil-up. X = mole fraction of low boiler in liquid, Y = mole fraction of low boiler in vapor. CP = control plate.

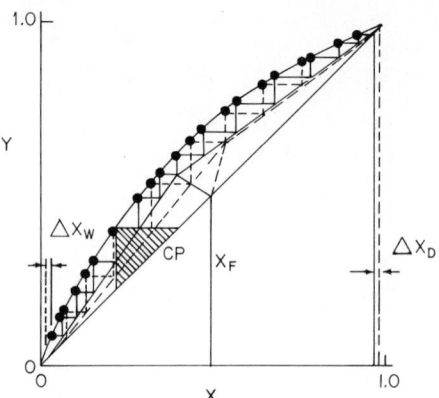

Fig. D-26. Effect of lowered feed enthalpy on a toluene-benzene separation operating with constant reflux and composition control of boil-up. X = mole fraction of low boiler in liquid, Y = mole fraction of low boiler in vapor. CP = control plate.

McCabe-Thiele construction is just as informative and can be obtained in the absence of a digital computer or a plate-by-plate calculation.

Effect of Increased Feed on Distillate Control: Figure D-24 shows the effect of increased feed on a column operating with distillate-composition control of reflux and constant boil-up.

Since the boil-up remains constant and the feed is increased, the ratio of liquid to vapor flow in the stripping section must increase. Consequently, the slope of the lower-column operating line will increase and the bottoms quality will be degraded. Increasing the feed will increase the bottoms rate substantially and the distillate rate

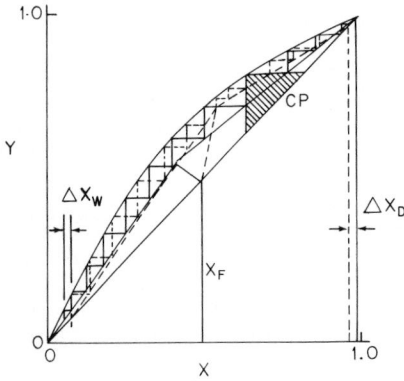

Fig. D-25. Effect of a reflux change under distillate-composition control. X = mole fraction of low boiler in liquid, Y = mole fraction of low boiler in vapor. CP = control plate.

slightly. The increase in distillate will be at the expense of reflux, so that the upper-column operating-line slope will be decreased. Consequently, the distillate quality also will be degraded. It is interesting to note that the column temperature profile below the control plate declines, while that above the control plate increases.

Effect of Decreased Feed Enthalpy on Distillate Control: Figure D-25 shows the effect of decreasing the feed temperature on the toluene-benzene separation operating with constant boil-up and distillate-composition control of reflux. The effects are similar to those resulting from increased feed. The ratio of total heat to feed declines. The slope of the lower-column operating line increases, and that of the upper-column operating line decreases, with a consequential degradation of separation. The column temperature profile shifts downward below the feed plate and increases above it.

Effect of Feed-Enthalpy Change on Bottoms-Composition Control: Figure D-26 shows the effect of feed-enthalpy change on the toluene-benzene separation operating with constant reflux and bottoms-composition control of boil-up. The lower enthalpy feed increases the liquid flow in the stripping section, and the boil-up is increased by the control to make up for the loss in feed enthalpy. The total heat supplied is therefore held unchanged or at worst decreased only slightly. Because a substantial fraction of the heat input is shifted from the feed plate to the bottom of the column, the separation actually improves slightly. The temperature profile above the control plate

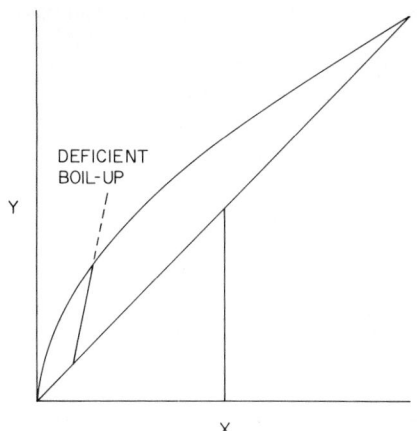

Fig. D-27. Illustration of the cause of flooding when the liquid- and vapor-handling capacity of a distillation column is exceeded. A deficiency of boil-up increases the slope of the lower operating line, thus calling for an unobtainable location of the lower operating line. X = mole fraction of low boiler in liquid, Y = mole fraction of low boiler in vapor.

shifts downward, but below the control plate it shifts to higher temperatures.

Failure of Separation: Any disturbance that increases the demand for reflux or boil-up can cause complete failure of separation if the limits of the column to handle the required liquid or vapor flow are exceeded or if the capacity of the boil-up

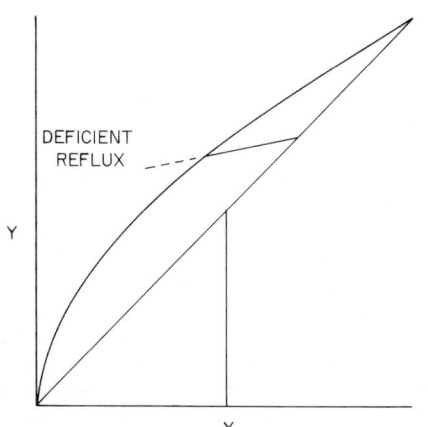

Fig. D-28. Illustration of the effect of a deficiency of reflux which decreases the slope of the upper-column operating line. X = mole fraction of low boiler in liquid, Y = mole fraction of low boiler in vapor.

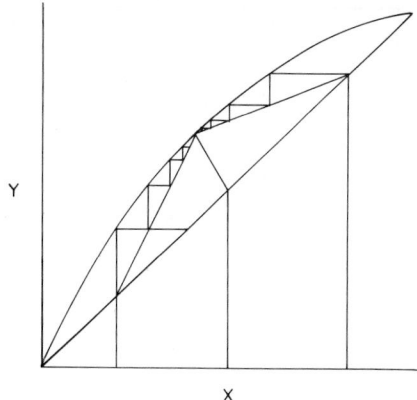

Fig. D-29. Illustration of the effect of an excessive feed rate when both the reflux and boil-up capabilities of a distillation column are overloaded. X = mole fraction of low boiler in liquid, Y = mole fraction of low boiler in vapor.

or reflux facilities is exceeded. Exceeding the liquid- and vapor-handling capacity causes flooding. Virtually no separation occurs. A deficiency of boil-up increases the slope of the lower operating line, and an unobtainable location of the lower operating line is called for, as illustrated in Fig. D-27. A deficiency of reflux decreases the slope of the upper-column operating line, with the similar result that an unobtainable location of the upper operating line is called for in the rectifying section, as shown in Fig. D-28. Figure D-29 shows the effect that an excessive feed rate can have when both the reflux and boil-up capabilities of a column are overtaxed. It is apparent that the control plate, whether in the top or bottom, would no longer be able to maintain its composition at the prescribed level.

Azeotropic Mixtures. An azeotropic mixture is one in which, over a certain composition range of two components, one component will be more volatile than the other, and after a point of convergence is exceeded, the volatility order of the two components will be reversed. Two systems of this type are shown in Fig. D-30: ethyl acetate–ethanol and chloroform-acetone. In the latter system, chloroform is less volatile than acetone up to a concentration of 66 mole % chloroform. Above this concentration, chloroform is the more volatile component. At the 66% point, both compounds have the same volatility; i.e., the vapor and liquid compositions are identical.

Scores of flow sheets in this volume indicate various kinds of distillation operations. Consult

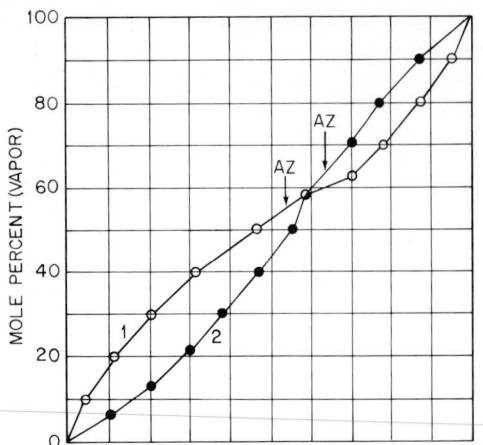

Fig. D-30. Vapor-liquid equilibrium data at 760 mm Hg for chloroform-acetone and ethyl acetate–ethanol systems. (1) ethyl acetate–ethanol, (2) chloroform-acetone. *AZ* = azeotrope.

the Subject Index and, in particular, see **Alkylation; Coal tar and derivatives; Lubricating oils; Petroleum; Petroleum processing;** and **Pyridine and derivatives.**

REFERENCES

Bogenstatter, G., and K. Hengst: Regelung von Destillationsklonnen, *Chem-Ing. Tech.*, vol. 31, no. 7, pp. 425–431, July 1959.

Buckley, P. S.: Override Controls for Distillation Columns, *Instr. Tech.*, vol. 15, no. 8, pp. 51–58, August 1968.

Gilliland, E. R., and C. M. Mohr: Transient Behavior in Plate-Tower Distillation of a Binary Mixture, *Chem. Eng. Prog.*, vol. 58, no. 9, pp. 59–64, September 1962.

Leva, M.: Film Tray Equipment for Vacuum Distillation, *Chem. Eng. Prog.*, vol. 67, no. 3, pp. 65–70, March 1971.

Lubowicz, R. E., and P. Reich: High Vacuum Distillation Design, *Chem. Eng. Prog.*, vol. 67, no. 3, pp. 59–63, March 1971.

Luyben, W. L.: Ten Schemes to Control Distillation Columns, *ISA J.*, vol. 13, no. 7, pp. 37–42, July 1966.

Nygren, P. G., and G. K. S. Connolly: Selecting Vacuum Fractionating Equipment, *Chem. Eng. Prog.*, vol. 67, no. 3, pp. 49–58, March 1971.

Oglesby, M. W., and D. E. Lupfer: Feed Enthalpy Computer Control of Distillation Column, *Control Eng.*, vol. 9, no. 2, pp. 87–88, February 1962.

Peister, A. M., and S. S. Grover: Dynamic Simulation of a Distillation Tower, *Chem. Eng. Prog.*, vol. 58, no. 9, pp. 65–70, September 1962.

Perry, R. H., and C. H. Chilton (eds.): "Chemical Engineers' Handbook," 5th ed., McGraw-Hill, New York, 1973.

Robinson, C. S., and E. R. Gilliland: "Elements of Fractional Distillation," 4th ed., McGraw-Hill, New York, 1950.

Rose, A., and E. Rose: "Distillation," Interscience-Wiley, New York, 1951.

Svrcek, W. Y., and H. W. Wilson: Case History of a Column Control Scheme, *Chem. Eng. Prog.*, vol. 67, no. 2, pp. 45–51, February 1971.

Distilled Spirits   (See **Ethyl alcohol;** and **Yeasts.**)

DNA (Deoxyribonucleic Acid)   (See **Peptides and proteins;** and **Polymerization.**)

Dolomite   (See **Ironmaking and steelmaking; Limestone;** and **Magnesium.**)

Domestic Waste Treatment   (See **Waste treatment.**)

Donors   (See **Catalysts.**)

Doré Metal   (See **Bismuth;** and **Tellurium.**)

Double Salt   (See **Salt.**)

Dough Mixers   (See **Mixing, pastes.**)

Downs cell   (See **Sodium.**)

Dowtherm Systems   (See **Energy systems for processes.**)

Drawing, Glass   (See **Glass.**)

Drugs   (Consult the **Subject Index.**)

Dry Cells   (See **Batteries.**)

Dry Cleaning   (See **Air pollution;** and **Perchloroethylene.**)

Drying, Gases   The assurance of end-product quality and the reliability of manufacturing operations frequently require the effective drying of gases, including various inert atmospheres, nitrogen, hydrogen, oxygen, and organic gases, such as chlorine-fluorine compounds (the Freons, etc.). The most commonly used gas, compressed air, requires drying and conditioning for trouble-free operation of pneumatic equipment, including tools, instruments, and controllers. In paint manufacture, dry inert gas is used to blanket agitation operations. Dry process gases, such as nitrogen or hydrogen, are used in metal-annealing operations. The manufacture of transistors requires blanketing with a dry gas in assembly operations. The degree of drying (defined for the purpose here as the removal of water vapor) varies, of course, from one application to the next.

Conditioning of Compressed Air: Because the basic principles of drying apply to most gases, a typical compressed-air drying installation is described here. The system shown in Fig. D-31

comprises the following major operations: (1) compression, (2) cooling (air, water, or refrigeration), and (3) desiccation (liquid or solid desiccant). In a typical untreated plant air system, the compressed air will contain numerous impurities, such as dust, dirt, chemical vapors, and moisture, that pass through the compressor intake filter. The untreated air also will contain oil and possibly sludge from the compressor (if an oil-lubricated compressor is used). Further, the air will contain heat from compression. Thus air dis-

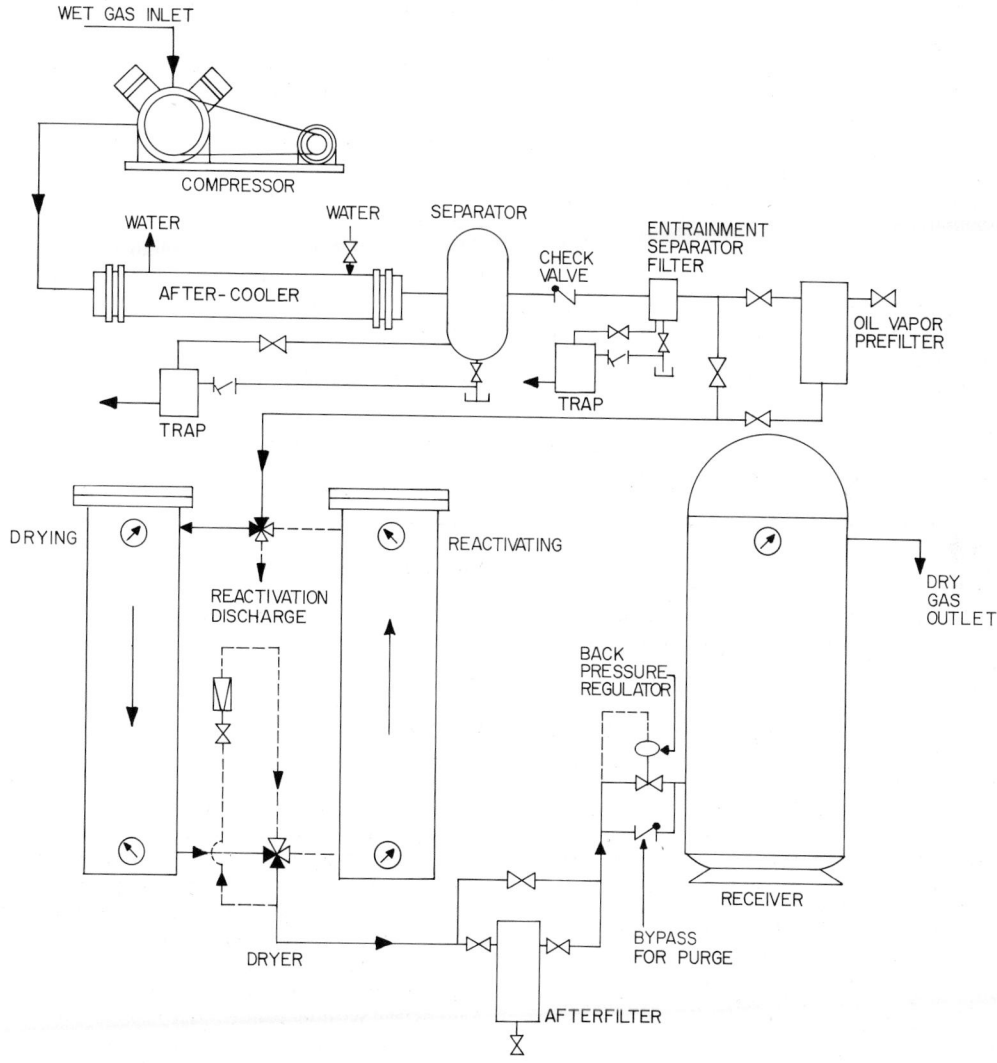

Fig. D-31. Compressed air drying and conditioning system.

charging from a compressor is not suitable for instrument or general plant air systems.

The compressor aftercooler (generally water-cooled) removes the heat of compression and also condenses the moisture present to the level of saturation. Generally, aftercoolers are designed to operate effectively within 10–15°F of the temperature of the cooling water used. Considering that air at 100°F when saturated contains almost twice as much moisture as saturated air at 80°F, the desirability of a low cooling temperature is evident. Maximum removal of moisture by an efficient aftercooler reduces the size and cost of subsequent desiccant-drying equipment.

Operating pressure also is an important factor in determining the moisture content of compressed gases inasmuch as saturated air at 105 psig contains approximately 20% less moisture than saturated air at 80 psig, assuming the same temperature condition. Therefore operating at the highest pressure consistent with plant design requirements will result in a smaller, more efficient, and more economical desiccant dryer.

The water separator collects the water condensed in the aftercooler, plus entrained oil, dust, rust, and other particulate matter. The accumulated condensate normally discharges through an automatic drain trap. Although a high-pressure receiver is used primarily to store the compressed air, the unit also will collect some moisture due to a reduction in air velocity. Should the receiver temperature become cooler than the air stored in it, further condensation of water vapor will occur. Frequently, the desiccant dryer is installed ahead of the receiver so that the receiver stores dried compressed air.

The need for further drying, as by a desiccant-dryer system, is demonstrated by the fact that 100 scf/min air at 100 psig and saturated at 100°F will add 7 gal water vapor each day to the compressed-air piping. In addition to the need for clean conditioned compressed air as previously described, the presence of water vapor in some climates may result in the freezing of critically important lines to the point where operations may have to be shut down.

Types of Dryers: Some of the factors that affect the selection of a drying system include (1) flow rate, (2) pressure, (3) temperature and saturation, (4) ambient temperatures, (5) final dew-point requirements, (6) available reactivation utilities, and (7) space limitations. Deliquescent dryers (dissolving desiccant types) and refrigeration-type dryers are not described here because these kinds of dryers only partially remove moisture from gases and, generally, additional drying equipment is required. The following systems use a solid, regenerable desiccant which *adsorbs* the moisture present. See also **Adsorption.** Desiccant regeneration usually is accomplished by the application of heat or purging or a combination of both procedures. Normally, dual drying towers are used for continuous service, allowing one tower to be on line while the other tower is being regenerated. The desiccants most frequently used are silica gel, activated alumina, and molecular sieves. Desiccant systems permit efficient dew-point performance in the range of −40 to −100°F (−40 to −73°C) and available in a wide range of capacities and pressure ratings (from atmospheric pressure up to 5,000 psig).

Purge-Type Dryer with Internal Regeneration Heating: Each of the drying towers shown in Fig. D-31 is filled with desiccant material. The air or other gas to be dried enters through a top manifold and passes through the tower that is on line. The dry gas, after passing through the desiccant bed, discharges from the bottom manifold into the process air or gas line. Regeneration usually is accomplished at essentially atmospheric pressure, utilizing embedded heaters (steam or electric) for heating the desiccant bed. A small portion of the dried product is purged through the tower being heated to sweep out the liberated moisture. Where the gas being dried is not suitable for purging, the design is modified to use an external purge source. Because of its simplified design, and economical operation, this is a basic arrangement for industrial applications.

Convection-Type Dryer with External Reactivation System: With the type of system shown in Fig. D-32, the external reactivation system consists of a heater (steam, electric, or gas), a blower, and a cooler. One captive volume of gas is recirculated by means of the external blower through the heater, desiccant bed, and into the cooler, where the liberated moisture is condensed and removed. Following the heating period, the blower circulates the captive volume of gas to cool the desiccant prior to going back into drying service. The closed system of regeneration prevents contamination of the process gas from outside sources and also provides a reactivation system that is completely independent of the process gas flow. A dryer of this type can operate 0–100% of its full-load rating without adjustment. Some convection dryers reactivate at atmospheric pressure; others reactivate at line pressure. This design generally

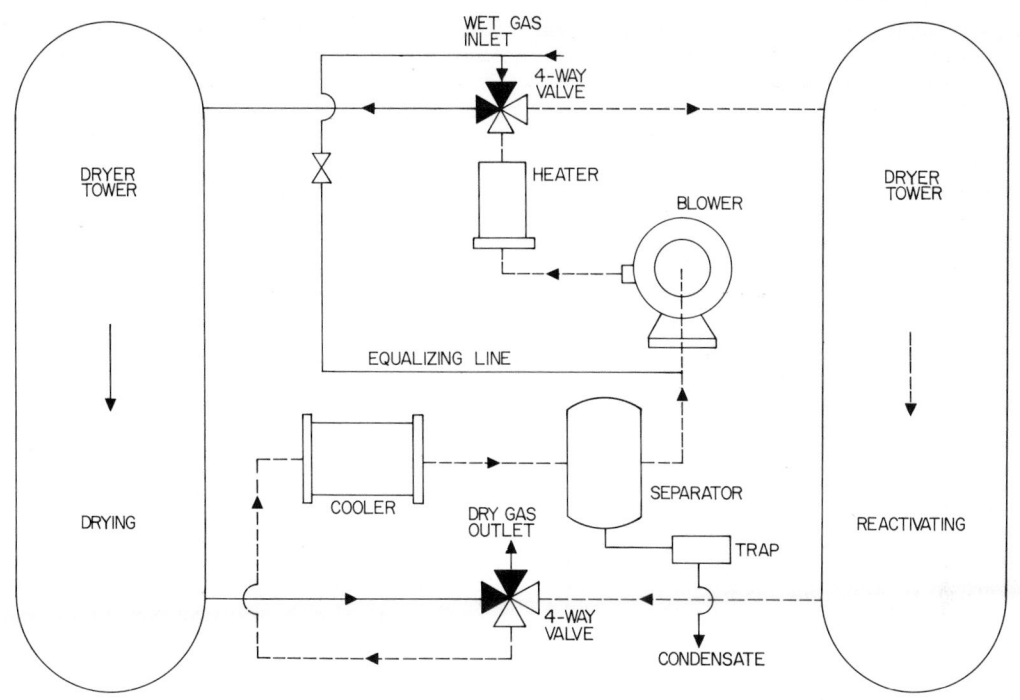

Fig. D-32. Convection-type dryer with external reactivation system. Dryer tower at left is on drying cycle; tower at right is on reactivating cycle. Dashed lines indicate piping used during reactivation.

is used where large flows are involved, where a closed reactivation system is desired, or when an independent reactivation system is desired. There are modifications of the design which eliminate the reactivation blower by diverting a portion of the process air for the reactivation requirements before being reintroduced into the tower on drying service. Generally, the blower-type design is the most efficient.

A modification of this design uses atmospheric air for regeneration. During reactivation, the blower circulates atmospheric air through the heater and then to the desiccant bed to liberate moisture adsorbed during the previous drying period. Upon leaving the desiccant bed, the hot moist reactivation air is vented to atmosphere. Because regeneration is not accomplished in a closed system, this type of dryer is not easily adaptable to handling gases other than air, unless contamination with air is not a problem.

Heatless-Type Dryer: The pressure swing or heatless dryer is similar to the purge-type unit described previously except that the embedded heaters are omitted and the design relies upon frequent tower reversals, where only a small quan-tity of moisture is adsorbed by the desiccant between reversals. After a short adsorption period, rapid depressurization lowers the vapor pressure of the adsorbed moisture, allowing the moisture to be desorbed from the desiccant and swept from the tower by a high purge rate of the dried product. Although no electric power or steam is required for regeneration heating, additional compressor horsepower is required because of the additional air required for reactivation purge. Generally, the heatless-type dryer is used for relatively small flow applications or where a plant has an excess of compressed-air capacity.

While compressed air or gas may seem complex when compared with the usual industrial utilities (water, fuel gas, or electricity), it must be emphasized that this utility is being produced on the plant site. In most instances, an adequate and economical overall compressed-air or gas system requires desiccant drying to assure optimum plant performance.

—Robert C. Frey, *The C. M. Kemp Manufacturing Company, Glen Burnie, Md.*

Drying, Paint (See **Air pollution;** and **Paint.**)

Drying, Solids One of the most frequently encountered chemical engineering unit operations for effecting a separation of two or more components from a starting material, drying generally signifies the removal, by evaporation, of a liquid from a solid, although gases also are dried. See **Drying, gases.** Usually, drying involves the processing of a material that, on a weight basis, contains more solid phase than liquid phase, and the majority of dryers are designed to handle this starting condition. However, there are dryers for very wet materials, such as sludges and slurries. In spray dryers, the starting material is very liquid; i.e., on a weight basis there is much more liquid phase than solid phase present, as represented, for example, by a solution or thin slurry or suspension. See also **Spray-drying.**

Mechanics of Drying: In drying, sufficient energy must be applied to the wet material to provide the heat of vaporization required to convert the liquid to a vapor along the surface of the material, as well as to cause the liquid phase to migrate in some fashion to the surface of the material, i.e., to flow out of capillaries and interstices of the solid material. When one considers that drying is applied to materials that range from large lumps of very dense substances to fine crystals and granules and powders on the bulk side, including also materials in the form of films and matted and webbed fibers, such as paper, the need for several basic dryer designs is evident. An efficient universal dryer suitable for the efficient handling of all kinds of starting materials is yet to be designed.

In methods other than dielectric heating, the heat applied by the dryer first flows to the outer, exposed surface of the material before reaching the interior of the solid. In dielectric heating, a higher temperature is produced within the material than on the surface, and hence drying commences from within. Compared with other drying methods, the use of dielectrically heated dryers is quite limited, due in large part to the higher costs of drying by the latter method. In the more conventional dryer designs, the heat required is provided by convection, conduction, radiation, or a combination of these methods. Freeze-drying is a special case which involves high vacuums and sublimation as well as very low temperatures. See also **Freeze-drying.**

Although several variables enter into designing and operating a dryer for maximum efficiency, the process essentially represents a time-temperature relationship. Within the limitations of the material being dried, drying proceeds, as may be expected, at a faster rate when the temperature is increased. The drying rate also increases when the vapor space immediately in contact with the surface of the solid material is subject to rapid movement and therefore to replenishment by less saturated air (or other gas) for carrying the vapors away.

The theory of drying is quite well developed. Theorists breakdown the drying process into two principal periods: the *constant-rate period* and the *falling-rate period.* The constant-rate period precedes the falling-rate period. During the constant-rate period, the rate of liquid removal per unit of drying surface is essentially constant and represents a condition where the surface of the material is saturated (fully wetted) and where moisture from within the solid is moved to the surface at a sufficient rate to maintain the surface saturation. During the constant-rate period, with no changes in the adjustments and control settings of the dryer that affect heat input or other variables, the drying rate will remain constant. The mechanism is similar to that of evaporation from a body of liquid wherein the temperature of the saturated surface remains constant.

During the falling-rate period, the drying rate at any given instant continues to decrease. The average moisture content of the solid at the time the constant-rate period ceases and the falling-rate period commences is known as the *critical moisture content.* This value has been established for a large number of materials and is very helpful for the design and application of drying equipment.

For purposes of analyzing the mechanics of the drying process during the falling-rate period, two situations are considered which essentially occur in sequence. First, the drying rate is reduced because the entire surface area of the solid no longer is saturated. It is self-evident that the drying rate of the unsaturated portion would be less than that for saturated portions. As drying continues, the saturated portions become smaller and ultimately disappear. This first phase of the falling-rate period, sometimes referred to as the *unsaturated-surface drying period,* is affected mainly by external conditions (dryer design and control settings) rather than by the nature of the solid. Second, the drying rate is affected by the movement of internal moisture to the surface of the solid. This second phase of the falling-rate period sometimes is referred to as the *internal-moisture movement period* and is predominantly a function of the nature of the solid material. Where materials are dried to low-moisture contents, this second period normally requires the largest percentage of total drying time.

For estimating total drying time, estimates of both the constant- and falling-rate drying times are needed. An approximation of total drying time for materials in which moisture movement is governed by capillary moisture flow is

$$\theta_t = \theta_c + \theta_f \qquad (1)$$

$$\theta_t = \frac{(W_o - W_c)\rho_s \lambda d}{h_t (t - t_s')_m}$$
$$+ \frac{(W_c - W_e)\rho_s \lambda d}{h_t (t - t_s')_m} \ln \frac{W_c - W_e}{W_t - W_e} \qquad (2)$$

$$\theta_t = B\left(\frac{W_o - W_c}{W_c - W_e} + \ln \frac{W_c - W_e}{W_t - W_e}\right) \qquad (3)$$

where

$$B = \frac{(W_c - W_e)\rho_s \lambda d}{h_t (t - t_s')_m} \qquad (4)$$

and

θ_t = total drying time, hr
θ_c = drying time for constant-rate period, hr
θ_f = drying time for falling-rate period, hr
W_o = initial moisture content, lb water/lb dry solid
W_c = average critical moisture content, lb water/lb dry solid
W_e = average equilibrium moisture content, lb water/lb dry solid
W_t = average moisture content at time θ_t, lb water/lb dry solid
ρ_s = bulk density of dry solid, lb/ft^3
λ = latent heat of vaporization at t_s', Btu/lb
d = depth of material on tray, ft
h_t = total heat-transfer coefficient given by Eq. (6), Btu/(hr)(ft^2)(°F)
$(t - t_s')_m$ = logarithmic mean temperature difference,

$$[(t_1 - t_s') - (t_2 - t_s')] \ln \frac{t_1 - t_s'}{t_2 - t_s'} \qquad (5)$$

t_1 = inlet-air temperature, °F
t_2 = exit-air temperature, °F
t_3 = solids temperature, °F

Frequently, particularly in tray-drying, heat arrives at the evaporating surface from the tray walls by conduction through the wet material. For this case, where both radiation and conduction are significant, the total heat-transfer coefficient * is

$$h_t = (h_c + h_r)\left[1 + \frac{A_u}{1 + d(h_c + h_r)/k}\right] \qquad (6)$$

where A_u = ratio of outside unwetted surface to evaporating-surface area
k = thermal conductivity of the wet material, Btu/(hr)(ft^2)(°F)(ft)
h_c = convection heat-transfer coefficient, Btu/(hr)(ft^2)(°F)
h_r = radiation heat-transfer coefficient, Btu/(hr)(ft^2)(°F)

The factor h_r must be corrected for emissivity of the surface. For insulated trays, the arithmetic average of inside and outside unwetted area should be used.

Classification of Dryers: A dryer may be termed *direct,* in which heat required for drying is applied through immediate contact between the wet material and the hot gases used, or *indirect,* in which the heat is transferred to the wet material through some intervening medium, such as a pipe or retaining wall. Further, a dryer may be designed for *batch* or *continuous* operation. Either direct or indirect dryers may be batch or continuous, as shown in the family tree of dryer designs given in Table D-3. Useful in the selection of a dryer for a given application is another classification which considers the characteristics of the materials being processed:

LIQUIDS: True and colloidal solutions and emulsions. Examples include inorganic salt solutions, extracts, milk, blood, waste liquors, and rubber latex.

SLURRIES: Pumpable suspensions, such as pigment slurries, soap and detergents, calcium carbonate, bentonite, clay slip, and lead concentrates.

PASTES AND SLUDGES: Filter-press cakes, sedimentation sludges, centrifuged solids, and starch are examples.

FREE-FLOWING POWDERS: Materials that are 100 mesh or less. They are relatively free-flowing in the wet state, but dusty when dry. Examples include centrifuged precipitates, pigments, clay, and cement.

GRANULAR, CRYSTALLINE, OR FIBROUS SOLIDS: Materials that are 100 mesh or larger. Examples

*Shepherd, Brewer, and Hadlock, *Ind. Eng. Chem.,* vol. 30, p. 388, 1938

TABLE D-3. *Classification of Dryers*

Direct Heating		Indirect Heating	
Continuous	Batch	Continuous	Batch
Tray dryers: Continuous metal belts, vibrating trays that use hot gases; also vertical turbodryers. *Sheeting dryers:* Continuous sheets of material go through dryer as taut sheets stretched on pin frames or as festoons. *Pneumatic conveyor dryers:* Often used where grinding also is involved. Material is moved by a stream of high-velocity gas at a high temperature to a cyclone collector. *Rotary dryers:* Material is conveyed and showered within a rotating cylinder through which hot gases flow. *Spray dryers:* For materials that can be atomized. See also **Spray dryers.** *Through-circulation dryers:* Material is supported by a continuous conveying screen. The hot gases are blown through the material. *Tunnel dryers:* Material is placed on trucks which move through a tunnel in which there is a flow of hot gases.	*Tray and compartment dryers:* Material is placed on trays which may be on permanent shelves or on portable trucks. Hot gases are blown across the trays. *Through-circulation dryers:* Material is placed on trays with screen bottoms. The hot gases are blown through the material.	*Cylinder dryers:* Material in the form of continuous sheets passes over and around cylinders which generally are heated with steam or hot water and which rotate. *Drum dryers:* Essentially for materials in liquid and slurry form. One or more heated drums dip into liquid, and drying occurs during part of rotation cycle. *Screw-conveyor dryers:* Essentially a conveyor within a heated, closed housing. Vacuum operation is possible. *Steam-tube rotary dryers:* Material passes through a long rotating cylinder in which steam, hot water, or other heating medium is contained within closed tubes. *Vibrating-tray dryers:* A continuous tray-dryer configuration in which heat is supplied by medium by way of conduction to material that is vibrated vigorously on trays.	*Agitated-pan dryers:* Material is placed in covered shallow cylindrical pans that are jacketed for heating. Agitator stirs material constantly. May be operated at atmospheric pressure or under vacuum. *Freeze dryers:* Involve combination of low temperature and high vacuum in which freezing, sublimation, and drying occur. See also **Freeze-drying.** *Vacuum rotary dryers:* Material is agitated within a stationary, horizontal shell under vacuum. The agitator also may be heated. *Vacuum tray dryers:* Material is placed on trays which are heated by conduction from shelves. Entire compartment is under a relatively high vacuum. No agitation of the material.

include rayon staple, salt crystals, sand, ores, and synthetic rubber.

LARGE SOLIDS (SPECIAL FORMS AND SHAPES): Pottery, brick, rayon cakes, shotgun shells, painted objects, lumber, and rayon skeins are examples.

CONTINUOUS SHEETS: Paper, impregnated fabrics, cloth, plastic sheets, and fiberboard are examples. See also **Pulp (wood) production and processing.**

DISCONTINUOUS SHEETS: Materials such as veneer, wallboard, photograph prints, leather, and foam-rubber sheets.

Dryer Design Criteria: Important factors to consider when designing or selecting a dryer include:

1. *Nature of the material* to be dried: (a) physical and chemical characteristics when wet and dry, (b) corrosiveness, (c) toxicity, (d) flammability, (e) particle size, and (f) abrasiveness.

2. *Drying behavior:* (a) type of moisture—bound, unbound, or in combination, (b) starting moisture content, (c) maximum allowable final moisture content, (d) maximum permissible drying temperature, and (e) estimated total drying time for various dryer designs.

3. *Nature of total processing operation:* (a) quantity of material to be handled per hour, (b) continuous or batch operation desired, (c) nature of the proc-

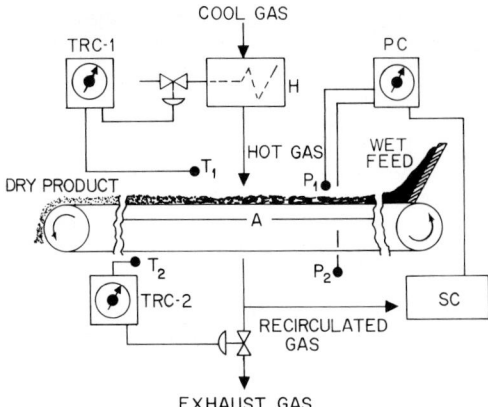

Fig. D-33. Traveling-screen dryer showing principal controls.
TRC-1 is a dry-bulb temperature controller. *TRC*-2 is a wet-bulb temperature controller. *PC* is a pressure-drop controller interlocked with *SC*, the belt-speed controller. *A* represents a typical controlled section of the traveling screen. *H* is a heater. Measurement of dry-bulb temperature T_1 of the hot gas controls the flow of hot medium to the heater. Wet-bulb temperature T_2 of the gas after leaving the screen and having passed through the wet material controls the amount of gas that is recirculated to the system. The belt speed is controlled from one section of the dryer only and is based on the pressure differential $P (= P_1 - P_2)$ across the moving screen.

essing just prior to drying operation, and (*d*) nature of subsequent processing operations.

4. *Characteristics desired of dried product:* (*a*) shrinkage permissible, (*b*) contamination allowance, if any, (*c*) tolerances of variations from final desired moisture content, (*d*) effects of overdrying on yield and acceptability of final product, (*e*) particle size and other physical characteristics of final product that may be altered during drying, (*f*) desired exit temperature of product, and (*g*) bulk density of final product.

5. *Recovery of materials,* including (*a*) dust and (*b*) solvents.

6. *Facilities requirements:* (*a*) space needed, (*b*) temperature, humidity, and cleanliness of raw air (other gas) required—or need for inert gas, (*c*) fuels available, (*d*) electric power required, (*e*) allowable noise, vibration, and dust, (*f*) allowable heat losses, (*g*) source of wet feed, and (*h*) exhaust-gas outlets.

7. *Control and instrumentation* requirements.

Direct-continuous Dryers: One drying section of a continuous *traveling-screen dryer* is shown in Fig. D-33. Dryers of this type are well suited to large-

scale operations involving granular, crystalline, and fibrous solids and may be adapted to a wide variety of solid shapes and forms. They are not suited to liquids and slurries. The controls on each zone of a multisection dryer of this type would be the same as shown, except that pressure *P*, the pressure drop through the material, will be measured in only one section for the control of belt speed. A belt-speed change may be required to hold a constant pressure drop through the material being dried if the consistency of the material or the throughput rate changes. The uncontrolled variable is feed rate.

Materials that are dried in *pneumatic conveying dryers* include clay, corn, gluten, coal, raw gypsum, various pharmaceuticals, silica-gel catalyst, various synthetic resins, and sewage sludge. A typical two-stage pneumatic conveyor dryer is shown in Fig. D-34. The system incorporates an ordinary single-stage dryer with a second stage containing a cage-mill disintegrator. The second stage ensures complete drying after thorough dispersion of lumps and agglomerates. If disintegration is required to disperse the wet feed, the stages can be reversed, or disintegration can be used in both stages. Depending upon the temperature sensitivity of the product, inlet-air temperatures range between 150°C (300°F) and 705°C (1300°F). Often, with a heat-sensitive solid, a high initial

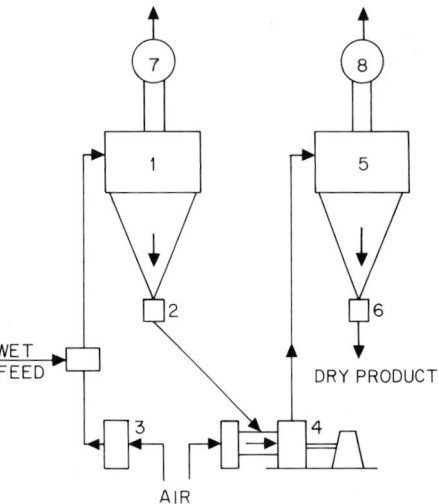

Fig. D-34. Two-stage air-stream and cage-mill pneumatic conveyor dryer. (1) Wet-stage cyclone, (2) air lock, (3) steam air heaters, (4) cage mill, (5) dry-stage cyclone, (6) air lock, (7) and (8) vent fans.

moisture content will permit a high inlet-air temperature. Evaporation of surface moisture takes place at essentially the wet-bulb air temperature. Until this is completed, by which time the air will have cooled appreciably, the surface moisture film prevents the solids temperature from exceeding the wet-bulb temperature of the air. Generally, dryers of this type are used for solids having initial moisture contents of 3–90% on a wet basis. The gas velocity in the conveying duct must be sufficient to convey the largest particles.

Direct-heat rotary dryers, as shown in Fig. D-35, are equipped with flights on the interior for lifting and showering the solids through the gas stream during passage through the cylinder. Although these flights may be continuous through the cylinder, they usually are offset every 2 to 6 ft to ensure more continuous and uniform curtains of solids in the gas. The shape of the flights depends upon the handling characteristics of the solids. When materials change characteristics during drying, the flight design may be changed along the length of the dryer. Spiral flights usually are provided in the first few feet at the feed end to accelerate forward flow. When cocurrent gas-solids flow is used, the flights may be omitted from the exit end to reduce entrainment of dry product in the exit

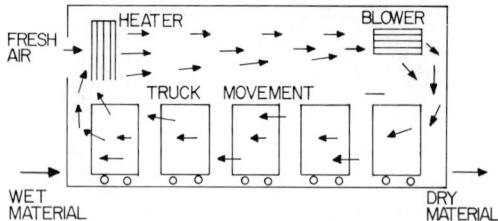

Fig. D-36. **Countercurrent tunnel dryer.**

gas. Showering of wet feed at the feed end of a countercurrent dryer frequently serves as an effective means for scrubbing dry entrained solids from the gas stream before leaving the cylinder. Knockers frequently are used to reduce clogging of flights. Countercurrent flow of gas and solids provides greater heat-transfer efficiency for a given inlet-gas temperature, but cocurrent flow may be used to better advantage for drying heat-sensitive materials at higher inlet-gas temperatures because of the rapid cooling of the gas during initial evaporation of surface moisture.

Continuous tunnel dryers of the type illustrated schematically in Fig. D-36, indicating countercurrent operation, involve the placing of the material to be dried on trays or trucks which move progressively through the tunnel in contact with hot gases. In some cases, each truck will occupy successive positions in the tunnel for a given period of time rather than moving at a constant slow rate through the tunnel. However, belt-type tunnel dryers are almost always fully continuous in operation. Air flow may be totally cocurrent, countercurrent (as illustrated), or a combination of the two flow schemes. Cross-flow designs also are used where the heating air flows back and forth across the trucks in series. When handling granular, particulate solids which do not offer high resistance to air pickup, a perforated belt conveyor may be used in which there is *through circulation* of gas to improve heat- and mass-transfer rates.

Direct-Batch Dryers: A widely used dryer in this category is the *tray dryer,* which usually consists of an insulated enclosure or compartment, equipped with fans and heating coils and suitable supports for the material. Tray dryers may be of the tray-truck or stationary-tray type. In the tray-truck configuration, the trays are loaded on trucks which are pushed into the dryer. In the stationary-tray configuration, the trays are placed directly on permanent racks within the drying compartment. The trucks may be fitted with flanged

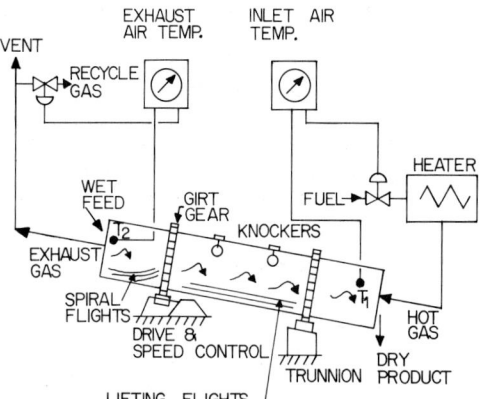

Fig. D-35. **Countercurrent direct-heat rotary dryer.** Numerous automatic control arrangements are possible. Hot-gas entrance temperature may be controlled from inlet T_1 through regulation of fuel to the gas heater. The amount of gas recycled may be controlled by measuring temperature T_2 at the wet end and controlling portion of gas vented. Material-moisture-measurement devices may be added. Wet-feed rate may be controlled as well as the rpm of the rotating cylinder. Selection of controls is largely dependent upon severity of load changes and alterations in the characteristics of the wet-feed material.

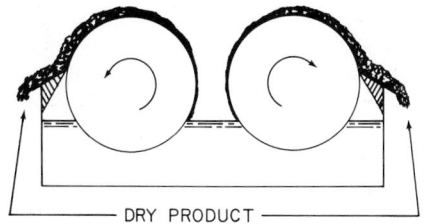

Fig. D-37. Double-drum dryer. One drum rotates clockwise, the other counterclockwise, each dipping into a pool of solids-containing liquid. Drying commences immediately upon the wet material being elevated from the liquid pool by the action of the heated, rotating rolls. The dry or partially dry product is scraped from the drum surfaces by doctor blades.

wheels to run on tracks or with flat swivel wheels. Monorail suspensions also are used. The trucks usually incorporate two tiers of trays with 18–48 trays per tier. Metal trays are preferred because of heat conductivity. Screen bottoms, where applicable, also promote faster drying rates. Circulation of air at velocities of 200–2,000 ft/min is desirable. Nonuniform air flow is one of the most serious problems in the operation of tray dryers. Air usually is circulated by propeller or centrifugal fans, the fan usually being mounted within or directly above the dryer. Total pressure drop through the trays, heaters, and ductwork generally does not exceed 1 to 2 in. of water. Air recirculation usually is 80–95%. Steam is the usual heating medium.

Indirect-continuous Dryers: Cylinder dryers are widely used for paper and other materials in continuous sheets. See **Papermaking.** *Drum dryers* are used principally for drying liquids and slurries and, where the material can be made to flow, as pastes and sludges. As shown in Fig. D-37, the drum dips into the liquid material and rotates at such speed that the material will be dried to its desired moisture content by the time the cylinder has rotated through so many degrees and the material has been picked off by a doctor blade. The use of twin drums, rotating in opposite directions, is common and greatly increases the capacity of the system. Drum dryers usually are heated with steam or hot water. Essentially the same design configuration, wherein the drum is refrigerated rather than heated, is used for cooling and solidification operations—as in separating wax from solvent in solvent dewaxing.

Steam-tube rotary dryers probably are used most widely of the indirect-heat rotary dryers. Steam-heated tubes running the full length of the cylinder are fastened symmetrically in one, two, or three concentric rows inside the cylinder and rotate with it. When handling sticky materials, one row of tubes is preferred. Lifting flights usually are inserted behind the tubes to promote agitation of the solids. Wet feed enters the dryer through a chute or screw conveyor. The product discharges through peripheral openings in the shell. These openings also admit purge air to sweep moisture or other evolved gases from the shell. In nearly all cases, the gas flow is countercurrent to solids flow. Steam is admitted to the tubes through a revolving steam joint, and condensate is removed by gravity continuously through the steam joint. Steam-tube dryers are used for continuous heating (or cooling) of granular or powdery solids which cannot be exposed to ordinary atmospheric or combustion gases. They are especially suited to fine dusty particles because of the low gas velocities required for purging the cylinder. In addition to drying, these units also may be applied to reacting and solvent-recovery situations.

Screw-conveyor dryers also permit continuous drying with indirect heat. Steam, hot water, or other heat-transfer media may be used, depending upon temperature required and sensitivity of product. Units of this type also are adaptable to continuous vacuum drying.

Indirect-Batch Dryers: Dryers in this category are particularly well suited to vacuum applications. A typical *vacuum rotary dryer* is shown in Fig. D-38. Vacuum operation is required where low solids temperatures must be maintained because of possible damage to product, or in some

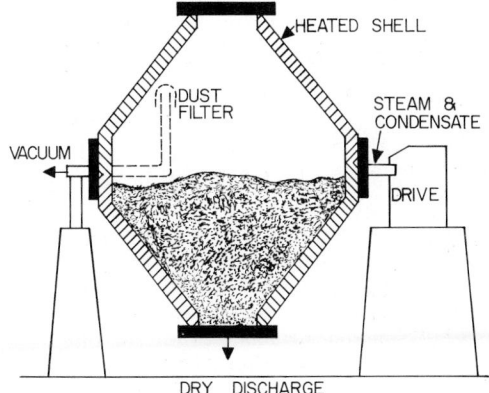

Fig. D-38. Vacuum rotary dryer of double-cone construction.

cases where air may tend to cause oxidation or an explosive condition. Often the latter situation arises where solvent must be removed from a material. In vacuum processing and drying, the objective is to create a large temperature-driving force between the jacket and the product. To accomplish this with fairly low jacket temperatures, the internal pressure must be reduced so that the liquid will boil at a lower vapor pressure. Few vacuum dryers operate below 10 mm Hg pressure unless of special design. Vacuum dryers usually are filled to 50–65% of the total shell volume. Agitator speeds are in the range 3–8 rpm.

Vacuum shelf dryers are indirect-heated batch dryers consisting of a pressuretight chamber, often constructed of cast iron or steel plate. Shelves within the chamber hold pans or trays of material to be dried. The shelves are hollow and permit the heat-transfer medium to heat the trays from underneath. Dryers of this type are used for drying pharmaceuticals and many other very temperature-sensitive or easily oxidizable materials. Dusty materials may be dried with negligible dust loss. Hygroscopic materials may be completely dried at a temperature below that required in atmospheric dryers. Equipment of this type also is used for freeze-drying processes, for metalizing furnace operations, and for the manufacture of semiconductor parts in controlled atmospheres. Shelf vacuum dryers usually operate in the range 1–25 mm Hg pressure. Drying cycles may require many hours.

Agitated-pan dryers consist of a fairly shallow, flat-bottomed cylindrical pan that is covered by a dished or conical cover. The bottom and walls of the pan are jacketed for heating steam or other medium. A central vertical shaft carries a slow-moving heavy-duty agitator which stirs the material in the dryer and moves it toward and away from the heat-transfer surfaces. The agitator usually is set to a very close clearance and may carry spring-loaded scrapers to clean the heated surfaces. Even when operated at atmospheric pressure, most agitated pan dryers are closed with a cover containing manhole sight glasses and an outlet through which heated vapors may escape by natural draft. Vacuum pan dryers have a dished or dome-shaped cover fitted with the vacuum connection. Typical units are 3–10 ft in diam and contain 15–300 ft^2 of heating surface. They generally are loaded about two-thirds full. Agitator speeds are 2–20 rpm. Agitated-pan dryers are required where a material must be agitated during drying. The units are fairly easy to clean, and thus are adaptable to handling successive batches of different materials. The units are most suitable for batch drying of difficult-to-handle materials for which continuous drying would not be economical, as when solvents must be vaporized from the solids and recovered; or drying must be under a high vacuum; or where evaporation, crystallization, and drying, with consequent large changes in physical properties, must be done in a single unit. Agitated-pan dryers are not suitable for materials which suffer particle-size degradation during drying, or substances that form into balls and case-harden.

Drying Oils   (See **Paint;** and **Vegetable oils.**)

Dusts   (See **Air pollution.**)

Dwight Lloyd Sinter Machine   (See **Cadmium; Copper production;** and **Zinc and zinc/lead smelting.**)

Dyestuffs and Dyeing (Textiles)   As described here, dyestuffs are compounds applied to textile substances (substrates) to produce a visual stimulation interpreted as color by the viewer. Dyes are as old as textiles themselves and predate written history. Old dyes, however, were natural products of either animal, vegetable, or mineral matter. In 1856 William Henry Perkin, while attempting to synthesize quinine, produced a lavender-colored substance, mauviene. Fortunately, he recognized the possibilities of this event, and the synthetic dyestuff industry was born. Now the vast majority of dyestuffs are synthesized organic compounds.

Color is a visual phenomenon. The eye sees color by virtue of interpreting to the brain impulses produced by reflected light rays. The specific color depends on the length of the reflected wavelengths. The dye molecules possess one or more special groups called *chromophores* which impart color to the textile substrate. Examples of these groups include the azo group, $-N=N-$; the thio group, $=C=S$; the nitroso group, $-N=O$; the carbonyl group, $=C=O$; the nitro group, $-NO_2$; and the azoxy group, $-N=N-$.
$$O$$

In addition to chromophores, organic-dye molecules contain auxochromes. These groups influence hue and intensity of colors. Examples of auxochromes include amino group, $-NH_2$; hydroxyl group, $-OH$; sulfonic group, $-SO_3H$; and substituted amino groups, $-N(CH_2)_2$ and

—$NHCH_3$. In addition to influencing the color, selected auxochromes also increase the solubility of dyestuffs.

For many years the presence of chromophores and auxochromes has been used as the basic explanation for the formation of color. The theory is still useful today, but in addition, developments in the field of spectroscopy have shed considerable new light on the behavior of organic compounds, including dyes. It is recognized that all organic compounds, whether or not they include chromophores in the molecular arrangement, absorb radiation. That some are colored is due simply to the fact that absorption bands are within the range of radiation visible to the human eye. These visible compounds do contain characteristic chromophore groups.

Dyes may be classified in several ways. The two most common are (1) classification according to chemical structure and (2) classification according to dyeing properties. The chemist uses the structural classification; the practical dyer and the technologist use dyeing properties as a basis for categorization. This procedure identifies the following groups: basic, or cationic; acid and premetalized; chrome and mordant; direct and developed direct; sulfur, azoic, vat, disperse, and reactive. Examples are as follows:

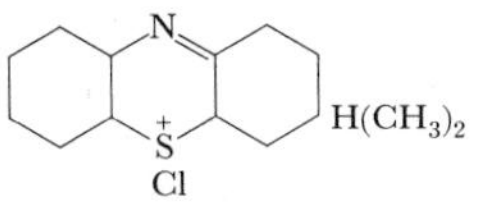

Basic dye: methylene blue

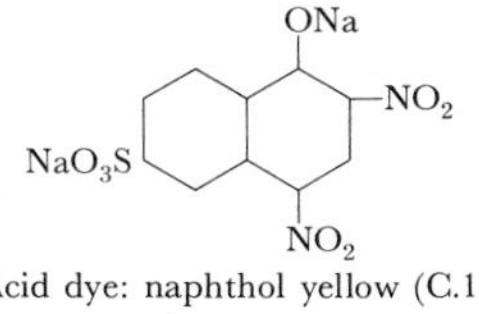

Acid dye: naphthol yellow (C.1.1)

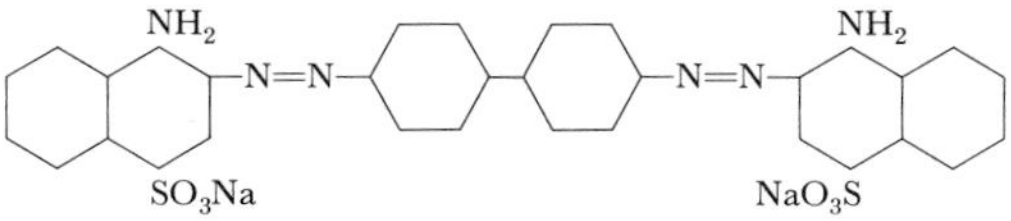

Direct dye: congo red (C.1.28)

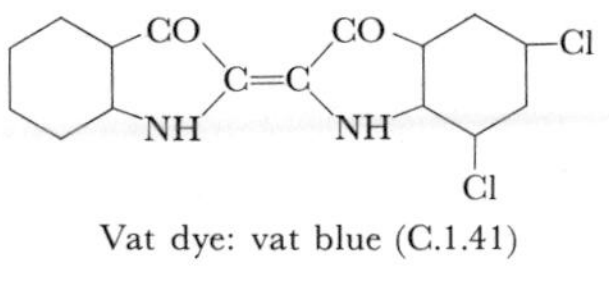

Vat dye: vat blue (C.1.41)

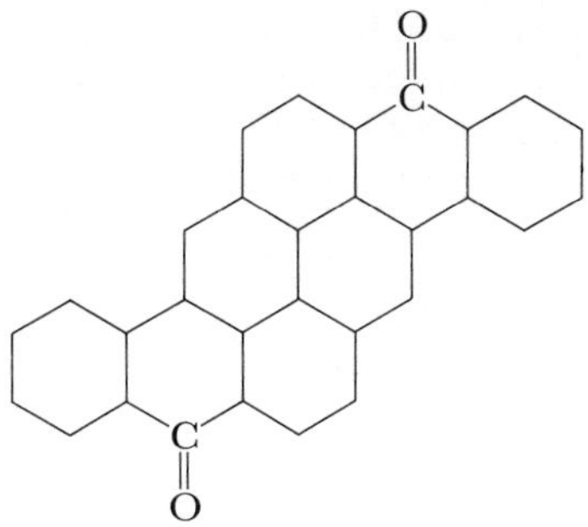

Vat dye: vat blue (C.1.41)

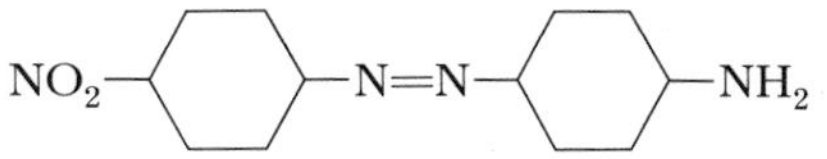

Disperse dye: disperse orange (C.1.3)

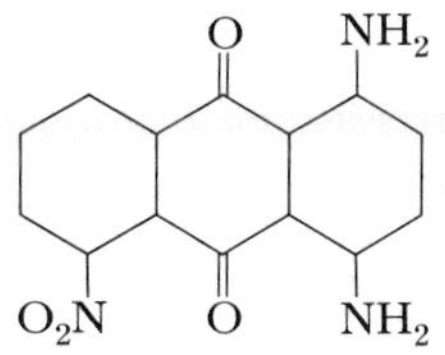

Disperse dye: disperse violet (C.1.8)

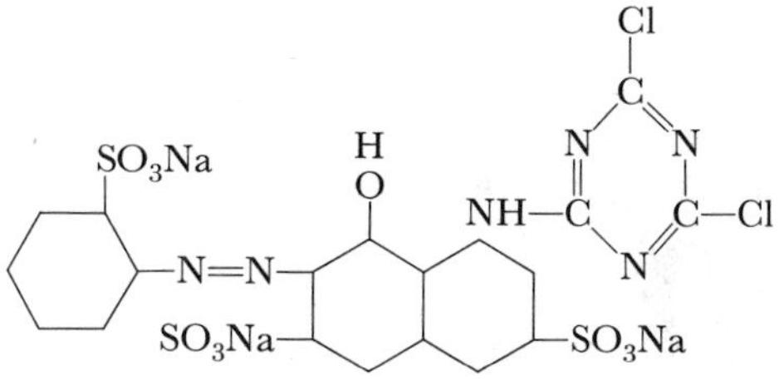

Reactive dye: red

The forces which anchor dyestuff molecules to fibers are complex. In general, three steps may be clearly identified:

1. Migration of the dyestuff from a solution to the fiber interface (substrate) followed by adsorption on the surface of the fiber

2. Movement or diffusion of the dye from the surface to the center of the fiber, where it locates around and between fiber molecules

3. The anchoring or attaching of the dye molecule to the fiber molecule by physical forces, hydrogen bonding, or actual covalent bonding with the fiber molecule

The movement of dye molecules from solution to the fiber is influenced by the presence of electropotential forces, by variation in temperatures, and by the degree of agitation of the solution and substrate. The relation between the ionic charge of a fiber and a dyestuff influences the feasibility of using a particular type of dye on specific types of fibers. Increased temperature tends to break dye micelles into small units, thus enabling the dye to attach itself to the fiber, and eventually move among the fiber molecules. Agitation is related to the assembly of dye molecules on the fiber substrate. Excessive agitation could hinder this assembly process, or it may move molecules rapidly enough to force greater contact.

Diffusion of the dye into the center of the fiber depends on relative size of the dye molecule and the pore opening of the fiber.

Attachment of the dye molecule to the fiber molecule occurs as a result of one or more of the following factors: (1) physical forces, (2) ionic or salt linkages, (3) hydrogen bonding, and (4) co-valent bonding.

Physical forces include the attachment of dyes by such forces as dye-molecule shape and size, the planar relation of the dye molecule, and the space or openings between fiber molecules. Dye molecules that are relatively linear and planar in shape tend to adsorb to fiber molecules. Where aggregation of dye molecules can occur within the fiber, the resulting dye molecule will be held due to size.

Ionic or salt linkages occur between protein fibers and anionic dyes that can form a link with the $-NH_3^+$ groups in fiber molecules.

Hydrogen bonding occurs primarily in cellulosic fibers. Dyes containing hydroxyl, amino, or azo groups can form hydrogen bonds with the hydroxyl group of the cellulose molecules. A critical factor in hydrogen bonding is the space (usually expressed in angstrom units) between groups capable of forming the bond.

Covalent bonding occurs when there is a definite sharing of electrons between atoms. This produces the most stable bond and is a characteristic of reactive dyes and some acid and some cationic dyes.

Dyeing Machines and Processes. Equipment used in applying dyes varies, depending on dyestuff, type of fiber, and form (physical shape) of the substrate. Some dyes require large proportions of liquid and require units capable of holding these large amounts of solution. Some fibers require pressure in order for dyeing to occur; thus containers that can be sealed and pressurized are needed. The form of the substrate plays a very important role. Loose fibers, yarns, or fabrics require different kinds of handling and machinery for effective dyeing.

Machines used in dyeing function in one of three ways:

1. Movement of the substrate (usually fabric in this case) through the dye liquor

2. Movement of the dye liquor through the substrate (may be used on fiber, yarn, or fabric)

3. Movement of both the substrate and the dye liquor (yarn and fabric)

Several process configurations are shown in Figs. D-39 to D-47. The essential requirements for any dyeing machine can be summarized as follows:

1. The machine should provide adequate movement for the dye to penetrate the substrate uniformly.

2. The movement must not be damaging to the substrate.

3. The equipment must be of materials that will not be damaged by the chemicals in the dye bath.

4. Uniform temperature control throughout the process is desirable.

5. It is desirable to have some source by which additional dye or other chemicals may be added without stopping the process.

Each type of dyestuff has particular problems associated with its dyeing procedure. Six of the most commonly used dyestuffs are discussed in relation to the dyeing operation. A term frequently used in discussion of dyes is *leveling*. This refers to the ease with which the dye colors the substrate uniformly, evenly, and to the degree desired.

Acid Dyes: Acid dyes are so named, first, because the original acid dyestuffs were applied in a bath containing either a mineral or an organic acid, and second, because the anion is the colored portion of the molecule. Most of the acid dyes are sulfonic acid salts. Protein fibers are frequently colored with acid dyes.

The solubility of acid dyes is high, and the dye bath exhausts to a maximum degree. Three general methods are used in applying acid dyes; these procedures depend upon the leveling and exhausting properties of the dye.

Method 1: This procedure is used for dyes that level readily, exhaust quickly, and require strong acid such as sulfuric. The bath (liquor) is formulated with water, sodium sulfate, and sulfuric acid. The liquor is raised to 40°C (105°F), the fiber, yarn, or fabric is entered, and the bath is run for 10–20 min. The dissolved dye is added,

Fig. D-39. Raw-stock dyeing. The dyeing of raw stock before spinning makes it possible to blend two or more different shades into a single yarn. Stock dyeing also has the advantages of eliminating leveling problems since any unevenness will be masked when the fibers are blended in the spinning process. Stock is dyed in open tanks and closed pressurized vessels (as shown). (*Fibers Division, Allied Chemical Corp.*)

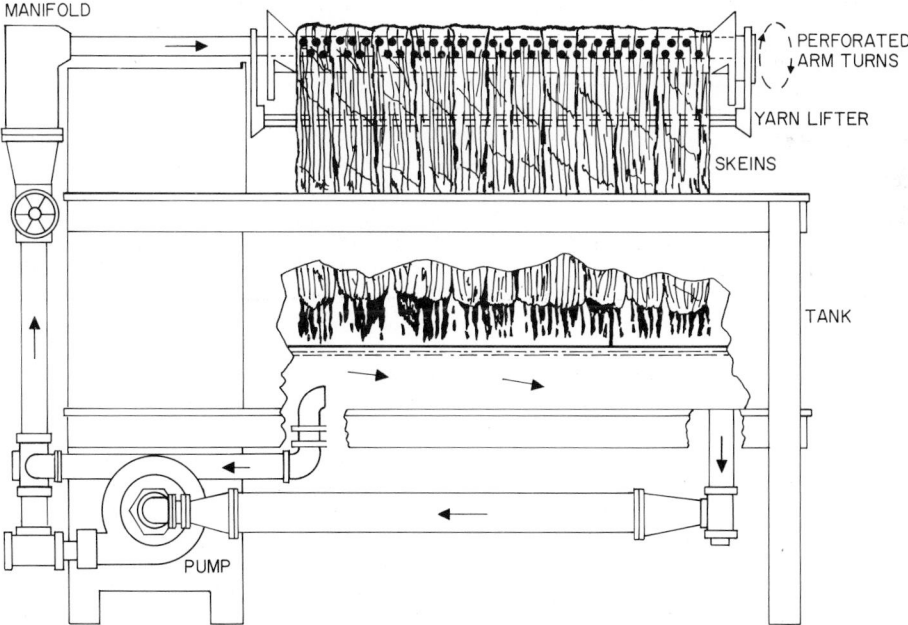

Fig. D-40. Skein dyeing (Cascade machine). Yarn hanks are suspended from perforated arms (stainless steel) located above an open tank which contains the prepared dye bath. Liquor from the bath is pumped from the tank through the arms and is distributed through the skeins, returning to the tank to be recirculated. Periodic rocking of the perforated arms assists in uniform distribution of the dyestuff.

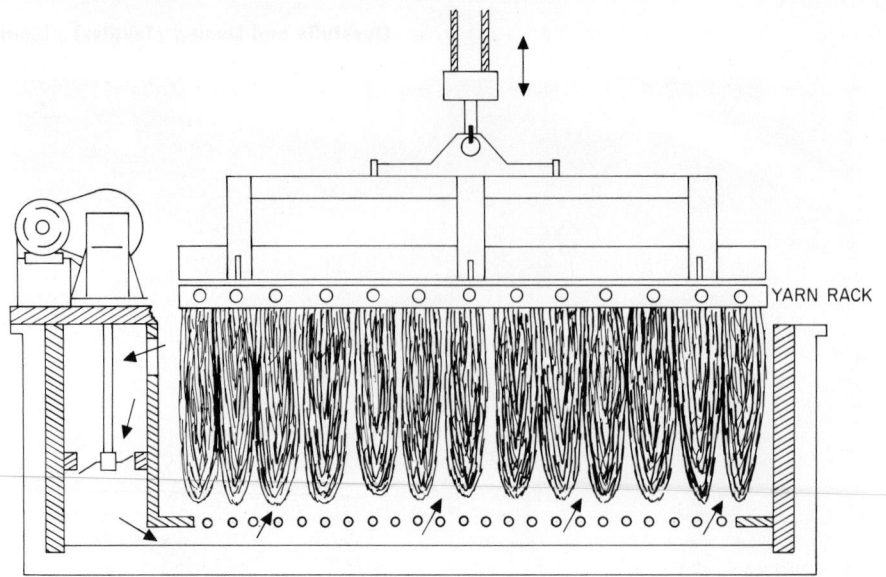

Fig. D-41. Skein dyeing (Hussong machine). A rack from which skeins are hung is lowered into the dye bath. A propeller circulates the dye through the skeins. In some machines, the cover may be closed to permit dyeing under pressure.

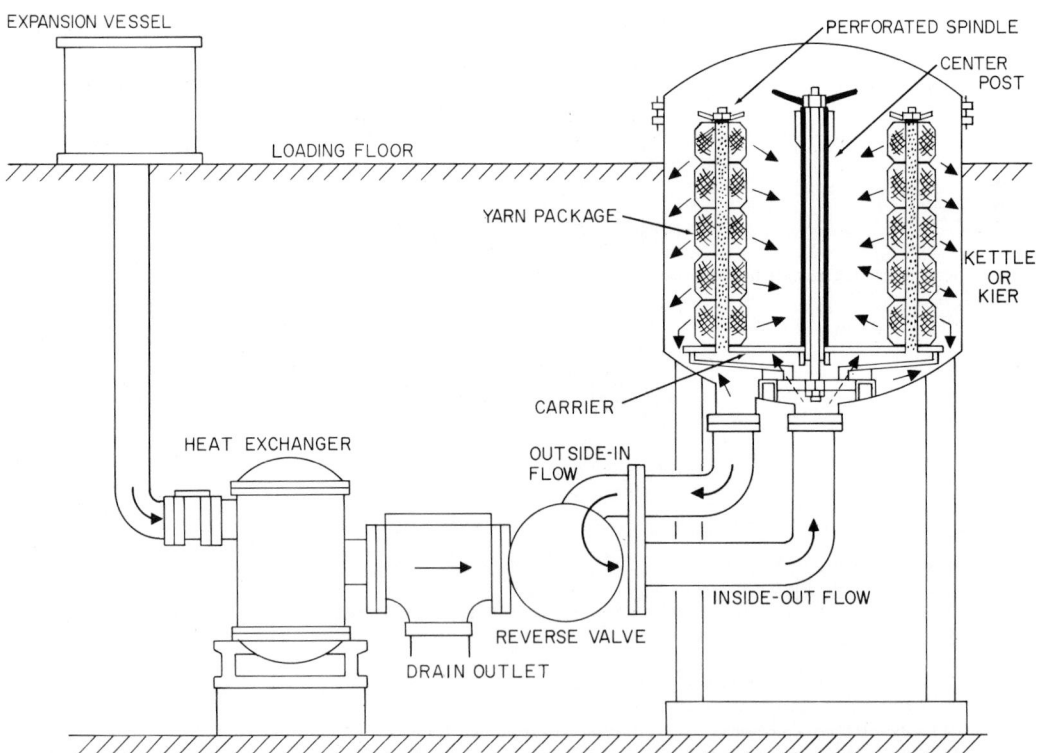

Fig. D-42. Package dyeing. Yarn packages are placed on vertical spindles of a portable carrier which is lowered into the dyeing vessel. The dye liquor is circulated continuously through the spindle perforations and through the packages, after which the flow pattern is reversed. These cycles are repeated until the desired shade is achieved.

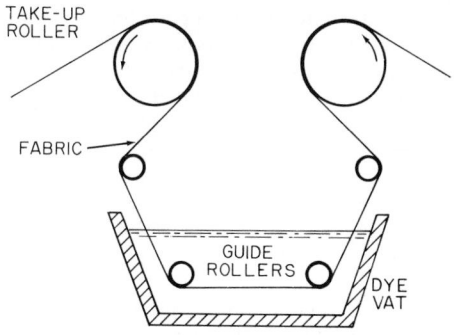

Fig. D-43. Fabric dyeing by jig. In the open jig illustrated, the fabric passes over and under a series of rollers through the dyebath. Usually, the direction of fabric travel will be reversed several times until the required depth of shade is achieved. When properly adjusted, there is a squeezing action to assist dye penetration into the fabric.

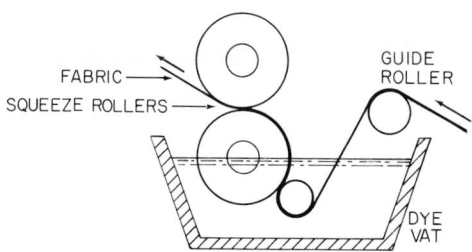

Fig. D-44. Fabric dyeing by padder. Squeeze rollers are used to press out the excess dye liquor. Where depth of color is required, several padders may be used in tandem.

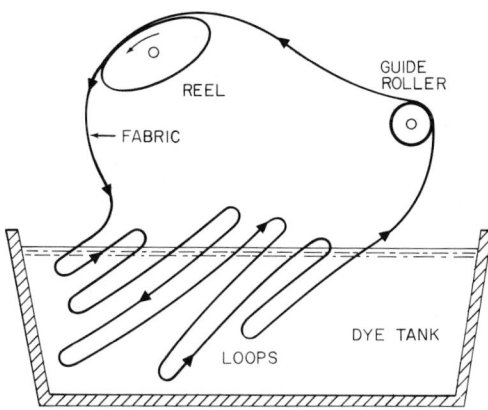

Fig. D-45. Dye beck. An endless band of fabric may be formed by sewing one end to another with fabric movement occurring as shown. Although the goods are in the tank during most of the cycle, the turning reel, by constantly moving the material, assures uniform distribution of the dye. Dye becks also process fibers in rope form.

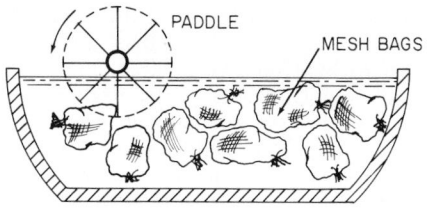

Fig. D-46. Paddle machine. Items such as finished garments, knit goods, and special fabrics are placed in loose-mesh bags and agitated in the dye bath by means of a paddle.

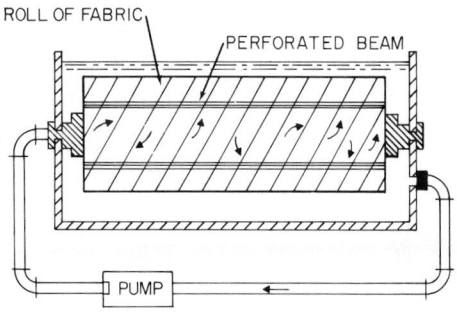

Fig. D-47. Beam-dyeing. In a unit known as the *Burlington beam dyer,* the dye is circulated continuously through a perforated beam on which is placed a roll of fabric. The process is better adapted to cloth with a relatively open structure than to closely woven fabrics.

the temperature is slowly raised to boiling, and dyeing is continued for 45–60 min. The pH of the dye liquor should be maintained between 2 and 3 for optimum results.

Method 2: The initial acidity is achieved by using organic acids such as acetic or formic. Dyes in this group do not level as easily as those in group 1. The dyebath is set with sodium sulfate, acetic acid, and dye. The temperature is raised to 37–48°C (100–120°F), the textile materials are entered into the dyebath, the temperature is raised to the boil, and dyeing continues for 30–45 min. Acetic or formic acid can be added as needed to increase exhaustion of the dye.

Method 3: This process is for the group of dyes that do not level well. The bath is set with sodium sulfate, ammonium acetate, acetic acid, and well-dissolved dye. The temperature is raised to 37–48°C (100–120°F), the materials are entered, and the temperature is raised slowly to the boiling point. Dyeing is continued for 60–90 min. Acetic acid may be added to exhaust the dyebath.

As stated, acid dyes are commonly used for protein fibers such as wool and silk. In addition, acid dyes are highly effective on nylon, modified acrylic fibers, and jute. These dyes are relatively colorfast to light on protein fibers and vary in colorfastness to laundering or dry cleaning. However, if properly applied, the acid dyes are very satisfactory for selected end uses of the protein-fiber fabrics. They are highly successful on the modified acrylic fibers which have had molecules grafted onto the acrylic fiber that react with the acid dyes.

Basic, or Cationic, Dyes: Basic, or cationic, dyes are so named because the dye molecules dissociate in water and the cation is the colored portion of the dye. The cation is attracted to some anionic sites in some fiber molecules and may form covalent bonds. Not all cationic dyes, nor all fibers, react to form this covalent bond; thus basic dyes are variable in their durability on the different substrates to which they are applied.

An outstanding characteristic of cationic dyes is their brilliance and intensity of color. Other properties include: not as soluble in water as some other types; may decompose at temperatures over $65°C$ ($150°F$). It is not recommended that basic dyes be applied simultaneously with acid or direct dyes because precipitation can occur. In general, basic dyes have low colorfastness on most fibers. On acrylic fibers cationic dyes exhibit relatively good colorfastness due to some covalent bonding reactions.

Basic dyes are applied from a dyebath with a pH of 5–6; when urea is used as a buffer, a pH of 7.5–8 is recommended. The temperature is held around $60-80°C$; tannic acid is used as an auxiliary. Retarders are commonly used with acrylic fibers to prevent uneven exhaustion.

Direct Dyes: Direct dyes constitute the largest group of dyes. They are water-soluble and easily applied to cellulose fibers. These dyes are frequently called *substantive colors*. Direct-dye molecules tend to be linear and will adsorb onto fiber molecules. In addition, hydrogen bonding and van der Waals forces help hold dye onto the substrate. Some aggregation of molecules occurs within the fiber which also lessens the migration of dye. Nonetheless, direct dyes are frequently of poor colorfastness and require special processing to improve fastness to washing or other water baths.

There are three general classes of direct dyes: class A, those that are self-leveling; class B, those that will level by the proper addition of salt; class C, those that require salt and additional temperature control to level.

Class A dye is made into a paste with cold water and a wetting agent (anionic or nonionic), and sufficient boiling water is added to form a uniform solution. This solution is added to the bath containing the required quantities of water and salt. The textile substrate is entered into the well-mixed dyebath at $40-50°C$ ($100-122°F$). The bath is raised to the boil over a period of 30–40 min, and dyeing continues at the boil for 45–60 min.

The method for class B differs only in that the salt is added in segments to promote leveling of color. The gradual addition of salt occurs during the time the temperature is being raised. For some dyes the salt addition may continue during the boiling period.

For class C dyes the salt is added after boiling commences and continues to be added slowly during the boiling period. It is necessary to control temperature rise so that the time to the boil is extended over a prescribed period.

Aftertreatment improves the colorfastness to laundry of direct dyes. Methods include formaldehyde aftertreatment, chrome aftertreatment, organic cationic fixing rinses, and development by diazotization.

Cationic fixing agents are used as final rinses. One to eight percent on weight of fabric (owf) of the agent is used in a rinse bath at $70°C$ ($160°F$). While these rinses tend to improve the washfastness, they frequently decrease the lightfastness of the dyes. They may alter the color.

The washfastness of direct dyes can be improved noticeably by rinsing in a solution containing 1–2% formaldehyde and 1–2% acetic acid. The dye substrate, usually in fabric form, is treated in the solution for 30 min at $60°C$ ($140°F$).

A few direct dyes can be improved in fastness to washing and bleeding by a chrome aftertreatment. The substrate is rinsed for 30 min at $60°C$ ($140°F$) in a solution containing either sodium dichromate or chromium fluoride plus acetic or formic acid.

The most successful method for improving washfastness of direct dyes is the development of the dye by diazotization. A developer such as beta-naphthol, *m*-phenylenediamine, phenol, phenylmethylpyrazolone, or resorcine is used to react with a free amino group on the dye molecule. The reaction should occur in situ on the fiber. The substrate is dyed according to a regular direct-dye method. The material is then diazotized for 20-30 min at $27°C$ ($80°F$) in a bath containing 1.5–3% sodium nitrite, 3–5% sulfuric acid, or 5–7% hydrochloric acid. The diazotized dyeings are developed for 20–30 min in a cold bath containing the developer recommended. The amount of developer used depends upon the type selected and the depth of color desired. Developing of direct dyes frequently changes the color because developer may add new chromophoric groups to the dye molecule during the chemical reaction that occurs.

Developed direct colors have improved colorfastness to laundry, but the colorfastness to light is frequently reduced. In general, direct dyes are inexpensive and easy to apply.

Disperse Dyes: Disperse dyes were developed first for cellulose acetate and were called *acetate dyes.* With the advent of the newer man-made fibers that are hydrophobic, scientists expanded the use of this group of dyes to include several other fiber groups, and the present term disperse found increased usage. Disperse dyes are not soluble in water, but they will form uniform dispersion in a water bath.

A standard theory states that disperse dyes, due to their lack of affinity for water, migrate toward and diffuse into the organic fiber and form a solid solution within the hydrophobic fiber. A newer theory states that the dye is transferred from the suspension to the fiber in molecular form. The pigments are soluble in water to a very minute degree, but in that state are highly substantive to the fiber. As the dye moves from the bath to the fiber, other dispersed molecules dissolve and in turn are absorbed by the fiber. In addition, there is evidence that hydrogen bonding and van der Waals forces increase dye retention.

Colorfastness of this group varies, depending on the fiber substrate. Major problems occur when some of these dyes on acetate and nylon are exposed to gas fumes and gas fading results, and when they encounter conditions that cause sublimation of the dye. The latter tends to occur when these dyes, on polyesters or acetates, are subjected to the temperatures involved in heat-setting processes.

Disperse dyes can be applied to fabrics under pressure. This is frequently needed for adequate dyeing of such fibers as polyester.

The processes for applying disperse dyes vary among fibers. They are applied from a water dispersion at temperatures appropriate to the fiber substrate. A typical application to nylon follows.

The dye liquor is set with a surface-active agent and a sequestering agent. The substrate is entered, and the liquid is circulated through the substrate for 10 min. Following this, the dye, well dispersed and diluted, is added; the temperature is raised to 93°C (200°F). The substrate is dyed for 20–40 min. The dye solution is drained, and the machine is filled with water and a small amount of sequestering agent. The substrate is rinsed in this solution for 10 min. The machine is again drained, and the material is rinsed in cold water.

When applying disperse dyes to polyester, the use of pressure is frequently required. The bath is prepared with a surface-active agent; the dye is added; the temperature is raised to 95°C (200°F); and the material is processed for 15 min. Acetic acid may be added to adjust to the pH recommended for the dye. The machine is sealed, and the temperature is raised very slowly to 120°C (248°F). The dyeing operation is continued for 1½ hr. The machine is cooled to 100°C (212°F) and opened; the material is then thoroughly rinsed.

Disperse dyes may be applied to polyesters by means of adding carriers to the dyebath. Care must be taken to ensure uniform solution of the carrier, or a spotty dyeing may result.

An important and more recent process used in applying disperse dyes is the *thermosol method.* In this technique the dye is padded into the substrate and it is subjected to dry heat. This method has several advantages: fabric moves on a continuous basis through the dye and the thermosoling range; carriers are not required; and fabrics are handled in open width, eliminating the possibility of unwanted creases.

Disperse dyes are used on many of the hydrophobic fibers. They give relatively good colorfastness and provide an adequate color range.

Vat Dyes: One of the best-known types of dyestuffs is the vat dye. Natural vat dyes were used in Egypt prior to 2000 B.C. A synthetic vat dye was developed in 1897, and since then a great number of vat dyes have been synthesized, and these have replaced natural vats. The original vat dyes included such well-known dyes as indigo and tyrian purple.

Vat dyes are organic compounds that are insoluble in water, but they possess the unique characteristic of reducing in an alkaline solution to a leuco form that is water-soluble and substantive to cellulose fibers.

These dyes are characterized by good to excellent fastness to laundering and, in general, good fastness to light.

Vat dyes are bonded to the cellulose molecule

by hydrogen bonding and van der Waals forces. After the dye is applied, it is oxidized to an insoluble form, and thus becomes trapped within the fiber.

Vat dyes can be categorized into four groups, depending on the method of application. Method 1 is used for dyes that require a high concentration of alkali and relatively high vatting and dyeing temperatures. No electrolyte is required for exhaustion of the dye. The temperature for vatting and dyeing is 60°C (110°F). The dye paste or powder is diluted with water at the correct temperature (vatting temperature); a wetting agent is added; the required amount of caustic soda is added; then the sodium hydrosulfite (reducing agent) is added slowly. The reduction process requires 10–15 min. When the bath is ready for dyeing, the goods are entered and processed at the required temperature for the recommended time.

In method 2 a moderate amount of alkali is required, lower temperature for vatting and dyeing are used, and salt (an electrolyte) is added to increase exhaustion. The salt is added to the dyebath as a last step. The temperature for vat reduction and for dyeing is 50°C (122°F). The general process is the same as for method 1.

The method 3 procedure differs from method 2 as follows. Slightly less alkali is used; the vatting temperature is 40°C (104°F) and the dyeing temperature is 20°C (68°F); the amount of salt (electrolyte) is considerably greater than for other methods.

Method 4 is not really a distinct method. Rather, it provides for those dyes that have individual requirements for dyeing.

Following the application of vat dyes, the substrate is oxidized to develop the color. To promote evenness and speed of oxidation, the dyed material is frequently rinsed in cold water containing an oxidizing agent such as hydrogen peroxide or sodium perborate. The material is then thoroughly soaped and rinsed, and dried.

Vat dyes are considered to have excellent colorfastness, but the actual degree of colorfastness depends upon the chemical formulation of each individual dye. Some do not perform as well as others. However, the vat dyes, in general, do possess colorfastness that is rated high.

Reactive Dyes: The most recent development in dyeing textiles has been the introduction of reactive dyes. These dyes are unique in that they form a covalent bond with the fiber molecule. This provides excellent washfastness and good to excellent fastness to other environmental factors. While these dyes are used primarily on cellulose fibers, some are being used successfully on silk, wool, nylon, and acrylic fibers.

Reactive dyes contain a chemical group that forms a bond with either a hydroxyl group in the cellulose molecule or with an amino group in the other fibers.

Although reactive dyes provide excellent colorfastness, they are somewhat more expensive than other dyestuffs, and in general require the addition of expensive dyeing assistants. Reactive dyes in solution cannot be stored because there is color loss and chemical breakdown.

These dyes may be applied by either continuous or noncontinuous methods. A typical example of noncontinuous methods would be dyeing in a winch-type machine. The substrate is loaded into the machine with the recommended amount of salt in the water. The liquor is heated to the required temperature, 40°C (105°F) or 60°C (140°F), and the predissolved dyestuff is added. When sodium bicarbonate is used, it is added at this time. The dyebath is run for 20–30 min and the recommended quantity of soda ash is added slowly; dyeing continues for 20–40 min. The goods are thoroughly rinsed, washed at the boil with soap, rinsed again, and dried.

Continuous methods are gaining in use because of the saving in labor. Excellent results occur. There are three general continuous methods: thermofixation, single-bath pad-steam fixation, and two-bath pad-steam fixation. In the thermofixation process the fabric is padded with a solution of dye and urea, wetting agent, and alkali. It is then dried in a temperature-controlled oven, rinsed, soaped, rinsed again, and dried.

The pad-steam method, single bath, pads the material with dye, wetting agent, and alkali; the fabric is dried and passed through steam for fixation of color; it is then rinsed, soaped thoroughly, rinsed again, and dried. All this is done on a continuous basis as the fabric passes through the machinery.

The only difference between bath methods 1 and 2 occurs between the first dry step and the steaming fixation. In bath 2 a second padding process is inserted in which strong alkali is padded into the material.

It should be emphasized that any dyestuff, regardless of type, comes from the dye manufacturer with suggestions for application. Most dye technologists use these suggestions, which usually save time, energy, and equipment and prevent inferior dyeing.

Two or more dyes that are of the same type and compatible in methods of application can be blended to produce color hues not obtained by a single dye.

Dyeing Blends: The dyeing of blends presents the dryer with several problems. These result from the fact that each fiber type reacts differently to dyestuffs and will accept certain classes of dyes only. A further problem is caused by the differences in time, temperature, and dyeing assistants needed for a selected fiber and a selected dyestuff.

Blends may be dyed in steps. That is, one fiber is dyed and the fabric is thoroughly rinsed; the second fiber is then dyed and the fabric rinsed. This can be repeated if more than two fibers are present that require different dye types which cannot be applied in a compatible bath.

Some dyes can be mixed and applied to the blend fabric in one step. This is possible when temperatures and time of application are compatible and when no adverse reactions occur between any of the dyeing assistants used. The reader is referred to technical bulletins prepared by both dyestuff and fiber manufacturers that provide recommended procedures for dyeing textile products.

Evaluation of Dyes: No colored textile product will be acceptable to consumers if it loses color easily and quickly under expected use and care. Thus dyes and dyed fabrics are subjected to various evaluation procedures to determine colorfastness to different environmental factors. These tests include colorfastness to washing, light and sunlight, acids, alkalies, perspiration, crocking (rubbing), chlorinated water, hypochlorite bleaching, chlorination, dry cleaning, pressing wet and dry, and gas fumes. Tests are clearly described in manuals from the American Association of Textile Chemists and Colorists and from the American Society for Testing and Materials. Evaluation scales are cited with each test procedure.

REFERENCES

Beech, W. F.: "Fibre-reactive Dyes," SAF International, New York, 1970.

Schmidlin, H. U.: "Preparation and Dyeing of Synthetic Fibers," Reinhold, New York, 1963.

Trotman, E. R.: "Dyeing and Chemical Technology of Textile Fibres," Griffon and Co., London, 1970.

Venkataraman, K.: "The Chemistry of Synthetic Dyes," Academic, New York (four volumes in print, two more to be published), 1952–1971.

Wilcock, C. C., and J. L. Ashworth: "Whittaker's Dyeing with Coal-Tar Dyestuffs," 6th ed., Textile Book Service, Metuchen, N.J., 1964.

—Marjory L. Joseph, *California State University, Northridge; Northridge, Calif.*

Dynamite (See **Explosives.**)

Dysprosium (Gk. *disprositos*, inaccessible.) **Dy** = 162.50 (at. wt.); 66 (at. no.). Dysprosium is the tenth of 15 elements in group III, period 6, generally shown as the rare-earth elements in a separate line below the main body of the periodic table; sometimes referred to as the *lanthanide series.*

Pure metallic Dy is silvery gray and retains its metallic luster in air at room temperature. It is stable up to 400°C and then oxidizes slowly up to 600°C. The metal is soft and can be worked by conventional tools and equipment to rod, ribbon, and foil.

The melting point of Dy is 1410°C; boiling point 2562°C; density 8.540 g/cm^3. It is the fourth most abundant of 15 metals in the lanthanide group. Estimated at three parts per million in the earth's crust, it ranks 42d of the 83 natural elements. This means that Dy is potentially more available than beryllium or tin.

Dysprosium has seven natural isotopes: 156, 158, 160 (2.3 w/o), 161 (18.9 w/o), 162 (25.5 w/o), 163 (24.8 w/o), and 164 (28.2 w/o).* Twelve artificial isotopes have been identified. The metal has a low acute-toxicity rating.

The thermal-neutron-absorption cross section of natural Dy is 940 barns per atom. This is the fourth highest absorption after gadolinium, samarium, and europium of the 83 available elements. The isotope 164 (28.2 w/o) absorbs at 2,700 barns per atom. None of the natural isotopes are radioactive.

Dysprosium has one valence of 3^+. Its oxide, Dy_2O_3, is very stable and does not react with moist air and hydrolize as is the case with lanthanum-cerium-group "light" rare earths. The white oxide powder melts at 2340 ± 10°C. The element is a strong cation (electropositive). Both the metal and oxide dissolve in mineral acids and react with inorganic and organic anions, forming a wide range of compounds. The characteristic color of the 3^+ ion in solution is yellow.

The reason Boisbaudran chose the name dysprosium, which means "hard to get at," is because

---

*This abbreviation is used in the rare-earth and related fields for weight percent.

of the successive discovery of new elements found in what was believed to be pure material. Dysprosium was the third from last of the 15 lanthanons to be identified (1886), over a century after the original mineral was noted. Dysprosium was separated from holmium, which, with thulium, had been found in erbium after, along with terbium and ytterbium, it had been separated from the "yttrium earths," a mineral now known as gadolinite. The name dysprosium—"hard to get at"—is no longer appropriate.

Dysprosium is the most plentiful of the "heavy" rare earths (at. nos. 65–71) when yttrium (no. 39) is excluded. In the known minerals, however, yttrium is present in concentrations at about 8 to 10 times the amount of dysprosium. Since minerals like apatite, euxenite, gadolinite, and xenotime are processed for their yttrium content, a plentiful supply of dysprosium (and other heavy lanthanides) is available as a coproduct. Fortuitously, the separation factor between yttrium and dysprosium in liquid-liquid organic and solid-resin organic ion-exchange systems is favorable. Excellent yields of both Y and Dy at high purity are obtained in ion-exchange systems that are under control.

Anhydrous DyF_3 is reduced by calcium in a sealed bomb reactor heated above the melting point of Dy.

See **Rare-earth elements and metals** for a detailed description of processing and a table of properties.

Uses: The four major natural isotopes of Dy, their high neutron-absorbing capability, and the stability of the metal makes Dy the logical material for detecting and measuring nuclear reactions and exposures. The use of Dy metal foil and Dy_2O_3 in *dosimeters* was the first application.

When Dy is used as a neutron sponge, no harmful decay radiations are emitted. Nor is helium generated (alpha particles), which would strain and eventually crack structural parts containing nuclear fuel. A stainless steel containing about 3% Dy_2O_3 is used in the shank of certain control rods for high-flux-beam reactors.

A dysprosium oxide–nickel cermet has been proposed to eliminate water cooling of nuclear-reactor control rods.

Dysprosium chemicals can be used as catalysts, although the relatively high cost has retarded its use. Dy_2O_3 on alumina polymerizes ethylene.

In mixed ferromagnetic "garnets," dysprosium oxide is used with yttrium and gadolinium oxides to obtain selected microwave properties and temperature-effect compensation. Dysprosium added to tantalum oxide for glass-phase semiconductor spark plugs is a potential large use.

Dysprosium oxide fluoresces yellow in glass under ultraviolet radiation. It could also be the activator for the yellow component of phosphor used in black-and-white TV.

This versatile element offers interesting properties when combined with other elements in the development of photoelectric, semiconducting, and thermoelectric materials.

REFERENCES

Overview No. 17 (application information including possible uses), Molybdenum Corporation of America, New York.

See also list of references under **Rare-earth elements and metals.**

—Joseph G. Cannon, *Molybdenum Corporation of America, White Plains, N.Y.*

EDTA (Ethylenediaminetetraacetic acid) (See **Chelating agents.**)

Eggs (See **Amino acids; Pasteurization;** and **Peptides and proteins.**)

Einstein's Law (See **Nuclear power plants.**)

Einsteinium (See **Chemical elements.**)

Elastomers An elastomer is a polymer* with elastic (or rubbery) properties and may be natural or synthetic. Natural rubber is a prime example of the first. See also **Rubber, natural.** Many high-molecular-weight materials having certain rubbery characteristics at room temperature are known as synthetic rubbers even though most of their chemical compositions differ from that of natural rubber. Since the synthetic elastomers of

---

*A polymer is a substance consisting of molecules which are, at least approximately, multiples of low-molecular-weight units, or monomers. Isoprene (2-methylbutadiene-1,3) is chemically C_5H_8, while polyisoprene is $(C_5H_8)_x$, where $x \geqq 2$ and normally is 1,000–10,000 for rubbers. See also **Butadiene; Isoprene;** and **Polymerization.**

commerce are popularly known as rubbers, this name will be used here where it seems appropriate.

Various synthetic rubbers were under development prior to 1940, mainly in an effort to find a rubbery material useful in places where natural rubber was inadequate. Though of value in various critical uses, none of these synthetics became of great importance to the rubber industry as a whole until most of the Western world's supply of natural rubber was cut off by World War II. At present the various synthetic rubbers fill 75–80% of the industrial and commercial requirements for elastomers in the United States.

Many rubbers normally classified as general-purpose rubbers are produced primarily for tires but also used for many other items on a smaller scale. Materials referred to as *specialty rubbers* find little if any use in tires but do contribute significantly to a variety of other products. A list of the major synthetic elastomers and principal use categories is given in Table E-1. It may appear that some of these rubbers duplicate each other and could be used interchangeably, but the following example will prove that this is not the case.

A tire tread made of a blend of styrene-

TABLE E-1. *Uses of General-Purpose and Specialty Elastomers*

Type	Percent of Total Consumption*
General-purpose elastomers:	
Styrene-butadiene rubbers (SBR)	50
Natural rubber	22
cis-Polybutadiene rubbers	10
cis-Polyisoprene rubbers	3
Specialty elastomers:	
Ethylene-propylene rubbers (EPDM)	<2
Butyl (IIR, copolymer of isobutylene and isoprene)	<3.5
Polysulfide rubbers	<0.5
Silicones	<0.5
Polyacrylates	<0.1
Fluorocarbons	<0.01
Neoprene	5
Nitrile rubbers (NBR)	2.5
Polyurethanes	0.5
Hypalon	0.5

Consumption of Specific Synthetic Rubbers

SBR . . . . Tires and tire products (68%), miscellaneous solid rubber uses (11%), footwear and shoe products (4.6%), miscellaneous latex uses (4.5%), foam-rubber products (3.7%), miscellaneous mechanical goods (3.7%), rubber hose and tubing (1.2%), other uses (each less than 1%): belts and belting, sponge rubber, proofed goods, wire and cable insulation, and adhesives

cis-Polybutadiene . . . Tires (92%), high-impact polystyrene (3.8%), belting (2.3%), and sponge-rubber underlay (1.9%)

Butyl . . . . Tire linings, tubes, curing bags, and off-road tires (77.7%); miscellaneous mechanical goods (11.2%); sealants, adhesives, and tapes (5.6%); hose (2.2%); seals (1.1%); coated fabrics (1.1%); and rolls (1.1%)

cis-Polyisoprene . . . . Tires and tire products (50%); mechanical goods (25%); miscellaneous uses, such as sponge, drug sundries, and sheet (19%); and footwear (6%)

EPDM . . . . Automotive parts (44.5%); hose (13.3%); tire development and tire products (sidewalls) (11.1%); appliance parts, excluding hose (6.7%); footwear (6.7%); wire and cable insulation (4.5%); O rings, seals, and gaskets (4.4%); miscellaneous uses (3.3%); belting (2.2%); rolls (2.2%); and proofed goods (1.1%)

Polysulfide . . . . Adhesives, sealants, and coatings (66.7%); miscellaneous uses (20.9%); rolls (8.3%); and hose (4.1%)

Silicones . . . . Electrical uses (41.4%), seals and gaskets (20.7%), electronic potting compounds (16.1%), sealants (16.1%), and miscellaneous uses (5.7%)

Polyacrylates . . . . . O rings, packings, and gaskets (80%); adhesives, sealants, and coatings (12%); and miscellaneous uses (8%)

Fluorocarbons . . . . . O rings, gaskets, and seals (81.6%); miscellaneous uses (8.8%); hose (4.8%); and proofed goods (4.8%)

Neoprene . . . . Mechanical goods (not elsewhere classified) and automotive uses (17.7%); miscellaneous uses (16.9%); latex (16.2%); wire and cable insulation (13.9%); hose (12.3%); adhesives, sealants, and coatings (7.7%); footwear and shoe products (3.8%); belts and belting (3.8%); O rings, packings, and gaskets (3.8%); coated fabrics (3.1%); and rolls (0.8%)

*SOURCES: *Chem. Week,* March 18, 1970 and *Rubber World,* February 1971.

butadiene rubber (SBR) and butadiene rubber (polybutadiene) wears longer than a tread made of natural rubber when tested under ideal conditions, and both materials lose tread-wear resistance rapidly as temperature increases. In actual practice, the blend generates more heat on flexing than natural rubber. Thus, the blend is used in small tires (automobiles), where lower temperatures can be maintained, and natural rubber is used in large, heavy-duty truck and bus tires, which run at appreciably lower temperatures and wear longer than if they were made of the blend. A hose made of SBR would conduct water just as well as a hose made of neoprene, but if increased fire and solvent resistance were required, neoprene might be preferred. Many other factors, e.g., low-temperature flexibility requirements, expected useful life of the product, and economics, enter in the choice of rubbers.

To add to the complexity, some plastics, such as polyvinylchloride (not usually regarded as an elastomer) can be plasticized to produce materials with certain elastomeric properties, sometimes referred to as *plastomers*. No sharp line of demarcation can be drawn between rubbers (elastomers) and plastics because for many items, e.g., common garden hose, certain plastics as well as rubbers can be used. In even more items they can be used together. In fact, most polymers, especially those in the categories considered here, are plastics (either crystalline and/or amorphous solids) at some temperature. As the temperature increases, they pass through an elastomeric stage and often a liquid stage before reaching their decomposition temperature.

Properties: Chemical composition determines solvent and fire resistance and, in combination with structure, determines high-temperature stability and low-temperature flexibility. Polar groups enhance resistance to oils and other hydrocarbon fluids or gases and often detract from resistance to water and water solutions. Many rubbers are necessary to fill a variety of needs, considering the wide range of flexible products used over a broad span of temperatures.

Elasticity, a property of paramount importance in products such as tires and flexible hose, is not always the prime criterion in selecting a given rubber. For example, significant quantities of rubbers are consumed in paints and adhesives. See also **Paints.**

Rubbers are distinguished from one another by the structure and/or chemical composition of the polymer molecules and various chemicals mixed or inadvertently associated with the polymer. An essentially pure polymer is rarely supplied for fabrication of consumer products. In extreme cases, such as a 50 pt oil masterbatch of emulsion SBR, the "rubber" to be compounded into a consumer product contains only about 60% polymer. It is normal practice to add about 1% polymer stabilizer, such as antioxidant, to all synthetic rubbers in their production processes. The microstructure of the polymer molecules, i.e., the orien-

tation of the atoms in the individual monomer molecules (segmers) in the polymer makes a large contribution to the physical properties and usefulness of the rubbers. Various polymerization processes yield microstructures unique for the particular combination of catalyst, initiator, solvent, and monomers. The most widely used diene monomers, e.g., butadiene and isoprene, can be combined in cis-1,4; trans-1,4; and 1,2 and 3,4 (isoprene only) structures in homo- and copolymers. Systems have been developed that yield polymers in which the segmers have essentially only one structure or, within broad limits, any particular given mixture of structures.

Production of Synthetic Rubbers: Two general processes, emulsion polymerization and solution polymerization, are used. In the emulsion* process, the monomers are dispersed as small droplets in a water solution of an emulsifier by agitation. After addition of catalyst (or initiator), polymerization occurs in emulsifier micelles, forming the latex particles to which the monomers migrate from the droplet reservoirs. The temperature is controlled by transfer of heat from or to the water medium surrounding the latex particles.

In the solution† process, from which water normally must be rigorously excluded to assure efficient use of the catalysts, the monomers are diluted with an organic solvent and polymerized by the addition of catalysts which are quite different from those used in emulsion processes. The organic solvent serves as the reaction medium as well as heat-transfer medium.

Two processes used in the manufacture of synthetic rubber are shown in Fig. E-1. Many variations of these processes are possible, depending upon the product itself and the form desired by the ultimate user. Homo- and copolymers can be prepared in both processes. The structure, molecular weight, and various properties are determined by the polymerization process used as well as by the catalyst and other ingredients in the polymerization system.

Styrene-Butadiene Rubbers. This family of SBR rubbers includes monomer ratios up to about 50% styrene. When more than 50% styrene is used, the resulting polymers generally are more like plastics than rubbers. The most widely used members of

---

*Suspension polymerization is similar in some respects to emulsion polymerization and is used for some plastics but normally not for rubbers.

†Bulk polymerization is a special form of solution polymerization and is seldom used for rubbers.

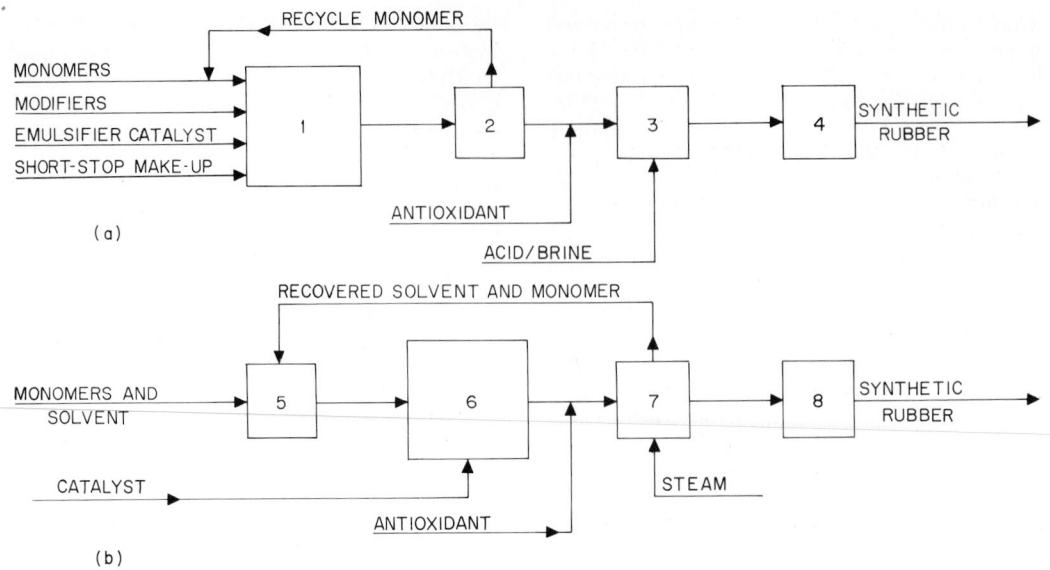

Fig. E-1. Processes for producing synthetic rubber. (a) Emulsion polymerization and (b) solution polymerization. (1) Reactor, (2) monomer recovery by steam distillation, (3) coagulation and washing, (4) drying and packaging, (5) monomer solvent drying, (6) reactor, (7) desolventizing and crumb dewatering, (8) drying and packaging.

the SBR family are those with about 25% styrene, which are polymerized in emulsion systems at 5–10°C. Although the major use for SBR is in tires, the SBR used in tread normally is different from that used in the sidewall or carcass. The best SBR for shoe soles, adhesives, and other products is also different. There is tremendous variation both in processing techniques and in the final products. They can be made at different temperatures, at different viscosities, on different emulsifiers, and in different solvents. They can be stabilized with different antioxidants, masterbatched with different oils and carbon blacks, and coagulated by different techniques. Each of these variables can be controlled to achieve desirable properties and to minimize undesirable characteristics for given end uses.

Emulsion SBR (E-SBR): Early plants for making SBR were designed for 50°C polymerization. When it was found that SBR made at a lower temperature is particularly desirable for tires, many plants were remodeled to produce large quantities of what sometimes is referred to as *cold SBR*. However, production of the higher-temperature, hot SBR has continued on a lesser scale for many specialty uses.

Typical hot and cold formulations are given in Table E-2. The polymerizations are carried out

TABLE E-2. *Emulsion SBR Polymerization Formulations*

	Hot SBR	Cold SBR
Ingredient, parts by weight:		
Water (deionized).	180	200
Fatty acid soap	5	4.5*
Styrene.	25	28
t-Dodecyl mercaptan . . .	0.35	0.2
Butadiene	75	72
Potassium persulfate . . .	0.3	
Redox initiator	. . .	†
Conditions:		
Temperature, °C	50	5
Time, hr	12	8
Conversion, %	75	60

*Usually a 50:50 mixture of fatty acid and rosin acid is used in the cold recipe to increase the fluidity of the soap solution and enhance the building tack of the rubber.

†A solution of $FeSO_4 \cdot 7H_2O$ (0.02 parts) chelated with the tetrasodium salt of ethylenediaminetetraacetic acid (0.04 parts) and sodium formaldehyde sulfoxylate (0.03 parts) is included in the soap solution, and cumene hydroperoxide (0.08 parts) is dissolved in a portion of the styrene for charging after the other ingredients have been mixed and cooled to polymerization temperature. Additional, equal quantities of the hydroperoxide may be required to maintain a high polymerization rate.

in steel reactors, often glass-lined or clad with stainless steel, equipped with an agitator, external jacket, and frequently internal coils for exchange fluids to effect temperature control. For large-scale production, the more efficient plants have groups of 6–10 reactors connected in series to form a continuous polymerization chain. The ingredients are pumped continuously into one end of the chain at a rate so that the desired conversion is obtained by the time the product, as latex, is discharged at the other end of the chain. Whether the process is batch or continuous, a *shortstop* ingredient is added to prevent further conversion. The latex is further processed to remove residual unpolymerized monomers and, after the addition of antioxidant, proceeds to coagulation equipment, where crumb is formed. The wet crumb is separated from the serum, dried, and baled for shipment.

Viscosity Specifications: Although rubbers normally are shipped as bales and handled as solids, they are very viscous liquids and much of their usefulness depends on their ability to flow during processing before conversion into elastic solids by the vulcanization or curing step. The raw or crude rubber must meet certain specifications, especially a viscosity limit, which is related to processibility and normally reported on a Mooney plasticity scale, abbreviated ML-4. The ML-4 specification for many rubbers is 50 ± 2. Some are produced at this ML-4; others, e.g., for oil masterbatching, are made at much higher plasticity so that the blend of rubber and oil will have an ML-4 of 50 ± 2. Liquid rubbers having a viscosity similar to molasses and useful in the preparation of solid propellants for rockets are too fluid to be measured by a Mooney test machine. At the other extreme, the higher the ML-4 the more difficult the rubbers are to process and the greater the power required for mixing in the compounding ingredients. Above ML-4 of 150, most rubbers crumble instead of flowing in the mixing operations. Thus, except in very special cases, rubbers and masterbatches are made with ML-4 less than 100 and most frequently in the 35–60 range.

The two most important factors controlling the ML-4 of emulsion rubbers are the level of chain-transfer agent (*t*-dodecyl mercaptan in the recipes cited) and conversion. Roughly, the ML-4 is inversely proportional to the mercaptan level and directly proportional to the conversion. It is also recognized that various quality features, particularly for emulsion SBR used in tires, go down as

conversion increases above about 65%. In order to produce the best quality SBR for tires, conversion is stopped at about 60%. SBRs for other purposes are often run to 75–80% conversion, and where the product is to be used as a latex 90–95% conversion is often practiced.

Molecular Structure: The polymerization of emulsion SBR is initiated by free radicals generated by the redox system in cold SBR and by persulfate or other initiator in hot SBR. The initiator rarely becomes involved in the molecules of the polymers and thus has essentially no control over their structure. Nearly all the molecules are terminated by fragments of the chain transfer agent, which in this case is a mercaptan. The majority of the polymer molecules may be schematically pictured as RSM_nH, where the RS represents the $C_{12}H_{25}S$ part of a dodecyl mercaptan molecule, M represents the monomers involved, n the degree of polymerization, and H a hydrogen atom formerly attached to the sulfur of a mercaptan molecule. In free-radical-initiated polymerizations of butadiene, either by itself to form homopolymers or with other monomers to form copolymers, the butadiene segmers will be about 18% 1,2; 16% cis-1,4; and 66% trans-1,4.

SBR Latex: A significant amount of SBR is consumed as latex for preparation of foam rubber, paints, adhesives, and fabric treatment. Normally latices with solids contents above 50%, often as high as 65–70%, are most suitable. The polymerization technique is basically the same as that already described, with recipe changes appropriate to materials involved and to the process used for achieving high solids.

The rubber in latices prepared for dry rubbers, as for tires, is dispersed in very small particles (less than 0.1 μm diam). A concentration above approximately 40% solids yields a very viscous, nonusable latex. Various techniques have been developed for producing larger particles which will yield fluid latices of high solids content. Use of less water and reduced emulsifier levels yields higher-solids, larger-particle-size latices directly in the reactor, but polymerization rates are very low and reaction times of 40–65 hr are not unusual. Chemical and mechanical methods have been developed for increasing the average particle size of the normally produced latices, after which, in a second step, they can be concentrated by creaming. The latex-concentration step is described under **Rubber, natural.** Concentration also can be effected by removal of part of the water by evaporation. For solids contents of about 55% or more, a two-step process normally is the most economical.

Styrene Butadiene Copolymers Prepared in Solution.

S-SBR (Solution SBR): One technique involves the copolymerization of styrene and butadiene by polar or ionic mechanisms to produce what is recognized as solution SBR. A wide range of styrene-butadiene ratios and molecular structures, i.e., the microstructure of the butadiene segmers and distribution of the styrene segmers, is possible. Copolymers with no chemically detectable blocks of polystyrene constitute a distinct class of solution SBRs and are most nearly like styrene-butadiene copolymers made by emulsion systems. As would be expected, these materials can be processed in the same manner as E-SBR and used as a direct replacement or in blends with the latter. Solution SBRs with terminal blocks of polystyrene (indicated by the notation S-B-S) have the properties of self-cured elastomers, which are processed by techniques suitable for thermoplastics and do not require vulcanization.

Lithium alkyls, such as LiC_4H_9, are used as the catalyst. The microstructure of the butadiene segmers is strongly influenced by the nature of the organic solvent used. The proportion of 1,2-butadiene segmers may be increased by the use of increased levels of ethers, such as diethyl ether and tetrahydrofuran, in the solvent. The formation of polystyrene blocks is controlled by variations in monomer-charging technique and conversion levels. Since temperature has not been found to affect quality as much as in emulsion systems, polymerizations are normally run at 60–70°C. The polymerization recipes are rather simple compared to those used in emulsions, consisting basically of only three ingredients: monomer at about 15% concentration in the organic solvent and catalyst.

Alfin Copolymer: This solution SBR is prepared by using a catalyst consisting of an alcoholate, such as sodium isopropoxide, allyl sodium, and sodium chloride. (The name *alfin* is derived from *al*cohol and ole*fin*.) Although alfin catalysts and polymers have been known since the mid-1940s, an effective method for controlling the polymer molecular weight is a relatively recent development. The microstructures of these polymers are very close to emulsion rubbers made from the same monomers.

Stereoelastomers. Stereospecific solution polymerization has received emphasis since the discovery of the complex coordination catalysts that yield polymers of butadiene and isoprene having highly ordered microstructures. The catalysts normally are mixtures of organometallic and transition-metal compounds. A typical example of triethylaluminum with titanium tetrachloride. Stereopolyisoprene with over 95% cis-1,4 structure closely resembles the elastomeric part of natural rubber. The solution polymers, as produced, have appreciably higher purity in that they are not mixed with protein or other nonelastomeric residues found in the natural product; nor do they contain electrolyte and emulsifier residues commonly present in emulsion rubbers. Low levels of materials such as antioxidants are added to improve their aging.

As indicated by Table E-1, the bulk of *cis*-1,4-polybutadiene is used in tires. The material seldom is used alone because of its poor processibility and poor traction. Blends with SBR yield tire treads that are appreciably better, overall, than can be made from a single rubber. The high purity of solution polybutadiene makes it a desirable replacement for emulsion polybutadiene in the preparation of high-impact polystyrenes. Since gum stocks of the high *cis*-1,4-polyisoprene have strength nearly as high as similar stocks of natural rubber, the natural material must face competition in connection with such items as cut elastic thread, golf balls, and shoes, as well as in products where carbon black is used, such as tires.

Stereospecific catalysts, also called *coordination catalysts,* generally have not been found useful in copolymerizations, although there is considerable research in this direction. *trans*-1,4-Polyisoprene is used in golf-ball covers, but uses for the other stereopolymers of the dienes, e.g., 1,2-polybutadiene, 3,4-polyisoprene, and the trans-1,4 polymers remain to be developed.

Butyl Rubber. Officially coded as IIR, this rubber is a copolymer of isobutylene and isoprene. The elastomers contain only 0.5–2.5 mole % isoprene, introduced to cause sufficient unsaturation to make the rubber vulcanizable. The polymerizations normally are carried out at −80 to −100°C in methyl chloride as solvent and with anhydrous aluminum chloride and a trace of water as the catalyst.

Butyl rubber, one of the earlier synthetic rubbers, lost out to emulsion SBR when the decision (during World War II) was made to scale up production of a synthetic material when the natural rubber supply was cut. Incompatibility of butyl with natural rubber and the difficulty with which it is cured contributed to the decision. It is claimed that the more recently developed chlorobutyl rubbers, containing 1.1–1.3% chlorine, overcome these difficulties.

The major use for butyl rubber for many years has been in inner tubes for tires. Development of tubeless tires for automobiles, however, adversely affected the growth rate for butyl-rubber production. Two characteristics make butyl rubber particularly useful in motor mounts and other vibration-damping applications, namely high energy absorption and low rebound. Since butyl rubber is practically free of double bonds, the material has high resistance to aging, a property which makes it useful in curing bags for tire production and in coatings exposed to weather.

Ethylene-Propylene Elastomers. Referred to as EPR, this material has limited usefulness because it cannot be vulcanized in readily available systems. The terpolymer EPT (or EPDM, where DM stands for a nonconjugated diene monomer) may have potential. The rubbers are made from low-cost monomers, have good mechanical and elastic properties, and show outstanding resistance to ozone, heat, and chemical attack. The cured rubbers are superior to butyl in dynamic resilience and are flexible to very low temperatures; their brittle point of about $-95°C$ is below that of SBR and natural rubber.

Acrylonitrile-Butadiene Rubbers (NBR). Commonly called *nitrile rubbers,* these are made by emulsion polymerization with recipes essentially the same as those used for emulsion SBR except for the monomers. Cold and hot NBRs at various acrylonitrile-butadiene ratios, covering the range of about 15:85 to 50:50, are commercially available in dry-rubber form as well as latex. Synthetic emulsifiers are often used in the polymerization to minimize discoloration and yield a more desirable rubber for certain uses.

The nitrile rubbers are noted for their solvent resistance, which increases with the acrylonitrile content. Unfortunately, the flexibility at low temperature decreases as the nitrile content increases, so that the 50:50 copolymer is very stiff at $-10°C$, becoming brittle at about $-20°C$.

The copolymerization of these monomers in emulsion is somewhat unusual in that a 33:67 ratio polymer tends to form no matter what the actual charge ratio is. In order to prepare a nearly homogeneous 15:85 polymer a much lower monomer ratio is charged initially and the remainder of the acrylonitrile is charged at an appropriate rate during the polymerization. Conversely, to make a uniform 50:50 NBR, a higher monomer ratio is charged initially and the remainder of the

butadiene during the polymerization. Being solvent-resistant rubbers, the nitriles are used for gaskets and oil and gasoline hoses and are blended with polyvinyl chloride for solvent-resistant electrical insulation and food-wrapping film. Nitrile latices are often used in treating fabrics that must withstand dry cleaning.

Neoprene. The generic name *neoprene* is used to designate a family of dry rubbers and latices, the first of which was introduced in 1932.* Neoprene is made by the free-radical-initiated polymerization of chloroprene in emulsion systems. Two general classes are prepared. The earlier, *sulfur-modified* type contains a low level of sulfur (up to about 2%), interpolymerized with chloroprene. The polymer, as prepared, is practically non-processible. However, thiuram disulfide, added to the latex, reacts with the polysulfide groups, causing polymer cleavage and thus yielding a lower-molecular-weight, processible neoprene. For the other class of neoprene, chain transfer agents, such as alkyl thiols (mercaptans), are added to the polymerization recipe to control the polymer molecular weight, the same as for E-SBR.

The different neoprenes in each class are made with variations in the polymerization conditions, i.e., the nature and amount of the modifying agents, initiators, emulsifiers, temperature, conversion, and polymer stabilization system. Variation by the copolymerization of other monomers is seldom resorted to because of the tendency of chloroprene to form homopolymers even in the presence of other available monomers.

Neoprene is noted for its combination of fire-retardant, good solvent-resistant, and high-temperature-stability properties compared to butadiene and isoprene homo- and copolymers. The chlorine in each segmer deactivates the adjoining carbon—carbon double bond, making it less sensitive to oxidative attack. Moreover, the low reactivity of neoprene's double bonds makes the normal sulfur vulcanization recipes used in SBR and natural-rubber compounds ineffective for the curing of neoprene compounds. Instead, metal oxides, such as ZnO and MgO, serve as curatives.

Polyurethanes. Polyurethanes probably constitute the most versatile family of polymers since its many members range from soft elastomers to rigid plastics in both solid and foam products. These materials are described under **Urethanes.**

Hypalon (Chlorosulfonated Polyethylene). This ma-

*By Du Pont and called Duprene.

terial is an example of the conversion of an already formed plastic, polyethylene, to an elastomeric product. Polyethylene in solution is treated with chlorine and sulfur dioxide to introduce approximately 1.3% sulfur and 29% chlorine into the polymer. Most of the chlorine is attached directly to the carbon atoms in the backbone of the polymer, and the remainder is in the form of sulfuryl chloride groups, $\cdot SO_2Cl$, through which cross-linking occurs in the curing step with metal oxides. Because of its good oxidation and ozone resistance, this material often is used in products exposed to weather, e.g., calendered stocks for lining ditches and ponds. See also **Polyethylene.**

Polysulfide Rubbers. These rubbers are commonly referred to as *Thiokol rubber.** Thiokols are prepared by the condensation polymerization of sodium polysulfides with a dichloro- (often blended with a trichloro-) organic compound. Thiokol A, the first type, was made from Na_2S_4 and ethylene dichloride. Varieties with a greater range of elastomeric properties are being made with other compounds, such as bis(chloroethyl)formal. The reaction is carried out in water, forming a latex-type product from which the rubber is isolated. Liquid polymers terminated in a low-molecular-weight mercaptan also are produced. Thiokols are noted for their very high resistance to organic solvents.

Polyacrylate Elastomers. The copolymerization of ethyl acrylate with the acrylate esters of higher-molecular-weight alcohols to form the acrylic rubbers normally is done in emulsion or suspension systems. Because of their solvent-resistant properties and stability at elevated temperatures (150–175°C), the acrylates are used in automatic-transmission gaskets for automobiles.

Silicone Elastomers. Commercially useful members of the family of silicone rubbers have alternating Si and O atoms for a backbone, and the members differ from each other mainly in the nature of the organic substituents on the Si atoms and the degree of polymerization. In the absence of double bonds in the backbone, the many forms of stereoisomers found in the unsaturated hydrocarbon rubbers lack counterparts in silicone rubbers. The chemical combination of organic and inorganic materials provides the silicone rubbers with useful properties over a very wide temperature range (approximately −70 to 225°C). Organic rubbers generally are stronger at room temperature but lose strength rapidly with increase in temperature. The silicones, having a lower temperature coefficient, often exceed the tensile strength of organic rubber at high temperature. The silicones are noted for their excellent dielectric stability and high resistance to oils, chemicals, and weathering. See also **Silicones.**

Fluoroelastomers. The most costly of the commercially available elastomers, fluoroelastomers can be used at over 300°C and have excellent resistance to aromatic solvents, acids, and alkalies. Most fluoroelastomers are used to make seals and gaskets for use in hot-liquid systems, aircraft, and missiles. Some fluoroelastomers are made by the emulsion copolymerization of perfluoropropylene and vinylidene fluoride; some are copolymers of chlorotrifluoroethylene and vinylidene fluoride; others are fluorosilicones.

See also **Petrochemical complex.**

—Glen E. Meyer, *The Goodyear Tire & Rubber Company, Akron, Ohio*

Electric Furnace   (See **Abrasives; Aluminum; Energy systems for processes; Iron- and steelmaking; Phosphoric acid; Phosphorus;** and **Silicon.**)

Electrodialysis   (See **Desalination; Peptides and proteins; Separation operations;** and **Water treatment.**)

Electrolytic Processes   (Consult the **Subject Index.**)

Electromagnetic Separation (Gases)   (See **Separation operations.**)

Electron   An electron is a subatomic particle in the shell structure of atoms. An electron carries one negative quantum of electric charge (1.6021×10^{-19} C), exactly the same as the value of positive electric charge carried by a proton. An electron has a mass of 0.00054876 atomic mass unit.

Electron Volt   An electron volt is a unit of energy equivalent to the amount of work required to transport an electron through a potential difference of 1 V.

Electronic Industry Materials   (Consult the **Subject Index.**)

Electrophilic Substitution   (See **Furan group.**)

---

*Made by Thiokol Chemical Corp. (United States).

Electrophoresis (See Peptides and proteins.)

Electroplating (Consult the **Subject Index**.)

Electrorefining (See **Cobalt**; **Copper production**; and **Nickel**.)

Electrostatic Painting (See **Paint**.)

Electrostatic Separation (Solids) (See **Beneficiation, ore**; and **Separation operations**.)

Elements (See **Chemical elements**.)

Elution Chromatography (See **Chromatography**.)

Embrittlement (See **Copper**.)

Emerald (See **Bauxite**; and **Beryllium**.)

Emery (See **Abrasives**.)

Emf Series of Metals The electromotive series arranges the metal elements in order of the amount of electromotive force (voltage) set up between metal and solution when the metal is placed in a 1 molal solution of any of its salts. Each metal is negative to those ahead of it in the list and positive to those following it.

TABLE E-3. *Electromotive Series*

Element	Ion	Electrode Potential, V at 77°F (25°C) for 1 m Metal-Ion Concentration
Magnesium	Mg^{++}	-2.34
Aluminum	Al^{3+}	-1.67
Zinc	Zn^{++}	-0.76
Chromium	Cr^{3+}	-0.71
Iron	Fe^{++}	-0.44
Cadmium	Cd^{++}	-0.40
Nickel	Ni^{++}	-0.25
Tin	Sn^{++}	-0.14
Lead	Pb^{++}	-0.13
Hydrogen	H^+	$0*$
Copper	Cu^{3+}	$+0.34$
Silver	Ag^+	$+0.80$
Palladium	Pd^{++}	$+0.83$
Mercury	Hg^{++}	$+0.85$
Platinum	Pt^{++}	$+1.2$
Gold	Au^{++}	$+1.42$

*Arbitrary.

Emissivity (See **Insulation, thermal**.)

Emulsifiers [See **Anionic surfactants (sulfurbearing)**; **Colloid systems**; **Defoaming agents**; **Detergents**; and **Surfactants**.]

Emulsion Polymerization (See **Elastomers**; **Polymerization**; and **Polyvinyl chloride**.)

Emulsion SBR Rubber (See **Elastomers**.)

Enamel (See **Paint**.)

Enargite (See **Arsenic**; and **Copper production**.)

Encapsulating (See **Epoxy resins**; and **Plastics**.)

Endothermic Reaction (See **Molecule**.)

Energy Balance (See **Distillation**; and **Energy systems for processes**.)

Energy Levels, Atomic (See **Atomic structure**; and **Nuclear power plants**.)

Energy Systems for Processes Fundamentally, a processing plant converts raw materials into finished products through a series of operations that require energy and, in some operations, the removal of energy. Although heat may be derived as the result of exothermic reactions within some processes, on balance most processes require rather large energy inputs. Even energy removal requires additional energy to pump cooling water and to activate refrigeration systems.

Processing plants derive their energy from two major sources: the combustion of fossil fuels and electric energy, procured from a local utility, or in some of the larger plants generated in part inside the plant. Full generation of all electric-power requirements usually is limited to plants situated in remote, underdeveloped locations, where there is no nearby utility to provide reliable service.

Material and Energy Balance. Basic to the design of an energy system for a processing plant is the determination of the energy balance, but before this can be formulated, a material balance must be constructed. As an example, the qualitative block diagram of a benzene hexachloride plant is shown in Fig. E-2. As the result of a careful step-by-step analysis of the flow diagram, a material-balance flow sheet of the type shown in Fig. E-3 will be prepared. Because the calcula-

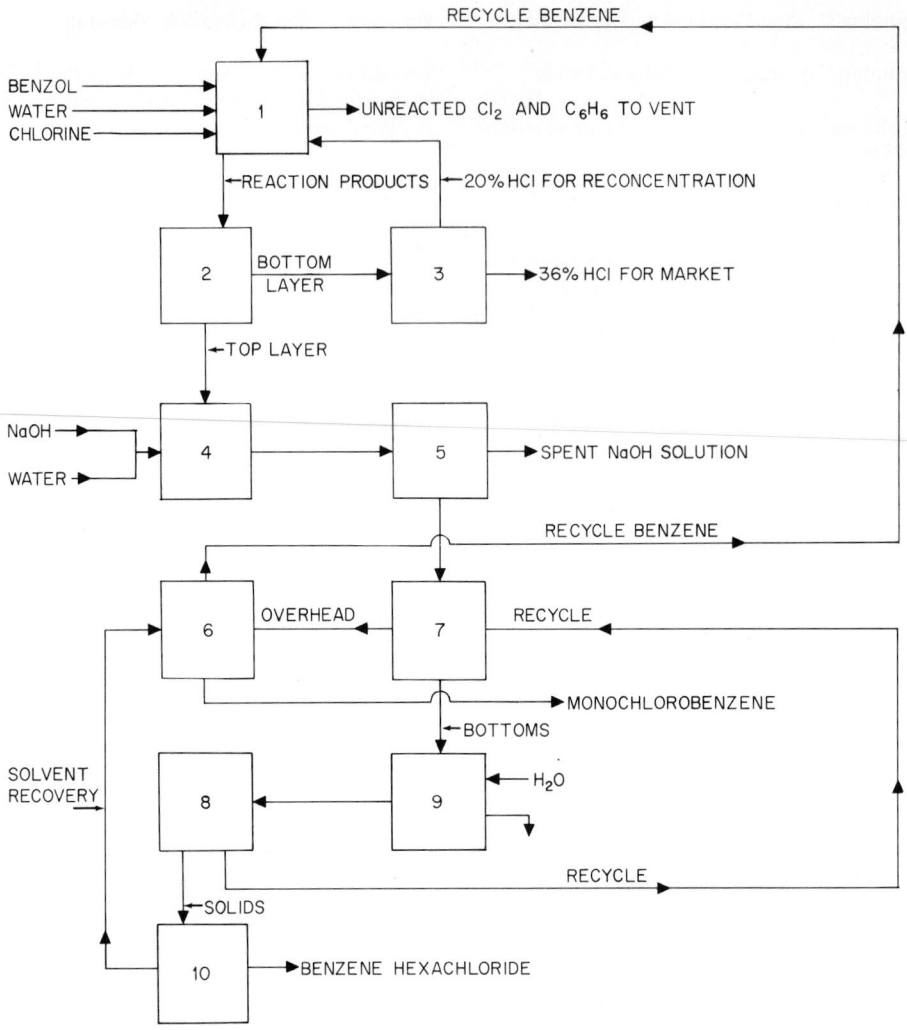

Fig. E-2. Qualitative block diagram of continuous process for making benzene hexachloride (BHC). (1) Reactor operated at 70°C ($C_6H_6 + 3Cl_2 \rightarrow C_6H_6Cl_6$), (2) separation by decantation (top layer consists of BHC and side-reaction products, such as crude monochlorobenzene; bottom layer consists of 24.6% aqueous HCl), (3) acid stripping still, (4) neutralization, (5) separation (top layer consists of mixed chlorinated organic products; bottom layer contains spent NaOH), (6) fractional distillation from which benzene is recycled, (7) flash distillation carried out at 100°C (overhead consists of C_6H_6 and C_6H_5Cl; bottoms consist of 56% $C_6H_6Cl_6$ plus C_6H_6 and C_6H_5Cl), (8) solids-liquid separation (liquid recycle consists of C_6H_6 and C_6H_5Cl), (9) chiller (product exits at 30°C), (10) vacuum drier from which solvent consisting of C_6H_6 *plus* C_6H_5Cl is recovered.

tions are complex and extensive, the process will be broken down into sections, as represented by the material-balance flow sheet of Fig. E-4. The energy balance for this same process section is shown in Fig. E-5. A continuous 24-hr/day operation for the plant is assumed. With this analysis, the basic requirements of the process for heating

and cooling are established. In addition to analyzing the heat content and needs of each materials stream entering and leaving the process, the energy requirements of all powered equipment, e.g., pumps, blowers, compressors, mixers, instruments, and controls, must be estimated.

This example is an oversimplification because

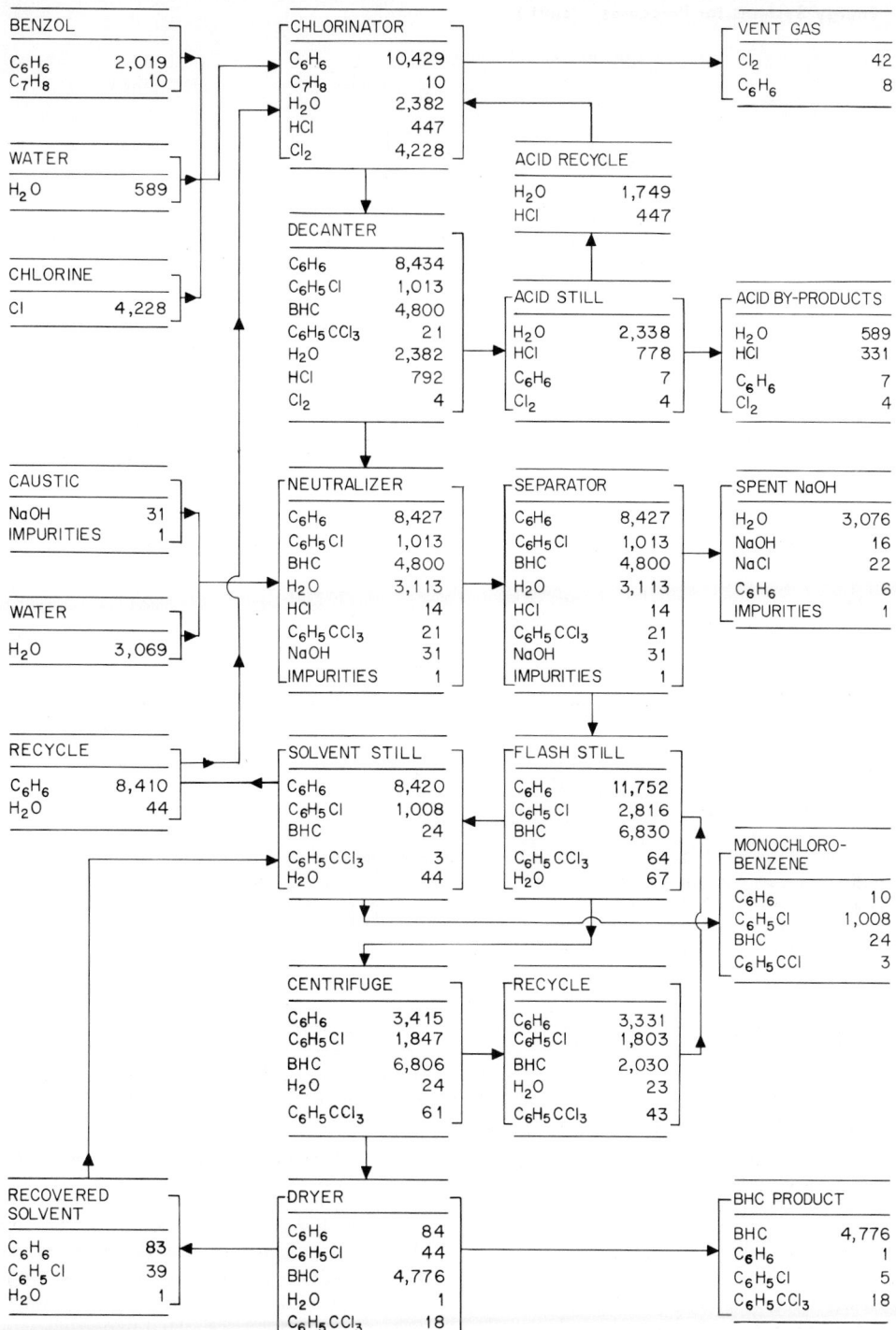

Fig. E-3. Material-balance flow sheet for benzene hexachloride process. Raw materials are shown in column at left, processing operations in middle column, and by-products in column at right. The figures indicated are weight units (in example, pounds per 24 hr of operation).

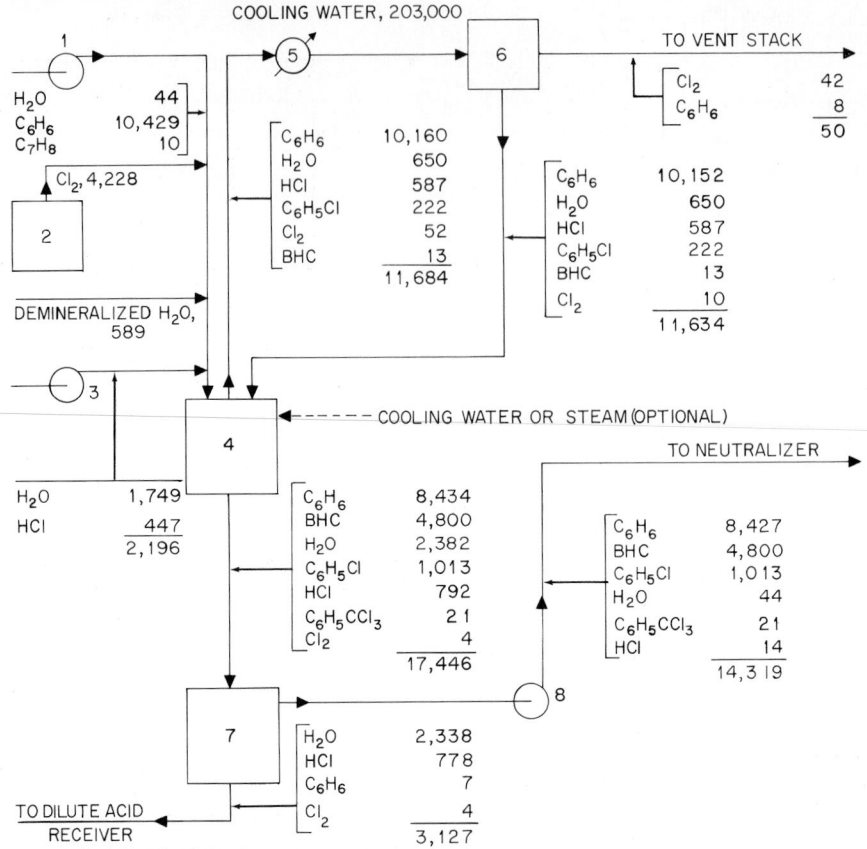

Fig. E-4. Material-balance flow sheet for a section of the benzene hexachloride process. (1) Benzol feed pump, (2) chlorine vaporizer, (3) acid-recycle pump, (4) chlorinator, (5) reflux condenser, (6) vent-gas separator, (7) acid still, (8) crude-product pump. The figures indicated are weight units (in example, pounds per 24 hr of operation). This type of material-balance summary is useful for simple processes but can become cumbersome for complex ones, for which a simple flow sheet usually is keyed to a tabular summary.

numerous other factors usually must be considered. How will this process relate from the point of view of energy with other processes already in operation at a given plant? How much investment should be made now in an energy system to accommodate a future expansion of this process? Planning a total-energy system for a plant is replete with technical-economic tradeoffs. See also **Ethylene.**

Energy Sources

Combustion of Fossil Fuels: The main source of heat energy for processing plants is the combustion of fossil fuels, notably the hydrocarbon fuels, such as fuel oil and natural gas. Coal remains a source of energy for some processes, particularly those of a metallurgical character, and of course coal remains a prime source of energy for the generation of electric power used in large amounts by processing plants. Many plants are designed to use multiple fuels, depending upon fuel availability and local pollution conditions. See **Fuels.**

Basically, the combustion equipment used is *direct-fired* or *indirect-fired.* In direct-fired equipment, the flame and/or products of combustion are in direct contact with the material being heated. Kilns are an example. See also **Cement.** In indirect-fired equipment, the flame and products of combustion are separated from contact with the material within the process, usually by metal or refractory walls. Kettles, vaporizers,

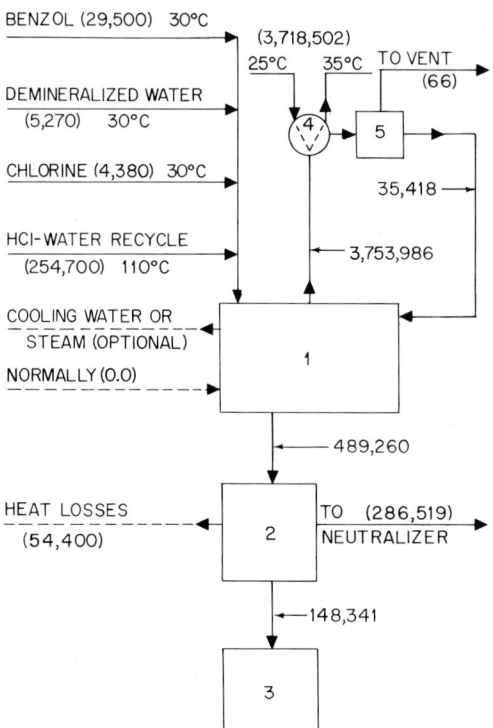

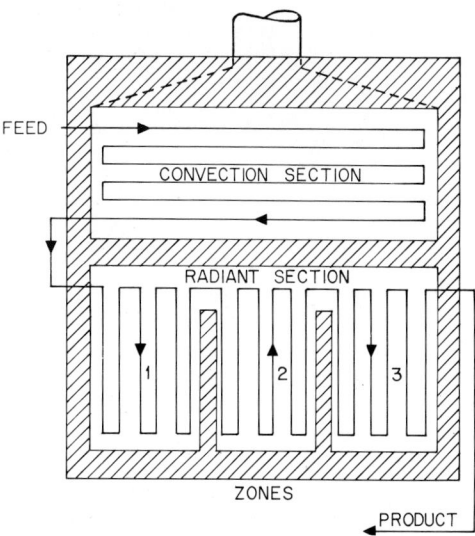

Fig. E-6. Multizone pyrolysis heater exemplifies an indirect-fired design. Feed enters the convection section and proceeds successively through three radiant sections, each zone-controlled, offering the advantages of heating-curve variation in processing various feedstocks and flexibility in control of product-yield distribution. (*Foster Wheeler Corporation.*)

Fig. E-5. Energy-balance flow sheet for a section of the benzene hexachloride process. (1) Chlorinator, (2) acid still, (3) dilute-acid receiver, (4) reflux condenser, (5) vent-gas separator. Figures are Btu/24 hr operation. Reference temperature is 25°C.

stills, furnaces of many types, and boilers are examples. A multizone pyrolysis heater is shown and described in Fig. E-6.

Electric Energy: Although some processing equipment can be operated on the basis of a direct thermal-mechanical conversion, e.g., a huge blower operated by a gas turbine, the great majority of processing equipment is operated directly by electric motors or indirectly in a pneumatic or hydraulic mode as the result of electrically driven compressors, pumps, and so on. In an important but relatively small segment of energized equipment electric energy is converted directly into heat. Electrical heating offers advantages to some operations (see Fig. E-7) including: (1) ease of obtaining high temperatures, (2) safety and convenience, (3) cleanliness in the absence of combustion by-products, (4) uniform heating and sometimes better control, and (5) general absence of oxidizing conditions. Electrical-resistance heaters

are particularly applicable in situations involving high pressures or high vacuums.

Electrochemical Systems: The application of electric energy directly to bring about chemical changes represents a fusion of chemical, metallurgical, mechanical, and electrical engineering. Electrochemical processes fall into three major areas: (1) in the melting, recovery, and refining of metals, e.g., the *electrolytic refining* of Cu, Pb, Ni, Sn, Ag, and Bi or in the production of elemental metals from *fused electrolytes,* as for Al, Mg, Na, Ca, Be, and K; (2) in the *electroplating* of metals and other materials, e.g., electrodeposition of latex; and (3) in the *electrolytic production* of important chemical materials, including anthraquinone, calcium cyanamid, carbon disulfide, caustic soda, chlorine, fluorine, hydrogen peroxide, hypochlorites, potassium hydroxide, sodium dichromate, and sodium perchlorate. Chemical technology also is concerned with the reverse situation, namely engineering products (batteries and fuel cells) for the conversion of chemical energy into electric energy. Also see the Subject Index, which lists the electrochemical operations described in this volume.

Nuclear Energy Systems: To the chemical and

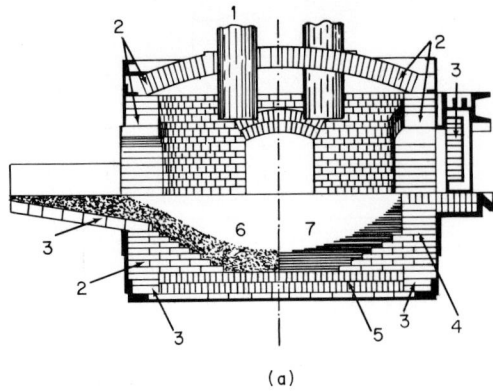

(a)

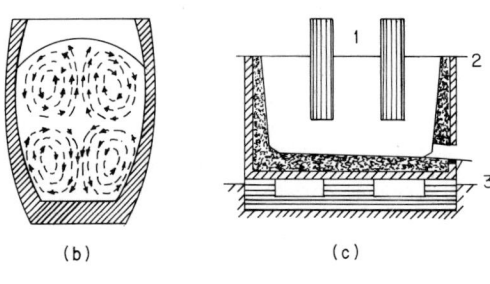

(b)   (c)

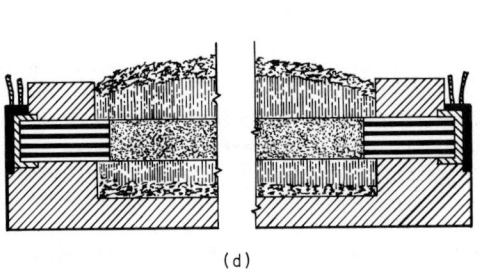

(d)

Fig. E-7.   Types of electric furnace. (a) Sectional view of three-phase arc furnace for producing ferrous alloys. Capacities range from 500-lb to 50-ton charges, requiring 250–10,000 kVA. Charge melts at approximately 1600°C (2912°F). (1) Electrodes, (2) silica brick, (3) fireclay brick, (4) silica or metalkase brick, (5) magnesite brick, (6) ground ganister mix, and (7) grain magnesite. Left-hand side of furnace is shown with an acid lining; right-hand side with a basic lining. (b) Stirring effects in a molten-metal mass that takes place in an induction-type furnace. (c) Open-type resistance furnace with movable electrodes for producing calcium carbide from lime and coke. (1) Electrodes, (2) upper floor level, (3) foundation and pouring level. (d) Open-top, buried-resistor furnace for making silicon carbide from sand and coke.

TABLE E-4. *Materials Commonly Used as Heat-Transport Fluids*

Fluid	Range of Application	
	Temperature, °F	Pressure, psig
Steam	200–1100	0–4500
Water	300–400	90–230
Dowtherm A . . .	450–750	0–145
Dowtherm E . . .	300–500	0–72
Oil	30–600	0
Molten salts . . .	290–1100	0
Mercury	600–1000	0–180
Flue gas or air . .	30–2000	0–100

process engineers, these systems are of interest from at least three vantage points: (1) the chemical design relating to the construction of nuclear power plants; (2) the availability of electric energy from nuclear power plants; and (3) the availability of radioisotopes from nuclear reactions. See **Nuclear power plants;** and **Radioisotopes.** The growing use of plasma to effect chemical reactions is described under **Plasma, chemical processing with.**

Heat Generation and Transport. Steam has many advantages and is the most widely used fluid for transporting thermal energy from generator to point of use. Table E-4 lists the most commonly used heat-transport fluids. When the desired temperature lies within 200–500°F, steam usually is the most effective heat-transport medium, being nontoxic, easy to handle, and relatively low in cost. Other thermal fluids include hot water; diphenyl diphenyloxide (Dowtherm A); o-dichlorobenzene (Dowtherm E); inorganic salt mixtures, such as 40% $NaNO_2$, 7% $NaNO_3$, 53% KNO_3, with a range 850–1100°F; mineral oils (range 30–600°F); flue gas or air; and, to a lesser extent, mercury. A combination vapor-and-liquid heating system that uses Dowtherm is shown in Fig. E-8.

Steam Systems: A conventional steam system for electric-power generation is shown in Fig. E-9. With the exception of certain metallurgical plants, the need for electric power is secondary to the requirement for thermal energy. The steam system shown in Fig. E-10 is much more suited to the needs of the processing plant because it provides for maximum utilization of turbine or engine exhaust for process heat needs. A third steam system—in fact, the most common except for the very large process plants—simply generates proc-

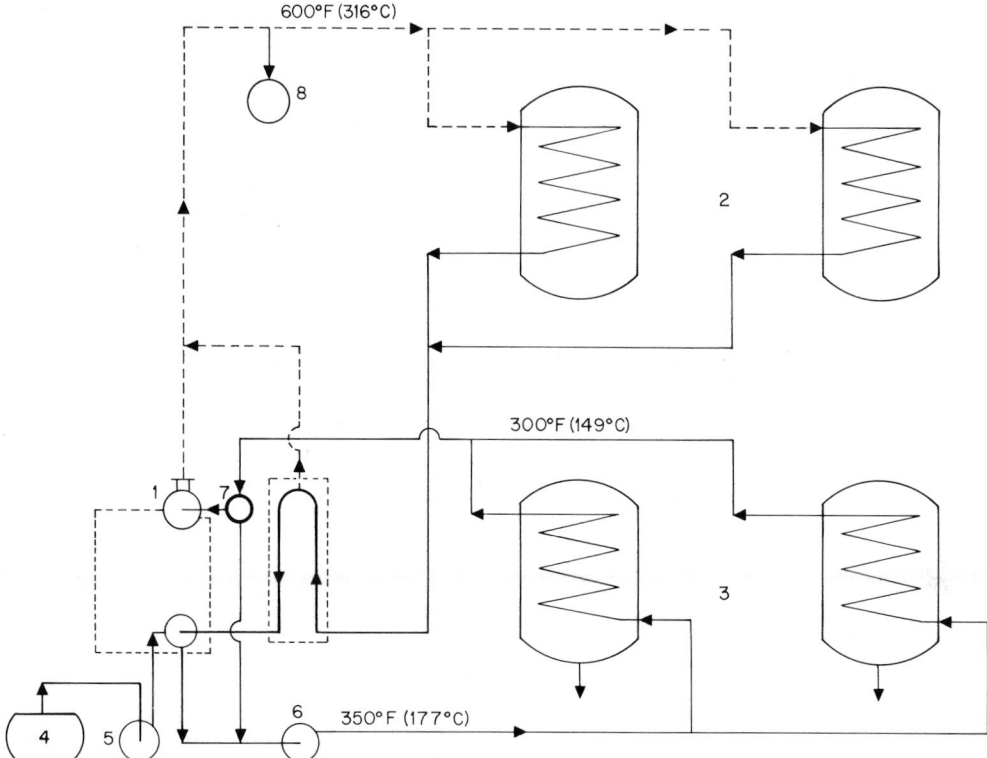

Fig. E-8.   Combination vapor-and-liquid heating system using Dowtherm. (1) Dowtherm vaporizer, (2) high-temperature vessels serviced by system, (3) low-temperature vessels serviced by system, (4) heat-transfer-medium storage tank, (5) charge pump, (6) circulating pump, (7) thermostatically controlled valve, (8) vent condenser. Dashed flow lines = vapor; solid flow lines = liquid.

ess steam, and no turbine is part of that system. The inclusion of electric-power generation in planning the total-energy system for a plant involves a thorough evaluation of plantwide thermal efficiency as well as a comparison of captive

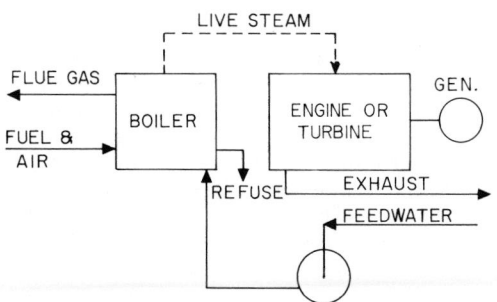

Fig. E-9.   Steam system with electric-power generation as the major objective.

power-generating costs with the rates and reliability of a local power utility. The decision frequently is strongly influenced by the availability of in-plant by-product fuels, such as bagasse, sawdust, coke-oven gas, or bark, which can be used in combination with purchased fossil fuels. In the usual case, where the objective is that of providing both electric power and process steam, the boiler generally delivers steam at high pressure to a turbine, the exhaust of which is delivered to a header for distribution to the process. It is evident that a two-purpose system of this kind has built-in limitations, i.e., that the needs for process steam and the needs for electric power must be in balance, a difficult balance to estimate and maintain. Of course, additional electric-power requirements can be met by the local utility, and thus the primary objective in designing a system of this kind is to err on the side of making certain that sufficient process steam will be produced. In a

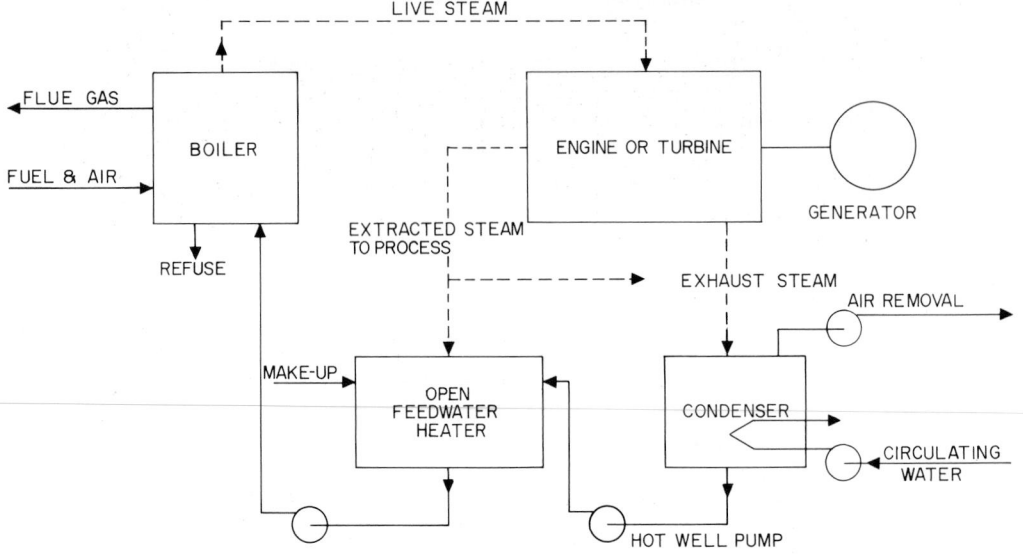

Fig. E-10. Steam system in which part of energy is used for generation of electric power and part for thermal energy to process.

system of this type, the process-steam portion sometimes is referred to as the *heat sink.* Starting with the major needs for thermal energy (the heat-sink portion), the designer will work back to ascertain whether it is most economic simply to provide a system that will provide those thermal needs or, in considering the needs for electric and mechanical energy as well and comparing the costs of purchased power to meet these needs, whether total costs will be less with a dual-purpose system.

The use of the term heat sink in connection with process energy systems can be misleading. The term has been used by electronics engineers to define a device, "usually a mass of metal that is added to equipment for the purpose of absorbing and dissipating heat (unwanted heat)." In process terms, the heat sink is the paramount objective and certainly not a means for dissipating unwanted energy.

Heat Transfer. In terms of heat exchange for recovering and recycling thermal energy, the shell-and-tube heat exchanger is the most common type. It can be used with liquid on both sides; gas on both sides; or liquid on one side and gas on the other. The need for liquid-liquid exchangers is greatest. There are several design configurations; common types are shown in Fig. E-11, along with names of components.

Many applications of heat exchangers in a

process are mandatory; i.e., a material strictly for processing reasons may have to be heated, cooled, or condensed, but other applications in connection with heat recovery and recycling offer the process engineer a chance to exercise some discretion. The initial relatively high cost of exchangers must be considered along with the additional pumping costs and maintenance to make certain that the investment in heat recovery will be justified on a long-term basis. Much work has gone into heat-exchanger design to improve their overall thermal efficiency and resistance to corrosion and abrasion and to lower the pressure drop they impose when inserted into a flow line. It is common to use different metals on the tube and shell side.

Heat exchangers perform many functions within a processing plant and consequently often are referred to by special names even though they remain fundamentally heat exchangers.*

Chillers: A chiller cools fluid to temperatures below those obtainable with ordinary cooling water by vaporization of a refrigerant. The fluid to be cooled is routed through the tubes while the low-boiling refrigerant vaporizes from a pool of liquid in the shell. This operation can be carried out in a kettle-type reboiler (Fig. E-11).

Partial Condensers: Many overhead vapors from

*The assistance of the Foster Wheeler Corporation in providing these definitions is gratefully acknowledged.

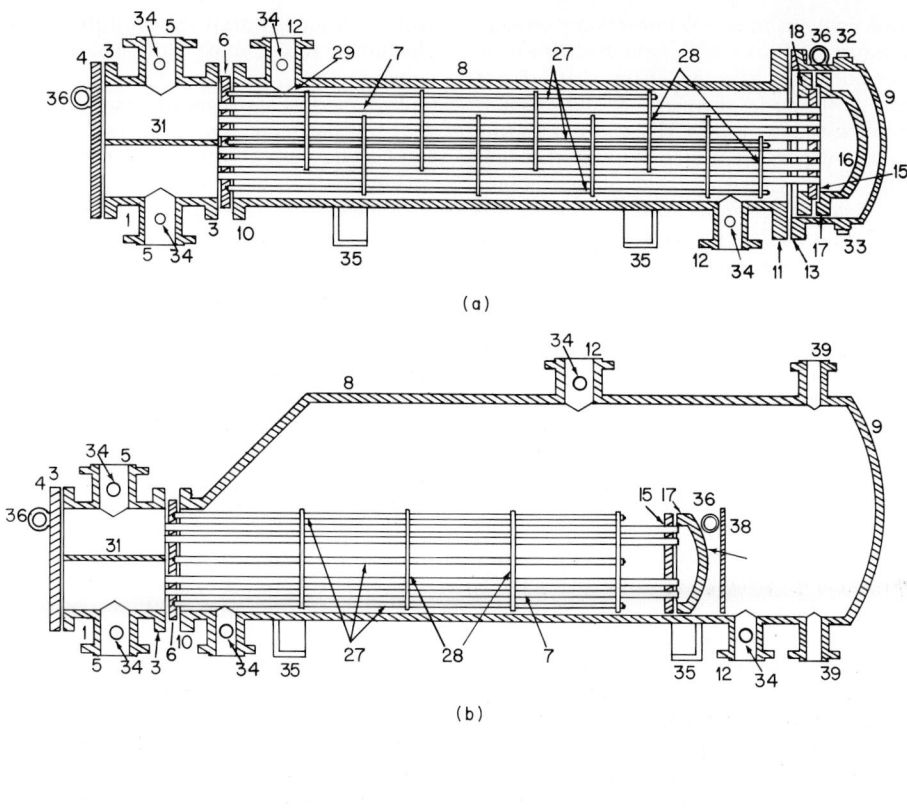

(a)

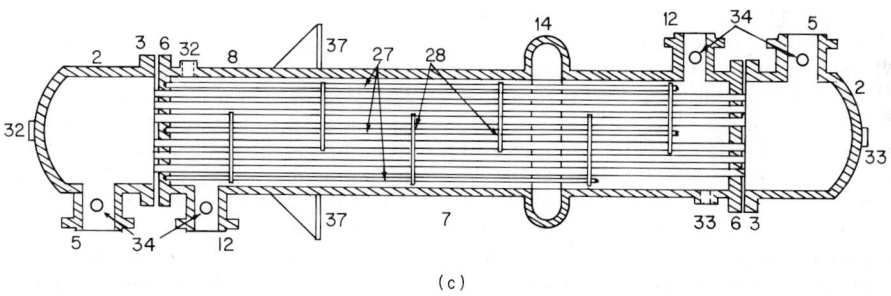

(b)

(c)

Fig. E-11. Three common types of heat exchangers. (a) Type AES, internal-floating-head exchanger (with floating-head backing device); (b) type AKT, kettle-type floating-head reboiler; and (c) type BEM, fixed-tube-sheet exchanger. (*Sketches adapted from specification diagrams from the Standards of Tubular Exchanger Manufacturers Association.*) (1) Stationary head, channel, (2) stationary head, bonnet, (3) stationary-head flange, channel or bonnet, (4) channel cover, (5) stationary-head nozzle, (6) stationary tube sheet, (7) tubes, (8) shell, (9) shell cover, (10) shell flange, stationary-head end, (11) shell flange, rear-head end, (12) shell nozzle, (13) shell cover flange, (14) expansion joint, (15) floating tube sheet, (16) floating-head cover, (17) floating-head flange, (18) floating-head backing device, (19) split shear ring, (20) slip-on backing flange, (21) floating-head cover, external, (22) floating tube sheet skirt, (23) packing-box flange, (24) packing, (25) packing follower ring, (26) lantern ring, (27) tie rods and spacers, (28) transverse baffle, (29) impingement baffle, (30) longitudinal baffle, (31) pass partition, (32) vent connection, (33) drain connection, (34) instrument connection, (35) support saddle, (36) lifting lug, (37) support bracket, (38) weir, (39) liquid-level connection.

distillation columns in petroleum-refinery services are a mixture of heavy and light hydrocarbons and noncondensable gases, i.e., gases that are not condensed at the outlet temperature and pressure of the condenser (air, H_2S, CH_4, and other light ends). These vapors are routed through the shell side while water is used as the cooling medium on the tube side of the unit. Condensation on the shell side begins at the saturation temperature of the heavy components and continues over a decreasing temperature range until part of the lighter components are condensed. Part of the existing liquid is sent back to the tower as reflux while the remainder is further refined or passes to the trim cooler and storage.

Trim Cooler: This unit condenses the last small remaining light-end vapors and cools the liquid to the ultimate storage temperature [often about 100°F (38°C)] by using cooling water. This cooling usually is not conducted in the main condenser because it would reduce column pressure.

Reboilers: These exchangers operate in conjunction with a distillation tower to vaporize enough liquid to assure vaporization of the overhead product. A hot process stream or steam may be used as the heating medium. Most reboilers are shell-and-tube exchangers located at the base of the tower. The vaporizing fluid is routed through the shell side of the exchanger, i.e., kettle reboiler (Fig. E-11).

Thermosiphon Reboiler: Flow of the vaporizing fluid depends upon the difference in static head between the column of liquid flowing from the tower to the reboiler and the partially vaporized column of liquid returning from the exchanger to the tower.

Forced-Circulation Reboiler: A pump is used to provide more positive circulation than available with the thermosiphon effect, e.g., in the vaporization of viscous fluids.

Vapor Heat Exchanger: This unit preheats a cool stream of process fluid by using heat from partially condensing vapor. The objective is to conserve heat and eliminate the requirement for a separate preheater. Caution should be exercised in employing this type of unit as an overhead condenser since the operation of the distillation tower will depend upon the constancy, i.e., flow rate and temperature, of the preheat stream.

Air-cooled Exchanger: As used in the petroleum industry, air-cooled exchangers normally comprise two headers joined by a horizontal bank of finned tubes. Usually two motor-driven fans located above (induced draft) or below (forced draft) the tubes are used to circulate the air over the finned surface. Induced-draft fans pull the air up across the tubes; forced-draft fans blow air across the tubes.

Exchanger: This is the general term for a unit that transfers heat from a warm fluid to a cooler fluid. The heat transferred is conserved by the process.

Steam Generator: A unit that produces steam for use throughout a plant, it often utilizes high-level heat available from burning such materials as tar and heavy oil.

Superheater: This unit heats vapor above the saturation temperature.

Waste-Heat Boiler: It generates steam and is similar to a steam generator except that hot gas or liquid produced by a chemical reaction (often combustion) is the heating medium.

Heat Transfer. The equation for calculating the required heat-transfer surface for a shell-and-tube exchanger is

$$A = \frac{Q}{U \, \Delta t}$$

where A = effective heat-transfer surface required based upon outside tube diameter, ft^2

Q = total heat to be transferred, Btu/hr

ΔT = mean temperature difference corrected for noncountercurrent flow, °F

U = overall heat-transfer coefficient, Btu/(hr)(ft^2)(°F)

The overall transfer coefficient is calculated by determining the value of the five resistances to the flow of heat: (1) resistance to heat transfer of the hot fluid, (2) resistance offered by the tube wall, (3) resistance of the cold fluid, and (4) and (5) the fouling resistance on each side of the tube.

Fouling resistance is a factor that accounts for the future deposition of dirt or products of corrosion on the heat-transfer surface, thereby increasing resistance to the flow of heat.

In calculating the heat-transfer coefficients consideration must be simultaneously given to the pressure drops of the fluids flowing either in the tubes or shell. The design must be consistent with the allowable pressure loss. At times the allowable drop is so low that the exchanger is designed just to accommodate the drop.*

---

*The calculation of heat-transfer coefficients and pressure drop for heat exchangers is well covered by D. Q. Kern, "Process Heat Transfer," McGraw-Hill, New York, 1950, and W. H. McAdams, "Heat Transmission," 3d ed., McGraw-Hill, New York, 1954.

Heat Storage. Inasmuch as there is no perfect insulator (see also **Insulation, thermal**), it is frequently necessary to store heat in rather large quantities in specially designed apparatus. Hot water is one of the easiest forms in which to store thermal energy that is immediately available. Electric energy and steam, on the other hand, essentially have to be generated on an as-needed basis. A particularly difficult problem is posed by, say, a blast furnace, where great amounts of a hot gas are required on a cyclic basis. To heat such huge quantities of air on a continuous, as-needed basis would be quite impractical with the present state of the art. The solution lies in the use of several stoves, as shown in Fig. E-12, which are quite large, often over 100 ft high and about 25 ft in diameter. The blast temperature of approximately 1000°F (528°C) is accomplished by preheating the stove checkerwork to a much higher temperature. The gas passing through the stove exhausts initially at 2000°F (1093°C). Mixing this with unheated air produces the required blast temperature. The stoves normally are heated for a period of 3 hr and are "on wind," or exhausting, for 1 hr. A similar system of checkerwork regenerators also is used in connection with glass-tank heat-storage systems. Pebble heaters (Fig. E-13) also are used for heating gases or removing heat from gases.

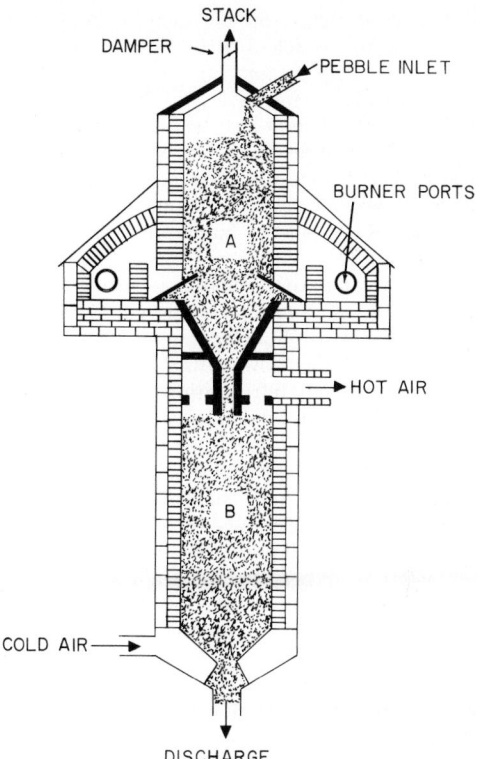

Fig. E-13. Pebble heater for heating steam to temperatures impractical in metallic units. Also used for heating air, hydrogen, methane, and other gases for processing purposes. In reverse, a pebble heater may be used to recover heat from hot gases. The pebbles are heated in top chamber A by direct contact with combustion gases and passed through a throat to lower chamber B, where heat is transferred to cool gases. The two chambers are maintained at the same temperature so that there will be no gas flow between them. An average cycle on the pebbles is 30–50 min.

Waste-Heat Removal. Most processing plants need to remove rather large quantities of low-value heat. Cooling often is effected through the use of conventional heat exchangers, i.e., to recycle heat energy through the system and thus keep energy costs to a minimum. The point is usually reached, however, where not all heat can be dissipated by a recovery device and thus must leave the process and be lost to it, usually to the air or to a stream or body of water adjacent to the plant. Excessive dumping of such heat is the cause of objectionable thermal pollution.

Evaporative cooling, where applicable, usually is the most economical way to remove excess heat.

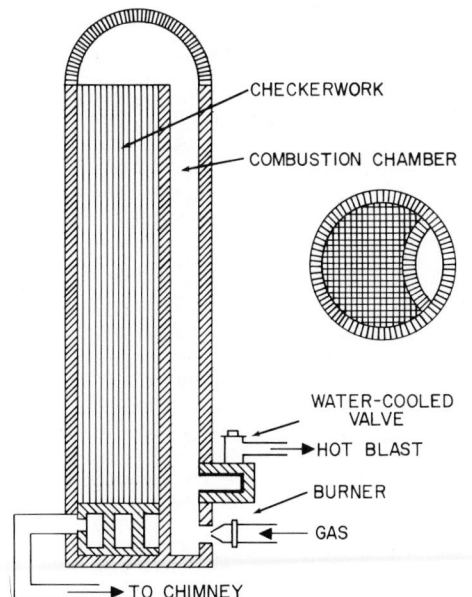

Fig. E-12. Blast-furnace stove for preheating large quantities of air.

The heat is picked up via heat exchangers by cooling water, which then is piped to a cooling tower or spray pond and sprayed to gain contact with the colder atmospheric air. The heat-transfer process involves latent heat transfer due to vaporization of a small portion of the water and sensible heat transfer due to the difference in temperature of the water and air. Approximately 80% of the heat transfer is due to latent heat and 20% to sensible heat.

When evaporative cooling is insufficient or impractical, refrigeration is required. Some processes, such as the liquefaction of gases, require temperatures much lower than can be achieved with conventional refrigeration methods. Temperatures below normal refrigeration minimums (Dry Ice melts at $-109.6°C$) are loosely referred to as *cryogenic*. See also **Helium;** and **Natural gas.**

Heat Transfer for Solid Materials. In processing plants it is often necessary to remove heat from a material in the liquid state in order to effect solidification. When the operation is done on a batch basis, the term *casting* usually applies, but when it is carried out continuously, the term *flaking* may be used. The reverse situation occurs when solids must be heated. For obvious reasons, conventional shell-and-tube heat exchangers are not well suited to these applications.

A rotating-shell device is shown in Fig. E-14. Application parameters are (1) low-range cooling [750°F (399°C) and lower] where the shell dips in water, (2) intermediate cooling [up to 1400°F (760°C)] where forced circulation of tank water is used, (3) primary cooling (above 1400°F) where water is copiously sprayed on the shell and the solids loading is light, (4) low-range heating (below steam temperature) where the shell is dipped

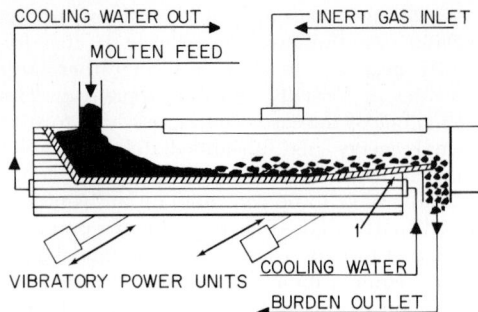

Fig. E-15. Vibrating-type heat-transfer equipment for batch solidification. Sometimes referred to as a *caster,* the device is widely used in certain industries. After cooling and solidification, intense vibratory action shatters cake into lumps; (1) is a liquid dam.

in hot water, and (5) high-range heating where heating is effected by tempered combustion gases or ribbon-type radiant-gas burners.

Vibrating-conveyor type equipment (Fig. E-15) sometimes is used. Water-cooled continuous metal belts offer another solution. For thick cake production, rotating-shelf units may be used. Most of this equipment can be considered in the general category of driers. See also **Drying, solids.** Equipment for melting solids is shown in Fig. E-16.

REFERENCES

Boelter, L. M. K., H. Cherry, H. A. Johnson, and R. C. Martinelli: "Heat Transfer Notes," McGraw-Hill, New York, 1965.

Bransom, S. H.: "Applied Thermodynamics," Van Nostrand Reinhold, Princeton, N.J., 1961.

Dossat, Roy J.: "Principles of Refrigeration," Wiley, New York, 1961.

Gröber, H., S. Erk, and V. Grigull: "Fundamentals of Heat Transfer," 3d ed., McGraw-Hill, New York, 1961.

Holt, Arthur D.: Heating and Cooling of Solids, *Chem. Eng.,* Oct. 23, 1967, pp. 145–166.

Irvine, Thomas F., Jr., and James P. Hartnett (eds.): "Advances in Heat Transfer," Academic, New York, 1970.

Kirkwood, J. G., and I. Oppenheim: "Chemical Thermodynamics," McGraw-Hill, New York, 1961.

Miller, Ryle, Jr.: Process Energy Systems, *Chem. Eng.,* May 20, 1968, pp. 130–148.

Norman, W. S.: "Absorption, Distillation, and Cooling Towers," Wiley, New York, 1962.

Perry, Robert H., and Cecil H. Chilton (eds.): "Chemical Engineers' Handbook," sec. 9 (Heat Generation and Transport), McGraw-Hill, New York, 1973.

Solberg, H. J., and O. C. Cromer: "Thermal Engineering," Wiley, New York, 1960.

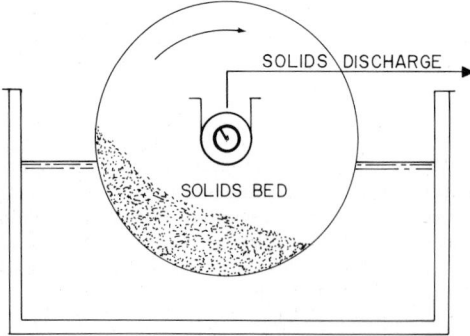

Fig. E-14. Plain rotating shell used for both heating and cooling. For high-range heating, tempered combustion gases may be used instead of water.

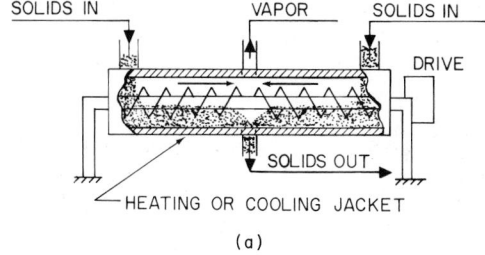

SOLIDS IN   VAPOR   SOLIDS IN

DRIVE

SOLIDS OUT

HEATING OR COOLING JACKET

(a)

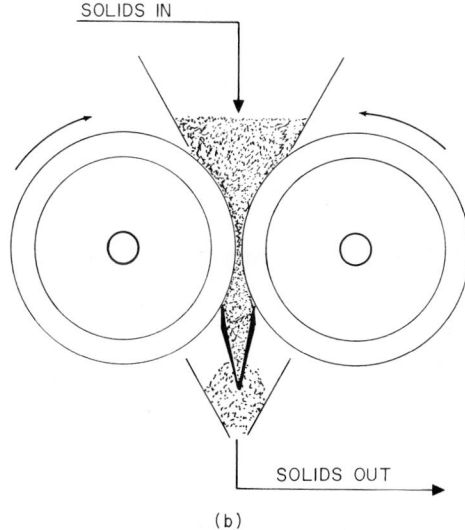

SOLIDS IN

SOLIDS OUT

(b)

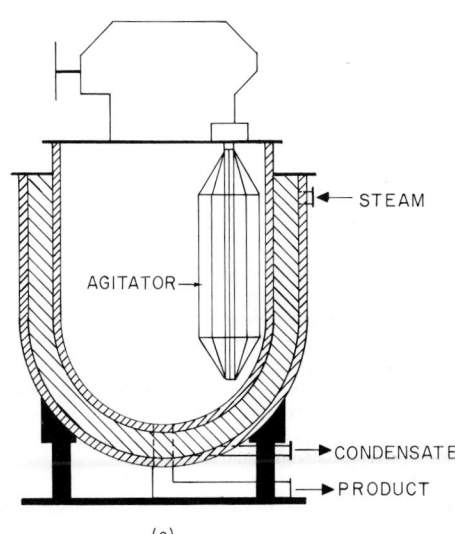

STEAM

AGITATOR →

CONDENSATE

PRODUCT

(c)

Enthalpy (See **Distillation.**)

Enzymes Enzyme (from the Greek meaning "leavened") designates scores of organic complexes that have a number of properties in common. A crisp classification of enzymes is much more difficult to construct than that of the much more easily defined chemical family, e.g., the acids, salts, or alcohols. The existence of enzymes has been known since the early 1600s because of their role in digestion and fermentation processes used to make alcohols and other industrial and commercial products. Only in comparatively recent years have some of the simpler enzymes been isolated. Urease was produced in crystalline form in 1926. Other enzymes that have been isolated include amylase, carboxypeptidase, chymopapain, ficin, lysozyme, papain, pepsin, and starch phosphorylase. As this list shows, many enzyme names end with the suffix -ase.

Enzyme complexes are generated by living cells; they function as catalysts in reactions that involve the metabolism of living organisms and thus play vital roles in biological chemistry and medicine. Expanding knowledge of the life processes thus depends upon furthering the investigation of enzymes—their constitution, structure, synthesis, and behavior. Just scanning the abridged compilation of enzymatic reactions given in Table E-5 suggests numerous applications for enzymes in the manufacture of industrial, medical, and commercial products.

Common properties of enzymes include (1) their predominant, established role as *catalysts* for several types of chemical reactions, often providing the means of effecting chemical conversions otherwise difficult and at lower rates of energy expenditure; (2) their *structure* and *constitution,* which suggest that most enzymes are simple or conjugated proteins; (3) their relatively high sensitivity to *environmental conditions,* such as temperature, pH,

Fig. E-16. Heat-transfer equipment used for solids handling. (a) Tank equipped with mixing ribbon spirals provides considerable agitation and is useful for melting or cooking dry powdered solids. Heat-transfer efficiency is only moderate because of the relatively deep beds of solid particles. (b) Double drum. Scraping knives may be engaged continuously or intermittently, depending upon the nature of the heated product. (c) Vertical agitated kettle. Although heat transfer through the jacket normally is quite poor, a kettle offers convenience in handling and cleaning and is particularly useful where batches of different materials must be processed frequently.

TABLE E-5. *Classification of Enzymes by Reactions Catalyzed*

Enzyme Group	Reactions and Examples
Hydrolases	Hydrolysis in the general manner of $R \cdot O \cdot R' + H \cdot O \cdot H \rightleftharpoons R \cdot O \cdot H + H \cdot O \cdot R'$
Carbohydrases	Hydrolysis of carbohydrates
Hexosidases	Hydrolysis of the $R \cdot O \cdot R'$ linkage in glycosides and disaccharides; major enzymes are maltase, invertase (sucrase or saccharase), lactase, and emulsin
Polyases	Hydrolysis of polysaccharides to disaccharides and to lower-molecular-weight polysaccharides; major enzymes include α- and β-amylases (which hydrolyze dextrin and starch), cellulase, and inulase
Esterases	Hydrolysis of esters and fats
Lipases	Hydrolysis of fats to glycerol and fatty acids
Phosphatases	Hydrolysis of esters of phosphoric acid to alcohols and phosphoric acid
Sulfatases	Hydrolysis of esters of sulfuric acid to an alcohol and sulfuric acid
Cholinesterase	Hydrolysis of acetylcholine to acetic acid and choline
Amidases	Hydrolysis of acid amides to ammonia and acids (*see also* **Amino acids**)
Glutaminase	Hydrolysis of glutamine to ammonia and glutamic acid
Arginase	Hydrolysis of arginine to ornithine and urea
Urease	Hydrolysis of urea to ammonia and carbon dioxide
Other important amidases include asparaginase, aspartase, hippuricase, histidase, oxidase, the prinamidases, transaminases, and tyramine oxidase	
Proteases	Hydrolysis of peptide linkages into proteins and peptides
Peptidases	Hydrolysis of peptides to α-amino acids; important peptidases include carboxypeptidase and aminopolypeptidase, which catalyze the hydrolysis of polypeptides to less complex peptides and amino acids
Dipeptidases	Splitting dipeptides to amino acids; protaminases and prolinase also members of this group
Proteinases	Hydrolysis of proteins to polypeptides; included in this group are pepsin, which is found in gastric juices; trypsin and chymotrypsin, in pancreatic juices; cathepsin, in animal tissues; ficin, in fig sap; bromelin, in pineapple juice; papain, in papaya; and rennin, which acts on casein (*see also* **Proteins and polypeptides**)
Nucleases	Hydrolysis of nucleic acid and nucleic acid derivatives
Nucleotidase	Hydrolysis of nucleotides to form phosphoric acid and nucleosides
Nucleosidase	Hydrolysis of nucleosides to the relevant bases and carbohydrates
Polynucleotidase	Hydrolysis of nucleic acid to nucleotides; this material is found in pancreatic and intestinal juices
Desmolases	Breaking products down further after hydrolysis to form H_2O and CO_2; in this process, the carbon-carbon links are destroyed
Carboxylases	Removal of carboxyl groups from acids; carboxylase, a member of this series, attacks pyruvic acid (pyroracemic acid, $CH_3COCOOH$) to produce CO_2 and acetaldehyde
β-Keto carboxylases	Formation of α-keto acids by attacking β-keto acids
Aldolase, or zymohexase	Splitting of C_6 monosaccharides into C_3 compounds; e.g., splitting fructose 1,6-diphosphate into phosphoglyceric acid and dihydroxyacetone phosphoric acid
Amino acid decarboxylases . . .	Formation of amines and carbon dioxide from action upon amino acids; although a zinc-protein enzyme, carbonic anhydrase, which forms CO_2 and H_2O by acting on carbonic acid, is considered a member of this group
Oxidases and reductases	Catalysis of oxidation-reduction systems
Dehydrogenases	Action on compounds that have a hydroxyl group to yield the corresponding carbonyl group; thus, hydroxy acids and alcohols are changed to keto acids and aldehydes; aldehydes are changed to acids
Lactic dehydrogenase	Converts lactic acid to pyruvic acid
Alcohol dehydrogenase	Converts ethyl and other alcohols to acetaldehyde and other aldehydes respectively

TABLE E-5. *Classification of Enzymes by Reactions Catalyzed* *(Continued)*

Enzyme Group	Reactions and Examples
Aldehyde dehydrogenase . . .	Forms acetic acid from aldehydes
β-Hydroxybutyric dehydrogenase	Yields acetoacetic acid from hydroxybutyric acid; both aldehyde dehydrogenase and β-hydroxybutyric dehydrogenase are found in liver
Isocitric dehydrogenase . . .	Forms oxalosuccinic acid from citric acid
Malic dehydrogenase	Yields oxalacetic acid from malic acid
Glucose dehydrogenase . . .	Changes glucose to gluconic acid
Robison ester dehydrogenase	Forms the corresponding phosphohexonic acid when it acts upon hexose 6-phosphate
Glycerophosphate dehydrogenase	Changes glycerophosphate to phosphoglyceric acid

Iron enzymes:

Cytochrome oxidase	Oxidizes cytochrome c (in reduced form) to cytochrome c (in oxidized form) plus H_2O; oxygen must be present in system
Catalase	Catalyzes breakdown of H_2O_2 into O_2 and H_2O; occurs in nearly all living cells
Peroxidase	Acts as catalyst in oxidizing ortho- and para-substituted phenols in presence of peroxygen compounds and hydrogen peroxide; occurs in milk, roots, and sprouting plants

Copper enzymes:

Tyrosinase	Oxidizes phenol compounds into quinones; identical with polyphenol oxidase (found in potatoes) and monophenol oxidase (found in mushrooms)
Ascorbic acid oxidase	Catalyzes formation of dehydroascorbic acid from ascorbic acid when oxygen is present

Yellow-enzyme group:

Triphosphopyridine nucleotide(TPN)– cytochrome c reductase	Catalyzes the reaction Cytochrome c + reduced coenzyme II → reduced cytochrome c + oxidized coenzyme II found in yeast and liver
Diphosphopyridine nucleotide(DPN)– cytochrome c reductase	Catalyzes the reaction Cytochrome c + reduced coenzyme I → cytochrome c + oxidized coenzyme I also found in yeast and liver
Xanthine oxidase	Participates in several oxidation-reduction actions: (1) oxidizes hypoxanthine to xanthine, (2) oxidizes xanthine to uric acid, (3) oxidizes reduced coenzyme I to oxidized coenzyme I, (4) changes aldehydes to acids; additional amino acid oxidases convert amino acids to keto acids, ammonia, and hydrogen peroxide
Haas' enzyme and Warburg's old-yellow enzyme	Both enzymes, found in yeast, participate in the reaction: reduced coenzyme II and reduced Yellow enzyme
Diaphorase	Participates in the reaction: reduced coenzyme I oxidized to oxidized coenzyme I and reduced diaphorase
Succinic dehydrogenase . . .	Changes succinic acid to fumaric acid
Hydrases (hydratases)	Participate in the addition or removal of water without causing hydrolysis; action occurs at the substrate level
Enolase	Forms phosphopyruvic acid by removing water from 2-phosphoglyceric acid
Aconitase	Forms aconitic acid by removing water from citric acid
Fumarase	Forms malic acid by adding water to fumaric acid
Glyoxalase	Forms lactic acid by adding water to pyruvic aldehyde (methyl glyoxal)
Mutases	Act at the substrate level to cause simultaneous reduction of one molecule and the oxidation of another; a mixture of alcohol dehydrogenase and aldehyde dehydrogenase also has this property
Phosphoglucomutase	Forms glucose 6-phosphate from glucose 1-phosphate

TABLE E-5. *Classification of Enzymes by Reactions Catalyzed* (*Continued*)

Enzyme Group	Reactions and Examples
Transfer enzymes	Participate in transfer of radicals
Transaminases	Function to put an amino group in α-keto acid, thus forming different keto and amino acids
Hexokinases (phosphate transfer enzymes)	Act as catalysts in reactions such as adenosine triphosphate (ATP) and glucose to yield adenosine diphosphate (ADP) and glucose 6-phosphate
Phosphorylases	Appear to invade the carbohydrate molecule with phosphoric acid (radicals), instead of water as in the case of a hydrolase, to form glucose 1-phosphate from phosphate and glycogen or starch
Isomerases	Fructose 6-phosphate is formed from glucose 6-phosphate by the action of phosphohexisomerase

and the presence of organic and inorganic materials; and (4) their origin from *living cells,* a striking parallel even in view of synthesis. The environmental tolerance of the enzymes closely parallels other substances associated with life processes: a relatively narrow temperature span with denaturation (deactivation) occurring at temperatures generally above 50°C (122°F) and greatly reduced activity well above the freezing point of water, a low tolerance to a pH below 4, minimal to no tolerance of certain organic solvents, such as alcohol and acetone, and destruction by numerous organic and inorganic substances.

Functional Classification: From the standpoint of their catalytic activity, enzymes may be roughly classified in accordance with the kind of reactions in which they participate: (1) the *hydrolases,* which catalyze hydrolysis in the general manner of $R \cdot O \cdot R' + H \cdot O \cdot H \rightleftharpoons R \cdot O \cdot H + H \cdot O \cdot R'$; (2) the *desmolases,* which catalyze reactions further breaking down products of hydrolysis to form water and carbon dioxide; (3) the *oxidases* and *reductases,* which respectively catalyze oxidation and reduction reactions; and (4) a miscellaneous category, including the *hydrases* (or *hydratases*), *mutases,* and *transfer enzymes,* which catalyze reactions not previously mentioned. As shown by Table E-5, each major classification includes several subgroups. Also see **Amino acids;** and **Proteins and polypeptides.**

Mixtures of Enzymes: The specific enzymes included in Table E-5 do not reflect some of the better known enzyme designations of the past because many enzymatic materials formerly considered to be single substances are in fact mixtures of two or more specific enzymes. Thus *zymase,* once thought to be the yeast enzyme for converting sugar into alcohol, is a complex mixture of several enzymes; *erepsin* comprises a mixture of

peptide-splitting enzymes; and *diastase* is a mixture of amylases. See also **Yeast.**

Enzymes differ from most inorganic catalysts in being very specific for the reactions they catalyze. As an example, in the hydrolysis of raffinose (a trisaccharide), an acid catalyst will yield glucose, fructose, and galactose; diastase will yield melibiose and fructose; emulsin will yield sucrose and galactose. The glucosidic linkages are hydrolyzed at approximately equal rates with the acid catalyst. On the other hand, the enzyme catalysts act on just one kind of linkage even though the difference in linkages is small. Acids may catalyze numerous compounds, including amides, acetals, and esters, whereas a given enzyme confines its action to a very specific compound or related group. Because of this behavior, mixtures of enzymes can be effective. See also **Starch.**

An exceptionally long list of references is given to demonstrate the extent to which enzymes function in life processes.

Besides their importance in life processes and the manufacture of industrial and pharmaceutical preparations, enzymes also have found use as ingredients of detergents. The attack of normal detergents on proteinaceous stains, e.g., blood, milk, food residues, and stains attached to carbohydrates, is considerably less effective than that on oils and greases. Unfortunately, some of these stains become more firmly bound into fibers during washing and when they are exposed to heat during drying. Certain enzymes tend to break down these proteins to water-soluble proteoses or peptones. To be useful in this role, the enzymes must obviously be compatible with the usual detergents, e.g., alkyl aryl sulfonates, fatty alcohol sulfates, and their ethoxylated derivatives. Of course, the enzymes must remain active at the relatively high pH values (8.5–9.5) of most washing compounds, and they must be stable under usual storage conditions.

See also **Detergents;** and **Pyridine and derivatives.**

REFERENCES

Barman, Thomas E.: "Enzyme Handbook," Springer-Verlag, New York, 1969.

Belding, Melvin E., and Seymour J. Klebanoff: Peroxidase-mediated Virucidal Systems, *Science,* vol. 167, pp. 195–196, Jan. 9, 1970. [Peroxidase (myeloperoxidase or lactoperoxidase), hydrogen peroxide, and a halide such as iodide, bromide, or chloride form a potent virucidal system that is effective against polio and vaccinia virus, particularly at a low pH. The peroxidase–halide–hydrogen peroxide system may contribute to the host defense against certain viral infections.]

Berlin, Richard D.: Specificities of Transport Systems and Enzymes, *Science,* vol. 168, pp. 1539–1545, June 26, 1970. (Selectivity of cells for exogenous compounds is enhanced by sequential action of membrane carriers and enzymes.)

Chakrabarty, Asit K., and Herman Friedman: L-Asparaginase-induced Immunosuppression: Effects on Antibody-forming Cells and Serum Titers, *Science,* vol. 167, pp. 869–870, Feb. 6, 1970. (Treatment of mice with L-asparaginase from *Escherichia coli* resulted in a marked suppression of the immune response, as assessed both cellularly and humorally. Suppression occurred only when the enzyme was injected together with the sheep erythrocytes used as antigen. There was little or no effect when the enzyme was injected before the antigen. Simultaneous injection of asparagine prevented suppression, an indication that the effect of the enzyme was due to depletion of an amino acid probably essential for normal lymphoid cell function during antibody production.)

Epel, Bernard, and Warren L. Butler: Cytochrome *a*3: Destruction by Light, *Science,* vol. 166, pp. 621–622, Oct. 31, 1969.

Filner, Philip, John L. Wray, and Joseph E. Varner: Enzyme Induction in Higher Plants, *Science,* vol. 165, pp. 358–367, July 25, 1969. (Environmental or developmental changes cause many enzyme activities of higher plants to rise or fall.)

Gibbs, Gordon E., and Guy D. Griffin: Beta Glucuronidase Activity in Skin Components of Children with Cystic Fibrosis, *Science,* vol. 167, pp. 993–994, Feb. 13, 1970. (As compared to that in normal children a decreased activity of β-glucuronidase, a lysosomal enzyme associated with mucopolysaccharide metabolism and with salt transport, has been detected in the epidermis and sweat-gland tissues of children with cystic fibrosis.)

Goldring, Irene P., Irving M. Ratner, and Leonard Greenburg: Pulmonary Hemorrhage in Hamsters after Exposure to Proteolytic Enzymes of *Bacillus subtilis,* *Science,* vol. 170, pp. 73–74, Oct. 2, 1970.

Greene, Martin L., James A. Boyle, and J. Edwin Seegmiller: Substrate Stabilization: Genetically Controlled Reciprocal Relationship of Two Human Enzymes, *Science,* vol. 167, pp. 887–889, Feb. 6, 1970.

Greenspan, Michael D., Claire H. Birge, Gary Powell, William S. Hancock, and P. Roy Vagelos: Enzyme Specificity as a Factor in Regulation of Fatty Acid Chain Length in *Escherichia coli,* *Science,* vol. 170, pp. 1203–1204, Dec. 11, 1970.

Hatfield, G. Wesley, and R. O. Burns: Ligand-induced Maturation of Threonine Deaminase, *Science,* vol. 167, pp. 75–76, Jan. 2, 1970. (The dimeric intermediate substructures of threonine deaminase, which are obtained by alkaline dialysis of the native tetrameric enzyme, are inactive after reassembly unless they are subsequently exposed to maturation-inducing ligands for which the enzyme possesses stereospecific binding sites. Ligand binding at these sites precedes the full maturation of the enzyme.)

Hymer, W. C., Andrea Mastro, and Elaine Griswold: Intestinal Enzymes: Indicators of Proliferation and Differentiation in the Jejunum, *Science,* vol. 167, pp. 1627–1630, Mar. 20, 1970. [Some intestinal enzymes were assayed which were related to (1) cellular proliferation, e.g., aspartate carbamoyltransferase, thymidine kinase, uridine kinase, and dihydroorotase; (2) cellular differentiation, e.g., lactase, invertase, maltase, alkaline phosphatase, and dipeptidase; and (3) lysosomes, e.g., β-glucuronidase, acid β-galactosidase, and acid phosphatase. These enzymatic determinations can be used to distinguish the crypt from the villus during healthy or diseased states.]

Kint, J. A.: Fabry's Disease: Alpha-Galactosidase Deficiency, *Science,* vol. 167, pp. 1268–1269, Feb. 27, 1970.

Kleinschmidt, Albrecht K., Joel Moss, and M. Daniel Lane: Acetyl Coenzyme A Carboxylase: Filamentous Nature of the Animal Enzymes, *Science,* vol. 166, pp. 1276–1278, Dec. 5, 1969.

Krishnamurty, Kotra V.: Carbonic Anhydrase in Seawater: Carbonato Complexes, *Science,* vol. 165, p. 929, Aug. 29, 1969.

Little, Brian W., and William L. Meyer: Ribonuclease-Inhibitor System Abnormality in Distrophic Mouse Skeletal Muscle, *Science,* vol. 170, pp. 747–749, Nov. 13, 1970. (Skeletal-muscle extracts from mice with muscular dystrophy contain severalfold higher than normal levels of free alkaline ribonuclease II activity and none of the free ribonuclease inhibitor normally present. This abnormal pattern is not seen in heart or liver extracts from dystrophic mice.)

Lloyd, K., and O. Hornykiewicz: Parkinson's Disease: Activity of L-dopa Decarboxylase in Discrete Brain Regions, *Science,* vol. 170, pp. 1212–1213, Dec. 11, 1970. (The activity of L-dopa decarboxylase was greatly reduced in the striatum, less so in the hypothalmus, and unchanged in the cortex of brains of patients with Parkinson's disease. However, it appears that even in the striatum enough activity remained to allow formation of dopamine from L-dopa in patients treated with large doses of L-dopa.)

Massaro, Edward J.: Horseshoe Crab Lactate Dehydrogenase: Tissue Distribution and Molecular Weight, *Science,* vol. 167, pp. 994–996, Feb. 13, 1970. [Lactate dehydrogenase from *Xiphosura* (*Limulus*) *polyphemus* is D(−) lactate specific. It does not use L(+)lactate, α-hydroxybutyrate, α-hydroxyvalerate, or α-hydroxy-isocaproate as substrate. In most tissues lactate dehydrogenase is composed of five isozymes with a molecular weight of 140,000 for each, as judged by gel-filtration chromatography. This suggests that the isozymes are tetramers comprising varying amounts of two physicochemically distinct subunits.]

Mitchison, J. M.: Enzyme Synthesis in Synchronous Cultures, *Science,* vol. 165, pp. 657–663, Aug. 15, 1969. (The patterns of enzyme synthesis suggest an ordered sequence of transcription throughout the cell cycle; linear enzymes; step and peak enzymes.)

Murphy, Robert C., Milica V. Djuricic, Sanford P. Markey, and K. Biemann: Tay-Sachs Disease: Generalized Absence of a Beta-D-N-Acetylhexosaminidase Component, *Science,* vol. 165, p. 698, Aug. 15, 1969.

Perry, Thomas O.: Dormancy of Trees in Winter, *Science,* vol. 171, pp. 29–35, Jan. 8, 1971. (Enzyme pattern in plants.)

Reed, Gerald: "Enzymes in Food Processing," Academic, New York, 1966.

Reik, Louis, Gary L. Petzold, Joan A. Higgins, Paul Greengard, and Russell J. Barrnett: Hormone-sensitive Adenyl Cyclase: Cytochemical Localization in Rat Liver, *Science,* vol. 168, pp. 382–384, Apr. 17, 1970.

Ryan, Wayne L., and Halvor C. Sornson: Glycine Inhibition of Asparaginase, *Science,* vol. 167, pp. 1512–1513, Mar. 13, 1970.

Scandalios, John G.: Alcohol Dehydrogenase in Maize: Genetic Basis for Isozymes, *Science,* vol. 166, pp. 623–624, Oct. 31, 1969.

Stormont, Clyde, and Yoshiko Suzuki: Atropinesterase and Cocainesterase of Rabbit Serum: Localization of the Enzyme Activity in Isozymes, *Science,* vol. 167, pp. 167–202, Jan. 9, 1970. (Zymograms reveal multiplicity of esterase isozymes in rabbit serum.)

Suelter, C. H.: Enzymes Activated by Monovalent Cations, *Science,* vol. 168, pp. 789–795, May 15, 1970.

Suggs, Joseph E., Robert E. Hawk, August Curley, and Elizabeth L. Boozer: DDT Metabolism: Oxidation of the Metabolite 2,2-bis(*p*-Chlorophenyl)ethanol by Alcohol Dehydrogenase, *Science,* vol. 168, p. 582, May 1, 1970. (A metabolite of DDT is a substrate of crystalline liver alcohol dehydrogenase. The oxidation of the substrate was detected spectrophotometrically.)

Tomkins, Gordon M., Thomas D. Gelehrter, Daryl Granner, David Martin, Jr., Herbert H. Samuels, and E. Brad Thompson: Control of Specific Gene Expression in Higher Organisms, *Science,* vol. 166, pp. 1474–1480, Dec. 19, 1969. (Expression of mammalian genes may be controlled by repressors acting on the translation of mRNA; enzyme induction in mammalian cells in continuous culture; role of enzyme turnover. Theory of enzyme induction.)

Turino, Gerard M., Robert M. Senior, and Bhagwan D. Garb: Serum Elastase Inhibitor Deficiency and α_1-Antitrypsin Deficiency in Patients with Obstructive Emphysema, *Science,* vol. 165, pp. 709–710, Aug. 15, 1969.

Whitt, Gregory S.: Homology of Lactate Dehydrogenase Genes: E Gene Function in the Teleost Nervous System, *Science,* vol. 166, pp. 1156–1158, Nov. 28, 1969.

Wuntch, Thomas, Raymond F. Chen, and Elliot S. Vesell: Lactate Dehydrogenase Isozymes: Kinetic Properties at High Enzyme Concentrations, *Science,* vol. 167, pp. 63–65, Jan. 2, 1970.

Yagi, Kunio (ed.): *Flavins and Flavoproteins,* Proc. 2d Conf., *Nagoya, Japan,* 1967, University of Tokyo Press, Tokyo.

Epichlorohydrin In the family of organic epoxides, epichlorohydrin (1-chloro-2,3-epoxypropane) is second only to ethylene oxide in quantity industrial use. This highly reactive compound has the structure

It is a colorless, clear, mobile liquid with a characteristic chloroform-like odor and the properties listed in Table E-6.

Reactions: Epichlorohydrin owes its industrial importance to its unique functionality. The molecule has two available reactive sites, the chlorine

TABLE E-6. *Physical Properties of Epichlorohydrin*

Molecular weight	92.53
Freezing point, °C	−57.1
Boiling point, °C	116.07
Density, d_4^{25}, g/cm^3	1.1750
Refractive index, n_D	1.4358
Flash point, Tag open cup, °C	33
Heat of fusion (calc.), cal/mole	2500
Heat of vaporization (calc.), cal/mole	8729
Vapor pressure (calc.):	

mm Hg	Temp., °C
1	−16.35
10	15.81
17	25.00
20	27.61
100	60.20
400	96.11

Viscosity, cP at 25°C	1.06
Solubility in water, g/100 g.	6.58
Water solubility in epichlorohydrin, g/100 g . .	1.47

atom and the epoxy group, which dominates the character of the compound. The three-membered epoxide ring is highly strained, and thus the bonds are weaker than linear ethers, resulting in a less stable molecule. This accounts for the ease with which epichlorohydrin can undergo acid-catalyst reactions and cleavage by bases.

Typical reactions of epichlorohydrin are shown in Table E-7.

Production: Epichlorohydrin is produced by chlorohydrination of allyl chloride:

$$H_2C\!=\!CHCH_3 + Cl_2 \rightarrow$$

Propylene

$$H_2C\!=\!CHCH_2Cl \xrightarrow{\text{HOCl}} ClCH_2CHClCH_2OH +$$

Allyl chloride   1,2-Dichlorohydrin

$$ClCH_2CHOHCH_2Cl \xrightarrow{\text{hydroxide}} \overset{O}{\overset{\diagup\!\!\diagdown}{CH_2\!-\!CHCH_2Cl}}$$

1,3-Dichlorohydrin   Epichlorohydrin

A production flow diagram is given in Fig. E-17.

During 1970 an estimated 350 million lb of epichlorohydrin was produced in the United States and of this amount approximately 60% was used in the manufacture of glycerin. Approximately 37% of the industrial use of epichlorohydrin was for the manufacture of epoxy resins.

Epoxy Resins: The most popular of these resins is the reaction product between epichlorohydrin and bisphenol A. See also **Epoxy resins.**

Epichlorohydrin-based Rubbers: The physical properties of polyepichlorohydrin rubber remain good over a wide range of temperatures. Permeability to gases is extremely low. The material also has good resistance to solvents, fuels, oils, and ozone. The good aging, high resiliency, and flexibility at low temperatures have led to applications in automotive and aircraft parts, seals and gaskets, hose, belting, wire, and cable jackets. See also **Elastomers.**

Wet-Strength Resins for the Paper Industry: Several wet-strength resins have gained acceptance and are of two classes.

Epichlorohydrin-modified Polyamides: The main advantage of this widely used class is that no alum or an acid medium is required (as with formaldehyde-based resins) for incorporating the resin into the cellulose pulp. These water-soluble resins are cationic and stable below a pH of 7. During the drying process, the resin cross-links and thus produces a paper with permanent wet-strength properties.

Addition of Epichlorohydrin to High-Molecular-Weight Polyalkylene Polyamines: This gives resins which perform like the epichlorohydrin-modified polyamides.

Ion-Exchange Resins: A stable, water-insoluble anion-exchange resin can be prepared by reacting epichlorohydrin with ethylene diamine or a similar amine. The reaction of epichlorohydrin with polymeric tertiary amines forms a strong-base

TABLE E-7. *Typical Reactions of Epichlorohydrin*

With	Reaction
Monohydric alcohols	$\overset{O}{\overset{\diagup\!\!\diagdown}{CH_2\!-\!CHCH_2Cl}} + ROH \rightarrow ROCH_2CHOHCH_2Cl$
Organic acids	$\overset{O}{\overset{\diagup\!\!\diagdown}{CH_2\!-\!CHCH_2Cl}} + RCOOH \rightarrow RCOOCH_2CHOHCH_2Cl$
Aldehydes	$\overset{O}{\overset{\diagup\!\!\diagdown}{CH_2\!-\!CHCH_2Cl}} + RCHO \rightarrow \overset{\overset{R\quad H}{\diagdown C \diagup}}{\overset{O \quad O}{\diagdown\,\diagup}} CH_2ClCH\!-\!CH_2$
Amines	$\overset{O}{\overset{\diagup\!\!\diagdown}{CH_2\!-\!CHCH_2Cl}} + RNH_2 \rightarrow RNHCH_2CHOHCH_2Cl$
Water	$\overset{O}{\overset{\diagup\!\!\diagdown}{CH_2\!-\!CHCH_2Cl}} + HOH \rightarrow CH_2OHCHOHCH_2Cl$

ion-exchange resin. A cation-exchange resin can be produced by the condensation of epichlorohydrin with polyhydroxyl phenols, followed by sulfonation. See also **Ion-exchange resins.**

Other Uses: Epichlorohydrin is used in a minor capacity in the *textile industry*. It has been used to modify the carboxy groups of wool, where the resulting product has increased durability and improved moth resistance. Epichlorohydrin also has been used in the synthesis of antistatic agents, wrinkle-resistant agents, and coating sizings. In the area of *bioproducts,* epichlorohydrin is an effective alkylating agent and has been evaluated as a fumigant against larvae and eggs of various insects. It is effective as a sterilizing agent and bacterial-growth inhibitor. Other applications include surface-active agents, inks, dyes, asphalt improvers, and corrosion inhibitors. Epichlorohydrin meets the requirements of a number of regulations administered (1971) by the U.S. Food and Drug Administration.

REFERENCES

Epichlorohydrin, "Chemical Economics Handbook," Stanford Research Institute, Menlo Park, Calif., 1970.

Epichlorohydrin: The Versatile Intermediate, Form 125-1005-68, The Dow Chemical Company, Midland, Mich., 1968.

Freuder, E., and C. D. Leake: The Toxicity of Epichlorohydrin, *Univ. Calif. Pub. Pharmacol.*, vol. 2, pp. 69–78, 1941.

Lee, H., and K. Neville: "Handbook of Epoxy Resins," McGraw-Hill, New York, 1967.

Morrison, R. T., and R. N. Boyd: "Organic Chemistry," 2d ed., pp. 887–892, Allyn and Bacon, Boston, 1966.

—Richard B. Yerman, *The Dow Chemical Company, Midland, Mich.*

Epoxidizing Agents (See **Peroxides, organic.**)

Epoxy Paints (See **Paints.**)

Epoxy Resins These thermosetting resins are classified as engineering plastics because of their excellent mechanical and electrical properties, dimensional stability, resistance to high temperatures and many chemicals, and strong adhesion to metal, glass, fibers, and many other materials.

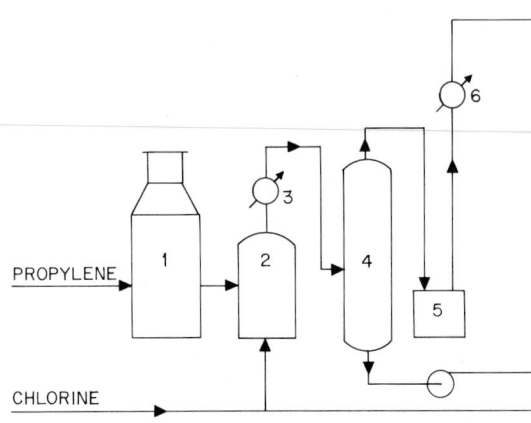

Chemistry and Types: The epoxy groups are three-membered rings with one oxygen and two carbon atoms. The most widely used conventional epoxy resins are manufactured by reaction of epichlorohydrin with a polyhydroxy compound, such as bisphenol A, in the presence of a catalyst. These epoxy resins are known as diglycidyl ethers of bisphenol A (bis-A). The structural formula is shown below.

These resins vary from low-viscosity liquids to high-melting solids, and the differences are brought about by changing the ratio of epichlorohydrin to bis-A used in their production. See also **Epichlorohydrin.**

Solid epoxy novolak resins, shown structurally

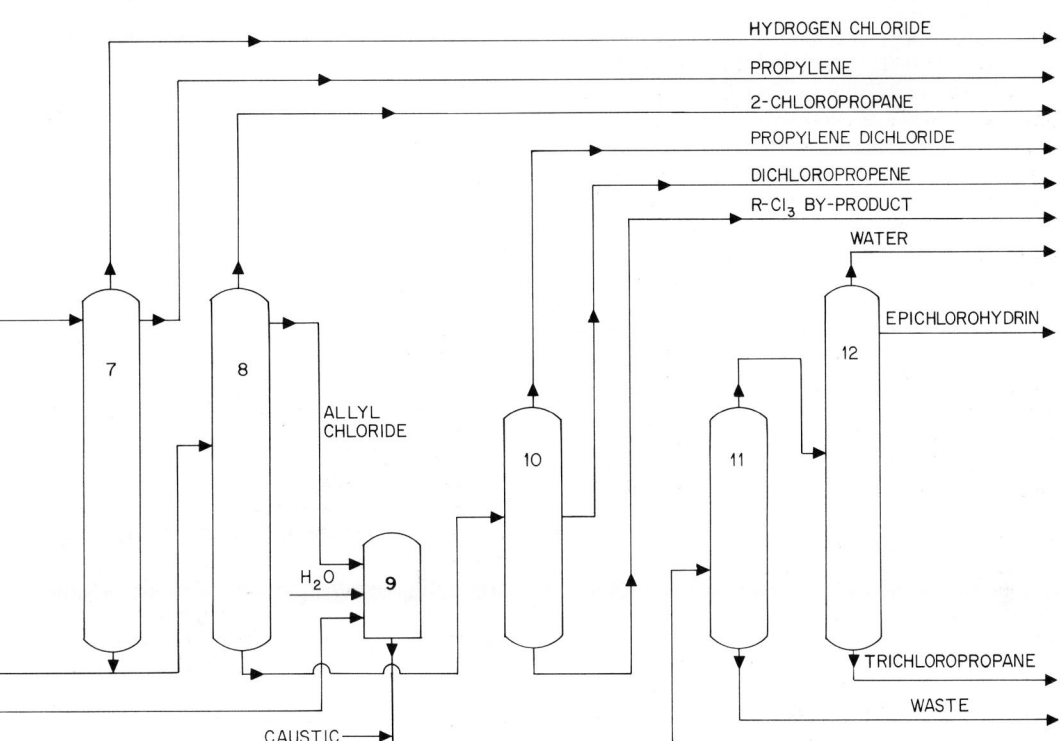

Fig. E-17. Schematic representation of materials flow in epichlorohydrin production. (1) Heater, (2), (9) reactors, (3), (6) coolers, (4) separation drum, (5) compressor, (7), (8), (10), (12) distillation columns, (11) hydrolyzer tower.

here, have better high-temperature properties than the conventional bis-A resins.

Of these, epoxy phenol novolak resins are the most important and are basically novolak resins whose phenoloic hydroxyl groups have been converted to glycidyl ethers. Their production is similar to bis-A resins. Epoxidized novolaks are used mainly in solid single-stage molding compounds and high-temperature laminating systems.

Liquid cycloaliphatic epoxies usually are produced by the peracetic acid epoxidation of cyclic olefins. The epoxide groups are attached directly to the cycloaliphatic ring. The cycloaliphatics have property advantages over conventional epoxies, including superior weatherability, arc and tracking resistance, and dielectric strength. Although cycloaliphatic epoxies can be formulated to have high heat-deflection temperatures (above 250°C), intermediate cure temperatures are sufficient.

Curing: Epoxy resins are not finished products but are reactive chemicals to be combined with other chemicals to yield systems capable of conversion to a predetermined thermoset structure. In general, the major producers of resins, hardeners, and other chemicals for epoxy systems do not supply finished compounds. Compounding is done by specialized firms and by some large epoxy users who operate captive facilities.

Epoxy resins can be cured by cross-linking agents, known as *hardeners,* or by catalysts which promote self-polymerization. When nearly all the reactive sites in the resin have been reacted, the system becomes a tough, infusible material and is considered cured. Selection of the proper curing agent for an application depends upon such system requirements as mixture viscosity, system mass, and temperature. Cured-system requirements, e.g., resistance to temperature and chemicals and electrical properties, also affect the selection of curing agent.

Typical cross-linking agents are primary and secondary aliphatic polyamines, such as diethylenetriamine, triethylenetetramine, tetraethylenepentamine, diethylaminopropylamine, and piperazines. Most of these materials are liquid chemicals of moderately low viscosity that can be blended with epoxy resins at room temperature. Aromatic polyamines are also used as hardeners.

Acid anhydrides and polybasic acids are also used extensively as epoxy-resin curing agents. Typical anhydride curing agents are methyl tetrahydrophthalic anhydride, hexahydrophthalic anhydride, and chlorendic anhydride. These anhydride hardeners and such polybasic acids as adipic acid are particularly effective with most cycloaliphatic epoxy resins. Polyamides and two-step phenolic resins that do not contain water can also be used as hardeners for some epoxy resins.

Catalysts added to epoxy resins permit the epoxy molecules to coreact but do not themselves serve as direct cross-linking agents. The catalysts commonly used are boron trifluoride, monoethylamine, dicyandiamide, and tertiary amines, e.g., triethylamine, dimethylaminomethylphenol, and benzyldimethylamine. Unlike most hardeners that must be used in stoichiometric ratios, catalyst concentrations may be varied within certain ranges to control curing rates. Elevated-temperature postcures are often required, particularly when tertiary amine catalysts are used.

Diluents: These materials are used to lower the viscosities of uncured epoxy-resin systems. Although hardeners or heat can increase flow, diluents are more effective. There are two types of diluents, those containing groups which react with epoxy systems and nonreactive diluents. The reactive type is preferred because nonreactive diluents (polyols or aromatic hydrocarbons) lower the performance of the cured epoxies. Typical reactive diluents are liquid glycidyl ethers. These materials reduce viscosity without excessive loss of properties. The combination of reduced viscosity or slower cure rate which results is often desirable, particularly in filled compounds. Without the use of a diluent in filled compounds, the addition of pigment and filler to the resin system would result in a mixture too viscous to pour, flow in the mold, or permeate wire coils.

Granular fillers, such as silica and calcium carbonate, may be dispersed in these resins and cur-

TABLE E-8. *Typical Properties of Epoxy Molding and Casting Resins*

	Molding Compounds		Casting Resins		
			Bisphenol A Type		
	Glass-Fiber Filler	Mineral Filler	No Filler	Silica Filler	Cycloaliphatics
Tensile strength, psi	14,000–30,000	5,000–7,000	4,000–16,000	7,000–13,000	10,000–20,000
Elongation, %	4	. . .	3–6.2	1–3	1.7–6.0
Tensile modulus, 10^5 psi .	30.4	. . .	3.5	. . .	5–9
Compressive strength, psi .	25,000–30,000	18,000–25,000	15,000–21,000	17,000–28,000	32,000–45,000
Flexural strength, psi . . .	20,000–26,000	10,000–15,000	13,300–21,000	8,000–14,000	15,000–32,000
Hardness, Rockwell M . .	100–108	101	80–110	85–120	85–120
Specific gravity.	1.8–2	1.6–2.06	1.11–1.4	1.6–2	1.10–1.12
Deflection temperature:					
°C.	204–260	149–260	121–260	121–288	66–288
°F.	400–500	250–450	115–550	160–550	150–550
Dielectric constant:					
60 kHz	4	4	4	4	3.9
60 MHz	5	5	5	5	4.5
Arc resistance, s	125–140	150–180	45–120	150–300	80–122
Water absorption, %. . . .	0.05–0.095	0.1	0.08–0.15	0.04–0.10	0.08–0.15

NOTE: Molding compounds and bisphenol A (silica filler) casting resin are self-extinguishing. The burning rate of the other compounds is slow.

ing agents to modify viscosity and reduce the peak exothermic temperature reached during cure. Fillers are used to increase thermal conductivity and reduce the thermal coefficient of expansion of the cured-resin system. Fibrous fillers, such as chopped glass fiber, asbestos, cotton, paper, synthetic fibers, and metal foil and fibers, may be added to increase strength and impact resistance.

Properties: The properties of epoxy molding and casting resins are summarized in Table E-8. In addition to their use in casting, potting, encapsulating, and molding applications, the good wetting and adhesion properties of epoxies permits bonding with reinforcing materials to produce high-strength composites. Epoxy-resin–glass-fiber laminates are commonly used in electrical printed circuits and in wing and fuselage structural members of jet aircraft, where they must maintain strength over wide temperature ranges. Filament-wound epoxy composites are used for rocket-motor casings for missiles. The properties of graphite and carbon fiber-reinforced epoxy laminates are given in Table E-9. These materials, with high strength, light weight, and heat resistance, are useful for many applications, including jet-engine turbine blades and missile parts.

Raw-Material Properties: Liquid epoxy resins of the bisphenol A type have a specific gravity of 1.5–1.17 although special grades run as high as 1.24. The most widely used resin has a viscosity of 11,000–13,500 cP.

Liquid cycloaliphatic resins show an apparent specific gravity of 1.16–1.18. The basic grade has a viscosity of 350–450 cP at 25°C.

Solid resins of the epoxidized novolak type have a specific gravity of 1.12–1.126 at 25°C and a softening point of 63–99°C (145–210°F) as determined by the ring-and-ball test.

Liquid hardeners for epoxy resins exhibit a wide range of specific gravities, from a low of

0.829–1.226. A widely used type (methyl tetrahydrophthalic anhydride) has a viscosity of 32–47 cP and a specific gravity of 1.204–1.224.

Users are warned to avoid skin and eye contact with most of these materials. Processors should avoid breathing vapors or mists given off. Protective clothing should be worn in a well-ventilated environment when handling most epoxies and hardeners. Finished, cured parts present relatively few toxicological problems.

Uses: Because of excellent electrical properties, epoxies are used widely in casting, potting, and encapsulation of a wide range of electrical and electronic parts. Since epoxies provide excellent adhesion and low shrinkage, the resin does not crack or separate from the parts during cure. The size of parts encapsulated, cast, or potted ranges from miniature coils and switches weighing a few grams to large motors and insulators weighing several pounds. Castings are made by pouring liquid compounds into molds and curing. In pottings, the epoxy systems are cast around parts, which are in containers or housings. Encapsulations are pottings with the containers removed and with the epoxy systems impregnated into the parts (see Fig. E-18).

Epoxy molding compounds are thermosetting systems made up of epoxy resins, hardeners, catalysts, and fillers or reinforcements. Colorants, release agents, and other additives may be used. These may be one-component products, containing all ingredients, or several components requiring mixing and dry blending before molding.

Molding compounds are a relatively recent growth area for epoxy resins. Originally, the B-stage formulations were highly unstable and required refrigeration. Single-stage epoxy molding compounds with room-temperature shelf stability are now available.

Single-stage epoxy formulations with long shelf life depend upon chemical-loaded molecular sieves (CLMS).* Known as synthetic zeolites, these sodium aluminum silicates are porous. Various catalysts, such as tertiary amines, are absorbed by the sieve in the pores and are locked in chemically. At low temperatures, the catalysts cannot react with the epoxy, thus providing indefinite shelf life. At the prescribed molding temperatures, the catalyst is released and reacts to provide a rapid cure. The proportion of amine absorbed by the molecular sieves is generally 15–20% by

TABLE E-9. *Properties of Epoxy-Graphite-Type Laminates*

	Longitudinal, psi	Transverse, psi
Flexural modulus . . .	41×10^6	1.2×10^6
Flexural strength	115,000	6,200
Compressive modulus .	44×10^6	1.1×10^6
Compressive strength .	97,000	27,000
Tensile modulus	44×10^6	0.9×10^6
Tensile strength	214,000	4,600
Shear strength	6,000	

*Developed by Linde Division, Union Carbide Corporation.

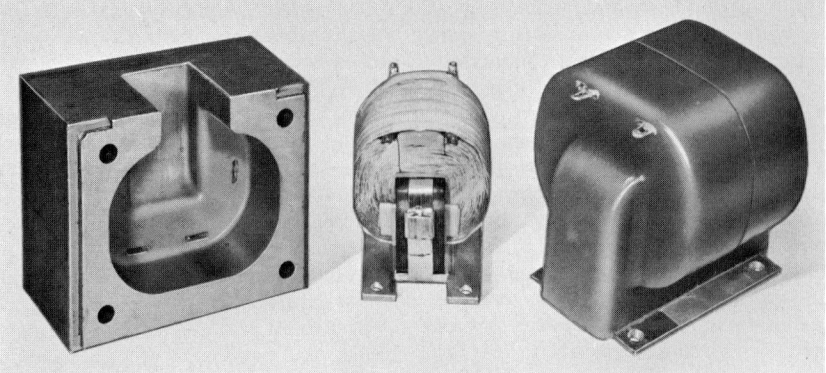

Fig. E-18.   Electrical component to be encapsulated in epoxy resin is shown in center. One half of the mold is shown at left. The final encapsulated component is shown at right. (*Union Carbide Corporation.*)

weight. Various amines, such as triethanolamine or piperidine, can be used. The shelf life is increased further by coating the absorption product of the molecular sieve and amine with a thermoplastic material which is compatible with the molding material.

Powder molding or coating is a growing technique for encapsulating electrical and electronic parts. In one method, the component to be encapsulated is heated, and an epoxy powder compound, suspended or fluidized by rising gas or air, melts in contact with the hot object, forming a coating. Powder compound can also be applied by flooding, flock guns, electrostatic guns, and electrostatic fluid-bed techniques. The powder technique has the advantage of high production rates and low unit costs compared with liquid encapsulants. Molds are not required, and waste due to overspray and premature gellation of liquid systems is eliminated. The excess powder can be reused. Use is limited to thin coatings on small electronic components having low heat capacities.

Epoxy-based *adhesives* are used for bonding dissimilar materials, e.g., plastics and metal, metal and wood, and rubber and ceramics. Only minimum pressure is necessary to effect a satisfactory bond. For these uses, epoxy adhesives are supplied as one- or two-part systems. One-part systems require curing at elevated temperatures; two-part systems can be cured at room temperature. Both have better properties when heat-cured. Some epoxy adhesive systems can withstand temperatures to 316°C (600°F) or higher, but curing at about 150°C (300°F) is required for these heat-resistant grades.

Epoxy resins also find wide application in *protective coatings*. The finishes can be formulated to resist many industrial chemicals and products, liquid foodstuffs, and other corrosive materials. Epoxy coatings include solvent and solventless systems. Solvent systems are room-temperature-cured or heat-cured (baked) finishes. The room-temperature-curing solvent systems include liquid or solid resins cured by hardeners and air-dry-curing liquid epoxy esters. The flexibility of epoxy-resin coatings makes them well suited for coating metal sheet for lithographed containers because the sheet can be coated while flat and then formed with no evidence of cracking. See also **Paint**.

—John J. Madden, *Union Carbide Corporation, New York.*

Erbium   (From Ytterby, a town in Sweden where a mineral containing the element was first identified.) **Er** = 167.26 (at. wt.); 68 (at. no.). Erbium is the twelfth of 15 elements in group III, period 6, generally shown as the rare-earth elements in a separate line below the main body of the periodic table, sometimes referred to as the *lanthanide series*. Er was discovered in 1843 by Carl Gustav Mosander in Stockholm.

Pure Er metal is silvery-gray, retains its metallic luster at room temperature, and is not affected by atmospheric gases or moisture. Massive Er metal, even when heated, oxidizes slowly compared to other elements in the series. Like the others, however, fine powder or chips of Er ignite and burn. Er metal is soft and can be cold-worked with conventional tools and equipment by using common annealing and size-reduction schedules. Intermetallic compounds are formed by Er with other metals, but they have not yet been investigated in detail. A recent study concluded that "erbium behaves mechanically very similarly to yttrium and in many respects like titanium and zirconium."

The melting point of Er is 1527°C; boiling point 2863°C; density 9.045 g/cm^3. Estimated at 2.8 ppm in the earth's crust, it ranks about the same as U and Be in potential availability.

Erbium has six natural isotopes: 162 (0.1 w/o*), 164 (1.6 w/o), 166 (33.4 w/o), 167 (22.9 w/o), 168 (27.1 w/o), and 170 (14.9 w/o). Twelve artificial isotopes have been prepared. The metal is not radioactive and has a low acute-toxicity rating. Its thermal-neutron-absorption cross section is 166 barns per atom, a relatively high figure and tenth of the natural elements.

The only valence known for erbium is 3$^+$. Its oxide, chemical compounds, and solutions are characterized by a reddish-violet color. Added to soda-lime glasses, Er_2O_3 imparts a distinctive pink color but no visible fluorescence when excited by ultraviolet. The element is a strong cation (electropositive) and forms compounds with most inorganic and organic anions. Both the metal and oxide are soluble in mineral acids. Many chemical compounds have been prepared and properties determined. Measurements confirm the *lanthanide contraction* resulting from the addition of one unpaired electron in the 4*f* subshell which has almost no effect on the valence. The eleven 4*f* electrons of Er record sharp absorption bands in the visible and ultraviolet ranges of the spectrum. This is the reason it was distinguished early from the other lanthanide metals in mixtures containing Y and Tb.

Er is found with the yttrium-heavy rare-earth elements. When minerals such as xenotime, certain types of apatites, or gadolinite are processed to obtain Y, the heavy lanthanide elements, including Er, are enriched in the remaining mixture. Organic ion-exchange and solvent-extraction processes then separate Er from the others. About 1 lb of Er can be recovered for every 15–20 lb of Y. The ion is usually precipitated from solution by oxalic acid and then calcined to Er_2O_3 having 99.9% purity.

Pure Er metal is produced by reducing $ErCl_3$ or ErF_3 with Ca or Li metal in a sealed-bomb

*This abbreviation is used in the rare-earth and related fields for weight percent.

reaction. The metal has a moderately high vapor pressure in air near its melting point. Therefore, melting and pouring should be completed rapidly, preferably in a positive pressure of inert gas. In vacuum Er volatilizes rapidly.

For a detailed description of processing and a table of properties, see **Rare-earth elements and metals.**

Uses: Prior to 1965, Er was considered a candidate metal for "burnable" poison in the start-up cycle of nuclear reactors. Other more available metals (at the time) were selected, however, and that application never developed into significant use. Because of its high absorption cross section and much greater availability since 1965, interest in Er and its compounds for nuclear control has revived.

The advent of color TV phosphors using yttrium oxide in 1965 made unprecedented amounts of high-purity Er compounds available.

Bell Telephone Laboratories has developed an erbium-activated phosphor which is coated on gallium arsenide diodes. The diode emits infrared waves, which are converted to visible light by the phosphor. By varying the energizing power and using a combination of rare-earth-activated phosphors, the primary colors of light can be obtained. Potential applications are in display panels and in color-picture production.

Erbium is one of the elements that provide the highest pulsed solid laser efficiencies in glass at room temperature.

A compound, erbium selenide, is a semiconductor with superior thermal stability. Er added to Y and Gd ferromagnetic "garnets" produces special temperature-effect compensation in microwave technology.

An outstanding use of Er is the absolute calibration of ionization gages that measure the degree of vacuum in outer space (and on earth). The erbium hydride–hydrogen system at a fixed temperature creates an extreme vacuum, making it possible to measure 10^{-4} to 10^{-11} torr with high precision.

Many research groups are studying the photoelectric, semiconducting, and thermoelectric properties of this element and its compounds for other new applications.

REFERENCES

Lundin, Charles E.: *9th Rare-Earth Res. Conf. Denver, Colorado, October* 1971, vol. 2, pp. 776–781, Denver Research Institute, University of Denver.

Overview 17, Molybdenum Corporation of America, New York, 1970.

Owen, C. V., and T. E. Scott: *J. Less Common Met.*, vol. 16, pp. 447–455, 1968.

See also list of references under **Rare-earth elements and metals.**

— Joseph G. Cannon, *Molybdenum Corporation of America, White Plains, N. Y.*

Erythromycin   (See **Antibiotics.**)

Essential Oils   (See **Terpenes.**)

Esters   Alcohols, much like metallic hydroxides, react with acids to form ethereal salts known as esters, for example, $CH_3OH + HOOCCH_3 \rightleftharpoons CH_3COOCH_3 + H_2O$, where CH_3COOCH_3, methyl acetate, is the ester. In the reversible reaction, esters are hydrolyzed by water to form an alcohol and an acid. To catalyze the hydrolysis of esters, sulfonic-type cation exchangers may be used. See also **Ion-exchange resins.** To prepare esters, the presence of water must be minimized. Dehydrating agents, such as $ZnCl_2$, HCl, or H_2SO_4, may be added to the mixture to hold the water and thus diminish hydrolysis (decomposition of the ester). Dibasic acids form two classes of esters.

A number of esters occur naturally in fruits, flowers, and plants, often contributing to the scent of the substances, e.g., amyl acetate (pear), methyl butyrate (pine), methyl salicylate (wintergreen), and isoamyl isovalerate (apple).

Esters and esterification are described particularly under **Acetate fibers; Alcohols, fatty (via hydrogenation); Amino acids; Anionic surfactants (sulfur-bearing); Aroma chemicals, Cellulose ester plastics, organic; Polymerization** (polyesterification); **Rayon; Surfactants; Vegetable Oils;** and **Waxes.**

Industrially, esters are used in explosives, plastics, photographic films, lacquers, rayon, paints, varnishes, and soaps and as intermediates. Several esters are high-tonnage chemicals, total production in the United States alone running in excess of 300 million lb annually.

Etching   (See **Hydrogen fluoride;** and **Waxes.**)

Ethane   (See **Ethylene; Natural gas;** and **Petrochemical complex.**)

Ethanol   (See **Ethyl alcohol.**)

Ethanolamines   The ethanolamines (mono-, di-, and triethanolamine) are hydroxyamines first reported by Wurtz in 1860 and first introduced commercially in the United States in 1928. The compounds are clear, viscous liquids at room temperature and white crystalline solids when frozen.

The physical properties of ethanolamines are given in Table E-10. Production of ethanolamines in the United States in 1968 was 221 million lb. The fatty acid derivatives of the ethanolamines are very important commercially. In addition to forming soaps, the ethanolamines also can form esters, amides, and esteramides. The ethanolamines have relatively low toxicity; they produce little primary skin irritation, although mild allergies have developed in a few individuals after prolonged and repeated contact of these substances with the skin.[1]

An area of great usefulness for the ethanolamines is scrubbing gases for removal of acidic

TABLE E-10. *Physical Properties of Ethanolamines*

Property	Ethanolamine, $NH_2CH_2CH_2OH$	Diethanolamine, $NH(CH_2CH_2OH)_2$	Triethanolamine, $N(CH_2CH_2OH)_3$
Boiling point, °C:			
At 750 mm Hg	171.0	*	360
50 mm Hg	100	187	246
10 mm Hg	69	153	201
Freezing point, °C	10.5	28.0	21.2
Refractive index n_D^{20}	1.4544	1.4747	1.4852
Apparent specific gravity, d_{20}^{20}	1.0179	1.019	1.1258
Vapor pressure at 20°, mm	0.36	0.01	0.01
Solubility in water, wt % at 20°	∞	96.4	∞
Flash point, open cup, °C	93	138	179

SOURCE: Alkanolamines and Morpholines, *Union Carbide Corp. Bul.* F-40332A, New York, 1960.
*Decomposes.

compounds. Hydrocarbon gases containing H_2S can be scrubbed with monoethanolamine, which combines with it by salt formation and effectively removes it from the gas stream. In plants synthesizing NH_3, H_2 and CO_2 are produced by the steam reforming of methane. H_2 can be obtained by countercurrently scrubbing the gas mixture in a packed or tray column with monoethanolamine, which absorbs the CO_2. The CO_2 can be recovered by heating the ethanolamine. See also **Ammonia; Hydrogen;** and **Reforming.**

The soaps of the ethanolamines are used widely as emulsifiers, shampoos, and textile-treating agents. The fatty acid amides of diethanolamines are useful as builders in heavy-duty detergents, in which alkyl aryl sulfonates are the surfactant compounds. See also **Detergents.** Triethanolamine, used as the alkali in photographic developing baths, promotes fine grain structure in the developed film. Ethanolamine also is used as a softening agent and as a humectant and plasticizing agent for leather coatings, textiles, and glues.

Production: Although the ethanolamines can be prepared by the reaction of ethylene chlorohydrin with NH_3, nearly all current commercial processes involve the reaction of ethylene oxide with NH_3, generally in aqueous solution. Typical of this type of process* is that shown in Fig. E-19 and described here.

*Developed by Halcon International, Inc.

In the production of ethanolamines, the ratio of mono-, di-, and triethanolamines (MEA, DEA, and TEA) is variable and can be modified by adjusting the NH_3–ethylene oxide ratio in the reaction and by recycling MEA and/or DEA to the reaction. Higher ammonia-oxide ratios favor high DEA and TEA yields and, conversely, lower ammonia-oxide ratios are used when maximum MEA production is desired.

The ethylene oxide and aqueous NH_3 feed (which contains both fresh gaseous NH_3 and recycle NH_3) and any recycle MEA and DEA are mixed and fed to the reactor. The reaction is noncatalytic and is carried out at sufficient pressure to prevent vaporization in the reactor. Essentially all of the ethylene oxide is consumed in the reaction. The reactor effluent is passed to an NH_3 stripper, where unreacted NH_3 is removed from the solution. The overhead vapors, containing NH_3 and water, are cooled and partially condensed and then sent to the NH_3-absorption column. The fresh NH_3 makeup feed also is fed into the NH_3-absorption column, and both streams are absorbed to form the aqueous NH_3 feed to the reactor.

The ammonia-free amines, in a water solution, are fed to an evaporation system, where the bulk of the water is removed. This water is recovered for use in the NH_3 absorber to produce the aqueous NH_3 feed to the reactor. The concentrated amines then are fed to a drying column to remove

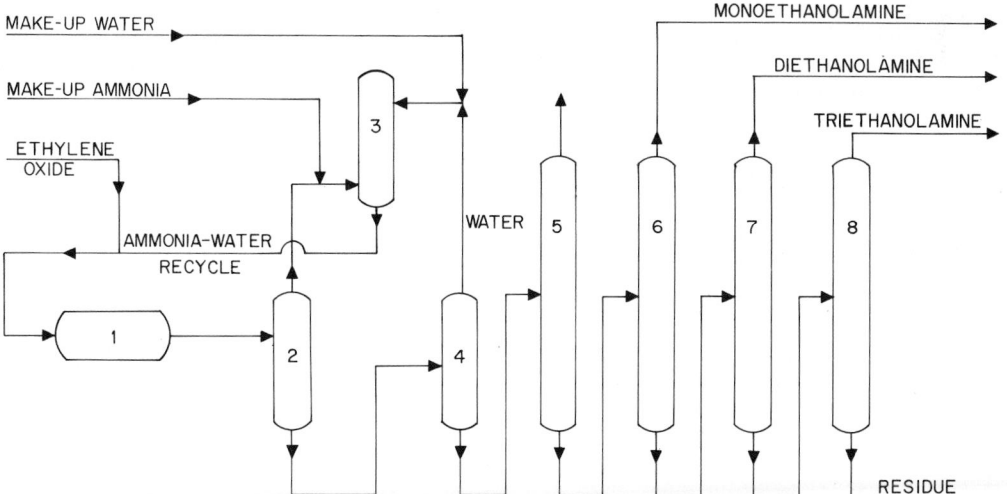

Fig. E-19.   Process for manufacture of ethanolamines. (1) Reactor, (2) stripper, (3) ammonia absorber, (4) evaporator system, (5) drying column, (6) monoethanolamine column, (7) diethanolamine column, (8) triethanolamine column. (*Developed by Halcon International, Inc.*)

the remainder of the water. The dried amines are separated in a series of distillation columns which produce the final products. The flows are evident from Fig. E-19.

Consumption: Approximately 40% of United States production of ethanolamines in 1968 went into the manufacture of surface-active agents, textiles, and cosmetics; 20% was used in gas treating; 14% was exported for various uses; 7% was used in the production of morpholine; and 19% went to miscellaneous applications, including the preparation of emulsion polishes and herbicides. Morpholine is a dye chemical of the acridine group, with the formula

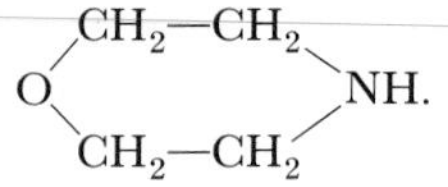

$$O \bigg\langle \begin{matrix} CH_2-CH_2 \\ CH_2-CH_2 \end{matrix} \bigg\rangle NH.$$

REFERENCES

1. Alkanolamines and Morpholines, *Union Carbide Corp. Bul.* F-40332A, New York, 1960.
2. "Chemical Economics Handbook," Stanford Research Institute, Menlo Park, Calif., 1970.
3. Ethylene Oxide, chap. 7 in S. A. Miller, "Ethylene and Its Industrial Derivatives," Benn, London, 1969.

—Alvin S. Cohan, *Scientific Design Company, Inc., New York*

Ethers   Structurally, the ethers are characterized by an oxygen linkage between two radicals in accordance with the general *chemical signature* $R \cdot O \cdot R$. The radicals of the linkage may be identical, as in dimethyl ether, $CH_3 \cdot O \cdot CH_3$, or diethyl ether, $C_2H_5 \cdot O \cdot C_2H_5$; or two different radicals may be involved, in which case the compound is a *mixed ether,* e.g., ethyl isopropyl ether, $C_2H_5 \cdot O \cdot C_3H_7$. Mixed ethers often are prepared from mixed alcohols. When both radicals are aliphatic, the ether may be termed an *alkyl ether* or an *alphyl oxide.* In the classic sense, such ethers are considered to be derivatives of the monohydric alcohols.

The general formula for saturated alkyl ethers is $C_nH_{2n+2}O$. Since each ether is isomeric with a saturated alcohol, the formula $C_4H_{10}O$ applies both to diethyl ether and to butyl alcohol. There are many isomeric ethers, commencing with $C_4H_{10}O$; e.g., methyl propyl ether is isomeric with diethyl ether, both having the formula $C_4H_{10}O$. Compounds that belong to the same family and differ only in the alkyl group present are termed *metameric.*

Phenols, which are structurally similar to the alcohols, also are capable of forming ethers. Anisole (methyl phenyl ether), $C_6H_5 \cdot O \cdot CH_3$, is an example of an aromatic ether. There are relatively few ethers with aromatic radicals on both sides of the oxygen link.

Thio ethers are structured like regular ethers but with a sulfur instead of an oxygen link: thus, $R \cdot S \cdot R$. Diethyl sulfide, $(C_2H_5)_2S$, is an example of a thio ether, while methyl ethyl sulfide, $CH_3SC_2H_5$, is an example of a mixed thio ether.

Nomenclature: As with the alcohols, several systems are used to designate the ethers. Some examples of equivalent terms follow:

p-Anethole. . . p-Propenyl anisole, 1-methoxy-4-propenylbenzene
Anisole. . . . . Methyl phenyl ether
Carbitol . . . . Diethylene glycol monoethyl ether
Diethyl ether   Ether, ethyl ether, ethoxyethane
Dimethyl ether . . . . . Methyl ether, methoxymethane
Ethylene oxide . . . . . Glycol oxide
o-Guiaicol . . . Methylorthohydroxyphenylene ether
Methyl n-propyl ether . . . . . Methoxypropane
Phenetol . . . . Ethyl phenyl ether

Profile of the Ethers: The specific properties of representative ethers are given in Table E-11. The following observations generally apply to most ethers:

1. Mobile, volatile, inflammable liquids lighter than water. An exception is dimethyl ether, which is a gas.

2. Quite inert chemically, resembling diethyl ether.

3. Not acted on by alkali metals or alkalis.

4. Do not react with dilute acids.

5. Decomposed if heated with strong acids, whereupon esters are yielded:

$(C_2H_5)_2O + 2H_2SO_4 \rightarrow$
Diethyl ether

$2C_2H_5 \cdot HSO_4 + H_2O$
Ethyl hydrogen sulfate

$CH_3 \cdot O \cdot C_2H_5 + 2HBR \rightarrow$
Methyl ethyl ether

$CH_3Br + C_2H_5Br + H_2O$
Methyl   Ethyl
bromide   bromide

TABLE E-11. *Properties of Representative Ethers*

Ether	Formula	Formula Weight	Sp gr	Mp, °C	Bp, °C
p-Anethole	$CH_3CH:CH\cdot C_6H_4OCH_3$	148.20	0.991	22.5	235.3
Anisole	$CH_3OC_6H_5$	108.13	0.990	−37.3	154–55
Carbitol	$C_2H_5O(CH_2)_2O(CH_2)_2OH$	134.17	0.990	...	201.9
Diallyl ether	$(CH_2:CHCH_2)_2O$	98.15	...	...	94
Diamyl ether	$(C_2H_5CH_2CH_2)_2O$	158.28	0.774	−69	190
n-Dibutyl ether	$(C_2H_5CH_2CH_2)_2O$	130.22	0.769	−98	142.4
Dibutyl ether (iso-)	$[(CH_3)_2CHCH_2]_2O$	130.22	0.762	...	122.5
sec-Dibutyl ether	$[C_2H_5(CH_3)CH]_2O$	130.22	0.756	...	121
Diethoxymethane	$CH_2(OC_2H_5)_2$	104.15	0.851	...	89
Diethyl ether	$(CH_3CH_2)_2O$	74.12	0.708	−116.3	34.6
Diethylene oxide	$O(CH_2CH_2)_2O$	88.11	1.035	12	101
Diethylene glycol	$(CH_2OHCH_2)_2O$	106.12	1.119	−6	245
Dimethyl ether	CH_3OCH_3	46.07	1.617	−138.5	−23.7
Diphenyl ether	$C_6H_5OC_6H_5$	170.20	1.073	27	259
Diphenylene oxide	$(C_6H_4)_2O$	168.18	...	86–87	287–8
Diphenylenemethane oxide	$(C_6H_4)(CH_2)(O)(C_6H_4)$	182.22	...	105	315
n-Dipropyl ether	$(C_2H_5CH_2)_2CO$	102.17	0.744	−122	91
Dipropyl ether (iso-)	$[(CH_3)_2CH]_2O$	102.17	0.725	−60	69
Ethyl furfuryl ether	$CH_3OCH_2C_4H_3O$	112.13	...	...	150
Ethyl isopropyl ether	$C_2H_5OC_3H_7$	88.15	0.745	...	54
Ethyl α-naphthyl ether	$C_{10}H_7OC_2H_5$	172.22	1.061	5.5	276.4
Ethyl n-propyl ether	$C_2H_5OC_3H_7$	88.15	0.732	−79	61
Ethylene oxide	$\langle(CH_2)_2\rangle O$	44.05	0.887	−111.3	13.5
Glycol ether	$(HO\cdot CH_2CH_2)_2O$	106.12	1.118	−10.5	244.8
o-Guiaicol	$CH_3O\cdot C_6H_4OH$	124.13	1.140	28.3	205
Methyl benzyl ether	$CH_3OCH_2C_6H_5$	122.17	0.987	...	174
Methyl furfuryl ether	$CH_3OCH_2C_4H_3O$	112.13	...	...	150
Methyl isopropyl ether	$CH_3OC_3H_7$	74.12	0.735	...	32
Methyl n-propyl ether	$CH_3OC_3H_7$	74.12	0.738	...	40
Phenetol	C_6H_5OH	94.11	0.967	−30.2	172
Propylene oxide	$CH_3(CHCH_2)O$	58.08	0.831	...	35

6. Chlorine and bromine form substitution products with ethers.

Reactions of diethyl ether may be summarized as follows:

1. With oxygen, in the presence of a flame or spark, an explosive reaction occurs yielding water and CO_2.

2. With water plus heat and an acid, such as H_2SO_4, ethyl alcohol is formed.

3. With concentrated H_2SO_4 plus heat, ethyl alcohol and ethyl hydrogen sulfate are formed.

4. With hydriodic acid gas, ethyl alcohol and ethyl iodide are formed.

5. With Cl_2 or Br, the corresponding halide-substitution product is formed.

6. With phosphorus halides, ethyl halide (2 moles) is formed.

7. With HNO_3, ethyl oxide is formed.

8. With chromic acid, ethyl oxide is formed. See also **Ethyl glycol ethers.**

Ethers generally make excellent solvents for fats, oils, resins, gums, alkaloids, and a number of other organic materials. Diethyl ether is still used in limited amounts as an anesthetic.

Ethoxide (See **Alcoholate.**)

Ethoxylation (See **Surfactants.**)

Ethyl Alcohol Ethyl alcohol (ethanol), CH_3CH_2OH, popularly known as a constituent of alcoholic beverages, is an important industrial chemical with volume requirements far surpassing that for potable use. Before 1930, all ethanol was produced by processes employing natural fermentation. In that year, the first plant synthesizing ethanol from ethylene started production. Since then, synthetically produced ethanol essentially has replaced fermentation alcohol in industrial markets.

Before 1948, synthetic ethanol was produced by variations of the indirect-hydration process, generally known as the *ethyl sulfate* or *sulfation-hydrolysis process,* which, although declining in use, remains a factor in current production. The first process* to synthesize ethanol commercially by the vapor-phase direct hydration of ethylene was introduced in 1948. In recent years, the direct-hydration route has dominated the manufacture of industrial ethanol. Significant in the conversion to direct hydration have been increased restrictions regarding environmental pollution, which are costly to observe in the acid-reconstituting step of the indirect-hydration process.

Physical Properties: Ethanol is a clear, colorless liquid with a mild characteristic odor. The major physical properties of pure ethanol, summarized in Table E-12, are strongly influenced by the hydroxyl group, which imparts polarity and hydrogen-bonding characteristics that are analogous to water and typical of low-molecular-weight alcohols. The association between ethanol molecules also is displayed by the highly nonideal behavior of ethanol in many solutions and by the numerous azeotropes formed. A selected list of binary azeotropes is given in Table E-13.

Chemical Properties: Ethanol is a primary alcohol and undergoes reactions typical of this group, namely esterification, dehydration, dehydrogenation, and oxidation.

Ethanol will react with inorganic and organic acids, acid anhydrides, and acid halides to form esters and water:

$$RCOOH + CH_3CH_2OH \rightleftharpoons$$
$$RCOOCH_2CH_3 + H_2O$$
$$CH_3CH_2OH + H_2SO_4 \rightleftharpoons$$
$$CH_3CH_2OSO_3H + H_2O$$

Ethyl alcohol can be dehydrated to form diethyl ether or ethylene:

$$CH_3CH_2OH \xrightarrow{\text{acid}} CH_2CH_2 + H_2O$$

$$2CH_3CH_2OH \xrightarrow{\text{acid}} CH_3CH_2OCH_2CH_3 + H_2O$$

Both products are formed, but conditions can be altered to promote the desired reaction.

*Shell Ethanol Process, Shell Development Company.

TABLE E-12. *Selected Physical Properties of Pure Anhydrous Ethanol*

Autoignition temperature, °C	390–430
Boiling point at 760 mm, °C	78.32
Change in boiling point dt/dp at 760 mm, °C/mm Hg	0.033
Coefficient of expansion per °C	0.0011
Critical temperature, °C	243.1
Critical pressure, atm	63.0
Critical volume, l/mole	0.167
Density, d_4^{20}, g/ml	0.7893
Dielectric constant at 20°C	25.7
Electrical conductivity at 25°C, Ω-cm	1.35×10^{-9}
Explosive limits in air, vol %:	
Lower	4.3
Upper	19.0
Flash point (Tag open cup), °F	60
Freezing point at 760 mm, °C	−114.1
Heat of combustion at 25°C, kcal/mole	328
Heat of formation at 25°C, kcal/mole	−66.36
Heat of fusion at freezing point, kcal/mole	−1.187
Heat of vaporization at boiling point, kcal/mole	9.30
Heat capacity at 25°C, cal/(g)(°C)	0.574
Molecular weight	46.07
Refractive index at 760 mm and 20°C	1.36143
Surface tension at 20°C, dynes/cm	22.2
Thermal conductivity at 20°C, (kcal)(m)/(hr)(m²)(°C)	0.15
Viscosity at 25°C, cP	1.078

TABLE E-13. *Selected Binary Azeotropes of Ethanol at 760 mm Hg*

Second Component	Ethanol, wt %	Azeotrope Bp, °C
Acetonitrile	56	72.5
Benzene	31.7	67.9
Butyraldehyde	60.6	70.7
Carbon disulfide	9	42.6
Carbon tetrachloride	16	65.0
Chloroform	7	59.4
Cyclohexane	31.3	64.8
Ethyl acetate	30.98	71.8
Ethyl acrylate	72.7	77.5
Heptane	49	70.9
Hexane	20.8	58
Isopropyl acetate	53	76.8
Isopropyl ether	17.1	64
Methylcyclohexane	47	72.1
Nitromethane	73.2	76.0
Pentane	5	34.3
Toluene	66.7	76.5
Trichloroethylene	27.5	70.9
Water	96	78.2

Acetaldehyde can be prepared by oxidation or dehydrogenation of ethanol, with reactions typically in the vapor phase over various metal catalysts:

$$CH_3CH_2OH + \tfrac{1}{2}O_2 \rightarrow CH_3CHO + H_2O$$
$$CH_3CH_2OH \rightarrow CH_3CHO + H_2$$

The hydrogen atom of the hydroxyl group can be replaced by an active metal to form ethoxides. In particular, sodium ethoxide can be prepared by the reaction between absolute ethanol and sodium or by refluxing absolute ethanol with anhydrous sodium hydroxide:

$$2CH_3CH_2OH + 2Na \rightarrow 2CH_3CH_2ONa + H_2$$

$$CH_3CH_2OH + NaOH \rightarrow CH_3CH_2ONa + H_2O$$

Other reactions involving the hydrogen atom of the hydroxyl group include the opening of an oxide ring to form glycol ethers:

$$CH_3CH_2OH + \overset{O}{\overset{\diagup \diagdown}{CH_2 - CH_2}} \rightarrow$$

$$CH_3CH_2OCH_2CH_2OH$$

Uses: Industrial ethanol is used as a solvent in toiletries, cosmetics, pharmaceuticals, and surface coatings and as a raw material chemically converted into other products. The major consumption of ethanol as a raw material in the manufacture of acetaldehyde is declining in favor of processes which convert ethylene to acetaldehyde directly. Other uses are in the manufacture of glycol ethers, ethyl chloride, amines, ethyl acetate, and vinegar.

Ethanol generally is purified and used as the azeotrope with water (95% ethyl alcohol), although anhydrous ethanol has become in greater demand, particularly for cosmetics and aerosols.

The distribution and use of ethanol in the United States is regulated closely by the federal government due to the taxes imposed on ethanol for beverage use. Any industrial utilization requires prior knowledge and approval by the government. Ethanol may be purchased as specially denatured alcohol, as proprietary or special industrial solvents, as completely denatured alcohol, and as tax-paid or tax-free pure alcohol.

Production

By Direct Hydration: Ethanol is produced in the direct-hydration process by the vapor-phase catalytic addition of water to ethylene according to the overall reaction

$$CH_2{:}CH_2 + H_2O \xrightleftharpoons{\text{cat.}}$$

$$CH_3CH_2OH + 19{,}000 \text{ Btu/lb mole}$$

A supported acid catalyst is commonly used. Since the conversion of ethylene to ethanol is limited to relatively low values by the thermodynamic equilibrium at practical operating conditions, a large recycle volume of unconverted ethylene is usually required. The temperature, pressure, water-ethylene ratio, and ethylene purity are some of the important variables affecting the amount of conversion. Small amounts of by-products are formed in other reactions, the primary side reaction being the dehydration of ethyl alcohol to form diethyl ether:

$$2C_2H_5OH \xrightleftharpoons{\text{cat.}} (C_2H_5)_2O + H_2O$$

The process* shown in Fig. E-20 is conveniently separated into three sections: (1) the reaction section, which produces crude ethanol; (2) the purification section, which produces purified 95 vol % ethanol; and (3) the dehydration section, which produces high-purity, water-free absolute ethanol.

In the reaction section, feed ethylene and water combine with recycle ethylene and are preheated to the desired reaction temperature. The hot vapors pass through the catalytic-bed reactor, where ethylene and water are converted to ethanol at yields of 97% or better. The reactor product vapors are cooled by heat exchange with the reactor feed and separated into liquid and vapor streams. Ethanol is recovered from the vapor stream by scrubbing with water. The aqueous ethanol solution from the scrubbing step, combined with the liquid product from the separator, forms the crude ethanol feed to the purification section. Gas from the scrubber is compressed and recycled to the reactor. If required, a small portion of the recycle gas is vented to limit the concentration of inerts in the reactor feed. The quantity of required process vent depends on the purity of the fresh ethylene feed.

The crude product is sent to the purification section for removal of light and heavy impurities. The product from this section is the 95 vol % ethanol-water azeotrope. This high-quality product has excellent odor characteristics, high permanganate time, and low trace-impurity content.

For anhydrous ethanol product, part or all of

*Shell Direct Hydration Process for the manufacture of ethanol, Shell Development Company.

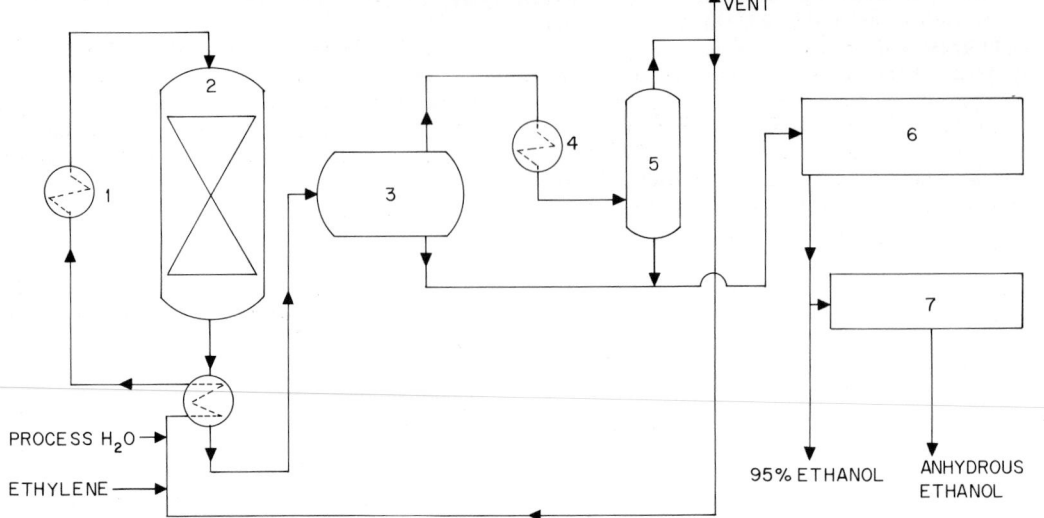

Fig. E-20.   Direct hydration of ethylene to produce ethanol. (1) Heater, (2) reactor, (3) separator, (4) cooler, (5) scrubber, (6) ethanol-purification section for light and heavies removal, (7) ethanol-dehydration section. (*Shell Ethanol Process, Shell Development Company.*)

the 95 vol % ethanol is fed to the dehydration section, where the water is removed by an azeotropic agent. Pure ethanol containing trace quantities of water is the resulting product. This high-quality product is suitable for such odor-critical applications as cosmetics and pharmaceuticals.

By Indirect Hydration: The preparation of ethanol by the sulfation-hydrolysis process can be generalized by the following four steps:

1. Absorption of ethylene in concentrated sulfuric acid to form mono- and diethyl sulfates:

$$CH_2{:}CH_2 + H_2SO_4 \rightarrow CH_3CH_2OSO_3H$$
$$2CH_2{:}CH_2 + H_2SO_4 \rightarrow (CH_3CH_2)_2SO_4$$

2. Hydrolysis of the ethyl sulfates to ethanol:

$$CH_3CH_2OSO_3H + H_2O \rightarrow$$
$$CH_3CH_2OH + H_2SO_4$$
$$(CH_3CH_2)_2SO_4 + 2H_2O \rightarrow$$
$$2CH_3CH_2OH + H_2SO_4$$

3. Recovery and purification of the crude ethanol

4. Reconcentration of the dilute sulfuric acid

The absorption is exothermic, and cooling is required. The rate of absorption is controlled by such variables as acid concentration, ethylene concentration, temperature, and pressure. In addition, the diethyl sulfate reacts with ethanol to form diethyl ether and more ethyl sulfate:

$$(CH_3CH_2)_2SO_4 + CH_3CH_2OH \rightarrow$$
$$CH_3CH_2OSO_3H + CH_3CH_2OCH_2CH_3$$

Diethyl sulfate is promoted by high ethylene-acid ratio, which in turn is promoted by increased ethylene pressure and concentration and increased acid concentrations. The diethyl sulfate not only causes an increase in ether formation but is also more difficult to hydrolyze than ethyl sulfate. The absorption products are hydrolyzed by dilution with water. Crude ethanol is steam-stripped from the dilute acid solution and commonly purified by distillation. The dilute acid is recycled to the absorber after concentration and removal of carbonaceous materials. The reconcentration of the dilute sulfuric acid represents a significant portion of the capital investment for this process.

By Fermentation Processes: These processes may use any agricultural raw material in which the carbohydrate is present as sugar or materials, such as starches, that can easily be converted to sugars. After the materials are in sugar form, yeast enzymes are added to ferment them to ethanol. Historically, blackstrap molasses has been the principal source of sugars converted to industrial ethanol in the United States. Molasses, a by-product of cane-sugar manufacture, contains about 50% sugars and is easily fermented. See also **Enzymes;** and **Yeasts.**

In the fermentation process molasses is diluted with water to a final sugar concentration of about 15 wt % and acidified to form the beginning mash. Yeast culture is added to the extent of about 3–10 vol %. The yeast contains two enzymes, invertase and zymase. Invertase converts sucrose into the invert sugars glucose and fructose:

$$C_{12}H_{22}O_{11} + H_2O \xrightarrow{\text{invertase}}$$

$$\underset{\text{Glucose}}{C_6H_{12}O_6} + \underset{\text{Fructose}}{C_6H_{12}O_6}$$

Sucrose

These sugars are converted by zymase into ethanol and carbon dioxide.

$$C_6H_{12}O_6 \xrightarrow{\text{zymase}} 2C_2H_5OH + 2CO_2$$

Acidity and temperature are carefully controlled. Nutrients are added as necessary to sustain yeast activity. The fermentation period is about 2 days. The final mash, called *beer,* usually contains up to 12% ethanol. The crude ethanol is recovered from the beer by steam distillation.

Other processes for the production of ethanol have been investigated, including the hydrogenation of acetaldehyde, hydrocarbon oxidation, hydration of diethyl ether, and hydration of ethylene by dilute acids. See also **Petrochemical complex.**

REFERENCES

Hatch, L. F.: "Ethyl Alcohol," Enjay Chemical Company, New York, 1962.
"Shell Ethyl Alcohol: A Guide for Industrial Users," Shell Chemical Company, Industrial Chemicals Division, Houston, Tex., 1970.
"Synthetic Ethanol and Isopropanol," *Stanford Res. Inst. Process Econ. Program, Rep.* 53, Menlo Park, Calif., 1969.

—D. E. Dodd, *Shell Development Company, Houston, Texas*

Ethyl Cellulose Ethyl cellulose is a versatile thermoplastic cellulose ether that is compatible with a wide variety of solvent systems, resins, oils, and plasticizers, allowing for diversity in the design of formulations for numerous and varied applications. Ethyl cellulose molding compounds possess a natural surface gloss which imparts a pleasing appearance to molded and extruded parts. Most significantly, molded ethyl cellulose has excellent toughness, flexibility, and shock resistance

over a temperature range from about −40 to +100°C.

Ethyl cellulose is prepared by reacting grade wood pulp or cotton linters having a high α-cellulose content with ethyl chloride and sodium hydroxide according to the following simplified reaction scheme, where R_{cell} is the cellulose radical:

Main reaction:

$$R_{cell}-OH + NaOH \rightleftharpoons$$
$$R_{cell}-OH \cdot NaOH \text{ (complex)}$$
$$R_{cell}-OH-NaOH \rightleftharpoons R_{cell}-ONa + H_2O$$
$$R_{cell}-ONa + CH_3CH_2Cl \rightarrow$$
$$R_{cell}-O-CH_2CH_3 + NaCl$$

Side reactions:

$$CH_3CH_2Cl + NaOH \rightarrow CH_3CH_2OH + NaCl$$
$$CH_3CH_2OH + NaOH \rightarrow CH_3CH_2ONa + H_2O$$
$$CH_3CH_2ONa + CH_3CH_2Cl \rightarrow$$
$$CH_3CH_2-O-CH_2-CH_3 + NaCl$$

Minor side reaction:

$$CH_3CH_2Cl + NaOH \rightarrow$$
$$CH_2{=}CH_2 + NaCl + H_2O$$

The side reactions account for over one-half the ethyl chloride consumed.

The structural formulas of cellulose and ethyl cellulose with complete (54.9%) ethoxyl substitution are

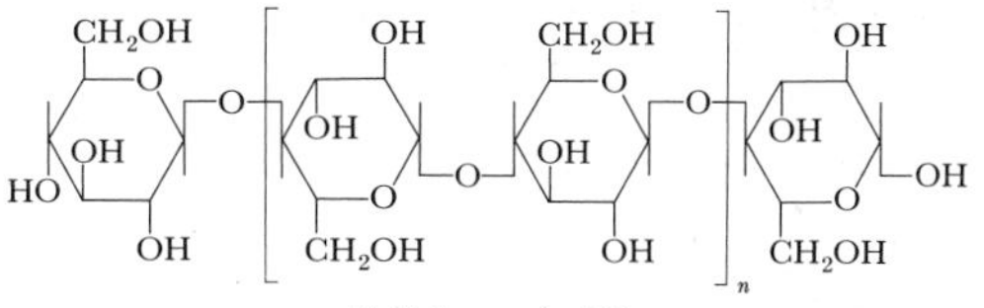

Cellulose, $n > 500$

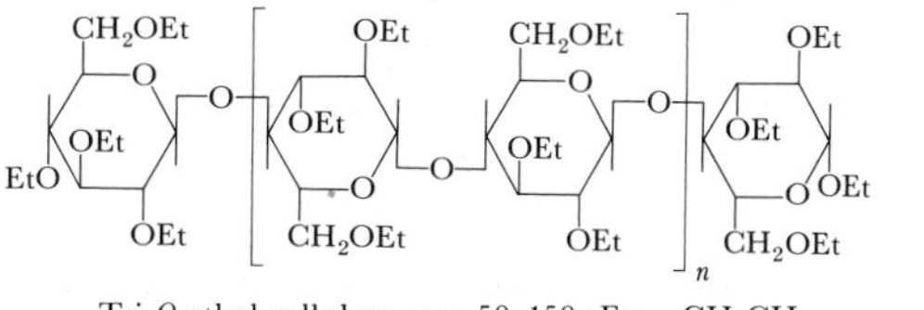

Tri-*O*-ethyl cellulose, $n = 50$–150; Et = CH_3CH_2

Manufacture: There are four basic steps.[1,2]

Preparation of Alkali Cellulose: This is accomplished by slurrying shredded cellulose into excess concentrated alkali prior to addition of ethyl chloride using the alkali solution as the suspending agent or by dipping the cellulose sheet into 50–75% alkali, allowing controlled caustic pickup. This is followed by cutting the alkali-cellulose sheet for loading to the reactor.

Reaction: The ethyl chloride and, if desired, a solvent are loaded into the reactor along with the alkali cellulose and heated to 110–150°C for a controlled time depending on the degree of ethylation desired. For a limited caustic pickup a second addition of sodium hydroxide is usually necessary during the reaction to complete the ethylation. Viscosity control is obtained by adding controlled amounts of air to the reactor or by exposing alkali cellulose chips to the atmosphere for a controlled time before reactor loading.

By-Product Recovery: The diethyl ether, ethanol, unreacted ethyl chloride, and solvent (if used) are flashed from the crude ethyl cellulose in the reactor or in a separate step and recovered by fractionation.

Washing and Drying: The water-insoluble ethyl cellulose is thoroughly washed to remove excess alkali, salt, and by-products. At some point in the wash system dilute acid is added for final pH adjustment. The product is then dried and packaged. In order to meet product specifications regarding average percent ethoxyl and viscosity, batch blending is usually necessary.

Properties: The types of ethyl cellulose produced are summarized in Table E-14. The material is available in 50-lb bags as an untreated granule usually smaller than 20 mesh or as plasticized pellets for extruding or injection molding. Ethyl cellulose is tasteless and nontoxic and has received clearance for use in foods and food containers. The density (1.14) is the lowest of any cellulosic plastic. The natural color of ethyl cellulose is colorless to light amber, but it can be formulated into a wide range of transparent, translucent, and opaque colors. The material is slightly hygroscopic and should be dried before molding. Absorption of immersed wafers under test varied from 8% (low ethoxyl) to 4% (high ethoxyl). Gas permeability is high compared with other thermoplastics. The transmission rate for O_2 through ethyl cellulose is about 1,000 cm^3-mil/100 in.^{2} in 24 hr. For water vapor, the rate is from about 50 (high ethoxyl) to about 100 (low ethoxyl) g/100 in.^{2} in 24 hr through a 1-mil film.

Compression-molding temperature ranges from 250–390°F (121–199°C) and pressure from 500–5,000 psi. Injection-molding temperature is 350–500°F (177–260°C) and pressure 8,000–32,000 psi. Tensile strength is 2,000–8,000 psi and flexural strength 4,000–12,000 psi. The dielectric constant at 60 Hz is 3.0–4.2.

Strong acids decompose the material, whereas weak acids and strong alkalis have only a slight effect. Weak alkalis do not attack the material. Ethyl cellulose is widely soluble in organic solvents.

End Uses: Ethyl cellulose is useful as a strippable coating for metal parts, in coatings for papers, in medicinal tablets, and in coatings for bowling pins. Also as a binder for inks and abrasives and in plastic formulations for molded parts. Although ethyl cellulose can withstand high temperatures for short periods, it should not be used where exposure to temperatures in excess of approximately 212°F (100°C) because of thermal degradation. Since the material also is susceptible to ultraviolet degradation, which is intensified by the presence of moisture, formulations subject to outdoor exposure are only moderately stable.

Ethyl cellulose is not compatible with other

TABLE E-14. *Commercial Types of Ethyl Cellulose*

Ethyoxyl Range		Viscosity Range, cP*	Uses
%	Degree of Substitution		
44.5–46.5	2.21–2.35	14–5,000+	Injection molding where heat-distortion resistance is desired; compared with other ethyl cellulose compositions, generally are harder and have higher melting ranges and better solvent resistance
47.5–49.5	2.43–2.59	4–300	Broad range of resin, solvent, and plasticizer compatibility
49.0+	2.53+	10–200	Useful where maximum water resistance and hydrocarbon compatibility is desired

*Determined as 5% solutions at 25°C in 4:1 toluene-ethanol.

thermoplastics, and therefore processing equipment must be thoroughly cleaned prior to use. The material can be processed with conventional injection molding or extruding equipment and requires no special adjustments or special materials of construction for the molds or extrusion die. In general, injection molding is performed within 10–15°F of the burning-point temperature of the plastic. Ethyl cellulose plastic parts can be joined together or to other materials by a number of adhesives.

Ethyl cellulose has been formulated into compositions with high impact resistance, high heat-distortion temperatures, and good processibility. Sheeting of ethyl cellulose is tough, flexible, and transparent yet sufficiently rigid to withstand rough handling. The improved properties and extensive flexibility in formulating the material for a specific use make ethyl cellulose very attractive for many new applications.

REFERENCES
1. Ethyl Cellulose, *Chem. Met. Eng.*, September 1945, p. 129.
2. Ott, Emil, and Harold M. Spurlin: "High Polymers," vol. 5, p. 913, Wiley, New York, 1954.
                    —K. L. Krumel, *The Dow Chemical Company, Midland, Mich.*

Ethyl Chloride   (See **Chlorine organics.**)

Ethylene   Although ethylene has virtually no direct end uses, it is the most important petrochemical feedstock, both in quantity and economic value. Ethylene is the basic feedstock for such building blocks as ethylene oxide, ethylbenzene, ethyl chloride, ethylene dichloride, ethyl alcohol, and polyethylene, all of which, in turn, are used in the production of a multitude of end products.

Ethylene is a colorless, flammable gas at standard temperature and pressure, has a molecular weight of 28.03, a normal boiling point of $-103.7°C$, a critical pressure of 49.98 atm, and a critical temperature of 9.5°C. Essentially all ethylene production is high-purity (99.9%). A typical specification for high-purity ethylene is shown in Table E-15.

Chemistry: As the first member of the *alkene family*, ethylene has the formula

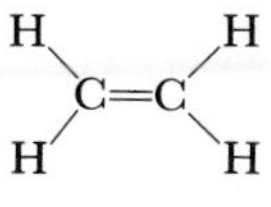

TABLE E-15. *Typical Specifications for High-Purity Ethylene*

Ethylene, %	99.90–99.99
Acetylene, mole ppm	<10
Oxygen, mole ppm	<5
Carbon monoxide, mole ppm	<5
Hydrogen sulfide, wt ppm	<2
Water, mole ppm	<10

SOURCE: S. A. Miller, "Ethylene and Its Industrial Derivatives," Benn, London, 1969.

Ethylene first was produced by the dehydration of ethyl alcohol over alumina:

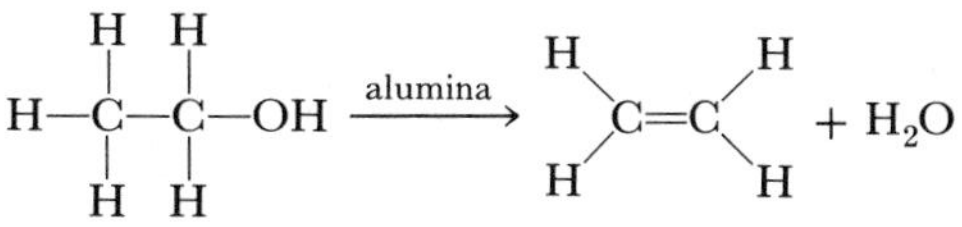

Ethylene now is commercially produced by steam cracking of ethane:

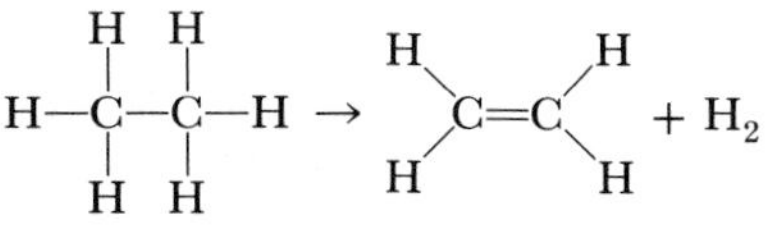

or propane:

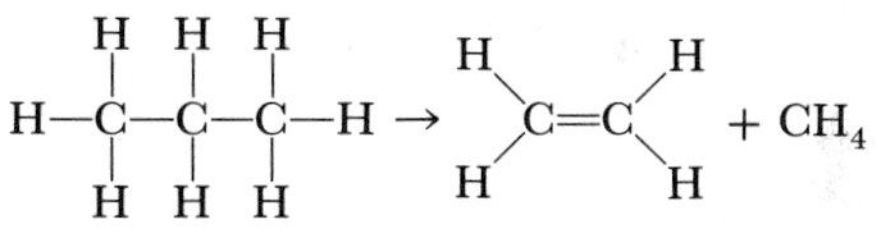

Ethylene is also produced from other paraffinic or naphthenic hydrocarbons. These reactions are highly endothermic (34,400 kcal/kg mole of ethane cracked at ∼900°C) and proceed in the direction shown at temperatures above approximately 620°C without a catalyst.

The importance of ethylene as a petrochemical feedstock derives from its versatility in reacting, forming a large number of chemical intermediates. The reactivity of the double bond, as well as the ability to homopolymerize and copolymerize with other monomers, accounts for this versatility. The most important reactions from a commercial viewpoint of ethylene are:

Polymerization to high- and low-density polyethylene:

$$nCH_2{=}CH_2 \rightarrow (-CH_2-CH_2)_n$$

Oxidation to ethylene oxide:

$$CH_2{=}CH_2 \rightarrow \overset{\displaystyle O}{CH_2{-}CH_2}$$

Alkylation of benzene to ethylbenzene for the production of styrene:

$$CH_2{=}CH_2 + \bighexagon \rightarrow \bighexagon{-}C_2H_3$$

Oxychlorination to ethylene dichloride for the production of vinyl chloride:

$$CH_2{=}CH_2 + 2HCl + \tfrac{1}{2}O_2 \rightarrow C_2H_4Cl_2 + H_2O$$

Hydration to ethanol:

$$CH_2{=}CH_2 + H_2O \rightarrow C_2H_5OH$$

Oxidation to acetaldehyde:

$$CH_2{=}CH_2 + \tfrac{1}{2}O_2 \rightarrow CH_3{-}\overset{\displaystyle O}{\underset{H}{C}}$$

Chlorination to ethyl chloride:

$$CH_2{=}CH_2 + HCl \xrightarrow[\text{cat.}]{\text{acidic}} CH_3{-}CH_2Cl$$

Major Uses: The consumption of ethylene in the United States has risen from 1 billion lb/yr in 1948 to 5×10^9 lb/yr in 1960 and more than 15×10^9 lb/yr in 1971. Total free-world consumption is estimated to be in excess of 37×10^9 lb/yr (1970).

The demand for ethylene has come primarily from its polymeric end uses. As shown by Table E-16, polyethylene is the principal ethylene derivative. Copolymers of ethylene with propylene and other monomers, although representing a small share of the market, grew rapidly in the late 1960s. Other derivatives that are ultimately processed to high polymers, such as ethylbenzene (polystyrene) and ethylene dichloride (polyvinyl chloride), have maintained a steady growth rate. Although the production of ethanol and ethylene oxide from ethylene has increased, percentagewise these derivatives have dropped from 50% of total ethylene production in 1960 to 26% in 1970 (United States).

The principal ethylene derivative, polyethylene, is produced in two basic forms, high-density poly-

TABLE E-16. *Consumption of Ethylene*

	Percentage of Total			
	United States		Western Europe	
Use	1960	1970	1960	1970
Polyethylenes	27	34	44	50
Ethylene oxide	30	19	26	15
Ethyl alcohol	21	7	10	3
Ethylbenzene	10	8	9	8
Ethylene dichloride .	7	11	5	16
Others	5	21	6	8

SOURCE: J. R. Lambrix, C. S. Morris, and H. J. Rosenfeld: The Implications of Heavy Oil Cracking, *Chem. Eng. Prog.*, November 1969.

ethylene ($\sim$0.96 g/cm^3) and low-density polyethylene ($\sim$0.92 g/cm^3). Low-density polyethylene finds its major end uses in films, coatings, and injection molding. High-density polyethylene is used extensively in blow and injection molding. In addition to its homopolymer, ethylene is finding increasing use as a copolymer with propylene, alkyl acrylates, acrylic acid, and vinyl acetate.

The second most important derivative of ethylene, ethylene oxide, is a basic petrochemical building block which is further reacted to a multitude of end products including ethylene glycols, ethanolamines, and glycol ethers. Approximately 60% of the ethylene oxide produced goes into the production of ethylene glycol.

Production: Ethylene was first prepared and identified in the eighteenth century by the dehydration of ethanol. On an industrial scale, ethylene has been produced by catalytic dehydration of ethanol, as a coproduct with acetylene in the high-temperature electric-arc cracking of hydrocarbons, by the hydrogenation of acetylene, and by recovery from coke-oven gases. However, in the industrialized nations today, virtually all ethylene is produced by the thermal cracking of hydrocarbons, either as a primary product in ethylene units or as a by-product of petroleum-cracking processes.

Feedstocks: Almost any naphthenic or paraffinic hydrocarbon heavier than methane can be steam-cracked to yield ethylene and varying quantities of coproducts.

In the United States, the preferred feedstock, which accounts for 75% of the ethylene production, has been ethane and/or propane recovered from natural gas or from the volatile fractions of petroleum. However, in recent years, as the avail-

TABLE E-17. *Raw Materials Used in 1970 for Producing Ethylene*

	Percentage of Total	
	United States	Europe
Ethane and propane cracking	75	6
Refinery gas	15	4
Liquid feedstocks . .	10	90

SOURCE: P. Collinswood, Ethylene in Europe, *AIChE 68th Annu. Meet., March 1971, Houston, Tex.*

ability of new natural-gas supplies which can be recovered economically has become less certain, producers have been turning to heavier petroleum fractions, such as gas oils, as feedstocks. Also affecting the selection of heavier feedstocks is the increasing value of coproducts such as propylene, butadiene, isoprene, cyclopentadiene, and aromatic gasolines, which are produced in greater quantities in the cracking of the heavier feeds. In Europe and Japan, however, the greatest quantities of ethylene have been produced by the steam cracking of light naphtha fractions (see Table E-17). Historically, the main reasons for this difference in feedstock preference are as follows:

1. With no indigenous supplies of natural gas in Europe and Japan, the transportation costs of ethane and propane made these feedstocks less attractive.

2. Because of the relatively low demand for motor gasoline and a high demand for middle distillates and fuel oils, the fractionation of crude produced an excess supply of light liquid feedstocks.

3. Due to the relatively low demand for motor gasoline in Europe and Japan, catalytic cracking of heavy fractions, which produces substantial quantities of propylene and butylene in the United States, has been practiced to only a limited extent, thus leaving a demand gap for these products. This gap has been filled by the propylene and butylene which is coproduced with ethylene in the cracking of naphtha.

Coproducts: As already mentioned, in the steam cracking of liquid hydrocarbons, large, variable quantities of coproducts are obtained. The principal coproducts in steam cracking are propylene, a C_4 stream rich in butadiene, a pyrolysis gasoline rich in aromatics, and other hydrocarbon streams suitable as fuel.

The coproducts produced vary in type and quantity with the feedstock and the severity or depth of cracking. In general, lighter feedstocks yield greater quantities of ethylene while heavier feedstocks yield greater quantities of coproducts.

As shown in Table E-18, in cracking ethane approximately 4 lb of premium-value coproducts (propylene, butadiene, and C_6–C_8 aromatics) is produced per 100 lb of ethylene whereas 118 lb of premium-value coproducts per 100 lb of ethylene is produced in the cracking of a typical heavy gas oil. Due to the high proportion of coproducts produced, a plant cracking liquid feedstock is

TABLE E-18. *Typical Product Distributions for Various Feedstocks*

Product	Feedstock, million lb/yr			
	Ethane	Propane	Naphtha	Heavy Gas Oil
Ethylene	1,000	1,000	1,000	1,000
Propylene, chemical grade	25	410	530	600
C_4's:				
Butadiene	20	50	160	220
Butylene, butane	13	30	150	200
Pyrolysis gasoline:				
C_5, 204°C nonaromatics	. . .	. . .	230	400
C_6–C_8 aromatics	10	80	500	360
Total C_5 and heavier	28	160		
Fuel oil	. . .	. . .	140	900
Off gas	164	680	540	450
Total products	1,260	2,410	3,250	4,130
Feed rate	1,260	2,410	3,250	4,130

SOURCE: J. R. Lambrix, C. S. Morris, and H. J. Rosenfeld, The Implications of Heavy Oil Cracking, *Chem. Eng. Prog.,* November 1969.

often referred to as an *olefins unit* rather than an *ethylene unit.*

The ratio of coproducts produced can also be altered by changing the mode of operation with the same feedstock. For example, a typical naphtha may yield anywhere from 1.0 lb of propylene per pound of ethylene (when cracked at low severity without ethane recycle) to 0.45 lb of propylene per pound of ethylene (when cracked at high severity with ethane recycle).

Cost of Production: An important reason for the rapid increase in ethylene consumption has been the steady decrease in the cost of this material while the cost of alternatives has increased. The selling price of ethylene in the United States declined from 5 cents per pound in 1960 to 4½ cents in 1965 to 3 cents or less in 1970.

This reduction has been achieved in part by advances in design technology and the reduction in energy costs through total integration of energy systems. The construction of larger units has also contributed to economy, decreasing the investment per unit of ethylene produced. The effect of capacity on production costs for various feedstock plants is shown in Fig. E-21. The size of a typical ethylene unit in the United States rose from 100–200 million lb/yr in 1960 to 1,000 million lb/yr in 1970.

Modern Cracking Processes: A number of processes for the pyrolysis of the various hydrocarbon

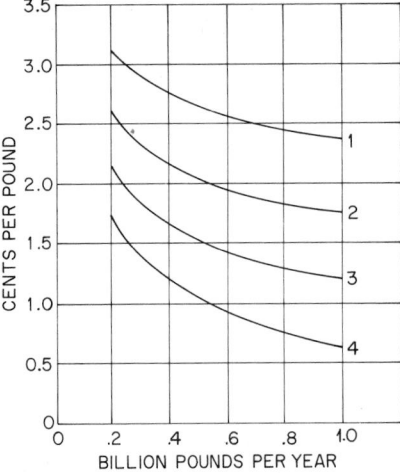

Fig. E-21. Effect of plant capacity on ethylene production costs for various feedstocks. (1) Ethane feed, (2) propane feed, (3) naphtha feed, (4) gas-oil feed. Based upon premium value of coproducts and typical raw-materials costs prevailing on the U.S. Gulf Coast in 1970.

feedstocks are used and although they differ in specific operating conditions, the basic schemes are quite similar. A flow diagram for a typical modern ethylene plant is given in Fig. E-22. The plant may be divided into three major process sections: (1) cracking heaters and effluent quench, (2) gas preparation, and (3) gas recovery.

Cracking Heaters and Effluent Quench: Final product yields and the degree of heat recovery and steam generation are controlled by the design of the cracking or pyrolysis section of the plant. Here, olefin-rich gas is produced by the controlled pyrolysis of hydrocarbons in the radiant coils of the cracking furnaces. Yield flexibility is provided by controlling the temperature, pressure, residence time, and time-temperature profile in the pyrolysis furnace.

Along with the hydrocarbon feed to the furnace, a quantity of dilution steam is added in a weight ratio to the hydrocarbon feed of 0.3 : 1–1 : 1, depending upon the molecular weight of the feed. Typically this steam-hydrocarbon ratio is about 0.3 : 1 for ethane, 0.5 : 1 for naphtha, and 0.5 : 1–1.0 : 1 for gas oils.

The addition of dilution steam serves two purposes: (1) Its primary function is to lower the partial pressure of the hydrocarbon, thus retarding the formation of excess tail gas (methane). (2) A secondary benefit is the reaction between the steam and the coke which forms on the furnace tubes, reducing the rate of coke buildup on the tubes, and thus increasing the run time of the cracking furnace.

The pyrolysis is carried out in a number of furnaces operating at approximately atmospheric pressure and consisting of a radiant section, a convective section, and a waste-heat section. The convective section serves to heat the hydrocarbon and steam feed to the point at which cracking begins (~600°C). In the radiant section, the endothermic heat of cracking and sensible heat required to raise the temperature from the point of incipient cracking to the desired furnace exit temperature (800–900°C) is added. The residence time of the cracking fluid in the radiant section of the furnace is between 0.2 and 0.7. In the waste section of the furnace the sensible heat of the flue gases is recovered, e.g., preheating boiler feedwater and superheating steam.

The hot pyrolysis gases from the furnaces are cooled from the exit temperature of 800–900°C to 350–450°C in a quench boiler to stop the cracking reaction. The exit temperature of the effluent gases from the quench boilers is maintained above the dew point to prevent high-boiling polymerized materials from coating out onto the tubes of the quench boilers. The heat removed from the furnace effluent gases in the quench boilers is used to generate high-pressure steam.

Pyrolysis gases are further cooled by direct-contact quenching with hot oil (for naphthas and gas oils) or water (for ethane and propane) in the pyrolysis-effluent fractionator. The design of this fractionator depends upon the feedstock, the products desired, and the energy requirements of the plant. For a naphtha or gas-oil feedstock, the vapors are quenched with hot fuel oil which recirculates over a baffled section (used because of the fouling which could occur in typical distillation trays). The heat removed from the vapors by the circulating hot oil is used to generate low-pressure steam and for other low-level heat users. The upper, trayed section of the fractionator separates the fuel oil in the pyrolysis effluent from the gasoline and lighter gases.

When the feedstock is ethane or propane, a water quench is normally used. The recirculating quench water recovers heat as a low-level heat source.

Gas Preparation: The cooled gases are then compressed to about 35 atm in a multistage centrifugal compressor with interstage cooling and liquid removal between stages. The condensed hydrocarbon liquids are sent to a gasoline stripper, where the C_4's and lighter are recovered overhead and recycled to the compressor circuit while a gasoline product is taken from the bottoms. (In ethane and propane cracking, the amount of C_5's and heavier is too small to be stripped out at this stage.) Vapors from the last stage of compression are cooled and pass to the caustic tower, where sulfur compounds and CO_2 are removed. For heavy gas-oil feedstocks, which contain an appreciable quantity of sulfur, a regenerable absorption system, such as that based on diethanolamine, is used prior to the caustic wash in order to minimize caustic consumption and the quantity of waste effluent which must be disposed of. After being washed with caustic, the vapors are scrubbed with water to remove entrained caustic.

Vapors leaving the water scrubber are passed to the drying system to remove moisture from the vapors in order to prevent freezing in the low-temperature-distillation section of the plant. Activated alumina or molecular sieve is generally used as a drying medium.

Gas Recovery: The effluent from the driers then flows to the cold section of the gas separation, where the vapors are chilled in a succession of *core-type* interchangers and refrigerated exchangers. An integrated, two-refrigerant cascade system is usually employed to refrigerate at a low horsepower. In the cold section a major portion of the vapor is condensed. The vapor left uncondensed at the lowest temperature level ($-134°C$) is a hydrogen-rich stream, which can be taken as a product or further purified. The liquids condensed out in this section, containing methane and heavier components, are pressured to the demethanizer, which strips the methane and remaining hydrogen from the C_2's and heavier. The methane-hydrogen overhead stream, commonly referred to as *tail gas,* from which ethylene has been rectified, is expanded and flows back through the core interchangers, where the refrigeration is recovered before passing to the fuel system. Where economically justifiable, expansion engines and turboexpanders may be used to obtain lower refrigeration temperatures and to recover energy.

The demethanizer bottom stream is pressured to the deethanizer, where the C_2's are fractionated overhead and the C_3's and heavier taken as bottoms. The deethanizer overhead, containing the acetylene, ethylene, and ethane, flows to the acetylene converter, where acetylene is hydrogenated to ethylene and ethane over a noble-metal catalyst. (With ethane or propane as feedstock the acetylenes are converted in the total raw-gas stream, prior to the fractionation section. This processing sequence is possible since butadiene, which is also hydrogenated in the acetylene converter, is not a significant coproduct in ethane-propane cracking.)

The converter effluent passes to a C_2 splitter, where pure ethylene product is fractionated overhead and ethane is removed from the bottom of the tower. The ethane is exported as a fuel or recycled to a cracking furnace for conversion to ethylene. Recovery of approximately 99% of the ethylene produced in the pyrolysis furnaces is achieved at purities as high as 99.98 mole %.

The deethanizer bottoms, containing C_3's and heavier materials, are pressured to the depropanizer, where they are separated into a C_3 overhead stream and a C_4 and heavier bottoms stream. The overhead product is sent to methyl acetylene converters, where methyl acetylene and propadiene are hydrogenated to propylene and propane over a noble-metal catalyst. Depending upon the feedstock and severity of cracking, the converter effluent may contain up to 93% propylene, which is suitable for direct export as chemical-grade pro-

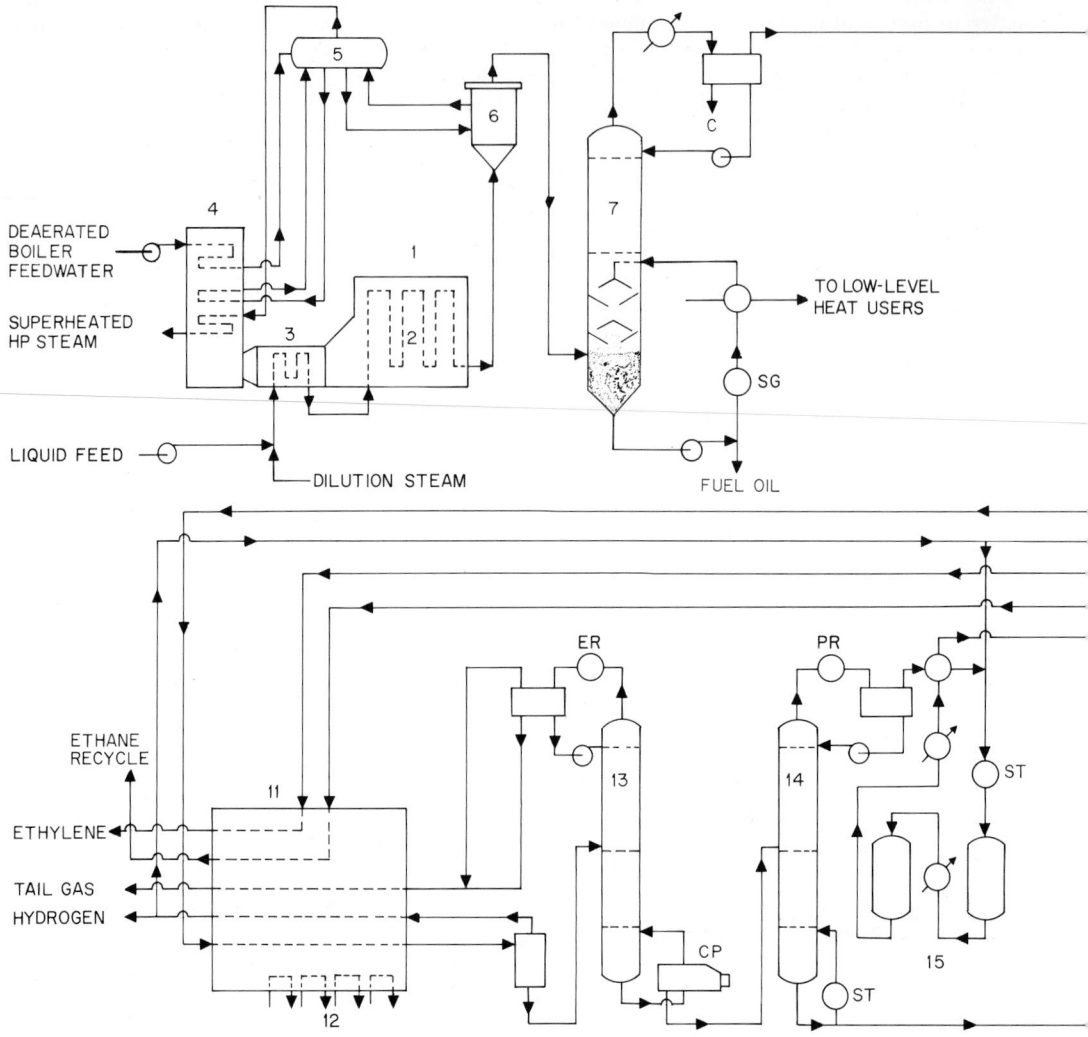

Fig. E-22. Ethylene process (naphtha cracking). (1) Pyrolysis furnace, (2) radiant section, (3) convection section, (4) waste-heat boiler, (5) steam drum, (6) quench boiler, (7) pyrolysis-effluent fractionator, (8) gasoline stripper, (9) caustic and water wash tower, (10) process gas drier, (11) demethanizer-feed heat exchanger, (12) multiple levels of ethylene and propylene refrigeration, (13) demethanizer, (14) deethanizer, (15) acetylene converters, (16) C_2 splitter, (17) depropanizer, (18) methyl acetylene converters, (19) C_3 splitter, (20) debutanizer. *C*, condensate; *CP*, condensed propylene; *ER*, ethylene refrigeration; *HP*, high pressure; *HW*, hot water; *LPG*, liquefied petroleum gas; *PR*, propylene refrigeration; *SG*, steam generation; *ST*, steam. (*The M. W. Kellogg Company, a division of Pullman Incorporated.*)

pylene. If polymer-grade propylene product (99 + %) is desired, or if the propylene content of the converter effluent is less than 92% propylene, the stream is sent to a C_3 splitter, where propylene product is fractionated overhead and an LPG product is taken from the bottom.

The C_4 and heavier stream from the depropanizer bottom flows to the debutanizer, where the C_4's are fractionated overhead. This stream, containing 40–60% butadiene, may be sent to an extraction unit for recovery of butadiene. The debutanizer bottoms, containing C_5's and heavier,

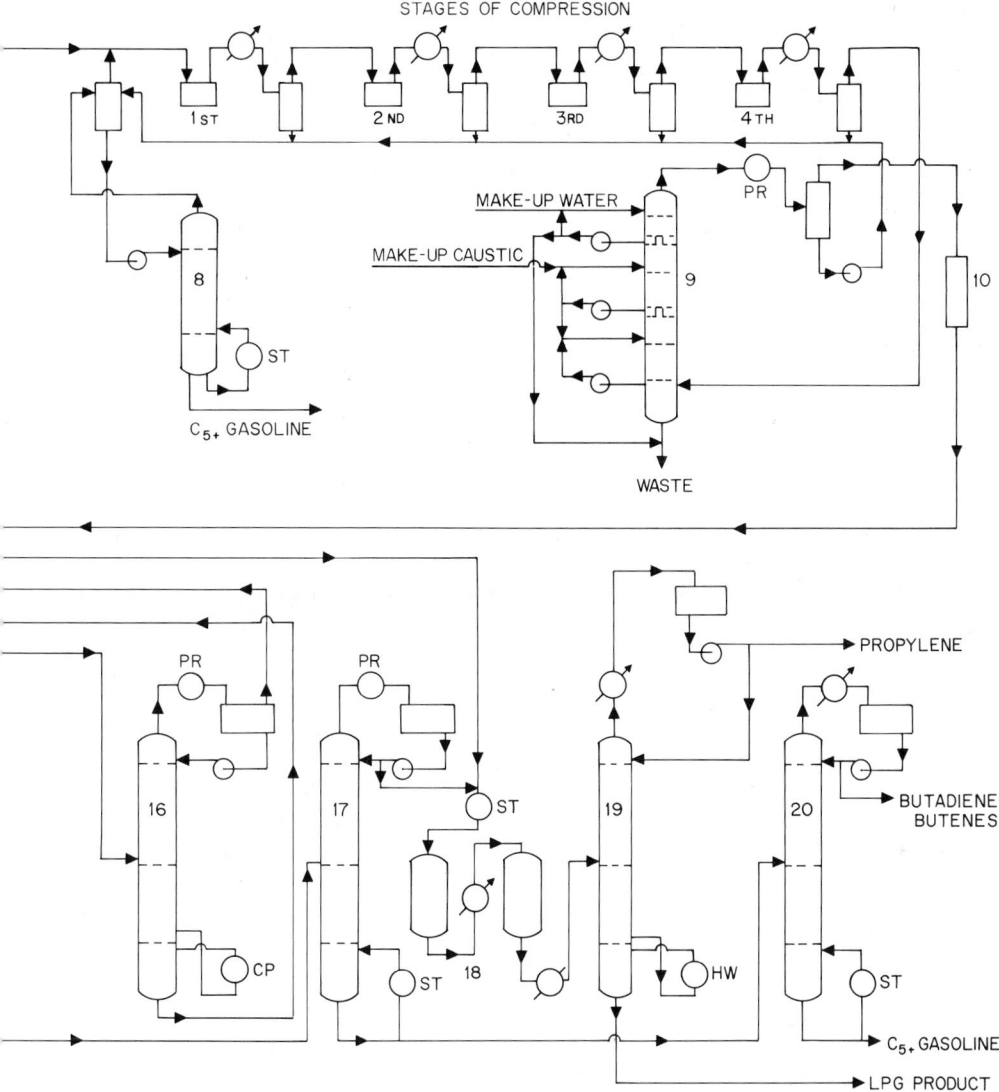

STAGES OF COMPRESSION

1ST 2ND 3RD 4TH

8

ST

C$_{5+}$ GASOLINE

MAKE-UP WATER

MAKE-UP CAUSTIC

PR

9

10

WASTE

PR

PR

16

17

ST

18

19

20

CP

ST

HW

ST

PROPYLENE

BUTADIENE-
BUTENES

C$_{5+}$ GASOLINE

LPG PRODUCT

are joined with the gasoline-stripper bottoms and flow to storage. This material, commonly referred to as *pyrolysis gasoline,* contains 50–70% aromatics. After hydrotreating, it is suitable as feed to an aromatics recovery unit or for blending in a gasoline pool.

Refrigeration Systems: The separation of light gases, such as hydrogen, methane, ethane, and ethylene, requires fractionation at low temperatures because the critical temperatures of these materials are below ambient temperature. The fractionation sections of ethylene units therefore require large cooling loads, which must be supplied by refrigeration rather than cooling water. The process refrigeration requirements are supplied by a propylene refrigeration system at the higher-temperature refrigeration levels ($+9$ to $-34°C$) and an ethylene system for lower temperature levels (down to $-100°C$). Both systems utilize centrifugal compressors and steam-turbine drivers. Ethylene and propylene are chosen as the refrigerants because of their ready availability and advantageous thermodynamic properties.

The propylene compressor is a three-stage machine, the pressure of the final-stage discharge being selected to allow cooling water to condense the discharge propylene vapors. Condensed propylene is then let down to various pressures depending upon the refrigeration levels required. The lowest level of propylene refrigeration normally chosen is the bubble point of propylene at just above atmospheric pressure, which allows operation of the entire refrigeration system above atmospheric pressure without danger of air leakage into the circuit.

The ethylene compressor is also a three-stage machine. The final-stage discharge pressure is selected to allow the ethylene to be condensed against the lowest level of propylene refrigeration. Condensed ethylene then is let down in much the same manner as the propylene, the lowest level being selected on the same basis.

A flow scheme showing a typical propylene-ethylene cascade-type refrigeration system for an ethylene unit is represented in Fig. E-23.

Energy Consumption: Although the ethylene plant is a large consumer of fuel, steam, and power, it contains the potential of being a large producer and (if the plants employ propane or liquid feedstocks) even an exporter of energy. To make maximum use of this potential, modern ethylene plants are designed to permit an integration of the energy available within the process with that required to operate the plant. The excess may be exported as fuel oil, tail gas, or steam.

For example, the high-level heat of the furnace effluent is recovered by the generation of high-pressure steam in the quench boilers. This steam may then be superheated and used as the motive force to the turbine drives of the process-gas refrigeration compressors. Lower-pressure process

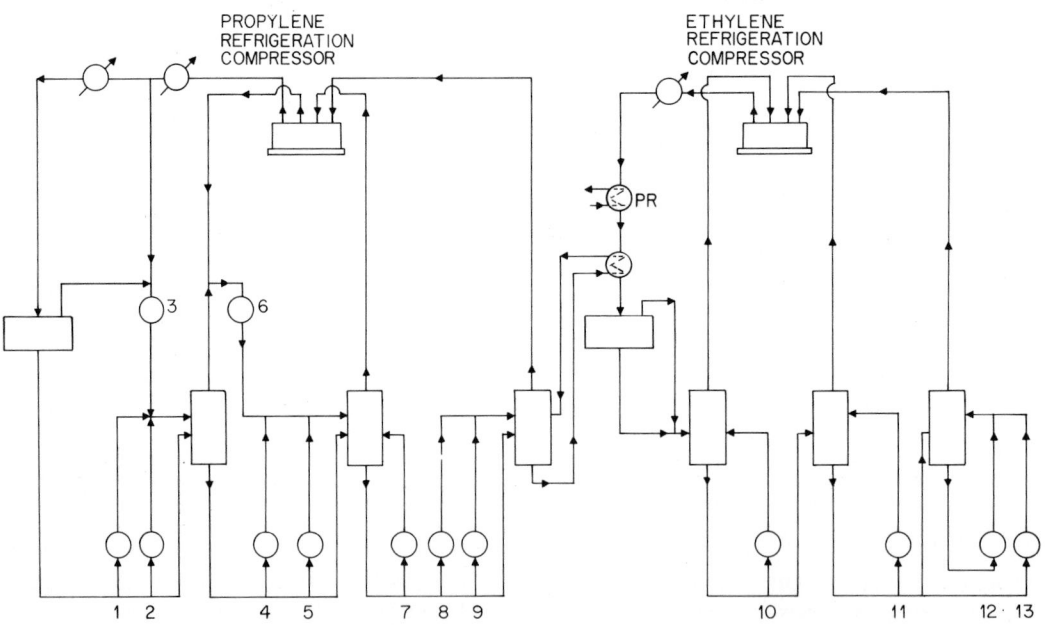

Fig. E-23. Cascade refrigeration system used in ethylene production. Exchangers are (1) drier–feed chiller, (2) (5), (8), (10), (11), (12) demethanizer feed, (3) demethanizer reboiler, (4) ethylene refrigeration desuperheater, (6) C$_2$ splitter reboiler, (7) deethanizer condenser, (9) C$_2$ splitter condenser, (13) demethanizer condenser. *PR*, propylene refrigeration. (*The M. W. Kellogg Company, a division of Pullman Incorporated.*)

steam may then be extracted from these turbines at the desired intermediate pressure levels. Low-level heat is recovered from the quench section of the pyrolysis-effluent fractionator to generate low-pressure and dilution steam as well as to preheat boiler feedwater. Even lower-level heat requirements, such as those of the demethanizer and C_2 splitter reboilers, are satisfied by exchange against condensing propylene refrigerant, thus lowering the cooling-water and horsepower requirements of the refrigeration system.

The fuel oil and tail gas produced in the cracking step are used as fuel for the pyrolysis furnaces. The quantity of fuel produced is normally in excess of that required for internal use except for ethane feedstock, where fuel must be imported to the unit to meet the energy demands.

Materials of Construction: Usually the feedstock or products of an ethylene unit do not require special materials of construction. An exception is gas-oil feedstocks, which may contain significant quantities of sulfur compounds. However, in several areas of the plant, special alloys are required because of extremes of temperature. In the pyrolysis furnace, tube-wall temperatures well above 1000°C occur due to the high heat-flux requirements of the modern short-residence-time furnace. Therefore, modern ethylene pyrolysis furnaces use 25:20 Cr-Ni steel alloy for the radiant tubes because of its superior high-temperature strength.

In the cold section of the plant special materials are also required to avoid the brittleness of carbon steels at low temperature. Killed steel, aluminum, and stainless steel are all used in this section of the plant, depending upon the temperature, service, and economics.

Distribution and Storage: Due to its physical properties, ethylene usually is moved at lowest cost as a vapor by pipeline, which may be a simple line connecting an ethylene-producing unit with an ethylene-consuming unit or a complex grid of piping extending over hundreds of miles and interconnecting numerous ethylene producers and consumers. Examples of such grids are the "spaghetti bowl," connecting ethylene producers with petrochemical plants along the U.S. Gulf Coast, and the European ethylene grid interconnecting 13 ethylene units in three western European countries over 700 mi of pipeline.

Along the Gulf Coast, fluctuations in the quantity of ethylene being consumed and produced in the units connected to the pipeline are handled by using huge salt domes as intermediate storage facilities. When the amount of ethylene being produced exceeds consumption, ethylene is automatically diverted from the pipeline to the salt domes, where it is stored in the dense phase at a pressure in excess of 100 atm. When demand for ethylene exceeds that being produced, ethylene flows from the salt domes, through driers, and back to the pipeline. Pressure in the domes is maintained by pumping in brine when the volume of ethylene is decreasing.

Where pipelines are not used, ethylene is usually shipped as a refrigerated liquid (−100°C) at approximately atmospheric pressure in insulated stainless-steel or aluminum trailers or tankers.

Storage of ethylene, other than in domes, normally is as refrigerated liquid in insulated tanks at close to atmospheric pressure. Small quantities of ethylene for laboratory use are shipped and stored as compressed gas in cylinders.

Other References to Ethylene: This important hydrocarbon is mentioned under numerous other topics throughout this volume; consult the Subject Index, and in particular see **Aldehydes; Alkylation; Ethylene oxide; Natural gas; Petrochemical complex; Petroleum; Polyethylene; Polymerization;** and **Thermal cracking.**

> —James E. Wallace, *The M. W. Kellogg Company, a division of Pullman Incorporated, Houston, Tex.*

Ethylene Alkylation Ethylene alkylation offers the refiner a high-octane blending stock in the front-end portion of the gasoline range. The properties of ethylene alkylate help to produce a balanced gasoline, especially in a lead-free gasoline economy.

The major component in ethylene alkylate is 2,3-dimethylbutane (2,3-DMB), which has a clear Research octane number of approximately 103.5. Typical octane numbers (on an unleaded and unblended basis) for alkylates produced from various olefins are compared in Table E-19. Thus, ethylene alkylation yields a product having a

TABLE E-19. *Typical Octane Numbers for Alkylate from Several Olefins*

Olefin	Octane Number	
	Research	Motor
Ethylene	99–101	92–93
Propylene	89–91	88–89
Butenes (1 and 2 mixed) . .	96–93	94–96
Amylenes	91–93	90–91

higher Research octane number than that obtained by conventional alkylation of other olefins. Additionally, because of its boiling range, ethylene alkylate provides front-end octanes. Ethylene alkylate therefore complements the higher-boiling traditional alkylates and reformates to provide an octane-balanced gasoline less subject to knocking under full-throttle acceleration or where maldistribution is a problem.

Chemistry: Ethylene and isobutane are reacted in the presence of $AlCl_3$ to form primarily 2,3-DMB:

$$H_2C{=}CH_2 + H_3C{-}\underset{H}{\overset{CH_3}{C}}{-}CH_3 \xrightarrow{\ AlCl_3\ }$$

$$HC_3{-}\underset{H}{\overset{CH_3}{C}}{-}\underset{H}{\overset{CH_3}{C}}{-}CH_3$$

The reaction product* typically is about 70 wt % 2,3-DMB (on a C_4-free basis). The remainder is primarily a mixture of C_5–C_8 isomers, with a small amount of C_9–C_{10} material. The side products result from cracking, isomerization, polymerization, and disproportionation reactions. The amount of *n*-butane in the alkylate depends mainly on isobutane feedstock purity but is typically 2–4%.

The reaction catalyst is a solution of free aluminum chloride in aluminum chloride complexed with hydrocarbon. Trace amounts of HCl are added as a reaction promoter. A thorough discussion of the chemistry, which is by a carbonium-ion mechanism, is presented by Schmerling.[1]

Process: With reference to Fig. E-24, dry isobutane and ethylene are reacted in the presence of the $AlCl_3$ catalyst to form primarily 2,3-DMB. Ethylene and fresh plus recycle isobutane are mixed ahead of the reactor. Fresh makeup $AlCl_3$ is brought into the system by solution in a slipstream of isobutane feed. The activity of the catalyst is maintained at a high level by passing a continuous slipstream through the catalyst-regeneration system.

Catalyst regeneration reduces the amount of $AlCl_3$ makeup to 10% or less of that normally necessary for nonregenerative catalyst operation. A small amount of spent catalyst is purged from

*Shell Ethylene Alkylation Process.

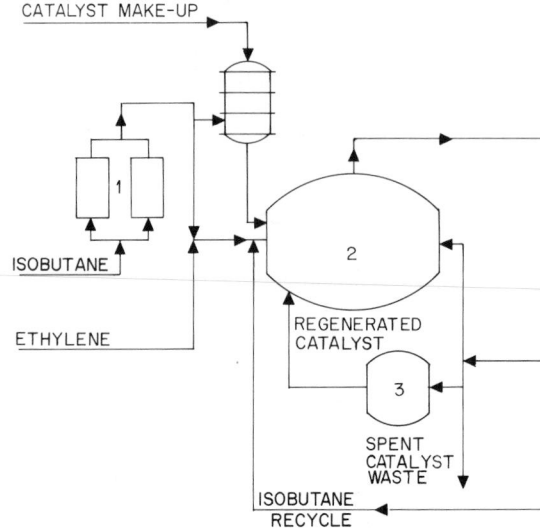

the system to prevent buildup of inert components.

Following the alkylation reaction, the two-phase liquid mixture is separated, and the denser catalyst phase is recycled back to the alkylation reactor. The lighter hydrocarbon phase containing the ethylene alkylate product flows to the fractionation and recovery section. In the fractionation section, the crude alkylate is recovered as a bottom product from the deisobutanizer. Isobutane is recovered as a side cut and is recycled back to the alkylation reactor. The deisobutanizer overhead is partially condensed, and the condensate is depropanized to provide additional isobutane recycle. Relatively pure propane, depending on plant feed composition, can be removed as a side cut from the depropanizer. Vent gas from the two columns is caustic-scrubbed to remove trace amounts of acid and normally is sent to the refinery fuel system.

REFERENCE

1. Schmerling, L.: Alkylation of Saturated Hydrocarbons, in G. A. Olah (ed.), "Friedel-Crafts and Related Reactions," vol. 2, pt. 2, pp. 1075–1097, Interscience-Wiley, New York, 1964.

—Ronald L. Dickenson and William S. Reveal, *Shell Development Company, Houston, Tex.*

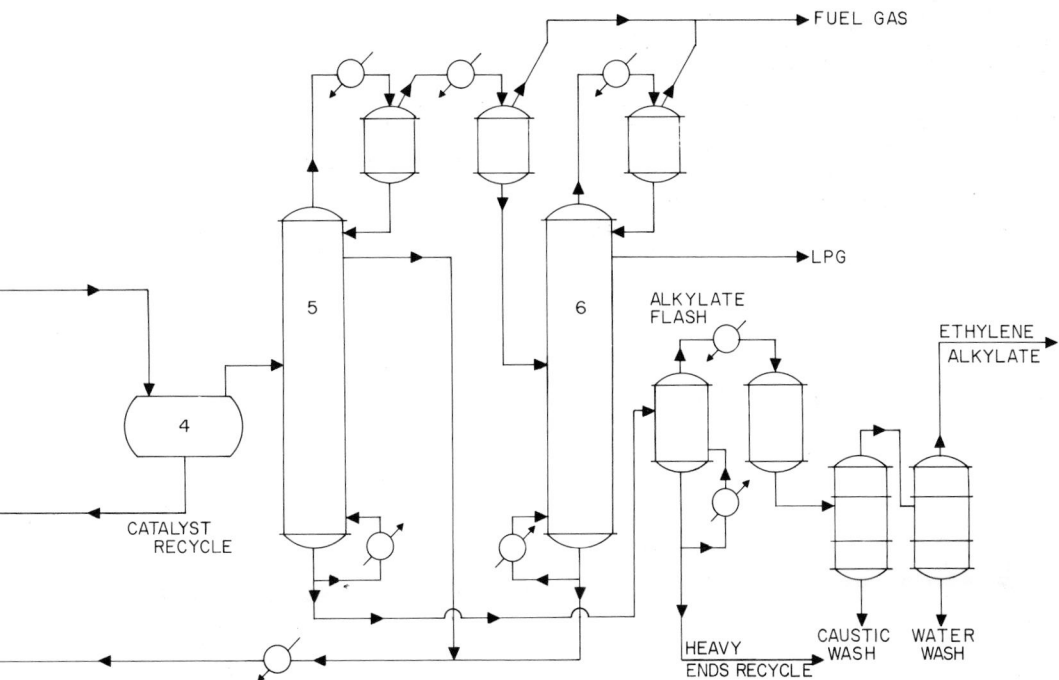

Fig. E-24. Ethylene alkylation process. (1) Driers, (2) reaction section, (3) regeneration section, (4) settler, (5) deisobutanizer, (6) depropanizer. (*Shell Development Company.*)

Ethylene Chlorohydrin (See **Chlorine organics.**)

Ethylenediaminetetraacetic Acid (EDTA) (See **Chelating agents;** and **Detergents.**)

Ethylene Dibromide (See **Bromine.**)

Ethylene Dichloride (See **Chlorine organics.**)

Ethylene Glycol Ethylene glycol, $HOCH_2CH_2$-OH, manufacturing traditionally has consumed the major share of ethylene oxide production. Historically, ethylene glycol has been used primarily as a permanent-type antifreeze for water-cooled internal-combustion engines. In the 1960s, ethylene glycol became a significant chemical intermediate in the manufacture of polyesters for fibers, films, and coatings. Other important uses are in deicing solutions, as a hydraulic fluid, and in the manufacture of low-freezing-point explosives and glycol ethers. In addition, di- and triethylene glycols are important coproducts normally produced during the manufacture of ethylene glycol.

The main end uses of diethylene glycol, $HOCH_2CH_2OCH_2CH_2OH$, have been in the manufacture of unsaturated polyester resins and polyester polyols for polyurethane-resin production and in the textile industry as a conditioning agent and lubricant for various natural and synthetic fibers. Other uses are as an extraction solvent in the petroleum industry, as a desiccant in natural-gas processing, in the production of triethylene glycol, and in the manufacture of certain plasticizers and surfactants. See also **Polyester fibers.**

Triethylene glycol, $HOCH_2CH_2OCH_2CH_2$-OCH_2CH_2OH, is used primarily in the dehydration of natural gas and as a humectant. Smaller amounts are used as a solvent in printing inks and in the manufacture of plasticizers, unsaturated polyester resins, and polyester polyols for polyurethanes.

Chemistry and Product Characteristics: The process* illustrated in Fig. E-25 includes provision for the production and purification of ethylene oxide

*Shell Ethylene Oxide–Ethylene Glycol Process.

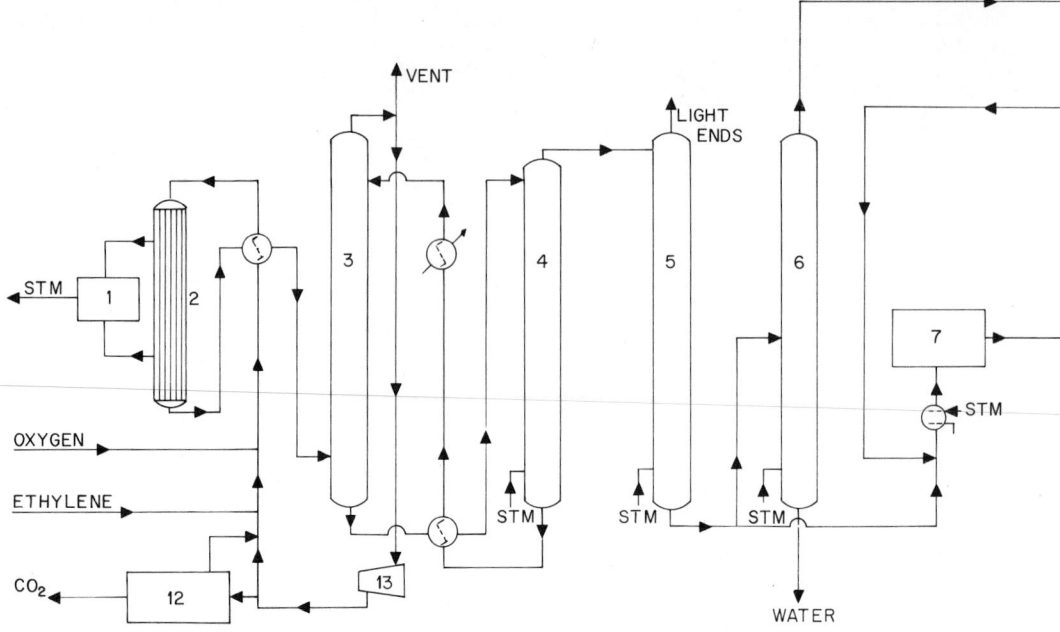

Fig. E-25. Ethylene oxide–ethylene glycol process. (1) Steam generator, (2) ethylene oxide reactor, (3) ethylene oxide absorber, (4) ethylene oxide stripper, (5) light-ends column, (6) ethylene oxide purification column, (7) glycol reactor, (8) glycol dehydrator, (9) ethylene glycol purification column, (10) diethylene glycol purification column, (11) triethylene glycol purification column, (12) carbon dioxide removal, (13) recycle compressor. (*Shell Development Company.*)

and ethylene glycol as well as the heavier glycol coproducts. The two sections usually are integrated to minimize overall manufacturing costs. Either section, however, can be designed and constructed independently of the other. In this process, ethylene oxide is produced by the direct oxidation of ethylene with purified O_2 over an Ag catalyst. Ethylene oxide, in turn, is thermally hydrolyzed to ethylene glycol, diethylene glycol, and triethylene glycol by the following series of reactions:

$$H_2C\overset{O}{\overgroup{-}}CH_2 + H_2O \rightarrow H_2\overset{\overset{\displaystyle H}{\overset{|}{O}}}{C}-\overset{\overset{\displaystyle H}{\overset{|}{O}}}{C}H_2$$

Ethylene oxide   Ethylene glycol

$$H_2\overset{\overset{\displaystyle H}{\overset{|}{O}}}{C}-\overset{\overset{\displaystyle H}{\overset{|}{O}}}{C}H_2 + H_2C\overset{O}{\overgroup{-}}CH_2 \rightarrow$$

$$HO-CH_2-CH_2-O-CH_2-CH_2-OH$$

Diethylene glycol

$$HO-CH_2-CH_2-O-CH_2-CH_2-OH + H_2C\overset{O}{\overgroup{-}}CH_2 \rightarrow$$

$$HO-CH_2-CH_2-O-CH_2-CH_2-O-CH_2-CH_2-OH$$

Triethylene glycol

Trace amounts of higher glycols also are formed. Water normally is kept in large excess in order to favor formation of ethylene glycol and to suppress the sequential reaction of ethylene glycol with more ethylene oxide to form the higher glycols.

This series of ethylene oxide hydrolysis reactions, which can be catalyzed by dilute strong aqueous acid or carried out (as in the process illustrated) in a noncatalyzed manner at high temperature and pressure, forms the basis of production of virtually all the mono-, di-, and triethylene glycols produced in the United States. A small amount of ethylene glycol is produced by the hydrogenation and hydrogenolysis of molasses, and earlier some ethylene glycol was produced by a process that used formaldehyde, carbon monoxide, and water for starting materials.

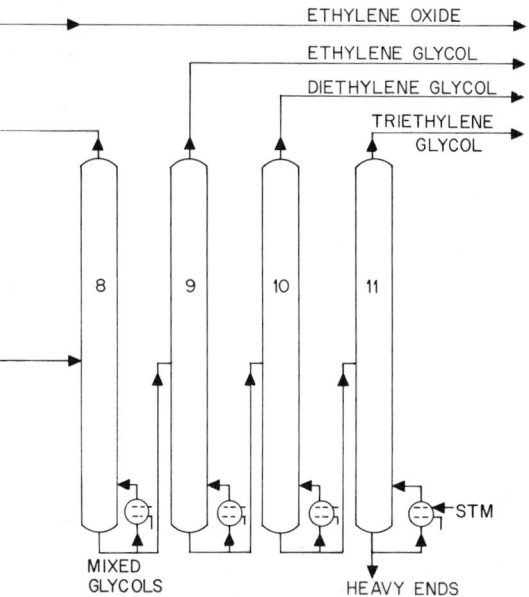

ETHYLENE OXIDE

ETHYLENE GLYCOL

DIETHYLENE GLYCOL

TRIETHYLENE GLYCOL

8 9 10 11

STM

MIXED GLYCOLS

HEAVY ENDS

Production Process: With reference to Fig. E-25, ethylene oxide (EO) is formed by direct oxidation of ethylene with O_2 over a rugged Ag catalyst. The heat of reaction is recovered in a steam-generation system. The EO is recovered from the reactor exit gas by absorption in water. The unreacted gases, except for a small vent stream, are compressed and after addition of ethylene and O_2 makeup feeds are recycled to the EO reactor. A portion of the recycle gas is treated to remove CO_2, formed in the reactor by a side reaction, from the system. The EO is recovered from the absorbant water by steam stripping. The aqueous EO then is condensed and stripped of light ends. The noncondensables are recovered and recycled. A portion of the stabilized EO is fed to the ethylene glycol reaction system. The remainder is charged to a final purification column, where high-purity EO product is made. There may be variations of the purification system. See also **Ethylene oxide.**

The stabilized EO stream is mixed with a large excess of water, preheated, and fed to the ethylene oxide reactor. The EO and water react under conditions of high temperature and pressure to form primarily ethylene glycol. The amounts of di- and triethylene glycols produced are, within practical limits, a function of reactant concentrations, which are manipulated to give the desired amount of each product. The crude glycols are dehydrated and then recovered individually as highly pure overhead streams from a series of vacuum-operated purification columns. The yield of glycols, based on EO charged to the glycol reactor, is in excess of 99.5%. Only a very small amount of heavy ends are formed, and mechanical losses can be held to a low level.

The mono-, di-, and triethylene glycols are colorless and essentially odorless stable liquids with low viscosities and high boiling points. Because they are members of a homologous series of dihydric alcohols, the three glycols have similar properties. Differences in their applications are due mainly to variations in their physical properties, some of which are given in Table E-20.

TABLE E-20. *Physical Properties of Glycols*

Property	Ethylene Glycol	Diethylene Glycol	Triethylene Glycol
Molecular weight	62.07	106.12	150.17
Boiling point at 760 mm Hg, °C	197.6	245.0	287.4
Vapor pressure at 20°C, mm Hg	0.06	<0.01	<0.01
Apparent specific gravity (20°C/20°C)	1.1155	1.1184	1.1254
Refractive index, n_D^{20}	1.4316	1.4472	1.4559
Viscosity at 20°C, cP	20.93	35.7	47.8
Flash point, Cleveland open cup, °F	240	290	330
Specific heat at 20°C, cal/(g)(°C)	0.561	0.500	0.525
Surface tension at 20°C, dynes/cm²	50.5	47.0	47.5
Heat of vaporization at 760 mm Hg, cal/g	191	83	99
Freezing point in air at 760 mm Hg, °C	−13	−8	−7.2
Water solubility	Complete	Complete	Complete

SOURCE: Glycols, *Shell Chemical Company Bull.* IC: 67-58, 1967.

REFERENCE

Curme, G. O., Jr., and Franklin Johnston: "Glycols," ACS Monograph Ser. 114, Van Nostrand Reinhold, New York, 1952.

—Ronald L. Dickenson, *Shell Development Company, Houston, Tex.*

Ethylene Glycol Ethers   The ethylene glycol ethers constitute a series of colorless liquids of high solvency obtained by reaction between an alcohol and ethylene oxide (EO). Various alcohols have been used in the reaction, including methyl, ethyl, isopropyl, *n*-butyl, and isobutyl, to form their respective ethers. The initial reaction product is the monoethylene glycol ether, which in turn can react with additional EO to form heavier products. The alcohol homologues, up to the butyl derivatives, generally are completely miscible with water and virtually all common solvents. They have mild, pleasant odors and, in general, have reasonably low viscosities and pour points and relatively high flash points. The chemical properties are those to be expected from their structures. The compounds can be oxidized to aldehydes and acids, dehydrated to vinyl ethers, esterified with the usual reagents to form a further useful series of solvents, and converted into the usual series of alcohol derivatives.

The major outlets for glycol ethers are jet-fuel deicing additives, surface-coating solvents, printing-ink solvents, and brake fluids. Monoglycol ethers are used primarily in the first application, whereas the di and tri products are used as a diluent for brake fluids.*

Chemistry: Ethylene oxide reacts with compounds possessing an active hydrogen atom. The reaction of EO with an alcohol is

$$ROH + \overset{O}{\overset{/\backslash}{CH_2-CH_2}} \rightarrow ROCH_2CH_2OH$$

Alcohol   Ethylene oxide   Mono product

The reaction is first order with respect to both reactants.

Depending upon the nature of the alkyl (R) group, the reaction proceeds readily with an acid or an alkaline catalyst. The reaction also proceeds at a much lower rate when uncatalyzed. A secondary alcohol reacts more slowly than a primary

*Production in the United States in 1969 approximated 0.5 billion lb.

alcohol, and a lower-molecular-weight alcohol reacts faster than a higher-molecular-weight alcohol. Since the first addition product of the reaction itself contains an active hydrogen, like the starting alcohol group, the reaction with EO is

$$ROCH_2CH_2OH + \overset{O}{\overset{/\backslash}{CH_2-CH_2}} \rightarrow$$

Mono product   Ethylene oxide

$$ROCH_2-CH_2-O-CH_2-CH_2OH$$

Di product

The reaction may continue further, giving products with three or four more equivalents of EO. The relative proportions of mono, di, tri, and heavier products formed depend upon the ratio of the alcohol to EO in the reactor feed, the type of catalyst used, the amounts of mono and di ether in the feed, and the nature of the alcohol.

Production Process: One process* for the production of ethylene glycol ethers (Fig. E-26) is functionally divided into a reaction section and a product-purification section. In the reaction section, EO is mixed with an excess-alcohol feed stream, which contains recycle alcohol from an alcohol-recovery column and fresh makeup alcohol. The combined stream is heated to reaction temperature and fed to the reactor. The preheat temperature depends upon the type of feed alcohol used and on the alcohol-EO ratio. The reaction is exothermic. The EO essentially is reacted to extinction in the reactor. The reactor pressure is controlled to keep the reactants in the liquid phase.

The reaction-product stream is sent to an alcohol-recovery column, where the excess alcohol is removed as top product and returned to the reactor. The crude glycol ethers, removed as column bottoms, are sent directly to the product-purification section.

In the product-purification section, the crude glycol ethers are separated into specification monoglycol ether, diglycol ether, triglycol ether, and heavier ethers by a train of vacuum-fractionation columns. The high-purity ethers are taken as overhead products. The heavy-residue stream

*Shell Ethylene Glycol Ethers Process. This process is in use by Shell Chemical U.K. Ltd., Carrington, England; Shell Nederland Chemie, N.V., Pernis, The Netherlands; Shell Chemical Company, Geismar, La.; and Texas Eastman Company, Longview, Tex.

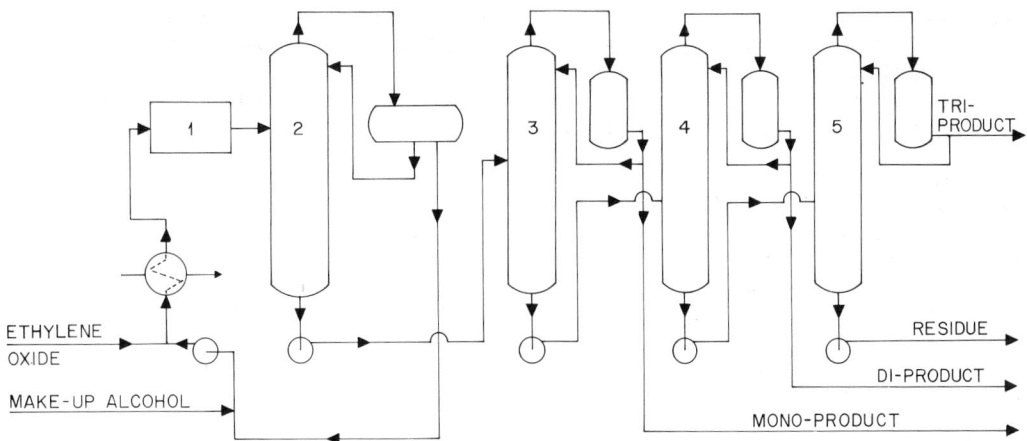

Fig. E-26. Ethylene glycol ethers process. (1) Reactor, (2) alcohol column, (3) mono column, (4) di column, (5) tri column. (*Shell Development Company.*)

from the bottom of the last column is sent to disposal.

In this process, the equipment is optimized for a multiproduct glycol ethers plant. Different alcohols may be selected as starting material. The process is quite flexible, and the product distribution for any alcohol may be changed easily to meet the market demands. Individual stream factors for each alcohol situation may be changed to allow even greater operating flexibility. Glycol ether product specifications from this process are given in Table E-21.

REFERENCES

"Chemical Economics Handbook," Stanford Research Institute, Menlo Park, Calif. 1969.
Curme, O., Jr. (ed.): "Glycols," ACS Monograph Ser. 114, Van Nostrand Reinhold, New York, 1953.
Miller, S. A. (ed.): "Ethylene and Its Industrial Derivatives," Benn, London, 1969.

—J. Yu, *formerly with Shell Development Company, Houston, Tex.*

Ethylene Oxide The direct-oxidation process for the production of ethylene oxide from ethylene

TABLE E-21. *Product Specifications of Typical Glycol Ethers*

Compound	Sp gr (20°C/20°C)	Distillation Range at 760 mm Hg	
		Initial Bp, °C	Dry Point, °C
Monoethylene glycol ethers:			
Methyl	0.966	123.5	125.0
Ethyl	0.931	134.0	136.0
Isopropyl	0.906	139.5	144.5
n-Butyl	0.902	167	173
Isobutyl	0.892	158	162
Diethylene glycol ethers:			
Methyl	1.023	192.0	200.0
Ethyl	0.990	198.0	204.0
n-Butyl	0.955	227.0	235.0
Isobutyl.	0.947	217	225

NOTE: All these compounds are miscible with water. They have a color rating (Pt-Co) of 10. All have an acidity of 0.01 wt %.

was discovered by the French researcher Le Fort and was first disclosed in patents of the Societé Française de Catalyse Generalisée. The essential feature of the process consists of the partial oxidation of ethylene in the presence of a metallic-silver catalyst. According to the original patents, operation advantageously takes place at 150–300°C under a pressure of at least 50 atm. The new procedure at once aroused lively interest. Since the time of its introduction, the direct-oxidation process has substantially eliminated the use of the older chlorohydrin process, which depended on the reaction of ethylene with aqueous hypochlorous acid to form ethylene chlorohydrin. The ethylene chlorohydrin was reacted with calcium hydroxide to form ethylene oxide and calcium chloride as a by-product.

In 1966, 90% of all of the ethylene oxide produced in the free world was made by direct oxidation (~2.1 million tons/yr total[1]). In 1970, it is probable that the percentage of the chlorohydrin-produced oxide did not exceed 1% of the total.*

Consumption of ethylene oxide in the United States was approximately 2.6 billion lb in 1968 and 3.2 billion lb in 1969. Most of the ethylene oxide, about 60%, is used to produce ethylene glycol. The largest use of ethylene glycol is as antifreeze, almost all the rest going to polyethylene terephthalate for manufacture of polyester fibers. About 12% of the ethylene oxide is used to produce surfactants; 8% goes to ethanolamines, used in detergents and soaps; 10% goes to polyethylene glycols, used in solvents, plasticizers, and lubricants; and about 9% goes to glycol ethers, used as solvents and jet-fuel additives. Important physical properties of ethylene oxide include molecular weight 44.05; boiling point 10.7°C at 760 mm Hg; freezing point −111.3°C; and specific gravity 0.8711 at 20°C.

The following reactions take place in the oxidation of ethylene with silver as catalyst:

$$CH_2{=}CH_2 + \tfrac{1}{2}O_2 \rightarrow \underset{\displaystyle O}{CH_2{-}CH_2} \qquad (1)$$

$$CH_2{=}CH_2 + 3O_2 \rightarrow 2CO_2 + 2H_2O \qquad (2)$$

*Of the total ethylene oxide thus produced by direct oxidation, roughly 35% was made by Union Carbide Corporation in its own plants using its own process; about 37% was made by licensees of Scientific Design Company, in some 90 or more plants using either the Scientific Design air or oxygen process; and the remaining 28% was made by firms using other processes (estimate as of mid-1969). See also **Petrochemical complex.**

Modern yields are about 70–72% of theory. Selectivity is highly sensitive to temperature, dropping as temperature rises. It is therefore essential that the commercial plant provide for efficient heat removal and careful temperature control.

Tubular reactors with fixed catalyst beds are preferred in industrial plants. Over the years, plant sizes have increased dramatically. In 1958, a plant producing 28,000 tons/yr of the product was considered large. Plants in some cases are now well over 0.25 million tons/yr and approaching 0.5 million tons/yr capacity.[1,2]

Only silver is effective as a catalyst with high selectivity; it is precipitated onto inert carriers to increase its life. The selection of the carrier appears to be of considerable importance because of the critical need for the removal of heat. Many investigations have been conducted into ways of retarding the undesired combustion reaction to form CO_2 by additions of inhibitors (ethylene dichloride, ethylene dibromide, alcohol, amines, or organometallic compounds). Despite the many compounds for which such usefulness is claimed, it seems likely that among them only ethylene dichloride and some polychloro aromatic compounds have been used industrially.[1]

The preparation of ethylene oxide using the direct-oxidation technique* is shown in Fig. E-27. While neither the schematic flow sheet nor the process description here bears a close correspondence to any modern operating ethylene oxide plant, both are sufficiently like them to provide a real sense of how such plants operate.

A mixture of ethylene, air, and return gas, in which the ethylene content is 3–5 vol %, is conducted under a pressure of 10–20 atm gage to a tubular reactor with fixed-bed silver catalyst. A heat-transfer agent makes it possible to adjust the reaction temperature in a narrow range of 220–280°C, in each case according to the activity of the catalyst. The gas leaving the reactor first gives up most of its heat content to the circulating gas in a heat exchanger and then enters a scrubbing tower, in which the ethylene oxide is scrubbed with water.

The unabsorbed gas, which contains unreacted ethylene and oxygen, is in part returned to the reactor and in part forwarded after renewed heating in a heat exchanger through a second reactor, where the oxidation is completed. The reactor is again connected to a scrubbing tower into which

*Scientific Design Company, Inc.

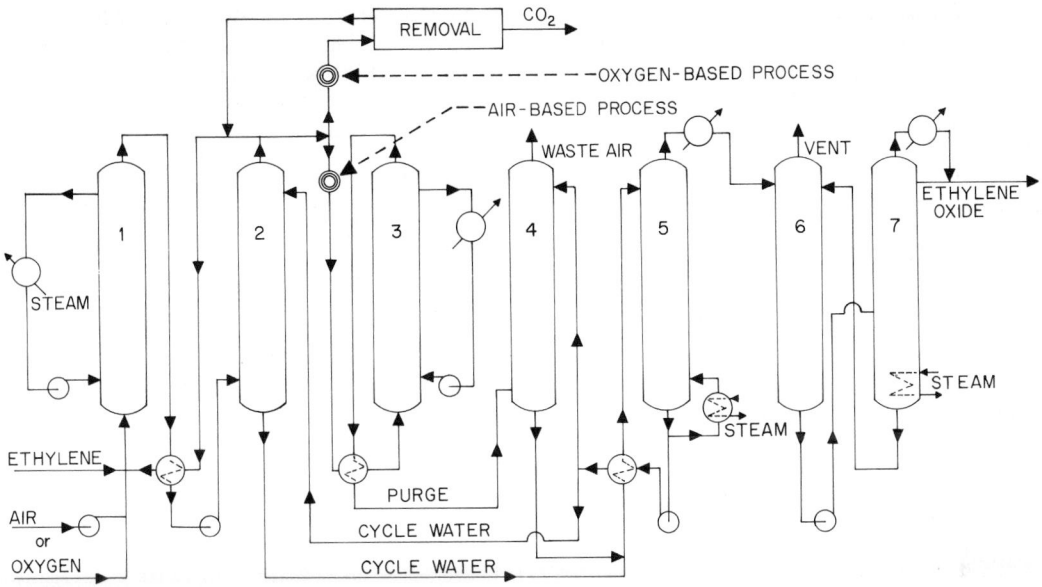

Fig. E-27. Schematic flow sheet of ethylene oxide production by direct-oxidation route. (1) Main reactor, (2) main absorber, (3) purge reactor, (4) purge absorber, (5) desorber, (6) stripper, (7) refiner. (*Scientific Design Company, Inc.*)

the cooled reaction gases enter. The ethylene oxide is absorbed in water, and the now almost ethylene-free gas can be purged.

The dilute aqueous solutions from the scrubbing towers are combined and flow to a stripping tower, in which ethylene oxide is expelled by heating and is fractionated to pure ethylene oxide in other distillation columns. The water is again conducted to the scrubbing towers after the expulsion of the ethylene oxide. The yield amounts to 70–72%; that is, about 100 kg of ethylene is necessary to produce 100 kg of ethylene oxide.

Aside from the increase in plant size, the change of greatest significance since the process was introduced has been the shift from the direct introduction of air to supply the oxygen needed for the reaction to the use of highly purified oxygen. One of the major differences between the older chlorohydrin process and Le Fort's direct oxidation lies in the fact that the older process, on a once-through basis, converts all the ethylene fed to the process almost completely. In contrast, direct oxidation operates best with a conversion below 50%. The admixture of the unconverted ethylene with large quantities of nitrogen (which must be purged from the reaction system) makes the recovery of the ethylene uneconomical and its subsequent conversion inefficient.

Accordingly, there developed the use of reactors in series (usually two), the last of which was operated to obtain not maximum oxide selectivity but maximum ethylene conversion. These purge reactors vastly improved the economics of the direct-oxidation process. Today, pure oxygen is frequently used as the source of oxidation gas, avoiding the introduction of large amounts of inert diluents which must be purged from the system and thus making possible an efficient recycle system (which requires only a minor purge stream) in which the ethylene introduced can be utilized almost completely.

Such a mode of operation reduces the importance of "per pass" conversion and makes it possible to select operating parameters which minimize the cost of producing the oxide. Of course, the oxygen system does require that the undesired by-product, CO_2, be removed; once again, however, developments in the chemical industry over the years make it relatively easy to meet such a requirement.

The process has also been improved in other ways. The use of higher pressures, better heat-transfer designs, better catalyst, improved kinetic design data, and the like have all made important contributions.[2]

Three factors arising from differences between

the air process and the oxygen process exert an effect on capital costs. The air process suffers a disadvantage in its requirement for a purge reactor and the associated purge absorber, neither of which is necessary in an oxygen process plant. This capital disadvantage is, however, offset by the oxygen-process requirement for both a CO_2 removal system and an oxygen manufacturing plant, which, of course, is an oxygen-nitrogen air-separation unit. Under the right circumstances, when purchased from a large facility serving a number of customers, the shared economies of a very large oxygen plant can be realized; in that event, the oxygen process shows a distinct economic advantage over the air process, especially for larger plants.

Some companies are converting an existing air-based plant to an oxygen-based plant.

REFERENCES

1. Landau and Lidov: Ethylene Oxide, in S. A. Miller (ed.), "Ethylene and Its Industrial Derivatives," Benn, London, 1969.
2. Landau, Brown, Saffer, and Porcelli, Jr.: Ethylene Oxide Economics: The Impact of New Technologies, *Chem. Eng. Prog.*, March 1968.
3. Faith, W. L., D. B. Keyes, and R. L. Clark: "Industrial Chemicals," 3d ed., p. 380, Wiley, New York, 1965.

—Alvin S. Cohan, *Scientific Design Company, Inc., New York*

Ethylene-Propylene Elastomers   (See **Elastomers.**)

Ethylene Terephthalate   (See **Polyester fibers.**)

Ethylene–Vinyl Acetate Copolymers   EVA copolymers are polyolefin materials approaching rubbery materials in softness and elasticity, yet they can be processed like other thermoplastics. They require no curing or plasticizer. Parts made from these copolymers have little or no odor, and their elasticity is permanent. EVA copolymers can be injection-, blow-, compression-, transfer-, and rotationally molded or extruded into film, sheeting, pipe, and profiles. Since EVA parts have good clarity and gloss, stress-crack resistance, good barrier properties, low-temperature flexibility and toughness, good adhesive properties, and good resistance to ultraviolet radiation, they offer some physical and processing advantages over polyvinyl chloride and rubber. The main limitation of available EVA copolymers is comparatively low resistance to heat and solvents. When heated, these resins soften at 160°F (71°C).

EVA copolymers are used principally in specialty applications, replacing plasticized polyvinyl chloride and rubber. The resins meet U.S. Food and Drug Administration requirements for use in direct contact with foods in food-processing machinery and in packaging applications.

The effect of resin density on physical properties of EVA copolymers is not the same as for polyethylene resins. Density of EVA resins is a function of the amount of comonomer added to the ethylene. It is also affected by the degree of crystallinity of the resin. Softening temperature, torsional stiffness, and modulus of elasticity are lowered as density increases, contrary to the behavior of polyethylene resins. The same is true of clarity. Increasing the density of EVA resins results in extreme transparency that cannot be obtained with polyethylene resins.

EVA copolymers are not attacked by alcohols, glycols, or weak organic acids. Chlorinated hydrocarbons, straight-chain paraffinic solvents, and benzene and its derivatives do attack the polymers to varying degrees.

—Richard E. Kowal, *U.S. Industrial Chemicals Co., a division of National Distillers and Chemical Corporation, New York*

Europium   (From the continent Europe.) **Eu** = 151.96 (at. wt.); 63 (at. no.). Europium is seventh of 15 elements in group III, period 6, generally shown as the rare-earth elements in a separate line below the main body of the periodic table, sometimes referred to as the *lanthanide series.* Eu was discovered in 1889 by Sir William Crookes, English chemist and physicist.

Pure Eu metal is unstable in air and is the most reactive of the lanthanide elements. Of all the elements only Ba, Rb, and Li are more electronegative than divalent Eu. This extremely reactive metal must be handled in an inert atmosphere. In vacuum Eu volatilizes at its melting point. The room-temperature form of Eu is body-centered cubic, which exists up to its melting point. The consolidated pure metal which is divalent, is very soft compared to most of the other rare-earth metals, and a number of other properties of Eu are similarly anomalous. Eu metal is not known to be toxic, but precautions should be used in handling and storage because of its high reactivity.

The melting point of Eu is 822°C; boiling point 1597°C; density 5.245 g/cm^3. Estimated at 1.2 ppm in the earth's crust, it is one of the least plentiful of the rare-earth elements although at this level Eu is several times more abundant than Sb, Bi, or Cd.

Two stable isotopes of Eu are found in nature, 151 (47.8 w/o*) and 153 (52.2 w/o). Sixteen artificial isotopes of Eu have been identified. When thermal neutrons are absorbed by natural Eu, ^{151}Eu forms ^{152}Eu, with a half-life of 13 years; ^{153}Eu forms ^{154}Eu, with a half-life of 16 years, which further decays to ^{155}Eu, having a half-life of 1.7 years. Since certain emissions from the decay of Eu isotopes are of extremely high energy, when large masses of Eu are concentrated into control rods, high activity may develop. This activity, combined with the relatively long half-life of the chain isotopes, is a desirable property in the design of some nuclear reactors.

Dilute mineral acids dissolve Eu rapidly, and the metal reacts slowly with water at room temperature. Sodium and ammonium hydroxide solutions and sodium nitrate solution slowly corrode Eu at room temperature. The metal is soluble in liquid ammonia, giving a blue solution, which is similar to the behavior of alkaline and alkaline-earth metals. Although the tripositive state of Eu^{3+} is more stable in solution, Eu is reduced to the divalent state, Eu^{++}, by zinc. Reduction of Eu in solution is a method used to separate and purify the element. Hydrated trivalent compounds are formed with many inorganic and organic anions. Divalent europous carbonate and chloride compounds, which are unstable in air, are precipitated from reduced solutions.

Pure europium oxide powder, Eu_2O_3, is almost white, having a pale pink hue in natural light. Its melting point is near 2050°C. The oxide is more dense (7.42 g/cm^3) than the metal (5.245 g/cm^3). Insoluble in water, Eu_2O_3 dissolves readily in mineral acids. Added to soda-lime glass at 1 w/o, Eu_2O_3 is colorless in natural light and gives strong visible orange-red fluorescence when activated by ultraviolet light. Water-soluble europium sulfate, contrary to normal behavior of chemicals, decreases in solubility with increasing temperature.

Analytical methods for detecting Eu in minerals and chemical solutions within 5 ppm accuracy are well known. The sharp absorption bands of trivalent and divalent Eu make it practical to use x-ray fluorescence, emission spectrography, and colorimetric procedures for analyzing micro and macro amounts of Eu.

*This abbreviation is used in the rare-earth and related fields for weight percent.

The principal source of Eu is the rare-earth fluocarbonate mineral bastnasite, found at Mt. Pass, Calif. This natural mixture of light rare-earth elements contains between 0.09 and 0.11 w/o Eu_2O_3 consistently. Minerals such as monazite, xenotime, and yttrium-heavy rare-earth concentrates sometimes show an equal or higher Eu_2O_3 content, but more extensive processing of these phosphate minerals is necessary, making them less attractive economically.

After the rare-earth content in bastnasite is acid-leached into solution, the Eu is enriched by a series of liquid-liquid organic ion-exchange solvent extractions. The adjacent elements, samarium (62) and gadolinium (64), are also enriched. By reducing the Eu to Eu^{++} with Zn in the enriched Gd-Eu-Sm solution and adding ammonium carbonate, europous carbonate, $Eu(CO_3)_2$, which has a distinctive green color, is precipitated. Europium oxide results when the carbonate is calcinated in air. World capacity to supply 99.99% Eu_2O_3 is about 25,000 lb/yr with existing plants from sources already developed. For a detailed description of mineral resources, processing, and a table of properties, see **Rare-earth elements and metals.**

Uses: Although the nuclear properties of Eu are among the most favorable for any natural element, its use by the atomic-energy industry is still very limited. When reactors were being designed in the 1950s for the propulsion of submarines and aircraft and for stationary power production, Eu compounds in control rods were explored intensively. At the time, it was determined that insufficient mineral resources existed to justify its use in reactors, economic considerations aside. Therefore, other more available and economic materials were selected.

During the 1960s, several small reactors were built using europium molybdate as the major control-rod component. These account for less than 5% of total Eu consumption per year as the Eu must be replaced at regular intervals. One reactor of special interest is installed at McMurdo Base to supply power to United States expeditionary teams stationed in the Antarctic. In that environment, maximum reliability, portability, and serviceability (replacement) was required. Quantities of Eu were sufficient to meet anticipated needs, and this installation plus others are now operating. Europium continues to be used in small amounts by nuclear-research laboratories. Since the supply of Eu is now assured for many years in the future, increasing use by the nuclear-energy industry is one of the most promising applications.

A major technological breakthrough and a turning point for the rare-earth industry occurred in June 1964. The research laboratory of General Telephone & Electronics, Inc. (GTE-Sylvania) announced the discovery of a new red phosphor for commercial television. Initially the phosphor was europium-activated yttrium orthovanadate, $Eu:YVO_4$. The brightness of this phosphor increases linearly with increased beam-current density, resulting in an immediate 40% improvement in light output. An equally important factor is the white body color of the phosphor, which increased the rate at which color TV screens could be produced, thereby lowering costs of production.

In order to meet the commercial demand for 99.99% pure Eu_2O_3, the rare-earth industry built new plants using organic-solvent extraction processes that had previously been evaluated only on a laboratory scale. At the same time, other plants were constructed to produce the required pure yttrium oxide, Y_2O_3, which because of its chemical and physical properties is usually considered a rare-earth element. See also **Yttrium.**

An average color TV set requires about $\frac{1}{2}$ g of Eu_2O_3 and 6 g of Y_2O_3. By 1970 the world market for color TV reached 10 million sets. The rare-earth industry is geared to produce double the needs for the 11,000 lb of pure Eu_2O_3 required. Most of the Eu_2O_3 in color TV is consumed by the United States and Japan. The European and Southern Hemisphere nations offer an even greater potential consumption.

The stimulus provided the rare-earth industry by color TV since 1965 has resulted in the development of other new phosphors using Eu in various host matrices. These new compounds are used in x-ray screens, high-intensity mercury-vapor lamps, general-purpose fluorescent lamps, neutron scintillators, and charged-particle detectors.

Ferromagnetic europium chalcogenides (selenides, tellurides, and sulfides) are used in optically read memory systems. Various forms of Eu are used in orthoferrites, magnetooptics, and beam (laser) addressable memory systems, all of which are under development for future computer and electronic switching applications.

REFERENCES

Overview 1, 3, 11, 28 (properties and application information including possible uses), Molybdenum Corporation of America, New York.

See also list of references under **Rare-earth elements and metals.**

—Joseph G. Cannon, *Molybdenum Corporation of America, White Plains, N. Y.*

Eutectic A eutectic is that particular mixture out of a possible combination of two or more mixtures of materials which has the lowest melting point. The term is used most commonly in connection with metal alloys. Eutectic mixtures also are important in freeze-drying. See also **Freeze-drying.**

Euxenite (See **Uranium;** and **Ytterbium.**)

Evaporation The escape of molecules of water vapor from the surface of a liquid to an adjacent gaseous mixture is termed evaporation. As a unit operation, evaporation is the largest volume producer of vaporized water from a liquid—in nature as well as in commercial practice. See also **Desalination.**

Compared with related unit operations, evaporation is the vaporization of a portion of a solvent in which a solute or solid is present. Distillation is another form of evaporation in which the vaporization of one liquid from another liquid occurs. Sublimation is the vaporization of a solid from other solids and its subsequent condensation. Drying is the removal of a liquid from a solid generally by vaporization. All these processes are related because the common principles associated with vaporization must be present.

Vaporization occurs if a liquid is in a space where the partial pressure of its vapor is less than the vapor pressure of the liquid at that temperature. This change in state occurs with a large absorption of heat, and if vaporization is to continue, heat must be supplied continuously. In addition, the vapor generated must be removed continuously to avoid the condition whereby the vapor pressure of the liquid becomes equal to the partial pressure of its vapor, in which case vaporization will cease.

Thus, in vaporization from oceans, lakes, or rivers, the radiation from the sun and convection of heat to the water supply the necessary heat to produce a change in state, and air currents provide the means to remove the vapor. In commercial evaporation practice, heat generally is provided by an external source, such as steam. Vapor removal is accomplished by condensation, either through the transfer of the heat in the vapor to the tubes inside of which a cooler liquid medium circulates

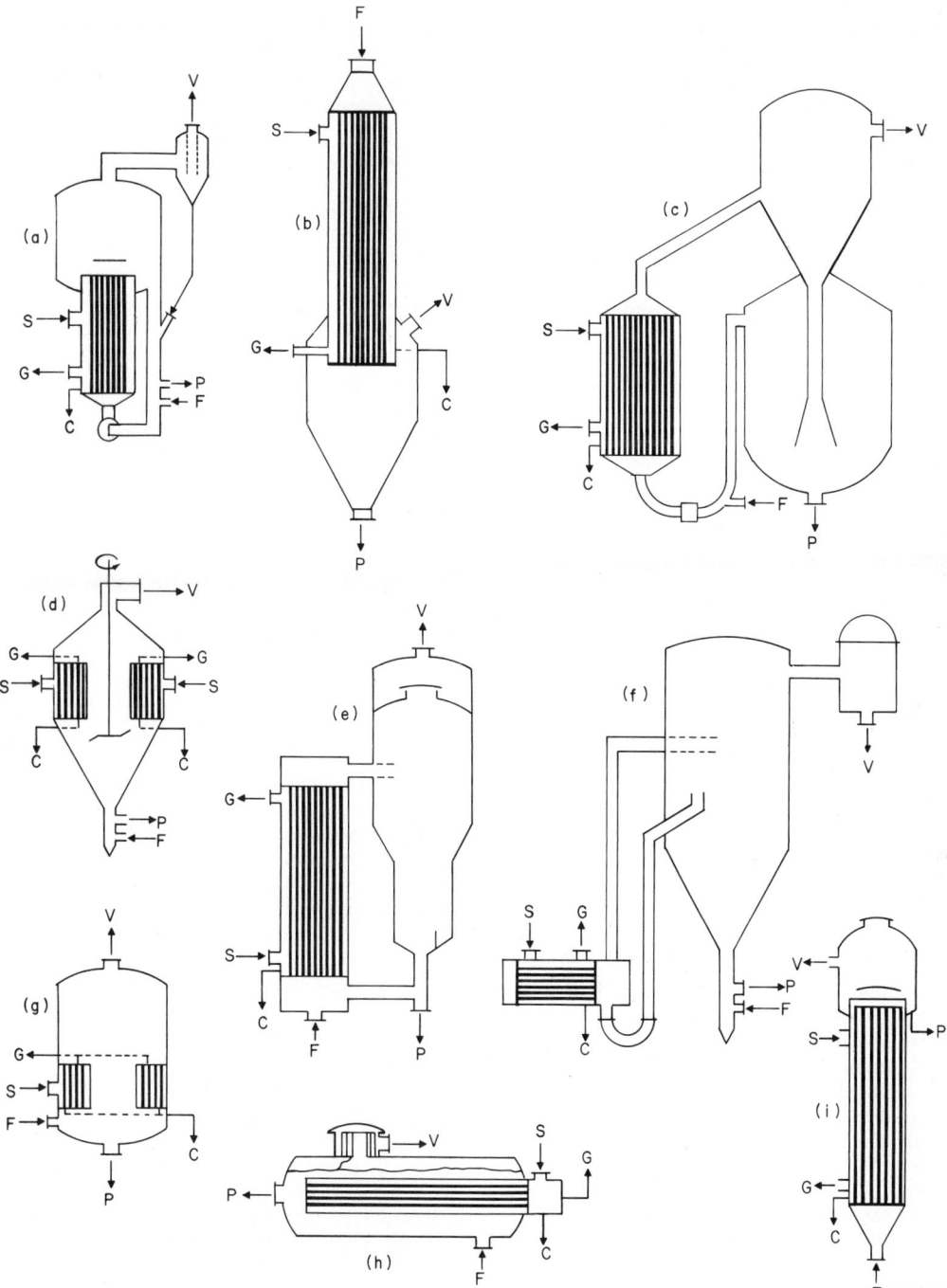

Fig. E-28. Principal types of evaporators shown schematically. (a) Forced-circulation, (b) falling-film, (c) Oslo-type crystal-lizer, (d) propeller calandria, (e) recirculating long-tube vertical, (f) submerged-tube, forced-circulation, (g) short-tube vertical, (h) horizontal-tube, (i) long-tube vertical. C, condensate; F, feed; G, vent; P, product, S, steam; V, vapor.

or by condensing the vapor in a stream of water.

Heat Transfer: The rate of heat transfer from an external heat source for promotion of vaporization is determined by a form of Newton's law:

$$Q = UA \ \Delta T$$

where Q = heat transferred, Btu/hr
U = overall heat-transfer coefficient, Btu/(hr)(ft^2)(°F)
A = heat-transfer area, ft^2
ΔT = overall temperature difference between points from and to which heat is flowing, °F

The overall heat-transfer coefficient U represents a means for accounting for all the factors affecting heat transfer, such as thermal conductivity, viscosity, specific heat, and density.

Designing evaporators for industrial processes involved a consideration of nonboiling as well as boiling heat-transfer coefficients for the promotion of vaporization. The equation developed theoretically by Nusselt (but usually known as the *Dittus and Boelter equation*) is often used in such design considerations:

$$\frac{hD}{K} - 0.023 \left(\frac{DG}{u} \right)^{0.8} \left(\frac{Cu}{K} \right)^{0.4}$$

where h = film coefficient, Btu/(hr)(ft^2)(°F)
D = tube ID, ft
K = thermal conductivity, (Btu)(ft)/(hr)(ft^2)(°F)
G = mass velocity per cross section of tubes, lb/(ft^2)(hr)
u = viscosity in consistent units (equal to cP × 2.42)
C = specific heat

This represents the basic equation used by the evaporator designer in optimizing the heat-transfer surface necessary to do specific work. Other and equally important design considerations must be taken into account for each type of evaporation process. Each process has liquids with different physical and chemical analysis and consequently different behavior.

Industrial Processes: Evaporation on a commercial basis is used extensively as a most economical means of recovery of the solute from a solvent (generally water). Operation of an evaporator with a condenser combined with an air ejector makes possible a combination of pressure-vacuum operation which increases the driving force ΔT and permits a greater total quantity of Btus to be transferred. The higher ΔT permits multiple-effect op-

eration, which reduces the amount of steam required, thus providing lower utility costs.

Types of Evaporators: Several major types of evaporators used in industry are shown schematically in Fig. E-28. A classification chart of the various types is given in Fig. E-29.

Pulp and Paper: The pulp and paper industry is the largest user of evaporation equipment and produces the largest quantity of evaporation. As one of the important stages in the recovery of chemicals in pulp mills, the principles of multiple-effect evaporation have been utilized more extensively than by any other industry. Sextuple and septuple-effect evaporators are common in the industry (see Fig. E-30).

In pulp mills, after the digestion system the pulp is leached with water, dissolving the chemical solids almost completely in various types of pulp-washing systems. It is the recovered liquid from the washing operation which is fed to the evaporator, generally at about 15% total dissolved solids. The evaporator removes much of the water and in doing so concentrates the liquid to 55–65% total dissolved solids. This solution generally is burned in the recovery furnace, where further evaporation is accomplished, the chemicals are recovered, and steam for mill purposes is generated.

This is a general description of part of the various processes in the pulp and paper mill industry. The actual process used depends on the type of paper produced and economic factors.

In the production of kraft papers, an alkaline digestive process is used, and the recovered chemicals from the washing system are concentrated through evaporation in a type of evaporator referred to as the long tube vertical (LTV). Such evaporators have tubes generally 20–36 ft long. The tube diameter commonly is 2 in. OD.

The LTV is a natural-circulation evaporator with vaporization occurring within the tubes. Unlike other evaporator designs, e.g., the forced-circulation, calandria, or basket-type evaporator, the liquor level is not controlled but establishes itself automatically.

In the alkaline pulping process the evaporator is manufactured with type 304 stainless-steel construction with stainless-steel tubes for the first and second effects and the balance of the evaporator constructed of steel.

In the acid pulping process, whether magnesium, magnesite, sodium or neutral sodium sulfite, generally a forced-circulation evaporator is used, since its higher liquor velocities result in more

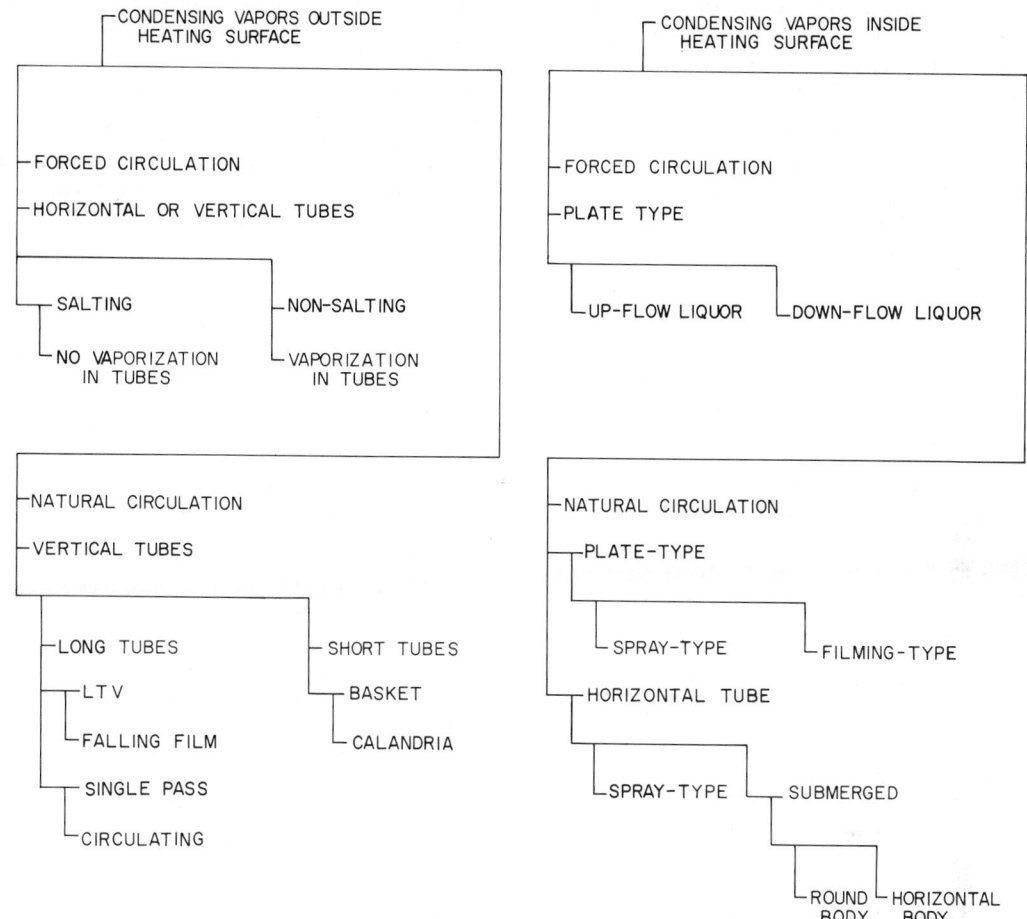

Fig. E-29. Classification of evaporators. LTV, long-tube vertical. In the classification of long tubes under the vertical-tube class, $L/D > 80$; for short tubes, $L/D < 80$.

on-time operation, because calcium sulfate present would scale the inside of the tubes in the first, second, and sometimes the third effect of a multiple-effect evaporator. The inverted solubility curve of calcium sulfate results in deposition of this material, which rapidly reduces the heat-transfer rate U and lowers the evaporator capacity.

The forced-circulation evaporator in this industry is stainless steel 316L or 317L in all parts in contact with liquor, vapor, or condensate. The tubes used are generally 2 in. OD, 18 BWG, 24 ft long. The circulation is provided by stainless-steel pumps, and velocities in the tubes are 6–12 ft/s. Design considerations include maintaining

sufficient hydrostatic head above the heating elements and low temperature rises through the heating element to avoid boiling at the inside of the tube wall to prevent scaling.

In this industry the viscosities of the liquor handled range from less than 1 cP at 15% total dissolved solids to approximately 180 cP at 65% total dissolved solids and at operating temperatures. Specific heats similarly are about 0.93–0.70 from unconcentrated to highly concentrated liquors.

Salt (NaCl) Production: The evaporation of salt brines requires extensive equipment. While many of the earlier salt evaporators were of the natural-circulation type and later modified with internal

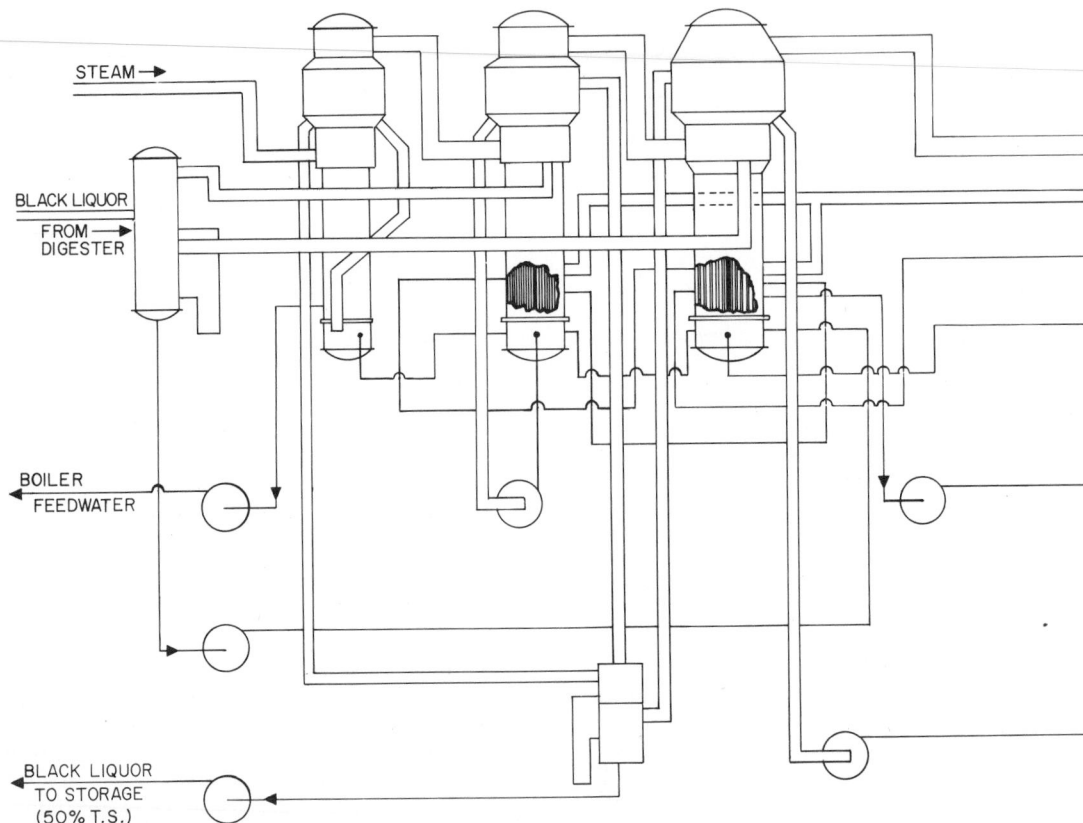

Fig. E-30. Multiple-effect kraft-mill heat-recovery evaporator. *T.S.*, total solids.

propellers to provide a degree of forced circulation, all modern evaporators are of the forced-circulation type. The largest salt evaporator on the North American continent is in Canada, where an installation has two quadruple-effect evaporators with vapor heads approximately 40 ft in diameter. The total capacity of the forced-circulation pumps exceeds 1.1 million gal/min. A salt evaporator in southern Italy is the largest single train of evaporators and includes a 40-ft-diam recompression evaporator body. Further details of sodium chloride evaporation are covered under **Salt (NaCl)**.

Caustic-Chlorine Production: Salt brines after puri-

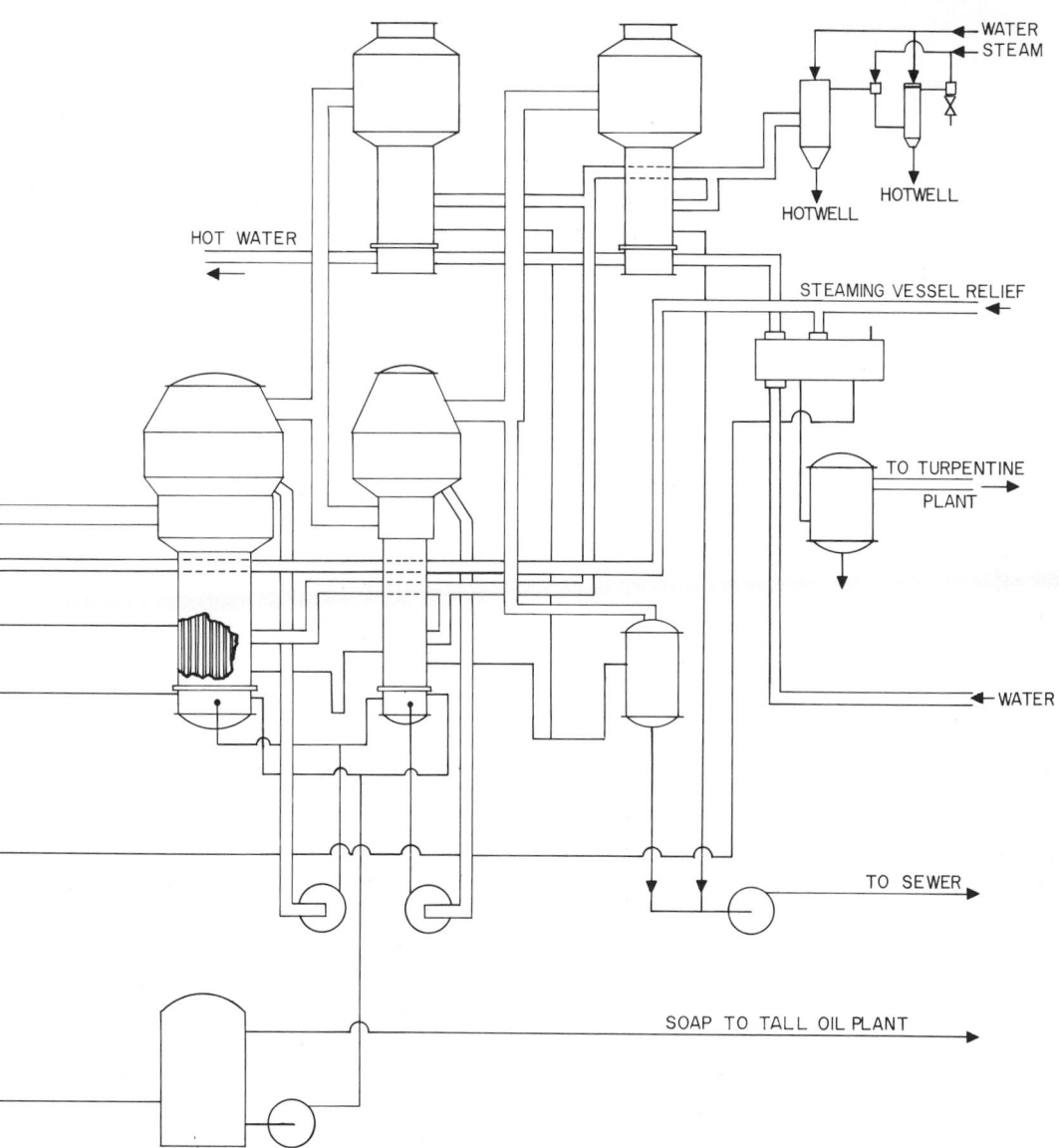

WATER
STEAM

HOTWELL

HOTWELL

HOT WATER

HOTWELL

STEAMING VESSEL RELIEF

TO TURPENTINE
PLANT

WATER

TO SEWER

SOAP TO TALL OIL PLANT

fication and other treatment represent a base for the production of caustic and chlorine. The use of mercury cells makes possible a concentrated caustic solution (50% NaOH) without evaporation. This process is used when the cost of mercury and power and other utilities makes it economical to do so. Concentration to 70% NaOH or to anhydrous caustic is performed in specially designed evaporators. See also **Caustic soda;** and **Chlorine.**

The electrolysis of purified brines in diaphragm cells results in the separation of chlorine and the production of a dilute caustic cell-liquor solution with an analysis of approximately 9–11% NaOH, 15–17% NaCl, and about 1–2% Na_2SO_4. Caustic solution is then concentrated to 50% NaOH. Subsequent evaporative stages may be followed, including concentration to the anhydrous.

The cell liquor is generally concentrated in specially designed triple-effect evaporators to approximately 50% NaOH, which has a specific gravity of about 1.51 at 180°F (82°C) and a boiling-point rise of approximately 80°F (26.7°C). During the concentration of the cell liquor, saturation with respect to NaCl is reached and salt is precipitated. The salt produced is thickened in elutriating legs or in settlers and then sent to centrifuges for removal of accompanying liquors and washing.

The concentrated 50% NaOH liquor is cooled in specially designed equipment to about 75°F, at which point a triple salt containing sodium sulfite, sodium chloride, and sodium hydroxide is precipitated and removed.

As with all salting-type evaporators, design considerations can have a considerable bearing on the on-stream time or operating cycles. An improperly designed evaporator can result in short operating cycles between boil-outs to remove salt encrustations on the heating surface or actual plugged tubes. Some evaporators now operating must be boiled out on 3- to 7-day cycles whereas properly designed evaporators may operate from 2 weeks to 2 months between boil-outs if the concentration of sodium sulfate in the caustic liquor is not too high.

The forced-circulation type of evaporator is generally used for concentration of cell liquor to 50% NaOH since it provides the best performance. The triple-effect caustic evaporator is generally of the construction indicated in Table E-22.

Phosphate Industry: The increasing use of fertilizers has resulted in major increases in the production of phosphoric acid, the details of which are described under **Phosphoric acid.** The modern phosphoric acid evaporator is of the forced-circulation type and consists of rubber-lined steel vapor heads, carbon-tubed heaters with the tubes cemented into carbon tube sheets, an Alloy 20 circulating pump, rubber-lined steel circulating piping, rubber-lined vapor piping, and rubber-lined direct-contact-type condensers. Rubber lining and carbon construction has proved most desirable, while the use of various metals, including stainless steels, Inconel, and others, has been minimal due to the highly corrosive liquor (see Fig. E-31).

Falling-Film Evaporator: In the industrial processes described thus far, evaporation is required as a means of providing a change in concentration or state for a specific purpose. The types of evaporators used in these illustrations represent the two most generally used in industry.

Another type of evaporator, the falling-film evaporator, is used on processes where the material is heat-sensitive and the time in contact with the heat-transfer surface must be as short as possible. This type is generally operated in single pass through the tubes without recirculation. The liquor is fed to the top of the tubes through individual tube orifices or through a perforated plate mounted above the tube with the perforation sized for a specific pressure drop to assure uniform feeding. The low ΔT permits multiple-effect operation at relatively low temperatures so that heat degradation with heat-sensitive materials is avoided.

Auxiliary Equipment

Pumps: Most forced-circulation evaporators are supplied with low-head axial-flow pumps of the single- or double-elbow type as the main circulating pump in the circulating system. Such pumps have been used with capacities as high as 100,000 gal/min per pump. It has been mentioned that one evaporator installation of the forced-circulation type has a total pumping capacity of well over 1 million gal/min. Other types

TABLE E-22. *Materials of Construction for Triple-Effect Caustic Evaporators*

	First effect	Second Effect	Third Effect
Vapor bodies	Solid nickel	20% nickel-clad	20% nickel-clad
Heaters:			
Tube side	Nickel	Nickel	Nickel
Shell side	Steel	Steel	Steel
Circulating piping	Nickel	Nickel	Nickel
Circulating pumps. . . .	Nickel	Alloy 20	Alloy 20
Vapor piping	Steel	Steel	Steel
Transfer piping.	Nickel	Nickel	Nickel

Fig. E-31. **Two-stage forced-circulation phosphoric acid evaporator.**

of evaporators generally employ various types of centrifugal pumps to pump liquids to the evaporator, as transfer pumps between effects, and to discharge from the evaporator system.

Condensers and Vacuum Equipment: Since almost all evaporators are operated at pressures above and below atmospheric pressure, provision must be made for use of condensers and air ejectors to operate the evaporator at 3–4 in. Hg vacuum abs. The condenser and air ejectors are installed so that the bottom point on this equipment provides an adequate barometric lag (34 ft at sea level). A condenser often used is the tubular-surface type, with the cooling water inside the tubes, condensing the vapor on the outside tubes. This type provides uncontaminated heated water and at the same time avoids dilution of pollutants that may be present in the vapor that is condensed. Because of this latter feature, surface condensers are gaining in use as antipollution laws become more stringent.

The direct-contact barometric condenser operates on the lowest terminal difference because of the intimacy of the condensing vapor with the cooling water. Thus this type uses considerably less cooling water than other types of condensers.

The multijet spray condenser is also a direct-contact condenser, but its use is diminishing with the greater advantages of the direct type or surface condensers.

Steam-jet air ejectors are used for removing noncondensable gases after the condenser. These are generally of the two-stage type and provided with precoolers of the barometric-spray type or surface-type intercondensers or after condensers.

Instrumentation: The successful operation of evaporator equipment is markedly assisted by adequate instrumentation properly located. Steam-pressure and flow controllers, liquor flowmeter, multipoint temperature recorders, boiling-point-rise density controller, and level controllers have become important requirements for good evaporator operation.

—R. E. Bergstrom, *Swenson Division, Whiting Corporation, Harvey, Ill.*

Exothermic Reaction (See **Explosives;** and **Molecule.**)

Explosives An explosive is a solid, gas, or liquid substance which can be made to release tremendous heat and pressure through a rapid, self-sustaining exothermic decomposition. Although explosives can be classified as mechanical, nuclear, or chemical, chemical explosives are by far the best known and most widely used both for commercial and military applications.

Classification: Commonly, there are two main classifications: (1) the low, or *deflagrating, explosives,* which function through burning processes characterized by rather slow, progressive reaction rates and pressure buildups that create a heaving action (typical decomposition of cellulose nitrate used in propellants: $C_{24}H_{30}N_{10}O_{40} \rightarrow 5N_2 + 10H_2 + 5H_2O + 11CO_2 + 13CO$); and (2) *detonating explosives,* which are distinguished by rapid chemical reactions, causing tremendously high pressure and brisance (shattering action) [decomposition of nitroglycerine: $4C_3H_5(ONO_2)_3 \rightarrow 12CO_2 + 10H_2O + 6N_2 + O_2$], and the extremely rapid rates of reaction often characterized by detonation waves that frequently obtain a velocity in excess of 20,000 ft/s.

Early Developments: Black powder (KNO_3 or $NaNO_3$, 75%; charcoal, 15%; and sulfur 10%), the first explosive developed by mankind, has been attributed alternately to Chinese and Egyptian ingenuity and undoubtedly predates the birth of Christ. Black powder was not introduced into western Europe until the thirteenth century and was apparently used first as a propellant. Black powder, which is a deflagrating explosive, was adapted for blasting purposes early in the 1600s, and its use as a blasting explosive for all types of construction blasting increased until it eventually was replaced by the introduction of Nobel's dynamite and the development of other high explosives and blasting agents.

Some of the earliest high explosives include *mercury fulminate,* $HgC_2N_2O_2$, discovered in the late seventeenth century, and nitrostarch, $C_{12}H_5(ONO_2)_{30}$, a sensitizing ingredient used in modern commercial explosives and discovered by Braconnot in 1832. In 1838, Dumas and Pelouse produced nitrocotton by treating cotton and paper with HNO_3 in the same manner that Braconnot treated starch with HNO_3. Sobrero first produced *nitroglycerin,* $C_3H_5(ONO_2)_3$, when he found that he could obtain a powerful high-explosive oil by treating glycerin with HNO_3. Nitroglycerin caused many tragic accidents and was not developed into a safe explosive until Nobel found that the hazards of handling the explosive oil could be reduced by utilizing a diatomaceous earth, kieselguhr, as an inert absorbent. After developing his first *dynamite,* Nobel continued to work on the refinement and improvement of energy combinations and introduced $NaNO_3$ and later NH_4NO_3 into his dynamite formulations. In 1875, while working with cellulose tetranitrate, Nobel mixed the collodion with nitroglycerin and developed blasting gelatin. Gelatine dynamite, a refinement of blasting gelatin, was produced by incorporating a portion of blasting gelatin into dynamite formulations. Development of the blasting cap used in conjunction with a safety fuse permitted positive, safe initiation of the dynamite.

Trinitrotoluene (TNT), $C_6H_2(CH_3)(NO_2)_3$, was first prepared by Wilbrand in 1863 but was not manufactured on a large scale until about 1900. Shortly thereafter, Germany recognized TNT as a replacement for *picric acid,* one of their standard military explosives. With the advent of World War I, the use of TNT became widespread, and its manufacture and use were limited solely by the availability of toluene. Based upon World War I performance, TNT was adopted as the basic standard military explosive by many countries and its use became widespread.

Pentaerythritol tetranitrate (PETN), $C(CH_2NO_3)_4$, prepared initially by Tollens in 1891, was not commercially available for use as an explosive until after World War I. The availability and relatively low cost of formaldehyde and acetaldehyde, as developed between World War I and II, permitted production of PETN on a large scale at a relatively low cost, and this in turn contributed to its use as an explosive.

Although cyclotrimethylenetrinitramine, or cyclonite (RDX), was discovered by Henning in 1899, its potential as a high explosive was not realized until about 1920. Used extensively during World War II as a component of bursting charges, RDX also served as a component of numerous cyclotols and plastic explosives.

RDX

Cyclotetramethylenetetranitramine (HMX), $C_4H_8N_8O_8$, often is formed in the manufacture of RDX but does not have the same explosive importance.

Ammonium nitrate, NH_4NO_3, or Glauber's salt, known as early as 1659, was not used extensively in explosive formulations until 1867, when it was introduced in Nobel's dynamite to replace a portion of the nitroglycerin. As previously noted, NH_4NO_3 was used later as an oxidizer in ammonia gelatins and other types of dynamites.

During World War I when supplies of TNT became rather critical due to toluene shortages, NH_4NO_3 was employed as a means of conserving TNT, and mixtures containing 80% NH_4NO_3 and 20% TNT or 50% NH_4NO_3 and 50% TNT (known as 80:20 or 50:50 amatols) were employed as military explosives for shells and bombs. *Amatols* were used later in World War II, especially by the Axis powers, to conserve supplies of TNT.

Even though NH_4NO_3 had been used in increasingly larger proportions in dynamite formulations since its original introduction, it was not until the mid-1950s that it gained widespread popularity as a blasting explosive, due to the Texas City, Tex., disaster of 1947, which highlighted the tremendous explosive power of NH_4NO_3, and the post–World War II emergence of low-cost NH_4NO_3 in prilled form. See also **Ammonia.**

Robert Akre combined prilled NH_4NO_3 and

carbon black in a 94:6 percentage ratio to produce an explosive suitable for blasting in open-pit strip mines and patented this explosive under the name Akremite. In order to obtain more homogeneous mixtures, field experiments were conducted with NH_4NO_3 in which liquid hydrocarbons were used to replace the carbon black of the original Akremite formulation. Typical reaction for the detonation of a 94:6 NH_4NO_3–fuel oil (ANFO) mixture is $3NH_4NO_3 + CH_2 \rightarrow 3N_2 + 7H_2O + CO_2 + 82$ kcal/mole.

Diesel fuel oil can be mixed with NH_4NO_3, to provide an explosive of rather consistent quality, and the use of ANFO explosives became widespread in the late 1950s and early 1960s. By the mid-1960s, this explosive accounted for over half of all the commercial explosives consumed in the United States.

The most serious deficiency of ANFO explosives was that NH_4NO_3 had no intrinsic water resistance and ANFO could not be used for wet blasting conditions. Initial efforts were made to enhance the water resistance of ANFO by coating the NH_4NO_3 prills or packing the ANFO in waterproof, sealed plastic tubes. These methods, however, were not very successful because the waterproof coating on the prills often affected the sensitivity of the NH_4NO_3 or prevented it from absorbing the fuel oil, and the packaging of ANFO in waterproof containers frequently produced explosive units with densities of less than 1.0 which would not sink in water. Attempts to increase the density of the ANFO by tamping or pressing were never quite successful.

The need for an NH_4NO_3-based explosive having the low cost and safety features of ANFO but with improved density and water resistance led Cook, Farnum, and others to the development of *slurry explosives*. Slurry explosives consist of oxidizers (NH_4NO_3 and $NaNO_3$), fuels (coals, oils, aluminum, other carbonaceous materials), sensitizers (TNT, nitrostarch, and smokeless powder), and water mixed with a gelling agent to form a thick, viscous explosive with excellent water-resistant properties. Slurry explosives are manufactured as cartridged units or are mixed on site in bulk and pumped into place at the point of use.

The *nuclear explosives* developed by the U.S. Manhattan Project during World War II have, for the most part, been maintained as classified information, although much work has been done in trying to adapt nuclear explosives for commercial applications in the postwar period.

Some recent explosive advancements include the development of hexanitrostilbene and various hydrazine compositions and specialized applications involving lasers. Other recent developments are described later.

Basic Types

Low, or Deflagrating, Explosives: These release their energy through rapid burning, the rate of which can be generally controlled. Black powder was initially produced for use in firearms, later employed as a burning train for blasting or safety fuse, and eventually bulk-packaged and used for general blasting work. The need for a less hygroscopic explosive which would also produce less smoke led to the development of the smokeless powders, which are used as propellants for both military and sporting arms. The introduction of modern rocket technology provides a large range of low explosives to meet the increasing need for specialized propellants.

High, or Detonating, Explosives: This class can be further grouped into categories of primary (initiating explosives) and secondary (noninitiating explosives).

Initiating, or Primary, Explosives: Types which can be readily detonated by heat, impact, or friction are generally used for the manufacture of initiating devices such as blasting caps, electric blasting caps, percussion caps, and delay initiating devices. Included in the list of initiating or primary explosives are mercury fulminate, silver fulminate, fulminate-chlorate mixtures, lead azide, silver azide, diazodinitrophenol, and lead styphnate.

Secondary, or Noninitiating, High Explosives: These include a number of organic and inorganic compounds which have been combined to produce desired properties of sensitivity, brisance, detonating velocity, and stability. Common examples are the dynamites, TNT, nitrostarch, cyclonite, tetryl, and ammonium picrate.

Blasting Agents: In addition, another group of explosives is characterized by their relative insensitivity. Because these compounds, which include nitrocarbonitrates and water slurries, must be initiated with a high-explosive booster to assure positive detonation, they are commonly referred to as blasting agents rather than as high explosives. Nitrocarbonitrates are a combination of oxidizers (NH_4NO_3 and $NaNO_3$) and carbonaceous materials (fuel oil, coal dust, pit meal, carbon black, and aluminum dust), while the water slurries usually are mixtures of an explosive sensitizer (TNT, smokeless powder, nitrostarch, or cyclotol), oxidizers (NH_4NO_3 and $NaNO_3$), fuels (coal dust, oil, and sulfur), and a gelling or thickening agent. A combination nitrocarbonitrate-water slurry is a blasting agent in which no high-explosive sensitizers are used, and the compound consists of oxidizers (NH_4NO_3 and $NaNO_3$), fuels (coal and oils), and aluminum, sulfur, HNO_3, or other sensitizing agents which are not explosives per se.

Properties and Testing of Explosives. Although there is no *ideal* explosive, some explosives, because of their intrinsic characteristics, are better suited for particular applications. A number of standard tests devised and developed by the U.S. Bureau of Mines, Bureau of Explosives, various government installations, arsenals, testing laboratories, and the commercial explosive manufacturers can be used to determine the physical properties of an explosive. Thus, by determining certain physical properties from test data, explosives can be comparatively evaluated.

Explosives to be employed for types of applications other than those considered purely experimental must have a relatively high degree of thermal and chemical stability so that they can be used or stored under conditions that may be far from ideal or favorable. High explosives of extremely limited thermal stability are unsuited for practically all types of commercial or military applications. To determine thermal stability, explosives are tested to determine the rate at which the product decomposes at some arbitrarily selected temperature, 75–100°C. Explosive samples subjected to the 75–100°C range for given periods of time are examined periodically and the rate of decomposition determined. Some explosives can be subjected to elevated temperatures for extended periods of time with little or no appreciable change while others decompose rather rapidly. Since most high explosives must be stored for varying periods of time (1–2 years is quite common), they must have a marked degree of chemical stability. Also, from both an economic and safety standpoint, the manufacturer of the explosive must be certain that the mixture can be stored for a relatively long period of time without fear of the ingredients reacting chemically either to induce the explosive to break down completely or become supersensitive. Prolonged storage tests are designed to determine chemical stability, and during the testing period, explosives are examined for change in pH, change in physical properties, evolution of gas, and other evidence that a chemical breakdown or a discernible change in chemical stability is taking place. Selected stabilizing ingredients are often incorporated in explosives to enhance chemical stability,

and in commercial dynamites calcium carbonate, talc, zinc oxide, chalk or marble dust are commonly used to preclude the presence of acidity and to maintain chemical stability.

Since the basic purpose of a high explosive is to detonate, the explosive must possess the sensitivity necessary for it to react predictably from the action of the initiating medium designed to produce detonation. To determine the detonation sensitivity of an explosive, standard test detonators of graded strength ratings from $\frac{1}{2}$ to 16 are utilized. The strength ratings are controlled by varying the type and weight of charge and equating the detonators to 90:10 mercury fulminate–potassium chlorate or lead azide–PETN detonators which specify the following charges:

Test Detonator, no.	Weight of Charge 90/10 ful. comp., g	Weight of Charge PETN-Lead Azide
$\frac{1}{2}$	0.26	
1	0.30	
$1\frac{1}{2}$	0.35	
2	0.40	
$2\frac{1}{2}$	0.47	
3	0.54	
$3\frac{1}{2}$	0.59	
4	0.65	
5	0.80	
6	1.00	
7	1.50	
8	2.00	
12	. . .	1.20 g PETN–0.35 g lead azide/ignition mixture
16	. . .	1.50 g PETN–0.35 g lead azide/ignition mixture

Grades of commercial explosives not cap-sensitive and most military explosives which cannot be detonated with a no. 6 commercial blasting cap must be tested with detonators of no. 8–16 strength in order to determine their sensitivity. Less sensitive explosives, i.e., nitrocarbonitrates (ANFO), explosive slurries, and nitrocarbonitrate slurries, necessitate the use of extremely large initiators of 3–10 g for ANFO mixtures to 50 g or larger for the less sensitive slurries in order to determine basic sensitivity. Normally, initiation test charges larger than 3 g are made of pentolite, composition B, or other high-strength explosives, which, in turn, are detonated by a blasting cap.

Obviously, since explosives must be handled during the manufacturing process, transported from point of manufacture to point of storage or use, and handled again in the end application, they must have an acceptable level of sensitivity to shock, friction, and impact to preclude premature detonation by dropping, bumping, or shaking, during the course of normal handling.

Friction sensitivity as determined by the pendulum friction apparatus consists of dropping a pendulum with a 20-kg steel- or fiber-faced shoe 1 m so that it swings across a grooved anvil into which a sample of explosive has been spread. The passing of the shoe over the explosive sample is recorded by the number of snaps, cracklings, ignitions, and explosions that occur. The steel-faced shoe will often produce snaps, crackles, or detonations in explosives that withstand the fiber-faced shoe.

Impact sensitivity can be expressed by denoting the minimum height of drop at which a standard test weight will detonate an explosive sample. Test weights of 2 and 10 kg are commonly used to determine minimum height of drop at which detonations will occur.

Shock sensitivity of an explosive can be gaged by subjecting a sample to the impact of a rifle bullet. In this test the caliber and type of bullet and the manner in which the explosive sample is contained or positioned during the test can be varied; e.g., the sample can be suspended freely in a thin-walled container, placed against a steel backing, or confined in a capped, steel pipe. Confinement generally enhances the probability that an explosive will detonate in the rifle-bullet test. Recent testing data show that some of the newer, smaller-caliber sporting rifles with high muzzle velocities offer a more severe test than the conventional .30 caliber rifle. Test results are expressed by the number of detonations or ignitions observed during testing sequence.

One of the properties of an explosive most commonly measured is its detonating velocity, the speed at which the detonation wave travels through the explosive, or the rate of chemical reaction of the detonation process. High-speed photographic testing has determined this reaction of the order of a few microseconds, showing that as the detonation wave progresses, the area behind the wave has undergone a chemical change while the area or material ahead of the detonation wave remains unchanged until the wave passes through it. Factors affecting detonation velocity include chemical composition, particle size, density, cross-sectional area, moisture content, method of initiation, and degree of confinement. Detonation velocity can be measured with high-speed streak cameras, spectrometers, or electric counters which record the make-break cycle the detonation wave causes in an electric circuit. For many years detonation velocity was determined by the D'Autriche method, a test that utilized speed-tested detonating cord. The detonating cord is cut to a convenient length, and the ends are inserted as probes spaced a predetermined distance apart in the explosive column. The center of the detonating-cord loop is placed at a mark inscribed on a lead plate. The explosive sample is then detonated, and as the detonation wave travels through the explosive column it initiates the ends of the detonating cord inserted as probes in the sample. The detonation waves traveling in both directions in the detonating cord collide and produce a line at the point of collision. Knowing the rate of the detonating cord and the distance the probe ends were separated in the explosive column, it is a simple matter to measure the distance between the marks on the lead plate and set up a proportion to calculate the detonation velocity of the explosive sample. This relationship is

$$V_d = \frac{D_c \times S_c}{2 \times D_p}$$

where V_d = detonating velocity of explosive tested, m/s

D_c = distance (spacing) of detonating cords, cm

S_c = speed of detonating cord, m/s

D_p = distance between marks on lead plate, cm

An explosive property closely allied to detonating velocity is brisance, or shattering power; an explosive that has a high detonating velocity will have extremely high shattering power because it releases its energy rapidly. Brisance is measured by a series of tests which determine the explosive's ability to crush sand, fragment steel pipe, or deform solid lead cylinders. The results of these tests are recorded as the weight of sand crushed, the number and size of pipe fragments produced, or the distance the cylinder is deformed.

Explosive power is a measure of the maximum pressure developed by detonation and the rate at which this pressure is released. Explosives that develop substantially large amounts of gas upon detonation but release it rather slowly are characterized by a heaving action. Typical of this type of explosive are the low-density, low-velocity ammonia dynamites. Explosives that generate extremely high pressures very rapidly include the straight and high-velocity gelatin dynamites and the military explosives, RDX, pentolite, and TNT. The ballistic pendulum used for. testing explosive power consists of a mortar (suspended as a pendulum from knife-edges) in which an explosive sample is detonated to expel a heavy shot from the mortar and cause the suspended mortar to recoil. The recoil of the mortar is proportional to the propulsive force of the explosive, and the arc of swing is registered in centimeters on a fixed scale and then compared and correlated with the arc recording of some standard explosive such as 50% straight dynamite or TNT.

Another important physical property of an explosive is its water resistance, or ability to perform efficiently when exposed to water or moisture or subjected to water pressure (see Fig. E-32). Some explosives are insoluble in water and can be used

Fig. E-32. Testing explosives underwater to determine water resistance and ability to accept initiation and detonate when subjected to water pressure.

underwater or subjected to wet storage conditions for indefinite periods of time without losing any effectiveness. Others are extremely hygroscopic and will absorb moisture from the atmosphere and lose their potency if not encased in an impermeable package. The water resistance of explosive formulations may be enhanced by the inclusion of a waterproofing agent such as nitrocellulose (gelatin dynamites). Sensitizing oils (nitroglycerin, ethylene glycol dinitrate, and dinitrotoluene) also affect water resistance, proportional to the amount of explosive sensitizer contained in the formulation. In recent years widespread use has been made of a number of gelling or blocking agents (guar gum, starch, bone meal, psyllium seed, and alum) to provide water resistance to explosives. Casting or pressing the explosives, when feasible, increases density and resistance to penetration of water and/or absorption of moisture. Water resistance can also be achieved to some extent by combining various explosives. For example, in the amatol type of explosives, NH_4NO_3, which is extremely hygroscopic and has very little water resistance, can be improved measurably by coating it with molten TNT.

Besides moisture or water attacking an explosive and rendering it ineffective, water *pressure* also has a marked effect on many explosives, and high heads of water can "dead press" explosives, some dynamites becoming insensitive or reacting erratically when subjected to extremely high heads of water. Loss of sensitivity, change in detonation velocity, or alteration of other characteristics due to the pressing effect of water pressure can often be prevented by incorporating certain ingredients in explosives which tend to resist densification and help maintain consistency. Minute, hollow spheres, or *microballoons,* of phenolic resin or glass are used extensively in high-velocity gelatins and seismograph explosives that must detonate consistently at high rates after being loaded under hundreds of feet of water for long periods of time.

Toxicity of an explosive depends to a large extent upon chemical composition although the physical state (finely divided material) can cause respiratory problems, dermatitis, or other complaints. Explosives containing liquid sensitizing oils which are volatile and/or toxic often exude them when subjected to high temperatures or poor storage conditions. Although many explosives, particularly those designed for military applications, can be handled with little regard to toxicity, others must be treated with extreme care under guarded conditions or be packed in special types of protective containers because of their toxic effects.

Whenever an explosive is detonated, gases (CO, CO_2, oxides of nitrogen, and steam or water vapor) are generated. Depending upon chemical composition, other gases including NH_3, NH_4Cl, H_2, and H_2S can also be evolved. Often, due to the conditions under which the explosive is to be used, the amount and type of gases or fumes liberated on detonation are of extreme importance. When an explosive is detonated in the open atmosphere, the fumes usually dissipate rapidly and cause no problem but under confined conditions it is imperative that an explosive have extremely good fume properties and not generate

long). The three fume classes set up under this system are as follows:

Fume Class	Cubic Feet Poisonous Gases per ($1\frac{1}{4} \times 8$) Cartridge of Explosive
1	< 0.16
2	0.16–0.33
3	0.33–0.67

The U.S. Bureau of Mines devised a fume classification for permissible explosives based on the liters of poisonous gas produced per $1\frac{1}{2}$ lb of explosive or the cubic feet of poisonous gas produced per pound of explosive. The Bureau of Mines Fume Classification also has three classes:

Fume Class	Liters Poisonous Gases per $1\frac{1}{2}$ lb of Explosive	Cubic Feet Poisonous Gases per 1 lb of Explosive
A	<53	<1.25
B	53–106	1.25–2.50
C	106–158	2.50–3.72

copious amounts of noxious or poisonous fumes. When the chemical composition is known, calculations can be made, based on the detonation reaction, to predict the amount and types of gases that will be liberated by an explosive. The U.S. Bureau of Mines and explosive manufacturers evaluate fume properties of an explosive by utilizing the Bichel gage, a large enclosed chamber in which a cartridge of explosive can be detonated and the resulting gases trapped and then released through a measuring or sensing apparatus. Used in conjunction with the Cranshaw-Jones, Burrell, and other analyzers, it determines the amount of CO, CO_2, oxides of nitrogen, and other gases generated by an explosive. Portable gas detectors can sample and evaluate the quantity and quality of the gaseous products resulting from an explosion.

For hygienic reasons, federal and state regulatory bodies frequently limit the amount of poisonous gases that can be liberated from an explosive intended for use where adequate ventilation is of prime consideration. To facilitate and standardize methods for measuring and rating the poisonous-gas properties of an explosive several fume classifications have been developed. The Institute of Makers of Explosives has established a fume-classification rating based on the cubic feet of poisonous gases produced per cartridge of explosive (standard cartridge $1\frac{1}{4}$ in. diam $\times$ 8 in.

Since the amount of poisonous gases or toxic fumes generated by an explosive involves all items that take part in the explosive reaction, the packaging or wrapping must also be considered when fume determinations are made. Heavy paper or cardboard or wrappings coated with paraffin, asphalt, or other carbonaceous materials normally release large amounts of CO when detonated. On the other hand, explosives that contain excessive amounts of O_2 carriers in their formulation or are deficient in fuel or carbon often liberate excessive amounts of the oxides of nitrogen. By chemical formulation it is possible to regulate the ingredients in a particular explosive in order to modify its fume characteristics. Thus, an explosive which releases excessive oxides of nitrogen may have its fume properties improved by the addition of carbonaceous materials which will combine with the excess oxygen and bring the oxygen balance closer to zero. Likewise, explosives which have a negative oxygen balance and tend to produce excessive CO can have their fume properties improved by increasing the quantity of oxidizers used in the formulation. Of course, fume properties can also be improved by modifying the amount of carbonaceous material (paraffin, type and thickness of paper, etc.) in the explosive package.

Manufacture. Because black powder can be initiated by flame or spark, its manufacture is considered rather hazardous, and manufacturing processes must be conducted so as to eliminate any possible incidents that might result from an accidental flame or spark. In making black powder it is essential to obtain an intimate mixture of the oxidizer (KNO_3 or $NaNO_3$), the charcoal, and the sulfur. To accomplish this, the KNO_3 or $NaNO_3$ is ground in a ball mill and the charcoal and sulfur pulverized together in another ball mill in order to ingrain the sulfur into the pores of the charcoal. Following the pulverization and incorporation of the nitrate and charcoal-sulfur mixtures the ingredients are transferred to a mixing building, where they are screened and mixed. The materials are dampened during the mixing cycle and then taken to a mill house, where giant mill wheels further increase the incorporation of the ingredients. The milled material is then pressed into cakes. The cakes of black powder are later broken up in a corning mill and the broken material screened to grain size. The screened material is tumbled for 6–12 hr in heated wooden drums to evaporate any moisture. When the moisture level reaches a prescribed level, the black powder is glazed with graphite and the drying cycle continued. The dried black powder is separated into standard granulations and packed into 25-lb kegs or bags.

Nitroglycerin can be manufactured in a batching process by nitrating glycerin with a mixed acid consisting of approximately 40.0% HNO_3, 59.5% H_2SO_4, and 0.5% H_2O. In the nitrating process 1 part by weight of glycerin is added to 4.3 parts of mixed acid, which is maintained at a temperature of 25°C or less. After all the glycerin has been added, the mixture is agitated and cooled to about 15°C. It is then run off into separating tanks and subjected to a drowning wash, which removes most of the dissolved acid. The nitroglycerin is then neutralized and transferred to storage tanks. When initially put into storage the nitroglycerin has a milky appearance because of its moisture content. With storage in a heated building it eventually becomes clear as the moisture content decreases to less than $1/2$%. In addition to the standard batching method just described, there are also several continuous processes. One continuous type is the Schmid process, which involves continuous nitration, separation of the nitroglycerin from the spent acid, and continuous neutralization and washing. Another more modern operation is the Biazzi process, which is considered the safest of all processes because it limits the amount of nitroglycerin in the system at any time. Due to its high degree of sensitivity, nitroglycerin is rather hazardous to handle; it is seldom used as an explosive by itself but is used as a sensitizer in other explosives such as dynamites and propellants.

Nitrocellulose is manufactured by subjecting cellulose (cotton linters, wood pulp, cotton fibers) to a mixed acid of approximately 25% HNO_3, 64% H_2SO_4, and 11% H_2O. The cellulose is dried to a moisture content of less than 1%, added to the mixed acid, and submerged below its surface. The mixture must be stirred constantly to prevent any overheating, and after being stirred for 30 min the cellulose is separated from the acid and dumped into cold water. The nitrocellulose must then be purified and subjected to a boiling process for up to 60 hr to eliminate any cellulose sulfate.

In the manufacture of nitrostarch, tapioca or cornstarch is nitrated with a mixed acid consisting of approximately 38% HNO_3 and 62% H_2SO_4. After nitration is complete, the nitrostarch is separated from the acid and the nitrated starch is purified by washing it in water and NH_3 until all traces of acid are removed. After washing the starch nitrate is separated from the water, filtered, and air-dried in a drying house at a temperature of approximately 40°C.

Dynamite, the most widely known commercial high explosive, is produced by combining a sensitizer (nitroglycerin, ethylene glycol dinitrate, or nitrostarch) with oxidizers (NH_4NO_3, $NaNO_3$) and fuels or carbonaceous materials (coals, sulfur, wood pulp, sawdust, starch, wax) and occasionally finely divided metals (atomized, granulated, or flake aluminum). In addition, depending upon its intended application, dynamite may contain gelatinizing and waterproofing materials, such as nitrocellulose, or water-swellable or blocking agents (guar gums, starch, or bone meal). To counteract possible acidity dynamites contain antacids ($CaCO_3$, ZnO, $NaHCO_3$, talc, chalk, or marble dust), which act as stabilizers to neutralize the acid and halt decomposition. The normal procedure for manufacturing dynamite consists of blending all the ingredients except the sensitizing agents into a *dope mix*. After a thorough blending cycle has been completed, the dope is combined with the sensitizing agents and run through a predetermined mixing cycle. The mixed or finished dynamite is then taken to a cartridging or pack house, where it is packed into cartridges, bags, or other types of containers.

As previously noted, primary explosives (lead azide, fulminate of mercury, lead styphnate, and diazodinitrophenol) are extremely sensitive to

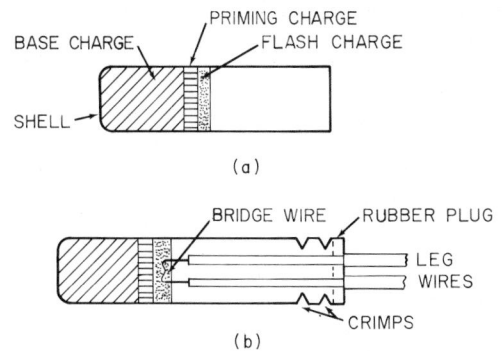

Fig. E-33. (a) **Typical blasting cap (fuse cap)**, (b) **typical electric blasting cap (bridge wire)**. The shell in either is usually made of aluminum or copper. The base charge is RDX or PETN. The priming charge is lead azide. The flash charge is lead styphnate. Diameter is approximately $\frac{1}{4}$ in. The length of (a) averages 1–1$\frac{1}{2}$ in., (b) 1–2 in. The leg wires of (b) are 20 to 22-gage copper wire.

heat, impact, and friction. For this reason, they are not used in large quantities in dynamites, bombs, or demolition explosives but are employed in the manufacture of initiating devices designed to initiate or detonate less sensitive secondary high explosives or explosive compounds. The two most commonly used initiating devices are the blasting cap and the electric blasting cap (Fig. E-33). The blasting cap is crimped to a length of safety fuse and is set off by the "spit" of the flame from the black-powder train of the fuse impinging upon the flash increment of the cap charge. The electric blasting cap is similar except that it uses a high-resistance bridge wire which is heated to incandescence by passing an electric current through the leg wires of the cap. Both blasting caps and electric blasting caps are mass-produced on complex sophisticated assembly lines which employ a high degree of automation and a refined system of quality control. Although special high-strength blasting caps can be manufactured by increasing the weight of the base charge in the cap, the commercial blasting cap is described as having a no. 6 strength rating.

Detonating cord is an exploding, flexible cord used in conjunction with various types of blasting caps for initiating high explosives. The detonating cord consists of a core load of PETN (pentaerythritol tetranitrate) or RDX (cyclotrimethylenetrinitramine) encased in plastic or textile coverings. The core loads of detonating cords are 4–400 gr of explosive per lineal foot, and the cord diameters are approximately $\frac{1}{16}$–$\frac{5}{8}$ in.

Trinitrotoluene, (TNT), because of its versatility, is probably the most widely known military explosive. It can be used for melt loadings and in the formulation of a number of binary explosives (amatol, tritonal, torpex, pentolite, tetrytol, ammonal, and various composition explosives, B and C). The availability of synthetic toluene has contributed substantially to the potential high production and low manufacturing cost of TNT. Trinitrotoluene is manufactured in one-, two-, or three-stage nitrations utilizing toluene and mixed acids as the raw materials. As the nitration processes are completed, the TNT is purified by washing with water and is then neutralized to eliminate any free acid.

PETN and RDX, explosives previously noted as being used as core loads for detonating cords, are also used by the military in the binary explosives, pentolite (PETN/TNT), composition B (RDX/TNT), and torpex (RDX/TNT/aluminum).

In the manufacture of PETN, pentaerythritol is treated with nitric acid, the nitrated mixture dumped into cold water, and the PETN precipitated on a filter. The PETN is again washed in water and then dissolved in acetone. The solution is filtered and the PETN precipitated by the addition of cold water. As the PETN is separated, it is caught on a filter and washed with water to remove the acetone. PETN is then held as a water-wet material until it is ready for use, at which time the water is removed in a drying operation.

For the manufacturing of RDX hexamethylenetetramine is obtained from the reaction of NH_3 and formaldehyde and nitrated. The nitrating process can consist of one utilizing only HNO_3 or another in which acetic anhydride and NH_4NO_3 are also employed. In the latter process about 10% of HMX is produced. Because of the two types of manufacturing processes there are two classes of RDX listed by the military: type A is made by the HNO_3 process and type B by the acetic anhydride process.

Use. The use of explosives can generally be divided into two general categories, military and commercial. A large proportion of military explosives consist of the propellant type employed as fixed ammunition for small arms and artillery and the rocket propellants developed during World War II, which are based upon nitrocellulose, nitroglycerin, or cellulose–diethylene glycol dinitrate mixtures. New propellant-type explosives are constantly being introduced for use in artillery rockets, jet-assisted takeoff mechanisms, and power-actuated devices. The latter include a number of automatic opening and closing systems and fastening, severing, and ejection devices used in jet aircraft.

Special binary explosives with high blast effect have been compounded by the military for use as warheads in conventional artillery and rocket missiles and bursting charges in aerial bombs, naval and land mines, naval torpedoes, and hand grenades. The shaped-charge principle, or Munroe effect, has also been utilized in demolition charges used for breaching concrete fortifications and in special types of armor-piercing rounds or antitank weapons such as the bazooka and 3.5 rocket launchers of World War II.

In the past TNT has been used by the military for general demolition work as pole and satchel charges. However, during and after World War II the military developed a number of RDX-based explosives, including the plastic explosives, composition C-2, C-3, and C-4, for general combat demolition work, composition B and torpex for melt loading of bombs, and military dynamite, a cartridged explosive similar in size and shape to commercial dynamite but with much safer handling and better storage characteristics.

Normally, military explosives designed for combat military operations must have a greater degree of stability, less sensitivity, and more versatility than explosives manufactured strictly for commercial applications. Military explosives are generally much more expensive than commercial explosives.

The mining, construction, and seismic-prospecting industries are the largest consumers of commercial explosives. Included in the mining industry is the mining of coal, the recovery of metallic and nonmetallic ores, and the quarrying of dimension stone and aggregate. In the construction field explosives are used where rock or other types of material must be moved or removed expeditiously and includes not only excavation for foundations, construction of grades for railroads and highways, and tunneling through mountains but also demolition work involved in razing stacks, demolishing buildings, and blasting foundations and other concrete structures to make way for alterations or new construction. In seismic prospecting or exploration, which is closely allied with the mining industry, special types of explosives are used on land and offshore to generate shock waves which can be recorded on seismographs to enable the geologist to plot underground contours and formations and establish probable locations of gas, petroleum, or mineral deposits.

The coal-mining industry has long been the single largest consumer of commercial explosives, and despite the decline in the use of coal as a domestic heating fuel, over 700 million pounds of explosives are still used annually in open-pit stripping and underground coal mines to satisfy coal demands. For underground coal mining, *permissible explosives*, special types of dynamite designed for use under gassy or combustible conditions, are employed. To be classified as a permissible, an explosive must be tested and approved by the U.S. Bureau of Mines. From experience gained in coal-mining applications, the use of permissible explosives has also been introduced into certain other mining operations where hazardous conditions would be created if conventional (nonpermissible) explosives were used.

In agriculture, dynamites are used for blasting stumps, clearing land, setting posts, breaking up hardpan, planting trees, digging wells, and blasting drainage and irrigation ditches.

Industry applies explosive technology to the forming, cladding, bonding, hardening, and welding of metals. These specialized explosive applications have simplified work and reduced costs by replacing conventional techniques in certain industrial processes. Extensive research is being conducted into these applications as well as the use of explosives for cutting, punching, riveting, and fastening metals and other materials and the development of explosive devices for controlling operations remotely by opening and closing valves, engaging switches, energizing relays, or disrupting power sources.

Safety. Explosives are manufactured under rigid control. The explosives industry has long enjoyed an enviable safety record. The trends in explosive manufacture are toward the development of less sensitive explosives which are much safer to manufacture and handle. Refined processing equipment, such as the Biazzi nitrator and the peristaltic pump, have permitted handling smaller quantities of highly sensitized explosives at any one time so that the danger of mass explosions involving large quantities of explosive materials is virtually eliminated.

Explosives are transported in accordance with regulations specified by the U.S. Department of Transportation, Interstate Commerce Commission, Department of Defense, and state and local boards. Other nations have similar regulations, varying in intensity and scope.

Fig. E-34. Truck-mounted mixing unit for on-site blending and pneumatic loading of ANFO mixtures. (*Amerind-MacKissic.*)

Before explosives can be offered for transportation, they must be tested and classified according to their degree of hazard. All explosives, commercial or military, must be stored in buildings that are designed and located specifically for the storage of explosives. These buildings, commonly called *magazines,* are normally fire-, bullet-, and theft-resistant. Explosive storage magazines often must be licensed by state or local authorities, and they must be located in accordance with the American Table of Distances, which prescribes the distance, based on storage capacity (pounds of high explosives) that a magazine must be positioned away from inhabited buildings, passenger railways, and public highways. This table also specifies the distance for magazine separation when more than one storage building is used. Because of stringent storage regulations, magazines generally are located in isolated areas and are frequently protected by natural or artificial barricades.

Advances. Research has been directed toward the perfection of *bulk placement systems* for explosives which will make it possible to transport all or a majority of the ingredients to the blast site as nonexplosives. At the site, the ANFO or the slurry ingredients are mixed and then blown or pumped directly into the drill hole (see Fig. E-34). The increased use of low-sensitivity explosives, such as ANFO and water slurries, has resulted in the need for special types of *initiating devices and boosters.* These generally are made of pentolite, composition B, torpex, or other grades of cast military-type explosives which possess the high initiating efficiency required to initiate relatively insensitive explosives. Because of the growing need for refinement in the field of *noise abatement and vibration reduction,* a number of special types of initiating devices have been developed, e.g., delay and millisecond-delay electric blasting caps, special types of detonating cord, detonating-cord delay units, special types of nonelectric blasting caps with detonating-cord delay assemblies, and an electric blasting device designed to facilitate delay or sequential initiation.

Probably the most controversial subject in explosive technology is the development of a *nuclear explosive* which can be used for commercial blasting purposes. Government projects aimed in this direction have included Rainier, Toboggan, and Plowshare. Objectives proposed include the creation of harbors with a single blast, excavation of new canals, and the exploitation of oil-shale production.

—Thomas P. Dowling, *Trojan-U.S. Powder, Division of Commercial Solvents Corporation, Allentown, Pa.*

Expression Expression is an operation to separate liquids from solids by compression under conditions that allow the liquids to be removed while retaining the solids between the compressing surfaces. Expression equipment can be grouped into batch and continuous presses. The major applications and characteristics of the principal types of expression equipment are summarized in Table E-23.

Batch Presses: Although this type is being replaced by the more efficient continuous presses because of relatively low capacities and high labor requirements, batch presses do provide a high-quality product and still are used extensively in the separation of oil from vegetable seeds and nuts.

Open Plate Press: The plates are equally spaced and are suspended from each other by linkages

TABLE E-23. *Operating Data for Expression Equipment*

Type of Press	Applications	Maximum Pressure, psig	Capacity Example	Feed Material, %	Discharge Cake, %
Plate.......	Fruit, olives, vegetable seeds, and nuts, particularly flaxseed	1,500–2,000	12 to 18 tons/day whole cottonseed	30–35 oil	5–10 oil
Box	Vegetable seeds and nuts, particularly cottonseed and peanuts	1,500–2,000	10 to 12 tons/day whole cottonseed	30–35 oil	5–10 oil
Cage.......	All oil seeds and nuts; practically any oily material, especially castor beans and copra	6,000–8,000	8 to 10 tons/day castor beans	35–50 oil	5–10 oil
Pot........	Cocoa butter and other fats not liquid at ambient temperature	3,000–6,000	6 to 10 tons/day cocoa butter	30–50 fat	6–10 fat
Screw:					
Low-pressure	Wood and pulp products; wastewater slurries; food and beverage products	100–1,000	400 bone-dry tons/day cooked wood chips	90–97 liquid	25–50 solid
High-pressure	Vegetable seeds and nuts; natural and synthetic rubber; rendered materials	10,000–40,000	200 bone-dry tons/day vegetable seeds	30–35 oil	3.5–5 oil
V-disk	Wood and pulp products, high-polymer resins, spent grains, food products, waste-water slurries, starch products	200–400	125 bone-dry tons/day kraft pulp	85–97 liquid	25–55 solid
Roll	Wood and pulp products; wastewater slurries; alkali cellulose; and sugarcane	5,000*	500 bone-dry tons/day wood pulp	92–98 liquid	30–55 solid

*Pounds per linear inch.

NOTE: The selection of expression equipment to fulfill a particular need depends upon the type and quantity of material to be processed, the consistency of the feed, and the desired consistency discharge of the cake. With these qualifications, this table is intended to provide a survey of the types of materials dewatered by expression equipment. The maximum pressures listed are those recommended by equipment manufacturers, and only one example of maximum capacity is given in the table. Because of their construction and operation, batch presses, can be used to process almost any quantity of material up to their maximum. The V-disk and screw press are available in sizes to express small amounts of material, but are not well adapted to intermittent operation; the smaller units often plug, giving the batch press an advantage in small-scale operations. Economics usually prohibits the use of the roll press where the amount of material to be dewatered is small. The consistency data given in the two right-hand columns refer to the specific capacity example.

which permit the entire assembly to become compressed (Fig. E-35). The oil-seed flakes are completely wrapped in press cloths and placed between the plates. Pressure is applied in stages, and the maximum pressure is maintained for 20–45 min to allow the oil to drain. A complete pressing cycle may require 30–55 min.

Box Press: This design is similar to the plate press except that boxes (2 in. deep) enclose the cake on two of the four sides. This simplifies folding the press cloth. Box presses have lower capacities than plate presses.

Cage Press: This is a closed press and consists of a finely perforated cylinder with a hydraulically operated ram (Fig. E-36). Press cloths usually are not required. Generally several pressings are performed on a given charge and a complete press often comprises several cages. These presses are particularly suited for seeds high in oil and low in fiber content and for high-pressure, low-temperature operation.

Pot Press: This is a special form of cage press in which the cage is replaced by a series of short, superimposed steam-heated pots. The walls of the pots are solid, and discharge takes place through perforated plates and filter mats in the bottom of

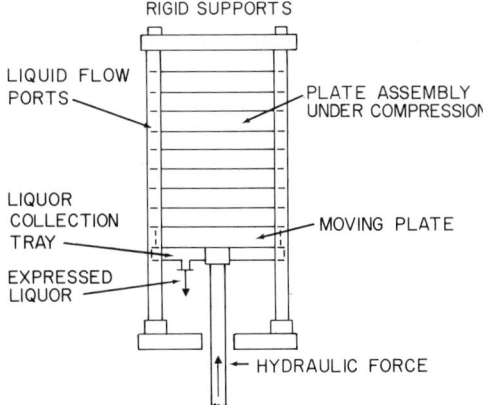

Fig. E-35. Plate press.

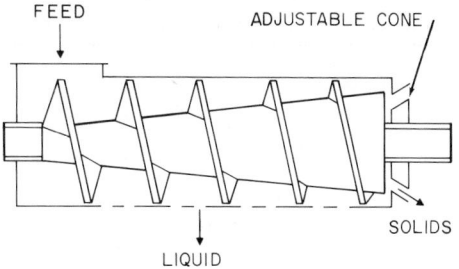

Fig. E-37. Continuous screw press. Screw pitch decreases in direction of flow.

each section. Generally a series of pots is used in each press, and the bottom of each pot serves as the ram for the pot below. Pot presses can be heated and can handle very soft, nonfibrous material, such as fruit pulp.

Curb Press: This design is similar to the cage press but operates at lower pressures with fewer drainage channels. It is used for expressing olive oil, fish oils, and other oils that do not require high pressures and for dewatering and recovering grease from garbage before incineration.

Continuous Presses: In these designs, material can be fed wet with a simultaneous discharge of dry cake.

Screw Press: This design consists of a continuous screw, or worm, which rotates in a cylindrical housing lined with perforated plate (Fig. E-37). The taper of the screw and/or the pitch decreases from the inlet to the discharge, exerting a powerful squeezing action. The degree of compression and capacity can be controlled by adjusting the area of the discharge port and by varying the rotational speed of the screw. For high-pressure applications, the continuous screw is replaced by a shaft fitted with a series of discontinuous worms,

and the perforated housing is replaced by a drainage barrel made up of bars that fit into a heavy bar housing.

The screw press effects a large saving in operating labor over batch presses and is used to mechanically dry a great variety of materials, as shown in Table E-23. Power consumption is high, and power dissipated in friction may raise the product temperature, necessitating an internal cooling system. Skilled operating labor is required, and maintenance costs can be high.

V-Disk Press: A typical design consists of two conical disks that face each other in a suitable casing (Fig. E-38). The disks are arranged so that as they rotate, they continuously converge from a point of maximum gap (feed) to a point of minimum gap. The rotating action of the disks carries the solid material from the feed point to the discharge opening. Once past the point of maximum pressure, the disks diverge to discharge materials from the outlet. Liquid passes through

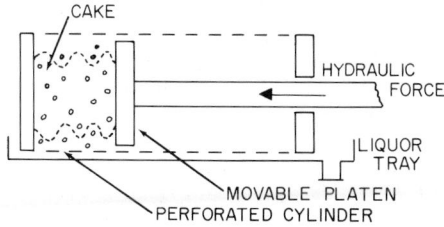

Fig. E-36. Cage press under compression.

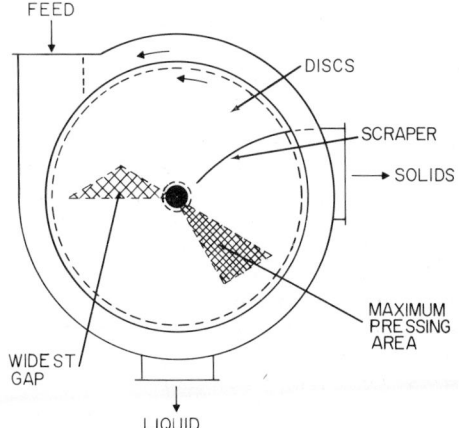

Fig. E-38. V-disk press.

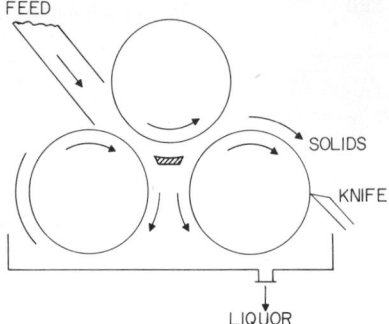

Fig. E-39. Continuous roller press.

snarls, which is important with certain materials.

Roll Press: This design (Fig. E-39) operates on the principle of the old-fashioned clothes wringer and is equipped with a minimum of two rollers but usually with three rollers, as illustrated. The material to be dewatered is fed into the nip of the first two rolls. The rotation of the rolls pulls the solids through, while the expressed liquid is collected in a trough below. To separate a greater quantity of liquid from the cake, the solids then are directed through another pressing cycle by a scraper blade. The feed material may be carried through the press on a continuous perforated belt, with the expressed liquid passing through the belt to a collection trough. In an alternate design, the press is equipped with perforated rollers, the interiors of which are maintained at reduced pressure to aid the flow of liquid away from the expressed cake. Although these presses have large capacities, they are more expensive than other types of continuous presses.

—William J. Murray, *Eimco Processing Machinery Division, Envirotech Corporation, Salt Lake City, Utah*

Extraction, Protein (See **Peptides and proteins.**)

Extractive Distillation (See **Distillation.**)

Extruding (See **Agglomeration; Mixing, pastes;** and **Plastics.**)

the screened surfaces of the disks and down through the discharge opening. The capacity and degree of compression can be controlled by changing the speed of rotation and by adjusting the width of the minimum gap. The capacity can also be affected by loading the press with a screw or other automatic feeder. Designs with hydraulic systems allow the disks to oscillate during operation to maintain constant pressure, thus permitting a relatively light construction inasmuch as the squeezing surfaces will move to accommodate changes in feed consistency. The synchronized movement of the rotating disks does not tear the solid materials and prevents the formation of fiber

F

Fabrics (Consult the **Subject Index.**)

Fabric Dyeing [See **Dyestuffs and dyeing (textile).**]

Faggoting (See **Iron and steel.**)

Fast Neutrons (See **Radioisotopes.**)

Fast Reactor (See **Nuclear power plants.**)

Fats [See **Alcohols; Alcohols, fatty (via hydrogenation); Detergents; Hydrogenation; Soaps; Surfactants;** and **Vegetable oils.**]

Fatty Acids [See **Alcohols, fatty (via hydrogenation); Carboxylic acids; Chlorine organics; Soaps;** and **Vegetable oils.**]

Feed Supplements (See **Amino acids; Pyridine and derivatives;** and **Urea.**)

Feedwater (See **Water treatment.**)

Feist-Benary Synthesis (See **Furan group.**)

Feldspar (See **Bauxite; Ceramics;** and **Glass.**)

FEP Resins (See **Fluoroplastics.**)

Ferberite (See **Tungsten.**)

Fermentation (See **Acetic acid; Amino acids; Antibiotics; Carboxylic acids; Enzymes; Ethyl alcohol; Pasteurization;** and **Peptides and proteins.**)

Fermium (See **Chemical elements.**)

Ferrite (See **Iron and steel.**)

Ferrites A ferrite is a powdered, compressed, and sintered magnetic material having high resistivity and consisting chiefly of ferric oxide combined with one or more other metals. The products consist of bivalent metallic oxides combined with Fe_2O_3 to form a crystal structure. Various mixtures of metallic oxides in conjunction with the iron oxide are calcined to produce ferrite powders. The powder also may be incorporated in a plastic or rubber matrix to produce flexible magnetic parts. The high resistance of ferrites makes eddy-current losses extremely low at high frequencies. Examples of ferrite compositions include nickel ferrite, nickel-cobalt ferrite, manganese-magnesium ferrite, yttrium-iron garnet, and single-crystal yttrium-iron garnet.

Ferrites resemble ceramic materials in production processes and physical properties. The dc resistivities correspond to those of semiconductors, being at least 1 million times those of metals. Magnetic permeabilities may be as high as 5,000 and dielectric constants in excess of 100,000. The Curie point is quite low, however, in the range 100–300°C. Saturation flux density generally is below 5,000 G. Ferrites provide design advantages over strip and powder cores for filter cores, deflection transformers, and yokes and in antenna rods, pulse transformers, delay lines, and waveguide elements. See also **Iron oxides, synthetic.**

Ferromanganese (See **Manganese.**)

Ferrous Chromite (See **Chromium.**)

Fertilizers Fertilizers are natural or manufactured materials containing plant nutrients in available form.[1,2,8] At least 16 elements* are essential to plant life. Carbon, hydrogen, oxygen, nitrogen, phosphorus, and potassium are needed in macro amounts. Calcium, magnesium, and sulfur are required in semimacro quantities, and the remainder—iron, manganese, silicon, cobalt, copper, and molybdenum—in trace amounts.[3,16] Carbon, hydrogen, and oxygen are obtained from ambient air and water, while a few plants, such as legumes, can use root bacteria to fix atmospheric nitrogen. Other elements are mostly assimilated from the soil via complex physicochemical root mechanisms.

*For example, the nutrients required to produce 150 bushels of corn, in pounds per acre required are N, 310; P, 120 as phosphate, 52 as phosphorus; Ca, 58, equivalent to 150 lb of agricultural limestone; Mg, 50, equivalent to 275 lb of epsom salt, or 550 lb of sulfate of potash-magnesia; S, 33; Fe, 3, equivalent to 15 lb of iron sulfate; Mn, 0.45; B, 0.10; Zn, Cu, and Mo in trace amounts; O, 10,200; C, 7,800; and H_2O, 3.225–4.175 tons, equivalent to 29–36 in. of rain. The effect of most fertilizer addition is asymptotic. For example, the yield of corn on an irrigated test farm shows that increasing the nitrogen applied per acre from 40 to 80 lb increases corn yield by 30 bushels; increasing N to 120 lb further increases the yield by 20 bushels; increasing N to 160 lb further increases the yield by 15 bushels; a further increase of N to 200 lb increases the yield by only another 10 bushels.

To ensure repeated healthy crops, farmers must maintain adequate soil-nutrient levels by appropriate fertilizer additions. Traditional low-analysis materials, such as manure and other animal or vegetable wastes, are usually in limited supply on large modern farms and have to be supplemented or replaced by high-analysis chemical fertilizers formulated to specific soil and crop needs. While some people express a preference for "naturally" fertilized crops, no alternative to chemical fertilizers as a way of meeting current world food needs of 3.5 billion people (estimated to reach 7 billion by the year 2000) is known.

Nitrogen Fertilizers. Studies by nineteenth century scientists established nitrogen as a primary plant nutrient and one of the key elements in organic fertilizers.[4,5] The first chemical nitrogen fertilizers were Chilean saltpeter, $NaNO_3$, which today finds a limited use as a tobacco topdressing, and ammonium sulfate produced as a coal-gas by-product. Ammonium sulfate still is applied in considerable quantities outside the United States. Early twentieth-century attempts at fixing nitrogen via electric arc led to the production of calcium cyanamide, $CaCN_2$, in the United States, and calcium nitrate via HNO_3 in Norway. The last $CaCN_2$ plant in the Western world closed in June 1971, but large tonnages of $Ca(NO_3)_2$ still are made in Europe. Today, virtually all nitrogen fertilizers are based on NH_3 synthesized from atmospheric nitrogen plus hydrogen derived from natural gas, oil, coal, lignite, or electrolysis of water (in decreasing order of use). See also **Ammonia.**

Ammonium Sulfate: $(NH_4)_2SO_4$ contains approximately 21% N. Initially it was produced by scrubbing coal gas with H_2SO_4, followed by evaporation, crystallization, separation, and drying. This method is still used to a limited extent, as is reacting H_2SO_4 with anhydrous NH_3. A major current source is caprolactam manufacture, which gives several tons of by-product sulfate per ton of caprolactam. Outside the United States, several large producers use the Merseburg reaction, by which natural or by-product gypsum is reacted with ammonium carbonate to yield ammonium sulfate:

$$CaSO_4 \cdot 2H_2O + (NH_4)_2CO_3 \rightarrow$$
$$CaCO_3 + (NH_4)_2SO_4 + H_2O - 3.9 \text{ kcal}$$

When properly prepared, the product is in the form of free-flowing, stable crystals.

Calcium Nitrate: $Ca(NO_3)_2$ contains approximately 15% N. This material was made initially by directly reacting limestone and HNO_3, a method still in limited use. $Ca(NO_3)_2$ currently is produced in large quantities as a by-product from nitrophosphate manufacture (described later). There are several commercial processes for air-prilled, oil-prilled, and crystalline products. $Ca(NO_3)_2$ is very hygroscopic, making suitable storage and shipping precautions imperative.

Ammonium Nitrate: NH_4NO_3 contains approximately 35% N. As shown in Fig. F-1, ammonium nitrate is produced by directly reacting NH_3 and HNO_3. Various commercial processes operating under vacuum and at atmospheric pressure or above are available: $NH_3 + HNO_3 \rightarrow NH_4NO_3 - 26\ kcal$. NH_4NO_3 also is made in large quantities by reacting calcium nitrate by-product from nitrophosphate plants with NH_3 and CO_2: $Ca(NO_3)_2 + 2NH_3 + CO_2 + H_2O \rightarrow 2NH_4NO_3 + CaCO_3$. It usually is produced as prills, granules, or solution, either alone or in conjunction with other nitrogen-containing liquids, e.g., urea and aqua NH_3. Solid and molten forms of NH_4NO_3 can be hazardous under certain conditions, e.g., when detonated and/or when organic matter is present. A popular explosive, ANFO, for example, is a mixture of NH_4NO_3 and fuel oil. Accordingly, suitable precautions are

necessary in production, storage, and handling. One way of minimizing danger is to produce nitro chalk, a mixture of NH_4NO_3 and $CaCO_3$ with a maximum nitrogen content of about 26%.

Precautions must also be taken against spontaneous slow burning as well as physical breakdown caused by volume changes induced by crystal transitions between forms IV and III, when storage temperatures are allowed to fluctuate in the 30–35°C range.[5] Small amounts of urea help to suppress self-burning, and certain alkali salts plus low free-moisture improve particle stability.[6] As NH_3NO_3 is hygroscopic, clay coatings and moisture-proof bags are necessary safeguards against spoilage in storage and transportation.

With reference to Fig. F-1, aqueous HNO_3, gaseous NH_3, and a conditioner solution (magnesite dissolved in HNO_3) are fed into a reactor to form NH_4NO_3 solution. The pH of the final NH_4NO_3 solution is adjusted by addition of NH_3. From the final neutralizer, the solution is pumped to a falling-film evaporator, where the water content is reduced to a maximum of 0.5%, necessary for the subsequent prilling step. The concentrated NH_4NO_3 solution is sent to a prilling tower, where it is finely distributed through spraying roses. The droplets descending in the tower are cooled with

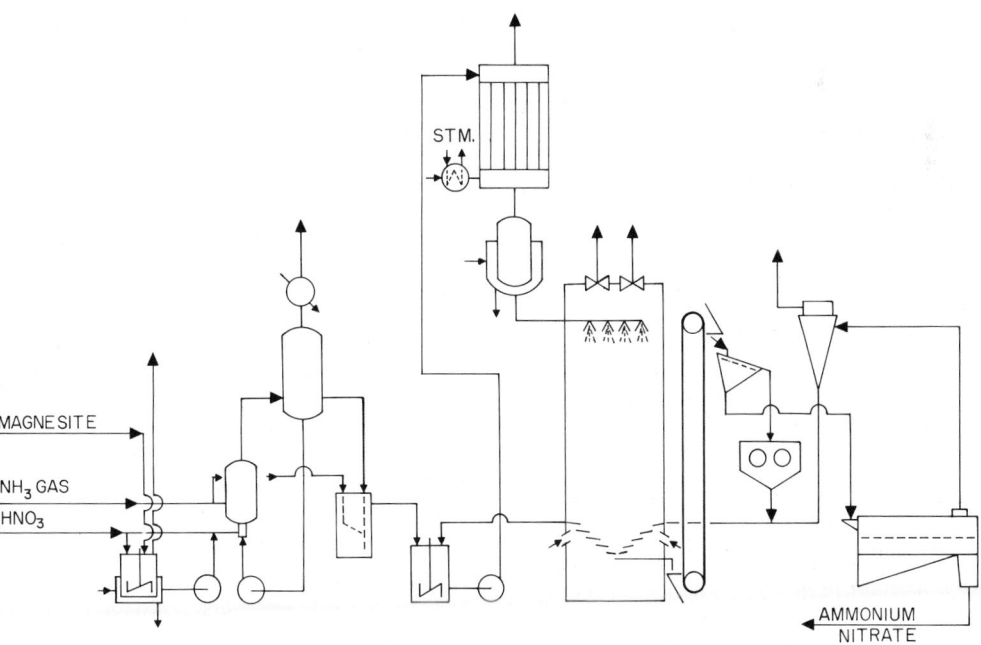

Fig. F-1. Production of ammonium nitrate. (*Hoechst-Uhde Corporation.*)

a countercurrent stream of air; the prills are withdrawn through an opening in the bottom of the tower for subsequent classifying. The oversize is crushed and dissolved in NH_4NO_3 solution together with the fines from the dust-collecting equipment. The normal-grain-size fraction is sent to a fluidized-bed cooler, where the product temperature is reduced by means of cold air to make the product suitable for storage.

Urea: $CO(NH_2)_2$ contains 46.65% N. First synthesized by Wöhler in 1828 from ammonium cyanate, urea has become the dominant nitrogen fertilizer in the last few years. Urea also is used increasingly as an animal feed and a raw material for melamine and urea-formaldehyde resins. The production of urea is described under **Urea.**

Trends in Nitrogen Fertilizers: In the United States, large quantities of liquid nitrogen fertilizers, such as anhydrous NH_3, aqua NH_3, ammonium salts, and many combinations, including some containing phosphate salts, are being used. Since all these materials are highly water-soluble, the possibility of ecological upset caused by nitrogen runoff has been raised. Fortunately, many soils rapidly fix ammonium ions, thereby checking waste and pollution. Controlled release of nitrogen can be obtained by reacting urea and formaldehyde to produce "ureaform" fertilizers, by coating urea with sulfur, or by using slowly soluble compounds containing nitrogen, such as isobutylidene diurea.[5,6] These materials are relatively costly and are sold principally to nonfarm markets.[15] Recent attempts to make controlled-release, high-analysis nitrogen fertilizers in combination with phosphorus show promise.[7]

Phosphate Fertilizers. In the mid-nineteenth century, Liebig and others showed that the traditional fertilizer properties of bones are due largely to a high phosphate content and that treatment with H_2SO_4 greatly increases effectiveness. Rapidly expanding fertilizer needs created a big demand for such chemical manures and led to an acute shortage of bones. This was overcome by the timely discovery of phosphate minerals in Florida and elsewhere, plus guano deposits in Peru. Today, bone and guano are mostly limited to special nonfarm uses and have been virtually replaced by fertilizers prepared from phosphate rock mined in many countries.[8,9]

An important property of phosphate fertilizers is *availability* to plants, which is largely a function of phosphate solubility in a specific soil. Many phosphate rocks consist of a clay-sand matrix

bearing an apatite mineral, $Ca_5(PO_4)_3R$, where R is usually fluorine but may be OH, CO_3, or Cl. Washing followed by screening and beneficiation yields pebbles and sandy concentrates often having the general composition $3Ca_3(PO_4)_2 \cdot CaF_2$ (about 30–40% P_2O_5). In water and alkaline and neutral soils, apatites and tricalcium phosphate, $Ca_3(PO_4)_2$, are highly insoluble, but they are moderately soluble in acid soils. Dicalcium phosphate, $CaHPO_4$, is readily soluble in acid soils and moderately so in water and alkaline and neutral soils, whereas monocalcium phosphate, $Ca(H_2PO_4)_2$, is soluble in water and all moist soils. Also affecting availability is the presence of Fe and Al phosphates, which are insoluble in water but soluble in weak acids.

In some countries total water solubility is demanded or is at a premium, which means that monocalcium and ammonium phosphates must be used. In others, slight solubility in water or appreciable solubility in weak acids such as citric is adequate, thus permitting fertilizers to include large amounts of dicalcium phosphate as well as phosphates of Fe and Al. Several nitrophosphate fertilizers and steel slags are in this category. Some humic and acid soils are able to assimilate ground phosphate rock without prior chemical treatment.

Single Superphosphate: Approximately 20% P_2O_5,[2,8,10] this is the oldest water-soluble phosphate fertilizer, and it still is produced in large quantities. The material is made by reacting ground phosphate rock and 70% H_2SO_4 in a batch den or on a continuous belt. A solid mass of monocalcium phosphate and gypsum is formed, which is cured by storing for several weeks before grinding and shipping. Gaseous compounds of fluorine and silicon are evolved and removed by water scrubbing. The empirical reaction is $3Ca_3(PO_4)_2 \cdot CaF_2 + 7H_2SO_4 \rightarrow 3Ca(H_2PO_4)_2 + 7CaSO_4 + 2HF$.

Wet-Process Orthophosphoric Acid: This is commercially 30–54% P_2O_5. The addition of H_2SO_4 to phosphate rock in amounts greater than needed to make single superphosphate produces orthophosphoric acid, H_3PO_4. Several reactions occur. Monocalcium phosphate reacts with H_2SO_4 to yield phosphoric acid, part of which reacts with more rock to make additional monocalcium phosphate. Qualitatively this is $Ca(PO_4)_2 \cdot CaF_2 + H_2SO_4 + H_2O \rightarrow H_3PO_4 + CaSO_4 \cdot H_2O + HF$.

This acid is a fertilizer intermediate and also is used to make detergent phosphates after purification.[4] Commercial processes are all based on violent agitation of the rock-acid slurry, followed by removal of $CaSO_4$ via filtration and evaporation of H_2O to obtain the desired concentration, up to 54% P_2O_5. Traditional processes separate the sulfate as impure dihydrate (gypsum), which usually is discarded. Some polymorphic P_2O_5 invariably is trapped in the filter cake, and this loss can be largely overcome by precipitating the sulfate as the hemihydrate and recrystallizing in dihydrate form, as is done in some recent processes of Japanese origin.[11] In this way, a gypsum suitable for wallboard, plaster, and cement manufacture can be produced simultaneously.

Triple Superphosphate: This is approximately 46–48% P_2O_5.[2,8] Acidulating phosphate rock with phosphoric acid produces concentrated or triple superphosphate, which is essentially monocalcium phosphate containing very little gypsum. The reaction is $Ca(PO_4)_2 \cdot CaF_2 + H_3PO_4 + H_2O \rightarrow Ca(H_2PO_4)_2 \cdot H_2O + HF$. Continuous processes are available for making powdered (run-of-pile) and granulated products. Triple superphosphate is used mostly to furnish P_2O_5 in mixed fertilizers

and is water-soluble. Its high analysis compared with single superphosphate offers savings in storage and shipping.

Ammonium Phosphates:[4-6] Several ammonium phosphates can be prepared, but only the mono and the di compounds are made for fertilizer purposes, alone or in combination with other salts. Numerous commercial processes are available whereby anhydrous NH_3 is reacted with H_3PO_4 and the resulting slurry is converted to solid form and dried. NH_3/H_3PO_4 ratios between 1 and 2 can be selected to produce various product grades. For example, for diammonium phosphate $2NH_3 + H_3PO_4 \rightarrow (NH_4)_2HPO_4$.

Wet-process acid is commonly used, but minor quantities are made from acid of electric-furnace origin. Since impurities in wet-process acid make crystallization difficult, the corresponding products are granulated. Mono- and diammonium phosphates based on furnace acid can be crystallized easily. The general process for making granular ammonium phosphate fertilizer is shown in Fig. F-2. Typical analyses are tabulated below:

Acid Used	Monoammonium Phosphate		Diammonium Phosphate	
	N, %	P_2O_5, %	N, %	P_2O_5, %
Wet-process	11	48	18	46
Electric-furnace . .	12	61	21	53

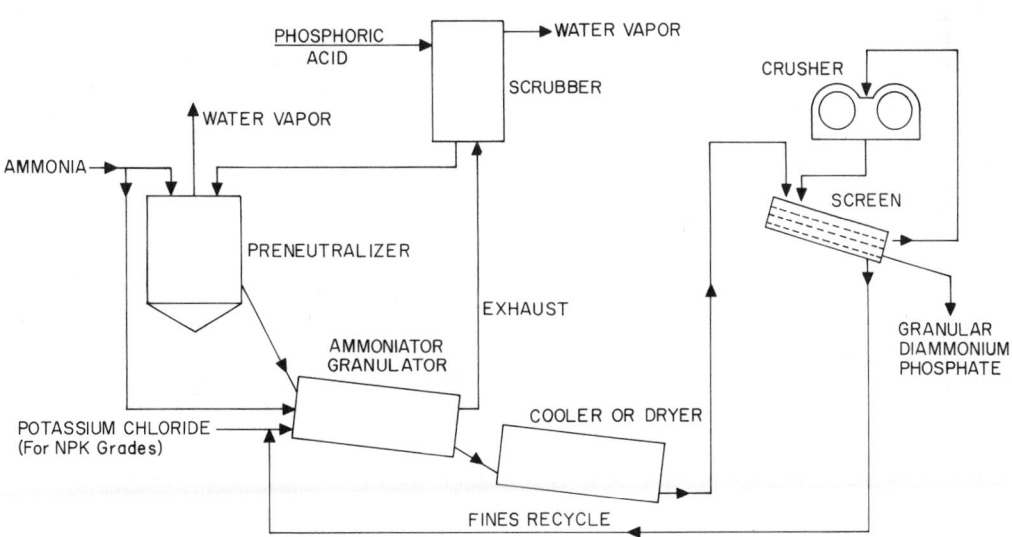

Fig. F-2. Granular ammonium phosphate fertilizer process.

Granulated diammonium phosphate made from wet-process acid is produced in much greater quantities than monoammonium phosphate and has become the most popular of all phosphate fertilizers, due to its high analysis, stability, and favorable economics. In conventional processes, neutralization is undertaken at atmospheric pressure. Recently, several monoammonium phosphate processes have become available in which NH_3 and acid are reacted under pressure, the slurry being flashed in a tower and converted to powder form by the heat of reaction.[5,6] The product is useful as an intermediate for making granulated mixed fertilizers.

Nitrophosphates:[2,5,6] Phosphate rock dissolves readily in HNO_3 to produce mixtures of calcium nitrate, phosphoric acid, and monocalcium phosphate, according to the amount of acid used. One example is $3Ca_3(PO_4)_2 \cdot CaF_2 + 18HNO_3 \rightarrow 4H_3PO_4 + CaH_2(HPO_4)_2 + 9Ca(NO_3)_2 + 2HF$.

If converted to solid form, the $Ca(NO_3)_2$ component would render the monocalcium phosphate unstable and make the product too hygroscopic for use. One method of overcoming these problems is to form calcium nitrate tetrahydrate crystals by chilling and then removing them by filtration or centrifuging. The calcium nitrate can be subsequently crystallized, prilled, and bagged or used to make ammonium nitrate. Ammoniation of the mother liquor produces a mixture of ammonium phosphate, dicalcium phosphate, and ammonium nitrate that can be concentrated and prilled or granulated. In terms of customary fertilizer nomenclature (percent N, P_2O_5, and K_2O), many product grades are possible according to raw-material ratios and process conditions, for example, 20-20-0. Adding potash before prilling or granulating yields formulations such as 15-15-15.

An alternative to $Ca(NO_3)_2$ removal is conversion in situ to less hygroscopic compounds. Several proprietary methods are available. Addition of phosphoric acid along with HNO_3, followed by ammoniation, yields dicalcium phosphate, ammonium phosphate, and ammonium nitrate. Alternatively, H_2SO_4 can be added with the HNO_3 to precipitate part of the calcium as $CaSO_4$, which, however, appears in the product and lowers the grade from 15-15-15 to perhaps 12-12-12. In other methods, ammonium or potassium sulfate or CO_2 is used to convert part of the $Ca(NO_3)_2$. In the latter case, phosphate in the product is in dicalcium form and is water-insoluble. The grade also is reduced by the presence of $CaCO_3$.

Other nitrophosphate processes developed or proposed incorporate circuits for converting $CaSO_4$ to $(NH_4)_2SO_4$ for reuse or decomposing $Ca(NO_3)_2$ to HNO_3 and lime for reuse and sale. Millions of tons of nitrophosphates are made annually in Europe, where less emphasis is placed on total phosphate water solubility. In the United States, because of ample availability of sulfur under normal conditions and a preference for rapid-acting, water-soluble phosphate fertilizers (especially on corn and grains), interest in nitrophosphates has been limited to date.

Nonorthophosphates:[5,6] Continued dehydration of 54% wet-process orthophosphoric acid removes remaining free water and then molecular water to yield the pyro acid:

$$2 \times \underset{\underset{OH}{|}}{\overset{\overset{O}{\|}}{HO-P-OH}} \rightarrow \underset{\underset{OH}{|}}{\overset{\overset{O}{\|}}{HO-P-O}} \underset{\underset{OH}{|}}{\overset{\overset{O}{\|}}{-P-OH}} + H_2O$$

Continued heating will remove more H_2O molecules until various insoluble compounds are formed. In practice, evaporation by submerged combustion or under vacuum is undertaken until a eutectic between ortho and pyro acids with a P_2O_5 content of about 72% is reached. This *superphosphoric acid* is shipped in insulated tank cars to market areas and ammoniated to produce liquid ammonium polyphosphate (APP) fertilizers, such as 10-34-0. These liquids possess strong sequestering properties and can keep impurities in solution, as well as salts of micronutrient metals that are insoluble derivatives of orthophosphoric acid. In addition to good stability in storage and shipping, some agronomic advantages are also claimed.

An alternative and more direct method of producing APP fertilizers[7,12] is to react NH_3 and wet-process acid under self-pressure and dissolve the melt in ammonia solution to produce 10-34-0. Addition of potash and a little clay makes possible nitrogen-phosphorus-potassium suspensions such as 13-13-13. Or, the melt can be granulated with care in a pug mill to produce solid 12-57-0. Addition of urea and potash, if required, makes granulation easier and makes solid grades such as 28-28-0 or 19-19-19 possible. However, the nonorthophosphate content is reduced from over 50% to 25%. Similar products having higher nutrient levels and nonortho contents can be made by using furnace acid instead of wet-process acid. In some cases, users may find that superior sequestering and suspension properties justify the higher cost. A recently disclosed simple way of producing polyphosphates by heating ammonium orthophosphates in NH_3 vapor shows promise.[12]

Potassium Fertilizers. Potassium, like phosphorus, was found in the early nineteenth century to be a major essential plant food and large deposits of sylvite, KCl, and carnallite, $KCl \cdot MgCl_2 \cdot 6H_2O$, were discovered at Stassfurt, Germany while drilling for halite, NaCl. This source soon replaced wood ashes and is still mined. Billions of tons of these minerals have been found in many countries, especially Canada, which now supplies most North American needs. Mining usually is by underground methods between 1,000 and 5,000 ft, although solution mining is also practiced in Canada. At the surface, ores are crushed, beneficiated, crystallized, and dried to produce commercial potash or, *muriate*, KCl, in various grades and particle sizes containing 60–62% KCl. Potash is also recovered from brines, the Dead Sea, and other sources.

Relatively small amounts of other potassium salts are used as fertilizer, generally for special purposes. Tobacco and some vegetables are adversely affected by high chloride concentrations, and K_2SO_4 or KNO_3 are preferred. K_2SO_4 is made in substantial quantities in Europe by the Mannheim process from potash and H_2SO_4 and in the United States and other countries by various exchange reactions between K, Na, and Mg salts. Some KNO_3 is produced in Chile from nitrate deposits and in the United States from potash and HNO_3 via proprietary techniques. Limited amounts of KOH are used in liquid-fertilizer formulations.

Potash usually is applied to the soil along with salts containing nitrogen and/or P_2O_5 in amounts prescribed by the agronomist.[3] One method is to combine crushed muriate with moist nitrogen and P_2O_5-containing compounds and granulate the mixture by conventional means.[2,5,8] Another is to dry-blend materials such as urea, diammonium phosphate, and potash and apply to the soil. The recent production of granular potash in sizes compatible with other fertilizer salts has made this possible by minimizing segregation in transit. The total water solubility of potash results in 100% initial K_2O availability in most soils. However, in clays and when rainfall is limited, excessive chloride buildup can be harmful. Controlled-release techniques or use of other potassium salts may become necessary in some areas.

Secondary Nutrients and Micronutrients. Heavy cropping and a decline in the use of single super-phosphate has created sulfur deficiencies in many areas of the world, including the United States. These deficiencies can be remedied by periodically applying gypsum or sulfur, directly or in conjunction with primary nutrients. Calcium levels are usually adequate and can easily be restored by lime or limestone dressings. Magnesium deficiencies are common and are corrected by applying Mg-containing minerals, such as kieserite or dolomite.[3]

Although micronutrients usually are needed only in ounces or a few pounds per acre, deficiencies soon lead to stunted and diseased crops and livestock.[16] Several of these trace elements have proved surprisingly difficult to add to the soil or plants in forms that can be assimilated readily. One way is to incorporate in fertilizers slowly soluble frits containing desired micronutrient compounds. Another way is to make organic chelates of metals, such as Fe, Cu, Co, B, and Zn. Superphosphoric acid and solid or liquid polyphosphates are also effective carriers for several micronutrients.[7,12]

REFERENCES

1. Collins, G. H.: "Commercial Fertilizers," 5th ed., McGraw-Hill, New York, 1955.
2. Sauchelli, V.: "Manual on Fertilizer Manufacture," Industry Publications, Caldwell, N.J., 1963.
3. McVickar, M. H., et al.: "Fertilizer Technology and Usage," pp. 269–340, Soil Science Society of America, Madison, Wis., 1963.
4. Sauchelli, V.: "Fertilizer Nitrogen," Van Nostrand Reinhold, New York, 1964.
5. Pratt, C. J., and R. Noyes: "Nitrogen Fertilizer Chemical Processes," Noyes Development Corp., Park Ridge, N.J., 1965.
6. Slack, A. V.: "Fertilizer Developments and Trends," Noyes Development Corp., Park Ridge, N.J., 1968.
7. Getsinger J., et al.: TVA 7th Demonstr., Dev. New Fert. Technol., Muscle Shoals, Ala., October 1968.
8. Sauchelli, V.: "Chemistry and Technology of Fertilizers," Van Nostrand Reinhold, New York, 1963.
9. Van Wazer, J.: "Phosphorus and Its Compounds," vol. 2, Interscience-Wiley, New York, 1961.
10. U.S. Department of Agriculture and Tennessee Valley Authority, various publications.
11. Slack, A. V., et al.: "Phosphoric Acid," vols. 1 and 2, Dekker, New York, 1968.
12. TVA 8th Demonstr., Dev. New Fert. Technol., Muscle Shoals, Ala., October 1970.
13. Gale, J., et al.: Fertilizer Situation, U.S. Department of Agriculture, Feb. 22, 1971.
14. Hicks, G. C., et al.: AID/TVA Fertilizer Intermediates for Use in Developing Countries, TVA, Muscle Shoals, Ala., 1970.
15. Schery, R. W.: Fert. Solutions, July–August, 1971.

16. Dennis, E.: "Micronutrients," National Fertilizer Solution Association, Peoria, Ill., 1971.

—Christopher J. Pratt, *Hoechst-Uhde Corporation, Englewood Cliffs, N.J.*

Fibers   Structurally, a fiber is a sinewy, threadlike object that may be described (1) as *long* and *thin* because the length of the fiber may be hundreds or even tens of thousands of times greater than the sectional dimension of the fiber, (2) as possessing *strength* to resist elongation and being pulled apart, (3) as exhibiting some *elasticity* to return to its original dimension after stretching, (4) as relatively *flexible,* particularly along its longitudinal dimension, and (5) as capable of *interlocking* or *mechanically bonding* with other fibers (like or unlike) to form a matrix of fibers that amplifies the foregoing characteristics. Although an individual fiber may be quite small (a cotton staple ranges from $\frac{1}{2}$ to $2\frac{1}{2}$ in. long with a diameter of less than $\frac{1}{1,000}$ in.), millions of fibers, properly arranged and working together, can provide the strength, size, and other properties that one normally associates with the qualities of metals and ceramics. Fibers possess many additional characteristics unobtainable from other forms of material. Thus, the fiber may be considered as a basic structural component of many materials.

Treating natural fibers and, in particular, the manufacture of synthetic fibers constitutes a major segment of the chemical industry. Many of the materials required for the manufacture of synthetic fibers are derived from petroleum. Fibers, along with synthetic films and plastic articles made from synthetic resins, account in large measure for the accelerated growth of the petrochemical industry. See also **Petrochemical complex.**

Nature of Fibers.   Natural organic polymers, which make up materials such as cellulose, resins, proteins, and starch, remained rather mysterious even to the organic chemist until instrumentation, exemplified by x-ray diffraction apparatus, the electron microscope, and ultracentrifuge, became available to probe the structure of giant molecules. Polymer chemistry did not get seriously under way until about 1920. It was found that these giant molecules are made up of a large number of units repeated in the structure. The term *polymer* derives from the Greek *poly,* "many," and *meros,* "part." The individual building blocks are termed *monomers.* Studies show that most polymers have the configuration of a long, flexible chain. Many natural polymers are partly crystalline and partly amorphous. The presence of a crystalline structure imparts rigidity, strength, resistance to heat, and insolubility to the material; whereas if the structure is predominantly amorphous, the material is elastic, soft and absorptive, and relatively permeable to fluids. Fibers, particularly synthetic fibers, are long-chain molecules constructed of many like or mixed monomers to form copolymers. The long, sinewy fibers from which materials are made possess the same long, stringy configuration in their inner, or molecular, structure. See **Polymerization.**

Since ancient times, man has taken advantage of the basic qualities of fibers as a material of construction. Fibers, both natural and synthetic, are found in *rigid structures,* such as wood or man-made fiberboard, where a complex of millions upon millions of fibers produces rigidity and strength comparable to conventional materials of construction. Research is accelerating in an effort to find means for providing greater rigidity and strength to synthetic polymers, but at present, it is in connection with *semirigid* and *nonrigid* materials, e.g., the limp fabrics required for apparel, cordage, carpeting, hose, tire cord, and paper products, that fibrous construction excells, and it is for these kinds of applications that the major tonnage of fibers produced (in terms of billions of pounds annually) is consumed.

Classification of Fiber Uses: Historically, certain fibers, such as the bast fibers made from the inner fibrous bark of the stems of various plants, have been used by primitive people to make rough cordage, huts, and rope suspension bridges, which may be referred to as *tying applications.* Nurserymen still use raffia for tying. Such fibers as broomcorn and broomroot have been used for centuries in what might be term *brush* applications. *Plaiting* and *rough-weaving fibers,* such as straw, palm leaves, bamboo, and rattan, have been used for basketry and furniture making. The tapa or kapa cloth of the Pacific Islands is a nonwoven fabric. Large tonnages of *filling fibers* are used as packing materials and as wadding and filling for upholstery, including the use of various grasses, reeds, and husks. For the foregoing applications, the natural fibers require little or no processing.

It is in connection with fibers for *spinning, webbing,* or *matting* that modern technology, particularly chemical technology, has played a large role. Spinning fibers are used to make yarn, which in turn is used in the manufacture of a vast array of textile products. Webbing or matting fibers are used to make unwoven materials, such as paper, paperboard, and some fabrics. Spinning and webbing fibers are of the greatest importance industrially and account for the greatest tonnage consumption of fibers. Webbing or matting fibers are described under **Pulp (wood) production and processing.** The remainder of this description pertains to the use of fibers by the textile industry.

Natural Fibers. There are three main sources of natural fibers: (1) animal fibers, e.g., wool, mohair, and silk, which chemically are complex proteins; (2) vegetable fibers, e.g., cotton, flax, and jute, which are predominantly cellulose; and (3) mineral fibers, e.g., asbestos. Cotton and wool are by far the most important natural fibers in terms of use.

Cotton: Despite the marked inroads of synthetic fibers into the textile industry, cotton continues as a major textile fiber, cotton consumption being expressed in billions of pounds per year. Cotton is one of the most versatile of all fibers and contributes its many good properties to blends with other fibers. Cotton fibers contain 88–96% by weight of cellulose, together with protein, pectin, sugar, and approximately 0.4–0.8% by weight of wax.

Special finishes for fabrics made from cotton have led to the introduction of both easy-care and stretch cotton fabrics. Postcured resin finishes, at the sacrifice of some abrasion resistance, permit permanently creased garments to be made of cotton. Although some synthetic fibers compete with cotton on a basis of technological superiority, the position of cotton as a major textile fiber is largely a question of economics. Research projects have been undertaken to eliminate much of the loss and damage that occur in harvesting and ginning, factors that have contributed to price instability. The relative advantages and limitations of cotton as a textile fiber, in comparison with other major fibers, are given in Table F-1.

Wool: The natural, highly crimped fiber from sheep, wool is one of the oldest fibers from the standpoint of use in textiles. Minute scales on the surface of the fibers allow them to interlock and are responsible for the ability of the fiber to *felt,* a phenomenon responsible for felt cloth and mill-finished worsteds. Crimpiness in wool is due to the open formation of the scales. Fine merino wool has 24 crimps per inch. Luster of the fiber depends upon the size and smoothness of the scales. The basic wool protein, keratin, comprises molecular chains that are linked with sulfur. When sulfur is fed to sheep in areas deficient of

sulfur, the quality of the wool improves. Wool fibers that fall below 3 in. in length are known as *clothing wool;* fibers 3–7 in. are referred to as *combing wools.* The wool-fiber diameter ranges from 0.0025 to 0.005 in.

Wool is known universally for its warmth, resilience, and characteristic hand (the feel of a fabric when handled). For normal apparel use, wool-fiber gradings range from coarse 40s up to fine 80s* (lamb's wool). Wool fabrics can be made shrinkproof and mothproof by means of special finishing treatments. Durable pleats can be put in by special chemical treatments. Coarse wools are used extensively in carpets, which are characterized by good resilience, ability to retain appearance, and ease of maintenance. See also **Peptides and proteins.**

Mohair: This very resilient hair is obtained from the Angora goat. The staple length ranges from 5–8 in., but Turkish fibers go up to 10 in. Mohair provides a characteristic crisp, resilient, and slightly scratchy hand to fabrics even when used in very low percentages with other fibers.

Silk: This fiber is extruded by the silkworm when it spins its cocoon. Silk is a fine, strong, lustrous fiber, the filaments of which are stuck together by a gummy substance called *sericin* that is normally removed by a special scour prior to dyeing. The remaining raw-silk fiber, known as *fibroin,* consists largely of the amino acid alanine, $CH_3CH(NH_2)CO_2H$, which can be synthesized from pyruvic acid. Silk is closely allied to cellulose and resembles wool in structure but contains no sulfur. Many varieties of silk depend upon the diet of the silkworm.

Flax: The bast fiber from the inner bark of the flax plant, *Linum usatatissimum,* is legally the only fiber that can be used in the manufacture of linen goods. Through trade usage, the term *linen* has incorrectly become associated with a particular fabric character or type without regard to fiber content. However, when reference is made to Irish linen, the fiber involved is always flax. Due to its limited production and the traditional hand labor involved, flax, in comparison with cotton, is quite expensive. Flax possesses most of the better qualities of cotton and in addition is more absorbent and stronger and quite resistant to at-

---

*Wools are classified by their fineness and coarseness in terms of spinning counts. A spinning count refers to the number of hanks of worsted yarn that a pound of a given grade of wool will spin. Wool classified as 60s will spin 60 times 560 yd, or 33,600 yd of worsted yarn.

TABLE F-1. *Properties Imparted to Fabrics by Major Fibers*

Property	Cotton	Wool	Flax	Silk	Acetate	Rayon (Viscose)	Nylon	Acrylic	Modacrylic	Polyester
Abrasion resistance	Good	Fair[a]	Fair	Fair	Poor	Fair	Excellent	Fair[a]	Fair-good[b]	Fair-good[b]
Hand	Excellent	Excellent	Excellent	Excellent	Excellent	Fair	Fair[c]	Very good	Good	Fair
Pilling resistance	Excellent	Good[d]	Fair	Good	Excellent	Excellent	Poor[a]	Fair[a]	Fair	Poor-fair[b]
Pressed-crease retention	Poor[e]	Poor	Very poor	Poor	Very poor	Fair	Fair	Very good	Good	Poor-good[b]
Safe ironing temperature: °F	425	300	450	300	350	375	300–350[b]	300–350[b]	225–275[b]	300–350[b]
°C	219	149	232	149	177	191	149–177	149–177	107–135	149–177
Stability to repeated launderings	Excellent[f]	Poor	Good	Good	Fair	Good[g]	Fair	Good	Fair	Excellent[h]
Strength	Good	Fair	Good	Good	Poor	Very good	Excellent	Fair-good[b]	Poor	Excellent
Sunlight resistance	Fair	Good	Fair	Poor	Fair	Good	Poor	Excellent	Excellent	Fair-good[b]
Colorfastness	Excellent	Good	Excellent	Good	Poor	Excellent	Good	Good	Good	Good
Wash and wear performance	Good[g]	Poor	Very poor	Poor	Fair	Fair	Fair	Good	Fair-good[b]	Excellent
Wrinkle resistance	Good[g]	Good	Poor	Good	Fair	Good[g]	Good	Good	Good	Excellent

[a] Good in carpets.
[b] Varies with manufacturer.
[c] Tends to be cold and clammy unless modified by texturing.
[d] Pills form, but tend to break off.
[e] Unless heavily resinated and postcured.
[f] If preshrunk.
[g] If resin-treated.
[h] Must be heat-set.

tack by bacteria and mildew. The average flax fiber is about 20 in. long.

Synthetic Fibers. The first major artificial or synthetic fiber for textile use was rayon, introduced about 1910 as a substitute for silk. The chemical composition of rayon is completely different from that of silk. Rayon thus typifies most reconstituted or synthetic fibers that perform almost as well as, and in many instances better than, natural fibers but which chemically and structurally are not exact duplicates of their natural counterparts. The production of synthetic fibers has grown rapidly, particularly since World War II. Qualities of synthetic fibers which are targets for continued scientific improvement include better moisture absorption; easy-care characteristics; soil and stain resistance; reduced electrical static and pilling; improved dyeability, reduction in flammability; resistance to light, thermal, and oxidative degradation; and higher tenacity and strength. Chemical technology is working in at least three directions (1) to introduce new, superior fibers, (2) to improve the characteristics of present synthetics, and (3) to develop methods for improving the natural fibers by chemical-treating processes.

Classification: Since the hundreds of chemical firms that produce synthetic fibers and the thousands of textile and apparel firms that consume them have persisted in using scores of trade names to identify fibers in their marketing, the nomenclature of synthetic fibers is unnecessarily complex and confusing. Some authorities define a synthetic fiber as a noncellulosic fiber of synthetic origin and consequently exclude rayon and acetate from their definition. Other authorities include rayon and acetate, along with nylons, polyesters, and acrylics in the total spectrum of synthetic fibers.

Because the physical and chemical properties of synthetic fibers cover a broad range, classification of the fibers on the basis of their chemical composition is the least complex. There is, of course, a twilight zone between elastomers and synthetic fibers; e.g., spandex, a fiber with a rubberlike quality (spandex is included in this description). For information on synthetic rubber and elastomers, see **Elastomers.**

The accepted generic names* for the various classes of synthetic fibers are used in the following descriptions.

Acrylic: A manufactured fiber in which the

*U.S. Federal Trade Commission (Textile Fiber Products Identification Act).

fiber-forming substance is any long-chain synthetic polymer composed of at least 85% by weight of acrylonitrile units, $-CH_2-CH-$. The char-

$$| \atop CN$$

acteristics of acrylic fibers are summarized, along with other major fibers, in Table F-2. See also **Acrylic fibers.**

Modacrylic: A manufactured fiber in which the fiber-forming substance is any long-chain synthetic polymer composed of less than 85% but at least 35% by weight of acrylonitrile units except when it qualifies as rubber.

Polyester: A manufactured fiber in which the fiber-forming substance is any long-chain synthetic polymer composed of at least 85% by weight of an ester of a dihydric alcohol and terephthalic acid, p-$HOOC \cdot C_6H_4 \cdot COOH$. See also **Polyester fibers.**

Rayon: A manufactured fiber composed of regenerated cellulose, as well as manufactured fibers composed of regenerated cellulose in which substituents have replaced not more than 15% of the hydrogens of the hydroxyl group. See also **Rayon.**

Acetate: A manufactured fiber in which the fiber-forming substance is cellulose acetate. Where not less than 92% of the hydroxyl groups are acetylated, the term *triacetate* may be used as a generic description of the fiber. See also **Acetate fibers.**

Saran: A manufactured fiber in which the fiber-forming substance is any long-chain synthetic polymer composed of at least 80% by weight of vinylidene chloride units, $\cdot CH_2 \cdot CCl_2 \cdot$. See also **Polyvinyl chloride.**

Azlon: A manufactured fiber in which the fiber-forming substance is composed of any regenerated naturally occurring proteins. See also **Peptides and proteins.**

Nytril: A manufactured fiber containing at least 85% of a long-chain polymer of vinylidene dinitrile, $\cdot CH_2 \cdot C(CN)_2 \cdot$, where the vinylidene dinitrile content is no less than every other unit in the polymer chain.

Nylon: A manufactured fiber in which the fiber-forming substance is any long-chain polyamide having recurring amide groups, $-C-NH-$, as an integral part of the polymer

$$\| \atop O$$

chain. See also **Nylon.**

Rubber: A manufactured fiber in which the fiber-forming substance is natural or synthetic rubber, including the following categories:

TABLE F-2. *Summary of Characteristics of Natural and Synthetic Fibers*

Fiber	Formula	Chemical Composition	Sp gr	Moisture Regain,* %
Cotton	$(C_6H_{10}O_5)_x$	Cellulose	1.54	7.0–8.5
Wool	$(C_{42}H_{157}O_{15}N_5S)_x$	Keratin	1.32	11–17
Rayon	HC— HCOH HOCH HCOH HC— HCH (O)	Regenerated cellulose	1.46–1.54	11–16.6
Acetate	[ring structure: H, OH, OCOCH$_3$, H, CH$_2$OCOCH$_3$]	Cellulose acetate	1.32	6
Triacetate	[ring structure: CH$_2$OCOCH$_3$, OCOCH$_3$, OCOCH$_3$]	Cellulose acetate with 92% or more of hydroxyls acetylated	1.3	3.2
Acrylic	$\begin{array}{cccc} H & CN & H & CN \\ -C- & C- & C- & C- \\ H & H & H & H \end{array}$	85% or more acrylonitrile	1.17–1.19	1.0–2.5
Modacrylic	$\begin{array}{cccc} H & H & H & H \\ -C- & C- & C- & C- \\ H & Cl & H & CN \end{array}$	35–84% acrylonitrile	1.30–1.37	0.3–4.0

*At 70°F 21.1°C and 65% relative humidity.   †Regular to medium tenacity.   ‡High tenacity.

Tensile Strength,* 1,000 psi	Thermal Effects	Resistance to Chemicals		
		Acids	Alkalis	Organic Solvents
60–120	Quite resistant to degradation by heat; after about 5 hr at 250°F (121°C), material yellows; decomposes above 300°F (149°C)	Cold concentrated acids and hot dilute acids disintegrate the material	Mercerizes, but without damage	Quite resistant to most solvents
17–29	Marked effects above 212°F (100°C); scorches at 400°F (204°C); chars at 570°F (299°C)	Hot H_2SO_4 destroys the material; otherwise quite resistant to acids	Strong alkalis destroy the material; attacked by weak alkalis	Quite resistant to most solvents
28–47† 58–88‡	Loss of strength at 300°F (149°C); decomposes at 350–400°F (177–204°C); burns readily, but does not melt	Acts like cotton	Loss of strength and swelling in strong alkalis	Resistant to most solvents
20–28	Loss of strength at 195–225°F (91–107°C); sticking point is 350°F (177°C); softens at 395–405°F (202–207°C); melts at 500°F (260°C); burns fairly slowly	Strong acids decompose material; dissolves in acetic acid	Resistant to weak alkalis; strong alkalis saponify the material	Dissolves in acetone and is softened by alcohol; affected by most solvents
18–23	Sticking point is 350–375°F (177–191°C); if heat-treated, sticking point is 465°F (241°C); melts at 575°F (302°C)	Acts like acetate	More resistant than acetate; good to pH of 9.8	Dissolves in acetone, chloroform, and methylene chloride; trichloromethylene causes swelling
30–62	Sticking point is 410–490°F (210–254°C); loses about 10% of strength after 10-hr exposure at 350°F (177°C); loses all strength after 325 hr	Resistant to most acids	Boiling, strong alkalis destroy the material; resistant to weak alkalis	Unaffected by common solvents
44–50	Stiffens and shrinks above 250°F (121°C); excess heat causes discoloration; does not support combustion	Resistant to most acids	Resistant	Generally unaffected, but warm acetone dissolves the material

Fiber	Formula	Chemical Composition	Sp gr	Moisture Regain,* %
Nylon 66	$-N-C-C-C-C-C-C-N-C-C-C-C-C-C-$ (with H and O substituents)	Polyamide based on hexamethylene diamine and adipic acid	1.14	4.0–4.5
Nylon 6	$-N-C-C-C-C-C-C-N-C-C-C-C-C-C-$ (with H and O substituents)	Polyamide based on caprolactam	1.14	4.5
Glass	SiO_2	Mostly silica	2.54	0
Polyester	$-O-C-C-O-C-$ (ring) $-C-$	Ester of a dihydric alcohol and terephthalic acid	~1.38	0.40–0.8
Spandex	$-O(CH_2)_4OCONH$ (ring with CH_3) $NHCOO(CH_2)_4O-$	Segmented polyurethane	1.21	0.75–1.3
Olefin	$-C-C-C-C-$ (with H and CH_3 substituents)	Polypropylene	0.90	<0.1
Fluorocarbon	$-C-C-C-C-C-$ (with F substituents)	Long-chain carbon molecules with available bonds saturated with fluorine	2.1	0

¶Regular filament.

Tensile Strength,* 1,000 psi	Thermal Effects	Resistance to Chemicals		
		Acids	Alkalis	Organic Solvents
65–85¶ 111–122‡	Sticking point is 445°F (230°C); melts at about 480°F (249°C); loss of 10% of strength in less than 1 hr of exposure at 350°F (177°C)	Resistant to weak acids; decomposed by strong mineral acids	Very little effect	Resistant but soluble in some phenolic materials and in 90% formic acid
60–85¶ 111–122‡	Slight discoloration if held above 300°F (149°C) for over 5 hr; melts at 415–430°F (213–221°C); Retains 95% of original tenacity after 24 hr at 330°F (166°C)	Similar to nylon 66	Very little effect	Similar to nylon 66
200–550	Strength tends to decrease above 600°F (316°C); material softens above 1350°F (732°C); nonflammable	Attacked by hot hydrofluoric and phosphoric acids; otherwise resistant	Affected by concentrated alkalis and by hot, weak alkalis	Not affected
81–88¶ 105–160‡	Sticking point is 455°F (235°C); melts at 480–550°F (249–288°C); up to 10% loss of strength after 18 hr at 350°F (177°C)	Disintegrates in concentrated H_2SO_4; otherwise resistant	Resistant; disintegrates slowly in boiling strong alkalis	Generally unaffected except by some phenolic materials
9–12	Sticking point is 345°F (174°C) and melts at about 445°F (230°C); deteriorates and yellows after exposure for 30 min to temperatures over 300°F (149°C)	Dilute HCl and H_2SO_4 cause yellowing	Resistant to weak alkalis	Resistant; dissolves in boiling cyclohexanone and dimethylformamide
35–80	Softens at 305–315°F (152–157°C) and melts at 325–340°F (163–171°C); tends to shrink above 265°F (130°C), depending on length of exposure	Quite resistant	Quite resistant	Dissolves in chlorinated hydrocarbons at elevated temperatures
14–38	Monofilament melts at 550°F (288°C); no degradation up to 400°F (204°C); vapors toxic	Essentially inert	Essentially inert	Only substances known to react with the fiber are alkali metals, fluorine gas, and chlorine trifluoride

1. The fiber-forming substance is hydrocarbon, such as natural rubber, polyisoprene, polybutadiene, copolymers of dienes and hydrocarbons, or amorphous (noncrystalline) polyolefins. See also **Butadiene; Elastomers; Isoprene;** and **Rubber, natural.**

2. The fiber-forming substance is a copolymer of acrylonitrile and a diene (such as butadiene) composed of not more than 50% but at least 10% by weight of acrylonitrile units. The term *lastrile* can be used as a generic description of fibers in this category. See also **Acrylonitrile.**

3. The fiber-forming substance is a polychloroprene or a copolymer of chloroprene in which at least 35% by weight of the fiber-forming substance is composed of chloroprene units, $-CH_2-C=CH-CH_2-$.

$$\begin{array}{c} | \\ Cl \end{array}$$

Spandex: A manufactured fiber in which the fiber-forming substance is a long-chain synthetic polymer comprising at least 85% by weight of a segmented polyurethane. See also **Urethanes.**

Vinal: A manufactured fiber in which the fiber-forming substances is any long-chain synthetic polymer composed of at least 50% by weight of vinyl alcohol units, $\cdot CH_2 \cdot CHOH \cdot$, and in which the total of the vinyl alcohol units and any one or more of the various acetal units is at least 85% by weight of the fiber.

Olefin: A manufactured fiber in which the fiber-forming substance is any long-chain synthetic polymer composed of at least 85% by weight of ethylene, propylene, or other olefin units except amorphous (noncrystalline) polyolefins qualifying as rubber.

Vinyon: A manufactured fiber in which the fiber-forming substance is any long-chain synthetic polymer composed of at least 85% by weight of vinyl chloride units, $\cdot CH_2 \cdot CHCl \cdot$.

Metallic: A manufactured fiber composed of metal, plastic-coated metal, or metal-coated plastic, or a core completely covered by metal.

Glass: A manufactured fiber in which the fiber-forming substance is glass. See also **Fiber glass.**

Anidex: A manufactured fiber in which the fiber-forming substance is any long-chain synthetic polymer composed of at least 50% by weight of one or more esters of a monohydric alcohol and acrylic acid, $CH_2 : CH \cdot COOH$.

Matrix: A biconstituent fiber of nylon and polyester polymers.

During the early growth of synthetic fibers in the 1940s and 1950s, producers of synthetic fibers largely concentrated on the introduction of general-purpose fibers in an effort to gain a broad market position. The more recent advances of polymer chemistry, coupled with much greater

TABLE F-3. *Dyes Commonly Used with Various Fibers*

Fiber	Dyes Used
Cotton	Direct, vat, azoic, basic, mordant, pigment, sulfur, and fiber-reactive
Wool	Acid, milling, chrome, mordant, vat, and indigo
Acetate	Readily dyed with disperse and azoic dyes; acid dyes are used for printing; solution-dyed material available
Rayon	Direct, vat, azoic, reactive, sulfur, and pigment types; processes similar to cotton; some rayons dye more slowly than cotton
Nylon	Materials have a strong affinity for all types of dyestuffs, including direct, acid, premetalized acid, chrome, and vat colors as well as newer complex types of dyes; for nylon 6/6, disperse, acid, and premetalized dyes usually preferred
Acrylic	Varies with manufacturer of fiber; usually disperse, basic, neutral, premetalized, acid, chrome dyes may be used; with one type of fiber, cationic and disperse dyes give a complete range of shades
Modacrylic	Disperse and cationic dyes with one type; neutral, premetalized, disperse, and basic or cationic with another type
Polyester	Varies with manufacturer of fiber; generally disperse and azoic dyes with carrier or heat for staple; disperse, dispersed and developed, and azoic combinations with carrier or heat for filament
Spandex	Good affinity for disperse, direct, selected acid, chrome, and premetalized dyes
Fluorocarbon	Cannot be dyed
Glass	Resin-bonded pigment systems; vat, acid, or chrome dyes will tint only
Polyethylene	Pigmented during manufacture
Polypropylene	Some pigmented during manufacture, but can be dyed with disperse, acid, and chelating dyes and certain vats, sulfurs, and azoics
Matrix	Generally acid and disperse dyes
Anidex	Disperse and selected acid and premetalized dyes

sophistication on the part of fiber consumers, have caused a trend in the direction of producing specialty synthetic fibers for specific applications. Where justified economically, much can be done to build desired properties directly into the molecular structure of the fiber. Blending two or more fibers has also opened up many opportunities for creating yarns and fabrics to end-use specifications. Dyes commonly used with synthetic fibers are summarized in Table F-3.

Nonwoven Materials. Fibers, in addition to their wide use in woven and knit goods, are finding increasing use in nonwoven materials, e.g., bonded fiber textiles, needle-punched fabrics, and fiber-reinforced paper fabrics. The nonwovens are a compromise between textile and paper products in terms of cost, production rates, and end-use characteristics. Nonwoven materials are not necessarily inexpensive substitutes for woven materials but more frequently are specialized products with characteristics engineered into them to meet specific end-use objectives. Nonwovens find use in apparel as interlinings and interfacings, as blankets, in carpet face and backing, as filter media, in medical applications where sanitary requirements warrant the use of disposable items, in tapes and ribbons, in wiping cloths, and in numerous household products, including curtains, diapers, napkins, shades, towels, and bookbinding.

Fibers also are finding application in oriented-fiber composites; in these structural materials nonwoven fibers are purposely oriented in a matrix to increase structural efficiency. Glass-filament-wound reinforced plastic tanks, pipe, and large vessels and structural components are examples where oriented-fiber composites provide high strength and a modulus of elasticity in relation to weight, which may be up to 3 times and up to 5 times that obtainable with conventional structural materials, including metals.

REFERENCES
Heimbold, Noreen C.: Manmade Fiber Chart: 1972, *Textile World,* McGraw-Hill., Atlanta, Ga.

Properties of the Manmade Fibers, 1972, *Textile Industries,* W. R. C. Smith Publishing Co., Atlanta, Ga.

Stanford Research Institute, Menlo Park, Calif. reports on synthetic fibers, oriented-fiber composites, nonwoven materials, and industrial textiles.

Fibers, Wood [See **Pulp (wood) production and processing.**]

Fiber Glass Fiber glass is simply glass in fibrous form. The material generally has properties simi-

lar to the glass from which it is made except that the tensile strength may be increased up to over 100 times that of the base glass. History records the use of strands of glass for decorating vases by the early Egyptians. The famed Venetian craftsmen had a limited knowledge of drawing glass fibers, but it was not until 1930–1940 that glass producers perfected a way to make fibers commercially.

Two forms of glass fiber are produced, *staple* or short-length fibers and *continuous* fibers. Staple fibers are used for thermal and acoustical insulation. Continuous filaments are used for yarn (as in fabrics), tire cord, and plastic reinforcements. Products using continuous filaments as reinforcements benefit from the high tensile strength of the filaments (as high as 300,000 psi compared with organic fiber strengths of less than 100,000 psi).

The base glass is made by heating raw materials, such as silica sand, limestone, dolomite, clay, boric acid, soda ash, and other minor ingredients, in a high-temperature furnace. See also **Glass.** Typical glass-fiber compositions are given in Table F-4. Fiber made from electrical-grade glass E is used most commonly for yarn, tire cord, and plastic reinforcement because of its high strength and electrical properties. Specialty glasses, although low in volume, fill important needs. S glass is a superior-strength glass primarily for defense applications, such as missile cases. C glass is more chemically resistant than E and is used for battery separator plates and chemical filters. Alkali glass A is used to some extent in the production of plastic reinforcement products.

Continuous-Fiber Products. In the *direct-melt* process (see Fig. F-3), raw materials are fed to a tank furnace to convert the mixture to glass. The glass flows to forehearths, which have platinum-alloy

TABLE F-4. *Typical Fiber-Glass Formulations*

Ingredient	Type of Glass, wt%				
	E	Insulating	A	S	C
SiO_2	54	63	73	64	65
Al_2O_3 . . .	14	5	1	24	4
MgO . . .	4	2	2	10	3
CaO	19	6	10	. . .	14
R_2O	0.5	16	14	. . .	8
B_2O_3	8	7	. . .	0–2	6
Fe_2O_3 . . .	0.3				
F_2	0.2	1			

NOTE: R = rare-earth element.

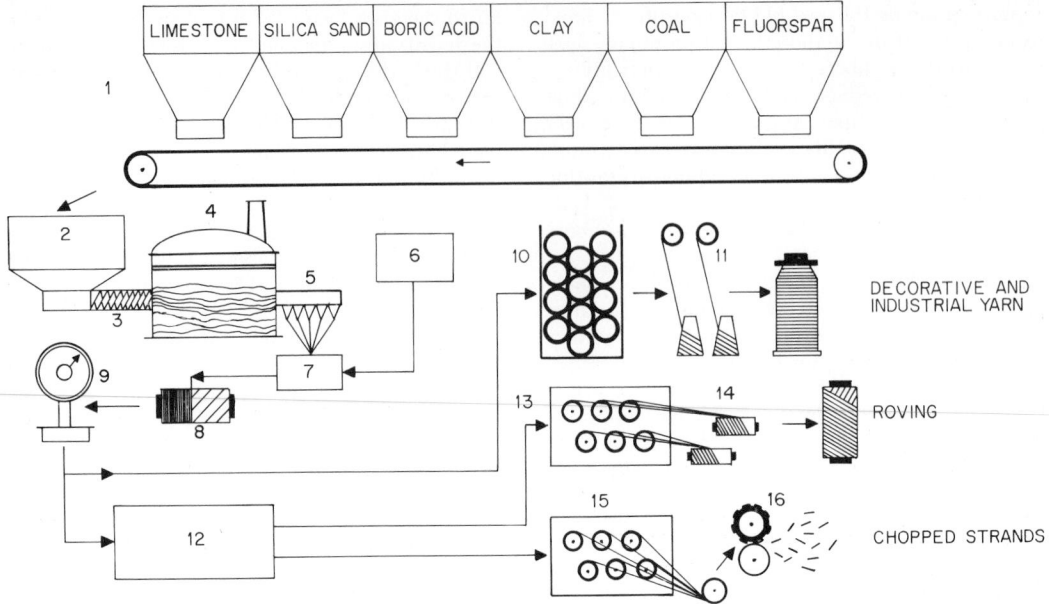

Fig. F-3. Direct-melt process for producing fiber glass. Raw materials (1) are automatically weighed and batched to mixer (2) prior to passing through screw feeder (3) to the glass melting tank (4). The molten glass flows to forehearths (5), at the bottom of which are platinum-alloy bushings or spinnerets. The latter are electrically heated and carefully temperature-controlled. Formulated binder material (6) is applied to the newly formed filaments (7) prior to high-speed winding (8). After weighing and inspecting (9), the wound filaments follow one of three paths in accordance with desired end product. In making decorative and industrial yarn, the fiber windings are placed in a conditioning room (10) prior to twisting (11). For the production of roving and chopped strands, the material from inspection operation (9) passes to an oven (12), where the filaments are heat-treated. This is followed by creeling (13) and roving winding (14) for production of roving. Following creeling (15), chopped strands (16) may also be made. There are several additional weighing and inspecting stations.

bushings or spinnerettes in the bottom. The bushings contain many holes, or orifices, each of which supplies a small stream of molten glass from which monofilaments are drawn. Mechanical attenuation, which produces a forming package, is accomplished by attaching the fibers to a rotating drum which turns up to 15,000 peripheral ft/min.

The *marble-melt process* consists of producing 1-in. marbles by a separate tank furnace. The marbles are then fed to a bushing unit, which is heated by electrical resistance. From this point, the process is identical to the direct-melt process.

Sizing: Because of the basic character of glass, the filaments are somewhat fragile and tend to abrade each other in close contact. A protective coating or sizing is necessary for the production, processing, and end use of all continuous-fiber glass products. Generally, a fiber-glass sizing or binder for textile or reinforcement products may contain (1) a film former, generally resinous in nature, that forms a strand or thread from grouped monofilaments; (2) a lubricant to aid in processing and end use of the fiber-glass product; and (3) additives to accomplish specified purposes, e.g., providing antistatic characteristics.

Sizings for plastics reinforcement also will have a coupling agent, such as a chrome complex, a silane, or combination of these two, to assure an interfacial bond between the glass surface and the resin matrix. Yarns for weaving normally have an oil-starch sizing. These coatings are applied before winding of the forming package.

Filaments ranging in number from 20 to 2,000 are then gathered together as a thread or strand before winding. As shown by Table F-5, the filaments are available in many diameters and are letter-designated. In continuous-fiber products, filaments range from designation B to P.

Continuous-filament products are designated by a letter-number system which specifies properties

TABLE F-5. *Designations of Filament Diameters*

Designation	Diameter × 10⁻⁵ in.
AAA	<3.0
AA . :	3.0–5.9
A	6.0–9.9
B	10.0–14.9
C	15.0–19.9
DE	23.0–27.9
G	35.0–39.9
H	40.0–44.9
K	50.0–54.9
P	70.0–74.9

important to end uses. For example, listing a strand as ECK67.5(200) 630 indicates that the material is made from E glass and is a C continuous fiber of K diameter; the strand contains 200 monofilaments and has a yield of 67.5×100 (or 6,750 yd/lb) and is coated with 630 binder. This particular binder is an oil-starch type.

Forming packages composed of wound strands normally are not supplied to industrial users without further processing. Strands are twisted and plied before being woven into fabric. A plied yarn, for example, is coated with a latex binder before being used as a tire-cord reinforcement.

End products made from continuous strand include fire-resistant curtains, reinforced tires and transmission belts, and many reinforced plastic items, such as boats, auto bodies, corrosion-proof pipe, roofing panels, and missile cases.

Staple-Fiber Products. Monofilament, short-length fibers are used for thermal and acoustical insulation, filtration, and cushioning. These products are made in basically three ways. In the high-temperature blast-jet process, 30-mil-diam fibers or rods first are produced by a bushing-type process. The coarse primary rods then are filamentized by a high-temperature, high-velocity blast burner. The blown mass of filaments is collected on a conveyor belt and bonded together by an inert thermosetting resin in a manner which creates many tiny air spaces throughout the material. The bonding process may be modified to produce flexible blankets, rigid board, or special molded shapes, such as pipe insulation. Coatings, facings, or jackets usually are applied for reflective, vapor-barrier, or decorative purposes.

In another process (replacing the high-temperature blast-jet process in some areas), a stream of molten glass is directed onto a rapidly rotating wheel which contains holes in its periphery. Centrifugal force directs glass through each hole to create fibers. A third process involves conversion of 2- to 6-in. chopped strand and textile-type yarns into separate and random monofilaments by a garnetting machine.

Properties of Staple Fibers: Fiber diameters range from AAA to G (Table F-5), with the largest volume in the range C–G. Thermal conductivity of glass-fiber products is influenced by fiber diameter, density or compactness of the fiber mass, and temperature conditions. Generally, thermal conductivity ranges between 0.20 and 0.80 (Btu)(in.)/(hr)(ft³)(°F). Temperature of applications ranges between near absolute zero to 593°C (1100°F), or to the softening point of the glass. Unbonded mat is used at extreme temperature conditions, whereas standard bonded insulation covers the range of −40 to 232°C (−40 to 450°F). Fiber-glass acoustical products are particularly good energy absorbers at the frequency levels of 500–2,000 Hz. Fiber diameter, density, and method of mounting control the absorbing characteristics.

—L. Dow Moore, *PPG Industries, Pittsburgh, Pa.*

Fibroin (See **Fibers;** and **Peptides and proteins.**)

Films (Consult the **Subject Index.**)

Films, Heat Conductance of (See **Insulation, thermal.**)

Filtration Filters used in chemical processing are classified into two distinct groups, continuous and intermittent. The principal types of filters in each of these categories are listed in Table F-6, grouped according to their primary function. The general form of the filter, the means of solids discharge, type of filter medium, and general applications are also given.

Continuous Filters. Filters that operate without interruption for weeks to months are classified as *continuous* units. The main types are the rotary drum and disk filters and various horizontal designs, represented by the circular-pan, tilting-pan, and traveling-belt types. All are operated primarily as vacuum units, although the rotary drum or disk filters are available in totally enclosed designs for pressure operation. Application of the latter generally is limited to special situations (volatile solids, toxic materials, and high temperature) because of their greater mechanical complexity and cost.

TABLE F-6. *Filter Classification for Continuous[a] and Intermittent Types*

Function			Type of Filter	General Form	Solids Discharge Means
Concentrate Solids		Clarifi-cation			
Dry Cake[b]	Slurry				

				Continuous		

Dry Cake[b]	Slurry	Clarification	Type of Filter	General Form	Solids Discharge Means
x		x	Rotary drum	Horizontal revolving drum	
			Basic unit	Medium attached	Air blow + scraper
					Scraper on medium
					Roll pickup
					Wire
			String	Medium attached	Strings over roll
			Precoat	Medium attached + filter aid	Scraper
			Cell-less		Scraper + blowback
			Traveling medium	Medium attached	Gravity over roll + scraper
			Belt	Medium moving	
			Coils	Medium moving	Gravity over roll + tine bar scraper
			Top feed	Medium attached	Scraper
			Internal feed	Medium attached	Pulsating air blow + scraper
			Totally enclosed	Medium attached	Scraper
			Pulp unit	Medium attached	Various (roll, etc.)
x		x	Rotary disk	Vertical revolving disks	
				Medium attached	Air blow + scraper
					Water jet
					Roll pickoff
x		x	Horizontal	Circular moving pan(s) or moving belt	
			Rotating pan	Medium attached	Scroll
			Tilting pan	Medium attached	Gravity (pan tipping)
			Traveling belt	Medium moving	Gravity over roll
	x	x	Filter-thickener Filter press	Plate and frame	Bleed-off recirculation
			Rotating disk	Vertical parallel disks	Back flush
			Leaf	Vertical parallel leaves	Mechanical shock
			Tubes	Vertical	Back flush
	x	x	Ultrafilter[d] Membranes	Cartridge or assemblies Sheet, leaf, tubes, or hollow fibers	Bleed-off recirculation

Driving Force			Type of Medium					Typical Applications
Grav-ity	Vac-uum	Pres-sure	Pre-coat	Paper	Textile Fabric[c]	Wov-en Wire	Other	
Continuous								
	x				x	x		Sewage sludge, various chemicals (titanium, lime mud, dyes, citrate, etc.) catalyst, flue dust
	x				x		x	Cane mud
	x				x			Clay, red mud, electric-furnace dust
	x				x			Zinc residue, calcium carbonate, peanut butter
	x				x			Starch, gluten, carbonates, antibiotics
	x	x	x			x		Juices, wines, antibiotics, petroleum
	x				x			Adipic acid, fibers, coal, ore, carbonation mud
	x				x	x		Sewage and paper-mill sludges, gluten, flue dust, starch and various chemicals
	x						x	Sewage and paper-mill sludges
	x				x	x		Salt and similar crystalline materials
	x					x		Metallurgical concentrates (iron ore, etc.)
	x	x			x	x		Petroleum dewaxing, solvent slurries, catalyst, hazardous materials
					x	x		Fiber washing, thickening and recovery (kraft, sulfite and other pulps)
	x	x			x			Metallurgical slurries (taconite, copper, lead, etc.) coal, cement
	x				x	x		Fiber recovery (paper-mill save-all)
	x				x			Cement, paper sludge.
	x	x			x	x		Rapid-settling solids (sand, coal, salt cake, gypsum), pulp fibers
	x				x			Gypsum (phosphoric acid)
	x				x			Medium and coarse solids, fibers
		x			x			Pigments, catalysts (continuous reactors)
	x	x			x			Metal hydroxides, coal refuse
		x			x			First carbonation juice (beet sugar)
		x			x			White liquor slurry (recausticizing)
		x					x	Electrocoating paints, enzymes, proteins, fine solids

Concentrate Solids — Dry Cake[b]	Slurry	Clarification	Type of Filter	General Form	Solids Discharge Means
				Function	
			Intermittent		
x		x	Leaf Stationary	Horizontal or vertical assembly Rectangular	Manual or automatic
			Rotating	Circular, horizontal	Air sluice, rotating leaves
			Traveling-medium	Medium on open mesh belt	With medium into hopper
				Medium between chambers	Scraper beyond end roll
			Tube	Vertical assembly	Vibration
			Filter press	Multiple plate and frame	Manual + automatic
			Nutsche	Vertical cylinder, depth medium	Manual or sluice
	x	x	Leaf Stationary	Vertical parallel leaves	Sluice
				Horizontal parallel leaves	Tipping
			Rotating	Circular parallel leaves	Sluice
			Tube	Vertical assemblies	Backwash
			Deep bed	Vertical or horizontal tank	Backwash
			Sand or coal	Fixed bed	Manual or automatic
				Moving bed	Continuous
			Plate	Stacked circular plates, vertical	Manual
				Stacked rectangular plates, horizontal	Tipping
		x	Pressure filter[e]	Assembly or elements in vertical tank	
			Cartridge		
			Replacement	Cylindrical disk or plates	Discard medium
			Permanent	Cylindrical disk or plates	Manual backwash
			Sheet or film	Single or multiple stack	Discard medium
			Bag or sock	Single or multiple	Discard medium

[a] Based on uninterrupted operation for weeks to months without shutdown for medium changing. Intermittent (or batch) type filters by contrast operate for much shorter periods (hours to several days or so).
[b] Containing residual moisture.
[c] Natural or synthetic fibers, woven or nonwoven.
[d] Also nonslurries where membrane functions to retain (and concentrate) certain higher molecular weight solubles, e.g., proteins.
[e] Including microfiltration.

Driving Force			Type of Medium					Typical Applications
Grav-ity	Vac-uum	Pres-sure	Pre-coat	Paper	Textile Fabric[c]	Woven Wire	Other	
						Intermittent		
		x			x	x		Red mud, contact clays, pigments, nickel catalysts
		x			x	x		Pharmaceuticals, clay and zinc slurries
x	x			x	x			Machine coolants, cutting oils, sludges
		x			x			Machine coolants, cutting oils, sludges
		x				x	x	Chemicals wastes
		x		x	x			Pigments, dyes, carbon clay, various chemicals
x	x	x					x	Chemicals, wastes (small volumes)
		x	x		x	x		Water, chemicals, beverages, oils, resins, soaps
		x	x		x	x		Water, chemicals, beverages, oils, resins, soaps
		x	x		x	x		Water, waste streams, sugar, chemicals
		x	x		x	x	x	Water, chemicals, beverages
		x						
		x					x	Process water, process liquors
		x					x	Process water, process liquors
		x	x	x	x	x		Industrial liquors, vegetable oils, organic chemicals
		x	x	x	x	x		Industrial liquors, vegetable oils, organic chemicals
		x						Numerous industrial chemicals
		x		x	x		x	Fluids, oils, water, paint, etc.
		x				x	x	Fluids, oils, water, paint, etc.
		x		x			x	Fluids, oils, water, paint, etc.
		x			x			Fluids, oils, water, paint, etc.

Rotary Drum Filters: This is the most versatile and widely used continuous filter in the chemical process industries. The design provides means for concentrating slurry solids to dry (moist) cakes, washing solubles from such cakes when required, and producing a clarified effluent. The continuous drum filter was introduced commercially in the United States by E. L. Oliver in 1908.

Many design configurations have been developed to utilize most effectively the basic principle for the varying filtration characteristics of slurries, ranging from extremely fine particles (1–10 μm) to very coarse particles (50–150 μm) and from thin, sticky cakes to thick, fairly dry sludges. These filters are used in some form in practically every process operation involving slurries where solids must be recovered from the liquor for further processing or disposal. It is estimated that over 25,000 rotary drum filters are in operation worldwide (see Fig. F-4).

As shown by Fig. F-5, a horizontally positioned drum rotates partially submerged in a slurry-containing vat. With the filtering surface at the drum periphery, the necessary driving force is secured by applying a vacuum through a central valve on the drum shaft (or trunnion) to individual compartments or sections that provide support and drainage for the filter medium. The filter cake forms while the sections are immersed. Upon emerging, further dewatering occurs as air passes through the cake, displacing a good portion of the mother liquor. For more effective removal of any soluble solids in the mother liquor, wash water may be applied before final dewatering. The dewatered cake then is discharged by cutting off the vacuum and applying a reverse air blow. As

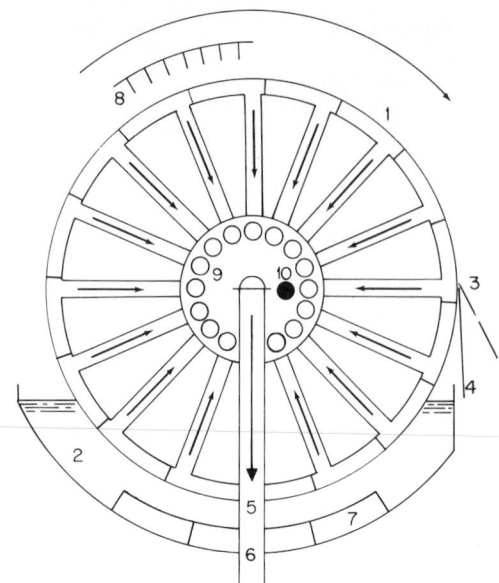

Fig. F-5. Basic elements of rotary drum filter. (1) Rotating drum, (2) slurry, (3) cake, (4) scraper deflector, (5) filtrate and air, (6) vacuum, (7) slurry agitator, (8) wash, (9) automatic valve, (10) air.

the cake separates from the filter cloth, it is deflected by a scraper blade to a discharge trough or conveyor below. Other means for cake discharge are described later.

The filter drum is compartmentalized at the periphery with a number of drainage sections (usually 12–24) individually sealed by division strips. Various types of drainage supports (wire mesh, plastic grids, ribs, and perforated plates) are used to support the filter medium and permit filtrate to flow laterally through the section to the outlet. Drainage piping in various arrangements (leading and/or trailing edge) provides for the flow of filtrate to the central valve on the filter trunnion.

The automatic valve provides the means for applying vacuum during the cake-forming, washing, and dewatering portion of the cycle while cutting off vacuum and permitting an air blow during cake discharge. The valve may be compartmentalized to permit separation of the wash liquor from the initial cake-forming filtrate. Filtrate separation, vacuum cutoff and cut-in as well as air blow are accomplished by movable bridges mounted in the annular channel of the valve.

One or more filtrate outlets from the automatic

Fig. F-4. Basic rotary drum filter. *(Dorr-Oliver Inc.)*

valve are connected to a vacuum source, usually a vacuum pump (either dry or wet type) except for certain applications, e.g., wood pulp, where it is customary to employ a barometric leg 20–30 ft high. When a dry vacuum pump is used, the two-phase stream of air and filtrate must flow to a receiver, where the air and liquid are separated, the air outlet at the top going to the vacuum pump for removal.

The filter medium (usually a natural or synthetic woven fabric) is fixed to the drum by applying a helical-path wire winding or steel strapping or by caulking into each division strip, permitting replacement of a single panel instead of the entire cover.

In operation, the drum revolves continuously at rather low speeds: 2–10 min/revolution for many applications, up to 4–5 rpm for free-draining wood pulp. Slurry is maintained in the vat at the desired level, for example, 25% submergence, with a slow-speed oscillating agitator to prevent settling and minimize classification. Vacuum applied to the outlet line creates the pressure differential causing the slurry to deposit solids on the filter medium and producing a cake of varying thickness depending on the permeability, applied vacuum, and drum speed.

When it is desirable to reduce the mother liquor in the cake to a low level, e.g., recovery is of chemical value or an undesirable constituent is removed, wash water is applied by sprays located at about the 10 o'clock position of the drum. The applied wash acts to displace the residual cake liquor and is referred to as *displacement washing*. It is more efficient than straight dilution due to a minimum mixing of the wash with the mother liquor. Following the wash operation, vacuum is retained so that the cake can be dewatered to provide low cake moisture.

The rotary vacuum filter operates continuously until blinding of the filter cloth affects performance, e.g., capacity, cake moisture, or washing results, sufficiently to necessitate changing the medium. Replacing the filter cloth usually involves a downtime of 4–8 hr.

The performance of a rotary vacuum filter is a function of two types of variables, those related to the machine and those applying to the process slurry. The basic machine operating variables are drum speed, vacuum, and submergence. Functional design characteristics may limit or enhance performance but essentially are fixed for the particular design. For example, filtrate flow rates are subject to hydraulic limitations imposed by the design of the grid supporting the filter medium, depth of grid, length of drainage section, sizes and locations of drainage piping, and flow restrictions in the automatic valve. Other design aspects include feed distribution, agitation, and discharge means.

The maximum capacity of any filter design, however, is achieved when the drum speed is at the maximum beyond which the cake is too thin for discharge. Thus, the combination of the maximum drum speed, submergence, and vacuum will provide the maximum output. This is true even if a cake-washing step is involved since the wash ratio remains the same due to the decrease in cake thickness with increasing drum speed.

The nature of the slurry, however, and the permeability of the cake-forming solids essentially govern the output of any filter. The solids concentration of the slurry and the viscosity of the mother liquor affect the cake-forming rate, varying directly with solids concentration and inversely with viscosity. Slurry temperature is significant only as it affects viscosity and the extent of flashing with increasing vapor pressure and higher vacuums.

The combined effect of machine and slurry variables determines the output of the rotary vacuum filter. The relationship of these basic variables (D. B. Purchas) is given by

Output (dry solids per unit time)

$$= \left(\frac{2WP^{1-s}T_f}{\eta r_o T_c}\right)^{0.5}$$

where W = weight of dry solids per unit volume of slurry
P = pressure differential
s = cake-compressibility factor (0–0.9)
T_f = cake-form time*/cycle
η = viscosity (liquor)
r_o = specific cake resistance at zero pressure
T_c = cycle time per revolution

Advantages: These include continuous operation until the filter medium requires changing (several weeks to months), adaptability to many types of slurries, effective washing with minimum dilution, dry-cake discharge (moisture content ranging from 20 to 70%), minimum of operating attention, and low maintenance.

Disadvantages: Installation costs sometimes are high, depending upon location, especially of aux-

---

*A function of drum submergence and valve bridging in relation to total cycle time.

iliaries (vacuum pump, receiver, filtrate pump, and cake conveyor) and extent of piping and electrical work required. The periodic filter-medium changing results in 2–6 hr downtime, depending on the method of attachment. Thin, sticky cakes present discharge problems, and variations in filtering characteristics of solids can affect capacity, cake moisture, and washing results.

Limitations: The filters generally are limited to slurries containing over 1% solids (except for precoat type) with moderate settling rates because rapid-settling materials present pickup problems and reduced capacities if classification occurs. Very slow filtering materials, such as slimes, cannot be handled unless applicable to precoat operation. High filtrate clarities usually are not obtainable except with precoat type. Slurry temperatures must be limited to around 71–88°C (160–190°F) due to flashing problems, depending on vacuum used (5–25 in. Hg). Thus, volatile solvents are not handled except in a pressure arrangement.

Size Ranges: Various diameters and face lengths provide a broad range of filtration areas of 10–1,500 ft² or more in diameters up to about 14 ft and lengths up to 34 ft or more. These larger units are used on wood-pulp or special metallurgical applications, including brown-stock washing, and uranium, gold, and copper applications. Some units as large as 18 ft diam have been field-assembled for special metallurgical applications. The primary materials of construction are mild steel and stainless alloys. Rubber-covered drums are obtainable in certain sizes. All-plastic units (polyester or polyvinyl chloride) are available in smaller sizes (4–6 ft diam).

Variations: A number of modifications of the basic rotary-drum unit have been devised for specific applications or to provide improved performance with respect to filtrate clarity or filter-medium life. Several of these forms are listed in Table F-7, along with advantages, disadvantages, limitations, and typical applications.

String-Discharge Filter: One of the earlier modifications provides a means for discharging thin ($\frac{1}{16}$–$\frac{1}{8}$ in.) and somewhat sticky cakes. A series of endless strings $\frac{1}{2}$–1 in. apart are wrapped around the drum and two external rolls. The strings lift the cake off the drum at approximately 60° beyond top center. When the cake and embedded strings reach the discharge roll, the strings separate from the cake as their direction is reversed and the cake moves on to discharge by gravity. The amount of inactive area depends on the point where the strings leave and return to the drum surface, but generally it is slightly greater than the conventional drum unit with a blow discharge.

Top Feed: Since a conventional drum filter cannot pick up rapid-settling solids, feeding on top of the drum was devised especially for crystal materials where hot-air drying was also needed to reduce the moisture content to a low level. This led to a combination dewatering and drying unit.

For hot-air drying, the filter is enclosed and heated air introduced so that cake moistures of 3% or less can be obtained. The slurry is fed near the top of the drum (40° before top center) through a distributor (adjustable riffle board) and is confined between the end flanges of the drum and flexible seals on the feedbox. The cake is discharged at several points, for example, 4, 6:30, and 9 o'clock positions, by means of three scrapers removing $\frac{1}{4}$-in.-thick layers with each cut when glazing from the hot air occurs (as with salt) and reduces porosity. Wash can also be applied, but wash separation is not practical. The product discharges into a hopper below. This unit overcomes crystal breaking and dusting that occur when centrifuges and rotary driers handle products such as salt.

Vacuum Precoat: The need for high filtrate clarity with difficult-to-filter solids forming very thin, sticky cakes often gelatinous or colloidal in nature led to the development of the continuous vacuum precoat unit, a major modification of the basic rotary filter. Instead of the conventional filter fabric, it employs a precoat of diatomaceous earth or perlite (expanded volcanic ash) several inches or more in thickness. A thin cake formed on the precoat is continuously removed by a knife-edge scraper set to shave a thin layer (around 0.001–0.003 in.) with each revolution. The depth of cut is set slightly more than the actual penetration of solids in order to minimize precoat consumption and obtain the longest possible operating cycle before the precoat is all consumed. It is now customary to use precoats 4–6 in. thick to obtain longer cycles. With precoat cuts 0.0004–0.009 in. and drum speeds 0.3–3.0 min/revolution, operating cycles can vary from as little as 8–12 hr up to a week or more. The length of cycle is determined by

$$T_c = \frac{D_{pc} N_d}{C}$$

where T_c = time of cycle, min
D_{pc} = depth of precoat, in.
N_d = drum speed, min/revolution
C = precoat cut, in.

TABLE F-7. *Variant Forms of Rotary Vacuum Filters*

Types	Advantages	Disadvantages and Limitations	Typical Applications
String discharge	Discharges thin cakes if not too sticky; suited to slow filtering solids	Cannot always discharge sticky cakes; string maintenance; limited to relatively thin cakes; medium changing awkward and time-consuming	Gluten, starch, antibiotics
Top feed	High capacity for dewatering coarse crystals and solids; designed for hot-air drying	Limited washing capability; fines (especially under 5 μm) must be minimal	Salt, iron concentrates, bone char, sodium sulfate
Vacuum precoat	High clarity for difficult-to-filter fine or gelatinous solids at low concentrations	Cost of precoat material; relatively short cycles (12–48 hr); solids disposal	Juices, wines, antibiotics, lube oil, slop oil
Traveling coil.	Long-life medium (stainless coils), essentially non-blinding, continuous medium washing off filter; no blowback	Effluent high in solids; poor fines retention	Sewage sludge, paper-mill sludge
Traveling belt	Filter-medium blinding minimized by continuous washing, thinner cakes dischargeable; no blow-back; easier medium changing	Irregular clarities due to leaks at edges and joining seam, shorter effective cycle, higher cost, more maintenance	Sludge, organic and inorganic solids
Cell-less with internal shoe	Ability to discharge thin cakes, higher capacity especially with free-filtering materials; automatic valve, individual drum sections, and agitator eliminated; also external receiver	Limited sizes (up to 140 ft²); some leakage from air blow to vacuum side; maintenance on internal shoe; slurry recirculation disperses flocs	Crude adipic acid, cellulose and polyester fibers
without internal shoe	Simpler design, low cost; eliminates blowback	Periodic replacement of product precoat; lacks flexibility; generally limited to single-stage washing	Coal, ore, flue dust, sludge, first-carbonation mud (beet sugar)
Internal feed	High dewatering capacity for rapid settling solids; no slurry vat, simplified construction and operation	Limited face lengths (usually not exceeding diameter); unsuited for washing	Iron concentrates, various metallurgical slurries
Enclosed drum	Permits handling slurries with volatile solvents and hazardous chemicals in inert- or active-gas atmospheres	High cost, extensive auxiliaries, and complex operation	Petroleum dewaxing, polymers, insecticides, radioactive materials

At the end of the cycle, the filter is precoated again, requiring 30–60 min. A suitable woven wire fabric (usually 24 × 110 mesh) or a plastic monofilament (76 × 68 count) covers the drum to retain the precoat and permit the cake to build up to the desired thickness, usually from a 1% slurry. The knife advance is then set through mechanical adjustment, although a hydraulic system is sometimes used in European designs. An automatic stop is provided when the precoat cake has been reduced to about $\frac{1}{4}$ in. thickness.

Traveling Medium Units: In the conventional rotary drum unit, the filter medium is affixed to the drum. Plugging or blinding occurs progressively, and eventually replacement of the medium is necessary. Recognizing that continuous cleaning of the medium could extend its useful life, a moving or traveling medium that is washed externally, off the drum, was introduced in the early 1950s.

Coil Filter: One of the first commercially successful versions, the coil filter employs a series of parallel helically formed coils. Filtrate flows between the coils with the cake forming on top. The coils leave the drum at about the 1 o'clock position with the lower layer separating from the top layer. The cake is carried by the top coils to the discharge roll, where it breaks away from the coils and falls by gravity to a solids conveyor below. The coils are subjected to continuous washing by spray nozzles and are then returned to the drum. Blinding is essentially eliminated. Since the stainless coils are rugged, their life is generally guaranteed up to 10 years (or 41,600 operating hours).

Belt Filter: The use of fabric in place of metal coils or woven wire was introduced in the mid-1950s in an effort to lengthen filter-medium life by external washing on a continuous basis and to provide improved clarities over metal media. The general arrangement of external rolls, filter-medium wash trough and sprays, take-up roll, and other accessories is shown in Fig. F-6.

The filter medium is supplied in a single length to be closed by zipper or clipper belt lacing, or a suitable sewing machine is provided. Depending

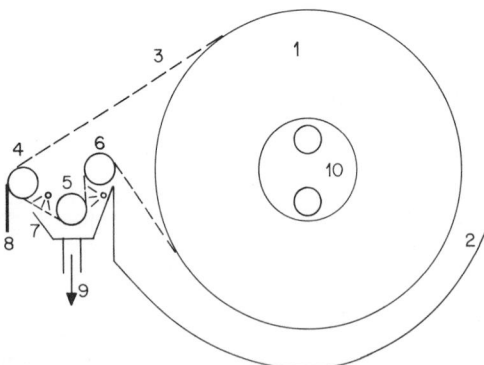

Fig. F-6. Rotary belt filter. (1) Filter drum, (2) feed tank, (3) filter medium, (4) discharge roll, (5) adjustable roll, (6) return roll, (7) wash trough, (8) cake deflector, (9) separate discharge to recycle or drain, (10) valve.

on the application and the effectiveness of the continuous washing, the life of the filter medium ranges from 5–6 weeks to 6 months or more. While various mechanical devices have been used for belt tracking, currently most designs utilize some type of edge gripper that engages a rubber extrusion fastened to each side of the filter-medium belt.

Pulp Filters: The rotary-drum principle as a means for washing and dewatering solids was applied to various pulp and paper operations in the early 1930s to take advantage of continuous processing. To handle pulp fibers, major design modifications were necessary due to the much greater hydraulic requirements and the problems related to discharging pulp sheets of varying character. The principal applications for vacuum drum units in the pulp and paper industry are (1) brown-stock washing (kraft, sulfite), (2) bleach washing, and (3) deckering (thickening).

In brown-stock washing, the spent pulping liquor must be recovered from the digested pulp with a minimum of dilution water used in the washing operation. This is accomplished by means of a multistage (2–4) rotary-drum washing system utilizing countercurrent washing. Vacuum is provided by barometric legs rather than conventional vacuum because of the large capacities involved. Feed to the washers is at a low consistency (around 1%) requiring a flow of 5,000 gal/min for a 300 tons/day mill. Washers are available to handle up to 1,000 tons/day or more.

In bleach washing, a similar washer is provided although the function differs in that the washing on each stage must be sufficiently complete to minimize carry-over to the next stage.

Deckering in the pulp and paper field is applied to dilute suspensions (0.5–1%) after screening or centrifugal cleaning to dewater the stock to 12–14% consistency for high-density storage. Vacuum washers are often employed for this service and follow the brown-stock washing operation after either subsequent screening or centrifugal cleaning. See also **Pulp (wood) production and processing.**

Internal-Feed Drum: An unusual arrangement for a rotary drum filter* is shown in Fig. F-7. Filtration takes place on the inside of the drum surface. Thus the feed is to the inside, obviating the need for the external slurry vat necessary with the conventional rotary drum. Drainage supporting grids are mounted between division strips,

*Commonly known as the Dorrco filter.

Fig. F-7. Internal-feed rotary drum filter. (*Dorr-Oliver Inc.*)

each section draining at the end by means of external piping to the filter valve. A slurry pool is maintained in the bottom to a depth of several inches. As the formed cake approaches top center, vacuum is cut off and a reverse, pulsating blow is gently applied to discharge the cakes to a chute below or sometimes a conveyor. The drum rests on external rolls and is driven by a variable-speed drive.

Totally Enclosed Filter: When the slurry comprises volatile or toxic substances, an open-type vacuum

filter is not suitable. A completely enclosed rotary-drum unit can be employed for a broad range of special applications involving various solvents and hazardous materials such as insecticides, carcinogens, and radioactive products. In addition, it can be designed to safely accommodate various atmospheres such as hydrogen, acetone, or similar vapors. Typical applications along with the various types of totally enclosed units are listed in Table F-8. For such applications, the basic rotary-drum unit is totally enclosed with a flanged top cover for access. Depending on the service, the outer shell may be designed for hood pressures of only a few inches H_2O (usually 5–10) or up to 50 psi working pressure.

When pressures only slightly over or below atmospheric are needed to prevent air-in leakage or vapor egress, a vacuum pump is used to produce 20–22 in. Hg and the necessary differential pressure. Inert gas (usually CO_2 or N_2 and CO_2) can be supplied to the hood by a blower and the same atmosphere employed for the blow on cake discharge. The cake is removed by a scraper, a scroll below providing the means for discharging through a single center outlet. When employed for dewaxing applications where methyl ethyl ketone is the solvent, the solvent-diluted lube-oil feed is chilled to $-45°C$ ($-50°F$) or lower. See also **Lubricating oils.**

For applications requiring above atmospheric pressure only, e.g., propane dewaxing, the hood enclosure must be designed for working pressures up to 50 psi. Consequently, it is of very heavy construction (wall thicknesses up to $1\frac{1}{2}$ in.), and cost is accordingly much greater. The rotary

TABLE F-8. *Types and Applications of Totally Enclosed Rotary-Drum Filters*

Type of Unit	Operating Pressure	Filter Medium	Typical Solvent	Application
Pressure	Up to 50 psi	Fabric	Propane	Dewaxing
			Water (121°C; 250°F)	Saturated solutions
	Up to 30 psi	Precoat	Water (fermentation broth)	Antibiotics
			Ether, toluene, hexane	Dibasic acid salts
Pressure or vacuum (vaportight)	2–10 in. Hg above or below atm	Fabric	Methyl ethyl ketone	Dewaxing
			Methanol	Polymers
	Vacuum side 20–22 in. Hg	Precoat	Oil (hot)	Clay
			Alkylated phenol	Clay

drum may be enclosed in a vertical-tank pressure vessel for simplifying pressure-tank design.

When the slurry to be filtered requires a precoat medium for reasons of clarity and/or cake discharge, a precoat-type unit is available. It follows the same basic design as the vacuum precoat unit except that the cake discharges by scraper to a scroll that expels the solids as a plug. Units are available in diameters of 3–8 ft.

Cell-less Rotary Drum: In the conventional rotary vacuum filter, the drum is sectionalized at the periphery so that the automatic valve on the drum shaft can provide the operational control (vacuum on or off) required. However, in the cell-less, or single-cell, design both the automatic valve and the individual collecting sections are eliminated, along with the internal piping. Practically the entire drum surface ~90% is used for filtering, washing, and dewatering. A unique feature is an internal shoe that permits an air blow for discharging the cake.

The inside of the drum acts as a receiver for separating air and filtrate, the latter being removed by an internal siphon extending through the shaft. This feature permits the operation of the filter with thin cakes, so that high drum speeds (up to 26 rpm) and substantially greater capacities result. It can be employed on both slow- and rapid-filtering materials although it has limited flexibility due to the elimination of the valve and individual section control. It is available in an open-type design or completely enclosed for vaportight or pressure operation up to 150 psi and temperatures of 177°C (350°F). Materials of construction are steel, stainless, nickel, Hastelloy, Monel, and rubber-covered. The United States version* is shown in Fig. F-8.

A cell-less unit without internal shoe is also in

*Known as the Bird-Young filter.

Fig. F-8. Cell-less rotary drum filter. (*Bird-Young filter, Bird Machine Co.*)

use (especially in Europe) due to the lower cost. Without the air-blow discharge, it must be operated as a precoat unit using the product for precoating and a conventional scraper to continuously peel off a thin layer of cake. Eventually (after 8–16 hr of operation), the product precoat becomes sufficiently plugged with fines to require renewing. The filter is stopped, and after the old precoat is dropped, a new precoat is applied. The filter is generally available in cast iron, steel, stainless steel, and rubber-covered.

Rotary-Disk Filters: The rotary-disk vacuum filter (originally known as the American filter) is an extended-area version of the rotary-drum unit. A series of parallel disks provides vertical filtering surfaces in place of the cylindrical surface on the rotary drum. Thus, for a given floor space, the rotary-disk unit provides substantially more filter area (up to 3.5 times) than a drum type.

The vertical disks are mounted on a shaft (trunnion) containing a number of drainage channels that connect to the automatic valve. Each disk comprises a number of individual sectors (usually 8–10) connected to the main shaft and held in place by tie rods, the outlet mating with the corresponding opening in the center shaft channel and gasket-sealed. The sectors are designed with a series of radial ribs or similar drainage means to provide for filtrate flow to the outlet. A perforated plate mounted over the ribs serves as support for the filter medium, which is bonded to the sector face (as with wire fabric) or made into bag form (as with woven fabrics), the bag being fastened around the outlet nozzle.

Since a sector must be completely immersed during the cake-forming portion of the cycle, the slurry depth is maintained at about 40% submergence. Vacuum applied through the rotary valve from the main filtrate outlet causes cake to form on the immersed sectors, and dewatering of the cake takes place after the sector emerges from the slurry and prior to discharge.

The cake is dislodged by applying a slight air blow (sometimes a sudden air blast referred to as a *snap blow*) that billows out the filter fabric and permits the scraper to peel away the cake while minimizing abrasion of the fabric. The slurry vat is crenelated to provide an open space, or chute, between the disks so the cake can be discharged to a conveyor below.

Since many metallurgical slurries, e.g., taconite, tend to settle rapidly, the slurry vat is designed to accommodate a paddle agitator to maintain the solids in suspension. For some applications, the agitator can be omitted.

With the filtering surface in a vertical position, cake washing for the most part is impractical and usually not attempted. For this reason, the disk filter is primarily a dewatering unit.

Normally, disk filters are furnished in mild steel and cast iron for general metallurgical applications and similar noncorrosive slurries. Where corrosion protection is required, stainless alloys (304 and 316 grades) are used. For pulp applications, sectors are offered in both stainless and plastic, the latter primarily to reduce weight for greater ease in handling and lower cost.

The rotary-disk filter is used primarily for dewatering where large tonnages or flows are involved and washing is not required. The greater area provided by the disk configuration over the drum (a factor of 2.25–4, depending on size) is an advantage in reducing plant space and installation costs, principally where multiple units are employed.

The disk design, however, imposes a number of functional restraints and thus limits flexibility compared to the rotary drum. The high submergence required to immerse the sector reduces the amount of the cycle available for drying; the ratio of dry to form time is 1:1, compared to 2:1 or more on the drum unit. Aside from the shortened dry cycle, the vertical position of the filtering surface precludes practical cake washing. Cake-discharge means are also limited with this configuration, the deflector-scraper generally predominating where cakes are $3/8$ in. or more. Rolls have been used for cement and similar applications but are expensive and somewhat troublesome.

Applications: The metallurgical field has provided the bulk of the applications, particularly in iron-ore processing and metal concentrates. Filtration rates cover a broad range: cement, 10–15 dry $lb/(ft^2)(hr)$; coal, 50–80; copper concentrate, 50–120; hematite, 200–300; and taconite, 300–500.

Disk Save-all: The disk filter also has found wide application in the pulp and paper field for recovering fiber from paper-machine white waters. Here, high flow rates must be handled to recover fiber present to the extent of 5–60 lb/1,000 gal. Flow rates per unit area are governed by the stock freeness and are 1–5 $gal/(ft^2)(min)$. The largest save-alls (12 ft diam × 24 disks—over 5,000 ft^2 filter area) can handle upward of 6,000 gal/min, depending on the application. Sweetener stock from the machine chest must be added to enable the unit to form an adequate sheet for discharge as well as to minimize effluent solids (usually less than 0.25 lb/1,000 gal) in the clear filtrate. Barometric legs rather than vacuum pumps are employed to produce the necessary vacuum. When the disk unit is employed as a save-all for fiber recovery from pulp- and paper-mill white waters, the pulp sheet is discharged by directing a small spray at the uppermost corner of the sector.

Size Ranges: Rotary-disk vacuum units are available in a range of sizes 4–12 ft diam with 1–20 disks, providing areas from a little over 100 ft^2 to 4,200 ft^2.

Horizontal Filters: When the filtering surface operates in a horizontal plane, the filter is designated as a horizontal type but identified usually by its characteristic form, e.g., stationary pan, tilting pan, or traveling-belt type. The characteristics and applications of these three main types are summarized in Table F-9.

Horizontal vacuum filters are employed primarily for rapid-settling solids, especially those which cannot be handled successfully on rotary drum or disk filters. They are also advantageous where cake washing is required, as displacement washing with high efficiency is readily achieved. Cake discharge varies with the design—gravity for the belt and tilting-pan types and mechanical plowing by scroll on the stationary-pan unit.

A broad range of sizes is available, the tilting-pan design providing the largest areas in a single unit. It is employed extensively in the fertilizer industry in filtering gypsum from phosphoric acid. The size range available for the various types of units is shown in Table F-10.

Stationary Pan: A typical stationary-pan vacuum filter (see Fig. F-9) comprises a rotating pan made up of a series of wedged-shaped filtering sections, a slurry-feed distributor, wash headers, spiral conveyor for cake discharge, and an automatic filter valve located beneath the pan to serve the usual functions of a continuous filter (applying or cutting off vacuum and air blow, collection and/or separation of filtrates). The design is shown schematically in Fig. F-10.

The filter cake is contained by an outer steel rim at the pan periphery. For discharge a spiral conveyor lifts the cake over the rim, leaving a thin heel cake. A modified form of the conventional pan* utilizes a flexible rubber belt for the rim. At the discharge point, the rim belt is moved away from the pan so that the cake can be readily discharged off the edge and the filter medium thoroughly washed. Materials of construction are steel, stainless steel, nickel, and Monel.

*Ucego.

TABLE F-9. *Advantages, Disadvantages, and Typical Applications of Horizontal Vacuum Filters*

Type	Advantages	Disadvantages and Limitations	Typical Applications
Stationary pan	Broad application for coarse, rapid-settling solids; stage washing; simplicity of operation; good accessibility	Size range generally to 24 ft diam (444 ft² — max., 1,000 ft²); scroll discharge leaves heel (1–4 in.) sometimes disadvantageous; filtrate separations from stage washing not always sharp	Caustic salt, sand, cotton linters, hardwood pulp, gypsum, aluminum hydrate, salt cake, potash, borax, catalyst, polymers, cottonseed meal
Tilting pan	Especially suited for very large capacities, e.g., 2,000 tons/day gypsum; complete cake discharge; filter medium washing after each cycle; stage washing with ready separation of filtrates	High cost; large floor space required; maintenance relatively high	Gypsum in phosphoric acid manufacture (including hemihydrate)
Traveling belt	Broadly applicable to coarse, rapid-settling solids for efficient stage washing, dewatering, and gravity discharge of cake, requires less floor space	Vacuum usually not above 15 in. Hg due to leaks at sealing edges; high maintenance costs with rubber drainage belts when used; size limitation (max. 460 ft² active area)	Salt, aluminum hydrate, pigments, crystallized materials, food processing

TABLE F-10. *Size Ranges of Continuous Horizontal Vacuum Filters*

Type	Diameter or Length, ft	Size	Active Filtration Area, ft²
Stationary pan	3–24 diam	18–20 sections	4–500*
Tilting pan	27–72 diam	12–30 pans	100–1,800
Traveling belt	4–40 long	1–12 ft wide	2–460

*Designs up to 1,000 ft².

Fig. F-9. Stationary-pan filter. (*Dorr-Oliver Inc.*)

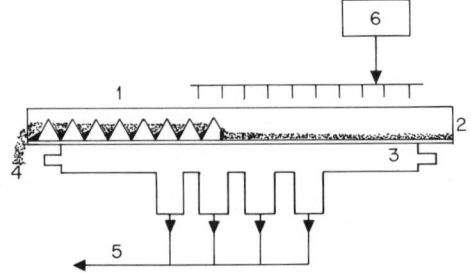

Fig. F-10. Schematic representation of horizontal pan filter. (1) Feed, (2) filter pan, (3) filter valve, (4) cake discharge, (5) filtrate, (6) wash water.

The pan filter can be employed for simple dewatering applications (sand and coal) or for washing in a three-stage countercurrent system, e.g., the filtration of gypsum from phosphoric acid. When volatile solids are involved, the entire unit can be hooded for pressures of 3–4 in. H_2O. Pressure units up to 30 psi have been designed but only in small sizes for a few applications.

Slurry is fed through a distributor toward a feed baffle to achieve a uniform slurry pool. A reverse air blow at the same point serves to dislodge the unremoved cake (referred to as *heel*) and remixes with the incoming slurry. This novel feature tends to minimize medium blinding and maintains capacity. One, two, or three stages of wash may be applied from suitably located wash headers depending on the cake dewatering characteristics. The washed and dewatered cake is removed by a scroll conveyor set at a close clearance ($\frac{1}{8}$–$\frac{1}{4}$ in.) to the filter medium. The residual cake is referred to as the heel. Vacuums up to 25 in. Hg can be obtained. Filter medium (woven wire or synthetic fabric) is affixed to the individual sections by caulking in the section grooves.

Tilting-Pan Filter: The tilting-pan filter design is a circular arrangement of trapezoidal pans which rotate about a center island housing an automatic valve to control vacuum application and filtrate separations. The individual pans are mechanically rotatable about their horizontal axis for complete cake discharge. The filtrate outlet line serves as the axis for rotation and connects to the radial outlet from the valve through a flexible coupling. The pans ride on rollers. At the discharge point, a series of cams on the trackway causes the pan to turn 180° and discharge its cake. Sprays thoroughly wash the filter medium before the pan is returned to its normal position.

Feed is applied through a distributor over the length of the pan. Wash is similarly applied at various points to provide three-stage countercurrent washing as the individual filtrates are diverted by the automatic valve to corresponding receivers.

Tilting-pan filters are available in sizes up to 1,800 ft^2 and have primarily been employed in phosphoric acid filtration (filtering gypsum), where large amounts of acid must be handled (up to 2,000 tons/day). The large units are mechanically complex but practical, as they function generally without excessive maintenance and permit efficient three-stage countercurrent washing with minimum dilution of the filtrate. Gravity discharge of the cake eliminates problems of medium wear, while washing after each cycle minimizes blinding and lengthens cloth life.

Horizontal Traveling Belt: The horizontal belt filter comprises a combination of a filter medium and a drainage belt traveling over a series of fixed vacuum boxes, as shown schematically in Fig. F-11. The filter medium in the form of a belt is carried by a rubber transmission belt driven by the front drum roll. Depending on the design, the transmission belt may be fabricated with suitable transverse ribs and outlets to provide channels for drainage of filtrate to the vacuum chambers underneath, or individual drainage members of rubber or plastic may be attached to the transmission belt.

Feed is applied by means of a distributor. Side dams along the filter belt contain the slurry until dewatering and washing are completed. Cake discharges by gravity as it passes over a small roll located beyond the large drive roll. On the return side below, the filter medium is washed by sprays applied to both sides.

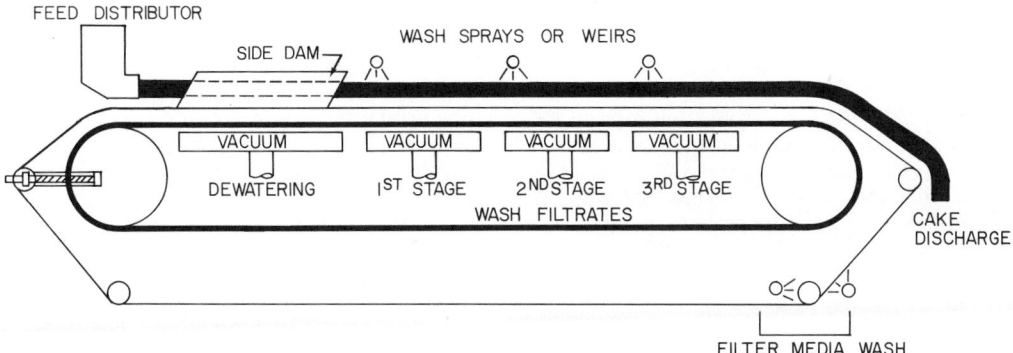

Fig. F-11. Horizontal traveling-belt filter.

This type of filter is well suited to coarse, rapid-draining solids, especially where efficient washing is required, either single or multistage. Sharp separation of filtrates can be attained. The filter can also be applied to finer materials where washing is required.

Filter-Thickeners: When a filter is employed to thicken slurry and at the same time produce clarified filtrate, it is classed as a filter-thickener. In this respect, it is duplicating the principal function of a sedimentation-type thickener. Application of a pressure differential accelerates the process compared to gravity settling and thus permits substantial reduction in the size of equipment required. It does so, however, at the expense of increased maintenance and operating costs, depending on the complexity of the device and the periodic replacement and/or cleaning of the filter medium due to blinding. Filter thickeners have enjoyed very limited success, except for their application to the beet sugar field (first-carbonation mud) in Europe. The basic types are the filter press, stationary and rotating leaves, rotating disk, and multitube.

The *filter-press* type utilizes a special plate with spiral channels that permit the slurry to traverse the entire area before exiting. By maintaining channel velocities of 5–15 ft/s solids deposition on the filter medium is minimized. As the slurry flows over the porous medium, a portion of the liquor flows through the medium to the opposing collecting plate. With continued recirculation of the slurry, thickening occurs, or if desired, the unit may be employed for dilution washing by continuously adding wash water to the main slurry tank. Available sizes range to 36-in.-wide plates with maximum area of about 360 ft². Its advantage rests in the simplicity of the plate-and-frame design along with a relatively low cost. Disadvantages are the labor time for changing the medium and the tendency for solids to deposit on the medium if velocities are insufficient.

The *stationary-leaf design*[*] was introduced in Europe for filtering first-carbonation mud in the beet sugar field. The unit operates at pressures up to 1 kg/cm² and on a relatively short cycle (20–30 min). Sludge is discharged periodically (every two cycles) from the bottom. As the slurry level falls, air is sucked into each leaf via the filtrate-effluent line and thus provides an air backwash of the medium (air purge) that is claimed to have a beneficial cleaning effect. The chief advantage is low cost and simplicity of operation; chief disadvantages are the discontinuous cycle, periodic cloth changing, and initial filtrate turbidity. More recently, a curved four-sided vessel design for high-pressure operation (to 2 kg/cm²) has been introduced.[†]

The *rotating-leaf* filter-thickener has been applied also to thickening and clarifying of first-carbonation mud, principally in European beet sugar mills. It comprises a series of vertical leaves hinged and rotating about a center axis in a pressure vessel. Each leaf outlet is connected to an internal automatic valve so that flow can be cut off during cake discharge. The latter is accomplished by mechanically holding one leaf back as the others are rotating. Sudden releasing causes the leaf to move forward rapidly and hit a mechanical stop, and the resulting shock dislodges the cake. Sludge is collected in a bottom extension of the main pressure vessel. The advantage of this design is the high ratio of filtering surface to tank volume and continuous operation. Its disadvantages lie in its mechanical complexity.

The *rotating-disk* filter-thickener employs a series of parallel disks mounted on a central shaft. The disks are made up of individual sectors, the same as used on rotary-disk vacuum filters. In essence, the unit is simply a rotary-disk filter completely submerged. The cake formed in deliquoring of the slurry is discharged by back-flushing each sector when it reaches its lowermost position. Sludge is collected in the cone bottom of the tank and periodically withdrawn. Units are furnished for either pressure or vacuum operation, the latter requiring only an open tank. The automatic valve is the same type as used on rotary-disk filters. Its advantage primarily rests in the use of proved components (disk and valve). Disadvantages are high cost and relatively low ratio of filtering area to volume compared to other units and the time consumed in changing media. Back-flushing flat filter surfaces also presents potential difficulties.

The *tube-type* filter-thickener comprises a multitube arrangement in a pressure vessel. Continuous operation is achieved by sectionalizing the drumhead and employing an enclosed rotating automatic valve that directs back-washing liquor (clarified) to each compartment on the desired time cycle in order to dislodge the cake and clean the medium. Cake settles to the bottom, where pumping out is regulated by a density sludge-level control device. This type of unit has been em-

---

[*]Grand Port.

[†]DDS-Copenhagen.

ployed to a limited degree for clarifying and thickening lime-mud slurry in recausticizing, replacing the conventional thickener. Its principal advantages are continuous operation, high effluent clarity, and greatly reduced space requirements. Disadvantages are periodic shutdowns for medium cleaning (acid) and changing and greater sensitivity to process changes that affect filterability of solids.

Intermittent Filters. Filters whose continued operation is limited by their solids-collecting capacity generally have been classified as batch-type units but are more appropriately designated as intermittent filters. Their cycle must be interrupted periodically when the solids accumulation has to be discharged or the filter element discarded. Filters in this broad category include pressure and vacuum leaf, tube or cylindrical element, filter press, deep-bed media, and permanent or disposable media.

Membrane Filters. The use of membrane filtration techniques to retain smaller particles than is possible on conventional filter cloth, e.g., microfiltration, ultrafiltration, and reverse osmosis, is covered under **Membrane filters.** See also **Water treatment.**

REFERENCES

Flood, J. E., H. F. Porter, and F. W. Rennie: Filtration Practice Today, *Chem. Eng.*, June 20, 1966, pp. 163–181.

Perry, R. H., and Cecil H. Chilton (eds.): "Chemical Engineers' Handbook," 5th ed., sec. 19, pp. 87–101, McGraw-Hill, New York, 1973.

Purchas, D. B.: "Industrial Filtration of Liquids," Leonard Hill, London, 1967.

Smith, W. C., and R. C. Giesse: Filtration Equipment Design, *Ind. Eng. Chem.*, vol. 53, no. 7, pp. 538–545, July 1961.

—Harry V. Miles, *Dorr-Oliver Incorporated, Stamford, Conn.*

Finishes (See **Insulation, thermal;** and **Paints.**)

Fire Retardants (See **Acrylic fibers; Bromine; Insulation, thermal;** and **Urethanes.**)

Fission Fission is a process of splitting an atomic nucleus into two or more nuclei of lighter elements, with the release of substantial amounts of energy. Fissionable materials include ^{235}U and ^{239}Pu. Fission products usually are highly radioactive.

Fission Products (See **Nuclear power plants.**)

Fixation (See **Ammonia; Fertilizers;** and **Nitrogen.**)

Fixed Beds (See **Adsorption.**)

Flaking (See **Agglomeration; Detergents;** and **Soaps.**)

Flames (See **Fuels;** and **Oxygen.**)

Flame Retardants (See **Acrylic fibers; Bromine; Insulation, thermal;** and **Urethanes.**)

Flares (See **Pyrotechnics.**)

Flashing (See **Distillation.**)

Flavors (See **Terpenes.**)

Flax (See **Fibers.**)

Flaxseed (See **Expression.**)

Flint, Sparkling (See **Lanthanum.**)

Flocculation (See **Acrylamide polymers; Liquid-solids separations; Paint;** and **Waste treatment.**)

Flotation Flotation is a means of separating a relatively small particle from a liquid medium. The particle may have a specific gravity greater than, less than, or the same as the liquid from which it is floated. The particulate floated may be a solid, an emulsified liquid, or both.

There are essentially two fundamental requirements in any flotation process: (1) a gas bubble and particle must come in contact with each other, and (2) the particle must have an affinity for the bubble. All advancement in the art of flotation from the first recorded patent by Haynes in 1860 has been toward these ends.

To achieve the first requirement, various methods of bubble production and particle agitation have been used. Over the years, in mineral flotation, these have resolved themselves into two basic methods:

1. *Dissolved gas, impeller agitation.* Gas under pressure is sparged into the bottom of a vessel in which an impeller mixes the rising bubbles with the agitated particles.

2. *Self-induced gas, impeller agitation.* The impeller is so positioned in the liquid that it inspirates ambient gas into the liquid as bubbles. These

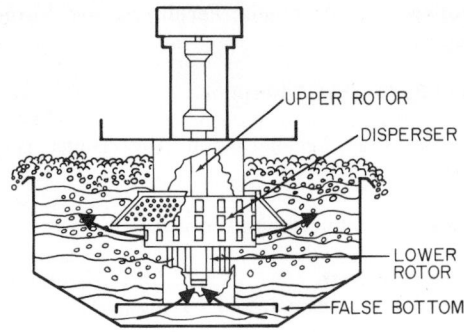

Fig. F-12. Sectional schematic of flotation cell. Upper portion of rotor draws air down the standpipe for thorough mixing with pulp. Lower portion of rotor draws pulp upward through rotor. Disperser breaks air into minute bubbles. Larger flotation units include false bottom to aid pulp flow. (*WEMCO Division, Envirotech Corporation.*)

bubbles are brought into contact with the agitated particle at the impeller's most dynamic zone (see Fig. F-12). In recent years, this has been the accepted method, brought about by advances in agitator mechanism design.

Reagents have been developed to satisfy the second requirement. Reagents are classified in five basic categories: (1) collection, (2) conditioning, (3) levitation, (4) frothing, and (5) depressant. Within each of these categories a number of reagents accomplish even more specific functions.

Frothers* generally are chemicals whose molecules contain both a polar and a nonpolar group. Recent developments include frothers which are completely miscible with water. The express purpose of a froth is to carry mineral-laden bubbles for a period of time until the froth can be removed from the flotation machine for recovery of its mineral content. Typical frothers are alcohols, pine oils, cresylic acids, eucalyptus oils, and camphor oils, which are slightly soluble in water. Soluble frothers in common use are represented by alkyl ethers and phenyl ethers of propylene and polypropylene glycols.

Collectors are reagents which selectively coat the particles to be floated with a water-repellant surface that will adhere to air bubbles. Collectors generally are classified as cationic, anionic, or nonionic. Examples include the xanthates, dithi-

*Definitions of frothers, collectors, and depressants were prepared by J. Wesley Burgess, Francis Carroll Moran, and Gailen T. Vandel, Fluor Utah Engineers & Constructors, Inc., San Mateo, Calif. and particularly reflect the practice of flotation in minerals beneficiation.

ophosphates, thiocarbonilides, and thionocarbonates, all of which are anionic collectors for sulfides; fatty acids and soaps, which are anionic collectors for nonsulfides; and amine salts, which are cationic collectors for nonsulfides.

Reagents used as depressants are chiefly inorganic salts, which compete with the collector for position on the sulfide surface. This permits the separation of one sulfide mineral from another. For example, in an alkaline solution, the addition of NaCN prevents flotation of sphalerite and pyrite by xanthates but not of galena, thus producing a higher grade of galena concentrates. The cyanide solution does not permanently affect the flotability of sphalerite, as it can be floated by adding $CuSO_4$ and xanthate.

Activators are reagents which alter the surface of a sulfide so that it can absorb a collector and float. The most widely used activator is $CuSO_4$. For example, xanthate as a collector will not readily float sphalerite, but the addition of $CuSO_4$ to the pulp changes the surface of the sphalerite particles to CuS. Xanthate then will readily float the activated sphalerite, as it behaves similar to CuS.

Uses: Although flotation was developed as a separations process for minerals processing and applies to the sulfides of Cu, Pb, Zn, Fe, Mo, Co, Ni, and As and to the nonsulfides, such as phosphates, NaCl, KCl, iron oxides, limestone, feldspar, fluorite, chromite, tungstates, silica, coal, and rhodochrosite, flotation also applies to nonmineral separations. The process is now being used effectively in waste disposal, particularly in connection with petroleum waste-water cleanup. For example, in one installation, a flotation process† with a rated capacity of 10,300 bbl/day of petroleum waste water is charged with a feed that contains an oil content of 75–115 ppm and discharges an effluent with an oil content reduced to 2–4 ppm. Horsepower requirements for the unit are 20.5.

REFERENCES
Booth, Robert B.: Froth Flotation, *American Cyanamid Co., Mineral Dressing Dept., Pub.* 21, New York, 1967.
"Flotation Fundamentals," The Dow Chemical Company, Midland, Mich., 1968.
Gaudin, A. M.: "Flotation," 2d ed., McGraw-Hill, New York, 1957.

—John R. Meidinger, *WEMCO Division, Envirotech Corporation, Sacramento, Calif.*

†Depurator, WEMCO Division, Envirotech Corporation.

Flotator Clarifiers (See **Sedimentation.**)

Flue-Gas Treatment (See **Absorption, acidic gases; Air pollution;** and **Sulfur dioxide removal.**)

Fluid Catalytic Cracking Introduced during World War II, fluid catalytic cracking progressively displaced the earlier thermal-cracking processes to a very large extent. See also **Thermal cracking.** Early catalytic-cracking processes used a fixed-bed cyclic system. That arrangement has been displaced by fluid catalytic-cracking units, which account for a capacity of 3.993 million bbl/day of fresh feed out of a total of 4.684 million bbl/day in the United States in 1970. The remaining capacity is made up of moving-bed catalytic-cracking units. The total free-world catalytic-cracking capacity is about 6 million bbl/day.

Catalytic cracking is used mainly to create gasoline, C_3/C_4 olefins, and isobutane primarily by the selective decomposition of heavy distillates. Because the cracking reactions are directed by specially prepared catalysts, the gasoline produced contains substantial proportions of high-octane-number hydrocarbon components, such as aromatics, branched paraffins, and olefins. Since the cracking reaction proceeds in accordance with the carbonium-ion mechanism, there are relatively minor amounts of fragments lighter than C_3 in the products. See also **Catalysts.** This result contrasts with the decomposition of hydrocarbons in thermal cracking by the free-radical mechanism, in which relatively large proportions of fragments lighter than C_3 are produced. This difference is illustrated by Table F-11, in which the typical proportions of C_1, C_2, C_3, and C_4 hydrocarbon fragments contained in the products of the two processes are shown. Some hydrogen is produced by both processes in varying amounts.

Cycle Oil: Catalytic cracking also produces a substance known as *cycle oil.* This is the distillate which boils above gasoline. A part of the cycle oil can be considered as a synthetic, cracked material that boils between the end point of the gasoline and the initial boiling point of the feedstock. The heavier portion of the cycle oil that falls within the boiling range of the feedstock represents the more refractory, uncracked components of the feed which are predominantly aromatic in nature. The cycle oils, withdrawn as net products from catalytic-cracking operations, are useful as components in heating oils, feedstocks to hydrocracking units, and for blending with heavy residuals to reduce viscosity; highly aromatic cycle oils are appropriate feeds for carbon-black manufacture.

Catalysts: Materials used as catalysts in modern catalytic-cracking units generally are crystalline in nature and sometimes are referred to as *zeolitic catalysts,* since they are modified hydrated alumina silicates. They have proprietary compositions. These catalysts, introduced in the early 1960s, offer improved stability over the powders, pellets, extrudates, and beads, which were synthetic amorphous silica-alumina composites or specially treated natural clays. The petroleum industry in the United States alone consumes approximately 100,000 tons/yr of catalysts in cracking units.

The catalysts used by fluid units are spray-dried microspheres. The average particle size of the equilibrium catalyst inventory in a fluid unit is typically 60 μm diam with nominally 10% by weight of the particles being smaller than 40 μm and 10% being larger than 105 μm. In circulation within the catalyst system, the microspheres gradually are reduced in size by formation of fines which leave the unit in the regenerator combustion-gas stream. Fresh catalyst is added to make up for these losses. For many units, such replacement turns out to be close to the economic optimum to sustain the catalyst activity and selectivity.

Fluid Catalytic-cracking Process: As shown by Fig. F-13, a typical fluid catalytic-cracking unit comprises (1) reactor, (2) regenerator, (3) main fractionator, (4) air blower or compressor, (5) spent-catalyst stripper, (6) catalyst recovery equipment, including (a) cyclones internal in the reactor and regenerator, (b) slurry settler, (c) an optional electrostatic precipitator, and (7) gas-recovery unit.

The feedstock, which may be preheated by exchange or in some cases by a fired heater, along with recycle from the fractionation section meets

TABLE F-11. *Typical Proportions of C_1, C_2, C_3, and C_4 Hydrocarbon Fragments in Products of Thermal and Catalytic-cracking Processes*

	Mole Percent in C_1–C_4 Fraction				
	C_1	C_2	C_3	C_4	Total
Thermal cracking . .	32	21	24	17	100
Catalytic cracking . .	12	11	29	48	100

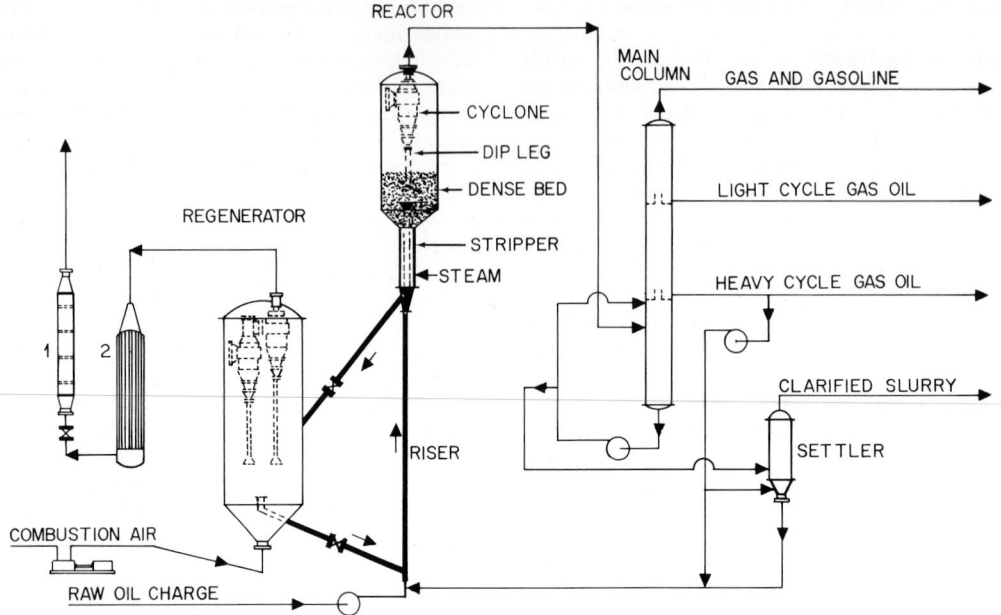

Fig. F-13. Fluid catalytic-cracking process. (1) Pressure-reducing orifice chamber, (2) flue-gas steam generator. The gas and gasoline from the main column go to the gas-concentration plant. (*Universal Oil Products Company.*)

a controlled stream of hot, regenerated catalyst. The resulting mixture of vaporized oil and catalyst ascends in the riser at a velocity such that the catalyst particles are suspended more or less discretely in a *dilute phase*. A large portion of the cracking occurs in the riser. Upon reaching the reactor, the linear velocity of the hydrocarbon vapors is reduced to such an extent that most of the catalyst settles out to form a *dense-phase* bed, the amount or height of which can be controlled by the valve in the spent-catalyst line leading to the regenerator. The remainder of the desired conversion is accomplished as the hydrocarbon vapors pass through the dense bed.

Above the dense bed in the reactor is another dilute phase of suspended fine catalyst particles, most of which are separated from the cracked vapors by one or more stages of cyclones in the upper portion of the reactor vessel. The separated particles are returned to the dense phase through a dip leg. The cracked vapors, carrying a minor concentration of catalyst fines, pass to the main fractionator, where they are distilled into several products. The column overhead product is separated by flashing into gas and unstabilized gasoline streams. These overhead products are routed to a gas-recovery unit, in which there can be

produced a debutanized gasoline, a C_3/C_4 cut as feed for alkylation, and a fuel-gas stream containing C_2 and lighter components. The steam used for stripping the spent catalyst and cycle oil is removed as water from the overhead separator (not shown).

The total cycle oil can be resolved in the main fractionator into a light cycle oil, which usually boils between 205 and 345°C (401 and 653°F), for ultimate use as a heating oil or diesel fuel. The light cycle oil generally is steam-stripped to remove light ends for flash-point control. The heavy cycle oil, boiling above 345°C (650°F) and removed as a side cut from the column, generally is recycled to the reactor, as shown in Fig. F-13. When the conversion is 75% or more, the heavy material is recycled to extinction, although some may be withdrawn as a net product. In catalytic cracking, the percent conversion of the feedstock has been reckoned by tradition as 100 minus the volumetric percentage yield of the total net cycle-oil product (100% minus the sum of light cycle oil, heavy cycle oil, and clarified slurry oil).

The column bottoms contain whatever catalyst fines escaped recovery by the cyclones. Generally, a stream is recycled to a higher point in the column to wash catalyst fines from the incoming vapors and then in most cases through heat exchange to heat colder liquid streams or to generate steam. A portion of the column bottoms stream is routed to a settler, in which catalyst fines are separated for return to the reactor as a concentrated settler bottoms stream. The settler overhead is a catalyst-free clarified oil which is withdrawn as a net product, generally in an amount up to about 5 vol % of the feedstock. This is a highly aromatic, refractory product that resists cracking. The clarified oil is useful as a cutter stock to reduce viscosities of heavy fuel oils.

The spent catalyst, which is the dense bed in the reactor, descends by gravity into a stripper where hydrocarbon vapors that are adsorbed, or within the interstices of the catalyst particles, are removed by a countercurrent flow of steam. The stripped catalyst descends through a control valve into the regenerator, in which the catalyst deposit is burned off. Air is introduced by a centrifugal compressor. The residual carbon on the regenerated catalyst is about 0.2 wt % of the catalyst.

As in the reactor, the dilute phase above the dense bed contains suspended fine catalyst particles, most of which are removed by two stages of cyclones and returned to the dense phase through dip legs. The combustion products, traditionally termed *flue gas,* are virtually free of oxygen and contain nearly equal amounts of CO and CO_2, along with a certain amount of steam formed by the combustion of a low-hydrogen-content catalyst deposit, brought into the regenerator as occluded stripping steam and as humidity in the combustion air.

The flue gas that leaves the regenerator through a pressure-control valve can be routed through a variety of auxiliary equipment, such as a heat exchanger for steam generation (shown in Fig. F-13). A CO boiler (not shown) is sometimes used, in which steam is generated by the combustion of the CO in the flue gas and other external fuel. Or the flue gas may be routed through an electrostatic precipitator for final removal of catalyst particulates where air pollution is a critical matter. In some instances, the cleaned flue gas may go to gas turbines for pressure letdown and the recovered energy used for driving electrical generators.

The fluid catalytic-cracking unit is operated as a heat-balanced system. The chemical heat of combustion of the catalyst deposit supplies (1) the heat required to vaporize and crack the oil coming to the riser, (2) the heat required to raise the temperature of the air to the flue-gas temperature,

and (3) the heat required to make up losses due to radiation. If the amount of catalyst deposit burned, the amounts and temperatures of flowing liquid and gas streams, and the temperature of the two dense catalyst beds are known, heat balances can be written for the reactor and regenerator and solved for the rate of catalyst circulation required to satisfy the heat balance. A unit processing 40,000 bbl/day of feedstock will circulate about 5 million lb/hr of catalyst between the reactor and regenerator in a typical instance. A typical unit is shown in Fig. F-14.

Process Variables: Because several process variables can be manipulated, the fluid catalytic-cracking process is quite flexible, with an ability to change the relative amounts (and to some extent the quality) of the products made and to process a wide variety of feedstocks. All the variables, including those described briefly here, are interacting and can be manipulated to vary the feedstock conversion, the relative amounts of gas, gasoline, and distillate fuels produced, the olefin content of the C_3 and C_4 fractions, and to some extent the octane numbers of the gasoline. Some of the major variables are as follows.

Fig. F-14. Fluid catalytic-cracking unit. (*Universal Oil Products Company.*)

*Weight Hourly Space Velocity.** This factor can be changed within limits by controlling the amount of catalyst in the dense bed of the reactor and is accomplished by altering the position of the slide valve in the spent-catalyst line to the regenerator. With higher inventories of catalyst in the reactor dense phase the conversion can be increased to a certain extent. Because of "back-mixing" effects in the dense phase, the conversion efficiency is not as high as it is in the riser, in which the catalyst and hydrocarbon vapors flow more or less concurrently.

Reaction-Zone Temperatures (Riser and Reactor): These temperatures are varied and controlled largely by adjusting the rate of flow of hot regenerated catalyst relative to the rate of flow of the combined feed (c/o ratio). Control is effected by adjusting the position of the slide valve in the regenerated-catalyst line leading from the regenerator to the riser. Higher temperatures in the reaction zone have the effect of (1) increasing conversion, (2) making the products more olefinic, and (3) at a given conversion rate reducing the ratio of gasoline to gas produced. Reaction temperature also can be increased by raising the preheat temperature of the hydrocarbon combined feed entering the riser. By raising the hydrocarbon preheat temperature, a given reaction temperature can be achieved at a lower c/o ratio.†

Residence Time: Although more of a design factor than an operating variable, conversion can be increased by raising the residence time in the riser.

Amount and Type of Recycle from Main Column to Riser: This is a manipulative variable. At low conversions (approximately 50%), generally the only material recycled to the reactor is a small amount of settled slurry concentrate that is carried to the reactor by a diluent stream of heavy cycle oil or feedstock. At high conversions (approximately 90%), heavy cycle oil is recycled to extinction. The amount recycled relative to fresh feedstock depends primarily on the activity of the catalyst. The cut temperature between the light and heavy cycle oil can also be varied, depending on the end-point specification desired for the light-cycle-oil net product. Increasing the recycle ratio of cycle oil to fresh feedstock tends to increase conversion at a given temperature because the c/o ratio based on fresh feedstock thereby is increased.

*Pounds per hour of oil feed to the reactor per pound of catalyst inventory in the reactor (and riser).

†Pounds per hour of catalyst circulation per pound per hour of oil feed to reactor.

Type and Condition of Catalyst: Compared with the older, amorphous catalysts, the zeolitic catalysts are more stable, and, other conditions being the same, these newer catalysts maintain a higher equilibrium activity. Thus, a much higher proportion of the total conversion is obtainable in the riser at higher gasoline-producing efficiency. The trend in the design of new catalytic-cracking units is to accomplish the total conversion in one or more risers at low recycle ratios.

With a given catalyst, the catalyst-activity level can be controlled by the rate at which fresh catalyst is added relative to the total inventory in the unit. If activities higher than those obtainable simply by adding fresh catalyst in an amount to replace the losses are desired, equilibrium catalyst may be withdrawn deliberately from the unit inventory so that greater amounts of fresh catalyst can be added. Rather complex economic studies are required to determine the feasibility of this practice.

Operating Pressure: This is a design factor and generally is in the range of 20–30 psig in the reactor and somewhat higher in the regenerator.

Stripping-Steam Rates Relative to Catalyst-Circulation Rate: These rates are held at values beyond which no additional benefit is obtained in terms of stripping hydrocarbons from the spent catalyst.

Metals Content of Feedstock: Metals such as nickel and vanadium in the feedstock accumulate on the catalyst and tend to cause the production of hydrogen under certain conditions. Since excessive hydrogen formation overloads the gas compressor and gas-recovery facilities, the metals content of feedstocks is watched carefully and held at a minimum.

Variations in Process Design: Among fluid units, two principal variations from the type shown in Fig. F-13 are noted in modern designs: (1) the reactor and regenerator vessels are situated side by side with catalyst transfer from vessel to vessel being made through two U-bends, an arrangement that permits units to be designed with a minimum height and number of thermal expansion joints, and (2) the two vessels are situated one above the other, with catalyst transfer being made in vertical internal lines.‡

Moving-Bed Units: In these units, the reactor is situated above the regenerator (referred to as the *kiln*). The catalyst particles, generally macrobeads, descend by gravity through the reactor and

*Model VI^SM, Esso Research & Engineering Co., and Orthoflow^SM, The M. W. Kellogg Co.

kiln and are returned to the top of the reactor by a pneumatic lift system. Catalyst-to-oil ratios generally are lower than in the fluid units, although in the latest designs heat-balance operation is utilized, thus eliminating the necessity for removing regeneration heat from the kiln by a cooling system. Nearly 700,000 bbl/day of fresh feed are handled by moving-bed units as contrasted with nearly 4 million bbl/day handled by fluid units (United States).

See also **Cracking; Petroleum;** and **Petroleum processing.**

—Melvin J. Sterba, *UOP Process Division, Universal Oil Products Company, Des Plaines, Ill.*

Fluidized-Bed Operations Processing operations involving gas-solids contacting have been markedly improved since the introduction of solids fluidization in the mid-1940s. One of the first applications was in connection with the gasification of coal, followed closely by the use of fluidization principles in the development fluid catalytic-cracking units in the petroleum industry. See also **Fluid catalytic cracking.** Fluidized-bed systems are used for several purposes, effecting reactions, heat exchange, and adsorption. See also **Iron- and steelmaking.**

The process of fluidizing converts a bed of solid particles into an expanded suspended mass that resembles a boiling liquid. This mass has a zero angle of repose, seeks its own level, and assumes the shape of the containing vessel. As in a vessel designed for boiling a liquid, space must be provided for vertical expansion of the solids and for disengaging splashed and entrained material. The usual shape is a vertical cylinder. The total cross-sectional area is determined by the volumetric flow of gas and the allowable or required fluidizing velocity of the gas at operating conditions. In some cases, the smallest permissible flow of gas is used; in other cases, the greatest flow of gas. The minimum velocity or mass flow rate is best determined by test in equipment where visual observations of the action of the bed can be made until experience with production units for a given application can be gained. The maximum flow generally is determined by the carry-over, or entrainment, of solids, and this is related to the dimensions of the disengaging space (cross-sectional area and height). Bed height is determined by (1) space-time yield, (2) gas-contact time, (3) space required for internal heat exchangers, (4) solids-retention time, and (5) additional geometric and stoichiometric factors.

In nearly all cases, the use of fluidization requires the use of a fluidized-bed *system* rather than an isolated piece of equipment. The arrangement of components of a system used in cases where the flow of solids is small, as is generally true in noncatalytic uses or in catalytic units where there is little or no deactivation of the catalyst, is shown in Fig. F-15. A catalytic-type unit like that found in petroleum-cracking operations where large quantities of solids flow into and out of the reactor and to and from the catalyst regenerator is shown in Fig. F-16.

The major parts of a fluidized-bed system are (1) reaction vessel, including the fluidized-bed portion, the disengaging space or freeboard, and the gas distributor, (2) solids feeder or flow control, (3) solids discharge, (4) dust separator for the exit gases, and (5) gas supply. Instrumentation, of course, is also an important consideration in the design of a fluidized-bed system.

The size of solid particles which can be fluidized varies greatly from less than 1 μm to 2.5 in. It is generally believed that particles distributed in size between 65 mesh and 10 μm are the best for smooth fluidization. Large particles cause instability and result in slugging or massive surges. Small particles (less than 10 μm) even though dry frequently act as if damp, forming agglomerates or fissures in the bed or spouting. Adding finer-sized particles to a coarse bed or coarser-sized particles to a bed of fines generally results in better fluidization.

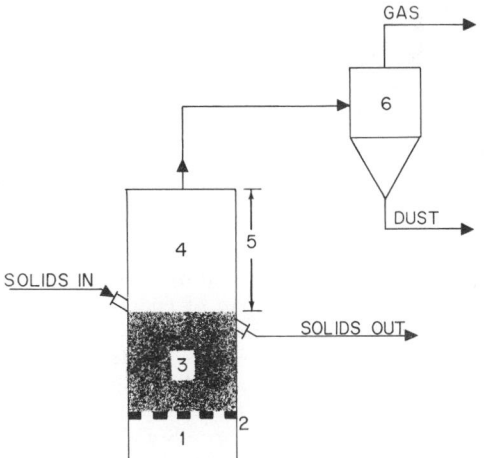

Fig. F-15. Noncatalytic fluidized-bed system. (1) Plenum chamber, (2) gas distributor or constriction plate, (3) fluid bed, (4) disengaging space, (5) freeboard, (6) dust separator.

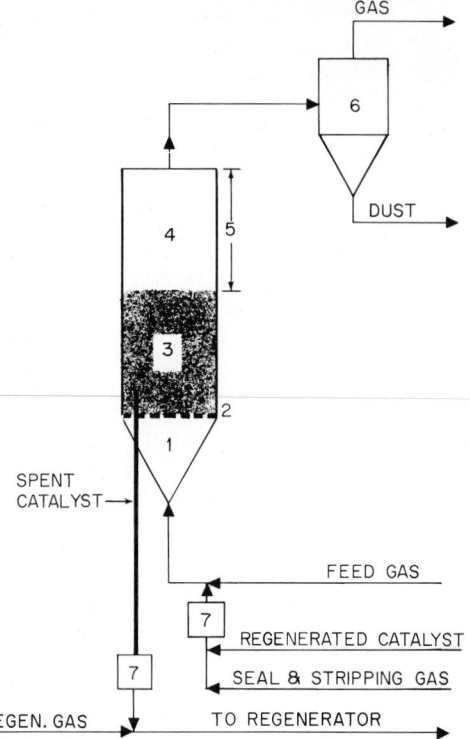

GAS

DUST

SPENT
CATALYST →

FEED GAS

REGENERATED CATALYST

SEAL & STRIPPING GAS

REGEN. GAS   TO REGENERATOR

Fig. F-16. Catalytic fluidized-bed system. (1) Plenum chamber, (2) gas distribution plate, (3) fluid bed, (4) disengaging space, (5) freeboard, (6) dust separator, (7) solids-flow control valves.

The upward velocity of the gas usually is between 0.5 and 10 ft/s. This velocity is based upon the flow through the empty vessel and frequently is referred to as the *superficial velocity.*

Either exothermic or endothermic reactions may be carried out in fluid beds. Exothermic reactions include roasting metal sulfides, coal- and oil-shale combustion and burning carbon adhering to cracking catalysts, where heat must be extracted from the fluid bed by direct or indirect heat exchange. Examples of endothermic reactions carried out in fluid beds include burning lime, calcining phosphate rock, the decomposition of ferric chloride or ferrous sulfate, and the reduction of metal oxides. HCN is produced in an electrically heated fluid bed. See also **Hydrogen cyanide.**

Heat* is often passed to the process from a

*Excerpted from L. Reh, Fluidized Bed Processing, *Chem. Eng. Prog.*, vol. 67, no. 2, February 1971, with permission of publisher.

combustion occurring within the fluid bed. Improved thermal efficiency can be achieved for such an internal combustion, especially if the excellent cooling effect of the fluid bed is put to use as control of a high theoretical combustion temperature, and to recover the sensible heat of roaster gas and solids. In connection with a process† for the calcination of aluminum trihydrate to alumina in a highly expanded bed, the heart of the process is the furnace (see Fig. F-17). Calcination proceeds in the furnace at 2000–2200°F (1093–1204°C) within an expanded fluid bed circulating via the return cyclone and a fluid-bed immersion seal.

The firing is achieved by direct oil injection into the lower furnace section, in which a zone of increased solids concentration is formed by dividing the fluidizing air into primary and secondary air. The near-stoichiometric combustion is carried out in the expanded bed, free of soot, without superheating and with CO contents of less than 0.5% at the furnace outlet and with O_2 contents of less than 1% in the flue gas. Primary and secondary air have been preheated to 840–940°F

†Vereinigte Aluminum Werke AG (VAW) and Lurgi, GmbH.

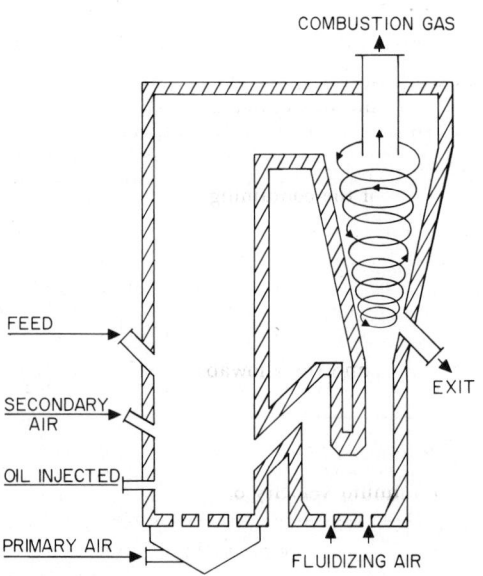

COMBUSTION GAS

FEED

EXIT

SECONDARY
AIR

OIL INJECTED

PRIMARY AIR

FLUIDIZING AIR

Fig. F-17. Circulating fluid-bed calciner with direct oil injection. Preheated hydrate enters feed tube at left. Oxide product is discharged to cooler from exit at right. [*Along lines of developments by Vereinigte Aluminum Werke AG (VAW) and Lurgi, GmbH.*]

(449–504°C) by cooling the discharged alumina in indirect and direct heat exchange. In the pilot plant for this process, gas velocities of more than 10 ft/s were run in the fluid-bed furnace of 3.28 ft ID and 26 ft inside height. The means solids concentration is approximately 6–12 lb/ft^3 of furnace volume, but because of internal solids recirculation, there is a substantial concentration distribution to be observed as a function of the furnace height. Temperature uniformity is comparable with that of conventional fluidized beds, although the gas velocities are several times higher.

The advantages of the principle of the highly expanded fluid bed over conventional beds can be gaged from the fact that a furnace with a 22 ft diameter is required at approximately equal heat utilization in a conventional multistage fluidized-bed furnace for a throughput capacity of 280 tons/day of alumina, whereas the 560 tons/day plant with barely 13 ft inside furnace diameter requires only two-thirds of the diameter and one-third of the grate areas. This system is tailored to fine-grained solids (20–250 μm) which have the innate advantages of high surface area, high rates of heat and mass transfer, and absence of internal diffusional resistance to chemical reaction. Prior to the concept of the highly expanded fluid bed, operating at unusually high fluidization velocities, such materials were limited to rather low fluidization velocities by the desire to limit entrainment of material from the bed.

Fluorapatite   (See **Phosphoric acid.**)

Fluorescent Whitening Agents   (See **Detergents.**)

Fluoridation   Fluoridation is the process of adding a fluoride-containing compound to water for the purpose of preventing dental caries. The introduction of approximately 1 ppm of fluoride to drinking water reduces the incidence of tooth decay in children by as much as 60% compared with a similar group of children who do not drink fluoridated water.

Experimental treatment of the water supply in a few American cities began in 1945. Favorable results were reported, and extensive publicity was generated, resulting in a rapid increase in the number of water supplies being fluoridated. In the United States, some 84 million persons are supplied with fluoridated water. Fluoridation is practiced in many countries throughout the world.

The compounds commonly used for fluoridation are sodium fluoride and sodium silicofluoride in the dry crystalline or powdered form and hydrofluosilicic (fluosilicic) acid in liquid form. The powdered compounds usually are applied by means of a dry chemical feeder with an integral solution tank or by a positive-displacement metering pump, which feeds a prepared solution of known concentration. Dry chemical feeders used for fluoridation normally are the gravimetric, loss-in-weight type, which are mounted on a scale permitting the dosage of the fluoride to be based upon unit weight rather than the unit volume of the chemical compound. Loss-of-weight feeders provide a relatively accurate measurement of the amount of chemical fed over a given period of time, thus limiting the possibility of overfeeding. The rate of feed can be set manually by adjusting the rate at which a motor-driven lead screw retracts a poise along the scale beam. For automatic pacing of chemical feed in proportion to flow, a signal from a flow transmitter controls the on-off time of the drive motor, usually over a 1-min cycle.

Fluoridation with dry chemicals involves the use of heavy feeding equipment, large storage area, and the frequent handling of bagged chemical with consequent hazard due to objectionable and noxious dust. Use of fluosilicic acid eliminates the inherent difficulties of dry chemical feeding. In large installations, the acid is stored in corrosion-resistant tanks and fed directly from the storage tank by a metering pump into the filtered water. In small installations, the drum in which the acid is shipped may be used as the storage tank. The drum generally is mounted on a scale and the weight checked periodically. Because of its unlimited solubility, fluosilicic acid can be used in very small plants, whereas the limited solubility of the dry chemicals presents limitations. Acid feed is proportioned to water flow by automatically adjusting the pump-stroke length or motor speed by a signal from a flow transmitter.

Since concentrations of fluoride in excess of 1.5 ppm may cause mottling of tooth enamel, careful control must be exercised to see that the desired dosage of 1.0 ppm is maintained. Most water departments or companies take daily samples from the plant effluent and taps in the distribution system. Fluoride concentration may be determined by a colorimetric laboratory analysis or by electrochemical methods, the latter being adaptable to continuous monitoring for control purposes.

—Leo J. Carroll, *Fischer & Porter Company, Warminster, Pa.*

Fluorine (From *fluor,* a mineral, from L. *fluere,* to flow.) **F** = 18.9984 (at. wt.); 9 (at. no.). Fluorine appears in group VII of the periodic table (halogens). The element has only one stable isotope, ^{19}F. At normal temperatures and pressures the diatomic fluorine molecule is a pale yellow-brown gas having a characteristic halogen odor. At $-188°$C fluorine condenses to a yellow-green liquid, and at $-219.62°$C it freezes to a yellow solid. A transition occurs at $-227.61°$C with an accompanying color change to a white solid. See Table F-12.

Like the other halogens, fluorine has seven valence electrons. All of the group exhibit high electronegativity, fluorine being the most electronegative element in the periodic table. See Table F-13. For this reason fluorine cannot be prepared by chemical reduction or replacement techniques. The high electronegativity of the element accounts for many of the chemical properties of fluorine and its compounds. In general, the properties of the halogens change steadily with increasing electronegativity. Unexpectedly, fluorine has a lower electron affinity than chlorine. This low value has been related to the low dissociation energy of the fluorine molecule (indicative of a weak F—F bond).

The size of the fluorine atom is small compared with the other halogens (one-fourth the size of chlorine). Since the fluoride ion, F$^-$, is similar in size to the oxide ion, the steric effects associated with chlorine and other halogens are therefore not found with fluorine. Owing to an increase in van der Waals forces in going from fluorine to iodine, the boiling points of the halogens increase with increasing atomic weight.

Fluorine is the most reactive element. It is the strongest oxidizing agent known. Under suitable conditions, fluorine combines with all of the elements except argon and neon. Oxygen reacts with fluorine endothermically under an electric dis-

TABLE F-12. *Physical Properties of Fluorine*

Molecular weight	38.00
Boiling point, °C	-188
Melting point, °C	-219.62
Triple point at 1.66 mm, °C	-219.62
Transition temperature, °C	-227.61
Density:	
Gas at 0°C, g/l	1.696
Liquid at boiling point, g/cm^3	1.108
Vapor pressure liquid at $-193.9°$C,	
atm	0.5000
Surface tension liquid at $-193.26°$C,	
dynes/cm	14.81
Critical temperature, °C	-129.2
Critical pressure, atm	55
Heat capacity:	
At constant pressure C_p,	
cal/(g mole)(K):	
Gas (actual) at 25.00°C	7.52
Liquid at $-223.16°$C	11.792
Solid at $-253.15°$C	2.240
At constant volume C_v, cal/g mole:	
Gas at 29.94°C	5.535
Ratio C_p/C_v	1.360
Heat of vaporization, cal/g mole	1564.
Heat of fusion, cal/g mole	122
Heat of transition, cal/g mole	173.90
Entropy of vaporization, cal/(g mole)	
(°C)	18.38
Entropy of fusion, cal/(g mole)(°C) . . .	2.28
Entropy of transition, cal/(g mole)	
(°C)	3.82
Viscosity, cP:	
Gas at 0°C	2.093×10^{-4}
Liquid at $-192.26°$C	0.275
Thermal conductivity gas at 0°C,	
cal/(s)(cm)(K)	5.92×10^{-5}
Index of refraction gas at 0 K and	
5890 Å	1.000214
Dielectric constant, liquid at boiling	
point	1.43
Oxidation potential, V	-2.85
Ionization potential, V	17

TABLE F-13. *Comparison of Fluorine with Other Halogens*

Halogen	At. No.	At. Wt.	Electronic Configuration	Electronegativity, Alfred-Rochow Scale	Electron Affinity, kcal/mole
F	9	18.9984	$1s^2, 2s^2, 2p^5$	4.10	79.6
Cl	17	35.453	[Ne] $3s^2, 3p^5$	2.83	83.3
Br	35	79.909	[Ar] $4s^2, 4p^5$	2.74	77.5
I	53	126.9044	[Kr] $5s^2, 5p^5$	2.21	70.6
At	85	(210)*	[Xe] $6s^2, 6p^5$	1.96	

*The mass number of the isotope of longest known half-life.

charge. With all other elements with which it combines fluorine reacts exothermically. Fluorine reacts with almost all inorganic compounds except fluorides in their highest valence state. Organic compounds react vigorously with fluorine (in some instances explosively). The only completely fluorine-resistant organic compound is carbon tetrafluoride, CF_4. The heats of reactions involving fluorine are always high. In some cases, a substantial amount of energy (activation energy) may initially be needed before even a highly exothermic reaction will proceed. Many reactions, however, occur spontaneously even at low temperatures upon exposure of the reactant to fluorine. The presence of easily oxidizable contaminants may initiate the reaction of fluorine and a less reactive material. Whether or not a spontaneous reaction occurs depends upon such factors as the initial temperature and pressure of the system, the thermoconductivity if the reactant is a solid, the surface-area–mass ratio, and the kinetic or steric effects.

Fluorine forms both ionic and covalent compounds, forming in most instances univalent bonds or uninegative ions. Fluorine also occurs in bridged compounds and may have a coordination number of 2 or more, for example, $K^+[(C_2H_5)_3Al \cdot F \cdot Al(C_2H_5)_3]^-$. Unlike other halogens, fluorine does not have positive oxidation states. Fluorine compounds may be extremely stable (SF_6) or extremely reactive (HF, halogen fluorides). Fluorine reacts very vigorously with metals, the resistance to fluorine attributed to some metals (nickel, copper, Monel) being due to the formation of a nonvolatile, nonreactive metal fluoride film on the surface of the metal. This film when intact retards further corrosion of the metal. The rate of reaction of fluorine and a metal depends to a large extent on the physical state of the metal.

Fluorine is estimated as the nineteenth most abundant element. Three minerals are important as commercial sources of fluorine: fluorite, CaF_2; fluorapatite, $CaF_2 \cdot 3Ca_3(PO_4)_2$; and cryolite, $3NaF \cdot AlF_3$. The most important industrial source for the production of hydrofluoric acid is fluorite (fluorspar). Fluorine compounds are described in Table F-14.

Isolation and Production: In 1886, Henri Moissan, a French chemist, first isolated fluorine in significant quantities. He employed electrolytic methods, using anhydrous hydrogen fluoride distilled from potassium acid fluoride, KHF_2, as the electrolyte, a platinum U-shaped cell, and plati-

num electrodes. The difficulties seemingly inherent in producing fluorine discouraged research for many years following Moissan's preparation. In 1919, Argo, Matthers, and coworkers simplified and increased the yield of the process. Other workers, such as Lebeau and Damiens in 1925–1927, improved the method for producing fluorine. World War II spurred research into better production methods. Fluorine was needed in order to separate from [238]U the fissionable isotope [235]U used in the atomic bomb. Also, work was conducted to develop compounds resistant to fluorine (highly fluorinated organic compounds) for use in diffusion plants. Until 1942, a commercial-scale cell for fluorine production had not been developed. In 1942, the Office of Scientific Research and Development (OSRD) issued contracts to several companies for the study of fluorine production by various methods. The laboratory cells in use at that time were small, had a low yield, required considerable attention, and had a short operating life.

Fluorine Cell: The main parts and construction materials of a typical fluorine cell are (1) a cathode integral with a mild-steel cell body, (2) a carbon anode, (3) a steel cell head including a Monel skirt, and (4) a Monel screen diaphragm. The purpose of the skirt is to prevent mixing of the hydrogen gas and fluorine gas formed. The diaphragm retains any carbon fragments broken from the anode and also helps to isolate the gases. The electrolyte used in a medium-temperature fluorine-generating cell is a fused mixture of KF and 2HF. Pure hydrofluoric acid, HF, is a nonconductor of electricity, whereas a solution of KF in HF is a good conductor. The temperature and electrolyte composition must be controlled. Since the hydrogen fluoride content of the electrolyte decreases during electrolysis, hydrogen fluoride must be added at intervals to the cell. The overall reaction is $2HF \rightarrow H_2 + F_2$, with hydrogen liberated at the cathode and fluorine at the anode.

Handling of Fluorine: In 1970, the American Conference of Governmental Industrial Hygienists adopted a threshold limit value (TLV) of 0.1 ppm for fluorine. This value refers to the airborne concentration of fluorine and represents conditions under which it is believed that nearly all workers may be repeatedly exposed day after day without adverse effects. Contact with the skin produces burns that may not be felt for some time, depending upon the degree of exposure. The burns are both thermal and chemical in nature, resulting from the heats of reaction of fluorine and skin and

TABLE F-14. *Properties of Representative Fluorine Compounds*

Compound	Formula	Formula Weight	Sp gr	Mp, °C	Bp, °C
Aluminum fluoride	AlF_3	83.97	3.07	1040	
Aluminum fluoride (fluellite) . . .	$AlF_3 \cdot H_2O$	101.99	2.17	120	250
Aluminum fluosilicate	$Al_2(SiF_6)_3$	480.12			
Aluminum sodium fluoride	$AlF_3 \cdot 3NaF$	209.96	2.90	1000	
Ammonium bifluoride	NH_4HF_2	57.05	1.21		
Ammonium fluoborate	NH_4BF_4	104.86	1.85		
Ammonium fluosilicate	$(NH_4)_2SiF_6$	178.14	2.01		
Ammonium fluotitanate	$(NH_4)_2TiF_6$	197.98			
Antimony ammonium fluoride . .	$Sb(NH_4)_2F_5$	252.84			
Antimony trifluoride	SbF_3	178.76	4.38	292	
Antimony pentafluoride	SbF_5	216.76	2.99	7	149.5
Arsenic fluoride	AsF_5	169.93	. . .	−80	−53
Arsenous fluoride	AsF_3	131.93	2.66	−8.5	63
Barium fluoride	BaF_2	175.36	4.83	1280	2137
Beryllium fluoride	BeF_2	47.02	1.99	800	
Bismuth fluoride	BiF_2	266.00	5.32		
Bismuth oxyfluoride	$BiOF$	244.00	7.5		
Boron fluoride	BF_3	67.82	. . .	−127	−101
Boron potassium fluoride	BKF_4	125.92	2.56	530	
Bromine trifluoride	BrF_3	136.92	2.49	8.8	135
Bromine pentafluoride	BrF_5	174.92	2.47	−61.3	40.5
Cadmium fluoride	CdF_2	150.41	6.64	1100	1758
Calcium fluoride	CaF_2	78.08	3.18	1360	
Calcium fluosilicate	$CaSiF_6$	182.14	2.66		
Carbon tetrafluoride	CF_4	88.00	. . .	−80	−15
Cerous fluoride	CeF_3	197.13	6.16	1324	
Cesium fluoride	CsF	151.81	3.59	684	1250
Cesium fluosilicate	Cs_2SiF_6	407.68	3.37		
Chlorine fluoride	ClF	54.46	1.62	−154	−100.8
Chlorine trifluoride	ClF_3	92.46	1.77	−83	11.3
Chromic fluoride	CrF_3	109.01	3.8	>1000	
Chromous fluoride	CrF_2	90.01	4.11	1100	>1300
Cobaltic fluoride	CoF_3	115.94			
Cobaltous fluoride	$CoF_2 \cdot 2H_2O$	132.97	4.43		
Cupric fluoride	$CuF_2 \cdot 2H_2O$	137.60	2.93		
Cuprous fluoride	Cu_2F_2	165.14	. . .	908	1100
Hydrogen fluoride	HF	20.01	0.987	−92.3	19.4
Ferric fluoride	FeF_3	112.84	3.18		
Ferrous fluoride	$FeF_2 \cdot 8H_2O$	237.96	4.09	100	
Fluorobenzene	C_6H_5F	96.04	1.024	−41.2	86
o-Fluorobenzoic acid	FC_6H_4COOH	140.04	. . .	122	
p-Fluorobromobenzene	FC_6H_4Br	174.95	. . .	. . .	152.5
Fluorodichloromethane	$CHFCl_2$	102.92	1.413	. . .	8.9
Fluoroiodobenzene	FC_6H_4I	221.96	. . .	. . .	182.4
o-Fluorotoluene	$CH_3C_6H_4F$	110.05	1.00	<−80	114
Fluorotrichloromethane	$CFCl_3$	137.37	1.53	. . .	23.7
Fluoroform	CHF_3	70.01			
Lead fluoride	PbF_2	245.22	8.24	855	1290
Lithium fluoride	LiF	25.94	2.29	870	1676
Lithium fluosilicate . . . . :	$Li_2SiF_6 \cdot 2H_2O$	191.97	2.33	100	
Magnesium fluoride	MgF_2	62.32	3.0	1396	2239
Magnesium fluosilicate	$MgSiF_6 \cdot 6H_2O$	274.47	1.79		
Manganous fluoride	MnF_2	92.93	3.98	856	
Manganese sesquifluoride	MnF_3	111.93	3.54		
Mercuric fluoride	HgF_2	238.61	8.95	645	650
Mercurous fluoride	HgF	219.61	8.73	570	

TABLE F-14. *Properties of Representative Fluorine Compounds*   (*Continued*)

Compound	Formula	Formula Weight	Sp gr	Mp, °C	Bp, °C
Mercuric oxyfluoride	$HgF_2 \cdot HgO \cdot H_2O$	473.24	. . .	100	
Molybdenum hexafluoride	MoF_2	210.00	2.55	17	35
Molybdenum oxyfluoride	$MoOF_4$	188.00	3.0	98	180
Nickel fluoride	NiF_2	96.69	4.63		
Nitrogen trifluoride	NF_3	71.01	1.54	−216.6	−120
Nitrosyl fluoride	NOF	49.01	. . .	−134	−56
Phosphorus trifluoride	PF_3	88.02	. . .	−160	−95
Phosphorus pentafluoride	PF_5	126.02	. . .	−83	−75
Phosphorus thiofluoride	PSF_3	120.08	1.63	<-20	265
Platinum difluoride	PtF_2	233.23			
Potassium fluoborate	KBF_4	125.92	2.50	500	
Potassium fluomanganite	K_2MnF_6	247.13			
Potassium fluoride	KF	58.10	2.48	880	1500
Potassium fluosilicate	K_2SiF_6	220.26	3.08		
Potassium fluostannate	$K_2SnF_6 \cdot H_2O$	328.92	3.05		
Potassium fluosulfonate	$KFSO_3$	138.16	. . .	311	
Silicon tetrafluoride	SiF_4	104.06	. . .	−77	−65
Silver fluoride	AgF	126.88	5.85	435	
Sodium fluoride	NaF	42.00	2.79	980	1700
Sodium difluoride	$NaHF_2$	62.00			
Sodium fluosilicate	Na_2SiF_6	188.05	2.76		
Strontium fluoride	SrF_2	125.63	2.44	1190	
Sulfur monofluoride	S_2F_2	102.12	1.5	−105.5	−99
Sulfur tetrafluoride	SF_4	108.06	. . .	−124	−40
Sulfur hexafluoride	SF_6	146.06	1.91	−50.8	−63.8
Stannic fluoride	SnF_4	194.70	4.78	. . .	705
Stannous fluoride	SnF_2	156.70			
Zinc fluoride	ZnF_2	103.38	4.84	872	

fluorine and water and the production of hydrofluoric acid, a solvent for protein.

Protective equipment used by personnel handling fluorine includes chemical goggles, gauntlet gloves, and neoprene rubber apron. In atmospheres containing low concentrations of fluorine (yet above the TLV), gas masks with cannisters approved by the Bureau of Mines for acid or acid gas–organic vapor should be used. In high concentrations, a self-contained Bureau of Mines–approved breathing apparatus with rescue harness and lifeline or a supplied-air respirator should be used. In very high concentration fluorine atmospheres, no protective equipment yet available is sufficient to provide protection, and the contaminated area must be evacuated.

Because of the extreme oxidizing character of fluorine, all containers, equipment, and piping must be of suitable materials and designated for fluorine service; they must be passivated before they can be used. The main steps involved in passivation are thorough cleaning, degreasing with a nonaqueous solvent, purging, and filling with increasing concentrations of fluorine. Passi-

vation is necessary to remove any easily oxidized material, such as grease, paint, pipe dopes, metal oxides, or metal filings. This procedure also allows a metal fluoride film to form on the surface of the treated surface, thus deterring further corrosion of the metal.

Uses: Fluorine is used in the production of fluorides (sulfur hexafluoride, uranium hexafluoride, cobalt fluoride, antimony fluoride, and halogen fluorides), in the synthesis of fluorocarbons, and as an oxidizer for rocket fuel.

Specifications and Shipping: The specification for technical grade fluorine is normally 98% minimum purity, with a typical analysis as follows:

Fluorine, % . . . . >99
Oxygen, % 0.35
Nitrogen, % 0.2
Carbon tetrafluoride, ppm 800
Sulfur hexafluoride, % 0.1
Hydrogen fluoride, % 0.2
Carbon dioxide, ppm 500
Water, ppm . . . . <1
Dew point, °F . . . . <−105

Fluorine is generally shipped as a nonliquefied compressed gas in seamless steel or nickel cylinders. Special permits are required from the Department of Transportation and the Bureau of Explosives for the transportation of liquid fluorine. Multijacketed dewars are used to contain the product. All fluorine containers must bear the D.O.T. flammable-gas label in addition to any other warning label, D.O.T. regulations stipulate that fluorine cylinders must not be charged to pressures exceeding 400 psig at 70°F and must not contain more than 6 lb of gas.

REFERENCES

Mellor, J. W., and G. D. Parkes: "A Comprehensive Treatise on Inorganic and Theoretical Chemistry," vol. 2, suppl. 1, Wiley, New York, 1962.

Rudge, A. J.: "The Manufacture and Use of Fluorine and Its Compounds," Oxford, London, 1962.

Slesser, C., and S. R. Schram: "Preparation, Properties, and Technology of Fluorine and Organic Fluoro Compounds," McGraw-Hill, New York, 1951.

—Mary E. Pisklak, *Industrial Gas Division, Air Products and Chemicals, Inc. Allentown, Pa.*

Fluorite   Also known as *fluorspar,* fluorite is a mineral having the chemical formula CaF_2 (a halide), belonging to the isometric crystal system (hexoctahedral class), and possessing a face-centered cubic structure. The calcium ions are located at the corners of the cube and in the center of each face, and the fluorine atoms are located at the center of the eight cubes considered to compose the unit cell. The theoretical fluorine content of CaF_2 is 48.9%. See Fig. F-18.

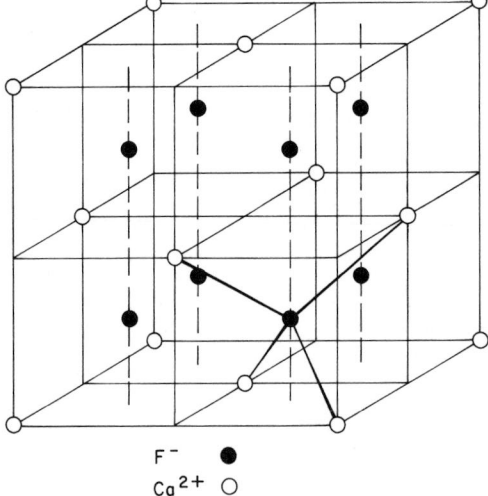

F^- ●
Ca^{2+} ○

Fig. F-18. Crystal structure of fluorite, CaF_2.

The mineral exhibits perfect octahedral cleavage. Its melting point is approximately 1330°C. On Mohs' scale of hardness (1–10) it has a relative value of 4 and can be scratched by a knife. The mineral has a vitreous luster and is transparent to translucent. The color of the mineral varies widely and is not homogenous throughout a particular specimen. Common colors are violet, green, yellow, pink, and colorless. Other colors such as blue, brown, and black have been observed. Some varieties of fluorite have the ability to emit light while being exposed to direct radiation or to an electric discharge in a vacuum tube. This phenomenon of fluorescence derives its name from fluorite. The word fluorite itself is derived from the Latin *fluere,* to flow. Besides fluorite and fluorspar, the mineral has been known as fluores, fluss, flusspol, glasspot, and spath fusible. The mineral occurs in crystals or clear masses. It is found as a vein material, being either the chief constituent or a gangue. As a gangue it is especially associated with the ores of lead, silver, and zinc. It also occurs with minerals such as a calcite, $CaCO_3$, quartz, SiO_2, barite, $BaSO_4$, and others. Fluorite is generally found in pneumatolytic deposits, sedimentary rocks, and as a sublimation product in connection with volcanic rocks. Countries having sizable deposits are United States, England, Canada, Germany, Russia, France, Italy, Spain, Tunisia, Morocco, South Africa, Korea, China, and Australia. The most significant deposits in the United States are located in Illinois and Kentucky. Fluorite is economically important as a flux in metallurgical industries because of its ability to reduce the viscosity of slags. Fluorite is treated with sulfuric acid in the commercial production of hydrofluoric acid, the chief raw material in the electrolytic manufacture of fluorine. The mineral is also used in ceramics.

—Mary E. Pisklak, *Industrial Gas Division, Air Products and Chemicals, Inc., Allentown, Pa.*

Fluoroelastomers   (See **Elastomers.**)

Fluoroplastics   Resins of a general paraffinic structure in which all or some of the hydrogen atoms have been replaced with fluorine are termed fluoroplastics. Major classes of these materials are *fluorocarbons,* notably polytetrafluoroethylene (TFE) and fluorinated ethylene propylene (FEP); *chlorofluorohydrocarbons,* notably polychlorotrifluoroethylene (CTFE); and *fluorohydrocarbons,* including polyvinylidine fluoride, VF_2, polyvinyl fluoride (PVF), and copolymers of fluorinated and halogenated ethylenes.

The practical service temperature, chemical resistance, friction properties, and electrical characteristics improve when the fluorine content is increased.

Both TFE and FEP resins are available in several forms, including film, filament, and dispersions. These resins are highly resistant to attack by chemicals and quite stable at high temperatures and maintain their toughness at low temperatures. The principal properties of these resins are summarized in Table F-15. The resins have a low dielectric constant, low dielectric loss, low coefficient of friction, and very desirable antisticking characteristics, the latter being a great advantage when these materials are used in thermal applications, e.g., the lining of vessels subjected to reasonable temperatures.

Because TFE resins have a high melt viscosity, they cannot be processed in conventional molding and melt-extrusion equipment. The processes used are similar to the techniques for making powder-metal parts. Unlike TFE, FEP is a true thermoplastic, and conventional molding and fabrication methods can be used. Both types of resins can be modified through the addition of filler and reinforcing materials. Applications for these resins fall into the following principal categories: (1) thermal systems, as vessels for heating and ablative shields; (2) electrical and electronic applications, including coatings for wire and cable; (3) components concerned with the handling of fluids, such as gaskets, packings, valve, pipe, pump, and hose linings; and (4) mechanical applications, including bearings as well as laminated parts involving metal, rubber, or other plastics. See also **Chlorofluorocarbons.**

Fluorosilicones (See **Silicones.**)

Fluorspar (See **Cryolite;** and **Hydrogen fluoride.**)

Foams (Consult the **Subject Index.**)

Foam Breakers (See **Defoaming agents;** and **Separation operations.**)

Foods, Fortification of (See **Amino acids.**)

Food Materials and Processes (Consult the **Subject Index.**)

Forging (See **Iron- and steelmaking.**)

Formaldehyde Over half of the methyl alcohol (methanol) produced in the world is converted into formaldehyde, HCHO, which thus represents a high-tonnage industrial chemical.

Chemistry: This compound is the lightest in the series of aliphatic aldehydes. In pure monomeric form, formaldehyde is a colorless, highly irritating, pungent gas, slightly heavier than air (vapor density 1.067). The compound is inflammable, igniting at 300°C, and forms explosive mixtures with air, the upper and lower explosive limits for gaseous formaldehyde being 73 and 7% by volume in air respectively. Formaldehyde is soluble in water and can be readily cooled to the liquid or solid phase. However, in both the liquid and gaseous phases, formaldehyde polymerizes rapidly at normal temperatures and below and hence is not commercially available as a gaseous monomer.

Formaldehyde gas (100%) has a molecular weight of 30.03. The melting point is $-118°C$ and the boiling point is $-19°C$. The specific gravity $(-20°C)$ is 0.815.

The two most important polymers of formaldehyde are trioxane and paraformaldehyde. Trioxane has a definite chemical composition, being a cyclic trimer, of formula $(CH_2O)_3$. It has a molecular weight of 90.05, melting point 61–62°C, boiling point 114.5–115°C, and a specific gravity

TABLE F-15. *Properties of Polytetrafluoroethylene and Fluorinated Ethylene Propylene*

	TFE	FEP
Dielectric constant, 60 and 10^6 Hz	2.1	2.1
Arc resistance, s	>300	>300
Continuous service temperature:		
°C.	260	205
°F.	500	400
Specific gravity	2.13–2.20	2.14–2.17
Water absorption, %	<0.01	<0.01
Melting point:		
°C.	. . .	274–295
°F.	. . .	525–563
Heat-deflection temperature:		
At 66 psi:		
°C	121	70
°F	250	158
At 264 psi:		
°C	56	51
°F	132	124
Thermal conductivity $(-200$ to $+360°F)$, $(Btu)(in.)/(hr)(ft^2)(°F)$ . .	1.7	1.4

(in the molten state) of 1.17 (65°C/20°C). Trioxane occurs as colorless crystals and in the pure state has an odor similar to that of chloroform. Stable at ordinary temperatures, trioxane can be distilled without decomposition.

Trioxane is combustible and also very volatile at ordinary temperatures. Its flash point is 37.8°C (100°F), and the upper and lower explosive limits are 28.7 and 3.5% by volume in air.

Paraformaldehyde is a linear polymer, $HO(CH_2O)_nH$, where n may vary between 8 and 50. This material occurs as a white powder or granules with a formaldehyde odor. Paraformaldehyde contains about 91–97% by weight of formaldehyde and melts within the range of 120–160°C.

Commercial Grades: Formaldehyde usually is manufactured as an aqueous solution. The original formaldehyde solutions contained 37% by weight of formaldehyde with 3–15% methyl alcohol, the latter acting as a stabilizer against paraformaldehyde deposition. The trend in recent years has been toward production of concentrated (for example, 50%) solutions containing as little methyl alcohol as necessary (0.5–1%). Paraformaldehyde deposition in such solutions is avoided in storage at elevated temperatures together, in some cases, with addition of special stabilizing chemicals.

Other commercial forms of formaldehyde are paraformaldehyde and trioxane. Both materials are normally prepared from the aqueous solution and therefore represent more costly forms. Examples of commercially available aqueous solutions are given in Table F-16. The acidity of a formaldehyde solution can be reduced to any desired level by passing it through an ion-exchange resin. See also **Ion-exchange resins.** Deacidification of formaldehyde is practiced widely, and suitable automatically controlled equipment is available.

Uses: By far the largest use for formaldehyde is in the manufacture of urea phenol, melamine, and acetal resins. These uses account for approximately 60% of the formaldehyde produced. The main outlet for formaldehyde is in production of urea resins. Urea-formaldehyde resins are showing an annual growth of about 6%; melamine-formaldehyde resins, about 5%; acetal resins, about 10%. These materials are described elsewhere in this volume (consult the Subject Index).

Pentaerythritol, $C(CH_2OH)_4$, as a market for formaldehyde has been growing at a rate of 3–4% per annum. Pentaerythritol is used in the manufacture of alkyd resins, pentaerythritol tetranitrate, additives for lubricants, and resin esters. Hexamethylene tetramine, made from formaldehyde and ammonia, is an important derivative and finds application in explosives manufacture and as a resin-curing agent.

Miscellaneous uses of formaldehyde include the manufacture of ethylene glycol, urea-formaldehyde fertilizers, acrylic esters, trimethylol propane for urethanes, textile-treating agents and tetrahydrofuran for elastomeric fibers, polyurethane elastomers, and as a solvent for synthetic and natural resins. Nitrilotriacetic acid (NTA) and isoprene manufacture may represent future major growth for formaldehyde use.

Commercial Production: All commercial formaldehyde is produced initially in the form of aqueous solution, and over 90% is made from methyl alcohol. Other feedstocks are methane, hydrocarbon gases, and dimethyl ether, but these materials are not in general use for this purpose.

Processes utilizing methyl alcohol feedstock fall into two categories: (1) *silver-catalyzed*, where formaldehyde is formed by a combination dehydrogenation-oxidation process:

TABLE F-16. *Properties of Commercial Aqueous Solutions of Formaldehyde*

	Formaldehyde, wt %			
	37	37	45	50
Methanol, wt %	1	5	1	1
Acidity as formic acid, wt %	0.01	0.01	0.01	0.01
	0.05	0.03	0.05	0.05
Iron, max., ppm	1	1	1	1
Color, max. APHA	10	10	10	10
Boiling point:				
°F	210	208	211	211
°C	98.9	97.8	99.4	99.4
Density at 18°C, g/ml	1.113	1.101	1.135	1.150

$$CH_3OH \rightarrow HCHO + H_2 \quad \text{(endothermic)}$$
$$H_2 + \tfrac{1}{2}O_2 \rightarrow H_2O \quad \text{(exothermic)}$$

and (2) *oxide-catalyzed,* where formaldehyde is formed by direct oxidation of methyl alcohol:

$$CH_3OH + \tfrac{1}{2}O_2 \rightarrow$$
$$HCHO + H_2O \quad \text{(exothermic)}$$

At present, most formaldehyde is made by the silver-catalyzed process (either gauze or crystals). The oxide processes are relatively recent developments.

In production-cost terms, there seems little to choose between the two routes. The better silver-catalyzed processes have the advantages of low initial capital outlay, lower power consumption, low catalyst-related charges, simplicity, and very large single-stream capability. On the debit side, silver processes require steam when distillation is necessary, but this cost can be minimized by proper heat-recovery measures. The oxide-catalyzed process can only claim the advantage of being steam exporters (excess of steam) and not requiring distillation to produce low-methanol products. However, capital charges, catalyst, and power consumption are higher. The relatively complicated reactor design, incorporating a thermostabilization (Dowtherm) loop, appears more

than equivalent in operating terms to a distillation unit. In addition, the maximum available single-stream capacity is lower. Allowing equal methanol conversion efficiencies, the respective advantages and disadvantages of each process virtually cancel each other out on cost terms. The choice of process for a particular location, therefore, is not straightforward.

Representative of modern silver technology is the silver-crystal-catalyzed process shown in Fig. F-19. The complete process* consists of two sections: (1) synthesis, yielding products containing finite amounts of (3–15%) methanol, and (2) distillation, required when the product must have a low methanol content.

In the synthesis section, vaporized methanol, together with air and steam preheated to 100°C, is passed over a thin bed of silver-crystal catalyst, where formaldehyde is formed by dehydrogenation of methanol. The reaction, being endothermic, requires heat, which is provided automatically by burning some of the hydrogen produced in the dehydrogenation reaction.

*Developed and operated by Imperial Chemical Industries Limited and available through Davy Powergas Ltd. Commercial plants are operating in the United Kingdom, United States, Scandinavia, Australia, New Zealand, and South Africa.

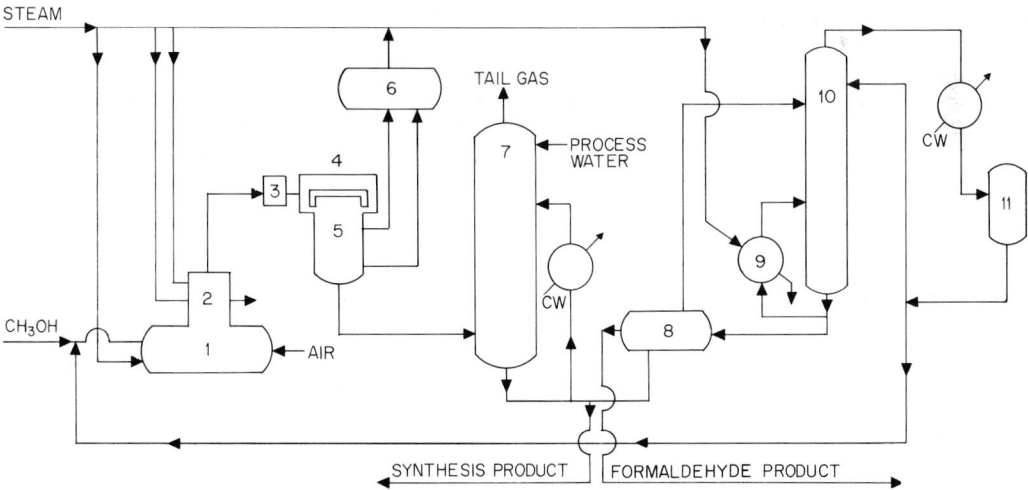

Fig. F-19. Formaldehyde process using a silver crystal catalyst in which formaldehyde is formed by a combination dehydrogenation-oxidation with methanol as feedstock. The final product contains up to 50% formaldehyde and less than 1% methanol. The synthesis product contains 42% formaldehyde and 15% methanol. (1) Vaporizer, (2) gas heater, (3) flame trap, (4) converter, (5) waste-heat boiler, (6) steam drum, (7) absorption tower (tail gas to tail-gas boiler or atmosphere), (8) feed preheater–bottoms cooler, (9) reboiler, (10) methanol stripping column, (11) reflux drum. *CW*, cooling water. (*Imperial Chemical Industries Limited; Davy Powergas Ltd.*)

The synthesis converter is mounted directly above the waste-heat boiler and is integral with it. This allows rapid quenching of the product gases and keeps side reactions to a minimum, thereby maximizing conversion efficiency at around 93%. The reaction products subsequently are condensed and absorbed to produce a methanolic formalin solution. The methanolic formalin from the absorber, in certain cases, can be the end product. However, if a low-methanol (less than 1%) product is required, the solution usually is fed to distillation, where excess methanol is taken off overhead and recycled directly to the vaporizer in the synthesis section. The final product, containing less than 1% methanol, is withdrawn from the base of the column. A wide range of products is obtainable from the process in the form of clear or faintly cloudy aqueous solutions containing 37–52.5 wt % of formaldehyde and less than 1% methanol. Single-stream units based upon this process are available in a range of capacities up to 136,000 tons of product per year (37% basis). This production capacity is directly comparable with the largest single-stream formaldehyde unit in operation.

The silver catalyst is easily replaced and cheaply regenerated. Downtime for catalyst replacement is about 36 hr.

Oxide catalysts used in the second route to formaldehyde production generally consist of a mixture of oxides of molybdenum, iron, and vanadium.

—D. G. Sleeman, *Davy Powergas Ltd., London, England*

Formalin   (See **Formaldehyde.**)

Formamide   (See **Amides and imides.**)

Formula Weight   (See **Molecular weight.**)

Foundry Compounds   [See **Silicates (soluble).**]

Fourdrinier   (See **Papermaking and finishing.**)

Fractionation   (See **Distillation.**)

Fractionation, Electrolytic   (See **Ion-exchange resins.**)

Fractionation, Protein   (See **Peptides and proteins.**)

Fragrances   (See **Aroma chemicals;** and **Terpenes.**)

Francium   (Derived from *France*.) **Fr** = 223* (at. wt.); 87 (at. no.). The element 87, visualized by Mendeleev as occupying the last position of the alkali metals in the periodic classification, was discovered in 1939 by Marguerite Perey, collaborator of Marie Curie. Previously, the search for this element in nature had been the object of many unsuccessful efforts. The history of its discovery begins, in fact, in 1913. J. A. Cranston studied the alpha activity of radiothorium, ^{228}Th, in the sources of mesothorium 2, $^{228}_{89}$Ac, a beta emitter, in order to establish the descendance of these two radioelements. Cranston observed that freshly purified mesothorium 2 emits a weak alpha radiation (approximately 1 for every 3,000 beta disintegrations), the intensity of which decreases with the period of mesothorium 2. Consequently, if ^{228}Ac disintegrated partially by alpha emission, the element 87 should have been produced. However, the search for it in radioactive ores carried out by O. Hahn and G. Hevesy in 1926 failed, perhaps because of the short half-life of approximately 2 min of the francium isotope involved, ^{224}Fr.

Cranston's results were confirmed by M. C. Guében in 1932. All the while, in 1914, S. J. Meyer, V. Hess, and F. Paneth had observed the emission of alpha particles by ^{227}Ac, but a doubt persisted because these particles could be attributed to the presence of protactinium traces or to Ac daughters. M. Perey proved a partial alpha disintegration (1.38%) of ^{227}Ac by chemically isolating the corresponding daughter, a beta emitter with a half-life of 22 min, having the expected properties of an ekacesium. This was the first isotope of element 87 to be identified and also the most important; it was called AcK. In 1946, M. Perey chose for element 87 the name *francium* (symbol, Fr), which was adopted by the International Chemical Union in 1949.

Isotopes: At present 22 isotopes are known, mass numbers 203–224. All are radioactive. The isotopes 222–224 are beta emitters (>99%). The lighter isotopes disintegrate by alpha emission. For isotope 212, electron capture (56%) competes with beta emission (44%) (see Fig. F-20). Three isotopes have half-lives longer than 10 min: ^{212}Fr 19.3 min, ^{222}Fr 14.8 min, and ^{223}Fr 22.1 min. In contrast, the half-life of ^{215}Fr is very short, <0.5 μs. The discontinuity in the variation of half-lives from ^{212}Fr to ^{215}Fr is associated with the

---

*Mass number of the isotope of longest known half-life.

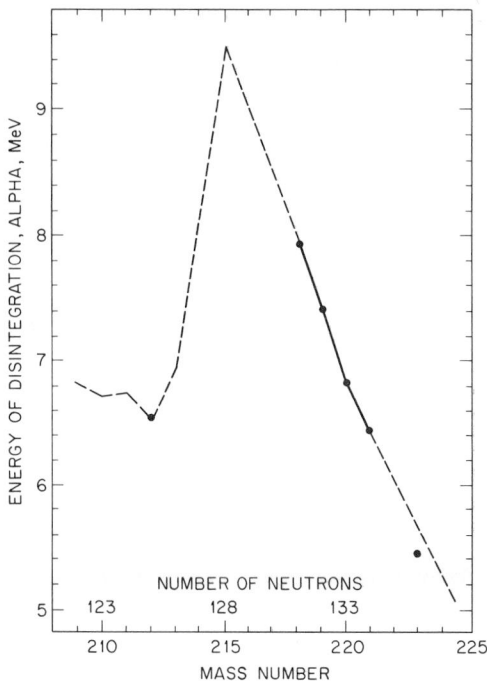

Fig. F-20. Energy of alpha disintegration as a function of the mass number of francium isotopes. Experimental data are indicated by dots and solid lines; dashed line indicates predictions.

sharp augmentation of instability of the alpha emitters when the number of neutrons becomes greater than the magic number 126 (corresponding to a closed shell). This instability is maximum for $N = 128$, that is, for the isotope of mass $128 + 87 = 215$. See Fig. F-20.

A systematic study of the radioactive properties of Fr isotopes leads to the conclusion that no isotopes with half-lives longer than those already known should exist. Thus, Fr is the most unstable of the first 101 elements of the periodic chart. The nonexistence of a stable or long-lived isotope explains the failures of chemists who, before the discovery of AcK, tried to identify a homologue of cesium in various ores and seawater by nonradiochemical methods.

The "discoveries" of *russium, alcalinium, moldavium,* and *virginium* were announced but remained without confirmation. The isotopes occurring in nature are found in thorium and uranium ores in which they are continually formed by disintegration chains that start with ^{232}Th ($4n$ family), ^{237}Np ($4n + 1$ family), and ^{235}U

($4n + 3$ family). These are, respectively, ^{224}Fr ($T_{1/2} \simeq 2$ min, β^-), ^{221}Fr($T_{1/2} = 4.8$ min, α) and ^{223}Fr (AcK) ($T_{1/2} = 22.1$ min, β^-: 1, 15 MeV, α $6.10^{-3}\%$, γ). One ton of natural uranium contains approximately 3.8×10^{-3} g of ^{223}Fr, corresponding to 0.17 μCi and 10^{-17} g of ^{221}Fr.

Preparation of Isotopes: AcK, the isotope with the longest half-life, can be prepared by extraction from the natural or artificial sources of ^{227}Ac (half-life 21.7 years). It reaches equilibrium with its parent radioelement in approximately 3 hr. AcK, like the other isotopes, can be artificially produced. The heavy isotopes are formed by irradiation of uranium or thorium by protons of high energy, whereas the lighter isotopes can be obtained by nuclear reactions induced in Au, Tl, or Pb targets by heavy ions. Examples include:

$$^{205}_{81}\text{Tl} \, (^{12}_{6}\text{C}, 5n) \, ^{212}_{87}\text{Fr}$$
$$^{197}_{79}\text{Au} \, (^{22}_{10}\text{Ne}; \, \alpha, 3n) \, ^{212}_{87}\text{Fr}$$

Because of the short lives of Fr isotopes, the properties of the element can be studied only at tracer-level concentrations, i.e., on the order of $10^{-15} \, M$.

Properties: As the higher homologue of cesium, Fr is an alkaline element. Its electronic structure consists of closed K, L, M, and N shells plus $5s^2$, $5p^6$, $5d^{10}$, $6s^2$, $6p^6$, $7s^1$.

The regularity of the variation of the physicochemical properties with the atomic number of the elements of the alkaline group makes it possible to estimate by extrapolation certain characteristics of Fr. These include:

Atomic radius, Å 2.8
Ionic radius, Å 1.8
Atomic ionization potential, eV 4
Density 2.4
Melting point, °C 20
Boiling point, °C 640

Francium, like its homologues, shows only one oxidation state ($+1$). Its electropositivity is close to that of cesium ($E_o = 2.923$ versus the normal hydrogen electrode). It forms an amalgam at more negative potentials than cesium.

The radiochemical work done on Fr confirmed the expected close analogy between its chemical properties and those of cesium, the ionic radius of which is not very different (1.65Å), as it results from the lanthanide contraction.

Like its homologues, Fr lacks any tendency to form complexes. Its salts are not isomorphic with the salts of sodium, the ionic radius of which is notably different, but they cocrystallize with those of cesium.

Coprecipitation is almost quantitative with double salts of cesium, e.g., chloroplatinate, chlorobismuthate, chloroantimonate, chlorostannate, the double cobaltinitrite of sodium and cesium, the silicotungstate, and $Cs_3Bi_2I_9$; but coprecipitation with simple salts of cesium, such as the perchlorate, picrate, iodate and tartrate, is only partial. In fact, the solubility of the double alkaline salts decreases with the atomic number whereas the solubility of the simple salts increases.

E. K. Hyde established that one of the best coprecipitating agents for Fr is free silicotungstic acid, precipitated by saturating the solution with gaseous HCl. In order to separate Fr, the precipitate is then dissolved in water and Fr adsorbed on a column of cationic Dowex 50 resin, which does not retain nonionized silicotungstic acid. Next, Fr is rapidly eluted with a small volume of concentrated HCl. Fr can also be coprecipitated with triheteropolyacids, e.g., vanadium phosphotungstic acid. It is easy to separate potassium from rubidium and cesium, the separation of rubidium and cesium is hard, and even more difficult is the separation of cesium and francium.

Like the other alkaline elements, francium remains in solution when other elements are precipitated as hydroxides, carbonates, fluorides, sulfides, chromates, and sulfates.

In the presence of sodium tetraphenylborate, Fr is extracted from its solutions at a pH of 9 by nitrobenzene. Its extraction from a very dilute sodium medium can also be carried out with dipicrylamine in nitrobenzene solution. Under these conditions, francium is much better extracted than cesium, and the separation factor can be as high as 7.

The separation of rubidium, cesium, and francium can be accomplished by chromatography on cation-exchange resins or on mineral exchangers, such as zirconium molybdate and zinc ferrocyanide. The distribution coefficients increase with atomic number, according to the variation of ionic radius.

The volatilization of francium chloride deposited on platinum begins at $110 \pm 15°C$ in vacuo and at $225 \pm 50°C$ at atmospheric pressure. At higher temperatures, the volatilization is complete and requires less than 1 min. However, it is not possible to separate francium and cesium by volatilization of their chlorides. According to theoretical estimates, the difference between the volatilities of francium and cesium halides would be maximum for the iodides and minimum for the fluorides.

Much radiochemical work on Fr has had as an objective the separation of AcK from the natural or artificial sources of ^{227}Ac or the separation of the Fr isotopes formed in targets irradiated with charged particles.

In her first experiments, M. Perey extracted AcK from a mixture of rare earths and from actinium in equilibrium with its disintegration products by treating a boiling HCl acid solution of the material with an excess of Na_2CO_3. Actinium and the rare earths, as well as RdAc (^{227}Th), AcX (^{223}Ra), AcB (^{211}Pb), and AcC (^{211}Bi), precipitate as the hydroxides or carbonates. AcC″ (^{207}Tl) is subsequently eliminated by coprecipitation with $BaCrO_4$. The filtrate contains AcK (^{223}Fr) radiochemically pure. AcK also can be rapidly separated from actinium and its daughters by paper chromatography, using a solution of $(NH_4)_2CO_3$ as eluting agent.

Silicotungstic acid has been used as a carrier for the separation of AcK as well as in the separation of Fr isotopes produced by synthesis. This rapid and very selective method makes it possible to obtain Fr solutions free from all foreign material except cesium, which also can be formed in the irradiation of certain targets.

Uses: ^{223}Fr (AcK) is utilized for the measurement of ^{227}Ac. The latter emits only a beta radiation of very weak energy ($E_{max} = 46$ keV) and an alpha radiation of low intensity (1.38 alpha particles per 100 disintegrations) and is therefore frequently determined by measuring the activity of one of its daughters. Among these, AcK (beta emitter, $E_{max} = 1.15$ MeV), presents advantages because it is easy to isolate and achieves equilibrium with ^{227}Ac in approximately 3 hr, whereas the equilibrium of the active deposit is reached only after an accumulation of several months.

It has been shown that Fr fixes itself in induced sarcomas in rats. This property offers the possibility of early diagnosis of certain types of cancers without risk to the organism because of the short half-lives of ^{223}Fr and ^{212}Fr.

REFERENCES

Bagnall, K. W.: "Chemistry of the Rare Radioelements," Butterworth, London, 1957.

Hyde, E. K.: Astatine and Francium, *J. Chem. Educ.*, vol. 36, pp. 15–21, 1959.

Lavrukhina, A. K., and A. A. Pozdnyakov: "Analytical Chemistry of Technetium, Prometheum, Astatine and Francium," Science Publishers, London, 1970.

Pascal, P. (ed.): "Traité de chimie minérale," vol. 3, pp. 131–141, Masson, Paris, 1957.

Perey, M.: Thesis, University of Paris, 1946.

The Radiochemistry of Francium, *Natl. Acad. Sci. Nuc. Sci. Ser.* 3003, 1960.

—G. Bouissières, *The University of Paris*

Frasch Process (See **Sulfur**.)

Free Radical (See **Acrylonitrile-butadiene-styrene resins; Antioxidants; Elastomers; Polymerization; Radicals; Vegetables oils;** and **Vinyl ester resins.**)

Freeze-drying Traditional methods of drying are unsatisfactory for preserving certain substances containing water. The increasing demand for preserving food products that can eventually be reconstituted to their original natural condition and the need for drying biological materials without any chemical, physical, or enzymatic changes have resulted in the development of freeze-drying, or *lyophilization.*

Freeze-drying is basically a technique for removing moisture from a wet material by bringing it to the solid state and subsequently subliming it. In most applications, water is the constituent to be removed. Usually, the wet material, whether it is a suspension, solution, or wet solid, is frozen at atmospheric conditions; then the ice is transformed into vapor and removed. The remaining dried material is a spongy mass of approximately the same size and shape as the original frozen mass; it has excellent stability, can easily be reconstituted in cold water, and possesses flavor and texture often indistinguishable from the original substance.

Whenever a wet material is freeze-dried, three basic operations are involved: (1) freezing, (2) sublimation (primary drying), and (3) desorption (secondary drying). These operations comprise several fundamental processes which occur simultaneously, and the conditions governing the rate of each process must be optimized for a specific product in order to achieve a satisfactory rate of drying.

Freezing: The materials processed by freeze-drying are usually complex mixtures of water and numerous substances. By cooling such materials to a temperature below 0°C (32°F), pure ice crystals separate out at first, and eventually the entire mass becomes rigid due to the formation of eutectics (Fig. F-21). Most food products and biological materials solidify completely at a temperature

in the range of −15 to −73°C (−5 to −100°F). At solidification of the entire mass, all the free water has been transformed into ice. Only a small quantity of the original water, the bound water, remains fixed in the internal structure of the material.

The rate of drying and the quality of the finished product are affected by the size, shape, and size distribution of the ice crystals formed during freezing and by the homogeneity of the frozen mass. It is therefore important to carry on the freezing or ice-crystallization operation under controlled conditions. Both the rate of cooling and, where possible, the degree of turbulence should be optimized for each product to arrive at a satisfactorily frozen mass. Slow freezing rates, resulting in large ice crystals, may have injurious biochemical effects on certain substances. Rapid freezing, resulting in small ice crystals, may cause undesirable color and texture changes in many food products.

Sublimation (Primary Drying): The frozen material can be subjected to ice sublimation at atmospheric pressure or under vacuum (less than 4.6 mm Hg). The vacuum technique is the most economical at the present state of the art.

Sublimation of ice crystals can be considered as comprising two fundamental processes, heat transfer and mass transfer. Heat is supplied to the ice crystal to sublime it, and the generated water vapor is transferred out of the sublimation interface. Thus, sublimation is rate-limited by resistances to heat and mass transfer within the specimen being dried.

As the sublimation interface recedes in the specimen (Fig. F-22), the dry layer presents a resistance to the flow of water vapor and a pressure difference must exist between the ice interface and the surface of the dry layer. A large pressure difference will facilitate high mass-transfer rates; however, the maximum allowable sublimation temperature at which no melting will occur and the cost of the vacuum equipment restrict this driving force to a limited range. In practice, the maximum allowable temperature and corresponding pressure at the sublimation interface is in the range of −9.4 to −40°C (+15 to −40°F) and 2,000 to 100 μm respectively.

The rate of heat input to the frozen specimen is a function of the operating-vacuum method of heat transfer and properties of the dried product. The operating vacuum determines the pressure difference and, in turn, the rate of mass transfer, which must be in balance with the rate of heat

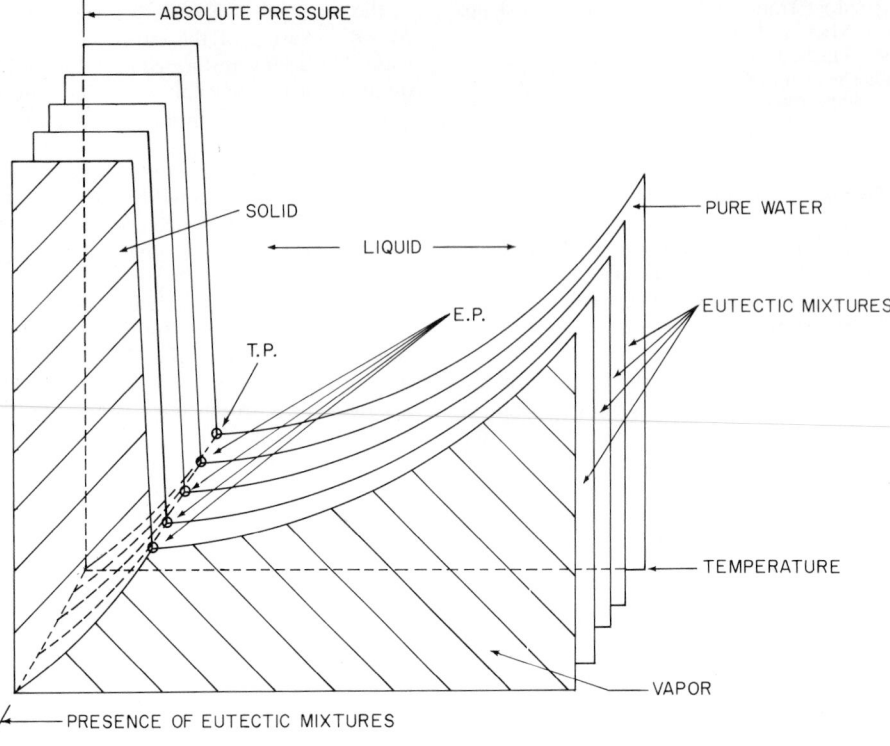

Fig. F-21. Eutectic phase diagram in freeze-drying process.

input; otherwise, either melting will occur at the sublimation interface and the purpose of freeze-drying will be defeated, or the sublimation temperature will decrease and the cost of processing will increase.

The heat required for sublimation (1200 Btu per pound of ice) can be supplied by conduction, radiation, electric resistance, microwave, or infrared heating. Figure F-22 shows three methods of heat input that have been investigated extensively. Depending on the method of heat transfer, the temperature gradient between the sublimation interface and the heat source is limited by the maximum temperature which can be tolerated on the surface of the dry layer or frozen mass. For radiation, the dry layer should not be heated to the point where charring and decomposition occur. For conduction, melting of the frozen mass in contact with the heating element should be avoided.

In most commercial applications, conditions are such that the rate of sublimation is controlled by heat transfer. Thus, developing techniques for

improving the heat-input rate is the objective of many investigations.

Desorption (Secondary Drying): Upon completion of sublimation of the ice crystals, final dehydration is carried out to remove the bound water which did not crystallize out during freezing and is bound by adsorption phenomena to the dried product. The product temperature is increased to 26.7–49°C (80–120°F), and under high vacuum the bound water and oxygen are removed from the dried product. The rate of desorption is considerably slower than sublimation. Although, the bound water is only 5–10% of the total water in many substances, the secondary drying may require 25–35% of the total drying time.

Drying Rates: Drying a frozen specimen proceeds initially at a constant rate with rapid evolution of water vapor. As the sublimation interface recedes within the product, water-vapor evolution decreases. This is the falling-rate period. When only bound water remains within the cellular structure of the product, the desorption period begins. During the constant-rate period, the sub-

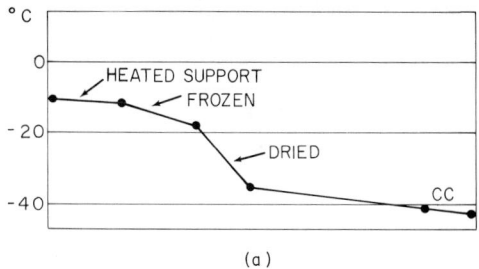

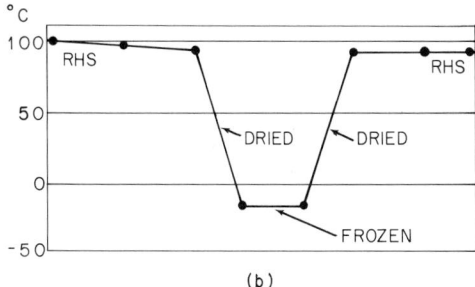

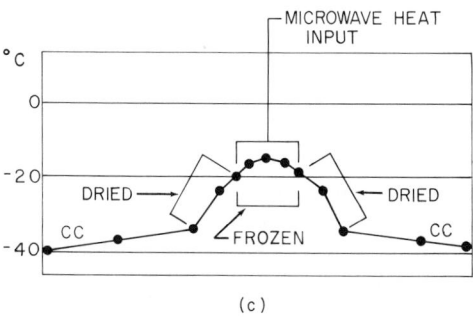

Fig. F-22. **Heat-input methods for freeze-drying processes.** (a) Conduction, (b) radiation, (c) microwave. *CC,* cold condenser; *RHS,* radiant-heat source.

limation rate can be expressed in terms of the heat of sublimation of ice and the heat-rate equation:

$$\text{Rate of sublimation} = \frac{UA\,\Delta T}{\Delta H_{\text{ice}}}$$

The overall heat-transfer coefficient U depends on the properties of the dry product and method of heat transfer. The heat-transfer rate A is influenced by the mechanical design of the heating elements and the conditioning of the frozen mass. The temperature gradient ΔT is limited by the maximum allowable temperatures at the sublimation interface and dry-layer surface. In the constant-rate period, the first one-half to two-

thirds of the drying cycle, 80% of the water is removed.

Processes and Equipment. In addition to the three fundamental operations discussed above, the freeze-drying process involves several other operations necessary to achieve an economically feasible system for large-scale manufacturing. The general commercial process comprises (1) preparation of the material, (2) freezing, (3) conditioning of the frozen mass, (4) drying, i.e., sublimation and desorption, and (5) conditioning of the product.

Preparation of the Material: It is not always economically feasible to subject a product in its original state to freeze-drying. One or more operations may be required to prepare the product. Wet solids, such as fruits and meats, are usually ground or sliced to facilitate drying by increasing the surface and reducing the thickness. Comestible liquids, such as coffee extract and fruit juices, and solutions in general are preconcentrated in order to minimize the water to be removed by sublimation and, in turn, to reduce the processing cost or to ensure that the final product will not be fragile. Some biologicals must be sterilized prior to freeze-drying. Chemical additives are used to concentrate dilute solutions which otherwise cannot be preconcentrated by such techniques as evaporation, freeze-concentration, or reverse osmosis and, in certain cases, to protect living organisms against the damaging effects of freezing or overdrying.

Freezing: The freezing operation can be accomplished by several techniques:

1. *Vacuum Cooling:* The material freezes itself by the evaporation of water as it is subjected quickly to high vacuum.

2. *Direct Contact:* The material is immersed in a cold liquid or in a stream of cold air or inert gases.

3. *Indirect Contact:* The material is frozen on cold surfaces.

Freezing the material is accomplished in the vacuum chamber where drying takes place in a separate piece of equipment or if the frozen mass requires some conditioning prior to drying.

Vacuum cooling is normally accomplished in the drying chamber. Its advantage is that large quantities of water are removed rapidly and no prior refrigeration is required. On the other hand, volatile flavor components are removed which might affect the quality of the final product. Also, removing water from the outer layer prior to complete freezing makes the cellular structure of certain materials collapse and subsequent drying and reconstitution may be inhibited. Therefore, this freezing technique has very limited application and is not recommended.

In direct-contact freezing, wet solid materials are placed in cold chambers and sprayed directly or immersed in cold air, inert gases, or liquid refrigerants such as Freon-12. Solutions or slurries may be frozen by spraying them in a cold stream of gas or liquid. Poor control of the rate of freezing limits these techniques to cases where quick freezing is desirable. Indirect-contact freezing is generally carried out in trays placed on refrigerated shelves inside the vacuum chamber.

A combination of indirect- and direct-contact equipment is in common industrial use. Trays containing the material to be dried may be placed on refrigerated shelves in a cold chamber and blasted with cold air, or inert gas or freezing belts may be used. Freezing belts in cold rooms are the most advanced technique for freezing solutions on a continuous basis. The solution is poured on one end of a moving refrigerated metallic belt and continuously removed at the other end as a frozen layer. By controlling the temperature of the surrounding air and the refrigerant temperature along the length of the belt, excellent control of the freezing rate is maintained.

Conditioning of the Frozen Mass: Materials frozen in a bulky or block form, meat, and solutions frozen on a belt or trays require further processing before drying. Granulation or slicing of such materials increases the available surface and minimizes the resistance to heat and mass transfer during drying. Standard size-reduction devices operating in cold chambers at about $-46°C$ ($-50°F$) are utilized.

Substances which do not freeze into a rigid solid at low temperatures, such as fruit juices, may be subjected to a devitrification treatment to avoid the soft-glass structure detrimental to optimum drying.

Drying: The conditioned frozen material to be dried is placed in a vacuum chamber, where sublimation and desorption of water occur. A schematic diagram of the vacuum chamber with the minimum associated equipment, water-removal and vacuum pumps, is shown in Fig. F-23. As soon as the chamber has been evacuated and the optimum vacuum has been reached (0.5–0.05 mm Hg), heating is applied so that the ice sublimes. For large-scale production of foods, the combination of conduction and radiation that results from circulating a hot fluid through coils or plates has been the most satisfactory method for heating.

A variety of vacuum-chamber designs have been suggested, depending on the plant capacity, product characteristics, and method of heat transfer. For laboratory or pilot-plant operations, most commercially available equipment is satisfactory. For large installations, custom engineering and fabrication are dictated to assure optimum performance with a specific product. The three vacuum chambers commonly used may be classified as batch, semicontinuous, and continuous.

The batch units are invariably cylindrical shelf-type driers equipped with heating and cooling coils or plates. Most major freeze-drying installations have adopted the batch unit with stationary heated shelves. Designs for better heat input, i.e., spikes or expanded metal sheet that penetrate the frozen mass or movable heated shelves compressing the frozen material, have met with limited acceptance because of the increased cost and complexity of the equipment. Overall drying cycles of 5–10 hr have been achieved with many food products.

The semicontinuous units are long cylindrical tunnels; trays with the frozen material are continuously conveyed through a series of heated zones. Interlocks are used in both ends for proper vacuum control. The frozen material is heated along the length of the tunnel with each zone maintained at a different temperature and is removed as a completely dried product.

Continuous units are designed to move the frozen material continuously on the heated surface and transfer it through a series of zones in which the temperature and vacuum are maintained at different levels. The frozen material is fed via interlocks in the chamber. Vibrators or other mechanical means are used to maintain the product in continuous motion.

During the constant-rate drying period, the temperature of the heat source (radiation) is 93–149°C (200–300°F) for food products; thus, high heat-input rates are achieved. This temperature is reduced to 51–66°C (125–150°F) in the falling-rate and desorption periods to avoid charring and decomposition of the dried products.

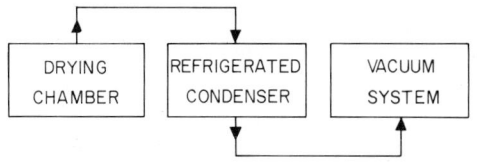

Fig. F-23. Elements of freeze-drying process.

Average drying rates of $0.5–1.6 \, lb/(ft^2)(hr)$ have been reported for food products.

The drying process is discontinued when the residual moisture content is sufficiently low to ensure good preservation of the specific product. For biological and food materials, the residual moisture is on the order of 1 and 3% respectively.

Water Removal: The three common methods for removing the water vapor are condensers, direct- and indirect-contact; desiccants, e.g., calcium chloride and zeolites; and vacuum pumps. The indirect-contact refrigerated condenser offers the optimum arrangement for water removal on an industrial scale. The condensing surface can be located in the drying chamber or in a separate chamber. The water vapor condenses and forms an ice layer on the cold surface and subsequently is removed, either intermittently by melting or continuously by scraping.

Vacuum Pumps: The function of these pumps is to evacuate the drying chamber quickly without allowing the prefrozen material to melt and thereafter to reduce the pressure progressively to the desired vacuum and maintain it at this level by removing the noncondensable gases. The vacuum equipment can be either an oil-sealed rotary vacuum pump or a multistage steam-ejector system.

Conditioning of the Product: The high porosity and low moisture content of the freeze-dried product require that the vacuum be broken and packaging be done under a dried inert-gas blanket to prevent oxidation during storage and maintain the low moisture content. Carbon dioxide or nitrogen are in common use for food products. For biological materials argon is sometimes preferred.

Commercial Applications. In addition to the manufacturing of pharmaceuticals, the freeze-drying process has been used for preserving many relatively inexpensive food products such as shrimp and orange juice. Currently, the major application in the food industry is coffee. It is estimated that the worldwide annual production of freeze-dried coffee is 45–50 million lb.

The process involves preparation of a coffee extract with 20–25% solids, clarification of the extract, freeze concentration of the extract to 30–40% solids, freezing extract to a completely frozen mass at -25 to $-43°C$ (-13 to $-45°F$), granulation of the frozen mass, sublimation of the ice at a vacuum of approximately $200 \, \mu m$ Hg abs., and drying the final product to a moisture content of 1–3%. For batch driers, the overall drying cycle is on the order of 6–8 hr.

REFERENCES

"ASHRAE Guide and Data Book, Applications for 1966–1967," ASHRAE, New York, 1967.

Colson, S., and D. B. Smith (eds.): "Freeze Drying of Foodstuffs," Columbine, London, 1963.

Meryman, H. T. (ed.): "Cryobiology," Academic, New York, 1966.

Noyes, R.: "Freeze-drying of Foods and Biologicals," Noyes Development Corporation, Park Ridge, N.J., 1968.

Pintauro, N.: "Soluble Coffee Manufacturing Processes," Noyes Development Corporation, Park Ridge, N.J., 1969.

Pintauro, N.: "Soluble Tea Production Processes," Noyes Data Corporation, Park Ridge, N.J. 1970.

Rey, L. (ed.): "Aspects théoriques et industriels de la lyophilisation," Hermann, Paris, 1964.

Rey, L. (ed.): "Advances in Freeze-drying," Hermann, Paris, 1966.

Tuomy, J. M.: "Freeze-drying of Foods for Armed Services," *Natick Lab. Natick, Mass., U.S. Army Food Lab. Tech. Rept.* 70-43-FL, February 1970.

—Neophytos Ganiaris, *Struthers Scientific and International Corporation, New York*

Freezing Process, Seawater   (See **Desalination.**)

Friedel-Crafts Reaction   [See **Anionic surfactants (sulfur-bearing); Catalysts; Chlorine organics; Coal tar and derivatives;** and **Isomerization.**]

Froth Flotation   (See **Flotation.**)

Fruits   (See **Aroma chemicals;** and **Pasteurization.**)

Fuels   Fuel is any matter used to produce heat by combustion. The combustion of interest here is the combination of a fuel and oxidizer at a rate sufficiently high to maintain itself. Slow oxidation, e.g., rusting, and fast oxidation, e.g., that found with pyrophoric, explosive, and incendiary materials, are not described here. Of concern here is the burning of fuels to liberate heat or to consume waste materials (incineration).

Although coal, oil, and gas are the primary industrial fuels, there are other combustible materials of commercial interest, including garbage, wastepaper, sawdust, bagasse, and many waste gases, including CO and H_2. These materials depend on the plant itself, and their use normally requires custom engineering. A material may be in excess in one plant and thus used as a fuel for combustion whereas another plant may be in short supply of the same material as an important process raw material. This is particularly true of

TABLE F-17. *Classification of Fuels*

Physical State	Natural or Primary	Manufactured or Secondary
Solid	Anthracite coal Bituminous coal Lignite Peat Wood	Coke Charcoal Briquettes
Liquid	Petroleum	Tar Petroleum distillates Petroleum residuums Alcohols Colloidal fuel
Gaseous	Natural gas	Illuminating gas Water gas Oil gas Producer gas Blast-furnace gas Acetylene

SOURCE: R. T. Haslam and R. P. Russell, "Fuels and Their Combustion," McGraw-Hill Book Company, New York, 1926.

gases, such as H_2. It is not uncommon for such materials to be piped between different kinds of plants. See also **Petrochemical complex.** A classification of fuels is shown in Table F-17. Combustion processes and problems differ markedly, depending upon whether the fuel is a solid, liquid, or gas. Heating values for various gases are given in Fig. F-24.

Besides infrared heat energy (usually the major

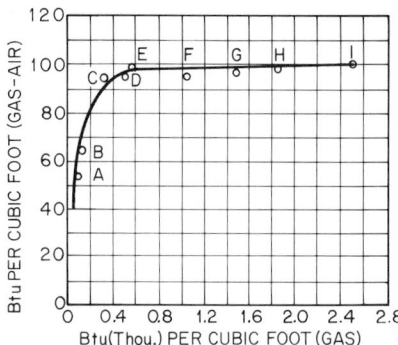

Fig. F-24. Btu of gas-air mixtures versus Btu of gas for various commercially produced gases. Conditions: 60°F (15.6°C), 30 in. Hg, H_2O saturated. *A*, blast-furnace gas; *B*, producer gas; *C*, blue water gas; *D*, coke-oven gas; *E*, coal gas; *F*, Sandusky, Ohio, natural gas; *G*, McKean County, Pa., natural gas; *H*, Follansbee, W. Va., residual gas; *I*, Follansbee natural gas.

goal of industrial combustion processes), there are other effects, e.g., the production of ultraviolet radiation, visible light, pulsating direct current, and noise. Use is made of the ultraviolet in flame-detection equipment, and the electrical phenomenon is the principle upon which some types of flame-detection equipment function.

Selection of Fuels and Combustion Systems: The choice of fuel for a particular application may depend upon many considerations but occasionally will be determined by a single overriding factor, such as *availability*. When there is a choice, other factors to consider include (1) economics, (2) the size of the process installation, (3) the interface of the fuel system with personnel, (4) pollution, and (5) compatibility with the product and process.

Economics: Traditionally the most powerful consideration is overall process cost. A thorough economic study should be made of each fuel situation, including fluctuations in the price of fuels, availability, and cost of the various fuels in the quantities required for the particular use at the site under consideration. Technology has evolved good methods of handling almost any fuel desired, but the process designer finds that geographic, social, and legal factors contribute to his decision as well.

Size of Process Installation: Small installations* tend toward the use of natural gas because this fuel (1) is readily adaptable to automatic control, (2) is easily distributed to point of use, (3) does not require protection from the weather, and (4) generally does not require any storage facility. Frost, which forms during sustained periods of extremely cold weather, reduces the capacity of the lines. When the purchased gas supply is interruptable, a backup system consisting of liquefied propane or butane may be required, including control equipment for switchover, blending, and/or air-gas-ratio equipment. Also, steam tracer lines may be necessary to prevent recondensation after vaporization has been achieved.

As the process size becomes larger, light fuel oils generally are considered (distillate oils). They require storage capacity and access for delivery, usually by truck. Occasionally, they require mild heating to keep the viscosity down and generally

*In order to picture this continuous spectrum as four bands, the following rough boundaries are offered: small, 0–1 million Btu/hr; second, 1–10 million Btu/hr; third, 10–100 million Btu/hr; and large, 100 million Btu/hr and up.

demand fuel-air-ratio equipment of greater complexity than gas.

As the process becomes larger still, residual oils are considered. These require more extensive storage facilities, since, due to the larger quantities used, delivery is frequently made by railroad tank car or barge. In these cases, heating for pumping is usually required at the storage tank and again near the point of use, so that the viscosity can be kept low enough for pumping and atomization. Here again the fuel-air-ratio equipment is more complicated.

At the largest process size, e.g., large steam-generating plants or electric-power-generating plants, consideration will be given to solid fuels. Unloading and storage facilities are required as shipment is made by railroad car or barge. If the climate cycle includes below-freezing weather, facilities will be needed to thaw frozen loads so they can be dumped out of railroad cars. Some protection may be required for part of the stored material so that it can be recovered from storage in freezing weather. Also, care must be taken to prevent spontaneous combustion in coal piles, particularly when much fine material is present. A system of conveyors is needed to bring the coal to the stoker or pulverizer. Extensive provision for fuel-air ratio and for removal of accumulated ash must be made.

The reason for this broad fuel choice and use follows the economics of the situation; i.e., gas or light oil (which requires the least capital investment) is usually the costliest on the cents per Btu basis, while residual oil and solid fuels are usually the cheapest on a cents per Btu basis but require an operation of considerable size in order to absorb the required capital investment to the point where economic advantages can be realized. With increasing complexity of the storage, processing, and handling of the fuel, additional labor and maintenance costs should be expected.

Groups of fuel-using processes like those found in a large plant usually mean that fuels can be purchased on a combined basis, which permits installation of handling and storage facilities for the less costly fuels. Therefore, the choice for any subsequent additions may be predetermined.

Interface of Fuel System with Personnel: A growing awareness of the extreme hardship sometimes suffered by workmen in contact with fuel-fired processes is making comfort a larger and larger consideration in the choice of fuels for industrial processes. Much can be done and has been done to protect workmen from disagreeable conditions, e.g., air-conditioned crane cabs, air-conditioned operator pulpits, improvement in ventilation, and laborsaving devices such as stokers on coal-fired equipment. However, there are conditions where the fumes, heat, and noise from a process force the designer to consider carefully the consequences of the fuel system. Consideration for the workman leads to a minimization of waste gases dumped into the work area as well as eliminating the toxic effects of incomplete combustion and the annoyance of particulate matter in the air. Solution of these problems installation by installation on a micro scale automatically helps alleviate the worldwide problems of air pollution on a macro scale.

Primarily, in this category, sulfur oxide products, SO_x, which are difficult to eliminate, must be considered.

Pollution: The four main categories of products or results of fuel combustion are outlined in Table F-18.

Carbon dioxide and *water vapor* are colorless, odorless, tasteless gases which are of little concern in small quantities but present problems in large quantities. It is feared that large amounts of CO_2 will change the ability of the atmosphere to absorb solar heat and may therefore have a long-term effect on the average temperature level in areas where the CO_2 level is high.

Water vapor is a normal part of our atmosphere, but occasionally the white plume resulting from its condensation becomes a traffic hazard in industrial areas or is mistaken by well-meaning laymen for an example of air pollution.

Nitrogen oxides are formed in detectable amounts

TABLE F-18. *Categories of Products and Results of Fuel Combustion*

Group 1. Results of complete (perfect) combustion
    a. Carbon dioxide, CO_2
    b. Water vapor, H_2O
    c. Nitrogen oxides, NO_x
Group 2. Results of incomplete (imperfect) combustion
    a. Carbon, C
    b. Carbon monoxide, CO
    c. Unburned hydrocarbons, CH_x
Group 3. Transient results of the combustion process
    a. Heat (rise in temperature)
    b. Noise
    c. Odor
Group 4. Lasting results of the combustion process
    a. Sulfur oxides, SO_x
    b. Particulate matter (dust or ash)
    c. Particulate matter (soot)

in normal combustion at temperatures above 2800°F (1583°C). The presence of excess air seems to increase the quantities formed. Holding for a short time between 2600 and 2800°F (1427 and 1538°C) will decrease the amount of NO_x in the flue gases. This topic is discussed in further detail under **Air pollution.**

All three members of group 2 (Table F-18) appear at about the same time when the combustion air is insufficient or the temperature is too low or the time is too short for complete combustion. A technical reevaluation of the specific problem usually will reveal the cause of incomplete burning. In some instances, a catalyst can be used to bring the reaction to completion.

Heat is usually the prime purpose of combustion, but frequently after doing useful work at high thermal head, heat is exhausted from the process in the form of hot waste gases, which contribute to the overall temperature of the environment. The rumble, roar, or hiss of industrial combustion is usually related to the physical structure or container and generally can be minimized by physical modification of the structure or using sound-absorbing materials. While incomplete, or *quenched,* combustion will produce an aldehyde odor, most industrial-odor problems come from the process and not the combustion. However, if sulfur is in the fuel, its presence can usually be detected.

The presence of sulfur in fuels is difficult to deal with, as it is undesirable in the atmosphere in all its forms. As sulfur is not destroyed by oxidation, the quantity in the products of combustion is directly proportional to the amount of sulfur in the fuel being used. All the fuels mentioned can carry sulfur, but usually it is easier to buy sulfur-free light oil or gas than it is to buy sulfur-free coal or residual fuel oil. However, this varies considerably from area to area. Sulfur removal is the object of much research work. See also **Sulfur dioxide removal.**

For items in group 4 (Table F-18), bag filters and/or electrostatic precipitators are usually required to remove very fine particulate matter from waste gases whether it comes from the combustion or the process. Ash is produced not only by coal but also by heavy fuel oils. See also **Separation operations.**

While soot is particulate, its removal can be accomplished by combustion adjustment or a high-temperature incinerator.

Compatibility with Product and Process: It might be argued that this criterion should be ranked most important, and in some cases it is; but usually the incompatibility can be overcome. For example, when solid fuels were available but gaseous fuels were needed, there arose the producer-gas science, in which coal is partially burned and partially distilled to produce a dirty but useful combustible gas. When coke was required and a coking coal was available, coke ovens produced a very useful coke-oven gas as well as the coke for metallurgical processes. Before the widespread distribution of natural gas, a great variety of gas-generating equipment was devised using coal or oil as the raw material.

When food products are involved, care must be exercised with heavy and solid fuels to prevent any change of taste. These problems are usually solved by firing the fuel at some central location to generate steam, which is used in heating, cooking, and pasteurizing. If flue gases will mix with the product and separation will be required, consideration must be given to the possibility that the gaseous and particulate matter in the combustion gases might contribute (favorably or unfavorably) to the end product, e.g., in large cement kilns.

When the product is exposed to the combustion gases but not mixed, the purity of the product must be considered; e.g., glass tanks which are overfired. Similar considerations are necessary when hot products of combustion are used to fluidize fine powders or when the combustion products are actually bubbled through the product, as in the submerged combustion melting of glass or sodium silicate. It must be remembered that products of combustion produce an environment which (while often neutral with respect to the product) can be adjusted to be oxidizing or reducing. Heating copper metal is particularly interesting at this point, as it oxidizes or reduces readily as the amount of air supplied is varied on either side of the stoichiometric combustion requirement.

Fuel Value: Almost all fuels are characterized by heating value per unit volume or unit weight such as Btu per cubic foot, Btu per pound, or kilocalories per cubic meter. Usually these numbers are designated as HHV (higher heating value) (gross heating value) or LHV (lower heating value) (net heating value). The difference is the latent heat of water vapor formed. With one exception, in all heating calculations for temperatures above 212°F (100°C) either value can be used as long as the calculation is consistent. However, it will be noted that an efficiency calculation based on higher heating value will produce a numerically lower result than the same calculation based on the lower heating value of the fuel. See Table F-19.

An important aspect of combustion arises when a very lean fuel (low heat content per unit weight or volume) is offered for burning. In considering this problem reference is made to the later discussion on incineration, because both the fuel and the air may be high in inerts (nitrogen, carbon dioxide, and water vapor), which must be raised above the combustion temperature for the mixture to burn. It will be found that 60 Btu/ft^3 of mixture is about the lower limit (see Fig. F-24). It will be recognized that in addition to waste-stream gases, this lean condition can be reached with too much excess air; but no matter what the cause, if the mixture must be consumed, the temperature can be kept high enough for the chemical energy to be released by adding heat to the combustion air or by adding enough of a richer fuel (and sufficient oxygen).

Major Factors Influencing the Combustion Process. Manipulating combustion for useful purposes requires four basic actions: (1) planning for good burning, (2) gasifying and mixing with the oxidizer, (3) raising the fuel to ignition temperature, and (4) maintaining the burning temperature.

Planning for Good Burning: The problem is to bring fuel and oxidizer together in an economical manner to do useful work. Some fuels are not ready for use without preparation, as anyone who has attempted to build a fire with wet logs knows. Wood may have to be dried, coal may have to be crushed, and oil may have to be heated and filtered before a successful combustion process can be carried on.

Gasifying and Mixing with Oxidizer: Since gases mix readily, this step is easily accomplished by pumps, venturi mixers, or other equipment. Oil may require heat to vaporize, energy to atomize, or both before it can be mixed with air. Solid fuels generally require crushing if the object is direct mixing with air; a traveling grate or stoker may be needed to support the coal in a hot zone during gasification and mixing with air.

Mixing with air is not a random business but requires a reasonable matching of the fuel need. Insufficient air will produce a fuel-rich mixture which will generate free carbon, CO, and unburned hydrocarbons. Excess air will depress maximum temperature and tend to favor high NO$_x$ in waste gas if the process temperature is high enough (over 2800°F, 1583°C).

Some gas-combustion systems mix the air and the gas completely before burning. Some oil and coal systems add the air in stages until burning is complete.

Raising to Ignition Temperature: Systems depend upon a hot chamber or burner tile, a pilot light, or self-piloting after ignition is accomplished, but some method of raising the combustible mixture to its ignition point is necessary (Table F-20) before the burning process will begin.

Maintaining the Burning Temperature: The combustion system is put together to liberate heat. Fuel must be supplied at a rate equal to the rate of heat loss or the temperature will drop below ignition temperature and combustion will cease.

The process requirements for heat include heat of reaction, heat of vaporization, wall losses, opening losses (radiation), and flue-gas losses, as well as the sensible-heat requirement of the product.

Open-Flame and Furnace Applications. There are two types of combustion problems, the open flame and the confined flame. The latter is used most often in industrial applications where the combustion takes place within a furnace. Major factors to consider include: temperature, time, and turbulence, and the burners and furnaces in which combustion occurs.

Temperature, Time, and Turbulence: There are interesting latitudes and limitations with regard to burning in a chamber. Causing fuel and air to combine in a hot chamber is not difficult if the chamber is *large enough* (enough time) and *hot enough* and provides *enough mixing*. Under this unusually easy circumstance, almost any crude system for introducing the fuel and air might be used. For example, two pipes side by side might be suitable for introducing gas and air (one story has it that regenerative glass tanks were once fired by hanging a bucket filled with fuel oil with a hole punched in the bottom over the port of the furnace so that hot air from the regenerator vaporized and burned the fuel oil as it entered the furnace as a solid stream).

Burners and Furnaces: Of major importance are characteristics of the flame, i.e., whether luminous or clear; piloting and backfiring; handling of hot gases from the burner; the burning rate of the fuels; the burning volume; and whether the installation is for one or more fuels.

Luminous or Clear Flame: Fuel and air introduced at relatively low velocity in parallel streams mix slowly and generally produce a long, luminous, soft flame. At the other extreme, fuel and air premetered, premixed, and injected at relatively

TABLE F-19. *Combustion Constants**†

No.	Substance	Formula	Molecular Weight	lb/ft^3	ft^3/lb	Sp gr (air=1.000)	Btu per ft^3 Gross (high)	Btu per ft^3 Net (low)	Btu/lb Gross (high)	Btu/lb Net (low)	Req. O_2 (mole)	Req. N_2 (mole)	Req. Air (mole)	Flue CO_2 (mole)	Flue H_2O (mole)	Flue N_2 (mole)	Req. O_2 (lb)	Req. N_2 (lb)	Req. Air (lb)	Flue CO_2 (lb)	Flue H_2O (lb)	Flue N_2 (lb)
1	Carbon‡	C	12.01	....	....	......	....	....	14.093	14.093	1.0	3.76	4.76	1.0	....	3.76	2.66	8.86	11.53	3.66	....	8.86
2	Hydrogen	H_2	2.016	0.0053	187.723	0.0696	325	275	61.100	51.623	0.5	1.88	2.38	....	1.0	1.88	7.94	26.41	34.34	....	8.94	26.41
3	Oxygen	O_2	32.000	0.0846	11.819	1.1053																
4	Nitrogen (atm.)	N_2	28.016	0.0744	13.443	0.9718																
5	Carbon monoxide	CO	28.01	0.0740	13.506	0.9672	322	322	4.347	4.347	0.5	1.88	2.38	1.0	....	1.88	0.57	1.90	2.47	1.57	....	1.90
6	Carbon dioxide	CO_2	44.01	0.1170	8.548	1.5282																
	Paraffin series:																					
7	Methane	CH_4	16.041	0.0424	23.565	0.5543	1013	913	23.879	21.520	2.0	7.53	9.53	1.0	2.0	7.53	3.99	13.28	17.27	2.74	2.25	13.28
8	Ethane	C_2H_6	30.067	0.0803	12.455	1.0488	1792	1641	22.320	20.432	3.5	13.18	16.68	2.0	3.0	13.18	3.73	12.39	16.12	2.93	1.80	12.39
9	Propane	C_3H_8	44.092	0.1196	8.365	1.5617	2590	2385	21.661	19.944	5.0	18.82	23.82	3.0	4.0	18.82	3.63	12.07	15.70	2.99	1.63	12.07
10	n-Butane	C_4H_{10}	58.118	0.1582	6.321	2.0665	3370	3113	21.308	19.680	6.5	24.47	30.97	4.0	5.0	24.47	3.58	11.91	15.49	3.03	1.55	11.91
11	Isobutane	C_4H_{10}	58.118	0.1582	6.321	2.0665	3363	3105	21.257	19.629	6.5	24.47	30.97	4.0	5.0	24.47	3.58	11.91	15.49	3.03	1.55	11.91
12	n-Pentane	C_5H_{12}	72.144	0.1904	5.252	2.4872	4016	3709	21.091	19.517	8.0	30.11	38.11	5.0	6.0	30.11	3.55	11.81	15.35	3.05	1.50	11.81
13	Isopentane	C_5H_{12}	72.144	0.1904	5.252	2.4872	4008	3716	21.052	19.478	8.0	30.11	38.11	5.0	6.0	30.11	3.55	11.81	15.35	3.05	1.50	11.81
14	Neopentane	C_5H_{12}	72.144	0.1904	5.252	2.4872	3993	3693	20.970	19.396	8.0	30.11	38.11	5.0	6.0	30.11	3.55	11.81	15.35	3.05	1.50	11.81
15	n-Hexane	C_6H_{14}	86.169	0.2274	4.398	2.9704	4762	4412	20.940	19.403	9.5	35.76	45.26	6.0	7.0	35.76	3.53	11.74	15.27	3.06	1.46	11.74
	Olefin series:																					
16	Ethylene	C_2H_4	28.051	0.0746	13.412	0.9740	1614	1513	21.644	20.295	3.0	11.29	14.29	2.0	2.0	11.29	3.42	11.39	14.81	3.14	1.29	11.39
17	Propylene	C_3H_6	42.077	0.1110	9.007	1.4504	2336	2186	21.041	19.691	4.5	16.94	21.44	3.0	3.0	16.94	3.42	11.39	14.81	3.14	1.29	11.39
18	n-Butene	C_4H_8	56.102	0.1480	6.756	1.9336	3084	2885	20.840	19.496	6.0	22.59	28.59	4.0	4.0	22.59	3.42	11.39	14.81	3.14	1.29	11.39
19	Isobutene	C_4H_8	56.102	0.1480	6.756	1.9336	3068	2869	20.730	19.382	6.0	22.59	28.59	4.0	4.0	22.59	3.42	11.39	14.81	3.14	1.29	11.39
20	n-Pentene	C_5H_{10}	70.128	0.1852	5.400	2.4190	3836	3686	20.712	19.363	7.5	28.23	35.73	5.0	5.0	28.23	3.42	11.39	14.81	3.14	1.29	11.39
	Aromatic series																					
21	Benzene	C_6H_6	78.107	0.2060	4.852	2.6920	3751	3601	18.210	17.480	7.5	28.23	35.73	6.0	3.0	28.23	3.07	10.22	13.30	3.38	0.69	10.22
22	Toluene	C_7H_8	92.132	0.2431	4.113	3.1760	4484	4284	18.440	17.620	9.0	33.88	42.88	7.0	4.0	33.88	3.13	10.40	13.53	3.34	0.78	10.40
23	Xylene	C_8H_{10}	106.158	0.2803	3.567	3.6618	5230	4980	18.650	17.760	10.5	39.52	50.02	8.0	5.0	39.52	3.17	10.53	13.70	3.32	0.85	10.53
	Miscellaneous gases:																					
24	Acetylene	C_2H_2	26.036	0.0697	14.344	0.9107	1499	1448	21.500	20.776	2.5	9.41	11.91	2.0	1.0	9.41	3.07	10.22	13.30	3.38	0.69	10.22
25	Naphthalene	$C_{10}H_8$	128.162	0.3384	2.955	4.4208	5854	5654	17.298	16.708	12.0	45.17	57.17	10.0	4.0	45.17	3.00	9.97	12.96	3.43	0.56	9.97
26	Methyl alcohol	CH_3OH	32.041	0.0846	11.820	1.1052	868	768	10.259	9.078	1.5	5.65	7.15	1.0	2.0	5.65	1.50	4.98	6.48	1.37	1.13	4.98
27	Ethyl alcohol	C_2H_5OH	46.067	0.1216	8.221	1.5890	1600	1451	13.161	11.929	3.0	11.29	14.29	2.0	3.0	11.29	2.08	6.93	9.02	1.92	1.17	6.93
28	Ammonia	NH_3	17.031	0.0456	21.914	0.5961	441	365	9.668	8.001	0.75	2.82	3.57		1.5	3.32	1.41	4.69	6.10		1.59	5.51
29	Sulfur‡	S	32.06	....	....	......	....	....	3.983	3.983	1.0	3.76	4.76	1.0 (SO_2)	....	3.76	1.00	3.29	4.29	2.00 (SO_2)	....	3.29
30	Hydrogen sulfide	H_2S	34.076	0.0911	10.979	1.1898	647	596	7.100	6.545	1.5	5.65	7.15	1.0 (SO_2)	1.0	5.65	1.41	4.69	6.10	1.88 (SO_2)	0.53	4.69
31	Sulfur dioxide	SO_2	64.06	0.1733	5.770	2.264																
32	Water vapor	H_2O	18.016	0.0476	21.017	0.6215																
33	Air	....	28.9	0.0766	13.063	1.0000																

Column groupings: columns "Req. O_2/N_2/Air" and "Flue CO_2/H_2O/N_2" (mole) fall under "For 100% Total Air, moles/mole Combustible or ft^3/ft^3 Combustible"; the (lb) set falls under "For 100% Total Air, lb/lb Combustible."

SOURCE: R. H. Perry and C. H. Chilton (eds.): "Chemical Engineers' Handbook," 5th ed., McGraw-Hill, New York, 1973.

* From American Gas Association.

† All gas volumes corrected to 60°F and 30 in. Hg dry.

‡ Carbon and sulfur are considered as gases for molal calculations only.

TABLE F-20. *Ignition Temperatures (at Atmospheric Pressure) of Gases, Liquids, and Solids*

| | Ignition Temperature | | | |
| | In Air | | In Oxygen | |
Substance	°F	°C	°F	°C
Hydrogen, H_2	1076–1094	580–590	1076–1094	580–590
Carbon monoxide, CO . .	1191–1216	644–658	1179–1216	637–658
Methane, CH_4	1202–1382	650–750	1033–1292	556–700
Ethane, C_2H_6	968–1166	520–630	968–1166	520–630
Propane, C_3H_8	. . .	. . .	914–1058	490–570
Ethylene, C_2H_4	1008–1018	542–547	932– 966	500–519
Acetylele, C_2H_2	763– 824	406–440	781– 824	416–440
Hexane,* C_6H_{14}	909	487	514	268
Decane, $C_{10}H_{22}$	865	463	396	202
Benzol, C_6H_6	1364	740	1224	662
Toluol	1490	810	1026	552
Phenol	1319	715	1065	574
Aniline	1418	770	986	530
Methyl alcohol	. . .	. . .	1031	555
Ethyl alcohol	1036	558	797	425
Propyl alcohol	941	505	833	445
Isopropyl alcohol	1094	590	954	512
n-Butyl alcohol	842	450	725	385
Amyl alcohol	768	409	734	390
Ethyl ether	649	343	352	178
Glycerin	932	500	777	414
Acetone	1292	700	1054	568
Sugar	725	385	712	378
Cylinder oil	783	417	608	320
Pennsylvania crude	601	367	468	242
Gas oil	637	336	518	270
Kerosene	563	295	518	270
Acetaldehyde	365	185	284	140
Benzaldehyde	356	180	334	168

SOURCE: R. T. Haslam and R. P. Russell, "Fuels and Their Combustion," McGraw-Hill, New York, 1926.

*Ignition temperatures for hexane and for other substances in the remainder of table are for dry air and dry oxygen.

high velocity generally produce a very short, hot, clear flame. Examples of industrial burners go from one extreme to the other.

Piloting and Backfiring: On the burner side of the combustion, consideration must be given to the burning rate of the fuel-air mixture so that enough fuel is burned at a piloting point, on an ignition tile, or on a burner block for the flame to be under control and firmly anchored at the appropriate location. With inadequate anchor, the flame will "blow off," and with excess piloting a premixed fuel and air may burn back into the piping supplying the burner at low firing rates.

Hot Gases from the Burner: To make a good application, it must be remembered that once a burner is chosen and is operating properly, there is a column of hot gases issuing from the burner or burner block. When this column of hot gases is visible, one is inclined to think of it as flame and try to avoid letting it touch the work being heated, but in clear-flame burners, where this column of hot gases is not visible, it must not be forgotten that the hot gases are there, and precautions must be taken in the placement of burners or work to avoid the impingement of this hot gas column on the work. Not all burners are the same. Some are designed specifically to spread their hot gases along the wall in which they are mounted, and in this case the work may be relatively close; other burners are designed to keep the hot gases intact, and with these the work must be kept fairly well removed.

Burning Rate of Fuels: Also affecting the length of the flame in burner design and the burning volume required is the burning rate of the fuel. Hydrogen and carbon monoxide burn relatively rapidly, while more complex compounds or mixtures, e.g., methane or oil, burn relatively slowly.

Burning Volume: On the furnace side of the burner, volume is required in order to give the combustion time to reach completion. In slow-burning applications, as much as $50\ ft^3/(million\ Btu)(hr)$ is required while high-velocity well-mixed fuel and air burn so rapidly that less than $0.006\ ft^3/(million\ Btu)(hr)$ is required.

In general this combustion volume should not be pressed to the minimum, as additional firing will cause burning to take place beyond the furnace chamber or in the stack.

One or More Fuels: Interest in precise placement of heat in chambers of limited size and lower temperatures has led to a considerable science in devices for metering, mixing, and piloting fuels. Some burners are very specialized to one particular fuel and firing condition, and others are ingeniously contrived to burn several kinds of fuels under a wide variety of conditions.

The discussion so far has centered around the use of one fuel at a time, but sometimes more than one fuel is burned simultaneously. One commercial example is atomization of fuel oil with natural gas. In this case, it is obvious that sufficient oxidant (air) to consume the fuel values (primarily carbons and hydrogens) in both fuels must be provided. A more subtle but equally important example is known as *incineration*.

Incineration. Incineration is usually considered when solids, liquids, or gases which contain some fuel value are to be thrown away by oxidizing them and dumping the products of combustion into the atmosphere. While this concept may have passed its peak of commercial and esthetic popularity, some understanding of its proper execution is in order.

Besides providing adequate oxygen for the carbon, hydrogen, and even sulfur values present, the reaction must be given adequate *time, temperature,* and *turbulence*. There must be adequate heat to pay off all the losses, including vaporizing water, and to get the reaction to the combustion temperatures; there must be adequate turbulence to mix the reactants thoroughly so that oxygen is exposed to all parts of the burning material and heat and temperature are conveyed to all parts of the reaction; and finally there must be adequate time for

the reaction to go to completion and not be quenched by contact with cold walls or cold atmosphere.

Of course, after performing the combustion reaction completely, sulfur oxides, nitrogen oxides, particulate matter, and in some cases water vapor may have to be removed. There is some indication that even the temperature of the hot gases being dumped into the atmosphere may have to be limited in the future.

Combustion-System Enhancement

Preheating: Economy dictates that one extract as much heat as possible from the combustion reaction and while still achieving the temperature required. In a simple illustration, merely maintaining a vessel at a given temperature requires exhausting the products of combustion at some temperature higher than that at which the box is maintained. If, however, some other process requirement can utilize heat at waste-gas temperature, some design options are available.

Preheat the Combustion Air: Often it is convenient and useful to use preheated air for combustion. Depending upon the equipment available, this air may have temperatures ranging from a few hundred degrees to $2000°F$ ($1093°C$). This adds sensible heat to the process, shortens the flame, and raises the maximum flame temperature.

Preheat the Fuel: Usually this is not fruitful unless the fuel is very lean, as preheating must be maintained below the cracking temperature of the fuel and the weight of the fuel is relatively small with respect to the weight of the air used in the combustion reaction.

Preheat the Product: Before introducing the product into the main chamber of the furnace, it may be preheated. A good example is a continuous furnace where the hot gases flow counter to the work flow.

Generation of Steam: When steam is required in a plant, it is always good practice to consider this additional function of a combustion system in the design phase rather than later fitting the system for steam generation.

Use of Oxygen: Higher flame temperatures, faster burning, and shorter flames will be produced when oxygen is used. Because of the relatively high costs of oxygen, only special circumstances warrant its use. Oxyacetylene cutting torches and fires for forming laboratory glassware are good examples.

Reforming: Should the size of the process warrant it, a gain in thermal efficiency can be achieved by steam-hydrocarbon reforming, using waste heat to break the constituents to H_2 and CO and preheat them before entry into the combustion chamber. This approach generally is confined to very large processes operating above 1500°F (816°C). See also **Reforming.**

Combustion Control Systems. Four factors are of major concern in a combustion control system: (1) temperature, (2) fuel-air ratio, (3) furnace pressure, and (4) safety.

Inspirator burners (venturi) pump air in proportion to fuel flow. Premix systems must embody ratio equipment. Most large systems meter and control the fuel and air, adjustment of controls being tied to flue-gas samples. The control of furnace pressure is very desirable to minimize air infiltration or "sting-out."

Safety controls are of high priority to all plant operators, insurers, municipalities, and in fact everyone closely associated with combustion processes. In the United States, Factory Mutual, Factory Insurance Association, Underwriters Laboratories, and various standards associations, as well as individual plants, evolve and disseminate advice, restrictions, or approvals with regard to the use of flammable and combustible materials and combustion safety equipment. On the face of it, the problem is simple and obvious: avoid the accumulation of combustible mixtures particularly in confined spaces. Prudent use of commercial safety equipment will reduce the hazard, while carelessness and neglect of reasonable safety precautions will certainly increase it.

One particularly insidious hazard is that resulting from the use of fuels that produce heavy vapors (specific gravity greater than 1), e.g., propane, butane, and diesel fuel. Each year newspapers describe destructive fires on small boats due to the failure to ventilate the engine space properly prior to start-up.

REFERENCES

American Gas Association: "Gas Engineers Handbook," Industrial Press, New York, 1965.

Haslam, R. T., and R. P. Russell: "Fuels and Their Combustion," McGraw-Hill, New York, 1926.

Jost, Wilhelm: "Explosion and Combustion Processes in Gases," McGraw-Hill, New York, 1946.

Lewis, B., and G. von Elbe: "Combustion, Flames, and Explosions," Academic, New York, 1961.

McAdams, W. H.: "Heat Transmission," 3d ed., McGraw-Hill, New York, 1954.

Perry, R. H., and C. H. Chilton (eds.): "Chemical Engineers' Handbook," 5th ed., McGraw-Hill, New York, 1973.

—Charles G. Bigelow, Jr., *Selas Corporation of America, Dresher, Pa.*

Fuels, Nuclear   (See **Nuclear power plants.**)

Fuel Additives   (Consult the **Subject Index.**)

Fuel Cell   A fuel cell, like a battery, converts chemical energy directly into electric energy. A fuel cell and a battery are similar in many respects. Both devices operate silently with no moving mechanical parts and generate little or no noxious fumes. Both devices generate a direct current rather than an alternating current and both devices are well adapted to uses where portability is important. Thus, in considering the potential use of the fuel cell as an energy source, there is a tendency to regard the fuel cell as a "superbattery." This is evidenced by the successful fuel cell applications to date, principally in spacecraft, space probes, and special military vehicles where the fuel cell has replaced more conventional batteries.

The primary feature that distinguishes the fuel cell from a battery is the ability of the fuel cell, within practical operating limitations, to produce electric power continuously just as long as the raw reactive ingredients are supplied to the cell. Thus, unlike a primary battery, the fuel cell is not limited to the generation of a given quantity of electric power because of fixed quantities of starting ingredients. The fuel cell also is unlike a secondary battery since the basic chemical reactions are not reversible in a practical sense and, hence, a fuel cell is not rechargeable by periodically reversing the chemistry through the application of an external source of electric current. Although fuel cells and batteries are similar in output characteristics and in several aspects of their construction, it is more realistic to consider the fuel cell in the sense of a continuous chemical processing plant, consuming raw materials at a constant rate to produce electricity as its principal aim but, as is the case of most chemical processing plants, the producing of by-products as well. It is in the terms of possible scale-up to large-size electric current producing units that fuel cells, unlike batteries, are of interest as potential major sources of electric current.

Considering the worldwide energy shortage, fuel cells are fundamentally attractive because theoretically they are very efficient power sources. Whereas steam-powered electric generators have nearly reached their peak of attainable efficiency (approximately 40% fuel conversion to useful energy) and power systems based upon internal-combustion engines are from 20–25% efficient, the fuel cell offers theoretical operating efficiency percentages in the high nineties. Much research remains, however, before the fuel cell can be considered a serious threat to conventional power sources, notably the requirement to develop far better catalysts to accelerate and promote reactive efficiency and to find ways to scale-up experimental or essentially smaller, portable device designs to large-scale generating units.

The concept of the fuel cell dates back to the early work of Ludwig Mond (Germany) and Carl Langer (Britain) in 1855 who built rudimentary devices which they termed "fuel cells." The work of Mond and Langer was spurred by an earlier demonstration (1839) by the English chemist, Sir William Grove, who showed that the production of hydrogen and oxygen by the electrolysis of water could be reversed, namely, that by bringing hydrogen and oxygen together in a suitable apparatus, water could be formed in a controlled fashion with the production of an electric current as a by-product.

Although researchers over the years have been attracted by the potential of the fuel cell because of its theoretically large efficiency as a power source, the fuel cell essentially remained a curiosity until urgent power needs of the space program required turning to nonconventional power sources. For example, the power needs of large, manned space vehicles would have required several tons of nickel-cadmium batteries. These, of course, were ruled out immediately because of the high energy-to-weight ratio in the kilowatt range needed. Nor were solar cells, which were satisfactory for unmanned satellites and probes, acceptable because of their comparatively low power output and the complexities inherent in orienting the cells to the sun. Thus, the space program gave large impetus, both in the Unites States and Soviet Union, to intensive fuel-cell research. The first fuel cells used were aboard the Gemini flights. Fallout from the space program, coupled with serious concern over the energy shortage, has greatly accelerated current fuel-cell research programs. Nevertheless, many critical design problems remain, particularly in terms of practical, large-scale units.

Hydrogen-Oxygen Cells: Most research to date has been directed at the reaction first proposed by Sir William Grove. As in the case of other fuel cells, the hydrogen-oxygen cell consists of an anode and a cathode, separated by a medium containing an electrolyte. Gaseous hydrogen (the fuel) is fed to and permeates a porous anode. By catalytic action, the hydrogen atoms are stripped

of their electrons which then flow by an external circuit (hence a source of electric current that can be tapped) to the cathode. Hydrogen ions (hydrogen atoms minus their electrons) flow through the ion-conducting medium (the electrolyte) to the cathode. Oxygen (purified or as an ingredient of air) is fed to the cathode. Each oxygen atom readily accepts two electrons and two ions to form a molecule of water. Thus, all steps in the reaction are completed.

Alkaline electrolytes (usually KOH) have been favored to date because the cathodic reduction of O_2 proceeds with less polarization in such solutions than in the case of neutral or acidic media. A disadvantage of an alkaline electrolyte is the accumulation of carbonates in the electrolyte resulting from any CO or CO_2 present in the raw hydrogen. Also, if a cell is operated on air (instead of purified O_2) for long periods, any CO_2 present in the air causes similar carbonate buildup. At least one commercial cell utilizes an ion-exchange membrane which is the equivalent of an acid electrolyte. The electrodes used to date in successful cells include carbon, nickel, platinum, and palladium—all of a porous nature to which catalysts are added to accelerate electrode reactions. Just in considering the hydrogen-oxygen cell, the choice of electrode materials, catalysts, and operating pressures and temperatures demonstrates the extensive research required to improve cell operation.

Theoretically, the standard potential for a hydrogen-oxygen cell is 1.23 V at 25°C. In designs to date, generally the terminal voltage does not exceed 1.05 V even under open-circuit conditions. Under load conditions, the voltage rarely exceeds 0.9 V and may be as low as 0.75 V. Investigations show that a considerable portion of the voltage loss occurs at the cathode due to the formation of H_2O_2 as an intermediate product during the reduction of O_2.

The reactions of an alkaline hydrogen-oxygen cell are

Anode: $2H_2 + 4OH^- \rightarrow 4H_2O + 4e^-$
Cathode: $O_2 + 2H_2O + 4e^- \rightarrow 4OH^-$
Cell: $2H_2 + O_2 \rightarrow 2H_2O$

Hydrazine-Oxygen Cells: Considerable research has gone into the development of fuel cells which use hydrazine rather than hydrogen as the fuel. Generally, a 65% (weight) hydrazine, N_2H_2, solution is fed directly to the anode compartment. Carbon electrodes, a caustic electrolyte, and suitable catalysts are used. Except for special applications, the comparatively high cost of hydrazine is a severe disadvantage of this configuration. Some efforts have been directed toward the substitution of ammonia for hydrazine, but to date such cells have been weakened by excessive polarization problems. The reactions of the hydrazine-oxygen cell are

Anode:

$$N_2H_4 + 4OH^- \rightarrow N_2\uparrow + 4H_2O + 4e^-$$

Cathode:

$$O_2 + 2H_2O + 4e^- \rightarrow 4OH^-$$

Cell:

$$N_2H_4 + O_2 \rightarrow N_2\uparrow + 2H_2O$$

Sodium Amalgam–Oxygen Cells: Aside from the disadvantages of the relatively high costs for the raw materials (sodium and mercury) and the costs entailed in the preparation of sodium amalgam, the sodium amalgam–oxygen cell possesses certain advantages, including a relatively high cell voltage (1.5 V at 150 mA/cm^2 of anode cross section) and an attractive power output as compared with the hydrogen-oxygen cell. The liquid sodium-mercury amalgam is allowed to flow over a vertically mounted steel plate in the anode chamber. The oxygen cathode is similar to that used in hydrogen-oxygen cells. The reactions of the cell are

Amalgamation:

$$4Na + 4x\ Hg \rightarrow 4Na(Hg)_x$$

Anode:

$$4Na(Hg)_x \rightarrow 4Na^+ + 4x\ Hg + 4e^-$$

Cathode:

$$O_2 + 2H_2O + 4e^- \rightarrow 4OH^-$$

Cell:

$$4Na + O_2 + 2H_2O \rightarrow 4NaOH$$

It is interesting to note that this cell generates its own electrolyte (NaOH) and consumes rather than generates water. Fortunately, seawater without prior demineralization can be used as makeup water. The cell also has an attractive operating temperature range of 20–80°C. The possible use of such cells in connection with chlorine-caustic production plants may prove attractive.

Hydrocarbon-fueled Cells: Particularly in connection with the development of fuel cells as possible major sources of electric power, both stationary and portable, much research is being directed to the utilization of hydrocarbons as the fuel rather than hydrogen. The advantages of the

readily available raw materials and the possible conversion of these materials into useful energy without first going through a heat cycle are evident.

Generally, the approach to date has been to use porous metal electrodes which are in contact with a fused alkali metal carbonate electrolyte. The latter obviously connotes high-temperature operation (in excess of 500°C). A cell of this type, operating on propane for example, involves the following reactions:

Anode:

$$C_3H_3 + 10CO_3^{-2} \rightarrow 13CO_2 + 4H_2O + 20e^-$$

Cathode:

$$5O_2 + 10CO_2 + 20e^- \rightarrow 10CO_3^{-2}$$

Cell:

$$C_3H_3 + 5O_2 \rightarrow 3CO_2 + 4H_2O$$

Because O_2 and CO_2 are cathode reactants, the maintenance of an acceptable output voltage requires that these materials be supplied to the cathode. Although CO_2 is available from the anode, capturing it from this source increases the complexity of the system. Therefore, an external source of CO_2 generally is considered the most desirable approach.

Operation of fuel cells on hydrocarbon feed and O_2 in which very concentrated phosphoric acid electrolytes and platinum or platinum-alloy electrodes are used have been reported, but to date the operating voltages have been excessively low (0.3 V or less). Research also is proceeding in the use of gaseous hydrocarbon-water and methanol-water mixtures. Most alcohol-consuming cells developed to date yield reasonable power densities when operated with caustic electrolytes in the temperature range of 40–100°C, but the accumulation of carbonates within the cells is excessive.

Fuel Oil   (See **Fluid catalytic cracking.**)

Fuller's Earth   (See **Adsorption.**)

Fulminate of Mercury   (See **Explosives.**)

Fumaric Acid   Fumaric acid is the trans isomer of maleic acid. It is similar in many respects to maleic acid and often is used interchangeably. Fumaric acid usually is produced in conjunction with maleic anhydride.

Since the early 1960s, fumaric acid has seen its major growth in its use as a food acidulant, competing with the commonly used citric acid. Because of fumaric acid's low cost and the low concentration required, food acidulants now account for about 30% of the consumption of fumaric acid. Low-moisture pickup makes it suitable for use in gelatin desserts and powdered food mixes, e.g., pie filling and milk-based puddings, to provide greater storage stability. Soft drinks constitute the largest market for food acids. In this field, fumaric acid ranks third after citric and phosphoric acids.

Unsaturated polyester resins consume about 24% of the fumaric acid produced. In this category, fumaric acid competes with less expensive maleic anhydride, but it is claimed that fumaric acid has better heat resistance and makes harder resins. Fumaric acid also is used in paper-size resins, surface coatings, and plasticizers. United States capacity for fumaric acid in 1970 was about 75 million lb.

Fumaric acid has been produced as a by-product of maleic anhydride. See also **Maleic anhydride.** Benzene is oxidized to maleic acid, which is dehydrated to maleic anhydride; this is purified and hydrated to maleic acid, which is catalytically isomerized to fumaric acid.* In another process, maleic acid is catalytically isomerized to fumaric acid, bypassing the production of maleic anhydride. The resulting fumaric acid is crystallized, washed, and dried. In a related manner, malic acid can also be produced. See also **Isomerism; Paint.**

—Robert Merims, *Scientific Design Company, Inc.,*
New York

Fungicides   (Consult the **Subject Index.**)

Furan Group   The furans are members of that class of heterocyclic compounds having in common a nucleus which is a five-membered ring containing four atoms of carbon and one of oxygen. In the so-called *simple furans,* this nucleus occurs as a fully unsaturated monocycle:

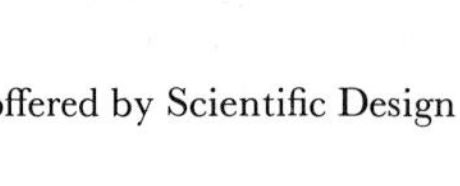

Furan
I

In the *condensed furan* series the nucleus is fused to at least one more cyclic compound. Typical struc-

*Processes offered by Scientific Design Company, Inc.

tures for condensed furans in which the other ring system is benzenoid are

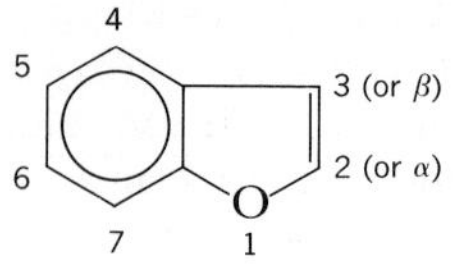

Benzofuran
II

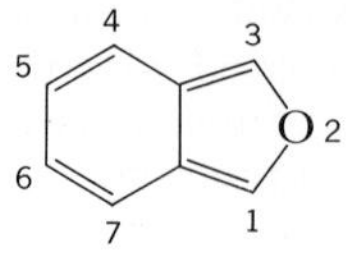

Isobenzofuran
III

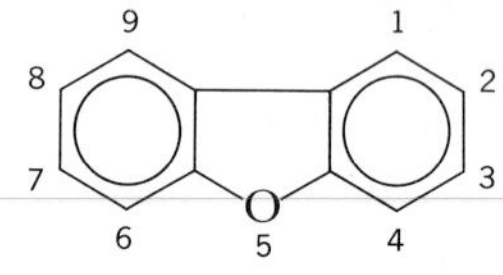

Dibenzofuran IV

The parent fully unsaturated monocycle is called *furan;* its completely saturated counterpart, *tetrahydrofuran.* The trivial names are firmly implanted and not likely to be displaced by the respective conventional (IUPAC) names oxole and oxolane.

Substituents on the furan ring are designated by the numbering system shown in I; the Greek letter notation is employed in the earlier literature. Because of asymmetry conferred by the hetero atom, the α and β positions are different. Thus, two isomeric monosubstituted products, e.g., 2- and 3-methylfuran, are possible. The number of isomers obtainable on further substitution can be determined logically from I.

In addition, certain terms have been used to designate radicals derived from furan, such as 2-furyl (V), 2-furfuryl (VI), 2-furoyl (VII), and 2-furfurylidene (VIII); likewise for the β (or 3) series.

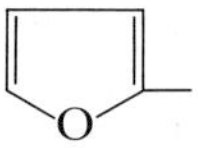

V

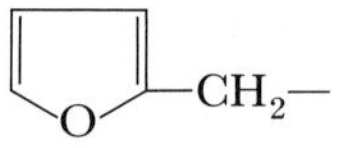

VI

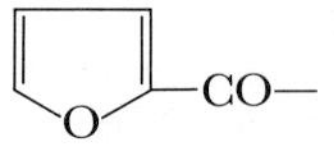

VII

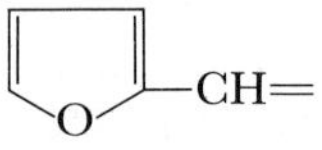

VIII

In the condensed-ring systems, trivial names such as benzofuran and isobenzofuran are preferred to the more definitive conventional names benzo[*b*]furan and benzo[*c*]furan respectively; the letter in the brackets designates which side of the furan nucleus I is fused to the benzene ring. By convention, polycyclic systems are numbered around the entire periphery, omitting the ring junctions. Position numbers for some condensed furans are illustrated in structures II–IV.

Except for those few which occur in constituents of essential oils, the fully unsaturated monocyclic furans are synthesized. The major methods involve cyclization reactions of open-chain compounds.

Dehydration of Sugars: The industrially significant furfural (precursor of other commercially important furans) is produced by acid-catalyzed dehydration of pentoses. Similarly, 5-methylfurfural is obtained from methylpentoses and 5-hydroxymethylfurfural from hexoses.

Paal-Knorr Synthesis: A general method for making furans is the acid-induced cyclization of 1,4-dicarbonyl compounds (visualized as occurring via the dienol). 2,5-Dimethylfuran is thus prepared from acetonylacetone and furan (in poor yield) from succindialdehyde.

Feist-Benary Synthesis: Another route to furans involves the condensation of an α-halocarbonyl compound with a β-ketoester. Ethyl 2,5-dimethyl-3-furancarboxylate, for example, is synthesized by reaction of α-chloroacetone with ethyl acetoacetate in the presence of pyridine.

The furan nucleus is shown in I as a conjugated diene and cyclic vinyl ether. Physical studies indicate that the ring is planar, with bond lengths between those normal for single- and double-bonded C—C and C—O bonds. This configuration requires delocalization of four π electrons from the carbon atoms and two paired electrons from the oxygen atom. The resulting sextet of delocalized electrons confers some aromatic character to the furan nucleus, which is confirmed by an observed stabilization energy of 25 ± 1 kcal/mole (the stabilization energy of the benzene ring is 37 ± 1 kcal/mole).

In agreement with the physical data, furans are less aromatic and hence more reactive than analogous benzene compounds. They undergo a greater variety of addition reactions (conjugated diene), are more susceptible to cleavage to open-chain forms (vinyl ether), and undergo substitution reactions more readily.

Through resonance, centers of high electron density are established at the 2 and 5 positions of the ring. Consequently, electrophilic substitution proceeds with great facility and predominantly in an α position if available. Orientation

follows the typical pattern of the benzene series; e.g., on nitration, 3-methylfuran yields 2-nitro-3-methylfuran (ortho) and 3-furancarboxaldehyde yields 2-nitro-4-furancarboxaldehyde (meta).

The introduction of substituents which interact electronically with the ring markedly affects the balance between aromatic and aliphatic character. For example, furan and its homologues participate as dienes in the Diels-Alder reaction, forming cyclic adducts by 2,5 addition of the dienophile. Furan types such as the halofurans, alkoxyfurans, furfuryl esters and ethers, and furfural diacetate react similarly, but furans containing electron-withdrawing substituents, e.g., furfural, furoic acid, and nitrofurans, demonstrate a considerably enhanced aromaticity in that they fail as dienes, even with the strongest dienophiles. Furthermore, whereas furan and its homologues are cleaved on acidic hydrolysis to open-chain dicarbonyl compounds (usually with some resinification), the negatively substituted furans are stabilized to cleavage by acids. However, the furan ring is deactivated by electron-withdrawing substituents and subject to nucleophilic attack. 2-Nitrofuran, particularly resistant to acids, is easily cleaved by strong alkali.

There are no analogs of phenols or aromatic amines in the furan series. No free furanol or furylamine exists; reactions which could lead to them invariably revert to the tautomeric dihydro derivative (or its decomposition product). For example, hydrolysis of 2-acetoxyfuran does not provide 2-furanol; instead, cleavage to β-formylpropionic acid occurs (presumably via 3-butenolide, the tautomer of 2-furanol).

Two types of partially saturated monocyclic furans are possible, the 2,3- (IX) and the 2,5-dihydrofurans (X).

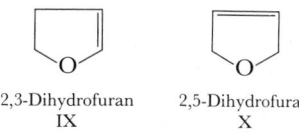

2,3-Dihydrofuran   2,5-Dihydrofuran
IX   X

The 2,3-dihydrofurans are enol ethers of γ-hydroxycarbonyl compounds, into which they readily revert on acid hydrolysis. The 2,5-dihydrofurans are prepared by acid-catalyzed cyclodehydration of appropriate *cis*-2-butene-1,4-diols; they are less sensitive to acids than the 2,3-dihydrofurans. A special class of 2,5-dihydrofurans is obtained on 2,5 addition of methanol or acetic acid, under controlled oxidative conditions, to furans free of electronegative substituents. These 2,5-dihydrofurans are cyclic acetals or acylals of unsaturated 1,4-dicarbonyl compounds to which they are readily converted on mild (acidic) hydrolysis. 2,5-Dimethoxy-2,5-dihydrofuran, for example, is hydrolyzed to malealdehyde.

Some tetrahydrofurans, i.e., fully saturated monocyclic furans, are prepared by cyclization of saturated 1,4-halohydrins and 1,4-glycols. These tetrahydrofurans are cyclic ethers. Other tetrahydrofurans are obtained by catalytic hydrogenation of furans and dihydrofurans. The hydrolytic stability of these tetrahydrofurans is governed by the substituents present in the α positions. (Tetrahydrofurans which comprise sugars in the furanose form are not considered here.)

Monocyclic Furans. Commercially available monocyclic furans include furan, tetrahydrofuran, methylfuran, methyltetrahydrofuran, furfural, furfuryl alcohol, and tetrahydrofurfuryl alcohol. Selected physical constants for the pure compounds are given in Table F-21.

Furan (furfurane, oxole), C_4H_4O, is a colorless, highly flammable, low-boiling liquid with a pronounced ethereal odor. It is miscible with most organic solvents, including aliphatic and aromatic hydrocarbons, chlorinated hydrocarbons, alcohols, esters, ethers, and ketones; the solubility of furan in water is ~1% at room temperature. Furan forms an azeotrope with water (1.2% H_2O) boiling at 30.5°C.

Furan is manufactured by catalytic decarbonylation of furfural. It has also been prepared by decarboxylation of 2-furoic acid. Furan is a potentially hazardous chemical and repeated exposure to the liquid or vapor should be avoided. Furan has a low flash point (−35°C, Tag closed cup) and its upper and lower flammability limits are 2.3 and 14.3% by volume in air, respectively. On exposure to air, furan forms an unstable peroxide by addition of oxygen across the 2,5 positions; due caution should be exercised whenever furan is to be distilled.

In the absence of catalysts and oxygen, furan is quite stable to heat. Pyrolysis to carbon monoxide, hydrocarbons, and hydrogen takes place on passage through a quartz tube at 670°C or in contact with nickel at 360°C. Maleic anhydride is formed on vapor-phase oxidation (air) over vanadium pentoxide. Furan reacts with ammonia in the presence of alumina at 400–450°C to give pyrrole; thiophene is formed with hydrogen sulfide and selenophene with hydrogen selenide under comparable operating conditions. Furan is hydrogenated to tetrahydrofuran over nickel catalysts at moderate temperature and pressure. Under more vigorous conditions, as well as with other catalysts, hydrogenolysis occurs, leading to *n*-butanol or butadiene; in the presence of steam, 1,4-butanediol is formed.

Because furan is sensitive to acids, electrophilic-substitution reactions are performed under carefully controlled conditions (see Table F-22 for reagents and typical products). Some of these reactions proceed via addition-elimination; e.g., the intermediate 5-acetoxy-2-nitro-2,5-dihydrofuran is obtained by the action of fuming nitric acid and acetic anhydride on furan before pyridine is added to strip off acetic acid (2,5) to form 2-nitrofuran. Principal metallic derivatives of furan are the mercurials (see Table F-22), 2-furyllithium from furan and butyllithium, and 2-furylsodium from furan and phenylsodium.

Tetrahydrofuran (tetramethylene oxide, oxolane), C_4H_8O, the fully saturated monocycle, is a colorless liquid with a characteristic etherlike odor and

TABLE F-21. *Physical Properties of Typical Furan Compounds*

Compound	Molecular Weight	Mp, °C	Bp, °C	Density, d_4^{20}	Refractive Index, n_D^{20}
Furan	68.076	−85.6	31.3	0.9378	1.4214
Tetrahydrofuran	72.108	−108.5	66	0.888	1.4073
Methylfuran	82.103	−88.68	~64	0.913	1.4320
Methyltetrahydrofuran	86.135	−136	80.2	0.854	1.4025*
Furfural	96.086	−36.5	161.7	1.1598	1.5261
Furfuryl alcohol	98.102	−14.63†	170	1.1285	1.4868
Tetrahydrofurfuryl alcohol	102.134	<−80	178	1.0543‡	1.4520

*n_D^{25}.

†Stable crystalline form; metastable form, −29°C.

‡d_{20}^{20}.

TABLE F-22. *Electrophilic-Substitution Reactions of Furan*

Reaction	Reagents	Principal Product*
Acylation	Acetic anhydride, BF_3	2-Acetylfuran
Alkylation.	Ethylene, BF_3	2-Ethylfuran
Bromination	Dioxane dibromide	2-Bromofuran
Chlorination	Cl_2, CH_2Cl_2 (below 0°C)	2-Chlorofuran†
Coupling	p-Chlorophenyldiazoacetate	2-(p-Chlorophenyl) furan
Formylation	(1) HCN, HCl; (2) H_2O	Furfural
Mercuration	$HgCl_2$, CH_3CO_2Na	2-Furylmercuric chloride
Nitration	(1) $CH_3CO_2NO_2$; (2) pyridine	2-Nitrofuran
Sulfonation	Pyridine-SO_3	2-Furansulfonic acid

*Some 2,5-disubstituted products are also formed.
†At low chlorine-furan ratio.

miscible with most common solvents including water. It is used as a solvent for polyvinyl chloride, polyvinylidene chloride, and some polyurethane resins in coating, film casting, and adhesive applications. Tetrahydrofuran is also an excellent medium for Grignard reactions and metal hydride reductions. Tetrahydrofuran forms an azeotrope with water (5.3% H_2O) boiling at 64°C.

Tetrahydrofuran is manufactured by catalytic hydrogenation of furan. It is also produced by cyclodehydration of 1,4-butanediol (obtained on hydrogenation of 1,4-butynediol from the reaction of acetylene with formaldehyde). Tetrahydrofuran has a low flash point (−14.5°C, Tag closed cup), and its flammability limits are 1.8–11.8% by volume in air. On exposure to air, tetrahydrofuran forms an explosive peroxide; commercially available tetrahydrofuran is stabilized to prevent accumulation of peroxide. More than trace amounts of peroxide should be destroyed before distilling tetrahydrofuran.

Tetrahydrofuran hydroperoxide is obtained in high yield at 25°C by passing air (oxygen) through tetrahydrofuran containing cobalt or nickel salts. Heating above room temperature decomposes the hydroperoxide to γ-butyrolactone and 2-tetrahydrofuranol (cyclic acetal of 4-hydroxybutyraldehyde). Oxidation of tetrahydrofuran with nitric acid yields succinic acid. Tetrahydrofuran reacts with ammonia over alumina (gas phase, 400°C) to give pyrrolidine; with hydrogen sulfide, tetrahydrothiophene is produced in high yield. Reaction of tetrahydrofuran with chlorine under mild conditions results in 2,3-dichlorotetrahydrofuran as the main product. With a free-radical initiator, tetrahydrofuran adds to maleic anhydride, forming tetrahydrofuryl-succinic anhydride.

Cleavage of the tetrahydrofuran ring affords a variety of products. With gaseous hydrogen chloride (room temperature; desiccant) 1,4-dichlorobutane is produced, together with some 4-chlorobutanol and di(4-chlorobutyl) ether. Acyl chlorides provide esters of 4-chlorobutanol, and anhydrides give esters of 1,4-butanediol. Carbon monoxide reacts with tetrahydrofuran in the presence of cobalt or nickel carbonyl (formed *in situ*), yielding δ-valerolactone and adipic acid. Most noteworthy of the ring-opening reactions is the cationic polymerization of tetrahydrofuran to poly(tetramethylene ether) glycol, $HO[(CH_2)_4O]_nH$, which, with appropriate isocyanates and extenders, is used in the manufacture of polyurethane elastomers (cast and thermoplastic systems), elastic fibers, and flexible foams.

Methylfuran (2-methylfuran, sylvan), 2-$C_4H_3O \cdot CH_3$, is obtained by catalytic hydrogenation (hydrogenolysis) of furfural (over copper chromite) or furfuryl alcohol (over copper). It is a colorless, flammable, ethereal liquid with a low flash point (−27°C, Tag closed cup). Methylfuran is miscible with most organic solvents; solubility in water is only 0.3% at 20°C. It forms an azeotrope with water (3% H_2O) boiling at 57–59°C. As with furan, precautions should be taken in handling and when distilling methylfuran.

Generally, the reactions of methylfuran are quite similar to those of furan. In the presence of sulfur dioxide, methylfuran adds to α,β-unsaturated carbonyl compounds; e.g., reaction with methyl vinyl ketone leads to (5-methyl-2-furfuryl)acetone. Methylfuran condenses with aldehydes or ketones (acid catalyst) to yield difurylalkanes; with formaldehyde, di(5-methyl-2-furylmethane) is obtained in low yield. Closely related is the Mannich (aminomethylation) reaction; using formaldehyde and dimethylamine hydrochloride, *N,N*-dimethyl-5-methyl-2-furfurylamine is readily prepared. Hydrogenation (hydrogenolysis) products of methylfuran include 2-methyltetrahydrofuran (Ni, 120°C), 2-pentanol (Pt, acetic acid), 2-pentanone (Ni, 225°C) and 1-hydroxy-4-pentanone (Pd, aqueous acetone).

Methyltetrahydrofuran (2-methyltetrahydrofuran), 2-$C_4H_7O \cdot CH_3$, is a colorless, mobile liquid with etherlike odor and low flash point ($-11°C$, Tag closed cup). Methyltetrahydrofuran is miscible with most organic solvents; with water it exhibits inverse solubility, i.e., its solubility in water decreases as the temperature increases (22.1% at 0°C, 13.9% at 25°C). It forms an azeotrope with water (10.6% H_2O) boiling at 71–72°C. Like most saturated ethers, it forms peroxides readily on contact with air; the usual precautions should be taken.

In most of its reactions methyltetrahydrofuran behaves much like tetrahydrofuran. Ring cleavage with unsymmetrical reagents, however, leads to a mixture of isomers; e.g., acetyl bromide yields 1-bromo-4-acetoxpentane (primary halide) and 1-acetoxy-4-bromopentane (secondary halide). The ratio of isomers depends largely on temperature, being 88% primary bromide at 121°C and only 39% at 50–60°C. All attempts thus far to prepare the homopolyether by cationic ring-opening polymerization of methyltetrahydrofuran have failed. Catalytic dehydration over a boron phosphate catalyst at 350°C provides piperylene (1,3-pentadiene) in high yield.

Furfural (2-furaldehyde, 2-furancarboxaldehyde), 2-$C_4H_3O \cdot CHO$, is a liquid aldehyde with a characteristic, pungent almondlike odor. Colorless when freshly distilled, it turns brown in contact with air. Furfural is miscible with most of the common organic solvents except saturated aliphatic hydrocarbons; its solubility in water is 8% at 20°C. Major applications for furfural are as a selective solvent in petroleum and rosin refining, in the extractive distillation of butadiene from other C_4 hydrocarbons, as a solvent and ingredient in the manufacture of bonded phenolic products, e.g., abrasive wheels and brake linings, and as a chemical intermediate in the production of other furans and tetrahydrofurans.

Furfural is manufactured from pentosan-rich agricultural residues such as corncobs, cottonseed hulls, oat hulls, rice hulls, and bagasse. The process involves digesting the raw material with dilute acid and removing the furfural continuously by steam distillation. The binary system furfural-water is nonideal, and at atmospheric pressure a minimum-boiling azeotrope (65% H_2O) distilling at 97.9°C is formed.

Thermally, furfural is quite stable in the absence of oxygen and specific catalysts; other than darkening in color, many hours heating at 230°C is necessary to produce detectable changes in physical properties. Furfural is not subject to extensive degradation by dilute acids except on long exposure at high temperature. In the presence of oxygen, furfural undergoes autoxidation leading to acidic ring-cleaved products (inhibited by small amounts of tertiary amines); catalytic vapor-phase oxidation provides maleic acid.

Furfural undergoes most of the reactions typical of aromatic aldehydes, complicated at times with side reactions stemming from the reactive ring. It gives a bisulfite complex, forms acetals, reacts with Grignard reagents, condenses with active methylene compounds, e.g., acids, anhydrides and esters, aldehydes and ketones, nitroparaffins, and affords the usual nitrogen derivatives (such as hydrazones). Furfural can be reduced to furfuryl alcohol, oxidized to furoic acid, and decarbonylated to furan. Sodium cyanide catalyzes dimerization of furfural to furoin (the furan analog of benzoin). In an unusual diene reaction, furfural (as dienophile) adds 2 moles of butadiene to form 2,3:4,5-bis(2-buteno)tetrahydrofurfural. Reaction of furfural with aqueous ammonium salts at ~150°C gives 3-pyridol in low yield. Aqueous bromine oxidizes furfural to mucobromic acid (α,β-dibromo-β-formylacrylic acid).

Furfuryl alcohol (2-furanmethanol, 2-furylcarbinol), 2-$C_4H_3O \cdot CH_2OH$, is manufactured by the continuous vapor-phase hydrogenation of furfural at relatively low pressure. The alcohol is a colorless liquid soluble in water and most organic solvents except the aliphatic hydrocarbons. On storage, furfuryl alcohol becomes gradually less soluble in water due to formation of bimolecular condensation products, a process accelerated by heat and/or acids but retarded by small amounts of bases. One use for furfuryl alcohol is as a dye solvent or dispersant in the textile industry.

Furfuryl alcohol is readily converted to resins in the presence of acidic materials. The reaction is highly exothermic, demanding careful control of catalyst concentration and temperature. The resins are thermosetting; when fully cured they are heat-stable and chemically resistant. Such resins are used in the production of corrosion-resistant mortars and (with asbestos, glass fiber, or other reinforcing material) in the manufacture of ducts, castings, pipes, tanks, and reactors for corrosion-resistant service. Other resins, based on furfuryl alcohol, urea, and formaldehyde, are widely used as foundry-core binders.

Under carefully controlled conditions furfuryl alcohol undergoes the typical reactions of a primary alcohol, i.e., esterification, etherification, and oxidation; neutral or basic reaction media are essential to avoiding resinification. Esterification is best achieved via ester interchange or the Schotten-Baumann reaction. The methyl ether is prepared using dimethyl sulfate in strong alkali; the ether behaves like a diene, forming an adduct with maleic anhydride. In dilute alcoholic acid, furfuryl alcohol undergoes an unusual ring cleavage to the corresponding ester of levulinic acid, e.g., with methanolic hydrogen chloride, methyl levulinate is obtained.

Tetrahydrofurfuryl alcohol (2-tetrahydrofuranmethanol), $2\text{-}C_4H_7O \cdot CH_2OH$, is manufactured by low-pressure vapor-phase hydrogenation of furfuryl alcohol. The saturated alcohol is a colorless, high-boiling liquid with a mild, pleasant odor; it is completely miscible with water and the common solvents. Tetrahydrofurfuryl alcohol is a solvent for esters and ethers of cellulose, vinyls, and other polymers, and as such is of interest in the leather, textile, and paint and varnish industries. The tetrahydro alcohol is readily esterified using conventional techniques; the esters are used as plasticizers and as carriers for stabilizers in vinyl resins. Tetrahydrofurfuryl oleate, for example, improves the low-temperature flexibility of vinyl films.

When tetrahydrofurfuryl alcohol is heated to ~270°C over alumina, it undergoes dehydration and ring expansion to 2,3-dihydropyran, a major chemical intermediate for 1,5-difunctional open-chain aliphatic compounds. Tetrahydrofurfuryl alcohol is cleaved by acidic reagents; e.g., 1,2,5-triacetoxypentane is obtained using acetic anhydride (ZnCl$_2$ catalyst). With ammonia over a foraminate nickel-aluminum catalyst (210°C), the alcohol is readily converted to tetrahydrofurfuryl amine; over alumina-type catalysts (500–600°C) the major product is pyridine. Tetrahydrofurfuryl alcohol is readily converted to tetrahydrofurfuryl chloride (thionyl chloride, pyridine); the chloride, a β-haloether, is cleaved by sodium to 4-pentenol.

Condensed Furans. The properties of the more important condensed furans are quite similar to those of the monocyclic furans when allowance is made for the stabilizing effects of aromatic rings. Benzofuran (II) is less sensitive to acids than furan, yet reagents such as sulfuric acid or alumi-

num chloride do cause polymerization. Compounds in the isobenzofuran series are much less stable than those in the benzofuran group, as might be inferred from the extended conjugation shown in structure III. Whereas 1,3-dihydrobenzofuran (phthalan, XI) is known, the unsaturated parent, isobenzofuran (III), is not. This is as would be expected since the fully aromatic benzene ring should confer considerable stability to the phthalan molecule (cf. structures III and XI). Thus it is not surprising that nearly all the known unreduced isobenzofurans have aryl substituents in the 1 and 3 positions.

Phthalan
XI

Benzofuran (benzo[*b*]furan, coumarone), C_8H_6O, is a constituent of coal tar, from which it is obtained in admixture with indene (the 1-methylene carbocyclic analog). Pure benzofuran is a colorless liquid, boiling point 173–175°C, $d_4^{22.7}$ 1.0913, $n_D^{22.7}$ 1.5645, soluble in alcohol, ether, and ethylene glycol and insoluble in water. It has been synthesized directly by catalytic dehydrogenation (and cyclization) of 2-ethylphenol. Strong acids convert it to resins, which have been used in the manufacture of varnishes and lacquers.

Although benzofuran is generally more stable than furan, electrophilic substitution reactions do require the same special conditions listed in Table F-23 for furan. When benzofuran is passed through a hot tube together with benzene, phenanthrene is produced; with naphthalene, the product is chrysene. Sodium cleaves benzofuran to sodium 2-ethynylphenoxide; ozone converts it into a mixture of salicylaldehyde and salicylic acid. Catalytic hydrogenation provides coumaran, the 2,3-dihydro analog. Benzofuran adds halogen to give the 2,3-dihalo addition product; the dibromo adduct yields 2-bromobenzofuran on heating and 3-bromobenzofuran when treated with alcoholic potassium hydroxide.

Dibenzofuran (diphenylene oxide), $C_{12}H_8O$, a solid, melting point 86–87°C, boiling point 287°C, is soluble in most common solvents but insoluble in water. It is a constituent of coal tar, presumably formed during the coking process. It is synthesized in nearly quantitative yield by dehydration of 2,2′-dihydroxybiphenyl and has been obtained in low yield on heating phenol with lead oxide. The furan ring in benzofuran is very stable, generally unaffected by acid or alkali; however, fusion with alkali cleaves the ring to 2,2′-dihydroxybiphenyl.

Dibenzofuran is halogenated, sulfonated, and acylated in the 2 position. Nitration affords some 2-nitrodibenzofuran, but the principal products are the 3-nitro and the 3,8-dinitro analogs. Mercuration (mercuric salts) and lithiation (butyllithium) occur in the 4 position of dibenzofuran. No direct route to 1-substituted dibenzofurans is known.

REFERENCES

Acheson, R. M.: "Introduction to the Chemistry of Heterocyclic Compounds," 2d ed., chaps. 3 and 4, Wiley, London, 1967.

Dolique, R.: Hétérocycles pentatomiques avec un atome d' oxygéne, in V. Grignard, G. duPont, and R. Licquin (eds.), "Traité de chimie organique," vol. 18, Masson, Paris, 1945.

Dunlop, A. P., and F. N. Peters: "The Furans," ACS Monograph 119, Reinhold, New York, 1953.

Elderfield, R. C., and T. N. Dodd, Jr.: Furan, chap. 4 in "Heterocyclic Compounds," vol. 1, Wiley, New York, 1950.

Stevens, T. S.: Compounds Containing a Five Membered Ring with One Hetero Atom, Oxygen or Sulfur, chap. 3 in E. H. Rodd (ed.), "Chemistry of Carbon Compounds," vol. IVA, Elsevier, Amsterdam, 1957.

—Edward Sherman, *The Quaker Oats Company, Barrington, Ill.*

Furfural   (See **Furan group.**)

Furnaces   (Consult the **Subject Index.**)

Furoic Acid   (See **Carboxylic acids.**)

Fuses   (See **Explosives.**)

Fusible Alloys   (See **Bismuth.**)

Fusion, Nuclear   Nuclear fusion is the process of combining light nuclei into heavier nuclei and usually is accompanied by a relatively large release of energy. Nuclear fusion often refers to a thermonuclear reaction, where hydrogen nuclei combine to form helium nuclei (hydrogen-bomb reaction). The associated loss in mass is released as energy. Extremely high temperature and pressure are required.

G

Gadolinite (See **Erbium**.)

Gadolinium (Named for Johan Gadolin, a Finnish chemist.) **Gd** = 157.25 (at. wt.); 64 (at. no.). Gadolinium is the eighth of 15 elements in group III, period 6, generally shown as the rare-earth elements in a separate line below the main body of the periodic table, and sometimes referred to as the *lanthanide series*. Discovered in 1880 by J. C. G. Marignac, a Swiss chemist, Gd was named for Gadolin, who identified the original mineral.

Pure Gd metal has a silvery-gray metallic luster which is slow to tarnish in the normal atmosphere. The pure metal is soft and malleable and can be fabricated with ordinary tools and equipment if processing temperatures are held below 150°C. Chips and machine turnings of Gd are mildly pyrophoric. Storage and handling of the metal coated with a light film of high-flash-point oil is recommended.

The melting point of Gd is 1312°C; boiling point 3266°C; density 7.898 g/cm³. Gadolinium metal has the unique property among metals of being ferromagnetic at room temperature. The Curie temperature T_c of Gd is 16°C (61°F), below which it is attracted to iron, cobalt and nickel.

Estimated at 5.4 ppm in the earth's crust, Gd is one of the less abundant metals but is potentially more available than tin, tantalum, or tungsten.

Gadolinium has seven natural isotopes, 152 (0.20 w/o),* 154 (2.16 w/o), 155 (14.68 w/o), 156 (20.36 w/o), 157 (15.64 w/o), 158 (24.96 w/o), and 160 (22.00 w/o). Eleven artificial isotopes have been identified. The element shows no natural radioactivity and has a low acute-toxicity rating.

The natural isotopic mixture of Gd has the highest thermal-neutron-absorption cross section (46,000 barns) of all elements, about 10 times higher than the next two elements, samarium and europium (5,800 and 4,300 barns), both of which are also lanthanides. Only two natural isotopes adsorb thermal neutrons, however, [155]Gd and [157]Gd, about 31% of the total weight. These are separated by a low-cross-section isotope, so that no chain relationship exists. Therefore the use of Gd is limited to applications such as start-up and shutdown, where rapid burnout is required in high-temperature reactors.

Dilute mineral acids react with Gd vigorously,

---

*This abbreviation is used in the rare-earth and related fields for weight percent.

545

forming solutions from which the chloride, nitrate, sulfate, or acetate can be prepared. The metal is almost inert to strong bases and boiling water. Compounds are formed with most inorganic anions and many organic anions. As a strong electropositive cation, Gd metal is an active reducing agent for metals, including Fe, Mn, Cr, Sn, Pb, and Zr.

Gadolinium has one valence of 3 and forms only tripositive compounds with anions. Its oxide, Gd_2O_3, is white and has a melting point of about 2350°C. The oxide dissolves in strong mineral acids, and the solutions are colorless. Pure Gd_2O_3 is relatively inert to moisture and CO_2 absorption from the air.

The minerals xenotime, monazite, and gadolinite and residues from uranium mining supply most of the requirements. Some Gd is recovered from bastnasite as a coproduct with Eu and Sm. Separation of Gd from the other chemically similar lanthanides is by liquid-liquid or solid-liquid organic ion-exchange processes. Commercial plants have been installed that can produce 50,000 lb/yr of pure Gd_2O_3 or chemicals. The metal is produced by the calcium reduction of anhydrous GdF_3 or $GdCl_3$ with subsequent vacuum remelting to obtain the purity level required. For a detailed description of resources and processes and a table of properties see **Rare-earth elements and metals.**

Uses: Interest in Gd because of its high thermal-neutron-absorption cross section stimulated laboratory research after 1945. It was one of the elements considered in the 1950s for control rods in nuclear reactors for propulsion of aircraft and submarines. Mineral sources that contained Gd and chemical- and isotope-separation processes were relatively obscure during this period. Commercial and military reactors were designed using other materials. During the 1960s Gd was alloyed in stainless steel and fabricated into tubing and structural parts for nuclear-reactor control rods. Although the use of Gd in nuclear reactors is less than 2 tons/yr, applications in atomic-energy control represent an expanding use of the element.

One of the first established uses of Gd discovered in the 1960s is in gadolinium iron garnets, $Gd_6Fe_5O_{12}$ (GIG), used for microwave frequency control. The crystalline structure of this solid-state-reacted oxide compound makes it useful in circulators, isolators, and bandpass filters in electronic circuitry.

Gadolinium oxide is used as the host matrix in the red phosphor for color television, activated by europium, another of the lanthanides. A gadolinium oxysulfide, Gd_2O_2S, phosphor is an x-ray-image intensifier requiring less x-ray dosage during medical examinations. Gadolinium is also used with yttrium and lanthanum activated by cerium in a phosphor for single-gun beam-indexing flying-spot scanning cathode-ray tubes.

The ferromagnetic property of Gd when alloyed with cobalt, cerium, iron, and copper ($Co_{3.5}CuFe_{0.5}Ce$) in permanent magnets imparts a desirable negative temperature coefficient of magnetic saturation. A magnetic glass containing about 5 w/o Gd_2O_3 has been produced.

Continued applied research into the unusual properties of this element and its many compounds should lead to further applications.

REFERENCES

Nesbitt, E. A.: "Cast Permanent Magnets of the Co_5RE Type with Mixtures of Cerium and Samarium," Bell Telephone Laboratories, Murray Hill, N.J., 1969.

Overview 27 (application information including possible uses), Molybdenum Corporation of America, New York.

—Joseph G. Cannon, *Molybdenum Corporation of America, White Plains, N.Y.*

Galena   (See **Arsenic; Flotation;** and **Lead.**)

Gallium   (L. *Gallia,* France.)   **Ga** = 69.72 (at. wt.); 31 (at. no.). Gallium was predicted by Mendeleev's periodic table and was discovered by François Lecoq de Boisbaudran in 1875 by observing the spectral lines (4172 and 4033) which appeared in a specimen of zinc blende from the Pyrenees. Ga is a member of group IIIB of the periodic table, along with B, Al, In, and Tl. There are two stable isotopes, ^{69}Ga and ^{71}Ga, and eight unstable isotopes, 64–68, 70, 72, and 73. See also **Chemical elements.**

Ga, with a melting point of 29.78°C (85.6°F), is the only metal except Hg, Cs, and Rb which can be liquid near room temperature. Considering this low melting point, Ga boils at a relatively high temperature of 2040°C (3760°F). Some other important physical properties of Ga are given in Table G-1.

Gallium is about as abundant as Pb and As (15 ppm) in the earth's crust. The only known deposit of the mineral gallite, $CuGaS_2$, in southwest Africa, contains up to 1% Ga. Zn, Cu, and Pb minerals contain small quantities of Ga (0.005–0.02%). The most important source is Al ore (bauxite) which contains up to 0.01% Ga.

Chemical Properties: Ga reacts as a trivalent

TABLE G-1. *Some Physical Properties of Pure Gallium*

Spectral lines, Å	4172.1, 4033.0
	2943.6, 2874.2
Density, g/cm³:	
Solid at 20°C	5.90
Liquid:	
At 29.8°C	6.095
300°C	5.905
1100°C	5.445
Surface tension of liquid, dynes/cm . . .	592
Viscosity of liquid, P:	
At 97.7°C	1.612
1110°C	0.578
Expansion upon solidifying, %	3.2
Hardness at 20°C, Mohs' scale.	1.5–2.5
Thermal conductivity at 30°C,	
(cal)(cm)/(s)(cm²)(°C)	0.07–0.09
Coefficient of linear expansion:	
Near 20°C, in./°C	18×10^{-6}
68°F, in./°F.	10×10^{-6}
Latent heat of fusion, cal/g	19.2
Heat of vaporization, cal/g	950–1020
Electrode potential at 20°C (standard	
hydrogen voltage), V.	−0.56

metal in most cases although there are compounds in which gallium is mono- or divalent. Like aluminum, it forms an oxide, Ga_2O, on the surface of the metal, which protects it from further oxidation. The metal becomes more easily oxidized as the temperature is raised and becomes quite reactive at 1000°C.

Gallium is not attacked by water under normal conditions; however, it can be reacted with water under high temperature and pressure. Gallium is only slowly attacked by concentrated acids. Aqua regia attacks gallium rapidly. Compounds of gallium can be formed with As, Se, P, S, Sb, and B.

Ga itself can be very corrosive to most metals because it diffuses rapidly into the crystal lattices of the metals. A trace quantity of Ga on an Al sheet or plate results in immediate embrittlement because Ga diffuses to the grain boundaries and separates them. Metals which tend to resist Ga are Ta, W, Mo, and Nb. Ga forms alloys with most metals at temperatures between its melting point and 600°C. Examples include Ba, Cu, Au, In, Fe, Pb, Li, Mg, Mn, Hg, Ni, Pd, Pt, Ag, Na, Pr, Tl, Sn, Ti, Ur, V, Zn, and Zr.

Halides of Ga include gallium trifluoride, which is an ionic salt of high melting point and low solubility, similar to AlF_3. The other trihalides are soluble and strongly hydrolyzed. $GaCl_3$ is a white solid (melting point 78°C). $GaCl_2$ can be

made by heating $GaCl_3$ in the presence of metallic Ga at a controlled temperature. At higher temperatures, $GaCl_2$ dissociates. GaI_2 dissociates into the tri- and monoiodide. GaCl can be made by heating Ga in the presence of Ar containing 1% Cl_2.

Oxides of Ga can be made by heating Ga in the presence of O_2. The heat of formation is −258 kcal for Ga_2O_3. If Ga_2O_3 is heated in the presence of Ga at 500°C in a vacuum, a black Ga_2O oxide is formed. The hydroxide, $Ga(OH)_3$, is amphoteric and very similar to $Al(OH)_3$. The solubility is lowest at a pH of 7.

Gallium nitride, GaN, is formed by reacting Ga with NH_3 at 1000°C or Ga_2O_3 in NH_3 at 480°C. The nitride is insoluble in water and acids. Gallium sulfide, Ga_2S_3, can be formed by heating Ga in the presence of sulfur vapor. Gallium sulfate, $Ga_2(SO_4) \cdot 16H_2O$, is formed by a reaction with H_2SO_4 and water.

Organic salts and organometallic compounds of complex nature include thiocarbamates, xanthates, oxalates, and alkoxides. Trialkylgalliums, R_3Ga, can be formed from gallium trihalides and aluminum alkyls when KCl is present; or they may be produced by heating Ga metal with mercury dialkyls. Like aluminum trialkyls, they are spontaneously inflammable either as liquids or solids and are hydrolyzed by water.

Ga and its compounds are not highly toxic. For rabbits and rats, the LD_{100} is approximately 100 mg of gallium per kilogram.

Uses: Perhaps the most notable use of Ga is in the electronics industry. Gallium arsenide, produced by reacting H_2 and As vapor with Ga_2O_3 at 600°C, has high electron mobility, 88,000 cm²/(V)(s), and its high-temperature stability is important for manufacturing diodes, laser diodes, and electroluminescent diodes. Since Ga is available in purities of 99.9999999+ % pure, it is used in the production of semiconductors. Gallium arsenide is used in solar batteries to convert solar energy into electricity in aerospace projects. Ga is used as an activator in phosphors and luminous paints. Other uses include dental amalgams, medicinals, arc rectifiers, as a sealant in vacuum systems, in transistors, and organic synthesis. Ga remains liquid over the largest temperature range of any metal and has a low vapor pressure at high temperatures. The metal expands 3.1% on solidifying and therefore should not be stored in glass. Ga wets glass to form a specular mirror.

Isolation and Purification: The original extraction by Lecoq de Boisbaudran was from zinc blende. Ga had little or no use in that period. Interest in the element was shown in 1916 in the United States, where it was distilled from Zn. Beginning in the 1930s, Ga was recovered as a by-product along with Ge from germanite, from copper-containing schists in Germany, and from flue dust from certain English coals. After World War II, commercial exploitation of Ga was made possible by recovering the element from the sodium aluminate used in the extraction of Al from bauxite by multiple concentration steps. In one process, $Ca(OH)_2$ is mixed with the sodium aluminate solution from the Bayer process. See also **Bauxite.** The Ga/Al ratio ranges from 1 : 3,000 to 1 : 8,000. Calcium aluminate is precipitated and filtered out, thus producing a filtrate that is enriched with Ga. The filtrate is agitated with CO_2, precipitating aluminum hydroxide and leaving an enriched gallate-in-caustic solution with a Ga concentration of approximately 0.2 g/l.

This solution is sufficiently enriched with Ga as gallate(III) to economically justify the electrolytic deposition of the Ga into a Hg cathode cell to further concentrate the Ga. The mercury-containing Ga is transferred from the cell to a reaction vessel, where the Ga is dissolved out of the Hg with boiling NaOH solution in the presence of Fe, which acts as a catalyst. At this point, the Ga concentration is approximately 80 g/l. The Hg is returned to the electrolytic cell to continue the concentrating electrolysis. The enriched solution is filtered to remove Fe and other entrapped oxides or particulate matter. The filtered solution now is ready to be electrolyzed to give metallic Ga, using a stainless-steel cathode in place of a Hg cathode. The Ga is easily removed from the stainless-steel cathode by raising the temperature to the melting point of Ga (29.78°C; 85.6°F). The Ga thus is ready for final purification by one of several processes: (1) crystallization as monocrystals, (2) chemical treatment with acids or O_2 at high temperatures, or (3) repeat resolution in pure boiling NaOH and reelectrodeposition.

Purification by these processes can lead to 99.99999% purity. The purity is monitored by spectrographic techniques or by electrical-resistivity tests near absolute zero.

REFERENCES

Adams, G. B., Jr., H. L. Johnston, and E. C. Kerr: *J. Am. Chem. Soc.,* vol. 74, p. 4784, 1921.

De La Breteque, P.: Swiss AIAG Process, *J. Metals,* November 1956, pp. 1528–1529.

Popp, E., A. Hejja, and J. Oevger: Extraction der gallium métallique des bauxite hongraises, *Acta Tech. Sci. Hung.,* vol. 14, no. 12, pp. 56–76, 1956.

Richards, T. W., and Sylvester Bayer: *J. Am. Chem. Soc.,* vol. 43, p. 274, 1921.

Roeser, W. F., and J. I. Hoffman: *J. Res. Natl. Bur. Stds.,* vol. 13, p. 673, 1934.

Spells, K. E.: *Proc. Phys. Soc. (Lond.),* vol. 48, pp. 299–311, 1936.

Wilkinson, W. D.: Properties of Gallium, *AEC Argonne Natl. Lab. Chicago,* ANL-4109, 1948.

—S. John Sansonetti, *Reynolds Metals Company, Richmond, Va.*

Galvanic Cells   (See **Batteries.**)

Galvanized Steel   (See **Conversion coatings;** and **Zinc.**)

Gamma Globulin   (See **Peptides and proteins.**)

Gamma Ray   A quantum of electromagnetic radiation emitted by a nucleus as the result of a quantum transition between two energy levels of the nucleus, gamma rays usually have energies between 10 keV and 10 MeV, with shorter wavelengths than x-rays. Gamma rays are more penetrating than alpha and beta particles and are not affected by magnetic fields. The ability of gamma rays to penetrate through considerable thicknesses of materials forms the basis for the use of gamma radiation from a radioactive source in various forms of instruments, including thickness gages and tank- and bin-level measurement devices. See also **Radioactive isotopes.**

Garnet   (See **Abrasives; Aluminum; Ceramics; Neodymium;** and **Yttrium.**)

Garnierite   (See **Nickel.**)

Gas, Artificial   (See **Methane;** and **Synthesis gas.**)

Gas Chromatography   (See **Chromatography.**)

Gas-Gas Separations   (See **Separation operations.**)

Gas-Liquid Separations   (See **Separation operations.**)

Gas Oil   (See **Petroleum processing.**)

Gas-Solid Separations   (See **Separation operations.**)

Gaseous Diffusion (See **Separation operations.**)

Gasification The process shown in Fig. G-1 produces synthesis gas (either H_2 plus CO or H_2 plus N_2) or hydrogen by the continuous noncatalytic partial oxidation of any liquid or gaseous hydrocarbon feed, ranging from the heaviest fuel oils, or even bitumen, to natural gas. Since the product gas may be readily desulfurized, high-sulfur fuels are acceptable feedstocks. The oxidant may be oxygen, air, or oxygen-enriched air. The process is flexible with respect to operating pressure and product-gas composition. A carbon-removal system results in a soot-free product gas. The small amount of separated carbon can be recycled to extinction.

Essentially the process * consists of the partial

*Shell Gasification Process (SGP). Over 90 reactors in operation at approximately 30 locations throughout

oxidation of hydrocarbons under pressure in the presence of steam with limited oxygen. Control of the hydrocarbon, oxygen, and steam feed rates yields a product as high as 96 vol % H_2 plus CO, with small amounts of CO_2, CH_4, N_2, and Ar (with some H_2S and COS if the feedstock contains sulfur). For chemical synthesis gas, e.g., methanol or oxo chemicals, the ratio of CO to H_2 can be adjusted, within limits, by choice of operating conditions and feedstock. Further adjustment of product-gas composition also is possible using a catalytic converter to shift CO and steam to H_2 and CO_2 by the water-gas shift reaction, followed by removal of CO_2 and other acid gases (H_2S, COS) by solvent absorption. See also **Absorption, acidic gases.**

A crude hydrogen stream can thus be produced

the world with an aggregate design capacity of over 1 billion scf/day of H_2 plus CO.

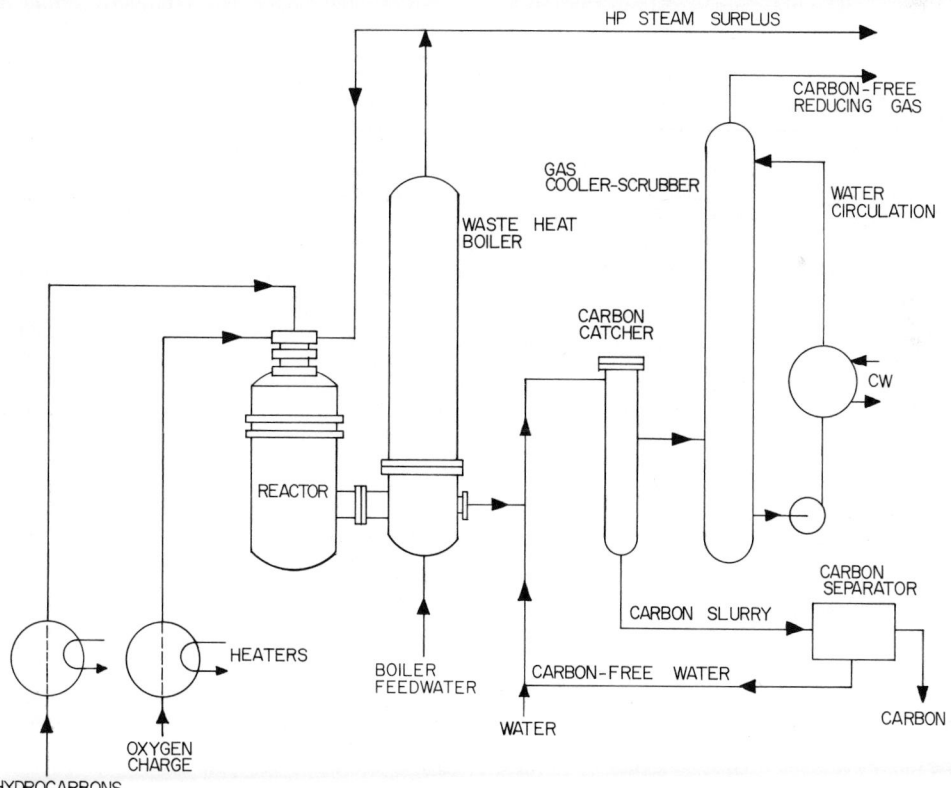

Fig. G-1. **Process for producing synthesis gas** (H_2 **plus** CO, **or** H_2 **plus** N_2). [*Shell Gasification Process (SGP), Shell Development Company.*]

for refinery or chemical-plant use. With air as the oxidant in the process there is, of course, a large concentration of nitrogen in the product gas. When ammonia synthesis gas (H_2 plus N_2) is desired, it is usually more economical to use O_2 from an air-separation plant for partial oxidation, with water-gas shift of the CO to H_2 and CO_2 followed by removal of CO_2 and then addition of N_2 from the air-separation plant to the product gas. This avoids energy losses in compressing, heating, and cooling a large volume of N_2 in the gasification process. Table G-2 shows typical results for gasification of three different feedstocks using oxygen as oxidant.

With reference to the simplified flow diagram (Fig. G-1), preheated hydrocarbon and oxygen are fed to the reactor through a burner of special design which provides very effective mixing. Homogeneous mixing of the reactants is vitally important to ensure uniform flame temperature. Typical flame temperatures range between 1300 and 1500°C. The hot gases pass from the refractory-lined reactor into a waste-heat boiler of

unique design; most of the steam generated is available for export from the process. This results in a high thermal efficiency for the process. From the boiler, the gas flows through a carbon catcher and a cooler-scrubber, where circulating water cools the gas and removes soot particles. Free-carbon content of the clean gas is typically only about 1 ppm. The carbon separator, also of special design, removes carbon from the water slurry. The carbon can be recovered separately or recycled to the reactor by slurrying it with the hydrocarbon feed.

The feedstock flexibility of the process makes it possible to switch feeds as process economics dictates. For example, relatively inexpensive high-sulfur fuel oil can replace more expensive natural gas as feed. Switching feeds usually requires only minor changes (such as feed pumps and burner nozzles). Increasingly stringent pollution controls can be accommodated, even for high-sulfur fuels which could not be burned in conventional boilers, by relatively simple desulfurization of the product gas, using existing pack-

TABLE G-2. *Typical Results for the Gasification of Various Feedstocks with Oxygen**

Characteristics and Conditions	Gasification Feedstock		
	Heavy Fuel Oil	Naphtha (liquid)	Natural Gas
Operating pressure,† psig	450	450	450
Preheat temperatures:			
Feedstock:			
°F	457	77	457
°C	236.1	25	236.1
Oxygen:			
°F	457	457	457
°C	236.1	236.1	236.1
Steam:			
°F	475	475	475
°C	246.1	246.1	246.1
Oxygen to reactor (as 100% at 95% volume purity), scf	12.7	14.9	16.6
Steam to reactor (sat. 475°F; 236.1°C), lb	0.40	0.35	0.05
Dry gas produced, scf	49.9	54.3	63.2
CO plus H_2 produced, scf	46.4	50.8	60.1
(CO plus H_2)/O_2 (as 100% at 95% volume purity), scf/scf	3.66	3.41	3.61
Composition of dry gas, vol %:			
H_2	46.1	51.6	60.9
CO	46.9	41.8	34.5
CO_2	4.3	4.8	2.8
CH_4	0.4	0.4	0.4
N_2 plus Ar	1.4	1.4	1.4
H_2S plus COS	0.9	70 ppm	
Delivery pressure of crude gas, psig	415	415	415

*Basis, 1 lb of feedstock to reactor.
†Process can be designed to operate at higher or lower pressures.

aged processes. In addition to synthesis gas or hydrogen, the process can be adapted to produce town gas, CO, or reducing gas for ore-reduction plants.

See also **Ammonia; Hydrogen;** and **Synthesis gas.**

—C. J. Kuhre, *Shell Development Company, Houston, Texas.*

Gasoline   (See **Alkylation; Cracking; Fluid catalytic cracking; Hydrocracking; Isomerization; Petroleum; Petroleum processing;** and **Thermal cracking.**)

Gate Mixers   (See **Mixing, pastes.**)

Gay-Lussac Law   (See **Ideal gas.**)

Gelatin   (See **Amino acids; Colloid systems; Peptides and proteins;** and **Starches.**)

Gelatinization   (See **Colloid systems;** and **Starch.**)

Gels   [See **Colloid systems;** and **Silicates (soluble).**]

Gems, Artificial   (See **Abrasives; Bauxite;** and **Ceramics.**)

Genetic Information   (See **Amino acids;** and **Peptides and proteins.**)

Geometric Isomerism   (See **Isomerism.**)

Geraniol   (See **Terpenes.**)

German Silver   (See **Copper.**)

Germanium   (L. *Germania,* Germany.) **Ge** = 72.59 (at. wt.); 32 (at. no.) Germanium is a very brittle metal (metalloid) with a bright, shiny surface, silvery in appearance. It is frangible, not ductile, and has a hardness of 6.25 on Mohs' scale. Crystalline structure is diamond cubic. Index of refraction is 4.068–4.143 and volume resistivity $60 \times 10^6 \, \mu\Omega$-cm at $25°C$; ionization potential 7.88 eV; valence 2 and 4; melting point $958°C$; boiling point $2700°C$;* specific gravity 5.35.

Ge is found in group IVA (along with C, Si, Sn, and Pb) and in period 4 of the periodic table. Discovery is attributed to Winkler (1886) although the element was predicted by Mendeleev

*Also reported up to $2830°C$.

(1871), who called it *ekasilicon.* Ge exists as five natural isotopes, 70, 72–74, and 76; there are seven radioactive isotopes, 67–69, 71, 75, 77, and 78. See also **Chemical Elements.**

Sources: The primary source of Ge is flue dust from the zinc industry. It can be obtained from reduction of oxide and sulfide ores. The principal ore is *germanite,* a Cu ore occurring in Southwest Africa and containing up to 20 different elements, about 45% Cu, 30% S, and 6–9% Ge as well as 1% Ga. The sulfide ore is *renierite,* occurring in Zaïre (formerly Belgium Congo) and containing up to about 8% Ge. A rare Pb-Ag ore, *ultrabasite,* also contains Ge. Small amounts of Ge are found in lepidolite, sphalerite, and spodumene. Some English coals contain up to 1.6% germanium oxide, GeO. The Ge metal of 99.9+ % purity required for electronics devices is produced by passing an ingot slowly through an induction heater, where soluble impurities pass along through molten zones and are cut off at the end of the ingot. Electronics users further zone-refine the metal.

Ge is insoluble in H_2O and HCl. The metal is soluble in aqua regia and H_2SO_4 and reacts with HNO_3 to form GeO_2. The metal combines directly with the halogens, but the products are readily decomposed by H_2O. Although insoluble in NaOH, Ge dissolves with incandescence in fused alkalis. There are two principal oxides, GeO (-ous), which is readily soluble in HCl or NaOH, and GeO_2 (-ic), which is slightly soluble in H_2O and acids. When fused with a fixed alkali hydroxide, GeO_2 forms a water-soluble germanate. The sulfides, GeS (red) and GeS_2 (white), are slightly soluble in H_2O, insoluble in HCl or H_2SO_4, but soluble in aqua regia and alkali sulfides and hydroxides. Chemically, the profile of Ge lies between Si and Sn. The hydroxides are amphoteric.

Electronics Uses: The major use of Ge is in solid-state electronic devices, such as transistors, which can be used as amplifiers and oscillators. The nerve center of a transistor, whether of Ge or Si, demands special skills to produce a high-purity material successfully, uniformly, and in sufficient quantity. Refined single crystals are cut up into thousands of tiny pellets. The transistor is a three-element control device which utilizes a small current to control a larger one. It requires no heated filament and usually operates with as little as one-millionth the power a vacuum tube needs for the same task.

Other Uses: Although as little as 0.35% Ge in Sn will double the hardness of Sn or provide greater strength and hardness to Al and Mg alloys,

Ge is not used to any extent for this purpose because of high cost. An alloy of Ge and Au (up to about 12% Ge) has been used for soldering jewelry. Melting point of the alloy is 359°C. Some Ge salts are used in glass to increase refraction characteristics.

Ge-Si alloys are under intensive investigation for use in thermoelectric generators.* Ge and Si form a continuous mixed-crystal series, but the production of homogeneous crystals is very difficult. Some of the advantages claimed for these materials are better thermoelectric qualities above 600°C, an improved efficiency per unit weight factor, and virtually no corrosion or decomposition.

Getters   (See **Cesium; Hafnium; Lanthanum; Rubidium;** and **Zirconium.**)

Ghatti   (See **Gums.**)

Gibbsite   (See **Bauxite.**)

Gilding Brass   (See **Copper.**)

Glass   Glass is an inorganic product of fusion which has cooled to a rigid solid without undergoing crystallization. It is a solid. It may be transparent, translucent, or opaque, and it may be colored. The chemical composition and corresponding properties may vary over a wide range. Glass will support a load, and may be shaped, broken, or cut. It is much like other solid materials, and yet it is unique.

Its uniqueness becomes obvious when it is examined on a submicroscopic level. Most solids have regular, orderly patterns for the arrangement of atoms, molecules, and ions, but glassy materials are highly disordered. There is some short-range order in glass, but beyond one or two atoms or ions the ordering may be described as random. Thus on a submicroscopic level, glassy solids look more like liquids than solids.

Since glasses do not have ordered structures with correspondingly specific bonding energies between rows, stacks, planes, or discrete ions, they do not have definite melting points. When a glassy material is heated, it softens slowly and transforms to the liquid state. Crystalline solids

generally transform from a solid to a liquid at a single specific temperature, the melting point. On cooling, a material that has a tendency to crystallize to solid will do so at the same temperature at which it transformed to a liquid. When a glass is cooled from a high temperature, it becomes increasingly viscous in a manner which is related to the inverse of the temperature until it becomes a rigid solid again. Thus a specific temperature where melting or freezing takes place cannot be found for a glass; i.e., glass does not have a melting point.

Most glasses can be made to crystallize if they are subjected to the right conditions of temperature and rate of cooling, which suggests that the glassy state is like a supercooled liquid. This is not borne out by measurements of density and other volume properties, which do not decrease in a linear manner as glass is cooled below its crystallization temperature.

Why does a glass form as a melt is cooled through a crystallization temperature in some cases and not in others? It is simply a question of whether the melt can be cooled through the temperature range of maximum crystal growth rate faster than the crystals can grow. Thus table salt cannot be formed as a glass, but sand, or SiO_2, can be. The maximum crystal growth rate is normally just below the melting point of the material, but materials that tend to form glasses are much more viscous at these temperatures. For example, in the extreme cases of salt and sand, the differences in viscosities at their respective melting points is about eight orders!

The two-dimensional drawing in Fig. G-2 shows SiO_2 in the ordered, or crystalline, and in the random, or glassy, state to illustrate the difference on a submicroscopic scale. Figure G-3 shows how the volume properties of a material would respond to temperature if they could be prepared as a glass, a supercooled liquid, or crystalline material.

Most glasses are composed of inorganic oxides, and most commercial glasses contain SiO_2 as their major constituent, but there are organic glasses and elemental metallic glasses. Glass is typically hard and brittle, and exhibits a conchoidal fracture. It is transparent or translucent in the visible portion of the spectrum.

Types of Glasses. A wide range of glass products exists, each type having special properties. The properties of glass are determined primarily by chemical composition, and since the composition may be varied almost infinitely, there are many thousands of different glasses. However, they may

---

*H. A. Herrmann and W. H. Dietz, The Methods of Making GeSi Alloys, Their Properties and Their Use in Thermoelectric Generators, in D. H. Collins (ed.), "Power Sources," Pergamon, Oxford, 1969.

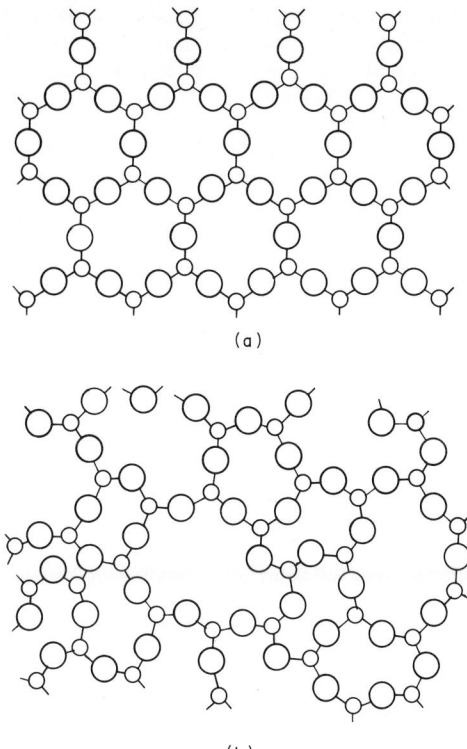

(a)

(b)

Fig. G-2. SiO$_2$ in (a) **crystalline** and (b) **glassy state.** The small circles represent Si and the large circles, O.

be generally classified into soda-lime-silica glasses; lead glasses; borosilicate glasses; and a number of special glasses, including solder glasses, laser glasses, silica glass, glass-ceramics, and colored glasses. These types essentially bracket the commercial glasses.

Soda-Lime-Silica Glasses: This is the most important group in terms of tonnage melted and variety of use. The combination of silica sand, soda ash, and limestone produces a glass that is easily melted and shaped and has good chemical durability. The raw materials are indigenous to most areas of the world and inexpensive. Soda-lime glasses are particularly suited to automatic-machine-forming methods and are the basis for most of the bottle-, sheet-, and window-glass industry. Very small amounts (often less than 1% of the total batch) of alumina, magnesia, boric oxide, and other chemicals are added to act as stabilizers and to increase durability.

Lead Glasses: The glasses of this group, composed basically of silica sand and lead oxide, have a high refractive index and high electrical resistivity. Potash is present as a significant constituent in most of these glasses. The slow rate of increase in viscosity with decrease in temperature makes lead glass particularly suitable to hand fabrication. The amount of lead may vary considerably, even up to 92% lead oxide; it is a more expensive glass, as the raw materials are relatively expensive and special care is needed in melting to avoid bubbles and seeds. Glasses of this type are used in high-quality art and tableware and for special electrical applications.

Borosilicate Glasses: This group of glasses is basically a combination of silica sand with boric oxide and soda ash. The glasses have excellent chemical durability and electrical properties, and their low thermal expansion yields a glass with a high resistance to thermal shock. High durability makes them ideal for demanding industrial and domestic use, such as chemical laboratory ware, cook ware, and pharmaceutical ware. These glasses were developed in the early part of this century to cope with the problem of cold rain on hot railway-signal lights.

Special Glasses

Solder Glasses: These glasses have low softening

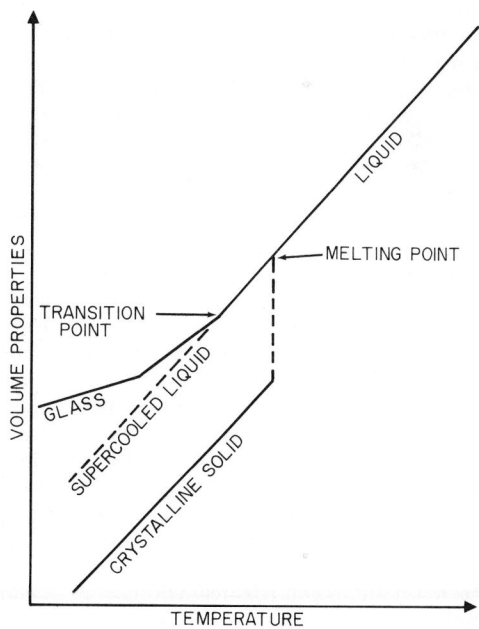

Fig. G-3. The volume properties of glass as opposed to crystalline solids as a function of temperature.

and annealing temperatures together with expansion characteristics which permit them to be used as intermediate glasses in making seals between two glass surfaces, between a glass and a metal, or between two ceramic surfaces. In fact, solder glass might be described as a high-grade glass glue. Normally, sealing temperatures are well below the annealing temperature of the glass being sealed, and there is little permanent effect on the glass parts being joined. The major constituents of these glasses include lead oxide, boric oxide, and zinc oxide.

Laser Glasses: Glass has various characteristics which make it an ideal laser host material. Its random structure permits broad emission and absorption bands, which provide higher efficiency, more energy storage, and greater energy per pulse than any other material. In addition, most lasing ions are easily soluble in the glass, and rods, fibers, or disks of any size and of high optical quality are easily fabricated. Of the several rare-earth ions which have been made to lase in a glass host, only neodymium has received commercial application. When a neodymium glass lases, it emits light at a rather fixed wavelength of 1.06 nm.

Silica Glass: A glass composed of silicon dioxide as the only constituent has a very high softening temperature and a very low thermal expansion. It is costly to make and fabricate because temperature in excess of 1800°C is required to manufacture it. However, its refractory character coupled with its very high resistance to thermal shock makes it ideal for special laboratory equipment, windows in high-temperature environments, and instruments.

Glass-Ceramics: Glass-ceramics are formed as glasses and subsequently heat-treated to produce a crystalline material. Since they are crystalline bodies, they are not glass, but the product is formed in the same manner as conventional glass. However, heat treatment causes a controlled nucleation and crystallization to form a crystalline ceramic. This family of materials is based on glasses whose major constituents are magnesium oxide, lithium oxide, aluminum oxide, and silicon dioxide. The crystalline phase or phases and their morphology control the properties of the materials, but the starting chemical composition and the heat treatment determine which crystalline phases will result. This new family of materials, the result of recent research efforts, has found applications as household cooking ware, reflective optics substrates, chemical processing components, and cooking-stove tops.

Colored Glasses: Nearly all glasses can be colored by adding one or more colorants to the batch in correct amounts. Production of some colors requires, or is enhanced by, the state of oxidation of the coloring agents and the atmospheres in which the glasses are melted. Table G-3 indicates the colors obtainable, colorants used, and chemical states required or utilized.

While the preceding paragraphs describe several classes of glass, within each class there can be infinite composition variations to fit the exact requirements of the user. Table G-4 shows typical composition ranges for commercial glasses.

Manufacturing Processes. Glass products are many and varied, and glass compositions range rather widely, depending on the desired products. Figure G-4 shows a typical cross section of a glass manufacturing facility. Raw-materials weighing, mixing, charging, and melting are common requirements regardless of the forming operation that is to follow. Most melting furnaces have a primary melting area, followed by a refining or homogenizing section, which is connected to the forming

TABLE G-3. *Colors Obtainable, Colorants Used, and Chemical States of Colorants Required or Utilized for Colored Glass*

Glass Color	Coloring Agent	Chemical State
Red	Cadmium sulfide, cadmium selenide	Reduced
	Cuprous oxide	Reduced
	Gold (metal)	
Yellow	Cerium oxide with titanium oxide	
Yellow-green . .	Chromic oxide	Oxidized
Blue-green . . .	Iron chromite	Reduced
Blue	Cobalt oxide	
Purple	Neodymium oxide	
Gray	Nickel oxide with titanium oxide	
Black	Copper, cobalt, nickel, and iron oxides in combinations of two or more	
Amber	Iron sulfide	Reduced
Flint (or colorless)	Selenium and cobalt oxide*	Oxidized

*Selenium and cobalt are used in flint glass to add red and blue hues in amounts only sufficient to balance the green hue resulting from iron oxide present as impurity in most naturally occurring raw materials. The intended result is an even light transmission over the whole visible spectrum.

TABLE G-4. *Composition by Weight Percent of Commercial Glasses*

	Soda-Lime-Silica Glass				Borosilicate Glass	Laser Glass	Solder Glass	Lead Glass	Glass-Ceramics
	Containers	Plate and Window Glass	Tableware	Fiber Glass Fabrics and Insulation					
SiO_2	70–74	71–74	71–74	65–74	70–82	61–69	0.5–16	35–70	62–70
Al_2O_3	1.5–2.5	1–2	0.5–2	2–4.5	2–7.5	0–5	0.1–4	0.5–2.0	17–22
B_2O_3				3–5.5	9–14		7–20		
Li_2O									3–5
Na_2O	13–16	12–15	13–15	8–16	3–8	12–24		4–8	
K_2O				0–1				5–10	
CaO	10–14	8–12	5.5–7.5	5–16	0.1–1.2	3–10			0–5
MgO			4.0–6.5	3–5.5					0–7
BaO					0–2.5		0–4		
ZnO							7–62		
PbO							4–77	12–60	
CuO							0–10		
Nd_2O_3						1–6			
CeO_2						0.1–1			
F_2							0–2		
ZrO_2 and TiO_2									3–10

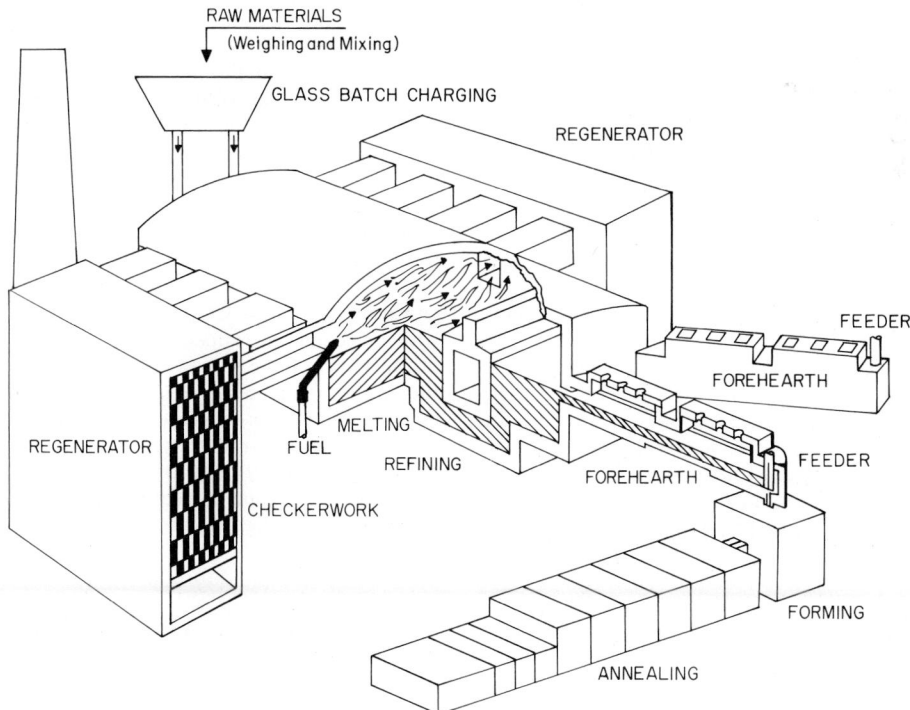

Fig. G-4. A simplified cross section of a continuous glass-manufacturing facility. The flames produce glass temperatures as high as 1650°C (3000°F). The regenerators and direction of firing are reversed every 20–30 min. The melting and refining sections are separated by a submerged refractory wall.

operation by channels called *feeders*. Although fiber glass is not passed through an annealing furnace after it is formed, most other glass products are annealed to relieve stresses caused by uneven cooling during and immediately after forming.

It is apparent that although there are several similar steps in all glass-manufacturing processes, the forming operations are the most diverse.

Batch Preparation: This begins with the selection, procurement, and storage of an adequate quantity of the raw materials. Selection is made on the basis of the oxides which each material contains and will provide to a glass and on the basis of purity and grain size. Naturally occurring raw materials are used wherever possible for economy, e.g., silica sand, limestones, feldspars, borates, soda ash, boric acid, potash, and barium carbonate. The prescribed quantities of these raw materials, depending on their chemical composition, are measured carefully and mixed together to provide a homogeneous batch. Such mixing is done on an intermittent or a continuous basis, depending on the volume of batch needed to charge the furnaces. The batch is conveyed by a variety of means to the furnaces but always in such a way that segregation is avoided. The importance of grain size of the various raw materials becomes evident in preventing dusting and/or segregation.

Furnaces: A variety of furnaces are used in the industry to melt the batch to produce glass. They must all accomplish the two purposes of confining the heat to the necessary area and containing the melted glass within the furnace. Crucibles or pots are sometimes used to contain the batch and the melted glass, in which cases the furnace merely retains heat; however, tank furnaces (Fig. G-4) are far more common. They are so constructed that the lower portion contains the glass and the superstructure retains the heat and provides combustion space for the fuels used. "Day" tanks are used in some instances where the operation is intermittent and the quantity of glass is small. The great majority of glass produced is melted in continuous furnaces, which are charged initially with batch and cullet (broken-up pieces of previously melted glass) which are melted, filling the tank to a specified depth, sometimes up to 66 in. Thereafter, batch and cullet are charged continuously at a rate equal to that at which the molten glass is withdrawn from the working end.

Continuous tank furnaces are designed to provide for a separate melter section and a refiner or conditioning section. The melting end is maintained at the necessary high temperatures to accomplish the melting and chemical reactions of the batch materials. The *refining*, or conditioning, section retains the glass long enough for it to cool to the necessary lower working temperatures.

Glass-melting furnaces are built of refractory materials of various types which will withstand the severe conditions to which they are exposed. The lower portion of the melter section, for instance, must be of the highest quality to withstand the corrosive action of the glass as well as the high temperatures used. Some sections may use lower-quality refractories because the temperature or corrosion conditions are not as severe.

Fuels used in today's furnaces in the United States are natural gas or oil. The fuel is fed to burners that project flames over the surface of the glass. Nearly all continuous furnaces utilize regenerators, which reclaim a portion of the heat from the exhausting combustion gases. Although some glass is melted entirely by the use of electric power, it is still too expensive in most areas to permit complete replacement of oil or gas. When electric power is used to augment the fossil fuels, it is called *electric boosting*.

For the areas that do have sufficiently low-cost electric power, the furnaces are constructed with conventional bottoms but with superstructure only adequate for initial heat-up. They depend on a blanket of batch floating on the surface of the glass to retain the heat within the tank that is provided by the submerged electrodes. Fresh batch is added to the blanket at a rate equal to the rate of melted glass withdrawn.

Melting: This provides the mutual solution of the oxide materials at high temperatures to yield a homogeneous liquid. Temperatures may range from $1427°C$ to over $1593°C$ ($2600–2900°F$), depending on the glass composition. Water vapor, entrapped air, and CO_2 are given off, some of which become entrapped in the glass, resulting, initially, in a foamy mass. As the melt moves to the higher-temperature regions, the viscosity is lowered and the gases escape. Deliberate hot spots enhance the natural convection currents, promoting homogeneity. More modern furnaces utilize bubblers, which introduce controlled pulses of air through the furnace bottom, further enhancing convection. This is particularly valuable for increasing temperatures near the tank bottom in melting those glasses which are more opaque to infrared radiation. A recent innovation has been the use of submerged burners introducing combustible mixtures through the furnace bottom which burn in passing upward through the molten glass.

The glass is essentially free from bubbles (or *seeds*) when it reaches the end of the melting chamber. It then passes under *floaters* in some furnaces, or through submerged throats in most, to the so-called *refining section* (more properly, the conditioning section). Here the refining or conditioning consists of allowing the glass to increase to a more usable viscosity level and to dissolve the remaining tiny seeds, or gaseous inclusions.

Furnaces supply glass to up to eight forming machines. Forehearths or alcoves serve to channel the glass to the individual machines or machine locations and to further change the temperature and viscosity.

Forming Operations: These are many and varied, involving two, three, or four major steps. The first is a further temperature conditioning to place the glass in the exact viscosity range, sometimes wide but often quite narrow, suitable for the selected primary forming operation. The second step is the primary forming itself, followed usually, but not always, by an annealing step. Single or multiple secondary operations may ensue. Only the major forming processes of drawing, pressing, blowing, and casting will be discussed.

Drawing is one of the simpler forming methods by which thousands of tons of window glass and millions of feet of rod and tubing are produced annually. Drawing window glass frequently utilizes a rectangular refractory frame, called a *debiteuse,* placed on the surface of the conditioned glass. It has a slot roughly 4–8 in. wide and 8 ft or more long through which the glass is pulled vertically. The width and length of the slot in the debiteuse, together with the drawing speed, aid materially in controlling the width and thickness of the sheet. The upward draw may continue until the sheet is nearly cold, when it can be stored and cracked off in suitable lengths, or it may be bent over a large roller at nearly the last moment it will withstand bending and conveyed horizontally into the annealing lehr.

Glass tubing may be drawn vertically in a manner similar to that for window glass. Another common method is the *Danner process,* in which a suitable stream of glass is flowed onto a conical rotating mandrel supported with its small end downward and its axis at a suitable angle to the horizontal. The tubing is drawn from the small end, through which sufficient air is blown to retain the desired cross section of the tubing. Drawing continues horizontally over rollers until the tubing can be cracked off in lengths at the cold end.

Plate glass may be formed by flowing the molten glass over the lip of the discharge end of the furnace between a set of large water-cooled rollers and then pulling it away by means of driven rollers. The resulting sheet is up to 1 in. or more thick and 10–12 ft wide. A more recent development is the *float-glass process,* which forms the glass into a sheet by floating it on a bath of molten metal such as tin. The glass flowing onto the bath of tin is pulled across the surface and cooled to the temperature at which it is rigid while still on the molten metal. The outstanding advantage of this process is that it produces a plate of glass both surfaces of which require no further polishing.

Modern methods of pressing, blowing, and casting usually involve an intermediate step, the formation of a suitable charge of glass, or *gob,* for the ensuing operation. The most common method involves a gob feeder located at the end of the forehearth. This consists of a bowl, or spout, kept full of glass by flow from the forehearth and having an orifice in its bottom and a refractory tube suspended in the bowl over the spout. The tube may be lowered to shut off the flow of glass or raised to permit flow at a selected rate. A refractory plunger operates vertically inside the tube. It provides a pumping action on its upstroke, momentarily restraining the flow of the glass. Its downstroke forces the accumulated glass out of the orifice, where it is sheared off. The result is a charge of glass, called a gob, of controlled size which is delivered to the forming machine by gravity.

Pressing, or press-forming, operations normally are used for relatively shallow, heavy-walled products. Pressing is accomplished by means of a metal mold (usually iron or steel), a ring which is centered on top of the mold, and a plunger which is forced into the mold through the ring. The mold shapes the exterior of the product, the ring the top, and the plunger the interior. A pressing machine may have many molds mounted on its circular rotating table, a ring for each mold or, more commonly, a single ring mounted on the same mechanism as the plunger, and a single plunger. After a gob is charged into the mold, the machine indexes one station under the plunger and the plunger moves down into the mold, dwells momentarily, then retracts. It is noteworthy that the plunger action flows the glass into the mold cavity rather than stamping out the product by a quick movement. Since considerable heat is removed from the glass by the plunger, it is cooled with water internally. The product remains in the mold for about half the revolution of the press table before removal to allow it to cool below its deformation temperature. The molds may be cooled by forced air.

Blowing methods work best for deep products and frequently must be used for thin-walled items. A common procedure, called the *blow and blow,* involves two steps, of which the first is shaping the glass charge into a form called a *blank* or *parison.* Gob-fed machines receive the gob in the parison mold, where it is shaped into a cylinder about two-thirds the height of the bottle. The finish, or top, of the bottle is formed in the same operation at the bottom of the mold by action of a small plunger entering the mold from below and delivering a puff of air. A transfer mechanism holding the parison by the completed finish then swings and inverts it into a second mold for the second step, blowing the glass into its final shape. A cross section of the molds shows this process in Fig. G-5.

Fig. G-5. Two newly made white-hot catsup bottles are lifted by the neck-rings onto a conveyor that takes them to the annealing lehr. In the rear are shown two red-hot parisons of glass hanging ready for the forming molds to close about them. The blow and blow process is used in this application. (*Owens-Illinois Charlotte, Mich., facility with a capacity of approximately 30 billion bottles and jars per year.*)

The Owens process employs vacuum to charge the glass into the blank or parison mold. Here, a blank mold dips into a shallow pot of molten glass, a vacuum is applied, and a charge of viscous glass is pulled into the blank mold. The finish is formed simultaneously at the top of the blank. This blank or parison is subsequently transferred into the blow mold, where the bottle is blown into its final form (Fig. G-6).

In another modern machine, the glass flows downward from an orifice in a continuous stream which passes between rollers that flatten it into a ribbon with alternate thick and thin spots. The ribbon is picked up by a horizontally moving support in which voids coincide with the thick portions of the ribbon. Blow heads on an endless belt operating from above the ribbon provide puffs of air to aid in producing a bulbous sagging in the thick portion of the ribbon. After sufficient sagging, molds on an endless belt close around the sagging glass from below, and air from the blow heads blows the glass into the shape of the mold. After the molds open, the product, frequently bulbs or Christmas ornaments, can be cracked off the ribbon. A very recent development utilizes this so-called ribbon machine to produce thin-walled bottles at a much faster rate than older methods.

Casting is usually restricted to two types of operations. The first involves the simple pouring of molten glass into molds. Examples include such massive shapes as the borosilicate mirror blank for the Mt. Palomar telescope and the large glass-ceramic mirror blanks for observatories in Australia and South America. The molds are specially constructed of refractory materials.

The second type of casting is *spin casting,* in which a gob from a gob feeder is fed into the bottom of a metal mold supported so that it can be rotated rapidly or spun on its vertical axis. The centrifugal force thus generated causes the glass to flow up the inclined sides of the mold, producing a conical shape. The initial movement of the glass is aided by insertion of a conical plunger into the glass at the bottom of the mold when spinning is begun. Mold speeds of up to 1,600 rpm are attained within 1 s. The funnel portion of television tubes is produced by this method.

Annealing: As with most substances on cooling, the temperature differential between the surface and interior layers of a piece of glass establishes temporary stresses, and the higher this differential the greater the stresses. Fracturing can occur when the stresses exceed the tensile strength of the

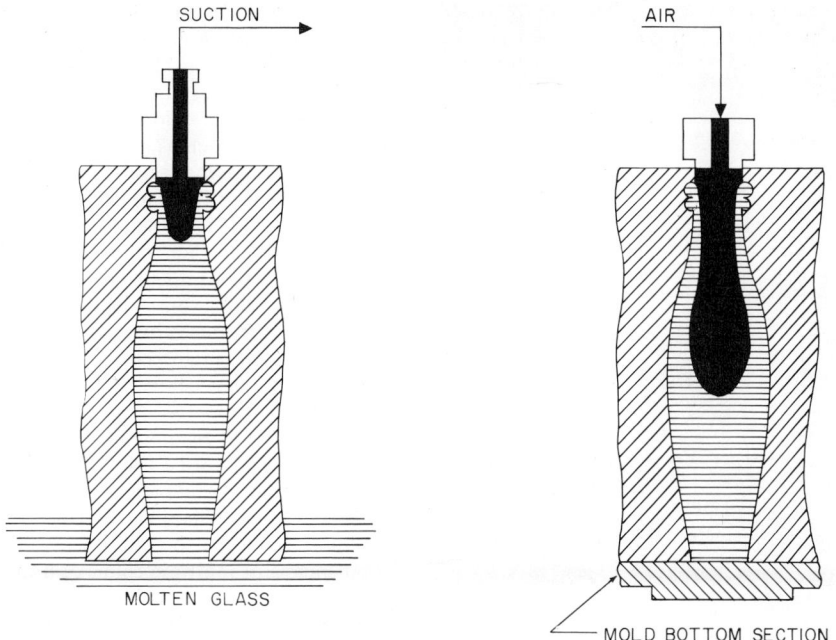

SUCTION

MOLTEN GLASS

AIR

MOLD BOTTOM SECTION

Fig. G-6. Owens process. The blank mold is dipped (left) into the surface of the molten glass, where it is filled by a vacuum suction. As the mold is lifted from the glass, a knife cuts off the glass and closes the mold. The blank mold (right) opens, and a puff of air is introduced to shape the parison before transferring it to the blow mold, where it is blown to its final shape.

glass. Permanent stresses can be avoided by carefully controlled cooling from a little below the annealing point to the strain point. This is the annealing range. Thereafter, the rate of cooling need only be such that the temporary stresses do not exceed the tensile strength of the glass. Glass manufacturers have learned to take advantage of these phenomena (see Fig. G-7).

Annealing immediately follows glass-forming operations. In continuous processes, the ware is placed on an endless belt, which carries it through the lehr, a tunnel in which the temperature is carefully controlled. Temperature of the ware is raised initially to near the softening point, then lowered slowly through the annealing range and thereafter at a more rapid rate to the point where it can be packed or stored. The process is designed to result in the degree of permanent stresses desired. Optical glass must be annealed very thoroughly to produce an essentially distortion- and strain-free lens; however, some stresses can be tolerated or become beneficial to most other products. Small rods and tubing, for instance, are strong enough because of their regular cross sec-

tion to require no annealing, while tempered glass has uniformly controlled stresses to increase its mechanical performance.

Secondary Operations: Lampworking is one of the many and varied operations utilized to produce glassware following the initial forming. The materials used are rod and tubing, which are softened in the flame of burners and shaped or blown as desired.

Grinding and *polishing* are important steps in many glass-manufacturing processes. Use of a sequence of increasingly finer gradations of abrasives, usually ending with jewelers' rouge or cerium oxide powder for polishing, produces the desired results. Optical lenses, prisms, and reflective optics parts are prominent examples. The plate-glass industry has used long lines of grinding and polishing equipment, but the glass produced by the float process has replaced much ground and polished plate glass.

Bending procedures are utilized to produce shapes otherwise difficult to fabricate, e.g., automotive windshields. They are produced by placing the flat pieces of proper shape and size on

Fig. G-7. Newly blown gallon jugs are aligned carefully for movement into the annealing oven, or lehr. After passing through the flames, the containers move into the lehr, in which the temperature first is raised and then lowered gradually in order to remove stresses and strains induced in the manufacturing process. An average trip through a lehr (about 100 ft in length) requires about 1 hr. (*Owens-Illinois.*)

molds and exposing them to temperatures above the softening point. The glass takes the shape of the mold by sagging or slumping with or without assistance from mold parts contacting the glass from above. Temperatures are maintained sufficiently low and the mold material is such that the surface of the glass is unaffected.

Laminating to produce safety-glass parts, as for automotive windows, is a common practice. A sheet of resin such as polyvinyl butyral is placed between properly sized sheets of glass and the whole exposed to slightly elevated temperatures and pressures to bond the glass tightly to the resin.

Coating of glass products such as containers is quite common, the objective being to protect the container from abuse to which it is subjected in handling during filling and shipping. A coating which is not visible, can be labeled, protects the surface, and provides lubricity is required and usually calls for a two-layer coating such as tin or titanium oxide, followed by a lubricious coating such as polyethylene. The oxide coatings are obtained by subjecting the hot container to a vapor of chloride which oxidizes to the oxide. Some glass containers such as aerosols have rather thick coatings of a polymer resin which provide protection as well as esthetic appeal. Thick opaque or translucent oxide and metallic coatings are sometimes used to provide attractive color effects or light protection. Many precision optical lenses are coated with thin vapor-deposited layers which reduce the light losses by reflection from the surface, and some architectural glass is coated to provide attractive colors and reflect undesirable infrared radiation.

Decorating glass or glassware is an old art that takes many and varied forms. Cutting, grinding, and mechanical or chemical polishing or etching are well known. Opaque, translucent, and transparent enamels can be applied by silk screens or other means in multiple colors and in almost any pattern. Low-melting vitreous enamels have been used for many years, and when properly fired, they provide good durability. More recently, organic polymers have been substituted for the vitreous enamel. They are not quite as durable as vitreous enamels, but they do not require high curing temperatures.

Tempering is the direct reverse of annealing; i.e., high permanent stress is induced in the glass. Rapid cooling or quenching is applied to the glass surfaces at a temperature slightly below the softening point, placing the surfaces in a high degree of compression while the balancing tensile forces are confined to the interior. Since glass always breaks in tension, very considerable strength is incorporated. Typical products are glass doors, windows, goggles, and even spectacles. Tempering must be the final step in the production line. Other products can be strengthened by judicious control of the degree of annealing if their shapes permit it.

Sealing glasses to each other or to other materials must take into account the thermal expansion-and-contraction characteristics. Many glasses have thermal-expansion properties which allow them to be sealed to metals, but each metal usually requires a different glass composition. Solder glasses are used to seal two pieces of glass to each other, two pieces of metal, or a piece of metal and a piece of glass. The glass seals on light bulbs and vacuum tubes are examples of commercial glass-metal seals, while color TV tubes are sealed together with solder glass at a temperature at which the phosphors are not degraded.

See also **Ceramics**.

—Earl D. Dietz, *Owen-Illinois, Inc., Toledo, Ohio*

Glass, Fiber   (See **Fiber glass**.)

Glasses, Liquid-Silicate   [See **Silicates (soluble)**.]

Glauconite   (See **Potassium**.)

Glazes   (See **Ceramics**.)

Globulins   (See **Peptides and proteins**.)

Glow-Discharge Chemistry   (See **Plasma, chemical processing with**.)

Glucose   (See **Cellulose ester plastics, organic; Ethyl cellulose; Rayon;** and **Starch**.)

Glue   (See **Adhesives**.)

Glutamic Acid   (See **Amino acids;** and **Peptides and proteins**.)

Glutamine   (See **Amino acids;** and **Peptides and proteins**.)

Glutathione   (See **Peptides and proteins**.)

Glutelin   (See **Peptides and proteins**.)

Gluten   (See **Starch**.)

Glycerides   [See **Alcohols, fatty (via hydrogenation); Anionic surfactants (sulfur-bearing); Soaps;** and **Vegetable oils**.]

Glycerin   (See **Polymerization; Soaps;** and **Vegetable oils**.)

Glycerol   (See **Alcohols; Epichlorohydrin;** and **Soaps**.)

Glycine   (See **Amino acids;** and **Peptides and proteins**.)

Glycols   (See **Ethylene glycol**.)

Glycol Ethers   (See **Ethylene glycol ethers**.)

Glyoxal   (See **Aldehydes**.)

Gold (OE. *geolu*, yellow.) (L. *aurum*, gold) **Au** = 196.967 (at. wt.); 79 (at. no.). Gold occurs in group IB, period 6 of the periodic table. It is the only yellow metal. Gold is relatively soft and is the most malleable and ductile of all metals. The purity or fineness of gold is expressed in karats (K). Pure Au is 24 K.

The melting point is 1063°C; boiling point 2970°C; and density 19.32 g/cm^3 at 20°C. Gold has 19 isotopes, 185–203. Only one of these (200) is stable. Gold has a face-centered cubic lattice and no allotropes. The metal is highly resistant to attack by acids and will not dissolve in any one of the common acids but will dissolve in aqua regia and in selenic acid, H_2SeO_4. O_2 and H_2S have no effect on Au, but both Cl_2 and Br_2 will react to form soluble halides. Gold is also dissolved in alkaline cyanide solutions. It is the only metal that forms no oxide film on its surface in air at ordinary temperatures. Even though very malleable, its elongation is not especially high (about 30% in the cast state). Gold leaf can be made so thin that it will transmit light. A summary of the more important physical and mechanical properties is given in Table G-5.

Gold is found chiefly as the free metal scattered through gravel (*placer gold*) or disseminated in veins of quartz (*vein gold*). Small quantities also are found in lead and copper sulfide ores. Nuggets of native gold, varying in size from that of a tiny

TABLE G-5. *Selected Physical and Mechanical Properties of Gold*

Thermal conductivity (20–100°C), (cal)(cm^2)(cm)(s)(°C)	0.70
Specific heat (20–100°C), cal/(g)(°C)	0.030–0.031
Electrical resistivity (20–100°C), $\mu\Omega$-cm	2.3–3.0
Temperature coefficient of resistivity (20–100°C) per °C	0.0039
Coefficient of linear thermal expansion (20–100°C) $\times 10^6$	14.2
Tensile strength, psi:	
Cast	18,000
60% cold-worked	32,000
Elongation, %:	
Cast	30
60% cold-worked	4
Hardness, Brinell:	
Cast	33
60% cold-worked	58
Modulus of elasticity, psi:	
Cast	10,800,000
60% cold-worked	11,500,000

pebble to a mass weighing as high as 248 lb, have been found. In a combined state, gold occurs in the mineral sylvanite, a telluride of gold and silver, (AuAg) Te$_2$, a rich ore found in Colorado. The bulk of the gold ores contain very little gold (about 5–15 g/metric ton). Some of the richest ores found in Africa contain 20–30 g/metric ton. Almost all countries produce some gold, but the leader, by far, is South Africa, followed by the Soviet Union and Canada and far behind by the United States, Australia, Ghana, and Rhodesia.

The treatment of gold ores involves (1) grinding, amalgamation, and/or cyanidation of ores containing coarse free gold and (2) very fine grinding, flotation, roasting, and amalgamation and/or cyanidation of ores containing gold telluride or sulfide. These processes produce an impure gold metal containing considerable silver and some copper plus other base metals. This is purified by melting and oxidizing the base metals or by melting and chlorinating (Miller process), which removes the base metals and silver. The silver-containing oxidized gold is purified by the electrolysis of gold chloride solutions containing an HCl solution (Wohlwill process). Here the anode is the alloy (Au-Ag) and the cathode is pure Au. The gold deposits then on the cathode, and the silver forms silver chloride and remains as a deposit about the anode.

The chief use of gold is as a standard of value for money. From 3400 B.C. to about A.D. 1930, it was the real basis for all coinage form. However, gold's role now is in backing up the complex international credit systems throughout the world.

The largest commercial use of gold is in the form of jewelry, and at the present time, most jewelry is made by the lost-wax process, which dates back to 3000 B.C. or earlier. Most of these applications employ the karat golds which contain 10 and 14 K (and less commonly 18 K) of gold (41.7, 58.3, and 75.0 wt % of gold, respectively). These alloys are of two general types. The red, yellow, and green golds are composed of Au, Cu, and Ag. These may contain Zn as well as minor amounts of deoxidizers and/or grain refiners. A wide variety of karat values of other color shades can be made by varying the composition ranges within this basic ternary system. The second widely used class is the white karat golds, which are produced in two basic alloy types. These are the original Au-Ni-Zn-Cu (18 K) and the Au-Cu-Ni-Zn (10 and 14 K) alloys, and of later development, Au-Pd-Ag-Cu and Au-Cu-Ni-Pd-Ag alloys. The last two are generally 14 and 10 K alloys, and

at 10 K, their color is probably better than that of the original white golds. Of course, the latter group is much more expensive.

Considerable brazing is done by jewelry manufacturers, and the solders may be of a lower karat content than the alloy being brazed. Usually, they contain much more Ag and Zn than the alloys themselves.

The pink golds derived mainly from the quinary system of Au-Ag-Cu-Ni-Zn, are essentially red golds which are whitened by the addition of Ag, Ni, and Zn.

The use of gold in the electrical, electronic, and other industrial fields has grown considerably in past years. The electrical and thermal conductivity, resistance to oxidation, and ease with which gold is electroplated make it an excellent coating for electrical contacts, particularly in metalized ceramics for use in electron tubes, transistors, and other electronic components because gold does not migrate into the ceramic as silver does. As a coating, too, its thermal characteristics make an excellent heat shield for such applications as space capsules.

Gold is used extensively in many solders and brazing alloys, ranging from the low-melting eutectics of Au with Ge, Si, and Sn to Au-Cu, Au-Ni, and Au-Pd-Ni alloys. The latter brazing materials have the ability to withstand long use at high temperatures and are particularly applicable to jet-engine fabrication.

Gold is also used in dentistry, but this application has declined in recent years. Gold alloys, such as Au-Ag-Cu with varying amounts of Pt and Pd, are used for restorations and for bridges, inlays, and partial dentures. These are cast with much more precision than jewelry, and have in fact, replaced wrought gold wire in many of these dental appliances. Gold wire, now used principally in orthodontic and prosthetic appliances, is a complex alloy containing Au, Pt, Pd, Ag, Cu, Ni, and Zn.

Compounds: Gold exhibits valences of 1 and 3 and generally forms two series of compounds, aurous (1$^+$) and auric (3$^+$). The aurous compounds generally are quite unstable and decompose readily when heated (moderate temperatures). Aurous oxide, Au$_2$O (purple), results when a solution of AuBr is treated with KOH in slight excess and subsequent boiling of the resulting mixture. Heated above 205°C, the oxide decomposes into the metal and O$_2$. Auric hydroxide, Au(OH)$_3$, can be formed by treating AuCl$_3$ with KOH. If the hydroxide is dried over CaCl$_2$, water is lost and auric oxide, Au_2O_3, formed. The latter decomposes if heated over 250°C. The Au oxides and hydroxides are insoluble in H_2O but are soluble in HCl and the halogen acids: $4HCl + Au(OH)_3 \rightarrow HAuCl_4 + 3H_2O$.

The aurous salts decompose in H_2O. Auric chloride and auric bromide are soluble in H_2O, but auric iodide is decomposed by H_2O. In H_2O, aurous sulfide, Au_2S, forms a colloidal solution. The sulfide is insoluble in dilute acids but will dissolve in alkali polysulfides and cyanides. Auric sulfide, Au_2S_3, will decompose in H_2O but is insoluble in the common acids. It will dissolve in aqua regia.

If NH_4OH or $(NH_4)_2CO_3$ is added to a solution of $AuCl_3$, a yellow precipitate (fulminating gold) will be formed, $AuN_2H_3 \cdot 3H_2O$, which is quite explosive. The sensitivity is increased by preparing the compound in the presence of a fixed alkali. Gold may be precipitated from acid or alkaline solutions by formaldehyde or other aldehydes, providing a good means for separating Au from Cu, Sb, Hg, Zn, Pb, Mn, Sn, As, and Pt. Au also is precipitated from acid solution by dimethylglyoxime. Hydrazine will precipitate Au from acid or alkaline solutions. The complex alkali cyanides of Au can be obtained by treating the oxides with NaCN. Au_2O yields $NaAu(CN)_2$, and Au_2O_3 yields $NaAu(CN)_4$.

Many Au compounds are decomposed by light; all are decomposed by heat.

REFERENCES

Smithells, Colin J.: "Metals Reference Book," vol. 3, Plenum, New York, 1967.
Wise, Edmund M. (ed.): "Gold: Recovery, Properties and Applications," Van Nostrand Reinhold, New York, 1964.

—Robert J. MacDonald, *Handy & Harman, Fairfield, Conn.*

Gold Number (See **Colloid systems.**)

Graft Copolymers (See **Polymerization.**)

Grainer Process [See **Salt (NaCl).**]

Graining (See **Separation operations.**)

Gram Mole [See **Mole (or mol).**]

Granulation (See **Soaps.**)

Graphite (See **Carbon; Coal tar and derivatives; and Iron and steel.**)

Gravity Blender (See **Mixing and blending, solids-solids.**)

Gray Iron (See **Iron and steel.**)

Grease Additives (Consult the **Subject Index.**)

Green Liquor [See **Pulp (wood) production and processing.**]

Grignard Reagents Grignard reagents are organomagnesium halides, usually represented by the formula RMgX, where R is any organic radical and X is a halogen. They are extremely important organic reagents due to their reactivity with a wide range of organic compounds as well as inorganic halides.

In 1899 Barbier prepared dimethylheptenol by reacting methyl iodide, dimethylheptenone, and magnesium in ethyl ether. Barbier's student Victor Grignard, studying the mechanism of this reaction in 1900, found that the reaction proceeds in two stages: (1) reaction of magnesium and an alkyl halide to form the alkyl magnesium halide and (2) reaction of the latter with a compound containing a carbonyl group to form a new carbon-carbon bond. Stage one signaled the first Grignard reagent. As a result of his research, Grignard was awarded the Nobel Prize in Chemistry in 1912, and chemistry was given its most versatile organic reagent. In the next 35 years Grignard and other researchers found that almost all alkyl and aryl halides react with magnesium in ether to form Grignard reagents, the aryl and vinyl derivatives proceeding with more difficulty.

Grignard reagents are prepared by slow addition of an organic halide to a suspension of magnesium in a suitable solvent, the two most commonly used being ethyl ether and tetrahydrofuran. Other solvents, e.g., the cyclic ethers, hydrocarbons, and nitrogen-containing solvents such as tertiary amines and substituted amides, have also been employed. The high reactivity of the Grignard makes it essential to exclude water and oxygen from the reaction. To activate the magnesium (slow to react due to the oxide film formed on exposure of magnesium to air) and thus initiate the reaction, the following procedures may be used: (1) addition of a small amount of iodine, ethylene dibromide, or methyl iodide, (2) crushing some of the magnesium after a small amount of halide has been added, (3) addition of a small amount of the Grignard reagent to the ether to ensure anhydrous conditions.

TABLE G-6. *Typical Reactants, Equations, and Products of Grignard Reagents*

Reactants	Equation	Products
Carbon dioxide	$RMgX + CO_2 \rightarrow RCO_2MgX \xrightarrow[H_2O]{HX} RCO_2H + MgX_2$	Carboxylic acids
Carboxylic acids	$R'COOH + RMgX \rightarrow R'CO_2MgX \xrightarrow{H_2O} R'R_2COH$	Tertiary alcohols
Acid halides .	$RCOX + 2R'MgX \rightarrow RR'COH$ $RCOX + R'MgX \rightarrow RCOR'$ (reverse addition)	Tertiary alcohols Ketones
Acid esters* . .	$RCO_2R' + 2R''MgX \rightarrow RR''_2COH + R'OH$	Tertiary alcohols
Aldehydes† . .	$RMgX + 'R'CN \rightarrow RR'C{=}NMgX$ $RR'C{=}NH \xrightarrow{H_2O} RR'CO$	Secondary alcohols
Ketones	$RMgX + R'COR'' \rightarrow RR'R''COMgX \xrightarrow{H_2O} RR'R''COH$	Tertiary alcohols
γ-Lactones . .	$\begin{array}{c} CH_2{-}CH_2 \\ \mid \qquad \mid \\ CH_2 \quad C{=}O \\ \diagdown O \diagup \end{array} + CH_3MgI \rightarrow HOCH_2CH_2CH_2(CH_3)_2COH$	Glycols
Nitriles	$RMgX + R'CN \rightarrow RR'C{=}NMgX \xrightarrow{H_2O} RR'C{=}NH \xrightarrow{H_2O} RR'CO$	Ketones
Metal halides .	$RMgX + SnCl_4 \rightarrow RSnCl_3 + R_3SnCl + R_2SnCl_2 + R_4Sn$	Organometallic compounds
Oxygen	$2RMgX + O_2 \rightarrow 2ROMgX \xrightarrow{H_2O} ROH + HOMgX$	Alcohols and phenols
Water	$RMgX + H_2O \rightarrow RH + Mg(OH)_2 + 2X^- + [Mg \cdot 6H_2O]^{++}$	Hydrocarbons
Hydrogen halides	$RMgX + HX' \rightarrow RH + X'MgX$	Hydrocarbons
Sulfur	$RMgX + S \rightarrow RSMgX \xrightarrow{H_2O} RSH$	Mercaptans
Ammonia . . .	$RMgX \xrightarrow[\text{decomposes}]{NH_3} RH$	Hydrocarbons

*Formic acid esters yield secondary alcohols or aldehydes.
†Formaldehyde yields primary alcohols.

In the mid-1950s Henry Normant and Hugh Ramsden, independently of each other, discovered that less reactive halides, such as chlorobenzene and vinyl chloride, readily form a Grignard reagent when tetrahydrofuran is used as the solvent. Tetrahydrofuran has since become an important solubilizing and complexing agent in Grignard synthesis generally since it allows higher reflux temperatures and greater safety than ethyl ether.

Di-Grignard reagents, XMgRMgX, while valuable in organic synthesis, are difficult to prepare. The most important include $BrMg(CH_2)_4MgBr$ and $BrMg(CH_2)_5MgBr$, which are used in the synthesis of heterocyclic compounds, and di-Grignards of o-bromoiodobenzene, useful in the synthesis of o-phenylene tertiary diphosphines.

The constitution of the Grignard reagent has come under considerable study since Grignard's original representation, RMgX. Schlenk and Schlenk suggested an equilibrium $2RMgX \rightleftarrows R_2Mg + MgX_2$; Dessy concluded that the structure was a $R_2Mg \cdot MgX_2$ complex; and Rundle, in an x-ray diffraction study of crystalline phenylmagnesium bromide, hypothesized the structure as $C_6H_5MgBr \cdot 2(C_2H_5)_2O$. There is still doubt as to the exact structure, but for practical purposes RMgX is used.

Grignard reagents serve as starting materials for a wide variety of compounds containing all types of functional groups. They are used in the synthesis of motor fuels, vitamins, hormones, pharmaceuticals, synthetic perfumes, organometallic compounds, and insecticides. They react with most organic functional groups with the possible exception of ethers, tertiary amines, and organic halides (with difficulty). Some typical reactants, with the equation and eventual products, are given in Table G-6. See also **Lithium;** and **Silicones.**

REFERENCES

Ashby, E. C.: *Q. Rev.,* vol. 21, pp. 259–285, 1967.
Coates, G. E., and K. Wade: "Organometallic Compounds," 3d ed., vol. 1, pp. 76–97, Methuen, London, 1967.
Kharasch, M. S., and O. Reinmuth: "Grignard Reactions of Nonmetallic Substances," Prentice-Hall, New York, 1954.
Nesmeyanov, A. N., and K. A. Kocheshkov: "Methods of Elemento-organic Chemistry," vol. 2, World, New York, 1967.
Walker, F. W.: "Grignard Compounds, Compositions and Mechanisms of Reaction," Ph.D. thesis, Georgia Institute of Technology, 1969.
Yoffe, S. T., and A. N. Nesmeyanov: "A Handbook of Magnesium-organic Compounds," Pergamon, London, 1957.

—Marguerite K. Moran, *M & T Chemicals Inc., Rahway, N.J.*

Grinding (See **Glass; Size reduction;** and **Soaps.**)

Grinding, Abrasive (See **Abrasives.**)

Grizzly (See **Screening.**)

Guar (See **Gums.**)

Gums Generally gums may be regarded as complex carbohydrates and usually are classified in terms of their sources, such as (1) exudates from plants and trees, e.g., gum arabic, ghatti, karaya, and tragacanth; (2) extracts of marine plants, e.g., agar, Iceland moss, Irish moss, and algin; (3) extracts of seeds, e.g., guar and locust; and (4) extracts of vegetables, e.g., pectin. Gums particularly occur in plants of arid regions. They usually are tasteless and odorless in commercial form. A number of gums are soluble in water; others unite with water to form a mucilaginous product.

Several gums are described briefly in alphabetical order.

Agar-agar is the dried mucilaginous material extracted from seaweed or marine algae and is the sulfuric acid ester of a linear galactan. The material is used mainly as a culture for bacteria and fungi. Containing salt, agar-agar sometimes is used instrumentally to form an electrical connection with solutions.

Algin, derived from several species of kelp, is a complex anhydride of an aldose sugar acid with all carboxyl groups free and all aldehyde groups conjugated. It is used in the preparation of ice cream, frozen desserts, candy, cake icings, paints, rubber, and pharmaceutical products.

Arabic (gum arabic), also known as *acacia gum,* is the exudation from a small tree, *Acacia arabica* (Africa). The substance has a molecular weight of about 240,000 and is a mixture of the Ca, Mg, and K salts of arabic acid in a complex of the saccharides arabinose, galactose, mannomethylose, and open-chain glucuronic acid. Gum arabic is used in adhesives, for thickening inks, in textile coatings, drug and cosmetic emulsions, and as a binder for pharmaceutical tablets, where it serves as a disintegrating agent to hasten tablet solubility. The gum also prevents the crystallization of sugar in candy glazes. A synthetic substitute for gum arabic is glyceryl monostearate.

Benzoin is a balsam derived from several species of *Styrax* trees (eastern Mediterranean) and is used in medicines, perfumes, and incense. Benzoin gum once was a source of benzoic acid.

Ghatti gum is derived from the tree *Anogeissus latifolia* (India) and is twice as effective as gum arabic as an emulsifier but less adhesive. The gum is used in textile finishing, particularly in India.

Guar is derived from the seed of the guar plant, *Cyanopsis tetragonoloba* (Pakistan, Texas, and Arizona) and is a polysaccharide with a straight chain of mannose units and one galactose group on alternating mannose units. The proportion of mannose to galactose is 4:1. The gum contains about 6% protein. Guar is more than 6 times as effective as starch in thickening power and is used for upgrading starches. Derivatives of guar gum are available that will stiffen gels even up to a water content of 99%.

Karaya gum is derived from the *Sterculia* tree (India) and also is known as *Indian gum* and *hog tragacanth.* The chief constituent is galactan. Uses are about the same as those for tragacanth.

Kauri gum is a fossil gum (New Zealand and New Caledonia) used in enamels and varnishes to increase body and enhance elasticity and hardness. The material also is used in adhesives and linoleum production.

Locust-bean gum, also known as *carob flour,* is obtained by milling bean kernels of locust trees (tropical America, Africa, and Mediterranean area), principally the *Ceratonia siliqua* (Spain, Cypress, Spain). The gum contains galactose and mannose in a complex polymer and may by regarded as a polysaccharide or complex sugar. The ratio of mannose to galactose is $3:1$, and the gum contains about 6% protein, somewhat similar in these respects to guar. Locust-bean gum is used for coating textiles; as a thickener and binder in glues, pastes, and latex; in leather finishes; sizings for yarn and paper; jellies; dog foods; and tobacco.

Okra gum comes from the pods of okra, *Hibiscus esculentus,* a plant of the cotton family. Pods sometimes are called *gumbos.* The substance is used as a thickening and stabilizing agent for foods and pharmaceutical products, also in electroplating baths for brightening Ni, Ag, and Cd plates.

Tragacanth gum is derived from the shrub *Astragalus gummifer* (eastern Mediterranean and Iran) and is used for adhesives and mucilage, leather dressings, textile printing, and as a general emulsifying, thickening, and suspending agent for foodstuffs, drugs, cosmetics, adhesives, and textile finishes. The chief constituent is galactan.

See also **Resins;** and **Rubber, natural.**

REFERENCES
Brady, George S.: "Materials Handbook," 10th ed., McGraw-Hill, New York, 1971.
Whistler, Roy L. (ed.): "Industrial Gums: Polysaccharides and Their Derivatives," Academic, New York, 1959.

Gum Rubber   (See **Rubber, natural.**)

Gummite   (See **Uranium.**)

Gun Metal   (See **Tin.**)

Gypsum   Sometimes there is confusion between *gypsum,* a commonly occurring mineral with the chemical formula $CaSO_4 \cdot 2H_2O$, and its useful product of partial dehydration, $CaSO_4 \cdot \frac{1}{2}H_2O$. This may be traceable to the ancient Greeks, who used the same word to designate both materials. Although there is some controversy about the discrete phases in the $CaSO_4$-H_2O systems, the identification given in Table G-7 generally is recognized.

Resources: The mineral gypsum is virtually unlimited in supply, although commercially desirable grades are not readily available in many areas. Major deposits in the United States occur in New York, Ohio, Michigan, Indiana, Iowa, Texas, Oklahoma, Nevada, and California. About 35% of the gypsum consumed in the United States is imported from Canada and Mexico and processed in seaboard plants. Total United States consumption in recent years has been 15–20 million tons/yr. Within a few years, consumption is expected to increase to 20 million tons/yr.

Properties: Gypsum is a soft rock occurring in sedimentary layers of a few feet up to 100 ft thick. Color varies, and gypsum can be white, gray, almost black, pink, or brown, depending upon the nature and amount of impurity present. Gypsum was originally precipitated from saline waters and

TABLE G-7. *Common Designations and Properties of Various* $CaSO_4$-H_2O *Systems*

Chemical Formula	Common Designations	Properties
$CaSO_4 \cdot 2H_2O$	Calcium sulfate dihydrate, rock gypsum, chemical gypsum, alabaster (white, fine grained), selenite (translucent, platey), satin spar (fibrous), land plaster (pulverized gypsum)	All forms (natural, synthetic, recrystallized) thermodynamically and crystallographically equivalent; habit may be needles, plates, or prisms
$CaSO_4 \cdot \frac{1}{2}H_2O$	Calcium sulfate hemihydrate, calcined gypsum, stucco, plaster of paris, molding plaster, gypsum plaster, chemical hemihydrate	Alpha and beta types exist depending on conditions of calcination; alpha is more stable, crystalline, lower energy; beta is less stable, disordered, higher energy
$CaSO_4$	Anhydrite I: high-temperature anhydrite	Produced by high temperature ($>1830°F$, $1000°C$) calcining. Contains free CaO
	Anhydrite II: insoluble anhydrite, inactive anhydrite, dead-burned gypsum, chemical anhydrite, mineral anhydrite	Produced by calcining at $480–1830°F$ ($250–1000°C$). Relatively inert; reactivity depends on calcining time-temperature relationship and particle size
	Anhydrite III: soluble anhydrite, active anhydrite, dehydrated hemihydrate	Produced by low-temperature ($350–500°F$, $177–260°C$) dehydration of hemihydrate. Avidly reacts with water and moist air to form hemihydrate.

frequently lies adjacent to anhydrite and lime-stone strata. These minerals are commonly found as impurities in gypsum. Acceptable quality gyp-sum rock is at least 80% $CaSO_4 \cdot 2H_2O$; good quality gypsum rock is at least 90% $CaSO_4 \cdot 2H_2O$. Limestone, anhydrite, and silica impurities do not detract from the usefulness of gypsum except as inert diluents. Other impurities, such as hygro-scopic clays and water-soluble salts, are deleterious to many gypsum products, and only a few tenths percent can be tolerated. Mining methods are similar to those for coal. Underground seams up to 500 ft deep are mined by the room-and-pillar method. Major sites in the western United States, Mexico, and in the Canadian maritime provinces are quarried or strip-mined.

Chemical Gypsum: Sometimes termed *by-product* or *synthetic gypsum,* this material is precipitated in various chemical processes, e.g., the manufacture of H_3PO_4. Types of chemical gypsum may be designated by the nature of the Ca salt from which the $CaSO_4$ is derived (phosphogypsum, fluorogypsum, chlorogypsum, and citrogypsum). Impure, by-product $CaSO_4$ results from new sys-tems for limestone scrubbing of SO_2 from power-plant stack gases. Except in agriculture, chemical gypsum has not been used widely in the United States. In Japan, where good mineral gypsum is scarce and costly, chemical gypsum has been up-graded to make an acceptable raw material for building products. Satisfactory chemical gypsum requires control of residual chloride, fluoride, and phosphate to less than 0.5%.

Processing: For thousands of years, it has been known that gypsum exposed to moderate heat gives off three-fourths of its water of hydration and that the resulting *stucco,* when mixed with water, forms a putty which can be shaped for artistic or structural purposes into articles that set and become strong, rigid, and durable.

Modern processing of gypsum rock involves crushing, grinding, and calcining to make gypsum plaster (see Fig. G-8). Usually, gypsum is calcined in indirect-heated kettles, which may be operated batchwise or continuous. A typical batch is 15 tons of 90% <100-mesh gypsum, which is heated to 300–350°F (149–177°C) $1\frac{1}{2}$–4 hr, depending upon the nature of the raw material and desired quality of the stucco. Release of steam from the calcining gypsum causes the kettle to boil vigor-ously. A gypsum kettle is a novel processing de-vice, an autogenous fluidized-bed reactor. Quality of the stucco is a function of heat flux, calcining temperature, reaction time, presence of natural or

added salts, and postcalcination treatment such as cooling and grinding. Gypsum is also calcined in direct-fired rotary kilns, flash driers, heated screw conveyors, and open pans.

In all of the foregoing systems, gypsum is de-hydrated at essentially atmospheric pressure in surroundings of unsaturated steam. The resulting stucco, predominantly beta hemihydrate, is char-acterized by high water demand (over 50 ml of H_2O per 100 g of stucco to make a pourable slurry). Gypsum calcined under controlled pres-sure in liquid water or saturated steam, in the presence of certain crystal modifiers, yields alpha hemihydrate, which has a low water demand (30–50 ml of H_2O per 100 g of stucco). Gypsum can also be dehydrated in hot, salty water or in mineral acids to form alpha hemihydrate or in-soluble anhydrite, depending upon temperature and salt or acid concentration.

Alpha hemihydrate can develop very high strength after set because of the low mixing-water demand and resulting high density of the set mass. However, after set at equal dry density, alpha hemihydrate is no stronger than beta hemi-hydrate. Lane[4] has shown that betahemihydrate readily disperses or disintegrates in water into a multitude of very fine particles; alphahemihy-drate is not dispersible, even with high shear mixing, explaining its relatively low water de-mand to obtain workable fluidity.

Plaster: The essential property of gypsum plas-ter, which permits economical fabrication of building products, is that the setting time can be precisely controlled from a few minutes up to many hours. This is accomplished by addition of fractional percentages of *accelerators,* typically water-soluble salts, such as K_2SO_4, or finely ground gypsum, and *retarders,* modified organic substances (casein, glue, blood, hair, and hoof meal), certain acids (citric, boric, and phosphoric), and salts of these acids. Accelerators are thought to provide additional nuclei for crystallization; retarders are thought to form protective colloids or insoluble salts which block water access to the plaster particle. Combinations of retarders and accelerators are frequently employed to achieve a controlled rate of reaction.

Wallboard: By far the largest single use of gyp-sum is to produce gypsum wallboard, which con-sists of a core of gypsum sandwiched between two layers of paper. Gypsum wallboard is the most commonly used material for lining the interior walls and ceilings of buildings. Its outstanding attributes are fire resistance, dimensional stability,

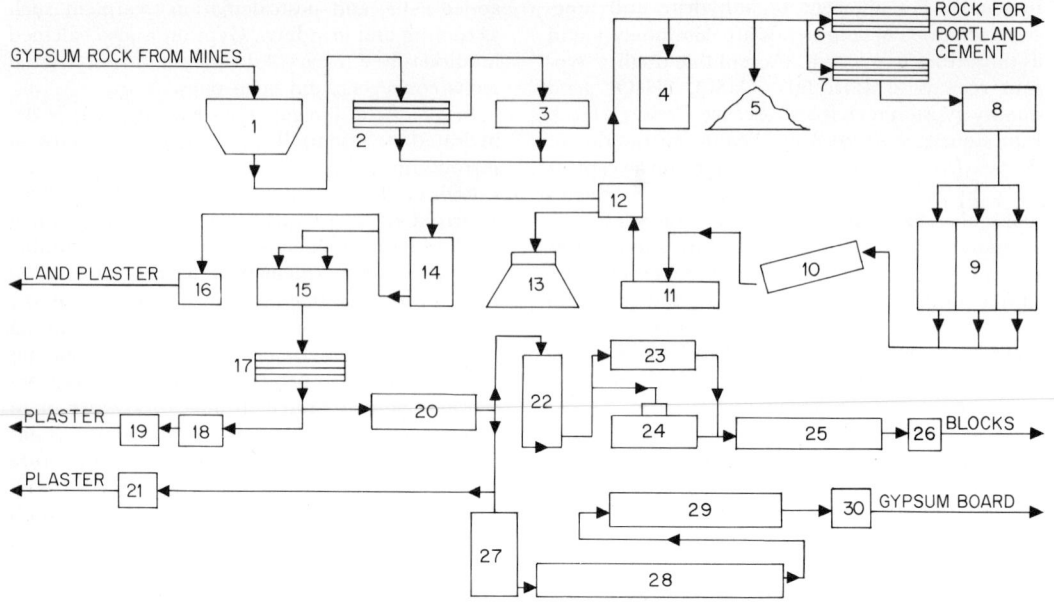

Fig. G-8. Flow diagram of a typical gypsum mill. Gypsum rock from the mine is crushed (1) and then screened (2). Oversize rock is passed through a hammer mill (3) where it is combined with the undersize for storage in a silo (4) or a stockpile (5). A portion of the crushed gypsum rock is further screened (6) and then sold for use in portland cement. The portion of rock to be used in the gypsum mill is further screened (7) and passed to a bulk-rock bin (8). Depending upon production requirements, the rock then is conveyed to a series of bins (9) from which it is drawn for processing in a rotary drier (10). The dried rock is further crushed in a hammer mill (11). An air separator (12), operating in conjunction with the hammer mill, separates the fines, which are then conveyed to a bin (14). Oversize material is sent to a grinding mill (13) for further pulverizing and recycling. Material out of bin (14) after packaging (16) is marketed as *land plaster*. The remaining portion of material from bin (14) passes to a kettle and hot pit (15), after which it is screened (17). A portion of this material is passed through a series of stone mills (18), packaged (19), and sold as plaster. The other portion of material from screen (17) passes to a tube mill (20) for further reduction, after which it goes directly into gypsum-block or gypsum-board production or is packaged as another grade of plaster (21). In block manufacture, the material from the tube mill passes to a storage bin (22). Blocks may be hand-molded (23) or machine-molded (24). In either case, the blocks pass through block kiln (25), after which they are packed (26) and readied for shipment. In gypsum-board manufacture, material is conveyed from tube mill (20) to a stucco silo (27), which supplies the board machine (28). Once the board has been formed and cut into appropriate lengths, the boards pass through the board kiln (29), after which they are packed (30) and readied for shipment. Any particular gypsum mill will display some variations in the flow and processing of materials.

easy workability, and low cost. Typical gypsum wallboard is $\frac{1}{2}$ in. thick, 48 in. wide, and 8–20 ft long and weighs 1.8 lb/ft^2. It is made continuously on board machines operating at 25–300 lineal ft/min. (See Fig. G-9.) Foamed plaster slurry is mixed and discharged upon a moving web of paper. The edges of the bottom paper are scored and folded so that the slurry is completely contained between that sheet and the top paper, which is laid upon the slurry. Thus, the paper surfaces not only provide strength and paintability to the finished board but are also a continuous mold within which the gypsum is cast. Board

thickness, which may be $\frac{1}{4}$–1 in., is controlled by the master roll and smoother bars. The board proceeds down a long rubber belt conveyor and within 5 min after forming sets hard enough to be cut to length and dried.

Making a fluid board slurry requires water much in excess of that required for plaster hydration; about 0.5 lb H$_2$O per pound of dry board must be removed by drying. This is done in multideck kilns, where heat is provided indirectly by steam coils or directly by products of combustion. Drying takes about 1 hr at air temperatures up to 600°F (316°C). The bond between the

Fig. G-9. In the manufacture of gypsum board, the continuously formed board travels down a long line for edge forming, setting, and cutting. (*United States Gypsum Company.*)

paper and gypsum core is preserved by addition of about 0.5% of acid-modified starch to the gypsum slurry. Water diffusion during drying leads to migration and deposition of soluble starch at the paper-core interface.

At the board slurry mixer, a separately generated foam is mixed with stucco and water to reduce dry board density to 40–50 lb/ft^3.

Fibers are added to provide crack resistance and/or fire resistance. Water-repellent chemicals may be added to the board core and/or the paper surface. Decorative and functional finishes can be factory-applied.

Gypsum board is attached to wood or steel framing members with nails, screws, or adhesives. It can easily and quickly be cut to size with a knife, not requiring sawing. Conventional gypsum board is limited to areas where it is not exposed to water, since gypsum is slightly soluble in water (about 2 g/l), the paper is weakened by water, and creep and erosion can occur. A special technology has evolved in connection with finishing the joints between gypsum boards to yield monolithic wall and ceiling surfaces. Gypsum boards are made in Australia and Germany with-

out paper surfaces by heavily loading the gypsum with reinforcing fibers.

In recent years, the market for gypsum board (dry wall) has increased while the use of gypsum plaster (wet wall) has decreased, due largely to economic factors. Gypsum plaster is formulated with set control additives to delay set for several hours. An aggregate of sand, perlite, or vermiculite is added along with sufficient water to make a mix that can be troweled or pumped. The gaged plaster is applied by trowel or spray over a water-absorptive base (gypsum lath), a mechanically perforated base (expanded metal), or properly treated masonry surface. Keene's cement, $CaSO_4$ (II), is a finishing plaster to provide especially hard, white, and water-resistant walls.

Other Uses of Gypsum: Industrial plasters find a number of unusual applications—dental plasters for making tooth impressions, orthopedic plasters for immobilizing broken bones, pottery plasters for molds for clay, oil-well cements, permeable plasters for casting nonferrous metals, art and statuary casting, lamp bases, patching and grouting compounds, insulating-brick production, and pattern and model making in the automotive and aircraft industries. Many of these applications require close control of set time and set expansion, usually achieved by proprietary methods. Water-reducing additives and reinforcing resins and cements may be added to achieve compressive strength of over 15,000 psi.

Aside from gypsum building products, the largest use of gypsum is in portland-cement manufacture. About 5% of gypsum is added to the cement clinker before grinding. The level of SO_3 addition should be optimized with respect to clinker quality and composition. Addition of gypsum helps to prevent undesirable false set and can increase early strength of the cement. Some cement producers favor a mixture of gypsum and anhydrite for optimum control of cement set. See also **Cement.**

In agriculture, gypsum performs as a soil conditioner, provides a source of available calcium and sulfate, helps retain organic nitrogen in the soil, and does not add acidity or alkalinity. Gypsum enjoys wide use in areas where soils are deficient in sulfur. Gypsum may be incorporated in mixed fertilizers and in animal feeds. See also **Fertilizers.**

Fine, white gypsum (terra alba) or dead-burned gypsum is used as a functional filler in paper and plastics and as an extender for TiO_2. Pure calcium sulfate, meeting U.S. Food and Drug specifications, is added to bread and other bakery products and finds use in beer production and as a pharmaceutical-tablet diluent. The Japanese use calcium sulfate in making *tofu*, a soybean curd.

Sulfur and H_2SO_4 can be derived from gypsum should adequate supplies not be available from normal sources. Plants are operating in Europe which make portland cement and sulfuric acid from gypsum or anhydrite. In the Muller-Kuhne process, gypsum is mixed with quantities of clay and silica necessary to make cement along with coke to reduce $CaSO_4$ to CaO. With equipment much the same as for portland-cement production, SO_2 is driven off and converted to sulfuric acid by the contact process. Other reducing agents and processing schemes may be employed to make S, H_2S, SO_2, H_2SO_4, CaS, and cement or lime from gypsum. $(NH_4)_2SO_4$ is made from gypsum by the Merseberg process $[CaSO_4 \cdot 2H_2O + 2NH_3 + CO_2 \rightarrow CaCO_3 + (NH_4)_2SO_4 + H_2O]$.

Dehydrated hemihydrate is a desiccant used in both the laboratory and industry. Glass-batch gypsum provides SO_3 for glassmaking. A major problem in desalinization of water and in opera-tion of heat exchangers and boilers is deposition of $CaSO_4$ scale on heat-transfer surfaces. Scien-tific investigation of this problem has contributed significantly to the literature on $CaSO_4$ chemistry.

REFERENCES

Edinger, S. E.: The Chemistry of Gypsum and Its Dehy-dration Products, U. S. Dept. Comm., *Bull. NTIS,* pp. 203–308, Sept. 1971.

Hansen, W. C., and J. S. Offutt: "Gypsum and An-hydrite in Portland Cement," 2d ed., U.S. Gypsum Company, Chicago, Ill., 1969.

Kelly, K. K., J. C. Southard, and C. T. Anderson: Thermodynamic Properties of Gypsum and Its Dehy-dration Products, *Bur. Mines Tech. Pap.* 625, 1941.

Lane, M. K.: Disintegration of Plaster Particles in Water, *Rock Prod.,* vol. 71, no. 3, p. 60 and no. 4, p. 73, 1968.

Schroeder, H. J.: Mineral Facts and Problems (Gypsum), *Bur. Mines Bull.* 650, 1970.

—John M. Summerfield, *United States Gypsum Company, Des Plaines, Ill.*

Gyratory Screens (See **Screening.**)

H

Haber-Bosch Process (See **Ammonia.**)

Hadfield Steel (See **Manganese.**)

Hafnium (L. *Hafnia*, Copenhagen.) **Hf** = 178.5
(at. wt.); 72 (at. no.). Although hafnium was not
discovered officially until 1923 by D. Coster and
G. C. de Hevesy, evidence of its existence was
reported by Urbain in 1911. Hafnia is the Latin
name for the city where it was officially discovered. Until about 1950, the element was referred
to as *celtium* in France.

Hafnium, a metal, occurs in group IVB of the
periodic table, along with Ti and Zr, and is in
period 6 of the table. There are eleven stable
isotopes (170–180) and one (181) radioactive isotope with a half-life of 46 days. See also **Chemical elements.**

Hf usually occurs with zirconium, and in fact
the chemical similarity between Hf and Zr is so
remarkable that the presence of Hf in chemicals
and minerals of Zr accounts for its rather late
isolation. Zircon, the principal commercial Hf-
bearing ore, contains 2–10% Hf. The ratio of Hf
to Zr in the earth's crust is estimated to be about
0.02, or about 4 ppm of Hf in the earth's crust.

Certain varieties of zircon, such as malacon, cyrto-
lite, and alvite, contain much higher ratios of Hf.
Only one mineral, the Sc-bearing thortveitite,
found in Scandinavia, is reported to contain more
Hf than Zr. Baddeleyite, pegmatite, monazite,
and zerkelite are other Hf-bearing minerals.

Many of the physical and mechanical properties
of Hf, e.g., heat capacity, thermodynamic functions, magnetic susceptibility, superconductive
transition temperatures, surface tension, modulus
of elasticity, and hardness, vary markedly with the
differences of Zr and interstitials present in Hf.
The form and method of production also vastly
affect Hf measurements, many of which have to
be extrapolated.

Depending upon Zr content, the density of
Hf = 13.09–13.36 g/cm^3; melting point = 1975–
2222°C; and temperature coefficient of electrical
resistivity at 20°C = 32.1–35.5 $\mu\Omega$-cm. The boiling point of Hf = 5400°C. The alpha form of Hf
is close-packed hexagonal; the beta form is body-
centered cubic. Linear chemical expansion coefficient (0–1000°C) = 5.9 × 10^{-6}/°C. Thermal
conductivity at 50°C = 0.0533 cal/(s)(cm)(°C).
Latent heat of vaporization = 72/kcal/mol.
Specific heat (25–100°C) = 6.25 cal/(mol)(°C).

Superconductivity transition temperature = 0.375 K.

Chemical Properties: These are almost analogous to Zr and also very similar to thorium, the similarities being a result of the size of the respective atoms and their electron distribution. Although the standard valence of Hf is 4, unstable di- and trihalides have been produced by the reduction of the tetrahalides. Hf is very resistant to attack from acids and bases. However, if NH_4F is added to any of the acids, a rapid disintegration occurs. Hf has somewhat better resistance to air oxidation at elevated temperatures than Zr and also good corrosion resistance to high-temperature high-pressure water. Of all the Hf compounds, perhaps the most interesting is hafnium carbide, which has the highest known melting point of any compound, 3890°C (7030°F). As finely divided metal powder ($<$325 mesh), Hf is pyrophoric and is usually wetted down. The standard potential for Hf ions in aqueous solution is 1.57 V.

Some of the common Hf chemicals produced are the acetate; boride, HfB_2; bromide, $HfBr_4$; carbide, HfC; chloride, $HfCl_4$; fluoride, HfF_4; hydride, HfH_2; iodide, HfI_4; nitride, HfN; oxide, HfO_2; oxychloride, $HfOCl_2 \cdot 8H_2O$; oxynitrate, $HfO(NO_3)_2 \cdot xH_2O$, silicide, $HfSi_2$; sulfate, $Hf(SO_4)_2 \cdot xH_2O$; and sulfide, HfS_2.

Extraction, Separation, and Production: Hf and Zr usually are extracted together, the zircon sand generally being broken down by carbiding or carbonitriding, followed by chlorination. The resultant mixture then is dissolved with a complexing agent and introduced into a liquid-liquid extraction process. In the system of countercurrent solvent-extraction columns perfected by the U.S. Bureau of Mines and Iowa State College, the Hf is stripped preferentially by the organic fraction and subsequently removed with a H_2SO_4 solution, precipitated as the hydroxide, calcined, and rechlorinated to form fairly pure $HfCl_4$. In addition to solvent extraction, fractional crystallization of the double fluorides is also used and generally results in a higher grade of Hf having a low Zr content. Commercial quality Hf produced by liquid-liquid extraction contains 1–4% Zr.

Most Hf metal is produced by the Kroll process, whereby $HfCl_4$ is reduced with Mg under an inert atmosphere. The resultant mixture of Hf sponge and $MgCl_2$ then is vacuum-distilled to separate the materials. A modified Kroll process, using Na, has the advantage of the lower cost of sodium. Sodium amalgam can also be used, since it seems to afford less stringent temperature and pressure requirements and fewer impurities are introduced to the metal during the reduction process. The iodide hot-wire or van Arkel–de Boer process then is used as a refining method for producing workable ductile Hf. Electrorefining, arc and induction melting, and zone refining may also be used.

Uses: World annual production of Hf metal is reported to be under 100 tons, with the United States, France, and the Soviet Union probably accounting for the majority. Potential commercial applications have been rather limited due primarily to a sparse supply and high costs. The chief use of Hf has been used as a control material in water-cooled nuclear reactors because of its resistance to radiation damage thanks to its ability to absorb neutrons above thermal energies. Hf actually is much more effective than may be indicated by its thermal-neutron cross section, which is 105 $\pm$ barns per atom for 2,200 m/s neutrons, and scattering cross section of 8 $\pm$ 2 barns per atom. Even under intense radiation, the isotopes formed usually have a rather high thermal-neutron absorption. It is for this reason that the control worth of Hf changes rather slowly.

Hf also is a useful flux-depressor material in water-cooled reactors. Flux depressors are placed strategically in a reactor to absorb neutrons, thereby decreasing the peaks in neutron flux. Hf is used successfully as a filament in gas-filled incandescent light bulbs, as a cathode in x-ray tubes, and as an alloying agent to strengthen tungsten and molybdenum filaments and electrodes used in high-pressure discharge tubes. Like Ti and Zr, Hf is an extremely effective getter material in vacuum tubes, particularly when used with Ba or Si. Hf also has been used in rectifying and high-pressure discharge tubes. Alloying small percentages of Hf with nichrome substantially increases the lifetime of these electrical-resistance heating elements. Hf has been successfully alloyed with many of the transition and refractory metals. Numerous patents have been issued, suggesting the possibility of Hf as an important alloying agent.

—Melvin Blum, *Atomergic Chemetals Co., Division of Gallard-Schlesinger Chemical Mfg. Corp., Carle Place, Long Island, N.Y.*

Half-Life   The average time required for one-half the atoms of a sample of a radioactive substance to lose their radioactivity by decaying into stable atoms is called the half-life. See also **Radioactive isotopes.**

Half-Life, Peroxide (See **Peroxides, organic.**)

Halogens (See **Bromine; Chemical elements; Chlorine; Fluorine;** and **Iodine.**)

Hammer Mill (See **Size reduction.**)

Hard Lead (See **Antimony;** and **Lead.**)

Hard Water (See **Water treatment.**)

Hardening (See **Iron and steel.**)

Hausmannite (See **Manganese.**)

Health Standards, Water (See **Water treatment.**)

Heat Balance (See **Distillation; Energy systems for processes; Ethylene;** and **Fluid catalytic cracking.**)

Heat Exchangers (See **Energy systems for processes.**)

Heat of Combustion (See **Fuels.**)

Heat Storage (See **Energy systems for processes.**)

Heat Transfer (See **Insulation, thermal.**)

Heat Treating (See **Iron and steel;** and **Pasteurization.**)

Heavy Hydrogen (See **Deuterium; Deuteron; Heavy water;** and **Tritium.**)

Heavy Water Also known as deuterium oxide, D_2O, heavy water is water in which the hydrogen of the water molecule consists entirely of the heavy-hydrogen isotope having a mass number of 2. Density of heavy water is 1.1076 at 20°C. Heavy water has been used as a moderator in nuclear reactors as well as a coolant. See also **Deuterium; Deuteron; Nuclear power plants;** and **Radioisotopes.**

Helium (Gk. *helios,* sun.) **He** = 4.0026 (at. wt.); 2 (at. no.). Helium was discovered by an Englishman, Sir Joseph Norman Lockyer, in 1868, while he was studying the sun. He noted a mysterious element in the ruddy gaseous layer surrounding the sun and between it and the corona, thus giving rise to the naming of this element. At the time of Lockyer's discovery, a French astronomer,

P. J. C. Janssen, reported on his investigations of the sun. The Lockyer and Janssen findings reached the French Academy of Science on the same day, and both men were credited with the discovery of He.

Later, work by W. F. Hillebrand, of the U.S. Department of the Interior, and Sir William Ramsey located the element on the earth. With the discovery of helium-bearing natural gas by H. P. Cady in Kansas in 1905 and subsequent investigations of physical properties by many scientists, most notably, Kamerlingh, Onnes, Keesom, McLennan, and Seibel, the aura of mystery around He began to disappear. The recovery of He from natural gas in 1918 made it a practical possibility to exploit the several unusual properties of the element.

Helium is a colorless, tasteless, and odorless gas with a normal boiling point of 4.2144 K and a specific gravity of 0.124 at 4.2144 K. It does not have a triple point and can be solidified only by applying high pressure to the liquid phase. Normal He can be transformed into He II by lowering its boiling temperature to 2.18 K (lambda point) under vacuum. At this temperature, a remarkable phenomenon occurs. The fluid stops boiling, and the physical properties are drastically altered. Under these conditions, He II is a better conductor of heat than all known metals, and it loses all its viscosity. He II can easily flow through the minutest porous restriction and can climb up and over the outside walls of a vertical, upright, open-mouthed container placed in the liquid-He bath. This extraordinary behavior continues until the He II liquid level in the container is the same as in the bath.

The lightest member of the noble gases, He has a molecular structure such that its two protons are balanced by two electrons and hence is chemically inert. No chemical reactions involving He are known. The isotopes of He are described under **Chemical elements.**

During the Federal Helium Conservation Program crude He production in the United States was approximately 3.5 billion scf/yr, but with disbanding of the program production is declining. Crude helium also has been produced in Canada, and He-bearing natural gases have been found in the North Sea, Algeria, and the Soviet Union. No production figures are available.

In *space exploration,* He is used in ground-support equipment, in the propellant tanks of liquid-fuel missiles as a compressed gas which expands and takes the place of the fuel as the latter is con-

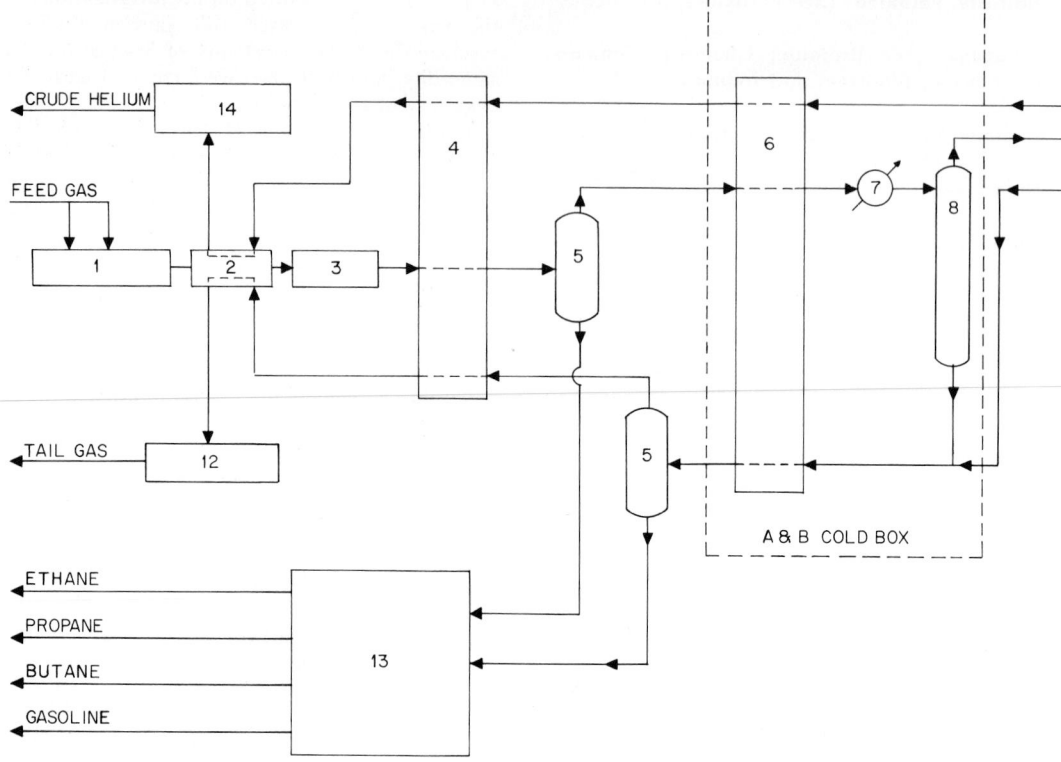

Fig. H-1. Helium recovery from natural gas. Propane, methane, and nitrogen refrigeration systems are used. (1) Feed-gas compression, (2) coolers, (3) drier, (4) heat exchanger, (5) flash drums, (6) heat exchanger, (7) methane cooler, (8) helium stripper, (9) helium cold-end heat exchanger, (10) helium fractionator, (11) nitrogen-methane tower, (12) tail-gas compression, (13) fractionation section, (14) helium compression. (*The M. W. Kellogg Company, a division of Pullman Incorporated.*)

sumed, to test complex rocket fuel and control systems, and in communication satellites to provide the low temperatures required by sensitive electronic systems. In *medicine,* He is mixed with O_2 to provide a breathing atmosphere for patients with respiratory ailments and mixed with explosive anesthetics for operating-room safety. In *industry,* He is utilized to provide an inert gaseous shield for arc welding, for leak detection in pressure and vacuum systems, to grow transistor crystals of Si and Ge, in the production and fabrication of Ti and Zr, in shock-tube tests, to fill the space between lenses in optical instruments, as the carrier gas in some chromatographic gas-analysis equipment, as a liquid bath for cryotrons and masers which permits them to reach productivity maximum, as a refrigeration medium for supplying the low temperature necessary for super-conducting electrical equipment, in lasers, and as a diluent gas in deep-sea diving applications. In *nuclear science,* helium is used as a heat-transfer medium in gas-cooled nuclear reactors. In *aeronautics,* He is used as a lifting gas for airships and for balloons used in meteorological programs.

Grade A helium can be compressed and shipped as a dense-phase gas by truck or rail. The liquefied grade A helium can be transported in specially designed containers by truck, rail, or air.

Production: Normal helium is recovered from natural gas* by a series of processing sequences shown in Fig. H-1. (1) Two natural-gas streams enter the plant at varying pressure levels of 470–550 psig, undergo compression to approximately

*National Helium Company, Liberal, Kans. (850 million ft³/day).

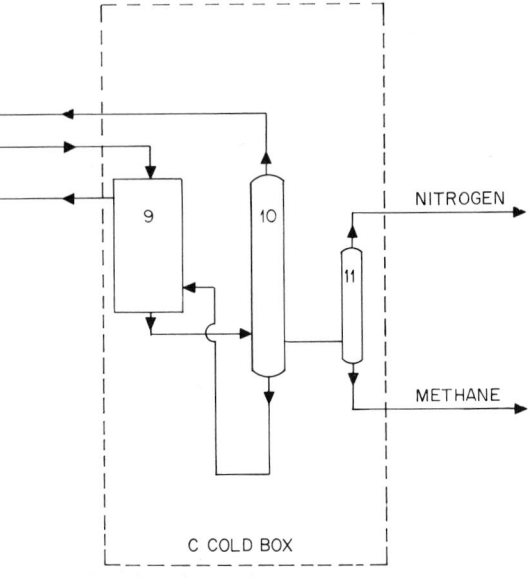

NITROGEN

METHANE

9

10

11

C COLD BOX

operating at temperatures far below ambient conditions. (7) At −101°C, the streams, approximately 95% liquefied, enter two helium strippers, where liquid and vapor fractions are separated initially by flashing. The liquid fractions from the flash are stripped within the towers by countercurrent flow with vapors generated in reboilers. (8) The −101°C vapors from the strippers, having about 3% He concentration, are combined and enter a third cold box *C*, where they are cooled to about −120°C (−185°F) through heat exchange with an effluent product stream, followed by methane refrigeration. (9) The −120°C stream then enters a fractionator. Dissolved He is removed in the stripping section and concentrated in the rectifying section. Top-tray vapors are cooled with liquid nitrogen refrigerant at −170°C (−275°F). The vapors at this temperature form the He product and are transferred to storage. The entire process sequence takes only 11 min and recovers approximately 99% of the He in the feed.

Crude He produced by the foregoing extraction facilities can be processed further to make grade A He gas (99.995% He). This product can then be liquefied and stored in specially designed cryogenic storage vessels. A typical process sequence for purification of grade A He includes several steps. (1) Crude He and air are compressed and directed into a deoxo unit. (2) The trace H_2 in the effluent from the compressor is catalytically reacted with the O_2 contained in the feed air to form water. (3) The water is removed in a fixed-bed drier, and the effluent gas leaves with a water dew point of less than −73.3°C (100°F). (4) The hydrogen-free gas is chilled and partially condensed by liquid nitrogen and cold returning He vapor. The gas then is sent to a separator. (5) Liquid-vapor separation is accomplished; the liquid effluent from the separator, containing almost all the nitrogen, is revaporized against the feed and vented to the atmosphere. The He-rich vapor effluent is directed into a nitrogen-absorber system. (6) In the nitrogen-absorber system, trace nitrogen is removed; the effluent grade A He from this system is heat-exchanged against the feed and directed into grade A He gas storage or to a liquefaction unit.

650 psig, and then are combined into a single stream. (2) The feed gas is cooled from 43 to 10°C (110–50°F) through heat exchange with effluent plant products. (3) Water (8 lb/million scf) is removed in fixed-bed driers, where the feed is dried to a dew point approaching −73°C (−100°F) to prevent hydrate formation and plugging in subsequent processing. (4) The gas is chilled to −45.6°C (−50°F) at 615 psig through further heat exchange with product streams supplemented by propane refrigeration, after which the gas is sent to a feed separator. (5) In the separator, the chilled liquid, representing 7% (weight) of the feed stream, is removed to prevent heavy-hydrocarbon plugging at colder levels. The feed stream undergoes heat exchange with the feed gas and enters a liquid-propane-gas (LPG) recovery unit, where propane, butane, and natural gasoline are recovered. (6) The chilled vapors from the separator are split into two equal streams and enter parallel cold boxes *A* and *B*, where they are cooled to −101°C (−150°F) through heat exchange with effluent cold-box streams assisted by methane. A cold box is a structure which houses the equipment necessary for a particular cryogenic sequence. The box serves as an outer barrier for the bulk insulation used to minimize the heat influx to the equipment

REFERENCES
Deaton, W. M.: *Cryogenic Engineering News,* May 1968.
M. W. Kellogg Company, *Kelloggram,* No. 3, 1963 series, Houston, Tex.
Kropschot, R. H., B. W. Birmingham, and D. B. Mann:

Technology of Liquid Helium, U.S. Bureau of Standards, Monograph 111, Oct. 1968.

Scott, R. B.: "Cryogenic Engineering," Van Nostrand Reinhold, Princeton, N.J., 1959.

Seibel, C. W.: "Helium, Child of the Sun," Univ. Press of Kansas, 1968.

—Henry E. Duckham, Jr., *The M. W. Kellogg Company, A Division of Pullman Incorporated, Houston, Tex.*

Hematite   (See **Iron ores; Iron oxides, synthetic;** and **Iron- and steelmaking.**)

Hemicellulose   [See **Pulp (wood) production and processing.**]

Hemoglobin   (See **Iron;** and **Peptides and proteins.**)

Herbicides   (See **Arsenic;** and **Pyridine and derivatives.**)

Hessite   (See **Tellurium.**)

Heterocumulenes   (See **Urethanes.**)

Heterocyclic Compounds   A ring-type compound (one or more rings) in which the atoms in the ring framework or nucleus are *not* all the same is termed a heterocyclic compound. Whereas all the atoms in the benzene ring are carbon (hence it is a *homocyclic compound*), pyridine and furan are examples of heterocyclic compounds, pyridine containing five carbon atoms and one nitrogen atom in its nucleus and furan four carbon atoms and one oxygen atom in its nucleus. Ring compounds that contain only carbon atoms in the nucleus are also called *carbocyclic compounds*.

Heterogeneous Process   (See **Catalysts.**)

Hexachloroplatinic Acid   (See **Platinum metals and platinum.**)

Hexamethylenediamine   [See **Cyclohexanol/ Cyclohexanone (KA oil);** and **Nylons.**]

High-Bulk Fibers   (See **Acrylic fibers.**)

High-Temperature Chemistry   (Consult the Subject Index.)

Histidine   (See **Amino acids;** and **Peptides and proteins.**)

Histone   (See **Peptides and proteins.**)

HMX (Cyclotetramethylenetetranitramine)   (See **Explosives.**)

Holmium   (L. *Holmia,* Stockholm.) **Ho** = 164.93 (at. wt.); 67 (at. no.). Holmium is the eleventh of 15 elements in group III, period 6, generally shown as the rare-earth elements in a separate line below the main body of the periodic table and sometimes referred to as the *lanthanide series*. Holmium was discovered in 1879 by Per T. Cleve and named for his native city.

Pure Ho is a silvery-gray metal which is very slow to tarnish or oxidize at room temperature in normal atmosphere. Even at elevated temperature, bulk Ho oxidizes slowly. The hot metal gradually deteriorates in 10-torr vacuum by reacting with water vapor, CO_2, NH_3, and hydrocarbons. The metal is soft and can be formed with conventional tools and equipment.

The melting point of Ho is 1472°C; boiling point 2695°C; density 8.780 g/cm³. Estimated at 1.2 ppm in the earth's crust, it is one of the least abundant elements although it ranks ahead of Sb, Cd, Bi, and Hg. Little is known about its alloys or intermetallic compounds.

Holmium has 1 natural isotope, ^{165}Ho, and 18, artificial isotopes. The metal is not radioactive and has a low acute-toxicity rating.

The only valence known for Ho is 3^+. Since Ho is strongly cationic (electropositive), it will react and form compounds with acids, halogens, and most organic anions. Its pure salts are characterized by pink-orange colors.

Holmium is extracted from apatite, xenotime, and yttrium-heavy rare-earth minerals, by organic ion-exchange techniques. The ion is precipitated from an acid solution by oxalic acid and calcined to an oxide. Purity of 99.9% or better is readily attained. The anhydrous fluoride or chloride compounds, HoF_3 and $HoCl_3$, are reduced by calcium or lithium in sealed-bomb reactions. Subsequent vacuum distillation to purify the metal is generally required.

See **Rare-earth elements and metals** for processing details and a table of properties.

Uses: Applications of Ho are just beginning to reach the commercial stage. Many pounds of pure compounds are available as a result of processing minerals to obtain yttrium.

Bell Telephone Laboratories has found that Ho added to orthoferrites being developed for computer memory and logic circuits improves the mobility and control of isolated magnetic bubble domains.

Bell Labs has made Ho-containing phosphors, which are coated on gallium arsenide diodes. These solid-state devices are able to shift invisible wavelengths of light into visible light by varying the energizing power.

Holmium is one of the ions that show the highest pulsed-solid-laser efficiencies in glass at room temperature. Applications for the coherent beam of light from lasers are in their infancy.

The photoelectric, semiconductor, and thermoelectric properties of Ho and its many compounds are being evaluated in research laboratories that study the fundamental properties of new materials.

REFERENCES

Molybdenum Corporation of America: *Overview* 17 (application information including possible uses), New York, 1970.

See also references under **Rare-earth elements and metals.**

—Joseph G. Cannon, *Molybdenum Corporation of America, White Plains, N.Y.*

Holocellulose [See **Pulp (wood) production and processing.**]

Homocyclic Compounds (See **Heterocyclic compounds.**)

Homogeneous Process (See **Catalysts.**)

Homogenization Homogenization is a process for reducing the size of particles in a liquid. Such a reduction of particle or globule size in a mixture of two immiscible liquids makes an emulsion possible. If an emulsifying agent is present, a more stable emulsion can be produced and coalescence of the dispersed phase prevented. The homogenizer is also used to produce dispersions by reducing the particle size with solid-in-liquid mixtures. As in the preparation of an emulsion, a dispersing agent is needed to maintain a homogeneous mixture.

The homogenizer consists of a high-pressure positive-displacement pump and an adjustable orifice. The pump is a piston or plunger type, usually consisting of three plungers, although some homogenizers are made with five or even seven plungers. The cylinder for each plunger has an inlet and discharge valve. The plunger pump must push the product through the homogenizing valve (adjustable orifice). For two-stage homo-genization, two valves arranged in series are used.

When homogenizing milk, the size of the fat globules is reduced and more surface of the fat is exposed, resulting in a stable emulsion which does not allow separation, or a cream layer. Dispersion of the small fat globules in the protein results in a soft curd, which is easily digested. When homogenizing ice cream mix, viscosity can be controlled, and because the fat globule size is reduced, overrun is easier to control in the final product without the risk of churning the fat. Homogenization of tomato juice and tomato catsup makes a uniform particle size and desired viscosity possible and prevents separation or reduces its rate. The same principles apply to baby foods. Salad dressings (not mayonnaise) are homogenized to prepare a stable emulsion of oil, starch, egg, vinegar, and flavoring with a viscosity similar to that of mayonnaise. Flavor concentrates containing oils are homogenized to produce a stable emulsion. Cosmetics are homogenized to produce a smooth body cream while controlling viscosity.

A typical homogenizing valve consists of a seat and plug of very hard abrasion-resistant materials, such as Stellite. The seating surfaces must be lapped smooth and be parallel. In operation, the plug is spring-loaded against the seat. Spring compression is adjusted so that when the product flows, energy in the form of pressure is required to lift the plug. While many products can be homogenized at pressures below 3,000 psig, machines are made to develop pressures in excess of 8,000 psig. Another type of valve uses a compressed cone of stainless-steel wire inserted into a socket, the product being homogenized by flowing between the wires.

Theories: There are several theories of what actually breaks up the particles in the homogenizer: (1) as the product enters the area between the lapped surfaces, it is suddenly accelerated to velocities as high as 30,000 ft/min (at 5,000 psig). When acceleration is this sudden, the particle (especially the liquid particle) is stretched or elongated to the point of breaking; (2) at this high velocity, there are shear forces between layers of liquids under flow that break up particles; and (3) Loo et al.[3] present evidence supporting cavitation as the primary cause of homogenization. When the pressure energy is converted into velocity energy, the vapor pressure of the product exceeds product pressure, resulting in the formation of vapor cavities which collapse upon leaving the valve at higher pressures. This collapsing, or implosion, of cavitation exerts tremendous force, breaking up the particles. Most homogenizers are designed to incorporate one or more of the foregoing principles.

Milk Products: The homogenization of milk is generally accomplished with one valve at pressures between 1,500 and 2,000 psig and at temperatures between 130 and 180°F (54–82°C). At these temperatures, the fats are fluid, permitting the pressure to reduce all globules to sizes less than 2 μm, with the majority at 1 μm. Because of the low fat content (3–5%) in whole milk and the quantity of natural emulsifiers, the globules adsorb sufficient emulsifiers to cover the increased surface and prevent agglomeration.

When homogenizing milk products, such as ice cream, having a higher percentage of fats (12–16%) to form a stable emulsion, the natural stabilizer is often insufficient to cover the increased surface thoroughly. Consequently, the globules are attracted to each other and form clumps. A second-stage valve operated in series with the first-stage valve will disperse the clumps, allowing greater opportunity for more emulsifier adsorption on the surface. With the clumps present, there is an increase in viscosity. The second stage therefore provides control of viscosity, which is important in ice-cream freezing.

The homogenizer is controlled by adjusting and maintaining a desired pressure across the valve (s). The pressure sensor must be heavily damped because of pulsations of the plunger-type pump. The sensor may be a dial pressure gage or equipped with recording and automatic controlling features. Force on the valve can be attained by compression of a spring by a threaded handle or direct force of compressed-air or hydraulic-fluid pressure.

REFERENCES

1. Becher, Paul: "Emulsions: Theory and Practice," Van Nostrand Reinhold, New York, 1965.
2. Jennes, R., and S. Patton: "Principles of Dairy Chemistry," Wiley, New York, 1959.
3. Loo, C. C., W. L. Slatter, and R. W. Powell: A Study of the Cavitation Effect in the Homogenization of Dairy Products, *J. Dairy Sci.*, vol. 33, p. 672, 1950.
4. Trout, G. M.: "Homogenized Milk," Michigan State University Press, East Lansing, 1950.
5. Grade "A" Pasteurized Milk Ordinance, U.S. Dept. of Health, Education and Welfare, Washington, 1965.

—Norbert L. Miller, *Cherry-Burrell Corporation, Cedar Rapids, Iowa*

Homologous Series   The general formula for the family of fatty acids is $C_nH_{2n}O_2$. The simplest compound in the series contains one carbon atom ($n = 1$) and is CH_2O_2 (formic acid). The next highest member of the series is $C_2H_4O_2$ (acetic acid). These two compounds differ by the structural increment of CH_2. A series like this is termed a *homologous series,* and any compound in the series is said to be a *homologue* of any other compound in the series.

Homologous series are particularly important in organic synthesis because generally a compound can be converted to the next higher or lower member of the series by appropriate reactions that will add or take away the CH_2 increment respectively. For example, a higher homologue in the fatty acid series can be created:

$$CH_3COOH \rightarrow CH_3CHO \rightarrow CH_3CH_2OH \rightarrow$$
Acetic acid   Acetaldehyde   Ethyl alcohol

$$CH_3CH_2I \rightarrow CH_3CH_2CN \rightarrow CH_3CH_2COOH$$
Ethyl iodide   Ethyl cyanide   Propionic acid

Or a next lower homologue can be created:

$$CH_3CH_2COOH \rightarrow CH_3CH_2CONH_2 \rightarrow$$
Propionic acid   Propionamide

$$CH_3CH_2NH_2 \rightarrow \quad CH_3CH_2OH \rightarrow CH_3COOH$$
Ethylamine   Ethyl alcohol   Acetic acid

For examples of other homologous series, see **Alcohols; Carbon compounds;** and **Hydrocarbons.**

Homopolar Bond   (See **Molecule.**)

Homopolymers   (See **Polymerization.**)

Hoopes Cell   (See **Aluminum.**)

Hormones   (See **Peptides and proteins;** and **Rubber, natural.**)

Huebnerite   (See **Tungsten.**)

Hussong Machine   [See **Dyestuffs and dyeing (textile).**]

Hutchinsonite   (See **Thallium.**)

Hydantoin Process   (See **Amino acids.**)

Hydrate Process, Seawater   (See **Desalination.**)

Hydration Processes   [See **Acrylamide polymers; Catalysts; Ethyl alcohol; Lime; Molecule;** and **Silicates (soluble).**]

Hydrazine Hydrazine, N_2H_4, is a colorless liquid with a freezing point of 1.4°C and a boiling point of 113.5°C, a specific gravity of 1.011 at 15°C referred to water at 4°C, and a formula weight of 32.05. The compound is soluble in all proportions with water and soluble in alcohol. Important derivatives include dihydrazine sulfate, $(N_2H_4)_2 \cdot H_2SO_4$, a white crystalline compound soluble in water and containing 37.5% available hydrazine, and hydrazine hydrate, $N_2H_4 \cdot H_2O$, a colorless liquid miscible with water.

Hydrazine is a tonnage chemical with several uses, e.g., a propellant for rockets (yields exhaust products at high temperature and of low molecular weight), a strong reducing agent, and a blowing agent for foamed rubber. Other uses are found in the manufacture of chemicals and plastics.

Modern processes for making hydrazine essentially are of a proprietary nature. Earlier processes involved urea as a raw material. Ammonia oxidation is now widely used, one two-step process, consisting of the following reactions:

Step 1:

$$NH_3 + NaOCl \rightarrow NH_2Cl + NaOH$$

Step 2:

$$NH_3 + NH_2Cl + NaOH \rightarrow$$
$$N_2H_4 + NaCl + H_2O$$

A mercury cell or caustic soda of equivalent purity is used to prepare the high-grade hypochlorite required for step 1. The sodium hypochlorite solution is mixed with an ammonia solution (20–28%) to accomplish step 1. A high mole ratio of ammonia to hypochlorite and low reaction temperatures are maintained. High yields of hydrazine from step 2 are favored by rapidly heating the reactants to elevated temperatures while maintaining a high mole ratio of ammonia to hypochlorite in liquid contact. Various agents, such as glue, gelatin, ethylenediamine tetraacetic acid, high alcohols, or formaldehyde, are used as inhibitors to minimize undesirable side reactions that would reduce the hydrazine yield through the formation of ammonium chloride and nitrogen: $N_2H_4 + 2NH_2Cl \rightarrow 2NH_4Cl + N_2$. Most manufacturers use the purest ammonia and hypochlorite obtainable in at least a 20:1 mole ratio, at temperatures between 120 and 180°C and pressures of 200–450 psi. The dilution rate is optimized between the cost of evaporation and equipment investment at a high dilution versus the loss of raw materials at higher concentrations.

The pressurized product stream from the hydrazine reactor passes to a low-pressure column, where the NH_3 is stripped from solution and recovered in an absorber and then recycled with fresh makeup to the chlorination reactor. A fractionator receives the stripped hydrazine solution and concentrates it to a nearly saturated salt solution with a N_2H_2 content 5 to 6 times greater than the original reaction liquor. Bottoms from the fractionator are fed to an evaporator that removes the by-product, NaCl, and boils off the hydrazine-water phase, which is sent to the fractionators.

In another process, chlorine, ammonia, and sulfuric acid, along with methyl ethyl ketone, are used as the charge. Products include hydrazine hydrate, hydrazine sulfate, ketazine, and dialkyldiazacyclopropane.

See also **Ketones;** and **Nuclear power plants.**

Hydrazones (See **Aldehydes;** and **Ketones.**)

Hydrocarbons A hydrocarbon is a compound comprising solely carbon and hydrogen. Because of the ability of a carbon atom to share one, two, or three valences with another carbon atom, and because of the great affinity between hydrogen atoms and available carbon bonds, the number of hydrocarbon compounds is very large.

Natural gas and petroleum are major natural sources of hydrocarbons. See also **Natural gas; Petrochemical complex;** and **Petroleum.** Hydrocarbons from these sources serve as major starting ingredients for thousands of organic chemicals, including those for manufacturing most of the synthetic adhesives, fibers, films, finishes, plastics, and resins. Coal also is a major source of hydrocarbon materials, particularly of the aromatic hydrocarbons, such as benzene, naphthalene, and anthracene. See also **Coal tar chemicals.**

Classification. There are nine major classes of hydrocarbon compounds from the standpoint of structure, which are summarized in Table H-1. The properties of representative hydrocarbons in each class are summarized in Table H-2. See footnote to Table H-1 regarding nomenclature.

Paraffins: Usually regarded as the simplest of all hydrocarbons is methane, CH_4 (see also **Methane**), which is a saturated compound in that all available carbon bonds are taken by hydrogen atoms. Methane is the building block of a long series of similarly structured compounds known as the *paraffins*. The term paraffin derives from the Latin *parum affinis*, meaning "small or slight affinity" and pertained originally to one of the earlier known hydrocarbons, namely paraffin wax. In the series of *normal* (straight-chain) paraffins, the compounds may be regarded as two

TABLE H-1. *Major Classes of Hydrocarbons**

Class	Formula of Homologous Series	Examples
Chain Structures		
Saturated, all single carbon bonds: Paraffins	C_nH_{2n+2}	Methane, CH_4; isobutane, $(CH_3)_3CH$
Unsaturated: One double carbon bond: Olefins	C_nH_{2n}	Ethylene, $H_2C\!:\!CH_2$; isobutylene, $(CH_3)_2C\!:\!CH_2$
One triple carbon bond: Acetylenes	C_nH_{2n-2}	Acetylene, $HC\!:\!CH$; ethylacetylene, $CH_3CH_2C\!:\!CH$
Two double-carbon bonds: Diolefins	C_nH_{2n-2}	Propadiene, $H_2C\!:\!C\!:\!CH_2$; bivinyl, $H_2C\!:\!CH\!\cdot\!CH\!:\!CH_2$
Two triple carbon bonds: Diacetylenes	C_nH_{2n-6}	Butadiyne, $HC\!:\!CC\!:\!CH$; bipropargyl, $HC\!:\!CCH_2CH_2C\!:\!CH$
Single-Ring Structures, Saturated, All Single Carbon Bonds		
Cycloparaffins	C_nH_{2n}	Cyclopropane, $(CH_2)_3$; cycloheptane, $(CH_2)_7$
Single- or Multiple-Ring Structures, Unsaturated, with One or More Double Carbon Bonds		
Cycloolefins	C_nH_{2n-2}	Cyclobutene, $CH_2\!\cdot\!CH_2CH\!:\!CH$
Cyclodiolefins	C_nH_{2n-4}	Cyclohexadiene, C_6H_8
Benzenoids		
Six-carbon ring structure, unsaturated, three single carbon bonds and three double carbon bonds per ring: Single rings	C_nH_{2n-6}	Benzene, C_6H_6; toluene, C_7H_8; cymene, $C_{10}H_{14}$
Chain-bonded double rings	C_nH_{2n-14}	Biphenyl, $C_{12}H_{10}$; diphenylmethane, $C_{13}H_{12}$
Interlocked double rings	C_nH_{2n-12}	Naphthalene, $C_{10}H_8$
Interlocked triple rings	C_nH_{2n-18}	Anthracene, $C_{14}H_{10}$
Mixed chain-ring structures: Paraffin-benzenoid . . .	. . .	Diphenylmethane, $(C_6H_5)_2CH_2$
Olefin-benzenoid	. . .	Phenylethylene (styrene), $C_6H_5CH\!:\!CH_2$
Acetylene-benzenoid . .	. . .	Phenylacetylene, $C_6H_5C\!:\!CH_2$

*Two systems for naming hydrocarbons are used widely in the literature: (1) the older *paraffin-olefin-acetylene* et al. names and (2) the newer *alkane, alkene,* and *alkyne* designations. The older names persist in engineering and commercial literature, while the newer designations are gaining acceptance in scientific treatises. Equivalents are

Paraffins	Alkanes	Acetylenes	Alkynes
Cycloparaffins	Cycloalkanes	Diacetylenes	Alkadiynes
Olefins	Alkenes	Benzenoids	Arenes
Diolefins	Alkadienes	Paraffin-benzenoids	Mixed alkane-arenes
Cycloolefins	Cycloalkenes	Olefin-benzenoids	Mixed alkene-arenes
Cyclodiolefins	Cycloalkadienes	Acetylene-benzenoids	Mixed alkyne-arenes

TABLE H-2. *Properties of Representative Hydrocarbons**

Hydrocarbon	Formula	Formula Weight	Sp gr	Mp, °C	Bp, °C
Paraffins					
Methane	CH_4	16.03	0.558†	−184	−161.4
Ethane	CH_3CH_3	30.05	1.049†	−172	−88.3
Propane	$CH_3CH_2CH_3$	44.06	1.558†	−189.9	−44.5
n-Butane	$CH_3(CH_2)_2CH_3$	58.08	0.601	−135	0.6
Isobutane	$(CH_3)_3CH$	58.08	0.603	−145	−10.2
n-Pentane	$CH_3(CH_2)_3CH_3$	72.09	0.631	−131.5	36.2
t-Pentane	$(CH_3)_4C$	72.09	. . .	−20	9.5
Isopentane	$(CH_3)_2CHCH_2CH_3$	72.09	0.621	−159.7	28
Hexanes:					
β-Methylpentane	$(CH_3)_2CH(CH_2)_2CH_3$	86.11	0.654	. . .	60
β,γ-Dimethylbutane	$(CH_3)_2CHCH(CH_3)_2$	86.11	0.668	−135.1	58.1
β-Ethylbutane	$CH_3CH(C_2H_5)_2$	86.11	0.676	. . .	64
n-Methylpentane	$CH_3(CH_2)_4CH_3$	86.11	0.660	−94.3	69
β,β-Dimethylbutane	$(CH_3)_3C(C_2H_5)$	86.11	0.649	−98.2	49.7
n-Heptane	$CH_3(CH_2)_5CH_3$	100.12	0.684	−90.5	98.4
n-Octane	$CH_3(CH_2)_6CH_3$	114.14	0.704	−56.5	125.8
n-Nonane	$CH_3(CH_2)_7CH_3$	128.16	0.718	−51	150.6
n-Decane	$CH_3(CH_2)_8CH_3$	142.17	0.730	−32	174
n-Undecane	$CH_3(CH_2)_9CH_3$	156.19	0.741	−26.5	197
n-Dodecane	$CH_3(CH_2)_{10}CH_3$	170.20	0.785	−31.5	214
n-Tridecane	$CH_3(CH_2)_{11}CH_3$	184.22	0.757	−6.2	234
n-Tetradecane	$CH_3(CH_2)_{12}CH_3$	198.23	0.765	5.5	252.5
n-Pentadecane	$CH_3(CH_2)_{13}CH_3$	212.25	0.772	10	270.5
n-Hexadecane	$CH_3(CH_2)_{14}CH_3$	226.27	0.775	20	287.5
n-Heptadecane	$CH_3(CH_2)_{15}CH_3$	240.28	0.778	22.5	303
n-Octadecane	$CH_3(CH_2)_{16}CH_3$	254.30	0.777	28	317
n-Nondecane	$CH_3(CH_2)_{17}CH_3$	268.31	0.777	32	330
n-Eicosane	$CH_3(CH_2)_{18}CH_3$	282.33	0.778	38	205
n-Heneicosane	$CH_3(CH_2)_{19}CH_3$	296.34	0.778	40	215
n-Docosane	$CH_3(CH_2)_{20}CH_3$	310.36	0.778	44.4	317.4
Hexacontane	$CH_3(CH_2)_{58}CH_3$	843.58	. . .	101	
Olefins					
Ethylene	$CH_2:CH_2$	28.03	0.978†	−169.4	−103.8
Propylene	$CH_3CH:CH_2$	42.05	1.498†	−185.2	−47
α-Butylene	$C_2H_5CH:CH_2$	56.10	0.6	−130	−5
β-Butylene	$CH_3CH:CHCH_3$	56.10	. . .	−127	3
Isobutylene	$(CH_3)_2C:CH_2$	56.10	. . .	. . .	−6
n-Amylene (α)	$CH_3CH_2CH_2CH:CH_2$	70.08	. . .	. . .	40
Isoamylene (α)	$(CH_3)_2CHCH:CH_2$	70.08	. . .	. . .	25
n-Amylene (β)	$CH_3CH_2CH:CHCH_3$	70.08	0.651	−139	36.4
Isoamylene (β)	$(CH_3)_2C:CHCH$	70.08	0.668	−124	38.4
Hexene	$C_4H_9CH:CH_2$	84.09	0.683	−98.5	64.1
Tetramethylethylene	$(CH_3)_2C:C(CH_3)_2$	84.09	0.712	. . .	73
Heptene	$CH_3(CH_2)_4CH:CH_2$	98.18	. . .	. . .	99
Decylene	$CH_3(CH_2)_7:CH_2$	140.16	0.763	. . .	172
Acetylenes					
Acetylene	$CH:CH$	26.02	0.906†	−81.8	−83.6
Methylacetylene	$CH_3C:CH$	40.03	0.660	−104.7	−27.5
Ethylacetylene	$C_2H_5C:CH$	54.05	0.668	−130	18.5
n-Propylacetylene	$C_3H_7C:CH$	68.06	0.722	−95	40
Isopropylactylene	$(CH_3)_2CHC:CH$	68.06	0.685	. . .	29.3

Hydrocarbon	Formula	Formula Weight	Sp gr	Mp, °C	Bp, °C
Diolefins					
Allene	$CH_2:C:CH_2$	40.03	. . .	−146	−32
1,2-Butadiene	$CH_3CH:C:CH_2$	54.09	. . .	. . .	18.5
1,3-Butadiene	$CH_2:CHCH:CH_2$	54.09	0.621	−108.9	−4.4
Isoprene	$CH_2:CHC(CH_3):CH_2$	68.06	0.679	−120	34
Biallyl	$CH_2:CHCH_2CH_2CH:CH_2$	82.14	. . .	. . .	60
Diacetylenes					
Butadiyne	$CH:CC:CH$	50.02			
Bipropargyl	$CH:CCH_2CH_2C:CH$	78.05	0.805	−6	85
Cycloparaffins					
Cyclopropane	C_3H_6	42.08	0.720	−126.6	−34
Cyclobutane	C_4H_8	56.10	0.703	−50	11.5
Cyclopentane	C_5H_{10}	70.13	0.745	−93.3	49.5
Cyclohexane	C_6H_{12}	84.09	0.779	6.5	81.4
Cycloheptane	C_7H_{14}	98.18	. . .	. . .	118
Cycloolefins					
Cyclobutene	$CH_2CH_2CH:CH$	54.05	0.693	. . .	3
Cyclopentene	$CH:CHCH_2CH_2CH_2$	68.06	0.775	. . .	−45
Cyclohexene	$CH_2CH_2CH_2CH_2CH:CH$	82.14	0.810	−103.7	83.3
Cycloheptene (suberene)	$CH:CHCH_2CH_2CH_2CH_2CH_2$	96.09	0.811	−12	118.1
Cyclodiolefins					
Cyclopentadiene	$CH_2CH:CH:CH$	66.05	0.805	. . .	42.5
1,2-Cyclohexadiene	C_6H_8	80.12	. . .	. . .	78.5
1,3-Cyclohexadiene	C_6H_8	80.12	. . .	. . .	80.5
1,4-Cyclohexadiene	C_6H_8	80.12	. . .	. . .	85.5
Limonene	$C_{10}H_{16}$	136.23	0.842	−96.9	177
Benzenoids					
Benzene	C_6H_6	78.11	0.879	5.5	80.1
Toluene	C_7H_8	92.13	0.866	−95	110.8
o-Xylene	C_8H_{10}	106.16	0.881	−25	144
m-Xylene	C_8H_{10}	106.16	0.867	−47.4	139.3
p-Xylene	C_8H_{10}	106.16	0.861	13.2	138.5
Ethylbenzene	C_8H_{10}	106.16	0.867	−94.4	136.2
Cumene (isopropylbenzene) .	C_9H_{12}	120.19	0.862	−96.9	152.5
p-Cymene (*p*-methyliso-propylbenzene)	$C_{10}H_{14}$	134.21	0.857	−73.5	176.5
Naphthalene	$C_{10}H_8$	128.16	1.145	80.2	217.9
Biphenyl	$C_{12}H_{10}$	154.20	0.992	69.5	254.9
Diphenylmethane	$C_{13}H_{12}$	168.23	1.001	26.5	265
Diphenylenemethane (fluorene)	$C_{13}H_{10}$	166.08	. . .	116	295
Bibenzyl	$C_{14}H_{14}$	182.25	0.978	52.5	284
Anthracene	$C_{14}H_{10}$	178.22	1.25	217.5	341
Phenanthrene	$C_{14}H_{10}$	178.22	1.179	99.5	340
Triphenylmethane	$C_{19}H_{16}$	244.32	1.014	93.4	359
Tetraphenylmethane	$C_{25}H_{20}$	320.16	. . .	282	431
Tetraphenylethane	$C_{26}H_{22}$	334.17	1.182	209	383

*See footnote to Table H-1 regarding nomenclature.
†Air.

methyl groups, one at each end of the chain, with a specific number of CH_2 (methylene groups) in between to form a long continuous chain:

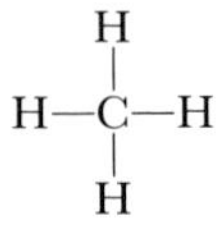

Methane 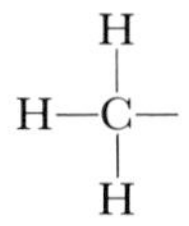   Methyl group

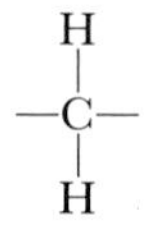

Methylene group 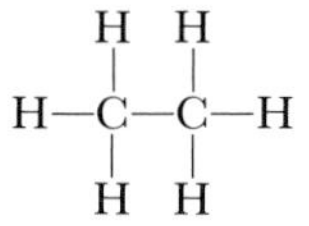   Ethane

In ethane, of course, this compound simply is two methyl groups linked together. The chains build up as follows:

$$CH_3-CH_2-CH_3 \qquad CH_3-CH_2-CH_2-CH_3$$
Propane   Butane

$$CH_3-(CH_2)_x-CH_3$$

Very long chains are known, including hexacontane, $CH_3(CH_2)_{58}CH_3$. The general formula for the homologous series of paraffins, C_nH_{2n+2}, can be compared with the formulas for other hydrocarbon series in Table H-1.

Starting with the butanes, branching or forking of chains occurs, and hence as the number of carbons increases, the number of isomers increases. See also **Carbon compounds.**

The paraffins with 4 or fewer carbons all are

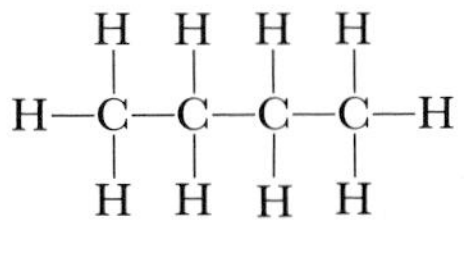

n-Butane 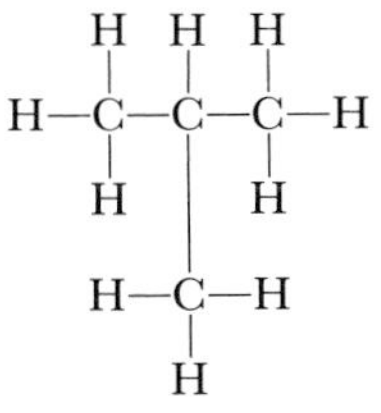   Isobutane

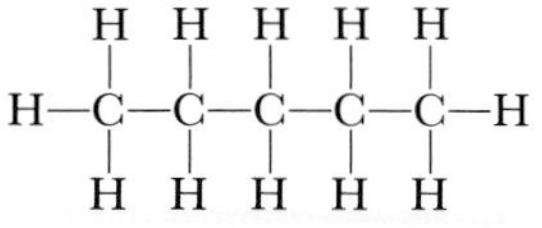

Pentane 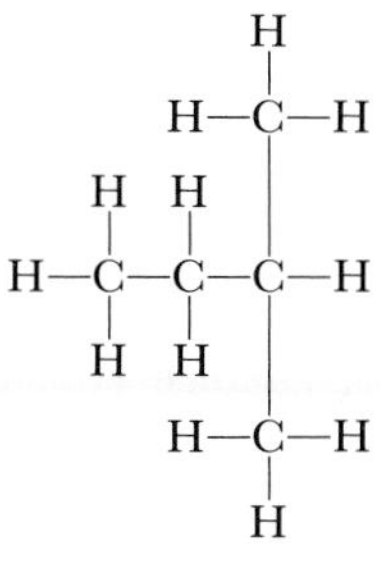   Isopentane

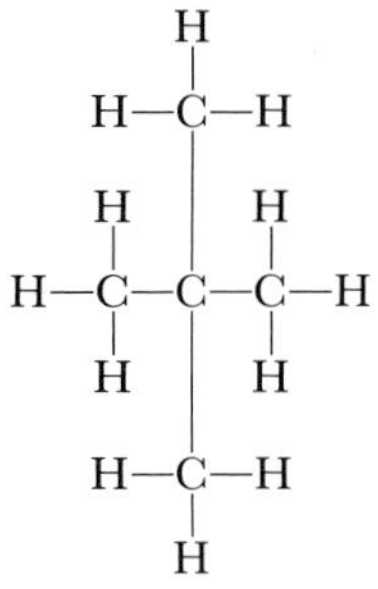

Tetramethylmethane

gases. Paraffins with 5–6 carbons are liquids at standard conditions, and all compounds with 17 or more carbons are solids. The change in boiling point and melting point with increasing carbon content is evident from Table H-2. The paraffins are quite stable and at normal temperatures do not react with powerful oxidizing agents, such as potassium permanganate or chromic acid, or with alkalis, sodium, fuming sulfuric acid, or nitric acid. The paraffins do react readily with chlorine to form many chlorine derivatives, several of which, e.g., methyl chloride, ethyl chloride, chloroform, and carbon tetrachloride, are important high-tonnage commercial chemicals. The paraffins are miscible with many organic liquids but not with water.

Olefins: The simplest of the olefins is ethylene, C_2H_4, containing two carbon atoms joined by a double bond and hence an *unsaturated* compound. Ethylene is one of a series of structurally similar hydrocarbons. However, unlike the paraffin chains, where the bonding of carbon atoms is consistent throughout the chain, the olefins have *only one* double bond regardless of how many carbons there may be in the total chain. As the number of carbons increases in the olefin chain, numerous isomers appear, just as with the paraffins.

The olefins form polymers under proper conditions of temperature and pressure and account for

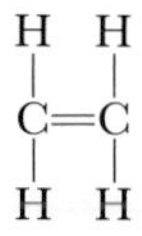

Ethylene,   β-Isoamylene
C_2H_4 (trimethylethylene), C_5H_{10}

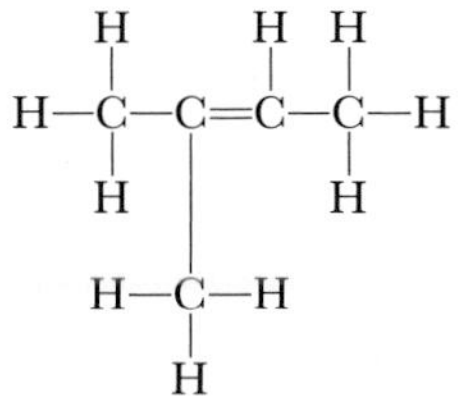

many high-tonnage synthetic products. See also **Polyethylenes;** and **Polymerization.**

With exception of ethylene, propylene, and the butylenes, which are gases, the higher olefins are liquids, with the boiling point rising with an increasing carbon count. Because of their double carbon bond, olefins are considerably more reactive than the paraffins, combining readily with chlorine, bromine, hydrogen bromide, sulfuric acid, and hypobromous and hypochlorous acids to form saturated additive products. Olefins can be converted into paraffins by hydrogenation in the presence of a catalyst. Hypobromous and hypochlorous acids convert olefins into bromo-hydrins and chlorohydrins, respectively:

$$CH_2:CH_2 + HOBr \rightarrow CH_2Br \cdot CH_2OH$$

Oxidation of olefins yields glycols:

$$CH_2:CH_2 + O + H_2O \rightarrow CH_2(OH) \cdot CH_2OH$$

The very heavy demands for ethylene are met in large part by cracking refinery gases and other refinery products—ethane, propane, butanes, pentanes, natural gasoline, naphtha, kerosine, and gas oil. Feedstocks are cracked in the presence of steam in tubular pyrolysis furnaces. Steam-generation pressures range from 400 to 1,800 psi. The uncondensed gases are compressed and treated with caustic soda and water, after which the treated gas may be dried with alumina or a molecular sieve. The fractionation sequence may vary to suit feedstocks but typically includes depropanization or deethanization prior to demethanization. Ethylene and ethane are separated by low-temperature fractionation. There are many variations of ethylene production processes. See also **Ethylene.**

Acetylenes: Acetylene, C_2H_2, is the simplest compound in the *acetylene series* of hydrocarbons, containing two carbons joined by a triple bond and hence an unsaturated compound. Like ethylene, acetylene is an anchor compound for several structurally similar compounds in the series (see Table H-2). Note that regardless of the number of carbons in the chain, only *one* triple bond appears in an acetylene. Up to a carbon count of

HC≡CH

Acetylene

H—C—C≡C—H (with H above and below the C)

Allylene
(methylacetylene)

Crotonylene
(dimethylacetylene)

12 ($C_{12}H_{22}$), the acetylenes are gases or very volatile liquids with a peculiar, distinct odor. These compounds readily form metallic compounds, such as silver acetylide, C_2Ag_2, and copper acetylide, C_2Cu_2. The acetylenes are easily oxidized and changed to products that contain a smaller number of carbons than the original reactants. They combine with ozone to form ozonides.

Acetylene gas (see also **Acetylene**) is an important starting compound in the creation of various synthetic compounds. Generally commercial production of acetylene starts with feedstocks made up of refinery products, ranging from naphtha cuts to middle distillates or, in some processes, from ethane to gas oil. Usually acetylene and ethylene are produced concurrently by high-temperature pyrolysis. The condensed hydrocarbons from the pyrolysis furnace are treated to remove carbon dioxide and hydrogen sulfide, after which the gases are dried and cooled. Hydrocarbons heavier than C_3 are generally recycled to the reactor. In one process, a highly selective solvent is used to remove the acetylene with a product recovery of 99.9% purity.

Diolefins: A *diolefin* contains two double bonds in the chain, as typified by isoprene and biallyl. Diolefins are important in the synthesis of elastomers and fibers. Isoprene is a high-tonnage product. See also **Isoprene.**

Isoprene (methylbutadiene)   Biallyl (1,5-hexadiene)

Butadiene, $CH_2:CH \cdot CH:CH_2$, also called bivinyl, vinyl ethylene, erythrene, and pyrrolylene, is another important diolefin, often produced commercially by the dehydrogenation of butane from natural gas or petroleum. Butadiene is also made from alcohol. This colorless gas is used in the production of synthetic rubber, nylon, latex paints, and resins. See also **Butadiene;** and **Elastomers.**

Diacetylenes: A *diacetylene* contains two triple bonds in the chain, as typified by bipropargyl, an important intermediate isomeric with benzene.

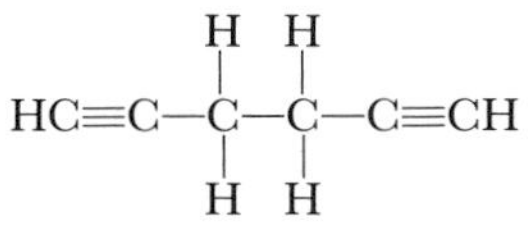

Bipropargyl (1,5-hexadiine)

Diacetylene, $CH:C \cdot C:CH$ (butadiine or butadiyne), also is in this class of hydrocarbons. Vinyl acetylene, $CH:C \cdot CH:CH_2$, is an interesting related compound.

Cyclohydrocarbons: There are many ring-type hydrocarbon compounds (see also **Carbon compounds**), which range from the three-carbon ring of cyclopropane, C_3H_6; the four-carbon ring of cyclobutane, C_4H_8; the five-carbon ring of cyclopentane, C_5H_{10}; to the six-carbon ring of cyclohexane, C_6H_{12}. These compounds all are paraffinic; i.e., they contain single bonds and are saturated. Cyclohexane, in particular, is a very important petrochemical. See also **Cyclohexanol/cyclohexanone (KA oil).**

There are several process variations, but most commercial processes involve the catalytic hydrogenation of benzene.

Camphane, $C_{10}H_{18}$, the parent material of the camphor group, is classified as a *dicycloparaffin*. This substance, also identified as dihydrobornylene, comprises a bridged ring. α-Pinene, camphene, and the terpenes also are classified as *dicycloolefins*. See also **Terpenes.**

There are several cycloolefins, including the isomeric forms of cyclohexadiene, C_6H_8, and limonene, $C_{10}H_{16}$.

Benzenoids: Benzene-type hydrocarbons, termed *benzenoids,* include several structural forms:

1. Six-carbon ring with three single bonds and three double bonds between carbons, typified by

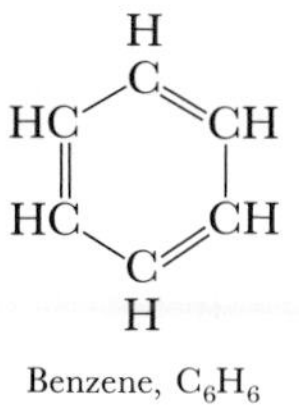

Benzene, C_6H_6

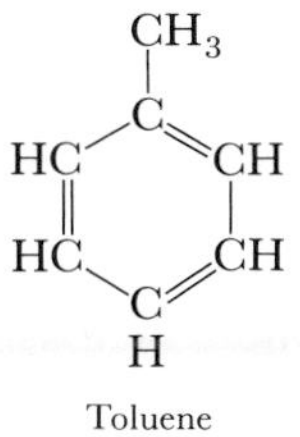

Toluene
(methylbenzene or phenylmethane)

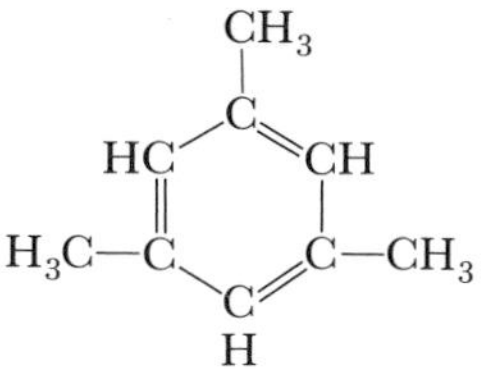

Mesitylene
(1,3,5-trimethylbenzene)

2. Two six-carbon rings that are chain-bonded, as typified by

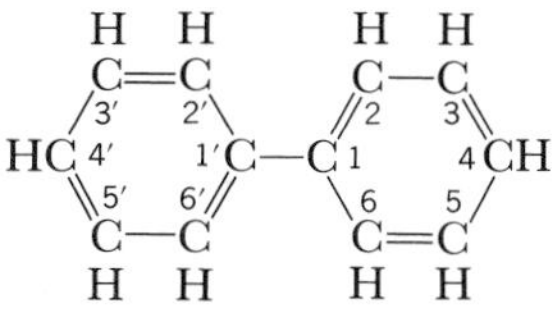

Biphenyl, $C_{12}H_{10}$

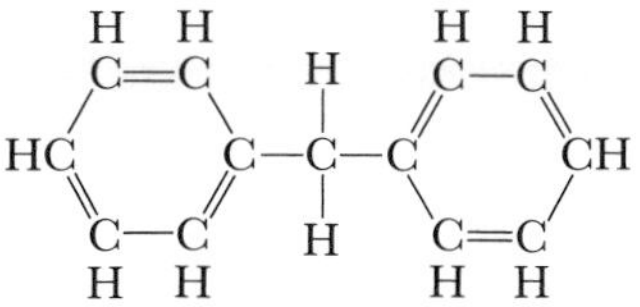

Diphenylmethane,
$C_{13}H_{12}$

3. Two or more six-carbon rings that are bonded at two carbons between rings, typified by

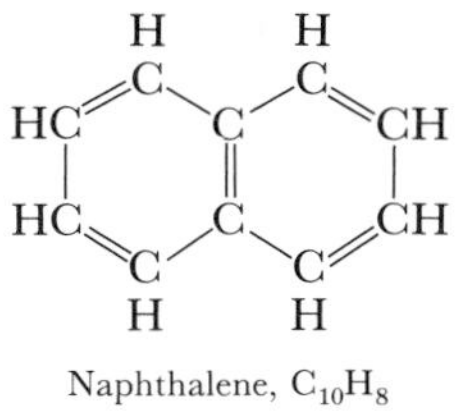

Naphthalene, $C_{10}H_8$

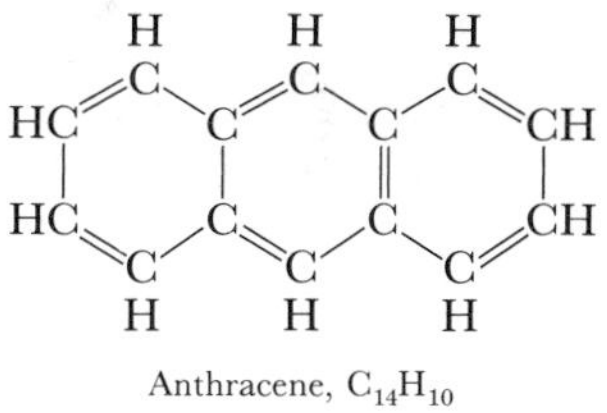

Anthracene, $C_{14}H_{10}$

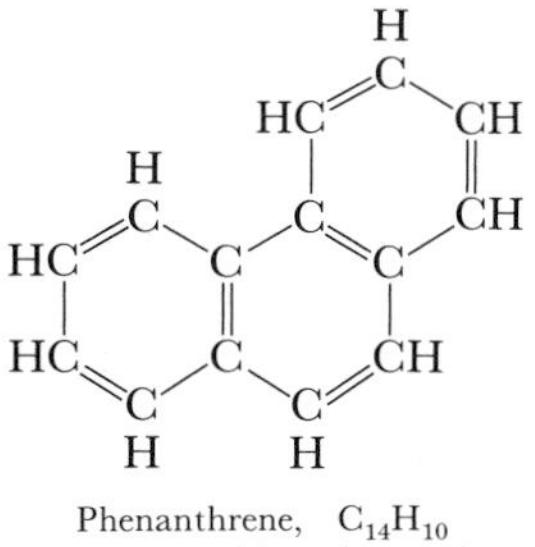

Phenanthrene, $C_{14}H_{10}$
(isomeric with anthracene)

Benzene is probably the most important of the benzenoid hydrocarbons because of the hundreds

TABLE H-3. *Representative Benzenoid Hydrocarbon Derivatives*

Compound	Formula	Formula Weight	Sp gr	Mp, °C	Bp, °C
C$_6$H$_5$ Compounds					
Chlorobenzene	C$_6$H$_5$Cl	112.50	1.540	140.7	
Bromobenzene	C$_6$H$_5$Br	156.96	1.497	−30.6	156.2
Iodobenzene	C$_6$H$_5$I	203.97	1.832	−31.4	188.2
Iodoxybenzene	C$_6$H$_5$IO$_2$	235.97	. . .	238	
Benzyl chloride.	C$_6$H$_5$CH$_2$Cl	126.51	1.103	−39	179.4
Benzal chloride.	C$_6$H$_5$CHCl$_2$	160.96	1.295	−17.4	214
Benzotrichloride	C$_6$H$_5$CCl$_3$	195.41	1.38	−4.8	220.7
Nitrobenzene	C$_6$H$_5$NO$_2$	123.05	1.207	5.7	210.9
Aniline	C$_6$H$_5$NH$_2$	93.06	1.022	−6.2	184.4
Acetanilide	C$_6$H$_5$NHCOCH$_3$	135.08	1.211	114.2	303.8
Phenyl Isothiocyanate	C$_6$H$_5$N:CS	135.11	1.135	−21	218.5
Methylaniline	C$_6$H$_5$NHCH$_3$	107.08	0.986	−57	195.7
Ethylaniline	C$_6$H$_5$NHC$_2$H$_5$	121.09	0.963	−63.5	204.7
Dimethylaniline	C$_6$H$_5$N(CH$_3$)$_2$	121.09	0.956	1.67	193.5
Diethylaniline	C$_6$H$_5$N(C$_2$H$_5$)$_2$	149.13	0.934	−34.5	216.3
Diphenylamine.	(C$_6$H$_5$)$_2$NH	169.09	1.159	53	302
Benzylamine	C$_6$H$_5$CH$_2$NH$_2$	107.08	0.980	. . .	184
Phenylhydrazine	C$_6$H$_5$NHNH$_2$	108.08	1.098	19.6	243.5
Diazoaminobenzene	C$_6$H$_5$N$_2$NHC$_6$H$_5$	197.11	. . .	96	
o-Aminoazobenzene	C$_6$H$_5$N:NC$_6$H$_4$NH$_2$	197.11	. . .	302	
Azoxybenzene	C$_6$H$_5$N:N(C$_6$H$_5$)O	198.09	1.246	36	
Azobenzene.	C$_6$H$_5$N:NC$_6$H$_5$	182.09	1.203	67	297.4
Nitrosobenzene	C$_6$H$_5$NO	107.05	. . .	68	59
Benzenesulfonic acid	C$_6$H$_5$SO$_3$H	185.14	. . .	46	
Phenol.	C$_6$H$_5$OH	94.05	1.072	41	182
Phenyl ethyl ether	C$_6$H$_5$OC$_2$H$_5$	122.08	0.965	−30.2	172
Phenyl acetate (phenetol) . .	C$_6$H$_5$OCOCH$_3$	136.06	1.078	. . .	195.5
Benzyl alcohol	C$_6$H$_5$CH$_2$OH	108.06	1.046	−15.3	205.8
Benzaldehyde.	C$_6$H$_5$CHO	106.05	1.046	−56	179.5
Benzoin	C$_6$H$_5$COCH(OH)C$_6$H$_5$	212.09	. . .	133	344
Acetophenone	C$_6$H$_5$COCH$_3$	120.06	1.026	19.7	202.3
α-Benzophenone	C$_6$H$_5$COC$_6$H$_5$	182.08	1.108	48.5	306
Thiophenol	C$_6$H$_5$SH	110.17	1.074	. . .	168.5
Diphenyl sulfide	(C$_6$H$_5$)$_2$S	186.14	1.119	. . .	293
Benzoic acid	C$_6$H$_5$COOH	122.05	1.266	121.7	249.2
Ethyl benzoate	C$_6$H$_5$COOC$_2$H$_5$	150.08	1.047	−34.6	213.2
Benzoyl chloride	C$_6$H$_5$COCl	140.50	1.211	−0.8	197.2
Benzamide	C$_6$H$_5$CONH$_2$	121.06	1.341	130	290
Benzonitrile.	C$_6$H$_5$CN	103.05	1.000	−13.1	190.7
Phenylacetic acid	C$_6$H$_5$CH$_2$COOH	136.06	1.228	76.6	265.5
α-Phenylpropionic acid	C$_6$H$_5$CH(CH$_3$)COOH	150.08	. . .	. . .	265
Isocinnamic acid	C$_6$H$_5$CH:CHCOOH	148.06	. . .	. . .	57
Mandelic acid	C$_6$H$_5$CH(OH)COOH	152.06	1.361	118	
C$_6$H$_4$ Compounds					
o-Chlorotoluene	C$_6$H$_4$ClCH$_3$	126.51	1.080	−35.1	159.4
o-Nitrotoluene	C$_6$H$_4$(CH$_3$)NO$_2$	137.06	1.168	−10.6	222.3
o-Nitroaniline.	C$_6$H$_4$(NO$_2$)NH$_2$	138.06	1.442	71.5	
p-Nitrosodimethylaniline . . .	C$_6$H$_4$(NO)N(CH$_3$)$_2$	150.09	. . .	85	
p-Sulfanilic acid	C$_6$H$_4$(NH$_2$)SO$_3$H	173.18	. . .	280	
o-Nitrophenol.	C$_6$H$_4$(NO$_2$)OH	139.05	1.447	45	214.5
Resorcinol	C$_6$H$_4$(OH)$_2$	110.05	1.285	110	276.5
Quinol	C$_6$H$_4$(OH)$_2$	110.05	1.332	170.5	286.2
o-Salicylaldehyde.	C$_6$H$_4$(OH)CHO	122.12	1.153	−7	196.5
Anisaldehyde	C$_6$H$_4$(OCH$_3$)CHO	136.06	1.123	2.5	247

TABLE H-3. *Representative Benzenoid Hydrocarbon Derivatives* (*Continued*)

Compound	Formula	Formula Weight	Sp gr	Mp, °C	Bp, °C
		C_6H_4 Compounds (*Continued*)			
Quinone.	$C_6H_4O_2$	108.03	1.31	115.7	
Anthranilic acid	$C_6H_4(NH_2)COOH$	137.13	. . .	144.5	
Phthalic acid	$C_6H_4(COOH)_2$	166.05	1.593	191	
Phthalic anhydride	$C_6H_4(CO)_2O$	148.03	1.527	130.8	284.5
Phthalimide	$C_6H_4(CO)_2NH$	147.05	. . .	238	
Isophthalic acid	$C_6H_4(COOH)_2$	166.05	. . .	330	
Terephthalic acid	$C_6H_4(COOH)_2$	166.05			
Salicylic acid	$C_6H_4(OH)COOH$	138.05	1.443	159	
Methyl salicylate.	$C_6H_4(OH)COOCH_3$	152.06	1.184	−8.6	223.3
Phenyl salicylate	$C_6H_4(OH)COOC_6H_5$	214.08	1.250	43	173
Acetylsalicylic acid	$C_6H_4(COOH)CO_2CH_3$	180.06	. . .	133.5	
p-Anisic acid	$C_6H_4(OCH_3)COOH$	152.06	1.385	184.2	280
Anthraquinone	$C_6H_4(CO)_2C_6H_4$	208.06	1.438	285	379.2
Alizarin	$C_6H_4(CO)_2C_6H_2(OH)_2$	240.06	. . .	290	430
Phenanthroquinone	$(C_6H_4)_2(CO)_2$	208.06	1.405	207	360
o-Tolylacetic acid	$C_6H_4CH_3CH_2COOH$	150.08	. . .	88.5	
o-Tolylchloride	$C_6H_4CH_3CH_2Cl$	140.53	. . .	. . .	199

upon hundreds of derivatives related to it. Just as methane may be regarded as the anchor material for the vast numbers of *aliphatic* compounds, so benzene is the starting substance for an extremely large number of *aromatic* compounds. Benzene is a by-product of coke ovens and is made synthetically from petroleum. In one process, toluene or C_8 aromatics are catalytically dealkylated to produce benzene. In addition to its use as a chemical raw material, benzene is an excellent solvent for waxes, resins, rubber, and other organic materials. Benzene also is an ingredient in various fuel blends. See also **Benzene.**

Toluene also is a by-product from coke ovens and also occurs in petroleum. It may be produced in quantity by dehydrogenation of petroleum fractions. In addition to its use as a raw material for numerous synthetic materials, toluene is an excellent solvent and is used in some fuel blends to improve octane rating. See also **Toluene.**

Xylenes (or dimethylbenzenes) are three in number and named ortho, meta, and para according to the position of the methyl groups. See also **Carbon compounds.** The xylenes are inflammable, mobile, pleasant-smelling liquids and are present in coal tar as a mixture of the three isomers. Separation is effected by fractional distillation. The xylenes also can be produced synthetically from petroleum products. In addition to use as solvents, xylenes also provide the base for a number of interesting derivatives for use in or-

ganic synthesis. *p*-Xylene is the prime starting material in one process for the production of terephthalic acid and other compounds required for a number of plastics, including the polyphenylene oxides, polyethylene terephthalates, and polyparaxylylenes (parylenes). See also **Petrochemical complex.**

Naphthalene, a white solid, is one of the heavy distillates from coal tar and is also produced by the catalytic dealkylation of alkylnaphthalenes. Refined naphthalene is used as an insect repellent. Technical grades are used for making dyestuffs, synthetic resins, coatings, and tanning agents.

Anthracene is contained in coal tar (see also **Coal-tar chemicals**) and is a colorless solid with a melting point of 218°C. It forms numerous derivatives, including anthraquinone, which are very important dyestuffs and intermediates. See also **Dyestuffs and Dyeing (Textiles).**

The benzenoid hydrocarbons are generally quite reactive, with numerous products formed by the substitution of halogens, nitro groups, and other radicals and elements for the hydrogens at various positions in the basic ring. Some of the major derivatives are listed in Table H-3.

Hydrochloric Acid Hydrochloric acid is an aqueous solution of hydrogen chloride gas, HCl, and is commercially available in several concentrations and purities. This acid is a high-tonnage chemical with many hundreds of uses both inside and outside the chemical industry.

Reagent-grade acid contains approximately 37% HCl (range = 36.5–38%) by weight and is a perfectly clear, water-white solution. Maximum limits on impurities commonly are NH_4, 0.003%; As, 0.000001%; free Cl_2, 0.0001%; heavy metals, Pb, 0.0001%; Fe, 0.00002%; SO_4, 0.0001%; SO_3, 0.0001%; and residue after ignition, 0.0005%.

Commercial-grade HCl acid usually is 20°B (31.45% HCl; specific gravity = 1.16), 18°B, and 22°B. The commercial varieties generally are slightly yellow because of impurities, such as dissolved Fe. Fuming HCl contains about 37% HCl with a specific gravity of 1.194. When 3 parts of HCl are mixed with 1 part of HNO_3, the solution is known as *aqua regia*, which dissolves many materials unaffected by either of the acids alone. Aqua regia dissolves gold and platinum.

Hydrogen chloride is a colorless gas, heavier than air, weighing 1.63911 g/l at standard conditions. When solidified, HCl melts at −111°C and boils at −85°C. Critical pressure is 83 atm; critical temperature is 51.3°C. The gas is quite soluble in H_2O, accounting for the high concentration of acid solutions obtainable. The gas has a suffocating effect when breathed and is poisonous.

The binary mixture of HCl and H_2O is a constant-boiling mixture at 110°C (pressure of 760 mm Hg). If any aqueous solution of HCl is boiled for a period of time, the temperature will rise gradually to 110°C and thereafter the temperature of the boiling mixture will remain constant. The constant-boiling mixture contains 20.24% HCl by weight with a density of 1.10. The composition of this mixture is so reliably reproducible that it can be used as a standard in quantitative analysis. The constant-boiling mixture results from the fact that a dilute solution of HCl evolves H_2O faster than HCl whereas a concentrated solution evolves HCl faster than H_2O. From solutions that contain less than 20.24% HCl, pure H_2O can be recovered in the distillate by continued fractional distillation, but it is not possible to recover pure HCl. On the other hand, pure acid can be recovered by fractionally distilling a mixture containing more than 20.24% HCl.

The partial pressure of H_2O in solutions of HCl for several concentrations and pressures is given in Table H-4. The specific heat of HCl solutions for several temperatures and concentrations is given in Table H-5.

Sources: Hydrochloric acid can be prepared by reacting H_2SO_4 with NaCl. The exothermic reaction occurs in two steps:

Step 1: $NaCl + H_2SO_4 \rightarrow NaHSO_4 + HCl$
Step 2: $NaCl + NaHSO_4 \rightarrow Na_2SO_4 + HCl$

Once formed, the HCl is absorbed in H_2O. The controls used in a system for manufacturing HCl acid of constant concentration by the absorption of anhydrous HCl gas in H_2O are shown in Fig. H-2. The gas is the uncontrolled primary flow

TABLE H-5. *Specific Heat of Hydrochloric Acid Solutions*

Mole % HCl	Temperature, °C				
	0	10	20	40	60
	Specific Heat, cal/(g)(°C)				
0.0	1.00				
9.09	0.72	0.72	0.74	0.75	0.78
16.7	0.61	0.605	0.631	0.645	0.67
20.0	0.58	0.575	0.591	0.615	0.638
25.9	0.55	. . .	. . .	. . .	0.61

from a production unit, with H_2O as the related secondary flow. The H_2O is introduced into an absorption tower as a spray and mixes intimately with gas entering the top of the tower. For all variations in the flow of the gas, there is an exact equivalent amount of H_2O to be added. The secondary controller automatically controls the valve to follow exactly the demands of the ratio signal from the primary transmitter.

Rather large quantities of HCl form as byproducts of the chlorination of organic compounds.

Anhydrous HCl gas also is available in steel cylinders under a pressure of 1,000 psi. Boiling point is 85.03°C.

Hydrochloric acid dissociates readily in H_2O and is one of the most active acids. HCl reacts

TABLE H-4. *Partial Pressure of Water in Hydrochloric Acid*

Concentration HCl, %	Temperature, °C					
	0	20	40	60	80	100
	Pressure, mm Hg					
6	4.18	15.9	50.6	139	333	715
10	3.84	14.6	47.0	130	310	677
20	2.62	10.3	33.3	93.5	230	510
30	1.26	5.41	18.4	53.5	136	310
38	0.53	2.51	9.5	29.6	72.5	182

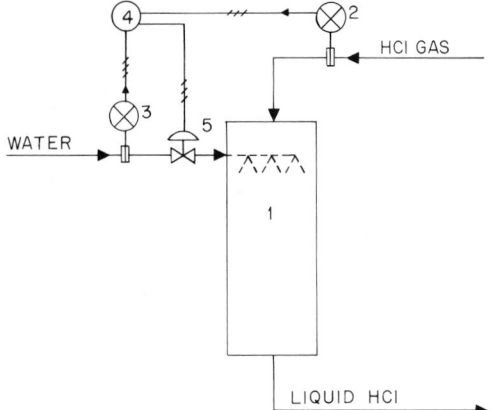

HCl GAS

WATER

LIQUID HCl

Fig. H-2. Ratio control of water to HCl gas to form constant-concentration hydrochloric acid. (1) Absorption tower, (2) primary flow transmitter that measures the uncontrolled (wild) flow of incoming HCl gas and transmits this intelligence to (4); (3) secondary flow transmitter that measures the controlled flow of incoming water and transmits this information to (4); (4) flow-ratio receiver-controller which, via control valve (5), accurately modulates the incoming water to assure production of liquid HCl of constant concentration. Product concentration can easily be altered by adjustment of (4). If desired, (4) can be calibrated in any convenient engineering terms, e.g., percent HCl by volume or by weight or specific gravity.

with many materials to form chlorides, which are all soluble except for AgCl, HgCl, CuCl, TlCl, and PbCl (latter soluble in hot H_2O). Generally, HCl is the most effective solvent for metallic oxides. Most common oxides and hydroxides will dissolve in HCl, but the aforementioned elements form insoluble chlorides. Other exceptions are some oxides which have been ignited, such as Cr_2O_3, Fe_2O_3, Al_2O_3, NiO_2, and SnO_2. Even these, with the exception of Cr_2O_3, will ultimately dissolve after long boiling with HCl.

Uses: A major use of HCl is to clean and prepare metals for other coatings. Significant quantities of HCl are used in the recovery of Zn from galvanized-iron scrap. Large quantities are used in the production of chloride chemicals. Before electrolytic methods for the production of Cl_2 were developed, large quantities of HCl were a source of Cl_2. Although nearly all Cl_2 is now produced from common salt by conventional electrolytic methods, economic circumstances occasionally justify the electrolysis of HCl to produce Cl_2 and H_2 and modern processes and equipment are available for this reaction. Inasmuch as the demand for HCl has not increased at the same rate as for alkalies, excess HCl can be converted into Cl_2. The installation of an HCl electrolysis plant should be considered wherever HCl is a by-product.

Hydrochlorination (See **Trichloroethylene.**)

Hydrocracking This process differs from catalytic cracking in using a different catalyst and an environment of hydrogen at total pressures of 800–2,500 psig. Since the accumulation of carbonaceous deposit on the catalyst is extremely slow, process periods on-line range from several months to a year or more. In some cases, the need to burn off the catalyst deposit is not the chief cause for shutdown.

Generally, the hydrocracking processes can accommodate a wider range of feedstocks than catalytic cracking. Feedstocks include not only heavy distillates but also solvent extracts from residuals that contain several parts per million organometallic complexes. In some designs, residual oils can be processed economically if the metals content is not too high.

Products: Hydrocracked products differ from those made by catalytic cracking in that they are not olefinic. The gasoline products are not as high in octane number. Although the C_5/C_6 fraction can be blended directly into the gasoline pool, the C_7+ naphtha usually is used as a feed to catalytic reforming, since it is high in naphthene content. The distillate products that are heavier than gasoline are not so aromatic as those from catalytic cracking. Thus the distillates are appropriate components for jet fuels and other uses where low aromatic content is a requirement. All hydrocracked products are low in sulfur. The light hydrocarbons have a predominance of branched isomers. The C_4 fraction is a valuable alkylation feed because of its high isobutane content. Hydrocracking is particularly applicable for the production of specialty products, e.g., low-pour-point low-sulfur-content diesel fuels, jet fuels, LPG, high-viscosity-index lube bases, and a wide variety of low-sulfur-content fuels.

A variety of proprietary catalysts are used and are supplied by the process licensors or through formulations furnished by the licensors.

Fixed-Bed Units: The fixed-bed process is one of two basically different techniques of hydrocracking, the other being the ebullating bed. Fixed-bed processes can be classified by the num-

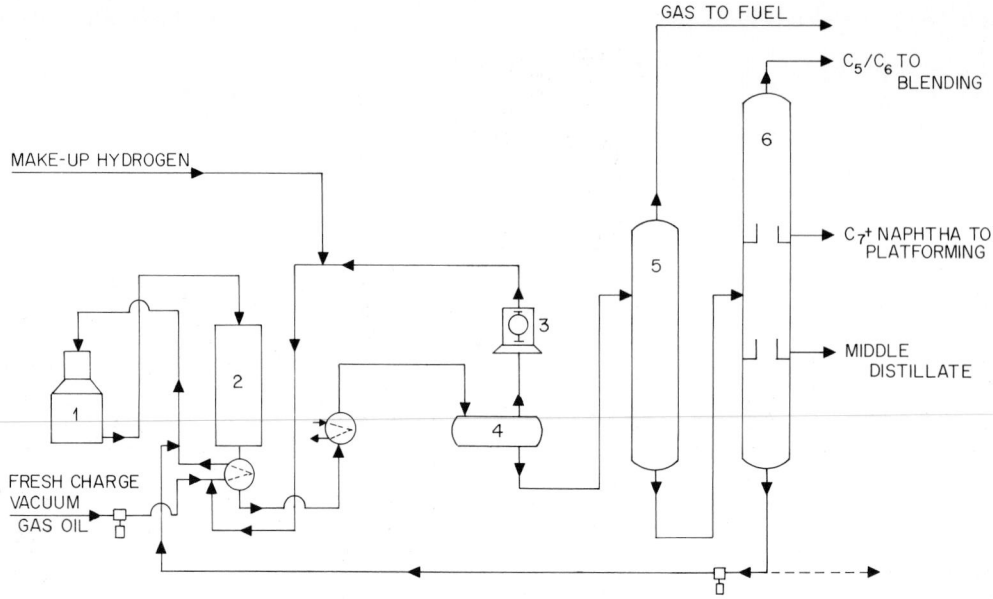

Fig. H-3. Hydrocracking process. (1) Heater, (2) reactor, (3) compressor, (4) separator, (5) debutanizer, (6) splitter column. (*UOP Isomax, Universal Oil Products Company.*)

ber of reactors used and the configuration of flow with respect to the reactor sequence and fractionation of reactor products. In the simplest arrangement, one reactor is followed by fractionation of the reactor effluent into net products and a recycle stream to be cracked further. A process of this type is shown in Fig. H-3. In another design, two reactors are used in series, each reactor employing a different catalyst, with the entire effluent from the first reactor being processed in the second reactor. Another two-reactor system, with different catalysts in each reactor, employs fractionation of the effluent from the first reactor to remove certain products before the remainder is processed in the second reactor (see also Fig. H-4).

Figure H-3 illustrates a single-reactor unit* designed to hydrocrack a heavy vacuum gas oil into gasoline and light distillate products. In this particular arrangement, the combined feed—comprising fresh feedstock, recycle and makeup hydrogen, and recycled unconverted distillate—is preheated by exchange with reactor effluent before it is brought to the desired reactor inlet temperature in the heater.

*UOP Isomax process.

The hydrocarbons, which are cracked in a down-flow, fixed-bed reactor, are cooled and brought to a drum, in which the recycle hydrogen is separated. The equilibrium liquid phase from the separator contains (in solution) some hydrogen as well as H_2S and NH_3 (formed from the sulfur and nitrogen in the feed) and net hydrocracked light hydrocarbons. These dissolved, light materials are removed in a debutanizer.

The stabilized liquid then is led to a fractionator for splitting into a C_5/C_6 fraction, which is suitable for blending into finished gasoline pools, and into a naphtha, which in most refining situations is used as a feedstock to catalytic reforming to improve its octane number. In some instances, the naphtha may be suitable as a blending component in finished gasoline pools.

A middle distillate is shown in Fig. H-3 as a net product. This material is suitable for use as a blending component for jet fuel, kerosine, diesel fuel, and a variety of heating fuels. The end point of the middle distillate can be adjusted to suit the specific purpose for which it is produced. The remainder of the reactor effluent is withdrawn as a higher-boiling column bottoms stream, which is recycled to the reaction zone for further hydrocracking.

Fig. H-4. Hydrocracking unit. (*UOP Isomax, Universal Oil Products Company.*)

In some designs, the relative positions of the debutanizer and the splitter (as shown in Fig. H-3) are reversed, so that only the light gasoline overhead from the first column is stabilized in the second column. Some designs in which relatively large amounts of light hydrocarbons are formed employ two stages of flashing of the reactor effluent prior to further separation by fractional distillation. The secondary flash helps reduce the vapor load in the first fractionator and leads to a more economical overall design. In some instances, the heat of reaction is high enough to make it practical to employ quenching at intermediate points in the reactor and thus keep the temperature profile of the reactor more nearly isothermal.

When the sulfur content of the feedstock is especially high, H_2S removal from the recycle hydrogen stream may be economical or H_2S removal from a slipstream of the recycle may be used, especially if H_2S recovery is routinely performed in the refinery.

In a unit of the type described, the relative proportions of gasoline and light distillate may be varied, either (1) by changing the operating conditions that influence the conversion per pass or

(2) by choice of catalyst. Some catalysts selectively convert heavy distillates into light distillates without further converting the primary product into gasoline; other catalysts convert any distillate selectively into gasoline; and still others are capable of hydrocracking naphthas (particularly) into LPG fragments, that is, C_3 and C_4.

Ebullating-Bed Reactor: In this design, the upward linear velocity of hydrogen and hydrocarbons is sufficiently high to expand the bed of catalyst particles and thus induce a condition of continuous, random motion. The upward linear velocity is not so high, however, that catalyst particles will be carried out of the reactor with the hydrocracked products.

Two catalyst sizes and configurations can be used. When a powdered catalyst is used, the recycle hydrogen and hydrocarbon feed is sufficient to ebullate the catalyst bed. In another design, where extruded catalyst particles are used, an internal ebullating pump is used to recirculate clear liquid from above the top of the expanded bed downward through a draft tube and upward through the bed. Provisions are made in the ebullating-bed design to withdraw spent catalyst and replace it with fresh catalyst at such rates that the catalyst inventory in the reactor can be maintained at a constant performance level—or at a new level if desired.

Since the ebullated catalyst beds are in a state of agitation, isothermal conditions are realized. Back mixing of hydrocracked products does occur. To compensate for diminished hydrocracking selectivity due to back-mixing effects, more than one reactor is used in series to obtain the benefit of staging.

Process Variables: Reactor temperature is the principal process variable that can be manipulated to alter the performance of the process. With other conditions constant, an increase in reactor temperature will increase the hydrocracking conversion but with some tendency to reduce the molecular weight of the cracked products. In addition to adjusting the inlet temperature to a fixed-bed reactor by changing the degree of heater firing, the operator has control over the temperature profile through the reactor by varying the amount of recycle-gas quench at intermediate points in the reactor bed. The unit design, of course, must provide this capability.

Hydrocracking catalysts slowly lose activity due to the accumulation of a carbonaceous deposit. In most instances, the hydrocracking conversion is maintained at a constant level by slowly increasing the reactor temperature to compensate for this decline in catalyst activity. Eventually, a reactor temperature is reached at which the hydrocracking efficiency has diminished to the point where such adjustments no longer are practical and where it is economical to restore the activity of the catalyst by burning off the carbonaceous deposit. For very heavy residual stocks, this point may be reached within a few months. For the lighter distillate stocks, one or more years of processing time may elapse before catalyst regeneration becomes necessary in fixed-bed units.

Space velocity (the rate of hydrocarbon feed per unit of catalyst inventory) is largely a design factor used to determine the reactor size for a given unit. At otherwise constant conditions, the lower the space velocity (or the greater the amount of catalyst in the reactor per unit rate of feed), the higher the hydrocracking conversion will be. Thus, the designer must make a judicious determination of the optimum reactor size for each unit, taking into account the interacting effects of many factors. In a given unit, the space velocity can be varied somewhat during a run by changing the feed rate within the limits of the equipment. A change of this type would be dictated by the overall economics governing the operation at a particular time.

Reactor pressure is a design factor. Units normally are operated within a narrow region of the design pressure. Higher pressures in practical ranges, at otherwise constant conditions, have the effect of increasing the rate of the hydrocracking reaction. At a given recycle hydrogen-to-oil ratio, increasing pressures tend to diminish the rate of accumulation of catalyst deposit with the net effect of prolonging run lengths between regenerations. But since equipment costs increase with higher design pressures, there is an optimum operating pressure for each processing requirement. As a general rule, the lighter feeds are hydrocracked in plants designed to operate at lower pressures than the heavier and dirtier feedstocks.

Hydrogen-to-oil ratio may be increased and, at a given pressure, will diminish the rate of deposit accumulation. This approach is costly, however, in terms of compressor size, compressor operating costs, and heating and cooling duties. Thus, along with design pressure, the engineering involves the judicious selection of an optimum recycle-to-oil ratio.

Growth: Hydrocracking capacity has been growing steadily since the late 1960s, at the relatively rapid rate of about 35% per year. It was estimated that in early 1970 the hydrocracking capacity in the United States was 603,000 bbl/ stream day, or about 13% of the total catalytic cracking capacity.

See also **Cracking; Fluid catalytic cracking; Petroleum;** and **Petroleum processing.**

—Melvin J. Sterba, *UOP Process Division, Universal Oil Products Company, Des Plaines, Ill.*

Hydroformylation   (See **Alcohols.**)

Hydrogels   (See **Colloid systems.**)

Hydrogen   (Gk. *hydro genes,* water-forming.) **H** = 1.00797 (at. wt.); 1 (at. no.). Hydrogen, the simplest and lightest element, is considered to have been the primordial substance from which all other elements in the universe developed. Stars were formed by gravitational forces from a rotating mass of hydrogen; the resultant high temperatures led to the fusion reaction converting hydrogen to helium, releasing thermal energy (as in the sun), and leading to the formation of the rest of the elements found on earth. Jupiter, Saturn, and Uranus still have substantial amounts of free hydrogen, whereas free hydrogen has long since escaped from the earth's lower atmosphere. However, at an altitude of 1,000 mi above the earth, hydrogen atoms are more abundant than either nitrogen or oxygen.

Hydrogen was identified by Cavendish in 1766 and named by Lavoisier in 1783. In 1931, Urey discovered a second isotope, deuterium, with mass 2; a third isotope, tritium, mass 3, was prepared synthetically in 1934 by Rutherford, Oliphant, and Harteck. Libby then detected the presence of tritium in water. Hydrogen, mass 1 (protium), and deuterium, mass 2, are stable isotopes, while tritium, mass 3, is radioactive, with a half-life of 12.26 years.

Although the abundant hydrogen isotope protium is the simplest known atom, it forms two diatomic molecules, namely, *ortho-hydrogen,* in which the two atomic nuclei spin in the same direction, and *para-hydrogen,* in which the nuclei spin in opposite directions. While the equilibrium composition of hydrogen gas is 75% ortho at ambient temperature, it changes to 99.8% para in the liquid state. The transition from ortho- to para-hydrogen is exothermic (168 cal/g), so that the heat released is more than enough to revaporize liquid hydrogen (heat of vaporization 107 cal/g). Recognition of the existence of the ortho-para transition and the development of catalysts to equilibrate the liquid during liquefaction have made possible the large-scale production, use, and storage of liquid hydrogen.

Hydrogen molecules dissociate to atoms endothermally at high temperatures (heat of dissociation about 103 cal/g mole), in an electric arc, or by irradiation. This property is used to effect atomic-hydrogen arc welding, in which hydrogen gas is dissociated by an ac electric arc between two tungsten electrodes, the hydrogen atoms recombining at the metal surface to provide the heat required for welding.

The properties of hydrogen are given in Table H-6.

Actual and potential uses for hydrogen can be predicted by inspection of its properties. Its low density, 7% that of air, plus its high thermal conductivity, 6.7 times that of air, have led to its use as a coolant in large rotating electrical equipment. The low density reduces windage friction losses to less than 10% those with air, while its high thermal conductivity and heat capacity permit more efficient heat transfer, the result being an overall increase in generator efficiency of as much as 1%. The high heats of reaction of hydro-

TABLE H-6. *Properties of Hydrogen*

Melting point, K	13.96
Heat of fusion at 14.0 K, cal/g	14.0
Boiling point at 1 atm, K	20.39
Heat of vaporization at 20.4 K, cal/g	107
Density, g/cm^3:	
Solid at 4.2 K	0.089
Liquid at 20.4 K	0.071
Critical temperature, K	33.3
Critical pressure, atm abs.	12.8
Critical volume, cm^3/mole	65.0
Critical density, g/cm^3	0.031
Heat of transition ortho to para at 20.4 K, cal/g	168
Specific heat:	
At constant pressure C_p, cal/g:	
Liquid at 17.2 K	1.93
Solid at 13.4 K	0.63
0–200°C	3.44
At constant volume C_v (0–200°C) cal/g	2.46
Ratio C_p/C_v (0–200°C)	1.40
Gas density at 0°C and 1 atm, g/l	0.0899
Gas specific gravity (air = 1.0)	0.0695
Gas thermal conductivity at 25°C, (cal) (cm)/(s)(cm^2)(°C)	0.00044
Gas viscosity at 25°C and 1 atm, cP	0.0089
Coefficient of thermal expansion per °C	0.00356
Heat of combustion at 25°C, kcal/g mole:	
Gross	63.3174
Net	57.7976
Heat of formation of HF at 25°C, kcal/g mole ΔH	−64.2
Flammability limit %:	
In oxygen	4–94
In air	4–74

gen with oxygen or fluorine, plus the low molecular weights of the product gases, have made hydrogen a prime fuel for rocket propulsion, since rocket thrust increases directly with the temperature and inversely with the molecular weight of the exhaust gases. Liquid hydrogen and oxygen were used in the second- and third-stage Saturn engines in the Apollo moon flights. The low atomic weight of hydrogen has made it the preferred propellant for nuclear rockets, in which nuclear emission provides heat for exhausting hydrogen gas at high temperatures.

Occurrence: Hydrogen is the ninth in abundance by weight of the elements in the earth's crust (third in number of atoms), most of it being found in water, which contains 11.2% hydrogen. It is present in acids, bases, hydrocarbons, all living organisms, and most organic compounds. Natural hydrogen contains about 0.015 mole % of deuterium and infinitesimal traces of tritium.

Chemical Reactions: Hydrogen is a colorless, odorless, tasteless gas, unreactive at ordinary temperatures in the absence of catalysts. It will react with oxygen at room temperature in the presence of a platinum catalyst, or explosively with a flammable mixture in the presence of a flame or spark, forming water. Hydrogen also reacts with sulfur, selenium, and tellurium.

It reacts with all the halogens, forming acids. The reaction is violent and exothermic with fluorine, mild and endothermic with iodine. Hydrogen reacts directly at elevated temperatures with most alkali and alkaline-earth metals to form hydrides, decomposable by water. Many metal oxides, halides, and sulfides react with hydrogen at elevated temperatures to form the free metals.

In the presence of Pt, Pd, or Ni catalysts, hydrogen reacts with various organic compounds to saturate double or triple bonds and to convert aldehydes, ketones, and esters to alcohols.

Manufacture: The catalytic reaction of hydrocarbons with steam (steam reforming) has become the most widely used process for producing hydrogen commercially. Hydrocarbons, from methane to petroleum naphtha, are vaporized, mixed with steam, and passed over a Ni catalyst at temperatures of 1200–1800°F (649–982°C), producing carbon oxides and hydrogen. A second widely used process is the noncatalytic partial oxidation of hydrocarbons, exemplified by two processes,*† which can utilize any liquid hydrocarbon, including crude oil and residual fuel oil, and hence can be used in locations where light hydrocarbons are unavailable or expensive. Preheated hydrocarbon, steam, and oxygen are fed through a burner to a refractory-lined combustion chamber, in which they react at temperatures of 2300–2700°F (1260–1482°C) to form carbon oxides and hydrogen. The two processes are similar chemically but differ in mechanical features. Most plants of the one type* use a direct quench for cooling reactions products, while plants of the other type† use indirect cooling in waste-heat boilers.

Steam reforming and partial oxidation have both been operated at pressures of 600 psig or higher, which permits more effective heat recovery and reduces compression costs. The partial-oxidation process has one important advantage, namely, the ability to utilize any available pumpable hydrocarbon, including those with high sulfur contents, but this is offset by the requirement

*Texaco process.
†Shell process.

for an air-liquefaction plant to supply the oxygen used in the process. Another advantage of partial oxidation is that no catalyst is needed. However, carbon removal from the product gas is required, an operation superfluous with steam reforming.

Other hydrogen manufacturing processes include catalytic partial oxidation of hydrocarbons, the steam–water-gas process, the steam-iron process, the electrolysis of water, ammonia dissociation, the steam-methanol process, and the thermal dissociation of hydrocarbons. Before 1940, the steam–water-gas process dominated hydrogen manufacture, but it has since been supplanted by the steam reforming of hydrocarbons. The steam–water-gas process uses coke as raw material, reacting it with steam to produce water gas, a mixture of CO and H_2. This is then reacted with steam over an iron oxide catalyst to convert the CO to CO_2 and H_2, which then is purified to produce H_2. The high cost of coke, the large steam requirement, and cyclic operation at atmospheric pressure have rendered this process obsolete.

The steam-iron process produces H_2 by reacting steam at high temperature with reduced iron oxide to give H_2 and then re-reducing the iron oxide with a reducing gas, such as water gas or producer gas, in a cyclic operation at atmospheric pressure. This process also is obsolete, although attempts have been made to modernize it with fluid-bed techniques.

The electrolysis of water to produce H_2 (and by-product O_2) has been utilized commercially where low-cost electricity is available, as in Canada and Norway. The electricity required amounts to 140 kWh or more per cubic foot of H_2 produced. Recent studies on high-efficiency, high-pressure electrolytic cells, together with the potential availability of low-cost electric power from nuclear reactors, have renewed interest in this process.

Catalytic partial oxidation of hydrocarbons, in which hydrocarbons up to and including petroleum naphtha are reacted with steam and O_2 over a Ni catalyst, is in commercial use mainly to produce synthesis gas for NH_3 or methanol. See also **Synthesis gas.** Like steam reforming, this process can be operated under pressures up to 30 atm or more. The necessary equipment is simpler than for steam reforming, since the catalyst chamber is not externally heated, but an O_2 supply is required. See also **Gasification.**

The steam-methanol and NH_3-dissociation processes have been used only in relatively small

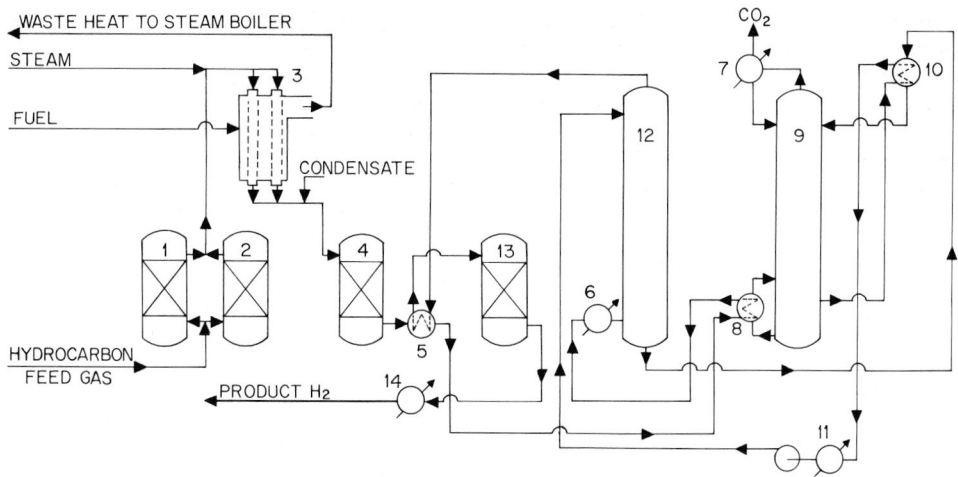

Fig. H-5. Production of hydrogen by steam reforming of hydrocarbons (1) and (2) Carbon drums, (3) reformer, (4) carbon monoxide converter, (5)–(8), (10), (11), and (14) exchangers, (9) amine regenerator, (12) carbon dioxide absorber, (13) methanator. (*C & I/Girdler Corporation.*)

installations for H_2 production, where the low cost of the equipment compared with steam reforming outweighs the cost of ammonia or methanol versus the cost of naphtha or other hydrocarbons.

Large quantities of H_2 are recovered as a by-product in certain oil-refining operations, principally catalytic dehydrogenation processes to convert aliphatic and naphthenic hydrocarbons to aromatics. Formerly, much of this by-product H_2 was burned as fuel or utilized for NH_3 synthesis, but present refining processes have such large hydrogen requirements for hydrodesulfurization and other H_2-consuming processes that all by-product H_2 is used and, in fact, often supplemented by H_2 produced by steam reforming or partial oxidation.

Purification: The crude H_2 produced by steam reforming or partial oxidation contains impurities, such as CO_2, CO, methane, carbon, H_2S, and water vapor. Carbon monoxide is converted to CO_2 and H_2 by reaction with steam (water-gas shift reaction), followed by CO_2 removal by various absorption processes, with trace amounts of carbon oxides being hydrogenated to methane (methanation). Methane in small amounts (1% or less) is unobjectionable in certain uses of H_2, such as NH_3 or methanol synthesis, but must be removed in other cases, e.g., where liquid hydrogen is desired. Methane may be removed by cryogenic means, by adsorption, or by re-reforming with steam. Carbon is removed by water-washing or filtering. Hydrogen sulfide is removed by ab-

sorption or adsorption processes. Water vapor may be removed by absorption, adsorption, or refrigeration. Synthetic zeolites (molecular sieves) have been used in adsorption processes for removing carbon oxides and water vapor from hydrogen.

Large quantities of extremely pure H_2 (99.9%) are produced by subjecting impure H_2 to diffusion through palladium films. At a temperature of about 600°F (316°C), molecular hydrogen will dissociate to atoms on palladium, the atoms then diffusing through the palladium and recombining on the opposite surface to yield pure H_2.

Production and Uses: World production of H_2 in 1970 was estimated as more than 3 trillion ft^3 or 8 million tons, about two-thirds of this being used in manufacturing NH_3. Petroleum refining operations and methanol manufacture were the next largest users. United States production of H_2 in 1970 for uses other than those just mentioned amounted to about 57 billion ft^3 out of a total production of 1.85 trillion ft^3.

The uses for H_2 other than NH_3, petroleum refining, and methanol include the space-exploration program, in which liquid hydrogen has become an important rocket fuel; oxyhydrogen and atomic-hydrogen welding; the hydrogenation of various organic compounds, such as vegetable and animal oils; and the reduction of many metallic oxides (Fe, Cu, Ni, Co, W, and Mo). Hydrogen chloride is produced by the direct reaction of H_2 and Cl_2.

A simplified flow sheet of H_2 production by the

steam reforming of hydrocarbons is shown in Fig. H-5. The feed gas is desulfurized by activated carbon, mixed with steam, reacted over a Ni catalyst in an externally heated reformer at 1600°F (871°C), quenched to 450°F (232°C) with condensate, passed over a CO conversion catalyst, cooled by heat exchange with gas en route to the methanator and monoethanolamine solution, further cooled to 120°F (49°C), passed through the CO_2 absorber, in which CO_2 is absorbed in monoethanolamine solution,* then through the heat exchanger and into the methanator at 500°F (260°C), where CO and CO_2 are hydrogenated to methane. The product H_2 then is cooled.

See also **Ammonia; Gasification; Hydrocracking; Hydrogenation; Hydrotreating; Isomerization; Methyl alcohol; Petrochemical complex;** and **Synthesis gas.**

—R. M. Reed, *C & I/Girdler Incorporated, Louisville, Ky.*

Hydrogen Arsenide   (See **Arsenic.**)

Hydrogen Bonding   (See **Molecule.**)

Hydrogen Bromide   (See **Bromine.**)

Hydrogen Chloride   (See **Hydrochloric acid.**)

Hydrogen Cyanide   Hydrogen cyanide, HCN, is a colorless gas which dissolves in water to form hydrocyanic acid (sometimes called *prussic acid*). The melting point = −14°C; boiling point = 26°C; critical temperature = 183.5°C; critical pressure = 50 atm. Density = 0.20 g/cm³; specific gravity = 0.697 at 18°C. Index of refraction of liquid HCN = 1.254. Formula weight = 27.03. There are two isomeric forms, HCN and HNC (inferred from its derivatives), which form *cyanides* and *isocyanides,* respectively.

HCN is soluble in water, alcohol, and ether in all proportions. Commonly, HCN is marketed as water solutions of 2–10 wt % strength. For processes, more frequently HCN is generated as required, and thus storage and handling are minimized. The very poisonous nature of HCN and derivatives is well known. Both gaseous and liquid HCN have a characteristic, subtle fragrance like bitter almonds.

HCN burns with a red-blue flame, producing

*Girbotol process.

CO_2, N_2, and H_2O. The aqueous solutions of HCN decompose slowly to give ammonium formate: $HCN + 2H_2O \rightarrow HCOONH_4$. Storage of aqueous solutions in the dark slows this decomposition. Chemically, HCN is classified as a very weak acid. Peaches, apricots, bitter almonds, cherries, and plums contain HCN derivatives in their kernels, sometimes in combination with glucose and benzaldehyde as a glucoside (amygdalin).

Au, Ag, Hg, and Pb are dissolved by solutions of NaCN with the absorption of O_2. Cu, Ni, Fe, Zn, Al, and Mg are similarly dissolved with the evolution of H_2. Most metal cyanides, such as NaCN, KCN, $Ca(CN)_2$, AuCN, and $Hg(CN)_2$, are soluble. AgCN and CuCN are insoluble. See also **Gold;** and **Silver.** The CN group enters into complex compounds, such as $K_4Fe(CN)_6$ and $Cu_2Fe(CN)_6$. There are several related acids, hydroferrocyanic acid, $H_4Fe(CN)_6$; hydroferricyanic acid, $H_3Fe(CN)_6$; cyanic acid, HCNO; and thiocyanic acid, HCNS.

The esters of HCN, termed *nitriles* or *alkyl cyanides,* include acetonitrile, $CH_3 \cdot CN$ (also called methyl cyanide); propionitrile, $C_2H_5 \cdot CN$ (ethyl cyanide), and acrylonitrile, $CH_2 : CHCN$. The latter are important compounds in the preparation of various synthetics. See also **Acrylonitrile.** The isonitriles, carbylamines, or isocyanides are isomeric with the corresponding cyanides.

Preparation: As an important raw material for several synthetics, HCN is prepared in significant tonnage quantities. Of course, preparation of many HCN derivatives does not require the use of HCN as a starting material. The most popular commercial process for HCN production in recent years has been the Andrussow process, in which a mixture of ammonia, methane, and air is subjected to partial combustion over a Pt catalyst in the range 900–1000°C, in accordance with the reaction $NH_3 + CH_4 + 1\frac{1}{2} O_2 + 6N_2 \rightarrow HCN + 3H_2O + 6N_2$. The yield of NH_3 generally is 55–60%.

In the special process configuration* shown in Fig. H-6, HCN is prepared from a mixture of (1) methane (refined or natural gas containing 50–100 vol % CH_4), and (2) ammonia. HCN (99.5 wt %) is produced with H_2 as a second product in accordance with the endothermic reaction $CH_4 + NH_3 \rightarrow HCN + 3H_2$. A Pt catalyst is used; reactor temperatures range from 1200 to 1300°C.

*BMA process, Degussa.

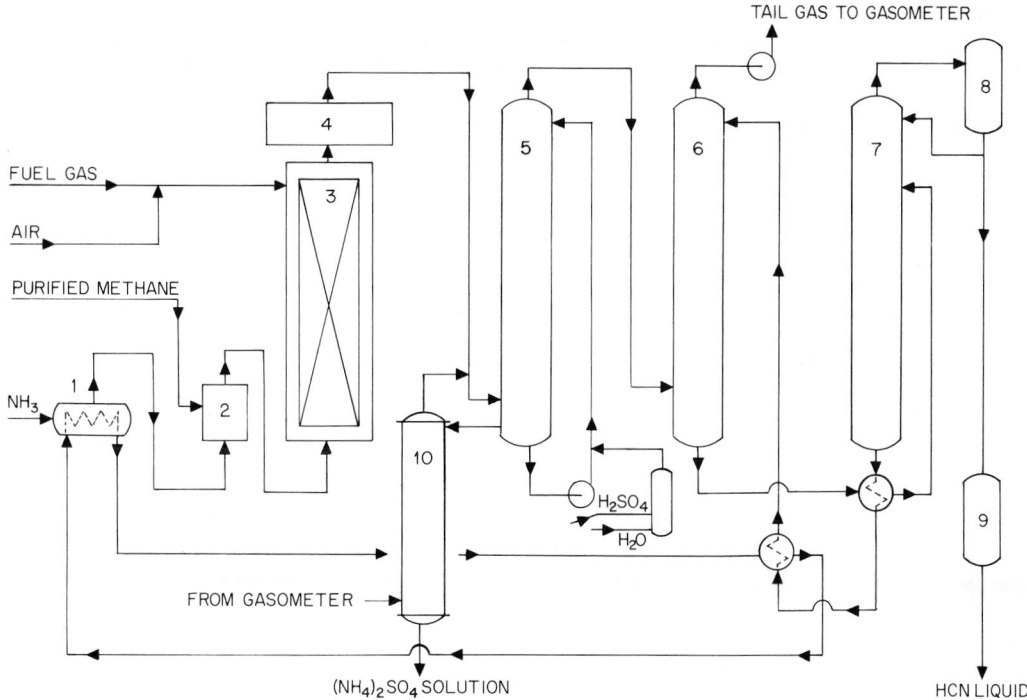

TAIL GAS TO GASOMETER

FUEL GAS

AIR

PURIFIED METHANE

NH_3

FROM GASOMETER →

$(NH_4)_2SO_4$ SOLUTION

H_2SO_4

H_2O

HCN LIQUID

Fig. H-6. Production of hydrogen cyanide with hydrogen as a second product. (1) Vaporizer, (2) mixer, (3) reactor, (4) cooler, (5) NH_3 absorber, (6) HCN absorber, (7) HCN rectification, (8) condenser, (9) aftercooler, (10) HCN stripper. (*Highly schematic representation of a type of process designed by Degussa.*)

After purification of the CH_4 and vaporization of the NH_3, the pressure of the gases is reduced, whereupon they are distributed over reactor tubes contained within a heated furnace. Product gases (volume percentages are roughly 22.8% HCN and 71.2% H_2) and unreacted gases (2.4% CH_4, 2.7% NH_3, and about 0.9% N_2) are immediately quenched upon exiting from the reactor. Unreacted NH_3 is removed with (dilute) H_2SO_4. HCN dissolved in the $(NH_4)_2SO_4$ is stripped with tail gas. Liquid HCN is produced by washing the HCN (free of NH_3) with H_2O. This solution is rectified to the 99.5 wt % HCN concentration. The H_2 produced goes to other processes or is used as a fuel. It is claimed that about 90% of the CH_4 and 84% of the NH_3 are converted to HCN. About 10% of the NH_3 may be recovered as $(NH_4)_2SO_4$.

In another process,* methane and propane are used as the hydrocarbon sources:

*Fluohmic process, Shawinigan Chemicals Ltd.

$$C_3H_8 + 3NH_3 \rightarrow 3HCN + 7H_2$$

or

$$CH_4 + NH_3 \rightarrow HCN + 3H_2$$

It is claimed that 85–90% yields are obtained on both raw materials. This process utilizes an electrically heated fluidized-bed reactor (reactor temperature 1510°C), a technique which may be applicable to other similar endothermic processes.

Uses: In addition to the broad use of HCN in the preparation of numerous chemical products and intermediates, HCN sometimes is used as a disinfectant (as a gas or as HCN-impregnated cellulosic disks). Cyanides are used in ore processing and metal treating. See also **Nitrogen.**

Hydrogen Fluoride Hydrogen fluoride is a colorless fuming liquid or gas (boiling point = 19.5°C) at atmospheric pressure. Although the formula weight for HF is 20.01 (calculated), its apparent molecular weight varies widely with temperature and pressure. For example, the molecular weight of saturated vapor at 19.51°C is 78.24 and at 100°C is 49.08. This is caused by strong hydrogen bonding between the molecules so that considerable polymerization occurs, resulting in large departures from ideal behavior in both liquid and gaseous phases. Thus, the boiling point of HF is abnormally high in the series of halogen acids: HF, +19.5°C; HCl, −85°C; HBr, −65°C; and HI, −36°C. The mechanism of the polymerization is as yet unknown; both ring- and chain-type structures have been postulated.

Other important physical properties include melting point = −83.37°C; density, liquid at 25°C = 0.9576 g/cm³, saturated vapor at 25°C = 3.553 g/l; vapor pressure (at 25°C) = 17.8 psia; critical temperature = 188°C; critical pressure = 941 psia; critical density = 0.29 g/cm³; heat of fusion at melting point = 46.93 cal/g; heat of vaporization (at boiling point at 1 atm) = 1609 cal/20.01 g; viscosity at 25°C = 0.26 cP; surface tension at 18.2°C = 8.6 dynes/cm; refractive index (5893 Å, at 25°C) = 1.1574; and dielectric constant at 0°C = 83.6.

Hydrogen fluoride undoubtedly was made by the early glass etchers. A. S. Marggraf (1764) showed that the active etchant was a gas. C. W. Scheele (1771) established that a new acid had been discovered. H. Davy (1814) proved that the acid contained a new element, fluorine, which finally was isolated (1886) by H. Moissan. Anhydrous hydrogen fluoride was first prepared (1856) by E. Frémy. The first commercial shipment of anhydrous acid was not made until 1936.*

Chemical Properties: The chemistry of hydrogen fluoride is characterized by the great strength of the hydrogen-fluorine bond, the strong tendency to hydrogen bonding, and the single oxidation state of fluorine. The molecule is one of the most stable diatomic molecules known, with a free energy of formation of −65.0 kcal per formula weight. Anhydrous hydrogen fluoride is a strong acid in the proton-donor sense and exhibits a solvent effect on many substances often accompanied by chemical reaction. It reacts with all metals below hydrogen in the replacement series except those which form an insoluble protective coating such as Al, Mg, Fe, and Ni. It reacts with many oxides and hydroxides to give water and the fluorides. In general, in the absence of protective coatings it will react with or displace any anion other than fluorine itself from a compound. Fluo-

*By Sterling Products Company (merged with Pennwalt in 1939).

rides of some metals, such as Sb, Pb, Cr, Mn, Ag, and Co are important fluorinating agents in combination with HF for organic compounds.

HF has a very strong affinity for water, so that no chemical substance will dry it; instead it reacts or complexes with the drying agent. It can be dried completely only by removal of the water by electrolysis. It is highly soluble in water, forming aqueous HF acid solutions which act as weak acids. This behavior anomaly has been attributed to extensive polymerization of the HF due to hydrogen bonding, and the phenomenon has complicated the determination of the ionization constant for HF. The calculated thermodynamic ionization constant is 6.16×10^{-4} based on measurements in $0.001\ M$ solutions.

With organic compounds, HF exhibits many interesting reactions. It is a powerful solvent for nitrogen-, oxygen-, and sulfur-containing compounds which are proton-accepting. Generally these solutions are conducting, due to proton transfer, and this is the basis for the electrochemical fluorination processes used to produce highly fluorinated compounds.

HF is also a powerful dehydrating agent and can char wood and paper. With other oxygen-containing compounds, it produces dehydrated polymerized products. When the removal of water is beneficial to a reaction, the dehydrating power of HF is very effective in promoting the reaction.

HF has also been recognized as a polymerization agent for unsaturated compounds and as an alkylation catalyst for both aliphatic and aromatic hydrocarbons.

It is, of course, a strong fluorinating agent, and will react with any compound that contains a readily replaceable halogen to form the corresponding fluoride. It adds to unsaturated compounds under proper conditions to form saturated fluorides. Finally, its aqueous solutions are capable of numerous hydrolysis reactions with proteins, cellulose, and esters.

Manufacture: Hydrogen fluoride (anhydrous or aqueous) is manufactured by the reaction of calcium fluoride, i.e., acid-grade fluorspar, with strong H_2SO_4 in a heated reactor. The acid-grade fluorspar typically is a finely ground product containing a minimum of 97.5–98% CaF_2, a maximum of 1% SiO_2, 0.1% H_2O, 0.05% S, and the remainder primarily $CaCO_3$. The H_2SO_4 should be not less than 96% (more dilute is corrosive to steel) but preferably not over 98%. The ratio of acid to spar, the temperature of the reacting mass,

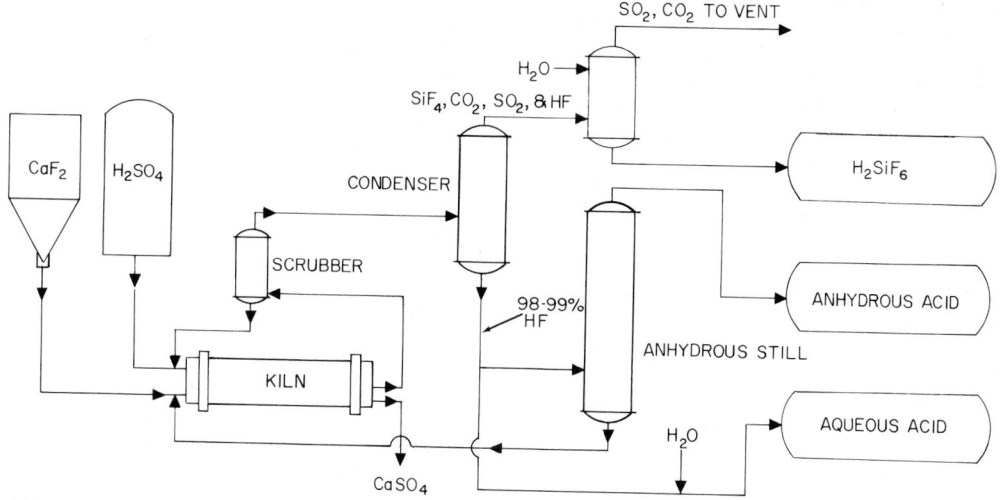

Fig. H-7. Hydrogen fluoride manufacturing process.

the time allowed for the reaction, and the degree of mixing of the acid and spar are additional important parameters in attaining maximum yield of HF. Also, silica is a highly objectionable contaminant since each pound will consume the equivalent of 2.6 lb of CaF_2 to form silicon tetrafluoride. The latter on absorption with HF in water results in an additional loss of HF due to the formation of fluosilicic acid:

$$CaF_2 + H_2SO_4 \rightarrow CaSO_4 + 2HF$$

$$4HF + SiO_2 \rightarrow SiF_4 + 2H_2O$$

$$2HF + SiF_4 \rightarrow H_2SiF_6$$

A flow diagram of a manufacturing process is given in Fig. H-7. The fluorspar and H_2SO_4 are fed into a heated horizontal reactor at 200–250°C and thoroughly mixed; the HF gas evolved is freed of dust and sulfuric acid fumes in some form of scrubber under slight vacuum and then collected in a condenser system as essentially 99% HF. This strong acid is then distilled to yield anhydrous acid of 99.9% purity. The spent $CaSO_4$ residue in the kiln is removed through an air-lock device. The noncondensible gases (CO_2, SiF_4, SO_2, and some HF) are scrubbed to recover additional HF and then absorbed in water to form fluosilicic acid (30–35% strength).

Bulk shipments of anhydrous HF are in tank cars of 42 or 22 tons capacity. HF also is available in steel cylinders of 200 or 100 lb capacity. Aqueous 70% acid is manufactured by dilution of strong (98%) acid, which is supplied in tank cars of 30 tons capacity or in steel drums (polyethylene-lined) of 450 and 260 lb capacity. Laboratory packages of 48, 60, and 70% acid are available in polyethylene bottles or carboys.

Uses: The major uses of HF are (1) in the manufacture of fluorinated organics as refrigerants, aerosol propellants, specialty solvents, and high-performance plastics (polytetrafluoroethylene, polyvinylidene fluoride, and polychlorotrifluoroethylene), (2) in the manufacture of aluminum fluoride and synthetic cryolite for the aluminum metal industry, and (3) in the manufacture of atomic-energy feed materials. Other significant uses are as an alkylation catalyst in the petroleum industry and as a pickling acid in the stainless-steel and nonferrous-metals industry. See also **Alkylation.**

In addition to the traditional use in etching and polishing glass, it is also used for the preparation of fluorides and fluoborates, for removal of sand from metal castings, for the formulation of laundry soaps and stain removers, for the preparation of special organic compounds, dyes, and pharmaceuticals, and in the manufacture of elemental fluorine.

Specifications and Analysis: Anhydrous HF is one of the purest chemicals in regular commercial distribution (as is the 70% aqueous acid made from it). A typical analysis shows 99.95% HF,

0.015% H_2SiF_6, 0.003% SO_2, 0.005% H_2SO_4, and 0.02% H_2O.

Hazards: Anhydrous HF should be regarded as a liquefied gas under pressure. All forms (liquid or vapor) and the normally used aqueous solutions of HF are extremely corrosive to the skin, eyes, mucous membranes, and lungs, resulting in painful burns and possible deep-seated ulceration. The proper use of protective clothing, goggles, face shields, and respirators is important in handling this chemical and a full knowledge of first-aid treatment is mandatory. Reference should be made to detailed instructions provided by manufacturers' data sheets or the Manufacturing Chemists Association Data Sheet (see references).

REFERENCES

Hydrofluoric Acid (Anhydrous and Aqueous), Chemical Safety Data Sheet SD-25, Manufacturing Chemists Association, Inc., Washington, D.C., 1957.

Kausch, Oscar: "Flussaure, Kieselflussaure und deren Metallsalze," Enke's Bibliothek für Chemie und Technik, vol. 24, Enke, Stuttgart, 1936.

"Mellor's Comprehensive Treatise on Inorganic and Theoretical Chemistry," vol. 2, suppl. 1, pt. 1, Wiley, New York, 1962.

Methods of Analysis for Anhydrous Hydrofluoric Acid, Procedures Recommended by Manufacturing Chemists Association, 1944 *Ind. Eng. Chem., Anal. Ed.*, vol. 16, pp. 483–486, 1944.

Simons, J. H.: "Fluorine Chemistry," vol. 1, pp. 225–259, Academic, New York, 1950.

—Paul A. Munter, *Pennwalt Corporation, King of Prussia, Pa.*

Hydrogen-Ion Concentration   [See pH (hydrogen-ion concentration).]

Hydrogen Peroxide   Hydrogen peroxide, H_2O_2, in pure anhydrous form is a viscous, colorless, clear

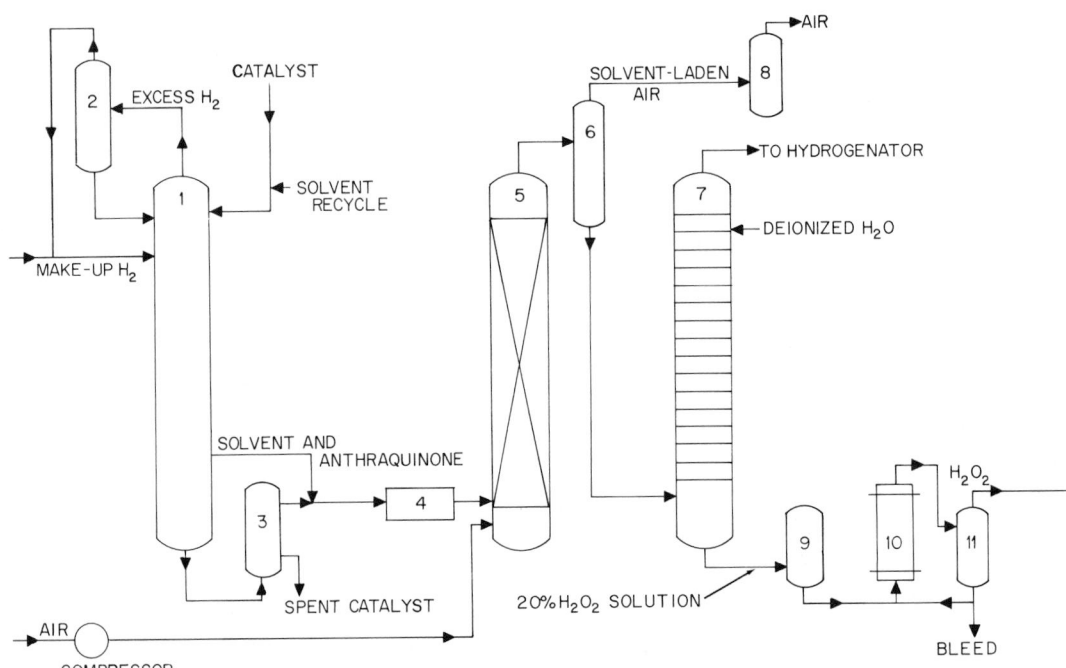

Fig. H-8. Autoxidation process for manufacture of hydrogen peroxide. Hydrogenation-oxidation cycle takes place in two identical units operating in parallel up to the extraction stage, where the two streams mix. H_2O_2 is extracted from the organic solvent by countercurrent contacting with deionized water. The main attraction of autoxidation processes over the electrolytic process is lower utility (electric) requirement as well as lower steam requirement because there is no hydrolysis of persulfate compounds. Further, in the electromechanical process, cell size is the limiting design factor; a plant can only be scaled up by adding more small electrolytic units. (1) Hydrogenator, (2) separator, (3) filter, (4) cooler, (5) oxidizer, (6) separator, (7) extractor, (8) activated-charcoal absorber, (9) organics removal, (10) vaporizer, (11) separator, (12) distillation column, (13) jet-spray condenser, (14) vaporizer, (15) separator, (16) distillation column, (17) jet-spray condenser.

liquid with a specific gravity of 1.44, a freezing point of $-0.89°C$, and a boiling point of $151.4°C$ at 760 mm Hg pressure. Formula weight is 34.02. When heated above $100°C$, pure H_2O_2 may explode violently.

Hydrogen peroxide is soluble in water in all proportions. It also is soluble in alcohol and ether but not in hydrocarbons. The chemically pure (cp) grade is a solution of 90% H_2O_2 and 10% H_2O and has a specific gravity of 1.39. This concentration contains 42% active O_2 by weight; 1 volume yields 410 volumes of O_2. H_2O_2 solutions are tonnage materials supplied commercially in various strengths, ranging from 3% H_2O_2 up to 35% H_2O_2. The commercial grades offered for oxidation and bleaching reactions usually contain 27.5–35% H_2O_2. H_2O_2, along with sodium peroxide, Na_2O_2, and several other compounds, is considered a powerful oxidant. See also **Peroxides, organic.**

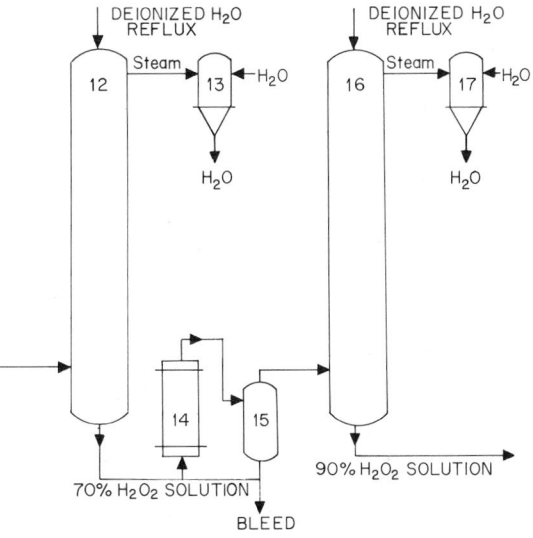

70% H_2O_2 SOLUTION

BLEED

90% H_2O_2 SOLUTION

Hydrogen peroxide solutions tend to decompose with time, a process that is retarded by storage at low temperatures in lightproof containers; sometimes an organic substance, such as acetanilide, is added to retard degradation. In addition to use in oxidizing and reducing reactions on an industrial level, H_2O_2 solutions (usually 3%) have

for many years found medicinal applications because of their antiseptic properties. H_2O_2 has been used to supply oxygen in various fuel mixtures, particularly for rockets and torpedoes, where it is used as an oxidizer in liquid bipropellant systems or as a monopropellant through controlled catalytic decomposition. These applications accounted for a marked increase in production of H_2O_2 during the 1960s.

A large portion of H_2O_2 production goes into bleaches for cotton, wool, and groundwood pulp, as well as for preparation of hair bleaches. Chemical manufacturing also consumes large amounts of H_2O_2 since the highly reactive H_2O_2 molecule is well suited for oxidation, epoxidation, and hydroxylation reactions. H_2O_2 also is used as a gas source in foaming rubber and plastics.

The reducing actions of H_2O_2 tend to be confined to very easily reduced substances, such as $KMnO_4$ and Ag_2O. The use of H_2O_2 to restore the colors of old paintings is an interesting application, by which the black PbS, resulting from years of exposure to the environment, is converted back to the original white lead sulfate.

Producers of H_2O_2 have long debated the relative merits of the electrochemical and chemical autoxidation processes for its manufacture. For many years, the traditional production route involved the electrolysis of aqueous solutions of H_2SO_4, $KHSO_4$, or NH_4HSO_4. In nonelectrolytic processes,* the charge materials may be an alkylated anthraquinone, quinone, and hydroquinone solvents, plus hydrogen, air or oxygen, water, and a catalyst (usually Ni, Pd, or Pt). Claimed conversion in these processes is approximately 90% of theoretical. The product is a solution of H_2O_2 (15–70%) in water. Anthraquinone contained in the solvent is hydrogenated (at about $40°C$ and 1–3 atm pressure) and reduced to hydroquinone (*p*-dihydroxybenzene) (step 1). The resulting hydroquinone solution then is oxidized with air or O_2 (step 2). R in the following equations may be a radical such as ethyl or tertiary butyl.

Step 1: $C_6H_4{:}(CO)_2{:}C_6H_3R + H_2 \xrightarrow{\text{cat.}}$
$$C_6H_4{:}(COH)_2{:}C_6H_3R$$

Step 2: $C_6H_4{:}(COH)_2{:}C_6H_3R + O_2 \longrightarrow$
$$C_6H_4{:}(CO)_2{:}C_6H_3R + H_2O_2$$

*FMC Corp. in the United States and Laporte Chemicals in the United Kingdom.

Theoretically, only hydrogen, atmospheric oxygen, and water are consumed in the process. A main difficulty is finding a solvent that will minimize side reactions during hydrogenation while at the same time dissolving both the hydrogenated and oxidized forms of the organic compound. Among solvents mentioned in the literature are benzene-methanol-cyclohexanol mixtures and primary and secondary nonyl alcohols. Because impurities in H_2O_2 cause spontaneous catalytic decomposition, strictest purity precautions are required. These precautions make H_2O_2 one of the purest chemicals in commercial production.

In some plants, aluminum is used extensively because its surface is passive enough to prevent peroxide decomposition in the vessels during processing. The aluminum equipment and piping may be pickled after fabrication prior to use to ensure chemical inactivity. Some of the older process plants (particularly German) employed the more costly enameled steel vessels. One modern process is shown schematically in Fig. H-8.

Hydrogen Sulfide   Hydrogen sulfide (molecular weight 34.08) at normal temperature and atmospheric pressure exists as a colorless gas having the offensive odor characteristic of rotten eggs. At $-82.9°C$, it liquefies (see Table H-7). It is a

TABLE H-7. *Physical Properties of Hydrogen Sulfide*

Molecular weight	34.08
Boiling point, °C	−59.6
Melting point, °C	−82.9
Triple point at 0.23 atm, °C	−85.5
Density:	
Gas at 21.1°C, g/l	1.43
Liquid at boiling point, g/ml	0.993
Specific gravity:	
Gas at 15°C (air = 1)	1.1895
Liquid, d_4^{60}	0.96
Critical temperature, °C	100.4
Critical pressure, atm	88.9
Critical density, g/cm³	0.349
Expansion ratio, liquid at boiling point to	
gas at 21.1°C	1:674
Solubility in water of gas at 26.7°C wt. % . .	0.32
Specific heat of gas at constant pressure at	
21.1°C, cal/(g mole)(°C)	8.2
Heat of vaporization, cal/g mole	44.63
Heat of fusion cal/g mole	568
Viscosity of gas at 0°C, cP	0.01166
Autoignition temperature, °C	260
Flammable limits in air, vol %	4.0–44

flammable gas and may explode upon ignition. Hydrogen sulfide has a relatively low ignition temperature (260°C) and a moderately wide flammability range (4.3–44% by volume) and burns with the liberation of heat (approximately 6230 cal/l at 15.6°C). The gas is also highly toxic. It is especially dangerous because it may paralyze the olfactory nerves, thus preventing detection of its characteristic odor. See also **Air pollution; Ethanolamines;** and **Sulfur dioxide removal.**

Since it is a mild reducing agent, hydrogen sulfide is oxidized under suitable conditions by such compounds as chlorine, oxygen, sulfur dioxide, and sulfuric acid. The halogens (fluorine, chlorine, bromine, and iodine) react with hydrogen sulfide to form the corresponding halogen acid. Upon ignition, hydrogen sulfide burns, yielding sulfur dioxide and water. It enters into reactions with many organic compounds. With solutions of heavy metals (silver, lead, copper, manganese), hydrogen sulfide forms metal sulfides. This reaction is employed in the selective separation of metals and in qualitative analyses and is responsible for the tarnishing of silver. In aqueous solutions, hydrogen sulfide hydrolyzes to a weak acid, accounting for its markedly increased corrosive action in the presence of moisture.

The gas occurs naturally as the decomposition product of metal sulfides and albuminous matter. It occurs in the air of mines (*stink damp*), mineral springs, and sewers. It is a by-product of many chemical processes, such as those involving viscose rayon, synthetic rubber, petroleum products, dyes, and leather. On a laboratory scale, it is generated by treating a sulfide with an acid or by heating thioacitamide, $CH_3C(:S)NH_2$. Industrially, depending upon the quantity and the purity needed, hydrogen sulfide is prepared by one of the following general reactions:

Sulfur and hydrogen:
$$S + H_2 \rightarrow H_2S$$

Sulfur and an alkali:
$$4S + 2NaOH + H_2O \rightarrow 2H_2S + Na_2S_2O_3$$

Sulfide and an acid:
$$2NaHS + H_2SO_4 \rightarrow 2H_2S + Na_2SO_4$$

By-product hydrogen sulfide is converted into sulfur or sulfuric acid. In the analytical laboratory, the gas is a well-known reagent. Commercially it is used to prepare sulfides (sodium sulfide, sodium hydrosulfide) and sulfur-containing organic compounds (thiophenes, mercaptans, and organic sulfides). It is a precipitating agent for the removal of copper, cadmium, and titanium from spent catalysts. It is also used in the production of extreme-pressure lubricants and oils and rare-earth phosphors for color TV tubes. See also **Amino acids.**

Hydrogen sulfide may be supplied by on-site generators or in cylinder quantities in steel cylinders (approved by Department of Transportation) as a liquefied gas under its own vapor pressure of 252 psig at 70°F. Department of Transportation regulations require a red flammable gas label on hydrogen sulfide containers. Quantities supplied in cylinders generally range from 0.5 to 175 lb of gas. Large containers with a capacity of 1,100 lb are also available.

—Mary E. Pisklak, *Industrial Gas Division, Air Products and Chemicals, Inc., Allentown, Pa.*

Hydrogen Sulfide Removal (See **Absorption, acidic gases.**)

Hydrogenation Generally considered as the reaction, i.e., addition, of hydrogen with other materials, hydrogenation occurs throughout the chemical and process industries. The term *reduction* is applied to reactions where oxygen is withdrawn from a compound and also where hydrogen is added. Sometimes both processes occur.

As early as 1897, Sabatier reduced nickel oxide in a stream of gaseous hydrogen at about 300°C and then used the finely divided Ni as a catalyst to react hydrogen with unsaturated organic materials at a temperature of about 175°C. Nickel, platinum black, palladium black, copper metal, copper oxide (Adkin catalyst), nickel oxide, aluminum, and other materials subsequently have been developed as hydrogenation catalysts. Temperatures and pressures of the process have been increased to improve yields. For example, the hydrogenation of methyl ester to fatty alcohol and methanol occurs at about 3,000 psig and 290–315°C (550–600°F). Hydrotreating to improve the quality of liquid hydrocarbon fuels occurs in fixed-bed reactors at pressures ranging from 100 to 3,000 psig, depending upon the design of the proprietary process.

Of the products of hydrogenation processes, hydrogenated oils are among the better known. Vegetable and fish oils are hardened or solidified by catalytic hydrogenation. Partial hydrogenation clarifies some oils and makes them odorless. Fatty oils, such as oleic acid, are converted into stearic acid by hydrogenation. Coconut oil, pea-

nut oil, and cottonseed oils can be made to appear, taste, and smell like lard; or they can be made to resemble tallow. Synthetic shortenings consist of hydrogenated oils. Generally, hydrogenated oils have higher melting points and lower iodine values than the natural untreated oils.

For numerous examples of hydrogenation in this volume see **Alcohols, fatty (via hydrogenation); Hydrotreating; Soaps;** and **Vegetable oils;** also consult the Subject Index under Hydrogenation.

Hydrology (See **Water treatment.**)

Hydrolysis (See **Acetate fibers; Alcohols; Cellulose ester plastics, organic; Ketones; Peptides and proteins; Silicones; Soaps; Urea;** and **Vegetable oils.**)

Hydrometallurgy (See **Cobalt;** and **Nickel.**)

Hydrophilic Substances [See **Anionic surfactants (sulfur-bearing); Colloid systems; Detergents;** and **Surfactants.**]

Hydrophobic Substances [See **Anionic surfactants (sulfur-bearing); Colloid systems; Defoaming agents; Detergents;** and **Surfactants.**]

Hydrosols (See **Colloid systems.**)

Hydrotreating Generally understood to include a variety of applications in which the quality of liquid hydrocarbon streams is improved by subjecting them to mild or severe conditions of hydrogen pressure in the presence of a catalyst, hydrotreating may be regarded as a rather specialized kind of *hydrogenation.* See also **Hydrogenation.** The primary purpose of hydrotreating is to selectively convert to a desirable material or eliminate from the system one or more unwanted materials in the feedstock. As shown by Table H-8, the use of hydrotreating is extensive, being involved in over 30% of the crude refined in the United States.* Although the catalysts and technique were known earlier, the availability of a ready supply of by-product hydrogen from catalytic reforming accelerated the use of hydrotreating in the early 1950s. See also **Reforming.**

Applications of hydrotreating are numerous,

*On a free-world basis, hydrotreating processes represent a capacity of approximately 10 million bbl/day (1970).

TABLE H-8. *Applications for Hydrotreating Processes**

Material Treated	Capacity, bbl/day
Catalytic reformer feeds.	2,155,965
Middle distillates	1,055,490
Catalytic cracking feeds and recycle.	263,645
Naphtha saturation	246,600
Lubricating-oil stocks	135,700
Heavy gas oil.	92,900
Residual desulfurization (and other feeds) .	102,200
	4,052,500

*United States, 1970.

and the feedstocks processed range from light fractions of gasoline to heavy residual stocks, as evidenced by the objectives of hydrotreating which include (1) pretreatment of naphtha feeds for catalytic reforming units; (2) desulfurization of distillate fuels; (3) improvement of burning quality of jet fuels, kerosines, and diesel fuels; (4) improvement of color, odor, and storage stability of various fuels and petroleum products; (5) pretreatment of catalytic cracking feeds and cycle oils by removal of metals, sulfur, nitrogen, and reduction of polycyclic aromatics; (6) upgrading of lubricating-oil quality; (7) purification of light aromatic by-products from pyrolysis operations; and (8) reduction in sulfur content of residual fuel oils.

Reactions: Some of the reactions commonly employed in hydrotreating processes include:

1. Removal of sulfur from its organic combinations in various types of sulfur compounds by hydrodesulfurization to form H_2S.

2. Removal of nitrogen as NH_3 from its organic combinations.

3. Hydrogenation of diolefins and olefins to paraffins or naphthenes.

4. Hydrogenation of monoaromatics to naphthenes to improve the burning quality of certain fuels.

5. Hydrogenation of polycyclic aromatics so that only one aromatic ring remains in the molecule; or, if desired, all the aromatic rings can be saturated.

6. Removal of oxygen from its organic combinations as H_2O.

7. Decomposition and removal of organometals, e.g., arsenic compounds in naphthas, by retention of these metals on the catalyst. Vanadium and nickel can be removed from gas oils intended as feedstocks for catalytic cracking.

H_2S, NH_3, and H_2O are removed from the hydrotreated liquid product by stripping in the stabilization section of the unit.

Hydrotreating Process: As shown by Fig. H-9, which is a generalized representation of the majority of hydrotreating processes, the essential components are (1) heaters and heat-exchange equipment; (2) the fixed-bed reactor section, which contains the catalyst and operates at pressures ranging widely from 100 to 3,000 psig, depending on the requirement of the treatment; (3) a gas-liquid separation section; (4) the hydrogen recycle system; and (5) a liquid product stripper or stabilizer.

Liquid feed is preheated by exchange with the reactor effluent and brought to a controlled reactor-inlet temperature in a fired heater. A recycle hydrogen stream joins the feedstock. The amount of the hydrogen recycle stream is in excess of that required for the chemical reactions in order to suppress the accumulation of deactivating carbonaceous deposits on the catalyst. Some of the cold recycle hydrogen may be brought into the reactor at intermediate points in the reaction zone to serve as a heat-absorption medium, thereby making the temperature profile through the reactor more nearly isothermal than it would be without the cold-gas quench.

To provide hydrogen consumed by the reactions and by solution losses, a stream of fresh makeup hydrogen is brought into the system either before or after the recycle gas compressor and in sufficient quantity to maintain the desired pressure in the unit. In many hydrotreaters, particularly in units used to pretreat naphthas for catalytic reforming, the output of hydrogen from the reformer is sufficient and can be routed directly through the hydrotreater without need for a recycle compressor. In these cases, the once-through hydrogen separated from the cooled reactor effluent liquid is released from the hydrotreater by pressure control and routed to other units where it can be used or to the refinery fuel-gas system.

The cooled reactor effluent stream is brought to a separator vessel, where the recycle or net hydrogen is removed. The liquid is routed to a stripper or stabilizer, which functions to remove H_2, H_2S, NH_3, H_2O, and light hydrocarbons dissolved in the separator liquid. When relatively large amounts of gases are dissolved in the separator operated at plant pressure, there is an advantage in routing the liquid from the primary separator to another separator maintained at a lower pressure. In this latter unit, a portion of

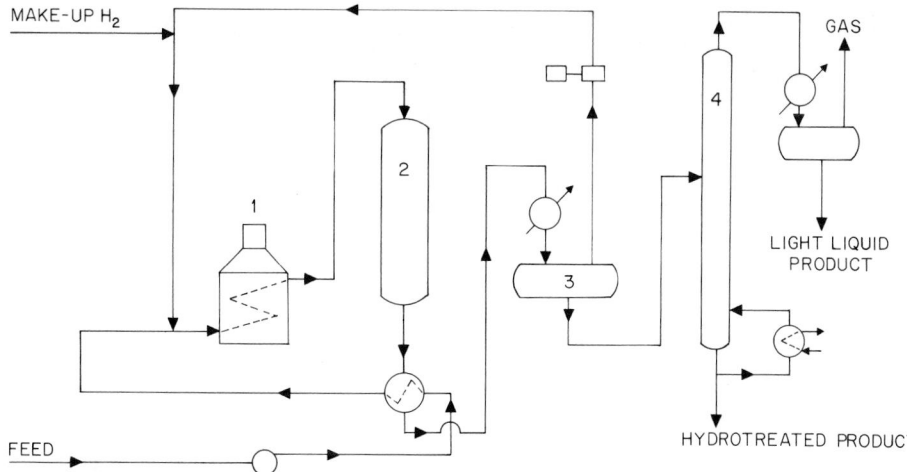

Fig. H-9. A representative hydrotreating unit. (1) Heater, (2) reactor, (3) separator, (4) stripper.

the dissolved gases is flashed from the liquid and removed at this point of lower pressure, thus relieving the load on the stabilizer. The stabilized hydrotreated liquid, free of dissolved, unwanted contaminants, is routed to subsequent processing or to product fuel blending.

Process Variables: The variables usually associated with fixed-bed catalytic units are common to most hydrotreating processes and include (1) reaction temperature, (2) pressure, (3) space velocity, and (4) hydrogen-to-hydrocarbon ratio. Other design factors are the reactor geometry, internal equipment, the mode of introducing temperature-quenching media into the reaction zone, and the manner of handling the reactor products.

The kinetics of all of the reactions involved in hydrotreating is improved as the *reaction-zone temperature* is increased. As the catalyst slowly deactivates during the course of a run, the reactor-inlet temperatures are raised by the operator to maintain a constant product quality from the standpoint of its content of residual sulfur or other unwanted material. Depending upon the nature of the feed, operating temperatures as low as 150°C (300°F) or as high as 455°C (850°F) are employed, but the usual operating range is in the region of 315°C to about 425°C (600–800°F).

Generally, hydrotreaters intended to process light feeds are designed to operate at *pressures* lower than those which will process higher-boiling feedstocks. Units that are designed to hydrogenate aromatics will run at higher pressures than those which are intended only to desulfurize or to remove nitrogen from feeds of comparable boiling range. Because high partial pressures of hydrogen suppress the formation of carbonaceous catalyst deposits, longer processing runs are obtainable where higher pressures are used— although at a higher cost.

Space velocity can be varied by the operator only by changing the feed rate within the limits of the ability of the equipment to handle these rates or by changing the loading of catalyst in the reactor(s) from one run to another. As space velocity is decreased (lower feed rate per unit volume of catalyst), other factors being constant, the extent of all of the reactions will increase if, of course, they have not already reached completion or chemical equilibrium. Since contaminants are more easily removed from light feeds, these units are designed to operate at higher space velocities than when heavier fractions are involved.

The net hydrogen produced by the reformer is sufficient for most units which pretreat naphthas prior to catalytic reforming so that the hydrogen can be passed once-through with the hydrocarbon feed through the naphtha hydrotreater and then cascaded to other hydrotreaters in the refinery. For heavier feeds and more difficult situations, a sufficiently high hydrogen partial pressure in the reaction zone must be maintained to aid the hydrogenation reactions and to suppress catalyst deposit formation. The designer employs a combination of pressure and hydrogen recycle rate to obtain the desired optimum hydrogen partial pressure.

On-stream processing periods for hydrotreaters range from several months to a year or more. Most catalysts are regenerated in situ by controlled combustion of the carbonaceous deposit with diluted air, using the hydrogen-recycle system to circulate the inert-gas–air mixture. Some regenerations are accomplished with a once-through steam-air mixture. Catalysts may be removed after regeneration and screened to remove fines before returning to the reactor.

Process Variations: There are over 25 different proprietary hydrotreating processes. Hydrocarbon feedstocks are processed in once-through operation, as shown in Fig. H-9, but in some instances there are significant modifications, involving such items as reactor temperature quenching, secondary flashing of reactor products, H_2S removal from the H_2 recycle gas, reactor internals for flow distribution, and two-stage reactor systems.

—Melvin J. Sterba, *UOP Process Division, Universal Oil Products Company, Des Plaines, Ill.*

Hydroxyalkyl Acrylates and Methacrylates The hydroxyalkyl acrylates and methacrylates are hydroxy functional esters of acrylic and methacrylic acid. Included in this summary are the following vinyl monomers:

$$CH_2\!\!=\!\!CHCOOCH_2CH_2OH$$

2-Hydroxyethyl acrylate (HEA)

$$CH_2\!\!=\!\!CHCOOCH_2\underset{\underset{\displaystyle OH}{|}}{C}HCH_3$$

2-Hydroxypropyl acrylate (HPA)

$$CH_2\!\!=\!\!CHCOOCHCH_2OH$$
$$\underset{\displaystyle CH_3}{|}$$

2-Hydroxy-1-methylethyl acrylate (HPA) (2-1)

$$CH_2\!\!=\!\!C(CH_3)COOCH_2CH_2OH$$

2-Hydroxyethyl methacrylate (HEMA)

$$CH_2\!\!=\!\!C(CH_3)COOCH_2\underset{\underset{\displaystyle OH}{|}}{C}HCH_3$$

2-Hydroxypropyl methacrylate (HPMA)

$$CH_2\!\!=\!\!C(CH_3)COOCHCH_2OH$$
$$\underset{\displaystyle CH_3}{|}$$

2-Hydroxy-1-methylethyl methacrylate (HPMA)

Some of the important properties of these monomers are given in Table H-9.

The most important property of the hydroxyalkyl acrylates is their ability to undergo free-radical addition polymerization to form high-molecular-weight polymers. These hydroxy functional monomers impart various properties to the polymeric system in which they are incorporated. By far the majority of the applications deal with their use in coatings and films. In most cases the monomers have been utilized to incorporate functionality into a polymeric structure, thereby allowing it to be cross-linked with suitable reagents. The most popular commercial method is to use substituted melamines, such as hexakis(methoxymethyl)melamine. As a class, the thermosetting acrylic resins offer performance superior to that obtainable with older established resins in a variety of applications in the field of industrial finishes. Typical finishes include automotive topcoats, and appliance and coil coatings. See also **Paints.**

In addition to industrial finishes prepared by solution-polymerization techniques, the hydroxyalkyl acrylates and methacrylates have been incorporated in latexes for the main purpose of preparing coatings of various types. Additional uses for one or more of these hydroxy monomers include the preparation of radiation-curable coatings, dispersing agents, embedding media for electron microscopy, inks, hydrophilic gels for medical uses, and water-soluble polymers for a number of different purposes, including paper additives. Reference 3 discusses the preparation, polymerization, and applications of these materials in considerable detail.

Since the hydroxy monomers usually contain appreciable amounts of acrylic or methacrylic acid, the monomers should be stored in stainless-steel, aluminum, or glass containers. Storage tanks should be of mild steel with a 5-mil coating of a baked phenolic resin. Although they are inhibited by the methyl ether of hydroquinone, there is evidence that acrylates and methacrylates polymerize under the influence of light. Therefore, it is suggested that these monomers be stored in the dark. Also, for proper action of the inhibitor, O_2 must be present in the monomer. The shelf life is suggested to be over 1 year in the absence of initiators or exposure to high temperatures or ultraviolet light.

Although the hydroxyalkyl acrylates and methacrylates can be prepared by a number of different routes, the method used commercially is the reac-

TABLE H-9. *Properties of the Hydroxyalkyl Acrylates and Methacrylates*

Property	Monomer			
	HEA	HPA	HEMA	HPMA
Molecular weight	116.06	130.08	130.08	144.10
Hydroxy ($\cdot$OH) content, theoretical percent of pure compound	14.65	13.08	13.08	11.80
Boiling point, °C [mm Hg]	65 [2]	90 [12][a]	85 [5]	82–84 [3][b]
Density, g/ml at 20°C	1.1076	. . .	1.0712	1.0316[b]
Flash point, Cleveland open cup, °C	104	100[c]	108	121[d]
Refractive index, 20°C	1.4545	. . .	1.4520	1.4439[b]
Glass transition temperature, °C (polymer)	−15	−7[c]	55	72[b]
Alfrey and Price copolymerization parameters:				
Q	0.85	0.87[c]	0.93	0.79[c]
e	0.68	0.75[c]	0.40	0.20[e]

[a] This value is for 2-hydroxypropyl acrylate.
[b] This value is for 2-hydroxypropyl methacrylate.
[c] This value is for a 75:25 mixture of 2-hydroxypropyl acrylate and 2-hydroxy-1-methylethyl acrylate.
[d] T. Alfrey, Jr., and C. C. Price, *J. Polym. Sci.,* vol. 2, p. 101, 1947.
[e] This value is for a 75:25 mixture of 2-hydroxypropyl methacrylate and 2-hydroxy-1-methylethyl methacrylate.

tion of ethylene or propylene oxide with the appropriate ethylenically unsaturated carboxylic acid.

$$CH_2{=}CRCOOH + CH_2CH_2 \rightarrow$$
$$\overset{\diagdown\,\diagup}{O}$$
$$CH_2{=}CRCOOCH_2CH_2OH$$

$$CH_2{=}CRCOOH + CH_2CHCH_3 \rightarrow$$
$$\overset{\diagdown\,\diagup}{O}$$
$$CH_2{=}CRCOOCH_2CHCH_3 +$$
$$\underset{OH}{|}$$
$$CH_2{=}CRCOOCHCH_2OH$$
$$\underset{CH_3}{|}$$

These reactions are carried out in the presence of a catalyst to promote ester formation and an inhibitor to retard polymerization. Typical catalysts are tertiary amines and their quaternary salts, Cr(III) ion, aluminum chloride, and Fe(III).

The hydroxyalkyl acrylates and methacrylates may be irritating to the eyes and skin as well as being absorbed through the skin. Since there is some evidence that HEA and HPA may cause allergenic responses in an occasional susceptible person, skin contact should be avoided.

See also **Acrylic plastics.**

REFERENCES

1. Dow Chemical Company: *Tech. Bull.* 114-284-70, Midland, Mich. (issued periodically).

2. Rohm and Haas Company: *Tech. Bull.* SP-216, Philadelphia, (issued periodically).
3. Nyquist, E. B.: R. H. Yocum, (ed.), in "Functional Monomers: Their Preparation, Polymerization and Applications," Dekker, New York, 1973.

—Edwin B. Nyquist, *The Dow Chemical Company, Midland, Mich.*

Hydroxycarboxylic Acids (See **Carboxylic acids.**)

Hydroxylamine (See **Aldehydes; Amines; Ketones; and Radicals.**)

Hygiene, Chemical (Consult the **Subject Index.**)

Hypalon (See **Elastomers.**)

Hyperon A hyperon is an unstable, subatomic particle the mass of which is between that of a neutron and a deuteron. A deuteron consists of a neutron and a proton.

Hypersorption (See **Adsorption.**)

Hypobromous Acid (See **Bromine.**)

Hypochlorites (See **Chlorine.**)

Hypochlorite Bleaches (See **Chlorine;** and **Detergents.**)

I

Ice Cream (See **Pasteurization.**)

Ideal Gas Boyle (1662) observed that "at constant temperature, the volume of a gas is inversely proportional to the pressure." Gay-Lussac (1801) discovered that "at constant pressure, the volume of a gas is directly proportional to its absolute temperature." A corollary to these two laws is that "at constant volume, the pressure of a gas is directly proportional to its absolute temperature." An ideal gas (which does not exist) would follow the foregoing laws precisely. Hydrogen and helium come the closest of the gases to meeting these requirements. Consequently, numerous empirical equations have been developed to portray the behavior of gases. The law of Gay-Lussac also is referred to as *Charles' law*. See also **Avogadro's number.**

Ideal Solution In an ideal solution, there is no special force of attraction between the components, and upon mixing there is no change in internal energy. Thus, in simply mixing two ideal-solution components, such as diluting one liquid with another, the nature of the liquids does not change by virtue of the mixing. Further, no heat is given off or absorbed. The volume of the resulting solution exactly equals the total of the volumes of the starting liquids. The resulting properties, including vapor pressure, fluidity, and refractive index, are weighted averages of the original liquids in proportion to the amount of each present.

Few liquids, in practice, yield ideal solutions. Toluene and benzene solutions closely approach the ideal. When the two substances are mixed, there is a very slight cooling effect. Compounds that are closely related structurally tend to remain essentially unchanged when mixed.

The concept of the ideal solution is fundamental to the thermodynamic study of mixtures.

Ignition Temperature (See **Fuels.**)

Ilmenite (See **Titanium;** and **Zirconium.**)

Imides (See **Amides and imides.**)

Impeller Mixers (See **Mixing, fluids.**)

Impingement Separators (See **Separation operations.**)

Incineration (See **Fuels.**)

Indium (From its indigo-blue emission lines.) **In** = 114.82 (at. wt.), 49 (at. no.). It has two stable isotopes with mass numbers of 103 (4.23%) and 115 (95.77%). Indium is located in group IIIA of the periodic table, which includes the elements boron, aluminum, gallium, and thallium. In the elemental form indium exhibits a tetragonal crystal structure.

Under ordinary conditions, indium is a silvery white metal which exists in only one stable crystalline form. It is one of the softest metals and can easily be cut with a knife. It is highly ductile and quite malleable. Like tin, it cries on bending. It is a superconductor at 3.37 K.

Some of its common physical properties are listed in Table I-1. See also **Chemical elements.**

Discovery: Indium is a relatively modern metal, discovered in 1863 by F. Reich and T. Richter as they were making a spectrographic examination of sphalerite ore for thallium. Instead of detecting thallium, they observed prominent indigo-blue spectral lines, which they attributed to a new element. This new element was named indium in view of the color prominence of its emission lines.

Occurrence: Indium is classified as a rare element, its crustal abundance being of the order of 0.1 ppm. Little is known of its geochemistry. It is distinctly chalcophilic in nature but may not be exclusively concentrated in sulfide minerals. There is apparently no reference relating to the observation of characteristic indium minerals. It is often found with sphalerite, in which it is believed to substitute for zinc, and also with cassiterite, wolframite, chalcopyrite, germanite, and coal. The major recovery of indium is as a byproduct from the treatment of zinc ores.

Recovery: Several schemes have been developed for the recovery of indium, and the choice depends on the product in which the indium is present. From a zinc calcine or zinc oxide fume it is recovered by acid leaching with dilute sulfuric acid, which removes the bulk of the zinc. The residue is then treated with a stronger sulfuric acid solution to dissolve the indium. The indium is precipitated from solution as the hydroxide. The precipitate is purified, and the indium is recovered by dissolving the precipitate in acid and sponging it out with zinc. The sponged metal may be further purified by dissolving it in hydrochloric acid. The solution is purified by adding barium chloride to remove the sulfates and hydrogen sulfide to remove the heavy metals. The indium is recovered by precipitation with aluminum or by electrolysis.

It can be recovered from the dross resulting from the smelting of lead-zinc concentrates by first treating it to form a complex slag containing indium. The slag is crushed and ground and the copper is removed by flotation. The tailings are reduced with coke, resulting in a lead bullion. This material is cast into anodes and treated electrolytically to produce a lead-tin cathode and indium-rich slimes. The slimes are roasted with sulfuric acid, and the indium is leached out with water. The solution is purified and the indium removed from the solution by sponging with zinc or aluminum. It is further refined by electrolysis.

When the metal is present in a crude zinc-lead metallic product, it is treated in the molten state with lead chloride and sodium chloride to form a chloride slag containing the indium. The slag is leached with dilute sulfuric acid, and the indium is sponged from solution with zinc dust. The indium sponge is melted, and the residual zinc is removed by chlorination. The indium may be refined further by electrolysis.

Chemical Properties: In its chemical behavior indium may exhibit valences of 1^+, 2^+, and 3^+. Its primary valence is 3^+; in some of its reactions it resembles zinc and in others aluminum and tin.

In the elemental form indium is quite stable to atmospheric exposure. It is attacked by mineral acids, the reactivity being more rapid with heat. It is relatively inert to alkalis and water. It will burn in air or oxygen with a blue flame to form indium sesquioxide, In_4O_6, more commonly referred to as indium trioxide.

Some of the indium compounds of interest follow:

Indium trioxide, In_2O_3, may be formed by burning indium in air or oxygen or by the decomposition of one of its salts, e.g., hydroxide, carbonate,

TABLE I-1. *Selected Physical Properties of Indium*

Density, g/cm³	7.31
Atomic volume, cm³/g atom	15.7
Melting point, °C	156.6
Boiling point, °C	2000
Specific heat at 20°C, cal/(g)(°C)	0.057
Thermal conductivity at 0°C, (cal)/(s)(cm³)(°C)	0.206
Coefficient of thermal expansion; in./(in.)(°C)	24.8×10^{-6}
Electrical resistivity at 22°C, Ω-cm	8.8×10^{-6}

or nitrate. It is pale yellow. In the amorphous form the oxide is acid-soluble; however, in the crystalline form it is insoluble in acids. Other oxides of indium have been reported, such as In_2O, InO, and In_3O_4.

Indium hydroxide, $In(OH)_3$, is formed by precipitating an indium salt with a base. It is a white gelatinous precipitate.

Indium sulfate, $In_2(SO_4)_3$, is produced by dissolving indium in sulfuric acid and evaporating the solution to dryness. It is a white crystalline solid which is deliquescent.

Indium sulfides, In_2S, InS, and In_2S_3, can be formed by direct combination of the metal with sulfur. In_2S_3, most common, is precipitated with H_2S from a weak acidic solution of an indium salt. The color of the sulfide, which depends on the conditions of preparation, varies from red to yellow.

Indium nitrate, $In(NO_3)_3 \cdot 3H_2O$, can be prepared by dissolving the metal or oxide in nitric acid and evaporating to dryness. Heating the trihydrate to $100°C$ will decompose it to the monohydrate.

Indium phosphate, $InPO_4$, is prepared by dissolving the metal or oxide in a suitable acid and precipitating it from solution with sodium phosphate.

Indium carbonate, $In_2(CO_3)_3$, is formed by dissolving the metal or oxide in a suitable acid and precipitating it from solution with ammonium or sodium carbonate.

Halides of indium are formed by the direct combination of the elements. Some properties of the halides are listed in Table I-2.

The mono- and dichlorides are unstable in water solutions and decompose to yield elemental indium and indium trichloride.

Although organic indium compounds are known, information on their preparation is sparse.

TABLE I-2. *Some Properties of Indium Halides*

Compound	Color	Mp, °C	Bp, °C
InBr	Red brown	220	662*
InBr$_2$	Pale yellow	235	632*
InBr$_3$	White to yellow	436	*
InCl	Yellow or dark red	225	550
InCl$_2$	White	235	550–570
InCl$_3$	White	586	
InI	Brown red	351	711
InI$_2$		212	
InI$_3$	Yellow	210	
InF$_3$		1170	>1200

*Sublimes.

Indium is quite soluble in many molten metals. It combines with nonmetallic elements to form such compounds as indium arsenide, indium antimonide, indium sulfide, indium selenide, indium telluride, and indium phosphide, all of which have interesting semiconductor properties.

Economy and Uses: Current free-world statistics on the production of indium are not available. In 1968, it was estimated at about 2 million troy ounces. Because of the association of indium with zinc ores, its availability can be influenced by the demand for zinc.

One of the early uses of indium was in the production of aircraft bearings, which were made from silver with a thin overlay of lead onto which indium was plated to improve corrosion resistance.

With the advent of semiconductor technology, indium became an important *p*-type element for doping germanium diodes and transistors. It also formed an interesting class of semiconductor compounds known as III-V compounds, which include InAs, InSb, and InP. As the oxide, it is used in electroluminescent panels.

Indium is alloyed with bismuth, lead, tin, and cadmium to form a variety of low-melting-point *fusible alloys* which can be used as forming alloys or solders. A eutectic alloy of indium-tin is a good solder for glass-to-glass or glass-to-metal seals. Copper-silver-indium and copper-gold-indium alloys are useful as brazing alloys, with a melting range of 700–800°C.

An alloy of silver, indium, and cadmium is used for control rods in nuclear reactors.

When indium is combined with mercury and thallium, the resulting eutectic alloy has a freezing temperature of $-63°C$ and is useful for switches, seals, and thermometers in low-temperature, arctic, and stratospheric applications.

Hygienic Considerations: Elemental indium has been handled for a number of years under normal industrial conditions without any effect on general health or incidence of dermatitis.

—S. C. Carapella, Jr., *American Smelting & Refining Company, South Plainfield, N.J.*

Inert Atmosphere   (See **Nitrogen.**)

Inert Gases   (See **Chemical elements.**)

Ingot Iron   (See **Iron and steel.**)

Inhibitors   (See **Antioxidants.**)

Initiators, Reaction (See **Peroxides, organic.**)

Injection Mixing (See **Mixing, fluids.**)

Injection Molding (See **Plastics.**)

Inks (See **Iron oxides, synthetic;** and **Waxes.**)

Inner Transition Elements (See **Rare-earth elements and metals.**)

Insecticides (Consult the **Subject Index.**)

Insulation, Thermal Thermal insulation has been defined as a material (or a mixture of materials or a composite structure) designed to reduce heat flow between boundary surfaces at specified temperatures. The effectiveness of a thermal insulation is judged on the basis of thermal conductivity, which depends on the physical structure of the insulation.

Consideration of thermal insulation for industrial use is divided for design purposes into high-temperature service and low-temperature service, depending upon the field of use. Industrial high-temperature insulation usually encompasses temperatures between ambient and 2000°F (1093°C). Refractory insulation design is used with higher temperatures. Industrial low-temperature insulation falls in the range of ambient to −100°F (−73.3°C). Cryogenic insulation design is used at lower temperatures.

Economic Thickness of Insulation. The costs involved in insulating are large enough to warrant careful selection of the type and quantity of insulation material. Economic thickness is the minimum annual value of the cost of heat loss plus the cost of insulation, i.e., that thickness of a selected insulation which will provide minimum heat loss while paying for itself within the assigned time period. Economic thickness can be estimated from established data, using tables provided by insulation manufacturers.

Heat loss is not always the main reason for insulating. Other factors may have equal or more importance, e.g., process control or improvement, prevention of condensation, personnel protection, fire protection, and noise reduction.

Fundamentals of Heat Transfer. Heat can flow through an insulation by several different mechanisms: (1) *solid conduction* through the materials making up the insulation and conduction between individual components of the insulation system across areas of contact; (2) *gas convection* in void spaces contained within the insulation material; and (3) *radiation* across these void spaces and through the components of the insulation.

Because these heat-transfer mechanisms operate simultaneously and interact with each other, it is not possible to superimpose the separate mechanisms to obtain an overall thermal conductivity. Since the thermal conductivity of an insulation is not strictly definable analytically in terms of variables such as temperature, density, and physical properties of the component material, it is useful to refer to an *apparent* thermal conductivity which is measured experimentally during steady-state heat transfer.

Overall Heat Transfer: In most of the steady-state heat-transfer problems encountered in practice, more than one of the heat-transfer modes is involved simultaneously. Therefore it is convenient to combine the various heat-transfer coefficients into an overall transmission coefficient in order to calculate the total heat transfer from the terminal temperatures.

The methods of calculating the overall coefficient of heat transmission require knowledge of the thermal conductivity and thickness of homogeneous elements, the thermal conductance of nonhomogeneous elements, surface conductances on both sides of the construction, and the conductances of any air spaces. Procedures for calculation may be found in Ref. 2.

A recommended practice for determination of heat gain or loss and surface temperatures of insulated pipe and equipment systems by use of a computer program is being developed by ASTM Committee C-16.

It must be emphasized that conductivity and conductance values are sometimes published in different units. Care must be taken in all calculations to determine that units are consistent. See Table I-3.

Heat-Transfer Symbols: U = overall coefficient of heat transmission or thermal transmittance (air to air); the time rate of heat flow usually expressed in Btu per (hour) (square foot) (Fahrenheit degree temperature difference between air on the inside and air on the outside of a wall, floor, roof or ceiling). The term is applied to the usual combinations of materials and also to single materials, such as window glass, and includes the surface conductance on both sides. This term is frequently called the U *value*.

k = thermal conductivity; the time rate of heat flow through a homogeneous material under steady-state conditions through unit area per unit

TABLE I-3. *Conversion Factors Used in Heat-Transfer Calculations*

Coefficients of Heat Transfer					
Btu/(hr)(ft²)(°F)	kcal/(hr)(m²)(°C)	cal/(s)(cm²)(°C)	W/(cm²)(°C)	W/(cm²)(°F)	hp/(ft²)(°F)
1	4.88	1.355×10^{-4}	5.68×10^{-4}	2.04×10^{-3}	3.94×10^{-4}
0.205	1	2.79×10^{-5}	1.16×10^{-4}	4.18×10^{-4}	8.07×10^{-5}
7380	36,000	1	4.19	15.1	2.01
1761	8,592	0.239	1	3.697	0.694
490	2,392	0.0664	0.278	1	0.193
2538	12,390	0.344	1.44	5.18	1

Thermal Conductivities				
Btu ft/(hr)(ft²)(°F)	Btu in./(hr)(ft²)(°F)	cal cm/(s)(cm²)(°C)	W cm/(cm²)(°C)	kcal m/(hr)(m²)(°C)
1	12	4.13×10^{-3}	0.0173	1.49
0.0833	1	3.44×10^{-4}	1.44×10^{-3}	0.124
242	2905	1	4.19	361
57.8	694	0.239	1	86.1
0.671	8.05	2.77×10^{-3}	0.0116	1

temperature gradient in the direction perpendicular to an isothermal surface. Its unit is Btu inch per (hour) (square foot) (Fahrenheit degree). Materials are considered homogeneous when the value of the thermal conductivity is not affected by a change in thickness or in area within the range normally used in construction. Some materials are not isotropic with respect to thermal conductivity. Care should be taken that the test method used is suitable for the particular material and gives a value of conductivity applicable to the intended use.

C = thermal conductance; the time rate of heat flow expressed in Btu per (hour) (square foot) (Fahrenheit degree average temperature difference between two surfaces). The heating, air-conditioning, and refrigerating engineer, however, dealing largely with compound walls with parallel surfaces, makes considerable use of the term *unit conductance,* or conductance per unit area. The average temperature is one which adequately approximates that obtained by integrating the temperature of the entire surface. The term is applied to specific materials as used, either homogeneous or heterogeneous, for the thickness or construction stated, not per inch of thickness.

The conductance of an air space depends on the temperature difference, the height, the depth, the position, character, and temperature of the boundary surfaces. Since the relationships are not linear, accurate values must be obtained by test and not by computation. The space must be fully described if the values are to be meaningful.

f = film or surface conductance; the time rate of heat exchange by radiation, conduction, and convection of a unit area of a surface with its surroundings. The surroundings must involve air or other fluid for radiation and convection to take place. The value is usually expressed in Btu per (hour) (square foot of surface) (Fahrenheit degree temperature difference). Subscripts i and o are usually used to denote inside and outside surface conductances respectively.

e = emissivity; the ratio of the total radiant flux emitted by a surface to that emitted by an ideal blackbody at the same temperature.

E = effective emissivity; the combined effect of the surface emissivities e of the boundary surfaces of an air space; the boundaries are assumed to be parallel and of large dimensions compared to the distance between them.

r = surface reflectivity; the ratio of the radiant flux reflected by an opaque surface to that falling upon it.

R = thermal resistance; the reciprocal of a heat-transfer coefficient, as expressed by $U, C, f,$ or e. Its unit is Fahrenheit degrees per (Btu) (hour) (square foot). For example, a wall with a

U value of 0.25 would have a resistance value of $R = 1/U = 1/0.25 = 4.0$.

Selecting Insulation. The following factors must be considered: (1) *thermal efficiency.* How much heat loss or gain can be tolerated? Insulations with a low k value have good efficiency. (2) *Operating temperature.* What is the maximum or minimum temperature? Some insulations can be used only up to 170°F (77°C); others are suitable for use beyond 2000°F (1093°C). (3) *Structural strength.* Will the insulation be required to carry any load or be subject to damage? (4) *Moisture resistance.* Will the insulation be used in a moisture-laden atmosphere? The water resistance of insulation ranges from poor to excellent. (5) *Chemical resistance.*

Types of Thermal Insulation Materials

Loose Fill and Cement: Loose-fill insulation consists of powders, granules, or nodules, which are usually poured or blown into cavities or other spaces. Insulating cement is a loose material mixed with water to obtain plasticity and adhesion, and troweled or blown wet on the surface and dried in place to serve as insulation. Both loose fill and insulating cement are especially suited for covering uneven and irregular surfaces. Application is by trowel or spray.

Flexible and Semirigid: Materials with varying degrees of compressibility and flexibility, generally blanket, batt, or felt insulation, are available in sheets and rolls of many types and varieties, both organic and inorganic. Coverings and facings may be fastened to one or both sides to serve as reinforcing, vapor barriers, reflective surfaces, or surface finishes. These coverings include combinations of laminated foil or plastic and paper, wire mesh, or metal lath. Thickness and shape of insulation may be of any dimension conveniently handled, although standard sizes are generally used. Application is by banding, impaling on welded pins or studs, or with adhesive.

Rigid: These materials are available in rectangular dimension called *block, board,* or *sheet,* preformed during manufacture to standard lengths, widths, and thicknesses. Insulation for pipes and curved surfaces is supplied in half-sections or segments with the radii of curvature to suit all standard sizes of pipe and tubing and greater diameters up to several feet. Application is by bonding, wiring, impaling on welded pins or studs, or with adhesive.

Reflective: Reflective material is available in sheets and rolls of single or multilayer construction and in preformed shapes with integral air spaces. Application is by mechanical means. Emissivity and reflectivity are the measure of effectiveness. The installed system must include uniform-width still-air spaces. See Table I-4.

Formed-in-Place: These are available as liquid components or expandable pellets, which may be poured or sprayed in place in order to form rigid or semi-rigid foam insulation. Fibrous materials mixed with liquid binders may also be sprayed in place.

Low-Temperature Service: Table I-5 lists materials suitable for low-temperature service, showing the temperature range of service for each material. In recent years cellular plastics have gained acceptance as low-temperature insulation. Other widely used materials include cellular glass and the mineral fibers. Cork and animal or vegetable fibers are now seldom used as industrial insulating materials. Because of the importance of moisture resistance for low-temperature service, emphasis on selection of material is changing.

TABLE I-4. *Emissive and Reflective Characteristics of Materials and Surfaces**

Material	Emissivity and Absorptivity	Reflectivity
Aluminum:		
Foil	0.02–0.05	0.95–0.98
Polished	0.02–0.045	0.955–0.98
Paint	0.40–0.65	0.35–0.60
Brass:		
Natural surface . .	0.06–0.07	0.93–0.94
Polished	0.05	0.95
Brick	0.93	0.07
Chromium, polished .	0.08	0.92
Copper:		
Polished	0.04	0.96
Oxidized black . .	0.78	0.22
Enamel, white	0.92	0.08
Glass, polished	0.95	0.05
Ice	0.96	0.04
Iron:		
Polished	0.242	0.758
Rusty	0.85	0.15
Monel, polished . . .	0.02–0.05	0.95–0.98
Nickel, polished . . .	0.11	0.89
Paint, ordinary	0.85–0.95	0.05–0.15
Paper:		
White	0.94	0.06
Roofing	0.91	0.09
Plaster	0.93	0.07
Wood	0.90–0.92	0.08–0.10

*Based on surface temperature in the range 70–100°F (21.1–38°C).

High-Temperature Service: High-temperature materials, listed in the lower portion of Table I-5, are grouped by their upper temperature limits. Insulations can be grouped by composition, shape, and end use. These classifications are somewhat arbitrary since there is no hard-and-fast separation and considerable overlapping occurs. Temperature limit seems to be the most practical way of listing. It is possible for these materials to be used over a range of temperature from ambient to the maximum shown. In choosing insulation materials, thermal properties undoubtedly come first in making comparisons. Low conductivity values are important, and it is assumed that the material will not disintegrate at temperatures up to the maximum shown.

Cryogenic Insulation Systems: A multilayer insulation system consists of many layers of alternate radiation-reflecting shields separated by low-conductivity spacers. This assembly is placed perpendicular to the flow of heat. Each layer contains a thin, low-emissivity radiation shield enabling the layer to reflect a large percentage of the radiation it receives from a warmer surface.

TABLE I-5. *Thermal Properties of Insulating Materials**

Low-Temperature Insulations

Temperature Range		Material Classification	Density, lb/ft^2	Thermal conductivity k, Btu in./(hr)(ft^2)(°F)
°C	°F			
−170 to 99°C	−275 to 210	Plastic foams, polyurethane . .	1.8 to 2.2	0.09 at −200°F; 0.17 at 100°F
−129 to 80	−200 to 175	Polystyrene	1.0 to 4.0	0.26 at 40°F; 0.28 at 75°F
−40 to 71	−40 to 160	Polyvinyl chloride.	4.5 to 26	0.26 at 75°F
−40 to 93	−40 to 200	Cellular rubber	3.5 to 20	0.24 at 25°F; 0.30 at 100°F
−46 to 871	−50 to 1600	Expanded silica (perlite)	4.0 to 10	0.33 at 0°F; 0.38 at 75°F
−240 to 427	−400 to 800	Glass (cellular)	9.0 to 18	0.36 at 25°F; 0.42 at 100°F
−157 to 982	−250 to 1800	Mineral fibers (rock, glass, or slag)	0.5 to 10	0.27 at 25°F; 0.30 at 100°F

High-Temperature Insulations

Accepted Maximum Temperature		Material Classification	Density, lb/ft^2	Thermal conductivity k, Btu in./(hr)(ft^2)(°F)
°C	°F			
1260	2300	Alumina-silica ceramic fiber	3 to 12	0.31 at 300°F; 0.82 at 1000°F
1204	2200	Fibrous potassium titanate	15 to 18	0.33 at 300°F; 0.52 at 1000°F
1038	1900	Diatomaceous silica (bonded with clay and asbestos)	24	0.67 at 300°F; 0.75 at 1000°F
982	1800	Mineral fiber (rock and slag)	16 to 24	0.59 at 400°F; 0.75 at 1000°F
871	1600	Expanded silica (perlite)	4 to 10	0.33 at 0°F; 1.13 at 1000°F
649	1200	Calcium silicate	11 to 13	0.41 at 300°F; 0.60 at 700°F
649	1200	Asbestos (amosite with binder)	15 to 18	0.37 at 200°F; 0.72 at 900°F
538	1000	Felted glass fiber (no binder)	4½	0.35 at 300°F; 0.71 at 800°F
427	800	Glass (cellular)	9 to 18	0.55 at 300°,F; 1.04 at 800°F
316	600	85% magnesia (magnesium carbonate with asbestos fiber)	11 to 12	0.44 at 300°F; 0.46 at 400°F
271	520	Processed Gilsonite†	40 to 44	0.6 to 0.8
204	400	Glass fiber (organic binder)	½ to 3	0.35 at 300°F
982	1800	Mineral-fiber cement (rock, slag, or glass)	24 to 30	0.55 at 200°F; 0.80 at 600°F
649	1200	Mineral-fiber fill (rock, slag, or glass)	10 to 12	0.26 at 100°F; 0.65 at 600°F

*Values of temperature, density, and conductivity are approximate. For specific design problems consult the insulation manufacturer for recommended design data.

† This material is used for underground insulation.

The radiation shields are separated from each other to attenuate heat transfer by radiation, to decrease heat transfer by solid conduction, and (with evacuation) to decrease heat transfer by gas conduction. When the designer selects materials for multilayer insulation, he tries to reduce each of the three possible heat-transfer modes.

For bulk storage of liquefied gases, where the economics of multilayer insulation may not be acceptable and where the effect of overall heat gain through the insulation system may be compensated for by refrigeration techniques, other insulation systems are in use. These include double-wall tanks having the annulus filled with fibrous or granular loose-fill insulation material and evacuated. Another system of tank storage uses tanks formed in the ground, utilizing frozen earth. This construction involves prefreezing a cylindrical shell of earth (up to 100 ft thick) in water-bearing strata to the required depth, excavating the earth from inside, and covering the hole with a ground-level gastight aluminum roof suitably insulated. The prefreezing takes about 6 months. The ice barrier so formed prevents leakage of the refrigerant and also serves as insulation.

Another design makes use of a concrete outer containment vessel buried in the ground. Mass insulation, ordinarily a cellular plastic, is applied to a suitable thickness on the interior of the concrete vessel. The primary containment for the liquefied gas is a stainless-steel inner tank composed of prefabricated components of stainless steel.

For cryogenic transfer piping under certain conditions it may be unwise to insulate. For example, a liquid-nitrogen transfer tube may be a bare pipe of thin-walled steel. The layer of frost which quickly covers such a siphon when in use provides adequate insulation. If powder or fiber insulation protected the pipe, it could become saturated with water from the atmosphere after one transfer and so lose its effectiveness. This could be avoided by providing an impermeable outer vapor barrier, but as the air that condenses on the outside of the cold liquid-nitrogen line will be rich in oxygen, the high flammability hazard of organic materials in contact with the condensed oxygen could result in explosions and fires.

Insulation Accessories: A considerable part of installed insulation cost covers the accessory materials and the application labor cost. Accessory materials include fasteners, both mechanical and adhesive; finishes, both exterior and interior; vapor-barrier coatings; jackets; weather-barrier coatings; joint sealants; lagging adhesives; membranes; and flashing compounds. The function of these accessories is to assist in the application and permanence of the installation and to protect the installed insulation from a variety of environmental factors which otherwise could affect performance and service life.

Finishes: Weather-barrier finishes for thermal insulation should meet specific design criteria including (1) resistance to deterioration by weather exposure; (2) resistance to specific environmental factors, such as water, chemicals, and mildew; (3) ability to hold up well within specified temperature limits; (4) ability to prevent accumulation of water within the insulation by water-vapor flow, which can change the heat-flow rate; (5) low flame spread, i.e., insusceptibility to ignition and consequent surface spread of flame; (6) availability in a variety of colors for certain applications: appearance, solar-heat load, and process identification; (7) compatibility with long-term service conditions, e.g., black finishes may cause softening of plastic foams from solar-heat gain; and (8) the necessary mechanical strength.

Mastic Finishes: Where mastic finishes are to be specified, it should be recognized that application has a significant effect on performance. In selecting mastic finishes, the following criteria should be evaluated: (1) application consistency, resistance to deformation or flow; (2) coverage; (3) flammability; (4) drying time; (5) shrinkage; (6) insulation compatibility; and (7) application temperature range. Selected properties of several types of mastic finishes are summarized in Table I-6.

The process temperature is first among the design factors to be considered in selecting an appropriate finish. If the process temperature is below ambient temperature for any significant time, it is essential that a vapor-barrier finish be applied to limit the migration of water vapor from the atmosphere into the insulation. Subsequent condensation and possible freezing of water within the insulation will increase the rate of heat flow through the insulation and can cause physical deterioration of the installed insulation. The use of insulation materials of low permeance should be considered in designing the vapor-barrier system.

On hot insulation, it is general practice to apply finishes which are sufficiently permeable to allow trapped water to escape as vapor without blistering the finish. Such finishes are known in the trade as *breathers*.

TABLE I-6. *Properties of Mastic Finishes*

Service Properties	Desirable Performance	Asphalt Emulsion Mastics	Asphalt Cutback Mastics	Copolymer Emulsion Mastic
Outdoor durability	10 years (min.)	5 years	10 years	5 years (min.)(est.)
Specific environmental resistance	Suitable for industrial petrochemical environment	Suitable	Suitable	Suitable
Temperature limits	−50 to 180°F (−46 to 82°C)	0 to 180°F (−18 to 82°C)	0 to 180°F (−18 to 82°C)	−50 to 180°F (−46 to 82°C)
Fire hazard (ASTM E84). .	Flame spread not exceeding 50	75 to 150	100 to 200	25
Water-vapor permeance rate (ASTM E96 and C355)	Vapor-barrier type: 0.5 perm max.; breather type: 3.0 perms, min.	1 to 3 perms	0.00 to 0.01 perm	3 perms freshly applied; changing to 0.5 within a few months
Color	Available in variety of colors	Black only	Black and aluminum	White and colors
Insulation compatibility . . .	Compatible with all types	Usually compatible	Usually compatible	Compatible
Stress-strain properties . . .	Suitable for industrial petrochemical environment	Poor to fair	Fair	Suitable (good)

NOTE: 1 perm = 1 grain per ft^2-hr-in.-Hg (test conditions must be stated).

Sheet materials are often used as weather barriers in hot-insulation service. Their unsealed joints permit the escape of water vapor from the insulation. Sheets are used primarily on tanks and piping where relatively smooth contours and uninterrupted areas make application economical. Impermeable sheet materials are not effective as vapor barriers unless all joints are permanently sealed. This can be accomplished with sealants. In critical designs adequate sealing of metal jackets requires field-welding techniques, the high cost of which usually dictates the choice of a mastic or coating as the vapor-barrier finish. The difference between mastics and coatings is one of thickness. Coatings are generally applied to a wet thickness of 15–30 mils. Mastics are semifluid materials applied in thicknesses of 30 mils and more. For heavy-duty outdoor protection of thermal insulation in industrial atmospheres, coatings have a limited life. It is good practice to use a thicker (mastic) finish for long-time protection of the insulation investment. Mastic finishes have the advantage of being applicable over rough and porous insulation surfaces, making them smoother and more pleasing in appearance than the thinner coatings.

REFERENCES

1. Ede, A. J.: "Introduction to Heat Transfer," Pergamon, Oxford, 1967.
2. American Society of Heating, Refrigerating and Air-Conditioning Engineers: "ASHRAE Handbook of Fundamentals," New York, 1967.
3. Glaser, P. E.: Thermal Insulation Systems, NASA SP-5027, Washington, 1967.
4. Tye, R. P. (ed.): "Thermal Conductivity," Academic, New York, 1969.
5. Malloy, J. F.: "Thermal Insulation," Van Nostrand Reinhold, New York, 1969.
6. "How to Determine Economic Thickness of Insulation," National Insulation Manufacturers Association, New York, 1961.

—Wayne P. Ellis, *Foster Division, Amchem Products, Inc., Ambler, Pa.*

Insulin (See **Peptides and proteins.**)

Intermediate An intermediate generally is considered to be a material (usually a chemical compound) that occurs somewhere in a chemical manufacturing process between the introduction of the basic raw materials and the creation of the final end products. When two or more separate chemical reactions are involved, the intermediate may be the product of one of the *between reactions* and serve as charge material for a subsequent reaction. For example, in the manufacture of aromatic polyester, several materials and reactions are required. The fundamental raw materials are nitric acid, xylene, methanol, and ethylene glycol. In one reaction, *p*-xylene and nitric acid yield terephthalic acid. The terephthalic acid then is esterified with methanol using sulfuric acid as a catalyst to yield dimethyl terephthalate. The dimethyl terephthalate then undergoes an ester interchange with ethylene glycol which yields bis(β-hydroxyethyl) terephthalate, later condensed to polyethylene terephthalate. This low-molecular-weight polymer then is polymerized to a high-molecular-weight polyethylene terephthalate. In this operation, terephthalic acid and dimethyl terephthalate can be regarded as intermediates. In some instances, a producer will procure intermediate materials from the outside rather than produce them in his own plant, particularly in the pharmaceutical and dye industries. Thus, a number of intermediates are high-tonnage items of commerce. Some intermediates are of low-tonnage requirement, and consequently the economics of materials and production sometimes favors the outside producer who can supply several users. A representative list of intermediates would include *o*-aminophenol-*p*-sulfonic acid, 2,6-dichloro-4-nitroaniline, 4-sulfophthalic acid, *o*-tolidine dihydrochloride, diphenylmethane, diphenylacetaldehyde, methyl cyclopentylphenylglycolate, and 2,3-dichloro-5,6-dicyanobenzoquinone. See also **Petrochemical complex.**

Iodine   (Gk. *ioeides,* violet.)   I = 126.9044 (at. wt.); 53 (at. no.). Iodine is a soft, black to violet orthorhombic crystalline solid of a metallic luster (although not metallic) with a melting point of 113.5°C and boiling point of 184.35°C. Thin crystals have a brownish-red appearance. Iodine can be precipitated as a dark-brown powder. The vapor is a beautiful violet color. Upon heating at normal pressures, solid iodine sublimes to the vapor form. Liquid iodine is red by transmitted light and brown by reflected light. Specific gravity = 4.93 at 20°C. Lattice constants are $a = 4.777$, $b = 7.251$, and $c = 9.773$. Specific heat = 0.052 cal/(g)(°C) at 20°C. Heat of fusion = 14.2 kcal/g, or 25.6 Btu/lb. Thermal conductivity = 10.4×10^{-4} (cal)(cm)/(s)(cm²)(°C) at 20°C. Electrical resistivity = 1.3×10^{15} $\mu\Omega$-cm at 20°C. Ionization potential = 10.44 eV. Valence values are 1, 5, and 7.

Iodine is a member of the halogen family of elements and is found in group VIIA (along with F, Cl, Br, and At) and in period 5 of the periodic table. Iodine has one stable isotope, 127, and 14 unstable isotopes, 122–126 and 128–136. The half-life of ^{129}I is very long, 1.72×10^7 years. See also **Chemical elements.**

Iodine was discovered in France by Courtois (1812) while studying kelp. There are several natural sources of iodine, including seawater (1,000 tons of iodine in 1 mi³), seaweed, salt brines, waste water of some oilwells, and Chilean saltpeter (once the primary source).

Iodine is only very slightly soluble in water but is soluble in alcohol, ether, benzene, chloroform, and several other organic compounds. Iodine is quite soluble in solutions of iodides. The molecular formula below 600°C is I_2; above 1500°C, dissociation is virtually complete, and the molecule comprises a single atom of iodine. Vapor pressure of solid iodine is 0.305 mm Hg at 25°C.

At warm temperatures, iodine reacts slowly with Pb and Ag but much more rapidly with most other common metals. In chemical reactions, iodine may play the role of oxidant (becoming I^-) or reductant (becoming IO_3^- or IO_4^-). Iodine oxidizes ferrocyanides to ferricyanides and reacts with HNO_3 to form HIO_3 and NO. H_3PO_2 and H_3PO_3 are oxidized to H_3PO_4. Iodine reacts with H_2S to form S and HI in the presence of moisture. H_2SO_3 is oxidized to H_2SO_4. Tetrathionates are produced from thiosulfates: $2Na_2S_2O_3 + I_2 \rightarrow Na_2S_4O_6 + 2NaI$. Iodine reacts directly with Cl_2 to form iodine chloride, ICl. With water present, HCl and HIO_3 are formed. Bromine also unites directly with iodine to form iodine bromide. The formation of a polyiodide, KI_3 (KII_2), results when iodine is combined with soluble iodides, for example, KI. Iodine usually is detected by the color of its solutions (violet in CCl_4 and yellow in H_2O). A blue color is formed when iodine is exposed to a cold solution of starch.

Representative compounds of iodine are given in Table I-7.

Uses: For scores of years, iodine has been used as an effective antiseptic. It is marketed commercially as tincture of iodine, which is a 7% alcohol solution of iodine in a 5% solution of KI. This form produces a mild burning of the skin and a stain. In another form 2% iodine is contained in an oil-water emulsion also containing lecithin. This concentration does not burn, and the mild stain washes off easily. There are several patented iodine-containing medicines, including deter-

TABLE I-7 *Representative Iodine Compounds*

Compound and Formula	Formula Weight	Sp gr	Mp,* °C	Bp, °C	Solubility in H₂O
Iodine monobromide, IBr	206.84	4.4	42	116	Decomposes
Iodine monochloride, ICl	162.38	3.2	27	97	Decomposes to HIO₃
Iodine pentafluoride, IF₅	221.92	3.5	−8	97	Decomposes
Iodine dioxide, IO₂	158.92	4.2	d 75	. . .	Decomposes to HIO₃
Iodine pentoxide, I₂O₅	333.84	4.8	d 300	. . .	Quite soluble
Sodium iodate, NaIO₃	197.92	4.3	d	. . .	Fairly soluble
Sodium iodide, NaI	149.92	3.7	651	1300	Quite soluble
Potassium iodate, KIO₃	214.02	3.9	560	d	Fairly soluble
Potassium periodate, KIO₄	230.02	3.6	582	. . .	Slightly soluble
Potassium iodide, KI	166.02	3.1	773	1330	Quite soluble
Potassium iodochloride, KCl · ICl₃	307.85	1.8	d	. . .	Decomposes
Triiodomethane (iodoform), CHI₃	393.80	4.1	s 119	. . .	Practically insoluble
Benzyl iodide, C₆H₅ · CH₂I	217.99	1.7	24	93	Insoluble
Hydriodic acid, HI	127.93	5.66	lq −51	−35	Very soluble

*d decomposes; lq liquid; s sublimes.

gents. Iodoform, CHI_3 (triiodomethane), made by reacting acetone or ethyl alcohol with sodium hypoiodite, once was a very popular antiseptic but now is used only in limited quantities. A number of iodo compounds are used as reagents and intermediates in organic syntheses. The diazo reaction is used to introduce iodine into benzenoid compounds. Iodine is used in tablet form for sterilizing drinking water in small quantities and causes less odor and taste than chlorine. Iodine is essential to proper cell growth in man and animals and exists in all cells of a normal body, the largest concentration being in the thyroid gland. Iodized table salt contains a small quantity of an iodine salt as a preventive of goiter and associated glandular disturbances. Iodine also is a supplement for cattle feeds. Iodine chemicals find use in photography and printing processes.

The McKechnie-Seybolt process for the production of vanadium metal is essentially the calcium reduction of vanadium pentoxide in the presence of iodine. The entire reaction is carried out in a steel bomb initially at about 700°C. The reaction produces vanadium metal, lime, and calcium iodide. An iodide process also is used in the production of high-purity zirconium.

Ion-Exchange Resins   Ion-exchange resins are insoluble solid acids or bases which have the property of exchanging ions from solutions. During the ion-exchange reaction, the ion-exchange resins are converted into insoluble acids, bases, or salts. Cation-exchange resins contain fixed electronega-tive charges which interact with mobile counterions having the opposite, or positive, charge. Anion-exchange resins have fixed electropositive charges and exchange negatively charged anions. Ion-exchange resins are three-dimensional macromolecules or insoluble polyelectrolytes having fixed charges distributed uniformly throughout the structure.

Most ion-exchange resins are used in fixed-bed processing equipment for softening and deionizing water. Equations for the removal of sodium chloride from water are:

Exhaustion, or service, step:

$$NaCl + RSO_3H \rightarrow HCl + RSO_3Na \quad (1)$$

Regeneration step:

$$2RSO_3Na + H_2SO \rightarrow Na_2SO_4 + 2RSO_3H \quad (2)$$

Equations (1) and (2) illustrate the reversible exchange of sodium ions for the hydrogen ion from the sulfonic cation-exchange resin. When the resin is depleted of hydrogen ions, it is regenerated with a dilute (5%) solution of H_2SO_4 [Eq. (2)]. The duration of the *service step* is usually a matter of hours; the *regeneration step* takes about 30 min.

The removal of NaCl from water (deionization) is completed by passage of the cation-exchanger effluent through a bed of anion-exchange resin.

Exhaustion, or service, step:

$$HCl + ROH \rightarrow HOH + RCl \quad (3)$$

Regeneration step:

$$RCl + NaOH \rightarrow ROH + NaCl \quad (4)$$

The effluent [Eq. (3)] which is free of NaCl is deionized water. The resin is prepared for the next service cycle by treatment with a 5% NaOH solution [Eq. (4)].

Most ion-exchange resins used by industry today are manufactured from uniform spheres of styrene-divinylbenzene (DVB) copolymers having diameters 0.3–1.0 mm (20–50 mesh, U.S. standard screens). The copolymer beads are formed by *pearl polymerization* and converted to ion-exchange resins by a second processing step. Sulfonic-type cation-exchange resins are made by sulfonation of the copolymer beads at elevated temperatures. Strong-base anion-exchange resins are produced by means of chloromethylation and amination of the copolymer spheres. A typical bead-form styrene-DVB anion exchanger is shown in Fig. I-1.

Desirable properties of ion-exchange resins include (1) complete insolubility in water and solvents to prevent imparting tastes, odors, or color bodies to the solution being treated; (2) high exchange capacity per volumetric unit, with high regenerant efficiency; (3) rapid and complete exchange with counterions; (4) good chemical stability to prevent degradation by oxidizing and reducing agents; (5) resistance to osmotic shock to prevent loss in use by physical breakdown; and (6) low initial cost.

Early Development: The first ion-exchange resins were described by Adams and Holmes,[1] a water-treatment expert and polymer chemist respectively, of the British Chemical Research Laboratory (1935). These ion-exchange resins were condensation products of phenol and formaldehyde. The granular-type cation-exchange resin contained sulfonic groups, and the anion exchanger contained aromatic amine groups. They

Fig. I-1. Photomicrograph of strong-base anion-exchange resin (40×).

are termed *strong-acid* and *weak-base* ion exchangers. A number of condensation-type ion-exchange resins were manufactured during 1935–1945. The first commercial deionization system was installed in 1939.

The next important step in ion-exchange resin technology was the synthesis of sulfonated styrene-DVB cation exchangers.[2] Commercial quantities of strong-base styrene-DVB anion exchangers appeared in 1948.[3] The first anion exchangers, the weak-base type, removed only strong mineral acids from water, such as HCl and H_2SO_4. The strong-base materials remove all acids, thus paving the way for production of water of equal or better quality than distilled water and at a much lower cost. The combination of the styrene strong-acid and strong-base exchange resins in a single tank (the mixed-bed deionizer), commercialized in 1949[4], produces water containing just a few parts per billion of dissolved salts and at a very low operating cost. The mixed-bed process produces ultrapure water from most fresh-water supplies at a fraction of the cost of distillation. This is the basic method used for central-station high-pressure boilers (5,500 psig) in the power industry and for applications in the electronics, chemical, and pharmaceutical industries.

Classification: A classification of ion-exchange resins by type, active exchange group, and configuration of the active group on the polymer is given in Table I-8.

The chemical behavior of ion-exchange resins is governed by the nature of the active exchange groups. The acid or basic strength of ion-exchange resins is determined by means of an acid-base titration. Strong-acid and strong-base ion-exchange resins have titration curves similar to H_2SO_4 and NaOH respectively. Weak-acid and weak-base ion-exchanger titration curves are very close to those of CH_3COOH and NH_4OH respectively.

The hydrogen-form strong-acid and the hydroxyl-form strong-base anion exchangers convert a solution of a neutral salt into the corresponding acid and base, while the ion-exchange resins are converted to the salt form.

$$RSO_3H + NaCl \rightarrow RSO_3Na + HCl \quad (5)$$
$$ROH + NaCl \rightarrow RCl + NaOH \quad (6)$$

Weak-acid and weak-base ion exchangers react with strong and weak bases and acids but do not split neutral salts.

$$RCOOH + NaOH \rightarrow RCOONa + HOH \quad (7)$$
$$RNH_3OH + HCl \rightarrow RNH_3Cl + HOH \quad (8)$$

TABLE I-8. *Classification of Ion-Exchange Resins*

Type	Active Group	Typical Configuration
Cation-Exchange Resins		
Strong acid	Sulfonic acid	⬡—SO_3H
Weak acid	Carboxylic acid	～CH_2CHCH_2～ 　　　│ 　　$COOH$
Weak acid	Phosphonic acid	⬡—$PO(OH)_2$
Anion-Exchange Resins		
Strong base	Quaternary ammonium	⬡—$CH_2N(CH_3)_3Cl$
Weak base	Secondary amine	⬡—CH_2NHR
Weak base	Tertiary amine (aromatic matrix)	⬡—CH_2NR_2
Weak base	Tertiary amine (aliphatic matrix)	—$CHCH_2NCH_2$ 　│　　　│ 　OH　　CH_2 　　　　　│

Ion-exchange reactions are generally reversible and are analogous to reactions which occur in solution. When a cation-exchange resin with A as its counterion is in a solution containing B cations, the reaction is

$$R^-A^+ + B \rightarrow R^-B^+ + A^+ \qquad (9)$$

where R is the cation-exchange resin.

After equilibrium is established according to the law of mass action, the reaction is

$$K_C = \frac{[\underline{B^+}][A^+]}{[\underline{A^+}][B^+]} \qquad (10)$$

A bar under the ion represent the ion in the resin phase; the absence of the bar indicates the ion in solution; the brackets indicate activities. Activity coefficients of ions in the resin phase cannot be precisely determined, and K_C is not constant with change in ionic concentration. The value K_C is considered a *selectivity coefficient* rather than an equilibrium constant. K_C is a useful measure of ion affinity. Values of K_C generally follow this order for strong-acid and strong-base ion exchangers: (1) divalent ions are preferred over monovalent ions, and (2) higher-molecular-weight ions are preferred over lower-molecular-weight ions of equal valence.

Applications: All ion-exchange-resin applications are based upon the same fundamental principles. Applications may be assigned to one of five categories: (1) transformation of ionic constituents, (2) removal of ionic impurities, (3) concentration of ionic substances, (4) fractionation of ionic substances, and (5) miscellaneous.

Transformation of Ionic Constituents: Softening water with the sodium form of a cation-exchange resin is the prime example of transformation of ionic constituents. Calcium and magnesium ions (hardness) occur in all freshwater supplies, forming objectionable scale and precipitates in boilers, laundries, and home appliances. These ions react with soap to form Ca and Mg stearates and reduce the effectiveness of detergents. The softening process is accomplished by passing the hard water through a vessel containing the Na form of cation-exchange resin. When the Na ions on the resin are depleted by exchange with Ca and Mg ions, the resin is regenerated with a 10% solution of NaCl and rinsed, and the softening cycle is repeated. Today's styrene-DVB cation exchangers have outstanding stability for this application, there being many examples of resins in use for as long as 20 years. See also *Water treatment.*

Other examples of this type of reaction are the conversion of the antibiotic streptomycin sulfate to its corresponding chloride by means of anion exchange, the exchange of Na ions in milk for the K ion, and the conversion of Na_2CrO_4 to H_2CrO_4 by cation exchange. The latter process is used extensively in the plating industry to concentrate H_2CrO_4 from rinse waters, with subsequent reuse of a toxic chemical and reuse of the rinse water in what might be termed a *closed system.*

Removal of Ionic Impurities: The major use of various combinations of ion-exchange resins under this category is the deionization of water for many purposes. Municipal and industrial water supplies contain dissolved salts, such as Ca, Mg, $NaHCO_3$, chlorides, and sulfates, which must be removed before use. Deionized water must be used in supercritical boilers to prevent scale formation. Deionization is also used to remove dissolved silica from boiler feedwater and condensate. At operating pressures above 1,000 psig, silica is carried over with the steam and condenses on turbine blades in power plants, causing a marked reduction in efficiency. Condensate purification at flow rates of 50 gal/(ft²)(min) of ion-exchange-resin bed area is used in most power plants having high-pressure boilers to reduce the Na, Fe, Cu, and silica concentrations to less than 50 ppb total contaminants.

A number of aqueous solutions of organic and inorganic chemicals are purified commercially. Dissolved salts, acids, bases, and color bodies are removed by ion-exchange resins (see Table I-9).

Concentration of Ionic Constituents: Ion exchange is successfully applied to concentrate electrolytes from dilute solutions with subsequent elution by a more concentrated regenerant solution to obtain a more concentrated solution of the electrolyte. An example is the recovery of H_2CrO_4 from rinse waters in the metal-finishing industry. The rinse waters are passed through a two-bed strong-base

TABLE I-9. *Chemicals Purified by Ion Exchange*

Chemical	Materials Removed	Ion-Exchange Method*
Formaldehyde . .	Formic acid	AE
Methanol	Ammonia	CE
Glycerin	Salts, acids, color	MB
Sorbitol	Salts, color	MB
Gelatin	Salts, color	MB
Sugars (sucrose, dextrose, lactose)	Salts, acids, color	MB
Citric acid	Acids, salts, color	CE, AE
Uranium	Ionic impurities	AE
Chromic acid . . .	Heavy-metal ionic impurities	CE
Copper	Ionic impurities	CE

*AE anion exchange; CE cation exchange; MB multibed ion exchange.

deionizer, and the deionized water is recycled to the plating system. The H_2CrO_4 is recovered as Na_2CrO_4, which is converted to H_2CrO_4 by treatment with the hydrogen form of a cation exchanger. Similar exchange processes are used to recover heavy and noble metals, such as Cu, Ni, Pt, and Au.

Fractionation of Electrolytes: Ionic species with opposite charges can be separated with either cation- or anion-exchange resins. The separation of ionic species of the same charge is possible if differences exist in acidic or basic strength, valence of the ion, or ionic radius. Examples of fractionation of electrolytes practiced on a commercial scale include (1) removal of a strong acid from an organic acid, e.g., sulfuric acid from citric acid; (2) ion-exchange chromatography to produce pure rare earths from a mixture; and (3) concentration of copper and cobalt from dilute solutions and their fractionation by using a carboxylic-type cation exchanger, an example of concentration and fractionation done at the same time.

Miscellaneous Applications: In the four categories just discussed the exchange of ions is common to all applications. It should be stressed that ion-exchange resins are reactive but insoluble acids, bases, and salts. These properties are used to advantage on an industrial scale for the adsorption of acidic and basic gases from gas streams. Gases which form an acid or a base with water can be removed by cation- or anion-exchange resins. Examples are SO_2, NH_3, CO_2, and H_2S.

Ion-exchange resins have been used as catalysts for a number of years, some of the advantages being (1) that catalyst-free products are obtained by means of simple filtration, (2) that catalyst can be reused for a number of cycles, (3) that continuous production is possible by passage of the solution through a bed of the material, and (4) that side reactions usually are kept to a minimum. Examples of ion-exchange resin catalysts are (1) sucrose inversion by means of the hydrogen form of a sulfonic-type cation exchanger, (2) ester hydrolysis with sulfonic-type cation exchangers, and (3) epoxidation of fats and oils with RSO_3H-type cation exchangers.

Ion Exclusion: Another process that uses ion exchange resins without exchanging ions is *ion exclusion,* as reported by Bauman.[5] This process uses an ion-exchange resin having high exchange capacity which excludes free electrolytes from the inner phase. Low-molecular-weight soluble nonelectrolytes distribute themselves equally between the resin and solution phases. If the solution of an electrolyte and a nonelectrolyte is passed through a column of a sulfonic-type cation exchanger (such as Dowex 50-W, 50–100 mesh) whose exchangeable ions are the same as in the electrolyte, the nonelectrolyte will not be retarded and the electrolyte, or salt, will be *excluded.* The effluent from such a column can be collected as a relatively pure salt solution, followed by a solution of the mixture and then by a pure product cut. The middle cut usually is recycled. Ion exclusion has the advantage of separating nonionic materials from ionic species without the use of chemicals for regeneration.

Examples of separations that are possible by ion exclusion are:

Ionic	Nonionic	Resin
HCl	Acetic acid	RSO_3H
Salt	Ethanol	RSO_3Na
Salt	Glycerin	RSO_3Na
Salt	Sucrose	RSO_3Na

The ion-exclusion process is particularly suited to sugar processing, e.g., sucrose recovery from molasses.

REFERENCES

1. Adams, B. A., and E. L. Holmes: *J. Soc. Chem. Ind. (Lond.),* vol. 54, pp. 1-6T (1935); British Patents 450,308 and 450,309 (1936); U.S. Patent 2,151,883 (1939).
2. D'Alelio, F. G.: U.S. Patent 2,366,077 (1944)(to General Electric Company).
3. McBurney, C. H.: U.S. Patent 2,591,573 (1952)(to Rohm and Haas Company).
4. Stromquist, D. M., A. C. Reents, and M. E. Veerman: U.S. Patent 2,771,424 (1956)(to Illinois Water Treatment Company).
5. Bauman, W. C.: U.S. Patent 2,684,331 (1954)(to The Dow Chemical Company).

—A. Curtis Reents, *Techni-Chem, Inc., Cherry Valley, Ill.*

Ion Exclusion   (See **Ion-exchange resins.**)

Ion Pair   A pair of ions resulting from the splitting of an atom or molecule into two charged fragments.

Ionization   [See **Amino acids; Atomic structure; Molecule;** and **pH (hydrogen-ion concentration).**]

Ionomer Resins   One of the later-developed synthetic resins, ionomer resins contain ionized carboxyl groups that form ionic cross-links in the intermolecular structure of these materials. The

Fig. I-2. Luncheon-meat packaging operation in which ionomer-resin film and three other plastic films combine to make the package. (*Du Pont.*)

TABLE I-10. *Properties of Various Grades of Ionomer Resins**

Grade†	Sp gr	Tensile Strength, 1,000 psi	Yield Strength, 1,000 psi	Elongation, %	Heat-Deflection Temperature at 66 psi		Refractive Index	Dielectric Constant (100 kHz–1 MHz)	Water-Vapor Transmission, g/(24 hr) (100 in.²)(mil)(100°F) at 90% RH
					°F	°C			
General-purpose film	0.94	5.0	1.8	450	113	45	1.51	2.4	1.5–2.0
Industrial film	0.95	4.0	1.8	450	...	...	1.51	...	1.5
Extrusion coating	0.94	4.8	2.4	400	...	...	...	...	2.0
Laminating	0.95	4.0	1.8	450	110	43	...	2.4	1.4
Sheet and blow molding . .	0.95	4.5	2.3	350	114	46	...	...	1.4
General-purpose molding . .	0.94	3.4	1.6	400	112	44	...	2.4	
High-clarity molding	0.95	4.2	2.0	350	108	42	1.51		
Electrical	0.96	3.8	2.4	450	110	43	...	2.36	

*DuPont Surlyn.
†All grades are slowly attacked by acids, show a high resistance to bases, swell slightly upon exposure to hydrocarbons, and are resistant to ketones and esters. All grades have a high resistance to vegetable and animal oils and a good resistance (below 50°C) to mineral oil.

special properties of the ionomer resins result from the ionic interchain forces that are clustered between the long-chain molecules of the polymer's structure. As expected, the resins have properties usually associated with a cross-linked structure. Of interest, however, is the fact that these polymers can be processed like other thermoplastic resins at conventional temperatures. Particularly desirable properties include good flexibility, adhesion and oil resistance, toughness, and high transparency. Their main limitation is the rather low upper-use temperature range (71–82°C, 160–180°F), beyond which there is an increase in creep rate and rapid decrease in modulus.

The materials are available in several grades, as shown by Table I-10, which lists key properties. Major applications of ionomer resins include film lamination and coextrusion to form composite structures for packaging food products. Shown in Fig. I-2 is a luncheon-meat vacuum-packaging operation* that combines an ionomer-resin film with polyester film for the flexible web of the package and polyvinyl chloride and polyvinylidene films for the rigid web, an interesting illustration of how several plastic materials can be combined into one product.

Iridium   (Gk. *iris,* rainbow.) **Ir** = 192.2 (at. wt.); 77 (at. no.). Iridium is a member of the family of platinum metals, group VIII of the periodic table. The melting point = 2442°C; boiling point = 5300°C; density at 20°C = 22.54 g/cm^3.

There are 2 stable isotopes of Ir, 191 and 193. There are 10 unstable isotopes, 187–190, 192, and 194–198. The element was identified and named by Tennant (England) in 1804.

Metallic Ir is not attacked by any mineral acid unless it is very finely divided. When fused with Na_2O_2 or an alkaline oxidizing flux, water-soluble iridates(IV) are formed. The finely divided metal is oxidized by air or O_2 at red heat to the dioxide, which decomposes into its elements at higher temperature. The valences of Ir are 1–6, the 3 and 4 valences being most common.

Iridium black is only slightly soluble in aqua regia. When fused with alkalies and alkaline nitrates or Na_2O_2, the metal is converted to an acid-soluble form. The metal at red heat reacts to a small extent with O_2, S, and P. At elevated temperature, the metal is attacked by Cl_2 and F_2. When fused with NaCl and treated with Cl_2, the

water-soluble sodium hexachloroiridate(IV), Na_2IrCl_6, is formed.

Iridium(III) hydroxide is a yellow-green or blue-black compound soluble in alkali and insoluble in water. It is made by adding KOH to a solution of potassium hexachloroiridate(III), K_3IrCl_6, in an inert atmosphere. When the trihydroxide is heated, a mixture of iridium(IV) oxide and the metal is formed. Iridium(III) oxide, Ir_2O_3, is made by fusing potassium hexachloroiridate(IV) with Na_2CO_3 and then leaching the mixture with water. At about 1100°C, both the oxide and the hydroxide decompose into the metal and O_2. When a solution of Ir is heated with $NaBrO_3$ at a pH of approximately 6, the dark-blue precipitate $Ir(OH)_4$ or $IrO_2 \cdot H_2O$ is formed. This water-insoluble compound, when heated to 350°C in N_2, loses its H_2O and is converted to the black oxide, IrO_2.

When iridium(III) chloride is heated in Cl_2 at 773–798°C, iridium(I) chloride is formed. The copper-red crystals are insoluble in acids and alkalies. The compound sublimes in Cl_2 at 790°C and decomposes into Cl_2 and metallic Ir. Iridium(II) chloride is stable from 763 to 773°C. The brown crystals are insoluble in H_2O, acids, and alkalies. Iridium(III) chloride is an insoluble green compound made by reacting the elements at about 600°C. The reaction is catalyzed by CO. Iridium(III) chlorides are prepared by reducing the corresponding iridium(IV) chlorides with oxalate or SO_2. Iridium(III) bromide is made by dissolving iridium(III) hydroxide in HBr. The blue solution yields olive-green crystals of $IrBr_3 \cdot 4H_2O$. When heated, the anhydride is formed. The triiodide is formed in analogous fashion as a trihydrate. Iridium(IV) chloride can be made in solution by the action of Cl_2 or aqua regia on ammonium hexachloroiridate(III). The relative insolubility of ammonium hexachloroiridate(IV), $(NH_4)_2IrCl_6$, makes it useful in the purification of Ir. The compound may be reduced in H_2 to the metal. The analogous sodium salt is very soluble, and the potassium salt is relatively insoluble.

Iridium(VI) fluoride is made from the element at 300–400°C. The bright yellow solid melts at 44°C and boils at 53°C. Potassium hexafluoroiridate(V) also has been prepared. Iridium fluoride can be made by heating the hexafluoride with the metal in a sealed tube at 150°C or heating it with glass above 200°C, at which temperature the glass reduces it to the tetrafluoride. This yellow solid melts at 106–107°C and boils above

300°C. Iridium(III) fluoride is formed by reducing the tetrafluoride with glass for 12–18 hr at 430–450°C.

Iridium(II) sulfide is formed by burning the metal in sulfur or by heating a higher sulfide at 700°C in N_2. This black solid is insoluble in H_2O, acids, and aqua regia. Iridium(III) sulfide, Ir_2S_3, is formed as a brown-black insoluble precipitate by passing H_2S through a hot acidic solution of an iridium(III) chloride. The precipitation is usually not quantitative. This amorphous black solid is not attacked by HNO_3 but is slowly dissolved by aqua regia or fuming HNO_3. The brown insoluble iridium(IV) sulfide, IrS_2, is partly formed by treating a tetravalent Ir solution with H_2S; when it is prepared in this way, some iridium(III) sulfide also is formed by reduction. Iridium(III) sulfate can be formed by dissolving iridium(III) hydroxide in H_2SO_4 in the absence of air. Trivalent iridium forms numerous cationic and anionic complexes in which it has a coordination number of 6. The amines are extremely stable and, once formed, difficult to destroy. Tetravalent Ir also forms complex ions, but to a lesser extent.

Further information concerning the sources, refining, and uses of Ir will be found under **Platinum metals and platinum.**

—Lionel E. Simmons, *Simmons Refining Company, Chicago, Ill.*

Iron (OE., *iver.*) **Fe** (L. *ferrum*) = 55.847 (at. wt.); 26 (at. no.). Pure iron is a silvery-white metal capable of taking a fine polish. Iron is found in group VIII in the periodic table, along with Ru, Os, Co, Rh, Ir, Ni, Pd, and Pt. There are nine isotopes of iron, 52–60. Four of these isotopes have fairly short half-lives, from 8.9 min (53) to 2.94 years (55), while 54, 56, 57, and 58 are fully stable. The half-life of ^{60}Fe is approximately 3×10^5 years.

There are three allotropic forms of iron: (1) *alpha iron,* which exists below 769°C (1416°F); (2) *gamma iron,* between 906 and 1404°C (1663 and 2559°F); and (3) *delta iron,* between 1404 and 1536°C (2559 and 2797°F).

Pure iron has a melting point of 1536°C (2797°F) and a boiling point of approximately 3000°C (5432°F). The density of iron is 7.874 at 68°F. Iron has valence numbers of 2^+ (-ous) and 3^+ (-ic). Iron, the hardest of the ductile metals, is surpassed only by Co and Ni in tenacity. Pure iron is attracted by a magnet but does not retain magnetism. Permanent-magnet materials are made of iron alloyed with Si, Co, Mn, Cr, Al, Cu,

TABLE I-11. *Properties of Iron*

Atomic radius, cm	1.71×10^{-8}
Atomic volume, cm³/g atom	7.1
Boiling point:	
°C	3000 ± 150
°F	5430 ± 270
Crystal structure at 25°C	fcc
Lattice edge, cm	2.8664×10^{-8}
Density:	
Liquid at 1564°C, g/cm³	7.00
lb/in.³	0.253
Solid at 20°C, g/cm³	7.874
lb/in.³	0.284
Electrical conductivity, volume, % of annealed Cu at 20°C	17.75
Electrical resistivity, $\mu\Omega$-cm:	
At 0°C	8.9
20°C	9.71
Temperature coefficient of electrical resistance (0–100°C)	0.65×10^{-4}
Electrochemical equivalent, mg/(s)(A) abs.:	
Fe^{++}–Fe	0.1929
Fe^{3+}–Fe	0.2893
Electrode potential (standard hydrogen scale) at 25°C, V	-0.44
Electron arrangement	2, 8, 14, 2
Electronic structure	$1s^2, 2s^2, 2p^6, 3s^2$ $3p^6, 3d^6, 4s^2$
Emissivity at 0.65 μm, %	40
Heat of combustion, cal/g atom	88,355
Ionization potential, V:	
First	7.90
Second	16.18
Third	30.65
Latent heat of fusion, cal/g	65.5
Latent heat of vaporization, cal/g	1598
Magnetic susceptibility at 820°C, g⁻¹	$1,000 \times 10^6$
Mechanical properties at 25°C (99.9% Fe):	
Brinell hardness	82–100
Percent elongation	30–40
Yield strength, psi	10,000–20,000
Tensile strength, psi	30,000–40,000
Melting point, °C	1536
Modulus of elasticity, psi	28.5×10^6
Modulus of rigidity, psi	11.64×10^6
Poisson's ratio	0.28
Reflectivity for light from tungsten filament (1% carbon), %:	
2500Å	38
Specific heat at 25°C, cal/(°C)(g atom)	6.55
Surface tension at 1550°C, dynes/cm	1,835–1,865
Thermal conductivity at 0°C, (cal)(cm)/(s)(cm²)(°C)	0.18
Thermal expansion, linear coefficient:	
cm/(cm)(°C)	11.76×10^{-6}
in./(in.)(°F)	6.53×10^{-6}
Average coefficient of linear expansion at 77°F:	
cm/(cm)(°C)	12.3×10^{-6}
in./(in.)(°F)	6.83×10^{-6}
Thermal-neutron-absorption cross section, barns	2.53
Valence	2, 3
Viscosity, cP:	
At 1743°C	4.45
1390°C	7.85

TABLE I-12. *Properties of Representative Iron Compounds*

Compound, Formula, and Formula Weight	Sp gr	Characteristics
Ferric acetate (basic) $FeOH(C_2H_3O_2)_2$ 190.89	. . .	Textile and leather dyeing; wood preservative; medical uses
Ferric ammonium oxalate $(NH_4)_3Fe(C_2O_4)_3$ 373.96	1.78	Monoclinic green crystals
Ferric chloride $FeCl_3 \cdot 6H_2O$ 270.20	. . .	Brownish yellow, very deliquescent crystals; melts at 37°C; boils at 280–285°C; used in photography; as an oxidizing agent; as a reagent in organic synthesis; very corrosive
Ferric citrate $FeC_6H_5O_7 \cdot 3H_2O$ 298.93	. . .	Reddish-brown scales; soluble in water; used for reproduction papers
Ferric nitrate $Fe(NO_3)_3 \cdot 6H_2O$ 349.96	. . .	Cubic crystals; melts at 35°C; soluble in water
Ferric oxalate $Fe_2(C_2O_4)_3$ 899.61	. . .	Amorphous; decomposes at 100°C
Ferric oxide (hematite) Fe_2O_3 159.68	5.26	Reddish brown to black; melts at 1565°C; insoluble in water; a major iron ore and pigment
Ferrous acetate $Fe(C_2H_3O_2)_2 \cdot 4H_2O$ 245.95	. . .	Needles; very soluble in water; solutions sometimes referred to as *iron liquor;* used as a mordant in dyeing
Ferrous bromide $FeBr_2$ 215.67	4.64	Green-yellow crystals; very soluble in water; decomposes before melting
Ferrous carbonate (siderite) $FeCO_3$ 115.84	3.8	Gray; decomposes before melting; practically insoluble in cold water but soluble in a solution containing CO_2; a major iron ore
Ferrous chloride $FeCl_2 \cdot 4H_2O$ 198.82	1.93	Monoclinic, blue-green crystals; deliquescent; quite soluble in H_2O; very corrosive
Ferrous hydroxide $Fe(OH)_2$ 89.86	3.4	Hexagonal pale green or white amorphous; decomposes before melting; very slightly soluble in H_2O
Ferrous iodide FeI_2 309.68	. . .	Hexagonal gray; melts at 177°C
Ferrous iodide $FeI_2 \cdot 4H_2O$ 381.74	2.87	Gray-black crystals; deliquescent; very soluble in cold water; decomposes in hot H_2O
Ferrous nitrate $Fe(NO_3)_2 \cdot 6H_2O$ 287.95	. . .	Rhombic green; decomposes before melting at 60.5°C; soluble in H_2O
Ferrous oxalate $FeC_2O_4 \cdot 2H_2O$ 179.87	2.28	Rhombic pale yellow; decomposes before melting at 160°C; very slightly soluble in H_2O

TABLE I-12. *Properties of Representative Iron Compounds*   (*Continued*)

Compound, Formula, and Formula Weight	Sp gr	Characteristics
Ferrous sulfate $FeSO_4 \cdot 7H_2O$ 278.01	1.9	Monoclinic blue-green; melts at 64°C ($-6H_2O$); boils at 300°C ($-7H_2O$); soluble in H_2O; source of ferrous compounds, disinfectant, water purifier, wood preservative, weed killer; used in inks, textiles, and leather processing
Iron carbide Fe_3C 179.52	7.4	Cubic gray; major microconstituent of steel; melts at 1837°C; insoluble in H_2O
Iron tetracarbonyl $Fe(CO)_4$ 167.84	1.99	Dark-green, lustrous crystals; decomposes before it melts at 140–150°C; insoluble in H_2O
Iron-ferricyanide (Turnbull's blue) $Fe_3[Fe(CN)_6]_2$ 591.47	. . .	Dark blue; decomposes before melting; insoluble in H_2O; coloring ingredient
Potassium ferricyanide $K_3Fe(CN)_6$ 329.19	1.89	Cubic colorless; granular white; deliquescent; very poisonous; decomposes before melting; moderately soluble in H_2O
Potassium ferrocyanide $K_4Fe(CN)_6 \cdot 3H_2O$ 422.33	1.89	Monoclinic lemon yellow; melts at 70°C ($-3H_2O$); decomposes before boiling; moderately soluble in H_2O

Ba, Ag singly or in combination. Rare-earth metals are also used in magnetic materials with iron.

Iron, the discovery of which predates recorded history, is important as an extremely versatile material of construction and engineering. Besides its use for these purposes in relatively pure form, such as malleable and wrought iron, there are literally thousands of iron-base alloys, including the steels. See also **Iron and steel;** and **Iron- and steelmaking.**

Iron is rarely found in native form except in meteorites. The principal iron ores are hematite, limonite, magnetite, and siderite. See also **Iron ores.**

The important physical properties of iron are summarized in Table I-11.

In addition to the ores from which iron is extracted, iron forms several compounds of commercial importance. The properties of representative iron compounds are summarized in Table I-12.

Iron is used as a catalyst and as a reducing agent and precipitates the free metal from solutions of Ag, Au, Pt, Hg, Bi, and Cu. Most iron salts are decomposed by heat. Ignition in air changes ferrous compounds to the magnetic oxide, and ferric compounds ignited on charcoal or by

a reducing flame are similarly changed to the magnetic oxide.

Iron reacts with egg white to form iron albuminate and with peptone to form iron peptone, both compounds finding medical use. Iron occurs in hemin, $C_{33}H_{32}O_4FeCl$, and in hematin, $C_{33}H_{32}O_4N_4FeOH$. The formation of these compounds provides a very sensitive test for blood. Iron compounds are used as reducing agents for several organic reactions. For example, ferrous chloride reacts with aniline and water to yield ferrous hydroxide and aniline hydrochloride: $FeCl_2 + 2C_6H_5NH_2 + 2H_2O \rightarrow Fe(OH)_2 + 2C_6H_5NH_2 \cdot HCl$.

—S. C. Desai, *Davy Ashmore International Ltd., Stockton-on-Tees, England*

Iron and Steel   The use of chemically pure iron essentially is limited to powder metallurgy and chemical applications where iron is used as a catalyst and for the production of certain ferrous and ferric chemicals. The composition of the principal forms of pure iron is given in Table I-13. Ingot iron, classified among pure irons, does have structural and mechanical uses described later.

Commercially, iron generally is used in the form of steels and cast irons, which contain iron as the major ingredient in association with carbon and other elements. Depending on the content of various elements and the mechanical and heat treatments given, these iron-base alloys exhibit the following properties in varying degrees: (1) brittleness (impact strength), (2) cohesive strength, (3) compressive strength, (4) creep, (5) fatigue, (6) ductility, (7) hardness, (8) malleability, (9) shear strength, (10) yield strength, (11) torsional strength, (12) electrical conductivity, (13) thermal conductivity, (14) thermal stability (change of properties with temperature), (15) thermal expansion, (16) corrosion resistance, (17) magnetic properties, and (18) heat treatability.

Because of the wide range of properties available by varying compositions and/or heat treatment, the low cost of production compared with other metals, and the ease of forming and machining into desired shapes, steels and cast irons are the most widely used structural and engineering materials in commerce and industry.

Steels and cast irons embrace such a wide variety of products with greatly differing properties that it is not possible to arrive at strict definitions. So descriptions here are limited to major families of these important materials. For practical purposes these broad definitions can be applied:

Steel: Iron which is malleable in at least one range of temperature and also is (1) cast into an initially malleable mass or (2) is capable of hardening greatly by sudden cooling or (3) both. Iron containing less than 1.7% carbon with or without other elements generally shows these properties.

Alloy Steels: Steels which owe their properties chiefly to an element other than carbon.

Cast Iron: Carbon content of the iron is such that it is not malleable at any temperature. This is generally the case when the carbon content is more than 2.2%.

Malleable Iron: Iron when first made is in the condition of cast iron and is made malleable by subsequent treatment without fusion.

Alloy Cast Irons: Irons which owe their properties chiefly to the presence of an element other than carbon.

With modern iron- and steel-production techniques ferrous products can be custom-produced to highly specific requirements. The two major classifications depend on the condition in which the products are used. In *cast products,* the basic shape or starting shapes are achieved by pouring molten metal into molds, close to or approximating the final shape and frequently followed by machining operations. Both cast irons and steels castings can be produced. In *wrought products* the basic shapes are achieved by hot and/or cold working of large pieces, usually beginning with blocks and producing the required shapes, e.g., bars, rods, wire, sections, strip, axles, or tubes, by forging, extrusion, rolling, and drawing. Generally only steels can be worked into shapes, thanks to their malleability.

The physical properties of ferrous products can be altered by cold working or heat treatment, which effect changes in the microstructure.

The *iron-carbon diagram* to some extent explains the role played by carbon in ferrous products. Fig. I-3 shows the relationship between carbon content and temperature and provides key information on microstructure and heat treatment.

Pure iron when cooled solidifies at about 1536°C (2797°F) as delta iron having a body-centered-cubic lattice structure and changing allotropically to gamma iron with face-centered-cubic lattice below 1404°C (2559°F). It is nonmagnetic. On further cooling gamma iron changes to alpha iron, with body-centered-cubic lattice. At 768°C (1415°F) a nonallotropic change occurs making the alpha iron strongly magnetic, with marked changes in the electrical resistance, rate of thermal expansion, and specific heat. These changes occur in the reverse order if pure iron is heated instead of cooled.

Carbon dissolves in molten iron and forms iron

TABLE I-13. *Composition of Pure Irons*

Iron Type	Ingredients, Percent of Total								
	C	Mn	P	S	Si	Cu	O_2	N_2	Total Impurities
Ingot	<0.020	<0.020	0.005±	0.020±	trace	0.04±	Some	0.004±	
Electrolytic	0.006	. . .	0.005	0.004	0.005	. . .	Some		
Carbonyl	0.0004	. . .	. . .	. . .	. . .	. . .	<0.01		
Hydrogen-purified .	0.005	0.028	0.004	0.003	0.001	. . .	0.003	0.0001	0.024

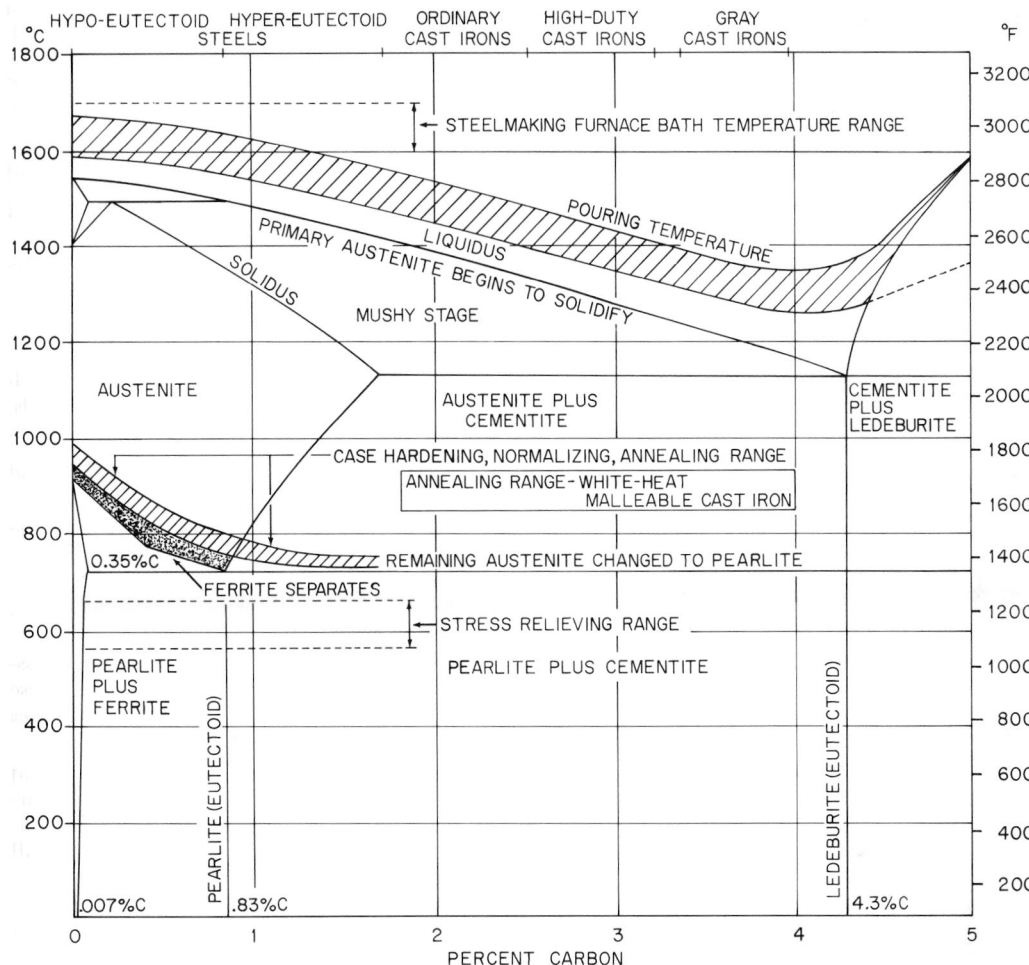

Fig. I-3. Iron-carbon equilibrium diagram.

carbide. With increasing carbon content the liquidus temperature, i.e., the melting point of iron, is progressively lowered, the eutectic point (lowest melting temperature) being 1130°C (2066°F) at 4.3% carbon. The solidus temperature is similarly lowered to 1130°C (2066°F) up to 1.7% carbon. The resulting transformations and the products formed after cooling the iron-carbon alloy below the solidus temperature depend on the carbon content, the temperature, and the rate of cooling. A brief explanation is given for some of them.

Austenite: An allotropic form of gamma iron with carbon in solid solution. It transforms to other products on cooling below 723°C (1333°F). The products depend on the rate of cooling.

Austenite containing only carbide is not stable at ordinary temperatures and therefore cannot be completely retained by quenching. However, the stability can be increased by adding certain alloying elements.

Ferrite: Practically pure iron, which can exist in magnetic alpha-iron form in iron, with up to 0.83% carbon. It exists at room temperature and up to 910°C (1670°F) in the absence of carbon. The upper limit of its existence is lowered progressively to about 723°C (1333°F) as the carbon content increases up to 0.83%. It cannot dissolve carbon, is soft and ductile, and has poor abrasive resistance.

Cementite: Iron carbide, Fe_3C, containing 6.67%

carbon. It is hard, brittle, and crystalline and is precipitated when austenite cools.

Pearlite: A eutectoid consisting of a laminated structure of ferrite and cementite. It is formed by transformation of austenite on cooling. Fineness or coarseness of laminated structure is governed by the rate of cooling. The lamellar arrangement of ferrite and cementite produces a very tough structure and is responsible for the mechanical properties of the steels.

Graphite: The free or uncombined carbon generally found in cast irons. Because it occurs as flakes, cast irons are easily machinable in spite of their resistance to abrasion.

Except for almost pure iron, e.g., wrought iron and ingot iron, the production techniques used mean that irons and steels contain, in addition to iron and carbon, some Si, Mn, P, S, and trace elements, e.g., Cu, Sn, As, and Sb. However, the principal ferrous products can be broadly classified into three main groups depending on the carbon content of the iron: (1) pure irons, (2) steels, and (3) cast irons. The major commercial products in each of these categories are described next.

Pure Irons. Since ingot iron and wrought iron contain only traces of carbon and other elements and therefore consist almost entirely of ferrite, they are extremely ductile.

Ingot Iron: This material is the only relatively pure iron made on a tonnage basis as strip, sheet, wire, galvanized sheet, and rail sections. Ingot iron, as distinguished from electrolytic pure iron, is made by some steel producers in open-hearth furnaces using melting practices roughly equivalent to those used in making steel. The composition of representative ingot iron is C, usually $<0.020\%$; Mn, 0.020%; P, $\sim0.005\%$; S, $\sim0.020\%$, with traces of Cu, O_2, and N_2.

Annealed ingot irons show a compressive strength in the form of elastic limit of 19,400 psi, proportional limit of 19,200 psi, and a yield strength of 20,600 psi. When ingot iron is cold-drawn to a Brinell hardness number of 100–130, it shows 50% of the machinability of the standard B1112 steel employed in the ASM machinability test. Typical applications for ingot irons are contact rails, radio-speaker cores, burial vaults, and culverts. In general, ingot iron is more resistant to corrosion than soft steel, has improved tensile properties at elevated temperatures, and has greatly improved electrical conductivity. The improvement in properties is difficult to evaluate numerically because extremely small amounts of impurities have a major effect on some of the properties.

Wrought Iron: This is a commercial form of iron produced by hand puddling, mechanical puddling, the Aston or Byers, busheling, or faggoting process. Because of the low temperature employed, the product is in pasty or semifused state and contains slag in dispersed state. When hammered, the metal grains elongate and weld together, part of the slag being squeezed out and part remaining intermingled with iron. This produces the characteristic fibrous structure of wrought iron.

The presence of slag does not greatly reduce the resistance of wrought iron to fatigue so long as the applied stress is longitudinal, but it markedly increases the corrosion resistance of wrought iron compared with steels. When wrought iron is exposed to corrosive media, it is quickly coated with a film of oxide, or rust. As the corrosion proceeds, more oxide is formed, but the slag filaments begin to function as rust resistors. The initial oxide film is fastened or pinned to the surface of the metal by the slag filaments.

Depending upon the process of manufacture, wrought iron composition is C, 0.06–0.08%; Mn, 0.015–0.045%; Si, 0.101–0.183%; P, 0.062–0.115%; S, 0.009–0.015%, and slag 1.20–2.85 wt%. Tensile strength is 34,000–54,000 psi. Brinell hardness numbers, 100–110.

Alloy wrought irons have been made successfully with Ni, Mo, Co, and a combination of these elements. The presence of up to 3.5% Ni raises the elastic limit and the tensile strength substantially.

Steels. Steel is essentially an alloy of iron and carbon with or without other elements. The carbon content of steels usually does not exceed 1.7%. Plain carbon steels contain, in addition to C, small amounts of Si, Mn, P, and S, derived from the raw materials and fuel used in the steelmaking process. Within limits, Si and Mn are beneficial and are purposely added both as deoxidants and additions. Since S and P are deleterious elements, their content is usually kept as low as possible except in free-cutting steels, where S is intentionally added.

Trace elements like Sn, Sb, Cu, and As are harmful. Small amounts of Al, Ti, Zr, and V can be present or are added as deoxidizing and/or grain-refining agents.

A number of elements are deliberately added to the steel, including some of the above, which are fortuitously present in small quantities, in amounts to confer certain desired properties. The elements thus used to produce alloy steels are Cr, Cu, Ni, Mo, W, Zr, V, Co, Pb, Cb, (Nb), Mn, Si, and S.

Irrespective of the process and technique adopted for making steel, the final product from the furnace is liquid steel of the desired composition. Depending on the deoxidation state of steel, which is related to carbon content and deoxidant addition, reaction between iron oxide and carbon proceeds, affecting the characteristics of the solidified product.

Killed steel is produced in several ways. Involved is the use of several deoxidizing elements, which act with varying intensities, the most common being Si and Al, which remove oxygen by forming solid oxides because of their affinity for oxygen is greater than that of carbon. The reaction forming carbon and oxygen gas is suppressed, and the killed steel lies quiet in mold when poured, shrinking on solidification, with formation of a conical cavity usually known as a *pipe*. Because of their greater uniformity and soundness killed steels are used for forging, carburizing, heat treating, and other applications.

Rimming steel is produced by leaving sufficient oxygen to react with carbon to evolve bubbles of CO. These conditions are best achieved in steels with low carbon and manganese contents by controlled addition of deoxidants. The effervescing action of CO evolution produces a pure outer skin of ferrite and a tougher inner core containing carbon, impurities, and inclusions on solidification of the poured block. The sandwichlike macrostructure is retained during subsequent shaping operations. Thin sheets intended for deep drawing or deep pressing, used in the manufacture of car bodies and domestic appliances, are made from rimming steel, because they give a smooth surface with adequate strength. In practice they usually contain less than 0.15% C, less than 0.50% Mn, and only traces of Si. Rimming steels are also used for forgings requiring smooth surfaces.

Sealing the top end of the ingot after pouring the rimming steel produces *capped steels*. Big-end-down bottle-top molds are used and after a small addition of Al are sealed by using a heavy metal cap. Capped steels, however, may also be cast in open-top molds, substituting for the metal cap an addition of Al or ferrosilicon, on top of the molten steel, to cause the steel on the surface to lie quietly and solidify rapidly. Steels of this type are used for producing sheet, strip, skelp, tin plate, wire, and bars.

A *semikilled or balanced steel* is produced by adjustment of the Si and Al added to low-carbon steel before teeming. It has a character intermediate between the killed and rimming steels. It is deoxidized less than the killed steel, leaving enough oxygen to react with carbon and form blowholes to compensate entirely or in part for the shrinkage that accompanies solidification. Semikilled steels generally have carbon content within the range of 0.15 to 0.30% and are used for less severe drawing and pressing than rimming steel and for structural shapes, plates, and merchant bar.

Effect of Elements on Steel Properties: Since the effect of a single element on steelmaking practice and properties depends on the effects of other elements, this interrelationship must be considered when a composition is evaluated.

Carbon: The effects of carbon can be seen by examining the iron-carbon diagram (Fig. I-3). The microstructure of steels containing less than 0.83% C, known as *hypoeutectoid steels,* when slowly cooled consists of pearlite and ferrite; of eutectoid steel containing 0.83% C entirely of pearlite; and those containing more than 0.83% C, known as *hypereutectoid steels,* pearlite and cementite. Each increase in the carbon content of the steel increases the hardness and tensile strength of steel in the *as-rolled* or *normalized* condition up to 0.83% C. Above this figure, the effect is less pronounced. The maximum hardness attainable after quenching also increases with the carbon content up to about 0.60% C. Very slight additional hardness can be secured above this figure. The strength of quenched and tempered steels depends upon the tempering temperature. Ductility decreases as the carbon content increases, and weldability is impaired above certain levels.

Manganese: In small amounts Mn occurs in many iron ores and is therefore present in most iron and steels. Further Mn is generally added to bring the amount to between 0.5 and 1.0%, and it contributes to both strength and hardness but to a lesser degree than carbon. Conversely, Mn lowers ductility, but to a smaller extent than carbon. Mn is beneficial to surface quality in all carbon ranges, particularly in resulfurized steels. It also increases the rate of carbon penetration during carburizing.

When Mn is specified within the limits of 1.65–2.10%, a steel qualifies as an alloy steel. Mn contributes to strength and is of major importance in increasing hardenability (the depth of hardness penetration after quenching); 13% Mn steel finds a special application as a wear-resistant steel. Chemical compositions requiring high Mn content combined with low C content necessitate the use of special low-carbon ferromanganese because the regular grades of ferromanganese contain appreciable amounts of C. See also **Manganese.**

Phosphorus: Some P is present in all irons and steels, and it is usually desirable to keep the level as low as possible. It increases the strength and hardness to approximately the same extent as carbon in steels which are normally used in the hot-rolled condition. In some types of steel, high P content is undesirable because it decreases ductility and impact toughness. Phosphorus is particularly undesirable in the higher-carbon steels because of excessive loss of ductility. Phosphorus promotes machinability in the lower-carbon steels and, along with copper, improves resistance to atmospheric corrosion. In acid open-hearth and acid bessemer processes, P is not removed in refining and necessitates higher permissible specification limits.

Sulfur: Present in varying amounts in all irons and steels, S is derived from the ores and the fuels used in their production. It is beneficial to machinability but detrimental to surface quality, particularly in low-C and low-manganese steels. It also decreases transverse ductility and impact resistance but has only a slight effect on longitudinal properties. Weldability decreases with increasing S content. Additions of 0.2–0.4% S are made in free-cutting steels to improve machinability.

Silicon: Like Mn, Si occurs in all steels naturally or as a result of additions during steelmaking. Since Si is one of the principal deoxidizers used in steelmaking, the amount of Si present varies with the type of steel. Rimmed and capped steels contain no significant amounts. Si promotes the adherence of Zn coating on hot-dipped galvanized wire. It is less effective than Mn in increasing strength and hardness. In low-carbon steels, Si usually is detrimental to surface quality, and this condition is more pronounced with the resulfurized grades.

When specified within the limits of 0.60-5.00% Si qualifies a steel as an alloy steel. Increasing the Si content increases the resiliency of steel for spring applications. Si raises the critical temperature for heat treatment. Increasing the Si content promotes the susceptibility of steel to decarburization. Very low carbon steels with 0.6-5.00% Si have a low hysteresis loss and a high electrical resistance and are used as transformer steels.

Aluminum: Used to deoxidize steels and obtain fine grain size, Al is sometimes used to obtain nonaging characteristics and to prevent the recurrence of stretcher strains in sheets and strip. Added in amounts approximating 1%, Al promotes nitriding properties (surface hardening by means of nitrogen-bearing gases at high temperatures).

Copper: In the small amounts used in carbon steel, Cu has a minor effect on mechanical properties. It is beneficial to atmospheric corrosion resistance when present in amounts exceeding 0.20%. In appreciable amounts, Cu is detrimental to hot-working operations. Cu adversely affects forge welding and is detrimental to surface quality but does not affect seriously arc or acetylene welding. Inasmuch as Cu is not removed by the conventional steelmaking processes, it is becoming increasingly difficult to hold within low limits.

Nickel: Except for manganese Ni is the most commonly used alloying element. Appreciable amounts up to 5% result in higher-strength steels with improved shock resistance. Ni counteracts the brittleness that develops in most pearlite steels at subnormal temperatures, lowers the critical temperature of steel, widens the temperature range for successful heat treatment, and promotes corrosion resistance. The use of Ni in quantities greater than 5% results in stainless and heat-resistant steels, described later.

Columbium (Niobium): This is used to stabilize chromium and stainless steels by the addition of amounts up to 1%. Additions of about 0.02% have been found to increase the yield point of medium-carbon steels by up to 50% without any loss of weldability.

Tungsten: Added to steels in amounts up to 20% in the production of high-speed steels, W greatly improves the hardness of a steel, and this hardness is maintained at high temperatures. Additions of smaller amounts are made to hot-working steels.

Zirconium: This element is added in small amounts to high-chromium steels to improve their machinability.

Cobalt: This gives cutting efficiency to high-speed steels and is a constituent of heat-resisting steels, as it confers resistance to creep and scaling.

Chromium: Cr is used in constructional steels primarily to increase hardness, improve hardenability, and promote the formation of carbides. Cr steels are relatively stable at elevated temperatures and have exceptional wear resistance. The use of Cr in stainless and heat-resistant steels is described later.

Molybdenum: This element has a major effect on increasing hardenability and a strong effect on increasing the high-temperature tensile and creep strengths (the slow deformation of steel under stress at elevated temperatures) of alloy steels. Steels that contain Mo are considered less susceptible to temper brittleness.

Vanadium: A strong deoxidizing agent, V promotes a fine austenitic grain size. The amount used in constructional steels is about 0.03–0.25%, although larger quantities are used in tool steels. Hardenability of medium-carbon steels is increased with a minimum effect on grain size with V additions of about 0.04–0.05%. Above this content, the hardenability decreases with normal quenching temperatures. However, the hardenability can be increased with the higher V contents by increasing the austenizing temperatures.

Titanium: In pearlitic steels Ti acts as a deoxidizer. In amounts of 0.02–0.05%, Ti increases the yield point of plain-carbon steels. Weldability is promoted without the need for normalizing.

Boron: Added to increase hardenability, B is effective only when added to fully killed steels. A few thousandths of 1% of B normally remains in the steel, and boron steels are evaluated by increased hardenability rather than chemical content. Boron intensifies the hardenability characteristics of elements which already are present in the steel and makes possible alloy conservation when used with steels containing small amounts of alloying elements. Boron is effective when used with low-carbon alloy steels, but its effectiveness diminishes as the carbon increases.

Classification of Steels: Various systems are in use for classification of steels, based on composition, physical properties, or certain special properties. These classifications often overlap and are too intermingled to provide a single general classification for all the types of steel. Steels in general use in commerce and industry can be broadly classified into two main categories: plain carbon steels and alloy steels.

Plain Carbon Steels: Mostly containing iron and carbon with small amounts of other elements, these represent the most important group of engineering materials. They constitute about 95% of the total world crude-steel output, which goes into the production of castings, forgings, structural shapes, plates, sheets, strip, tubes and pipes, rods and bars, wire, tools, wheels, tires, axles, and rails.

Plain carbon steels are classified as hypoeutectoid or hypereutectoid steels, depending on whether the carbon content is above or below 0.83% eutectoid composition. However, generally they are classified as (1) low-carbon steels, in which the carbon content is below 0.20%, (2) medium-carbon steels, with a carbon content between 0.20 and 0.50%, and (3) high-carbon steels with a carbon content above 0.50%.

The average physical properties of plain carbon steels depend on the carbon content. As shown in Fig. I-4, the tensile strength, yield strength, and hardness increase with increasing carbon content; elongation, reduction in area, and impact values show a marked decrease. As mentioned earlier, these properties depend upon the microstructure. By heat treatment it is possible to alter the microstructures significantly, with corresponding variations in the physical properties.

Alloy Steels: The American Iron and Steel Institute define alloy steels as follows: "By common custom alloy steel is considered to be alloy steel when the maximum of the range given for the content of alloying elements exceeds one or more of the following limits: manganese 1.65 percent; silicon 0.60 percent; copper 0.60 percent; or in which a definite range or a definite minimum quantity of any one of the following elements is specified or required within the limits of recognized field of constructional alloy steels: aluminum, boron, chromium up to 3.99 percent, cobalt, columbium [niobium], molybdenum, nickel, titanium, tungsten, vanadium, zirconium, or any other alloying element added to obtain a desired alloying effect." As an exception some stainless steels contain more than 4.0% chromium.

High-strength low-alloy steels, constructional alloy steels, stainless steels, heat-resisting steels, electrical steels, and alloy tool steels are the alloy steels in commercial use.

High-Strength Low-Alloy Steels: The objective of these alloys is twofold: higher mechanical properties and greater resistance to atmospheric corrosion than obtainable with structural-grade carbon or copper-bearing steels. Frequently steels in this category are proprietary to a producer, who gives a specific steel a trade name. All high-strength low-alloy steels meet chemical limitations of a maximum of 0.22% C; 1.25% Mn; and 0.05% S. In addition, P, Cu, Si, Ni, Cr, Mo, V, Al, Ti, and Zr sometimes are added, singly or in combination, to improve strength, toughness, corrosion resistance, weldability, and other properties. In comparison with structural-grade carbon steels, high-strength low-alloy steels may have a resistance to atmospheric corrosion 4–6 times greater, show a savings in weight because of higher

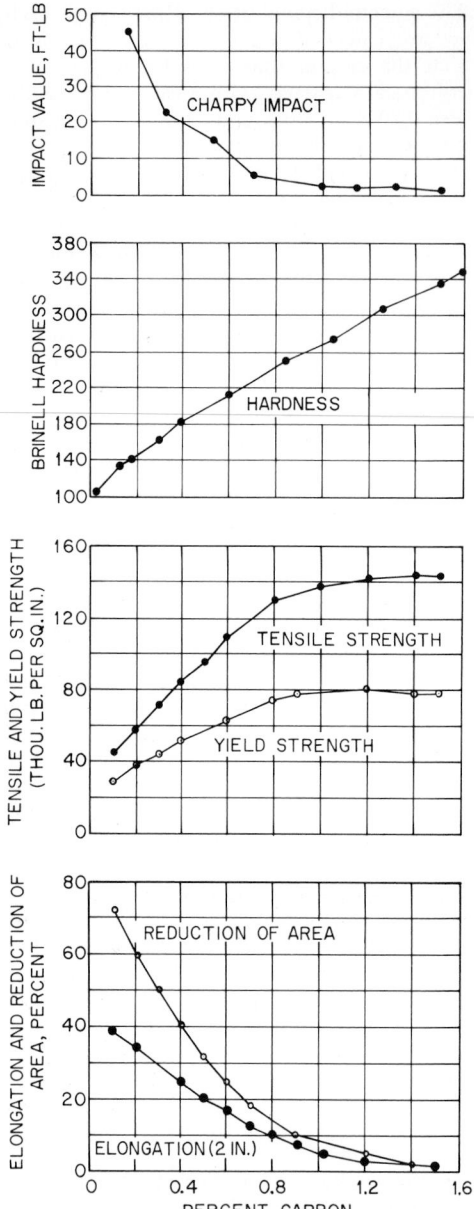

Fig. I-4. Average mechanical properties of as-rolled, 1-in.-diam bars of plain carbon steels versus carbon content. (*Based upon observations of Nead, Lagenberg, Wahlberg, Roberts-Austen, and Cowland.*)

strength levels, and offer improved weldability, greater resistance to wear and abrasion, improved paint adherence, and increased strength at moderately high temperatures. For a representative steel in this category, the tensile strength will approximate 70,000 psi for a $\frac{1}{2}$-in.-thick section and have a yield point of approximately 50,000 psi. Improved impact properties also are claimed for some of these alloys.

Constructional Alloy Steels: These steels, which form a major part of the tonnage of alloy steels produced, are used mostly in the automotive and aircraft industries. The alloying additions have an indirect effect on the steel properties by modification of the heat-treatment characteristics. By modification of microstructure it is possible to obtain desirable physical properties over a much wider range of sizes and sections than in plain carbon steels. These alloy steels are normally quench-hardened and tempered with or without carburizing.

Stainless Steels: Stainless steels comprise a group of ferrous alloys characterized by a high degree of resistance to chemical attack, a property sometimes referred to as *passivity*. This property results when iron is alloyed with at least 11% chromium. Higher Cr and the addition of Ni and other elements further improve corrosion resistance. A steel with 12% Cr will stain but will not show progressive rusting in normal atmospheres. Usually a steel with 18% Cr will not stain in normal atmospheres but may discolor in heavy industrial areas. The addition of 8% Ni makes an 18% Cr steel stain-resistant in all but the worst of atmospheres. The addition of Mo and other elements further enhances corrosion and heat resistance. The effect of Cr content on resistance to boiling 65% HNO_3 is given in Table I-14.

TABLE I-14. *Influence of Chromium Content on the Resistance of Low-Carbon Steel to Boiling 65% Nitric Acid**

Chromium, %	Average Rate of Corrosion, Inches of Penetration per Month
4.5	12.900
8.0	0.144
12.0	0.010
18.0	0.0025
25.0	0.00063

*Reprinted by permission from Charles L. Mantell, Editor-in-Chief, "Engineering Materials Handbook," McGraw-Hill Book Company, New York, 1958.

Although the iron-chromium alloys and their resistance to corrosion were known in England and France 130 years or more ago, the phenomenon of passivity was not recognized until 1910 by Borchers and Monnartz in Germany. Following this discovery, commercial stainless steels were developed almost simultaneously in Germany, England, France, and the United States. Besides resistance to corrosion, stainless steels have other desirable properties which have broadened their field of application. The martensitic alloys have the ability of hardening deeply in heavy sections, while the austenitic stainless steels are amenable to deep drawing and exhibit resistance to oxidation and creep at elevated temperatures.

Martensitic Types: These are chromium-iron alloys with Cr in the lower range (12–17%) and have a wide range of carbon. Their main characteristic is an ability to harden by heat treatment in a manner similar to carbon steels. Martensitic types 403, 410, 414, 416, 420, 431, 440A, 440B, and 440C are hardenable steels because they have a high carbon-to-chromium ratio that is necessary to cause transformation of structure upon cooling from over 815°C (1500°F). The best mechanical and corrosion-resisting properties are found where these steels are in the hardened condition. Tensile strengths obtainable are 70,000–105,000 psi for annealed steels and 125,000–200,000 psi for hardened steels. Brinell hardness, in the hardened condition, can range between 140 and 600. The martensitic steels are ferromagnetic. They are especially suitable for hot working and forging, and the lower-carbon types can be cold-worked without difficulty. These steels generally show the best machining characteristics of the stainless steels. Because they are air-hardening, care must be exercised in cooling after welding and hot working to avoid excessive stresses and possible cracking. These steels are well suited for most moderately corrosive conditions, including water, weather, and some chemicals, and are particularly valuable in applications requiring high strength, hardness, and resistance to abrasion and both wet and dry corrosion.

Ferritic Types: These stainless steels are chromium-iron alloys with higher Cr (18–30%) and lower carbon content. The microstructure is predominantly ferritic, and the steels are not hardenable by heat treatment. The ferritic steels are ferromagnetic and have a relatively low coefficient of thermal expansion, less than that of the austenitic types and about the same as carbon steel or slightly less. Ferritic types offer good resistance to oxidation and corrosion and often are selected for high-temperature service in applications which require intermittent heating and cooling because they tend to retain the scale or oxide coating which has formed. Type 430 is the basic steel in this group. Type 405 is a low-chromium nonhardenable type that can be used for services in which type 410 (martensitic) may be used but for which a nonhardening steel is required. Type 430F is a free-machining variation of type 430 (martensitic). Type 446 is an extremely high chromium stainless and heat-resisting steel designed for high-temperature, relatively low-stress applications, especially where service is intermittent or where sulfur compounds are present.

Austenitic Types: These are iron-chromium-nickel alloys (8–30% Cr, 6–20% Ni). The alloy content is sufficiently high to retain austenite at room temperature. The principal characteristics of this class are high ductility of the austenite, work-hardening ability, good corrosion resistance, and superior high-temperature properties. These alloys are termed *austenitic* or 18-8 *steels*. Type 302 is the basic steel in this group.

The austenitic steels are inherently tough and well adapted for fabrication by deep drawing. They can be welded easily and soldered by proper techniques. Type 303 is a special free-machining grade. Tensile strength in the annealed condition is considerably higher than that of mild steel, approximately 90,000 psi with a yield strength of about 35,000 psi.

Heat-resisting Steels: Steam-generating boilers, turbines, internal-combustion engines, pressure vessels, distillation equipment, and furnaces operated at high temperature must be made of materials which can retain the necessary physical properties at elevated temperatures. Creep, rupture strength, oxidation resistance, and corrosion resistance can be improved by the addition of alloying elements. In applications where temperatures exceed 538°C (1000°F) Mo is used in conjunction with Cr. Chromium (2%) imparts protection against oxidation up to 621°C (1150°F); 10–14% Cr is required for temperatures up to 760°C (1400°F), stainless steel being used for higher temperatures. Steels containing 25% Cr and 20% Ni or 27% Ni are used for service in temperature range 816–1093°C (1500–2000°F).

Electrical Steels: High permeability, high electrical resistance, and low hysteresis loss are the properties desired in construction of electrical equipment, e.g., power transformers, rotors and stators of motors and generators, and communications equipment.

These properties are obtained by adding 0.6–5.0% Si to a relatively carbon-free steel. The amount of Si added depends on the duty requirements.

Alloy Tool Steels: Tools and dies used in engineering industries require (1) hardness at room temperature, (2) abrasion resistance, (3) red hardness, (4) resistance to distortion, (5) hardenability, (6) toughness.

Low-alloy, manganese-containing oil-hardening, shock-resisting, high-carbon high-chromium, high-speed, hot-die, Ni-Cr-Mo tool steels are some of the innumerable varieties. The choice of tool or die steel depends on the application.

Cold Working: The term *cold-worked* refers to steel that is cold-drawn or cold-rolled. Mechanical properties are significantly affected by cold working, e.g., increase in tensile strength, yield strength, torsional strength, hardness, and wear resistance. Cold working usually is accompanied by some decrease in elongation and reduction in area. The effects of cold work depend upon such factors as chemical composition, cross section, method of steel production, and type of thermal treatment. By a suitable combination of these factors, cold-worked steel is produced having distinctive characteristics not readily obtained by other methods of manufacture. Some cold-worked steels have mechanical properties comparable to those of heat-treated bars. The increase in yield strength resulting from cold working is proportionately greater than the increase in tensile strength and, from an engineering standpoint, is important when yield strength is used as a working basis for design. One of the best known characteristics of cold-worked steel bars is improved machinability, particularly in the low-carbon steels. In automatic-screw-machine work, the accuracy of size and section of cold-finished bars, compared with hot-rolled bars, tends to minimize collet troubles, and the freedom from scale increases tool life. The ratio of yield strength to tensile strength is an influencing factor on machinability. The high yield-strength ratio resulting from cold drawing minimizes the plastic flow of metal during machining, permitting much of the tool energy to be utilized in shearing the chip.

Heat Treatment: Only a few highlights of heat-treating processes can be given here. The mechanical properties of most hot-rolled and cold-worked steels can be modified to suit end-use applications by heat treatment. In a broad sense,

heat treatment may be defined as an operation or series of operations involving the heating and cooling of steel in the solid state for the purpose of obtaining desired end properties of a metal.

In the conventional sense, heat treatment involves three basic operations: (1) *heating* the steel above the critical range to approach a uniform solid solution (austenite), (2) *hardening by quenching* in oil, water, or air to induce the formation of martensite (the hardest microconstituent of steel), and (3) *tempering by reheating* to a temperature below the critical range to secure the desired combination of strength and ductility. Martensitic stainless steels are hardened by quenching in air, water, or oil, depending upon the type of steel, followed by proper tempering.

In general, three types of steel do not respond to this type of heat treatment: (1) those which contain very low amounts of carbon, (2) austenitic types for which the critical ranges are below room temperature, and (3) ferritic stainless steels. Products which usually can be supplied in the quenched and tempered condition include carbon and alloy steel plates; carbon, alloy, and martensitic stainless-steel bars; hot-rolled alloy steel sheets; alloy-steel tubular products; and carbon-steel wire.

Normalizing: This operation consists of heating to a suitable temperature above the critical range, followed by cooling to below that range in still air. For wire, normalizing consists in heating to a suitable temperature above the critical range. Hot-rolled products are subject to nonuniformity in properties due to variables in mill practice, cross section, and mass. Normalizing promotes uniformity of structure and alters end properties beneficially. Products which can be normalized include (1) carbon, alloy, and high-strength low-alloy steel hot-rolled bars, (2) carbon, alloy, and high-strength low-alloy steel hot-rolled plates, (3) carbon and alloy semifinished steel, (4) carbon, alloy, and high-strength low-alloy steel hot-rolled sheets, (5) carbon, alloy, and high-strength low-alloy steel hot-rolled and cold-rolled strip, (6) carbon and alloy steel tubular products, and (7) carbon and alloy steel wire.

Annealing: Regular annealing for carbon steels, alloy steels, and martensitic and ferritic stainless steels consists in maintaining the steels at a temperature in or near the critical range, followed by cooling at a predetermined rate or cycle. Austenitic stainless steels generally are annealed by holding at appropriate temperatures and rapidly cooling to minimize the precipitation of those carbides which may adversely affect corrosion resistance. The purpose of annealing is to obtain softness; improve machining, forming, or shearing; reduce stress; improve or restore ductility; and modify other properties. Annealing generally is performed on stainless and heat-resisting steels and usually can be performed on the same products as those listed under normalizing.

Box annealing consists in annealing the steel in a suitable metal container to protect the steel from objectionable oxidation. In some cases, annealing is conducted in a reducing atmosphere. *Spheroidize annealing* consists in prolonged heating at an appropriate temperature followed by slow or cyclic cooling to produce a globular condition of the carbide. This treatment produces a structure that may be desirable for machining or cold-forming, cold-drawing operations, or for the effect it will have on subsequent heat treatments. *Stress relieving* is the process of reducing internal stresses by heating to a temperature below the critical range and holding long enough to equalize the temperature throughout the piece. *Patenting* consists in continuous heating of individual strands to above the critical range, followed by comparatively rapid cooling. Applicable only to wire rods and wire, patenting makes the product sufficiently tough to withstand severe distortion or drawing without actual or incipient breakage. *Isothermal annealing* consists of heating to the correct temperature above the critical for proper austenitizing, followed by rapid cooling to a suitable temperature and holding long enough to complete the transformation. The proper temperature and holding time for each composition can be deducted from isothermal transformation diagrams (called *S curves* because of their frequent resemblance to the letter S).

Case hardening is a process of surface hardening involving a change in the composition of the outer layer of an iron-base alloy, followed by appropriate thermal treatment. Typical case-hardening processes are carburizing, cyaniding, carbonitriding, and nitriding.

Carburizing is a process of case hardening in which carbon is introduced into a solid iron-base alloy by heating above the transformation temperature range while in contact with a carbonaceous material which may be a solid, liquid, or gas. Carburizing is frequently followed by quenching to produce a hardened case.

In *cyaniding,* the metal is heated in a cyanide salt. In *nitriding,* the metal is heated in an atmosphere of ammonia or in contact with nitrogeneous

material. Surface hardening by this method can be produced by absorption of nitrogen without quenching.

Temper refers to a condition produced by mechanical or thermal treatment marked by characteristic structure and mechanical properties. A given alloy may be in the softened or annealed temper, or it may be cold-worked to the hard temper or further to spring temper. Intermediate tempers produced by cold working (rolling or drawing) sometimes are called *quarter hard, half hard,* and *three-quarter hard.*

Cast Irons. In making liquid iron, carbon is generally used as a reducing and/or thermal agent and dissolves in the iron. Cast irons are alloys of iron and carbon, with or without other elements, usually containing 2.4 to 4.5% carbon.

The carbon in cast iron can occur wholly as graphite, as in soft gray iron, or wholly as carbide, as in chilled cast irons, or partly as iron carbide and partly as free graphite, which is the form in most commercial cast irons.

Pig Iron: This material is produced by reducing iron ore in a blast furnace, electric smelter, or low-shaft furnace. At one time molten iron produced in the blast furnace was allowed to flow from a central launder into a sand bed to produce small blocks. The arrangement resembled a sow and her pigs, from which the term pig iron is derived. The predominant use of pig iron is as a raw material for steelmaking. See also **Iron- and steelmaking.** However, rather large tonnages of pig iron also are used, along with scrap and coke, for melting in foundry cupolas to produce iron castings. A representative foundry pig iron will have the following composition: C, 3.00–4.50%; Si, 1–4%; P, 0.04–0.10%; and Mn, 0.20–1.50%. When pig irons are used to make other materials and their identity is lost, the physical properties are unimportant.

Castings made from foundry pig irons alone, without alloying ingredients, have Brinell hardness of 100–500, tensile strength of 10,000–40,000 psi, and a tendency to grow, i.e., increase in volume after repeated heating to temperatures between 455 and 900°C (850–1650°F); for ordinary castings, a 5% growth is extreme.

Gray Iron: Where no special alloying materials are added, the final cast material is known as *gray iron,* with tensile strengths ranging from 30,000 to 45,000 psi and Brinell hardness 187–235. The composition of a representative gray iron is C, 3–3.35%; Si, 2–2.4%; S, up to 0.12%; P, up to 0.2%; and Mn, 0.6–0.7%. Much of the carbon in gray iron is graphitic, and when the metal is fractured, the graphite flakes impart a gray appearance to the surface. Gray iron is used for many machinery applications because of its fatigue resistance. Also, because of a low sensitivity to surface notches, gray iron can be used for rotating and oscillating parts where keyways may be required. The toughness of gray iron makes it suitable for crankshafts, cams, gears, and for components of precision machinery. Gray iron performs well in certain corrosive environments, e.g., cast-iron pipe.

A series of cast irons may be obtained, ranging from those with all the carbon in graphitic form to those with a substantial portion of the carbon in combined form. The variation of properties is wide because of the result of the hardening effect of the carbides. Graphite gray iron is soft and readily machinable and has high damping capacity. In tensile strength, ductility, and impact, gray iron is much inferior to steel. Cast irons with a high percentage of combined carbon, as carbide, are brittle and hard, have high wear resistance, and are difficult to machine.

Silicon in the composition promotes the decomposition of cementite into free iron and graphite. Slow cooling aids graphite formation. Rapid cooling aids the formation of cementite. Rapidly cooled irons, often referred to as *chilled cast irons,* have a white surface. Over a range of composition, it is desirable to keep the Si content up to about 2% so that the amount of combined carbon in gray cast iron is less than 1.7 eutectoid percentage. Sulfur functions in a manner opposite to that of Si by stabilizing the carbide and thereby tending to "chill" the iron. Manganese prevents this if present in amounts twice that of S, plus 0.2% combined with the S, forming manganese sulfide. The P present is almost entirely combined with Fe and C to form the steadite eutectic.

Ductile Iron: Also called *nodular graphite iron, sperulitic iron,* and *spheroidal graphite iron,* ductile iron has a structure in which the graphite occurs not in flakes but as tiny balls, or spherulites. This structure also differs from the compacted aggregates of malleable iron. The presence of a few hundredths of a percent of Mg, Ce, or rare earths introduced into the molten iron just before casting causes the spherulitic graphite particles to form during solidification. With Mg or Ce additions, the final S content of the iron must be below 0.015% for the treatment to succeed.

The tensile strength of ductile (nodular) iron is 55,000–120,000 psi, and, also depending upon composition, the Brinell hardness numbers range

from 121 to 273. Although ductile iron lacks impact resistance comparable with that of steel, ductile iron is applicable for many severe uses involving impact, e.g., housings, gears, valves, and steel-mill rolls. The good resistance to fatigue makes ductile iron well suited for critical components of high-speed machinery, such as crankshafts. Ductile iron resists growth and oxidation up to temperatures of 810°C (1490°F), thus making it suitable for such applications as coke-oven doors, incinerator gates, sintering-machine pallets, and glass-forming molds.

Malleable Iron: This alloy, consisting mainly of iron and carbon, is hard and brittle as cast but is rendered tough and ductile by a subsequent heat-conversion process. White-iron castings, of which malleable iron is made, are produced in greensand molds from metal melted in an air furnace of the acid-hearth reverberatory type. The metal may have been melted originally in a cupola, but then subsequent refinement before casting must take place in an air or electric furnace for the proper control of the elements, so that the composition will permit conversion into malleable iron by subsequent heat treatment. The heat conversion is accomplished in an annealing furnace, in which the castings are heated for some time, first at about 870°C (1600°F) to break up the massive carbides and subsequently at a temperature of about 705°C (1300°F) to convert the last of the combined carbon to ferrite and temper carbon.

Standard malleable iron differs from gray cast iron in that it has ductility and substantial resistance to impact. Malleable iron differs from cast steel in that all its carbon is in the free form and therefore it is more easily machined than cast steel. The range of composition of standard malleable iron is C, 2.00–2.65%; Si, 0.90–1.40%; Mn, < 0.55%; P, < 0.18%; and S, < 0.18%.

Historically, malleable iron dates back to 1722, when the French scientist Réaumer first experimented with heating small iron castings to a bright redness for many days and noted the characteristics of the resulting malleable iron. Malleable iron, commercially available only in the cast form, is rather widely used in such products as railroad rolling stock, construction and farm machinery, electrical equipment, and automotive and truck parts.

High-Silica Cast Irons: In these materials, the Si content may be as high as 14.5%. These irons have a Brinell hardness of about 520, possess good corrosion and abrasion resistance, and find application in pumps, valves, ejectors, fans, jets, pipe, fittings, heat exchangers, tanks, tank sections, and kettles for severe service in the chemical process industries.

Alloy Cast Irons: Alloying elements such as Ni, Cu, Cr, Mo, and Ti are added in amounts of up to about 20%. These alloyed cast irons are used widely by the steel, chemical, and automotive industries for applications where resistance to wear, heat, and corrosion along with high strength, rigidity, and damping capacity are the properties required.

—S. C. Desai, *Davy Ashmore International Ltd.,*
Stockton-on-Tees, England

Iron Coagulants (See **Waste treatment**.)

Iron Ores Iron-ore reserves of the world total about 250 billion tons. In addition, over 500 billion tons are presently regarded as commercially nonexploitable. Use of large ore carriers has lowered freight costs, making some of the ore reserves previously considered unsuitable available for large-scale mining operations. Additional new sources of iron ore include Western Australia, Brazil, Chile, Liberia, and new locations in the Soviet Union. The geographical breakdown of these figures is Europe (principally West Germany, France, and Russia), 41%; North America, 23%; Asia (principally China, India, and Korea), 16%; South America (principally Brazil), 11%; Africa, 4%; with the remainder including Australia and Central America. Of course, the world-reserves picture bears no direct resemblance to current major sources.

Present important worked deposits include (1) Kerch oolitic limonite (Crimea, Russia), (2) Salzgitter limonite and hematite (Germany), (3) minette limonite and hematite (France, Germany, and Luxemburg), (4) blackband ironstones (British Isles), (5) Siegerland siderite (Germany), (6) Clinton hematites (Alabama), (7) Wabana oolitic hematites (Newfoundland), (8) Minas Gerais hematite (Brazil), (9) Krivoi Rog hematites (Ukraine, Russia), (10) Bihar, Orissa, and Bastar hematites (India), (11) Labrador hematite (Quebec and Labrador), (12) Lake Superior taconites, jaspilites, hematites, and magnetites (Michigan, Wisconsin, Minnesota, and Ontario), (13) Cerro Bolivar and El Pao hematites (Venezuela), (14) Kirunavaara magnetite (Sweden), and (15) hematites (Australia).

Annual world iron-ore production is in excess of 700 million tons, compared with 500 million tons as recently as 1960. Total production of iron ore in the Western Hemisphere is estimated to be over 200 million tons, with the United States accounting for about 90 million tons and Canada 50 million tons. Production of iron ore in Russia is approaching 200 million tons. Dependence of major steel nations on imported iron ore is France, 15%; United States, 36%; United Kingdom, 56%; West Germany, 84%; Japan, 92%; and Italy, 93%.

The increasing need for iron has accelerated two major developments: expansion of iron-ore pellet-plant production in those industrialized nations where supplies of high-grade ores have diminished and an intensive search for new supplies of high-grade ore. Shipment of iron ore as a slurry in ocean-going tankers is a relatively recent development in which iron-ore concentrates are pumped aboard a tanker as a slurry containing about 75% solids. After the slurry has settled, most of the water is pumped off, leaving a nonshifting cargo of 92% solids. The ore is reslurried on arrival with high-pressure water jets. Thus, iron ore can be moved in tankers rather than conventional bulk-carrying ships, and delivery via offshore pipeline pumping eliminates the need for dockside unloading equipment.

A large number of minerals contain iron but only a few are used as commercial sources of iron. As shown in Table I-15, important iron-bearing minerals can be grouped according to their chemical composition. Oxide minerals are the most important sources, followed by carbonates, sulfides, and silicates.

Beneficiation of Iron Ores: The term *beneficiation* encompasses all the methods used to process ore to improve its chemical or physical characteristics in ways that will make it a more desirable feed for the blast furnace. Such methods include crushing, screening, blending, grinding, concentrating, classifying, and agglomerating.

Concentration operations include jigging, flotation, and magnetic separation. These operations are described elsewhere in this book (see the Subject Index). Agglomerating processes are of growing importance in the processing of iron ores. The blast furnace is a countercurrent gas-solid reactor in which the solid charge materials are moving downward while the hot reducing gases are flowing upward. The best possible contact between the solids and the reducing gas is obtained with a permeable burden which permits not only a high rate of gas flow but also a uniform gas flow with minimum channeling of the gas. The primary purpose of agglomeration is to improve burden

TABLE I-15. *Principal Iron-bearing Minerals*

Oxides:

Magnetite, Fe_3O_4, 72.36% Fe; dark gray to black; occurs in igneous, metamorphic, and sedimentary rocks; strongly magnetic; when Ti content is 2–15% or more, the ore is termed *titaniferous magnetite;* specific gravity 5.16–5.18

Hematite, Fe_2O_3; 69.94% Fe; steel gray to dull or bright red; associated with vein deposits; igneous; metamorphic; and sedimentary rocks; also a product of the weathering of magnetite; common varieties are *crystalline, specular, martite* (pseudomorphic after magnetite), *maghemite* (magnetic ferric oxide), *earthy, ocherous,* and *compact;* specific gravity 5.26

Ilmenite, $FeTiO_3$, 36.8% Fe; associated in small amounts with magnetite; usually mined as a source of Ti with Fe recovery as a by-product

Hydrous oxides:

Limonite, the term for a group of mixtures of goethite and lepidocrocite; *goethite,* $HFeO_2$, 62.9% Fe; yellow or brown to nearly black; *lepidocrocite,* $FeO(OH)$; both ores occur with other iron oxides and in sedimentary rocks; limonites important sources of iron throughout the world

Carbonate:

Siderite, $FeCO_3$, 48.2% Fe; white to greenish gray and brown; commonly contains varying amounts of Ca, Mg, and Mn; these ores are usually calcined before feeding to the blast furnaces; frequently contain enough lime and magnesite to be self-fluxing; specific gravity 3.83–3.88

Silicates:

Chamosite, stilpnomelane, greenalite, minnesotaite, grünerite; many silicate minerals contain small amounts of iron associated with other bases, but comparatively few have iron as the principal base; they often have a rather complex chemical formula, with specific gravities higher than 2.8, and occur in various shades of green or black; iron silicates not important as a source of iron ore, but of interest as a primary source of oxide iron ores, which form through weathering or hydrothermal oxidation of the silicate minerals; have a wide distribution in sedimentary rocks and metamorphic iron formations

Sulfides:

Pyrite, FeS_2, 46.6% Fe; pale brass yellow; occurs in sedimentary, metamorphic, and igneous rocks and in veins; also known as *iron pyrites;* specific gravity 4.95–5.10

Marcasite, FeS_2, 46.6% Fe; pale brass yellow; occurs in limestones, clays, and lignite deposits; also known as *white iron pyrites;* differs from pyrite in crystalline structure and is less stable chemically

Pyrrhotite, FeS, 69.4% Fe; bronze yellow to copper red; often considered an indicator of Ni deposits because of its common association with NiS; iron sulfides sometimes mined as a source of sulfur but usually for other valuable elements, such as Cu, Ni, Zn, Au, and Ag; production and by-production of iron oxide is of growing importance as an iron source

permeability and gas-solid contact and thereby reduce blast-furnace coke rates and increase the rate of reduction. A secondary consideration is lessening of the amount of fine material blown out of the blast furnace into the gas-recovery system.

A good agglomerate for blast furnace use should contain 60% or more of iron, a minimum of undesirable constituents, a minimum of material less than $\frac{1}{4}$ in. in size, and a minimum of material larger than 1 in. The agglomerate should be strong enough to withstand degradation during stockpiling, handling, and transportation to the furnace so that it arrives at the furnace skip with a minimum of approximately 85–90% of material over $\frac{1}{4}$ in. In addition, the agglomerate must be able to withstand the high temperature and the degradation forces within the furnace without slumping or decrepitating. The agglomerate should also be reasonably reducible so that it can reduce at a satisfactorily high rate in the blast furnace. At present, there is less definite knowledge of the preferred shape, the most suitable size within the $\frac{1}{4}$- to 1-in. range, minimum strength required, or most desirable mineralogical structure.

Four principal types of agglomerating processes have been developed: (1) sintering, (2) pelletizing, (3) briquetting, and (4) nodulizing. To date, sintering and pelletizing have proved to be the most important.

Sinter is an agglomerate made from small particles of iron-bearing materials fused or fritted together at a high temperature. In the *sintering process* this high temperature is achieved by burning carbon in the form of coke breeze or other forms in the sintering-machine feed mix. A certain amount of flux can also be added in the feed mix to eliminate flux charging partially or completely in the subsequent ironmaking operations. The flexibility of the process permits conversion

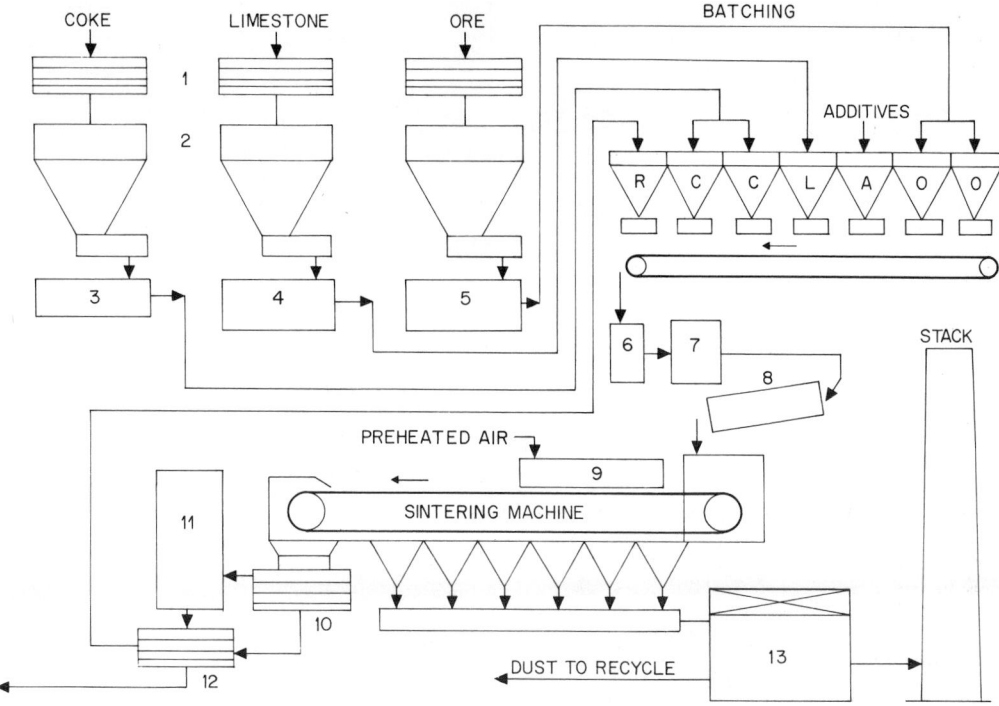

Fig. I-5. Continuous iron-ore sintering process. *A*, additives, *C*, coke fines, *L*, limestone flakes, *O*, ore fines, *R*, return fines. (1) Raw-material screens, (2) raw-material storage hoppers, (3) rod mill to reduce size of coke, (4) limestone crusher, (5) ore crusher, (6) mixing drum, (7) raw-sinter mixer, (8) rerolling drum, (9) burner hood, (10) sinter screening (hot), (11) sinter cooler, (12) sinter screening (cold), (13) electrostatic precipitator. (*Along lines of process developed by Metallgesellschaft A. G.*)

of a variety of materials, including naturally fine ores; ore fines from screening operations, flue dust, and ore concentrates; and other iron-bearing materials of very small particle size into a granular, relatively coarse form well suited for the blast furnace. A continuous sintering process is shown in Fig. I-5. A traveling grate conveys a bed of ore fines or other finely divided iron-bearing material intimately mixed with approximately 5% of a finely divided fuel such as coke breeze or anthracite. Near the head or feed end of the grate, the bed is ignited on the surface by gas burners, and as the mixture moves along the traveling grate, air is pulled down through the mixture to burn the fuel by downdraft combustion. As the grates (or pallets) move continuously over the wind boxes toward the discharge end of the strand, the combustion front in the bed moves progressively downward. This creates sufficient heat and temperature (1313–1480°C, 2400–2700°F) to sinter

the fine ore particles together into porous, coherent lumps. Sinter plants with a suction area of up to 5,200 ft² (400 m²), strand width of 17 ft (5 m), and production capacity of about 20,000 tons/day can be constructed.

Pelletizing differs from sintering in that agglomeration of the material is carried out prior to heat treatment by forming a "green" unbaked pellet or ball (sometimes referred to as *glomerule*), which is then hardened by heating. The iron ores to be pelletized are finely ground to present an adequate surface area for the formation of green balls and mixed with water and a binding agent. Bentonite, normally used as a binder, aids the balling operation, increases the strength of the green balls, promotes good surface properties, and preserves the ball strength during drying. The rolling action necessary to bring the moist particles together is usually achieved in a drum, cone, or disk with retaining lip. A small amount of fine solid fuel

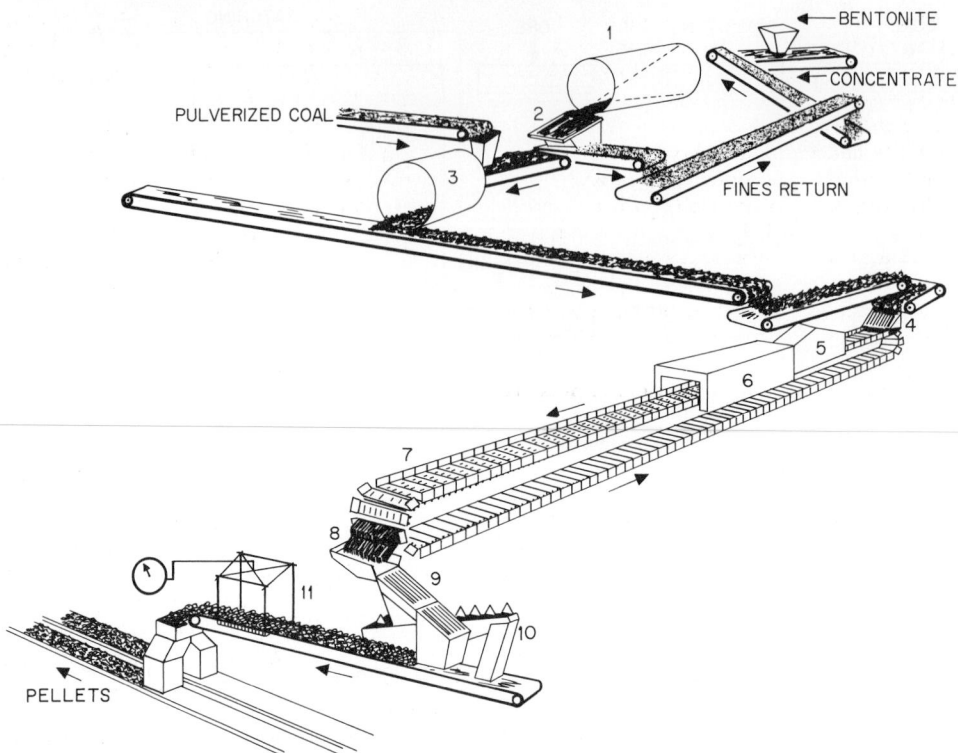

Fig. I-6. Pelletizing process in which a traveling grate is used. (1) Balling drum, (2) vibrating screen, (3) coal-coating drum, (4) vibrating screen, (5) drying hood, (6) ignition furnace, (7) pelletizing machine, (8) vibrating feeder, (9) vibrating screen, (10) classifier, (11) scale.

may be added to the pellet mix or coated on the pellets to supply part of the heat. Oxidation of a pelletized magnetite concentrate to hematite during the firing step may also supply a significant portion of the process heat requirements. Optimum moisture content for good pellets depends upon the fineness of the concentrate and use of additives and usually is in the 9.5–12% range. Bentonite is the principal additive. Others, such as soda ash, limestone, or dolomite, are sometimes used to improve pellet strength.

Pellet hardening by heating is carried out by various techniques, the common ones being (1) traveling-grate process, (2) combined grate and rotary kiln, and (3) shaft furnace. Figure I-6 shows a typical traveling-grate pelletizing process.

—S. C. Desai, *Davy Ashmore International Ltd., Stockton-on-Tees, England*

Iron Oxides, Synthetic Iron oxide pigments were probably the first inorganic coloring material used

by man. The colors are nontoxic, and their permanence is attested by the age-old articles and pictures colored by these pigments. Synthetic iron oxides were developed to improve the colors for printing ink, paint, and building materials. They surpass every prime pigment except white (TiO_2) in volume of use.

Synthetic yellow, brown, red, and tan colors are bright and clean in hue and well controlled in pigmentary properties, yet still carry the low cost associated with earth colors. The covering power has been increased to as high as $1,200 \text{ ft}^2/\text{lb}$ of oxide as measured by standard paint-industry tests. This is about 3 times greater than the value obtained by the best whites.

Development of the ferromagnetic properties of iron oxide has made these materials highly desirable for the production of magnetic components for the electronics industry. Recording tape is made from gamma ferric oxide, usually produced from synthetic yellow iron oxide. Ceramic magnets, *ferrites,* are produced from pure red iron oxides by reacting them at elevated temperatures with other metallic oxides. Ferrites with special magnetic properties can be synthesized to meet custom specifications. The magnets show excellent stability and reproducibility.

Several methods for producing iron oxides synthetically were used in the eighteenth and nineteenth centuries, but it was not until the twentieth century that most of today's techniques were developed.

Yellow iron oxide, similar in composition to the mineral goethite, is hydrated ferric oxide, $Fe_2O_3 \cdot H_2O$. It is produced synthetically by precipitation from an iron salt solution. The salts are generally by-products of other processes, e.g., spent steel-pickling liquor and titanium dioxide manufacture (copperas).

One widely used method for manufacturing synthetic yellow iron oxides is the Penniman-Zoph process, in which a hydrated ferrous oxide seed or nucleating particle is put into a tank containing a solution of ferrous sulfate and selected grades of scrap iron. Heat and compressed air (O_2) are then introduced, causing the iron oxide to grow onto the seed according to the reactions

$$FeSO_4 + 2NaOH \rightarrow Fe(OH)_2 + Na_2SO_4$$
Seed formation

$$4Fe(OH)_2 + O_2 \rightarrow$$
$$2Fe_2O_3 \cdot H_2O + 2H_2O$$
$$4FeSO_4 + O_2 + 6H_2O \rightarrow$$
$$2Fe_2O_3 \cdot H_2O + 4H_2SO_4$$
$$H_2SO_4 + Fe \rightarrow FeSO_4 + H_2$$

The scrap metal acts as the neutralizing agent for the sulfuric acid formed and consequently is an additional source of ferrous sulfate. The oxidation reactions which form $Fe_2O_3 \cdot H_2O$ must be carried out under carefully controlled conditions. Because of the small particle size, very light yellows are obtained first. As the reactions continue, particle size increases and the shades of yellow become deeper and redder. The reaction is terminated when the desired particle size as indicated by color is reached. The final precipitate is then carefully washed, dried, and packaged.

Black and brown synthetic iron oxides have the general chemical formula $FeO \cdot Fe_2O_3$. The ferrous constituent, FeO, is present in amounts varying from a trace to approximately 22%. When 19–22% FeO is present, the resulting black iron oxide is often called *synthetic magnetite.* This form of iron oxide displays the familiar ferromagnetic properties. Smaller percentages of FeO result in iron

oxide characterized by various shades of brown.

Both black and brown oxides are obtained by precipitating ferrous hydroxide from an iron salt solution. Oxidizing is rigidly controlled to obtain the desired ratio of FeO to Fe_2O_3. Black iron oxide can also be produced by chemical reduction of synthetic red iron oxide using specific atmospheres. Pure brown iron oxide can also be produced by physically blending red, yellow, and black synthetic iron oxides. For some applications, carbon black may be used in place of black iron oxide to produce nonmagnetic colors. These pigments vary in chemical analysis and differ in color characteristics from those produced by direct precipitation.

Red iron oxide, Fe_2O_3, ferric oxide or *synthetic hematite,* is produced by several synthetic methods, each having a specific influence on particle shape, surface area, and chemical purity. Within each method, variations in color ranging from light tile red to deep maroon may be produced. This synthetic red iron oxide is by far the most stable form. It is not affected by weather or weak acids or alkalis and does not change color with high temperatures. Other iron oxides usually change to the red (Fe_2O_3) form when subjected to such extreme conditions.

One of the methods of producing red iron oxide involves the dehydration of synthetic yellow iron oxide in furnaces using controlled temperatures and oxidizing conditions. Low-temperature processing retains the acicular characteristics of the starting yellow oxide and very high temperature processing will sinter the particles, causing shortening and rounding of the acicular crystals. Red iron oxides produced by this method are often termed *ferrite reds.*

In Europe a large quantity of synthetic red iron oxide is produced by oxidizing the FeO portion of Fe_3O_4 (black iron oxide) under controlled heating and oxidation conditions. Since this reaction can be self-sustaining, it is very difficult to control. The final product retains the particle shape of the original cubic crystal.

Roasted copperas reds are red iron oxides produced by the calcination of ferrous sulfate, $FeSO_4 \cdot 7H_2O$, sometimes referred to as *copperas* or *melanterite.* This method involves a two-stage calcination and generally produces spherical or nodular particles. This product is characterized by low surface area. The copperas solution, derived either from spent pickling liquor from steel mills or from titanium dioxide production, is saturated with pure iron and purified through recrystallization. The product $FeSO_4 \cdot 7H_2O$ is very pure and can meet the most stringent quality requirements. Because these iron oxides contain extremely low quantities of copper, manganese, and other heavy metals, they are ideally suited for applications involving food packaging, cosmetics, and rubber.

Synthetic reds are also produced by direct precipitation from solutions of ferrous salts. After World War II several patents were issued for the production of red iron oxide using these methods. The seed, similar to that used to produce yellow iron oxide, is reacted in tanks where the pH is very carefully controlled. As the particles grow, the color developed ranges from a light tile red to dark maroons. The particle shape of the precipitated red iron oxide is spherical, resulting in a low vehicle demand. The pigments are more uniform in particle size and shape than those from earlier processes. They are characterized by brightness of color but contain the impurities inherent in untreated precipitated products.

It should be apparent from these brief descriptions of manufacturing methods that the pigment color depends upon the finished particle size. Light yellow oxides, commonly described as the *light lemon* shades are characterized by an average particle size of about 0.2 μm when tested by sedimentation methods. Darker yellow or lemon shades average about 0.4 μm, while orange and dark orange materials range from 0.5 to 1.0 μm. Hiding-power tests show that 1 gal of paint containing 1 lb of lemon iron oxide will cover, or hide, 375 ft^2 of background. Other yellow oxides vary slightly from this value in a descending direction since the lemon-yellow iron oxide exhibits the optimum particle size for maximum hiding power.

The hue and strength of synthetic red iron oxide are also influenced by particle size. Red oxides for pigmentary applications range in particle size as shown in Table I-16.

Ferric oxide can exist in two different crystal structures, which exhibit different magnetic behavior. The alpha, Fe_2O_3, which is the common and stable form, is nonmagnetic. Carefully controlled reduction-oxidation processes are used to convert red Fe_2O_3 or yellow $Fe_2O_3 \cdot H_2O$ into gamma ferric oxide, which is highly magnetic. Acicular gamma ferric oxide, which is made from yellow iron oxide, is used extensively in the magnetic-recording industry. The needlelike particles permit a greater degree of orientation and higher permeability than is possible with nonacicular particles.

Uses: Many of the uses described briefly here are shared with natural iron oxides, but the synthetics are used whenever their special properties are required or they are economically feasible.

Catalysts: Due to the high surface area and excellent chemical stability red iron oxides are used in catalysts for sulfur removal from certain petrochemicals. Iron oxide–containing catalysts are also used in styrene production, liquid-hydrogen production, and solid rocket fuels.

Cement Products: Cement and concrete products are being produced in colors and there is increasing use in this steadily growing market because of the outstanding lightfastness and alkali resistance and relative ease of dispersion of iron oxides.

Electronics: Ferrite manufacture requires iron oxides of controlled uniformity and high purity. Ferrites are used in computer-memory cores, hi-fi speakers, color TV, and other electronic products.

Fabrics: Permanent colors are created in coated fabrics such as wall coverings, artificial leather, and recreational products. Iron oxides are also used to pigment the actual textile fibers.

Flooring: Good alkali resistance enables the iron oxide colors to withstand strong-alkali cleaning compounds and therefore makes them suitable for pigmenting flooring and wall-covering materials. These include the newer synthetic-fiber materials as well as the more familiar vinyl, asphalt, rubber, and linoleum sheet and tile products.

Inks: Textile-printing, engraving, rotogravure, and magnetic printing inks use the synthetic oxides.

Leather Finishes: Permanent iron oxides are used as pigments to make russet, brown, and tan shades in leather finishes.

Paints and Protective Coatings: All types of paints and protective coatings require synthetic iron oxides because of their lightfastness, chemical inertness, high opacity, and pleasing color tones either alone or in combination with organic pigments. Ease of processing, insolubility in water and organic solvents (good bleed resistance), high ultraviolet absorption, and infrared reflectance are all obtained at low cost. For these reasons, the paint and protective-coatings industry continues as a large user of iron oxides, both natural and synthetic. An interesting aspect of the infrared-reflecting properties of iron oxides is their use in camouflage coatings to prevent detection by aerial infrared photography.

Paper: Iron oxide pigments are used for beater coloring and surface coating of many types of paper products.

Plastics: Virtually any plastic material—rigid,

TABLE I-16. *Properties of Synthetic Iron Oxides*[a]

Type	Color[b]	Chemical Composition, %	Oil Absorption[c]	Particle Shape	Average Particle Size, μm	Hiding Power, ft²/lb	Sp gr	pH
Hydrated ferric oxide	Pigment yellow 42, CI 77492	$Fe_2O_3 \cdot H_2O$ 98+	32–52	Acicular	0.1–1.0	300–375	4.04–4.09	5.7–6.5
Dehydrated yellow (ferrite) reds	Pigment red 101, CI 77491	Fe_2O_3 98+	54–76	Acicular	0.4–2.9	1,000–600	4.46–4.58	5.0–6.0
Roasted copperas reds	Pigment red 101, CI 77491	Fe_2O_3 99+	13–26	Spheroidal	0.3–3.7	1,200–700	5.18	5.5–7.5
Precipitated reds	Pigment red 101, CI 77491	Fe_2O_3 97+	22–24[d]	Rhombohedral	0.2–1.5	1,000–800	4.90	6.0–7.0
Precipitated browns[e]	Pigment brown 6, CI 77491	$FeO \cdot Fe_2O_3$: FeO 1.6–9.2 Fe_2O_3 90.1–98.0	24–30	Cubic	0.2–0.8	1,000	4.69–4.77	8.0–9.0
Precipitated black[e]	Pigment black 11, CI 77499	$FeO \cdot Fe_2O_3$: FeO 22.0–25.6 Fe_2O_3 72.5–77.0	18–22	Cubic	0.2–0.8	900	4.95	7.0–9.0
Gamma iron oxide[e]	Pigment brown 11, CI 77499	Gamma Fe_2O_3 98+	24–28	Cubic	0.3–1.2	950	4.70	6.5–7.5
	Brown	Gamma Fe_2O_3 98+	30–60[d]	Acicular	0.4–0.8	…	4.45–4.60	2.5–7.5
Venetian reds[f]	Pigment red 101, CI 77491	$Fe_2O_3 + CaSO_4$: Fe_2O_3 40 $CaSO_4$ 60	18–20[d]	Spheroidal	0.75–3.0	300–400	3.45	4.0–5.0

[a] Values are typical ranges for synthetic iron oxides shown.
[b] CI = Colour Index published by the Society of Dyers and Colourists, Bradford, Yorkshire, England. Coordinated in the United States by the American Association of Textile Chemists and Colorists, Lowell, Mass. The United States secretary of the Colour Index is at the AATCC Technical Center, Research Triangle Park, N.C. 27709.
[c] Gardner-Coleman method unless otherwise indicated, ASTM D-1483-60.
[d] Spatula rub-out method, ASTM D-281-31.
[e] Have magnetic characteristics.
[f] Venetian reds are produced by the simultaneous calcining of ferrous sulfate or copperas with hydrated calcium oxide. The end product is an intimate combination of ferric oxide and calcium sulfate. They have been widely used as a pigment for barn paints.

flexible, reinforced, plastisol, organosol, or foam—can be pigmented with some type of iron oxide or closely related product such as a ferrite. The chemical purity and inertness of synthetic iron oxides are an advantage where storage stability is a factor.

Polishing: Iron oxide rouge has been used for centuries to polish metal and jewelry. It is also used in producing plate glass, mirrors, and optical lenses.

Recording Tape: The recording-tape industry is a large and growing user of synthetic iron oxides.

Roofing Granules: Coatings composed of a binder, such as sodium silicate pigmented with iron oxides, are used to coat roofing granules. The coating helps protect the base material, which is usually asphaltic, from embrittlement and rapid breakdown by ultraviolet light.

Rubber and Synthetic Elastomers: Rubber soles and heels, silicone-rubber stocks, coated fabrics, wire insulation, matting, home products, gasket stocks, and brake linings are only a few of the products in this broad classification colored with synthetic iron oxides. Roasted-copperas synthetic red iron oxides are widely used in rubber because of their low Cu and Mn content.

REFERENCES

American Society for Testing and Materials: "Annual Book of Standards," Philadelphia, Pa.
Color Stability of Red Iron Oxide Pigments, *Phila. Soc. Paint Technol. Off. Dig.*, vol. 29, no. 394, November 1957.
Gaynes, Norman I.: "Formulation of Organic Coatings," Van Nostrand Reinhold, New York, 1967.
National Bureau of Standards: "Organic Coatings," Washington, 1968.
Payne, Henry F.: "Organic Coating Technology," vol. 2, Wiley, New York, 1961.
Penn, W. S.: "PVC Technology," Maclaren, London, 1962.
U.S. Patents 360,967; 1,327,061; 1,368,748; 2,357,096; 2,716,595; 2,785,991; 2,866,686; 2,939,767.

—Carl W. Fuller, *Cities Service Company Incorporated, Trenton, N.J.*

Iron Phosphate   (See **Conversion coatings.**)

Iron Pyrite   (See **Arsenic.**)

Iron- and Steelmaking   Iron ores and ferrous scrap are the main iron-bearing raw materials used for producing iron and steel. Iron ore is smelted to produce liquid iron or, alternatively, reduced to give a solid product known as *sponge iron*. Liquid iron also can be made by melting ferrous scrap.

Steel is made by refining liquid iron by melting scrap and/or sponge iron. The liquid steel is cast into blocks and mechanically worked to obtain usable products. Liquid iron and steel can also be cast directly into the required shapes. Figure I-7 indicates the principal commercial processes which can be considered for various products.

Ironmaking. Iron is made mostly by (1) smelting run-of-mine or beneficiated iron ore in a blast furnace, low-shaft furnace, or electric smelter to give a liquid product; (2) reducing run-of-mine or beneficiated iron ore by direct reduction processes to produce sponge iron; and (3) melting ferrous scrap in a cupola, electric furnace, or fuel-fired furnace.

Chemistry of Iron-Ore Reduction: Since the most common iron ores contain iron in the form of oxides, iron is obtained by reduction of these oxides by a suitable reducing agent. The three main oxides of iron are tabulated.

Oxide	Fe, %	O, %
Hematite, Fe_2O_3	69.94	30.06
Magnetite, Fe_3O_4	72.36	27.64
Wüstite, FeO	77.78	22.22

The reducing agents of industrial importance are carbon, carbon monoxide, hydrogen, and hydrocarbons, e.g., methane. The reduction by carbon monoxide is exothermic and by carbon, hydrogen, and hydrocarbons endothermic. Hematite can be reduced in a reasonable period of time by either carbon monoxide or hydrogen over the entire operating temperature range. Magnetite can be similarly reduced by carbon monoxide; however, it cannot be completely reduced by hydrogen at temperatures above 600°C, where wüstite is formed. The reaction rate in the early stages of reduction is considerably higher with hydrogen than with carbon monoxide. The reaction rate with hydrocarbons is slow, and therefore they are reformed by cracking and partial oxidation before use as reductants.

The iron oxides form dark-gray porous sponge retaining the size and shape of the original material when reduced at temperatures below 1000°C (1832°F). A temperature of 950–1000°C (1742–1832°F) is necessary to achieve complete reduction in a reasonable length of time unless the ore is of fine size. Above 1000°C (1832°F) the product begins to soften and forms a pasty porous mass at about 1250°C (2192°F). The reduced iron

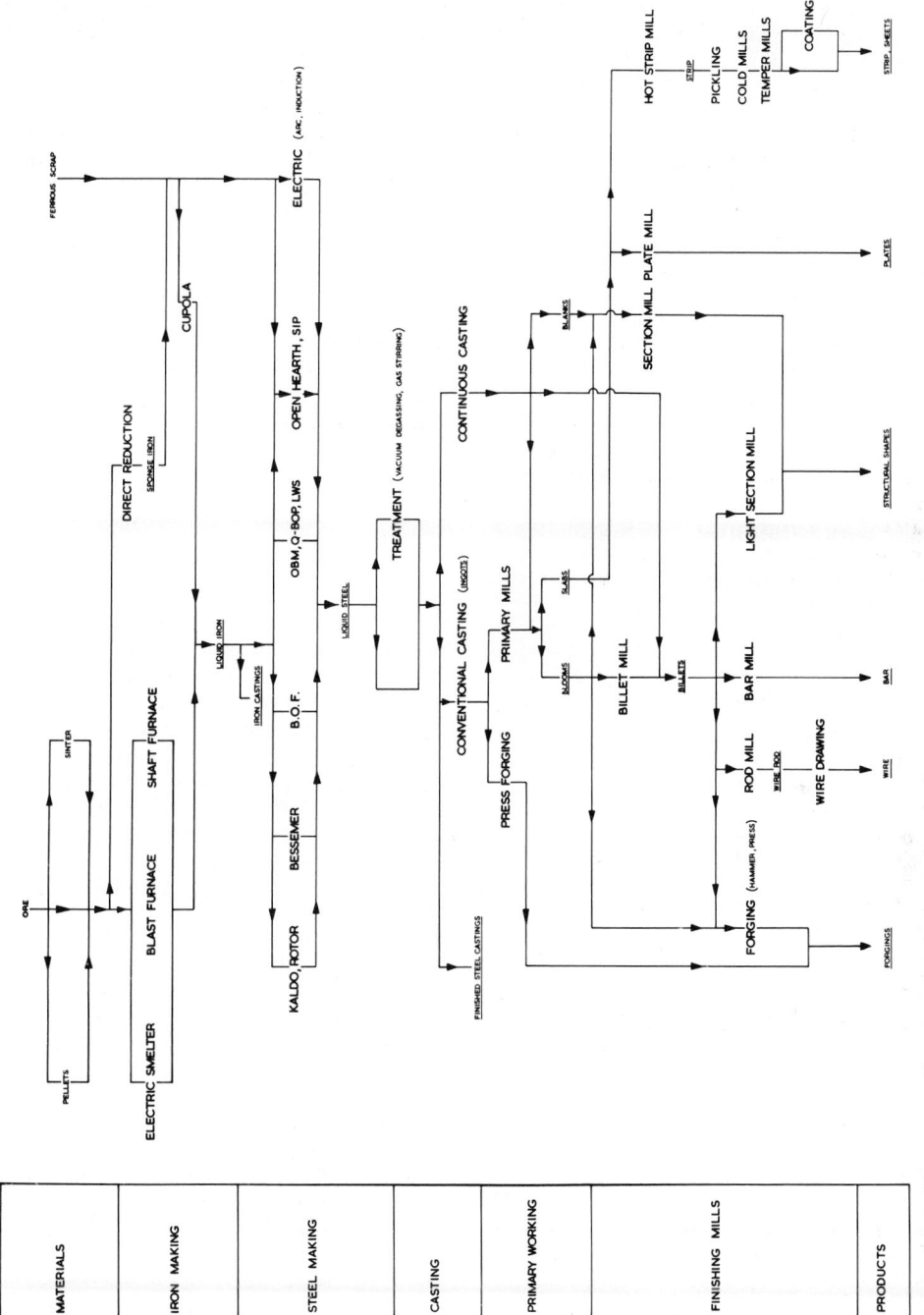

Fig. I-7. Principal processes and process routes in iron- and steelmaking.

647

Fig. 1-8. Modern blast furnace. (*Davy-Loewy Ltd., Sheffield.*)

absorbs carbon, if present, above 1300°C (2372°F) and fuses and melts to liquid iron between 1300 and 1500°C (2372–2732°F), depending on the carbon content.

Blast Furnace: Most of the iron produced in the world is made in a blast furnace (Fig. I-8), a tall refractory-lined furnace using the countercurrent-flow principle to achieve high efficiency. The raw materials, which include iron ore as sinter or pellets, are charged at the top along with coke as a reducing and thermal agent and limestone for fluxing the gangue material. Hot-air blast introduced at the bottom end burns the coke to heat, reduce, and melt the charge as it descends. The liquid iron and slag collect in the furnace hearth and are tapped at regular intervals through separate tapholes. The iron-ore smelting process is continuous, but if necessary the furnace can be damped down for short periods.

The waste gas from the combustion of coke, which contains about 28% CO and has a calorific value of about 90 Btu/ft^3 (800 kcal/m^3), is collected from the top end of the furnace by a downcomer pipe, cleaned to remove dust particles, and used as a fuel.

Blast furnaces vary in capacity from 100 to 10,000 tons/day with hearth diameter 9–46 ft and height 50–150 ft. The trend is toward large furnaces to reduce the capital and operating costs.

The general layout of a blast-furnace plant is shown in Fig. I-9. The blast furnace proper consists of a bottom, hearth, bosh, stack, and top. The raw materials are charged from the top by a skip hoist or belt-conveyor system via a double bell-and-hopper arrangement. The air required for combustion is supplied by turboblowers and preheated in hot-blast stoves lined with refractory-brick checkerwork. Usually three stoves are provided per furnace operated alternatively on a regenerative principle; i.e., one stove provides the blast while the other two are being heated up. Cleaned blast-furnace gas is used with coke-oven gas or oil for heating the stoves. The hot blast is supplied by a gas main to a bustle pipe, which encircles the bosh and distributes it to water-cooled tuyeres situated below the bosh for injection into the furnace. The metal is tapped into refractory-lined open-top torpedo or Kling-type ladles and transported to the pig-casting machine or to the steel plant.

New developments which have improved blast-furnace performance include the use of sized and prepared burden, high top pressure, higher blast temperature, fuel injection through tuyeres, reducing-gas injection in bosh, and oxygen enrichment of blast.

A few blast furnaces are operated with charcoal instead of coke, but their capacity is limited to a maximum of about 300 tons/day by the low crushing strength of charcoal.

Low-Shaft Furnace: Iron production in blast furnace requires a coarser burden to permit flow of reducing gases without channeling and good-quality abrasion-resistant high-strength metallurgical coke. For finer raw materials and low-grade coke or lignite the low-shaft furnace has been developed, which is very similar to the blast furnace in design and operation. Low-shaft furnaces are circular or oval in cross section, the oval shape allowing greater hearth area without increasing the required depth of penetration of the blast supplied through the tuyeres. Although this type of furnace was projected as being suitable for small-scale iron production, only a few have actually been installed.

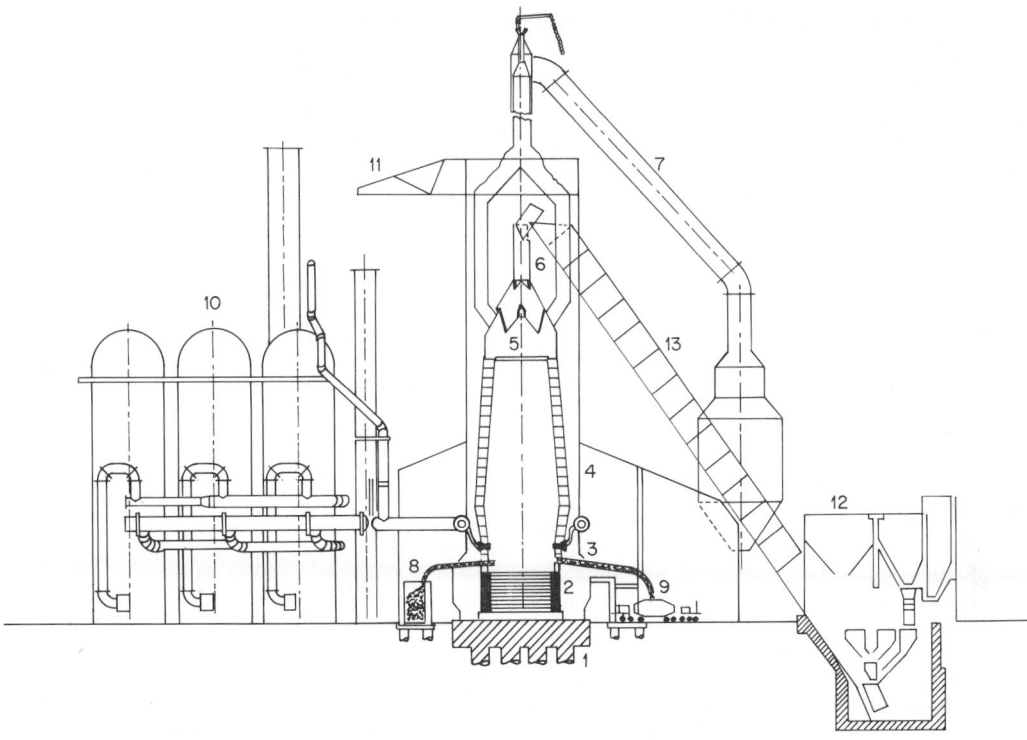

Fig. I-9. Cross section of a representative blast-furnace plant. (1) Foundations, (2) hearth, (3) bosh, (4) stack, (5) large bell, (6) small bell, (7) downcomer, (8) slag to slag pit, (9) iron to torpedo car, (10) hot-blast stoves, (11) outrigger, (12) raw-material bunkers, (13) skip hoist.

Electric Smelter: The electric iron-smelting furnace uses electric energy to provide heat, a function performed by coke in a blast furnace. The most widely used furnaces are of the submerged-arc type, developed by Tysland and Hole in Norway and using the Söderberg continuous self-baking electrodes. The furnace is circular or rectangular in cross section with transformer ratings of up to 60,000 kVA. Low-grade coke is used as the reducing agent. The electric furnace is used only when cheap power is available. By preheating and prereducing the iron-ore charge, power and the coke consumption for producing iron are lowered and production capacity increased. Rotary kilns or shaft furnaces are used for this purpose.

Direct Reduction Processes: A number of processes have been developed for partial or complete reduction of iron ore as an alternative to the blast furnace. Rotary and stationary kilns and furnaces, reverberatory furnaces, retorts, fluid-bed reactors, pot furnaces, and jet smelting, together with a variety of fuels and reductants, e.g., lignite, coal, coke, char, fuel oil, tar, and various gases, have been tried singly or in combination. Many of the processes have become obsolete or have not progressed beyond pilot-plant stage. Some have found only limited application under certain local conditions. Several direct-reduction processes* have been developed to semicommercial or production units. They are used to make a highly reduced product containing mostly metallic iron and little gangue material so that it can be used as a substitute for ferrous scrap to provide iron in steelmaking operations.

The SL/RN process employs a rotary kiln using solid reductant. The standard process based on the use of high-grade pellets or lump ores and anthracite operates as follows. Iron ore or pellets,

*SL/RN; HyL; Purofer; Midland-Ross; Fior; and U.S. Steel.

anthracite, dolomite, or limestone, and return coal are fed together into a rotary kiln. Shell burners supplied with air and gas or oil control the temperature. The aim is to maintain a constant temperature over the maximum possible kiln length in order to achieve a high utilization of the kiln volume. A uniform temperature of about 1100°C (2012°F) is obtained over approximately 60% of the kiln length. Coal can be injected from the discharge end. After leaving the reduction kiln, the charge is passed through a gastight seal into a water-cooled drum. It is cooled to a temperature below 100°C (212°F) to prevent reoxidation of the sponge iron. The cooler discharge consists of coarse- and fine-grained sponge iron, return coal (char), coal ash, and the desulfurizing agent. The grain size of the kiln feed is adjusted so that the major part of the sponge iron can be separated by ordinary screening. The fine-grained sponge iron is removed by low-intensity magnetic separators. The return coal is separated from the coal ash and the desulfurizing agent by screening to remove the fraction below 1 mm, comprising the desulfurizing agent, which is fed in a grain size < 1 mm, and the major part of the coal ash.

The HyL is a batch-cyclic process in which rich lump-iron ores or pellets are reduced in a fixed-bed reactor by reducing gas produced by steam reforming of natural gas or other hydrocarbons. The reducing gas, prepared by catalytic conversion of methane and steam using Ni as a catalyst, contains CH_4, 5%; CO, 13%; CO_2, 8%; and H_2, 74% by volume.

Four reactors are employed, each normally passing through the following four steps in an overall 12-hr cycle: (1) removal of cold sponge iron and loading with fresh iron ore or pellets, (2) preheating and secondary reduction with partially spent reducing gas from another reactor, (3) primary reduction to sponge iron, and (4) cooling the sponge iron with fresh, cool reducing gas and controlled deposition of carbon if desired.

The Purofer and the Midland-Ross are both continuous processes using shaft furnaces charged from the top with iron ore or pellets and discharging reduced product at the bottom. Reducing gas is generated by reforming natural gas or methane-rich gas made from naphtha. The hot reducing gas flows countercurrent to the descending charge.

The Fior and the U.S. Steel processes are also continuous direct-reduction processes using reformed natural gas or hydrocarbons as reducing agents in fluid-bed reactors. Depending on the form in which it comes from the mine or concentrator, iron ore may need to be dried and ground. It is then charged into a fluid-bed reactor, where it is heated and contacted with the reducing agent. Reduced ore may be briquetted or used as fines.

In the direct-reduction processes using gaseous reductants the iron ore or pellets are not mixed with any other materials, but in processes using solid reductant some ash and char is produced, and some desulfurizing agents, e.g., limestone or dolomite, are charged. These materials must be separated from the discharged reduced sponge iron.

Cupola and Other Furnaces: The cupola is the most economic and most common melting unit for producing iron. It is used widely for making iron for casting in iron foundries and occasionally to supplement iron needed in steelmaking furnaces. Cupola output is 2–75 tons/hr, and they are operated from a few hours daily to 2–3 months at a time. Whereas a blast furnace is a reducing unit, cupola only melts ferrous scrap and cold pig iron or prereduced sponge iron charged in alternate layers with coke. Limestone is added to flux the ash from the coke and form slag. The cupola is a vertical cylindrical shaft furnace working on the countercurrent-flow principle to heat and melt charge as it descends. Heat is supplied by almost complete combustion of coke by air injected through tuyeres near the hearth zone. Many of the cupolas are equipped for hot-blast supply through recuperators. The molten iron collects in the hearth and is removed continuously or intermittently through a taphole.

Reverberatory air furnaces or rotary furnaces fired with oil, gas, or pulverized coal and electric-arc furnaces are also used for melting ferrous scrap or cold pig iron to produce iron.

Steelmaking. Steel can be made by using (1) liquid iron as the main constituent of the charge, (2) steel scrap as the main constituent of the charge, (3) prereduced sponge iron, and (4) a mixture of liquid iron, scrap, and sponge iron.

A number of processes are available using external fuel in the form of gas, oil, and electric power or chemical heat produced by exothermic reactions of oxidation of metalloids, e.g., C, Si, Mn, and P, contained in the charge material to make steel.

Open-Hearth Process: The open-hearth process has been in commercial use for more than 100 years. The refractory-lined furnace, which can be of stationary or tilting type, consists of a shallow hearth enclosed by walls with charging doors on one side and a taphole on the other; it has a skewed roof and is fired from either end. Regenerative checkers are provided at either end to heat fuel and/or air before burning over the hearth. By switching firing from alternate ends at regular intervals and providing the right kind of flame, steelmaking temperature is achieved in the furnace. The main feature of the open-hearth process is its versatility in handling a variety of raw materials to make most grades of steels. Charges containing 100% scrap, 100% hot metal, or scrap and hot metal in intermediate ratios can be used to make steel. The acid open-hearth process imposes limitations on the S and P contents of the metallic charge, as they cannot be removed during refining. The basic open-hearth process can handle almost any type of metallic charge.

The Ajax process, which uses modified tilting-type open-hearth furnaces, can refine 100% hot-metal charges with oxygen.

Bessemer Process: In the bessemer process liquid iron is refined in a bottom-air-blown converter, which is a refractory-lined pear-shaped cylindrical vessel open at the top to permit charging of materials and allow escape of gases produced during blowing. The acid bessemer process requires iron low in P but high in Si. The basic bessemer process, also known as the *Thomas process,* can refine high-phosphorus iron.

The bessemer process can melt only 5–10% scrap, as a considerable amount of heat is wasted in heating nitrogen contained in the air. Also the nitrogen content of the steel produced is high. To solve these problems oxygen enrichment of air blast was tried but gave rise to tuyere refractory problems. Oxygen-steam or oxygen–carbon dioxide mixture is used instead of air to produce low-nitrogen steels.

Basic Oxygen Process: The basic oxygen steelmaking process has become a significant contributor to the steel output of the world, its outstanding success being due to its ability to produce high-grade steels at rapid rates, thereby reducing operational and capital costs compared with the other steelmaking processes based on the use of liquid iron.

Molten pig iron is refined to steel by top-blowing oxygen at high pressure onto the surface of the metal through a water-cooled lance contained in a tilting furnace, as shown in Fig. I-10. Oxidation of C, Si, Mn, and P provides the heat required for converting the molten iron into steel. The excess heat generated allows up to 30% scrap in

Fig. I-10. A 300-ton-capacity basic oxygen furnace. (*Davy Ashmore International Ltd., Stockton-on-Tees.*)

the charge. The conventional basic oxygen process can refine iron containing up to 0.3% P into most grades of steel. If the P content is higher than this amount, a modified version of the process which uses injection of powdered lime with the oxygen stream or double slagging practice is necessary. Figure I-11 shows a typical basic oxygen steel plant.

OBM, Q-BOP, LWS, and SIP Processes: Recently, the OBM and Q-BOP processes have been developed which use bottom blowing of oxygen and a shielding hydrocarbon through tuyeres in the bottom of a converter vessel. The endothermic dissociation of hydrocarbon by its cooling effect prevents excessive refractory wear in the tuyere area. This results in a substantial increase in the life of the bottom refractory plug. The OBM and Q-BOP processes are essentially the same except OBM generally refers to converters that refine high-phosphorus pig iron and Q-BOP refers to converters that refine low-phosphorus pig iron. Natural gas, propane, or LP gas are widely used as the gaseous hydrocarbon shield injected through an outer concentric gap around the central oxygen tuyere.

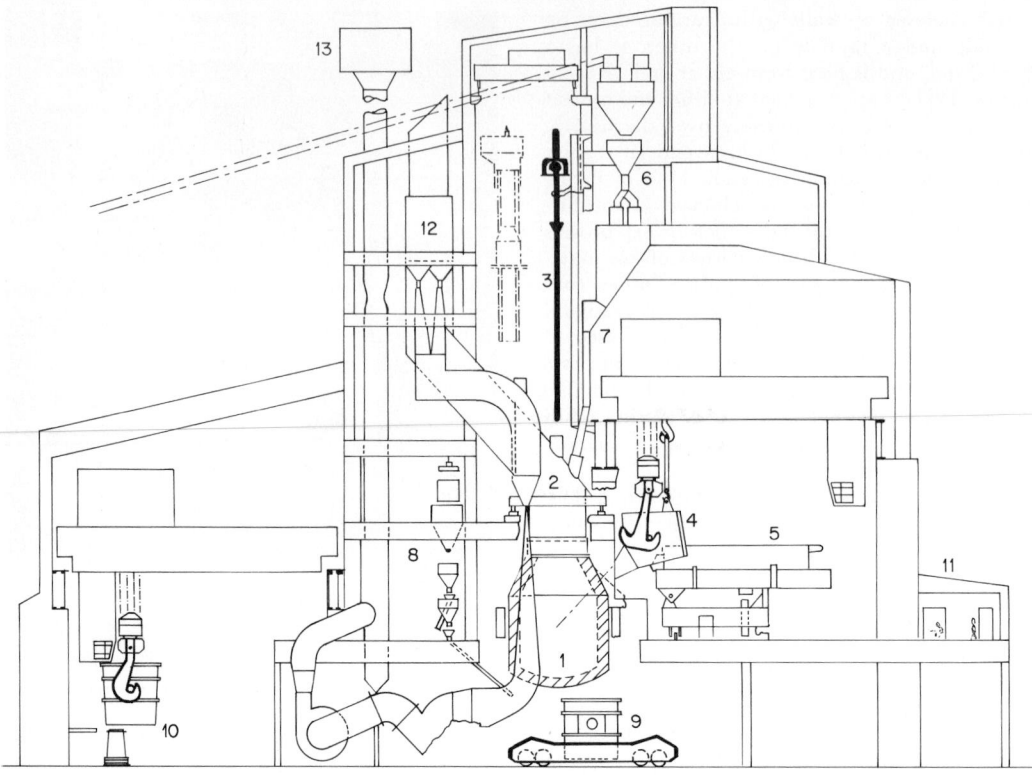

Fig. I-11. Section through basic oxygen steel plant. (1) Basic oxygen furnace, (2) fume hood, (3) oxygen lance, (4) hot-metal charging ladle, (5) scrap-charging car, (6) flux-addition system, (7) flux chute, (8) alloy-addition system, (9) steel-ladle transfer car, (10) ingot casting, (11) control pulpit, (12) gas-cleaning system, (13) exhaust stack.

A similar development, preferably using fuel oil as a hydrocarbon shield, is termed the LWS process.

The OBM tuyere is successfully inserted in the bottom of an open hearth furnace for injecting oxygen into the metal bath for refining. This new steelmaking technique is known as SIP (Submerged Injection Process).

Kaldo and Rotor Processes: The Kaldo process, developed in Sweden, was initially considered as a strong competitor to the basic oxygen steelmaking process. In this process the refining of molten iron to steel is carried out in a tilted pear-shaped basic-lined converter (Fig. I-12) rotating at speeds up to 40 rpm. Oxygen is blown at an oblique angle to the metal bath through a water-cooled lance. Most of the CO produced by the oxygen and carbon reaction is burned inside the converter vessel, the heat so generated being absorbed by the rotating vessel and transferred to

the bath, thus giving high thermal efficiency with a consequent increase in the amount of scrap the process can consume (up to 40%). The rotation of the vessel affords more accurate control over the bath composition and temperature, although the refining time is somewhat longer than for the basic oxygen process. The Kaldo process is very flexible in its ability to refine different types of iron. Both low- and high-phosphorus irons can be treated successfully for producing various grades of steel.

The investment costs for a Kaldo plant are higher than for a basic oxygen steelmaking plant. The refractory consumption is high and the lining life low, which reduces the operating availability of the furnace. For these reasons the process has not been very widely applied.

The Rotor process uses a long cylindrical horizontal steelmaking vessel, rotating slowly at 1–5 rpm. Oxygen is injected by two lances, one with

Fig. I-12. Charging scrap into a Kaldo furnace. (*Davy Ashmore International Ltd., Stockton-on-Tees.*)

high-purity oxygen into the bath and the second with low-purity oxygen to burn the CO evolved by the refining reaction to CO_2 within the vessel. The process has characteristics similar to those of the Kaldo process, but the long narrow shape of the vessel causes difficulties in charging scrap.

Electric Furnace: Direct-arc electric furnaces are commonly used for making steel. The furnace consists of a tilting cylindrical bowl-shaped hearth with three graphite electrodes inserted vertically through the roof. Three-phase supply is connected to the electrodes via a transformer. The furnace is charged through the door or from the top by using a swing-type roof. An arc is struck between the charge and the electrodes to supply heat. The rate of heat input is high since the temperature of the arc is about $3400°C$ $(6152°F)$. It is possible to operate the furnace under oxidizing, reducing, or neutral conditions, thus making it very suitable for the production of high-grade steels. The furnace can be tilted to tap steel and remove slag.

The electric steelmaking process is comparable to the open-hearth process in its ability to use raw materials and produce various grades of steel. Even though a few electric-furnace steel plants are operated with liquid iron in the charge, steel scrap and prereduced pellets are the most common raw materials.

Acid practice is confined to small foundry units. Basic electric-furnace steelmaking has shown spectacular growth in recent years. Improvements in furnace design and increase in the power input have resulted in lowering capital and operating costs. With ultrahigh-power-input operation and transformer ratings of up to 100,000 kVA it is possible to produce common grades of steel, requiring single slag practice only, in up to 300-ton heat sizes in less than 3 hr.

Steelmaking techniques have been developed for continuous charging of prereduced sponge iron as a substitute for steel scrap.

Current of high or medium frequency is passed through a coil surrounding a refractory crucible containing the charge in an *induction furnace,* used mostly for making special steels.

Other Processes: Attempts are being made to develop new steelmaking processes, but none have yet achieved large-scale commercial recognition.

In the *spray steelmaking* process liquid iron is poured through a tundish and refined continuously into steel by injecting powdered lime and oxygen tangentially from a ring onto the surface of the metal stream. The fuel-oxygen-scrap (FOS) *process* utilizes a vessel similar in shape to the electric-arc furnace but with greater height-to-diameter ratio. The roof is removable to permit rapid charging of scrap. Instead of the electrodes used in the arc furnace, the heat is supplied by an oxyfuel burner inserted through a central opening in the roof.

The *Wocra* and the *Irsid* processes are based on continuous melting and/or refining technique. The *cyclosteel* and *jet-smelting* processes attempted to make liquid iron or possibly steel by flash smelting of iron ore.

Casting of Liquid Steel: Whichever process of steelmaking is adopted, liquid steel is tapped from the steelmaking furnace into a ladle, a refractory-lined cylindrical container with trunnion attachment for crane lifting to transport steel. The bottom of the ladle is generally fitted with stopper-rod nozzle or sliding-gate nozzle for pouring steel. Lip-poured ladles are occasionally used.

Deoxidizing, recarburizing, or alloying additions are made to the ladle during steel tapping from the furnace to adjust the final composition of the steel to the desired specification. *Vacuum* or *gas-stirring* treatment is sometimes used before casting steel.

By treating liquid steel under vacuum it is possible to reduce the amounts of H_2, N_2, O_2, and harmful nonmetallic inclusions, resulting in improved physical properties and quality of steel. Removal of H_2 from certain forging steels eliminates the need of lengthy heat treatment lasting 2–3 weeks in subsequent processing. Methods developed for vacuum degassing of steel before casting are (1) ladle degassing in a chamber, (2) stream degassing by pouring from ladle to ladle or ladle to ingot, (3) vacuum-lifter or circulation degassing, (4) mold degassing, (5) combination of arc heating and degassing, (6) vacuum-furnace degassing, and (7) using a consumable electrode under vacuum.

Inert gases, e.g., Ar or N_2, are bubbled through liquid steel to equalize temperature and improve steel quality. This gas-stirring treatment uses a hollow stopper rod or porous refractory bottom plug in the ladle.

After tapping from the steelmaking furnace, vacuum degassing, or stirring treatment liquid steel is teemed into molds as ingots, continuously cast or pressure-poured into semifinished shapes, or poured into molds as steel castings. Generally an overhead traveling crane is used for transporting the ladle containing liquid steel for pouring.

In *conventional casting-pit practice* steel is teemed from the ladle into iron ingot molds of square, rectangular, polygonal, or round cross section to solidify as blocks. The iron molds may be placed

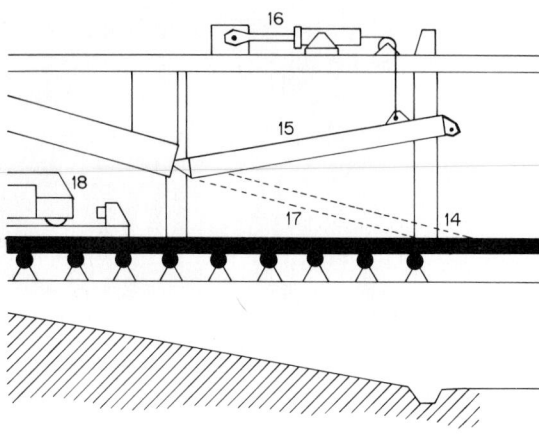

on the floor or on special casting bogies. The ingots can be top-poured directly into individual molds or bottom-poured simultaneously into a cluster through a trumpet-and-runner arrangement. Hot tops are used on molds to counteract shrinkage during solidification of killed steels. After teeming the molds are stripped and ingots charged into soaking pits for heating for subsequent processing or allowed to go cold for placing in stockyard.

To eliminate the need for heavy rolling-mill equipment for primary reduction of ingots, more and more use is being made of *continuous casting* (Fig. I-13) to pour liquid steel directly into semifinished shapes, e.g., slabs, blooms, blanks, or billets. Steel from the ladle is poured via a tundish into a water-cooled copper mold. As casting begins, the bottom of the mold is sealed with a dummy bar onto which the steel solidifies.

Solidified cast product is continuously removed through a direct spray-cooling withdrawal roll and cutoff system, maintaining a desired molten metal level in the copper mold. After cutoff the cast product is discharged onto a cooling bank. The following types of single or multistrand con-

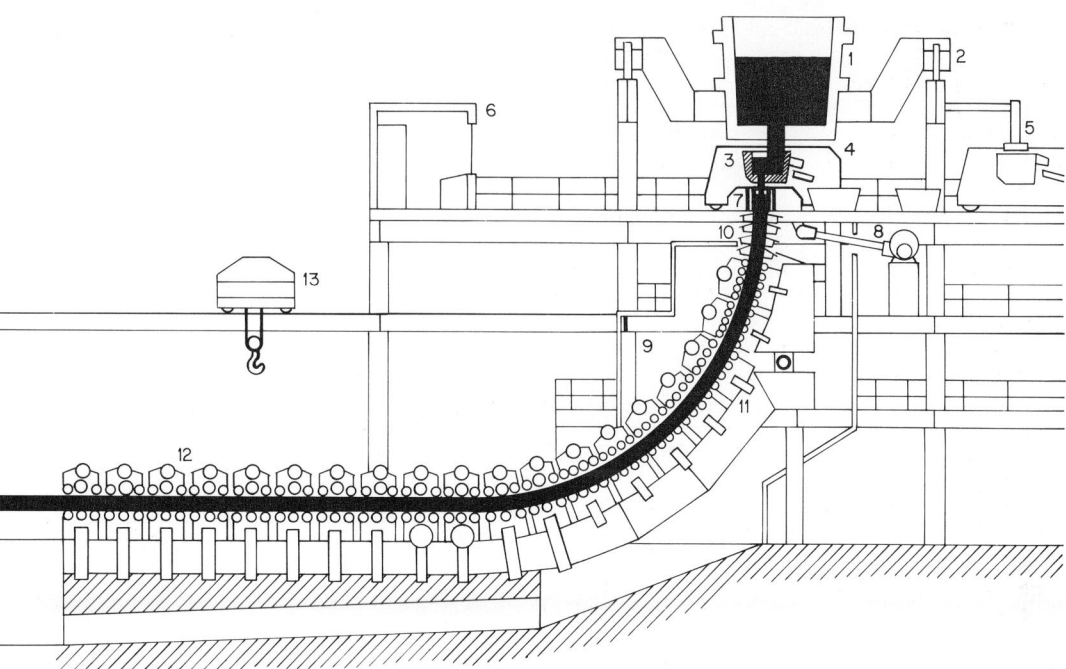

Fig. I-13. Continuous slab-casting machine. (*Concast AG, Zurich.*) (1) Ladle, (2) ladle car, (3) tundish, (4) tundish car, (5) one of the tundish preheating stations, (6) control room, (7) mold, (8) mold reciprocating drive, (9) secondary cooling zone, (10) cooling plates, (11) roller segments, (12) extended roller-segment zone, (13) auxiliary hoist for roller-segment maintenance, (14) tiltable dummy-bar head, (15) dummy-bar storage, (16) dummy-bar storage-cradle elevating mechanism, (17) cast strand, (18) cutting station.

tinuous-casting machines are available: (1) vertical mold with vertical cutoff, (2) vertical mold with bending rolls and horizontal cutoff, (3) curved mold with bending and horizontal cutoff, and (4) machines with direct strand-reduction units to reduce cross section by rolling before cutoff.

In the *pressure-pouring process* (Fig. I-14) the molten metal is forced up through a refractory tube into a mold by means of compressed air. Two systems are used to cast a number of molds in succession. Either the ladle is placed in a stationary airtight pressure chamber and molds moved over the chamber, or, alternatively, the molds are stationary and the pressure chamber containing the ladle is transported underneath the molds. The height to which the liquid steel is raised depends upon the air pressure applied, and the rate of pouring is determined by the rate of air-pressure increase.

Continuous casting and pressure pouring both give increased yield from liquid steel to semi-

finished product compared with conventional casting into ingots. The capital costs for the overall plant are also lower, due to elimination of stripping, mold preparation, soaking pits, and primary rolling mills.

Some products which have intricate shapes or cannot be formed by mechanical working can be obtained by pouring molten steel in a mold of the desired shape and allowing the steel to solidify in the mold cavity. Sand molds are most commonly used in this method of producing *steel castings*.

Shaping Steel: The ultimate aim of any iron- and steelmaking operation is to produce a usable product. Except for a small tonnage of castings, most steel is cast first as ingots or semifinished products and then mechanically worked into the desired shapes and sizes. This processing reforms the cast structure and improves the physical properties of the steel.

Normally, steel is heated until it is plastic enough for mechanical working. For heating ingots and semifinished products, the *soaking pit*

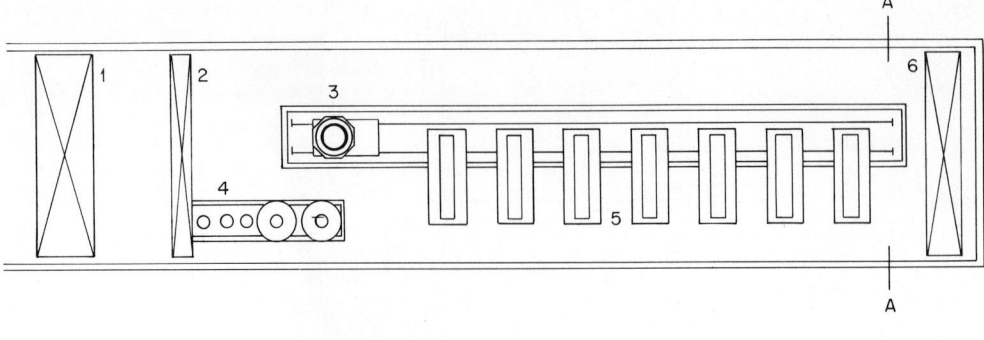

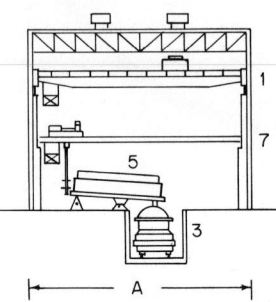

Fig. I-14. Pressure-pouring plant. Fixed-mold and moving-tank layout. (1) Ladle crane, (2) tube crane, (3) mobile pouring tank, (4) tube preheater, (5) molds, (6) slab crane, (7) mold coating machine.

and the *reheating furnaces* are used respectively. The heating temperature varies in the range 1150–1350°C (2102–2462°F), and many types of furnaces are available. The selection is determined by the grade of steel, shapes and sizes of the cast product, the output rate, and the operating temperature level desired.

The three main methods of forming steel by hot working are forging, extrusion, and rolling.

Forging: Forging steel consists in working it into a finished shape by hammering or pressing. Forged products include axles, crankshafts, rolling-mill rolls, boiler drums, turbine rotors, and components of cars and other machinery. In *hammer forging* the deformation of the red-hot steel block which rests on the anvil takes place under the action of repeated blows of the heavy part of the hammer, called the *ram*. Steam or compressed air is the usual operating medium, but other methods may be used. If intricate shapes are required, the ram and the anvil may be fitted with detachable dies shaped like the final product in each half. This method, called *drop forging* or

stamping, is usually employed when precise dimensions and a large quantity of items of one pattern are required.

In *press forging* (Fig. I-15) the heated steel block is formed into shape in a hydraulic press by a steady squeeze which penetrates through the entire thickness of the forging.

Extrusion: In the hot-extrusion process the heated piece of metal is placed in a chamber. A high pressure applied from one end by means of a hydraulically operated ram causes the metal to flow through a restricted orifice at the other end to produce desired shapes, e.g., rounds, squares, and hexagons. A die and a mandrel are used to produce tubes. Because of its high cost the process is applied on a limited scale for stainless and high-alloy-steel products.

Rolling: More than 90% of the world steel production is processed by rolling, which forms the final stage of most of the iron- and steelmaking operations. Rolling mills (Fig. I-16) are used to produce various semifinished and finished products, e.g., blooms, billets, slabs, rails, beams, channels, angles, rounds, squares, plates, sheets, and strip.

The rolling operation consists of passing the material between two rolls revolving at the same peripheral speed in opposite directions. The rolls have smooth or grooved surfaces, and the gap between them is smaller than the height of the entry material. The rolls grip the material and during its passage through them effect a reduction in the cross-sectional area with corresponding increase in the length. Multiple passes through one or more sets of rolls produce the final shape from a large block.

Various types of rolling mills and roll arrangements are employed in steelworks depending on the product desired. If the starting point is a large steel ingot, primary heavy rolling is done in the *blooming* and/or *slabbing mills* to produce blooms or slabs. Blooms range in size from 5×5 in. upward. The products which originate from a bloom are so numerous that only a few can be mentioned, e.g., rails; heavy, medium, and light sections, which are the structural shapes used for bridge and steel-framed building construction, window framing, and steel partitions; bars in a variety of cross sections, from which nuts, bolts, shafts, and machinery parts are produced; rod, which is drawn into wire and which is also used

Fig. 1-15. A 2,000-ton forging press. (*Davy-Loewy Ltd., Sheffield.*)

Fig. 1-16. A 42 × 102 in. blooming and slab mill. (*Davy-Loewy Ltd., Sheffield.*)

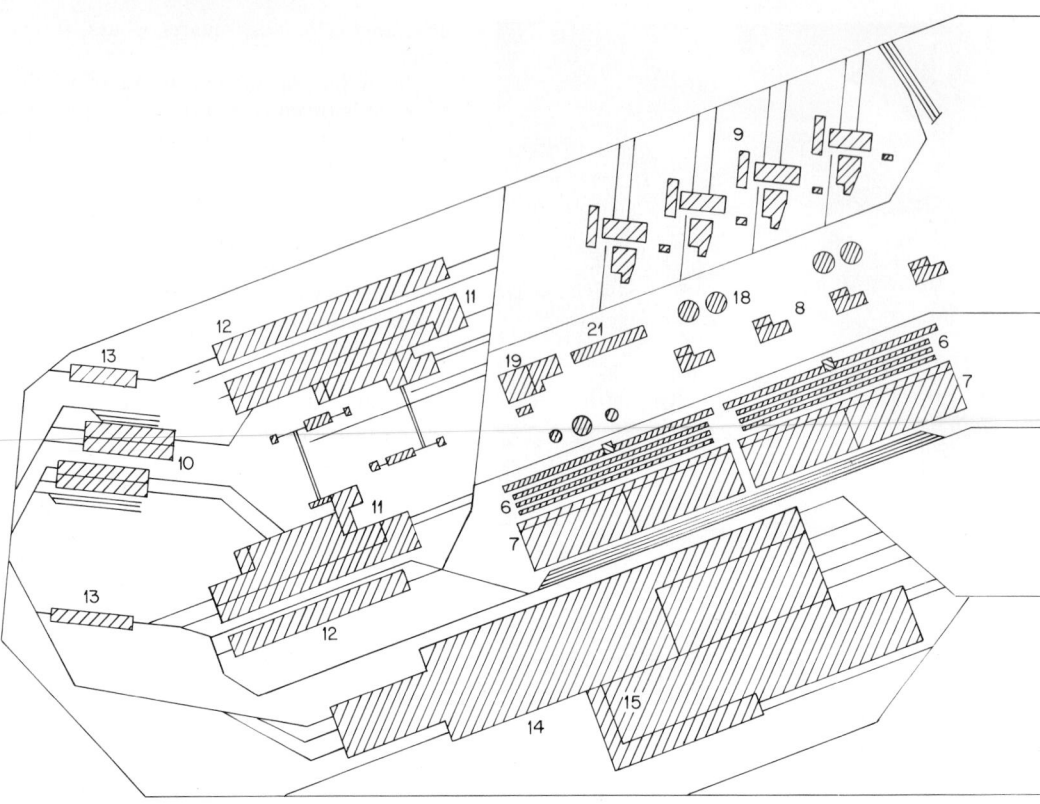

Fig. I-17. Layout of a modern integrated iron- and steelworks, capacity 8–10 million ingot tons per year. (1) Ore terminal, (2) ore stock, (3) coal stock, (4) ore-preparation plant, (5) sinter plant, (6) coke ovens, (7) by-products, (8) blast-furnace bins, (9) blast-furnace plant, (10) scrap stock, (11) steel plant, (12) mold preparation, (13) ingot stripper, (14) small section mill, (15) section and billet mill, (16) strip mill, (17) plate mill, (18) Dorr thickeners, (19) power station, (20) central administration office, (21) central laboratory, (22) central engineering and maintenance shops, (23) central stores area, (24) central garage, (25) central oil farm. A total facility of this type is extensively interlaced with several miles of railway track, only the major links of which are shown schematically on this diagram.

in reinforced-concrete work; and narrow strip used to manufacture articles such as wheel rims, razor blades, tubes, and pipes.

Blooms obtained by rolling or continuous casting are further processed in rail and structural mills, heavy beam mills, medium section mills, billet mills.

Billets produced by rolling in a billet mill or directly by continuous casting are usually 2–5 in. square in cross section. *Narrow strip mill* is used to roll strip up to 18 in. wide. *Rod and bar mills* produce light sections, bars, and rods. Rods can be drawn into wires.

Plates, which are used in a variety of industries, e.g., heavy engineering, shipbuilding, refineries, chemical plants, bridge building, and railway rolling stock, are produced by direct rolling of ingots or slabs produced from slabbing mills or from continuous casting machines in a *plate mill.*

Steel sheets and strip are extensively used in the manufacture of motor cars, refrigerators, washing machines, cookers, and many other domestic and industrial products. Slabs produced in a slabbing mill, a continuous casting machine, or pressure-poured molds are hot-rolled in the *hot strip mill,* descaled by pickling in an acid solution, and cold-rolled and tempered in the *cold-reduction* and *temper mills.* The cold-rolled strip may be sold in coil form or be taken to a side-trimming and sheet-shearing line, where it is cut into lengths of sheet to meet customer requirements.

Tin cans used as containers in the food, drink, and other industries are made from tin plate, i.e.,

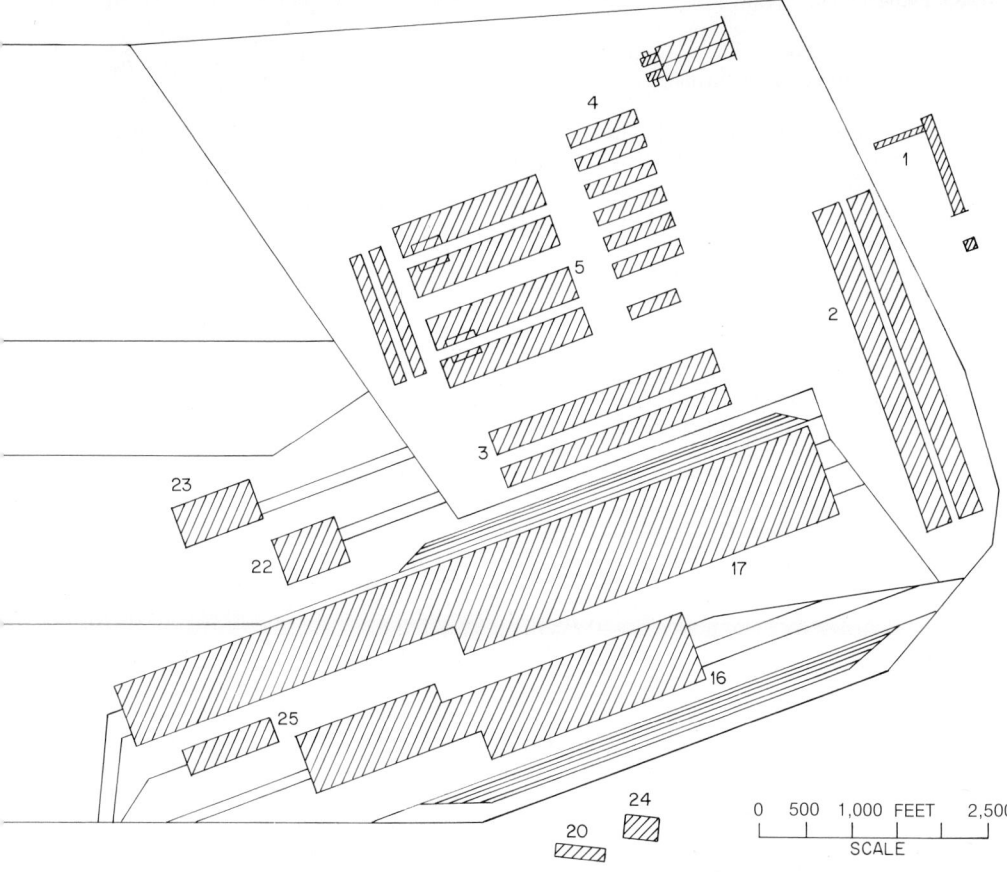

<table>
</table>

sheet steel thinly coated on each side by hot-dipping or the electrolytic method. The processing steps consist of pickling hot strip and passing it through the cold and the temper mills, annealing and tinning lines. Zinc, terne, aluminum, or plastic coating can be applied to obtain corrosion resistance.

Steel tubes and pipes, made by a variety of processes, can be divided into seamless and welded tubes. The rotary forge process or extrusion is used for seamless-tube production; welded tubes are made by the hydraulic-weld, electric-fusion-weld, continuous-weld, and electric-resistance-weld processes.

Iron- and Steelworks. The iron- and steel-production industry is highly capital-intensive. The facilities require integration of a number of process and plant units involving heavy capital investment. Iron- and steel-production plants can be classified as follows:

1. *Nonintegrated steelworks,* in which the plant consists only of rolling mills and finishing facili-ties. Billets, blooms, slabs, or ingots are imported from outside and processed in the steelworks.

2. *Semi-integrated steelworks,* in which a steelmaking plant is installed in addition to the rolling and finishing facilities. This type generally includes open-hearth or electric-arc furnaces for making steel by scrap melting from imported sponge iron. A few steelworks use blast-furnace iron transported in torpedo ladles from iron works located up to about 50 mi away.

3. *Integrated iron and steelworks* consists of all facilities, starts with iron ore as the main raw material, and goes up to the rolling and finishing operations.

An integrated iron- and steelworks consists of raw-materials unloading and stockyard, coke ovens, sinter or pelletizing plant, blast furnaces, steelmaking and casting plant, rolling mills, and finishing facilities. In addition water-treatment facilities, power plant, repair and maintenance workshops, oxygen plant, lime kilns, fuel-storage tanks, and laboratories are included. Figure I-17

659

shows the layout of a typical large-scale integrated iron- and steelworks based on blast-furnace iron-making.

Since direct-reduction processes for producing sponge iron are finding increasing application, integrated works using the direct-reduction route for making sponge iron from iron ore and melting it in electric-arc furnaces are also installed.

The production capacities of a steelworks range from a few hundred tons to 8–10 million tons/yr. As more and more new sources of cheap high-grade iron ores are developed, the tendency is to build large integrated iron- and steelworks on coastal sites. By installing a large blast furnace, basic oxygen furnace, and rolling-mill units the capital investment and the production costs per ton of product are substantially reduced.

On-line process and production control by computers is used increasingly because of advances in continuous measurement and analysis techniques.

—S. C. Desai, *Davy Ashmore International Ltd.,*
Stockton-on-Tees, England

Irrigation Water   (See **Desalination.**)

Isenthalpic Process   A process in which there is no change in enthalpy is termed an isenthalpic process. See also **Isentropic process.**

Isentropic Process   A process in which there is no change in entropy is termed an isentropic process. See also **Isenthalpic process.**

In processes of gas separation at low temperatures (below $-200°F$), the process of liquefaction is required. A major question is that of the most effective liquefaction method. Essentially, this involves refrigeration or heat pumping at low-temperature levels. Three methods have proved practical: (1) vaporization of a liquid; (2) using the Joule-Thomson effect in gases; and (3) causing expansion of a gas in an engine that is required to do external work. The methods have been used both separately and in combination.

The lowest temperature that can be effected by method 1 is 63 K, the triple point of nitrogen, inasmuch as there is no more volatile fluid that will condense at this temperature.

Although the self-cooling produced on expanding a gas such as air from a high pressure to 1 atm through a throttle is relatively small (method 2), use of an efficient heat exchanger to make it accumulative permits temperatures within a few degrees of absolute zero to be reached.

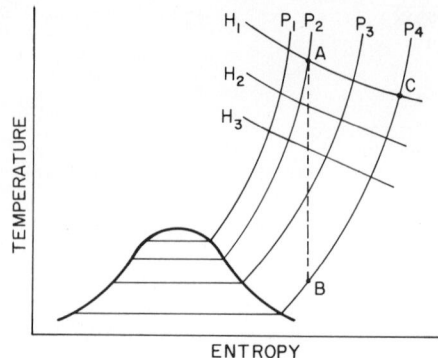

Fig. I-18. Comparison of the cooling effects in isentropic and isenthalpic expansions. H_1, H_2, and H_3, constant-enthalpy lines; P_1, P_2, P_3, P_4, constant-pressure lines.

Method 3 is probably the most important of the three inasmuch as it offers the best economy for a large-scale process. By utilizing expansion with external work (*isentropic* expansion at the limit) instead of *isenthalpic* expansion, the amount of cooling for a given pressure difference is much greater. This is shown diagrammatically by Fig. I-18, where $T_A - T_B$ is the *isentropic* cooling and $T_A - T_C$ is the *isenthalpic* cooling for adiabatic expansions between the same pressure limits.

Isobaric Process   A process that is conducted at constant pressure is termed an isobaric process. The term *isopiestic process* also is used.

Isobutane   (See **Alkylation; Isomerization;** and **Petroleum processing.**)

Isobutylene   (See **Elastomers.**)

Isochoric Process   A process that is conducted at constant volume is termed an isochoric process. The term *isometric process* also is used.

Isocyanates   (See **Urethanes.**)

Isocyanic Acid   (See **Urea.**)

Isocyanurate   (See **Urethanes.**)

Isoelectric Point   (See **Amino acids; Colloid systems; Peptides and proteins.**)

Isoleucine   (See **Amino acids;** and **Peptides and proteins.**)

Isomers (See **Carbon compounds; Isomerism;** and **Isomerization.**)

Isomerism When two or more chemical compounds contain the *same elements* in exactly the *same numbers,* the compounds are said to be *isomeric.* A compound of this type is termed an *isomer* or *isomeride.*

Normal butane, $CH_3CH_2CH_2CH_3$, and isobutane, $(CH_3)_2CHCH_3$, both with the general formula C_4H_{10}, are isomers. Normal butane is an open, straight chain of four carbon atoms, whereas in isobutane one of the carbons lies in a short branch from a main chain of three carbon atoms. See also **Carbon compounds.**

As the number of carbon atoms in a compound increases, the possibility for branches and even subbranches also increases. Thus, compounds with high carbon counts usually have several isomers—if not in fact, certainly in theory.

Isomers need not be as similar as the examples of normal butane and isobutane, where both compounds are saturated open chains. Two isomers of rather different structures are exemplified by bipropargyl (1,5-hexadiine) and benzene:

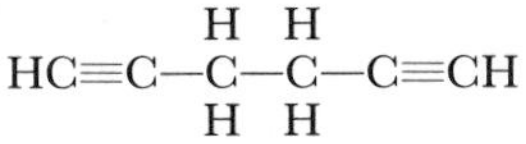

$$HC\equiv C-\overset{\overset{\displaystyle H}{|}}{C}-\overset{\overset{\displaystyle H}{|}}{\underset{\underset{\displaystyle H}{|}}{C}}-\overset{}{\underset{\underset{\displaystyle H}{|}}{C}}-C\equiv CH$$

Bipropargyl, C_6H_6'

Open, straight-chain structure
Aliphatic compound
Three single carbon bonds; two triple carbon bonds
Unstable; oxidized readily
Bp, 85°C; formula weight, 78.11

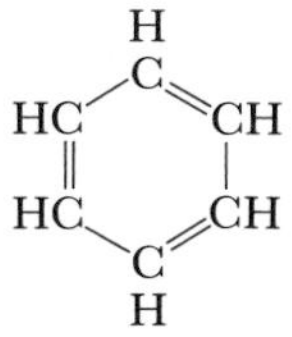

Benzene, C_6H_6

Ring structure
Aromatic compound
Three single carbon bonds; three double carbon bonds
Stable; oxidized with difficulty
Bp, 80.1°C; formula weight, 78.11

Benzene and its homologues have numerous isomers; for example, there are three trimethylbenzenes. Cumene (isopropylbenzene) also is isomeric with these compounds:

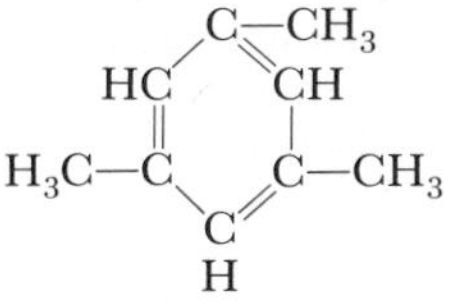

Mesitylene
(Symmetrical, 1,3,5-trimethylbenzene)
Sp gr: 0.863
Bp: 164.6°C

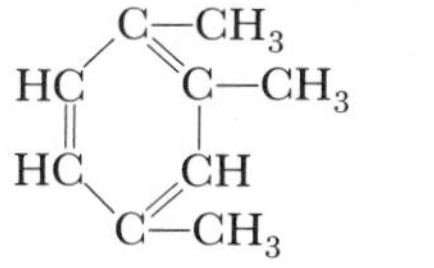

Pseudocumene
(1,2,4-trimethylbenzene)
Sp gr: 0.876
Bp: 169.8°C

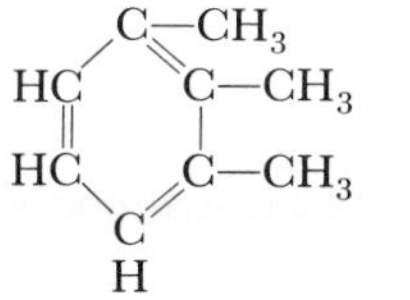

Hemimellitene
(1,2,3-trimethylbenzene)
Sp gr: 0.895
Bp: 176.5°C

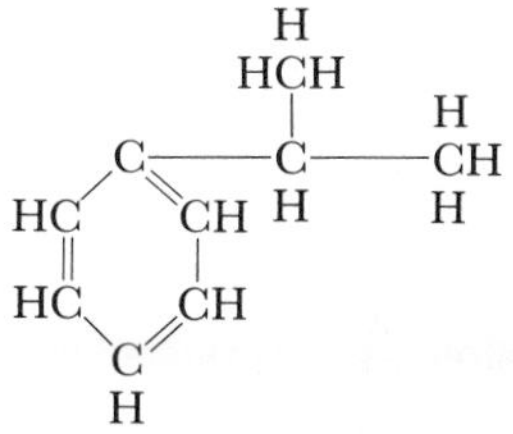

Cumene
(Isopropylbenzene)
Sp gr: 0.864
Bp: 153.4°C

The study of the many isomers of benzene and its homologues was a major step in establishing benzene as a symmetrical structure.

Geometric Isomerism: Normally, in depicting the structure of a chemical compound, the symbolism is not critical; i.e. a given atom or radical that is connected to a carbon atom can be shown above, below, to the right, or to the left of the carbon atom without affecting the analogous representation of the compound. Some compounds, however, are geometric isomers and the formula must reflect this fact.

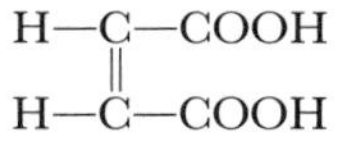

Maleic acid

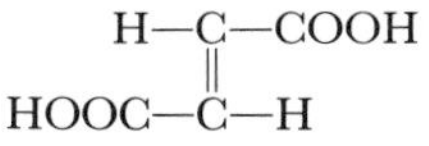

Fumaric acid

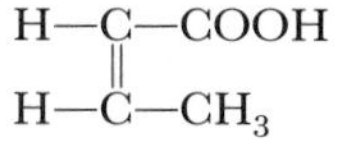

Isocrotonic acid
cis

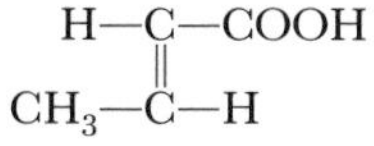

Crotonic acid
trans

When identical atoms or groups are in juxtaposition, the compound is designated as the cis (L., on this side) form; whereas the term trans is used to designate the form where the like atoms or groups are on opposite sides of the structure. *Trans*- and *cis*-polybutadiene are further examples of geometric isomers. See also **Elastomers.**

Stereoisomerism: Stereoisomerism is isomerism in space, i.e., three-dimensional isomerism in contrast to plane isomerism. The behavior of certain isomers presented puzzling effects that required explanation. As early as 1874, van't Hoff and Le Bel developed the concept of stereoisomerism in attempting to explain the behavior of lactic acid (α-hydroxypropionic acid), $C_3H_6O_3$.

α-Hydroxypropionic acid differs from one of its isomers (a plane isomer) in the following manner:

$$
\begin{array}{cc}
 & H \\
 & | \\
CH_3 & H-C-OH \\
| & | \\
H-C-OH & HCH \\
| & | \\
COOH & COOH \\
\text{Alpha form} & \text{Beta form}
\end{array}
$$

α-Hydroxypropionic acid also exists in two other basic forms: (1) *dextro* lactic acid, which rotates the plane of polarized light to the right, and (2) *levo* lactic acid, which rotates the plane of polarizing light to the left. A mixture of these two forms is ordinary lactic acid, which does not rotate the plane of polarized light. Ordinary lactic acid is termed dextrolevo lactic acid. Levo is abbreviated *l*; dextro *d*; and dextrolevo *dl*.

Graphic representations of chemical formulas, at best, are rough approximations of atomic arrangements and bonding and, for most substances, assist in clarifying and understanding the macro mechanics of a chemical structure and reaction. It is only reasonable to recognize, however, that

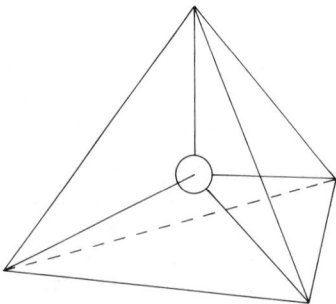

Fig. I-19. Pivotal carbon atom of compound (asymmetric atom) can be visualized as being in the center of a regular tetrahedron (equilateral pyramid).

molecules operate within three dimensions and that the compounds do not exist as the thin disk-like structures without depth which two-dimensional shorthand tends to depict.

Visualize (Fig. I-19) the pivotal carbon atom of the compound (the asymmetric atom) as being in the center of a regular tetrahedron (equilateral pyramid) with its valence bonds extending in four directions equidistantly to the four corners of the pyramid. Instead of duplicating the pyramid structure, the three-dimensional valence lines of the two lactic acids may be shown:

$$
\begin{array}{cc}
CH_3 & CH_3 \\
| & | \\
\,\,\,C-OH & \,\,\,C-H \\
H\diagup & HO\diagup \\
HO-C=O & HO-C=O
\end{array}
$$

$$
\begin{array}{cc}
CH_3 & CH_3 \\
| & | \\
H-C-OH & HO-C-H \\
| & | \\
HO-C=O & HO-C=O
\end{array}
$$

Tartaric acid also exhibits stereoisomerism. As with plane isomerism, the number of carbon atoms in a compound increases the possibilities for the formation of optical isomers. Saccharic acid, $COOH(CHOH)_4COOH$, for example, may exist in 10 optically isomeric forms. Several polyhydric alcohols, carbonates, and carbohydrates also exhibit extensive optical activity.

When a process is designed specifically to make isomeric changes, it is termed *isomerization*. See also **Isomerization.** Stereoisomerism also is discussed under **Amino acids.**

Isomerization   Isomerization can be defined as the rearrangement of the structural configuration of a molecule without changing its molecular weight. Although structural changes of this type occur in other processes, e.g., catalytic reforming and cracking, this description considers only processes in which isomerization is the principal reaction.

In petroleum refining, isomerization processes are used to change the structural configuration of C_4 paraffins (*n*-butane) into isobutane in order to supplement other sources to provide enough isobutane for alkylation with olefins in the production of motor fuel. C_5 and C_6 paraffins are isomerized to the more highly branched structures to improve their antiknock ratings. Isomerization also is applied to a much lesser extent to C_8 aromatic hydrocarbons.

In the United States at the beginning of 1970, the isobutane-producing capacity of C_4 isomerization units amounted to 43,000 bbl/stream day. The C_5/C_6 isomerization capacity was about 35,000 bbl/stream day. Any significant increase in the production of unleaded motor fuel will accelerate the need for C_5/C_6 isomerization. Isomerization first was used extensively in the early 1940s because of the demand for isobutane in the manufacture of alkylate needed for aviation gasoline. Among the components in motor gasoline, the one having the lowest unleaded octane number is the light, straight-run gasoline, which is composed generally of the C_5 and C_6 paraffins native in the crude. The octane numbers of these native paraffin fractions can be improved markedly by isomerization.

Table I-17 lists the octane numbers and normal boiling points of the several paraffin isomers in the C_5 and C_6 fractions arranged in order of their boiling points. A typical, light, straight-run gasoline would contain (in addition to the paraffins listed) minor amounts of cyclopentane, methylcyclopentane, cyclohexane, and benzene and would have a clear Research Method octane rating of 68–70. The C_5 paraffin fraction would contain about 60% n-pentane. The n-hexane content of the C_6 paraffin fraction would be about 48%, with the high-octane dimethylbutanes representing only 4% of the straight-run C_6 paraffin fraction. Thus, the very low octane n-paraffins are predominant in these two light, straight-run fractions.

An ideal situation would exist if a catalyst could be found to convert the paraffins completely to the highest-octane-number structures, i.e., to con-

TABLE I-18. *Equilibrium Vapor-Phase Compositions at Two Temperatures for C_4, C_5, and C_6 Paraffin Hydrocarbons*

Component	Approximate Composition, wt %	
	200°F (93°C)	400°F (204°C)
Butanes:		
Isobutane	75	56
n-Butane	25	44
Pentanes:		
Isopentane	86	70
n-Pentane	14	30
Hexanes:		
Methylpentanes	43	48
2,2,-Dimethylbutane .	43	29
2,3-Dimethylbutane . .	9	9
n-Hexane	5	14

vert, say, to isopentane and 2,3-dimethylbutane. Unfortunately, thermodynamic equilibrium limits the extent of conversion possible, and this extent is highly dependent upon the temperature at which the conversion occurs. The equilibrium is more favorable to the appearance of a greater proportion of the more highly branched isomers at low temperatures. Thus, it is of great interest to devise a catalyst that has sufficient activity to provide a high rate of isomerization at a low temperature. In a similar way, low reaction temperatures favor higher proportions of isobutane in the C_4 system at equilibrium. The influence of temperature on these equilibria in the vapor phase is illustrated by Table I-18, which reveals that equilibria at low temperatures favor the appearance of the high-octane-number structures of the C_4, C_5, and C_6 paraffins.

With respect to the C_5 and C_6 fractions, the refiner is interested in octane numbers rather than in isomer distribution. For the C_5 fraction, the Research Method clear octane numbers corresponding to equilibrium compositions at 200°F (93°C) and 400°F (204°C) are 89 and 84 respectively. These octane numbers for the C_6 equilibrium mixtures are 82 and 76 at 200 and 400°F respectively. Thus, an improvement of five to six octane numbers can be realized at equilibrium by dropping the temperature by 200°F in this region.

Catalysts: The early catalysts used during the 1940s for butane isomerization were of the expendable Friedel-Crafts type, e.g., aluminum chloride on a support or dissolved in molten antimony trichloride. These systems were promoted

TABLE I-17. *Characteristics of C_5 and C_6 Paraffin Isomers*

Paraffin Isomer	Bp		Approximate Research-Method Octane Number, Clear
	°F	°C	
Isopentane	82.0	27.8	93
n-Pentane	98.0	36.7	62
2,2-Dimethylbutane .	121.5	49.7	93
2,3-Dimethylbutane .	136.4	58.0	104
2-Methylpentane . . .	140.5	60.3	73
3-Methylpentane . . .	145.9	63.3	74
n-Hexane	155.7	68.7	30

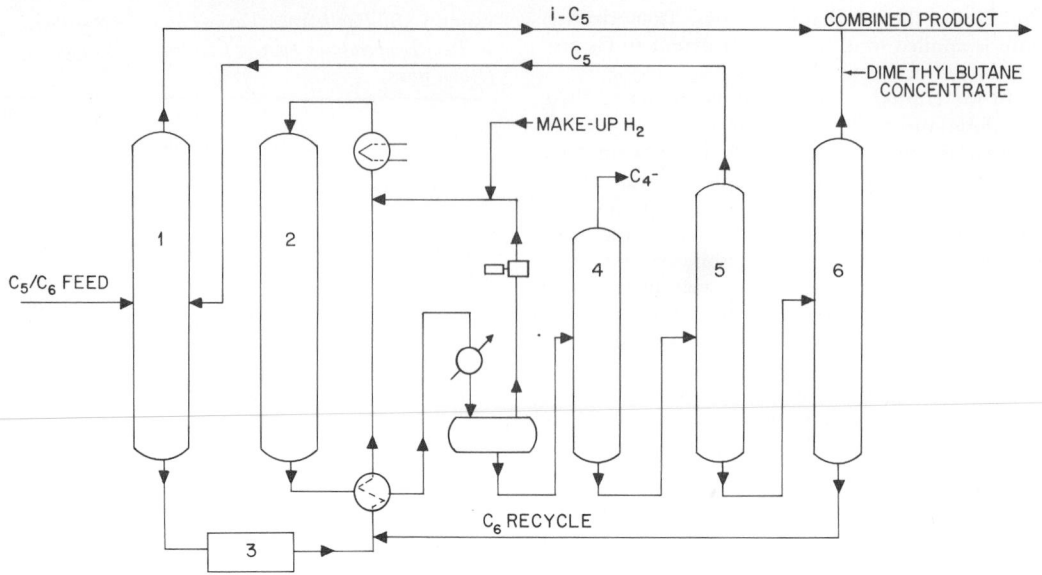

Fig. I-20. C$_5$/C$_6$ isomerization unit. (1) Deisopentanizer, (2) reactor, (3) drier, (4) stabilizer, (5) depentanizer, (6) deiso-hexanizer. (*UOP Penex Process, Universal Oil Products Company.*)

with HCl and operated in the general range 200–300°F (93–149°C), and the reaction environment was rather corrosive. More recently developed catalysts may be classified as *hydroisomerization* types in that they perform in the presence of a hydrogen atmosphere in order to minimize formation of carbonaceous deposits, which tend to deactivate the catalyst. The newer catalysts are usually a supported noble metal employed in fixed-bed reactors. The catalysts have a service life of several years.

Isomerization Process: One isomerization process[*] is shown in Fig. I-20. This unit is arranged to process a C$_5$/C$_6$ mixture with fractionation facilities to provide for the recycling of both *n*-pentane and *n*-hexane. A desulfurized C$_5$/C$_6$ blend first is fractionated to remove the native isopentane as a net product. The deisopentanizer bottoms are desiccant-dried before being joined by *n*-hexane recycle and brought to reaction temperature by heat exchange and suitable preheating. Before entering the reactor, the combined feed stream is joined by hydrogen recycle gas, which functions to suppress catalyst-deposit formation.

The fixed-bed reactor effluent is cooled and passed to a high-pressure separator. Gas from the

[*]UOP Penex unit.

separator, along with a small quantity of dried makeup hydrogen, is recycled to the reactor. The separator liquid is stabilized as a next step to remove any C$_4$ and lighter hydrocarbons that may be introduced with the makeup hydrogen, plus a very minor amount of light hydrocarbons formed by hydrocracking in the reactor. Hydrogen dissolved in the separator liquid is also removed by the stabilizer.

The next fractionator in series receives the stabilized liquid, from which it separates an equilibrium isopentane–*n*-pentane mixture that is routed back to the deisopentanizer for separating the isopentane as a net product. Thus, the *n*-pentane content of the feed is converted entirely to isopentane in the flow arrangement shown.

As the final step in the fractionation sequence, the hexane fraction is separated into a dimethyl-butanes concentrate as a net overhead product and a *n*-hexane-rich bottoms stream to be recycled for the further isomerization of the *n*-hexane and methylpentanes. With economically practical fractionation, the methylpentanes split between the overhead and bottoms of the deisohexanizer column. For the C$_5$ fraction, the boiling points of the two isomers are far enough apart to make a relatively clean split economically feasible. For the C$_6$ fraction, the greater number of isomers and the bunching of some of their boiling points preclude precision separations in columns having a reasonable number of plates.

Organic chloride promoter is added continuously and is converted to HCl in the reactor, but since the catalyst functions with only parts per million of the promoter, it is not necessary to provide separate facilities to recover and recycle the HCl. The HCl leaves the system by way of the stabilizer overhead product, which is treated with caustic before use as a refinery fuel. Because the system is dry and the concentration of HCl is low, the environment is noncorrosive. Carbon steel construction is permissible.

Process Variations: Apart from differences in plant design which employ either the Friedel-Crafts or the hydroisomerization catalysts, most process variations involve different recycle configurations. Generally, C_4 isomerization units are integrated with alkylation units so that a single deisobutanizer is common to both units. However, the C_4 isomerization unit can have its own deisobutanizer and deliver relatively pure isobutane as a net product.

Once-through processing of a typical C_5/C_6 (68–70 octane number) straight-run fraction results in a product having a Research Method octane number (clear) of about 83. By recycling the unconverted *n*-pentane, *n*-hexane, and most of the methylpentanes, as shown in Fig. I-20, a product having an octane number of about 93 would result. Obviously, any octane number between 83 and 93 could be produced, depending on the amount and quality of the equipment installed to separate the reactor effluent into net product and recycle streams.

Process Variables: Temperatures employed in the isomerization of the light paraffin hydrocarbons vary with the hydrocarbon being processed and especially with the type of catalyst used. In modern systems, the temperatures are relatively low, the reactor temperatures falling within the range 200–400°F (93–204°C). Because the influence on reaction equilibria is, in all cases, most favorable in the direction of lower temperatures, it is practical to operate units at the lowest possible temperature consistent with available catalyst activity. Low-temperature operation also minimizes undesirable hydrocracking side reactions.

Hydrogen recycle is used to minimize catalyst deposits, although the amount of hydrogen recycled relative to the hydrocarbon feed is less than in higher-temperature processes, such as catalytic reforming. Processes employing the fixed-bed hydroisomerization-type catalysts do not practice in situ catalyst regeneration.

The *operating space velocity and temperature* must be carefully balanced so that the isomerization reaction is brought just up to chemical equilibrium as it leaves the reactor. Chemical driving forces that push the reaction beyond this point will tend only to hydrocrack the reactants or products and lead to low yields of the desired isomers.

Generally, the operator has control over the *condition of the feedstock* with respect to the presence of materials other than the paraffin hydrocarbons undergoing isomerization. Moisture is the most important impurity to minimize. The absence of sulfur compounds results in better catalyst performance and life. For C_5 and C_6 isomerization, minor amounts of C_7 in the feed are tolerable, so that precision fractionation for feed preparation is not necessary. However, some hydrocracking of the C_7 occurs under the conditions selected to optimize C_5/C_6 isomerization. Also, some processes can tolerate the presence of normal amounts of benzene in native C_6 fractions while other processes may require a separate prehydrogenation step to saturate the benzene in the feed before it enters the isomerization section.

See also **Catalysis; Cracking; Hydrocracking;** and **Petroleum processing.**

—Melvin J. Sterba, *UOP Process Division, Universal Oil Products Company, Des Plaines, Ill.*

Isoprene Isoprene (2-methyl-1,3-butadiene) has been identified as the building block in natural rubber, which accounts for the growing interest in isoprene as an industrial chemical. The monomer (see Table I-19) is very reactive because of the conjugated double bond. The availability of high-purity monomer and the ability to control the degree of stereospecificity and the mole-weight distribution during polymerization make polyisoprene and copolymers, such as butyl rubber, an important part of polymer technology. In comparison with natural rubber, isoprene quality is more uniform, and the price is stable. See also **Butadiene; Elastomers; Polymerization;** and **Rubber, natural.**

From Tertiary Amylenes: The production of isoprene from tertiary amylenes contained in catalytically cracked C_5 gasolines requires three process units: (1) a tertiary-amylenes recovery unit, (2) a dehydrogenation unit, and (3) an isoprene-purification unit. The production rate of isoprene by this route is approximately 350 million lb/yr.

TABLE I-19. *Physical Properties of Isoprene, 2-Methylbutene-1, and 2-Methylbutene-2*

	Isoprene	2MB1	2MB2
Mole weight	68.11	70.13	70.13
Normal boiling point:			
°F.	93.32	88.09	101.42
°C.	34.07	31.16	38.57
Specific gravity (60°F/60°F)	0.6861	0.6557	0.6676
Coefficient of expansion at 60°F per °F	0.00086	0.00090	0.00087
Heat capacity (liquid at 1 atm), Btu/(lb)(°F)	0.5245	0.5266	0.5119
Heat of vaporization at 1 atm, Btu/lb	153	156	161

The tertiary amylenes, 2-methylbutene-1 (2MB1) and 2-methylbutene-2 (2MB2), are converted into isoprene by catalytic dehydrogenation. The product of the recovery unit contains about 90 wt % tertiary amylenes. Appropriate C_5 gasoline feedstocks are available from catalytic cracking which contain 28–30 wt % tertiary amylenes (2MB1 and 2MB2). Other components in such a fraction are saturates, other olefins, and cyclic compounds. Replacement of high-alumina cracking catalysts by zeolitic catalyst causes a reduction of the tertiary-amylene content of the C_5 gasoline fraction to 16–20 wt %, and for this more dilute feed prefractionation is economically beneficial. The physical properties of 2MB1 and 2MB2 are also included in Table I-19.

Recovery of Tertiary Amylenes: In the process* shown in Fig. I-21, the recovery of the tertiary amylenes is effected by acid absorption. About

*Shell process for tertiary-amylene recovery (Shell Development Company).

75% of the feed amylenes are recovered. The C_5 feed is extracted with aqueous acid, usually in multiple absorption stages. The lean-hydrocarbon fraction from the absorption step is treated by caustic washing and water washing to remove residual acid. This fraction usually is blended into gasoline stocks. Tertiary amylenes are extracted from the fat acid in a reversion step with a solvent. The fat solvent, after caustic and water washing, is stripped to recover the tertiary-amylene concentrate. In the absorption and reversion sections, the control of the dispersion and settling of the acid-hydrocarbon mixtures is important. This requires careful design of the mixers and settlers. In addition, proprietary additives are introduced which modify the properties of the emulsions. A typical feed composition is:

Wt %

Saturated C_5	45.4
Tertiary amylenes	28.9
Other C_5's.	25.7

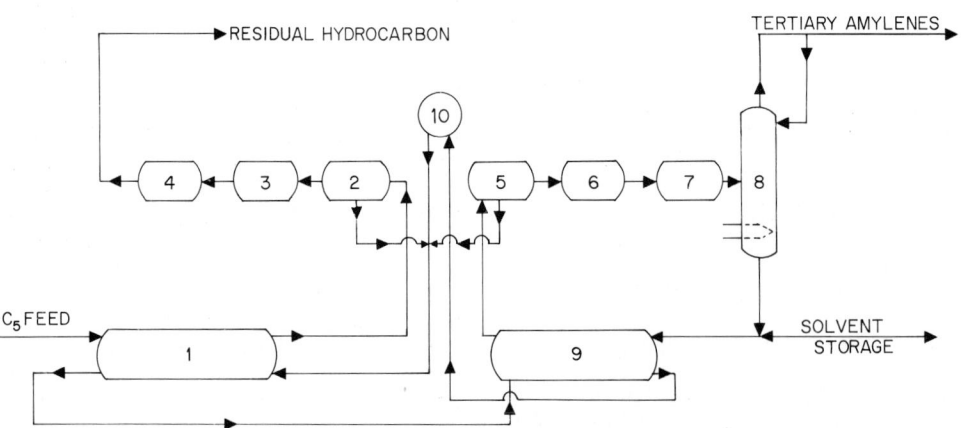

Fig. I-21. Process for recovery of tertiary amylenes from catalytically cracked C_5 gasolines. (1) Absorption, (2) surge tank, (3) caustic wash, (4) water wash, (5) surge tank, (6) caustic wash, (7) water wash, (8) solvent stripper, (9) reversion, (10) lean-acid surge. (*Shell Development Company.*)

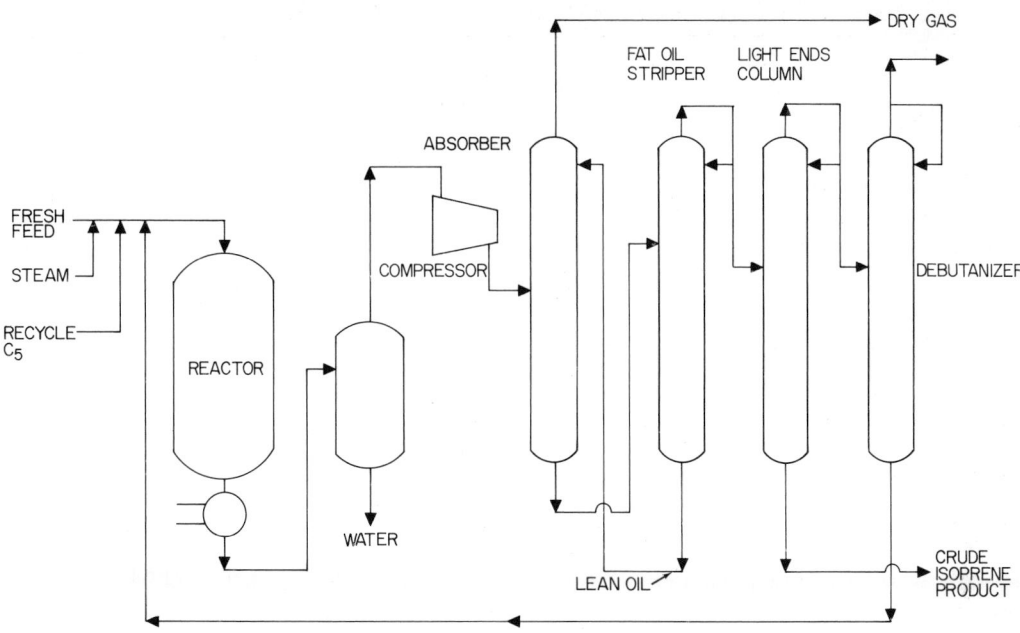

Fig. I-22. **Process for isoprene production by dehydrogenation of amylenes.** (*Shell Development Company.*)

Most of the plant is built of carbon steel because at process temperatures the acid is noncorrosive. More corrosion-resistant alloys are used in a few key parts of the process. Since the recovery of tertiary amylenes essentially is a once-through process, the unit can be located apart from the subsequent dehydrogenation and isoprene-recovery units.

Dehydrogenation of Amylenes: In the process* shown in Fig. I-22, isoprene is obtained from C_5 gasoline fractions by the dehydrogenation of tertiary amylenes (2MB1 and 2MB2 previously described) and 3-methylbutene-1 to isoprene. The preferred amylenes content (2MB1 and 2MB2 plus 3-methylbutene-1) of the reactor feed is 70 wt % or greater. The amylene-dehydrogenation and isoprene-purification units normally are located together to minimize the cost of shipping intermediate recycle streams. In this case, the reactor feed consists of a tertiary-amylenes concentrate and a recycle of unreacted amylenes from the isoprene recovery and purification unit. Through the reaction step, the overall molar yield of isoprene is approximately 75% of the amylenes

*Shell process for isoprene production by dehydrogenation of amylenes (Shell Development Company).

in the fresh feed. The remainder of the feed is rejected as a fuel gas and a C_4 fraction.

With reference to Fig. I-22, the fresh tertiary amylene feed and recycle C_5's are combined with steam and passed to the catalytic reactor. The important reaction is the dehydrogenation of amylenes to isoprene. However, some C_4's, C_2's, CO_2, and H_2 also are produced. Conversion and selectivity to isoprene are strongly influenced by the reaction temperature. Heat recovery from the reactor effluent provides steam for other parts of the process. The C_4's and C_5's are recovered from the reactor-effluent absorber-stripper section. The dry gas from the absorber usually serves as fuel gas. The stripper overhead product is processed for light-ends removal, mainly C_4's, which are fed to the debutanizer. The debutanizer recovers a C_4 fraction overhead and a bottoms product which is recycled to the reactor. The bottoms product from the light-ends column is the crude isoprene product from the dehydrogenation unit. Equipment for this process,† with the exception of furnaces for feed and steam superheating, is constructed of carbon steel.

†Although other dehydrogenation catalysts have been described in the literature, there are no equivalent commercial processes available (1972).

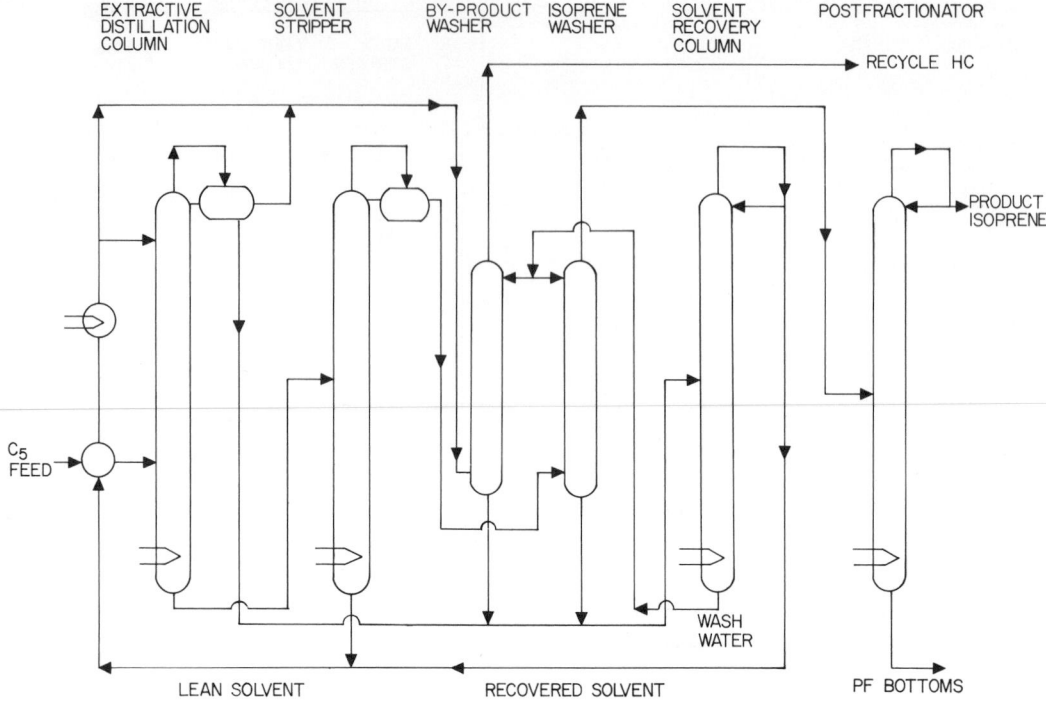

Fig. I-23. Process for isoprene recovery using acetonitrile as the extractive solvent. (*Shell ACN process, Shell Development Company.*)

Recovery: The crude isoprene, produced as just described, normally is fed to a process* of the type shown in Fig. I-23. In this process, high-purity isoprene may also be recovered from a mixed C_5 stream by extractive distillation, using acetonitrile, $CH_3 \cdot CN$, as the extractive solvent. Although product specifications depend on the user, a typical set of specifications for high-purity isoprene product is shown in Table I-20. Of these, the specifications for cyclopentadiene, acetylenes, and carbonyls are most restrictive.

With reference to the schematic flow sheet of Fig. I-23, the C_5's are extractively distilled with the acetonitrile solvent for by-product-hydrocarbon removal. The isoprene concentrate is recovered in the solvent stripper. Both the by-product hydrocarbon and the isoprene concentrate are washed to recover acetonitrile, which is returned to the solvent-circulation loop. The crude isoprene is purified in a postfractionator. Postfractionator bottoms usually are recycled to the tertiary-amylenes dehydrogenation system.

The acetonitrile solvent is not corrosive, and the

*Shell ACN (acetonitrile) process for isoprene recovery.

entire unit is built of carbon steel. Acetonitrile is a low-boiling solvent which leads to low temperatures in the solvent-stripping section, resulting in reduced fouling and high yields due to low polymerization losses. Also, the low operating temperature tends to reduce steam consumption by reducing sensible-heat pickup by the solvent.

The lower molecular weight of acetonitrile and

TABLE I-20. *Typical High-Purity Isoprene Specifications*

Component	Amount (All Maximum except Isoprene)
Isoprene, percent (wt.)	99.5
2-Methylbutene-2, percent (wt.)	0.2
Other pentenes, percent (wt.)	0.2
Cyclopentadiene, ppm (wt.)	3
Piperylenes, ppm (wt.)	100
Nonisoprene diolefins, percent (wt.)	0.05
Total acetylenes, ppm (wt.)	30
Peroxides, ppm (wt.)	10
Isoprene dimer, percent (wt.).	0.1
Nonvolatile residue, percent (wt.)	0.02

TABLE I-21. *Physical Properties of Acetonitrile*

Mole weight	41.05
Normal boiling point:	
°F	176.2
°C	80.1
Specific gravity (60°F/60°F)	0.7876
Viscosity 68°F, cP	0.38

the resulting lower molar volume tend to reduce the volume of solvent which must be circulated and thus tends to reduce equipment sizes. The low viscosity of acetonitrile (0.38 cP) tends to increase distillation-column tray efficiencies and thereby to reduce column height and cost. Because of the higher vapor pressure of acetonitrile, solvent cleanup under subatmospheric pressure is avoided. This reduces the likelihood of O_2 leakage into the system, minimizing its resulting increase in fouling tendency. The physical properties of acetonitrile are given in Table I-21.

Other Processes: Different routes for the recovery and synthesis of isoprene include (1) similar extractive distillation, using solvents other than acetonitrile, such as n-methylpyrrolidone and dimethyl formamide,[*] and (2) the synthesis of isoprene with subsequent recovery (in various stages of development and commercialization), including the following:

Reaction of isobutene and formaldehyde[†]

Reaction of acetone and acetylene:[‡]

[*]BASF and GEON (Japan) respectively.
[†]Soviet Union.
[‡]SNAM Progetti.

Dimerization of propylene:[§]

—A. E. Handlos, *Shell Development Company, Houston, Texas*

Isoprene Rule   (See **Terpenes.**)

Isopropyl Alcohol   Isopropyl alcohol (isopropanol), $(CH_3)_2CHOH$, one of the lower aliphatic alcohols, is a colorless, inflammable, mobile liquid, soluble in water, alcohol (ethyl), and ether (diethyl). Isopropanol has a molecular weight of 60.09, a specific gravity of 0.786 (20°C/20°C) a melting point of −89.5°C, and a boiling point of 82.5°C. The compound forms explosive mixtures with air, the lower explosive limit being 2.5 vol % isopropanol in air. Maximum allowable concentration is 400 ppm (volume). The flash point (closed cup) of isopropanol is 12°C, while the normal ignition temperature is 399°C.

Grades and Uses: Commercial isopropanol is available in solutions of concentrations 99, 95, and 91%. Since by far the greatest consumption of isopropanol (55–60%) is in the production of acetone, the economics of isopropanol production is tied closely to that of acetone. In recent years, considerable quantities of by-product acetone, particularly from the cumene-phenol process, have become available. While acetone, in its traditional role as a solvent in the organic chemicals industry, has been meeting considerable competition from other solvents, other uses of acetone are showing a marked increase, thus tending to balance the situation. Particular instances are in the production of bisphenol A (for epoxy and polycarbonate resins), methyl methacrylate, and methyl isobutyl ketone (solvent for vinyl resins).

In addition to its source for acetone, early uses for isopropanol were in rubbing alcohols and as an antistalling agent in winter-grade motor fuels. Although its use as an antistalling agent has virtually disappeared from the United States market, the practice still holds in Europe and could reappear in the United States as tetramethyl lead and tetraethyl lead are replaced in gasoline. Iso-

[§]Goodyear.

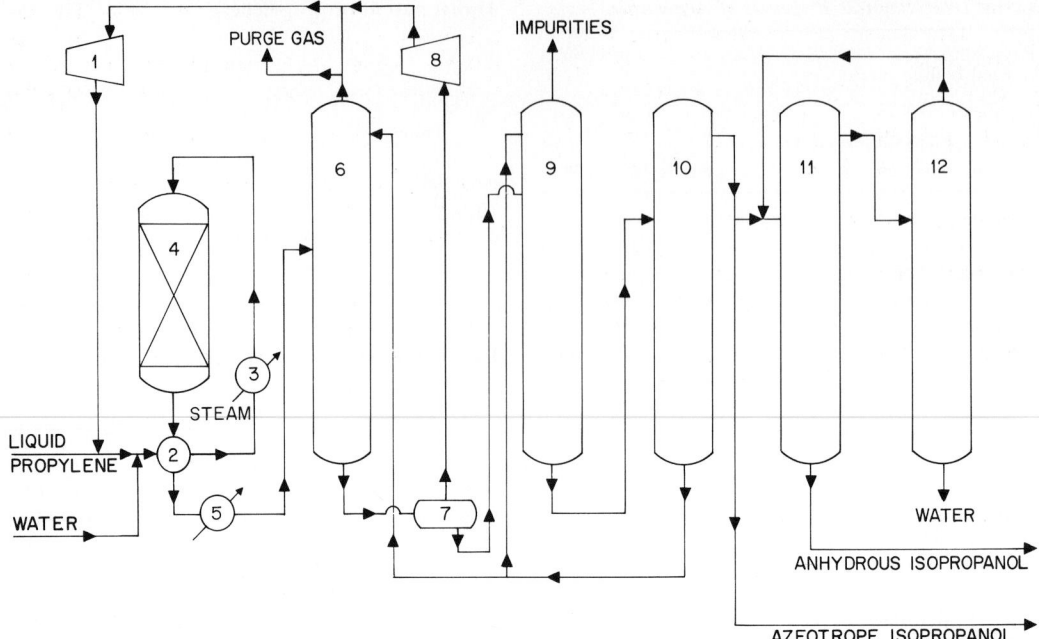

Fig. I-24. Direct-hydration process for production of isopropanol. (1) Recycle-gas compressor, (2) feed-effluent-heat exchanger, (3) preheater, (4) reactor, (5) cooler, (6) recycle-gas scrubber, (7) flash drum, (8) vent-gas compressor, (9) wash column, (10) constant-boiling mixture column, (11) drying column, (12) solvent-recovery column. (*Veba-Chemie, A. G., Germany; Davy Powergas Ltd., London.*)

propanol also is used as a solvent for oils, gums, shellac, and synthetic resins and in the production of isopropyl acetate and herbicidal esters and xanthates. Isopropanol also has a range of miscellaneous uses in the production of drugs and pharmaceuticals.

Production: The two basic commercial methods in use are (1) absorption of propylene in H_2SO_4 to form alkyl hydrogen sulfate, followed by hydrolysis of the ester, and (2) direct hydration with water, using a catalyst. An inherent disadvantage in the type 1 process is the need to handle H_2SO_4. Further, the type 1 processes yield little more than 70% isopropanol on feed propylene. Selection of the process is affected by local conditions, but most modern plants use the direct-hydration technique.

A leading process* of the direct-hydration type is illustrated in Fig. I-24, in which propylene feed, mixed with demineralized water and recycle pro-

*Veba-Chemie, A. G., Germany; Davy Powergas Ltd., London.

pylene, is heated to reaction temperature in a preheat train before passing through the catalyst bed. The reactor effluent is cooled and scrubbed with water in a high-pressure scrubber to remove products of reaction from the recycle gas, which then can be returned to the reactor. The aqueous product from the scrubber base is drawn off and flashed down to essentially atmospheric pressure to recover the dissolved propylene, which is returned to the reaction section. The crude isopropyl alcohol then is purified by extractive distillation with water and the dilute product rectified to produce the azeotrope, which contains 87.7 wt % isopropyl alcohol. The azeotrope then can be dried by azeotropic distillation with benzene in a two-column system.

All direct-hydration processes can be represented by

$$C_3H_6 + H_2O \rightarrow C_3H_7OH + \text{heat}$$

The equation shows that the reaction is favored by high pressure and low temperature. Because of rate considerations, however, the reaction is carried out commercially at elevated temperatures. The optimum conditions for the process illustrated are in the range 180–260°C and 25–45 atm. The conversion per pass is limited by the thermodynamic equilibrium for the reaction. See also **Petrochemical complex.**

—N. W. Browne, *Davy Powergas Ltd., London*

Isotactic   (See **Polymerization.**)

Isothermal Process   A process in which the temperature remains constant is termed an isothermal process. The isothermal compression of a gas is an example. When a gas is compressed, its temperature will rise unless heat is removed. If heat is removed during compression so that the gas temperature remains constant, the process is termed *isothermal compression*. If the temperature is allowed to rise, the process is termed *adiabatic compression*.

Isotone   Atoms of different elements (and hence different atomic numbers) that have the same number of neutrons are termed *isotones*.

Isotope   Atoms of the same element that differ in their number of neutrons are termed isotopes. See also **Chemical elements;** and **Radioactive isotopes.**

J

Japan Wax   (See Waxes.)

Jarosite   (See Potassium.)

Jaw Crusher   (See Size reduction.)

Jet Fuel   (See Alkylation; and Petroleum processing.)

Jet Mixers   (See Mixing, fluids.)

Jewelry Bronze   (See Copper.)

Jigging   (See Beneficiation, ore; and Separation operations.)

Jojoba   (See Waxes.)

K

Kainite   (See **Potassium.**)

Kaldo Process   (See **Iron- and steelmaking.**)

Kanamycin   (See **Antibiotics.**)

Karaya   (See **Gums.**)

Keene's Cement   (See **Cement;** and **Gypsum.**)

Kellner Cell   (See **Chlorine**)

Keratin   (See **Fibers;** and **Peptides and proteins.**)

Kerosene   (See **Fluid catalytic cracking; Petrochemical complex; Petroleum;** and **Petroleum processing.**)

Ketones   Structurally, the ketones are characterized by a *carbonyl, C:O,* linkage between two radicals in accordance with the general chemical signature $R \cdot CO \cdot R'$. The radicals of the linkage may be identical, as in the case of dimethyl ketone (acetone), $CH_3 \cdot CO \cdot CH_3$; or two different radicals may be involved, as in the case of methyl ethyl ketone, $CH_3 \cdot CO \cdot C_2H_5$. Compounds of the latter type sometimes are referred to as *mixed ketones*. When both radicals are *alkyls* (aliphatic), the ketone may be referred to as an *alphyl* ketone. In the classical sense, such ketones are considered as derivatives of the secondary alcohols.

The general formula for saturated alphyl ketones is $C_nH_{2n}O$, and it is interesting to note that each ketone is isomeric with the aldehyde that contains the same count of carbon atoms. Thus, acetone, C_3H_6O, is isomeric with propaldehyde, C_3H_6O.

Aromatic alcohols also are capable of forming ketones. Acetophenone (phenyl methyl ketone), $C_6H_5 \cdot CO \cdot CH_3$, is an example. This is a mixed *aryl* (aromatic) and *alphyl* (aliphatic) ketone. Fully aromatic (*diaryl*) ketones also exist, e.g., benzophenone, $C_6H_5 \cdot CO \cdot C_6H_5$.

Because of the presence of the carbonyl group, ketones exhibit certain behavioral similarities with the aldehydes. Just as the ethers represent a first stage of oxidation of the alcohols to acids, the ketones, like the aldehydes, represent a mid-stage of oxidation.

Nomenclature: The common or trivial names of the ketones are made up by following the designation of the alphyl or aryl group with the word

TABLE K-1. *Properties of Representative Ketones*

Ketone	Formula	Formula Weight	Sp gr	Mp, °C	Bp, °C
Acetone	CH_3COCH_3	58.08	0.792	-94.6	56.5
Acetophenone	$CH_3COC_6H_5$	120.14	1.033	20.5	202.3
Acetophenone acetone. . .	$C_6H_5COCH_2CH_2COCH_3$	176.30	. . .	. . .	162
Acetylacetone.	$CH_3COCH_2COCH_3$	100.11	0.975	-23	137
Acetylcarbinol	CH_3COCH_2OH	74.10	1.802	-17	146
Anthraquinone	$C_6H_4:(CO)_2:C_6H_4$	208.20	1.438	286	379–381
Anthrone	$C_6H_4{<}{CO \atop CH_2}{>}C_6H_4$	. . .	. . .	154	
Benzophenone	$C_6H_5COC_6H_5$	182.21	1.083	48.5	305.4
Benzoylacetone.	$C_6H_5COCH_2COCH_3$	162.19	. . .	81	262
Camphor	$C_{10}H_{16}O$	152.23	0.999	178–179	209.1
Carone	$C_{10}H_{16}O$	152.23	. . .	. . .	210
Carvone.	$C_{10}H_{14}O$	150.21	0.961	. . .	230
Cycloheptanone	$CH_2(CH_2)_5CO$	112.17	. . .	. . .	180
Cyclohexanone	$CH_2{<}(CH_2CH_2)_2{>}CO$	98.14	0.947	-45	155–156
Cyclopentanone	${<}(CH_2CH_2)_2{>}CO$	84.11	0.948	-58.2	129–130
Diacetyl	$CH_3COCOCH_3$	86.09	0.990	-3	88
Dibenzyl ketone	$(C_6H_5CH_2)_2CO$	210.26	. . .	34–35	330.6
Diethyl ketone	$(C_2H_5)CO$	86.13	0.816	-42	101.7
Dihexyl ketone	$(C_6H_{13})_2CO$	198.35	. . .	30.5	
Diphenylene ketone	$C_6H_4COC_6H_4$	180.21	. . .	84	342
n-Dipropyl ketone	$(C_2H_5CH_2)_2CO$	114.18	0.822	-32.6	144.2
Ethyl isopropyl ketone . .	$C_2H_5COCH(CH_3)_2$	100.17	0.830	. . .	114
Ethyl phenyl ketone	$C_2H_5COC_6H_5$	134.11	1.010	21	218
Ethyl *n*-propyl ketone . . .	$C_2H_5COCH_2CH_2CH_3$	100.17	0.818	. . .	124
Isodiamyl ketone.	$[(CH_3)_2CHCH_2CH_2]_2CO$	170.29	0.821	14.6	228
Isodibutyl ketone.	$[(CH_3)_2CHCH_2]_2CO$	142.23	0.805	. . .	168.1
Isodipropyl ketone	$[(CH_3)_2CH]_2CO$	114.18	0.806	. . .	123.7
Lauryl ketone	$(C_{11}H_{23})_2CO$	338.62	0.809	69	
Menthone.	$C_{10}H_{18}O$	154.12	0.897	. . .	207
Methyl ethyl ketone	$CH_3COC_2H_5$	72.10	0.805	-85.9	79.6
Methyl isopropyl ketone .	$CH_3COCH(CH_3)_2$	86.13	0.809	-92	93
Methyl *n*-nonyl ketone .	$CH_3(CH_2)_8COCH_3$	170.29	0.828	13.5	228
Methyl *n*-propyl ketone . .	$CH_3COCH_2CH_2CH_3$	86.13	0.812	-77.8	102
Methyl vinyl ketone	$CH_3COCH:CH_2$	70.09	0.836	. . .	81
Michler's ketone	$[(CH_3)_2NC_6H_4]_2CO$	268.35	. . .	174	360*
Palmitone.	$(C_{15}H_{31})CO$	239.42	0.795	. . .	83
Stearone.	$(C_{17}H_{35})_2CO$	506.44	0.793	. . .	88

*Decomposes.

ketone; thus, the designation methyl ethyl ketone, where the methyl and ethyl groups are attached to the C:O. If the groups are identical, the designation may be *di*methyl ketone. In another system, the ketone may be named after the alcohol from which it can be derived or the acid to which it can be oxidized; thus, such terms as acetone and propione.

Profile of the Ketones: Specific properties of representative ketones are given in Table K-1. The following observations apply to most ketones:

1. Neutral, volatile, mobile liquids below C_{11}; higher-carbon compounds are solids.

2. Reasonably agreeable odor.

3. The lower-carbon ketones are quite soluble in water; solubility decreases with rise in formula weight.

4. The higher-carbon ketones are practically insoluble in water, but most ketones are miscible with alcohol and ether.

5. Specific gravity rises gradually to about 0.83 as the formula weight increases.

6. As the number of carbon atoms increases, the boiling point rises

Chemical Reactivity: Although essentially of theoretical importance, it is interesting to note that the ketones, like the aldehydes, are unsaturated because of the double bond between the C and O of the carbonyl group. It should be emphasized that the hydrogen atom that is directly attached to the carbonyl group is not easily displaced. The ketones are readily acted on by reducing agents to form secondary alcohols. Catalytic reduction is used widely for the commercial production of certain compounds. Upon reduction, acetone yields isopropyl alcohol and pinacol. Secondary alcohols are not the only product of the reduction of ketones, varying amounts of ditertiary alcohols (pinacols) also being produced. Ketones are much more stable than aldehydes and consequently do not combine readily with alcohols, generally do not combine with ammonia at ordinary temperatures, and do not reduce alkaline solutions of metals. Unlike aldehydes, ketones do not undergo polymerization at ordinary temperatures.

The methyl ketones and some the cyclic ketones form crystalline additive compounds with solutions of sodium bisulfite, a valuable property in the purification of some ketones. Ketones combine with hydroxylamine to yield *ketoximes.*

$$(CH_3)_2CO + H_2N \cdot OH \rightarrow$$
Acetone   Hydroxylamine

$$(CH_3)_2C:N \cdot OH + H_2O$$
Acetoxime
(dimethylketoxime)

With hydrazine, ketones react to form *hydrazones.*

$$(CH_3)_2CO + H_2N \cdot NHC_6H_5 \rightarrow$$
Acetone   Phenylhydrazine

$$(CH_3)_2C:N \cdot NH \cdot C_6H_5 + H_2O$$
Acetone hydrazone

The ketones react with semicarbazide to form *semicarbazones.*

$$(CH_3)_2CO + NH_2 \cdot CO \cdot NH \cdot NH_2 \rightarrow$$
Acetone   Semicarbazide

$$(CH_3)_2C:N \cdot NH \cdot CO \cdot NH_2 + H_2O$$
Acetone semicarbazone

When treated with magnesium amalgam, ketones, after hydrolysis, yield a 1,2-glycol in a reaction known as the *pinacol-pinacolone rearrangement.* In the presence of sulfuric or hydrochloric acid, ketones can be made to undergo a cyclic trimerization, involving the loss of water. Upon oxidation, ke-

tones decompose to form two acids, each containing fewer carbon atoms than the originating ketone. Many acids can be formed by the oxidation of mixed ketones. Ketones combine with hydrogen cyanide to form *cyanohydrins.*

$$(CH_3)_2CO + HCN \rightarrow (CH_3)_2C(OH) \cdot CN$$

an important reaction because cyanohydrins readily hydrolyze to form hydroxy acids.

Ketones react with hydrazine, $NH_2 \cdot NH_2$, to produce azines.

Phosphorus pentachloride and pentabromide react with the ketones to form dihalogen derivatives of the paraffins in which the oxygen atom of the carbonyl group is replaced by two hydrogen atoms. See also **Peroxides, organic.**

Acetone is produced (1) by passing acetic acid vapor over heated lime, giving calcium acetate, which breaks down into acetone and calcium carbonate, $CH_3 \cdot CO \cdot O \cdot Ca \cdot OOC \cdot CH_3 \rightarrow CH_3 \cdot CO \cdot CH_3 + CaCO_3$; (2) by fermentation of starch, such as maize, along with the production of butyl alcohol; (3) by the oxidation of cumene to cumene hydroperoxide, which subsequently decomposes into acetone and phenol; and (4) by direct oxidation of propylene in which air is used (see Fig. K-1).

In the propylene process, a catalyst solution comprising copper chloride and small quantities of palladium chloride is used. The reaction is $CH_3CH:CH_2 + \frac{1}{2}O_2 \rightarrow CH_3COCH_3$. The palladium chloride is reduced to elemental Pd and HCl during the reaction. Cupric chloride causes reoxidation. During the catalyst-regeneration cycle, the cuprous chloride is reoxidized. The reaction temperature is approximately $100°C$ under moderate pressure.

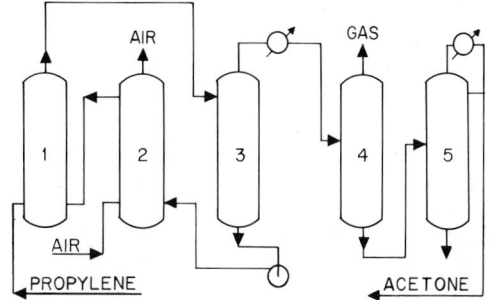

Fig. K-1. Production of acetone by direct oxidation of propylene with air. (1) Reactor, (2) regenerator, (3) crude-acetone separator, (4) degasser, (5) still. (*Highly schematic representation of a type of process designed by Hoechst-Uhde Corp.*)

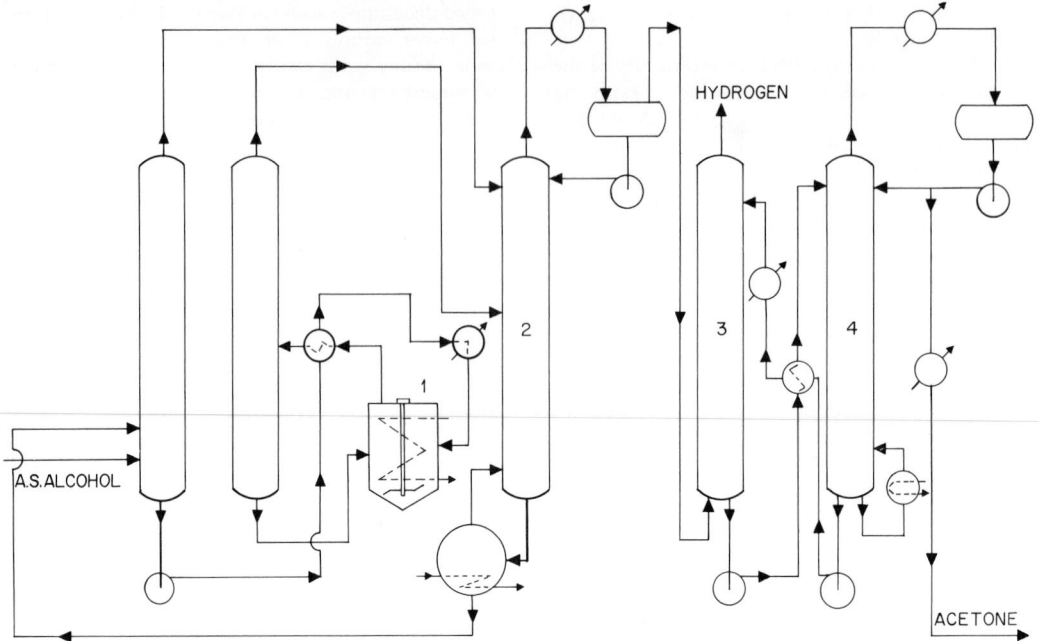

Fig. K-2. Production of ketones from corresponding secondary alcohols. Charge stock is an anhydrous secondary alcohol (ASA). (1) Reactor, (2) fractionator, (3) water absorber, (4) stripper. (*Highly schematic representation of a type of process designed by Institut Français du Pétrole.*)

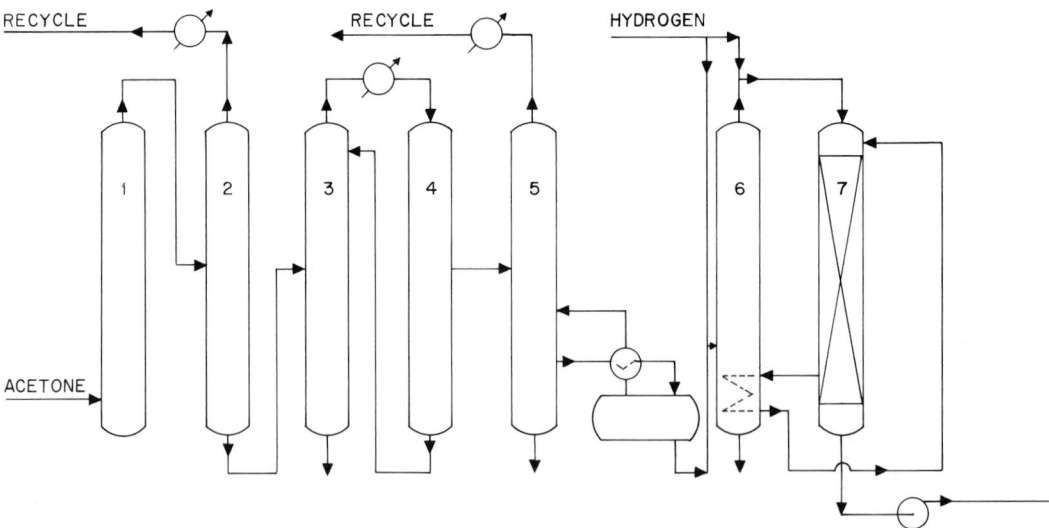

Fig. K-4. Production (simultaneous) of methyl isobutyl ketone (MIBK) and methylisobutylcarbinol (MIBC) with acetone as a charge stock. (1) Condensation column, (2) diacetone alcohol–concentration column, (3) dehydration column, (4) concentration column, (5) acetone recovery column, (6) mesityl oxide vaporizer, (7) hydrogenation reactor, (8) concentration column, (9) low-boiling-point column, (10) MIBK column, (11) MIBC column. (*Highly schematic representation of a type of process designed by Kyowa Hakko Kogyo Co., Ltd.*)

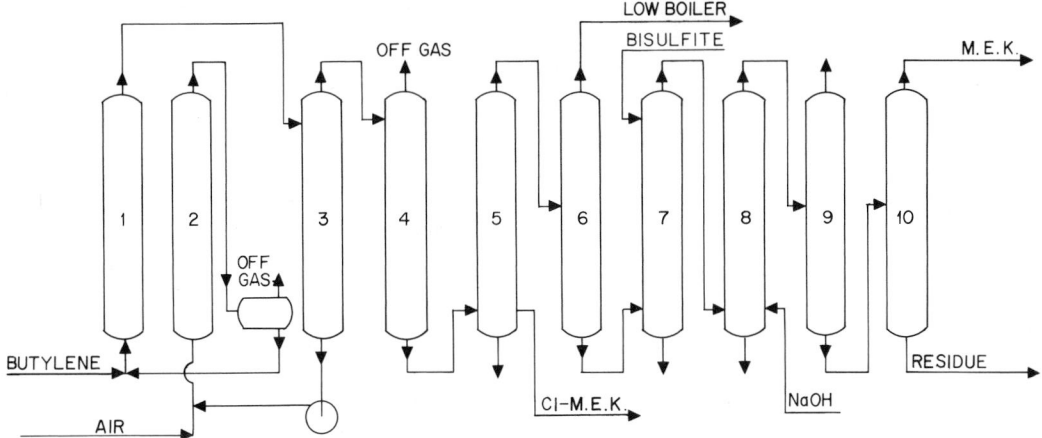

Fig. K-3. Production of methyl ethyl ketone (MEK) by direct oxidation of butylene with air. (1) Reactor, (2) oxidizer, (3) crude-product separator, (4) separator, (5) water and Cl–MEK separator, (6) low-boiler separator, (7) bisulfite extractor, (8) NaOH extractor, (9) water removal, (10) still. (*Highly schematic representation of a type of process designed by Hoechst-Uhde Corp.*)

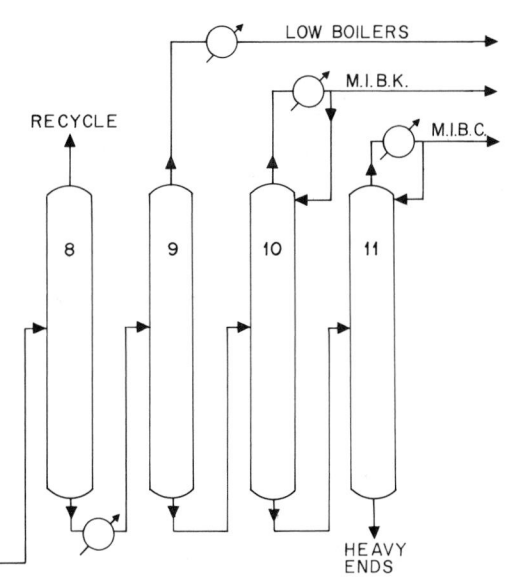

Acetone and many other ketones, according to the starting material used, can be produced by the process of Fig. K-2. High-purity hydrogen is a by-product. The process is versatile because, with appropriate changes, the same plant can be used at one time to produce acetone from isopropanol and at another time, e.g., methyl ethyl ketone from secondary butyl alcohol. The process may be designated as a liquid-phase dehydrogenation occurring at comparatively low temperatures (150°C). High molar yields (99.5% and better) are obtained.

Methyl ethyl ketone can be produced by the direct oxidation of butylene in which air is used (see Fig. K-3). This process uses a copper chloride–palladium chloride catalyst solution, similar to the acetone process of Fig. K-1. The reaction is $C_4H_8 + \frac{1}{2}O_2 \rightarrow CH_3COC_2H_5$. The action of the catalyst and regeneration occur as previously described. This reaction also takes place at about 100°C at moderate pressure. The methyl ethyl ketone–water mixture is treated with sodium bisulfite and caustic soda, followed by a distillation stage to yield the pure methyl ethyl ketone (MEK).

Methyl isobutyl ketone can be produced by using acetone as a charge stock, as shown in Fig. K-4. First, the acetone is changed to diacetone alcohol (DAA) by condensation (under pressure) with an alkaline catalyst. This may be a material, such as $Ca(OH)_2$ or $Ba(OH)_2$, which is slightly soluble in water. The reaction is exothermic, and thus cooling is required. In the next step, mesityl oxide

(MSO) is produced by dehydration of the DAA. An acid catalyst is used, and the temperature is moderate (100–120°C). The DAA condensate is concentrated to approximately 80% while the unreacted acetone is separated and recovered. In forming the MSO from DAA, the DAA partially decomposes into acetone, and thus a distilling stage is required to separate and recover the acetone. In a following step, the MSO is hydrogenated to yield both methyl isobutyl ketone (MIBK) and methylisobutylcarbinol (MIBC), thus requiring later fractionating stages to separate the two products. The ratio of MIBK/MIBC produced varies with temperature and hydrogen mole ratio. An increase in these variables tends to favor greater yields of MIBC. Thus, the producer has some control over the relative amounts of end products formed and can vary them in accordance with the market. The hydrogenation reactions are $(CH_3)_2C:CHCOCH_3 + H_2 \rightarrow (CH_3)_2CHCH_2 \cdot COCH_3$, methyl isobutyl ketone; and $(CH_3)_2CHCH_2COCH_3 + H_2 \rightarrow (CH_3)_2CHCH_2 \cdot CHOHCH_3$, methylisobutylcarbinol.

Uses: The ketones are used widely as reactants and intermediates for the manufacture of numerous synthetic materials, including resins. They also find extensive use as solvents.

Related Compounds: A *ketose* is a polyhydric ketone. Monosaccharoses are open-chain polyhydroxy aldehydes or ketones, also referred to respectively as *aldehyde sugars* (aldoses) and *ketone sugars* (ketoses). Fructose, for example, is a ketose. Dihydroxyacetone, $HO \cdot CH_2 \cdot CO \cdot CH_2 \cdot OH$, is a very simple ketose. *Ketonic acids* contain both a carbonyl and a carboxylic group and have reactive properties of both an acid and a ketone. Examples of monobasic ketonic acids are pyroracemic acid,

$CH_3 \cdot CO \cdot CO_2H$, acetoacetic acid, $CH_3 \cdot CO \cdot CH_2 \cdot CO_2H$, and levulinic acid, $CH_3 \cdot CO \cdot CH_2 \cdot CH_2 \cdot CO_2H$. *Ketonic hydrolysis* occurs, for example, when the ethyl ester of acetoacetic acid hydrolyzes to form acetone, carbon dioxide, and ethyl alcohol.

Ketone-Alcohol (KA oil) [(See **Cyclohexanol/ Cyclohexanone (KA oil)**.]

Ketonic Acids (See **Carboxylic acids;** and **Ketones.**)

Ketoses (See **Ketones.**)

Ketoximes (See **Ketones.**)

Kettle Soap (See **Soaps.**)

Killed Steel (See **Iron and steel.**)

Kilns (See **Bauxite; Cement;** and **Lime.**)

Kinin (See **Peptides and proteins.**)

Kneaders (See **Mixing and blending, solids/ solids.**)

Kraft Pulping Process [See **Pulp (wood) production and processing.**]

Krennerite (See **Tellurium.**)

Kroll Process (See **Beryllium;** and **Hafnium.**)

Krypton (See **Chemical elements.**)

L

Lacquer (See **Paint.**)

β-Lactams (See **Antibiotics.**)

Lactic Acid (See **Isomerism.**)

Lactic Ferment (See **Carboxylic acids.**)

Laminates (Consult the **Subject Index.**)

Lampblack (See **Carbon.**)

Langbeinite (See **Potassium.**)

Lanolin (See **Waxes.**)

Lanthanides (See **Chemical elements; Lanthanum;** and **Rare-earth elements and metals.**)

Lanthanum (Gk. *lanthanein,* to escape notice.) **La** = 138.91 (at. wt.); 57 (at. no.). Lanthanum is the first of 15 elements in group IIIA, period 6, generally shown as the rare-earth elements in a separate line below the main body of the periodic table and sometimes referred to as the *lanthanide series.* La was discovered in 1839 by C. G. Mosander in Stockholm.

Pure La metal retains its silvery-gray metallic luster only briefly because of the rapid formation of white oxide powder on surfaces exposed to air. The oxide itself is hygroscopic and tends to spall from the metal, allowing fresh surfaces to oxidize and leading eventually to complete reversion of the metal to an oxide unless it is protected. Metal chips and powders are quite pyrophoric. La metal will spark and burn when abraded. The pure metal is soft, comparable to tin, and easily workable with ordinary tools and forming equipment. In all situations La metal should be protected from the atmosphere by a coating of high-flash-point oil or stored in leakproof containers sealed after purging with an inert gas.

The melting point of La is 920°C; boiling point 3457°C; density 6.166 g/cm^3; it is thirty-fifth in abundance of 83 elements in the earth's crust, low toxicity rating.

La has 2 natural isotopes, ^{139}La (99.9 w/o*) and ^{139}La (0.1 w/o), and 19 artificial isotopes. The small amount of ^{138}La (0.1 w/o) in some minerals is mildly radioactive with a long half-life (10^{10}–10^{15} years). The La atom is always trivalent

*This abbreviation is used in the rare-earth and related fields for weight percent.

679

and is colorless, meaning that it has no absorption bands in the visible, ultraviolet, or near-infrared regions of the spectrum. La is the only element that has no $4f$ electrons. La metal becomes a superconductor below 6 K, the only rare-earth element having this property. See also **Rare-earth elements and metals.** La is the second most plentiful element found with the light rare-earth, or cerium-group, elements, ranging from 25 to 35% of the mixture. It is separated from other rare-earth elements by ion-exchange processing after acid leaching of bastnasite or monazite minerals. The pure metal is produced commercially by (1) electrowinning of La from La_2O_3 in a molten fluoride electrolyte, (2) electrolysis of La from fused anhydrous $LaCl_3$, or (3) metallothermic reduction of lanthanum fluoride, LaF_3, by calcium in a reactor sealed from the atmosphere.

Dilute mineral acids readily dissolve La metal. Concentrated mineral acids and acetic and formic acid dissolve La_2O_3. La metal reduces magnesia, alumina, zirconia, and the oxides of transition metals and the rare-earth elements at elevated temperatures. The hydrogen electrode potential of La^{3+} is $+2.522$ V. It has the largest atomic (and ionic) size of the trivalent elements and combines with most organic and inorganic anions.

The chemical nature of this element is revealed by its name, which means to escape notice; 36 years after cerium was believed to be in pure form, La was discovered in a partially decomposed cerium nitrate. Even today there are no wet chemical or spectrographic procedures to identify small amounts of La in other compounds; analysis relies upon x-ray fluorescence and nuclear-activation techniques.

Uses: As a component in mixed rare earths, La is consumed in mischmetal for sparking flints, burned in the cores of carbon electrodes for high-intensity lighting, and used in optical-glass polishing compounds. There is no question of its ability to create a hot spark in lighter flints, but its contribution to illumination and its effectiveness in glass polishing are questionable.

Pure lanthanum oxide, which melts at 4190°F (2310°C), has been used for many years as a component in the best grades of optical glass to lower dispersion of light and to improve the index of refraction. The hygroscopic nature of the oxide has limited its application as a refractory material even though it ranks eleventh of the most refractory metal oxides.

Pure lanthanum oxide is used in electronic capacitors, thermistors, and electroluminescent powders. It is a host matrix for fluorescent phosphors.

The largest use of La with other rare earths is in molecular-sieve catalysts for cracking crude petroleum. Over 10 million lb of lanthanum rare-earth chloride are consumed yearly in catalysts for oil refining to increase yields and efficiencies.

Lanthanum metal and its alloys find broad use in metallurgy. While La does lack sufficient mechanical strength or favorable properties for singular use, it has a very high affinity for oxygen, sulfur, nitrogen, and hydrogen. As a component in mischmetal, it is used to scavenge gases from molten metals and in processing vacuum tubes and lamps.

Cobalt-base alloys which specify La metal in the range 0.08–0.15% show increased resistance to hot corrosion and oxidation.[1]

Another intermetallic compound, $LaNI_5$, exhibits a remarkable ability to absorb and desorb large amounts of hydrogen at room temperature. The density of hydrogen absorbed by $LaNi_5$ under 2.5 atm at room temperature is 7.6×10^{22} atoms/cm³, nearly twice as high as in liquid hydrogen.[2,3]

REFERENCES
1. Herchenroeder, R. B.: Cobalt Base Alloy, U.S. Patent 3,418,111 (Stellite Division, Cabot Corporation), December 24, 1968.
2. van Vucht, F. A., Kuijpers, and Bruning: "Reversible Room-Temperature Absorption of Large Quantities of Hydrogen by Intermetallic Compounds," *Philips Res. Rep.* 25, pp. 133–140, August 1969.
3. Wiswall, R. H., and J. J. Reilly: Metal Hydrides for Energy Storage, *Pubn.* BNL-16889, Brookhaven National Laboratories, Upton, N. Y.

—Joseph G. Cannon, *Molybdenum Corporation of America, White Plains, N.Y.*

Laser Glass   (See **Glass;** and **Neodymium.**)

Lateritic Ores   (See **Cobalt;** and **Nickel.**)

Latex   (See **Acrylonitrile-butadiene-styrene resins; Adhesives; Elastomers; Paint; Resins;** and **Rubber, natural.**)

Laundry Compounds   (See **Detergents.**)

Lauric Acid   (See **Vegetable oils.**)

Lawrencium   (See **Chemical elements.**)

Leaching (See **Separation operations;** and **Urea.**)

Lead (OE. *lead.*) **Pb** (L. *plumbum*) = 207.21 (at. wt.); 82 (at. no.). Lead is a heavy, soft, malleable metal with a bluish cast. The metal surface oxidizes in normal atmospheres to a soft, gray patina. Pb is in group IV of the periodic table in the subgroup with Ge and Sn. There are four naturally occurring isotopes which, in order of abundance, are 208, 206, 207, and 204. There also are eleven unstable isotopes, 200–203, 205, and 209–214. See also **Chemical elements.** Lead is one of the metals sometimes termed *prehistoric.* A lead figure found at Abydos (Dardanelles) dates to about 3800 B.C.

Lead melts at 327.4°C, boils at 1740°C, and at 20°C has a density of 11.35 g/cm³. Pb has a thermal conductivity of 0.082 (cgs units) in the range (0–100°C); a mean specific heat of 0.0310 cal/(g)(°C) over 0–100°C; a resistivity of 20.6 $\mu\Omega$-cm; a temperature coefficient of resistivity of 3.36×10^2 over 0–100°C; and a coefficient of expansion of 29.0×10^6 over 0–100°C. The crystal is face-centered cubic with an edge length of 4.950 Å. The atomic radius of 1.75 Å is derived from the crystal structure.

Sources: Over 95% of the production of mined Pb is from primary ores in which galena, PbS, is predominant. The galena occurs in ore bodies containing 3–30% Pb. Galena is one of the most widely distributed sulfide minerals and occurs most frequently along with sphalerite, ZnS. The Pb-Zn ores contain recoverable amounts of Cu, Ag, Sb, and Bi. The major deposits being worked are in Australia, the United States, Canada, Mexico, Peru, Yugoslavia, and the Soviet Union.

Galena crystallizes in the isometric hexoctahedral system, commonly cubic or cubooctahedral, less often octahedral. Carbonates and sulfates are the chief secondary lead minerals formed by weathering. Cerussite, $PbCO_3$, is formed by the action of groundwater on galena. Anglesite, $PbSO_4$, is derived from galena which has been contacted by sulfate solutions generated by the oxidation of sulfide minerals. Properties of these minerals are summarized in Table L-1. Galena is decomposed by dilute HNO_3 with the separation of S and formation of $PbSO_4$. Some specimens effervesce, giving off H_2S.

Production: The major steps in the processing of lead ore to pig lead for market are shown in Fig. L-1. The ore is crushed, wet-ground, and classified (ore dressing) to at least 90% less than 200 mesh. Flotation agents and conditioning agents are added to aid the separation of the sulfide ore from the gangue. The ore attaches itself to the froth bubbles and is floated off to thickeners, where it is dewatered. The concentrates containing 45–60% Pb, 0–15% Zn, a few ounces of Au, up to 50 oz of Ag per ton, up to 3% Cu and 0.4% As, and as much as 2% Sb, along with 10–30% S, are the smelting feed. The concentrates are roasted (Dwight-Lloyd sintering machine) to lower the S content because PbS is not reduced by either carbon or CO at blast-furnace temperatures. The sinter, along with limestone and coke, is fed into the blast furnace. The lead-bullion output of the furnace is refined to marketable products, while the by-products are recovered by appropriate methods (also indicated by Fig. L-1). See also **Zinc and zinc/lead smelting.**

Lead Alloys: Pb is alloyed with other elements to improve the physical properties, to lower the melting point, to improve the corrosion resistance to specific chemicals, or to attain some other specific property. Sn, As, Sb, Bi, Cd, Ag, and Ca are added singly or in combination to obtain specific properties.

Antimony, the most common alloying agent with Pb, forms a eutectic alloy at 11.1% Sb, which melts at 252°C. Antimonial lead alloys with 3.5–7% Sb are used as grids, connectors, and posts

TABLE L-1. *Properties of Principal Lead Ores*

	Galena	Cerussite	Anglesite
Formula	PbS	$PbCO_3$	$PbSO_4$
Lead, %	86.6	77.5	68.3
Hardness, Mohs' scale	2.5–2.75	3–3.5	2.5–3
Luster	Metallic	Admantine to vitreous, resinous	Admantine to vitreous, resinous
Streak	Lead gray	Colorless to white	Colorless to white
Color	Lead gray	Colorless to white	Colorless to white
Density, g/cm³	7.58 ± 0.01	6.55 ± 0.02	6.38 ± 0.01

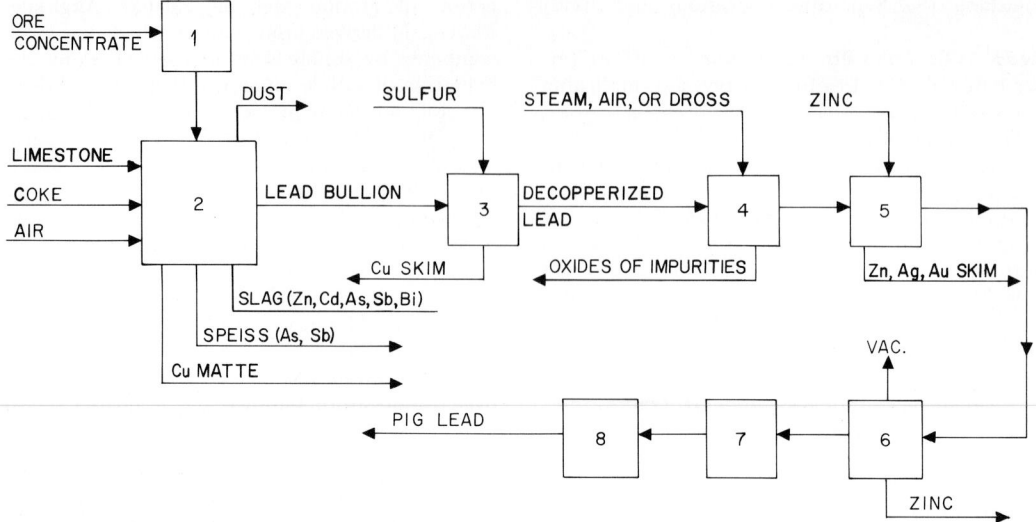

Fig. L-1. Schematic representation of materials flow in production of pig lead from ore concentrate. (1) Sintering, (2) lead blast furnace, (3) drossing kettle, (4) softening furnace, (5) desilverizing kettle, (6) dezincing kettle, (7) casting kettle, (8) molds. Dust fumes from the blast furnace go to the bag house. After separation, some of the bag-house dust is returned to the furnace. By-product metals shown are all forwarded to special metal-recovery processes.

in lead acid storage batteries. Antimonial lead is stronger than pure Pb and lends itself to automated production of grids, in addition to being more stable for battery applications. Up to 0.75% As may be added to antimonial lead to further improve the strength and precipitation-hardening characteristics.

Antimony (6% by weight) is added to Pb to yield an alloy with improved physical properties called *hard lead*. Since all Pb alloys are relatively weak structurally, it is common practice to support them with stronger materials. Sheet-lead tank linings may be used in wood, steel, or concrete tanks. Lead pipe may be loosely lined into steel or copper pipe or homogeneously bonded to it. Valves and fittings are available in hard lead or as lead-lined steel.

Corroding lead is a designation used in the industry to describe Pb refined to a high degree of purity.

Chemical lead signifies the undesilverized Pb produced from southeastern Missouri ores.

Acid copper lead is made by adding Cu to fully refined Pb.

Common desilverized lead designates fully refined desilverized lead.

Chemical and acid copper lead generally are used for chemical construction purposes.

Lead is resistant to chemical attack due to the formation of a protective coating on the surface which is both adherent and impervious. The coating, or film, is a chemical combination of Pb and the particular corrosive agent to which it has been exposed, that is, $PbSO_4$ (exposure to H_2SO_4) or $Pb(H_2PO_4)_2$ (exposure to H_3PO_4). Lead is used extensively in the construction of chemical plants, especially where H_2SO_4 is handled. Rolled sheet, extruded pipes, castings, and other forms are available in four grades as described in ASTM Spec. B-29. The compositions of these four grades are given in Table L-2. The mechanical properties of Pb and Pb alloys are given in Table L-3.

Lead is available as sheet 0.0117–1.0 in. thick. Pipe is produced from $\frac{1}{4}$ to 12 in. ID with the wall thickness varying with pressure rating. Welding bars, wire, pig metal, shot, and special castings and extrusions also are available. Lead has a high density and is relatively limp, contributing to its excellent sound-attenuation properties. Its high density also accounts for its use as a gamma-ray shielding material. Sheet, shot, bricks, and castings are all used for shielding where space is restricted. Corrosion resistance to a wide variety of atmospheres accounts for the use of Pb as a sheathing on aerial, underground, and submarine power and telephone cables. Being

TABLE L-2. *Compositions of Major Grades of Lead*

Constituent	Corroding Lead	Chemical Lead	Acid Copper Lead	Common Desilverized Lead
Silver:				
Max.	0.0015	0.020	0.002	0.002
Min.	. . .	0.002		
Copper:				
Max.	0.0015	0.080	0.080	0.0025
Min.		0.040	0.040	
Together, max.	0.0025	0	0	0
Arsenic, antimony, and tin				
together, max..	0.002	0.002	0.002	0.005
Zinc, max.	0.001	0.001	0.001	0.002
Iron, max.	0.002	0.002	0.002	0.002
Bismuth, max.	0.050	0.005	0.025	0.150
Lead (by difference), min.	99.94	99.90	99.90	99.85

soft, malleable, and easily welded, Pb sheet can be formed into an impermeable membrane for waterproofing.

Chemical lead with 6% Sb (hard lead) has good corrosion resistance to a variety of chemicals and is available in all forms, including sheet, pipe, valves, fittings, and anodes. An alloy of 0.06–0.08% Ca has physical properties similar to 4% antimonial lead. The melting point is essentially that of pure Pb. Tin alloys with Pb in all proportions. The eutectic alloy, 61.9% Sn with 38.1% Pb, melts at 183°C. The alloys containing 2–63% Sn are used widely as solders. Low-tin alloys are used for sealing the seams of food cans, while the higher-tin alloys (low melting) are used for electronic soldering. Sn-Pb alloys (15–20% Sn) are hot dip-coated onto steel and copper for improved corrosion resistance. Roofing, gasoline tanks, oil filters, and electronic chassis are made of terne plate. See also **Tin.** An alloy of 7% Sn and 93% Pb is used as an insoluble anode for electroplating chromium.

Chemical Activity: Generally, Pb does not dissolve in dilute acids except in the presence of an ample supply of atmospheric oxygen, due in part to the fact that H_2 is evolved on pure Pb only at a considerable overvoltage. In many cases, Pb is protected from solution by the formation of an insoluble coating on the surface, which protects the surface from further dissolution. The stability of Pb in moderately concentrated H_2SO_4 is important for its use in accumulators and for the manufacture of H_2SO_4 by the chamber process. See also **Sulfuric acid.** Lead is practically inert to HCl, but HNO_3 dissolves it readily because of the strong oxidizing power of the acid. Compact lead metal (as opposed to finely divided metal) is attacked only superficially by atmospheric O_2 but alloys readily with other metals.

Compounds: Because of the relative inertness of the metal as a raw material for the manufacture of lead chemicals, most inorganic Pb compounds are prepared from the oxides, nitrate, or carbonate.

Lead Monoxide: Aside from tetraethyl lead, lead monoxide, PbO, is the most important commercial compound of lead. It exists in two polymorphic forms, a reddish to reddish-yellow tetragonal

TABLE L-3. *Mechanical Properties of Lead and Lead Alloys*

Lead Type	Tensile Strength, psi	Elongation, %	Brinell Hardness
Pure (99.9+%).	1,750–1,900	55	3.5–4
Chemical:			
Rolled	2,190	50	4.5
Cast.	2,765	50	
6% antimonial:			
Rolled	4,500	50	7.5
Cast.	6,700	22	11.8

form (alpha PbO, litharge), n_{Li}^{20} 2.67, 2.74, stable at ordinary temperatures, and a metastable yellow orthorhombic crystalline form (beta PbO, massicot), n_{Li}^{20} 2.51, 2.61, 2.71. The transition temperature from the alpha to the beta form is about 488°C. The melting point of alpha PbO is about 888°C, but it is appreciably volatile below its melting point.

A crude form of lead monoxide, containing up to 40% lead metal, is commonly manufactured by oxidizing molten lead in a blast of air. The purer grades are produced by heating pig lead (or the crude lead oxide) in reverberatory furnaces at controlled temperatures. The desired crystalline form can be obtained in the furnace by precise temperature control or by boiling lead hydroxide with caustic soda at specific concentrations.

PbO reacts readily with common acids to form corresponding lead salts and dissolves in alkali to form plumbites or hydroxyplumbites. Reducing agents such as carbon, hydrogen, or carbon monoxide heated with lead monoxide readily reduce it to metal.

The major use of lead monoxide is for the preparation of storage-battery plates. For this application the high free-metal grades are employed. Pure grades are used for the preparation of inorganic lead salts. Various grades of litharge are employed as a starting material in the production of leaded glasses, glazes, varnishes, cements, chrome pigments, red lead, lead soaps, and greases. It is also employed as a flux in porcelain painting and as an activator in rubber compounding.

Red Lead: In terms of consumption, red lead, Pb_3O_4, also known as *minium,* is the second most important of the lead oxides. It is practically insoluble in water but dissolves in molten potassium nitrate. It turns dark when heated but is restored to its original color on cooling. Red lead may be considered to be lead(II) orthoplumbate, $Pb_2(PbO_4)$, and not merely a mixture of PbO and PbO_2, as borne out by the difference in thermal decomposition pressures between red lead and PbO_2 and by the distinct crystal structure of Pb_3O_4.

Red lead is formed as a brilliant red powder when finely divided lead monooxide is heated in air to about 500°C. A minium of brighter orange color is produced by starting with an oxide obtained from the decomposition of lead carbonate or nitrate.

Red lead is decomposed by dilute HNO_3 or HCl into lead dioxide, PbO_2, and the corresponding lead(II) salt.

A major use of red lead, together with other oxides, is in the preparation of positive storage-battery plates. See also **Batteries.** The compound is used extensively as a pigment in oil-based paints for protecting iron substrates against corrosion. The material also is used for luting joints in steel plates and tubes and is the preferred starting material for the manufacture of lead ferrites, $PbO \cdot 6Fe_2O_3$, used as permanent magnets.

Lead Dioxide: Although PbO_2 is sometimes called *lead peroxide,* since no H_2O_2 is formed in the decomposition of PbO_2 by acids, it is not a true peroxide. PbO_2 is a strong oxidant, evolving O_2 when treated with H_2SO_4 or when heated alone to 290°C or above. PbO_2 is fairly soluble in HCl and in a mixture of HNO_3 and H_2O_2.

Lead dioxide is produced commercially by treatment of an alkaline red lead slurry with chlorine. It is also obtained by the electrolytic oxidation of lead salts, by the action of dilute nitric acid on red lead, or by the oxidation of lead acetate with bleaching powder.

When heated with acids, PbO_2 tends to give the corresponding salt of the divalent Pb(II) ion because of the instability of the Pb(IV) ion. When heated with strongly basic oxides, lead dioxide combines with them to form plumbates.

Lead dioxide is the principal active constituent in the positive plate for lead storage batteries. The lead dioxide is formed when a charge is impressed on the plate holding the battery paste consisting of lead oxide, water, and sulfuric acid. The negative-plate paste changes to sponge lead during the process.

Combined with easily combustible substances such as sulfur or red phosphorus, lead dioxide is also used in the manufacture of matches and pyrotechnic materials. As an oxidant it is used in the manufacture of dyes, chemicals, and rubber substrates.

REFERENCES

Hansen, M.: Constitution of Binary Alloys, in "Metals Handbook," 8th ed., American Society for Metals, Metals Park, Ohio, 1970.

Hoffman, W.: "Lead and Lead Alloys," Springer-Verlag, Berlin, 1970.

Liddell, D. M.: "Handbook of Non-ferrous Metallurgy," McGraw-Hill, New York, 1945.

Palache, C., H. Berman, and C. Frondel: "Dana's System of Mineralogy," 7th ed., Wiley, New York, 1963.

Smithells, C. J.: "Metals Reference Book," Plenum, New York, 1967.

—Elbert J. Minarcik (lead and galena) and William D. Lang (chemical compounds), *NL Industries, Inc., Hightstown, N.J.*

Lead Arsenate   (See **Arsenic.**)

Lead Azide   (See **Explosives.**)

Lead Glass   (See **Glass.**)

Lead Styphnate   (See **Explosives.**)

Leaded Brass   (See **Copper.**)

Leaf Filter   (See **Filtration.**)

Lebeau's Method   (See **Beryllium.**)

Lehr   (See **Glass.**)

Lepidolite   (See **Germanium;** and **Lithium.**)

Leucine   (See **Amino acids;** and **Peptides and proteins.**)

Leucite   (See **Potassium.**)

Leucoxene   (See **Titanium dioxide.**)

Levorotatory Compounds   (See **Isomerism.**)

Lignin   [See **Pulp (wood) production and processing.**]

Lignite Coal   (See **Carbon;** and **Coal tar and derivatives.**)

Lime   The term *lime* includes a variety of chemicals manufactured from limestone or derived from chemical processes which utilize Ca compounds. According to the composition of the parent limestone, lime may be designated as *high-calcium lime* or *dolomitic lime.* Both *quicklimes,* CaO and CaO·MgO, and *hydrated limes,* CaO·H₂O, Ca(OH)₂·MgO, and CaO·MgO·2H₂O, are conventionally called lime. Therefore, precise terminology requires complex wording, e.g., dolomitic quicklime to denote CaO·MgO.

The various lime oxides and hydroxides are the lowest cost and most widely used sources of alkali for the chemical and metallurgical industries. In 1971, United States merchant lime sales were 12.37 million tons. Total lime consumption (including limes processed within captive facilities) was 19.63 million tons. Very little lime is imported or exported by the United States. Over 80% of the lime used in the United States is by the chemical and related industries, mostly as quicklime. About 10% is dead-burned dolomite

and less than 10% goes into construction uses, mostly as hydrate. Worldwide production of lime is about 100 million tons, the United States and the Soviet Union each producing about 20% of the total.

Some of the general physical properties of limes are given in Table L-4. All quicklimes are crystallized in the cubic system. Crystallites vary greatly in size, some being so fine as to appear amorphous. Degree of porosity also varies greatly, depending upon the structure of the limestone and the severity of its calcination. This property is measured best by specific surface area, which by the BET* method typically is 1–2 m^2/g. Specific surface area also is a reliable index of the reactivity of quicklime. Reactivity with water is measured directly by the temperature rise resulting from slaking (ASTM C-110).

All varieties of lime are but slightly soluble in water and most solvents. Thus, its performance in most liquid-solid reaction systems is diffusion-rate-controlled, and particle size, porosity, and surface area are important parameters. Solubility in water is only slightly affected by addition of other inorganic compounds. However, solubility may be increased a hundredfold in sugar solutions. Solubility in sucrose is used in determining the available-lime index (ASTM C-25). Available lime represents the total free lime in a lime product and is a measure of effective purity or active CaO. Available lime may be lowered by the presence of inert impurities in the limestone, incomplete calcination of the stone, or carbonation of the lime from exposure to air and moisture.

Both quicklime and hydrated lime are stable chemical compounds. However, quicklime reacts quickly with liquid water and slowly with moist air to form the hydrate. Hydrated lime will very slowly react with CO_2 and moisture in air to form carbonate.

Manufacture: The basic processes are calcination and hydration. Starting with high-calcium limestone, the reactions are

$$CaCO_3 + heat \rightleftharpoons CaO + CO_2 \qquad (1)$$
$$CaO + H_2O \rightleftharpoons Ca(OH)_2 + heat \qquad (2)$$

If dolomitic limestone is used, the reactions are

$$CaCO_3 \cdot MgCO_3 + heat \rightleftharpoons$$
$$CaO \cdot MgO + 2CO_2 \qquad (3)$$
$$CaO \cdot MgO + H_2O(l) \rightleftharpoons$$
$$Ca(OH)_2 \cdot MgO + heat \qquad (4a)$$

or

*Brunauer, Emmett, Teller method.

TABLE L-4. *Comparison of General Physical Specifications for High-Calcium and Dolomitic Limes[a]*

Type and Formula	Mol. Wt.	Sp gr		Dissociation Temperature, °C	Mp, °C[b]	Hardness, Mohs' Scale	Max. Solubility, g/l	Specific Heat, kcal/(g)(°C)	Neutralization Value[c]
		True	Apparent						
High-calcium limestone, CaCO$_3$	100.08	2.65–2.75	...	898[d]	d	2–4	0.013	0.203[e]	100
High-calcium quicklime, CaO	56.08	3.34–3.40	2.0–2.2	...	2570	2–3	1.31[f]	0.177[e]	178
High-calcium hydrate, Ca(OH)$_2$	74.10	2.3–2.4	...	540	d	...	1.76[f]	0.260[e]	135
Dolomitic limestone, CaCO$_3$·MgCO$_3$	184.40	2.75–2.90	...	725[d]	d	3.5–4	0.32[g]	0.215[h]	108
Dolomitic quicklime, CaO·MgO	96.40	3.5–3.6	2.06–2.29	...	2460	2–3	1.76[f]	0.185[e]	190
Dolomitic monohydrate, type N, Ca(OH)$_2$·MgO	114.42	2.7–2.9	...	540	d	...	1.76[f]	0.242[e]	175
Dolomitic double hydrate, type S, Ca(OH)$_2$·Mg(OH)$_2$	132.44	2.4–2.6	...	350–540	d	...	1.76[f]	0.273[e]	151

[a] Data from Robert Weast, Editor-in-Chief, "Handbook of Chemistry and Physics," 49th ed., Chemical Rubber, Cleveland, 1969; R. S. Boynton, "Chemistry and Technology of Lime and Limestone," Wiley, New York, 1966; and "International Critical Tables," McGraw-Hill, New York, 1926.
[b] d = decomposes.   [c] Neutralizing value as equivalent CaCO$_3$; which equals 100, and assuming each of the materials listed to be 100% pure.
[d] At 760 mm pressure in 100% CO$_2$ atmosphere.   [e] At 0°C.   [f] At 10°C.   [g] At 18°C.   [h] At 37.8°C.

$$CaO \cdot MgO + 2H_2O(g) + pressure \rightleftharpoons$$
$$Ca(OH)_2 \cdot Mg(OH)_2 + heat \quad (4b)$$

High-calcium limestone dissociates at 1650°F (899°C) in 100% CO_2 atmosphere at 1 atm pres. Under similar conditions, dolomitic limestone dissociates over 1340–1650°F (727–899°C). The heat of reaction required to convert $CaCO_3$ to CaO is about 2.8 million Btu per ton of CaO. In practice, heat input may vary from 4 to 10 million Btu per ton of lime. Calcination of limestone particles proceeds by a receding-surface mechanism. To attain reasonable rates of heat transfer into the center of rock or pebble-sized stone, operating temperatures in lime kilns are 1800–2300°F (982–1260°C). Reaction rate is increased and opportunity for recarbonation of the oxide is decreased by rapid removal of CO_2 from the kiln.

Except for old mixed-feed vertical kilns, all lime kilns operate with countercurrent flow of raw material and heat. Modern lime kilns utilize coolers to preheat air by recuperating heat from the hot quicklime. Lime kilns are fired directly with coal, oil, or gas. The type of fuel affects productivity and lime quality.

Two major types of lime calciners are the rotary and the vertical kiln. In the United States, rotary kilns are standard equipment for lime calcination, while in Europe vertical kilns are most widely used. Rotary kilns typically have higher output, up to 600 tons/day, and lower labor cost. Vertical kilns can be designed for higher fuel efficiency and lower capital investment. Vertical kilns normally require at least 3 × 6 in. rock feed, although some special vertical kilns are available which handle down to $3/4$-in. stone. Rotary kilns can handle down to $1/4$-in. stone. In general, lime made in a rotary kiln is of higher quality than lime made in vertical kilns. Production cost depends mainly upon stone and fuel costs and efficiency.

Many special types of lime kilns are available which have some advantages over both the rotary and vertical kilns. The latter require limestone rock that is hard enough to withstand handling through a moving-bed processing system. Many varieties of limestone tend to decrepitate into a powder during calcining, creating handling problems. One type of kiln which alleviates this problem is the moving-grate kiln. For instance, in the Calcimatic rotary-hearth kiln, the lime bed is quiescent during calcining. Controlled heat input is provided by multiple burners and firing zones. Finely crystalline limestone that cannot be maintained in pebble size can be calcined in the Fluo-

Solids kiln. Here the stone is pulverized to about 8–65 mesh and calcined in a multistage fluidized-bed reactor. Precise control of temperature and residence time is possible, and high-quality lime is produced. However, fluidized-bed calcination is usually not feasible with hard limestone because of high grinding cost. Precipitated calcium carbonate in the form of sludge or filter cake is calcined in rotary, fluid-bed, or multiple-hearth furnaces.

Lime quality is significantly dependent upon raw materials and conditions of calcination. Paramount variables are purity and texture of the limestone; size and uniformity of size of kiln feed; type of fuel, combustion control, and kiln-gas analysis; and time and temperature of calcination. In general, *soft-burned* highly reactive lime is produced by low-temperature calcination (1800–2000°F, 982–1093°C); *hard-burned* low-reactivity lime is produced by high-temperature calcination (2200–2400°F, 1204–1316°C) or too long a calcination. Attempts to push kiln productivity above design output can result in lime that contains both over- and underburned material. The latter, called *core,* is unconverted limestone and may range up to 5% by weight in poor-quality lime. In physical form, quicklime products are graded as lump, pebble, ground, and pulverized (see Fig. L-2). Some fine quicklime is pelletized

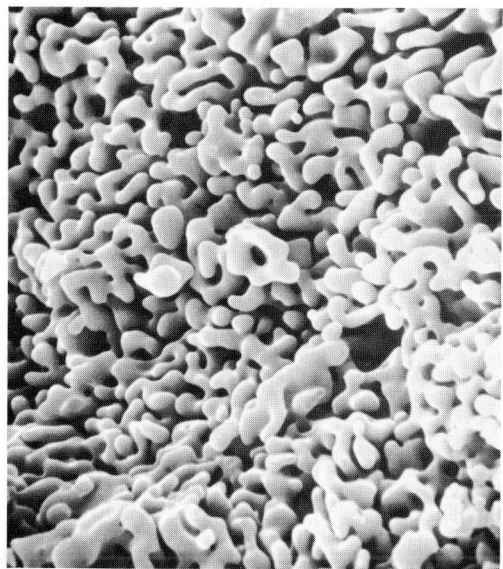

Fig. L-2. Electron scanning micrograph of calcined clamshells (quicklime). Magnification, approximately 2,400×.

for easy handling. Almost all quicklime is shipped in bulk.

Quicklime can be hydrated by the *dry* or *wet* method. In the dry method only enough water is added to chemically combine with the oxide and to compensate for losses due to evaporation. In the wet method, called *slaking,* an excess of water is added, resulting in a slurry or suspension of calcium hydroxide in water. For some applications it is desirable to slake with a lower water-to-solids ratio to yield a paste or a putty.

Dry hydration of high-calcium quicklime proceeds rapidly with evolution of 490 Btu/lb. Dry hydration of dolomitic quicklime at atmospheric pressure results in dolomitic single hydrate (type N hydrated lime per ASTM C-6). Dry hydration of dolomitic quicklime at 25–100 psi steam pressure converts the MgO component, giving the double hydrate (type S hydrated lime per ASTM C-206). Hydration is carried out in batch or continuous hydrators, which are essentially mixers that keep the crushed quicklime agitated while water is added.

A very substantial temperature rise occurs upon slaking quicklime, depending upon the water-lime ratio and the reactivity and purity of the quicklime. With insufficient water violent boiling occurs. Both wet and dry hydration result in physical changes in the lime particle. Hydrated lime is much finer than pulverized quicklime, and the specific surface area of the lime may increase tenfold upon hydration. Dry hydrated lime is usually whiter and purer than quicklime since coarse, heavy inerts can be removed in the hydrator system. Dry hydrate is a sticky powder having low bulk density and poor dry flowability.

Uses: The applications of lime encompass almost all sectors of technology. The oldest use is as a *structural material,* including masonry mortars, wall plasters, sand-lime brick, and soil stabilization. See also **Gypsum.** Double-hydrated dolomitic lime or specially processed high-calcium lime mixed with gypsum plaster is troweled on interior walls or ceilings to provide a hard, white finished surface. It is mixed with cement and sand to make exterior plaster or stucco. Mason's mortar, used to lay up bricks or blocks, usually contains lime. The lime provides plasticity, water retention, and easy troweling. *Sand-lime bricks* are made in large volume in Europe and Russia but not in the United States. About 10% hydrated lime is mixed with graded sand and water, pressed into shape, and put into autoclaves for 4–8 hr at 300–400°F (149–204°C). The reaction product, calcium silicate, results in a strong, white brick. A similar, lightweight, cellular product is made by introducing a blowing agent, usually Al powder, into the lime-sand-water mix. Large units, such as blocks, roof decks, and wall slabs typically are formed, having a density of 25–50 lb/ft^3.

In Japan, some portland cement is made by separately calcining limestone, mixing the quicklime with sand and clay, and clinkering and grinding this composition. A patent describes a nonclinkering process for manufacture of a cement by very fine grinding of quicklime and an aluminosilicate. See also **Cement.**

Stabilization of clay-bearing soils for construction is a rapidly growing use for lime. Typically, 5–7% hydrated lime is added to soil that has been scarified 6 in. deep. Water is added, and the lime, soil, and water are mixed. The soil is compacted by rolling. After curing for several days, the soil is ready to be surfaced. Because of the agglomerating and cementing action of the lime with clay and soil, the resulting base is more stable and resistant to moisture and temperature changes.

In *steel production,* the displacement of the open-hearth process by the basic oxygen steelmaking process has increased lime requirements fivefold, up to about 140 lb of lime per ton of steel. Lime is used as a flux. See also **Iron- and steelmaking.**

Dead-burned dolomite, formed by calcining dolomite at about 3000°F (1649°C) to convert MgO to periclase, is used as a *refractory.* Dolomitic quicklime sometimes is used in place of dolomitic limestone in *glassmaking.* Quicklime is used in the *Bayer process* for making alumina from bauxite. See also **Bauxite.** Large tonnages of lime are used in the *sulfate process for making paper.* See also **Pulp (wood) production and processing.** The use of lime to produce *calcium carbide* for making acetylene is declining. In the Dow process (seawater) for making *magnesium metal,* lime reacts with $MgCl_2$ in the seawater to precipitate $Mg(OH)_2$. See also **Magnesium.** Lime also is used in other magnesium-production processes. *Calcium metal* also is made from lime by reducing CaO with coke. See also **Calcium.**

Lime is employed in preparation of *insecticides* (calcium arsenate and lime-sulfur and bordeaux mixtures), *calcium hypochlorite* for bleaching paper pulp, *ethylene glycol* (permanent antifreeze) by the chlorohydrin process, citric acid, and various calcium-based salts (phosphate, stearate, citrate, tartrate, sulfonate, and benzoate). In the *sugar industry,* lime is used to purify the crude-sugar solution by precipitating impurities, which are filtered out. The remaining solution is carbonated to remove residual lime as calcium carbonate. Since both CaO and CO_2 are needed, sugar producers operate captive lime plants, often regenerating CaO from the $CaCO_3$ sludge.

Lime has some limited use as a *desiccant,* since quicklime avidly takes up moisture to 24% of its original weight. Hydrated lime absorbs CO_2 and prevents deterioration of apples during storage.

Lime is used in *water treatment,* sanitary engineering, and pollution control. In water softening, hydrated lime is added to remove temporary hardness or in the lime-soda process to remove permanent hardness. Addition of excess lime to water and retention for 24–48 hr purifies water by killing bacteria and by removing phenol. Lime acts as a coagulant for removing suspended solids by sedimentation from both municipal and industrial water. See also **Water treatment.** Dolomitic lime removes silica from boiler feedwater due to silica absorption by $Mg(OH)_2$. Lime is widely employed for acid neutralization in industrial wastes (fluoride effluent from phosphate plants and pickle liquors from steel plants) and mine wastes (acid coal-mine water).

REFERENCES

Allsman, Paul L.: "Minerals Yearbook," vols. 1 and 2 (Lime), U.S. Bureau of Mines, Washington, 1968.

Boynton, R. S.: "Chemistry and Technology of Lime and Limestone," Interscience-Wiley, New York, 1966.

United States Gypsum Company, Chemicals Division, Chicago, Ill.: "Profitable Lime Management," 1970.

—John M. Summerfield, *United States Gypsum Company, Des Plaines, Ill.*

Limestone   Limestone is a rock containing chiefly calcium carbonate and variable quantities of magnesium carbonate. Production of limestone in the United States (1970) is approximately 630 million tons/yr. Limestone is classified as *high-calcium* ($CaCO_3$ containing 5% or less $MgCO_3$); *dolomitic* ($CaCO_3 \cdot MgCO_3$, usually with over 35% $MgCO_3$); and *magnesian* (predominantly $CaCO_3$ but containing 5–35% $MgCO_3$). High-calcium limestone occurs as two minerals, *calcite* and *aragonite.* The former is the ordinary stable form, having a specific gravity of 2.71, Mohs' hardness of 3.0, and rhombohedral crystal structure. Aragonite is $CaCO_3$ precipitated from salty water either by inorganic processes or by the formation of marine shells. Aragonite is relatively unstable, slowly transforming to calcite. Aragonite has a specific gravity of 2.93 and Mohs' hardness 3.5–4.0, and crystallizes in the orthorhombic form. Dolo-

mite is a chemical compound, not a physical mixture, resulting from contact of Mg-bearing waters with calcite. Dolomite has a specific gravity of 2.8–2.9, Mohs' hardness of 3.5–4.0, and a rhombohedral crystal structure. All limestones evolve CO_2 and bubble in dilute HCl, providing a simple means of mineral identification. Dolomite reacts only in hot dilute HCl, while calcite will decompose in cold dilute HCl.

Limestones vary greatly in color and texture. Texture ranges from dense and hard limestone, e.g., marble or travertine, which can be sawed and polished for use as decorative stone, to soft, friable forms, e.g., chalk and marl. Chalk is an extremely fine-grained white limestone, while marl is an impure deposition product containing clay and sand. Texture, hardness, and porosity appear to be functions of degree of cementation and consolidation during formation of these materials. Color variation is due to impurities. Impurities may be deleterious, e.g., sulfur and phosphorus in stone for metallurgical processes, or harmless or beneficial, e.g., argillaceous impurities in portland cement rock.

Over 90% of all limestone is quarried (see Fig. L-3); the balance is mined underground. Although limestone occurs widely, good chemical- and metallurgical-grade limestone is scarce in many areas. Along the coast, oyster- or clamshells are dredged as a source of $CaCO_3$. Limestone normally is processed through a series of crushing, screening, and grinding operations. Because of transportation costs, the proximity of limestone sources to users is essential.

The major use of limestone is in *construction* (asphalt filler, road stone, riprap, and bituminous aggregate), representing about 60% of total consumption. Portland cement consumes about 15%, lime manufacture about 5%, agricultural liming about 5%, and use as a blast-furnace flux in iron production about 5%. Other applications include ground carbonate fillers for paint, paper, plastics, and adhesives; incorporation in animal feeds; as a fertilizer filler; in glass and refractories manufacture; and for acid neutralization. See also **Fertilizers.** Limestone is the source of the cheapest alkali for the chemical industry and a leading candidate for control of SO_2 pollution from industrial combustion processes. Lime also is used to neutralize acidic mine waters.

Precipitated $CaCO_3$ is produced in numerous chemical processes. See also **Lime.** In many cases, it is economical to dry and calcine the by-product

Fig. L-3. Limestone quarry at Genoa, Ohio. (*United States Gypsum Company.*)

to regenerate CaO or Ca(OH)$_2$. Some precipitated CaCO$_3$ is made to specific particle size and shape, whiteness, and purity for use as a functional filler for paper coatings, paint, and polymers. These products command a premium price compared with pulverized limestone fillers.

—John M. Summerfield, *United States Gypsum Company, Des Plaines, Ill.*

Limonite (See **Iron ores;** and **Nickel.**)

Linalool (See **Terpenes.**)

Linear Alkyl Sulfates [See **Anionic surfactants (sulfur-bearing); Detergents;** and **Surfactants.**]

Linerboard [See **Pulp (wood) production and processing.**]

Linkages (See **Carbon compounds;** and **Polymerization.**)

Linnaeite (See **Cobalt.**)

Linoleic Acid (See **Paint;** and **Vegetable oils.**)

Linolenic Acid (See **Vegetable oils.**)

Linseed Oil (See **Paint;** and **Vegetable oils.**)

Lipoproteins (See **Peptides and proteins.**)

Liquation (See **Tin.**)

Liquefied-Gas Storage (See **Insulation, thermal.**)

Liquefied Petroleum Gas (LPG) (See **Fuels; Hydrocracking;** and **Petroleum processing.**)

Liquid Chromatography (See **Chromatography.**)

Liquid Crystals A very limited number of organic compounds which are liquids at room temperature and atmospheric pressure exhibit *cybotaxis,* that is, under x-ray examination they show diffraction patterns and other characteristics that normally are expected only from solid crystalline substances. From these studies, it is inferred that the molecules in these materials are arranged in accordance with specific spatial configurations. As a class, these liquids are known as *cybotactic liquids.*

A much more limited family of materials, known as *liquid crystals,* possess the qualities of cybotactic liquids, but exhibit even more striking crystalline characteristics. Studies indicate that large groups of molecules in liquid crystals maintain their mobility (as in a true liquid) but, nevertheless, also retain a form of structural relationship—a loosely held-together spatial geometry. This underlying structure imbues the liquid crystal with many of the optical properties of a true crystal.

Although cybotactic liquids and liquid crystals have been known to scientists for many years, only during the past decade have serious investigations been made. These studies have been motivated largely because of the potential applications for liquid crystals in a variety of commercial electronic and optical devices.

At the present state of the art, liquid crystals fall into three classifications: (1) *smectic liquid crystals* in which the molecules appear to be arranged in rather precise layers where the longer axes of the molecules are perpendicular to the plane of the layers; (2) *nematic liquid crystals* in which the molecular arrangements are somewhat less ordered, where the long axes of the molecules are parallel, but where the molecules are not always arranged in precise layers; and (3) cholesteric liquid crystals in which the long axes of the molecules lie in the plane of the layers, but where the layers are very thin. Cholesteric crystals derive their name from the fact that many of these compounds contain cholesterol.

Because of the potential commercial value of these materials, the identification of specific new compounds frequently is withheld from the literature. Among the better known examples are *p*-ethyl azoxybenzoate (smectic), *p*-azoxyanisole (nematic), and cholesterylnonanoate, cholesterylchloride, and cholesteryloleate (all cholesteric).

A common experimental configuration which may become a building block for electronic and optical devices is the *liquid crystal cell*. These cells still are in an early stage of development. A drop of a liquid-crystal substance is sandwiched between two parallel plates of glass. A thin coating (tin oxide) on the inside plate surfaces ensures a uniform electric field across the cell. Tiny spacers are used between the glass plates to seal and to maintain a uniform spacing of from 5 to 25 µm. Electrodes then are attached to each glass plate. The cells may be square or circular and from a fraction of an inch to many inches across, depending upon the intended application.

This construction has the characteristics of a parallel plate capacitor (about 200 pF/cm^2) with the liquid-crystal substance as the dielectric. With no electric potential applied across the cell, the sandwiched material is quiescent, essentially transparent. As a dc voltage or low-frequency (60 Hz) ac voltage is applied, the sandwiched material becomes opaque. One researcher reports that the applied voltage must be on the order of $5 \times 10^3 \text{ V/cm}$, or about 6 V for a 0.5-mil thick cell. The opaque state is not due to a chemical reaction, but rather it results from the liquid becoming turbulent, causing it to scatter rather than transmit light. The cell thus becomes an optoelectric switching device. The phenomenon has led to the term *dynamic scattering*.

The concept of dynamic scattering postulates that the molecules of the liquid crystal cannot align with their dipole moment in the direction of the field when an electric field is applied to a thin film of the material. Rather, the axes of the molecules remain at some angle with respect to the electric field. The off-axis alignment of the molecular axis and electric field is essential for the scattering mode. Ions traveling in the liquid crystal under the influence of the field disrupt the normal pattern of the molecules, with some molecules being forced to line up in the directions of the local field around the ion. This gives rise to regions of discontinuity in the distribution of the molecules. These regions of discontinuity also result in regions of changing index of refraction for incident light on the liquid. Thus, the formation of light-scattering centers and resulting overall opacity.

It is evident that a basic device of this type can be used either in a transmissive or reflective mode. Devices designed for use of the transmissive mode have been termed *electronic windows*. Inasmuch as the ability of the liquid-crystal cell to transmit or reflect light is a function of the amount of emf applied to the electrodes, such windows, in conjunction with a photocell, can control the desired amount of light that may reach a receptor in an electrooptical circuit. Suggested uses have included automatically adjustable greenhouse windows, aircraft and automobile windshields, and camera apertures.

Reflective-type cells already are finding application in numeric counters wherein the digital numbers desired to "light" for a display are in the form of liquid-crystal cells. Upon activation for the reflective mode, the number(s) is seen by the viewer because of reflected ambient light. An all-electronic clock with no moving parts also can be constructed in which the reflective-type cells are used.

Color Switching: One of the most intriguing aspects of certain liquid crystals is the ability to obtain color without the use of a dye. In complex electronic data displays, color coding is highly desirable. One researcher reports development of a crystal mix that changes color from red to green when 150 V are applied to the cell. At lower voltage levels, the color change is not so pronounced (yellow-red hues). Still other researchers have added dyes to liquid-crystal materials. Combinations of photochromic and pleochroic dyes have been used. The materials exist in two forms, each exhibiting a different absorption spectrum. When a liquid crystal made of these materials is irradiated with light at the proper wavelength, a reversible conversion of one form to another is produced. Thus, the absorption spectra of the cell (hence color) are changed. The availability of such devices also leads to the possibility of constructing optical color-switching devices as well as voltage color-switching devices.

Liquid-Gas Separations (See **Separation operations**.)

Liquid-Liquid Dispersions (See **Mixing, fluids**.)

Liquid-Liquid Separations (See **Separation operations**.)

Liquid-Phase Reactions (See **Catalysts**.)

Liquid-Solid Mixing (See **Mixing, fluids**.)

Liquid-Solids Separations The need to separate liquids from solids occurs frequently in chemical and metallurgical processing. Complete separation of the solids and liquids is not always required, the degree of separation being dictated by the requirements of the process and giving rise to several kinds of liquid-solids separation equipment. Further, the nature of the liquids and solids determines the separation equipment most appropriate for specific conditions.

Because the major liquid-solids separation processes are described elsewhere in this volume, the objective of this description is to assist in the initial evaluation of a liquid-solids separation problem, emphasizing the relative advantages and limitations of the major operations for particular starting conditions and desired end results. An overall correlation is given in Table L-5.

Factors Influencing Separation. The major characteristics of liquid-solids mixtures (often termed *slurries*) which affect the efficiency of separation, the degree of separation achievable, and the selection of specific separations equipment include (1) particle-size distribution, (2) concentration of feed suspended solids, (3) specific gravity of solids and liquids, (4) chemical and physical properties of the materials, (5) surface and shape of particles, and (6) desired specifications of the liquids and solids once separated. Initial and operating costs for various types of separation equipment, of course, also are important. See Table L-6.

Particle-Size Distribution: The average size of the solids and the size distribution are of major importance. From Stokes' law (see also **Sedimentation**), the factor gD^2 is a gravity or weight factor and is the essential driving force for sedimentation, classification, and centrifugation. It also plays an important part in screening. Stokes' law does not apply in filtration except for homogeneous solids suspension in the filter tank. The effect of particle-size distribution on the major liquid-solids operations may be summarized as follows:

Screening: Coarse materials are the easiest to separate. Separation becomes more difficult as the near-mesh fraction increases.

Classification: The principles are similar to screening except for the coarse fraction, where $1/4$-in. or larger material is difficult to separate. Classification at less than 200 mesh is difficult by gravity but can be achieved by cyclones.

TABLE L-5. *General Characteristics of Liquid-Solids Separators*

Driving Force	Unit Operation	Manner of Separation	Mechanism	Characteristics of Solids	Characteristics of Liquid
Gravity	Sedimentation	Bulk solid/liquid	Hydraulic	Pumpable	Clear
	Screening	Separation of coarse solids from fine solids plus water	Fixed opening	Drained	Dirty
	Classification		Hydraulic	Fluid pulp	Dirty
Mechanical	Cycloning		Hydraulic	Fluid pulp	Dirty
	Centrifugation		Hydraulic	Fluid pulp	Dirty to Clear
			Permeable medium	Dewatered	
	Filtration	Bulk solid/liquid	Permeable medium	Dewatered	Clear
Thermal	Drying	Evaporate or lower viscosity	Evaporation or hydraulic	Dewatered	Vapor

TABLE L-6. *Separation Methods versus Product Characteristics**

Separation Method	Solids Characteristics	Liquid Characteristics
Thickening	40–75% solids (all particle sizes)	100–400 ppm of extreme fines
Screening	80–90% coarse solids	5–20% fine solids
Classifying	75–85% coarse solids	5–30% fine solids
Cycloning	40–80% coarse solids	5–30% fine solids
Filtration	80–95% solids (all particle sizes)	Less than $\frac{1}{2}$% of extreme fines
Centrifugation	70–95% solids	Less than $\frac{1}{2}$–20% of fines and extreme fines
Drying.	95–100% solids (all particle sizes)	Vapor and sometimes extreme fines

*The *degree of separation* required is influenced by the value of the materials being handled, the size distribution, processing problems, and contamination of the liquid. *Screening* always requires an undersize stream containing fines and gives excellent recovery on the oversize material. Dissolved solids can be recovered by washing the oversize material so long as the particles are reasonably coarse and free-draining (1 mm or larger). *Classifying* usually is not used as a final dewatering or separation operation. *Filtration* provides the highest recovery of all sizes, especially in the range below 100 mesh. It also provides the highest potential recovery of dissolved solids by single- or multiple-stage displacement washing. *Centrifugation* will produce high recovery for solids larger than 100 mesh and can produce good dissolved solids recovery. High-strength liquor concentrations, as well as high-solids concentrations, can be achieved, particularly on coarse solids.

Sedimentation: Coarse materials are the easiest to separate. However, a top size limitation must be considered because the torque requirements for the raking mechanism increase substantially with particle size. Top size for materials is

Mesh

Coal 14–28
Sand 40–65
Iron-ore concentrate 65

Centrifugation: Again, the coarse materials are the easiest to separate. Extremely fine, low-specific-gravity materials are difficult to separate.

Filtration: Filters usually operate best with solids of uniform particle size. Top size is limited by that which can be maintained in suspension in the filter tank. Since filtration is based upon the flow of fluids through capillaries and not upon Stokes' law, filtration acts on all particle sizes. Thus, fine particles can be handled effectively.

Drying: The coarser materials facilitate drying because generally less liquid is associated with the coarser particle due to the smaller relative surface area. In drying, fine materials may create dust problems.

Flocculation: Since particle size is very important in all liquid-solids separations, the possibility of adjusting or altering the particle size by flocculation should always be considered. Flocculation generally is used where the separation would not be possible without it, e.g., the filtration of sewage sludges. Since mechanical or chemical flocculation adds to operating costs, even though floccula-

tion may increase capacity, change the liquid content of the principally solids stream, or influence solids recovery, the additional cost versus efficiency and savings must be justified.

In some cases where loadings on existing equipment equal or exceed a proper design base, flocculants can be used to increase equipment capacity and thus delay the need for additional equipment. This is an economic consideration in balancing capital cost versus operating cost. The design of any flocculation system is important to achieving the most effective use of the energy or chemicals introduced. Since flocculation depends upon the number of collisions, time, and surface or electrical-charge characteristics of the materials, these factors become important design criteria. Because many floc structures are quite fragile, sheer stresses generally should be avoided to maximize the effect of flocculation. Due to the high-force area within a centrifuge, flocculation of centrifuge feeds has produced mixed results. Some success has been achieved by introducing chemicals within the bowl of the centrifuge itself.

Concentration of Feed Suspended Solids. Increasing the concentration of feed solids generally will increase equipment capacity because this action reduces the overall quantity (volume) of material to be handled. However, in classification, the mobility of the feed stream is important to ensure that relative movement between particles is not hindered by excessive collisions. In this case, higher solids concentration will hinder classification efficiency. Concentration of feed suspended solids affects liquid-solids operations as follows.

Screening: The feed stream must be mobile so that the fine particles can move to the screen and pass through, rather than be carried off with the bulk of oversize material. The effectiveness of separation also depends upon the split of the liquid phase, since the very fine particles ($<10\ \mu m$) will follow the liquid proportionately. As an example, if 20% of the liquid follows the oversize and 80% follows the undersize, 20% of the material smaller than 10 μm will follow the oversize and 80% will follow the undersize. Thus, the more dilute the feed stream, the greater the proportion of the undersize stream, allowing fewer fines to carry through with the oversize.

Classification: Dilution is important in classification to prevent hindered movement of the particles. The fines also report in at least the ratio of the liquid phase (as described for screening). Since the undersize stream will contain solids that produce some degree of hindrance, this factor must be considered in control of the separation point.

Sedimentation: Where Stokes' law applies (unhindered settling), the settling rates of the particles is constant. The degree of thickening required decreases as feed concentration increases. Thus, the capacity increases as the feed concentration increases. However, in most sedimentation or thickening applications, the solids must pass through a hindered settling condition described by the Coe-Clevenger equation:

$$A = \frac{1.33(F - D)}{Rg_1}$$

where A = unit area, ft²/ton of dry solids/24 hr
F = weight ratio of liquids to solids for settling rate R
D = weight ratio of liquids to solids in discharge
R = settling rate with an initial dilution F, ft/hr
g_1 = specific gravity of liquid

As feed dilution F decreases, the sizing basis decreases, but the settling rate also changes due to the increased hindered condition. Thus, the two changing factors may offset each other, and increasing the feed solids concentration will increase capacity only where unhindered settling is concerned. However, in many cases, increased feed solids will not significantly influence the capacity of the unit. In some cases, diluting the feed will increase the capacity. In situations of this type, experience with the same or similar materials is a paramount consideration.

Centrifugation: When a centrifuge is operated as a classifier, the foregoing observations apply with regard to the mobility and hindrance of particle movement. However, for separation, the higher the feed concentration, the more capacity a centrifuge will have. It will also trap more solids within the bed of the unit. Since the centrifuge must swirl the entire feed stream, the energy required for a higher feed concentration for the same amount of solids will be less than for a lower concentration. Maximum concentration is affected mainly by prefeed and feed-handling considerations.

Filtration: The rate of filtration increases as the feed solids concentration increases due to increased cake permeability and less liquid passage. High feed solids concentration may increase cake moisture slightly due to the formation of a less compact cake. If the filter rate is limited by cake washing or moisture reduction, increased feed solids may not influence design rate in any case.

Drying: From an economic standpoint, less liquid to be evaporated requires less heat. Feed concentration also depends upon handling characteristics required for effective feeding of the drier.

Flocculation Factors: As the solids concentration increases, intimate, uniform distribution of flocculating agents throughout the slurry becomes more difficult. The resulting floc may differ significantly from that produced in the same slurry in a more dilute condition.

Specific Gravity of Solids and Liquid. From Stokes' law, it is apparent that sedimentation, classification, and centrifugation all depend upon the relative difference between the specific gravity of the solids and that of the liquor for effective separation. Thus, in these operations, as the specific gravity of the solids approaches that of the liquid, the separation becomes more difficult. Filtration and screening are not directly influenced by the specific gravity of either phase.

Chemical and Physical Properties of the Liquid

Viscosity: In all cases, viscosity influences capacity, ease of separation, and the degree of liquid elimination from the principally solids stream. The lower the viscosity (which is inversely proportional to temperature), the better the separation. But, in a few cases, higher viscosity allows classification at larger mesh of separation than with water, due to the decreased settling rate of a given size particle.

Surface Tension: With the exception of screening, surface tension generally is not important. In screening fines, a greater degree of liquid or material may adhere to the surface of the particles and carry with it a higher proportion of the extreme fines, resulting in a larger proportion of fines being carried with the oversize.

Interfacial Tension: Adjustments or modification of interfacial conditions between solids and liquid can cause hydrophobic or hydrophilic conditions affecting moisture content of the principally solid stream. Generally, this modification is costly and not economically practical, but it may have application where high-cost materials are involved. An example of this would be to introduce an oil under high-shear agitation into a water-solids suspension, causing the solids-water interphase to change to a solids-oil-water interfacial system.

Dissolved Solids: Soluble materials follow the liquid phase in identical proportions, and this, in many cases, fixes the limitations for dissolved-solids removal by washing or dilution in liquid-solid separation equipment. Often, it is desirable to remove dissolved solids from the principally solid stream by rinsing or washing with fresh or recycled wash liquor. It is important to distinguish between dilution washing and displacement washing. In dilution washing, the solubles-free wash liquor is mixed intimately with the principally solids stream to dilute the concentration of solubles in the original liquor. Then, by further separation, the amount of dissolved solids remaining with the solids stream is reduced. The maximum effectiveness of this washing is limited by the relative concentrations and volumes of mother and wash liquors.

In displacement washing, the solubles-free wash liquor is passed through the principally solids stream in an attempt to achieve ideal plug flow by displacing mother liquor with the wash liquor. The maximum effectiveness of this type of wash is limited by the relative quantity of wash liquor applied, the soluble concentration of the wash liquor, and the efficiency of achieving theoretical displacement of the mother liquor. Thus, the degree of solubles removed by displacement washing is significantly greater than that removed by dilution washing.

The effect of dissolved solids on the major liquid-solids operations may be summarized as follows.

Sedimentation and Classification: Pumping or handling of the underflows limits the concentration of dry solids. Classifiers can concentrate dry solids but are limited to the coarser solids. Thus, liquid removal is limited, and solubles recovery necessitates further washing. Classifiers are poor washers due to ineffective mixing caused by the tendency

of wash liquor to roll over the surface instead of actually passing through the solids bed. Dilution of the mother liquor occurs, as separation of strong and weak liquors cannot be achieved. Also, addition of wash liquor into the classifier causes more difficult control of the separation point.

Centrifugation: In solid-bowl machines, washing will tend to wash out some fine solids because the wash commonly is applied at the area of centrifugal force. Although some degree of displacement washing is achieved, the wash does not pass uniformly through the entire bed. Separation of mother liquor and weak liquor cannot be achieved. For the screen-type centrifuge, washing is more effective but also causes some degree of dilution. Strong and weak liquors can be separated although washing will also tend to wash fine solids into the centrate.

Filtration: Filters operate entirely with a displacement wash. Wash liquors can be separated effectively from the strong liquor, and on many machines countercurrent washing can be used. This is important in applications where maximum-strength liquor from the washing system is critical to process economics.

Effect of pH: In many cases, the pH of the feed stream may affect the flocculant characteristics of the slurry and consequently the efficiency of the liquid-solids separation.

Surface and Shape of Particles: In some cases, surface or porosity may produce a lower specific gravity of the solids. Surface and shape of particles affect the major liquid-solids operations as follows.

Screening: Generally there is no influence from the standpoint of surface, but shape is very important. Screening is usually based on the second largest dimension of the particle to be separated. The screening job becomes more difficult as the difference in dimensions increases.

Classification: As the particles become less spherelike, the separation becomes more difficult. As the sphericity decreases, the resistance to passage of the particle through the medium increases.

Sedimentation: Particle shape is important because design generally is based upon, or limited by, the smallest particle size to be removed.

Centrifugation: Particle surface influences only pumping, plugging, and handling problems. The shape generally is not important except relative to handling and discharge of the principally solid stream out of the unit. Porous materials may tend to retain liquid and require a higher driving force for liquid removal and for effective washing.

Filtration: Shape influences this operation only

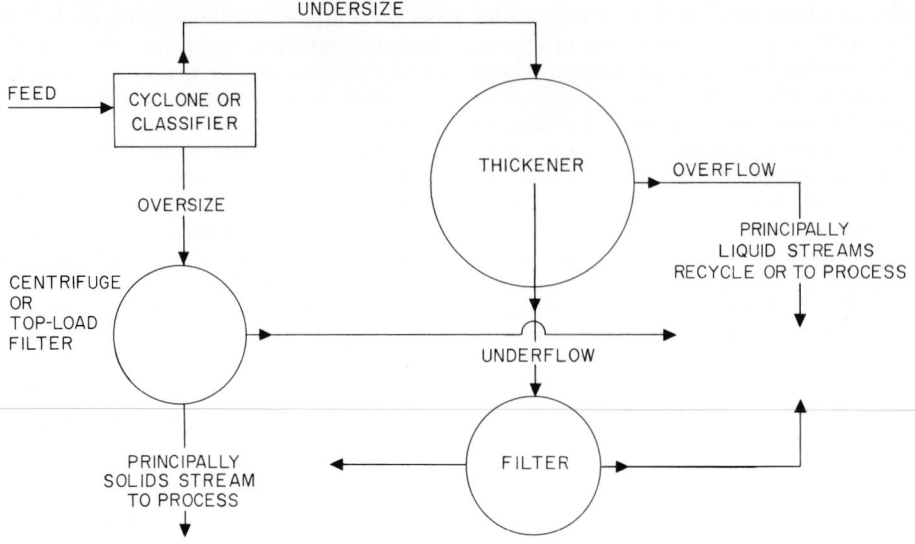

Fig. L-4. Typical liquid-solids circuit in which classification and separation equipment are combined.

in the case of extreme fines ($<$10 μm) even if the material is flocculated. Bentonite clay, which is a very fine platelet material, is a typical example. Porosity of particles hinders moisture reduction and wash effectiveness.

Drying: Shape and surface of particles essentially affect only handling but may influence feed method and conditions necessary to achieve uniform spreading of the feed for intimate heat contact. Porosity of the solids may hinder or reduce drying efficiency.

Influence of Total System on Liquid-Solids Separations. A liquid-solid separation device is influenced by conditions upstream and downstream. The device accepts a feed stream and produces a principally solid stream and a principally liquid stream. Refinements, such as washing, can be included in the total process.

Feed-Stream Factors: Pertinent considerations include the following.

The Necessity of Prior Thickening of the Feed Stream: A centrifuge acts on the total feed stream, and prior thickening may significantly reduce the energy requirement. In filtration, prior thickening may not be necessary for operation but may be desirable to reduce filter size. Both thickeners and classifiers are effective thickening devices. Classifiers have the advantage of separating fine solids, such as required in upgrading a concentrate.

Combination Operations: In some cases, a stream

may be handled best by a combination of classification and separation equipment. A typical circuit of this type is shown in Fig. L-4.

Flocculation: Flocculation can often be helpful in making difficult separations possible. Flocculation of thickener feeds usually is performed by adding chemicals to the feed before it enters the feed well, possibly in the feed launder on influent pipe. For filter applications, the flocculant usually is introduced in a special mixing-contactor prior to feeding into the filter tank. In centrifugation, the use of flocculants is questionable due to the high shear forces produced in the machine.

Precipitation and Crystallization: These steps may precede a separation operation, where they permit adjustment of particle size and character for optimum liquid-solid separation.

Processing Liquid-rich Streams: In any separation operation where solids will pass the separating device in the liquid-rich stream, some means of handling the stream is essential. If the stream is recirculated upstream, the fine solids returning to the separator will tend to build up in the recirculating system. To maintain operating effectiveness of the separator, the solids must be removed from the system. If the solids are allowed to build up, the feed to the separator eventually will change in character (become much finer) and the removal effectiveness and capacity of the separator will be seriously limited.

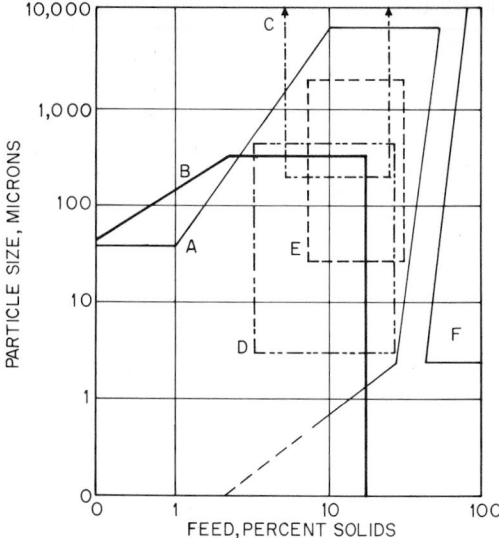

Fig. L-5. Areas of application for liquid-solids separation equipment. *A,* Filters and centrifuges, by varying the design and separation technique, can be used over very wide ranges of solid particles and concentrations. In general, the system must have sufficient flow to allow uniform feeding of the machine. The quantity of liquid required usually is a function of particle size and specific gravity of the particles. *B,* Sedimentation systems generally are restricted to feed solids finer than 35 mesh and 9–40% solids. *C,* Screens find the bulk of their application above 100 mesh size and with slurries ranging from 7 to 25% solids. *D,* Cyclones usually are used for applications between 5 μm and 28 mesh and 4–35% solids. *E,* Classifiers normally are applicable from 325 to 14 mesh and 9–40% solids. *F,* Driers are applicable to a wide range of particle sizes. Generally, it is desirable to minimize the liquid content of the system before feeding the drier. The amount of dewatering which can be accomplished economically normally is a function of particle size. Most continuous driers are subject to high dust losses if extreme fines are present as discrete particles. By agglomerating the fines into porous nodules, however, it is possible to dry very fine materials without excessive dust loss.

In many cases, a separator is used to dewater underflow from a preceding thickener with the filtrate or centrate being returned to the thickener for recovery of any solids which may pass the separator. If a significant amount of solids is allowed to pass the separator, they will be relatively finer than the representative feed solids and will eventually build up in the thickener to the point of overloading the machine. Although the problem may appear to be in the thickener, the

separating device actually may be the cause by recirculating fines which are not being removed from the circuit.

Processing Solids-rich Streams: Recirculation within the process of the solids-rich stream, unlike the liquid-rich stream, usually is not practiced. The solids-rich stream normally will proceed to the following step in the process.

The material balance around the separator is extremely important on each particle size and should include consideration of any breakage through the separator. In filtration, this is not too critical, except that recovery of all the fines may reduce the rate and increase cake moisture content. However, in centrifugation, this is an important factor, especially where a large amount of fines is present or if high solids recovery is important. In sedimentation, as the underflow concentration is increased, the size and torque requirements of the unit also increase. The underflow concentration is limited by the pumping and handling characteristics of the sludge.

Equipment Applications. The areas of application for liquid-solids separation equipment, in terms of percent solids in the feed and particle size, are summarized graphically in Fig. L-5.

Costs: Generalized cost figures for various cate-

TABLE L-7. *Generalized Cost Comparison of Liquid-Solids Separation Devices**

Type of Device	Cost Dollars per Ton per Day	
	Solids	Liquid
Thickener	200	50
Screen	400	200
Classifier	275	150
Cyclone	250	75
Filter	1,500	1,000–2,000
Centrifuge	2,000	1,500–3,000
Drier	3,100	14,000

*Values are given on the basis of 10 years of operation. By taking a mean figure of capital cost for each type of equipment and starting materials and final-product specifications and operating costs (in terms of dollars per ton per day of solids or liquids treated) and then writing off the equipment charges over a 10-year period and accumulating the operating charges, these generalized costs were developed. From this particular tabulation, it would appear that the lowest-cost method would be the various sedimentation devices, while the most expensive would be the drier. Obviously, these values are strongly influenced by the write-off time assumed. A long write-off period favors the machines that have high capital investment and low operating costs.

gories of liquid-solids separation equipment are listed and explained in Table L-7. The generalized cost tabulation points up the possibility of stagewise eliminating liquid from a slurry. Although one of the lower-cost systems may not produce product of the required specifications, it may provide an economical method of preparing the feed stream for the more costly system.

See also **Centrifuging; Drying, solids; Filtrations; Screening; Sedimentation; Separation operations;** and **Waste treatment.**

—L. L. Palm, *Eimco PMD Division, Envirotech Corporation, Salt Lake City, Utah*

Litharge   (See **Lead.**)

Lithium   (Gk. *lithos,* stone.) **Li** = 6.94 (at. wt.); 3 (at. no.). Discovered by the young Swedish chemist, Johann August Arfvedson, (1817) working in Berzelius' laboratory in Stockholm, lithium was named by Berzelius. It is the first member in group Ia of the periodic system, which includes the other alkali metals. Like its congeners, Li is electropositive, and the lithium atom is readily oxidized, losing its single valence electron to form a univalent ion. Compared with the other alkali metals, Li has the lowest density (lightest of *all* solid elements), the smallest ionic radius, the highest ionization potential and electronegativity, and the highest heat capacity. In general, Li is the least reactive, although there are exceptions, e.g., its reaction with C, N, and H.

In each group of the periodic table, the first member is peculiar, and thus Li differs from the other alkalies much more than they differ from one another. The compounds of Li are in general analogous to those of the other alkalies but often exhibit anomalous properties, in accordance with the diagonal relationship in the periodic table, whereby Li shows marked similarity to Mg and, in certain respects, to the alkaline-earth metals. This is noted in the low aqueous solubility of the carbonate, phosphate, oxalate, and fluoride and in the high solubility of the halides (other than the fluoride). Other such similarities include reaction with O_2 to form the normal oxide instead of peroxides, direct combination with N and C, the low thermal stability of the carbonates, high solubility of the alkyls in hydrocarbons, and high solvation of the ion as indicated by the formation of hydrates and ammoniates.

Many of the unusual properties of Li and its compounds can be attributed to the small radius of the Li ion. In the alkali group, the ionic radii

TABLE L-8. *Some Physical Properties of Lithium*

Stable isotopes:	
^{6}Li:	
Atomic weight	6.01703
Natural abundance %	7.6
^{7}Li:	
Atomic weight	7.01823
Natural abundance, %	92.4
Atomic radius, Å	1.33 Å
Ionic radius, Å	0.60 Å
Crystal structure (-195 to $+180°$C)	bcc
a_0 at 20°C, Å	3.5
Hardness, Mohs' scale	0.6
Density at 20°C, g/cm^3	0.534
Tensile strength, psi	84
Surface tension at 180°C, ergs/cm^2	398
Viscosity, mP:	
At 183.4°C	5.918
700°C	2.89
Melting point (triple point) (ΔV on	
melting = $+1.5\%$)	180.54
Heat of fusion, cal/g	103.2
Boiling point at 760 mm Hg, °C	1340
Heat of vaporization, cal/g	5100
Specific heat at 25°C, cal/g	0.849
Thermal conductivity (181°C)	
cal/(s)(cm)(°C)	0–105
Electrical resistivity at 20°C, $\mu\Omega$-cm	9.446
Ionization potential, eV	5.37
Standard electrode potential at 25°C, V	-3.045

range from 0.60 Å for Li^+ to 1.69 Å for Cs^+. This results in a high positive charge density for Li^+, which in turn accounts for the high degree of hydration (or solvation) of the Li ion. Thus, hydration converts the smallest alkali ion to the largest, and as a result, aqueous solutions of Li salts often exhibit abnormal properties. The small size of the Li atom (or ion) is an important factor in the behavior of Li in alloys and in glass and ceramic compositions. Finally, the low equivalent weight of Li is often advantageous, from both economic and weight considerations. Physical properties are summarized in Table L-8.

Uses: Until World War II, the uses for Li and its compounds were minor. Subsequently, due to the special and valuable properties of Li products, many important applications were found, which resulted in the exceptional growth of the lithium industry. Li metal is the starting material for the production of the important group of organolithium compounds. The metal also is used for the preparation of the hydride and nitride. Li metal is used as a deoxidizer, desulfurizer, and degasifier for the treatment of several molten

TABLE L-9. *Characteristics and Uses of Important Inorganic Lithium Compounds*

Compound, Formula, and Formula Weight	Mp, °C	Other Properties and Uses
Lithium carbonate Li_2CO_3 73.89	726°	Soluble in H_2O (1.5% wt at 0°C); used in glass, enamel, and ceramic formulations; in electrowinning of aluminum; in manufacture of other Li compounds; in treatment of manic-depressive psychoses
Lithium hydride LiH 7.95	686.4	Insoluble in nonreactive solvents; reacts vigorously with H_2O, with NH_3 to form the amide; used to produce $LiAlH_4$ and other double hydrides; production of $LiNH_2$; lightweight source of H_2 (1 lb gives 45 ft^3 H_2); also lightweight shield for thermal neutrons
Lithium hydroxide monohydrate $LiOH \cdot H_2O$ 41.96	471 (LiOH)	Soluble in H_2O; dehydrates on moderate heating; used in preparation of Li soaps for multipurpose greases, in manufacture of various Li salts, as additive to electrolyte of alkaline storage batteries; LiOH is efficient, lightweight absorbent for CO_2
Lithium chloride LiCl 42.39	608	Soluble in H_2O; good solubility in alcohols; highly hygroscopic; forms four hydrates; used in production of Li metal, as component of brazing fluxes for Al and Mg, in dehumidification systems, in molten-salt technology, as additive to electrolyte of dry cells for low-temperature use, in low-freezing fire-extinguishing solutions
Lithium bromide LiBr 86.84	550	Soluble in H_2O; good solubility in alcohols; highly hygroscopic; forms four hydrates; chief use is in absorption-refrigeration air-conditioning systems in which H_2O is the refrigerant; in recycling the latter, strong LiBr brine is used to absorb water vapor
Lithium fluoride LiF 25.94	848	Soluble in H_2O (slight); used in glass and enamel formulations, as component of welding and brazing fluxes, in electrowinning of Al, in molten-salt chemistry

metals, especially Cu and Cu alloys. Li is a component of the X2020 Al-Li alloy (1% Li), a structural alloy with improved high-temperature strength; the LA141 Mg-Li alloy (14% Li), an extremely light structural alloy; and Ag-Li alloys for fluxless brazing. Lithium metal shows promise for use in lightweight high-energy primary and secondary batteries. Lithium also shows potential for use as a coolant and heat-transfer liquid in nuclear reactors and as a tritium source for controlled-fusion reactors.

The uses and characteristics of Li compounds (inorganic) are summarized in Table L-9.

Occurrence and Sources: Trace amounts of Li are found in most rocks, as well as in many brines and spring waters. The Li content of the earth's crust currently is estimated at 10–20 ppm. Ocean water contains about 0.1 ppm, whereas some natural lake brines and brine wells may have several hundred parts per million. Lithium concentrations are also found in pegmatites, which are coarse-grained igneous rocks usually composed largely of quartz, feldspar, and mica but often containing minerals rich in some of the less abundant elements.

The commercially important lithium-rich minerals found in pegmatites are *spodumene,* $LiAlSi_2O_6$, with 8.0% Li_2O; *petalite,* $LiAlSi_4O_{10}$, with 3.5–4.5% Li_2O; and *lepidolite,* a complex mica with 3–4% Li_2O. Lithium pegmatites are widely distributed throughout the world. Spodumene is the only important Li ore mineral in North America, with extensive deposits in North Carolina and several Canadian provinces. As mined, spodumene ore usually carries only 1–2% Li_2O, due to associated gangue. Prior to extraction, the ore is concentrated to 5–6% Li_2O, usually by froth flotation. In addition to their major use for the production of Li chemicals, these three ores are used as direct additions to special glass and ceramic compositions that require a low content of color-forming elements, such as Fe and Mn.

The natural brines, as a Li source, are becoming increasingly important. Since 1938, Li products have been produced in moderate amounts from the brine of Searles Lake, Calif., as a by-product of potash and borax operations. Large-scale production of Li chemicals from the Li-rich brine at Clayton Valley, Nev., began in 1966.

Extraction Processes: A large number and variety of processes have been proposed for the extraction of Li from its silicate ores. Most of them can be classified as (1) base-exchange methods, in which the ore is reacted with an alkali sulfate or chloride, either at high temperature or autoclaved with an aqueous solution, to form a soluble Li salt, (2) the alkaline process, involving the reaction of the ore with limestone or lime at appropriate temperatures followed by water leaching of LiOH, and (3) the acid process, in which the ore, after high-temperature treatment, is heated at a moderate temperature with H_2SO_4 followed by water leaching of Li_2SO_4. The base-exchange processes are now of minor importance. The alkaline process can be used on all three of the silicate ores, whereas the acid process is limited to spodumene and petalite.

In the *alkaline process for spodumene,* a slurry of ground spodumene and limestone is fed into a long, coal-fired rotary kiln and discharged at 1900°F. The reaction products are calcium silicates, lithium aluminate, and lime. The clinker is wet ball-milled and the leach solution finally separated from the solids by a classifier and counter-current washer-thickeners. In leaching, lithium aluminate reacts with the lime to form insoluble calcium aluminate and a dilute solution of lithium hydroxide. The latter is concentrated in a triple-effect evaporator-crystallizer, where lithium hydroxide monohydrate is crystallized and separated from the mother liquor by centrifugation.

In the *acid process for spodumene,* the spodumene concentrates are heated in a long, gas-fired kiln to about 1075–1100°C, converting the natural α-spodumene to the more reactive β-spodumene. The kiln discharge is cooled, ball-milled, and mixed with 93% H_2SO_4 in moderate excess over that equivalent to the lithium. This mixture is heated in a kiln to about 250°C, resulting in a reaction in which hydrogen ion replaces Li in the mineral, forming Li_2SO_4 but leaving the ore residue essentially unattacked. Water leaching, neutralization with ground limestone, and filtration yield an impure alkali sulfate solution. Following removal of Ca and Mg, and concentration to about 20 wt % Li_2SO_4, soda ash is added to precipitate Li_2CO_3, which is removed by centrifugation. Glauber's salt, $Na_2SO_4 \cdot 10H_2O$, is recovered from the filtrate and converted to the anhydrous salt as a by-product.

Recovery of Li from natural brines is different and normally results in LiCl as the primary product.

Manufacture of Lithium Metal: In 1818, Sir Humphry Davy first prepared Li metal in minute amount by the electrolysis of lithium oxide; this was followed by preparation in gram amounts by R. Bunsen and A. Matthiessen in 1850 by the electrolysis of LiCl. The first commercial production was in Germany during World War I for limited alloying use. Moderate production in the United States began in 1930 and increased greatly during and after World War II.

Although Li metal can be prepared from its compounds by chemical reduction with such metals as aluminum, commercial production always has employed the electrolysis of the fused chloride. Following the work of Guntz (1893), the electrolyte is a low-melting mixture of LiCl and KCl. The cell body is a heavy, mild-steel box (8 × 6 × 5 ft deep) with thermally insulated sides. Heat is supplied at the bottom by a gas flame. Five graphite anodes, each 8 in. diam and 6 ft long, are arranged vertically. The cathodes are mild steel. Chlorine, discharged at the anodes, is collected and removed. Dc power at 6–6.5 V is used. The current efficiency is about 80%. The electrolyte is 55 wt % LiCl and 45% KCl. The temperature of the cell is maintained at 450–475°C (above the melting point of the mixture 430°C).

In operation, droplets of molten Li formed at the cathode rise and collect in a pool at the surface, where it is periodically skimmed and poured into steel ingot molds. Pure LiCl cell feed is added after each skimming. The purity of commercial Li metal is 99.8% or better, with metallic impurities less than 0.1%. The metal is commercially available as cast ingots (various sizes), extruded wire, ribbon, or rod, and as sand and dispersions, the latter in the 10- to 30-μm range.

Inorganic Chemistry and Compounds: The reaction of Li with water is less vigorous than that of the other alkalies, and ignition does not normally occur unless the metal is finely divided. Reaction with alcohols is slower, forming the alkoxide and hydrogen. Exposure to moist air results in rapid tarnishing, with the initial formation of LiOH, followed by nitridation. Dry air is unreactive, even at the melting point of Li.

Reaction with N_2 to form Li_3N proceeds slowly at room temperature and requires the presence of moisture. Li reacts with H_2 at elevated temperature to form the relatively stable LiH. Li is oxidized by CO_2 at elevated temperature. Reaction with NH_3 yields the amide, $LiNH_2$. Inorganic acids react violently, except for cold, concentrated H_2SO_4. Also see Table L-9.

Organolithium Compounds: These compounds are uniquely versatile reagents which have in many cases superseded Grignard reagents in the synthesis of pharmaceutical intermediates. Their greatest application is as catalysts for the anionic polymerization of dienes and vinyl aromatics. This versatility is due to a combination of a high degree of reactivity, good stability, and excellent solubility in a variety of solvents.

The most important organolithium compounds are *n*-butyllithium, *sec*-butyllithium, dilithioisoprene adduct, methyllithium, and lithium acetylide.

n-Butyllithium and *sec*-butyllithium are liquids with a low viscosity and a high boiling point. Both compounds decompose above 100°C to butenes and lithium hydride. Both are highly soluble in hydrocarbon solvents, such as *n*-hexane and benzene. In both pure and solution form, these compounds are highly sensitive to O_2 and moisture and, like all organolithium compounds, normally are handled under dry nitrogen. In addition to use as catalysts in the preparation of elastomeric products, they are used as metalating agents for the preparation of a variety of pharmaceutical intermediates.

Dilithioisoprene adduct is another highly useful polymerization catalyst. Having two carbon-lithium bonds per molecule, this catalyst (as a benzene solution) is used to prepare low-molecular-weight polybutadienes with thermal hydroxy groups, useful intermediates in elastomeric polyurethane technology.

Methyllithium, phenyllithium, and lithium acetylide are solid organolithium compounds, generally soluble only in highly polar media, such as ethers and tertiary amines. Lithium acetylide must be complexed with ammonia or ethylenediamine to preserve its stability for any length of time. These three reagents are used in attaching methyl, phenyl, and ethynyl groups to organic substrates for the preparation of pharmaceutical intermediates.

REFERENCES

Bach, R. O., C. W. Kamienski, and R. B. Ellestad: Lithium and Lithium Compounds, in "Kirk-Othmer Encyclopedia of Chemical Technology," 2d ed., vol. 12, Wiley, New York, 1967.

"Gmelins Handbuch der Anorganischen Chemie," 8th ed., Lithium, Syst.-Nr. 20, Verlag Chemie, Weinheim, 1920; suppl. vol., Syst.-Nr. 20, 1960.

Houben-Weyl: "Methoden der Organischen Chemie," 4th ed., vol. 13/1, "Metallorganische Verbindungen Li Na K Rb Cs Cu Ag Au," Thieme, Stuttgart, 1970.

Schreck, A. E.: Lithium: A Materials Survey, *U.S. Bur. Mines Inf. Circ.* 8053, 1961.

—R. B. Ellestad and C. W. Kamienski, *Lithium Corporation of America, Bessemer City, N.C.*

Lithium Silicates [See **Silicates (soluble).**]

Locust Bean Gum (See **Gums.**)

Loellingite (See **Arsenic.**)

Lorandite (See **Thallium.**)

Low Brass (See **Copper.**)

Low-Shaft Furnace (See **Iron- and steelmaking.**)

Lubricants, Synthetic Synthetic lubricants are a broad range of compounds derived from chemical synthesis rather than from petroleum or oils of animal or vegetable origin. Since they are man-made under controlled conditions, synthetic-base fluids usually consist of pure materials or of mixtures of pure materials. Petroleum and animal- and vegetable-oil lubricant base stocks, on the other hand, are complex mixtures of compounds which may vary in composition depending on the source of the base stock and the degree of refining.

Synthetic lubricants were developed primarily as a result of the discovery of the lubricating properties of various nonpetroleum fluids; the increase of performance requirements, which exceeded the capabilities of petroleum lubricants; and wartime shortages of petroleum. Considerable work was needed to identify the components of petroleum oil responsible for their lubricating properties. This led, in 1934, to the preparation of synthetic hydrocarbons for lubricants. Further work on this type and on ester-based lubricants was carried out in Germany before and during World War II in order to overcome petroleum shortages and shortcomings, specifically, high-performance aircraft-engine lubricants and low-temperature automotive and ordnance lubricants for use in winter operations.

After the war, much work was done in Great Britain and the United States on ester-based lubricants for turbojet and turboprop aircraft.

TABLE L-10. *Classes of Synthetic Lubricants*

Class	General Formula	Suppliers*
Synthetic hydrocarbon	$[-CH_2-CH_2-CH_2-]_n$	2, 11, 13
Diester	$R-O-\overset{\displaystyle O}{\overset{\|}{C}}-(CH_2)_n-\overset{\displaystyle O}{\overset{\|}{C}}-O-R$	10, 11, 16
Neopentyl polyol ester	$C-(CH_2-O-\overset{\displaystyle O}{\overset{\|}{C}}-R)_4$	7, 9, 10
Silicone	$\left[\begin{array}{c} R \\ \| \\ -Si-O- \\ \| \\ R' \end{array}\right]_n$	4, 6, 17
Polyether (polyalkylene glycols)	$\left[\begin{array}{c} R \\ \| \\ -CH_2-CH-O- \end{array}\right]_n$	3, 17
Phosphate ester	$R-O-\overset{\displaystyle O}{\underset{\underset{R'}{\|}}{\overset{\|}{P}}}-O-R''$	12, 15, 16
Silicate ester	$Si-(O-R)_4$	1, 12
Fluorinated compounds (fluorocarbons)	$CF_3-(CF_2)-CF_3$	4, 5, 8
Poly aromatics (polyphenyls)	$\left[\begin{array}{c} \text{⬡—⬡} \end{array}\right]_n$	12, 14

*1, Chevron; 2, Conoco; 3, Dow; 4, Dow Corning; 5, Du Pont; 6, General Electric; 7, W. R. Grace; 8, Halocarbon; 9, Hercules; 10, Humble; 11, Mobile; 12, Monsanto; 13, Shell; 14, Shell Chemical; 15, Stauffer; 16, Tenneco; 17, Union Carbide.

Types: Examples of synthetically produced fluids used as lubricants are synthetic hydrocarbons, carboxylic acid esters, silicones, polyethers (polyalkylene glycols), phosphate esters, silicate esters, highly fluorinated compounds, and polyaromatics (polyphenyls and polyphenyl ethers). These compounds are identified in Table L-10.

Synthetic-base fluids are usually compounded with additives in order to improve or modify one or more properties. Some types of fluids are more responsive to additive treatment than others, and their performance can be greatly improved.

Additives are usually grouped by the effects they produce as antioxidants, antiwear, and extreme-pressure additives, rust and corrosion inhibitors, and viscosity-index improvers. Levels typically are 0–15%. Properties of compounded lubricants are shown in Table L-11.

In addition, gelling agents, either organic (soaps and dyes) or inorganic (clays and silicas), can be used with synthetics to produce greases, which generally reflect the properties of their base fluids.

Applications: As a class, synthetic lubricants were first thought of as lubricants to be used only in severe situations where normal lubricants are incapable of performing and considered a last resort, primarily because of their generally higher cost compared to standard lubricants. In some cases, this is still justified, especially when the lubricant is expected to withstand extremes in temperature or is exposed to harsh environments, strong oxidants or nuclear radiation. However, there is a growing awareness that certain types may economically replace petroleum-based lubricants (even though they may require higher initial expense) in order to get longer equipment life, reduced downtime, and lower long-term lubrication costs. In many original equipment applications, changing to a synthetic may increase the

TABLE L-11. *Properties of Fully Compounded Lubricants* *

Base Oil Type	Lubricity and Antiwear†	Fluid Range	Viscosity Index	Additive Response of Base Oil	Oxidation Stability	Thermal Stability	Hydrolytic Stability	Fire Resistance	Compatibility with Petroleum Lubricants	Compatibility with Paints, Plastics, and Elastomers	Cost
Conventional petroleum	G	F	F	FG	F	F	E	P	E	FG	Very low
Synthetic hydrocarbons	G	G	G	VG	G	G	E	P	E	G	Medium
Esters	VG	VG	VG	E	G	FG	G	F	VG	FP	Medium
Silicones	P	E	E	P	G	G	VG	F	P	G	High
Polyethers (polyalkylene glycols)	G	G	VG	G	F	F	G	F	F	F	Medium
Phosphate esters	E	F	F	G	FG‡	F	FP‡	VG	G	P	Medium
Silicate esters	F	E	E	F	F	VG	P	F	F	F	High
Highly fluorinated compounds	G	F	P	P	E	VG	VG	E	P	G	Very high
Polyaromatics (polyphenyls and polyphenyl ethers)	F	P	P	F	G	E	E	F	G	F	Very high

*E, excellent; VG, very good; G, good; FG, fairly good, F, fair, FP, fairly poor, P, poor.

†For metals, particularly steel on steel.

‡Degradation products are strongly corrosive.

SOURCE: Reprinted from D. R. Fairbanks, M. H. Knapp, and A. K. Lazarus, Synthetic Lubricants, *Machine Design*, July 1969, with permission of copyright owner.

initial cost of the product slightly but will result in lower operating costs to the consumer and fewer warranty complaints to the manufacturer.

A major application for synthetic lubricants today is the lubrication of turbojet engines used in commercial and military aircraft. These engines require lubricants with good temperature-viscosity relationships, low volatility, and good load-carrying capability. Petroleum lubricants are unequal to the task. Turbojet engines use either diester-based oils meeting MIL-L-7808G specifications or neopentyl polyol esters meeting MIL-L-23699 specifications. Both are formulated to provide lubrication at bulk oil temperatures of 300–400°F (149–204°C). The neopentyl ester provides higher-temperature and greater load-carrying capabilities.

Synthetics are now being used as direct replacements for petroleums. Where petroleum lubricants do an adequate job, synthetics often do an outstanding job. An example is their recent introduction as air-compressor lubricants for rotary-vane, reciprocating and helical-screw units. In one type of unit, the lubricant change interval with petroleum lubricants was every 500 hr to prevent lacquering and sludge buildup. Even with this cycle it was necessary to shut down for overhauling every 2,000 hr. With a diester-based lubricant, the same units operate on a 4,000-hr cycle, and even at this extended period there is little evidence of lacquering or sludging. Other advantages over petroleum oils are lower operating temperatures and a lower oil-consumption rate.

Other synthetics with a variety of applications include phosphate esters for use in fire-resistant hydraulic fluids, glycols for use in brake fluids, fluorosilicones for use in gas compressors, and polyphenyls for radiation resistance.

REFERENCES

Braithwaite, E. R.: "Lubrication and Lubricants," Elsevier, Amsterdam, 1967.
Fairbanks, D. R., M. H. Knapp, and A. K. Lazarus: Synthetic Lubricants, *Mach. Des.,* July 1969.
Gunderson, R. C., and A. W. Hart: "Synthetic Lubricants," Van Nostrand Reinhold, New York, 1962.

—Malcolm H. Knapp, *Intermediates Division, Tenneco Chemicals, Inc., Piscataway, N.J.*

Lubricant Additives   (See **Carbon;** and **Molybdenum.**)

Lubricating Oils   Lubricating oils are used to provide a film between the moving parts of machines and engines to prevent wear with little or no loss of power. The starting raw material for lubricating oil, as for other petroleum products, is crude oil. It should be stressed, however, that lubricating oils simply are not by-products of gasoline and other petroleum fractions. Lubricating oil manufacture requires crude to be processed strictly with lubricating oil as the objective. By-products are refined asphalts, resins, and waxes, which must be separated from the lubricating oil stocks. For these reasons many refiners consider the manufacture of lubricating oils much more complex than that of fuels.

As shown by Fig. L-6, the conventional steps in lubricating oil manufacture are pretreatment of the crude oil charge, as required, followed by distillation of the crude in two steps (an atmospheric tower and a vacuum tower), deresining or deasphalting (as required by the nature of the crude oil charge), dewaxing, solvent extraction, finishing (which may include filtration), and blending, including mixing various additives with the final lubricating oil. The recovery and refining of asphalts, resins, and waxes are important to the overall economy of lubricating oil manufacture.

The chemical composition of lubricating oils is exceedingly complex, the number of carbon atoms varying from approximately 20 to 70. Well-refined lubricating oils contain very little olefinic unsaturation but do contain some aromatic unsaturation. The compounds contained in lubricating oils include paraffins, cycloparaffins (with also multi-ring or condensed compounds), and aromatics (with also multi-ring or condensed aromatic ring compounds).

Waxes generally are paraffin compounds, both straight and branched, and also contain 3–25% cycloparaffins, depending upon the source of the crude oil.

Petroleum resins are hydrocarbons of very high molecular weight containing small amounts of oxygen, sulfur, and nitrogen compounds, which can be found in bridge compounds or in ring compounds. The hydrocarbons include the paraffin, cycloparaffin, and aromatic types in varying amounts and configurations.

Asphalt is physically made up of brown solids called *asphaltenes,* i.e., asphaltic resins of high molecular weight, viscous compounds with a degree of unsaturation and oils. The asphaltenes are believed to contain condensed aromatic ring compounds with oxygen, sulfur, and nitrogen in ring compounds or in bridge positions.

Pretreatment of Crude: In order to remove in-

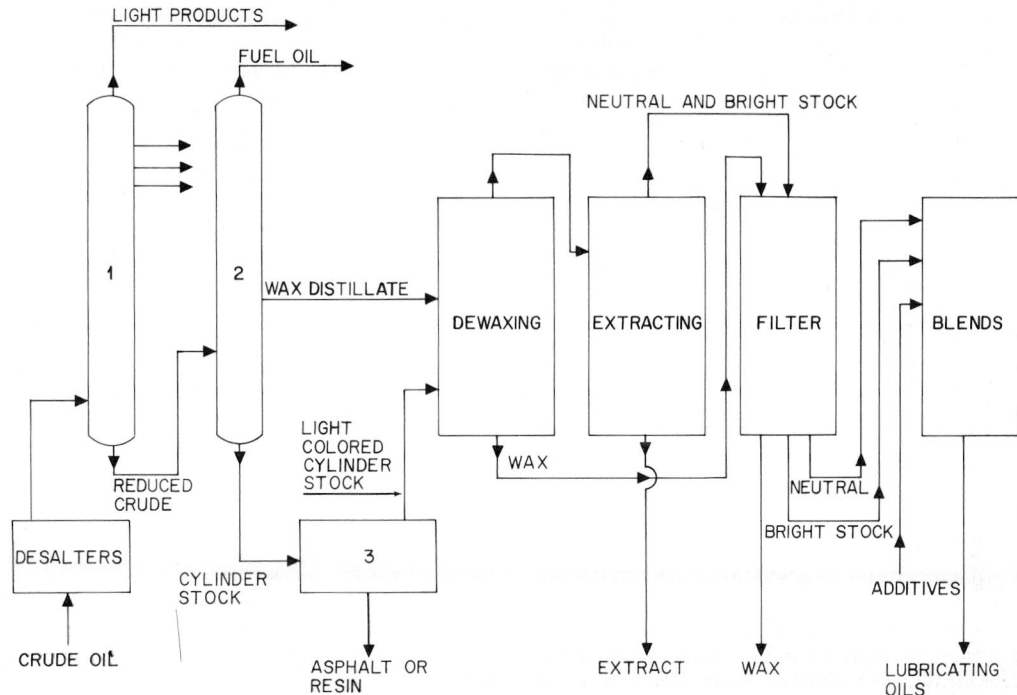

Fig. L-6. Representative lubricating-oil manufacturing processes. (1) Atmospheric tower, (2) vacuum tower, (3) deresining or deasphalting.

organic salts from the crude oil charge to the crude-distillation unit, chemical or electrostatic desalting is used. Salts in the crude oil cause fouling of heat exchangers, corrosion in the distillation units, and increased coking of the furnaces.

Distillation: With further reference to Fig. L-6, crude oil, after pretreatment, is charged to the atmospheric tower, where the crude is separated into light products, which are taken off overhead, and several fractions in side streams. The products removed from the atmospheric tower are straight-run gasoline, naphtha, water-white distillate, and fuel oil, which are sent to other parts of the plant for further processing. The bottoms of the atmospheric tower, or reduced crude, are charged to the vacuum tower. Reduced crude contains the lubricating oil base and a very small portion of heavy fuel oil to be used as reflux for the vacuum tower.

The prime object in the manufacture of lubricating oil is the initial separation of the light products and the separation of wax distillate and cylinder stock without any decomposition or cracking of the lubricating fractions; thus, a vacuum distillation unit is used to separate the wax distillate and cylinder stock at a lower temperature.

The vacuum tower produces some fuel oil overhead, which is sold as a separate product or sent to another area of the refinery for further processing and blending. The two main products from the vacuum tower are wax distillate (with a boiling range of approximately 675–950°F, 357–510°C), which is taken off about mid-column, and cylinder stock, which is the bottoms product. Both streams contain desirable lubricating oil constituents as well as desirable by-products. The wax distillate is charged directly to the dewaxing unit. The vacuum tower bottoms, or cylinder stock are charged to a deresining or deasphalting unit.

Basic stocks for lubricating oil manufacture are wax distillate and cylinder stock, although at the crude unit many refiners make a long residuum from which a certain grade of motor oil can be made.

Deresining or Deasphalting: The character of

these operations depends upon the type of crude used. Asphalts, which are contained in asphaltene crudes, can be constituted into different properties by further distillation or by air blowing. Resins occur in paraffin or low-asphaltic cylinder stocks.

Propane is used as a solvent and at different temperatures and ratios causes asphalts or resins to separate from the oil due to a difference in solubility. This process requires a tower or towers for the separation of the oil and the asphalt or resin. Other equipment is used for the further separation of the resins and recovery for reuse of the propane from the various solutions of oil and resins or asphalts.

Dewaxing: Wax is probably the most troublesome product in the manufacture of lubricating oil. Its presence in lubricating oil prevents free movement at lower temperatures; however, once removed, wax is the most valuable product in refining.

Solvent dewaxing is the common approach to wax removal and is advantageous in that both light and heavy stocks can be alternately charged to the solvent dewaxing unit. Usually, methyl ethyl ketone (MEK) and an aromatic solvent, such as toluene, are used. The MEK causes the wax in the oil to crystallize, and the toluene is used to dissolve the oil. The principles of the process are those of manufacture and removal of wax by solvent injection and chilling. The solvent mixture, at a carefully controlled temperature, is added in measured amounts to either the light or heavy wax-containing stocks at points in the chilling to produce proper crystallization of the wax. The oil, wax, and solvent mixture is chilled further to the temperature that will produce the desired pour point of the dewaxed oil. This operation is accomplished in scraped-tube double-pipe heat exchangers. Ammonia or propane is used as a refrigerant, and cold filtrate is pumped through the exchanger to maximize the heat exchange.

The chilled mixture is then filtered on continuous rotary-vacuum filters, where a constant level of chilled feed mixture is maintained. Inside the filter, a drum covered with cloth is rotated through the mixture of wax, solvent, and oil. Wax from the mixture forms a cake on the outside of the drum and is washed continuously with chilled solvent to displace the oil in the wax cake. Before the wax cake is removed, it is dried and loosened with chilled flue gas. Both the wax solution and the oil solution are distilled for removal of solvent (to be reused) and to provide solvent-free wax and oil. Thus the two products are a wax-free oil and an oil-free wax.

The lubricating oil made after wax removal from wax distillate is neutral stock, which is a little less in viscosity than an SAE 10 motor oil. The lubricating oil made after wax removal from cylinder stock is bright stock, which is a little more viscous than an SAE 70 motor oil.

Solvent Extraction: To upgrade or improve the quality of neutral or bright stock, some refiners find it necessary to solvent-refine for the removal of constituents, e.g., aromatic compounds, that cause low viscosity indexes. Solvent extraction can be performed before or after dewaxing, but most solvent extraction is performed after wax removal in order to prevent any interference from wax in the charge oil. Such solvents as chlorex, nitrobenzene, phenol, furfural, or benzene and sulfur dioxide are used. The solvent is contacted with the charge oil in countercurrent treating towers, and the solvent and oil are mixed. The improved oil (or raffinate) and solvent are taken overhead from the treating tower, and unsaturated material or extract and solvent are removed from the bottom. Solvent is removed from both the raffinate and extract in recovery equipment and reused. Yielded are the improved lubricating oil of higher viscosity index and extract.

Filtration: Most refiners use deresining or deasphalting, dewaxing, and solvent extraction with filtration. See also **Adsorption.** Bauxite is the common filtering medium or adsorbent for the removal of asphaltic and resinous undesirables or simply for light filtration or finishing. The bauxite is placed in a vertical steel tank, and the oil is permitted to gravitate through the bauxite in the filter. A screen in the bottom head of the filter prevents the bauxite from being removed from the filter with the oil. When the bottom outlet oil stream reaches a predetermined color, the bauxite (after washing with naphtha and steaming) is regenerated by burning off the adsorbed material and reused. The number of regenerations for bauxite, space velocity, and temperature of filtration are the main factors in determining the efficiency of the filtration process. Instead of filtration for color improvement and impurity removal, some refiners use hydrogen and a catalyst.

A relatively new process for increasing the quantity and quality of lubricating oil is hydrocracking. By using a catalyst and hydrogen, a heavier, lower-viscosity-index stock can be transformed into a lighter high-viscosity-index lubricating oil and valuable by-products. Charge stock to the process can be cylinder stock from the crude unit or deresined or deasphalted oils.

Blending and Additives: In blending neutral and

bright stock, additives are used for many purposes. Additive blending is done after bauxite filtration, since filtering will remove the additives. Additives are used according to the severity of the operating conditions for which the lubricating oil is intended and the quality of the base lubricating oil. Naturally, a high-quality lubricating oil base requires fewer additives than a lower quality base. Many additives are available, several of which are multipurpose. The manufacturer must be certain that the additives used are compatible with each other and with the lubricating oil bases.

Several common additives and their function in lubricating oils are:

Antioxidant to reduce oxidation or thermal degradation of an oil

Corrosion inhibitor

Rust inhibitor

Extreme-pressure additive in gear oils to react with metallic surfaces under high pressure to control frictional properties and reduce wear

Antifoam

Antiwear, to reduce wear by forming a tighter and thicker film

Viscosity index improver to make oil more suitable for use over a greater temperature span

Pour-point-temperature depressant

Detergent to lessen deposition of harmful deposits

Dispersent to provide dispersion of products of degradation and combustion

Antisquawk to supply a noise suppressor for high-speed automatic-clutch application.

Antichatter to improve the surface frictional properties of special clutches

Classification: In order that an engine manufacturer may specify the proper oil to be used in his product, a lubricating oil manufacturer may know what blends to produce, and the consumer may know what oil to use in a given machine, classification and specification of oils is mandatory.

Through the cooperation of the American Petroleum Institute (API), the American Society for Testing and Materials (ASTM), and the Society of Automotive Engineers (SAE), an Engine Service Classification System was established for diesel and gasoline engine service conditions. Nine engine classifications as guides are given in *API Bull.* 1509, 6th ed., January 1971, obtainable from the API.

—Harry A. Fennick, *Pennzoil Company,*
Oil City, Pa.

Luminescent Materials (Consult the **Subject Index.**)

Lutetium (L. *Lutetia,* village on the present site of Paris.) **Lu** = 174.97 (at. wt.), 71 (at. no.). Lutetium is the last of 15 elements in group III, period 6, generally shown as the rare-earth elements in a separate line below the main body of the periodic table and sometimes referred to as the *lanthanide series.* Lu was discovered in 1907 by the French chemist Georges Urbain and named for his native city.

Pure lutetium is silvery-gray and retains its metallic luster in normal atmosphere indefinitely. A few fundamental properties (elastic modulus and thermal expansion) have been measured only recently. Hardness of the metal is about 3 times that of cerium. Very little is known about its workability or alloying behavior. Calculations and extrapolations from known data for Lu and other lanthanide elements indicate no anomalies in the physical, mechanical, or chemical properties of the element.

The melting point is 1663°C; boiling point 3395°C; density 9.835 g/cm^3. It is the least abundant of the lanthanide elements, estimated at 0.5 ppm in the earth's crust. Even at this level, however, it is potentially more plentiful than mercury, cadmium, or any of the precious metals. Lutetium is found in combination with other heavy rare earths in relatively minor amounts. One reason it is so difficult to identify is that because the Lu 4f subshell is completely filled with 14 electrons, it has no absorption bands in the visible or near-ultraviolet region of the spectrum to allow easy identification by spectrographic equipment. By chemical methods, it is equally difficult to identify because the three valence electrons, 5d, 6s^2, are the same for the other 14 elements in the lanthanide series (57–71).

Lutetium has 2 natural isotopes, ^{175}Lu (97.4 w/o*) and ^{176}Lu (2.6 w/o), and 14 artificial isotopes. Natural ^{176}Lu is radioactive, with a half-life of 2.2 × 10^{10} years. Lutetium, although not studied as extensively as the other lanthanides, is classified as having a low acute-toxicity rating.

The valency of Lu is 3$^+$. Its oxide, Lu$_2$O$_3$, is white, which is characteristic of the many chemical compounds it forms. The metal and oxide are soluble in mineral acids, giving colorless solutions.

When minerals such as xenotime, certain types of apatite, and yttrium-heavy rare-earth residues from uranium ores are processed for yttrium, the remaining rare earths include lutetium. Before 1965, the small amount of Lu available had been

*This abbreviation is used in the rare-earth and related fields for weight percent.

recovered at great expense from monazite, the source for thorium which was processed mainly for the light rare-earth elements. Enriched mixtures of heavies containing Lu are available today in relatively large amounts. Based on the established uses for yttrium, several thousand pounds per year of lutetium could be produced from this source at much lower cost.

Processing to obtain pure Lu is by organic ion exchange using synthetic resins or by organic liquid-solvent extraction. Purity of 99.9+% is assured by these methods. The ion is precipitated from an acid solution with oxalic acid. The oxalate is calcined to Lu_2O_3.

Lutetium metal is produced by the reduction of anhydrous $LuCl_3$ or LuF_3 using calcium or lithium in a sealed-bomb reaction.

See **Rare-earth elements and metals** for a more detailed description of processing and a table of known properties of lutetium.

Uses: The possibility that lutetium would be available in sufficient quantity at prices low enough to warrant research expenditures has been evident only since 1965. A reasonable assumption is that when commercial applications are found for Lu, it will be as available as europium, another of the less abundant elements in the lanthanide group. Since so little is known about Lu, it is likely to be several years before this assumption can be tested.

Bell Telephone Laboratories has successfully proved the feasibility of manipulating isolated magnetic bubble domains in lutetium orthoferrites for computer memory and storage functions.

The elements lanthanum, gadolinium, and yttrium (always associated with the lanthanides) have a common characteristic with lutetium. They are all noncoloring, exclusively trivalent metallic elements because of their electronic structures. Important new uses have been discovered for Y, La, and Gd since 1960. With the knowledge that the oxides of these elements are highly effective as host matrices for luminescent phosphors, the electrical and electronic properties of lutetium are now being studied in many laboratories.

REFERENCES

Molybdenum Corporation of America: *Overview* 17 (application information including possible uses), New York, 1970.

Tonnies, J. J., K. A. Gschneidner, Jr., and F. H. Spedding: Elastic Moduli and Thermal Expansion of Lutetium Single Crystals from 4.2 to 300°K, *J. Appl. Phys.*, vol. 42, August 1971.

See also references listed under **Rare-earth elements and metals**.

—Joseph G. Cannon, *Molybdenum Corporation of America, White Plains, N.Y.*

Lutidines   (See **Pyridine and derivatives.**)

Lye   (See **Caustic soda.**)

Lyophile   (See **Colloid systems; Freeze-drying;** and **Peptides and proteins.**)

Lyophobe   (See **Colloid systems.**)

Lysine   (See **Amino acids;** and **Peptides and proteins.**)

Lysozyme   (See **Peptides and proteins.**)

M

Macrolides (See **Antibiotics.**)

Macromolecules (See **Polymerization.**)

Magnesium (Gk., *Magnesia,* a district in Thessaly.) **Mg** = 24.312 (at. wt.); 12 (at. no.). Magnesium is a silvery-white metal probably best known for its lightness. With a specific gravity of only 1.74 and its ability to be strengthened by alloying with other metals, Mg long has been recognized as the lightest structural metal.

Magnesium is one of the alkaline-earth metals and appears in group IIA of the periodic table between Be and Ca. Mg forms many compounds and has a single valence of 2. There are three stable natural isotopes of Mg, 24–26, and three radioactive isotopes, 23, 27, and 28. Mg is extremely abundant in the earth's crust, being the eighth most abundant chemical element and the sixth most abundant metallic element. Of the structural metals, Mg is the third most plentiful, exceeded only by Al and Fe. See also **Chemical elements.**

The melting point of Mg = 650°C; boiling point = 1110 ± 10°C; and critical temperature (calculated) = 1867°C. The atomic volume is 14.0 cm^3/g atom. The electron arrangement in free atoms is (2)(8)2. Mg has a crystalline structure that is close-packed hexagonal, and so do the common alloys except those containing Li in amounts of 11% or more, which have a body-centered cubic structure. The accepted values for lattice parameters of pure Mg are 5.199 Å for c_0 and 3.2023 Å for a_0. Consequently, the axial ratio c/a is 1.6236. Other physical constants of Mg are listed in Table M-1.

The history of magnesium dates back to 1695, when an English physician and botanist, Nehemiah Grew, evaporated the mineral waters from Epsom, England, and obtained the crystalline product commonly known as epsom salt, MgSO$_4$. For many years, magnesia was confused with lime until 1754, when Joseph Black was able to show that they were different substances. The chemical identity of magnesia was not established, however, until 1808, when Sir Humphry Davy pointed out that it was the oxide of a new metal which he named *magnium.*

One of Davy's experiments to isolate the metal in magnesium alba consisted of passing potassium vapors over magnesium oxide heated to a red heat and then extracting the Mg with Hg. In another

709

TABLE M-1. *Some Physical Constants of Magnesium*

Density, g/cm³:	
At 20°C	1.74
650°C, solid	1.64
Liquid	1.57
700°C.	1.54
Volume contraction, 650° (liquid) to 650° (solid), %	4.2
Linear contraction, 650° (solid) to 20° (solid), %	1.8
Electrical:	
Resistivity, $\mu\Omega$-cm:	
At 20°C	4.46
300°C	9.5
600°C	17.0
650°C (liquid)	28.0
900°C	28.0
Temperature coefficient at 20°C, $\mu\Omega$-cm/°C	0.017
Conductivity at 20°C (annealed Cu standard), %:	
Mass	198
Volume	38.6
Thermal:	
Flame temperature (theoretical), °C . .	4850
Coefficient of expansion, in./(in.)(°C):	
At 20–100°C	0.0000261
20–200°C	0.0000271
20–300°C	0.0000280
20–400°C	0.0000290
20–500°C	0.0000299
Specific heat, cal/(g)(°C):	
At 20°C	0.245
300°C	0.275
650°C, solid	0.325
Liquid	0.316
Conductivity, (cal)(cm)/(s)(cm²)(°C) . .	0.37
Diffusivity (at 20°C), cm²/s	0.87
Heat of combustion, cal/g mole	145,000
Latent heat of fusion, cal/g	88±2
Latent heat of vaporization, cal/g . . .	1260±30

experiment, he is said to have electrolyzed moistened $MgSO_4$ using Hg as the cathode. In both instances, he obtained an amalgam, and he proved that magnesia alba was the oxide of a new metal. It has never been firmly established, however, that Davy actually obtained Mg in pure metallic form. The French scientist A. Bussy generally is credited with having first isolated metallic Mg. In 1828, he obtained small globules of the metal by fusing $MgCl_2$ with potassium. The first production of metallic Mg by electricity was announced by Michael Faraday (1883). Robert Bunsen (1852) developed the first electrolytic cell to produce Mg metal.

Chemical and Metallurgical Uses: The alloys of Mg have received much attention for their role in aerospace and numerous commercial applications where lightness is important. Not so well known and certainly not so well publicized are the many chemical and metallurgical uses of primary Mg. It is a very active metal and as a consequence has many chemical and metallurgical uses. It is a powerful deoxidizer and desulfurizer and, in addition, will combine chemically with most nonmetallic elements. Mg has been used for many years in the well-known Grignard reaction for the production of certain organic chemical compounds. See also **Grignard reagents;** and **Magnesium organics.**

Because of its relatively high position in the electromotive series of metals, Mg can reduce many metals from their compounds. As a consequence, it is used in the production of such metals as beryllium, hafnium, titanium, uranium, yttrium, and zirconium.

Mg forms alloys with most nonferrous metals, and its presence can have a significant influence on properties. Table M-2 summarizes these effects on the properties of several common metals. At the present time, the greatest consumption of Mg for alloying with the metals listed in the table is in the manufacture of aluminum- and zinc-base alloys. In copper- and nickel-base alloys, the primary function of Mg is to deoxidize and desulfurize melts of these metals and their alloys although there are some beneficial effects due to alloying action.

Production: Mg is very prevalent in the earth's crust and is a constituent of more than 150 minerals. The element also is found in seawater, bitterns, and subterranean brines and salt beds. Only a few of the minerals, however, are important commercially as sources for Mg metal or its compounds, dolomite probably being utilized to the greatest extent. More than half the world's production of metallic Mg today comes from seawater.

Two types of magnesium production processes are in use, electrolytic and thermal. The former accounts for about 80% of the world's production. The remaining 20% is produced by the ferrosilicon process in various countries. Currently, all production of primary Mg in the United States is by the electrolysis of $MgCl_2$, which breaks the compound down into its components of magnesium and chlorine.

In one process,* seawater from the Gulf of

*The Dow Chemical Company (world's largest producer of primary Mg).

TABLE M-2. *Summary of Magnesium Use in Various Alloys*

Base Alloy	Effect of Mg on Properties	Remarks
Aluminum . . .	Increases mechanical properties and resistance to corrosion; facilitates heat treatment	Remelting can reduce Mg content, which should be replaced by adding pure Mg to casting pot or ladle
Copper	Permits age hardening and improves tensile strength	In copper-alloy melting practice, Mg is used primarily as a deoxidizer in such alloys as Cu-Ni-Zn and leaded brasses and bronzes
Lead	Increases strength, hardness, and resistance to creep	Alloyed only in very small amounts; Mg also used as a debismuthizer in refining primary Pb
Nickel	In combination with carbon forms an age-hardenable alloy	As with Cu alloys, principal use of Mg is to deoxidize and desulfurize melts including pure Ni, Ni-Cr, and Ni-Cu alloys
Tin	Increases tensile strength and hardness	Effect on strength is dramatic, but too much Mg can reduce corrosion resistance and ductility
Zinc	Improves dimensional stability and reduces intergranular corrosion of zinc die castings; in zinc sheet, Mg refines the grain and increases hardness and creep strength	Also used in zinc-base bearing metal and in zinc-alloy metalworking dies

Mexico is pumped into large settling tanks, where it is treated with lime (roasted oystershells can be used). The lime causes the Mg in the seawater to precipitate as the insoluble hydroxide. Next comes a filtration step, in which the magnesium hydroxide is filtered out and then reconverted into a slurry with fresh water. Treatment with HCl converts the magnesium hydroxide into magnesium chloride. The resulting solution is evaporated, leaving a residue of $MgCl_2$, which then is dried and fed to the electrolytic cells. The electrolysis yields molten Mg, which is removed from the cells and cast into ingots. The Cl_2 gas goes back into the process as HCl to react with magnesium hydroxide.

Alloys: Pure metals must be alloyed with other metals to give them the necessary strength and other qualities required for structural applications. Some common Mg alloys are listed in Table M-3, along with their chemical compositions, principal characteristics, and typical uses. The forms in which these Mg alloys are commonly produced and key properties are given in Table M-4.

Structural uses for Mg alloys are many and varied but usually have one quality in common—lightness. Consequently, Mg alloys have been used in aircraft and space vehicles, dockboards, hand trucks, containers and materials-handling equipment, chain saws, portable electric and pneumatic tools, levels and other hand tools, luggage, sporting goods, tooling jigs and fixtures—in fact, wherever weight reduction is desired.

Magnesium alloys in their various forms can be expected to continue filling a real need in products which must be constructed with a minimum of weight to overcome inertia, conserve power, and reduce worker fatigue. A very important consideration in weight reduction of manually handled equipment is the safety factor. Lighter-weight equipment means easier handling, fewer injuries, and a reduction in lost time due to accidents.

Inorganic Compounds: Some of the more important magnesium inorganics are summarized in Table M-5.

REFERENCES

Ball, C. J. P.: The History of Magnesium, *J. Inst. Metals* (*Lond.*), vol. 84, pt II, pp. 399–411, 1955–1956.

Beck, A.: "The Technology of Magnesium and Its Alloys," 2d ed., trans. of "Magnesium and seine Legierungen," Hughes, London, 1941.

Church, F. L.: Magnesium: Starting to Make Out in Mass Markets, *Mod. Met.*, vol. 22, no. 6, pp. 57–81, July 1966.

Comstock, H.: "Magnesium and Magnesium Compounds: A Materials Survey," U.S. Bureau of Mines, Washington, 1963.

Emley, E. F.: "Principles of Magnesium Technology," Pergamon, London, 1966.

TABLE M-3. *Chemical Composition, Characteristics, and Uses of Magnesium Alloys*

Alloy	Nominal Chemical Composition, %							Characteristics	Uses
	Al	Mn	Rare Earths	Th	Zn	Zr	Other		
AZ31B . . .	3.0	...	...	...	1.0	...	...	Moderate strength, good formability, dent resistance, and weldability	Sheet, plate, extrusions and forgings; general purpose wrought alloy
AZ91B . . .	9.0	...	...	...	0.6	...	...	Good strength and castability	Most popular die-casting alloy for portable tools, business machines, and vehicles
AZ91C . . .	8.7	...	...	...	0.7	...	...	Good castability, pressure tightness, and weldability; moderate strength	General-purpose sand and permanent-mold casting alloy
HK31A . . .	...	...	...	3.0	...	0.7	...	Good short-time, elevated-temperature properties; weldable without stress relief; low microporosity in cast form	Sand and permanent-mold castings, sheet and plate for aerospace applications at 400–700°F (204–371°C)
HM21A . .	...	0.6	...	2.0	...	...	...	Very stable at elevated temperatures; good creep strength and formability; weldable without stress relief	Sheet, plate, and forging for aerospace applications at 400–800°F (204–427°C)
HM31A . .	...	1.2 min.	...	3.0	...	...	...	Excellent properties at elevated temperatures; weldable without stress relief	Extrusions for aerospace applications at 400–800°F (204–427°C)
QE22A . . .	...	...	2.0 didymium	...	...	0.7	2.5 Ag	Superior tensile yield strength plus excellent creep and fatigue strength	Castings for aerospace applications up to 500°F (260°C)
ZK60A . . .	...	...	...	...	5.7	0.5	...	High strength, good toughness, good spot-weldability, limited arc weldability	Highly stressed parts especially in aerospace and military applications; appreciable use as a forging alloy

TABLE M-4. *Typical Room-Temperature Mechanical Properties of Magnesium Alloys*

Form	Alloy	Temper	Section Thickness, in.	Tensile Strength, ksi*	Tensile Yield Strength, ksi*	Elongation in 2 in., %	Compressive Yield Strength, ksi*
Sand and permanent-mold castings	AZ91C	T4	½-in. separately cast test bars according to ASTM procedure	40	12	14	12
	AZ91C	T6		40	19	5	19
	HK31A-T6	T6		32	15	8	15
	QE22A-T6	T6		40	30	4	30
Die castings	AZ91B	F	Separately cast test bars	33	22	3	22
Extruded bars, rods and shapes	AZ31B	F	0.250–1.499	38	29	15	14
	HM31A	T5	<1 in.²	44	38	8	27
	ZK60A	T5	<2 in.²	53	44	11	36
Sheet and plate	AZ31B	H24	0.016–0.249	42	32	15	26
	AZ31B	H24	0.250–0.374	40	29	17	23
	HK31A	H24	0.016–0.249	38	30	9	23
	HM21A	T8	0.016–0.249	35	23	11	19
	HM21A	T8	0.250–0.500	37	27	12	23
Forgings	AZ31B	F	†	38	28	9	12
	HM21A	T5	†	34	22	9	16
	ZK60A	T6	Up to 3	47	39	11	25

*ksi = 1,000 psi.
† Section thickness limits have not been established.

TABLE **M-5.** *Properties of Some Important Inorganic Magnesium Compounds*

Compound and Formula	Formula Weight	Color	Crystalline Form	Sp gr	Mp, °C
Magnesium acetate, $Mg(C_2H_3O_2)_2 \cdot 4H_2O$ ..	214.47	White	Monoclinic prisms	1.454	80
Magnesium aluminate (spinel), $MgO \cdot Al_2O_3$	142.26	Colorless	Cubic	3.6	2135
Magnesium ammonium chloride, $MgCl_2 \cdot NH_4Cl \cdot 6H_2O$	256.83	White	Rhombic	1.456	195
Magnesium ammonium phosphate (struvite), $MgNH_4PO_4 \cdot 6H_2O$	245.44	Colorless	Rhombic	1.715	100*
Magnesium ammonium sulfate (boussingaultite), $MgSO_4(NH_4)_2SO_4 \cdot 6H_2O$	360.62	Colorless	Monoclinic	1.72	>120
Magnesium carbonate (magnesite), $MgCO_3$	83.43	White	Trigonal	3.037	350*
Magnesium carbonate (nesquehonite), $MgCO_3 \cdot 3H_2O$	138.38	Colorless	Rhombic	1.852	100
Magnesium carbonate (basic) (hydromagnesite), $3MgCO_3 \cdot Mg(OH)_2 \cdot 3H_2O$	365.37	White	Rhombic	2.16	*
Magnesium chloride (chloromagnesite), $MgCl_2$	95.23	Colorless	Hexagonal	2.325	712
Magnesium chloride (bischofite), $MgCl_2 \cdot 6H_2O$	203.33	White	Monoclinic	1.56	118*
Magnesium hydroxide (brucite), $Mg(OH)_2$ ..	8.34	White	Trigonal	2.4	*
Magnesium oxide (periclase, magnesia), MgO	40.32	Colorless	Cubic	3.65	2800
Magnesium perchlorate, $Mg(ClO_4)_2$	223.23	White	Deliquescent	2.60	*
Magnesium peroxide, MgO_2	56.32	White	Powder	.....	275†
Magnesium pyrophosphate, $Mg_2P_2O_7$	222.60	Colorless	Monoclinic	2.598	1383
Magnesium potassium chloride (carnallite), $MgCl_2 \cdot KCl \cdot 6H_2O$	277.88	Colorless	Rhombic, deliquescent	1.60	265
Magnesium potassium sulfate (picromerite), $MgSO_4K_2SO_4 \cdot 6H_2O$	402.73	Colorless	Monoclinic	2.15	72*
Magnesium silicofluoride, $MgSiF_6 \cdot 6H_2O$ ...	274.48	Colorless	Trigonal	1.788	*
Magnesium sodium chloride, $MgCl_2 \cdot NaCl \cdot H_2O$	171.70	Colorless			
Magnesium sulfate, $MgSO_4$	120.38	Colorless	....	2.66	1185
Magnesium sulfate (epsom salt, epsomite), $MgSO_4 \cdot 7H_2O$	246.49	Colorless	Rhombic	1.68	70*

*Decompose. †Explodes.
NOTE: Dolomite, $MgCO_3 \cdot CaCO_3$, is a type of limestone and described under **Limestone.**

Shear Strength, ksi*	Bearing Strength, ksi*	Brinell Hardness	Density, lb/in.³	Mp °F	Mp °C	Thermal Conductivity at 20°C (68°F), (cal)(cm)/(s)(cm²)(°C)	Electrical Resistivity at 20°C (68°F), μΩ-cm
17	60	53	0.0653	1105	596	0.11	16.2
20	75	66	0.0653	1105	596	0.13	12.9
22	61	55	0.0648	1200	649	0.22	7.7
...	...	78	0.0655	1020	549	0.25	6.8
20	...	67	0.0653	1105	596	0.12	14.3
19	56	49	0.0642	1160	627	0.18	9.2
22	62	63	0.065	1121	605	0.25	6.6
26	79	82	0.0660	1175	635	0.29	5.7
29	77	73	0.0642	1160	627	0.18	9.2
28	72	...	0.0642	1160	627	0.18	9.2
26	67	57	0.0648	1200	649	0.27	6.1
19	63	56	0.0642	1202	650	0.33	5.0
...	67	...	0.0642	1202	650	0.33	5.0
19	70	55	0.0642	1160	627	0.18	9.2
18	36	50	0.0642	1202	650	0.33	5.0
...	...	...	0.0660	1175	635	0.30	5.7

Gross, W. H.: "The Story of Magnesium," American Society for Metals, Metals Park, Ohio, 1949.

Roberts, C. S.: "Magnesium and Its Alloys," Wiley, New York, 1960.

—W. H. Gross, *The Dow Chemical Company, Midland, Mich.*

Magnesium Carbonate   (See **Limestone.**)

Magnesium Organics   The only commercially significant organomagnesium compounds are the well-known and widely used Grignard reagents (see **Grignard reagents**). Magnesium dialkyls and diaryls have been prepared in pure form, but they do not share the ease of preparation or the versatility in organic synthesis of the Grignards.

Compounds of the R_2Mg type (where R is alkyl or aryl) may be prepared by (1) reaction of magnesium metal with the appropriate diorganomercury compound, preferably in ether, (2) dioxane precipitation of RMgX and MgX_2 from ether solutions of Grignard reagents, or (3) reaction under pressure at about 100°C of Mg metal with an olefin in the presence of hydrogen or magnesium hydride or of magnesium hydride with an olefin in the presence of ethers.

Diorganomagnesium compounds which have been prepared include methylenemagnesium, dimethylmagnesium, diethylmagnesium, dipropylmagnesium, diisopropylmagnesium, dibutylmagnesium, di-*t*-butylmagnesium, diamylmagnesium, ethylisopropylmagnesium, diethylmagnesium sulfate, divinylmagnesium, dibutenylmagnesium, dicyclohexylmagnesium, bis(methylcyclopentadienyl)magnesium, bis(1-methylbutyl)magnesium, bis(2-methylbutyl)magnesium, dicyclopentadienylmagnesium, dihexynylmagnesium, diphenylmagnesium, bis(triphenylmethyl)magnesium, dicinnamylmagnesium, bis(phenylethynyl) magnesium, and dibenzylmagnesium.

Magnesium dialkyls are practically nonvolatile white solids which are insoluble in organic solvents but easily soluble in ether. They are spontaneously flammable in air or CO_2. Dimethylmagnesium is stable to at least 220°C, but the higher homologues decompose readily between 175 and 200°C with evolution of hydrocarbon. Methylenemagnesium, an amorphous brown solid, is incapable of solvation. Dicyclopentadienylmagnesium, melting point 176°C, decomposes at 300°C, is soluble in ether, tetrahydrofuran, and liquid NH_3 and reacts violently with H_2O, CS_2, CCl_4, CO_2, and air.

Diphenylmagnesium, a colorless solid insoluble in organic solvents but soluble in ether, decomposes at 280°C to give biphenyl.

The only reported uses of magnesium dialkyls and diaryls are as catalysts in the polymerization of olefins and in coating glass and metals with magnesium by decomposition.

—Marguerite K. Moran, *M & T Chemicals Inc., Rahway, N.J.*

Magnesium Oxide   (See **Lime.**)

Magnetic Materials   (See **Cobalt; Ferrites; Gadolinium; Iron oxides, synthetic; Lanthanum; Oxygen; Praseodymium; Rare-earth elements and metals;** and **Samarium.**)

Magnetic Separation, Solids   (See **Beneficiation, ore;** and **Separation operations.**)

Magnetite   (See **Iron ores;** and **Iron- and steelmaking.**)

Mahogany Soaps   [See **Anionic surfactants (sulfur-bearing).**]

Maillard Reaction   (See **Amino acids.**)

Maleic Anhydride   Maleic anhydride is used primarily in making polyester resins, an end use accounting for about half the market for the product. Fumaric acid, insecticides (such as Malathion), maleic hydrazide, and alkyd resins also consume comparatively large quantities of maleic anhydride. In the United States, the production of maleic anhydride increased more than threefold in 1957–1967, and in 1971 it was about 300 million lb. See also **Isomerization; Paint.**

Maleic anhydride was first produced commercially* in 1933, using the catalytic vapor-phase oxidation of benzene, a process based on investigations by Weiss and Downs first described in 1920. Current commercial production of maleic anhydride is almost entirely by direct air oxidation of benzene and, to a minor extent, by the air oxidation of the butenes. Catalytic air oxidation of benzene† is described here.

In this process, benzene is mixed with air, preheated, and passed through a multitubular, cooled catalytic reactor, where the benzene is oxidized to maleic anhydride. A number of side reactions

---

*National Aniline Division of Allied Chemical Corporation.

†Scientific Design Company, Inc.

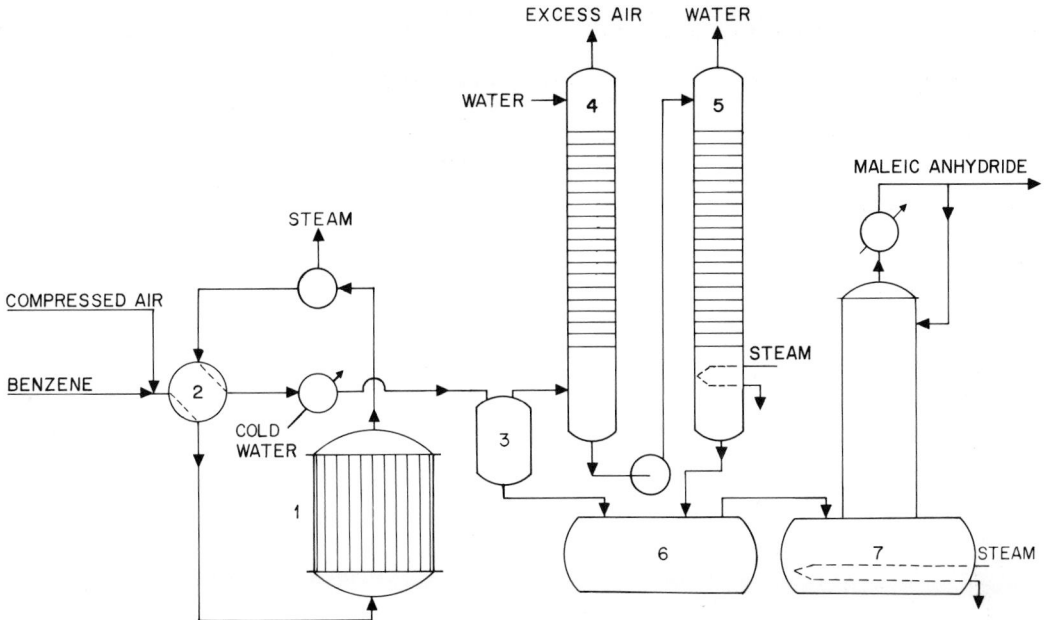

Fig. M-1. Maleic anhydride and fumaric acid process. Fumaric acid and malic acid are produced from maleic anhydride by processing in an isomerizer (not shown). Principal components of the process include (1) tubular catalytic reactor, (2) heat exchanger, (3) separator, (4) scrubber, (5) dehydrator, (6) crude-maleic surge tank, (7) refining column. Cold water is supplied to the coolers shown. (*Scientific Design Company, Inc.*)

occur, the most important being the oxidation of benzene to CO_2 and H_2O. Reaction temperatures are in the 350–450°C range (see Fig. M-1).

The reactor effluent gases then pass through three heat exchangers in series. The first generates steam, the second preheats the feed gas and also cools the effluent gases, and the last further cools the reactor effluent gases by using cooling water.

The cooled gases then pass to a water scrubber, where maleic anhydride is absorbed as maleic acid. The acid solution is then dehydrated to crude maleic anhydride, purified further by refining, and finally pelletized and bagged.

The favorable economics of this process can be attributed mainly to the high sensitivity of the vanadia-based fixed-bed catalyst, which has demonstrated a long operating life in numerous plants. In addition to high yield and product quality, the process, including the recovery and purification system, is characterized by easy operability. This is significant, considering the high temperatures employed, the use of fused salt for heat transfer in the reaction section, the tendency of maleic anhydride to solidify, and the corrosive-

ness of the product. As with most high-volume chemicals, the trend in maleic anhydride production is to larger plants.

The equipment in a maleic anhydride plant lends itself to scale-up. The economic advantages of a large plant are consequently more attractive than those of a small one. The equipment is such that a single train can usually be employed for most plants that are commercially viable in most parts of the world today. In the case of the largest plants, there may be a requirement for duplicate reactor trains.

A centrifugal compressor supplies the air for oxidizing the benzene. Machines of this type, which would be adequate for a plant of 75,000 tons/yr capacity, are readily available. Axial compressors provide 3 times the capacity of a centrifugal machine. Thus a single machine will meet all the requirements of a maleic plant contemplated in the foreseeable future.

The maleic anhydride product is recovered and purified in tray columns. There is no need for duplicate columns in any of the recovery or purification units in the process described here, even for the largest plant foreseeable in the next few years.

The reaction is carried out in a large tubular reactor of the fixed-tube sheet type. Shipping considerations limit reactor size though this is not a problem from a process standpoint.

With conventional reactor designs mechanical and shipping considerations sometimes increase the cost of large reactors to a point where the economic advantage of a single large reactor train compared with two parallel trains disappears or is very slight—despite recent advances in mechanical design which permit reductions in reactor tube sheet and flange thickness. Thus a very large plant might have more than one reactor train.

A graded temperature zone reactor design* has recently been developed which improves yield. Catalyst improvements* have also been realized.

—Robert Merims, *Scientific Design Company, Inc., New York*

Malleable Iron   (See **Iron and steel.**)

Malt   (See **Yeasts.**)

Manganese   (Gk. *Magnes,* a city in Magnesia.) **Mn** = 54.9380 (at. wt.); 25 (at. no.). As early as 1740, Pott demonstrated that although it contains no iron the mineral pyrolusite exhibits magnetic qualities. Recognized earlier by Scheele and others as a new element, Mn was first isolated by Gahn (1774) by reduction of the oxide with carbon.

Of the five isotopes of manganese, only one, ^{55}Mn, is stable. The pure metal may exist in one of four allotropic forms. The alpha form is stable at room temperature. The unit cell of alpha Mn contains 58 atoms, which are in at least four different valence states. The gamma form, which changes to the alpha form at ordinary temperatures, can exist at lower temperatures. Unlike the alpha form, which is harder and more brittle than iron, gamma Mn is soft, flexible, capable of being cut, and easily bent. Mn belongs to group VIIA in the first transition series, other members of the group being technetium and rhenium.

Manganese may have the oxidation numbers -3, -1, 0, $+1$, $+2$, $+3$, $+4$, $+5$, $+6$, and $+7$. The more important Mn compounds are those which contain Mn^{++} and Mn^{7+}. Mn^{4+} is significant because of manganese dioxide, MnO_2. Compounds containing Mn^+ are stable in acids but

*Scientific Design Company, Inc.

TABLE M-6. *Properties of Manganese*

Melting point, °C	1245
Boiling point, °C	2150
Density at 20°C, g/cm^3	7.43
Specific heat at 25.2°C, cal/g	0.115
Latent heat of fusion, cal/g	63.7
Linear coefficient of expansion (0–100°C)	22×10^{-6}
Hardness, Mohs' scale	5.0
Compressibility	8.4×10^{-7}
Solidification shrinkage, %	1.7
Latent heat of vaporization at boiling point, cal/g atom	53,700
Standard electrode potential, V	1.134
Magnetic susceptibility (cgs)	8×10^{-6}

are oxidized in bases. Mn^{7+} is found only in oxy compounds. Manganese in the zero oxidation state exists in bimetallic compounds. Mn^{3-}, Mn^-, and Mn^+ are found in organometallic complexes. See Table M-6.

Ores: The most common Mn ores are pyrolusite, MnO_2; braunite, Mn_2O_3; hausmannite, Mn_3O_4; and rhodochrosite, $MnCO_3$. Mn also exists as the sulfite, sulfate, silicate, and tungstate, which are of little industrial value. Mn metal can be obtained from oxide ores by reduction with C, Al, Mg, or Na; or by electrolysis. World production of Mn outside the United States is almost entirely from the oxides.

The Soviet Union is the largest producer of Mn ore, other large producers including the Republic of South Africa, Brazil, India, and China. Relatively little of the large deposits of Mn ore in the United States is suited to commercial production. Slag from open-hearth furnaces is a potential source of Mn but processing is not economically feasible with present techniques. Large quantities of manganese nodules lie on the ocean floors, particularly in the Pacific, and constitute an important potential source. They contain as much as 24% Mn.

Uses: Although Mn has a variety of uses, the metal is employed principally in metallurgy, applications which consume about 90% of the Mn produced. The steel industry, where Mn acts as a deoxidizer or cleanser in molten metal, is the largest consumer. Mn also combines with sulfur in the molten metal to remove sulfur from the final product. In certain stainless steels, Mn is used as an alloying agent in place of the more costly Ni. Mn also is used in alloys with other metals, such as Al, Sb, and Cu.

The chemical industry consumes the remaining Mn. Since the element is essential to plant and animal life, compounds containing it are used in fertilizers and feed. A number of Mn salts are used as catalysts in organic reactions. Various organometallic compounds containing Mn are used as herbicides and as agents for improving combustion. Mn is an important component in ferrites used in the electronics industry. The metal also is contained in soaps used as paint driers.

Manganese dioxide, MnO_2, is used as an oxidizing agent in the production of numerous chemicals, including dyes and hydroquinone. MnO_2 also has been used in the manufacture of chlorine and iodine. Common uses for MnO_2 include its role as a depolarizer in battery dry cells, in making a black enamel for pottery, and in water treatment. Both Mn metal and manganous oxide, MnO, are used as coatings on welding electrodes. Other Mn compounds of importance include manganese phosphate, used in metal surface treatments; manganese chloride, used in dyeing cotton; manganous sulfate, used in calico printing; and sodium and potassium permanganate, important in chemical analytical work and as disinfectants.

Ferromanganese: The principal form of Mn used in metallurgical work is ferromanganese, which contains about 80% Mn and 20% Fe. Ferromanganese generally is produced in a blast furnace or an electric-arc furnace. High-purity Mn metal is also prepared by aqueous electrolysis.

There are no rigid ore requirements for ferromanganese produced in a blast furnace. A mixture of different ores generally is used, the mix proportioned to meet the specifications of the final alloy. Low-silica ores are preferred because they reduce the slag volume. The phosphorus content of the ore also should be low in order to maintain a low phosphorus content in the alloy. For the electric-furnace process, the charge consists of Mn ore, coke, and limestone mixed in proper proportions. The quantity of silica present determines the extent of losses to the slag, basic slag giving the lowest losses. Recoveries usually are about 85%. In the production of high-purity Mn, the ore is first roasted to produce MnO, after which it is leached with H_2SO_4 to form manganese sulfate. The solution then is neutralized to precipitate iron and aluminum. Other impurities are removed as sulfides. After treatment, the solution is electrolyzed, yielding 99.94% pure metal. A high-purity ferromanganese also is produced by a fused-salt electrolysis.

Metallurgical Applications: Electrolytic (high-purity) Mn often is used as a deoxidizing agent and also is a constituent in nonferrous metals to improve strength, ductility, and hot-rolling properties. Manganese-base alloys containing 72% Mn, the balance being Cu and Ni, are used as a component in bimetals for switching relays because of their very large temperature coefficient of expansion. Alloys containing 60–80% Mn, with the balance of Cu, are used because of their high vibration-damping properties.

Low-carbon ferromanganese is added to steel when the carbon content is critical. Standard ferromanganese, which contains 7% C and 74–78% Mn, is used for adding Mn to steel, both as an alloy and in deoxidation. *Spiegeleisen,* which is 10–23% Mn and 4–5% C and was first employed in 1856 by Robert Mushet, is used in alloys and as a cleaning agent. Silicomanganese is an effective blocking agent used to stop the reaction of carbon and oxygen in steel. Hadfield steel, developed in 1888, contains about 13% Mn and is useful wherever hard materials are handled because the metal becomes harder with shock and impact. Mn also is replacing Ni in the 200 series of stainless steels, developed in an effort to obtain more economical austenitic materials.

Manganese Inorganics: A number of chemical processes have been developed to upgrade Mn ore which produce an intermediate Mn compound. These intermediates usually are free of most siliceous matter. Although these processes were designed to convert the compound to an oxide for use in metallurgical applications, the purity of the compounds often renders them suitable for commercial use.

The properties and uses of several Mn inorganic compounds are summarized in Table M-7.

The ammonium carbonate process* is the only such upgrading process that has reached commercial application. The high-grade manganese carbonate produced is sold to the chemical industry. The process involves reducing the ore to MnO by roasting with gases rich in CO as the initial step. The calcine is then ground and leached in an aqueous solution containing 18 moles of NH_3 and 3 of CO_2. The resulting product is decomposed to yield $MnCO_3$ and NH_3.

Processes of lesser importance include the chloride processes, the sulfur oxide processes, and bacterial leaching. The first type involves a reduction of the ore to MnO, followed by leaching with

*Developed by Manganese Chemicals Corp.

Compound, Formula, and Formula Weight	Density	Characteristics and Uses
Inorganic Compounds		
Manganese borate MnB_4O_7 210.18	. . . .	Pale red; insoluble in H_2O; used as a varnish drier
Manganese carbonate $MnCO_3$ 114.94	3.125	Pink; insoluble in H_2O; decomposes slowly in air, leaving hydrated oxides containing higher-valence Mn
Manganous chloride $MnCl_2 \cdot 4H_2O$ 197.91	2.01	Rose-colored monoclinic crystals; prepared by action of HCl on Mn, carbonate, or oxide; melts at 58°C; boils at 200°C
Manganese dihydrogen phosphate $Mn(H_2PO_4)_2$ 249.31	. . . .	Used in phosphatizing iron and steel surfaces; made by dissolving carbonate in H_3PO_4
Manganese sulfate $MnSO_4$ 150.99	3.235	Pink crystals in hydrated form; hydrates contain 1, 4, 5, or 7 H_2O; made by treating lower-valence Mn compounds with H_2SO_4; melts at 700°C; decomposes at 850°C
Manganic fluoride MnF_3 111.93	3.54	Red crystals; prepared in anhydrous form by treating MnI_2 with F_2; reverts to MnF_2 on heating
Manganese nitrate $Mn(NO_3)_2 \cdot 6H_2O$ 287.04	1.82	Colorless needles; prepared from $MnCO_3$ or Mn and dilute HNO_3; melts at 25.8°C; boils at 129.5°C
Manganous oxide MnO 70.93	5.18	Found in nature as manganosite; isomorphous with oxides of calcium, magnesium, and cadmium; may be prepared by reduction of higher oxides with reducing gas, such as CO and H_2; obtained by thermal decomposition of $MnCO_3$ or MnC_2O_4 in absence of air
Manganous hydroxide $Mn(OH)_2$ 88.95	3.258	Appears in nature as pyrochroite; very low solubility in H_2O; near-pure material practically white
Manganomanganic oxide Mn_3O_4 228.79	4.70	Occurs in nature as mineral hausmannite; ferromagnetic oxide; most stable of all the Mn oxides; conveniently prepared by decomposing $MnCO_3$ in air in high-temperature environment or by oxidation of MnO
Manganese sesquioxide Mn_2O_3 157.86	4.81	Exists in dimorphic forms; occurs in nature as bixbyite; prepared by thermal decomposition of MnO_2
Manganese sesquioxide hydrated MnO(OH) 87.94	. . . .	Belongs to group of trivalent compounds known as *oxyhydroxides* with compositions intermediate between oxide and hydroxide; prepared by oxidation of slurry of manganous hydroxide with air, H_2O_2, or Cl_2
Manganese dioxide MnO_2 86.93	5.026	Pyrolusite, beta MnO_2, is the form most commonly referred to; polymorphic; color varies from black to metallic gray depending on particle size
Manganese heptoxide Mn_2O_7 221.86	>1.84	Volatile anhydride of permanganic acid; stable at −5°C; begins decomposing to dioxide at 0°C; prepared by adding powdered $KMnO_4$ to concentrated H_2SO_4 cooled in a freezing mixture
Permanganic acid $HMnO_4$ 119.04	. . . .	Unstable; very soluble in cold H_2O; decomposes in hot H_2O

Compound, Formula, and Formula Weight	Density	Characteristics and Uses
Inorganic Compounds		
Potassium manganate K_2MnO_4 197.13	. . . .	May be produced in the laboratory by boiling $KMnO_4$ with strong KOH; soluble in 10% KOH with rapid decomposition; decomposes in pure H_2O or dilute acids; an intermediate in the commercial preparation of $KMnO_4$
Potassium permanganate $KMnO_4$ 158.03	2.703	Rhombic, dark-purple crystals; upon heating, liberates O_2 at 240°C; soluble in H_2O; somewhat soluble in methanol, glacial acetic acid, acetone, and pyridine; one of the most important Mn compounds because it is irreplaceable in many processes
Manganese dihydrogen phosphate $Mn(H_2PO_4)_2$ 249.31	. . . .	Used in phosphatizing iron and steel surfaces; can be made by dissolving the carbonate in phosphoric acid
Manganese sulfate $MnSO_4$ 150.99	3.235	Pink crystals in hydrated form with 1, 4, 5, or 7 H_2O, almost white in anhydrous state; monoclinic tetrahydrate obtained commercially; commercial grades are one of the most important Mn compounds; can be prepared by treating almost any lower-valence Mn compound with H_2SO_4
Organic Compounds		
Manganous acetate $Mn(C_2H_3O_2)_2 \cdot 4H_2O$ 245.08	1.589	Pale-red crystals soluble in H_2O, methanol, and ethanol; anhydrous salt prepared by heating $Mn(NO_3)_2$ with acetic anhydride; commercially made by treating the hydroxide, carbonate, manganous oxide, or electrolytic Mn with acetic acid, used as an oxidation catalyst in organic reactions
Manganic acetate $Mn(C_2H_3O_2)_3 \cdot 2H_2O$ 268.10	. . . .	Light-brown crystals, prepared readily by oxidizing manganous acetate with either Cl_2 or permanganate; used as starting material in preparation of other manganic compounds
Manganous acetylacetonate $Mn(C_5H_7O_2)_2$ 253.16	. . . .	Light-tan colored crystals; decomposes above 180°C; also forms dihydrate; may be prepared by mixing solutions of manganese chloride and acetylacetone and then precipitating the product by addition of ammonia or alkali-metal hydroxide to pH 6–8
Manganic acetylacetonate $Mn(C_5H_7O_2)_3$ 352.27	. . . .	Brown to black crystalline powder; relatively stable; insoluble in H_2O; soluble in organic solvents; prepared by addition of solutions of manganous sulfate, potassium permanganate, acetylacetone, and ammonium hydroxide; pH of result mixture should be maintained at 5.5; promoted as a catalyst for several organic reactions
Biscyclopentadienylmanganese $(C_5H_5)_2Mn$ 209.15	. . . .	Pyrophoric in air; reacts with H_2O, aqueous acids, and bases; may be prepared by several methods, including reacting $MnCl_2$ with C_5H_5MgBr followed by heating; used in gas plating in vacuo of heated metal surfaces; derivatives have antiknock properties
Manganous ethylenebisdithio-carbonate $(CH_2NHCS_2)_2MN$ 265.30	. . . .	Yellow powder; fungicide; prepared by adding a 14% NH_3 solution and a 69% $(CH_2NH_2)_2$ solution to CS_2, diluting the mixture, neutralizing it with acetic acid, and combining with a 20% $MnCl_2$ solution; resulting product insoluble
Methylcyclopentadienylman-ganesetricarbonyl $CH_3C_5H_4Mn(CO)_3$ 218.09	. . . .	Light amber liquid; insoluble in H_2O; soluble in organic solvents; can be made by at least six methods, e.g., reacting biscyclopentadienylmanganese with CO; used as an antiknock agent

hydrochloric acid. Relatively pure $MnCO_3$ is precipitated from the leach solution by the addition of sodium carbonate. There are several variations of this procedure. The sulfur oxide processes involve passing sulfur dioxide gas through a slurry of low-grade ore to produce a $MnSO_4$ solution. There also are several modifications of this process, including different processing temperatures. The last type involves the use of bacteria found near manganese ore plants which have the ability to dissolve manganese oxides in solutions of pH 5–6 by the slow addition of H_2SO_4. The only requirement other than the organisms is a nutrient solution. The extraction of manganese as a sulfate is on the order of 71.7–99.9% depending on the ore. The action of the bacteria is not completely understood.

Manganese Organics: The principal use of the organic compounds of Mn is in the catalysis of organic reactions. Because of their ability to catalyze the oxidation and polymerization of oils, many organomanganese compounds have been widely utilized as driers and soaps. Mn organics also are used in agriculture as fungicides and in petroleum products as antiknock agents. The properties and uses of several Mn organic compounds also are summarized in Table M-7.

REFERENCES

Kirk, R. E., and D. F. Othmer: "Encyclopedia of Chemical Technology," vol. 12, pp. 887–905; vol. 13, pp. 1–55, Wiley, New York, 1967.

Lange, Norbert A.: "Handbook of Chemistry," 10th ed., pp. 73–74, McGraw-Hill, New York, 1961.

Weast, Robert C.: "Handbook of Chemistry and Physics," 51st ed., Chemical Rubber, Cleveland, 1970.

—Joseph C. Koziar and J. Y. Welsh, *Chemetals Division, Diamond Shamrock Chemical Company, Baltimore, Md.*

Mannheim Process   (See **Fertilizers.**)

Marble   (See **Limestone.**)

Margarine   (See **Vegetable oils.**)

Marl   (See **Limestone.**)

Martensitic Alloys   (See **Iron and steel.**)

Mass Number   The mass number of an atom is the total number of nucleons in the nucleus. Further,

$$A = N + Z$$

where A = mass number

N = number of neutrons in nucleus

Z = atomic number

Mass number also is defined as that whole number which is closest in value to the atomic mass with the latter quantity expressed in atomic mass units. See also **Chemical elements.**

Mass Polymerization   (See **Polymerization.**)

Mass Transfer   (See **Absorption;** also consult the **Subject Index.**)

Mastic Finishes   (See **Insulation, thermal.**)

Material Balance   (See **Distillation;** and **Energy systems for processes.**)

Matrix Fibers   (See **Fibers.**)

McCabe-Thiele Diagram   (See **Distillation.**)

Medical Materials   (Consult the **Subject Index.**)

Melamine   Melamine, $C_3N_3(NH_2)_3$, is a white solid with a melting point of 355°C and solubility in water of 32 g/100 g at 20°C. As the triamide of cyanuric acid, melamine resembles an amide more than an amine. In water it acts as a weak base and forms solids that can be recovered.

2,4,6-Triamino-*s*-triazine

Most melamine produced is condensed with formaldehyde or other aldehydes to form resins that have excellent resistance to heat, water, and many chemicals as well as good electrical properties and surface hardness. Accordingly, protective and decorative laminates account for about 45% of consumption; molding compounds, about 30%; textile resins, 9%; coatings, 7%; and paper-treating resins and various adhesives, 9%.

Demand for melamine in the United States (1970–1971) was approximately 125 million lb/yr with a projected growth of about 5% annually to 1975.

Production: First prepared by Liebig (1834), melamine was produced for many years from calcium cyanamide, which was converted to dicyanodiamide and then to the trimer melamine. This route has largely been replaced by synthesis from urea via several proprietary methods based upon the following steps. Urea is thermally decomposed endothermically into a cyanic acid–ammonia gas mixture

$$CO(NH_2)_2 \rightarrow$$
$$HCNO + NH_3 + 780 \text{ kcal/kg of solid urea} \quad (1)$$

Cyanic acid gas is thermally decomposed exothermically into a melamine–carbon dioxide vapor

$$6HCNO \rightarrow C_3N_3(NH_2)_3 \rightarrow$$
$$3CO_2 - 710 \text{ kcal/kg of melamine} \quad (2)$$

The overall reaction, which is endothermic, can be written

$$6CO(NH_2)_2 \rightarrow C_3N_3(NH_2)_3 + 3CO_2 + 6NH_3$$

In view of the large amounts of CO_2 and NH_3 evolved during reactions (1) and (2), this process is best undertaken in conjunction with urea manufacture, which permits off gases to be usefully recycled. Synthesis can be undertaken at low or medium pressures with a catalyst or without a catalyst if high pressure is used.

In the process* shown schematically in Fig. M-2, fertilizer-grade urea from an adjacent plant is melted and sprayed into a bed of catalyst "sand" (an aluminosilicate) fluidized with NH_3 gas and heated by a circulating molten-salt system. Reactions (1) and (2) are allowed to occur in the same reactor vessel, even for plants with annual capacities in the 70 to 80 million lb/yr category, thereby saving heat requirements and capital costs. Melamine vapor plus other reaction products and NH_3 carrier gas pass from the reactor to a quench tower, where they are cooled against returned mother liquor to below the sublimation temperature of melamine. Product crystals are washed out of the gas and collect in the base of the tower. This suspension is thickened in a liquid cyclone, concentrated to about 45 wt %, and expanded by steam in a stripper to about 1 atm abs. to remove most of the dissolved gases.

To make a product that will meet the purity standards required by the resin industries, the crude melamine crystals are dissolved in returned mother liquor to separate condensation impurities,

*Stamicarbon process (Dutch State Mines).

such as "melan" and "melon", which remain as solids. Neutralization with CO_2 gas precipitates hydrolyzation by-products ammelide and ammeline, and the addition of activated carbon removes colored contaminants. These solids are separated in a filter, and pure melamine is recovered by vacuum crystallization, thickening, and centrifuging. The product is air-dried, ground, and sent to storage. Ammonia in the stripper off gas is separated from the inerts and returned to the reactor as carrier gas. Other off gases are converted to ammonium carbamate and returned to the adjoining urea plant. Optimum synthesis temperatures for the process are said to be 330–450°C, and corresponding operating pressures are about 7 atm abs.

Process requirements per metric ton of melamine are:

Urea, kg	3,100
NH_3, liquid, kg	460
CO_2, kg	30
Catalyst, kg	8
Activated carbon, kg	2
Filter aid, kg	17
Steam, kg, at 31 atm, gage	5,100
At 11 atm, gage	3,000
Cooling water ($\Delta t\, 8°C$), m³	650
Power, kWh	500
Fuel (gas), kcal	3.4×10^6

By-products per metric ton of melamine are concentrated carbamate solution at 18 atm abs. and 100°C with the following composition:

	kg
NH_3	1,400
CO_2	1,150
H_2O	850
	3,400

Product quality of the melamine produced may be summarized as follows:

Condition	White, crystalline powder, free from physical impurities
Analysis:	
Melamine, wt %	99.8 min.
Ash, wt %	0.01 max.
Moisture, wt %	0.05 max.
Fe, ppm	1.0 max.
Appearance of the melamine-formaldehyde resin solution at 40°C	Clear

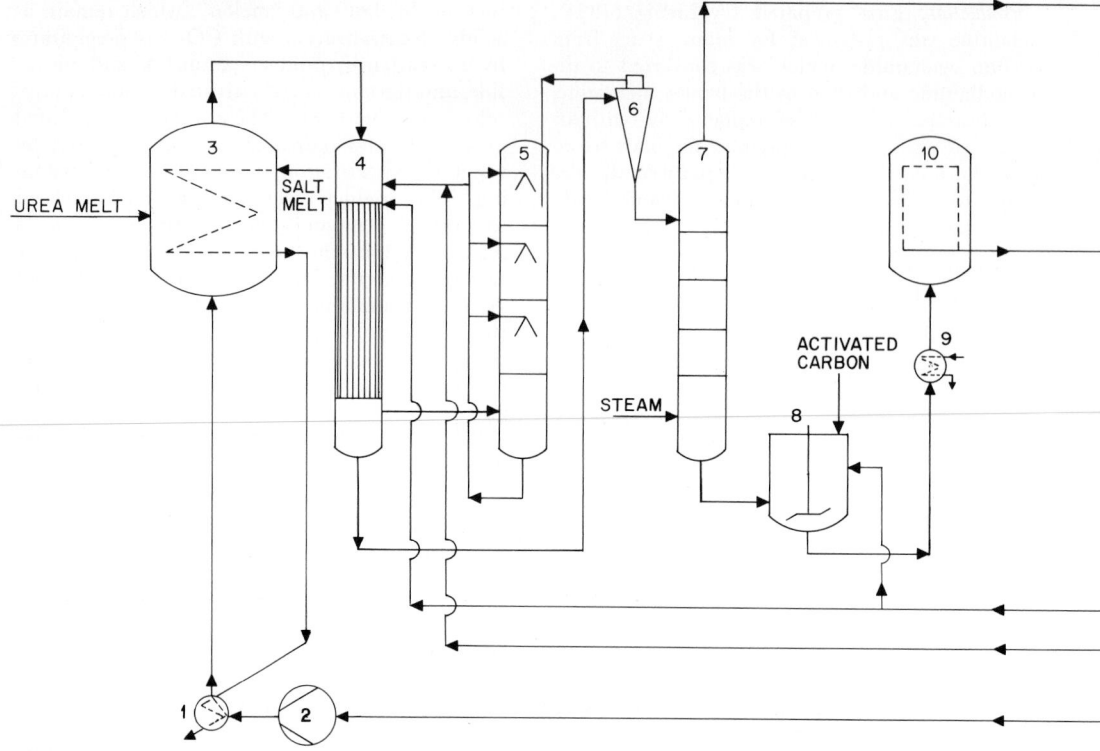

Fig. M-2. Melamine production process. (1) NH$_3$ preheater, (2) NH$_3$ compressor, (3) reactor, (4) saturator cooler, (5) quench-gas scrubber, (6) thickener cyclone, (7) stripper, (8) dissolving vessel, (9) solution preheater, (10) clarifying filter, (11) crystallizer, (12) vapor condenser, (13) prethickener, (14) centrifuge, (15) drying conveyor, (16) separator, (17) mill, (18) product scale, (19) washing column, (20) NH$_3$ condenser, (21) desorber, (22) absorber, (23) inert-gas scrubber.

APHA color of the filtered
    solution at 25°C. 20
Bulk density, kg/l 0.6
Packed density, kg/l 0.8–1.0

In another process,* separate reactors are used for reactions (1) and (2), which are undertaken at atmospheric pressure. A fluidized sand bed heated by a molten-salt system and a fixed catalyst bed are used respectively. Gaseous melamine, NH$_3$, and CO are separated, and a crystalline product recovered in circuits similar to those described in the Stamicarbon process. However, NH$_3$ and CO$_2$ are usually recovered not as a carbamate but as ammonium carbonate or as gaseous NH$_3$ and CO$_2$, which are normally recycled to an adjacent urea plant. A flow sheet is given in Fig. M-3.

Process requirements per metric ton of mela-

*Osterreichische Stickstoffwerke (OSW) process.

mine and for total decomposition of off gases to NH$_3$ and CO$_2$ are:

Urea, kg 3,250
Ammonia. 0
CO$_2$. . . . . Minor quantities
Catalyst . . . . Minor quantities
Steam needed at 13 atm
    gage, kg 5,400
Cooling water, 25°C
    ($\Delta t = 10$°C), m^3. 657
Power, kWh 811
Fuel, kcal . . . . 4.2 × 10^6

By-products per metric ton of melamine are:

	kg
Ammonia	935
Carbon dioxide	1,330
Waste water.	2,500
Steam condensate	2,900

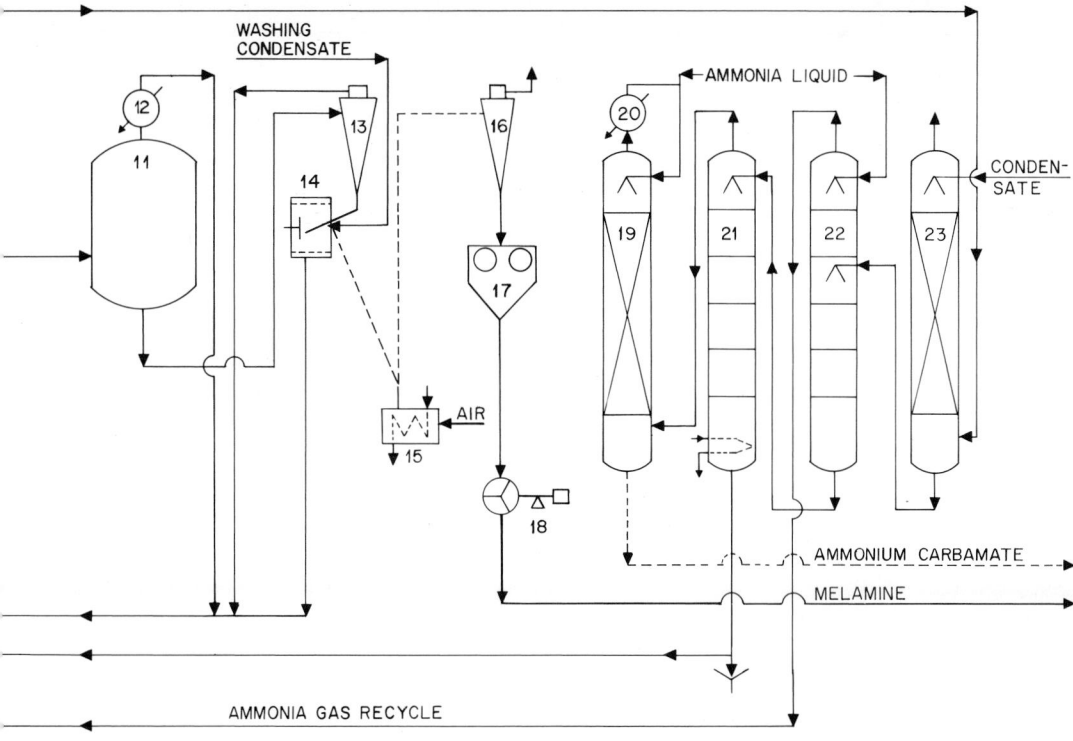

Product quality of the melamine produced by this process may be summarized as follows:

Condition	White, crystalline powder

Analysis:

Melamine, wt %	99.9
Ash, wt %	0.04
Moisture, wt %	0.1
APHA formalin test	15
Bulk density, kg/l	0.6

In still another process,* molten urea from an adjoining plant is compressed to about 100 kg/cm² and delivered to a high-pressure wash tower, where melamine vapor contained in the off gas from the synthesis reactor is absorbed by the urea. The urea melt is fed to the reactor, along with NH₃ gas, which is preheated to about 400°C and compressed to 100 kg/cm². Heat is supplied to the reactor by a molten-salt system, and the urea decomposes into an aqueous melamine solution. Off gas from the reactor, after being scrubbed by incoming molten urea to recover melamine, is sent to an absorber and returned to the urea plant as a carbamate solution at about

*Nissan.

200°C and 100 kg/cm². Product melamine passes from the reactor into a pressurized quencher and is cooled by aqueous NH₃.

After separation of part of the NH₃ in a stripper, the liquor is filtered, prior to recovery of the melamine by crystallization. The product is centrifuged, dried, pulverized, and sent to storage.

NH₃ released in the process is absorbed, liquefied, and recycled to the synthesis reactor. By returning the reactor off gases to the urea plant as a carbamate solution at about 100 kg/cm², integration with urea production and overall economics are said to be improved. Major material costs are substantially the same as for other urea-based methods, but since this is a high-pressure process, no catalyst is used and appreciable savings in capital and operating costs are claimed, compared with methods based on lower pressures. High yields and purity standards are also said to result.

—Christopher J. Pratt, *Hoechst-Uhde Corporation, Englewood Cliffs, N.J.*

Melamine-Formaldehyde Resins (See **Amino resins.**)

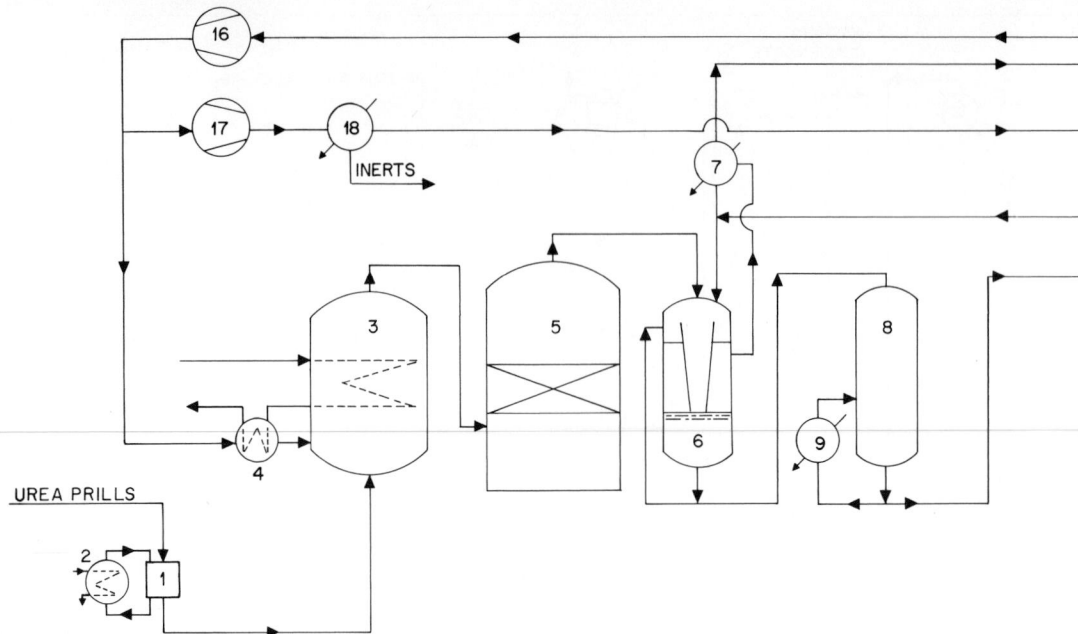

Fig. M-3.   Melamine production process. (1) Melt vessel, (2) circulating heater, (3) decomposer, (4) NH_3 preheater, (5) contact furnace, (6) separator, (7) recooler, (8) collecting vessel, (9) circulating cooler, (10) centrifuge, (11) drier, (12) sieve, (13) mill, (14) scale, (15) first separation column, (16) and (17) NH_3 compressors, (18) condenser, (19) second separation column, (20) third separation column, (21) mother-liquor heater, (22) hydrolyzing column.

Melting Operations   (See **Energy systems for processes;** and **Iron- and steelmaking.**)

Melting Points   (See **Chemical elements.**)

Melting Tanks   (See **Glass.**)

Membrane Filters   Since the mid-1950s, a new technology has evolved around efforts to develop filtration media which will retain smaller particles than conventional filter cloth. This work has led to the development of membrane structures for use as filtration media which retain particles in the micrometer range and below. Further developments produced even tighter membranes capable of retaining large dissolved molecules and colloids. Finally, very tight membrane structures were made which can retain simple sugars and inorganic salts, such as sodium chloride.

Membrane separations can be classified according to the molecular or particle size which can be retained: microfiltration, ultrafiltration, and reverse osmosis.

Microfiltration: This method encompasses the filtration of particles 10–0.1 μm. The membrane consists of a number of pores which pass directly through the membranes. The pores are uniform in size and occupy approximately 80% of the membrane volume. Although the membranes often are made of cellulosic material, they also may be made of synthetic polymers. Typical operating pressures are 10–100 psig.

Microfilters normally are used to remove very small quantities of fine suspended particles and function as screens when fluids flow through the membrane. All particles larger than the filter pores are retained on the filter surface. Although most small particles pass through the filter, some are retained by the filter by random entrapment and adsorption.

Relatively high flux rates [1–10 gal/(ft^2)(min)] are typical, but the open-pore structure of the microfilter means that it tends to plug quite readily. Therefore, this type of filtration is a batch operation, i.e., the membranes with the retained particles must be removed from service periodically and new membranes inserted.

Ultrafiltration: This method encompasses the

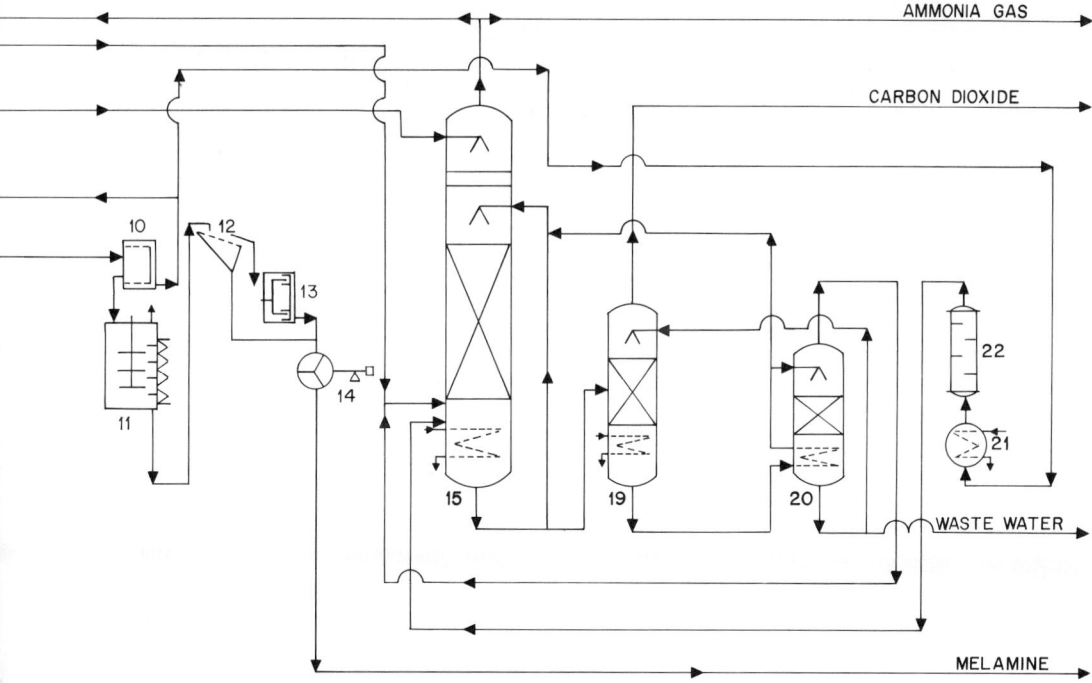

separation of large dissolved molecules categorized as macromolecules or colloids. In this size range, molecules usually are classified in angstrom units according to their size. Ultrafiltration includes sizes ranging from 10 to 1000 Å. Ultrafiltration membranes are anisotropic; i.e., they have a very thin skin, which is supported on a spongy sublayer of membrane material. The thin skin is the working part of the membrane, and separations take place at the skin surface. Flow of the solvent phases through the membrane skin is predominantly by a pore-flow mechanism (Fig. M-4). However, instead of a uniform pore structure, the membrane skin has a plurality of small, irregular passageways.

The separation of macromolecules takes place

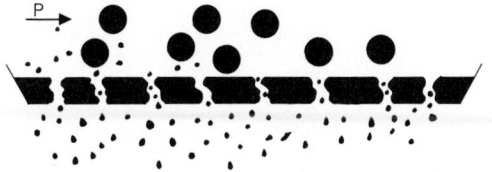

Fig. M-4. Schematic representation of pore flow in ultrafiltration.

at the upstream side of the membrane. A dry cake never is obtained. Substantial velocity is maintained across the membrane surface (2 to 10 ft/s) to prevent accumulation of retained substances. At pressures of 10–50 psig, membrane flux rates of 5–20 gal/(ft²)(day) are typical, and membrane life is a matter of years. Normally, the membranes are made of cellulose acetate or synthetic polymer.

Reverse Osmosis: This method separates inorganic salts and simple organic compounds under pressure. The size of the species usually is considered in terms of molecular weight. Reverse osmosis can be defined as the retention of solutes of molecular weight below 500. Solvent transport through reverse-osmosis membranes is substantially diffusive in nature (Fig. M-5). These membranes also are anisotropic (thin-skinned, overlaying a porous substructure). In all probability, the skin contains no pores.

If water is the solvent phase, the water passes through the membrane by true diffusion. At some point in its passage through the membrane, the solvent water actually becomes a part of the membrane water structure. As low-molecular-weight solutes exhibit osmotic pressure when con-

TABLE M-8. *Useful Ranges of Various Membrane Processes*

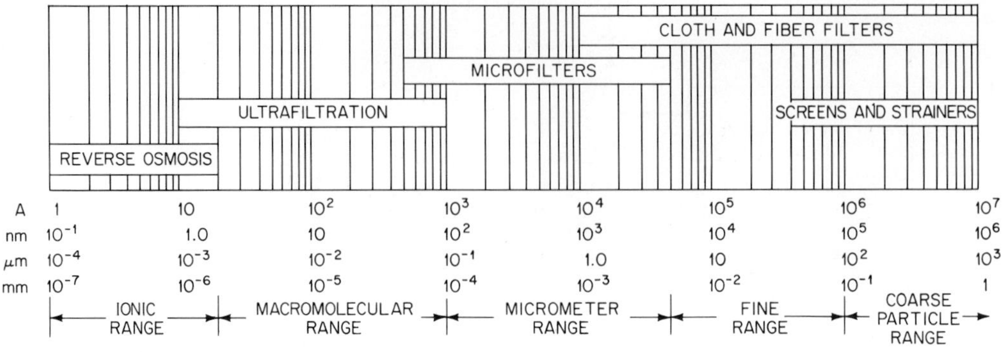

Reproduced by permission of copyright owner, Dorr-Oliver Inc.

centrated, system pressure in excess of the osmotic pressure of the concentrated solutes must be applied to create a pressure driving force.

In reverse osmosis, system pressures of 200–1,500 psig generally are used. Accumulation of retained solutes at the membrane surface (concentration polarization) occurs, and flow conditions generally are maintained over the surface of the membrane to control accumulation. Flux rates usually are 5–15 $gal/(ft^2)(day)$, and membrane life is a matter of years. Reverse-osmosis membranes normally are made of cellulose acetate or some synthetic polymer. They must be kept moist.

Ranges of Membrane Processes. The size range of separation of each technique is given in Table M-8.

Ultrafiltration: Industrial ultrafiltration equipment first appeared commercially about 1965. This low-pressure membrane unit operation has found increasing use in industrial macromolecular separation and concentration. The method provides a means for continuous separation which

does not involve a phase change. Separations can be made economically at room temperature or even below where required.

The skin and porous substructure of the membrane, previously described, are illustrated in Fig. M-6. Some of the standard ultrafiltration membranes commercially available are listed in Table M-9. Since the membranes are polymeric substances, they do not make sharp molecular-weight cutoffs. The molecular shape and size significantly affect the separation achieved. In almost all cases, it is necessary to run small-scale membrane-selection studies for prior determination of the best membrane for a particular use.

Operation of the ultrafiltration process is shown in Fig. M-7. A pressurized feed solution flows over the skin surface of a supported membrane. Under pressure, solvent and low-molecular-weight solutes pass through the membrane while larger macromolecules are retained in the system. The retained materials tend to collect on the surface of the membrane, forming a gel layer which limits the flux rate. To minimize the thickness of this gel layer, ultrafiltration systems are designed so that flow sweeps across the membrane surface. A

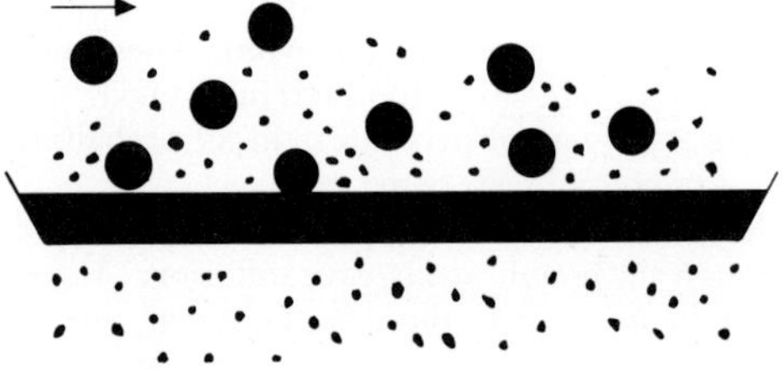

Fig. M-5. **Schematic representation of molecular diffusion in reverse osmosis.**

TABLE M-9. *Ultrafiltration Membranes*

Designation	Protein Molecular-Weight Cutoff	Temperature Limit, °C	pH Range
UM-10 . .	10,000	50	2–10
FP-12 . . .	12,000	60	2–10
XP-24 . . .	24,000	70	1–11
PM-30 . . .	30,000	200	1–11
HFA-300 .	45,000	40	2–8

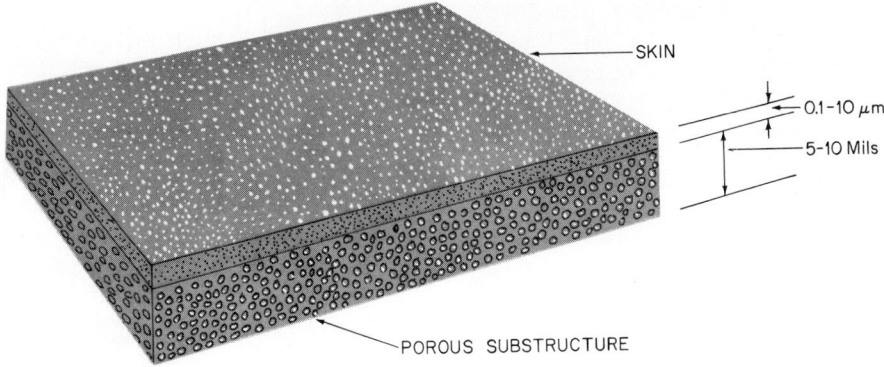

Fig. M-6. Anisotropic ultrafilter.

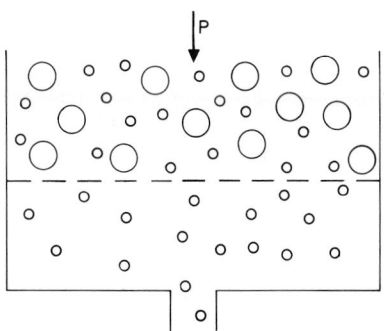

Fig. M-7. Operation of the ultrafiltration process. Pressure forces solutes of a preselected molecular weight through the permeable membrane while solutes not satisfying the criteria are held at the membrane surface.

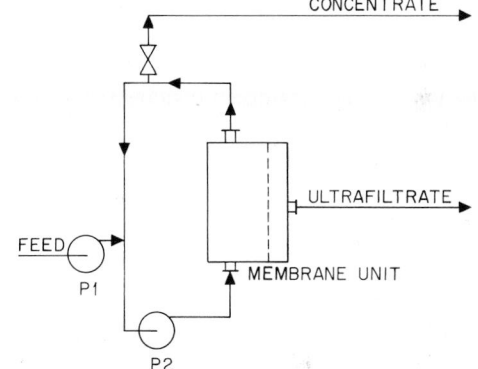

Fig. M-8 Ultrafiltration process flow. P_1-pressurization pump, P_2-recirculation pump.

typical process flow diagram is shown in Fig. M-8. A pressurization feed pump feeds a recirculation loop, which includes a recirculation pump and the ultrafiltration modules. The recirculation provides flow through the membrane modules to maintain velocity across the membrane surfaces. The feed pump provides the required 30- to 50-psi system pressure.

A modular concept is used to build ultrafiltration systems. The modules are connected in parallel and series arrays to provide the required area. The membrane modules must be designed to satisfy a number of different considerations for successful operation. Since a relatively large amount of membrane area is required to process a given flow, the membrane module should be compact to reduce the space required. Because it is desirable to have a significant velocity across

the surface of the membrane, the modules must be constructed so that flow can be pumped across the membrane surface without any undue loss of head. Since solids are present in most process streams, the modules must be constructed to accommodate suspended solids.

There are two basic module designs in use, the tube and the parallel-leaf design. One configuration of the tubular design is shown in Fig. M-9. The tubes usually are $\frac{1}{2}$ or 1 in. diam, and the membrane is supported on the inside of a porous tube. The porous tube must be able to withstand the system pressure. The tubes (6 ft to 8 ft long) are connected in various series or parallel arrays to build up the required membrane area.

The parallel-leaf module consists of a number of parallel sheets of porous backup material covered with membrane. The sheets are spaced about

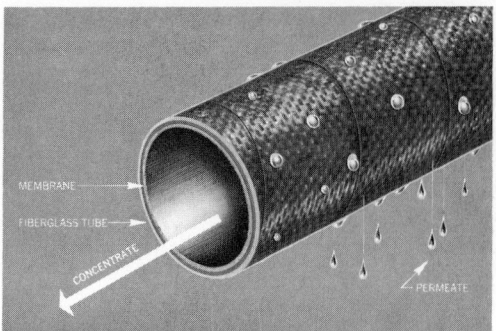

Fig. M-9. Tubular-type membrane. (*Osmotik tube, Calgon-Havens.*)

Fig. M-10. Ultrafiltration modules of parallel-leaf design. (*Standard Iopor module, Dorr-Oliver Inc.*)

$\frac{1}{8}$ in. apart. The parallel-leaf design is shown in Fig. M-10. Three 20-ft^2 ultrafiltration cartridges form a housing containing 60 ft^2 of membrane area. Flow through the module passes between the leaves in the cartridges, providing the required velocity at the surface of the membranes. Ultrafiltrate is removed through one edge of each sheet and collected. As shown in Fig. M-11, the modules are connected directly together to build a large system.

Application: Industrial ultrafiltration is used for many macromolecular separations, including the separation and retention of proteins, starches, complex organic compounds, and colloidally dispersed substances, such as clays, pigments, latex particles, microorganisms, and pharmaceuticals.

One of the results of processing milk to make cheese is a substantial quantity of whey, which often is regarded as waste. Normal cheese whey may contain 0.5% protein, 0.5% ash, and 5% lactose, the remainder being water. Ultrafiltration of cheese whey separates the protein from the lactose, the protein being retained by the membrane and thus concentrated. The concentrated soluble protein is a valuable food product. The lactose in the ultrafiltrate can be crystallized.

Ultrafiltration systems have been used widely to process paint from electrodeposition tanks, retaining the pigment and resin solids while allowing the inorganic salts, water, and excess solubilizer to pass into the ultrafiltrate. The systems also are used to process rinse waters, concentrating the retained paint solids. The loss of paint solids

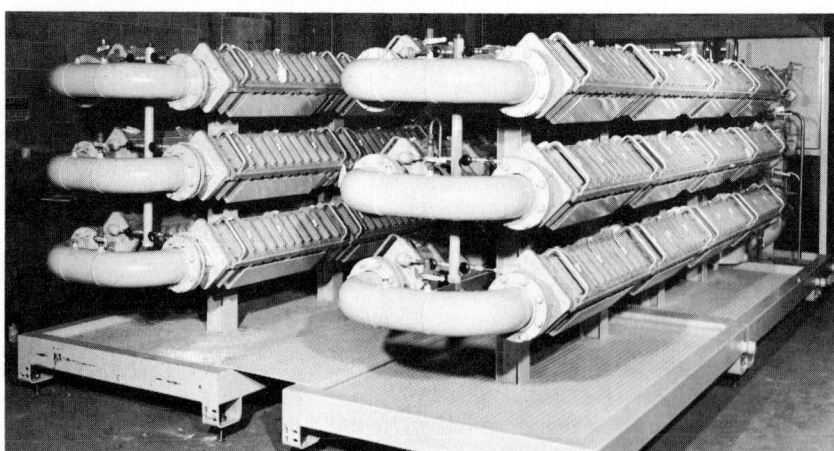

Fig. M-11.   System made up of ultrafiltration modules of parallel-leaf design for industrial application. (*Iopor Ultrafiltration System, Dorr-Oliver Inc.*)

due to drag-out thus is eliminated. Systems also have been used in the concentration and purification of various enzymes from fermentation processes. The means is provided for concentrating the enzymes without any application of heat, which tends to destroy enzyme activity. The enzyme (a protein) is retained on the ultrafiltration membranes, while the water and small organic and inorganic impurities pass through.

Ultrafiltration also has application in the treatment of sewage. Air is bubbled into raw sewage to grow activated sludge, which is passed over the membranes. The quality of the ultrafiltrate far exceeds current requirements. The membranes retain all bacteria and viruses, and thus the ultrafiltrate contains zero coliform count. See also **Water treatment.**

REFERENCES

Forbes, F.: Ultrafine Filtration for Electrophoretic Painting, *Prod. Finish.*, November 1970.
Friedlander, H. Z., and R. N. Rickles: Membrane Technology, II, *Anal. Chem.*, vol. *37*, pp. 27A–68A, 1965.
Merson, R. L.: Reverse Osmosis in Food Processing, *Natl. Canners Ass. and USDA Symp., Albany, Calif.*, Jan. 23, 1969.
Michaels, A. S.: New Separation Technique for the Chemical Process Industries, *Chem. Eng. Prog.*, vol. *64*, no. 12, p. 31, 1968.
Porter, M. C., and A. S. Michaels: Membrane Ultrafiltration, *Chem. Tech.*, January 1971.
Sourirajan, S.: "Reverse Osmosis," Academic, New York, 1970.
Stavenger, P. L.: Putting Semipermeable Membranes to Work, *Chem. Eng. Prog.*, vol. 67, no. 3, p. 30, 1971.

—Robert A. Fiedler, *Dorr-Oliver Incorporated, Stamford, Conn.*

Mendelevium   (See **Chemical elements.**)

Mercaptans   (See **Polymerization;** and **Sulfur.**)

Mercuric Acetate   (See **Catalysts.**)

Mercury (Planet Mercury.) **Hg** (L. *hydrargyrum*.) = 200.59 (at. wt.); 80 (at. no.). Mercury is a lustrous, silvery liquid with a slight blue tinge that is liquid at normal temperature. The only other element that is liquid near this range is gallium.

Hg is a member of group IIB, along with Zn and Cd, and of period 6 of the periodic table. The melting point of $Hg = -38.87 \pm 0.02°C$; boiling point $= 357°C$; and density is 13.55 g/cm^3 at 20°C. There are seven stable isotopes of Hg,

196, 198–202, and 204, and seven unstable isotopes, 192–195, 197, 203, and 205. Other important mechanical and physical properties of Hg are summarized in Table M-10. See also **Chemical elements.**

Hg was mined over 2,300 years ago. Mercury is mentioned in Aristotle's writings. Before Bartolomé de Medina's process for silver recovery by amalgamation with Hg (1557), Torricelli's use of Hg in a barometer (1643), and Fahrenheit's Hg thermometer (1714), mercury played the role of a curiosity rather than a useful element.

Mercury, also known as *quicksilver,* ranks tenth in quantity in world output of nonferrous metals, but the unusual and desirable combination of properties of Hg—including liquidity at ordinary temperatures, high density, uniform volume expansion, electrical conductivity, ability to alloy readily, high surface tension, chemical stability, and toxicity of its compounds—gives Hg an importance industrially and militarily far out of

TABLE M-10. *Some Properties of Mercury*

Specific heat (20°C), cal/(g)(°C) . . .	0.033
Heat of fusion, kcal/g	2.7
Thermal conductivity at	
20°C, (cal)(cm)/(s)(cm^2)(°C)	0.0201
Electrical resistivity at 0°C, $\mu\Omega$-cm . .	94.1
Crystal structure	Rhombohedral
Lattice constant at 20°C, a, Å . . .	2.999
Ionization potential, eV	10.434
Vapor pressure, mm Hg:	
At 126.2°C	1
184.0°C	10
261.7°C	100
290.7°C	200
323.0°C	400
Viscosity, cP:	
At −20°C	1.85
0°C	1.68
20°C	1.55
100°C	1.21
200°C	1.01
Surface tension, dynes/cm:	
At 20°C	465
112°C	454
300°C	405

proportion to the small size of the mercury-producing industry.

Sources: Of 25 minerals known to contain Hg, the chief source is the red sulfide, cinnabar, HgS, which contains 86.2% Hg and 13.8% S. Cinnabar deposits are relatively shallow and confined to areas of late Tertiary orogeny and volcanism.

Despite widespread occurrences of cinnabar, deposits of commercial importance are found in only a few countries, China, Italy, Mexico, the Philippines, Peru, Spain, the Soviet Union, the United States, and Yugoslavia. Measured, indicated, and deduced world reserves of Hg ore are an estimated 4 million flasks, with the United States accounting for about 300,000 flasks. Most of the reserve is in the principal mercury-producing mines of the world.

Depending on the type of deposit, Hg ore is mined by surface or underground methods. With either method, Hg mining is a comparatively small-scale operation because the deposits are characteristically small and irregular in size. With the less costly surface methods, ore averaging 3 lb of Hg per ton may be mined profitably, whereas the underground methods may require a grade of 10 lb/ton or more. Beneficiation of Hg ore, other than crushing to a suitable size, is seldom done, but Hg ore may be upgraded by hand sorting, screening, jigging, tabling, or flotation. Of the various processes, flotation is the most efficient, and concentrates containing 25–50% Hg with a recovery of about 90% have been produced commercially.

Roasting is the conventional process for extracting Hg from its ores and concentrates ($HgS + O_2 \rightarrow Hg + SO_2$). It is essentially a distillation process in which the Hg ore is heated in a mechanical furnace or retort to vaporize the Hg, followed by cooling and condensation of the vapor to liquid metal. Recovery of Hg is high, averaging about 95% for furnace plants and 98% for retort installations. In addition, the product—prime, or virgin, mercury—averages about 99.9% purity, which is satisfactory for virtually all uses. Dirty or contaminated Hg is cleaned and refined by filtering, oxidation, or acid leaching of the impurities or by distillation.

Virgin mercury is shipped in 76-lb wrought-iron and spun-steel flasks, 4–7 in. diam with an average height of 12 in. In establishing the flask at the 76-lb figure, Spanish standards were used because that country has been the world's leader in Hg production for centuries. The Spanish apothecaries' quintal of 100 apothecaries' pounds was in use when Spain adopted the metric system. One Spanish quintal of 100 apothecaries' pounds would equal 76.07 lb avdp (U.S.), and the present Spanish metric weight of 34.5 kg equals 76.06 lb. The accepted standard of 76 lb 1 oz per flask would equal 76.063 lb avdp (U.S.); the extra ounce was dropped, and the even 76-lb unit was adopted.

Secondary mercury is reclaimed from waste products, such as dental amalgams, sludges, used batteries, and other Hg-bearing materials, in retorts similar to those used for treating primary materials. The quantity of Hg recovered from secondary material in the United States is significant, approaching 15–20% total domestic production.

World mine output of Hg is dominated by Italy and Spain. United States imports of Hg from Spain traditionally have represented close to one-half of total Hg imports; nearly 20% comes from Italy, approximately 17% from Mexico, and 10% from Yugloslavia; the remainder is made up of domestic production and imports from Japan and the Philippines. Import-export patterns and prices of Hg fluctuate widely, prices ranging from about $300 per flask to well over $500 per flask in a given year. Fluctuations arise from the great adverse effect of slight overproduction as well as wide swings in use. The United States government and some others have stockpiling policies to maintain ample reserves of the metal during times of emergency and to contribute to some degree to price stability.

Grades: USP mercury conforms to U.S. Pharmacopeia specifications. Technical mercury is a grade generally used for industrial purposes. Triple-distilled mercury conforms to American Dental Association and National Formulary requirements, and reagent grade conforms to American Chemical Society specifications. These and other forms of Hg produced from prime virgin or scrap metal usually are packaged in small containers (from 10 lb to 4 oz) of earthenware, glass, or plastic.

Toxicity: Poisoning may occur in mining and extracting metallic Hg and in any industry in which Hg is used, as a result of handling or exposure to its vapors. Mercury is capable of being subdivided into very minute particles. The finely divided dark-gray powder form, for example, can be readily obtained by shaking Hg with grease, chalk, sugar, ether, and a number of other substances. When handled and changed to a highly divided form, the chances of poisoning are greatly increased because of lessened awareness of the metal's presence. All mining and handling areas should be properly ventilated with employment of respirators and masks. Suspected contaminated surfaces should be sprayed with various compounds to render the Hg inert. Much remains to be learned about formation of simple mercury organic compounds that may become water- or airborne and persist in the environment for long

TABLE M-11. *Uses of Mercury*

Field of Use	Percent of Total Consumed by the Uses Listed*	Trend if any
Electrical apparatus	28.3	
Electrolytic preparation of chlorine and caustic soda	20.2†	Up
Paint, antifouling, and mildewproofing	18.4	Up
Industrial and control instruments	8.1	Down
Pharmaceuticals	7.8	Down
Agriculture	7.3	Down
Dental preparations	3.3	Down
General laboratory uses . .	2.1	Down
Catalysts	1.9	
Paper and pulp manufacture.	1.4	Down
Amalgamation	1.2	

*United States. Total annual consumption of Hg in the United States about 80,000 flasks.

†Can go as high as 30–35% during period of heavy construction of new chlorine–caustic soda installations because of heavy initial Hg requirements for cells.

periods. Until more information is available, there will be continued stress on the safe handling and disposal of Hg and on the search for effective substitutes for the metal.

Uses: The major uses of Hg are given in Table M-11. A very important application during recent years has been as a cathode in the electrolytic preparation of chlorine and caustic soda. Although consumption of Hg per installation is small, large quantities are required for the initial installation. A large expansion of this use essentially accounted for the meteoric rise in Hg prices during the mid-1960s.

Significant quantities of Hg are used in electrical apparatus, e.g., lamps, arc rectifiers, batteries, and switches; and in instruments, including thermometers and barometers. Very substantial quantities of Hg also are used in dental preparations, as catalysts, and in general laboratory applications. Compounds of Hg are used extensively in insecticides, fungicides, and bactericides for agricultural and industrial purposes; the latter applications will undoubtedly be subjected to severe scrutiny because of the poisonous nature of these materials. Pharmaceuticals, the control of slime in pulp and paper mills, and protective paints also use Hg in somewhat smaller amounts in the form of Hg compounds. An expanding use of Hg is foreseen in connection with catalysts and new amalgam metallurgical techniques. Frozen

Hg patterns for precision casting also are an interesting potential.

For agricultural and industrial purposes, copper and numerous organic chemicals may be substituted for certain Hg compounds. In the pharmaceutical field, sulfa drugs, iodine, and various antiseptics and disinfectants have made inroads on mercury chemicals. The use of metal powders, porcelain, and plastic materials are preferred to Hg amalgams in many dental uses. Lead azide, diazodinitrophenol, and other organic initiators serve the same function as mercury fulminate. The most difficult of mercury's properties to duplicate are its high specific gravity, fluidity at room temperatures, and excellent electrical conductivity coupled with its other properties. The use of mercury as a heat-transfer medium in boilers has largely been abandoned.

Chemistry: Hg readily dissolves a number of metals to form amalgams but not Fe or Pt. The metal is not oxidized in air at ordinary temperatures but converts slowly to HgO or HgO_2 with continued heating near the boiling point. Mercurous oxide, a black powder, is formed when mercurous salts are reacted with fixed alkalis. Since the formation of HgO is slow when heated with air, the oxide is made on a large scale by heating a mixture of $Hg(NO_3)_2$ and Hg or of $Hg_2(NO_3)_2$ alone until the emission of NO_2 is completed. The oxide also can be prepared by treating a mercuric salt with a fixed alkali. The peroxide, HgO_2, results when $Hg(NO_3)_2$ is treated with an excess of perhydrol at a cool temperature.

The most effective solvent for Hg metal is HNO_3, but a small quantity of HNO_2 should be present. Ferric ion retards and manganous and sodium ions speed up the reaction. Hg is attacked by sulfur to form HgS, but dry H_2S does not react with Hg. The sulfide is formed when Hg is reacted with $Na_2S_2O_3$. Dilute or concentrated H_2SO_4 do not attack Hg. However, hot concentrated H_2SO_4 will form the sulfate, Hg_2SO_4, with generation of SO_2 and formation of H_2O if the Hg is in excess. If the acid is in excess, $HgSO_4$ is formed.

Mercury forms both Hg^+ (-ous) and Hg^{++} (-ic) salts. The Hg^+ salts are converted to the Hg^{++} salts by powerful oxidizing agents although they are stable in air at ordinary temperatures. Some of the Hg^{++} salts, although also permanent in air, are reduced to Hg^+ salts or even metallic Hg with relative ease. Most Hg salts are insoluble in water, particularly if no free acid is present to prevent hydrolysis and the consequent precipitation of a basic salt. $HgCl_2$ is fairly soluble in H_2O (67.6 g/l), whereas $HgBr_2$ is less than moderately soluble and HgI_2 is only very slightly soluble.

Of all the metals, Hg has the greatest affinity for sulfur with the possible exception of Pd. Free Hg will precipitate Ag, Au, and Pt from solution and also will reduce Hg^{++} to Hg_2^{++}. Hg is precipitated from solution by numerous reducing agents, such as Pb, Sn, Bi, Cu, Cd, Al, Fe, Zn, H_3PO_2, and H_2SO_3. In laboratory reactions, stannous chloride is often used as the reducing agent.

Over the years, a number of interesting and useful organic mercury compounds have been produced, and some still find commercial and medical application. These include acetoxymercuri-2-ethylhexylphenolsulfonic acid (Mertoxol), which contains about 40% Hg and is used as an antiseptic. The formula is estimated to be $CH_3COOHg \cdot C_6H_2OH \cdot C_8H_{17} \cdot SO_3H$. Sodium 2,3-dihydroxy-3,5-(dihydroxymercuri)benzophenone-2'-sulfonate is mixed in equal parts with ammonium 2,4-dihydroxybenzophenone-2'-sulfonate (plus sodium acetate and H_2O) to form Meroxyl. The combined mixture contains about 28% Hg and also has antiseptic uses. The familiar Merthiolate, Metaphen, and Mercurochrome all are mercury organic compounds used as antiseptics, disinfectants, and germicides. Merthiolate, for example, is $C_2H_5HgS \cdot C_6H_4 \cdot COONa$, that is, the sodium salt of ethylmercurithiosalicylic acid. Mercurochrome (also Merbromin) is $C_{20}H_8Br_2HgNa_2O_6$, that is, the disodium salt of 2,7-dibromo-4-hydroxymercurifluorescein. Mercurophen is $NaOC_6H_3(NO_2) \cdot HgOH$, that is, sodium hydroxymercuri-o-nitrophenolate. Mercarbolide, $C_6H_4OH \cdot HgCl$, or 2-hydroxyphenylmercuric chloride, is used in soaps as an antiseptic ingredient. Used as a preservative in certain commercial preparations and as an antiseptic is phenylmercuric nitrate (basic phenylmercuric nitrate), $C_6H_5HgOH \cdot C_6H_5HgNO_3$. The foregoing compounds and numerous others are generally known in the pharmaceutical field as *mercurials*.

Some of the simpler mercuric organic compounds (not all previously understood in depth) are of current concern because of their persistence as poisonous pollutants. Hg is present in a number of noncomplex organometallic compounds, including mercurous formate, $HgCHO_2$; mercuric oxalate, HgC_2O_4; diphenylmercury, $(C_6H_5 \cdot C_6H_4)_2Hg$; chloromercuriphenol, $o\text{-}C_6H_4OHHgCl$; dibenzylmercury, $(C_7H_7)_2Hg$; $(C_2$-$H_5)_2Hg$; dimethylmercury, $(CH_3)_2Hg$; and phenylmercuric cyanide, C_6H_5HgCN.

Mercury Cell (See **Caustic soda**; and **Chlorine**.)

Mercury Fulminate (See **Explosives**.)

Mesityl Oxide (See **Ketones**.)

Mesitylene (See **Hydrocarbons**.)

Meson A meson is an elementary subatomic particle having a rest mass intermediate in value between the mass of an electron and that of a proton. Mesons are found in cosmic rays and are produced in high-energy nuclear reactions. Mesons may have a positive or negative charge or be neutral. The average lives of mesons are always shorter than 1 μs. Mesons differ in masses and spins and can in some cases change from one kind to another. Known mesons include μ mesons, π mesons, and τ mesons.

Mu meson (also referred to as a *muon*), mass = 210 times mass of electron. Carries a positive or negative charge.

Pi meson (also referred to as a *pion*), mass (π^0 meson) = 264 times mass of electron; π^+ meson = 273 times mass of electron; π^- meson = 273 times mass of electron.

Meta Compounds (See **Carbon compounds**.)

Metals (Consult the **Subject Index**.)

Metal Pretreatment (See **Conversion coatings**.)

Metallocenes (See **Osmium**; and **Ruthenium**.)

Metallurgical Processes (Consult the **Subject Index**.)

Metaphosphoric Acid (See **Fertilizers**; and **Phosphoric acid**.)

Metasilicate [See **Silicates (soluble)**.]

Metering Pumps (See **Mixing, fluids**.)

Methacycline (See **Antibiotics**.)

Methane Methane, CH_4, is the principal constituent of natural gas from the oil fields, averaging 75% by weight but rising to nearly 99% in some cases, e.g., the gas from the Pennsylvania fields.

Some Kentucky gas, however, has a methane content as low as 23%. Typical pipeline gas from several fields, free of carbon dioxide, hydrogen sulfide, and water vapor, contains about 78% methane, 13% ethane, 6% propane, 1.7% butane, and 0.6% pentane, the remaining small fraction consisting of gases higher in the paraffin series. Consequently, methane may be classified as a major fuel. See also **Fuels;** and **Natural gas.**

Other names for methane include *marsh gas* (due to decomposition of natural organic matter in bogs and low damp places), *firedamp* (in coal mines), and methyl hydride. Methane is odorless, colorless, and lighter than air (0.55) and has a melting point of $-182.6°C$ and a boiling point of $-161.5°C$. The heating value of methane (995 Btu/ft^3 at 60°F and 30 in. Hg pressure) is low when compared with ethane (1730 Btu), propane (2465 Btu), and butane (3200 Btu).

Natural gas is a raw material for numerous synthetic products. Consequently, methane is an important starting ingredient for many processes. Since it usually is not necessary to isolate and purify the methane, it tends to lose its identity in a list of basic petrochemicals. See also **Petrochemical complex.** The conversion or use of methane is not always complete, and hence methane is a major ingredient of some stack waste gases.

The presence of CH_4 in various feedstocks makes possible the creation of synthesis gas: $CH_4 + H_2O \rightarrow CO + 3H_2$. Synthesis gas (the percentages of CO and H_2 vary according to the end product to be made) is used widely in the production of ammonia, oxo chemicals, and methanol. See also **Synthesis gas.** Hydrogen cyanide (in one process) is prepared from methane and ammonia at a temperature of about 1250°C in the presence of a platinum catalyst: $CH_4 + NH_3 \rightarrow HCN + 3H_2$. The methane content of the charge stock (natural or refined gas) varies from 50 to 100%. A major outlet for methane and other light paraffins is in the production of olefins. Methane also is used as a raw material in controlled oxidation to produce acetylene.

Although there has been a strong trend for many years toward the replacement of artificially made fuel gases by natural gas, large quantities of artificial gas still are produced, variously named *artificial gas, producer gas, coal gas, water gas, manufactured gas,* and *town gas.* Coal and petroleum materials are the primary sources of carbon, and steam is the source of hydrogen. There are several processes operating under various pressures and temperatures. Artificial gases of these kinds contain a high methane content.

In addition to its value as a fuel and basic chemical, methane is of particular interest because CH_4 is the anchor compound of the aliphatics, all of which are considered to be derivatives of methane. See also **Aliphatic compounds; Carbon compounds;** and **Hydrocarbons.**

Methanol   (See **Methyl alcohol.**)

Methionine   (See **Amino acids;** and **Peptides and proteins.**)

Methoxide   (See **Alcoholate.**)

Methyl Alcohol   Methyl alcohol (methanol), the lightest of the series of aliphatic alcohols, is a clear, colorless, inflammable, volatile, and very mobile liquid. The compound is poisonous as liquid or vapor and is soluble in water, alcohol (ethyl), and ether (diethyl ether) in all proportions.

The molecular weight of methanol is 32.04, and its specific gravity is 0.792 (20°C/4°C). The melting point is $-97.6°C$, and the boiling point is 64.6°C. Methanol is slightly denser than air, having a vapor density of 1.11 (air = 1). The upper and lower explosive limits (percent by volume in air) are 36.5 and 6.0 respectively. The maximum allowable toxic concentration in air is 200 ppm by volume. The self-ignition temperature is 470°C, and the flash point (closed cup) is 11°C.

Commercial Grades: Three grades* may be defined as follows: *grade A,* synthetic, 99.85% by weight (solvent use); *grade AA,* synthetic, 99.85% by weight (hydrogen and carbon dioxide generation); and *grade C,* wood alcohol (denaturing use).

Uses: Synthetic methanol is one of the major raw materials of the organic chemical industry. Methanol has economic stability and a steady growth rate owing to the low costs of production and diversity of applications. Nearly all the methanol producers also make formaldehyde, which is the main end use (more than 50%) of methanol. The other main end uses are dimethyl terephthalate, methacrylates, methylamines (for resins, herbicides, and fungicides), methl halides (for silicones, tetramethyl lead, butyl rubbers, paint removers, photographic films, aerosol propellents, and degreasing compounds), acetic acid,

_____

*U.S. Federal Specification O-M-232e (July 20, 1968).

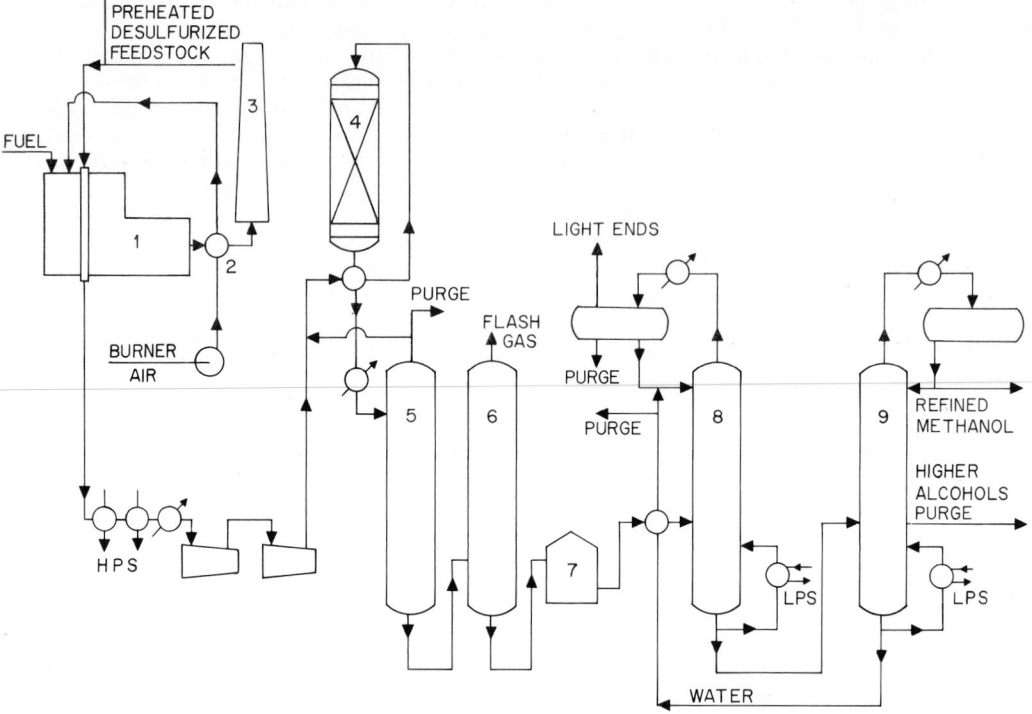

Fig. M-12. Low-pressure methanol-production process. (1) Burner and superheater, (2) air preheater, (3) stack, (4) methanol converter, (5) separator, (6) flash vessel, (7) crude storage, (8) topping column, (9) refining column. *HPS*, high-pressure steam; *LPS*, low-pressure steam. (*Imperial Chemical Industries, Ltd.*)

and solvents. Future wide use of methanol as a fuel may occur. See also **Amino acids.**

Commercial Production: The most recent advances in methanol synthesis are the low- and intermediate-pressure processes of the type shown in Fig. M-12. The synthesis step of this process* relies upon a copper-based catalyst, which gives good yields of methanol at pressures of 50 and 100 atm. These pressures are substantially below those of the 250–350 atm required by earlier processes. The high catalyst activity allows the synthesis reaction to take place at a relatively low temperature of 250–270°C. As a result, methanation is avoided, and by-product formation is lower, giving increased process efficiency.

The development of this low-pressure technology has caused a major reassessment of the economics of methanol production. The energy required to compress the synthesis gas from its production pressure to the synthesis unit is re-

*Developed by Imperial Chemical Industries, Ltd.

duced by a factor between 2 and 3. The lower synthesis pressure allows the exclusive use of centrifugal compressors in plants with capacities as low as 15 million gal/yr. Small producers find attractive the savings in investment, operating, and maintenance costs made possible by low-pressure operation. Plants range in capacity from 15 to 200 million gal/yr.

The low-pressure methanol process consists of three basic operations: (1) synthesis-gas preparation, (2) methanol synthesis, and (3) methanol purification.

Synthesis gas is prepared by the steam reforming or partial oxidation of a liquid or gaseous hydrocarbon feedstock or by direct combination of CO_2 with purified hydrogen-rich gases. See also **Reforming;** and **Synthesis gas.** Economic considerations usually favor the steam-reforming route for a naphtha or natural-gas feedstock. In this instance, desulfurized feedstock is preheated, mixed with superheated steam, and reacted over a conventional catalyst (normally nickel-based) in a multitubular reformer. The reformer usually is operated at between 15 and 30 atm and at a tube outlet temperature of 840–900°C. The reforming conditions are chosen to give the most economic overall production costs. Methane slip (amount of unconverted methane) usually is greater than for conventional high-pressure synthesis processes, since the cost of compressing the additional methane is less significant with the low-pressure process. With a naphtha feedstock, an almost exact stoichiometric ratio of carbon oxides to hydrogen in the synthesis gas is achieved, but when natural gas is the feedstock, there is an inherent deficiency of carbon. Established practice for many years has been to add CO_2 from an external source in preparing a stoichiometric synthesis gas. Development of the low-pressure process has shown that this addition of CO_2 is not required and that, depending upon the cost of CO_2, production of methanol from natural-gas feedstock alone is economic.

After heat recovery and cooling, the synthesis gas is compressed to the required synthesis pressure and passed into the synthesis loop at the suction of a circulator. The circulator, which boosts the pressure of the circulating gases to make up the total loop pressure drop, also is a centrifugal machine. Feed-gas preheating is carried out by heat exchange with the hot gases leaving the converter. Synthesis takes place in a hot-wall converter over the low-pressure methanol-synthesis catalyst at 250–270°C. Temperature control of the converter is effected by injecting cold gas at appropriate levels in the catalyst bed, using specially developed distributors which provide excellent gas mixing while allowing free passage of the catalyst for easy charging and discharging. After leaving the converter and passing through the feed-gas preheater, the converted gases are cooled, and crude methanol is condensed and separated from the uncondensed gases, which are recycled with makeup synthesis gas to the converter. A continuous gas purge is taken from the synthesis loop in order to remove an accumulation of inert gases. This purge is recycled to the synthesis-gas preparation section as reformer fuel. The crude methanol is reduced in pressure before passing forward to the methanol-purification section, where methanol of the required purity is produced by conventional distillation methods.

—P. G. Robinson, *Davy Powergas Ltd.,*
London, England

Methyl Bromide (See **Bromine**.)

Methyl Chloride (See **Chlorine organics**; and **Silicones**.)

Methyl Esters [See **Alcohols, fatty** (via hydrogenation); and **Vegetable oils**.]

Methyl Ethyl Ketone (See **Ketones**.)

Methyl Group (See **Hydrocarbons**.)

Methyl Isobutyl Ketone (See **Ketones**.)

Methyl Methacrylates (See **Hydroxyalkyl acrylates and methacrylates**; and **Polymerization**.)

Methylene Chloride (See **Chlorine organics**.)

Methylene Dichloride (See **Chlorine organics**.)

Methylene Group (See **Hydrocarbons**.)

Methylethyl Lead (See **Chlorine organics**; **Petroleum**; and **Petroleum processing**.)

Methylfuran (See **Furan group**.)

Methylisobutylcarbinol (See **Ketones**.)

Micelles (See **Surfactants**.)

Microbiology (See **Amino acids**; **Antibiotics**; and **Polymerization**.)

Microcline (See **Potassium**.)

Microcrystalline Paraffins (See **Waxes**.)

Microfiltration (See **Membrane filters**.)

Milk (See **Amino acids**; **Homogenization**; **Membrane filters**; **Pasteurization**; and **Peptides and proteins**.)

Miller Process (See **Gold**.)

Milling (See **Size reduction**; and **Soaps**.)

Mineral Beneficiation (See **Beneficiation, ore**; and **Flotation**.)

Mineral Fibers (See **Fibers**.)

Mineral Oils Mineral Oil is the title currently used in the official compendia (U.S. Pharmacopeia, National Formulary) for products formerly known as White Mineral Oil, Liquid Petrolatum, or Liquid Paraffin. The products are petroleum distillates, of specified viscosity and specific gravity ranges, which have been exhaustively refined, usually with oleum or SO_3, to remove reactive compounds. Each compendium has its own monograph describing Mineral Oil. The monographs are similar in their definitions of the products, tests for purity, and the effect of solid paraffins. They differ in the ranges allowed for viscosity and specific gravity. The capitalized term, Mineral Oil, is used here to designate products that meet the purity and quality standards of the official compendia and of the Food Additives Regulation CFR 121.1146.

There has long been considerable confusion in the nomenclature of many petroleum products, including Mineral Oil. The common term mineral oil can refer to any oil of mineral origin, and the indefinite word oil signifies only that a substance is liquid at some prevailing temperature and lacks miscibility with water. The capitalized form, Mineral Oil, used in the official compendia, is the result of a change in federal law calling for use of the simplest generic names for drugs. In USP XVI, and before, the title of the monograph was Liquid Petrolatum, with synonyms including White Mineral Oil. White Mineral Oil and White Oil have been the most commonly used names for many years. Even these terms are misleading, however, because the product is not white but colorless. Some oils of other than mineral origin have been called White Oils, and some mineral oils that are nearly colorless but not of medicinal quality have also been referred to as White Oils.

The official name for the pharmaceutical or drug product as defined by the official compendia is Mineral Oil, USP, or Light Mineral Oil, NF. Undoubtedly the name White Mineral Oil will continue to be widely used. It is the title used in Food Additives Regulations CFR 121.1146 to designate the highest quality product of this type. The White Mineral Oil of CFR 121.1146 is the only type permitted for direct use in food processing.

Classification: One simple classification breaks these products down into (1) Light or NF Mineral Oils with viscosity at 100°F (38°C) below 37 cSt and specific gravity at 25°C/25°C between 0.818 and 0.880; and (2) Heavy or Extra Heavy USP Mineral Oils with viscosity at 100°F above 38.1 cSt and specific gravity at 25°C/25°C between 0.845 and 0.905. A classification based on com-

position is more useful but more difficult to make. Such a classification attempts to designate a Mineral Oil as naphthenic or paraffinic. No commercially produced Mineral Oil is either entirely naphthenic or entirely paraffinic. All Mineral Oils are mixtures in varying proportions of (1) naphthenic or saturated-ring hydrocarbons, either a single ring or two or more rings fused together, (2) isoparaffinic, or branched open-chain, hydrocarbons, and (3) normal, or straight-chain, paraffinic hydrocarbons. While there are isoparaffinic and normal paraffinic molecules in all Mineral Oils, certainly the percentage of normal paraffinic molecules is low. Most molecules contain both naphthene rings and normal or isoparaffinic side chains. Thus when reference is made to a naphthenic or paraffinic type of Mineral Oil, it means only that the oil is more naphthenic or more paraffinic than another, reference oil or that the Mineral Oil is on one side or the other of some arbitrary dividing line based on composition data or on some physical or chemical characteristics. In earlier work such classifications were based almost entirely on empirical relationships such as those of Waterman, Van Nes, Van Western, and many others. Today information on composition can be obtained more directly by analyzing samples in the mass spectrometer or nuclear-magnetic-resonance spectrometer. The information obtained, however, is still subject to interpretation of the data.

Comparisons of Mineral Oils of equivalent viscosities show the paraffinic type to be lower in specific gravity, refractive index, and volatility and higher in average molecular weight than the corresponding naphthenic type. For many years most Mineral Oils of approximately 100 Saybolt Seconds Universal (SSU) at 100°F and lower viscosity were manufactured from paraffinic-type distillates because they were considered more desirable than the naphthenic type, especially from an odor and taste standpoint. At the same time the higher-viscosity Mineral Oils were made from naphthenic-type distillates because it was uneconomic to process the higher-viscosity paraffinic distillates due to formation of very viscous sludge and consequent slow and inefficient production. More recently, with new technology, oils of both types are available over a rather wide viscosity range. See Table M-12.

A naphthenic type of Mineral Oil may be preferred over a paraffinic type of similar viscosity where aqueous emulsions are involved because the higher specific gravity may aid in forming a more

TABLE M-12. *Comparison of Typical Inspections of Naphthenic and Paraffinic-Type Mineral Oils of Equivalent Viscosities**

	Naphthenic Type			Paraffinic Type		
	1	2	3	1	2	3
Viscosity, SSU:						
At 100°F	129	184	212	128	189	210
210°F	40.7	44.2	46.3	41.9	45.6	48.2
Specific gravity (60°F/60°F)	0.8762	0.8778	0.8735	0.8550	0.8660	0.8602
Pour point, ASTM, °F	< -25	< -25	. . .	10		
Flash point, ASTM, °F	380	375	420	410	440	460
Refractive index, n_D^{20}	1.4762	1.4770	1.4755	1.4680	1.4731	1.4724
Waterman Analysis:						
Carbon atoms in naphthene rings, %	43.0	42.5	36.0	28.5	31.0	24.0
Average number of naphthene rings	2.6	2.9	2.6	2.0	2.5	2.0
Carbon atoms in paraffinic						
structure, %	57.0	57.5	64.0	71.5	69.0	76.0
Average molecular weight	373	408	435	422	442	481

*Oils compared have comparable viscosities at 100°F.

stable emulsion, especially if the emulsion is in critical balance. With modern technology, however, and the extensive variety of emulsifiers available, the specific gravity of the Mineral Oil need not be a factor in the preparation of the desired emulsions. In other applications a paraffinic type Mineral Oil may be preferred over a naphthenic type of similar viscosity because of its lower volatility. The paraffinic-type Mineral Oils have been considered to have a more pleasant feel on the skin than the naphthenic type. Naphthenic Mineral Oils show a somewhat wider or higher range of solvency than corresponding paraffinic types.

Production: Mineral oils are manufactured from distillate fractions from all types of crude petroleum. Fractions from paraffin-base and mixed-base (mid-continent) crudes used for processing are the *neutral oils* obtained from the wax-distillate portions of the crudes by removing the paraffin-type wax which is present. The neutral oils commonly used extend from about 50 SSU at 100°F viscosity to 300 SSU and higher. They often are solvent-extracted as well as dewaxed. The oil selected is usually treated in a batch operation carried out in a tank with a conical bottom called an *agitator,* where small increments of oleum (about 10% by weight of the oil) are added as *dumps.* After a series of dumps, or a *round,* the reaction products are drawn off and the treated oil is neutralized with soda ash or caustic. In the treatment some of the oleum or SO_3 adds to the aromatic compounds in the oil to form sulfonic acids. Neutralization of these acids yields the oil-soluble petroleum sulfonates, which are extracted from the oil with an aqueous solution of isopropyl or other alcohol and then further refined into an article of commerce. After the oil has received the required amount of oleum treatment, and after the last of the sulfonate has been extracted, moisture and other volatile products are removed by heating and blowing dry steam or air through the oil. Finally the oil is contacted with, or percolated through, an absorbent such as bauxite or fuller's earth to give a Mineral Oil of medicinal quality. In modifications of the described procedure, SO_3 may be used in the place of oleum, or the treatment with oleum may be carried out on a continuous basis. Also hydrofining or hydrogenation may be used to change the reactive substances in the charge stock or partially or wholly remove them. By saturating the unsaturated and aromatic compounds in the oil rather than removing them, hydrogenation might be expected to give Mineral Oils a higher than usual specific gravity.

Properties of these Mineral Oils, e.g., minimum chemical reactivity and lack of odor, taste, and color, along with the inherent properties of oils lead to a myriad of uses, beginning in the pharmaceutical and cosmetic industries and spreading into a wide range of industrial applications. Industrial uses include food processing, e.g., lubrication of dough dividers in bakeries; plastics, as a secondary plasticizer in styrene polymers; textiles, as a base for synthetic-fiber lubricants; and many others.

REFERENCES

Fiero, George W.: Purity of White Mineral Oil, *Ann. Allergy,* May 1965.

Franks, Arthur J.: White Mineral Oil and Petrolatums in Cosmetics, *Soap, Perfum. Cosmet.,* March-April, 1964.

Meyer, Erich: "White Mineral Oil, Petrolatum and Related Products," Chemical Publishing, New York, 1968.

Waterman, H. I.: "Correlation between Physical Constants and Chemical Structure," Elsevier, Amsterdam, 1958.

—Charles Steenbergen, *PENRECO Inc., Butler, Pa.*

Mining   (Consult the **Subject Index.**)

Mischmetal   (See **Cerium; Lanthanum; Neodymium;** and **Rare-earth elements and metals.**)

Mixing, Fluids   Processes frequently require mixing two or more liquids or gases. Equipment designs are markedly determined by the phases of the materials involved, there being no universal approach to dispersion operations. The terms *mixing, blending, commingling,* and *compounding* essentially are synonyms for fundamentally the same processing objective, namely that of *dispersing* one material throughout another.

Liquid-Liquid Mixing Systems. Whether the liquids involved in these systems are miscible or immiscible has a decisive bearing on equipment design. Major applications where liquids must be mixed include (1) blending of liquid products, as exemplified by the blending of various liquid fuels to obtain end products of desired specifications, (2) liquid-liquid extraction systems, (3) the mixing of liquids to bring about chemical reactions where the liquids per se may be the reactants or where their solutes may be the reagents, and (4) in the preparation of emulsions in which one liquid is dispersed throughout another in such finely divided form that later separation does not occur.

Line Blending: Particularly for the blending of miscible liquids, centrifugal pumps in which two liquids are fed to the suction side of the pump find wide use. Digitally controlled systems are used in which two or more liquid ingredients are carefully flow-controlled in accordance with specifications governing proportions, which sometimes vary from one customer to the next. If there are numerous specifications to meet, a batch system is used. If the blending involves a standard mixture for sale or use within the plant in large quantity, a continuous system may be preferred. Continuous systems are used where the blend is pumped directly into a pipeline. If the blended liquids go to a blending-storage tank, and if there is any tendency for the liquids to separate in time, some form of slow mechanical stirring apparatus or pneumatic agitation is provided.

If the mixing action of the pump may be insufficient, one of numerous designs of jet mixers, injectors, or orifices and mixing nozzles may be used. Jet mixers depend upon the impingement of one liquid on the other to obtain a dispersion. One liquid is pumped through a small nozzle or orifice into the flowing stream of the other liquid. In an injector, the flow of one liquid is induced by the flow of the other to produce an aspirating effect. In the orifice mixer, both liquids are pumped through constrictions in a pipe, the pressure drop of which is partly utilized to create the dispersion. As many as 20 orifice plates may be used. Valves, too, may be considered adjustable orifice mixers.

Thus, with the exception of the instrumentation and control system, it is evident that for many liquid-liquid blending problems, existing process equipment (tanks, pumps, and valves) can be configured into an effective mixing system, as contrasted with the need for specialized vessels and equipment for other phase dispersions. Continuous in-line or batch blending systems to handle two or more components are readily engineered by using regular process pumps with special flowmeters and valves. Flowmeters are available which provide digital electric signals which can be fed to relatively simple counter-computers to maintain any desired ratio between various input flows. If sufficient turbulence is not present in the line, special mixing devices of the type previously mentioned may be required; or the flows from the electronic blending system may be piped to a blending tank equipped with a paddle, air agitation, or circulating pump to effect the additional mixing required. A fully automated electronic battery of quantity control valves and pulse-generating equipment with temperature compensation (to correct for mass flow because of density changes with temperature) is shown in Fig. M-13. The proper proportioning of two or more liquids to a blending system also may be effected through the use of volumetric metering pumps, which are particularly useful for adding small quantities of one liquid into a large volume of a main fluid stream. In such cases, the main stream is measured by an orifice-plate, venturi, positive-displacement, or other appropriate flowmeter, which generates a control signal to throttle

Fig. M-13. Fully automatic electronic liquid-blending system. (*Rockwell.*)

the action of the metering pump. Or, in an open-loop arrangement, several metering pumps may be preset to given rates, and so long as the pressure on the suction and discharge side of the metering pumps is not altered, the desired proportions of fluids will be maintained.

Homogenizers: Most homogenizers function by passing the product under pressure between closely clearing but relatively fixed surfaces. The high velocity, hydraulic shear, pressure release, and impact rend the dispersed phase into a very fine state of subdivision on the order of 1 μm diam. See also **Homogenization.**

Liquid-Solid Mixing Systems. Two types of processing occur where liquids and solids must be mixed: (1) dissolving solids to form solutions, sometimes requiring considerable agitation because of relatively low solubilities, and (2) preparing slurries and suspensions. Generally, the same kinds of mixing equipment can be used for both functions. The mixing mechanism and the vessel or tank in which mixing takes place should be engineered as a total system. Some of the criteria to be considered include providing uniformity of vessel content, particularly where chemical reactions are concerned, and producing a final mixture that can be pumped easily.

Impeller Mixers: There are two main classes of impellers, the *axial-flow* impeller, in which the blade makes an angle of less than 90° with the mixer axis, several configurations of which are shown in Fig. M-14, and *radial-flow impellers* which have blades parallel to the axis of the drive shaft. The smaller configurations with multiple blades are referred to as *turbines,* whereas the larger, slower impellers may be called *paddles.* The anchor-type impeller (Fig. M-15) also is of this type. Some radial-flow impellers are shown in Fig. M-16. Baffles frequently are installed in mixing tanks to improve the process. The power and mixing mechanism may be mounted permanently to the mixing vessel (top, bottom, or side entry), or the entire mixing mechanism may be portable. Thus, there is a large range of design combinations and capacities, from laboratory mixers to tanks that hold many thousands of gallons.

Injection Mixers: Although not suitable for suspending free-settling solids, introducing compressed air to effect a mixing action can be used in the production of high-density slurries, where settling rates are relatively slow. The Pachuca tank exemplifies an air-mixing system. Air is injected in the center of a draft tube, causing the pulp to rise in the tube and to flow down in the annular space around the tube. The arrangement is shown schematically in Fig. M-17. Some Pachuca tanks utilize an impeller at the base of the draft tube to provide circulation instead of using air.

Gas-in-Liquid Dispersions. Dispersion of a gas in

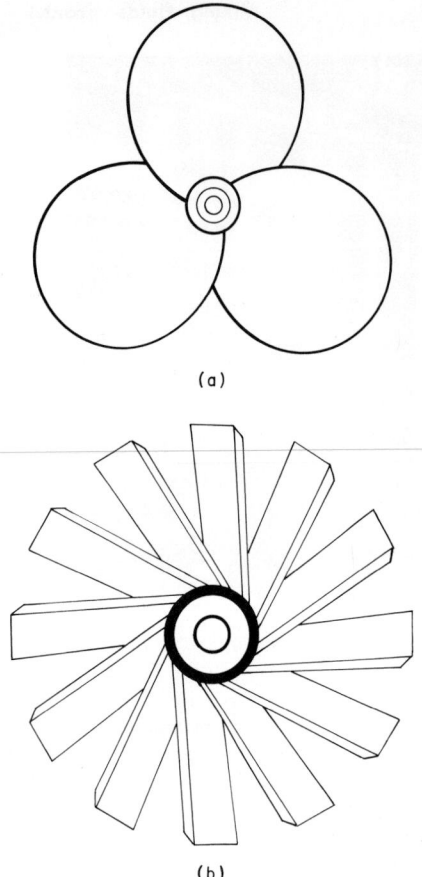

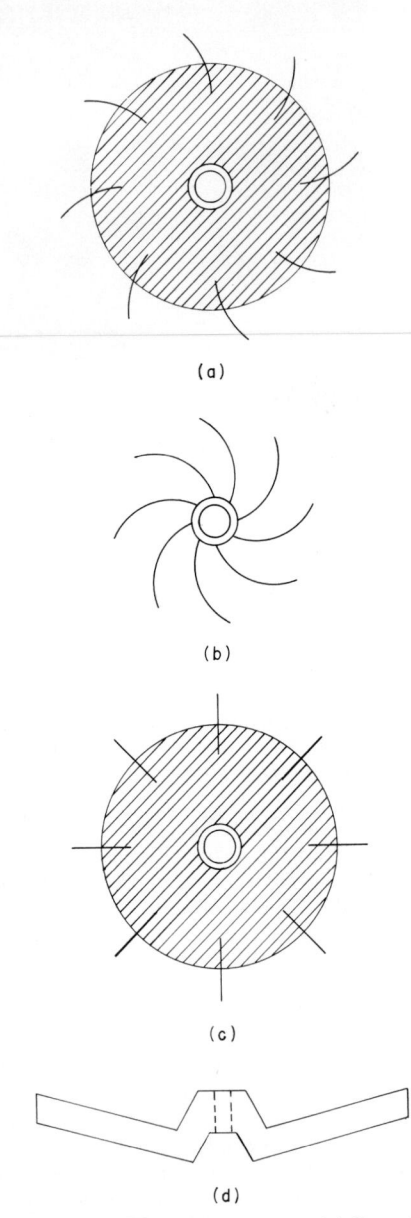

(a)

Fig. M-14. Types of axial-flow impellers. (a) Marine type, (b) fan turbine.

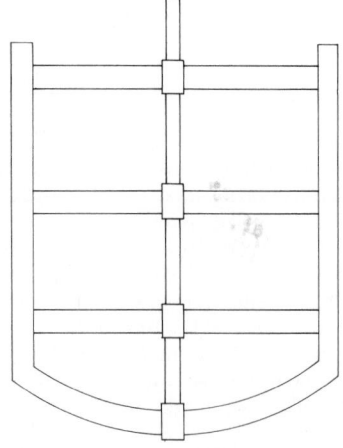

Fig. M-16. Types of radial-flow impellers. (a) Curved-blade turbine, (b) spiral backswept turbine, (c) flat-blade turbine, (d) paddle.

Fig. M-15. Anchor-type impeller used in horseshoe mixer.

740

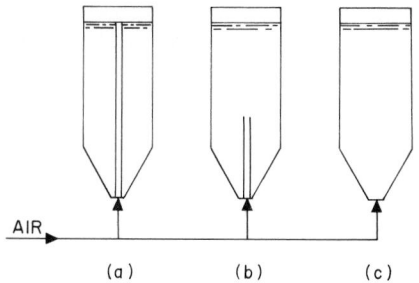

AIR

(a)　　　(b)　　　(c)

Fig. M-17. Arrangements of a Pachuca tank. (a) Full-center column, (b) stub-column, (c) free air-lift tank.

a liquid usually has one or more of the following objectives: (1) agitation, where bubbling air into a process vessel causes a mixing action, thus obviating the need for mechanical means (this method is well suited to liquids of low volatility and not affected by air); (2) foam production, as in the production of foam plastics, elastomers, and glass; fire-fighting froths; and various food products; and (3) gas-liquid contacting, where a solution of the gas is desired or a gas is to be reacted chemically with a solution.

Spargers: These devices essentially are perforated pipes or plates with orifices usually ranging from $\frac{1}{8}$ to $\frac{1}{2}$ in. diam. They are widely used as agitators. For obtaining much finer bubble formation, porous septa in the form of plates, tubes, or disks may be used. These devices are well suited for gas absorption, as in sewage and waste aeration tanks and sulfite, pulp and paper, and pharmaceutical waste-treatment systems.

Precipitation: If a gas is precipitated from a supersaturated solution, the result is usually a fine dispersion of bubbles throughout the liquid. This approach is used in the making of cellular rubber. Uncured natural or synthetic rubber is heated and saturated with an inert gas at very high pressure. Before vulcanization, the pressure is released, allowing liberation of the gas dissolved in the rubber to form sponge rubber. Pressure saturation, followed by flashing and bubble precipitation, is also used in some flotation and thickening units.

Generation: Essentially, this is a chemical method in which a finely divided, suspended material (sometimes called the *blowing agent*) is decomposed to form a gas which produces well-dispersed bubbles. This method is used in producing cellular elastomers and plastics. Sodium and ammonium bicarbonate, ammonium nitrate, calcium carbonate, and certain organic compounds make good blowing agents. This action is akin to that of leavening agents used by bakers.

Dispersers: A mechanical disperser induces its own air supply to the mixture.

Liquid-in-Gas Dispersions. Many industrial requirements for dispersing a liquid in a gas (often air) include atomizing fuels for combustion, as in oil burners, air washing, humidifying, and spraying cooling water in towers and ponds, and spraying paint and other finishes. Devices for spraying insecticides and other agricultural chemicals require similar equipment. In processing, liquid-in-gas dispersions are important in spray drying, in spray towers for washing and absorption, in scrubbers for gas-liquid contacting, and for distributing liquids over packed beds.

Depending essentially upon the size of the liquid drop dispersed, the dispersion may be classified as free drops, a spray, a mist, or a fog. Generally, a spray is a mechanically produced dispersion comprising relatively large drops and usually quite unstable. Mists and fogs, normally associated with atmospheric phenomena, are usually formed by condensation rather than atomization. A small discrete portion of liquid, essentially spheroidal in shape, is termed a *drop.* Drops in the μm range and below are known as *droplets.* Drop formation and stability depend upon temperature, pressure, and composition of the liquid.

Nozzles: The nozzle is the predominant device for creating liquid-in-gas dispersions and may be one of three principal types: (1) pressure nozzles, which break the liquid into drops by impacting the liquid under pressure with the atmosphere or another jet or fixed plate; (2) the rotating nozzle or spinning atomizer, in which the fluid at low pressure is fed to the center of a disk or cup that rotates at high rates and centrifugal force breaks the fluid into drops; and (3) gas-atomizing nozzles, in which a high-velocity gas impinges on the fluid to break it into drops. There are many nozzle designs. Hollow-cone nozzles are the most commonly used. Considering all design variations of hollow-cone nozzles, the orifices range from 0.02 to 2 in. diam with discharge rates ranging from 0.01 gal/min or less to 200 gal/min.

Gas-Gas Dispersions. Gases tend to mix readily without the need for special equipment. Usually, the fans, blowers, compressors, and other equipment required for handling gases also provides excellent means for effecting good mixing of two or more gases. Where mixing must be thorough and fast, as in the case of an oxyhydrogen torch, jet mixers are used. In gas-gas systems, the greater concern usually relates to proportioning a quantity of one gas to that of another. This requires precision flow measurement and control. For example, in most inert-gas generators, air and gas are mixed in a blower, the proportion of gas and air being controlled automatically by a mixing valve located at the suction of the blower.

Fluid Mechanics of Mixing. Any circulation mixer creates two conditions within the mixing vessel, circulation of the fluid and fluid shear. Impeller mixers also supply energy to circulate the fluid. The power P consumed by an impeller is related to the volumetric circulation rate Q (also called pumping capacity) and the velocity head H from the impeller by

$$P = \rho Q H \tag{1}$$

where ρ is the fluid density. The pumping capacity of an impeller is defined as the volumetric flow rate normal to the impeller discharge area. The pumping capacities of a geometrically similar series of impellers are given by

$$Q \propto N D_a^3 \tag{2}$$

where N is the speed and D_a the impeller diameter. The impeller head H is proportional to the square of the velocity of the fluid leaving the impeller blades, which in turn is related to the pumping capacity Q. Equation (1) shows that many different combinations of impeller speed and impeller diameter will give the same power consumption. For example, in a baffled tank, the power consumed by any type of impeller turning at high speed in a low-viscosity liquid is given by

$$P \propto N^3 D_a^5 \tag{3}$$

If this is combined with Eqs. (1) and (2), the ratio of pumping capacity to impeller head at constant power consumption becomes

$$\frac{Q}{H} \propto D_a^{8/3} \tag{4}$$

This leads to one of the basic principles of impeller operation. A large impeller running at a slow speed gives a large circulating capacity and a low fluid shear rate, while a small impeller running at high speed gives a high fluid shear rate and a low total circulating capacity. The power input may be distributed in different ways by the choice of the ratio of impeller size to tank size.

The fluid velocity leaving the tips of the impeller establishes a fluid shear rate in the tank.

This fluid shear rate is related to a fluid shear stress. Some processes are especially sensitive to the actual velocity of the fluid, in terms of both its direction and magnitude, while other systems are more influenced by the fluid shear stress in the system. It is important to distinguish between these two conditions.

If a process depends upon the pumping capacity of the mixer and the average fluid shear rate in the vessel, the ratio of flow to fluid shear rate is usually controlling. In these systems, there usually is an optimum ratio of flow to fluid shear rate. The shear rate (or velocity gradient) produces shear stresses throughout the fluid in the tank. With nonnewtonian fluids in particular, the fluid shear rate is a key factor. With laminar flow, the shear stress can be calculated from the shear rate if the viscosity is known. With turbulent flow, however, this is not true. Turbulent shear stress results from the behavior of transient random eddies, including large-scale eddies which decay to small eddies or fluctuations. The scale of the large eddies depends on the tank size and is different in different systems. Small eddies, on the other hand, appear to be similar in small and large systems. Small eddies dissipate energy primarily through viscous shear. Since their behavior is almost independent of tank size, processes which depend on this effect have similar characteristics in both small and large tanks. The shear rate in the fluid is much higher near the impeller than it is near the tank wall. The difference is greater in the large tanks than it is in the small ones.

The chief criterion of the type of flow is the tank Reynolds number, defined by

$$N_{\mathrm{Re}} = \frac{N D_a^2 \rho}{\mu} \tag{5}$$

where N = speed
D_a = impeller diameter
ρ = fluid density
μ = absolute viscosity

Since N_{Re} is dimensionless, it is important to use the same unit of time in calculating both N and μ. Flow in the tank is turbulent whenever N_{Re} is greater than 10,000. Thus, viscosity alone is not a valid indication of the type of flow to be expected. Between Reynolds numbers of 10,000 and approximately 10 is a transition range, in which flow is neither laminar nor fully turbulent. At Reynolds numbers below 10, flow is laminar.

Mixing, Pastes Pastes include creams, greases, doughs, pulps, and muds and are not necessarily sticky or tacky. Pastes may be mixtures of solids and liquids or of two or more heavy liquids. Paste-mixing equipment is used widely in the rubber, paint, baking, candy, glue, ceramic, adhesives, grease, cosmetics, and chemical industries. Colloid chemistry enters into the preparation of many of these products.

High-velocity impellers are not applicable to most paste-mixing operations because the impellers simply bore a hole in the mass without mixing. Because of the flow properties, impact and shear are required very close to the mixer blades. Thus, many paste mixers feature small clearances between the rotating blades and the body of the mixer and baffle bars when used. Paste properties usually dictate rather slow shearing action. If, during the mixing operation, the material can be folded, stretched, and compressed, a more uniform paste mix will usually result.

Change-Can Mixers: These devices are appropriately named because the ingredients are contained in cans which are portable and economic to use. The mixing head is permanently located. Mixing is effected in a batch manner, one can at a time. There are two major designs: (1) the can is fastened into a fixed position, and the mixer blades turn, usually in a combined rotary and planetary motion to continuously contact the inside walls of the can as well as the center; (2) the blades are fixed, and the can is rotated by a turntable, as shown in Fig. M-18. The two arrangements can be combined. Together with shakers, these mixers are widely used for paints.

Gate Mixers: In these devices, a flat, rotating structure of horizontal and vertical bars cuts the paste at different levels and at the tank wall, where stationary bars sometimes are fastened to give points of intensive shear. Slow speed is required to prevent the entire mass from rotating within the tank. Among materials processed in gate mixers are coatings, starch pastes, paints, and sizes.

Kneaders are used for pastes as well as for solids-solids mixes and are described under **Mixing and blending, solids-solids,** as are *muller mixers* and *pug mills.* Crutchers are described under **Soap.**

Banbury Mixer: This mixer (Fig. M-19) also produces a kneading action and is used predominantly in the rubber and plastics industries. Much friction is produced in the confined space,

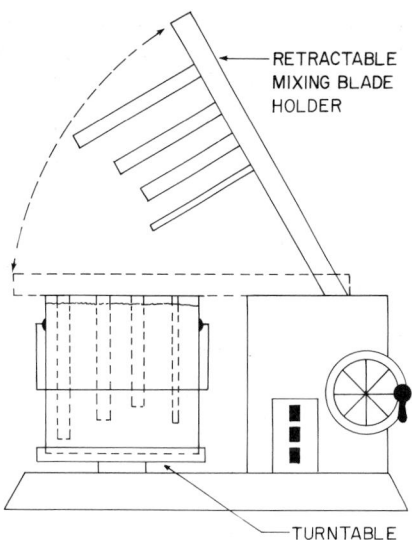

RETRACTABLE MIXING BLADE HOLDER

TURNTABLE

Fig. M-18. Schematic representation of change-can mixer with rotating table and retractable, fixed mixing blades.

and, therefore, cooling is required. See also **Rubber, natural.**

Roll Mills: In these devices, two or more rolls turn at different speeds, between which the materials to be mixed are passed. A kneading, tearing, stretching, folding, shearing action is produced. Roll mills are well suited for applications where an intimate mixture of a solid with a liquid is desired, as in printing-ink manufacture. The units also are used for the heaviest type of work in which mixing is possible, such as mixing fillers into rubber and blending rubber stocks. The rolls are often corrugated to afford a better grip on the material.

Mixer-Extruders: In the type of machine shown in Fig. M-20, material is fed as a dry solid, is fluxed in the barrel to form a paste, and then resolidified at the discharge. Action in the barrel is one of shearing, rubbing, and kneading. One continuous screw or two screws rotate in the closely fitting barrel. The work of the screw is augmented by forcing the material through breaker screens and around breaker disks just before the material is forced out through the nozzle. Because the mixing zone is of small cross section, and because the material motion is essentially in one direction, very little volume blending occurs. Thus, the ingredients must be preblended

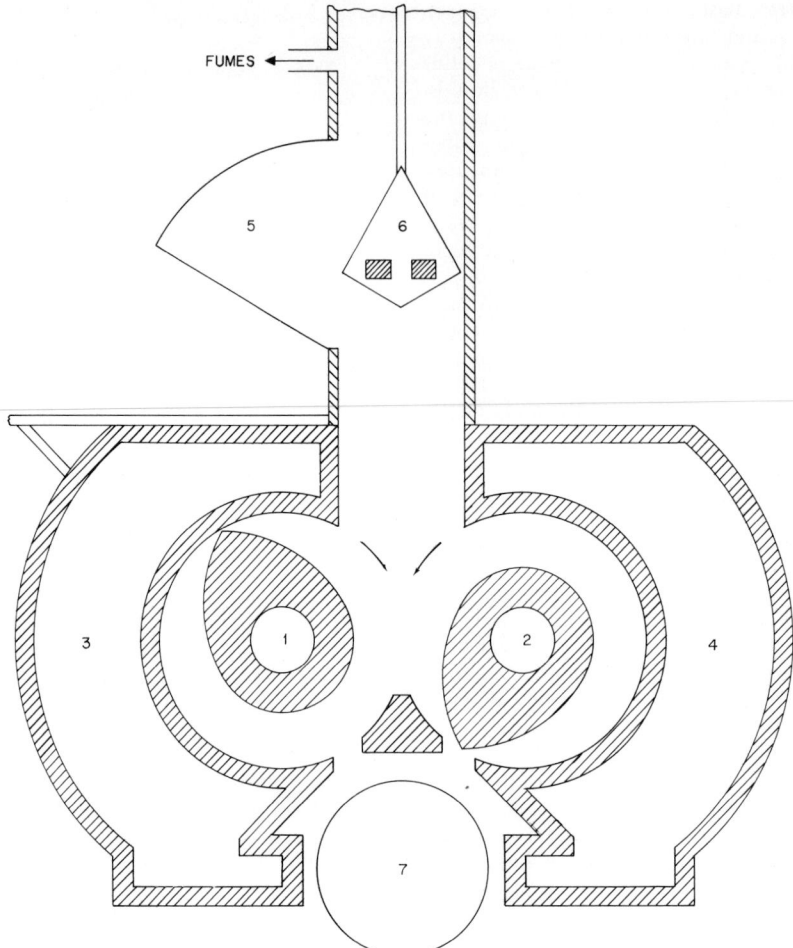

FUMES

Fig. M-19. Highly schematic sectional representation of Banbury mixer. (1), (2) Specially shaped rotors cored for cooling water or steam, (3), (4) spray-cooled chambers, (5) feed hopper, (6) floating weight, (7) sliding discharge door. (*Farrell Corporation.*)

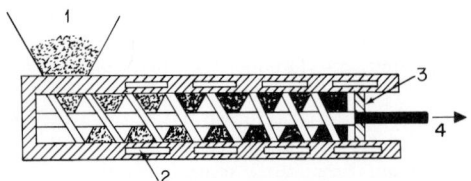

Fig. M-20. Mixer-extruder. (1) Charge stock, (2) heating or cooling chambers in extruder jacket, (3) die, (4) extruded product.

before entering the machine. These machines are used for the extrusion of soft chemical and food mixes which do not require fluxing, as well as for the extrusion of hard plastics, some of which must be fluxed at temperatures above 400°F (204°C). Wires can be covered and shapes of intricate cross section can be produced. Plastic resins also are blended in extruders to form pellets for later press and injection molding.

Mixing and Blending, Solids-Solids There is a tendency to consider the terms *mixing* and *blending* as interchangeable. Mixing may be defined as the commingling or dispersion of two or more ingredients of varying proportions so that the resulting product or mixture shows unit portions or fractions having the same percentages of each ingredient as charged. Blending, on the other hand, is best described as a homogenization of a product to provide uniformity and to reduce variations in such process variables as bulk density, viscosity, color, and melt index. Blending usually is handled in large volumes prior to packaging or bulk load-out for shipment. There is an increasing demand for greater control and better accuracy of mixing and hence more sophisticated procedures for analyzing the resulting product. No one mixer or blender is best for all applications.

Solid mixing and blending operations generally are classified into two broad operations, *batch* or *continuous*. In batch operation, the ingredients are charged to a mixing vessel in some predetermined order and formulation, held within the mixing vessel under agitation for a period known as the *retention time,* then discharged after the desired degree of uniformity is achieved. During mixing, no portion of the charge is discharged from the unit. Continuous mixing or blending allows for the operation to be carried on-stream with the retention time determined by the size of the unit and the length of time required for given particles to pass from inlet to discharge. In continuous operations, accuracy of mixing depends upon accuracy of the feeding or proportioning equipment used to charge the several ingredients to the unit.

Unmixing: The same random motion that is imparted to particles during mixing also serves to make unmixing possible. This can be explained in two ways: (1) the electrostatic charges on the surface of the particles are unbalanced by the frictional effects of the particles as they rub against one another. This introduces intensified charges on certain particles. (2) This same motion gives the particles sufficient mobility for them to be free to segregate themselves. Once a uniform distribution is achieved, no further benefit can be derived from further random motion of the particles. From this point on, if the particles are susceptible to unmixing from electrostatic-charge accumulation, further motion will only undo what already has been accomplished in the mixer. The problem arises of knowing when this optimum time occurs and when to cease mixing. This is determined from experience with a particular formula and equipment, coupled with analytical testing.

As demands for quality control become more rigid, the methods of checking uniformity and accuracy of mixing and blending are becoming more complex. At one time, the accepted standard was a visual check, e.g., the spatula or smear test. The present tendency is to express the uniformity in terms of standard deviation, or statistical measure of how the concentration of one constituent varies from the true concentration. Testing procedures include chemical analysis, spectrographic analysis, microscopic examination, and even the use of radioactive tracers.

Selecting Equipment: Many factors influence the selection of the optimum mixer or blender. Information required by the manufacturer of this equipment to determine the most effective design for a given application is summarized in Table M-13.

Basic Types of Equipment. Because of the multitude of different mixer and blender designs available, details on all types cannot be given. A few representative types are described.

Rotary Drum: This unit operates on the principle of tumbling to achieve mixing of the dry materials. Drum-type mixers have the axis of rotation horizontal to the center of the drum.

TABLE M-13. *Checklist for Selecting Mixing and Blending Equipment*

Feeding
Number and identification of ingredients
Percentage of each
How ingredients are fed to mixer

Material Characteristics	
Bulk density of each ingredient	
Particle-size distribution	
Particle shape or flowability	
Moisture or other liquid content	
Special properties:	
Heat sensitivity	Explosiveness
Abrasiveness	Fibrous nature
Friability	Tendency to smear
Volatility	Presence of
Static electric charge	agglomerates

Mixing Rate	
Batch	*Continuous*
Batch size, ft^3	Rate, lb or ft^3/hr
Batches/hr	Retention time
Charging and discharging	required
time available	Method of feeding

TABLE M-13. *Checklist for Selecting Mixing and Blending Equipment* (*Continued*)

Heating or Cooling Required	
Product temperature required	Fluid temperature
	Fluid pressure
Inlet or feedstock temperature	ASME code required
Specific heat of each ingredient	
Type of heat-exchange fluid	

Discharge
Location and number of parts
Manual or air-operated
Height off floor

Product Specifications
Mixing or blending accuracy required
Control standards
Method of sampling
Statistical analysis to be used in evaluation

Materials of Construction
Carbon steel, stainless steel, or other alloys
Interior surface finish or polish
Weld grinding or finish
Special sanitation codes or requirements

Electric Current Characteristics
Voltage, frequency, and phase
Enclosures required (TEFC, explosion-proof)
Starting equipment (NEMA standard)

Special Considerations
Special inlets or covers
Vent connections
Special gaskets
Special shaft seals, i.e., air-purged
Liquid-addition metering and control equipment
Structural-steel supports or framework
Location of motor and drive
Type of drive guards (special local or state codes)

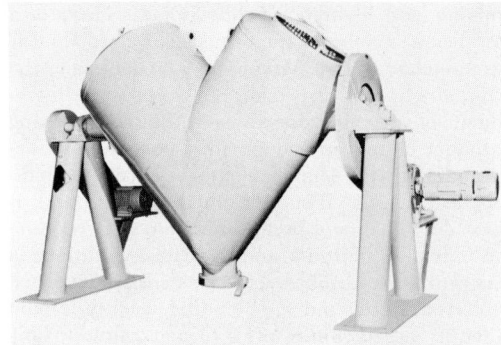

Fig. M-21. Twin-shell blender. (*The Patterson-Kelley Co., Inc.*)

Often the drum is provided with internal baffles or helical plates to improve the cross flow. This limitation of good cross flow along the axis is perhaps the chief disadvantage of this type of unit, and therefore it is not generally recommended for precise mixing or where cleanout is a problem. Sometimes inclining the drum on the axis improves the mixing action without necessitating baffles. Intake and discharge of material takes place through one opening in the end of the drum.

Double Cone: In this design, two cones are joined to a relatively short cylindrical section, and the axis of rotation is centrally located on the cylindrical portion. The cones normally have a 90° included angle or a 45° discharge angle, which is adequate for discharge of most materials. This type of blender gives a much better cross-sectional flow than the horizontal drum since the blender contains no flat spots and all motion is rolling. In addition there is an interfolding action of the material as it transfers from one cone to the other. It is general practice not to load these units to more than 65% of the total volumetric capacity; hence, the units must be significantly oversize to accomplish a given capacity. As the unit is normally free of baffles and other internal obstructions, it is fairly easy to clean.

Twin Shell: This unit, a variation of the tumbling mixer, is formed from a cylinder bisected at an angle and joined together to form a V shape (see Fig. M-21). It provides a nonsymmetrical shape about the axis of rotation, which is located to provide equal loading throughout a full rotation. This unit combines the blending action of the inclined cylinder with the intermeshing action that occurs when two inclined cylinders combine their flow. Many units are furnished with an intensifier bar, which breaks up minute agglomerates. The V shape permits the use of large access openings on each leg and allows for a central discharge port, completely discharging all contents and giving excellent cleanout.

Muller or Pan: Essentially this unit consists of a circular pan with wide wheels mounted on ad-

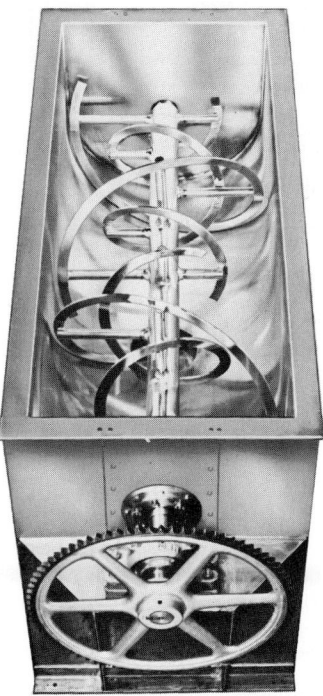

Fig. M-22. Horizontal ribbon mixer viewed from above. The gear drive is in the foreground, and the ribbon is mounted in a U trough. (*Sprout, Waldron & Co.*)

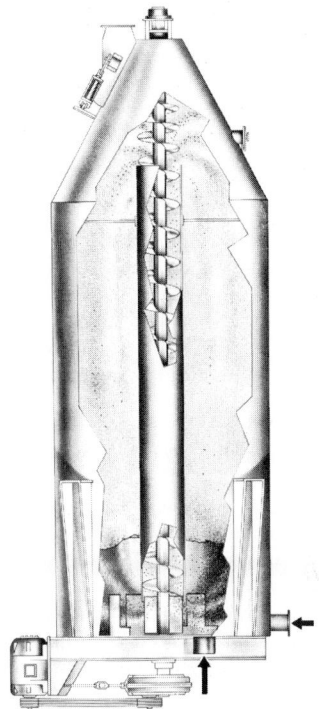

Fig. M-23. Vertical-screw mixer. (*Sprout, Waldron & Co.*)

justable axles connected to a central vertical drive shaft. Plows or scrapers fitting closely to the bottom of the pan direct the material into the path of the mulling wheels. The pan is stationary or is rotated in the direction opposite to the muller wheels. This action—actually a spatula device—has the added advantage of breaking up caked materials or incorporating oils or pigments into solids. It has been compared to the ancient mortar and pestle. Because units of this type do not provide for precision mixing, have high requirements, and result in a high degree of particle-size attrition, they are highly specialized equipment.

Horizonal Ribbon: This unit, probably the oldest and most widely used mixer in industry, consists of a stationary U-trough shell (cylindrical troughs also are quite common) which is usually 2 to 3 times as long as it is wide and which is fitted with a longitudinal shaft on which are mounted arms supporting a combination of spiral ribbons, paddles, or other configurations (see Fig. M-22). Good cross flow of material is achieved by having the

outer ribbon move the material in one direction and the inner one in the opposite direction. This arrangement also prevents material buildup on one end of the unit. Units may be jacketed for heating or cooling. With the addition of liquid-spray nozzles, they can be used for mixing small quantities of liquids or resins. The discharge opening may be at the center of the trough or at one of the ends. The end discharge has proved more effective in terms of mixing efficiency and time to achieve good mixing.

Mixing uniformity in this type of unit is very good, and mixing time is short to moderate, but degradation of particle size is greater than experienced in tumbling units. Cleanout is somewhat more difficult because of the many internal ribbons, ribbon arms, and other obstructions. Power requirements are somewhat higher than normal because the entire volume of material is constantly being agitated. However, the unit has greater versatility than any other unit available, which accounts for its great popularity.

Vertical Screw: This unit (Fig. M-23) consists

of a vertical tank with a hopper bottom in which a centrally located spiral screw picks material up from the bottom of the blending tank and discharges it from the top of the screw, cascading it across the entire cross section of the mixing tank. Here the material moves by gravity to the bottom of the vessel, where it is picked up and recirculated. Recirculation continues until the desired mixing is achieved. Units are relatively low in initial cost and have a distinct advantage where large volumes are required. They are particularly adaptable for the blending of pellets, granules, and other free-flowing materials. Materials which tend to bridge or cake should not be used in this type of unit. Power requirements are low in relation to the volume because energy is expended only in moving that quantity upward through the volume encompassed by the screw itself. The balance of the movement throughout the mixer is by gravity. Cleanout is difficult, and mixing uniformity is not as satisfactory as that provided by the ribbon blenders or any of the tumbling types.

In a variation of the screw blender, the rotating screw also orbits around the inside wall of the mixing vessel, thus moving material from bottom to top as well as around the periphery of the tank. Although this design offers a more intense mixing action, it requires increased maintenance because of the complexity of the drive.

Double-Arm Kneader, or Sigma Blade: This design is used where products are of high viscosity, such as plastics and rubber. The basic design consists of a rectangular trough with curved bottom, forming two half-cylinders and a saddle (Fig. M-24). The blades, usually sigma type, revolve toward each other and either overlap or rotate tangentially. This overlapping action provides a cutting as well as a kneading action.

Pug Mill: This unit consists of a longitudinal trough containing two parallel shafts with short heavy-duty paddles. As the material moves gradually from one end of the trough to the other, the paddles cut and knead the material. Materials handled usually are heavy clays and similar moist materials that require relatively high energy inputs.

Mixing by Air: Mixing is achieved with a mixing head located at the bottom of a vertical tank and containing a manifold, a series of nozzles, and a cone valve. Compressed air enters the mixing tank through the manifold and a specially designed Laval nozzle. Filters are installed to prevent fine material from passing through the noz-

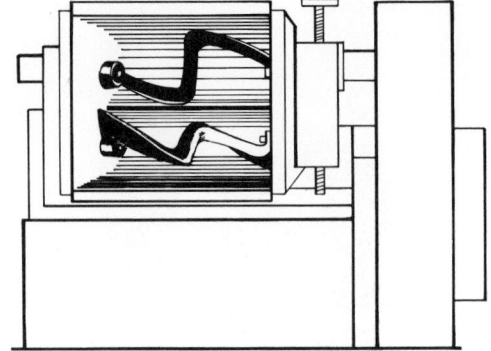

Fig. M-24. Highly schematic representation of a type of medium-weight kneader-mixer manufactured by the J. H. Day Co., Inc.

zles and into the manifold when the air is not flowing. The cone valve acts as a discharge gate to be opened at the completion of the mixing cycle and has been designed to prevent the segregation of the particles of varying size and density during discharge.

The mixing cycle usually consists of a series of intermittent air blasts. The air enters the mixer through the nozzles and the mixing head at a velocity near the speed of sound. The nozzles are positioned at an angle to cause the material to spiral upward along the side of the tank (see Fig. M-25). Materials in the center of the mixer flow down toward the mixing head to complete the circulation of material. By adjusting the duration of the air blasts, this vortex of material can be completely controlled. This unit offers excellent mixing uniformity in a very short time. Materials of varying densities and particle size can be mixed without segregation. The mixing interior is free from obstructions, and cleanout is complete. The air which is used for mixing can also be used to facilitate discharge, particularly with materials which otherwise tend to bridge. Initial cost is relatively high compared to other units of the same volume. Extremely short mixing times, however, permit use of smaller batch sizes, and initial cost can be put in perspective with other batch blenders.

Double-Agitator Mixer: This unit essentially is a variation of a pug mill. It is lighter in construction and can be used on a continuous basis. Paddles are adjustable to vary retention time and intensity of action. Units can be furnished without a solid bottom or with a drop bottom to facilitate cleanout.

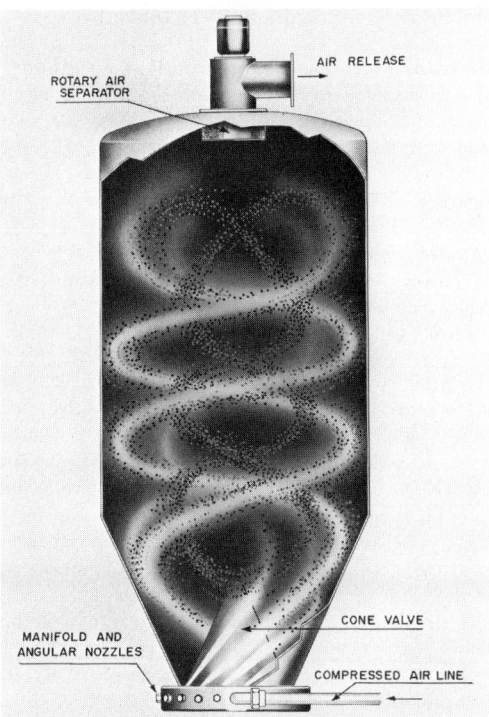

Fig. M-25. Cross section of air mixer. (*Sprout, Waldron & Co.*)

Fig. M-26. Continuous blender utilizing gravity tumbling and short residence time. (*The Patterson-Kelley Co., Inc.*)

Special Tumbling: A variation of the tumbling blender is the unit consisting of a series of V-shaped compartments, all smooth and accessible without screens, baffles, flights, or impellers (Fig. M-26). This is known as a *zigzag blender*. Moisture can be added through a dispersion bar in the first compartment, where liquids are injected through annular orifices. The material is moved back and forth at each V junction during each revolution of the blender. A portion of the material goes upstream, the balance down. A slight incline is generally most effective. The material discharges by gravity from the last compartment.

Continuous Screw: This unit is a variation of the standard screw conveyor which incorporates a cut and folded flight. The result is mixing in addition to conveying action. This is the least sophisticated of the mixer designs and generally is used where mixing uniformity is not of prime importance. Equipment investment is low, and the dual advantage of conveying and mixing sometimes eliminates a second step in the process and thus addi-

tional space and investment in special mixing equipment.

Attrition Mills: Often referred to as *disk mills* and used in conjunction with other premixing equipment for the intimate blending of powders, these mills consist of two vertical disks and horizontal shafts with adjustable disk clearance. Special plates permit intensive blending with a minimum of grinding action and hence low horsepower requirements. Basically these are the same as the mills used for size reduction but with a change in the configuration of the replaceable disks. Throughput rates per horsepower are high.

Gravity Blender: This unit consists of a main blending chamber with a series of downspouts in a spiral pattern around the periphery of the chamber. Material to be mixed is charged into the chamber, after which particles are withdrawn through the various downspouts in the center of the chamber into a collecting hopper beneath the main vessel. As the particles enter the downspouts at different levels and from different locations of the blending chamber simultaneously, the cumulative mix entering the bottom hopper at any instant represents a composite of the contents. In addition to the blending upon convergence, some mixing takes place within the bottom collector. The unit is adaptable for either batch or continuous operation. In a batch system the product is recycled until an acceptable blend is obtained. In a continuous operation the product may be conveyed continuously from the blender, either to a second and similar blender or to other process equipment. This unit offers low capital investment for the given mixing or blending volume. It uses only the horsepower required by the recirculation system (none for the blending itself). It is adaptable to vessels of extremely large volume. Existing storage vessels may be converted by the installation of downcomers and the secondary collecting chamber. For most applications where any significant degree of uniformity is required, two or three recycles are usually necessary. Beyond that, the improvement is not consistent with the added investment.

REFERENCES

Fischer, J. J.: Solid-Solid Blending, *Chem. Eng.*, Aug. 8, 1960.

Parker, Norman H.: Modern Theory and Practice on the Universal Operation of Mixing, *Chem. Eng.*, June 8, 1964.

Rathmell, C.: Granular Solids Mixing, *Chem. Eng. Prog.*, April 1960.

Staff Feature Report: Mixing Trends, *Chem. Process*, June 1965.

"Survey of Mixing Equipment in Chemical Processing Plants," Putman, Chicago, 1961.

—Kenneth R. Sterrett, *Sprout, Waldron & Company, Inc., Muncy, Pa.*

Modacrylics   (See **Acrylic fibers;** and **Fibers.**)

Moderators, Nuclear   (See **Nuclear power plants.**)

Mohair   (See **Fibers.**)

Molal Concentration   The molal concentration of a substance in solution is the number of moles of the substance contained in 1,000 g of *solvent.* For example, 0.5 molal solution of sodium chloride in water contains 0.5 × (gram molecular weight of NaCl = 58.454), or 29.227 g of salt in 1,000 g of water. See also **Molar concentration;** and **Normal concentration.**

Molar Concentration   The molar concentration of a substance in solution is the number of moles of the substance contained in 1,000 ml of *solution.* For example, 0.5 *M* solution of sodium chloride is made by putting 0.5 × (gram molecular weight of NaCl = 58.454), or 29.227 g of salt in a vessel, adding water, and thoroughly dissolving the NaCl until a total volume of 1,000 ml results. When solution density is known, molar concentration can be converted into molal concentration and vice versa. Molar solutions sometimes are called *formal* solutions, but the term is better avoided because of possible confusion with *normal* solutions. See also **Molal concentration;** and **Normal concentration.**

Molding   (Consult the **Subject Index.**)

Mole (or Mol)   A mole (or mol) is an amount of a substance, in specified mass units, equal to the molecular weight of that substance. A *gram mole* or *gram-molecular mass* of hydrogen, H_2, for example, weighs 2 × (atomic weight of H) in grams, or 2.016 g, a gram mole of carbon dioxide, CO_2, weighs 1 × (atomic weight of C) + 2 × (atomic weight of O) in grams, or 12.011 plus 31.998 = 44.009 g. A *pound mole* of ammonia, NH_3, weighs 1 × (atomic weight of N) + 3 × (atomic weight of H) in pounds, or 17.0306 lb. See also **Avogadro's number.**

Mole Fraction   The mole fraction of a substance in a system is the number of moles of that substance divided by the total number of moles in the system. In an ideal-gas mixture, the mole fraction is numerically equal to the volume fraction. The *volume fraction* of a component in a mixture is equal to the volume of that component at the total pressure and temperature of the mixture divided by the volume of the mixture at the same temperature and pressure.

The mole fractions, N_A and N_B, respectively, of components A and B of a binary solution are:

$$\text{Mole fraction of } A = N_A = \frac{n_A}{n_A + n_B}$$

$$\text{Mole fraction of } B = N_B = \frac{n_B}{n_A + n_B}$$

where n = number of moles of specific component present.

It is evident that the mole fraction of A plus the mole fraction of B must equal one or, if expressed as a percentage (mole percent), 100. In the case of three or more components, the denominators of the foregoing expressions must reflect the additional moles present.

Consider a solution containing 50 g of methanol in 1,000 g of water. What is the mole percent of each?

$$\text{Moles of } CH_3OH = 50/32 = 1.562$$
$$\text{Moles of } H_2O = 1,000/18 = \underline{55.556}$$
$$\text{Total moles} \dots\dots\dots\dots 57.118$$

Mole percent CH_3OH
$$= 1.562/57.118 \times 100 = 2.735\%$$
Mole percent H_2O
$$= 55.556/57.118 \times 100 = 97.265\%$$

The same solution expressed in terms of weight percentage is:

Weight percent CH_3OH
$$= 50/1,050 \quad = 4.762\%$$
Weight percent H_2O
$$= 1,000/1,050 = 95.238\%$$

In nuclear chemistry, mole fraction sometimes is used to express the number of atoms of a given isotope in an isotopic mixture as a fraction of the total number of atoms of that element in the mixture.

Mole Volume Under definite conditions, a mole of gas will occupy a definite volume regardless of the nature of the gas. This definite volume is termed the mole volume. At a pressure of 760 mm Hg and a temperature of $0°C$, a gram mole of gas will occupy 22.41 liters. This observation also applies to a mixture of gases. A pound mole of gas will occupy 359 ft^3 at a pressure of 760 mm Hg and a temperature of $32°F$.

Inasmuch as a volume of 1 g mole of gas at any given pressure and temperature contains the same number of molecules of gas regardless of how many gases are in the mixture, the percent by volume of any given gas also equals both the percent pressure exerted by that gas and the mole percent of that gas. Mole percent $\times$ volume percent = pressure percent.

Molecular Distillation (See **Separation operations**.)

Molecular Sieve (See **Adsorption; Epoxy resins;** and **Peptides and proteins**.)

Molecular Weight The molecular weight of a substance is the total of the atomic weights of all atoms contained in a molecule of the substance. Generally, molecular weight is synonymous with *formula weight*. In stoichiometric calculations, care must be exercised to ensure that the molecular formulas of the compounds correspond perfectly with the compounds as represented in the equations of reaction. Hydrogen and most other gases, except the inert gases, are diatomic, and thus the molecular weight is twice the atomic weight. Serious errors result if the water of hydration of a compound is overlooked. For example, the molecular weight of anhydrous magnesium fluosilicate, $MgSIF_6$, is 166.38. The formula weight of the same substance with six molecules of water of hydration, $MgSIF_6 \cdot 6H_2O$, is 274.47.

Molecule Classically, a molecule is defined as the smallest subdivision of a substance that can exist by itself and still possess the *complete properties* of the substance. This traditional approach to the structure of matter is rather easy to comprehend in terms of chemical compounds formed of two or more different atoms, but the molecular definition and concept become less obvious where pure metals, for example, are concerned. In fact, in numerous treatises that describe the structure of pure metals the term molecule never appears. Although tremendous advances have been made in our understanding of the solid state, this realm still poses many unanswered questions, including the application of the traditional molecular concept.

The classical definition of a molecule can also be misleading in terms of the inert gases, such as He and Ar. Chemically, these gases are inactive. They do not form compounds. Unlike such gases as H_2 and O_2, which form molecules comprising two atoms and thus symbolized with a subscript 2 (diatomic molecules), the inert gases do not appear to unite in pairs or other atomic groupings and hence are symbolized without subscripts. The molecular theory is brought into play by the statement that such gases comprise monatomic molecules.

A frozen inert gas is probably the simplest of all solids. The fact that these gases, upon cooling, first form liquids and then solids shows that the atoms are held together by a weak force, identified as the *van der Waals attraction*. The theoretical explanation is complex, but these forces essentially result from a shifting of position of the negatively charged cloud of orbital electrons with reference to the positively charged nucleus of the atom, causing an electrical imbalance and hence a weak force that attracts other atoms.

Range of Molecules: Molecules range in size and complexity from the very small hydrogen, H_2, molecule, with a diameter of 2.4×10^{-8} cm, to the giant molecules of the natural and synthetic polymers. In addition to using the term *diatomic* with reference to gaseous molecules, such as H_2 and N_2, which contain two atoms per molecule, the term also applies to molecules that comprise two different atoms, e.g., silver bromide, AgBr, or hydrogen chloride, HCl. There are numerous *triatomic* molecules (three atoms), e.g., nitrogen dioxide, NO_2, and water, H_2O. Molecules that contain four atoms include hydrogen peroxide, H_2O_2, and ammonia, NH_3. Five-atom molecules are exemplified by methane, CH_4, and chloroform, $CHCl_3$. Examples of common chemical substances with molecules containing from six to many tens of molecules could also be given. Glycerol contains 14 atoms; mannitol 26 atoms; dibenzylaniline 40 atoms; common table sugar (sucrose) 45 atoms, while polymeric substances like nylon contain thousands. A few of the giant molecules have been investigated by electron microscopy although most molecules are far too small to be observed directly by instrumental means. That some materials, such as certain oils, form a monomolecular film if spread on the surface of water affords a key to direct measurement. Some of these films have been found to be about 1×10^{-7} cm thick.

Electron Transfer: The mechanism of molecule formation from atoms can be explained in a rather crude fashion which at least hints at the phenomenon of compound formation. At best, these explanations are analogies rather than accurate depictions of what really occurs.

The manner in which electrons surround the atomic nucleus is described under **Atomic structure.** The chemical activity (ability to form molecules with other atoms) stems mainly from the electrons in the outer structure of the atom, frequently termed *valence electrons,* which under proper conditions are available to participate in electron transfers from one atom to the next or in sharing between atoms.

Table A-57 under **Atomic structure** indicates the number of electrons contained in each shell of the elemental atom. Some elements just begin to fill their outer shells, while others have nearly filled outer shells. In some instances, e.g., the inert gases, the outer shells are completely filled (2 electrons in the K shell for He; 8 electrons in the M shell for Ar; and 8 electrons in the N shell for Kr). This condition imparts great stability, i.e., resistance to chemical reaction. Thus, only through the application of great *activation energy* e.g., from electron bombardment or the effects of electric-glow discharge, will these elements be altered.

This same table indicates that Na, with its total of 11 electrons, has inner shells K and L filled but only one electron in its M shell, which, as the table indicates, can accommodate several additional electrons. This solitary electron of Na, then, is vulnerable to attraction by external forces because the internal electrical attraction which maintains a balanced condition is less when one or a few electrons exist in an unfilled shell than when the shell is filled or nearly filled. Sodium is said to have a valence of 1 (Na^+), a positive valence because once the Na atom transfers this lone electron, the atom no longer is electrically neutral

(with a perfect balance of charges between electrons and protons) but has a net positive electric charge of 1.

The table also indicates that Cl, with its total of 17 electrons, has inner shells K and L filled and has 7 out of a possible 8 electrons in the first M subshell. Thus, the electrical-force balance of the Cl atom is such that, unlike the Na atom, there is no tendency for Cl to part with its 7 outer electrons. Instead Cl tends to attract an electron from outside to fill up the vacancy. In this case, Cl also is said to have a valence of 1, but of opposite sign (Cl^-) to that of Na^+ because, in incorporating an additional electron, the Cl atom has 18 electrons with only 17 protons to balance them and hence has a net negative electrical charge of 1. Thus, Cl has a valence of 1^-.

The atomic electrical-force conditions are well prepared for an electron to be transferred from a Na atom to a Cl atom, the result being the formation of a molecule of NaCl.

In the formation of molecules and hence chemical compounds by the electron-transfer phenomenon, the process is not limited to the transfer of one electron. Some elements have positive valences as high as 7^+ and negative valences as much as 4^-. Some elements have both positive and negative valences.

Electron Transfers in Solution: When an atom acquires or loses one or more electrons without any resulting compound formation (as frequently happens in solutions of stable compounds), the atom with the additional electron is termed a *negative ion.* Thus, when NaCl dissociates in water, the Cl atom is changed to a Cl^- ion. Similarly, the Na atom because an Na^+ ion. A substance that behaves in this manner is termed an *ionic salt.* While ions generally are thought of in terms of gases and liquid solutions, solid NaCl, for example, may be considered as being held together by ionic forces. In other words, a molecule of NaCl could be regarded as a binding together of one Na^+ ion and one Cl^- ion.

Electron Sharing: Instead of transferring electrons from one atom to the next, certain atoms *share* electrons. Carbon and hydrogen have a strong tendency to share, or *pair,* electrons. Carbon has a valence of $2+$, $4+$, and $2-$, indicating its great flexibility to form compounds and accounting, in part, for the hundreds of thousands of carbon compounds that may be formed. How one atom of carbon and four atoms of hydrogen unite to form a molecule of methane can be depicted by

$$
\begin{array}{c}
\text{H} \\
\text{H} : \overset{\cdot\cdot}{\underset{\cdot\cdot}{\text{C}}} : \text{H} \\
\text{H}
\end{array}
$$

where the dots indicate, for each pair, an electron from the hydrogen atom and an electron from the carbon atom's L shell. This type of bond is termed a *covalent bond* or *homopolar bond* and is fundamental to organic chemistry. Germanium and silicon also have four electrons outside an inert shell and thus can behave somewhat like carbon.

When both electrons in the bond are supplied by one atom, the result is termed a *dative* or *semipolar bond*. For example, in the oxidation of a tertiary amine, both electrons required to form a bond with oxygen come from the nitrogen atom. Sharing of electrons is not limited to one pair, but triple covalent bonds can be formed, as in the case of nitrogen.

In still another form of bonding termed *hydrogen bonding,* hydrogen, with its single electron, can form two bonds. The dimer of formic acid, $HCOOH_2$, is the result of hydrogen bonding of two molecules of formic acid, HCOOH:

Two molecules of formic acid   Formic acid dimer

The hydrogen bonding is indicated by the dotted lines. Formation of hydrates by the water molecule also involves hydrogen bonding.

Metals: In the solid metallic state, each metal atom is considered to be surrounded closely by a large number of similar atoms with but a few electrons in the outer shells. This permits the electron clouds to overlap, the loosely held electrons in the outer orbits being fully shared without regard to association with individual atoms. Thus, the bond in metals is nonspecific, and hence many dissimilar metals can form alloys or easily join one with another. Together with this loose overlap of outer-shell electrons, there is very close packing down to and including the inner-shell electrons. This general description of metallic structure accounts for many of the characteristic properties of metals: (1) freedom of movement of electrons accounts for excellent electrical conductivity; (2) the large number of free electrons tends to absorb and reradiate energy freely, giving metals their luster and opacity; (3) the ability of free electrons to transfer thermal energy accounts for excellent heat conductivity; and (4) the close

packing helps to explain the toughness of metals.

Molecular Combinations: Many molecular combinations (compound formation) require *activation energy* for their formation. A small spark, for example, is required to cause the explosive combination of hydrogen and oxygen to form water. In this instance, only a small amount of energy is required to activate a few of the atoms because once the reaction starts, it becomes fully *exothermic;* i.e., sufficient energy is released to cause all available H and O atoms to unite. Once started, this form of reaction is termed a *chain reaction;* i.e., it is self-perpetuating so long as materials are available. This type of chain reaction should not be confused with a nuclear chain reaction. A reaction that requires energy continuously to keep it going is termed an *endothermic* reaction.

When compounds, such as water, are formed with the release of a lot of energy, it generally follows that they are quite stable. Thus, the products of an explosion usually are quite stable. The reverse situation also holds. Acetylene, C_2H_2, for example, requires a lot of energy for formation from C and H atoms. Acetylene is unstable and is ready to release that energy under the proper conditions. The high temperature of the oxyacetylene flame is explained on this basis by three energy-releasing actions: (1) C_2H_2 dissociates into carbon and hydrogen, releasing energy; (2) carbon combines with oxygen to form CO_2, liberating further energy; (3) concurrently, the hydrogen combines with oxygen to form water and to release still more energy.

REFERENCES

Kauzmann, W.: "Quantum Chemistry," Academic, New York, 1970.

McWeeny, R., and B. T. Sutcliffe: "Methods of Molecular Quantum Mechanics," Academic, New York, 1969.

Pask, J. A. (ed.): "An Atomistic Approach to the Nature and Properties of Materials," Wiley, New York, 1967.

Ziman, J. M.: "Principles of the Theory of Solids," Cambridge, London, 1964.

Molybdenum (Gk. *molybdos,* lead.) **Mo** $= 95.94$ (at. wt.); 42 (at. no.). Molybdenum is a silver-gray metallic element of group VI of the periodic table. Credit for its discovery goes to Scheele (Sweden) in 1778. Despite the derivation of the name, any similarities between Mo and Pb are extremely superficial. The melting point of Mo, for example, is 2610°C, higher than all but three metallic elements. The boiling point $= 4800$°C and density $= 10.2 \, g/cm^3$ at 20°C. There are seven stable isotopes, 92, 94–98, and 100, and five unstable isotopes, 90, 91, 93, 99, and 101.

The principal ore is molybdenite, MoS_2; powellite, $Ca(MoW)O_4$, and wulfenite, $PbMoO_4$, are of lesser importance. A typical molybdenite contains on the order of 0.3% Mo. Mo is recovered by crushing, grinding, and flotation operations that produce a concentrate containing approximately 90% MoS_2.

Approximately 85% of all Mo produced in the free world is used as an alloying agent in iron-base alloys, of which alloy steels, stainless steels, tool steels, superalloys, and alloy cast irons are the most significant. Of the remaining Mo production, roughly 72% is diverted to Mo chemicals, including purified MoS_2 for lubricant applications.

The starting material used in the production of most Mo chemicals is either molybdic oxide or ammonium molybdate. Two grades of molybdic oxide are produced, a technical grade, manufactured by roasting molybdenite concentrate, and a pure grade, made by sublimation of technical oxide. Ammonium dimolybdate is made by dissolving pure oxide in NH_4OH, filtering out the insolubles, and crystallizing the filtrate.

Chemistry: Several factors contribute to the complex chemistry of Mo, most important being the variable valence of the element in compound formation. Compounds are known with valence states of 0, 2^+, 3^+, 4^+, 5^+, and 6^+. In addition, disproportionation occurs to transform compounds of a single valence into mixtures containing several valence states, and coordination numbers change readily with slight modification of conditions. Mo has enormous complexing power, particularly in the higher valences. Ions such as the molybdate readily aggregate or disaggregate, depending on pH, to yield mixtures of polyions by polymerization-condensation reactions.

Oxides: As many as 11 oxides of Mo have been reported, but most are metastable and prepared only by heating stoichiometric mixtures of oxides and metal. The most common and most stable compounds are the dioxide and the trioxide. Reduction of the trioxide in hydrogen at 300–400°C proceeds smoothly to the dioxide. Reduction continues to metallic Mo at higher temperatures. Some physical and chemical properties of the common oxides are given in Table M-14.

Molybdenum trioxide, or molybdic oxide, is the most important compound. Most known molybdenum compounds are prepared directly or indirectly from the trioxide.

TABLE M-14. *Properties of Molybdenum Dioxide and Trioxide*

	MoO_2	MoO_3
Molecular weight	127.95	143.93
Color	Red brown	White
Density, g/ml	6.47	4.69
Melting point, °C	Dissociates	795
Sublimation temperature, °C	1100	700
Boiling point, °C	. . .	1155
Heat capacity at 25°C, (cal)/(mole)(°C)	13.20	17.93
Solubility:		
H_2O	Insoluble	Very slight
HCl	Insoluble	Moderate
HNO_3	*	Moderate
H_2SO_4	Slight	Soluble
HF	Insoluble	Moderate
Alkalies	Very slight	Soluble

*MoO_2 is converted into MoO_3 in concentrated HNO_3.

Sulfides: Molybdenum disulfide, in addition to being the predominant naturally occurring compound, is important industrially for its lubricating properties. The molecular structure consists of two layers of sulfur atoms separated by a layer of molybdenum atoms. The weak S—S bonds provide planes that can be sheared at low levels of force. Values as low as 0.017 have been measured for the coefficient of friction of MoS_2. Low friction values are retained over a temperature range from -200 to 400°C, the latter being the temperature at which oxidation is first observed. In vacuum, MoS_2 is stable to 1370°C, where dissociation begins.

Commercial utilization of the disulfide as a lubricant is in the form of a dry film or as an additive to grease or oil. The use of MoS_2 as a filler in nylon is described under **Nylons.**

Molybdenum disulfide has been used as a catalyst in a variety of hydrogenation-dehydrogenation reactions. This application has involved deposition on a carrier as well as free MoS_2 in a micronized form.

Other compounds that have been observed are the sesquisulfide, Mo_2S_3, and the trisulfide, MoS_3. The sesquisulfide is a gray-black amorphous compound that can be prepared by the direct reaction of the elements in sealed, evacuated tubes or by the thermal decomposition of MoS_2. The trisulfide is a brown to black amorphous compound formed by heating or acidifying ammonium tetra-thiomolybdate, $(NH_4)_2MoS_4$.

Several thiomolybdates have been observed, and their properties are similar to the corresponding sulfur-free molybdates.

Halides: Molybdenum pentachloride, $MoCl_5$, is the most important halide. The compound is generally prepared by the chlorination of the metal powder. The reaction is exothermic, and the reactants must be pure and dry to obtain a satisfactory product. Other preparation methods have used the trioxide or the sulfide as the molybdenum contributor reacting with CCl_4, hexachloropropene, or thionyl chloride.

Molybdenum pentachloride has limited solubility in chlorinated solvents. It reacts with a number of polar solvents such as ethers, alcohols, aldehydes, ketones, esters, amines, acids, and acid anhydrides. It also reacts to a limited extent in aromatic and aliphatic hydrocarbons; it is extremely sensitive to small concentrations of air and water. Oxychlorides, hydroxy chlorides, or other oxygen-containing compounds are formed, depending on the specific environment.

Molybdenum pentachloride has been used to catalyze a number of polymerization reactions, e.g., olefins, vinyl monomers, norbornenes, trioxane, ethylene, vinylcyclohexane, cyclopentene, and butadiene. It is a convenient compound to use in the preparation of vapor-phase coatings of molybdenum on metallic or ceramic substrates.

Many additional halides and oxyhalides of Mo are known, representing a wide range in stability. The highest member of each series ($MoCl_5$, $MoBr_4$, MoF_6, and MoI_3) is generally used to prepare the lower halides by reaction with Mo metal or by reduction in hydrogen or a hydrocarbon.

Molybdates: An extremely wide range of molybdates is known, including ammonium molybdates of a variety of compositions. Ammonium polymolybdate (known also as ammonium dimolybdate and molybdic acid 85%) is the highest-purity Mo compound available commercially. It is not unusual to obtain a product purity of 99.97% by reduction of this compound to the metal powder.

Ammonium molybdate is used as a source of high-purity MoO_3 for deposition on support materials for catalyzing hydrogen-treating and hydrocracking processes. It is used in electroplating baths and as a laboratory reagent for determining phosphates, arsenates, and lead.

Sodium molybdate, the simplest water-soluble molybdenum compound, is prepared by dissolving MoO_3 in an excess of caustic soda. The anhydrous material has the composition Na_2MoO_4; the crystals $Na_2MoO_4 \cdot 2H_2O$.

It has recently been determined that molybdenum is an important element in the life processes of both plants and animals. Because of high water solubilities, the sodium and ammonium molybdates are important in fertilizers and diet supplements. Large increases in crop yields have been obtained in molybdenum-deficient soils by the application of molybdates as fertilizers, foliar sprays, or seed coatings. Many vegetables and even some trees respond to molybdate applications.

Mo is now recognized as an essential micronutrient in animals. A minimum daily requirement for human beings has been established in the Soviet Union. A very recent development has been the discovery of the importance of Mo in promoting dental health, and molybdenum-containing dentifrices are now marketed in Europe.

Sodium molybdate, an important chemical in the pigment industry, is used in the manufacture of molybdate-chromate orange pigments and numerous phosphomolybdic acid–organic pigments. It is also used as a condensation catalyst for phthalocyanine pigments, as a corrosion inhibitor in glycol-based antifreeze, and in synthetic cutting fluid. Its use in aqueous solutions is important when freedom from toxicity and staining are required.

Soluble molybdates react with certain organic compounds to produce dyes for furs and hair which are reasonably colorfast.

Many other molybdates have technological significance. Zinc molybdate is an important white nontoxic pigment with excellent corrosion-inhibiting properties. Lithium molybdate is used as an additive for porcelain enamel coatings. Molybdates of Fe, Co, and Ni are used extensively in the petroleum and chemical industries to catalyze hydrogenation, desulfurization, denitrification, and hydrocracking. Lead molybdate is used in the application of vitreous designs to glass bottles.

Many elements can act as central atoms in a wide variety of heteropolymolybdates.

Hexacarbonyl: Molybdenum exhibits an oxidation state of 0 in the hexacarbonyl, $Mo(CO)_6$, generally prepared by reacting $MoCl_5$ with CO in ether at pressures well above atmospheric and temperatures of 20–100°C. The crude product can be purified by sublimation or distillation.

Because of the zero oxidation state, molybdenum hexacarbonyl is subject to attack by oxidizing agents; it has limited solubility in most agents because of the difficulties of solvent association with its strongly hexacoordinate molecule. The hexacarbonyl is stable in air and water at room temperature but decomposes when heated to approximately 150°C. It forms a number of organic and organometallic derivatives. Nitrogen-, phosphorus- and oxygen-coordinated complexes are numerous. The hexacarbonyl is useful in the vapor deposition of Mo films on metals and ceramics. By adjusting deposition conditions, molybdenum carbide coatings can be formed. The hexacarbonyl has been used as a catalyst for a number of hydrocarbon reactions, e.g., the disproportionation of olefins, the cyclization of ethylene, the isomerization of 1-olefins, and the epoxidation of propylene. The hexacarbonyl has been studied as an additive for gasoline and it is claimed that in leaded gasoline it suppresses preignition resulting from lead deposits.

Organic Compounds: Molybdic oxide, soluble molybdates, molybdenum hexacarbonyl, and certain halides form complex compounds with a great variety of organic oxygen, nitrogen, and sulfur compounds.

Oxygen-coordinated compounds include acetylacetonates, oxalates, carboxylates, alkoxides, phenoxides, and organic chlorides. Nitrogen-coordinated compounds include a number of organic chlorides and organic molybdates. Sulfur-coordinated organic compounds include dialkyldithiocarbamates, dialkyldithiophosphates, cysteine complexes, and α-diketone complexes. Some mixed oxygen-nitrogen donor complexes are known.

Examples of industrial applications of organic compounds include molybdenum acetylacetonate as a catalyst for the polymerization of ethylene and the formation of polyurethane foam; pyrogallol-molybdate complexes as dyes; molybdenum oxalate in certain photochemical systems; and molybdenum dithiocarbamate as a lubricant additive.

Metallurgical Uses: Many alloy steels contain Mo because of its beneficial effects on hardenability, toughness, cold-formability, and weldability. Although the majority of these steels contain less than 1% Mo, this use accounted for over 59 million lb of Mo in the free world in 1970. An interesting development has been a high-strength low-alloy steel containing manganese, molybdenum, and columbium (niobium) for use in line pipes in extreme cold weather.

Stainless steels produced in 1970 contained about 27 million lb of Mo. The element contributes to the corrosion resistance, elevated-temperature strength, and weldability of many basic types of stainless steel. Ferritic stainless steels containing 18–26% Cr and 1–2% Mo now challenge the traditional 18-8 types for many applications.

Molybdenum added to tool steels (15 million lb in 1970) results in better hot strength and improved resistance to softening and thermal cycling effects. Many of the early tungsten grades have been replaced by Mo grades because of improved properties, better price stability, and lower metal density.

The foundry industry used approximately 11 million lb of Mo in 1970 to improve the strength and abrasion resistance of cast iron. An additional 6.75 million lb went into superalloys for high-temperature environments, e.g., jet engines. A new Ni-base superalloy containing 18% Mo has a significantly higher melting point, together with lower density and lower coefficient of thermal expansion, than the superalloys now in use.

REFERENCES

Climax Molybdenum Company: Heteropoly Compounds of Molybdenum and Tungsten, *Bull.* Cdb-12a, November 1969; Properties of Molybdenum Dioxide, *Bull.* Cdb-1a, August 1960; Properties of Molybdic Oxide, *Bull.* Cdb-1, August 1969; Properties of Molybdenum Disulfide, *Bull.* Cdb-5a, February 1962; Properties of Simple Molybdates, *Bull.* Cdb-4, October 1962; Properties of Molybdenum Hexacarbonyl, *Bull.* Cdb-13a, April 1970; Properties of Molybdenum Pentachloride, *Bull.* Cdb-3a, June 1969; Evaluation of Zinc Molybdate vs. Basic Lead Silico Chromate, *Publ.* C-45, December 1970; Organic Complexes of Molybdenum, *Bull.* Cdb-9, suppl. 7, September 1970.

Hampel, C. A.: "Rare Metals Handbook," 2d ed., Van Nostrand Reinhold, New York, 1961.

Industrial Applications of Molybdenum Chemicals, *J. Agric. Food Chem.*, vol. 3, August 1955; *Ind. Eng. Chem.*, vol. 47, August 1955.

Molybdenum, *Min. Annu. Rev.*, June 1971.

—Robert Q. Barr, *Climax Molybdenum Company, New York*

Monazite (See **Cerium; Gadolinium; Lanthanum; Lutetium; Neodymium; Praseodymium; Rare-earth elements and metals; Samarium;** and **Ytterbium.**)

Monoammonium Phosphate (See **Fertilizers.**)

Monochlorobenzene (See **Chlorine organics.**)

Monochlorostyrene Monochlorostyrene is a vinylic monomer which readily enters into poly-

TABLE M-15. *Properties of Monochlorostyrene*

Property	Isomer		
	Ortho	Meta	Para
Molecular weight	138.6	138.6	138.6
Boiling point, °C	187.5	190	191.5
Melting point, °C	−64.3	−74.9	−14.9
Specific gravity (25°C/4°C)	1.0956	1.090	1.0818
Refractive index n_D^{25}	1.55988	1.5598	1.56364
Flash point, °C, Cleveland open cup .	74	74	71
Heat of polymerization, cal/g	116	. . .	114
Polymer T_g, °C	102	. . .	115

merization reactions to yield additional polymers. Of the three possible isomers of this monosubstituted styrene, only the two with the chlorine atom in the ortho or para position are found in the commercially produced material. Some pertinent properties of the three monochlorostyrene isomers are given in Table M-15.

$$CH=CH_2$$

Monochlorostyrene

Monochlorostyrene is miscible with most common organic solvents, including methanol, acetone, benzene, and hexane. It is insoluble in water.

A number of synthetic routes have been used to prepare monochlorostyrene. Two examples are the *dehydration of chlorophenethyl alcohol*

$$\text{(1)}$$

and the *catalytic dehydrogenation of chloroethylbenzene*

$$\text{(2)}$$

Monochlorostyrene readily polymerizes with itself and with a large number of other unsaturated monomers to produce high-molecular-weight polymers. In general the monomer behaves much like styrene and has been successfully incorporated into many polymer systems in place of styrene. Bulk, solution, suspension, and emulsion techniques have been used for polymerization.

The Alfrey-Price copolymerization parameters for *o*-monochlorostyrene are $Q = 1.28$ and $e = -0.36$ and for *p*-monochlorostyrene are $Q = 1.03$ and $e = -0.33$. Table M-16 lists the reactivity ratios r_1 and r_2 for the copolymerization of these two monochlorostyrene isomers and several common monomers.

Commercially monochlorostyrene is used as the reactive diluent in unsaturated polyester resins and in the preparation of fire-retardant thermoplastic molding materials. Unsaturated polyester

TABLE M-16. *Copolymerization Reactivity Ratios of Monochlorostyrene Monomers*

M_2	M_1			
	o-Monochloro styrene		*p*-Monochloro styrene	
	r_1	r_2	r_1	r_2
Acrylic acid.	0.74	0.38	0.62	0.38
Acrylonitrile	1.22	0.07	1.04	0.09
1,3-Butadiene.	0.61	0.91	0.55	1.09
Butyl acrylate	1.52	0.10	1.29	0.13
Methyl methacrylate .	1.32	0.43	1.04	0.54
Styrene	1.50	0.55	0.88	0.66
Vinyl chloride	23.8	0.03	19.8	0.04
Vinylidene chloride . .	4.49	0.13	3.72	0.17

resins formulated using monochlorostyrene or mixtures of monochlorostyrene and styrene cure faster than the conventional all-styrene diluted resins. These differences in cure times depend upon the nature of the polyester alkyd; however, resins diluted with a 50:50 monochlorostyrene-styrene mixture generally cure $1\frac{1}{2}$ to 5 times faster than the same resins diluted with styrene. Because monochlorostyrene contains 25.6% chlorine, it is often used in producing fire-retardant polyester resins which depend on chlorine- or bromine-containing components to achieve their low flammability. In this application the monomer is generally used in conjunction with a halogen-containing alkyd and a synergist such as antimony oxide.

The low flammability of polymonochlorostyrene, the homopolymer of monochlorostyrene, has been responsible for the polymer's use in self-extinguishing molding materials. This thermoplastic will burn in air when ignited; however, when it is compounded with a small amount of a fire-retarding synergist such as antimony oxide (1–3%), it will not continue to burn and is classified as self-extinguishing. The higher heat-distortion temperature (102 versus 87°C by ASTM D648 264-psi test) and the lower gas permeability of polymonochlorostyrene compared to polystyrene are other significant differences between these two polymers.

Commercially produced monochlorostyrene is a 60:40 mixture of the ortho and para isomers and is supplied containing 100 ppm of the polymerization inhibitor t-butylcatechol. Since dissolved O_2 is necessary for the inhibitor to function, the monomer should be stored under an air blanket.

Monochlorostyrene does not form explosive mixtures with air at room temperature. Storage temperature should not exceed about 27°C (80°F) and preferably should be kept below 21°C (70°F). Monochlorostyrene has been shown to have low acute oral toxicity and it is not likely to present a hazard from ingestion under the conditions of normal industrial operation.

—J. L. Brewbaker, *The Dow Chemical Company, Midland, Mich.*

Monoethanolamines   (See **Ethanolamines.**)

Monosodium Glutamate   (See **Amino acids.**)

Mooney Plasticity Scale   (See **Elastomers.**)

Morpholine   (See **Ethanolamines;** and **Paint.**)

Mortar   (See **Cement.**)

Moving Beds   (See **Drying, solids.**)

Mulling   (See **Mixing and blending, solids-solids.**)

Mullite   (See **Ceramics.**)

Muntz Metal   (See **Copper.**)

Muscovite   (See **Potassium.**)

Myosin   (See **Peptides and proteins.**)

Myrcene   (See **Terpenes.**)

Myristic Acid   (See **Vegetable oils.**)

N

Naphtha (See **Acetylene; Cracking; Ethylene; Hydrocracking; Hydrotreating; Petrochemical complex; Petroleum; Petroleum processing;** and **Reforming.**)

Naphthalene Naphthalene, $C_{10}H_8$, sometimes called *tar camphor*, is a colorless, crystalline flaked solid with the familiar odor of moth balls. Physical properties include: melting point $= 80.2°C$; boiling point $= 217.9°C$, subliming slowly at room temperature; specific gravity $= 1.145$ at $20°C$ (referred to water at $4°C$); very slightly soluble in water (0.003 part in 100 parts at $25°C$); moderately soluble in alcohol (9.5 parts in 100 parts at $20°C$); very soluble in ether and benzene. The compound burns with a smoky flame. The structural formula for naphthalene, consisting of two joined benzene rings, is given under **Carbon compounds.**

A good commercial grade of naphthalene (called 78° naphthalene, referring to the melting point) is approximately 96% pure. Refined naphthalene is available in flakes, pellets, and balls. Use of naphthalene as an insect repellent has dwindled in recent years as other materials, e.g., *p*-dichlorobenzene, have displaced it. About 75%

of the naphthalene produced is used in the production of phthalic anhydride, although this position is threatened by petroleum-derived xylene. Naphthalene is used as an intermediate for a variety of products, including 1-naphthol and 1-naphthylmethylcarbamate insecticides and 2-naphthol tanning agents. Chlorinated naphthalene ranges from low-viscosity oils through waxlike solids to resinlike solids. These materials are highly resistant chemically, have good dielectric strength, and are used as solvents for fireproofing and waterproofing fabrics and as coatings for electric cable. Naphthalene crystals have been used in very small quantity in photomultiplier tubes as gamma-ray detectors.

In the United States (1968) total naphthalene production was 900 million lb, of which a little over one-third was produced from petroleum, largely by dealkylation of methylnaphthalenes. The remainder was derived from coal tar, as described under **Coal tar and derivatives.** The dealkylation of alkylnaphthalenes may be effected thermally or catalytically. In the process shown in Fig. N-1, a thermal, noncatalytic hydroalkylation route is followed. The charge stock may be light catalytic gas oils, reformate bottoms, aro-

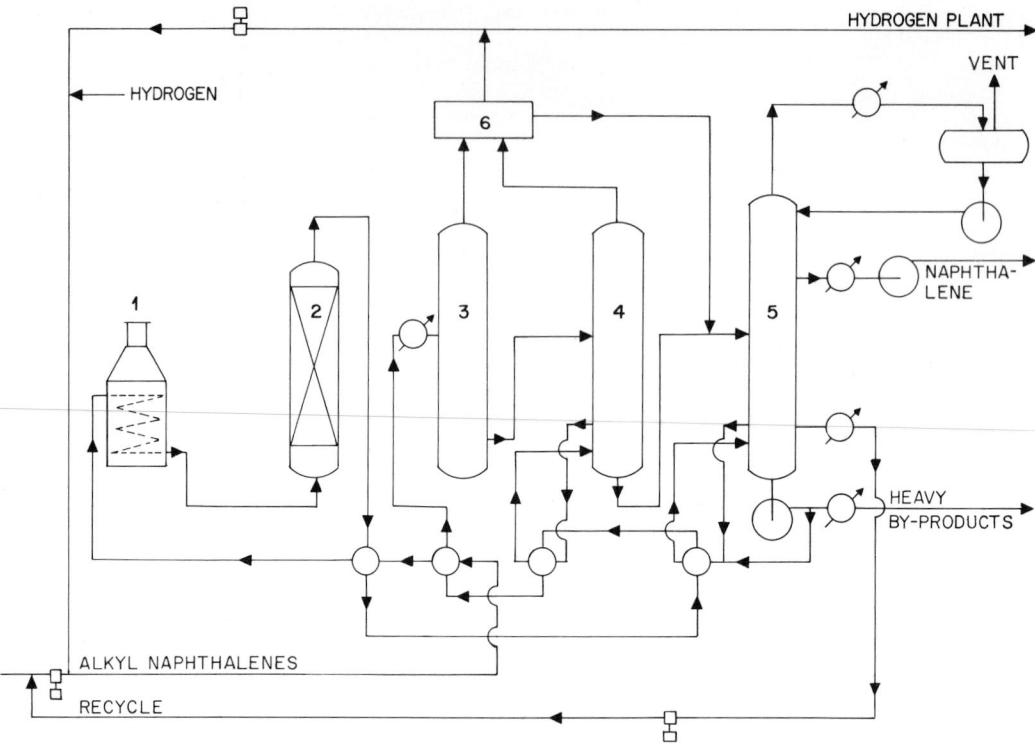

Fig. N-1. Thermal, noncatalytic hydroalkylation route for producing naphthalene. (1) Heater, (2) reactor, (3), (4) strippers, (5) fractionator, (6) vapor recovery.

matic extracts, or coal-tar fractions, plus hydrogen. In a mixed-feed operation, both naphthalene and benzene may be produced at the same time. The yield of naphthalene depends upon the amount of double-ring compounds present in the charge stock.

Naphthalene differs from benzene, in which monosubstitution products exist in one form only, in that it yields two isomeric monosubstitution products; e.g., there is only one nitrobenzene, regardless of how the material is prepared, but naphthalene yields α- and β-nitronaphthalene.

Important derivatives of naphthalene include naphthalene tetrachloride, $C_{10}H_8Cl_4$; α- and β-nitronaphthalene, $C_{10}H_7NO_2$; α- and β-naphthylamine, $C_{10}H_7NH_2$; α- and β-naphthol, $C_{10}H_7OH$. Mono- and disulfonic acids also can be derived from naphthalene. Upon sulfonation, naphthalene yields naphthalene α- and β-sulfonic acids, $C_{10}H_7 \cdot SO_3H$. In theory, 14 isomeric naphthylamine monosulfonic acids, $C_{10}H_6(NH_2) \cdot SO_3H$, can be obtained, 7 from α-naphthylamine

and 7 from β-naphthylamine. Used in dye manufacture, 1-4-naphthylamine-monosulfonic acid, or naphthionic acid, results from the treatment of α-naphthylamine with H_2SO_4. When naphthalene itself or mono- and disubstitution products of naphthalene are oxidized under proper conditions, α-naphthaquinone, $C_{10}H_6O_2$, can be formed. When α-amino-β-naphthol is oxidized, β-naphthaquinone can be formed. Tetrahydronaphthalene, $C_{10}H_{12}$ (tetralene), can be formed by the reduction of naphthalene with H_2 over a Ni catalyst. Other derivatives include hydrindene, C_9H_{10}; indene, C_9H_8; α- and β-hydrindone, C_9H_8O; and acenaphthene, $C_{12}H_{10}$.

Naphthyl (See **Radicals.**)

Natural Gas Natural gas is a mixture of methane, ethane, propane, and other paraffinic hydrocarbons, along with H_2S, CO_2, N_2, He, and traces of other elements and compounds. Large naturalgas deposits have been discovered throughout the

TABLE N-1. *Analysis of a Sample of Natural Gas* *

Component	Mole %
Methane	76.2
Ethane	6.4
Propane	3.8
n-Butane	1.3
Isobutane	0.8
n-Pentane	0.3
Isopentane	0.3
Cyclopentane	0.1
Hexane + hydrocarbons	0.3
Nitrogen	9.8
Oxygen	Trace
Argon	Trace
Hydrogen	0.0
Hydrogen sulfide	0.0
Carbon dioxide	0.2
Helium	0.45

*Panhandle natural gas field (Texas).

world, most notably in the United States, Canada, the Soviet Union, and the Middle East. Natural gas is found underground at various depths and pressures and in solution with crude-oil deposits.

The constituents of natural gas vary with geographic location. No one composition can be considered typical. The analysis of a gas sample taken from the Panhandle natural gas field in Texas is given in Table N-1. Since some portions of the earth lack natural gas or consume more than is locally available, this valuable raw material commonly is imported either by gas-transmission lines or in a liquefied state.

Natural gas is most widely known as a fuel and because of its clean-burning characteristics minimizes air pollution. Recent technological advancements in steam-methane reforming (see also **Reforming**) have made natural gas a valuable feedstock for the production of ammonia, the main building block for agricultural fertilizers. In the United States, approximately 11 million short tons of NH_3 are produced each year by this process. See also **Ammonia;** and **Fertilizers.**

Natural gas is also used as a feedstock for the production of hydrogen, methanol, and other organics. See also **Ethylene; Hydrogen; Methyl alcohol;** and **Petrochemical complex.**

Natural-Gas Processing: Normally, natural gas is processed to recover components heavier than methane. Depending upon the constituents and percentage recovery required, two general techniques are available: an absorption-oil process for low ethane recoveries and a cryogenic process for high ethane recoveries.

The basic *absorption process* is illustrated in Fig. N-2. Rich natural gas is fed into the bottom of an abosrption tower, where the gas is contacted countercurrently by a lean presaturated absorption oil. The lean tail gas exits from the top of the absorber and is directed back to the pipeline. The lean-oil circulation rate and temperature are determined by the product spectrum and the required percentage recovery. High recovery percentages require large lean-oil circulation rates. If ethane recovery is desired, the lean oil must be refrigerated. Hence the feed gas must be dehydrated to prevent plant freeze-ups.

The absorption oil removes the desired hydrocarbons from the incoming gas stream and then is directed into a demethanizing tower, where absorbed methane is removed from the absorption oil by fractionation and exits from the top of the tower. The methane is further refrigerated, combined with regenerated lean oil, and sent to a presaturator. The vapor phase containing mostly methane is then combined with the tail gas from the absorber overhead. The presaturated liquid is used for reflux in the demethanizer and as lean oil to the absorber.

The bottoms of the demethanizer are directed into the lean-oil still, where the desired hydrocarbon components are separated from the lean oil. The overhead vapors containing ethane and heavies are directed into a fractionating train, where the desired product spectrum is obtained. The bottoms liquid then is recycled through a number of heat exchangers back to the presaturator.

Reboiling heat can be supplied by hot oil or steam. Tower reflux can be generated by using refrigeration, water, or air coolers. Gas turbines can be used to drive the necessary rotating equipment.

The basic *cryogenic process* is shown in Fig. N-3. A compressed and dried natural-gas stream is directed into a cryogenic processing unit, where its temperature is progressively lowered, with the subsequent effect of condensing the desired hydrocarbons. Various temperature-level liquid dropouts are situated in the cold box to optimize the hydrocarbon liquid recovery. One such level is maintained by propane at $-37°C$ ($-35°F$). Ethane or ethylene refrigerant can be used to stabilize the lowest-temperature-level dropout.

The condensed hydrocarbons are separated from the vapor-liquid mixture in the knockout

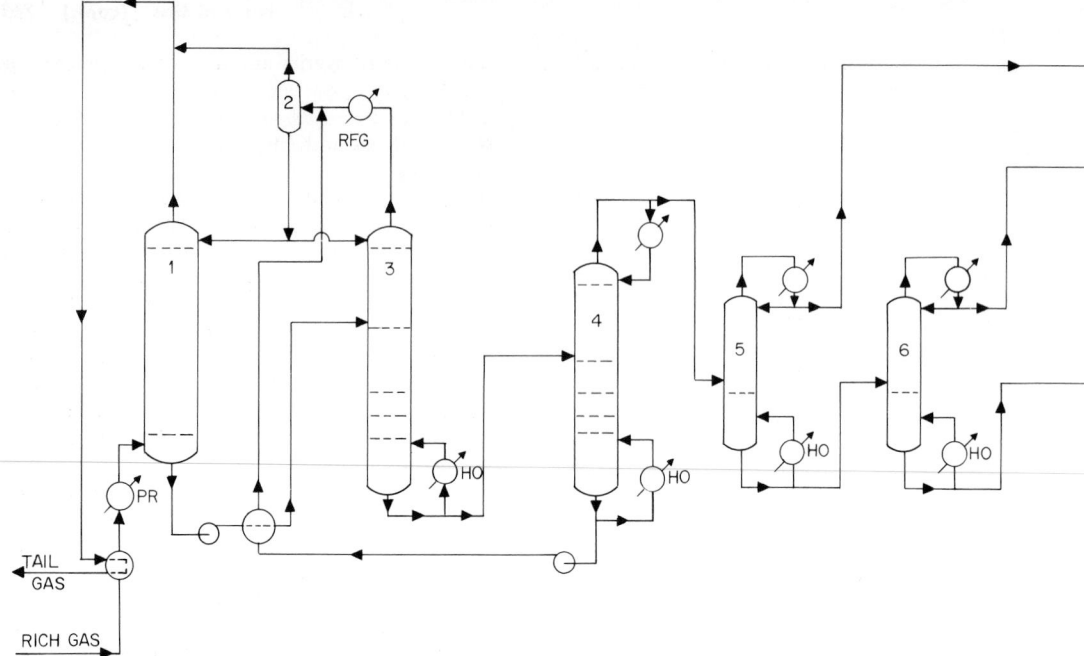

Fig. N-2. Representative absorption-oil process for recovery of ethane and heavier hydrocarbons from natural gas. (1) Absorber, (2) presaturator, (3) demethanizer, (4) lean-oil still, (5) deethanizer, (6) depropanizer, (7) debutanizer, (8) C₄ splitter. *RFG,* refrigerant, *PR,* propane refrigerant, *HO,* hot oil. (*The M. W. Kellogg Company, a division of Pullman Incorporated.*)

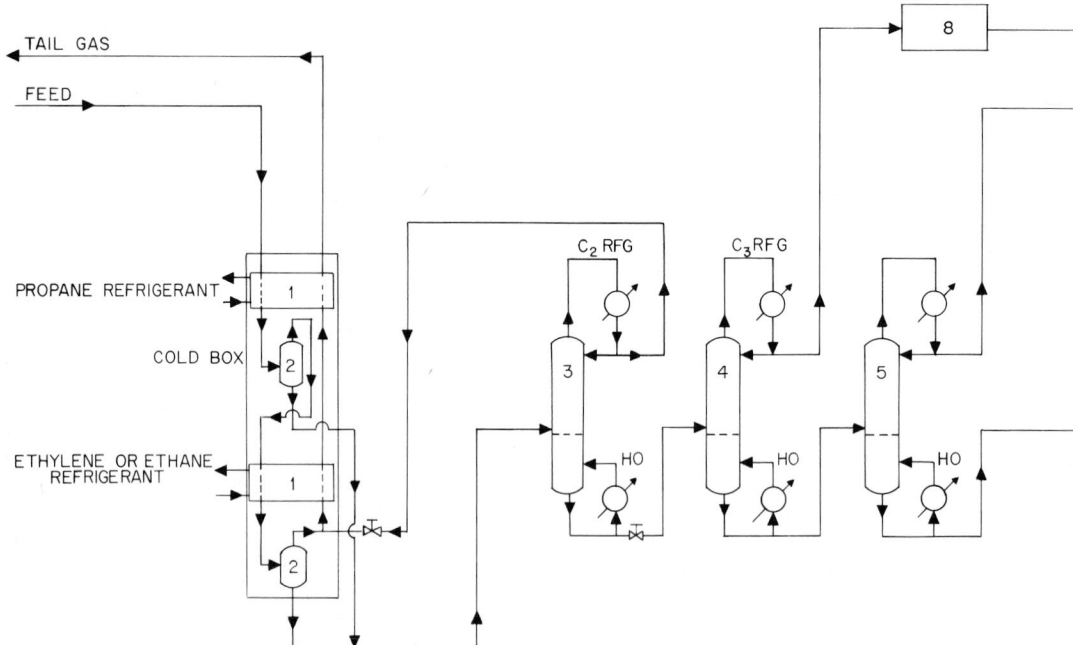

Fig. N-3. Cryogenic recovery of hydrocarbons from natural gas. (1) Exchangers, (2) knockout pots, (3) demethanizer, (4) deethanizer, (5) depropanizer, (6) debutanizer, (7) C₄ splitter, (8) carbon dioxide removal. *RFG,* refrigerant, *HO,* hot oil. (*The M. W. Kellogg Company, a division of Pullman Incorporated.*)

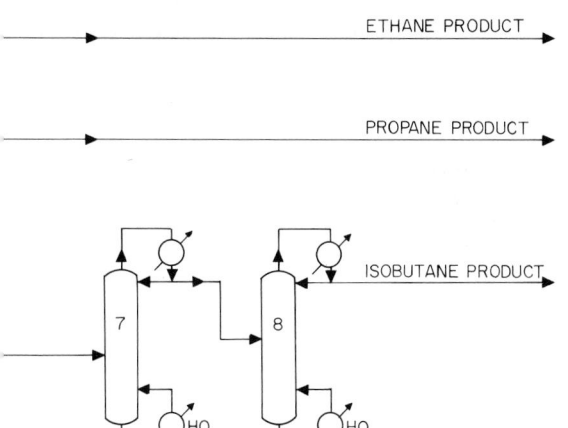

ETHANE PRODUCT

PROPANE PRODUCT

ISOBUTANE PRODUCT

BUTANE PRODUCT

NATURAL GASOLINE PRODUCT

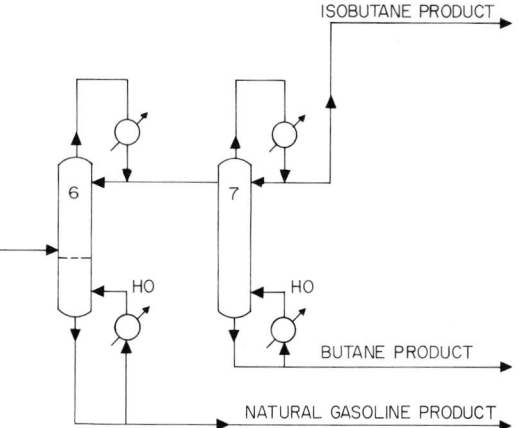

ETHANE PRODUCT

PROPANE PRODUCT

ISOBUTANE PRODUCT

BUTANE PRODUCT

NATURAL GASOLINE PRODUCT

drums provided. The vapor effluent from the knockout drums (the Btu-controlled tail gas) is heat-exchanged against the incoming feed and exits from the cold box at essentially ambient temperature.

The condensed hydrocarbon phases are directed to the high-pressure demethanizer, where the methane is fractionated from the product hydrocarbon stream. The cold high-pressure methane vapor is isenthalpically expanded and directed into the heat exchangers, where it is combined with the cold tail gas. The bottoms of the demethanizer contain the recovered hydrocarbons and subsequently are fractionated into the desired hydrocarbon products, as shown by the flow diagram.

If the feed contains CO_2 and it is considered a contaminant, the CO_2 is removed from the ethane product stream. The refrigeration compressors may be driven by a gas turbine. Waste heat is recovered by a circulating hot-oil system, which in turn supplies reboiler heat to the fractionating train. Fractionating-tower condenser duties are handled by refrigeration or air coolers, depending upon the towers involved. See also **Cryogenic processes.**

—Henry E. Duckham, Jr., *The M. W. Kellogg Company, A Division of Pullman Incorporated, Houston, Tex.*

Natural Rubber   (See **Rubber, natural.**)

Naumannite   (See **Selenium.**)

Neat Soap   (See **Soaps.**)

Nematic Crystals   (See **Liquid Crystals.**)

Neodymium   (Gk. *neo*, new and *didymous*, occurring in pairs.) **Nd** = 144.24 (at. wt.); 60 (at. no.). Neodymium is the fourth of 15 elements in group III, period 6, generally shown as the rare-earth elements in a separate line below the main body of the periodic table and sometimes referred to as the *lanthanide series.* Nd was discovered in 1885 by C. A. von Welsbach in Vienna.

Pure Nd metal has a silvery gray metallic luster, which dulls on exposure to moist air at ambient temperatures. When impurities are low in the metal, it is soft and malleable and can be processed with ordinary tools and equipment. The metal is pyrophoric; processing and storage under a nonreactive coolant, inert gas, or vacuum is required.

The melting point of Nd is 1016°C; boiling

point 3068°C; density 7.004 g/cm^3. Estimated at 28 ppm in the earth's crust, Nd is potentially more abundant than cobalt, niobium, or lead.

Neodymium has seven natural isotopes, 142 (27.09 w/o*), 143 (12.14 w/o), 144 (23.83 w/o), 145 (8.29 w/o), 146 (17.26 w/o) 148 (5.74 w/o), and 150 (5.63 w/o). Seven artificial isotopes have been identified. ^{144}Nd is mildly radioactive with a half-life of 10^{10}–10^{15} years. The metal has a low acute-toxicity rating.

Dilute mineral acids dissolve Nd readily. The metal reacts slowly with cold water. This element forms chemical compounds with most inorganic and many organic anions. Brittle intermetallic compounds are formed with most of the elements to the right of Fe in the periodic table. Properties of the many possible alloys need to be defined and evaluated.

Neodymium has one valence of 3$^+$. Its oxide, Nd$_2$O$_3$, melts at 2270°C and is hygroscopic. Pure Nd$_2$O$_3$ is light blue and is soluble in strong mineral acids and formic and acetic acid. Solutions of Nd^{3+} and its many chemical compounds are characterized by a distinctive red-violet color. It has eight narrow absorption peaks in the ultraviolet, visible, and infrared regions of the spectrum. Anhydrous Nd$_2$O$_3$ (7.24 g/cm^3) is denser than Nd metal (7.004 g/cm^3); and has a larger coefficient of thermal expansion. Added to soda-lime glass up to 5 w/o, Nd$_2$O$_3$ produces colors from light violet to red-purple; Nd does not produce any visible fluorescence under ultraviolet excitation, however, as some of the other lanthanide elements do.

Neodymium is the third most plentiful of the light lanthanide elements. By 1839 cerium and lanthanum had been identified as separate elements in the mineral named *cerite earths* found near Bastnäs, Sweden. The balance of the mineral was called *didymium*. In 1880, another element, gadolinium, was separated from didymium. In 1885, Welsbach separated didymium into two components, naming them neo- (new) and praseo- (green) dymium.

The minerals bastnasite and monazite, which contain 15–25% Nd, supply all requirements. Separation of Nd from the other chemically similar elements is by liquid-liquid or solid-liquid organic ion-exchange processes. Commercial separation plants have been constructed that can produce several hundred thousand pounds of

*This abbreviation is used in the rare-earth and related fields for weight percent.

Nd$_2$O$_3$ or chemicals a year. The metal is produced by electrolysis of fused anhydrous NdCl$_3$ or the electrolytic reduction of the oxide in molten NdF$_3$. For a detailed description of resources and processes and a table of properties see **Rare-earth elements and metals.**

Uses: Neodymium, as one of the three major components in light rare-earth mixtures, has been used since late in the nineteenth century in mischmetal, a pyrophoric metal known as *lighter flint*. It is also used in the mixture burned in the cores of arc carbons which produce the intense light required for motion-picture projection lamps. A mixture of rare earths containing Nd is also used to polish optical glass although it is questionable whether Nd is an effective element in the application.

Since early in the 1960s the largest consumption of Nd is by the petroleum-refining industry, which uses mixtures of rare-earth elements that have proved to be the most efficient catalysts for cracking crude oil.

The first application for elemental Nd was to color glass. The range of pure violet to purple colors obtainable makes Nd useful for a variety of art glass objects and tableware. It is also used in sunglasses, in goggles to protect welders' eyes, and in decorative fiber optics. In glass at 3–5 w/o Nd exhibits dichroic properties; i.e., it shows two or more colors depending upon the light under which the glass is viewed.

One of the first solid laser materials discovered in the 1960s was a Nd-doped glass, later developed into Nd-doped single-crystal yttrium–aluminum oxide garnets (Nd:YAG) which have the highest pulsed-laser efficiency operating at room temperature. The strong narrow absorption peaks of Nd in the visible spectrum and the high-temperature stability of its oxide assure its use as broader applications for lasers are discovered.

Classical didymium, which contains about 75% Nd and 25% Pr, is a metallurgical additive to extend the operating temperature range of cast magnesium-alloy parts used in aircraft and satellites.

Another commercial use of Nd$_2$O$_3$, developed in 1970, is in barium titanate capacitors. Additions of Nd$_2$O$_3$ up to 30 w/o increase the dielectric strength of these electronic components over a wider temperature range. Nd is a sensitizer in special phosphate-type phosphors, which have application in coded inks used for automatic sorting by the postal service and financial institutions.

The optical and magnetic properties of this

element offer a fertile field for further research and development.

See references under **Rare-earth elements and metals.**

—Joseph G. Cannon, *Molybdenum Corporation of America, White Plains, N.Y.*

Neon (See **Chemical elements.**)

Neoprene (See **Elastomers.**)

Neptunium (See **Chemical elements.**)

Network Polymers (See **Polymerization.**)

Neutralization Value (See **Vegetable oils.**)

Neutrino A neutrino is a subatomic particle which carries no charge and which has a rest mass described as *vanishingly small*. These particles do not react readily with other matter.

Neutron A neutron is an elementary subatomic particle located in the nucleus of an atom. The neutron carries no electric charge and has a mass of 1.008986 amu (atomic mass units). Further

$$N = A - Z$$

where N = number of neutrons in nucleus
A = mass number
Z = atomic number
Neutrons, along with protons, make up the nuclei of atoms.

Type	Description
Delayed	Released from fission products following the initial fission
Fast	Energy in excess of 0.1 MeV
Intermediate . .	Energy between 100 eV and 0.1 MeV
Slow	Energy under 100 eV
Thermal.	Energy and velocity determined by surrounding temperature; at 21°C *average energy* = 0.025 eV and *probable velocity* = 2,200 m/s

Neutron flux is the number of neutrons that will pass through a unit volume in a unit of time.

See also **Nuclear power plants;** and **Radioactive isotopes.**

Neutron Moderation (See **Nuclear power plants;** and **Radioisotopes.**)

Newtonian Substance (See **Colloid systems.**)

Nickel (G. *kupfernickel,* Old Nick's copper.) **Ni** = 58.71 (at. wt.); 28 (at. no.). Nickel exists in nature in five stable isotopes, of which the two most common are ^{58}Ni, accounting for 67.7% of the earth's nickel deposits, and ^{60}Ni, with 26.2%. The rarer forms are ^{61}Ni, 1.25%; ^{62}Ni, 3.66%, and ^{64}Ni, 1.6%. One of the most abundant elements in the universe, ranking eleventh or twelfth, nickel ranks twenty-fourth among constituents of the earth's crust, making up a scant 0.008% of the whole. Nickel is in group VIII of the periodic table, along with palladium and platinum.

White in color, refined Ni is strong, tough, ductile and is further characterized by a high degree of resistance to heat, corrosion, and abrasion. Thus, Ni is one of the most important alloying elements. It has a density of 8.908 g/cm^3 at 20°C, a melting point of 2647°F (1452°C), and a boiling point of 5252°F (2900°C). Although Ni is magnetic, these properties are lost when the metal is heated above 345°C.

Early Use and Isolation: Ni alloys have been used by man, albeit unknowingly, since he first learned to use metals. Meteorites are believed to be prehistoric man's earliest source of Ni, which he used to forge implements and weapons that had to be strong and tough. *Kupfernickel,* a seventeenth-century word meaning Old Nick's copper, was used by Saxon miners to describe ores that were difficult to smelt. Ni was isolated and identified in 1751 by Axel Fredric Cronstedt (1722–1765), a chemist with the Swedish Department of Mines, from a sulfide ore known as *los* or *gersodorffite,* NiAsS. Cronstedt carried out a series of experiments to determine the new metal's properties and named the element in 1754, when he published Continuation of Results and Experiments on the Los Cobalt Ore.

Sources: For more than 100 years after Cronstedt's work, the major source of Ni was Germany and Scandinavia, and the major use was as a plating material. In 1865, vast deposits of lateritic (also called oxide) nickel ores were discovered in New Caledonia, which for the next 40 years was the world's primary source of the metal. In 1883, even larger deposits (sulfide) were discovered in Sudbury, Ontario, which has been the world's largest source of Ni since 1905.

The most common sulfide nickel-bearing mineral is pentlandite, $(FeNi)_9S_8$, a bronze-yellow mineral that contains about 34% Ni. Pentlandite usually is found in sulfide ores together with large amounts of pyrrhotite, $Fe_{n-1}S_n$, and significant amounts of chalcopyrite, $CuFeS_2$. Sulfide ores

TABLE N-2. *Major Nickel-bearing Ores*

Ore	Formula (Idealized)	Nickel, %	Color	Crystal System
Arsenate:				
Annabergite	$Ni_3As_2O_8 \cdot 8H_2O$	29.40	Apple green	Monoclinic
Arsenides:				
Gersdorffite	NiAsS	35.42	Steel gray	Isometric
Niccolite	NiAs	43.92	Copper red	Hexagonal
Maucherite.	$Ni_{11}As_8$	51.85	Platinum gray	Tetragonal
Rammelsbergite	$NiAs_2$	28.15	Tin white	Orthorhombic
Silicate and oxide:				
Garnierite	$(Ni,Mg)_6Si_4O_{10}(OH)_8$	Up to 47	Green gray	Amorphous
Nickeliferous limonite.	$(Ni,Fe)_2O_3 \cdot nH_2O$	Low		
Sulfides:				
Heazlewoodite	Ni_3S_2	73.30	Bronze yellow	Isometric
Millerite	NiS	64.67	Brass yellow	Hexagonal
Pentlandite.	$(Fe,Ni)_9S_8$	34.22	Bronze yellow	Isometric
Polydymite	Ni_3S_4	57.86	Steel gray	Isometric
Siegenite	$(Co,Ni)_3S_4$	28.89	Steel gray	Isometric
Violarite	$(Ni,Fe)_3S_4$	38.94	Violet gray	Isometric

thus contain large amounts of Fe and Cu along with Ni, as well as varying amounts of Co and precious metals. The largest known reserves of sulfide ores are in Canada and the Soviet Union. Other reserves are in Australia, Finland, the Republic of South Africa, and Rhodesia.

About four-fifths of the world's known Ni reserves are contained in oxide and silicate ores located mainly in tropical and subtropical regions. The most common lateritic minerals are garnierite, found in New Caledonia, and limonite, found in Cuba. Other major lateritic deposits exist in Australia, Indonesia, the Philippines, Central and South America, the United States, and the Soviet Union. Lateritic ores lie closer to the earth's surface than sulfide ores and thus are easier to mine, but since the Ni is dispersed in a lateritic ore rather than concentrated as it is in a sulfide ore, lateritic ores are more difficult to extract.

Major Ni ores are listed in Table N-2.

Processing: Ni in sulfide ores occurs as a distinct mineral and can be separated by beneficiation methods, e.g., flotation and magnetism, after the ore has been crushed and ground. The enriched ore then is refined. Because the Ni content is dispersed throughout lateritic ores, the entire ore must be treated by one or a combination of three basic refining steps—hydrometallurgy, pyrometallurgy, or vapometallurgy—in order to extract the nickel.

The process used on sulfide ores in the Copper Cliff, Ont. smelters of The International Nickel Company of Canada, Limited is shown in Fig. N-4. The matte or concentrate products of sulfide ores then must be refined to obtain pure metallic nickel. The two most common refining processes are the electrolytic process and the Mond (or carbonyl) process. The electrolytic process, shown in Fig. N-5, yields Ni of more than 99.9% purity, and permits the efficient recovery of other metals in the sulfide ores. The carbonyl process, used at International Nickel's refinery at Clydach, Wales, is shown in Fig. N-6. In this process, CO converts impure Ni powder to nickel carbonyl gas, which is decomposed into pure Ni pellets. International Nickel expanded its production in mid-1973 with the installation of a carbonyl process facility at Sudbury, Ontario. This plant has a capacity of 100 million lb of nickel pellets as well as 25 million lb of nickel powders for the North American market.

Forms and Uses: The three most widely used forms of primary Ni are (1) electrolytic cathode sheets, which vary in size from 1 in.^{2} to 1 yd^2, (2) pellets produced by the decomposition of nickel carbonyl, and (3) ferronickel. The price and purity of electrolytic and pellet forms are the same. Pellets are traditionally favored in Europe, whereas electrolytic Ni is preferred in North America. Ferronickel is widely used for the manufacture of stainless steel. Ferronickel has a Ni content of 24–48%. Ni also is available in the form of powder, briquettes, ingots, and shot. Nickel oxide sinters also are produced for applications where pure nickel is not required.

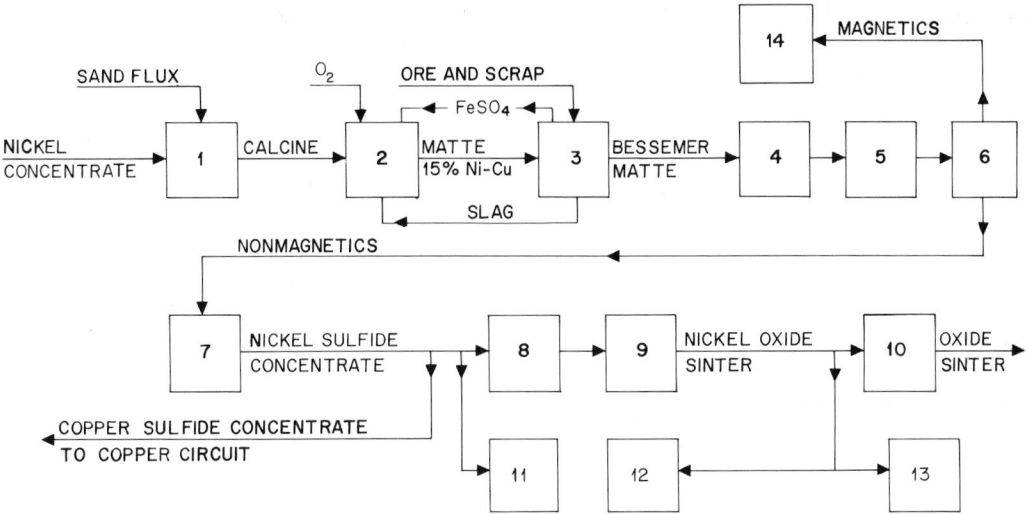

Fig. N-4. Generalized representation of nickel circuit for concentrating sulfide ores at Copper Cliff smelter. (1) Roaster, (2) reverberatory furnaces, (3) converter furnaces to which high-grade lump ore and scrap are fed, (4) controlled cooling molds, which receive bessemer matte (77% Ni; 22% Cu sulfides), (5) breaking, crushing, and grinding, (6) magnetic separation, (7) matte flotation, (8) pelletizer, (9) fluid-bed roasters, (10) chlorinator circuit, (11) Port Colborne refinery (sulfide anodes), (12) Port Colborne refinery (metal anodes), (13) Clydach refinery, (14) precious-metal recovery. (*The International Nickel Company of Canada.*)

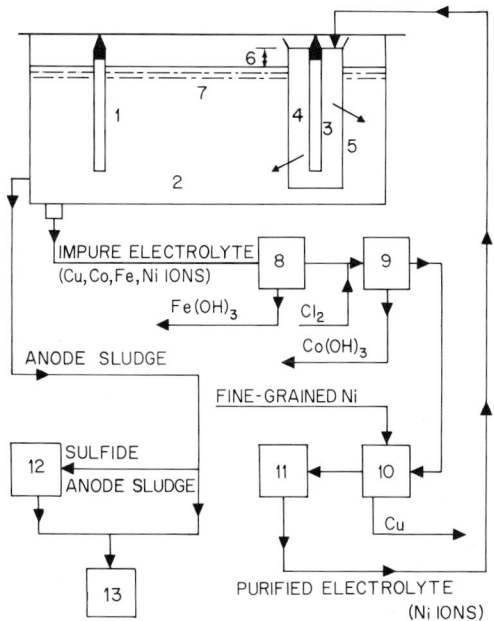

Fig. N-5. Generalized electrorefining materials flow for production of nickel. (1) Impure anode, (2) impure electrolyte, (3) cathode, (4) pure electrolyte, (5) canvas compartment, (6) hydrostatic head, (7) electrolytic tank, (8) precipitation of Fe, (9) precipitation of Co, (10) precipitation of Cu, (11) pH adjustment, (12) recovery of S, (13) precious-metal recovery. (*The International Nickel Company of Canada.*)

Over half of all Ni consumed is for stainless steels and high-nickel alloys. Other major uses are plating, constructional alloy steels, iron and steel castings, in copper and brass products, and in coinage. The largest single use is in austenitic stainless steels, which contain 3.5–22% Ni and 16–26% Cr, depending upon grade. The presence of Ni stabilizes the austenite and enhances the ductility of the steel. Ni and Cr both contribute to the corrosion-resistant properties of stainless steel. Ni in amounts up to 9% adds strength, hardness, and toughness to a large number of alloy steels. A 9% Ni steel remains stable at low temperatures and will handle liquefied gases. Low-nickel steels containing 0.5–0.7% Ni are strong, ductile, and tough and are used for parts of automobiles, power machinery, and construction equipment. There are several thousand nickel-containing alloys, ranging from 0.02% hardenable silver alloy to 99% malleable nickel. See Table N-3.

Finely divided Ni can dissolve up to 17 times its volume of hydrogen, a property that makes it suitable for catalyzing hydrogenation and other processes. Ni also finds application in alkaline storage batteries.

Compounds: Most Ni salts are pale green crystals, yellow in the anhydrous form. The borate, carbonate, cyanide, ferrocyanide, ferricyanide, oxalate, phosphate, and sulfide of nickel all are insoluble. Several Ni salts form soluble compounds with ammonium hydroxide. $NiCl_2$ is sol-

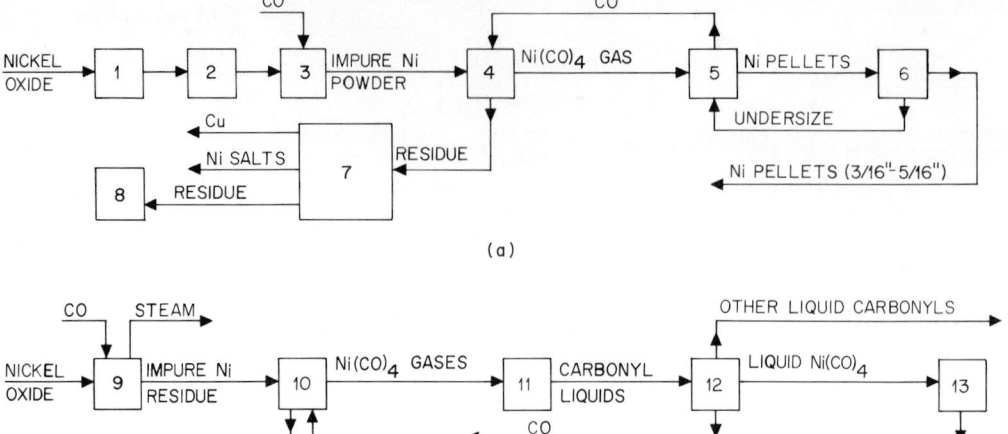

Fig. N-6. Generalized materials flow of Mond (or carbonyl) process for refining nickel. (a) Atmospheric carbonyl process; (b) pressure carbonyl process. Representation is based upon Clydach refinery. (1) Grinding, (2) rotary-hearth roasters, (3) reducers, (4) volatilizers, (5) decomposers, (6) screen, (7) leaching, (8) precious-metals recovery, (9) reducer, (10) volatilizer operating under pressure, (11) condenser, (12) distillation column, (13) vaporizer, (14) decomposer, (15) precious-metals recovery with processing for Ni, Co, Cu, and precious metals. (*The International Nickel Company of Canada.*)

TABLE N-3. *Physical Properties of Wrought Nickel and Representative Nickel Alloys*

		Alloys						
	Wrought Nickel	Duranickel 301	Monel 400	Hastelloy B	Hastelloy F	Inconel 600	Incoloy 800	Illium G
Approximate composition, %:								
Ni	99	93.9	66.0	63.5	45.5	72	32.5	56.0
Cu	0.25	0.05	31.5	. . .	. . .	0.5	0.75	6.5
C	0.15	0.15	0.12	0.05	0.05	0.15	0.10	
Fe	. . .	0.15	1.35	5.0	20.5	8.0	45.6	
Al	. . .	4.5						
Ti	. . .	0.5						
Co	. . .	. . .	. . .	2.5	2.5			
Cr	. . .	. . .	. . .	1.0	22.0	15.5	21.0	22.5
Mo	. . .	. . .	. . .	28.0	6.5	. . .	. . .	6.5
V	. . .	. . .	. . .	0.3				
W	. . .	. . .	. . .	. . .	1.0			
Cb + Ta	. . .	. . .	. . .	. . .	2.0			
Density, lb/in.³	0.321	0.298	0.319	0.334	0.295	0.304	0.290	0.310
Melting range, °F	2615–2635	2550–2620	2370–2460	2408–2462	2350	2500–2600	2475–2525	2290–2440
Specific heat, Btu/lb	0.109	0.104	0.102	0.0907	0.1025	0.106	0.12	0.105
Electrical resistivity, $\Omega/(cmil)(ft)$	57	255	307	811	673	620	595	734
Modulus of elasticity in tension, 10^6 psi	30.0	30.0	26.0	30.8	29.0	31.0	28.0	31.0
Poisson's ratio	0.31	0.31	0.32	. . .	0.305	0.29	0.30	0.29
Thermal conductivity at 70°F, $(Btu)(in.)/(hr)(ft^2)(°F)$	420	165	151	. . .	<100	103	80	94
Coefficient of thermal expansion, 70–600°F, $\mu in./(in.)(°F)$	8.0	7.7	8.8	6.4	8.7	7.9	9.0	

TABLE N-4. *Properties of Representative Nickel Compounds* *

Compound	Formula	Formula Weight	Sp gr	Mp, °C	Bp, °C
Nickel acetate	$Ni(C_2H_3O_2)_2$	176.74	1.80	d	
Nickel ammonium chloride	$NiCl_2 \cdot NH_4Cl \cdot 6H_2O$	291.19	1.65		
Nickel orthoarsenate	$Ni_3(AsO_4)_2$	453.93	4.98		
Nickel carbonate	$NiCO_3$	118.69	. . .	d	
Nickel carbonyl	$Ni(CO)_4$	170.69	1.32	−25	43
Nickel chloride	$NiCl_2$	129.60	3.55	s	973
Nickel cyanide	$Ni(CN)_2$	110.71	. . .	200	d
Nickel dimethylglyoxime	$Ni[(CH_3)_2(CNO)_2H]_2$	288.83	. . .	s	
Nickel ferrocyanide	$Ni_2Fe(CN)_6 \cdot 11H_2O$	527.44	. . .		
Nickel fluoride	NiF_2	96.69	4.63		
Nickel formate	$Ni(CHO_2)_2 \cdot 2H_2O$	187.74	2.15	d	
Nickel fluosilicate	$NiSiF_6 \cdot 6H_2O$	308.84	2.13		
Nickelic hydroxide	$Ni(OH)_3$	109.71	. . .	d	
Nickelous hydroxide	$Ni(OH)_2$	92.71	4.1	d	
Nickel nitrate	$Ni(NO_3)_2 \cdot 6H_2O$	290.80	2.05	56.7	136.7
Nickel orthophosphate	$Ni_3(PO_4)_2 \cdot 7H_2O$	492.22	. . .		
Nickel sulfate	$NiSO_4$	154.75	2.12	d	
Nickel monosulfide	NiS	90.75	4.60	797	
Nickel tetrapyridine fluosilicate	$Ni(C_6H_5N)_4SiF_6$	516.94	2.31		

*d decomposes; s sublimes.

uble in alcohol, and $Ni(NO_3)_2$ is soluble in dilute alcohol. The properties of several representative Ni compounds are given in Table N-4.

REFERENCES

Boldt, Joseph R., and Paul Queneau: "The Winning of Nickel," Longmans, Toronto, 1967.

Howard-White, H. F.: "Nickel: An Historical Review," Van Nostrand, Princeton N.J., 1963.

Nickel: Canada and the World, *Min. Rep.* 16, Queen's Printer, Ottawa, 1968.

Nickel and Its Alloys, *U. S. Natl. Bur. Stand.: Monogr.* 106, Washington, 1968.

Nickel Silver (See **Copper.**)

Nicotinic Acid (See **Pyridine and derivatives.**)

Ninhydrin Reaction (See **Amino acids.**)

Niobium (Gk. *Niobe,* daughter of Tantalus.) **Nb** = 92.91 (at. wt.); 41 (at. no.). This element, first observed by C. Hatchett (1801) while examining a mineral found in Connecticut, also is commonly referred to as *columbium,* so named by Hatchett. The name was selected because of the original discovery of the element in the United States (Columbia = United States). The symbol of columbium is **Cb.**

Later, European chemists essentially rediscovered the element and renamed it *niobium* because of its close association with tantalum and, in so doing followed the traditions established by Berzelius in favoring the naming of elements from ancient mythology rather than using names of places. A study of the list of the over 100 elements (see also **Chemical elements**) shows, of course, that the naming of the elements has been anything but consistent. Further, the naming of the elements has been subject to controversy and disagreement in a number of instances, including some of the most recent discoveries, e.g., atomic numbers 101 and 103.

Even though in some official circles, the name of this element now is formally accepted as niobium with the symbol Nb, the use of columbium persists throughout much of the literature. This element is described in detail under **Columbium.**

Nitration (See **Explosives;** and **Nitric acid.**)

Nitric Acid Nitric acid, HNO_3, has been, first, a curiosity of the alchemists and, later, an important industrial commodity in the 1,000 years of its recorded history. The acid was known to the alchemists as aqua fortis (strong water) or aqua valens (powerful water). It was of particular interest to these early experimenters because it dissolved metals including Cu and Ag. The early chemists also were fascinated by the fact that addition of sal ammoniac (ammonium chloride) gave aqua regia (royal water), which dissolves Au as well as Ag

Nitric acid is a colorless liquid with a specific gravity of 1.503 at 25°C, a freezing point of −41.6°C, and a boiling point of 86°C. This 100% acid is not entirely stable and must be prepared from its azeotrope (constant-boiling mixture) by distillation with concentrated H_2SO_4.

Reagent-grade concentrated HNO_3 is a water solution containing about 68% HNO_3 by weight. This strength corresponds to the constant-boiling mixture of the acid with water, which is 68.4% HNO_3 by weight at atmospheric pressure and boils at 121.9°C.

Nitric acid is completely miscible with water. It forms two solid hydrates, $HNO_3 \cdot H_2O$ and $HNO_3 \cdot 2H_2O$, with corresponding melting points of approximately −38 and −18.5°C.

Nitric acid is a strong acid and a powerful oxidizer. In dilute solution it is almost completely ionized to H^+ and NO_3^- ions and behaves like a strong acid.

With organic compounds, HNO_3 may act as a nitrating agent, as an oxidizing agent, or simply as an acid. The classic example of nitration is its reaction with benzene or toluene in the presence of concentrated H_2SO_4 to form nitrobenzene or trinitrotoluene (TNT). An example of oxidation properties is in the oxidation of cyclohexanol by HNO_3 to produce adipic acid, an intermediate of nylon. Behaving like an acid, it forms nitroglycerin by esterification of glycerol in the presence of concentrated H_2SO_4.

One peculiar property of HNO_3 is its ability to passivate some metals such as iron and aluminum. This property is of significant industrial importance since modern processes for producing the acid depend on it. Modern suitably formulated stainless-steel alloys are usefully resistant to nitric acid through a wide range of conditions. The acid's passivity or the metal's resistance to attack is attributed to the formation of a protective oxide layer on the surface of the metal.

United States production of HNO_3 now exceeds 5 million tons/yr. About 75% is used in the manufacture of agricultural fertilizers, largely in the form of NH_4NO_3. Some 15% is used in explosives (nitrates and nitro compounds), and about 10% is consumed by the chemical industry. As the red fuming acid or as nitrogen tetroxide HNO_3 is used extensively as the oxidizer in propellants for space rockets and other missiles.

Production: Three commercial methods have been developed for nitric acid production: (1) the reaction between sulfuric acid and sodium nitrate, (2) the thermal combination of oxygen and nitrogen in the air, and (3) the catalytic oxidation of ammonia and absorption of the gaseous products in water.

Large-scale production of HNO_3 based on the reaction between sulfuric acid and sodium nitrate obtained from crude Chilean saltpeter was started about 1825. This process dominated the field for almost 100 years until it was superseded by the ammonia oxidation process in the early 1920s.

Production of HNO_3 by the thermal combination of oxygen and nitrogen in the air has been accomplished by the Birkeland-Eyde process, developed in Norway about 1900, and by the Wisconsin thermal process, developed at the University of Wisconsin in the 1930s and successfully demonstrated by the U.S. Army in the 1950s.

Today, all commercial production is based on the catalytic oxidation of ammonia and absorption of the gaseous products in water. This process, developed by the German chemist Wilhelm Ostwald, was based on earlier work by the French chemist C. F. Kuhlmann. By the Ostwald process, HNO_3 is produced in a three-stage operation: ammonia is oxidized to nitric oxide, the nitric oxide is further oxidized to nitrogen dioxide, and the gases are absorbed in water to yield HNO_3, according to the simplified equation

$$4NH_3 + 5O_2 \rightarrow 4NO + 6H_2O$$
$$2NO + O_2 \rightarrow 2NO_2$$
$$3NO_2 + H_2O \rightarrow 2HNO_3 + NO$$

The nitric oxide formed in the last equation returns to the gas phase, is reoxidized to nitrogen dioxide, and reabsorbed. These reactions are highly exothermic. In reality, numerous complex reactions occur in connection with this process.

In a manufacturing plant, air is preheated, mixed with superheated ammonia vapor, and reacted catalytically over a gauze composed of 90% Pt and 10% Rh at a temperature of 800–960°C and operating pressures between atmospheric and 120 psig. The reaction produces nitrogen dioxide, NO_2, and nitric oxide, NO. The latter is oxidized to NO_2 in the reaction train. The NO_2 actually exists in equilibrium with its dimer, N_2O_4. This equilibrium mixture, sometimes referred to as *nitrogen peroxide,* is absorbed in water in a cooled absorber tower to form HNO_3 at a strength of 55–60% HNO_3.

The first plants were operated at atmospheric

or low pressures. It was soon found that higher pressures would both reduce capital costs and increase operating efficiency. Higher pressures favored increased oxidation of the NO, improved absorption, and higher acid strength. This led to the development of the Du Pont pressure process using up to 110 psig. Several proprietary HNO_3 processes are offered.

The plant illustrated in Fig. N-7 has a capacity of 500 tons/day and a high overall conversion efficiency, averaging at least 95%. Most conventional nitric acid units convert 90–92.5% of the nitrogen in ammonia to the acid. The plant shown makes extensive use of alloy steel and thoroughly mixes the air and ammonia before entry to the converter. The main heat exchangers are butted together to form a single unit.

A centrifugal compressor, driven by a 4,000-hp steam turbine and a tail-gas turbine type of expander, supplies air for the feed stream. The steam turbine is used for starting the compressor and for supplemental power during normal operation.

The compressed air stream, preheated to 500°F (260°C) joins the vaporized ammonia feed in a double mixing operation based on orificing. Air (90 vol %) passes axially through a pipe section, while ammonia (10 vol %) enters radially into the stream through other orifices for more mixing. This ammonia-air mixture then enters a converter and flows through a Pt-Rh gauze catalyst, where the NH_3 stream ignites and burns very rapidly to NO and H_2O at about 1650°F ($\cong$900°C) and 9 atm.

Hot reaction products and excess air leaving the converter pass through a row of heat exchangers that provide air preheating, tail-gas heating, by-product steam for the 4,000-hp turbine, and export steam. A filter then removes any stray catalyst particles the process stream may have picked up from the gauze.

The heat exchangers, along with the ammonia converter and catalyst filter, are butted against each other, forming a compact unit that permits high heat recoveries as well as savings in space, piping, and exchanger heads. This arrangement is claimed to promote the oxidation of NO to NO_2 before the process stream undergoes condensation.

Partially cooled process gas from the heat-exchanger system passes into a cooler-condenser of special shell-and-tube design, where some weak acid is formed as soon as by-product water from the ammonia oxidation undergoes condensation and absorbs a portion of the nitrogen dioxide.

This weak acid is separated and introduced to a middle tray of the absorption tower while the gas fraction enters at the bottom. In the absorber (a 72-ft-high column fitted with cooling coils and 37 trays), NO_2 reacts with water to form HNO_3. The stripped process gas, which is now essentially nitrogen, leaves as a wet overhead (tail gas) while HNO_3 of about 58% concentration is withdrawn at the bottom.

After a demisting operation, the tail gas is heated against steam and then passes through the shell side of the heat exchangers that cool the reaction products. To abate air pollution, the tail gas next goes through a catalytic burner that uses natural gas or ammonia-plant waste gases as fuel. This combustion reduces the residual nitrogen dioxide so that the final tail gas is a colorless stream consisting mostly of N_2, O_2, and CO_2. Use of the combustor has the advantage of increasing the tail-gas temperature for more efficient power recovery by the turbine expander.

Now purified, the gas stream can be safely directed to the turbine expander for power recovery to assist in driving the main compressor. A steam superheater (not shown in Fig. N-7) is installed between the catalytic burner and the expander to reduce the tail-gas temperature to the level required for proper operation of the turbine-gas expander.

The catalytic burner, which reduces NO_2 contamination of the atmosphere, also increases the life of the expander because tail gas has no HNO_3 carry-over that could eventually strip out the turbine blades. See also **Air pollution.**

Handling: Safety precautions include (1) avoiding contact with the acid and using protective equipment, particularly for eye protection (safety goggles with side shields or cup-type rubber-framed goggles); (2) avoiding breathing the fumes given off when HNO_3 dissolves metals or oxidizes organic compounds (the fumes, often red-brown, are toxic and can cause serious delayed effects without immediate discomfort); and (3) flushing any spills with copious quantities of water and disposing immediately, with caution, or any wood or other cellulosic product that has been exposed to the acid (although HNO_3 itself does not burn, it oxidizes organic matter, such as paper, cloth, or wood, to form a highly flammable material).

REFERENCES
Bingham, E. C., Jr.: Compact Design Pays Off at New Nitric Acid Plant, *Chem. Eng.,* May 23, 1966.
Chilton, Thomas H.: "Strong Water," MIT, Cambridge, Mass., 1968.

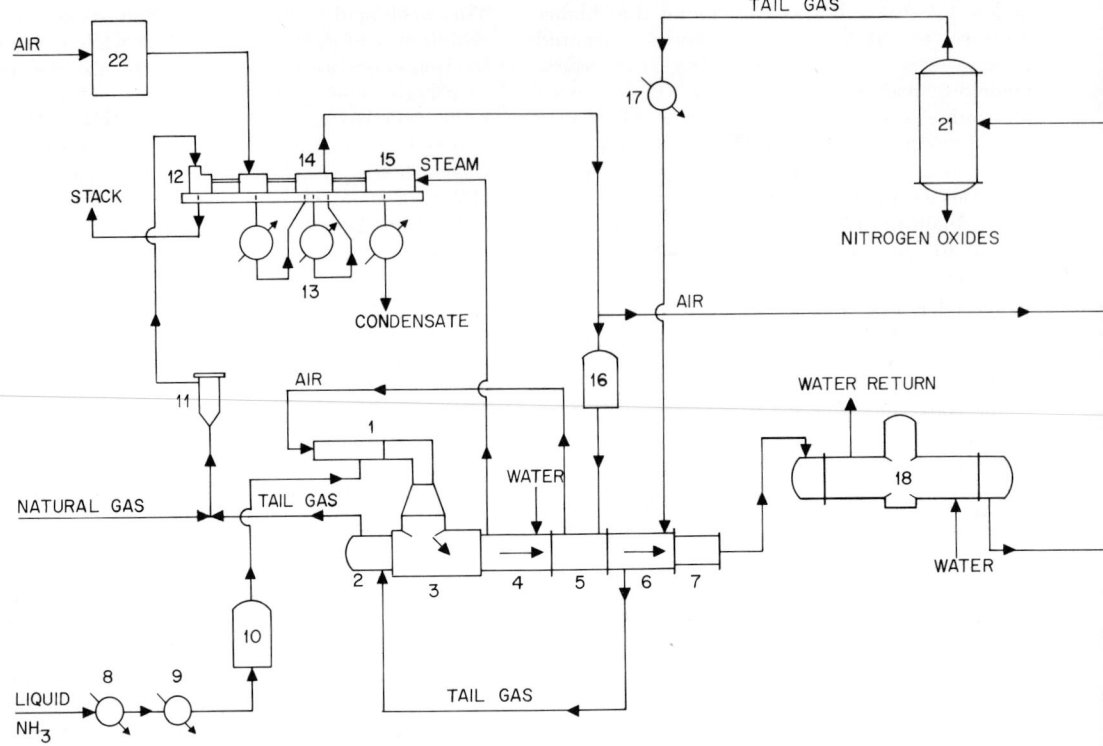

Fig. N-7. Schematic representation of materials flow in a 500 tons day nitric acid plant. (1) Mixer, (2) turbine gas heater, (3) converter, (4) waste-heat boiler, (5) air preheater, (6) tail-gas heater, (7) filter, (8) vaporizer, (9) superheater, (10) ammonia filter, (11) tail-gas burner, (12) expander turbine, (13) intercoolers, (14) axial centrifugal compressor, (15) steam turbine, (16) air filter, (17) tail-gas preheater, (18) cooler-condenser, (19) weak-acid separator, (20) absorber tower, (21) entrainment separator. Air from (5) introduced to mixer (1) is at 260°C (500°F) and at a pressure of 135 psia. Burned tail gas from (11) consists of N_2, O_2, and CO_2. Tail gas enters tail-gas preheater (17) at 38°C (100°F). The bottoms from entrainment separator (21) consist of dissolved nitrogen oxides, which flow to the acid sump.

Pratt, Christopher J., and Robert Noyes: "Nitrogen Fertilizer Chemical Processes," Noyes Development Corp., Pearl River, N.Y., 1965.

—Edward C. Bingham, *Farmers Chemical Association, Inc., Harrison, Tenn.*

Nitrides (See **Aluminum; Iron and steel;** and **Nitrogen.**)

Nitriding (See **Iron and steel.**)

Nitrile Rubbers (See **Elastomers.**)

Nitrilotriacetic Acid (NTA) (See **Chelating agents,** and **Formaldehyde.**)

Nitrocotton (See **Explosives.**)

Nitrogen (Gk. *nitron,* niter.) **N** = 14.0067 (at. wt.); 7 (at. no.). Nitrogen, N_2, is a colorless, odorless, tasteless, nontoxic gas at standard conditions with a density = 1.1649×10^{-3} g/cm³ at 20°C (0.04209×10^{-3} lb/in.³ at 68°F). The melting point = -210.0 ± 0.1°C; boiling point = 195.8°C; specific gravity = 1.026 at -252.5°C and 0.808 at -195.8°C; the critical temperature = -147.1°C; the critical pressure = 33.5 atm. Thermal conductivity = 0.000060 (cal)(cm)/ (s)(cm²)(°C) at 20°C. Heat of fusion is 172 cal/mole at the melting point; heat of vaporization is 1336 cal/mole at the boiling point.

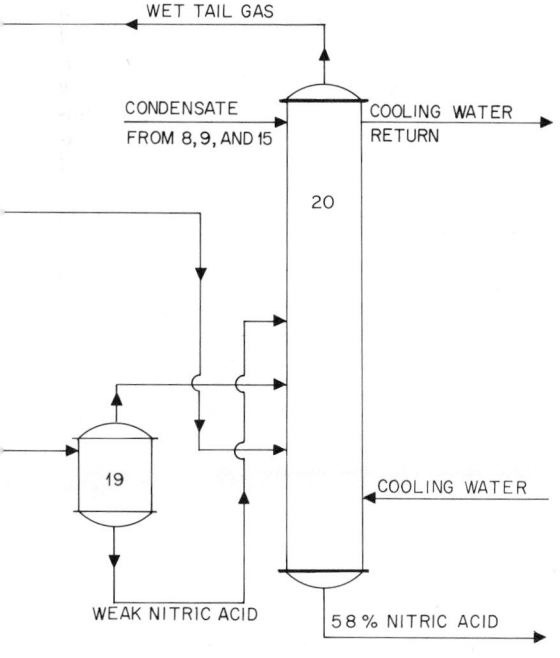

WET TAIL GAS

CONDENSATE
FROM 8, 9, AND 15

COOLING WATER
RETURN

20

COOLING WATER

19

WEAK NITRIC ACID

58% NITRIC ACID

N_2 is quite inert and will not attack free elements under ordinary conditions. When nitrogen is mixed with O_2 and subjected to an electric discharge, nitrogen peroxide is formed. Nitrogen will combine directly with several metals when ignited with them to form nitrides, such as Li_3N and Mg_3N_2. Halogens when combined with nitrogen produce very explosive compounds. The ionization potential of nitrogen is 14.54 eV.

N_2 is slightly soluble (less than O_2) in cold H_2O (2.35 parts N_2 in 100 parts H_2O at 0°C) with solubility decreasing with rise in temperature (1.55 parts N_2 in 100 parts H_2O at 20°C). N_2 is only slightly soluble in alcohol and is practically insoluble in most other known liquids.

Nitrogen occurs in group VA and period 2 of the periodic table. There are two stable isotopes of nitrogen, 14 and 15, and four unstable isotopes 12, 13, 16, and 17. See also **Chemical elements.**

Nitrogen, although the predominant component of the earth's atmosphere, is estimated as the twenty-eighth element in terms of terrestrial abundance with an average of 46.3 g/metric ton

of terrestrial matter. Dry atmospheric air, disregarding pollutants, contains approximately 78.09 vol % and 75.54 wt % N_2.

Nitrogen was not identified as an element until it was isolated by Daniel Rutherford (Scotland) in 1772. Lavoisier (France) further confirmed nitrogen as a gaseous element in 1776. Nitrogen, like oxygen, is an essential ingredient to practically all forms of life and consequently is very important in foods and fertilizers. See also **Amino acids; Fertilizers;** and **Peptides and proteins.** Nitrogen serves as an important diluent of O_2 in the atmosphere and thus controls natural burning and respiration processes. Removal of N_2 from hydrocarbon streams is described under **Hydrotreating.**

Compounds: There are too many inorganic and organic nitrogen compounds to discuss in full here. A partial list of families of nitrogen-bearing compounds would include alkaloids, amides, amines, cyanides, cyanogens, diazo compounds, hydrazines, imides, nitrates, nitrides, nitrites, nitriles, oximes, purines, pyridines, and ureas (consult the Subject Index).

Ammonia, NH_3, ranks first among industrial chemicals in terms of value and second in terms of tonnage production, worldwide production exceeding 42 million short tons/yr. See also **Ammonia.** *Urea,* $NH_2 \cdot CO \cdot NH_2$, is important in fertilizers, animal feeds, and synthetic resin production and approximates 15 million tons/yr of production and is rapidly growing. Among the chemicals used as fertilizers, nitrogen is present in urea in the greatest proportion as a form available to plant life. See also **Fertilizers;** and **Urea.** *Hydrogen cyanide,* HCN, which forms hydrocyanic acid when dissolved in H_2O, is a high-tonnage nitrogen-bearing chemical, the salts and derivatives of which are used widely as intermediates in organic syntheses. See also **Hydrogen cyanide.**

Nitric acid, HNO_3, is produced in high tonnages (United States production is estimated in excess of 5 million tons/yr) and is a basic inorganic chemical used for hundreds of applications as a reactant, a solvent, and an oxidant. Although several nitrates occur in nature, e.g., $NaNO_3$ from Chile and Bolivia, and KNO_3, $Ca(NO_3)_2$, and $Mg(NO_3)_2$, HNO_3 is an important starting ingredient for other nitrates. In a process known as nitrogen fixation, numerous plants and bacteria have the ability to convert elemental nitrogen of the air into compounds used in the further construction of proteins and other compounds found in living tissues. The term *fixation* is also used in connection with the synthetic manufacture of nitrogen-bearing fertilizers. The production of calcium cyanamide, $CaCN_2$, was the object of early efforts to achieve synthetic nitrogen fixation. In a reverse process, some plants and bacteria break down nitrogeneous materials to form elemental nitrogen or ammonia, with the characteristic odors of putrefaction.

Nitrogen oxides, NO_x, are considered serious air pollutants in a number of industrial areas. See **Air pollution.** *Nitric oxide,* NO, is heavier than air and is a colorless gas (although it forms brownish-red fumes in contact with air), toxic, and somewhat soluble in H_2O. Upon cooling to $-153.6°C$, the gas becomes a colorless liquid. Critical temperature is $-92.9°C$; critical pressure 64.4 atm. In the presence of air, NO is easily converted to the more stable nitrous oxide. Nitric oxide does not combine directly to form an acid with H_2O. *Nitrous oxide,* N_2O, is a colorless gas with a somewhat sweet taste and odor. Critical temperature is $36.5°C$; critical pressure, 71.66 atm. The gas also is known as *laughing gas* because of the mild exhilaration it produces when breathed, and at one time was used rather widely as a shallow anesthetic. The gas supports combustion in the same manner as air. Other oxides of nitrogen include nitrogen dioxide, NO_2; nitrogen tetroxide, N_2O_4; and nitrogen sesquioxide, N_2O_3.

Nitrous acid, HNO_2, is formed when N_2O_3 is added to water. Freshly prepared nitrous acid is blue but quickly fades with the emission of brown fumes: $3HNO_2 \rightarrow HNO_3 + 2NO + H_2O$. It is classified as a moderately strong acid.

Manufacture: Aside from its elemental function as a diluent of atmospheric air, nitrogen is predominantly important in the form of its hundreds of compounds, the preparation of which usually does not require the isolation of N_2 in the elemental, gaseous form. Pure nitrogen is a by-product of the manufacture of pure O_2 from air. See also **Oxygen.**

Nitriding of certain alloy steels is a process in which the metals are heated to 900–1200°F (482–649°C) in an atmosphere of partially dissociated NH_3. Thus, nitrogen plays a role, but the pure form is not required. The metals are case-hardened by the formation of nitrides. Surface hardening is produced by the absorption of nitrogen without quenching. *Cyaniding* is a process of case hardening of an iron-base alloy by the simultaneous absorption of carbon and nitrogen by heating the metal in a cyanide salt. Cyaniding usually is followed by quenching to produce a hard case. *Powder metallurgy* also utilizes NH_3 (dissociated) atmospheres in certain instances.

For several applications, N_2 makes an excellent inert atmosphere, e.g., in the gaseous insulation of transformers. Inert atmospheres also are used widely in connection with electric-furnace operations. The basis of an artificial atmosphere is the exclusion of oxygen (air) from the heating chamber by the substitution of some other gas or mixture of gases. Artificial atmospheres are classified as active, or process, atmospheres and inactive, or protective, atmospheres. N_2 is one of three gases (also H_2 and CO) used as protective atmospheres and serves three purposes: (1) it slows the reactions of any active gases present; (2) it lowers the partial pressure of each active gas; and (3) it reduces the inflammability of any active gases present. Usually, nitrogen constitutes a large part of the volume of the protective atmosphere. Commercial nitrogen contains more or less O_2, water vapor, and CO_2 in small amounts but sufficient to cause slight oxidation. The addition of small amounts of a reducing gas, such as methane, corrects these deficiencies.

Nitrogen Balance   (See **Fertilizers;** and **Peptides and proteins.**)

Nitrogen Fertilizers   (See **Fertilizers.**)

Nitrogen Group   (See **Chemical elements.**)

Nitrogen Oxide Pollutants   (See **Air pollution;** and **Nitrogen.**)

Nitroglycerin   (See **Explosives.**)

Nitrophosphates   (See **Fertilizers.**)

Nitrostarch   (See **Explosives.**)

Nobelium   (See **Chemical elements.**)

Nonionic Surfactants   (See **Detergents;** and **Surfactants.**)

Nonwovens   (See **Fibers;** and **Rayon.**)

Normal Concentration   A *normal solution* contains one *gram-equivalent weight* of the substance whose normality is identified in one liter of solution. The equivalent weight of a substance is that weight which will involve, in a chemical reaction, one atomic weight of hydrogen or that weight of any other element or portion of a substance which, in turn, will involve in reaction one atomic weight of hydrogen.

For example, the chlorine atom of NaCl also is found in hydrochloric acid, HCl, in combination with one hydrogen atom. Therefore, the gram-equivalent weight of NaCl is 58.454, the same as its gram-molecular weight. A one normal solution (abbreviated 1 N) of NaCl contains 58.454 g of salt per liter of solution.

For a given solution, the molar and normal concentration values are equal only when the gram-molecular and gram-equivalent weights are equal. Examples where these values are not equal include sulfuric acid, H_2SO_4, which contains two active hydrogen ions and whose gram-equivalent weight is therefore one-half its gram-molecular weight; phosphoric acid, H_3PO_4, which contains three active hydrogen ions and whose gram-equivalent weight is therefore one-third its gram-molecular weight; and calcium hydroxide, $Ca(OH)_2$, which contains two active hydroxyl ions (each equivalent to a hydrogen ion) and whose gram-equivalent weight is therefore one-half its gram-molecular weight.

Normalizing (See **Iron and steel.**)

Nozzles (See **Mixing, fluids.**)

NTA (Nitrilotriacetic acid) (See **Chelating agents;** and **Formaldehyde.**)

Nuclear Explosives (See **Explosives.**)

Nuclear Power Plants Nuclear power stations generate a significant segment of electric-power requirements. In the United States alone, as of December 1970, there were 104 civilian power reactors operating or on order, having an electrical generating capacity of 86,794 MW, with 8,000 MW operational. The total electrical generating capacity in the United States at that time was 341,000 MW operational.

The application of nuclear-reactor systems to commercial power-generating stations began in December 1957 with the operation of Shippingport Atomic Power Station, which operates in the Duquesne Light Company grid, supplying power to the Pittsburgh metropolitan area.

Energy Release from Fission of Heavy Nuclides. In a nuclear power plant, energy is generated from the fission of heavy nuclides, as contrasted with conventional electrical generating plants, where energy is generated from the chemical combustion of fossil fuels. The heat generated in conventional or nuclear plants is transferred to a working fluid, and from this point on the two types of plants are essentially similar.

In nuclear fission, the nucleus of a heavy atom is split into two or more fragments. The reaction is initiated by the absorption of a neutron. One typical reaction is

$$^{235}_{92}U + ^1_0n \rightarrow ^{137}_{56}Ba + ^{97}_{36}Kr + 2\,^1_0n + \Delta E \quad (1)$$

In this reaction, a ^{235}U atom absorbs one neutron, becomes unstable, and subsequently fissions into two fission fragments plus two neutrons. This is just one of the ways in which ^{235}U might fission, many others being possible. The number of neutrons produced in a fission reaction is usually 2 or 3. The excess neutrons produced by the fission reaction provide the means of self-sustaining the chain reaction.

Only three nuclides, ^{233}U, ^{235}U, and ^{239}Pu, are fissionable by neutrons of all energies and sufficiently stable to have practical significance. Of these, only ^{235}U occurs in nature. Natural uranium contains 0.7% ^{235}U and 99.3% ^{238}U. The other two fissionable nuclides are produced artificially from ^{232}Th and ^{238}U, respectively, by neutron capture followed by two stages of radioactive decay. Since the probability that a neutron will initiate a fission reaction in these nuclides is essentially an inverse function of neutron velocity, it is desirable to slow down the high-energy neutrons (1-2 MeV) liberated in the fission reaction. The most important slowing-down mechanism is elastic scattering on elements of low mass number. Materials like light and heavy water, beryllium oxide, and graphite are purposely placed in reactors to slow down, or *thermalize,* the neutrons to an energy of about 0.025 eV. These materials serve as *moderators* in a nuclear reactor. Certain nuclides are, to some extent, fissionable by high-energy (greater than 1 MeV) neutrons only. ^{232}Th and ^{238}U fall into this category. These nuclides, however, cannot support a chain reaction because the probability of fission is small even for neutrons with energies in excess of 1 MeV.

The energy of a nuclear reaction can be computed from the change in mass between reactants and products according to Einstein's law:

$$\Delta E = \Delta mc^2$$

For example, the mass difference in this equation is $\Delta m = 0.2058$ amu; therefore, $\Delta E = 931$ MeV/amu $\times$ 0.2058 *amu* $= 191.6$ MeV. The average amount of energy released in the various fission reactions is about 200 MeV. This energy is distributed in the fission process as follows:

	MeV
Kinetic energy of fission fragments	165
Radioactive-decay energy	23
Kinetic energy of neutrons	5
Prompt gamma-ray energy	7
	200

The energy of a chemical reaction, approximately 3–4 eV, is much lower than that of the nuclear reaction. Hence, the fission of ^{235}U yields 2.5 million times as much energy as the combustion of the same weight of carbon. The fission of 1 g/day of uranium results in a power production of 1 MW.

The importance of fission, from the standpoint of energy production, lies in two facts: (1) a large amount of energy is released in the fission reaction; and (2) the production of excess neutrons permits a chain reaction. These two circumstances make it possible to design nuclear reactors in which self-sustaining chain reactions occur with the continuous release of energy. Nuclear fission is not the only energy-releasing nuclear reaction. The fusion of light nuclides, like hydrogen, into heavier elements is also an energy-producing process. The amount of energy released in nuclear fusion is even larger than the energy release of nuclear fission.

Nuclear-Reactor Types. The three basic components of a nuclear reactor are the fuel, the coolant, and the moderator. The selection of these three components specifies the reactor type. The third item, the moderator, is not a necessity. While in some designs there is an advantage in slowing down the neutrons, in other designs the majority of fissions are produced by fast neutrons. Reactors in which most of the fission reactions are induced by thermal neutrons due to the use of moderator materials are called *thermal reactors*. Similarly, if most of the fission is initiated by high-energy fast neutrons, the reactor is a *fast reactor*. Fast reactors do not use moderator materials.

Nuclear fuels can be grouped into two broad categories. In the first category are the three nuclides which are fissionable by neutrons of all energies, namely ^{233}U, ^{235}U, and ^{239}Pu. These three nuclides are usually called *fissile nuclides*. The second category contains ^{232}Th and ^{238}U, which are *fissionable nuclides*. Thermal reactors use fissile nuclides as fuel, while fast reactors are designed to burn fissionable materials. In fast reactors, only a small portion of the ^{232}Th and ^{238}U are fissioned directly. A larger portion of these materials is converted into ^{238}U and ^{239}Pu, respectively, through neutron absorption. Thus, this reactor type not only consumes fuel but also produces (breeds) new fuel material; hence the term *breeder reactor*. Breeding is possible in thermal reactors also, but to a lesser extent. The fuel material in a fast reactor must contain a significant amount (about 10%) of one of the fissile materials. The remainder of the fuel must have a high mass number in order to avoid slowing down the neutrons. The natural reserves of fissionable materials are more than 100 times greater than the reserves of fissile materials. Consequently, from the viewpoint of utilization of available energy resources, fast reactors are of great importance.

Most of the power reactors in service or under construction today are thermal reactors. The many types of thermal reactors depend on the selection of the coolant and moderator. A majority of the plants built in the United States use water as coolant and moderator material. There are two types of water reactors, *pressurized water reactors* (PWR) and *boiling-water reactors* (BWR). Great Britain played a pioneering role in the development of *gas-cooled* (CO_2) *and graphite-moderated reactors* and is the largest manufacturer of gas-cooled reactors. Canada specialized in reactors cooled and moderated by heavy water. The few fast reactors in operation or under construction today are basically used as demonstration plants. Economically, they cannot compete with thermal reactors at the present time. Fast reactors are the type of the future. They use either materials with high mass number (liquid metals or sodium) or gases as coolant.

Pressurized Water Reactor. A typical pressurized-water-reactor system (PWR) is shown in Fig. N-8. Heat generated in the nuclear core is removed by water (reactor coolant) circulating at high pressure through the primary circuit. The water in the primary circuit both cools and moderates the reactor.

The heat is transferred from the primary to the secondary system in a heat exchanger, or boiler, thereby generating steam in the secondary system. Typical operating conditions are summarized in Table N-5. The steam produced in the steam generator, a tube-and-shell type heat exchanger, is at a lower pressure and temperature than the primary coolant. Therefore, the secondary portion of the cycle is similar to that of the moderate-pressure fossil-fueled plant.

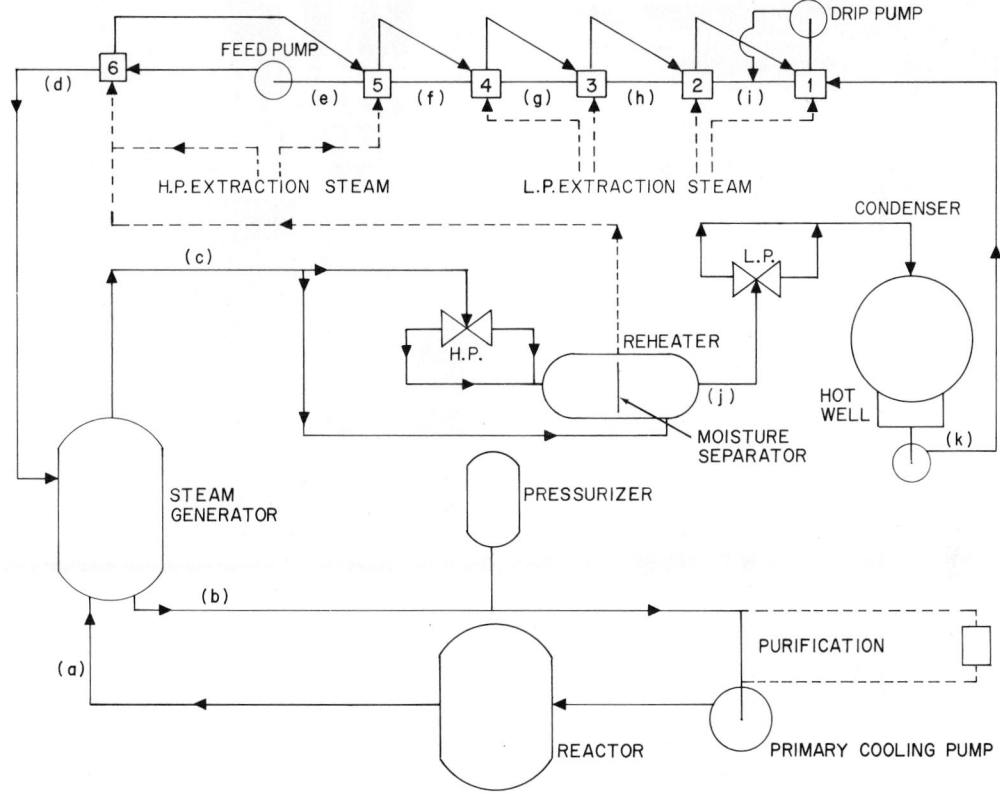

Fig. N-8. Representative pressurized-water-reactor cycle. Conditions prevailing in the system: (a) 2,100 psi, 605°F (319°C); (b) 550°F (288°C); (c) 735 psi, 509°F (265°C); (d) 435°F (224°C); (e) 380°F (193°C); (f) 317°F (158°C); (g) 265°F (130°C); (h) 210°F (99°C); (i) 170°F (77°C); (j) 475°F (246°C); (k) 2 in. Hg abs. pressure, 100°F (38°C). *HP*, high pressure; *LP*, low pressure.

In boiling-water or direct-cycle systems, steam is generated in the core and is delivered directly to the steam turbine.

Reactor Components: The reactor is composed of a core or assemblage of fuel elements, control rods, coolant, and moderator (Fig. N-9).

Heat energy is generated in the *reactor core,* which contains an array of fuel assemblies. Typical cores are rated at approximately 1,500–3,500 MW$_t$ (megawatt thermal) with corresponding electrical output of 500–1200 MW$_e$ (megawatt electric). An 800-MW$_e$ core is approximately a right circular cylinder, having an equivalent diameter of 136 in. and an active height of 137 in. and is made up of 217 fuel assemblies. About 80 metric tons of uranium are used in the reactor core. The core is completely open, shrouded only at the outer periphery.

TABLE N-5. *Typical Operating Conditions for PWR Systems*

	Approximate Range
Primary system:	
Temperature:	
°F	550–610
°C	288–321
Pressure, psig	2,000–2,250
Secondary system:	
Steam temperature:	
°F	510–525
°C	266–274
Pressure, psig	700–900
Power ratings, MW:	
Thermal	1500–3500
Electrical	500–1200

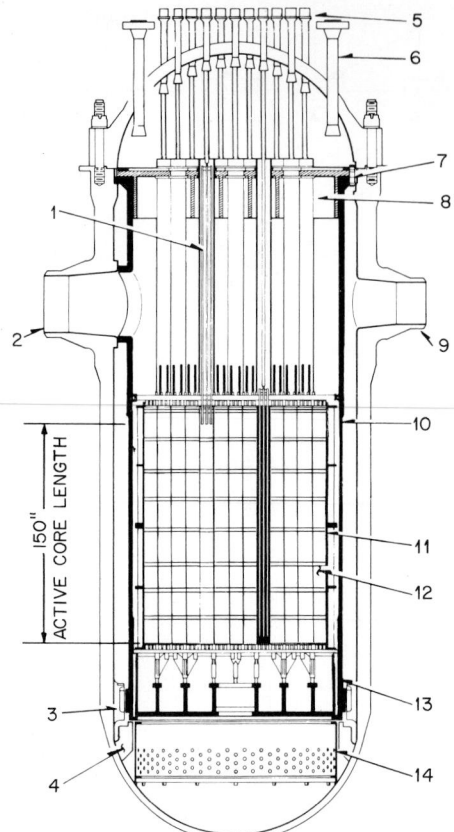

Fig. N-9. Reactor arrangement. (1) Control-element assembly fully withdrawn, (2) 42-in.-ID outlet nozzle, (3) snubber, (4) core stop, (5) control-element drive mechanism, (6) instrumentation nozzle, (7) alignment pin, (8) upper guide structure, (9) 30-in.-ID inlet nozzle, (10) core-support barrel, (11) core shroud, (12) fuel assembly, (13) core-support assembly, (14) flow skirt.

Each assembly is made up of 176 fuel rods supported within a rigid frame consisting of stainless-steel upper and lower end fittings, Zircaloy fuel-rod spacer grids, and five Zircaloy axial guide tubes which will accept a control element assembly (control rod). The overall length is about 155 in., and the cross section is approximately 8 × 8 in. Each assembly weighs about 1,300 lb (Fig. N-10).

Each fuel rod contains sintered UO_2 pellets approximately 0.4 in. diam by 0.6 in. long. The Zircaloy 4 cladding tube is 0.026 in. thick. This material has good corrosion behavior in the reactor environment and low neutron-absorption

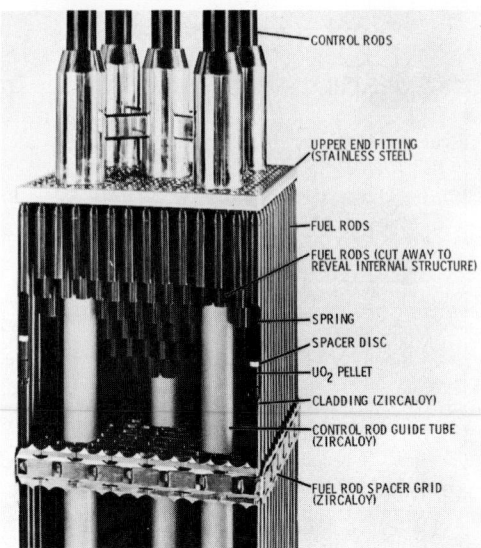

Fig. N-10. Cutaway model showing construction details of a PWR fuel assembly. (*Combustion Engineering, Inc.*)

cross sections. The tube length accommodates the fuel-pellet column and a fuel and gas expansion space. A diametral gap between the pellets and the cladding provides for differential expansion between the cladding and fuel to limit clad strain.

Control rods are located in selected fuel assemblies and contain materials possessing high neutron-absorption cross sections. Their relative position in the core regulates the rate of power generation. The control materials commonly used are boron as the carbide, B_4C, or the alloy Ag-In-Cd (Fig. N-10).

The *coolant,* or core heat-removing medium, is water. A typical 800-MW_e PWR coolant-system flow is $\cong$300,000 gal/min. The coolant temperature at the core inlet is 550°F. The coolant temperature at the core exit is 600°F. The coolant water also serves as *moderator* in a PWR.

Fuel Management: To demonstrate fuel management, an example core containing 217 fuel assemblies divided into three batches has two batches with 72 fuel assemblies each and the third with 73 assemblies. At equilibrium burnup conditions, each fuel batch will remain in the core for three cycles so that every assembly will receive approximately the same burnup.

The core volume is divided into two radial zones. At the beginning of an equilibrium burnup cycle, the outer zone, consisting of one-third the core volume, contains fresh fuel while the inner zone, consisting of two-thirds the core volume, contains a mixture of two fuel batches, one-third and two-thirds depleted. At the end of the cycle, the fully exposed batch of fuel is removed from the central core. The fuel batch in the outer zone, which by this time has reached approximately one-third the rated exposure, is now moved to the central zone. Fresh fuel is placed in the outer zone and the core is ready to operate over the next burnup cycle. The initial core, to simulate the equilibrium core, uses three different ^{235}U enrichments. The burnup in the equilibrium cycle is about 10,000 MW/day per metric ton of uranium; discharge burnup after three cycles is about 30,000 MW/day per metric ton of uranium.

The fissile material consumed during an operating cycle varies by reactor type. Current PWRs consume one-half to two-thirds of the beginning ^{235}U in a normal 30,000 MW/day per metric ton of uranium burnup. Plutonium, also a fissile material, is produced and partially consumed in the cycle. It is produced by neutron capture in ^{238}U and decays to long-lived ^{239}Pu:

$$^{238}U + {}_0^1 n \rightarrow {}^{239}U \xrightarrow{\beta} {}^{239}Np \xrightarrow{\beta} {}^{239}Pu \quad (2)$$

Fuel with a beginning-of-life enrichment of 3.5 w/o* ^{235}U will produce about 1 w/o Pu, of which 30% is consumed in fission reactions. The fuel discharged from the reactor has a residual of fissile material which is recovered for future use.

The fuel material is capable metallurgically of burnup in the region of 60,000–100,000 MW/day per metric ton of uranium. The limiting factor is the ability of the cladding to withstand the volume increase of the fuel, which is burnup-dependent. Another factor which limits burnup is the increase in fission-gas release at higher burnup. This can be accommodated by an ade-

*This abbreviation is used in the rare-earth and related fields for weight percent.

quate plenum design (free volume) within the fuel.

Reactivity Control: Nuclear reactors are designed to sustain controlled chain reactions in the reactor core so that the total number of neutrons in each successive generation is the same. Due to the constant neutron population, the number of fissions per unit time and therefore the power production are constant. PWRs operating at a steady power level require very little control. They are inherently stable. Control is required when power-level changes are made (short-term control) and to compensate for the burnup of the fuel during a cycle (long-term control).

Control of nuclear reactors is achieved by inserting or withdrawing from the reactor core neutron-absorbing materials. When neutron-absorbing materials are inserted into the core, the neutron population and energy production decrease. Similarly, withdrawal of these materials results in an increase in power production. Due to the very short lifetime of neutrons in a reactor ($\cong 30$ μs) reactors respond quickly to control. The rate of power-level change will depend on the *multiplication factor k*, defined as the ratio of the present neutron population to that of the previous neutron generation. The worth of control materials is measured in terms of *reactivity* ρ, defined as

$$\rho = 100 \frac{k-1}{k}\% \quad (3)$$

Thus, while the reactor operates at a steady power level, the multiplication factor is unity and the reactivity is zero. A multiplication factor greater than unity (positive reactivity) represents power increase; a multiplication factor less than unity (negative reactivity) represents power decrease.

PWRs use control materials in three different forms, as shown in Table N-6. The control rods are adjusted according to need to compensate for short-term changes, thus maintaining the power

TABLE N-6. *Types of Control Methods Used in Pressurized Water Reactors*

Control Method	Purpose	Material	Reactivity Worth, Approx. %
Control rods	Power maneuvering, shutdown	Boron carbide, silver-indium-cadmium	8.0
Burnable poison	Power shaping, compensation for fuel burnup	Boron or aluminum	4.0
Soluble poison	Compensation for fuel burnup, shutdown	Boric acid	10–15

level. The movement of the rods is usually initiated by an automatic controller. Power maneuvering is also accomplished by control-rod movement. A certain fraction of the control rods is always kept in the *full-out position,* providing the means for fast reactor shutdown if needed.

Burnable poisons are employed as fixed poison rods in the fuel assemblies. They accomplish two purposes simultaneously, namely compensate for fuel burnup and help shape the power distribution within the core. Early in core life, when the new fuel contains more fissile material, the burnable poisons absorb neutrons, thus holding the reactivity and power down. By the end of cycle, when a large portion of the fissile material is gone due to fuel burnup, the burnable poisons are not present. Through reactions induced by neutron absorption, they are converted into neutral materials with small neutron-absorption cross sections. Also, in the beginning of the cycle, when new fuel is introduced into the core, the local variations in power density are relatively high. Judicious locating of the burnable-poison rods makes more uniform power generation possible. Since power output is limited by the hottest portion of the core, uniform power distribution results in higher power output for the same core size.

Water reactors are characterized by large reactivity changes over a core lifetime, long-term changes due to fuel burnup and shorter-term changes associated with power maneuvering. When control rods are used to compensate for these changes, their changing pattern of insertion introduces substantial and time-varying distortions of the spatial power distribution in the reactor core. Unless the control-rod system is designed and used with great care, these distortions may correspond to highly nonuniform power-density distributions. The adoption of *chemical shim* control (reactivity control by boric acid dissolved in the reactor water) for PWRs has largely bypassed this problem and thus has been a source of major improvement in power distribution. In principle, with chemical shim, there is no need for control rods in the core during power operation, although they must be held in the withdrawn position as a means of rapid shutdown.

In practice, however, the adjustment of the boric acid concentration, which is done by a feed-and-bleed system, is cumbersome and expensive for short-term reactivity adjustments. The system is used most beneficially for compensating the excess reactivity incorporated for fuel burnup. This amounts to 10–15% reactivity for a core

employing a three-batch refueling scheme. Since the early PWR plants were intended for base-load operation, the other control requirements during power operation are small and these reactors will operate almost as rodless reactors.

The limitations of chemical-shim control are quite apparent. In addition to the problem of short-term reactivity adjustment, they include a practical limit on total capability for reactivity compensation. The limit is imposed by the positive component of the moderator-temperature coefficient of reactivity, which increases approximately in proportion to the boric acid concentration, and which may cause undesirable dynamic characteristics if it becomes large enough to outweigh the negative component and yield a net positive coefficient. The use of chemical shim is expected to remain an important feature of the PWR technology, but the importance of control rods is expected to increase with time. In order to control core reactivity, boric acid is controlled much like any additive or treatment chemical in power operations. Boric acid dissolved in the primary coolant controls only the slowly changing reactivity effects. The mechanism by which this occurs is the absorption of neutrons by the ^{10}B isotope of natural boron. Natural boron contains $\cong 20$ atm % ^{10}B, which has a very high thermal neutron cross section of $\cong 3{,}840$ barns. The reaction is

$$^{10}\text{B}(n,\alpha)\,^{7}\text{Li} \tag{4}$$
$$^{10}\text{B}(n,2\alpha)\,^{3}\text{H} \tag{5}$$

Reaction (5), although substantially less probable, does produce tritium activity which must be analyzed and accounted for in waste handling.

Reaction (4) results in ^{7}Li production of about 0.1 ppm/day at a boron concentration of 1,070 ppm (^{10}B 200 ppm) in a plant having rated power of 2,500 MW$_\text{t}$. The effect of alkali-metal input is to partially neutralize the boric acid and increase the coolant pH (water quality is discussed later).

Reactor Coolant System: The reactor coolant system is designed to remove heat from the reactor core and transfer it to the secondary system by the forced circulation of pressurized water. The major components of the system are the reactor vessel; heat transfer loops, each containing a steam generator and a reactor coolant pump; a pressurizer connected to one of the reactor-vessel outlet pipes; and associated piping.

Reactor vessels are presently designed and fabricated in accordance with the ASME Boiler and

Pressure Vessel Code Section III, class A. The vessels are fabricated largely from SA-533, grade B, class I (nickel-modified Mn-Mo) steel. All surfaces in contact with primary coolant are clad with austenitic stainless steel of Ni-Cr-Fe alloy. A typical reactor vessel for an 800 MW$_e$ plant is approximately 14 ft in diameter and has a height of approximately 41 ft.

The nuclear steam-supply system utilizes steam generators to transfer the heat generated in the reactor to the secondary system. Figure N-11 is typical of the steam generators in an 800-MW$_e$ plant (two are required). The steam generators, vertical U-tube heat exchangers, operate with the reactor coolant on the tube side and the secondary coolant on the shell side. All surfaces in contact with reactor coolant are Ni-Cr-Fe alloy or austenitic stainless steel.

Reactor coolant enters the steam generator through an inlet nozzle, flows upward through U-tubes, and leaves through two outlet nozzles. The steam produced in the shell side flows upward through the moisture separators and leaves the steam generators through the steam outlet nozzle at the top. Specially designed moisture-separating equipment limits the moisture content of the steam to a maximum of 0.20%. The steam-generator secondary-side materials can tolerate normal water chemistry associated with fossil-fueled boilers.

The pressure of the reactor coolant system is controlled by the pressurizer, where steam and water are maintained in thermal equilibrium. During full-load operation, the pressurizer volume is almost evenly divided between saturated water and saturated steam. Steam is formed by energizing immersion heaters in the pressurizer or is condensed by a subcooled pressurizer spray as necessary to maintain operating pressure and limit pressure variations due to plant-load transients.

Materials of Construction: The materials of construction listed in Table N-7, typical of modern PWRs, consist of 300 series stainless-steel primary-system surfaces, Zircaloy-clad fuel elements, and Inconel 600 steam-generator tubing. With the possible exception of core materials, conventional construction materials are used throughout the reactor coolant and steam-condensate cycle.

Water Quality: High-purity water is needed in nuclear applications, principally to preclude fouling of heat-transfer surfaces, corrosion, and the creation of radioactive materials through nuclear-activation processes. Corrosion can impair material integrity or increase corrosion-product

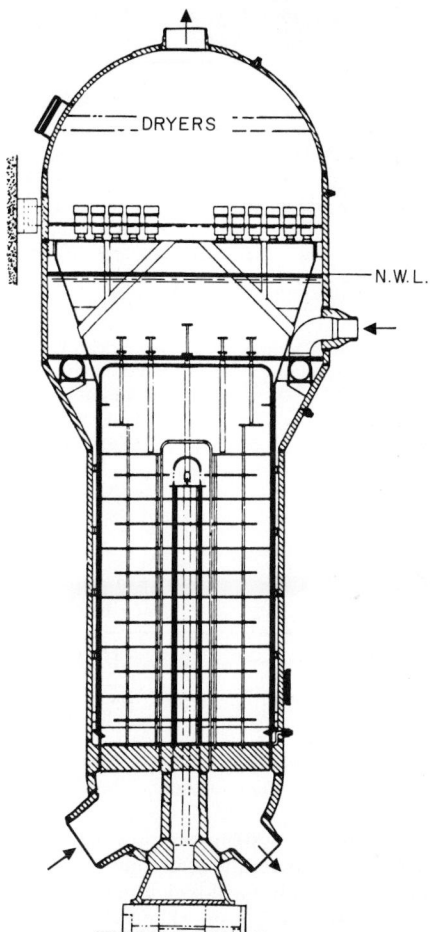

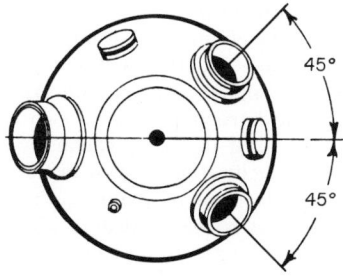

Fig. N-11. Steam generator for an 800-MW$_e$ nuclear steam-supply system. Two such generators are required. The lower portion of the generator is 13 ft, 8 in. diam; upper portion, 19 ft, 11 in. diam. Overall unit height is 62 ft, 7 in. *NWL,* normal water level.

TABLE N-7. *Materials Exposed to Coolant in a Typical Reactor Coolant System*

Reactor vessel:
Vessel cladding . . . . Weld-deposited SS*
Vessel internals . . . . 304 SS
Instruments and control-rod
  drive nozzles . . . . Ni-Cr-Fe alloy–600†
Fuel cladding and guide
  springs . . . . Zircaloy 4
Control-rod drives . . . . . . 304 and 410 SS
Pipe cladding main-loop pipes . 304L SS
Surge and spray piping . . . . . 316 SS
Steam generator:
Bottom-head cladding . . . . Weld-deposited SS*
Tube-sheet cladding . . . . . Weld-deposited
                                  Ni-Cr-Fe alloy–600†
Tubes . . . . Ni-Cr-Fe alloy–600†
Divider plate . . . . 410 SS
Pumps:
Casing . . . . 316 SS
Internals . . . . 304 SS
Pressurizer:
Cladding . . . . 304 SS* and Ni-Cr-Fe
                                  alloy–600†
Heaters . . . . Ni-Cr-Fe alloy–600†

*Equivalent to type 304 stainless steel as deposited.
†Inconel 600.
SOURCE: Combustion Engineering, Inc.

concentrations, which result in activated residues on the primary-system surfaces, filters, and waste systems.

In the primary system, the presence of the reactor neutron flux results in unique chemical conditions not found in fossil-fueled plants. This flux has two major effects on chemical control: (1) impurities can be made radioactive and (2) radiation can induce normally nonoccurring chemical reactions to take place. These conditions create the need for maintaining certain unique controls over the water chemistry in the nuclear sector of the cycle, e.g., the elimination of impurities from the primary water and the inclusion of certain additives to suppress undesirable radiation-induced chemical reactions. These two requirements will be met if the water chemistry is held to the typical specifications shown in Table N-8. The water must be equal in quality to the effluent of a mixed-bed ion-exchange demineralizer, with boric acid added for reactivity control.

To assure this water quality in the reactor primary coolant, makeup water must be maintained to equivalent specifications. The typical water-quality specification for reactor-coolant makeup water is shown in Table N-9. One common im-

TABLE N-8. *Primary Coolant Chemistry*

Minimum specific resistivity
  prior to additives, MΩ-cm 0.5
Normal pH:
  At 573°F (301°C) 5.3–9.5
       77°F (25°C) 4.5–10.2
Hydrogen at 77°F, cm^3/kg (STP) . . 10–50
Oxygen at 77°F, ppm . . . . <0.1
Halides, ppm:
  Chloride . . . . <0.15
  Fluoride . . . . <0.10
Additives, ppm . . . . <25 of Li^7, K,
                                            or NH_3
Total solids, prior to additives, ppm   0.5
Maximum boric acid concentration,
  nominal, ppm:
  At 573°F (301°C) 0–9,800
       77°F (25°C) 2,280–9,800

purity which can be introduced is the chloride ion. The potential for stress corrosion of austenitic stainless steels in the presence of chloride and oxygen is well documented. Since austenitic stainless steel is used extensively in construction of PWR systems, it is obvious that chloride and oxygen must be rigidly controlled. The specification for chloride and oxygen is as follows:

$$mg/l$$

Chloride . . . . <0.15
Dissolved oxygen (temperature >150°F) . <0.10

Hydrazine is employed during start-up operations to ensure that oxygen is not present to induce the stress-corrosion reaction. Oxygen reacts with hydrazine as follows:

$$N_2H_4 + O_2 \rightarrow N_2 + 2H_2O \qquad (6)$$

In the absence of dissolved oxygen, N_2H_4 provides a sink for dissolved O_2 by maintaining metal oxides at their lower oxidation states.

TABLE N-9. *Reactor-Coolant Makeup-Water Chemistry*

	Specifications	Operating Limit
Specific resistivity, MΩ-cm		≥0.5
Total solids, ppm		≤0.5
pH at 77°F (25°C)		6.0–8.0
Dissolved N_2 and O_2		Nondeaerated
Halides, ppm:		
Chlorides		0–0.15
Fluorides		0–0.1

The reaction rate is temperature-dependent and is extremely rapid in the range 150–250°F (66–121°C). A competing decomposition reaction also accelerates as the temperature increases. At $\cong 400°F$ (204°C), hydrazine decomposes rapidly, and chemical deoxygenation of water is no longer practical. At power, oxygen can enter the coolant system in aerated makeup water. This oxygen is immediately scavenged by dissolved hydrogen in a reaction induced by the reactor radiation.

In a PWR water has been shown to undergo various equilibrium reactions which are normally unexpected for the thermal conditions of the coolant. These reactions generally can be summarized as follows:

Water decomposition and formation:

$$H_2O \rightleftarrows H_2 + \tfrac{1}{2}O_2 \qquad (7)$$

Ammonia synthesis and decomposition:

$$3H_2 + N_2 \rightleftarrows 2NH_3 \qquad (8)$$

Nitric acid synthesis:

$$2H_2O + 2N_2 + 5O_2 \rightarrow 4HNO_3 \qquad (9)$$

Nitric acid reduction:

$$2HNO_3 + 5H_2 \rightarrow N_2 + 6H_2O \qquad (10)$$

The kinetics of these reactions is quite rapid during reactor operation, and many intermediate products are postulated before formation of the end products. Since these reactions are dependent on the reactor flux, they are first-order with respect to the reactor power level and the concentrations of reactants.

It is apparent from Eqs. (7) to (10) that the presence of free hydrogen can shift the equilibrium of these reactions to preclude the excess decomposition of water or formation of nitric acid. For this reason, PWRs operate with an excess of hydrogen (added as hydrogen gas).

Hydrogen-concentration control in the current generation of PWRs is accomplished by the chemical and volume control system (CVCS), unique to PWRs, which allows direct and continuous control of water quality in the primary coolant on a completely automatic basis. The CVCS is essentially a side-stream loop on the primary coolant system (PCS). It permits continuous bleeding and feeding of water from PCS, through purifying ion exchangers and filters, and into a gas-pressurized water reservoir called the *volume-control tank*. Purified water from this tank is charged back into the reactor at the same rate as the letdown flow, thereby stabilizing reactor coolant volume. The simplicity of the bleed and feed process permits the addition of concentrated boric acid to increase boron concentrations or the replacement of the reactor coolant with makeup water free of boric acid to reduce boron concentrations. The hydrogen concentration in the coolant is controlled by maintaining a hydrogen overpressure on the volume-control tank.

The presence of boric acid fixes the lower pH specification of 4.5. Boric acid is slightly ionized at reactor operating temperatures, so that it does not greatly influence the fluid pH. Therefore, the addition of very small amounts of strong- or weak-base constituents greatly influences the pH at high temperature.

There are several benefits arising from operation at adjusted pH: (1) the general corrosion of system materials is reduced, and (2) core reactivity has been reported to be affected by changes in coolant pH in large commercial PWRs.

In general, the metal corrosion rates from typical PWR materials are low, and tests have shown that they stabilize after 200 days. Since an increase in pH reduces the corrosion rate, the coolant chemistry is adjusted to obtain this effect.

Radioactivity. Nuclear steam-supply systems employing PWRs are indirect cycle systems. The primary coolant, which picks up radioactivity from a number of sources, is separate from the water, which serves as working fluid for the steam turbine, and the steam generator serves as a barrier between the two fluids, making it possible to hold the off-site release of radioactivity to a very low level.

Because of its closed-cycle design, it is easy to control radioactivity in a PWR. Holdup in the system provides for decay of short-lived radioisotopes. Other fission-product activities are continuously and efficiently removed from the water by the ion exchangers in the CVCS.

Reactor primary coolant contains two basic categories of radioactivity, fission products and activation products. The fission products are due to the presence of tramp uranium on the surface of fuel cladding or leakage from minor fuel-cladding defects. The activation products are due to neutron activation of (1) the moderator, (2) contaminants introduced into the system with makeup water, or (3) corrosion products released from system materials. Corrosion products are a major source of long-lived activation products and so are important contributors to radiation levels on plant surfaces and long-lived radioactivity in discharged coolant (waste water).

Fission Products. Fission products from two basic sources may be present in the reactor coolant: (1) Uranium contamination on the surface or in the metal matrix of fuel cladding and core structural materials is essentially the background or base activity and has been present in all reactors. (2) Penetration or minor leakage in the cladding which encapsulates the fuel has been observed in reactors and does not impede safe continued operation of the plant.

The system design provides for operation of the core with a significant fraction of the fuel cladding being defective. Routine monitoring of fission products provides the operator with trends in coolant activity and a history of the relationship between various activity species. The information on activity and ratios of activity of specific nuclides is useful for assessing changes in the source of fission products.

One method widely used for routine monitoring of fission products is the radiochemical analysis of ^{131}I and ^{133}I. The relationship between these isotopes is a function of the release mechanism, i.e., recoil from surface or a cladding interstitial uranium atom produces a different ratio than diffusion from a defective fuel rod. Both mechanisms may be operating and releasing fission products to the coolant simultaneously. However, the ratio between the two iodine isotopes reveals which release mechanism was predominant.

Activation of Water. The isotopes ^{13}N, ^{16}N, ^{17}N, and ^{18}F result from neutron activation of oxygen isotopes occurring naturally in the light-water moderator. The activity levels are a function of the reactor flux and system volume:

Nuclide	Half-Life	Reaction	Target Source
^{16}N	7.13 s	$^{16}O(n,p)^{16}N$	Water
^{17}N	4.14 s	$^{17}O(n,p)^{17}N$	Water
^{18}F	109.8 min	$^{18}O(p,n)^{18}F$	Water
^{13}N	9.99 min	$^{16}O(p,\alpha)^{13}N$	Water

The level of ^{16}N activity determines the secondary shielding requirements for the reactor plant. This isotope (half-life 7.13 s) emits beta particles having maximum energy of 3.32, 4.39, and 10.49 MeV and gamma rays of 6.18 and 7.1 MeV, ^{17}N, which emits delayed neutrons, must also be considered in designing plant shielding and sample-line delay times (transit time from reactor to sample point) to protect personnel during sampling.

A typical decay curve of radioactive reactor-coolant contaminant activities is shown in Fig.

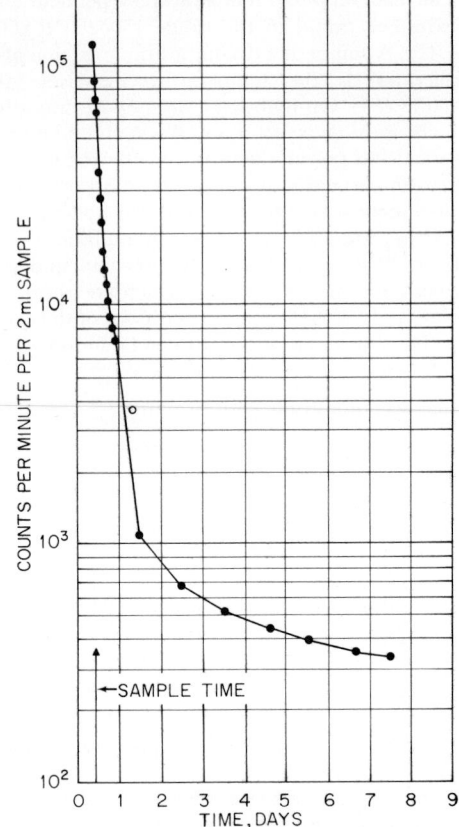

Fig. N-12. Gross gamma decay of reactor coolant (Shippingport, Pa., installation). A 2-ml liquid sample is used; $1\frac{3}{4} \times 2$ in. NaI (Tl) crystal; $\frac{5}{8}$-in.-diam. well; counter-bias at 50 KeV.

N-12. The gamma activity, after a sample has decayed about 15 min, is due principally to ^{13}N (half-life 10 min) and ^{18}F (half-life 1.87 hr). The samples were taken with the reactor operating at steady-state full power. Normally, the fission-product contribution to gross gamma activity is very small.

Table N-10 lists the major nonvolatile activities present in the Shippingport plant reactor coolant and shows the approximate percent of total gamma activity 1 hr after sampling. The activity is diminished very rapidly by a factor of 10,000 in a few hours until a long-half-life residual of corrosion-product activity remains. Volatile activities are the noble-gas fission products xenon and krypton and ^{41}Ar. ^{40}Ar enters the system with

TABLE N-10. *Shippingport Atomic Power Station, Nonvolatile Gamma Activity, Core 1 Seed 3*

| Nuclide | Half-Life | Activity at Sampling Time | | Percent of Total Gamma Radiation at 1 hr |
		Disintegrations/ (min)(ml)	μCi/ml	
^{13}N	10.1 min	1.82×10^4	8.19×10^{-3}	0.3
^{187}W	24 hr	2.22×10^3	9.00×10^{-4}	2.9
^{24}Na	15 hr	1.92×10^3	8.64×10^{-4}	1.8
^{18}F	1.87 hr	2.8×10^4	1.26×10^{-2}	21.3
^{138}Cs	32 min	1.03×10^4	4.68×10^{-3}	3.1
^{56}Mn	2.58 hr	5.65×10^3	2.54×10^{-3}	4.5
^{54}Mn	300 days	3.43×10^2	1.09×10^{-4}	0.2
^{64}Cu	12.8 hr	6.4×10^4	2.88×10^{-2}	39.2
^{131}I	8 days	4.2×10^1	1.89×10^{-5}	
^{133}I	21 hr	1.03×10^3	4.67×10^{-4}	14.8
^{134}I	52 min	1.00×10^4	4.5×10^{-3}	
^{135}I	6.7 hr	1.92×10^2	8.64×10^{-5}	
Percent of gross gamma activity at 1 hr accounted for				88.1

air dissolved in makeup water and becomes activated to ^{41}Ar.

Corrosion-Product Activation. Some of the corrosion products generated by the primary-system materials are released to the coolant and are transported to the core surfaces. This causes the metallic constituents to become activated, resulting in radioactive oxide deposits. Formation of corrosion products in a PWR results in a concentration of about 0.05 mg/l in the circulating coolant.

Generally, the corrosion products correspond in composition to the system materials. In a modern PWR, the available surface areas in contact with reactor coolant and their respective corrosion rates are as given in Table N-11. These corrosion and wear products are released from the surface of the plant piping and components and transported by the circulating coolant. A small percentage of these oxides is removed from the circulating coolant by the ion exchanger in the CVCS. The metal

oxides of interest are those which are deposited on the core, where they become activated by core radiation, the degree of activation depending upon the residence time on the core.

These activated metal oxides are then transported by the circulating coolant. Again, a small percentage will be removed by the CVCS ion exchanger, but some will redeposit on the plant surfaces as well as on the core. When they deposit on the plant surfaces, a radiation-level increase on these surfaces will result and will be reflected in the radiation levels of the primary system. Subsequent operation will result in release of deposits back to the coolant, which can again redeposit on the core to become further activated. The nuclides which contribute most to these radiation levels are shown in Table N-12.

The deposits of irradiated corrosion products found on components generally have not seriously impaired equipment maintenance. They do require that maintenance be planned and that con-

TABLE N-11. *Surface Areas in Contact with Reactor Coolant and Their Respective Corrosion Rates*

Material	Corrosion Rates, mg/(dm^2)(mo)	Area, ft^2	%	Annual Metal Release, g/yr
Stainless steel.	2	26,000	10	5,860
Inconel 600	4	154,000	60	68,500
Zircaloy	Small	54,000	30	Small
		234,000	100	

NOTE: This totals $\cong$ 165 lb/yr of metal or 205 lb of metal oxide.

TABLE N-12. *Irradiated Corrosion Products Found in a PWR*

Nuclide	Half-Life	Nuclear Reaction	Target Source
^{51}Cr	27.8 days	$^{50}Cr(n,\gamma)^{51}Cr$	Steel and Inconel
^{54}Mn	312 days	$^{54}Fe(n,p)^{54}Mn$	Steel and Inconel
^{59}Fe	45 days	$^{58}Fe(n,\gamma)^{59}Fe$	Steel and Inconel
^{58}Co	71 days	$^{58}Ni(n,p)^{58}Co$	Steel and Inconel
^{60}Co	5.24 yr	$^{59}Co(n,\gamma)^{60}Co$	Stellite, steel
^{64}Cu	12.9 hr	$^{63}Cu(n,\gamma)^{64}Cu$	17-4 PH steel
^{97}Zr	65 days	$^{94}Zr(n,\gamma)^{95}Zr$	Zircaloy
^{181}W	130 days	$^{180}W(n,\gamma)^{181}W$	Steel
^{187}W	24 hr	$^{186}W(n,\gamma)^{187}W$	Steel

tamination and personnel exposure be considered in this planning.

REFERENCES

Abbott, W. E.: New Design Features for Large PWR, *Nuclex-69, Basle,* 1969.

Abbott, W. E., F. Bevilacqua, and F. J. Hanzalek: Design and Operating Features of Large PWR Systems, *Am. Power Conf. Chicago,* 1968.

Chernock, W. P.: Materials for Water Cooled Reactor Cores, Seminar on Nuclear Materials, Brooklyn Polytechnic Institute, 1968.

Cohen, P.: "Water Coolant Technology of Power Reactors," ANS Monograph, Gordon and Breach, New York, 1969.

Dickinson, N. L., et al.: An Experimental Investigation of Hydrazine-Oxygen Reaction Rates in Boiler Feedwater, *Proc. Am. Power Conf., 1957,* vol. 19, pp. 692–706.

Dietrich, J. R.: Operating Considerations in Fueling of the PWR, *Nuclex-69 Ind. Colloq., Basle,* 1969.

El-Wakil, M. M.: "Nuclear Power Engineering," McGraw-Hill, New York, 1962.

Graves, R. H.: "Chemistry and Waste Management at the Connecticut Yankee Atomic Power Plant," National Association of Corrosion Engineers, Houston, Tex., 1969.

Lechnick, W., J. A. Martucci, and R. D. Scherer: Shippingport Operations, Duquesne Light Company, Sponsor, DLCS 36403, Operational Chemistry, pt. V, chap. 1, 1963.

Miller, D. A., and P. E. C. Bryant: "Corrosion and Coolant Chemistry Interactions in Pressurized Water Reactors," National Association of Corrosion Engineers, March 1970.

Weisman, J., and S. Bartnoff: The Saxton Chemical Shim Experiment, World Conf. on Atomic Power, WCAP-3269-24, July 1965.

Wells, A. F.: "Structural Inorganic Chemistry," 3d ed., Oxford, Fair Lawn, N.J., 1962.

Zinn, W. H., and J. R. Dietrich: Progress in Nuclear Steam Generating Systems Employing Pressurized Water Reactors, *XII Nucl. Congr. Rome,* 1967.

—John A. Martucci and Zolian R. Rosztoczy, *Combustion Engineering, Inc.. Windsor, Conn.*

Nucleation (See **Crystallization.**)

Nucleic Acids (See **Peptides and proteins;** and **Polymerization.**)

Nucleon A nucleon is a subatomic, nuclear particle; the name applies to a proton or to a neutron.

Nucleotide Sequence (See **Peptides and proteins;** and **Polymerization.**)

Nuclide A nuclide is an atom that differs from all other atoms in *both* the mass number and the atomic number. The designation is useful because of the existence of two or more isotopes for the majority of elements and because isotopes of several elements may have the same mass number.

The atomic number of an element is distinctive, i.e., exclusive to any given element. Thus, no element other than magnesium can have an atomic number of 12. However, magnesium has three stable and three radioactive isotopes, each with a different mass number (23–28). Several other elements have isotopes with mass numbers between and including mass numbers 23 and 28, e.g., Ne, Na, Al, Si, and P. Thus, to pinpoint a specific atomic configuration of a specific element, i.e., a nuclide, both the atomic number and the mass number must be specified. Including the elements through nobelium (at. no. 102), there are hundreds of nuclides.

See also **Chemical elements;** and **Nuclear power plants.**

Nutrition (See **Amino acids; Fertilizers; Peptides and proteins;** and **Urea.**)

Nylons The first nylon developed (type 6/6) was discovered in 1938 by W. H. Carothers. Since that time, nylons have filled an important role for industry and the consumer in various formulations, shapes, and forms, e.g., oriented fibers, which are subsequently processed into fabrics, fishing line, and other monofilament uses; injection-molded nylons, used as bearings, gears, and other parts subjected to wear and impact; extruded nylon tubing and hose, used in large quantity because of its chemical inertness, high strength, and flexibility; oriented nylon strip used as strapping for packaging, displacing traditional steel strapping; and heavy cast-nylon parts, frequently used in the textile, papermaking, and bottle-handling fields.

Most nylons exhibit a combination of high melt

point, high strength, impact resistance, wear resistance, chemical inertness, and a low coefficient of friction. The physical and chemical properties of several types of nylon are summarized in Table N-13.

Types: Type 6/6 and type 6 nylons are the most broadly used, dominating the field of textile fibers and in many respects having similar properties. Nylon 6/10, a lower-strength material produced in less volume, is used for industrial applications requiring improved moisture stability and high dielectric strength. It also has a lower melting point, lower specific gravity, and higher cost than types 6/6 and 6. Nylons 11 and 12 are relatively new specialties. Although of growing importance, they are used in rather low volume. They are very flexible and have a low order of moisture absorption; they are therefore preferred where consistent properties are needed in the presence of moisture, as well as chemical inertness, flexibility, and, in certain cases, transparency. Other nylons are of comparatively little commercial significance compared with these five.

Formulation: Types 6/6 and 6/10 are formed by the condensation of diamines with dibasic organic acids into linear chains containing amide groups. Types 6, 11, and 12 are self-condensed amino acids.

Type 6/6:

$$NH_2(CH_2)_6NH_2 + HOOC(CH_2)_4COOH \rightarrow$$
Hexamethylenediamene   Adipic acid

$$[NH(CH_2)_6NHCO(CH_2)_2CO]_n + H_2O$$
Polyhexamethyleneadipamide

Type 6/10:

$$NH_2(CH_2)_6NH_2 + HOOC(CH_2)_8COOH \rightarrow$$
Hexamethylenediamene   Sebacic acid

$$[NH(CH_2)_6NHCO(CH_2)_8CO]_n + H_2O$$
Polyhexamethylenesebacamide

Type 6:

$$NH(CH_2)_5CO \rightarrow [NH(CH_2)_5CO]_n$$
ε-Caprolactam   Polycaprolactam

Type 11:

$$NH_2(CH_2)_{10}COOH \rightarrow [NH(CH_2)_{10}CO]_n$$
Aminoundecanoic acid   Polyaminoundecanamide
$$+ H_2O$$

Type 12:

$$NH(CH_2)_{11}CO \rightarrow [NH(CH_2)_{11}CO]_n$$
Laurolactam   Polydodecanolactam

Besides the basic nylons, a variety of copolymers can be manufactured, some of which are commercially available. Nylons and nylon copolymers can be blended to form alloys with specifically tailored properties.

Nylons can be modified by the addition of certain plasticizers, fillers, reinforcements, and stabilizers. Ordinarily, nylons used for injection molding, such as type 6/6, have relatively low molecular weights (on the order of 15,000 to 20,000). High molecular weights are available to provide higher melt viscosity for nylon resins which are to be extruded into tubing or shapes. The molecular weight of nylon generally is determined by a relative-viscosity test, as described in ASTM D-789.

Resin Forms and Properties: Nylon resins usually are supplied in the form of cylindrical or rectangular diced pellets. The specific gravity of the natural resin ranges from 1.02 (nylon 12) to 1.15 (nylon 6/6). Fillers and reinforcements increase the specific gravity. The practical limit of glass reinforcement of about 40% results in a specific gravity of approximately 1.41. Melt points range around 300–500°F (150–250°C) and vary with nylon type over this range. For a given nylon formulation, the melt point normally will not vary more than ±10°F (±5.6°C) from lot to lot. Most commercial nylon molding resins are nontoxic. If a large amount of residual monomer is retained in the resin, as can occur with certain unextracted formulas, the material should not be in prolonged contact with food because of the possibility of monomer leaching.

Shipping and Storage: Nylon resins must be shipped and stored in a manner to exclude moisture. Nylons have a propensity to absorb moisture, and if the moisture is absorbed before the resin is melted for further processing, several difficulties can result, e.g., bubbling, surface imperfections, and degradation. If the resin has absorbed moisture, these defects in processing are very apparent on visual inspection. Unless the resin can be predried before use in extruders or injection-molding machines, nylon resins should be ordered in moisture-proof, sealed cans. If drying is necessary, this can be accomplished with circulating 180°F (82°C) dry air. Drying cycles may range up to 5 hr.

If predrying is available before further processing, other packaging methods can be considered, e.g., bags, fiber drums, large fiber packs which contain 1,000 or 1,200 lb of resin, and bulk rail-car shipments.

TABLE N-13. *Physical Properties of Certain Nylons*

	Type 6/6	Type 6	Type 6/10	Type 11	Type 12	30% Glass-reinforced Type 6/6	2.5% MoS_2-filled Type 6/6	Direct-polymerized Cast Nylon
Tensile strength, 1,000 psi	11.8	11.8	8.5	6.9–9.2	6.1–9.25	19.5–21.5	10–14	11–14
Elongation, %	60	200	85	180–380	250–400	5–10	5–150	10–50
Tensile yield stress, 1,000 psi	11.8	11.8	8.5	5.7–6.4	3.85–6.4			
Flexural modulus, psi $\times 10^5$	4.1	3.95	2.8			10–12.5	4–5	3.5–4.5
Tensile modulus, psi $\times 10^5$	4.2	3.8	2.8	2.1	0.6–1.7	10–12.5	4.5–6	
Rockwell hardness, R scale	118	119	111	108	106	118–120	110–125	112–120
Tensile impact, ft-lb/in.²	76		160			65–90	50–180	80–100
Impact strength, Izod, ft-lb/in. of notch	0.9	1–2	0.9	1.4–4.4	2.0–5.5	1–2	0.5–1	1–2
Deflection temperature under load at 264 psi:								
°F	185	147	135	118	120–131	470–490	200–470	300–425
°C	85	70	57	47	49–55	243–254	93–243	149–219
Dielectric constant, 60 Hz	4.0	3.8	3.9		4.2	4.0		3.7
Water absorption, 24-hr, %	1.5	1.6	0.4	0.4	0.25	0.65–0.75	0.5–1.4	0.6–1.2
Specific gravity	1.13–1.15	1.13	1.07–1.09	1.03–1.04	1.01–1.02	1.34–1.37	1.14–1.18	1.15–1.17
Melting point:								
°F	482–500	420–435	405–430	367	352–356	482–500	482–500	420–440
°C	250–260	216–224	207–221	186	178–180	250–260	250–260	216–227

Resistance

To strong acids	Not recommended*
To strong bases	Good*
To hydrocarbons	Excellent*
To chlorinated hydrocarbons, aromatic and aliphatic alcohols	Good

*All types

NOTE 1: Most of the nylon resins in this table are used for injection molding, and test values are determined from standard injection-molded specimens. In these cases, a single typical value is listed. Exceptions are MoS_2-filled nylon and direct-polymerized (cast) nylon, which are largely sold in various sizes of semifinished stock shapes. The range of values listed results from tests on various forms and sizes produced under varying processing conditions. The value ranges for types 11 and 12 reflect the different levels of plasticizer available in these materials.

NOTE 2: Since single values apply only to standard molded specimens and properties will vary in finished articles of different sizes and forms produced by various processes, values in this table should be used for comparison and preliminary design considerations only and not for final-design requirements. For final-design purposes the manufacturer should be consulted for test experience with the actual form being considered. None of the values listed should be used for specification purposes.

NOTE 3: All tests are conducted by the appropriate ASTM method on specimens which are dry as molded (0.2% moisture max.) unless otherwise indicated.

Fillers and Reinforcements: Several types of materials are used with the various nylons to modify the material for special properties, applications, or cost levels. The most common reinforcement is glass fiber. Nylon molding resins reinforced with glass fiber are commonly used for the manufacture of injection-molded gears and semistructural components in automotive applications and appliances. Glass reinforcements greatly increase the tensile strength and modulus. Most of the other beneficial properties of nylon are maintained, except that the wear resistance is not as good as with unfilled resin. The resistance to distortion at elevated temperatures is a particular improvement afforded by glass-fiber reinforcement. The most common filler level is 30% although there are commercial applications for filler levels ranging between 10 and 50%.

Molybdenum disulfide is widely used to improve the lubricity, strength, and thermal properties of nylon, and it improves the wear resistance of all common nylons. Nylon molding resin used for difficult mechanical applications often contains glass and molybdenum disulfide, the latter acting as a lubricant and nucleating agent which promotes a high level of crystallinity in finished parts, thus improving their strength and stability.

Less extensively used specialty fillers include small quantities of polytetrafluoroethylene to reduce the coefficient of friction, glass spheres to improve the stability and strength, carbon to improve resistance to ultraviolet light, and asbestos to improve the strength and reduce the cost. By working with this group of fillers and reinforcements in various levels, the properties of nylon can be tailored to a specific application or requirement.

Environmental Effects: Nylons require modification or stabilization to improve their resistance to certain environmental effects. Unstabilized nylon is degraded by ultraviolet light. The commonest and most reliable stabilizer for this purpose is approximately 2% well-dispersed carbon black, which has proved effective in the absorption of ultraviolet light. Actually, few critical nylon applications involve constant exposure to ultraviolet light. The nylons are considered quite weatherable, the only exception being their ultraviolet resistance. The nylons are considered adequate for outdoor applications if they are not exposed to direct sunlight.

Each nylon has a different degree of sensitivity to moisture and equilibrates at a different level. For example, type 6/6 nylon in certain amorphous sections equilibrates at 8.5% moisture when ex-

Fig. N-13. Segments of cast nylon are bolted into a cast-iron frame, and gear teeth are generated. The nylon gears are used on papermaking machinery, where they offer wear resistance and quiet operation.

posed to 100% relative humidity (RH). At 50% RH, it will equilibrate at 2.5% moisture. Nylons 11 and 12 are the most insensitive to moisture absorption. Type 12 nylon will absorb 1.4% at saturation when exposed to 100% RH and 0.7% at 50% RH. The moisture content in nylon manifests itself like a plasticizer; i.e., it makes the material more flexible. Some nylons also expand measurably as moisture is absorbed.

Nylons oxidize very slowly at 180°F (82°C), and oxidation becomes more rapid as the ambient temperature is raised. Nylons which are not heat-stabilized commonly are used at 200°F (93°C). For critical applications where temperatures above 200°F are anticipated, a heat-stabilized formulation should be used. All nylons are available in heat-stabilized grades. Heat-stabilized nylons can be used continuously to approximately 250°F (121°C), with short-time exposures up to 300°F (149°C).

Nylon is a crystalline polymer, the level of crystallinity varying with the method of processing and the rate of cooling from the melt. Rapidly cooled articles are amorphous and therefore more flexible and ductile, whereas slowly cooled articles are of higher crystallinity and therefore more stable, rigid, and strong.

Conversion and Fabrication: Because of the high-volume use of nylons in textiles, the largest-volume manufacturing process is the drawing and orientation of monofilaments. They are further processed like other textile fibers into finished fabrics and yard goods. Other monofilaments are processed into bristles for brushes.

The second most common method of conversion to finished articles is by injection molding, e.g., gears, cams, sliders, combs, and other high-volume consumer and industrial parts. Extrusions of tubing, rod, strip, and complex profiles are readily available. Tubing is also extruded from nylon and then braided with nylon monofilament to improve the burst strength of the pressure tubing. A nylon sheath is then extruded over the braid to make an all-nylon hose which is superior in certain respects to rubber hose. When high performance, quality, and close tolerances are required, nylon parts are machined from extruded mill shapes using standard metalworking techniques. Machined nylon parts of almost any complexity and size are available. If large complex parts are required, nylon castings are available. Nylon cast shapes are produced at atmospheric pressure by anionic polymerization of ϵ-caprolactam. Massive castings up to several hundred pounds have been made (see Fig. N-13).

Nylon film, generally a type 6, is made in high volume for boil-in-the-bag food packaging, as well as see-through windows in packages, packing-slip envelopes, and packages requiring high strength and durability.

—D. D. Carswell, *The Polymer Corporation, Reading, Pa.*

Nytrils (See **Fibers.**)

O

Octane Number (See Alkylation; Cracking; Fluid catalytic cracking; Isomerization; Petrochemical complex; Petroleum; and Petroleum processing.)

Octanol (See Alcohols.)

Odor Chemicals (See Aroma chemicals.)

Oils (See Mineral oils; Petroleum; and Vegetable oils.)

Oleandomycin (See Antibiotics.)

Olefins (See Alkylation; Cracking; Ethylene; Fluid catalytic cracking; Hydrocarbons; Hydrocracking; Isomerization; Petrochemical complex; Reforming; and Thermal cracking.)

Oleic Acid (See Defoaming agents; Paint; Vegetable oils; and Waxes.)

Oleoresins (See Resins.)

Oleum (See Sulfuric acid.)

Oligomer (See Polymerization.)

Oligopeptides (See Peptides and proteins.)

Olive Oil [See Anionic surfactants (sulfurbearing); and Vegetable oils.]

Open-Hearth Furnace (See Iron- and steelmaking.)

Optical Bleaches (See Detergents.)

Optical Materials (Consult the Subject Index.)

Ore Beneficiation (See Beneficiation, ore; and Iron ores.)

Organic Chemistry (See Carbon compounds.)

Organic Waste Treatment (See Waste treatment.)

Organometallic Compounds (Consult the Subject Index.)

Organosols (See Paint.)

Orpiment   (See **Arsenic**.)

Ortho Compounds   (See **Carbon compounds**.)

Orthoclase   (See **Potassium**.)

Orthophosphoric Acid   (See **Fertilizers;** and **Phosphoric acid**.)

Orthosilicates   [See **Silicates (soluble)**.]

Osmium   (Gk. *osme*, odor.) **Os** = 190.2 (at. wt.); 76 (at. no.). Osmium is a member of the family of platinum metals, group VIII of the periodic table. The melting point = 3000°C; boiling point = 5500°C; density at 20°C = 22.47 g/cm³. Os is the heaviest of the naturally occurring elements. There are seven stable isotopes of Os, 184, 186–190, and 192, and six unstable isotopes, 182, 183, 185, 191, 193, and 194. Os was identified by Tennant (England) in 1803.

Finely divided Os oxidizes in air, producing the poisonous and volatile tetroxide. The compact metal is not attacked by nonoxidizing acids. The finely divided metal dissolves in fuming HNO_3, aqua regia, and alkaline hypochlorite solutions. When fused with Na_2O_2 or KNO_3 and KOH, the metal is converted to the corresponding water-soluble osmate, K_2OsO_4. The brown or black insoluble osmium(IV) oxide, OsO_2, can be made by heating Os with a limited amount of O_2 or with osmium(VIII) oxide. This compound forms a brown to black-blue dihydrate that can be prepared by reducing a solution of the tetroxide or by hydrolyzing a solution of sodium hexachloroosmate, Na_2OsCl_6.

Osmium(VIII) oxide, the most important compound, is formed in one of the reactions unique to the platinum metals. Its ease of formation and volatility make it useful in a purification step for the refining or analysis of Os. The tetroxide is readily formed by heating the metal in air or distilling an osmium-containing solution from HNO_3. Although an aqueous solution of osmium(VIII) oxide is neutral to litmus, it is a weak acid with first dissociation constant of about 8×10^{-13}. Osmium(VIII) oxide is soluble in water, alcohol, and ether. The compound is widely used as a stain for tissues. When an alkaline solution of osmium(VIII) oxide is reduced with alcohol or KNO_2, an osmate(VI) is formed. Potassium osmate(VI) is formed by adding an excess of KOH to such a solution, resulting in the precipitation of violet crystals of $K_2OsO_4 \cdot 2H_2O$.

The osmate(VI) ion is probably better written as $OsO_2(OH)_2^{--}$.

Osmium(II) chloride can be prepared by heating osmium(III) chloride in vacuum at 500°C. This dark brown compound is insoluble in HCl or H_2SO_4. HNO_3 or aqua regia oxidize it to the tetroxide. Osmium(III) chloride is best made by decomposing ammonium hexachloroosmate(IV), $(NH_4)_2OsCl_6$, in a Cl_2 stream at 350°C. The brown hygroscopic powder sublimes above 350°C, and at about 560°C it disproportionates into the tetrachloride and dichloride. Osmium(IV) chloride is formed from the elements at 650–700°C. The black compound slowly dissolves in water, eventually forming the dioxide. The free acid, H_2OsCl_6, is stable in solution and can be made by refluxing osmium(VIII) oxide with HCl and alcohol. The ammonium salt can be precipitated by adding NH_4Cl to such a solution. This salt is reduced to the metal when heated in H_2. The potassium salt is well known. Both are brownish-red solids yielding orange solutions in water.

Recent studies have established the reaction product of Os metal and F_2 at 300°C to be the hexafluoride, OsF_6. This yellow volatile solid had previously been described as an octafluoride. Osmium(VI) fluoride melts at 33.4°C and boils at 47.5°C. OsF_6 can be reduced to a pentafluoride and a tetrafluoride. The pentafluoride is a blue-gray crystalline solid that melts at 70°C to a green viscous liquid and boils at 226°C. The tetrafluoride distills at about 290°C. Potassium hexafluoroosmate(V), $KOsF_6$, can be made by reacting KBr, osmium(IV) bromide, and bromine trifluoride. The white powder dissolves in water to form a colorless solution that hydrolyzes to yield some osmium(VIII) oxide. On addition of 1 equiv of KOH to a fresh solution, an orange color develops, O_2 is evolved, and yellow crystals of potassium hexafluoroosmate(IV), K_2OsF_6, form.

Os forms many complexes with nitrite, oxalate, carbon monoxide, amines, and thio ureas. The latter are important analytically. Osmium forms the interesting aromatic "sandwich" compound, osmocene. A *metallocene* is described under **Ruthenium**.

Further information on sources, refining, and uses of Os will be found under **Platinum metals and platinum**.

—Lionel E. Simmons, *Simmons Refining Company, Chicago, Ill.*

Osmosis   When solutions of different concentration are separated by a membrane, e.g., a thin sheet of collodion, they diffuse through it from the more concentrated solution to the less concentrated solution. This process is called *osmosis*. The pressure that would have to be applied to the solution to nullify this process is termed *osmotic pressure*.

Osmosis, Reverse (See **Membrane filters;** and **Water treatment.**)

Ovalbumin (See **Peptides and proteins.**)

Ovens, Coke (See **Coal tar and derivatives.**)

Oxamide (See **Amides and imides.**)

Oxidation (See **Air pollution; Alcohols; Antioxidants; Catalysts; Chlorine; Ethylene; Ethylene oxide; Fuels; Iron- and steelmaking;** and **Waste treatment.**)

Oximes (See **Aldehydes;** and **Ketones.**)

Oximino (See **Radicals.**)

Oxo Process In this process (oxonation), an olefin is reacted with water gas (a mixture of CO and H_2) in the presence of a catalyst (usually cobalt)

to form an aldehyde that contains one more carbon than the olefin feed. Thus, if propylene is the feedstock, the reaction is

$$CH_3CH:CH_2 + (CO + H_2) \rightarrow$$

Propylene   Water gas

$$CH_3CH_2CH_2CHO + (CH_3)_2CHCHO$$

n-Butyraldehyde   Isobutyraldehyde

The oxo process usually is the beginning step of an alcohol synthesis and is followed by an aldol condensation step and a hydrogenation step to produce an alcohol (in this example, 2-ethylhexanol). See also **Alcohols.**

Purity of the feed usually is relatively high (92% or better propylene in this example); it is scrubbed free of sulfur compounds and any CO_2 that might interfere with the reaction. The CO and H_2 are of 98–99% purity and are fed to the reactor in a 1:1 mole ratio. The exothermic reaction takes place in the liquid phase at pressures in the range of 200–300 atm and at a temperature of 120–150°C.

In Fig. O-1 the overhead mixture from the reactor containing aldehydes and some unreacted material is cooled and separated by high- and low-pressure flashing stages. Unreacted gas is recycled from each of these separators to the oxo reactor. Most of the CO catalyst is separated

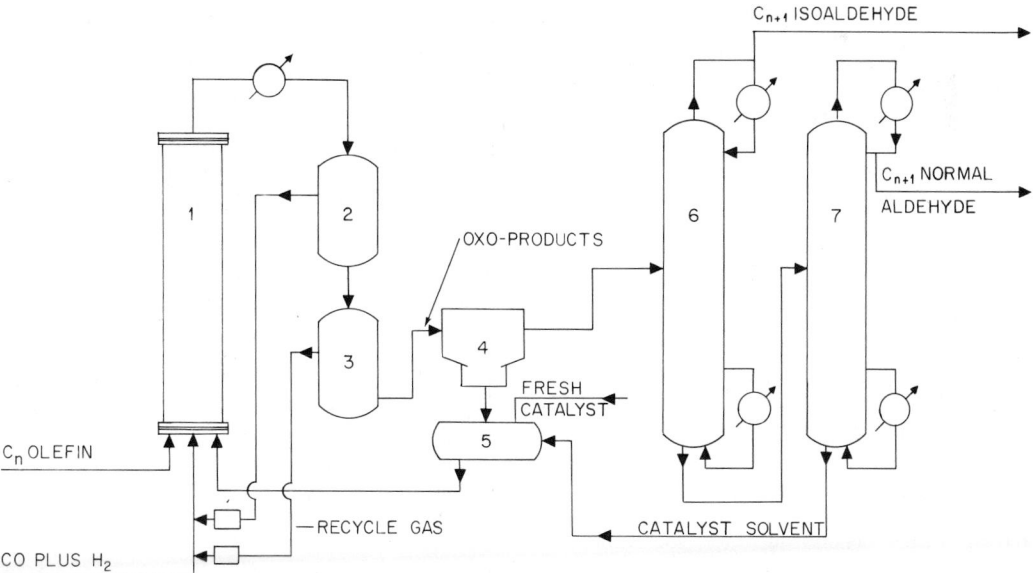

Fig. O-1. Oxo process for converting olefins into aldehydes. (1) Oxo reactor, (2) high-pressure separator, (3) low-pressure separator, (4) catalyst separations, (5) catalyst regeneration, (6), (7) aldehyde separation.

continuously from the liquid phase and regenerated to be returned to the reactor along with whatever fresh catalyst may be required. In the example given here, the ratio of n-butyraldehyde to isobutyraldehyde produced is approximately 4:1, but it is claimed that the proportion can be controlled between 3.8:1 and 4.8:1. See also **Gasification**; and **Synthesis gas.**

Oxyacetylene Welding   (See **Oxygen.**)

Oxychlorination   (See **Chlorine; Ethylene; Trichloroethylene;** and **Vinyl chloride monomer.**)

Oxygen   (F. *oxygène,* coined by Lavoisier and intended to mean acidifying.) **O** = 15.9994 (at. wt.); 8 (at. no.). Oxygen, O_2, is a colorless, slightly magnetic, odorless gas at standard conditions with a density of 1.3318 × 10^{-3} g/cm^3 at 20°C (0.048118 × 10^{-3} lb/in.^{3} at 68°F). The melting point is −218.9°C, boiling point −183.0°C, specific gravity 1.14 at −183°C, critical temperature 118.8°C, and critical pressure 49.7 atm. Thermal conductivity is 0.000059 (cal)(cm)/(s)(cm^2)(°C) at 20°C. Heat of fusion is 106 cal/mole at the melting point; heat of vaporization is 1629 cal/mole at the boiling point. Ionization potential is 13.614 eV. Oxygen is the most electronegative of all elements with the exception of fluorine and combines directly or indirectly with all elements except the rare gases (and fluorine with difficulty).

O_2 is slightly soluble in cold H_2O (4.89 parts O_2 in 100 parts of H_2O at 0°C), solubility decreasing with rise in temperature (2.6 parts O_2 in 100 parts H_2O at 30°C; 1.7 parts O_2 in 100 parts H_2O at 100°C). O_2 is only slightly soluble in alcohol. Molten silver will dissolve approximately 10 times its volume of O_2 and then easily give up the O_2 on cooling.

Oxygen occurs in group VIA and period 2 of the periodic table. There are three stable isotopes of oxygen, 16–18, and three unstable isotopes 14, 15, and 19. See also **Chemical elements.** There are three known allotropic forms of oxygen: (1) the ordinary oxygen of the atmosphere, with two atoms per molecule, O_2; (2) ozone, O_3, with three atoms per molecule; and (3) the rare, very unstable, nonmagnetic, pale-blue O_4, which breaks down readily into two molecules of O_2.

Pure ozone (Gk. *ozein,* to smell) has a characteristic, rather penetrating odor, sometimes noticed near electrical equipment. It is formed in the atmosphere by lightning and during the evaporation of water, particularly sea spray. Ozone can be prepared by passing dry, cold O_2 through an electric discharge. Although O_3 is more efficient as an oxidant than O_2, relatively few economically feasible industrial uses have been found for it. Ozone and ozonides now are identified as objectionable air pollutants, occurring under inversion and other climatic conditions over industrial areas. Ozone reacts with olefins to form ozonide addition compounds. Ozone has a serious deleterious effect on certain elastomers. See also **Antioxidants; Elastomers; Peroxides, organic; Rubber, natural;** and **Water treatment.**

Oxygen leads all other elements in terms of terrestrial abundance, averaging 466,000 g/metric ton, or 46.6% of all terrestrial matter. Water on earth comprises 88.8% O_2 by weight. Minerals and rocks average 44–48% O_2 by weight. Atmospheric air, disregarding pollutants, contains approximately 21% O_2 at sea level (20.98% by volume; 23.15% by weight of dry air). Lunar rocks returned by Apollo 11 averaged 39.9 ± 1.1% O_2.

Despite its abundance, oxygen was not identified until 1774 by Priestley in England (as dephlogisticated air) and Scheele in Sweden independently. Lavoisier (France) established the elemental nature of oxygen during the same period and is responsible for its name. Nearly all known species of living things require oxygen in some form, either free or chemically bound.

At one time, the term *oxidation* meant merely a reaction in which a substance combines with oxygen and of course a large percentage of oxidation reactions, such as combustion (see also **Fuels**), fit the old definition very well. In the modern sense, however, oxidation refers to the loss of one or more electrons from the outer shell of an atom, and hence oxidation may take place in the complete absence of oxygen. For every oxidation reaction there is an accompanying reduction reaction.

Oxidation in which oxygen is the oxidant proceeds over a wide span of reaction rates, ranging from the very slow processes of rusting and most other forms of corrosion, through the combustion of fuels, to the extremely fast reactions found with pyrophoric, explosive, and incendiary materials.

In a broad sense, the term *corrosion* applies to the destructive alteration of a metal or alloy by chemical reaction with any substance. The mechanisms by which these reactions occur are varied and complex. Oxygen and other oxidizing agents may affect the corrosion rates of metals in two ways: (1) they can increase the rate of attack by supporting the cathode reactions; (2) oxidizing agents may substantially reduce or eliminate cor-

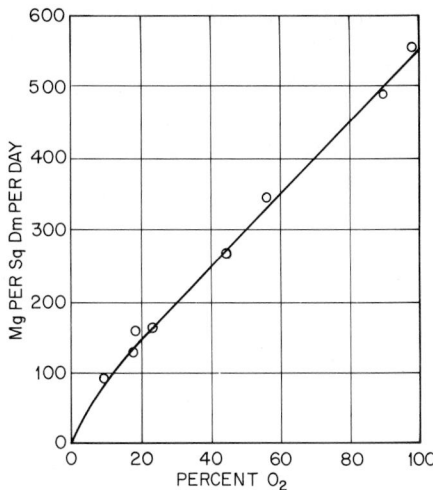

Fig. O-2. Influence of oxygen concentration in the saturating atmosphere on the corrosion of Monel metal in 5% H_2SO_4 at room temperature.

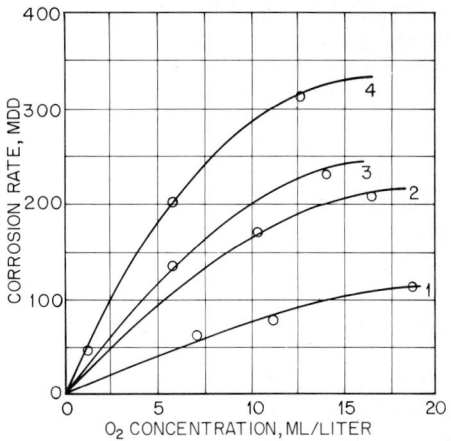

Fig. O-3. Effect of velocity and oxygen concentration on corrosion of steel in water at 23°C (73.4°F) during a 5- to 7-day test. (1) Velocity 0.0 ft/s, (2) 3 ft/s, (3) 0.5 ft/s, (4) 1.2 ft/s; mdd, milligrams of metal corroded per square decimeter per day.

rosion by forming protective films on the metal surface. In still other situations variations in concentration of oxidizing agents on a metal surface can lead to severe localized corrosion. The influence of oxygen concentration in two cases is shown in Figs. O-2 and O-3.

Compounds: Hundreds of oxygen-bearing compounds are described in this volume. The majority of elements ignited in oxygen combine readily with it. According to their valence, some elements form two or more series of oxides. Usually a lower oxide can be oxidized to a higher oxide. A number of ores evolve O_2 to form lower oxides or reduce to elemental form. The ubiquitous and versatile nature of oxygen applies equally to organic and inorganic compounds.

Manufacture: Although O_2 can be produced by the electrolysis of H_2O, high-tonnage O_2 production generally uses air as a raw material. These processes fall into two classifications, liquid-oxygen processes, in which O_2 is fractionally distilled from liquid air, and gaseous-oxygen processes.

Because of the relatively high energy costs of compressing and refrigerating involved in O_2 production by these means, a score or more process designs have been developed and tested, some being abandoned over the years. An idea of the alternatives which face the process designer can be gathered from scanning the methods available

specifically in the area of producing refrigeration for these processes: (1) Joule-Thomson effect only; (2) Joule-Thomson effect plus auxiliary refrigeration with an ordinary liquid-vapor cycle at moderate or high temperature levels, i.e., relative to liquid-air temperature; (3) Joule-Thomson effect plus approximately reversible expansion of the air or products in an expander; (4) refrigeration essentially due only to approximately reversible expansions of auxiliary fluid or fluids, e.g., helium, through expanders, i.e., processes in which the fluid remains entirely in the gas phase; (5) refrigeration essentially due only to auxiliary fluid or fluids operating in liquid-vapor cycles, i.e., the cascade process; (6) refrigeration essentially due only to approximately reversible expansion of air or products in an expander, i.e., the low-pressure process; and (7) processes using an auxiliary nitrogen-liquefaction cycle. For further information pertaining to cryogenic processes of the type used in producing O_2, see **Ethylene; Helium;** and **Natural gas.**

Uses: Listing the applications of the scores of compounds of O_2 is well beyond the scope of this volume. Air is a major chemical raw material, predominantly because of its O_2 content. Frequently, to create the chemical concentrations required and to avoid transporting nitrogen (78.03% by weight of dry air is nitrogen), O_2 in a much purer form is used as a process raw material. The costs of pure or nearly pure O_2 often affect the selection of a process design. For example, there are several ways to synthesize ammonia. In the noncatalytic partial-oxidation processes designed to produce H_2 from a wide range of hydrocarbon liquids, the hydrocarbon feed is oxidized. The required O_2 usually is supplied by an air-separation plant from which N_2 also can be used as feed for the synthesis gas. The process has advantages over some other approaches, but the need for high-purity O_2 can be considered a disadvantage.

Significant quantities of purified O_2 are used in welding and cutting metals. In oxyacetylene welding, the combination of O_2 and C_2H_2 in correct proportions yields a flame with a temperature of approximately 6300°F (3482°C). This combustion reaction releases 1433 Btu per cubic foot of gases used. Combination of high temperature and heat content enables the flame to melt or fuse locally many metals and alloys. O_2 with a minimum purity of 99.5% is required.

Oxyhydrogen flames are used for welding light-gage aluminum and magnesium alloys and for underwater cutting. The flame temperature is somewhat lower than that of oxyacetylene flame, but the same purity of O_2 is required.

If iron is heated to about 1500°F (816°C) and contacted with high-purity O_2, the metal will burn rapidly with evolution of much heat. Theoretically, the process is self-supporting; in actual practice, continuous preheating is required, often by oxyacetylene or oxyhydrogen flames. O_2 purity is very important for cutting applications; a reduction from 99.5 to 99.0% purity of the O_2 will reduce cutting efficiency by over 10%.

Basic oxygen steelmaking has upped the tonnage requirements for pure O_2. The basic oxygen process gets its name from the procedure of blowing nearly pure O_2 at high rates into a bath of molten iron and scrap contained in a pear-shaped vessel. The oxygen combines with carbon and other unwanted elements in the bath, burning them out and converting the charge to steel. The basic oxygen furnace (BOF) is highly efficient and can produce raw steel in much less time than the open-hearth furnace. See also **Iron- and steelmaking.**

A milestone in United States steelmaking was reached late in 1970, when monthly BOF output surpassed that of the open hearth for the first time. Worldwide basic oxygen steelmaking now exceeds 300 million short tons of raw steel per year.

Relatively small quantities of oxygen are used in medical applications. Cryogenic, aerospace, and missile applications also have increased the demand for pure O_2.

Oxygen Group   (See **Chemical elements.**)

Ozokerite   (See **Waxes.**)

Ozone   (See **Elastomers; Oxygen; Peroxides, organic; Rubber, natural;** and **Waste treatment.**)

P

Paal-Knorr Synthesis (See **Furan group.**)

Package Dyeing [See **Dyestuffs;** and **Dyeing (textile).**]

Packaged Process Plants (See **Process plants, packaged.**)

Packaging (Consult the **Subject Index.**)

Packed Columns (See **Absorption.**)

Padding [See **Dyestuffs and dyeing (textile).**]

Paint Paint is a very broad word used to describe many different types of materials used for decorative and protective coating of various surfaces. In this description, the word *paint* is used to describe all types of pigmented coatings, such as enamels, lacquers, latex coatings, and alkyd flat wall paints.

Classification: The type of *binder,* such as alkyds, vinyls, and epoxy, may be the basis for classifying paints. Since most binders are used in several different kinds of paint, however, paints also may be classified according to the properties of the product or *end use.* Alkyd enamels, for instance,

are gloss paints with good abrasion resistance and good cleanability, while alkyd flat wall paints are characterized by a very low sheen and inferior abrasion resistance and cleanability.

Paints used as the final coat on a surface are referred to as *finish coats* or *topcoats.* Paint applied before the topcoat is called an *undercoat.* Undercoats often are classified according to use. *Fillers* are undercoats used to fill holes, pores, or irregularities to provide a uniform surface for the topcoat. *Primers* are used to aid the adhesion of the topcoat to a surface and to prevent absorption of the topcoat into a porous surface. Primers can also be used to prevent corrosion of metals that are to be painted. *Surfacers* are highly pigmented undercoaters used to make a surface more uniform and give adhesion to the final coat. Surfacers often are formulated so that they can be sanded smooth before the topcoat is applied. *Sealers* are clear or pigmented materials applied to a surface to prevent some materials in the surface to be painted, e.g., a dye, from migrating into the topcoat.

Paints also can be classified into two very broad groups, those which use water as the primary liquid of the paint and those which use other liquids, such as hydrocarbons or aliphatic or oxygenated compounds. Many of the same polymers and pigments can be used in both types of paint. The primary difference is that most non-water-dilutable paints are *solution paints,* where the liquid is a solvent in which the polymer is dissolved. *Water-based paints,* however, are primarily latex paints, where the polymer particles are a discontinuous phase and water is the continuous phase. *Organosols,* however, consist of a resin dispersed in a liquid other than water. There also are water-based paints which have a water-soluble binder instead of a latex dispersion.

In addition to the pigment, binder, and liquid, a paint also may contain many additives, such as defoamers, thickeners, flow agents, catalysts, wetting agents, and plasticizers to improve various properties of the paint. The selection of the pigments, binder, and additives depends, of course, upon the properties desired in the paint to be made.

Manufacture: A primary concern in the manufacture of paint is the dispersion of pigments in the liquid portion of the paint.

Pigments used in paint contain aggregates of fine particles, which are held together by strong forces of attraction. Work is required in the form of high shear or attrition applied to the pigment-liquid slurry in order to reduce the pigments to their ultimate particle size as determined by the manufacturer of the pigment. Considerable work may also be required to wet the surface of the pigment and to displace any air or moisture that may be absorbed on the surface by the liquid. Wetting agents such as soaps or detergents in the water phase of latex paints are often used to help wet and disperse pigments.

After the pigments are dispersed, they must be prevented from coming together again and flocculating, or forming soft lumps of pigment and liquid. The nonpolar binder of a solvent-thinned paint will usually serve to envelope the pigment particles and prevent the electrostatic charges on the surface of the pigments from causing flocculation. In dispersions of water-based paints surfactants in the water phase will neutralize the charges on the pigment surfaces, and protective colloids, such as water-soluble cellulosic polymers, are introduced into the water phase to prevent flocculation after dispersion. In practice, few pigments are dispersed to their ultimate particle size, and paints usually contain many aggregates and flocculants.

Since some pigments are harder to disperse than others, different types of dispersing equipment or mills are used. Dispersing pigments in a liquid is often called *grinding* although there is very little reduction in size of the original pigment particles during the dispersion operation of paint manufacture.

The high-speed stone mill consists of a stationary carborundum stone and a high-speed rotating stone. Pigment pastes are passed between these stones; the distance between the stones can be varied for more or less shearing action. These mills are suitable for high production rates of paints fairly easy to disperse, e.g., architectural paints, where very fine dispersion is not required.

Roller mills consist of steel rollers rotating in opposite directions at different speeds. The pigment-liquid paste is passed between the rolls, which can be adjusted to different clearances. Three-roll mills are the most widely used in the paint industry. Roller mills have relatively slow production rates and require skilled operators but are capable of producing fine dispersion.

Heavy-duty dough mixers, consisting of two roughly S-shaped blades which overlap and rotate in opposite directions, are sometimes used to disperse very heavy pastes.

Ball and pebble mills consist of large cylindrical steel tanks which rotate around a horizontal axis. The mill is partly filled with steel or porcelain balls or pebbles and the material to be dispersed. Baffle bars are usually added to the sides of the tank to help lift the balls for better dispersion. Steel balls are more efficient because of their greater density but cannot be used to produce white paints. Ball mills require little attention after they have been charged and are capable of producing good dispersion. See also **Size reduction.**

The sand mill consists of a cylinder containing coarse sand as a grinding medium. The pigment paste to be dispersed is fed into the mill, and rotating impeller disks driven by a vertical shaft impart a circulation pattern to the sand-paste mixture. The difference in velocity between the particles near the surface of the impellers and the rest of the material develops a high shear action to disperse the pigments. A coarse screen allows the pigment slurry to pass through the mill while retaining the sand in the mill. Pigment slurrys can be passed through these mills for continuous operation. Production rates of sand mills can be fairly good, and dispersion is quite good. These mills are often used for high-quality industrial finishes.

The high-speed disperser consists of a tank containing a circular impeller driven at high speed by a vertical shaft. Dispersion of the pigment-liquid mixture is achieved by high shear action developed near the surface of the impeller. High-speed dispersers are used where very fine dispersion is not required or whenever the pigments will disperse easily in the liquid. The production rate is very high, and this type of equipment is used to manufacture most architectural paints.

After the pigment is dispersed and stabilized in the liquid, it is usually transferred to a thin-down tank, equipped with slower agitation, where the paint is adjusted to the desired viscosity and shaded. Materials can be added to the paint in the thin-down tank that are not required in the grinding operation and can be added with slow agitation. The binder of a solvent-thinned paint is usually added during the grinding operation; the binder of a latex paint is usually added in the thin-down tank. Thin-down tanks are usually on a floor below the grinding equipment to take advantage of gravity flow, and the finished paint is usually dropped to a floor below the thin-down tanks for straining and filling.

Application: In addition to the familiar brush and roller methods, paints can be applied with air or airless spray equipment; electrostatic, hot, or steam spraying; use of aerosol packaging; dip, flow, and electrodeposition coating; roller coating machines; and powder coating.

Although most architectural paints are applied with a brush or roller, much paint is now being applied by professional painters with compressed air or airless spray equipment. With airless spray equipment the paint is atomized by forcing it through a very small orifice under very high pressure.

With *electrostatic spraying* the atomized paint is attracted to the conductive object to be painted by an electrostatic potential between the paint and the object. Very little paint is lost with this process, and irregular objects can be coated uniformly.

Hot spray application consists of heating the paint so that it is more fluid and higher-solids paints can be applied. With *steam spraying,* steam is used to atomize the paint.

Two-component spray equipment consists of two material lines leading to the spray gun so that two materials, e.g., an epoxy and a catalyst, can be mixed in the gun just before application.

Aerosol is a method of packaging paint in a can containing a compressed gas so that the paint can be atomized through a small orifice opened by a push button.

Many different methods are used for industrial application of paint, including most of the spray methods.

Dip application is a simple method where objects to be coated are suspended from a conveyor chain and dipped into a large tank containing the paint. This method is often used for undercoating objects where paint uniformity and appearance are not important.

In *flow coating* the paint is allowed to flow over the object to be painted, which is usually suspended from a conveyor. This process is similar to dip coating but is used where the object, e.g., a bed spring, is too large for a tank.

Electrodeposition consists of depositing a paint on a conductive surface from a water bath containing the paint. The negatively charged paint particles are attracted to the object to be coated, which is the anode when an electric potential is applied. Paint can be applied to very irregular surfaces at very uniform thickness with little loss of paint. The system is limited to one coat of limited film thickness, and equipment cost is high.

Roller coating machines are used to apply paint to one or both sides of flat surfaces, e.g., fiberboard or tin plate. The thickness of the coating can be controlled by the clearance between a doctor blade and the applicator rolls. Decorative effects, such as wood-grain patterns, can be applied with these machines. Flat sheets of wood, fiberboard, or metal and rolls of fabric, paper, or metal can be coated with these machines.

In *powder coating* paint in a dry powder form is applied on the surface of a heated or electrostatically grounded object to be coated. Following powder application, the object is heated to fuse and cure the coating.

Drying and Curing: There are two main mechanisms by which a wet paint applied to a surface becomes a dry solid coating. Some paints, e.g., lacquers and most latex paints, dry only by the evaporation of the liquids in the paint. The polymer binder is completely cured when the liquids evaporate, and no chemical change is required to harden the polymer. In latex paints, the latex binder consists of very small particles of solid polymer separated by water, which is the continuous phase. When the water evaporates, the polymer particles touch each other and fuse together, or coalesce, into a continuous paint film. Pigment particles are also dispersed in the water phase, and the dry paint film consists of a mixture of pigment and polymer particles fused together. If the latex particles are so hard that they will not fuse together when the water evaporates, coalescents must be added, e.g., carbitol acetate or dibutyl phthalate.

The other broad method of drying takes place by a chemical reaction to cross-link a soft polymer after the liquids have evaporated to produce a hard paint film. Some paint binders contain unsaturated compounds, such as linseed or soya drying oils, which react with oxygen in the air to form solid polymers. Other materials, such as isocyanates, react with water vapor in the air in order to polymerize.

Some materials, e.g., epoxy resins, must be cured by reaction with curing agents such as amines and polysulfide resins. The majority of these paints are supplied as two-package systems to separate the resin from the curing agent until just before the paint is applied.

Oxidative polymerization is usually aided by polymerization catalysts called *driers*. Lead octoate, cobalt naphthenate, or manganese isodecanoates are often used. Heat applied to the paint film is also used as an aid to polymerization reactions.

Solvent Paints. The many different types of vehicles, e.g., alkyds, chlorinated rubber, acrylics, and nitrocellulose, used in solvent paints will be discussed under specific types of paints.

Because of its high refractive index, titanium dioxide is the primary pigment used in solvent paints to give opacity to white and light-colored paints. Among other white pigments used is zinc oxide for hardness, drying, and mildew resistance of exterior paints. See also **Titanium dioxide.**

Extender pigments, such as talc, silica, mica, clay, and calcium carbonate, are used to control many different properties of a paint (hardness, gloss, settling, rheological properties, and rust prevention). Since the solvent evaporates and contributes nothing to the dried film, the least expensive combination of solvents can be used which will dissolve the polymer, give the desired viscosity, and conform to air-pollution standards.

Antiskinning agents are often used to prevent oxidation and surface hardening of the paint resin in the can. Oxime compounds, e.g., methyl ethyl ketoxime, and substituted phenols, e.g., guaiacol, are often used.

Scratch- and marproof additives, such as amino resins, wax, polyethylene and cellulose derivatives, are sometimes added to harden the paint film or develop a lubricated surface.

Flooding and floating is the separation of one or more pigments from the rest of the paint at the surface of the film, usually due to a difference in densities of the pigments. Antiflood and antifloating agents such as silicones can be used to correct this problem.

Additives which will attach to the surface of the pigments and form a pigment-vehicle bond for better grinding or dispersing are often used. Compounds which contain both hydrophilic and lipophilic groups, e.g., morpholine, methyl ethyl ketoxime, and soya lecithin, are used so that the hydrophilic part will attach to the pigment and the lipophilic part will attach to the vehicle.

Flow or leveling agents are used to correct irregularities or defects in applied paint films. These defects can be caused by differences in concentration between the surface and the interior of the paint during evaporation of the solvent, over-polymerized particles in the vehicle, an unclean surface, pigment flocculation, or solvent mixtures with a wide range of evaporation rates. Some of the materials used are silicone resins and unsaturated organic acids.

Drying-Oil Paints: Drying oils, which have been used for years in paint binders or vehicles, are naturally occurring materials usually of vegetable origin, i.e., liquid triglycerides with three molecules of long-chain fatty acids to each glycerin molecule. The majority of these acids have 18 carbon atoms, many of which are unsaturated. The oils differ greatly in their drying properties, which depend upon the degree of unsaturation of the acids. Linseed, safflower, soya, tall oil, cottonseed, tung, and oiticica contain fairly large percentages of unsaturated acids, e.g., oleic, linoleic, linolenic, eleostearic, and licanic acids.

Drying oils are often used in exterior architectural paints because of their durability, ease of application, and moderate cost. Disadvantages are slow drying and poor chemical resistance.

Alkyd Paints: Alkyds are polyester resins made from polybasic acids and polyhydric alcohols. Glycerol and pentaerythritol are often used for the polyalcohol, and phthalic anhydride and maleic acid are often used for the polycarboxylic acids. All the oils previously discussed are also used in alkyds by converting the fatty acid oils into monoglycerides and then reacting with a dibasic acid such as phthalic anhydride:

Phthalic anhydride   Fatty acid

Alkyd resins vary greatly in their properties because of the many different oils, alcohols, and acids that can be used to make them.

Alkyds have faster drying, better gloss retention, and better color than oils. Most unmodified alkyds have low chemical and alkali resistance. Alkyds can be modified with rosin esterified in place of some oil acids. Phenolic resins, such as *o*- or *p*-phenylphenol, can also be used in order to produce greater hardness and better chemical resistance.

Styrene and vinyl toluene are also used to modify alkyds for faster dry, better hardness, and toughness.

Copolymers of silicones and alkyds are often used for stove or heater finishes or for coating articles which may be subjected to heat up to about 450°F (232°C). They have good adhesion, hardness, flexibility, toughness, exterior durability, and resistance to solvents, acids, and alkalis.

Acrylic monomers can be copolymerized with oils to modify alkyd resins for fast dry, good initial gloss, adhesion, and exterior durability.

Aromatic acids, e.g., benzoic or butylbenzoic, may be used to replace part of the fatty acids for faster air dry, high gloss, hardness, chemical resistance, and adhesion.

Vinyl Paints: Vinyl chloride–vinyl acetate copolymers in solution form are often used in paints. These paints are resistant to alkali and organic acids, alcohols, oils, and aliphatic hydrocarbons. They can be dissolved in ketones, esters, and chlorinated hydrocarbons; have good water resistance, toughness, and flexibility; are nonflammable; and have good clarity and good exterior durability. Vinyl copolymers often are used in paints for railway car finishes and organosols.

Epoxy Paints: Epoxy resins are prepared from epichlorohydrin CH_2CHCH_2Cl and a dihydroxy

$$\underset{O}{\overset{\diagup\diagdown}{}}$$

compound, usually a biphenol. See also **Epoxy resins.**

Two reactions are involved in the polymerization: condensation to eliminate HCl and addition reactions to open epoxide rings along the chain to produce hydroxyl groups. The polymer has epoxide rings at each end and hydroxyl groups along the chain, which ensure good adhesion to polar surfaces such as metals. There are two types of epoxy resins, *catalyzed types* and *epoxy esters*.

Catalyzed epoxies must be converted to useful products by reaction with curing agents, e.g., amines, polyamide resins, polysulfide resins, anhydrides, metallic hydroxides, or Lewis acids.

Most of these materials are supplied as two-package systems to separate the materials until just before application. Polymer curing takes place by reaction of a curing agent with epoxide rings to cross-link the polymer. Paints made from these polymers have excellent chemical resistance and hardness and are often used for maintenance coatings, trade sales specialties, and industrial finishes. Chalking with exterior exposure and the two-package system limit the use of epoxy finishes.

The epoxy resin can also be reacted with drying oils or fatty acids to produce epoxy esters, which cure by air drying or heat. Paints made with epoxy esters do not have as good chemical and solvent resistance as catalyzed epoxies, but they are superior to oils and alkyds in this respect. They also exhibit chalking on exterior exposure.

Acrylic Paints: Acrylic resins can be divided into: thermoplastic and thermosetting types. Acrylic resins used in paints are mono- or copolymers of acrylic acid or methacrylic acid esters. Some of the common monomers are methyl methacrylate, butyl methacrylate, methyl acrylate, butyl acrylate, ethyl acrylate, and 2-ethylhexyl acrylate. See also **Hydroxyalkyl acrylates and methacrylates.**

Thermoplastic resins become soft when heated and reharden when cooled. Paints made with these resins have excellent exterior durability and excellent gloss and color retention. They are non-yellowing and can have good resistance to stain. Heating after application allows these paints to reflow, producing excellent appearance. Thermoplastic acrylic finishes are primarily used for automotive and product finishes. They have the disadvantages of poor adhesion and high price for both polymer and solvents.

Thermosetting acrylic resins have at least one monomer belonging to the acrylic family which will react with itself or other resins at elevated temperatures to cross-link in order to cure. In addition to the acrylic monomers previously listed, acrylonitrile, acrylamide, styrene, and vinyl toluene are often used in these polymers.

Polymers which react to cross-link primarily because of hydroxyl groups are usually combined with an epoxy resin; those which react mainly with carboxyl groups usually are combined with an amine resin. Thermosetting acrylic paints, which are hard and stain-resistant and have high gloss, are often used for appliance finishes. Tough flexible finishes can be formulated for coil coatings.

Cellulosic Paints: Cellulosic polymers, such as

nitrocellulose, ethyl cellulose, ethyl hydroxyethyl cellulose, cellulose acetate, and hydroxyethyl cellulose, can be dissolved in solvent mixtures which are primarily oxygenated liquids, such as esters, ketones, ethers, and alcohols. Coatings made with these polymers dry only by loss of solvent, which is often referred to as *lacquer drying*. Cellulosic lacquers must often be plasticized with compounds such as dibutyl phthalate or butylbenzyl phthalate. Compatible resins, e.g., ester gum or acrylic or maleic resins, are often used to improve gloss, adhesion, and elongation.

Cellulosic lacquers have very fast dry but low film build. They are used extensively for furniture finishing and automobile refinishing.

Chlorinated-Rubber Paints: Chlorinated rubber is made by chlorinating natural rubber, is soluble in aromatic hydrocarbons, and is used to make paints with excellent chemical resistance, water resistance, and fast-drying qualities. The polymer is rather brittle and must be plasticized with materials such as drying oils and chlorinated biphenyl. The disadvantage of these paints is poor resistance to organic solvents and oils; exterior exposure is fairly good. They are used primarily where good chemical or water resistance is required, as in industrial maintenance.

Styrene-butadiene copolymer resins are very similar to chlorinated rubber in solubility and other physical properties.

Polyester Paints: Polyesters are unsaturated thermosetting polyester resins similar to those used for reinforced plastic. Although alkyds can be considered unsaturated polyesters, this term has been reserved for resins which have unsaturated compounds in the backbone of the polymer. These resins are made by reacting unsaturated dibasic acids, e.g., maleic anhydride, citraconic anhydride, fumaric acid, itaconic acid, phthalic anhydride, and adipic acid, with polyhydric alcohols, e.g., propylene glycol. Styrene or some other aromatic vinyl monomer is added to the polyester resin, which is then solubilized and made into a paint. Inhibitors, e.g., hydroquinone, are added to prevent premature polymerization in the can; organic peroxides or some other catalyst must be added to initiate polymerization of the styrene monomer and the polyester resin for curing, which is often carried out at elevated temperatures.

Polyester finishes are very hard, tough, resistant to solvents, and fairly heat-resistant. They are often used for furniture finishes. Adhesion of these paints is often poor.

Phenolic Paints: Phenolic resins as used in coatings are primarily made from phenol and para-substituted phenols reacted with formaldehyde to form methylol groups on the phenol ring.

Phenol   Formaldehyde   Methylol phenol

Condensation polymers are then produced by reacting these groups with phenol. Phenolic coatings have fast dry, high build, and good resistance to moisture and chemicals. Their poor initial color and tendency to yellow after application limit their use. Phenolic coatings are often used for baked can coatings, and oil-modified phenol-aldehyde finishes are often used for marine finishes and aluminum paints.

Polyurethane Paints: Polyurethanes are based upon reactions of isocyanates, RNCO. Urethane coatings have excellent solvent and chemical resistance, abrasion resistance, hardness, flexibility, gloss, and electrical properties. They are, however, rather expensive, and the aromatic isocyanates yellow after application. See also **Urethanes.**

Polyisocyanates such as toluene diisocyanate react with hydroxylated drying oils to produce resins analogous to alkyds and epoxy esters. Finishes made with these urethane oils air-dry by oxidation of the unsaturated oils.

Polyhydroxy materials can be reacted with isocyanates with an excess of the isocyanate so that the polymer will contain NCO groups, which react with moisture in the air after the coating is applied to cross-link the polymer for curing. Moisture-cured urethanes are difficult to pigment since the pigments must be completely dry and nonalkaline. They are used primarily for clear coatings.

Phenol will react with an isocyanate to block the isocyanate or prevent it from reacting with hydroxylated materials in the system. When this type of coating is heated to 150°C after application, the phenol volatilizes from the film, leaving the isocyanate free to react with the hydroxyl-bearing resin to cure the film.

Two-package systems are also used where a

catalyst, e.g., a tertiary amine, is added to cross-link the polymer.

Coumarone-Indene Paints: Coumarone-indene resins which are derived from coal tar are used widely to make aluminum paints since they aid leafing of the aluminum and minimize gas formation. They have a yellow color, however, and only fair durability except in aluminum paints.

Indene   Coumarone

Bitumen Paints: Bitumen resins are also used for aluminum paints as well as for maintenance and insulating coatings because of their low cost. They are dark in color and have low solvent resistance and only fair durability except with aluminum.

Other Materials for Solvent Paints: Urea-formaldehyde, melamine-formaldehyde, and other triazine-formaldehyde resins are all hard, glossy, colorless, brittle, and chemically resistant thermosetting polymers used to modify basic coating vehicles, such as alkyds, acrylics, and vinyls. Although rosin is sometimes used in making vehicles, such as maleic rosin alkyds, for use in consumer products or industrial finishes because of their moderate cost, it is being replaced gradually by other materials. Silicones sometimes are used to modify alkyds and polyesters for better exterior durability.

Latex Paints. Rutile titanium dioxide is the primary pigment used in latex paints to obtain opacity, or hiding, in white or pastel paints. Semichalking grades are used for interior paints, chalk-resistant grades for exterior paints, and fine-particle-size grades for semigloss paints. Zinc oxide is sometimes used for exterior paints to help prevent mildew, but care must be taken in using this pigment because of its chemical reactivity.

A number of pigments, such as calcined clay and delaminated clay, can be used to advantage in latex paints as titanium dioxide extenders to increase opacity. These pigments have large surface areas due to irregular surfaces or fine particle size, and the latex vehicle will not cover all the pigment surface when the paint film is dry, leaving entrapped air in the film. The interfaces of air with pigment and vehicle increase the light refraction of the film and thus the opacity. Good

hiding can be obtained with these pigments at a low cost, but the paint film often becomes porous and difficult to clean.

Most of the common extenders, e.g., mica, calcium carbonate, clay, talc, silica, and wollastonite, can be used in latex paints. Since these pigments vary in particle size, shape, hardness, color, surface treatment, and water demand, they can affect viscosity, flow, gloss, color, cleanability, scrubbability, enamel holdout, uniformity of appearance, and even opacity to some extent. Extender pigments are selected to obtain the desired properties for each type of paint. Slightly soluble ammonium phosphate compounds are used as the primary pigment in intumescent fire-retardant paints.

Surfactants are used in latex paints to help wet and disperse pigments, emulsify liquids, and function as defoamers. These materials have a balanced polar-nonpolar structure, which in water-base paints is usually referred to as a *hydrophile-lipophile balance.* The chemical composition of surfactants can vary greatly, and they are usually only classified into anionic, cationic, and nonionic types. Anionic surfactants (aryl alkyl sulfonates, sulfosuccinic acid esters, soaps, water-soluble amines, and sulfonated oils) and nonionic surfactants (partial esters of polyhydric alcohols with long-chain carboxylic acids, long-chain alcohols with free hydroxyl groups, and ethers of polyhydric alcohols with long-chain fatty alcohols) are used primarily in latex paints.

The more hydrophilic water-soluble surfactants are used to wet and disperse pigments. Surfactants with a lipophilic chain to dissolve in a polar liquid and a hydrophilic group to dissolve in water are used to emulsify varnishes or oils in latex paints. Surfactants which are not water-soluble are used as defoamers. See also **Surfactants.**

Thickeners or protective colloids are used in latex paints to produce the desired viscosity and help stabilize emulsions and pigment dispersions. Water-soluble protein or casein dispersions and cellulosic polymers (carboxymethyl, hydroxyethyl, and methyl cellulose) are the most commonly used. Soluble polyacrylates, starches, natural gums, and inorganic colloidal materials have also been used.

Protective colloids can affect many properties of a paint, such as washability, brushability, rheological properties, and color acceptance. Since latex paints are susceptible to bacterial attack, they should contain preservatives. Several different types of preservatives can be used: phenolic, mercuric, arsenic, or copper compounds, formaldehyde, and certain quaternary chlorinated compounds. Some of these compounds are chemically active, and some are toxic, facts which must be considered in selecting a preservative.

Polymers Used in Latex Paints: Many of the polymers used in the plastics industry and in solution coatings previously described can also be obtained in latex form. An advantage of using polymers in this form is that high-molecular-weight fully cured polymers can be made to flow well whereas in solution form they would have high solution viscosities.

Latex paints dry fast and are easy to apply; equipment can be cleaned with water; and there is no fire hazard or air pollution during application. A latex paint does not lose its liquid gradually with a gradual increase in viscosity during drying as a solution paint does because of the inversion of the two-phase system when most of the water has evaporated. Sometimes this can produce application or flocculation problems. The tendency of latex polymers to coalesce when frozen and the fact that they seldom are made with more than 55% solids also can be disadvantages.

The main types of latex polymers used in latex paints are styrene-butadiene, vinyl homo- or copolymers, and acrylic polymers or copolymers.

Styrene-Butadiene Latex: This low-cost material can produce paints with good color acceptance, low temperature coalescence, toughness, adhesion, cleanability, and good sealing properties. The styrene-butadiene copolymer remains a little soft after the water has evaporated from the film, but the polymer contains some unsaturated groups which air-oxidize to cross-link the polymer and harden the film. This oxidation will cause the polymer to yellow slightly. Styrene-butadiene paints are used primarily for interior wall applications.

Vinyl Latexes: These materials are usually copolymers of a vinyl monomer, e.g., vinyl acetate or vinyl chloride, and some other monomer, e.g., ethylene or vinylidene chloride, acrylonitrile, methyl acrylate, methyl methacrylate, fumaric acid, maleic anhydride, or styrene.

The various vinyl latexes have different properties depending upon the monomers used, the particle size of the latex, and the surfactant system. There are, however, a few general properties of vinyl latexes. They dry fast and are fully cured when the water evaporates from the paint; are

nonyellowing; have good adhesion and good scrubbability; and may have good flow and leveling properties. Some vinyl latexes do not have very good freeze-thaw resistance, low-temperature coalescence, or color acceptance. Vinyl copolymers are usually plasticized with liquids such as dibutyl phthalate, butyl cellosolve acetate or tributyl phosphate. Vinyl latex paints are used primarily for interior and exterior architectural surfaces.

Acrylic Latexes: These materials are polymers or copolymers which are composed primarily of monomers of the acrylic family, e.g., methyl methacrylate, butyl methacrylate, methyl acrylate, and 2-ethyl hexylacrylate. Other monomers, e.g., styrene or acrylonitrile, can be polymerized with acrylic monomers.

Acrylic latexes vary greatly in their physical properties, according to the monomers used, the particle size, and the surfactant system of the latex. In general, acrylic latexes are cured by loss of water only, do not yellow, have good exterior durability, are tough, and may have good abrasion resistance.

Acrylic polymers are fairly expensive, and some latexes do not have very good color compatibility. Acrylic latex paints can be used for concrete floors, interior flat and semigloss finishes, and exterior surfaces.

Other Latex Materials: Hydrocarbon resins or drying oils are also available in emulsion or latex form, but these materials are used primarily to modify the properties of the three major types of latex just described. Thermosetting or catalyzed polymers can be made in latex form for industrial paints. To date, no large application has been developed for them. Alkyd varnishes or drying oils sometimes are emulsified in water-base paints to modify the properties of the latex, e.g., plasticizing it, or to add desired properties to the paint, e.g., adhesion to exterior chalky surfaces.

Water-soluble Paints. Water-soluble paint binders, such as egg albumin, gum arabic, and casein, have been used for many years. Most of these materials have serious limitations, including water sensitivity and poor durability. Many of the synthetic polymers now used in solvent or latex paints can be solubilized in water. Carboxylic, hydroxyl, epoxy, or amine groups on a polymer in conjunction with coupling solvents, such as alcohols, alcohol ethers, or glycol ethers, are the primary mechanisms by which resins are solubilized.

Maleic or fumaric acids can be reacted with drying oils to produce resins with some carboxy groups which can be solubilized with ammonia or amines. Alkyds can be solubilized by leaving a reactive carboxylic group on the resin instead of terminating the reaction with a monobasic acid or drying-oil acid. See also **Fumaric acid;** and **Maleic acid.**

$$CH_3\overset{\overset{\displaystyle CH_2OH}{|}}{\underset{\underset{\displaystyle CH_2OH}{|}}{C}}COOH$$

Dimethylol propionic acid, $CH_3C(CH_2OH)_2 \cdot COOH$, and trimellitic anhydride have been used for this reaction. A styrene–alkyl alcohol copolymer esterified by a fatty acid and reacted with maleic anhydride to provide solubility has been used as a water-soluble paint vehicle. Amine-solubilized polyesters, acrylics, epoxy esters, and phenolics can also be prepared. Water-soluble resins are usually solubilized by volatile alkalis, such as amines, to prevent the paint resin from remaining water-soluble after application.

Water-soluble resins have been used for air-dry and low-bake industrial primers and finishes, coatings applied by electrodeposition, and architectural semigloss paints. Contamination and mechanical and chemical stability can be a problem with these resins. See also **Silicates (soluble).**

—Ralph S. Armstrong, *The Sherwin-Williams Co., Cleveland, Ohio*

Palladium (Asteroid Pallas.) **Pd** = 106.4 (at. wt.); 46 (at. no.). Palladium is a member of the family of platinum metals, group VIII of the periodic table. The melting point of Pd = 1552°C; boiling point = 3980°C; density at 20°C = 12.02 g/cm³. There are six stable isotopes of Pd, 102, 104–106, 108, and 110, and seven unstable isotopes, 100, 101, 103, 107, 109, 111, and 112. Pd was discovered by Wollaston (England) in 1804.

Pd has some similarities with both Ni and Ag and many with Pt. Pd dissolves more readily in acids than any other member of the platinum group of metals. In aqua regia, the metal dissolves quickly. Even the compact metal dissolves slowly in HCl. In finely divided form, it is quite soluble in all acids. When heated in air at red heat, the monoxide, PdO, is formed. Pd is similarly converted to the dihalides under the same conditions when it is exposed to F_2 or Cl_2. The metal is not affected by H_2S.

The black compound, palladium(II) oxide is formed by fusing palladium(II) chloride with NaNO₃ at 600°C and then leaching out the salts with water. This strong oxidizing agent is easily reduced to the metal by H_2. The compound is insoluble in water and acids, including aqua regia. The hydroxide, $Pd(OH)_2$, is made by the hydrolysis of palladium(II) nitrate. The compound is soluble in acids, and water is evolved on heating, but even at 500–600°C some water still remains. At this temperature, the compound starts to lose O_2.

Palladium(III) oxide, P_2O_3, is made as a hydrate by careful oxidation of a solution of palladium(II) nitrate either by anodic oxidation or ozone treatment at −8°C. This unstable brown powder reverts to the monoxide in about 4 days. When heated, the compound loses water and may explode as it changes to the monoxide.

Palladium(II) chloride is formed by direct combination of the elements at 500°C. It is the only stable solid chloride over 500–1500°C. The red crystals are partly soluble in water and completely soluble in HCl. The fraction insoluble in water is probably a polymer. Palladium(II) chloride also is the product obtained by evaporation of a solution of Pd in HCl. Palladium(II) bromide can also be made from the elements.

When KI is added to a solution of palladium(II) chloride, an insoluble diiodide is precipitated. The dark red-black crystals are soluble in excess iodide with formation of the tetraiodide complex ion. Palladium(II) iodide evolves iodine at 100°C, the decomposition to the elements being complete at 330–360°C. The black compound, palladium(III) fluoride, is made by direct combination of the elements. On reduction, the brown difluoride is formed.

Divalent Pd forms many planar complexes with a coordination number of 4. The tetrachlorides are quite soluble. When a solution of palladium(II) chloride is oxidized with chlorite or chlorate ion, Pd(IV) is formed, which has a coordination number of 8. The addition of NH_4Cl to such a solution precipitates ammonia hexachloropalladate(IV) as a red compound. It is somewhat less stable than the platinum analog.

The soluble yellow-brown palladium(II) nitrate is formed by dissolving finely divided Pd in warm HNO_3 and then crystallizing the compound from this solution. The analogous sulfate is similarly formed from H_2SO_4. It crystallizes as a red-brown dihydrate. Both these compounds easily hydrolyze.

Palladium(II) sulfide is precipitated as a brown powder by adding H_2S to a solution of palladium(II) ion. When this sulfide is heated with sulfur at 400°C, the insoluble disulfide is formed. The excess sulfur can be extracted with CS_2 to yield the gray-black crystalline palladium(IV) sulfide. This compound is not soluble in single acids but is soluble in aqua regia.

Some Pd complexes are important analytically or in the refining of Pd. The yellow dimethylglyoxime compound is quantitatively precipitated from a HCl solution of palladium(II) chloride by the addition of an alcoholic solution of dimethylglyoxime. Palladium(II) has a great affinity for nitrogen-containing ligands. The di- and tetramine find use in refining.

Further information on the sources, refining, and uses of Pd will be found under **Platinum and Platinum metals.** Pd also is mentioned under **Aldehydes; and Ketones.**

—Lionel E. Simmons, *Simmons Refining Company, Chicago, Ill.*

Palmitic Acid (See **Vegetable oils;** and **Waxes.**)

Pan Filter (See **Filtration.**)

Paper Chromatography (See **Chromatography.**)

Papermaking and Finishing The basic principle of papermaking has remained virtually unchanged since paper was invented in China some 2,000 years ago. A thin sheet or mat is formed by draining water from a suspension of vegetable (cellulosic) fibers through a fine screen, subsequently submitting it to pressure, and finally drying by heat. Unlike other felted mats, in which the fibers are held together simply by friction or by adhesives, paper possesses natural bonding, primarily through hydrogen bonds, in the contact area between fibers. The extent of this bonding has a marked effect upon the physical properties of the paper.

In the beginning of the nineteenth century, the advent of machinery capable of making paper in a continuous web, instead of individual sheets, marked the transition of papermaking from a hand craft to a mass-production process.

Wet End: A modern paper machine begins with a flow spreader, or distributor, conveying a dilute fiber suspension (0.1–1% fibers) to a headbox which delivers a jet of the suspension or slurry through a slice (sluice) across the full width of the machine, almost 400 in. in some large machines. In the headbox, the fibers are dispersed, and the flow is rectified as well as possible so that the jet is delivered onto a moving endless fine-mesh wire screen with uniform composition, flow rate, and velocity. The pressure in the headbox and its slice opening are adjusted so that the jet velocity matches the speed of the wire screen (up to 3,000 ft/min for newsprint); the proper stock flow per unit width corresponds to the desired basis weight* of the paper.

The dispersion of fibers in the headbox is brought about by subjecting the slurry or suspension to shear stresses, usually with turbulence. In the rectifier-roll type of headbox, rotating perforated rolls extending across the width are used. Turbulent or orderly shear or both are also induced by causing the slurry to flow through appropriately shaped passages.

In a more recent development,† the suspension flows through several converging channels formed by thin plastic sheets. These channels terminate in a thickness of 0.2 in. or less close to the point where the jet emerges from the slice. The shear stresses associated with a fine-scale turbulence imposed by these relatively long, thin multiple channels lead to very effective fiber dispersion and a relatively stable jet (free from large-scale disturbances). The headbox is a much more compact structure than a conventional headbox with rectifier rolls.

As shown in Fig. P-1a, the most common type of paper machine is the Fourdrinier, in which the moving wire screen is in the form of an endless conveyor belt stretched between two large rolls. The roll situated under the headbox slice is called the *breast roll*. The roll located generally at the end of the straight wire run is the *couch roll*. Drainage of the slurry through the wire screen is induced by several types of driving forces. In the early, slow-speed machines, the principal force was gravity. Later, the hydrodynamic action of table rolls, which support the wire and rotate with it, began to play an important part in drainage as speed increased. More recently, foils came into use, i.e., rigid, stationary, hydrodynamically shaped elements which support the wire and exert a pumping action through the wire screen. Other means are perforated or slotted boxes with vacuum over which the wire runs. When only water

---

*Basis weight is weight per unit area and varies with grades and sizes of papers. For example, newsprint has a basis weight of 32 lb, which means that the weight per unit area is 32 divided by 3,000 ft^2 (the reference area), or slightly over 0.01 lb/ft^2.

†Beloit Converflo headbox.

is drained, they are called *wet boxes*. When applied toward the dry end of the wire screen, they also suck air through the wet paper mat and are called *suction boxes*. Vacuum boxes may also be fitted inside a perforated couch roll and on special machines inside the breast roll. On all modern Fourdriniers a forming board located close to the breast roll is used to scrape off the water, drained initially by gravity, from the bottom of the wire.

Most of these drainage elements, especially the table rolls, exert some positive pressure pulses near their contact with the wire before drainage occurs by virtue of their suction power. These upward pulses, acting in combination with turbulent forces in the suspension flow, disturb the suspended fibers and the paper mat being formed. The disturbances are beneficial in some instances because they break up fiber flocs, or clumps, which have not been completely dispersed in the headbox or have subsequently reformed on the wire. On the other hand, if these disturbances lead to larger-scale redistribution of fibers on the wire, they are detrimental to the uniformity of the final paper. The more perfect the dispersion of fibers and the more quiescent the flow in the jet issuing from the headbox, the less desirable these disturbances become. The pressure-suction sequence taking place at the drainage elements contributes in an important way to the distribution of the fine fiber fragments (fines) and nonfibrous additives (such as pigments) through the thickness of the paper.

A relatively recent and important development in paper forming is the twin-wire former* (Fig. P-1*b*). In this type of machine, the fiber suspension is confined between two wire screens, and water is removed through both wires either simultaneously or alternately. This two-sided drainage leads to greater symmetry of distribution of fines and other nonfibrous particles through the thickness of the sheet. Besides the other drainage-inducing devices which may also be used with twin-wire formers, there are two added drainage forces, centrifugal force and the hydraulic pressure between the two converging wires under tension while traveling around rolls, curved surfaces, and other drainage devices. A significant feature of twin-wire forming is the elimination of the free surface of the fiber-water suspension while the sheet is being formed. This greatly reduces the larger-scale disturbances (waves, streaks, and jumps) which occur at higher speeds on Four-

drinier wires. Similar principles are used in some tissue-forming devices and multi-ply paperboard machines to be discussed later. It appears that twin-wire forming, in general, and particularly in combination with the headbox† previously described has attractive potential for producing more uniform paper and board for almost all grades with high-speed operations in more compact systems than Fourdriniers.

Not all the fiber and other solid materials are retained by the forming wire. For this reason and because so much water is used in the papermaking process the *white water* removed in the sheet-forming process is recirculated in the overall system. A large part of it is added directly to the high-consistency stock and fed back to the headbox, while a small portion goes into a *save-all* device, which recovers much of the solids from the white water. These extracted fibers and other solids are returned and added to the suspension. The clarified water is used in showers for cleaning wires and felts and other purposes so that only a small amount of the reused water eventually is discharged.

Press Section: At the end of the forming system, the paper web is transferred from the wire to a press *felt*, a fine-textured woolen or synthetic fabric. At this point, the web contains about 4 or 5 parts water to 1 part solids. The wet paper web and one or more press felts pass through two or more press-roll nips, where water is squeezed out. Pressing also compacts the paper mat. This increases the potential interfiber contact areas where bonds will be formed.

The early *plain press* used a pair of metal and rubber-covered solid rolls. The expressed water had to flow out of the nip in the upstream direction, parallel to the paper web, as in the old washing-machine wringer. Nip pressures were then limited by the damage to the wet web (*crushing*) caused by this lateral flow. Providing closely spaced small holes in the surface of the softer roll next to the felt alleviated this situation. In the suction press, the drilled roll became a perforated shell with an internal vacuum box, allowing greater control over the temporary retention and subsequent removal of water from the holes and in some cases also aiding the dewatering of the felt and the sheet by sucking air through them just before and/or after the nip.

Later development work led to the *fabric press*, in which the felt contacting rolls are wrapped with

---

*Beloit Bel Baie Former.

†Beloit Converflo headbox.

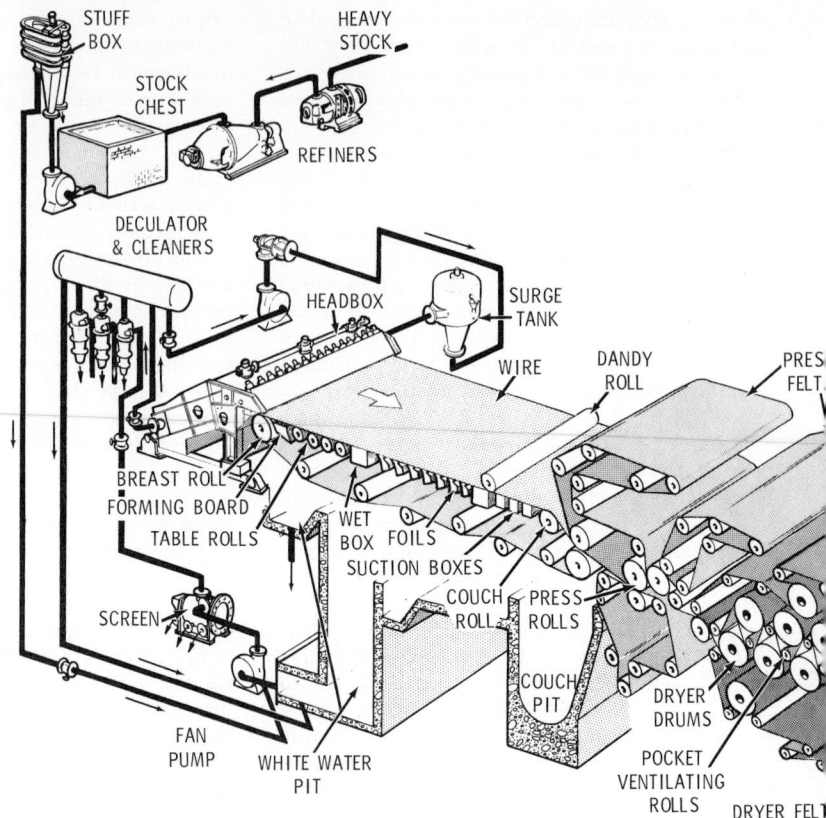

Fig. P-1(a). Fourdrinier machine for producing printing-grade paper. (SOURCE: *Beloit Corporation.*)

a relatively coarse mesh and incompressible fabric. In another development,* the felt contacting rolls have narrow, closely spaced circumferential grooves. In both types the lateral flow of the expressed water is virtually eliminated. With these later types of presses, called *transversal-flow presses,* nip pressures up to 800 pli (pounds per

*Beloit Ventanip press.

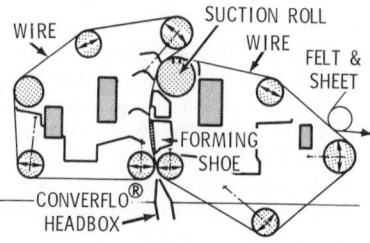

Fig. P-1(b). Twin-wire former. (*Bel Baie Former, Beloit.*)

lineal inch) for paper grades and up to 1,200 pli for heavier board grades have been used at high speeds without crushing. The press rolls are not perforated, and no vacuum is required.

While the development of the modern presses has achieved high performance with simple constructions, the remaining problems of flow resistance and web rewetting leave room for improvement. It is generally recognized that mechanical removal of water is much cheaper than drying.

Driers: After water removal by pressing has been done to the extent which is practical with present technology, the paper web leaves the press section with $1\frac{1}{2}$–2 parts of water to 1 part fibers. Most of this remaining water, down to 5–10%, must be removed by evaporative drying. In the most common method of paper drying, the paper web is passed over a series of staggered cast-iron drums (Fig. P-1*a*) internally heated by condensing steam at pressures ranging up to approximately

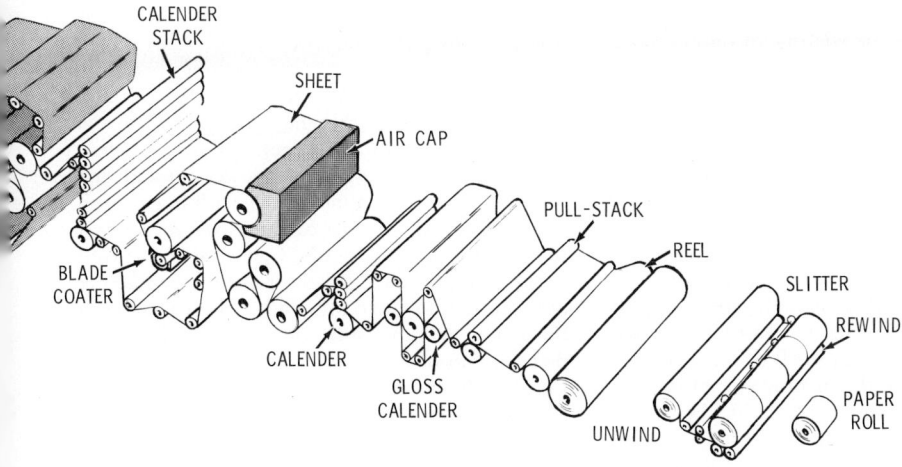

CALENDER STACK

SHEET

AIR CAP

PULL-STACK

REEL

SLITTER

REWIND

BLADE COATER

CALENDER

GLOSS CALENDER

UNWIND

PAPER ROLL

150 psig. The paper web is held in contact with the rotating drums by means of drier felts under tension. The diameter of the drier drums is typically 5 or 6 ft. There may be as many as 100 of them in heavyweight paperboard machines.

An important recent development is the introduction of ventilating devices which blow air of controlled temperature and humidity through the drier felts into the spaces between adjacent driers, where the air is confined by the sheet and felt runs. These pocket ventilating systems, together with greater control of the flow patterns within the drier hood which usually encloses the entire drier section, have led to significant improvements in cross-machine uniformity of paper drying. This results in paper and board of improved suitability for modern high-speed converting and printing operations.

Relatively recent developments in paper drying include high-velocity hot-air impingement; radiant heating, usually from natural-gas-burning devices; and the use of electric energy for *dielectric drying* in the lower-megacycle-frequency ranges and for microwave drying in the higher-frequency ranges. These developments were particularly motivated for drying coated paper. However, air impingement is also widely used for drying tissue, as an adjunct to the Yankee drier, discussed later. The gas-fired radiant driers have found considerable use in drying coating, but their wider application has been inhibited by the low thermal efficiency associated with present inability to use the residual heat in the very hot exhaust gases. Radio-frequency drying, while having the unique advantage of being able to generate heat within the paper web or coating layers, has not attained wide use because of the high cost of the equipment and of the converted electric energy.

Size Press and Coaters: Many printing grades of paper and paperboard are coated with an aqueous suspension of pigments (such as clay) in adhesives (such as starch) to provide a smoother surface, control the penetration of inks, and improve the pick resistance, appearance, brightness, and opacity. These and other materials are also applied, such as *functional coatings,* to provide such features as water resistance, pressure sensitivity for carbonless copying, and a wide variety of other properties. The appropriate materials may be added to the papermaking furnish during some stage of stock preparation (called *internal sizing*). Application of sizing or coating to one or both surfaces of the formed and dried sheet, rather than as internal sizing, simplifies the sheet-forming process and provides better control of surface properties.

The principal methods of surface coating may be classified as roll, blade, and air-knife coating, according to the method used to apply and control the final coating-layer thickness and smoothness. The size press may be considered as the simplest type of roll coater. In this method, generally used for application of low-viscosity solutions or suspensions, the paper web is passed through the nip of a pair of rolls while the ingoing side of the nip is kept flooded with the sizing fluid. For more viscous compounds, the roll coater may involve several rolls for each side of the sheet, film splitting and leveling out of layer thickness accompanying the successive transfer of coating material from one roll to another. The coating is applied to the sheet by the last roll, similar to the roller application of paint.

With the blade and air-knife coaters, excess material is first applied to the web by a roll which dips into a pan kept filled with the coating mixture and a variety of other means. The blade coater removes the excess by troweling the surface of the sheet with a flexible blade as the web travels through the nip between the blade and a roll. The air knife, usually used with less-viscous compounds having less than 50% solids, removes the excess by blowing at the wet coated surface with a high-velocity jet of air from a thin slot. Blade

coating fills in the valleys of the paper surface to achieve a level, smooth surface, while the air knife leaves a coating layer of uniform thickness.

A recently developed coater* simultaneously coats and smooths both surfaces of the paper web by running it down through the nip between a blade and a roll while maintaining two puddles, one between the web and the roll and the other between the flexible blade and the web, thus eliminating the necessity for two coating stations.

After sizing or coating, the solvent, usually water, must be removed from the coating by evaporative drying. With some coating formulations and paper grades, drying can be done on ordinary steam-heated drums without damage to the coated surface, particularly if the surface of the first drum is smooth (sometimes chromeplated). However, it is often desirable to do the initial drying with air impingement or radiant heating.

Surface coating can be done on the machine as a step in the paper-machine operation, as shown in Fig. P-1*a* or as a separate off-machine operation.

Calenders and Winders: Nearly all paper grades are calendered after they have been dried to the desired final moisture content. Ordinary calendering involves passing the paper web through one or more nips between metal rolls with linear pressures up to 500 pli. The calendering process flattens out the paper structure by virtue of the high pressure and "irons" the sheet with a burnishing action due to a small amount of sliding between the paper and the rolls, which results from the deformation of the sheet as it passes through the nip. Calendering causes bulk reduction, which often is not desired, and surface smoothing, which is. The results strongly depend upon moisture content, calender-roll temperature, roll pressure, and speed.

In supercalendering, an off-machine operation, the calender rolls consist of alternating chilledsteel and paper-filled rolls, i.e., paper disks clamped on a steel shaft. These roll fillers have to be replaced periodically. Very large loads are used, up to 2,500 pli. The increased burnishing action associated with deformation of the relatively soft paper roll and the very high roll pressures impart a smoother, glossier surface to the web than ordinary calendering with all-metal rolls. This type of calendering is frequently used on coated sheets to provide a glossy coated surface.

In gloss calendering, an alternative to super-

*Billblade.

calendering, the paper web is passed over a steam-heated cast-iron drum, on which it is pressed and burnished by (usually two) rubber- or plastic-covered rolls. Since the pliable materials used for covering the pressure rolls are much more durable than the supercalender-roll fillers, the gloss calender is suitable for on-machine operation. Gloss calenders were introduced for use on board grades, for which nip pressures of about 500 pli are used. They are also used on paper grades with nip pressures up to 1,000 pli. The hot surface of the drum makes it possible to achieve the smoothing and glossing action with less pressure and less reduction in bulk, although not quite as much gloss is achieved as with a supercalender.

One of the basic problems associated with calender and other loaded rolls on wide paper machines is their deflection. Varying the roll diameter across the width, i.e., crowning the roll, provides only a predetermined correction which is not adjustable with changing loads or for other reasons. Recently roll constructions have been developed in which a stationary center shaft supports the rotating shell with a force distribution which is uniform across the width. In one such design, one-half of the annular space between the rotating shell and center shaft is filled with oil under pressure with appropriate seals. In one system,* this force transmission is accomplished by a hydrodynamic bearing that pushes against the inside of the rotating shell.

Since most modern papermaking and converting operations are high-speed, continuous processes, the paper web is handled in rolls. At the reel, the web is wound up into a full machine-width roll. When it reaches a certain diameter, the roll is removed from the paper machine, unwound, slit, and rewound on a rewinder into rolls of the proper width and diameter for shipment.

Other Types of Machines: Fourdrinier machines are used for making almost all grades of paper and board. To achieve the various basis weights and other required properties and to accommodate the various types of pulp furnish which vary considerably in ease of drainage, certain adjustments of configuration are needed, particularly in the length of the Fourdrinier wire, the number and types of drainage elements, and the number of drums in the drier section.

The cylinder machine, invented at about the same time as the Fourdrinier, consists of a rotating cylindrical mold covered with a wire screen and

*Beloit Controlled Crown Roll.

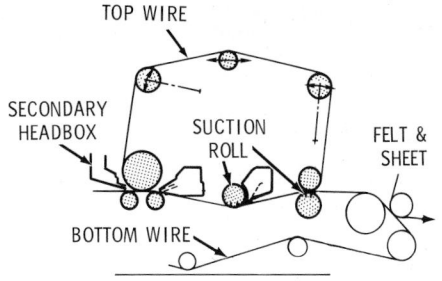

Fig. P-1(c). **Two-wire forming device used for manufacture of multi-ply board at high speeds.** (*Inverform, Beloit.*)

partially submerged in a vat. The stock flows into the vat, and a mat is formed on the cylinder under a hydraulic head difference between the stock level in the vat and the white-water level inside the cylinder. The wet mat is picked up by a felt running through the nip between a couch roll and the cylinder. The cylinder machine is used for making multi-ply board, employing several vats in series. Because of speed and other limitations, the cylinder machine is becoming obsolete.

In recent years, new types of formers have emerged, of which four examples are illustrated. One system† has already been described.

The Inverform‡ (Fig. P-1c) is the first two-wire forming device used for the manufacture of multi-ply board at much higher speeds than the old cylinder machine. With this device, after a base sheet is laid down on the bottom wire, as in the Fourdrinier machine, secondary headboxes and *top-wire* dewatering units permit the successive delivery and draining of additional layers. Natural bonding holds the separate layers together, resulting in a single, thick sheet of paperboard.

The Hydraulic Former‡ (Fig. P-1d) illustrates an advanced example of a cylinder machine. Stock is delivered to a roof-confined forming zone over a wire-wrapped open cylinder. This provides large and easily adjusted driving forces for drainage and makes the machine versatile for a wide range of heavy board grades and weights. Several units generally are used in series to form a multi-ply board. The forming cylinders may be fitted with internal suction boxes for added capacity.

The Crescent Former§ (Fig. P-1e) illustrates

† Beloit Bel Baie Former.
‡ Beloit.
§ Kimberly-Clark.

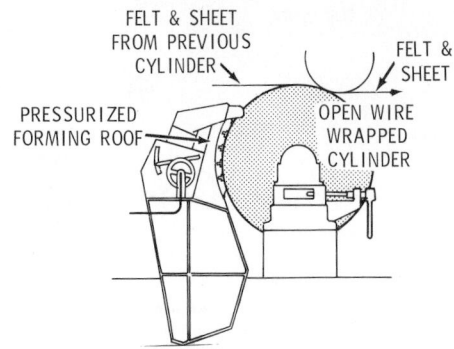

Fig. P-1(d). Hydraulic former, an advanced example of a cylinder machine. (*Hydraulic Former, Beloit.*)

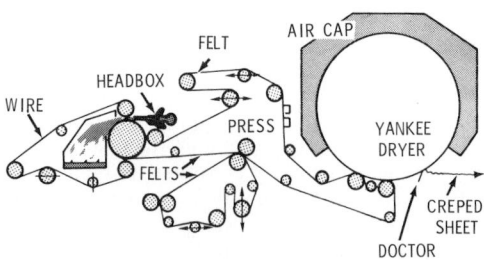

Fig. P-1(e). Two-wire forming where one "wire" is a felt. (*Crescent Former, Kimberly-Clark.*)

another variation of two-wire forming, where one "wire" is a felt. This device is used primarily for tissue grades at very high speeds (5,000 ft/min or more). The illustration shows the unique drying system used for sanitary grades, such as tissue and towelling. The thin web carried by the felt is pressed by a pressure roll onto a large (12–18 ft diam) steam-heated cast-iron cylinder called a *Yankee drier,* which often is equipped with a high-velocity air-impingement hood (air cap). Natural and added adhesives cause the web to stick to the Yankee drier. At about 5% moisture, the sheet is scraped off the Yankee by a doctor blade. This doctoring, called *creping,* sharply bends and compresses the sheet in its own plane, giving it a soft, bulky texture.

Some manufacturers now are using through-air drying for tissue, i.e., sucking or blowing hot air through the paper web. Through-air drying replaces some or all of the water removal normally accomplished by pressing and by the Yankee drier, leading to an improvement in the bulkiness of

tissue. The introduction of through-air drying, in combination with high-velocity hot-air impingement, also has been used on an experimental basis for newsprint.

Paper-Machine Drives and Control Systems: The basic paper-machine speed must be controlled with reasonable accuracy ($\pm0.5\%$). Of more importance is the control of the differential speed between sections in order to accommodate shrinkage of the paper and to maintain proper tensions in the tender wet web of paper as it is carried through the machine. The differential speed between sections, i.e., the wire, each press, and three or four drier sections, is usually controlled within $\pm0.1\%$. This is accomplished with mechanical drives using differential gearing or tapered pulley-and-belt systems or with electronic systems, e.g., dc motors with feedback control of armature voltage or, more recently, solid-state controlled-frequency synchronous-motor systems.

In addition to the drive speeds, many other variables and conditions must be controlled or taken into account in the papermaking process. These include flow rates, pressures, and temperatures, which, in principle, can be controlled by conventional, autonomous analog controllers or simply by the manual setting of valves. But there are also unavoidable variations in the inputs to the process, the environment which affects it, the condition of the process equipment, and sheet breaks. Furthermore, limitations in the accuracy of available sensor devices lead to deviations from control set points which are detected not directly but by their effect on other measured variables.

For these reasons control of the papermaking process is not simply a matter of independent control of individual variables at set points corresponding to some unique set of optimum values but requires compensatory interaction of variables in order to achieve acceptably constant properties of the paper produced. The control relationships involve a wide range of response times, transport lags, and dead times; long-term trends as well as rapid variations; and, in some parts of the process, nonlinear relationships.

Under these conditions, stable control with reasonable tolerance of variations is impossible with analog systems without considerable intervention of human operators and is difficult even then. Therefore, the much greater capabilities of direct digital-computer control systems to automate the control of papermaking are being investigated with considerable interest.

Data-logging and alarming computer systems

have been installed in several mills, being in some, the first step toward a supervisory computer control system in which the computer drives the set points of analog controllers. Direct digital-computer control has been applied to a few paper machines.

See also **Pulp (wood) production and processing.**

—Robert A. Daane, *Paper Machinery Division, Beloit Corporation, Beloit, Wis.*

Paper-Pulp Processing [See **Pulp (wood) production and processing.**]

Para Compounds (See **Carbon compounds.**)

Paraffins (See **Alkylation; Cracking; Fluid catalytic cracking; Hydrocarbons; Hydrocracking; Isomerization;** and **Reforming.**)

Paraffin Waxes (See **Lubricating oils;** and **Waxes.**)

Paraformaldehyde (See **Formaldehyde.**)

Parkes Process (See **Bismuth.**)

Particulate-Matter Removal (See **Air pollution; Filtration;** and **Waste treatment.**)

Parylene Polymers Parylene is the generic name for members of a polymer series* produced by vapor-phase deposition. The basic member of the series, parylene N, is poly(*p*-xylylene) and is used where its primary dielectric property is required. For its toughness and barrier properties, parylene C, poly(chloro-*p*-xylylene) is the most extensively used member.

Formation: Parylene can be formed as a continuous, adherent film as thin as 1000 Å (about 0.004 mil). The process is not line of sight as is a vacuum metalization, but coats all surfaces of the substrate evenly. The coating deposits uniformly on sharp edges, complex shapes, and in deep narrow recesses. Below 300°F (149°C), no solvent for any commercial member of the parylene series of polymers is known.

Under normal conditions, adsorption of the monomeric xylylene on a substrate leads to immediate polymerization. Deposition rates and efficiencies are improved at lower substrate temperatures. Even without external cooling devices, the

*Introduced by Union Carbide Corporation in 1965.

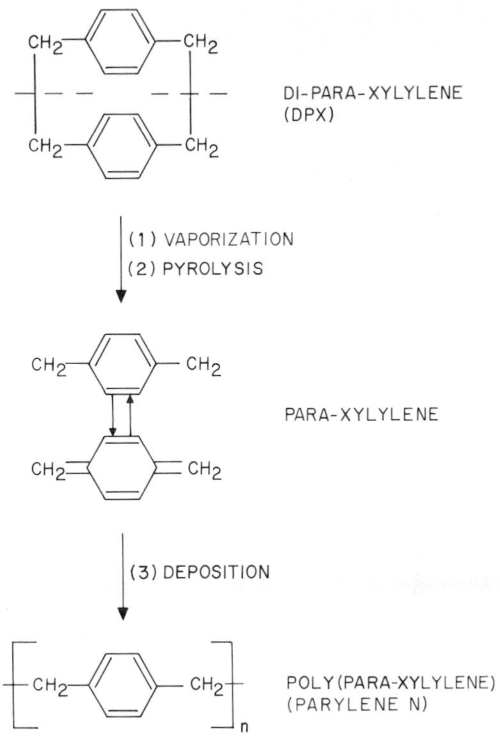

Fig. P-2. **Major steps in production of parylene N.** Vaporization is carried out at about 150°C (302°F) and 1 mm Hg pressure. The pyrolyzer or cleavage zone occurs at approximately 650°C (1202°F) with process under pressure of about 0.5 mm Hg. The deposition zone occurs at about 25°C (77°F) and pressure of 0.1 mm Hg.

substrate temperature is never increased by more than a few degrees during the deposition process.

Unlike most plastics, parylene is not produced and marketed as a polymer. Although it is a linear polymer and, therefore, categorized as a thermoplastic, it is not practical to melt, extrude, mold, or calender parylene, and it cannot be applied from solvent systems.

The parylenes are generated by the consumer who is supplied the appropriate dimer. He then forms parylene coatings on his substrates in specially designed, but not particularly complex, vacuum equipment. This equipment generates monomeric xylylene from the dimer.

A schematic representation of production of parylene N is shown in Fig. P-2.

Properties: The key properties of parylene N and parylene C are summarized in Table P-1. Short-term (1,000-hr) service *temperatures* in air for paryl-

TABLE P-1. *Properties of Parylenes**

Physical and mechanical:	Parylene N	Parylene C
Tensile strength, psi	6,500	10,000
Yield strength, psi	6,100	8,000
Elongation to break, %	30	200
Yield elongation, %	2.5	2.9
Density, g/cm³	1.11	1.289
Coefficient of friction, static and dynamic	0.29	0.25
Water absorption, 24 hr, %	0.06	0.01
Index of refraction, n_D^{23}	1.661	1.639
Thermal:		
Melting point, °C	405	280
Linear coefficient of expansion, 10^{-4} (cal)(cm)/(s)(cm²)(°C)	~3	
Electrical:		
Dielectric strength, V/mil at 1 mil (short-time)	7,000	5,600
Corrected to ⅛ in.	700	590
Volume resistivity at 23°C and 50% RH, Ω-cm	1×10^{17}	6×10^{16}
Surface resistivity at 23°C and 50% RH, Ω-cm	10^{13}	10^{14}
Dielectric constant: 60 Hz	2.65	3.15
1 kHz	2.65	3.10
1 MHz	2.65	2.95
Dissipation factor:		
60 Hz	0.0002	0.020
1 kHz	0.0002	0.019
1 MHz	0.0006	0.013
Typical barrier properties:		
Gas permeability (24 hr at 23°C and 1 atm), cm³-mil/100 in.²:		
Nitrogen	7.7	1.0
Oxygen	39.2	7.2
Carbon dioxide	214	7.7
Hydrogen sulfide	795	13
Sulfur dioxide	1,890	11
Chlorine	74	0.35
Moisture-vapor transmission (24 hr at 37°C and 90% RH), g-mil/100 in.²	1.6	0.5

*Physical, mechanical, thermal, and electrical properties according to appropriate ASTM method.

enes N and C are about 200 and 240°F (93 and 116°C) respectively. These materials perform better in inert atmospheres. After 1,000 hr at 410°F (210°C), elongation of parylene N decreases to about 5%, density increases to about 1.13, and tensile strength increases to about 11,000 psi. Modulus of elasticity remains unchanged at 350,000 psi; dielectric strength remains at about 700 V/mil, dissipation factor at 0.0002, and dielectric constant at 2.65. Parylene is *transparent.* Visible light transmission is limited only by fresnel interfacial reflection effects. Parylene N can be considered a primary *dielectric* because of its low dissipation factor, which varies only slightly with frequency. As regards *chemical resistance,* parylene resists attack and is insoluble in all organic solvents up to 300°F (149°C). Parylene C can be

dissolved in chloronaphthalene at 345°F (174°C), and parylene N is slightly soluble at the solvent boiling point of 505°F (263°C). Both polymers resist permeation by all solvents except aromatic hydrocarbons. They also are unaffected by stress-cracking agents, such as Hostepal, Igepal, and lemon oil.

Parylene is an excellent barrier to gas and water vapor. Parylene C is approximately equal to polyethylene terephthalate in gas permeability, and its water-vapor transmission rate is extremely low. Typical barrier properties are given in Table P-1.

Applications: Parylenes are used widely as conformal coatings for electronic components and circuit boards. One coat of parylene 0.6 mil thick will cover the most complex circuit board free of pinholes, providing moisture protection with virtually no change in electrical properties.

Pellicles or membranes of parylene as thin as 250 Å, mounted on hoops or frames, can be vacuum-metalized and are used in optics and electronics as beam splitters, x-ray windows, nuclear-radiation measuring devices, micrometeoroid detectors, and low-mass fast-response thermocouples.

Surface replicas can be formed by depositing a parylene film on a model; the duplicated surface on the parylene film can then be studied by electron microscopy.

Since parylene is not affected by biological fluids, devices coated with the material can safely be placed in the biological environment.

Chambers for coating objects with parylene can be constructed of almost any size.

—W. F. Beach, *Union Carbide Corporation, Bound Brook, N.J.*

Passivation   (See **Fluorine; Nitric acid;** and **Titanium.**)

Pastes   (See **Adhesives.**)

Paste Mixers   (See **Mixing, pastes.**)

Pasteurization   Louis Pasteur introduced pasteurization in 1864 to prevent acid, bitter, and ropy spoilage in wines and later recommended the process to prevent spoilage of beer. Pasteurization, which usually consists of heating a product at a temperature below the boiling point of water, destroys vegetative cells of many forms of bacteria that cause spoilage but usually not bacterial spores.[3] Pasteurization is used with many food products, particularly those with a pH below 4.5, to retard bacterial spoilage.

The legal definition (United States) of pasteurization defines the process as a time-temperature relationship, with a considerable margin of safety, to destroy *Mycobacterium tuberculosis.* Since about 1950, the use of ultra-high-temperature pasteurization of dairy products has been widely practiced, at temperatures above those legally required, with the objective of improving the keeping quality of milk by killing many of the spoilage organisms present as well as the pathogenic types.

Pasteurization also finds wide application in the canning industry and in connection with fermentation products, e.g., beer, wines, and various industrial products. Nonheat methods of "pasteurization" have been attempted to eliminate or reduce spoilage, e.g., irradiation (including ultra-violet), high pressure, fluctuating pressure, centrifugal force, addition of chemicals, and ultrasonic vibration, but none of these methods are used commercially to an extent comparable with pasteurization.

Technical Basis: The thermal death of bacteria can be assumed to be a first-order exponential reaction mechanism, particularly below 250°F (121°C). Once a certain minimum killing temperature is obtained, the time required at elevated temperatures reduces tenfold for a given specific temperature rise. The number of degrees Fahrenheit required to reduce the time tenfold for equal bacterial kills is referred to as the *Z value.* Many organisms exhibit a *Z* value close to 18, but the value may range from 2 to 36. A knowledge of the *Z* value of the organisms present in a product is very useful in designing and selecting optimal pasteurization equipment where time, temperature, throughput, and configuration of processing (batch or continuous) all determine the effectiveness and economics of the process.

The pasteurization time-temperature parameters for a number of representative products are given in Table P-2.

Equipment and Systems: To obtain the thermal exposure required, the product may be heated, held, and cooled in or with (1) the basic container (jar, bottle, or can), (2) jacketed vessels, and (3) continuous heat exchangers with holding tubes. Heat-exchanger configurations may be (1) internal tube, (2) shell and tube, (3) plate, (4) steam injection with flash evaporative cooling, (5) scraped surface, or (6) a combination of these designs. Other methods, including dielectric and microwave heating, have been applied in a limited way.

The trend is toward continuous heat exchangers to minimize thermal damage to the product and to improve productivity. Because of the relationship of the rates of bacteriological kill (first-order logarithmic reaction) to chemical or thermal changes (first-order reaction), a considerable safety factor from the bacteriological-kill standpoint can be designed into the process without significantly affecting the chemical quality of the product.

The equipment used varies from the very simple, with minimal control, where the product is *heated in a vat* and filled hot, held, and cooled in the container or *heated continuously* and filled hot, to the rather elaborate system that employs either vat or continuous methods, provided with numerous safety devices, interlocks, and automatic-diversion devices to ensure adequately processed products.

TABLE P-2. *Pasteurization Time-Temperature Parameters for Representative Products*

Product and Conditions	°F	°C	min	s	Heating Method
Milk, to be bottled and stored	143	61.7	30	15	Vat
(refrigerated) at a temperature below	161	71.7	. . .	15	Continuous
45°F (7.2°C)	191*	88.3	. . .	1	Ultrahigh con-
	194*	90.0	. . .	0.5	tinuous method
	201*	93.9	. . .	0.1	for indirect heat
	204*	95.5	. . .	0.05	exchangers*
	212*	100.0	. . .	0.01	
Ice-cream mix[3]	150	65.5	30	. . . .	Vat
	166	74.4	. . .	15	Continuous
For ultrahigh continuous method for indirect heat exchangers, same parameters as for milk					
Liquid egg products[8] (to be held refrigerated):					
Whole eggs	140	60.0	3.5	. . .	Continuous
	133	56.1	30.0	. . .	Batch
Salt whole eggs (2% or more salt added)	146	63.3	3.5	. . .	Continuous
	144	62.2	6.2	. . .	Continuous
Sugar whole eggs (2–12% sugar added)	142	61.1	3.5	. . .	Continuous
	140	60.0	6.2	. . .	Continuous
Plain yolk	142	61.1	3.5	. . .	Continuous
	140	60.0	6.2	. . .	Continuous
	135	57.2	30.0	. . .	Batch
Sugar yolk (2% or more sugar added)	146	63.3	3.5	. . .	Continuous
	144	62.2	6.2	. . .	Continuous
	139	59.4	30.0	. . .	Batch
Salt yolk (2–12% salt added)	146	63.3	3.5	. . .	Continuous
	144	62.2	6.2	. . .	Continuous
	139	59.4	30.0	. . .	Batch
Fruit juices and nectars[6] (for hot fill or aseptic packaging):					
pH 3.9–4.1	205	96.1	. . .	30	Continuous
4.1–4.2	210	98.8	. . .	30	Continuous
4.2–4.3	216	102.2	. . .	30	Continuous
4.3–4.4	221	105.0	. . .	30	Continuous
3.9 or less (including purees, tomato catsup and chili sauce)	190	87.7	. . .	30	Continuous
	235–265	112.8–129.4	†	Several	Continuous
Tomato juice, puree, sauce, and pizza sauce[6]	245	118.3	. . .	30	Continuous
Tomato paste and fruit concentrates (17% solids and above)[6]	200	93.3	. . .	30	Continuous
	190	87.7	. . .	. . .	Vat
Fermented products:					
Beer	160	71.1	. . .	20	‡
Still red wines	185–190	85.0–87.7	up to 1.0	. . .	‡
Vinegar, in kettle	150	65.5	. . .	. . .	‡
Hot-filled into bottle	150–160	65.5–71.1	. . .	. . .	‡
Continuously in bulk	140–150	60.0–65.5	. . .	. . .	Ref. 4

*Applicable to all milk and milk products (whole milk, including Vitamin D and fortified skim milk, low-fat milk, chocolate milk and drink, cream, ice-cream mix, eggnog, and concentrated milk) normally pasteurized in plate-type heat exchangers. Because some lethality does occur during the heating and cooling portion of the process, ultrafast heat exchangers, such as steam injection with flash cooling, cannot be used with these parameters.[7]

†Sufficient to inactivate pectic enzymes.

‡In addition to killing the spoilage bacteria, the heat also causes various beneficial chemical changes, e.g., coagulating certain colloids and hastening aging of certain wines.

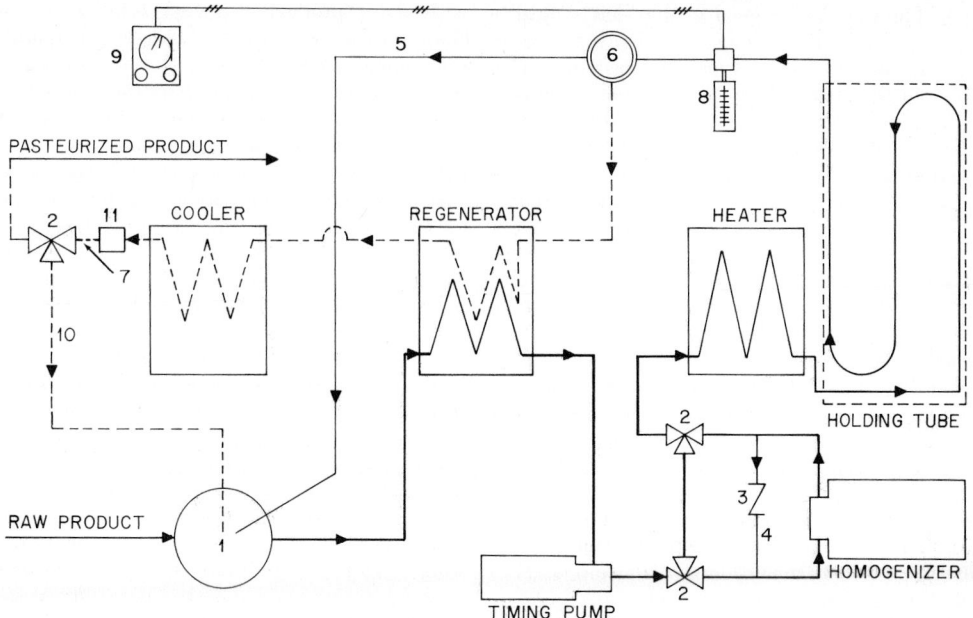

Fig. P-3. HTST (high-temperature, short-time) pasteurizer with homogenizer of larger capacity than timing pump. (1) Raw-product constant-level tank, (2) three-way bypass valves, (3) sanitary check valve, (4) recirculating line, (5) diversion line, (6) diversion valve, (7) horizontal line that must be at least 12 in. above any raw product in the system, (8) indicating thermometer, (9) recorder-controller, (10) bypass line, (11) vacuum breaker. (*Cherry-Burrell Corporation.*)

Several factors dictate the time-temperature relationship required, including (1) the intended use of the product, (2) the type of container, (3) the shelf or storage life required, (4) whether hot-fill or other packaging techniques are used, (5) special objectives, such as inactivation of enzymes, (6) the pH and solids content of the product, and (7) the A_w level* associated with the organisms.

Typical System: The numerous products, problems, and resulting equipment and instrumentation and control-system configurations required are evident. Only one typical system will be described here, as shown in Fig. P-3. The important process functions which must be controlled and interlocked are (1) pasteurization temperature, (2) diversion of product if the pasteurizing temperature should fall below the low limit, (3) holding time and system functions affecting the holding time, such as vacuum chambers, pumps, and homogenizer, (4) concentration or

---

*A_w (water activity) describes the moisture available to the organism(s) and is defined as the vapor pressure of the solution (of solutes in water in most foods) divided by the vapor pressure of the solvent (often water).

dilution in vacuum chambers, particularly if steam is added, and (5) pressure in heat exchangers, particularly after the holding tube and the flow-diversion valve (FDV) to ensure that any possible leakage will occur from the pasteurized side to the raw side rather than vice versa.

In addition, the system must be arranged to prevent contamination of the pasteurized product, either from the raw product or from utilities, e.g., vacuum-chamber condenser cooling water.

The pasteurizing temperature is controlled indirectly by controlling the temperature of the hot water, vacuum steam, or pressure steam used as the heat-transfer medium in the final heater section. Initially, the system is started on water, and often a chemical sanitizing agent is used to inactivate any contaminants which may be on the surface of the equipment. The system then is flushed with pasteurized water and switched to the product. The pasteurizer temperature at the discharge of the holding tube is recorded. Should this temperature fall below the low allowable temperature, the FDV diverts the product back to the raw surge tank. An event marker, actuated by the FDV, records that a diversion has occurred. The FDV is designed so that if it should leak in the diverted position, it will be to atmosphere and not to the forward, or pasteurized, portion of the system. Once the product temperature has been reestablished, the valve switches to the forward flow position. Resanitization of the valve is accomplished by the product temperature. In this example, it is assumed that other downstream segments of the system have not been contaminated as the system has remained closed.

All controls and instrumentation are explicitly defined by the regulatory agency (in the United States, the U.S. Public Health Service Grade A Pasteurized Milk Ordinance).[9]

Generally, the other important process functions, such as the addition and flash evaporation of added steam, are closely controlled and interlocked with the main operation of the pasteurizing section. This is accomplished with a ratio controller or temperature-differential controller. Temperatures just before steam injection and immediately after flash evaporation are controlled to prevent dilution or concentration of the product. The physical arrangement of the system is made to provide drainage to the raw side of any nonprocessed product. Check valves and cutouts are included to prevent backflow of the product so that flow rates above those for which the system is designed do not occur. The holding time is measured initially at a rate within the limitations of a positive-displacement timing pump. This pump is sealed, and if a new pump is required or the speed of the system and pump are increased in the future, the holding tube must be retimed and the new pump resealed. Holding times are calculated and field checked by using a salt injection unit with a conductivity cell in conjunction with a timing device. Holding-tube diameters are calculated for varying capacities based upon turbulent flow when the product is of low viscosity, e.g., most dairy products, or laminar flow when the products are of high viscosity, e.g., most egg products.

Holding times for typical dairy products can be considerably shorter than for products having higher viscosities or containing particles because heat penetration and equilibrium can be assured in shorter times with the lower-viscosity products. Some products are held at the pasteurizing temperatures for longer periods purposely to eliminate the need for higher temperatures, particularly if the product may boil or become damaged at the higher temperature.

A more recently approved pasteurizer for use with dairy products employs tubular heat exchangers and operates at much higher temperatures and pressures than the conventional plate-type unit. This system sterilizes the products and cools them aseptically while at the same time satisfying pasteurization requirements. The primary difference in this system lies in the location of the FDV and the very short holding times. The FDV is located at the discharge of the final coolers to remove it from the high-pressure–high-temperature area. Thermocouple electronic instrumentation is used throughout the system.

Because the system assures complete resanitation of all possible contaminating areas after a temperature failure, more accurate control of the pasteurizing system is possible; certain regulatory groups favor this approach over the conventional one. If aseptic components, such as the FDV, heat exchangers, and other valves, are used, the system can be used for the production of "commercially sterile" products, which is necessary if the pasteurizing system is to eliminate spoilage organisms.

REFERENCES

1. Amendment to 3-A Accepted Practices for the Sanitary Construction, Testing and Operation of High Temperature Shorttime Pasteurizers (Revised Effective January 22, 1967), *J. Milk Food Technol.*, vol. 32, p. 187, May 1969.
2. Amerine, M. A., and W. V. Cruess: "The Technology of Wine Making," Avi, Westport, Conn., 1960.
3. Elliker, P. R.: "Practical Dairy Bacteriology," McGraw-Hill, New York, 1949.
4. Frazier, W. C.: "Food Microbiology," McGraw-Hill, New York, 1958.
5. Hall, C. W., and G. M. Trout: "Milk Pasteurization," Avi, Westport, Conn., 1968.
6. "Hot Fill–Hold–Cool Procedures for Various High Acid Products in California," National Canners Association, Western Research Laboratories, Berkeley, Calif., 1969.
7. Reed, R. B., R. W. Dickerson, Jr., and H. E. Thompson, Jr.: Time-Temperature Standards for the Ultra-high Temperature Pasteurization of Grade "A" Milk and Milk Products by Plate Heat Exchange, *J. Milk Food Technol.*, vol. 3, p. 72, 1968.
8. Inspection of Egg and Egg Products: Notice of Proposed Rule Making, *USDA Consum. Mark. Serv. Fed. Regist.*, 1971.
9. Grade "A" Pasteurized Milk Ordinance—1965: Recommendation of the U.S. Public Health Service, Washington, 1965.
10. Van Den Bogaert, X., S. Hassan, and F. Quesada: Flash Pasteurization and Aseptic Bottling, *Tech. Q.*, vol. 5, p. 218, Stamford, Conn., 1966.

—V. R. Carlson, *Cherry-Burrell Corporation, Cedar Rapids, Iowa*

Peanut Oil (See **Expression;** and **Vegetable oils.**)

Pearl Polymerization (See **Ion-exchange resins.**)

Pearlite (See **Iron and steel.**)

Peat (See **Carbon.**)

Pebble Heater (See **Energy systems for processes.**)

Pebble Mill (See **Paint;** and **Size reduction.**)

Pegmatities (See **Tungsten.**)

Pelletizing (See **Agglomeration; Catalysts;** and **Iron ores.**)

Penicillin (See **Antibiotics;** and **Peptides and proteins.**)

Penniman-Zoph Process (See **Iron oxides, synthetic.**)

Pentachlorophenol (See **Coal tar and derivatives.**)

Pentaerythritol (See **Chlorine organics; Formaldehyde;** and **Polymerization.**)

Pentaerythritol Tetranitrate (See **Explosives.**)

Pentasodium Tripolyphosphate (STP) (See **Detergents.**)

Pentlandite (See **Nickel.**)

Pentolite (See **Explosives.**)

Pepsin (See **Enzymes;** and **Peptides and proteins.**)

Peptides and Proteins Compounds made up of two or more amino acids covalently bound in an amide linkage comprise the peptides. The characteristic amide linkage, in which the carboxyl group of one amino acid joins with the amino group of the next amino acid, is called a *peptide bond*. A peptide is a chain of amino acid residues and provided that the chain is not circular or blocked at either of the ends has an N-terminal amino acid, bearing a free amino group, and a C-terminal amino acid, bearing a free carboxyl group.

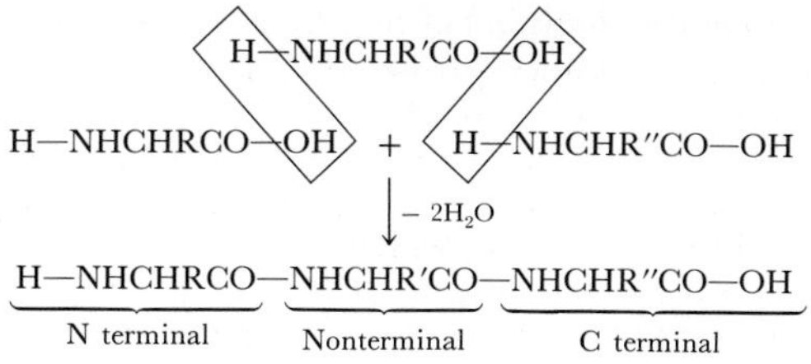

$$H-NHCHRCO-NHCHR'CO-NHCHR''CO-OH$$

N terminal | Nonterminal | C terminal

A particular peptide is named according to the sequence of amino acids contained in it, beginning with the N terminal and moving toward the C terminal. All the residues prior to the C terminal are named as acyl substituents, e.g., valylglycylalanine if in the diagram, $R = CH(CH_3)_2$, $R' = H$, and $R'' = CH_3$. A shorthand notation is used to represent the structure of a peptide, e.g., H-Val-Gly-Ala-OH, where abbreviations for each amino acid, usually expressed by the first three letters of the name, are arranged with the prefix H, denoting the amino terminal (N terminal), and the suffix OH, denoting the carboxyl terminal (C terminal). The official abbreviations suggested by the tentative rules of the International Union of Pure and Applied Chemistry and the International Union of Biochemistry are in Table P-3.

TABLE P-3. *Abbreviations and Genetic Codon for Amino Acids*

Amino Acid	Abbreviation*	Genetic Codon†
Glycine	Gly	GGPu, GGPy
Alanine	Ala	GCPu, GCPy
Valine	Val	GUPu, GUPy
Leucine	Leu	CUPu, CUPy, UUPu
Isoleucine	Ile (Ileu)	AUPy
Serine	Ser	UCPu, UCPy, AGPy
Threonine	Thr	ACPu, ACPy
Aspartic acid	Asp	GAPy
Glutamic acid	Glu	GAPu
Asparagine	AsN (AspN)	AAPy
Glutamine	GlN (GluN)	CAPu
Methionine	Met	AUG
(Half cystine	CyS)	
Cysteine	CySH	UGPy
Phenylalanine	Phe	UUPy
Tyrosine	Tyr	UAPy
Tryptophan	Trp (Try)	UGG
Arginine	Arg	CGPu, CGPy
Lysine	Lys	AAPu
Histidine	His	CAPy
Proline	Pro	CCPu, CCPy

*These are the abbreviations for amino acid residues in a peptide chain. Valylglycylalanine is formulated as H-Val-Gly-Ala-OH. Earlier abbreviations are shown in parentheses.

†Genetic code for translating the nucleotide sequence of mRNA (A, adenosine; G, guanosine; C, cytidine; U, uridine, Pu, A or G, and Py, C or U) into the amino acid sequence of the peptide. The sequence of ···UUUCCC··· is translated into the sequence of ···Phe-Pro···.

Peptides consisting of two, three, six, and eight amino acid residues are known respectively as di-, tri-, hexa-, and octapeptides. Such peptides of lower molecular weight are termed *oligopeptides.* Peptides consisting of 10 or more amino acid residues and of molecular weight in the range of $1\text{-}5 \times 10^3$ are termed *polypeptides.* Proteins, which are of central importance in all biological systems, are also peptide in nature. This fundamental concept of protein chemistry was first proposed by Emil Fischer, father of protein chemistry, early in the twentieth century. No sharply defined line separates the large polypeptide from the small protein. Insulin (hormone protein), protamine, and some components of histone (basic proteins of chromosomes) are examples of small proteins. Nearly all proteins are composed of amino acid residues, more than 100 in number, and their molecular weight is as large as 10^4 up to 10^7. Some values for the molecular weight of a few proteins are listed in Table P-4.

TABLE P-4. *Molecular Weights of Selected Proteins*

Protein	Molecular Weight $\times 10^3$
Insulin.	6
Ribonuclease (pancreas)*	13
Lysozyme (eggwhite)*	15
Chymotrypsinogen*	21
Ovalbumin*	43
Serum albumin*	66
Hemoglobin	68
Gamma globulin (IgG)	160
Fibrinogen	340
Urease.	460
Thyroglobulin	640
Myosin	850
Hemocyanin, octopus	2,800
Snail	8,900
Tobacco mosaic virus	40,000

*Protein consisting of a single peptide chain.

Some proteins are composed of a single peptide chain; others, especially of higher molecular weight, consist of two or more peptide chains. Proteins of huge molecular weight (millions) are enormous aggregates of protein subunits, each of which may not be so large (molecular weight $= 1.5\text{—}10 \times 10^4$ in most cases). Independent peptide chains constituting a protein molecule are frequently held by the disulfide bridges of cystine residues. For a single chain, these bridges may bring together two quite distant points in terms of the linear amino acid sequence, forming a large loop structure:

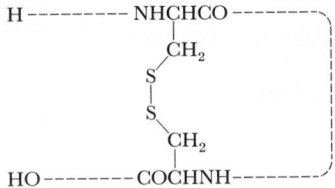

Of the more than 200 amino acids which have been found in living organisms, only 20 α-amino acids of the L configuration listed in Table P-3 serve as the building units for proteins and the related peptides. Their peptide linkages are α-peptide between the α-carboxyl group and the α-amino group. The amino and carboxyl groups on the side chain are free. These 20 amino acids occur in widely varying proportions in different proteins. Some proteins are completely lacking in one or more of them. In addition, several other amino acids occur only in certain specific proteins. Hydroxyproline, for example, has been found only in collagen and elastin, which are proteins of animal connective tissue, and in gelatin derived from collagen. These exceptional amino acids are derived secondarily from each of the common protein constituents (proline in these cases), which have been incorporated into peptide chains according to the genetic codon. See also **Amino acids.**

Peptides

Naturally Occurring: Nearly all living organisms appear to contain lower peptides of various kinds, which can be classified as proteinous and nonproteinous peptides. The former are the α-peptides of amino acids which are normal protein constituents. A large number of peptides of this type have been isolated in vitro through the partial hydrolysis of proteins, especially since the advent of chromatographic techniques. In vivo enzymic cleavages of proteins which result from tissue autolysis, bacterial action, digestive processes, and the like undoubtedly give rise to a large variety of peptides of similar nature. Such peptides have been detected in blood plasma in small amounts, and in liver and other tissues, as well as in urine and in secreting fluids in somewhat larger amounts.

Nonproteinous: These substances are characterized by the structure containing, in part, an amino acid or linkage not found in normal proteins. The amino acid β-alanine, $NH_2CH_2 \cdot$

CH₂COOH, is present as a component of carnosine (β-alanyl-L-histidine) and anserine (β-alanyl-1-methyl-L-histidine), which are found in vertebrate muscle in a rather large concentration. The γ-glutamyl linkage is most common in nonproteinous peptides.

$$-NHCHCH_2CH_2CO-$$
$$\overset{|}{COOH}$$

Glutathione, widely distributed in plant and animal cells and known to perform a large number of biological functions, has a structure of γ-L-glutamyl-L-cysteinylglycine.

Many of the antibiotics produced by bacteria are peptide in chemical nature and contain amino acids of the unnatural configuration. Gramicidin S, a main component of the antibiotics of *Bacillus brevis,* is a cyclic decapeptide, (-L-ornithyl-L-Leu-D-Phe-L-Pro-L-Val-)₂, having no terminal residues. Bacitracin, produced by *B. lichenoformis,* is a cyclic hexapeptide, -D-ornithyl-L-Ile-D-Phe-L-his-L-Asp-L-Lys, with two branches, one peptide branch from the β-carboxyl group of aspartyl residue and the other toward the ε-amino group of lysyl residue. Further examples are tyrocidin from *B. brevis,* polymixin from *B. polymixa,* and subtilin from *B. subtilis.*

A number of strains of *Streptomyces* also produce peptide derivatives of antibiotic activity. Actinomycin, having antitumor activity and rather high toxicity, is an example. Phalloidin and amanitin are toxic heptapeptides isolated from a toadstool, *Amanita phalloides.* These substances contain hydroxyproline of the unnatural configuration and a thioether bridge in their cyclic peptide structure. The most familiar antibiotic, penicillin, from *Penicillium chrysogenum,* also is a peptide derivative. The bicyclic structure, including β-lactam and thiazolidine, originates from the L-cysteinyl-L-valine portion by enzymic condensation and conversion of the configuration at the valyl residue. Other examples of cyclic compounds derived from peptides are ergotoxine, an ergot alkaloid, and luciferin, which causes the bioluminescence of the firefly.

Biologically Active Peptides in Animals: A number of peptide hormones have been isolated from higher vertebrates, some of which are listed in Table P-5. They are proteinous peptides and include tripeptides, e.g., the thyrotropin releasing factor (TRF), some oligopeptides, e.g., oxytocin and vasopressin, and large polypeptides, e.g., adrenocorticotropic hormone (ACTH). Some of their N-terminal groups are blocked with acetyl,

TABLE P-5. *Some Peptide Hormones from Higher Vertebrates*

Hormone	Number of Amino Acid Residues
Peptide hormones:	
TRF (hypothalamus)	3
Oxytocin (neurohypophysis)	9
Vasopressin (neurohypophysis)	9
α-MSH (neurohypophysis)	13
β-MSH (neurohypophysis)	22
ACTH (neurohypophysis)	39
Gastrin (stomach mucosa)	17
Secretin (duodenum)	27
Glucagon (pancreas)	29
Insulin (pancreas)*	51
Kinin in blood:	
Andiotensin	8
Bradykinin	9
Kallidin	10
Toxic peptides:	
Apamin (bee venom)	18
Melittin (bee venom)	26
Erabutoxin (snake venom)*	62

*Usually classified as a protein.

as in α-melanocyte–stimulating hormone (MSH), or with pyroglutamyl residue, as in TRF and gastrin. Some of their C terminals also are masked with a simple amide, as in TRF, α-MSH, and gastrin.

Blood kinin is a biologically active peptide derived from proteins normally present in the plasma but biologically inert. Angiotensin, which has pressor activity, for example, is a split product of the plasma protein andiotensinogen. Formation of insulin from proinsulin in the pancreas also is achieved by such limited proteolysis:

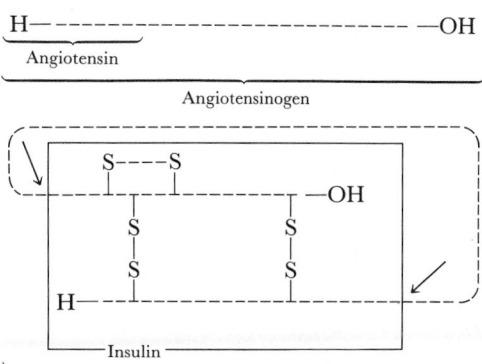

The arrows indicate the split position during limited proteolysis.

Another type of biologically active peptides are the toxic components in venom of snakes, bees, scorpions, and spiders. Apamin and melittin in bee venom are proteinous peptides of medium size, but since the toxic peptides in snake venom are of much higher molecular weight, they are frequently classified as protein. These toxic peptides are rich in basic amino acids and in cystine, which gives the disulfide loop in their single peptide chain.

Determination of Amino Acid Sequence: Sequence can be determined by partially cleaving the chain, identifying each fragment, and then fitting the individual segments into a unique sequence for the original chain. The procedure is far from trivial for a polypeptide. The first successful determination was not completed until 1955, when the complete primary structure of insulin was elucidated by F. Sanger, who introduced novel techniques for the identification of the N-terminal residues and other factors. The N-terminal amino acid usually is determined by one of three methods.

1. The peptide is arylated with 1-fluoro-2,4-dinitrobenzene (Sanger's reagent), yielding the 2,4-dinitrophenyl (DNP) derivative of the N-terminal residue. Complete hydrolysis, extraction of the DNP–amino acid into an organic solvent, and chromatographic identification of the amino acid involved reveal the nature of the N terminal. Partial hydrolysis, isolation of the DNP-peptide fragments, and determining the constituents reveal the amino acid sequence from the N terminal.

2. The second method is the phenylthiocarbamylation of the N terminal with phenylisothiocyanate (Edman's reagent). Treatment of this product (I) with acid results in the cyclization of the N-terminal residue to phenylthiohydantoin (II) and the liberation of the second residue as a new N terminal (III).

$$H\text{—}NHCHRCO\text{—}NH\text{-----------}$$

$$\downarrow C_6H_5N\text{=}C\text{=}S$$

$$C_6H_5NH\text{—}CS\text{—}NHCHRCO\text{—}NH\text{----------}$$

(I)

$$\begin{array}{c}CS\\C_6H_5\text{—}N\qquad NH\end{array} + H\text{—}NH\text{-----------}$$
$$CO\text{—}CHR$$

(II)   (III)

By repeating phenylthiocarbamylation and cyclization stepwise and identifying (II) formed at each step, the N-terminal amino acid sequence is revealed, provided, of course, that the peptide chain is composed of α-amino acids.

3. The use of exopeptidase also is applicable to terminal amino acid determination of proteinous peptides consisting of α-L-amino acids. Aminopeptidase requires the presence of a free α-amino group for activity and hence splits only the N terminal from the peptide chain, resulting in the unmasking of the next residue. By following the kinetics of the progressive release of amino acids from a peptide, information about the N-terminal sequence can be obtained. In a similar way, an enzyme carboxypeptidase, which requires a free α-carboxyl group for activity, can be employed to determine the C-terminal sequence.

The chemical method used for the C-terminal determination is a reaction with anhydrous hydrazine, which breaks up the peptide chain and converts all amino acid residues (except the C terminal, whose carboxyl is not involved in the peptide linkage) to the aminoacyl hydrazides. The C-terminal amino acid can be separated from the hydrazides, identified, and assayed (Akabori's method).

The amino acid sequence of oligopeptides is revealed only by combining these methods, but the sequence determination of a long peptide chain requires cleavage into small fragments of manageable size. Pure proteolytic enzymes or chemical reagents which permit a split at specific peptide bonds are used to yield a peptide mixture that is not so complicated. The isolation of each segment so produced and analyses of their sequence reveal the amino acid sequence on each portion of the original chain.

Prior to the sequence determination, all the disulfide bridges of cystine are generally cleaved, converting the residue to cysteine derivatives, and two or more peptide chains (if formed by this procedure) are separated. The locations of disulfide cross-linkages are established separately, after the amino acid sequence of each chain has been established.

Synthesis: Peptides can be obtained through the condensation of carboxyl, A = —CHRCO—OH, and amino, H—NHCHR′— = B, groups:

$$H\text{—}(A)\text{—}OH + H\text{—}(B)\text{—}OH \rightarrow$$
$$H\text{—}(A)\text{—}(B)\text{—}OH \quad (1)$$

A definite sequence can be constructed by repeating the condensation with individual amino acid residues following the sequence (stepwise elongation) and in some cases by the condensation between the two preformed peptide fragments (fragment condensation). In these condensations, the amino and carboxyl groups which are not to participate in the reaction must be blocked:

$$H—(A)—OH \rightarrow X—(A)—OH \qquad (2)$$

$$H—(B)—OH \rightarrow H—(B)—OY \qquad (3)$$

$$X—(A)—OH + H—(B)—OY \rightarrow$$
$$X—(A)—(B)—OY \quad (4)$$

$$X—(A)—(B)—OY \rightarrow H—(A)—(B)—OH \qquad (5)$$

instead of a simple formula as shown in (1).

The blocking groups, X and Y, should be readily introduced, stable to the condensation reaction, and removed selectively from the accomplished peptide:

$$X—(A)—(B)—OY \rightarrow$$
$$H—(A)—(B)—OY \text{ or } X—(A)—(B)—OH \quad (6)$$

for the further elongation of peptide chain toward the N- or C-terminal side.

Amino groups can be protected with a great range of reagents. N-Alkoxycarbonylation with benzyl or t-butyl chlorocarbonate is most widely used, as the blocking group is readily decomposed by treating with HBr in glacial acetic acid. The N-benzyloxycarbonyl can be removed by hydrogenolysis, and N-t-butoxycarbonyl by trifluoroacetic acid. Other examples for amino blocking are tosylation with p-$CH_3C_6H_4SO_2Cl$, tritylation with $(C_6H_5)_3CCl$, and phthaloylation or trifluoroacetylation with the corresponding acid anhydride. These four protecting groups can be removed to yield a free amino group, respectively with sodium in liquid ammonia, with trifluoroacetic acid, with hydrazine, and with dilute alkali.

Carboxyl groups are generally protected through formation of an ester. Once the peptide bond is formed, the ester linkage is hydrolyzed selectively in a dilute alkaline solution at room temperature. The benzyl ester may be split by hydrogenation; methyl or ethyl esters may be converted to acyl hydrazides, without hydrolysis, by treating with hydrazine and then with nitrous acid to azide, for the further condensation with the next amino component (see equations at top of next column).

The azide method is a general procedure for the stepwise elongation of a peptide chain toward the C terminal.

If a peptide involves amino acids with side

$$X—(A)—OH + H—(B)—OCH_3$$
$$\downarrow \text{condensation}$$
$$X—(A)—(B)—OCH_3$$
$$\downarrow NH_2NH_2$$
$$X—(A)—(B)—NHNH_2$$
$$\downarrow HNO_2$$
$$X—(A)—(B)—N_3 + H—(C)—OCH_3$$
$$\downarrow -N_3H$$
$$X—(A)—(B)—(C)—OCH_3$$

chains that may react during condensation, the problem of protection becomes increasingly difficult. A great range of reactive groups on side chains—amino, carboxyl, thiol, hydroxyl, and so on—must be adequately blocked. Thus, their blocking must be stable to unmasking of the α-amino or α-carboxyl block for stepwise condensation and be readily removed at the final stage, leaving the accomplished peptide moiety intact. The successful synthesis of oxytocin by V. du Vigneaud was the first example of inventiveness in solving these special problems of peptide synthesis, although simple peptides with many more amino acid residues had already been synthesized by E. Fischer. Hydrofluoric acid, proposed by S. Sakakibara, is an excellent reagent for removing these blocking groups, especially in peptides of larger size and complexity.

Formation of the peptide bond from amino and carboxyl components may be accomplished through activation of the latter, as acyl halides, anhydrides, azides, or esters. The last two are used most frequently because the amino acid residue concerned scarcely racemizes during the reaction. p-Nitrophenyl ester of alkoxycarbonylamino acid is a typical example of an active ester which permits the elongation of a peptide chain toward the N terminal:

$$t\text{-BuO}—CO—(B)—OC_6H_4NO_2 + H—(C)—OR$$
$$\downarrow -p\text{-nitrophenol}$$
$$t\text{-BuO}—CO—(B)—(C)—OR$$
$$\searrow HBr/AcOH$$
$$t\text{-BuO}—CO—(A)—OC_6H_4NO_2 + H—(B)—(C)—OR$$
$$\downarrow$$
$$t\text{-BuO}—CO—(A)—(B)—(C)—OR$$

Carbodiimide, based on the following reaction, is the most widely used for the condensation:

$$RN{=}C{=}NR \xrightarrow{R'COOH} RNHC{=}NR \xrightarrow{H_2NR''}$$

with R'COO attached above the central carbon.

$$R'CO{-}NHR'' + RNH{-}CO{-}NHR \quad (7)$$

Automation of Peptide Synthesis: The synthesis of higher peptides is a tedious process, involving many steps. For example, 20 steps for oxytocin; 80 steps for ACTH, and 250 steps for ribonuclease. Nearly all the steps, however, are a cyclic repeat of the condensation reaction and the unmasking of the α-protecting group from the condensate. Automation is possible if the conditions for the two reactions and the purification procedures of each intermediate are standardized. Recently, R. B. Merrifield has devised a solid-phase method of peptide synthesis, in which (1) the C-terminal amino acid is initially attached to a resin particle as an ester; (2) the reaction cycle for the stepwise elongation of the peptide chain toward the N terminal is repeated on the amino component attached to the resin, for example, R = resin in Eq. (7); and (3) the completed peptide chain is liberated from the resin particle. Purification of reaction intermediates in (1) and (2) can be achieved simply by filtering and washing the resin particle chemically modified at each step. The reactions proceed readily under conditions that are not so affected by the size and nature of the peptide chain concerned as the usual liquid-phase reaction is because the reactants attach to the same resin matrix of high molecular weight. Thus, the synthetic procedures are markedly simplified, and the time required for one cycle is shortened perhaps from a day to an hour. By using the solid-phase method, manually or as an automated peptide synthesizer, blood kinins, insulin, and their chemical analogs and some proteins, e.g., ribonuclease and trypsin inhibitor, have been synthesized successfully.

Polyamino Acids: Polypeptides obtained by a single polymerization are known as *polyamino acids*. Polyamino acids of the $[NH(CH_2)_nCO]_x$ type, where $n = 6$ (nylon 6), 3 (nylon 7), 4 (nylon 9), and 10 (nylon 11), are important as synthetic fibers as well as plastics and are prepared by polycondensation of the corresponding amino acids. Poly-α-amino acids are most conveniently synthesized from *N*-carboxy anhydrides formed in the reaction of an amino acid with phosgene:

$$H_2NCHRCOOH \xrightarrow{Cl_2CO} \underset{CO-O}{\overset{CHR}{HN\diagdown CO}} \xrightarrow{-CO_2}$$

$$\left(NHCHRCO\right)_x$$

From a mixture of two or more different anhydrides, copolymers can be prepared.

The properties of these polymers vary in a wide range with their amino acid constituents as well as the degree of polymerization. A high degree of polymerization can be achieved by the proper choice of reaction temperature, solvent, and the nature and concentration of catalyzer on each reactant of a given concentration. Polyamino acids so obtained have been used widely as simple models for understanding the relationship between protein structures and their properties and may lead to potential materials for fibers, plastics, and other synthetics. See also **Polymerization.**

Proteins. Proteins are peptides of high molecular weight found universally in the cells of living organisms and in such biological fluids as blood plasma. Proteins are exceedingly diverse in properties and functions. Some are relatively inert fibers (the fibroin of silk; the keratins of wool, hair, feather, or horn; or the collagens of tendons and connective tissue) which play a structural role in the animal organism. Other proteins are readily soluble in water or dilute salt solutions, e.g., the proteins in eggwhite, blood plasma, or cytoplasm. Since most of their molecules are almost spherical in shape, these substances often are called *globular protein* in contrast with *fibrous protein*.

Proteins are indispensable components of all biological systems. They serve functions in cellular structure, catalysis, regulation, and transportation. Intracellular organization and particles, such as cellular or nuclear membrane, mitochondria, the Golgi apparatus, various reticula, and the like, are constructed with *lipoproteins*. Contractile elements of muscles and flagella or for intracellular movements also are proteins. All the enzymes, whether dissolved in cytoplasm or bound to cellular structures, are protein in nature. Many hormones are proteins or peptides. The chromosomes, which carry the genetic information in the cell nuclei, are highly complex nucleoproteins. Some proteins on cell membranes are concerned in the active transport through the membrane. Blood contains several proteins responsible for transportation. For example, hemoglobin in red cells is an effective oxygen carrier. Immunoglobulins in blood plasma protect the body from a number of infections. Indeed, proteins are intimately concerned with virtually all physiological events. It is appropriate, then, that the word *protein* is derived from the Greek word meaning "first" or "most important."

Classification: Many proteins have been isolated. Each protein is named with reference to the function, source, characteristic constituent, or other factors. In 1907, a combined committee, representing the American Society of Biological Chemists and the American Physiological Society, proposed a formal classification of proteins into three major categories: (1) simple proteins; (2) conjugated proteins; and (3) derived proteins. The last classification embraces all denatured proteins and hydrolytic products of protein breakdown and no longer is considered as a general class.

The simple proteins are defined as those proteins composed only of a peptide chain and yielding α-amino acids and their derivatives on complete hydrolysis. Simple proteins were further classified, on the basis of their solubilities, into seven groups:

1. Albumin, soluble in salt-free water
2. Globulin, insoluble in salt-free water but soluble in dilute salt solutions
3. Glutelin, insoluble in all neutral solvents but soluble in dilute acid or alkali
4. Prolamine, insoluble in water and in absolute alcohol but soluble in 70–80% alcohol
5. Albuminoid, highly insoluble in all neutral solvents
6. Histone, soluble in water, insoluble in ammonia, and coagulated by heating; rich in basic amino acids but containing a wide variety of other amino acids
7. Protamine, soluble in water and in ammonia and not coagulated by heating; contains almost exclusively arginine, together with only a few kinds of other amino acids

It has gradually become clear that the foregoing classification is not completely satisfactory and by no means rigid. The term *albuminoid* is no longer used, the category now being called *scleroprotein* or *fibrous protein,* the two principal subclasses of which are collagen and keratin. Histone and protamine are specific proteins of basic character in cell nuclei. Glutelin and prolamine also are specific proteins in plant grain.

In contrast, albumin and globulin are widely distributed in almost all living materials. Globulin may be divided into two subclasses, the less soluble euglobulin and the more soluble pseudoglobulin, but there appears to be no distinct boundary between euglobulins and pseudoglobulins or between pseudoglobulin and albumin.

Conjugated proteins are defined as a complex of proteins with nonproteinous moieties that are an integral part of the molecule. The protein part is known as apoprotein. According to the nature of components attached to apoprotein, conjugated proteins may be classified into metal proteins, phosphoproteins, glycoproteins, nucleoproteins, lipoproteins, mucoproteins, hemoproteins, chromoproteins, and so on. In some cases, two or more types of nonproteinous components are involved independently. Thus, ovalbumin contains an oligosaccharide and two phosphates, each of which links to the single peptide chain.

Purification: Basically, the problems concerned with the purification of proteins consist in the choice of a starting material, the extraction of proteins from tissues, mycelia, or cells, and the isolation of the desired component from the extract. Each case presents somewhat different problems, depending on the properties of the protein to be isolated and the nature of the source material used. With a biologically active protein, it is possible to use the specific activity as an index of alterations occurring at each fractionating step. When no such index is available, electrophoretic patterns may give useful information on the protein components involved.

The isolation of a pure protein from a highly complex mixture of proteins and nonproteinous components of living material usually is a formidable task. The protein of interest may constitute only 1% or even less of the dry material weight of the starting material. Much of the remainder may consist of other proteins, some of whose properties closely resemble those of the desired product. In addition, the decided instability of most proteins imposes limitations on the conditions that can be used during isolation. In general, fractionations must be achieved using aqueous solutions as solvents, keeping the temperature as low as possible, and avoiding extremes of pH, excessive agitation, and foaming.

Extraction: An efficient extraction of proteins usually is accomplished by destroying the cellular organization of the tissue by grinding in a homogenizer, food blender, or other like equipment or by exposing the material to ultrasonic vibrations. If the protein is bound to a particular portion of the cell, pretreatment with a lipid solvent or detergent may be necessary to free it. Acetone powder, prepared by grinding the source material with a large volume of acetone in a food blender and filtering off the acetone layer, frequently has been used as a convenient method of enzyme preparation. In some cases, it is advisable to isolate the particular fraction of the cell as an initial purification step.

The choice of solvent for extraction also is important. Insulin in pancreas has been extracted successfully with 65–75% alcohol at a pH below 7. The alcohol facilitates the extraction of insulin but minimizes the solution of other proteins. The acid prevents the undesirable action of proteolytic enzymes. Once in solution, any combination of several fractionation methods may be applied—precipitation, chromatography, gel filtration, electrophoresis, or adsorption. Dialysis and lyophilization are the most common procedures for obtaining preparations of the final or intermediate products in a dry and salt-free state.

Fractional Precipitation: The solubility of most proteins is a sensitive function of pH, ionic strength, temperature, and concentration of organic solvent. By appropriate selection of these factors, a series of precipitated fractions can be obtained in which the desired component is concentrated. The solubility of proteins, because they are high-molecular electrolytes with many valences, is markedly affected by neutral salts. In accordance with the concentration of salt, a protein may be salted in or salted out.

Salting In: Salting in, the increase of solubility caused by salt of lower concentration, is especially notable in globulins, which can be precipitated from a salt solution by dialysis against water or by adding large volumes of water, leaving the other proteins in solution. Some proteins may crystallize out almost in the pure state, e.g., crystallization of hemoglobin by electrodialysis of horse hemolysate and of hemocyanine by high dilution of octopus hemolymph. With certain water-insoluble proteins, such as myosin in muscle, purification is better achieved by extracting the more soluble components with water, leaving the desired material in the residue. This then can be extracted with other solvents for further purification.

Salting Out: High concentration of neutral salts tends to decrease the solubility of proteins, resulting in precipitation in most cases. The concentration necessary for such salt-out procedures varies widely from one protein to another. Adding salt stepwise and centrifuging the precipitates off at each step will give a series of arbitrary fractions. The desired protein thus will be in one or two of the adjacent fractions. For example, three fractions rich in fibrinogen, globulins, and albumin are precipitated in this order by stepwise salting out of blood plasma.

The effect of salts on protein solubility is a function not only of the molar concentration but also of the charge on the ions. Thus, the salt effects are described best in terms of ionic strength, defined by

$$\text{Ionic strength } \frac{\Gamma}{2} = \frac{1}{2} \sum c_i Z_i^2$$

where c_i is the concentration of the ith ionic species and Z_i is its charge. At a given pH and temperature, the change in solubility of a protein with change in salt concentration can be approximated by the simple relationship

$$\log S = \beta - K_s \frac{\Gamma}{2}$$

where S = solubility
K_s = salting-out constant
β = hypothetical solubility of a protein at zero ionic strength

The constant K_s depends upon the nature of the protein and the salt used but not on the pH and temperature. The value of β, on the contrary, strongly depends on pH and temperature. Thus, two types of fractional precipitation are possible, K_s fractionation and β fractionation. The former is usually achieved with a stepwise increase of salt concentration at a given pH and at a given temperature, the latter by changing the pH or the temperature at a given salt concentration.

Isoelectric-Point Precipitation: Protein solubility varies markedly with pH and is at a minimum at the isoelectric point. By raising the salt concentration and adjusting the pH to the isoelectric point it is often possible to obtain a precipitate considerably enriched in the desired protein and to crystallize it from a heterogenous mixture. For example, ovalbumin and serum albumin can be obtained in crystalline form in essentially two steps. The globulins in eggwhite or serum are removed as precipitates by adding ammonium sulfate to 0.4 saturation at a neutral pH. The supernatant fluid then is adjusted to a pH of 4.7, the isoelectric point of these albumins. The ammonium sulfate concentration gradually is increased until a slight opalescence develops. Addition of seed crystals is frequently all that is necessary to initiate crystallization. Lysozyme, a minor component of eggwhite, can be crystallized out directly from the starting material when sodium chloride of 5% concentration is added and adjusted to a pH of 9.5.

Precipitation with Organic Solvents: Materials such as ethanol and acetone sharply reduce the solubility of most proteins. At room temperature, denaturation proceeds rapidly in the presence of these reagents. However, these solvents can be used successfully for protein fractionation if the operation is carried out at a temperature near or below 0°C and at reagent concentrations below certain critical values. A number of proteins have been crystallized by this method: urease of the jack bean (the first crystallization of an enzyme by Sumner, in 1926) and Taka-Amylase of *Aspergillus oryzea* from aqueous acetone and hemoglobin from aqueous ethanol.

Another outstanding example of this technique is in the systematic separation of the components of human plasma by methods developed by E. J. Cohn, J. T. Edsall, and coworkers. By appropriate selection of pH, ionic strength, temperature, and ethanol concentration, a series of precipitated fractions, which represent to a considerable degree the purification of the major component, can be obtained.

Electrophoresis: Because one of the distinctive features of proteins is their difference in charge (see Table P-6), and because electrophoretic mobilities are principally a function of charge, electrophoresis is one of the best methods for separating protein mixtures. Immunoelectrophoresis, a combination of gel electrophoresis of high resolution with immune precipitation reaction of high specificity, has been applied successfully to the study of protein mixtures as well as protein homogeneity. The zone method, employing a support-

ing medium of high capacity, also may be useful for preparation. After the electrophoresis, the supporting medium can be divided into a number of separate zones, and the proteins in each zone can be eluted out separately. Typical supports for protein purification include cellulose, starch gels, starch blocks, and polyacrylamide gels.

Chromatography: More frequently, protein ions have been separated by ion-exchange chromatography. Ion exchangers of hydrophilic backbones, usually cellulose or chemically modified dextrane,* are the most satisfactory for this purpose, and their diethylaminoethyl (DEAE), carboxymethyl (CM), and sulfoethyl (SE) derivatives are used most commonly. Such columns are developed by washing with buffers of increasing ionic strength and/or changing values of pH as appropriate. Properly used, the resolving power of these columns is great, and very substantial purifications frequently can be obtained. Nearly all the major enzymes—amylase, lipase, and several proteolytic enzymes, for example—can be isolated from pancreatic juice directly by a single chromatographic step. The capacity of the column is fairly high, permitting handling gram quantities of protein.

The selective adsorption of proteins onto certain materials and the selective elution of proteins following adsorption can also be used as a purification procedure. Common materials for the adsorption of proteins include calcium phosphate gel, alumina gel, and kaolin.

Gel Filtration: Protein purification can be achieved by gel filtration through a column of chemically modified dextrane,* agarose, or polyacrylamide gels that are available in a variety of pore sizes. The fractional elution of proteins from the column is due to the action of gel matrix as a molecular sieve, which sorts out molecules according to size. The method can also be used in studies of molecular weight, association, or dissociation of proteins.

Structure and Properties: The proteins essentially are polymers of α-amino acids joined in an α-peptide linkage. As mentioned earlier, only 20 different amino acids of the L configuration are found as common constituents. The properties of a protein are determined in part by its amino acid composition. For example, the net charge at any given pH is largely a function of the relative number of dibasic and dicarboxylic amino acids present in the protein structure.

TABLE P-6. *Isolectric Point of Some Proteins*

Protein	Isolectric Point (pH)
Pepsin	1
α-Casein	4.0
Serum albumin	4.7
Insulin	5.4
Fibrinogen	5.8
Conalbumin (transferrin)	6.8
Hemoglobin	7.1
β_2-Globulin	7.3
Ribonuclease	9.5
Lysozyme	10.7
Clupeine	12.1

*Sephadex.

Many of the characteristic properties of a protein depend upon the integrity of the conformation of the peptide chain. The biological activity of enzymes and protein hormones, for example, can be destroyed by mild procedures which do not break any of the peptide linkage. Heating, exposure to weak acids or alkalies, solution in the presence of urea, guanidine, organic solvents, detergents, or heavy-metal ions—all are procedures that can profoundly alter the conformation of the peptide chain in a protein molecule. Any such disorienting process is called *denaturation*. See also **Enzymes.**

Most globular proteins denature readily, decreasing their solubility. Heat coagulation of proteins, e.g., those of eggwhite, is a familiar example of denaturation. Proteins which crystallize readily in the native state become incapable of crystallization after denaturation. An increase in viscosity of solution and in susceptibility to proteolysis and changes in optical rotation also accompany denaturation.

Primary Structure: Protein structure and properties may be considered from several levels of structural organization. These levels are termed *primary, secondary,* and *tertiary structures.* The primary structure of a protein is defined by the amino acid sequence along the peptide chain and corresponds to the usual structural formula written for an organic compound. The sequence in which the 20 different amino acids are linked in any given protein is highly specific and characteristic for that protein. The number of possible permutations in even so small an enzyme as pancreatic ribonuclease (124 residues) is astronomic, namely 20^{124} permutations! However, it is now rather well established that the cell of a given species actually makes only one possible sequence as the enzyme.

This specificity of protein biosynthesis is a consequence of the precise and particular arrangement of the nucleotide residues in the relevant RNA molecule. As this arrangement is derived from the nucleotide sequence in the parent DNA molecule of cell nuclei, it follows that the amino acid sequence of protein in a particular cell is directed by the types of DNA concerned. Thus, the biosynthesis of protein may be considered as "a translation of the nucleotide sequence of DNA into the amino acid sequence of peptide chain."

The nucleotide sequence carrying the genetical information on a given protein is initially transcripted into a messenger RNA (mRNA) under the action of DNA-dependent RNA-polymerase. The enzyme generates RNA from four nucleoside triphosphates,* ATP, CTP, GTP, and UTP, only in the presence of a single-stranded DNA as primer. The RNA generated has the nucleotide sequence completely complementary to that of primer DNA. mRNA, or *template* RNA, is then transported from cell nuclei to cytoplasm and associates with ribosome. The complex polysome is the enzymic system for protein biosynthesis.

The amino acids for protein biosynthesis have to be activated and conjugated with transfer RNA (tRNA) by the action of amino acid–activating enzymes

$$NH_2CHRCOOH + tRNA + ATP \rightarrow$$
$$NH_2CHRCO—tRNA + AMP + pyrophosphate$$

Their actions are highly specific for any individual amino acid and for amino acid acceptor, tRNA.

Formation of a peptide bond on the polysome proceeds as a transfer reaction of amino acid residue from aminoacyl tRNA to the peptide chain, which is initiated at the N terminal and grows toward the C terminal by the stepwise addition of amino acid:

$$Peptide—COOH + NH_2CHRCO—tRNA \rightarrow$$
$$peptide—CO—NHCHRCOOH + tRNA$$

The active amino acid donors are aminoacyl tRNA bound to mRNA whose nucleotide sequence decides the binding sequence of individual aminoacyl tRNA. The tRNA with its amino acid attached links with the appropriate and complementary site of the mRNA in polysome, through three nucleotide units. For example, the nucleotide sequence ···UUUCCC··· permits the binding of phenylalanyl tRNA (at UUU) and then of prolyl tRNA (at CCC). As the process continues with other aminoacyl-tRNA molecules through the mRNA chain, individual amino acids are brought together in the correct sequence.

The genetic codon for translating the nucleotide triplet is given in Table P-3.

Proteins that perform functionally identical tasks in different species may exhibit subtle differences in their primary structure. Such species differences are mainly due to the replacement of individual amino acids, and the greater the species difference the greater the structural difference. For example, the α-chains of ape, monkey, horse, and chicken hemoglobin differ from human

*Adenosine, cytidine, guanosine, uridine triphosphates.

hemoglobin respectively at only the 1–2,* 5–7,* 17, and 34 residues, out of the total 141 residues in each peptide chain. In spite of these species differences, proteins of identical function still possess several peptide segments of the identical or quite similar sequence in common.

Secondary Structure: A long peptide chain can assume an almost infinite number of different conformations. This multiplicity of conformational possibilities is due to the rotation about covalent bonds other than those in the peptide linkage, each of which is fairly rigid, fixing the position of C=O and N—H on a plane with a trans configuration:

Any ordering of the otherwise flexible peptide chain resulting from the formation of hydrogen bonds between the C=O and N—H of the peptide backbone, is termed *secondary structure.* L. Pauling and R. B. Corey established a set of criteria for formation of the most stable secondary structure. Acceptable structures that meet these requirements fall into two broad classes; helical structures and sheet structures.

The most typical of the latter class are pleated sheets, as shown in Fig. P-4. Fibrous proteins, such as silk (fibroin) and wool stretched in steam (β-keratin), as well as polyamino acids oriented unidirectionally in fibers or films, are composed mainly of this spatial structure. Their x-ray diffraction shows an axial spacing of 6.5–7 Å and a backbone spacing of 4.7 Å in common, together with a side-chain spacing which varies according to their spatial size. The antiparallel pleated-sheet structure also is involved in some globular proteins as a hydrophobic segment. The arrangement in a hydrophilic molecule of spherical shape is depicted in Fig. P-6c.

The most important helical structures are those of polyproline and collagen and the α helix (Fig. P-5). The configuration of an α-helix is characterized by a translation along the helical axes of 5.4 Å per turn, which is made for every 3.6 residues. Thus, the translation per residue is near 1.5 Å, and the rotation per residue is 100°. Nearly

*= difference among the species.

all the globular proteins have an α-helical region of various length interspersed with a region of random coils (irregular conformation). The content of the helical part is estimated to be 25% in β-lactoglobulin, 42% in lysozyme, and 77% in myoglobin and hemoglobin. Polyamino acids in solution also are present as a helix or as a random coil, depending mainly upon the nature of the solvent and the side chain of the polymer.

The trans conformation from random coil to helix or pleated sheet can be measured by the hydrogen-deuterium exchange method or by infrared spectrometry because the hydrogen-bond formation of N—H···O=C results in lowering the reaction velocity of H-D exchange:

$$>N-H + D_2O \rightarrow >N-D + DHO$$

The absorption spectrum near 1650 cm^{-1} (stretching mode of C=O) and 1535 cm^{-1} (deformation mode of N—H) also are altered. The infrared dichroism is another important factor, as the hydrogen bonds in an α-helix are oriented parallel to the helical axis and those in the pleated sheet are perpendicular to the fiber axis.

As may be seen from Fig. P-5, an α-helix is an asymmetric structure, and the denaturation, i.e., the conformation change to random coil, is accompanied by an increase of the levo rotation. The helical structure in collagen, on the contrary, denatures with a marked decrease in the levo rotation. Studies of optical rotation and, more precisely, of optical rotatory dispersion have frequently been made to estimate the helical content in the protein molecule and the disorder during the denaturation process.

Tertiary Structure: In addition to the backbone of α-peptide bonds and the hydrogen bonds between their C=O and N—H, just described, a protein molecule may be linked in other ways, covalently and noncovalently. The most important covalent linkage is the disulfide bridge joining two cysteine residues at distant points along a peptide chain or two cysteine residues between the two distinct peptide chains. These disulfide linkages are formed after the construction of the peptide chain and constrain the three-dimensional conformation of the chain. In the lysozyme molecule, for example, the single peptide chain of 129 amino acids is folded as a structure of a reasonably compact and roughly ellipsoid shape by four disulfide bridges between the cysteine residues 6 and 127, 30 and 115, 64 and 80, and 76 and 94, as shown in Fig. P-6.

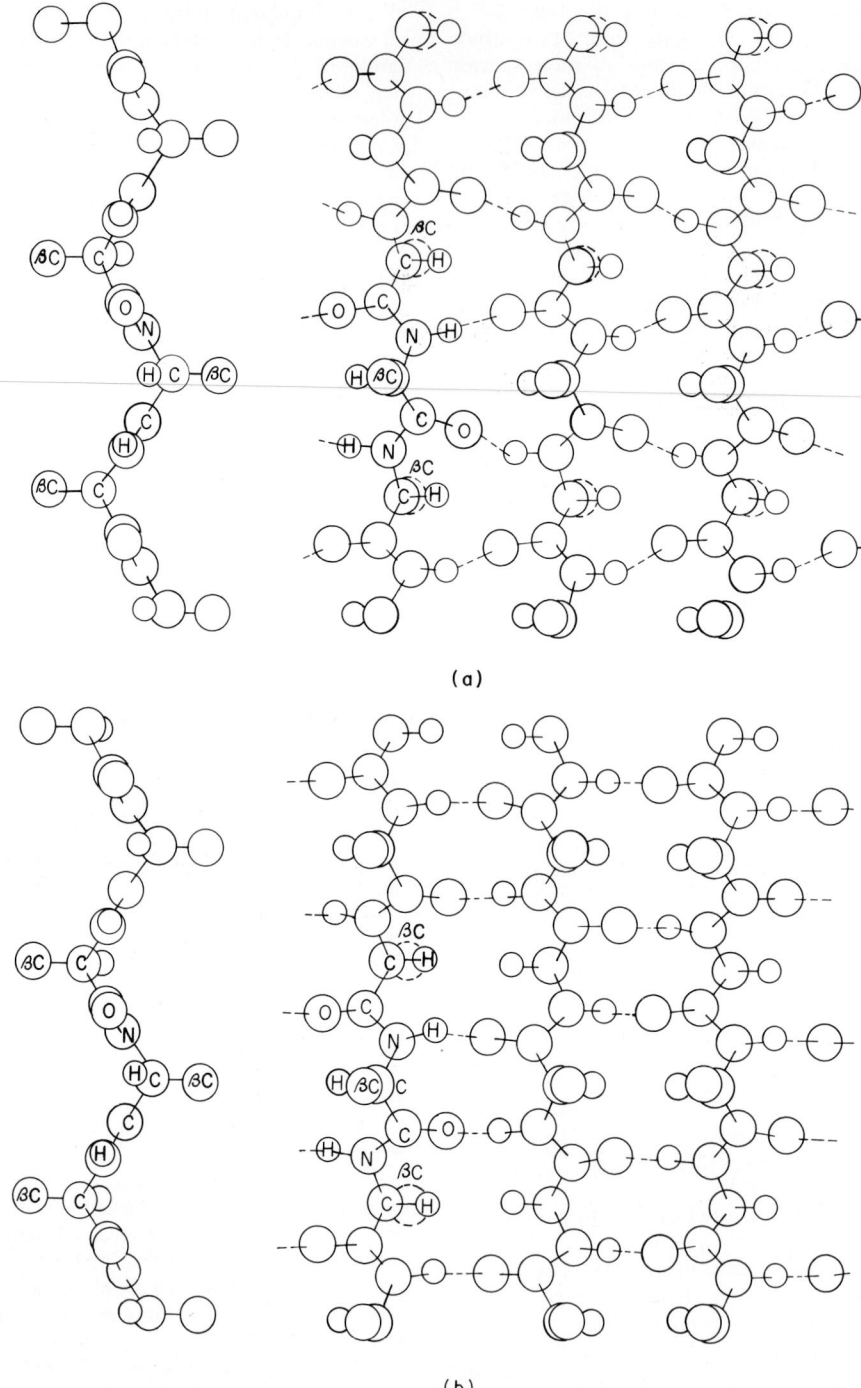

(a)

(b)

Fig. P-4. Representation of the (a) parallel and (b) antiparallel pleated-sheet structure for polypeptide chains.

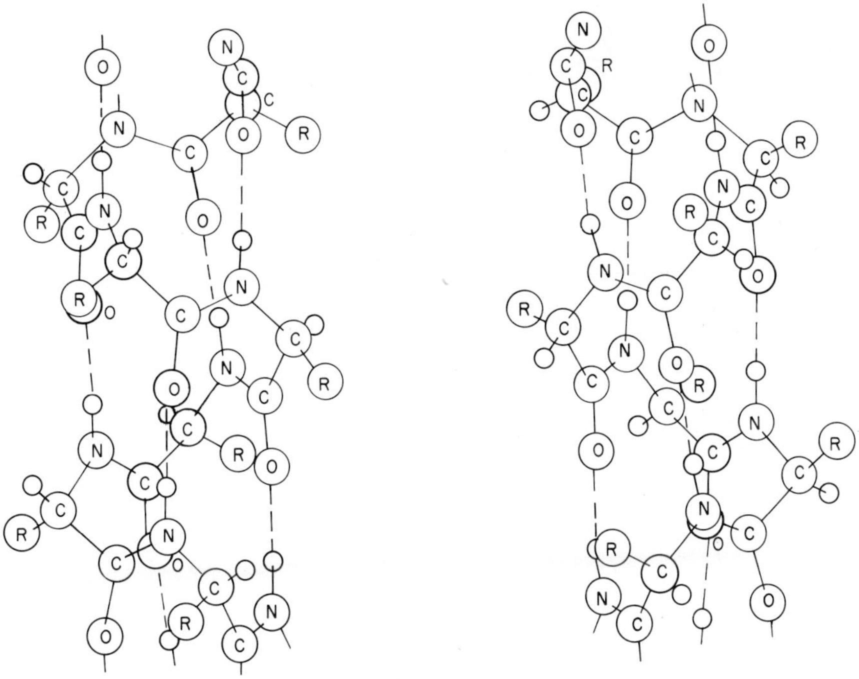

(a)   (b)

Fig. P-5. (a) Left-handed and (b) right-handed α-helical forms of a peptide chain containing L-amino acids.

As the figure shows, the conformational structure of lysozyme is very complex, particularly in the region bounded by residues 35 and 80. Thus, the molecule contains several segments of α-helix and a segment of antiparallel pleated sheet. These segments might themselves assume various conformations without any side-chain interactions. Any ordering of otherwise flexible segments resulting from the interactions between the side chains is known as tertiary structure.

Such interactions may include stable covalent bonds as well as noncovalent bonds of much less stability. Most common of the latter type are hydrogen bonds, van der Waals interactions, charge-transfer forces, salt linkages, and so on. Tertiary structure of proteins may also be affected by the formation of a metal complex, such as Zn^{++} in insulin, Fe^{++} in hemoglobin, and Cu^+ in hemocyanin; or by an association with nonproteinous compounds as coenzymes and substrates in enzymes.

Three-dimensional structure of proteins is determined most precisely by x-ray diffraction methods, although the hydrodynamic behavior of protein solutions, such as viscosity, frictional coefficient, or light scattering, has yielded information concerning the general topography of the macromolecule, the size (molecular weight), and shape. Following the guidelines established in the work on myoglobin by M. F. Perutz and J. C. Kendrew, spatial structures of several proteins (hemoglobin, cytochrome c, lysozyme, chymotrypsinogen, and others) have been established recently. X-ray studies of other proteins are in progress.

Metabolism and Nutrition: Individual protein in a living organism has its own life-span—4 months for hemoglobin and only 1 week or so for serum albumin in the normal human body. The aged proteins are digested by proteolytic enzymes of tissue, such as cathepsin. An appreciable part of the recovered amino acids may be available for the biosynthesis of new proteins, but another part is catabolized and the nitrogen is excreted as urea in mammals, uric acid in birds, reptiles, and insects, or ammonia in organisms of lower classes.

An animal is said to be in nitrogen balance if its daily intake of nitrogen is just balanced by its daily excretion. For the maintenance of nitrogen balance and for growth, the human organism

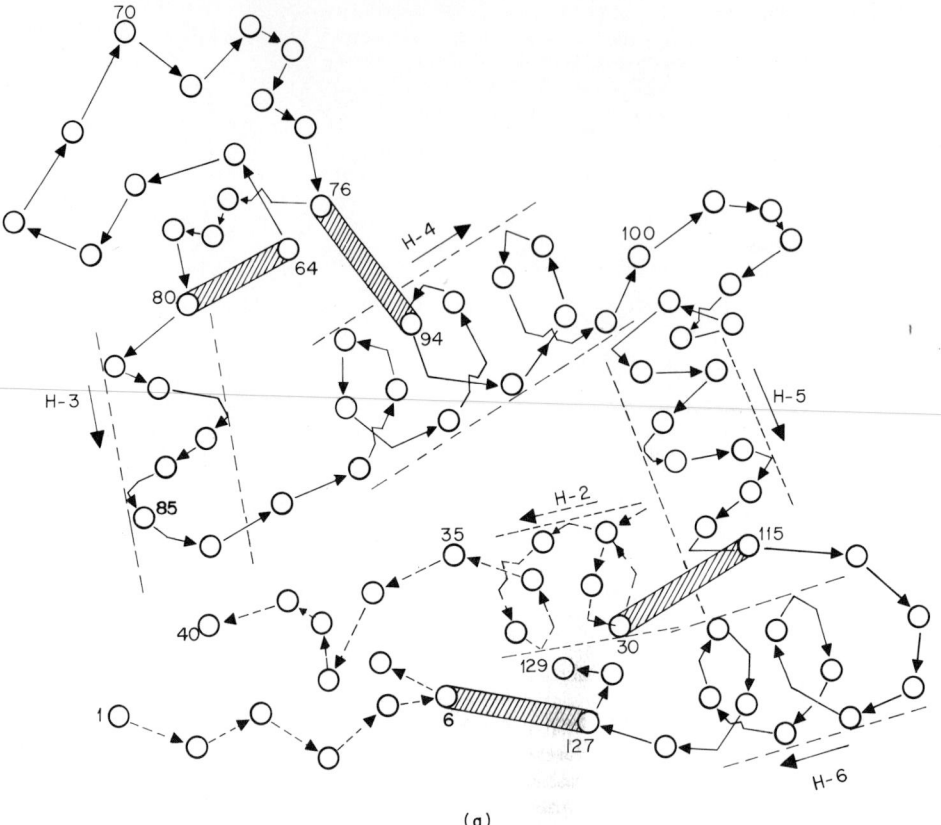

(a)

Fig. P-6. Schematic representation of the peptide-chain conformation of lysozyme. The shaded rectangles represent —CH_2S—SCH_2— bridges (S—S), and the numbers refer to individual residues, counting from the N terminal. (a) The C-terminal portion from 64 to 129 and the four disulfide bridges are shown together with two neighboring segments, 1–7 and 9–40. The S—S between 30 and 115 is before the helical segment *H-2*.

needs a daily intake of 70–80 g of proteins. Based upon this value, the current world population needs 3×10^7 tons of animal proteins and 10×10^7 tons of plant proteins annually. A 50% increase of this production will have to be achieved by the beginning of the next century to feed the expanding world population.

All living organisms, without exception, can synthesize their own proteins from amino acids. In terms of the ability to carry out the de novo synthesis of amino acids, however, there are wide variations among different organisms.

Plants can synthesize all the amino acids from nitrogen in the form of ammonium salts or nitrate

and a few other simple compounds. The annual production of cereal and vegetable proteins so assimilated from inorganic nitrogen over the world is estimated at present to be approximately 10×10^7 tons: 4×10^7 as wheat, 2×10^7 as rice, and 1×10^7 as corn plus other sources.

Some microorganisms, such as lactic bacteria, require preformed amino acids for growth, as they lack some synthetic ability. This requirement has served as the basis for a widely used method of bioassay for a given amino acid content of food and various media. On the other hand, some microorganisms often do very well with ammonium sulfate and carbohydrate as the sole sources

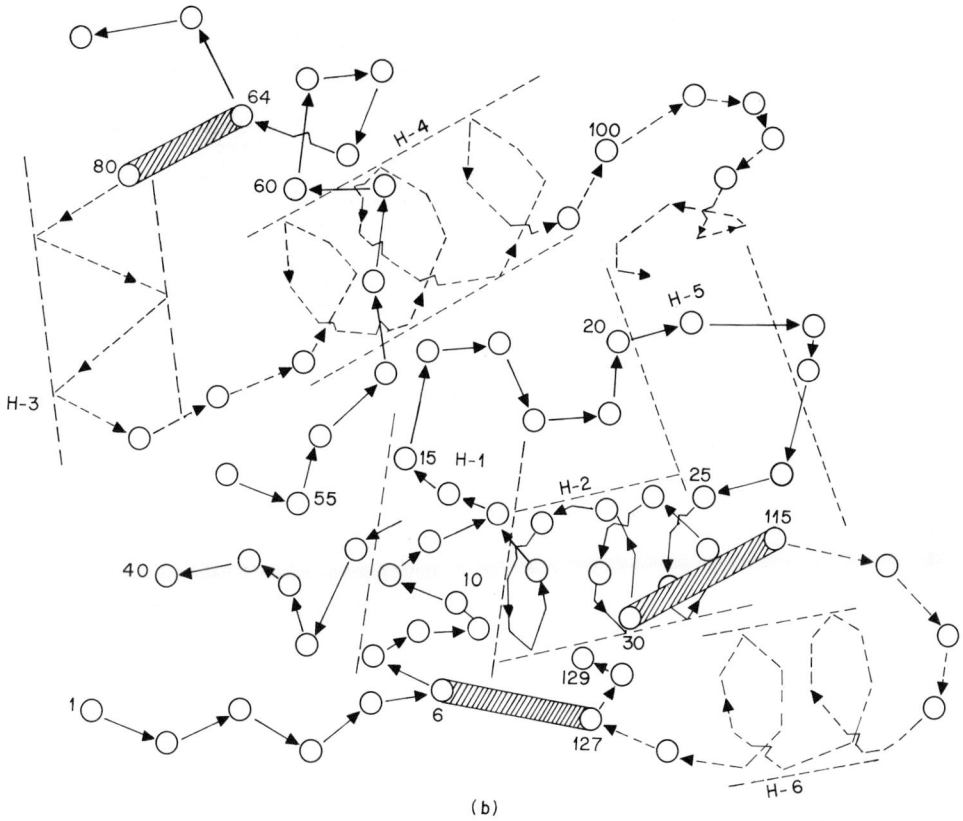

Fig. P-6. (cont.) Schematic representation of the peptide-chain conformation of lysozyme. (b) The N terminal portions, *H*-1 and 20–25, are present behind *H*-2 and *H*-5, respectively. The 55–64 portion is over the helical segment, *H*-4.

of nitrogen, sulfur, and carbon and, in some cases, accumulate particular amino acids in a process known as *amino acid fermentation.*

Animal organisms can scarcely assimilate inorganic nitrogen into body protein without the effective cooperation of intestinal flora, as in ruminants. Thus, the human organism requires daily the intake of 70–80 g of protein as the source of amino acids. More than half of the protein-constituent amino acids, however, can be derived from other amino acids by their own enzymic reactions. Thus, amino acids are classified as *essential* or *nonessential.*

The classical studies of W. C. Rose have established that, for the young rat, leucine, isoleucine, valine, lysine, methionine, phenylalanine, tryptophan, threonine, histidine, and arginine are essential amino acids. The first eight have been found

to be essential for all animal species so far studied, including the silkworm, *Bombix mori.*

In contrast with the young rat, the human organism can synthesize sufficient histidine for optimal growth and sufficient arginine for the maintenance of nitrogen balance. Thus, arginine is half essential for man, i.e., nonessential for adult and essential for infant. The dietary requirement for protein and for particular amino acids is a function of several variables, including the presence of metabolically related substances in the diet. For example, tyrosine exerts a sparing effect on phenylalanine, as does glutamic acid on arginine, or cystine on methionine. Thus, a proper composition of amino acids reduces the requirement of dietary proteins.

In addition, amino acid requirements vary with the physiological state of the animal, with age,

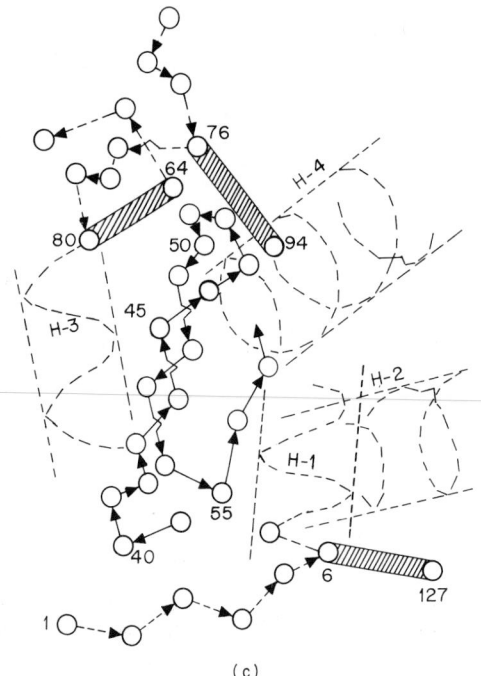

(c)

Fig. P-6. (cont.) Schematic representation of the peptide-chain conformation of lysozyme. (c) The pleated-sheet structure (40–55) is present in a hole surrounded by the helical segments, *H*-1, *H*-2, *H*-3, and *H*-4.

and perhaps with the nature of the intestinal flora.

The standard-reference protein proposed by the Food and Agricultural Organization (FAO) in 1957 contains the following essential amino acids in the ratios indicated:

Percent of crude protein

Isoleucine	4.2
Leucine	6.2
Lysine	4.2
Methionine	2.2
Phenylalanine	2.8
Threonine	2.8
Tryptophan	1.6
Valine	5.0

Such a distribution of amino acids in dietary proteins can be obtained by taking both animal and plant proteins at a proper ratio, 1:3–4. Plant proteins are low in price but markedly deficient in some essential amino acids. Thus, their protein efficiency is low without any additional supply of

the deficient amino acids. Enrichment of human food and animal feed with free amino acid(s), lysine, methionine, threonine and/or tryptophan, in place of animal proteins, has been done successfully.

Appreciably excess amounts of amino acids may be acceptable without any undesirable effects provided their compositions are proper. Excess intake of a particular amino acid, however, may be injurious, as observed in a rat that feeds on eggwhite proteins with threonine or isoleucine added in a high concentration. Such an undesirable effect is termed *imbalance of amino acids* and is not observed on eggwhite protein-rich diet or on the protein diet to which both threonine and isoleucine have been added.

The effects of amino acids on nutrition are also discussed under **Amino acids.**

Protein Industry: A number of proteins or crude materials containing particular proteins have been of wide use in human life over thousands of years, e.g., the keratins of wool, feather, or horn; fibroin of silk; collagenous tissues as leather; proteins in milk, wheat, soybean, egg, and many other natural substances. Wool and silk are still important textile materials. Tanning hides to form leather is a well-known protein process known to the ancients. Making cheese from milk casein and flavor seasonings from plant or fish proteins are also very old processes. Gelatin derived from collagen has been used widely in food processing and as an adhesive material (glue) and still is important to the photographic-film and food industries. Glutein in cereal is a protein which plays an important role in baking and from which glutamic acid has been extracted. Other examples of protein in connection with food processes include those in meat (myosin), egg (ovalbumin), rice (oryzenin), soybean (glycinin), and corn (zein).

Yeast or bacteria grown on ammonium salts and carbohydrates have been used successfully as animal supplements because the cell is rich in proteins, together with vitamins. Current research is focused on replacing carbohydrates with hydrocarbons as the carbon source. Various kinds of microorganisms, such as *Pseudomonas, Micrococcus* and yeast, can utilize normal paraffins, crude oils, gas oil, or natural gas for their growth. The petroprotein so produced from petroleum is an interesting source of protein, as these organisms fed on hydrocarbons, grow very rapidly at a high density compared with higher plants or domestic animals and the yield of proteins is fairly high (see Fig. P-7).

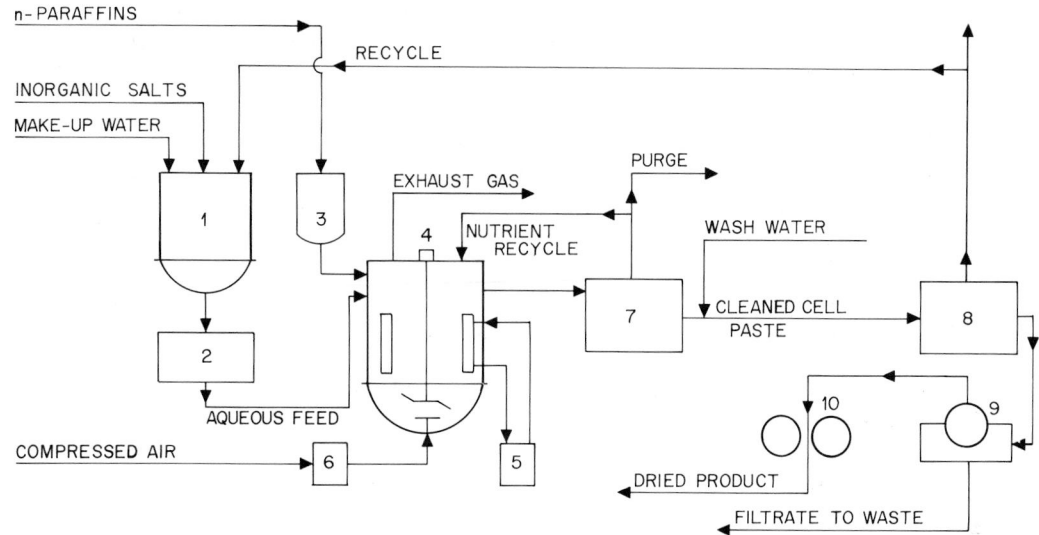

Fig. P-7. Schematic representation of materials flow in a continuous, single-cell protein-production process from petroleum. (1) Medium-preparation vessel, (2) medium sterilizer, (3) feed vessel, (4) fermenter, (5) refrigeration unit, (6) air filter, (7) separator, (8) desludging separator, (9) vacuum filter, (10) drum dryer. The n-paraffins represent 90% or more of the feed. The cleaned cell paste from the separator contains approximately 20% solids. (*Daniel I. C. Wang, Chem. Eng., Aug. 26, 1968.*)

REFERENCES

Anfinsen, C. B., Jr., J. T. Edsall, and F. M. Richard (eds.): "Advances in Protein Chemistry," vol. 25, Academic, New York, 1970.

Florkin, M., and R. H. Stotz (eds.): "Comprehensive Biochemistry," vols. 7 and 8, Elsevier, Amsterdam, 1963.

—Shiro Akabori, *Consultant, Ajinomoto Co., Inc. and President of the Protein Research Foundation, Osaka;* and Kazuo Satake, *Consultant, Teijin Ltd., Tokyo*

Perchloroethylene Perchloroethylene(tetrachloroethylene), $Cl_2C:CCl_2$, is a chlorinated hydrocarbon solvent. It has a high molecular ratio of chlorine to carbon, making it nonflammable and nonexplosive. Perchloroethylene rapidly dissolves vegetable, animal, and mineral oils, greases, tars, and some waxes and resins. Current production in the United States is about 0.5 billion lb annually. The most common use is in dry cleaning, for which it has been used as a solvent since 1934. One of the primary advantages of perchloroethylene as a dry-cleaning solvent is its nonflammability.

Perchloroethylene also is used extensively in industry as a degreasing solvent for liquid and vapor cleaning of metals. The compound is used in metal cleaning in preference to other chlorinated hydrocarbons when its high boiling point is advantageous. This property permits a longer and more thorough rinsing action. Perchloroethylene is especially applicable for removal of high-melting pitches and waxes and for cleaning parts excessively contaminated with oil.

The compound also is used as a chemical intermediate and in the manufacture of blended safety solvents. It is available for use by the pharmaceutical industry for extraction purposes. Perchloroethylene also is used in the manufacture of fluorocarbons.

Production: Perchloroethylene can be produced by reacting chlorine with a hydrocarbon such as methane, ethane, or propylene. HCl is formed from the hydrogen in the hydrocarbon material. The reaction between chlorine and the hydrocarbon gives off a large amount of heat. Some of the heat of the reaction must be removed or absorbed by inert materials, or the temperature in the reaction vessel will become so high that carbon and other undesirable products will be produced. Because of this, carbon tetrachloride vapors are fed into the reactor along with chlorine and hydrocarbon to absorb some of the heat of reaction and control the temperature in the reactor vessel at the desired levels.

Hot gases leaving the reactor are cooled by contacting them with liquid carbon tetrachloride and perchloroethylene in a quench column. The heat removed from the gases as they are cooled produces carbon tetrachloride and perchloroethylene vapors. These vapors, along with the gases and vapors from the reactor, flow out the top of the quench column and through a series of condensers which return the carbon tetrachloride and perchloroethylene to a liquid state. Some of this condensate is then returned to the quench column to cool the reactor gases, and some is returned to be evaporated into the reactor.

The crude stream containing perchloroethylene is taken from the middle of the quench column and fed to distillation columns. Here the perchloroethylene is separated from other chlorinated hydrocarbon products which have been produced in the reaction. After distillation, the perchloroethylene is further purified by removing any traces of chlorine or acid which may remain. After purification, perchloroethylene streams are fed into check tanks. The perchloroethylene is then sampled for a complete laboratory analysis before it is pumped to storage tanks.

Properties: Ordinary steel tanks generally are satisfactory for storing perchloroethylene. Where there is excessive moisture, stainless-steel or lined ordinary steel tanks will minimize rust formation.

Perchloroethylene is moderately toxic by inhalation, by prolonged or repeated contact of the liquid with the skin or mucous membranes, or when taken by mouth. The most important effect of vapor inhalation is narcosis, or anesthesia, which begins when concentrations in excess of 200 ppm are inhaled. Other effects are irritation of the eyes, nose, and skin. Although it is not highly injurious to the liver, liver injury has been reported in man following excessive exposure.

Key physical properties of perchloroethylene are summarized in Table P-7.

—Harold D. DeShon, *The Dow Chemical Company, Midland, Mich.*

TABLE P-7. *Key Physical Properties of Perchloroethylene*

Molecular weight	165.85
Boiling point at 760 mm:	
°C.	121.2
°F.	250.2
Steam distillation point at 1 atm:	
°C.	87.7
°F.	189.9
Solvent-water ratio, by weight	5.32:1
Latent heat of vaporization	
(boiling point):	
cal/g	50.0
Btu/lb	90.0
Specific gravity (20°C/4°C)	1.623
Evaporation rate (ether = 100)	9
Solubility in H_2O at 25°C, g/100 g . .	0.015
Dielectric constant, liquid at 25°C . .	2.365
Flammability	Nonflammable
Underwriters' Laboratory rating . .	0
Threshold limit value (TLV), ppm . .	100

Periodic Law   In 1869, Dmitri Mendeleev developed a matrix which demonstrated that the chemical elements can be placed in groups having similar physical and chemical behavioral patterns, based upon the increasing atomic numbers of the elements. This was a discovery of great significance and one that has contributed much to the advancement of theoretical chemistry. Mendeleev found, however, that his tabular arrangement of the then known elements would hold only if certain positions in his table were left vacant, and he therefore predicted elements yet to be discovered and confirmed, notably gallium, scandium, and germanium. For scandium, termed *ekaboron* by Mendeleev, the properties predicted were strikingly similar to those later determined after the element was discovered by the Swedish chemist Lars Fredrik Nilson, in 1879. Fortunately, Mendeleev lived to witness the fulfillment of his predictions.

Several earlier attempts to classify the elements into some systematic arrangement had been made, including the work of de Chancourtois (1862) and Newlands (1863). Only 1 year after Mendeleev's discovery, Lothar Meyer, a German chemist, independently demonstrated periodicity of atomic volumes, which Meyer defined as the resulting figure when atomic weight is divided by the element's specific gravity in the solid state.

With intervening advancements in theoretical chemistry and physics, aided by improved instruments and analytical procedures, the original Mendeleev periodic chart has undergone numerous modifications and refinements. One version is given in Table P-8. See also **Atomic structure.**

Peroxides, Organic   Organic peroxides contain the ·OO· group with at least one of the peroxy oxygens attached to an organic radical. Several types of organic peroxides are shown in Table P-9.

TABLE P-8. *Periodic Table of Chemical Elements*

GROUP

PERIOD	I A	II A	III B	IV B	V B	VI B	VII B	VIII			I B	II B	III A	IV A	V A	VI A	VII A	O
1	1 H																	2 He
2	3 Li	4 Be											5 B	6 C	7 N	8 O	9 F	10 Ne
3	11 Na	12 Mg											13 Al	14 Si	15 P	16 S	17 Cl	18 Ar
4	19 K	20 Ca	21 Sc	22 Ti	23 V	24 Cr	25 Mn	26 Fe	27 Co	28 Ni	29 Cu	30 Zn	31 Ga	32 Ge	33 As	34 Se	35 Br	36 Kr
5	37 Rb	38 Sr	39 Y	40 Zr	41 Nb	42 Mo	43 Tc	44 Ru	45 Rh	46 Pd	47 Ag	48 Cd	49 In	50 Sn	51 Sb	52 Te	53 I	54 Xe
6	55 Cs	56 Ba	57 La	72 Hf	73 Ta	74 W	75 Re	76 Os	77 Ir	78 Pt	79 Au	80 Hg	81 Tl	82 Pb	83 Bi	84 Po	85 At	86 Rn
7	87 Fr	88 Ra	89 Ac	104														

Metals

Nonmetals

RARE EARTHS

58 Ce	59 Pr	60 Nd	61 Pm	62 Sm	63 Eu	64 Gd	65 Tb	66 Dy	67 Ho	68 Er	69 Tm	70 Yb	71 Lu
90 Th	91 Pa	92 U	93 Np	94 Pu	95 Am	96 Cm	97 Bk	98 Cf	99 Es	100 Fm	101 Md	102 No	103 Lw

TABLE P-9. *Principal Types of Organic Peroxides*

Peroxide Type	Structure
Hydroperoxides:	
Alkyl hydroperoxides	$ROOH$
Organometallic hydroperoxides	$R_mM(OOH)_n$
Peroxides:	
Dialkyl peroxides	$ROOR'$
Alkyl peroxymetalloids	$R'_mM(OOR)_n$
Diorganometallic peroxides	R_nMOOMR_n
Peroxyacids:	
Peroxycarboxylic acids	$R(CO_3H)_n$
Peroxysulfonic acids.	RSO_2OOH
Peroxyesters:	
Alkyl peroxycarboxylates	$R'(CO_3R)_n$
OO-Alkyl *O'*-alkyl' monoperoxycarbonates	$R'OC(O)OOR$
Dialkyl diperoxycarbonates	$C(O)(OOR)_2$
Alkyl peroxycarbamates	$\diagdown$NC(O)OOR
Alkyl peroxysulfonates	$R'SO_2OOR$
OO-Alkyl *O',O'*-dialkyl' monoperoxyphosphates	$(R'O)_2P(O)OOR$
Diacyl peroxides:	
Diacyl (aroyl) peroxides	$RC(O)OOC(O)R'$
Dialkyl peroxydicarbonates	$ROC(O)OOC(O)OR$
OO-Acyl *O'*-alkyl monoperoxycarbonates	$RC(O)OOC(O)OR'$
Acyl alkyl(aryl)sulfonyl peroxides	$R'SO_2OOC(O)R$
Di[alkyl(aryl)sulfonyl] peroxides	RSO_2OOSO_2R

α-Oxy and α-peroxy hydroperoxides and peroxides

$$\text{contain the grouping} \quad \underset{\diagup}{\overset{\diagup}{C}}\begin{matrix}OO{-}\\\\O{-}\end{matrix}$$

Excluding the peroxy acids, commercially available organic peroxides find over 90% of their application in the polymer industry, where they are used as *initiators* for the free-radical polymerizations of vinyl monomers and as *curing* or *crosslinking agents* for resins, elastomers, rubber, and polyolefins. Although used in small amounts, organic peroxides are not true catalysts because they do take part in the reaction and are not recovered.

Organic peroxides decompose (sometimes explosively) upon heating or irradiation to generate two free radicals:

$$ROOR' \xrightarrow[h\nu]{\Delta \text{ or}} RO\cdot + \cdot OR'$$

The structure of R and R' determines the temperature at which decomposition occurs, which can vary from below room temperature to above 150°C. The decomposition rates are expressed in half-lives and largely determine the application area. The more important commercial organic peroxides are listed in Table P-10 in order of increasing thermal stability, i.e., their 10-hr half-life temperatures.

The peroxy acids, often prepared in situ, are primarily used as epoxidizing and bleaching agents.

Many, but not all, organic peroxides are also sensitive to shock and/or friction, which can cause explosions, especially with the pure or concentrated peroxide solutions. Commercial products are formulated to exclude such hazards by incorporating solvents, water, and/or other additives as necessary.

The decomposition of many organic peroxides can be accelerated to much lower temperatures (even below room temperature) in the presence of certain compounds such as transition-metal salts and amines. For example, dibenzoyl peroxide will decompose explosively on contact with dimethylaniline. Such activity is used to advantage in the polymer industry, e.g., polyester thermosets are cured at room temperature using the

TABLE P-10. *Some Commercial Organic Peroxides*

Peroxide	Structure	10-hr Half-Life, °C*	Major Applications†
Acetyl cyclohexanesulfonyl peroxide	cyclo-$C_6H_{11}SO_2OOC(O)CH_3$	31	1a
Diisopropyl peroxydicarbonate	$[(CH_3)_2CHOC(O)O\cdot]_2$	35	1a, b, d, f, 2d
Di(2,4-dichlorobenzoyl) peroxide	$[2,4\text{-}Cl_2C_6H_3C(O)O\cdot]_2$	54	1a, c, e, 2a, c
t-Butyl peroxydivalate	$(CH_3)_3CC(O)OOC(CH_3)_3$	55	1a–e
Didecanoyl peroxide	$[CH_3(CH_2)_8C(O)O\cdot]_2$	61	1a, b, e
Dilauroyl peroxide	$[CH_3(CH_2)_{10}C(O)O\cdot]_2$	62	1a, b, d, e
Dipropionyl peroxide	$[CH_3CH_2C(O)O\cdot]_2$	64	1b
Diacetyl peroxide	$[CH_3C(O)O\cdot]_2$	69	1b, e, 2a
Dibenzoyl peroxide	$[C_6H_5C(O)O\cdot]_2$	72	1a, c, d, e, 2a, c, 4
t-Butyl peroxyisobutyrate	$(CH_3)_2CHC(O)OOC(CH_3)_3$	79	1b
OO-t-Butyl O-isopropyl monoperoxycarbonate	$(CH_3)_3COOC(O)OC_3H_7\text{-}i$	99	1, 2a, 3b
2,5-Di(benzoylperoxy)-2,5-dimethylhexane	$[C_6H_5C(O)OOC(CH_3)_2CH_2\cdot]_2$	100	1c, 2a
t-Butyl peroxyacetate	$CH_3C(O)OOC(CH_3)_3$	102	1b, c
t-Butyl peroxybenzoate	$C_6H_5C(O)OOC(CH_3)_3$	105	1b–e, 2a, c
Dicumyl peroxide	$[C_6H_5C(CH_3)_2O\cdot]_2$	113	1c, e, 2a, 3a–c
2,5-Di(t-butylperoxy)-2,5-dimethylhexane	$[(CH_3)_3COOC(CH_3)_2CH_2\cdot]_2$	119	1b–d, 2a, c, 3a
Di(t-butyl) peroxide	$[(CH_3)_3CO\cdot]_2$	126	1b–f, 2a–c
2,5-Di(t-butylperoxy)-2,5-dimethylhexyne-3	$[(CH_3)_3COOC(CH_3)_2C\equiv]_2$	128	1b–d, 2a, 3a
p-Menthane hydroperoxide		133‡	1, 2a, b
2,5-Dihydroperoxy-2,5-dimethylhexane	$[HOOC(CH_3)_2CH_2\text{—}]_2$	154‡	1, 2a
Cumene hydroperoxide	$C_6H_5C(CH_3)_2OOH$	158‡	1c, e, g, 2a, b
Diisopropylbenzene hydroperoxide		‡	1, 2a, b
t-Butyl hydroperoxide	$(CH_3)_3COOH$	172‡	1c–e, 2a
Ketone peroxides Methyl ethyl ketone peroxides	Mixtures containing various amounts of peroxidic structures, e.g., 	¶	2a

TABLE P-10. *Some Commercial Organic Peroxides*   *(Continued)*

Peroxide	Structure	10-hr Half-Life, °C*	Major Applications†
Cyclohexanone peroxides	$-(CH_2)_5-$	¶	1e, 2a
Peroxyacetic acid	CH_3CO_3H	. . .	4, 5

*Half-lives were determined using 0.2 M benzene solutions.

†(1) Initiator for vinyl monomer polymerizations and copolymerizations: (*a*) vinyl chloride; (*b*) ethylene; (*c*) styrene; (*d*) vinyl acetate; (*e*) acrylics; (*f*) fluoroolefins; and (*g*) butadiene-styrene; (2) curing agent: (*a*) thermoset polyesters; (*b*) styrenated alkyds and oils; (*c*) silicone rubbers; and (*d*) poly(allyl diglycol carbonate); (3) cross-linking agent: (*a*) polyethylene; (*b*) ethylene-propylene rubbers; and (*c*) styrene-butadiene rubbers; (4) bleaching agent; (5) epoxidizing agent.

‡Hydroperoxides are generally used with reducing agents (redox systems) in emulsion polymerization systems.

¶These peroxides are used with activators, and half-life data have little significance.

dimethylaniline–dibenzoyl peroxide or a methyl ethyl ketone peroxide–cobalt naphthenate combination, and hydroperoxide–iron salt combinations are used to prepare styrene-butadiene rubber near 0°C.

Organic peroxides are also used in a variety of other applications. They are generators of free radicals and have been used as free-radical catalysts and/or reactants in many organic syntheses. Di-*t*-butyl peroxide, for example, is an excellent source of methyl free radicals:

$$\left[(CH_3)_3CO\right]_2 \cdots \xrightarrow{\Delta} 2(CH_3)_3CO\cdot \longrightarrow$$

$$\text{acetone} + CH_3\cdot$$

This peroxide is also used as an ignition accelerator for diesel fuels.

Diacyl peroxides and peroxy esters are useful bleaching agents and pharmaceutical additives or intermediates. Dibenzoyl peroxide is still the preferred bleaching agent for flour. In organic synthesis, peroxy esters and peroxydicarbonates are excellent acyloxylating agents in the presence of copper salts.

Cumene hydroperoxide is used in the manufacture of phenol and acetone:

$$C_6H_5C(CH_3)_2OOH \xrightarrow{H^+}$$

$$\left[\begin{array}{c} CH_3 \\ | \\ C_6H_5O\overset{}{C}+ \\ | \\ CH_3 \end{array} + H_2O\right] \rightarrow C_6H_5OH + CH_3\overset{\overset{O}{\|}}{C}CH_3$$

Organic peroxides, especially hydroperoxides, play an important role in biological processes involving enzymes such as peroxidase, catalase, and hydroperoxidase.

Organic peroxides are synthesized from four sources of peroxy oxygen: (1) molecular oxygen (including air), (2) ozone, (3) hydrogen peroxide (including alkali and other metal salts), and (4) other organic peroxides.

Many substances react with oxygen (or air) by a process called *autoxidation* to produce organic peroxides:

$$R\text{—}H + O_2 \rightarrow ROOH \text{ (and sometimes ROOR)}$$

When sulfur dioxide is also present, the process (called *sulfoxidation*) produces peroxysulfonic acids:

$$R\text{—}H + SO_2 + O_2 \rightarrow RSO_2OOH$$

Aldehydes are readily autoxidized to peroxycarboxylic acids:

$$R(O)CH + O_2 \rightarrow RCO_3H$$

Autoxidations are usually initiated by light or some other free-radical source.

Ozone reacts with unsaturated compounds to produce a variety of α-oxy and α-peroxy hydroperoxides and peroxides, depending upon conditions used:

Most organic peroxides are prepared from hydrogen peroxide, which reacts with a variety of substances to form many of the peroxide types shown in Table P-9. Hydroperoxides and dialkyl peroxides are usually prepared from olefins and/or alcohols in the presence of strong acids:

$$ROH + H_2O_2 \xrightarrow{H^+} ROOH \text{ and/or } ROOR$$

Carboxylic acids form peroxycarboxylic acids:

$$RCO_2H + H_2O_2 = RCO_3H + H_2O$$

Anhydrides and acyl halides (including sulfonyl chlorides and chloroformates) give diacyl peroxides:

$$2R(O)CCl + H_2O_2 \xrightarrow[-2HCl]{base} R(O)COOC(O)R$$

Aldehydes and ketones react with hydrogen peroxide to form various mixtures of α-oxy and α-peroxy hydroperoxides and peroxides:

$$\text{C=O} + H_2O_2 \rightleftharpoons \text{HOCOOCOH} +$$

$$\text{HOOCOOCOOH} + \text{HOCOOCOOH} + \cdots$$

Organometallic amines, halides, hydroxides, or oxides react with hydrogen peroxides or hydroperoxides to form organometallic peroxides or hydroperoxides:

$$R_3SiCl + H_2O_2 \xrightarrow[-HCl]{amine}$$

$$R_3SiOOH \text{ or } R_3SiOOSiR_3$$

Alkyl hydroperoxides react with the same reagents that react with hydrogen peroxide to form the analogous alkyl peroxy derivatives. Thus with olefins or alcohols, unsymmetrical dialkylperoxides are formed:

$$R'OH + ROOH \xrightarrow{H^+} ROOR'$$

Anhydrides and acyl halides produce peroxy esters:

$$R'(O)CCl + ROOH \xrightarrow[-HCl]{base} R'(O)COOR$$

Aldehydes usually give α-hydroxyperoxides (hemiperoxyacetals), HOCHOOR, while ketones give *gem*-peroxides (diperoxyketals):

$$\text{C=O} + 2ROOH \xrightarrow{H^+} \text{C(OOR)}_2$$

Unsymmetrical diacyl peroxides are prepared from peroxy acids and acyl halides or anhydrides:

$$R'(O)CCl + RCO_3H \xrightarrow[-HCl]{base} R'(O)COOC(O)R$$

Diacyl peroxides can be hydrolyzed to peroxycarboxylic acids while alkyl peroxycarboxylates are saponified to alkyl hydroperoxides:

$$R'(O)COOR \xrightarrow{saponification} R'CO_2H + ROOH$$

Alkyl peroxymetalloids are easily hydrolyzed to alkyl hydroperoxides, while alcoholysis of alkyl peroxysilanes has given unsymmetrical dialkyl peroxides:

$$(ROO)_4Si + 4R'OH \rightarrow 4ROOR' + Si(OH)_4$$

REFERENCES
Criegee, R.: Herstellung und Umwandlung von Peroxyden, in Houben-Weyl, "Methoden den organischen Chemie," 4th ed., Thieme, Stuttgart, 1952.
Davies, A. G.: "Organic Peroxides," Butterworth, London, 1961.
Edward, J. O.: "Peroxide Reaction Mechanism," Interscience-Wiley, New York, 1962.
Hawkins, E. G. E.: "Organic Peroxides," Van Nostrand Reinhold, New York, 1961.
Lucidol Division, Pennwalt Corporation: "Lucidol Organic Peroxides," Buffalo, N.Y., 1970.
Swern, D. (ed.): "Organic Peroxides," Interscience-Wiley, New York, 1970.

—C. S. Sheppard, *Lucidol Division, Pennwalt Corporation, Buffalo, N.Y.*

Peroxygen Bleaches (See **Detergents.**)

Pesticides (See **Arsenic; Maleic anhydride;** and **Pyridine and derivatives.**)

Petalite (See **Lithium.**)

Petrochemical Complex In a generic sense, *petrochemicals* are those chemicals which are derived in whole or in part from petroleum or natural gas constituents. Certain petrochemicals are separated in their final product form directly from petroleum without undergoing changes in chemical composition. Other petrochemicals may experience several intermediate steps in their synthesis before becoming final useful products. This fact has led to the term *petrochemical intermediate* to designate chemical compositions at stages between one or more raw materials and the final products of commerce.

The term *petrochemical complex* applies to an

assortment of aggregated processing units which are engaged in the conversion of several raw materials derived from petroleum into a variety of petrochemical intermediates or final products. The complex usually is associated with, is near, or in some fashion receives its raw materials largely from a petroleum refinery which produces a diversity of fuels as its primary products. Most of the raw materials used by a petrochemical complex arise as by-products or coproducts from the petroleum refinery. Propylene from catalytic cracking or aromatics from catalytic reforming are examples. Or the complex per se may generate some of its own raw materials, such as ethylene and hydrogen, from natural gas components. On a worldwide basis, 80% of the feedstocks for petrochemicals originates from petroleum refinery gases and liquids, with natural gas as the second most important raw material.

The principal end-use products of the modern petrochemical industry include the following broad classifications: (1) synthetic elastomers, (2) plastics and resins, (3) synthetic fibers and films, (4) detergents, (5) solvents, paint vehicles, and plasticizers, (6) agricultural chemicals, such as fertilizers, pesticides, and herbicides, (7) automotive chemicals, including antifreeze agents and lead alkyls, and (8) pharmaceuticals.

Progress in Petrochemicals: The original, primary sources of raw materials for the organic chemical industry were coal, coal tar, and other substances of animal or vegetable origin, occurring in nature or agriculturally grown. The impetus toward petroleum and natural gas sources began during World War II, when the demand for aromatics rose more rapidly than the ability to supply aromatics as by-products of the coal-coking industry, which was tied to steelmaking. It is significant that the production of former natural sources of materials, including coal mining and natural-rubber plantations, were essentially labor-intensive enterprises, whereas the petroleum refinery, involving continuous and automated operations achieved through high capital investment, is not so sensitive to labor costs. Thus, as labor costs increased in other areas, an economic incentive triggered the transition from conventional raw materials to petroleum sources.

The period following World War II saw intensive development of the petrochemical industry. Petroleum replaced coal as a main source of aromatic chemicals, and the petrochemical industry took over a large part of the production of alcohols formerly secured only from fermentation of carbo-

hydrates. World War II also introduced the era of synthetic polymer substitutes for inorganics, such as metals and glass, and for many other natural substances, including leather, wood, rubber, fibers, glues, waxes, and gums. The large-scale production of these new substances necessitated supplies of raw materials far in excess of those available from refinery off-gases. The additional requirements for olefins (primarily ethylene) began to be supplied by the steam pyrolysis of light hydrocarbons recovered from natural gas. A parallel phenomenon was the rapid rise in demand for ammonia and its derivatives for use in fertilizers on a worldwide scale. The steam reforming of methane to produce synthesis gas for ammonia manufacture was practiced wherever natural gas was available.

During the decades that followed World War II, numerous factors motivated a continuing, rapid expansion of petrochemical activity. Expanded knowledge of surface phenomena led to the discovery and development of a variety of synthetic detergents, adhesives, and water repellants. A better understanding of huge molecules and polymerization led to whole new families of materials, such as plastics, fibers, foams, rubber, and coating materials. Accelerated scientific research, coupled with a growing dependence upon petroleum as a new source of raw materials, had a dramatic effect on the economics of certain traditional industries, including natural-rubber production in the Far East, the effect of synthetic detergents on the market for conventional soaps, the effect of synthetic nitrates on the former Chilean monopoly, and the effects of synthetic fibers on natural materials, such as Australian wool, worldwide cotton production, and Japanese silk.

Of large significance was the synthesis of aromatics by catalytic reforming and the development of processes for the inexpensive extraction from the reformates of the building-block aromatics, benzene, toluene, and the C_8 aromatic family. To adjust the imbalance between benzene and toluene, processes were developed for the dealkylation of toluene (and C_8 aromatics) to produce benzene. Thus, dependence upon coal as a source of raw materials for the synthetic chemical industry no longer prevailed.

Interlocking Production Complex: In order to illustrate the possible arrangement of a number of processing units that might constitute a petrochemical complex, the flow sheet of Fig. P-8 has been devised. It shows a particular combination of processes that receives certain raw materials from a petroleum refinery and converts them mainly into petrochemical intermediates. The flow sheet is not complicated by the inclusion of additional processing units for converting the intermediate building blocks into finished products. Many petrochemical complexes which are associated with refineries carry their processing sequence to about the stage of intermediate manufacture, as shown in Fig. P-8.

Processing arrangements within petrochemical complexes can be quite similar on a worldwide basis, since the variety of intermediate or finished products is rather common wherever petrochemical manufacture occurs on a large scale. Certain raw-material sources can differ from one nation to another, however, as exemplified by the fact that in the United States catalytic-cracking units provide almost all of the propylene and butylene requirements, but because catalytic cracking is employed to a much lesser extent outside the United States, petrochemical manufacturers must provide these olefinic raw materials, along with ethylene, largely by the pyrolysis of naphthas. These pyrolysis units also provide an abundant amount of their light-aromatic raw-material needs.

The present trend is for petrochemical complexes to extend their manufacture farther toward the finished product. Particularly among the larger organizations, the petrochemical complex produces and markets a great number of finished products. There is no typical petrochemical complex inasmuch as each differs from the others both in the variety of raw materials it digests and in the number and kind of products (intermediate or finished) it markets.

The hypothetical complex of Fig. P-8 receives the following materials from a petroleum refinery:

1. Ethane, propane, or a light naphtha for olefin manufacture

2. A C_6–C_8 cut from a catalytic reformer as a source of aromatics

3. A hydrotreated kerosine as a source of n-paraffins for detergent manufacture (see also **Hydrotreating**)

4. Propylene from a catalytic cracking unit (see also **Cracking**)

5. Hydrogen as a by-product from catalytic reforming and other manufactured sources (see also **Reforming**)

In addition, an input of natural gas is shown and designated for the manufacture of ammonia and its derivatives. The flow sheet is devised so that the primary products from the first step in the conversion of raw materials just listed interact with each other to a great extent, a practice that is quite common to those complexes which carry their processing sequence to at least one step beyond the first. At the same time, these complexes (as shown on the flow sheet) market their primary products to other firms for further subsequent syntheses.

A steam-pyrolysis unit is shown for the conversion of ethane, propane, or a light naphtha primarily into ethylene, but with propylene, butylene, butadiene, and an aromatic concentrate shown as products. With ethane as the feedstock, the primary product is ethylene. Propane can be made to yield both ethylene and propylene in varying ratios. Light naphtha yields a spectrum of light olefins and a coproduct which is rich in light aromatics.

The flow sheet shows the ethylene as a feed for ethylbenzene manufacture within the complex and as a net product for sale and subsequent conversion to such important petrochemicals as polyethylene, ethanol, acetic acid, ethylene oxide, and vinyl chloride. In 1970, 65% of the ethylene was produced by cracking ethane or propane; 26% was obtained from refinery off-gases; while 9% was produced by cracking liquid feeds. During 1970, 34% of the ethylene was converted into polyethylene, 19% into ethylene oxide, 11% into ethylene dichloride, and 8% into styrene.

Propylene produced by steam pyrolysis is shown supplanted by propylene from the refinery catalytic cracking unit and (on the flow sheet) is to be used within the complex for the manufacture of cumene as well as being a net product for sale. Of the propylene used in the United States for petrochemicals, 88% originates from refineries; the remaining 12% is produced by steam pyrolysis. Further conversion of propylene (not shown on the flow sheet) results in such important products as polypropylene, acrylonitrile, propylene oxide, glycerin, isopropanol, and acetone.

The C_4 olefins are shown in Fig. P-8 as net products. Butadiene condensed with styrene finds its way largely into SBR rubber, while isobutylene is converted into butyl rubber. Normal butane is used for the manufacture of alcohols, plastics, maleic anhydride resins, and methyl ethyl ketone.

The aromatic concentrate, shown as a product from the pyrolysis unit when light naphtha is the feedstock, contains mostly benzene, toluene, and C_8 aromatics, but associated with it are mono- and diolefins. It is common practice to hydrotreat

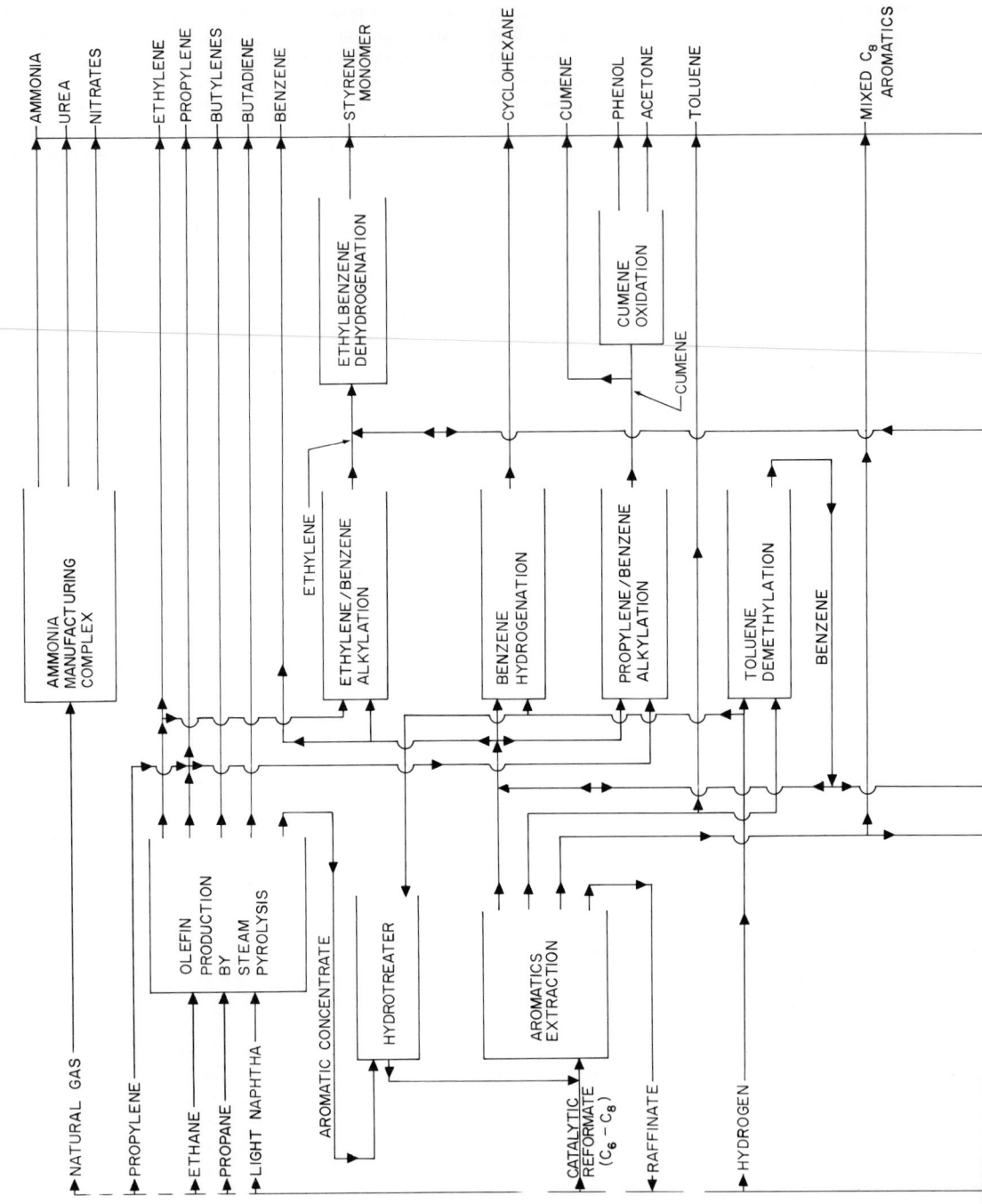

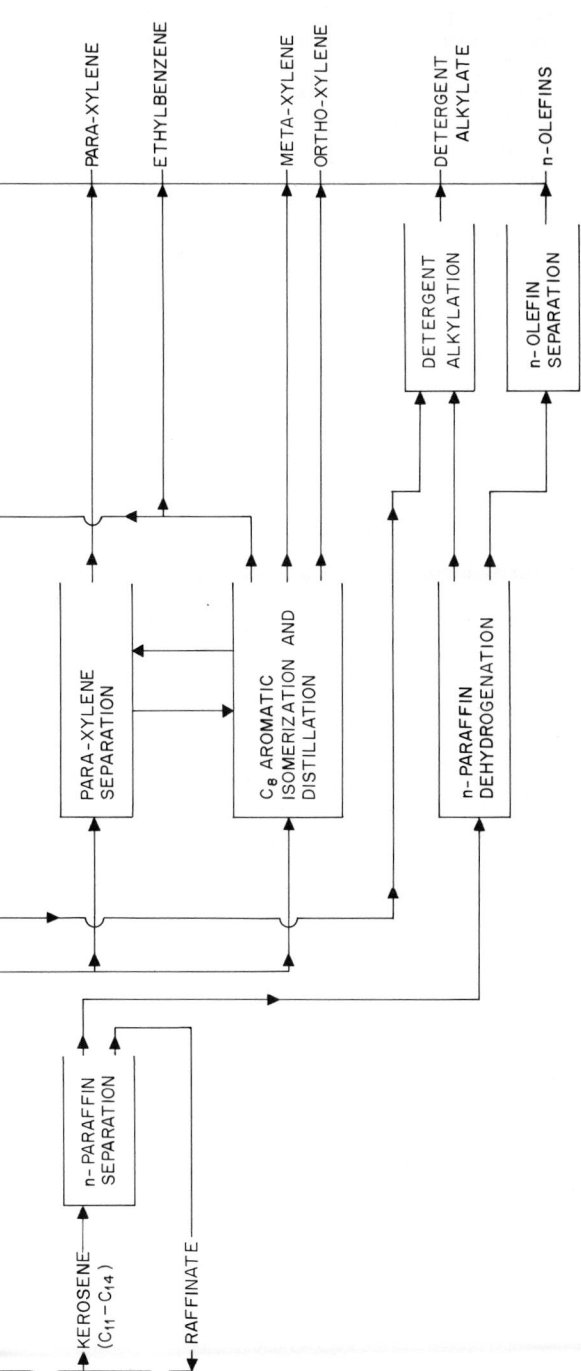

Fig. P-8. Interlocking processes and flow of materials in a petrochemical complex. (*Copyright* © 1973 *Universal Oil Products Company.*)

this stream (shown on the flow sheet) to saturate the olefins and to purify it of sulfur and nitrogen compounds before it is routed to an extraction unit for recovery of the contained aromatics in pure form.

An aromatics extraction unit* is shown yielding pure benzene, toluene, and C_8 aromatics as extracts from a C_6–C_8 catalytic reformate cut and the hydrotreated pyrolysis unit coproduct. The benzene product is used, as shown within the flow-sheet complex, for ethylbenzene manufacture, for hydrogenation to produce cyclohexane, for cumene manufacture, for the production of detergent alkylate, and as a net product. These are all quite common secondary processing steps for benzene conversion to the next intermediate in many petrochemical complexes. Benzene derivatives, of course, are many in number and include rubber products, polystyrene plastics, fibers, detergents, surface coatings, aspirin, weed killers, insecticides, dyes, sulfa drugs, solvents, films, and resins.

The extracted toluene is shown on the flow sheet as routed to demethylation for benzene manufacture within the complex and as a net product for sale. Among the derivatives of toluene in subsequent processing are explosives, solvents, plastics, dyestuffs, pharmaceuticals, surface-coating vehicles, and food preservatives.

The extracted C_8 aromatic family would contain the distribution of isomers in a typical catalytic reformate as shown by Table P-11, arranged in the order of increasing normal boiling points.

The mixture is useful as a solvent and as a high-octane-number blending component for motor fuels. For chemical syntheses, however, it is necessary to separate the isomers in relatively pure form. From their boiling points, it is noted

that it is possible to separate the ethylbenzene or the o-xylene by precision fractional distillation. On a commercial scale, however, these columns are huge, and this technique is not practical in separating the p- and m-xylenes.

Until 1971, p-xylene was separable from the mixture (from a practical standpoint) only by fractional crystallization on the basis of its high melting point relative to its isomers. The first unit of a novel technique† for the continuous adsorptive separation of p-xylene in pure form and with high recoveries from its isomers was put into commercial operation in 1971. Isomerization techniques‡ also are available to bring any C_8 aromatic mixture to an equilibrium composition. The combination of separation and isomerization operations, appropriately integrated, can convert a C_8 aromatic mixture into predominantly any one or any desired distribution of isomers. The flow sheet of Fig P-8 shows all four isomers as net products in pure form, available for further synthesis. Figure P-8 also shows the separated ethylbenzene being routed to a dehydrogenation unit for styrene production or as a net product from ethylbenzene alkylation. The direction of flow will be determined by varying market demands.

The usual next step in the conversion of o-, m-, and p-xylenes is to phthalic anhydride and isophthalic and terephthalic acids respectively. These materials, in turn, can be further converted into alkyd resins and polyesters, useful in surface coatings and fibers as end products.

Figure P-8 also shows a sequence of processing steps to produce a biodegradable detergent alkylate and n-olefins, starting with a hydrotreated kerosine fraction containing, typically, C_{11}–C_{14} hydrocarbons. The primary processing step is the separation of the n-paraffins in pure form from the

---

*Typified by a Sulfolane (proprietary name) or Udex^{SM} unit (both processes licensed by UOP Process Division, Universal Oil Products Company).

†Parex^{SM} process, licensed by UOP Process Division, Universal Oil Products Company.

‡Isomar^{SM} process, licensed by UOP Process Division, Universal Oil Products Company.

TABLE P-11. *Isomers of C_8 Aromatic Family*

Isomer	Percent	Normal Bp		Mp	
		°F	°C	°F	°C
Ethylbenzene	18	277.1	136.1	−138.9	−94.9
p-Xylene	18	281.0	138.3	+55.9	+13.3
m-Xylene	44	282.4	139.1	−54.2	−47.9
o-Xylene.	20	291.9	144.4	−13.3	−25.2

kerosine mixture by molecular-sieve adsorption.* The raffinate from the separation process normally is returned to the refinery, where it finds its way into fuel blends.

The *n*-paraffins are catalytically dehydrogenated as a next step.† Because of chemical equilibrium, the product is a mixture of *n*-olefins and unconverted *n*-paraffins. A biodegradable detergent can be prepared by alkylating the olefins with benzene (as shown) and the *n*-paraffins returned to the dehydrogenation step for further conversion.

Similarly, if *n*-olefins are desired as a net product, they can be separated in pure form by a continuous-adsorptive technique‡ and the *n*-paraffin stream returned to the dehydrogenation unit for further conversion.

In the hypothetical complex of Fig. P-8, benzene is shown as:

1. Alkylated with ethylene to produce ethylbenzene, which subsequently is dehydrogenated to a styrene monomer as a net product. A number of petrochemical complexes include further processing to produce a styrene polymer for sale as a net product, and in some instances molded articles are manufactured as finished products.

2. Hydrogenated to cyclohexane, shown as a net product on the flow sheet but converted in subsequent steps, mainly into nylon.

3. Alkylated with propylene to form cumene, which is oxidized within the complex to phenol and acetone.¶

About 90% of the benzene produced in the United States is converted into styrene, cyclohexane, phenol, and detergent alkylate. Thus, the complex flow sheet in Fig. P-8 represents a processing arrangement that accounts for most of the important outlets for benzene.

Many petrochemical enterprises are involved in the manufacture of ammonia. Natural gas is shown on the flow sheet as a raw-material feed to a steam reforming system that provides the synthesis gas for the production of ammonia. Although not detailed, the ammonia complex is shown to produce urea and nitrates as derivatives, so that a diversified package of agricultural fertilizer components is a typical net product for an enterprise of this nature. See also **Ammonia; Fertilizers;** and **Urea.**

*Molex[SM].
†Pacol[SM].
‡Olex[SM].
¶ Unisir[SM]; all processes (UOP Process Division, Universal Oil Products Company) are typical.

Interdependence of Plants and Material Resources:
It is evident from Fig. P-8 that the several processing units are integrated in such fashion that the relative amounts of products can be varied over wide ranges as desired or as demanded by economic forces—acting in the marketplace, in the area of raw-material supply, or within the complex itself. An ability to shift product distribution is characteristic of petrochemical complexes largely because of the diversity of the processing units and their interlocking nature.

Petrochemical entities of the kind described have the choice of expanding their operations in at least three different directions:

1. Processing units can be added to increase the number of intermediate products for sale; e.g., units might be added to manufacture chlorobenzenes or nitrobenzenes; the butylenes or butadiene might be appropriately polymerized.

2. Intermediates already being made can be further processed in the direction of the finished final product; e.g., the complex depicted on the flow sheet might produce polystyrene from the styrene monomer, convert phenol to phenolic resins, or produce acrylonitrile from ammonia and propylene. Because of wider profit margins which generally result from considerably higher unit prices for the finished product than for the intermediate, there often is an economic incentive in this direction.

3. The raw-material situation may be improved. This usually is accomplished by acquiring a small refinery or by joint ventures between a refiner and a petrochemical manufacturer. Joint ventures seem to be especially attractive because each partner has a special contribution to make to the combined operation in terms of technology, markets, and financial matters.

Still another popular route for expansion is for the petroleum refiner to diversify his operations in the direction of undertaking petrochemical manufacture. Most frequently this is done by performing petrochemical manufacture within the framework of the operation and business structure of the refinery itself. A petrochemical division or subsidiary may be formed as expansion proceeds. This is an easy entry for the refiner because he has captive raw materials and expertise in catalysis and automated continuous-flow processing. Further, the refiner is accustomed to the tradition of high capital investment in equipment per employee, a characteristic of the petrochemical industry. In addition, petroleum refiners are research- and development-oriented with a

TABLE P-12. *Important Petrochemicals†*

Acetic acid	Hydrazine
Acetone*	Hydrogen
Acrylonitrile	Isobutylene*
Adipic acid*	Isophthalic acid
Ammonia	Isoprene
Amylenes*	Isopropyl alcohol
Benzene	Maleic anhydride
Benzoic acid*	Mesitylene*
Bisphenol A*	Methane
Butadiene	Methyl alcohol
Butylene*	Methyl chloride*
Butyraldehydes*	Methyl ethyl ketone (MEK)*
Caprolactam	Methylisobutylcarbinol (MIBC)*
Cumene	Methyl isobutyl ketone (MIBK)*
Cumene hydroperoxide (CHP)*	Monochlorostyrene
Cyclododecatriene (CDT)*	Naphthalene
Cyclohexane*	Natural gas
Cyclooctadiene (COD)*	Perchloroethylene
Cymene*	Petroleum
Diacetone alcohol (DAA)*	Phenol
Dimethylformamide (DMF)*	Phthalic anhydride
Dimethylmetadioxane (DMD)*	Potassium terephthalate (TPAK)*
Dimethylterephthalate (DMA)*	Propane*
Epichlorhydrin	Propylene*
Ethane*	Propylene oxide
Ethanolamines	Styrene
Ethyl alcohol	Synthesis gas
Ethylbenzene*	Terephthalic acid (TPA)*
Ethylene	Toluene
Ethylene glycol	Tolyene diisocyanate (TDI)*
Ethylene oxide	Urea
Formaldehyde	Vinyl chloride
Fumaric acid	Vinyl cyclohexane (VCH)*
Glycerin*	Xylenes*
Glycol ethers	Xylene diamine*
Hexamethylenetetramine*	

†More detailed information pertaining to unstarred materials is listed alphabetically in this volume. If an asterisk follows the name, consult the Subject Index. See also the following alphabetical entries: **Acetals; Acetate fibers; Acrylics; Acrylic fibers; Acrylonitrile-butadiene-styrene resins; Adhesives; Alkyds; Allylics; Amino resins; Cellulose; Cellulose ester plastics; Elastomers; Epoxies; Ethyl cellulose; Fibers; Fluoroplastics; Ionomers; Nylons; Paint; Parylenes; Phenolics; Phenoxy resins; Plastics; Polycarbonates; Polyester fibers; Polyethylenes; Polyimides; Polyphenylene oxides; Polypropylenes; Polystyrenes; Polysulfones; Polyvinyl chloride; Rayon; Resins; Silicones; Urethanes; Vinyl chloride monomer;** and **Vinyl esters.**

background of accelerated discoveries of new and less costly processing methods—desirable attributes for the petrochemical industry.

Important petrochemicals mentioned here and several not included in this description are listed in Table P-12. Some of these materials are described in more detail elsewhere in this volume, as noted. Of course, even this rather long list of petrochemicals is not fully representative of the hundreds of compounds that could be so classi-fied. For excellent insight into the inner workings of the petrochemical industry, see **Ethylene.**

—Melvin J. Sterba, *UOP Process Division,*
Universal Oil Products Company,
Des Plaines, Ill.

Petrolatum, Liquid   (See **Mineral oils.**)

Petroleum   The definitions of petroleum (also variously referred to as *crude oil* or *rock oil*) are numerous, but may be summarized as follows: an *oily, flammable liquid of widely varying viscosity, possessing a unique, characteristic heavy odor, varying in color from yellow to dark reddish-brown or black but usually exhibiting a distinct greenish fluorescence,* believed to have been formed from marine and semimarine near-shore sediments by geological action (precise mechanism unknown) over millions of years.

This extremely important natural raw material is found throughout the world, from deserts to arctic regions to the continental shelves. Source configurations range from surface oozes, to subsurface seeps and tar sands and pits, to rock strata which may be buried from a few hundred feet to several miles deep. Petroleum is produced in commercial quantities on virtually every land mass of significant size on the earth with the present exception of Antarctica and Greenland and their adjacent offshore areas. Crude petroleum is a major source of energy worldwide.

Basic Classes: Petroleums are nonuniform, highly complex mixtures of paraffinic, naphthenic, and aromatic hydrocarbons. Small amounts of sulfur and even smaller amounts of nitrogen and oxygen compounds usually are present. The terms *paraffinic, aromatic,* and *asphaltic-base* (or *naphthenic*) are applied by the petroleum industry to designate the most prevalent class of chemical constituents found in the crude oils from various localities. Similarly, the terms *sour* and *sweet* are used. Sour crudes contain sulfur and have an unpleasant, sometimes sickening, odor of garlic or rotten eggs. The odorous sulfur usually exists in the form of mercaptans or hydrogen sulfide. Sweet crudes contain very little sulfur and have comparatively pleasant odors.

The oils derived from so-called *oil shales* are not truly petroleum, although petroleumlike products can be made from them by specialized chemical processing. Shales are sedimentary rocks which have a relatively high content of a bituminous substance named *kerogen* and 30–60% organic matter and fixed carbon. Kerogen, although not a definite chemical compound, yields an oily substance when heated (retorted) in the absence of air. Extraction of oil shale with ordinary solvents produces no oil, and their solubility in solvents is scanty. This evidence supports the conclusion that the "oil" is the result of a chemical change, the thermal cracking or fragmenting of the molecules that make up kerogen, i.e., pyrolysis.

Representative oil shales yield 20–50 gal per ton of crude oily material when retorted. The resulting oil is of a relatively unsaturated or olefinic character in comparison with a natural petroleum. The oil may be significantly high in sulfur and nitrogen compounds and possess a highly objectionable odor. The oil may be refined by hydrotreating in conjunction with other petroleum refining processes.

Natural gas is not formally defined as a component of crude petroleum, although natural gas commonly exists in the same geological formations, often directly in contact with crude petroleum. Sometimes the pressure of the natural gas helps drive the crude petroleum to the surface. Natural gas also may exist alone as a *dry gas.* Natural gas is a mixture of light paraffin hydrocarbons, beginning with methane (usually the principal constituent) and extending through butane. Varying proportions (up to 15% or more) of nitrogen, carbon dioxide, hydrogen sulfide, and occasionally small amounts of helium may also be in the mixture. Small amounts of heavier hydrocarbons may also be present, especially if the gas is found associated with crude petroleum. See also **Helium;** and **Natural gas.**

Natural Gasoline, or *casinghead gasoline,* is the propane-and-heavier-hydrocarbons content of an as-produced natural gas, particularly when the gas is geologically associated with crude oil. It is extracted from the gas by compression, absorption, or other processing. By convention, all of it must vaporize at a maximum temperature (end point) of 375°F (191°C). There are 24 different commercial grades, defined by Reid vapor pressure and the percent evaporated at 140°F (60°C).

Commercial grades of liquefied petroleum gas (LPG), as well as propane and butanes, result from tailoring the raw natural gasoline to meet the various grade specifications. LPG also results from denuding any natural gas of contained hydrocarbons heavier than ethane to prevent condensation of liquids in natural-gas pipelines under conditions of high pressure and/or cold temperatures.

Originally, it was found that natural gas produced with crude petroleum tended to condense partially when placed under pressure, subjected to low temperature, or both. This occurred at the wellhead if the gas was vented (because expansion cooled the gas) or collected at low places in the piping. The liquid was called *casinghead* or *drip gasoline* and was used without special refining in early motor fuels to secure the desired volatility and other properties. It was too "wild" (actually boiling in the vehicle tank and causing frost to form) to be used directly in automobile engines of that day until it had been first exposed to the atmosphere and the light ends—largely propane and butanes—allowed to evaporate by weathering.

Natural gasolines generally are subjected to various refining procedures (see **Isomerization; and Reforming**) to improve the octane rating before they are used in modern automotive gasolines. Generally the natural-gasoline plant has removed most of the propane and butanes and sells them to the refiner and the chemical plant as separate products of specific characteristics or as LPG for use in areas beyond the gas mains.

Gas condensate also is found in various geological formations throughout the world and particularly in parts of Louisiana and the southwestern United States. These formations are important sources of light petroleum fractions (up to and including kerosine) and natural gas, particularly in Louisiana and Texas. The material may be considered as a natural gas containing 10% or less of an unusual light crude petroleum (approximate gravity 50–60° API). Huntington* describes the substance as one in which *all* the components exist as a single-phase fluid—neither a gas nor a liquid, actually—as the result of the particular conditions of high temperature (often substantially above 150°F; 66°C) and high pressure (1,500 psi and up) which exist in these generally deeply buried formations.

Gas condensate reservoirs cannot be produced economically by merely allowing the pressure to dissipate in pushing the fluid to the surface because the resulting pressure reduction in the reservoir would cause the thermodynamic phenomenon of retrograde condensation, a condition in which most of the heavier hydrocarbons (including some of the butanes) would condense (liquefy) and become so dispersed on the grains of the formation that bringing the material to the surface would not be economically feasible. Instead, these fields are operated through cycling plants by producing the fluid, denuding it of condensables, and recompressing the "dry" methane and ethane for return to the formation to maintain pressure. The heavy hydrocarbons are marketed as LPG, gasoline, kerosine, and allied products. When all of the heavier components have been recovered from the formation fluid, the reservoir is produced in a conventional manner as a natural-gas field.

Composition of Crude Oils: Petroleum oils vary

*R. L. Huntington, "Natural Gas and Natural Gasoline," pp. 42–47, McGraw-Hill, New York, 1950.

considerably in composition, even when closely associated geographically. In some areas of the United States, for example, crude oils near the surface may have quite different chemical compositions from those in deeper strata. Depth alone, however, does not correlate significantly with composition.

Analyses of typical crude oils found in representative areas of the United States are given in Table P-13. It may be generalized that crudes found in the East and Midwest are predominantly sweet and paraffinic; those found along the Gulf Coast usually are naphthenic; those occurring in the inland Southwest are sour and naphthenic; and those found along the West Coast are asphaltic. The waxy, sweet paraffinic oils found in Pennsylvania first became prominent because of the high quality of lubricating oils and greases that could be made from them. The severe stresses imposed by the bearings and close-fitting reciprocating surfaces of modern machinery led to the development of refining processes and the discovery of additive materials whereby many other crude oils also can be transformed into excellent lubricants. Today, even the Pennsylvania oils require special refining and the use of additives in order to meet modern quality specifications.

Analyses of some crude petroleums found outside the United States are given in Table P-14, which illustrates the variety of crudes existent but is not intended to give a typical or representative picture of worldwide petroleum source compositions.

Significance of Composition: Although greatly abridged in analytical details, the tabular information provides a basis for understanding the importance of composition differences, particularly for the types of processing operations used in converting various crudes to useful end products. Interpretation of some of the key variations would include the following:

API Gravity: This parameter (API = American Petroleum Institute), expressed in "degrees," is mathematically related to specific gravity and usually is determined by a hydrometer. The specific gravity of water (arbitrarily defined as unity) is 10.00 when expressed as degrees API. As used in Tables P-13 and P-14, API gravity usually, although not infallibly, indicates the gasoline and kerosine contents of the crude. As an example, the Mississippi, Texas, New Mexico, and Louisiana crudes have API gravities between approximately 35 and 40; and so do the Arabian, Iranian, and Colombian crudes. In checking the distilla-

TABLE P-13. *Analyses of United States Crude Oils*

	McComb, Mississippi		Southwest Texas		East Texas	Wyoming (Sour)	New Mexico	N. Kenai Penin., Alaska	San Ardo, California		Ospelousas, Lousiana		Velma, Oklahoma
Gravity, °API	40.7		36.5		39.1	17.9	37.5	25.9	13.3*		38.2		29.1
Distillation, °F: Type	D 86¶	UOP 76§	D 86	UOP 76	D 86	D 86	D 86	D 86	UOP		D 86	UOP 76	D 86
IBP (initial boiling point)	152	152	158	158	125	306	118	160	IBP 180	0.3†	127	127	145
5% over	192	192	208	208	191	408	162	196	180– 380	1.6†	206	206	224
10	224	224	244	244	211	476	224	220	380– 550	16.1†	256	256	482
30	344	344	461	461	355	633	354	514	550– 650	10.6†	438	438	659
50	504	506	700	770	539	675	530	660	650– 900	17.3†	545	575	712
70	655	703	760+	1075	702	710	626	690	900–1000	6.0†	672	720	748
90	760+	977	...	...	760	732‡	728	710‡	1000+	48.1†	760+	905	748+
EP (end point)	...	1062	...	...	760+	732+	734+	710+	...	...	...	978	...
% recovered	98.5	95.5	98.5	90.5	97.0	89.0	93.0	80.0	...	...	98.5	96.0	95.5
% bottoms	1.6	4.5	1.8	9.5	2.8	11.0	5.6	...	...	...	0.7	4.0	...
% coke, wt.	...	...	...	...	...	...	...	...	...	...	...		...
% recovered At 400°F	37.0	...	32.0	...	34.5	4.5	35.0	...	1.9*		26.5		22.5
525°F	53.0	...	48.0	...	54.0	15.0	55.0	...	17.9*		46.0		33.5
572°F	60.0	...	54.0	...	...	20.0	...	...	...		54.0		39.0
Total sulfur, wt %	0.07		0.45		0.2	3.33	1.0	1.036	1.93*		0.08		1.13
Reid vapor pressure, psi	4.6		3.4		...	0.2	8.8	4.2	...		8.4		4.0
Pour point, °F	60		30		55	–5	25	...	...		40		<–30
Bottom sediment and water, vol %	0.10		0.10		0.1	0.3	0.3	2.5	...		0.15		0.2
Conradson carbon residue, wt %	0.79		1.74		...	0.6	...	...	...		0.52		...
Salt (as NaCl) lb/1,000 bbl.	4		<0.5		31	6.3	14.0	76.0	1.9		5.0		78.0
Gasoline, vol%	35.5		32.0		29.0	9.1	37.8	14.4	16.1		26.1		22.3
Kerosine, vol %	18.1		12.1		10.1	...	...	18.0	10.6		18.9		17.3
Diesel fuel, vol %	14.6		38.0		13.8	14.0	41.2	18.4	23.3		22.9		8.5
Gas oil, vol %	28.1		12.6		(47.1)	30.7	20.8	22.3	48.1		27.9		31.9
Asphalt (bottoms), vol %	3.7		5.3		...	39.9	...	25.7	...		4.1		20.0
Metals (in gas oils), ppm: Nickel	0.06		...		...	...	...	...	0.15		...		...
Vanadium	0.08		...		...	...	...	...	<0.1		...		...

*Calculated.
†% over in indicated boiling range.
‡Cracked at 80% over.
¶Designates data from Method of Test D 86, Committee D-2 on Petroleum Products and Lubricants, ASTM.
§Designates data from UOP Laboratory Test Methods for Petroleum and Its Products, No. 76, Universal Oil Products Company.
Copyright © 1973 Universal Oil Products Company.

TABLE P-14. *Analyses of World Crude Oils*

	Arabian	Minas, Central Sumatra, Topped		Putomayo, Colombia		Gulf Nigeria		Zulia, Venezuela		Iran		Kuwait	
Gravity, °API	30.0	35.3		35.0		34.7		25.2		36.6		31.5	
Distillation, °F:													
Type	D 86	D 86*	UOP 76†	°F	Vol %	°F	Vol %	Hempel‡	Vol %	Hempel‡	Vol %	Hempel‡	Vol %
IBP (initial boiling point)	77	173	594	IBP 400	34.1	IBP 140	6.3	IBP 122	0.1	IBP 122	1.5	IBP 122	2.8
5%	160	216	662	400–500	9.3	140–170	1.8	122–167	1.2	122–167	2.8	122–167	2.5
10	231	246	699	500–650	20.3	170–310	16.8	167–212	2.3	167–212	4.5	167–212	3.2
20	287	295	750	650–750	9.0	310–520	26.5	212–257	3.7	212–257	6.3	212–257	4.2
30	...	341	792	750–900	11.4	520–680	19.3	257–302	3.9	257–302	7.1	257–302	4.3
50	330	427	890	900+	17.2	680+	30.9	302–347	4.1	302–347	5.6	302–347	4.5
70	435	497	1042	...	...	...	...	347–392	3.6	347–392	5.4	347–392	4.0
90	452	575	Cracked	...	...	...	...	392–437	3.4	392–437	5.5	392–437	4.0
95	526	619	...	...	...	...	...	437–482	4.7	437–482	7.4	437–482	4.2
EP (end point)	526	639	1042+	...	...	...	...	482–527	6.0	482–527	2.6	482–527	5.5
								527–583	2.2	527–583	6.6	527–583	2.6
								583–633	5.1	583–633	5.6	583–633	7.0
								633–687	4.9	633–687	5.1	633–687	4.4
								687–738	5.2	687–738	5.9	687–738	4.8
								738–790	7.3	738–790		738–790	5.3
% recovered	...	99.0	72.5	101.3		101.6		57.7		77.3		63.3	
% residue	...	1.0	27.5	...		...		42.1		22.5		34.8	
Total sulfur, wt %	3.05	0.2		0.49		0.16		1.69		1.12		2.62	
Reid vapor pressure, psig	3.8												
Pour point, °F	−33	0		45		20							
Gasoline, vol %	29.1	11¶		34.1		24.9		<5		5		<5	
Kerosine, vol %	16.0	16¶		9.3		26.5		18.9		32.2		25.5	
Gas oils, vol %	12.5	14¶		40.7		19.3		14.1		18.3		13.7	
Residuum, vol %	42.4	59¶		17.2		30.9							
Metals in gas oils, ppm:													
Vanadium	0	...		25§		0.7§							
Nickel	0	...		11		5.1§							
Iron	3			...									
Salt, lb/1,000 bbl	12			Trace		5							

*IBP to 650°F.
†On 650°F + bottoms.
‡Bureau of Mines Hempel, vol % at stated cut points.
§Estimated.
¶In crude oil.
Copyright © 1973 Universal Oil Products Company.

tion figures, it will be noted that the gasoline content [that fraction boiling below about 400°F (204°C)] of these crudes ranges from about 25% to over 35% by volume. The kerosine portions of such "light" crudes also are usually high. In contrast, Wyoming sour crude with an API gravity of 17.9 will be noted to contain but 6% gasoline and about 40% asphalt. California crude has an even greater content of residuum and almost no gasoline.

Sulfur Content: The amount of sulfur in crude is important in terms of handling the crude within the refinery and the undesirable effects of sulfur in finished products. High-sulfur crudes require special materials of construction for refinery equipment because of their corrosiveness. Certain refinery processes require desulfurization of sour charge stocks prior to use as a feedstock, not only because of their corrosiveness but also because of the effect of sulfur-bearing compounds on expensive catalysts. From the standpoint of the consumer, sulfurous gasoline has an unforgettably offensive smell unless specially sweetened (see also **Hydrotreating**) and may corrode the fuel system and engine parts as well as polluting the atmosphere after it has been burned. United States government specifications and those of most states limit the sulfur content of gasoline to 0.15%, which probably will be further reduced in the future. Other petroleum products also are adversely affected when they have a high content of sulfur.

Other factors of importance in the tabular data include:

DISTILLATION RANGE, which indicates what fractions and how much of each fraction is present.

POUR POINT, defined as the lowest temperature at which the material will pour and a function of the oil's composition in terms of waxiness and bitumen content.

SEDIMENT AND WATER, a measure of dirt and other foreign matter as well as water.

SALT CONTENT, not confined to NaCl but usually interpreted in terms of it; undesirable because of the tendency to obstruct fluid flow, to accumulate as an undesirable constituent of residual oils and asphalts, and a tendency of certain salt compounds to decompose when heated and cause corrosion of refining equipment.

METALS CONTENT, the heavy metals vanadium, nickel, and iron tend to accumulate in the heavier gas oil and residuum fractions where the metals may interfere with refining operations, particularly by poisoning catalysts. The heavy metals also contribute to the formation of deposits on heated surfaces in furnaces and boiler fireboxes, leading to premature failure of equipment, interference with heat-transfer efficiency, and increased maintenance.

Diversity of Processing Routes: Fortunately, even though there is a wide variation in crudes, processing operations, including pretreating units, are available to handle a reasonable variety of charge stocks within a given refinery. It is important to stress that a given petroleum refinery may process a number of crudes with varying characteristics over a period of time.

The initial step in refining a crude petroleum or a gas condensate liquid is to separate it into fractions by first distilling at or near atmospheric pressure. That portion boiling above approximately 343°C (650°F) and called *reduced crude* is subjected to vacuum flashing or vacuum distillation if several fractions are desired (as in manufacturing lubricating oils). See also **Petroleum processing.**

Light gases, straight-run gasoline, kerosine, and distillate fuel oils are the primary products of atmospheric distillation, as shown in Table P-15, but generally are unsuited for direct use without further processing. Subsequent processing steps are dictated in part by the amount and quality of the respective fractions, equipment available in the refinery, and the particular markets that the individual refiner supplies.

The products from vacuum distillation can be used directly if only heavy fuel oils and pitch are to be made; otherwise, the distillables are taken as column overhead and side cuts for further processing. The residuum is cut back with lighter oils to form a salable heavy fuel oil (as for ship's bunkers). Alternatively, it may be thermally cracked or hydrocracked.

Refiners vary their processing schemes according to the season of the year. For example, about midsummer most United States refiners start reducing their gasoline production by adjusting cracking operations to produce more fuel oils in anticipation of winter needs. Some proprietary hydrocracking processes can have special capability designed into them to produce all gasoline under certain conditions and to maximize fuel oils under other conditions.

Table P-16 shows the production of crude oil and stocks of crude and refined products during

TABLE P-15. *Derivation and Use of Major Petroleum Products*

Product	Refining Processes Employed	Typical Uses
Light gases	Distillation of crude petroleum	Chemical manufacturing, gasoline manufacturing, fuels (LPG)
Gasoline	Distillation of crude petroleum Cracking of heavy fractions (thermal; catalytic; hydrocracking; coking of pitch, residues, etc.) Reforming (catalytic and thermal) Polymerization of light olefins Isomerization of C_5 and C_6 paraffins Alkylation of olefins with isoparaffins	Automotive and aircraft fuels, solvent, chemical manufacturing, illuminant, cooking fuel
Kerosine	Distillation of crude petroleum Cracking of heavier fractions (thermal, catalytic, etc.)	Jet-aircraft fuel, illuminant, cooking and space-heating fuel, solvent
Gas oils	Distillation of crude petroleum (atmospheric and vacuum) Cracking of heavier fractions (thermal and catalytic, etc.) Hydrocracking of residual oils Vacuum distillation	Domestic and light industrial fuels, diesel fuels, chemical manufacturing, gasoline manufacture, solvents, road oils
Lubricating oils	Distillation of crude petroleum (atmospheric and vacuum) Hydrocracking special residual oils Solvent refining of residual oils	Various lubricating oils, pharmaceutical white oils, sources of waxes and petrolatums, petroleum jelly, asphalt, lubricating greases
Residua	Vacuum distillation of petroleum Distillation of synthetic petroleum made by cracking	Manufacturing of fuel oils and gasoline, chemical manufacturing, source of coke, asphalt
Coke	Thermal cracking of residuums and pitches	Fuel, metallurgy, industrial electrodes

the first quarter of 1971. Other countries produce and store a different spectrum of products suited to their markets.

TABLE P-16. *United States Petroleum Statistics, First Quarter 1971*

	Daily Average, bbl	Total Stocks, bbl
Crude petroleum:		
Produced	9,848,000*	275,094,000†
Refined	10,496,000	
Motor gasoline	5,410,000	225,590,000
Aviation gasoline	43,000	4,972,000
Jet fuel:		
Naphtha	281,000	6,733,000
Kerosine	576,000	22,031,000
Kerosine fuel oil	146,000	20,248,000
Distillate fuel oils	2,465,000	117,163,000
Residual fuel oils	648,000	47,222,000

*Includes 470,000 bbl/day of lease condensate. Worldwide production average, January–March 1971, 48,412,900 bbl/day, of which noncommunist areas produced 40,538,900 bbl/day.
†Includes 21,154,000 bbl of imported petroleum.

Since requirements for petrochemicals and for gasoline, for example, often compete for the same hydrocarbon molecule, it is common for certain process units in some refineries to be run *blocked operation.* For a specified period, the unit is adjusted to produce the maximum of aromatic hydrocarbons, say, the product being stored and withdrawn as needed. The unit then is readjusted to make the desired quality of gasoline-blending component and run for the required time, when the unit is returned to making aromatics. Segregated storage and the unit's charge capacity are coordinated so that alternating service of some refinery units has no undesired effect upon overall refinery operation. See also **Petrochemical complex.**

Chronology of Petroleum Industry Progress: Various problems have arisen to plague the refining industry, and it is interesting to see how some of them were solved (Table P-17).

—David P. Thornton, Jr., *Universal Oil Products Company, Des Plaines, Ill.*

TABLE P-17. *Chronology of Petroleum-Industry Progress*

Problem	Solution
1. Better gasoline quality and greater quantity needed	1. Thermal cracking processes (about 1910)
2. Need to improve odor and stability of gasoline and kerosine	2. Refining with chemical solutions and synthesis and use of oxidation inhibitors; started in late 1920s
3. Better gasoline quality and greater quantity needed	3a. Discovery of tetraethyl lead (1921) b. Polymerization of light olefins to make "poly" gasoline by catalysis (mid-1930s) c. Catalytic cracking invented and improved (late 1930s)
4. Combat grade aviation gasoline testing above 100 octane needed for World War II	4. Alkylation of light olefins with light isoparaffins by catalysis; discovered in 1932; commercialized in early 1940s
5. More aromatic hydrocarbons needed, especially toluene for TNT; benzene, toluene, and other high-octane aromatics needed for combat grade aviation gasoline and for chemical synthesis	5a. Catalytic reforming to make toluene from petroleum naphthas (early 1940s), using non-noble metal catalyst b. Extractive distillation of toluene from reformate with phenol and other materials (early 1940s) c. Extraction with SO_2, suggested in 1907 to purify kerosine, applied to secure aromatics from reformate (early 1940s) d. Alkylation of propylene with benzene using solid H_3PO_4 catalyst to make cumene (early to mid-1940s)
6. Butadiene needed for synthetic rubber in wartime	6. Thermal and catalytic processing applied to petroleum distillates, "quicky" butadiene program (early 1940s)
7. More isobutane for alkylation in wartime aviation-gasoline program	7. Isomerization of *n*-butane (early 1940s)
8. Improve quality of straight-run gasoline	8. Catalytic reforming using noble-metal catalyst (1949)
9. Remove catalyst poisons and sulfur compounds from gasoline and naphtha	9. Catalytic hydrotreating (early 1950s)
10. Increase supplies of pure aromatic hydrocarbons and aromatic concentrates	10. Liquid-liquid solvent extraction processes using aqueous glycols and improved contacting means (1952)
11. Purify kerosines and light and heavy distillates	11. Modified catalytic hydrotreating (mid-1950s)
12. Improve quality of light hydrocarbons used in gasoline	12. New catalytic isomerization processes using noble-metal catalysts, converting C_4, C_5, and C_6 *n*-paraffins to isoparaffins (mid-1950s)
13. Increase production of light fuels and gasoline; reduce production of heavy fuels	13. Development of catalytic hydrocracking processes having great flexibility (1959–1960)
14. Ethylbenzene for styrene manufacture	14. Catalytic alkylation process developed, uniting benzene directly with dilute ethylene in refinery gases (1958)
15. Separation of normal paraffins from mixtures with isoparaffins	15. Molecular sieves used as solid adsorbants (1959, but not commercialized until late 1960s)
16. Increase benzene supply and decrease toluene	16. Hydrodealkylation of toluene; produce naphthalene from alkyl naphthalenes (early 1960s)
17. Synthesize cyclohexane for nylon	17. Catalytic hydrogenation of benzene (early 1960s)
18. Inprove quality of heavy fuel oils	18. Hydrodesulfurization of heavy fuels, also by hydrocracking (mid-1960s)
19. Biodegradable synthetic detergents	19. Development of processes of dehydrogenate *n*-paraffins to *n*-olefins and alkylate benzene with them (mid-1960s)
20. Increase production of *p*-xylene	20. Isomerization of C_8 aromatics to *p*-xylene (late 1950s)
21. Improve supplies of individual pure xylene isomers	21. Adsorptive separation of *p*-xylene in high yield and purity, making possible separation of other isomers by precise fractionation (early 1970s)

Petroleum Processing   Although no two petroleum refineries are alike, a typical integrated refinery that uses currently representative refining processes is shown schematically in Fig. P-9. In the interest of simplicity, the diagram shows only operations concerned with the production of fuels and omits the manufacture of lubricating oils, waxes, solvents, road oils, asphalt, petrochemicals, and other nonfuel products. See also **Lubricating oils; Petrochemical complex;** and **Petroleum.**

The processing units basic for the manufacture of fuel products in the refining industry* include (1) crude distillation, (2) catalytic reforming, (3) catalytic cracking, (4) catalytic hydrocracking, (5) alkylation, (6) thermal cracking, (7) hydrotreating, and (8) gas concentration. For further details on some of these processes, see **Alkylation; Cracking; Fluid catalytic cracking; Hydrocracking; Hydrotreating; Reforming;** and **Thermal cracking.** This immediate description concentrates on the interrelationship and fundamental function of these basic processing units.

Petroleum refineries also use many auxiliary systems, e.g., treating units to purify both liquid and gas streams, waste-management and pollution-control systems, cooling-water systems, units to recover H_2S from gas streams and convert it into elemental sulfur or H_2SO_4, electric-power stations, steam-producing facilities, and provisions for storage of crude oil and products. Many of these topics are covered elsewhere (consult the Subject Index).

Crude-Oil Distillation: To minimize corrosion of refining equipment, a crude-oil distillation unit generally is preceded by a *desalter,* which reduces the inorganic salt content of raw crudes. Salt concentrations vary widely (from nearly zero to several hundred pounds, expressed as NaCl/1,000 bbl). The crude unit functions simply to separate the crude oil physically, by fractional distillation, into components of such boiling range that they can be processed appropriately in subsequent equipment to make specified products.

Although the boiling ranges of these components (or fractions) vary between refineries, a typical crude unit will resolve the crude (as shown by Fig. P-9) into the following fractions:

A. By distillation at atmospheric pressure
   1. A light straight-run fraction, consisting primarily of C_5 and C_6 hydrocarbons but

---

*Based upon United States refinery practice. Outside the United States, the gasoline-creating processes, e.g., catalytic cracking, alkylation, and catalytic hydrocracking in the refineries, are less common.

also containing any C_4 and lighter gaseous hydrocarbons dissolved in the crude
   2. A naphtha fraction having a nominal boiling range of 200–400°F (93–204°C)
   3. A light distillate with boiling range of 400–650°F (204–343°C)
B. By vacuum flashing
   1. Heavy gas oil having a boiling range of 650–1050°F (343–566°C)
   2. A nondistillable residual pitch

In the atmospheric-pressure distillation section of the unit, the crude oil is heated to a temperature at which it is partially vaporized and then introduced near, but at some distance above, the bottom of a distillation column. This cylindrical vessel is equipped with numerous trays through which hydrocarbon vapors can pass in an upward direction. Each tray contains a layer of liquid through which the vapors can bubble, and the liquid can flow continuously by gravity in a downward direction from one tray to the next one below. As the vapors pass upward through the succession of trays, they become lighter (lower in molecular weight and more volatile), and the liquid flowing downward becomes progressively heavier (higher in molecular weight and less volatile). This countercurrent action results in fractional distillation, or separation of hydrocarbons based on their boiling points. A liquid can be withdrawn from any preselected tray as a net product, the lighter liquids, e.g., naphtha, from trays near the top of the column, and the heavier liquids, e.g., diesel oil, from the trays near the bottom. The boiling range of the net product liquid depends on the tray from which it is taken. Vapors containing the C_6 and lighter hydrocarbons are withdrawn from the top of the column as a net product, while a liquid stream boiling higher than about 650°F (343°C) is removed from the bottom of the crude distillation column. See also **Distillation.**

This bottom liquid stream, called the *atmospheric residue,* is further heated and introduced into a vacuum column operated at an absolute pressure close to 50 mm Hg maintained by the use of steam ejectors. In this vacuum column, a flash separation is made to produce the heavy gas oil and the nondistillable pitch products previously described. Although the vacuum column contains certain internal hardware to minimize the entrainment of pitch in the rising vapors and to aid in heat transfer between vapor and liquid, it is more nearly a chamber in which vapor and liquid are separated by a single-stage flash than a fractional-distillation column.

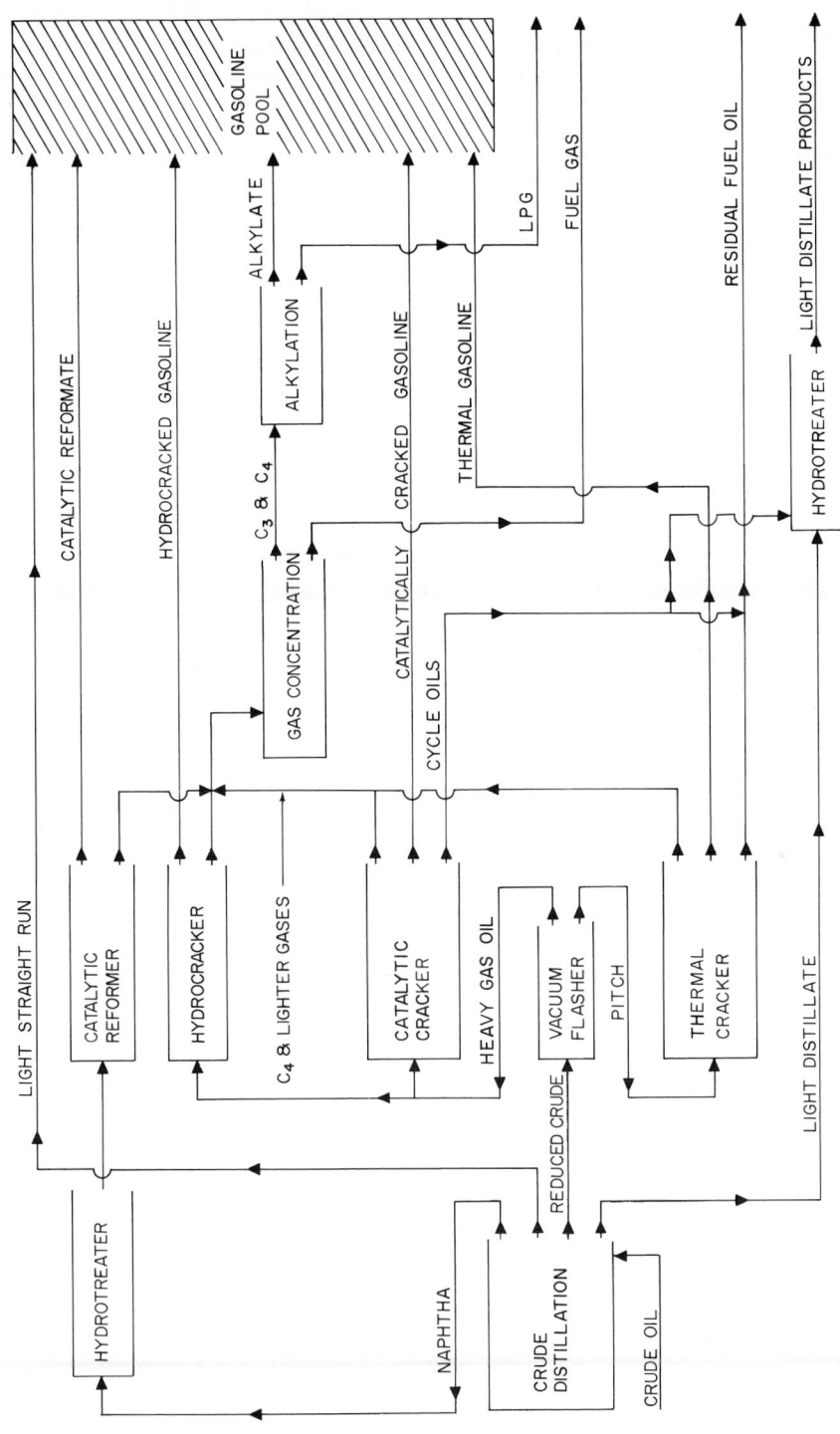

Fig. P-9. Simplified representation of materials flow in an integrated refinery for manufacturing fuels. *(Copyright © 1973 Universal Oil Products Company.)*

The crude oil and atmospheric residue are brought to their desired temperatures in tubular heaters. Oil is pumped through the inside of the tubes contained in a refractory combustion chamber fired with oil or fuel gas in such manner that heat is transferred through the tube wall in part by convection from hot combustion gases and in part by radiation from the incandescent refractory surfaces.

Light Straight-Run Gasoline: This fraction generally contains all hydrocarbons lighter than C_7 in the crude and consists primarily of the native C_5 and C_6 families. This light fraction is stabilized (not shown in Fig. P-9) to remove the C_4 and lighter hydrocarbons which are routed to a central gas-concentration unit for further resolution. The stabilized C_5/C_6 blend usually contains odorous mercaptans, which are treated for odor improvement before delivery to the refinery gasoline pool. See also **Hydrotreating.**

Of the components in modern gasoline pools, the light straight-run fraction has the lowest octane number* (antiknock rating). Its unleaded octane number, in a typical case, will be just under 70, while the unleaded octane number for the entire refinery pool (on a United States average basis) will be about 89. The light straight-run fraction has a good octane-number response to the additions of lead alkyls. Isomerization is also used to improve its octane rating. See also **Isomerization.**

Naphtha
Catalytic Reforming: The chemical composition of the naphtha fraction, and therefore its octane number, varies with the crude source, but in an average case it will be in the range of 40–50 octane. To become a suitable component for blending into finished gasoline pools, its octane number must be raised by changing its chemical composition. Nearly all refineries of the world accomplish this change by catalytic reforming. See also **Reforming.**

Practically all naphtha feedstocks to catalytic-reforming units are hydrotreated first to prolong the processing life of the reforming catalyst. An important by-product of catalytic reforming is hydrogen, which is used in hydrotreating and whatever hydrocracking may be practiced in the refinery. In some cases, supplementary hydrogen is produced by the steam reforming of natural gas or light naphtha cuts.

*In this description, Research Method octane numbers are used.

Heavy Gas Oil
Catalytic Cracking: The primary function of catalytic cracking is to convert into gasoline those fractions having boiling ranges higher than that of gasoline. An important secondary function is to create light olefins, such as propylene and butylenes, to be used as feedstocks for motor-fuel alkylation and petrochemical production. Isobutane, a necessary reactant for the alkylation process, also is an important product of catalytic cracking.

Although the principal feedstock is the gas oil separated from the crude by vacuum distillation, this feed often is supplemented with portions of light distillates and with distillate fractions resulting from thermal coking operations. These options are not shown in Fig. P-9.

For practical reasons, the conversion of distillate feedstocks to lighter materials is not carried to completion. The remaining, uncracked distillates (cycle oils) are used as components for domestic heating fuels (generally after hydrotreating) and to blend with residual fractions to reduce their viscosity to make acceptable heavy fuel oil, as shown in Fig. P-9. In some refineries, cycle oils are hydrocracked to complete their conversion to gasoline.

Unleaded octane numbers of catalytically cracked gasolines fall in the range of 89–93. After treatment for odor control, they are blended directly into the refinery gasoline pool.

Hydrocracking: In a sense, hydrocracking is complementary and supplementary to catalytic cracking, in that hydrocracking processes over a catalyst in a hydrogen environment heavy distillates and, in some instances, cycle oils which are impractical to convert completely in catalytic cracking units. The main product is gasoline or jet fuels and other light distillates.

An important secondary product is isobutane. See also **Hydrocracking.**

Generally, the C_5/C_6 fraction is blended into the gasoline pool. In some instances, the heavier portion of the gasoline is also blended into the gasoline pool; in other cases, this portion may be reformed first to improve its octane number. The flow diagram of Fig. P-9 shows only heavy gas oil as a feedstock and the entire liquid product as gasoline routed directly to the refinery gasoline pool even though the aforementioned options are widely performed in various combinations.

Pitch
Thermal Cracking: The pitch, as produced by most vacuum-flashing units, is too viscous to be

marketed as a heavy fuel oil without further treatment. In some refineries, the pitch is processed further in a thermal cracking unit (*visbreaking*) under relatively mild conditions to reduce its viscosity. In many instances, the thermal cracking does not reduce the viscosity sufficiently, and, as shown in Fig. P-9, additional viscosity reduction is obtained by blending in a required amount of catalytic cycle oil to produce marketable residual fuel oil.

In certain situations, it is more economical to process the pitch in a thermal coking unit, from which the main products are gasoline, distillates, and coke. The gasoline from a coking unit is handled as previously described. The coke is useful, after calcination, for electrode manufacture where it meets certain purity specifications, but the coke is used principally as a metallurgical coke or as fuel. Distillates from thermal coking operations may be used as feedstock for catalytic cracking, or the lighter distillates may be routed to the refinery distillate product pool after hydrotreatment.

A few refiners obtain additional feedstock for catalytic cracking or hydrocracking operations by the solvent extraction of the vacuum pitch, usually with propane as the solvent. The extract is relatively free of organometallic compounds and highly condensed aromatic structured hydrocarbons. Thus, the extract is suitable for handling by catalytic units. Extracted pitch is subsequently processed in thermal units or converted to asphalts.

The small amount of thermal gasoline which is made as a by-product is routed, after treatment, to the gasoline pool or to catalytic reforming through a hydrotreating unit because its octane number is relatively low.

Hydrotreating: As a processing tool, hydrotreating has numerous applications in a refinery, where its principal function is to purify, cleanse, and improve the quality of the feedstock. The process employs hydrogen and a catalyst. See also **Hydrotreating.** The use of hydrotreating for pretreating naphthas prior to catalytic reforming has already been mentioned.

Figure P-9 shows all of the crude light distillate and the net catalytic cycle oil being hydrotreated in a single block before being routed to the refinery light distillate pool. In certain cases, the light distillate in the crude may be sufficiently low in sulfur content to bypass hydrotreating; in other cases, only a portion of the stream is hydrotreated to remove native sulfur compounds. Some refiners hydrotreat portions of their catalytic cracking feeds, particularly if they originate from thermal operations or if they are inordinately high in sulfur content.

Desulfurization is also an objective in the production of low-sulfur residual fuel oils. Reduced crudes which are especially high in sulfur (of the order of 4% or more) can be brought to sulfur levels of the order of 1% by vacuum flashing, hydrodesulfurizing the overhead vacuum-distilled gas oil, and blending the gas oil with a very low sulfur content with the untreated pitch to obtain a reconstituted low-sulfur fuel oil.

Gas Concentration: The gas-concentration system, as shown in Fig. P-9, collects gaseous product streams from various processing units and physically separates the components to provide, in the usual case, a C_3/C_4 stream as a feedstock for alkylation and a C_2 and lighter stream that is almost always used in its entirety to supply process heat requirements within the refinery.

Hydrogen sulfide is removed from gas streams in which it occurs by selective absorption in liquid solutions (usually organic amines). The H_2S released from the rich solution is converted by further processing into elemental sulfur or H_2SO_4.

Alkylation: In motor-fuel refineries, the alkylation units produce a high-quality paraffinic gasoline by the chemical combination of isobutane with propylene and/or butylenes. A small amount of pentenes also are alkylated. The alkylation is accomplished with the catalytic aid of hydrofluoric or sulfuric acid to produce a gasoline having octane numbers in the range of 93–95. See also **Alkylation.**

Propane and *n*-butane associated with the olefins in the feedstocks are withdrawn from alkylation units as by-products. A portion of the *n*-butane is routed to the gasoline pool to adjust the vapor pressure of the gasoline to a level which permits prompt and easy starting of engines. The remainder of the *n*-butane and the propane are available for LPG (liquefied petroleum gas), a clean fuel which is easily distributed even to remote points as bottled gas for heating purposes.

Average Refinery Output: Although the number of products manufactured by the refining industry* is extensive, it is convenient to state the out-

*Table P-18 applies to the United States. On a free-world basis (not including the United States in these figures), overall refinery gasoline yield based on crude is 20–25%, or half that in the United States; while the residual fuel-oil yield is about 35%, or close to 3 times that in the United States.

put on a simplified basis by classifying it into a few major groups according to their general boiling ranges. A simplified statement is given in Table P-18, in which the items correspond with the product classes shown in Fig. P-9.

TABLE P-18. *Refinery Output of Major Products* *

Product	Vol % of Crude Processed
Liquefied petroleum gas (LPG)	3
Gasoline.	48
Distillates	33
Residual fuel oils	13
Petroleum coke.	2†

*United States (1970).
†Weight percent of crude.

Gasoline is the predominant product of refineries in the United States, distillates being the next largest class of fuels produced. In the 33% distillate fuel figure are included such products as kerosine and solvents which are not large in volume. Thus, the dominant products in this class are domestic heating oil, diesel fuel, and jet fuel.

In the residual-fuel-oil classification of Table

TABLE P-19. *Average Gasoline-Pool Composition* *

Component	Vol %
Catalytically cracked†	37
Catalytic reformate	32
Alkylate.	13
Straight-run‡.	13
Thermally cracked	5

*United States (1970).
†Includes 1.5% hydrocracked.
‡Includes 8% butanes for vapor-pressure enhancement.

P-18, are arbitrarily included road oils (0.5%), asphalts (3.5%), and waxes and lubricating oils (2%). Thus, the residual fuel oil itself amounts to about one-half of the 13% figure shown. The actual consumption of residual fuel oil in the United States is more than twice that produced by the domestic refining industry, a deficiency made up by imports, largely from South America.

Refinery Gasoline Pool: The composition of the average total gasoline pool (United States, 1970) with respect to the types of gasoline used to make the blend is approximated in Table P-19. The components routed to the gasoline pool in Fig. P-9 correspond with the items that appear in Table P-19, from which it will be noted that 82% of the United States gasoline is produced by catalytic cracking, catalytic reforming, and alkylation, the three major processes used by the United States refining industry.

The total gasoline pool can be resolved into the 40% premium and 60% regular grades (as marketed in early 1970) having octane numbers and lead contents as shown by Table P-20. At that time, Research octane ratings of the premium and regular grades were nominally 100 and 94.5, respectively, as sold at the filling-station pumps. The corresponding clear Research octane numbers of these two grades were 92.5 and 87, as produced by the refiner's processing units and before the indicated lead additions were made to enhance their octane ratings to marketable levels.

The octane numbers of the premium- and regular-grade gasolines and the proportions of the two grades sold have changed historically to accommodate the octane-number requirements of the mixture which constitutes the car population at a given time. These figures for the last three decades are shown in Table P-21. It may be noted that during the past three decades there has been an appreciable increase of about 20 octane numbers in the antiknock quality of both grades of

TABLE P-20. *Octane Ratings of Premium and Regular Gasoline Grades* *

Grade	Vol % of Pool	Lead Content g/U.S. gal	Octane Number			
			Research Method		Motor Method	
			With Pb	Unleaded	With Pb	Unleaded
Premium	40	2.8	100.0	92.5	92.5	83.5
Regular	60	2.3	94.5	87.0	86.5	77.5
Average for total pool	...	2.5	96.5	89.0	89.0	80.0

*United States (early 1970).

TABLE P-21. *Historical Trend of Premium and Regular Gasoline**

	1940	1950	1960	1970
Premium gasoline, % of total sales.	10	28	31	40
Research octane numbers:				
Premium grade	80	90	99	100
Regular grade	74	83	92	94.5

*United States.

TABLE P-22. *Capacities of Important Refining Processes,** *Percent of Crude Processed*

Process	1930	1940	1950	1960	1970
Thermal cracking	42	50	37	16	13
Catalytic polymerization†	. . .	1.3	1.2	1.4	<1
Catalytic cracking	. . .	3	23	36	37
Alkylation†	. . .	. . .	1.7	4	6
Catalytic reforming	. . .	. . .	1	19	21
Catalytic hydrocracking	. . .	. . .	. . .	. . .	5

*United States.
†Product as percent of crude.

gasoline. Also, there has been a marked trend upward in the proportion of premium-grade gasoline used. To accommodate these changes, there have been distinct alterations in refining techniques.

Capacities of Refinery Processing Units: Capacities of the important refining processes in relation to time are indicated by Table P-22. Before about 1935, all refining was performed by thermal processes. In chronological succession, catalytic cracking began on a commercial scale in the late 1930s and alkylation in the early 1940s. Catalytic reforming of the type employing noble-metal catalysts was introduced in 1949, and catalytic hydrocracking appeared in the early 1960s. The increasing application of catalysis to petroleum refining and the accompanying decrease in thermal processing are quite apparent.

—Melvin J. Sterba, *UOP Process Division, Universal Oil Products Company, Des Plaines, Ill.*

Petzite (See **Tellurium.**)

Pewter (See **Antimony;** and **Tin.**)

pH (Hydrogen-Ion Concentration) All water solutions of acids and bases owe their chemical activity to their relative hydrogen- and hydroxyl-ion concentration. In water, the equilibrium product of the hydrogen- and hydroxyl-ion concentrations is a constant, 10^{-14}, at 22°C. The pH scale is uniquely related to water at this temperature. By definition,

$$pH = \log \frac{1}{\text{hydrogen-ion concentration, moles/l}}$$

When the concentrations of H^+ and OH^- in pure water at 22°C are equal, the H^+-ion concentration must be 10^{-7}; thus, the pH = $\log (1/10^{-7})$ = 7.0. As shown by Fig. P-10, the pH scale covers the range of both acid and alkaline solutions. Pure water is neither acidic nor basic. Acid solutions increase in strength as the pH values fall below 7. The scale is not linear with concentration, a change in 1 unit of pH representing a tenfold change in the effective strength of acid or base; i.e., a solution of pH 3 is 10 times as strong an acid solution as a solution of pH 4.

It is important to recognize that pH measures only the concentration of hydrogen ions actually *dissociated* in solution and not the total acidity or alkalinity, a fact which accounts for the observed pH change in pure water with temperature. As the water temperature increases, the amount of dissociation increases and the quantity of hydrogen and hydroxyl ions is increased equally. Since pH is related to the concentration of H^+ ions alone, the pH actually decreases although the water is still neutral. Therefore, unless the relationship between dissociation constant and temperature is known it is not possible to predict the pH of a solution at a desired temperature from

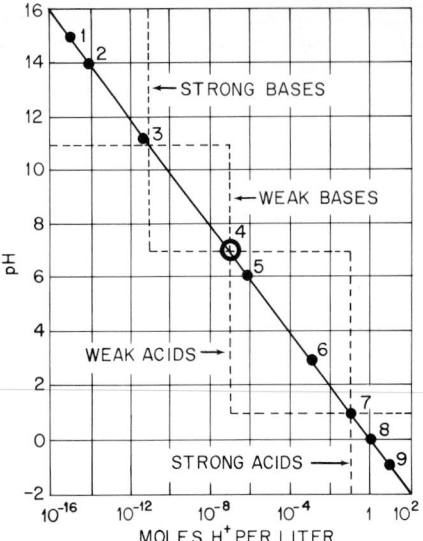

Fig. P-10. Relationship between pH and hydrogen-ion concentration in moles H⁺ per liter. (1) pH of 10.0 M NaOH, (2) 1.0 M NaOH, (3) 0.1 M Na_2CO_3, (4) neutral, the pH of pure water at 22°C (71.6°F), (5) the pH of pure water at 100°C (212°F), (6) 0.01 M acetic acid, (7) 0.1 M HCl, (8) 1.0 M HCl, (9) 10.0 M HCl.

a known pH reading at some other temperature.

pH meters generally cover a range of 0–14 pH units, but it is possible to make measurements beyond these limits in very concentrated solutions. The pH values and characteristics of various materials are given in Table P-23.

Two methods are in general use for the direct measurement of pH in chemical systems, chemical indicators and potentiometric instruments, i.e., pH meters. The color of certain compounds in solution depends upon H⁺-ion concentration. Indicators generally are weak acids and bases and their salts. When the salt of a weak acid is different in color from the nonionized acid, the resultant color of the solution will depend upon the ratio of the concentration of the two forms. The pH range in which the color change takes place with an acid indicator, for example, depends upon the ionization constant of the acid. When the ratio of acid form to salt is 1, then $K = $ H⁺; this is the midpoint of pH range of color change. pH determinations are made with indicators by adding a small quantity of indicator solution to the sample and comparing the color with that of a color standard. When good color standards are

available in steps of 0.2 pH unit and observations are made in a comparator by a skilled observer, a change of 0.1 pH unit can be detected. Instrumented comparators for making these observations eliminate the effects of human error and bias. For rough measurements, the colorimetric method is convenient and inexpensive. Turbid and colored solutions, of course, cannot be observed with accuracy, and many indicators are not stable in strongly oxidizing or reducing solutions.

The basic measuring system for electrometric determinations consists of (1) a pH-responsive electrode, (2) a reference electrode, and (3) a potential-measuring instrument. The hydrogen electrode is a standard for such measurements, but is not used in practical plant systems. Essentially, the hydrogen electrode is employed by bubbling hydrogen gas past a wire or foil which is able to catalyze the reaction $H^+ + e \rightleftarrows \frac{1}{2}H_2(g)$ and thus establish an equilibrium between molecular hydrogen and the hydrogen ions. The metal usually is platinum which has been pretreated to provide a platinum-black catalyzing surface.

It was not until the development of the glass electrode that pH measurement became a simple, reliable tool for all types of uses (see Fig. P-11). The possibility that a thin glass membrane of special composition could develop a potential was described as early as 1909 by the German chemist Fritz Haber, but little progress was made until

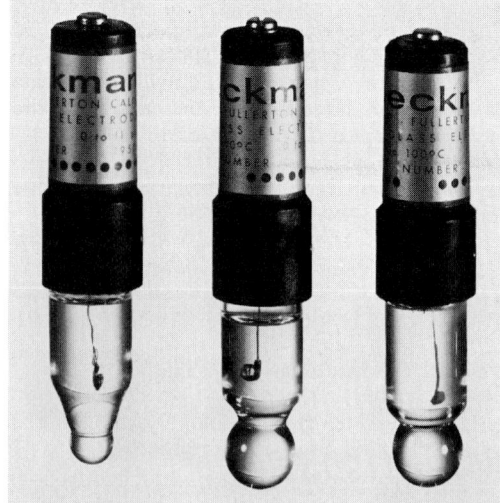

Fig. P-11. Configuration of glass electrode. (*Beckman Instruments, Inc.*)

TABLE P-23. *pH Values and Characteristics of Various Materials*

Compound or Material	Total Normality	Effective H$^+$ (as Normality)	pH (at 25°C)
Acids and Common Acidic Solutions			
Hydrochloric acid	1.0	0.8	0.1
	0.1	0.083	1.08
	0.001	0.001	3.00
Sulfuric acid	1.0	0.48	0.32
	0.1	0.68	1.17
Acetic acid	1.0	0.0043	2.37
Lemon juice	. . .	0.01–0.0063	2.0 –2.2
Acid fruits	. . .	10^{-3}–3 × 10^{-5}	3.0 –4.5
Vegetables, including melons	. . .	10^{-5}–10^{-7}	5.0 –7.0
Jellies, fruit	. . .	10^{-3}–3 × 10^{-4}	3.0 –3.5
Fresh milk	. . .	3 × 10^{-7}–2.2 × 10^{-7}	6.50–6.65

Compound or Material	Total Normality	Effective OH$^-$ (as Normality)	pH (at 25°C)
Bases and Basic Solutions			
Sodium hydroxide	1	0.57	13.73
	0.1	0.071	12.84
Ammonia (10% NH$_3$)	5.9	0.006	11.8
	0.1	0.0018	11.27
Lime water [Ca(OH)$_2$ sat.]	. . .	0.04	12.4
Trisodium phosphate (Na$_3$PO$_4$·10H$_2$O, 2%) . .	. . .	0.009	11.95
Blood plasma, human	. . .	. . .	7.3–7.5

Temperature, °C	pH
Water at Various Temperatures	
0	7.472
22	7.00
25	6.998
50	6.631
100	6.13

the mid-1920s. Developments since that time have accelerated to the point where both laboratory and continuous-process pH measurements are commonplace. The mechanism of the electrode is complex. The glass electrode produces a predictable potential directly related to the H$^+$-ion concentration of the solution in which it is immersed. The electrode responds in a predictable fashion throughout the normally accepted 0–14 pH range, developing 59.2 mV/pH unit at 25°C, consistent with the classical Nernst equation. Unlike earlier types of pH electrodes (quinhydrone and antimony), the glass electrode is not influenced by oxidants or reductants in the solution. Glass electrodes have resistances varying from a few megohms to a thousand or more megohms, but with electronic circuitry this poses no problem.

The second or reference electrode in a pH measurement system simply is required to complete the circuit. A good stable reference electrode (1) must produce a predictable potential compatible with the glass measuring electrode, (2) must be linear with respect to temperature change, and (3) must be simple to apply.

The mercury–mercurous chloride (calomel) electrode once was widely used. Commonly used now is the silver–silver chloride reference electrode.

The benefits of properly applied pH measurements may be summarized as follows: (1) many chemical reactions can be controlled better, resulting in greater processing efficiency, better quality of the ultimate product, and sometimes a safer process, (2) records are provided to assure management that optimum processing conditions have prevailed, (3) often the only practical means for controlling liquid waste products and their effects on treatment systems and streams is by way of pH measurement, and (4) in some cases, continuous pH measurement has permitted processes to be converted from a batch to a continuous basis, e.g., chemical-fertilizer production and several food manufacturing processes.

—Thomas J. Kehoe, *Process Instruments Division,*
Beckman Instruments, Inc., Fullerton, Calif.

Pharmaceuticals   (Consult the **Subject Index.**)

Pharmacokinetics   (See **Antibiotics.**)

Phenol   Phenol, C_6H_5OH, sometimes called carbolic acid or hydroxybenzene and discovered by Runge in 1834, is a colorless, needlelike crystalline solid with a characteristic, rather sweet odor. The compound is quite corrosive (caustic) to the skin and is poisonous. Physical properties include melting point 42–43°C; boiling point 181.4°C; specific gravity 1.071 at 25°C (referred to water at 4°C); slightly soluble in water (8.2 parts in 100 parts at 15°C); completely miscible with alcohol, ether, glycerol, and chloroform. Phenol absorbs water and turns a reddish color upon exposure to air.

Technical grades of phenol contain 82–84% and 90–92% phenol. Liquid phenol is prepared by melting the crystals and adding water. For many years, phenol has been known as an effective antiseptic and disinfectant. About one-half of the phenol produced is used to make phenol-formaldehyde resins. See also **Phenolic resins;** and **Phenoxy resins.** Other important uses are as an intermediate in the production of bisphenol A (for epoxy resins), alkyl phenols, and caprolactam (nylon production). Several of these compounds are described elsewhere in this volume (consult the Subject Index).

In the United States (1968), total phenol pro-

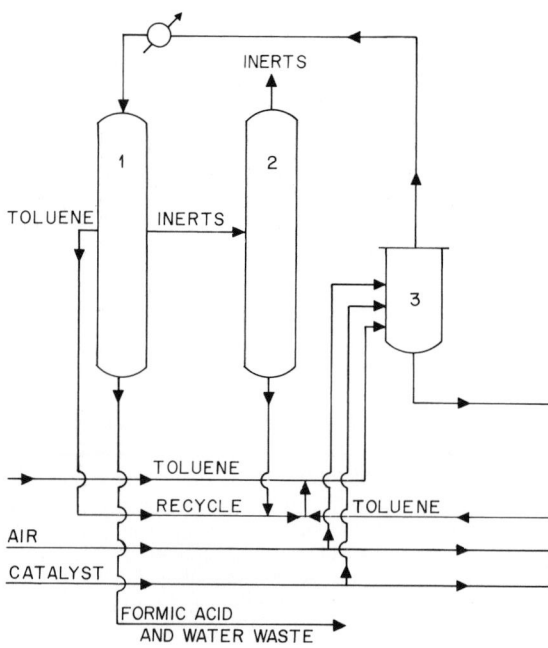

duction was just slightly under 1.5 billion lb. Only 34 million lb was derived from coal tar, which once was the primary source of phenol. The production of phenol from coal tar is described under **Coal tar and derivatives.** Synthetic phenol has been produced in constantly increasing amounts since 1940 and accounted for 1.463 billion lb in the United States in 1968.

Synthetic phenol processes include the following:

1. A two-step process in which toluene is oxidized to benzoic acid, followed by another oxidation step converting the benzoic acid to phenol (see Fig. P-12). Air and cobalt catalyst are charged to a first reactor (120–175°C and 30 psig). The reactor bottoms, comprising a mixture of benzoic acid and toluene, are distilled to produce pure benzoic acid. In the second reactor, the benzoic acid is mixed with steam, air, and a catalyst (copper and magnesium salts) at 220–245°C and atmospheric pressure to form phenol. Vapors from the reactor, comprising benzoic acid, water, and phenol, are transferred to a

distillation column to remove crude phenol, which goes to a second distillation column for final purification. Both the toluene and benzoic acid are recycled.

2. Benzene is converted to monochlorobenzene with HCl and air: $C_6H_6 + HCl + \frac{1}{2}O_2 \rightarrow C_6H_5Cl + H_2O$, followed by hydrolysis of the monochlorobenzene to form phenol: $C_6H_5Cl + H_2O \rightarrow C_6H_5OH + HCl$. The three operations involved are oxychlorination, hydrolysis, and fractionation.

3. Cumene is oxidized to cumene hydroperoxide (CHP), which is followed by decomposition of the CHP into phenol and acetone. The reaction is exothermic and carefully temperature-controlled to minimize production of side products. Phenol represents 69% and acetone 43% by weight of the cumene charged to the process. Utility-grade phenol and acetone (ASTM Spec. D 329) are produced.

4. High-purity phenol is produced from benzene or cyclohexane in a reaction whereby fresh and recycle hydrogen, mixed with benzene feed,

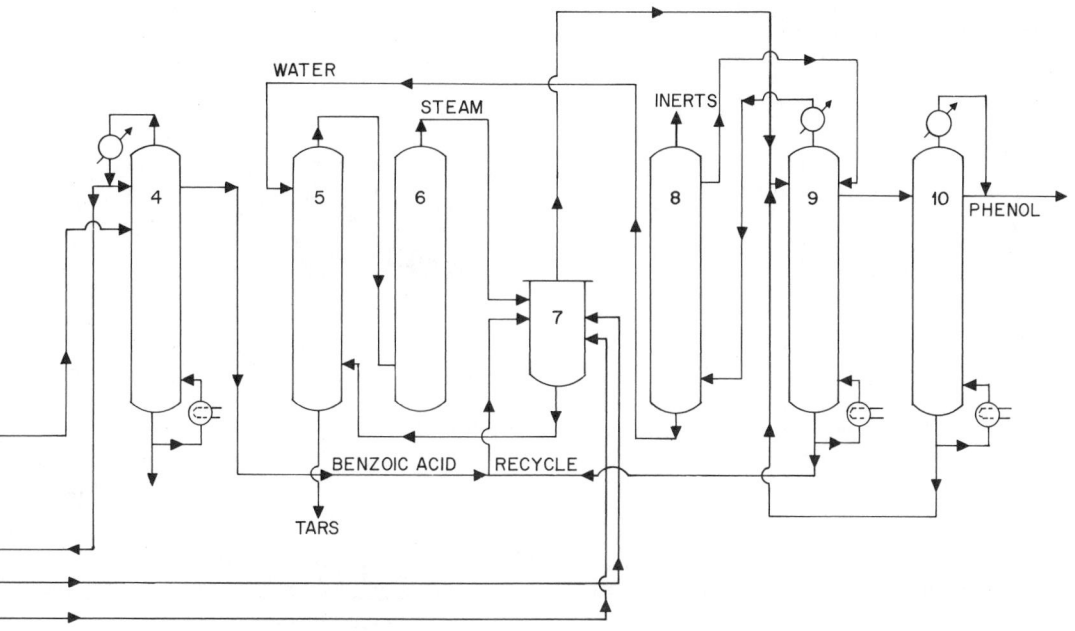

Fig. P-12. Production of phenol involving oxidation of toluene to benzoic acid, followed by another oxidation step to convert benzoic acid to phenol. (1) Decanter, (2) absorber, (3), (7) reactors, (4) still, (5) extractor, (6) evaporator, (8) decanter, (9), (10) stills.

pass to a hydrogenation reactor. The reaction occurs in a single stage.

Derivatives and homologues of phenol include phenyl methyl ether (anisole), $C_6H_5 \cdot O \cdot CH_3$; phenyl ethyl ether (phenetole), $CH_3CO \cdot O \cdot C_6H_5$; the nitrophenols, $C_6H_4(NO_2) \cdot OH$; picric acid (trinitrophenol), $C_6H_2(NO_2)_3 \cdot OH$; catechol, $C_6H_4(OH)_2$; resorcinol, $C_6H_4(OH)_2$; quinol (hydroquinone), $C_6H_4(OH)_2$; pyrogallol (pyrogallic acid), $C_6H_3(OH)_3$; phloroglucinol, $C_6H_3(OH)_3$; thiophenol (phenyl mercaptan), $C_6H_5 \cdot SH$; and diphenyl sulfide, $(C_6H_5)_2S$.

Alkylated phenols, such as *t*-butylphenol, $C_{10}H_{14}O$; *t*-amylphenol, $C_{11}H_{10}O$; and nonylphenol (a mixture of monoalkyl phenols with side chains of random-bracketed alkyl radicals in the molecule) are used in the production of oil-soluble phenol-formaldehyde resins. Bisphenol A, $(CH_3)_2C(C_6H_4OH)_2$, is a starting ingredient of several synthetics. Phenylphenol is used in cosmetics to prevent bacterial deterioration and to control odors.

See also **Petrochemical complex.**

Phenolic Antioxidants (See **Antioxidants.**)

Phenolic Carboxylic Acids (See **Carboxylic acids.**)

Phenolic Resins Resins that are prepared from reacting phenol with formaldehyde often are simply termed *phenolics*. Resins of this type are among the oldest of the synthetic plastic materials, dating back to 1909, when Leo Hendrik Baekeland developed bakelite, which found rapid and wide acceptance by a growing electrical industry as a substitute for shellac and other natural resins. Phenolics have been high-tonnage plastic materials for many years, and although scores of newer synthetics have competed with them for many applications, thanks to improved molding compounds and processing equipment, the phenolics have remained a major synthetic thermosetting resinous material. See also **Plastics; and Resins.**

The process for making phenol-formaldehyde resins is relatively simple and involves mixing a 37–40% formaldehyde, HCHO, solution with phenol, C_6H_5OH, in the presence of ammonia as a catalyst and heating to approximately 80–90°C (175°–195°F), when a resinous mass is formed. The water is boiled off, and the resinous product is allowed to cool in thin layers. The reaction can be arrested by dilution with a solvent, e.g., acetone

or alcohol. In another process, the formaldehyde and phenol are mixed (without ammonia) and heated slowly in a digester to a temperature ranging from 140 to 160°C (284–320°F) over a period up to 3 hr. Water is bled off as steam. The thin resinous material then is allowed to cool, after which it may be mixed with various modifying agents and fillers to provide a wide range of final properties after curing under pressure and heat.

Coatings and adhesives consume about 75% of phenolic resins, the other 25% going into molded products. Phenolic adhesives and laminates are widely used in construction products. Molded products have excellent dielectric strength, impact strength, and dimensional stability as well as good resistance to heat, acids, alkalis, and water. See also **Paints.**

Phenolic molding compounds are divided into six major categories: (1) general-purpose compounds, usually two-stage resin types in which the materials are preheated prior to molding and curing; molding temperatures range from 150 to 200°C (300–390°F) and molding pressures from 2,000 to 15,000 psi, the higher pressures being required for transfer molding; (2) nonbleeding compounds (two-stage) widely used for bottle-cap closures; (3) heat-resistant compounds (two-stage) containing a mineral filler and applicable to such products as appliance handles and bases and heater plugs, where service up to 260°C (500°F) may be required; (4) impact compounds, sometimes glass-filled, for use where strength and heavy wear are requirements e.g., electrical hand tools, gun handles, instrument housings, and heavy-duty electrical components; (5) electrical compounds (usually two-stage) where low electrical-loss characteristics are needed, e.g., automotive ignition parts, aircraft electrical parts, and communications gear; (6) special-purpose compounds, particularly where contact with water and detergents requires exceptional corrosion resistance, e.g., soap dispensers, sink sprays, shower heads, and chemical pump housings and impellers. Machining of phenolic parts is not recommended for shapes that cannot be molded. Wood flour, as a filler, provides the best machinability properties. Fabric and cotton fillers are more prone to chipping, and mineral fillers are unsatisfactory where machining is required.

Representative uses of phenolics include (1) as reinforcement ingredients for acrylonitrile-butadiene and styrene-butadiene synthetic rubber, natural rubber, and neoprene, where phenolics

TABLE P-24. *Properties of Phenolic Molding Compounds*

Type	Color	Density, oz/in.³	Heat-Deflection Temperature		Impact Strength, ft-lb/in. of notch	Flexural Strength, psi	Tensile Strength, psi	Compressive Strength, psi	Hardness, Rockwell M	Dielectric Constant
			°F	°C						
General-purpose	Brown or black	0.780–0.815	330–340	166–171	0.25–0.31	9,000–11,000	6,500–7,000	25,000–32,000	70–95	5.0–6.2
Nonbleeding	Brown or black	0.798	370	188	0.28–0.30	10,000–10,500	7,000–7,500	30,000	77–82	6.5
Heat-resistant	Black	0.896–0.954	300–360	149–182	0.26–0.28	8,500–9,500	5,500–6,500	25,000–28,000	112	7.0–12.8
Impact	Black	0.792–1.069	300–600	149–316	0.60–15.0	10,000–20,000	6,000–7,000	16,500–30,000	87–113	5.6–6.8
Electrical	Black	0.792–0.971	275–300	135–149	0.26–0.36	9,500–10,000	5,000–6,500	24,000–30,000	105–115	5.2–10.0
Special-purpose.	Black	0.786–0.809	300–320	149–160	0.20–0.48	8,000–10,000	4,500–7,000	26,000–30,000	107–114	7.4–11.0

increase stiffness, hardness, and resistance to heat abrasion and chemicals; (2) in nitrile rubber cements for better adhesion and heat resistance; (3) in molded shapes, along with wood particles, e.g., croquet balls, shuffleboard disks, and toilet seats; (4) in brake linings, clutch facings, and other friction parts as a binder for asbestos and other dry ingredients, greatly improving mechanical strength; (5) in insulating materials, e.g., mineral-wool batts and glass-fiber insulation, where moisture and heat resistance are paramount; (6) in treating and impregnating paper sheets, tubes, and other forms, e.g., battery separators, oil and air filters, and counter tops.

The properties of representative phenolic molding compounds are summarized in Table P-24.

Phenoxy Resins   These high-molecular-weight thermoplastics are derived from bisphenol A and epichlorohydrin. Although chemically similar to epoxy resins, the phenoxies contain no epoxy groups, their molecular weight is much higher, and they are essentially linear polymers. The principal advantages of the phenoxies are an excellent dimensional stability and creep resistance, low mold shrinkage (on the order of 0.003 to 0.004 in./in.), and very good processability. Maximum continuous use, however, should not exceed approximately 77°C (170°F). They have excellent colorability, both with pigments and dyes. Phenoxies generally are resistant to chemicals except for aromatic and polar solvents and esters. They resist staining by most household products except alcohol and materials containing phenol. Phenoxy resins are injection-molded at temperatures between 193 and 271°C (380–520°F). Extrusion temperatures run from 204 to 218°C (400–425°F) for sheets and somewhat higher for thicker sections. Phenoxies can be joined by heat-sealing or bonded by adhesive-grade phenoxy dissolved in an appropriate solvent. They also can be bonded (heat) to metal parts. The properties of phenoxy resins are summarized in Table P-25.

Phenyl   (See **Radicals.**)

Phenylacetic Acid   (See **Carboxylic acids.**)

Phenylalanine   (See **Amino acids;** and **Peptides and proteins.**)

Phenylhydrazine   (See **Aldehydes;** and **Ketones.**)

Phosgene   (See **Chlorine organics;** and **Polycarbonates.**)

Phosphates   (See **Detergents; Fertilizers;** and **Phosphoric acid.**)

Phosphating   (See **Conversion coatings.**)

Phosphines   (See **Amines.**)

Phosphine Oxides   (See **Surfactants.**)

Phosphors   (See **Dysprosium; Erbium; Europium; Gadolinium; Holmium; Lanthanum; Neodymium; Promethium; Terbium; Yttrium;** and **Zinc.**)

Phosphoric Acid   The name phosphoric acid commonly refers to orthophosphoric acid, H_3PO_4. Anhydrous orthophosphoric acid is a white, crystalline solid, which melts at 42.35°C. It forms a hemihydrate, $2H_3PO_4 \cdot H_2O$, which melts at 29.32°C. Although it is possible to produce almost any desired concentration, it is common practice to supply the material as a solution containing from 75% H_3PO_4 (melting point = 17.5°C) to 85% H_3PO_4 (melting point = 21.1°C).

When phosphoric acid is heated to temperatures above about 200°C, water of constitution is lost. A series of acids is formed by the dehydration, ranging from pyrophosphoric acid, $H_4P_2O_7$, to metaphosphoric acid, $(HPO_3)_n$. Salts of the dehydrated acids are used for the preparation of certain types of liquid fertilizers and are present in many detergents in percentages up to 40%. The dehydrated acids can form water-soluble complexes with many metals, such as calcium.

One, two, or three of the hydrogens in phosphoric acid may be neutralized. When one hydrogen is replaced with sodium, the product is slightly acid, while replacement of all three hy-

TABLE P-25. *Properties of Phenoxy Resins.*

Impact strength, ft-lb/in. of notch	2.5
Tensile strength, psi	9,500
Flexural strength, psi	14,000
Dielectric constant	4.1
Specific gravity	1.182
Heat-distortion temperature at 264 psi:	
°C	86
°F	188
Hardness, Rockwell R	123
Water absorption (24-hr immersion), %	0.13

drogens produces a highly alkaline product. The three compounds are:

pH of 1% Solution

Monosodium phosphate, NaH_2PO_4 . . . . . .4.0
Disodium phosphate, Na_2HPO_4 . . . . .8.3*
Trisodium phosphate, Na_3PO_4 . . . . 12.0

*Not fully certain.

Other phosphorous acids of little commercial significance are hypophosphorous acid, H_3PO_2; orthophosphorous acid, H_3PO_3; and pyrophosphorous acid, $H_4P_2O_5$.

Since phosphorus is essential for plant and animal life, large quantities are used, either directly or indirectly, for food production. Bones contain high percentages of tricalcium phosphate.

Uses: Total production of H_3PO_4 in the United States amounts to about 5 million tons of contained P_2O_5, while world production is estimated to be close to 10 million tons. Large quantities of phosphatic fertilizers are manufactured from phosphate rock by direct means, e.g., normal superphosphate, thus avoiding the production of H_3PO_4 as a separate material. See also **Fertilizers.**

About 80% of United States production is wet-process acid, most of which is subsequently converted into calcium phosphates or ammonium phosphates for fertilizer. Another major use is the production of sodium and potassium phosphates, either ortho or dehydrated, for use in detergents and cleaning and boiler compounds. Phosphoric acid is used as such for metal-surface treatment and in soft drinks. Other uses include the preparation of calcium phosphates for food supplements, phosphates for use in baking, and ammonium phosphates for fireproofing. Phosphates are also included in other materials, e.g., insecticides, lubricating-oil additives, toothpaste and other polishing compounds, and certain types of glass.

Raw Materials and Manufacture: The major sources of H_3PO_4 are mineral deposits of phosphate rock. Mining operations are extensive in the United States, the Mediterranean area, and Russia, with smaller operations in many other countries. Known deposits can supply world requirements for many centuries.

The major constituent of most phosphate rocks is fluorapatite, $3Ca_3(PO_4)_2 \cdot CaF_2$. Commercial rocks contain 30–38% P_2O_5 plus a variety of impurities. Of these, iron, aluminum, magnesium, silica, carbon dioxide, sodium, potassium, and sulfates may be present in appreciable quantities.

Two major methods are utilized for the production of phosphoric acid from phosphate rock. The *wet process* involves the reaction of phosphate rock with sulfuric acid to produce phosphoric acid and insoluble calcium sulfates. Many of the impurities present in the phosphate rock are also solubilized and retained in the acid so produced. While they are of no serious disadvantage when the acid is to be used for fertilizer manufacture, their presence makes the product unsuitable for the manufacture of phosphatic chemicals.

In the other method, the *furnace process,* phosphate rock is combined with coke and silica and reduced at high temperature in an electric furnace, followed by condensation of elemental phosphorus. Phosphoric acid is produced by burning the elemental phosphorus with air and absorbing the P_2O_5 in water. The acid produced by this method is of high purity and suitable for all uses with little or no treatment.

Wet Process: There are a number of proprietary processes for the manufacture of phosphoric acid. Figure P-13 is a flow sheet for a modern plant, which can be designed for capacities of up to 1,000 tons of P_2O_5 (1,400 tons of H_3PO_4) per day in a single unit. The basic reaction in this process is

$$3Ca_3(PO_4)_2 \cdot CaF + 10H_2SO_4 + 20H_2O \rightarrow$$
$$10CaSO_4 \cdot 2H_2O + 6H_3PO_4 + 2HF \quad (1)$$

Numerous side reactions also occur.

Phosphate rock and sulfuric acid, together with recycled weak liquors, are carefully metered to a large, stirred reactor, providing 4–8 hr retention. Conditions in the reactor are carefully controlled to maintain preselected conditions. Typical control levels are as follows:

Temperature, °C 77–83
Acid concentration, % . . . . 29–32 P_2O_5
    (40–44% H_3PO_4)
Solids, % 34–40
Dissolved sulfates, % . . . . . 1.5–3 (H_2SO_4)

Temperatures are controlled by removing the excess heat of reaction with a vacuum cooler or by blowing air through the phosphoric acid slurry.

The slurry, which contains precipitated gypsum, is sent to a filter, commonly a large, horizontal rotary-pan-type unit. The first filtrate is sent to further treatment or use. The gypsum is washed with water in several countercurrent steps, and weak liquor is returned to the reaction stage.

For most uses, the acid requires further concentration, normally done in vacuum evaporators,

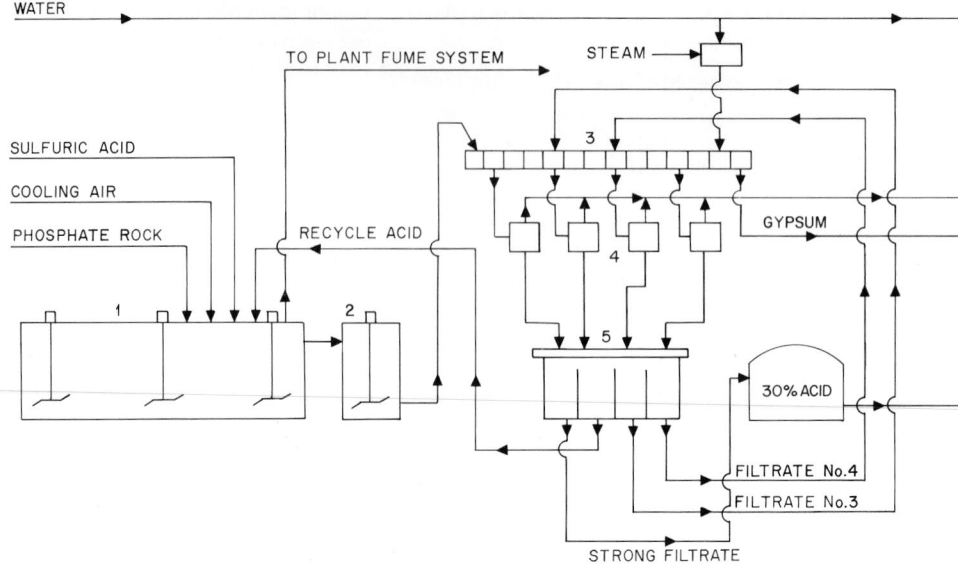

Fig. P-13. Phosphoric acid production and concentration by the wet process. (1) Digester, (2) filter feed tank, (3) filter, (4) filtrate receivers, (5) filtrate tank, (6) condenser, (7) concentrator.

operating at 2–5 in. Hg abs. Merchant-grade acid for shipping is generally concentrated to about 54% P_2O_5 (75% H_3PO_4). See also **Evaporation.**

The highly corrosive nature of the ingredients requires corrosion-resistant materials. Rubber-lined steel is frequently used for vessels and large piping. Reactors are sometimes constructed of concrete, with a rubber or mastic corrosion barrier. Carbon brick is often used to protect linings. Pumps, agitators, and other metal parts are stainless steel. Various types of plastic may be used for piping, small vessels, and other specialized equipment.

Effluents and gypsum disposal pose problems. Fluorine is evolved at various steps in the process, and scrubbers are required to reduce release to the atmosphere to acceptable levels. In the United States this is usually about 20 lb/day for large plants. Gypsum is frequently piled in diked areas or dumped into abandoned mines. In a few cases, it may be disposed of in rivers or in the ocean.

Waste water from the plant is heavily contaminated with fluorine, phosphates, sulfates, and other compounds. It is commonly impounded in large ponds, where a portion of the contaminants

may precipitate or be lost by other processes. The cooled effluent from the ponds is recycled to the production unit. Any excess water must be treated with lime before it can be allowed to enter streams.

Recent developments include plants designed to precipitate the calcium sulfate in the form of the hemihydrate instead of gypsum. In special cases, hydrochloric acid is used instead of sulfuric for rock digestion, the phosphoric acid being recovered in quite pure form by solvent extraction. Solvent-extraction methods have also been developed for the purification of merchant-grade acid, which normally contains impurities amounting to 12–18% of the phosphoric acid content. Processes for recovering part of the fluorine in the phosphate rock are in commercial use.

Material and other requirements for the production of 1 ton of P_2O_5 in the form of acid containing 75% H_3PO_4 are shown in Table P-26.

Electric-Furnace Process: This process is more expensive than the wet process but has the advantage of producing acid of high purity. A single electric furnace may have a capacity of up to 300 tons/day of P_2O_5, but many installations consist of several furnaces.

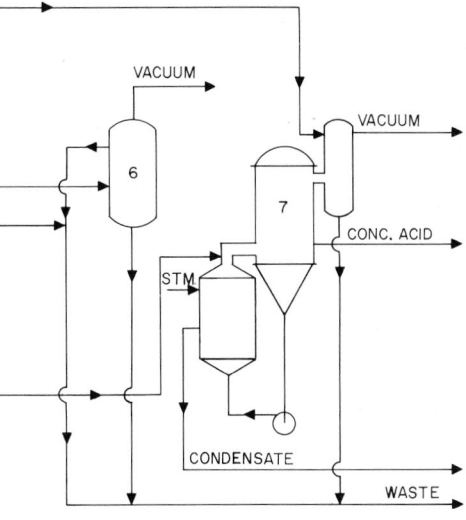

TABLE P-26. *Materials and Energy Requirements for Production of 1 Ton of P$_2$O$_5$ by Wet Process*

Phosphate rock, tons	3.0–3.5
Sulfuric acid, tons	2.6–3.0
Electric power, kWh.	100–140
Water, gal	20,000–30,000
Steam, lb	4,500–5,000
Waste gypsum, CaSO$_4$·2H$_2$O, tons . . .	4.6–5.2

Figure P-14 is a simplified flow sheet of an electric-furnace plant. A mixture of coke, silica, and phosphate rock is formed into nodules by heating in a nodulizing kiln, and the resulting lump material is transferred to the electric furnace, where it is heated with an electric current introduced by means of graphite electrodes. The entire charge is melted, and elemental phosphorus is volatilized. The slag is tapped off intermittently, while the phosphorus vapor is condensed.

The liquid phosphorus is then sent to the acid plant and burned with air to form P$_2$O$_5$. The acid anhydride is absorbed in water to form phosphoric acid.

The reactions involved are basically

$$2Ca_3(PO_4)_2 + 6SiO_2 + 10C \rightarrow$$
$$P_4 + 10CO + 6CaSiO_3 \quad (2)$$
$$P_4 + 5O_2 \rightarrow 2P_2O_5 \quad (3)$$
$$P_2O_5 + 3H_2PO \rightarrow 2H_3PO_4 \quad (4)$$

Typical requirements for the production of 1 ton of P$_2$O$_5$ in the form of phosphoric acid are shown in Table P-27. The by-product slag is usable for light-weight building aggregate, road construction, and other minor uses.

Storage and Handling: H$_3$PO$_4$ is a strong acid, and contact with body or tissues may result in

TABLE P-27. *Materials and Energy Requirements for Production of 1 Ton of P$_2$O$_5$ by Electric-Furnace Process*

Phosphate rock, tons	3.6–4.2
Coke, tons	0.6–0.7
Silica, tons	0.7–1.3
Carbon electrodes, lb	20–30
Water, gal	8,500–20,000
Electric power, kWh.	5,500

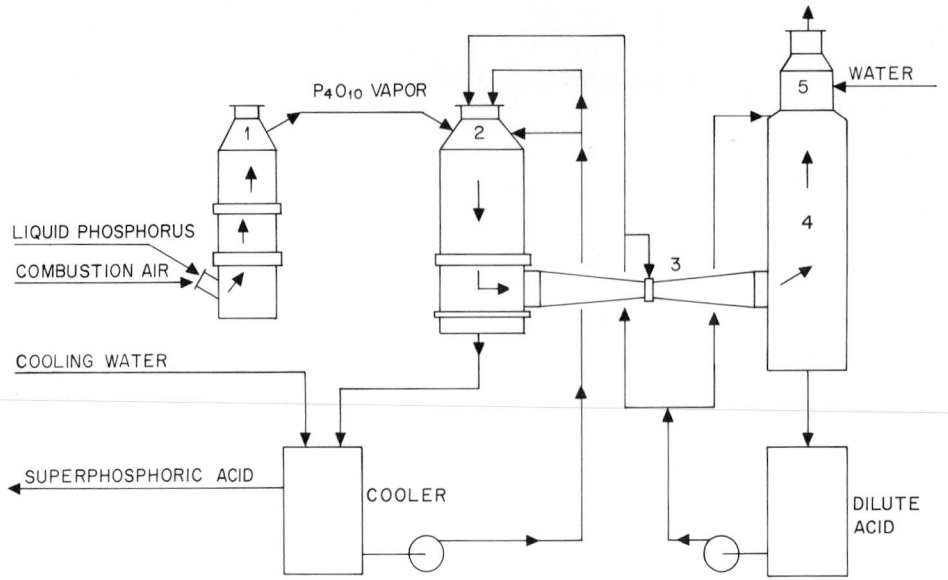

Fig. P-14. Phosphoric acid production by the electric-furnace process. Production of elemental phosphorus is not shown (see Fig. P-15). (1) Combustion chamber, (2) hydrator, (3) venturi scrubber, (4) separator tower, (5) mesh demister.

severe burns. Precautions similar to those used for handling sulfuric acid should be employed. Phosphoric acid exhibits no peculiar or systemic toxicity, and any damage is attributed to its corrosive nature.

Phosphoric acid is commonly shipped and stored in rubber-lined equipment, including tank trucks, tank cars, and specially designed ships. Stainless-steel and polyethylene containers are also used. Wet-process acid, which contains fluorine, may be corrosive to glass or coatings containing silica.

REFERENCES

Fertilizer Manual, *UN Pub.* 67.11.B1, 1967.
Phosphatic Fertilizers: Properties and Processes, *Sulphur Inst. Tech. Bull.* 8, 2d ed., Washington 1966.
Phosphorus: Properties of the Element and Some of Its Compounds, *TVA Chem. Eng. Rep.* 8, 1950.
Van Wazer, John R.: "Phosphorus and Its Compounds," Interscience-Wiley, New York, 1961.

—William A. Lutz, *Consulting Engineer, Louisville, Ky.*, and Enrico Pelitti, *C & I/Girdler Incorporated, Louisville, Ky.*

Phosphoric Anhydride   (See **Fertilizers.**)

Phosphorus   (Gk. *phosphorus,* light-bearing.) **P** = 30.98 (at. wt.); 15 (at. no.). Phosphorus is a nonmetallic element in group VA, period 3, of the periodic table. Atomic volume (yellow) = 16.96 cm^3/g atom; specific gravity, solid (yellow) = 1.83; (red) = 2.20. Melting point = 44.1°C; boiling point = 280.5°C. Usual coordination numbers are 3 or 4, but 1–6 are known.

There is one stable nuclide, 31, with a nuclear spin of 0.5 bohr. There are four radioactive isotopes, 29, 30, 32, and 34. See also **Chemical elements.** Several allotropic forms[1] exist, the most common being the familiar alpha, white (or yellow) with a density of 1.82 g/cm^2 and soluble in CS$_2$, benzene, and ether. This form slowly reverts to stable, crystalline, red phosphorus(IV). The reversion is rapid upon heating in presence of a catalyst,[2] such as I, S, Se, or Na. Above 580°C, phosphorus reverts to the yellow form. Critical temperature of liquid P is 675°C; critical pressure is 80 atm.

The P atom contains two electrons in the inner K shell, eight in the L shell, and five in the outer M shell. This structure determines its properties and chemical behavior. The molecule is believed to contain four P atoms, in the shape of a tetrahedron, each atom sharing an electron from adjacent atoms (coordinate covalence = 3) to form a stable K shell of eight electrons. Above 800°C, dissociation into a two-atom molecule occurs. The element is twelfth in abundance and is widely distributed in mineral and organic combination but not in free form. Major minerals contain $Ca_5X(PO_4)_3$, where X = F, CO_3, or OH.

Discovery of phosphorus is attributed to the chemist Hennig Brandt (Germany) in 1669, who distilled urine with sand and coal.[3] It is produced industrially as a highly toxic, inflammable, pale-yellow liquid or solid and in smaller quantities as the less poisonous red solid.

Manufacture: Elemental phosphorus is continuously produced by thermally reducing the tricalcium phosphate in phosphate rock, carefully mixed with coke and silica.[4] P_2 vapor and CO evolve and pass to a condenser to separate the phosphorus as a liquid which is filtered before storage and use. The CO is burned to supply much of the heat needed for feed preparation and about 25 million Btu per short ton of P_4 can be recovered in this way (Fig. P-15). Periodically, ferrophosphorus alloy and calcium silicate slag are tapped from the furnace. The process is strongly endothermic, and in addition to the heat of combustion of coke thermal requirements of 11,000–13,000 kWh per short ton of P_4 are supplied via electric arc.[5] Small blast furnaces have been used to a limited extent, and techniques based on fluidized beds, rotary kilns, and plasma jets have been proposed.

Although numerous intermediate and side reactions occur, the overall reaction, which takes place at atmospheric pressure and 1300–1500°C can be represented by

$$Ca_3(PO_4)_2 + 5C + 3SiO_2 \xrightarrow{\Delta}$$
$$P_2 + 5CO + 3Ca \cdot SiO_3 - 700 \text{ kcal}$$

Maximum release of P_4 from the phosphate and longer furnace life are said to occur with a SiO_2/CaO weight ratio of about 0.8, as this gives minimum melting-point eutectics in the systems $CaO\text{-}SiO_2$ and $Ca_3(PO_4)_2\text{-}SiO_2$.

The basic P_4 electric patent (United Kingdom) was issued to Readman in 1888, and a 100-kW unit was built in England in 1893, followed by a 1,500-kW furnace in the United States at Niagara Falls in 1896. Modern furnaces are in the 70- to 75-MW category, with three-phase self-baking electrodes 4 ft or more in diameter, each carrying loads of 50–60 kA. A 70-MW furnace will produce 44,000 short tons/yr of P_4, equivalent to 100,000 tons of P_2O_5 as acid. Careful feed preparation is imperative for smooth and economical furnace operation[4] and, as most phosphate rock is available only as a sand, agglomeration by incipient fusion to form $\frac{1}{2}$- to 2-in. lumps is also necessary. As for all electrothermal processes, low-cost power is essential and is often a deciding factor in plant location, as customary industrial rates are usually prohibitive.

Production Statistics and Uses: Worldwide phosphorus production capacities are estimated at about 1.5 million short tons, the United States representing an installed capacity of about 700,000 short tons of this total figure. Up to 1969, growth in the United States was 5.5%/yr to 629,000 short tons, but this fell to 597,000 short tons in 1970[6] due to concern over ecological effects of phosphate detergents. This, plus dwindling supplies of Tennessee brown rock and recent availability of processes for purifying wet-process

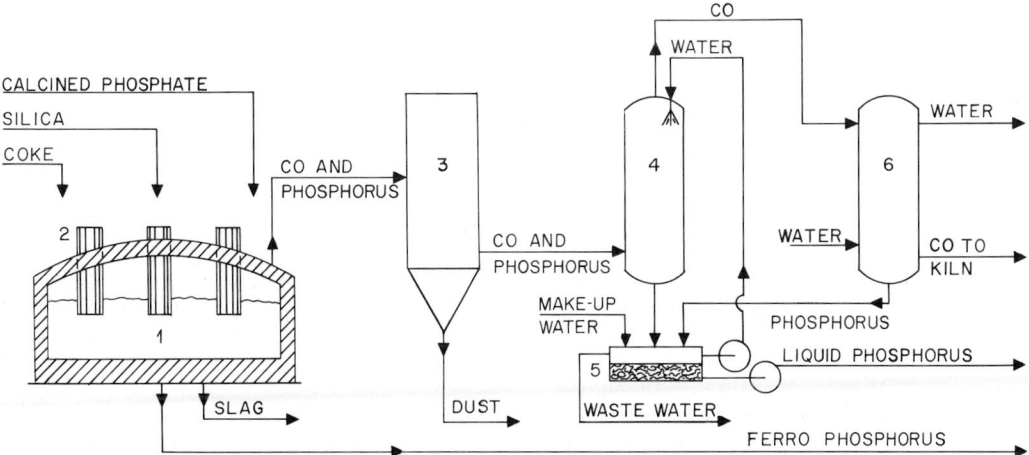

Fig. P-15. Schematic representation of materials flow in the manufacture of elemental phosphorus. (1) Electric furnace, (2) electrodes, (3) electrostatic dust precipitator, (4) phosphorus condenser, (5) phosphorus sump, (6) tubular condenser.

phosphoric acid, makes future growth uncertain. Until recently, eight producers operated 35 furnaces with capacities from 5,000 to over 30,000 short tons/yr, but several are now idle and more closings are anticipated. Production elsewhere in the world is expected to continue growing at about 10% annually for the next few years.

More than 80% of the P_4 produced in the United States is converted into thermal acid by burning with air and absorbing the P_2O_5 in water under controlled conditions.[4] (In 1969, 878,400 short tons of P_2O_5 as acid were made, compared to 3,136,600 by wet processes.) In recent years, 50% of the total P_4 production has been converted into detergent phosphates (which is likely to change), 14% to liquid fertilizers, and 16% to food, water-treatment, pharmaceutical, and miscellaneous chemicals, via the acid. The remaining 20% has been directly converted to alloys, pyrotechnics, and organic intermediates for oil and fuel additives, plasticizers, and pesticides. For phosphorus produced and consumed as fertilizer materials, see **Fertilizers;** see also **Phosphoric acid.**

Compounds: All but a few of the many thousands of phosphorus compounds known contain tri- or quadrivalent P linkages in which the outer *M* shell has a stable eight-electron structure.

Oxygen Derivatives: The phosphorus oxidation mechanism is uncertain and could be a stepwise progression from P_4 to P_4O_{10} via addition of O or a series of reactions in which PO is formed at each stage, that is, $P_4 + O \rightarrow PO + P_3$; $P_3 + O \rightarrow PO + P_2$, etc.

When phosphorus is burned in a free supply of air, phosphorus pentoxide, P_2O_5 (or P_4O_{10}), a key compound, is formed. This polymorphic molecule of four tetravalent P atoms and ten O atoms exhibits at least three crystalline and two amorphous forms, according to temperature. The best known oxides and oxy acids are:

Oxide	*Oxy Acid*
Trioxide, P_2O_3 . .	Hypophosphorus acid,H_3PO_2
Tetroxide, P_2O_4. .	Phosphorus acid,H_3PO_3
Pentoxide, P_2O_5 .	Hypophosphoric acid,$H_4P_2O_6$
+ 3H_2O	Orthophosphoric acid,2H_3PO_4
+ 2H_2O	Pyrophosphoric acid,$H_4P_2O_7$
+ H_2O	Metaphosphoric acid,2HPO_3

The tribasic ortho acid forms three series of salts, e.g., NaH_2PO_4, Na_2HPO_4, and Na_3PO_4. These and corresponding calcium salts are of major commercial importance in fertilizers, detergents, and pharmaceuticals.

Other Inorganic Compounds: With hydrogen, PH_3 (phosphine) and P_2H_4 (phosphoretted hydrogen) can be formed. The former yields phosphonium salts with dry halogen hydracids (PH_3 + HX $\rightarrow PH_4X$). The latter can exist as several liquid and solid varieties. With halogens, and according to the ratios used, trivalent PX_3 and pentavalent PX_5 compounds result, except for the iodides,[7] which are PI_3 and P_2I_4. Direct reaction between P_4 and the gaseous halogen is frequently employed. (Use of liquefied halogens can be dangerous.) Addition of water to the halides produces oxyhalides, for example, $PCl_5 + H_2O \rightarrow POCl_3$ + 2HCl. With sulfur, cautious reactions with phosphorus yield a series of sulfides, P_2S_5, P_4S_7, and P_4S_3. Many of these inorganic compounds have important industrial uses, either directly or as intermediates.

Organic Compounds: Most of these compounds are based on inorganic PCl_3, $POCl_3$, P_2S_5, or P_2O_5 plus a suitable intermediate.[8] For example, ester intermediates such as alkyl phosphoryl chlorides can be made by addition of primary alcohols to $POCl_3$. Triaryl phosphate plasticizers and gasoline additives such as tricresyl phosphate (TCP) can be prepared from PCl_5 and an appropriate phenolic compound. Alkyl diaryl phosphates can be made from $POCl_3$ and the corresponding phenols. Numerous thiophosphate esters containing PS plus an ethyl or methyl group and a substituted aryl group and based on $PSCl_3$ or P_2S_5 have been developed for pesticide purposes. Dialkyl dithiophosphates prepared with the aid of P_2S_5 find use in flotation-agent, oil-additive, and insecticide manufacture. True organophosphorus compounds containing a C—P bond, however, are more difficult to prepare than esters. Some syntheses used for pesticide production have not been disclosed. Other organic phosphorus compounds, as yet synthesized to very limited degrees, are those found in every life process.[3,9] These include the phosphoglycerides needed for fermentation, adenosine phosphates required in photosynthesis and muscle activity, and the complex phosphorus-containing groups identified in the nucleotides.

REFERENCES
1. Bell, C., et al.: "Modern Approaches to Inorganic Chemistry," 2d ed., Butterworth, London, 1966.
2. Van Wazer, J.: "Phosphorus and Its Compounds," vols. 1 and 2, Interscience-Wiley, New York, 1961.
3. Farber, E.: History of Phosphorus, *U.S. Natl. Mus. Bull.* 240, pap. 40, pp. 177–200, Smithsonian Institution, Washington, 1965.

4. Hartlapp, G.: "Phosphoric Acid," pt 2, pp. 927–982, Dekker, New York, 1968.
5. Hoechst-Uhde Corporation: Phosphorus Technical Bulletin, Englewood Cliffs, N.J., 1969.
6. United States Department of Commerce, Bur. of the Census, Washington, D.C., 1971.
7. Partington, J. R.: "A Text-Book of Inorganic Chemistry," p. 615, Macmillan, London, 1931.
8. Chadwick, D. H. and R. S. Watt: Manufacture of Phosphate Esters and Organic Phosphorus Compounds, in Ref. 2.
9. Katchman, B. J.: Phosphates in Life Processes, in Ref. 2.

—Christopher J. Pratt, *Hoechst-Uhde Corporation, Englewood Cliffs, N.J.*

Phosphorus Pentachloride (See **Aldehydes;** and **Ketones.**)

Photochemistry and Photolysis Chemical change brought about by exposure of substances to light is termed *photochemistry*. Where the change involves chemical decomposition of the radiated material, the process is termed *photolysis*. The term light in this context includes visible light and ultraviolet radiation. See also *Photosynthesis.*

Few kinds of molecules are activated by visible light because the energy in a quantum of visible light is insufficient to match the dissociation-energy requirements. Notable exceptions are the molecules of chlorine, bromine, and iodine. The fact that these substances are strongly colored indicates that they absorb light. As a consequence, certain halides are widely used in photography. The dissociation energy for Br_2 is approximately 48 kcal/g mole of quanta. Visible light with a wavelength of about 600 nm is sufficient to bring about this dissociation. The dissociation energy of Cl_2 is about 60 kcal/g mole of quanta and visible light of a wavelength of approximately 490 nm can cause this dissociation. The dissociation energy of I_2 is even less—approximately 37 kcal/g mole of quanta and, hence, its dissociation can be effected with infrared radiation of a wavelength of approximately 780 nm.

A slight shift of the radiant-energy spectrum to the ultraviolet greatly increases the number of possible photochemical reactions. The largest number of reactions occur in the near- and middle-ultraviolet regions, between 400 and 250 nm wavelengths. The dissociation energy of HBr is approximately 88 kcal/g mole of quanta, equivalent to radiation in the near-ultraviolet region of about 375 nm wavelength. Hydrogen chloride and the diatomic molecules S_2 and H_2 all have dissociation energies of about 103 kcal/g mole of quanta, thus falling between the middle-ultraviolet and far-ultraviolet regions (about 275 nm wavelength). Radiation of about 240 nm wavelength in the far-ultraviolet region is required for the dissociation of O_2 (approximately 120 kcal/g mole of quanta) and a radiation of about 190 nm wavelength (just beyond the far-ultraviolet region and into the vacuum ultraviolet region) is required to dissociate NO which has a dissociation energy of approximately 150 kcal/g mole of quanta.

From the foregoing examples, it will be evident that the energy content of a quantum of radiation (or photon) is directly proportional to frequency and inversely proportional to wavelength. A gram-mole of quanta carries 6.06×10^{23} photons and is the number of photons needed to dissociate a gram-mole of a diatomic molecule were the quantum yield unity.

Many photochemical reactions occur in steps and often at great speed. Molecular species with lifetimes in terms of microseconds are not uncommon and, as instrumental techniques improve, it is likely to be found that some of these reactions occur in terms of nanoseconds and even picoseconds.

Upon exposure to a quantum of light, a molecule generally will react in one of four ways in absorbing that energy: (1) the most common reaction is that of converting the new electronic excitation into atomic vibrations within the molecule—this vibrational energy ultimately is passed to the surrounding environment as heat; (2) the excited molecule may fluoresce, i.e., emit or reradiate energy of a wavelength somewhat longer than the exciting energy; (3) the molecule may be chemically transformed (the basis of photochemistry); or (4) the molecule may be ruptured, in which case the process is termed photolysis. The molecules are said to be photolysed.

Probably the best understood and the most extensive example of photolysis is the production of ozone, O_3, in the upper atmosphere, a reaction critical to life on earth because ozone acts as a filter of the middle- and far-ultraviolet radiations which destroy living organisms. Regular oxygen, O_2, absorbs solar ultraviolet radiation with a wavelength of 190 nm. The released oxygen atoms may combine with oxygen molecules present to form ozone, or the freed oxygen atoms may recombine to form O_2. Thus, there is a continuing combination of processes in dynamic equilibrium, i.e., the synthesis and the photolysis of ozone.

In the early 1900s, it was observed that a quantum yield of close to 1 million molecules of HCl resulted when pure H_2 and Cl_2 were exposed to ultraviolet radiation. This observation was in contradiction to the postulate that the quantum yield should be unity. This was later explained by Max Bodenstein (1912) who proposed that actually a chain reaction rather than one reaction occurred. The reaction now is considered to take place in the following steps:

$$(1) \quad Cl_2 + energy \ (330 \ nm) \rightarrow Cl + Cl$$
$$(2) \quad Cl + H_2 \rightarrow HCl + H$$
$$(3) \quad H + Cl_2 \rightarrow HCl + Cl$$
$$(4) \quad Cl + H_2 \rightarrow HCl + H$$
$$(5) \quad H + Cl_2 \rightarrow HCl + Cl$$
$$(6) \quad Cl + H_2 \rightarrow HCl + H$$
$$(7) \quad H + Cl_2 \rightarrow HCl + Cl$$
$$(8) \quad Cl + H_2 \rightarrow \ . \ . \ .$$

The explanation lies with the fact that some of these reactions involve an excess of energy, whereas others involve a deficit. Thus, an overall energy balance is maintained and there is no conflict with the original postulate.

Much investigation has been made of the photolysis of acetone. This reaction yields primarily carbon monoxide and ethane. The current explanation of the mechanics of this reaction (as when acetone is radiated with ultraviolet energy of 310 nm) is that first two free methyl radicals and one free acetyl radical are formed, along with the formation of carbon monoxide. In a second step, the two methyl radicals combine to form ethane. This, of course, is an oversimplification of what occurs. The lifetime of the free radicals may be only a few ten-thousandths of a second and, thus, laboratory measurements are difficult. The advent of flash spectroscopy now makes it possible to record the existence of such free radicals during their brief existence. The concept of the *triplet state,* stemming from early work done by G. N. Lewis and collaborators at the University of California at Berkeley (late 1930s to early 1940s) also is contributing to an understanding of these kinds of reactions. For further details, consult references listed at the end of this entry.

Much progress has been made in the photochemical studies of plastics and other synthetics, including a photochlorination process* involving the exposure of polyvinyl chloride to chlorine in

*B. F. Goodrich Chemical Co.

the presence of ultraviolet radiation. This procedure produces a plastic material with a much higher heat-distortion temperature, making it possible to use the material for hot-water piping and other elevated-temperature applications for which the untreated material is entirely unsuited. (See also **Anthracene; Caprolactam; Polymerization;** and **Silver;** and consult Subject Index for other examples of photochemical processes.)

REFERENCES

Green, A. E. S. (ed.): "The Middle Ultraviolet: Its Science and Technology." Wiley, New York, 1966.

Kosar, Jaromir: "Light-sensitive Systems: Chemistry and Application of Nonsilver Halide Photographic Processes," Wiley, New York, 1965.

Oster, Gerald, and Nan-Loh Yang: Photopolymerization of Vinyl Monomers, *Chem. Rev.,* vol. 68, no. 2, pp. 125–151, March 25, 1968.

Oster, Gisela K., and H. Kallmann: Energy Transfer from High-lying Excited States, *J. de Chim. Phys.,* vol. 64, no. 1, pp. 28–32, January 1967.

Porter, George: Flash Photolysis and Some of Its Applications, *Science,* vol. 160, no. 3834, pp. 1299–1307, June, 21, 1968.

Photoelectric Materials (Consult the **Subject Index.**)

Photographic Chemicals (Consult the **Subject Index.**)

Photosynthesis Photosynthesis is the chemical process by which plants capture and convert solar radiant energy essential to their growth and maturation. Artificial sources of radiation, within certain limits, also activate photosynthesis.

Broadly stated, light excites selected, complex, radiation-absorbing molecules within the plant structure. The responses of these molecules as they relate to the capture and conversion of energy to sustain the life cycle of a plant are closely related to the molecular response mechanisms which make *vision* in animals possible. The processes of photosynthesis and vision both exhibit *photoperiodism,* i.e., the reaction to the cycle of night and day. Photoperiodism in terms of photosynthesis does not simply involve the daily turning on and turning off of external radiant energy sources, but also is concerned with much more subtle and continuous molecular alterations within plants as brought about by the gradual seasonal changes in the relative length of day and night. In a sense, the molecular processes are imbued with a form of memory that is involved in the further programming of molecular change.

The molecular alterations involved in photosynthesis also require a supply of carbon dioxide and water and they function best within the rather narrow temperature range of 10–35°C. Practical experience has taught much concerning the needs and conditions required by photosynthesis. For example, the minimum rate of photosynthesis (threshold) occurs when the radiation energy input drops below 1% of full sunlight intensity. Carbon dioxide constitutes only about 0.03% of the earth's atmosphere, yet this concentration has proved to be sufficient for the maintenance of healthy development and growth of plants. However, it is believed that an increase in CO_2 concentration would increase the rate of photosynthesis. In addition to CO_2 and H_2O, numerous other substances (nitrogen, sulfur, phosphorus, to name a few) are required for plant health. Knowledge of the total chemistry of plant behavior, including how the fundamental process of photosynthesis interacts with other chemical changes within plants, still is in a comparatively early phase of development. Biochemists, molecular biologists, and scientists in associated fields, however, have made significant progress in this area, particularly during the last two decades.

The overall chemical reaction of photosynthesis is

$$6CO_2 + 6H_2O \rightarrow C_6H_{12}O_6 + 6O_2$$

Thus, the energy absorbed is stored in plants in the form of carbohydrates and many other compounds derived from these carbohydrates. More complex sugars, such as sucrose, are built up from the simpler molecules of the hexoses.

In vascular plants, photosynthesis takes place mainly in the leaves. Carbon dioxide from the air diffuses into the intercellular spaces of the leaves and then, upon meeting water in the mesophyll cells, dissolves. In most plants, water is absorbed by the root structure from the soil from which it is transported to the leaves. The CO_2-H_2O solution then diffuses to the surface of the chloroplasts where photosynthesis occurs.

Chlorophyll molecules are the main energy receivers in a plant. There are several chlorophylls, differing by virtue of their side groups. A molecule of chlorophyll *a* is shown above right.

Only a brief, oversimplified description of photosynthesis is included here. There still remain differing theories as to what actually occurs during photosynthesis. For further details, consult the references listed at the end of this entry.

At the start, the impact of visible light excites an electron in the chlorophyll molecule, raising

Phytol   Chlorophyllin

the normal energy level of the electron to a higher level. It is estimated that within less than a hundred-millionth of a second, the excited electron returns to its normal state. The reversion, of course, could take the form of the reemission of visible light, but this would not advance the process of photosynthesis. Instead, it is theorized that the reversion takes place in a number of steps, during which process the energy needed for photosynthesis is passed along a chain of molecules. Although a small part of the energy received is reemitted as light of a slightly longer wavelength (hence, the dark red light characteristic of chlorophyll when it fluoresces), the greater portion of the received energy is molecularly transferred to a recipient molecule that ultimately effects the photochemical synthesis. This energy, according to one concept, causes two endoergic reactions (energy-absorbing reactions): (1) the phosphorylation of the nucleotide adenosine diphosphate (ADP) to form adenosine triphosphate (ATP), and (2) the reduction of the coenzyme triphosphopyridine nucleotide (TPN) to its reduced form TPNH. There remains the carbon reduction cycle of photosynthesis in which the CO_2 is assimilated. This cycle does not require radiant energy, but utilizes the ATP and TPNH produced by the first two steps. The carbon reduction cycle is known to be extremely complex, involving several carbohydrates and other complex compounds. One of the simplest of these is 3-phosphoglyceric acid (PGA) with the formula $POCH_2 \cdot CHOH \cdot COOH$, where P signifies a phosphate group. In one of the earliest of the chain of reactions in photosynthesis, it is theorized that PGA reacts with ATP and TPNH to form an aldehyde of PGA, i.e., 3-phosphoglyceraldehyde (a triose). This latter material, in turn, yields many other substances, directly or by reactions with other triose phosphates which have been formed by isomerization. It is further theorized that five molecules of the aforementioned aldehyde of PGA, as the result of a complex series of reactions, form three molecules of a pentose phosphate (ribulose-1, 5-diphosphate). The latter may combine with an additional molecule of CO_2 in still another revolution of the cycle. The end result, which is much better understood, is that three molecules of CO_2 and H_2O form a 3-carbon compound, but in so doing at least eleven different phosphorylated sugars have been produced as intermediates.

REFERENCES

Allen, Mary Belle (ed.): Comparative Biochemistry of Photoreactive Systems, *Symp. Comparative Biology of the Kaiser Foundation Research Institute,* The Kaiser Foundation Research Institute, Richmond, Calif., 1960.

Bonner, James, and J. E. Varner (eds.): "Plant Biochemistry," Academic, New York, 1966.

Chargaff, Erwin, and J. N. Davidson (ed.): "The Nucleic Acids," Academic, New York, 1955.

Davidson, J. N., and Waldo E. Cohn (ed.): "Progress in Nucleic Acid Research and Molecular Biology," Academic, New York, 1963.

Godwin, T. W. (ed.): "Chemistry and Biochemistry of Plant Pigments," Academic, New York, 1965.

Johnson, Frank H., Henry Eyring, and Milton J. Polissar: "Kinetic Basis of Molecular Biology," Wiley, New York, 1954.

Phthalic Anhydride   (See **Petrochemical complex.**)

Phthalimide   (See **Amides and imides.**)

Pickling   (See **Conversion coatings;** and **Iron- and steelmaking.**)

Picolines   (See **Pyridine and derivatives.**)

Picric Acid   (See **Explosives.**)

Pierce-Smith Converter   (See **Copper production.**)

Pig Iron   (See **Iron and steel;** and **Iron- and steelmaking.**)

Pigments   (See **Antimony; Cadmium; Iron oxides, synthetic; Paint;** and **Titanium dioxide.**)

Pinacol-Pinacolone Rearrangement   (See **Ketones.**)

Pinacols   (See **Ketones.**)

Pinene   (See **Terpenes.**)

Piperidine   (See **Pyridine and derivatives.**)

Pitch   (See **Coal tar and derivatives;** and **Petroleum processing.**)

Pitchblende   (See **Uranium.**)

Plant Synthesis   (See **Amino acids.**)

Plasma, Blood   (See **Peptides and proteins.**)

Plasma, Chemical Processing with   Descriptions assigned to plasmas have ranged from a simple "ionized gas" to "the fourth state of matter." For this discussion, a plasma is considered to be partially ionized gas with enough energetic ions and electrons to produce a highly reactive environment. Such a medium can be employed as an energy source to drive both homogeneous and heterogeneous chemical reactions. Although conventional heat sources are adequate for many endothermic reactions, plasmas may offer special advantages relative to reaction kinetics, specificity, or purity of products. For reactions requiring very high temperature and enthalpy, a plasma may be the only means available to achieve economically acceptable reaction rates.

One reason for the wide diversity of definitions of plasma is the extreme variation of physical characteristics, which depend on how the plasma is generated. Various generators are described in a number of literature sources.[1] In relatively simple terms, plasmas have been classified as (1) thermal plasmas or arcs, (2) cold plasmas or glow discharges, and (3) hybrid plasmas such as the corona discharge.[2,3] Since each has unique properties, they will be briefly described with a few selected applications. The various types of plasmas and their characteristic current-pressure relationship are illustrated schematically in Fig. P-16.

Thermal Plasmas: The typical average temperature of an arc is generally in excess of 8000 K, and it may be much higher. Historically, such systems have been used for applications requiring intense heating such as cutting, welding, cladding, and spheroidizing. Recently, a number of chemical processes have been described.

Any chemical mixture reaching the temperature of an arc is dissociated into atoms. For the moment let us consider a perfect arc with every atom at the same high temperature. The chemical

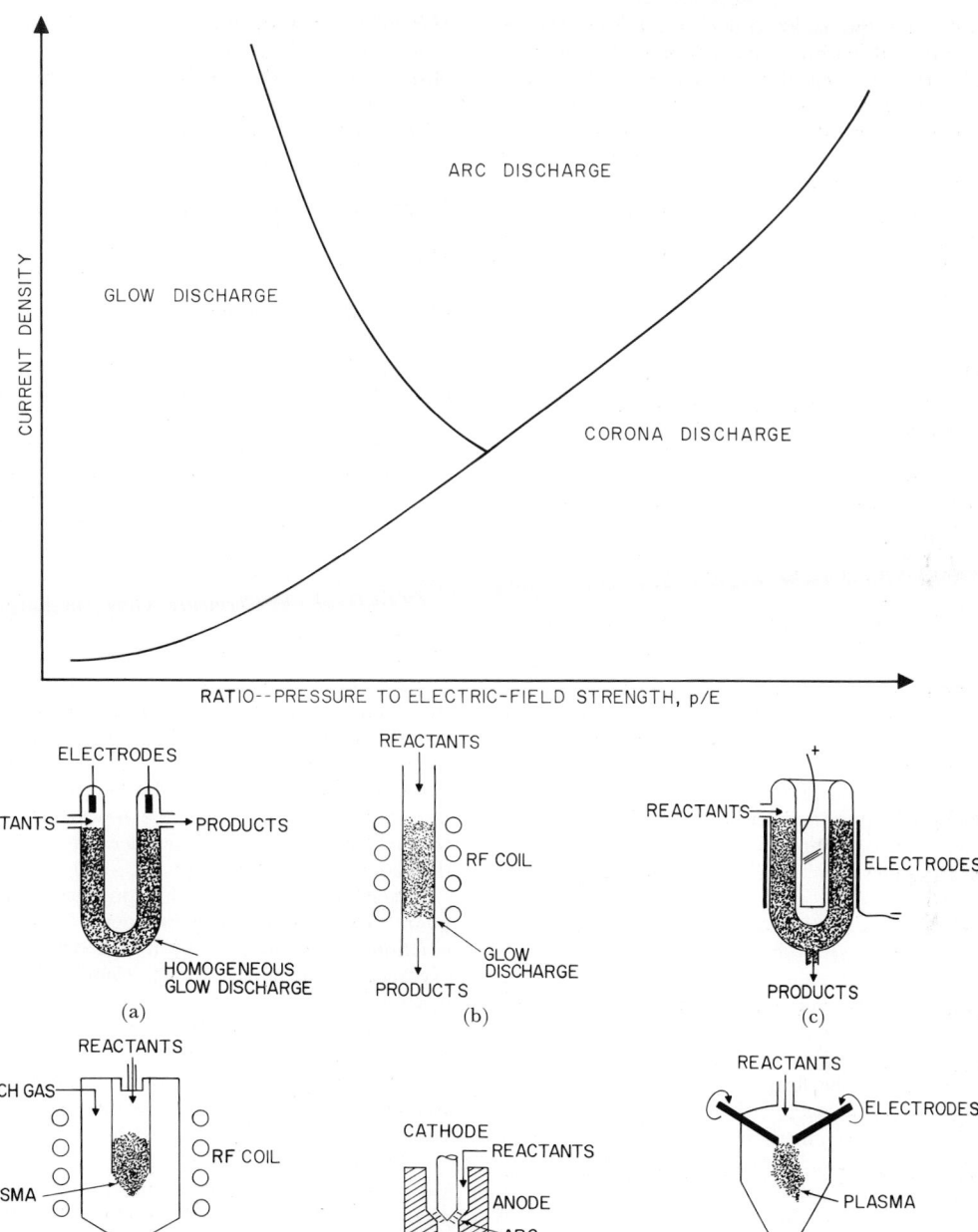

Fig. P-16. Operating regions of various types of plasma environments. (a) Glow discharge, ac or dc. Electrodes may be tungsten or platinum. (b) Radio-frequency coil, 200 kHz–20 MHz. (c) Corona discharge (ozonizer discharge), 15–30 kV, 60 Hz–10 kHz, may be applied to electrode surfaces which handle about 1 W/cm^2. (d) Induction plasma, radio-frequency coil, 200 kHz–20 MHz. (e) Plasma jet, ac or dc. Cathode is tungsten; anode is copper. (f) Carbon-arc furnace. (*Adapted from Ref. 3.*)

products obtained from such an arc depend on how the various atoms recombine and on the rate of cooling or quenching applied to them. If these atoms are allowed to cool slowly, the reactions that ensue are governed by the laws of thermodynamics and can be predicted with some accuracy. If these hot atoms are quenched rapidly at a rate greater than 10^6 K/s, the resulting chemistry is primarily controlled by the kinetics of the various atom-recombination reactions. Often unpredicted and highly desired results are produced.

Generally, three types of arcs have been used to promote chemical reactions: (1) plasma jet, dc or ac, with water-cooled nonconsumable electrodes, (2) the carbon arc, using electrodes that are consumed, and (3) the induction plasma, where energy is supplied at radio frequencies (10^4–10^7 Hz) with the elimination of electrodes.[4,5] The chemistry of these systems has been reviewed.[1,2,6–12]

The carbon arc is generally the simplest and cheapest system. Increased capital investment and operating cost of the induction plasma system are offset by certain unique advantages. Reactive gases such as O_2, H_2, or Cl_2 can be efficiently and reliably heated to over 3000 K. Under these conditions, the lifetime of electrodes in the dc plasma jet is just a few hours. Since the velocity of gas through the dc and ac plasma jet is high, the residence time in the hot zone is severely shortened. This is important when considering gassolid reactions or the heating of solid particles with the plasma. In contrast, the residence time in the induction plasma is 5 to 10 times greater and with no electrode problems.

Both homogeneous and heterogeneous reactions have been observed in arc plasmas. In addition, a third category, dissociation, merits attention.

Homogeneous Reactions: Typical reactions of this type include the formation of nitric oxide from oxygen and nitrogen,[13] the preparation of hydrogen cyanide from methane and nitrogen,[14] and the partial oxidation of hydrocarbons by oxygen atoms from a quenched oxygen plasma. Typical products are formaldehyde from ethylene and oxygen and methanol from methane and oxygen.[15] Interest in ultrafine particles of refractory carbides and nitrides has led to investigation of the reaction of gaseous transition-metal chlorides with methane or nitrogen.[16]

Heterogeneous Reactions: Reactions between solids and gases, which represent the prime candidates for plasma processing, suffer from the inherent difficulty of obtaining sufficient exposure of molecules within solid particles to the reactive gas.[17] This difficulty is heightened in plasma chemistry because of the generally short exposure times. Incomplete reaction of the solid species is generally the result, but encouraging work has been performed in investigations of the reaction of hydrogen with coal to form acetylene[18] and the preparation of cyanogen from carbon and nitrogen.[19]

Dissociation: Significant interest has been shown in the general phenomenon of bond breakage and rearrangement, which occurs with minerals subjected to the temperature of the arc. Many otherwise intractable minerals are made reactive to rather mild processing conditions by pretreatment in plasma. This area has recently been reviewed.[20] A plasma process based on the dissociation of zircon and its subsequent treatment to produce zirconium dioxide and sodium silicate has been commercialized.[21] Figure P-17 shows the flow sheet for this process.

Glow-Discharge Chemistry: Glow discharge exists at pressure below 100 torr and generally is operated around 10 torr. The mean free path for electrons at low pressures is quite large, and since the electrons are mobile, energy is selectively fed into them rather than to the heavier atoms. Thus, the gaseous plasma is not in thermal equilibrium. The gas temperature may be several hundred degrees, whereas the electrons can possess energies equivalent to an electron "temperature" of many thousand degrees.

In general, glows are used to produce workable quantities of atoms for subsequent downstream reactions. The low pressure allows the atoms to exist for several hundred milliseconds. Atoms of hydrogen, oxygen, chlorine, boron, nitrogen, bromine, and iodine have all been successfully generated in a cold plasma.

Perhaps the most significant use of cold plasmas from an industrial standpoint is the surface treatment of polymers known as *casing*[22,23] (*c*ross-linking by *a*ctivated *s*pecies of *in*ert *g*ases). Physical properties such as wettability and adhesive strength can be markedly altered by subjecting the surface to a cold plasma of argon. Polymer films can also be produced, and several patents[24,25] have been granted that deal with this technique. Application of cold plasmas to textile treatment has been discussed.[2]

Corona Chemistry: A corona discharge is produced around high-voltage electrodes situated in a high-impedance environment. An example is

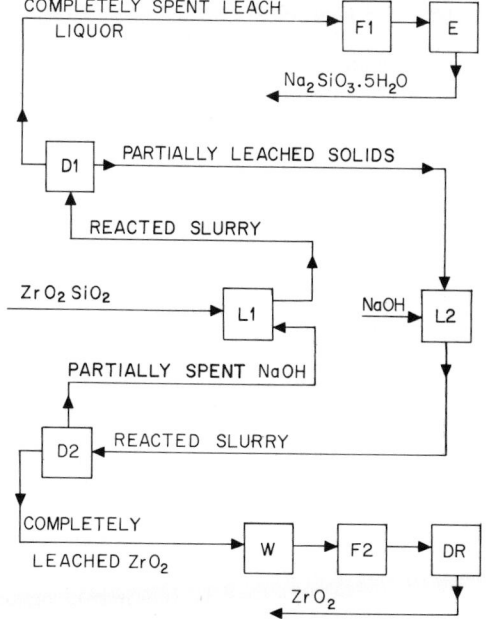

Fig. P-17. Schematic representation of materials flow for production of zirconium dioxide and sodium silicate from dissociated zircon. $L1$, leach 1, which is carried out at 143°C (290°F); $L2$, leach 2, carried out at same temperature; $D1$, decant 1; $F1$, filtration, which is carried out at approximately 80°C (175°F); E, evaporation; $D2$, decant 2; W, washing; $F2$, filtration; DR, drying. On the basis of 1 lb mole of feedstock, the following materials balance prevails: 183 lb of ZrO_2SiO_2 is charged to the process; 160 lb of 50% NaOH is charged to leach 2; 122 lb of ZrO_2 is produced; 212 lb of $Na_2SiO_3 \cdot 5H_2O$ is produced.

the glow around high-tension cables on a humid day. Usually, for chemical applications, the reactant mixture is separated from the electrodes by an insulating barrier, and energy is supplied at frequencies of $60-10^4$ Hz. A nonequilibrium filmy discharge is produced in the vicinity of the high-voltage electrode. It should correctly be called a condensed spark discharge or ozonator discharge. Nevertheless, the term corona discharge is used by chemists.

This type of chemical system is completely different from the arc. Industrial production of ozone is carried out by using corona-discharge techniques. The corona contains a very small quantity of highly energetic electrons. It operates above 100 torr and frequently at atmospheric pressure. Corona discharge is used to promote

condensation reactions and polymerization of unsaturated organic molecules. The mechanism can be explained as follows. Collision with an electron will tend to open up a double bond or fracture a C—H bond. Several activated molecules will then combine to form a condensation product, and the molecular chain will grow. As it grows, it presents a greater probability of being hit by another highly charged electron that will fracture it. Thus, an asymptotic balance is achieved which prevents further chain growth. Coffman and Browne[26] have reviewed corona chemistry.

Silanes and siloxones are produced in a corona. In addition, the viscosity of unsaturated organic liquids can be increased when they are treated with hydrogen or nitrogen in a corona discharge. An ingenious addition to corona condensation chemistry is the use of a selective liquid absorbent to remove the product from the discharge volume when it would normally be a gas. A group at Imperial Chemical Industries in England has produced hydrazine in a corona discharge and found that yields are significantly improved when the product is removed by absorption with ethylene glycol.[27]

REFERENCES

1. Ibberson, V. J., and M. W. Thring: Plasma Chemical and Process Engineering, *Ind. Eng. Chem.*, vol. 61, no. 11, p. 48, 1969.
2. Bradley, A., and J. D. Fales: Prospects for Industrial Applications of Electrical Discharge, *Chem. Technol.*, April 1971, p. 232.
3. Dundas, P. H., and M. L. Thorpe: Economics and Technology of Chemical Processing with Electric-Field Plasmas, *Chem. Eng.*, vol. 76, no. 14, p. 123, June 30, 1969.
4. Thorpe, M. L.: Induction Plasma Heating: System Performance, Hydrogen Operation and Gas Core Reactor Simulator Development, NASA CR-1143, August 1968.
5. Thorpe, M. L., and L. W. Scammon: Induction Plasma Heating: High Power, Low Frequency Operation and Pure Hydrogen Heating, NASA CR-1343, May 1969.
6. Warren, I. H., and H. Shimizu: Applications of Plasma Technology in Extractive Metallurgy, *Can. Min. Met. Bull.* 58, p. 551, 1965.
7. Goldberger, W. M.: Trends in High-Temperature Chemical Processing, *Chem. Eng.*, Mar. 14, 1966, p. 173; Mar. 28, 1966, p. 125.
8. Korman, S.: High Intensity Arcs, *Int. Sci. Technol.*, June 1964, p. 90.
9. Gibson, J. O., and R. Weidman: Chemical Synthesis via the High Intensity Arc Process, *Chem. Eng. Prog.*, vol. 59, p. 53, 1963.

10. Kana'an, A. S., and J. R. Margrave: Chemical Reactions in Electric Discharges, *Adv. Inorg. Chem.*, vol. 6, 1964.
11. Jolly, W. L.: The Use of Electrical Discharges in Chemical Syntheses in H. B. Jonassen and A. Weissberger (eds.), "Technique of Inorganic Chemistry," vol. 1, Interscience-Wiley, New York, 1963.
12. Vurzel, F. B., and L. S. Polak: Plasma Chemical Technology: The Future of the Chemical Industry, *Ind. Eng. Chem.*, vol. 62 no. 6, p. 8, 1970.
13. Eremin, E. H., S. S. Vasilev, and N. I. Kobozev: Investigation of the Oxidation of Nitrogen in a High-Frequency Glow Discharge, II, *Zh. Fiz. Khim.*, vol. 9, p. 48, 1937.
14. Grosse, A. V., H. W. Leutner, and C. S. Stokes: "Plasma Jet Chemistry," Temple University Research Institute, Philadelphia, 1961.
15. Denis, M.: Electric Discharge Apparatus for Chemical Reactant Flowing Gases, U.S. Patent 3,308,050, Mar. 7, 1967.
16. Neuenschwander, E. J.: Herstellung und Charakterisierung von ultrafeinen Karbiden, Nitriden und Metallen, *J. Less Common Met.*, vol. 11, p. 365, 1966.
17. Szekely, J., and J. W. Evans: Studies in Gas-Solid Reactions, II, *Met. Trans.*, vol. 2, p. 1699, 1971.
18. Avco Arc Coal Process, Phase I Feasibility Report, Office of Coal Research, Washington, August 1968.
19. Leutner, H. W.: The Production of Cyanogen from the Elements, Using Plasma Jet, *Ind. Eng. Chem. Proc. Des. Dev.*, vol. 1, p. 1966, 1962.
20. Charles, J. A., G. V. Davies, R. M. Jervis, and G. Thursfield: Processing of Minerals in an Induction-coupled Plasma Torch, *Trans. Inst. Min. Met.*, vol. 79, p. C54, 1970.
21. Wilks, P. H., C. L. Grant, P. Ravinder, R. J. Downer, P. A. Pelton, and M. L. Talbot: The Commercial Production of Submicron ZrO_2 via Plasma, *AIChE Ann. Meet.*, *San Francisco*, November 1971.
22. Hansen, R. H., and H. Schonhor: A New Technique for Processing Low Surface Energy Polymers for Adhesive Bonding, *J. Polym. Sci.*, pt. B, *Polym. Lett.*, vol. 4, p. 203, 1966.
23. Bamford, C. H., and J. C. Ward: The Effect of High Frequency Discharge on the Surface of Solids, *Polymer*, vol. 2, p. 277, 1961.
24. Coleman, J. H.: Polymerizing Method and Apparatus, U.S. Patent 3,068,510, Dec. 18, 1962.
25. Coleman, J. H.: Polymerizing Method and Apparatus for Carrying Out the Same, U.S. Patent 3,069,283, Dec. 18, 1962.
26. Coffman, J. A., and W. R. Browne: Corona Chemistry, *Sci. Am.*, vol. 205, p. 91, June 1965.
27. Thornton, J. D., W. D. Charlton, and P. L. Spedding: Hydrazine Synthesis in a Silent Electric Discharge, *153d Natl. Meet. ACS, Miami*, April 1967.

—Merle L. Thorpe, *Ionarc Smelters Ltd. and Humphreys Corporation, Bow, N.H.*; Clarence L. Grant, *University of New Hampshire, Durham, N.H.*; and Phillip H. Wilks, *Ionarc Smelters Ltd., Bow, N.H.*

Plaster   (See **Gypsum**.)

Plaster of Paris   (See **Gypsum**.)

Plastics   *Plastic* generally refers to an article that has been formed from a *resin* through the application of pressure or heat or both. Derivation of the word stems from the plastic nature of many of the starting materials rather than from the characteristics and properties of the formed end products, the majority of which are relatively hard, nonflowing, dimensionally stable, and quite *nonplastic* in nature.

Conventionally, the manufacture of the resinous raw materials is considered a part of the chemical industry, whereas the further compounding of the base resins with softeners, solvents, plasticizers, fillers, and other various additives and the ultimate forming processing (such as molding and extruding) constitute the plastics industry. The line of demarcation is not sharp. Some resins are relatively easy to manufacture from readily available commercial chemicals and thus may be made in house by a plastic manufacturer. At the other extreme, either because of complexities in forming the final end products or for proprietary marketing reasons, the chemical manufacturer may elect not to sell the starting resins but supply only end-stock items, e.g., bars, tubes, rods, sheets, and films, to end users. Such unilateral approaches generally are difficult to maintain over long periods, particularly when a large demand develops for a new material.

Classification: Important to the production and end-use characteristics of plastic products is the thermal behavior of the resinous ingredients. There are two main categories, *thermoplastic resins,* which can be heated and softened, cooled and hardened limitless times without undergoing a basic alteration, and *thermosetting* resins which do not tolerate thermal cycling and which attain new, irreversible properties when once set at a temperature which is critical to each resin. A thermosetting resin cannot be resoftened and reworked after molding, extruding, or casting.

Major thermoplastic resins include (1) acrylonitrile-butadiene-styrene (ABS) resins, (2) acetals, (3) acrylics, (4) cellulosics, (5) chlorinated polyethers, (6) fluorocarbons [polytetrafluoroethylene (TFE), polychlorotrifluoroethylene (CTFE), fluorinated ethylene propylene (FEP)], (7) nylons (polyamides), (8) polycarbonates, (9) polyethylenes (including copolymers), (10) polypropylenes (including copolymers), (11) polystyrenes, and (12) vinyls (polyvinyl chloride).

Major thermosetting resins include (1) alkyds, (2) allylics, (3) the aminos (melamine and urea), (4) epoxies, (5) phenolics, (6) polyesters, (7) silicones, and (8) urethanes.

The basic chemistry, method of manufacture, and major properties of the final end products of the foregoing resins are described elsewhere (consult the Subject Index).

Fabrication of Plastic Products: Final plastic end products generally result from two basic approaches and sometimes a combination of them: (1) the product is constructed from basic stock forms (rods, bars, sheets, and so on), which are machined and handled much as metals are handled; and (2) the end products are made directly from the raw ingredients. A rough analogy with metals might be making a valve body from a large precast bar or ingot by hogging out the desired shape and using machining methods to achieve the product or, in contrast, casting the valve body by pouring molten metal into a mold. Plastics are particularly adaptable to the latter approach, which accounts in large measure for the very high production rates that can be attained in plastic-products manufacture.

Plastics may be formed into basic stock shapes and almost innumerable end configurations highly specific to given requirements by several methods.

Casting: The raw resinous materials in monomer or modified-monomer form (sometimes referred to as *syrups*) are poured into molds that usually are carefully temperature-controlled. Subsequent heat treating or annealing may be required to complete polymerization and provide stress relief. Nylons, silicones, epoxies, acrylics, polyesters, and styrenes commonly are cast in this manner. The technique is used for potting and encapsulating as well as for the formation of basic shapes, e.g., cylinders, tubes, rods, slabs, and sheets. The labor costs and scrap rates of casting usually are relatively high. The method further is limited by the availability of suitable starting formulations. Casting often will provide better strength and optical characteristics than extruding or molding. The method is well suited to making short-run items, such as prototypes, because molds can be made from relatively inexpensive materials. Frequently, in casting, fillers and reinforcing fibers are added to the starting formulation to improve specific properties. These materials may include glass roving, mat, or cloth, fuller's earth, wood fibers, flock, and a number of oxides which perform as extenders.

Film Casting: In this process, the plastic raw materials are dissolved in solvents, after which the liquid material is spread out over polished drums (large diameter) or smooth belts, which continuously convey the formed film through curing chambers, where the solvents are evaporated and recovered. Films of vinyl, polyvinyl alcohol, acetates, polysulfones, and modified acrylics are fabricated in this manner. In addition to basic films, special configurations for packaging, liners, decals, laminates, and enclosures may be produced by film casting. The method allows the use of numerous additives that might thermally degrade in other forming methods. Film casting is well adapted to forming tough products with high tensile, elongation, and tear factors as well as excellent clarity, and ease of heat sealing. Film-casting equipment is costly and thus is limited to high-production runs.

Compression Molding: This process usually involves placing a partially formed thermosetting resin into a temperature-controlled cavity. Through the application of heat and pressure on the closed mold, the plastic material fills the cavity when softened. The method is applicable to virtually all thermosetting resins and is used occasionally with TFE fluorocarbons and vinyls. Compression molding is well suited to making large parts, e.g., body panels for automobiles, washing-machine parts, furniture parts, and various housings. There usually is little waste, and finishing costs are low. The method is not useful for intricate parts or where tolerances of ±0.005 in. or less are required.

Injection Molding: This process is quite similar to that used in the diecasting of metals. The thermoplastic starting material (usually in granular form) is carefully heated in a cylinder until it becomes plastic, after which it is forced into a temperature-controlled mold. The method is used for practically all thermoplastic materials except TFE fluorocarbons. Injection-molding production rates are high, and quite intricate parts can be produced. Metallic inserts may be added. Dimensional accuracy is quite high. Because of high tooling costs, the method is not well suited to short runs. Part size is limited to available standard machinery. The chilling of material, particularly when forced into thin sections, sometimes causes incomplete parts. Difficulties are experienced with some kinds of holes and with undercuts.

Transfer Molding: Essentially, this is a two-step injection-molding process in which the raw material first is heated to plasticity in a separate (transfer) chamber before being fed to the plunger. The method has advantages in connection with thin sections. Production speeds are high. The method is used mainly with phenolics and melamines, although it is applicable to polyesters, silicones, alkyds, ureas, and diallyl phthalates. Rather elaborate, costly molds are required. Size usually is limited to parts containing about 16 oz of material. Waste can run high because there is no recycling of sprues and culls.

Cold Molding: This process is similar to compression molding except that the materials are molded by pressure only and then placed in baking ovens for curing. The molds do not require heating or cooling. The process is mainly used for phenolics in making such items as utensil handles and electrical-insulation parts.

Blow Molding: A thin cylinder, known as a *parison,* first is produced by extrusion. The parison then is positioned in a split mold and pneumatically forced against the mold faces. The process is used with polypropylene and polyethylene for bottles, various housings, and packaging configurations. Acetals, butyrates, ABS, vinyls, styrenes, and nylon are similarly treated. The process resembles that used in the manufacture of certain glass products. See also **Glass.** Blow molding is well adapted for making large, thin-walled parts almost free of strains, thus reducing stress cracking and warping. The molds are relatively simple to construct and are low cost, an advantage for short production runs. The process normally is limited to tubular and hollow parts where dimensional tolerances of $\pm 5\%$ are acceptable.

Slush or Rotational Molding: A slurry consisting of the plastic raw material (often polyethylene or vinyl) and large amounts of plasticizers is poured into a mold that is heated and rotated. The process results in very little shrinkage, good control over thickness by regulation of the temperature and length of time in which the slurry is in contact with the mold, and relatively low mold costs. Rates of production are relatively slow, and the choice of starting formulations is limited.

Extruding: A majority of stock plastic shapes are made in this way. Thermoplastic materials are heated in a plasticizing cylinder and by means of a rotating screw are forced through a die to provide the desired cross section. The process is well suited for sheets, rods, bars, and tubes. A variation of the process is used for extruding coatings of soft plastic materials over other materials. Almost any profile can be imparted to the product, but of course variations in profile are limited to one direction. The tooling costs are low compared with injection molding. Thickness of the material can be controlled quite precisely. Production rates are high.

Calendering: The starting formulations are warmed and formed into a doughlike mass and then fed between large heated rolls. These rolls work and shape the material to the correct caliper. Additional rolls sometimes are used to impart a textured or patterned surface effect, as for a simulated leather. Calendering is used with cellulosics, styrenes, and vinyls with close control over tolerances. Where excessive plasticizer is required, the material may become brittle and crack and shrink with age.

Thermoforming: This process starts with a sheet of material which is heated before being drawn down onto a form or mold (by vacuum, positive pressure, or mechanically). Most thermoplastic materials can be thermoformed, with the exception of nylon, acetals, and fluorocarbons. The process is well adapted to the high rate production of low-cost parts that have relatively large surface areas. Tooling generally is low cost, and the draws can be quite deep. There is virtually no limitation on part size. Labor costs run higher than for injection molding. Thermoforming does not permit undercuts.

Sintering: Granules of the raw plastic first are compacted under pressure to the approximate desired dimension, after which they are fused together by the application of heat. Once fused, the material may be further formed by heat and pressure. The process is mainly used for TFE resins although nylons are processed in this way for some uses. Lubricants can be incorporated into porous nylon parts by this method. TFE parts made this way can be quite dense, with excellent mechanical and electrical properties. Tooling and production costs run high. There is a limited selection of starting formulations. The process is not adaptable to very thin sections or tight angular variations in cross-sectional thickness.

Plastic Assembly Methods: The principal methods for assembling plastic parts include:

- A. Mechanical
    1. Self-tapping screws
    2. Threaded inserts
    3. Press fitting
    4. Snap fits
    5. Cold heading
- B. Thermal and chemical
    1. Heat joining
    2. Spin welding
    3. Hot-gas welding
    4. Cementing

Heat joining simply results from applying heat to the parts (usually of similar materials) and then pressing them together to form a fusion-type bond. *Spin welding* is particularly suited for joining parts of circular cross section. One part is rotated in relation to the other while the two free ends to be joined are pressed against each other. The frictional heat causes melting at the interface, which is allowed to solidify by cooling while the parts still are under pressure. Rotational speeds range from 20 ft/s to 5,000 rpm, and the pressure during spinning normally is about 200 psi. Polypropylene and polyethylene resins may be spin-welded at pressures of 50–100 psi. Although this method is applicable to almost all thermoplastics, any given weld usually is confined to parts of the same material.

In *hot-gas welding,* a hot-gas torch and filler rod are used, and the welds appear much like those of metal welds produced by electric-arc welding. The diameter of the filler rod is usually $\frac{1}{16}$–$\frac{3}{16}$ in. The tensile strength of a butt weld will approach 90% of the strength of the parent material. A welding temperature of approximately 540°F (282°C) is required for polyethylene, while a temperature of about 630°F (332°C) is required for acetals. The hot gas usually is nitrogen. The method has a relatively low cost. Ultrasonic welding techniques also are employed.

Cementing techniques usually are tailored to specific plastic materials and are of three main types: (1) solvent, (2) dope, or (3) chemical cements. In solvent cementing the surfaces to be joined are dissolved by an appropriate solvent to produce soft mating edges. Upon application of appropriate pressure and allowing the solvent to evaporate, the two treated parts are bonded together with no foreign substances present. A dope type of cement operates like a straight solvent cement. A chemical cement contains resins which polymerize upon application to form a strong bond. All cementing techniques can be used with thermoplastics, but chemical cements apply only to thermosetting resins. Most cements should be cured to develop a high strength in the joint. During the curing period, the parts must retain their alignment under appropriate pressure.

Relative Advantages of Plastics and Metals: Be-

Property	Comment
Weight	Plastics light; metals heavy
Strength	Metals generally considerably stronger, particularly at high and low temperatures
Chemical resistance	If other characteristics are acceptable, a plastic material, specifically selected, can offer greater resistance to corrosion from chemicals
Shock and vibration	Plastics often outperform metals
Absorption of vibrations, including acoustic range	Plastics usually excell
Transparency and translucency	Not obtainable with metals
Lubrication	Metals usually require fairly complex lubricating systems, whereas some plastics are self-lubricative
Color	A large number of plastics can be produced in a variety of colors, whereas most metals require postfabrication finishing
Hardness	Metals generally excell
Dimensional stability	Many plastics tend to absorb moisture and solvents and thus undergo dimensional changes
Effects of radiation	Some plastics undergo degradation from ultraviolet radiation
Flexibility	Even the most rigid plastics have more resiliency than metals
Electrical properties	Most plastics are very good insulators; consequently there is no contest with metals where this is a factor
Thermal insulation	Most plastics are excellent thermal insulators
Wear and abrasion resistance	Plastics generally are superior to metals, particularly where corrosion is also a problem
Flammability	Some plastics support combustion when ignited; often this objection can be overcome through the effective use of fillers and additives
Production methods	Particularly where final products can be directly molded or extruded, the equipment costs are lower; hence cost per piece is lower
Strength-weight ratio	Although plastics (unless effectively reinforced) have less strength than metals, they often provide greater strength on a weight basis because of their low specific gravity

TABLE P-28. *Comparison of Candidate Plastics and Metals for Given Uses*

Application	Plastic	Metal or Other Material
Heavily stressed mechanical components, such as gears, cams, couplings, and rollers	Polycarbonates, fabric-filled phenolics, nylons, and TFE-filled acetals	Brass, steel, and cast iron
Chemical and thermal equipment	Epoxy-glass, polypropylene, high-density polyethylene, polyvinylidene fluoride, chlorinated polyether, fluorocarbons	Titanium, stainless steels, columbium (niobium), other sophisticated metals and alloys
Containers, ducts, housings, and shrouds	Acrylonitrile - butadiene - styrene, high-density polyethylene, modified acrylics, epoxy-glass, cellulose acetate butyrate, polypropylene, and high-impact styrene	Die-cast metals; stamped or cast magnesium or aluminum, and formed steel
Electrical parts	Polyesters, silicones, allylics, aminos, polyphenylene oxides, polycarbonates, alkyds, phenolics, and epoxies	Glass and ceramics

cause of the tremendous number of plastic materials as well as metal alloys available, comparative generalizations are difficult and sometimes misleading. The tabulated observations (page 885) should not be interpreted too literally because these frequently are exceptions to these generalizations. See also Table P-28.

Laminated Plastics: These materials are made of several plies of sheet materials, which are coated or impregnated with a thermosetting resin, known as the *binder*, and then bonded permanently together by the application of pressure and heat.

Phenolic resins are widely used with a number of base materials, e.g., paper, cotton fabric, asbestos paper and fabric, nylon cloth, and continuous-filament glass fabric. Melamine, silicone, and epoxy resins also are used in laminates. Metals also enter into composite laminates. Some x-ray shields are made from lead sheets bonded between paper-base laminates; nonmagnetic beryllium copper is used in laminates to provide good thermal conduction; stainless-steel-clad laminates are used to enhance corrosion resistance; laminates with silver or gold facings are used for electrical

TABLE P-29. *Characteristics of Unreinforced and Reinforced Plastics*

Plastic	Cost Ratio, Reinforced/Unreinforced	Tensile Strength, 1,000 psi	Tensile Modulus, 10^5 psi	Shear Strength, 1,000 psi	Deformation, 4,000-psi load, %	Water Absorption, 24 hr, %	Sp gr	Volume Resistivity, Ω-cm $\times 10^{15}$	Dielectric Constant at 60 Hz
Polystyrene	5X								
Unreinforced		9	4	—	1.6	0.03	1.05	10	2.6
Reinforced		14	12	9	0.6	0.07	1.28	36	3.1
Polypropylene	3X								
Unreinforced		5	2	4.6	—	0.01	0.90	17	2.3
Reinforced		7	5	4.7	6	0.05	1.05	15	—
Polyamide	2.4X								
Unreinforced		12	4	10	2.5	1.5	1.14	450	4.1
Reinforced		30	—	14	4	0.6	1.53	2.6	4.5
Polycarbonate	1.6X								
Unreinforced		9	3	9	0.3	0.3	1.2	20	3.1
Reinforced		20	17	12	0.1	0.09	1.53	1.4	3.8
Acetal	2.4X								
Unreinforced		10	4	9.5	—	0.2	1.4	0.6	—
Reinforced		13	8	9.1	1	1.1	1.7	38	—

contacts where the metals provide high electrical conductivity and the laminate provides insulation and strength.

Reinforced and Filled Plastics: Certain properties of plastic materials, e.g., impact resistance, strength, resistance to crazing and cracking, shrinkage, and brittleness, can be enhanced by adding fibrous materials to the raw plastic formulations. Glass and asbestos fibers are widely used. Such fillers as talcs, silicates, clays, asbestos fines, paper, and carbonates are also added to plastics to improve appearance, processibility, and resistance to crazing, cracking, and shrinkage. In designing plastic formulations, certain fundamental relationships apply: (1) increased flexibility usually means a loss of chemical resistance, sometimes weatherability, and almost always hardness; and (2) increased hardness almost always also increases brittleness and loss of impact strength and resilience. A comparison of unreinforced and reinforced materials is given in Table P-29.

Plasticizers   (Consult the **Subject Index.**)

Plastomers   (See **Elastomers.**)

Plate Columns   (See **Absorption;** and **Distillation.**)

Plate Glass   (See **Glass.**)

Plate Press   (See **Expression.**)

Platinum Metals and Platinum   (Sp. *platina*, little silver.) **Pt** = 195.09 (at. wt.); 78 (at. no.). Plati-num is one member of a family of six elements (group VIII of the periodic table), called the *platinum metals,* which almost always occur together. Before the discovery of the sister elements, the term platinum was applied to an alloy with Pt as the dominant metal, a practice that persists to some degree even today. The major properties of the platinum metals are given in Table P-30. See also **Iridium; Osmium; Palladium; Rhodium;** and **Ruthenium.**

Occurrence: These metals occur in both primary and secondary deposits. The primary deposits are generally associated with Ni-Cu sulfide ores. The Sudbury ores of Canada and the deposits of the Bushveld complex of South Africa are of this type. Native platinum occurs as a primary deposit in the Ural Mountains of the Soviet Union and also in the Choco district of Columbia. Weathering and erosion of these deposits have resulted in the formation of secondary, or placer, deposits of native Pt in riverbeds and streams. One nugget of Pt found in the Urals weighed over 25 lb. Most of the world's platinum comes from Canada, the Soviet Union, and South Africa. Minor amounts have been found in Alaska, Columbia, Ethiopia, Japan, Australia, and Sierra Leone.

Because of their unique properties and in spite of their high initial cost, the platinum metals find many applications in industry. Since used platinum metals retain a large portion of their initial value, many scrap materials are a major source of recoverable platinum metals. Practically every application of platinum generates scrap in some form which is eventually returned to the platinum

TABLE P-30. *Some Physical Properties of the Platinum Metals* *

Property	Ruthenium	Rhodium	Palladium	Osmium	Iridium	Platinum
Atomic volume, cm^3/ g atom	8.29	8.27	8.88	8.43	8.54	9.09
Atomic radius	1.336	1.342	1.373	1.350	1.355	1.385
Crystalline form	hcp	fcc	fcc	hcp	fcc	fcc
Lattice parameter, Å:						
a	2.7041	3.804	3.8902	2.7341	3.8389	3.9310
b	4.2814	. . .	. . .	4.3197		
Thermal conductivity at 20°C, (cal)(cm)/						
(s)(cm^3)(°C)	. . .	0.21	0.168	. . .	0.14	0.166
Electrical resistivity at 0°C, $\mu\Omega$-cm	7.2	4.5	10.8	9.5	5.3	10.6
Thermal expansivity, °C × 10^6 at 20°C . .	9.6	8.3	12.4	6.6	6.6	9.0
Hardness, Mohs' scale	6.5	. . .	4.8	7.0	6.5	4.3
Specific heat, cal/g atom at 20° C	0.057	0.059	−0.0584	−0.031	0.031	0.031
Heat of fusion, kcal/mole	6.1	5.2	4.0	7.0	6.3	4.7
Heat of vaporization, kcal/mole	135.7	118.4	90	150	134.7	122

*Melting point, boiling point, and density for these elements are given under the alphabetical description of the element.

refiner for recycling. Although there are ample mine reserves, they soon would be depleted without constant scrap recycling.

Refining Processes: The refining procedures are a good introduction to the complex chemistry of the platinum metals. Some of these methods are still the best analytical techniques available for the separation of the metals. South African ore is smelted to form a Cu-Ni matte containing small amounts of the platinum metals (0.18%). The matte is melted, cast into anodes, and electrolytically dissolved. The contained Cu is deposited at the cathode, the Ni remains in the H_2SO_4 electrolyte, and the Pt metals are contained in the anode slimes. The resulting Cu is refined and the $NiSO_4$ solution purified and crystallized. The anode slimes are treated by roasting to remove sulfur and leached with dilute H_2SO_4 and air to remove Cu and Ni. The leached slimes are treated with aqua regia. The aqua regia solution is evaporated to concentrate the solutions and expel the excess HNO_3. The residue from this treatment contains Rh, Ir, Ru, Os, and Ag. The solution contains Pt, Pd, and Au.

Pt is first removed by precipitating as ammonium hexachloroplatinate, NH_4PtCl_6, by the addition of a saturated solution of NH_4Cl. The precipitate is washed, dried, and calcined to form platinum sponge about 98% pure. The sponge is purified by redissolving in aqua regia and evaporating the solution to dryness with NaCl. The resulting sodium hexachloroplatinate is dissolved in H_2O and boiled with $NaBrO_3$ to convert impurities, such as Ir, Rh, Pd, and base metals, to valence states which produce readily filterable hydroxides. The Pt left in solution is free of impurities. It is then treated with NH_4Cl, and the pure ammonium hexachloroplatinate precipitate is calcined at 1000°C to pure Pt sponge.

The first aqua regia solution is treated with $FeSO_4$ to precipitate the gold. Pd is precipitated by oxidizing the solution with HNO_3 and adding NH_4Cl. Ammonium hexachloropalladate is formed (analogous to the Pt compound). This salt is purified by dissolving in NH_4OH, filtering off the impurities, and reprecipitating the Pd by the addition of HCl. The insoluble complex $Pd(NH_3)_2Cl_2$ is formed, which when calcined and reduced in H_2 yields pure Pd sponge.

The insolubles from the first aqua regia treatment are fused with a flux of litharge, soda ash, borax, and carbon in a gas-fired furnace at 1000°C for 1 hr. A lead phase is formed which collects all the rare metals. The slag phase fluxes alumina, silica, and some base metals.

The melt is poured into ingots and cooled, the two phases separating. The slag is discarded. The lead portion is heated with HNO_3, which dissolves the Pb and Ag. The Pb is precipitated as a sulfate and then the Ag as a chloride. The residue is treated with concentrated H_2SO_4 at 300°C. Rh will dissolve, leaving Ir, Ru, and Os as insolubles. The Rh solution is treated with Zn powder, precipitating an impure Rh. The impure Rh is heated in an atmosphere of Cl_2. Many impurities form volatile chlorides at this temperature and are expelled. Rh forms a polymeric trichloride, which is insoluble in aqua regia. The rhodium trichloride is digested in aqua regia for several hours, then filtered dried, and calcined, yielding a commercial grade of Rh sponge.

The residue insoluble in H_2SO_4 is fused with Na_2O_2, poured into thin slabs, and cooled. Ir is oxidized in the fusion to IrO_2, which is insoluble in H_2O. Ru and Os form soluble sodium salts and are separated from the Ir by filtration. The insoluble IrO_2 is dissolved in aqua regia, and ammonium hexachloroiridate is precipitated by the addition of NH_4Cl. Calcining yields pure Ir sponge.

The filtrate from the dissolution of the Na_2O_2 fusion contains $NaRuO_4$ and $NaOsO_4$. Ethyl alcohol is added to the solution, causing the precipitation of RuO_2, which is separated by filtration.

The Ru is purified by distilling with Cl_2. Volatile ruthenium tetroxide is collected. A saturated solution of NH_4Cl is added, causing the precipitation of ammonium hexachlororuthenate. The precipitated salt is calcined in H_2 yielding commercial Ru sponge.

The filtrate from the alcohol precipitation of Ru contains the Os. The solution is neutralized with HCl and is treated with powdered Zn, reducing the Os to the metallic state. Osmium tetroxide is formed by roasting the impure Zn in a current of O_2. The volatile OsO_4 is trapped in an aqueous solution of KOH. Ethanol is added to the solution, precipitating potassium osmate, which is mixed with an excess of NH_4Cl and calcined in an atmosphere of H_2. The resulting Os sponge is leached to remove KCl, leaving a commercial-grade Os sponge.

The refining of secondary scrap follows much the same procedures with minor variations. For example, solid metallic Pt and especially the Rh and Ir alloys of Pt are very difficult to dissolve in aqua regia. Therefore, the scrap generally is alloyed with Cu, Ni, Pb, or Zn before dissolution with acids.

Uses: These metals, in various forms, currently are used as catalysts for a wide variety of reactions. Products include high-octane gasoline, HNO_3, H_2SO_4, HCN, vitamins, antibiotics, H_2O_2, cortisone, alkaloids, and fuel-cell chemicals. These catalysts also are used to remove trace impurities, e.g., acetylene in ethylene or O_2 in H_2, or noxious constituents of partial combustion, e.g., automobile exhausts. Although substitutes are being sought, Pt is by far the best catalyst for pollution control of auto exhausts. See also **Catalysts.**

The corrosion resistance of the Pt metals has made the Pt crucible and the Pt electrodes commonplace laboratory tools. The glass industry makes use of large amounts of Pt and its alloys for manufacturing very pure glass. Synthetic fibers often are extruded through spinnerettes made of Pt alloys. The large use of Pt metals in dental and medical devices, in jewelry, and for decorative purposes is based on the corrosion resistance and general appearance of these metals.

Because of their high melting points and stability, Pt alloys have found applications in thermocouples, resistance thermometers, potentiometer windings, electrodes, insoluble anodes, high-temperature furnace winding, crucibles that can withstand corrosive materials at high temperature, and generally as materials of construction that will not contaminate products at very high temperatures. Often Pt and Pd are alloyed with Rh, Ir, Ru, or Os to increase their strength, hardness, and corrosion resistance.

Platinum metals, in particular Pd, find extensive use in the electrical industry. Most of these metals are used as contacts, particularly in telephone relays, where their resistance to oxidation and sulfidization results in circuits of reliability and stability. Alloys of Pt find use as grids for electronic tubes, in electrodes for aircraft spark plugs, for contact metal in printed and solid-state circuits, and in pressure-rupture disks.

Platinum. The melting point of Pt is 1769°C, boiling point, 4530°C, and density at 20°C 21.45 g/cm^3. There are five stable isotopes, 192, 194–196, and 198, and seven unstable isotopes, 188–191, 193, 197, and 199. Chavaneau succeeded in preparing malleable Pt and patented the process in 1783. Wollaston (England) is credited with isolating pure Pt in 1803. Julius Caesar Scaliger wrote in 1557 of a metallic substance in the mines of Mexico and Panama which cannot be melted in the Spanish furnaces.

Compounds: The best known and most abundant of the platinum-group metals, Pt forms many di-

and tetravalent compounds. The latter valence is more common and more stable. Pt in compact form is inert to all mineral acids except aqua regia. Under oxidizing conditions, fused alkalies will attack Pt to some extent. Molten halides, carbonates, and sulfates have little effect on the metal. Concentrated boiling H_2SO_4, fused cyanides, and fused alkaline sulfides will attack the finely divided metal. Pt is vigorously attacked by Cl_2 at elevated temperatures. In hot aqua regia or HCl containing chlorate ion or H_2O_2, the metal slowly dissolves, yielding a solution of hexachloroplatinic acid, H_2PtCl_6.

Platinum(II) hydroxide is made by adding KOH to a solution of platinum(II) chloride. The unstable black powder is easily oxidized by air and must therefore be handled in an inert atmosphere. In hot alkali or HCl, it disproportionates into the platinum(IV) compound and the metal. Very careful dehydration results in the formation of a gray powder that approaches the composition of platinum(II) oxide. Platinum(II) oxide can also be made by combining the elements at 420–440°C at an O_2 pressure of 8 atm.

When a solution of hexachloroplatinic(IV) acid is boiled for some time with NaOH, all the chloride ions are replaced by hydroxide ions. The resulting sodium hexahydroxoplatinum, $Na_2Pt(OH)_6$, is soluble in the basic solution, but it can be precipitated as hexahydroxoplatinic(IV) acid, $H_2Pt(OH)_6$, by the addition of acetic acid. The hydroxide ions of the salt are replaced by the corresponding ions of mineral acids when the compound is dissolved in acid. Hexahydroxoplatinic acid can be dehydrated to yield compounds corresponding to the tri-, di-, and mono-hydrate of platinum(IV) oxide. The last water molecule cannot be removed without some destruction of the dioxide.

Brown-black, insoluble, anhydrous platinum(IV) oxide is made by fusing hexachloroplatinic(IV) acid with $NaNO_3$ at about 500°C. The alkali salts are washed out with H_2O to free the fine insoluble residue of platinum(IV) oxide. This compound is known as *Adam's catalyst.*

When Pt is heated to 500°C in the presence of Cl_2, yellow-green, insoluble platinum(II) chloride is formed. At a pressure of 1 atm of Cl_2, the compound is stable from 435 to 581°C. It can also be made by heating hexachloroplatinic(IV) acid in Cl_2 at about 500°C. Platinum(II) chloride is soluble in HCl as tetrachloroplatinic(II) acid. It forms many salts that are water-soluble. These salts can be made by reducing a hot solution of the corresponding hexachloroplatinate(IV) with oxalic acid or SO_2. Platinum(III) chloride has a narrow range of stability. It can be made by contacting Pt or a platinum chloride with 1 atm of Cl_2 at 364–374°C. This dark-green to black compound is practically insoluble in cold concentrated HCl but does dissolve on warming, forming a mixture of tetrachloroplatinic(II) and hexachloroplatinic(IV) acids. Anhydrous platinum(IV) chloride is very difficult to prepare. This brown soluble solid can be made by heating hexachloroplatinic(IV) acid in Cl_2 at 360°C. The most common Pt compound, hexachloroplatinic(IV) acid, is readily made by dissolving Pt in aqua regia, followed by several evaporations with additional HCl to destroy nitrosyl compounds. The acid crystallizes as a hexahydrate. It is difficult to stop the evaporation at just this point, and slight local overheating causes excess loss of water. The sodium salt is quite soluble, and the compound is resistant to hydrolysis in basic solution, allowing the bromate hydrolysis to precipitate base metals and other Pt metals as their hydroxides. The Pt remains in solution. The insolubility of ammonium hexachloroplatinate(IV) often is used in refining Pt. Its slight solubility can be overcome sufficiently by mass action to allow its use as a gravimetric procedure for the determination of Pt. This yellow compound decomposes at red heat, yielding pure Pt sponge. The insolubility of the potassium salt is used for the gravimetric determination of potassium.

A series of di-, tri-, and tetrabromides is well known. Platinum(II) iodide is precipitated as a black insoluble compound by the addition of 2 equiv of iodide to a hot solution of platinum(II) chloride. The black, insoluble, graphitelike substance, platinum(III) iodide, is made by combining the elements in a sealed tube at 350°C.

In contrast with Pd, Pt does form a Pt(IV) iodide. When a concentrated solution of hexachloroplatinic(IV) acid is treated with a hot solution of KI, this brown-black substance is precipitated. The compound is somewhat unstable and light-sensitive. It dissolves in excess KI to form the complex salt, also rather unstable.

Pt forms a nonvolatile tetrafluoride, a pentafluoride, and a volatile hexafluoride. The dark red PtF_6 melts at 56.7°C and is very reactive. It even reacts with O_2 at 21°C to form dioxygenyl hexafluoroplatinate(V), O_2PtF_6.

When sulfur and Pt sponge are ignited, some platinum(II) sulfide is formed. The naturally occurring mineral is called *cooperite*. When heated in air or H_2, the products are metallic Pt and S.

Platinum(IV) sulfide can be made by heating ammonium hexachloroplatinate(IV) or Pt and S at 650°C. When precipitated by H_2S from chloroplatinic acid, the compound may exist as $PtS_2 \cdot H_2S$.

Divalent and tetravalent Pt probably form as many complexes as any other metal. The platinum(II) complexes are numerous with N_2, S, halogens, and C. The tetranitritoplatinum complexes are soluble in basic solution. Tetranitritoplatinum(II) ion is formed when a solution of platinum(II) chloride is boiled, at about neutral pH, with an excess of $NaNO_3$. The ammonium salt may explode when heated. Generally, platinum-metal nitrites should be destroyed in solution. They never should be heated in the dry form. Platinum(II) complexes most often have a coordination number of 4. Many compounds have been prepared with olefins, cyanides, nitriles, halides, isonitriles, amines, phosphines, arsines, and nitro compounds.

Platinum(IV) has a coordination number of 6. It forms complexes with halides, nitrogen and sulfur compounds, and other donors but to a lesser extent than platinum(II).

REFERENCES
Beamish, F. E.: "The Analytical Chemistry of the Noble Metals," Pergamon, London, 1966.
DeMent, Jack, and H. C. Drake: "Rarer Metals," Chemical Publishing, New York, 1946.
Griffith, W. P.: "The Chemistry of the Rarer Platinum Metals," Interscience-Wiley, New York, 1967.
Hampel, Clifford A.: "Rare Metals Handbook," Van Nostrand Reinhold, New York, 1954.
Hoke, C. M.: "Refining Precious Metal Wastes," Metallurgical Publishing, New York, 1940.
Kolthoff, I. M., and P. J. Elving (eds.): "Treatise on Analytical Chemistry," vol. 8, Interscience-Wiley, New York, 1963.
"Mellor's Comprehensive Treatise on Inorganic and Theoretical Chemistry," vols. 15 and 16, Longmans, London, 1937.
Smith, E. A.: "The Sampling and Assay of the Precious Metals," Griffin, London, 1947.
Wise, E. M.: "Palladium: Recovery, Properties, and Use," Academic, New York, 1968.

—Lionel E. Simmons, *Simmons Refining Company, Chicago, Ill.*

Plodding   (See **Soaps.**)

Plutonium   (See **Chemical elements;** and **Nuclear power plants.**)

Plywood   [See **Adhesives;** and **Silicates (soluble).**]

Pneumatic Mixers   (See **Mixing and blending, solids/solids.**)

Polishes   (See **Waxes.**)

Polishing   (See **Glass.**)

Pollution, Air   (See **Air pollution.**)

Pollution, Water   (See **Waste treatment.**)

Polonium   (After Poland, M. Curie's birthplace.) **Po** = 218.2 (at. wt.); 84 (at. no.). Polonium was discovered by Pierre and Marie Curie in 1898, in the pitchblende from Joachimsthal. It is the first element whose existence was demonstrated by radiochemical methods. Pierre and Marie Curie, having observed that the radioactivity of uranium ores was higher than that of uranium itself, came to the conclusion that this was due to the presence of traces of strongly radioactive, hitherto unknown elements in the ores. Chemical fractionation of pitchblende led to part of the radioactivity being concentrated with the sulfides which precipitated in acid medium. The existence of a new element, a homologue of tellurium, was thus established.

All polonium isotopes are radioactive. Among the 27 that have been identified (mass numbers ranging from 192 to 218), 7 exist in nature as members of the thorium or of the uranium series:

Half-life

^{210}Po (radium F). . .   138.3 days, α emitter
^{211}Po (actinium C') .   0.52 s, α emitter
^{212}Po (thorium C'). . 3×10^{-7} s, α emitter
^{214}Po (radium C') . . 1.637×10^{-4} s, α emitter
^{215}Po (actinium A) . . 1.83×10^{-3} s, α and β emitter
^{216}Po (thorium A) . .   0.158 s, α emitter
^{218}Po (radium A)   . .   3.05 min, α and β emitter

The longest half-lives are those of ^{208}Po (2.93 years) and of ^{209}Po (103 years).

The most important isotope is the one discovered by P. and M. Curie, ^{210}Po (RaF) which belongs to the uranium-radium series (cf. Table P-31). It decays by alpha emission to stable lead 206 (RaG) with a half-life of 138.3 days. The energy of the alpha particles is 5.3 MeV, which corresponds to a range of 3.87 cm in air at 15°C and 760 mm Hg (formation of 152,000 ion pairs), and of 22 μ in aluminum. The alpha radiation is practically pure as it is only accompanied by a weak gamma radiation, 1.25 (0.8 MeV) photons per 10^5 disintegrations. One curie of ^{210}Po (3.7×10^{10} disintegrations per second) corresponds to 222.2 μg of the element and constitutes

TABLE P-31.   *Uranium Radioactive Family*

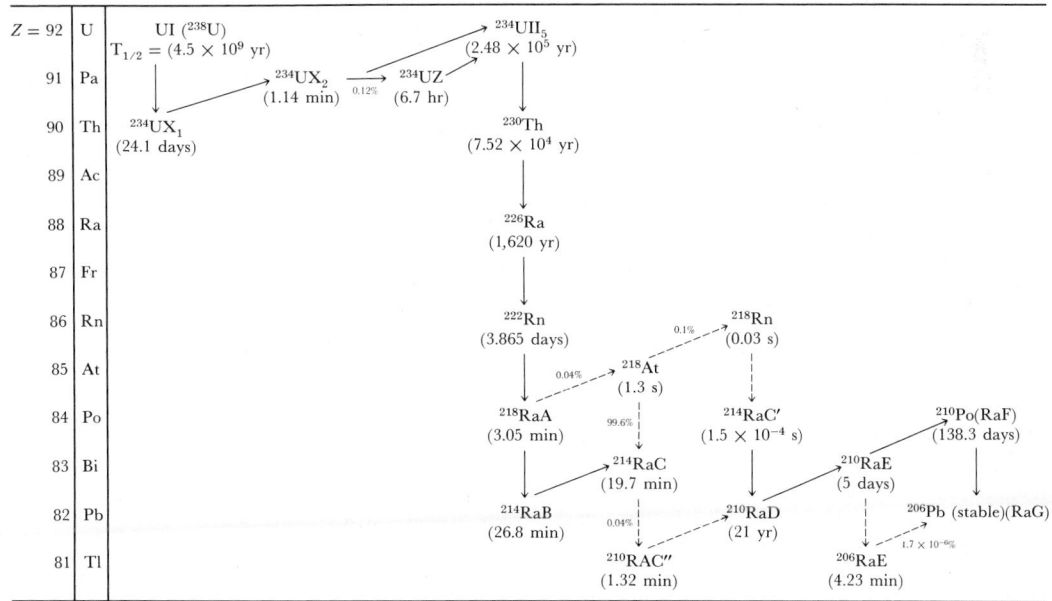

a source of heat yielding 27.224 cal/hr (0.032 W). Marie Curie and André Debierne have carried out a direct evaluation of the Avogadro number by measuring the volume of helium accumulated after the emission of a known number of α particles (helium nuclei) by a ^{210}Po source.

The abundance of ^{210}Po in uranium ores at radioactive equilibrium is 7.6×10^{-11} g/g uranium. The direct extraction of polonium from such ores is therefore a laborious task. The natural source generally resorted to for preparing trace amounts of ^{210}Po is radium D which can be recovered by acid washing of aged radon needles. Radium D decays with a half-life of 21 years to RaE and then to RaF (see the uranium-radium series). The separation of Po from RaE and RaD can be achieved readily by spontaneous deposition of polonium from dilute hydrochloric, nitric, and acetic media onto silver foil. This is subsequently dissolved in nitric acid and polonium separated from silver by precipitating the latter as the chloride.

Milligram amounts of ^{210}Po are at present produced by neutron irradiation of bismuth in a nuclear reactor

$$^{209}_{83}\text{Bi}(n,\ \gamma)^{210}_{83}\text{Bi}(\text{RaE},\ T_{1/2} = 5\ \text{days}) \longrightarrow\ ^{210}_{84}\text{Po}$$

The thermal neutron cross section of bismuth is 0.020×10^{-24} cm^2. Polonium can be separated from irradiated bismuth by distillation at 750–850°C. Alternatively, the bismuth target can be dissolved in hydrochloric acid and polonium precipitated by reduction to the metal with stannous chloride. The other isotopes are produced by irradiating bismuth or lead with neutrons, deuterons or helium ions, or lighter targets with heavy ions

$$^{116}_{48}\text{Cd} + ^{84}_{36}\text{Kr} \rightarrow\ ^{197}_{84}\text{Po} + 3\,n$$

Polonium was first artificially produced in 1936 by Livingood who irradiated bismuth with 5.4 MeV deuterons:

$$^{209}_{83}\text{Bi} + d \rightarrow p +\ ^{210}_{83}\text{Bi} \xrightarrow{\beta-}\ ^{210}_{84}\text{Po}$$

Very pure metallic polonium can be obtained in milligram amounts by decomposing the monosulfide PoS at 275°C in vacuo and subliming the metal at 450–500°C. Polonium is a noble metal; the standard potential of the couple Po/Po^{IV} in nitric medium is estimated to be $E_h^\circ = +0.77$ volt. It is spontaneously deposited onto silver, nickel, and less noble metals from dilute hydrochloric, nitric, or acetic solutions; a cathodic de-

posit of polonium is also obtained by electrolysis of its solutions, both acid and alkaline.

For quite a long time, the physico-chemical properties of polonium were exclusively investigated by radiochemical methods applied to 10^{-11}–10^{-6} g samples. Experiments on a milligram scale became possible when ^{210}Po was produced by neutron irradiation of bismuth. However, the interpretation of the data is difficult owing to important radiolytic effects which result from the high specific activity of this isotope (4,500 Ci/g). Isotopes 208 and 209 would present a substantial advantage from this point of view, if they could be prepared in macroquantities, as their half-lives are considerably longer.

In agreement with its electronic structure (completely filled K, L, M, and N shells, $5s^25p^65d^{10}6s^26p^4$), polonium belongs to group VIB of the periodic table. As the higher homologue of tellurium and a neighbor of bismuth, it shares a certain number of properties with these.

The metal is soft, silver grey, melts at 254°C and boils at 962°C. It has two allotropic forms: form α is stable at room temperature, is cubic, has a density of 9.196 at 36°C, and corresponds to an atomic radius of 1.69 Å; form β is rhombohedral, appears at about 36°C, and has a density of 9.398°C at 39°C.

More than a hundred lines have been identified in the visible and ultraviolet emission spectrum of polonium. The most intense are those at 2450, 2558, 3003, 4170, and 4493 Å. The ionization potential of the neutral atom is estimated to be 8.43 eV.

The most stable oxidation state of polonium in aqueous solution is IV; the element also shows the VI, II, and −II (and possibly a very unstable III) oxidation states. The metal is oxidized by air at room temperature and develops a superficial layer of yellow oxide. It forms the tetrahalides by reacting with chlorine and bromine. Metallic polonium dissolves in 2 N HCl yielding a red solution which becomes yellow owing to the oxidation of Po(II) to Po(IV). It is sparingly soluble in dilute nitric acid (1.9×10^{-5} mole/l in 0.1 N HNO$_3$) but dissolves readily in concentrated nitric acid. With fuming sulfuric acid it forms the red compound PoSO$_3$ which is very unstable and decomposes to yield PoO, black.

Polonium salts are very easily hydrolyzed. In the absence of a complexing agent Po^{4+}, PoO^{2+}, and PoO(OH)$_2$ appear successively with increasing pH. Formation of basic salts and of insoluble hydroxide causes colloidal behavior in neutral and in dilute acid or alkaline media. The hydroxide dissolves in concentrated solutions of sodium or potassium solutions, forming polonites (Na_2PoO_3, K_2PoO_3) analogous to the tellurites.

Polonium enters into a variety of complexes; the hexachloropolonites and the hexabromopolonites of the alkali elements are isomorphous with the corresponding hexachloro- and hexabromotellurites.

Sulfur dioxide and hydrazine in acid medium reduce polonium to the valence 2 which is unstable. Stronger reducing agents such as stannous chloride, titanous chloride and hypophosphorous acid in hydrochloric medium, or hydrazine in alkaline medium precipitate the metal. Hydrogen sulfide also acts as a reducing agent and causes the precipitation from dilute hydrochloric solution of a black monosulfide which is insoluble in ammonium sulfide. At tracer concentration, polonium is oxidized to the valence 6 by dichromate and by ceric salts. The standard potential of the couple $Po(IV)/Po(VI)$ in $6 N$ HCl is about $+1.5$ V. A very pure peroxide is deposited at a gold anode from nonreducing solutions.

Some polonium compounds are very volatile even at room temperature. Such is the case of the hydride PoH_2, polonium-dimethyl, and the carbonyl.

Polonium 210 is usually estimated by measuring its radioactivity. Quantities of the order of 100 or more millicuries can be estimated calorimetrically.

Like other alpha emitters, ^{210}Po presents a high degree of radiological hazard. When ingested or inhaled it accumulates in the spleen, kidneys, liver, and blood cells. The maximum permissible body burden is 4.5×10^{-12} ($0.02 \mu Ci$).

Polonium 210 has been utilized frequently in nuclear physics as a source of virtually pure alpha radiation. Mixed with beryllium, it constitutes a source of neutrons (free from gamma radiation), produced according to the reaction

$$^9Be + \alpha \rightarrow {}^{12}C + n$$

The discovery of artificial radioactivity by Irene and Frédéric Joliot-Curie in 1934 is an outcome of their investigations on the bombardment of light elements, boron, magnesium, aluminum, with the alpha particles emitted by a strong source of ^{210}Po.

The ionization of air by the radiation from polonium sources has been utilized for eliminating the accumulation of static charges on insulating materials. The heat liberated by the radioactive decay of polonium may be converted into electric energy at low voltage.

REFERENCES
Bagnall, K. W.: "Chemistry of the Rare Radioélements," Butterworth's Scientific Publications. Londres, 1957.
Curie, Marie: "Traité de Radioactivité," Ed. Hermann, Paris, 1935.
Haissinsky, M.: le polonium, *Act. Sci.*, no. 157, Ed. Hermann, Paris, 1937.
Haissinsky, M.: Polonium, Nouveau Traité de Chimie Minérale, vol. 13, Ed. Masson, Paris, 1961.
Joliot-Curie, I.: "Radioéléments naturels," Ed. Hermann, Paris, 1946.

—G. Bouissières, *The University of Paris, France*

Polyacrylamides   (See **Acrylamide polymers.**)

Polyacrylate Elastomers   (See **Elastomers.**)

Polyallomers   Polyallomers are highly crystalline polymers produced from propylene and ethylene. Polyallomers have a density of 0.902 g/cm^3, with a softening point ranging from 54 to 60°C (130–140°F). Heat-deflection temperature at 264 psi ranges from 46 to 60°C (115–140°F); at 66 psi ranges from 66 to 99°C (150–210°F). Brittleness temperature ranges from -23 to -40°C (-10 to -40°F). The yield strength falls between 2,900 and 4,200 psi; hardness ranges from 50 to 85 Rockwell R; tensile impact strength from 170 to 250 ft-lb/in^2. The dielectric constant of polyallomers is 2.3. They are thermoplastic. Crystallinity is exhibited to a degree normally associated only with polymers made from one monomer. The polyallomers exhibit properties that differ both from the homopolymers and the copolymers made from propylene and ethylene via other polymerization routes.

Polyallomers possess the characteristic milky color of the polyolefins but in thin sections are much clearer than the other polyolefins. Polyallomers exceed linear polyethylene in moldability, flow characteristics, mold shrinkage, and stress-crack resistance. Formulations of these resins can be injection-molded, extruded, and thermoformed. Injection-molded products include fishing-tackle boxes, typewriter cases, molded cowl panels for autos, shoe lasts, bowling-ball bags, and threaded container closures.

Polyamides   (See **Urethanes.**)

Polyamino Acids   (See **Peptides and proteins.**)

Polybasic Acids (See **Carboxylic acids.**)

Polycarbonates In the chronology of synthetic resins, polycarbonates are among the more recent thermoplastics in commercial importance although pilot production dates back to 1957. Polycarbonates may be defined as linear, low-crystalline, high-molecular-weight (=18,000) polymers in which the linking elements are carbonate radicals.

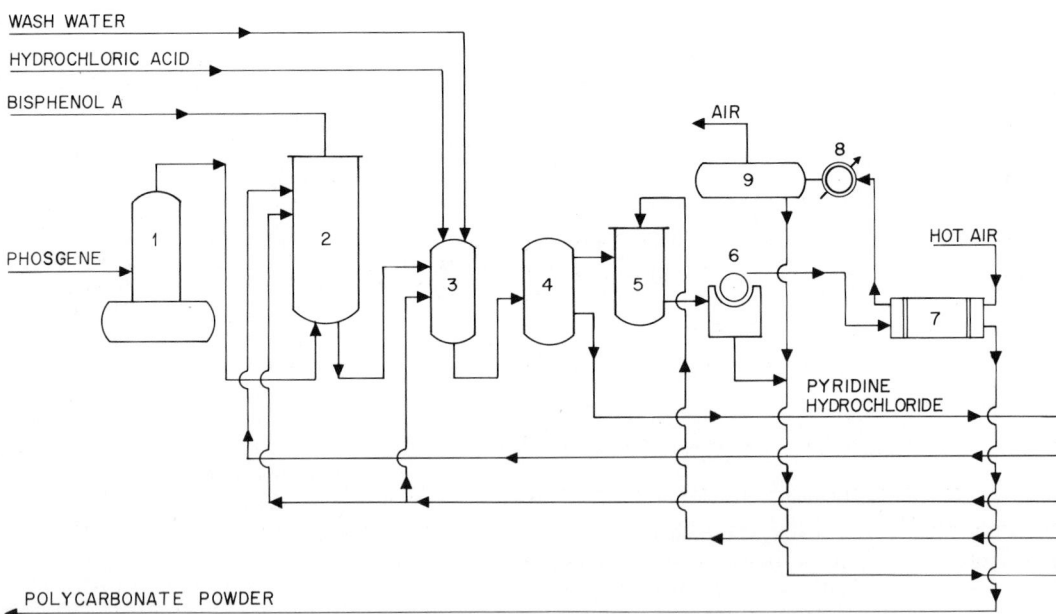

Two processes are followed in producing polycarbonates: (1) ester exchange between a carbonate diester and a dihydroxy aromatic compound and (2) the phosgenation of a dihydroxy aromatic, shown in Fig. P-18. Bisphenol A [2,2-bis(4-hydroxyphenyl)]; phosgene, $COCl_2$; pyridine, C_6H_5N; and a solvent (undisclosed) enter a polymerization reactor. The reacting mixture is carefully maintained at a temperature of about 40°C (or slightly lower) (104°F) for a residence time of from 1 to 3 hr. Variables that may be regulated include residence time and the proportions and temperatures of the entering ingredients. Feed purities, of course, affect the reaction. The reaction mixture comprises the polycarbonate polymer, pyridine hydrochloride, and unreacted pyridine in solvent.

The entire mixture proceeds (via polyethylene piping) to a wash tank, where water wash and HCl are added, the acid reacting with the residual pyridine. To lower the viscosity, more solvent may

Fig. P-18. Manufacture of polycarbonates by the phosgenation of a dihydroxy aromatic. (1) Phosgene vaporizer, (2) polymerization reactor in which polymer solution is made, (3) water wash tank(s), (4) decanter (polymer solution is top layer; pyridine hydrochloride solution is bottom layer), (5) precipitation tank where action of organic antisolvent produces a polycarbonate slurry, (6) filter, which yields a polycarbonate cake and a solvent-antisolvent solution, (7) drier, which produces polycarbonate powder for blending and extrusion, (8) condenser for drier volatiles, (9) separator which returns solvent-antisolvent solution to solvent-recovery section, (10) caustic mix tank, (11) solvent stripper, (12) azeotropic distillation column, which produces a water-pyridine azeotrope (top) with a sodium

be added at this point. By decantation, the polymer solution is separated from the pyridine hydrochloride solution. An antisolvent (undisclosed) is added to the polymer solution to precipitate the polycarbonate in slurry form. This slurry is filtered, after which the solid polycarbonate is dried to a powder.

The aqueous pyridine hydrochloride solution coming from the decanting operation is mixed with a caustic solution, allowing removal of chloride ions as NaCl. Any solvent present is stripped in a column prior to the caustic treatment. The stripped solution proceeds to an azeotropic-distillation column, where NaCl solution is removed as bottoms. The overhead is an azeotrope of about 43% water and 57% pyridine. The condensed azeotrope is combined with a breaking agent (undisclosed), and this mixture proceeds to a pyridine distillation column, which produces

water-free pyridine as overhead. The bottoms are treated to recover the breaking agent. The process includes a recovery system for both the solvent and antisolvent.

Properties: As shown by Table P-32, polycarbonate resins have good thermal stability and impact strength over a wide range, -168 to $+121°C(-275$ to $+250°F)$ and thus can be used continuously, under load, up to the higher temperatures stated as well as having a high resistance to creep or cold flow. Because of their high ductility, polycarbonates yield rather than shatter under high impacts. Dimensional stability is evident from their behavior in boiling water, which does not cause dimensions to alter by more than 0.001 in./in. Polycarbonates appear to be unaffected by ordinary humidity changes. Although generally unaffected by oils, greases, and acids, polycarbonates are dissolved by chlorinated hy-

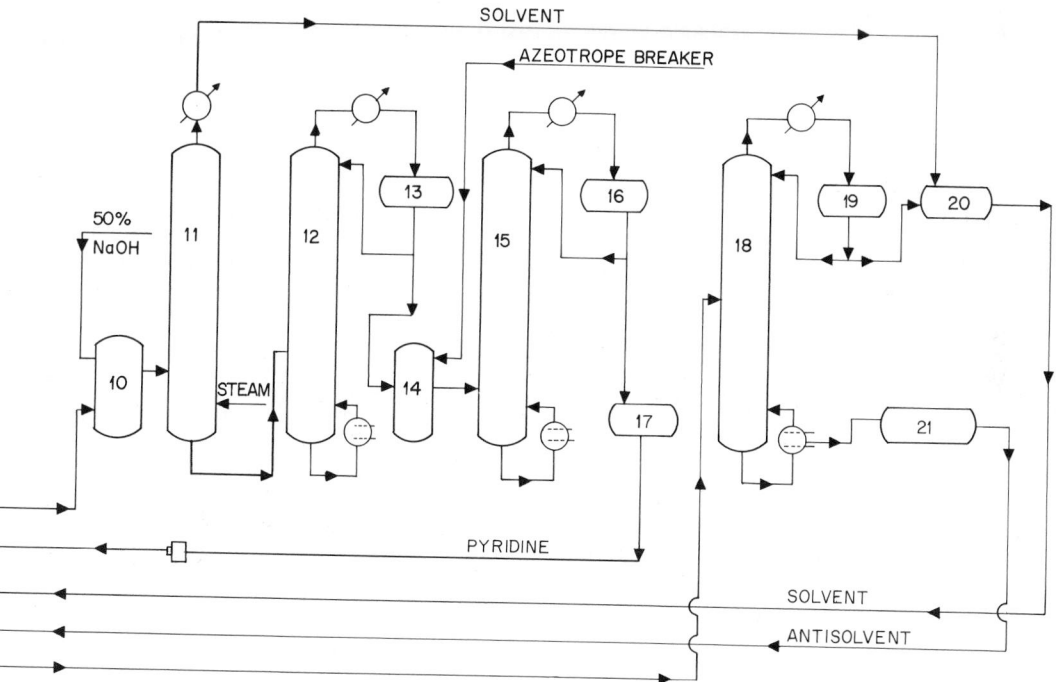

chloride (bottoms product) for disposal, (13) condenser, (14) mixing tank, to which is added an azeotrope breaker, (15) pyridine column, which produces dry pyridine (top) and bottoms for further processing to recover azeotrope breaker, (16) condenser, (17) pyridine storage drum, (18) solvent recovery column, which separates solvent (top) from antisolvent (bottoms), (19) condenser, (20) solvent storage drum, (21) antisolvent storage drum. It will be noted that columns (11), (12), and (15) and associated vessels constitute the pyridine-recovery portion of the process, whereas column (18) and associated equipment constitute the solvent-recovery portion of the process.

TABLE P-32. *Properties of Polycarbonates*

	Unreinforced	Reinforced with 40% Glass Fibers
Tensile strength, psi	9,000	20,000
Impact strength, ft-lb/in. of notch	16	4–4.7
Flexural strength, psi	12,000	26,000
Compressive strength, psi	11,000	19,000
Coefficient of thermal expansion per °F	3.0×10^{-5}	0.7×10^{-5}

drocarbons and attacked to some degree by many aromatic solvents, ketones, and esters. Methylene chloride can be used to bond polycarbonate parts.

Among products made from polycarbonate resins are injection-molded portable hand tools, small appliances, camera parts, pump impellers, light globes, safety helmets, air-conditioner housings, and electrical connectors. The raw material is available in pellet form over a wide range of colors. As with most other raw resin materials, a variety of end products are available as the result of blending, filling, and reinforcing.

Polycondensation Reactions (See **Polymerization.**)

Polyelectrolytes (See **Waste treatment.**)

Polyester Fibers Polyester fibers are defined as synthetic fibers containing at least 80% of a long-chain polymer composed of an ester of a dihydric alcohol and terephthalic acid. The first polyester fiber to be commercialized was prepared from the ester in which the dihydric alcohol was ethylene glycol and is the material used in the largest quantity by the textile industry. The only other ester to achieve commercial importance is 1,4-dimethyldicyclohexyl terephthalate, which represents a small fraction of total polyester production.

Properties of Poly(Ethylene Terephthalate): As produced for fiber manufacture, this material is a white solid with a molecular weight of 10,000–25,000. The basic chemical structure is

$$HOCH_2CH_2O\overset{O}{\underset{}{C}}{-}\!\!\left\langle\bigcirc\right\rangle\!\!{-}\overset{O}{\underset{}{C}}{-}O{-}\!\!\left[CH_2CH_2O\overset{O}{\underset{}{C}}{-}\!\!\left\langle\bigcirc\right\rangle\!\!{-}\overset{O}{\underset{}{C}}{-}O\right]_n\!\!{-}CH_2CH_2OH$$

If rapidly quenched from the melt, the material is substantially amorphous, but if it is cooled slowly or if the amorphous material is reheated to 95–180°C, it will crystallize to an extent that depends upon time and temperature. The crys-

talline material has a sharp melting point (related to crystallite size, type, impurities, and molecular weight) of between 250 and 265°C. Again, depending upon the degree of crystallinity, the density varies between 1.33 and 1.45. The glass-transition temperature T_g is in the range 70–80°C.

The polymer as manufactured generally contains a few percent of diethylene glycol (DEG) units, $\cdot OCH_2CH_2\cdot O \cdot CH_2CH_2O \cdot$, built into the polymer chain because of a side reaction. The higher the DEG content, the lower the melting point and the lower the melt viscosity for a given molecular weight and temperature. In addition to DEG, minor compounds, such as catalyst residues, oxidation and thermal degradation inhibitors, and optical brighteners and toners, may be found in fiber-grade polymer. TiO_2 also is added as a delustering agent in amounts up to 2%. The basic polymer may be modified by forming copolyesters, containing up to 15% of a comonomer, selected to provide basic dyeability and/or deeper-disperse dyeability, or other specific properties.

Manufacture: The original process, still in wide use for making the polymer, employs dimethyl terephthalate (DMT) and ethylene glycol as raw materials. More recently, a process using direct esterification of terephthalic acid (TPA) with ethylene glycol has gained increasing acceptance because of the greater availability of highly purified TPA.

With either process, the first step is the preparation of the intermediate diester, bishydroxyethyl terephthalate (bisHET), which then is further condensed to the polymer (page 897).

Reaction (1) takes place under conditions of an excess of glycol, high temperatures, and one or more catalysts. The mole ratio of glycol to DMT is 2:2.5, and temperatures in the vicinity of 200°C are used. The methanol produced, of course, is distilled off, thus forcing the reaction to the right. Hundreds of different catalysts have been claimed in the patent literature; most of them are oxides

$$CH_3O\overset{\displaystyle O}{\overset{\|}{C}}-\underset{}{\bigcirc}-\overset{\displaystyle O}{\overset{\|}{C}}-OCH_3 \; +$$

DMT

$$2HOCH_2CH_2OH \rightarrow$$

$$HOCH_2CH_2O\overset{\displaystyle O}{\overset{\|}{C}}-\underset{}{\bigcirc}-\overset{\displaystyle O}{\overset{\|}{C}}-OCH_2CH_2OH \; +$$

BisHET

or

$$2CH_3OH \quad (1)$$

$$HO\overset{\displaystyle O}{\overset{\|}{C}}-\underset{}{\bigcirc}-\overset{\displaystyle O}{\overset{\|}{C}}OH \; + \; 2HOCH_2CH_2OH \rightarrow$$

TPA

$$HOCH_2CH_2O\overset{\displaystyle O}{\overset{\|}{C}}-\underset{}{\bigcirc}-\overset{\displaystyle O}{\overset{\|}{C}}-OCH_2CH_2OH \; +$$

BisHET

$$2H_2O \quad (2)$$

or organic salts of alkali (especially lithium), alkaline-earth, or transition metals. A typical catalyst would be a manganese acetate–antimony oxide mixture, at a few hundred parts per million.

Although Eq. (1) shows bisHET as the sole reaction product, in reality a certain amount of condensation takes place and significant quantities of dimer, trimer, and tetramer are formed. The reaction product is more properly designated *prepolymer.*

The mole ratio of glycol to TPA generally is considerably lower in reaction (2) than in the DMT process, being about 1.3 to 1.5. This means that there is more low-molecular-weight polymer in the prepolymer produced. The reaction is believed to be carried out at a temperature of 200–250°C to accelerate the reaction. At these temperatures, increased pressures (2–10 atm) are required to avoid loss of ethylene glycol (boils at 197°C). The water produced may be distilled off under these conditions with minimum loss of glycol. No catalysts are necessary with this system, but they may be added at this point to assist in the polycondensation stage.

Using either the DMT or TPA system, the resulting prepolymer is next exposed to conditions which will lead to polycondensation by raising the temperature to about 280°C while simultaneously reducing the pressure to 0.5–1.0 mm Hg. The reaction is

$$nHOCH_2CH_2O\overset{\displaystyle O}{\overset{\|}{C}}-\underset{}{\bigcirc}-\overset{\displaystyle O}{\overset{\|}{C}}OCH_2CH_2OH \rightarrow$$

$$HOCH_2CH_2O\overset{\displaystyle O}{\overset{\|}{C}}-\underset{}{\bigcirc}-\overset{\displaystyle O}{\overset{\|}{C}}O\left[CH_2CH_2O\overset{\displaystyle O}{\overset{\|}{C}}-\underset{}{\bigcirc}-\overset{\displaystyle O}{\overset{\|}{C}}-O\right]_{n-1}CH_2CH_2OH \; +$$

$$(n-1)HOCH_2CH_2OH$$

The ethylene glycol is distilled off under these conditions of temperature and pressure. A time cycle of 4–6 hr is required to reach the desired molecular weight, which is monitored by the viscosity of the molten polymer as revealed by power consumption of the stirrer used in the reaction vessel. When the desired molecular weight is reached, the reaction mass is discharged into cold water to quench it and then passed to a dicer or granulator, or it is transferred while still molten directly to the spinning unit.

The foregoing reaction can be carried out with many variations. The early processes were batch, but many of the newer plants use continuous operation. Continuous operation can be profitably connected directly to spinning, thus obviating the necessity for dicing and remelting. Many different catalysts have been described in the patent literature, and the choice of catalyst may be dictated by time and temperature desired, amount of DEG and color desirable in the product, and thermal and hydrolytic stability of the product. If a delustered yarn or fiber is wanted, as is usually the case, a suspension of TiO_2 is added to the reaction mass prior to polycondensation. Other additives may be oxidation and thermal/degradation inhibitors, optical brighteners, and toners, all of which are present to varying degrees in commercial polyester fibers. The major steps of a continuous process for making polyester resin are shown in Fig. P-19.

Spinning: The basic process for making polyester fibers from the polymer is called *melt spinning,* i.e., heating the polymer above its melting point, forcing it through small holes in a metal plate, and then quenching the molten stream as it issues from the holes by means of a current of cool air.

In modern spinning plants the polymer is heated and conveyed to the spinning head by means of a melt extruder. If the polymer is initially in the form of quenched chips, it must be

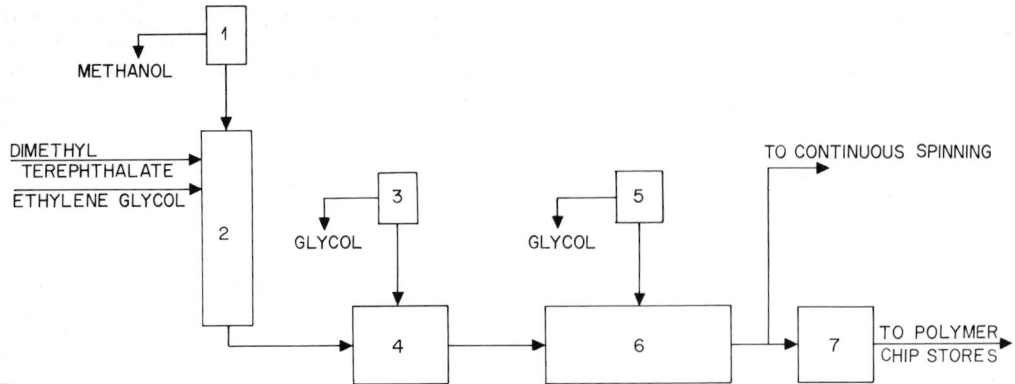

Fig. P-19. Continuous production of polyester via the DMT process. (1) Condenser, (2) ester-interchange column, (3) medium-vacuum source, (4) polymerization and glycol strip, (5) high-vacuum source, (6) polycondensation reactor, (7) quench and dicer.

thoroughly dried (moisture content <0.01%) before melting or the molten resin will degrade by ester hydrolysis during the spinning process. The extruder accepts the dried chips from a hopper and melts them by the application of heat electrically through the walls of the extruder and mechanically by friction between the softened, highly viscous chips. At the end of the extruder, where the chips have melted and the temperature is 282–299°C (540–570°F), the root of the screw increases abruptly applying pressure to the melt to force it to the metering pump, which is next in line (see Fig. P-20). The metering pump delivers a precise flow rate to the filter and spinnerette to ensure accurate denier control of the filaments. The molten polymer in its passage from extruder to jet has a viscosity of the order of 1,000 P, more or less, depending on molecular weight and temperature of the melt. Pressures in the area range from 1,000 to 5,000 psi. The filter consists essentially of a cup containing purified sea sand, although other media, e.g., ground glass, metal screens, and porous metal filters, have been used. The spinnerette is made of stainless steel with a number of holes ranging from 14 to several hundred, depending upon the filament count of the yarn being spun. The holes have diameters varying from 0.008 in. to about 0.025 in., according to the denier being spun. The depth of the holes is 1.3–3 times the diameter.

On emerging from the spinnerette, the filaments, aided by gravity, pass down through a vertical spinning tube (10–20 ft long) or chamber to the take-up device. Just below the spinnerette,

the filaments are cooled by a transverse flow of air from the quench chamber, which solidifies them. As the filaments emerge from the spinning tube, they are brought in contact with a ceramic wheel, which is rotating in a bath of finish. The finish is applied to afford lubrication to the filaments during further textile processing and to serve as an antistatic application to prevent the buildup of electrostatic charges on the filament inasmuch as polyester is a markedly hydrophobic material. In the final step, the yarn is wound on a tube or spool at speeds of 500–1,500 m/min. This speed is greater than the velocity at the jet by a factor of 10 or 100, producing *jet stretch*, which is an important parameter in determining the quality of the spun yarn.

The spun yarn is weak and highly extensible because the polymer molecules are randomly oriented. To impart strength and dimensional stability the yarns must be drawn at temperatures above the glass-transition temperature of the material by pulling the yarn between two *godet* wheels, the second of which is rotating at a speed 3–6 times as fast as the first. The higher the draw ratio, i.e., the ratio of the two speeds, the more oriented the molecules become and the stronger the yarn. The two godets may be heated, and the yarn may pass over a hot plate between the two. Another technique is to snub the yarn around a heated ceramic pin set between the two godets. The heat applied, as well as the mechanical heat generated during drawing, induces crystallization in the yarn, which locks the molecules in their oriented position, thus conferring dimensional

POLYMER CHIPS

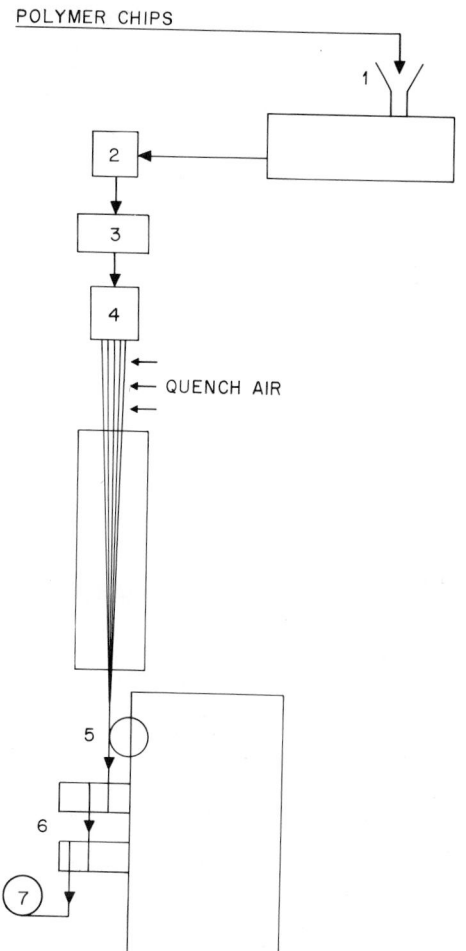

QUENCH AIR

Fig. P-20. Process for melt spinning polyester filaments.
(1) Melt extruder, (2) pump, (3) sand filter, (4) spinnerette,
(5) finish wheel, (g) godets, (7) take-up.

the first. The tow may be heated between the rolls
by a hot platen, infrared heater, or hot-water
bath. After drawing, the tow is passed through
a mechanical crimper and then a heat-setting
oven, which further crystallizes the fibers so that
they will retain the crimp in later operations.
After heat setting, the tow is cut into staple of
lengths varying from $1\frac{1}{2}$ to 6 in. depending on the
end product desired, and conveyed to a baler,
which compresses the staple and binds it.

Properties: The physical properties of a textile
fiber depend on the molecular weight of the polymer, the degree of crystallinity, and the degree of
orientation of the molecules with respect to the
fiber axis. By virtue of the chemical nature of
poly(ethylene terephthalate) it is possible to spin
the fiber in an amorphous, relatively unoriented
form and then control the orientation and crystallinity of the final product by controlling the
variables of temperature, speed, and draw ratio
in the drawing and heat-setting process. This
means that a wide variety of physical properties
can be imparted to the fibers to adapt them to
particular end uses in the textile market. For
example, a staple fiber designed for low-pilling
fabrics may have a tenacity of 2.5 g/denier at 43%
elongation and have been spun from a polymer
of molecular weight 10,000. On the other hand,
a high-tenacity yarn for tires or other industrial
purposes may have a tenacity of 7.0 g/denier at
20% elongation with a molecular weight of
25,000. Almost any value of mechanical properties between these two extremes can be made and
is available on the commercial market, with the
general limitation that the higher the tenacity, the
lower the ultimate elongation. In addition, the
shape of the stress-strain curve can be widely
varied. This is useful in matching the stress-strain
curve of the polyester to that of other fibers with
which it is often blended, e.g., cotton, wool, and
high-wet-modulus rayon.

The glass-transition temperature T_g of the polymer, another important property of textile fibers,
is the temperature above which extensive rotation
of large segments of the molecules takes place,
thus relieving stresses imposed at the molecular
level. The T_g of poly(ethylene terephthalate) is
75–100°C, depending on the degree of crystallinity. When a polyester fabric is held in a flat
state and heated above T_g, the stresses in the
molecules introduced in drawing and weaving are
relaxed and the fabric tends to remember and
return to the flat geometry. Thus, apparel made

stability against further heating. The maximum
degree of drawing is set by the preorientation in
the spun yarn, determined by the jet stretch.

In the manufacture of staple, the yarn is not
collected on spools at the spinning machine but
gathered together with a large number of other
yarns into a bundle called *tow.* The tow is collected in large boxes or cans and transported to
the drawing area. Here the tow is withdrawn
from the box in the form of a wide, flat band,
which is drawn between two sets of heavy-duty
rolls, the second traveling at a higher speed than

from fabric heat-set in this manner tends to resist wrinkling and mussing. In addition, severe wrinkling can be removed by tumble drying the fabric during laundering at temperatures above the T_g. These properties place polyesters in the forefront of the easy-care fabrics.

The major disadvantage of polyester as a fiber-forming material is its very low moisture retention. Under normal, everyday conditions of temperature and relative humidity, polyester will absorb only a few tenths of 1% moisture, compared with 9–12% for cellulosic materials and wool. A fabric with low moisture-sorption capacity has an uncomfortable feel, since perspiration is not absorbed from the skin but tends to spread through the interstices of the fabric, which feels cold and clammy. As a result, polyester is often blended (in staple form) with cotton, wool, or rayon to give fabrics with the strength and easy-care characteristics of polyester and the moisture absorption of cellulose or wool for comfort. The percentage of polyester in such blends varies from 50 to 65%.

This low sensitivity to moisture and the general chemical inertness of polyester make these fibers more difficult to dye than cotton, rayon, nylon, or acetate fiber. Dyeing is done with the class of dyestuffs known as *disperse dyes,* insoluble pigments which are dispersed in water before being applied to the fabric. The pigments are soluble in the organic material of the fibers and penetrate the organic phase at high temperatures. In order to get deep dye shades it is necessary to autoclave the dyebath and raise the temperature to 120°C or to use organic *carriers,* i.e., substances such as chlorinated phenols which swell the polyester and permit easier penetration of the fiber. In a type of dyeing called Thermosol the dye dispersion is laid onto the fabric, after which the fabric is raised to very high temperatures (177–204°C, 350–400°F), where the dye diffuses into the fiber matrix.

This inherent difficulty in dyeing poly(ethylene terephthalate) has led to the commercial development of two basic variations of the fundamental polyester molecule, the so-called *deep-dyeable* and *cationic* or *basic dyeable* polyesters. They are both copolymers. In deep-dyeable polyester a few mole percent of a high-molecular-weight comonomer is incorporated into the molecule, interfering somewhat with the crystallization of the fibers, leading to a more open structure of the molecular network, and permitting a higher degree of penetration of the disperse dyestuff. This enables deeper dye shades to be obtained, often without the use of carriers. In cationic dyeable polyesters, the comonomer contains dye sites such as sulfonate groups to permit the fiber to be dyed with basic dyestuffs which will not dye ordinary or deep-dye polyesters.

Types: The two main classes are continuous-filament yarns and short-cut fiber, called *staple.* A wide range of deniers is available in continuous-filament yarns, varying from very fine deniers of about 20 up to 2,000 for heavy industrial yarns. The denier of a yarn is the weight in grams of 9,000 m and thus is a measure of fineness. The number of filaments in these yarns ranges from about 7 for the 20-denier yarns up to 384 for the heavy material.

Staple fiber is produced in sizes ranging from 1.5 to about 15 denier per filament. The finer deniers are used in making blends with cotton and rayon for apparel, while the coarser-denier yarns generally are used in making carpets. Staple lengths vary from $1\frac{1}{4}$ to 6 in.

Fiber with no added delustrant is designated as *clear. Bright* fiber has about 0.1% TiO_2; *semidull* fiber has about 0.25% TiO_2; and *dull* fiber may have up to 2% TiO2. Other variations in physical properties and dyeing characteristics include optically brightened, high-modulus, high-shrink, high-tenacity, low-pilling, deep-dyeable, and cationic-dyeable fibers.

—James P. Dux, *American Viscose Division, FMC Corporation, Marcus Hook, Pa.*

Polyester Polyols   (See **Urethanes.**)

Polyester Resin Paints   (See **Paint.**)

Polyesterification   (See **Polymerization.**)

Polyethylene   Polyethylene is a thermoplastic material available in a wide variety of formulations with such useful properties as toughness at temperatures ranging from -57 to $+93$°C (-70 to $+200$°F); stiffness, ranging from flexible to rigid; and excellent chemical resistance. The plastic can be fabricated by all thermoplastic processes.

Formulations are classified primarily by density (specific gravity) of the resin: (1) ASTM type I, specific gravity 0.912–0.925, variously designated as low-density, regular, conventional, or high-pressure polyethylene; (2) ASTM type II, specific gravity 0.925–0.940, commonly referred to as intermediate-density or medium-density polyethylene; and (3) ASTM type III, specific gravity 0.940–0.965, commonly termed high-density, linear, or low-pressure polyethylene. Within each density classification, products with different melt indexes are available.

Chemical Composition: Polyethylene is formed from the polymerization of ethylene under specific conditions of temperature and pressure and in the presence of a catalyst, according to the simple equation

$$
\begin{array}{c}
\underset{\substack{| \\ H}}{\overset{\substack{H \\ |}}{C}} = \underset{\substack{| \\ H}}{\overset{\substack{H \\ |}}{C}} \xrightarrow[\text{cat.}]{\text{pressure}} \left[-\underset{\substack{| \\ H}}{\overset{\substack{H \\ |}}{C}} -\underset{\substack{| \\ H}}{\overset{\substack{H \\ |}}{C}} -\underset{\substack{| \\ H}}{\overset{\substack{H \\ |}}{C}} -\underset{\substack{| \\ H}}{\overset{\substack{H \\ |}}{C}} - \right]_n
\end{array}
$$

The reaction is exothermic and may form polymer from a molecular weight of 1,000 to well over 1 million. The high-pressure process, which normally produces types I and II, uses oxygen, peroxide, or other strong oxidizers as catalyst. Pressure of reaction ranges from 15,000 psi to over 40,000 psi. The polymer formed in the high-pressure process is highly branched, with side branches occurring every 15–40 carbon atoms on the chain backbone. Crystallinity of this polyethylene is approximately 65%. Amorphous content of the polymer increases as the density is reduced.

The low-pressure process normally produces type III, but variations of this process are known to produce types I and II also. Catalysts used in the low-pressure process vary widely, but the most frequently used are metal alkyls in combination with metal halides or activated metal oxides. Reaction pressures normally fall within 50–200 psi. Polymer produced by this process is more linear in nature, with branching occurring about every 1,000 carbon atoms. Linear polyethylene is approximately 85% crystalline.

Copolymers of ethylene with other monomers, e.g., butene-1, hexane, ethyl acrylate, vinyl acetate, and acrylic acid, have been formulated to develop such specific properties as environmental-stress-crack resistance and improved flexibility or toughness. Chlorinated polymers and others with ultrahigh molecular weights are in development to extend the property range of polyethylenes from extremely rigid to elastomeric.

Uses: Polyethylene products include injection-molded housewares, containers, automotive parts, institutional seating components, and toys; extruded film, sheet, electrical-cable jacketing, pipe, tubing, and coatings; blow-molded bottles; and rotationally molded tanks, carboys, and toys.

Properties

General: The principal properties of compression- and injection-molded polyethylene are summarized in Table P-33. Tensile strength, hardness, chemical resistance, surface appearance, and flexural modulus increase with an increase in density (from type I through type II to type III).

Optical: Ethylene is translucent to opaque white in thick sections, opacity increasing with density. The average refractive indexes are type I, 1.51; type II, 1.52; and type III, 1.54. Relatively clear film can be extruded from polyethylene, especially if it is quenched rapidly. The plastic accepts pigmentation readily. Most coloring is performed using dry-blend techniques. Color-dispersion devices are required to ensure a thorough mixing of resin and pigment.

Frictional: Coefficients of friction for polyethylene are in the middle range for plastic materials. Values are from 0.30 for type III on itself or on steel to 0.06 for type I on itself or on steel.

Mechanical: These properties vary with density and melt index. Low-density polyethylenes are flexible and tough; high-density products are quite rigid and have creep resistance under load. Toughness is the primary mechanical property affected by melt index, with lower-melt-index polyethylenes having greater toughness. Under loads, polyethylene is subject to creep, stress relaxation, or a combination of both.

Electrical: Excellent dielectric characteristics at all frequencies and high electrical resistivity have made polyethylene one of the most important insulating materials for wire and cable. Typical electrical properties are given in Table P-34.

Thermal: At no-load conditions, polyethylene has good heat resistance. However, small loads can cause distortion at relatively low temperatures. Dimensional stability of polyethylene is fair to good. Dimensional changes caused by crystallization during cooling usually occur in a nonuniform pattern, resulting in warpage. Warpage is encountered most often in high-density molding resins. Most shrinkage occurs within 48 hr after fabrication and for type I and type II materials is 0.01–0.03 in./in. Shrinkage in type III materials is 0.015–0.05 in./in. The thermal expansion of polyethylene is approximately linear up to the melting point. The rate of expansion is 9.5×10^{-5} for type I; 8.0×10^{-5} for type II; and 7×10^{-5} for type III. Expansion and contraction can cause a high stress buildup if a polyethylene

TABLE P-33. *Typical Properties of Molded Polyethylene*

Polyethylene	Melt Index	Yield Tensile Strength, psi	Yield Elonga-tion, %	Ultimate Tensile Strength, psi	Ultimate Elongation, %	Modulus in Tension, psi
Compression-molded:						
Type I:						
Sp gr 0.916*.	0.3	1,500	. . .	2,000	550	16,000
	2.0	1,300	17	1,700	500	13,000
	20.0	1,250	20	1,400	520	15,000
Sp gr 0.920*.	2.0	1,500	16	1,700	500	20,000
Type II, sp gr 0.926†. .	2.0	1,800	15	1,500	170	31,000
	20.0	1,900	14	1,400	150	30,000
Type III, sp gr 0.960†	0.7	4,000	. . .	2,200	400	140,000
	1.0	3,800	. . .	2,000	400	140,000
	5.0	4,100	. . .	1,900	200	140,000
Injection-molded:						
Type I, sp gr 0.916* . .	2.0	1,600	. . .	2,500	150	18,000
	20.0	1,100	. . .	1,400	150	14,500
Type II, sp gr 0.926† . .	5.0	1,800	. . .	2,000	150	31,000
	20.0	1,700	. . .	1,800	125	30,000
Type III, sp gr 0.960 . .	1.0	3,400	. . .	2,300	500	130,000
	5.0	3,700	. . .	2,200	400	135,000

*Test at 20 in./min.
†Test at 2 in./min.

part is fastened to a material with a much lower thermal coefficient. Slotted holes or other means of allowing the plastic to move are recommended. Thermal conductivity of types I and II is 2.3 (Btu)(in.)/(hr)(ft²)(°F). The value for type III is 3.4. Specific heat of all types is 0.55 Btu/(lb)(°F).

Environmental: Rupture of molecular bonds by external and internal stress in the presence of certain compounds is referred to as *environmental stress cracking*. Small molecular fractures in the amorphous regions propagate until visible cracks appear. In time, the part may fail. Chemical agents which accelerate stress cracking in polyethylene include detergents; aliphatic and aro-matic hydrocarbons; soaps; animal, vegetable, and mineral oils; ester-type plasticizers; organic acids; and aldehydes, ketones, and alcohols. There is no adequate test for stress cracking. The most practical data are obtained from end-use tests of fabricated parts. Low-melt-index polyethylene has better stress-crack resistance than high-melt-index material of the same density made by the same process. The more flexible, low-density material develops a lower internal stress at a given deflection, resulting in improved stress-crack resistance.

Deterioration occurs in uncolored polyethylene exposed to weather. Ultraviolet light causes photoactivated oxidation. Satisfactory weathering for-

TABLE P-34. *Typical Electrical Properties of Polyethylene*

Property	Type I	Type II	Type III
Dielectric strength, short-time, ¼-in. specimen, V/mil . . .	460–700	460–650	450–500
Volume resistivity, Ω-cm	10^{-15}–10^{-16}	10^{-15}–10^{-16}	10^{-15}–10^{-16}
Dielectric constant at 10^3 and 10^6 Hz.	2.25–2.35	2.25–2.35	2.25–2.35
Dissipation factor:			
At 10^3 Hz	0.0002	0.0002	0.0002
10^6 Hz	0.0002	0.0002	0.0003
Loss factor:			
At 10^3 Hz	0.0003	0.0003	0.0004
10^6 Hz	0.0003	0.0003	0.0007
Arc resistance, s	150	150	150

mulations contain 2–2.5% of well-dispersed carbon black and stabilizers. The carbon black prevents ultraviolet-light penetration.

Unmodified polyethylenes are flammable and are classified in the slow-burning category by the National Board of Fire Underwriters. Burning rate is approximately 1–1.5 in./min. Self-extinguishing formulations are available.

Chemical Resistance: At room temperature, polyethylene is insoluble in practically all organic solvents, although softening, swelling, and environmental stress cracking can occur. At high temperatures, some concentrated acids and oxidizing agents chemically attack polyethylene. Above 60°C (140°F), the material becomes increasingly soluble in aliphatic and chlorinated hydrocarbons. Chemical resistance increases slightly as density is increased.

Adsorption and Permeability: Polyethylene is water-resistant and is a good water-vapor barrier. Less than 0.1% of water is absorbed in a 2-in.-diam, 1/8-in.-thick disk of polyethylene in 24 hr. Transmission of other gases is high compared with that of other plastics. Polyethylene is not satisfactory for retention of vacuum.

Fabrication: The principal methods used are injection molding, blow molding, film and sheet extrusion, pipe extrusion, and thermal-fusion processing.

Injection molding: Polyethylene of all melt indexes is fabricated on screw-type molding machines. High-melt-index polyethylenes are used for fast-cycle thin parts without severe physical-property requirements. Low-melt-index resins are used for parts having heavier wall sections with impact-strength or stress-crack-resistance specifications. Good mold-temperature control is essential.

Blow Molding: Low-melt-index resins are best for blow molding since during processing the polyethylene is extruded downward as a free tube which must have sufficient melt strength to avoid stretching with its own weight. Furthermore, most applications require a resin with good stress-crack resistance.

Film and Sheet Extrusion: Polyethylene film is extruded either as blown tubing or cast flat film for use in packaging and industrial-film applications. Sheet can be extruded with standard techniques. The extruder must be capable of high heat input and of developing high shear rates. The roll stack must have sufficient cooling capacity to cool the sheet quickly and uniformly.

Pipe Extrusion: Extruded pipe is made on conventional equipment. Most pipe is made in coils for easy handling. Lower-melt-index polyethylenes are used to produce pipe with good long-term burst resistance.

Thermal Fusion: Polyethylene is well suited for this process. Large tanks and drums, drum liners, and small boats are made this way. Rotational molding is often used to produce completely closed, hollow parts from polyethylene in powder form.

Secondary Fabrication Methods: Standard machining methods can be used on polyethylene, but frictional heat should be minimized. Solvent-type adhesives are not satisfactory because of the chemical inertness of polyethylene. Rubber-base contact adhesives are used occasionally. For best adhesion, the surface of the plastic should be oxidized. Polyethylene parts can be welded by conventional thermoplastic methods.

Thermoforming: High-density resins are used extensively in thermoforming. This method requires a low-melt-index product to avoid sagging during the heating step. Long heating cycles are necessary because of the low thermal conductivity and high softening temperature of polyethylene.

Decorating: Polyethylene parts are decorated by silk screening, hot stamping, or dry offset printing. For satisfactory printing, the surface must be oxidized by hot air, flame, chlorination, sulfuric acid–dichromate solution, or electronic bombardment. Hot-air or flame methods are used with molded parts; flame or electronic methods with film. Inks specially made for polyethylene give best results. Roll-leaf hot stamping does not require pretreatment of the surface.

Design: Because of high mold shrinkage, parts must be carefully designed to minimize warpage. Wall cross-sectional thicknesses should be uniform throughout the part. Large flat areas should be avoided. Corners should be radii rather than square. Stiffing ribs should be <80% of the thickness of the wall to which they are attached. Thermoformed parts require liberal radii and draft angles. Slight undercuts can be incorporated when a female mold is used. Dimensional variations in a part made of polyethylene are difficult to predict. In general, greater tolerances should be allowed than with the more rigid plastics. See also **Petrochemical complex.**

—B. A. Geisert, *The Dow Chemical Company, Freeport, Tex.*

Polyethylene Wax (See **Waxes.**)

Polyhalite (See **Potassium.**)

Polyimide Resins These resins are not of a single chemical composition but comprise a family of organic polymers. Starting ingredients may be an anhydride and a diamine to form a polyamide acid, which then is imidized to form the polyimide resin. Particularly at high temperatures, the mechanical and electrical properties of molded polyimides are superior to most other engineering plastics. Powder-metallurgy techniques or conventional thermoset methods of fabrication are used in producing final polyimide parts.

Polyimide film is used as wire and cable wrap and for insulation and slot liners for motors, printed-circuit backing, and other electrical applications because of its combination of mechanical and electrical properties, particularly at elevated temperatures. Shrinkage is only 3.5% at 400°C (752°F), an important factor where magnetic tape and printed circuits must be exposed to high ambient temperatures. The film also is attractive in aerospace applications because of its resistance to ionizing radiation, which exceeds that of other organic films.

Molded polyimide parts can be used continuously in air up to 260°C (500°F) and in an inert atmosphere up to 316°C (600°F). Polyimide parts also perform satisfactorily at cryogenic temperatures. Selected properties of typical molded polyimides are given in Table P-35.

Polyisocyanates (See **Urethanes.**)

Polymerization Polymerization is the joining together of small molecules to form larger molecules. When only a few molecules are so combined, yielding a polymer with a molecular weight of a few hundred or a thousand, the product is called an *oligomer*. An oligomer may contain two units (a *dimer*), three units (a *trimer*), four units (a *tetramer*), and so on. If the molecular weight is very large, say, 50,000 or 1,000,000, the product is called a *high polymer* or a *macromolecule*. The word *polymer* is derived from the Greek *poly,* many, and *meros,* parts.

Giant polymeric molecules are important constituents of all living organisms, as illustrated by the proteins (muscle, skin, hair, tendon, enzymes, hemoglobin, and antibodies), the nucleic acids (genetic material of all cells), the carbohydrates (starch and cellulose), and many other natural materials. From earliest days, man has used these naturally occurring polymers as materials for clothing, shelter, and tools. Cotton, wool, silk, wood, leather, and rubber are examples. However, an understanding of the molecular *structure* of polymers and the ability to *synthesize* useful polymers has come only in modern times.

The German chemist Emil Fischer began to study starch, polypeptides, lignin, cellulose, and

TABLE P-35. *Properties of Molded Polyimides*

	Type SP-1*	Type SP-2*
Dielectric constant at 10^5 Hz and 73°F (22.8°C) . . .	3.4	7.6
Deflection temperature at 264 psi:		
°F.	~680	~680
°C.	~360	~360
Specific gravity.	1.43	1.51
Tensile strength, psi:		
At 73°F (22.8°C)	13,000	9,000
482°F (250°C).	6,600	6,000
600°F (316°C).	5,200	5,000
Elongation, %:		
At 73°F (22.8°C)	7–9	4–6
472°F (250°C).	6–8	3–5
Compressive strength, psi:		
At 73°F (22.8°C)	40,000	32,000
482°F (250°C).	20,000	13,000
Flexural modulus, 1,000 psi:		
At 73°F (22.8°C)	450	540
482°F (250°C).	290	370
600°F (316°C).	260	325
Coefficient of friction, dynamic, air, unlubricated . . .	0.20–0.61	0.06–0.35

*Du Pont designations for specific materials.

rubber in the late 1890s. A breakthrough in the synthesis of polymers occurred in 1909, when the Belgian-American chemist, Leo Baekeland, trying to ascertain the constitution of sticky, resinous deposits that formed on the chemical glassware he was using for handling phenol and formaldehyde, found that the gummy material turned hard and became transparent with the application of heat. The resulting material had excellent electrical, chemical, and mechanical properties. Thus, the first thermosetting plastic, phenolformaldehyde (phenolics), named *bakelite,* arrived when there was a great demand for a material to construct electrical apparatus, although the material also found many other uses. Urea-formaldehyde resins appeared shortly thereafter.

Shortly after World War I, the central concept of polymer chemistry, that of a very long covalent chain, was established, and several commercial synthetic polymers followed shortly. Since World War II, polymer chemistry has become a predominant element in industrial organic chemistry, and today many categories of synthetic materials are produced in large quantities: elastomers, fibers, films, coatings, and plastics of many types. Most of these materials are described elsewhere in this volume.

Many different kinds of polymerization reactions can be used to prepare high-molecular-weight linear and network polymers, of which the most important are (1) polycondensation reactions, (2) addition polymerization of compounds containing a C=C bond, (3) ring-chain conversions, and (4) combinations of the foregoing.

Polycondensation Reactions. Many organic functional groups undergo condensation reactions which join together two molecules, often with the elimination of H_2O or some other small molecule. For example, a carboxylic acid, $R \cdot COOH$, can condense with an alcohol, $R' \cdot OH$, to form an ester, $R \cdot COO \cdot R'$:

$$RCOOH + R'OH \rightarrow RCOOR' + H_2O$$

An amine can condense with an acid to yield an amide:

$$RCOOH + R' \cdot NH_2 \rightarrow RCONH \cdot R' + H_2O$$

Acid chlorides react even more readily than carboxylic acids:

$$R \cdot COCl + R'OH \rightarrow R \cdot COO \cdot R' + HCl$$

An isocyanate, $R \cdot NCO$, can react with an alcohol to form a urethane or with an amine to form a urea, without the elimination of a small molecule:

$$R \cdot NCO + R'OH \rightarrow R \cdot NHCOO \cdot R'$$
$$R \cdot NCO + R'NH_2 \rightarrow R \cdot NHCONH \cdot R'$$

In principle, any of these condensation reactions can be used to produce polymers by using *bifunctional* starting materials. If adipic acid, $HOOC \cdot CH_2CH_2CH_2CH_2 \cdot COOH$, is reacted with ethylene glycol, $HO \cdot CH_2CH_2 \cdot OH$, the product is a linear polyester and the reaction is termed *polyesterification:*

$$HO{-}CH_2CH_2{-}OH + HOOC{-}(CH_2)_4{-}COOH$$

$$+ \ HO{-}CH_2CH{-}OH \rightarrow$$

$$---O{-}CH_2CH_2{-}\overset{\overset{O}{\|}}{C}{-}(CH_2)_4{-}\overset{\overset{O}{\|}}{C}{-}CH_2CH_2{-}O---$$

In the same way, a dibasic acid and a diamine can be condensed to form a polyamide, and a diisocyanate can be condensed with a diol to yield a polyurethane.

These polycondensation reactions proceed in a stepwise manner (in striking contrast to addition polymerizations, which are usually chain reactions). The functional groups react randomly, forming high-molecular-weight polymers only at very high conversion. At 50% conversion of an equimolar mixture of ethylene glycol and adipic acid, where half of the functional groups are reacted and half are unreacted, the number fractions of various species are as tabulated below.

Times Reacted	Formula	Fraction
None	$HOOC \cdot (CH_2)_4 \cdot COOH + HO \cdot CH_2CH_2 \cdot OH$	0.5
Once	$HOOC \cdot (CH_2)_4 \cdot COO \cdot CH_2CH_2 \cdot OH$	0.25
Twice	$HOOC \cdot (CH_2)_4 \cdot COO \cdot CH_2CH_2 \cdot OOC \cdot (CH_2)_4 \cdot COOH +$ $HO \cdot CH_2CH_2 \cdot OOC \cdot (CH_2)_4 \cdot COO \cdot CH_2CH_2 \cdot OH$	0.125
Three times	$HOOC \cdot (CH_2)_4 \cdot COO(CH_2)_2OOC \cdot (CH_2)_4 \cdot COO \cdot (CH_2)_2 \cdot OH$	0.0625

Only at much higher conversions will the poly-condensation product be a high polymer. If the size of a particular polyester chain is represented by n (the number of fragments it would hydrolyze to), the number average value of n is related to the fractional conversion f by

$$\bar{n} = \frac{1}{1 - f}$$

In order for $\bar{n}$ to equal 100, f must be 0.99, that is, 99% conversion.

It follows that not all condensation reactions can produce high-molecular-weight linear polymers but only those which can be driven cleanly to very high conversions. Further, the reactants must be present in exact stoichiometric amounts in order for high-molecular-weight polymers to form.

A minor variation in the above pattern is the use of an unsymmetrical bifunctional molecule instead of two symmetrical bifunctional molecules. For example, an ω-amino acid can undergo polycondensation to yield a polyamide:

$$H_2N—(CH_2)_n—COOH \rightarrow$$

$$\overset{O}{\overset{\|}{\cdots C}}—NH—(CH_2)_n—\overset{O}{\overset{\|}{C}}—NH—(CH_2)_n\cdots$$

Network Polymers by Polycondensation. Polycondensation reactions involving bifunctional starting materials yield linear polymer chains. If trifunctional or tetrafunctional molecules are included, a polycondensation reaction can yield branched chains and three-dimensional networks. Examples of multifunctional alcohols are glycerin (trifunctional) and pentaerythritol (tetrafunctional).

$$HO—CH_2—\underset{\underset{OH}{|}}{CH}—CH_2—OH$$

Glycerin

$$HO—CH_2—\underset{\underset{CH_2}{|}\atop{\underset{OH}{|}}}{\overset{\overset{OH}{|}\atop\overset{CH_2}{|}}{C}}—CH_2—OH$$

Pentaerythritol

The higher the proportion of multifunctional reactants to bifunctional reactants, the more tightly

cross-linked the resulting network polymer will be. In contrast to linear polymers, network polymers are insoluble and infusible.

When polyfunctional molecules are used to form network polymers by polycondensation, the requirement that the condensation reaction proceed cleanly to very high conversion can be relaxed. Wastage of a few functional groups by side reactions does not prevent the formation of very large molecules in this case.

The basic condensation involved in the formation of bakelite is

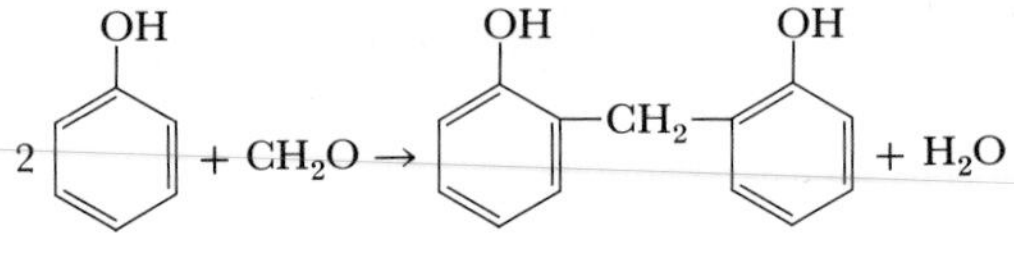

This condensation can take place at both ortho and para positions, leading to three-dimensional network structures. Another important family of thermosetting plastics are the urea-formaldehyde resins.

Addition Polymerization of Vinyl and Diene Monomers. Many unsaturated organic molecules can undergo addition polymerization to yield high-molecular-weight polymers. The simplest example is ethylene, which polymerizes to polyethylene:

$$CH_2{=}CH_2 \rightarrow \cdots CH_2CH_2CH_2CH_2CH_2CH_2\cdots$$

Some other important monomers for addition polymerization are:

$$CH_2{=}CH—CH_3 \qquad CH_2{=}CH—\overset{O}{\overset{\|}{O}}CCH_3$$

Propylene   Vinyl acetate

$$CH_2{=}CH—COOCH_3 \qquad CH_2{=}\overset{\overset{CH_3}{|}}{C}—COOCH_3$$

Methyl acrylate   Methyl methacrylate

$$CH_2{=}CH—C_6H_5 \qquad CH_2{=}CHCl$$

Styrene   Vinyl chloride

$$CH_2{=}CCl_2 \qquad CH_2{=}CH—CN$$

Vinylidene chloride   Acrylonitrile

Many 1,3-dienes, the simplest being butadiene, $CH_2{=}CH—CH{=}CH_2$, can undergo addition polymerization. Depending on the polymerization conditions, the diene monomer units in the polymer chain may be 1,2 or 1,4, or a mixture of these.

$$-CH_2-\underset{\underset{CH=CH_2}{|}}{CH}- \qquad -CH_2-CH=CH-CH_2-$$

1,2-Butadiene unit   1,4-Butadiene unit

Mechanisms of Addition Polymerization: There are several distinct mechanisms by which an unsaturated monomer can be polymerized to form an addition polymer. Three of these are indicated below:

Carbonium-ion:

$$Chain-CH_2\underset{\underset{X}{|}}{\overset{\overset{H}{|}}{C}}{}^+ + CH_2=CHX \rightarrow$$

$$chain-CH_2-\underset{\underset{X}{|}}{\overset{\overset{H}{|}}{C}}-CH_2-\underset{\underset{X}{|}}{\overset{\overset{H}{|}}{C}}{}^+$$

Free-radical:

$$Chain-CH_2-\underset{\underset{X}{|}}{\overset{\overset{H}{|}}{C}}\cdot + CH_2=CHX \rightarrow$$

$$chain-CH_2-\underset{\underset{X}{|}}{\overset{\overset{H}{|}}{C}}-CH_2-\underset{\underset{X}{|}}{\overset{\overset{H}{|}}{C}}\cdot$$

Carbanion:

$$Chain-CH_2-\underset{\underset{X}{|}}{\overset{\overset{H}{|}}{C}}:^- + CH_2=CHX \rightarrow$$

$$chain-CH_2-\underset{\underset{X}{|}}{\overset{\overset{H}{|}}{C}}-CH_2-\underset{\underset{X}{|}}{\overset{\overset{H}{|}}{C}}:^-$$

These are all *chain* reactions, in contrast to the stepwise polycondensation reactions. Carbonium ions, free radicals, and carbanions are all very reactive with unsaturated monomer molecules. A growing chain with an active end group rapidly adds a monomer molecule, lengthening the chain by one unit and regenerating the active end. Multiple repetition of this addition step quickly leads to the formation of a high-polymer molecule. Chain growth is finally stopped, either by some termination reaction which destroys the active chain end (the usual case) or by depletion of monomer. Carbonium-ion polymerization is initiated by strong acids; carbanion polymerization by strong bases; and free-radical polymerization by easily cleaved molecules, such as peroxides, or by ultraviolet light.

In addition to the above mechanisms, many unsaturated monomers can be polymerized by various complex catalyst systems, such as $TiCl_4$ plus aluminum alkyls, sometimes referred to as *Ziegler catalysts.* In these systems, both the active growing chain end and the monomer molecule are *complexed* with the catalyst. In some cases, the growing chain can be considered to be a carbanion, bound as a ligand to a metal atom of the catalyst, with the monomer molecule bound as a ligand to the same metal atom. This fixed geometrical relationship between the growing chain end and the monomer molecule results in a high degree of stereochemical control over the structure of the polymer formed.

When an olefinic monomer, $CH_2=CHX$, is polymerized, the substituent X can extend above or below the plane of the extended zigzag polymer chain, corresponding to a *d* or *l* configuration of the carbon atom in question. A sequence of monomer units may be identical (*isotactic*), perfectly alternating (*syndiotactic*), or random (*atactic*), with respect to this stereochemical configuration:

Isotactic:

$$-CH_2-\underset{\underset{X}{|}}{\overset{\overset{H}{|}}{C}}-CH_2-\underset{\underset{X}{|}}{\overset{\overset{H}{|}}{C}}-CH_2-\underset{\underset{X}{|}}{\overset{\overset{H}{|}}{C}}-CH_2-\underset{\underset{X}{|}}{\overset{\overset{H}{|}}{C}}-$$

Syndiotactic:

$$-CH_2-\underset{\underset{X}{|}}{\overset{\overset{H}{|}}{C}}-CH_2-\underset{\underset{H}{|}}{\overset{\overset{X}{|}}{C}}-CH_2-\underset{\underset{X}{|}}{\overset{\overset{H}{|}}{C}}-CH_2-\underset{\underset{H}{|}}{\overset{\overset{X}{|}}{C}}-$$

Atactic:

$$-CH_2-\underset{\underset{X}{|}}{\overset{\overset{H}{|}}{C}}-CH_2-\underset{\underset{X}{|}}{\overset{\overset{H}{|}}{C}}-CH_2-\underset{\underset{H}{|}}{\overset{\overset{X}{|}}{C}}-CH_2-\underset{\underset{X}{|}}{\overset{\overset{H}{|}}{C}}-$$

The complexed catalyst systems can enforce a high degree of stereochemical regularity (isotactic or syndiotactic), in contrast to the first three mechanisms, which normally yield rather atactic polymers. This subtle structural difference can be of tremendous practical importance. Isotactic polypropylene, for example, is a highly crystalline polymer with excellent physical properties; whereas atactic polypropylene is a soft amorphous material of little value.

A few monomers can be polymerized by any of the four addition polymerization mechanisms; styrene is one example. On the other hand, many monomers are limited to one or two of the possible mechanisms. Some of the more important commercial processes are classified below with regard to mechanism.

Process	*Mechanism*
Low-density (branched) polyethylene	Free radical
High-density (linear) polyethylene	Complexed catalyst
Isotactic polypropylene	Complexed catalyst
Polystyrene	Free radical
Vinylidene chloride copolymers	Free radical
Polyvinyl chloride	Free radical

Copolymerization and Copolymers. A copolymer is a polymer molecule which contains two (or more) *different* kinds of monomer unit; the process of forming such a molecule is called *copolymerization*.

From any given pair of monomers, a wide variety of copolymers is possible, the properties depending on the relative *amounts* of each monomer and the *arrangement* of the two kinds of unit in the copolymer molecule. Some of the possible types of copolymer are shown in Fig. P-21.

When a mixture of two vinyl monomers undergoes free-radical addition polymerization, the product is normally a *random* copolymer. Random copolymers can also be produced by polycondensation, e.g., reaction of ethylene glycol with a mixture of adipic and succinic acids.

In some cases of addition copolymerization, the free-radical end group on the growing chain has a strong preference for adding the *other* monomer as the next unit; in such cases, the preferred product is a strongly *alternating,* rather than random, copolymer.

Block copolymers are most readily prepared by the *carbanion* mechanism. In the absence of water and other terminating agents, carbanions are very long-lived species. Polymerizing carbanion chains of one monomer can grow until the monomer is depleted, and then a different monomer can be supplied to yield blocks of the second kind, after

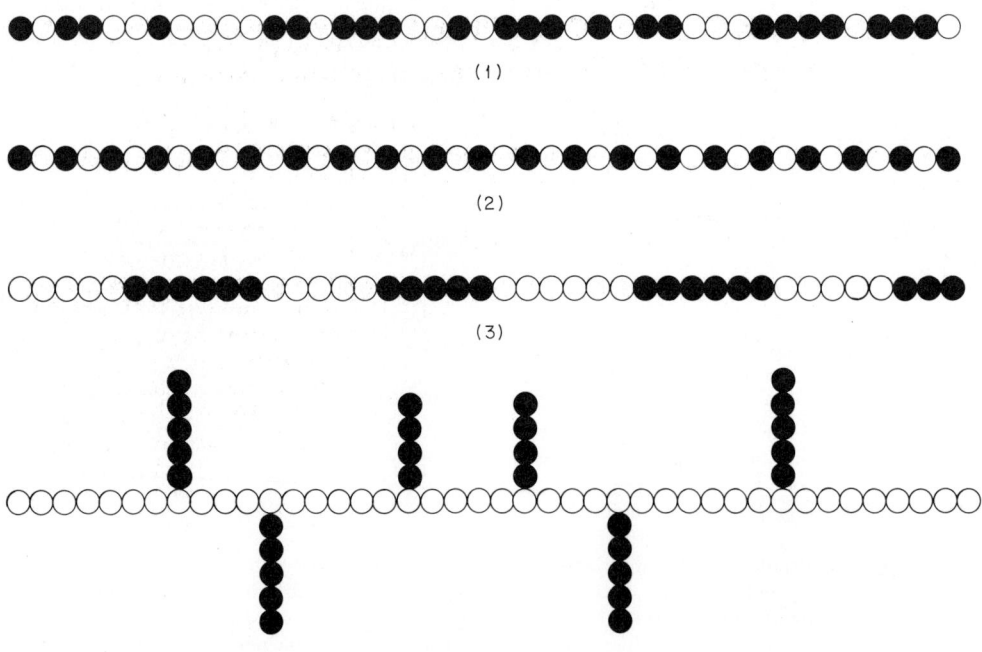

Fig. P-21. Copolymers in the form of polymer chains made up of two kinds of monomers represented by black and white spheres. Four types of formation are shown: (1) random, (2) alternating, (3) block, (4) graft.

which the first monomer is again introduced, and so on.

Graft copolymers can be prepared by first polymerizing one monomer, then adding a second monomer and polymerizing it (by free-radical mechanism) in the presence of the first polymer. Chain transfer with the first polymer results in the growth of branches (second monomer) attached covalently to the backbone polymer.

Copolymerization processes greatly broaden the scope of available polymeric materials, copolymers playing roles analogous to those of *alloys* in metallurgy. As with alloys, the *properties* of a copolymer may be intermediate between those of the corresponding homopolymers or they may be new and completely distinctive.

Network Polymers by Addition Polymerization: Since the addition polymerization of a vinyl monomer, $CH_2=CHX$, yields long chains, the corresponding polymerization of a divinyl monomer yields network polymers. Typical divinyl monomers are divinyl benzene and ethylene diacrylate

Divinyl benzene

$$CH_2=CH-COO-CH_2CH_2-OOC-CH=CH_2$$

Ethylene diacrylate

If styrene monomer containing 1 mole % divinyl benzene undergoes addition polymerization, the product is an insoluble, infusible network polymer, with a cross-link occurring on the average at every 50 units.

Portion of network polymer,
showing a cross-link between two chains

Clearly, the higher the proportion of divinyl monomer in the monomer mixture, the more closely knit the resulting network polymer will be.

Polymerization Processes (Free Radical Mechanism). Carbonium ions and carbanions react rapidly with water; consequently, any polymerization reactions which proceed by these mechanisms must be carried out under anhydrous conditions. On the other hand, water is rather inert toward most organic free radicals; as a result, the basic free-radical polymerization mechanism has been developed into a wide range of specific processes, some of which involve water as a carrier medium.

The principal process used in the manufacture of polystyrene is *mass* polymerization. Styrene monomer is heated to an elevated temperature, and free radicals are initiated thermally. Provision must be made for heat transfer to remove the exothermic heat of polymerization (approximately 17 kcal/mole). The polymerization is carried to a high conversion, and the molten product is devolatilized to remove residual monomer.

Some polymers are prepared by *solution* polymerization in an inert solvent. Upon completion of the reaction, polymer must be separated from solvent; this can be done by precipitation and filtering or by evaporating the solvent, e.g., by spray drying.

In *suspension* polymerization, liquid monomer is dispersed in water in the form of tiny droplets in a stirred kettle. Usually a polymerization initiator, such as a peroxide, is dissolved in the monomer phase. Upon heating, polymerization takes place in the monomer droplets in a manner essentially identical with mass polymerization; the continuous, low-viscosity water phase facilitates heat transfer to the kettle jacket. The product is a dispersion of small polymer particles, which can easily be separated from the water phase by centrifugation or filtering.

Emulsion polymerization also employs water as a continuous phase, but its mechanism is more involved than that of suspension polymerization. A simple illustration of emulsion polymerization is provided by the following recipe:

	Parts
Water	60
Styrene monomer	40
Soap	3
Potassium persulfate	0.5

The ingredients are stirred and heated. The styrene monomer disperses as an emulsion, stabilized by part of the soap. Some of the soap exists in the form of *micelles* (small association clusters), which imbibe some styrene monomer. The water-soluble persulfate ion spontaneously cleaves to form two radical-ions, which initiate polymerization in the monomer-swollen soap micelles. The tiny polymer particles so formed inbibe more monomer and grow larger. The much larger monomer droplets serve as a supply reservoir, feeding the growing polymer particles by diffusion through the aqueous phase.

A typical commercial emulsion polymerization process is much more complicated than the foregoing illustration. For example, an early recipe for styrene-butadiene rubber (GR-S) contained the following ingredients:

	Parts
H_2O	60
Butadiene	25
Styrene	8.3
Rosin soap	1.6
Daxad 11	0.03
Dodecyl mercaptan	0.08
Cumene hydroperoxide	0.03
Dextrose	0.3
$K_4P_2O_7$	0.06
$FeSO_4 \cdot 7H_2O$	0.05
KCl	0.16
KOH	0.03

Free-radical initiation is by a redox system involving hydroperoxide, ferrous ion, and dextrose. The mercaptan is a chain-transfer agent, provided to control the molecular weight.

Ring-Chain Conversions. Under proper reaction conditions, many small ring molecules can be opened up and strung together to form linear polymer chains, an important example being the production of nylon 6 from caprolactam monomer:

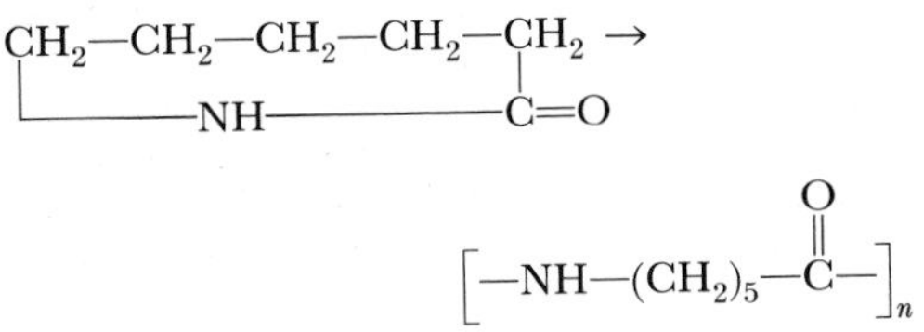

$$CH_2\!-\!CH_2\!-\!CH_2\!-\!CH_2\!-\!CH_2 \rightarrow$$
$$\underset{\displaystyle -\!\!-\!NH\!-\!\!-\!\!-\!\!-\!\!-\!C\!=\!O}{\rule{0pt}{0pt}}$$

$$\left[-NH-(CH_2)_5-\overset{\displaystyle O}{\overset{\displaystyle \|}{C}}-\right]_n$$

Another example is the polymerization of elemental sulfur. At room temperature the stable form of sulfur is the cyclic molecule S_8. When heated above 159°C, these ring molecules open up and string together to form linear polymeric chains.

Certain three-membered ring molecules are particularly susceptible to polymerization, e.g., ethylene oxide, ethylene imine, and ethylene sulfide:

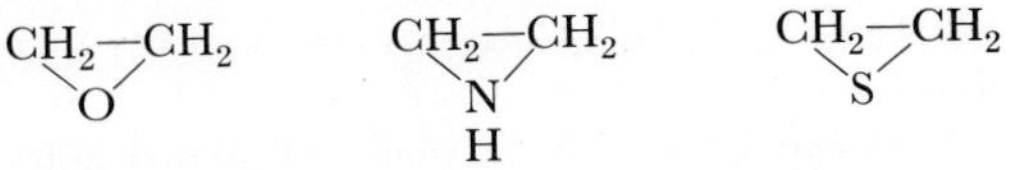

Ethylene oxide   Ethylene imine   Ethylene sulfide

Because of the strain in the three-membered ring, these molecules are much more reactive than ordinary ethers, amines, and sulfides. Both acidic and basic catalysts cause ethylene oxide to polymerize to polyethylene oxide:

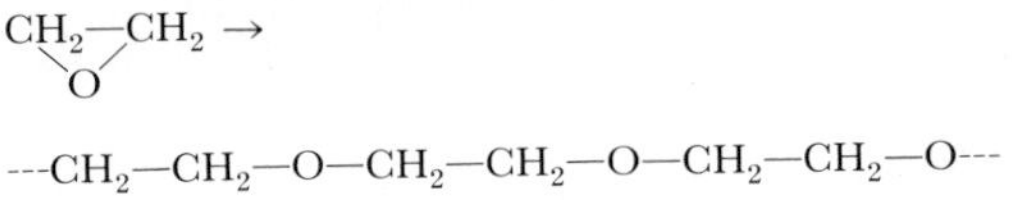

$$\underset{\displaystyle O}{CH_2\!-\!CH_2} \rightarrow$$

$$-\!\!-\!-CH_2\!-\!CH_2\!-\!O\!-\!CH_2\!-\!CH_2\!-\!O\!-\!CH_2\!-\!CH_2\!-\!O\!-\!-\!-$$

A small molecule which contains two or more of these highly reactive three-membered rings can be readily converted to a three-dimensional network polymer. An important monomer of this type is the diglycidyl ether of bisphenol A:

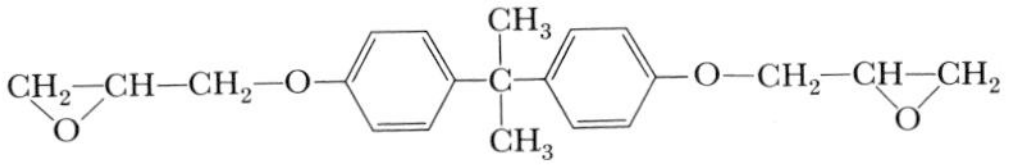

A whole class of polymeric materials, the *epoxy resins,* can be prepared by polymerization of this monomer, using a wide variety of catalysts and coreactants.

Combinations of Polymerization Reactions. While most commercial polymers are produced by one or another of the above types of polymerization reaction, many useful polymeric materials are formed by the sequential use of two different types of reaction; e.g., a low-molecular-weight polyester can be formed by polycondensation of an unsaturated dibasic acid, such as maleic acid, $HOOC\!-\!CH\!=\!CH\!-\!COOH$. This condensation polymer contains double bonds and can subsequently be copolymerized with a vinyl monomer, such as styrene (addition polymerization).

The final product is an insoluble, infusible three-dimensional polymer. Alternatively, the addition polymerization can precede the polycondensation; a vinyl monomer containing some condensable side group can be subjected to addition polymerization or copolymerization, and the resulting polymer used as an ingredient of a polycondensation. Addition polymerization can also be combined with ring-opening polymerization, as for glycidyl methacrylate,

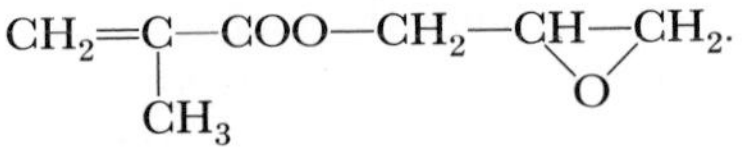

$$CH_2\!=\!\underset{\displaystyle CH_3}{\overset{\displaystyle |}{C}}\!-\!COO\!-\!CH_2\!-\!\underset{\displaystyle O}{CH\!-\!CH_2}.$$

Biological Polymerization Reactions (Biopolymers).
Organic polymers of many kinds play important roles in all living organisms. Of particular importance are proteins, nucleic acids (DNA and RNA), and polysaccharides.

Proteins are condensation polymers of α-amino acids:

$$H_2N—CH—COOH$$
$$\mid$$
$$R$$

α-Amino acid

$$---NH—CH—CO—NH—CH—CO—NH—CH—CO---$$
$$\mid \qquad\qquad \mid \qquad\qquad \mid$$
$$R \qquad\qquad R' \qquad\qquad R''$$

Protein

There are 20 different amino acid monomers, which differ in the nature of the substituent R. A given protein molecule is a *specific-sequence* copolymer of these different amino acids. See also **Peptides and proteins.**

The polymerization of these α-amino acid monomers to form a protein molecule involves two difficulties not faced in synthetic polymer chemistry: (1). protein molecules, in an aqueous medium, are thermodynamically unstable relative to their hydrolyzates (the monomers); (2). the monomer units must be strung together in the precise *sequence* characteristic of the given protein.

A nucleic acid is an alternating copolymer of a sugar unit and a phosphate unit, with a heterocyclic nitrogen base attached to each sugar residue:

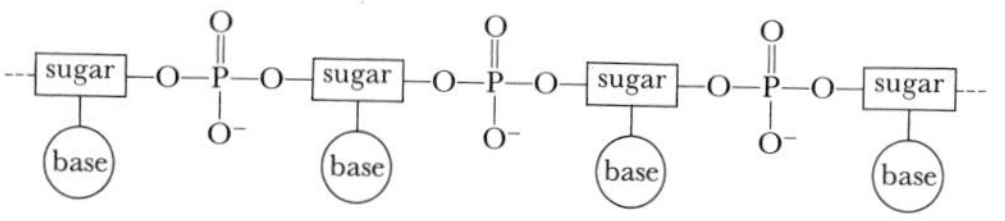

In RNA, the sugar is ribose, and the base is adenine, cytosine, guanine, or uracil. In DNA, the sugar is deoxyribose, and the base is adenine, cytosine, guanine, or thymine. The *primary structure* of a nucleic acid designates the precise sequence of bases along the chain. Cellular DNA, which constitutes the genetic material of the cell, is in the form of a two-stranded helix. The primary DNA chain is accompanied by a complementary companion chain hydrogen-bonded (base to base), with adenine always paired with thymine and cytosine always paired with guanine (above right). As with proteins, the biosynthesis of nucleic acids involves two difficulties: the polymers are thermodynamically unstable relative to their hydrolyzates, and the polymerization must produce a chain with a specific sequence of bases.

The general pattern of biosynthesis of these

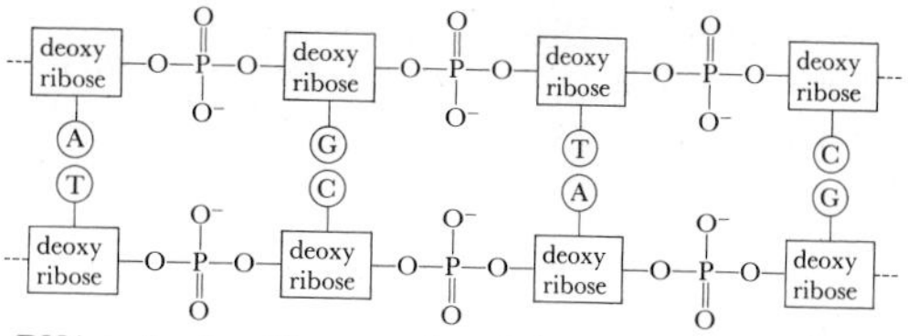

DNA molecule with companion chain; this ribbon is twisted into the form of a double helix

polymers is believed to be essentially as follows. The genetic substance, DNA, is able to replicate itself, thus opening the way to cell division. DNA can also transcribe its message onto a RNA chain, i.e., control the RNA synthesis. The RNA, in turn, diffuses out of the cell nucleus and controls the polymerization of a protein molecule. The primary structure (amino acid sequence) of the protein is directly controlled by the base sequence of the messenger RNA and indirectly by the base sequence of the nuclear DNA. (A *triad* of three successive bases codes for one amino acid unit.)

$$DNA \rightarrow RNA \rightarrow protein$$

The mechanism of the DNA replication is indicated schematically in Fig. P-22.

Required for this polymerization reaction are (1) the old DNA double helix, (2) a mixture of the four monomers in the form of *high-energy triphosphates* (overcoming the thermodynamic barrier to polymerization of the monophosphates themselves), and (3) an enzyme, *DNA polymerase,* which *catalyzes* the reaction.

The reaction product consists of a new DNA primary chain and a new companion chain identical in primary structure with the originals.

The controlled polymerization of messenger RNA proceeds in a similar fashion, the growing end of the RNA molecule being coordinated with a local section of the primary DNA chain (locally disrupted from its companion chain). Required for RNA formation are (1) the guiding DNA, (2) the four RNA monomers in the high-energy triphosphate form, and (3) an enzyme, *RNA polymerase.*

Controlled protein synthesis is somewhat more complicated. Briefly, the 20 amino acid monomers are present, each attached by a *high-energy bond* to an appropriate small transfer RNA molecule. Matching a triad of bases on the tRNA with the triad of bases on the mRNA chain ensures the selection of the correct amino acid, and an enzyme catalyzes the transfer of the activated monomer unit from the tRNA to the growing protein chain.

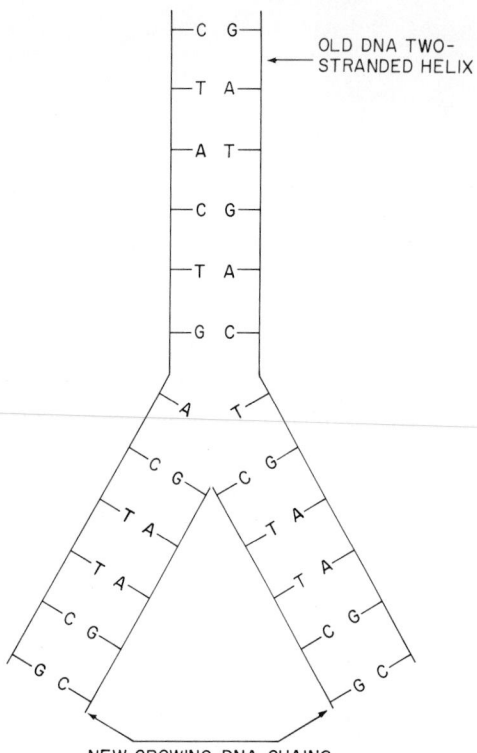

OLD DNA TWO-STRANDED HELIX

NEW, GROWING DNA CHAINS

Fig. P-22. Replication of DNA shown schematically. A = adenine, C = cytosine, G = guanine, T = thymine.

The overall process of biopolymerization is thus complex and subtle. Nucleic acids are necessary in the synthesis of proteins; and proteins (the enzymes) are required in the synthesis of nucleic acids. See also **Enzymes.**

Other specific aspects of polymerization will be found under **Acrylic fibers; Acrylic plastics; Elastomers; Ethylene; Hydrogen fluoride; Ion-exchange resins; Paint; Rayon; Rubber, natural;** and **Silicones.** Also consult the **Subject Index** under Polymerization.

REFERENCES

Billmeyer, F. W.: "Textbook of Polymer Science," 2d ed., Interscience-Wiley, New York, 1971.
Morgan, P. W.: "Condensation Polymers," Interscience-Wiley, New York, 1965.
Proskauer, E. S., E. H. Immergut, and C. G. Overberger: "Perspectives in Polymer Science," Interscience-Wiley, New York, 1966.
Sorenson, W. R., and T. W. Campbell: "Preparative Methods of Polymer Chemistry," 2d ed., Interscience-Wiley, New York, 1968.

Watson, J. D.: "Molecular Biology of the Gene," Benjamin, New York, 1965.

—Turner Alfrey, Jr., The *Dow Chemical Company, Midland, Mich.*

Polyoxyethylene Compounds (See **Surfactants.**)

Polyparaxylene (See **Parylene polymers.**)

Polypeptides (See **Peptides and proteins.**)

Polyphenylene Oxides The polyphenylene oxides are relatively new thermoplastic materials based upon oxidative-coupling technology. The structures have a large degree of symmetry, lack strong polar groups, and incorporate a rigid phenylene oxide backbone. One of the most important properties of these materials is the high heat-deflection temperature coupled with a very low brittleness temperature, giving this series of resins a practical temperature range of about 360°C (650°F). Hydrolytic stability is another outstanding characteristic. Parts made from polyphenylene oxide can be steam-sterilized repeatedly with no significant change in properties, accounting for their wide use in the medical field. The materials can be used in all aqueous environments, including dilute bases, acids, and detergents. They will withstand the action of many organic chemicals but are not recommended where aliphatic hydrocarbons, ketones, and esters are involved. For example, both chloroform and toluene soften or dissolve polyphenylene oxide and consequently, can be used for solvent cementing. The materials can be injection molded, blow molded, extruded, and thermoformed and also can be machined. The properties of polyphenylene oxides are summarized in Table P-36.

TABLE P-36. *Properties of Polyphenylene Oxide Resins*

Specific gravity	1.06
Impact strength, ft-lb/in. notch	1.5–1.9
Tensile impact strength, ft-lb/in.^{2}	350–450
Tensile strength at 125°C (257°F), psi	6,000
Flexural strength, psi	14,500
Hardness, Rockwell R	120
Heat-deflection temperature at 264 psi:	
°C	190
°F	375
Brittleness temperature:	
°C	−170
°F	−275
Dielectric constant	2.58

Polyphosphates   (See **Detergents.**)

Polypropylene   Polypropylene thermoplastic materials have a good balance of properties and economics, which has resulted in their use in the production of film, fibers, and a broad variety of injection-molded and extruded shapes. Chemical and heat resistance (augmented by good mechanical properties obtainable with reinforcing additives) are the important attributes of polypropylenes.

The properties of the base resin are attributable to the stereoregularity of the propylene monomeric units in the polymer structure. These units are arranged in the polymer chain so that their respective spatial configurations are in some particular order. Polymers evidencing an orderly arrangement along the individual polymerical chain are said to be *tactic*. In polypropylene resins, the structure is said to be *isotactic* because the monomeric units are positioned in a duplicating pattern on the same side of the polymer chain.

The isotactic arrangement is significant in polypropylene because it relates to the development of a crystalline molecular structure. Properties are in turn directly related to the degree of crystallinity in the processed polymer. Consequently, the manufacturing process is regulated to result in the maximum yield of isotactic material.

Manufacture: Polypropylene is produced by polymerization of propylene monomer obtained by cracking petroleum products, including natural gas and light oils. The C_3 fraction contains both propylene and propane. Since propane adversely affects the function of the catalyst and properties of the product, subsequent distillation step is employed to realize essentially pure propylene. Water also destroys the catalyst and must be removed.

The polymerization reaction is carried out in the presence of Ziegler catalyst. Transition-metal salts and organometallic compounds are used to produce the former. Characteristically, the catalyst may be formed in situ or as required in a separate vessel. Preformed catalysts, by contrast, are made outside the reactor at some earlier time.

This approach utilizes metal salts such as chromium oxide, molybdenum oxide, cobalt molybdate, and other molybdates. A dispersion of the catalyst system in a hydrocarbon solvent and the monomer are added to an agitated reactor maintained at approximately 60°C. The polypropylene polymer forms around the catalyst particles. Following formation, the polymer granules are washed with acidified alcohol solution to decompose residual catalyst.

Finally, the granules are washed free of the alcohol solvent and dried. As produced, the polymer is in the form of a finely divided white powder. A variation of the conventional polymerization process involves the introduction of ethylene for the manufacture of copolymers. Adding ethylene groups to the polymer chain improves impact properties.

The variety of polypropylene products now commercially available results from blending stabilizer systems and fillers with the polymer granules. A marketable product is formed by subsequent melt mixing, extrusion, and pelletizing. Stabilizer systems include those especially compounded for extended exposure to heat, for resistance to extraction in chemical solutions, and for production of film or fiber. Mechanical properties are enhanced by addition of talc, asbestos, and glass fibers. The base resin has a specific gravity of 0.90 and a melting point of 166°C (330°F). Because the natural resin is semitranslucent and milky-white in color, pigmentation is readily accomplished. The addition of fillers reduces color intensity. Polypropylene materials are usually supplied in pellets. Limited quantities of the polymer are used in the powder form. Such material is involved when the resin is compounded by the processor.

Processing and Applications: Polypropylene resins are processed by injection-molding, extrusion, and blow-molding techniques. Minor quantities are used in laminating, rotational molding, electrostatic coating, and fluidized-bed coating. Parts produced by injection molding include a broad variety of appliance, automotive, and electrical components.

The introduction of polypropylene carpeting and synthetic turf has resulted in major growth of extruded fibers. Blow molding is utilized to form bleach and detergent bottles. Polypropylene provides the chemical resistance and rigidity required of these containers.

Rotational-molding and electrostatic-coating techniques utilize granular polypropylene. The chemical resistance makes the resin useful for forming and coating chemical tanks.

Polypropylene resins lend themselves to a variety of postfabrication processes, e.g., hot-gas welding, butt welding, and spin welding. The resin can be decorated by hot stamping, painting, vacuum-metallizing, and electroplating.

Properties: The important properties of the polypropylene polymer are chemical resistance, heat resistance, electrical behavior, and processibility. Impact resistance of the homopolymer is low, particularly at low temperatures. When copolymerized with ethylene or with additions of rubber, impact resistance is markedly improved. Load-bearing qualities, as reflected by the flexural modulus, may be increased 3–5 times by addition of talc, asbestos, or glass-fiber reinforcing agents. Resistance to degradation with exposure to elevated temperature, ultraviolet radiation, and chemical solutions is imparted by the introduction of selected stabilizer systems during compounding.

The specific gravity of the homopolymer and copolymer is 0.90; of the 40% asbestos or 40% talc-reinforced resins, 1.24; of the 20% glass-reinforced resin, 1.04. Tensile strength of the homopolymer is 5,000 psi; of the copolymer, 4,000 psi; of the 40% asbestos- and 40% talc-reinforced resins, 4,800 psi; and of the 20% glass-reinforced resin, 7,000 psi. The percent elongation at yield for the homopolymer is 200; for the copolymer, 400; and 5 or less for the reinforced resins. Flexural modulus for the homopolymer is 180×10^3 psi; for the copolymer, 135×10^3 psi; for 40% asbestos-reinforced resin, 650×10^3 psi; for 40% talc-reinforced resin, 575×10^3 psi; and for 20% glass-reinforced resin, 550×10^3 psi. Deflection temperature at 66 psi for the homopolymer is 93°C (200°F); for the copolymer, 88°C (190°F); for 40% asbestos-reinforced resin, 143°C (290°F); for 40% talc-reinforced resin, 140°C (284°F); and for 20% glass-reinforced resin, 152°C (305°F). Shore D hardness for the homopolymer is 72; copolymer, 70; and reinforced resins, 77. Dielectric strength (short-time 0.125 mV/mil) is 650 for both homopolymer and copolymer and 450 for the reinforced resins. Volume resistivity for the homopolymer and copolymer is 10^{17} Ω-cm and just a little less for the reinforced resins.

—John S. Houston, *Amoco Chemicals Division, Standard Oil of Indiana, Naperville, Ill.*

Polystyrenes   Polystyrene is a water-white thermoplastic produced from coal tar and petroleum gas. Mechanical properties of this material can be altered by addition of modifying agents such as rubber (for impact strength), methyl or α-styrene (for heat resistance), methyl methacrylate (for light stability), and acrylonitrile (for chemical resistance). High-heat materials are produced as copolymers. The amount of alloying monomer varies with the heat resistance required. Polystyrene has excellent electrical properties, good thermal and dimensional stability, resistance to staining, and low cost. The material can be reheated and remolded, although excessive heat causes degradation. Polystyrene is not compatible with many common solvents and thus has a tendency to stress crack under load. Exposure to ultraviolet rays causes yellowing and loss of mechanical strength.

General-purpose polystyrenes are used for knobs, cocktail glasses, compact cases, indoor light shields (with light-stabilizer formulations), and a variety of other applications. High-heat polystyrenes are used in appliance parts, tape reels, oriented polystyrene film, table-model radios, and containers. Impact high-heat materials also are used for appliances. Medium-impact grades are used for parts requiring good rigidity and moderate impact strength, including containers, closures, and packaging. High-impact polystyrenes are used in many appliance parts, such as breaker strips, refrigerator inner doors, refrigerator inner cabinets, hot-drink cups, and structural foams. Super high-impact polystyrenes are used in toys, some appliance parts, and numerous small-volume applications.

Chemistry. The polymerization of styrene is a chain reaction which proceeds readily by all known polymerization techniques with the formation of water-white polymers. This reaction can be shown schematically as

Styrene   Polystyrene

The exact nature of the beginning and end of such a polymer chain is not certain. In general, the polymer can be characterized by its average degree of polymerization, i.e., the value of n, or more precisely by the distribution of n values. This polymerization reaction proceeds by heat alone, for which the heat of polymerization is 17.4 ± 0.2 kcal/mole at 26.9°C with or without catalysts.

Styrene also will polymerize in the presence of various inert materials, such as solvents, fillers, dyes, pigments, plasticizers, rubbers, and resins. Moreover, it forms a variety of copolymers with other mono- and polyvinyl monomers.

It is a matter of general observation that with styrene the polymerization-rate curves will exhibit three distinct phases, the nature of which can be determined by the polymerization conditions and the purity of the monomer: (1) an initial slow period at the beginning of the reaction, known as the *induction period,* which appears to be associated with the presence of an inhibitor or other impurity in the monomer; (2) a period of relatively rapid polymerization, which persists almost to the end of the reaction, and for which the rate is exponentially dependent upon temperature; and (3) a final slowing down in rate as the reaction approaches completion and the monomer becomes exhausted. This effect is particularly apparent at low temperatures with relatively impure monomers.

Production. The basis for this description is a plant that will produce either general-purpose polystyrene or impact polystyrene.

General-Purpose Polymer Process: In this process (Fig. P-23), styrene is fed continuously into the three-stage, stirred polymerization train, where approximately 90% conversion is attained. The solution containing 90% polymer then is pumped to a devolatilizer, where the residual unreacted styrene is vaporized, condensed, and recycled continuously back to the first-stage polymerizer. The hot polystyrene melt flows through a feeder at the conical bottom of the devolatilizer into a pelletizing extruder and is pelletized, cooled, dried, and screened. The polystyrene pellets then are air-conveyed to a storage bin.

*High-Impact Polymer:** Synthetic rubber is chopped into small pieces and fed continuously into a dissolver, along with styrene monomer. This rubber-styrene solution may contain small amounts of lubricants or flow agents. The mixture is polymerized to obtain about 35% conversion of the styrene and rubber feed in the first reactor. The reactor solution, fed to the top of the second polymerization column of the series, exits from the bottom and is fed to the top of the third reactor. The 85–90% polymerized product, from the bottom of the third polymerizer, is heated before being pumped into the devolatilizer to recover and recycle the unreacted monomer. The finishing

*Description based on U.S. Patents 2,694,692 and 3,243,481 issued to The Dow Chemical Company.

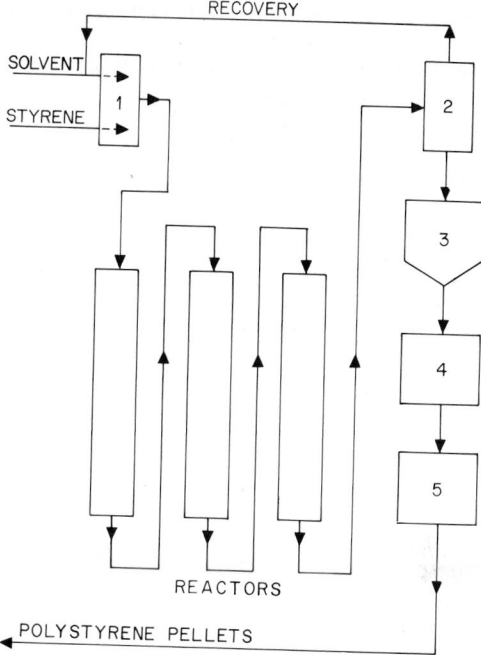

Fig. P-23. Block diagram of continuous solvent process for styrene polymerization. (1) Mixer, (2) devolatilizer, (3) extruder, (4) cooler, (5) cutter. (*The Dow Chemical Company.*)

operation is similar to that described for general-purpose polystyrene.

General Properties

Density: Specific gravity of polystyrene at 73°F (22.8°C) varies from 1.04 to 1.09, depending on color. Density varies slightly with pressure, but for practical purposes, the plastic is noncompressible.

Heat-Resistance: Deflection temperatures range from about 160 to 210°F (66–99°C), depending upon the formulation. Continuous resistance to heat for polystyrene is usually 140–175°F (60–80°C). Time and load have a significant influence on the useful service temperature of a part.

Toxicity: Polystyrene is nontoxic when free from additives and residuals. It has no nutritive value and does not support fungus or bacterial growth.

Dimensional Stability and Moisture Absorption: Dimensional stability of polystyrene resins is excellent. Mold shrinkage is generally about 0.0045 in./in. The low moisture absorption (about 0.02%) allows fabricated parts to maintain dimensions and strength in humid environments.

Permeability: Gas permeability of polystyrene to oxygen and nitrogen is shown in Fig. P-24. Moisture transmission is greater than that of most polyethylenes.

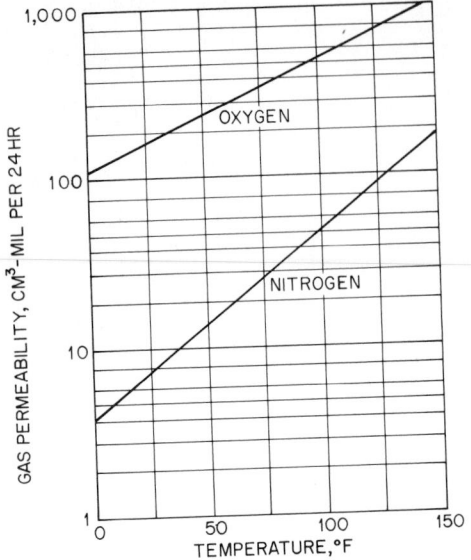

Fig. P-24. Gas permeability of polystyrene in oxygen and nitrogen, 100 in.² at atmospheric pressure.

Transparency and Optical Properties: General-purpose polystyrene is water white, and transmission of visible light is about 90%. Modifiers reduce this property, and translucence results. The refractive index is about 1.59; critical angle about 39°.

Polystyrene molecules do not have the same optical properties in all directions. When molecules become oriented in a given direction during fabrication, a double refraction occurs and a birefringence effect can be observed if the part is examined through a polarized lens under a polarized light source. Injection moldings often exhibit birefringence in a random pattern. This can be beneficial if the birefringence is in the direction of load.

Abrasion Resistance: Polystyrene is a hard plastic which can be marred and scratched more readily than the relatively resilient materials.

Solubility: Polystyrene is soluble in most aromatic and chlorinated solvents. Typical solvents are benzene, carbon tetrachloride, trichloroethylene, carbon disulfide, tetralone, dioxane, acetic ester, methyl ethyl ketone, toluene, perchloroethylene, pyridine, ethyl benzene, and styrene monomer. Polystyrene generally is insoluble in alcohols, such as methanol, ethanol, normal heptane, and acetone. It is insoluble in water.

Mechanical Properties. Typical mechanical properties of polystyrene are given in Table P-37. The long-term load-bearing strength of most polystyrene materials is only about one-third to one-fourth of the typical tensile strengths given in Table P-37. Design strength of most commercial polystyrenes is between 800 and 1,500 psi for loads that are to be held longer than 2 weeks.

Fabrication. Polystyrene materials can be readily fabricated by many processes common to most thermoplastics, including injection molding, extrusion and orientation, and compression molding. Extrusion is almost as important as injection molding in the manufacture of polystyrene parts. The material is easily extruded because of the wide temperature range which may be used without danger of material degradation. Stock and die temperatures usually are between 350 and 450°F (177–232°C). Compression molding is seldom used for anything other than large parts with heavy cross sections or for parts made in small volume.

Embossed surfaces improve scratch resistance and light-diffusion properties. Laminates impart

TABLE P-37. *Properties of Compression-molded Polystyrene*

	General-purpose	Rubber-modified
Tensile strength, psi	4,000–7,000	2,000–5,000
Compressive strength, psi	11,000–16,000	4,000–9,000
Flexural strength, psi	10,000–17,000	4,000–12,000
Impact strength, Izod, ft-lb/in.		
of notch	0.020–0.40	0.5–4.0
Hardness, Rockwell M	65–80	60
Ultimate elongation, %	0.8–2.0	5–60
Tensile strength of elasticity, psi	400,000–500,000	200,000–400,000

Fig. P-25. Injection-molding machine for producing high-impact polystyrene items.

gloss, preprinted decoration, or chemical resistance. Heat polishing or flame polishing often is done by means of infrared heaters. A gloss as high as 90% (as measured on a 60° glossmeter) can be attained without significantly reducing mechanical properties. However, if the treated side of a heat-polished sheet is overheated, the mechanical properties can be drastically impaired.

Benefits produced by annealing polystyrene parts include (1) lower internal stresses, (2) improved mechanical strength, and (3) higher heat-distortion temperature. Usual annealing temperature for molded parts is 10°F (5.5°C) below the practical heat-distortion temperature.

Thermal forming has become one of the most important fabrication techniques for polystyrene. Extruded sheet can be formed into complex shapes by using prestretch, plug-assist, or vacuum-forming techniques.

Injection Blow Molding: After 30 years of development, injection blow molding is becoming a major process for converting polystyrene into bottles, jars, and other types of open containers (see Fig. P-25). Injection blow molding can be used to process polystyrene which, because of low-melt strength, is difficult to control in free parison extrusion and is difficult to trim without cracking in extrusion blow molding. Injection molding is capable of inducing preferred orientation in a blown article to a greater degree than has been possible in extrusion blow molding.

Foam Extrusion: Impact polystyrene for quality furniture components and cabinetry permits fast production of highly detailed reproductions of structural and nonstructural wood-grain parts having intricate shapes. Improvements in foam injection-molding processes, coupled with equipment changes and more effective blowing agents, have made it possible to reproduce structural polystyrene foam parts with smooth integral-skin surfaces. Both general-purpose and impact polystyrene resins, coupled with nucleator and blowing agent, can be directly extruded into foam sheet for use in egg cartons, meat trays, and disposables for single-use applications (cups, plates, and bowls) by hospitals, cafeterias, and schools.

—J. Dennis Griffin and John Y. Glass, *The Dow Chemical Company, Midland, Mich.*

Polysulfide Rubbers (See **Elastomers.**)

Polysulfones Compared with other engineering thermoplastics, polysulfones have an exceptionally high service temperature, retaining their physical properties from −101°C to over 149°C (−150 to 300°F). Acceptance of this relatively new synthetic material can be attributed to a good balance of mechanical, electrical, heat-resistant, and self-extinguishing properties. Parts of a variety of sizes, shapes, and wall thicknesses can be produced by a number of commercial processes. Some of the uses include electrical components, e.g., coil bobbins, circuit-breaker components, housings of various types and battery cases. Electronic applications include connectors, carriers, TV components, tube bases, flexible film, and circuit boards. Other uses are found in the appliance, automotive, and aircraft industries.

Synthesis: The commercially available bakelite polysulfone is prepared by the nucleophilic substitution reaction between the sodium salt of 2,2-bis(4-hydroxyphenyl)propane and 4,4'-dichlorodiphenyl sulfone. The sodium phenoxide end groups are reacted with methyl chloride to terminate the polymerization. This controls the molecular weight of the polymer and contributes to thermal stability.

Structure: The unique feature of the chemical structure of this polymer is the diaryl sulfone grouping. This is a highly resonating structure, in which the sulfone group tends to draw electrons from the phenyl rings. The resonance is enhanced by having oxygen atoms para to the sulfone group. Oxidation, by definition, is a loss of electrons. Having the electrons tied up in resonance imparts excellent oxidation resistance to the polymer. Also, the sulfur atom is in its highest state of oxidation.

The high degree of resonance has two additional effects: it increases the strength of the bonds involved and fixes this grouping spatially into a planar configuration. This provides rigidity to the polymer chain, which is retained at high temperatures.

The excellent thermal stability of polysulfones is verified by thermal gravimetric analysis, which shows it to be stable in air up to 500°C; this, coupled with high oxidation resistance, provides very good melt stability for molding and extrusion.

The ether linkage imparts some flexibility to the polymer chain, giving inherent toughness to the material. In common with all tough, rigid thermoplastic polymers, polysulfone has a second

low-temperature glass transition at −101°C (−150°F), which is assigned to the ether linkages.

In these polysulfones, the linkages connecting the benzene rings are hydrolytically stable. Therefore, the polymers are resistant to hydrolysis and to aqueous acid and alkaline environments.

Thermal Properties: Heat-deflection temperature for polysulfone is 174°C (345°F) at 264 psi. The maximum service temperature is generally considered to be around 160°C (320°F), at which point the flexural modulus is about 310,000 psi, compared with 390,000 psi at room temperature. At low temperatures, polysulfone remains tough and maintains useful properties at as low as −101°C (−150°F). Creep under continuous load is low, even at elevated temperatures.

Dimensional Stability: The coefficient of thermal expansion for polysulfone is 3.1×10^5 in./(in.) (°F) from −5.6 to +149°C (−22 to +300°F). Mold shrinkage is 0.007 in. Thermal cycling and exposure to moisture induce little dimensional change. Under such conditions as 28 days exposure to water at 22.2°C (72°F) or cycling between water at 60°C (140°F) and air at 149°C (300°F), the linear dimensional change remained less than 0.001 in./in.

Adhesive Characteristics: An adhesive grade of polysulfone provides high bond strengths on aluminum, stainless steel, and other metals. Tensile lap shear-strength values of 3,500 psi at room temperature have been obtained with aluminum. This exceeds the Mil-A-5090D, type II requirement. Creep under a load of 1,600 psi for 8 days is less than 0.00005 in., compared with 0.015 in. allowed under the specification. Peel strengths as high as 50 lb/in. have been achieved. Bonding is generally carried out by applying a primer coat of polysulfone in a suitable solvent, baking to release the solvent, inserting a polysulfone film between the surfaces to be bonded, and applying pressure at a temperature of 260°C (500°F) or higher.

Mechanical Properties: Polysulfone is a high-strength, high-modulus material. The notched Izod value is moderate, 1.3 ft-lb per inch of notch, but this value holds substantially constant at temperatures as low as −40°C (−40°F) and for thicknesses of $\frac{1}{8}$–$\frac{1}{2}$ in.

Electrical Properties: Typical electrical data for polysulfone illustrate its excellent dielectric characteristics as well as the relative constancy of values over a wide range of temperatures and even after immersion in water. Dielectric strength (short-time) of a 10-mil sheet is 2,200, 2,250, and

1,600 V/mil at 25, 100, and 160°C (77, 212, and 320°F) respectively.

Environmental Properties: Polysulfone exhibits exceptional chemical resistance to all materials except polar solvents. Solutions of inorganic acids and alkalies, alcohols, and aliphatic hydrocarbons do not affect the material. Polysulfone swells and partially dissolves in esters and ketones and is soluble in chlorinated hydrocarbons. Stress-crack resistance is good, specimens having survived for over 500 hr in a 2% detergent solution at 60°C (140°F) and 1,500 psi stress. Under most conditions, polysulfone absorbs less than 1% moisture. Polysulfone is inherently flame-resistant and self-extinguishing without requiring modifiers. The material has been approved for use in articles in repeated contact with food and for handling potable water.

Resin Properties: With a specific gravity of 1.24, polysulfone weighs about 0.7 oz/in.³ or 77 lb/ft³. The resin is shipped in 50-lb polyethylene bags as free-flowing pellets. The resin must be thoroughly dried before melt-processing. A reasonable quantity of dried reground material may be mixed with virgin resin without affecting performance. Natural grades of polysulfone are transparent and light amber in color. A broad range of transparent and opaque colors can be supplied as well as glass filled grades.

Processing: Polysulfone can be fabricated with conventional equipment by extrusion, injection molding, blow molding, or thermoforming. It differs from other thermoplastics only in that higher stock temperatures must be maintained.

—R. K. Walton, *Union Carbide Corporation, Bound Brook, N.J.*

Polyunsaturates (See **Vegetable oils.**)

Polyurethanes (See **Urethanes.**)

Polyvinyl Butyral (See **Adhesives.**)

Polyvinyl Chloride The polymer of vinyl chloride, polyvinyl chloride (PVC), and copolymers of vinyl chloride and such materials as vinyl acetate are produced by three basic processes.

Mass Process: Vinyl chloride is polymerized in vinyl chloride as a carrying medium. The polymerization is stopped at a low conversion, and the polymer is separated from the residual monomer, which is recycled. The product is a granular material.

Emulsion Process: The vinyl chloride is emulsi-fied with water by use of emulsifying agents. The polymerization is carried to a high degree of conversion with a small amount of recovered monomer recycled to the process. The product is obtained as an emulsion or spray-dried to produce a very fine powder.

Suspension Process: Vinyl chloride is suspended as small droplets of monomer in water, and the polymerization is carried to a high degree of conversion. A small amount of monomer is recovered and recycled. The granular product is obtained by centrifugation and drying.

While a wide range of products can be made by the first two methods, they are generally limited to specialty products which are difficult or impossible to make by the suspension process. The largest proportion of PVC is produced, particularly in the United States, by the suspension process because it requires the least capital investment and has the lowest operating cost while still producing products required for the largest number of applications.

The process described here is typical of the batch suspension technique used by the largest number of producers. The schematic diagram (Fig. P-26) indicates the features of the process. Vinyl chloride, which is a vapor at room temperature, is received as a liquid in pressurized tank cars. If inhibited, the inhibitor is removed by contact with caustic, followed by distillation. The reactors, which are in parallel, are stirred, jacketed, glass-lined pressure vessels. Process water is purified by demineralization and charged to the reactors. Catalysts, suspending agents, and other polymerization chemicals, as needed, are charged to the reactor followed by vinyl chloride. The mixture is then heated by hot water in the reactor jacket to bring the contents to reaction temperature, at which point the reaction becomes exothermic. Temperature is then controlled by cooling water circulated through the jacket.

Product type determines time of reaction. After the reaction is complete, minor quantities of unreacted vinyl chloride are recovered by vacuum stripping, condensed, distilled, and then recycled. The product in the reactor is a slurry, which is pumped to agitated storage tanks, where it is held with other batches for the drying process. The slurry is then pumped to a continuous horizontal centrifuge, where the process water is separated and sewered. The wet filter cake drops into a rotary drier, where it is dried by hot air, which also conveys the end product to a collector. The dried resin is screened to remove oversize particles

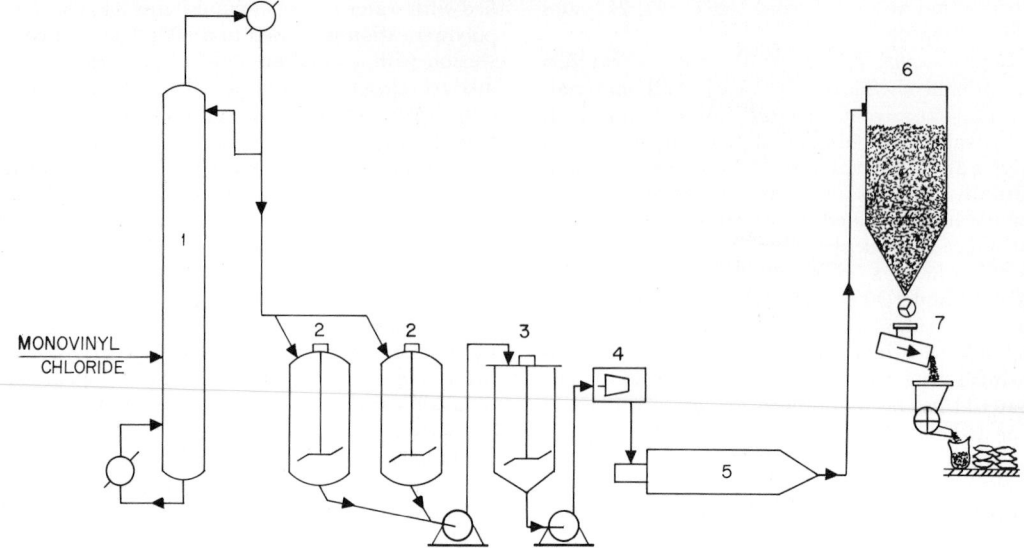

Fig. P-26. Polyvinyl chloride production by the suspension process. (1) Purification, (2) reactors, (3) blend tank, (4) centrifuge, (5) drier, (6) resin collector, (7) weighing and bagging. (*Scientific Design Company, Inc.*)

and then is bagged in 50-lb paper bags or transferred to silos for bulk shipment.

The process yields approximately 95% prime resin. The materials of construction are primarily stainless steel or glass-lined equipment to ensure a clean product.

See also **Stabilizers (PVC).**

—Robert M. Brown, *Scientific Design Company, Inc.,*
New York

Porcelain Enamel   (See **Ceramics.**)

Porosity   (See **Catalysts.**)

Porphyry Coppers   (See **Copper production.**)

Portland Cement   (See **Cement.**)

Positron   A positron is a subatomic particle that has the same mass as an electron but carries an equal and opposite charge. A positron has the ability to annihilate an electron, whereupon the masses of the positron and electron are converted into gamma radiation.

Pot Press   (See **Expression.**)

Potash   (See **Fertilizers;** and **Potassium.**)

Potassium   (NL. *potassa,* K_2O, from English potash.) **K** (NL. *Kali,* from Arabic *gily*) = 39.102 (at. wt.); 19 (at. no.). Potassium is a soft, silvery-white metal in group IA of the periodic table (alkali metals). Both physically and chemically it is similar to sodium and the other alkali metals.

Naturally occurring isotopes of potassium and their abundance in the earth's crust are ^{39}K 93.08%; ^{40}K 0.0119%; and ^{41}K 6.91%. The characteristics of all isotopes, including eight artificial isotopes, ranging in half-life from 6.7 ns to 22.4 hr, are given under **Chemical elements.** The naturally occurring isotope of mass 40 is radioactive, with a half-life of 1.3×10^9 years. The products of the decay are ^{40}Ca and ^{40}Ar. The decay to argon has permitted estimates of the age of rocks and mineral formations.

The physical properties of K are summarized in Table P-38. Metallic K is miscible with sodium in all proportions. The eutectic, 77.2% K, 22.8% Na, has a freezing point of $-12.3°C$. The sodium alloys containing 40–90% K are liquid at room temperature. K is also soluble in mercury, liquid ammonia, aniline, and ethylenediamine.

The K atom has one valence electron and exhibits only the 1+ valence. The inorganic salts of K are completely ionized. K is one of the most chemically reactive metals, is a powerful reducing agent, and is capable of reacting with the oxides

TABLE P-38. *Major Properties of Potassium**

Melting point, °C	63.7
Boiling point, °C	760
Density at 20°C, g/ml	0.86
Crystal structure	bcc
Lattice constant, Å	5.32
Atomic radius, Å	2.35
Cationic radius, Å	1.33
Electrode potential, V	2.93
Specific heat (liquid), cal/g	0.1422 + 0.000668t (°C)
Heat capacity at 200°C, cal/g	0.189
Thermal conductivity at 200°C, (cal)(cm)/(s)(cm²)(°C)	0.107
Latent heat of fusion, cal/g	14.6
Latent heat of vaporization, cal/g . . .	496

*Data from Refs. 1, 2, 3, 9, and 10.

and salts of other metals and with many organic compounds. Some of the specific chemical reactions of K are summarized in Table P-39.

Manufacture of the Metal: The first reported preparation of the metal was by Sir Humphry Davy in 1807, by the electrolysis of fused KOH. This procedure has been used subsequently for limited commercial production of the metal, but the method never has been entirely satisfactory.

K cannot be prepared by simple electrolysis of aqueous solutions of its salts because hydrogen, rather than potassium, is liberated at the cathode. However, a saturated aqueous solution of the chloride can be electrolyzed by use of a mercury cathode to yield a potassium amalgam, from which the K metal can be separated by distillation.

In 1949, a patent was issued for production of potassium[4] by a process involving the reaction $Na + KCl \rightleftarrows K + NaCl$. By ingenious application of engineering principles, the inventors were able to devise an economical process for continuous production of Na/K mixtures with a potassium content of up to or greater than 99.5%. In this process, vaporized Na metal is made to rise in a stainless-steel packed column, countercurrent to a descending stream of molten KCl. K is replaced by Na, with the result that a mixture of Na and K vapors is formed. In the upper part of the column, the metal vapors are separated by fractionation. Control of the conditions permits control of the Na/K ratio of the product.

Potassium also has been prepared by heating potassium carbonate with carbon, by the thermite reduction of aluminates,[5] and by reaction of KCl with CaC_2. An Italian patent based on the latter reaction reports K yields of greater than 70% (CaC_2 basis).[6]

Uses: Although K could be used for many of the purposes for which sodium is used in quantity, its much higher cost has limited its use to those applications where it has unique advantages. Na production exceeds K production by a factor of more than 1,000.

Most of the metallic K produced is used for preparation of the superoxide, KO_2, an ingredient of gas-mask canisters, and in the form of the NaK alloy, as a heat-exchange liquid. Use of the vaporized metal in turbines has been reported. The Na alloy has been used in magnetohydrodynamic power generation, as a catalyst, and for removal of O_2, H_2O, and CO_2 from inert-gas systems.

TABLE P-39. *Some Reactions of Potassium*

With:	Reaction
Water	Reacts rapidly, liberating H_2 and forming KOH in solution; reaction can be violent, with ignition of H_2
Air or oxygen	Reacts rapidly; can ignite spontaneously to give the superoxide, KO_2, if oxygen is in excess, or the monoxide, K_2O, if oxygen is deficient
Halogens	Reacts rapidly; detonates in contact with liquid bromine
Chlorinated hydrocarbons . . .	Forms explosive mixtures
Carbon	Reacts at elevated temperatures
Hydrogen	Reacts at elevated temperatures
Carbon dioxide	Reacts at elevated temperatures to give CO, C, and K_2CO_3
Helium, argon	No reaction
Nitrogen	No reaction
Saturated hydrocarbons	No reaction
Carbon monoxide	Forms explosive carbonyl
Sulfur	In liquid NH_3 solutions, reacts to give K_2S_2 and K_2S_4
Ammonia (liquid)	In presence of catalysts (metals), reacts to give amides and H_2

Handling Hazards: K can ignite spontaneously in air and reacts violently with water, liberating hydrogen, which can ignite spontaneously. A potassium fire gives off a dense smoke of potassium oxides, which present a considerable hazard to fire fighters because of their caustic nature. Protection to eyes and respiratory system is mandatory for persons approaching such a fire. These hazards are compounded by the formation of potassium superoxide, KO_2, a powerful oxidant that can be dangerous in contact with many substances. Materials useful in fighting a potassium fire include dry sodium chloride, soda ash, calcium carbonate, silica, certain ternary eutectic compounds, and Met-L-X powder. Carbon tetrachloride must *not* be used. Potassium never should be allowed to contact any part of the body or clothing.

For shipment, K is packed under an inert gas, such as argon or nitrogen, or under a hydrocarbon oil. Since it is a flammable solid, U.S. Interstate Commerce Commission regulations require shipping containers to be marked with a yellow label.

Occurrence of Potassium; Potash Production: K is the seventh most abundant element in the earth's crust, constituting about 2.5%. Because it is so reactive, K is never found in the elemental state. Potash is the name rather loosely applied to K compounds in general, particularly those of commercial significance. The name is derived from the fact that potassium carbonate is found in the ashes remaining after combustion of wood, kelp, and other vegetable matter, until relatively recently, the most important sources of this chemical.

Although the major basic potash chemical today is KCl, the K content of potash souces is generally expressed in terms of K_2O. Most of the potash produced today is derived from mineral deposits formed by evaporation of ancient lakes and seas which had become enriched in K salts leached from the soil, a process enhanced by the fact that most simple inorganic K compounds are highly soluble in water.

Although the most economically significant K occurrence is in the form of solid deposits, very large quantities of K salts are found concentrated in certain bodies of water, most notably the Great Salt Lake, Utah; Salduro Marsh, Utah; the Dead Sea, between Israel and Jordan; and Searles Lake, in California. These brines are used for commercial potash production.

Important naturally occurring K minerals include:

Sylvite, KCl
Sylvinite, KCl/NaCl mixture
Carnallite, $KCl \cdot MgCl_2 \cdot 6H_2O$
Kainite, $MgSO_4 \cdot KCl \cdot 3H_2O$
Langbeinite, $K_2SO_4 \cdot 2MgSO_4$
Polyhalite, $K_2SO_4 \cdot MgSO_4 \cdot 2CaSO_4 \cdot 2H_2O$
Jarosite, $K_2Fe_6(OH)_{12}(SO_4)_4$
Alunite, $K_2Al_6(OH)_{12}(SO_4)_4$
Leucite, $K_2O \cdot Al_2O_3 \cdot 4SiO_2$
Microcline, $K_2O \cdot Al_2O_3 \cdot 6SiO_2$
Orthoclase, $K_2O \cdot Al_2O_3 \cdot 6SiO_2$
Muscovite, $K_2O \cdot 3Al_2O_3 \cdot 6SiO_2 \cdot 2H_2O$
Biotite, $H_2K(Mg,Fe)_3 \cdot (Al,Fe)(SiO_4)_3$.

The major workable potash mineral deposits are located in Stassfurt, Germany (sylvinite, carnallite, kainite); Alsace (sylvinite); New Mexico (polyhalite, sylvinite, langbeinite); Saskatchewan, the Soviet Union, Spain, and Poland (sylvinite, carnallite); Italy (volcanic deposits, leucite); Atlantic Seaboard, United States (greensands, glauconite, $KFe(SiO_3)_2 \cdot H_2O$); and Utah (alunite).

The total reserve of recoverable potassium (as K_2O) in the world is estimated at 53×10^9 tons. Of this reserve, Canada has 37%; Russia, 49%. German (East and West) reserves constitute most of the remainder, but the United States, French, and Spanish reserves are substantial (0.3–0.5×10^9 tons each). There also are significant reserves in Great Britain, Italy, Australia, Zaïre (formerly Belgian Congo), and Israel. Some of the saline-water reserves, although not heavily exploited as yet due to high cost, are substantial. For example, the Dead Sea contains 1.5–2×10^9 tons (as K_2O) of potassium.

In 1965, world potash consumption was 15 million tons.[12] Although production capacity was somewhat larger than consumption, production facilities were able to operate at a high percentage of capacity, and prices generally were firm. Since that period, however, world capacity has increased out of proportion to demand, with an overcapacity of 50% reported in 1969.

Compounds: The characteristics and uses of major potassium compounds are summarized in Table P-40. See also **Fertilizers.**

REFERENCES

1. Kolthoff, I. M., and Philip J. Elving: "Treatise on Analytical Chemistry," pt. II, vol. 1, pp. 306–460, Interscience-Wiley, New York, 1961.
2. "Kirk-Othmer Encyclopedia of Chemical Technology," 2d ed., vol. 16, pp. 361–400, Wiley, New York, 1968.

TABLE P-40. *Characteristics and Uses of Major Potassium Compounds*

Name, Formula, and Formula Weight	Characteristics and Uses
Potassium aluminum sulfate $Al_2(SO_4)_3 \cdot K_2SO_4 \cdot 24H_2O$ 948.78	Colorless crystals; soluble in water; made by mixing solutions of K_2SO_4 and $Al_2(SO_4)_3$, followed by crystallizing; used for water purification, fur treatment, leather tanning; in photography, pickle making, dyeing and fireproofing textiles; manufacture of artificial stones, statuary, pigments, color lakes, vegetable glue, marble and porcelain cements, matches, medicines; in taxidermy
Potassium bicarbonate $KHCO_3$ 100.12	White, granular solid; soluble in water; begins to decompose at temperatures over $60°C$, although not as rapidly as the sodium analog; used in dry fire extinguishers, detergents, fertilizers, and pharmaceuticals; as a neutralizing agent in leather tanning, a dusting agent in rubber and plastics, and in textile finishing
Potassium carbonate K_2CO_3 138.21	White, granular solid; soluble in water; marketed as the anhydrous product, as the hydrate, and in solution; very hygroscopic; alkaline reaction; made by reaction of CO_2 and KOH liquor; in addition to use as a starting material for the manufacture of other potash chemicals, such as phosphates, silicates, and persulfates, it is used in the manufacture of glass, particularly TV picture tubes, soaps, and chocolate, and as a CO_2 absorber; also used in vat dyeing and textile printing, and in titanium enamels, photographic chemicals, boiler compounds, and electroplating baths
Potassium chlorate $KClO_3$ 122.55	Colorless crystals; soluble in water; a powerful oxidizing agent; made by electrolysis of KCl solution or reaction of calcium chlorate with KCl in solution; used in manufacture of matches, pyrotechnics, explosives, percussion caps, medicines, dyes, and paper; due to its oxidizing power, requires care in handling
Potassium chloride KCl 74.56	Also called potassium muriate; colorless or white crystalline material; soluble in water; occurs widely in natural deposits and saline waters; used directly as a fertilizer; starting material for manufacture of metallic K and many K compounds; some KCl is used as a dietary salt substitute when Na intake must be limited
Potassium chromate K_2CrO_4 194.20	Yellow crystals; soluble in water; nonhygroscopic; made by reaction of KOH with potassium bichromate; used as a corrosion inhibitor, as a component of certain inks, ceramics, fluxing and metal coating compounds, and photographic chemicals; textile and dye production, leather finishing
Potassium cyanide KCN 65.12	White solid; deliquescent; soluble in water; when acidified, liberates HCN gas; made by reaction of carbon and NH_3 with molten K_2CO_3; used in metallurgy, electroplating, extraction of gold from ores, pesticide fumigant, in photography and analytical chemistry, engraving, lithography; *very poisonous,* either by direct ingestion or inhalation of HCN gas produced upon acidification
Potassium dichromate $K_2Cr_2O_7$ 294.19	Orange-red crystals; soluble in water; stable and nonhygroscopic; good oxidizing agent; made by reaction of sodium dichromate with KCl; used in manufacture of safety matches, pyrotechnics, pigments, colored glass; used as a laboratory reagent, in blueprint developing, and in wood preservation
Potassium hydroxide KOH 56.11	Also called caustic potash; a nearly white solid; soluble in water; a strong alkali; very hygroscopic; made by electrolysis of aqueous KCl, either in diaphragm cells or cells having a Hg electrode; marketed as the dehydrated solid (90–92% KOH) and as a liquor (45–50% KOH); used in manufacture of soaps, chemicals, drugs, dyes, alkaline batteries, adhesives, fertilizers, and alkylates; for purifying industrial gases and petroleum products, scrubbing HF; in asphalt emulsions, adhesives; as a drain-pipe cleaner; because of caustic reactive nature, *should be handled carefully*
Potassium iodide KI 166.01	Colorless crystals; soluble in water; in solution, forms complex iodides with elemental iodine; made by (1) reaction of KCl and HI, followed by distillation of HCl, (2) reaction of iodine with KOH or K_2CO_3 in solution, (3) reaction of ferrous/ferric iodide with K_2CO_3, and (4) evaporation of natural brines; used in photography, medicine, preservation of fruit, chemical analysis, and organic synthesis; in preparation of iodized salt and supplement in nutrition

TABLE P-40. *Characteristics and Uses of Major Potassium Compounds* (*Continued*)

Name, Formula, and Formula Weight	Characteristics and Uses
Potassium nitrate KNO_3 101.11	Colorless crystals; soluble in water; powerful oxidant; generally made by (1) mixing $NaNO_3$ and KCl in solution, followed by removal of $NaCl$ by precipitation, then crystallization of KNO_3 or (2) reaction of KCl with HNO_3; one form of the latter process yields Cl_2 as a by-product; used as a fertilizer, in manufacture of matches, explosives, pyrotechnics, glass, and medicines; also as a food preservative, color fixative and pickling agent for meat; rocket-fuel oxidizer, and in the heat treatment of steel
Potassium permanganate $KMnO_4$ 158.03	Purple crystals; soluble in water and H_2SO_4; a strong oxidizing agent; made by reaction of Mn ore (MnO_2) with KOH and O_2, followed by reaction of the intermediate K_2MnO_4 with CO_2; alternatively, the K_2MnO_4 in solution can be electrolyzed; used in metal treatment, as a disinfectant, in the manufacture of medicines and pharmaceuticals (including vitamins), and other chemical manufacture; as an analytical reagent, for bleaching paper pulp, and in water purification; *requires care in handling* due to its strong oxidizing character
Potassium sulfate K_2SO_4 174.27	Colorless crystals; soluble in water; made by reaction of H_2SO_4 with KCl, or from kainite ore directly by solution and fractional crystallization; used for manufacture of potash alum and glass, as a fertilizer, and in medicines; almost 90% of demand is for fertilizer; total United States demand (1968) was 253,000 tons (K_2O basis)
Potassium superoxide KO_2 71.10	Yellow solid; strongly alkaline and a powerful oxidizing agent; reacts with water to liberate O_2 and give a strongly basic solution containing hydroperoxide; made by reaction of K metal with O_2; chief use is in breathing apparatus, where moist air being exhaled liberates oxygen while CO_2 is simultaneously absorbed by the KOH formed in the reaction between H_2O and KO_2; *must be handled with care* because of strong oxidizing character and high alkalinity

3. Hey, D. H. (ed.); "Kingzett's Chemical Encyclopaedia," 9th ed., Van Nostrand, New York, 1966.
4. Jackson, C. B., and R. C. Werner: U.S. Patent 2,480,655, Aug. 30, 1949 (to Mine Safety Appliances Co.).
5. Isabaev, S. M., et al.: *Izv. Vysshikh Uchebn. Zaved., Tsvetn. Met.*, vol. 9, no. 4, pp. 39–42, 1966 Russian; *Chem. Abstr.*, vol. 66 col. 5010s.
6. Perugini, Giancarlo: Italian Patent 632,506, Jan. 29, 1962; *Chem. Abstr.*, vol. 61, col. 2773f.
7. Sittig, Marshall: "Inorganic Chemical and Metallurgical Process Encyclopedia," Noyes Development Corporation, Park Ridge, N.J., and London, 1968.
8. Bretschneider, O., et al.: U.S. Patent 2,920,951, Jan. 12, 1960.
9. "Mellor's Comprehensive Treatise on Inorganic and Theoretical Chemistry," vol. 2, suppl. 3, Wiley, New York, 1963.
10. Lange, Norbert A.: "Handbook of Chemistry," 10th ed., McGraw-Hill, New York, 1967.
11. Adams, Samuel S.: pp. 1–21 in *Role of Potassium in Agriculture, Proc. Symp.*, 1968; *Chem. Abstr.*, vol. 70, col. 59593u.
12. *Chem. Week*, Nov. 19, 1969, p. 36.

—John G. Benson, *Allied Chemical Corporation, Industrial Chemicals Division, Solvay, N.Y.*

Potassium Hydroxide (See **Caustic potash.**)

Potassium Hypochlorite (See **Chlorine.**)

Potassium Nitrate (See **Fertilizers.**)

Potassium Silicates [See **Silicates (soluble).**]

Potassium Sulfate (See **Fertilizers.**)

Potting (See **Epoxy resins.**)

Pound Mole [See **Mole (or mol).**]

Powder Metallurgy (See **Iron; Iron and steel;** and **Tungsten.**)

Powder Paints (See **Paint.**)

Power, Nuclear (See **Nuclear power plants.**)

Praseodymium (Gk. *prasios*, green, and *didymous*, occurring in pairs.) **Pr** = 140.91 (at. wt.); 59 (at.

no.). Praseodymium is the third of 15 elements in group III, period 6, generally shown as the rare-earth elements in a separate line below the main body of the periodic table and sometimes referred to as the *lanthanide series*. Pr was discovered in 1885 by Baron G. A. von Welsbach in Vienna.

Pure praseodymium metal has a silvery-gray metallic luster, which dulls rapidly due to the reaction with moisture and oxygen, forming a complex nonadherent oxide. When nonmetallic impurities such as nitrogen, hydrogen and oxygen are low, Pr metal is soft and workable with conventional tools and equipment. Processing requires an inert atmosphere or vacuum, and storage under a nonreactive liquid is recommended. Fine Pr powder, chips, and turnings are pyrophoric, burning at a red heat and consuming the metal completely.

The melting point of Pr is 934°C; boiling point 3512°C; density 6.772 g/cm^3. Estimated at 8.2 ppm in the earth's crust, Pr is potentially 4 times more available than tin, tantalum, or beryllium.

Only one isotope, ^{141}Pr exists in nature, but 14 artificial isotopes have been identified. This element shows no radioactivity and has a low acute-toxicity rating.

Dilute mineral and halogen acids dissolve Pr metal easily. This element forms compounds with most inorganic and many organic anions. Intermetallic compounds of Pr with most of the elements to the right of iron in the periodic table have been observed. The properties of the many compounds of Pr still need to be defined and evaluated, however.

Praseodymium has multivalent properties, and its oxide system is very complex. Studies show that Pr reacts with oxygen to form a large number of phases in the composition range PrO_x, $1.5 \leq x \leq 2.0$. The accepted commercial formula for the oxide is Pr_6O_{11} (in which $x = 1.83$). This complex oxide has a characteristic black color. Depending on the conditions under which the oxide is prepared (calcined), however, the formula PrO_2 is yellow-green, which decomposes at 350°C to Pr_2O_3, which is brown-black. Praseodymium oxide dissolves in concentrated mineral acids, and its solutions are predominately trivalent and green in color. Added to soda-lime glass, Pr_6O_{11}, up to 5% w/o* produces a yellow-green color. The glass

*This abbreviation is used in the rare-earth and related fields for weight percent.

shows a very weak visible fluorescence under ultraviolet excitation.

Praseodymium is the fourth most plentiful of the light lanthanide elements. Only cerium and lanthanum had been identified as separate elements by 1839 in the mineral called *cerite earths* found near Bastnäs, Sweden. The balance of the mineral was called *didymium*. In 1880 an element, gadolinium, was separated by the Swiss chemist, J. C. G. Marignac, from didymium. Then, in 1885, von Welsbach in Germany succeeded in separating the two components in didymium naming them *neo-* (new) and *praseo-* (green) *dymium*.

The minerals bastnasite and monazite, which contain 4 to 8% Pr, supply all requirements. Separation of Pr from the other chemically similar elements is by liquid-liquid or solid-liquid organic ion-exchange processes. Commercial separation plants have been constructed that can produce over 100,000 lb/yr continuously. Praseodymium metal is produced by the continuous electrolysis of Pr_6O_{11} in a molten fluoride electrolyte or a calcium reduction of PrF_3 or $PrCl_3$ in a sealed-bomb reaction. The metal, oxide, and chemicals are available in several grades to the highest purity levels. For a detailed description of resources, processes, and a table of properties, see **Rare-earth elements and metals.**

Uses: Praseodymium, as one of the components in light rare-earth mixtures, has been used since late in the nineteenth century in mischmetal, a pyrophoric alloy known as *lighter flint*. The mixture of Ce, La, Nd, and Pr as fluorides and oxides is burned in the cores of arc carbons, which produce the intense light required for motion-picture projection lamps and searchlights. A mixture of the rare-earth oxides containing Pr is also used to polish optical glass.

Natural mixtures of the lanthanons containing about 5% Pr are used in petroleum-cracking catalysts. A concentrated mixture containing about 10 w/o Pr (with 30% w/o Nd and 60 w/o La) used by the petroleum-refining industry in a catalyst for cracking crude oil is the largest commercial use of Pr and other lanthanide elements.

One of the first applications for elemental Pr was to color glass. A range of clear yellow to green colors makes Pr useful for a variety of art-glass objects and tableware. It is also used in sunglasses, in goggles to protect welders' eyes, and optical filters.

A praseodymia-zirconia yellow stain is used by the ceramic-tile industry. The stability of this pure color during firing of the glaze is the result of reacting 3–5 mole % Pr_6O_{11} with zirconia.

A most important metallurgical use of Pr is as an intermetallic compound with cobalt, $PrCo_5$. This compound has unsurpassed permanent-magnet properties. Outstanding are its resistance to demagnetization (coercive force H_c) and magnetic saturation value (intrinsic induction B_i) which has the highest theoretical energy product $(BH)_{max}$ of the rare-earth cobalt alloys. The $PrCo_5$ is first crushed to particles about 20 μm (in inert gas or liquid). The powder is then pressed into shape in a magnetic field which orients the crystals in a uniform direction. Using appropriate sintering additives, with control of time and temperature, a dense, stable magnetically receptive solid is obtained that has a Curie temperature T_c of 612°C and a theoretical energy product of 36.0 MGOe.

Praseodymium mixed with other lanthanide elements like samarium, lanthanum, cerium, and yttrium and alloyed with cobalt constitutes a "new family of permanent magnet materials." These magnets are produced on a commercial scale for traveling-wave tubes, electric watches, hearing aids, and gyroscopes. New systems for power generation, transportation, and material handling are being designed to take advantage of their unusual properties.

The uses of praseodymium in magnets, glass ceramics, and catalysts already discovered warrant further investigation of the many possible compounds and potential applications of this element.

REFERENCES

Becker, J. J.: Permanent Magnets, *Sci. Am.*, December 1970, pp. 92–98.
Molybdenum Corporation of America: *Overview* 1, 3, and 25 (application information, including possible uses), New York (published periodically).

    —Joseph G. Cannon, *Molybdenum Corporation of America, White Plains, N.Y.*

Precipitation   (See **Colloid systems; Detergents; Iron oxides, synthetic;** and **Mixing, fluids.**)

Precipitator   (See **Air pollution;** and **Separation operations.**)

Precoating Filters   (See **Filtration.**)

Prefabricated Process Plants   (See **Process plants, packaged.**)

Preservatives   (Consult the **Subject Index.**)

Presoak Compounds   (See **Detergents.**)

Pressing   (See **Glass.**)

Prilling   (See **Explosives; Fertilizers;** and **Urea.**)

Primary Waste Treatment   (See **Waste treatment.**)

Process   (See **Chemical engineering.**)

Process Industries   The term *process industries* is rather loosely used to describe those industries which depend heavily upon chemical and metallurgical technology. Manufacturing of all types tends to fall into two major classifications: (1) *discrete-piece manufacturing systems,* in which individual, identifiable (discrete) pieces of solid materials are handled and such pieces may be cut, formed, machined, and otherwise shaped and ultimately fastened together in some fashion to form a simple or complex final assembly. Machinery, transportation equipment, electrical apparatus, communications equipment, appliances, garments, tools, utensils, instruments, and furniture and fixtures typify the products that result from discrete-piece manufacturing systems. In plants of this kind an inventory can be taken by counting pieces. (2) In *fluid-bulk processing systems* gases, liquids, and bulk (sometimes fluidized) solids are handled. In these systems the discrete pieces are molecular in dimension in terms of the gases and liquids handled and range from colloidal, submicroscopic particles up to fairly large rocks in terms of the solids handled, but even in the case of large solid objects, individual identity of pieces seldom is important. In a fluid-bulk plant, an inventory is *not* taken by counting pieces except at the very end of the process, where products are packaged and containerized in some fashion.

Although some industries are predominantly discrete-piece or fluid-bulk in their manufacturing nature, there are relatively few industries that fall simply into one category or the other. The chemical and petroleum industries come about as close as any to typifying a fluid-bulk processing industry, and the garment industry is essentially a pure discrete-piece manufacturing industry. Where an industry is predominantly fluid-bulk in nature, it qualifies under the approximate definition as a *process industry*. In the literature, one also will find reference to the *chemical process industries*. Within the rather loose framework of definition that prevails, process industry and chemical process industry may be considered synonymous.

TABLE P-41. *Process Industries and Important Branches*

Industry and Products	Standard Industrial Classification Code	Industry and Products	Standard Industrial Classification Code
Chemicals and petrochemicals:		Candy and other confectionary products	2071
Alkalies	2812	Carbonated beverages	2086
Chemicals, industrial		Cereals	2043
Inorganic, nec	2819	Cheese	2022
Organic, nec	2818	Chewing gum	2073
Chemicals and chemical preparations, nec	2899	Chocolate and cocoa products	2072
Chlorine	2812	Cigarettes	2111
Coal-tar crude chemicals	2814	Cigars	2121
Dyes, dye (cyclic) intermediates, and		Coffee roasting	2095
organic pigments (lakes and toners)	2815	Cookies and crackers	2052
Gases, industrial	2813	Corn (wet milling)	2046
Gum and wood chemicals	2861	Fish and seafood	
Rubber, synthetic (vulcanizable elastomers)	2822	Canned and cured	2031
Drugs and medicines:		Frozen	2036
Biological products	2831	Flavoring extracts and syrups	2087
Medicinal chemicals and botanicals	2833	Flour and other grain-mill products	2041
Perfumes, cosmetics, and other toilet		Flour (blended and processed)	2042
preparations	2844	Food preparations, nec	2099
Pharmaceutical preparations	2834	Fruits, jams, jellies, and preserves (canned)	2033
Explosives and fireworks:		Fruits	
Ammunition, artillery	1921	Dried and dehydrated	2034
Ammunition, nec	1929	Frozen	2037
Small-arms	1961	Pickled	2035
Explosives	2892	Fruit juices (frozen)	2037
Matches	3983	Gelatin and glue	2891
Ordnance and accessories, nec	1999	Ice, manufactured	2097
Fats and oils:		Liquors, distilled and blended	2085
Animal and marine fats and oils (except		Macaroni, spaghetti, vermicelli, and	
grease and tallow)	2095	noodles	2098
Candles	3984	Malt	2083
Cottonseed oil milling	2091	Malt liquor	2082
Fatty acids	2894	Meat packing	2011
Grease and tallow	2094	Milk	
Margarine	2096	Condensed and evaporated	2023
Shortening, table oils, and other edible fats		Fluid	2026
and oils, nec	2096	Rice milling	2044
Soybean-oil milling	2092	Salad dressing	2035
Vegetable-oil milling (except cottonseed and		Sugar	
soybean oil)	2093	Beet	2063
Fertilizers and agricultural chemicals:		Cane, except refining	2061
Agricultural chemicals, nec	2879	Refining	2062
Agricultural pesticides	2873	Tobacco (stemming and redrying)	2141
Fertilizers	2871	Tobacco products (except cigarettes and	
Mixing only	2872	cigars)	2131
Fibers and textile-related products:		Vegetables, canned	2033
Cellulosic synthetic fibers	2823	Dried and dehydrated	2034
Cotton finishing	2261	Vegetable juices, frozen	2037
Dyeing and finishing of textiles, nec	2269	Vegetable sauces and seasonings	2035
Silk and synthetic-fiber finishing	2262	Wine	2084
Synthetic organic fibers (except cellulosic)	2824	**Leather and associated products:**	
Wool dyeing and finishing	2231	Artificial leather and other impregnated and	
Wool scouring	2297	coated fabrics (except rubberized)	2295
Foods, beverages, and other consumable products:		Oilcloth	2295
Brandy and brandy spirits	2084	Tanning and finishing	3111
Bread	2051	**Lime and Cement:**	
Butter	2021	Cement, hydraulic	3241

nec, not elsewhere classified

Industry and Products	Standard Industrial Classification Code	Industry and Products	Standard Industrial Classification Code
Concrete, dry mixture	3272	Sand and gravel	1442
Concrete-curing compounds	2899	Silver	1044
Concrete plants	3531	Soda ash	1474
Gypsum products	3275	Sulfur	1477
Lime	3274	Talc, soapstone, and pyrophyllite	1096
Metallurgical and metal products:		Titanium	1093
Aluminum (primary smelting and		Tungsten	1064
refining)	3334	Uranium and radium	1094
Batteries		Vanadium	1094
Storage	3691	Zinc	1031
Primary (wet and dry)	3692	Paints, pigments, and allied products:	
Blast furnaces (ferrous), including coke		Inorganic pigments	2816
ovens	3312	Paints, varnishes, lacquers, and enamels	2851
Coatings, engravings, and allied services,		Putty, caulking compounds, and allied	
nec	3479	products	2852
Copper (primary smelting and refining)	3331	Petroleum refining and hydrocarbons:	
Electrometallurgical products	3313	Asphalt felts and coatings	2952
Electroplating, plating, polishing,		Crude petroleum	1311
anodizing, and coloring of metals	3471	Hydrocarbon processing	1321
Enameled iron and metal sanitary ware	3431	Liquefied petroleum gas	1321
Lead (primary smelting and refining)	3332	Lubricating oils and greases	2992
Nonferrous metals, primary smelting and		Natural gas, processing	1321
refining, nec	3339	Transmission	4922
Secondary smelting, refining, and		Paving mixtures and blocks	2951
alloying	3341	Petroleum products, refined (chemicals,	
Zinc (primary smelting and refining)	3333	fuels, and solvents)	2911
Mining and ore beneficiation:		Pipelines	
Aluminum	1051	Crude petroleum	4612
Anthracite coal	1111	Pipelines, nec	4619
Asphalt and bitumins	1494	Refined petroleum	4613
Barite	1472	Rubber products:	
Bituminous coal	1211	Footwear	3021
Borates	1474	Reclaimed rubber	3031
Clay, bentonite	1452	Rubber products, nec	3069
Fire	1453	Tires and inner tubes	3011
Clay, nec	1459	Soap, cleaning, and polishing products:	
Copper	1021	Cleaning preparations	2842
Feldspar	1456	Detergents	2841
Fluorspar	1473	Finishing agents	2843
Fuller's earth	1454	Polishing preparations	2842
Gold	1042	Sanitary preparations	2842
Gypsum	1492	Soap	2841
Iron	1011	Sulfonated oils	2843
Kaolin and ball clay	1455	Surface-active agents	2843
Lead	1031	Stone, clay, glass, and ceramic products:	
Lignite coal	1212	Abrasives	3291
Limestone	1422	Asbestos products	3292
Manganese	1062	Bathroom fixtures	3261
Mercury	1092	Bricks	3251
Metallic minerals, nec	1099	Clay structural products	3251
Mica	1493	Dishes and chinaware	3261
Nonmetallic minerals, nec	1499	Earthenware	3263
Peat	1498	Glass	
Potash	1474	Flat	3211
Pumice and pumicite	1495	Pressed and blown	3229
Salt	1476	Glass containers	3221

nec, not elsewhere classified

TABLE P-41. *Process Industries and Important Branches* (*Continued*)

Industry and Products	Standard Industrial Classification Code	Industry and Products	Standard Industrial Classification Code
Insulating board.	2661	Utilities:	
Insulation		Manufactured gas	4925
Molded asbestos	3293	Power generation	4911
Plaster	3275	Sewage treatment	4952
Mineral products nec.	3295	Water treatment.	4941
Mineral wool	3296	Wood, pulp, paper, and board products:	
Packing, boiler, pipe, and steam	3293	Building-paper and building-board mills . .	2661
Pottery products nec	3269	Linoleum.	3982
Refractories		Paper coating and glazing	2641
Clay	3255	Paper mills, except building-paper mills . .	2621
Nonclay	3297	Paperboard mills	2631
Tile, ceramic floor and wall	3253	Pencils	2952
		Pulp mills	2611
		Wood preserving	2491

nec, not elsewhere classified

A number of industries start their manufacturing in a bulk-fluid way and end up with discrete-piece manufacturing operations. Ferrous and nonferrous metals production, for example, begins with bulk ores, followed by numerous fluid-bulk operations before discrete, countable ingots, bars, tubes, and other specific forms emerge. The glass industry essentially is a bulk-solids industry until the point where containers are formed. The textile and pulp and paper industries both essentially start with the processing of fluid-bulk materials but create discrete products for final shipment. Even the chemical and petroleum industries embrace discrete-piece handling in the final packaging and containerizing of products.

In a number of industries, one manufacturing operation tends to be an interface between fluid-bulk handling and discrete-piece handling. This operation involves the handling of very long, continuous lengths of solid materials. The bleaching and dyeing of long lengths of textiles; the rolling and forming of long rails, pipes, bars, and other shapes in the metals industries; the shaping and treating of long lengths of plate glass; the forming and processing of long lengths of films and other plastic shapes; and the manufacture of paper and other webbed materials are examples of *continuous-length* handling operations. From an engineering, instrumentation, and control viewpoint, these continuous-length operations all tend to pose similar technical problems that usually fall midway between the chemical engineering

approach to fluid-bulk processing and the mechanical engineering approach to discrete-piece manufacturing.

Sometimes, the term *continuous* is closely associated with the process industries. Although it is true that continuous, 24 hr/day production typifies petroleum and some segments of chemical production, one also finds numerous batch processes in industries that essentially are fluid-bulk in character. Thus, the term *continuous* must be used with discretion.

Although rather arbitrary, as will be understood from the foregoing description, the list of industries in Table P-41 is generally accepted by business planners as comprising the process industries. The Standard Industrial Classification code number (as adapted by the Executive Office of the President, U.S. Bureau of the Budget, 1967) for each industry is included because census and other statistical information is frequently reported by these code numbers.

Process Plants, Packaged Packaged process plants are preassembled by the manufacturer and shipped to the plant site essentially complete, as contrasted with plants in which individual equipment items are supplied unassembled, with all piping, instrumentation, insulation, and painting done on the plant site.

Several kinds of process plants have been available as packaged units for a number of years, e.g., air-dehydration units employing solid desiccants;

ammonia dissociators, in which ammonia is converted to a mixture of hydrogen and nitrogen; and inert-gas generators, in which a suitable fuel, such as natural gas or distillate fuel oil, is burned in air to generate flue gas. The gas is then scrubbed with monoethanolamine solution to remove CO_2, leaving an inert gas consisting principally of nitrogen. Other early packaged process plants were shop-preassembled gas-purification units,* used to remove H_2S or CO_2 from natural gas by scrubbing with monoethanolamine solution.

During World War II, a number of packaged hydrogen-manufacturing units built† for military use produced H_2 for barrage balloons and other uses by reacting methanol with steam to form CO_2 and H_2, the CO_2 being removed by treatment with monoethanolamine solution. These units were completely preassembled and required only a supply of methanol and water for operation.

The next major application of the packaged-plant concept was in the development of CO_2 units used to produce liquid CO_2 and Dry Ice. CO_2 is obtained by burning fuel oil, and the heat generated is used to strip CO_2 from the monoethanolamine solution used for CO_2 absorption. The pure CO_2 gas thus obtained is scrubbed with $KMnO_4$ solution to remove odorous impurities and then compressed and liquefied. The liquid CO_2 may be charged into cylinders or converted

*Girbotol
†Girdler Corporation.

into solid Dry Ice by the addition of a Dry Ice press. Typical plant sizes produce 150–300 lb/hr of CO_2.

Another important packaged-plant application has been for CO_2 removal from the air in nuclear submarines. These units use the same monoethanolamine scrubbing process and are completely shop-assembled and rigorously tested for performance, noise levels, and dependability before installation in submarines.

A packaged plant* has been built for the continuous production of sodium hypochlorite bleach by reacting Cl_2 and $NaOH$. The unit, with a capacity of 9,000 gal/hr, occupies a space 15 ft long × 8 ft high × 7 ft wide. Air-liquefaction plants, including completely automated units for producing high-purity nitrogen (99.999%) have been supplied.† A packaged unit‡ for producing polyvinyl chloride, which includes a naphtha-cracking unit with a capacity of 25 million lb/yr of ethylene, has been supplied. A 500-W portable electric generator§ for the U.S. Army combines a hydrogen-fired fuel cell developed at Fort Monmouth with a hydrogen unit in which kerosene is reformed with steam to produce the hydrogen feed for the fuel cell.

A number of packaged hydrogen units, utilizing steam reforming of natural gas instead of metha-

*The Dow Chemical Company.
†Linde and Air Products.
‡Chemical Systems Inc.
§Pratt and Whitney.

Fig. P-27. Portion of packaged hydrogen plant mounted on permanent skid. (*C & I/Girdler Incorporated.*)

nol, were built.* One of these units is shown partially in Fig. P-27. On the skid is the CO_2-removal equipment for a 600,000 ft³/day H_2 plant. In addition to providing support during shipment, the skid eliminates the need for separate foundations for all the equipment mounted on it.

The packaging concept also has been used in the fertilizer industry for the production of ammonia and urea. A typical application† is a complete NH_3 manufacturing plant with capacities of 60–100 tons/day. Equipment, piping, structural steel, and instrumentation are completely shop-assembled on a number of skids (usually 16–20), sized for shipment by rail, truck, or ship. Interconnecting the skids is the only work required at the plant site and can be carried out in 1 or 2 months. Processwise, packaged NH_3 plants are conventional, utilizing natural gas or liquid hydrocarbons, e.g., propane, butane, or naphtha, as feed material. These plants prove economical in isolated locations which must serve a relatively small market and where the higher operating cost implicit in small-capacity operation is offset by high freight costs for NH_3 delivered from major producing centers.

—R. M. Reed and E. Pelitti, *C&I/Girdler Incorporated, Louisville, Ky.*

Prolamine (See **Peptides and proteins.**)

Proline (See **Amino acids;** and **Peptides and proteins.**)

Promethium (Gk. *Prometheus,* who stole fire from heaven and gave it to man, thereby teaching him many useful arts and sciences.) **Pm** = 147 (at. wt.); 61 (at. no.). The discovery was announced and the name of this element suggested by J. A. Marinsky, L. E. Glendenin, and C. D. Coryell in 1947 at the Oak Ridge (Tenn.) National Laboratory.

Promethium is the fifth element in group III, period 6, generally shown as the rare-earth elements in a separate line below the main body of the periodic table. The existence of element 61 was suggested by Mendeleev in 1869 and predicted by Moseley in 1914. During 1923-1926, scientists in Illinois and in Italy both claimed identification of element 61 from x-ray examination, but positive proof was missing. Pool and Quill first observed radioactivity from element 61

*C&I/Girdler.
†Ammopac (C&I/Girdler).

in 1938. Milligram quantities of ^{147}Pm were first chemically separated from uranium nuclear-reactor fission products in 1945. Promethium was identified in pitchblende from Zaïre (formerly Belgian Congo) in 1966. The content reported was about $4.5 \times 10^{-6}\%$ resulting from fission of ^{235}U and ^{238}U. The man-made radioactive isotope with the longest half-life is ^{145}Pm (18 years).

The few physical properties of Pm metal that have been reported mainly concern its potential as a fuel source. In pure form ^{147}Pm is silvery-white, relatively soft, and can be cast or machined into complex shapes. Because of radioactivity, all work with Pm must be carried out in a shielded glove box. The melting point of Pm is 1124°C; the estimated boiling point is 2460°C; density 7.22 ± 0.02 g/cm³. Other properties that may have been determined remain classified by the U.S. Atomic Energy Commission (AEC) or are proprietary.

F. Weigel, of Munich, is credited with being the first to make Pm metal in 1963 by the reduction of promethium fluoride, PmF_3, with lithium metal in vacuum.

The nuclear properties of man-made promethium isotopes have been studied in great detail. Some 15 to 18 Pm isotopes have been identified (140–156), most with very short half-lives. ^{147}Pm extracted from uranium or plutonium reactor wastes is by far the most important and predominant. This isotope can also be produced by neutron irradiation of neodymium. ^{147}Pm has a half-life of 2.62 years, decays primarily by beta emission (0.225 MeV), and gives a specific power of 0.333 W/g; its disintegration rate is 928 Ci/g. The pure ^{147}Pm isotope is important because it decays by beta emission only, at a sufficiently low energy level compared to most fission products to require only light to moderate shielding.

Two other isotopes, ^{146}Pm and ^{148m}Pm, are produced from the reactor wastes. Even though these are associated in only small amounts, they both produce penetrating gamma radiation which must be eliminated if the outstanding properties of ^{147}Pm as a heat source and a power source are to be achieved.

Extensive research at AEC laboratories and private concerns has made the processing of ^{147}Pm practical. The product, which is available in kilogram quantities from the AEC, is >99.9% pure Pm_2O_3. In 1970 the price amounted to $560 per watt. By 1990 it is projected that about 1,000 kg/yr of ^{147}Pm will be available from the processing of reactor wastes.

The key to separating the beneficial isotope, ^{147}Pm, from other isotopes is an aging process. Figure P-28 shows the formation, typical amounts, half-life, and decay products. Radioactive wastes that have been stored for at least $3\frac{1}{2}$ years are selected for chemical processing.

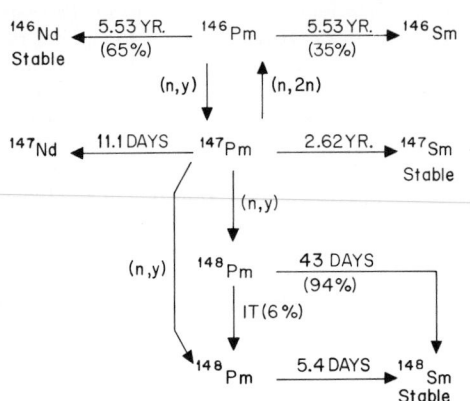

Fig. P-28. Methods of forming promethium isotopes.

After other waste products, spent uranium, and plutonium are separated, the rare earths produced by the reactor are put into solution and processed by solid-liquid organic ion-exchange methods. See **Rare-earth elements and metals.**

Promethium metal is prepared by calcium or magnesium reduction of PmCl$_3$ in a bomb reaction, as shown in Fig. P-29.

The beta-emitting isotope, ^{147}Pm, is used to activate a luminescent phosphor. Hundreds of tiny inert beads, called Microspheres,* containing the isotope mixed with phosphor provide a long-lived, reliable green light to assist astronauts in docking and maneuvering in outer space. Microspheres and disks were used to guide the lunar modules in the Apollo moon landings.

The same principle is used to provide failproof switch-tip light sources inside the dark space capsule. The phosphor microspheres, emitting four different colors, are embedded in holes drilled in the ends of instrument-panel flip and toggle switches.

Commercial applications of the power provided by ^{147}Pm are being developed in such devices as betavoltaic cells for surgical implant with heart pumps and pacemakers. Properly designed ^{147}Pm beta energy sources cannot cause harmful biologi-

*3M Company trademark.

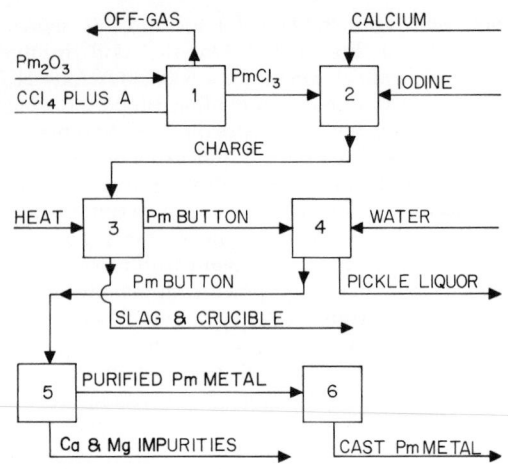

Fig. P-29. Schematic representation of materials flow in promethium metal preparation. (1) Chlorinator which operates between 550° and 600°C (1022–1112°F), (2) blender, (3) reduction vessel, (4) pickling tank, (5) vacuum distillation, (6) casting chamber.

cal exposure of any significance under any imaginable circumstances.

REFERENCES

Deonigi, D. E., R. W. McKee, and D. R. Hoffner: Isotope Production and Availability from Power Reactors, Pacific Northwest Laboratory, Richland, Wash., BNWL-716, July 1968.

Fullman, H. T., and H. H. Van Tuyl: Promethium Technology: A Review, *Isot. Radiat. Technol.*, vol. 7, no. 2, Winter 1969–1970 (contains 72 references).

Gschneidner, K. A., Jr.: *Rare-Earth Inf. Cent. News*, vol. 2, no. 1, March 1967.

Lahr, T. N.: Use of Radioluminescent Materials in Space, *Isotop. Radiat. Technol.*, vol. 7, no. 1, Fall 1969.

—Joseph G. Cannon, *Molybdenum Corporation of America, White Plains, N.Y.*

Prontosil   (See **Amides and imides.**)

Propaldehyde   (See **Aldehydes.**)

Propane   (See **Ethylene;** and **Hydrocarbons.**)

Propellants   (See **Chlorofluorocarbons;** and consult the **Subject Index.**)

Propionate Plastics   (See **Cellulose ester plastics, organic.**)

Propylene (See **Alkylation; Amino acids; Ethylene; Hydrocarbons; Petrochemical complex;** and **Propylene oxide.**)

Propylene Oxide Propylene oxide has been produced since the 1920s by the chlorohydrin process, starting with propylene and chlorine. The propylene chlorohydrin formed is dehydrochlorinated with calcium hydroxide, producing propylene oxide, with $CaCl_2$ as a waste product. The reaction is similar to the early chlorohydrin process for ethylene oxide. As new direct-oxidation plants for ethylene oxide were built, the old plants were converted to produce propylene oxide. See also **Ethylene oxide.**

A new process* for producing propylene oxide now used commercially is based on the original discovery that hydroperoxides can be made to react with olefins in the presence of selected catalysts to give high yields of both alcohols and epoxides. With propylene as the olefin, selectivities to propylene oxide can be made better than 95%. Hydroperoxides, e.g., those obtained from ethylbenzene, diethylbenzene, and cumene, can be used in the reaction with propylene, leading to propylene oxide and the corresponding alcohol. For example, a plant starting with propylene oxide and ethylbenzene would produce propylene oxide and styrene (by dehydration of the corresponding alcohol). The process has no waste by-products.

The United States capacity for propylene oxide is in the vicinity of 1.7 billion lb/yr, nearly all of which is represented by the newer and more direct process.

Polypropylene glycol and polyether polyols used in the manufacture of flexible urethane foams account for 53% of the propylene oxide made; propylene glycol, used as an intermediate for unsaturated polyester resins, for softening cellophane, and as brake and other functional fluids, accounts for 29%. Dipropylene glycol, used in polyester resins, plasticizers, and printing inks, accounts for 5%. Another 5% is used for making glycol ethers used as solvents, leaving 8% for miscellaneous applications.

—Alvin S. Cohan, *Scientific Design Company, Inc., New York.*

Protactinium (See **Chemical elements.**)

Protamine (See **Peptides and proteins.**)

*Oxirane Corporation.

Protective Atmosphere (See **Nitrogen.**)

Protective Coatings Coatings on materials are used primarily for two reasons, protection against the environment to which the material is exposed and for decorative purposes. Use of color for coding (as with piping) tends to combine these two objectives. In most applications of coatings, particularly architectural, a coating will be selected with both objectives in mind.

There is a fine line of demarcation between the typical coating applied in liquid form, which upon drying and curing becomes a thin film or lining, and the thicker linings which are applied with adhesives. A next step, of course, in equipment protection is metal cladding or refractory linings in furnaces and other high-temperature equipment.

Liquid coatings, including enamels, lacquers, latex coatings, and paints, are described under **Paint.** Consult the Subject Index for other types of protective coatings and linings.

Estimates of total plant-construction costs in the chemical industry that go into painting vary from 2 to 3%. The most costly phase of industrial painting generally is surface preparation, representing about 50% of the total cost. Painting costs generally are made up of two-thirds for labor and one-third for materials. There are many tradeoffs involved in determining the best coating or lining for a specific installation, including quality versus the life of the coating or lining, particularly in view of very high labor content. An excellent summary of some of these factors is given in *Chem. Eng. Prog.,* vol. 66, no. 8, pp. 31–52, August 1970.

Protective Colloid (See **Colloid systems;** and **Paint.**)

Proteolysis (See **Peptides and proteins.**)

Protium Protium is *light,* or *regular,* hydrogen, with a mass of 1.008145 atomic mass units. The isotopic abundance of protium is 99.9851%.

Proton A proton is an elementary subatomic particle located in the nucleus of an atom. It has one positive quantum of electric charge $(1.6021 \times 10^{-19}C)$ and a mass of 1.007596 atomic mass units. The number of protons contained in the nucleus of any given atom is expressed by the atomic number Z for that atom.

Pseudoplastic Fluids (See **Colloid systems.**)

Pug Mill (See **Mixing and blending, solids-solids.**)

Pulp Filter (See **Filtration.**)

Pulp (Wood) Production and Processing Pulps can be defined as fibrous products derived from cellulosic fiber-containing materials and used in the production of hardboard, fiberboard, paperboard, paper, and molded-pulp products. With suitable chemical modification, pulps can be used in the manufacture of rayon, cellulose acetate, and other familiar products. Pulps can be produced from any material containing cellulosic fiber; but in North America, wood is the predominant source of pulp, and this discussion is concerned only with the production and processing of wood pulp.

Wood is a cellular substance chemically composed of roughly 70% holocellulose, 25% lignin, and 5% water. These percentages are based on oven-dry wood. Also present are small quantities of soluble organic extractives (in a solution of 1 vol 95% alcohol and 2 vols benzene). The chemical composition and physical character of wood vary from species to species, within species grown in different geographical locations, and within a given tree, depending upon the location of the fiber cell in the tree. Both lignin (noncarbohydrate) and holocellulose (carbohydrate) are polymeric substances. Holocellulose is composed of approximately 70% alpha cellulose and 30% hemicellulose, the long-chained alpha cellulose being characterized by nonsolubility in alkali; whereas the shorter-chained hemicellulose is alkali-soluble, the degree depending upon the alkali concentration. Lignin concentration in wood substance is greatest in the middle lamella (the zone around each individual fiber cell), decreasing in concentration through the cross section of the fiber, and reaching a concentration of about 12% at the inner layer of the fiber adjacent to the fiber cavity, or lumen. It is the middle-lamella material (lignin and hemicellulose) that cements the fiber cells together, thus giving rigidity to the fibrous wood structure.

The objective of wood pulping is to separate the cellulose fibers one from another in a manner that preserves the inherent fiber strength and to remove as much of the lignin, extractives, and hemicellulose materials as required by pulp end-use considerations. Wood pulp to be used for the manufacture of hardboard, for example, requires only the removal of water-soluble wood sugars and sufficient fiberization, i.e., separation of fibers, to permit effective felting of the fibers in a sheet-forming operation. In a subsequent operation in which the felted fiber sheet is subjected to high pressure and heat, the lignin in the fiber mass softens and flows, ultimately acting as a bonding agent cementing the fibers together into a coherent hardboard. At the other extreme, wood pulp to be used for rayon manufacture must be of a high alpha-cellulose content (~88–93%), have extremely low amounts of noncarbohydrate material, and be well fiberized to permit uniform reactions during chemical processing.

Pulping Processes. Wood is converted to pulp by mechanical and chemical actions which constitute the pulping process; their selection depends upon the type of wood supply available and the pulp qualities desired. Pulps can be characterized on the basis of the unbleached pulp yields achieved by the pulping process used, i.e., the yield of oven-dry (OD) pulp obtained from oven-dry debarked wood.

In Fig. P-30, five major types, or classes, of pulps are related to pulp yield ranges normally considered to define each class of pulp. Pulp yield is a direct indication of degree of chemical action (delignification and chemical attack on carbohydrate and other nonligneous material). Also indicated in Fig. P-30 are the degrees of defibration effected by chemical and mechanical action utilized to produce the pulp, although this representation is not strictly correct. For example, in producing a full chemical pulp, wood chips are subjected to chemical action (digestion or cooking) in a pressure vessel; when digestion is completed, the cooked and softened chips retain the same physical form as the raw chips originally charged to the digester but separate into essentially discrete fibers as a result of mechanical action occurring upon sudden release of the chips from the pressure vessel into a receiving tank, which ordinarily is at atmospheric pressure.

At the other extreme, no chemicals are used in the production of mechanical pulp, and defibration is effected by subjecting wood to a mechanical grinding or attrition action; in this instance, the defibration is aided by some small degree of chemical change and solubilization of wood substance occasioned by heat generated by the grinding operation.

The pulps listed in Fig. P-30 are characterized on an unbleached basis as produced by processes conventionally called pulping processes. In many

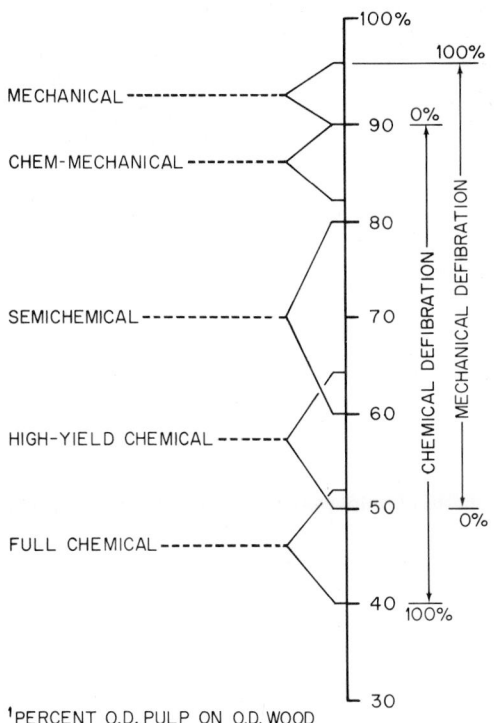

UNBLEACHED PULP YIELD[1]

MECHANICAL

CHEM-MECHANICAL

SEMICHEMICAL

HIGH-YIELD CHEMICAL

FULL CHEMICAL

100%

90

80

70

60

50

40

30

100%

0%

CHEMICAL DEFIBRATION

MECHANICAL DEFIBRATION

0%

100%

[1]PERCENT O.D. PULP ON O.D. WOOD

Fig. P-30. Characterization of wood pulps on a yield basis.

largely been superseded by the kraft process, which is characterized by its use of sodium hydroxide and sodium sulfide as active delignification agents in the chip-cooking phase of the process.

Chip-digestion parameters are digester pressure and temperature, digestion time to and at maximum temperature, amount of active alkali used per unit weight of OD wood (percent active alkali), percentage ratio of sulfide to active alkali (percent sulfidity), and weight ratio of cooking liquor (including chip moisture) to OD wood weight. No two kraft pulp mills use the same set of parameter values; such values must be frequently adjusted, even within a given mill, because of variations in incoming wood and pulp-quality requirements.

Kraft processes are applicable to nearly all species of wood, and effective means of recovering spent cooking chemicals for recycle in the process have been developed. Some sodium and sulfur losses do occur and are replenished in the cooking-liquor system by adding sodium sulfate at the recovery boiler, where it is converted to sodium carbonate and sulfide. In order to maintain a proper sulfur-to-sodium ratio in the recovered chemicals, other chemicals such as sodium carbonate, sodium sulfite, and sulfur are sometimes used for chemical makeup.

In contrast to the highly alkaline (pH 11–13) kraft processes, sulfite pulping processes are acidic in nature and are of two general types. (1) The acid sulfite processes utilize calcium, sodium, magnesium, or ammonium bisulfite in combination with free or excess sulfur dioxide as cooking chemicals (pH 1.7–2.3). (2) Bisulfite processes use sodium, magnesium, or ammonium bisulfite (pH 3.5–5.5) for chip digestion.

Several sulfite processes are multistage and use various combinations of acid sulfite and bisulfite cooking stages and can even use the alkaline kraft cook as one of the multistages. Although spent calcium acid sulfite cooking liquor can be incinerated, there is no recovery of calcium or sulfur. The sodium and magnesium bases can be recovered with or without sulfur recovery, and spent ammonium base liquor can be burned with recovery of sulfur as an option.

High-yield chemical pulps can be produced by the soda, kraft, or sulfite processes in which chemical use and digestion time and/or temperature are suitably reduced to effect a milder cook than used for full chemical pulps. Mechanical defibrators are used to complete the separation of

instances these pulps must be further treated chemically to remove residual lignin, hemicellulose, and color bodies before they can be considered suitable for use in specific applications. This further treatment is called *bleaching,* and the bleaching operation is actually an extension of the pulping process.

Customarily, pulping processes and bleaching processes are considered separately, although the choice of bleaching process is highly dependent upon the pulping process used to produce the pulp which is to be bleached. With this distinction between pulping and bleaching in mind, it will be understood that the pulping processes to be briefly described here pertain only to the production of unbleached pulps.

The soda, kraft, and sulfite pulping processes are used to prepare full chemical pulps. The soda process, which uses sodium hydroxide as the cooking chemical for delignification purposes, has

wood fibers not accomplished by the chemical action.

Semichemical pulps are usually prepared by the neutral sulfite semichemical (NSSC) process, although modifications of the full chemical processes can also be used. Active pulping chemicals are (in the sodium-base NSSC process) sodium sulfite buffered with sodium bicarbonate (pH 7.0–9.0) and (in the ammonium-base NSSC process) ammonium sulfite with ammonium hydroxide used as a buffer. Defiberization is usually accomplished by attrition mills of the disk type.

Chemimechanical pulps are produced by processes in which roundwood or chips are subjected to impregnation, usually at elevated temperatures at atmospheric pressure, with weak solutions of pulping chemicals, such as sulfur dioxide, sodium sulfite, or sodium hydroxide, followed by mechanical defibration. The mild chemical action softens wood lignin and promotes easier defibering with less fiber damage than the purely mechanical processes.

Wood Pulping Operations. The preceding discussion of pulps and pulping processes should serve as a background for the various operations involved in the preparation of wood pulp. The pulping system of a typical kraft linerboard mill, as indicated in the simplified flow diagram in Fig. P-31, is illustrative of that required for the preparation of both full and high-yield chemical pulps. Linerboard normally is two-layered, the base, or primary sheet, being formed from a high-yield chemical pulp (50–54% yield) and the top, or secondary sheet, being formed from a full chemical pulp, either unbleached (48–50% yield) or

bleached (46–48% unbleached yield) laid upon the wet primary sheet on the sheet-forming wire.

Pulp, paper, and paperboard mills are characterized by high capital investment costs and use of high-tonnage and rugged but precisely engineered machinery capable of continuous operation with minimum maintenance. Modern kraft linerboard mills, for example, are designed to produce boards at 400–1,000 tons/day at installed mill costs, including woodlands, of $60,000–$80,000 per daily ton of board. Indications of machinery sizes will be given in following paragraphs.

Wood-Chip Preparation: As indicated in Fig. P-31, pulping operations begin with receipt of wood at the mill site. Pulpwood is supplied in log form (roundwood) or chips in accordance with specifications set by the pulp mill. Roundwood is usually received with bark on and in lengths and diameters suitable for proper handling in the wood-preparation equipment at the mill. It has been customary for mills to specify multiple lengths of pulpwood, that is, 4 and 8 ft as standard receipts, but a recent trend has been to the procurement of tree-length logs, up to 70 ft in length, either exclusively or in combination with short logs. Another trend has been to the use of chips already prepared, except perhaps for final screening, by independent suppliers or by satellite wood yards operated by the pulp mill itself.

Although linerboard mills formerly used only softwoods (coniferous) for pulping, continued improvements in pulping and board-making technology have permitted the inclusion of up to 20% or more of hardwoods (deciduous) in the wood furnish to the mill, with improved utilization of

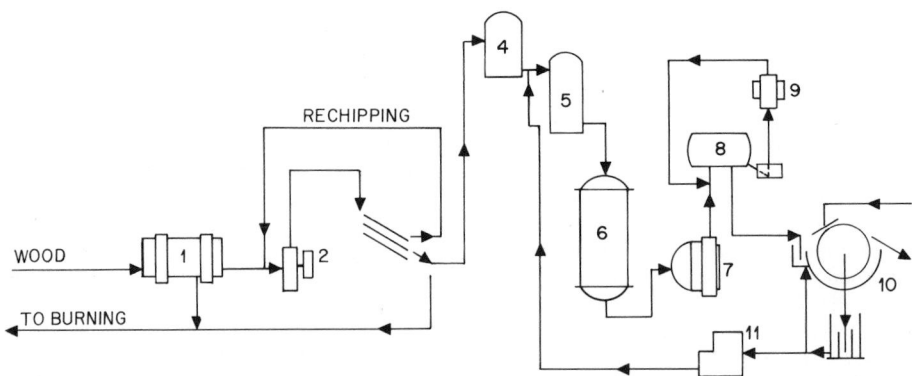

Fig. P-31. Simplified flow diagram of kraft pulp mill. (1) Debarking, (2) chipping, (3) screening, (4) steaming, (5) impregnating, (6) digesting, (7) fibrilizing, (8) screening, (9) fiberizing, (10) washing, (11) chemical recovery.

woodlands as a beneficial result. Softwood and hardwood species are processed through the chipping operation and stored separately; they are blended into the digester and cooked together, or they are processed separately and the respective pulps blended just ahead of the board machine.

Former practice was to store pulpwood receipts in either a debarked or unbarked condition in stacks or random piles in the woodyard and to reclaim the yard wood for processing into chips just a few hours in advance of chip needs at the digester. An increasing practice is to convert the wood into chips immediately after pulpwood receipt and to place the chips, usually by belt or air conveyance, in chip piles built up on concrete or asphalt pads. Separate piles are provided for softwood and hardwood chips, and storage capacities of 40,000 cords or greater can be maintained.

Dwell time of chips in the outside chip storage (OCS) varies, depending upon the geographic location of the mill and wood-supply conditions, but normally chips are reclaimed within 1–3 months after initial placement in order to avoid undue deterioration from fungus or insect attack.

Pulp logs are conveyed to the debarking area, where they are cut to proper length, if necessary, and sorted. Accepted logs are mechanically fed into one end of a large horizontal, cylindrical drum consisting of three or more sections constructed of spaced steel plates, channels, or bars mounted in carrying rings and supported on trunnions and driven by ring gears or suspended from an overhead structure by heavy chains one or more of which are motor-driven. This *barking drum*, as it is called, rotates at a speed of 5–8 rpm, and as the logs tumble about in passing from the intake to the discharge end of the drum, bark removal is effected by the logs rubbing against each other or against the bars or plates constituting the drum shell. Provision is made for immersion of the logs in water in one or more of the drum sections to assist the barking action and wash the wood.

A barking drum is usually 12 ft diam and 40–70 ft long and will deliver well-barked softwood at an average rate of from 0.5–0.8 cord/hr per foot of drum length and hardwood at about one-half that rate. The motor-drive requirement is roughly 3 hp connected per foot of drum length.

Bark removal can also be accomplished by use of a *ring barker*, in which logs feed through rotating knurled rolls, knives, and hammers, peripherally mounted and arranged to contact the bark regardless of wood diameters, or *hydraulic barkers*

which employ high-pressure (1,500 psi) water jets for bark stripping. These two types are usually used for debarking long logs and can be arranged for log feed through the barkers directly into a chipper.

Bark removed from pulpwood logs is collected, shredded in a hog or hammer mill, and used as fuel in steam boilers, where it contributes about 9000 Btu per pound of dry solids. If incineration is not practical, bark is used as landfill or further processed into a mulch.

Debarked wood is conveyed to a chipper for conversion into chips of proper length for chemical treatment in a subsequent cooking operation; chip lengths of $\frac{1}{2}$–1 in. are conventional, depending upon the wood species and pulping process. A chipper is essentially a cast-steel disk, rotated on a horizontal shaft, on the face of which knives are mounted radially; the distance the cutting edge of these knives projects from the faceplate of the disk is adjustable and determines the length of the chip. Logs are fed to the chipper through an angular spout opening in a hood which encloses the disk. The shearing action of the rotating knives on the log bearing upon a bedplate (bedknife) set at the bottom of the spout both cuts the wood and pulls the log to the surface (wear plate) of the disk for further shearing action.

Wood chips of a length determined by the knife setting but of varying widths (chip cards) pass through slots behind the knives to the backside of the disk, where the cards are broken or reduced in width. Chips drop through an open bottom of the chipper-hood enclosure or, if the disk is equipped with vanes on its circumference, are blown from a totally enclosed hood to a cyclone separator. Disk diameters are 50–120 in.; number of knives, 4–15; and spout diameters, 20–35 in. in the most common sizes of disk chippers. As an example of drive-motor requirement and capacity, a 112-in., 15-knife, 28-in.-diam feed-spout chipper for softwoods is driven at a speed of 360 rpm by a 2,000-hp synchronous motor and has a capacity of 120 cords/hr.

Wood chips discharging from the chipper contain varying amounts of dirt, wood dust, slivers, and oversize chips, which must be removed to facilitate the digestion operation. Separation of acceptable chips from reject material is accomplished by screening directly after the chipping operation or following reclaim from bulk chips storage (silo or OCS). Chipper discharge is conveyed to a surge bin of ample size (1,000–1,500 ft^3) to level output irregularities and to provide sufficient chip volume for uniform feed to conveyors which transfer the chips to the screens or bulk chip storage for later feed to the screens. A triple-deck vibratory screen is conventionally used for chip screening, requiring about $4\,ft^2$ of top screen area per cord per hour and consisting of two screening sections and a pan, mounted one atop the other in an oscillating frame and slightly declined from the horizontal.

Chips are fed to the top screen and move down the screening surface, with oversize chips and slivers discharging at the lower end of the top screen and acceptable chips, dust and dirt passing through the top screen to the lower screen. Accepted chips discharge from the end of the lower screen, and dust and dirt pass through the screen to the bottom pan and discharge to a bin from which the rejects are conveyed for combining with bark for incineration or other end use. Slivers and oversize chips are rechipped in a small disk chipper or hammer mill and returned for rescreening. Accepted chips are conveyed to screened chip storage ahead of the digestion operation.

Chip Digestion: The chip-digestion, or cooking, operation is accomplished in a digester, either batch or continuous. Essentially a chip digester is a large vessel provided with suitable raw-chip feed and cooked-chip discharge ports and equipped with means for heating and maintaining its contents to and at a specified temperature for the required length of time. Batch digesters are vertical, stationary cylindrical pressure vessels into which chips and cooking liquor are charged and in which liquor is constantly moved, either by percolation within the digesters aided by direct addition of steam for heating purposes or by continual withdrawal of liquor through screened ports and reintroduction of the liquor, after heating in heat exchangers, onto the top (and sometimes into the bottom) of the chip mass within the vessel. Modern batch digesters are typically $4{,}000$–$6{,}000\,ft^3$ in volume, with height-to-diameter ratios of 3.5–5.5 and per-cook pulp capacities of 10–20 tons.

Continuous digesters have been developed as part of the highly successful effort to convert pulp and papermaking from a series of strictly batch operations into an integrated series of continuous operations. A number of successful types of continuous digesters range from horizontal and inclined tube (single or multiple) designs, in which the chip charge is moved through the digester by mechanical screw or bucket conveyors, to vertical digesters, in which chip movement is effected by

gravity in downflow digesters or by a lifting mechanism in upflow digesters. For purposes of illustration, the operation of a continuous down-flow digester is outlined here.

Screened chips are conveyed from storage to a chip-supply bin in the digester house. If hardwood and softwood chips are to be cooked together, they are blended by weight proportion during this transfer. Chips feed by gravity from the bin to a chip meter, either a twin-screw or a multipocket rotary feeder, the speed of which determines chip and cooking-liquor flow rate to the digester and pulp discharge rate.

Metered chips drop to a low-pressure rotary feeder valve, through which the chips are introduced into a steaming vessel maintained at a pressure of about 15 psig, where the chips are preheated, air is expelled from the chip interior, and chip moisture content is leveled in preparation for impregnation with cooking liquor. Steamed chips are fed concurrent with liquor feed, by rotary valve or screw, into an impregnation zone—a separate vessel or the upper portion of the digester itself—which is maintained at a temperature about $20°F$ ($5.6°C$) lower than that in the digester cooking zone ($340°F$) ($171°C$) into which the impregnated chips enter after about a 20-minute dwell time in the impregnation zone.

Since cooked chips are continually being removed from the bottom of the digester, chips pass downward in the digester, replacing those discharged; time of passage through the cooking zone is normally 90–120 min. Liquor in the cooking zone is withdrawn for heating and is recirculated, whereas the liquor introduced with chips in the impregnation zone simply moves down with the chips into the cooking zone. As cooked chips reach the bottom zone of the digester, they are plowed to a central well in the bottom of the digester while being mixed with filtrate from the pulp washer for cooling and dilution purposes. Cooked chips cooled to a temperature of about 250–$260°F$ (121–$127°C$) are forced by digester pressure through a discharge pipe leading from the well through an adjustable orifice valve and thence through a blowline to a receiving vessel called a *blow tank*, usually maintained at atmospheric pressure. Flash steam, noncondensable gases generated during the cook, and volatile material are recycled to the chip-steam vessel. Mechanical forces exerted in the transfer of chips from the digester to the blow tank effect fiberization of the chips, the degree of which depends upon cooking conditions. The fibrous material collected in the blow tank can now be called pulp, and separate blow tanks are normally used to collect the several types of pulp produced alternately in the digester.

The objective of the digester operator is to produce a pulp characterized by a specific *kappa number*, defined as the number of milliliters of $0.1\ N$ $KMnO_4$ solution consumed per gram of moisture-free pulp under standard test conditions, with results adjusted to be equivalent to a 50% consumption of the permanganate in contact with the pulp. The kappa number is a measure of the oxidizable wood substance left in the pulp after all water-soluble material has been washed from it and, for a given wood sample, is directly relatable to lignin content and yield of pulps produced from that wood source by means of a given pulping process over a fairly narrow yield range, that is, 46–56%.

Some mills have experimentally determined the correlation between kappa number and yield of various pulps produced from their particular wood mixes and thus are able to predict with a fair degree of accuracy the pulp yield being obtained at any given time, rather than having to use inventory figures for monthly yield determinations. The pulp is sampled as close to the digester discharge valve as possible for control purposes, and the digester operator adjusts cooking conditions if necessary to maintain the pulp kappa number at a desired level, which is different for each type of pulp produced.

Pulp Screening and Washing: Pulp (brown stock) discharged to the blow tank is in admixture with black liquor, a water solution of spent and residual cooking chemicals and dissolved wood substance, and is at a consistency of from 10–18%. The term *consistency* has a meaning peculiar to the pulp and paper industry and refers to the percentage ratio of washed, dry (either oven- or air-dried) fiber to total fiber slurry weight. The fiber bundles left in the pulp after blowing must be fiberized, i.e., separated into discrete fibers, and the black liquor removed in order for the pulp to be refined (a conditioning of individual fibers) and formed into a fiber sheet on the board machines.

Pulp is diluted with filtrate from the pulp washer to a consistency of about $4\frac{1}{2}\%$ in the lower portion of the blow tank and fed to fibrilizers, which serve the purposes of metal trapping, fiber-bundle breaking, rough screening, and pumping. As the pulp enters the unit through a large inlet chamber, the flow velocity diminishes, allowing pieces of metal and other heavy debris to fall into a trash box, which is emptied periodically. A rotating screen and impeller assembly at the far end of the inlet chamber provides a pumping action which draws the pulp into the machine, through the screen plate, and discharges it to further processing. Large fibrous material is retained on the screen plate until broken up by a stationary scraper blade to a size permitting passage through the screen perforations ($\frac{3}{4}$-in.-diam holes).

Stock from the fibrilizers (two or more units are normally used at each blow tank) is pumped through an in-line consistency regulator, which controls a dilution valve immediately ahead of the fibrilizers and provides a fine control to the dilution effected at the blow tank, to multiple units of sealed centrifugal screens. Before entering the screens, the stock is diluted to a consistency of about $1\frac{1}{2}\%$ with washer filtrate. A centrifugal screen essentially consists of a rigidly mounted, cylindrical, perforated-metal basket with a rotor which turns inside the screen basket. Stock is admitted at the feed end of the basket, and acceptable fiber passes through the perforations into the space between the basket and the sealed screen housing and discharges by gravity flow to the washer headbox.

Oversize material which will not pass through the perforations is moved along the axis of the screen by the action of the rotor until it reaches the reject-outlet end of the basket. Washer filtrate dilution liquor is admitted to either side of the reject outlet to sluice acceptable fiber away from the reject flow. Rejected fiber passes through an adjustable reject orifice and is pumped to a secondary screen, identical in construction to the primary screens except that smaller perforations are used. Accepts from the secondary screen discharge to the washer headbox, and reject fiber is piped to subsequent processing.

The amount of reject fiber discharging from prewasher screens depends upon the degree of cooking or yield of pulp being produced. A 48% yield pine pulp, for example, could contain 7% (by weight) reject fiber, whereas a 52% yield pulp produced from the same wood furnish could contain up to 16% reject fiber. For economic reasons it is essential to utilize this reject fiber. Modern practice is to fiberize the reject material by mechanical action and to return it to the pulp system, usually as a somewhat lower-grade (higher-yield) pulp than that from which it was obtained. Fiberization can be accomplished either in attrition mills of the disk type or in twin-screw fiber conditioners, where the rubbing action of the disk or screw surfaces and between the fibrous material itself effects the fiber separation, preferably without cutting or shortening the fiber length. This mechanical fiberization is never 100% efficient, and the stock must be returned to the pulp system at a point ahead of the prewasher screens.

Removal of the black liquor from screened brown stock is usually accomplished on rotary-drum vacuum filters, arranged for multistage countercurrent washing, as illustrated in Fig. P-32. A line of four vacuum washers (drum dimensions $11\frac{1}{2}$ ft diam and 20 ft face width) can readily remove 98% of the black liquor solids contained in brown stock being washed at a rate of 600 oven-dry tons of pulp per day. Screened pulp entering the headbox of the first-stage washer has purposely been excluded from contact with air from the time it left the blow tank in order to avoid generating foam, which interferes with screening and washing operations.

In the headbox the pulp is diluted with first-stage filtrate to a consistency of about 1.0% and gravity-fed to the first-stage washer vat, in which the washer drum rotates at a speed of 1–5 rpm at about a 50% submergence. As the drum rotates in the vat, the liquor drains from the fiber slurry through the face of the drum, which is covered with a 30- to 40-mesh wire supported by deck sections, into drainage channels within the drum structure and through a trunnion pipe to a drop leg leading to a filtrate tank. As the liquor drains from the pulp due to the pressure-head differential between the vat content and drum channels and a vacuum induced in the drainage channels by the drop leg (or sometimes by a vacuum pump), a pulp mat is formed on the face wire. This fiber mat is permitted to drain for a short period of travel as the drum surface leaves the vat and is then contacted with a flow of filtrate from the next (second-stage) washer, distributed by a series of shower nozzles or weir boxes across the upcoming drum surface.

As the fiber-covered drum surface passes downward just beyond the vertical centerline of the drum, the vacuum is released so that the fiber mat can be easily removed from the drum-face wire by a takeoff roll or doctor blade (air or steam). Vacuum is applied again after the clean drum surface has submerged into the vat a short distance and a more gentle and uniform initial mat deposit has been formed by gravity filtration than would have been obtained under full vacuum. Application of vacuum is controlled by a multiport valve located in the circumference of the drum or in the trunnion. Discharged pulp is repulped with filtrate from the second-stage washer, and the washing cycle is repeated as many times as the number of drums provided (three or four stages of washing are usual for kraft pine

Fig. P-32. Line of three brown stock washers (9 ft 6 in. diam $\times$ 16 ft long) equipped with multiport circumferential valve. Discharge end in foreground. (*Improved Machinery Inc.*)

pulp) using a counterflow of wash liquor with fresh water entering the washing system as shower water on the last-stage washer.

Washed pulp discharged from the final-stage washer at a consistency of 14–20% is diluted to about 12% consistency and pumped by a positive-displacement pump to a high-density pulp storage tank or diluted to $4\frac{1}{2}$% consistency and transferred by a centrifugal pump to low-density pulp storage. Normally, separate washer lines are used for each type of pulp produced, and these pulps are retained in separate storage tanks until sent to the papermill for refining, final screening, and board making.

Refining is accomplished by disk mills, as previously mentioned, equipped with different plate designs or patterns than those used for defibration. During the refining operation, cellulose fibrils, which wind spirally around the fiber at various positions in its cell wall, are loosened, the cell wall swells due to water absorption, and the fiber is conditioned for sheet formation and interfiber bonding in the paper- or board-making operation.

Chemical Recovery: Black liquor washed from the pulp is used as first-stage washer filtrate, for pulp dilution as indicated, and excess filtrate at a soluble-solids content of 12–15% and is sent to multiple-effect evaporators and concentrated to a 45–50% soluble-solids content. This liquor is further concentrated to a 62–65% solids content in direct-contact evaporators of the cascade or cyclone type or in indirect-contact tube-type heat exchangers and burned in recovery boilers for the generation of process steam.

Sodium and sulfur compounds in black-liquor solids are converted in the burning operation into sodium carbonate and sodium sulfide, which leave the furnace as a smelt. The smelt is mixed with recycled water in a dissolving (smelt) tank to form green liquor, so termed because insoluble impurities impart a green color to the liquor. These impurities are removed in a green-liquor clarifier in the form of dregs, which are then washed to remove soluble chemicals and sent to landfill areas.

Clarified green liquor is causticized with lime to convert sodium carbonate into sodium hydroxide. The resulting white liquor containing sodium hydroxide and sodium sulfide plus insoluble calcium carbonate and excess lime is clarified in tray clarifiers; clarified white liquor is returned to the digester as cooking chemical, and lime mud (clarifier underflow) is washed and burned in a rotary lime kiln or fluidized-bed reactor to form calcium oxide for recycling in the causticizing operation.

Economic and environmental control factors dictate that chemical and heat values of black-liquor solids be carefully conserved, and the recovery system of the modern kraft pulp mill has developed into a highly sophisticated system with still more improvement in efficiency continually being sought.

Bleaching: As previously mentioned, bleaching wood pulp is an extension of the pulping process. Conventional cooking chemicals are not specific for lignin and other noncarbohydrate constituents of wood but also attack the carbohydrate content of wood substance. Bleaching has long been an integral part of the production of pulp for fine paper manufacture, but only recently has the operation been included in linerboard mills, where bleached pulp can be used for formation of the entire board sheet or simply as a top liner constituting 12–20% of the total sheet weight.

Bleachable pulps are produced by cooking wood chips to a point (yield of 48% or less) where further cooking would remove excessive amounts of carbohydrate material and then subjecting washed and screened brown stock to a series or sequence of treatments with chemicals more specific for the removal of noncarbohydrate material than cooking chemicals. A typical bleaching sequence for a kraft pine pulp, for example, would be, in pulp-mill parlance, *C-E-H-D;* translated, this means that pulp is subjected to treatment with chlorine (C), followed by an extraction (E) with sodium hydroxide, followed by a calcium or sodium hypochlorite (H) treatment, with a final treatment with chlorine dioxide (D). These chemical treatments are carefully controlled with respect to pulp consistency, chemical concentration, reaction temperature and pH, and time and are conducted in a series of reaction towers with pulp washing between each treatment stage.

Oxygen may be used in an *O-C-E-D* sequence, which would permit recycling of the oxygen-stage washer effluent back to the pulp mill for eventual combination with black liquor being sent to the recovery plant.

REFERENCES
MacDonald, Ronald G., and John N. Franklin (eds.): "Pulp and Paper Manufacture," 2d ed., 3 vols., McGraw-Hill, New York, 1969.
Rydholm, Sven A.: "Pulping Processes," Interscience Wiley, New York, 1965.

Rydholm, Sven A.: Continuous Pulping Processes, *Tech. Ass. Pulp Pap. Ind. Publ. STAP 7, Spec. Tech. Ass.* 1970.

—Henry F. Szepan and Dunbar G. Terry,
*Improved Machinery Inc. (subsidiary of Ingersoll-Rand),
Nashua, N.H.*

Pumice (See **Abrasives.**)

Purines Purine, the parent compound of the purine group, was first prepared by the German chemist Emil Fischer, about 1900. Fischer also determined the formulas of a number of purine derivatives. Some purines may be considered as complex cyclic diureides derived from one molecule of hydroxy dibasic acids and two molecules of urea.

Purine   Uric acid (2,6,8-trioxypurine

Xanthine (2,6-dioxypurine)   Theobromine(3,7-dimethylxanthine)

Caffeine(1,3,7-trimethylxanthine)   Guanine(2-amino-6-oxypurine)

Uric acid is the oxidation product of complex nitrogen-containing compounds of living organisms and is present in the urine, blood, and muscle juices of carnivorous animals and excreta of birds, terrestrial reptiles, and insects. Hippuric acid is secreted by herbivorous animals.

Although of scientific importance, the purines have no major industrial or commercial use. Caffeine (theine), also classified as an alkaloid, finds limited use in some pharmaceuticals, it occurs in tea leaves (2–5%), in coffee beans (0.75–1.75%), and in cola leaves (1–2.5%) and consequently is an ingredient of several prepared hot and cold beverages. Theobromine occurs in cocoa and cola beans and in small amounts in tea leaves but not in coffee. There is no bromine in the compound, which is a more powerful stimulant than caffeine. A derivative, dimethylxanthine, is used in medicine. Citrated caffeine, made

by reacting citric acid with caffeine, contains about equal proportions of anhydrous caffeine and citric acid and is used in some pharmaceutical preparations. Guanine is the chief constituent of pearl essence, which is used for making imitation pearls and certain plastic lacquers. It can be derived from iridescent fish scales, notably those of the sardine and herring; 100 tons of herring will yield 1 ton of scale, which in turn will yield 1 lb of pearl essence. Synthetically produced pearl essences (basic lead carbonate and lead monohydrogen phosphate) also have the desired multiple reflectivity. Xanthine is present in small amounts in the blood, liver, and urine, also tea leaves, beet juice, and sprouting seeds.

Pyocyanase (See **Antibiotics.**)

Pyridine and Derivatives Pyridine and derivatives of pyridine occur widely in nature as components of alkaloids, vitamins, and coenzymes. These compounds are of continuing interest to theoretical, physical organic, and biochemistry. They have many commercial uses, e.g., herbicides and pesticides, pharmaceuticals, feed supplements, solvents and reagents, and chemicals for the polymer and textile industries.

Structure and Nomenclature: The pyridine group consists of six-membered, heterocyclic, aromatic compounds with one nitrogen atom in the ring. The parent compound of this group is pyridine (I) with ring positions numbered as shown.

Alternate denotations of the 2, 3, and 4 positions in the ring are α, β, and γ respectively. Pyridine is a hygroscopic, colorless liquid with a characteristic unpleasant smell. The compound, a tertiary amine, is a somewhat stronger base than aniline and readily forms quaternary ammonium salts.

The behavior of pyridine in substitution reactions can be understood on the basis of its resonance structures (I–I*d*) and on the basis of the electron-density distribution at the various ring positions as derived from molecular-orbital-theoretical calculations. An example of the published π-electron density distribution is shown in II. The resonance energy of pyridine is 35 kcal/mole (versus 39 kcal/mole for benzene).

Electrophilic substitution occurs at the 3 and 5 positions but usually requires drastic conditions because the species actually being attacked is a pyridinium ion. For example, nitration of pyridine with KNO_3 and concentrated H_2SO_4 at 300°C gives a 15% yield of 3-nitropyridine. Electrophilic substitution in the pyridine ring is facilitated by the presence of electron-donating substituents.

Nucleophilic substitution occurs in the 2, 4, and 6 positions of pyridine under relatively mild conditions. As an example, amination of pyridine with sodium amide in *N,N*-dimethylaniline at 180°C gives 2-aminopyridine in good yield.

Homolytic (free-radical) substitution may occur in any of the 2 to 6 positions of pyridine. Thus, the reaction of pyridine with benzenediazonium salts gives a mixture of 2-, 3-, and 4-phenylpyridine.

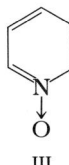

III

Many pyridine derivatives which are difficult to make directly from pyridine are readily accessible starting from pyridine *N*-oxide (III), made by oxidation of pyridine with hydrogen peroxide in acetic acid. As but one example, the nitration of pyridine *N*-oxide gives 4-nitropyridine *N*-oxide in high yield. Reduction of the *N*-oxide to the parent pyridine nucleus is readily effected by hydrogenation or reagents such as PCl_3 or triphenyl phosphine.

Selected physical properties of pyridine and some of its simple derivatives are summarized in Table P-42. Trivial names for the methylpyridines are the *picolines;* the dimethylpyridines are the *lutidines;* and the trimethylpyridines (and in older literature the ethyldimethylpyridines) the *collidines.* The refractive indices for these alkyl pyridines and for pyridine itself fall in the range $n_D^{20} \sim 1.50$–1.51.

Production of Pyridine and Important Homologues

By-Products of Coke Manufacture: In United States practice, coking of coal is done almost exclusively by high-temperature (>800°C) processes. For a long time, the major source of the pyridines was the chemical-recovery coke oven. The volatiles produced in the coke oven are only partially condensed. The noncondensed gases are passed through a scrubber (the ammonia saturator) containing H_2SO_4. After removal of crystals, $(NH_4)_2SO_4$, a solution of $(NH_4)_2SO_4$ and pyridinium sulfates is obtained and treated with NH_3 to liberate the contained pyridine bases (~70% is pyridine itself).

The balance of the pyridine bases is extracted from the crude coal tar, i.e., the condensed, main portion of the volatilization products from coking. The crude tar contains ~ 0.1–0.2% pyridine bases. In the tar-distillation plant, the crude tar is submitted to continuous dehydration then continuous distillation. The tar acids (phenolics) and tar bases (pyridines and quinolines) are recovered by caustic wash and acid wash, respectively, of certain distillation cuts. The selection of fractions so treated may vary at different plants, but the general principles are the same. The chemical oil fraction of boiling range ~130–196°C/atm contains most of the pyridine, picolines, lutidines, and other polyalkyl pyridines. As an example of plant practice, this particular chemical-oil fraction is first washed with dilute NaOH to remove phenolics (tar acids) and then extracted with dilute H_2SO_4. The acid wash is steamed to remove the entrained neutral oils, the liquor is saturated with a salt such as Na_2SO_4 to minimize the solubility of the bases, and the pyridine bases are liberated (*sprung*) with aqueous NaOH by bringing the liquor to pH ~9. The layer of bases is then separated and dried by azeotropic distillation with benzene. The crude bases contain ~35–40% pyridine and ~10% 2-picoline, the balance being higher-boiling tar bases. Final workup is by fractionation. Fractions commercially available include denaturing-grade pyridine, boiling point 115–160°C; 2° pyridine, boiling point 113.5–115.5°C; refined α-picoline, boiling point 128–130°C; "refined" mixed picolines, boiling point 141–145°C; "refined" 2,4-lutidine (~60% 2,4-lutidine), boiling point 156–160°C; "refined" collidines (~80% 2,4,6-collidine), boiling point 169–173°C; and 15–18 grade bases, boiling point 150–180°C. The isolation of pure components

TABLE P-42. *Selected Properties of Pyridine and Derivatives*

Compound	Bp at 760 mm, °C	Fp, °C	Density, d_4^{20}	pK_a in water at 25°C
Pyridine.	115.2–5.3	−41.67	0.9819	5.17
2-Picoline	129.4	−66.87	0.9455	5.97
3-Picoline	144.1	−18.22	0.9564	5.68
4-Picoline	145.4	3.64	0.9546	6.02
2,3-Lutidine	163–164	−15.22	. . .	6.57
2,4-Lutidine	158.4	−64.00	0.9332	6.63
2,5-Lutidine	159–160	−15.54	. . .	6.40
2,6-Lutidine	144.0	−6.12	0.9237	6.72
3,4-Lutidine	163.5–164.5	−11.04	. . .	6.46
3,5-Lutidine	170–171	−6.50	. . .	6.15
2,3,4-Trimethylpyridine	192–193			
2,3,5-Trimethylpyridine	186.8			
2,3,6-Trimethylpyridine	172.8	−11.5	0.9142	
2,4,5-Trimethylpyridine	189.8			
2,4,6-Trimethylpyridine	170.3	−44.2		
3,4,5-Trimethylpyridine	211.9			
2-Methyl-5-ethylpyridine	178.3	−70.3	0.9208	
Pyridine *N*-oxide	138–140[a]	68–69	. . .	0.79
2-Chloropyridine	170	. . .	1.5330	
2-Bromopyridine	192	. . .	1.5658[b]	
2-Pyridinealdehyde	181	. . .	1.126[c]	
3-Pyridinemethanol	142–143[a]	. . .	1.5440[d]	
2-Aminopyridine	211	57.5		
3-Cyanopyridine	204–205	50–51		
4-Cyanopyridine	. . .	78.5–80		
Picolinic acid	. . .	136.5–138		
Nicotinic acid	. . .	235–236.5	. . .	4.76
Nicotinamide	. . .	131–132	. . .	10.6, 13.5
Isonicotinic acid	260[a,e]	319		
Isonicotinic acid hydrazide	. . .	171–172		
Quinolinic acid (pyridine-2,3-dicarboxylic acid)	. . .	190[f]		
Isocinchomeronic acid (pyridine-2,5-dicarboxylic acid	. . .	254[f]		
Dipicolinic acid (pyridine-2,6-dicarboxylic acid	. . .	226–227[f]		
2-Vinylpyridine	159	. . .	0.976	
4-Vinylpyridine	64–67[g]	. . .	0.979[d]	
2-Methyl-5-vinylpyridine	181.2[h]	. . .	0.958	
Piperidine.	106	−7	0.08615	11.1

[a]At 15 mm.
[b]At 30°C
[c]At 18°C.
[d]At 25°C.
[e]Sublimes.
[f]Decomposes.
[g]At 15–17 mm.
[h]At 160 mm.

from the mixed-picoline fraction and the higher-boiling fractions requires relatively costly processing, e.g., the formation and separation of addition compounds, azeotropic distillation, or selective extractions; these purifications are little practiced. **See also Coal tar and derivatives.**

Synthetic Methods of Manufacture: Due to rising demand, production of the pyridine bases by large-scale syntheses passed the volume of tar bases extracted from coal tar in the 1960s. United States capacity for synthetic pyridine, the picolines, and 2-methyl-5-vinylpyridine (MEP) by

1970 was estimated at 35–40 million lb/yr. All these products can be made by condensation reactions of aldehydes and ammonia.

When acetaldehyde and ammonia in a 3:1 mole ratio are fed over dehydration-dehydrogenation catalysts such as PbO or CuO on alumina, ThO_2 or ZnO or CdO on silica-alumina, or CdF_2 on silica-magnesia at 400–500°C/atm, an equimolar mixture of 2- and 4-picolines can be obtained in 40–60% yields. When a mixture of acetaldehyde, formaldehyde, and ammonia in about 2:1:1 mole ratio is passed over such catalysts, pyridine and 3-picoline are produced; their ratios are usually 1/0.7, but the amount of pyridine can be increased by changes in the feed.

A major producer* of pyridine bases has reported a pyridine–3-picoline synthesis by reacting acetylene, methanol, and ammonia in a fluidized-bed reactor at 400°C, using catalysts consisting of Zn, Cu, or Cd salts on alumina-silica-magnesia supports.

The lowest-cost synthetic pyridine base, 2-methyl-5-ethylpyridine, is made in a liquid-phase process from paraldehyde (derived from acetaldehyde) and aqueous ammonia in the presence of ammonium acetate at ~1,500–2,800 psig and 220–280°C in 70–80% yield; 2- and 4-picoline are minor by-products.

In the synthetic processes, mixtures of products are often obtained, and variation in the supply-demand balance of the alkyl pyridine isomers has led to much research on processes which might alleviate such imbalances, including development of the catalytic hydrodealkylation of alkyl pyridines as well as the alkylation of pyridine.

Pyridine Derivatives and Uses

Herbicides: A major outlet for pyridine is in the manufacture of the desiccant herbicides and aquatic weed killers, such as 1,1'-ethylene-2,2'-dipyridilium dibromide (IV)† and 1,1'-dimethyl-4,4'-dipyridilium dichloride (V)‡ or dibromide or dimethylsulfate. 2-Picoline is the source of 2-chloro-6-trichloromethylpyridine (VI)¶, which is useful as a fertilizer additive for reduction of nitrogen losses in the soil due to bacterial oxidation. 2-Picoline also is the starting material for the production of 4-amino-3,5,6-trichloropicolinic acid (VII)§, a powerful broad-spectrum herbicide for broad-leaved plants.

Pesticides: 2-Picoline is a component of 1-[(4'-amino-2'-n-propyl-5'-pyrimidinyl)methyl]-2-picolinium chloride hydrochloride (VIII),* a broad-spectrum coccidiostat. A newer coccidiostat is 3,5-dichloro-4-hydroxy-2,6-lutidine (IX).† The acaridicide O,O-diethyl-O-(3,5,6-trichloro-2-pyridyl) thiophosphate (X)‡ is used to control ectoparasites. Di(n-propyl) isocinchomerate (XI)¶, used in fly repellents, is made by oxidation of 2-methyl-5-ethylpyridine and esterification of the isocinchomeronic acid obtained. Nicotine (XII) sulfate,§ used as an agricultural insecticide, as an external parasiticide, and as an anthelmintic, is obtained by extraction of tobacco wastes and not by synthesis.

Pharmaceuticals: There are several pyridine and piperidine derivatives with widely varying, often multiple, drug action. A few examples are described briefly here.

A number of *antihistamines* contain the pyridine moiety in their structure, as exemplified by chlorpheniramine maleate (XIII, 2-[*p*-chloro-α-(2-dimethylaminoethyl)benzyl]pyridine acid maleate), doxylamine succinate (XIV, 2-[α-(2-dimethylamino)ethoxy-α-methylbenzyl]-pyridine acid succinate), and pyrilamine maleate [XV; 2-(2-dimethylaminoethyl-2-*p*-methoxybenzyl) aminopyridine acid maleate]. These products are synthesized, e.g., from the appropriate benzylpyridines or aminopyridines.

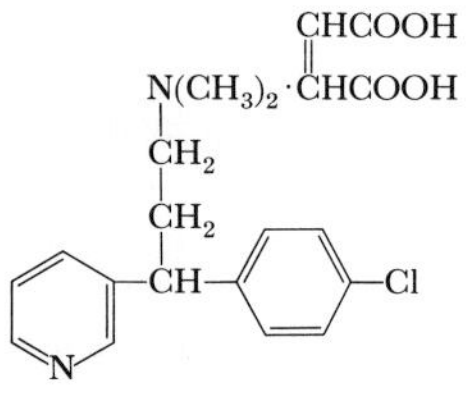

XIII

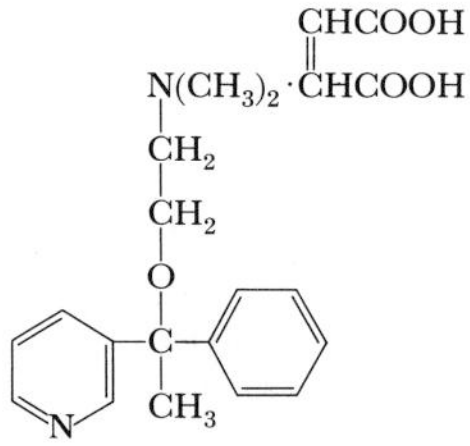

XIV

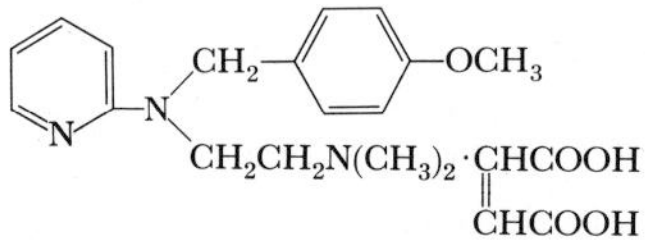

XV

Cetylpyridinium chloride (XVI) is used as a germicide and antiseptic, e.g., in mouthwashes; it is made by quaternization of pyridine with cetyl chloride.

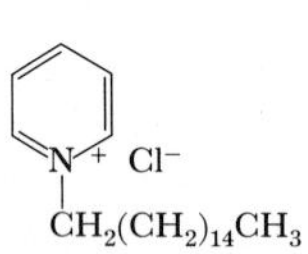

XVI

Isonicotinehydrazide (XVII, isoniazid) is an important antitubercular drug made by oxidation of 4-alkylpyridine (or 2,4-lutidine) or by hydrolysis of 4-cyanopyridine to isonicotinic acid (pyridine 4-carboxylic acid) and reaction of an ester or the acid chloride of the latter with hydrazine.

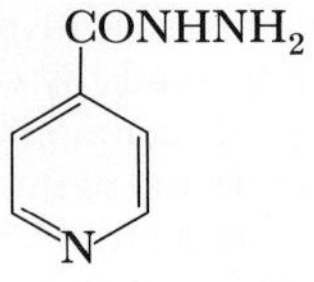

XVII

Meperidine hydrochloride (XVIII,* 1-methyl-4-carbethoxy-4-phenylpiperidine) is an important narcotic and analgesic; made not from piperidine but by ring-closure reactions of appropriate precursors.

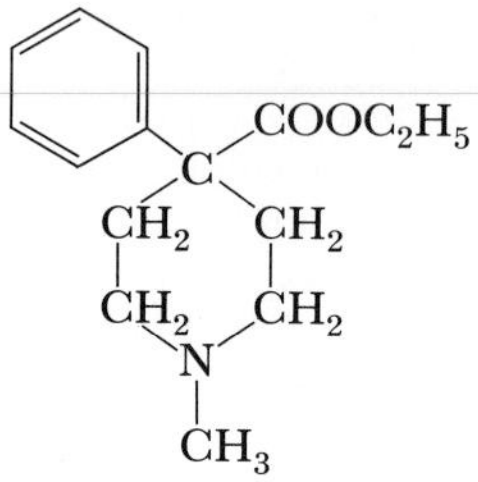

XVIII

Nicotinic acid (XIX, pyridine-3-carboxylic acid; or niacin) and *its amide* (XX, niacinamide) are members of the vitamin B group, used as additives for flour and bread enrichment. Other important uses for niacin are in the treatment or prevention of pellagra, as a vasodilator (niacinamide does not have this effect) as such or as fructose-1,3,4,6-tetranicotinate (nicofuranose), for example, as an agent to control cholesterol levels in the blood and organs, and as a feed additive (about 70% of the total current United States consumption of over 4 million lb/yr of niacin went into animal-feed enrichment). Nicotinic acid is made by the oxidation of 3-picoline, 2-methyl-5-ethylpyridine (the isocinchomeric acid, XXI, produced is partially decarboxylated), or quinoline (the intermediate quinolinic acid, XXII, is also partially decarboxylated) with H_2SO_4 in the presence of selenium dioxide at ~300°C, or with HNO_3 (as the dilute acid under pressure or as its mixture with sulfuric acid) at ~185–250°C, or by electrochemical oxidation. Nicotinic acid can also be made from 3-picoline by catalytic ammoxidation to 3-cyanopyridine (XXIII) followed by hydrolysis. Nicotinamide is prepared by partial hydrolysis of the nitrile (XXIII) or by amination of nicotinic acid chloride or its esters.

*Demerol.

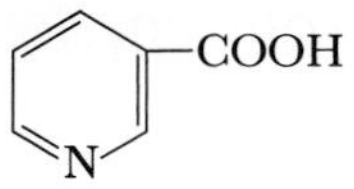

XIX

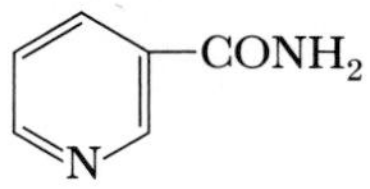

XX

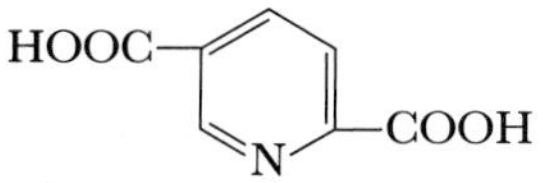

XXI

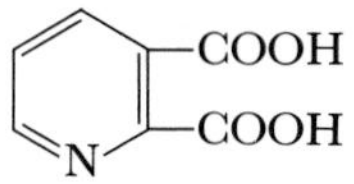

XXII

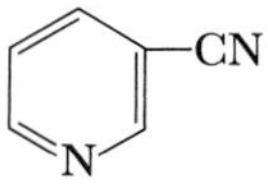

XXIII

Nikethamide (**XXIV**,* *N,N*-diethylnicotinamide) made by reaction of nicotinic acid esters or the acid chloride with diethylamine, is a respiratory and heart stimulant, used beneficially against overdoses of barbiturates and morphine.

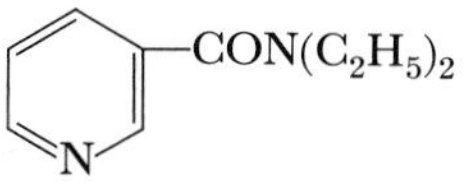

XXIV

Pipadrol(**XXV**,α,α-diphenyl-2-piperidinemethanol) is a central-nervous-system stimulant, made by condensation of 2-pyridylmagnesium chloride with benzophenone and catalytic hydrogenation of the pyridine ring of the resultant carbinol.

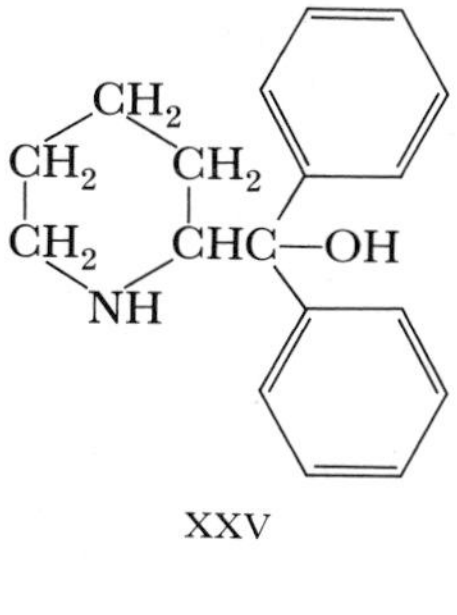

XXV

Piperocaine hydrochloride [**XXVI**, *d,l*-(2-methyl-piperidino)propyl benzoate hydrochloride] is used as a local anesthetic. It is made by reaction of 2-methylpiperidine with 3-chloropropyl benzoate.

*Coramine.

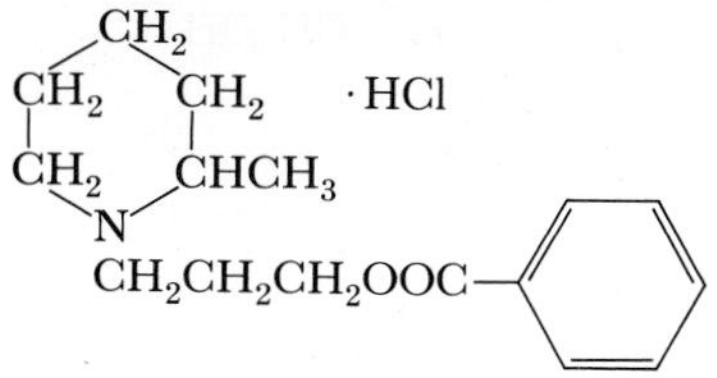

XXVI

Pyrithione (**XXVII**,† 2-mercaptopyridine *N*-oxide) exists in equilibrium with *N*-hydroxy-2-pyridinethione and is a fungicide and bactericide, prepared by reaction of 2-chloropyridine *N*-oxide with sodium hydrosulfide and sodium sulfide. The zinc salt of pyrithione is used as a component of antidandruff shampoos and as a bactericide in soap and detergent formulations.

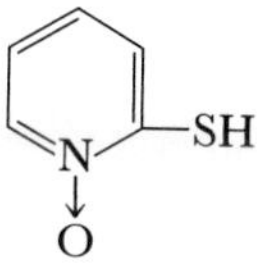

XXVII

Sulfapyridine (**XXVIII**, 2-sulfanylamidopyridine) is used to treat dermatitis herpetiformis; it has veterinary applications against pneumonia, shipping fever, and foot rot of cattle. It is made by condensation of 2-aminopyridine with the appropriate sulfonyl chloride.

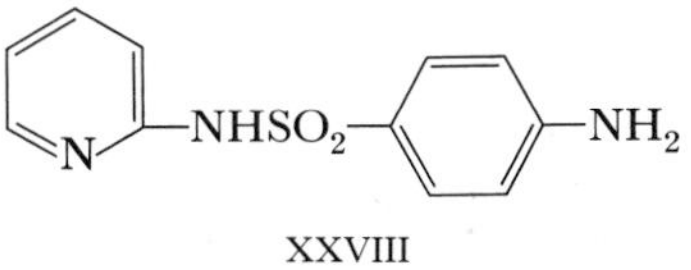

XXVIII

Vitamin B$_6$ [**XXIX**, 2-methyl-3-hydroxy-4,5-di(hydroxymethyl)pyridine or pyridoxol] is one of several related pyridine derivatives (the pyridoxins) with vitamin B$_6$ activity, widely occurring in nature as such or as their phosphate esters. For example, grains and meats are relatively rich in vitamin B$_6$. Commercial production is by synthesis starting for example with the base-catalyzed condensation of cyanoacetamide and ethoxyacetylacetone. Pyridoxol is used in the treatment of nutritional (vitamin B$_6$ complex) deficiencies in man and animals.

†Omadine.

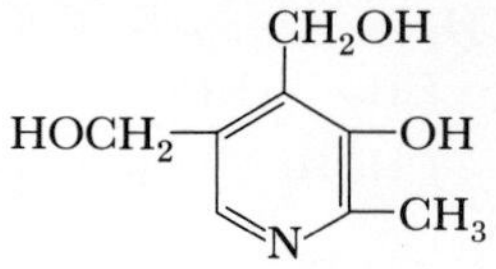

XXIX

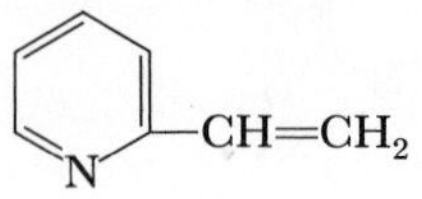

On a commercial scale *piperidine* (XXX, hexa-hydropyridine) is prepared by the catalytic hydrogenation of pyridine, e.g., with Ni catalysts, at ~1,000–2,000 psig at 150–200°C or under mild conditions with noble-metal catalysts. Pyridine derivatives can be similarly reduced to substituted piperidines. The major uses of piperidine and derivatives are in the pharmaceutical field (examples above) and as rubber-vulcanization accelerators, e.g., piperidinium pentamethylenedithiocarbamate (XXXI).*

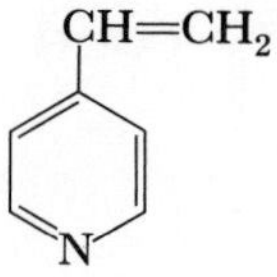

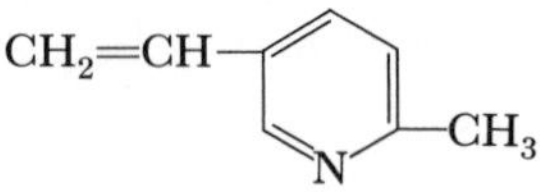

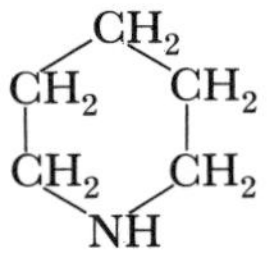

XXX

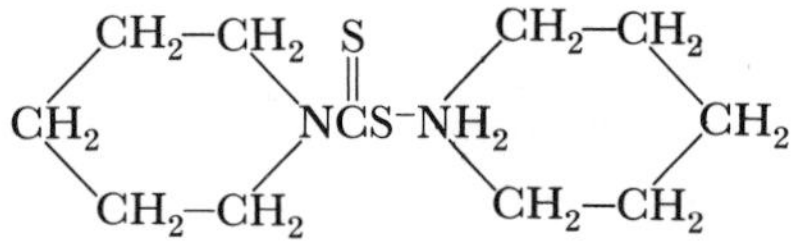

XXXI

Textile Chemicals: *Stearamidomethylpyridinium chloride* (XXXII) is used in waterproofing textiles. It is made by reacting pyridine hydrochloride with stearamide and formaldehyde. *Vinylpyridines* are used as components of acrylonitrile copolymers to improve the dyeability of polyacrylonitrile fibers.

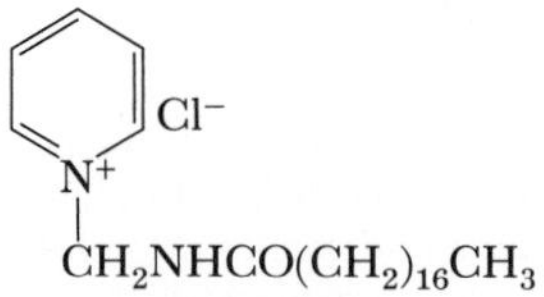

XXXII

Vinylpyridines: The commercially important products are 2-vinylpyridine (XXXIII), 4-vinylpyridine (XXXIV), and 2-methyl-5-vinylpyridine (XXXV).

*Accelerator 552.

XXXIII   XXXIV

XXXV

XXXIII and XXXIV are made by the reaction of 2- or 4-picoline, respectively, with formaldehyde to give the 2- or 4-(2-hydroxyethyl)pyridine, which is catalytically dehydrated to the vinylpyridine, usually in the vapor phase over an alumina catalyst. 2-Methyl-5-vinylpyridine (XXXV) is produced by dehydrogenation of 2-methyl-5-ethylpyridine (MEP) over catalysts such as Fe_2O_3-Cr_2O_3-K_2O on alumina. The technique of oxydehydrogenation of the ethylpyridines has also received attention.

The major uses of the vinylpyridines are in the production of acrylic fibers of improved dyeability via copolymers containing $\leq 5\%$ vinylpyridine and as vinylpyridine-butadiene-styrene terpolymer latices (~10–20% vinylpyridine content) used as components of tire-cord dips which improve the bonding of fiber to rubber. On rayon, nylon, and polyester cords the latex pickup is about 5%, whereas on the recently introduced fiber-glass cords up to 20% latex pickup is reported.

Pyridine hydrochloride is used in the manufacture of polycarbonates. See also **Polycarbonates.**

—Hans Dressler, *Koppers Company, Inc.,*
Monroeville, Pa.

Pyrite   (See **Iron ores;** and **Sulfur.**)

Pyrochlore   (See **Uranium.**)

Pyrolusite   (See **Manganese.**)

Pyrolysis   (See **Acetylene; Energy systems for processes; Ethylene; Petrochemical complex;** and **Thermal cracking.**)

Pyrometallurgy   (Consult the **Subject Index.**)

Pyromucic Acid   (See **Carboxylic acids.**)

Pyrophosphoric Acid (See **Fertilizers; and Phosphoric Acid.**)

Pyrotechnics Pyrotechnics are physical mixtures of chemical elements and/or compositions formulated so that they ignite readily and through exothermic reaction produce light, sound, or smoke. Pyrotechnics are used primarily for signaling, illuminating, marking, or creating special effects, a basic or apparent difference being whether they are designed for commercial or military applications. Commercial pyrotechnics include fireworks, railway signal torpedoes, highway flares, signaling torches, and emergency or rescue illuminators. Battlefield illuminating flares, warning or trip flares, signal flares, tracer ammunition, smoke compositions, and artillery simulators are common military pyrotechnics.

Pyrotechnic compositions generally consist of finely divided materials pressed into some particular shape or form. The most important parts are the fuels and oxidizing agents, common fuels being sulfur, charcoal, phosphorus, zirconium, calcium silicide, and powdered forms of aluminum, magnesium, and other metals and alloys. Barium, sodium, potassium, and strontium nitrates; potassium and ammonium perchlorates; barium, lead, and strontium peroxides; and oxides of iron, lead, copper, and silver are some of the

TABLE P-43. *Typical Compositions for Military Signal Flares*

Component	Ingredients for Colored Signals, %		
	White	Green	Red
Aluminum	15	—	—
Magnesium	17	15	40
Barium nitrate	55	60	—
Potassium perchlorate . .	—	—	20
Strontium nitrate	5	—	30
Hexachlorobenzene . . .	—	20	5
Copper dust	—	3	—
Asphalt	5	—	5
Linseed oil	3	—	—
Castor oil	—	2	—

oxidizing agents employed. In addition to oxidizers and fuels, pyrotechnics usually contain color intensifiers, which form color bands in the flame spectrum during decomposition; retardants to control or slow the burning rate of the fuel-oxidizer mixture; and binders or waterproofing agents when necessary.

Typical compositions for military signal flares are shown in Table P-43.

—Thomas P. Dowling, *Trojan–U.S. Powder, Division of Commercial Solvents Corporation, Allentown, Pa.*

Q

Quantum Number   (See Atomic structure.)

Quartz   (See Abrasives; and Silicones.)

Quaternary Ammonium Compounds   (See Amines; and Pyridine and derivatives.)

Quaternary Ammonium Silicates   [See Silicates (soluble).]

Quicklime   (See Lime.)

Quicksilver   (See Mercury.)

Quinoline   (See Coal tar and derivatives.)

R

Racemization (See **Amino acids;** and **Isomerism.**)

Radiation Shielding (See **Carbon; Dysprosium; Erbium; Lead; Nuclear power plants; Radioisotopes; Tungsten;** and **Zirconium.**)

Radiation Vulcanization (See **Rubber, natural.**)

Radicals A combination of atoms that frequently participate in chemical reactions as a group, behaving in this respect like an element, sometimes may be referred to as a *radical*. There are hundreds of inorganic and organic groups like this, some of which are listed in Table R-1.

For many years the evidence available indicated that radicals do not exist in the free state, or if they do, only for very short periods of time. In 1900, however, Gomberg discovered the *free radical* triphenylmethyl. This constituted a breakthrough in this area of chemistry, but unfortunately, for many years after, these pursuits were channeled largely into dye chemistry. Relatively recent studies* point to the existence of long-lived, stable

---

*George A. Olah, Stable Carbonium Ions in Solution, *Science,* vol. 168, June 12, 1970. Free radicals also are discussed under **Polymerization.**

carbonium ions in very acid solutions. A new area of carbonium-ion chemistry is opening up, aided by nuclear-magnetic-resonance spectroscopic analytical methods and by the use of highly acidic solvent systems leading to the formation and study of stable carbonium ions. Crystalline forms of these ions now are contemplated. Efforts are under way to formulate an appropriate system of nomenclature, which is becoming increasingly important as additional carbonium ions are identified. Currently, for example, the trimethylcarbonium ion may also be referred to as *t*-butyl cation.

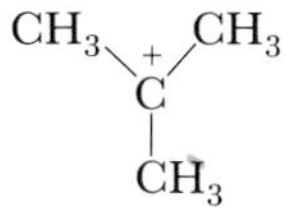

$$CH_3 \underset{\underset{\displaystyle CH_3}{|}}{\overset{+}{C}} CH_3$$

Trimethylcarbonium ion

Single elements or groups of two or more, depending upon valence, also may be referred to as radicals, e.g., chloride, Cl; sulfide, S; and nitride, N.

TABLE R-1. *Representative Groups that Frequently React in Concert*

Group	Formula	Formula Weight	Valence
Inorganic			
Aluminate	Al_2O_4	117.92	2
Ammonium.	NH_4	18.04	1
Antimonate			
Meta	SbO_3	169.72	1
Pyro	Sb_2O_7	355.43	4
Arsenate	AsO_4	138.88	3
Arsenite.	AsO_3	122.89	3
Borate			
Meta	BO_2	42.79	1
Per	BO_3	58.78	1
Tetra	B_4O_7	155.17	2
Bromate.	BrO_3	127.88	1
Carbonate	CO_3	59.98	2
Chlorate.	ClO_3	83.40	1
Per	ClO_4	89.41	1
Hypo	ClO	51.44	1
Chloroaurate	$AuCl_4$	338.69	1
Chloroplatinate	$PtCl_6$	407.67	2
Chlorostannate	$SnCl_6$	331.27	2
Chromate	CrO_4	115.96	2
Cyanate	CNO	42.01	1
Cyanogen (cyanide)	CN	26.02	1
Dichromate	Cr_2O_7	215.93	2
Fluosilicate	SiF_6	142.09	2
Hydrazine	H_2NNH_2	32.06	1
Hydroxyl	OH	17.00	1
Hydroxylamine.	H_2NOH	33.03	1
Iodate	IO_3	174.87	1
Iron cyanide	$Fe(CN)_6$	211.97	3
			4
Manganate	MnO_4	118.90	1
Molybdate	MoO_4	159.90	2
Nitrate	NO_3	61.98	1
Nitrite.	$\cdot NO_2$	45.99	1
Nitroso	NO	30.00	1
Phosphate, meta	PO_3	78.94	1
Ortho.	PO_4	94.93	3
Pyro	P_2O_7	173.87	4
Selenate.	SeO_4	142.92	2
Silicate	SiO_3	76.06	2
Stannate	SnO_3	166.66	2
Meta	Sn_5O_{11}	769.34	2
Sulfate	SO_4	96.02	2
Sulfite.	SO_3	80.03	2
Sulfonic	SO_2OH	81.04	1
Thiocyanate	CNS	58.08	1
Thiosulfate	S_2O_3	112.09	2
Titanate.	TiO_3	95.87	2
Tungstate	WO_4	247.81	2
Uranate	UO_4	301.99	2
Vanadate, meta	VO_3	98.91	1
Zincate	ZnO_4	129.33	2

TABLE R-1. *Representative Groups that Frequently React in Concert*   (*Continued*)

Group	Formula	Formula Weight	Valence
Organic			
Acetate	$C_2H_3O_2$	59.03	1
Acetyl	CH_3CO	43.04	1
Aldehyde	CHO	29.01	1
Allyl	$CH_2:CHCH_2$	41.08	1
Amino	NH_2	16.03	1
Amyl			
Iso	$(CH_3)_2CHCH_2CH_2$	71.16	1
n	$CH_3(CH_2)_3CH_2$	71.16	1
Azoxy	N_2O	44.01	1
Benzal	C_6H_5CH	90.13	2
Benzoate	$C_7H_5O_2$	121.10	1
Benzoyl	C_6H_5CO	105.11	1
Benzyl	$C_6H_5CH_2$	91.14	1
Benzylidene, see benzal			
Butyl			
Iso	$(CH_3)_2CHCH_2$	57.13	1
n	$CH_3CH_2CH_2CH_2$	57.13	1
Butyryl	C_3H_7CO	71.10	1
Cacodyl	$As(CH_3)_2$	105.00	1
Capryl	$CH_3(CH_2)_6CO$	127.22	1
Carbonyl	CO	28.00	2
Carboxyl	$COOH$	73.00	1
Cinnamate	$C_9H_7O_2$	147.14	1
Citrate	$C_6H_5O_7$	189.04	3
Decyl	$CH_3(CH_2)_8CO$	155.28	1
Diazo	$C_6H_5N_2$	105.13	1
Diphenyl	$C_6H_5 \cdot C_6H_5$	154.22	2
Ethoxy	OC_2H_5	45.06	1
Ethyl	C_2H_5	29.07	1
Ethylene	CH_2CH_2	28.06	2
Ethylidene	CH_3CH	28.06	2
Formate	CHO_2	45.00	1
Formyl	$CH:O$	29.01	1
Furfuryl	$C_4H_3OCH_2$	81.09	1
Glyceryl	C_3H_5	41.08	3
Heptyl	$CH_3(CH_2)_5CH_2$	99.22	1
Hexyl	$CH_3(CH_2)_4CH_2$	85.19	1
Imino	NH	15.02	2
Isopropyl	$(CH_3)_2CH$	43.10	1
Lactate	$C_3H_5O_3$	89.05	1
Lauryl	$CH_3(CH_2)_{10}CH_2$	169.37	1
Malate	$C_4H_4O_5$	132.03	2
Malonyl	$C_3H_2O_2$	70.03	1
Methine (methenyl)	CH	13.02	3
Methoxy	OCH_3	31.03	1
Methyl	CH_3	15.04	1
Naphthyl	$C_{10}H_7$	127.17	1
Nonyl	$CH_3(CH_2)_7CH_2$	127.28	1
Octyl	$CH_3(CH_2)_6CH_2$	113.25	1
Oleate	$C_{18}H_{33}O_2$	281.49	1
Oxalate	C_2O_4	87.98	2
Oxalyl	C_2O_2	56.00	1
Oximino	NOH	31.01	2
Palmitate	$C_{16}H_{31}O_2$	255.45	1
Phenyl	C_6H_5	77.11	1
Phthalate	$C_8H_4O_4$	164.08	2
Propargyl	$CH:CCH_2$	39.06	1

TABLE R-1. *Representative Groups that Frequently React in Concert* (*Continued*)

Group	Formula	Formula Weight	Valence
Organic			
Propionyl	C_2H_5CO	57.07	1
Propyl	$CH_3CH_2CH_2$	43.10	1
Salicylate	$C_7H_5O_3$	137.09	1
Stearate	$C_{18}H_{35}O_2$	283.51	1
Succinyl	$C_4H_4O_2$	84.06	1
Tartrate	$C_4H_4O_6$	148.02	2
Tolyl	$CH_3C_6H_4$	91.14	1
Vinyl	$CH_2:CH$	27.05	1
Xylyl	$C_6H_4(CH_3)CH_2$	105.17	1

NOTE: The *alkyl* or *alphyl* radicals, including methyl, ethyl, and propyl and sometimes called the alcohol radicals, are considered to be derived when one atom of hydrogen is removed from the paraffin molecules, such as methane, ethane, and propane. The *acyl* radications, including acetyl, propionyl, and butyryl, are sometimes called the acid radicals and are considered to be derived when OH is removed from the fatty acid molecules, such as acetic acid, propionic acid, and butyric acid. The *aryl* radicals comprise the aromatic series, including phenyl, naphthyl, and tolyl.

Examples of a radical appearing on both sides of a chemical equation are:

Hydroxyl: $2NiCl_2 + 4K\underline{OH} + H_2O_2 \rightarrow$
$$4KCl + 2Ni(\underline{OH})_3$$

Acetate: $Fe(C_2H_3O_2)_3 + 2H_2O \rightarrow$
$$2H\underline{C_2H_3O_2} + Fe(OH)_2(\underline{C_2H_3O_2})$$

Sulfate: $Mn\underline{SO_4} + 2Na_2CO_3 + O_2 \rightarrow$
$$2CO_2 + Na_2\underline{SO_4} + Na_2MnO_4$$

Nitrite: $Co(\underline{NO_2})_2 + 2H\underline{NO_2} \rightarrow$
$$H_2O + NO + Co(\underline{NO_2})_3$$

Nitrate: $Hg_2(\underline{NO_3})_2 + H_2S \rightarrow$
$$2H\underline{NO_3} + HgS + Hg$$

Cyanogen: $Cd(\underline{CN})_2 + 2K\underline{CN} \rightarrow$
$$K_2Cd(\underline{CN})_4$$

$$H_3PO_4 + 12(NH_4)_2MoO_4 + 21HNO_3 \rightarrow$$
$$(\underline{NH_4})_3\underline{PO_4} \cdot 12MoO_3 + 21(\underline{NH_4})\underline{NO_3} + 12H_2O$$

In the last example three radicals (phosphate, ammonium, and nitrate) take part in a single reaction.

To qualify as a radical, a combination of atoms need not always behave in concert but simply exhibit a frequent chemical behavior pattern. For example, in the following reaction, the manganate radical, MnO_4, does not remain intact but is reduced to manganese dioxide, MnO_2.

$$2KMnO_4 + 3MnSO_4 + 2H_2O \rightarrow$$
$$K_2SO_4 + 5MnO_2 + 2H_2SO_4$$

Radioisotopes Isotopes that are distinguishable from other species of atoms with the same atomic number by radioactive transformation are known as radioactive isotopes or radioisotopes. They are useful as sources of radiation and as tracers, a use that arises because they exhibit substantially the same chemical behavior as the stable species of isotopes while emitting radiation which makes it easy to determine their location and identity. Table R-2 summarizes the principal types of radiation emitted by radioisotopes and gives important properties of these radiations.

Some form of instrumentation is almost always associated with the practical uses of radioisotopes. Industrial process and research and medical applications are among the most important areas using such instrument systems. Radioisotopes are used experimentally to determine flow rates of sewers, rivers, and large pipes; to locate leaks; to measure wear; to determine pathways in photosynthesis; to establish CO_2 fixation rates; and for numerous other applications in industry, medicine, and the life sciences. As of 1972, there were more than 10,000 installations of nuclear gages in the United States. These gages utilize radiation from sealed radioisotope sources to measure level, density, thickness, moisture content, and composition. Thousands of other radioisotope instrumentation procedures are used in medical, biochemical, and other research laboratories.

Important terms dealing with radioactivity, e.g., isotope, half-life, radioactivity units, curie, roentgen, and neutrons, are described in their proper alphabetical location in this volume. See also **Atomic structure; Chemical elements; and**

TABLE R-2. *Principal Types of Radiation Emitted by Radioisotopes*

Type	Symbol	Description	Rest Mass (O = 16)	Charge (Electron = 1)	Range in Air, cm	Ion Pairs per Centimeter in Air
Alpha particles	α	Nuclei of He atoms	4	+2	2–9 (for 3–10 MeV)	30,000–70,000 (varies with distance from source)
Beta particles	β	Electrons ejected from a nucleus	$\dfrac{1}{1,840}$	−1	160–2,000 (for 0.5–5 MeV)	150–40 (for 0.1–5.0 MeV)
Gamma rays	γ	Electromagnetic radiations produced only in nuclear processes	None	None	15,000 $\frac{1}{2}$ value thickness (for 1.5 MeV)	$\frac{1}{100}$ number of pairs produced by same energy β

NOTE: X-rays resulting from filling nuclear shells and bremsstrahlung generated by the deceleration of beta particles are two additional sources of electromagnetic radiation produced by radioisotopes that find use in instrumentation and research. None of the radiations described in this table produce detectable amounts of radioisotopes.

Nuclear power plants and consult the Subject Index.

Inverse-Square Law: Radiation emitted by radioisotopes is uniformly distributed in all directions in space; thus the number of particles, or quanta, passing through a unit volume at any point distant from the source varies inversely as the square of the distance from the source, or

$$I = \frac{I_0 r_0^2}{r_1^2}$$

where I_0 = radiation intensity at distance r_0 from source
I = radiation intensity at distance r_1 from same source

This equation neglects absorption effects and assumes a point source; it is most useful for gamma radiation, where the source-to-detector distance usually is much larger than any source or detector dimension.

Radiation Absorption and Measurement of Density, Level, and Thickness: Measurement of density, level, and thickness all depend upon a determination of the number of radiations per unit time penetrating the sample, e.g., a pipe filled with slurry and producing a measurable signal in the detector. An increase in the amount of matter between the source and the detector *usually* results in a decrease in the signal.

The exponential nature of the attenuation of beta,* x-,† or gamma radiation is indicated by the

*Beta radiation has a finite range, while in theory x- and gamma rays are exponentially attenuated.

†X-radiation of each energy exhibits sharp changes in absorption coefficient for certain absorbers.

relationship

$$\frac{I}{I_0} = Be^{-\mu\rho t}$$

where I_0 = initial radiation intensity
I = radiation intensity through absorbing material
B = a buildup factor, depending on energy and collimation of the source and on ρ and t; B accounts for radiation scattered or changed in direction by interactions that do not stop the radiation
μ = absorption coefficient, dependent upon composition of absorbing material
ρ = specific gravity of absorbing material
t = thickness of absorbing material

Optimum performance and accuracy are obtained for an absorption gage when the signal change is maximized for a given process or product change. Signal change is at a maximum for $\mu\rho t = 1$. For thickness gaging where ρ and t are fixed, proper selection of the source permits optimizing $\mu\rho t$. For density gaging, the pipe diameter or path length t can be selected to optimize $\mu\rho t$.

Since all material between source and detector can contribute to scattering (a change in direction of the radiation) and absorption, errors in absorption-type gage readings can occur due to changes in air density, pipe-wall thickness, and solids deposits. Gages must be regularly zeroed and calibrated. Simple solid standards usually are provided for calibration.

Scattering can produce a 180° change in direction of the incident radiation; measurement of this *backscattered* radiation, which is sensitive to density and atomic number of the scattering material, provides a means of measuring the density of compacted soils with a surface reflection gage. Backscattering also permits measuring the weight per unit area of a sheet on a roll or backup plate when the sheet, e.g., rubber, has a different atomic number than the roll, e.g., iron.

Neutron Moderation and Moisture Measurement: If a source of fast neutrons (Sb-Be, Pu-Be, ^{252}Cf) is placed on or in a medium containing hydrogen, some of the neutrons will be slowed down (moderated or thermalized) by collisions with the hydrogen atoms; the number of slow neutrons per unit area per unit time can be determined with a detector selectively sensitive only to slow neutrons, e.g., a ^{3}He- or BF$_3$-filled proportional or Geiger-Müller detector. The relative response of the detector will provide a measure of the hydrogen content of the medium surrounding or adjacent to the source. If water is the only hydrogen-containing variable constituent of the medium (or the principal one), the technique can be used to measure moisture content.

Gages are affected by changes in dry bulk density and to a lesser extent by composition changes of the medium. Calibration at regular intervals is essential.

Neutron moisture gages are used for highway, water content of snow, foundation, and agricultural purposes and are finding applications as process control instruments.

Radioisotope X-Ray Fluorescence Spectrometry: Radioisotope x-ray fluorescence is a relatively recent instrumentation application resulting from the availability of sealed radioisotope sources, x-ray detectors, and portable electronics. Commercially available portable instruments are in use for analysis and identification of alloys, geophysical prospecting, mineral processing, and regulating the thickness of coatings in tinning and galvanizing processes.

The instruments include a radioisotope source (alpha, bremsstrahlung,* or x-ray), a sample holder, a detector for the excited or fluorescent x-rays, a high-voltage supply for the detector, and an amplifier-analyzer-scaler-timer, or rate meter. Filters sometimes are used in front of the detector to filter out unwanted x-rays. Proportional, scintillation, and semiconductor detectors are all used

*The deceleration of electrons (beta rays) or positrons in matter produces x-rays just like those produced in x-ray tubes. When produced by braking (deceleration), these x-rays are called bremsstrahlung from the German word for brake.

in various types of fluorescence gages. In principle, radiation from the sealed source penetrates the surface layers of the sample, which can be solid, liquid, powder, or suspension, and excites characteristic x-rays for some elements in the sample. A selection can be made of which elements will be excited by changing the source and thus the excitation energy. Problems in using x-ray fluorescence result from inhomogeneity of samples, matrix effects, particle-size effects, heterogeneity, and rough-surface phenomena.

Statistical Nature of Isotopic Radiation: The inherent statistical nature of isotopic radiation gaging systems results from the fact that radioisotopes do not decay and emit radiation continuously but in a random manner. Because the disintegration rate is random in time, the actual number of radiations emitted in a given unit of time is related to the average value by a probability function.

The relationship between the actual or observed number of events (the measured signal) and the average becomes closer in terms of percent deviation as larger numbers of radiations are measured. This can be attained by increasing the size of the source (number of millicuries), improving the geometry (moving source closer to detector), improving the efficiency of the detector, or increasing the time for the measurement.

In most systems, the detector responds to background radiation due to cosmic rays and naturally radioactive materials (mostly potassium, radium, thorium, and their decay products) as well as to radiation from the source. Statistical variations in this background radiation produce unwanted variations in the response of the system; background effects are minimized by shielding the detector with iron or lead and by using sources that produce radiation intensities several orders of magnitude above the background. In making specific gravity measurements of liquids or slurries, temperature changes can produce major signal-generation changes. Air bubbles, grease deposits, or pipe-wall corrosion can also cause major errors.

Industrial Applications for Radioisotope Measurements

Pressure Measurement: The interaction of alphas with a gas (or solid) results in the formation of positive and negative ions, which can be collected and measured as electric current. Since the number of ions produced in a gas by alpha particles from a source of fixed size depends on the density and composition of the gas, measurement of ionization can be used as a method of measuring pressure or composition. This principle is used in a number of vacuum gages.

Level Measurement: The ability of gamma radiation to penetrate steel walls of pressure vessels and brick linings of cupolas and the absorption or scattering of this gamma radiation by material inside these vessels permits the level of solids or liquids to be measured without the use of floats or other internal detection devices. In the radiation-absorption technique, the source of gamma rays, such as radium or ^{60}Co, is placed on one side of the vessel. The radiation-detecting device is placed on the opposite side, or a number of sources are placed in wells and the detector is located opposite the sources. The principle is applied in a number of design configurations.

Thickness and Coverage Measurement: For very thin materials (under 5 mg/cm^2, 0.0009 in. of Al), the absorption of alpha radiation is used to provide a measure of weight per unit area. For 0.5–1,000 mg/cm^2 (0.00009–0.16 in. of Al), beta radiation, as emitted by ^{90}Sr, ^{85}Kr, ^{14}C, and other radioisotopes, is used for measuring weight per unit area. For thicknesses above 1,000 mg/cm^2, bremsstrahlung or gamma sources are used.

Wear Measurement: Gears, bearings, piston rings, and distributor points have been irradiated to produce radioisotopes directly in the irradiated materials. The radioactive part then is placed in the equipment whose wear rate is to be measured. Piston rings have been placed in test engines which are operated with varying cylinder temperature, type of lubricant, and quantity of sulfur in the gasoline. The amount of material worn from the ring is determined by counting radiations from the radioactive material in the lubricating oil. The same principle has been applied in many other ways, including measurement of the wear of a tungsten carbide die, rubber tires, floor waxes, and automotive protective coatings.

Uniformity of Mixing: It is possible to use radioisotopes to determine the uniformity of mixing of liquids or dry solids. Radioactive ^{24}Na in NaCl as the tagged compound has been used to determine the uniformity of mixing various chemical products. Catalysts tagged with ^{51}Cr, ^{46}Sc, and ^{144}Ce have been used to determine mixing patterns in fluid catalytic cracking units; 1 mCi of a 1-meV gamma emitter uniformly mixed in 500 tons of catalyst will produce a net counting rate of 275 cpm/kg* above a background of

*The abbreviation cpm stands for counts per minute.

200 cpm when 1-l catalyst samples are counted with a 2-in.-thick, 1$\frac{3}{4}$-in.-diam NaI scintillation counter. The mixing and degree of short circuiting in model sewage-settling tanks has been studied using radioactive tracers.

Flow Measurement: In a peak-timing technique, a gamma emitter such as ^{60}Co or ^{124}Sb is injected quickly at a point close to the section of pipe in which the velocity is to be determined. The time of passage of the peak of the tracer wave is determined using two detectors, e.g., Geiger-Müller counters or scintillation detectors, located a known distance apart and external to the pipe. In a second method (dilution method), a radioactive tracer is bled into the line at a known rate. Measured downstream, the concentration in the line is inversely proportional to the flow rate.

Diffusion and Other Measurements: Radioisotopes are used for measuring extremely slow velocities like those obtaining in diffusion in metals, porous bodies, liquids, or gases. For measuring self-diffusion, e.g., cobalt in cobalt or silver in silver, radioisotopes make measurements possible that cannot be done by other techniques. Radioactive cobalt has been plated on a cobalt bar which is then heated for the desired time and the surface counted to determine the depth of diffusion as a function of time and temperature.

Tritiated water is used to study the permeability of thin, flexible plastic sheets by clamping a septum of the wrapping over a dish of tritiated water solidified by gelatin. Methane is passed over the upper surface of the septum and into a counter. The counter pulses resulting from the tritium diffused in the methane are measured.

Radioisotopes have been used for studying flow patterns of underground water supplies or tracing cross flow between oilwells. Water tagged with ^{131}I has facilitated the study of the subsurface flow of water used in secondary oil recovery to determine the path, velocity, and carrying strata of the water. In pipeline operations, to minimize loss of different products at the terminus, an oil-soluble ^{124}Sb compound can be injected into the line at the time a change in product is being made. Pumping stations along the route have radiation detectors located ahead of the station, which pick up the arriving interface and automatically operate controls for appropriate handling.

Licensing and Safety. In the United States, the Atomic Energy Commission originally controlled the use of all by-product materials produced in nuclear reactors, leaving uncontrolled the naturally occurring radioisotopes, such as radium and polonium, and isotopes produced in accelerators. In recent years, a number of states have established their own radiation-safety operations, usually under the department of health or equivalent, and have assumed the issuance of licenses and inspection of radiation sources, including radium and polonium. States with such practices have become known as the *agreement states*.

The radioisotopes used in sources usually are furnished in doubly sealed, stainless-steel containers welded closed, often called *encapsulated sources*.

Federal and state regulations require verification of the integrity of such sources by a simple leak or wipe test, which involves wiping the source with a moistened piece of filter paper (maintaining a safe distance from the source) at from 6-month to 3-year intervals and then checking the wipe for radioactivity. Most source suppliers offer a wipe-test service at a reasonable cost. Unshielded sources never should be handled except with tongs.

Alpha, beta, x-ray, and gamma radiation

TABLE R-3. *Properties of Common Radioisotope Radiation Sources*

Radioisotope Sources	Radiation of Interest	Half-Life	Energy of Radiation	Principal Use
Alpha:				
^{210}Po	α	138 days	5.3 MeV	Static elimination; thin film thickness gaging
^{226}Ra	α	1,620 years	5.15 MeV	Vacuum gage
Beta:				
^{14}C	β	5,700 years	0.155 MeV	Thin plastic gaging
^{85}Kr	β	10.8 years	0.695 MeV	Light paper and plastic gaging
^{90}Sr–^{90}Y	β	28 years	0.61 and 2.18 MeV	Heavy paper, thin metal, rubber gaging
X-ray and bremsstrahlung:				
^{3}H	Bremsstrahlung	12.46 years	17.6 keV and down	S and Pb analysis
^{90}Sr–^{90}Y	Bremsstrahlung	25 years	2.18 MeV and down	Steel, Cu, Al sheet
^{57}Co	X-ray	0.74 years	122 keV 136 keV 14 keV	X-ray fluorescence analysis
^{210}Pb	X-ray	22 years	47 keV 11–13 keV	X-ray fluorescence analysis
^{241}Am	X-ray	458 years	60 keV 14–21 keV	X-ray fluorescence analysis
^{238}Pu	X-ray	86 years	12–20 keV	X-ray fluorescence analysis
^{55}Fe	X-ray	2.7 years	6 keV	X-ray fluorescence analysis and sulfur analysis
^{109}Cd	X-ray	1.3 years	88.2 keV 22.2 keV	X-ray fluorescence analysis
Gamma ray:				
^{137}Cs	γ	33 years	0.66 MeV	Level, density, thick plate gaging
^{60}Co	γ	5.2 years	1.2 and 1.3 MeV	Level, density, thick plate gaging
^{226}Ra and daughter	γ	1,620 years	Up to 0.8 MeV	Level, density, thick plate gaging
Neutron:				
^{226}Ra–Be	Neutrons*	1,620 years	Up to 13 MeV	Moisture and density gaging
^{210}Po–Be	Neutrons	138 days	Up to 11 MeV 4 MeV av	Moisture and density gaging
^{124}Sb–Be	Neutrons*	60 days		Moisture and density gaging
^{239}Pu–Be	Neutrons	24,360 years		Moisture and elemental gaging
^{252}Cf (spontaneous fission source)	Neutrons*	2.65 years	2.3 MeV av 6.0 MeV max.	Activation analysis, moisture measurement
^{241}Am–Be	Neutrons	475 years		Moisture measurement

*Also emits gamma radiation.

TABLE R-4. *Values of Relative Biological Effectiveness (RBE)* *

Type of Radiation	RBE
X- and gamma	1
Beta rays	1
Alpha rays	20
Fast neutrons†	10
Thermal or slow neutrons	5

*Since the same number of roentgens (R) from various types of radiation produces different amounts of body damage, the roentgen equivalent man (rem) is used in stating allowable radiation exposure values; rem = R × RBE, where RBE has the values given in this table. The danger of exposure also varies with the parts of the body involved.

Permissible Exposure (rems/Calendar Quarter)	
Whole body; head and trunk; active blood- forming organs; lens of eyes; gonads	1.25
Hands and forearms; feet and ankles	18.75
Skin of whole body	7.5

The dose to the whole body when added to the accumulated occupational dose must not exceed

$$\text{MPD} = 5(N - 18) \text{ rem}$$

where MPD = maximum permissible accumulated dose, rem

N = person's age

†Neutrons having energies of 0.1–10 MeV. Above 10 MeV, the RBE increases rapidly.

sources *do not* cause other materials to become radioactive. Neutrons can make materials radioactive, but with their low intensity and relatively short irradiation times, the neutron sources used in instruments produce no measurable residual radioactivity.

Radioisotope Radiation Sources. Important properties of common radioisotope radiation sources are given in Table R-3. Values of relative biological effectiveness are given in Table R-4.

—Jerome Kohl, *Department of Nuclear Engineering, North Carolina State University, Raleigh, N.C.*

Radium (L. *radius*, a ray) **Ra** = 226* (at. wt.); 88 (at. no.). The discovery of radium by Pierre and Marie Curie and their collaborator Gustave Bémont followed the discovery of polonium by several months in 1898. The existence of this new element, close to barium in its chemical behavior, was established by following the study of the fractionation of the pitchblende of Joachimsthal. It was possible to observe its concentration with

*Mass number of the isotope of longest known half-life.

barium owing to its emitting radiation. This property is the origin of the name radium. The discovery, based on radiochemical data, was confirmed by Demarçay who observed in sufficiently enriched fractions of radium-containing barium a line of wavelength 3815 Å which was due to no other known element and the intensity of which increased with the radioactivity during the fractionations by fractionated crystallizations. The isolation some years later of pure radium chloride permitted Marie Curie to effect a determination of the atomic weight and to fix uniquely the place of radium in the periodic table as a homologue of barium.

Today, 25 isotopes, all radioactive, have been identified (mass numbers between 206 and 230). Of these, four exist in nature as daughter products of uranium and of thorium, and are:

	Half-Life
^{223}Ra (actinium X)	11.41 days, α-emitter
^{224}Ra (thorium X)	3.64 days, α-emitter
^{226}Ra (radium)	1622 years, α-emitter
^{228}Ra (mesothorium 1)	6.7 years, β-emitter

The isotope ^{226}Ra generally called radium, which is the one that P. and M. Curie discovered, belongs to the natural radioactive family of uranium (see Table R-5). Radium 228 is a daughter of ^{232}Th. These two isotopes are the most abundant in nature. They are present in the minerals of uranium and thorium, respectively, in radioactive equilibrium in the proportions by weight: ^{226}Ra/U = 3.4 × 10^{-7} and ^{228}Ra/Th = 4.7 × 10^{-10}. The quantity of radium in the lithosphere can be estimated at approximately 3–10^7 metric tons. The radium content is on the order of 10^{-12} g/g in rocks, only 3 × 10^{-13} g/g in rocky meteorites, and less if not none in iron meteorites. In rivers and in the oceans, its concentration is on the order of 10^{-13} g/l. It is present in plants in the proportion of approximately 10^{-14} g/g and in animals 10^{-15} g/g.

Both isotopes, ^{226}Ra and ^{228}Ra, have sufficiently long half-lives to be extracted directly from uranium and thorium minerals. This extraction was very rapidly done on an industrial scale from uranium minerals: pitchblende, autunite, carnotite, betatite to satisfy the needs of medicine and technology. The first refinery for the preparation of radium was created in France in 1904 by Armet de Lisle. Other refineries were then installed in Austria, England, Germany, and the United States. But since 1922, the majority of the radium production was secured by Belgium which oper-

TABLE R-5. *Uranium Radioactive Family*

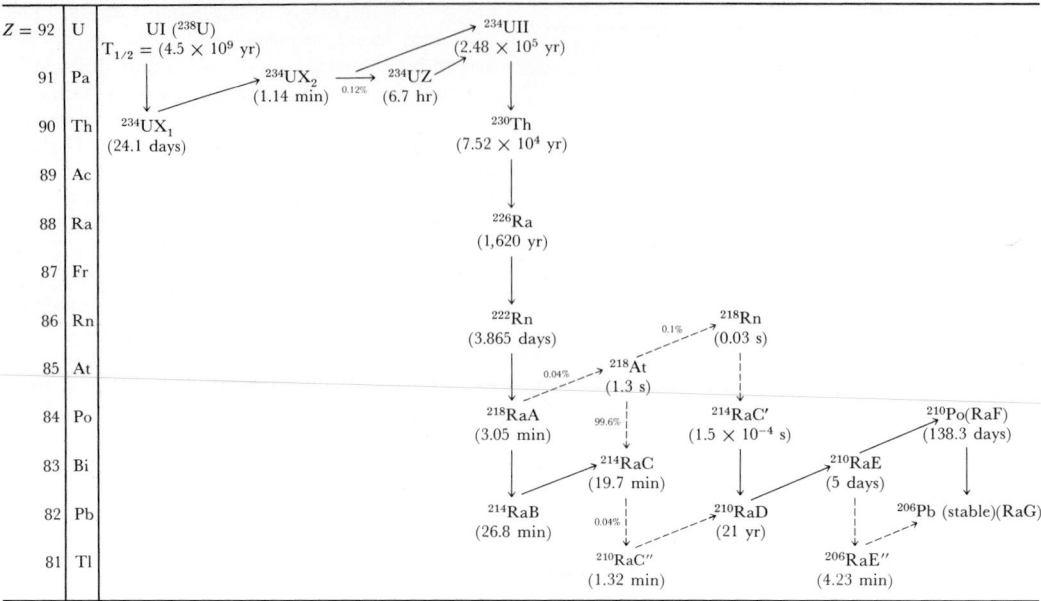

ated the Oolen pitchblende refinery of the Congo. Before World War II, the world store of radium rose to about 800 g. Since the use of uranium as a nuclear fuel, radium has become a subproduct of the preparation of uranium. Artificial radioactive elements generally have supplanted the former major applications of radium.

The principle of processing uranium minerals was established by P. and M. Curie. After acid attack, radium can be precipitated in the sulfate state by being carried down by the sulfate of barium and/or lead, elements usually present in the minerals, and the sulfates of which are isomorphs of $RaSO_4$. Then, the lead sulfate is made soluble by the action of a sodium chloride solution; then the radium-containing barium sulfate is converted into carbonates by boiling with a sodium carbonate solution. The radium can be separated from the barium by fractional crystallization of the chlorides, the bromides, or preferably the chromates. The coefficient of fractionalization at 20°C is, respectively, 4.5, 10, and 15 for the aforementioned salts. The radium-barium separation also can be effected on cation resin, the radium being absorbed more than the barium. However, this method is not convenient for large quantities of radium whose radiation changes the resin.

Radium 226 disintegrates by alpha emission of 4.78 MeV giving rise to a radioactive gas, radon 222. With its half-life of 1,622 years, 1 g of this radioelement emits 3.62×10^{10} alpha particles per second. This rate of disintegration was taken by definition to be equal to 1 Ci until 1950. Since that date, in order to be free from the precision with which the half-life of ^{226}Ra can be determined, the definition of the curie has been the quantity of any radioelement whose number of disintegrations per second is 3.700×10^{10}. P. Curie and A. Laborde were the first (1903) to observe the release of heat by radium salts. For 1 g of radium in balance with its daughters, this release of heat, which is due to the disintegration from the kinetic energy of alpha, beta, and gamma rays, is close to 140 cal/hr.

According to its position in group II of the periodic table, the electron structure of radium is: $[Rn] + 7s^2$.

The chemical properties of radium are very close to those of barium, its homologue. Indeed, their ionic radii are close to the point of the lanthanide contraction, Ra^{++}: 1.39 Å; Ba^{++}: 1.33 Å. Radium is the most electropositive of the alkaline-earth elements. The normal potential, calculated from the pair Ra^{++}/Ra, is $E_H = -2.92$ V. The ionization potentials corresponding to the extraction of the first and second valence electron are 5.262 and 10.099 eV, respectively.

The first preparation of the metal was performed by M. Curie and A. Debierne by electrolyzing a chloride solution of 106 mg on a mercury cathode. It forms an amalgam which is distilled under reduced hydrogen pressure to isolate the metal. It can also be obtained in an impure form by thermal decomposition of the nitride $Ra(N_3)_2$ between 180–250°C in a vacuum. F. Wiegel and A. Trinkl (1967) prepared the metal by reduction of 100 to 200 g of RaO by aluminum under a deep vacuum.

The metal is a brilliant white when it has just been prepared, but blackens in air for it probably forms a nitride. The element melts at 700°C. An isomorph of barium, radium crystallizes in the cubic-centered system: $a_0 = 5.148 \pm 0.015$ Å. For a coordination number 8, the atomic radius is 2.23 Å, and the specific weight is 5.50 g/cm^3.

As all the alkaline-earth elements, radium has only a 2$^+$ valence. Radium salts are isomorphs of the corresponding barium salts. They are almost all white, but color on ageing under the effect of their own radiation into yellow or violet. They spontaneously emit a blue light and color a flame carmine red. The sulfate, carbonate, chromate, and iodate are extremely insoluble in water—less soluble than those of barium. Specifically, 100 g of water at 20°C dissolves only 2.1×10^{-4} g of radium sulfate, making it the least soluble of all the sulfates. The precipitation of radium sulfate is inhibited by the presence of ethylenediamine tetraacetic acid, which complexes the radium. The very slight solubility of the chloride and of the nitrate of radium in concentrated solutions of the corresponding acid is a property that facilitates certain separations.

If in weighable quantity, ^{226}Ra can be determined by gravimetry. However, its radioactivity is usually measured, or that of one of its daughters, in radioactive equilibrium. Thus, the radium contained in a sealed ampul can be determined by comparison with a standard by measuring the penetrating gamma rays of the derivative radium C (^{214}Bi) in equilibrium after an accumulation of approximately one month. Marie Curie (1911) prepared an international standard filed with the International Bureau of Weights and Measures in Paris which consisted of a glass ampul containing 21.99 mg of anhydrous radium chloride. This standard was replaced in 1934 by a new international standard of 22.23 ± 0.02 mg of RaCl, manufactured by O. Hönigschmid. In a thin layer, radium is determined by the measurement of its alpha radiation of 4.78 MeV. Finally, a very sensitive method permitting determination of 10^{-14}–10^{-16} g of Ra in samples of several grams consists of liberating radon gas. This is possible provided the sample can be put into solution or melted so as to effect the liberation of radon. It is allowed to accumulate for several hours in an ionization chamber or a scintillation counter and thus the radioactivity of this gas can be measured.

The toxicity of radium is due to the emission of alpha, beta, and gamma rays by the element and its derivatives. In an organism, radium primarily is fixed to the bone marrow and the skeleton. The maximum quantity tolerable in the human body is 10^{-9} g. The release of the emanation considerably increases the dangers of handling radiferous preparations.

Some of the former applications for radium, now essentially replaced by less costly and often less dangerous radioelements, include curietherapy and gammatherapy, preparation of luminous paints through incorporation of the luminescent zinc sulfide, and production of neutrons mixed with beryllium powder.

Radium can be used as a raw material for the synthesis of weighable quantities of ^{227}Ac through irradiation with thermal neutrons in a nuclear reactor. The cross section of the process ^{226}Ra (n, γ) ^{227}Ra $\xrightarrow{\beta}$ ^{227}Ac is 23 barns (23×10^{-24} cm^2).

REFERENCES

Bagnall, K. W.: "Chemistry of the Rare Radioelements," Butterworth, London, 1957.

Bouissières, G.: "Radium, dans le Nouveau traité de Chimie Minérale," Institut de Physique Nucléaire, l'Université de Paris, 1958.

Joliot-Curie, I.: "Radioéléments naturels," Ed. Hermann, Paris, 1946.

—G. Bouissières, *The University of Paris*

Rare-Earth Elements and Metals Fifteen elements (at. no. 57–71 and also usually including 39) make up this series. The series also is known as the lanthanon, lanthanide, or lanthanoid elements; the cerium and yttrium group elements; the fraternal fifteen; inner transition elements; reactive and refining elements; and mischmetal.

Symbol	Mixtures of	Empirical* Mol. Wt.
RE or Ln	Rare-earth elements	140
RE$_2$O$_3$·Ln$_2$O$_3$,REO	Rare-earth oxides	164, 328
REM or MM	Rare-earth metals	140

*Based on the average distribution of rare-earth elements found in minerals.

These elements are in group III, period 6, and generally are shown in a periodic system of their own below the periodic table. Because their names are unfamiliar to many, their symbols are given along with a guide to pronunciation:

Element	Symbol	Atomic No.	Pronunciation
Cerium	Ce	58	se′ri-um
Dysprosium . . .	Dy	66	dys-pro′si-um
Erbium	Er	68	er′bi-um
Europium	Eu	63	u-ro′pi-um
Gadolinium . . .	Gd	64	gad′o-lin′i-um
Holmium	Ho	67	hol′mi-um
Lanthanum . . .	La	57	lan′tha-num
Lutetium	Lu	71	lu-te′shi-um
Neodymium . . .	Nd	60	ne′o-dim′i-um
Praseodymium . .	Pr	59	pra-seo-dim′i-um
Promethium . . .	Pm	61	pro-me′thi-um
Samarium	Sm	62	sa-ma′ri-um
Terbium	Tb	65	tur′bi-um
Thulium	Tm	69	thu′li-um
Ytterbium	Yb	70	i-tur′bi-um
Yttrium	Y	39	it′ri-um

Discovery and Development: In 1787, an unusual black mineral, identified some years later as containing yttrium and *rare-earth oxides,* was noted by C. A. Arrhenius in a quarry near Ytterby, Sweden. By 1869, when Mendeleev proposed the periodic ordering of the elements, only 4 of these elements had been discovered; his theory said that 12 more should exist. References contain a plethora of accounts of scientists in several European countries who contributed to their eventual identification. After 1907, when lutetium was isolated, only element 61 remained to be found; 40 years later, in 1947, scientists at the Atomic Energy Commission (AEC) laboratory at Oak Ridge, Tenn., produced atomic number 61 from uranium fission products and named it promethium, Pm. No stable isotopes of Pm have been found in the earth's crust.

By 1947, however, AEC scientists had discovered that when a uranium atom fissions, the rare-earth elements are important fission products. This created an intense interest within the AEC to develop fast and efficient methods of separating these elements from their minerals and from each other. Laboratories at Ames, Iowa, and Oak Ridge, Tenn., soon showed that separation by organic ion-exchange resins is practical.

Until that time, knowledge of the individual group III, period 6 elements was limited to the research laboratory.

These 15 elements are alike in that they share a valence of 3 and have a very strong affinity for oxygen. It was soon established, however, that they almost form a periodic system of their own. Differences in the fundamental properties of each element such as atomic (or ionic) size and valence were discovered (see Table R-6).

Natural mixtures of RE elements have been used commercially since early in the century. Mischmetal (MM) is the source of the hot spark in lighter flints. Mixed RE fluorides are burned in the cores of carbon electrodes to create the intense sunlike illumination required by motion-picture projectors and searchlights. Mixed RE oxides (REO) are used to grind and polish almost all optical lenses and TV faceplates.

Soon after 1947, it was discovered that REM effectively controls the shape of carbon in normally brittle cast iron, resulting in ductile or nodular iron, a new metallurgical application at the time.

During the 1950s, interest in several of the pure elements (Eu, Gd, Dy, Sm, Er) was stimulated because they have the highest thermal-neutron-absorption properties. Yttrium metal was fabricated into tubing and mill products because it is almost transparent to thermal neutrons and has a unique stability at high temperature in contact with liquid uranium, potassium, and sodium. Nuclear aircraft and submarine propulsion programs were the main impetus for these efforts.

Early in the 1960s mixtures of the RE were incorporated with synthetic molecular-sieve catalysts, resulting in increased petroleum-refining efficiency. In 1964, a new red phosphor for color television was discovered. Relatively large quantities of highly purified europium and yttrium oxides were needed as commercial color TV production started immediately.

Permanent magnets, having properties several times superior to any other known materials, were invented in 1967. The metals praseodymium, yttrium, samarium, lanthanum, and cerium are precisely alloyed with five atoms of cobalt, $RECo_5$. This new family of permanent-magnet materials is expected to alter many aspects of power generation and electronic communication.

During the 1960s, the RE metals were established as reactive and refining metals in the iron and steel industry. As alloying elements, lanthanum and yttrium improve the high-temperature oxidation and corrosion properties of superalloys. REM is more effective than calcium, magnesium and aluminum in refining ferrous and nonferrous metals.

The Institute for Atomic Research sponsors a Rare-Earth Information Center at Iowa State University, Ames, Iowa, which provides a comprehensive service to science and industry by cataloging the vast amount of technical information generated about these elements each year.

Occurrence and Sources: RE mineral sources exist in many parts of the world; the overall potential supply is unlimited. As a group they rank fifteenth in abundance, somewhat more plentiful than zinc. Cerium, yttrium, lanthanum, and neodymium rank twenty-eighth, thirtieth, thirty-first, and thirty-fifth of the 83 naturally occurring elements in the earth's crust, each more available than lead, tin or molybdenum.

RE minerals are generally classified as sources for *light* (La through Gd) or *heavy* (Y plus Tb through Lu) elements. Most minerals, however, contain some of all 15 elements. (See Table R-8 for typical mineral distributions.)

Until 1964, monazite, a thorium-RE phosphate, $REPO_4Th_3(Po_4)_4$, was the main source for the RE elements: Monazite is the source for thorium and is often a coproduct with rutile, ilmenite, and zircon (heavy mineral sand) mining operations. Australia, India, Brazil, Malaysia and the United States are active sources. India and Brazil supply a mixed RE-chloride compound after thorium is removed chemically from monazite.

Bastnasite, a rare-earth fluocarbonate mineral, $REFCO_3$, is a primary source for light RE. Since 1965, an open-pit resource at Mt. Pass, Calif., has supplied about two-thirds of world requirement for REO. Proven reserves at this source are 2.5 million tons REO content.

The main source for yttrium and heavy-RE is a by-product of uranium mining in the Elliott Lake region, Ontario. Some xenotime, found in Malaysia, is processed in Japan and Europe.

Processing: Crushed and finely ground bastnasite containing about 70% REO is roasted under oxidizing conditions to convert soluble trivalent Ce compounds to insoluble tetravalent CeO_2. The roasted product is leached with HCl, which dissolves the remaining RE (La, Pr, Nd, Sm, Eu, Gd), leaving behind a concentrated cerium product. The solution is passed through liquid-liquid organic solvent-extraction (SX) cells, resulting in a primary separation of La-Nd-Pr from Sm-Eu-Gd. Further SX separates a pure lanthanum solution and a concentrated Nd-Pr solution, which another SX circuit separates. Europium is reduced to a divalent state in solution and precipitated. A final SX system separates and purifies gadolinium and samarium. Pure elements are usually precipitated as oxalates and calcined to oxides.

Natural flotation-concentrated bastnasite at 70% REO may be directly chlorinated in the presence of carbon and heat to produce anhydrous $RECl_3$. A caustic soda metathasis reaction to exchange the fluocarbonate, hydrochloric acid leach, and evaporation produces a hydrated $RECl_3 \cdot 6H_2O$. The chlorides of the mixed RE are produced in large quantity and used for the production of mischmetal, in catalysts, and as feedstock for solvent extraction and ion-exchange separations.

Monazite, containing about 55% REO and 5% thorium, is treated by either of two methods: finely ground particles are (1) leached with hot H_2SO_4, which dissolves thorium and RE, leaving an insoluble phosphate residue, or (2) reacted with hot caustic (NaOH), which dissolves the phosphate, creating a solution of trisodium phosphate which may be recovered as a by-product. The thorium and RE-hydrate cake is then dissolved in H_2SO_4. Thorium sulfate is selectively precipitated by pH adjustment. Separation of the other RE in solution is usually completed by selective absorption on ion-exchange resins and elution from ion-exchange columns.

After thorium is removed from the H_2SO_4 solution, the RE remaining are precipitated using sodium hydroxide forming a double salt, $NaRESO_4 \cdot xH_2O$, known as *pink salts* because of their characteristic color. This salt is dissolved in HCl, treated to remove impurities, and evaporated until the hydrated $RECl_3 \cdot 6H_2O$ can be cast.

The Canadian yttrium-heavy RE concentrate is leached with HNO_3, which brings all RE into solution. SX separates yttrium from the other heavy RE, each of which can eventually be separated by further SX. Xenotime is leached with hot H_2SO_4 and separation of yttrium and the heavy RE completed in ion-exchange columns.

The liquid-liquid organic-solvent extraction cycle is complete within 5–10 days and is a continuous process. The resin ion-exchange cycle requires 60–90 days and is a batch process. Both processes result in pure RE oxides and chemicals.

Metal Production: Commercial production of mischmetal is produced by electrolysis. The most widely used process starts with dehydrated RE chloride produced from monazite or bastnasite. It is important to keep RE oxychlorides from forming during the dehydration process, which is done in air, in vacuum, or by reverse hydrolysis with ammonium chloride.

TABLE R-6. *Summary of Atomic and Thermal Properties of Rare-Earth Elements**

Atomic number Symbol Element	39 Y Yttrium	57 La Lanthanum	58 Ce Cerium	59 Pr Praseodymium	60 Nd Neodymium	61 Pm Promethium
Estimated abundance:						
ppm	33	30	60	8.2	28	0
g/ton	28–70	5–18	20–46	3.5–5.5	12–24	0
Atomic constants:						
Atomic weight, (CN = 12)	88.91	138.91	140.12	140.91	144.24	(145)
Metallic radius, Å, (CN = 12)	1.773	1.877	(+3) 1.846 (+4) 1.672	1.828	1.822	1.809
Volume, cm³/g atom.	18.99	22.53	(+3) 21.43 (+4) 15.92	20.81	20.60	20.24
Density,						
g/cm³	4.457	6.166	6.773	6.772	7.004	7.220
lb/in.³	0.162	0.223	0.245	0.245	0.253	0.261
Crystal structure at 25°C	hcp	dhcp	fcc	dhcp	dhcp	dhcp
Unpaired 4f electrons	0	0	1	2	3	4
Number of isotopes:						
Natural	1	2	4	1	7	0
Artificial	14	19	15	14	7	15–18
Lattice constants, Å:						
a	3.650	3.772	5.1601	3.6725	3.659	3.65
c	5.741	12.144	—	11.8354	11.799	11.65
Ionic radius, Å:						
+2						
+3	0.893	1.061	1.034	1.013	0.995	0.978
+4			0.92	0.90		
Color of 3⁺ ion (in solution)	Colorless	Colorless	Colorless	Green	Reddish violet	Pink
Electronegativity	1.177	1.117	(+3) 1.123 (+4) 1.43	1.130	1.134	1.139
Absorption bands, 3⁺ ion, Å	None	None	2105 2220 2380 2520	4445 4690 4822 5885	3540 5218 5745 7395 7420 7975 8030 8680	5485 5680 7025 7355
Thermal properties:						
Melting point:						
°C	1520	920	798	934	1016	1124
°F	2768	1688	1468	1713	1879	2055
Boiling point at 1 atm.:						
°C.	3338	3457	3427	3512	3068	(2460)
°F	6040	6255	6201	6354	5554	(4460)
Heat of fusion ΔH_f, kcal/g atom .	2.724	1.482	1.305	1.646	1.705	(1.94)
Heat of sublimation ΔH_s at 25°C, kcal/g atom .	101.287	103.084	101.146	85.286	78.507	(64)
Heat capacity C_p at 25°C, cal/(g atom)(°C) . . .	6.34	6.48	6.44	6.56	6.56	(6.50)
Coefficient of expansion, per °C × 10⁻⁶	10.8	4.9	8.5	4.8	6.7	(9.0)
Nuclear properties:						
Thermal neutron capture, barns/atom	1.31	8.9	0.73	11.6	50	

*Table compiled by Molybdenum Corporation of America, White Plains, N.Y. (Joseph G. Cannon); edited by Rare-Earth Information Center, Institute for Atomic Research, Iowa State University, Ames, Iowa (Karl A. Gschneidner, Jr. and N. Kippenhan). Data from S. R. Taylor, Abundance of Chemical Elements in the Continental Crust: A New Table, *Geochim. Cosmochim. Acta,* vol. 28, pp. 1273–1285, 1964; E. T. Teatum, et al., Compilation of Calculated Data Useful in Predicting Metallurgical Behavior of Elements in Binary Alloy Systems, *Univ. Calif., Los Alamos Sci. Lab. Rep.* LA-4003, pp. 11–12, Dec. 24, 1968; Clifford A. Hampel, "Rare Metals Handbook," 2d ed., chaps. 1 and 35, Reinhold Company, New York, 1961; O. A. Songina, "Rare Metals: Scandium, Yttrium, Lanthanide

62 Sm Samarium	63 Eu Europium	64 Gd Gadolinium	65 Tb Terbium	66 Dy Dysprosium	67 Ho Holmium	68 Er Erbium	69 Tm Thulium	70 Yb Ytterbium	71 Lu Lutetium
6.0	1.2	5.4	0.9	3.0	1.2	2.8	0.5	3.0	0.5
4.5–7	0.14–1.1	4.5–6.4	0.7–1	4.5–7.5	0.7–1.2	2.5–6.5	0.2–1	2.7–8	0.8–1.7
150.35	151.96	157.25	158.92	162.50	164.93	167.26	168.93	173.04	174.97
	(+2) 2.041							(+2) 1.939	
1.802	(+3) 1.798	1.801	1.783	1.775	1.767	1.758	1.747	(+3) 1.741	1.735
								(+2) 24.82	
19.95	(+2) 28.93	19.91	19.30	19.03	18.78	18.49	18.14	(+3) 17.98	17.79
7.536	5.253	7.898	8.234	8.540	8.780	9.045	9.318	6.972	9.835
0.272	0.190	0.285	0.298	0.309	0.317	0.327	0.337	0.252	0.356
rhom	bcc	hcp	hcp	hcp	hcp	hcp	hcp	fcc	hcp
5	6	7	6	5	4	3	2	1	0
7	2	7	1	7	1	6	1	7	2
11	16	11	17	12	18	12	17	10	14
3.626	4.580	3.636	3.604	3.592	3.578	3.559	3.5375	5.483	3.505
26.18	—	5.782	5.698	5.655	5.626	5.595	5.558	—	5.553
1.11	1.09						0.94	0.93	
0.964	0.950	0.938	0.923	0.908	0.894	0.881	0.869	0.858	0.848
			0.84						
Yellow	Pale pink	Colorless	Almost colorless	Yellow	Pink	Reddish violet	Green	Colorless	Colorless
1.145	(+2) 0.98	1.160	1.168	1.176	1.184	1.192	1.200	(+2) 1.02	1.216
	(+3) 1.152							(+3) 1.208	
3625 3745	3755 3941	2729 2733	3694 3780	3504 3650	2870 3611	3642 3792	3600 6825	9750	None
4020		2754 2756	4875	9100	4508 5370	4870 5228	7800		
					6404	6525			
1073	822	1312	1356	1410	1472	1527	1545	815	1663
1963	1512	2394	2473	2570	2682	2781	2813	1499	3025
1791	1597	3266	3223	2562	2695	2863	1947	1194	3395
3256	2907	5911	5833	4644	4883	5185	3537	2181	6143
2.061	2.202	2.403	2.580	2.643	2.911	4.757	4.025	1.830	4.457
49.257	42.5	95.347	93.374	70.038	72.330	76.086	55.787	36.473	102.245
7.06	6.48	8.86	6.91	6.72	6.49	6.71	6.46	6.39	6.40
10.4	32	6.4	7.0	8.6	9.5	9.2	11.6	25.0	12.5
5,800	4,300	46,000	46	940	64	160	125	37	80

and Actinides," chap. 6, trans. from Russian (1970), 3d ed. (1964), U.S. Dept. of Interior and The National Science Foundation, Washington, D.C.; Karl A. Gschneidner, Jr., "Solid State Physics," vol. 16, "Physical Properties and Interrelationships of Metallic and Semimetallic Elements," pp. 275–426, Academic Press, New York, 1964; Clifford A. Hampel, "The Encyclopedia of the Chemical Elements," Reinhold Company, New York, 1968; R. Hultgren, R. L. Orr, and K. K. Kelley, supplement to "Selected Values of Thermodynamic Properties of Metals and Alloys," Wiley, New York, 1963; Data from Department of Mineral Technology and Lawrence Radiation Laboratory, The University of California, Berkeley, Calif. (data and revisions published periodically).

TABLE R-7. *Summary of Mechanical, Electrical, and Oxide Properties of Rare-Earth Elements**

Atomic number Symbol Element	39 Y Yttrium	57 La Lanthanum	58 Ce Cerium	59 Pr Praseodymium	60 Nd Neodymium	61 Pm Promethium
Mechanical properties:†						
Yield strength:						
kg/mm^2	37.2	19.0	11.6	20.2	16.1	N.A.
1,000 psi	53.0	27.0	16.5	28.8	22.9	N.A.
Elongation, %	25	8	24	10	11	N.A.
Tensile strength:						
kg/mm^2	46.0	22.5	15.4	21.9	21.1	N.A.
1,000 psi	65.5	32.0	21.9	31.2	30.0	
Vickers hardness, 10-kg load, kg/mm^2	38	37	24	37	35	N.A.
Elastic properties (values in parentheses estimated):						
Compressibility, cm^2/kg × 10^{-6}	2.68	4.04	4.10	3.21	3.00	(2.78)
Shear modulus, kg/cm^2 × 10^6	0.263	0.152	0.122	0.138	0.148	(0.170)
Young's modulus, kg/cm^2 × 10^6	0.661	0.387	0.306	0.332	0.387	(0.430)
Poisson's ratio	0.258	0.288	0.248	0.305	0.306	(0.278)
Electrical properties at 25°C:						
Resistivity, μΩ-cm	53.0	56.8	75.3	68.0	64.3	300.0
Hall coefficient, V-cm/(A)(Oe) × 10^{12}	−0.77	−0.8	+1.81	+0.71	+0.97	N.A.
Work function, eV	3.23	3.3	2.84·	2.7	3.3	(3.07)
Magnetic properties:						
Moment, theoretical for 3$^+$ ion, Bohr magnetons .	0	0	2.5	3.6	3.6	N.A.
Susceptibility, emu/g atom × 10^6	191	101	2430	5320	5650	N.A.
Curie temperature, °C	None	None	None	None	None	N.A.
Néel temperature, °C	None	None	−260.6	None	−253	N.A.
Metal oxide:						
Formula	Y$_2$O$_3$	La$_2$O$_3$	CeO$_2$	Pr$_6$O$_{11}$	Nd$_2$O$_3$	Pm$_2$O$_3$
Color.	White	White	Buff	Black	Light blue	White
Molecular weight	225.81	325.82	172.12	1021.79	336.48	342.
Melting point:						
°C	2410	2300	2800	2040	2310	2350
°F	4370	4172	5072	3704	4190	4262
Density, g/cm^3	5.03	6.58	7.22	6.83	7.31	7.60

*Table compiled by Molybdenum Corporation of America, White Plains, N.Y. (Joseph G. Cannon); edited by Rare-Earth Information Center, Institute for Atomic Research, Iowa State University, Ames, Iowa (Karl A. Gschneidner, Jr. and N. Kippenhan). Data from S. R. Taylor, Abundance of Chemical Elements in the Continental Crust: A New Table, *Geochim. Cosmochim. Acta*, vol. 28, pp. 1273–1285, 1964; E. T. Teatum, et al., Compilation of Calculated Data Useful in Predicting Metallurgical Behavior of Elements in Binary Alloy Systems, *Univ. Calif., Los Alamos Sci. Lab. Rep.* LA-4003, pp. 11–12, Dec. 24, 1968; Clifford A. Hampel, "Rare Metals Handbook," 2d ed., chaps. 1 and 35, Reinhold Company, New York, 1961; O. A. Songina, "Rare Metals: Scandium, Yttrium, Lanthanide and Actinides," chap. 6, trans. from Russian (1970), 3d ed. (1964), U.S. Dept. of Interior and The National Science

The mixed RE chloride is fused in an iron, graphite, or ceramic crucible with the aid of electrolyte mixtures made up of potassium, barium, sodium, or calcium chlorides. Figure R-1 shows a schematic of a reduction cell.

Carbon anodes are immersed in the molten salt. Electrically, the crucible is connected as the cathode. As direct current flows, molten mischmetal builds up in the bottom of the crucible. These cells may be operated in semicontinuous production by ladle removal of the metal into molds, where it solidifies under a protective layer of electrolyte. The solidified electrolyte is readily separated from the cast metal after cooling.

The electrolysis-bath temperature is 800–900°C. Usual operation is 2,000–2,200 A at 12–15 V. The reaction products are mainly gaseous chlorine and carbon compounds, which are drawn off and absorbed in a suitable system to control air contamination.

A recent commercial practice for producing mischmetal and 99.9% pure cerium, lanthanum,

62 Sm Samarium	63 Eu Europium	64 Gd Gadolinium	65 Tb Terbium	66 Dy Dysprosium	67 Ho Holmium	68 Er Erbium	69 Tm Thulium	70 Yb Ytterbium	71 Lu Lutetium
11.8	N.A.	27.4	N.A.	38.0	22.6	28.8	N.A.	6.7	N.A.
16.2		39.0		47.0	32.1	41.0		9.5	
3	N.A.	8	N.A.	6	5	11	N.A.	6	N.A.
12.7	N.A.	39.7	N.A.	43.6	26.4	35.6	N.A.	7.3	N.A.
18.0		56.5		62.0	37.5	50.8		10.4	
45	17	57	46	42	42	44	48	21	77
3.34	6.66	2.56	2.46	2.55	2.47	2.39	2.47	7.39	2.39
0.129	(0.060)	0.227	0.233	0.259	0.272	0.302	(0.310)	0.071	(0.345)
0.348	(0.155)	0.573	0.586	0.644	0.684	0.748	(0.770)	0.182	(0.860)
0.352	(0.286)	0.259	0.261	0.243	0.255	0.238	(0.235)	0.284	(0.233)
92.0	81.3	134	116	91	94	86	79	28	59
−0.2	N.A.	−4.48	−4.3	−2.7	−2.3	−0.34	−1.8	+3.77	−0.54
3.2	(2.54)	(3.07)	(3.09)	(3.09)	(3.09)	(3.12)	(3.12)	(2.59)	(3.14)
1.5	3.5	7.95	9.7	10.6	10.6	9.6	7.6	4.5	0
1275	33,100	356,000	193,000	99,800	70,200	44,100	26,100	71	17.9
None	None	+17	−53	−185	−254	−253	(−241?)	None	None
−258	−165	None	−43	−97	−143	−188	−216	None	None
Sm_2O_3	Eu_2O_3	Gd_2O_3	Tb_4O_7	Dy_2O_3	Ho_2O_3	Er_2O_3	Tm_2O_3	Yb_2O_3	Lu_2O_3
Cream	Pale pink	White	Dark brown	Cream	Cream	Rose	Light green	White	White
348.70	351.92	362.50	747.69	373.00	377.86	382.52	385.87	394.08	397.94
2330	2330	2400	2340	2390	2395	2400	2390	2410	2470
4226	4226	4352	4244	4334	4343	4352	4334	4370	4478
7.11	7.29	7.61	7.87 (Tb_2O_3)	8.16	8.41	8.65	8.90	9.21	9.41

Interrelationships of Metallic and Semimetallic Elements," pp. 275–426, Academic Press, New York, 1964; Clifford A. Hampel, "The Encyclopedia of the Chemical Elements," Reinhold Company, New York, 1968; R. Hultgren, R. L. Orr, and K. K. Kelley, supplement to "Selected Values of Thermodynamic Properties of Metals and Alloys," Wiley, New York, 1963; Data from Department of Mineral Technology and Lawrence Radiation Laboratory, The University of California, Berkeley, Calif. (data and revisions published periodically).

†Highest reported value for metal at room temperature after 10–50% reduction in area or annealed or as-cast; purity unknown.

N.A.—not available.

praseodymium and neodymium metals is by electrowinning from the oxide in a molten fluoride bath. The electrolyte mixture is about 50% of REF_3, 25% BaF_2, and 25% LiF. This mixture is heated externally in a carbon crucible. When molten at about 1000°C, carbon anodes are lowered into the electrolytic. A molybdenum or tungsten cathode is also submerged (see Fig. R-2).

Mixed REO, CeO_2, La_2O_3, Pr_6O_{11}, or Nd_2O_3 are fed to the cell at regular intervals. The process uses the principle that oxides are soluble to some extent in the fluorides of the RE elements. The cell operates at 8–10,000 A and 4–6 V. The metal ion coalesces at the submerged cathode, and metal is tapped from the bottom of the molten bath at regular intervals. The anodes are consumable, and the reaction product is CO_2. The purity of the metal produced depends on the purity of the oxide fed to the cell.

Properties: There is a greater difference in the properties of individual RE metals than of their oxides or chemicals (see Table R-7). Properties

TABLE R-8. *Typical Distribution of Active Mineral Sources of Rare Earths*

Reported as Oxides	Xenotime, Malaysia, %	U Residues, Canada, %	Monazite, Australia, %	Bastnasite, California, %
Lanthanum	0.5	0.8	20.2	32.0
Cerium	5.0	3.7	45.3	49.0
Praseodymium . .	0.7	1.0	5.4	4.4
Neodymium . . .	2.2	4.1	18.3	13.5
Samarium	1.9	4.5	4.6	0.5
Europium	0.2	0.2	0.05	0.1
Gadolinium	4.0	8.5	2.0	0.3
Terbium	1.0	1.2		
Dysprosium	8.7	11.2		
Holmium	2.1	2.6		
Erbium	5.4	5.5	2.0	0.1
Thulium	0.9	0.9		
Ytterbium	6.2	4.0		
Lutetium	0.4	0.4		
Yttrium	60.8	51.4	2.1	0.1
	100.0	100.0	100.0	100.0

SOURCE: NMAB *Rep.* 266, October 1970.

of the hydrated oxides and compounds show great similarity with respect to chemical processing.

The RE elements can be grouped into four families according to the arrangement of their $4f$ electron shells:

1. Y, La, Gd, and Lu are always trivalent and optically neutral (colorless). The $4f$ shells are empty (Y and La), full (Lu), or contain the maximum number of unpaired electrons (Gd).

2. Nd, Dy, Er, Ho, and Tm are also always trivalent but are optically and magnetically active. All have several unpaired $4f$ electrons and exhibit a variety of distinctive colors and electronic properties.

3. Sm, Eu, and Yb are trivalent but prefer to become divalent under certain conditions.

4. Ce, Tb, and Pr are sometimes trivalent but prefer to add oxygen and stabilize at a higher valence between 3 and 4.

As the RE elements add one electron to the $4f$

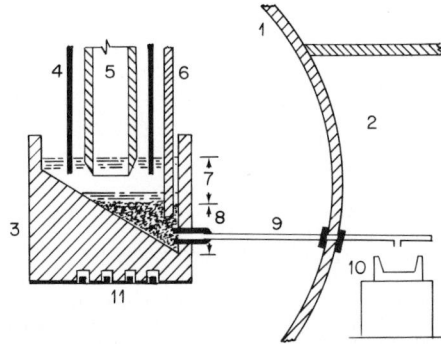

Fig. R-2. **Electrowinning reduction of the U.S. Bureau of Mines for production of pure rare-earth metals and mischmetal.** (1) Main chamber, (2) casting chamber, (3) graphite crucible, (4) cathode, (5) anode, (6), molybdenum rod, (7) liquid electrolyte, (8) molten cerium metal, (9) tapping pipe (molybdenum), (10) casting ladle, (11) heating elements.

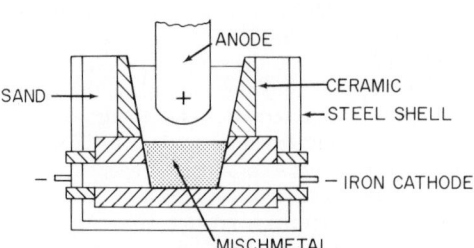

Fig. R-1. Electrolytic reduction cell for mischmetal.

shell (increase in atomic number), they exhibit the *lanthanide contraction*. This is shown in the table of properties, for, as the number increases, the ionic size decreases (the atom is compressed). Most of the physical and chemical properties plot on a smooth line. Europium (at. no. 63) and ytterbium (at. no. 70) are exceptions, however; they deviate in many properties because they are divalent in the metallic state.

Each element in the group is strongly cationic when compounding. Individual and mixed-RE cations combine with most inorganic anions and compound anions. Organic anion compounds of the RE are even more extensive. An infinite variety of materials seems possible; the properties of most still need to be defined.

REM additions strongly influence physical, electrical, and corrosion properties of many nonferrous metals and ferrous alloys. When used in alloys, they are considered *reactive* *e*lements. When used to getter oxygen, sulfur, hydrogen, and nitrogen, they are considered *refining* *e*lements.

New separation techniques developed within the last 25 years, i.e., liquid-liquid and solid-liquid organic ion exchange, make it possible to obtain each individual element in the form and purity needed. The price trend of pure RE elements and mixtures is lower as markets expand and diversify.

RE are classified as having a low acute-toxicity rating. The light RE metals are pyrophoric, the heavies and yttrium less so. Precautions must be taken to handle and store fine powders which can ignite spontaneously. Bulk RE metal forms a protective oxide layer rapidly but can ignite if heated in air to 160–200°C. The light metals are usually stored with a protective oil film, under an inert liquid, or in vacuum. The heavies and yttrium are stable in air at room temperature.

See also individual descriptions of each of the rare-earth elements.

REFERENCES

Becker, Joseph J.: Permanent Magnets, *Sci. Am.*, December 1970, pp. 92–100.

Callow, R. I.: "The Industrial Chemistry of the Lanthanons, Yttrium, Thorium and Uranium," Pergamon, London, 1967.

Eyring, L. (ed.): "Progress in the Science and Technology of the Rare Earths," Macmillan, New York, vol. 1, 1964; vol. 2, 1966.

Gschneidner, K. A.: "Rare Earth Alloys," Van Nostrand Reinhold, New York, 1961.

Gschneidner, Karl A., Jr.: Rare Earths: The Fraternal Fifteen, U.S. Atomic Energy Commission, Oak Ridge, Tenn., December 1964.

Hampel, Clifford A.: "The Encyclopedia of the Chemical Elements," Van Nostrand Reinhold, New York, 1968.

Hirschhorn, I. S.: Commercial Production of Rare Earth Metals by Fused Salt Electrolysis, *J. Met.*, vol. 20, pp. 19–22, March 1968.

Molybdenum Corporation of America: *Overview* 1–28, New York, 1968–1971.

Parker, John G., and Charles T. Baroch: The Rare-Earth Elements, Yttrium and Thorium, U.S. *Bur. Mines Publ.* IC8476, 1971.

Seitz, Frederick, and David Turnbull: "Solid State Physics," vol. 16, Academic, New York, 1964.

Songina, O. A.: Rare Metals, Izdatestvo "Metallurgiya," Moscow 1964; trans. J. Schmorak, 1970; U.S. Dept. of Commerce, Clearinghouse for Federal Scientific and Technical Information, Springfield, Va. 22151.

Spedding, F. H., and A. H. Daane (eds.): "The Rare Earths," Wiley, New York, 1961.

Stevenson, P. C., and W. E. Nervik, "The Radiochemistry of the Rare Earths, Scandium, Yttrium and Actinium," Lawrence Radiation Laboratory, Livermore, Calif., February 1961.

Strnat, K. S.: The Recent Development of Permanent Magnet Materials Containing Rare Earth Metals, *Tech. Rep.* AFML-RE69-299, University of Dayton, Dayton, Ohio, June 1970.

Topp, N. E.: "The Chemistry of the Rare-Earth Elements," American Elsevier, New York, 1965.

—Joseph G. Cannon, *Molybdenum Corporation of America, White Plains, N.Y.*

Raschig Ring (See **Absorption.**)

Rayon Unlike the synthetic fibers derived from synthetic polymers, rayon is produced from the natural polymer *cellulose.** The most widely used process for manufacturing regenerated cellulose fibers is the viscose-rayon process, which depends upon the solubilization of cellulose through the sodium xanthate ester. Another process of less commercial importance is the cuprammonium rayon process, which is based on solubilization of cellulose in ammonia through copper complexes. Viscose rayon is the oldest commercial man-made fiber, the process having been discovered in 1891 by Cross and Bevan in England. The growth and popularity of rayon are accounted for by many factors: (1) cellulose remains the cheapest polymer for fibers; (2) cellulose has the inherent advantage of moisture absorption and excellent dyeability; and (3) rayon-fiber properties can be varied widely to meet the requirements of many end products.

Chemically, cellulose is a linear polysaccharide having the basic chemical building unit of β-an-

---

*The U.S. Federal Trade Commission in 1952 defined rayon as "man-made textile fibers and filaments composed of regenerated cellulose, and yarn, thread, or textile fabric made of such fibers and filaments." Previously all cellulose-base, man-made fibers were referred to as rayon and included cellulose acetate.

hydroglucose which is illustrated in the structural formula for cellulose:

$$\beta\text{-Anhydroglucose Unit}$$

The degree of polymerization (DP) for cellulose is equivalent to $n + 2$ according to the structural formula. The average DP for some cellulose products is cotton 2,500, chemical cellulose (dissolving-grade wood pulp) 600–1,500, rayon 250–800. In all cellulose products, there is a distribution of molecular weight, and in rayon the molecular weight may cover a DP range of 50–2,000 units (8,000–320,000 molecular weight).

The cellulose molecule bristles with hydroxyl groups, OH^-, which accounts for the anionicity of rayon as well as its hydrophilic properties. Hydrogen bonding of the hydroxyl sites of adjoining cellulose chains accounts for the intermolecular forces and the consequent lateral and longitudinal strength of the cellulose structure. Since hydrogen bonds may be disrupted by water, which causes a swelling of the cellulose structure, rayon fibers are much weaker wet than dry. See also **Cellulose ester plastics, organic.**

The chemical reactivity of cellulose depends largely upon the hydroxyl groups present on cellulose. The normal properties of rayon can be modified through etherification or esterification of hydroxyl sites. Cross-linking reactions, which are commonly employed in durable-press treatments for cotton and rayon fabrics, depend on etherification of hydroxyl sites.

Structure: Like natural cellulose fibers, rayon consists of a network of morphological units consisting of fibril assemblies. Fibrillar interlinkages of molecules or bundles of molecules of lower order and higher accessibility tie the fibril assemblies together.

Although there are limitations to the control of the rayon fiber structure, the variables of the rayon process permit altering (1) the molecular length of the fiber-forming polymer molecule, (2) the degree of crystallinity, or perfection of order, (3) the crystallite size, (4) the size of the fibrillar aggregates which form the network structure, (5) the degree of orientation of the fibrillar aggregates and the interlinking molecular elements, and (6) the uniformity of packing, called the *lateral order distribution.*

The fibril assemblies present in rayon fibers are approximately 40 Å wide and have an irregular length, perhaps 300–1000 Å. From these data, it may be supposed that the cellulose molecular chains within the fibril consist of sequences of 60–200 glucose units. These molecular chains in the fibril may take on a parallel configuration and become so highly ordered as to form crystallites, while in adjacent areas the disorder of the molecules gives rise to amorphous structure. Many investigators have shown that a given cellulose chain may pass through both ordered and disordered regions.

The physical properties and some of the chemical properties of rayon are determined by the relative amounts of crystalline and amorphous cellulose and also by the size of the crystallites. Since the proportion of crystalline material is lower in rayon than in natural cellulose fibers, rayon fibers are more accessible and consequently more reactive chemically, exhibiting greater dye affinity and higher moisture regain.

Regenerated cellulose fibers have a gross morphology known as the *skin-core structure.* Differentiation of the skin and core structure, a measure of lateral order distribution, is found in cross sections of rayon fibers which have been subjected to suitable staining techniques, where dye receptivity and fastness are a measure of the cellulose crystallite sizes. The skin-core differentiation procedure may show the cross section as two phases, an outer area (skin) and an inner area (core) representing fine and coarse crystalline regions respectively.

The cross-sectional shape of viscose rayon fibers varies from round to kidney-bean, highly serrated, dumbbell, and elongated shapes. The shape of the cross section is controlled by the spinning-coagulation conditions and the shape of the spinnerette hole (see Fig. R-3). The cross-sectional shape of the fiber influences (1) the longitudinal and lateral tensile properties, (2) the packing properties of fibers in yarns and the consequent covering properties of fabrics, (3) the bending moduli, and (4) the optical properties.

Types of Rayon Fiber: Rayon is produced as *filament yarns* and *staple* and *tow.* Similar spinning processes are used for the filament and staple fibers of a given type (see Table R-9).

Physical Properties: A summary of typical properties for various types of rayon filament yarns and staples is given in Table R-10.

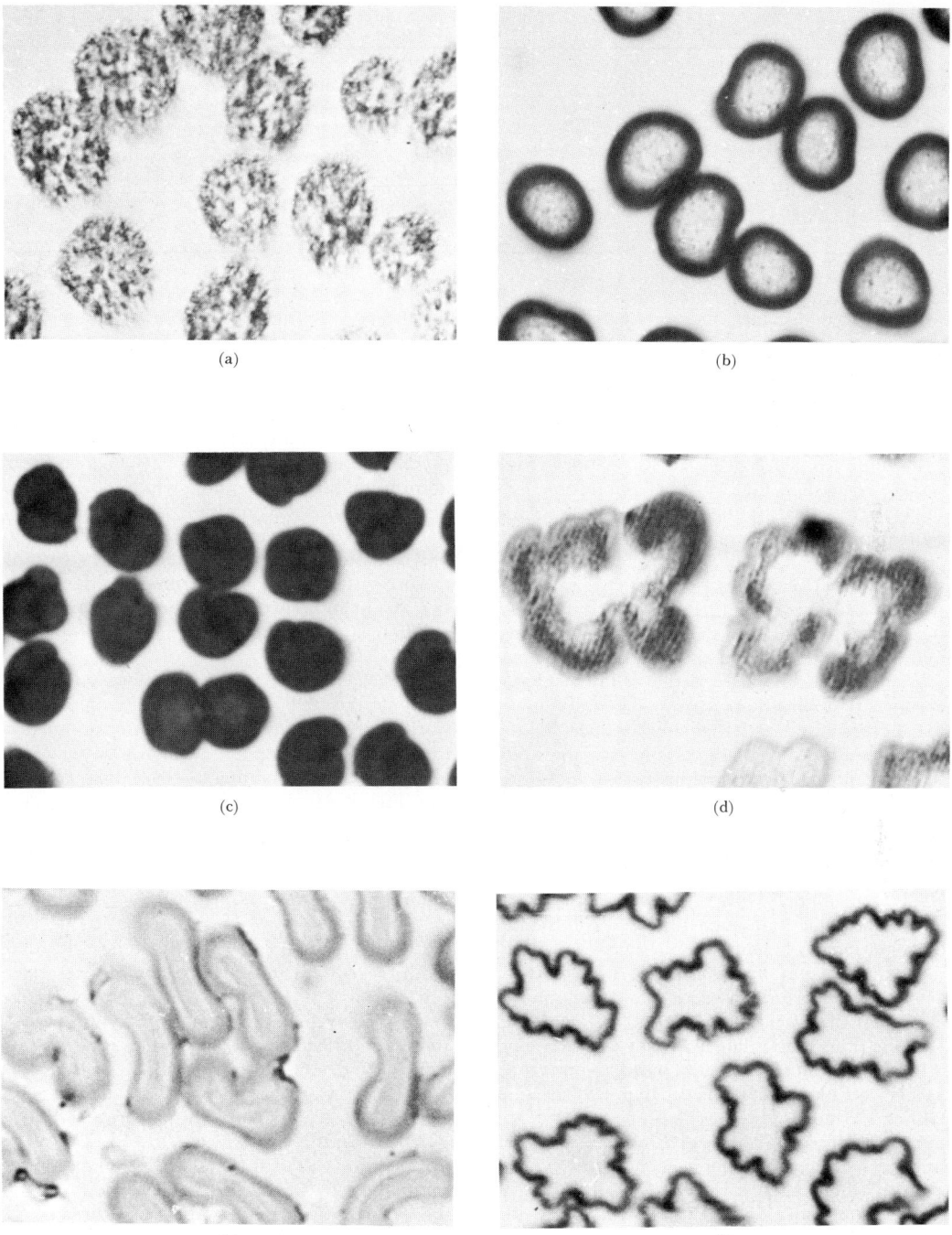

Fig. R-3. Typical rayon-fiber cross sections. (a) Regular rayon, skin-core structure, (b) high-wet-modulus rayon, skin-core structure, (c) crimped rayon, broken-skin structure, (d) high-tenacity rayon, all skin, (e) elongated cross section, (f) fire-retardant rayon.

TABLE R-9. *Types of Rayon Fiber*

Filament		Staple	
Description	Yarn Denier/Filament	Description	Denier/Filament
Regular tenacity, textile	75–100/10–150	Regular	0.75–40
Medium tenacity, textile	150–900/40–350	Crimped	1.5–15
High tenacity, textile	600–900/450–720	High-strength	1.5–3
Industrial.	1,100–3,300/980–3,000	Extra-strength	1.5–3
		High-wet-modulus . . .	1.0–5.5

Tenacity, conditioned and wet, can be varied over wide ranges as a function of the spinning process. The high-tenacity and high-wet-modulus rayons are used for apparel and domestic and industrial fabrics where high strength is a requirement. Unlike the synthetic fibers, rayons are very hydrophilic, which is reflected in relatively high *moisture regain* and high *water* of imbibition, i.e., the water retained by the fiber after soaking and centrifuging under standard conditions, reported as a percentage of the dry fiber weight. The relatively low wet tenacities for rayon fibers are attributed to the hydrophilicity of the rayon structure. Water plasticizes the amorphous regions; hydrogen bonds between the cellulose chains are weakened; and slippage of the cellulose chains occurs with application of stress. The ratio of wet to conditioned strength is greater for fibers having high crystallinity and orientation, e.g., the high-tenacity and high-wet-modulus rayons.

Uses: Filament and staple rayons are used in virtually every area where textile fibers find application. The largest consumption of rayon staple is in wearing-apparel fabrics, where it may be used 100% or combined with cotton, wool, polyester, nylon, and other man-made fibers. When used in combination with hydrophobic synthetic fibers, rayon contributes its hydrophilic properties to achieve the moisture absorption and distribution characteristics essential to comfortable clothing. Durable-press fabrics largely are designed around cellulose-polyester blends, and when rayon is used as the cellulose component, high-performance fabrics having unique *hand** properties are obtained. Filament rayon is used in such fabrics as linings, suiting, gabardines, bengalines, voiles, crepes, failles, taffetas, velvets, and plush. Filament rayon finds wide application in draperies, upholstery, and automobile seat covers, where

*The esthetic property of a fabric which defines its compliance, softness, and fullness.

durability combined with vibrant color is required.

Industrial uses of rayon include automobile tires, power-transmission belts, conveyor belts, and industrial hose, where strength and chemical and thermal stability are important. Sanitary hospital products use rayon because of its high moisture absorption, high purity, high whiteness, and freedom from lint. Rayon also is a principal fiber in the field of nonwovens, which are made by dry and wet-laid (paper) systems. Rayon is adaptable to both systems. Rayon contributes to the properties sought in nonwovens, including moisture absorption, strength, softness, nonelectrostatic qualities, and low cost.

Production Process: The viscose-rayon process accounts for 99% of the rayon manufactured worldwide. The rayon manufacturer obtains the raw material, cellulose, in the highly purified form of a pulp sheet containing 87–98% alpha cellulose (see Fig. R-4). Pulp in bulk or sheet form is treated with 17–19% NaOH to form alkali cellulose (an operation termed *steeping* or *mercerization*). Excess NaOH solution is pressed out, reducing the soluble part of the pulp, hemicellulose, and resinous impurities.

The alkali cellulose (30–35% cellulose; 15–17% alkali) is reduced to fluffy crumbs in a shredder. The crumbs are aged for a few minutes to more than 30 hr (depending upon end results desired and raw materials) under controlled conditions to reduce the degree of polymerization of the cellulose. The properly aged alkali cellulose next is reacted with CS_2 in a churn, or baratte. This is a batch operation with about a 2-hr cycle. The CS_2 esterifies the alkali cellulose, forming cellulose xanthate crumb, a deep-yellow product.

The cellulose xanthate crumb is charged with NaOH solution to a mixer or dissolver and is dispersed uniformly under controlled temperature, using agitators and attritors. The cellulose xanthate solution is called *viscose* and is an orange

TABLE R-10. *Physical Properties of Rayons*

Property	Filament Yarns			Staple Fibers			
	Regular Textile	High Tenacity	Cuprammonium	Regular Textile	High Tenacity	High-wet Modified Type	Modulus Polynosic Type
Tenacity, g/denier:							
Conditioned	1.5–2.9	3.0–5.0	1.8–2.3	2.0–2.9	3.0–4.5	4.5–5.0	4.0–6.0
Wet	0.7–1.9	1.9–3.2	1.1–1.3	1.2–1.9	2.0–2.8	2.8–3.3	3.0–4.8
Elongation, %:							
Conditioned	15–30	10–25	10–17	16–30	20–30	17–20	8.0–13
Wet	20–40	14–32	17–33	22–40	18–32	19–22	10.0–15
Wet modulus at 5% elongation, g/denier	. . .	. . .	. . .	3	3.0–5.0	10–12	15–21
Loop strength, g/denier*	1.0–1.7	1.8–3.6	2.3–3.2	0.3–1.0	1.2	1.2–1.5	0.6–1.0
Elastic modulus, g/denier	75	125	68				
Specific gravity	1.50–1.52	1.49–1.51	1.52–1.54	1.50–1.52	1.49–1.51	1.52	1.52–1.53
Moisture regain at 70°F and 60% RH, %	12.5	12.5–13.5	12.2	13.0	12.5–13.0	11.8	11.5–12.0
Water of imbibition, % of dry weight	90–100	70–80	90	110–120	75–85	75	65–70
Weight loss in 6% NaOH† at 20°C, 10 min	. . .	. . .	. . .	15–18	15	8	4–6

Thermal properties . . . . Thermal stability adequate for general textile applications; ironing temperatures of 400°F (204°C) not degradative with short exposure; rayon loses strength above 300°F (149°C); chars and decomposes at 350–400°F (177–204°C); long exposure to high temperatures causes yellowing (oxidative damage); does not melt or stick at elevated temperatures

Crimp . . . . Relatively large proportion of rayon staple is produced by a chemical crimp process, contributing bulkiness to yarn and fabrics and resulting in interfiber cohesion which improves processing efficiency; chemical crimp is result of macro-structure control, where the rayon is produced with a broken-skin-core type cross section; crimp frequency varies from 10–30 crimps per inch for $1\frac{1}{2}$ denier per filament rayon to 6–15 crimps per inch for 15 denier per filament rayon

Light resistance . . . . Photocellulose formed on exposure to light; rayon exhibits tensile-strength loss, depending upon temperature, humidity, and time

Biological resistance . . . . Sensitive to mildew, which causes discoloration and tensile-strength loss; Encountered only in very moist climates; may be prevented with mildewcides

Luster . . . . Rayons are made in bright and dull luster; dull fiber made by dispersing TiO_2 or other opacifying agents into spinning solution; opacifying agent (0.75–2.5%) may be included, depending upon end use

Color . . . . Normally produced as white or translucent fiber; color pigments and dyestuffs in viscose spinning solution yield solution-dyed rayons, recognized for their high degree of color uniformity and excellent fastness to laundering, light, perspiration, seawater, bleaching, dry cleaning, and cracking

Dyeing properties . . . . Similar to cotton; affinity for direct, sulfur, naphthol, vat, and basic dyestuffs; direct dyestuffs most commonly used

Chemical properties . . . . Rayon subject to degradation through hydrolysis when exposed to hot dilute or cold concentrated mineral acids; alkalies cause rayon to swell in critical concentrations (9% NaOH < 25°C); structure is disrupted with final disintegration of fiber; unaffected by most organic solvents and usual dry-cleaning agents; although strong oxidizing agents degrade rayon, fiber is not damaged by hypochlorite or peroxide bleaches when applied as recommended

Flammability . . . . Rayon is flammable and must be treated with flame retardants to meet federal flame-retardant fabric requirements in the United States imposed, for example, on special end products; flame-retardant chemicals used by rayon producers are based upon halogenated organic phosphonates; also phosphonitrillates

*A measure of the transverse strength, toughness, and abrasion resistance. Optimized with fibers having high tenacity and high elongation properties.

†Related to the fine structure, orientation, crystallinity, and degree of polymerization of the rayon. High resistance to caustic predicts better wet performance (long laundry life) and higher residual strength after mercerization.

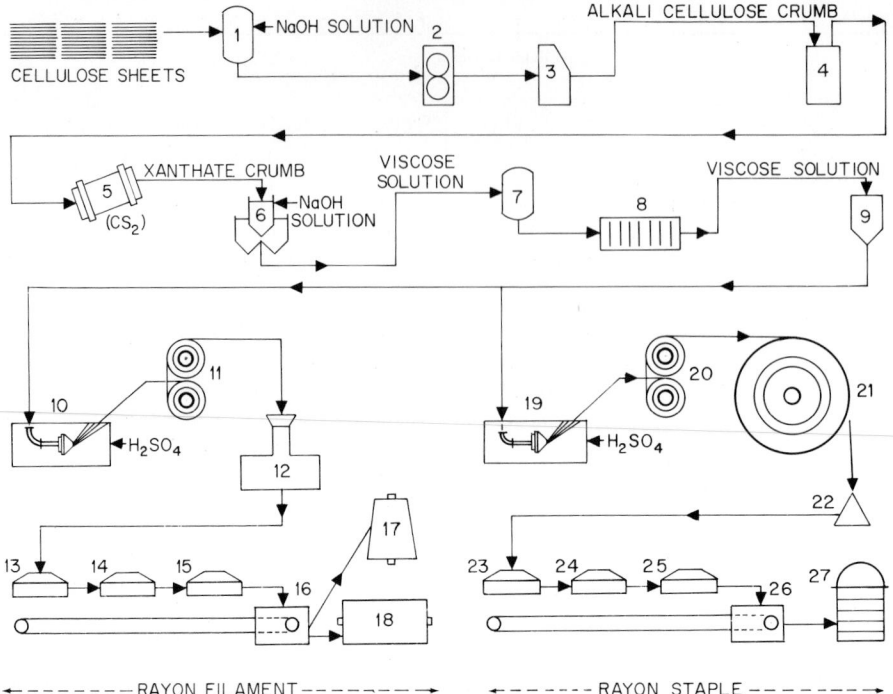

Fig. R-4. Rayon manufacturing process. Cellulose sheets are conveyed to continuous steeping tank (1), which contains a caustic soda solution, after which the material passes to squeeze rolls (2) and a shredder (3). Alkali cellulose crumb then passes to aging bin (4), from which it is conveyed to a mechanical churn (5) containing carbon disulfide. The xanthate crumb produced here then passes to a mixing tank (6), which contains a caustic soda solution. The resulting product (viscose solution) then passes to an aging tank (7) and then through filters (8), whereupon it is piped to spinning tank (9). At this point, depending upon whether rayon filament or rayon staple is being made, the process splits. The first step in filament production is the spinning machine (10), which incorporates a sulfuric acid bath. Filaments then pass through godet wheels (11) and then to the spinning box (12). The material then is conveyed through a series of hooded operations—(13) desulfurizing, (14) bleaching, (15) washing, (16) drying. The material then may go to coning (17) or beaming (18). Spinning is the first operation in staple production (19), followed by passage through godet wheels (20) and then to a tow wheel (21). The filaments then are cut (22), after which the material is processed essentially the same as filament by (23) desulfurizing, (24) bleaching, (25) washing, and (26) drying, followed by baling (27). (*FMC Corporation, American Viscose Division.*)

liquid containing 5–10% cellulose and 5–10% NaOH. The viscose solution is prepared for spinning by (1) deaerating under vacuum, (2) aging or ripening, which permits a chemical rearrangement and improved distribution of the xanthate groups along with partial decomposition of the cellulose xanthate, and (3) filtering to remove foreign debris and gels.

Viscose ready for spinning is pumped to the spinning machine, where it is blended with delustrants, color pigments, or other chemicals for specialty fibers. The viscose then is accurately metered to each spinning position, where it passes

through a precious-metal alloy spinnerette, which is alkali- and acid-resistant. As the viscose leaves the spinnerette, it enters into a coagulation-regeneration bath of dilute H_2SO_4, $ZnSO_4$, and Na_2SO_3. The temperature of the bath may be varied between 20 and 60°C, depending upon the desired regeneration rate. Coagulation of the filament occurs, along with dehydration of the swollen gel fiber and decomposition of the cellulose xanthate. The partially regenerated fiber is stretched between wheels called *godets*, and frequently the freshly created fiber is plasticized and further regenerated by applying hot, dilute acid baths to the fiber as it passes through the stretch zone.

Filament yarns, after stretching, may be collected in centrifugal spinning boxes, or they may be purified and dried on continuous spinning-processing machines. Box-collected yarns are purified in cake form by processing the yarn with desulfurizing, bleaching, washing, and lubricant-finish baths. Finally, the yarn is dried. In the staple process, bundles of fibers, or tows, are stretched and regenerated and then are fed to a cutter, which reduces the tow to staple having nominal lengths of 1–6 in. The staple fibers are sluiced onto a conveyor, where they form a dense blanket which is purified with processing baths as described for filament yarn. After drying, the staple rayon is opened and baled.

—Joseph W. Schappel, *FMC Corporation, American Viscose Division, Marcus Hook, Pa.*

RDX (Cyclotrimethylenetrinitramine) (See **Explosives.**)

Reactions, Catalytic (See **Catalysts.**)

Reactions, Plasma-generated (See **Plasma, chemical processing with.**)

Reactive Dyes [See **Dyestuffs and dyeing (textile).**]

Reactor, Nuclear (See **Dysprosium; Erbium;** and **Nuclear power plants.**)

Reaeration (See **Waste treatment.**)

Realgar (See **Arsenic.**)

Reboilers (See **Distillation;** and **Energy systems for processes.**)

Recrystallizer Process [See **Salt (NaCl).**]

Rectification (See **Distillation.**)

Recycling (See **Agglomeration; Gold; Mercury; Platinum metals and platinum;** and **Silver.**)

Red Brass (See **Copper.**)

Red Lead (See **Lead.**)

Reduction, Metals (See **Cadmium; Hydrogen;** and **Iron- and steelmaking.**)

Reflectivity (See **Insulation, thermal.**)

Refluxing (See **Distillation.**)

Reforming The purpose of reforming is to rearrange or reform the molecular structure of hydrocarbons, particularly with the objectives of upgrading naphthas having poor antiknock characteristics to premium-quality motor fuels or producing aromatics, notably benzene, toluene, and C_8 aromatics, from selected naphtha fractions.

The chemical reactions involved in reforming the hydrocarbon molecules in naphtha include:

1. Dehydrogenation of naphthenes to aromatics. This reaction is kinetically rapid and proceeds nearly to completion at conditions employed in modern processes.

2. Dehydrocyclization of paraffins to form aromatics.

3. Isomerization of paraffins to more highly branched isomers to the extent permitted by chemical equilibrium. Another important isomerization reaction is the conversion of five-carbon naphthene rings to six-carbon rings, which upon dehydrogenation form aromatics. See also **Isomerization.**

4. Hydrocracking of heavy paraffins in naphthas to paraffins of lower molecular weight. Control of the hydrocracking reactions is needed to minimize the formation of fragments lighter than C_5, which would not be included within the boiling range of gasoline. See also **Hydrocracking.**

Of the hydrocarbon types residing in naphthas and the reformed product, aromatics have the highest octane ratings, and *n*-paraffins have very low ratings. Naphthenes are classed as being intermediate, and isoparaffins as high in octane ratings. As the reactions proceed, the reformed product will contain increasing concentrations of aromatics and decreasing concentrations of heavy paraffins with a resulting increase in octane rating. These chemical reforming reactions are accelerated and directed by noble-metal catalysts in an environment of hydrogen under moderate pressures. Catalyst fouling and consequent loss in activity is inhibited by the application of adequate hydrogen partial pressure in the reaction zone.

Trends in Development: The upgrading of naphthas by thermal cracking started about 1930. In this noncatalytic process, olefins were formed by cracking and dehydrogenation of paraffins, and aromatics native in the naphtha feed were concentrated by cracking paraffins to light gaseous hydrocarbons. A very small amount of aromatics was synthesized. All these reactions tended to improve the octane number of the naphtha, but liquid-product yields and octane-number values were distinctly lower than those obtainable by modern catalytic processes. During the 1940s, toluene for TNT manufacture was produced by a fixed-bed catalytic reforming process.* This process used a non-noble-metal catalyst to promote the dehydrogenation of C_7 naphthenes contained in an appropriate naphtha fraction. Presently, all naphtha upgrading and aromatics production from naphtha fractions are accomplished by processes using noble-metal catalysts† (see Table R-11).

Catalysts: The noble-metal catalyst introduced‡ in the late 1940s contained platinum on a support and originally was in the shape of a cylindrical tablet. Presently, the most common forms are spheres and extrudates. Reforming catalysts have experienced progressive improvement since their

---

*Developed largely by Standard Oil Development Co. (now Esso Research & Engineering Co.), the process was named Hydroforming^{SM}. The use of thermal reforming has since been discontinued.[1]

†Platforming^{SM} introduced by Universal Oil Products Co. in 1949.[2]

‡By Universal Oil Products Co.

TABLE R-11. *Catalytic Reforming Capacities**,†

Proprietary Name of Process	bbl/stream day
Platforming	1,286,760
Ultraforming	297,190
Englehard	249,740
Powerforming	211,800
Mobil	194,800
Houdriforming	148,450
Catforming	81,700
Others	305,780
	2,776,220

*United States (1970). The total catalytic reforming capacity shown in this table represents 22% of the crude capacity of United States refiners. Since the average naphtha content of crudes processed in the United States is close to this figure of 22%, it is significant to note that about all the available straight-run naphthas were being catalytically reformed as of the report date. This table is reproduced from *Oil and Gas Journal*, April 6, 1970, by permission.

†Catalytic reforming capacity for the free world is about 6 million bbl/day (1970).

---

introduction, particularly as a result of careful attention to the formulation, the purities of raw materials, and the details involved in the many steps of manufacturing the catalysts. Major improvements were realized in terms of the activity, selectivity, stability, and physical properties of the catalysts.

A new generation of reforming catalysts that contain another metal, e.g., rhenium, in addition to platinum appeared in 1968. These bimetallic catalysts exhibited such exceptional stability characteristics that it became possible to employ them at substantially diminished operating pressures. Under these conditions, marked increases in yields of reformate of a given octane number, of hydrogen, and of aromatics were obtainable— and still with good stability of catalyst activity and selectivity.

The Catalytic Reforming Process: The essential components of a particular reforming process§ are shown in Fig. R-5 and include (1) reactors which contain the catalyst in fixed beds, (2) heaters to bring the naphtha and recycle gas to reaction temperature and to supply heats of reaction, (3) product cooling system and a gas-liquid separator, (4) hydrogen-gas recycle system, and (5) stabilizer to separate light hydrocarbons dissolved in the receiver liquid. See also Fig. R-6.

Practically all naphtha feedstocks to catalytic reforming units are hydrotreated (see also **Hydrotreating**) to remove nonhydrocarbons which would adversely affect the stability of noble-metal reforming catalysts from the standpoint of their activity and selectivity. Materials removed include S, N, O, and organic compounds of As and Pd; all are catalyst poisons.

The catalyst is contained as a fixed bed in three or more separate adiabatic reactor vessels with feed and hydrogen-recycle-gas preheating before the first and reheating between the subsequent reactors. Because of the rather large endothermic heats of the dehydrogenation reactions, there is a substantial drop in temperature of the flowing and reacting stream, particularly in the first reactor, where the rapid naphthene dehydrogenation occurs. Therefore, the effluents from the first and second reactors are reheated to bring them to the proper inlet temperature for the subsequent reactor. Often, the charge heater and the interheaters are contained in the same furnace housing.

---

§Platforming, Universal Oil Products Co. The original of the family of processes using noble-metal catalysts and the most extensively used process in this group.

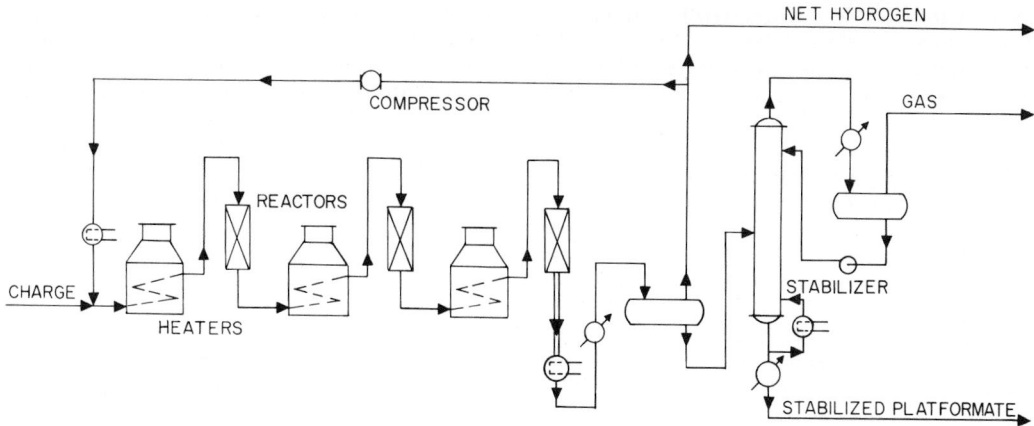

Fig. R-5. Platforming catalytic reforming process. (© 1973 *Universal Oil Products Company.*)

Effluent from the last reactor in the train is cooled and led to a receiver in which the product mixture is separated into a liquid and a gas stream. Most of the separated gas stream (largely hydrogen) is compressed and recycled to the reactors to provide the protective hydrogen partial pressure in the reaction environment. A net hydrogen-rich product stream is withdrawn from the system by pressure control, as shown in Fig. R-5.

The receiver liquid which contains dissolved light hydrocarbons is routed to a fractionator to produce a stabilized reformate* suitable for blending into finished gasoline. This liquid is generally free of hydrocarbons lighter than C_5. The C_4 and lighter hydrocarbons, separated as an overhead product stream by the stabilizer fractionator, usually are routed to a gas-concentration system within the refinery.

Process Variations: Most variations in catalytic reforming processes involve frequency of catalyst regeneration. At relatively high pressures and ratios of hydrogen recycle gas to naphtha feed, the catalyst is fouled and deactivates rather slowly, continuous-processing runs varying from a few months to more than a year. At the end of this uninterrupted processing period, the reactor inlet temperature needed to maintain the desired reformate octane number may have reached the limit of heater capabilities or the catalyst selectivity may have diminished to a point where it is economic to terminate the run and restore the activ-

*This stream from a Platforming unit is known as Platformate[SM].

Fig. R-6. UOP Platforming process at Coastal States' Petrochemical Company refinery, Corpus Christi, Tex. Three reactors are stacked, with a fourth to the side. In this installation, the catalyst is withdrawn, regenerated, and continuously returned to the reactors in a closed system. (© 1973 *Universal Oil Products Company.*)

ity or selectivity of the catalyst by an in situ regeneration.

Regeneration usually is performed by burning the carbonaceous deposit accumulated on the catalyst with air diluted with combustion gases, using the gas-recycle compressor to circulate the air-combustion-gas mixture through the reactor system at a controlled burning temperature. This regeneration procedure can restore the catalyst nearly to its performance characteristics at the beginning of the prior processing cycle. An alternative procedure, which avoids the downtime for in situ regeneration, is to unload the spent catalyst and reload the reactors with fresh material. This spent catalyst usually is returned to the supplier for recovery of the precious-metal content.

At the other extreme is the process designed to operate at distinctly lower pressures and ratios of hydrogen recycle gas to naphtha, at which higher reformate yields of a given octane number are obtainable. In this approach, the catalyst fouls and deteriorates much more rapidly, and thus frequent catalyst regenerations are necessary. These plants usually are provided with an additional reactor which is manifolded to the other reactors and appropriately valved to allow one reactor to be on regeneration while the others continue to process naphtha feed. A reactor can be removed from processing service for regeneration as frequently as once per day. Although higher reformate yields are obtainable by use of this *swing-reactor* design concept, the units tend to be more costly because of the additional reactor, manifold piping, and continuous-regeneration auxiliaries.

Between these two extremes are designs and operating techniques that can perform at any intermediate regeneration frequency most economically suited to a particular refining situation.

Process Variables: Apart from the catalyst itself and the mechanical design of the unit and its reactors, four major process variables affect the conduct of the reforming reactions and performance results: (1) temperature in the reaction environment, (2) space velocity (or feed rate per unit of catalyst), (3) reaction pressure, and (4) recycle rate of hydrogen-rich gas per unit of naphtha feed. In a sense, the catalyst itself is a process variable in that significant improvements have been made in its performance during the history of the process and it is not difficult to change the catalyst loading in the reactors from one run to another.

As *temperatures* in the catalyst zone are increased, the reaction rates and extents of the reactions

mentioned earlier increase in a given system so that the overall result is the production of a reformate with an increased octane number. There is flexibility with respect to temperature distribution among the three reactors in that the inlet temperature to each reactor can be controlled independently. As the catalyst slowly deactivates, it is common operating practice to increase reactor inlet temperatures during the processing life of the catalyst to maintain the production of a reformate having a constant octane number. Should it be desired to change the octane number to satisfy variable blending requirements in a refinery, it is usual practice to vary the reactor inlet temperatures appropriately and thus change reformate quality.

Space velocity may be defined as the rate of liquid naphtha feed per unit of catalyst in the reactors. Both volumetric and weight units are used in the industry. Space velocities may be expressed as barrels of liquid (at 60°F) naphtha feed per hour per barrel of catalyst in the reactors (liquid hourly space velocity); or as pounds of naphtha feed per hour per pound of catalyst contained in the reaction zone (weight hourly space velocity). As the space velocity is lowered, the reactions proceed to a greater extent if they are not limited by chemical equilibrium, and the reformate octane number increases.

Space velocity is used more often as a design factor than as an operating variable. Reforming units which are expected to produce reformates having high octane numbers generally are designed for relatively low space velocities, to contain relatively more catalyst per unit of feed rate than units designed for lower reformate octane numbers. In a given reforming unit, space velocity can be used as an operating variable by changing the naphtha feed rate only within the capacity limits of the pumps, heaters, exchangers, and other equipment that handle the feed, recycle gas, and products. Space velocity can also be changed by varying the amount of catalyst loaded in a given unit from one run to another, but only within the limits permitted by the reactor design.

Pressure and hydrogen recycle also are important variables. Although pressure is used primarily as a design factor, a given plant can be operated over a limited range of pressures that are permitted, for example, by the strength of equipment at high values and by the capacity of the recycle-gas compressors at low values. Diminishing reaction pressure improves dehydrogenation equilibrium and suppresses hydrocracking reactions. Thus, higher yields of reformate of a given octane number are obtained at lower pressures. Hence, there is an economic incentive to operate at lower pressures. However, at a given ratio of hydrogen recycle to naphtha feed, the partial pressure of hydrogen in the reaction environment is decreased at the lower operating pressure. Consequently, more rapid fouling and activity decline of a particular catalyst is experienced.

REFERENCES

1. Davidson, R. L.: An Appraisal of Catalytic Reforming, *Pet. Process.*, vol. 10, no. 8, pp. 1174–1175, August 1955.
2. Bland, W. F.: Platforming: Operating Results and Design, *Pet. Process.*, vol. 5, no. 4, pp. 351–355, April 1950.

—Melvin J. Sterba, *UOP Process Division, Universal Oil Products Company, Des Plaines, Ill.*

Refractories (Consult the **Subject Index.**)

Refractory Pitch (See **Coal tar and derivatives.**)

Refrigerants (Consult the **Subject Index.**)

Refrigeration Systems (See **Chlorofluorocarbons; Cryogenic processes; Energy systems for processes; Ethylene; Helium; Insulation, thermal; Natural gas;** and **Oxygen.**)

Regenerator (See **Adsorption; Drying, gases; Fluid catalytic cracking;** and **Reforming.**)

Reinforced Plastics (See **Fiber glass;** and **Plastics.**)

Relative Biological Effectiveness (See **Radioisotopes.**)

Resins Although natural resins are carbon compounds that contain oxygen and hydrogen, this does not distinguish a resin because the generalization also applies to many other organic compounds. Practically, the word *resin* implies certain physical characteristics. Natural resins generally are viscous and sticky liquids and semiliquids that gradually harden when exposed to air, becoming amorphous, brittle solids. The term *resin* may be applied to either stage—before or after oxidation. Natural resins generally are yellowish in color and insoluble in water but are quite soluble in CS_2 and many other organic solvents, e.g., benzene, ether, and alcohol. Synthetic resins possess many of these same properties.

Sources of Natural Resins: Secretions from plants, particularly those which appear on the surface of the plant when it has been cut or injured, are the principal source of natural resins. The resinous substance protects the plant against microorganisms entering the wound and against loss of sap from the wound. Commercially, natural resins are obtained by deliberately cutting the bark of trees and directing the exudate into collecting buckets.

Turpentine, obtained in this manner from various coniferous trees, constitutes a large industry in many parts of the world. The longleaf pine, the mountain pine, and the slash pine are the principal sources in the United States. Bordeaux turpentine is obtained in France from the cluster pine; Scotch turpentine is obtained from the Scotch fir; Strassburg turpentine from the fir (*Abies*); Swiss turpentine from the stone pine; and Venetian turpentine from the European larch. True turpentine is considered to be Canada balsam, which is obtained from the bark of *A. balsamea*. The exudation contains 75–90% resin, the remainder being an oil. Crude, commercial turpentine, although it has numerous uses as a solvent and in medicine, is of value principally as a source of resins. See also **Terpenes.**

Rosin is obtained from turpentine by distilling off the oil of turpentine. Also known as *colophony*, rosin is an amber, brittle, translucent, and hard resin with a specific gravity of approximately 1.08, a softening point of 80°C, and a melting point which varies from 120 to 135°C, depending upon the specific source of the material. Rosin is widely used in making paints and varnishes, in sizing paper, as a flux for soldering, and in the manufacture of soaps and linoleum. In some products, rosin is used along with more costly resins to reduce costs.

Natural resins fall into three categories: (1) *hard resins,* e.g., rosin, copal, mastic, amber, and sandarac, which are odorless, tasteless, brittle, and hard and fracture like glass; (2) *oleoresins,* e.g., copaiba, dragon's blood, and the balsams, which are amphorous solids containing essential oils and very sticky; and (3) *gum resins,* e.g., as benzoin, myrrh, frankincense, and asafetida, which contain gums.

Latex, obtained from the rubber tree (*Hevea brasiliensis*) in the same manner as turpentine, is not to be confused with a resin. It is an emulsion which contains resins, gums, alkaloids, proteins, starches, tannins, oils, and sugars. See **Rubber, natural.**

Synthetic Resins: A large variety of synthetic resins provides the base for many materials of construction with properties that technically or economically are unobtainable from metals, ceramics, wood, animal fibers, plant fibers, and other natural substances. *Structures,* with a spectrum of mechanical, electrical, and chemical properties (see **Plastics**); *fibers,* with qualities superior to those of cotton, wool, and silk (see **Fibers**); thin *films,* with strength, flexibility, and other properties superior to paper; *adhesives,* with improved tenacity, ease of application, toughness, and resistance to aging; *coatings and finishes,* with greater resistance to corrosion, effects of sunlight and weather, faster drying times, and convenience of use (see **Paints**)—these are some of the improvements that synthetic resins have contributed to materials. Not to be overlooked is the simplification of fabrication and assembly (and hence often lower costs) that synthetics based upon resins provide. By adding several dimensions to available materials, the synthetic-resin industry has become one of the major segments of the chemical industry. See also **Petrochemical complex.**

Early Synthetic Resins: The production and use of synthetic resin started its rise to a major industry in the late 1930s just before World War II and, of course, was strongly influenced by the needs of war when many of the previously conventional materials of construction, including metals, were in short supply. Development of synthetic resins and plastics, however, dates back to the mid-1800s. One notable early development stems from the economics of ivory, then widely used for the musical keys of instruments, for billiard balls, buttons, and decorative items. In an attempt to win the substantial reward offered by a manufacturer of billiard balls, John Wesley Hyatt developed celluloid, which he produced by subjecting cellulose nitrate of low nitration (pyroxylin), plasticized with camphor, to high-pressure working. This product immediately found numerous uses and, considering the economy of that period, became a rather high-tonnage product.

One of the first truly synthetic resins was developed by Leo Hendrik Baekeland, a Belgian-American inventor, in 1909. The material (bakelite) was derived from phenol and formaldehyde and found its first wide acceptance by the electrical industry, which had hitherto depended upon shellac and other natural resins because of their resistance to heat and electricity. Urea-formaldehyde resins came along a little later. These tough, transparent resins had many of the properties of glass but were easily scratched and became dull upon long exposure to moist air. To overcome these faults, fillers (now used extensively in plastics), e.g., wood flour, paper, and cotton, were developed. Methyl methacrylate (Lucite and Plexiglas) first appeared in 1937. Cellulose acetate butyrate (Tenite II) appeared in 1938, and the first of the polyamide resins (nylon) went into large-scale production in 1940. Since then well over a score of major families of resins have been developed.

Because the pace of introduction of new resins has been high, and because the industry has continued to grow at a rate far in excess of the general economy, there is a tendency to consider synthetic resins as fairly new to the scene, but as we have just seen, synthetic resins are now well into their second century. Synthetic resins have progressed from the time when they were looked upon as substitutes (often associated with the belief that their qualities and values were inferior to those of the natural materials) to their present stature, where synthetics are regarded as materials in their own right, each with its own set of relative advantages and limitations. Because of the availability of so many different kinds of synthetic resins, problems of materials selection are indeed complex. The distinction between thermoplastic and thermosetting resins is described under **Plastics.** Natural resins derived from pitch are described under **Coal tar and derivatives.** Resins designed for ion exchange are described under **Ion-exchange resins.** Numerous synthetic resins are described in this volume; consult the Subject Index. For the general chemistry of synthetic resins, see also **Polymerization.**

Reverse Osmosis   (See **Desalination; Membrane filters;** and **Water treatment.**)

Reynolds Number   A Reynolds number N_{Re} is any of several dimensionless quantities of the form $LV_{\rho/\mu}$ which are all proportional to the ratio of inertial force to viscous force in a flow system. Here L is a characteristic linear dimension of the flow channel, in feet; V is linear velocity, in feet per second; ρ is fluid density, in pounds per cubic foot; and μ is fluid viscosity, in pounds per foot per second. The *critical* Reynolds number corresponds to the transition from turbulent flow to laminar flow as the velocity is reduced. Its value depends upon the channel geometry, being 2,000–3,000 for circular pipe.

—Donald F. Boucher and George E. Alves, from "Perry's Chemical Engineers' Handbook," 4th ed., McGraw-Hill Book Company, New York, 1963, by permission.

Rhenium (G. *Rhine.*) **Re** = 186.2 (at. wt.); 75 (at. no.). Rhenium is a heavy transition metal that forms an extensive series of compounds covering the widest range of valences of any element. Re is in group VIIB of the periodic table, falling below manganese and technetium. It has two natural isotopes, 185 and 187, and eight artificial isotopes, 177, 178, 180, 183, 184, 186, 188, and 189. See also **Chemical elements.**

Because of its high melting point (3180°C), high modulus of elasticity (68.0 × 10⁶ psi at 20°C), and its heat-stable crystalline structure, Re is an excellent refractory metal. Other important physical properties include density (20°C) = 21.04 g/cm³; boiling point = 5900°C; specific heat at 27°C = 0.0320 cal/(g)(°C), 0.0354 at 500°C, 0.0405 at 1500°C, and 0.0434 at 2000°C; linear coefficient of thermal expansion 6.12 μin./(in.)(°C) (25–500°C); and electrical resistivity 17.5 $\mu\Omega$-cm at 0°C, 25.4 at 100°C, 80.5 at 1100°C, and 106.5 at 2100°C. Re is paramagnetic; the magnetic susceptibility is 0.366 × 10⁻⁶ emu. The lattice type is hexagonal close-packed. Lattice constants at 20°C are a_0 = 2.760 Å and c_0 = 4.458 Å. The atomic radius (12 coordination) is 1.38 Å; ionic radius (Re^{7+}) = 0.56 Å. The thermionic work function at Richard constant of 52 A/(cm²)(deg²) is 4.80 eV; photoelectric work function is 4.66 ± 0.01 eV. Cross section for thermal neutrons is 86 barns per atom. Ionization potentials in volts are first, 7.8; second, 13.1; third, 26.0; fourth, 37.7; fifth, 51; sixth, 64; and seventh, 79.

Rhenium was discovered by the German scientists W. Noddack and I. Tacke in 1925, when it was first detected in platinum and columbite (niobite) ores. A few milligrams were produced in 1927 and the first gram in 1928. Although Re was produced in the United States, Europe, and Russia in small quantities starting in the early 1930s, production of significant commercial quantities did not begin until the early 1960s as the result of advanced recovery techniques and greater demand.

Re occurs as a minor constituent (100 ppb) of molybdenite-bearing porphyry Cu ores mined in substantial volume in the United States and South America and generally is associated with the Mo mineralization. Currently, Re is recovered from by-product molybdenite, where it is concentrated to levels of 300–1,000 ppm, as the result of the roasting of MoS_2 to MoO_3. At the elevated temperatures of this process, Re is oxidized and volatilized as rhenium heptoxide, Re_2O_7, and is recovered from the flue gases by wet scrubbing and chemical separation techniques as relatively crude ammonium perrhenate. Re metal is produced by reducing purified ammonium perrhenate with H_2.

Chemistry: Re differs considerably from Mn and Tc, its neighbors in the periodic table, in both chemical and catalytic properties, just as tungsten differs from chromium. For example, metallic Mn is rather reactive, slowly decomposing water and reacting with dilute acids. In contrast, Re metal is relatively inert. It does not react with water or nonoxidizing acids. For Mn, +2 is the most stable oxidation state. $KMnO_4$ in the +7 state is well known as a powerful oxidizing agent. On the other hand, Re is most stable in the +7 state and lacks oxidizing power, as evidenced by low electrochemical potentials. However, both elements form stable complexes in the lower valence states. The common coordination numbers of both metals are 4 and 6. However, Re readily forms complexes with a high coordination number, but technetium does so only occasionally. With the halides, Re forms a stable heptafluoride and pentachloride, but Tc does not. Similarly, $ReCl_6$ is more stable than $TcCl_6$. On the other hand, $TcCl_4$, unlike the corresponding Re compound, is very stable and sublimes without decomposition.

In many respects Re occupies a position intermediate between W and the platinum metals. Re has well-defined anhydrous oxides in the +3, +4, +6, and +7 oxidation states similar to the high-valent oxides of W, Os, and Ir. However, platinum oxides exist only in the +2, +3, and +4 states. A somewhat similar situation occurs in the sulfides. Like W, Re forms numerous oxo cations and oxy anions; the number of such ions falls off from Os to Pt. In a variety of its carbon compounds, Re again is similar to the elements from W to Ir but quite different from Pt.

A unique feature of Re chemistry is the unusually high solubility of the heptoxide in water and oxygenated solvents. By contrast, the oxides of W and most of the Pt metals have very low solubilities. (Osmium tetroxide, the single exception, is moderately soluble in water.)

In summary, Re is similar in some respects to Mn and Tc, which fall in the same group. In other respects, it is similar to its neighbors in groups VI and VIII, W and the Pt metals. Nevertheless, rhenium's unusual combination of properties from both a chemical and metallurgical point of view makes it unique in the periodic table.

Re has found important applications in catalysts for reforming in conjunction with platinum, in selective hydrogenation, and in other chemical reactions. The most common processes in which Re is used or has been tested and used as a catalyst include alkylation, dealkylation, dehydrochlorination, dehydrogenation, dehydroisomerization, enrichment of water, hydrocracking, hydrogenation, oxidation, and reforming. The outstanding property of Re catalysts is their high selectivity, particularly in hydrogenation reactions. Re also displays unusually high resistance to such catalyst poisons as N_2, S, and P. In activity Re generally surpasses W, Mo, Co, and other metals usually employed as oxides and sulfides. Re approaches Ni, Pd, Pt in catalytic applications.

The perrhenate ion with an ionic weight of 250 is one of the heaviest simple anions available in readily soluble salts. It can be used as a precipitant for K^+ and certain other heavy univalent ions, as a precipitant for such complex ions as $Co(NH_3)_6^{++}$, for the separation of alkaloids and other organic bases, and for fractional crystallization of the rare earths.

Use in Alloys: In addition to the pure metal, Re is highly desirable as an alloying addition with other refractory metals such as W and Mo. The addition of Re greatly enhances the ductility and tensile strength of these metals and their alloys, even after heating above the recrystallization temperature. A prime example is the complete ductility exhibited by a Mo-Re fusion weld.

Re and its alloys are gaining rapid acceptance in nuclear reactors, semiconductors, electronic-tube components, thermocouples, gyroscopes, miniature rockets, electrical contacts, thermionic converters, and other commercial and aerospace applications.

Metallurgical Properties: In addition to the high modulus of elasticity previously given, Re is attractive from the standpoint of workability, weldability, and machinability. Re is easily formed at room temperatures but must be cold-worked to avoid hot shortness. Unusually rapid work hardening requires frequent intermediate anneals in an inert or reducing atmosphere. A ductile to brittle transition temperature does not exist. The recrystallization temperature ranges between 1400 and 1600°C; the stress-relieving temperature lies between 1200 and 1400°C. Re can be welded by inert-gas or electron-beam methods, using caution to protect against oxidation. Welds are ductile at ambient temperatures. Electrical-discharge machining (EDM), electrochemical machining (ECM), abrasive cutting, or grinding should be used. Re is very difficult to machine with carbide tools and other conventional methods.

REFERENCES

Davenport, W. H., V. Dollonitsch, and C. H. Kline: Advances in Rhenium Catalysts, *Ind. Eng. Chem.*, vol. 60, 1968.

Peters, J. E.: Recent Developments in Rhenium and Rhenium Alloy Powder Metallurgy, *Jl. Met.*, April 1961.

Sutulov, A.: Molybdenum and Rhenium Recovery from Porphyry Coppers, University of Concepcion, Chile, 1970.

—Michael B. Whiteman, *formerly with Cleveland Refractory Metals, Division of Chase Brass & Copper Co., Incorporated, subsidiary of Kennecott Copper Corporation, Solon, Ohio*

Rheopectic Fluids   (See **Colloid systems.**)

Rhodium   (Gk. *rhodon*, rose.)   **Rh** = 102.91 (at. wt.); 45 (at. no.). Rhodium is a member of the family of platinum metals, group VIII of the periodic table. The melting point of Rh is 1960°C; boiling point 4500°C; and density at 20°C 12.41 g/cm^3. There is one stable isotope of Rh, 103, and seven unstable isotopes, 99–101 and 104–107. Rh was discovered by Wallaston (England) in 1803.

Compact Rh is almost insoluble in all acids at 100°C, including aqua regia. Hot concentrated H_2SO_4 will slowly dissolve the finely divided metal. When alloyed with 90% or more of Pt, it is soluble in aqua regia. The metal is attacked by fused bisulfates. Rh is soluble in molten Pb. This is the basis of the classic separation of Rh and Ir.

Rh compounds exhibit valences of 2, 3, 4, and 6. The trivalent form is by far the most stable. When Rh is heated in air, it becomes coated with a film of oxide. Rhodium(III) oxide, Rh_2O_3, can be prepared by heating the finely divided metal or its nitrate in air or O_2. The rhodium(IV) oxide is also known. Rhodium trihydroxide may be precipitated as a yellow compound by adding the stoichiometric amount of KOH to a solution of $RhCl_3$. The hydroxide is soluble in acids and excess base. When the freshly precipitated

$Rh(OH)_3$ is dissolved in HCl at a controlled pH, a yellow solution is first obtained in which the aquochloro complex of Rh behaves as a cation. The hexachlororhodate(III) anion is formed when the solution is boiled for 1 hr with excess HCl. The solution chemistry of $RhCl_3$ is often very complex. Two trichlorides of Rh are known. The trichloride formed by high-temperature combination of the elements is a red, crystalline, nonvolatile compound, insoluble in all acids. When Rh is heated in molten NaCl and treated with Cl_2, Na_3RhCl_6 is formed, a soluble salt that forms a hydrate in solution. Rhodium(III) iodide is formed by the addition of KI to a hot solution of trivalent Rh.

Rhodium(III) sulfate exists in yellow and red forms. If $Rh(OH)_3$ is dissolved in cold H_2SO_4, the product is the yellow form, in which the sulfate is ionic. If this solution is evaporated in hot H_2SO_4, the product is a red nonionic sulfate. When Rh is treated with F_2 at 500–600°C, RhF_3 is slowly formed. This compound is practically insoluble in water, concentrated HCl, HNO_3, H_2SO_4, H_2F_2, or NaOH.

If a solution of $RhCl_3$ is treated with $NaNO_2$, the very soluble sodium hexanitritorhodate(III), $Na_3Rh(NO_2)_6$, is formed. The solubility of this compound in alkaline solution makes it useful for refining, as many base metals are precipitated as their hydroxides under these conditions. The analogous ammonium and potassium salts are relatively insoluble.

When H_2S is passed into a solution of a trivalent Rh salt at 100°C, the hydrosulfide, $Rh(SH)_3$, is formed. This black precipitate is insoluble in $(NH_4)_2S$. Rh forms many complexes with NH_3, amines, cyanide, chloride, bromide, and numerous polynitrogen and polyoxygen chelating agents.

Further information concerning the sources, refining, and uses of Rh will be found under **Platinum metals and platinum.**

—Lionel E. Simmons, *Simmons Refining Company, Chicago, Ill.*

Ribbon Mixers (See **Mixing and blending, solids-solids.**)

Ribonuclease (See **Enzymes;** and **Peptides and proteins.**)

Rice (See **Amino acids.**)

Ricinoleic Acid (See **Vegetable oils.**)

Ricordite (See **Tellurium.**)

Rimming Steel (See **Iron and steel.**)

Ring Compounds (See **Carbon compounds.**)

Ring-Chain Conversions (See **Polymerization.**)

RNA (Ribonucleic Acid) (See **Peptides and proteins;** and **Polymerization.**)

Roasting (Consult the **Subject Index.**)

Rock Gypsum (See **Gypsum.**)

Rock Salt [See **Salt (NaCl).**]

Rod Mill (See **Size reduction.**)

Roentgen A roentgen (R) is that amount of gamma or x-radiation which will produce 2.083×10^9 ion pairs (one electrostatic unit of charge) per cubic centimeter of free air at a temperature of 0°C and a pressure of 1 atm. One roentgen corresponds to a radiation field dissipating 83.8 ergs per gram of air, which dissipates approximately 93.8 ergs per gram of body tissue. Ionization in tissue is a measure of physical damage; thus allowable radiation exposures for human tissue are expressed in roentgens. Since the same number of roentgens from various types of radiation produces different amounts of body damage, a measure called *roentgen equivalent man* (rem) is used in stating allowable radiation exposure values. See Table R-4 under **Radioisotopes.**

Roll Crusher (See **Size reduction.**)

Roll Press (See **Expression.**)

Roller Mill (See **Mixing, pastes; Paint;** and **Size reduction.**)

Rolling Mill (See **Iron- and steelmaking.**)

Rosin (See **Paint;** and **Resins.**)

Rotary Disk Filter (See **Filtration.**)

Rotary Drum Filter (See **Filtration.**)

Rotary Drum Mixer (See **Mixing and blending, solids-solids.**)

Rotor Process (See **Iron- and steelmaking.**)

Rubber, Natural Natural rubber is the name applied to the polymer *cis*-polyisoprene obtained chiefly from the *Hevea brasiliensis* tree.* Originally, the tree grew wild in the Amazon valley, but during the last part of the nineteenth century, it was planted in well-organized plantations in tropical lands of the Far East and later in Africa (see Table R-12). The average rubber tree stands about 40–50 ft high. For optimum growth, a tropical climate having 80 in. or more of annual rainfall is required.

Rubber comes from the tree as a milky white fluid, which is a colloidal suspension of rubber in a liquid consisting mostly of water. The tree is tapped by well-trained workmen using a sharp-edged tool, and the cutting action goes at an angle of 30° from top left to bottom right. It is important that the rubber latex-bearing cells be cut but that the blade not wound the inner cambium layer, as this would harm the tree. A cup is hung below the cut to collect the white milklike latex,

*It is alleged that English scientist Joseph Priestly observed that the material could be used for rubbing out lead pencil marks and thus gave the material the name *rubber*. (*Introduction to the Theory of Perspective,* Joseph Priestly, 1835)

which contains about 35% rubber, the rest being water, protein, resins, organic materials, and other plant substances.

The yield of the tree can be increased by applying to the bark immediately below the tapping cut chemicals such as 2,4-dichlorophenocyacetic acid (2,4-D) and 2,4,5-trichlorophenocyacetic acid (2,4,5-T). Recently it has been found that Ethrel (2-chloroethylphosphonic acid)† applied to the tree increases rubber production through the release of ethylene, a natural plant hormone. These chemicals are applied in high-viscosity liquid forms, usually mixed with palm oil or other diluents. These yield stimulants appear to function by an anticoagulation action. Under normal conditions after latex begins to come from a tree, a clotting action takes place which decreases the flow and eventually stops it completely. These chemical stimulants decrease the coagulation action and allow the liquid latex to flow longer.

Presently the average annual yield for Malaysia is 600 lb/acre, which includes high-yielding trees of 1,500 lb/acre and many of the older, lower-yielding clones. The processing of crude rubber is described later.

†P. D. Abraham, Field Trials with Ethrel, Rubber Research Institute of Malaya Conference, July 1970, preprint F-1.

TABLE R-12. *Natural Rubber Exported (1,000 Long Tons)*

Area	1940	1960	1965	1968	1970	1971
Malaysia	547	775	919	1,114	1,304	1,356
Indonesia	543	587	695	729	755	804
Other Asian countries and Oceania	283	400	458	487	484	490
Africa	16	149	155	167	207	190
Tropical America	26	15	14	7	10	10
Total	1,415	1,926	2,241	2,504	2,760	2,850

SOURCE: *Rubber Statist. Bull.,* vol. 27, no. 4, January 1973, Secretariat of International Rubber Study Group, Brettenham House, 5–6 Lancaster Place, London WC.

NOTE: Estimated World Consumption of Natural Rubber

	Percent
Total western Europe	28.5
United Kingdom	6.5
West Germany	6.5
United States	21.0
Eastern Europe (including U.S.S.R.)	15.0 or less
Other areas	35.5

Inasmuch as stockpile quantities of natural rubber exist in several countries, production and consumer quantities do not always match. The selling price of natural rubber varies widely (from 12–89 cents per pound during 1945–1970); 29½ cents (August 1969); 18 cents (October 1970); 29 cents (March 1973).

Properties: Chemically, natural rubber is *cis*-polyisoprene and has a broad molecular-weight distribution ranging from several million to 100,000.

$$\underset{-CH_2}{\overset{H_3C}{\diagdown}}C=C\underset{CH_2-CH_2}{\overset{H\quad H_3C}{\diagup\diagdown}}C=C\underset{CH_2-CH_2}{\overset{H\quad CH_3}{\diagup\diagdown}}C=C\underset{CH_2^-}{\overset{H}{\diagup}}$$

Natural rubber is soluble in practically all aromatic and aliphatic hydrocarbons and particularly in halogenated hydrocarbons. When cements and solvent adhesives are made using natural rubber, methyl ethyl ketone (MEK) frequently is used to reduce viscosity. Although MEK is not a solvent, it tends to disperse large molecular particles, resulting in lower-viscosity dilution. Crude rubber is decomposed by heat and can be cyclized at 250°C. It can be easily hydrogenized and reacts readily with halogens. Two products developed from the latter reaction are chlorinated rubber, used as a paint base for concrete, and rubber hydrochloride, used to make a moisture-resistant and heat-sealing transparent film called Pliofilm.

The stress-strain properties of natural rubber are the best of all the elastomeric polymers. In vulcanized films made by the latex process, the tensile strength may exceed 6,000 lb/in.^{2}, and ultimate elongation is as high as 700% or more.

Natural rubber is readily attacked by O_2. Cu and Mn, which if present in amounts greater than the specified 0.001%, greatly accelerate oxidation. There are, however, naturally occurring antioxidants in natural rubber which help preserve it until vulcanization. All vulcanized natural-rubber products contain added antioxidants to ensure satisfactory life.

Rubber burns quite readily and generates more than 10,000 cal/g. The specific gravity of rubber is 0.934, a property utilized in concentrating natural-rubber latex by the centrifuge process. The serum, which is mostly water and has a specific gravity of about 1.0, tends to separate readily from the rubber. The liquid concentrated latex is used in making foam rubber, dipped goods, adhesives, and carpet backing for nonwoven carpets. An industry has developed around this application, which involves spreading foamed latex to the underside of carpeting, making an integral carpet-foam system.

Compounding and Vulcanization: Crude rubber in the raw state has few applications with the exception of crepe soles for shoes. To make commercial rubber products, the material must be mixed with a variety of chemicals and vulcanized into desirable end shapes. Charles Goodyear discovered in 1839 that adding sulfur to rubber and heating the mixture greatly enhances the physical properties of rubber. The material no longer becomes tacky in warm weather and in cold weather it does not become brittle. The material is much tougher, and the quality of products made this way results in service for a much longer period of time. In addition to sulfur, which cross-links the large rubber molecules and makes it a giant organic molecule, ZnO, organic accelerators, antioxidants, reinforcing pigments, and other processing aids are used in compounding rubber for useful vulcanized products.

In the manufacture of rubber products, crude rubber is masticated on a two-roll mill or in an internal mixing machine, called a *Banbury,* where heat and mechanical mixing reduce the rubber to a very viscous plastic mass. Antioxidant such as an alkylated diphenylamine,

$$C_8H_{17}\!-\!\!\bigcirc\!\!\underset{}{\overset{H}{-N-}}\!\!\bigcirc\!\!-C_8H_{17}\quad(2\text{ parts by}$$

weight), two accelerators, one a primary type such as benzothiazyl disulfide (1 part by weight) and the other a secondary one such as tetramethyl-thiuram disulfide (0.1 part by weight), activators of cure, such as stearic acid (2 parts by weight) and zinc oxide (5 parts by weight)—all based on 100 parts by weight of rubber—are added to the plastic mass and thoroughly dispersed in it. The compounded rubber is passed through the two-roll mill to make it into $\frac{1}{4}$-in. slabs, which are then cut into 8×4 in. dimensions, placed in a flat mold so that there is a slight excess in volume, and subjected to pressure and 140°C temperature for 30 min. After this vulcanization step, the mold is opened and the slab of hot rubber is removed. The next day the sample is tested and has developed a tensile strength of 3,850 lb/in.^{2}, elongation of 680%, stress at 500% elongation of 420 lb/in.^{2}. This is known as a *pure gum compound,* and this type of rubber can be used for hot-water bottles, water-hose stock, crude tire tubes, and any application requiring high stress-strain (tensile and elongation) properties.

The function of the antioxidant is to improve service life of the product against such well-known degrading agents as oxygen, light, and nitroso compounds. One theory is that the antioxidant selectively reacts with the degrader, slowing down its reaction with the rubber molecule, which would result in scission and eventually poorer physical properties. During recent years, considerably aggravated by air pollution, degradation of vulcanized rubber by small quantities of ozone (a few parts per million) in the air has become a serious problem. Ozone has little noticeable effect on unstretched rubber, but even under slight stretch it causes cracks in the surface which grow perpendicularly to the direction of extension. Hundreds of different antioxidants and antiozonants are employed, amine and phenol complexes being the basis of most. See also **Antioxidants.**

Accelerators act as catalysts of vulcanization, but unlike most catalysts they undergo chemical change during the reaction. Benzothiazyl disulfide is one of the oldest types, going back to 1925, but it still accounts for the greatest use in the industry today. Besides the thiazole types, other popular accelerators are sulfenamides; aryl guanidines; dithiocarbamates, extremely fast accelerators used mostly in latex compounds; and thiurams, also very fast and often used as a secondary accelerator to hasten the vulcanization rate. Accelerators also contribute to improved aging properties of the end product.

Stearic acid is an activator of vulcanization, as is zinc oxide, both reaching to form zinc stearate, which enhances the activity of the organic accelerators. Zinc stearate is impractical to add directly to the rubber because its slippery lubricating nature makes it difficult to mix in.

With an accelerated system a simple network structure with dialkenyl mono- and disulfide cross-links and conjugated triene units as main-chain modifications is obtained:

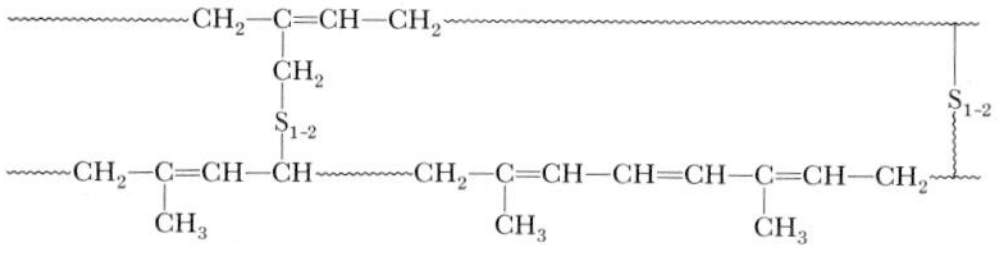

With an unaccelerated sulfur–natural-rubber system, the poor cross-linking efficiency results in sulfur being incorporated into the rubber network as long polysulfide cross-links, cyclic monosulfides, and vicinal cross-links, which are very close together and act physically as a single cross-link:

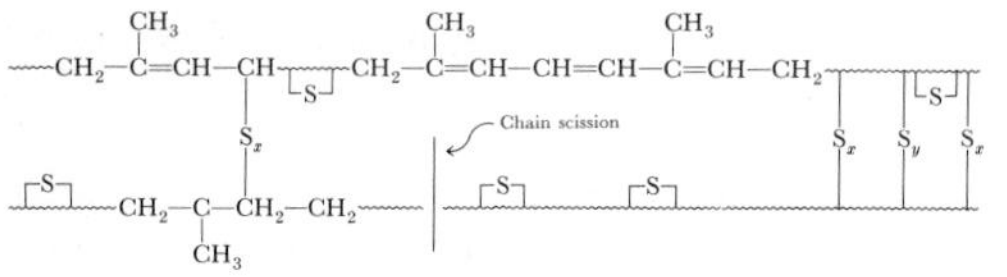

It is theorized that between the complex network structure of the unaccelerated system and the simpler network structure of the accelerated system, structures made up of the two models represent natural-rubber vulcanizates made at various times and temperatures of cures, with different reactant concentrations, and showing the effects of other variants.

At any given degree of cross-linking, the tensile strength is highest with polysulfide bonds. High elongation at break is obtained by slightly decreasing the cross-linking action. If lower elongation is required, slightly excessive cross-linking is used, usually accompanied by higher tensile strength. Vulcanization of rubber decreases its solubility in solvents, and this property frequently is used as a qualitative measure of cure.

Vulcanization by sulfur accounts for practically all the commercial products. However, peroxide types of curing systems may be used, especially for some of the synthetic rubbers. Radiation vulcanization has been under laboratory investigation for the past 15 years and is now entering commercial application.* The energy is supplied by ionizing radiation, which produces free radicals and cross-links the polymers. The physical properties of the vulcanizate are similar to those obtainable with sulfur.

Carbon black is the major reinforcing pigment, not only for natural rubber but for practically all the synthetic rubbers. As much as 40–50 parts by weight, based upon 100 parts of rubber, is used in all tire-tread compounds. Carbon black greatly increases tensile strength at low elongations (modulus) and results in longer-wearing tires. Colloidal silica contributes some reinforcing properties to rubber but not to the same degree as carbon black.

Uses: Thousands of flexible products requiring top performance characteristics are made of natural rubber, e.g., the huge earthmover tires, truck tires, airplane tires for large jet aircraft, bridge supports, and surgeons' gloves. The treads of most passenger-car tires in the United States consist mostly of styrene-butadiene synthetic rubber because of lower cost and the lower temperature buildup during use. Because of its excellent high- and low-temperature properties, many products used in the arctic and tropical areas of the world are made from natural rubber; it is not suitable for applications where there is contact with naph-

*K. H. Morganstern, Radiation Vulcanization, *Rubber Age,* vol. 103, no. 3, March 1971.

tha, e.g., gasoline hoses, because the material swells. Almost all elastic bands are made from natural rubber. Because of its excellent tack properties, the material is used in solvent and latex form as the base for adhesives.

World production (1969) of natural rubber was over 2.9 million long tons; of synthetic rubbers, over 4.5 million long tons, making a total of over 7.5 million long tons. Because natural rubber has numerous properties giving it both use and economic advantages for many applications, and thanks to new methods for tapping collection and built-in antioxidants, the continued use of natural rubber as a major elastomer is assured. See also **Elastomers;** and **Petrochemical complex.**

Processing Raw Materials: Crude latex is assembled in a rubber-producing plant connected with the estate and is converted into one or more of several types of crude rubber; or it is stabilized and concentrated into a higher-solids latex. About 92% of the natural rubber shipped is in dry form; the remainder is shipped as concentrated liquid latex.

Smoked sheet is a common type of crude rubber, and the price of rubber usually is quoted on ribbed standard smoked sheet 1 (RSS 1). This material is made by coagulating fresh latex with formic acid in long horizontal tanks. The coagulum is run through a series of smooth metal rolls with the clearance being decreased from one set to another, an arrangement that squeezes out the serum and densifies the wet rubber. Water is run over the wet coagulum to wash out nonrubber materials. The last unit consists of ribbed rolls which imprint ribbed markings on the sheet. After drying in air for a few hours, the sheets are hung in a drying shed at 40–50°C until dry. Originally, wood was used for drying in smokehouses. Modern installations use efficient drying tunnels. Sheets are inspected by holding them over a strong light to determine clarity, color, presence of dirt, and other factors. The material is classified 1, 2, and on down to lower grades. The sheets are piled up and squeezed in a baling machine to form 250-lb bales of 19 × 19 × 24 in. (480 × 480 × 600 mm) dimensions.

Crepe, another popular type of commercial rubber, is of two major classes, pale crepe and thick blanket crepe. Pale crepe is made by adding dilute $NaHSO_3$ to latex to inhibit discoloration and softening during processing. Formic acid is used as the coagulant. The wet coagulum is passed through rolls with longitudinal grooves which give the rubber a crepelike appearance.

Water running over the surface cleans out dirt and other nonrubber materials. Sheets are hung up to dry in circulating warm air. The quality of pale crepe is assessed on its whiteness and how well the finished rubber appears. This is a premium product and commands a higher price than RSS 1. Blanket crepes are of lower quality and are made from wet slabs obtained from small holders. These are creped, dried, and baled. Other types of crepe are made from coagulum left in collection cups and from dried skin remaining from the tapping incision. In addition to collecting latex, a tapper collects all dried and coagulated rubber that remains from the round before. This last type of crepe is a high-grade rubber but not as high as pale crepe or RSS 1.

Granular Forms: Several types of natural rubber are processed in granular form, the most popular of which is *Hevea* crumb from Malaysia. The wet rubber coagulum is treated with a small amount of noncompatible oil (usually castor oil) and passed through creping rolls, which shred the product into fine crumbs. The material is washed and because of its finely divided state easily dried, resulting in a high-grade, completely unoxidized crude rubber. The rubber is pressed into 75-lb bales and marketed at a premium price, usually a few cents above standard RSS 1. Other methods for making granular rubber consist in mechanically cutting or chopping the wet coagulum or sheets.

Extrusion Process: In this process, wet coagulum collected from small estates is taken to a central place and fed into an extrusion machine having a two-stage drying system which drives off the water. The rubber produced looks like popcorn and is easy to bale. This process offers the small holders an opportunity to upgrade their rubber and obtain a better price.

Grading on Technical Standards: Standard Malaysian Rubber (SMR) was introduced in 1965 and is a fast-growing natural-rubber class. SMR is graded on technical standards rather than visual properties, as is done with smoked sheet and crepes. Dirt, ash, nitrogen content, volatile matter, plasticity, change in plasticity on aging, and color standards have been established for the various SMR grades. An additional advantage is the packaging of SMR in smaller, more uniformly wrapped bales similar to those used for synthetic rubbers.

—Thomas H. Rogers, Jr., *Goodyear Tire & Rubber Company, Akron, Ohio*

Rubber, Synthetic (See **Elastomers.**)

Rubidium (L. *rubidus*, red.) **Rb** = 85.48 (at. wt.); 37 (at. no.). Rubidium is one of the alkali metals, a member of group IA, period 5, of the periodic table. It was discovered by Bunsen and Kirchhoff (1861) in the mineral lepidolite by means of the spectroscope. In 1861, Bunsen obtained the free element by electrolysis of its fused salts.

Two isotopes of Rb occur in nature, ^{85}Rb and ^{87}Rb, the latter being radioactive. Several unstable Rb isotopes with mass numbers from 81 to 92 or higher have been obtained artificially. See also **Chemical elements.**

Rubidium is a silvery-white, soft metal closely related to K but more volatile. Rb also has some similarity to Li and can be cut with a knife. The crystal structure is body-centered cubic. Principal physical properties are given in Table R-13.

Chemical Properties: Rb distills and forms a blue vapor when heated to a red heat in the absence of O_2. Rb is soluble in liquid NH_3 and insoluble in hydrocarbons. Rb is more readily oxidized than K. Upon exposure to air, a gray-blue oxide rapidly forms. This film is a mixture of the monoxide, Rb_2O, the peroxide, Rb_2O_2, and the *superoxide,* RbO_2. Rb reacts with water, even in the form of ice, at $-108°C$ to form rubidium hydroxide. At room temperature, the reaction is violent, the evolved H_2 igniting. Rb dissolved in liquid NH_3 combines with ozone. Rb unites directly with Br_2 and Cl_2 with production of a flame. A summary of Rb compounds is given in Table R-14.

Although only extremely small quantities are involved, Rb gas-cell systems now serve as secondary time standards, like quartz-crystal oscillators, because they must be referenced to more accurate systems. Rb systems have a characteristic resonance at 6,835 MHz and are useful because, unlike other atomic-frequency standards, they require little power and are relatively compact. Portable Rb atomic clocks weighing as little as 44 lb and occupying about 1 ft^3 were introduced by the U.S. Army Electronics Command in 1963. These clocks operate on standard 110-V current, the 24-V output of military vehicles, or both. The clocks are used to set precise radio broadcasting frequencies, to synchronize radar nets, and to assist in the accurate tracking of missiles and satellites. See also **Cesium.**

Classified as an absolute-type magnetometer, Rb-vapor instruments were developed in the United States in 1958 by government scientists, including those of the Coast and Geodetic Survey and the National Bureau of Standards. In operation, the Rb-vapor magnetometer utilizes an Rb lamp, which is mounted in the tank coil of a radio-frequency oscillator. After collimating and filtering, the Rb light is circularly polarized and passed through a Rb-vapor cell, then focused on a sensitive photocell. Various combinations of amplifier parameters and Rb isotopes allow considerable range in the measurement of ambient magnetic fields. Since the total world range is 15,000–80,000 gammas, any system capable of this span is usable anywhere on earth.

Production: Rb is obtained from a by-product of potassium extraction, known as artificial carnallite, which contains about 0.02% RbCl. Rubidium and cesium can be obtained as precipitates of silicomolybdate. Molybdenum is removed by volatization as oxychloride when a current of hydrogen chloride is passed through. Alcohol is added to the residue to form an acid solution. Potassium and rubidium are then precipitated out. Cesium antimony chloride is precipitated alone by using antimony chloride; after repeating the silicomolybdate step, potassium-free Rb is obtained. To yield a pure metal, the fused hydride is subjected to electrolysis.

Alloys: Rb forms alloys with Ag, Au, K, and Na, as well as amalgamating quickly with Hg. In the system Na-Rb there is a eutectic, but no compounds or solid solutions are formed. K and Rb are completely miscible in the solid state, while Cs and Rb form an uninterrupted series of solid solutions. These alloys are used to eliminate the last traces of air from vacuum tubes (known as *getters*) and for obtaining high vacuums in other communications devices.

Several Rb compounds are available commercially, including the bromide, carbonate, chloride, fluoride, hydroxide, and iodide. Rubidium has an excellent potential for use in ion propulsion en-

TABLE R-13. *Some Physical Properties of Rubidium**

Melting point, °C	38.8
Boiling point, °C	688
Density, g/cm^3, at 20°C	1.53
Mean specific heat (0–100°C), cal/(g)(°C)	0.085
Resistivity at 20°C, $\mu\Omega$-cm	12.5
Coefficient of lineal thermal expansion (0–100°C)/°C	9×10^{-6}

*Additional properties can be found in Ref. 3.

TABLE R-14. *Some Rubidium Compounds and Their Formation*

Oxygen compounds:

Rb combines with O_2 very vigorously and forms a variety of compounds; Rb oxides react with H_2O to form RbOH, liberating O_2, except for Rb_2O, which forms only RbOH

Monoxide, Rb_2O (yellow, cubic); sp gr 3.72

Dioxide or peroxide; Rb_2O_2

Dioxide or peroxide; Rb_2O_2

Superoxides; Rb_2O_3 (trioxide); Rb_2O_4 (tetraoxide)

Halogen compounds:

The halides of Rb generally are obtained by reacting RbOH or Rb_2CO_3 with the corresponding halogen acid (example: $Rb_2CO_3 + 2HCl \rightarrow 2RbCl + H_2O + CO_2$)

RbBr (cubic); refractive index 1.5528; sp gr 3.35; soluble in H_2O

$RbBr_3$ (rhombic)

RbCl (cubic); refractive index 1.4936; sp gr 2.80; moderately soluble in H_2O

RbF (white, cubic); refractive index 1.396; sp gr 3.56; quite soluble in H_2O

RbI (colorless, cubic)

Several crystalline hydrates and acid salts have been obtained among the fluorides, namely $2RbF \cdot 3H_2O$, $RbF \cdot 3H_2O$, $RbF \cdot 2HF$, $RbF \cdot 3HF$

The characteristic behavior of Rb in forming poly- and complex halides is shown by several compounds; $RbBr_3$, RbI_3, RbI_7, RbI_9, $RbBr_2I$, RbBrICl; for example, RbI_3 can be obtained by dissolving RbI in sulfamic acid: $3RbI + 2NH_2SO_3H + H_2O \rightarrow RbI_3 + Rb_2SO_4 + (NH_4)_2SO_3$

Oxyhalides:

Rb readily combines with the oxyhalide ion giving compounds:

$RbBrO_3$ (bromate) sp gr 3.68, slightly soluble in H_2O

$RbClO_3$ (chlorate) (trimetric) sp gr 3.19; slightly soluble in cold H_2O; moderately soluble in hot H_2O

$RbClO_4$ (perchlorate) (rhombic) sp gr 2.9; practically insoluble in cold H_2O; moderately soluble in hot H_2O

$RbIO_4$ (periodate) (colorless, tetragonal) sp gr 3.918; practically insoluble in cold H_2O

Sulfur compounds:

Sulfides: Rb forms a series of sulfides, some of which are hydrated: Rb_2S (red, deliquescent) sp gr 2.912; very soluble in H_2O; other sulfides include $Rb_2S \cdot 4H_2O$ (sulfide tetrahydrate), Rb_2S_3 (trisulfide), Rb_2S_4 (tetrasulfide), and Rb_2S_5 (pentasulfide)

Sulfates:

Rb_2SO_4 (rhombic) refractive index 1.5133; sp gr 3.613; soluble in H_2O

$RbHSO_4$ (bisulfate)

$Rb_2S_2O_7$ (pyrosulfate)

Borates:

Rb generally exists in combination with boron as borates; $RbBO_3 \cdot H_2O$ (perborate); $Rb_2B_4O_7$ (tetraborate); $RbBO_2$ (metaborate); $RbBO_4 \cdot \frac{1}{2}H_2O$; as well as the borohydride, $RbBH_4$, which is extremely stable

Other compounds:

Some uncommon salts of Rb include Rb_3PO_4, RbSCN, RbCNO, Rb_2WO_4, $RbMoO_4$, $RbReO_4$, $RbTaO_3$, and $RbNO_3 \cdot HNO_3$

Rb also forms the following, the majority of which are difficultly soluble compounds and can be used for the detection and eventual separation of Rb from accompanying elements; e.g., Rb_2SiF_6, $RbCl \cdot CoCl_2 \cdot 2H_2O$, Rb_2PtCl_6, $RbPu(NO_3)_6$, $Rb_2O \cdot GeO_2 \cdot 12MoO_3 \cdot 9H_2O$

Rubidium forms a number of mixed ferrocyanides; Rb also forms organic compounds, but the known organic and organometallic compounds of Rb are few

gines (space vehicles) because, like cesium, Rb is easily ionized and has a reasonably high atomic weight. The potential of Rb and Cs for use in the generation of electricity via the magneto-hydrodynamic principle also is good.

REFERENCES

1. Perelman, F. M.: "Rubidium and Caesium," trans. R. G. P. Townsdrown, Pergamon-Macmillan, New York, 1965.

2. Simons, Eric N.: "Uncommon Metals," Hart Publishing, New York, 1967.

3. Smithells, Colin J.: "Metals Reference Book," 4th ed., Plenum, New York, 1967.

—Technical Staff, *Kawecki-Berylco Industries, Inc., Reading, Pa.*

Ruby (See **Bauxite**.)

Ruthenium (L. *Ruthenia,* Russia.) **Ru** = 101.1 (at. wt.); 44 (at. no.). Ruthenium is a member of the family of platinum metals, group VIII of the periodic table. The melting point is 2250°C; boiling point 4900°C; density at 20°C is 12.32 g/cm^3. There are seven stable isotopes, 96, 98–102, and 104, and five unstable isotopes, 95, 97, 103, 105, and 106. Ru was discovered by Claus (Germany) in 1844.

The chemistry of Ru is still poorly understood. The existence of at least eight valence states, coupled with the tendency to complex with many ions, often results in the presence of several different complexes in a given solution.

Ru metal is quite refractory. It is not significantly soluble in any single acid; even aqua regia has little effect. At room temperature, the metal does not react with O_2, but when heated in air, a film of the dioxide appears. The metal is insoluble in fused sulfates. Molten alkali slowly dissolves the metal. The rate of attack is rapid under oxidizing conditions, and a molten mixture of NaOH and Na_2O_2 will readily dissolve the metal.

The finely divided metal is soluble in hypohalites if an excess of alkali is present. At red heat, the metal combines with Cl_2 to form the dichloride. Ruthenium(VIII) oxide is formed when an alkaline ruthenium solution is treated with a strong oxidant, such as chlorine, or bromate ion when the Ru is in acid solution.

Ruthenium(III) hydroxide is formed by the action of alkali on a solution of ruthenium(III) chloride. It is easily oxidized by air to the tetravalent state. The dioxide, RuO_2, forms when the metal is heated in air. Hydrous ruthenium(IV) oxide can be precipitated by adding alcohol to a less than 3 *M* NaOH solution of ruthenium(VIII) oxide, followed by boiling. Above 3 *M* NaOH, complete reduction is not obtained. The hydrous oxide that is soluble in concentrated HCl tends to occlude impurities.

The only known octavalent Ru compound is the tetroxide, RuO_4, which exists in a yellow and a brown form. The volatile and poisonous tetroxide melts at about 25°C and sublimes readily. It may explode in contact with oxidizable substances or when heated above 100°C. It is formed by distillation from either an alkaline or acid solution under strongly oxidizing conditions. The tetroxide is moderately water-soluble. When dissolved in alkali, it initially forms a green solution of heptavalent perruthenate of the form $MRuO_4$, which further reduces to the orange ruthenate

M_2RuO_4. The reduction to the hexavalent state is quicker in strong alkali. The ruthenates also are made by fusing finely divided metal with a mixture of alkali hydroxide and nitrate or peroxide.

Anhydrous ruthenium(III) chloride, $RuCl_3$, is made by direct chlorination of the metal at 700°C. Two allotropic forms result. The trihydrate is made by evaporating a HCl solution of ruthenium(III) hydroxide to dryness or reducing ruthenium(VIII) oxide in a HCl solution. The trihydrate, $RuCl_3 \cdot 3H_2O$, is the usual commercial form. Aqueous solutions of the trihydrate are a straw color in dilute solution and red-brown in concentrated solution. Ruthenium(III) chloride in solution apparently forms a variety of aquo- and hydroxy complexes. The analogous bromide, $RuBr_3$, is made by the same solution techniques as the chloride using HBr instead of HCl.

Ruthenium(III) iodide, RuI_3, is a black, insoluble compound precipitated by the addition of iodide ion to a solution of $RuCl_3$.

Tetravalent ruthenium chloride, $RuCl_4$, and the hydroxychloride, $Ru(OH)Cl_3$, are intermediate products when $RuCl_3$ is prepared by evaporating the tetroxide in HCl. When the hydrooxychloride in hot HCl is treated with Cl_2, it is converted to the tetrachloride. The anhydrous tetrachloride also is known. The tetrabromide and tetraiodide have not been isolated; attempts to prepare these compounds result in the formation of the respective trihalides.

The only pentavalent Ru compounds known are the fluorides; RuF_5 is made by combining the elements. The compound melts at 107°C and boils at 313°C. The salt, $NaRuF_6$, was recently made by mixing $RuCl_3$ with NaCl and treating the mixture with BrF_3.

Ru forms many complex ions. The nitrosyl compounds are frequently encountered by accident due to the great affinity of Ru for the nitrosyl group. Ruthenium(III) nitrosylchloride, $Ru(NO)Cl_3 \cdot 4H_2O$, is a by-product of most solutions of $RuCl_3$ in aqua regia or solutions containing HNO_3. It also is present in HCl solutions resulting from a KOH and nitrate fusion of the metal. The chloride and bromide are respectively raspberry and violet in solution. Alkaline chlorides form complex salts of the type $M_2Ru(NO)Cl_5$, which can be crystallized from solution. A black gelatinous precipitate of the nitrosylhydroxide, $RuNO(OH)_3$, is slowly formed when a solution of the nitrosylchloride is heated with a strong base. A series of nitrato and nitro derivatives of nitrosylruthenium also have been described and separated.

It is generally accepted that the disulfide is the only certain sulfide of Ru. It is formed by the action of H_2S on a solution of Ru or from the elements at about 1000°C. When ruthenium(IV) sulfide is treated with HNO_3, the sulfate is formed.

Ruthenium dicarbonyl dichloride, $Ru(CO)_2Cl_2$, is formed when $RuCl_3$ is heated above 210°C in the presence of CO. It is a yellow, insoluble, volatile compound. The bromine and iodine analogs are similarly formed.

When finely divided Ru metal is heated at 180°C under 200 atm of CO, ruthenium pentacarbonyl, $Ru(CO)_5$, is formed.

Ruthenium forms a large number of complex ions with amines.

Recently, a new group of organometallic sandwich compounds, called *metallocenes,* has been discovered. Ruthenocene is made in about 50% yield by reacting $RuCl_3$ with cyclopentadienylsodium in tetrahydrofuran. After refluxing and distilling the solvent, the light yellow crystals of ruthenocene are sublimed. The compound, $Ru(C_5H_5)_2$, undergoes a large number of substitution reactions typical of aromatic systems.

Further information concerning the sources, refining, and uses of Ru will be found under **Platinum metals and platinum.**

—Lionel E. Simmons, *Simmons Refining Company, Chicago, Ill.*

Rutile (See **Titanium; Titanium dioxide;** and **Zirconium.**)

S

Saccharides (See **Starch.**)

Saccharin (See **Toluene.**)

Saccharomyces (See **Yeasts.**)

Safe Handling of Chemicals (Consult the **Subject Index.**)

Salad Oil (See **Vegetable oils.**)

Salt A salt is generally defined as a substance that yields ions, excluding substances that yield hydrogen ions (acids) and hydroxyl ions (bases). Often salts are classified on the basis of their parent acid and base, i.e., (1) a salt of a strong acid and a strong base; (2) a salt of a strong acid and a weak base; (3) a salt of a weak acid and a strong base; and (4) a salt of a weak acid and a weak base. Sodium chloride, formed by reacting the strong base sodium hydroxide with the strong acid hydrochloric acid, is in category (1). When a salt of this type is dissolved in water, any recombination of ions present simply results in the formation of sodium hydroxide and hydrochloric acid in equal proportions, and thus an aqueous solution of sodium chloride is neutral and if the

water is pure will have a pH of 7. Salts in the other three categories *hydrolyze* when dissolved in water; i.e., they do not form a neutral solution but one that is acidic or basic. An example is provided by sodium acetate, the salt of a weak acid (acetic acid) and a strong base (sodium hydroxide). When sodium acetate is dissolved in water, it gives an alkaline reaction that is the result of hydrogen ions from the water being bound up by the undissociated acetic acid, thus leaving hydroxyl ions in excess. The hydrolysis of ferric chloride produces an acid reaction inasmuch as hydroxyl ions are removed from solution in the form of insoluble ferric hydroxide, leaving hydrogen ions in excess. Similarly, ammonium chloride hydrolyzes to provide an acidic reaction as the result of the hydroxyl ions being removed from the water in the form of ammonium hydroxide, leaving hydrogen ions in excess. Thus, hydrolysis of salts occurs when the ions of the water (hydrogen or hydroxyl or both) are removed by the ions of the salt either to form insoluble or undissociated substances. See also **pH (hydrogen-ion concentration).**

A significant percentage of important high-tonnage industrial chemicals are salts, and several of them are described elsewhere in this volume (consult the Subject Index). Generally, inorganic salts are designated by the anions which they contain, e.g., as chlorides, which can be considered as reaction products of hydrochloric acid, nitrates (nitric acid), sulfates (sulfuric acid), phosphates (phosphoric acid), sulfites (sulfurous acid), nitrites (nitrous acid), acetates (acetic acid), and hypochlorites (hypochlorous acid). The prefix normally is simply the cation, as *sodium* chloride, *potassium* nitrate, and *magnesium* sulfate.

In addition to the direct reaction between an acid and a base to produce a salt, a salt can also be formed by displacing the hydrogen of an acid by a metal. Thus, zinc dissolves in hydrochloric acid to form hydrogen and the salt, zinc chloride.

The alums are examples of *double salts,* which include the compound known as alum, $KAl(SO_4)_2 \cdot 12H_2O$, or potassium aluminum sulfate. Other alums include $NaCr(SO_4)_2 \cdot 12H_2O$, or sodium chrome alum; and $NH_4Fe(SO_4)_2 \cdot 12H_2O$, or ferric ammonium alum. From the foregoing, it is obvious that aluminum sulfate, $Al_2(SO_4)_3 \cdot 18H_2O$, known as *papermakers' alum,* is not a true alum.

Because salts yield ions, they are electrolytes and form the basis for numerous electrochemical operations.

As shown by Table S-1, the solubility in water of salts varies widely. For example, the great majority of ammonium salts are readily soluble; a large percentage of chlorides are soluble; while most oxalates, phosphates, and sulfides are essentially insoluble.

Advantage can be taken in chemical processing of the so-called *salting-out* effect when a salt solution is added in certain situations. For example, the presence of a dissolved salt reduces the solubility of gases in water. Although the degree of

TABLE S-1. *Relative Solubilities of Principal Inorganic Salts*

	Aluminum	Ammonium	Antimonous	Barium	Bismuth	Cadmium	Calcium	Chromium	Cobalt	Cupric	Ferric	Ferrous	Lead	Magnesium	Manganese	Mercuric	Mercurous	Nickel	Potassium	Silver	Sodium	Stannic	Stannous	Strontium	Zinc
Acetate															MS		SS			MS					
Arsenate	I		I	AI	I		AI	AI	I	I	I		I	AI	AI	I	AI	I		I		I		SS	I
Arsenite			I						I	I	AI			AI		AI	I	AI		AI				SS	
Borate				AI	SS									AI	AI					AI				AI	
Bromide													SS			SS	I			I					
Carbonate				AI	I	AI	AI		I	AI			AI	I	AI	AI	I	AI		AI				AI	AI
Chlorate																									
Chloride													MS			MS	I			I					
Chromate			I	AI		SS	AI		AI	MS			I			SS	SS	AI		AI				SS	SS
Cyanide				MS					I			I	SS		AI		AI			I					AI
Ferricyanide				I	I				I	I			I	AI		I		I		I					
Ferrocyanide			SS	I	I				I	I	I	I	I	AI		I		I		SS		AI	AI		I
Fluoride	SS			SS		SS	AI	AI	MS	AI	AI	SS	AI	AI	SS	MS	MS	MS		MS				AI	MS
Hydroxide (oxide)	I		AI	MS	I	I	SS	I	I	I	I	I	AI	AI	I	I	I	I		AI		I	I	SS	I
Iodate		MS		AI	AI	SS			SS	SS				AI		AI	AI			AI				MS	SS
Iodide			SS		AI								AI			I	I			I		MS	SS		
Nitrate																									
Oxalate	AI	MS		AI	AI	AI	I	SS	I		AI		AI	I	AI	AI	I	I	I	AI	MS			AI	I
Phosphate	I			I	I	I	AI	I	I	I	I	I	AI	I	AI	I	I	I	I	I		I		I	I
Silicate	I			I		I	I	I	I		I	I	I	I	I		I							I	I
Sulfate				I			SS						AI			MS	AI			SS				AI	
Sulfide			I	MS		I	I	SS	I	I	I	I	I		I	I		I		I		I	I		I
Sulfite				AI	AI	AI			AI				AI	AI	MS		AI			AI			AI	AI	
Thiocyanate				SS						AI				MS		AI	I			I					

	Rare or nonexistent compound
	Soluble (over 50 g/l)
MS	Moderately soluble (10-50 g/l)
SS	Slightly soluble (1-10 g/l)
AI	Almost insoluble (0.01-1 g/l)
I	Insoluble (less than 0.01 g/l)

the effect of salting out varies from one salt to the next, the decrease in solubility is the same for different gases for a given salt. Theoretically, the presence of the salt reduces the role of water as a solvent for the gases because a portion of the water is required for hydration of the salt. The salting-out effect is used in the manufacture of soaps, application of dyes, and the precipitation of albumins (using ammonium sulfate). See also **Peptides and proteins; and Soaps.**

Salt (NaCl) Sodium chloride is produced in virtually every country of the world although in some countries production is sufficient only to fill local needs. The 10 largest salt-producing countries* are the United States, mainland China, the Soviet Union, West Germany, France, the United Kingdom, India, Italy, Canada, and Mexico. In Europe, salt is produced from bedded deposits found in the north and by solar evaporation in the south. Salt from South America, Asia, Africa, and the Middle East is mainly solar salt, except for some rock salt in China, Pakistan, northern India, and certain areas of Africa. Several of the key physical properties of salt are given in Table S-2.

Salt occurs as bedded or domed deposits in 28

*"U.S. Bureau of Mines Minerals Yearbook," Washington (annual).

TABLE S-2. *Some Properties of Sodium Chloride* *

Mineral name	Halite
Chemical symbol	NaCl
Molecular weight	58.4428
Composition:	
% Na	39.342
% Cl	60.658
Specific gravity	2.165
Crystal habit	Isometric system, cubes
Hardness, Mohs' scale	2.5
Refractive index, n_D^{20}	1.554
Melting point:	
°C	800.8 ± 0.5
°F	1473.4 ± 0.9
Boiling point:	
°C	1413
°F	2575.4
Specific heat, cal/(g)(°C) . . .	0.204
Solubility in water,	
g/100 g H_2O:	
At 0°C	35.8
100°C	39.1

*Extensive additional tabulations have been made by Kaufmann.[1]

states of the United States and in several provinces of Canada. Michigan, New York, Texas, Ohio, Louisiana, and Kansas account for the bulk of the production of rock salt in the United States. This excludes solar salt production at Great Salt Lake, Utah, and on the west coast. Salt deposits were formed in past geologic ages by evaporation of impounded salt water under desert conditions. Two types of deposit are found: (1) essentially horizontal stratified beds, as originally laid down, and (2) a plug, or dome, formed by superincumbent weight on deep-lying horizontal beds, causing upward flow toward the surface through zones of weakness.

All rock salt is basically sodium chloride, but because of the manner in which it was formed, physical characteristics vary. Commercial rock salt ranges in purity from about 97% (Kansas) to about 99% (Louisiana) and averages about 98% for the northern United States. By far the most significant impurity in rock salt is the mineral anhydrite (calcium sulfate, $CaSO_4$). Dolomite, quartz, calcite, and iron oxides are found in minor quantities. For example, rock salt from New York state averages 98.24% NaCl; 1.48% $CaSO_4$; 0.05% $CaCO_3$; 0.06% $MgCO_3$, 0.11% SiO_2; 0.04% Fe_2O_3; and 0.02% H_2O. Of these ingredients, only the Fe_2O_3 and the SiO_2 are truly insoluble. The mineral impurities are quite inert and possess no toxic or otherwise undesirable properties. No bacteria exist in rock salt.

Rock salt is produced by dry mining a deposit of salt underneath the earth's surface in a manner similar to coal mining. Rock salt is not purified in any way but is crushed and screened to commercial sizes and marketed both in bulk and bagged forms.

Evaporated-Salt Production: This form of salt is made by evaporating water from brine to form salt crystals. Not all evaporated salt is the same. There is evaporated *granulated salt,* in which each crystal is a tiny cube, and there is *grainer, or flake, salt,* which is irregular in shape, frequently thin and flaky, and unusually soft.

The conventional process for producing evaporated salt starts with rock salt in its natural underground formation (see Fig. S-1). Holes are drilled into these salt deposits, water is pumped into them to dissolve the salt, and the brine is brought to the surface for refining. In this operation all the insolubles are left behind in the well. The brine is then at least partially purified by the addition of chemicals to remove hardness and dissolved gases. The resulting semipure brine is evaporated in multiple-effect vacuum pans, where the salt crystallizes as perfect cubes of sodium chloride. The same kind of brine can be evaporated in open pans to produce grainer salt, having crystals of a characteristic hopper structure. Purified salt is manufactured conventionally by taking the same brine and subjecting it to intensive chemical purification before feeding it to the vacuum pans for crystallization.

Granulated or vacuum-evaporated salt is produced by boiling brine at less than atmospheric pressure in large tightly sealed evaporators, called *vacuum pans,* installed singly or in multiple units.

The underlying principle of the process is the lowering of the boiling point of the brine by decreasing the pressure of the vapor above the liquid. With the multiple-effect vacuum-pan system, each vacuum pan acts not only as an evaporator but also as a boiler, producing heated vapor or steam for boiling in the next succeeding pan and as the condenser for the pan immediately preceding. To illustrate the foregoing, the hot vapor produced by the evaporation in the first vacuum pan enters the heating element of the second pan. This vapor, though not as hot as the live steam used in the first pan, is hot enough to boil the brine because of the lower pressure in the second pan. Pressure in succeeding pans is similarly lowered by the condensation of vapor, aided by vacuum pumps.

When evaporation has concentrated the brine to the point of saturation, salt begins to crystallize in perfect cubic grains. The size of the crystals is controlled by the rapidity of evaporation, depending in turn on the degree of vacuum, temperature, and agitation. When the crystals have grown to the proper size, they drop to the bottom of the pans. See also **Evaporation.**

From the salt legs at the bottom of the pans the salt is drawn continuously in the form of slurry from each of the pans and is pumped to a wash tank, where additional washing is done to remove the cocrystallized anhydrite and to displace the calcium and magnesium chlorides present in the mother liquor which wets the crystal. The slurry is then pumped to a drier-filter, from which the salt is discharged with a moisture content of approximately 0.1% at a temperature of 200°F (93°C). It then passes through a rotary cooler, where the remaining traces of moisture evaporate off. This cooler also removes traces of dust from the surface of the crystal, giving it polish. From the cooler the salt is conveyed to a series of vibrating screens, where it is graded, and stored in the

proper bins for bulk shipment or feeding to the bagging machines.

Grainer Process: In this process, grainer salt is produced by surface evaporation of brine in flat pans open to the atmosphere. Steam pipes a few inches above the tank bottom supply the heat. Due to evaporation at the brine surface, the salt crystals form on the surface and are held there by surface tension. They grow laterally, first forming thin flakes; then, as they grow, they tend to sink and develop into hollow pyramids with thin sides and relatively heavy bottom points floating point down. These are known as *hopper crystals.* Eventually they sink and are scraped to one end of the pan, removed, dried, and screened. The fragile hoppers are broken up during handling, giving the name *flake salt.*

The Alberger process is a modification of the grainer process, a portion of the feed being a hot slurry formed when superheated brine under pressure is passed through a purifying graveler and flashed to ambient temperature.

Recrystallizer Process: This process was designed to make high-purity evaporated granulated salt from rock salt and takes advantage of the differential solubility of anhydrite with respect to salt. It also features lower steam costs, which make it attractive even when the feed is solar salt, and it serves as an outlet for waste rock salt, either fine in screening or low in purity. This method does not replace the multiple-effect vacuum-pan process, for which the raw material is brine from wells rather than solid crystals of salt.

At temperatures close to the boiling point of brine, salt solubility increases with temperature while the solubility of calcium sulfate in brine decreases. Based on this difference, the fundamentals of the recrystallizer process are simply explained. Rock salt is fed at a controlled rate into a slurry tank, where it is mixed with relatively cool recycle brine saturated with respect to salt and undersaturated with respect to calcium sulfate. It is raised to a high temperature by direct injection of steam, which makes it undersaturated with respect to salt and saturated with respect to calcium sulfate. The brine flows through a saturator, where complete dissolution of the salt takes place. Since calcium sulfate is already saturated in the brine at the higher temperature, any additional calcium sulfate that is fed in with the raw salt is prevented from dissolving. The brine is then clarified, and all the undissolved calcium sulfate and sludge is removed. The hot clear brine is allowed to cool, and the pressure is reduced.

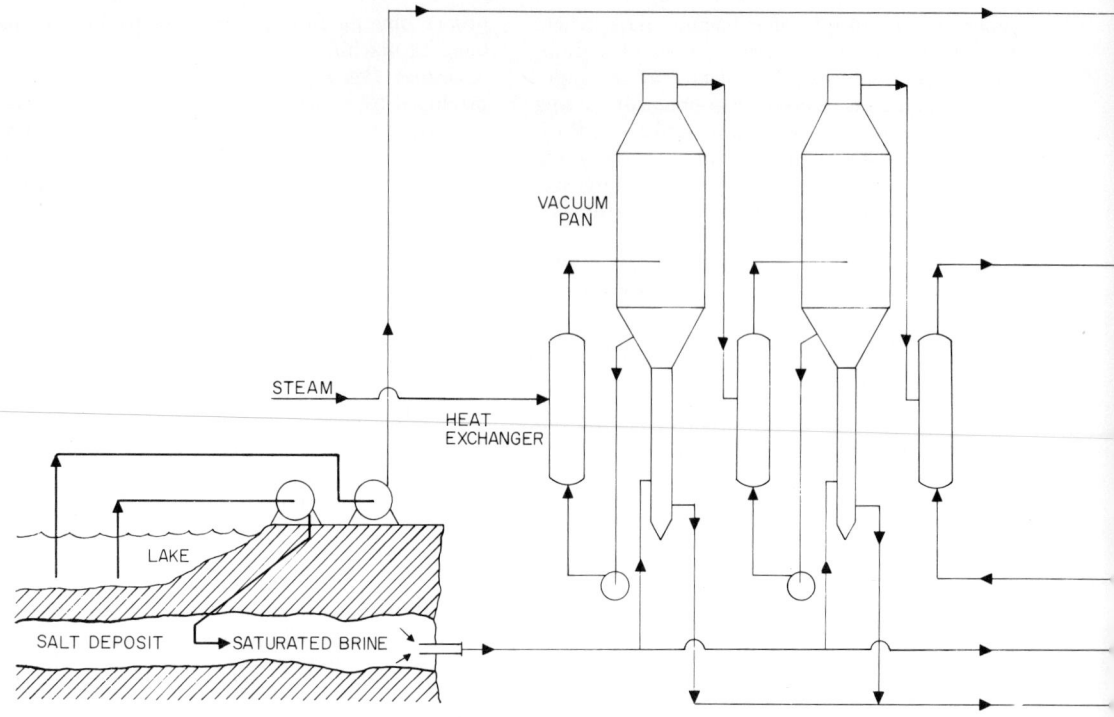

Fig. S-1. Multiple-effect vacuum pans used for salt production.

Under these conditions the brine is supersaturated with respect to salt, and the salt recrystallizes. Since the calcium sulfate is undersaturated at the lower temperature, it remains in solution and thus almost completely separated from the newly created crystals of salt. The salt is withdrawn and pumped to a top-feed rotary filter-drier, where the crystals are dewatered and dried.

Solar Salt: Because successful operation of a solar-salt facility depends upon favorable climatological conditions, it can be used only in certain areas. Principal operations in the United States are in California and at the Great Salt Lake. The brine is pumped first to concentrating ponds, where much of the water is evaporated. Some impurities are precipitated, and saturated brine is then transferred to crystallization ponds, where most of the salt is precipitated in a relatively pure state.

Evaporation takes place at the surface and produces flakes and hoppers in the same way as grainer salt, but the flakes are much larger and coarser. The salt accumulates on the bottom of the lagoons. Bitterns containing magnesium chlo-

ride, magnesium sulfate, potassium chloride, and other salts native to the raw brine are drawn off and disposed of or processed further to recover these salts. The sodium chloride is harvested by scraping machinery and is then washed, dried, and screened.

Uses: Several basic chemicals are produced directly from NaCl as a raw material: soda ash, calcium chloride, caustic soda, sodium, sodium sulfate, sodium bisulfate, hydrochloric acid, sodium cyanide, and sodium hypochlorite. A high percentage of the salt used in the United States is produced by the users directly in captive facilities. The salt usually is produced in the form of brine and carried by pipeline to various facilities for the manufacture of other chemicals. The chemical industry consumes two-thirds of the salt produced, most of this salt going to electrolytic plants to produce the aforementioned compounds. The remainder of the total salt tonnage (dry salt) finds many other industrial, highway, farm, and home uses.

Additives: Granulated salt has a tendency to absorb moisture. In its dry form the salt is quite

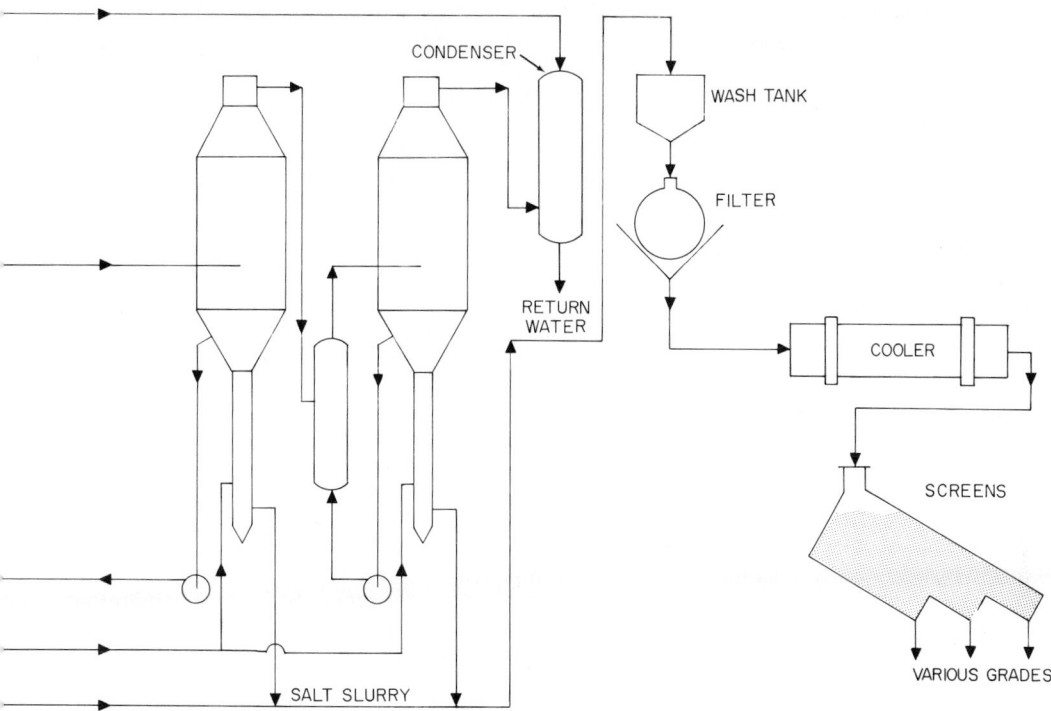

free-flowing, but as soon as it absorbs a certain amount of moisture, the granules tend to stick to each other. The addition of upward of 0.5% of a talclike compound, such as sodium silicoaluminate, calcium phosphate, magnesium carbonate, or calcium silicate, tends to protect the salt from the adverse effects of moisture and render it free-flowing even when slightly wet. These materials are called *fillers*.

Salt also has a tendency to cake, which occurs when the salt dries out after having previously absorbed moisture. The moisture is present as saturated brine, and when it evaporates, the brine recrystallizes, forming new salt crystals. These new crystals normally act as a cement to bind together the mass of granulated salt. Sodium ferrocyanide (yellow prussiate of soda) has a unique effect in preventing the cementing action. With as little as 13 ppm by weight of NaCl, the cementing action is stopped. The use of 13 ppm of sodium ferrocyanide in salt for human consumption is approved by the U.S. Food and Drug Administration (CFR 21-121.1032). Many special-purpose additives also are used in NaCl.

Commonly, KI is added to table salt to prevent goiter, and trace minerals are added for animal nutrition.

REFERENCE
1. Kaufmann, D. W.: "Sodium Chloride," Van Nostrand Reinhold, New York, 1960.

—C. E. MacKinnon, *United Salt Corporation, Houston, Texas*

Salting Out [See **Anionic surfactants (sulfur-bearing); Crystallization; Peptides and proteins; Salt;** and **Soaps.**]

Samarium (From samarskite, a mineral named after Col. M. von Samarski, a Russian inspector of mines.) **Sm** = 150.35 (at. wt.); 62 (at. no.). Samarium is the sixth of 15 elements in group III, period 6, generally shown as the rare-earth elements in a separate line below the main body of the periodic table, sometimes referred to as the *lanthanide series*. Sm was discovered in 1879 by François Lecoq de Boisbaudran, a French chemist, who named the element from the mineral samarskite, found in the Ural mountains.

Pure metallic samarium retains a silvery gray luster in dry air but is only moderately stable in moist air, forming an adherent oxide coating. This element is the only one in the lanthanide series that crystallizes in the rhombohedral form at room temperature. At 917°C it transforms to body-centered cubic form. When nonmetallic impurities, particularly oxygen and hydrogen, are at very low levels, samarium metal is soft and malleable, but it must be fabricated in inert gas or with a protective cladding. Vacuum processing is unsuitable because of samarium's high vapor pressure at elevated temperatures. Chips of samarium metal and fine powder are pyrophoric, igniting spontaneously in air and burning at 150–180°C. Storage in leakproof containers purged of air or under an inert liquid is recommended.

The melting point of Sm is 1073°C; boiling point 1791°C; density 7.536 g/cm^3; vapor pressure at 1535°C is 1.9×10^2 mm Hg. Estimated at 6.0 ppm in the earth's crust, it is the fifth most plentiful of the lanthanon elements, about 3 times the abundance of tantalum or tin. The metal has a low acute toxicity rating.

Samarium has seven natural isotopes: 144 (3.16 w/o*), 147 (15.07 w/o), 148 (11.27 w/o), 149 (13.84 w/o), 150 (7.47 w/o), 152 (26.63 w/o), and 154 (22.53 w/o). Eleven artificial isotopes have been identified. Natural ^{147}Sm is a very weak alpha emitter with a half-life of 2.5×10^{11} years. The thermal-neutron-absorption cross section at 0.025 eV of the Sm isotope mixture is 5,800 barns, making it the second highest (after gadolinium) of the elements found in nature. ^{149}Sm has a cross section of about 40,000 barns but is separated by low-cross-section isotopes, so that no chain reaction exists during neutron absorption, resulting in a fast burnout rate.

Samarium dissolves readily in dilute mineral acids and some organic acids. Many intermetallic compounds of Sm are formed with elements to the right of iron in the periodic table. With the exception of cobalt-samarium the properties of most of these metallics need to be defined and evaluated. As a strongly cationic element, Sm is an active reducing agent for the oxides of other elements as well as carbon monoxide and carbon tetrachloride. Since Sm has a tendency to absorb

*This abbreviation is used in the rare-earth and related fields for weight percent.

one outer valence electron into its $4f$ shell, it sometimes exhibits alloying properties similar to the other trivalent lanthanides, and at other times it is like the alkaline-earth metals. While most of the measured and calculated properties of Sm metal fall at points on a curve expected of a trivalent rare-earth element, there are some anomalies. Compared to other metals in the series, Sm has a low boiling point and a high vapor pressure. Its crystal structure is rhombohedral instead of hexagonal close-packed. The metal cannot be prepared by reduction with calcium, barium, or lithium or by electrolysis; only divalent chloride or fluoride salts are formed.

Samarium oxide, Sm_2O_3, has a cream-yellow color and is relatively stable in air; melting point 2350°C. It is soluble in concentrated mineral acids and insoluble in hot water. Solutions of Sm are yellow and predominately trivalent. Sm forms trivalent chemical compounds with most organic anions and many inorganic anions and divalent compounds with chlorine and fluorine. Divalent samarium is identified by intense absorption bands in the ultraviolet spectrum using spectrographic techniques. Other sharp spectral lines make Sm analysis in micro and macro amounts reliable by x-ray fluorescence and atomic absorption. Soda-lime glass containing 1 w/o of Sm_2O_3 produces a strong orange fluorescence when excited by ultraviolet light but is practically colorless in natural light.

Sources of Sm are the minerals monazite and bastnasite, which contain about 4.5 and 0.5% Sm_2O_3 respectively. Present demands for Sm are met by separation as a coproduct with europium and gadolinium from these minerals. Another potential source which contains 4.1% Sm_2O_3 is the "heavy" rare-earth residues from uranium mining in Canada. The Sm content can be separated as a coproduct with Y_2O_3, adding substantially to existing supplies. Separation of Sm from the other chemically similar elements is by liquid-liquid or solid-liquid organic ion-exchange processes. Commercial separation plants have been installed that can produce 50,000 lb/yr or more of pure Sm_2O_3 and chemical compounds.

The production and availability of Sm metal presents a different problem from that of the other light rare-earth metals. The salts and oxide of Sm do not reduce to metal using calcium, barium, or lithium, nor have electrolysis processes been discovered; instead very stable divalent compounds of Sm are formed. The most effective reducing agent for Sm_2O_3 is lanthanum (La) metal, also a rare earth. An excess of La is mixed with Sm_2O_3, and the mixture is heated in a tantalum crucible in high vacuum. The Sm metal as it is reduced volatilizes and is collected as a powder or sponge on a cooled copper or tantalum condenser plate. Subsequent remelting of the Sm in a positive argon or inert atmosphere and casting into a graphite mold produces pure metal depending on the feed material and process control.

For a detailed description of resources and processes and a table of properties see **Rare-earth elements and metals.**

Uses: Samarium-gadolinium alloyed with aluminum is sometimes used in the form of spheres and cylinders by the nuclear industry in safety devices to absorb neutrons for short periods. The nuclear properties of Sm in cermets, intermetallics, and chemical forms present one of the most promising future uses of this element.

Small amounts of Sm_2O_3 added to borosilicate glass tubing encase lanthanum borate glass rods which are drawn into fine fibers. The outside surface of the fibers containing Sm increases the transmission of light through the flexible glass fibers known as *fiberoptics*. Small quantities of Sm_2O_3 are also used in optical-glass filters.

Another recent application for Sm is in coded inks to be used by the postal system and financial institutions for automatic sorting. Samarium is an activator in phosphate-type phosphors, producing a strong narrow emission in the near-infrared region of the spectrum.

The first established use of Sm of commercial significance is the permanent-magnet alloy $SmCo_5$. This intermetallic compound is one of several rare-earth compounds discovered by the U.S. Air Force Materials Laboratory, Dayton, Ohio and recognized as a new family of permanent-magnet materials. The permanent magnetic strength already achieved by $SmCo_5$ is 4 times higher than with any previously developed material. Permanent magnets are produced by methods similar to those used for alnico, platinum-cobalt, and ferrites. Samarium is first precisely alloyed with cobalt, then crushed to micron-sized powder, pressed in a magnetic field to orient the particles, sintered in an inert atmosphere, ground to tolerance, and magnetized. Several tons a year of pure Sm metal are consumed in permanent magnets for traveling-wave tubes, electric watches, gyroscopes, and hearing aids. Designs using $SmCo_5$ and other rare-earth metals which will further enhance magnetic properties

and lower the costs of these magnets are rapidly reaching commercial production in loudspeakers, meter movements, material-handling devices, motors, and generators.

The discovery of practical uses since 1965 for Sm, which has a number of unique physical and chemical properties, justifies additional research and development efforts.

REFERENCES

Becker, J. J.: Permanent Magnets, *Sci. Am.,* December 1970.

Molybdenum Corporation of America: *Overview* 1, 3, and 28 (application information including possible uses), New York. Published periodically.

Strnat, Karl J.: The Recent Development of Permanent Magnetic Materials Containing Rare-Earth Metals, *Tech. Rep.* AFML-TR-69-299, University of Dayton, Dayton, Ohio, June 1970.

—Joseph G. Cannon, *Molybdenum Corporation of America, White Plains, N.Y.*

Sand [See **Glass;** and **Silicates (Soluble).**]

Sand Mill (See **Paint.**)

Sanitary Sewage (See **Chlorine;** and **Waste treatment.**)

Sanitation Agents [See **Anionic surfactants (sulfur-bearing); Chlorine; Detergents; Surfactants.**]

Saponification (See **Soaps;** and **Vegetable oils.**)

Saponification Value (See **Vegetable oils;** and **Waxes.**)

Sapphire (See **Bauxite;** and **Ceramics.**)

Saran (See **Fibers.**)

Saturated Compounds The term saturated frequently is applied to carbon compounds in which the atoms of carbon are fully combined with other atoms, so that a chemical change can be effected only by substitution and not by simple addition. Methane exemplifies a saturated compound in that all four available carbon bonds are taken by hydrogen atoms. Ethylene, on the other hand, contains a double carbon bond, which, under proper conditions, can be made available to accept additional hydrogen or other atoms by elimination of the double carbon bond:

$$CH_2{:}CH_2 + H_2 \rightarrow CH_3CH_3$$

Unsaturated   Saturated
(ethylene)   (ethane)

Compounds with double and triple carbon bonds therefore are *unsaturated*. Compounds that have only single carbon bonds are *saturated*.

The product of a reaction in which a saturated compound is changed by substitution (not addition) is known as a *substitution product*:

$$CH_4 \;+\; Cl \rightarrow \qquad CH_3Cl \qquad + \; Cl \rightarrow$$

Saturated   Monosubstitution product
(methane)   (methyl chloride)

$$CH_2Cl_2 \qquad + \; Cl \rightarrow \qquad CHCl_3$$

Disubstitution product   Trisubstitution product
(methylene dichloride)   (chloroform)

$$+ \; Cl \rightarrow \qquad CCl_4$$

Tetrasubstitution product
(carbon tetrachloride)

Substitution products are often termed *derivatives*.

Save-all   (See **Filtration**.)

Scandium   (L. *Scandium*, Scandinavia.) **Sc** = 44.956 (at. wt.); 21 (at. no.). Mendeleev predicted the existence of this element in 1869, called it *ekaboron*, and accurately foretold many of its properties—atomic weight, oxide formula and characteristics, and reactions of its salts and their hydrolysis. Nilson first extracted a small amount of the oxide from euxenite and gadolinite in 1879 and called it *scandia*. He was followed in the same year by Cleve, who had larger and purer quantities of the oxide and was able to prepare and describe many of its compounds and identify it with Mendeleev's ekaboron. The element was then found widespread in nature; by 1910, Eberhardt had determined its presence in more than 800 mineral species. The chemistry of the element attracted the attention of many workers, but not until 1937 was Sc obtained in metallic form by Fischer, Bruniger, and Grieneisen, who electrolyzed a fused-salt mixture of the chloride with alkali chlorides. Later, Petru and Loriers separately reduced the halides with calcium metal, and this has remained the basic method of preparation.

Scandium metal is relatively soft (Brinell hardness 125) and has a silvery luster which rapidly oxidizes in air. Table S-3 lists the salient physical characteristics. The existence of a face-centered cubic allotrope, although given here, has not received absolute recognition. The element is highly reactive and combines readily with water, air, acids, halogens, and chalcogenides. Several carbides have been reported, but the most stable one is the monocarbide, ScC, of hexagonal habit; similarly with the nitride, ScN.

Scandium has no naturally occurring radioactive isotopes although some 15 (^{40}Sc–^{50}Sc) have been prepared artificially with half-lives ranging from 0–18 s to 84 days; these decay by β^+, β^-, and γ emission with energies of 0.18–4.9 MeV.

With an atomic number of 21, Sc has the

TABLE S-3. *Physical Properties of Scandium*

Crystal type:	
Alpha	hcp Mg
Beta	fcc Cu
a_0, Å:	
Alpha	3.309
Beta	4.541
c_0, Å:	
Alpha	5.273
Beta	
Radius (CN = 12):	
Alpha	1.641
Beta	1.605
Density:	
Alpha	2.985
Beta	3.19
Metallic valence	3
Electronegativity	1.27
Work Function, eV	3.23
Melting point, °C	1538 ± 20
ΔH_f, kcal/mole	4.2
Vaporization, kcal/mole:	
ΔH_s	80.8
ΔH_v	78.6
Boiling point, °C	2900
Vapor pressure (log $P_{mm} = A/T + B$):	
A	17,200
B	8.31
Entropy, cal/(mole)(°C)	8.3
Heat capacity ($C_p = A + Bt + Ct^2$):	
A	6.00
$B \times 10^3$	1.1
Magnetic susceptibility, emu/mole	8.08×10^6
Magnetic moment	0.14
Hardness, Brinell	125
Ionization potential, eV:	
First	6.54
Second	12.80
Third	24.75
Fourth	73.9

atomic structure $[Ar]3d^14s^2$ and is diamagnetic, its residual paramagnetism increasing slightly with decreasing temperature from $7.0–8.2 \times 10^{-6}$ esu over 292–90 K. Most recent work on the magnetic susceptibility of scandium has given the value of 4.76 units after correcting for a diamagnetic susceptibility of 25.8×10^{-6} esu.

Solutions of Sc salts have no visible absorption spectra. Arc and spark spectra show numerous lines, however, with abundant multiplets. ScO lines predominate at longer wavelengths.

Sources: The geochemistry of scandium has been studied by many workers, and Borisenko has provided an excellent survey of available data. Terrestrial abundance of the element has been determined to be $5–6 \times 10^{-4}\%$, making it more prolific than antimony, bismuth, silver, and gold and cosmically equal to beryllium, boron, strontium, tin, selenium, and tungsten. Typically disperse, Sc appears mainly in basic and ultrabasic rocks. Sc_2O_3 concentrations in seawater of 4×10^{-5} g/ton and in fresh water of 10 mg/l have been found. Because the scandium ion readily forms both anionic and cationic complexes, its widespread dissemination is easily understood and, because of its chemical similarity to the lanthanide series, a considerable number of scandium-bearing minerals are, essentially, sources of the lanthanides. Sc is never found without the lanthanide elements, particularly those of atomic number greater than 64, and some association with yttrium also is seen because of this symbiosis.

Only three true Sc minerals exist, thortveitite, a silicate; and sterrettite and kolbeckite, both phosphates. Apart from these, only wiikite and bazzite, complex niobates (columbates) and silicates respectively, are known to contain more than 1% Sc. Because of the magnitude of its deposits, davidite, having a concentration of 0.02% Sc_2O_3, is a major source of the element.

Sc is usually extracted from these sources together with the lanthanide group and uranium by acid extraction or alkali roasting. Oxalate precipitation then yields a concentrate suitable for separation by ion-exchange techniques. Following crude separation, refinement of the process will yield scandium oxide having a purity of 99.99% or better. Solvent-extraction techniques (thiocyanate in ether, etc.) have also been applied but do not compare economically with the ion-exchange approach. Compounds of scandium are usually prepared from the oxide, and the metal is now customarily prepared by reduction of the fluoride,

ScF_3, by calcium metal followed by purification by distillation at 1700°C and 10^{-7} mm Hg. By this means metal of better than 99.99% purity has been obtained.

Chemistry: In aqueous solution, the Sc ion has a triple positive charge, and its ionic size has been given as 0.81 Å. Recent studies show, however, that the simple Sc^{3+} ion rarely exists, the solutions being highly polymerized and hydrolyzed, hydroxy-bonded structures being the rule rather than the exception. In this context Sc shows greater similarity to Al, Y, Ga, In, and Tl than to the lanthanide series. No evidence has been presented for a valence in solution other than 3^+.

In compound formation, Sc closely follows the properties of Al, Y, and La, as would be expected from the periodic table. The white oxide, Sc_2O_3, is more basic than either yttria or lanthana. It has the cubic Tl_2O_3 structure with 16 moles of Sc_2O_3 in the unit cell with a 9.79 Å and Sc—O distances 18 Å. The oxide is readily soluble in mineral acids. Scandium hydroxide, $Sc(OH)_3$, can be precipitated from solution by alkali hydroxides but dissolves in excess to form alkali scandiates, thus showing similarity to Al but not Y. With ammonium hydroxide hexamminoscandium ions are obtained, $[Sc(NH_2)_6]^{3+}$. In the absence of alkali ions, the water solubility of scandium hydroxide is $\sim 5 \times 10^{-7}$ g mole/l. On ignition at 600°C the hydroxide converts to the oxide.

Simple (normal) scandium carbonate does not exist, but double carbonates are precipitated of the type $Na_5Sc(CO_3)_4 \cdot xH_2O$ and $NH_4Sc(CO_3)_2 \cdot xH_2O$.

The normal scandium oxalate differs from those of the lanthanide elements to a marked degree. Precipitation is always incomplete although water solubility of the oxalate is about 60 mg/l. Soluble complex oxalates are obtained with alkali oxalates. The oxalate decomposes at 635°C to the oxide.

Scandium halides are readily prepared by dissolution of the oxide in the corresponding acid. Only the fluoride is insoluble, and all are white. Again, double compounds such as K_3ScF_6 are formed. The perchlorate, $Sc(ClO_4)_3 \cdot 6H_2O$, is very hygroscopic and undergoes extensive polymerization in solution. The nitrate, $Sc(NO_3)_3 \cdot 4H_2O$, crystallizes readily but is deliquescent and soluble in ether and other organic solvents. Scandium sulfate, $Sc_2(SO_4)_3 \cdot 5H_2O$, again forms double salts, of which the potassium duplex is much less soluble in ether than the sodium or ammonium salt.

Several organometallic complexes of scandium have been prepared, notably the acetyl acetonates, and cyclopentadienyls, of which the latter are quite volatile.

The Sc ion appears to have no function in human metabolism, but when ingested may migrate to the bony structure to effect a partial replacement of calcium.

Analysis for and of scandium, both qualitative and quantitative, is not a simple matter and is best confined to spectroscopic techniques—x-ray, arc, spark, atomic absorption, or neutron activation. Wet methods are highly unspecific and unreliable, and there exists no specific unequivocal chemical test for the presence of Sc.

Limited Applications: Except possibly in exotic light sources, where the iodide appears to enhance luminosity, no real industrial application currently exists for Sc or its compounds. Earlier suggestions that the metal might enhance or replace aluminum in the aircraft industry were not fulfilled because of the metal's reactivity and high price. Some minor inclusions of Sc are made in substituted yttrium iron garnets for specific electronic purposes, but it is unlikely that this use will be of any magnitude.

REFERENCES

Borisenko, L. F.: "Scandium: Its Geochemistry and Mineralogy," Consultants Bureau, New York, 1963.

Daane, A. H.: p. 261 in E. V. Kleber (ed.), "Rare Earth Research," Macmillan, New York, 1961.

Vickery, R. C.: "Chemistry of Yttrium and Scandium," Pergamon, London, 1960.

Vickery, R. C.: "Analytical Chemistry of the Rare Earths," Pergamon, London, 1961.

Vickery, R. C.: Scandium, Yttrium and Lanthanum in A. F. Trotman-Dickenson (ed.), "Comprehensive Inorganic Chemistry," Pergamon, Oxford, 1973.

—R. C. Vickery, *Hudson Laboratories, Hudson, Fla.*

Scavengers (See **Cesium; Hafnium; Lanthanum; Rubidium;** and **Zirconium.**)

Scents (See **Aroma chemicals.**)

Scheelite (See **Tungsten.**)

Schweitzer's Reagent (See **Cellulose.**)

Scrap-Metal Processing (See **Iron- and steelmaking; Gold; Mercury; Platinum metals and platinum;** and **Silver.**)

Screening Numerous processes require the separation of solid particles of various dimensions for a variety of reasons, including the separation of over- and undersize materials from final products for market and similarly of raw materials to assure that the particle size is optimal for subsequent processing equipment. In the case of separating materials into two size categories, the screen essentially is a go no-go gage. In more complex separation operations, several size categories may be separated, but each stage operates on the go no-go principle.

The screening surface may comprise woven wire, silk, or plastic cloth; perforated or punched plate; grizzly bars; or wedge wire sections, depending upon the particle sizes involved and the abrasiveness and other properties of the materials to be separated. Wire cloth generally is specified by mesh, i.e., the number of openings per linear inch, counting from the center of any wire to a point exactly 1 in. distant. Or, the specification may be by the dimension of the clear opening (space) between the wires, given in inches or millimeters. The former method usually prevails for cloth 2 mesh and finer, while the latter method applies for space cloth with $\frac{1}{2}$-in. opening and coarser (see Table S-4). The open area of square-mesh wire cloth can be determined by

$$P = \frac{O^2}{(O + D)^2} \times 100 = (OM)^2 \times 100$$

where P = percentage of open area
O = size of opening
D = diameter of wire
M = mesh

Screens may be divided into five main types: (1) grizzlies, (2) revolving screens, (3) shaking screens, (4) vibrating screens, and (5) oscillating screens.

A *grizzly screen* consists of a set of parallel bars held apart by spacers to provide some predetermined opening. Grizzlies are used before primary crushers in rock- and ore-crushing plants to remove fines before the crushing operation. Except for flat grizzlies, which are located on top of ore and coal bins and under unloading trestles for retaining occasional pieces of large material, stationary grizzlies are installed at a slope of 20–50°. These devices are used for scalping or rough screening of dry material at 2 in. and coarser and are unsatisfactory for moist and sticky materials. Where applicable, stationary grizzlies require no power and little maintenance. Adjustment of the openings is very difficult, and separations may not

TABLE S-4. *U.S. Sieve Series and Tyler* Equivalents, ASTM E-11-70*

Sieve Designation		Sieve Opening		Nominal Wire Diameter		Tyler Screen Scale Equivalent Designation
Standard†	Alternate	Millimeters	Inches (Approximate Equivalents)	Millimeters	Inches (Approximate Equivalents)	
125 mm	5 in.	125	5	8	.3150	
106 mm	4.24 in.	106	4.24	6.40	.2520	
100 mm	4 in.‡	100	4.00	6.30	.2480	
90 mm	3½ in.	90	3.50	6.08	.2394	
75 mm	3 in.	75	3.00	5.80	.2283	
63 mm	2½ in.	63	2.50	5.50	.2165	
53 mm	2.12 in.	53	2.12	5.15	.2028	
50 mm	2 in.‡	50	2.00	5.05	.1988	
45 mm	1¾ in.	45	1.75	4.85	.1909	
37.5 mm	1½ in.	37.5	1.50	4.59	.1807	
31.5 mm	1¼ in.	31.5	1.25	4.23	.1665	
26.5 mm	1.06 in.	26.5	1.06	3.90	.1535	1.050 in.
25.0 mm	1 in.‡	25.0	1.00	3.80	.1496	
22.4 mm	⅞ in.	22.4	0.875	3.50	.1378	0.883 in.
19.0 mm	¾ in.	19.0	0.750	3.30	.1299	0.742 in.
16.0 mm	⅝ in.	16.0	0.625	3.00	.1181	0.624 in.
13.2 mm	0.530 in.	13.2	0.530	2.75	.1083	0.525 in.
12.5 mm	½ in.‡	12.5	0.500	2.67	.1051	
11.2 mm	7⁄16 in.	11.2	0.438	2.45	.0965	0.441 in.
9.5 mm	⅜ in.	9.5	0.375	2.27	.0894	0.371 in.
8.0 mm	5⁄16 in.	8.0	0.312	2.07	.0815	2½ mesh
6.7 mm	0.265 in.	6.7	0.265	1.87	.0736	3 mesh
6.3 mm	¼ in.‡	6.3	0.250	1.82	.0717	
5.6 mm	No. 3½¶	5.6	0.223	1.68	.0661	3½ mesh
4.75 mm	No. 4	4.75	0.187	1.54	.0606	4 mesh
4.00 mm	No. 5	4.00	0.157	1.37	.0539	5 mesh
3.35 mm	No. 6	3.35	0.132	1.23	.0484	6 mesh
2.80 mm	No. 7	2.80	0.111	1.10	.0430	7 mesh
2.36 mm	No. 8	2.36	0.0937	1.00	.0394	8 mesh
2.00 mm	No. 10	2.00	0.0787	0.900	.0354	9 mesh
1.70 mm	No. 12	1.70	0.0661	0.810	.0319	10 mesh
1.40 mm	No. 14	1.40	0.0555	0.725	.0285	12 mesh
1.18 mm	No. 16	1.18	0.0469	0.650	.0256	14 mesh
1.00 mm	No. 18	1.00	0.0394	0.580	.0228	16 mesh
850 μm	No. 20	0.850	0.0331	0.510	.0201	20 mesh
710 μm	No. 25	0.710	0.0278	0.450	.0177	24 mesh
600 μm	No. 30	0.600	0.0234	0.390	.0154	28 mesh
500 μm	No. 35	0.500	0.0197	0.340	.0134	32 mesh
425 μm	No. 40	0.425	0.0165	0.290	.0114	35 mesh
355 μm	No. 45	0.355	0.0139	0.247	.0097	42 mesh
300 μm	No. 50	0.300	0.0117	0.215	.0085	48 mesh
250 μm	No. 60	0.250	0.0098	0.180	.0071	60 mesh
212 μm	No. 70	0.212	0.0083	0.152	.0060	65 mesh
180 μm	No. 80	0.180	0.0070	0.131	.0052	80 mesh
150 μm	No. 100	0.150	0.0059	0.110	.0043	100 mesh
125 μm	No. 120	0.125	0.0049	0.091	.0036	115 mesh
106 μm	No. 140	0.106	0.0041	0.076	.0030	150 mesh
90 μm	No. 170	0.090	0.0035	0.064	.0025	170 mesh
75 μm	No. 200	0.075	0.0029	0.053	.0021	200 mesh
63 μm	No. 230	0.063	0.0025	0.044	.0017	250 mesh
53 μm	No. 270	0.053	0.0021	0.037	.0015	270 mesh
45 μm	No. 325	0.045	0.0017	0.030	.0012	325 mesh
38 μm	No. 400	0.038	0.0015	0.025	.0010	400 mesh

*W. S. Tyler, Incorporated Screening Division, a Subsidiary of Combustion Engineering, Inc., Mentor, Ohio.

†These standard designations correspond to the values for test-sieve apertures recommended by the International Standards Organization, Geneva, Switzerland.

‡These sieves are not in the fourth root of 2 Series, but they have been included because they are in common use.

¶These numbers (3½–400) are the approximate number of openings per linear inch but it is preferred that the sieve be identified by the standard designation in millimeters or micrometers; 1,000 μm = 1 mm.

be complete. The vibrating grizzly is simply a bar grizzly that is mounted on eccentrics so that the entire assembly moves and thus is less likely to clog than the stationary versions.

Revolving screens, now largely displaced by vibrating screens, consist of a cylindrical frame surrounded by wire cloth or perforated plate, open at both ends, and inclined at a slight angle. Material is delivered at the upper end, and oversize material is discharged at the lower end. Desired product falls through the wire cloth openings. The devices usually revolve at slow speeds, 15–20 rpm. The capacity is not large, and efficiency is low. Sometimes these units are referred to as *trommel screens.*

Shaking screens consist of a rectangular frame which holds wire cloth or perforated plate. The unit is slightly inclined and suspended by loose rods or cables or supported from a base frame by flexible flat springs. The frame is driven with a reciprocating motion. The material is fed at the upper end and is advanced by the forward stroke of the screen, while the finer particles pass through the openings. Screens of this type require low headroom and low power. Maintenance costs are usually high. The capacity is low compared with that of inclined high-speed vibrating screens.

Vibrating screens are widely used where large capacity and high efficiency are required. They provide accurate sizing, good capacity per square foot, and low maintenance per ton, and require relatively little space. They are of two main types, mechanical and electrical. Mechanical vibration is effected by an eccentric or unbalanced shaft, although other designs are used, including rotating unbalanced weights. Electrically vibrated screens are particularly useful in the chemical industry, where they can handle many light, fine, dry materials and metal powders from about 4 down to 325 mesh. An electromagnet provides low-amplitude vibration (1,500–7,200 vibrations/min).

Oscillating screens are characterized by low-speed (300–400 rpm) oscillations in a plane essentially parallel to the screen cloth. They are used for materials from $\frac{1}{2}$ in. to 60 mesh. Some light, free-flowing materials can be separated at 200–300 mesh. Silk cloths often are used. *Reciprocating screens* are used widely in chemical processing for handling fine separations even down to 300 mesh. They handle dry, light, or bulky materials, light metal powders, powdered foods, and granular materials and are not intended for handling heavy tonnages, such as rock and gravel. An eccentric

under the screen supplies oscillation, ranging from gyratory (about 2 in. diam) at the feed end to reciprocating motion at the discharge end. Frequency is 500–600 rpm. The screen is inclined about 5°, setting up secondary high-amplitude vibration. Further vibration is caused by balls bouncing against the lower surface of the screen cloth. *Gyratory screens* are boxlike devices, round or square, with a series of screen cloths nested atop one another. Oscillation, supplied by eccentrics or counterweights, is in a circular or near-circular orbit. In some designs, a supplementary whipping action is set up. Most gyratory screens also have an auxiliary vibration caused by balls bouncing against the lower surface of the screen cloth.

Where solid particles in a mixture of solids and liquid require separation, the operation may be termed *wet classification* if methods *other than* screening are used. In general, the two products resulting are a partially drained fraction containing the coarse material (termed the *sand*) and a fine fraction along with the remaining portion of the liquid medium (termed the *overflow*). Equipment of this type is described under **Sedimentation.**

Screens   (See **Waste treatment.**)

Screw Mixers   (See **Mixing and blending, solids-solids.**)

Sealers   (See **Adhesives; Glass;** and **Paint.**)

Seawater   [See **Bromine; Desalination; Iodine; Magnesium; Salt (NaCl); Titanium;** and **Water.**]

Sebacic Acid   (See **Carboxylic acids;** and **Nylons.**)

Secondary Acetate   (See **Acetate fibers;** and **Cellulose ester plastics, organic.**)

Sedimentation   Sedimentation may be defined as the removal of solid particles from a liquid stream by gravitational force. The operation is effected by slowing the velocity of a feed stream in a large-volume tank so that gravitational settling can occur. Generally, the feed enters a baffled configuration (feed well) which is located at the center of the tank and at the level of the liquid surface. The solids (underflow) settle to the bottom of the tank for removal by rakes or other means, while the clear liquids overflow the tank perimeter.

Sedimentation is divided into two functional operations: (1) *thickening,* where the primary purpose is to increase the concentration of suspended solids of the feed stream, i.e., to *remove liquids,* and (2) *clarification,* where the purpose is to remove fine-sized particles and to produce a clear effluent, i.e., to *remove solids.* Some equipment does both, and the dividing line between thickening and clarification is not always sharp.

The feed stream to a thickener or clarifier usually is termed the *feed* or *influent.* Clarified liquid from the unit is called *overflow supernatant* or *effluent.* Settled solids may be termed *underflow, pulp, sludge,* or *mud,* depending upon the industry in which the operation is performed and the specific purpose of the operation. Solids removal may be a batch operation, as in the case of disposal lagoons, and range from hand removal to intermittent clamshell crane removal systems. Continuous, mechanically unassisted removal is possible by means of settling cones or tanks, which have sharp bottom apex angles (45–60°), but their size and capacity are limited by the angle of the cone. Consequently, the most widely used continuous sedimentation devices combine a tank with a shallow sloped bottom and a motor-driven bottom-raking mechanism. These mechanisms can be categorized by the particulate size of product which they can accept, ranging from the hydroseparator, for handling very coarse particles, to the solids-contact clarifier, for handling very fine particles in feed streams containing less than 0.5% solids. Thickening requires a raking removal mechanism with a higher torque capacity than clarification. The latter operation frequently requires special features to aid coagulation of fine particles.

Fundamental Operation of a Continuous Thickener. The major vertical zones within a continuous thickener are shown in Fig. S-2. Zone *A* is a clear supernatent, essentially free of solid particles in most applications except in the hydroseparator, described later. Zone *B* consists of pulp of feed density which is in a state of *unhindered settling,* often in a state of flocculation (or coagulation). Zone *C* is a transition zone between unhindered settling and the beginning of compression of the solid particles. Zone *D* is the compression zone, where the weight of solids and the action of raking arms (not shown in Fig. S-2) forces the liquid from spaces (interstices) between the solid particles.

The area of the thickener must be sufficient to allow the slowest settling particle to reach the discharge of the unit before its associated liquid overflows the tank. The settling rate is not the same for all zones of the thickener. Thus, sufficient space must be provided to allow the particles to pass through each zone. The upflow rate of the liquid must not be greater than the settling rate of the slowest particle.

Major Design Factors. The design of thickeners and clarifiers represents a combination of applied theory and the results of years of experience with numerous types of slurries. The fundamental starting factors are (1) the total daily throughput required, (2) the nature (complete specifications) of the influent, (3) the specifications of the clear liquid to be discharged, and (4) the specifications of the solids discharge. Sedimentation equipment usually must be designed to operate efficiently over a period of many years and is further characterized by the need to handle very large volumes of materials. This is contrasted with most other process equipment, where downtime of several days per year is expected and tolerated. Thus, in addition to design size parameters (tank diameter and depth to accommodate total throughput and the actions of the various vertical zones as previously described), the reliability of the rake-driving mechanism is extremely critical.

Because there are few exactly duplicate situations, the detailed design of sedimentation equipment normally is preceded by a series of tests on the particular slurry to be handled. Numerous testing procedures (long-tube settling experiments, Coe-Clevenger, and Kynch procedures, and others) are used. Description of these procedures is beyond the scope of this volume.

Types of slurries: For equipment design and sizing purposes, including equipment diameter and depth, feed slurries generally fall into two categories: (1) *type A slurries,* which settle with a clear line of demarcation (subsidence) between clear (supernatent) liquor and the underflow pulp, and (2) *type B slurries,* which gradually clarify without any line of demarcation. Type A slurries further can be divided into *slimes* and *sands.*

The most important factors that influence the settling rate of slurries are (1) feed dilution, (2) size and shape of particles, (3) specific gravity of solids and liquid, (4) viscosity of liquid, and (5) temperature of slurry. In a free-settling slurry with sufficient dilution so that particles do not interfere with one another in settling, the rate of particle settling is expressed as a form of Stokes' law, but for practical purposes the formula cannot be used for sizing equipment. If flocculants are used, these materially affect the sizing and testing

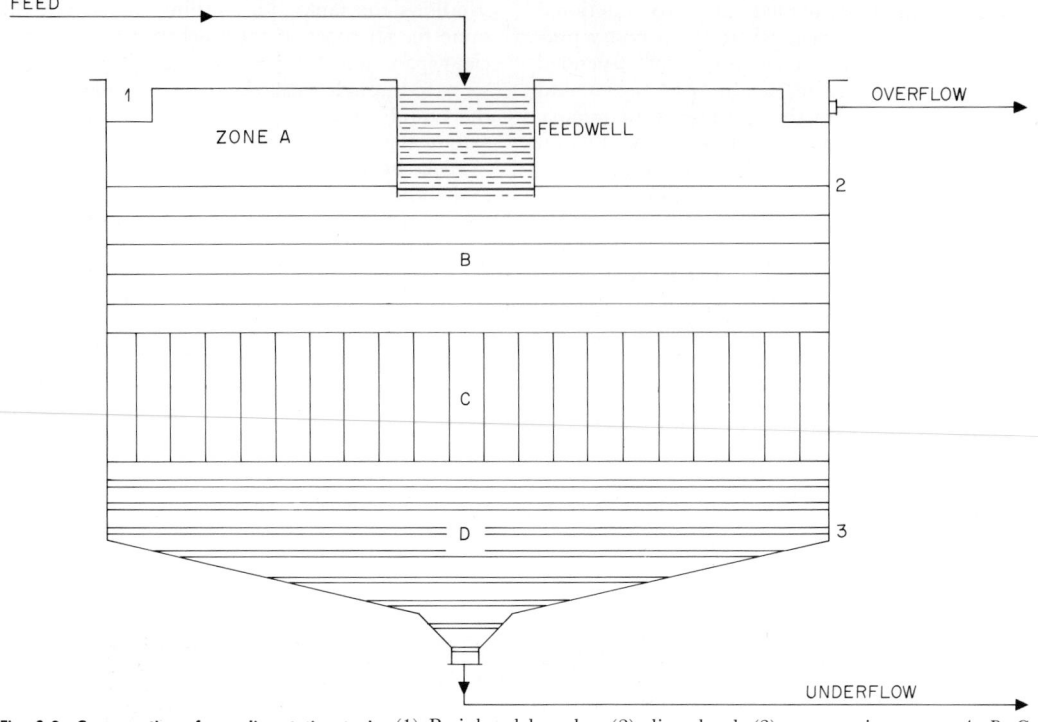

Fig. S-2. Cross section of a sedimentation tank. (1) Peripheral launder, (2) slime level, (3) compression zone. *A, B, C,* and *D* are major zones.

methods used. See also **Liquid-solids separation;** and **Waste treatment** for a description of floccu-lants.

Thickener Mechanisms. Selection of equipment is based upon the area required for settling, the amount of solids in the feed per unit time, the feed dilution, the specific gravity of the solids and liquid, and other sizing parameters. The required duty of the mechanism is classified according to (1) unit area loading in square feet per ton per 24 hr, (2) underflow dilution in percent solids, (3) screen analysis of solids in percent of 200 mesh, (4) specific gravity of solids, and (5) prior experi-ence. See Table S-5.

Applications must be analyzed carefully for establishing torque levels. When torque require-ments are established, thickener horsepower is calculated from the raking-speed requirement. Raking speed varies with the type of material handled. Rake speed (arm-tip speed) usually is 20–25 ft/min for slow-settling materials, 25–35 ft/min for average metallurgical materials, and up to 50 ft/min for concentrates and crystal-

line materials. Rake speeds for clarifiers and solids-contact units are 10–20 ft/min.

Continuous Circular-Type Mechanisms: There are three varieties, which differ in the way the mechanism is supported and driven:

1. *Bridge-supported Thickeners.* These units are built to a maximum of 150-ft diam. The drive mechanism is supported by a bridge that spans the tank. The drive turns raking arms and the center-discharge cone scrapers through a center shaft. The arms move settled solids to the center, where they are withdrawn through a conical opening.

2. *Center-Column-supported Thickeners.* These units are built in sizes of 100–500 ft diam. The drive is supported on a center column that rests on the bottom of the tank. A cage located around the center column supports and turns the raking arms. Settled solids are removed through a dis-charge trench which encircles the center column. See Fig. S-3.

3. *Center-Column-supported Traction-driven Thicken-ers.* These units resemble those described under

TABLE S-5. *Slurry Characteristics versus Sedimentation-Equipment Types*

Equipment Type	Capability of Particle-Size Separation	Feed Solids, sp gr	Solids in Feed, Wt %	Solids in Underflow, Wt %
Hydroseparator.	<4 mesh; 65 to 200-mesh particles go to overflow with dispersants; 5-μm separations possible	1.5–6.0	4–20	20–60
Thickener.	<35 mesh	1.1–6.0	2–20	5–60
Combination clarifier-thickener. . . .	<150 mesh	1.0–1.5	<2	1–5
Clarifier	<100 mesh (usually)		<2	1–5*
Solids-contact unit. . . .	<100 mesh (usually)	1.0–1.5	<½	1–5*
Flotator	<200 mesh (usually)	<1.1	<½	†

*Depends on application.

†Underflow generally limited to a small amount of grit. Float concentration, 1–5%.

2 except that they are driven by a motorized carriage that rides on the tank wall. The units usually are in excess of 200 ft diam. Cautions to be observed in specifying a unit of this type include (1) high cost of wall for supporting traction drive, (2) possible operating difficulties in cold weather, and (3) lack of availability of a practical lifting device to relieve feed solid surges and torque overloads.

Thickener Drive Units: The drive unit always

Fig. S-3. Thickeners used in connection with recovery of magnesia from seawater. Each of the 250-ft-diam thickeners shown handles 36 million gallons per day (25,000 gpm) for thickening magnesium hydroxide and producing an overflow of high clarity which then is returned to the Pacific Ocean at this California location. (*Dorr-Oliver Incorporated.*)

is torque-rated at levels considerably above normal expected loadings. Momentary overloads caused by process upsets require a high factor of safety. The drive unit normally is supplied with an overload alarm and automatic cut-out. High loads may be caused by overthickened pulp, oversize solids in the feed, surges of dense materials, or sloughing of materials accumulated on tank sides or rake arms.

Rake-Arm Design: The rake mechanism moves settled solids to the discharge point and in so doing thickens the sludge by creating paths in the compression zone which allow escape of trapped liquids. Bridge-supported and center-column-supported thickeners commonly utilize two long arms with the option of two short raking arms, the latter added when necessary to rake the inner area. Traction-driven thickeners have one long arm and three short arms.

Arms are available for a variety of tank-bottom configurations, including a design which conveys material to the tank periphery, where it is withdrawn. Processes which produce a thixotropic sludge (one that is sensitive to the size and speed of the raking structure) require special arms in which the truss is elevated above the sludge bed. This design also is useful when large-scale loads can be expected to accumulate on the raking structure. When a hard-packing material is encountered, spikes or serrations are used on the bottom of the rake blades.

Hinged rake arms are reappearing on mechanisms [see Fig. S-4(a) and (b)]. Updated designs involve wire rope suspensions and driving members for the arms. This design is particularly well suited to pulps which build up scale loads in the arms and other applications where the addition of an expensive mechanical lift mechanism cannot be justified.

Feed, Overflow, and Underflow: Feed to the sedi-

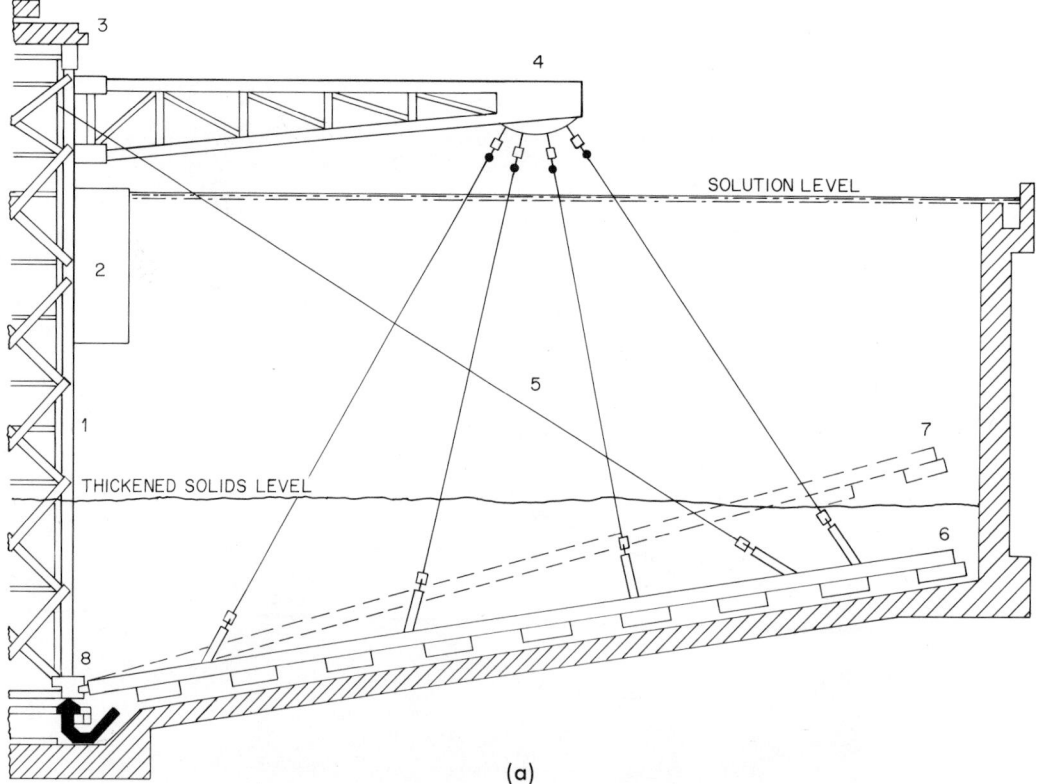

(a)

Fig. S-4. (a) Thickener employing drive truss and cable-supported rake arm. (1) Center column, (2) feed well, (3) drivehead, (4) drive truss, (5) cables, (6) tubular rake arm and blades in normal position, (7) in up position, (8) special arm hinge. (*Dorr-Oliver Inc.*) (b) Bridge-supported thickener employing cable-supported rake arm. (1) Center

mentation unit usually is introduced at the center and baffled by a variety of feed wells. The feed well ideally provides a quieting zone from which flow to the tank will be as nearly laminar as possible. Overflow of clarified effluent usually is handled by a peripheral launder located at the upper tank edge. V-notch overflow weirs help to distribute the internal flow patterns evenly within the tank.

For underflow, with low-specific-gravity solids that do not cause plugging or with dilute concentrations, buried or inaccessible sludge lines can be used. These provide the lowest installed cost of all underflow arrangements. Usually several sludge lines are employed, with one of the lines as an emergency bypass. Water for back flushing should also be available.

The peripheral discharge is also a low-cost underflow arrangement, which allows a flat-bottom tank to be placed on compacted oiled sand thus

reducing installation cost. The design, however, requires an outward raking mechanism, which in turn requires more torque to move a given amount of solids to the discharge point and thus limits this approach to small units.

The most common underflow technique is a tunnel to allow complete accessibility to the discharge point, valves, and associated piping. Although the most costly approach, the tunnel provides positive, flooded suction and consequently is used on difficult pumping applications. When hot or corrosive slurries are handled, the tunnel often is extended across the entire tank diameter for increased safety.

Center-shaft or column pumping lowers installation costs and can be used where ground conditions prohibit a tunnel. Variations include (1) a special room below the thickener with access through the center column, (2) a submergible pump in a flooded center column, (3) a pump at

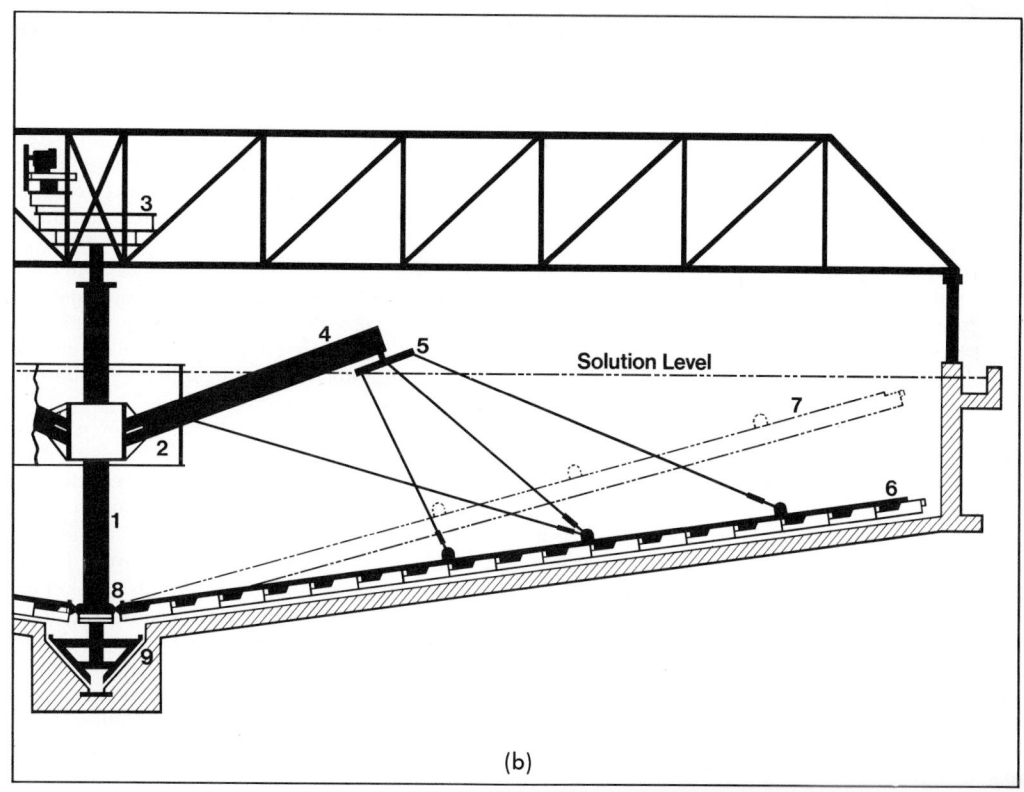

(b)

shaft, (2) feed well, (3) drivehead, (4) drive arm, (5) tension equalizing cable connection, (6) tubular rake arm and blades in normal position, (7) in raised position, (8) hinge connection, (9) guided discharge cone scraper. (*EIMCO, Envirotech Corporation.*)

the bottom of a "dry" column with the pump inlet connected to an inlet port at the base of the column, (4) pumping via the center shaft on units where this is practical, and (5) mounting the pump on the bridge with the inlet line in the center column. The latter approach is the most common but has the slight disadvantage of requiring a vacuum prime; and the critical head requirements of the pump require careful selection.

Hydroseparators. Applied principally in metallurgical and less frequently in chemical processing, hydroseparators usually are considered to be classifiers because water added to the unit allows classification of the finer particles removed in the overflow. There also is a definite thickening of coarse solids in the underflow. Thus, a hydroseparator properly should be considered as a combination classifier and thickener.

Clarifier Mechanisms. Gravity clarifiers are supplied to produce a relatively clear overflow, generally of a type B slurry or pulp as previously described. These units are used mainly for industrial and domestic waste and water clarification. The sludge collected usually is smaller in volume and less dense than in thickener applications, so that construction usually is restricted to light-duty (torque) operation. Except for the rectangular clarifier, the concept and layout are similar to those of the circular thickener.

Clarifiers usually include (1) adjustable squeegees fitted to the raking blades to clean the tank bottom and to prevent accumulation of materials that may become septic and (2) surface skimmers attached to the rake arms which deposit floating materials in a scum box at the tank periphery. Due to the wide range of particles in a waste stream, underflow pumps are selected for their nonclogging properties. Low-head, recessed-impeller centrifugal pumps usually are preferred. Flocculants are not often used.

Clarifier-Thickener. The clarifier-thickener is a variation of the basic clarifier. The unit clarifies dilute suspensions to the same degree as a standard unit and in addition achieves underflow concentrations 2–3 times that available from a conventional clarifier. Additional thickening capacity is obtained with a deep sludge pump adjacent to the center column. Here, additional detention time allows increased sludge concentration with only a slight increase in torque by utilizing the extended-post arrangement. So that the sludge does not become septic, approximately 10% of the final effluent is recirculated and added just above the sludge pump. This gives sufficient oxygen to keep the sludge fresh. The clarifier-thickener unit is built with diameters of 30–200 ft. See Fig. S-5.

The suction-type unit is another variation in which solids are removed by suction tubes rather than standard rakes. The unit is used primarily as a final clarifier in activated-sludge treatment plants. Fresh sludge always is available because the unit withdraws sludge across the entire tank bottom continuously. This is accomplished by suction ducts on the rake arms. Radial squeegees provide positive solids movement to the openings.

Solids-Contact Units. These are considered the most efficient for chemical mixing, flocculation, and clarification in a single tank. The configuration is shown in Fig. S-6. The solids-contact clarifier has a combination drive that turns the rake mechanism at a very slow speed and, by means of a concentric shaft, rotates a turbine in the reaction well at a higher speed. Previously settled solids are contacted with the influent and chemicals by the pumping action of the turbine. Because of good mixing and high solids concentration, flocculation is improved. The mixture flows from the reaction zone to the outer clarification portion of the unit, where the flocculated particles settle out. The rake arms move these settled solids to the center, where they are pumped into the recirculation drum. A portion of the solids is removed by an automatic desludging arrangement.

Solids-contact clarifiers have found application in the water-treatment field for turbidity and color removal, lime softening, and tertiary waste treatment. The units are often used where coagulation and flocculation are required before sedimentation.

Flotators. The operation of a flotator is similar to a clarifier and is designed for removal of fine colloidal solids and emulsified oils in addition to normal settleable solids. Dissolved air in a pressurized feed stream forms a mass of fine bubbles, which adhere to the suspended matter, floating it to the surface. The floating solids are skimmed to a scum trough, while settling material is collected in the conventional way by the mechanism rakes. Flotators normally are available in sizes from 6 to 70 ft diam. Costs are approximately 10–15% higher than clarifiers, excluding cost of the pressurization system. See also **Waste treatment.**

Comparative Equipment Costs. Sedimentation equipment usually is found to be the lowest-cost method for obtaining bulk separation of liquids

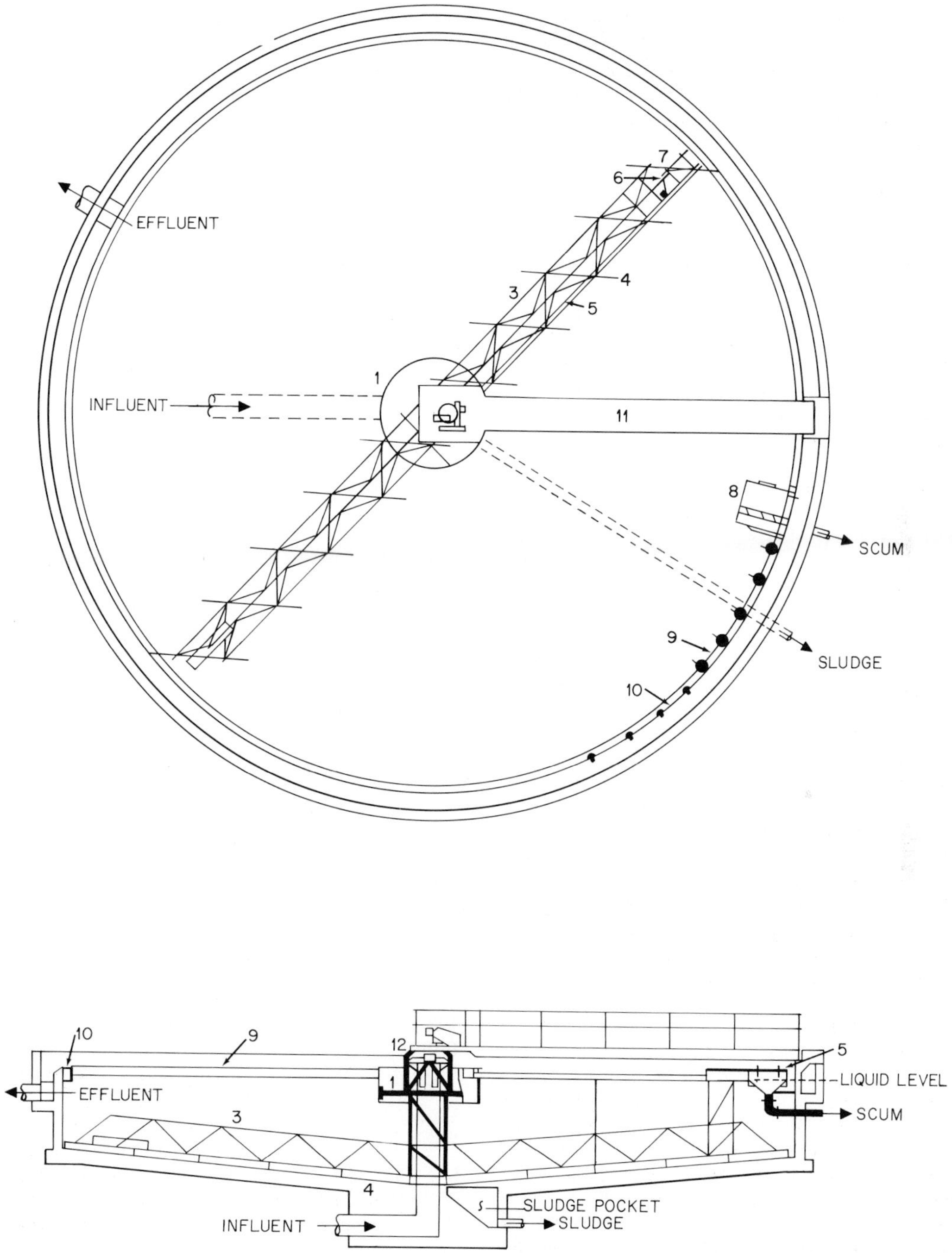

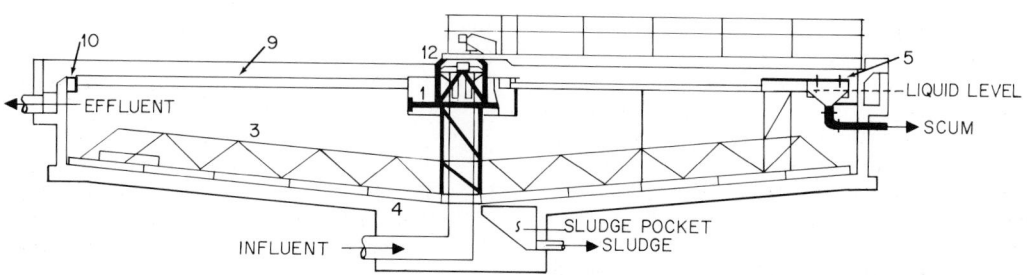

Fig. S-5. Clarifier-thickener. (1) Feed well, (2) drive platform, motor, and drive control, (3) arm, (4) blades and squeegees, (5) skimmer, (6) hinged skimmer, (7) scum-skimming device, (8) scum box, (9) baffle, (10) weir, (11) walkway, (12) cage. (*Envirotech.*)

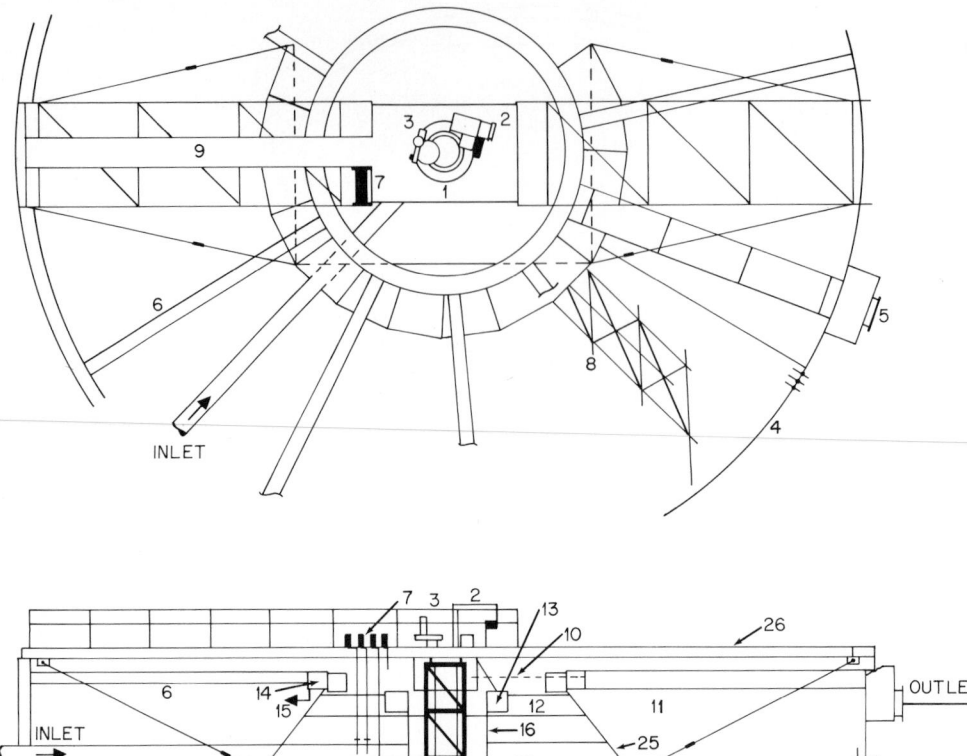

Fig. S-6. Solids-contact unit. (1) Operating platform, (2) turbine drive motor, (3) rake drive motor, (4) outlet radial, (5) outlet pocket, (6) radial launders, (7) chemical piping, (8) blades, (9) walkway, (10) maximum water level, (11) outlet launder, (12) baffles, (13) turbine, (14) annular collection launder, (15) auto launder drain, (16) recirculation drum, (17) center column, (18) rotating drum, (19) cage, (20) rake arm, (21) sludge pump, (22) desludging, (23) sludge pipe, (24) sludge sump scraper, (25) reaction well, (26) superstructure. (*Envirotech.*)

from solids in relatively dilute feed streams. A cost comparison of the various types of mechanisms in the usual applied size ranges is given in Fig. S-7 (page 1013). Curve *A* shows approximate costs of thickeners. The curve is not exact, inasmuch as there are cost variations at a given diameter for components such as lifting devices, higher torque drives, and special arm designs. Costs in the chart are based on steel mechanisms, steel tanks and bottoms to 100 ft diam, and steel tanks with concrete bottoms over 100 ft diam. They include erection plus normal site preparation (surveying, excavation, reinforcing steel, underflow tunnel, and backfill). The costs do not include pumps, piping, electrical work, or instrumentation.

Clarifier costs are proportional to diameter as shown in curve *B* of Fig. S-7. These costs are based on a steel mechanism, steel weir and baffle, steel skimmer and scum box, and concrete sides and bottom. The costs include erection and normal site preparation but exclude pumps, piping, electrical work, and instrumentation.

—L. L. Palm, *Eimco PMD Division, Envirotech Corporation, Salt Lake City, Utah*

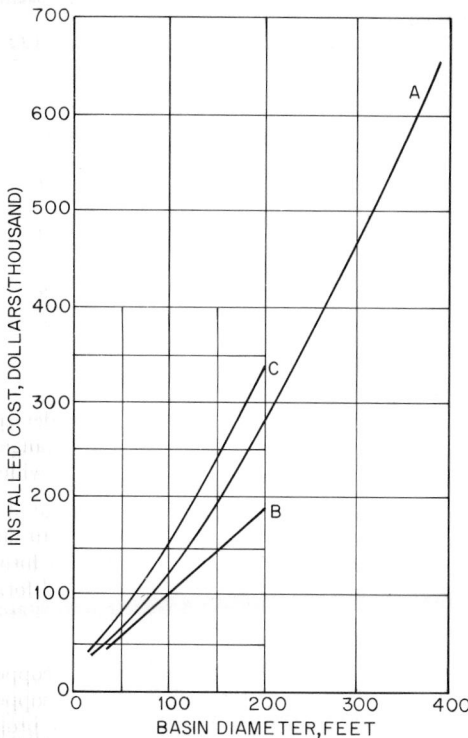

Fig. S-7. **Comparative sedimentation-equipment costs.** Installed cost versus basin (mechanism) diameter: A thickener, B clarifier, C solids-contact clarifier.

Seismograph Explosives (See **Explosives.**)

Selenite (See **Gypsum.**)

Selenium (Gk. *selene,* moon.) **Se** = 78.96 (at. wt.); 34 (at. no.). Selenium exists in several allotropic forms and resembles sulfur in this respect, although the allotropes of Se are not so well defined as those of S. Se is a member of the oxygen-sulfur family of elements and is found in group VIA (along with O, S, Te, and Po) and in period 4 of the periodic table. There are six natural isotopes, 74, 76–78, 80, and 82, and seven radioactive isotopes 72, 73, 75, 79, 81, 83, and 84. See also **Chemical elements.** Valences are 2, 4, and 6.

Amorphous selenium is red and softens in the range of 40–50°C, achieving complete fluidity only at the melting point, 217°C, but showing elasticity at 70°C. At the latter temperature, Se can be drawn into small wires (exhibiting 100–200% elongation). The stretched wires, if cooled below 70°C, do not contract until warmed above 70°C. The specific gravity is 4.26.

Vitreous, glassy, brownish-black, Se is very brittle and exhibits a conchoidal fracture. It has dielectric properties, being electrified by friction. The material is red in thin layers. The specific gravity is 4.28. This material results from the rapid cooling of liquid Se. Vitreous Se does not appear to have a precise melting point. The thermal conductivity is 0.000293–0.000328 (cal)(cm)/(s)(cm^2) (°C) at 25°C.

The *metallic,* or hexagonal, crystalline form of Se is the most stable and considered to be the normal form of Se. This form results when amorphous Se is heated slowly in the range 90–220°C. Specific gravity = 4.79; melting point = 217°C. It is an excellent electrical conductor, and like germanium, crystalline PbS, silicon carbide, silicon, and cuprous oxide, metallic Se is classified as a semiconductor, where conduction is by electrons and electron holes and extremely sensitive to impurities. Metallic Se has a resistance of:

°C	0	20	60	100	170
Ω	66	49.1	23.8	8.0	2.0

Ion potential = 9.750 eV. Heat of sublimation = 219.4 cal/g at the melting point. Hardness is 2.0 on Mohs' scale. Dielectric constant = 6.6 at 10^8 Hz.

Se also exists in a monoclinic crystalline form of deep red color with a specific gravity of 4.46 and a melting point in the range 170–180°C. This form can be obtained from solutions in carbon disulfide.

All forms of Se change to the metallic gray form when heated above the transition point.

Se is odorless and tasteless, but Se vapor has a distinctly disagreeable odor. Metallic Se is insoluble in H_2O but will dissolve in cold, concentrated H_2SO_4 to yield a green solution, which upon dilution with H_2O produces metallic Se. Aqua regia and HNO_3 attack Se to form selenous acid, H_2SeO_3.

Se was discovered by Berzelius in 1817. The element has been found in cavities of Vesuvian lavas and in some shales. For example, the volcanic tuff of Wyoming contains about 150 ppm Se; the black shale of Idaho contains Se up to about 1 lb/ton. Se occurs in numerous minerals but chiefly in *eucairite,* (Ag, Cu)$_2$Se, *naumannite,* Ag$_2$Se, and *zorgite,* (ZnCu)$_2$Se. Before the obsolescence of the lead-chamber process for the manu-

TABLE S-6. *Representative Selenium Compounds* *

Compound and Formula	Formula Weight	Sp gr	Mp, °C	Bp, °C	Solubility in H_2O
Selenium monobromide, Se_2Br_2	318.23	3.6		227 d	Decomposes
Selenium monochloride, Se_2Cl_2	229.31	2.9	−85	130 d	Decomposes
Selenium tetrachloride, $SeCl_4$	221.03	3.8	305 s	288 d	Decomposes
Selenium hexafluoride, SeF_6	193.20	3.3	−39	−34	Decomposes
Selenium dioxide, SeO_2	111.20	4.0	340	s	Soluble
Selenium trioxide, SeO_3	127.20	3.6	120 d		Very soluble
Selenium oxybromide, $SeOBr_2$	255.03	3.6 lq	42	217 d	Decomposes
Selenium oxychloride, $SeOCl_2$	166.11	2.4	8.5	176	Decomposes
Selenium oxyfluoride, $SeOF_2$	133.20	2.7	4.6	124	Decomposes
Selenium monosulfide, SeS	111.26	3.1	118 d		Insoluble
Selenic acid, $H_2SeO_4 \cdot H_2O$	162.99	2.6	26	205	Very soluble
Selenous acid, H_2SeO_3	128.98	3.0	d		Very soluble

*d, decomposes; s, sublimes; lq, liquid.

facture of H_2SO_4, lead-chamber muds were an important source of Se. Currently, a prime source is copper refining, where blister copper anodes, for example, contain from 0.03–0.14% Se. Copper-refinery slimes may contain 8–9% Se, and commercial recovery may reach 0.66 lb of Se per ton of Cu. Se also is recovered from residues when iron sulfide ores are roasted for H_2SO_4 production.

Se forms one principal oxide, SeO_2 (-ous), and other variously reported oxides, such as Se_2O_3, Se_3O_4, and SeO_3. SeO_2 dissolves in water to form selenous acid, H_2SeO_2. Selenic acid, H_2SeO_4, also exists and is readily soluble in H_2O. Several salts (selenites and selenates) are formed with the alkaline-earth metals—all insoluble in H_2O. Selected representative compounds of Se are given in Table S-6.

Selenium is well known and widely used because of its photosensitive properties. In the Se photoelectric cell, an emitter and a collector are contained within an evacuated, sealed container. The emitter is a metal surface covered with a thin layer of Se. The cell functions because of the change in electrical resistance of Se upon exposure to light. The response is proportional to the square root of incident energy. The spectral sensitivity lies in the red portion of the spectrum. An external source of emf is required, but the resistance is low and can be readily amplified. In the selenium photovoltaic cell, a metallic surface is covered with a thin film of vitreous or metallic Se. A transparent film of another metal, such as Pt, goes over the Se. This cell generates its own emf, while its internal resistance decreases with increasing irradiation. The response is essentially proportional to the incident energy. These devices are not unduly sensitive to temperature changes.

Se also exhibits unipolar conduction, which qualifies it as an ac rectifier. A representative unit may comprise a Ni or Ni-plated steel or an aluminum disk with a thin layer of Se applied to one face. The Se may be applied in vitreous form, which converts to the metallic or hexagonal form upon heat treatment. The Se layer then is coated with a thin layer of a low-melting alloy, e.g., one in the Cd-Bi-Sn system.

Se is added to stainless steel and to copper alloys to increase machinability. Selenium copper combines free-machining and hot-working properties with high electrical conductivity. Available in rod form, selenium copper is widely used to make copper-to-glass seals. Se is used in glass manufacture as a decolorizer to counteract green ferrous shades. Sodium selenite is used for the production of clear red glass and for making red enamels. From 1 to 3% Se has been found to increase the abrasion resistance of vulcanized rubber. Se also is used in photographic and printing reproduction chemicals.

Although Se in elemental form is not considered poisonous, many Se compounds are quite toxic. For example, hydrogen selenide, H_2Se, is a toxic gas of unpleasant odor that is chemically similar to arsine, stibine, and hydrogen telluride. Soils containing Se are toxic to man and animals.

Selenium Copper   (See **Copper**.)

Semicarbazides   (See **Aldehydes**.)

Semicarbazones (See **Aldehydes;** and **Ketones.**)

Semiconductors (See **Arsenic; Catalysts; Erbium; Gallium; Germanium; Holmium; Indium; Selenium;** and **Silicon;** also consult the **Subject Index.**)

Semipolar Bond (See **Molecule.**)

Separation Operations One of the commonest problems in processing materials is that of separating one or more materials from another or several other materials. From the standpoint of the phases represented by the materials present in a mixture, there are six fundamental combinations, as shown by Table S-7. One gas may have to be separated from another gas or a whole mixture of gases; a liquid may have to be separated from a solid; or a solid from a gas; and so on. Separation operations usually are not so clear-cut as to involve only two different materials. Further, separations often do not effect a full and precise separation of one material from the next but require several stages of separation or the return of some of the product to the separating equipment to obtain a better separation and puri-

fication of the components. In a mixture of materials, usually one material is of greater economic value than the remaining materials and may, in essence, be the target of separation. The target substance may be present in large or small amounts compared with the total mixture. In other instances, several materials may be of relatively the same value.

Major separation operations are given in Table S-8. These operations also are described briefly here with reference to more extensive details elsewhere in the volume. Although some chemical substances are added to the mixture to be separated in some separation operations, additives participate principally in a physical way. Operations that bring about separation essentially by chemical reaction are not covered by Table S-8.

Solid-Solid Separations. These operations range from the separation of large pieces of rock as they exit from a coarse crusher, where it is desired to classify the pieces by size for subsequent handling, to the purification of crystalline materials by sublimation.

Screening: This operation simply involves the use of a screen with specific size openings or any other form of matrix with uniform openings

TABLE S-7. *Separation Operations Classified on Basis of Phases of Components in Starting Mixture*

Liquid-Gas	Gas-Gas	Gas-Solid
Gravity settling	Low-temperature distillation	Gravity settling
Impingement separation	Absorption	Impingement separation
Cyclone separation	Adsorption	Cyclone separation
Foam breaking	Gaseous diffusion	Electromagnetic separation
Drying	Thermal diffusion	Filtration
Boiling	Electromagnetic separation	
Condensation	Chromatographic separation	
	Condensation	

Liquid-Liquid	Liquid-Solid	Solid-Solid
Distillation:	Sedimentation	Size classification
Rectification	Thickening	Screening
Azeotropic	Clarifying	Dense-medium separation
Extractive	Filtration	Jigging
Molecular	Centrifugation	Tabling
Extraction:	Drying	Magnetic separation
Solvent	Spray drying	Electrostatic separation
Precipitative	Evaporation	Flotation
Dialysis	Crystallization	Sublimation
Electrodialysis	Prilling	Freeze-drying
Thermal diffusion	Graining	
	Leaching	
	Expression	

TABLE S-8. *Classification of Physical Separation Operations*

Operation	Principle
Essentially Solid-Solid	
Size classification	Centrifugal force separates particles of different sizes
Screening	Particle versus aperture size; also applies to slurries
Classifying	Differential settling rates due to density difference; water added if not originally present
Dense-medium separation	Layering due to specific-gravity differences between light and heavy components and dense media; subsequent magnetic separation of dense-medium particles; dense medium comprises water and heavy magnetic particles
Jigging	Differential settling rates due to density differences
Tabling	Differential momentum due to density differences
Magnetic separation	Magnetic attraction
Electrostatic separation	Electrically charging particles before separation
Flotation	Adherence of light-density material to air bubbles
Sublimation	Passage from solid to gas phase without forming liquid phase
Freeze-drying	Special form of sublimation
Essentially Liquid-Solid	
Thickening	Gravitational settling with object of concentrating
Clarifying	Gravitational settling with object of purifying
Filtering	Trapping solid particles on surface of membrane or in pores of medium; filter aids sometimes used
Centrifuging	Heavy solid particles propelled through liquid or filter medium by centrifugal force
Drying, solids	Evaporation of volatile solvent or of liquid entrained in solid
Spray drying	Evaporation of atomized solution in stream of hot gas
Evaporation	Vaporization of solvent from solution or slurry
Crystallization	Formation of supersaturated solution followed by cooling
Prilling	Combination of spray drying of concentrated solution and crystallization
Leaching	A form of solvent extraction
Expressing	Squeezing liquid out of pores of solid
Essentially Gas-Solid	
Gravity settling	Gravitational force attracts heavier particles
Impingement separating	Inertia of solid particles is greater than gas
Cyclone separating	Gravitational settling force is replaced by centrifugal force
Bag filtering	Trapping solid particles by cake formed on bag
Electrical precipitation	Exposure of solid particles to gas ions
Air filtering	Trapping solid particles by filter medium
Essentially Gas-Gas	
Low-temperature distillation	Differential vapor pressures of liquid gases
Absorption	Selective solubility of a gas in a solvent
Adsorption	Affinity of certain solids holds gases
Gaseous diffusion	Differential diffusibility of gases
Thermal diffusion	Large molecules tend to go to cooler medium
Electromagnetic separation	Principle of mass spectrometer
Gas centrifuging	Differential action of centrifugal force on light and heavy molecules
Chromatographic separation	Differential migration of substances through sorptive media
Condensing	Widely divergent boiling points of components

TABLE S-8. *Classification of Physical Separation Operations* (*Continued*)

Operation	Principle
Essentially Liquid-Liquid	
Distillation	Differential vapor pressures of components
Rectification (fractional distillation)	Return of portion of condensed vapors from distillation to enrich rising vapors in one component
Azeotropic and extractive distillation	Addition of third component alters volatility of one of the two components to effect sharper separation
Liquid-liquid extraction	Selective solubility; improved by refluxing
Dialysis	Small molecules tend to diffuse through semipermeable membrane
Electrodialysis	Anions and cations permeate separate membranes
Thermal diffusion	Large molecules tend to go to cooler medium
Chromatographic separation	Differential migration of substances through sorptive media
Essentially Liquid-Gas	
Gravity settling	Heavier liquid particles tend to settle by gravity
Cyclone separating	Differential action of centrifugal force on liquid particles and gas
Impingement separating	Inertia of liquid particles is greater than gas
Foam breaking, mechanical*	Mechanical or acoustic interference with foam
Thermal*	Hot surface near foam tends to vaporize liquid content of foam
Drying, gases	Affinity of certain solids for vapors (often water)
Boiling	Partial separation by driving off gas along with some liquid vapor
Condensing	Heated liquid vapors condense; some gases will not condense

*Wherever practical, chemical addition agents (antifoam agents) are used to prevent or reduce foam formation. Agents include fatty acid esters, metallic soaps, and hydrocarbons. See also **Defoaming agents.**

through which pieces smaller than the openings will pass through (the *undersize* or *minus material*) while the pieces which are larger than the openings will be held back (the *oversize* or *plus material*). Although the principle is exceedingly simple, practical problems include the selection of suitable screening materials to withstand the corrosion and abrasion of the mix, the provision of means to keep all pieces or particles in motion and thus available to pass through the apertures without clogging, and the careful selection of aperture (mesh) size to effect the desired separation as efficiently as possible. See also **Screening.**

Wet Classifying: This operation involves the separation of solid particles in a mixture of solids and liquid into fractions in accordance with particle size or density by means other than screening. Generally, the two products that result are (1) a partially drained fraction that contains the coarse material (the *sand*) and (2) a fine fraction, together with the remaining liquid termed the *overflow*. Classifying is carried out in a pool of fluid pulp contained in a large tank arranged to allow the coarse solids to settle out, whereupon they are removed by gravity, mechanical means, or induced pressure. See also **Sedimentation.**

Dense-Medium Separation: This operation is useful for the separation of solid particles of differing densities. In essence, a liquid suspension of finely-divided high-gravity solids is prepared. Ores of different densities, for example, when exposed to such a suspension will tend to separate by rising or settling in the liquid suspension. Many types of solids have been used to obtain a high-gravity medium, but the magnetic solids (ferrosilicon and magnetite) are used in most cases. These solids, alone or in combination, can provide a suitable dense medium over a gravity range of 1.25–3.4. Dense-medium separation is applicable to any ore in which the valuable component has an appreciable gravity difference from the gangue components. In coarse-ore heavy-medium separation plants, the limiting bottom size of dense-medium feed is 10 mesh, and the upper size limit is 12 in. The magnetic particles of the dense medium subsequently are removed by a magnetic separator.

Jigging: In this operation, a pulsating stream of liquid flows through a bed of materials of different specific gravities, causing the heavy material to work down to the bottom of the bed and the lighter material to rise to the top. This is a very old operation used for concentrating heavy mineral from the lighter gangue. Costs to construct are low, but power and water consumption are high. The process is widely used for the concentration of coal. See also **Beneficiation, ore.**

Tabling: This is a concentration process whereby a separation between two or more minerals is effected by flowing a pulp across a riffled plane surface inclined slightly from the horizontal, differentially shaken in the direction of the long axis, and washed with an even flow of water at right angles to the direction of motion. A separation between two or more mineral deposits depends mainly on the difference in specific gravity between the effective specific gravity (specific gravity of mineral minus specific gravity of water) of the valuable and waste materials. Tables treat metallic ores effectively in size ranges from 6 to 150 mesh but can be used to treat lighter materials, such as coal of a considerably larger size. Dry tables also are used. The shaking motion is similar except that the direction of motion is inclined upward from the horizontal, and instead of water acting as the medium of distribution a blast of air is driven through a perforated deck. Tables also are used for selective flocculation or agglomeration of grains of one mineral in an aggregate by the addition of an agglomerating agent. See also **Beneficiation, ore.**

Magnetic Separation: This operation is used to remove tramp iron and for the concentration and purification of a magnetic material from a nonmagnetic material.

Electrostatic Separation: If one or more of the materials in a granular mixture can receive a surface electric charge on or before entering an electrostatic field, the grains of that material will be repelled from one of the electrodes and attracted toward the other, depending upon the sign of the charge on the grain. By causing such grains to fall into separate chutes, a separation or concentration results. The major electrification mechanisms of separation or beneficiation of solid substances fall into three groups: (1) contact electrification, (2) electrification by conductive induction, and (3) electrification by ion bombardment. In this operation, all particles to be separated must be completely surface-dry. The surrounding humidity must be less than 35%. Electrostatic separation methods are used to separate such materials as feldspar from quartz; concentrating potassium feldspar from sodium feldspar; columbite, ilmenite, and cassiterite from gangue materials; and in the separation of ilmenite, rutile, leucoxene, and zircon.

Flotation: In froth flotation, ore is suspended in water at a pulp density of 15–35% solids by means of mechanical or air agitation. The surfaces of specific mineral particles are treated with substances called *promoters* or *collectors,* which render those particles air-avid and water-repellent. With vigorous agitation and aeration in the presence of a frother, a layer of froth or foam forms at the top of the flotation machine. The air-avid minerals become attached to air bubbles and rise to the surface, where they collect in the froth and are skimmed off. Undesired minerals are depressed or rendered nonfloatable either by leaving their surfaces unaltered by collector adsorption or through the use of additional modifying agents. In glass-sand flotation, iron-bearing minerals are floated off in the froth, while high-grade silica materials are in the overflow. A large percentage of Cu, Pb, Zn, Mo, Sb, and Ni are produced from ores that are concentrated first by froth flotation. See **Flotation.**

Sublimation: In this operation, a solid is converted to a gas without passing through the liquid phase. The complete operation involves going from a solid to a gas and back to a solid. In addition to purifying solid materials, sublimation sometimes is used to introduce a vaporized solid material into a gaseous stream for further processing. Also, where corrosion or product decomposition are problems at distillation temperatures, sublimation may be a preferred method because of the lower temperatures involved.

Freeze-drying: This operation is a special form of sublimation in which a solid material containing moisture is freed of that moisture by first freezing the moist solid and then removing the water directly as vapor from the solid state. A high vacuum is required because the vapor pressure of ice is considerably below atmospheric pressure. Freeze-drying is used extensively in the food and drug industries. See also **Freeze-drying.**

Liquid-Solid Separations. There are many equipment configurations for separating liquids from solids, necessitated by the wide variety of materials to be handled, which range from a light slurry containing a very small percentage of solids that must be filtered out to a slightly moist solid that must be converted to a bone-dry condition. Liquid-solid separations are among the most common operations in the process industries. The fundamental considerations in the design and selection of equipment in this category are given under **Liquid-solid separations.** The major operations are defined briefly here.

Sedimentation: This is a rather general term for

an operation wherein suspended solids are removed from a liquid by gravitational settling. The two primary forms of sedimentation equipment are (1) thickeners and (2) clarifiers. The term *decanting* sometimes is used to designate sedimentation. See also **Sedimentation.**

Thickening: The principal objective of thickening is to increase the concentration of the feed stream. The mechanical continuous thickener, equipped with sludge-raking arms, is the most common type. Usually the operation is performed in cylindrical tanks. The sludge collection system and removal system are designed to move the settled material continuously across the tank floor to a discharge point. Feed enters through a central feed well designed to distribute the flow to the basin. Clarified liquor overflows into a launder around the periphery. Thickened sludge, raked toward the center by a slowly revolving mechanism, enters a central collecting trough or cone and is discharged through a spigot or removed by a sludge pump. See also **Sedimentation.**

Clarifying: The main objective of clarifying is to free solids from a relatively dilute stream. These units operate on the basis of gravity sedimentation and utilize a raking mechanism as in a thickener. See also **Sedimentation.**

Filtering: This operation may be defined as the separation of undissolved, particulate, suspended solids from a fluid mixture by passage of most of the fluid through a septum or membrane that retains the solids *on* or *within* itself. The mixture to be separated is called the *feed slurry;* the fluid that passes through the septum is termed the *filtrate.* When the separated solids accumulate in amounts that visibly cover the medium, they are called the *filter cake.*

The numerous designs of filters can be classified by (1) the *driving force,* i.e., the manner by which the filtrate is induced to flow through the septum, usually effected by a hydrostatic head (gravity), by above-atmospheric pressure applied upstream of the septum, or by a vacuum applied downstream; (2) the *function,* i.e., whether the filtrate or the cake is the product of greater value; (3) *operating cycle,* i.e., batch or continuous; (4) the *nature of the solids,* i.e., relatively large, easily filterable particles versus minute particles that may require ultrafiltration; and (5) *filtration mechanism.* In one major type, the filtered solids are stopped at the surface of the medium and pile up to form a cake; in the other type, the solids are trapped within the pores or body of the medium. See **Filtration.**

Centrifuging: In this operation, centrifugal force is employed to replace the weaker force of gravity used in settling and in some filtering operations. The mix to be separated is whirled about a center point. Centrifugal force causes sedimentation of heavy solid particles through a layer of liquor or filtration of a liquid through a bed of porous solids held inside a perforated rotating cylinder. Centrifugal force is used in several other separating operations. See also **Centrifuging.**

Drying: Unless otherwise specified, drying generally refers to the removal, by evaporation, of a liquid from a solid, although gases also are dried. Usually, drying involves the processing of a material that, on a weight basis, contains more solid phase than liquid phase. The majority of driers are designed to handle this starting condition. In *spray driers,* the starting material is very liquid; i.e., on a weight basis, there is more liquid phase than solid phase present, as represented by a solution, thin slurry, or suspension.

In all but dielectric heating, the heat required is directed first against the outer, exposed surface of the material before reaching the interior of the solid. In dielectric heating, a higher temperature is produced within the material than on the surface. In the more conventional drier designs, the heat required is provided by convection, conductance, radiation, or a combination of these methods. Although several variables enter into designing and operating a drier for maximum efficiency, the operation essentially represents a time-temperature relationship. Within the limitations of the material being dried, drying proceeds, as one would expect, at a faster rate when the temperature is increased. There are many drier-design configurations, batch and continuous, with direct or indirect heat transfer to the wet solid. See **Drying, solids;** and **Spray drying.**

Evaporation: In this operation, solvent is removed as a vapor from a solution or slurry. The vapor may or may not be recovered, depending upon its economic value. The end product may be a solid, but the transfer of heat in the evaporator must be to a solution or to a suspension of the solid in liquid if the apparatus is to be classified as an evaporator and not as a drier. Evaporators are similar to stills or reboilers or distillation columns except that no attempt is made to separate the components of the vapor. See **Evaporation.**

Crystallization: This operation permits the removal of a solute from a supersaturated solution in the form of crystals. Supersaturation usually is reached by starting with a hot saturated solution and gradually cooling. The rate of cooling affects crystal formation, slow cooling tending to favor the formation of larger crystals. Fractional crystallization permits the removal of one solute from two or more solutes by taking advantage of the differences in solubilities. While crystallization qualifies as a separation operation, the function of crystallization can go beyond that of separating materials. Crystallization is excellent for making pure, attractive, and concentrated materials for marketing. The operation is not confined to liquid solutions but is also applicable to crystallizing metals from the molten state, as in the zone refining of metals. See **Crystallization.**

Prilling: This operation is a combination of crystallization and spray drying and is used extensively in the production of urea and ammonium nitrate. See also **Explosives; Fertilizers;** and **Urea.**

Graining: Evaporators that directly produce crystals are referred to as crystallizing evaporators. When supersaturation is developed by evaporation rather than by cooling, the operation may be referred to as graining.

Leaching: This operation is a form of solvent extraction. In leaching, a solute is removed from a solid by using a selective liquid solvent. A simple form of leaching apparatus is a tank in which the solvent and solids are agitated and then allowed to separate, either in the same vessel or in a separate settling tank. Leaching is applied widely in ore processing, as in the case of extracting bauxite with sulfuric acid. See also **Urea.**

Expressing: In this operation, a liquid is separated from a two-phase solid-liquid system by compression. The liquid literally is squeezed out of the solid. In some applications, e.g., dewatering paper, expression is competitive with drying. Mechanical removal of water usually is less costly than thermal methods. In vegetable-oil processing, expression and solvent extraction are competitive. Expression is less thorough but may yield both oil and meal products of higher quality. In the recovery of juice from sugar cane, expression on three-roll mills is combined with solvent extraction using water, in a series of alternating or simultaneous operations. See **Expressing.**

Gas-Solid Separations. The removal of solid dispersoids from gases predominates in this category of materials separation. Major applications include dust collection and smoke abatement. The dust may be a product of economic value, e.g., that collected from the exit of a drier or smelter. The spray drying of milk, eggs, and soap and the manufacture of high-purity zinc oxide and carbon black make extensive use of dust collectors. Dust may be collected for internal air-cleaning operations e.g., where pharmaceuticals, photographic film, electronic, and optical products may be poisoned or damaged by the presence of dust. Dusts often present an explosion hazard, as typified by flour dusts from milling and bagging operations, or a health hazard, in the case of siliceous and metallic dusts. Dusts generally are deleterious to equipment and are removed, e.g., the removal of dust from engine air intakes.

Chromatographic separation of gas-solid systems is used principally (1) on a laboratory or pilot plant scale, and (2) in connection with chemical analysis. See also **Chromatography.**

Gravity Settling: Possibly the simplest type of dust collector is the gravity settling chamber, in which the gas velocity is reduced to permit the dust to settle by the action of gravity. Gravity collectors generally are built in the form of long, empty, horizontal, rectangular chambers with an inlet at one end and an outlet at the side or top of the other end. For a given volumetric air-flow rate, the collection efficiency depends upon the total plan cross section of the chamber and is independent of the height. The height must be sufficiently large to ensure that the gas velocity in the chamber is not high enough to cause reentrainment of separated dust. Horizontal shelves within the chamber improve separation. Such chambers are difficult to clean. The pressure drop through a settling chamber is small, consisting primarily of entrance and exit losses. Such chambers occasionally are used on natural-draft exhausts from kilns, but they are gradually being replaced by low-pressure drop cyclones or other more compact equipment.

Impingement Separators: These devices are based upon the principle that when a dust-laden gas impinges on an object, the true fluid or gas will tend to be deflected around the object whereas the dust particles, because of their greater inertia, will tend to collect on the surface of the object (called the *target*). Generally, impingement collectors are operated at pressure drops of 0.1–1.5 in. H_2O and are limited to removing dusts that are predominantly larger than 10–20 μm diam. Rappers sometimes are used to shake the collected dust from the collecting bodies. Usually these units can be adapted to existing flues and ducts and may be used at elevated temperatures provided that the dust does not become tacky. In some designs, circulating-water films are used to keep the elements clear.

Cyclone Separators: In these devices, the dust-laden gas enters a cylindrical or conical chamber tangentially at one or more points and leaves through a central opening. The dust particles, by virtue of their inertia, tend to move toward the outside separator wall, from which they are led into a receiver. In essence, a cyclone is a settling chamber in which gravitational force is replaced by centrifugal acceleration. This latter force may range from 5 times that of gravity (in large-diameter, low-resistance cyclones) to as much as 2,500 times that of gravity in small, high-resistance units. Compared with other collection equipment, cyclones are relatively low in cost. They generally are applicable where particles of over 5 μm diam are involved, although with less efficiency particles down to 3 μm diam sometimes are handled.

Mechanical Centrifugal Separators: In these units, a powered rotating motor is used to provide the force to whirl the particles from the gas rather than depending entirely on gas velocity as a cyclone does. The units are compact and offer an advantage where large numbers of individual collectors may be required.

Bag Filters: In this equipment, the dust-laden gases are passed through a woven-fabric bag which collects the dust, allowing the gas to pass through the pores of the bag. The efficiency of these units is low until sufficient dust-laden gas has been passed through the apparatus to precoat the fabric pores inasmuch as the pores usually are much larger than the dust particles to be removed. Normally the precoat layer forms in a matter of seconds to a few minutes, after which the efficiency of separation will approach 99%. Woven-fabric filters are limited to 1 to 8 ft^3/min per square foot of cloth because of the compacting action of higher velocities. Felt filters also are widely used. Felt as the filter medium permits somewhat higher velocities on some dusts without excessive penetration, since the filter cake plays a lesser role in the filtration than is the case with woven-fabric filters.

Electrical Precipitators: Equipment in this category operates on the principle that when particles suspended in a gas are exposed to gas ions in an electrostatic field, the particles will become charged and migrate under the action of the field. To avoid thick dust accumulation, perforated or slotted electrodes are sometimes used. Moisture or conductive salts may be added to increase the conductivity of the collected dust layer. Although equipment of this type involves a number of design problems, the state of the art is quite well advanced and the units are widely used, e.g., in cement mills, power plants, and processing plants of a great variety.

Air Filters: These devices are used for the elimination of atmospheric dust, and because in these applications the dust has no economic value, the concern is more with the quantity of dust present than its quality. Process dust concentrations may run as high as several hundred grains per cubic foot, although not usually exceeding 20 gr/ft^3. Atmospheric-dust concentrations generally are below 5 gr/1,000 ft^3. Thus, the burden of separation per volume of gas processed is much lower than for the units previously described. Air filters may be of the viscous or dry type. In the former, the filter medium is coated with a viscous material to retain the dust. The filter pad may consist of one of several materials, e.g., glass fibers, animal hairs, wood shavings, corrugated fiberboard, split wire, or metal screening. The pads are coated with mineral oil or other chemicals of high viscosity and flash point to acts as a dust holder. Some units can be cleaned and recoated; others are of the throwaway variety. Typical dry-filter media are sheets of cellulose pulp, cotton, felt, or spun glass. Felt filters generally are reconditioned by vacuum or dry cleaning. Inexpensive filter media usually are discarded when the maximum dust load has been reached.

Acoustic Collectors: High-intensity acoustic vibrations cause particle collisions and thus tend to flocculate fumes and mists, whereupon they can be collected by conventional apparatus. Generally, the optimum frequency is 1,000–10,000 Hz, below and above which effective flocculation does not occur.

Gas-Gas Separations. In separating gases, advantage is taken of properties that may distinguish the gas or gases to be separated out of two or more gases in a mixture, e.g., (1) vapor pressure, (2) solubility in a solvent, (3) affinity for a solid adsorbent, (4) diffusibility, (5) molecular and atomic-weight differences, (6) density, and (7) chemical properties.

Low-Temperature Distillation: By combining distillation with cryogenic techniques, advantage is taken of the differences in vapor pressures of two different gases at low temperatures, the process first requiring, of course, the liquefaction of the gases involved. Visualize a gas mixture comprising G_1 and G_2. Cryogenic cooling will reduce these gases to liquids, L_1 and L_2. Careful heating will allow L_1 to vaporize and separate from L_2.

Once G_1 has been separated from L_2, further heat can be applied to vaporize L_2 and thus collect G_2 as a separate fraction of the initial mix. See **Cryogenic processes; and Distillation.**

Absorption: A specific gas can be separated from a mixture of gases by contacting the mixture of gases with a solvent that will selectively dissolve the specific gas. For example, the solubility of NH_3 at a temperature of 30°C for a partial pressure of NH_3 of 260 mm Hg is 20 weights of NH_3 per 100 weights of H_2O. Reagents can be added to the solvent to increase solubility, and in some instances the gas reacts with the absorbent to form a loose chemical compound. Organic bases, such as mono-, di-, and triethanolamine, may be used to absorb CO_2 and H_2S. Often, the objective of the absorption operation is that of removing an objectionable gas from a gaseous stream. There are numerous nonaqueous solvents for gas absorption, including acetic acid, acetone, amyl alcohol, benzene, carbon tetrachloride, ethyl alcohol, nitrobenzene, and toluene. Gases typically absorbed from gaseous streams indlude CO, Cl_2, HCl (as in the production of hydrochloric acid), H_2S, CH_4, O_2, SO_2 and SO_3.

Gas-absorption operations usually are conducted in (1) a tower filled with irregular solid packing material, (2) an empty tower into which the liquid is sprayed, or (3) a tower that contains a number of bubble-cap or sieve plates. Usually, the gas and liquid streams flow countercurrently through the contacting apparatus in order to obtain the greatest rate of absorption. In some instances, gas-absorption operations are carried out in spray columns, wetted-wall columns, and stirred vessels.

The vapor-liquid equilibrium relations for the system must be known to (1) determine the amount of liquid required to absorb the requisite amount of the soluble component from the gaseous mixture or (2) determine the quantity of gas needed to strip the requisite amount of the volatile component from the liquid. The latter consideration refers to the inverse of gas absorption, namely *stripping* or *desorption,* in which it is desired to transfer a volatile component from a liquid mixture to a gas. See **Absorption.**

Adsorption: A solid adsorbent that has an affinity for a given gas or gases may be contained in a fixed bed or in a moving bed and thus used to effect gas separations. The simplest form of gas-solids contactor is a static bed, in which particulate solids are held in fixed positions, one particle resting upon another with no relative motion among particles. Gas is made to flow over, impinge upon, or flow through the voids among the particles. Contacting is confined to the interface between the gas phase and the solid phase. Channeling, a difficult condition to correct, may result from nonuniformities in bed packing or density and pose a serious operating problem. Fixed beds almost always require cyclic operations because of the difficulty of adding or removing solids during operation. However, as contrasted with moving beds, the solids in a static bed do not have to move throughout a portion of the system and hence there are fewer losses, e.g., from abrasion. Solid adsorbents include aluminas, aluminosilicates, diatomaceous earth, silica gel, carbon, anhydrous calcium sulfate, iron oxide, magnesia, and various organic resins.

In a moving-bed adsorption system, there is relative movement among the solid particles, and this exposes new surfaces for gas contacting, prevents short-circuiting, and reduces the formation of stagnant pockets. Although primarily used for drying or reaction applications, rotating vessels, such as kilns, calciners, and rotary driers afford a means of keeping a bed of solid particles in motion. In a fluidized-bed system the solid particles are supported by a distributor plate, through which the gas mixture is forced under pressure. With sufficient gas velocity, the solid particles are forced to move about in a random fashion in the fluid-bed portion of the equipment. In the disengaging space above, the particles free themselves from the gas stream and fall back into the lower portion of the vessel. Gas adsorption, of course, is only one application for an equipment configuration of this type. Fluidized beds are widely used for effecting chemical reactions, providing heat transfer, mixing solids with gases, drying solids and gases, heat treatment, and other operations involving contact between the gas and solid phases. See **Adsorption; and Fluidized-bed operations.**

Gaseous Diffusion: If a binary gaseous mixture, G_1, G_2, at a high pressure is passed over a microporous barrier, a fraction of the gas will diffuse through the barrier into a low-pressure discharge chamber and will be found to be richer in component G_1. This is known as *Knudsen diffusion.* The passage of gas mixtures through the barrier is governed by the unequal collision frequency of each molecular species upon the walls of the pores. Fast, so-called light molecules separate from slower, heavier molecules within the barrier. The Oak Ridge, Tenn., plant designed for the enrichment of $^{235}UF_6$ from the naturally occurring uranium hexafluorides that contained 99.3% $^{238}UF_6$ represented the first application of gaseous diffusion on a production basis. The molecular weight of the hexafluoride of ^{235}U is 349, whereas that of the hexafluoride of ^{238}U is 352. Inasmuch as the rate of diffusion of a gas is inversely proportional to the square root of density, the greatest separation factor for one stage of separation is $\sqrt{352/349}$, or 1.0043. Obviously, only part of the gas can diffuse through a given barrier, and hence the separation factor is less. Thus, the number of diffusion stages for the Oak Ridge plant was approximately 4,000, requiring a plant that covered several acres. Polymeric barriers also have been considered. It is possible that with improvements in the state of the art, gaseous diffusion may be considered seriously for the recovery of CO_2, He, and N_2 from natural gas.

Thermal Diffusion: Application of a thermal gradient to a homogeneous solution (gas or liquid) causes a concentration gradient, and this affords a means of separating materials. The logic behind thermal diffusion stems from the kinetic theory of gases and the cage model of liquids. If there is no marked size difference, heavier species tend to concentrate in the cold region. Where materials of identical molecular weight are involved, the larger molecules go to the cold region by virtue of their greater momentum. In the static mode, differential concentration can be established by eliminating convection currents that otherwise would tend to negate the effects of the applied thermal gradient. In the reflux method, hot and cold materials are flowed countercurrently. The reflux usually is provided by using the density gradient that results from the imposition of the temperature gradient. Equipment of this latter type sometimes is referred to as a *thermogravitational column* or a *Clusius-Dickel column.* Limited applications of thermal diffusion separations include those for concentrating dilute mixtures of isotopic gases. Equipment costs tend to be high and efficiencies low.

Electromagnetic Separation: The same principle that operates in a mass spectrometer can be applied on a larger scale for isotope separation in gases.

Gas Centrifuge: The principles of the centrifuge also apply to gas separation because in a vertical rotating cylinder the higher-molecular-weight components of a gas mixture tend to concentrate near the walls of the cylinder while the lower-molecular-weight materials congregate toward the axis. The separation factor of a gas centrifuge depends upon the difference between the molecular weights of the two components and not on the square root of the ratio of the molecular weights, as in a diffusion process. This factor is much more favorable for isotopic mixtures of the heavy elements than for those of the light elements.

Condensing: When the boiling point of one component of a gas mixture is very different from those of the other components, the condensation method can be used to separate that component from the mixture. For example, in separating helium from natural gas, the feed gas (which contains less than 2% He) is approximately 98% condensed to yield a remaining gas of crude helium and nitrogen with very little methane. In another example, in the commercial production of zinc, zinc oxide is reduced by carbon in a heated retort. The products of reaction are mainly zinc vapor and carbon monoxide. Zinc is separated from the CO by condensation in a splash condenser.

Liquid-Liquid Separations. As in other separations, liquid-liquid separations take advantage of differences in properties from one liquid to the next, including (1) vapor pressure, (2) solubility in a solvent, and (3) permeability through various membranes. Immiscibility, coupled with specific-gravity differences, also plays a role in liquid-liquid separations.

Chromatographic separation of liquid-liquid systems is used principally (1) on a laboratory or pilot plant scale and (2) in connection with chemical analysis. See also **Chromatography.**

Distillation: In this operation, the components of a liquid mixture can be separated by partially vaporizing the mixture and separately recovering the vapor and residue. As the operation proceeds, the lighter, more volatile components of the original mixture (*distilland*) concentrate in the vapor, while the less volatile components concentrate in the liquid residue (*bottoms*). Because of the similarity, it is well to emphasize that in distillation the vapors evolved are recovered by condensation whereas in evaporation, the vapor (usually water) is discarded, the solid residue being the prime target of the separation. Condensed vapors are termed the *distillate.* See also **Distillation.**

Molecular Distillation: This is a special form of distillation conducted at pressures of 1–7 μm in the laboratory and 3–30 μm in industrial installations. Compared with conventional laboratory vacuum distillations carried out between 1 and 10 mm Hg, this is a very high vacuum; 1 μm equals 0.001 mm

Hg. The other feature of the molecular still is that the condenser is located within a distance less than the mean free path of the evaporating molecules from the evaporator portion of the apparatus. Thus, although a molecule may return to the distilland many hundred times before reaching the exit of a conventional vacuum still, 50% of the molecules in a properly functioning molecular still reach the exit on their first try. Thus, efficiency is remarkably high. Because of the absence of convection due to ebullition and because high viscosities and high molecular weights may impede diffusion within the distilland, the surface of the distilland in a molecular still may not always represent the total liquid. Therefore, efficient molecular distillation requires the mechanical renewal of the surface film. This is achieved by vigorous agitation, as in the *stirred-pot still;* by employing a *falling film;* or by using centrifugal force, as in the *centrifugal* molecular still. Commercial installations of falling-film stills have achieved throughputs of many tens of liters per hour, whereas the centrifugal still is capable of outputs of several hundred liters per hour. Centrifugal stills usually are arranged in groups of from three to seven to permit fractionation by multiple redistillation. Among the uses are the separation of mono- and diglycerides for bread and paraffin wax for milk cartons; the distillation of plasticizers, fatty acid dimers, and synthetics; the distillation of vitamin A esters and intermediates; and the stripping of α, β, γ, and δ-tocopherols and sitosterols from vegetable oils.

Liquid-Liquid Extraction: This operation, also referred to as solvent extraction, is accomplished by treating a mixture of different substances with a selective liquid solvent. At least one of the components of the mixture must be immiscible or partly miscible with the treating solvent so that at least two phases are formed over the entire range of operating conditions. To be effective, one or more of the components must be dissolved from the mixture by the solvent preferential to the other components present. The solvent-rich phase that contains the preferentially dissolved component is termed the *extract layer,* and the residual phase formed by the undissolved component, or diluent, and usually containing some solvent is called the *raffinate* layer. Either layer may be at the top or bottom of the separating vessel, depending upon relative densities. Other forms of solvent extraction include leaching, washing, and precipitative extraction (*salting out*).

Liquid-liquid extraction is used in separating the components of condensed mixtures where vaporization methods, such as distillation and evaporation, are impractical because the substances to be separated (1) have comparable volatilities, (2) are relatively nonvolatile, (3) are heat-sensitive, or (4) have one component present in very small concentration.

Dialysis: In this operation, two solutes, L_1 (a low-molecular-weight substance) and L_2 (a high-molecular-weight substance) are separated by permitting the solutes to diffuse through a semipermeable membrane. The operation works best where there is a large difference in molecular weights because of the relative diffusion rates of the two solutes. Complete separation occurs only when one substance will not pass through the membrane pores. Various membrane materials are used, including cellophane, parchment paper, nitrocellulose (denitrated), polyvinyl chloride, and other microporous synthetic materials. Dialysis has been used for recovery of caustic soda from a mixture with hemicellulose and in the refining of beet sugar. Other applications include the recovery of colloidal organics in the production of pharmaceuticals and the recovery of mineral acids from water solutions in various metallurgical operations. Most modern dialyzers are of the continuous countercurrent type. A dialyzer frame is built along the lines of a plate-and-frame filter press.

Electrodialysis: In this operation, an emf is used to effect the transport of ionized materials through diaphragms that separate two or more solutions. In a simple case, visualize starting solution L_1 that contains ionizable solutes S_2 and S_3. The objective is to separate the solutes from the solvent. In the example, the solvent could be water and the solute could be sodium chloride. Two membranes are used in the apparatus, one an ion-exchange resin that is permeable to cations; the other an ion-exchange resin that is permeable to anions. Thus, S_2 will pass through one membrane; S_3 through the other. In this way, the solution becomes depleted of the solute, and separation is effected. Demineralization of water is one of the common applications of this operation. See also **Ion-exchange resins.**

Liquid-Gas Separations. Unlike the five combinations of phases just described, dispersions of gases in liquids and liquids in gases tend to be relatively unstable and commonly tend to separate from each other if allowed to remain undisturbed. In manufacturing systems, of course, where these dispersions are desirable for some processing reason, their later separation usually does not permit the time required for a natural, unassisted separation. Phase separation is also complicated in many cases, by the presence of various agents that have been added to the original dispersion to provide a degree of stability.

Liquid-gas mixtures are separated for a number of reasons: (1) product purity, e.g., air bubbles in paint; (2) equipment operating efficiency, e.g., the nuisance of foams and froths; (3) loss of valuable materials because of entrainment; (4) prevention of damage to equipment, e.g., the effects of an acid spray on fans and blowers; (5) prevention of atmospheric pollution.

The principal methods of liquid-gas separations are listed in Table S-8.

Gravity Settling: Probably the simplest way to separate liquid particles from a gas is to pass the gas stream at very low velocity through a large vessel. The gas may be passed vertically upward through the vessel, whereupon the larger liquid particles will settle out. Better separation can be obtained by allowing the gas to flow horizontally through a long, rectangular chamber, but even this arrangement is uneconomical when particles of less than 40 μm diam must be removed.

Cyclone Separators: As with other separations already described, centrifugal force may be used. Frequently, a cyclone separator designed primarily for dust elimination also can be used to eliminate liquid entrainment.

Impingement Separators: Because liquid droplets have greater momentum relative to gas flow, the droplets will impinge upon an obstructing surface and can be collected. Wire mesh is commonly used as an impingement target for entrainment separation.

Foam Breakers: Aside from a few special processing situations, e.g., flotation or the purposeful preparation of foam products, the formation of a foam or froth in a piece of process equipment is a nuisance and interferes with equipment efficiency. Possibly the most common approach to foam breaking is the use of chemical defoamer compounds. See **Defoaming agents.** When defoaming agents cannot be used because they contaminate the product, physical means are necessary and may be mechanical, electrical, or thermal in nature. The mechanical devices are quite simple, e.g., rotating breaker bars. The presence of a hot surface near a foam tends to destroy the foam. Essentially, a portion of the foam is evaporated, causing the acceleration of its breakdown. It is also known that electrical discharges tend to weaken or destroy films.

Gas Drying: In addition to industrial gases, e.g., nitrogen, hydrogen, oxygen, Freon, and the inert gases, which are almost always dried before packaging and delivery to end users, compressed air, widely used in chemical and other manufacturing plants, normally is dried to prevent corrosion and to assure top operating efficiency of pneumatically operated equipment, e.g., instruments, tools, and control valves. Many gas driers incorporate a combination of heating element and dessicant beds (silica gel, activated alumina, and molecular sieves). See **Drying, gases.**

Boiling: The solubility of a gas usually diminishes with increasing temperature although there are some exceptions. Boiling a liquid containing a dissolved gas lowers its solubility, and further the formation of liquid vapor rising through the boiling liquid helps to entrain and remove gas from the liquid.

Condensing: Visualize liquid L_1 containing gas G_2. In most cases, L_1 can be vaporized to G_1 to produce a mixture of G_1 and G_2. Where G_2 is a noncondensable gas, G_2 can be separated from G_1 in the condenser, where G_1 is returned to the liquid phase L_1.

Sequestering Agents (See **Chelating agents;** and **Detergents.**)

Serine (See **Amino acids;** and **Peptides and proteins.**)

Serum Albumin (See **Peptides and proteins.**)

Settling, Gravity (See **Separation operations.**)

Settling Tanks (See **Sedimentation;** and **Waste treatment.**)

Sewage Treatment (See **Waste treatment.**)

Shakers (See **Screening.**)

Shortening (See **Hydrogenation.**)

Siderite (See **Iron ores.**)

Siegenite (See **Cobalt;** and **Nickel.**)

Sieves (See **Screening.**)

Sieve Tray (See **Absorption.**)

Silanes (See **Plasma, chemical processing with.**)

Silica (See **Abrasives; Ceramics;** and **Silicon.**)

Silica Glass   (See **Glass.**)

Silicates (Soluble)   Soluble silicates are systems containing varying proportions of an alkali metal or quaternary ammonium ion and silica, the most common and commercially used soluble silicates being those of sodium and potassium. They can be produced over a wide range of stoichiometric and nonstoichiometric composition and are distinguished by the *ratio of silica to alkali.* This ratio is used to specify these systems, generally expressed as the *weight percent ratio* of silica to alkali-metal oxide (SiO_2/M_2O). Sometimes, particularly with lithium and quaternary ammonium silicates, the molar ratio is used.

Sodium silicates are widely used in all types of detergents and cleaning compounds as active alkaline builders. Soluble silicates have a long history as adhesives and cements. Sodium and potassium silicates are important bonding agents in a large variety of ceramic cement and refractory applications because of their heat stability and resistance to chemicals. Alkali-metal silicate bonds are used in high-temperature ceramic products for the fabrication of electrical components. Soluble silicates are used widely for pelletizing, granulating, and briquetting finely divided particles, e.g., ores, fertilizers, and clays. Sodium silicates have been of interest for many years as bonding materials for foundry mold and core compositions. Soluble sodium and potassium silicates have wide application in coatings of many different types since they adhere well to many different substrates, such as metal and paper. Sodium silicates are used to protect waterlines and tanks from corrosion. The ability to form sols and gels is one of the most interesting and characteristic properties of soluble silicates. Activated sols are used in water-clarification processes. Silica gels are used on a large scale as desiccants and as carriers for the production of petroleum-cracking catalysts, and they are used as raw materials for the production of zeolites. These and other applications are described later.

Manufacture: Sodium and potassium silicates generally are produced by fusion of pure sand with alkali-metal carbonate or alkali-metal sulfate and carbon in large open-hearth furnaces at temperatures of about 1300–1500°C. The glasses are ground and used in this form or dissolved in water to produce silicate solutions. Sodium and potassium silicate solutions are also produced by dissolving sand in sodium or potassium hydroxide solution at elevated temperatures and pressures (autoclaves).

Lithium silicate glasses are insoluble in water, but they can be produced by dissolving silica gel in, or mixing silica sols with, lithium hydroxide solutions.

Commercially, the anhydrous sodium metasilicate is made from the anhydrous melt. The salt also crystallizes rapidly from its aqueous solution at temperatures above 80–85°C. The sodium metasilicate pentahydrate melts incongruently at 72.2°C and is ordinarily prepared by allowing a solution of the five-hydrate concentration to crystallize completely.

Typical Properties: Sodium and potassium silicates are commercially produced in the form of glasses, crystalline and amorphous powders (hydrated and anhydrous), and aqueous solutions. Table S-9 gives representative compositions and properties. Only a few types of lithium (Table S-10) and quaternary ammonium silicates (Table S-11) are commercially available today.

The most important property of sodium and potassium silicate glasses and hydrated amorphous powders is their solubility in water. The dissolution of vitreous alkali is a two-stage process: (1) as the result of an ion-exchange process between the alkali-metal ions in the glass and the hydrogen ions in the aqueous phase, the aqueous phase becomes alkaline due to the excess of hydroxyl ions produced while a protective layer of silanol groups is formed on the surface of the glass; (2) a nucleophilic depolymerization similar to the base-catalyzed depolymerization of silicate micelles in water takes place.

In general, the solution rate depends greatly on the ratio, concentration, temperature (pressure), and particle size. Glasses more alkaline than a silica-to-alkali mole ratio of 2 dissolve fairly rapidly in hot or cold water. However, glasses more siliceous than the disilicate dissolve very slowly even in boiling water, although some alkali is leached out. The siliceous silicates must be dissolved by steam under pressure.

The rate of solution is greatly increased by prehydration of highly siliceous systems, as is the case with hydrous amorphous silicate powders, usually produced by spray drying soluble silicate solutions.

Most properties of glasses as a function of composition change along a smooth curve. The same is true for alkali-metal silicate glasses, but there is often a change in slope in the curve of property values of the sodium silicate glasses at about 60–70% SiO_2, the disilicate region. This is demonstrated by the density and viscosity of sodium silicate melts. Mixed glass systems of sodium and

TABLE S-9. *Typical Amorphous Commercial Sodium and Potassium Silicates*

Commercial Silicates	Weight Ratio* SiO_2/M_2O	Mole Ratio* SiO_2/M_2O	M_2O,* %	SiO_2, %	Softening Point,† °C	Flow Point,‡ °C	H_2O, %	Baumé 20°C	Sp gr d_{20}^{20}	Viscosity at 20°C, P	pH
Anhydrous Glasses											
Sodium	3.22	3.33	23.5	75.5	655	840					
	2.00	2.06	33.0	66.0	590	760					
Potassium . . .	2.50	3.92	28.3	70.7	700	905					
Hydrated Amorphous Powders											
Sodium	3.22	3.33	19.2	61.8	...	...	18.5				
	2.00	2.06	27.0	54.0	...	...	18.5				
Solutions											
Sodium	1.60	1.65	19.7	31.5	...	...	...	58.5	1.68	70	12.8
	2.00	2.06	18.0	36.0	...	...	...	59.3	1.69	700	12.2
	2.50	2.58	10.6	26.5	...	...	...	42.0	1.41	0.6	11.7
	2.88	2.97	11.0	31.7	...	...	...	47.0	1.49	9.6	11.5
	3.22	3.32	8.9	28.7	...	...	...	41.0	1.39	1.8	11.3
	3.75	3.86	6.8	25.3	...	...	...	35.0	1.32	2.2	10.8
Potassium . . .	2.50	3.93	8.3	20.8	...	...	...	29.8	1.259	0.4	11.30
	2.20	3.45	9.05	19.9	...	...	...	30.0	1.261	0.07	11.55
	2.10	3.30	12.5	26.3	...	...	...	40.0	1.381	10.5	11.70
	1.80	2.83	16.4	29.5	...	...	...	47.7	1.490	13.0	12.15

Crystalline Sodium Silicates

	Mole Ratio	Formula	Mp, °C	Density, g/ml
Orthosilicate · · · · · · · · · · · · · · · ·	0.5	$2Na_2O \cdot SiO_2$	...	2.5
Sesquisilicate pentahydrate · · · · · ·	0.67	$3Na_2O \cdot 2SiO_2$		
Metasilicate, anhydrous · · · · · · · ·	1.0	$Na_2O \cdot SiO_2$	1089	2.61
Pentahydrate · · · · · · · · · · · · · ·	1.0	$Na_2O \cdot SiO_2 \cdot 5H_2O$	72.2	1.749

*M represents Na or K respectively. †Viscosity reaches 4×10^7 P. ‡Viscosity reaches 10^5 P.

TABLE S-10. *Typical Commercial Lithium Silicate Solutions*

| SiO$_2$/Li$_2$O | | | | | | | |
Weight Ratio	Mole Ratio	Li$_2$O, %	Na$_2$O, %	SiO$_2$, %	Density, g/cm^3	Viscosity, cP	pH
9.4	4.7	2.2	. . .	20.7	1.17	. . .	10.6–10.8
9.6	4.8	2.1	. . .	20.0	. . .	4	11
11.8	5.9	1.6	. . .	18.8	1.18		
17.0	8.5	1.2	. . .	20.0	. . .	2.5	11
6.5*	4.1*	1.8	1.2	19.6	1.19	. . .	10.7

*SiO$_2$/(Li$_2$O + Na$_2$O).

potassium silicates exhibit viscosity lowering at intermediate and high temperatures. The viscosity may be more than two orders of magnitude below that of either pure silicate at the same temperature. Maximum decrease in viscosity occurs when K$_2$O constitutes about 30 mole% of the total alkali present.

The only soluble crystalline alkali-metal silicates commercially available today are the sodium silicates at mole ratios of 1:1 SiO$_2$/Na$_2$O or lower (see Table S-9). Of these, the commercially important ones are the anhydrous sodium metasilicate, Na$_2$SiO$_3$; the sodium metasilicate pentahydrate, Na$_2$SiO$_3 \cdot 5$H$_2$O (Na$_2$H$_2$SiO$_4 \cdot 4$H$_2$O); and the sodium sesquisilicate pentahydrate, Na$_3$HSiO$_4 \cdot$ 5H$_2$O (3Na$_2$O$\cdot$2SiO$_2 \cdot 11$H$_2$O). The commercial anhydrous forms of sodium sesqui- and orthosilicate are mixtures of sodium metasilicate with caustic soda.

Soluble silicate solutions can be completely identified by any two properties chosen from a group such as density, weight ratio, percent alkali, percent silica, and viscosity. Viscosity is one of the most important physical properties of soluble silicate solutions. Viscosities of the more siliceous sodium silicate solutions rise much more rapidly with increasing concentration than those of the more alkaline products.

Viscosities at constant solids contents for sodium silicate solutions at varying ratios show a pronounced minimum near a weight percent ratio of 2:1 SiO$_2$/Na$_2$O.

When sodium silicate solutions of intermediate ratios are concentrated to a thick gum, they become very sticky and tacky. This property is important in many of the adhesive applications for sodium silicates and is related to high cohesion and low surface tension but not primarily to viscosity.

Since dry films of sodium or potassium silicates

TABLE S-11. *Typical Commercial Quaternary Ammonium Silicate Solutions*

| SiO$_2$/(N$^+$R$_4$) Mole Ratio | SiO$_2$/(N$^+$R$_4$) | | NR$_4$, % | SiO$_2$, % | Sp gr d$_{20}^{20}$ | Viscosity at 20°C, P | pH |
	Mole Ratio	Wt % Ratio					
			Solutions				
0.5	1.2	0.37	54.5	10.0	1.24	10.7	13
4.5	10.8	2.8	21.5	30.0	1.23	0.2	11.5
7.5	18.0	5.5	11.0	30.0	1.26	0.2	11.2
9.5	22.4	7.0	12.9	45.0	1.41	0.6	11.1
12.5	30.0	9.0	10.0	45.0	1.40	0.6	11.1
			Hydrated Amorphous Powders*				
7.5	15.6	4.3	28.4	66.5			
9.5	19.7	5.9	24.0	71.0			

*Both products contain 5.0% H$_2$O; the time to complete solution at 25°C (50 g water) is 2 min maximum.

rehydrate and gradually dissolve, heat curing of silicate films or bonds is not sufficient to obtain complete insolubility. This can be accomplished only by chemical modification, e.g., reaction with multivalent metal ions which all form insoluble silicates, or by gelation, i.e., treatment with acids, acidic salts, or organic compounds which form acidic materials on hydrolysis in the alkaline medium. The organic systems include anhydrides, amides, and aldyhydes. Soluble silicates can also be insolubilized by reaction with amphoteric metal powders, e.g., aluminum, silicon, ferrosilicon, or calcium silicide, or by reaction with complex metal hydrides, e.g., $LiAlH_4$ or $NaBH_4$. Sodium aluminate, zincate, and plumbate also react with soluble silicate solutions to form gels.

The stability of soluble silicate solutions depends strongly on pH and concentration. The addition of acids and acid-forming compounds gives rise to the formation of silica gels.

Soluble alkali-metal silicate solutions are not compatible with most organic water-miscible solvents. The addition of alcohols and ketones causes phase separation into liquid layers, the silicate accumulating in the lower layer. A few organic systems, particularly such polyols as glycols, glycerins, sugars, and polyethylene glycols, are compatible and miscible with alkali-metal silicate solutions.

Uses: The major applications for soluble silicates were mentioned earlier. The specific functioning of the soluble silicates for each of several major applications will be described here.

Detergents and Cleaning Compounds: The silicates contribute significantly to the effectiveness of detergents and cleaners by maintaining a sufficiently high pH in the system through their unique buffering ability. They also perform certain basic functions, e.g., saponification of animal and vegetable oils and fats, emulsification of mineral oils, deflocculation of solid dirt particles, suspension of soil, prevention of redeposition of suspended dirt, and prevention of corrosion of soft metals (Sn, Al, Cu, and brass) from attack by other ingredients in the detergent mixture. See also **Detergents.**

Deflocculation: The deflocculation effectiveness of soluble silicates, i.e., the ability to disperse agglomerated masses into very fine particles which repel each other, is important in many applications beyond detergents, as in clay mining and refining, the production of clay slips, in ore flotation, oil refining and reclaiming, stabilizing drilling muds, and deinking wastepaper.

Adhesives and Cements: Sodium silicate adhesives are extensively used in combining fiberboard, corrugated paperboard, spiral tube winding, laminating metal foils, plywood bonding, and in the production of building and insulating boards. Advantages include good spreading and contact; wide control over the rate of set; and formation of strong, rigid bonds resistant to heat, moisture, vermin, and bacterial attack.

Bonding Agents: The selection of sodium versus potassium silicates for ceramic cement and refractory bonding applications depends upon performance requirements. Both compounds offer heat stability and resistance to chemicals. Dehydrated silicate films have good dielectric properties making the alkali-metal silicates well suited to the fabrication of electrical components. Conductivity minima can be obtained with ternary Na_2O-K_2O-SiO_2 glasses at about equimolar amounts of alkali-metal oxide.

Setting of the silicate in bonds for finely divided particles, as in ores and fertilizers, is accomplished by air drying or heating. In most cases, a water-resistant bond is required, which is obtained by heating at a very high temperature to form a glassy bond or by chemical setting. Almost any acidic or heavy-metal compound will react with soluble silicate to cause a chemical set. The reaction generally is instantaneous, so that these materials must be applied as an aftertreatment.

The commonly termed *carbon dioxide process,* which has gained wide acceptance in the foundry field, is based on forming a bond of the mold refractory by setting the silicate through brief exposure to CO_2. Use of soluble-silicate–ferrosilicon combinations as binders for foundry molds is one of the later developments.

Coatings: In addition to their adherence properties in coatings, as on metal and paper, soluble silicates are excellent vehicles for fire-retardant or zinc-rich corrosion-resistant paints for iron and steel.

Sols and Gels: Activated sols are used in water-clarification processes. Such active sols are used widely in papermills for increasing fiber and filler retention on the wire and for white-water treatment.

Other important applications of sodium and potassium silicates are in ore flotation as dispersant for siliceous gangue, slime dispersant, and surface conditioner in a variety of ores. The textile industry uses large volumes for stabilizing bleach baths and for fiber preparation. Potassium silicates are used in flux coatings for welding electrodes where sputtering is to be avoided and as a binder for the phosphor on the face of black-and-white television screens. Soluble silicates are used as impregnants of woods and porous metals, binders in abrasive wheels, in soil solidification, concrete-road curing and concrete hardening. The production of building products and insulating materials uses soluble silicates. They also are used as raw materials for the preparation of stable colloidal silica sols, metallic silicates, and finely divided silica hydrogels.

REFERENCES

Eitel, W.: "Silicate Science," Academic, New York, 1964.

Iler, R. K.: "The Colloid Chemistry of Silica and Silicates," Cornell University Press, Ithaca, N.Y., 1955.

Vail, J. G.: "Soluble Silicates," ACS Monograph 46, Chemical Catalog, New York, 1928.

Vail, J. G.: "Soluble Silicates," 2 vols., ACS Monograph 116, Van Nostrand Reinhold, New York, 1952.

Weldes, H. H.: Polysilicates as Detergent Builders, *Soap/Cosmetics/Chemical Specialties*, vol. 48 (4), pp. 72–78, 80, 96, 1972.

Weldes, H. H., and K. R. Lange: Properties of Soluble Silicates, *Ind. Eng. Chem.*, vol. 61, pp. 29–44, 1969.

—Helmut H. Weldes, *Philadelphia Quartz Company, Philadelphia, Pa.*

Silicomanganese   (See **Manganese**.)

Silicon   (L. *silex*, flint.) **Si** = 28.086 (at. wt.); 14 (at. no.). Silicon appears in group IVA of the periodic table. This element is not found in nature in the elementary state but is bound to oxygen. Bound to oxygen alone, Si appears as quartz and quartzites; bound to oxygen and alkalies, earth alkalies, aluminum, iron, or similar elements, silicon appears as silicates. See **Silicates (soluble)**.

In 1787, Lavoisier pointed out that SiO_2 is not one single element, as was then commonly believed, but an oxide of a hitherto unknown element. At the turn of the eighteenth century, Scheele, Davy, Gay-Lussac, and Thénard tried to isolate the element. In 1817, Berzelius found Si in a cast-iron melt and succeeded in isolating it in 1823 by reduction of potassium fluorosilicate with potassium. H. E. Sainte-Claire Deville (1854) and C. Winkler (1864) produced small laboratory quantities of Si. Around 1900, the excellent properties of Si as a deoxidizing agent for steel production were discovered, and soon ferrosilicon alloys using quartzite, coke, and iron pellets were developed and produced in electric refining furnaces of the type already producing calcium carbide. With this technique and raw materials of sufficient purity, a silicon concentration of up to about 98% can be obtained.

Pure Si is compact and macrocrystalline and is of a deep dark-gray color with a high gloss. Special production techniques also yield transparent yellow flakes of brown powder consisting of minute crystals.

In the periodic table, Si is located between C and Ge and, like its neighbors, has four valence electrons ($3s^2$, $3p^2$) and the structure of diamond (face-centered cube). The lattice constant is 5.43086 Å. Each Si atom is covalently connected with four other atoms, surrounding it tetrahedrally. The atomic distance (Si—Si) is 2.35 Å. In the third row of the periodic table, the left neighbor of Si is Al (a metal); the right neighbor is P (a typical nonmetal). Si, together with Ge and Sn—the latter in its gray modification with diamond structure—belongs to the remarkable

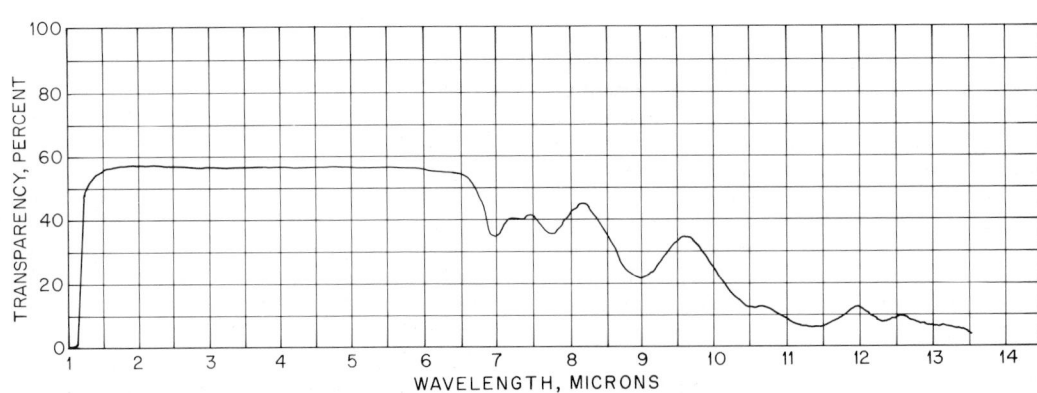

Fig. S-8. Transparency of high-purity silicon to infrared radiation.

group of electronic semiconductors. Si in its purest form, at 300 K, like carbon in diamond modification and like P, is practically an insulator. By adding heat energy (about 900 K) or traces of elements of the neighboring groups III or V (B, Al, P, As, and Sb), Si can be made electronically conductive. Since about 1950, Si and Ge have gained an extraordinarily important position in the electronics industry as semiconductors. Also, Si is infrared-transparent for wavelengths between 1.2 and 10 μm with a maximum transparency of about 58% (uncoated); thus Si is an important optical material (see Fig. S-8). The important properties of Si are given in Table S-12.

With nonmetals having greater electronegativities than Si, Si forms covalent compounds like Si_3N_4 and SiC. Compounds like SiO_2 and SiF_4, silanes, and chlorosilanes partly show polar bonding characteristics apart from the covalent ones. This becomes obvious by comparison with similar carbon compounds. While SiH_4 and the chlorosilanes are instantaneously hydrolyzed with a nucleophilic agent, such as OH^- (water and bases), the comparable carbon compounds remain completely resistant. Also, Si forms Si·Si chains only as hydrides Si_nH_{2n+2}, chains unknown for all other Si compounds.

Chemical Properties: Si in compact form is fairly resistant to acids, including HF. An iron alloy with 20% Si is useful as acid-resistant material for equipment in the chemical industry. At temperatures above 250°C, Si reacts with gaseous HCl to form $SiCl_3H$ and $SiCl_4$; with CH_3Cl in the presence of Cu and $CuCl_2$ to form alcylchlorosilanes, the base materials for silicones. See also **Silicones**. While Si forms SiF_4 with F at room temperature, Si reacts with Cl_2, Br, and I only at 350–650°C to form tetrahalides.

Dissolved in liquid metals, Si reacts vigorously with O_2 and its metal compounds, forming respectively SiO_2 and SiO, which are volatile at 1100°C; hence the use of Si as a deoxidizing agent in metallurgy. The heat of combustion for SiO_2 is 203 kcal/mole. Other relevant physical properties are given in Table S-12. When heated in open air or under O_2, Si forms a very thin surface coating of SiO_2, which serves as a protective layer against further oxidation. When Si and SiO_2 are melted together under vacuum, SiO is formed. This compound is stable only in the heated state and turns into a brown, glasslike substance on sudden cooling.

Even with diluted bases, Si reacts vigorously, developing H_2 and forming alkali silicates.

Silicon combines with metals (melting under inert gases, such as He and Ar) to form silicides and alloys. Of technical importance as deoxi-

TABLE S-12. *Some Important Properties of Silicon**

Electronegativity	1.74
Atomic radius, Å	1.33
Lattice constant, Å	5.43086
Si—Si bond distance, Å	2.35
Thermal properties:	
Melting point, °C	1412
Vapor pressure (melting point), torr	3×10^{-6}
Boiling point, °C	3145
Heat of fusion, kcal/mole	12.1
Heat of vaporization, kcal/mole	71
Specific heat, cal/(g)(°C)	0.18
Above melting point, cal/(g)(°C)	0.219
Thermal conductivity at 20°C,	
(cal)/(s)(cm)(°C)	0.3
Linear thermal expansion coefficient	
at 20°C, per °C	2.33×10^{-6}
Expansion on freezing, % volume	
increase	9
Critical temperature, °C	4920
Critical pressure, atm	1,450
Mechanical properties:	
Hardness:	
Mohs' scale	7
Vickers scale	1,000
Knoop scale	950–1,150
Elastic constant at 20°C, dynes/cm²:	
C_{11}	1.657×10^{12}
C_{12}	0.6393×10^{12}
C_{44}	0.7957×10^{12}
Young's modulus, dynes/cm²	1.9×10^{12}
Bulk modulus, dynes/cm²	6.04×10^{11}
Poisson's ratio	0.048–0.0403
Surface tension, dynes/cm	720
Electrical properties:	
Band gap ($\mathcal{E}_g$), eV:	
At 0 K	1.15
295 K	1.107
Variation of $\mathcal{E}_g$ with temperature,	
eV/°C	-2.4×10^{-4}
Intrinsic resistivity at 300 K, Ω-cm	2.3×10^5
Electron mobility at 20°C,	
cm²/(V)(s)	1,350
Temperature dependence,	
cm²/(V)(s)	$4.0 \times 10^9\ T^{-2}$
Hole mobility at 20°C, cm²/(V)(s)	480
Diffusion constant, cm²/s:	
For electrons	34
For holes	12
Dc dielectric constant	11.8
Optical properties:	
Index of refraction:	
At 1.35 μm	3.4975
3.00 μm	3.4320
5.00 μm	3.4223
Temperature coefficient of refractive	
index, per °C	3.9×10^{-5}
Transmission at 4.0 μm at 25°C, %	58
Reflection coefficient at 4.0 μm	0.301
Magnetic susceptibility at 293 K, g^{-1}	0.111×10^{-6}

*Isotopes of silicon are described under **Chemical elements**.

dizing agents are alloys with iron (ferro alloys), calcium, and aluminum. Copper-silicon (about 20% Cu) is important as a basis for alcylosilanes (silicones). Alloys with small percentages of Si in steel (transformer cores and construction steel) and in Al or Mg (corrosion-resistant alloys with high mechanical strength used in the aircraft industry and for combustion engines) are important to modern technology.

Production of Silicon for Alloys: The manufacture of raw steel uses Si at about 1.6–1.7 kg/ton. World production of ferrosilicon is estimated at 2.5 million tons (1970). Ferrosilicon with up to about 20% Si (as used for cast iron) can be produced in a conventional blast furnace. Higher concentrations of Si (45, 75, 90, and 98%) are obtained only in electric refining furnaces. Raw materials are pure quartzites. Impurities (Al_2O_3 and CaO) disrupt the melting process by formation of dross. Chemical coke is used as a reducing agent. For the high concentrations of Si (90–98%), highly ash-free petroleum coke and charcoal are required. For 45–75% alloys, Fe is added in the form of small pellets or chips.

Furnaces of 5–12 MW capacity have proved satisfactory. Söderberg electrodes of 1 m diam are positioned in the furnace (diam 5 m, depth 2 m) to form the corners of an equilateral triangle. For products with 98% Si, pure graphite or heat-treated rod electrodes are used. Several newly constructed furnaces are placed on a rotating pedestal, allowing a slow rotation (a few degrees per hour) to avoid the formation of dross. The furnaces* are emptied about every 70–90 min, after

*Furnace constructors include Demag Elektro-Metallurgie, Düsseldorf, Germany, and Elektrokemisk, Oslo, Norway.

cooling. The ingots are ground to size in accordance with end-use requirements. Material and energy requirements of these furnaces are given in Table S-13.

Calcium silicon alloy has a typical composition of 30–33% Ca, 60–64% Si, 3–5% Fe, 1–2% Al, 0.3–0.6% C, and <0.15% S and P. The same furnaces as for FeSi are used. Raw materials are quartz, CaC_2, Ca, coke, and charcoal. Energy requirements are about the same as for 98% Si. Calcium silicon is even more effective than FeSi as a deoxidizing agent for steel because of the presence of Ca. The melt and dross can be heated to higher temperatures, whereupon the dross viscosity becomes lower and thus residual dross is easier to handle during hot forming of steel at 1000–1100°C.

Silicon carbide, SiC, with an atomic distance for Si·C of 1.9 Å and a purely covalent bonding explains the extraordinary hardness of this compound (9.5 on Mohs' scale). Each Si atom is surrounded tetrahedrally by four C atoms, and, correspondingly, each C atom is surrounded by four Si atoms. Industrially produced by reduction of pure quartz (glass-sand) with low-ash petroleum coke in an electrical-resistance furnace (Acheson process), the technical grade has the form of light-green to black hexagonal crystals. The material is used widely as an abrasive for metals and as a ceramic raw material for dross-repellent linings. See also **Abrasives;** and **Ceramics.**

Hyperpure Silicon for Semiconductors: The purity necessary for semiconductor devices is a minimum of 10^5 Si atoms per atom of impurity.

The most widely used method for production of hyperpure Si starts sith $SiCl_4$ or trichlorosilane, $SiCl_3H$, either of which can be reduced with H_2

TABLE S-13. *Electric-Energy and Raw-Material Consumption for Producing Ferrosilicon*

	Ferrosilicon with			
	45% Si	75% Si	90% Si	98% Si
Quartz, kg/ton of product	1,000–1,100	1,650–1,750	2,300–2,600	2,700–3,000
Graphite, kg/ton of product	550–700	850–1,000	1,100–1,400	
Charcoal, kg/ton of product	. . .	150–200	200–300	1,500–2,000
Scrap iron, kg/ton of product . . .	600–700	180–210		
Electrodes, number	40–60	60–70	70–110*	110–130†
Electric energy, kWh	4,500–5,500	8,500–9,000	13,000–15,000	14,000–16,000

SOURCE: Silicium, in "Ullmanns Enzyklopädie der technischen Chemie," 3d ed., vol. 15, p. 684, Urban & Schwarzenberg, Munich and Berlin, 1964.
*Söderberg-Stampfmasse.
†Burned electrodes.

to form a compact deposition of Si on hot surfaces (800–1200°C). The basic requirement for obtaining high-purity Si is to start with the highest possible purity of all raw materials. The $SiCl_4$ and $SiCl_3H$ are freed from B and P compounds by fractional distillation and the use of absorbing agents. The H_2 used is purified to the highest degree by passing it, preferably under high pressure, through molecular sieves and other absorption agents at a low temperature (-190°C).

Gaseous $SiCl_4$ or $SiCl_3H$ with a high excess of H_2 is blown through heated quartz tubes. Si is deposited on the walls. In a more advanced process, the reactive-gas mixture is blown into quartz bell jars and deposited on filaments of Ta or W or on thin rods (3–6 mm) of hyperpure Si, heated by resistance or radio-frequency energy. The latter process yields polycrystalline rods of Si up to about 150 mm diam and about 1 m long.

Production of Single Crystals: For semiconductor applications, the polycrystalline Si has to be converted into single crystals of defined type of conductivity (n or p), tightly toleranced in range of resistivity, and of the highest possible crystallographic perfection.

Crucible free vertical float zoning, developed at the Bell Telephone Laboratories by W. G. Pfann, P. H. Keck, and M. J. E. Golay, and perfected by Siemens and Wacker-Chemie, may be used. The purification method is based on the fact that impurities, particularly those of metallic nature, have higher solubility in molten than in solid Si (see Table S-14). Boron, however, is practically irremovable by float zoning, whereas P, As, and O_2 can be evaporated when float zoning is done under vacuum.

Crucible pulling is an alternate method for producing single crystals of Si, particularly of lower resistivity (5×10^{-3}–5×10^1 Ω-cm), by pulling them out of a melt (Czochralski technique).

All techniques for producing single crystals of Si can be applied under protective gas (He, Ar) at normal or reduced pressure or under high vacuum (10^{-5} torr).

Commercial Forms of Hyperpure Silicon: The usual forms of very high-purity Si are:

A. *Polycrystalline material*
   1. Ingots for float zoning; 10–50 mm diam; 400–1,000 mm long.
   2. Cylindrical pieces for crucible pulling made to fit the dimensions of quartz crucibles used; 50–200 mm diam; weights from 200 g to several kilograms.

TABLE S-14. *Distribution Coefficients of Impurities in Silicon*

Element	K^*
B	8×10^{-1}
Al	2×10^{-3}
Ga	8×10^{-3}
In	4×10^{-4}
Ge	3.3×10^{-1}
Sn	1.6×10^{-2}
P	3.5×10^{-1}
As	3×10^{-1}
Sb	2.3×10^{-2}
Cu	4×10^{-4}
O	5×10^{-1}
S	1×10^{-5}
Fe	8×10^{-6}
Ta	1×10^{-7}

SOURCE: F. A. Trumbore, *Bell Sys. Tech. J.,* vol. 205, 1960; P. Aigrain and M. Balkanski, "Tables of Constants and Numerical Data," vol. 12, Pergamon, New York, 1961.

*

$$K = \frac{C_s}{C_l}$$

where C_s = concentration in solid silicon
C_l = concentration in liquid silicon

3. Polycrystalline material, due to its infrared transparency, can be used for lenses, prisms, and domes in infrared systems (fire detection, medical, space, and missile applications).

B. *Single crystals*
   1. Ingots, diam 5–85 mm.
   2. Slices, cut with industrial-diamond saws from the ingots with surfaces as cut or polished; slice diameters, a few millimeters to 85 mm; standard thickness from 100 μm.

A rough estimate (1970) indicates that about 600 tons of polycrystalline hyperpure silicon was consumed; not a high tonnage, but certainly a relatively high monetary value.

REFERENCES

Cotton, F. A., and G. Wilkinson: "Advanced Inorganic Chemistry," 2d ed., Wiley, New York, 1962.

Eljustin, W. P., J. A. Pawlow, and B. C. Lewin: "Ferrolegierungen," V. E. B., Verlag, Technik, Berlin, 1953.

Glang, R., and E. S. Wajda: "The Art and Science of Growing Crystals," Wiley, New York, 1966.

Haberecht, R. R., and E. L. Kern (eds.): "Semiconductor Silicon," The Electrochemical Society Inc., New York, 1969.

Hansen, M.: "Constitution of Binary Alloys," McGraw-Hill, New York, 1958.

Kane, P. F., and G. B. Larrabee: "Characterization of Semiconductor Materials," McGraw-Hill, New York, 1970.

Runyan, W. R.: "Silicon Semiconductor Technology," McGraw-Hill, New York, 1965.

"Ullmanns Enzyklopädie der technischen Chemie," 3d ed., vol. 15, pp. 678–692, Urban & Schwarzenberg, Munich and Berlin, 1964.

—Hans Striebel, *Wacker Chemie G.M.B.H., Munich, Germany*

Silicon Bronze   (See **Copper.**)

Silicon Carbide   (See **Abrasives;** and **Silicon.**)

Silicones   Silicones are semiorganic polymers with a quartzlike structure in which various organic groups are attached to the silicon atom. Varying the kind and number of organic groups produces a variety of materials ranging from liquids through gels and elastomers to rigid solids (resins).

A common commercial process for making silicones begins with quartz, SiO_2. The quartz is reduced by carbon to silicon in an electric-arc furnace, and the product is crushed to a fine powder. By contacting this powder with methyl chloride in the presence of a catalyst under proper conditions a chlorosilane is formed. This is known as the *direct process*. An alternate method of preparing the intermediate chlorosilanes involves reacting Grignard reagents with silicon tetrachloride. The organosilicon chlorides or chlorosilanes are purified by distillation and hydrolyzed with water to form cyclic tetramers of siloxane. This material, in turn, is reacted further in the presence of an end-blocking agent to form a polymer.

Chemistry: Basically, polyorganosiloxanes are characterized by a molecular backbone of alternating Si and O atoms with organic groups attached to the Si atoms. The type of organic group and the extent of cross-linkage between polymer molecules determine whether the silicone will be fluid, elastomeric, or resinous. See also **Silicon.**

The basic polydimethylsiloxane structure, shown here, can be modified and varied in many ways.

$$CH_3-\underset{\underset{CH_3}{|}}{\overset{\overset{CH_3}{|}}{Si}}-O\left[\underset{\underset{CH_3}{|}}{\overset{\overset{CH_3}{|}}{Si}}-O\right]_x\underset{\underset{CH_3}{|}}{\overset{\overset{CH_3}{|}}{Si}}-CH_3$$

For example, substituting a phenyl group for one of the methyl groups produces a new polymer or, as shown here, a copolymer:

$$CH_3-\underset{\underset{CH_3}{|}}{\overset{\overset{CH_3}{|}}{Si}}-O\left[\underset{\underset{CH_3}{|}}{\overset{\overset{CH_3}{|}}{Si}}-O\right]_x\left[\underset{\underset{CH_3}{|}}{\overset{\overset{C_6H_5}{|}}{Si}}-O\right]_y\underset{\underset{CH_3}{|}}{\overset{\overset{CH_3}{|}}{Si}}-CH_3$$

The phenyl groups extend the thermal stability of the polymer in both directions, making it a more serviceable material at both higher and lower temperatures than the polydimethylsiloxane polymer.

Introducing fluorine atoms along the chain in the form of a trifluoropropyl group has a profound effect on solubility. Fluorosilicones are insoluble in common aliphatic, aromatic, and chlorinated solvents. In contrast dimethylsiloxane polymers are completely soluble in these solvents. At the same time there is very little change in thermal properties. Fluorosilicone fluids exhibit improved lubrication characteristics as well. This combination of chemical inertness and good lubricity is valuable in many lubricant applications in the chemical process industry, e.g., a compressor lubricant. Fluorosilicone oils have greatly reduced maintenance costs on HCl and methyl chloride compressors.

The polymers discussed so far are completely reacted materials; i.e., they have no reactive sites and therefore are incapable of further reaction. However, a class of reactive siloxane polymers of intermediates is also available, containing silanol groups, $\cdot SiOH$, which react with alkyds, phenolics, epoxies, cellulosics, polyesters, and other organics containing hydroxyl groups.

Silicone Fluids: These materials are clear liquids available in viscosities of 0.65–1 million cSt or higher. The other properties of dimethylpolysiloxane fluids, for example, also vary widely. Specific gravity is 0.761–0.975 (at 25°C referred to water at 25°C). Their dielectric constants range from 2.18 (at 1,000 Hz and 23°C) to 2.82. Freezing points are −86 to −44°C (−123 to −47°F) and boiling points (at 760 mm Hg) range from 99.5°C to over 230°C (211°F to over 446°F). The refractive index (at 25°C) is 1.375–1.4035.

The fluids may be used alone, combined with organic thickeners to produce greases, or combined with inorganic fillers to produce water-repellent dielectric compounds. Applications include defoamers, release or parting agents, cosmetic or polish fluids, heat-transfer fluids, compressible fluids in liquid springs, and cell-control additives for foams. Special fluids containing silane hydrogen are effective for permanent water-repellency treatment of fabrics. See also **Lubricants, synthetic.**

Silicone Elastomers: The elastomers are essentially high-molecular-weight fluids which are generally compounded with inorganic reinforcing fillers (special silicas) and suitable curing agents (organic peroxides). After molding or extruding, they are vulcanized by heating to cross-link the linear molecules into a flexible three-dimensional structure. The resulting rubber has a multifold increase in strength properties, serviceable from $-90°C$ to over $250°C$.

Room-temperature-vulcanizing (RTV) elastomers constitute another category of silicone rubber products of increasing importance. These RTV rubbers have excellent weathering resistance and do not become embrittled with aging at ordinary temperatures. Their flexibility over a wide temperature range, ease of use and handling, and ability to adhere to many substrates results in wide application as a sealing medium. Many types of silicone RTV rubbers are made into flexible molds for casting plastic parts, e.g., furniture. These molds duplicate the most minute detail, are easy to use and durable, withstand shelf aging, and resist damage from repeated use and rough handling.

One-part, or one-component, RTV rubbers have found ready acceptance in the appliance, aerospace, automotive, consumer, and building industries as adhesives and sealants.

Resins: Silicone resins can be provided in varying ranges of flexibility. Flexible resins have been used as electrical insulation for coating and impregnating varnishes and as protective paint films. Rigid resins are more commonly used as binders in laminated composites and in molding compounds.

Laminated Composites: Silicone resins combined with heat-resistant reinforcements (glass fibers and cloth, asbestos, or mica) are used in a wide variety of electrical applications. These require an initial cure of $177°C$ followed by an aftercure through $193°C$, which develops optimum properties. High-pressure silicone laminates—sheets, tubes, or rods—can be machined in a manner similar to other laminates reinforced with abrasive materials. Coil forms, bobbins, generator and transformer insulation, switch spacers, terminal boards,

and slot wedges for electric motors are typical fabricated uses of laminates. When complex shapes have to be fabricated, some silicone resins can be laminated at low pressures (30–200 psi) in hydraulic presses or by autoclave and bag molding. Lay-up techniques are similar to those employed with organic resins. Filament-wound silicone-glass laminates can be produced with flexural strengths of 100,000 psi or more. Such laminates are used in making radomes, nose cones, and rocket-motor cases.

Molding Compounds: Silicone molding compounds are molded in the same way as organic thermosetting plastic. Compounds designed for encapsulation of electronic components can be molded at pressures as low as 150–1,000 psi. Those filled with glass fiber require higher molding pressures (5,000 psi). These materials are designed for applications demanding high impact resistance.

The molded plastic is highly resistant to environmental extremes; its heat-distortion point ranges above $400°C$; it is self-extinguishing and highly resistant to thermal shock. Encapsulated modules show low dielectric losses even at high frequencies, high operating temperatures, or after exposure to water. This reliability is due in part to the fact that these resins do not contain ionic or polar constituents.

Potting and Encapsulating: Solventless thermosetting silicone resins are used where electronic parts must be protected from mechanical and thermal shock, moisture, corrosion, and atmospheric ozone. Tough and flexible, these resins can be poured around units to be potted and set up at room temperature without exothermic heat. Portions of the resin can easily be cut away for replacement or repair of defective components. New material can be poured in and cured in place.

Another type of solventless resin designed for use as encapsulant is poured in place and, when heated, cures to a crystal-clear jellylike protective cushion. This dielectric gel has good electrical properties and is transparent, permitting visual inspection of potted parts.

Physical Properties: Principal physical properties which distinguish silicone polymers from organic counterparts are retention of properties over a wide temperature range, low surface tension, excellent release properties, extreme water repellency, outstanding electrical properties, and chemical and physiological inertness.

Temperature Resistance: Silicone polymers are relatively unaffected by high or low temperatures.

Silicone fluids perform over a temperature range of -73.3 to $260°C$. Compared to the viscosity of organic fluids, the viscosity and/or shear properties of silicone fluids change relatively little over a wide temperature range, permitting their applications in damping devices for instruments subjected to temperatures of -45.6 to $232°C$.

Thermal oxidative stability also makes silicones useful as release agents for molds operating at $232°C$. They do not carbonize and eliminate the need for frequent mold cleaning.

When used as binders for paints, silicone resins operate continuously at $316°C$ and intermittently at temperatures as high as $538°C$. Such paints exhibit excellent long-term aging and weathering properties.

Surface Tension: Silicone polymers possess unusually low surface tension. When silicones are soluble in a system, the surface tension of the entire system is reduced and the effect is similar to that of adding a surfactant to water. The system foams and wets more completely. The ability of silicones to support and maintain foam has great commercial utility in one-shot urethane-foam systems. On the other hand, when the polysiloxane is insoluble in the system, it acts as a defoamer. See also **Defoaming agents.**

Release Properties: Very few substances will stick to a silicone polymer. Highly polar surfaces are made nonpolar when dimethylsiloxane polymers orient themselves on a surface. Most materials do not adhere well to these nonpolar, low-energy surfaces. Application is made of this property in the release of rubber and plastic parts from metal molds. Release coatings for paper are also based on this property. Silicon-treated fabrics are more easily cleaned since stains and dirt do not stick so tenaciously.

Other Properties: The ability to depolarize surfaces produces extraordinary water repellency. Clothing, leather, bricks, and masonry become highly water-repellent when treated with silicones. Silicone polymers exhibit high dielectric strength and low dissipation factors and maintain good electrical properties in severe ambient conditions. Many silicones, particularly the dimethyl polymers, are *physiologically inert*. They have been engineered into devices to aid or replace human parts and can be implanted in the body permanently without ill effects. Such functional devices as heart valves, hydrocephalus shunt valves, and retinal reattachment units have been fabricated from these elastomers. Implants to replace such lost or ill-formed parts as breasts, ears, chins, and noses serve both a cosmetic and a psychological purpose.

—Myron Kin, *Dow Corning Corporation, Midland, Mich.*

Silizanes (See **Silicon;** and **Silicones.**)

Silk (See **Fibers;** and **Peptides and proteins.**)

Siloxanes (See **Silicon;** and **Silicones.**)

Siloxones (See **Plasma, chemical processing with.**)

Silver (OE. *seolfor.*) **Ag** (L. *argentum.*) $= 107.868$ (at. wt.); 47 (at. no.). Silver occurs in group IB, period 4, of the periodic table. It is a white, lustrous metal surpassing all other metals in several important properties, including electrical and thermal conductivity, optical reflectivity, ductility, and malleability, as well as being very stable chemically. If it were not for its scarcity, silver would be one of the most widely used of all metals. A summary of important physical and mechanical properties is given in Table S-15.

TABLE S-15. *Selected Physical and Mechanical Properties of Silver*

Thermal conductivity, (0–100°C) (cal)(cm)/(s)(cm^2)(°C)	0.999
Specific heat (20–100°C), cal/(g)/(°C)	0.053–0.054
Thermal expansion (0–100°C) per °C .	19.68×10^{-6}
Electrical resistivity at 20°C, $\mu\Omega$-cm .	1.59
Electrical conductivity (% IACS*) . . .	108.4
Temperature coefficient of resistance (0–100°C), per °C	0.004098
Diamagnetic properties, cgs units . . .	-0.181×10^{-6}
Surface tension (at 1268 K), dynes/cm	920
Mechanical properties of commercially pure silver (99.9$^+$ %):	
Tensile strength, psi:	
Annealed	25,000
Cold-worked, 50%	44,000
Elongation, %:	
Annealed	52
Cold-worked 50%	5

*International Annealed Copper Standard.

The melting point of Ag $= 960.8°C$; boiling point $= 2187–2259°C$; density $= 10.49$ g/cm^3 at $20°C$. Silver has two stable isotopes, 107 and 109. In addition, there are reported to be 25 less stable isotopes, ranging in half-life from 5 s to 253 days. See also **Chemical elements.** Silver has a face-centered cubic lattice and therefore has no allotropes.

The earliest records show that man in southern Europe and Asia Minor used Ag both for ornaments and as a medium of exchange as early as 4000–3000 B.C. The mining of Ag began about 500 B.C. in Greece and was said to be the chief support of Athens during the Persian Wars. From 1000 to 1 B.C., vast expansion of Ag production took place, coins were stamped, and Ag was gilded with gold amalgam. There are three important facets of the modern history of silver in western Europe and the United States: (1) development of silversmithing, (2) the political-monetary aspects of the metal, and (3) the growth of industrial uses of the metal.

Silver is widely distributed throughout the world. It rarely occurs in native form but is found in ore bodies as AgCl or, more frequently, as simple or complex sulfates. Almost two-thirds of the current Ag production in the United States is obtained as a by-product of base-metal ores of Cu, Pb, and Zn. Here, Ag usually is associated with the sulfide of the base metals. Mexico is currently the leading silver-producing country in the world, with an average production of about 44 million troy oz/yr. Mexico is followed by Peru, the United States, Canada, the Soviet Union, Australia, and Japan.

A great amount of Ag is recycled; in fact, in the United States, far more Ag is recycled than is mined. Much of this scrap comes from photographic film, the arts, electrical, and other industrial fields. In 1963, recovery of Ag from these sources resulted in about 95 million oz of Ag.

Different types of Ag-containing ore require different processing. In general, the ores are ground very fine and then concentrated by selective flotation treatments. Cu and Pb concentrates are smelted to yield Cu and Pb bullions, from which Ag subsequently is recovered by fire and/or electrolytic refining. Zn concentrates are commonly processed by leaching and electrolytic methods, although sometimes retorting practices are used in place of electrolysis.

The smelting and refining of secondary silver scrap differs somewhat from that of base-metal scrap. The high value of the feed material dictates the necessity for very careful preparation of a homogenous sample and close accuracy in analyzing for precious metals content. The highly complex nature of scrap, with respect to physical form, chemical composition, and grade, requires a complicated but highly flexible process flow sheet. The primary features of scrap recovery are very efficient and fast processing so as to minimize metal losses and tie up high-value materials.

Uses: In 1971, the industrial use of Ag in the United States was about 140 million oz. Of this, photography used 32%; electrical applications, 31%; jewelry and silverware, 17%; brazing alloys, 13%, and miscellaneous applications the balance of 7%.

The predominant place of silver salts as photographic receptors is not the result of any unusual primary sensitivity to illumination but is due to the fact that they undergo an unusual secondary amplification process called *development.* Silver salts, like the salts of many metals, when immersed in solutions of reducing agents, are changed to metallic Ag. The photographic system depends upon the fact that when certain mild reducing agents (called *photographic developers*) are chosen, the rate of reduction is increased many times if the silver salt crystals carry very small amounts of metallic Ag at the developed-crystal interface. The development process provides an amplification factor of up to 10^{11} to the effect produced by the original light exposure. Whereas new photographic or recording devices are being developed not involving silver, none yet approach the packing density of a fine-grained image possible using Ag. Thus it appears that Ag will be used in photography for many years to come.

Among the electrical uses for Ag are electrical contacts, printed circuits, and batteries. By far the greatest use is in electrical contacts, where the high electrical and thermal conductivities, as well as corrosion and oxidation resistance, of silver are major reasons. Although Ag has a strong tendency to weld under heavy currents, this is counteracted by alloying or by adding nonmetallic substances (such as cadmium oxide) to the Ag matrix. The use of silver–cadmium oxide and silver-tungsten materials in electrical contact applications is widespread. The alloys used to improve the wear resistance and to reduce the sticking tendency of silver include Ag-Au, Ag-Cu, Ag-Pd, and Ag-Pt. More complex alloys include Ag-Cu-Ni, Ag-Mg-Ni, Ag-Au-Cd-Cu, and Ag-Cd-Cu-Ni. Silver–cadmium oxide alloys are prepared either by combining Ag and Cd oxide by powder-metal techniques or by the internal oxidation of a wrought Ag-Cd alloy. Electrical alloys, which are impossible to combine by conventional melting, lend themselves to powder-metal fabrication. Such composite structures as silver-graphite, Ag-Fe, and Ag-W are good examples.

In silver batteries, the silver oxide–zinc secondary battery has found application where energy delivered per unit of weight and space are of prime importance. The major disadvantages lie in its high cost and relatively short life. Consequently, almost 90% of the silver-battery market is concerned with defense and space components. About 10 million oz of silver were used during 1965 for this purpose. See also **Batteries.**

Ag and Cu form a simple eutectic system with limited solid solubility. This system can absorb elements such as Zn, Cd, Sn, and In. These additions lower its melting temperatures while still preserving its identity as a eutectic. It can also absorb higher melting elements such as Ni or Pd, which raise its melting temperatures without destroying its identity as a eutectic. It can also absorb higher-melting elements such as Ni or Pd, which raise its melting temperatures but may improve its wetting characteristics, corrosion resistance, and strength at elevated temperatures.

Between the two World Wars, low-temperature Ag-Cu brazing alloys were found to be useful on Cu and its alloys and Fe and its alloys (including stainless steel). Silver solders or brazing alloys can make joints far stronger and more durable than common soft solder (such as Pb-Sn) alloys. They are used in most refrigeration systems to join Cu tubing. Also, extensive use is found in the assembly of automotive parts, military components, aircraft assemblies, and other hard-goods manufacture.

One Ag alloy containing about 70% Ag, 26% Sn, 3% Cu, and 1% Zn is used extensively by dentists in combination with Hg to fill cavities in teeth. The "amalgam" manufacturers supply dentists with is the alloy in the form of powder (filed, or more recently, atomized). This is mixed with mercury, using 8–5 parts of Hg to 5 of alloy, and the cavity is packed. In the cavity, a metallurgical reaction takes place in which the Ag-Sn compound in the alloy becomes a durable Ag-Sn-Hg compound.

Compounds: Silver oxide, Ag_2O, is formed when alkali hydroxides act on Ag salts or from heating the carbonate above 200°C. Ag_2O decomposes to metallic Ag and O_2 at 300°C. Black silver peroxide, Ag_2O_2, results from treating $AgNO_3$ with $Ag_2S_2O_3$. The peroxide is an active oxidizing agent. Ag forms the largest number of insoluble salts of all metals, approached only by Pb and Hg. AgCl is moderately soluble in HCl and chlo-

rides because of the complex argentichloride, H_2AgCl_3 and H_3AgCl_4, formed. AgCl is not attacked by HNO_3 but is soluble in NH_4OH. All Ag salts are soluble in NH_4OH except AgI and Ag_2S. All Ag salts are soluble in HNO_3 except AgCl, AgBr, AgI, $AgBrO_3$, $AgIO_3$, AgCN, and AgCNS. Alkali thiosulfates will dissolve almost all Ag salts, and the cyanides, including NaCN, will dissolve all Ag salts except Ag_2S. $AgNO_3$ is very soluble in H_2O but sparingly in alcohol and ether.

Ag should not be dissolved in NH_4OH in the absence of NH_4 salts because of the likely formation of *fulminating silver*, which is highly explosive (NAg_3 or $NAgH_2$).

Silver, its oxides, halides, and other salts play important roles in industrial chemistry. Silver is an excellent catalyst in oxidizing reactions such as in the production of formaldehyde from methanol and oxygen, ethylene oxide from ethylene and oxygen, and glyoxal from ethylene glycol and oxygen. Silver, its oxides, and its salts have oligodynamic properties, and silver was once used, before chlorination, to sterilize water. At present, the chief silver-based medicinal is Argyrol.

REFERENCES
Addicks, L.: "Silver in Industry," Van Nostrand Reinhold, New York, 1940.
Butts, A., and C. Coxe: "Silver: Economics, Metallurgy, and Use," Van Nostrand Reinhold, New York, 1967.
Smithells, Colin J.: "Metals Reference Book," vol. 3, Plenum, New York, 1967.

—Robert J. MacDonald, *Handy & Harman, Fairfield, Conn.*

Silver-Alloy Brazing   (See **Copper.**)

Silver Azide   (See **Explosives.**)

Silver Fulminate   (See **Explosives.**)

Sink-Float Concentration   (See **Beneficiation, ore;** and **Separation operations.**)

Sintering   (See **Agglomeration; Iron ores; Iron- and steelmaking;** and **Smelting.**)

Sintering, Plastic   (See **Plastics.**)

Size Enlargement   (See **Agglomeration.**)

Size Reduction   Size reduction implies a general process for reducing any given particle to a smaller size by cleavage or fracture of the existing structure of the particle and forming new surfaces. Some of the specific operations that fall within this general process class are variously termed *crushing, grinding,* and *pulverizing,* the specific term depending upon design and application variations as described briefly here. Size reduction is a very important process to most manufacturing industries that deal with solid materials in bulk form, ranging from heavy rock-type raw materials to very finely ground end products.

Basic Laws: Probably more than most unit operations, size reduction remains more of an art than a science. From a theoretical point of view, certain basic formulas or laws have been developed, e.g., Kirk's law, which states, "the work for a given reduction ratio is the same irrespective of the starting size." Rittinger's law states, "work is proportional to the surface produced," and, more recently, Bond postulated that "work is inversely proportional to the square root of the diameter of the particle."

Although these basic laws are valid in the overall sense and hence of value, the laws do not provide detailed answers to most specific size-reduction problems. Most manufacturers of grinding and pulverizing equipment have found it necessary to establish test laboratories in order to evaluate the performance of their equipment. There is no one universal and all-purpose unit that will do an optimum job of grinding or pulverizing all conceivable products. Numerous designs have been developed to process given materials to specific product requirements.

Forces at Work: In any given size-reduction process, there are combinations of forces applied. The situation is complex, and it is rare for only one such force to be utilized. These forces are (1) compression, (2) tension, (3) shear, (4) impact, and (5) attrition.

Tension is the primary cause of failure in most brittle materials; however, no practical size-reduction unit has been devised to apply a primary tensile force. Nevertheless, tension is a contributing factor in many crystalline materials, e.g., salt; certain minerals, e.g., limestone; and miscellaneous materials, e.g., glass. Pure compression applied to brittle materials will usually result in fractures or reduction. With more ductile materials, compression by itself will generally produce a flattening effect or even flaking. Therefore compression generally is combined with other forces such as rubbing or shear to produce the desired results.

Attrition is another term for rubbing. Attrition by itself seldom is selected, although this is the predominant action in some mills, such as disk mills. Even here there are other forces present, e.g., shear and impact.

Shearing action is predominant in cutting and chipping, e.g., rotary knife cutters, dicers, slitting machines, and chippers. The term implies simply that a particle is cut sharply and cleanly between two sharp edges, either singular or multiple.

Impact is the sudden application of a force by moving tools, e.g., hammers in a hammer mill. This force is always present to some extent in every size-reduction operation because of the effect of a moving particle impacting against a stationary surface or similar particles. The impact is therefore difficult to control in most unit machines and sometimes works against the desired results.

Physical Characteristics: Information required by the manufacturer of size-reduction equipment to determine the most effective design for a given application is summarized in Table S-16. Good design and application engineering requires a clear understanding of the user's requirements for the finished product as well as the starting materials and other production considerations.

Classification of Size-Reduction Operations: In general, size-reduction operations can be grouped into basic areas. The first is *primary crushing,* wherein large lumps or pieces of virgin material, typified by those from a mining operation, are involved. These may run to 5 ft in size and perhaps larger. *Secondary crushing* generally is limited to 6- or 8-in. size and most frequently is used as a precrushing stage prior to other grinding or pulverizing operations. *Grinding* is a broad term which refers to all size-reduction operations from roughly 1 in. down to a particle size of 100–150 mesh. On the other hand, *pulverizing* covers the range of 150 mesh downward, including the micrometer range and below. Grinding and pulverizing can each be subdivided into primary and secondary classifications. A summation of the range of application for each of these classes is given in Table S-17.

Jaw Crushers: These devices generally are used as a primary crusher because they are adaptable for handling large pieces of hard and semihard materials in large volume. For example, they are available with feed openings up to approximately 6×7 ft with capacities of approximately 1,000 tons/hr. Power requirements are high, ranging up to 300 hp in larger sizes. They are capable of producing products ranging from 2 to 12 in., with some of the product falling below this. These

TABLE S-16. *Checklist For Selection of Size-Reduction Equipment*

Feedstock dimensions
   Maximum lump or particle size
   Particle-size distribution
Bulk density
If sheet stock:
   Sheets or continuous rolls?
   Width and thickness of sheets
Physical characteristics
   Hardness (if possible, use Mohs' scale)
   Moisture content
   Liquids present, e.g., oils and resins
   Temperature sensitivity
   Corrosiveness
   Abrasiveness
Product requirements
   Particle-size range required
   Percentage of oversize tolerated
   Percentage of fines tolerated
   Special shape considerations
Production considerations
   Production rate, e.g., pounds per hour
   Sifting and recycling allowed?
   Inlet temperature of feedstock
   Precrushing allowed?
Electric-current characteristics
   Voltage, frequency, and phase
   Enclosures required, (e.g., totally enclosed or explosion-proof)
Special considerations
   Materials of construction
   Explosive characteristics
   Limitation on metallic pickup?
   Sanitary requirements
   Changes in characteristics with temperature
   Toxicity of product

crushers give a compression or squeezing action between two surfaces similar to the jaw of a vice. The movable jaw is pivoted at the top, allowing the greatest movement to the smaller lumps. The movement of the jaw is effected through a knuckle action by the rising and falling of a pitman, which is moved by an eccentric shaft. Vertical movement is communicated to the jaw by two toggles.

TABLE S-17. *Operating Ranges of Size-Reduction Equipment*

Operation	Feedstock	Product
Primary crushing . . .	Up to 60 in.	3–4 in. downward
Secondary crushing . .	6–8 in.	$\frac{3}{4}$–1 in. downward
Grinding	1 in. max.	10–150 mesh
Pulverizing	20 mesh max.	150 mesh and finer

One basic type is the Blake crusher illustrated in Fig. S-9. Other units, such as the Dodge, use a bottom movable pivoted jaw. Some units have curved jaw surfaces, which are reputed to reduce clogging. Others use two movable jaws pivoted at the top.

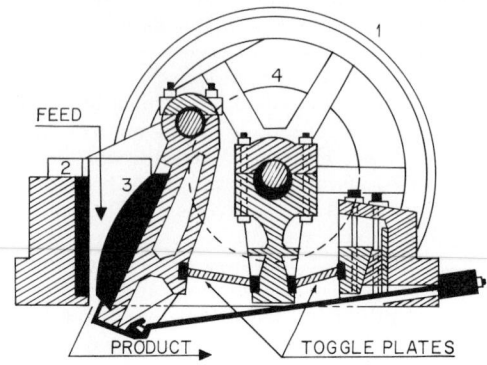

Fig. S-9. Blake-type jaw crusher. (1) Flywheel, (2) fixed-jaw plate, (3) swing-jaw plate, (4) pulley.

Gyratory Cone Crushers: As shown in Fig. S-10, these crushers are vertical-shaft units having a conical head or crushing element operating with an eccentric gyratory motion within an inverted

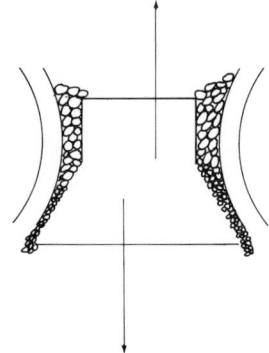

Fig. S-10. Highly schematic representation of the operating principle of a gyratory cone crusher.

telescopic member. The coarse feed enters at the top, where there is the greatest spacing between the two cone-shaped members; and the finer, crushed material leaves at the bottom, where the spacing between the members is much smaller. The conical head supported from the base is operated by an eccentric drive through gears and a countershaft, thus transmitting the gyratory motion. The upper frame is held in a fixed position by heavy springs, allowing relief from any choked condition caused by overfeeding. The material receives rapid impact as it passes through the crushing chamber. The openings between the crushing element and the stationary member may be decreased by raising the position of the rotating element with the subsequent reduction in size of material and capacity potential. These crushers are available with feed openings 2–10 in. wide. With the various sizes available, capacities range from 20 tons/hr of a $\frac{3}{8}$-in. product (requiring 25 to 30 hp) up to 600 tons/hr of $1\frac{1}{2}$ in. discharge product (requiring 250 to 300 hp). Some manufacturers combine a primary and secondary crusher by using a two-step crushing head, with the top portion of the head doing the coarse grinding and the lower portion further reducing the material.

Heavy-Duty Hammer Mills: The design principle of this unit is evident from Fig. S-11: pivoted

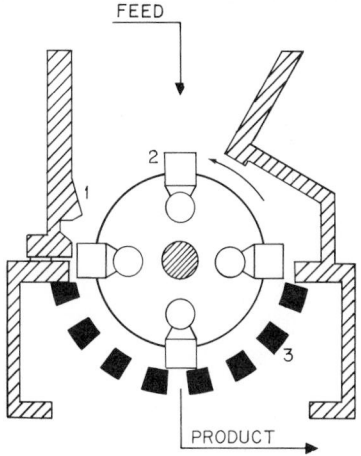

Fig. S-11. Highly schematic representation of a hammer crusher of a type made by Jeffrey Mfg. Co. (1) Breaker bar, (2) hinged hammer, (3) grate bars.

hammers mounted on a horizontal shaft impact the feed material against breaker plates. Material is fed in at the top throat and discharged by gravity through a grid of bar stock or a screen. Usually feed sizes are large and preclude the use of screw-controlled devices other than conveyor belts or bucket elevators. Feed sizes may be as large as 10–20 in. with power requirements varying from 5 to 500 hp. These machines will produce from a few tons per hour up to 500 tons/hr

and are adapted to minerals, ores, salts, ceramics, and slag.

For secondary crushers there are roll crushers, sawtooth crushers, and edge runners (pan crushers).

Roll Crushers: These may be divided into two classes, smooth and corrugated rolls. In either, the unit consists of rolls of the same diameter revolving toward each other at speeds which may be the same or at a differential. Smooth-roll crushers are generally designed to permit one shaft to rotate in fixed bearings while the other is in movable bearings. The distance between the rolls is adjustable, and powerful springs hold the movable roll to a preset clearance. Tension springs exert pressure on the rolls up to 6,000 lb per inch of roll face for light duty to as high as 40,000 lb per inch for heavy duty. Standard diameters for smooth rolls are 12–48 in., and the mills generally operate at speeds of 200–2,000 ft/min. With grooved or toothed rolls, only one roll may be used; although frequently two rolls are used.

Sawtooth Crushers: The crushing element of these units is a series of breaker plates with sawteeth, alternating with spacers and rotating at differential speeds on twin shafts, as illustrated in Fig. S-12. Rotors are normally approximately 10

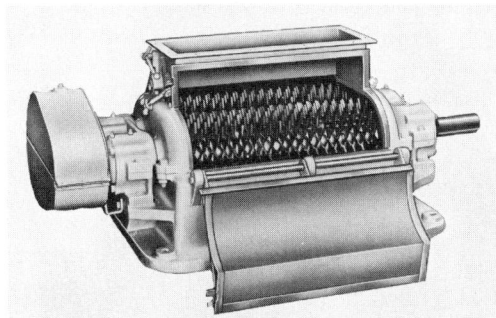

Fig. S-12. Sawtooth crusher. (*Sprout, Waldron & Co.*)

or $16\frac{1}{2}$ in. diam (tip of saw). Although these units involve a greater investment than horizontal or vertical crushers of comparable capacity, they have the advantage of being able to accept comparatively wide pieces, e.g., slabs, filter-press cake, and continuous sheets. In addition, they can produce a limited size range with a low percentage of fines, high capacity, and low power consumption, thanks to an action that stresses tearing rather than impact or compression.

Edge Runners: These mills are sometimes called

dry-pan mills. Although not used as frequently as in the past, they are still used for grinding clay in the ceramic industry and consist of rotating stones or wheels revolving in a circular pan or trough, the activating motion being a central shaft carrying an arm supporting the circular disk. The bottom pan may be rotating or stationary with the material discharged through grids on the pan periphery. These have pan diameters of 2–7 ft.

In grinding, the following units typify a few of the many available.

Attrition Mills: These units, or disk mills, as they are sometimes referred to, consist of the two vertical disks, horizontal shafts, adjustable clearance between these vertical disks, feed close to the shaft or the central point at one disk and discharge at the periphery (Fig. S-13). Frequently, attrition mills have one stationary and one rotating disk; in others, both disks rotate at differential speeds or in opposite directions. Each disk has interchangeable metal plates which may be flat-grooved, spiked, or toothed. These mills are best applied to materials which are reduced by a tearing or shredding action although the flat grinding plate is often used for granulating and fine grinding. Because of the frictional nature of the reduction, temperature buildup is usually more of a factor than in certain other types of mills, and heat-sensitive materials must be avoided.

Fig. S-14. Turbo mill. (*Pallmann Pulverizers Co., Inc.*)

A variation, but not an attrition mill per se, is the Pallman Turbo Mill (Fig. S-14). This mill uses attrition-mill plates but chiefly in the periphery rather than on the flat rotating surface. Reduction is primarily by a rotating arm which has blades in close proximity to these plates. The unit requires close control of air passage through the mill and can be used as a classifier.

Cage Mills: These mills consist of two or more concentric cages, alternate cages operating from separate horizontal shafts running in opposite directions. Both the cage mill and the attrition mill generate a considerable amount of air and usually require a products-collection system.

Rotary Cutters: These units consist of a rotor shaft carrying a series of rigid fly knives or blades,

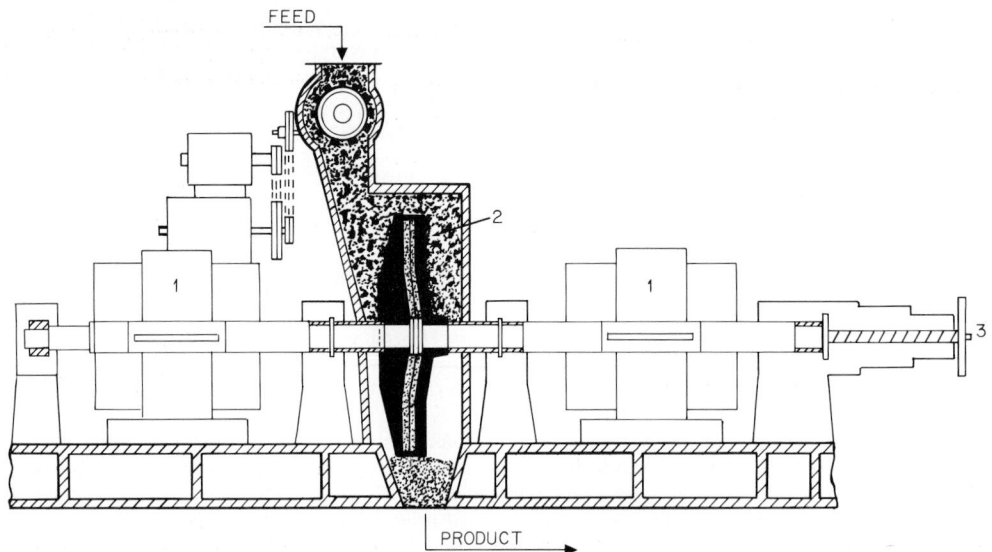

Fig. S-13. Highly schematic representation of a double-runner attrition mill of a type manufactured by Sprout, Waldron & Co. (1) Drive motors, (2) grinding disks, (3) handwheel to adjust clearance between grinding plates.

sometimes slightly spiral in configuration. The rotating knives operate within a series of stationary, or bed, knives, mounted in the housing of the cutter. A perforated plate or screen encloses 50–75% of the circumference of the grinding chamber, thus controlling the top particle size of the product (Fig. S-15). The product size is determined

Fig. S-15. Rotary knife cutter. (*Sprout, Waldron & Co.*)

by the screen openings; fines are controlled in part by the rotor speed, number of knife cuts per revolution, and clearance of rotor and bed knives. Typical uses of these cutters are plastic sheets, rag stock, polyethylene film, leather scrap, and wastepaper. A variation of the cutter is the dicer, a precision unit. One type of dicer has a set of spaced circular rotating cutter blades operating on a shaft parallel to the rotary cutter and uses no screens. A second set of blades provides action at right angles to the rotary knife cutter and delivers a uniform or diced product.

Hammer Mills: The hammer mill has already been described in principle in connection with primary crushing. In a smaller and lighter-duty version, the same principle is applied to grinding. Most hammer mills have 30–200 hp and generally are furnished with a perforated-sheet or wire-mesh screen.

Roller Mills: These units for grinding or coarse pulverizing are basically scaled-down versions of the large roll crushers used for secondary crushing. Generally the surface of the rolls is corru-

gated or grooved to control particle size and the rolls are smaller, perhaps 6, 9, or 10 in. diam at the most. They usually should not be fed with a feedstock coarser than $\frac{1}{2}$ or $\frac{5}{8}$-in. ring size.* They are often arranged in pairs of one, two, or three high and frequently as double mills with two pairs lying side by side (Fig. S-16). The advantage of the latter arrangement is that the mill can be combined with a sifter, removing the in-range material between reduction breaks and returning the oversize to the second pair of rolls. The mills are always furnished with a feeder-distributor which uniformly regulates the feed across the entire roll width. The power requirement on these mills is usually low, and they have the advantage of giving a high yield of limited-size-range product with a minimum of fines. They are not adaptable to materials which lend themselves to smearing, nor are they particularly good for abrasive materials.

Fig. S-16. Roller mill. (*Sprout, Waldron & Co.*)

Pulverizing is a broad term applied to very fine grinding and ultrafine operations where the product requirement is generally finer than 150 or 200 mesh. This encompasses the extremely fine ranges of the micrometer and smaller products.

Ball, Pebble, and Rod Mills: Basically all these mills consist of a rotating drum operating solely on a horizontal axis and filled partially with a free-moving grinding medium which is harder and tougher than the material to be ground (Fig.

*Ring size: For materials $\frac{1}{4}$ in. and finer, the particle size is designated by mesh size and characterized by a distribution of said sizes, e.g., "the screen analysis." Particles larger than $\frac{1}{4}$ in. are normally defined by "ring size." A 2-in. ring-size lump means that no lump of that dimension would exceed 2 in. and, hence, would pass through a 2-in. ring.

S-17). The tumbling action of the grinding medium crushes and grinds the material by a combination of attrition and impact. The medium may be a round metal ball operating in a drum with

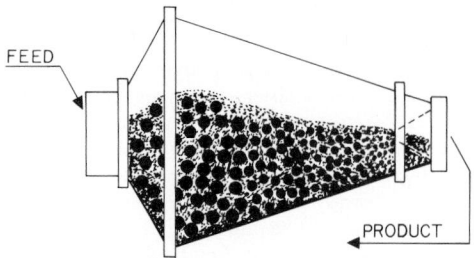

Fig. S-17. Highly schematic representation of operating principle of a cone-type pebble or ball mill.

a metal lining (ball mill). A pebble mill uses a nonmetallic medium, e.g., flint, pebbles, or even large pieces of the material being ground, and has a nonmetallic lining of flint or porcelain blocks. In the rod mills the grinding medium consists of metallic rods as long as the mill cylinder which rotate freely like balls or pebbles as the mill turns. Rod mills have metallic linings. Feed for a rotating drum varies by a maximum of $1\frac{1}{2}$-in. ring size downward, and these units can be operated either batch or continuous. Grinding is generally prolonged up to several hours to obtain the necessary fineness within the particle-size limits. Frequently some form of classification is used with return of oversize materials to the feed point of the system. By using classification, better control of particle size is achieved and a greater capacity horsepower is obtained. See also **Size classification.**

High-Speed Hammer Mills: These mills use peripheral speeds as high as 21,000 ft/min frequently with controlled feed rate. The material is impinged against a cover, smooth or corrugated. The product discharge is through a perforated plate or screen at the lower portion of the casing with relatively fine perforations. Factors governing the performance of these units include hammer-tip peripheral speed; type, sharpness, and clearance of hammers; feed rate; and screen-perforation openings. Hammers are generally mounted in balanced rows on a rotor and extend across the entire width of the unit, operating at close clearance (Fig. S-18). Except at low rotor speeds, screen-perforation openings have less effect on the discharged particle size than would be expected, because at elevated rotor speeds and close clearances the particle strikes the screen at

a tangent rather than perpendicular. These units are often used in conjunction with built-in high-speed mechanical classifiers.

Vertical-Shaft Roller, Ball and Bowl Mills: Basically these are a production-size mortar and pestle having a rotating shaft carrying hinged arms with terminal rollers or balls. As the rotor turns, the hinged arms are thrown out by centrifugal force against a vertical sidewall housing and the material is ground between the revolving rollers and the stationary sidewalls. Most of these mills incorporate a built-in air classifier.

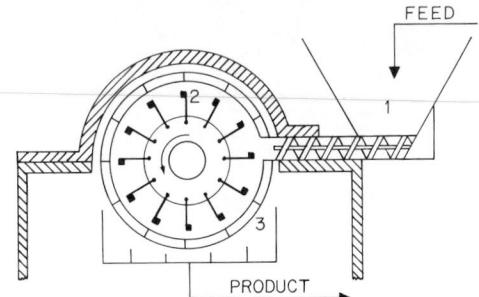

Fig. S-18. Highly schematic representation of a type of hammer-mill pulverizer. (1) Screw feed, (2) small T-shaped hammers, (3) product screen.

Fluid-Energy Mills: These mills differ from the types previously discussed in that they have no moving parts. Their source of power is air, steam, or other gases at high pressures. Since the use of these gases as a grinding medium is considerably more expensive than that of electric power, applications are limited to cases where ultrafine particle sizes are essential. If temperature is not a factor, steam is used as the source of power. Air is used on low-melting-point products such as insecticides and pharmaceuticals. Cooling action of the rapidly expanding air serves to discharge the ground material at approximately the same temperature as it was fed into the unit. The basic principle is that of two diametrically opposed high-pressured jets and the removal of particles of suitable fineness by an upward air classification (Fig. S-19). This has been sophisticated in certain mills by incorporating a vertical cylindrical grinding chamber in which a circulating charge of material to be ground is acted upon by a series of impinging jets introduced through orifices spaced around the periphery. Another version is the vertical tubular oval chamber with impinging jets at the lower portion and removal of the acceptable particles through openings toward the upper portion.

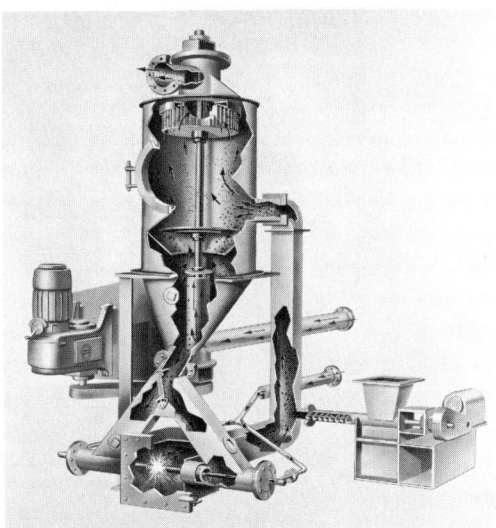

Fig. S-19. Fluid-energy mill. (*Majac, Incorporated.*)

REFERENCES

Galanty, H.: Size Reduction Theory and Practice, *ACS Ann. Meet.*, March 26, 1962.

Perry, R. H. and C. H. Chilton (eds.): "Chemical Engineers' Handbook," 5th ed., McGraw-Hill, New York, 1973.

Stern, A. L.: A Guide to Crushing and Grinding Practice, *Chem. Eng.*, Dec. 10, 1962.

—Kenneth R. Sterrett, *Sprout, Waldron & Company, Inc., Muncy, Pa.*

Sizing   (See **Fibers;** and **Fiber glass.**)

Slabbing Mill   (See **Iron- and steelmaking.**)

Slag   (See **Iron and steel.**)

Slow Neutrons   (See **Radioisotopes.**)

Sludge Disposal   (See **Waste treatment.**)

Sludge Filter   (See **Filtration;** and **Waste treatment.**)

Slurry Explosives   (See **Explosives.**)

Slurry Processing   (See **Filtration; Liquid-Solids separation;** and **Sedimentation.**)

Smaltite   (See **Cobalt.**)

Smectic Crystals   (See **Liquid Crystals.**)

Smelting   Smelting is the term applied to the production of a metal in elemental form from its ore by pyrometallurgical processes. Several examples of smelting processes are given, for example, under **Cadmium; Lead; Tin;** and **Zinc and zinc-lead smelting;** also consult the Subject Index.

Smithsonite   (See **Cadmium.**)

Smoke Signals   (See **Pyrotechnics.**)

Smokeless Powder   (See **Explosives.**)

Soaps   The meaning of soap varies with one's point of view. If only the chemical nature of the material is considered, soap is any salt of a fatty acid containing eight or more carbon atoms. Some of these salts are soluble in water; others are not. Because of the importance of its physical properties, it is customary to limit the definition of soap to fatty acids that exhibit surface-active properties, such as sudsing, detergency, and lowering of surface tension. Synthetic detergents are chemically unrelated to soaps.

Although soap has detergent properties, several factors distinguish soap from synthetic detergents, or syndets. Soap is made from natural fats and oils, while detergents fundamentally are produced from synthetic organic compounds. Soap forms a curdy, insoluble compound in hard water. Detergent surfactants work well in all kinds of water and do not form bathtub ring. Certain advantages have led to wide acceptance of detergents in place of soaps, particularly for heavy-duty cleaning (laundry, floors, woodwork). Soaps, however, remain popular for mild cleaning and particularly for personal use.

Function of Soap. The most important function of soap is to aid water in the cleaning process. Simple as this seems, it is in reality quite complex, primarily because of the varied nature of soils and materials encountered. Cleaning skin, fabrics, and hard surfaces with soap is as varied as the treatment of metals with acid before soldering, laboratory glassware with oxidizing agents, and clothing with cleaning fluid. Soils which soap commonly encounters are widely different—dust, soot, lubricating oils, foods of all kinds, perspiration, blood, and so on. That soap is effective over such a wide range is quite remarkable.

Manufacture of Soap

Ingredients: The primary materials used in the manufacture of bar soaps are natural fats and oils. Manufacturers today use a blend of top-quality animal tallow and coconut oil. These basic materials eventually are converted to their neutral salts by use of some alkaline material, such as sodium hydroxide. Additional, minor ingredients are added, e.g., sodium silicate or magnesium sulfate, to control alkalinity, odor, and aging stability.

Soapmaking Process: The basic process simply is that of splitting the fat stocks into fatty acid and glycerin, followed by separation and neutralization of the resulting fatty acids with alkali. Two systems are in common use, the kettle method and the continuous hydrolyzer.

Kettle Method: The pioneers used a simplified kettle process when they boiled animal fat and wood ashes (for alkalinity) for several hours in a large pot. The modern soap kettle has a capacity of 60,000–300,000 lb and is equipped for heating, settling, and blending the fats, alkali, salt, and water.

The kettle first is charged with fat and a sodium hydroxide solution. Then follows a sequence of heating, separating, and washing to convert the raw materials to *finished base soap* and to separate the impurities and by-products. The process normally takes several days for any single kettle. Although there have been improvements in handling and purification, such as continuous centrifugation, the basic process of saponifying fats directly with caustic remains unchanged.

Hydrolyzer Process: The development of continuous hydrolysis is the most important basic improvement in the processing of fats into soap since the early days of soapmaking. There are several advantages over the kettle process: (1) better quality soaps can be made from darker fats, (2) glycerin recovery is simplified because no salt is needed and the resulting finished glycerin is of higher quality, (3) a single hydrolyzer unit produces about the same quantity of soap as 10 kettles, thus effecting savings in manufacturing space and a reduction of in-process inventory, and (4) greater flexibility is possible in controlling the chemical and physical properties of the finished soap.

The hydrolyzing process consists essentially of (1) hydrolysis, (2) fatty acid distillation, (3) post-hardening (optional), (4) neutralization, and (5) glycerin recovery. The basic hydrolyzer process is shown in Fig. S-20.

Hydrolysis: Development of continuous hydrolyzing was the key step toward the continuous soapmaking process. In this reaction, fat and water react to form fatty acid and glycerin:

$$(RCOO)_3C_3H_5 + 3H_2O \rightleftarrows$$
$$3RCOOH + C_3H_5(OH)_3$$

where R is an alkyl of C_8 or larger. This equation represents the complete hydrolysis. Actually, the reaction takes place in a stepwise fashion, forming intermediate diglyceride and monoglyceride.

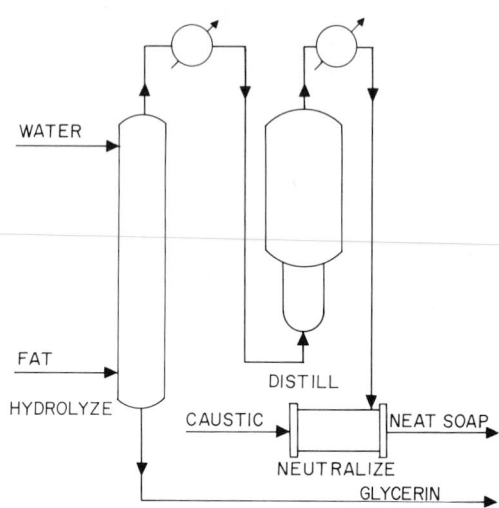

Fig. S-20. Basic hydrolyzer process used in soap manufacture.

The reaction can be accomplished only through intimate contact between water and fat molecules. High temperature makes it possible to dissolve an appreciable quantity of water in the fat phase and to obtain this intimate contact. At room temperature, water and fat are essentially insoluble. At elevated temperature, the solubility of water increases to 12–25%, depending upon the type of fat. At the higher temperatures, high pressures also are necessary to keep the water from flashing into steam.

The reaction is reversible. In order to make it proceed to the right, the proportion of water to fat can be increased or the glycerin can be removed. Removal of glycerin is used as the reaction-forcing method. The required combination of high temperature, high pressure, and continuous glycerin removal is accomplished in a countercurrent hydrolyzer column.

Fat stocks, blended in the proper formula, are mixed with dry zinc oxide catalyst. The mixture is maintained at about 212°F (100°C) to ensure dryness and to keep the catalyst in solution. Hot water for the hydrolysis reactions is put under high pressure by piston-type feed pumps with adjustable drives so that the rates and proportions of fat to water can be accurately controlled. The fat and water are heated to the hydrolyzing temperature by direct steam injection or by heat exchangers. The fats are pumped into the column near the bottom, and the water enters near the top. Thus, a countercurrent flow of water downward through rising fatty material is obtained.

The hydrolysis occurs in a two-phase reaction system. The fats and fatty acids flow continuously with droplets of water falling through them. Glycerin from the hydrolysis is dissolved in the excess water falling through the column. The rate-limiting factor is the transfer of glycerin into the water droplets. Zinc oxide catalyzes the reaction by forming zinc soap, which increases the glycerin transfer across the oil-water interface. Fresh water entering the column at the top reduces the glycerin to the lowest possible point, while a glycerin-water seat maintained at the bottom of the column (where the glycerin content is highest) prevents fat from washing out.

The fatty material passes upward through the column with about 99% completeness in splitting. The fatty acids, saturated with water, are discharged through an orifice into a flash tank. The dissolved water vaporizes, cooling the fatty acids and blanketing them with steam. The fatty acid contains the zinc soap catalyst and the remaining unsplit fat.

The column, pumps, and piping in contact with the hot fatty acid are made from corrosion-resistant stainless steel. The column is a hollow vessel, containing no baffles, trays, or packing material of any kind. The quality of the hydrolyzing operation is determined by the degree of split obtained on the fat. The fatty acid stream should contain very little free glycerin, if any.

Distillation: The second key step in continuous soapmaking is distillation. Originally, fatty acids made in hydrolyzers were acid-washed to split out the zinc soap and then bleached to improve color, but continuous distillation of the hydrolyzer fatty acids results in lighter soap from darker stocks at lower cost.

The fatty acids from the hydrolyzer are collected in the still feed tank and vacuum-dried to reduce moisture to low levels. Then they are flash-distilled at an absolute pressure of 2–5 mm Hg. The still bottoms are recirculated through heat exchangers back to the still to carry the heat necessary for vaporizing the fatty acids. The still bottoms, which contain the zinc soap catalyst and unsplit fat, are removed from the system, acidu-

lated to remove the zinc, and frequently used in animal feeds. The fatty acid vapors from the still pass to several water condensers in series. The condensed fatty acids drop to a surge tank for posthardening or directly for neutralization.

The two prime objectives of this process, maintenance of good odor and color in the distillate and proper bottoms yield, are achieved by effective control over vacuum, temperature, and distillation rate.

Posthardening: Not shown in Fig. S-20 is an optional further treatment of the fatty acids known as *posthardening*. This operation involves hydrogenation of some of the unsaturated carbon-carbon bonds on the fatty acid molecules. Originally, the purpose of this step was to improve color and odor. As such, the hardening was intended only to eliminate polyunsaturates, leaving the majority of the monounsaturates unaffected. A greater amount of hardening can be performed, however, to tailor some of the physical properties of the finished bar. Control of hydrogenation, although not delicate, is becoming more critical in modern soapmaking.

The fatty acids from distillation are heated and passed with a metered hydrogen supply through hardening tubes which contain a fixed bed of granular nickel catalyst where the hydrogenation takes place. The hardened fatty acids flow through a filter to remove traces of catalyst. The filtered stock drops to a flash tank, where excess hydrogen is removed. Hardening is controlled by temperature, pressure, hydrogen flow, residence time, and catalyst age. The fatty acids then are cooled for neutralization.

Neutralization: The saponification reaction between alkaline solutions and fatty acids is almost instantaneous:

$$RCOOH + NaOH \rightarrow RCOONa + H_2O$$

Each reactant is metered accurately into the neutralizer, where intimate mixing occurs and the reaction takes place. Soap from the neutralizer is discharged at about 200°F (93°C) to a blend tank equipped with agitation and recirculation to ensure uniform composition of the soap. This base soap (or *neat soap*) normally is stored in large kettles until required for subsequent processing into finished products, e.g., soap powders, granules, or toilet bars.

The characteristics of the neat soap are controlled easily by accurately governing the composition of the alkaline solution used. Normal hydrolyzer neat soap contains about 69% actual soap,

30% water, and less than 1% NaCl, plus other stabilizers. Neat soap is a uniform, translucent, white, viscous fluid at 180–200°F (82–93°C).

Glycerin Recovery: The glycerin water stream from the hydrolyzer is concentrated by evaporation, purified, and subsequently sold or used in other processes.

Varieties of Soap. Up to the point of the process just described, soapmaking is the same for all forms of the finished product, whether (1) a milled toilet bar,* (2) a floating bar,† or (3) a blown granule.‡ Only the ratios of the different types of fats and oils to the hydrolyzer or the alkaline agent used for neutralization will vary.

Milled Toilet-Bar Soap Manufacture. Milled toilet soap is a high-grade soap in which critical crystal-phase changes have been brought about through the use of mixers, milling rolls, and plodders. The milled soap is made by drying a good grade of neat soap to about 15% moisture content, breaking up the crystalline structure that develops during drying and cooling, plasticizing and converting a sufficient portion of the soap to a desirable phase condition, deaerating and compacting the resulting mass, and forming it into bars. Perfume, coloring matter, preservatives, and special additives frequently are incorporated prior to the milling operation. A milled bar is particularly hard, dense, and smooth, and it lathers freely without forming excessive soft soap on the surface of the bar.

Drying: Liquid base soap is dried from a 30% water liquid form to a solid of about 15% water content. If desired, some minor ingredients may be blended into the soap stream prior to drying. Methods of drying used in common practice are (1) chip drying, (2) atmospheric flash drying, and (3) vacuum flash drying.

Chip Drying: Sometimes called *ribbon drying,* this process involves spreading a thin layer of hot base soap on a large chilled drum which cools and firms up the soap. Drying is promoted primarily by the difference in water-vapor pressure between the soap chips and the air surrounding them. No attempt is made to increase drying rate by heating the soap itself.

Atmospheric Flash Drying: A tower similar to a synthetic-granules spray-drying tower is used. The heat for drying, however, is put into the soap by heating it under high pressure before flashing

*E.g., Camay.
†E.g., Ivory.
‡E.g., Duz.

it into the tower. During flashing, the pressure on the soap is abruptly relieved, and soap moisture flashes to steam. Air to the tower is used for cooling. The soap temperature as it enters the flashing nozzles determines the final moisture of the dried soap.

Vacuum Flash Drying: In this, the most recent technique, drying takes place in a vacuum vessel similar to an atmospheric tower but smaller. The soap is similarly heated before flashing but under less pressure, so that boiling (actually drying of the soap) occurs in the heat exchangers. Since there is boiling in the heaters, the moisture of the dried soap depends primarily upon soap flow rate, soap pressure, and steam pressure to the heater and to a minor extent on the absolute pressure in the vacuum chamber. The final temperature of the soap depends entirely upon the absolute pressure in the vacuum chamber.

Mixing: After drying, the soap noodles or flakes are mixed with all additional minor ingredients required by the final product formula. Mixing is done in batch processes or continuously.

Minor ingredients include dye, perfume, preservatives, deodorants, opacifiers, and special-purpose items, such as cold cream, cleansing cream, and skin conditioners. The type and proportion of these materials is largely what makes one brand of milled soap different from another. Not all brands include all categories of the minor ingredients.

In batch mixing, dried soap and minor ingredients are measured and dumped into a dry blender, where gross mixing occurs. The batch process is cumbersome and slow, and it is difficult to maintain uniform quality. Continuous mixing operations for improving economy and efficiency of mixing include precision metering pumps to spray the minor ingredients onto the soap noodles as they are pulverized and conveyed through the mixer. Although the minor ingredients constitute but a small portion of the total product, their effect on the physical properties, e.g., softness, resistance to cracking, lathering, and resistance to dissolving, are considerable.

Milling: The three objectives of milling are (1) thorough and intimate final mixing of the soap, perfume, and other ingredients without overheating, (2) crushing lumps of overdried soap and pulverizing them into pieces too small to appear as lumps or hard specks in the finished bar, and (3) conversion of a sufficient portion of the soap into the waxy, plastic phase by cold working.

Soap is milled by forcing it through a series of

rolls, thus subjecting it to a strong rubbing action. This cold working at the proper moisture content changes the crystalline structure or phase of the soap. Temperature control during milling is important. If the temperature is too low, the wrong crystal structure will be formed, resulting in soft soap or a hard, brittle structure prone to cracking. If the temperature is too high, the soap will become sticky and difficult to process further.

Another method for complete mixing and working uses multiple plodding and screening, in which the soap and minor ingredients are pushed together through finer and finer mesh screens.

Plodding: After milling, it is necessary to form the soap into a shape for making the final bar. This usually is accomplished with a plodder, which essentially is a large-size meat grinder with a barrel that terminates in a cone. The plodder functions to compact the pellets or flakes of soap into a solid mass, squeeze out any pockets of entrapped air, and extrude it as a firm, uniform, and continuous strip.

Operations which follow include cutting, stamping, wrapping, and packing.

Floating-Bar Soap Manufacture. Hydrolyzer (or kettle) soap made from the desired blends of fat and oil first is flash-dried to a moisture content of about 22%. It then enters a mechanical mixer called a *crutcher*, where it is thoroughly mixed with perfume, preservatives, and air. The amount of air controls the density of the final product, giving the bar a density of less than 1 and making it floatable.

From the crutcher, the mix goes to a freezer to reduce the temperature of the soap to the point where it will hold its shape when extruded. In the earlier steps, the soap mix is in liquid form. Rapid chilling is required to put it into a solid state. The machine is similar to a commercial ice-cream freezer, consisting of a horizontal cylinder surrounded by a jacket and housing a rotating shaft (mutator) on which scraping blades are mounted. The liquid soap mix from the crutcher is pumped into one end of the cylinder. A refrigerated brine solution is circulated through the jacket to chill the soap. The scraping blades on the mutator remove the chilled soap from the cylinder walls and maintain uniformity of the mix. The nose of the freezer is equipped with an oblong orifice through which the soap is extruded, after chilling, in the form of a continuous ribbon which has the same cross section as the final bar. There follows a series of cooling, storing, stamping, and packaging operations.

Granule-Soap Manufacture. For granule soap, soap from the hydrolyzer is conveyed directly to a spray-drying tower, where it is *blown* in the form of light-density granules. If an "unbuilt" product, e.g., one designed for fine-fabric washing, is being made, the soap plus small amounts of preservative and brighteners will be mixed in a crutcher and then blown. If a heavy-duty product for laundry service is desired, *builders* such as carbonates and silicates will be mixed with the base soap in the crutcher and then blown. Processing soap granules from the crutcher stage forward is essentially the same as for detergents. See also **Detergents.**

REFERENCES

Bailey, Alton E.: "Industrial Oil and Fat Products," Wiley, New York, 1950.
Cooke, Edward I.: "The Modern Soap and Detergent Industry," 3d ed., vol. 2, Technical Press, London, 1951.
Shreve, R. Norris: "Chemical Process Industries," 3d ed., pp. 543–559, McGraw-Hill, New York, 1967.

—George S. Speidel, III, *Bar Soap and Household Cleaning Products Division, The Procter & Gamble Company, Cincinnati, Ohio*

Soda Ash   Sodium carbonate, Na_2CO_3, commonly is referred to as soda ash. Chemically pure sodium carbonate has a formula weight of 106.00 and is a white powder with a refractive index of 1.535. Specific gravity = 2.533; melting point = 851°C; solubility in cold water = 7.1 parts in 100 parts of water and 48.5 parts in 100 parts of water at 104°C. Sodium carbonate is insoluble in alcohol and ether and decomposes before it reaches a boiling temperature under normal pressure.

Soda ash is a high-tonnage chemical raw material and approaches a production rate of 10 million tons/yr in the United States. About 40% of soda ash is used in glassmaking; about 35% goes into the production of sodium chemicals, such as sodium chromates, phosphates, and silicates; nearly 10% is consumed by the pulp and paper industry; with the remainder going into soaps and detergents and nonferrous metal refining.

Historically, soda ash is one of the oldest tonnage chemicals. The name stems from early methods of obtaining soda ash by burning seaweed. Leblanc invented the first process for preparing soda ash during the French Revolution. In the Leblanc process, sodium chloride is converted to sodium sulfate, which subsequently is heated with limestone and coke in a furnace in accordance with the reactions:

$$Na_2SO_4 + 2C \rightarrow Na_2S + 2CO_2 \qquad (1)$$

$$Na_2S + CaCO_3 \rightarrow Na_2CO_3 + CaS \qquad (2)$$

The Leblanc process largely was replaced by the ammonia, or Solvay, process in the mid-1800s, named for E. Solvay, who first used this method in a plant near Brussels. In the Solvay process, still widely used, carbon dioxide is passed through an ammonia-saturated solution of sodium chloride to form sodium bicarbonate, which is then calcined:

$$NH_3 + CO_2 + NaCl + H_2O \rightarrow$$
$$HNaCO_3 + NH_4Cl \quad (3)$$
$$2HNaCO_3 \rightarrow Na_2CO_3 + CO_2 + H_2O \quad (4)$$

There has been a recent trend, particularly in the United States, to go to the natural mineral *trona* as the raw material source. Trona, found in great abundance near Green River, Wyo., chemically is sodium sesquicarbonate, $Na_2CO_3 \cdot NaHCO_3 \cdot 2H_2O$. The natural ore is crushed and then dissolved in agitated tanks to form a concentrated solution of sodium sesquicarbonate. Most of the impurities in trona, such as boron oxides, calcium carbonate, silica, sodium silicate, and some shale rock, are insoluble in hot water. The insoluble matter is separated in large settling tanks, and further purification follows by passing the solution through pressure-leaf filters. The hot sesquicarbonate solution then is cooled in vacuum crystallizers, where fine needlelike crystals are formed. The crystals are removed from solution in basket centrifuges. The sesquicarbonate crystals are heated to about 240°C in rotary calciners, where carbon dioxide and bound water are released to form what is termed *natural soda ash*. Over one-fourth the soda ash production in the United States comes from trona. The purity of natural soda ash is quite high (99.88%) with the iron content as low as 7–9 ppm and chlorine content less than 0.02%. Because of its needlelike shape, natural ash flows freely without abrading or forming dust during handling and storage and assists glassmakers in obtaining uniform and homogeneous mixes.

Soda-Lime Silica Glass   (See **Glass**.)

Soda Pulping Processes   [See **Pulp (wood) production and processing.**]

Söderberg Cell   (See **Aluminum;** and **Silicon.**)

Sodium   (E. *soda.*) **Na** (L. *Natrium*) = 22.997 (at. wt.); 11 (at. no.). Sodium is a waxy, silvery white metal that melts at 97.83°C. The liquid boils at 882.9°C (1 atm pressure). Other key physical properties of the metal are given in Table S-18.

TABLE S-18. *Key Physical Properties of Sodium*

Melting point, °C	97.83
Triple point, °C	97.82
Boiling point, °C	882.9
Heat of fusion at melting point:	
cal/g	27.06
Btu/lb	48.7
Heat of vaporization at boiling point:	
cal/g	951
Btu/lb	1712
Density, g/cm³:	
Solid at 0°C	0.9721
Liquid at melting point	0.9268
Vapor pressure:	
At 300°C:	
mm Hg	0.014
psia	0.00027
At 900°C:	
mm Hg	866.4
psia	16.7
Heat capacity, cal/(g)(°C):	
Solid at 0°C	0.2866
Liquid at melting point	0.3201
Thermal conductivity, (cal)(cm)/(s)(cm²)(°C):	
Solid at 21.2°C	0.317
Liquid at 200°C	0.193
Surface tension at 250°C, dynes/cm	199.5
Electrical resistivity at 200°C, $\mu\Omega$-cm	13.18
Hardness, Mohs' scale	0.4
Heat-transfer coefficient, Btu/(hr)(ft²)(°F)	7000
Of NaK eutectic	3500
Crystal structure	fcc
Lattice constant at 20°C, Å	$a = 4.2820$
Closest approach of atoms, Å	3.708
Ionization potential, eV	5.138

Na is a member of group IA, period 3, of the periodic table and thus is one of the alkali metals, along with Li, K, Rb, and Cs. There is only one stable isotope, 23, and five known unstable isotopes, 20–22, 24, and 25.

Sir Humphry Davy first isolated sodium metal (1807) by the electrolysis of sodium hydroxide. The element remained a laboratory curiosity, however, until Oersted discovered (1824) that Na reduces aluminum chloride to produce pure aluminum metal. This led to efforts to develop a commercial process for the manufacture of Na, described later.

Chemical Behavior: Sodium, like the other alkali metals, has only one valence electron. All members of the group are highly electropositive. The inorganic compounds of Na are all ionic, or salt-like, and nearly all are soluble in water and other polar solvents. A few compounds, such as the hydride, amide, and carbide, are decomposed by water, often quite violently. To a great extent, the chemical properties of inorganic Na compounds reflect the characteristics of the anion. Na, in its inorganic chemistry, may be looked upon primarily as a carrier for anions. See Table S-19.

The organic compounds of sodium, in which Na is linked directly to carbon, have been known and used for many years. These compounds are readily made by reacting Na dispersed in an inert hydrocarbon with alkyl and aryl chlorides. Amylsodium, the sodium organic of the greatest commercial importance, is used extensively as an essential ingredient of various polymerization catalyst systems. The alfin catalyst, made by reacting sodium isopropoxide with amylsodium, is used in the production of butadiene-styrene rubber (Buna S). Amylsodium and titanium tetrachloride form a catalyst for the low-pressure polymerization of ethylene. Other catalyst systems containing amylsodium, as well as propylene sodium, cyclopentadienylsodium, and others, are useful in the polymerization of a variety of olefinic monomers.

Another important use for sodium organics is in the production of other metal organics. For example, triphenylbismuth is formed by the reaction

$$3C_6H_5Na + BiCl_3 \rightarrow (C_6H_5)_3Bi + 3NaCl$$

Similarly, many metal halides can be reacted with various sodium organics to form the corresponding metal organic.

Occurrence: Sodium is the sixth most abundant element in the earth's crust and is present as the chloride in seawater and in rock-salt deposits. Chili and Peru have deposits of sodium nitrate, and on the sites of dried-up lakes are found carbonates, sulfates, and borates of sodium. Sodium salts are present in nearly all natural brines and in plant and animal fluids. The important role of sodium in the functioning of living organisms has received much attention in recent years, re-

TABLE S-19. *Properties of Representative Sodium Compounds*

Compound and Formula	Formula Weight	Sp gr	Mp, °C	Bp, °C
Acetate, $NaC_2H_3O_2$	82.04	1.528	324	
Amide, $NaNH_2$	39.02	. . .	210	400
Benzoate, $NaC_7H_5O_2$	144.11			
Bicarbonate, $NaHCO_3$	84.01	2.20*	270	
Bisulfate, $NaHSO_4$	120.06	2.742	>315	†
Tetraborate, $Na_2B_4O_7$	201.27	2.367	741	
Bromide, $NaBr$	102.91	3.205	755	1390
Carbonate, Na_2CO_3	106.00	2.533	851	†
Chlorate, $NaClO_3$	106.45	2.490	248	†
Chloride, $NaCl$	58.45	2.163	800.4	1413
Chromate, Na_2CrO_4	162.00	2.723	392	
Cyanide, $NaCN$	49.02	. . .	563.7	1496
Fluoride, NaF	42.00	2.79	992	
Hydride, NaH	24.005	0.92	d.800	
Hydroxide, $NaOH$	40.00	2.130	318.4	1390
Hypochlorite, $NaOCl$	74.45	. . .	†	
Iodide, NaI	149.92	3.667	651	1300
Nitrate, $NaNO_3$	85.01	2.257	308	380†
Peroxide, Na_2O_2	77.99	2.805	†	
Tribasic phosphate, Na_3PO_4	163.97	2.537	1340	
Pyrophosphate, $Na_4P_2O_7$	265.95	2.45	988	
Metasilicate, Na_2SiO_3	122.05	. . .	1088	
Sulfate, Na_2SO_4	142.05	2.698		

* $-CO_2$.
† Decomposes.

sulting, as one example, in low-sodium diets to counteract certain ailments.

Production: In 1886, Castner produced Na by the reduction of caustic soda with carbon, but in this same year the inventor of the Hall process for aluminum production removed the only existing market for Na. Castner continued, however, to work on uses for Na and improved processes for its production; he also invented processes for the manufacture of sodium peroxide and sodium cyanide.

In 1895, a plant was built in Niagara Falls, N.Y. (because of its cheap electric power) for the manufacture of Na using the Castner process for the electrolysis of fused NaOH. In 1921, the Castner process was superseded by the Downs process, in which a mixture of fused NaCl and $CaCl_2$ is electrolyzed to produce Na. The Downs process is still in use, essentially unchanged except for increases in cell size and amperage.

The Downs cell, shown schematically in Fig. S-21, consists of a steel shell lined with refractory

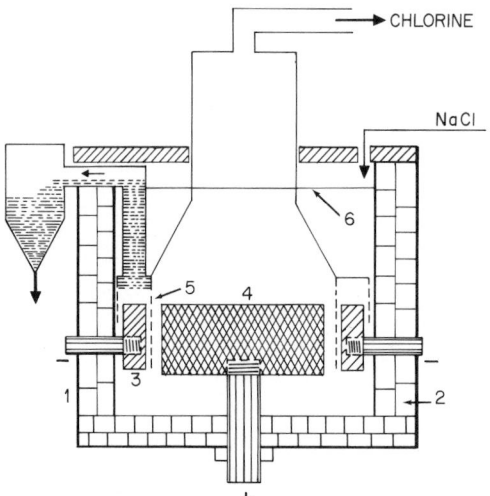

Fig. S-21. Schematic representation of Downs cell used for production of sodium metal. (1) Iron shell, (2) firebrick, (3) iron or copper cathode, (4) graphite anode, (5) iron-gauze diaphragm, (6) level of fused sodium chloride.

brick. In the original design, a single cylindrical graphite anode projected upward through the cell bottom. Modern cells have four anodes. The anodes are surrounded by a steel cathode, leaving an annular electrolysis zone. In four-electrode cells, the four cathode cylinders are joined together as a unit having heavy steel arms that

project through the sides of the cell for attachment of the bus bars. Suspended above the electrodes is a steel collector assembly. Chlorine gas, generated at the anodes, is directed into an inverted cone that projects above the electrolyte level and is connected to the chlorine-collection duct. Over the cathode is an inverted trough that directs the Na to a vertical riser pipe. Hanging from the collector assembly are steel wire-mesh diaphragms that extend down into the electrolysis zone and serve to prevent recombination of the Na and Cl_2.

The separate melting points of the electrolyte ingredients used in the Downs cell are NaCl 800°C and $CaCl_2$ 772°C. A mixture of 33.2% NaCl and 66.8% $CaCl_2$ is a eutectic which melts at 505°C. The composition actually used, however, contains 58–59% $CaCl_2$ and has an initial solidification temperature in the range 575–585°C. The presence of $CaCl_2$ in the cell results in the formation of some Ca along with the Na at the cathode. Calcium has only limited solubility in Na, and the excess Ca reacts with the NaCl: $Ca + 2NaCl \rightarrow 2Na + CaCl_2$.

The sodium, saturated with Ca, is cooled in the riser pipe. This reduces the solubility of Ca in Na, precipitating Ca, which falls back into the cell, where it reacts to form more Na. The Na that overflows at the top of the riser pipe contains 1% or less Ca. The Na is further purified by filtration at a temperature near its melting point, reducing the Ca content to about 0.05%.

The sodium cells operate at 7–8 V and are arranged in groups of 25–40 or more cells, connected in series. The current is 25,000–40,000 A or more. Dry, purified NaCl is fed continuously to the cells. $CaCl_2$ is added intermittently as required to maintain the cell-bath composition in the desired range.

Uses: Sodium has had many uses in industry and a number of potential uses. Old, abandoned uses have been replaced by new ones with the net result of a fairly steady increase in the amount of Na produced from 20 to 30 million lb/yr in 1920–1925 to more than 310 million lb in 1969. The growth in Na production generally has paralleled the growth in the manufacture of tetraethyl and tetramethyl lead, for which metallic Na is an essential raw material. Sodium metal also is used in the reduction of titanium in substantial quantities.

The use of Na as a heat-transfer medium has enjoyed a mixed reception. The metal has several attractive properties for this purpose, including a low density compared with most metals, contributing to low cost per unit volume and relative ease of pumping. In fact, the density of Na is only approximately one-half that of the more commonly used nitrate-nitrite heat-transfer salts. The vapor pressure of Na is relatively low even at temperatures as high as 550°C. The viscosity of molten Na is quite low. The heat capacity of Na metal is higher than that of most common metals in liquid form. The thermal conductivity of liquid sodium is 5–10 times greater than the conductivities of Hg or Pb and approximately 50 times greater than that of most organic heat-transfer media. Na is exceeded only by Ag, Cu, Al, and Au in terms of electrical conductivity. Because of this latter property, bus bars have been constructed from steel pipe filled with Na. The passage of an electric current through Na vapor at low pressures produces the characteristic yellow sodium light, which has enjoyed a mixed reception for use on commercial highways and for industrial lighting.

Metallurgically, Na has been used in the modification of aluminum-silicon alloys. Such alloys, normally coarse and brittle, can be transformed into fine-grained alloys with good casting properties through the addition of small amounts (a fraction of 1%) of Na. Na also has been used as a hardening agent in bearing metals. Na, when added with an alkaline-earth metal, such as Ca, increases the hardness of Pb appreciably. The German *Bahnmetal* is an alloy of this type. Because of its own low surface tension, Na can be added to other metals. When added to Pb, Zn, or Sn, improved adhesion and a smoother coating are claimed.

Handling: In general, plain carbon steel is suitable for handling sodium at the temperatures likely to be encountered in chemical processing, just above the boiling point. All-welded pipeline construction and bellows-sealed packless valves are commonly employed. In fighting fires involving sodium, a protective layer or blanket must be formed over the metal while it cools below its burning temperature in air. Conventional fire extinguishers containing water, carbon dioxide, or chlorinated hydrocarbons should not be used because they react violently with the metal. Preferred are soda ash, salt, and graphite, but they must be dry. Sand normally is not recommended because perfectly dry sand for emergency purposes is difficult to obtain. The extent of protective clothing required when handling sodium metal depends on the quantity of metal involved and the temperature of exposure. In manufacturing operations where temperatures are high, an apron, leggings, and a complete face covering should be used. At normal temperatures or where small quantities of the metal are involved (as in a laboratory), conventional protective gear of goggles and gloves usually is adequate.

REFERENCES
"Chemical Profiles: Sodium," Schnell Publishing Company, New York, 1969.
Hey, D. H. (ed.): "Kingzett's Chemical Encyclopedia," 9th ed., Van Nostrand Reinhold, New York, 1966.
Kolthoff, I. M., and Philip J. Elving: "Treatise on Analytical Chemistry," pt 2, vol. 1, Interscience-Wiley, New York, 1961.
Mantell, C. L.: "Electrochemical Engineering," 4th ed. of "Industrial Electrochemistry," McGraw-Hill, New York, 1960.
"Mellor's Comprehensive Treatise on Inorganic and Theoretical Chemistry," Wiley, New York, 1963.
Sodium Builds on Solid Base, *Chem. Week*, Aug. 19, 1967.

Sodium Bromide   (See **Bromine**.)

Sodium Chloride   [See **Salt (NaCl)**.]

Sodium Hydroxide   (See **Caustic soda**.)

Sodium Hypochlorite   (See **Detergents**.)

Sodium Phosphates   (See **Phosphoric acid**.)

Sodium Silicates   [See **Plasma, chemical processing with;** and **Silicates (soluble)**.]

Sodium Sulfite   [See **Ketones; Pulp (Wood) Production and Processing**.]

Softener, Water   (See **Detergents;** and **Water treatment**.)

Soil-removal Agents   (See **Detergents; Soaps;** and **Surfactants**.)

Soil Stabilizers   [See **Lime;** and **Silicates (soluble)**.]

Solar Cells   Photovoltaic cells which convert solar radiation directly into electric energy have been used extensively in space vehicles and probes and to a lesser extent as a source of electric energy for instruments and other small apparatus which must operate unattended in remote locations for relatively long periods. Early cells were of the *n*-type silicon doped with boron to produce *p*-on-*n* semiconductors. Later designs have used phosphorus as the dopant to produce *n*-on-*p* semiconductors.

The operation of the solar cell results from conversion of electromagnetic energy of incident radiation into carrier (electron-hole) pair generation across a *p-n* junction or within the silicon crystals which produce an emf that varies somewhat with the wavelength of the incident light. Approximately 0.2–0.5 V per cell is available and current output is proportional to the area of each cell. The current output of a cell with an area of 2 cm^2 is only about 0.05 A. Hence, very large numbers of cells are required for most applications. Further, for maximum utility, a means must be provided to orient the cells wherever possible so that the incident light will be perpendicular to the face of the cells.

The net efficiency of the solar cell is low (from 10 to 18%), but because of the interest in better utilizing solar energy as a way to alleviate energy shortages and pollution problems arising from fuel-burning generating facilities, schemes have been proposed that involve huge solar collectors in stationary earth orbit. In one proposal, a collector would be placed about 22,300 mi above the equator. A 5 × 5 mi panel of solar cells would intercept 8.5 × 10^7 kw of radiant solar power. This electric power would be converted into microwave radiation and beamed to earth. Upon receipt, the energy would be reconverted, resulting in about 10^7 kw, an amount sufficient to meet the power requirements of a very large city. The magnitude of such a project is demonstrated by the fact that it would require over 20 billion cells, if each had a receiving surface area of 1 in^2.

Solar Salt   [See **Desalination**; and **Salt (NaCl)**.]

Solation   (See **Colloid systems**.)

Solder Glass   (See **Glass**.)

Soldering   (See **Aluminum; Copper; Lead**; and **Tin**.)

Solidification Processes   (See **Energy systems for processes**.)

Solid-Solid Separations   (See **Separation operations**.)

Solid-State Electronic Materials   (See **Boron; Germanium; Selenium**; and **Silicon**; also consult the **Subject Index**.)

Solid-Waste Handling   (See **Agglomeration**.)

Sols   [See **Silicates (soluble)**.]

Solubility   [See **Crystallization; Defoaming agents; Salt; Salt (NaCl)**; and **Surfactants**.]

Soluble Silicates   [See **Silicates (soluble)**.]

Solution Polymerization   (See **Elastomers**; and **Polymerization**.)

Solvent Extraction   (See **Butadiene; Isoprene; Lubricating oils; Perchloroethylene**; and **Toluene**.)

Solvent Paints   (See **Paint**.)

Solvents   (See **Air pollution; Chlorine organics; Chlorofluorocarbons; Coal tar and derivatives; Perchloroethylene; Petrochemical complex; Pyridine and derivatives; 1,1,1-Trichloroethane**; and **Trichloroethylene**; also consult the **Subject Index**.)

Solvolysis   (See **Terpenes**.)

Sorption Operations   (See **Absorption**; and **Adsorption**.)

Soya Oil   (See **Paint**; and **Vegetable oils**.)

Soybeans   (See **Amino acids**.)

Space Velocity   (See **Fluid catalytic cracking; Isomerization**; and **Reforming**.)

Spandex   (See **Fibers**.)

Spargers   (See **Mixing, fluids**.)

Spectra, Atomic   (See **Atomic structure**.)

Spermaceti   (See **Alcohols**; and **Waxes**.)

Sphalerite   (See **Cadmium; Flotation; Germanium; Indium**; and **Lead**.)

Spheroidizing   (See **Iron and steel**.)

Spiegeleisen   (See **Manganese**.)

Spinning   (See **Acetate fibers; Acrylic fibers; Fibers; Fiber glass**; and **Polyester fibers**.)

Spodumene   (See **Ceramics; Germanium;** and **Lithium.**)

Sponge Metal   (See **Iron- and steelmaking;** and **Titanium.**)

Spray Drying   The spray drier is unique among driers in that it dries a finely divided droplet by direct contact with the drying medium (usually air) in an extremely short retention time (3–30 s). This short contact time results in minimum heat degradation of the dried product, a feature that led to the popularity of the spray drier in the food and dairy industries during its early development. Later, additional products and chemicals were added to the list of spray-dried materials as other advantages became apparent and as various types of spray driers became available. A representative list of products dried this way is given in Table S-20.

TABLE S-20. *Representative Spray-dried Products*

Substance	Water in Feed, %
Aluminum hydroxide	93
Aluminum sulfate	70
Animal blood	65
Coffee extract	50–70
Detergent product	40–64
Lignin	63
Magnesium carbonate	92
Magnesium chloride	52
Manganese sulfate	50
Pigments	73
Silica gel	95
Sodium sulfide	50
Tanning extract	45
Urea resin	60–70
Zinc sulfate	55

Mechanics: Drying from a particle generally takes place in two stages, the *constant-rate* and the *falling-rate period*. Since the particles are finely divided in the spray drier, the time for each of these stages is very short.

In the constant-rate period, evaporation takes place at the surface of the particle. The evaporation rate is controlled by the diffusion rate of the vapor through the surrounding air film. The primary drying force is the ΔT between the temperature of the surrounding air and the temperature of the particle, the latter generally considered to be at the wet-bulb temperature of the drying medium. In the constant-rate period, the diffusion rate of water within the particle to the surface of the particle is capable of being equal to the evaporation rate or greater. This allows a film of water to cover the entire surface of the particle.

When the diffusion rate of the water to the surface of the particle is no longer capable of keeping the particle surface wet, the diffusion rate becomes the controlling factor and the falling-rate period of drying is reached. During this stage, evaporation takes place at some plane within the surface of the particle. It is during the falling-rate period of drying that the particle temperature begins to rise above the wet-bulb temperature and approaches the temperature of the surrounding air.

As the vapor-liquor interface recedes within the particle, a solid phase begins to appear on the surface, forming a crust. Depending upon the properties of this crust and the temperature of the surrounding air, different types of dried particles are formed. There are two common types. (1) If the crust is rigid and porous, the shape of the particle is formed by the *initial* forming action since the vapor can diffuse through the particle, depositing solids on the inside wall of the particle, and air can diffuse into the particle to fill the void. (2) If the crust is skinlike, the particle can expand; expand, rupture, and collapse; or expand and shatter—all to varying degrees.

The wall thickness of the particle will depend upon the crust properties and solids concentration of the feed.

Controlling Factors: Considering the two different phases of drying, the design of a spray drier can be altered to give products of specific qualities, both in physical and chemical properties—limited, of course, by the material that is to be dried. Some of the qualities that can be controlled to a degree are (1) particle size, (2) particle distribution, (3) particle sphericity (shape), (4) particle wall thickness, (5) heat degradation of the product, (6) density, and (7) product moisture.

Atomization: Inasmuch as a spray drier operates by drying a finely divided droplet, the feed to the drier must be capable of being atomized sufficiently to ensure that the largest droplet produced will be dried in the retention time allowed. There are different requirements on the degree of atomization required to give the desired product. These factors include minimizing the fines and/or coarse fractions, controlling particle dryness, and controlling bulk density. All commercial atomizers, whether of the centrifugal-wheel, pressure-nozzle, or other type, will produce a particle-size distribution that follows a probability curve. As the total energy input increases, the average particle size will decrease and the particle-size distribution will improve; i.e., the spread between the largest and the smallest particles will be less.

Centrifugal-Wheel Atomizers: The wheel consists of a disk, or adaptation, which is rotated at a very high speed (1,700–50,000 rpm) (see Fig. S-22).

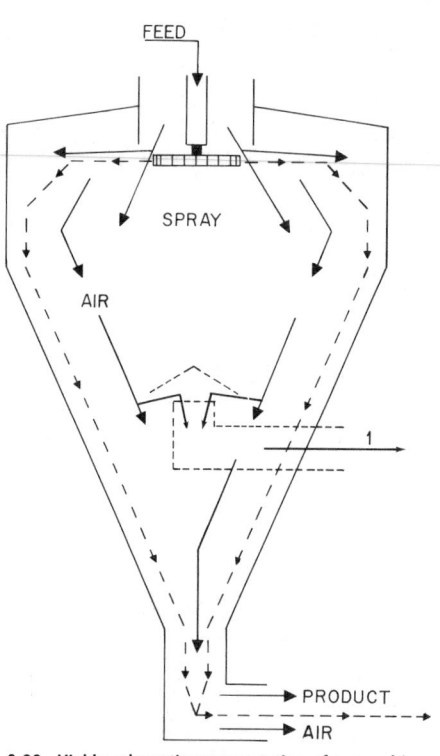

Fig. S-22. Highly schematic representation of spray drier with wheel atomizer. (1) Air outlet when drying chamber is used for initial separation.

Feed generally is introduced to the center, with centrifugal force dispersing the feed and throwing out a thin film to the periphery. As the film leaves the disk, it breaks up into a thread, which in turn forms droplets. The disk is located in the hot-air stream so that even though droplets are thrown toward the wall of the drier, the hot air travels cocurrently and dries the particle enough to prevent wall buildup upon contact. A spray drier with a wheel atomizer must be relatively large in diameter and shorter than a drier with pressure nozzles.

Pressure-Nozzle Atomizers: This system consists of an orifice placed after a fixed mechanism, called a core, swirl chamber, or whizzer, depending on the manufacturer. A high-pressure pump pumps the feed to the nozzle body at a pressure of 250–8,000 psig. The feed slurry is pumped through

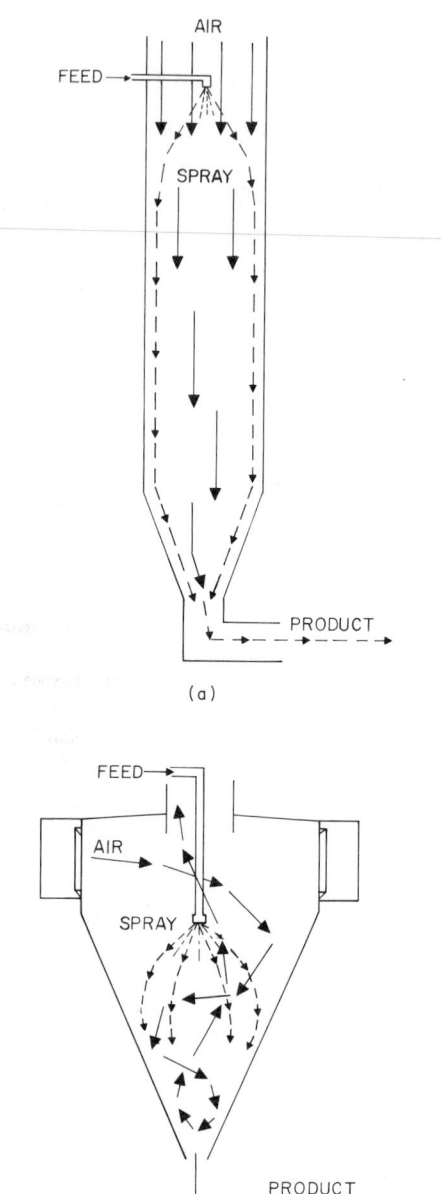

Fig. S-23. (a) Parallel-flow and (b) mixed-flow spray drier.

high-pressure piping to the whizzer, where a spin is imparted to the fluid before it enters the nozzle orifice. This results in a hollow-cone spray which throws droplets either cocurrent or countercurrent to the air flow.

The flow pattern is such that a cocurrent spray drier must be relatively long and small in diameter (see Fig. S-23a), while a countercurrent dryer is shorter and larger in diameter. A third type, sometimes referred to as a *mixed-flow drier* (see Fig. S-23b), uses an air pattern similar to a cyclone collector, i.e., the spray is introduced at the up-coming air stream (countercurrent) and the particles transfer to the air sweeping the wall (cocurrent).

Modifications: Spray driers are modified to use the properties of the product to best advantage during drying. Some of these features include:

Cool-Air Sweep: The use of ambient air introduced at the drier wall or cone to prevent dried particles that are heat-sensitive or that have low melting points from sticking on a hot wall also dilutes the exhaust air to lower the exhaust temperature artificially.

Cold-Wall Feature: A double wall on the spray drier accomplishes the same purpose as the cool-air sweep but to a lesser degree and without the dilution.

Separating Cone: Cools the powder to near ambient conditions and gently separates the powder from the drying medium.

—A. C. Patsavas, *Swenson Division, Whiting Corporation, Harvey, Ill.*

Spray Towers (See **Absorption.**)

Spreading Agents (See **Defoaming agents; Paint;** and **Surfactants.**)

Stabilizers (PVC) Of all the polymers in use today, polyvinyl chloride (PVC) is probably the most versatile and the least stable. This instability constitutes a crucial problem. As soon as un-protected PVC is processed, the effects of degradation become an immediate concern and steps must be taken to stabilize the polymer. The function of a PVC stabilizer is to protect the polymer against thermal and ultraviolet degradation. When PVC is exposed to heat and light, hydrogen chloride, HCl, is liberated, ultimately resulting in a conjugated-double-bond chromophoric structure. As degradation proceeds, the color of the plastic changes from white to yellow through tan and brown to reddish brown and finally to black. Since color formation usually precedes serious deterioration of physical properties of the polymer, one requirement for a good stabilizer is the ability to prevent discoloration during manufacture and use of a PVC compound (compound here means a blend of PVC, plasticizer, stabilizer, and filler).

While a great deal of investigative work has been carried out on the mechanism of PVC stabilization, no single theory is accepted by all authorities. The mechanisms of PVC stabilization will not be discussed in detail. In general, however, the stabilizer is considered to react with the liberated HCl and prevent further degradation of the polymer. Stabilizers are usually metal soaps or other materials capable of reacting with HCl.

A wide variety of compositions is offered as commercial PVC stabilizers. Some relate to specific needs of a particular consumer and some to general use. Most are combinations of two or more active ingredients which perform synergistically. Such combinations are available as powders and liquids. The basic components of PVC stabilizers may be classified broadly as follows: (1) compounds of Pb, Ba, Ca, Sr, Cd, and Zn, (2) auxiliary organic compounds, and (3) organotin compounds.

Lead compounds were among the earliest stabilizers. While they are quite effective in both initial and long-term heat stabilization, they have the disadvantage of being toxic, imparting opacity to PVC compounds, and having poor sulfide-stain resistance. The primary use for leads in PVC is in electrical applications such as wire coatings. The principal lead compounds for this use are tribasic lead sulfate, basic lead sulfate silicate, basic lead carbonate, and basic lead phthalate. Use levels usually range from 3–8 parts per 100 parts resin (PHR).

Barium and cadmium stabilizer systems, which are less toxic than lead, have lower cost, better sulfide-stain resistance, and good clarity, have become the major class of stabilizers. Often included in this classification because of similar performance are the compounds of Ca, Sr, and Zn. Compounds in this class show a wide variety of anions, e.g., octoates, benzoates, laurates, stearates, and substituted phenolates.

Barium compounds serve as the building blocks for the majority of the stabilizer systems. They are seldom used alone since PVC stabilized solely with barium compounds rapidly becomes yellow, amber, and then dark brown. Strontium performs in a similar fashion but because of cost is used to a limited degree.

Cadmium compounds used alone in PVC offer excellent initial color, but after a relatively short time at elevated temperatures they cause sudden, severe degradation and color development. Advantage is taken of this early activity of Ca compounds by using them in combination with Ba compounds which offer long-term stability, thus providing the balance required for stabilizing PVC.

Zinc compounds have activity similar to Ca compounds, but while Zn compounds are responsible for early stability, once color develops, these compounds cause rapid degradation of the polymer which is difficult to control. For this reason they are used in minor amounts, in combination with Ba and Ca compounds to further upgrade early color and improve the sulfide-stain resistance of these systems. Zinc compounds also find use as nontoxic stabilizer components in combination with Ca and Mg materials.

Calcium compounds function like barium-containing materials but with greatly reduced efficiency. As a result, calcium-based materials are used only to a limited degree in general-purpose stabilizers but are major components of nontoxic stabilizers.

Stabilizers containing Ba, Cd, Ca, and Zn compounds are generally used in PVC at levels of 1–4 PHR.

Organic phosphites may be used as an integral part of a liquid stabilizer, or they may be added directly to the PVC composition during compounding when solid stabilizer systems are used. Organic phosphites act as nonstaining antioxidants and may chelate the metallic salts formed during processing, resulting in greatly improved performance of the barium-cadmium class of stabilizers.

Polyhydric alcohols are used in the stabilization of rigid PVC, nontoxic systems, and vinyl asbestos compositions. They control the activity of the cadmium and zinc components, extending the heat stability and thus providing improved color hold along with long-term stability. Examples are pentaerythritol, mannitol, and sorbitol.

Nitrogenous compounds, e.g., melamine and dicyandiamide, are used only in vinyl asbestos compositions, quite often in combination with polyhydric alcohols and metallic compounds. They provide heat stability and retard water growth (dimensional increase due to immersion in water) of vinyl asbestos materials.

Epoxy plasticizers, derived from a soybean oil and isooctyl tallate, are a standard part of most PVC formulations primarily as a supplement to barium-cadmium stabilizer systems. The normal level of use is 3–8 PHR. They extend the life of the metal-stabilizer systems and offer some flexibility to the PVC compound. Liquid barium-cadmium stabilizers containing 40–60% epoxy plasticizers are currently available from most stabilizer suppliers.

Organotin compounds offer good overall stabilization, with excellent clarity and complete freedom from sulfide staining. Their disadvantages are relatively high cost (compared to barium-cadmium compounds) and, in some cases, poor light stability. Certain tin stabilizers comply with U.S. Food and Drug Administration requirements for specific food-packaging applications.

For a stabilizer to function efficiently it must react with HCl, disrupt double-bond formation, act as an antioxidant, react with impurities, and prevent the formation of degradation by-products.

Stabilizers cannot be expected to completely prevent the breakdown of PVC for an indefinite period, but they serve to retard degradation through the critical processing period and extend the useful life of the finished product.

—John W. McBroom, *Intermediates Division, Tenneco Chemicals, Inc., Piscataway, N.J.*

Stainless Steels   (See **Chromium;** and **Iron and steel.**)

Stannite   (See **Tin.**)

Staple   (See **Fibers.**)

Starch   Starch is the major reserve carbohydrate of many higher plants, where it is deposited in the form of insoluble granules in storage organs, such as seeds and tubers. Chemically, starch is a homopolymer of α-D-glucopyranoside of two distinct types. The linear polysaccharide, amylose, has a degree of polymerization (DP) on the order of several hundred glucose residues connected by α-D-(1 → 4)-glucosidic linkages. The branched polymer, amylopectin, has a DP on the order of 10,000 to 100,000 glucose residues. The segments between the branched points average about 25 glucose residues linked by α-D-(1 → 4)-glucosidic bonds, while the branched points are linked by α-D-(1 → 6) bonds. The structure and properties of these two polymers are described in Refs. 2 and 5.

Most cereal starches are made up of about 75%

amylopectin and 25% amylose molecules. Root starches, however, are slightly higher in amylopectin, while waxy corn and waxy milo starch contain almost 100% amylopectin. At the other extreme, high-amylose cornstarch and wrinkled-pea starches contain 60–80% amylose.

The molecules of amylose and amylopectin are synthesized by enzymes inside the living cell in plastids known as *amyloplasts* and are deposited as starch granules;[6] see also **Enzymes.** These granules are microscopic in size, diameters ranging from 3–8 μm for rice starch up to 100 μm for the larger potato-starch granules. Most cornstarch falls in a range of 5–25 microns. An experienced observer usually can identify the genetic origin of a sample of starch by the size and shape of the granules. The granules are insoluble in cold water but swell rapidly when heated to the gelatinization temperature range for that particular starch. As the granules swell, they lose their characteristic cross pattern under polarized light and imbibe water rapidly until they are many times their original size.

Upon continued heating or mechanical shear, the swollen granules begin to disintegrate, and the viscosity, having reached a maximum, begins to decrease. There usually are some granules and segments of granules, however, that do not completely disperse in aqueous systems even under the most stringent conditions.

As the partially solubilized paste is cooled, the hydrated molecules and segments of granules begin to insolubilize. In a dilute system (approximately 1%), the segments and molecules retrograde or precipitate. At higher concentrations, enough intermolecular and intersegment bonds form to fix the entire system into three-dimensional gel. The rigidity of this gel is affected by many factors, but the amylose content is perhaps the most significant. High-amylose starches when thoroughly cooked form very rigid gels. Waxy corn or waxy milo starch pastes form little if any gel structure when cooled.

While some wheat and potatoes are processed (United States), over 90% of all starch is produced from corn in what is called the *corn wet-milling industry*. In recent years, over 200 million bushels of corn (about 5% of the total corn crop) was converted into wet-process products.[4]

Corn Refining Process (See Fig. S-24.): The first phase of the process is essentially a mechanical separation of the starch from the protein, germ, fiber, and solubles that make up the balance of the corn kernel. The corn first is cleaned and then steeped for 36–48 hr in circulating warm water containing a small amount of sulfur dioxide. The steeping process removes some of the solubles and toughens the germ as well as softening the protein matrix surrounding the starch granules. The steepwater, containing much of the solubles, is drawn off as the first by-product. The softened kernels then go through the degerminating mills, which are designed for tearing the soft kernels into coarse particles, freeing the rubbery oil-bearing germ without crushing it, and loosening the hulls. The wet, macerated kernels are conveyed in an aqueous suspension to flotation tanks or centrifugal hydrocyclones, in which the germ, being lighter than the other components of the mixture, can be separated. This germ, which contains most of the oil in the kernel, is processed subsequently to recover the corn oil by expellers and solvent extraction. The germ residue is sold as animal feed as corn-germ meal or after recombination with other by-products of the process.

The remaining mixture of starch, protein or gluten, and hull is finely ground and washed through a series of screens to remove the hull from the starch and gluten. The hull then is combined with the protein steepwater to give corn gluten feed. The remaining suspension of gluten and starch is pumped to high-speed centrifugal machines which utilize the difference in specific gravity to separate the relatively heavier starch from the lighter gluten. After further processing, the protein-enriched fraction is sold as gluten meal for animal feeds. The separated starch still is in aqueous suspension and now is ready for washing and drying or further processing into modified starch products or dextrin or conversion into corn syrup or sugar. A 56-lb bushel of corn typically yields about 32 lb of starch at 12% moisture; 15 lb of feed and feed products at 10% moisture; and about 2 lb of oil. The remainder is water.

Conversion Operations: Over half of the starch produced (United States) is not dried but converted directly into syrup or dextrose (glucose) from the aqueous suspension. Hydrochloric acid is added to a pH of about 1.5–2.0, and the mixture is heated well above 100°C in continuous pressurized reactors. The acid catalyzes the cleavage of the glucosidic linkages and if carried to the end point, gives a solution of crude dextrose. In modern practice, the reaction usually is stopped well short of the end point by neutralizing the acid in order to minimize the formation of reversion products. This gives rise to a series of acid-converted syrups depending upon the extent of the

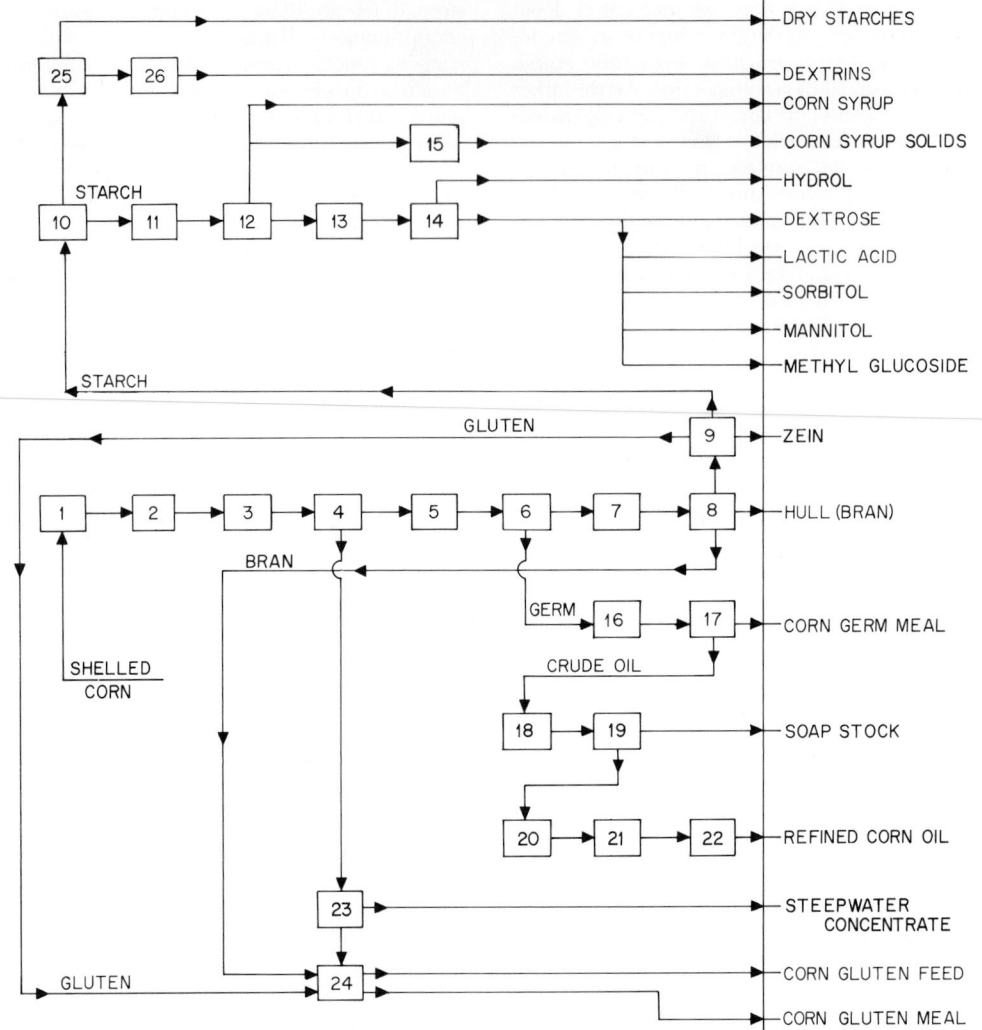

Fig. S-24. Schematic representation of materials flow in corn-refining process. (1) First corn cleaners, (2) storage bins, (3) second corn cleaners, (4) steep tanks, (5) degerminators, (6) germ separators, (7) grinding mills, (8) washing screens, (9) centrifugal separators, (10) starch-washing filters, (11) syrup and sugar enzyme or acid converters, (12) decolorizing and evaporating, (13) sugar crystallizers, (14) centrifugals, (15) drum or spray driers, (16) germ washing and drying, (17) oil extractors, (18) filters, (19) centrifugal separators, (20) bleaching and winterizing, (21) deodorizers, (22) filters, (23) steepwater evaporators, (24) feed driers, (25) starch driers, (26) dextrin roasters. (*Corn Refiners Association, Inc. Reproduced by permission.*)

acid hydrolysis. These syrups can be purified, concentrated, and sold; or they can be converted further with enzymes to give high-maltose syrups or a dextrose solution with fewer reversion products for better subsequent crystallization. In some cases, it is preferable to replace the acid-conversion step during the early stage of hydrolysis with

liquefying enzymes which even further reduce reversion products or give unusually low conversion syrups.

The degree of conversion generally is specified by the dextrose equivalent (DE), in which the reducing power of the syrup is expressed as a percent of the reducing power of pure dextrose.

Syrups are defined as those hydrolysates having a DE of 20 or over; those with a DE of less than 20 are classified as maltodextrins.

Chemical Modifications: While much of the remaining starch is dried with little if any treatment, a substantial portion is chemically modified to build in specific physical properties. In the main, these chemical treatments are controlled carefully so that the physical integrity of the starch granule will be undamaged. Otherwise, the cost of subsequent filtering and drying is too high.

One class of chemical treatment is designed to cleave the starch molecule and thus reduce the viscosity of the cooked paste. These treatments include acid thin boiling, oxidation, dextrinization (dry roasting), and enzyme conversion. The latter reaction takes place in the swollen granules and therefore usually is carried out in the user's plant.

A second class of chemical modification results in chemical derivatives through the hydroxyl group on the starch molecules. These linkages may be ethers, esters, or hemiacetals, and the substituent groups may terminate in anionic, cationic, and nonionic end groups. Since these end groups confer their particular properties to the starch polymer, it is possible to build in a large number of different physical and chemical properties with a small degree of substitution.

A third class of chemical treatment is obtained by cross-linking adjacent starch molecules by treatment with reagents having two or more reactive groups. This stabilizes the granule against disintegration and gives pastes that are highly resistant to breakdown under shear or high temperature. The preparation, analysis, and properties of the various modified starches are described in Refs. 7 and 8.

Tonnage Production: The quantities of shipments by the corn wet-milling industry are summarized in Table S-21. The food industry is the major user of these products, almost every processed food using one or more of the starch products. The food industry consumes almost all the syrup and dextrose produced as well as a substantial portion of the starch. The paper industry is a distant second user, followed by the textile and adhesive industries. Uses are described in detail in Refs. 3, 7, and 8.

REFERENCES

1. U.S. Bureau of the Census: *Census of Manufacturers: 1967,* Industry Series, Corn Wet Milling, SIC Code 2046.
2. Kerr, R. W.: "Chemistry and Industry of Starch," Academic, New York, 1950.
3. Liebenow, Robert C.: "The Corn Refining Industry," Corn Refiners Association, Washington, 1966.
4. USDA: "Agricultural Statistics," p. 33, Washington, 1970.
5. Whistler, R. L., and E. F. Paschall: "Starch Chemistry and Technology," vol. 1, Academic, New York, 1965.
6. Ibid., pp. 133–175.
7. Ibid., vol. 2, 1967.
8. Whistler, R. L.: "Methods in Carbohydrate Chemistry," vol. 4, "Starch," Academic, New York, 1964.
9. Sinclair, P. Michael: Enzymes Convert Starch to Dextrose, *Chem. Eng.,* August 30, 1965, pp. 90–92.

—Clifford H. Hullinger, *American Maize-Products Company, Hammond, Ind.*

Steam Cracking (See Ethylene.)

Steam Generation (See Energy systems for processes; and Nuclear power plants.)

Steam Reforming (See Fuels; and Hydrogen.)

Stearic Acid (See Carboxylic acids; Rubber, natural; Vegetable oils; and Waxes.)

Steel (See Iron and steel; and Iron- and steel-making.)

Stereoelastomers (See Elastomers.)

Stereoisomerism (See Isomerism.)

Sterilization (See Amino acids; and Pasteurization.)

TABLE S-21. *Tonnage Production of Starch Products**

Product	Million Pounds
Cornstarch	3,119.0
Corn syrup	2,777.9
Corn gluten feed	2,365.9
Corn sugar	1,227.9
Corn gluten meal	875.2
Other wet-process by-products	341.4
Dextrin (corn, tapioca, etc.)	168.7
Corn-syrup solids	127.1
Other starches	121.3
Steepwater concentrate	78.0

*United States (1967). Corn-oil tonnage (not included) ranked fourth in dollar value.

Steroidal Antibiotics   (See **Antibiotics.**)

Stibine   (See **Antimony.**)

Stibines   (See **Amines.**)

Stibnite   (See **Antimony.**)

Stoichiometry   The application of the fundamental laws of chemistry, physical chemistry, and physics to the heat, energy, and material balances of a chemical system and other chemical engineering calculations is referred to as stoichiometry.

Stone Mill   (See **Paint.**)

Storage Batteries   (See **Batteries.**)

Storage, Chemical   (Consult the **Subject Index.**)

STP   (Pentasodium Tripolyphosphate)   (See **Detergents.**)

Strecker Synthesis   (See **Amino acids.**)

Streptomyces   (See **Antibiotics;** and **Peptides and proteins.**)

Streptomycin   (See **Antibiotics.**)

String-Discharge Filter   (See **Filtration.**)

Stripping Columns   (See **Absorption;** and **Distillation.**)

Strontium   (From Strontian, a village in Scotland.) **Sr** = 87.63 (at. wt.); 38 (at. no.). Strontium is chemically similar to its neighbors, Ca and Ba in the IIA (alkaline-earth) series of elements. This white metal is relatively workable and very reactive. Discovered in compound form by T. C. Hope in 1792, its name derives from the mining region of Strontian in Scotland. Sir Humphry Davy in 1808 reduced the metal to an amalgam and R. Bunsen later produced it in relatively pure form by electrolysis of the chloride.

Because of its extreme reactivity, Sr is difficult to isolate in a high-purity form and to maintain in a high degree of purity during handling. Therefore, reported property data are often of limited accuracy. Properties as given in the literature are shown in Table S-22.

TABLE S-22. *Selected Physical Properties of Strontium*

Chemical valence	2^+
Isotopes (stable)	84, 86, 87, 88
Electron configuration	2, 8, 18, 8, 2
Density at 20°C, g/cm³	2.6
Melting point, °C	772 ± 2
Boiling point, °C	1372
Specific heat, kcal/g	0.176
Heat of fusion, kcal/g	23
Heat of vaporization, kcal/g	42.3
Electrical resistivity at 20°C, μΩ-cm . . .	13.4
Crystal form:	
Above 605°C	bcc
215–605°C	hcp
Below 215°C	fcc
Fcc lattice constant, Å	6.085
Closest approach of atoms, Å	4.31

Strontium* exists in the face-centered cubic form at room temperature. It transforms into hexagonal closed-packed form at 215°C and to body-centered cubic form at 605°C.

Sr is reduced rather readily from its oxide by metallothermic reduction with Al in equipment similar to that used for the reduction of Ca. The major difficulty in obtaining a relatively pure metal is the lack of availability of high-purity oxide.

Because of its similar properties, Sr would be useful in many applications where Ca is used, but there is no known commercial demand for the metal because of its relatively high cost.

The main minerals of strontium are the carbonate, strontianite, and the sulfate, celestite. Sr (as the sulfate or oxide) is used as an additive in the manufacture of some hard-magnet ferrites. Sr compounds are used in the manufacture of pyrotechnics. A radioactive isotope, ^{90}Sr, is a hazard from nuclear blasting activities because of its long half-life and tendency to be retained in the body.

REFERENCES

American Society for Metals: "Metals Handbook," 8th ed., vol. 1, Metals Park, Ohio, 1970.
Elliot, R. P.: "Constitution of Binary Alloys: First Supplement," McGraw-Hill, New York, 1965.
Hampel, C. A.: "Rare Metals Handbook," 2d ed., Van Nostrand Reinhold, New York, 1961.
Rashid and Kayser: *J. Common Met.,* vol. 25, pp. 107–108, 1971.

—Frank Emley, *Pfizer Metals & Composite Products, Pfizer Inc., Easton, Pa.*

*Produced in the United States by Pfizer Inc., New York, and in Canada by Dominion Magnesium Ltd.

Stucco   (See **Gypsum**.)

Styrene-Butadiene Rubber (SBR)   (See **Elastomers**.)

Sublimation   (See **Freeze-Drying**; and **Separation operations**.)

Submerged-Flame Process   (See **Acetylene**.)

Substitution Product   (See **Saturated compounds**.)

Succinimide   (See **Amides and imides**.)

Sulfanilamide   (See **Amides and imides**; and **Sulfur**.)

Sulfapyridine   (See **Pyridine and derivatives**.)

Sulfates   (See **Sulfur**.)

Sulfathiazole   (See **Sulfur**.)

Sulfation   [See **Anionic surfactants (sulfur-bearing)**.]

Sulfite Pulping Process   [See **Pulp (wood) production and processing**.]

Sulfoethyl Esters   [See **Anionic surfactants (sulfur-bearing)**.]

Sulfonamides   (See **Amides and imides**; and **Sulfur**.)

Sulfonation   [See **Anionic surfactants (sulfur-bearing)**; **Sulfur**; and **Surfactants**.]

Sulfone Group   (See **Polysulfones**.)

Sulfonic Acids   [See **Anionic surfactants (sulfur-bearing)**; and **Surfactants**.]

Sulfosuccinates   [See **Anionic surfactants (sulfur-bearing)**.]

Sulfur   (L. *sulfur*.) **S** = 32.07 (at. wt.); 16 (at. no.). Sulfur, one of the nine elements in group VI of the periodic table, is between oxygen and selenium in the odd series. Atomic volume, rhombic = 15.5 cm^3/g atom; monoclinic = 16.4 cm^3/g atom. Melting point, rhombic = 113°C; monoclinic = 119°C. Boiling point = 446.6°C. Density, rhombic = 2.07 g/cm^3 at 20°C; mono-

clinic = 1.96 g/cm^3 at 20°C. Bulk density = 85 to 90 lb/ft^3; powdered, 33–36 lb/ft^3. Angle of repose = 35°. Critical temperature = 1050°C; critical pressure = 116 atm. Sulfur is insoluble in water; soluble in CS_2. Valences are 2, 4, and 6.

Sulfur exists in two crystalline forms, rhombic (R) and monoclinic (M), plus several amorphous states, including an elastomer. Some 36 native crystalline types based on the stable rhombic form are known. Molecules in S_R contain eight atoms in the form of a twisted ring and joined to form unit cells of 16 rings[4] (see Fig. S-25). S_R changes

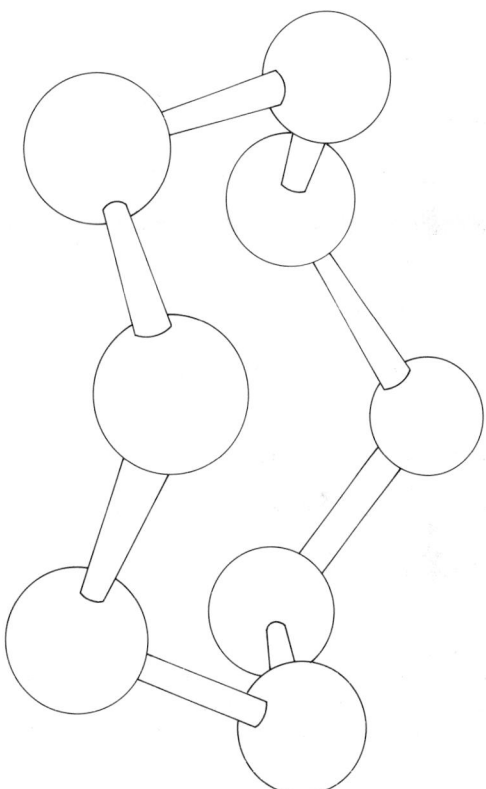

Fig. S-25. A sulfur octatomic molecule.

(reversibly) to S_M at 95.4°C with absorption of 2.99 cal/g. S_M contains six rings and becomes the stable variety up to the melting point of 115.2°C when pure. The viscosity-temperature relation of molten sulfur is unusual. Heating from 120 to 160°C decreases viscosity and darkens color. Above 160°C, viscosity abruptly increases up to 230°C, followed by a decrease up to the boiling

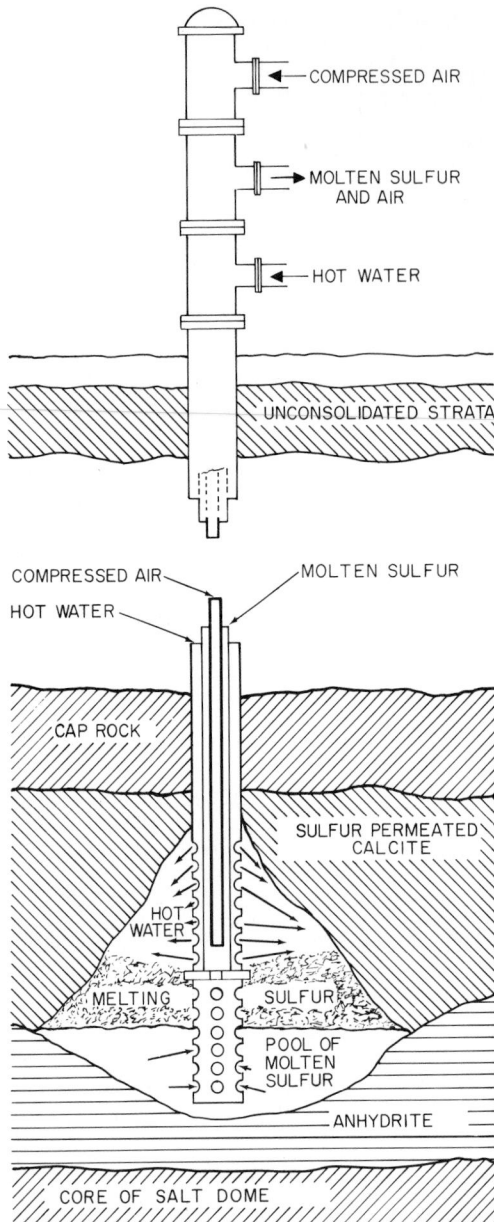

Fig. S-26. Mining salt-dome sulfur by the Frasch method. The process entails placing three concentric pipes inside a casing that lines a shaft drilled to the base of the sulfur deposit. Typical diameters for the pipes are 8, 4, and 1 in. Water heated to a temperature well above the boiling point is pumped down the space between the 4- and the 8-in. pipe and also, in the initial stage, down the 4-in. pipe. It flows through holes in the casing and

point. These changes are reversible and probably due to formation of long-chain polymers.[7] Rapid chilling of molten sulfur creates an elastomer. Above 900°C, sulfur molecules are mostly diatomic. Four stable isotopes, 32–34 and 36 are known, plus four radioactive isotopes, 31, 35, 37, and 38. ^{32}S predominates in nature, and ^{35}S is a useful tracer.[4]

Occurrence and Production: Earliest sources were elemental sulfur deposits in Sicily and other volcanic regions, but rising demands in the late nineteenth century for sulfuric acid led to the widespread use of pyrites, FeS, as a cheaper alternative. This was supplemented by SO_2 evolved in smelting other metal sulfides. Next came the development of the Frasch hot-water process[4,6] for mining elemental sulfur in U.S. Gulf Coast salt domes, which was later extended to Mexican domes and recently sedimentary deposits in Texas and Poland (see Figure S-26). Removal of organic and inorganic sulfur, H_2S, from Canadian sour natural gas by solvent scrubbing and conversion to elemental sulfur via the Claus reaction[6] has now become the major North American supply source and is also undertaken in France, West Germany, and other countries processing sour gas. These will soon be supplemented by elemental sulfur recovered via petroleum-product desulfurization, plus sulfur and H_2SO_4 resulting from flue-gas purification (see Table S-23). Enormous reserves of sulfur, little used to date, also exist as calcium sulfate minerals and in seawater. See also **Sulfur dioxide removal.**

World sulfur demand in all forms is expected to grow by at least 5% annually from a current level of over 40 million long tons to nearly 70 million by 1980. United States demand is likely to show a similar long-term growth, from about 11 million long tons to at least 15 million long tons by 1980.[12] No shortages in the next decade are foreseen; in fact a cumulative oversupply especially in North America may reach several million tons. This is largely due to removal and recovery of sulfur from Canadian sour natural gas, high-sulfur fuel oils, and stack gases rich in SO_2. These secondary supplies are unrelated to market demands and are overshadowing traditional sources. Worldwide, over 80% of all sulfur produced is

into the rock, where it melts the sulfur, which forms a pool at the bottom of the well and, under the pressure of the water, rises part way up the 4-in. pipe. Compressed air forced down the 1-in. pipe lightens the sulfur so that it rises to the surface.

TABLE S-23. *Typical Molten-Sulfur Analyses*

Constituent	Source	
	Sour Gas	Frasch Well
Sulfur	99.96%	99.85% (bright)
		99.45% (dark)
Free moisture.	0.025%	0.03%
Free carbon and hydrocarbons	0.0056%	0.1% (bright)
		0.25–0.5% (dark)
Selenium	<1 ppm	<2 ppm
Tellurium	0	<2 ppm
Arsenic	<1 ppm	<0.25 ppm
Chlorides	0	<0.001%
Ash	0.0036%	0.005% max.

converted to sulfuric acid, half of which is used to make fertilizers, especially phosphates.[6,12] The chemical, petroleum, and many other industries consume the remainder. Nonacid uses are principally in rubber compounding and production of SO_2, SO_3, pesticides, and pharmaceuticals. Today, most sulfur is transported in molten form and is moved in ocean-going vessels on a large scale.

Sulfur Compounds

Oxygen Derivatives: Several oxides and corresponding acids are known or suspected:

Oxide			Acid	
Mon-	SO	Monosulfurous	$(H_2SO_2)_n$
Sesqui-	S_2O_3	Hyposulfurous	$H_2S_2O_4$
Di-	SO_2	Sulfurous	H_2SO_3
Tri-	SO_3	Sulfuric	H_2SO_4
Hept-	S_2O_7	Perdisulfuric	H_2SO_8
Tetr-	SO_4	Permonosulfuric	H_2SO_5

also, thiosulfuric acid, $H_2S_2O_3$, and pyrosulfuric acid, $H_2S_2O_7$, as well as a range of thionic acids, from dithionic, $H_2S_2O_6$, to hexathionic, $H_2S_6O_6$. Most important industrial compounds are SO_2, SO_3, and corresponding acids.

Sulfur dioxide is easily produced by burning sulfur or mineral sulfides in air; by the action of H_2SO_4 on carbonaceous materials; or by adding strong acids to sulfites. It dissolves readily in water to produce H_2SO_3 and is a useful mild reducing and bleaching agent. It has two isomers and is dibasic, forming normal sulfites, X_2SO_3, and bisulfites, $XHSO_3$. Direct oxidation is not spontaneous and requires O_3 or the use of excess O_2 plus a catalyst. In industry, millions of tons are converted annually to SO_3 for sulfuric acid

manufacture,[8] usually via a vanadium oxide catalyst at 500–750°C: $2SO_2 + O_2 \rightarrow 2SO_3$. See also **Air pollution;** and **Sulfuric acid.**

The properties of SO_2 and SO_3 are given in Table S-24. Unless stabilized, liquid SO_3 forms gamma, beta, or alpha crystals at 17, 32.5, and 62.3°C, the latter being difficult to melt. A strong oxidizing and dehydrating agent, SO_3 reacts violently with water to produce H_2SO_4 and ignites readily combustible materials. SO_3 absorbed in concentrated H_2SO_4 is termed *oleum.*

Two series of sulfates exist, normal, R_2SO_4, and bi-, or acid, $RHSO_4$. Normal sulfates of alkali and heavy metals occur as natural minerals and can be made by action of dilute H_2SO_4 on the metal (strong acid may induce passivation) or by double decomposition, for example, $BaCl_2 + Na_2SO_4 \rightarrow BaSO_4 + 2NaCl$. Many sulfates form sulfides when heated with carbon or hydrogen, for example, $CaSO_4 + 4C \rightarrow CaS + 4CO$, or to SO_2 and O_2 when heated with HCl. Most derivatives of higher oxides have limited uses.

Hydrogen Derivatives: Sulfur forms a series of sulfides with hydrogen, from H_2S_2 to H_2S_6, most stable and important being the monosulfide, H_2S. This is a toxic gas at normal temperature and pressure prepared directly from H_2 and sulfur vapor at 500°C via a silica-alumina catalyst or by action of acids on metallic sulfides. It is a weak reducing agent and moderately soluble in water, forming sulfides with alkali metal, hydroxides, and then hydrosulfides (for example, KHS). The second hydrogen of H_2S has a very small dissociation constant, which allows selective precipitation of many metallic sulfides by choosing appropriate pH, as in analytical work and pigment manufacture.

TABLE S-24. *Properties of Sulfur Dioxide and Sulfur Trioxide*

Property	SO_2	SO_3
Melting point, °C	−75.5	16.8
Boiling point, °C	−10	44.8
Density at 16°C, g/cm³	2.264	1.95
Heat of formation, kcal/mole	70.9	22.5
Toxicity	Asphyxiating above 20 ppm	Toxic above 2 ppm
Critical temperature, °C	157.2	218.3
Critical pressure, atm	77.7	83.6

Halogen Derivatives: Numerous chlorine compounds are known, for example, SCl_4, SCl_2, S_2Cl_2, S_3Cl_2 and perhaps S_3Cl_4, the most familiar being the monochloride, S_2Cl_2. This is made industrially by directly heating sulfur and Cl_2 and also results as intermediate product when making CCl_4 from CS_2 and Cl_2. It is a stable, yellow liquid used in vulcanization and in chlorination and sulfonation of organic products. The vapor is toxic above 1 ppm. Sulfur monobromide, S_2Br_2, is a red liquid made by heating sulfur plus Br_2 under pressure. The hexafluoride, SF_6, is made by burning sulfur spontaneously in F_2, thus confirming the maximum valency (6) of sulfur. No direct compound of sulfur and I_2 apparently exists. Several oxyhalides known.

Another group of sulfur-halogen compounds can be made by reacting H_2SO_4 with PCl_4 to replace one or more OH groups by Cl and form chlorosulfonic acids, for example, $SO_2(OH)Cl$, which is also made industrially from oleum and dry HCl. Sulfuryl chloride, SO_2Cl_2, can be made directly from SO_2 and Cl_2 over activated carbon. These compounds find industrial uses as sulfonating and/or chlorinating agents, especially in organic syntheses.

Other Inorganic Derivatives: Under appropriate conditions, sulfur and carbon combine directly to form sulfides such as C_3S_2, CS, and CS_2. COS is produced by reacting CO and H_2S at high temperatures. Chlorination of such compounds produces toxic materials, e.g., thiophosgene, $CSCl_2$, and Cl_3CSCl, a mercaptan. The principal industrial compound is CS_2, a solvent made by heating sulfur vapor and charcoal in a retort or by reacting methane and sulfur vapor over a silica catalyst, followed by absorption in a suitable oil and distillation. Thiocyanates such as NH_4CS_2·NH_2, NaSCN, and KSCN have important industrial uses and can be made by reacting NH_3 and CS_2 or by heating appropriate cyanide salts in a

sulfur suspension. At least three direct compounds of sulfur and nitrogen are known; these appear to be nitrides, S_4N_4, S_5N_2, and S_2N. All are red liquids, S_4N_4 being dangerously unstable. Recently, a group of sulfur, phosphorus, and nitrogen compounds has been developed for possible agricultural purposes by reacting phosphorus, sulfur, and NH_3 in vapor phase. Typical products include phosphorus thiotriamide, $PS(NH_2)_3$, and phospham, PNNH.[14]

Organic Derivatives: The many valencies of sulfur plus its ability to form ring and chain complexes and to combine readily with carbon, hydrogen, and many negative elements lead to innumerable compounds. Many take part in essential biological processes and are components of proteins and amino acids found in wool, skin, grains, and scents. The role of sulfur as a vital plant food is becoming increasingly recognized. Many fossil fuels contain sulfur in both organic and inorganic combination. In organic synthesis, addition and oxidation reactions are readily possible:

$$RSR + C_2H_5Br \rightarrow R_2C_2H_5S + Br$$

$$RSX + O_2 \rightarrow \begin{matrix} R & & O \\ & \diagdown S \diagup & \\ X & & O \end{matrix}$$

Reductions via hydrogen, H_2S, and PCl_5 are usually straightforward, S—S bonds are mostly weak; S—C bonds are often stable and may break only via strong attack by mineral acids, molten cyanides, or heat. With compounds containing carbon, hydrogen, and oxygen, sulfur usually acts in a bivalent manner; and presence of a halogen may induce stability. Several comprehensive descriptions of organosulfur compounds are available.[9,10]

Sulfur dyes[11] form an important organosulfur group and are made by baking organic intermediates such as aromatic amines with sulfur and/or refluxing with a polysulfide, or for example, Na_2S, and removing excess sulfur with sodium sulfite or other means. Oxidation with air imparts insolubility. Thio ethers, $R—S—R$, prepared by alkylating mercaptans or adding Na_2S to an alkylation agent, are widely used as oil additives, elastomers, and antioxidants. Still important are the antibacterial sulfa drugs, such as sulfanilamide and sulfathiazole. One synthesis route is to chlorinate acetanilide, react with NH_3, and separate the N-acetyl group with an alkali. Mercaptans (thiols), with a general formula RSH, often resemble alcohols and are easily esterified but yield only one hydrogen on oxidation to form disulfides: $2CH_3SH + \frac{1}{2}O_2 \rightarrow CH_3SSCH_3 + H_2O$. Lower members, which are malodorous, are found in decomposing animal matter and some petroleum crudes; removal from thermally cracked stocks via sodium plumbite or alkalis improves odor and octane levels. Organic thiocyanates, $HSC\!:\!N$, and isothiocyanates $HN\!:\!C\!:\!S$ (mustard oils), have specific applications as intermediates and pesticides.

Attaching a sulfo group, $\cdot SO_2OH$, to a carbon or nitrogen atom, known as *sulfonation*, is a large-scale industrial process used in making detergents, emulsifiers, and intermediates. Methods include reacting with oleum or SO_3, catalytic oxidation with SO_2 and Cl_2, and treatment with sulfites. Sulfonation is a simple way of expanding the usefulness of many organic materials by adding a polar hydrophilic group. *Sulfation*, the addition of an $\cdot OSO_2OH$ group to a carbon atom or an $\cdot SO_2OH$ group to an oxygen atom, is another extensive industrial practice, e.g., the sulfation of long-chain saturated alcohols and fish-oil products. Typical methods include treatment with SO_3 or sulfuric or chlorosulfonic acid. Sulfation of alkenes with concentrated H_2SO_4 followed by addition of H_2O yields alcohols; for example, $C_2H_4 + HOSO_2OH \rightarrow C_2H_5OH + HOSO_2OH$. The vulcanization mechanism of sulfur treatment of rubber is described under **Rubber, natural.** The role of sulfur in amino acids is described under **Amino acids.**

REFERENCES

1. Meyer, B.: "Elemental Sulfur Chemistry," Interscience-Wiley, New York, 1965.
2. Nickless, G.: "Inorganic Sulphur Chemistry," American Elsevier, New York, 1968.
3. Partington, J. R.: "A Textbook of Inorganic Chemistry," Macmillan, London, 1931.
4. Texas Gulf Sulphur Company: "Sulfur Manual," New York, 1969.
5. Freeport Sulphur Company: "Sulfur Data Book," McGraw-Hill, New York, 1954.
6. Pratt, C. J.: Sulfur, *Sci. Am.,* vol. 222, no. 6, June 1970.
7. Wiwiorowski, T. K.: *Sulphur Inst. J.,* Summer 1968.
8. Duecker, W. W., and J. R. West: "Manufacture of Sulfuric Acid," Van Nostrand Reinhold, New York, 1959.
9. Suter, D. M.: "Organic Chemistry of Sulfur," Wiley, New York, 1944.
10. Fieser, L. F., and M. Fieser: "Advanced Organic Chemistry," Van Nostrand Reinhold, New York, 1961.
11. Venkataraman, K.: "Chemistry of Synthetic Dyes," vol. 2., Academic, New York, 1952.
12. Manderson, M. C.: *ACS Meet. Chic.,* 1970.
13. Pratt, C. J.: Sulfur Supply Pattern for the Future, *ACS Symp., N. Y.,* September 1969.
14. Getsinger, J., et al.: *TVA 7th Fert. Demonstr.,* Muscle Shoals, Ala., 1968.

—Christopher J. Pratt, *Hoechst-Uhde Corporation, Englewood Cliffs, N.J.*

Sulfur Dioxide Removal Stringent pollution regulations on sulfur oxide, SO_x, emissions has accelerated the search for processes to remove SO_2 from power plants and other industrial stack gases. The process chemistry for SO_2 removal presents a variety of choices, many of which have been investigated in recent years. All processes are still in a development phase. Several have been carried to the point of testing in large units tied to sulfur-emitting power or chemical plants. These processes fall into three categories: (1) chemisorption of SO_2 by solids or liquids, with regeneration to produce sulfur; (2) catalytic oxidation of SO_2 to produce H_2SO_4, and (3) physical adsorption of SO_2 on char with regeneration to produce concentrated SO_2. The process described here* falls in category 1. Detailed descriptions of other processes may be found in the literature, and a survey of the field can be obtained from the references.

Process Description: The process shown in Fig. S-27 uses a dry acceptor (cupric oxide on activated alumina) in a static bed. The acceptor is unique in that it allows acceptance and regeneration steps of equal duration to be carried out at substantially the same temperature in two identical reactors in swing operation. The net effect is that an SO_2-rich gas, free from O_2 and particulate matter, is produced from a flue gas lean in SO_2. The process typically achieves an overall sulfur removal efficiency of 90% and does not produce any waste

*Shell Flue Gas Desulfurization Process.

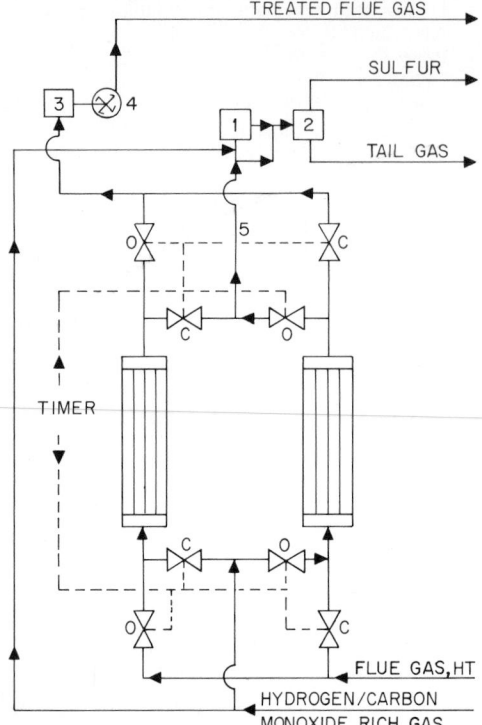

TREATED FLUE GAS

SULFUR

TAIL GAS

TIMER

FLUE GAS, HT

HYDROGEN/CARBON
MONOXIDE RICH GAS

Fig. S-27. Flue-gas desulfurization process with partial reduction and Claus unit for production of elemental sulfur. Valves are automatically controlled by a timing device. Diagram depicts a particular step in cycle. *C,* closed; *O,* open. Flue gas enters from waste-heat recovery, high temperature (*HT*). Tail gas is recycled to furnace or inlet flue gas. Treated flue gas goes to stack. (1) SO_2 reduction unit, (2) Claus unit, (3) waste-heat recovery, low temperature, (4) blower, (5) regeneration off gas. (*Shell Development Company.*)

products. Because of a special reactor design, removal of particulate matter from the flue-gas stream of an oil-fired heater or boiler is not required. The regeneration off gas can be used to produce either elemental sulfur or H_2SO_4.

Process Chemistry: The acceptance cycle begins with the regenerated acceptor in the form of copper on the activated-alumina carrier. Three reactions occur simultaneously. SO_2 is taken up from the flue gas by the acceptor according to the reaction

$$SO_2 + \tfrac{1}{2}O_2 + CuO \rightarrow CuSO_4 \qquad (1)$$

The formation of CuO from Cu, which is produced during the regeneration cycle, proceeds by

$$Cu + \tfrac{1}{2}O_2 \rightarrow CuO \qquad (2)$$

The regenerated acceptor may also contain some Cu_2S and carbon deposits, which will be oxidized in accordance with

$$Cu_2S + 2\tfrac{1}{2}O_2 \rightarrow CuO + CuSO_4$$
$$C + O_2 \rightarrow CO_2 \qquad (3)$$

All these reactions are fast, but reaction (1) is the slowest. Therefore, the SO_2 in the flue gas will be accepted by oxidized Cu in the reactor.

SO_2 will be accepted from the flue gas until the quantity of unloaded acceptor in the static bed becomes too small to ensure complete SO_2 removal. At this point, SO_2 will start to slip through, and the flow stream will then be directed to a bed of regenerated acceptor.

In the regeneration cycle, $CuSO_4$ releases the bulk of the accepted sulfur in the form of SO_2, upon contacting with reducing agents, such as H_2, CO, and light hydrocarbons. $CuSO_4$ is outstanding in this regard because other relevant metal sulfates are normally reduced to their sulfides at process temperature. With hydrogen and methane as reducing agents, the overall regeneration reactions are

$$CuSO_4 + 2H_2 \rightarrow Cu + SO_2 + 2H_2O$$
$$CuSO_4 + \tfrac{1}{2}CH_4 \rightarrow Cu + SO_2 + \tfrac{1}{2}CO_2 + H_2O$$

Unused CuO from the acceptance period is also reduced to Cu.

A third reaction in the regenerator is:

$$6Cu + SO_2 \rightarrow Cu_2S + 2Cu_2O$$

Hydrogen is an extremely active regeneration agent. Light hydrocarbons display lower regeneration rates, which increase with carbon number. However, hydrocarbons form some coke on the acceptor. The coke formation is insignificant with methane but increases with increasing carbon number.

Technical Considerations: Through development of a dry absorbent (CuO on activated alumina) operated at the same temperatures for acceptance and regeneration, this process avoids the disadvantages of wet processes (costly heat-exchange equipment constructed from corrosion-resistant materials; loss of thermal buoyancy of the flue gas; and waste-disposal problems) as well as the disadvantages of dry processes (regeneration temperature being considerably higher than acceptance temperature, with consequent need for heat exchange).

A cyclic process with static packed beds is used for contacting flue gas with the acceptor in the process illustrated. A continuous system requires the acceptor material to be moved and will inherently suffer from relatively complex solids-handling operations. A moving bed leads to high flue-gas pressure drop and hence to high compression costs. The dispersion of fine acceptor particles in the flue-gas stream would inevitably result in loss of acceptor by attrition and call for a very efficient separation of fine solid particles from the treated flue gas. In conventional form, a packed bed is unsuitable for the processing flue gas because of the inevitable rapid plugging by soot and ash. The plugging problem was solved in the process illustrated by developing a specially designed reactor which avoids solids deposition.

This process may be followed by the production of elemental sulfur from the SO_2-rich regeneration off gas. Many processing routes to elemental sulfur have been developed. The type and capacity of existing available processing units and the type of regeneration agent used in the sulfur removal process will determine the most economical route.

A typical solution will be H_2 or an H_2-CO-rich gas for regeneration if neither H_2S nor existing capacity of Claus unit is available.* The preferred route to sulfur here, denoted *partial reduction, Claus,* is to process two-thirds of the regeneration off gas through a reduction reactor for the conversion of SO_2 into H_2S and to feed this stream, after admixture with the unconverted one-third part of the regeneration off gas, to a Claus unit. An actual design may differ in several respects from that shown in Fig. S-27 to suit the requirements of the particular application.

REFERENCES

Arthur G. McKee and Company: Report 993, Systems Study for Control of Emissions, Primary Nonferrous Smelting Industry, vols. 1 and 2 (U.S. Dept. of Commerce Clearinghouse Reports PB 184 884 and PB 184 885), June 1969.

Bienstock, D., L. W. Brunn, E. M. Murphy, and H. E. Benson: Sulfur Dioxide: Its Chemistry and Removal from Industrial Waste Gases, *U.S. Bur. Mines Inf. Circ.* 7836, 1958.

Dautzenberg, F. M., M. E. Naber, and A. J. J. van Ginneken: The Shell Flue Gas Desulfurization Process, *68th Natl. Meet. AIChE, Houston, Texas,* March 1971.

Katell, S., and K. D. Plants: Here's What SO_2 Removal Costs, *Hydrocarbon Process.,* vol. 46, no. 7, p. 161, July 1967.

Slack, A. V.: Air Pollution: The control of SO_2 from Power Stacks, III, Processes for Recovering SO_2, *Chem. Eng.,* vol. 74, no. 25, p. 188, Dec. 4, 1967.

—W. L. Kaiser, *Shell Development Company, Houston, Texas*

Sulfuric Acid Sulfuric acid, H_2SO_4, has been a high-tonnage chemical for over 100 years. Chemically pure H_2SO_4 (formula weight 98.08) is a corrosive, colorless, oily, slightly viscous liquid with a refractive index of 1.429. The specific gravity is 1.839 referred to water at 15.5°C, and the melting point is 10.49°C. H_2SO_4 boils at about 237°C with partial decomposition. Some SO_3 passes off as vapor, and the boiling point rises until the acid has attained a strength of 98.3% H_2SO_4. The acid then distills over, unchanged, at a constant boiling point of 338°C. H_2SO_4 is completely miscible with water at all liquid temperatures but decomposes alcohol and most other organic solvents. Since the heat of solution of H_2SO_4 in water is very high solutions should be made by adding the acid to the water (to avoid spluttering) rather than vice versa.

Worldwide production of H_2SO_4 is estimated at 100 million tons/yr. United States production (1973) is estimated at 32 million tons. Phosphate fertilizers consume by far the majority of H_2SO_4 made (United States), accounting for 42% of the total. Other important uses include petroleum refining, 9%; production of $(NH_4)_2SO_4$, 6%; manufacture of pigments, 5%; manufacture of explosives, 3%; with five areas each accounting for 2%; alcohol manufacture, $Al_2(SO_4)_3$ production, copper-ore leaching, steel pickling, and uranium processing. The manufacture of detergents consumes 1% and miscellaneous chemicals and other uses, 24%. The main forms in which H_2SO_4 is available commercially are listed in Table S-25.

Early Processes: Dating back to the days of alchemy, H_2SO_4 has been made from natural sulfates, jointly distilling iron vitriol with sand, and by burning a mixture of KNO_3 and sulfur in a ladle suspended in a glass vessel partially filled with water. Baron von Liebig's work in fertilizer chemistry (circa 1850) led to the first large-scale commercial processes for making H_2SO_4, the lead-chamber process, so named be-

*The Claus process consists of a vapor-phase catalytic reaction in which a mixture of H_2S and SO_2 in the proper proportion produces elemental sulfur according to the reaction $2H_2S + SO_2 \rightarrow 3S + 2H_2O$. Because the reaction is equilibrium-controlled, one catalytic converter will recover only 70–75% of the sulfur in the gas. A second converter is normally used, and a sulfur recovery of 90–93% is possible. The elemental sulfur produced in these stages is condensed and drawn off to a sulfur pit. In the context of a flue-gas desulfurization process, the Claus unit tail gas containing the unrecovered sulfur may be recycled to the furnace or to the flue-gas desulfurization unit.

TABLE S-25. *Typical Commercial Grades of Sulfuric Acid*

Grade	Sulfuric Acid, or Equivalent	Sp gr Referred to H_2O at 15.5°C
Baumé:		
60°	77.67	1.7059
66°	93.19	1.8354
Percent:		
98	98.00	1.8437
Oleum,* %:		
20	104.50	1.915
30	106.75	1.952
40	109.00	1.983
65	114.63	1.992

*Oleum is a mixture of H_2SO_4 and free SO_3; 20% oleum contains 20% SO_3 and 80% of 100% H_2SO_4; 1 lb of this mixture is equivalent in acid value to 1.045 lb of H_2SO_4.

cause of the two or more very large lead-lined wooden chambers (resembling warehouses) into which a gaseous mixture of SO_2 and oxides of nitrogen was passed. This process now is essentially obsolete. Chamber acid never exceeded a concentration of 78% by weight. Modern H_2SO_4 plants depend upon the use of SO_2 mixed with air and the catalytic oxidation of SO_2 to SO_3.

Raw Materials: The main sources of SO_2 for H_2SO_4 manufacture in the United States are elemental sulfur, 73%; smelter gases from pyrites, 5%, from zinc, 4%, and from copper, 8%; spent and sludge acids, 7%; and H_2S, 3%. The utilization of Cu smelter gases (formerly wasted to the atmosphere) is increasing because of pollution regulations. Thus, some of the acid currently made from elemental sulfur may be replaced.

Types of Plants: H_2SO_4 plants are of two major types, sulfur-burning plants, which start with elemental sulfur, and metallurgical plants, which use SO_2 gas produced by the processing of sulfide ores. See also **Copper production.** The type of plant is selected on the availability and economics of the raw material. In addition, metallurgical plants often are included as part of a smelter complex to minimize atmospheric pollution (apart from economics).

Sulfur-burning plants have the advantages of lower capital and operating costs, greater flexibility with regard to operating rate and grades of acid produced, a credit for by-product steam, and no weak-acid effluent disposal problem.

Metallurgical plants are used for raw materials other than elemental sulfur. In addition to smelter gases, hydrogen sulfide and spent acids may also be used. Often these raw materials are

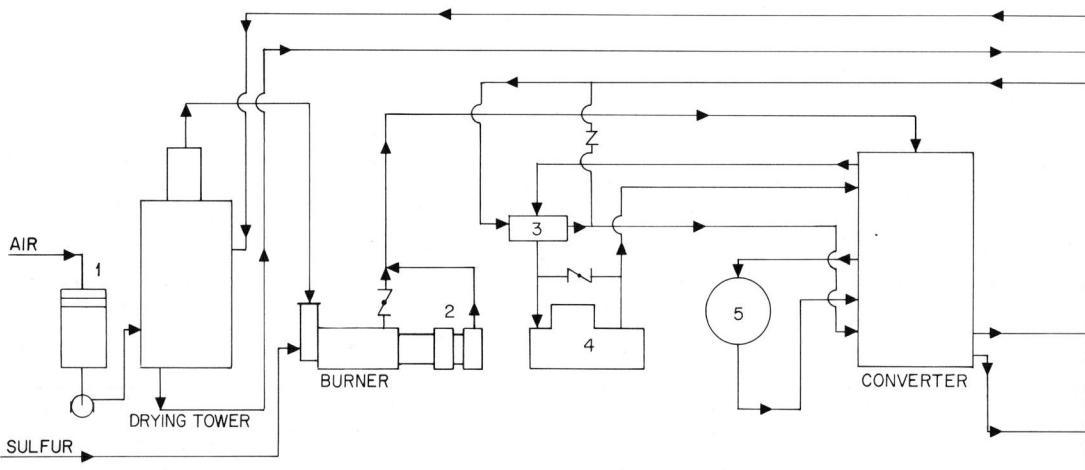

Fig. S-28. Sulfur-burning interpass-absorption sulfuric acid plant. (1) Filter, (2) waste-heat boiler, (3) heat exchanger, (4) waste-heat boiler, (5) steam superheater, (6) interpass heat exchangers, (7) economizer. (*Monsanto Enviro-Chem Systems, Inc.*)

available on a by-product basis and hence at low cost.

Metallurgical plants require a significant gas-purification system to remove solid, liquid, and gaseous impurities and excess water vapor. Typically, the purification system consists of a gas scrubber-cooler circulating a weak-acid solution. This is followed by an electrostatic mist precipitator. Water vapor is absorbed in a drying tower using H_2SO_4 manufactured by the process as the drying medium. A series of gas-to-gas heat exchangers is used to recover heat generated by the oxidation of SO_2 to SO_3 in the catalytic converter. This heat is used to preheat the incoming gas to reaction temperature.

The interpass absorption design (also known as *double catalysis*) may be applied to both the sulfur-burning and metallurgical types of plants. This design includes an additional absorbing tower for removal of SO_3 formed from SO_2 prior to the last stage of the converter. This removal of SO_3 shifts the equilibrium of the reaction in favor of oxidation of essentially all the SO_2 to SO_3.

With an interpass absorption design, it is possible for the yield on SO_2 to be 99.7–99.9% and for the SO_2 in the exit gas to be in the range of 100–350 ppm. The corresponding values for an efficient noninterpass plant are a 98% yield and 2,000 ppm SO_2 in the exit gas. The trend in plant design is toward interpass absorption to comply with existing or expected air-pollution regulations. A modern sulfur-burning interpass-absorption sulfuric acid plant is shown schematically in Fig. S-28. See also **Air pollution.**

Catalytic Converters: The catalyst in the converter vessel is in the form of small pellets and typically is arranged in four layers. Provision is made for removal of the heat of the reaction after each layer or stage. The usual catalyst contains vanadium pentoxide, V_2O_5, with a potassium promoter on a silica carrier. The catalyst may be used for a number of years with only a very moderate decrease in activity.

The rate of the reaction, $SO_2 + \frac{1}{2}O_2 \rightarrow SO_3$, in the presence of vanadium catalyst is negligible below 400°C. In practice, the temperature of the process gas stream entering the catalyst layers is maintained in the range 420–450°C. With typical process gas streams containing no more than 10% SO_2, the reaction kinetics is such that the maximum temperature in the catalyst rarely exceeds 600°C.

Handling: Sulfuric acid circulated in the manufacturing process is handled in cast-iron pipe, acid-resistant brick-lined steel vessels, Alloy 20 valves and pumps of acid-resistant cast-iron and

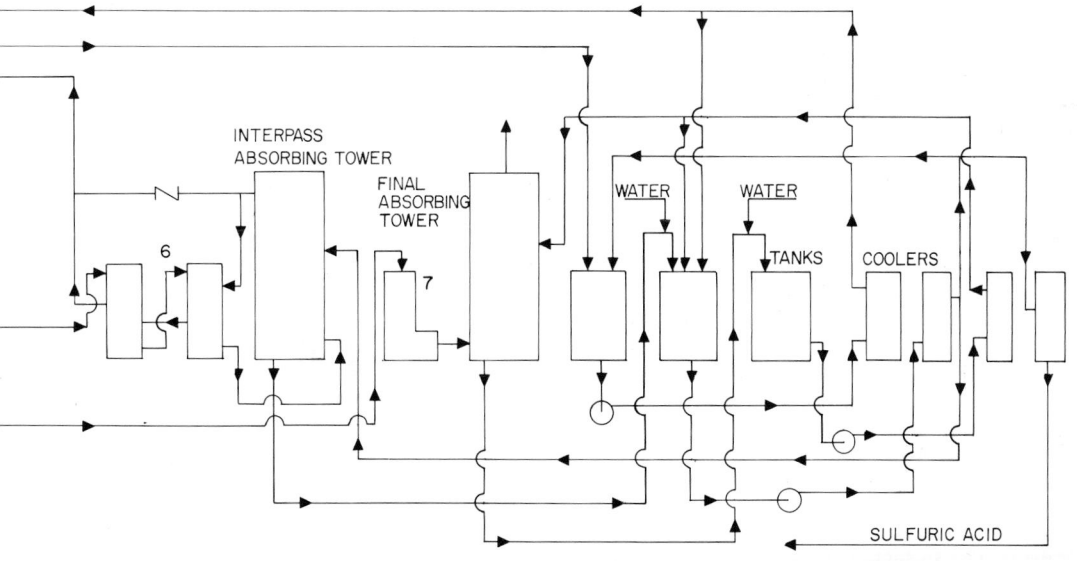

alloy parts. Oleum is handled in steel pipe and unlined steel vessels. Sulfuric acid at strengths greater than 60°Be (77.67%) and oleum are commonly stored at ambient temperatures in steel storage tanks. Shipment is made in steel tank trucks, tank cars (up to 90 tons), and barges (up to 2,000 tons). Loading and unloading lines are commonly of Schedule 80 steel, sized on the basis of low velocity (less than 3 ft/s). Pumps and valves in this service usually are of Alloy 20 construction or equivalent.

REFERENCES

Duecker, W. W., and J. R. West: "The Manufacture of Sulfuric Acid," Van Nostrand Reinhold, New York, 1959.

Lange, N. A.: "Handbook of Chemistry," 10th ed., p. 1367, McGraw-Hill, New York, 1967.

Parkes, G. D. (ed.): "Mellor's Modern Inorganic Chemistry," p. 505, Wiley, New York, 1967.

Stanford Research Institute: "Chemical Economics Handbook," Menlo Park, California, 1966.

—Paul J. Stuber, *Monsanto Enviro-Chem Systems, Inc., Chicago, Ill.*

Supercalendering (See **Papermaking and finishing.**)

Superconductors (Consult the **Subject Index.**)

Superphosphate (See **Fertilizers.**)

Surface-active Materials [See **Anionic surfactants (sulfur-bearing); Catalysts; Defoaming agents; Detergents; Soaps;** and **Surfactants.**]

Surfactants The term surfactant, derived from *sur*face-*ac*tive *agent*, technically defines a group of compounds which, when dissolved in water or another solvent, orient themselves at the liquid interface with other surfaces and effect a modification of the liquid properties at that interface. This modification generally is evidenced by increased solubilization, foaming, frothing, emulsification, wetting, dispersing, or penetration power of the solution.[1]

Surfactants usually are compounded with other ingredients, e.g., water softeners, corrosion inhibitors, and soil suspension agents, to formulate a detergent, i.e., cleaning agent. The surfactant promotes emulsification and dissolution of oily and greasy substrates from the surfaces being cleaned. The surfactant wets particulate matter so that upon application of mechanical energy, as in hand-scrubbing or washing-machine action, the soil is removed easily and dispersed in the cleaning solution. See also **Detergents.**

The largest-volume use of surfactants is in household cleaning products. In 1970, it is estimated that more than 750 million lb of surfactants were sold in granular household products. Including surfactants used in liquid detergents, a total of more than 1 billion lb/yr is used in the United States. This discussion centers on major household applications.

Structure: Molecularly, all surfactants have a common structure, being composed of a long hydrocarbon chain which is oil-soluble but water-insoluble, i.e., a *hydrophobic portion,* and a terminal group which is water-soluble, i.e., a *hydrophilic portion.* This end group confers water solubility on the entire molecule. The molecule may be depicted structurally as shown in Fig. S-29.

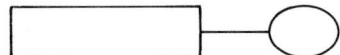

Fig. S-29. Diagrammatic representation of hydrophobic group (oil-solubilizing) shown at left and hydrophilic group (water-solubilizing) shown at right.

Surfactants achieve their unique effects in solutions because of the orientation of these molecules. At the solution interface with a second phase, the hydrophobic group is thrust outward and the hydrophilic group downward into the solution (see Fig. S-30). Close packing of molecules can be achieved with this geometry, leading to a higher concentration of surfactant at the interface than in the bulk solution. A marked reduction in surface tension results, which promotes improved penetration and oil emulsification. In common household practice today, wash

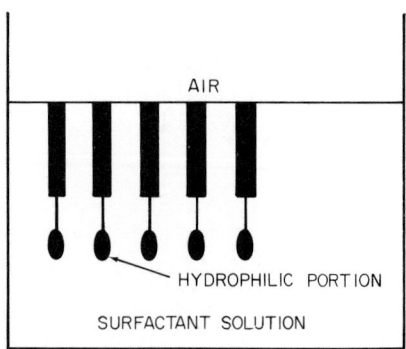

Fig. S-30. Orientation of surfactant molecules at air-solution interface.

solutions have a surface tension of 25–35 dynes/cm compared with pure water at 72.8 dynes/cm (20°C).

In the bulk of the liquid, groups of surfactant molecules aggregate to form micelles. Although the precise structure of micelles continues to be the subject of detailed physicochemical investigation, they may be considered to be a cluster of molecules in a spherical form as depicted in Fig. S-31. Energy considerations indicate that the spherical form is most likely. The hydrophilic ends of the molecule are oriented outward into the solution; the hydrophobic inward. These micelles are capable of holding oils and particulate matter which have been removed from surfaces in solution or suspension.

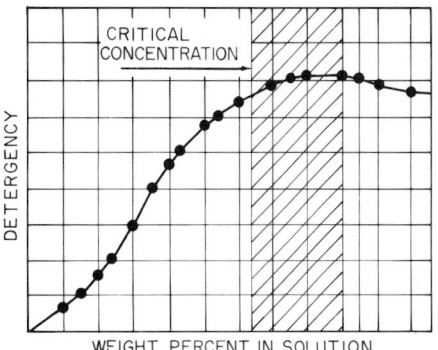

Fig. S-32. Variation of detergency with weight percent surfactant in solution.

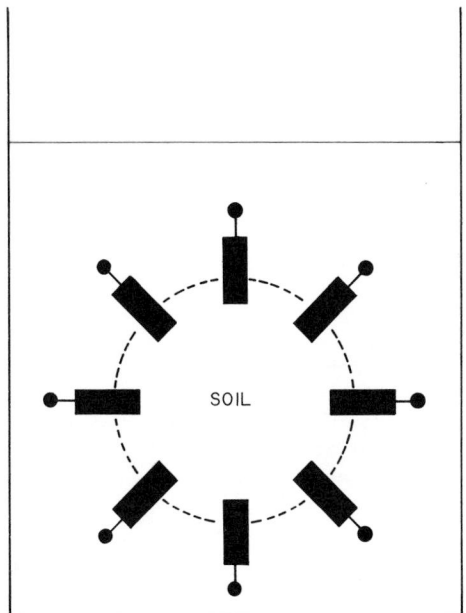

Fig. S-31. Diagrammatic representation of micelles of surfactant dissolved in a wash solution.

Surfactants exert their effect upon the solution in which they are dissolved at extremely low concentrations. Most of the fabric washing done in the modern home is with an aqueous solution containing less than 0.03 wt% (0.0001 mole%) surfactant. The optimum cleaning ability with a given surfactant is reached at a solution concentration called the *critical micelle concentration* (CMC) (see Fig. S-32).[2] This concentration varies with temperature, the presence or absence of other organic compounds and inorganic salts, and pH. Physically, the CMC closely coincides with the minimum surface tension of the system.

The major economic utilization of surfactants is for detergents. Literally thousands of compositions have been formulated and sold as powders, gels, solutions, and pastes. Each form and composition is designed for uniqueness in some application. For optimum detergency, the hydrocarbon hydrophobic tail of the molecule is in the C_{12}–C_{18} range. Compounds above C_{18} in chain length generally suffer from low aqueous solubility; those below C_{12} do not form effective micelles.

Classification: Chemically, surfactants may be classified as anionic, cationic, and nonionic. The nonionics include polyoxyethylene compounds, amides, and esters, semipolar, and zwitterionic.* Anionics and cationics ionize in aqueous solution and thus show migration properties under the influence of an electric field. Nonionics do not ionize, although a wide range in polarity does exist among the compounds in this grouping. Examples are given in Table S-26.

Anionic Surfactants: These are the least expensive and constitute the largest group of surfactants used commercially, particularly the dodecyl and tridecyl benzene sulfonates, the alkyl sulfates of C_{12}–C_{14} chain length primarily derived from coconut-oil sources, the alkyl sulfates of the C_{16}–C_{18} chain length from tallow sources, and the alkyl ether sulfates. This latter group achieves improved water solubility and mildness to skin through condensation of 3-12 moles of ethylene oxide on the fatty alcohol base prior to sulfation.

*From the German *Zwitter,* hybrid, i.e., exhibiting the properties of two different ions.

TABLE S-26. *Examples of Major Types of Surfactants*

Surfactant Type	Hydrophobic Group	Hydrophilic Group	Chemical Name or Description
Anionic	$C_{12}H_{25}O^-$	SO_3Na^+	Sodium dodecyl sulfate
Cationic	$C_{12}H_{25}-\overset{\overset{\displaystyle CH_3}{\mid}}{\underset{\underset{\displaystyle CH_3}{\mid}}{N}}-CH_3^+$	Cl^-	Trimethyl dodecyl ammonium chloride
Nonionic			
Polyoxyethylene type	$C_{12}H_{25}O$	$(CH_2-CH_2O)_{10}H$	Condensation product of 1 mole of dodecyl alcohol with 10 moles ethylene oxide
Semipolar	$C_{12}H_{25}-\overset{\overset{\displaystyle CH_3}{\mid}}{\underset{\underset{\displaystyle CH_3}{\mid}}{N^+}}\rightarrow O^-$		Dimethyl dodecyl amine oxide
Zwitterionic	$C_{12}H_{25}-\overset{\overset{\displaystyle CH_3}{\mid}}{\underset{\underset{\displaystyle CH_3}{\mid}}{N^+}}-(CH_2)_2COO^-$		Dimethyl dodecyl ammonium proprionate

The alkyl ether sulfates thus find major application in household liquid detergents intended for hand dishwashing.

Raw materials other than fats and oils are mainly of petroleum origin. Alkyl benzene is produced by catalytic alkylation of benzene with linear C_{12} and C_{13} alkyl chlorides using Lewis acids as catalysts. The linear C_{12} and C_{13} hydrocarbons are separated from feedstocks by properly sized molecular sieves and are chlorinated. Before 1965 the detergent industry used hydrocarbons of ~C_{12} chain length derived from a propylene tetramer, but the high degree of branching led to an unacceptably low rate of biodegradability in rivers and streams. The linear, more rapidly degradable type used today, is commonly referred to as LAS (*l*inear *a*lkylate *s*ulfonate).

Fatty alcohols and olefins constitute two other prime sources of surfactant raw materials. Both are produced commercially via Ziegler chemistry utilizing ethylene buildup on aluminum trialkyls. See also **Alcohol process, ALFOL.**

The hydrophilic portion of anionic surfactants is a sulfonate, sulfate, phosphate, or carboxylate; sulfates and sulfonates predominate. They are produced by reaction of the base hydrocarbon with H_2SO_4, 10–30% oleum, SO_3 in a stabilized liquid form, or chlorosulfonic acid, $HO \cdot SO_2 \cdot Cl$. The reactions are vigorously exothermic and must

be closely temperature-controlled to avoid charring.

Sulfonation:

Alkyl benzene   Alkyl benzene sulfonic acid

Neutralization:

Alkyl benzene sulfonic acid

Sodium alkyl benzene sulfonate (NaLAS)

See also **Anionic surfactants (sulfur-bearing).**[3]

Cationic Surfactants: These are primarily ammonia derivatives. The alkyl hydrophobe dissolved in water carries a positive charge and is strongly attracted to negative surfaces such as cellulose and other fibers, human hair, and metal. The predominant use of cationics is based upon this phenomenon. They are used as antistatic agents in hair rinses and as fabric softeners when deposited on fabrics such as cotton. In addition, unrelated to their surfactant activity, cationics are used as sanitizing agents because of a unique biocidal activity of some species.

Cationics cannot be formulated with anionics since a neutralization reaction destroys the activity of both. They can be formulated with nonionics.

Nonionic Surfactants: Since this type does not ionize in water, solubility is achieved either through the presence of oxygen in the molecule (as in ethylene oxide or propylene oxide hydrophylic linkages, terminal hydroxyl groups, or oxygen linked directly to nitrogen as in amine oxides) or via sulfonate, phosphate, or carboxylate terminal groups in the case of the zwitterionics.

Nonionic surfactants are relatively less affected by Ca^{++} and Mg^{++} salts than anionics since no precipitates are formed due to lack of ionization. They are more oil-soluble than the polar anionics or cationics and have a greater ability to dissolve oil and grease from soiled surfaces when used in a concentrated form. In dilute aqueous solutions, however, nonionics give only the same cleaning as anionics.

The base material for the majority of nonionics is an alkyl phenol or an aliphatic alcohol for ethoxylates or an alkyl amine for semipolar or zwitterionic compounds.

Ethoxylation involves the direct addition of highly reactive ethylene oxide to the alkyl phenol or aliphatic alcohol using either acidic or basic catalysis:

$$R-\!\!\!\bigcirc\!\!\!-OH + (CH_2CH_2O)_n \xrightarrow[OH^-]{\substack{H^+ \\ or}}$$

Alkyl phenol

$$R-\!\!\!\bigcirc\!\!\!-O\,(CH_2CH_2O)_nH$$

Ethoxylated alkyl phenol

$$ROH + (CH_2CH_2O)_n \xrightarrow[OH^-]{\substack{H^+ \\ or}}$$

Alcohol

$$RO\!-\!(CH_2CH_2O)_nH$$

Ethoxylated alcohol

In industrial practice the vigorously exothermic reaction is normally carried out batchwise using heat-exchange surfaces and the stepwise addition

of ethylene oxide to control the reaction temperature. Since ethylene oxide addition is a buildup reaction, a distribution of chain lengths results, the average being determined by the starting weight ratio of ethylene oxide to hydrocarbon.

Nonionic surfactants generally produce less suds than anionics.

The semipolar amine oxide type of nonionic surfactant has achieved considerable industrial importance in recent years. In alkaline detergent solutions it generates copious suds, is mild, is efficient at low concentrations, and does not interact with anionics. A high water solubility and hydroscopic nature limits the use of amine oxides to liquid detergents.

$$\underset{\substack{\text{Alkyl dimethyl} \\ \text{amine}}}{R-\overset{\textstyle CH_3}{\underset{\textstyle CH_3}{N}}} + H_2O_2 \rightarrow \underset{\substack{\text{Alkyl dimethyl} \\ \text{amine oxide}}}{R-\overset{\textstyle CH_3}{\underset{\textstyle CH_3}{N}}} \rightarrow O + H_2O$$

Phosphine oxides are analogous compounds containing phosphorus rather than nitrogen.

REFERENCES
1. Schwartz, A. M., and J. W. Perry: "Surface Active Agents," 3d ed., Interscience-Wiley, New York, 1949.
2. Shinoda, Ozo, et al.: "Colloidal Surfactants," Academic, New York, 1963.
3. Jungermann, Eric: "Cationic Surfactants," Dekker, New York, 1970.

—Richard D. Walker, *The Procter & Gamble Company, Cincinnati, Ohio*

Suspension Polymerization (See **Polymerization;** and **Polyvinyl chloride.**)

Sylvinite (See **Potassium.**)

Sylvite (See **Fertilizers;** and **Potassium.**)

Syndiotactic (See **Polymerization.**)

Syneresis (See **Colloid systems.**)

Synthesis Synthesis, in the chemical sense, is the process of building chemical compounds through a planned series of steps (reactions and separations).

Synthesis usually is implemented (1) when the desired compound is not present in natural materials from which it can be isolated; (2) when the compound cannot be easily obtained from reacting readily available materials in a few simple steps; or (3) although it is available within a natural complex, the economics of separation and purification are prohibitive or, particularly in biochemistry, too little of the natural material is obtainable.

Even more important, synthesis plays a key role in developing new, untried chemical structures which on paper appear to have properties that may be of great value, e.g., a new synthetic material, a new drug, or a new fuel. Chemicals by design from prior knowledge of related materials generally are created via the route of synthesis. Further, synthesis is fundamental to broadening the base of chemical knowledge. Sometimes unexpected results occur, i.e., compounds with unusual and often desirable practical properties.

Because of the hundreds of thousands of organic substances (already established but many still remaining to be "built"), organic synthesis predominates. Most of the synthetics (elastomers, fibers, and other polymers, coatings, films, adhesives, and numerous other products) that have appeared during the last few decades (many of which are described in this book) resulted from research involving organic synthesis.

Some of the early work in organic synthesis dealt with the creation of certain fatty acids and ketones. A few examples are given to provide an insight into the workings of synthesis.

In the following examples, only the main starting ingredients and products are shown. No attempt is made to indicate by-products or the conditions of the reactions involved:

A. Target compound: ethylpropylacetic acid, $(C_2H_5)(C_3H_7)CH \cdot COOH$

1. Acetic anhydride $\rightarrow$ ethyl acetate
   (+ alcohol)

2. Ethyl acetate $\rightarrow$ ethyl acetoacetate
   (sodium + dil. acids)

3. Ethyl acetoacetate $\rightarrow$ sodium derivative
   (+ sodium ethoxide)   of ethyl acetoacetate

4. Sodium derivative of ethyl acetoacetate $\rightarrow$
   (+ propyl iodide)

   ethyl ethylpropyl acetoacetate

5. Ethyl ethylpropyl acetoacetate $\rightarrow$
   (+ strong alc. potash)

   ethylpropylacetic acid

B. Target compound butyl acetone, $CH_3 \cdot CO \cdot CH_2 \cdot C_2H_4$

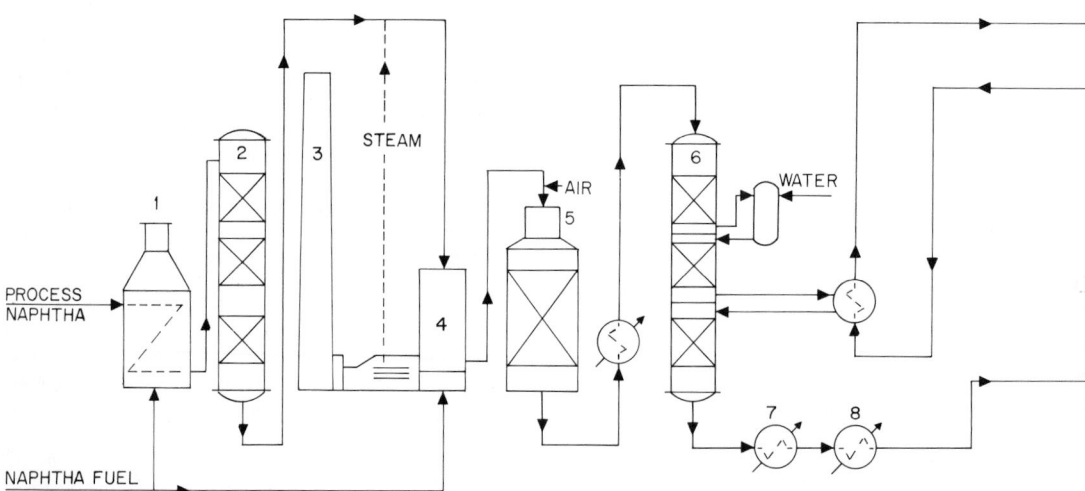

Fig. S-33. Process for production of synthesis gas for ammonia manufacture. (1) Naphtha vaporizer, (2) desulfurizer, (3) stack, (4) primary reformer, (5) secondary reformer, (6) carbon monoxide conversion, (7) feedwater heater, (8) reboiler, (9) carbon dioxide absorber, (10) air cooler, (11) reactivator, (12) methanator, (13) feedwater heater. (*Imperial Chemical Industries Limited; Davy Powergas Ltd.*)

1–3, as above

4.   Sodium derivative of ethyl acetoacetate →
  (+ butyl iodide)

        ethylbutylpropyl acetoacetate

5. Ethylbutylpropyl acetoacetate →
  (+ dil. alc. potash)

        butyl acetone

  C. Target compound *n*-valeric acid, $CH_3 \cdot CH_2 \cdot$
  CH_2CH_2COOH

1. Potassium chloroacetate →
  (+ potassium cyanide)

        potassium cyanoacetate

2.   Potassium cyanoacetate   →
  (+ alcohol + hydrogen chloride)

        ethyl malonate

3.   Ethyl malonate   →
  (+ sodium ethoxide)

        sodium derivative of ethyl malonate

4. Sodium derivative of ethyl malonate →
  (+ propyl iodide)

        ethylpropyl malonate

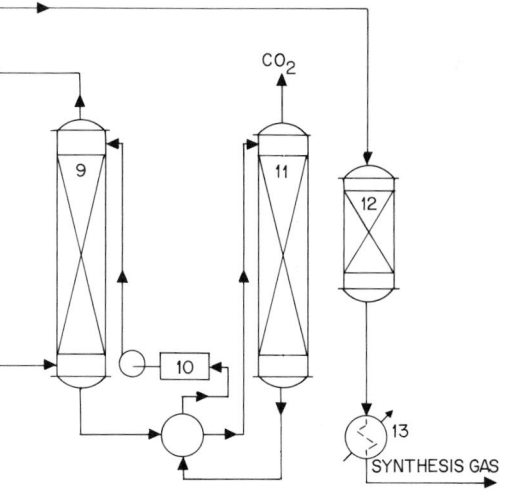

5. Ethylpropyl malonate → propylmalonic acid
  (+ alc. potash)

6.   Propylmalonic acid   → *n*-valeric acid
  (+ decomposition by heat)

Numerous examples of synthesis are give in this volume. Urea, the first organic compound to be synthesized is now produced in very high tonnage. See **Urea;** also see **Amino acids; Antibiotics;** and **Peptides and proteins** for examples of complex synthesis.

Synthesis Gas   Synthesis gas is the name given to mixtures of gases in suitable proportions for the production of synthesis products without adding further reactants. Typical examples are hydrogen and nitrogen for ammonia synthesis; carbon monoxide and hydrogen for methanol synthesis; carbon monoxide, hydrogen, and olefins for oxo alcohols. While the example used in this description is ammonia, similar principles apply to the production of the other synthesis gases.

Ammonia Synthesis Gas: The hydrogen for ammonia synthesis gas can be obtained from several sources in commercial quantities, including coke-oven–water gas; from water by electrolysis; by steam reforming of hydrocarbons; and by partial oxidation of hydrocarbon feedstocks.

Nitrogen can be added as pure nitrogen from an air-separation plant or by the introduction of air in some part of the process. As synthesis takes place at a very high pressure, it is desirable to generate gas at high pressure and thus reduce the compression cost. Hence, conventional coke ovens, which work at low pressure and have high capital costs, are no longer economical for this application. Likewise hydrogen production by electrolysis is very costly unless cheap electricity is available, together with an air-separation plant and a market for oxygen. Unless exceptional conditions are encountered, hydrocarbon feedstocks are the optimum choice. The process route based on these feedstocks involves the production of crude synthesis gas and subsequent purification of the gas to remove impurities which could be harmful to the synthesis catalyst.

The two most common routes available for the generation of crude synthesis gas are steam reforming or partial oxidation of a range of feedstocks from natural gas to crude oil. In both cases, the producer is faced with the problem of removing CO, CO_2, and sulfur.

In steam reforming plants, it is essential that

the sulfur content be reduced to less than 0.5 ppm before the reforming reaction because sulfur poisons the expensive catalyst used. *Sulfur reduction and removal* is achieved by a number of processes including:

1. *Activated carbon,* suitable for treating gaseous feedstocks containing organic sulfur and limited amounts of H_2S.

2. *Zinc oxide and iron oxide,* suitable for removal of H_2S. If certain organic sulfur compounds are present, it is necessary to employ a hydrodesulfurization catalyst in series to convert the compounds to H_2S, which then can be removed readily in beds of zinc oxide or iron oxide.

3. *Acid wash system,* suitable for liquid feedstocks with sulfur contents of up to 1,500 ppm with outlet sulfur content of 50 ppm minimum.

4. *Caustic wash,* suitable for use with gaseous hydrocarbons when it is required to remove small quantities of sulfur as H_2S to the low levels required.

5. *Amine washer,* most suitable for removal of H_2S to less than 1 ppm from gaseous feedstocks but also adaptable for use with liquid feedstocks.

6. *Hydrorefining,* suitable for treating liquid feedstocks and reducing sulfur contents up to 1,500 ppm. Several proprietary processes of this type are available. See also **Hydrotreating.**

The *steam-hydrocarbon reforming process* involves the reaction between a desulfurized hydrocarbon feedstock (natural gas or light distillates up to 220°C boiling point) and steam at elevated temperature and pressure over a catalyst. The reaction conditions range up to 850°C and 30 atm. The catalyst used, normally based on nickel, is packed in externally heated tubes suspended in a furnace box. The reaction products are H_2, CO, CO_2, and methane, together with undecomposed steam. These gases then are passed to a second reforming stage, together with air to add nitrogen. The secondary reactor is normally a refractory-lined vessel containing a slightly less active catalyst than that in the primary reformer. The secondary reaction is autothermic, and the outlet product consists of a mixture of gases, including hydrogen and nitrogen in such proportions that subsequent purification stages will give rise to the correct empirical ratio for NH_3 synthesis, that is, $3:1$ (see Fig. S-33, page 1076).

An alternative *partial-oxidation route* is noncatalytic and can handle a range of feedstocks from natural gas through to crude oil. See **Gasification.**

The first stage in the *purification* procedure normally involves reacting steam with CO over an iron oxide catalyst to produce more H_2 and CO_2. The CO conversion stage can incorporate two stages of high-temperature conversion or a high- and low-temperature stage, the latter taking place over a Cu-Zn-based catalyst. The choice is determined by the economics of a specific installation. After the CO conversion stage, the gases pass to a CO_2 removal stage, a continuous system in which the gas passes up a tower and meets a countercurrent stream of stripping solution, often activated caustic. The spent solution leaving the absorption stage passes to a regenerator, where CO_2 is removed by steam stripping and the solution is rendered usable for further absorption. The synthesis gas, stripped of CO_2, passes to a final purification stage to remove residual carbon oxides and water from the gas, after which the gas is ready to enter the synthesis loops.

For many years, CO was removed by a continuous copper-liquor wash. This process still is used, but methanation offers an alternative route. In methanation, the carbon oxides are reacted with H_2 over a catalyst to produce methane. The methane, which is not harmful to synthesis catalyst, can be purged from the synthesis loop and used as fuel.

See also **Ammonia; Desulfurization; Gasification;** and **Reforming.**

—G. M. Hampson, *formerly with Davy Powergas Ltd., London*

Synthetic Fibers (See **Fibers.**)

Synthetic Resins (See **Resins.**)

Synthetic Rubber (See **Elastomers.**)

T

Tableting   (See **Agglomeration.**)

Tabling   (See **Beneficiation, ore;** and **Separation operations.**)

Tall Oil   (See **Mineral oil; Paint;** and **Vegetable oils.**)

Tallow   [See **Alcohols; Alcohols, fatty (via hydrogenation); Detergents; Hydrogenation; Soaps; Surfactants;** and **Vegetable oils.**]

Tantalum   (*Tantalus,* king punished by being surrounded by food and drink that receded when he reached for it.) **Ta** = 180.95 (at. wt.); 73 (at. no.). Tantalum is a ductile metal that is easily fabricated. The high melting point of Ta (2996°C) classifies it as one of the refractory metals. It is a heavy, steel-blue metal which becomes platinum-white when polished. The crystal structure is body-centered cubic with no allotropic transformations. Tantalum 181 is a natural isotope. Isotope 182 has been used as a source of gamma rays. Ta is a member of group VB of the periodic table, along with vanadium and columbium (niobium), and is in period 6 of the table. See also **Chemical elements.**

Ta was discovered (1802) by Ekeberg of Sweden as a new element in yttrium minerals from Sweden. In pure form, Ta was first obtained by

TABLE T-1. *Some Physical Properties of Tantalum*

Crystallographic properties:	
Crystal structure	bcc A2
Lattice constant, Å	3.2959
Coordination number	8
Goldschmidt radius, CN-8, Å	1.426
Atomic and nuclear properties:	
Density at 20°C, g/cm³	16.62
	16.64
Atomic volume, cm³/g atom	10.9
Cross section, thermal neutrons, . . .	21.3
barns/atom	(absorption)
Thermal properties:	
Melting point, °C	2996
Boiling point, °C	5427
Specific heat at 0°C, cal/g	0.03322
Heat capacity,	
At 0°C, cal/mole:	6.024
25°C	6.08
Entropy at 25°C, cal/mole	9.90
Latent heat of fusion, cal/mole	7500
Latent heat of vaporization,	
kcal/g atom	180
Heat of combustion, cal/g	1380

1079

TABLE T-2. *Some Tantalum Compounds and Their Formation*

Suboxides:

The suboxides are formed at about 300–500°C in O_2 at 1 atm pressure; approximate composition, Ta_6O, Ta_4O, and Ta_2O

Oxides:

Two basic crystalline types of Ta_2O_5 are formed at various temperatures during calcination: at 700–1300°C, the beta form is prevalent; in excess of 1360°C, conversion to the alpha form occurs.

Ta_2O_5 (white, rhombic), sp gr 8.735 (61.2°C); insoluble in H_2O; soluble in fused $KHSO_4$; insoluble in alcohol

Ta_2O_4 (dark gray), insoluble in H_2O and alcohol

Ta_2O_2 (?) (brown powder), sp gr 7.35; insoluble in H_2O and HNO_3 + HF

Hydrous oxides:

Although hydrous oxides are known to exist, they appear to have no definite composition and can vary from $Ta(OH)_5$ and $TaO(OH)_3$ to $Ta_6O_{19}(OH)_n$ (8 + n)

Tantalates:

When anhydrous Ta_2O_5 is fused with alkali-metal hydroxides or carbonates, tantalates are formed; Rose described a number of tantalates in which the amounts of potash are related as 1:4:8:12 and in the case of soda as 1:4:6:18:24:32; most tantalates are insoluble in water, the exceptions being those alkali-metal tantalates having a base-acid ratio greater than 1

Water-soluble:

According to Marignac, only one clearly defined species is water-soluble, $4K_2O$, $3Ta_2O_5 \cdot xH_2O$; however, Jander and colleagues give evidence of the solubility of $7K_2O$, $5Ta_2O_5 \cdot 24H_2O$; a double tantalate also soluble is $Na_3K_5Ta_6O_{197}$

Anhydrous:

Among these, four species have been identified: $3K_2O \cdot Ta_2O_5$; $K_2O \cdot Ta_2O_5$; $K_2O \cdot 2Ta_2O_5$; $K_2O \cdot 5Ta_2O_5$

There also exist mixed oxides with metals other than alkali metals, including the series ATa_9O, where A = P, As, or V

Nitrates:

There is no simple nitrate of Ta, but the trinitratotantalum(V) oxide is known and is obtained by the reaction $TaCl_5 + 4N_2O_5 \rightarrow TaO(NO_3)_3 + 5NO_2Cl$

Phosphates:

The metaphosphate, $TaOPO_4$, is known, as well as the orthophosphate, $Ta_3(PO_4)_5$

Sulfato complexes:

The empirical formula for tantalum sulfate is $Ta_2(SO_4)_5$ ($Ta_2O_5 \cdot 5SO_3$) or $(TaO_2)_2S_5O_{16}$. Double sulfates of tantalum with $(NH_4)_2SO_4$ also have been described, e.g., $(NH_4)_2SO_4$, $Ta_2(SO_4)_5$, and $(NH_4)_3Ta(SO_4)_4$

Tantalum oxide fluorides:

Two oxide fluorides of Ta are known, $TaOF_3$ and TaO_2F

Tetrahalides:

TaF_4 to date has not been characterized; $TaCl_4$ (black crystals), $TaBr_4$ (large brown-black crystals), and TaI_4 (has two forms according to the temperature when prepared)

Trihalides:

TaF_3 (cubic structure), $TaCl_3$ (black crystals), $TaBr_3$ (graphitelike feeling on rubbing), and TaI_3 (characterization still uncertain)

Other halides:

$TaBr_5$ (yellow crystals), sp gr 4.67; decomposes in cold H_2O

$TaCl_5$ (yellow prisms), sp gr 3.68 (27°C); decomposes in cold H_2O

TaF_5 (tetragonal), sp gr 4.74; soluble in cold H_2O

Carbides:

TaC and Ta_2C (exists, but has no industrial importance)

Silicides:

Four silicides are well characterized: $TaSi_2$, Ta_5Si_3, Ta_2Si, and $Ta_{4.5}Si$

Borides:

Four known borides, TaB, TaB_2, Ta_3B_4, and Ta_3B_2

Nitrides:

Ta forms TaN, Ta_2N, and Ta_3N_5 (red, amorphous), insoluble in H_2O, soluble in HNO_3 + HF

Phosphides:

Two known P compounds: TaP and TaP_2

Sulfide:

Ta_2S_4 (dark gray), insoluble in H_2O, slightly soluble in HF and HNO_3, insoluble in HCl

Arsenides:

Two have been studied: TaAs and $TaAs_2$

Antimonides:

Three have been studied: Ta_3Sb (cubic), Ta_5Sb_4 (tetragonal), and $TaSb_2$ (monoclinic)

Pertantalic acid and salts:

Ta compounds, when treated with H_2O_2 in acid solutions, are converted to pertantalic acid, and the corresponding sodium salts have been isolated, sodium metapertantalate, $NaTaO_4 \cdot NaTaO_5 \cdot 13H_2O$ and sodium orthopertantalate, $Na_3TaO_8 \cdot H_2O$

Organic compounds:

$TaCl_5$ forms substitution derivatives with alcohols and phenols in which the metal atom is bonded through oxygen bridge atoms: $TaCl_5 + 5C_6H_5OH \rightarrow Ta(OC_6H_5)_5 + 5HCl$; $TaCl_5$ also reacts with acetylacetone and related diketones to form compounds such as $TaCl_2(OCH_3)_2 \cdot C_5H_7O_2$ and $Ta_2O(O_3PCH_3)_4$, tantalum methylphosphinate. Compounds in which Ta is bound directly to the C atom are extremely unstable.

Berzelius (1820) by heating potassium tantalofluoride with potassium. Important physical properties of Ta are summarized in Table T-1.

Chemical Properties: When Ta is exposed to the atmosphere, it forms a tough and impermeable Ta_2O_5 film which protects it from corrosion and is therefore highly resistant to chemical attack below about 150°C. It resists corrosion by all acids except fuming H_2SO_4, HF, and acid solutions containing fluoride ions. Ta reacts slowly with alkali solutions.

Ta absorbs H_2 readily, starting at room temperature, via cathodic electrolysis. Exposure to H_2 atmospheres above about 350°C also results in the formation of Ta hydrides (interstitial combinations), which are brittle. The H_2 can be removed by heating to about 800°C or higher in a vacuum. Ta is not generally attacked by the low-melting liquid metals in the absence of O_2 or N_2. Liquid Bi has little effect below 1000°C. At 1200°C, Ta is slightly attacked by Ca. Ta resists molten Ga at 450°C but not above 600°C. Ta is highly resistant to liquid Pb and Li up to 1000°C. Ag attacks Ta slightly at 1200°C. Molten Zn, however, attacks Ta readily above 450°C.

At 500–1000°C Ta oxidizes linearly at O_2 pressures between 10 mm Hg and 600 psi. The oxidation rate increases significantly from 600 to 800°C with O_2 pressures above 0.5 atm, and Ta oxidizes catastrophically at 1300°C. See Table T-2.

Production: Ta is commercially available as powder, ingot, bar, rod, wire, sheet, strip, foil, tubing, and expanded mesh. Most Ta is obtained from $(Fe, Mn)(Ta, Cb)_2O_6$ minerals, which are powdered and fused with caustic soda. Materials such as Mn and Fe are eliminated by the use of HCl. Potassium fluotantalate, K_2TaF_7, is crystallized out of the residue, leaving columbium (niobium) double fluoride, K_2CbOF_5. This Marignac separation process has been virtually superseded by the newer liquid-liquid extraction process, which starts out with digestion of the ore with aqueous HF.

Uses: The characteristics of high reliability, compact size, and environmental versatility make Ta attractive in the electronics industry, where Ta foil or porous pellets are used in electrolytic capacitors, emitters, and getters. Ease of fabrication, excellent corrosion resistance, and high heat-transfer coefficient encourage the use of tantalum in heat-transfer applications. Because its corrosion resistance is much like that of glass, Ta is used in conjunction with glass and glass-lined steel and for repairing glass-lined equipment.

The leading anticorrosion applications of Ta are in the manufacture of HCl and H_2O_2, in Cr plating baths, in Br_2 heaters and stills, in condensing C_2H_5Br, and in the preparation of fine chemicals.

Ta is used as susceptors and resistance heaters in high-temperature furnaces and for some nuclear-reactor parts.

Since Ta is inert to body fluids and is tolerated by the human body, it plays an important part in surgical applications. Tantalum can be placed in the skull or other parts of the body without rejection. Ta strips and screws are used to hold broken pieces of bone together, and Ta wire mesh is used for reinforcements, surgical staples, and braid for sutures.

Alloys: As an alloy addition, Ta is used in Ni- and Ni-Co-base superalloys for jet-engine and gas-turbine parts.

Ta-base alloys for aerospace structures and space power systems are noted for their high-temperature strength and stability. T111 (Ta-8W-2Hf), T222(Ta-10W-2.5Hf), and T811C (Ta-8W-1Hf-1Re-0.02C) are alloys with applications at temperatures exceeding 1600°C.

Small alloying additions of Zr to Ta increase tensile strength at room temperature and up to 1200°C. Mo, Hf, Re, V, and W also strengthen Ta at the 5% or more addition level. Ternary alloys of Ta with 30% Cb (Nb) and 5% V or Zr, or 10% Hf or W have room-temperature tensile strength of approximately 3 times that of pure Ta. An alloy of Ta with W has been used for springs with applications at high temperatures and high vacuum.

REFERENCES

Simons, Eric N.: "Uncommon Metals," Hart, New York, 1967.
Sisco, Frank T., and Edward Epremian: "Columbium and Tantalum," Wiley, New York, 1963.

—Technical Staff, *Kawecki-Berylco, Inc., Reading, Pa.*

Tape, Recording (See **Iron oxides, synthetic.**)

Tar Chemicals (See **Coal tar and derivatives;** and **Pyridine and derivatives.**)

Tartar Emetic (See **Antimony.**)

Tartaric Acid (See **Carboxylic acids;** and **Isomerism.**)

Tartronic Acid (See **Carboxylic acids.**)

Technetium   (Gk. *tech etos*, artificial.) **Tc** = 99 (at. wt.); 43 (at. no.). This element was discovered in 1937 by Perrier and Segré (France) as the result of experiments in which molybdenum was bombarded with deuterons. Element 43 was vacant in the periodic table for many years. An earlier claim to have found this element (calling it *masurium*) was false.

There is no isotope of technetium of sufficient stability to exist in nature. The longer-lived isotopes have half-lives in the order of 10^5–10^6 years. Technetium has, however, been detected in certain stars, and theories of stellar evolution and element synthesis must explain this fact.

The fission of 1 g of ^{235}U results in the formation of 27 mg of ^{99}Tc. Thus significant quantities of technetium have become available with the advent of nuclear power. Small quantities have been isolated, and larger amounts could be made available from this source if applications were developed. An isotope of importance in diagnostic medicine is produced as a decomposition product of radioactive molybdenum.

chlorine. The metal can be prepared by reducing the sulfide in hydrogen.

The principal application of technetium is as a diagnostic aid. Since technetium concentrates in the liver, it is valuable in radiological examination of that organ and in labeling the liver during examination of other organs. Soluble pertechnetates or sulfur colloids are used for this purpose. The ^{99}Tc isotope is generally produced from the decay of ^{99}Mo. Separation is by ion exchange or liquid-liquid extraction, the latter being preferred.

—Robert Q. Barr, *Climax Molybdenum Company,*
New York

Tellurium   (L. *Tellus*, earth.) **Te** = 127.60 (at. wt.); 52 (at. no.). Tellurium appears in group VIA of the periodic table (oxygen group) and has properties similar to those of sulfur and selenium. Its properties are more metallic than sulfur and selenium but less so than those of polonium. In the long period, iodine, which follows tellurium, has a lower atomic weight, due to the preponderance of heavier isotopes of tellurium:

Mass no.	120	122	123	124	125	126	128	130
%	0.08	2.46	0.87	4.61	6.99	18.71	31.79	34.49

As would be expected, information on the chemistry of technetium is scanty. In general, the chemistry is intermediate between manganese and rhenium and more closely resembles the latter. Valence states of 0, 2+, 4+, 5+, 6+, and 7+ are thought to exist. The 6+ state disproportionates to the 5+ and 7+ states within minutes, and the 5+ state disproportionates to the 4+ and 7+ states within hours. The 4+ state exists as the black dioxide and as the chlorine ion, $TcCl_6$. The dioxide is a much weaker oxidizing agent than MnO_2. The heptavalent state is the most stable. The heptavalent ion TeO_4^- in acid is a fair oxidizing agent. Tc^{++} ions tend to be stable in water solutions. The heptavalent oxide, formed on heating the metal above 400°C, is very hygroscopic and dissolves in water to give a pink solution with the characteristics of a strong monobasic acid. Hygroscopic crystals of $HTcO_4$ separate when the acid solution is evaporated. The corresponding ammonium salt, NH_4TcO_4, is stable at moderate temperatures.

Technetium forms a dark brown sulfide, Te_2S_7, which is insoluble in acids. The sulfide forms a volatile chloride when heated in the presence of

Tellurium exists in only one crystalline form, a hexagonal crystal structure having trigonal symmetry. Te is silvery gray in appearance and brittle. Table T-3 lists some common physical properties of the element. See also **Chemical elements.**

Discovery: Tellurium was first discovered in 1782 by Franz Müller von Reichenstein, an Austrian chemist who observed a peculiar phase in gold ores being mined in Transylvania. It was not until 1798 that it was characterized as a new element by Klaproth, who named it tellurium.

TABLE T-3. *Selected Physical Properties of Tellurium*

Density at 25°C, g/cm^3:
  Crystalline 6.24
  Amorphous 6.00
Atomic volume, cm^3/g atom 20.45
Melting point, °C . . . . 449.8 ± 0.05
Boiling point, at 760 mm, °C . . . . . 989.9 ± 4
Specific heat, cal/(g)(°C) 0.047
Thermal conductivity,
  (cal)(cm)/(s)(cm^2)(°C) 0.014
Coefficient of thermal expansion,
  in./(in.)(°C) . . . . 16.75 × 10^{-6}
Electrical resistivity at 25°C, Ω-cm . . 436,000 × 10^{-6}

Occurrence: Tellurium is a relatively rare element which ranks seventy-seventh in the order of terrestrial abundance, amounting to about 0.002 ppm.

Geologically, tellurium is found in significant amount in pyrrhotite magmas together with pentlandite and chalcopyrite. Selenium often occurs in close association with tellurium in these magmas. While the selenium is found substituted for the sulfur, the tellurium is usually disseminated as independent minerals consisting of tellurides of Au, Ag, Bi, Hg, Ni, Pb, Cu, or Pt. Light-metal tellurides usually are not found.

Some of the most common minerals of tellurium are hessite, Ag_2Te; petzite, Ag_3AuTe_2; calaverite, $AuTe_2$; krennerite, $(Au,Ag)Te_2$; altaite, PbTe; ricordite, Cu_4Te_2; and tetradymite, Bi_2Te_2S. Native tellurium has occasionally been found. There are no known rich mineralizations of tellurium from which it can be recovered. Since it is widely dispersed in nature, most of the tellurium of commerce is obtained as a by-product from the treatment of certain sulfidic ore deposits of Cu, Cu-Ni, and Pb. It is estimated that 80–85% of the tellurium recovered comes from the processing of Cu and Cu-Ni ores while the remainder is obtained from processing Pb ores.

Recovery: Tellurium is recovered principally from the treatment of complex insoluble slimes which result from the electrolytic refining of impure copper anodes. In addition to the tellurium the slimes contain Cu, Se, Ag, Au, and sometimes trace quantities of other precious metals. The slimes are decopperized and subjected generally to a slagging treatment with soda ash. In this operation, the Au and Ag are concentrated as an alloy or bullion called *doré metal,* while the Se and Te are collected in the soda slag in the form of water-soluble -ite and -ate compounds. After water leaching, the Te is separated from the Se as a tellurous acid precipitate by acidifying the leach solution with H_2SO_4 to a pH of 6.2. After filtering, the crude tellurous acid is redissolved in caustic, treated with sodium sulfide to precipitate impurities such as Pb, and reacidified to yield a purified tellurous acid which upon mild heating is dehydrated to TeO_2.

Elemental Te can be recovered from the TeO_2 by a reduction with finely divided carbon under a borax flux; by dissolving it in caustic and electrowinning the Te from solution; or by dissolving it in HCl or H_2SO_4 and reducing it to metal with SO_2.

Chemical Properties: In its chemical behavior, Te exhibits a valence of 2^- (Na_2Te), 2^+ ($TeCl_2$), 4^+ (TeO_2), and 6^+ (TeO_3). Many of its chemical properties resemble those of selenium and sulfur, with which it is classified as a chalcogen. It is more strongly amphoteric than either element. In the elemental form tellurium does not tarnish readily when exposed to the atmosphere for prolonged periods of time. It is attacked and dissolved by both H_2SO_4 and HNO_3 forming $TeSO_3$ and $2TeO_2 \cdot HNO_3$ respectively. Hydrochloric acid has no effect on tellurium. It is attacked by hot caustic alkali but is not affected by NH_4OH. In the molten state, it will readily oxidize to form TeO_2.

Hydrogen telluride, H_2Te, does not form readily by the direct combination of the elements since it is thermodynamically unstable above 0°C. It can be prepared by the action of a reducing acid on a compound such as Al_2Te_3. It is a gas and has a disagreeable odor. Proper precautions should be taken when handling hydrogen telluride since it is quite toxic.

Tellurium dioxide, TeO_2, is formed by the oxidation of molten tellurium or by the reaction of tellurium with HNO_3 and the thermal decomposition of the oxynitrate, $2TeO_2 \cdot HNO_3$. It is a white crystalline solid which becomes yellow on heating and melts to a dark clear-red liquid at 733°C. It is sparingly soluble in water, forming tellurous acid.

Tellurium trioxide, TeO_3, is formed by the dehydration of telluric acid. The alpha form is orange, and the beta form is grey.

Tellurium pentoxide, Te_2O_5, a bright yellow solid has reportedly been formed by prolonged heating of H_6TeO_6 at 406°C.

Tellurium monoxide, TeO, has been reported. A black phase may form when melting tellurium under conditions of partial oxidation and was thought to be TeO but has been found to be a mixture of elemental Te and TeO_2. While TeO exists in the vapor state, evidence supporting solid TeO is inconclusive.

Tellurous acid, H_2TeO_3, is a white solid which readily dehydrates to the dioxide. Very dilute solutions of tellurous acid result from the reaction of TeO_2 with water. It can be prepared by the acidification of an aqueous tellurite solution with dilute nitric acid or by hydrolysis in the cold of a tellurium tetrahalide.

Orthotelluric acid, H_6TeO_6, $H_2TeO_4 \cdot 2H_2O$, or $Te(OH)_6$, is produced by oxidizing the element, dioxide, or tellurites with strong oxidizing agents. Two modifications of the acid exist; an alpha, cubic form, and a beta, monoclinic form, which is the stable form at room temperature.

Polymetatellic and pertelluric acids also exist.

Tellurium will combine with the halogens to form the tabulated compounds. Of the alkali salts, sodium tellurite, Na_2TeO_3, and tellurate Na_2TeO_4, are most common.

Compound	Mp, °C	Bp, °C
TeF_4	129.6	193.8*
TeF_6	−34.6 †	−38.4‡
Te_2F_{10}	−33.7	59
$TeCl_2$	208	328
$TeCl_4$	224	390
$TeBr_2$¶	279	340
$TeBr_4$	363	414*
Te_2I_2		
TI_4	280§	

*Decomposes.
†At 1,500 mm.
‡At 945 mm.
¶Solid solution of Te in $TeBr_4$.
§In sealed tube.

Due to its electronegative character, tellurium readily forms tellurides with many metals. Some of the tellurides of commercial interest are bismuth telluride, Bi_2Te_3, for thermoelectric application, and cadmium telluride, CdTe, for solar cells and infrared applications. Both sulfur and selenium are completely miscible in molten tellurium. Compounds of tellurium and sulfur have been alleged to exist based on chemical preparations; however, these compounds must be considered to be elemental mixtures with a fortuitous stoichiometric composition.

Organotellurium compounds have been prepared, but there is no extensive literature comparable to that available for sulfur and selenium on the preparation and properties. Generally, organo compounds of tellurium are not thermally stable and have a tendency to oxidize at room temperature.

Economy and Uses: The 1970 free-world production of tellurium is estimated at about 450,000 lb. Because of its by-product nature, the availability of tellurium is related to economic factors which influence the demand for the primary metals from which it is recovered. Commercial tellurium is 99.7% pure. It is also offered in purities of 99.99 and 99.999 + %.

In metallurgical applications tellurium is used as an additive to improve the machinability of low-carbon steels, stainless steels, and copper. See also **Copper.** In cast iron it is used as a carbide stabilizer. It is also added to steel castings to control pinhole porosity. The addition of about 0.05% of tellurium to lead improves its physical properties. In chemical applications, it is used in rubber vulcanization as a curing agent and accelerator. It is used in primer fuses for explosives. It is added to glass and ceramics. Some applications as a catalyst have been found. In electronics it is used to produce compounds and materials for thermoelectric, solar cell, infrared-detector, and emitter applications. Small quantities are also used as a dopant for semiconductor devices.

Hygienic Considerations: The accepted concentration limit for an 8-hr daily exposure to Te and its compounds (except H_2Te) as dust or fume in air is 0.1 mg of Te per cubic meter of air. Exposure at this level may cause garlic breath. In addition to precautions taken for exposure to dust and fume, care should be taken to avoid accidental intake by handling tobacco products and food with unwashed hands. Exhaust ventilation should be used in processes emitting dust and fume. Elemental Te may be handled with bare hands. Precaution should be taken to avoid skin contact when handling soluble tellurium compounds.

—S. C. Carapella, Jr., *American Smelting & Refining Company, South Plainfield, N.J.*

Tellurium Copper   (See **Copper.**)

Tempering   (See **Iron and steel.**)

Terbium   (Sw. *Ytterby,* town in Sweden where the mineral containing the element was first identified.) **Tb** = 158.92 (at. wt.); 65 (at. no.). Terbium is the ninth of 15 elements in group III, period 6, generally shown as the rare-earth elements in a separate line below the main body of the periodic table and sometimes referred to as the *lanthanide series.* Tb was discovered in 1843 by Carl Gustav Mosander in Stockholm.

Pure terbium metal is silvery gray and relatively stable in moist air. The metal is malleable when provisions are made to prevent reaction with any of the gases or halogens. Little is known about its characteristics in alloys or intermetallic compounds.

The melting point of Tb is 1356°C; boiling point 3223°C; density 8.234 g/cm³. Estimated at 0.9 ppm in the earth's crust, it is one of the least abundant of the rare-earth elements. At this level, however, it is potentially more available than Sb, Cd, Bi, or Hg.

Terbium has 1 natural isotope, ^{159}Tb, and 17 artificial isotopes. The metal is not radioactive and has a low acute toxicity rating.

Although a valence of 3^+ for terbium is most common, this element forms compounds as a 4^+ ion and at intermediate levels. The oxide system is very complex. The accepted commercial formula is Tb_4O_7, but x-ray diffraction shows it to consist of two phases, $TbO_{1.713}$ and $TbO_{1.818}$. When Tb_4O_7 is dissolved in dilute acids, the Tb^{3+} ion appears in solution and a TbO_2 residue results. Tb_4O_7 and TbO_2 are dark brown—almost black—in color, while carefully prepared Tb_2O_3 is almost white.

Terbium is strongly cationic (electropositive) and forms compounds with most inorganic and organic anions. TbF_4 is a known stable compound.

Terbium is found in the yttrium, or heavy, rare-earth minerals. It is separated from the other elements by organic ion-exchange or solvent-extraction processes. Because of its low abundance relative to the other elements, it is recovered as a coproduct when minerals such as apatite or xenotime are processed for yttrium. Terbium metal is produced by calcium reduction of anhydrous TbF_3 in a sealed-bomb reaction. The oxides, compounds, and metal are available at 99.9 + % purity.

For a summary of the properties of terbium and a description of processing see **Rare-earth elements and metals.**

Uses: Since about 1965, when the europium-activated yttrium oxide red phosphor made color TV popular, research has accelerated rapidly, exploring the luminescent and electronic properties of the pure separated lanthanide elements. Previously there had been no known applications for Tb outside the research laboratory.

Terbium-activated lanthanum oxysulfide, $Tb:La_2O_2S$, is a phosphor now used as an image intensifier for x-ray screens.

Terbium-activated indium borate, $Tb:InBO_3$, phosphor emits an intense narrow green light 5450–5500 Å. This phosphor is suitable for information display systems under high ambient light conditions.

The next generation of color TV design may use terbium-activated yttrium silicate, $Tb:Y_2SiO_5$, or yttrium phosphate, $Tb:YPO_4$, green phosphors. These are high-efficiency post-deflection focused phosphors that eliminate the need for a shadow mask in a TV tube.

Terbium oxide can be used as a ceramic stain, but it does not color glass. In soda-lime glass, a small amount of terbium gives a strong green-blue fluorescence under ultraviolet light.

This element is being studied in great detail for its potential in photoconductive, semiconductor, and thermoelectric applications with other materials.

REFERENCES

Overview 17 (application information and possible uses), p. 17–3, and *Overview* 28, pp. 28–1–28–4, Molybdenum Corporation of America, New York, 1970 and 1971.

See also references listed under **Rare-earth elements and metals.**

—Joseph G. Cannon, *Molybdenum Corporation of America, White Plains, N.Y.*

Terephthalic Acid   (See **Polyester fibers.**)

Terne Plate   (See **Tin.**)

Terpenes   Essential oils have been obtained from plants for centuries by gentle heating or stream entrainment. As early as 1592, 60 oils were recognized. In the 1800s, scientists began identifying components and realized that many essential oils contained a class of related hydrocarbons with an empirical formula of $C_{10}H_{16}$. To this class of hydrocarbons they applied the name *terpenes*.[1,2] Typically, the terpenes are characterized by acyclic, monocyclic, and bicyclic structures:

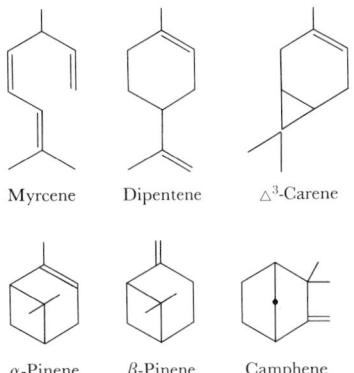

Myrcene   Dipentene   △³-Carene

α-Pinene   β-Pinene   Camphene

The nomenclature of these compounds is complicated by the continued general use of the unsystematic common names given them before their structures had been determined. To indicate that the terpene is composed of 2, 3, 4, or 6 five-carbon building blocks, this nomenclature uses the mono-, sesqui-, di-, or tri- prefixes. This section is concerned mainly with the monoterpenes.

The lack of systematic nomenclature for terpenes is well recognized and the subject of an ACS monograph.[4] Preferred nomenclature is suggested in this monograph, and additional details may be found in indexes to *Chemical Abstracts*.

At an early stage in investigations and identification of terpenes, it was recognized that the five-carbon olefin *isoprene* is a basic building block. This fact soon developed into the *isoprene rule,* which states that any terpene can be constructed from isoprene units. Although inexact, the rule is a good preliminary basis for determination whether a ten-carbon hydrocarbon falls into the class of terpenes.

Sources: The terpenes are distributed widely in nature and can be found in grasses, softwood trees, and occasionally hardwoods. Their biological function is obscure although some terpenes have fungistatic and biostatic activity and may serve to preserve the plant. The three sources of commercial quantities of turpentine include (1) gum turpentine, obtained by tapping the living tree; (2) steam-distilled wood turpentine, obtained by extraction from pine stumps; and (3) sulfate turpentine, obtained as a by-product of Kraft papermaking. Smaller amounts of terpenes are produced by extraction of other plant materials.

The sulfate pulp industry is the most important commercial source. During the process of pulping softwood by the Kraft process, a crude turpentine is produced as a steam distillate. This crude sulfate turpentine is condensed, processed to remove included impurities, and then fractionally distilled to produce α-pinene, β-pinene, dipentene, and pine oil as the major products. Typical compositions and properties are given in Table T-4. See also **Resins.**

Characteristics: As a class, the common acyclic, monocyclic, and bicyclic terpenes are mobile water-white liquids with boiling points generally between 150 and 180°C, which have piney or turpentine odors. The materials are stable under alkaline and neutral conditions and can be distilled readily at atmospheric pressure. The excellent solvent power of the terpene mixture turpentine led in the past to its extensive use as a paint solvent.

Chemistry: Chemically, terpenes are strongly influenced by their generally cyclic nature and their double bonds. The terpenes are sensitized to oxidation by their allylic hydrogen and olefin content. The widely different rates of oxidation are the result of the varied cyclic structures.

The most outstanding chemical reactivity of the terpenes is displayed in the presence of traces of mineral acidity. Typically a combination of double-bond isomerizations, ring expansions, ring openings, and carbon-chain isomerizations occur. Rapid and complex rearrangements and polymerizations are the result. Typical of these is the Wagner-Meerwein transformation of α-pinene to camphene.

α-Pinene

Camphene

The extremely specific rearrangements of terpenes and knowledge of their stereoisomeric changes have led to substantial theoretical interest and use of these olefins as model compounds for study of reaction mechanisms.

Most of the reactions that can be applied to olefins (hydrohalogenation, oxidation, polymerization, cracking) can also be carried out with terpenes, leading to a number of commercially useful products. α-Pinene, β-pinene, and dipentene, the most abundant terpenes, are converted into flavors and fragrances, resins, and insecticides. Synthetic flavors and fragrances are generally produced by thermal cracking of β-pinene to the acyclic terpene myrcene, hydrochlorination (hydrogen chloride and copper catalyst), and solvolysis of the hydrochloride to produce a variety of valuable oxygenated flavor components, such as nerol, geraniol, and linalool. A variety of other derivatives of these compounds are also made.[5]

Resins are produced from β-pinene and dipentene and various other terpene mixtures by cationic polymerization, using aluminum chloride as the catalyst.[3] These resins* have the unique quality of producing tack or stickiness in polymeric mixtures, such as natural rubber, and serve as the

*Arizona Chemical Company.

TABLE T-4. *Composition and Properties of Turpentine Products**

	α-Pinene	β-Pinene	Dipentene
Physical constants:			
Boiling point, °C	156.3–157.4	166.0–172.5	176.5–180.5
Refractive index at 20°C. . . .	1.4661	1.4774	1.4750
Specific optical rotation	+25.8	−21.8	−55.9
Composition by gas-liquid chromatography, %:			
α-Pinene	95.0	7.5	
β-Pinene	2.5	76.3	0.5
Dipentene	. . .	11.0	88.8
Camphene	2.5	1.7	
p-Cymene	. . .	0.5	5.6
Others	. . .	3.0	5.1

*Acintene distilled terpene fractions (a registered trademark of Arizona Chemical Company).

basic tackifying resin used in pressure-sensitive tapes of all kinds.

The largest single agricultural use of terpenes is as a starting material for preparation of terpene polychlorinates, such as toxaphene. These materials are produced by catalyzed chlorination of camphene or α-pinene–camphene mixtures. The products contain 50–70% chlorine and are used as insecticides. They find extensive use in cotton production for control of boll weevils, grasshoppers, and moths. The increasing utilization of the terpenes has stimulated studies pertaining to the synthetic production of turpentine materials.[6]

REFERENCES

1. Mayo, P. de: "The Chemistry of Natural Products," vol. 2, "Mono- and Sesquiterpeoids," Interscience-Wiley, New York, 1959.
2. Simonsen, J., and D. H. R. Barton: "The Terpenes," vols. 1–3, Cambridge University Press, London, 1932 and 1961.
3. Roberts, W. M., and A. R. Day: *J. Am. Chem. Soc.,* vol. 72, p. 1226, 1950.
4. American Chemical Society: Nomenclature for Terpene Hydrocarbons, *Adv. Chem. Ser.* 14, 1955. (For more detailed nomenclature see indexes to *Chemical Abstracts.*)
5. Guenther, Ernest: "The Essential Oils," vols. 1–3, Van Nostrand, Princeton, N.J., 1948–1949.
6. Meuly, W. C.: Synthetic Terpene Chemicals from Isoprene, *Am. Perfum. Cosmet.,* vol. 85, p. 123, 1970.

—H. G. Arlt, Jr., *Arizona Chemical Company, Stamford, Conn.*

Tetracyclines (See **Antibiotics.**)

Tetradecanol (See **Alcohols.**)

Tetradymite (See **Bismuth;** and **Tellurium.**)

Tetraethyl Lead (See **Chlorine organics; Petroleum;** and **Petroleum processing.**)

Tetramer (See **Polymerization.**)

Tetramethylene Oxide (See **Furan group.**)

Tetrasodium Ethylenediamine Tetraacetate (EDTA) (See **Detergents.**)

Tetryl (See **Explosives.**)

Textile Chemicals (Consult the **Subject Index.**)

Textile Dyeing [See **Dyestuffs** and **dyeing (textile).**]

TFE Resins (See **Fluoroplastics.**)

Thallium (Gk. *Thallos,* young shoot.) **Tl** = 204.39 (at. wt); 81 (at. no.). It has two stable isotopes with mass numbers of 203 (29.5%) and 205 (70.5%). Thallium is a member of the group IIIA elements, along with boron, aluminum, gallium and indium.

In the elemental form thallium exhibits two crystalline forms, hexagonal close pack and face-centered cubic. The ordinary form of the element is hexagonal close pack.

Unoxidized freshly prepared thallium metal has a silvery-gray luster, which upon standing in the atmosphere dulls to a blue gray and finally becomes encrusted with a black oxide. It is relatively soft and ductile. It is a superconductor below 2.38 K.

Some of its physical properties are listed in Table T-5. See also **Chemical elements.**

TABLE T-5. *Selected Physical Properties of Thallium*

Density, g/cm^3 . . . . 11.85
Atomic volume, cm^3/g atom 17.25
Melting point, °C 303
Boiling point, °C 1457
Specific heat at 20°C, cal/(g)(°C) 0.031
Thermal conductivity,
  (cal)(cm)/(s)(cm^2)(°C) 0.093
Coefficient of thermal expansion,
  in./(in.)(°C) . . . . 28 × 10^{-6}
Electrical resistivity, at 20°C, Ω-cm . . . . . 18 × 10^{-6}

Discovery: Thallium was discovered in 1861 quite unexpectedly when Sir William Crookes was using a spectrograph to examine a seleniferous residue of the production of H_2SO_4 from pyrite. Looking for tellurium, he observed bright green lines in the emission spectrum. Since the position of these lines was not characteristic of any known element, he believed he had discovered a new element, which he called thallium. Almost simultaneously a French researcher, A. Lamy, identified the same element.

Occurrence: The reported crustal abundance of Tl varies from 0.3 to 3.0 ppm. The geochemical properties are both lithophilic and chalcophilic. As a lithophilic element, it is found in association with potassium minerals such as feldspar and micas. As a chalcophilic element, it forms independent minerals. While Tl minerals are considered rare, some of the more common ones are crookesite, $(Cu,Tl,Ag)_2Se$; lorandite, $Tl_2S \cdot As_2S_3$; and hutchinsonite, $PbS(Tl,Ag)_2S \cdot As_2S_3$. The most important economic source of the element is from the treatment of lead and zinc concentrates. In the thalliferous lead concentrates, the Tl occurs as a substitute for the Pb in galena.

Recovery: Tl is collected in the flue dusts which result from smelting lead and zinc concentrates and which may contain Cd, In, As, Se, and Te. The flue dust may be treated in several ways. When it is handled as a by-product from the recovery of cadmium, it is leached with H_2SO_4, solubilizing both the Cd and Tl. Impurities are precipitated from solutions as sulfides or hydroxides. The solution is then electrolyzed to remove the cadmium. When the ratio of Tl to Cd in the electrolyte reaches 1:10, the cathodes are removed and new ones are inserted. After completion of the electrolysis, a final cathode is obtained which contains 5–20% Tl. This cathode metal is treated with boiling water and steam, dissolving the Tl as a hydroxide along with a trace of Cd. The solution is treated with Na_2CO_3 to remove the Cd. The thallium is precipitated from solution as the sulfide, which is then dissolved in H_2SO_4, from which Tl can be recovered electrolytically.

Chemical Properties: Like indium, which belongs to the same group, thallium exhibits valences of 1^+, 2^+, and 3^+. Compounds of thallium with a valence of 1^+ are the most numerous.

In the elemental form, thallium will oxidize slowly at room temperature in dry air. Its rate of oxidation increases rapidly with temperature. It will dissolve in water in the presence of oxygen to form thallous hydroxide. It is soluble in both H_2SO_4 and HNO_3. When attacked by HCl, it will form thallous chloride, which is insoluble.

Some common thallium compounds of interest follow:

Thallous oxide, Tl_2O, formed by the low-temperature oxidation of the metal or the decomposition of thallous hydroxide at 100°C

Thallic oxide, Tl_2O_3, produced from the oxidation of thallous oxide

Thallium tetroxide, Tl_2O_4, reported to be prepared electrolytically

Thallous hydroxide, TlOH, formed by dissolving thallous oxide in water or the metal in water containing oxygen

Thallic hydroxide, $Tl(OH)_3$, formed by adding NH_4OH to a solution of a thallic salt

Thallium halides, formed by the action of the respective halide acids on soluble thallium salts

Thallous sulfates, Tl_2SO_4, prepared by dissolving thallium or a thallous compound in H_2SO_4 and evaporating to dryness

Thallium sulfide, prepared by precipitating a soluble thallous salt from solution with H_2S

Thallous dichromate, $Tl_2Cr_2O_7$, formed by precipitating a soluble thallous salt from a slightly acidic solution with $K_2Cr_2O_7$

Thallous nitrate, $TlNO_3$, formed by dissolving the metal in HNO_3 and evaporating to dryness (three different crystalline modifications are said to exist)

Thallium metal alloys readily with many metals. Some limited liquid solubility of thallium in the molten state has been observed with the elements Cu, Al, Zn, Ni, Mg, and Se.

Economy and Uses: No statistics are available on the production of thallium, but its availability is sufficient to satisfy any new use of several thousand pounds.

The applications of thallium are quite limited. Until recently, the majority were based on its poisonous properties for use in the extermination of rodents and ants, thallium sulfate being used for this purpose.

In electronic applications, thallium-activated sodium iodide crystals are used in photomultiplier tubes. Thallium bromoiodide crystals will transmit infrared radiation, and crystals of thallium oxysulfide will detect infrared radiation. This combination of crystals has been employed for military communication systems.

The organic compounds thallous formate and thallous malonate are used in the preparation of heavy-liquid sink-float solutions, which are useful for the gravity separation of minerals.

Low-melting-point glasses based on mixtures of thallium, arsenic, sulfur, and selenium have been investigated for the encapsulation of semiconductor devices. Small quantities of thallium may be added to the counterelectrode alloy used for selenium rectifiers to improve the properties of the rectifier. The life of a tungsten-filament incandescent lamp is reputed to be increased by the addition of a thallium salt to absorb traces of oxygen. Thallium is used in special glasses to improve the optical properties and increase the refractive index.

Alloys of Pb and Tl have melting points above the constituent metals and are useful for applications in specialty fuses. A binary eutectic Hg-Tl, alloy freezing point $-60°C$, and the ternary eutectic Hg-Tl-In alloy, freezing point of $-63.3°C$, are useful for switches, seals, and thermometers in low-temperature, arctic, and stratospheric applications.

Hygienic Considerations: When handling thallium metal and its compounds, skin contact and inhalation of dusts and fumes should be avoided. Impervious gloves and aprons should be worn, and exhaust ventilation should be available for handling dusts and fumes. Although industrial poisoning by thallium is rare, proper precautions should be taken when handling.

—S. C. Carapella, Jr., *American Smelting & Refining Company, South Plainfield, N.J.*

Thermal Conductivity (See **Insulation, thermal.**)

Thermal Cracking Developed during the early part of the twentieth century, thermal cracking was the first of the major cracking processes. See also **Cracking.** Largely replaced by fluid catalytic cracking for increasing gasoline production and improving its quality, thermal cracking is used principally in coking and viscosity breaking. Both processes are used to convert nondistillable residues into more valuable products.

Thermal Coking: This process converts heavy residual stocks into gas, gasoline, distillates, and coke with the objective of maximizing the yield of distillates and minimizing the yields of gas, gasoline, and coke. The light distillates are used in domestic and industrial heating oils. The heavy distillates are appropriate feedstocks for catalytic cracking. Gasoline from thermal coking units can be blended into motor fuels after suitable treating or further processed for octane number improvement by catalytic reforming after purification by hydrotreatment. See also **Hydrotreatment;** and **Reforming.** Although only a by-product of the process, the petroleum coke is useful as a fuel in steam generation, and if it meets purity and other required specifications, the coke is one of the major raw materials for the manufacture of carbon electrodes used in the aluminum and other electrometallurgical industries.

Two types of thermal coking processes presently are used: (1) a continuous fluid coking technique and (2) a cyclic, semicontinuous process, variously named *delayed coking, decarbonizing,* or *low-pressure coking.* In the cyclic processes, the coke is alternately formed in, and removed from, a vessel or drum so that with two or more such manifolded drums in a unit, one drum can be filling while the others are being emptied. Installed capacities of delayed coking and fluid coking units are given in Table T-6.

TABLE T-6. *Thermal-Cracking-Process Capacities in the United States**

Process	Feed, bbl/day	Coke, tons/day
Delayed coking	727,900	30,435
Fluid coking	124,300	4,620
Viscosity breaking . . .	298,950	
Total	1,151,150	35,055

*1970.

A delayed coking unit consists of the three sections shown in Fig. T-1 plus coke removal and handling equipment. Ordinarily, the residual feedstock is charged to the fractionator, where it encounters the hot vapors from the coke drum.

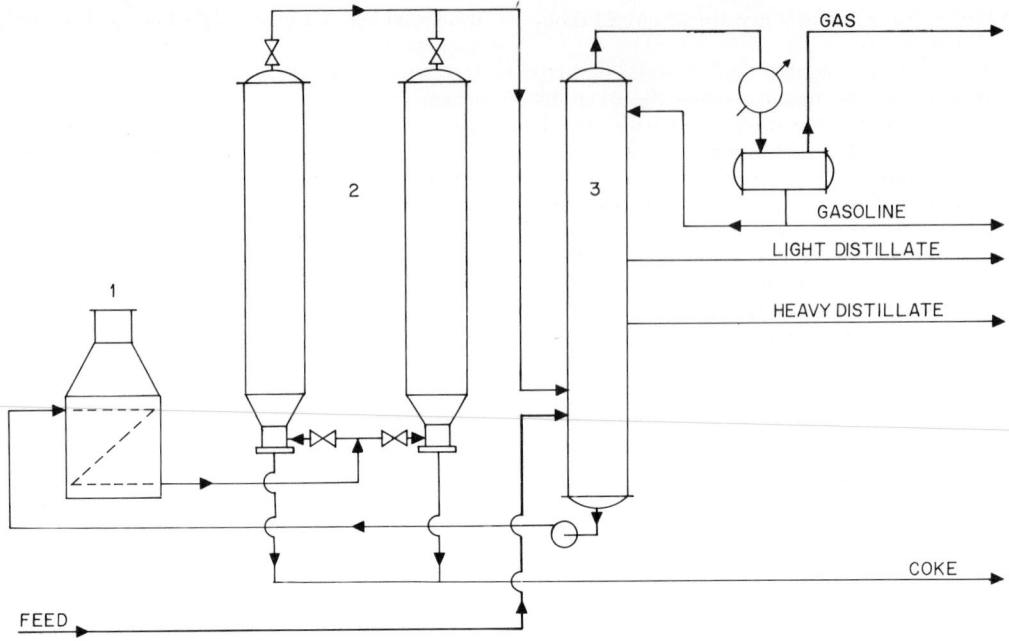

Fig. T-1. Delayed coking unit. (1) Furnace, (2) coke drums, (3) fractionator. (© 1973 *Universal Oil Products Company.*)

Here any light components are first flashed from the feed before the feed joins with recycle and is charged from the bottom of the fractionator as a combined feed to the furnace. The furnace quickly heats the charge to temperatures in the vicinity of 480°C (896°F). The heated, combined feed then is introduced to the bottom of one of two or more insulated vessels where the residence time is sufficient for the material to crack as the result of its contained heat and to form a solid coke residue. Simultaneously, lighter cracked products are evolved. Since the reaction in the drum is endothermic, the resulting temperature is in the region of 425°C (797°F). The coke remains in the drum, and the cracked products leave as vapors. The vapors are routed to the fractionation section, where they are separated into gas, gasoline, distillates of desired boiling ranges, and a heavy fraction, which is recycled to be cracked to extinction in the coking section of the unit.

When the coke accumulates to a predetermined level in one of the drums, the flow is switched to another empty, preheated drum so that operation of the furnace and fractionation sections is continuous. The filled drum just taken out of service is steamed to remove vapors, and water is ad-

mitted to cool the coke so that it can be removed safely. In modern designs, the coke is removed from the drum by hydraulic cutting, in which high-velocity water jets bore a hole downward from the top of the coke bed. Following this operation, laterally directed water jets remove coke by a cutting action from the bottom of the bed. The coke falls with the water into cars or is sluiced by water to storage.

In order to minimize the yield of coke, drum pressures are relatively low—in the general range of 10–70 psig. There can be a wide variation in yields of products, depending upon the nature of the feedstock and the boiling range of the liquid distillates produced.

The Fluid Coking[SM] process is performed in equipment similar to that used in fluid catalytic cracking (see also **Fluid catalytic cracking**) but operates by circulating small particles of coke as seed, which are formed by the process itself in one of two vessels to which the feed is charged. A small portion of the circulating coke stream is burned in the other vessel to raise its temperature enough to provide for the heat of reaction and feed preheat in the reactor vessel. The latter is maintained at temperatures of 480–565°C

(896–1050°F). The net coke-product particles can be transported to storage by a stream of air.

Other Thermal-cracking Operations: Viscosity breaking is used to reduce the viscosity of heavy residues by a mild thermal cracking with a minimum yield of gasoline. Feedstock is passed through a tubular furnace with a minimum of recycle. The slightly cracked products are separated by distillation into gas, gasoline, a light distillate with a nominal end point of about 345°C (650°F), and a fuel-oil residue having a considerably lower viscosity than the feedstock. The visbroken residue is often further reduced in viscosity by blending with catalytic cycle oils to produce a marketable fuel oil which meets viscosity specifications. Although visbreaking units consist basically of a cracking section and a separation section, a number of variations in flow arrangements exist.

Of the 12,650,000 bbl/day of installed crude refining capacity in the United States, the thermal-cracking capacity as shown by Table T-6 represents only 9% of the total. A minor amount of additional, miscellaneous thermal operations includes the cracking of gas oils. The extent of pyrolysis of a variety of feeds to produce olefins as petrochemical raw materials is illustrated by the production in 1970 of about 18 billion lb of ethylene.

Thermal cracking in the free-world nations amounts to about 10% of the crude refining capacity, a proportion not much different from that observed in the United States.

See also **Ethylene; and Petrochemical complex.**

—Melvin J. Sterba, *UOP Process Division, Universal Oil Products Company, Des Plaines, Ill.*

Thermal Diffusion   (See **Separation operations.**)

Thermal Insulation   (See **Insulation, thermal.**)

Thermal Plasmas   (See **Plasma, chemical processing with.**)

Thermal Reactor   (See **Nuclear power plants.**)

Thermonuclear Reaction   When the required activation energy for a nuclear-fusion reaction is supplied thermally, it is termed a thermonuclear reaction.

Thermoplastic Resins   (See **Plastics.**)

Thermosetting Resins   (See **Plastics.**)

Thiamides   (See **Amides and imides; and Sulfur.**)

Thickeners   (See **Sedimentation; and Waste treatment.**)

Thin-Layer Chromatography   (See **Chromatography.**)

Thio Ethers   (See **Ethers.**)

Thiokol Rubbers   (See **Elastomers.**)

Thiols   (See **Sulfur.**)

Thixotropic Fluids   (See **Colloid systems.**)

Thomas Process   (See **Iron- and steelmaking.**)

Thorium   (Thor, Norse god of thunder.) **Th** = 232.038 (at. wt.); 90 (at. no.). Thorium metal normally is a dark gray color, containing some oxide. Although stable at room temperature in air, the metal ignites and burns brilliantly upon application of heat, to form ThO_2. The metal will burn in chlorine, bromine, iodine, or sulfur at temperatures above 450°C and combines with hydrogen and nitrogen above 650°C. Th will alloy readily with several metals, including Al, Cu, Ni, and Zn. Th is more electropositive than Mg. A mixture of ThO_2 with 1% CeO_2 intensely heated will emit a very brilliant light, making it useful for gas mantles. Small quantities of Th also have been used in electric-light filaments. The element was discovered by Berzelius (Sweden) in 1828.

Thorium is found in monazite sand, which contains 3–9% ThO_2, depending upon source (Brazil, India, and in the Carolinas of the United States). After concentration, the ore is digested with strong H_2SO_4 to form the sulfate, which, after a series of reactions, yields thorium nitrate, the commercial form of the element. All thorium-containing substances are radioactive.

Th is a member of the transactinium earths, actinide series, described in detail under **Chemical elements.** Thorium, with a mass number of 232, captures slow neutrons and through a series of nuclear reactions forms fissionable ^{233}U. Consequently, Th has played an important role in nuclear chemistry and physics. There are eight unstable isotopes of Th, 227–234. The element has a specific gravity of 11.2, melting point of 1845°C, and boiling point greater than 3000°C.

Th dissolves in HCl but becomes passive in

HNO_3. The element is not affected by fusion with alkalis.

Th gained little recognition scientifically or commercially until the advent of the nuclear age.

Thorium Series (See **Chemical elements.**)

Threonine (See **Amino acids;** and **Peptides and proteins.**)

Thucholite (See **Uranium.**)

Thulium (Gk. *Thule*, the northernmost part of the habitable world or the most distant goal of human endeavor.) **Tm** = 168.934 (at. wt.); 69 (at. no.). Thulium is the thirteenth of 15 elements in group III, period 6, generally shown as the rare-earth elements in a separate line below the main body of the periodic table and sometimes referred to as the *lanthanide series.* Tm was discovered in 1879 by P. T. Cleve, a Swedish chemist.

Pure Tm metal retains a silvery-gray metallic luster, and corrosion is almost absent in air up to 200°C. The pure metal is malleable and can be worked with ordinary tools and equipment. It is soluble in dilute and concentrated mineral acids. The halogen gases react vigorously with Tm above 200°C to form trihalides. At elevated temperatures Tm combines with oxygen, sulfur, nitrogen, carbon, and hydrogen and forms intermetallic compounds with most metals to the right of iron in the periodic table.

The melting point of Tm is 1545°C; boiling point 1947°C; density 9.318 g/cm³; low toxicity rating. Estimated at 0.48 ppm in the earth's crust, Tm is the least abundant of the rare-earth elements. Even at this level, however, it is potentially more available than Sb, Cd, Bi, or any of the precious metals.

Tm has one natural isotope, ^{169}Tm, and 17 artificial isotopes. After bombardment of ^{169}Tm by thermal neutrons for 4 months, some radioactive ^{170}Tm (half-life about 129 days) is prepared. A complex conversion and radiation occurs, and subsequently ^{170}Tm decays by beta emission to ^{170}Yb. Continued high-flux bombardment creates ^{171}Tm, which has a half-life of 19 years. The ^{171}Tm decay proceeds by beta emission with no associated gamma-ray output. The $4f$ shell of Tm has 12 electrons, 2 of which are unpaired.

Thulium oxide differs from the lighter rare-earth oxides in that it has little or no reaction with moist air to form a hydroxide up to 200°C. The oxide dissolves easily in mineral acids and forms water-soluble hydrated salts. Aqueous solutions exhibit only the 3^+ ion. Nonaqueous media, however, indicate that Tm^{++} may exist, as revealed by the stoichiometry of TmI_2, TmH_2, TmS. Trivalent crystalline salts of Tm^{3+} are characterized by a green color.

Thulium is found with the yttrium, or heavy, rare-earth elements. Processing by organic liquid-liquid or solid-liquid ion-exchange techniques yields pure Tm compounds. Before the development of cation-exchange resins capable of separating the chemically similar rare earths (within the past 25 years), thulium was practically unavailable in pure form. Today, Tm is readily available because other heavy rare-earth elements, particularly yttrium, dysprosium, holmium, and erbium, are produced. The quantity available and the price of Tm depend on the markets for the other rare-earth elements; 2,000–3,000 lb/yr could be refined from the lanthanide elements remaining after pure yttrium is recovered.

Thulium metal is prepared by the direct reduction of thulium oxide by lanthanum metal at high temperature in vacuum. The vapor pressure of Tm near its melting point (1545°C) is quite high, allowing it to sublime and be condensed free of other metallic impurities. The trifluoride, TmF_3, is reduced by the metallothermic reaction with calcium in tantalum or tungsten crucibles. It is more difficult to obtain pure metal by this reaction because of the solubility of the crucible metals in molten thulium.

No commercial applications for Tm or its compounds are known. The artificial isotope ^{171}Tm is a potential safe energy source. The photoelectric, semiconductor, and thermoelectric properties of Tm and its many compounds in the near-infrared region of the spectrum are being studied.

—Joseph G. Cannon, *Molybdenum Corporation of America, White Plains, N.Y.*

Thyroglobulin (See **Peptides and proteins.**)

Thyroptropin Releasing Factor (TRF) (See **Peptides and proteins.**)

Tin (OE. tin.) **Sn** (L. *stannum*) = 118.70 (at. wt.); 50 (at. no.). Tin is a member of group IVB of the periodic table, falling between Ge and Pb and having a valence number of 2^+ or 4^+. Tin exists in two allotropic forms, the common soft white (beta) tin, which crystallizes in the body-centered tetragonal system, and the brittle gray (alpha) tin, which crystallizes in the diamond-type cubic system. The latter is formed when tin of high purity is exposed to temperatures below the transformation temperature of 13.2°C. The alpha tin converts back to white tin above the transformation temperature. Impurities in commercial grades of tin prevent the transformation of white to gray tin.

Tin has the largest number of naturally occurring stable isotopes of any of the elements; they are listed under **Chemical elements.**

Bronze articles with 10–15% Sn content date back to about 3500 B.C. A Sn-Pb alloy (now called *solder*) was reported by Pliny (A.D. 79). The tinplate industry did not begin until the fourteenth or fifteenth century in Bohemia. Saxony was the center for tinplating in the mid-seventeenth century. At present, the United States is the largest producer of tinplate.

Free-world consumption of primary tin (1970) is estimated at nearly 180,000 metric tons; in the United States the consumption of both primary and secondary tin by finished product is broken down* as follows: (1) tinplate 34.36%; (2) solder 27.43%; (3) copper alloys (brass and bronze) 18.49%, (4) white metals (babbitt, terne, type metal, pewter, britannia, jewelers' metal, and miscellaneous alloys) 7.90%; (5) tin chemicals 4.50%; (6) tinning 2.95%; (7) other products 4.37%. A large proportion of the tinplate consumed in the United States is absorbed by the canning industry. The nontoxic nature of tin, coupled with its high resistance to chemical attack by acids or other aqueous solutions, makes it an ideal protective coating for stronger materials having higher toxicity and less resistance to chemical attack, especially where these coated metals are used in food storage or processing. Tinned steel, copper, or brass is used in making food-processing machinery.

The tin coating is applied to steel by electroplating in a continuous process (now largely used in the United States) or by passing the cut sheet through a bath of molten tin. For brightness hot dipping the article in molten tin or *flow brightening* the dull, electrodeposited article by heating to the melting point by induction or in hot oil is employed. For a very thin film of tin, as on copper tubing, aluminum pistons, bearings, or brass pins, immersion tin coatings are used. Tin alloy coatings are used when the coatings must be harder, brighter, and more corrosion resistant. Alloys of Sn with Cu, Pb, Ni, Zn, Cd and ternary Sn-Cu-Pb coatings are commonly used. See also **Lead.**

*United States 1970.

Commercial tin alloys are of two classes, the *bronzes,* copper-base alloys containing Sn, and *white metals,* alloys of Sn with Sb, Pb, Bi, Cu, and other metals. The white metals include antifriction or bearing metals, printers' alloys, pewter, britannia metal, the solders and special alloys for use in castings in the chemical industry, battery plates, bullets, collapsible tubes and foil, and diecasting and other molds.

The principal types of tin bronzes are *phosphor bronzes,* with 5–10% Sn and small amounts of P, and the *gun metals,* which are tin bronze casting alloys modified with 1–6% Zn. The superiority of these alloys in elastic properties, bearing properties, and resistance to corrosion (atmospheric, fresh, and salt water), as well as to alternating stress and corrosion fatigue, makes the metals especially useful in marine and railway equipment, pumps, valves, pipe fittings, ship propellers, gears, and springs.

White bearing alloys (babbitts) are high-tin alloys containing more than 80% Sn and high-lead alloys containing up to 12% Sn and 70% or more Pb. These metals are used in machinery, including motors, compressors, and rolling-mill equipment.

Occurrence: Tin, present in the earth's crust to the extent of 40 g/ton, occurs in nine different minerals from two types of deposits: the highly dispersed and the most commercially significant ore, cassiterite (tinstone), SnO_2, and the complex sulfidic ores, a combination with the sulfides of base metals and pyrites. Cassiterite is heavily concentrated in bands and layers of varying thickness, forming economically valuable deposits, such as those found in southeastern Asia (Malaysia, Thailand, Indonesia, China). The complex sulfidic ores, found in lode deposits and economically significant only in Bolivia, are stannite, $SnS_2 \cdot Cu_2S \cdot FeS$; herzenbergite, SnS; teallite, $SnS \cdot PbS$; franckeite $2SnS_2 \cdot Sb_2S_3 \cdot 5PbS$; cylindrite, $Sn_6Pb_6Sb_2S_{11}$; plumbostannite, $2SnS_2 \cdot 2PbS \cdot 2(FeZn)S \cdot Sb_2S_3$; and canfieldite, $4Ag_2S \cdot SnS_2$. While the important tin-producing countries are Malaysia, Bolivia, Indonesia, Nigeria, Thailand, Zaïre (formerly Belgian Congo), and China, smaller quantities also are produced in the United Kingdom, Burma, Japan, Canada, Portugal, Spain, and Australia. Australia eventually may become a leading tin producer.

Tin is present in meteorites, in natural waters, in soil, in marine organisms and animals, in the tissues of edible animals, in the milk of lactating animals, and in small amounts in the organs of man. Until relatively recently, the presence of Sn in the human organism was not shown to be either beneficial or harmful. It is now suggested that Sn may be a hitherto unrecognized essential trace element in the growth of the mammalian organism.

Physical Properties: Tin, a soft, ductile, silvery-white, nontoxic metal, lends itself readily to rolling, stamping, extrusion, and spinning. The thin film of oxide formed on the surface of cast tin produces a slight yellowish color. Highly polished tin holds its brightness under both indoor and outdoor exposure. Tin differs from the common structural metals (Al, Cu, Fe) in its low melting point of 231.9°C. The interval between this low melting point and its high boiling point (2270°C) exceeds that for nearly all metals. Key physical properties of Sn are given in Table T-7.

Chemistry and Compounds: Tin forms stannous, Sn^{++}, and stannic, Sn^{4+}, compounds; complex salts of the stannite, M_2SnX_4, and stannate, M_2SnX_6, type; coordination complexes; organic

TABLE T-7. *Physical Constants of Tin*

Melting point, °C	231.9
Boiling point, °C	2270
Transformation temperature, °C	13.2
Heat of transformation, cal/g	4.2
Latent heat of fusion, cal/g	14.2
Specific heat, cal/g	
At 25°C	0.053
From 25°C to melting point	$0.0372 +$
	$0.053 \times 10^{-3}T$ K
From melting point to 1000°C	0.0615
Vapor pressure, mm:	
At 1000 K	7.4×10^{-6}
2000 K	30.6
2550 K	638
Density, g/cm³:	
White tin at 15°C	7.29
Gray tin at 13°C	5.77
Liquid tin at melting point	6.97
Viscosity, cP:	
At 250°C	1.88
800°C	0.87
Surface tension at 215°C,	
dynes/cm	685
Thermal conductivity at 0°C,	
(cal)(cm)/(s)(cm²)(°C)	0.150
Brinell hardness at 20°C	3.9
Latent heat of vaporization, cal/g	520 ± 20
Coefficient of linear expansion	
at 0°C	19.9×10^{-6}
Specific volume at 20°C	0.1395
Critical temperature, °C	3730
Critical pressure, atm	650

salts of tin, such as the tin soaps, in which tin is bonded through an element other than carbon; and organotin compounds, in which tin is bonded to at least one and up to four carbon atoms.

Being amphoteric, tin reacts with both strong acids and strong bases but remains relatively resistant to neutral solutions. In oxygen or dry air at ordinary temperatures, a thin oxide film develops on tin; moisture and heat accelerate this oxide formation. Tin does not form a compound by direct union with H_2, N_2, or CO_2. Tin is not oxidized by H_2O_2, nor is the latter decomposed catalytically by Sn. Even when heated, Sn is not attacked by gaseous NH_3. Halogens react readily with Sn to form stannic halides (with chlorine, bromine, and iodine reacting at normal temperatures and fluorine at 100°C). Tin is easily attacked by HI and HBr but less readily by HCl. It is feebly attacked by aqueous HF and slowly dissolved by aqueous HCl. Dilute H_2SO_4 reacts slowly on Sn in the cold; reaction takes place with hot concentrated H_2SO_4 to form stannous sulfate. HNO_3 reacts on tin in any concentration or at any temperature. With dilute HNO_3, soluble Sn nitrates are obtained; with concentrated HNO_3, Sn is oxidized to the insoluble hydrated stannic oxide.

Tin is attacked by aqueous solutions of the hydroxides and carbonates of Na and K, especially when warm. Stannates are formed by the action of strong NaOH or KOH solutions on Sn. A vigorous reaction results on heating tin and sulfur, with the formation of Sn sulfides. H_2S does not react with Sn at ordinary temperatures but does react at 100–400°C to form stannous sulfide. An aqueous solution of SO_2 attacks Sn, forming stannous sulfide. Molten tin combines with phosphorus to form a phosphide.

Inorganic Tin Compounds: These include all compounds in which tin is bonded to an element other than carbon, even though carbon may be present in the compound. The chlorides of Sn, stannous and stannic, are best prepared by reaction of Cl_2 on Sn metal. Double decomposition reactions of stannous chloride, stannous oxide, or stannous hydroxide with the appropriate reagents are generally used for preparing stannous salts. Stannic oxides reacts readily with acids but only slightly in aqueous media. Due to its refractory nature, cassiterite seldom is used as a raw material for preparing Sn chemicals. Metallic stannates are prepared by fusion of stannic oxide with the desired metal hydroxide or carbonate or by direct double decomposition.

The properties, preparation, and uses of the most commercially significant inorganic tin compounds are summarized in Table T-8.

Organotin Compounds: Most organotin compounds have a tetravalent structure of formula R_nSnX_{4-n} or R_4Sn, where R is an organic radical, X is an anion, and n may be any number from 1 to 4. Several divalent organotins have been reported, but their structures are questionable. Di- and polytins, such as $R_3Sn-SnR_3$, and compounds with other metals, such as R_3SnSiR_3, also are known. Unlike the organo derivatives of Si and Ge, present nomenclature of organotin compounds is based upon the use of tin as a suffix. Thus, $(C_4H_9)_4Sn$ is tetrabutyltin and $(C_4H_9)_2SnCl_2$ is dibutyltin dichloride. Although the first organotin (diethyltin diiodide) was synthesized by Frankland in 1849, these compounds did not attain any commercial significance until a century later. At present the most important organotins are the dibutyltin, tributyltin, dioctyltin, triphenyltin, and tricyclohexyltin derivatives. Production of organotins (1968) was estimated at 10,000 tons.

Tetraorganotins can be prepared by (1) alkylation of tin halides, especially stannic chloride, with Grignard reagents or alkyl lithium; (2) reaction of an organic halide with a tin-sodium alloy; (3) direct reaction of tin with an organic halide under special conditions (high temperatures, catalysts, irradiation); and (4) reaction of stannic chloride with alkyl aluminum compounds. All these processes are used commercially to some extent. For the Grignard route, ethyl ether is seldom used except possibly in catalytic amounts. Other organic solvents, e.g., tetrahydrofuran, toluene, and heptane, are normally used. Organotin halides are prepared by disproportionation of the tetraorganotin with stannic chloride or by partial alkylation of stannic chloride. The organotin halides thus prepared are converted into the oxides with alkali. The organotin halides and oxides are usually the intermediates employed in the syntheses of other organotin derivatives.

The largest use for organotins is as stabilizers for polyvinyl chloride (PVC), neoprene, and other polymers against degradation by light, heat, or oxygen. They are superior to other PVC stabilizers in their ability to produce colorless transparent flexible and rigid PVC products. Selected organotins have been sanctioned by the U.S. Food and Drug Administration (FDA) as being safe for use in PVC for food packaging. Other applications include their use as stabilizers in chlorinated polyethylenes, vinyl copolymers, silicones, and polyamides; as catalysts in many types of inorganic and organic reactions (condensation, esterification, halogenation, hydrogenation, oxidation, and polymerization) as well as curing catalysts for resins and rubbers and polymerization catalysts for polymers, e.g., polyurethanes, polyolefins, methacrylates and vinyls; as biocides, including insecticides, wood preservatives, fungicides, bactericides, antifouling agent in marine paints, molluscicides; as anthelmintics in veterinary medicine; in electroconductive coatings on glass; and as lubricant and fuel additives.

Diorganotin derivatives, particularly dibutyltin and dioctyltin, are most useful commercially as catalysts and stabilizers for resinous and polymeric systems. Dibutyltin diacetate and dibutyltin di(2-ethylhexoate) are used as catalysts in silicone manufacture. Polyurethane catalysts include dibutyltin di(2-ethylhexoate) and dibutyltin dilaurate. The largest volume of organotins, principally dibutyltin maleates, dibutyltin dilaurylmercaptide, and dibutyltin S,S'-bis(alkylmercaptoacetates), are used to stabilize PVC against the effects of heat, light, and oxygen. Organotins which have been approved by the FDA for use as stabilizers in PVC food packaging are dioctyltin maleate and dioctyltin S,S'-bis(isooctylmercaptoacetate). Nontoxic monobutyltin derivatives, i.e., monobutylthiostannonic acid anhydride, are also used as PVC stabilizers.

The most extensive application of triorganotin derivatives stems from their biocidal activity. Bis(tributyltin) oxide is used as an antifoulant for marine paint, as fungicides, bactericides, sanitizing agents, and wood preservatives. Bis(tributyltin) fluoride is used in ship-bottom paints; tributyltin benzoate and salicylate for control of fungi on textiles and leather; and tributyltin linoleate for control of fungi on plastics. Tricyclohexyltin hydroxide has received FDA approval for use as an insecticide on apples, citrus fruit, and pears. Triphenyltin hydroxide has FDA approval for use as an agricultural fungicide on sugar beet, carrot, pecan, and potato crops. Triphenyltin acetate, triphenyltin hydroxide, and tricyclohexyltin hydroxide are widely used in Europe as agricultural fungicides. Tributyltin chloride is an effective repellant for rodents.

In veterinary medicine dibutyltin dilaurate finds use as a poultry anthelmintic effective against *Eimeria tenella* and *E. necatrix* and as a control agent against worms of the type *Ascardia galli.*

TABLE T-8. *Properties, Preparation, and Uses of Selected Inorganic Tin Compounds*

Name, Formula, Formula Weight, and Physical Constants	Preparation, Uses, and Other Properties
Stannous oxide, SnO (134.70); mp = decomposes > 385°C; sp gr = 6.6	Blue-black lustrous crystals; soluble in acids and concentrated alkalies; obtained by precipitation of a SnO hydrate from a $SnCl_2$ solution with alkali and later treatment in H_2O (near bp and at constant pH); heated in air to give SnO_2; dissolved in acid, SnO yields stannous salts; chief uses: as intermediate in preparation of other stannous compounds; also used in manufacture of AuSn glass and CuSn ruby glass
Stannic oxide, SnO_2 (150.70); mp = >1600°C; sp gr = 6.6–6.9	White tetragonal crystals; prepared industrially by blowing heated air over molten Sn; by calcination of the hydrated oxide; or by burning finely divided Sn made by atomizing Sn via high-pressure steam; also, by high-temperature treatment of $SnCl_4$ with steam; hydrated SnO_2 (stannic acid) obtained by hydrolysis of stannates; alphastannic acid (orthostannic acid), the more highly hydrated and more soluble form, obtained from alkaline-metal stannate solutions by decomposing with acid or by decomposition of a $SnCl_4$ solution with alkali or excess H_2O; less soluble and less hydrated form, betastannic acid, made by heating, drying, or long standing of alpha-stannic acid; SnO_2 used as a ceramic opacifier, as ingredient in ceramic colors (chrome-tin pink and vanadium-tin yellow), as a polishing agent for marble and granite
Stannous chloride, $SnCl_2$ (189.61); mp = 246.7°C; bp = 623°C; sp gr = 2.6	White solid, soluble in H_2O, glacial CH_3COOH, alcohol, many organic solvents; available in anhydrous and hydrous forms; anhydrous salt obtained by treatment of molten Sn with Cl_2, hydrate by treatment of flaked Sn with HCl, followed by evaporation and crystallization or by cathodic reduction of a $SnCl_4$ solution; used in electrotinning of steel strip, immersion tinning processes; tin coating of sensitized paper; as a food additive, as catalyst in organic reactions, stabilizer for perfumes in soaps; sensitizing agent in plating of plastics and silvering of mirrors; as well-drilling mud additive; antisludge agent for oils
Stannic chloride, $SnCl_4$ (260.53); mp = −30.2°C; bp = 114°C; sp gr = 2.23	Colorless, fuming liquid, soluble in H_2O, many organic solvents; prepared by direct chlorination of Sn (molten Sn or suspended as finely divided Sn under $SnCl_4$); used in manufacture of blueprint and other sensitized papers, as a stabilizer of perfume in soap, in preparation of lube additives, as a catalyst (Friedel-Crafts), in treatment of nonconductive substrates, such as glass and plastics, to form conductive coating and thus improve physical properties; also, for treatment of hot-molded glass containers to enhance lubricity and prevent scratching defects; main use is preparation of inorganic Sn and organotin compounds
Stannous fluoride, SnF_2 (156.70)	White crystals, soluble in H_2O, alkali, and fluorides; prepared from SnO and HF; used mainly in toothpastes and dental preparations to prevent demineralization of teeth
Stannic fluoride, SnF_4 (194.70); mp = sublimes; sp gr = 4.78	White solid, soluble in H_2O
Stannous sulfide, SnS (150.76); two forms, see column at right	(1) Dark gray crystals, mp = 882°C, sp gr = 5.1; (2) black amorphous powder, mp = 1200°C; both soluble in concentrated HCl, alkali metal, or NH_4 polysulfide
Stannic sulfide, SnS_2 (182.82); decomposes; sp gr = 4.5	Golden leaflets, soluble in concentrated HCl, basic sulfides, alkali

TABLE T-8. *Properties, Preparations, and Uses of Selected Inorganic Tin Compounds* (*Continued*)

Name, Formula, Formula Weight, and Physical Constants	Preparation, Uses, and Other Properties
(a) Potassium stannate, $K_2Sn(OH)_6$ (298.88); (b) sodium stannate, $Na_2Sn(OH)_6$ (266.65)	Both colorless crystals, soluble in H_2O; produced in U.S. as product of alkaline detinning process; widely used in alkaline electroplating Sn baths; in immersion plating of Al; in plating Sn alloys; (2) also a stabilizer for H_2O_2; other stannates (Ba, Ca, Mg, Sr, Bi, Pb, Ni, Cu, Cd, Co, Fe) used in ceramic dielectrics
Stannous pyrophosphate, $Sn_2P_2O_7$ (411.44); mp = decomposes > 400°C; sp gr = 4.0	White powder soluble in HCl and $Na_4P_2O_7$
Stannous sulfate, $SnSO_4$ (214.77); mp = decomposes > 360°C	White crystalline powder soluble in H_2SO_4 and H_2O; obtained commercially by action of H_2SO_4 on $SnCl_2$ or Sn. Used in acid Sn plating bath for electrotinning steel strip, in immersion plating steel wire with Sn or Cu-Sn alloy coatings prior to drawing (liquor finishing)
Tin hydride, SnH_4 (122.72); mp = decomposes 145–150°C; bp = −52°C	Unstable gas
Stannous 2-ethyl hexoate, $Sn(C_8H_{15}O_2)_2$ (405.10); sp gr = 1.26	Clear, light-yellow viscous liquid, soluble in many organic solvents; used as a catalyst in curing silicone oils and production of polyurethane foams
Stannous oleate, $Sn(C_{18}H_{33}O_2)_2$ (681.61); sp gr = 1.06	Clear, light-yellow liquid, used as a catalyst for urethane and other polymers
Stannous oxalate, $Sn(C_2O_4)$ (206.71); mp = decomposes 280°C; sp gr = 3.56 (18°C)	White powder, soluble in hot concentrated HCl; used as a catalyst in esterification and coal hydrogenation
Stannous acetate, $Sn(C_2H_3O_2)_2$ (236.75)	Colorless crystal, soluble in CH_3COOH; useful as a catalyst and as a textile agent to promote exhaustion of dyes

NOTE: The tin salts of the carboxylic acids (octanoic, oxalic, tartaric, oleic, lauric, naphthenic, palmitic, stearic, formic, and gluconic acids) also are known as *tin soaps*, generally prepared by reaction of stannic or stannous oxide with the appropriate acid.

Tetraphenyltin is used as a scavenger for hydrogen chloride in dielectric fluids.

The properties of major organotins are summarized in Table T-9.

Metallurgy: Tin concentrates from alluvial deposits of almost pure cassiterite are smelted directly. Those from lode deposits, such as the Bolivian, must be upgraded before smelting. These complex ores, containing from 40–60% Sn, are treated by roasting, gravity concentration, and magnetic separation to remove the sulfidic minerals and other metals present.

Tin smelting is simply the reduction of cassiterite by carbon to yield Sn metal. To prevent loss of Sn from the high-tin-content slag, the smelting is carried out in three stages: (1) primary smelting in reverberatory or blast furnaces; (2) retreatment of slags, hardhead, and refinery dross; and (3) removal by refining of the residual impurities in the metallic Sn. In the modern reverberatory furnace, a charge containing dressed Sn ore, 15–20% anthracite, and smaller amounts of limestone, slag, sand, and by-products from the refining steps is heated with stirring for 10–12 hr at 1200–1300°C. The batch then is tapped into a settler, from which the slag overflows into cast-iron pots. Molten tin for further refining is cast from the settler bottom into 75-lb slabs. The first-run slag, containing 10–25% Sn, is crushed into ¾-in. lumps and resmelted at temperatures up to 1480°C in the presence of anthracite screening as a reducing agent, a limestone flux, and a metallic-iron catalyst for the reduction. The Fe-Sn compound formed is returned to the system, and the slag containing less than 2% Sn is discarded. Blast-furnace and electric smelting also are used, the latter on a large scale in Zaïre.

Refining: Raw tin from the smelting operations

TABLE T-9. *Physical Properties of Some Typical Organotins*

Compound	Formula	Sp gr d_{20}^{20}	Bp, °C (Except as Noted)	Mp, °C	Refractive Index n_D^{20}
Tetramethyltin	$(CH_3)_4Sn$	1.2905^a	76–77	−54	1.4415
Tetrabutyltin	$(C_4H_9)_4Sn$	1.0541	145^b	< −70	1.4727
Tetraoctyltin	$(C_8H_{17})_4Sn$	0.9605	268^b	. . .	1.4681
Tetraphenyltin	$(C_6H_5)_4Sn$	1.521	>420	229	
Dibutyltin dichloride	$(C_4H_9)_2SnCl_2$	1.36^c	135^b	43	1.499^c
Dibutyltin dilaurate	$(C_4H_9)_2Sn(OCOC_{11}H_{23})_2$	1.05^a	$>400°F^b$	. . .	$1,470^a$
Bis (tributyltin) oxide . . .	$[(C_4H_9)_3Sn]_2O$	1.1715^a	220–230^b 180^d	Freezes below −45	1.4880
Dioctyltin maleate	$(C_8H_{17})_2Sn\diagup^{OCOCH}_{\diagdown OCOCH}$	. . .	. . .	93–95	
Dioctyltin S,S′-bis(isooctyl-mercaptoacetate)	$(C_8H_{17})_2Sn(SCH_2COOC_8H_{17})_2$	1.06^a	Not distillable at 10 mm	Freezes below −15	
Tricyclohexyltin hydroxide .	$(C_6H_{11})_3SnOH$	. . .	. . .	195–197	
Triphenyltin acetate	$(C_6H_5)_3SnOCOCH_3$	1.84	. . .	122–124	
Hexamethylditin	$(CH_3)_3SnSn(CH_3)_3$	. . .	182	23	
Hexabutylditin	$(C_4H_9)_3SnSn(C_4H_9)_3$	1.148^a	198^b	. . .	1.5090^a
Dibutyldiphenyltin	$(C_4H_9)_2Sn(C_6H_5)_2$	. . .	154–157^e	. . .	1.5602^a

^{a}At 25°C. ^{d}At 2 mm.
^{b}At 10 mm. ^{e}At 1 mm.
^{c}At 50°C.

may be refined by electrolytic processes and further heat treatment, i.e., liquation and tossing. A purer grade of Sn is produced by electrolytic refining, but the high costs and the low demand for ultrapure grades of Sn dictate limited use. Taking advantage of the low melting point of Sn in comparison with the impurities contained, liquation permits separation. Tin is heated on the sloping hearth of a small reverberatory furnace to just above its melting point. Tin runs into a *poling kettle,* while the other higher-melting metals remain in the dross. Tossing (agitation of the molten Sn in the poling kettle with steam, compressed air, or poles of green wood) removes the remaining traces of impurities and produces pure tin (in excess of 99.8% Sn) in iron molds weighing 100 lb each.

Secondary Tin: Alloy scrap containing tin is reworked by a combination of liquation and smelting. Tinplate scrap is detinned by electrolytic or chemical means. Commercial installations employ (1) the alkaline electrolytic process, (2) the chlorine process, and (3) the alkaline chemical process. The latter, most widely used today, utilizes a caustic solution containing an oxidizing agent to remove both the Sn and the underlying Fe-Sn alloy from the steel. The resultant detinning solution can be treated by (1) crystallization to recover sodium stannate, (2) electrolysis to recover Sn metal, or (3) acidification with CO_2, H_2SO_4, or acidic gases to precipitate hydrated tin oxide.

In the United States (1970), there were about 85 secondary tin smelters but only 1 primary smelter. The principal tin smelters are located in the United Kingdom, Thailand, and Malaysia.

REFERENCES

Faulkner, C. J.: The Properties of Tin, *Tin Res. Inst., Publ.* 218, 1965.

Hedges, E. S. (ed.): "Tin and Its Alloys," Arnold, London, 1960.

Ingham, R. K., S. D. Rosenberg, and H. Gilman: Organotin Compounds, *Chem. Rev.,* vol. 60, pp. 459–539, 1960.

Luijten, J. G. A., and G. J. M. Van Der Kerk: "Investigations in the Field of Organotin Chemistry," Tin Research Institute, Washington, 1955.

Mantell, C. L.: "Tin: Its Mining, Production, Technology and Applications," 2d ed., Van Nostrand, Princeton, N.J., 1949.

Mellor, J. W.: "A Comprehensive Treatise on Inorganic and Theoretical Chemistry," vol. 7, pp. 276–483, Longmans, New York, 1957.

Moran, M. K.: bibliographies on Uses of Tin Compounds as Electroconductive, Abrasion Resistant, Refractive and Iridescent Coatings on Non-conductive Surfaces, *Bu.* CTG-1, 1959; Uses of Tin Chemicals as Catalysts, TC-2, 1962, and suppl. 1, 1969; Uses of Tin Chemicals as Stabilizers, ST-3, 1962; Uses of Stannous Chloride, SNC-1, 1964, M & T Chemicals Inc., Rahway, N.J.

Neumann, W. P.: "The Organic Chemistry of Tin," Wiley, New York, 1970.

Poller, R. C.: "The Chemistry of Organotin Compounds," Academic, New York, 1970.

Wright, P. A.: "Extractive Metallurgy of Tin," Elsevier, Amsterdam and New York, 1966.

—Marguerite K. Moran, *M & T Chemicals Inc., Rahway, N.J.*

Tires (See **Elastomers;** and **Rubber, natural.**)

Titanium (Gk. *Titan,* giant.) **Ti** = 47.88 (at. wt.); 22 (at. no.). Titanium is a relative newcomer to the family of engineering materials, its first significant use being in jet-engine components in the early 1950s. Pure Ti has about 56% the density of steel, but alloys can be formulated with strengths nearly equal to the strongest structural steel. The metal also exhibits very good resistance to corrosion, particularly in oxidizing environments.

Titanium appears in group IVB, period 4, of the periodic table. It is the fourth most abundant metallic element in the earth's crust and ninth most common element. There are eight isotopes of Ti, including five stable isotopes (46–50) with 48 the most abundant (73.99%) and three unstable isotopes, 44, 45, and 51. See also **Chemical elements.** Ti metal exists in two crystalline forms. Up to 1625°F (885°C), the structure is hexagonal close-packed; above that beta-transition temperature, Ti is body-centered cubic. See Table T-10.

Early Development: The discovery of Ti generally is credited to Gregor in 1789 and the naming of the element to Klaproth in 1795. Impure metal first was produced, followed by titanium tetrachloride; finally, in 1887, metal of 95% purity was produced by reduction of titanium tetrachloride with sodium. Little use was made of Ti until after 1860, when ferrotitanium was first used as an alloying element in steel and later as a deoxidizer in the production of steel.

The commercial production of titanium dioxide, TiO_2, as a white pigment was of major significance to metal development. Large-volume mining and concentration methods for recovery of mineral sands developed in the early 1900s and were later applied to Ti metal production. See also **Titanium dioxide.**

Van Aarkel's efforts in 1925 produced the first ductile Ti from the tetraiodide. In 1937, Kroll discovered that $TiCl_4$ could be reduced to Ti

TABLE T-10. *Some Properties of Commercially Pure Titanium*

Melting point:	
°F	3272
°C	1800
Boiling point:	
°F	5900
°C	3260
Density at 20°C:	
lb/in.^{3}	0.1628
g/cm^3	4.507
Specific heat at 25°C:	
Btu/(lb)(°F)	0.130
kcal/(g)(°C)	0.124
Linear coefficient of expansion (0–100°C), in./(in.)(°F) at 68°F	4.67×10^{-6}
Crystal structure:	
Below 1625°F	hcp
Above 1625°F	bcc
Resistivity at 68°F, $\mu\Omega$-cm	42
Modulus of elasticity, psi	16.8×10^6
Thermal conductivity at 400°F, (Btu)(ft)/(hr)(ft^2)(°F)	6.6
Chemical valence	4$^+$, 3$^+$, 2$^+$
Oxides (in order of decreasing stability)	TiO_2, Ti_2O_3, TiO

metal using Mg as a reducing agent under an inert-gas blanket. In 1946, the process was perfected and small-scale pilot production was initiated by the U.S. Bureau of Mines. Commercial production* started in 1948.

The young industry faced a significant marketing problem in that there were very few applications for the metal. To stimulate sponge production, the federal government negotiated several contracts in 1954 to procure sponge in excess of market requirements from the three producers in the United States. This stimulus led to the guaranteed availability of the metal and to ultimate market growth. Subsequent growth paralleled requirements for the metal in aircraft production. The base for use of Ti in aircraft has broadened, and new markets for the metal for corrosion-resistant equipment are significant.

Titanium Ores: Although Ti is a major constituent of a number of minerals, only rutile, TiO_2, is of commercial importance to the Ti metals industry. The sand dunes of Australia and Florida

*In 1947, Du Pont first made Ti sponge commercially available. In 1950, Titanium Metals Corporation of America was formed as a joint venture of National Lead and Allegheny Ludlum and became the first fully integrated ore-to-finished-metal producer.

are the most important commercial rutile sources. Australia supplies over 80% of the world's rutile requirements.

Ilmenite, $FeTiO_3$, is also found in rock and sand deposits but is little used commercially as a source for $TiCl_4$ and Ti metal. A number of projects under way in several countries to develop satisfactory ilmenite beneficiation methods may ultimately become an important source for the metal. Known reserves of ilmenite are 50 times those of rutile.

Large floating dredge concentrators are the most efficient method of mining the sand deposits. A large pond is formed to float the equipment. At the leading edge of the pond, cutter heads and hydraulic suction are used to bring crude sand through large pipes to the floating concentrator. The 1–4% heavy minerals in crude sand usually are concentrated by wet gravity methods, using tables, spirals, or sluices. Magnetic and electrostatic separation methods are applied to the gravity concentrate to produce several products, including rutile of about 95% TiO_2 content.

Processing Methods: Ti metal is produced commercially by reducing TiO_2 by either the magnesium or the sodium process. The description here applies to the techniques used by United States manufacturers of Ti metal.

Rutile of about 95% TiO_2 content and coke are charged into a continuous chlorinator. As Cl_2 gas is added, crude $TiCl_4$ is produced by an exothermic reaction. The reaction product vapors, including excess Cl_2, are passed through dust collectors, and the $TiCl_4$ is condensed and pumped to a settling tank to further remove suspended solids. The crude $TiCl_4$, a yellowish liquid, contains rutile, coke, $FeCl_3$, and about 2% other metal chlorides ($SiCl_4$, $SnCl_4$, VCl_4, $CrCl_3$, $CbCl_3$). The solids are removed by vaporizing and recondensing. The remaining metal chlorides are separated by appropriate chemical treatment and fractional distillation. Pure $TiCl_4$ is a colorless liquid.

Sponge Production: Magnesium metal pigs are charged into a steel reduction reactor, which then is welded shut to seal out the atmosphere. The reactor is evacuated, backfilled with inert gas, and then heated to 850°C to melt the Mg. $TiCl_4$ is added and reacts to form molten $MgCl_2$, Ti metal sponge, and excess Mg. The molten $MgCl_2$ is tapped from the bottom of the reactor and returned to Mg recovery, where Cl_2 and Mg are separated by electrolysis. The resulting Cl_2 is recycled to the chlorination step, and the Mg is cast into pigs for the tetrachloride reduction.

For *leached sponge,* the reactor vessel is cooled and then opened in a dry room and the Ti metal sponge is bored out. Normally, a titanium skull about $1-1\frac{1}{2}$ in. thick is left on the wall of the vessel to minimize pickup of Fe from the steel vessel during the reduction step. The sponge chips are separated and crushed to provide a uniform $\frac{5}{16}$-in. product. The sponge still contains both Mg metal and $MgCl_2$, which can be removed by acid leaching. Leaching with dilute aqua regia forms salts which are removed by water washing, both in continuous processes. The sponge is dried before further processing.

For *vacuum-distilled sponge,* a vacuum is applied to the hot reactor vessel immediately following the reaction. The Mg metal and chloride are vaporized and separated before the vessel is cooled.

The alternate process employs Na in place of Mg for $TiCl_4$ reduction. Liquid Na (above 100°C) and $TiCl_4$ are fed into a continuous argon-purged reactor where titanium dichloride and NaCl are produced. Additional Na is added in a higher-temperature batch reactor, where reduction to Ti metal is completed. After cooling to room temperature, the mixture is removed from the reactor, crushed, and dissolved in HCl. The Ti metal in powder form is recovered from the acid, water washed, and dried. The powder is formed into compacts for ingot production in much the same way as sponge from the Mg process.

Electrolytic Reduction: Methods for electrolytic metal production have been developed and may be used commercially after metal requirements exceed existing sponge capacity. The basic process employs a fused chloride salt electrolyte, graphite anodes, and metallic cathodes. $TiCl_4$ is fed into the cell, and the metal is deposited at the cathode. The cathodes are removed and cooled and the Ti separated from the cathode, followed by leaching, washing, and drying.

Ingot Production: Sponge of known analysis, crushed scrap, and alloy elements are blended to form a mixture of the required composition. The sponge is formed into compacts in a hydraulic press. The compacts are welded together to form a sponge electrode. Sponge electrodes are vacuum-melted into a single primary ingot. After cleaning, the primary electrode is double- or triple-vacuum melted to produce a high-quality finished ingot.

Mill Products: Titanium ingot is converted into mill products (billets, plates, strip, tubes, and wire) using conventional mill reduction practices. One important difference is that titanium's affinity for gases, such as O_2 and H_2, requires that high-temperature operations be conducted in vacuum or under an inert-gas blanket or that contaminated surfaces be removed from metal processed in air. In practice, a billet either must be ground or lathe-turned to remove scale after it is forged to size. Other products generally are chemically descaled and pickled. Vacuum annealing or inert gas is used on thin products to prevent contamination.

Alloys: Alloys are classified as alpha, alpha-beta, or beta, depending upon the phases present at room temperature. Alpha alloys generally result when the principal elements present are the alpha stabilizers, namely O_2, N_2, H_2, and C.

Alloy development has produced a class of superalpha alloys, which add substitution solid strengthening elements to alpha compositions to improve strength and stability in the higher temperature ranges.

Alpha-beta alloys and beta alloys contain increasing amounts of beta stabilizers, principally V, Mo, Fe, Cr, Mn, Ta, and Cb.

The alpha-beta class usually has higher room-temperature strength than alphas and can be heat-treated. Weldability varies with alloy additions. The annealed beta alloys exhibit poor thermal stability at temperatures over about 450°F (232°C) but have good formability and weldability. Beta alloys respond to aging heat treatments that precipitate some alpha and result in very high strength at room temperature.

Titanium is produced in all common mill forms. There are hundreds of experimental alloys, but only a few alloys are of major commercial importance. Table T-11 divides the alloys according to use. Commercially pure materials are designated by their ultimate strength, followed by letter *A*. Alloys are designated by the major alloy elements. The *ELI* suffix means extra low interstitial, or low gas levels that provide greater toughness.

Applications

Aircraft: The high strength-to-weight ratio and good atmospheric corrosion resistance of Ti alloys let to their use in airframe structural and engine components in the early 1950s. The earliest airframe applications used commercially pure Ti for simple nonstructural components, such as firewalls, where Al could not work, and a weight-savings advantage over steel was realized. Major structural components, such as main wing beams, now are constructed of Ti materials. In future aircraft, as friction-generated skin temperatures rise, making aluminum unsuitable, the use of Ti will increase.

In engines, Ti has been limited by strength and oxidation to the compressor. Most Ti-alloy development has been directed toward achieving higher creep resistance that can be used further back in the compressor or where greater engine temperatures are required for better performance.

Corrosion Resistance: Titanium's corrosion resistance results from a thin adherent oxide film that forms when the metal is exposed to the atmosphere, water, or any environment containing O_2. In some environments, such as dry Cl_2 or Br_2 gas, a very rapid (pyrophoric) reaction may take place. The addition of a small amount of water, however, results in good resistance to these environments.

Reducing acids, such as HCl and H_2SO_4, corrode Ti metal. The addition of HNO_3, other oxidizing agents, or positive metal ions (Cu, Ni, Fe) can result in complete passivity in the environment. Alloying with about 0.2% palladium also increases the resistance of Ti to reducing environments by providing a form of self-induced anodic protection to Ti. Anodic protection, using externally applied current, also is used to increase the range of Ti passivity.

Titanium is used widely in chemical process equipment. Most important commercially is its resistance to seawater at temperatures up to 250°F (121°C) under all conditions. Titanium also is used in the pulp and paper industry in chlorine and chlorine dioxide bleach-plant equipment, including pumps, washer screens, piping, and other equipment exposed to chloride-ion attack.

Titanium is used for marine-hardware components, including ball valves for naval vessels, pressure hulls or pressure bottles for deep-diving submersibles, and parts for high-performance sailboats.

Because of its resistance to seawater, titanium

TABLE T-11. *Major Titanium Alloys Classified by Application*

Airframe Alloys	Engine Alloys	Corrosion Alloys
Ti-75*A*	Ti-8Al-1Mo-1V	Ti-35*A*
Ti-SAl-2.5Sn	Ti-SAl-2.5Sn	Ti-50*A*
Ti-6Al-6V-2Sn	Ti-6Al-4V	Ti-65*A*
Ti-6Al-4V	Ti-6Al-2Sn-4Zr-2Mo	Ti-0.2Pd
Ti-7Al-4Mo	Ti-6Al-2Sn-4Zr-6Mo	Ti-2Ni
Ti-4Al-3Mo-1V		
Ti-8Mn		
Ti-13V-11Cr-3Al		

tubing is widely used as a heat-transfer surface for electric-power-generator steam-surface condensers and petroleum-refinery condensers where seawater cooling is used. Titanium tubing is used in multistage flash-distillation plants for converting seawater to fresh water. Titanium tubes also are used for evaporator heaters in the manufacture of caustic soda, salt, and corn syrup and for coolers for Cl_2 and HNO_3.

Titanium's corrosion resistance is of commercial value in anode or cathode applications for chemical or metal production. For example, Ti coated with appropriate noble metals is a standard material of construction, replacing graphite anodes in the manufacture of Cl_2. Titanium serves both as anode and cathode in bipolar cells for the manufacture of chlorates. Ti cathodes also are used as surfaces for plating Cu and other metals in electrowinning metal-recovery processes.

REFERENCES

Abkowitz, Stanley, John J. Burke, and Ralph H. Hiltz, Jr.: "Titanium in Industry," Van Nostrand Reinhold Company, New York, 1955.

Bachman, Robert Z., and Charles V. Banks: Critical Review of the Analytical Methods of the Titanium, Vanadium and Chromium Transition Element Groups, pt. 4 of "Analytical Chemistry," vol. 3, Pergamon, Oxford, 1963.

Codell, Maurice: "Analytical Chemistry of Titanium Metals and Compounds," Wiley, New York, 1959.

Comstock, George F.: "Titanium in Iron and Steel," Wiley, New York, 1955.

Elwell, W. T., and D. F. Wood: "The Analysis of Titanium, Zirconium, and Their Alloys," Wiley, New York, 1962.

Jaffee, R. I.: The Physical Metallurgy of Titanium Alloys, *Prog. Mater. Sci.*, vol. 4, 1953.

—James A. McMaster, *Titanium Metals Corporation of America, West Caldwell, N.J.*

Titanium Dioxide   Titanium dioxide is the principal white pigment of commerce, it has achieved a preeminence by virtue of its exceptionally high refractive index, negligible color, and great inertness. Total worldwide production of TiO_2 is estimated to be 1,600,000 metric tons (1970), of which approximately 700,000 short tons were produced in the United States. On a worldwide basis, about 60% of the TiO_2 produced goes into paint; 14% into paper; 12% into plastics, including floor coverings; 3% into printing inks; remaining uses include rubber, ceramics, roofing granules, and textiles.

Production Processes: There are two main processes for making raw pigmentary titanium dioxide, the sulfate process and the chloride process.

The sulfate route was the first process used commercially, starting about 1930 in Europe and the United States. It is a batch process, and nearly 80% of all TiO_2 produced today is made by this method. The newer process by the chloride route, a continuous process, was first put in commercial operation in the late 1950s. Most new plant capacities, announced or added during the mid-1960s, were employing the chloride process. Further advances in sulfate technology and a significant increase of chloride feed costs have made the economics of the process questionable. The trend reversed in recent years, and additional plant capacities announced by European producers were predominantly based on the sulfate process.

These two processes usually are followed by a finishing process, in which the pigmentary particles are deaggregated (if necessary) and then given a coating of hydrous oxides. Both the rutile and anatase crystal modifications are used as pigments. Rutile pigment is made by the chloride and sulfate processes, but anatase pigment is made only by the sulfate process. Both processes, as well as the finishing process, are illustrated schematically in Fig. T-2.

Chloride Process: The basic chemistry of the chloride process is:

Chlorination: $3TiO_2 + 4C + 6Cl_2 \rightarrow$
$$3TiCl_4 + 2CO + 2CO_2$$
Oxidation: $TiCl_4 + O_2 \rightarrow TiO_2 + 2Cl_2$

The chloride process requires a feedstock high in titanium and low in iron. Mineral rutile (95% TiO_2) is preferred, but leucoxene (65% TiO_2) can be used. Many routes for converting ilmenite to chloride feedstocks have been proposed, but few have yet been commercialized. The ore, mixed with coke, is chlorinated in a fluidized bed at about 900°C. The main product is titanium tetrachloride, but iron and other impurities are also chlorinated and must be removed by selective condensation and distillation. Selective reduction prior to the final distillation removes $VOCl_3$. The purified $TiCl_4$ is reacted with oxygen at temperatures above 1000°C in the presence of $AlCl_3$, which promotes the formation of rutile rather than anatase. Although the oxidation is exothermic, it is not self-sustaining and must be assisted by extra heat. This heat is supplied to either or both reactants through heat exchangers, by electrical dischargers, by the simultaneous combustion of CO, or by the use of a fluidized bed. The

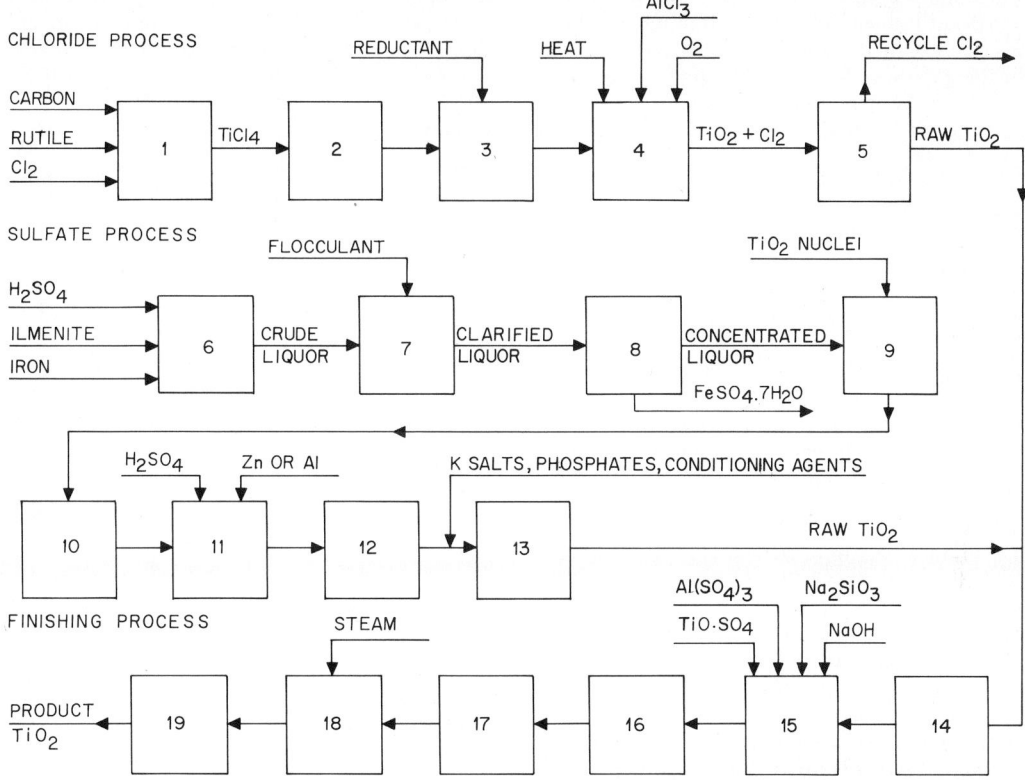

Fig. T-2. Schematic representation of materials flow in chloride and sulfate processes for manufacture of titanium dioxide. Finishing operations used with either process are identical and also are indicated. (1) Chlorination, (2) solids removal, (3) distillation, (4) oxidation, (5) cooling and solids preparation, (6) digestion, (7) clarification, (8) crystallization-concentration, (9) precipitation, (10) filtration, washing, and dewatering, (11) leaching, (12) filtration, washing, and dewatering, (13) calcination, (14) wet or dry milling, (15) coating with hydrous oxides, (16) filtration and washing, (17) drying, (18) fluid-energy milling, (19) packing final product TiO_2.

reactor must be designed to minimize the accumulation of solid products on the walls and burner parts and to give a product of the optimum crystal size (0.2 μm). Chlorine is regenerated in the oxidation step, and after cooling and separating it from the product it is recycled to the chlorinator. The raw titanium dioxide product is typically neutralized by washing in aqueous acid or alkali.

Sulfate Process: In this process, the feedstock is ilmenite (45–60% TiO_2) or a titanium-rich slag (70% TiO_2) obtained by electric smelting of ilmenite. The basic chemistry of the sulfate process is:

Digestion:
$$FeTiO_3 + 2H_2SO_4 \rightarrow$$
$$FeSO_4 + TiO \cdot SO_4 + 2H_2O$$

Precipitation:
$$TiO \cdot SO_4 + 2H_2O \rightarrow$$
$$TiO_2 \cdot H_2O + H_2SO_4$$
Calcination:
$$TiO_2 \cdot H_2O \rightarrow TiO_2 + H_2O$$

The ilmenite, or slag, first is digested with H_2SO_4, yielding a solid cake of titanium and iron sulfates, which is dissolved in water. To ensure that all the iron is in the Fe^{++} state, scrap iron is added to the solution until a small amount of Ti^{3+} is present. The solution is then clarified by an organic flocculant or by precipitated antimony sulfide. Some of the ferrous sulfate is crystallized from the solution (unnecessary when slag is used), which is then concentrated by vacuum distillation. The next stage, precipitation, is perhaps the most critical in the whole process. The concentrated liquor is nucleated, diluted with water, and boiled until about 96% of the titanium has been precipitated in the form of flocculated titanium dioxide (anatase) hydrate. The precipitate is filtered from the acid liquor and then leached under reducing conditions to remove most of the residual iron. Various *conditioning agents* are added, and the hydrate is then dried and calcined in a large rotary kiln having an exit temperature of about 900°C. The conditioning agents usually include a potassium salt and a phosphate and may also include zinc, antimony, and aluminum compounds. The purpose of these agents is to catalyze the formation of rutile from the anatase hydrate, to control the particle size of the product, and to improve other properties such as color, dispersibility, and photochemical stability. The raw product is then finished in a manner similar to that for the chloride process.

Finishing Process: In this process, the individual pigment particles are coated with a uniform layer of hydrous oxides in order to reduce their photocatalytic activity and to improve dispersibility. The raw chloride or sulfate product is usually milled before coating. This can be done dry in ring-roller mills or fluid-energy mills or wet in ball or sand mills. The milled, slurried product is coated by the successive addition of salt solutions, e.g., titanyl sulfate, aluminum sulfate, sodium silicate, and acids or alkalies. Many variations of this procedure are used to optimize the surface characteristics of the pigments for different applications. After coating, the pigments are filtered off, washed, dried, and fluid-energy-milled.

The several hundred commercial grades of pigmentary titanium dioxide vary in their crystal structure (anatase or rutile), particle size and shape, type of hydrous oxide coating, and content of other additives for specific applications. These pigments contain 80–99% TiO_2, the remainder being principally alumina and silica hydrates. Several nonpigmentary grades of titanium dioxide containing 99% TiO_2 are made for the glass, vitreous-enamel, welding-rod, and electroceramic industries.

—Oscar P. Muller, *Titanium Pigment Operations, NL Industries, Inc., New York*

TNT (Trinitrotoluene)   (See **Explosives**.)

Toluene   Toluene, C_7H_8, is a homologue of benzene in the C_nH_{2n-6} series of aromatic hydrocarbons. Also called toluol, methylbenzene, and methyl benzol, toluene is a colorless liquid with a distinc-

tive odor. It is flammable, burning with a smoky flame when ignited. Physical properties include formula weight 92.13; melting point −95°C; boiling point 110.8°C; specific gravity 0.866 at 20°C (referred to water at 4°C). It is practically insoluble in water (0.05 part in 100 parts at 16°C) and fully miscible with alcohol, ether, chloroform, and many organic liquids. Toluene fumes are poisonous. Unlike benzene, toluene is not easily purified by crystallization.

Chemically, toluene behaves much like benzene with exception of those properties that are due to the presence of the methyl group. In some respects, the methyl group gives toluene some of the behavior patterns of a paraffin. As an example, the hydrogen of the methyl group can be displaced by chlorine, which in turn can be displaced by

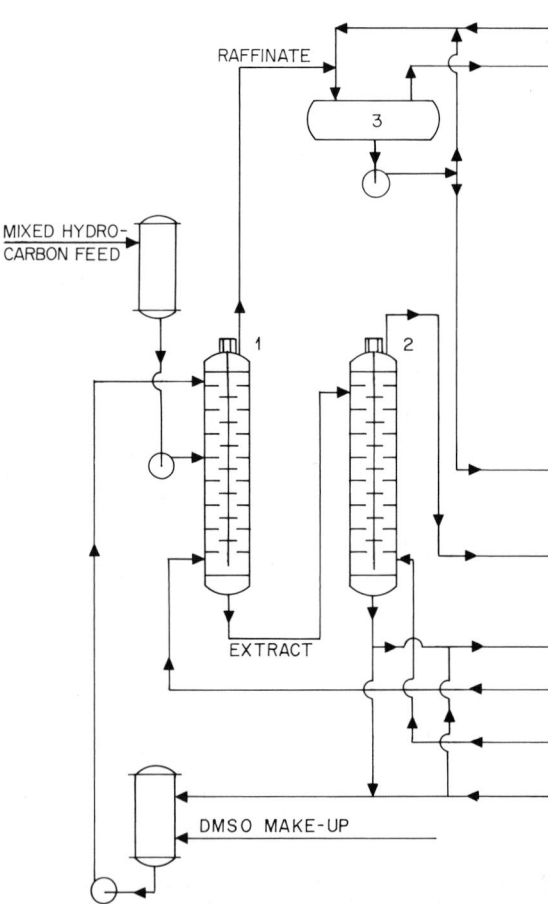

an amino or hydroxyl group. In this light, toluene may be regarded both as a derivative of benzene and as a monosubstitution product of methane.

The industrial pure grade of toluene distills between 108.6 and 112.6°C and is a water-white liquid with a specific gravity of 0.864–0.874 and a flash point of 2–5°C.

Production: United States production of toluene exceeds 2.5 million tons/yr, with over 95% coming from petroleum sources. The remainder is derived from coal gas and coal tar. See also **Benzene;** and **Coal tar and derivatives.** About one-half of the toluene produced is dealkylated and is a prime source of benzene. There is a small but growing demand for toluene in the production of diisocyanates. Use of toluene in fuels is diminishing.

Toluene is used in connection with several syn-

theses. Monochlorotoluene, $CH_3C_6H_4Cl$, is a solvent widely used for rubber and synthetic resins. Methyl cyclohexane, $C_6H_{11}CH_3$, produced by hydrogenating toluene, also is an effective solvent for oils, fats, waxes, and rubbers. Trinitrotoluene (TNT), $C_6H_2(CH_3)(NO_2)_3$, is a principal ingredient of many explosives. See also **Explosives.** Toluene reacts with concentrated H_2SO_4 to form *o*- and *p*-toluene sulfonic acids, $CH_3 \cdot C_6H_4 \cdot SO_3H$. Saccharin, the artificial sweetener, is derived from the ortho compound. An important antiseptic, chloramine T, is derived from the para compound. Toluene also is used as a charge stock in the production of benzoic acid and, in turn, in the manufacture of phenol. See also **Phenol.**

Some of the processes designed to produce benzene from aromatic stocks also can be adjusted

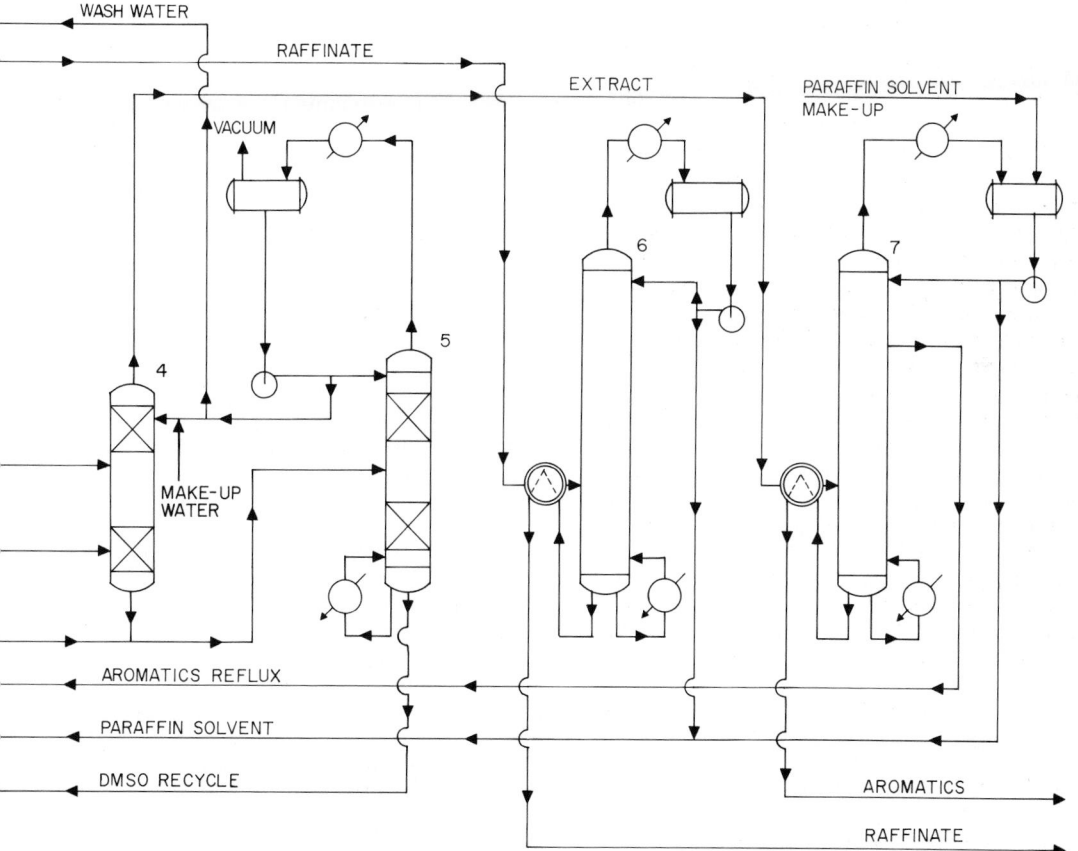

Fig. T-3. Aromatic-extraction process using DMSO as a solvent for producing toluene, benzene, and xylenes. (1) Aromatics extractor, (2) DMSO extractor, (3) raffinate wash, (4) extract wash, (5) DMSO distillation column, (6) raffinate distillation column, (7) aromatics distillation column.

to produce toluene, separately or simultaneously. A relatively recent aromatics-extraction process for producing toluene, benzene, and xylenes from mixed hydrocarbon stocks is shown in Fig. T-3. In the first extraction step, the aqueous dimethyl sulfoxide (DMSO) solution, used as the solvent, passes countercurrently against the mixed-hydrocarbon feed. A reflux of aromatic and paraffinic hydrocarbons is introduced at the bottom of the extractor. The reflux flow rates, reflux composition, and water content are chosen in accordance with the feed composition and the purity desired of the aromatic products. The extract phase (rich in aromatics) then enters the second extraction column, where it meets a countercurrent stream of paraffin solvent. This solvent displaces about 99% of the DMSO, which is recycled as a bottom stream to the first extractor. The raffinate from the first extractor and the overhead from the second extraction column are washed with water to separate out any entrained DMSO (about 1% and eventually recovered by vacuum distillation). See also **Distillation.**

A stream of light paraffins that can be used as an additional source of solvent for the DMSO extraction is provided by distillation of the raffinate. The major part of the paraffin solvent is recovered during distillation of the aromatic extract with some makeup added at the overhead condenser. The final product stream comprises benzene, toluene, and xylenes, which are separated subsequently into pure components.

DMSO is well known for its high solvent power and selectivity. It is to be noted that instead of separating the solvent-extract phase by distillation, this processes uses a second extraction step. The extraction steps take place at approximately room temperature.

See also **Petrochemical complex.**

Toluidines   (See **Amines.**)

Tolyl   (See **Radicals.**)

Tool Steels   (See **Iron and steel;** and **Tungsten.**)

Torpex   (See **Explosives.**)

Towers   (See **Absorption;** and **Distillation.**)

Town Gas   (See **Gasification;** and **Synthesis gas.**)

Toxins   (See **Antibiotics.**)

Tragacanth   (See **Gums.**)

Trans Compounds   (See **Isomerism.**)

Transactinium Earths   (See **Chemical elements.**)

Transfer Molding   (See **Plastics.**)

Transistors   (See **Gallium; Germanium;** and **Silicon;** also consult the **Subject Index.**)

Transport Systems, Heat   (See **Energy systems for processes.**)

Transuranium Elements   (See **Chemical elements;** and **Uranium.**)

Travertine   (See **Limestone.**)

Tray Columns   (See **Absorption;** and **Distillation.**)

Triacetate   (See **Acetate fibers;** and **Cellulose ester plastics, organic.**)

Triads   (See **Chemical elements.**)

1,1,1-Trichloroethane   Known as methyl chloroform and α-trichloroethane, 1,1,1-trichloroethane (1,1,1-Tri)

$$
\begin{array}{ccc}
 & H & Cl \\
 & | & | \\
H- & C- & C-Cl \\
 & | & | \\
 & H & Cl
\end{array}
$$

is one of a sequence of compounds in which chlorine atoms have replaced one or more hydrogen atoms on ethane or ethylene. Chemicals within this sequence vary distinctly in physical and chemical properties. For instance, 1,1,2-trichloroethane is a toxic liquid, boiling at 114°C, whereas 1,1,1-Tri boils at 74°C and is among the least toxic of the industrial solvents. Certain properties, however, can be forecast for the series. Each added chlorine atom reduces the volatility and the flammability of the chemical compared with the previous compound with one less chlorine atom. Thus, 1,1-dichloroethane has a flash point of 30°C and boils at 57°C, while 1,1,1-Tri is nonflammable. Molecules in this series containing $1\frac{1}{2}$ atoms of chlorine or more per carbon atom are nonflammable.

Some chemicals in the chlorinated ethane series are important as intermediates, including ethyl chloride to make tetraethyl lead and vinyl chloride and vinylidene chloride for polymer manufacture. Other members of this series are the chlorinated solvents. TCE, trichloroethylene, and perchloroethylene are the most important of this group. See also **Perchloroethylene;** and **Trichloroethylene.** These products have strong solvent action on organic chemicals, especially oils, greases, waxes, fats, and tars. These products, however, would be economically noncompetitive with alcohols, ketones, or petroleum solvents without the key advantage of nonflammability.

Still other members of the series are either flammable (1,1-dichloroethane, 1,2-dichloroethane, or *cis-* or *trans-*1,2-dichloroethylene) or highly toxic (*s*-tetrachloroethane, pentachloroethane, or hexachloroethane). The commercial importance of these latter chemicals is less, although they play important internal synthesis roles.

Early Chlorinated Solvents: The earliest commercialized chlorinated solvent, carbon tetrachloride, CCl_4, was broadly used both in metal cleaning and dry cleaning until its toxic hazard was recognized. This accelerated the research for safer solvents and led ultimately to 1,1,1-Tri for the room-temperature metal-cleaning uses of CCl_4.

TABLE T-12. *Some Properties of 1,1,1-Trichloroethane*

Boiling point:	
°F 165	
°C 74	
Freezing Point:	
°F . . . . −34	
°C . . . . −37	
Flash point . . . . None*	
Fire point . . . . None	
Threshold limit value, ppm . . . . 350†	
Vapor density 4.55	
Specific gravity 1.320	
Density, lb/gal 10.99	
Specific heat, Btu/(lb)(°F) 0.25	
Heat of vaporization, Btu/lb 102	
Viscosity at 25°C, cP 0.79	

*Under controlled laboratory conditions and a high-energy ignition source (10,000 V), the material can be ignited between the limits of 8–10.5% by volume in air at 25°C. These concentrations are most unlikely where proper ventilation is practiced.

†ANSI recommends a time-weighted average concentration of a maximum 400 ppm by volume in air for an 8-hr work day. Where the 400-ppm average is maintained, a maximum ceiling of 500 ppm is acceptable. Air concentrations not exceeding 800 ppm can be tolerated for intervals of less than 5 min, not more often than once every 2 hr if the two prior conditions are observed.

Perchloroethylene ultimately took over the dry-cleaning function.

Regnault discovered 1,1,1-Tri about 1840. Although inherently stable to light, oxidation, and most industrial metals, pure unstabilized TCE can react with aluminum. The development of a chemical inhibitor system to prohibit this reaction was critical to the successful displacement of CCl_4. An inhibitor system was first described in U.S. Patent 2,811,252. Some uses were recorded as early as 1947, but extensive commercial use depended on stabilization, which did not become available until 1954.

The properties of 1,1,1-Tri are summarized in Table T-12.

Manufacture: This is accomplished in a variety of ways, as described in the patent literature.* Variations occur in the raw materials, catalyst, gaseous or liquid phase, and chlorination or hydrochlorination reactions. Examination of the patents is suggested for greater detail. The key process reactions are summarized in the following equations.

$$CH_2{=}CCl_2 + HCl \xrightarrow{FeCl_3} CH_3{-}CCl_3 \qquad (1)$$

$$CH_3{-}CH_3 + Cl_2 \xrightarrow{CH_3{-}CH_2Cl} \begin{cases} CH_3{-}CCl_3 + \\ CH_2{=}CHCl + HCl \\ + CH_2{=}CCl_2 \end{cases} \qquad (2)$$

$$CH_3{-}CHCl_2 + Cl_2 \xrightarrow{CS_2} CH_3{-}CCl_3 + HCl \qquad (3)$$

$$CH_2{=}CHCl + HCl \rightarrow CH_3{-}CHCl_2 \qquad (4a)$$
$$\text{recycle}$$

$$CH_3{-}CHCl_2 + Cl_2 \xrightarrow[420{-}700°F]{} HCl + CH_3{-}CCl_3 \qquad (4b)$$

$$CH_2{=}CH_2 + 2Cl_2 \rightarrow HCl + CH_2Cl{-}CHCl_2 \qquad (5a)$$

$$CH_2Cl{-}CHCl_2 + NaOH \rightarrow CH_2{=}CCl_2 + NaCl \qquad (5b)$$

$$CH_2{=}CCl_2 + HCl \xrightarrow[FeCl_3]{liq.} CH_3{-}CCl_3 \qquad (5c)$$

Note: HCl is obtained from reaction (5a).

$$CH_3{-}CHCl_2 + Cl_2 \xrightarrow{90{-}170°C} CH_3{-}CCl_3 + HCl \qquad (6)$$

Obviously, numerous by-products are formed in most of these reactions. In other cases, as in Eq. (2), coproducts are formed in the same plant.

Industrial Cleaning Applications: 1,1,1-Tri was first successful as a replacement for CCl_4 to avoid the toxic hazard of that product. This application, which has continued to be a major use of 1,1,1-Tri, involves the removal of lubricating oils, greases, and tars in metal manufacturing and

*U.S. Patents 2,209,000, 3,304,337, 3,019,175, 3,059,035, 3,065,280, and 3,357,907.

maintenance operations. Cleaning of metal surfaces is accomplished by immersion in one or more tanks or by spraying, wiping, or brushing with the solvent. This manner of cleaning has been called *cold cleaning* to distinguish it from vapor degreasing (see **Trichloroethylene**). A wide variety of parts, from electronic components (including printed circuit boards) to massive parts for the automotive, aircraft, or railroad industries, are cleaned with 1,1,1-Tri.

Air-Pollution Factors: Although trichloroethylene has been the common solvent for vapor degreasing, 1,1,1-Tri is an important and growing vapor-degreasing solvent. Lower toxicity, lower boiling point, and more selective solvency (critically important in cleaning electric motors or printed circuits) have increased the use of 1,1,1-Tri in vapor degreasing. Recognition of air-pollution problems and the photoreactivity of trichloroethylene have stimulated interest in a substitute for it in vapor degreasing. First to control trichloroethylene emissions was the city of Los Angeles in the historic Rule 66. More recently, other cities, counties, and states have followed.

1,1,1-Tri is among the saturated halogenated hydrocarbons described as "virtually unreactive in the formation of oxidants" by the Environmental Protection Agency under the Clean Air Act of 1970. In the same reference, perchloroethylene is identified as non-oxidant-forming. However, 1,1,1-Tri has physical properties most similar to trichloroethylene, which permit it to be used in equipment designed for trichloroethylene. Thus 1,1,1-Tri has often been selected to comply with air-pollution regulations.

Both cold cleaning and vapor degreasing offer economical means for avoiding pollution problems. Proper design and operating practices permit minimum losses of 1,1,1-Tri from these systems. Soils removed by these processes can be concentrated for ease of disposal in a simple still, thanks to the nonflammability of 1,1,1-Tri, while the solvent is returned to the cleaning operation. The concentrated soils, solids, and oils, are nonvolatile and usually non-water-soluble. Consequently, the soil disposal does not contribute to air or water pollution.

Ultrasonic Cleaning: 1,1,1-Tri is a common ultrasonic cleaning agent. Cleaning action of the solvent is supplemented by sonically induced cavitation, producing strong physical cleaning. Available electricity is converted to approximately 20 Hz and directed to a crystal or magnetic transducer. The transducer expands and contracts with each cycle, producing strong shock waves in the solvent. These shock waves cause the momentary formation of low-pressure bubbles throughout the solvent which then collapse. The cavitation and collapse cause a strong scrubbing action, frequently used in the manufacture of electronic components and printed circuits, as part of a cold-cleaning or vapor-degreasing operation. Numerous other parts demanding ultracleanliness also use this technique.

Other Uses: Because of the properties previously described, 1,1,1-Tri finds numerous other uses, including adhesive formulations, mold-release carriers, clothing spot removers, laboratory solvent, aerosol solvent and vapor-pressure depressant, wig cleaners, asphalt extracting agent from paving aggregate, shoe polishes, and as a solvent to apply water and oil repellant to paper and textiles. 1,1,1-Tri is blended with other solvents to reduce their flammability or to provide added solvent properties. As little as 5% 1,1,1-Tri incorporated in metal-cutting compounds shows improved tool life and cutting speeds, particularly in drilling and tapping operations on hardened steels, Ti, Ta, and Mo. The material should not be used in cutting Al or other reactive metals. See also **Solvents**.

—Kenneth S. Surprenant, *The Dow Chemical Company, Midland, Mich.*

Trichloroethylene   Trichloroethylene, $CCl_2=CHCl$, the physical properties of which are summarized in Table T-13, is part of a family of chemicals including methyl chloride, methylene chloride, chloroform, carbon tetrachloride, per-

TABLE T-13. *Some Properties of Trichloroethylene* *

Boiling point:	
°C	87.2
°F	189
Freezing point:	
°C	−86.6
°F	−124
Flash point	None
Fire point	None
Vapor density	4.53
Specific gravity	1.459
Density, lb/gal	12.14
Specific heat, Btu/(lb)(°F)	0.22
Heat of vaporization, Btu/lb	101.6
Viscosity at 25°C, cP	0.54

*As a pure chemical, trichloroethylene is subject to decomposition from exposure to heat, light, metals, O_2, and reactive chemicals. Stabilizers are added to the compound to offset these deficiencies.

chloroethylene, and others that have unusual properties, particularly with regard to flammability. All members of this family contain one or more chlorine atoms per molecule, which causes it to have a lower vapor pressure than the corresponding hydrocarbon with one less chlorine and one more hydrogen atom. Each additional chlorine atom added to a molecule reduces its flammability. When the ratio of chlorine to carbon atoms equals or exceeds 1.5 : 1, the products become essentially nonflammable. Many of the commercially important members of this chemical family are liquids, essentially nonflammable, and exceptionally good solvents for organic chemicals, oils, greases, and tars. See also **Chlorine organics.**

Trichloroethylene is widely used as a solvent in vapor-degreasing applications, which account for about 95% of the production of this material in the United States. Other applications include caffeine extraction from coffee, wool-fabric scouring, adhesives, formulation of solvent blends, and nominal amounts for dry cleaning and as an anesthetic in childbirth.

Vapor Degreasing: This process, introduced in 1930, provides rapid and economical cleaning of greases, oils, tars, and waxes (used as manufacturing lubricants) from metal parts. In specially designed equipment, the solvent (most frequently trichloroethylene) is heated to a rapid boil. The resulting vapors rise in the equipment and are continuously condensed near the top of an enclosed chamber by cooling coils. The chamber thus becomes essentially filled with pure solvent vapors. The vapors remain distinct from the air above because they are 4.5 times heavier than air. The solvent condensate is collected in the condensate trough and routed through a water separator back to the boiling sump. Water is removed by gravity separation and decanted.

Metal parts are cleaned in this system by lowering them into the solvent vapors, where the vapors condense on the parts until they are heated to the vapor temperature. As a consequence, the parts are cleaned with essentially distilled solvent even though the solvent in the boiling sump may be contaminated with both soluble and insoluble soil. The time required to heat ordinary parts to vapor temperature usually is 2 min or less. Heat supplied to the parts during cleaning is sufficient to flash-evaporate the film of solvent left on the parts as they are withdrawn from the vapors. After this process, the parts are clean and dry and ready for painting, packaging, or assembly.

Vapor degreasing makes heavy demands on the solvent used. An ideal solvent for this use should have the following characteristics, and it is to be noted how well trichloroethylene meets these demands: (1) good solvent power; (2) inertness to industrial metals, (3) low flammability, (4) low toxicity, (5) low latent heat, (6) high vapor density, (7) chemical stability to heat and soils, (8) an intermediate boiling point (100–250°F, 37.8–121°C), (9) low cost, and (10) recoverability by distillation with stabilizers. While trichloroethylene satisfies the bulk of this market, both 1,1,1-trichloroethane and perchloroethylene are used widely as well. Specialty vapor-degreasing needs are fulfilled with methylene chloride and 1,1,2-trichlorotrifluoroethane.

Handling Properties: Although this compound is essentially nonflammable (with no fire or flash point as determined by standard methods, including Tag open-cup or Tag closed-cup methods), trichloroethylene vapors can be decomposed by exposure to open flame, hot metal surfaces, or welding. Decomposition products include phosgene, chlorine, and hydrogen chloride, all toxic and corrosive. The strong odors of Cl_2 and/or HCl may not serve as adequate warning. ANSI has suggested the following maximum exposures for people, assuming an 8-hr workday: (1) a time-weighted average concentration of 100 ppm or less in air; (2) a ceiling concentration of 200 ppm where 100 ppm average or less is maintained throughout the day; and (3) peak exposures of up to 300 ppm for not more than 5 min occurring once every 2 hr or less where other exposures comply with 1 and 2.

Manufacture: Three processes have been used, the acetylene process, direct chlorination, and oxychlorination. The acetylene process has historical precedence but has been abandoned due to the raw-material cost of acetylene compared with ethylene. In this process, acetylene is reacted with chlorine to produce 1,1,2,2-tetrachloroethane. This intermediate product then is dehydrochlorinated to produce 1,1,2-trichloroethylene:

$$CH{\equiv}CH + 2Cl_2 \rightarrow Cl_2HC{-}CCl_2H$$
$$Cl_2HC{-}CCl_2H \rightarrow Cl_2C{=}CClH + HCl$$

The direct-chlorination process uses ethylene and chlorine as raw materials. The initial step is chlorination of ethylene to 1,2-dichloroethane. This intermediate product is subjected to a complex series of dehydrochlorinations and chlorinations with the formation of several new intermediates, finally leading to trichloroethylene.

By-products are produced. Simplified equations may be given to describe the reactions:

$$CH_2{=}CH_2 + Cl_2 \rightarrow CH_2Cl{-}CH_2Cl$$
$$CH_2Cl{-}CH_2Cl + 3Cl_2 \rightarrow CHCl{=}CCl_2 + 3HCl$$

Oxychlorination uses ethylene as a starting material but derives the needed chlorine from the reaction of HCl and O_2. Water formed in this reaction causes severely corrosive conditions. Intermediates result from sequential chlorination, dehydrochlorination, and hydrochlorination. The overall reaction is

$$CH_2{=}CH_2 + 3HCl + 1\tfrac{1}{2}O_2 \rightarrow$$
$$CCl_2{=}CHCl + 3H_2O$$

Trichloroethylene may be shipped without U.S. Interstate Commerce Commission restrictions. The material is supplied in tank cars, tank trucks, compartmented tank cars or trucks, and 55-gal drums.

—Kenneth S. Surprenant, *The Dow Chemical
Company, Midland, Mich.*

Trickling Filter   (See **Water treatment.**)

Tricot Yarns   (See **Acetate fibers.**)

Triethanolamines   (See **Ethanolamines.**)

Triethylene Glycol   (See **Ethylene glycol.**)

Triglycerides   [See **Alcohols, fatty** (via **hydrogenation**); **Soaps**; and **Vegetable oils.**]

Trimerization   (See **Ketones**; and **Polymerization.**)

2,2,4-Trimethylpentane   (See **Alkylation.**)

Trinitrotoluene   (See **Explosives**; and **Toluene.**)

Trioxane   (See **Formaldehyde.**)

Triple Superphosphate   (See **Fertilizers.**)

Tripoli   (See **Abrasives.**)

Trisodium Nitrilotriacetate (NTA)   (See **Detergents.**)

Tritium   Tritium is the radioactive isotope of hydrogen having mass number 3 and is one form of heavy hydrogen, the other being deuterium. See also **Chemical elements.**

Trona   (See **Soda ash.**)

Tryptophan   (See **Amino acids**; and **Peptides and proteins.**)

Tumblers   (See **Mixing and blending, solids-solids.**)

Tung Oil   (See **Vegetable oils.**)

Tungsten   (Sw. *tung sten*, heavy stone.) **W** (G. *Wolfram*, meaning uncertain but influenced by wolf) $=183.85$ (at. wt); 74 (at. no.). Tungsten is characterized by a high melting point (3410°C), the highest of any metal and of any element except carbon, and by low vapor pressure, high hardness and density, and good electrical conductivity. Tungsten, classified as a refractory metal, is in group VIB, along with Cr and Mo, and in period 6 of the periodic table. Five isotopes, 180, 182, 183, 184, and 186, occur naturally. Unstable isotopes include 176–178, 181, 185, 187, and 188. See also **Chemical elements.**

The iron manganese tungstate mineral wolframite was first described about 1574 and was originally believed to be a mineral of Sn, with ores of which it is commonly associated. The origin *Wolfram* is obscure but may refer to the wolflike characteristic of tungsten, described by the early miners as devouring tin and causing low recoveries in the smelting operation.

The name tungsten, first used about 1758, is an adaptation of the Swedish and was applied to the mineral scheelite. In 1781, K. W. Scheele identified the acid constituent, tungstic oxide, of the calcium tungstate mineral scheelite, later named in his honor. J. J. de Elhuyer and F. de Elhuyer showed that tungstic oxide is a constituent of wolframite and are credited with first production of metallic tungsten in 1783 by carbon reduction of the oxide. In 1908, W. D. Coolidge obtained a patent for producing ductile tungsten wire. High-speed steels using tungsten were developed during World War I, and cemented tungsten carbide bonded with Co, in which tungsten carbide is liquid-phase-sintered with Co, was developed in the Krupp Laboratory in 1927.

Ores and Production: W never occurs in the free state. It is one of the rarer elements, constituting $7 \times 10^{-3}\%$ of the earth's crust, where it is twenty sixth in order of abundance among the elements. It is less plentiful than Sb, Ni, Sn, and Cr but more abundant than Cd, Zr, Hg, and Ag. With exception of tungstite, WS_2, tungsten always occurs as tungsten trioxide combined with oxides of Fe, Mn, or Ca and to a limited extent the oxides of Pb and Cu. W minerals are classified into two groups, wolframite and scheelite. The wolframite group is an isomorphous series consisting of hueb-nerite, wolframite, and ferberite, which are chemical variations of $(Fe,Mn)WO_4$. When the mineral contains 80% or more $MnWO_4$, it is called hueb-nerite; and when it contains 80% or more $FeWO_4$, it is called ferberite. The color varies from brown to black, the Mohs' hardness is 5–5.5, and specific gravity 7.1–7.5. The mineral scheelite, $CaWO_4$, has a specific gravity of 5.4–6.1, color ranging from brown to yellow to white, and Mohs' hardness 4.5–5. Pure scheelite exhibits a blue fluorescence under ultravoilet light, although contamination with powellite, $Ca(Mo,W)O_4$, can cause colors varying from white through gradations of yellow to flesh color.

Tungsten minerals are usually found in pegmatites, sills, and batholiths. Other minerals that accompany the tungsten minerals are those normally found in the same type of geological formations, and include cassiterite, quartz, feldspar, sulfides, arsenites, apatite, calcite, molybdenite, and bismuthinite. The most important W deposits in the world are, in order of decreasing magnitude, mainland China, United States, Korea, Bolivia, Portugal, Burma, Australia, Indochina, Spain, Argentina, Brazil, Russia, Peru, Zaïre (formerly Belgian Congo), Rhodesia, and Malaya. The most important producing states in the United States are California, Nevada, South Carolina, Idaho, and Colorado. W ore in the United States occurs underground and in low concentrations of 0.5–3% of WO_3, usually less than 1%. W ore is marketed on a short-ton unit basis, a unit being $\frac{1}{100}$ short ton, or 20 lb of WO_3, which contains 15.862 lb of W.

The ores are beneficiated to concentrates by crushing, grinding, ball milling, magnetic separation, flotation, tabling, roasting, leaching, and other upgrading methods.

The production of pure tungsten consists of extraction of tungsten from the ore, preparation of a pure ammonium paratungstate or tungstic oxide, reduction to metal powder, and conversion of the powder to massive metal by pressing, sintering, and mechanical working at elevated temperatures.

Wolframite ores can be digested with a strong solution of an alkali hydroxide, either under pressure or by boiling at atmospheric pressure. Alternately, the ore can be mixed with an alkali-metal carbonate or hydroxide and fused to form the water-soluble alkali-metal tungstate. The alkali tungstate is purified by recrystallization, leached with water, then converted into tungstic acid, H_2WO_4, by neutralization with HCl.

Scheelite is decomposed with HCl to form insoluble tungstic acid, which is then leached with NH_4OH in one process or with NaOH in another. Purification of the ammonium tungstate or the sodium tungstate solutions is usually performed to remove elements such as As, P, and Mo. Specific techniques are proprietary to the producers, but purification techniques include precipitation of As and P as magnesium arsentate and phosphate by additions of $MgCl_2$ (or NH_4Cl for treatment of sodium tungstate). Mo can be removed by the addition of Na_2S, followed by precipitation of W with $CaCl_2$ while the Mo remains in solution as sulfomolybdate, and the precipitated calcium tungstate is again digested with acid as before. Removal of impurities is also performed by proprietary liquid-liquid solvent-extraction processes.

Liquid ion exchange is used in some cases to convert the sodium tungstate to ammonium tungstate, which is subsequently crystallized as ammonium paratungstate, $5(NH_4)_2O \cdot 12WO_3 \cdot 11H_2O$. The ammonium paratungstate can be decomposed by heating to approximately 400°C to drive off the NH_3, resulting in a blue oxide, a slightly oxygen-deficient form of WO_3 which sometimes is assigned a chemical formula such as W_4O_{11}. Tungstic oxide, a yellow oxide, essentially pure WO_3, can be produced by a variation of the above chemical processes.

Chemistry: The valence of W in the maximum oxidized state is 6. In higher valences it has acidic properties, and in lower valences it has basic properties. Its position in the electromotive series depends upon the treatment of the surface and on the solution in which the tests are made, values varying between −0.7 and +0.7 V, probably due to the presence of a passivating oxide surface layer.

The oxides WO_2 and WO_3 are common commercial products. Mixtures of quadrivalent and hexavalent oxides form a complex system of the formula W_4O_{11}. Tungsten trioxide, WO_3, is the anhydride of the weak tungstic acid. Light yellow orthotungstic acid, H_2WO_4, is precipitated by addition of a strong acid to a hot solution of an alkali tungstate. Alkali thiotungstates are soluble, and acidification causes precipitation of part of the tungsten as tungsten trisulfide, WS_3. Tungsten disulfide, WS_2, is formed by reaction of sulfur with hot tungsten.

Tungstic acid has a tendency to form isopoly and heteropoly ions, which reflect in a series of orthotungstates, metatungstates, and paratungstates. Many heteropoly compounds have been described in which elements such as B, Si, P, and As serve as central atoms or nuclei. Phosphotungstic acid and silicotungstic acid are typical examples.

Tungsten forms compounds with all halogens. As a fluoride, tungsten exhibits a valence of 6; as a chloride or bromide, valences of 6, 5, 4, and 2; and as an iodide, valences of 4 and 2. Oxyhalides of W with a valence of 6 have formulas such as WOX_4 and WO_2X_2.

N_2 reacts very slowly with tungsten, and tungsten nitride, WN_2, forms only when W powder is subjected to NH_3 at elevated temperatures. No significant reaction has been noted between N_2 or NH_3 and massive W metal.

Binary Tungsten Hard-Metal Compounds: Such compounds as WC, W_2C, WB_2, and WSi_2, are characterized by their high hardness and melting points. These hard-metal compounds are usually formed by direct contact of a compound of C, B or Si with heated W powder, or they may be formed by a direct synthesis reaction. The carbides also can be formed by reduction of tungsten oxides with carbon. The tungsten carbides are the most important of the W hard-metal compounds. Tungsten carbide, WC, has a melting point of 2870°C, and ditungsten carbide, W_2C, has a melting point of 2860°C. Monotungsten carbide, WC, commonly called *tungsten carbide,* is usually prepared by chemical combination of W metal powder and finely divided lampblack and is formed at a temperature of 1450–1550°C. Also, WC can be produced by carbon reduction of ore by aluminothermic reduction or by forming the carbide in a molten menstruum bath.

Tungsten Powders: Different forms of W oxides have widely different physical properties and can be used to produce W powders with a range of physical properties. Thus, W powders having average particle sizes ranging from 0.5 to about 10 μm can be made using specific raw materials, temperature of reduction, and type of reducing agent. W raw materials can be reduced by carbon, hydrocarbon gases, or H_2, the purest metal being produced by H_2 reduction. The hydrogen-reduced material is used principally in pure, ductile metal and carbide applications, while carbon-reduced tungsten, which contains some residual carbon, is used for manufacture of cast tungsten carbide.

The tungsten powders are pressed in a mechanical press or by hydrostatic compaction to form green compacts. The pressed bars are presintered at about 1200°C to improve strength and vaporize some impurities. Then the presintered bars are sintered by induction or by passing a heavy electric current through them, thus heating the bars by their own resistance to a sintering temperature of about 2700°C while flowing H_2 through the sintering furnace. The square or round sintered ingots are worked to wire by swaging the bars through a series of progressively finer dies, with the material initially preheated to about 1600°C, then progressively reducing the temperature as the diameter is reduced finally to about 0.2 in. Wire drawing is carried out with conventional equipment. The larger diameters are drawn initially at about 800°C, and the drawing temperature is progressively lowered as finer sizes are produced. W sheet is produced by hot rolling rectangular sintered bars.

Metalworking and Properties: Fabrication and forming operations can be performed on W by working the material above its ductile-to-brittle transition temperature and observing other precautions to avoid brittle failures. Techniques for machining, cleaning, annealing, brazing, and welding have been developed. Shapes of W also can be made by plasma forming processes to make forms such as crucibles, tubes, and forging blanks. W also can be deposited on metal and other substrates by chemical vapor-deposition methods.

The coefficient of thermal expansion of tungsten is essentially linear with temperature over a wide range and is about 26% that of copper at room temperature. The electrical conductivity of tungsten is about one-third of that of annealed Cu but is higher than that of Ni, phosphor bronze, Hg, Pt, or Fe. The resistivity increase of W with temperature is essentially linear.

W exhibits exceptionally high strength and hardness at room temperature and elevated temperatures. Fine W wire reportedly has the highest tensile strength of all materials prepared by commercial production methods. In spite of its high density, 19.3 g/cm³, the strength-to-weight ratio of W at temperatures above about 1650°C is superior to that of all other metals tested.

Silicide-base coatings developed to protect W against oxidation at temperatures as high as about 1650°C are similar to coatings developed for Mo. These coatings are applied by casting, hot dipping, cementation, flame spraying, cladding, electrophoresis, electroplating, suspension methods, and vapor-phase reduction.

Principal Uses: About 50% of all W used is in the form of sintered tungsten carbides for cutting tools and wear-resistant parts. Wire for electrical lamps and metal shapes for aerospace and defense products consume 15%. High-temperature alloys and powder-metallurgy parts account for 14%. High-speed and other tools use 11%. All other uses total 10%.

The extreme hardness of tungsten carbide at ordinary and elevated temperatures makes it an excellent cutting material; in addition, tungsten carbide finds wide use in wear-resistant applications, e.g., dies for hot and cold working of wire, rod and tubing; mining tools; snow-tire studs; and ball-pen points. Tungsten carbide in the form of fine powder, 1–10 μm particle size, is bonded with cobalt for use in hard carbide tools and dies. Special-purpose carbide tools for machining steel and other hard materials generally contain varying portions of Ti, Ta, and Cb carbides along with tungsten carbide. Fused tungsten carbide finds extensive use in hard facing applications to resist wear.

Other uses for W include filaments for incandescent lamps and cathodes in electronic tubes; thermocouples; arc-lamp electrodes; electrodes in gas-shielded or atomic H_2 welding, including thoriated tungsten electrodes; electrochemical electrodes; contact disks for electrical switches; targets for x-ray tubes; cross hairs in telescopes; instrument springs; and glass-to-metal seals, where the coefficient of thermal expansion of W matches that of hard borosilicate glass. Ag-W and Cu-W compositions also are used in electrical contacts subject to severe arcing. W-Ni-Cu compositions (heavy-metal alloys containing about 7% Ni and 4% Cu) are used as shields or containers for radioactive materials, gyroscope rotors, counterweights in aircraft, and other uses where high density is required.

Nonferrous alloys of Co, Cr, and W, which are melted and cast, also are used as cutting tools, dies, and wear-resistant parts. W is used in steels to form stable carbides, strengthen ferrite, and refine the grain size and (chiefly in high-speed steels) to retain high hardness at elevated temperatures. Tungsten chemicals are used in textile dyes, inks, paints, enamels, and glass manufacture. Certain W compounds and derivative phosphors are used as luminescent light sources and in x-ray screens and television picture tubes.

REFERENCES

Chelius, J., and M. Schussler: *Mach. Des.,* 1970 Metals Reference Issue, vol. 42, chap. 19, pp. 87–88, Feb. 12, 1970.

Hampel, C. A. (ed.): "Rare Metals Handbook," chap. 30, Tungsten, Van Nostrand Reinhold, New York, 1967.

Stevens, R. F., Jr.: "Mineral Facts and Problems," U.S. Bureau of Mines, 1965.

Tungsten section, *Am. Met. Mark.,* Feb. 16, 1970.

Tungsten report, *Am. Met. Mark.,* Feb. 5, 1971.

—Mortimer Schussler, *Fansteel Inc.,*
North Chicago, Ill.

Turkey-red Oil [See **Anionic surfactants (sulfur-bearing).**]

Turpentine (See **Resins:** and **Terpenes.**)

Twitchellization (See **Vegetable oils.**)

Tyndall Phenomenon (See **Colloid systems.**)

Type Metal (See **Antimony.**)

Tyrosine (See **Amino acids;** and **Peptides and proteins.**)

U

Ultrafiltration (See **Membrane filters.**)

Ultrasonic Cleaning (See **1,1,1-Trichloroethane.**)

Unit Operation (See **Chemical engineering.**)

Unmixing (See **Mixing and blending, solids-solids.**)

Unsaturated Compounds (See **Saturated compounds.**)

Unsaturated Oils (See **Vegetable oils.**)

Uraninite (See **Uranium.**)

Uranium (Planet Uranus.) **U** = 238.04 (at. wt.); 92 (at. no.). Uranium metal is found in three allotropic forms. The alpha phase, stable below 668°C, is orthorhombic; the tetragonal beta phase exists between 668 and 774°C; at higher temperatures the body-centered-cubic gamma phase is stable. The gamma phase behaves most nearly like a true metal. The beta phase is brittle, while the alpha phase has many nonmetallic features in its crystallography.

Before the artificial production of elements, uranium was the element of highest known atomic number and atomic weight, and some difficulty was found in locating it in the periodic table of the elements, although in its chemical behavior it does show some resemblance to the elements of group VIB (Cr, Mo, and W) into which it might be expected to fall. Since the production of the transuranic elements, atomic numbers 93–103, these, together with actinium (89), thorium (90), protactinum (91), and uranium (92) have been placed in the actinide group of transition elements, similar in their mutual relations to the rare-earth group lanthanum (57) to lutetium (71). See Table U-l.

Discovery: Mining has been carried on at Joachimstal in Bohemia for centuries. Originally worked for Ag, which is found threaded and interleaved through a hard, black, and somewhat greasy-looking mineral, the ore also contains Ni and Zn.

For many years the black mineral was separated and discarded, but it was noted that after a day of exposure to sunlight a faint greenish-blue glow would be emitted in the dark. The mineral therefore received the name *pitchblende,* pitch from its appearance, and blende from a German word meaning dazzling or lustrous.

TABLE U-1. *Physical Properties of Uranium*

Isotopes (natural uranium), %:
^{238}U 99.28
^{235}U 0.72
^{234}U 0.005
Specific gravity . . . . 19.05 ± 0.02
Melting point, °C . . . . 1132 ± 1
Boiling point, °C 3818
Heat of fusion, kcal/mole 4.7
Vapor pressure at 1600°C, mm . . . . . . 10^{-4}
Electrical resistivity at 25°C, Ω-cm . . . . 35×10^6
Specific heat at 25°C 6.65

Investigating pitchblende, perhaps to establish the reason for the glow, the German chemist M. H. Klaproth (1743–1817) in 1789 isolated a yellow substance he thought to be the oxide of a new metal; by reducing this material with charcoal he obtained what he considered to be this metal, which he named uranium, after the planet Uranus, discovered by Herschel in 1781, which in turn was named after Uranus, Greek god of the heavens.

Klaproth's metal was probably the carbide or a mixture of carbide and oxide. The metal was not isolated until 1827 by Berzelius; in 1840 Peligot prepared fairly pure uranium by reducing a mixture of potassium and uranous chlorides with metallic sodium at red heat. In the 1880s Moissan used an electric furnace to reduce the oxide with sugar charcoal. Modern production methods use metallic Ca, Mg, or Na to reduce the tetrafluoride.

Radioactivity: In 1896, Henri Becquerel, who had obtained samples of pitchblende from Joachimstal to investigate the mystery of the emitted glow, placed a fragment near a supply of photographic plates. When the plates were developed, it was found that they were fogged; those nearest the samples had a hazy but unmistakable outline of the pitchblende that had rested on them. Repetition of the experiment under controlled conditions showed that pitchblende would cause fogging of plates even if they were covered with heavy paper, cloth, or thin metal. Pure uranium salts, however, were less effective than pitchblende. Investigation of this phenomenon by other workers, notably the Curies, led to the discovery of radium, polonium, and other elements, and to the first description of radioactivity. See also **Chemical elements; Nuclear power plants;** and **Radioisotopes.**

Chemistry: In its chemical compounds, uranium displays valences of 2–6; the compounds of valences 4 and 6 are the most stable and important.

The metal, even in massive form, is attacked by hydrogen at about 250°C with formation of the hydride, UH_3. At low pressures this compound decomposes into its elements above 400°C; the process is used for production of highly reactive uranium in fine powder form.

Oxygen and the halogens attack the metal; the precise composition of the compounds formed depends on the temperatures attained and reaction conditions; with fluorine mostly UF_6 is produced, with the other halogens mostly the tetrahalides. The halogen compounds ClF_3 and BrF_3 react vigorously, and the reactions can be controlled to give UF_6 as the sole product. Nitrogen or ammonia reacts with finely divided uranium at 300°C and with massive forms at higher temperatures. When uranium burns in air, mixtures of oxides and nitrides are formed. Both CO and CO_2 react at 750°C with formation of oxides and carbides.

The metal is readily attacked by HCl, HBr, or HI; attack by HF is limited by the formation of an insoluble fluoride coating on the metal surface.

Nitric acid also attacks the metal readily; the action of H_2SO_4 is slow but is assisted by oxidizing agents. The same statement applies to alkali attack. Boiling water attacks the metal slowly with liberation of hydrogen.

The dark-brown uranium dioxide, UO_2, pressed into pellets, is used as fuel in nuclear reactors; for this purpose a small amount of higher oxide, giving an O/U ratio of about 2.07, is always present, as this composition gives greater dimensional stability. The pure compound, produced by the reduction of higher oxides by hydrogen, ammonia, or hydrocarbons, is highly reactive, taking up oxygen from the air. It dissolves in acids to give uranous salts, which are rapidly oxidized by air on standing.

Uranium trioxide, UO_3, is an orange-yellow solid produced by careful thermal decomposition of hexavalent uranium compounds. The corresponding halides, UX_6, exist in the dry state but are decomposed by water to the uranyl compounds, UO_2X_2. Solution of UO_3 in sulfuric or nitric acids produces the uranyl compounds UO_2SO_4 or $UO_2(NO_3)_2$, which readily form complexes with excess acid in which the uranium is contained in the anion, for example, $UO_2(SO_4)_2^{2-}$ and $UO_2(SO_4)_3^{4-}$.

Produced during the concentration, refining, and purification of uranium from its ores, uranium diuranate (*yellow cake*) is a compound of uranium trioxide and ammonia of variable composition, usually represented as $(NH_4)_2U_2O_7$.

The oxide U_3O_8 is found in nature as the minerals uraninite and pitchblende. A dark-brown to black or greenish-black solid, it is produced by the ignition of most uranium compounds in air above 650°C; it starts to lose oxygen slowly above 900°C, ultimately forming UO_2.

The fluorides UF_4 and UF_6 are of importance. UF_4 (*green salt*) is the preferred raw material for the production of metallic uranium by calcium or magnesium reduction. It can be prepared from ammonium diuranate by calcination to UO_3, followed by reduction to UO_2 by hydrogen at 700°C and treatment with HF at about 450°C to convert it to UF_4. Continuous processes have been developed.

UF_4 is converted to UF_6 (uranium hexafluoride, or "hexa") by reaction with gaseous fluorine, ClF_3, BrF_3, or oxygen. Using ClF_3 at about 20 psig, the reaction can be carried out at room temperature; for oxygen higher temperatures are required, and 50% of the uranium is converted to nonvolatile UO_2F_2.

The separation of the isotopes ^{235}U and ^{238}U is based on the slight differences in diffusibility (diffusion process) or density (gas centrifuge) of their hexafluoride vapors. Both compounds are solids at room temperature and sublime unchanged at 56°C. A very large number of stages is needed in either process for concentration of the ^{235}U content above the 0.72% found in natural uranium, even to the figure of only 3–5% generally used in power-generation fuel.

Occurrence: Uranium is never found in nature as the elemental metal but occurs in combination with other elements in a wide variety of minerals. Although widely disseminated in the earth's crust in both igneous and sedimentary rocks, it is only rarely found in large concentrations. Its overall abundance in the earth's crust has been variously estimated at between 0.1 and 10 ppm; either figure makes it more abundant than gold or silver. Its presence in rocks is generally overlooked, because of the mainly low local concentration and because the uranium minerals as a class lack spectacular physical properties to call attention to them.

The principal free-world producers of uranium have been the United States, Canada, South Africa, Australia, and Zaïre (formerly Belgian Congo), with lesser quantities from France, Portugal, Sweden, Japan, and other countries. Various African countries have considerable potential, but owing to the current reduced demand for uranium, exploitation on a large scale appears improbable at present. Phosphate deposits in Florida and North Africa contain large amounts of uranium but at very low concentrations; at some future time it may become possible to recover it as a by-product.

Some of the principal uranium minerals are uraninite (uranium oxide, U_3O_8, in crystalline form), pitchblende (uraninite in massive form), brannerite (uranium and titanium oxide), coffinite (uranium silicate), autunite (hydrated uranium calcium phosphate), carnotite (hydrated uranium potassium vanadate), euxenite (uranium niobium titanate), pyrochlore (complex uranium niobium tantalum sodium oxide), thucholite (a mixture of hydrocarbons with uranium, thorium, and rare earths), gummite (hydrated uranium oxide), and becquerelite (hydrated uranium oxide). The last two, and many other hydrated and colored minerals, are generally found where weathering of primary minerals has occurred.

Production: Until it was required for weapons production, the only use of uranium was for the production of a lustrous black or alternately of a yellow color in ceramic glazes and glassware. Considerable quantities of ore were, however, mined in Bohemia during the twentieth century, and later in what was then the Congo, and Canada, but only for the purpose of recovering the radium associated with the uranium (to the extent of 1 part of radium to 3 million parts of uranium) for therapeutic and other purposes, the uranium being stockpiled or discarded. Much of this uranium was subsequently recovered when weapons demand arose; at about the same time, production of artificial radioactive isotopes eliminated the demand for radium, which is no longer produced.

Because of their general softness and the absence of striking physical properties, attempts to preconcentrate uranium ores by methods based on gravity, magnetic or electrical properties, or radioactivity, have not been generally successful, nor has flotation. The method of treatment has therefore generally involved treatment of the whole ore with an extractant and recovery of the uranium from solution after solid-liquid separation.

The extractant can be either an acid or an alkali. Extraction is improved by finer grinding of the ore, by increasing the concentration of extractant, and by the use of higher temperatures and oxidizing agents. The economics of variations in any one of these conditions or of any combination of them must be carefully assessed.

Separation of uranium from the inert material can be done by solid-liquid separation, filtration, or countercurrent washing, and uranium is recovered from the solution by treatment with ion-exchange resin or on solvent extraction. These processes can also be used without previous solid-liquid separation, but at some loss of efficiency. The uranium is reextracted into solution at a higher concentration than originally from either resin or loaded solvent and recovered from this solution by precipitation with suitable reagents. For original alkaline extractions, solid-liquid separation is always made and the uranium recovered by precipitation only.

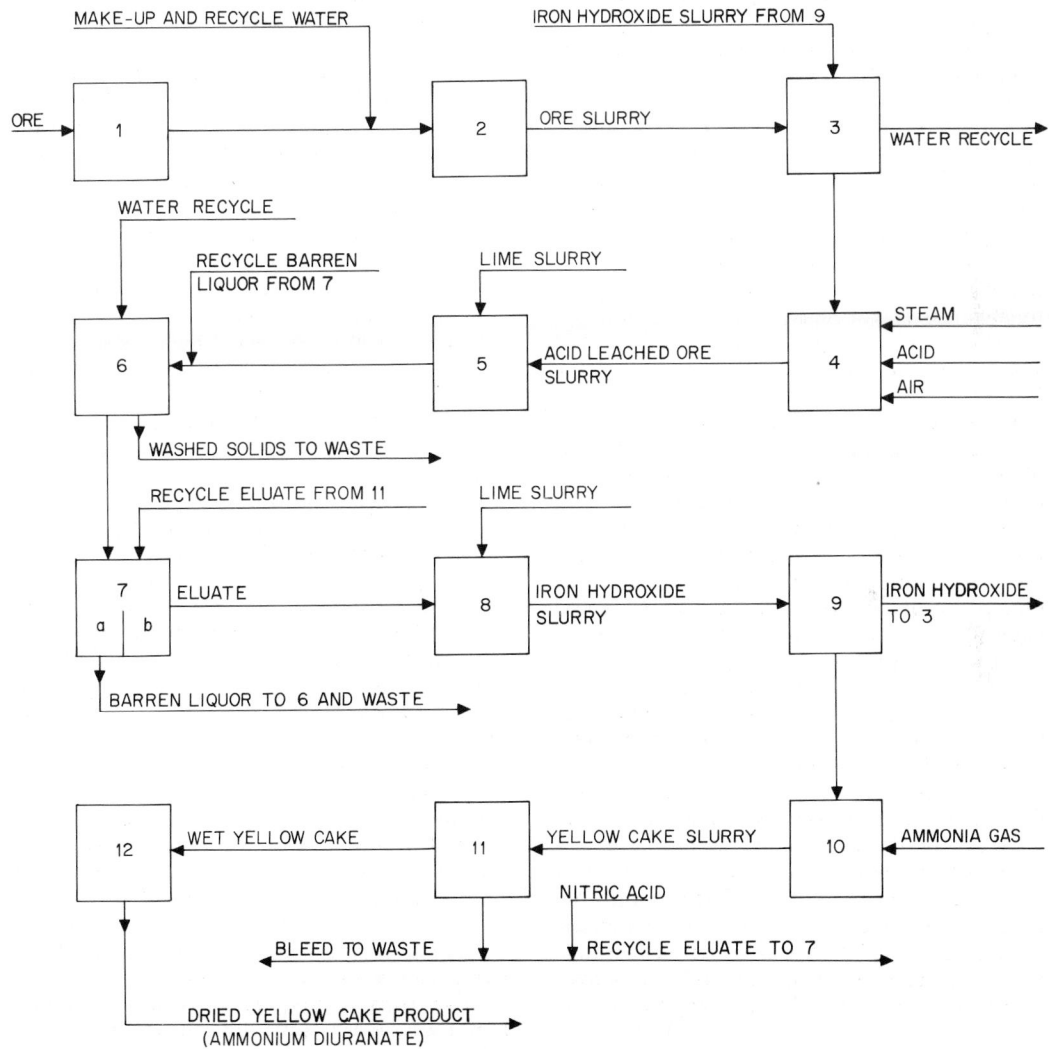

Fig. U-1. Schematic representation of materials flow for uranium concentration by acid leach and ion exchange. (1) Dry crushing, (2) grinding, (3) dewatering, (4) leaching, (5) partial neutralization, (6) solid-liquid separation and washing, (7) ion exchange (a) uranium absorption, (b) uranium elution, (8) iron precipitation, (9) solid-liquid separation, (10) uranium precipitation, (11) solid-liquid separation, (12) drying. Nitric acid is added after step (11) to adjust pH of recycle eluate.

Many more plants use acid than alkaline leaching. A fairly typical simple flow sheet using ion exchange after acid leach is shown in Fig. U-1.

Dewatering and liquid-solid separations (3), (6), (9), and (11) may be done by thickening or filtration; most usually by thickening followed by filtration. In operation (5), the pH of the leach slurry is raised to reduce its corrosive effect and to improve the ion-exchange operation on the uranium liquor subsequently separated. In ion exchange (7), resin contained in closed columns is alternately loaded with uranium and then eluted. The resin adsorbs the complex anions, such as $UO_2(SO_4)_3^{4-}$, in which the uranium is present in the leach solution. Elution is done with ammonium nitrate solution obtained by recycling the uranium filtrate liquor after pH adjustment. Some iron is adsorbed with the uranium and is eluted with it. Iron separation (8) is required because the iron hydroxide slurry is heavily contaminated with calcium sulfate and coprecipitated uranium salts. The slurry, therefore, is recycled to the dewatering stage (3). The washed solids from (6), the waste barren liquor from (7), and the uranium filtrate from (11) are combined, and the pH is raised to 7.5 by addition of lime slurry before the mixture is pumped to the tailings-disposal area.

The uranium concentrate produced by this process requires further refining before use in weapons or as fuel. This is done by dissolving the product in nitric acid, filtering, and treating the uranyl nitrate solution by solvent extraction methods.

—L. W. Vermeulen, *Rio Algom Mines Limited,*
Toronto, Ont., Canada

Uranium-Radium Series   (See **Chemical elements.**)

Urea   Urea, $NH_2 \cdot CO \cdot NH_2$, first discovered in urine by Rouelle in 1773 and identified as a pure crystalline organic compound in 1822, became famous in 1828, when Wöhler synthesized it from the inorganic compounds lead cyanate and ammonium hydroxide, thus proving for the first time that an organic compound could be produced outside a living organism. However, nearly 100 years passed before urea was manufactured in substantial quantities.

Properties: Urea has a formula weight of 60.06 and a nitrogen content of 46.65% by weight. It melts at 132.7°C (270.9°F) and has a specific gravity at 20°C of 1.335. Urea is very soluble in water, forming a 40% solution at 0°C and an 80% solution at 80°C. It is less soluble in methanol (18% at 20°C) and ethanol (5% at 20°C). Solubility in ammonia is 49% at 20°C; 76% at 50°C; and 90% at 100°C. Saturated aqueous urea solutions have viscosities ranging from 2 cP at 20°C to 1.7 at 50°C and 2.4 at 100°C.

Urea is a weak monoacid base, $K_B = 1.5 \times 10^{-14}$, forming stable salts, such as urea nitrate, $CO(NH_2)_2 \cdot HNO_3$, and urea oxalate, $2CO(NH_2)_2 \cdot H_2C_2O_4$. The various barbiturates used in medicine are derivatives of barbituric acid, $CO(NHCO)_2CH_2$, which is formed by reacting urea with malonic acid. Urethanes are formed from urea and alcohols (see also **Urethanes**). Urea reacts with formaldehyde to form compounds that are used as slow-release fertilizers (ureaform), adhesives, and plastics. Urea also reacts with H_2O_2 to yield a crystalline oxidizing agent.

Urea forms crystalline complexes (clathrates) with straight-chain alkanes, which has led to its use in the petroleum industry for separating straight- and branched-chain hydrocarbons. The hydrocarbon is recovered from the complex by adding water to dissolve the urea.

Urea hydrolyzes in acids or bases, yielding NH_3 and CO_2. Hydrolysis in aqueous solution is induced quantitatively by an enzyme (urease) obtained from jack beans or soybeans. Analysis of the NH_3 or CO_2 produced is used for the quantitative determination of urea.

Urea will sublime unchanged under vacuum at its melting point. Heating at atmospheric pressure above its melting point will cause urea to decompose, resulting in NH_3; isocyanic acid, $HNCO$; cyanuric acid, $(HNCO)_3$; biuret, $NH_2CONHCONH_2$, and other products. Biuret also is formed on heating aqueous urea solutions at atmospheric pressure. Biuret can be converted back into urea by heating at high pressure in the presence of excess NH_3. In the crystallization of urea from aqueous solutions, it has been found that the presence of about 5% biuret will change the urea crystals from long needles to short rhombic prisms, with improved handling properties.

Since heating of aqueous urea solutions normally occurs in manufacturing processes, the product will contain some biuret, this being 0.1% or less in crystals, 0.3% in prills made by remelting crystals, and up to 1% or more in prilled urea made from melt obtained by the evaporation of urea solutions.

Fertilizer-grade urea containing up to 1.5% biuret is satisfactory in most applications. However, for foliar application to citrus trees, pineapples, coffee, and cherry trees, the biuret content must be kept below 0.3%.

Pure biuret, manufactured from urea, has been found preferable to urea as a feed supplement for ruminants, since NH_3 is released from biuret more gradually than from urea. Thus, there is no danger of the animal's receiving an overdose of NH_3. When either biuret or urea is fed to cattle or sheep, enzymes in the rumen release NH_3; bacteria therein use it to form proteins, which are digested and utilized by the animal. See also **Fertilizers.**

When heated quickly to 350°C in a fluidized bed at atmospheric pressure, urea is decomposed to isocyanic acid and NH_3, which are passed over a catalyst at 400°C to form melamine, $C_3N_3(NH_2)_3$, the triamide of cyanuric acid. See also **Melamine**

Manufacture: Ammonia and CO_2 are reacted noncatalytically in an exothermic reaction to form ammonium carbamate, which is kept at an elevated temperature until water splits out, forming urea.

$$2NH_3 + CO_2 \rightarrow NH_2COONH_4$$
$$NH_2COONH_4 \rightarrow NH_2CONH_2 + H_2O$$

Manufacture is complicated by the extreme corrosiveness of the reaction mixtures, which has led to the use of reaction vessels lined with silver, lead, titanium, zirconium, or stainless steel. The high vapor pressure of ammonium carbamate at the elevated temperature (170–200°C) required for the dehydration step has required the use of pressures ranging from 160–250 atm. Under these conditions, a $2:1$ m mixture of NH_3 and CO_2 will be half converted to urea, the other half remaining as ammonium carbamate. The unreacted carbamate must then be separated from the urea and utilized elsewhere, e.g., in the production of $(NH_4)_2SO_4$, or recycled to the urea reactor. The use of an excess of NH_3 in the reactor feed will increase the conversion to urea to as much as 75% of the CO_2 feed, thus decreasing the amount of ammonium carbamate to be disposed of, but of course, at the expense of handling and reusing the excess NH_3.

Urea processes are classified as once-through, partial-recycle, or total-recycle processes. In a once-through process, unreacted ammonium carbamate in the reactor product is decomposed by heating the mixture, and the NH_3 evolved is utilized elsewhere. In a partial-recycle process, excess NH_3 and, in some cases, a portion of the unconverted CO_2 are returned to the reactor, with the balance of the unconverted NH_3 used elsewhere. In a total-recycle process, which makes up the greater part of current practice, all the unreacted ammonium carbamate is recovered and returned to the reactor.

There are four classes of total-recycle processes: gas-recycle, slurry-recycle, solution-recycle, and stripping processes.

Mixed gas recycle was used in Germany in 1920[*] in the first commercial plant making urea from NH_3 and CO_2. The NH_3-CO_2 mixture evolved on heating the urea reactor production solution was compressed and returned to the reactor. The gas mixture had to be kept hot to avoid formation of solid carbamate, and serious mechanical and corrosion problems resulted. Consideration has been given to hot mixed-gas recycle using centrifugal compressors.

Gas-recycle processes are in use in which NH_3 and CO_2 are separated by absorbing one or the other in a selective solvent, e.g., monoethanolamine solution for CO_2[†] or urea nitrate or ammonium nitrate for NH_3.[‡] The separated NH_3 and CO_2 are then compressed and returned as feed to the urea reactor. No water is fed to the reactor, but high energy is required for separating and recompressing the NH_3 and CO_2.

Slurry recycle was used,[¶] in which NH_3 and CO_2 evolved from the reactor product stream were recombined as a slurry of solid carbamate in oil and pumped back to the reactor. Patents have been issued on a process that would use a carbamate slurry in liquid NH_3 as reactor feed.

Solution recycle has been used most widely among the total-recycle processes.[§]

The flow sheet of a total-recycle urea plant utilizing solution recycle[||] is given in Fig. U-2. Liquid NH_3 and gaseous CO_2 ($4:1$ mole ratio), plus recycle aqueous ammonium carbamate solution, are fed to the urea reactor, which operates at about 185°C and 210 atm. The reaction products flow to the high-pressure decomposer. Vapors from the decomposer go to the high-pressure ab-

---

[*]I. G. Farbenindustrie.
[†]CPI-Allied Process.
[‡]Inventa process.
[¶]Pechiney process. Two plants constructed in early 1950s; none since.
[§]First used by Du Pont (1933) and Allied Chemical (1950) in the United States. Current processes include C&I/Girdler and Chemico in the United States; Lonza in Switzerland; Mitsui Toatsu in Japan; Montecatini in Italy; and Stamicarbon in Holland.
[||] C&I/Girdler process.

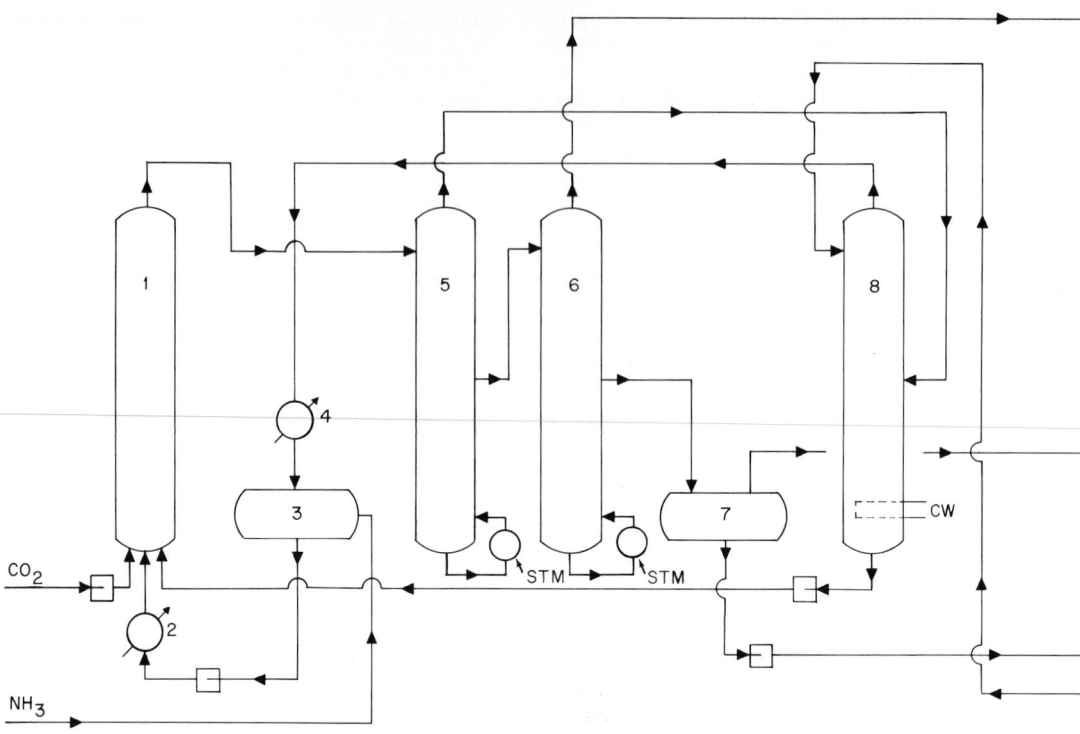

Fig. U-2. Urea manufacturing process utilizing solution recycle. (1) Reactor, (2) preheater, (3) ammonia receiver, (4) condenser, (5) high-pressure decomposer, (6) low-pressure decomposer, (7) gas separator, (8) high-pressure absorber, (9) low-pressure cooler, (10) gas condenser, (11) concentration equipment. *STM,* steam, *CW,* cooling water. (*C & I/Girdler Incorporated.*)

sorber, in which the CO_2 is absorbed as aqueous ammonium carbamate solution, with the excess NH_3 condensed and returned to the NH_3 receiver and the solution recycled to the urea reactor. The solution from the high-pressure decomposer flows to the low-pressure decomposer. NH_3 and CO_2 vapors evolved there flow to the low-pressure cooler to be absorbed as an aqueous solution, while the urea solution goes to a gas separator. Vapors from the separator, mostly water plus some NH_3 and CO_2, flow to a condenser to form the absorbent solution for the low-pressure cooler and high-pressure absorber. The 75% urea solution from the separator then is concentrated in evaporating equipment to a 99.5% (or more) melt. The melt is sprayed into a prill tower, in which the droplets are solidified into prills by a rising stream of cooling air. An alternate finishing process* is vacuum crystallization of the 75% urea solution,

followed by crystal drying and remelting to form prills containing 0.3% or less of biuret.

The prill tower, with its large airflow, great height, and attendant dust-recovery problems, has been eliminated in another process,† in which anhydrous urea melt from either an evaporator or a crystal melter is sprayed onto a bed of recycled fines in a rotating drum, thus coating the fine particles by accretion (onionskin effect) to produce dense spherical granules. Granule size can be regulated at any desired level, from the 2-mm diam typical of prills to the 5-mm diam preferred for forest fertilization, which cannot be achieved by prilling. All air used is passed through a wet scrubber to eliminate air pollution, with the liquid scrubber effluent being recycled.

Two stripping processes are in commercial use. In one,‡ CO_2 is used to strip carbamate out of

*Mitsui Toatsu process.

† C&I/Girdler-Cominco Spherodizer process.
‡ Stamicarbon (Holland).

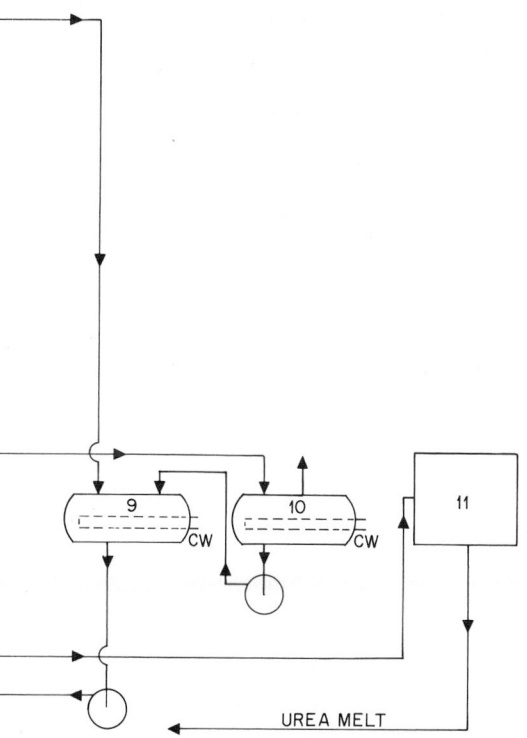

UREA MELT

the urea solution; the other process* uses NH_3 for this purpose. Both processes carry out the stripping in falling-film steam-heated carbamate decomposers operated at or near urea reactor pressure, with the evolved NH_3 and CO_2 recombined at this pressure and returned to the reactor.

Improvements in urea manufacturing technology are noted by comparing steam requirements (5 tons of steam per ton of urea) by the gas-recycle process (1951) to 0.8–0.9 ton of steam per ton of urea in modern facilities.

REFERENCES

Sauchelli, Vincent (ed.): "Fertilizer Nitrogen: Its Chemistry and Technology," pp. 247–294, Van Nostrand Reinhold Company, New York, 1964.

"Thorpe's Dictionary of Applied Chemistry," 4th ed., vol. 11, pp. 795–799, Longmans, London, 1954.

—R. M. Reed and E. Pelitti, *C & I/Girdler Incorporated, Louisville, Ky.*

*Snam Progetti (Italy).

Urea-Formaldehyde Resins   (See **Amino resins**.)

Urease   (See **Enzymes**; and **Peptides and proteins**.)

Urethanes   The urethanes are a conglomerate family of polymers in which formation of the urethane group, $\overset{\text{H}}{\underset{}{\text{N}}}{-}\overset{\text{O}}{\underset{}{\text{C}}}{-}\text{O}$, is an important step in polymerization. Because the urethane linkage usually is formed by reaction of hydroxyl and isocyanate groups, urethane chemistry is the chemistry of isocyanates. The high reactivity of isocyanates and knowledge of the catalysis of isocyanate reactions have made possible the simple production of diverse polymers from low- to moderate-molecular-weight liquid starting materials.

Isocyanate Chemistry: Isocyanates are among the class of heterocumulenes, compounds having adjacent double bonds between different atoms. Their electronic structure allows three resonance possibilities:

$$\left[R{-}\overset{-}{N}{-}\overset{+}{C}{=}O \leftrightarrow R{-}N{=}C{=}O \leftrightarrow R{-}N{=}\overset{+}{C}{-}\overset{-}{O} \right]$$

Compounds having active hydrogen atoms, i.e., those which will react with sodium, normally add across the $-N{=}C-$ double bond, the hydrogen atom becoming attached to the nitrogen and the remainder of the molecule to the carbonyl carbon.

$$R{-}N{=}C{=}O + HA \rightarrow R{-}\overset{H}{\underset{}{N}}{-}\overset{O}{\underset{}{C}}{-}A \quad (1)$$

The important isocyanate-hydroxyl reaction follows this pattern.

$$R{-}N{=}C{=}O + R'OH \rightarrow R{-}\overset{H}{\underset{}{N}}{-}\overset{O}{\underset{}{C}}{-}OR' \quad (2)$$

A urethane

Also important in polyurethane production is the isocyanate-water reaction, which both generates carbon dioxide to expand many urethane foams and forms substituted urea blocks in the polymer structure.

$$R{-}N{=}C{=}O + H_2O \rightarrow \left[R{-}\overset{H}{\underset{}{N}}{-}\overset{O}{\underset{}{C}}{-}OH \right] \rightarrow$$
$$RNH_2 + CO_2 \quad (3)$$

$$RNH_2 + RNCO \rightarrow R\overset{H}{\underset{|}{N}}-\overset{O}{\underset{||}{C}}-\overset{H}{\underset{|}{N}}-R \quad (4)$$

A substituted urea

The secondary reactions of isocyanates with urethanes to produce allophanates and with substituted ureas to produce biurets are important cross-linking reactions.

$$R-NH-CO-OR' + RNCO \rightarrow$$
$$R-\underset{|}{N}-CO-OR' \quad (5)$$
$$CO-NH-R$$

An allophanate

$$R-NH-CO-NH-R + RNCO \rightarrow$$
$$\overset{R}{\underset{|}{}}$$
$$R-NH-CO-N-CO-NH-R \quad (6)$$

A biuret

The reaction of isocyanates with carboxylic acids occurs in the formation of some polymers from polyesters having both terminal hydroxyl and carboxyl groups.

$$R-N{=}C{=}O + R'COOH \rightarrow$$
$$[R-NH-CO-O-CO-R'] \rightarrow$$
$$R-NH-CO-R' + CO_2 \quad (7)$$

An amide

An extension of this reaction to aromatic *o*-dicarboxylic acids produces an amide in a first stage and an imide on further heating.

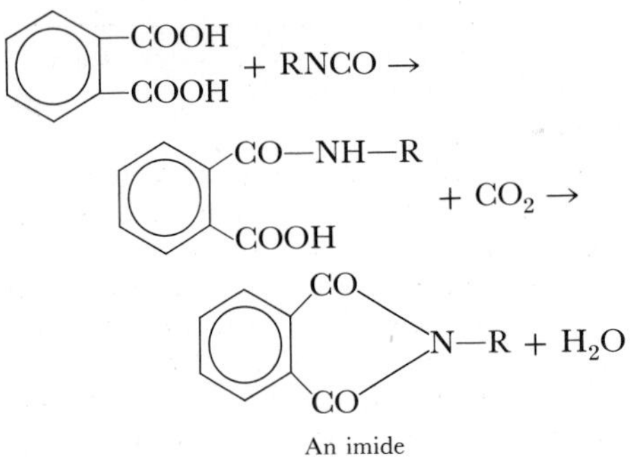

$$+ CO_2 \rightarrow$$

An imide

These reactions have been used to prepare amide-imide-containing polymers from systems based on trimellitic acid.

Homopolymerization of isocyanates occurs in some isocyanate-based systems and is promoted by basic anionic catalysts. Depending on the specific catalyst and temperature, cyclic dimers, cyclic trimers, linear *N*-substituted polyamides, or carbodiimides can be formed.

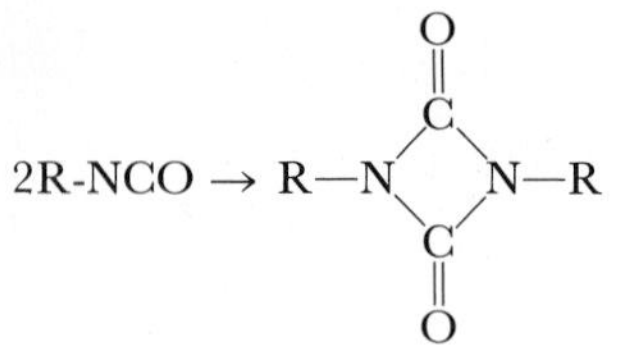

$$2R\text{-}NCO \rightarrow \quad (9)$$

Cyclic dimer
(uretidinedione)

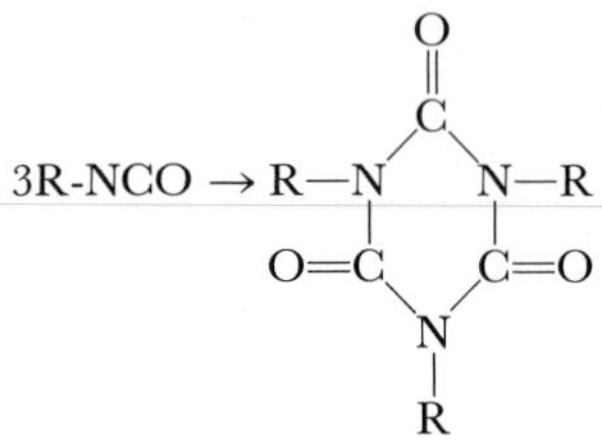

$$3R\text{-}NCO \rightarrow \quad (10)$$

Cyclic trimer
(isocyanurate)

$$nR\text{-}NCO \rightarrow \left(\overset{R}{\underset{|}{N}}-\overset{O}{\underset{||}{C}}\right)_n \quad (11)$$

N-substituted linear polyamide
(1-nylon)

$$2RNCO \rightarrow R-N{=}C{=}N-R + CO_2 \quad (12)$$

Carbodiimide

Dimer formation is a troublesome side reaction which can affect isocyanate stability but is sometimes used as a means of inactivating isocyanate groups for later regeneration by heating. The cyclic trimer or isocyanurate is more stable than the urethane and is the basis for preparation of flame-resistant, thermally stable polymers. Polycarbodiimides and the linear polyamides have no present commercial use.

Raw Materials: Common isocyanates used in preparing polyurethanes are listed in Table U-2. All are low-viscosity liquids at room temperature with the exception of 4,4'-diphenylmethane diisocyanate (MDI), a crystalline solid. The aromatic isocyanates are more reactive than the aliphatic isocyanates and are widely used in urethane foams, coatings, and elastomers. The cyclic structure of aromatic and alicyclic isocyanates contributes to molecular stiffness in polyurethanes.

Aliphatic isocyanates tend to form more flexible polyurethanes. Diisocyanates are used in preparing the more flexible, resilient types of urethane

TABLE U-2. *Isocyanates Used Commercially for Urethane Preparation*

Chemical Name	Common Name	Mp, °C or Room-Temperature State	Bp, °C
80/20 2,4-/2,6-Tolylene diisocyanate	80/20 TDI	11.5–13.5	120*
65/35 2,4-/2,6-Tolylene diisocyanate	65/35 TDI	3.5–5.5	120*
2,4-Tolylene diisocyanate	2,4 TDI	19.5–21.5	120*
Modified tolylene diisocyanates	Crude TDI		120*
4,4'-Diphenylmethane diisocyanate	MDI	37–38	194–199†
Polyisocyanates from aniline-formaldehyde condensates	Undistilled, crude, or polymeric MDI	Liquid	194–199†
1,6-Hexamethylene diisocyanate	HDI	Liquid	140–142‡
4,4'-Dicyclohexylmethane diisocyanate, mixed isomers.	. . .	Solid-liquid mixture	

*At 10 mm.
†At 5 mm.
‡At 21 mm.

foams and elastomers. Polyfunctional isocyanates provide higher cross-link densities in rigid urethane foams and solid polymers. Aliphatic and alicyclic isocyanates are most often used in coatings, where they produce urethanes which do not discolor when exposed to light.

Hydroxyl-terminated coreactants used with polyisocyanates can be classed as polyether polyols, polyester polyols, and natural products. The polyether polyols are used in the greatest volume because of low cost and a wide choice of types. Most are prepared by addition of 1,2-propylene oxide to hydroxyl- or amine-bearing compounds in the presence of strong base catalysts.

Choice of the initiating molecule and of the amount of propylene oxide to be added controls both the functionality and molecular weight of the polyol. 1,2-Propylene oxide normally adds to leave terminal secondary hydroxyl groups. Because secondary hydroxyl groups are less reactive toward isocyanates than primary hydroxyl groups, ethylene oxide is sometimes added to form activated polyols capped with primary hydroxyls.

Table U-3 indicates the variety of polyether polyols available commercially. All are liquids, although the viscosity is very high for the lower-molecular-weight highly branched types such as those based on sucrose. More expensive poly(1,4-tetramethylene ether) glycols are produced from tetrahydrofuran.

Hydroxyl-terminated polyesters used to prepare polyurethanes are produced by direct esterification of adipic, phthalic, and similar dicarboxylic acids with an excess of difunctional alcohols such

TABLE U-3. *Propylene Oxide–based Polyether Polyols*

Starting Compound	Functionality*
Water	2
Glycerin	3
Trimethylolpropane	3
1,2,6-Hexanetriol	3
α-Methyl glycoside.	4
Pentaerythritol	4
Sorbitol	6
Sucrose	8
Aniline	2
4,4'-Diphenylmethanediamine	4
Ethylenediamine	4
Diethylenetriamine	5

*Average number of hydroxyl groups per molecule.

as ethylene glycol, diethylene glycol, propanediols, and butanediols. Trifunctional reactants are added to produce chain branching when desired. Hydroxyl numbers of polyesters are in the range 50–450 with acid numbers usually below 10. A series of polyesters recommended for preparing urethane elastomers is produced by homopolymerization of ε-caprolactone to linear diols. Molecular weights of 500–2,000 are available. Several hydroxyl-containing natural products have been used as reactants in polyurethane production. Natural and modified castor oils have been included in foam and coating systems. Modified wood resins have been used as extenders in urethane foam systems. These resins vary but consist mainly of complex phenolic-type compounds and resin acids related to abietic acid.

Urethane Foams: Applications of flexible and rigid urethane foams, the best known forms of polyurethanes, are high-tonnage products.* Foam formulations contain isocyanates and polyols with suitable catalysts, surfactants for stabilization of foam structure, and blowing agents which produce gas for expansion. Urethane foam formulations are unique in consisting of low- to moderate-viscosity liquids which, when mixed in measured amounts, react, foam, and cure into a cellular mass. The largest volume of flexible urethane foam is used as a cushioning material and is produced from a formulation similar to the following:

	Parts *(weight)*
Polyether polyol, functionality 2–3, hydroxyl number 56	100
Water	3.5
Silicone copolymer stabilizer	0.1
Stannous octoate	0.2
Triethylene diamine	0.1
Tolylene diisocyanate	45

Reaction of polyether polyol and isocyanate is the primary polymerization reaction. Water also reacts with isocyanate to produce CO_2 blowing agent and substituted urea blocks which add to polymer strength.

Foam stabilizers which are block copolymers of poly dimethylsiloxane and mixed ethylene oxide–propylene oxide polymers are extremely efficient. Two catalysts are used to balance the isocyanate-hydroxyl and isocyanate-water reactions. Stannous octoate is the most stable and effective of a series of organometallic catalysts which are particularly active in promoting the isocyanate-hydroxyl reaction. Tertiary amines tend to favor the isocyanate-water reaction. The bicyclic structure of triethylenediamine, diazobicyclo[2.2.2]octane, is thought to account for its high activity. The amount of isocyanate used in foam formulations is approximately that required to react stoichiometrically with the available hydroxyl groups and water. An excess of 5–10% often is added to produce cross-links through secondary reactions.

Flexible urethane foams based on polyesters are used as clothing interlining and in other specialty applications. The polyesters have hydroxyl numbers of 50–60 and are frequently prepared from diethylene glycol, adipic acid, and sufficient trifunctional alcohol to give an average functionality of 2.4–3.0. Foam formulations are similar to that

*(United States, 1970) 650 million lb (flexible); 270 million lb (rigid).

given above for a polyether foam. Because polyesters of this type terminate in primary hydroxyl groups, less active catalysts, such as *N*-ethylmorpholine, are used. Surface properties of the polyester-foam mixtures also require different stabilizers, though of the same general type.

Rigid urethane foam formulations usually contain a nonreactive blowing agent, a polymeric isocyanate of the MDI type, and, in many instances, a flame retardant. A typical formulation could contain:

	Parts *(weight)*
Polyether polyol, functionality 4–8, hydroxyl number 450	100
Flame retardant	20
$CFCl_3$	35
Tertiary amine catalyst	2
Silicone copolymer stabilizer	1
Polymeric MDI-type polyisocyanate	115

Polyether polyols used in rigid urethane foams are the more branched lower-molecular-weight types. Flame retardants most often are liquid organic compounds which contain chlorine, bromine, and/or phosphorus. The compounds may not react with isocyanates and may bear hydroxyl groups which allow them to be built into the polymer structure.

The blowing agent, $CFCl_3$, commonly known as refrigerant 11, boils at 23.8°C, slightly above room temperature. Heat of reaction of isocyanate and hydroxyl groups vaporizes the liquid and produces a closed-cell–foam-cell foam structure containing $CFCl_3$ vapor. The low thermal conductivity of the vapor is responsible for the characteristic low thermal conductivity of rigid urethane foams.

Tertiary amines such as tetramethylguanidine, *N,N,N',N'*-tetramethylbutanediamine, triethylenediamine, and dimethylaminoethanol are used in most rigid urethane foam formulations. Stabilizers are similar to those used in flexible urethane foams.

A few rigid urethane foams are based on aromatic halogenated polyesters which contribute flame retardance, and other specialty types use carbon dioxide from the isocyanate-water reaction as the blowing agent. Crude TDI type isocyanates are chosen for some rigid foams, particularly in cost-conscious applications for which flame resistance is not a serious requirement.

Because the ingredients of urethane foam formulations are liquid, foam-production machinery consists of two or more metering pumps which feed reactants to a continuous mixer at controlled rates and in the desired proportions. The mixed output proceeds to a forming device, in which the foam rises and cures. Large continuous foam slabs are produced on long inclined conveyors lined with release paper. Rigid-foam panels are produced between two horizontal conveyors. Metal or reinforced plastic molds are used to form complex shapes. Mixed rigid-foam reactants are sometimes atomized by air or hydraulic pressure and can be sprayed onto uneven substrates.

Foam machines have throughput rates approaching 1,000 lb/min in flexible-foam-slab production and deliver as little as 2 lb/min in specialty molding operations. Larger machines are permanently installed. The smaller units, particularly those designed for field application of rigid urethane foams, are portable. In all instances, some control of raw-material temperatures is desirable.

Most urethane foams are produced by one-shot processes in which all raw materials are combined in a single step. In some specialized applications, there are advantages in prereacting the isocyanate and part of the polyol to form a prepolymer, which is then combined with the remaining reactants.

Urethane foams, both flexible and rigid, are most frequently made with densities of 1.2–2.0 lb/ft^3. Mechanical properties are high for the low density and vary exponentially with density. Low-density flexible polyether urethane foams have open cells, are resilient, and have low compression set under usual test conditions. The flexible polyester urethane foams are stiffer and less resilient. Both types of flexible foams have good chemical resistance but are swollen by strong solvents and hydrolyzed by strong acids and bases. Polyether urethanes are susceptible to oxidation and polyester urethanes to hydrolysis. The aging properties of both types are good, however.

Rigid urethane foams have closed cells and, when blown with CFCl$_3$, have the lowest thermal conductivity of any common insulation, approximately 0.11 (Btu)(in.)/(hr)(ft^2)(°F) when fresh, rising to approximately 0.15 when aged under conditions which allow air to diffuse into the cells. The mechanical strength, chemical resistance, and aging characteristics are excellent.

Higher-density flexible urethane foams produce energy-absorbing articles. Related semirigid or semiflexible urethane foams are often used in automotive crash pads and arm rests. High-density rigid urethane foams are used as molded structural parts of furniture and other products.

Urethane Elastomers: These materials are prepared from diisocyanates, polyester or polyether diols with molecular weights near 2,000, and low-molecular glycol or aromatic diamine chain extenders. The elastomers can be thermoplastic or thermosetting and are available as casting resins, millable gums, and thermoplastic resins. Although the reactants of some casting systems are processed in one step, most reactants are supplied as isocyanate-terminated prepolymers formed by reacting polyester or polyether diols with diisocyanates. This prepolymer then is cured by adding the glycol or diamine chain extender to form a thermoset polymer.

Thermoplastic urethane elastomers are similar in composition but are specially compounded to eliminate cross-linking and allow the resin to melt for injection molding and extrusion. Parts formed from some thermoplastic urethane resins can be postcured at higher temperatures to thermoset condition.

Millable gums are urethane elastomers of moderately high molecular weight which have the appearance of uncured rubber, can be processed on rubber-processing machinery, and are cured by rubber-curing agents.

Urethane elastomers have an outstanding combination of toughness, abrasion resistance, and elasticity. They can be formulated with hardness from 15 Shore A to 80 Shore D. Resistance to oxygen, oil, ozone, and other chemical agents is generally high.

Urethane Coatings: A wide selection of urethane coatings is prepared by dissolving polyurethanes or urethane reactants in appropriate solvents. These are described under **Paint.**

Physiological Properties: The urethanes must not be confused with *urethane* (ethyl carbamate), particularly with respect to physiological properties. Urethane polymers in general are physiologically innocuous, but isocyanates and strong amine catalysts used in their manufacture have histories of toxic effects in industrial applications.

REFERENCES

Buist, J. M., and H. Gudgeon: "Advances in Polyurethane Technology," Interscience-Wiley, New York, 1968.

Dombrow, B. A.: "Polyurethanes," 2d ed., Van Nostrand Reinhold Company, New York, 1965.

Saunders, J. H., and K. C. Frisch: "Polyurethanes: Chemistry and Technology," pts. I and II, Interscience-Wiley, New York, 1962 and 1964.

—John K. Backus, *Mobay Chemical Company, Pittsburgh, Pa.*

V

Vacuum Cooling (See Freeze-drying; Helium; and Natural gas.)

Vacuum Crystallizers (See Crystallization.)

Vacuum Degassing (See Iron- and steelmaking.)

Vacuum Filter (See Filtration.)

Valence (See Chemical elements; and Molecule.)

Valine (See Amino acids; and Peptides and proteins.)

Vanadium (From Vanadin, Norse goddess.) V = 50.942 (at. wt); 23 (at. no.). Vanadium is a member of group VB of the periodic table, along with columbium (niobium) and tantalum. The melting point is 1905°C and estimated boiling point 3310°C. The metal has two stable isotopes, 50 and 51, with 51 predominating (99.76%), and five unstable isotopes, 46–49 and 52. Vanadium is a hard metal and imparts both hardness and toughness to alloys with other metals. Key physical properties of vanadium are summarized in Table V-1.

1126

Vanadium was first reported by Andrés Manuel del Rio in 1801 and rediscovered by Nils Gabriel Sefström, a Swedish chemist, in 1830. It was not until about 1950, however, that the element could be produced in a sufficiently pure form and in large enough quantities to be studied as an engineering material. Since the early 1900s the primary use of vanadium has been as an alloying element for steels and irons. Some vanadium compounds are used in the chemical industry for

TABLE V-1. *Some Properties of Vanadium*

Atomic volume, cm³/g atom	8.5
Crystal structure	bcc
Lattice constants at 20°C, Å	$a = 3.033$
Closest approach of atoms, Å	2.627
Ionization potential, eV	6.743
Density:	
g/cm³	6.1
lb/in.³	0.22
Elastic modulus (room temperature), psi	~20,000,000
Specific heat (20–100°C), cal/g	0.12
Electrical resistivity at 20°C, μΩ-cm . . .	24.8
Thermal expansion (23–1100°C), μin./(in.)(°C)	8.3–10.9
Hardness Brinell (iodide V)	64

making oxidation catalysts and in the ceramic industry as coloring agents.

Occurrence: Ores containing small amounts of vanadium are widely distributed throughout the world. After nearly half a century as the world's major source, the asphaltite deposits of Minas Rigras, Peru, were phased out during the 1940s, and activity shifted to Colorado, Utah, and Arizona, where roscoelite and carnotite ores are mined for vanadium and uranium.

While vanadium still comes from the Colorado Plateau, as well as from Idaho, Chile, southwest Africa, Finland, Norway, and Venezuela, the current major sources are the titaniferous magnetites of South Africa and the Urals and the vanadiferous clays of central Arkansas.

World production has increased significantly since 1960. Production figures for 1970 were estimated at 28 million lb versus 14 million lb for 1960. Consumption of vanadium is largely in the form of ferrovanadium.

Extraction from Ores: Vanadium usually is recovered as a coproduct or by-product of other elements with which it is associated. Vanadium and uranium are extracted from carnotite and similar ores by acid leaching or by a roast-quench-leach process in which the ore is crushed to a size less than 10 mesh and roasted with salt in a multiple-hearth furnace to form the water-soluble sodium metavanadate, $NaVO_3$. The hot ore then is quenched in water or soda ash solution.

Uranium is precipitated first by neutralizing the leach liquor with H_2SO_4. After filtering off the uranium precipitate, vanadium is precipitated by adding more acid. The V precipitates as red cake, a complex sodium polyvanadate corresponding approximately to the hexavanadate, $Na_2H_2V_6O_{17}$. Vanadium and uranium also are extracted from carnotite by leaching the ground raw ore with H_2SO_4. Solvent extraction then is used to separate U and V and to enrich the solutions. When solvent extraction is used, V usually is precipitated as a low-soda red cake, approximating the composition $H_4V_6O_{17}$, and in some cases as ammonium metavanadate, NH_4VO_3.

Vanadium in iron ores is sometimes recovered by roasting and leaching the ground ore in a manner similar to that used for carnotite ore. In other cases, the ore is smelted to produce pig iron which contains the V, which is then recovered in a rich slag containing 5–25% V_2O_5 when the pig iron is converted to steel. The slag is then processed by a roasting and leaching procedure to extract the V in the form of red cake.

The vanadium contained in phosphate rock is recovered in ferrophosphorus during the electric-furnace smelting of phosphate rock to produce elemental P. The ferrophosphorus contains 3–7% V and is processed further by roasting and leaching to recover the vanadium.

Red cake contains 85–90% V_2O_5, 2–10% Na_2O, and about 3% water. Solution of red cake with soda ash forms sodium vanadate. Heating ammonium metavanadate drives off the ammonia and leaves a purified oxide, V_2O_5. For the production of alloys by smelting, red cake is fused and cast into flake form.

Ferrovanadium Reduction Methods: Ferrovanadium is produced by aluminum or silicon reduction of the oxide in the presence of iron in an electric-arc furnace. Aluminum reduction is commonly practiced. The reaction is exothermic, and little additional heat from the arc is required. Silicon reduction requires a two-stage reduction to achieve an efficient operation. Vanadium for metallurgical purposes is also produced by solid-state carbon reduction of vanadium pentoxide in a vacuum furnace. The product contains about 85% V, 12% C, and 2% Fe.

Vanadium-Metal Reduction Methods: The best known process for producing ductile vanadium metal now used in the United States is the calcium reduction of the oxide, as developed by McKechnie and Seybolt. This process consists of charging pure vanadium oxide, calcium metal, and iodine into a heavy-walled steel cylinder, taking special precautions to exclude moisture. The cylinder is sealed and evacuated, and the reaction between the ingredients is initiated by the application of heat. Within a short time, a sufficiently high temperature and pressure are reached to allow the molten droplets of vanadium to collect beneath the calcium oxide–calcium iodide slag and to form a single button, or regulus. A typical analysis of the product is 99.7% V, 0.10% O_2, 0.04% N_2, 0.008% H_2, 0.04% Fe, and 0.03% C.

Properties of Ductile Vanadium: In its pure form, V is soft and ductile. It can be hot- and cold-worked easily, but it must be heated in an inert atmosphere or in a vacuum because it oxidizes readily at temperatures above the melting point of its oxide (about 663°C, 1225°F). The strength of V is sensitive to interstitial impurities and varies from 30,000 psi in the purest form to 80,000 psi in the commercial form. This metal retains its strength unusually well at elevated temperatures. Vanadium has a relatively low cross section for neutron capture (5 barns), accounting for considerable interest in the nuclear field.

The resistance of V to HCl and H_2SO_4 is outstanding, and it withstands aerated saltwater attack better than most stainless steels. Vanadium cannot withstand HNO_3, either dilute or concentrated.

The density of V is 22% less than that of Fe and 28% more than that of Ti. The coefficients of electrical and thermal conductivity are both significantly higher than those of Ti.

Alloys: Most of the vanadium produced (about 80%) is consumed by the steel industry as an alloying agent. The formation of vanadium carbide when vanadium is added to steel is the basis for many of the unique properties imparted by V to steel. These carbides are extremely hard and wear-resistant; they do not coalesce readily but maintain a state of fine dispersion. A relatively small amount of V (0.06–0.10%) is soluble in the austenitic phase of steels. This small percentage markedly increases the ability of the steel to harden on rapid cooling. V also forms a stable nitride, and it can, in effect, lower the nitrogen content of steel.

Carbon and alloy steels consume more than half the V produced in the United States. Many plate, structural, bar, and pipe steels contain V to enhance strength and toughness. Higher strength for a minimum additional price over plain carbon steels without a corresponding decrease in ductility and weldability has accounted for this demand. V in the range of 0.02–0.08% is contained in many of these steels, together with higher Mn and often Cu. Most of these steels possess higher strength in the as-rolled condition, i.e., without heat treatment.

Sheet steels used for deep-drawing auto and home-appliance parts often contain V to suppress aging. In making such steels, ferrovanadium is added to rimming steels, providing a good, non-aging, deep-drawing steel at a lower cost than the aluminum-killed deep-drawing steel.

Vanadium has a promising future as a grain-size control agent in continuously cast steel, replacing aluminum, which to date cannot be used in this process. Many large steel forgings contain V in the range of 0.05–0.15%, where it acts as a grain refiner as well as for improving the mechanical properties of the forgings. In large steel castings and forgings a small percentage of V is unique in its favorable action of raising strength and ductility.

Tool steels are the third largest class of vanadium-containing steels. Almost all tool steels contain V, the amount ranging from 0.10 to 5.00%. Vanadium is required to ensure the retention of hardness and cutting ability at the elevated temperatures generated by rapid cutting of metals. Vanadium sometimes is used in cast iron to control the size and distribution of graphite flakes and to improve strength and wear resistance. The most widely used of the titanium-base alloys contains 4% V and 6% Al. A high-purity 40:60 V-Al alloy is produced for this purpose and accounts for a significant consumption of V.

Compounds: Vanadium forms many and frequently complicated compounds because of its variable valence. It has at least four oxidation states, 2^+, 3^+, 4^+, and 5^+. Vanadium is amphoteric, although predominantly basic in the lower oxidation states and acidic in the higher oxidation states. V forms derivatives from more or less well-defined radicals, such as VO^{++} and VO^{3+}. The oxygen acids readily form condensed acids. See Table V-2.

The most important compounds of V are vanadium pentoxide, V_2O_5, and ammonium metavanadate, NH_4VO_3. The pentoxide commonly is sold as the sodium salt of hexavanadic acid, $Na_2H_2V_6O_{17}$. Ammonium metavanadate is formed by adding excess NH_4Cl to an alkaline solution of V_2O_5. Pure (99.6%) V_2O_5 is made by calcining ammonium metavanadate.

Vanadium oxytrichloride is produced commercially for use as a catalyst in making ethylene-propylene rubber. Sodium metavanadate also is produced for use as an additive for alkaline amine solutions to reduce corrosive action in steel piping systems.

Many other compounds have been made on a smaller scale. Vanadyl sulfate and vanadyl chloride have had a rather regular demand. Compounds produced for study include the nitrates, acetates, oxalates, nitrides, carbides, bromides, iodides, and fluorides. Organometallic compounds include vanadyl linoleate, oleate, palmitate, phenolate, resinate, and stearate. Ammonium vanadyl tartrate has received attention as a hypocholesteremic agent.

The largest application of V compounds is the manufacture of oxidation catalysts for the chemical industry. Both ammonium metavanadate and vanadium pentoxide are used. Processes employing such catalysts include the manufacture of polyamides, such as nylon, the manufacture of H_2SO_4 by the contact process, the manufacture

TABLE V-2. *Representative Vanadium Compounds*

Compound	Formula	Formula Weight	Sp gr	Color and Form
Vanadium dichloride	VCl_2	121.86	3.23	Green, hexagonal
Vanadium trichloride	VCl_3	157.23	3.00	Pink, tabular
Vanadium tetrachloride	VCl_4	192.78	1.816	Red liquid
Vanadium dioxide	V_2O_2	133.90	3.64	Light-gray crystals
Vanadium trioxide	V_2O_3	149.90	4.87	Black crystals
Vanadium tetraoxide	V_2O_4	165.90	4.399	Blue crystals
Vanadium pentoxide	V_2O_5	181.90	3.357	Red-yellow, rhombic
Vanadyl chloride.	$(VO_2)Cl$	169.36	3.64	Yellow crystals
Vanadyl dichloride	$VOCl_2$	137.86	2.88	Green, deliquescent
Vanadyl trichloride	$VOCl_3$	173.32	1.829	Yellow liquid
Ammonium metavanadate	NH_4VO_3	116.99	2.326	Colorless crystals

of phthalic and maleic anhydrides, and various other organic oxidation reactions, such as anthracene to anthraquinone, alcohol to acetaldehyde, diphenylamine to carbazole, and sugar to oxalic acid.

For many years, V compounds have been used in the ceramics industry for glazes and enamels. Combinations of vanadium oxide and zirconia, silica, lead, zinc, tin, cadmium, and selenium produce various colors. Vanadium compounds are used in the production of aniline black for the dye industry.

REFERENCES

Bachman, Robert Z., and Charles V. Banks: Critical Review of the Analytical Methods of the Titanium, Vanadium and Chromium Transition Element Groups, pt. 4 of "Analytical Chemistry," vol. 3 Pergamon, Oxford, 1963.

Rostoker, William: "The Metallurgy of Vanadium," Wiley, New York, 1958.

—T. W. Merrill, *formerly with Foote Mineral Company, Exton, Pa.*

van Arkel–de Boer Process   (See **Hafnium.**)

van der Waals Attraction   (See **Adsorption;** and **Molecule.**)

Vanillin   (See **Aldehydes.**)

Vapometallurgy   (See **Nickel.**)

Vapor Degreasing   (See **Trichloroethylene.**)

Vaporization   (See **Evaporation.**)

Varnish   (See **Paint;** and **Resin.**)

Vat Dyes   [See **Dyestuffs and dyeing (textile).**]

V-Disk Press   (See **Expression.**)

Vegetable Fibers   (See **Fibers;** and **Peptides and proteins.**)

Vegetable Oils   Fats and oils are water-insoluble substances of vegetable or animal origin which consist predominantly of glycerol esters of fatty acids.

$$
\begin{array}{ll}
\text{H} & \\
\text{HC—OH} & \text{HOOCR} \\
\text{H—C—OH} + & \text{HOOCR}' \rightarrow \\
\text{H—C—OH} & \text{HOOCR}'' \\
\text{H} & \\
\text{Glycerol} & \text{Fatty acids}
\end{array}
$$

$$
\begin{array}{l}
\qquad\qquad \text{H} \\
\alpha\ \text{HC—OOCR} \\
\beta\ \text{HC—OOCR}' \\
\alpha\ \text{HC—OOCR}'' + 3H_2O \\
\qquad\qquad \text{H}
\end{array}
$$

Triglyceride

Most vegetable oils are mixed triglycerides since more than one kind of fatty acid is present. The type of fatty acid is the dominant factor influencing the properties of the vegetable oils, and to a lesser extent properties depend upon the α or β position of the attachment. Predominantly, the natural vegetable oils are composed of fatty acids with 18 carbon chains with varying degrees of unsaturation. The common fatty acids are summarized in Tables V-3 and V-4.

Reactions of the vegetable oils and their fatty acids may be divided into (1) reactions of the ester linkage of the ester or carboxyl end of the

TABLE V-3. *Properties of Saturated Fatty Acids*

Common or Other Name	Formula	Formula Weight	Mp, °C	Bp, °C at 760 mm	Density d_4^{70}	Refractive Index n_D^{80}	Neutralization Value*
Caprylic. . .	$C_7H_{15}COOH$	144.21	16.7	239.7	0.8615	1.4125	389.05
Capric. . . .	$C_9H_{19}COOH$	172.26	31.6	270.0	0.8531	1.4169	325.69
Lauric	$C_{11}H_{23}COOH$	200.31	44.2	298.9	0.8690	1.4230	280.08
Myristic . . .	$C_{13}H_{27}COOH$	228.36	53.9	326.2	0.8439	1.4273	245.68
Palmitic . . .	$C_{15}H_{31}COOH$	256.42	63.1	351.5	0.8414	1.4309	218.80
Stearic. . . .	$C_{17}H_{35}COOH$	284.47	69.6	376.1	0.8390	1.4377	197.23

*Neutralization value is defined as the number of milligrams of KOH required to neutralize 1 g of fatty acid.

fatty acid and (2) the reactive sites of the fatty chain.

Reactions of Fatty Carboxyls and Esters: The fatty acids can be reacted with glycerol or other hydroxyl-bearing compounds to form esters and water. Conversely, the triglycerides can be hydrolyzed to yield glycerol and fatty acids. Hydrolysis is effected at elevated temperature and pressure in a matter of minutes. A process called *Twitchellization* will hydrolyze an oil at low temperatures and pressure but at a much slower rate. During storage, an oil also may hydrolyze very slowly, resulting in an increase in acidity or acid value.

If an oil is split with alkali to yield glycerol and salt, or soap, of the alkali metal, the reaction is termed *saponification.* Soaps may also be made of Co, Pb, Mn, Ca, and Zr. Metal soaps are added to unsaturated oils to accelerate their oxidation, which converts these from liquid oils into rubbery insoluble materials.

Interesterification is one of the most important commercial processes. If a triglyceride oil is

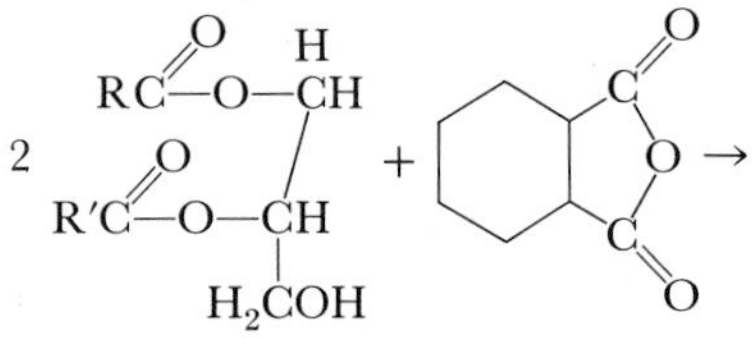

Phthalic anhydride

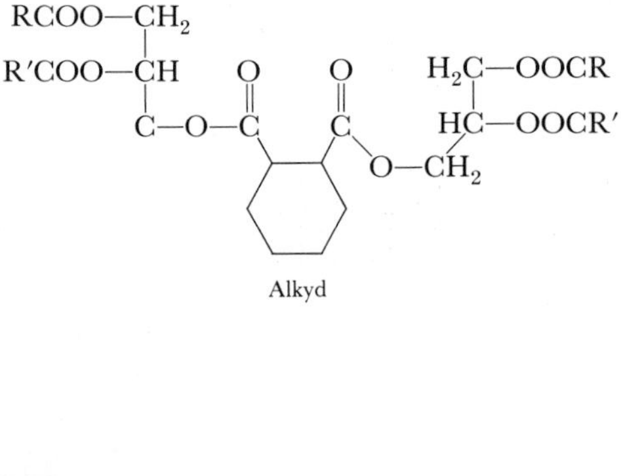

Alkyd

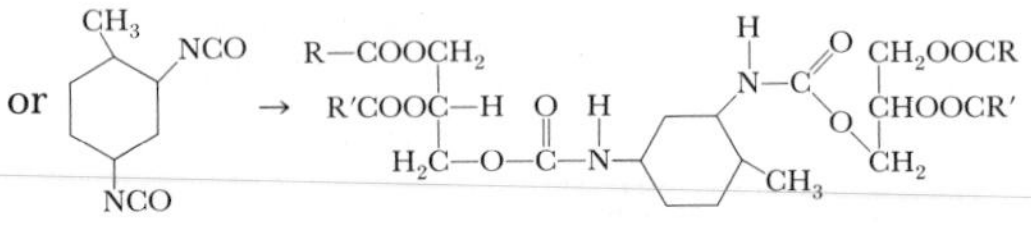

Toluene diisocyanate

Urethane

heated with additional polyol, e.g., glycerol or pentaerythritol, mixed partial esters are obtained. This reaction is known as *alcoholysis*. Similarly, fatty acids can be heated with a triglyceride to change the fatty acid composition of the triglyceride by a reaction called *acidolysis.*

The partial esters of the alcoholysis reaction are subsequently reacted with dibasic acids, usually o-phthalic anhydride to form alkyds. If the partial esters are reacted with diisocyanates, oil modified urethanes are produced.

The carboxyl group can be hydrogenated to yield fatty alcohols.

$$RCOOH + 2H_2 \rightarrow RCH_2OH + H_2O$$

The fatty acids react readily with ammonia or amines to form the corresponding soap. At elevated temperatures amides are formed.

$$RCOOH + NH_3 \rightarrow RCONH_2 + H_2O$$

The reaction may continue to form the corresponding nitrile.

$$RCONH_2 \rightarrow RCN + H_2O$$

Acid chlorides are prepared by the reaction of fatty acids, with PCl_3, PCl_5, $SOCl_2$, and $COCl_2$.

$$3RCOOH + PCl_3 \rightarrow 3RCOCl + H_3PO_3$$

*Reactions of the Fatty Acid Chain:** The dominant characteristic differentiating the fatty oils is the degree of unsaturation of the fatty acid sub-

*See A. J. Stirton, J. Turner, and R. W. Riemenschneider, *Oil Soap,* vol. 22, pp. 81–83, 1945, and J. E. Meyers, J. P. Kass, and G. O. Burr, *Oil Soap,* vol. 18, pp. 107–109, 1941.

TABLE V-4. *Properties of Unsaturated 18-Carbon Fatty Acids*

Common Name	Double Bond Position	Molecular Weight	Saponification Value[a]	Iodine Value[b]	Mp, °C	Bp, °C	Sp gr d$_4^{20}$	Refractive Index n_D^{20}	Source
Oleic	cis-9	282.45	198.63	89.9	16.0	234[c]	0.891	1.45823	Most common (olives and nuts)
Linoleic	cis-9, cis-12	280.44	199.5	181.0	−9.5	229[d]	0.9022	1.4699	Safflower, sunflower, soy, tall oils
Linolenic	cis-9, cis-12, cis-15	278.42	201.51	273.5	−11.3	231[e]	0.916	1.4800	Linseed and perilla oils
α-Eleostearic . .	cis-9, trans-11, trans-13	278.42	201.51	273.5	48–49	170[f]	0.903[g]	1.5112[h]	Tung oil
β-Eleostearic . .	trans-9, trans-11, trans-13	278.42	201.51	273.5	71.5	188[f]	0.8839[i]	1.5000[i]	Tung oil
Ricinoleic	cis-9, (+)-12 hydroxy	298.45	190.0	85.0	5.0	227[k]	0.9450	1.4716	Castor oil

[a] Defined as the number of milligrams of KOH required to saponify 1 g of fat or oil.
[b] Defined as the number of centigrams of iodine absorbed by 1 g of fat or oil.
[c] 15 mm.
[d] 16 mm.
[e] 17 mm.
[f] 1 mm.
[g] d$_4^{50}$.
[h] n_D^{50}.
[i] d$_4^{80}$.
[j] n_D^{80}.
[k] 10 mm.

TABLE V-5. *Oxidation Rates of Fatty Acids*

Acids	Double Bonds	Conjugation	100°C	20°C
Stearic..........	0	0	0.6	
Oleic	1, cis	0	6	
Linoleic........	2, cis	0	64	
Linolenic	3, cis	0	100	100
β-Eleostearic	3, trans	3	...	196
α-Eleostearic	3, cis, trans, trans	3	...	515

stituent. The vegetable oils frequently are classified as nondrying, semidrying, or drying oils. The first contain a higher content of saturated fatty acids and the last a higher content of unsaturated fatty acids. The degree of unsaturation of the oil or its fatty acid substituents affects its fluidity or melting point, density, refractive index, and reactivity with other chemical compounds.

The unsaturated oils can be hydrogenated with gaseous hydrogen by the use of catalyst of Ni, Pt, or Pd. Large quantities of oils are hydrogenated for use in food products.

The unsaturated oils are readily oxidized by air or chemical means.. The oxidation of an edible oil results in deleterious rancidity. Industrially, oils are oxidized by air or chemical means to produce many useful products. During the initial stages of oxidation the carbon α to the double bond is attacked together with a shift of the isolated double bonds to the more reactive conjugated position.

$$-CH{=}CH{-}CH_2{-}CH{=}CH{-} \xrightarrow{O_2}$$
$$-CH{=}CH{-}CH{=}CH{-}CH{-}$$
$$\underset{OOH}{|}$$

The relative rates of oxidation of fatty acids are given in Table V-5. By means of the free-radical mechanism, the oxidation continues to originate hydroxyl, epoxy, and ketoxy groups and to effect cission of the fatty chain and cross-linking by ester, ether, and hydrogen bonds. In a film of drying oil or resin the desired result is a flexible, tough, abrasion- and chemical-resistant coating which protects and beautifies the substrate. Besides participating in the curing of an oil-modified resin, the oil serves as a plasticizer to a resin which may be too hard and brittle by itself.

The unsaturated oils may be polymerized by heat and catalysts to produce viscous products useful in exterior house paints, flat wall paints, enamels, printing inks, and brake linings. The rate of polymerization is proportional to presence of conjugated unsaturation and the functionality of the unsaturated oil. The functionality is increased by increasing the amount of unsaturation and the functionality of the polyol to which the fatty acids are attached. Hence pentaerythritol esters polymerize much more rapidly than glycerol esters.

The effect of number and position of the double bonds of unsaturated esters on the rate of polymerization is shown in Table V-6.

TABLE V-6. *Polymerization Rates of Methyl Esters*

Diene		Triene	
Nonconjugated			
cis-trans	0.74		
cis-cis	1.0	cis-cis-cis	2.4
trans-trans	1.2		
Conjugated			
cis-trans	5.8	cis-trans-trans.....	170
trans-trans	26.0	trans-trans-trans ...	340

SOURCE: Paschke and Wheeler, *J. Am. Oil Chem. Soc.*, vol. 32, p. 471, 1955.

More rapid polymerization of unsaturated oils can be accomplished by isomerization of the double bonds from the isolated cis configuration to the conjugated trans configuration. Catalysts which have been found useful are alkali, Ni, Al, Si, SO_2 or anthraquinone compounds.

Even as unsaturated oils will homopolymerize, they will also copolymerize with the reactive vinyl monomers. Oils with conjugated double bonds copolymerize much more readily than oils with isolated double bonds. Copolymers of vegetable oils with styrene or vinyl toluene are items of commerce.

Castor oil is an important vegetable oil with 90% of its fatty acid composition consisting of one unusual fatty acid, ricinoleic acid, 12-hydroxyoctadec-9-enoic acid, $CH_3(CH_2)_5 \cdot CHOH \cdot CH_2 \cdot$ $CH:CH(CH_2)_7 \cdot COOH$. Castor oil is much more viscous than other triglyceride oils, and the viscosity of the oil changes less with temperature than other vegetable or mineral oils.

Because of viscosity and compatibility characteristics, castor and its derivatives are used for plasticization of polymers. The hydroxyl group of castor oil can be reacted directly with diisocyanates to give polyurethane foams, plastics, adhesives, or coatings.

The hydroxyl group of castor oil can be removed by acid catalysts in a process called *dehydration* since water is removed in the reaction. The removal of water from the ricinoleate chain includes the abstraction of a hydrogen from either the eleventh or thirteenth carbon atom. If the hydrogen is taken from the eleventh carbon atom, a reactive 9,11 conjugated system is formed. In a commercial unpolymerized dehydrated castor oil about 25–30% of the double bonds are conjugated.

Tall oil is listed with the vegetable oils in Table V-7. Large quantities are sold at low prices,

TABLE V-7. *Consumption of Vegetable Oils in the United States 1969*

Oil	Millions of lb
Castor	152
Coconut	732
Corn	441
Cottonseed	890
Linseed	194
Palm	148
Peanut	157
Safflower	89
Soy	5,948
Tall	1,269
Tung	31

SOURCE: United States Department of Commerce.

largely to the coatings industry. Tall oil is a misnomer since it consists of a mixture of fatty acids and rosin and is not a triglyceride oil. The refined rosin-free acids sell at a price approximately equal to the lower-priced drying and semidrying oils but are a low-cost source of fatty acids. Refined tall oil is about 50% oleic acid and 47% linoleic acid. The oil is a by-product of pine woodpulp-digestion process of the kraft paper industry.

Many other fatty acids occur in uncommon oils, primarily as triglycerides, which differ in carbon

chain length, position of the double bonds, or with reactive substituents, such as hydroxy, ketoxy, or epoxy groups. The compositional differences and analytical constants of common vegetable oils are summarized in Table V-8.

Uses of Vegetable Oils: The classification of vegetable oils as nondrying, semidrying, and drying is convenient because oils in a given grouping may be used in the same applications.

The nondrying oils have low unsaturation or contain less than 20% of linoleic acid. Coconut and olive oils are members of this group. Large quantities of these oils are consumed in cooking and salad oils, salad dressings, and similar edible products. Oils of greater unsaturation, e.g., soybean oil, may be used in similar products if the unsaturation is lowered by hydrogenation. The lower the unsaturation, the higher the melting point and hence the possible use in shortening, margarine, and bakery and confection products.

The nondrying oils are consumed in soap or surface-active products. Substantial quantities of inedible tallow, coconut, grease, and by-product fats and oils are used. These raw materials are converted into alkali-metal soaps or combined with ethylene oxide to make the nonionic surfactants and reach the consumer market as toilet or laundry bars, chips, flakes, washing powders, cleansers, and shaving creams.

The drying oils have greater unsaturation or greater quantities of fatty acids with higher unsaturation, e.g., linoleic acid, linolenic acid, or their isomers. The important oils of this group are linseed, tung, perilla, and dehydrated castor oil. The drying characteristics of these oils have no bearing on their water content but refer to the fact that the liquid mobile oil oxidizes, conventionally with driers, e.g., cobalt naphthenate, to solid insoluble films.

The drying oils are consumed in large quantities in paints, varnishes, caulking preparations, linoleum, plasticizers, lubricants, foundry-core sand binders, carbon electrodes, and brake linings.

Oils classified as semidrying are intermediate between the nondrying and drying oils in properties and contain about 40–60% linoleic acid. They are used in products of both classes depending on the market value of the oils and the cost of modification to make them suitable for the applications of either the drying or nondrying oils. Hence, soybean oil can be deodorized and inhibited with antioxidants, such as butylated hydroxytoluene, to make an edible salad oil or hydrogenated to make a plastic shortening. Con-

TABLE V-8. *Compositional Differences and Analytical Constants of Common Vegetable Oils*

Fat or Oil	Constants				Constituent Fatty Acids, g/100 g Total Fatty Acids						
	Sp gr (or Density) at 15°C	Refractive Index n_D^{40}	Iodine Value	Solidification Point, °C	Saturated			Unsaturated			
					Myristic	Palmitic	Stearic	Oleic	Linoleic	Linolenic	Other
Castor	0.961	1.4770	85	Turbid −12	[a]	[a]	[a]	7.4	3.1	. . .	87 ricinoleic
Coconut	0.924	1.4493	10	14 to 22	18.0	10.5	2.3	7.5	Trace	. . .	45.4 lauric
Corn	0.922	1.4734	123	−10 to −20	1.4	10.2	3.0	49.6	34.3		
Cottonseed .	0.917	1.4735	106	+12 to −13	1.4	23.4	1.1	22.9	47.8		
Linseed . . .	0.938	1.4782	179	−19 to −27	. . .	6.3	2.5	19.0	24.1	47.4	
Oiticica . . .	0.974	. . .	160	. . .	[b]	[b]	[b]	5.5	. . .	. . .	82.5 licanic
Olive	0.918	1.4679	81	−6.0	Trace	6.9	2.3	84.4	4.6	. . .	
Peanut . . .	0.914	1.4691	93	3	. . .	8.3	3.1	56.0	26.0	. . .	
Perilla	0.935	1.481[c]	195	. . .	[d]	[d]	[d]	7.0	39.0	46.0	
Safflower . .	0.900	1.462[e]	145	−13 to −18	. . .	6.4	3.1	13.4	77.0	0.2	
Soybean . . .	0.927	1.4729	130	−16	0.1	9.8	2.4	28.9	50.7	6.5	
Sunflower . .	0.923	1.4694	125	−17	. . .	5.6	2.2	25.1	66.2	. . .	
Tung	0.934	1.5174[c]	168	<17	[f]	[f]	[f]	4.1	8.5	. . .	82.0 eleostearic

[a] Total saturated 2.4.
[b] Total saturated 11.0.
[c] At 25°C.
[d] Total saturated 7.0.
[e] At 60°.
[f] Total saturated 5.5.

1134

versely, it may be modified with pentaerythritol and maleic anhydride to make a long oil alkyd performing much like linseed oil.

Of the commercially important oils, castor oil does not fit well into the above classifications. With its hydroxyl group on the twelfth carbon atom of the fatty chain, its properties and utilization are varied, though the hydroxyl group may be removed by dehydration to give an excellent nonyellowing drying oil. It is deodorized for medicinal (cathartic) and cosmetic use. It is used to make sulfonated oils, artificial leathers, plasticizers, lubricants, greases, hydraulic fluids, soaps, printing inks, dielectric fluids, perfumes, sebacic acids, and nylon.

See also **Expression** for description of how oils are obtained from sources.

REFERENCES

Eckey, E. W.: "Vegetable Fats and Oils," Van Nostrand Reinhold Company, New York, 1954.
Gunstone, F. D.: "Introduction to the Chemistry of Fats and Fatty Acids," Chapman & Hall, London, 1958.
Hilditch, T. P.: "The Chemical Constitution of Natural Fats," Wiley, New York, 1956.
Markley, K. S.: "Fatty Acids: Their Chemistry, Properties, and Uses," 5 vols., Interscience-Wiley, New York, 1960–1968.
Swern, D. (ed.): "Bailey's Industrial Oil and Fat Products," Interscience-Wiley, New York, 1964.

—Herbert M. Schroeder, *Spencer Kellogg Division of Textron Inc., Buffalo, N.Y.*

Vegetable Proteins (See **Amino acids;** and **Peptides and proteins.**)

Vehicles, Paint (See **Paint.**)

Vibrating Screens (See **Screening.**)

Vinegar (See **Acetic acid;** and **Pasteurization.**)

Vinyl Acetate (See **Polyvinyl chloride.**)

Vinyl Chloride Monomer The availability of low-cost raw materials and the versatility of the end product, polyvinyl chloride, has resulted in a highly competitive market for vinyl chloride monomer (VCM) and very efficient processes for its manufacture. See also **Polyvinyl chloride.**

VCM is a clear, colorless liquid with a boiling point of $-13.3°C$ ($+8°F$). Its flammability in air (at 4–22% by volume), along with its density, which is heavier than air, requires thorough safety precautions in handling.

The classic process for the production of VCM involves the catalytic hydrochlorination of acetylene. The reaction is generally carried out in tubes filled with activated carbon impregnated with mercuric chloride catalyst. The temperature normally varies from 100 to 280°C, the temperature being increased as the catalyst activity decreases with time. The high cost of acetylene in relation to the lower cost of other hydrocarbons, notably ethylene, has resulted in more economically attractive processes based on ethylene. Almost all of the acetylene-based plants have been phased out in favor of ethylene-based processes.

The recent advent of cheaper acetylene through hydrocarbon-cracking processes, e.g., the Wulff and Kureha processes, has allowed the continued use of the acetylene route in some instances. However, these processes produce coproduct ethylene in ethylene-acetylene ratios of $\frac{1}{2}$:1–8:1. Other processes in a complex must be tailored to fit these ratios of products.

The most versatile and economic route to VCM utilizes ethylene and chlorine or hydrogen chloride feedstocks through an intermediate product, ethylene dichloride (EDC). In the usual configuration, where elemental Cl_2 and ethylene are the feedstocks, EDC is produced by the direct chlorination of ethylene. The HCl produced as a by-product of the cracking of EDC to VCM is reacted with ethylene in the oxychlorination process to produce EDC for recycle. In this case, the direct chlorination and oxychlorination are of about the same capacity. If HCl is available from an external source, the oxychlorination process will be the larger and the direct chlorination process will be the proportionately smaller of the two processes. The three processes (see Fig. V-1) entail the following reactions:

Direct chlorination of ethylene:

$$C_2H_4 + Cl_2 \rightarrow ClC_2H_4Cl \qquad (1)$$

Thermal dehydrochlorination of EDC (EDC cracking):

$$ClC_2H_4Cl \rightarrow CH_2{=}CHCl + HCl \qquad (2)$$

Oxhydrochlorination of ethylene:

$$C_2H_4 + 2HCl + \tfrac{1}{2}O_2 \rightarrow$$
$$ClC_2H_4Cl + H_2O \qquad (3)$$

Direct Chlorination Process: The reactants are fed through distributors into liquid EDC, where the addition reaction takes place to form EDC. Substitution reactions yielding chlorinated by-

OXYHYDROCHLORINATION

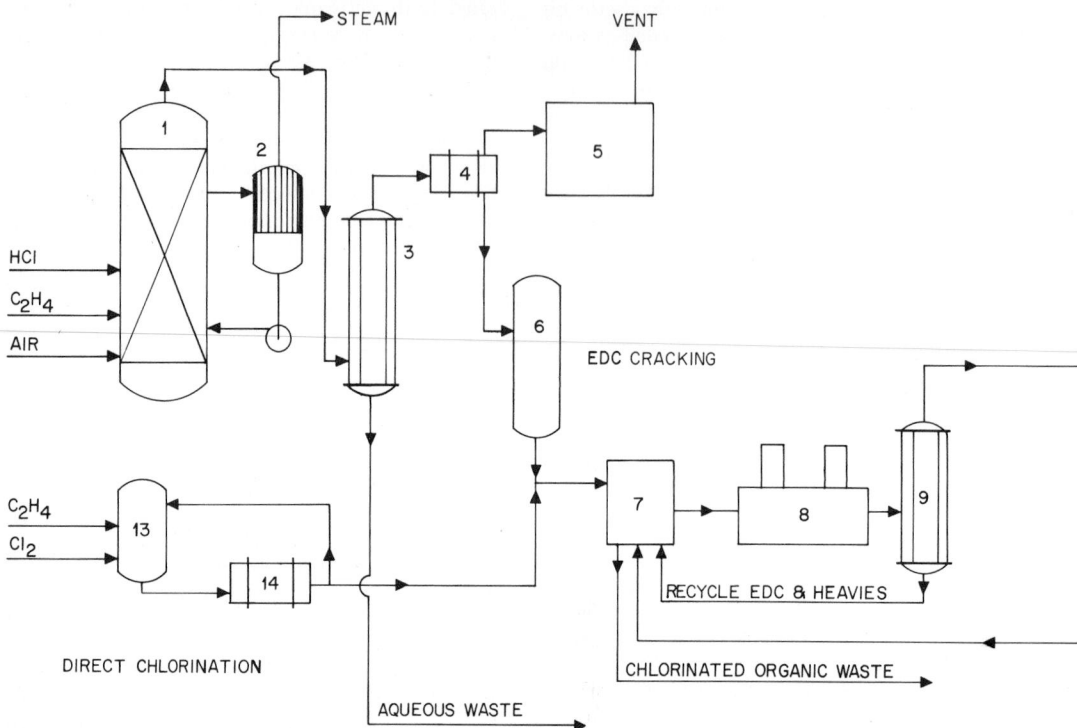

Fig. V-1. Process for making vinyl chloride monomer (VCM). (1) Oxy reactor, (2) steam-generation system, (3) quench column, (4) quench condenser, (5) ethylene dichloride (EDC) recovery, (6) crude EDC decanter, (7) EDC purification, (8) EDC cracking furnace, (9) quench column, (10) quench condenser, (11) anhydrous hydrochloric acid recovery column, (12) VCM recovery column, (13) direct chlorination reactor, (14) reactor cooler.

products and HCl are inhibited by the presence of ferric chloride, O_2, or air. The heat of reaction is removed easily by conventional cooling methods.

The crude EDC purity may be as high as 99.8%, the impurities being other chlorinated organics, HCl, ferric chloride, and Cl_2. The EDC then is combined with crude EDC from the oxyhydrochlorination unit, purified by distillation, and fed to the EDC cracking process.

The direct chlorination reactor and cooler can be of carbon-steel construction, but the washing equipment must be resistant to corrosion by HCl.

EDC Cracking Process: In the EDC cracking furnace, EDC is thermally cracked to yield VCM and HCl. The furnaces generally are direct-fired, using a suitable fuel gas or light distillate oil, such as kerosine. The reaction takes place inside hori-

zontal or vertical alloy tubes. Conversion usually is limited to 50–60% of the feed to limit coking or carbon deposition in the furnace tubes.

Furnaces with capacity of up to 140,000 metric tons/yr of VCM are in operation. Process pressures range up to 24 atm gage and exit temperatures usually fall within a temperature range of 500–550°C.

The hot furnace exit gases are cooled quickly in one of two types of quench columns. In a desuperheating quench, the gases are cooled to the dew point by evaporation of EDC and then partially condensed in a conventional condenser. In a condensing quench, the furnace exit material is cooled and partially condensed in the quench column, with the main process stream being fed from the bottom of the quench as a liquid to the HCl and VCM recovery columns. If low-pressure

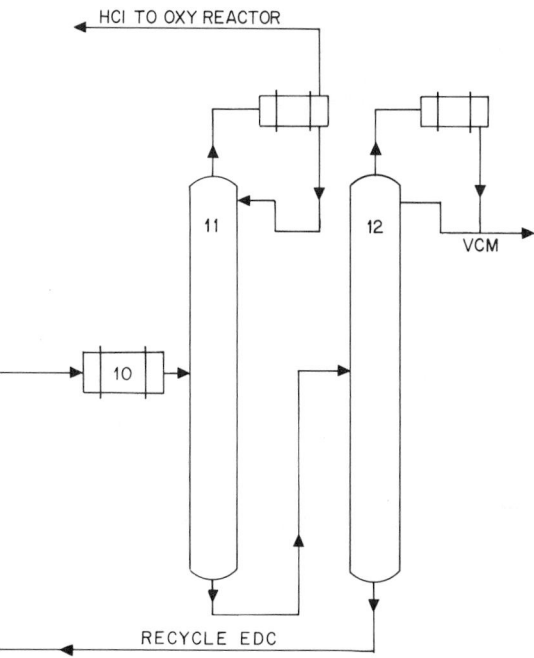

catalyst reactor may be used. Because of the presence of both HCl and product water, corrosion by HCl posed a serious problem in early development of the process. Special reactor design, using a fluid catalyst bed, has eliminated the problem, permitting the use of carbon-steel reactors.

The oxychlorination catalyst, basically cupric chloride impregnated on a fluid or fixed support, is highly reactive and selective and produces high conversion and high crude EDC purity in a one-stage reaction, using either a fluid or fixed catalyst bed. In a fixed bed, the catalyst must be replaced periodically due to loss of activity. In a fluid bed, the small continuous losses are made up by fresh catalyst to maintain an equilibrium activity and eliminate the need for replacing the entire charge.

The formation of EDC is believed to occur through a heterogeneous mechanism where Cl_2 from active cupric oxychloride sites adds to the ethylene double bond in a two-step addition reaction, thereby reducing the copper salt to cuprous oxychloride:

$$C_2H_4 + 2CuOCl_2 \rightarrow C_2H_4Cl_2 + 2CuOCl$$

The cuprous oxychloride subsequently is reactivated to the cupric state by the oxychlorination reaction:

$$2CuOCl + 2HCl + \tfrac{1}{2}O_2 \rightarrow 2CuOCl_2 + H_2O$$

The oxychlorination is highly exothermic. Cooling is accomplished either by direct generation of steam in reaction cooling coils or indirect generation of steam by circulation of a coolant through the reactor coils. The high thermal conductivity of fluid catalyst and the high heat-transfer coefficient between a fluid catalyst and the cooling surfaces, when combined with direct steam generation, allow design for production of over 1 lb of steam per pound of EDC produced at a pressure of up to 185 psig.

Unreacted HCl in the reactor effluent gas is absorbed in water in the quench column. The resulting hydrochloric acid stream is treated offsite before disposal. The scrubbed gas is condensed in a conventional shell-and-tube condenser. The EDC and water phases are separated in a decanter, the water phase passing offsite for treatment and the crude EDC phase being combined with crude EDC from the direct chlorination process for purification by distillation.

The aqueous effluent streams contain HCl, Cl_2, and EDC in solution. The acid materials can be neutralized with caustic, and the EDC can be stripped out and returned to the process. The

cracking is used, the VCM and EDC usually are removed from the quench vapor stream by condensation or absorption before the remaining HCl is compressed and fed to the oxychlorination unit. If higher-pressure cracking is used, the quench vapor stream can be fed directly to the distillation columns.

In the HCl recovery column, anhydrous HCl is recovered overhead by distillation with refrigerated reflux. This HCl must be purified to minimize the acetylene and vinyl chloride content which would oxychlorinate to produce overchlorinated by-products in the oxychlorination unit. In the VCM recovery column, EDC and other high-boiling impurities are separated by fractionation to produce high-purity VCM.

Oxychlorination of Ethylene: In this process for producing EDC, either a fluid catalyst bed or fixed

oxychlorination vent gas contains low levels of ethylene, CO_2, and CO. In some areas, catalytic burners or auxiliary fired burners are used to remove traces of organic materials by combustion. Of the chlorinated organic by-products, which are removed as the heads and bottoms of the purification columns, some can be converted through conventional processes to marketable solvents, such as perchloroethylene.

Alternately, the chlorinated by-products can be incinerated, the resultant HCl being recovered as concentrated hydrochloric acid or anhydrous HCl. In at least one instance, anhydrous HCl is recycled to the oxychlorination unit. See also **Petrochemical complex.**

—Lloyd Duwelius, *B. F. Goodrich Chemical Company, Cleveland, Ohio*

Vinyl Ester Resins The vinyl ester resins are a relatively recent addition* to thermosetting-polymer chemistry. Superficially, they are similar to unsaturated polyester resins insofar as they contain ethylenic unsaturation and are cured through a free-radical mechanism, usually in the presence of a vinyl monomer such as styrene. However, close examination of the chemistry and structure of the vinyl ester resin demonstrates several basic differences which lead to their unique characteristics.

Typical vinyl ester resin

Typical unsaturated polyester resin

Vinyl ester resins are manufactured through an addition reaction of an epoxy resin with an acrylic monomer, such as acrylic acid or the half-ester product of an hydroxyalkyl acrylate and anhydride.[2] In contrast, the polyester resins are condensation products of dibasic acids and polyhydric alcohols. The relatively low-molecular-weight precise polymer structure of the vinyl ester resins is in contrast to the high-molecular-weight random structure of the polyesters.

Of particular importance in describing the

difference between these two families of resins are the locations of the reactive unsaturation. In the polyester resin, these groups are located along the backbone of the polymer with terminal hydroxyl or carboxylic acid groups. In contrast, the vinyl ester resins contain no significant acidity but terminate in reactive vinyl ester groups. Because of the location of these reactive sites, the vinyl ester resins will homopolymerize as well as coreact with various vinyl monomers.

Resin Properties: Vinyl esters, because of their relatively low molecular weight and precise structure, can be characterized as low-viscosity, fast-wetting, consistent-reactivity products. Typical property profiles for some uncured vinyl ester resins are given in Table V-9. Properties of some cured resins are given in Table V-10. Typically, 6-month stability can be expected at 77°F (25°C) with decreased storage life under elevated temperatures. In general, anyone familiar with the proper storage and handling of unsaturated polyester resins and styrene monomer should have no difficulty with these materials.

Cure Mechanism: The free-radical cure mechanism of the vinyl ester resin is well understood. In most respects, it is similar to that of the unsaturated polyester resins. To initiate the curing process, it is necessary to generate free radicals within the resin mass. Organic peroxides are the most common source of free radicals. These peroxides will decompose under the influence of elevated temperatures or chemical promoters, e.g., organometallics or tertiary amines, to form free radicals. Generation of free radicals also can be effected by ultraviolet or high-energy radiation applied directly to the resin system. The free radicals thus formed react to open the double bond of the vinyl group. Once opened, the resin vinyl group is highly reactive and rapidly combines with several more vinyl groups available from both the unreacted resin and the monomer. This exothermic reaction is rapidly carried to completion forming a three-dimensional thermosetting network.

Applications: As might be expected from the wide variation in resin properties which can be built into the molecule, the vinyl ester resins find many applications. Chief among these are fiberglass-reinforced plastics where the inherent characteristics of the vinyl ester resin provide a price and performance advantage over other materials. The largest application at present is in the manufacture of corrosion-resistant reinforced plastic structures. Due to the reduced number of ester

*The first literature reference was a patent issued in 1962 for a tooth-filling compound.[1] Commercialization did not start until the late 1960s.

TABLE V-9. *Properties of Typical Uncured Vinyl Ester Resins**

Property	Standard Resin	Low-Viscosity Resin
Monomer type	Styrene	Styrene
Level, %.	45	45
Viscosity at 77°F (25°C), cP	550	200
Acid number	5	5
Specific gravity.	1.04	1.04
SPI gel time (1% benzoyl peroxide), min:		
At 180°F (82°C)	10	12
250°F (121°C)	1.4	1.5
Flash point (Tag open cup):		
°F	93	93
°C.	34	34

*The Dow Chemical Company.

groups within the resin structure, the corrosion-resistant vinyl ester resins are less prone to attack by hydrolysis than the bisphenol A–fumaric acid polyesters. In addition, the resilience of the vinyl ester resins (4–6% ultimate tensile elongation) results in a fabricated part which is less prone to damage during shipping, field erection, and service.

All fabrication methods commonly used in the manufacture of reinforced plastics can be used with the vinyl ester resins. In those applications, such as filament winding and bag molding, where fast wetting is important, significant increases in output can be realized. One technique which is

TABLE V-10. *Properties of Typical Cured Vinyl Ester Resins**

Clear-casting properties:	
Tensile strength, psi	12,000
Tensile modulus, psi	5.0×10^5
Ultimate elongation, %	5
Flexural strength, psi	18,000
Flexural modulus, psi.	4.5×10^5
Yield compressive strength, psi	17,000
Compressive modulus, psi	3.5×10^5
Deformation at yield, %	7
Heat-distortion temperature:	
°F	215
°C	101.7
Barcol hardness	35
Glass-reinforced laminate properties:	
Laminate thickness, in.	0.25
Fiber-glass content, %	30
Tensile strength, psi	19,000
Tensile modulus, psi	1.4×10^6
Flexural strength, psi	22,000
Flexural modulus, psi.	1.0×10^6

*The Dow Chemical Company.

particularly suited to the vinyl ester resin is in the manufacture of products using sheet molding compound (SMC). Here, the resin system (usually with a high filler loading) is combined in sheet form with the glass reinforcement and chemically thickened through the use of metal oxides, such as MgO. The SMC then is molded, usually in a matched-die molding operation, to give the desired final product, e.g., automotive parts, appliance housings, electrical structures, and panel configurations.[3,4] Vinyl ester resins diluted with vinyl toluene monomer are used in the production of high-temperature electrical laminating systems.

REFERENCES

1. Bowen, R. L.: U.S. Patent 3,066,112, Nov. 27, 1962 (assigned to Secretary of Commerce).
2. Bearden, C. R.: U.S. Patent 3,367,992, Feb. 6, 1968 (assigned to The Dow Chemical Company).
3. Jernigan, J. W.: U.S. Patent 3,466,259, Sept. 9, 1960 (assigned to The Dow Chemical Company).
4. Swisher, D. H. and D. C. Garms: U.S. Patent 3,564,074, Feb. 16, 1971 (assigned to The Dow Chemical Company).

—John H. Minsker, *The Dow Chemical Company, Midland, Mich.*

Vinyl Fibers (See **Fibers.**)

Vinyl Monomers (See **Hydroxyalkyl acrylates and methacrylates; Polymerization; Polyvinyl chloride;** and **Vinyl chloride monomer.**)

Vinyl Paints (See **Paint.**)

Vinylidene Chloride (See **Chlorine organics.**)

Vinyon   (See Fibers.)

Virus Separation   (See **Membrane** filters.)

Visbreaking   (See **Petroleum** processing; and **Thermal** cracking.)

Viscose Rayon   (See **Rayon**.)

Vitamins   (See **Alkaloids; Calcium;** and **Pyridine and derivatives;** also consult the **Subject Index**.)

Vulcanization   (See **Elastomers; Rubber, natural;** and **Tellurium**.)

W

Wagner-Meerwein Transformation (See **Terpenes.**)

Wallboard (See **Gypsum.**)

Washing, Pulp [See **Pulp (wood) production and processing.**]

Washing Compounds [See **Anionic surfactants (sulfur-bearing); Detergents; Soaps;** and **Surfactants.**]

Waste Gases (See **Air pollution; Fuels; Iron- and steelmaking;** and **Sulfur dioxide removal.**)

Waste Heat (See **Energy systems for processes.**)

Waste-Solids Handling (See **Agglomeration.**)

Waste Treatment Industrial waste treatment is an important problem in every industry which carries out liquid processing operations. If wastes are classified according to their pollutants, there are the following major categories: (1) suspended solids, (2) oils and greases, (3) organic matter, (4) dissolved metals, and (5) toxic chemicals. In Table W-1 the industries in which these pollutants are found, their characteristics, and the types of treatment used for their removal are summarized.

Suspended-Solids Removal. Under this general category, occasionally referred to as *primary treatment,* screening devices, gravity coagulation, flocculation, sedimentation equipment, and dissolved-air flotation systems are most common.

Gross Solids Removal: Large suspended solids and trash are removed by a screening device, such as a bar screen. Further removal of gross suspended solids can be accomplished on a revolving drum covered with a wire or cloth screen. The solids are retained on the cloth or drum surface as it rotates and are discharged separately from the liquid fraction, which passes through the screen. Variations in design consist of gyrating or table-type vibrating screens.

Sedimentation: This operation consists of the removal of suspended solid particles from a liquid stream by gravitational settling. Sedimentation may be divided into thickening and clarification. The primary purpose of thickening is to maximize the concentration of solids removed from the waste, while that of clarification is to minimize the solids concentration of the treated effluent. Although there is little difference in the appearance of the equipment used for thickening and

TABLE W-1. *Common Industrial Waste Treatment Methods and Applications*

Category of Pollutant	Characteristics	Type of Treatment	Industry
Suspended solids	Dense, rapid settling	Plain sedimentation	Mining, phosphate, steel mills, power plants (fly ash), beet-sugar processing (beet washing), pulp (hydraulic debarking), foundry
	Colloidal	Chemical coagulation followed by sedimentation or flotation	Pulp and paper, textile, petroleum and petrochemical, food plants, steel mills, mining, chemical plants
	Oily material or light-weight solids	Flotation (with chemical treatment if necessary)	Petroleum and petrochemical, laundry, meat packing, machining (cutting oil), aircraft or railroad-car washing, dairies, food plants
Organic matter	Vary with industry; some are easily oxidized biologically; others require special techniques	Trickling filter; activated sludge: conventional, high-rate, contact stabilization, aerobic digestion	Beet- and cane-sugar plants, dairies, meat packing, pulp and paper, canning, chemical plants, brewing, petroleum and petrochemical, tanneries
	Very strong organic wastes	Anaerobic treatment followed by aerobic treatment	
Dissolved metals	Generally cations of Al, Cr, Cu, Fe, or Zn in low-pH solution	Precipitate with lime, followed by sedimentation	Metal finishing, plating, rayon processing, steel mills, tanneries
	Chromates	Reduce with ferrous iron sulfate or sulfur dioxide; then precipitate with lime, followed by sedimentation	Tanneries, plating, metal finishing
Toxic materials	Cyanide (generally with metal complexes)	Oxidize with chlorine or hypochlorite; then treat with lime for precipitating metals	Plating, foundry

clarification, there is a definite difference in the process.

Clarifiers and thickeners (see also **Sedimentation** for equipment diagrams) are commonly used for primary treatment in waste treatment plants. They permit easy removal of 65% of the suspended solids (on the average) and 35–40% of the BOD_5 (5-day biochemical oxygen demand). These removals are not higher because since chemical coagulation typically is not used in primary treatment plants, the finely divided colloidal solids are not removed. Depending on the waste charac-teristics, greater removal could be experienced if chemicals were added to coagulate these solids.

Design rates for primary clarifiers on the average range between 500 and 1,400 gal/day per square foot of clarification area.

Thickening units in waste treatment plants **are** normally sized for average flows on the basis of overflow rate per unit area and detention time. Operating units of this type range from 5 ft to as large as 500 ft diam with flow rates typically ranging from a few gallons to hundreds of thousands of gallons per minute. Detention times vary from a few minutes to several days, depending on the application.

Clarifiers are utilized in biological treatment to concentrate activated sludge or trickling filter humus. Thickeners are used in virtually all filtration processes to concentrate the suspended solids to a filterable consistency prior to vacuum filtration.

Settling Tanks: According to the Ten State Standards (defined below), surface settling rates for primary tanks not followed by secondary treatment should not exceed 600 gal/(day)(ft^2) or plants having a designed flow of 1 million gal/day or less. The relationship between BOD removal and settling rate is shown in the Ten State Standards for Domestic Waste.

For intermediate settling tanks, the Ten State Standards recommend a settling rate not exceeding 1,000 gal/(day)(ft^2). These standards recommend that surface separation rates for final settling tanks not exceed 800 gal/(day)(ft^2), based upon significant flow periods.

Although these rates are applicable to sanitary sewage, they do not necessarily apply to industrial wastes because of the wide variation and settling characteristics of colloidal and suspended solids. For some wastes, a grit separator is needed for the removal of large or dense solids, but coagulation followed by dissolved-air flotation or sedimentation may be required instead of plain sedimentation.

Clarification of Water and Waste Water. These operations remove suspended solids and color. Particles that are retained on a glass filter mat or a 0.45-μm millipore filter are defined as suspended solids. Any material passing these filters is considered soluble.

The easiest method of clarification is to allow water to stand for a period of time or to reduce its velocity so that the readily settleable suspended solids are separated by gravity. The suspended solids remaining may eventually settle out but sometimes are so fine that they stay in suspension for an excessive length of time. Therefore, some technique other than simple gravity settling is necessary to remove these slowly settling or non-settling solids. The variations of suspended-solids removal with overflow rate is illustrated by Fig. W-1.

Flocculation can effectively agglomerate very fine particles and thus increase their settling rate for gravity removal. Flocculation is the process of bringing together fine particles so that they agglomerate. If flocculation alone is effective in

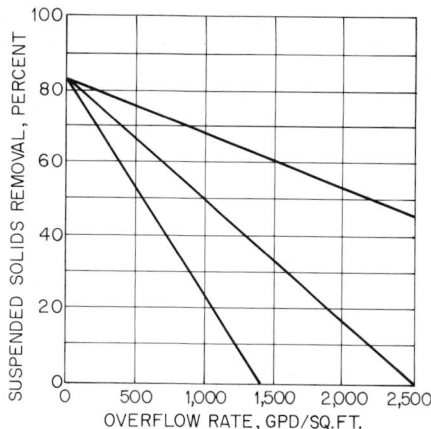

Fig. W-1. Variations of suspended-solids removal with overflow rate, based upon numerous sets of experimental data. The empirical relationship of one set of data is given by the upper line; another set by the lower line. The average is given by the middle line.

promoting agglomeration or self-coagulation, no chemical conditioning is required.

In most cases, the fine suspended solids are stable and will not agglomerate without chemical treatment to destabilize them so that they will stick together when flocculated. Coagulation is the method by which conditioning chemicals are rapidly dispersed in a liquid so that the characteristics of suspended particles are altered and agglomeration can occur. The coagulant is the conditioning chemical which changes the characteristics of the particles from being stable, or non-agglomerating, to unstable, or agglomerating.

Chemical Coagulants

Alum: The coagulant most often used for coagulation is alum, $Al_2(SO_4)_3 \cdot 18H_2O$. When alum is added to a suspension, the aluminum ion reacts with the alkalinity in the water to form polymeric aluminum hydrolysis species, more commonly called *alum floc*. For water treatment with low suspended solids (less than 50 mg/l) an alum dose of 10–50 mg/l is commonly used. In order to increase the settling rate of the slow-settling alum floc, a polymer or a weighting agent is added to aid in the clarification process. For clarification of waste water and phosphorus removal, alum doses of 50–300 mg/l have been used. Again, a polymer or a weighting agent is very effective in increasing the settling rate of the slow-settling alum floc. In general, for clarification using alum as the coagulant an overflow rate of 0.5–1 gal/(min)(ft^2) is used.

Iron: Iron salts, e.g., ferric sulfate, ferrous sulfate, and ferric chloride, have been used effectively as coagulants in both water and waste-water clarification. The general reactions and conditioning phenomena with iron salts are similar to those with aluminum salts. Iron salts have a wider pH range of application than alum. Normally, alum is most effective in a pH range of 5.5–7, ferric iron is effective at a pH of 5–9, and ferrous iron above 9. The dosage of iron salts for water and waste clarification and the overflow rates normally used are approximately the same as for alum.

Lime: Although lime is not considered an effective coagulant, it is used in water and waste-water treatment for clarification. Lime, $Ca(OH)_2$, does not produce a floc like that with aluminum and iron salts. It is added to water to react with bicarbonate alkalinity and phosphorus compounds and to adjust pH to cause precipitation of calcium carbonate and/or calcium hydroxyl apatite and magnesium hydroxide. Lime is used extensively to soften hard waters. Most hard water contains some magnesium. If the pH of the water is raised to 10.5 or above, flocculant magnesium hydroxide is formed. This assists in the clarification process. If the pH is not raised to 10.5, a coagulant such as alum or iron salt will normally be required in order to assure clarification.

In the treatment of a waste stream for clarification and phosphorus removal, a lime dosage of 100–500 mg/l is normally used. Only a small amount of softening is achieved by treating a domestic waste with lime. It has been observed that all the calcium added by the lime addition is removed, along with a small portion of calcium or magnesium from the waste stream. Normally, only about 10–20% softening is achieved even at dosages well over the stoichiometric amount required for theoretical softening of a waste stream, the reason being the interference of organics and phosphorus in domestic waste with the growth of $CaCO_3$ precipitates.

Polyelectrolytes: Polyelectrolytes are natural or synthetic long-chained (sometimes branched) organic macromolecules with a multiplicity of ionizable functional groups. When they are placed in water, the type of charge and degree of ionization of these functional groups determine the charge character of the polyelectrolyte. Cationic polyelectrolytes have an excess of positive sites (attracting anions) over negative sites (attracting cations) or only positive sites. Anionic polyelectrolytes have an excess of negative sites over positive sites or only negative sites. Nonionic polyelectrolytes have an equal distribution of positive and negative sites or no ionized sites.

The charge character and other important physicochemical properties result in polyelectrolytes being readily absorbed on the surface of suspended solids. Many types of polyelectrolytes have been successfully used as coagulants, but the interaction of polyelectrolytes and suspended solids which achieves coagulation is such that normally there is a relatively narrow dosage of polyelectrolytes which will effectively coagulate a given type and concentration of particles.

On the other hand, for aluminum and iron salts, there is a minimum dose at which coagulation is achieved but above which coagulation effectiveness is not lost.

Because of their specificity and nature of interaction with suspended solids, polyelectrolytes have not been widely used as coagulants. Depending on the type and concentration, say 20–200 mg/l of suspended solids, cationic polyelectrolyte doses of 5–40 mg/l might be required to achieve coagulation. There is no practical history of nonionic and anionic polyelectrolytes being used as coagulants. Polyelectrolytes should not be recommended as coagulants unless adequate jar tests are conducted to substantiate their effectiveness.

Polyelectrolytes are extensively used as flocculation aids, promoting growth and strength of floc particles. Experience has shown that nonionic and anionic polyelectrolytes are effective flocculation aids for coagulated particles. Dosages of 0.25–5.0 mg/l have been used to aid flocculation of coagulated suspended solids in concentrations of 20–200 mg/l.

Activated Silica Sols: Activated silica sols have been used both as coagulants and flocculation aids, chiefly the latter. The nature of their interaction with suspended solids is somewhat analogous to that of polyelectrolytes. Two features, however, differ: activated silica sol particles lack the long flexible chains of polyelectrolytes and therefore are considerably denser. Thus, activated silica is often referred to as a weighting agent which promotes faster settling of flocculant materials. It has been used as a nucleus for particle growth when clarifying waters with low suspended solids by alum or iron salt coagulation, flocculation, and gravity sedimentation. Dosages of activated silica sol for this application are about 20–60% of the alum dose used for coagulation. The activated silica sol is generally mixed in the water prior to alum addition, but when clarifying high-solids water, the activated silica is more effective when mixed into the water after the coagulant has been added. The optimum dose of activated silica is 0.5–8.0 mg/l.

Activated silica sols have also been used with and without alum to achieve clarification in lime water-softening plants. Approximately 0.5–5.0 mg/l may be required.

Coagulation reactions:

$$Al_2(SO_4)_3 + 3Ca(HCO_3)_2 \rightarrow$$
$$2Al(OH)_3 + 3CaSO_4 + 6CO_2$$
$$Al_2(SO_4)_3 + 3Na_2CO_3 + 3H_2O \rightarrow$$
$$2Al(OH)_3 + 3Na_2SO_4 + 3CO_2$$
$$Al_2(SO_4)_3 + 6NaOH \rightarrow 2Al(OH)_3 + 3Na_2SO_4$$
$$Al_2(SO_4)_3 \cdot (NH_4)_2SO_4 + 3Ca(HCO_3)_2 \rightarrow$$
$$2Al(OH)_3 + (NH_4)_2SO_4 + 3CaSO_4 + 6CO_2$$
$$Al_2(SO_4)_3 \cdot K_2SO_4 + 3Ca(HCO_3)_2 \rightarrow$$
$$2Al(OH)_3 + K_2SO_4 + 3CaSO_4 + 6CO_2$$
$$Na_2Al_2O_4 + Ca(HCO_3)_2 + 2H_2O \rightarrow$$
$$2Al(OH)_3 + CaCO_3 + Na_2CO_3$$
$$FeSO_4 + Ca(OH)_2 \rightarrow Fe(OH)_2 + CaSO_4$$
$$4Fe(OH)_2 + O_2 + 2H_2O \rightarrow 4Fe(OH)_3$$
$$Fe_2(SO_4)_3 + 3Ca(HCO_3)_2 \rightarrow$$
$$2Fe(OH)_3 + 3CaSO_4 + 6CO_2$$

Clarification Equipment. Using coagulation and flocculation equipment often significantly reduces the costs of installed equipment and operation. The type and size of mixing and feeding equipment has a significant effect on the results obtainable with coagulants. The simplest system consists of a tank equipped with a mixer. Such a setup provides the necessary contact between the coagulating chemical and the waste stream with sufficient detention time for the reaction. Sedimentation then occurs in a clarifier or thickener as previously described.

Recirculation of settled solids is beneficial in using chemicals to achieve optimum suspended solids removal with minimum chemical dosage. In some of the solids-contact units, a large portion of the precipitated solids are circulated by a low-shear high-pumping-rate turbine. A much higher concentration of solids is maintained in the reaction zone than would otherwise result in a single-pass reactor. Often, lesser amounts of chemicals are needed at this greater solids concentration to effect the desired removal rates. In addition, with the solids-contact phenomenon, the flocculi grow in size from the increased number of collisions with the finer particles so that the faster-settling solids result in a clearer overflow. The cone-and-impeller design of the solids-contact units is con-

tingent upon the particular waste being handled. See **Sedimentation** for a typical design. This same type of unit is used when precipitating and/or neutralizing chemicals are required. Units of this type are commonly sized at upflow rates of 0.5–2.0 gal/min per square foot of overflow area. The majority of these types of units allow the clarified overflow to pass through in $\frac{1}{2}$–4 hr although the solids can remain in the circulation circuit for as long as 2 weeks before being withdrawn. If not designed properly, inadequate or excessive mixing will result in ineffective treatment. The point of chemical addition is also important, as the degree of mixing prior to clarification determines in great part the completeness of the coagulation and flocculation.

The Ten States Standards* recommends peripheral speeds of 0.5–2.0 ft/s for flocculation used for water treatment and 1.5 for flocculation of sanitary sewage. It recommends that a flocculation basin be located as close as possible to a sedimentation basin and that the velocity of flocculated water be in the range of 0.5–1.5 ft/s to avoid deflocculation caused by turbulence in the pipelines.

The Ten States Standards has fairly general specifications for sedimentation basins. The velocity through settling basins should not exceed 0.5 ft/min; and the rate of flow over an outlet weir should not exceed 20,000 gal/day per foot of weir length.

The standards for solids contact units call for:
Not less than 30 min flocculation

*The Ten States Standards, is a shortened title of reports made by a General Committee with memberships from 10 states, namely Illinois, Indiana, Iowa, Michigan, Minnesota, Missouri, New York, Ohio, Pennsylvania, and Wisconsin, the exact title being "A Report of the Committee of the Great Lakes–Upper Mississippi River Board of State Sanitary Engineers on Policies for the Review and Approval of Plans and Specifications for Public Water Supplies." A similar report is issued for sewage treatment systems.

The standards are dynamic, i.e., frequently revised. As with any general standards, specific climatic conditions must be considered in the application of general rules; e.g., a trickling filter will not operate as well in extremely cold climates as in moderate climates.

Many of the other states have not set up such detailed rules and regulations but tend to adopt the Ten States Standards with slight modifications based upon their own requirements and climate. The Ten States Standards is viewed as the most comprehensive of all current approaches and is generally considered as the overall guidelines for waste treatment systems.

Not less than 2 hr retention for clarifiers or 1 hr for softeners

Weir loadings not exceeding:
20 gal/min (28,800 gal/day) per foot of weir length for softeners
10 gal/min (14,400 gal/day) per foot of weir length for clarifiers

Mechanical flocculators can be classified as rotary or reciprocating. There are reel-type units, a variety of paddle designs, and impellers or turbines. Reciprocating units include walking-beam and reciprocating agitators with sized blades or rotating paddles.

The reciprocating or vertical-shaft units avoid the need for underwater bearings, stuffing boxes, and dry wells. Solids-contact units frequently use variable-speed turbine agitators. These units have single mixing compartments, but short circuiting is minimized in some of the designs by recirculation in the reaction compartment.

Dissolved-Air Flotation Units. Flotation is most effective when the solids to be removed have a specific gravity close to that of water. If suspended solids have a specific gravity close to, but not more than, 1.0, they can be separated more readily by flotation than by sedimentation because of adherence of air bubbles. Most wastes which contain oils or greases are effectively clarified by flotation because of the specific gravity of the oils, regardless of whether they are separated as an oil phase or after being sorbed by a coagulant.

Flotation is effective when microscopic air bubbles released from solution adhere to suspended matter, oil droplets, or flocced particles, causing them to rise to the surface for separation as a float. Flotation is particularly effective for the removal of materials which, because they have a specific gravity close to 1.0, will not separate by gravity flotation or gravity sedimentation except at extremely slow rates. For many wastes, there will be a five- to sixfold reduction in area if dissolved air flotation is used instead of gravity separation. The savings in space and installed cost may more than offset the operating cost of pressurization. Flotation applications can generally be determined in the field by a simple sedimentation test. If a sample of raw waste contains solids that tend to float or remain in suspension in a graduated cylinder, or if after the addition of chemicals the resultant floc does not settle at a measurable rate, there is good possibility that such wastes can be handled most effectively by flotation. This can be confirmed by a bench-scale flotation test.

Total Flow Pressurization: This is the simplest method and is suitable and practical for flows up to about 5,000 gal/min. The entire flow must be passed through the pressurizing system and raised to a back pressure of 30–65 psig. A typical flow sheet depicting this arrangement is shown in Fig. W-2. Total flow pressurization is normally considered for applications on flows containing a large amount of unemulsified fats, grease, and oils.

Partial Flow Pressurization: This method is used where the flows are large and variable. The amount taken for pressurization by this method ranges between 25 and 50% of the total flow going to the flotation unit. It has the additional advantage that the nonpressurized flow can then reach the flotation unit by gravity. In this system there is always a pressurized stream added to the nonpressurized stream, ensuring flotation regardless of

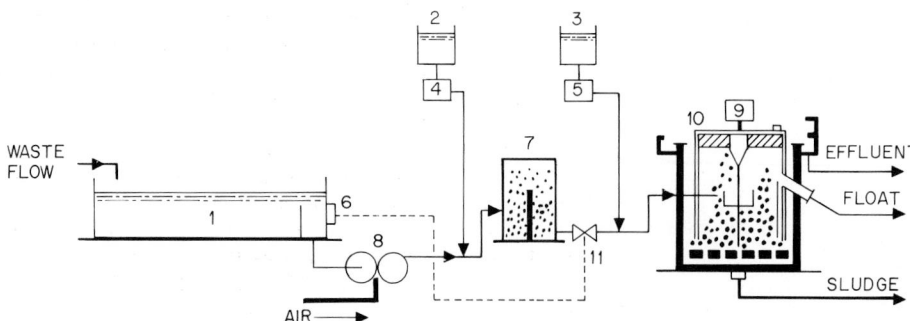

Fig. W-2. Dissolved-air flotation process. (1) Waste storage, (2), (3) chemical supply tanks, (4), (5) metering pumps, (6) liquid-level controller, (7) retention tank, (8) two-stage pressurizing pump, (9) drive unit, (10) float scrapers, (11) back-pressure valve.

the flow rate. This is a simple way of designing a flotation system to accommodate the variable forward flow.

Recycle Flow Pressurization: This method, also called pressurized recycle, is applicable where the partial flow pressurization is effective. It is based on recycling and pressurizing a portion of the effluent from the flotation unit. This arrangement is especially applicable where the waste contains large or unscreened particles or masses that would interfere with the proper operation of the pump or back-pressure regulator. However, this method has the limitation of adding to the total flow, thus requiring a larger flotation unit.

Sizing the Flotator Mechanism: One of the principal design criteria in the application of flotation is the surface overflow rate, just as it is an important consideration for the design of sedimentation equipment. The buoyant action of flotation causes the vertical separation to occur much faster than the same particles would settle under the influence of gravity alone. The overflow rate is based on the gross area inside the flotation compartment and is normally 1.5–3.0 gal/(min)(ft²) for most applications. These rates can be compared to sedimentation-type units, in which the surface overflow rate is ordinarily 0.3–1.0 gal/(min)(ft²).

Experience has shown that flotation will generally be effective if 3% by volume of a gas is applied, based on raw-waste flow. This is particularly true of oily wastes.

If the solids loading is high, or if flotation will be used for sludge thickening, the amount of gas should be calculated from the gas-solids (G/S) ratio. This has generally been expressed as the A/S ratio, which is defined as pounds of air released per pound of suspended solids in the influent waste. However, since other gases can be used for flotation, the ratio is more accurately expressed as the G/S ratio.

For clarification applications, a G/S ratio of 0.04 is frequently used, but the nature of the solids in the feed will determine the optimum ratio, which should be checked out in laboratory tests. This same relationship applies to the thickening of sludges, a 0.01–0.02 G/S ratio generally being used for activated-sludge thickening. (A typical circular flotation mechanism is shown under **Sedimentation.**)

Biological Treatment. Biological treatment of industrial wastes or sewage can be carried out on trickling filters or by an activated-sludge process. In either case, microorganisms are caused to grow under favorable conditions of oxygenation and nutrients (if required) with a balance maintained between the ingestion and digestion of the food on which the microorganisms live.

The food is the dissolved or suspended organic matter in the waste being treated. The food value, or strength, of the waste is normally expressed in terms of the 5-day biochemical oxygen demand (BOD_5). In the treatment processes, the microorganisms absorb this material and secrete enzymes to digest and utilize the absorbed material, thereby purifying the wastes.

The microorganisms consist of bacteria, protozoa, and other microscopic life which occurs in sewage, river mud, and soil. Given the proper amount of food, free oxygen, and a favorable environment, the organisms will grow and reproduce, forming the slime on trickling filters or the flocculant growths referred to as activated sludge.

The rate of microorganism growth is related to the balance between the food substrate and the microorganism population. A well-fed population will grow and reproduce rapidly; a starved population will not reproduce but will utilize its food to produce the energy required to stay alive. The waste products of the energy-producing function of bacteria (respiration) are carbon dioxide and water. A heavy loaded process will convert a major part of the food into new cells which will result in the maximum production of excess sludge to be wasted. In a sparsely loaded or starved process, there is very little excess growth.

Nutrient Requirements: Development of a healthy biological growth requires that the organisms be fed a balanced diet. Bacterial growth can be maintained with any waste containing BOD_5, providing certain essential elements are present. Some industrial wastes are deficient in nitrogen and phosphorus. Many high-carbohydrate wastes, e.g., from grain products, refining and paper-mill processes, and oil-refinery wastes, require the addition of these elements to maintain efficient treatment. Meat- or poultry-packing wastes, on the other hand, contain a high concentration of protein material and nitrogen in excess of basic requirements.

Nutrient requirements are often expressed in terms of the BOD-nitrogen-phosphorus ratio. Wastes having a ratio of 100:5:1 will normally ensure adequate nutrition.

Trickling Filtration: This is a process in which surfaces, e.g., rock, are coated with slime growth of bacteria and other microorganisms, which absorb and oxidize dissolved and organic matter in the same manner as activated-sludge organisms.

Crushed stone such as granite or limestone is usually used as the contact matter in the trickling filters, although plastic rings or corrugated sheets are frequently used.

The operation of a trickling filter can be summarized as follows:

1. An active biological growth forms on the stone or contact surface.

2. Dissolved and colloidal organic matter is absorbed by this growth.

3. The absorbed substances are digested by the microorganisms and oxidized or assimilated to promote further growth.

4. As the biological growth accumulates on the contact surface, it gets too heavy to adhere to the surface, sloughs off, goes out in the flow, and settles in the final clarifier.

Oxygen is supplied to the trickling filter by spraying the waste above the surface. In addition, the waste settling in the filter sets up a partial vacuum and draws air into the filter. The larger the surfaces, the greater the number of organisms that come into contact with the wastes to be purified and the higher the degree of purification.

Contact time in a trickling filter is relatively short compared with an activated-sludge process, and activated sludge provides an appreciably higher percentage of BOD removal. Trickling-filter effluent or final-clarifier effluent are frequently recirculated to increase the percentage of BOD removal.

Performance depends primarily on the ratio or recirculation to raw-waste flow. Where strong wastes are being treated, a two-stage filter plant with recirculation may be necessary to obtain the desired degree of purification.

Standard-rate filters, without recirculation, are designed to handle hydraulic loadings of 2–3 million gal/(acre)(day) with organic loadings of from 300–500 lb BOD_5/(acre-ft)(day) or 0.13–0.18 lb of BOD_5/(yd^3)(day). Reduction obtained through the complete plant, including the primary and final clarifiers, amounts to about 85% when treating normal domestic sewage. The BOD_5 removal percentage for standard-rate filters is shown in Fig. W-3.

With recirculation systems, surface loading of about 10 million gal/day is used with BOD loadings of 24,000 lb/acre-ft or 1.5 lb/(yd^3)(day) in the north; or 3,000 lb/(acre-ft)(day) [1.85 lb/(yd^3)(day)] in the south (United States).

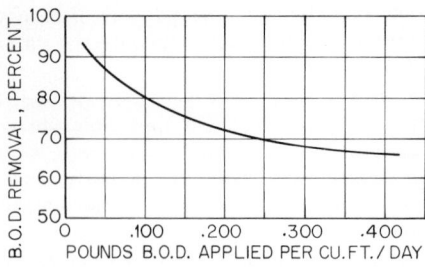

Fig. W-3. BOD_5 removal percentage for standard-rate filters.

BOD_5 reduction in high-rate filter plants is about 80–85% when treating normal domestic sewage.

The extent of purification without a high rate of recirculation is usually inadequate when treating strong wastes or wastes which are not readily purified. For this reason, an activated-sludge process is more widely used for industrial-waste treatment. In some cases, a trickling filter is used ahead of an activated-sludge plant to minimize shock loads, thereby providing more uniform and more complete treatment.

Activated Sludge Processes: Many conventional activated-sludge plants of the type depicted in Fig. W-4 have been designed on an empirical basis which provided 4–8 hr of aeration following primary treatment with sludge return from the final clarifier at a rate equal to about 25% of the raw-waste flow. Air supplied to the process varied from 0.5 to 2 ft^3 of air per gallon of sewage treated. Most of these plants operated successfully in a loading range of 20–40 lb of BOD_5 applied per day per 100 lb of mixed liquor-suspended solids (MLSS), with approximately 90% removal of BOD applied to the aeration unit. With primary treatment, the overall plant removal frequency is 95%.

BOD_5 loadings as high as 50 lb/day per 100 lb of MLSS have been handled without a decrease in efficiency in large well-operated plants. In smaller plants for part-time use or with unskilled operation, loadings of 20 lb of BOD_5 per day per 100 lb of MLSS are appropriate.

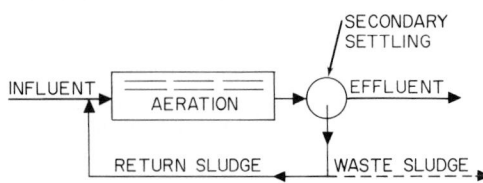

FIG. W-4. Conventional activated-sludge process.

With loadings of less than 10 lb of BOD₅ per day per 100 lb of MLSS the process is aerobic digestion, which has been used most successfully in the very smallest plants. At these loadings, BOD₅ removals of 85–90% are possible. If the excess sludge is wasted with the effluent, the lower range of BOD removals is obtained. For higher removals, separate periodic wasting of excess sludge is required.

Mixing in the aeration tank must be sufficient to cause vigorous circulation of the tank contents and prevent deposition at any point in the tank. Transverse velocities across the bottom of an aeration basin should exceed 0.5 ft/s to prevent deposition of solids.

Activated-Sludge-Process Modifications: Purification of sewage or industrial waste by activated sludge may be divided into two phases: (1) adsorption and absorption of the polluting material and (2) oxidation and digestion of the absorbed substances. It is now well known that the actual removal of pollutants from the liquid and their absorption by the activated sludge occurs within a very few minutes of contact between sewage and healthy activated sludge. The oxidation of the absorbed material takes much longer, a matter of hours. In the conventional process the two operations occur in the same aeration tank, absorption at the head end of the tank and oxidation at the downstream end. In this type of operation the oxygen demand of the mixed liquor is much greater at the head end of the aeration tank than at the effluent end. To meet the varying oxygen demand along the aeration tank air-diffusion equipment has sometimes been arranged to provide more air at the inlet end than at the effluent end and to gradually reduce the amount of air supplied along the length of the tank. The modification has been termed *tapered aeration.*

Another modification, introduced to distribute the oxygen demand uniformly along an aeration tank and eliminate the need for tapered aeration, is *step aeration.* In this process the sewage feed is introduced at several points along the aeration tank rather than at a single point at the head end. This modification serves to level the oxygen demand throughout the entire tank. The flow sheet for this modification is illustrated in Fig. W-5.

The present trend in aeration basin design is toward complete mixing. When mechanical aerators are used, the values of 0.75 HP to 1.0 HP/1,000 ft³ of aeration basin volume will meet a specification for ±10% suspended solids variation and no solids settling in the basin.

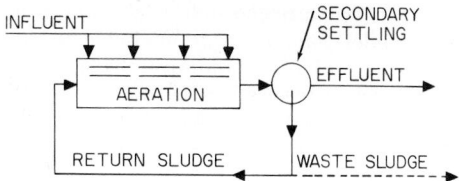

Fig. W-5. Step aeration process.

Complete mixing provides for rapid dilution of the incoming waste, thereby minimizing upsets due to shock loading.

Sludge Reaeration: Recognition of the two separate functions of an aeration tank has led to the development of several processes in which the functions of adsorption and oxidation are separated. The sewage is first contacted by activated sludge to remove the pollutants from the liquid; the sludge is then separated from the clarified liquid and reaerated to oxidize the absorbed material before returning it to contact fresh sewage again. The advantage of sludge reaeration has become apparent in many plants that have become overloaded as conventional activated-sludge plants but have been able to cope with increased loads by practicing sludge reaeration. Sludge reaeration permits more treatment in a smaller aeration tank by concentrating the required activated-sludge solids that are needed in the process into a smaller volume.

Sludge reaeration has been used for many years in various forms. The *contact-stabilization process,* illustrated in Fig. W-6, normally utilizes about a

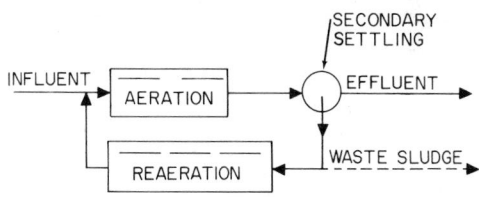

Fig. W-6. Contact-stabilization process.

30-min contact period followed by separation of the sludge and reaeration of the sludge for 4–8 hr. The brief contact period between activated sludge and waste provided by the contact-stabilization process, although giving a high degree of BOD₅ removal, does not always give as complete BOD₅ removal as might be obtained with longer contact periods.

The efficiency of removal by contact is related to some extent to the chemical nature of the waste material. Without considerable experience on the treatment of a particular industrial waste, bench-scale or pilot-plant tests are advisable to determine the treatability of the waste by these processes. Domestic sewage is in general amenable to treatment by sludge-reaeration processes, and 90% BOD_5 removal can normally be expected.

Extended aeration with 24 hr aeration for domestic sewage is usually not applicable in very cold climates, especially when a surface aerator is used. Also, aerobic digestion of waste sludge is not as effective in cold climates irrespective of the method of aeration. The activity of the organisms drops off considerably when liquid temperatures approach 40°F (4.5°C). Aerobic digesters can be heated, but this is not economical except perhaps at industrial installations where waste heat is available.

The recycle of activated sludge from the final clarifier to the aeration basin is usually expressed as a percentage of a sewage or waste flow. Generally, the practice is to use 25–50% of the average designed flow; however, if high MLSS are to be maintained, higher recycle ratios are required. Also, if the sludge has a high sludge volume (SVI), the recycle must be greater.

Aerated Lagoons: These stabilization ponds are widely used whenever space is available because of the lower first cost, lower operating cost, and simplicity of the biological treatment facilities.

If the suspended solids consist primarily of dispersed organisms and possibly some slow-settling "inert" material in the raw waste, the effluent may meet the requirements of the water-pollution authorities without final settling.

In other cases, a relatively large, unaerated pond can follow the aerated lagoon, where "polishing" of the effluent is accomplished. Due to the low oxygen demand, the normal surface absorption of oxygen is sufficient to maintain aerobic conditions. Any settleable solids will accumulate on the bottom. Anaerobic conditions may prevail on the bottom of such final basins, but the biological action is slow and confined to the bottom without causing odors or floating solids.

Where a higher-quality effluent with low suspended solids is required, aerated lagoons are followed by clarifiers with mechanical sludge-removal equipment. All settled solids may be wasted, or a large portion of them may be returned to the aerated lagoon, preferably under or near one or more of the surface aerators. The aerated lagoon then provides a higher degree of treatment because of the higher concentration of biological solids in suspension.

The BOD_5 reduction that can be accomplished with industrial wastes in an aerated lagoon can be quite variable and depends basically on the BOD_5 loading, the retention time, the degree of mixing, and the type of materials in the waste. Laboratory tests are required to determine the treatability of the waste and the BOD_5 removal obtainable in the absence of experience with that particular waste.

As a result of poor operating experience when sludge is allowed to remain too long in the final clarifier, efforts are made to remove sludge from the final clarifier as rapidly as possible. This has led to the use of bottom suction-removal mechanisms to remove sludge in a dilute state by suction through ports in rotating arms.

Particulate-Medium Filtration. Particulate-medium filters are used to remove oils and suspended solids in place of clarifiers if the concentrations are not high enough to cause frequent clogging of the filters. They are also used for polishing biological-treatment-plant or clarification-plant effluents.

The conventional sand filter, which has been used since about 1900, typically contains 24–30 in. of 0.50-mm siliceous sand supported on 14–18 in. of gravel above an underdrain system. These filters are capable of handling turbidities of 5–15 ppm without prohibitively short filter runs.

The dual-media filter subsequently was developed, which instead of having 24–30 in. of siliceous sand contains 12 in. of sand of 0.50-mm size, on top of which is placed 24 in. of approximately 1.0-mm anthracite coal. This provides a roughing filter on top of the sand filter and allows operation at rates up to about 5 gal/(min)(ft²) when handling turbidities in the range of 25 ppm.

The granular-media filter shown in Fig. W-7 is a gravity type which uses the hydraulic head of water stored in the filter to backwash it automatically. The filter is a vertical steel tank divided into three sections, backwash-storage space, filter-bed compartment, and filtered-water compartment under the underdrain.

The backwash-storage compartment holds the amount of water needed to properly backwash the filter bed. Because the backwash-storage space always fills to the same level, it gives a uniform backwash-water volume. The filter-bed compartment consists of a supporting plate on which are mounted distributor strainers which provide for uniform collection of filtered water and uniform distribution of backwash water. Because of the fine mesh screen in the strainers, no gravel bed is required.

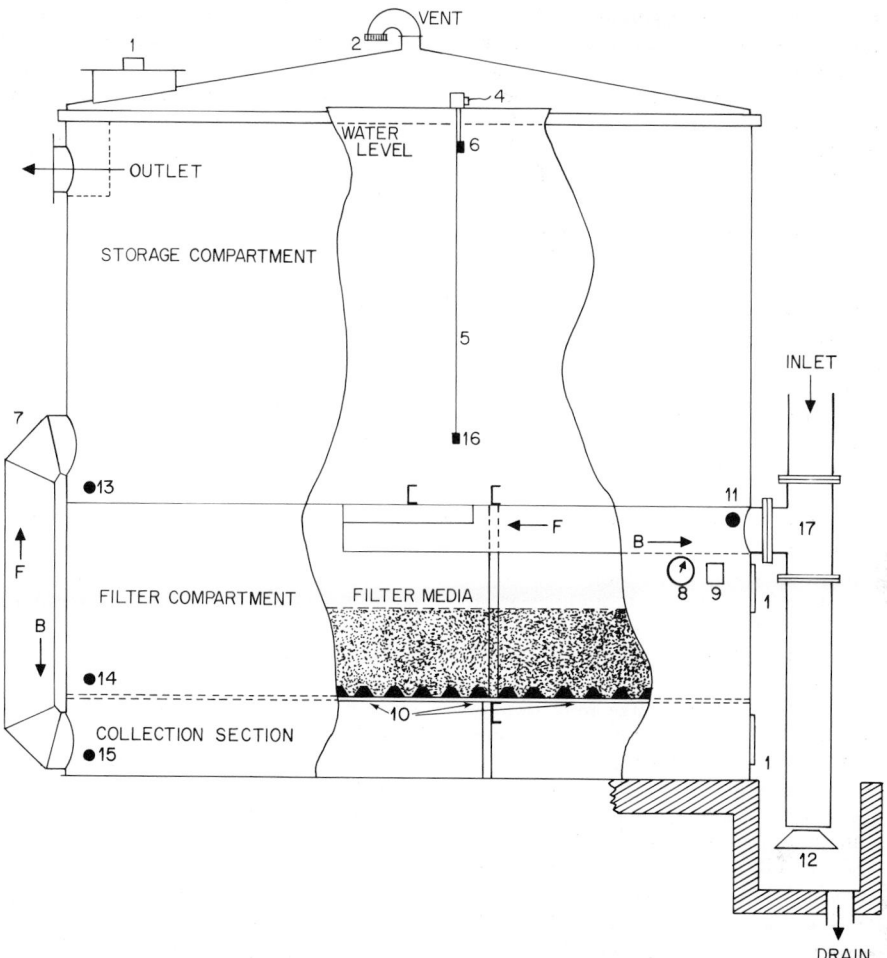

Fig. W-7. Granular-media filter with automatic controls and self-contained backwash system. (1) Manholes, (2) screen in vent line, (3) cover (optional), (4) electrode assembly, (5) probes, (6) reset, (7) transfer pipe(s), (8) head-loss gage, (9) switch assembly, (10) nozzles, (11) air-vent connection, (12) backwash throttle (adjustable), (13) drain (SVG) only, (14) sluicing connection, (15) drain, (16) backwash stop, (17) three-way valve. *F*, filter direction; *B*, backwash direction. (*Envirotech SVG filter.*)

Raw water enters the filter-bed section just above the sand bed and passes downward through the sand into the filtered-water compartment, whence it passes upward through an effluent pipe to the backwash section, which fills to the weir level in the outlet box.

When the head loss to the filter builds up to a predetermined maximum, an automatic backwash control repositions a three-way valve, cutting off the raw-water flow and opening the waste port so that the sand bed is washed with water from the storage compartment, flowing back through the collection compartment and up through the filter bed to waste.

High-Rate Deep-Bed Filtration: Deep-bed filters use relatively deep beds of medium-sized sand at filter rates of 8–10 gal/(min)(ft^2). Modifications of this design permit filtration rates of 15–18 gal/(min)(ft^2). These filters use a bed depth of up to 84 in., with either single or dual media, depending upon the application.

The filters must be capable of providing a filter

run of sufficient duration at the designed filtration rate without an excessive cost for waste water. If the filter medium is small, it may cause a short filter run and a high head loss. On the other hand, if the medium is too coarse, the filter will not produce a satisfactory effluent.

Backwashing: Washing a filter to remove the accumulated dirt is required with all media. With conventional sand filters, filter washing at 15–18 gal/(min)(ft^2) is the usual practice. Filter sweeps which apply a high-velocity water jet to the surface to avoid mud-ball formations are frequently used.

The use of an air-water backwash is becoming more general, the air providing a scouring action by the vigorous agitation in the bed, making it possible for the wash water to completely remove the dirt accumulated in the bed.

In deep-bed filters an air-water wash is mandatory because the removal of dirt from a 7-ft bed would require an excessive amount of wash water or would not keep the filter medium in a clean condition. Wash rates up to 40 gal/(min)(ft^2) and air applied at 4 ft^3/(min)(ft^2) are used.

Economic Considerations. The evaluation of deep-bed filters in place of clarifiers for any waste treatment application should take into account both first costs and operating costs. If filtration alone will provide a satisfactory effluent, the operating cost may favor the use of high-rate deep-bed filters. However, if the water contains colloidal solids or oil so that precoagulation is necessary, the filter installation must either be preceded by an efficient coagulation basin or the operating cost for chemicals to absorb oil and colloidal solids will be substantially higher. The disposal of the dirty wash water usually will require installation of a wash water clarifier for separating and thickening the solids removed in the filters.

Sludge Disposal. With exception of biological or chemical oxidation of organic matter, waste-disposal methods previously described serve only to concentrate the pollutants into sludge. Sludge disposal must be considered as part of any waste treatment application, and the following factors must be evaluated: (1) available space, (2) capital investment required, (3) operating costs, (4) atmospheric pollution, and (5) groundwater pollution.

If space is available, if the sludge is not putrescible, and if drainage will not contaminate underground water, sludge ponds are the simplest and least expensive means for sludge disposal. However, the evaluation should be made with long-range use in mind.

Alternative sludge-disposal methods include (1) thickening by sedimentation or flotation, (2) centrifugation, (3) vacuum filtration, (4) pipeline or tank-truck disposal to another location, and (5) incineration. The most satisfactory and economical method may involve a combination of the foregoing methods.

REFERENCES

See list under **Chemical engineering.**

—E. G. Kominek, *Eimco PMD Division, Envirotech Corporation, Salt Lake City, Utah*

Water   Water serves the chemically oriented industries in several major ways: (1) as a raw material for incorporation into final products without chemical change, (2) as a raw material for undergoing chemical change, (3) as a transport and conveyance medium with water acting as a solvent or carrier of solutions and suspensions in and out of reactions and physical-change operations, (4) as a heating and cooling medium over the wide temperature range from below normal freezing temperature (brine solutions) to those of superheated steam, (5) as an energy-storage medium, (6) as a gathering medium for waste products, (7) as a cleaning medium, (8) as a shield against heat and nuclear radiation (heavy water), (9) as a convenient standard in terms of temperature, density, viscosity, and other units important to the calibration of measuring and controlling equipment, and (10) with the exception of certain hazardous situations of which careful note must be taken, as a fire-fighting medium.

Conversely, water poses numerous problems for the process industries (1) by requiring removal when unwanted in raw materials and final products and from reactions when generated as an undesirable by-product, (2) by accelerating corrosion, (3) by interfering with electrical communications and equipment, (4) by freezing in pipelines and pneumatic systems, and (5) by offering no truly economic or convenient means for removing undesirable dissolved substances.

From the foregoing partial list it is apparent that chemical technology is concerned with practically all the properties of water, physical and chemical, and that these extend far beyond the capacity of one volume. Several water-oriented topics are covered in this book, however, including **Absorption; Chelating agents; Crystallization; Desalination; Detergents; Deuterium; Drying, gases; Drying, solids; Evaporation; Fluoridation; Freeze-drying; Ion-exchange resins; Mixing, fluids; Salt; Soaps; Spray drying; Surfactants;**

Tritium; Waste treatment; and **Water treatment.**
Tables W-2 to W-7 give selected key properties
of water.

*Special Behavior of Water:** The chemistry and
physics of water, of course, would require volumes
to report. The water molecule commonly is referred to as a polar molecule because the positive
and negative charges of the molecule are not arranged evenly around a center but distributed
asymmetrically to form positive and negative
poles. Greatly oversimplified, the water molecule
may appear as depicted schematically in Fig. W-8,
which shows the equilibrium position of the oxygen atom and the hydrogen atoms, i.e., the equilibrium position of the positive and negative
charges of the molecule. Consequently, the water

*So-called *polywater* is not described here because considerable further scientific investigation is required to
firmly establish or refute the existence of the claimed
new allotropic form of water ice (density, 2.3 g/cm³).

TABLE W-2. *Density of Pure Water at Atmospheric Pressure*

°C	g/ml	°C	g/ml	°C	g/ml
−13	0.997292	18	0.9986248	85	0.96865
−12	0.997292	19	0.9984346	90	0.96534
−11	0.997636	20	0.9982336	95	0.96192
−10	0.997935	21	0.9980221	100	0.95841
−9	0.008249	22	0.9978003	110	0.95099
−8	0.998501	23	0.9975684	120	0.94317
−7	0.998720	24	0.9973266	130	0.93494
−6	0.998950	25	0.9970751	140	0.92629
−5	0.999176	26	0.9968141	150	0.91721
−4	0.999380	27	0.9965437	160	0.90771
−3	0.999553	28	0.9962642	170	0.89776
−2	0.999673	29	0.9959757	180	0.88733
−1	0.999773	30	0.9956783	190	0.87639
0	0.9998676	31	0.9953722	200	0.86492
1	0.9999265	32	0.9950575	210	0.85290
2	0.9999678	33	0.9947344	220	0.84031
3	0.9999922	34	0.9954030	230	0.82712
4	1.0000	35	0.9950635	240	0.81330
5	0.9999919	36	0.9937159	250	0.79881
6	0.9999683	37	0.9933604	260	0.78368
7	0.9999297	38	0.9929970	270	0.76769
8	0.9998765	39	0.9926260	280	0.75063
9	0.9998092	40	0.9922473	290	0.73237
10	0.9997281	45	0.99024	300	0.71266
11	0.9996336	50	0.98807	310	0.69118
12	0.9995261	55	0.98573	320	0.66747
13	0.9994059	60	0.98324	330	0.64095
14	0.9992732	65	0.98059	340	0.61071
15	0.9991286	70	0.97781	350	0.57497
16	0.9989721	75	0.97489	360	0.52872
17	0.9988041	80	0.97183		

TABLE W-3. *Specific Volume of Water, cm³/g*

Pressure, kg/cm²	0°C	50°C	95°C
1	1.001	1.0121	1.0396
500	0.9772		
1,000	0.9568	0.9742	0.9985
1,500	0.9397	0.9583	0.9813
2,000	0.9249	0.9440	0.9662
3,000	0.8997	0.9202	0.9410
4,000	0.8796	0.8998	0.9195
5,000	0.8627	0.8825	0.9010
6,000	. . .	0.8669	0.8850
7,000	. . .	0.8531	0.8706
8,000	. . .	0.8408	0.8578
9,000	. . .	0.8297	0.8462
10,000	. . .	0.8193	0.8353
11,000	. . .	. . .	0.8527

molecule has a strong tendency to be oriented in
an electrical field. The dipole moment depends
upon the magnitude of the charge separation
within the molecule, and in the water molecule
this separation is large. Thus, water may be described as having an exceptionally large dipole
moment and consequently a large dielectric constant. On the basis of ascribing a dielectric constant of 1 for a vacuum, the dielectric constant
of water is 80; that is, in water two electrical
charges will attract or repel each other with only
one-eightieth as much strength as would be the
case in a vacuum. This accounts, at least in part,
for the remarkable ability of water to dissolve
substances, particularly materials whose molecules
are held together primarily by ionic bonding.

The bonding arrangements within the water
molecule also account for the exceptional cohesive

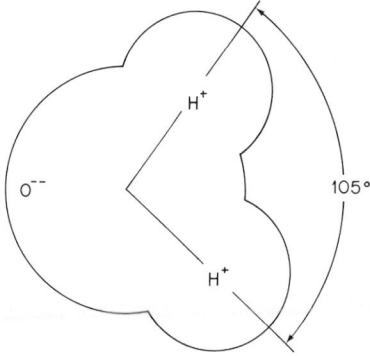

Fig. W-8. Arrangement of electrical charges in water molecule (schematic).

TABLE W-4. *Transition Parameters of Water and Ice* (*Fig. W-9*)

Pressure, kg/cm²	°C	$\triangle V$, cm³/g	Latent Heat, (kg)(cm)/g
I–liquid			
1	0	0.0900	3,410
1,130	−10.0	0.1122	2,900
2,115	−22.0	0.1352	2,390*
I–III			
2,115	−22.0	0.1818	220*
2,170	−34.7	0.1963	90*
Liquid–III			
2,115	−22.0	0.0466	−2,170*
3,530	−17.0	0.0241	−2,620*
I–II			
2,170	−34.7	0.2178	−430*
1,794	−75.0	0.2146	−380
III–II			
2,170	−34.7	0.0215	−520*
3,510	−24.3	0.0145	−720*
Liquid–V			
3,530	−17.0	0.0788	−2,660*
6,380	+.16	0.0527	−2,990*
III–V			
3,530	−17.0	0.0547	−40*
3,510	−24.3	0.0546	−40*
II–V			
3,510	−24.3	0.0401	680*
4,200	−34.0	0.0401	660
V–VI			
6,380	+0.16	0.0389	−10*
6,365	−20	0.0381	−10*
Liquid–V			
6,380	+0.16	0.0916	−3,000*
10,590	30.0	0.0663	−3,360
16,000	57.2	0.0478	−3,430
22,400	81.6	0.0330	−3.610*
VI–VII			
22,000	0.0	0.0567	−290
22,350	+40.0	0.0573	−60
22,400	81.6	0.0580	0*

*Triple point.

TABLE W-4. *Transition Parameters of Water and Ice* (*Continued*)

Pressure, kg/cm²	°C	$\triangle V$, cm³/g	Latent Heat, (kg)(cm)/g
Liquid–VII			
22,400	81.6	0.0910	−3,610*
28,000	124.1	0.0817	−4,840
34,000	161.1	0.0738	−5,650
40,000	192.3	0.0674	−6,550

*Triple point.

TABLE W-5. *Thermal Conductivity of Water*

Temp., °C	k, 10⁻⁵ W/(cm)(°C)	Temp., °C	k, 10⁻⁵ W/(cm)(°C)
0	554	160	682
10	576	170	680
20	598	180	676
30	615	190	672
40	630	200	666
50	643	210	659
60	654	220	652
70	665	230	644
80	671	240	635
90	676	250	624
100	680	260	614
110	684	270	602
120	686	280	590
130	687	290	576
140	686	300	564
150	685		

TABLE W-6. *Surface Tension of Water* (*against Air*)

Temp., °C	Surface Tension, dynes/cm	Temp., °C	Surface Tension, dynes/cm
−8	77.0	30	71.18
−5	76.4	40	69.56
0	75.6	50	67.91
5	74.9	60	66.18
10	74.22	70	64.4
15	73.49	80	62.6
20	72.75	100	58.9
25	71.97		

power exhibited in water's high surface tension and the outstanding ability of water to adhere strongly to a variety of materials (wetting). Bonding also accounts for the manner in which water crystals, e.g., snowflakes, are formed and for the maximum density of water (4°C), below which water assumes less dense forms, causing ice to

TABLE **W-7.** *Selected Properties of Pure Water*

Specific gravity at 100°C 0.958
Specific heat of liquid at 212°F,
  Btu/(lb)(°F) 1.005
Heat of vaporization, Btu/lb 970.2
Heat of fusion, Btu/lb 143.3
Cubical expansion coefficient 0.0024
Thermal conductivity liquid,
  (Btu)(ft)/(hr)(ft^2)(°F) 0.393
Diffusivity* (18°C), cm^2/s 0.0017

Viscosity		Dissociation Constant, $K \times 10^{14}$	
°C		°C	
0	1.7834	0	0.1139
10	1.3022	5	0.1846
20	1.0019	10	0.2920
30	0.7995	15	0.4505
40	0.6513	20	0.6809
50	0.5481	25	1.008
60	0.4687	30	1.469
80	0.3545	35	2.089
100	0.2813	40	2.919
		45	4.018
		50	5.474

Index of Refraction

Temp., °C	0.397 μm H	0.434 μm G	0.486 μm F	0.589 μm D	0.656 μm C
0	1.3444	1.3413	1.3380	1.3338	1.3319
20	1.3435	1.3404	1.3372	1.3330	1.3312
40	1.3411	1.3380	1.3349	1.3307	1.3290
80	1.332	1.3302	1.3270	1.3230	1.313

Velocity of sound in water, m/s:
  At 4°C 1,418.3
       25°C 1,493.2
Maximum permissible concentrations
of radioisotopes in water for continuous
exposure,† μCi/cm^3:
  β or γ emitter .10^{-7}
  α emitter . . . . .10^{-7}

*Diffusivity $= k/c\rho$, where $k =$ thermal conductivity, $c =$ specific heat, and $\rho =$ density.

†Values are reduced by a factor of 10 when applied to minors and to large populations. These are general values which are considered to be safe for limited exposure (for a few months) to any mixture of radioisotopes.

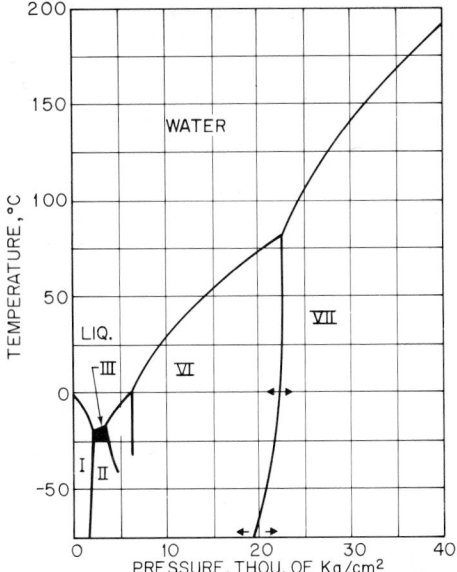

Fig. W-9. Phase diagram of water; see Table W-4.

float. Bonding is responsible for the exceptional heat capacity and exceptionally high latent heats of fusion and evaporation of water. See Fig. W-9.

Various observations over the years have led to continuing research into the chemical and physical behavior of water. Two unexpected observations are cited here briefly as examples.* It has been found that water sometimes will freeze in pipelines which carry natural gas at temperatures as high as 20°C (68°F), with resulting clogging by a slushy, snowlike material. Methane, the primary ingredient of natural gas, is very slightly soluble in water. It normally would be expected that methane, for example, would release much more energy in its dissolution in a substance such as hexane (in which it is readily soluble) than, say, in water because of its low solubility in water. Surprisingly, it was found that the dissolution of methane in water releases 10 times as much heat as that of methane in hexane. In another observation, it was noted that corn may show frost effects at temperatures of about 4°C (40°F), considerably above the normal freezing point. Investigation shows that water has a strong tendency to crystallize along the surfaces of protein molecules and, in special situations, water may expand abruptly to form ice and thus cause frost damage to living tissues.

Seawater: Three major purposes are served by seawater in terms of modern chemical and process technology: (1) as an advancing source of fresh water (see also **Desalination**); (2) as a heat-

*Arthur M. Buswell and Worth H. Rodebush, Water, *Sci. Am.*, April 1956.

exchange medium; and (3) as a source of materials, both inorganic and organic. Three-fourths of the surface of the earth is covered by seas and oceans, approximately 143 million mi^2. The volume of water in the oceans is estimated at over 323 million mi^3, but this quantity represents but $\frac{1}{4,500}$ the total volume of the earth. The average ocean depth is about 2.3 mi. The average height of the earth's land surface above sea level is only 0.5 mi. The greatest ocean depth is about 34,200 ft located in the Pacific Ocean just east of the Philippine Islands. Only about 4% of the ocean floor is below 18,000–30,000 ft.

A short ton (2,000 lb) of seawater contains approximately   55 lb NaCl,   2.54 lb Mg,   1.75 lb S, 0.8 lb Ca,   0.75 lb K,   0.125 lb Br,   0.025 lb Sr, 0.0009 lb I, 0.008 lb B, 0.0001 lb Pb, 0.00002 lb Cu, 0.00001 lb Zn, with lesser quantities of Ag, Au, and other elements. It is estimated that 1 mi^3 seawater contains approximately:

	Million tons
O	4,000
H	500
Cl	90
Na	50
Mg	6
S	4
Ca	2
K	2
Br	0.3
C	0.15

	Tons
Sr	40,000
B	22,000
Si	14,000
F	6,000
Ar	3,000
N	2,000
Li	900
Rb	550
P	320
I	230
In	90
Al, Fe, Mo, Zn (each)	50
Ba	30
As, Cu, Pb, Pr, Se, Sn (each)	15
Mn, V (each)	10
Ti	5
Ce, Co, Cs, Ni, Sb, Yt (each)	2
Ag, Bi, Kr, La, Ne, W, Xe (each)	1

	Pounds
Cd, Cr, Ge (each)	500
Ga, Hg, Sc (each)	250

Te	100
He, Nb (each)	50
Au	40

Water Problems: Major problems concerning water which require serious attention during the 1970s include (1) *pollution* (see **Waste treatment;** and **Water treatment**); (2) *heavy consumptive use,* which already is beginning to limit growth plans for cities and agriculture, particularly in the southwestern United States, (3) *evaporation losses* from reservoirs and storage ponds, particularly important in the arid Western and West Central sections of the United States, the control of which may be found to be more economic than developing new water sources; (4) *lowering water tables,* again of greatest concern to the Southwest and Pacific Coast, but also to some of the Central States near larger cities, and to some Middle Atlantic areas of the United States, a situation which is exerting considerable pressure now on the choice of irrigation areas; (5) *long-distance water-transmission systems,* which in the future may not be confined to the western United States; (6) *waste-water return to the oceans,* which can become the largest and least expensive potential secondary water source and a very attractive source for many industrial uses; (7) *salt-water encroachment,* which already is destroying some water sources and land, (8) *watershed trash vegetation,* the eradication of which can increase water yield, particularly in the Southwest, and (9) *storage of seasonal and flood flows,* which has been practiced for many years in most areas that are away from good lake or ground-water supplies but which will require extension as the water problem becomes more severe in less arid areas.

For the 1970s, the water-resources situation versus population-growth forecasts is summarized in the map of Fig. W-10.

The availability of water will continue to rank very high as a plant-location criterion for the large water users, including utility plants, chemical and petroleum processing plants, pulp and paper mills, and primary metal producers. Wherever water costs exceed approximately 20 cents per 1,000 gal, large-scale conservation programs become economically feasible. In some areas, this will cause large users to turn to low-grade sources for cooling and to reclaimed sewerage and other treated water for new process supply sources. For new plants, investment in water systems may be as much as 5% of total plant cost; for established plants as high as 10%.

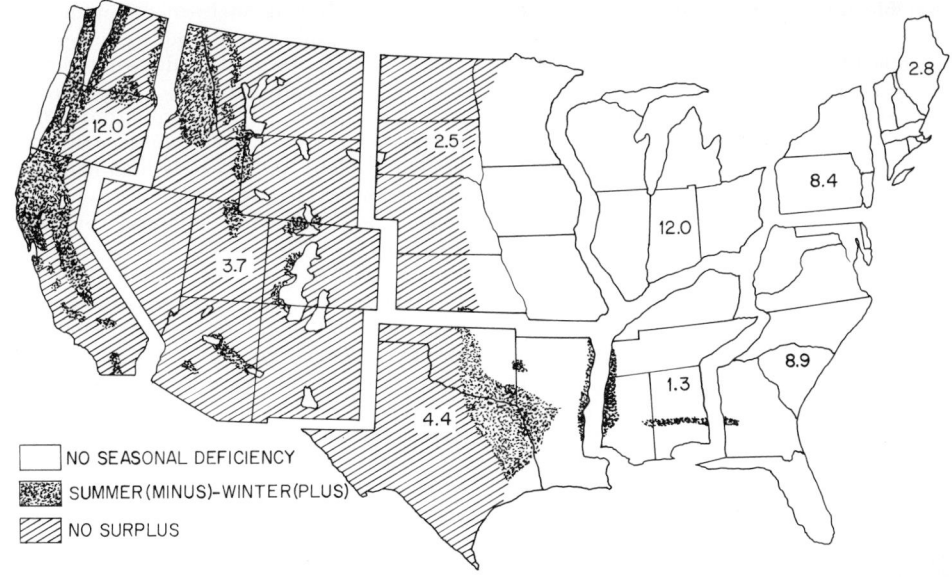

Fig. W-10. Water resources versus predicted population growth in the United States during the 1970s. Figures indicate population increase expected (millions of people).

REFERENCES

Allen, Leland C., and Peter A. Kollman: A Theory of Anomalous Water, *Science,* vol. 167, pp. 1443–1454, Mar. 13, 1970.

Babbitt, H. E., J. J. Doland, and J. L. Cleasby: "Water Supply Engineering," 6th ed., McGraw-Hill, New York, 1962.

Buswell, Arthur M., and Worth H. Rodebush: Water, *Sci. Am.,* April 1956.

Camp, Thomas R.: "Water and Its Impurities," Van Nostrand Reinhold, New York, 1963.

Castellion, G. A., D. G. Grabar, J. Hession, and H. Burkhard: Polywater: Methods for Identifying Poly-water Columns and Evidence for Ordered Growth, *Science,* vol. 167, pp. 865–867, Feb. 6, 1970.

Davis, S. Kenneth, and John Arthur Day: "Water the Mirror of Science," Anchor Books, Garden City, N.Y., 1961.

Delsemme, A. H., and A. Wenger: Superdense Water Ice, *Science,* vol. 167, pp. 44–45, Jan. 2, 1970.

Fair, Gordon Maskew, and John Charles Geyer: "Water Supply and Waste-Water Disposal," Wiley, New York, 1954.

Fair, Gordon Maskew, and John Charles Geyer: "Elements of Water Supply and Waste-Water Disposal," Wiley, New York, 1958.

Furnham, C. Fred (ed.): "Industrial Wastewater Control," Academic, New York, 1965.

Kamb, Barclay: Hydrogen-Bond Stereochemistry and "Anomalous Water": The "Polywater" Controversy Prompts a Close Look at Novel Kinds of H-Bonding, *Science,* vol. 172, pp. 231–242, Apr. 16, 1971.

Kurtin, S. L., et al.: Polywater: A Hydrosol?, *Science,* vol. 167, pp. 1720–1722, Mar. 27, 1970.

Linsley, R. K., and J. B. Franzini: "Water Resources Engineering," McGraw-Hill, New York, 1964.

Messmer, Richard P.: Polywater: Possibility of *p*-Electron Delocalization, *Science,* vol. 168, pp. 479–480, Apr. 24, 1970.

Rousseau, D. L., and S. P. S. Porto: Polywater: Polymer or Artifact?, *Science,* vol. 167, pp. 1715–1719, Mar. 27, 1970.

Steel, E. W.: "Water Supply and Sewerage," 4th ed., McGraw-Hill, New York, 1960.

Todd, David K.: "Ground Water Hydrology," Wiley, New York, 1959.

Wright, Forrest B.: "Rural Water Supply and Sanitation," 2d ed., Wiley, New York, 1956.

Water Gas   (See **Fuels;** and **Gasification.**)

Water Injection   (See **Air pollution.**)

Water Purification   (See **Ion-exchange resins.**)

Water Quality, Nuclear Reactor   (See **Nuclear power plants.**)

Water Reactors   (See **Nuclear power plants.**)

Water-soluble Paints   (See **Paint**.)

Water Treatment   Water is used by municipalities and industries in vast amounts. Water outranks all other raw materials. About 700 tons of water is used to make 1 ton of paper; 1,200 tons of water to make 1 ton of aluminum; and 250 tons of water for 1 ton of steel. The water, of course, is not consumed in these processes, but its physical form and its impurity level may be altered.

Water treatment chemistry and processes deal with the chemical or physical reactions of *small amounts of dissolved or suspended materials*. The unit of measurement commonly used is *parts per million* (ppm), which is equivalent to *milligrams per liter*. Thus, a surface water containing 200 ppm dissolved impurities is 99.98% water and 0.02% salts. It follows that water analyses and treatment processes are based upon specialized techniques designed for the determination and removal of trace quantities of materials.

Water Supplies and Their Impurities: Natural freshwater supplies are derived indirectly from the oceans. Clouds form by solar evaporation, and winds move the moisture overland to precipitate as rain, snow, or hail. The water flows over the surface or percolates into the ground. Excess water flows back to the oceans, completing the hydrological cycle.

Water supplies are classified as *surface supplies* and *underground waters*. Surface waters are rivers, lakes, creeks, ponds, and reservoirs. Wells and springs are the sources of underground water.

Water vapor in clouds is generally pure but gathers dust and gases. When it reaches the earth, suspended organic matter and soil turbidity are picked up. Minerals are leached from the soil and rocks, and organic matter is added from municipal and industrial wastes and decaying vegetation. Well waters usually are free of suspended materials and organic matter due to filtration through the earth. Surface waters are generally low in mineral content but relatively high in suspended and organic materials.

The major impurities in natural water supplies are listed in Table W-8. Most surface waters contain 50–250 ppm dissolved salts; well waters, 150–500 ppm. Water from the Great Lakes contains about 150 ppm total dissolved solids (TDS). Drinking Water Standards (1962) published by the American Public Health Association are given in Table W-9.

Water Analyses: The importance of an accurate and complete water analysis cannot be overemphasized. All water treatment processes are affected by variation in the dissolved and suspended impurities in the supply. Deep well waters generally have fairly constant impurity levels. Surface waters from rivers vary widely in mineral and turbidity levels.

Most water laboratories state the impurity levels as cations and anions in terms of calcium carbonate equivalents, or "as $CaCO_3$." Calcium carbonate is used as the common denominator because it has a molecular weight of 100, which facilitates calculations. In this ionic analysis method, total cations equal total anions, which simplifies prediction of the water analysis after various water treatment methods.

New analytical methods and instruments have made modern water treatment processes possible. Mineral content of waters is determined in many

TABLE W-8. *Major Impurities in Water*

Ionic and Dissolved		Nonionic and Undissolved	Gaseous
Cationic	Anionic		
Calcium	Bicarbonate ⎫	Turbidity, mud,	Carbon dioxide
Magnesium	Carbonate ⎬ Alkalinity	dirt, silt.	Nitrogen
Sodium	Hydroxyl ⎭	Organics*	Oxygen
Potassium	Sulfate	Color*	Hydrogen sulfide
Ammonium	Chloride	Microorganisms	Ammonia
Iron	Nitrate	Algae	Methane
Manganese	Phosphate	Bacteria	
	Silica	Oil	
	Fluoride	Metal oxides and	
	Organics*	hydroxides	
	Color*		

*Organic materials may be suspended or dissolved, ionic or nonionic.

TABLE W-9. *Standards for Drinking Water**

Constituent	Maximum of Concentration, ppm
Total solids	500
Chloride as Cl	250
Sulfate as SO$_4$	250
Nitrate as NO$_3$.	45.0
Calcium carbonate equivalents.	0.2

Alkalinity of Lime-Soda-treated Waters	Total Hardness + 35 ppm as CaCO$_3$

Fluoride at air temperature:	
From 50–53.7°F	0.9–1.7
To 79.3–90.5°F	0.6–0.8
Copper as Cu	1.0
Lead as Pb	0.05
Arsenic as As.	0.01
Selenium as Se.	0.01
Iron as Fe.	0.3
Silver as Ag	0.05
Manganese as Mn	0.01
Cadmium.	0.01
Zinc as Zn	5.0
Hexavalent chromium.	0.05
Barium	1.0
Phenolic compounds as phenol.	0.001
Cyanide (max. allowable 0.2)	0.01
Turbidity (silica scale)	10
Color (standard cobalt scale).	20
Alkyl benzene sulfonate (ABS).	0.5
^{226}Ra	3†
^{90}Sr	10 nCi/l†
Gross beta activity.	1,000 nCi/l†

*U.S. Public Health Service Standards (1962).

†Nanocuries per liter. A curie (Ci) is a unit of radio-activity defined as the quantity of any radioactive nuclide in which the number of disintegrations per second is 3.700×10^{10}. Measurements of the half-life of Ra indicate that the activity of 1 g of ^{226}Ra is slightly less than 1 Ci. A nanocurie is 10^{-9} Ci.

cases on a continuous basis by means of colorimetry, conductivity, and atomic adsorption. Organic matter is usually determined by oxidation methods and reported as chemical oxygen demand, oxygen consumed, or total organic carbon.

Water analyses are easily obtained in the United States. Important sources are the U.S. Geological Survey and the state geological departments or state water surveys. Water analysis methods are reviewed annually by *Analytical Chemistry,* published by the American Chemical Society.

Harmful Effects of Water Impurities: The first criterion of any water supply for human use is that it must be safe to drink. Fortunately, all harmful

bacteria are killed rapidly and inexpensively by means of chlorination. This method is universally used in the United States and most of the world. Chlorine gas is usually employed, but sodium hypochlorite is used occasionally for smaller installations. Most surface supplies must be chlorinated. Deep well waters are generally safe for drinking purposes, but most municipalities chlorinate these supplies to guard against surface contamination. See also **Chlorination;** and **Fluoridation.**

Harmful impurities in waters used in industries are classed as follows:

1. Scale deposits which form in boilers and other heat-exchange equipment. These deposits act as insulators and prevent efficient heat transfer, resulting in eventual tube and boiler failure.

2. Impurity carry-over. In steam, impurities carry over and foul turbines and processing equipment, causing failure of equipment or rapid loss of efficiency.

3. Corrosion of boilers, piping, and heaters by dissolved gases and acids in water supplies.

4. Spotting and staining of many industrial products, primarily in the metal finishing and plastics industries.

5. Contamination of electronic components, such as TV tubes, printed circuits, and transistors, by dissolved and suspended impurities resulting in rejects. The electronic industry is a large user of ultrapure water, requiring total impurity levels of less than 50 ppb.

6. Bad tastes and off colors in foods and beverages.

7. Waste of various chemicals such as soaps, detergents, and alkalis in laundering and dyeing operations.

Water Quality Specifications for Industrial Uses: Water quality requirements vary for different uses in industry over a wide range. For example, a plant may use seawater containing 35,000 ppm TDS for cooling purposes but require ultrapure water containing 0.05 ppm TDS (50 ppb) to manufacture its principal product. These limits of acceptable impurities have been determined for various industries by experience in product quality control and are, of course, variable to a degree. Table W-10 gives some water quality standards suggested for various industries by technical societies and committees and updated to accommodate new processes and products.

Table W-11 lists water treatment processes and the impurities they are designed to remove. Distillation is not covered, as this process is generally

TABLE W-10. *Suggested Water-Quality Tolerances for Various Industries*

Tolerances, ppm

Industry	Turbidity	Color	Taste and Odor	Hardness as CaCO₃	Alkalinity as CaCO₃	Iron	Manganese	Organics, as O₂ Consumed	Chlorides, as Cl	Total Solids	pH
Baking	10	10	Low	...	...	0.2	0.2	10	...	...	...
Carbonated beverages	0–2	10	None	250	50	0.2	0.2	10	...	850	7.0
Brewing	5–10	10	Low	100	75–100*	0.1	0.1	1	100	500–1,000	7.0
Distilleries, deproofing	1	1	None	0	1	0.02	0.1	1	1	2	...
Electroplating, final rinse, and bath makeup	1	5	Low	0	1	0.02	0.02	1	1	2	7.0
Food processing:†											
Canneries	5	5–10	None	25–50	25–50	0.1	0.1	...	...	850	7–8
Candy	5	5–10	None	0	...	0.2	0.2	...	...	100	7–8
Meat packing	5	5–10	None	0	...	0.2	0.2	...	...	500	7–8
Starch products	5	5–10	None	0	...	0.2	0.2	...	...	500	7–8
Ice manufacture	2	5	None	50	50	0.1	0.1	10	...	150	7–8
Laundering	1	5	Low	50	60	0.2	0.2	10	...	...	6.0–6.8
Mirror silvering	2	2	Low	0	...	0.1	0.1	1	1	2	6.8–7.2
Paper and pulp:						0.05	0.05				
Kraft	40	25	...	100	75	0.2	0.1	...	200	300	6.8–7.3
Sulfite	25	5	...	100	75	0.01	0.05	...	75	250	6.5–7.5
Fine	10	5	...	100	75	0.01	0.05	...	75	200	6.5–7.5
Pharmaceutical products‡	0	0	None	...	...	0	0	0.1	...	1	...
Textile:											
Cotton	1.5	5	...	10	75–100	0.05	0.05	...	...	200	6.8–7.2
Rayon	1.0	5	...	10	75	0.05	0.02	...	...	200	...
Wool	5.0	20	...	...	75	0.1	0.1	...	...	200	...

Quality Tolerances for Ultrapure Water

Industry	Turbidity	Color	Taste and Odor	Hardness as CaCO₃	Alkalinity as CaCO₃	Iron	Copper	Organics, as O₂ Consumed	Chlorides, as Cl	Resistivity, mΩ	pH
TV tubes	0	0	...	0	0	0.005	0.005	0.5	0	10	7.0
Semiconductor	0	0	...	0	0	0.005	0.005	0.1	0	18	7.0

*CaSO₄ desirable.

†Water must meet U.S. Public Health drinking water standards.

‡Must also meet U.S. Pharmacopeia standards; water for injection must be free of pyrogens.

TABLE W-11. *Methods of Removing Impurities from Water*

Process	Impurities Removed
Cold lime and hot lime-soda, softening-precipitation, settling and filtration	Calcium, magnesium, iron, manganese, turbidity, organic matter
Coagulation, using sludge contact clarifiers followed by filtration	Turbidity and suspended matter, oil, color, colloidal silica
Filtration: sand, multimedia, diatomaceous earth, and carbon filters	Turbidity and suspended matter, organic matter, iron, manganese
Breakpoint or superchlorination.	Organic matter, bacteria, tastes and odors, hydrogen sulfide
Oxidation (aeration) and precipitation; Cl_2 and alkali may be added.	Iron, manganese, and hydrogen sulfide
Degasification	Carbon dioxide, methane
Vacuum degasifier	Oxygen, nitrogen, carbon dioxide
Sodium-cycle cation exchange	Calcium and magnesium
Dealkalization by ion exchange, followed by aeration and pH adjustment	Bicarbonate, carbonate, and hydroxyl alkalinity
Deionization via ion exchange resins	All ionized salts, acids, or bases
Reverse osmosis	75–95% ionized salts from higher solids waters
Electrodialysis	Ionized salts from brackish waters down to approximately 300 ppm

too expensive to operate for freshwater treatment. Many of the water treatment methods are used in series operation to accomplish the desired degree of purification. For example, in the treatment of river waters, coagulation and filtration equipment is used prior to deionization. Removal of organic matter is done in the coagulation equipment by superchlorination.

Industrial Water Treatment Processes: One of the oldest methods of removing dissolved salts from water is the lime and lime-soda softening process. The hardness in water is due to the presence of calcium and magnesium ions. Most freshwater supplies contain about twice as much calcium as magnesium.

Lime and lime-soda softening is accomplished by adding a lime slurry and a solution of soda ash to the raw water in a solids-contact-type water treatment vessel. Hard waters contain *temporary* and *permanent* hardness. Temporary hardness refers to the calcium associated with the bicarbonate alkalinity. Upon heating, the temporary hardness precipitates as calcium carbonate. Calcium and magnesium chlorides and sulfates are not decomposed by heat, remain soluble, and are therefore termed permanent hardness. See also **Lime.**

The chemical reactions involved in temporary hardness with lime are

$$Ca(HCO_3)_2 + Ca(OH)_2 \rightarrow$$
$$2CaCO_3\downarrow + 2H_2O \quad (1)$$
$$Mg(HCO_3)_2 + 2Ca(OH)_2 \rightarrow$$
$$2CaCO_3\downarrow + Mg(OH)_2\downarrow + 2H_2O \quad (2)$$

Permanent hardness is removed by reaction with soda ash:

$$CaSO_4 + Na_2CO_3 \rightarrow CaCO_3\downarrow + Na_2SO_4 \quad (3)$$
$$MgSO_4 + Na_2CO_3 + Ca(OH)_2 \rightarrow$$
$$CaCO_3\downarrow + Mg(OH)_2\downarrow + Na_2SO_4 \quad (4)$$

This process lowers the total dissolved solids content equivalent to the solubility of calcium carbonate but has the limitation of leaving residual hardness and dissolved sodium salts. The chemical cost is low, but the overall cost may be high due to labor and space costs.

Coagulation processes designed to remove the suspended solids from surface waters utilize chemical additives to form hydrous precipitates to retain the turbidity. Alum, iron salts, activated silica, and polyelectrolytes are commonly used to clarify turbid waters. Sludge contact clarifiers are widely used. A typical clarifier used as a part of a metal-finishing waste disposal system is shown in Fig. W-11. The design for this type of equipment, used for lime softening and clarification, involves a mixing zone with the chemical feeds, a floc-forma-

Fig. W-11. Clarifier used as a part of a metal-finishing waste disposal system.

tion zone, and a sludge-blanket area. The clarified water is collected at the top of the unit, and sludge is discharged from the bottom outlet.

Effluents from lime softening or clarification units contain particulate matter which must be removed prior to further use. Sand or multimedia filters (quartz, anthracite, and garnet) are installed downstream of clarification units. Filters of this type reduce turbidity from 5–15 ppm to less than 1 ppm. Figure W-12 shows a battery of filters. This type of filtration equipment operates at a rate of 4–6 gal/(min)(ft²). When pressure drop across the filter reaches approximately 10 psi, the unit is backwashed, rinsed, and returned to service.

Dissolved and suspended organic materials in water supplies are removed by clarification and filtration with the help of oxidizing agents such as chlorine and ozone. Chlorine gas is used for larger installations and sodium hypochlorite solution for lower flows because of its convenience. As chlorine is added to water containing organic matter, a plot of chlorine added versus residual chlorine shows a steadily increasing residual followed by a sudden drop. This amount of chlorine is termed the *breakpoint*. Continued addition of

chlorine will give proportional chlorine residuals. Waters containing organic materials are usually chlorinated past the breakpoint to give about 1 ppm residual Cl_2.

Breakpoint chlorination equipment is normally followed by granular activated carbon filters operated at flow rates of 2–10 gal/min per square foot of bed area. The activated carbon removes residual chlorine, as strong oxidizing agents degrade ion-exchange resins located downstream. The activated carbon also removes residual organic matter by adsorption.

Aeration equipment is used extensively for the removal of heavy metals present in many well waters. Iron normally exists as soluble ferrous bicarbonate. Upon contact with air in a countercurrent aerator, the ferrous iron is oxidized to the ferric state. Ferric hydroxide, which precipitates, has a low solubility (0.23 ppm) and is easily removed by filtration. Manganese can also be oxidized and removed by this method at pH values of 9.5 +. This type of equipment is low in initial cost and economical to operate. Repumping is required from the catch tank located below the aerator.

Similar equipment is used for removal of free

Fig. W-12. Battery of filters in a water treatment system.

carbonic acid from hydrogen-cycle cation-exchange units. Bicarbonate alkalinity in fresh water is converted to carbonic acid in the cation exchanger by the reaction

$$Ca(HCO_3)_2 + 2RSO_3H \rightarrow (RSO_3)_2Ca + 2H_2CO_3$$

The carbonic acid content is lowered to about 10 ppm in a degasifier tower packed with Raschig rings or Berl saddles to provide a large surface area. The water flows by gravity in a downward direction while air is blown upward to remove the dissolved CO_2.

Water used for boiler feed must be free of oxygen and carbon dioxide. Boiler water treatment plants using the ion-exchange system contain a vacuum degasifier for removal of these dissolved gases. The vacuum is created by means of a steamjet eductor or, more commonly, a vacuum pump. CO_2 and O_2 residuals of 0.1 and 5 ppm are normally obtained.

The largest number of water treatment units are installed for water softening using the sodium form of a cation-exchange resin. These systems are used in homes, municipal softening systems, and most industries. The equipment consists of a vertical pressure vessel containing the ion-exchange resin, piping, valves, and a sequence timer for automatic operation. The softening tank contains carefully designed distribution systems to assure plug flow to attain uniform exhaustion of the cation-exchange resin. The common operating sequence is:

1. Service or softening cycle at flow rates of 5–25 gal/(min)(ft^2). The service time is usually 8 hr to a number of days. This varies with the amount of calcium and magnesium ions in the water being softened. This step continues until hardness appears in the effluent water, at which time regeneration occurs.

2. Backwash. The flow of water is reversed to expand the ion-exchange resin bed for regrading and removal of turbidity. Time is usually 10–30 min.

3. Regeneration or brining step. A 10–12% solution of NaCl is passed downward through the unit. The reaction is

$$2NaCl + (RSO_3)_2Ca \rightarrow 2RSO_3Na + CaCl_2$$

This step requires about 30 min. Salt dosages of 5–15 lb NaCl are used per cubic foot of exchange material.

4. Rinse. Excess NaCl is rinsed from the unit with hard water in a downward flow pattern. Time is approximately 30 min. The softener is then returned to its service cycle.

The same type of cycle is used for deionization equipment using cation- and anion-exchange resins.

Water containing calcium and magnesium alkalinity should not be used for boiler feed, as rapid scale formation and tube failures occur. The most common method of dealkalization involves the hydrogen form of a cation-exchange resin. The water is softened as the first step in the process, using sodium-cycle cation-exchange units previously described. The soft water passes through a pressure vessel containing the hydrogen form of a weak-acid cation exchanger. The dealkalization reaction is

$$2RCOOH + Ca(HCO_3)_2 \rightarrow$$
$$(RCOO)_2Ca + 2H_2CO_3$$

The carbonic acid is removed by passage through a degasification unit. Weak-acid cation-exchange resins have high exchange capacity and the additional advantage of efficient use of the acid regenerant, which may be either HCl or H_2SO_4.

Deionization of natural waters is the common process for removal of dissolved salts. Cation- and anion-exchange resins of many types are used by industry. The physical and chemical properties of ion-exchange resins are covered under **Ion-exchange resins.**

Deionization systems consist of three general types: (1) two-bed, weak-base anion exchanger, (2) two-bed, strong-base anion exchanger, and (3) mixed-bed.

Fig. W-14. Package-type mixed-bed deionizer. (*Illinois Water Treatment Company.*)

Fig. W-13. Two-bed deionizer.

Two-bed, weak-base anion exchangers remove essentially all ionized materials with the exception of carbonic and silicic acids. Strong-base anion exchangers, developed in 1948, incorporate a quaternary ammonium group in the matrix and will exchange the hydroxyl ion for the anions of all acids, weak or strong.

The cation-exchange vessel of two-bed deionizers contains a sulfonic cation-exchange resin which is regenerated with either HCl or H_2SO_4. Weak-base anion exchangers can be regenerated with dilute solutions of NaOH, NH_4OH, or Na_2CO_3. Strong-base anion exchangers require the strong base NaOH.

Most two-bed, weak-base ion-exchange units incorporate a degasifier for removal of H_2CO_3 from the deionized water, which is located between the cation- and anion-exchange units on a two-bed, strong-base system. Although the strong-base anion-exchange resin will remove H_2CO_3, degasifier operational costs are lower than the cost of sodium hydroxide regenerant. Figure W-13 shows a typical two-bed deionizer. This unit is manually operated, but many deionizers are automatic. The most common system uses air-operated valves controlled by solenoids and timers. Conductivity is normally used to signal the termination of a service cycle and to start the

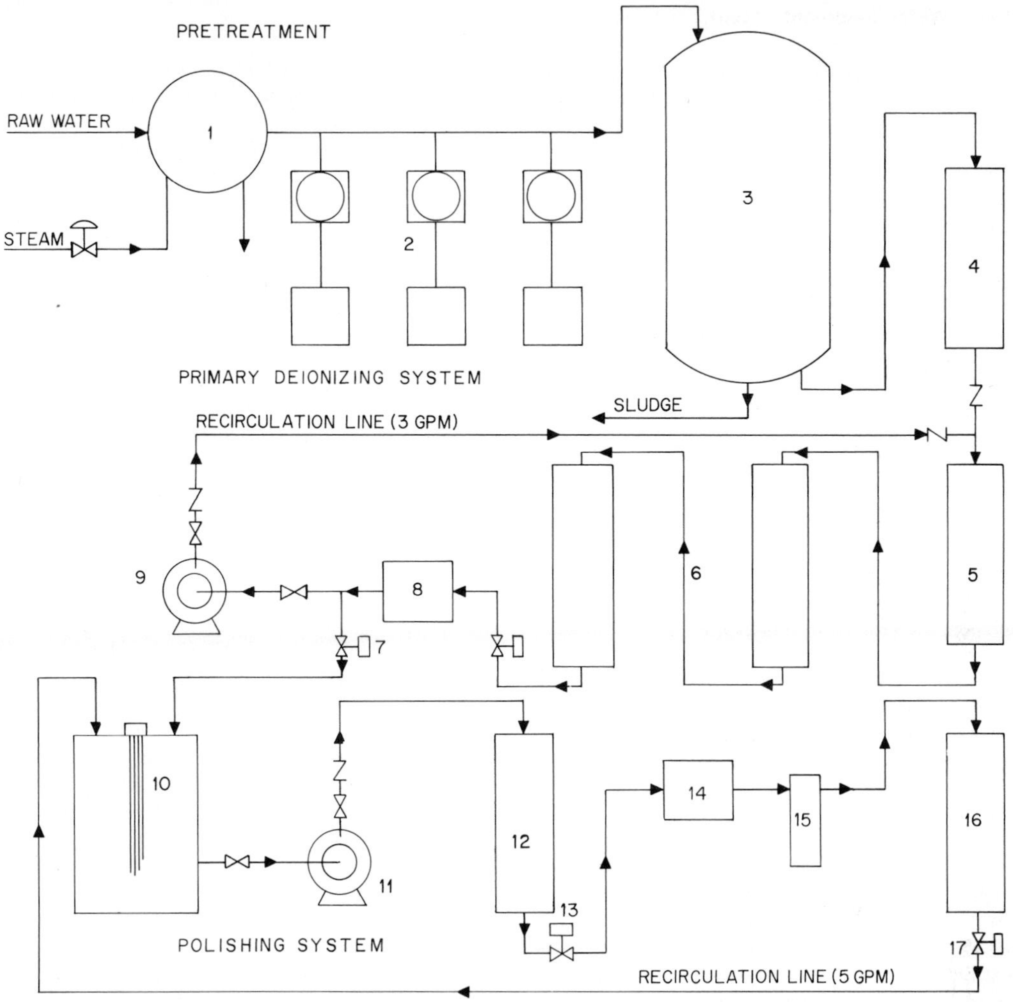

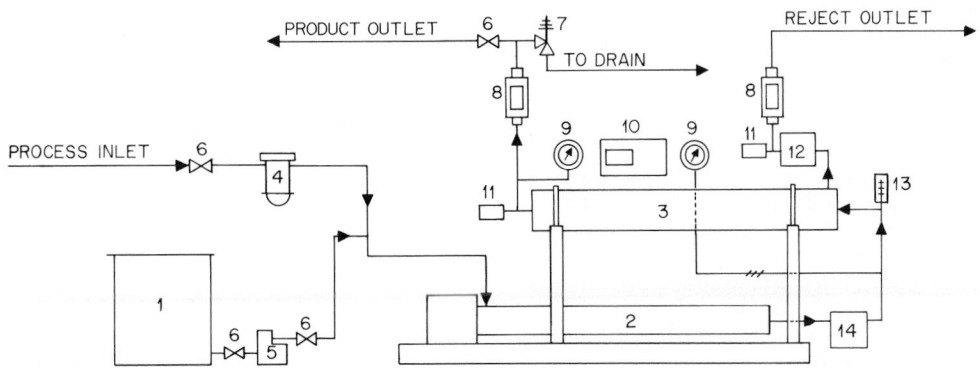

Fig. W-15. Flow sheet of a high-purity water system. (1) Heater, (2) chemical-solution pumps, (3) pressure clarifier, (4) sand filter, (5) carbon filter, (6) two-bed deionizer, (7) automatic valves, (8) ultraviolet sterilizer, (9) pump, (10) water storage tank with electrodes, (11) pump, (12) mixed-bed deionizer, (13) automatic valve, (14) ultraviolet sterilizer, (15) cartridge filter (0.5 μm), (16) nonregenerable deionizer, (17) automatic valve. (*Illinois Water Treatment Company.*)

Fig. W-16. Typical reverse osmosis system. (1) Chemical feeder (day tank), (2) pump, (3) permeator, (4) micron filter, (5) pump, (6) isolating valves, (7) relief valve, (8) flow indicator, (9) pressure gage, (10) conductivity monitor, (11) conductivity cell, (12) flow regulator, (13) thermometer, and (14) pressure regulator.

regeneration sequence. A variety of other instruments are used, such as flow indicators and recorders, temperature controllers, hardness and silica analyzers, and pH instruments.

Two-bed deionizers generally produce water containing 1–10 ppm $CaCO_3$ equivalents as ionized impurities. The use of the mixed-bed deionizer permits production of ultrapure water at a low operating cost. Water having a resistivity of 18 MΩ is produced routinely. Modern electronic and power generating plants could not operate economically without this high quality water. Figure W-14 shows a typical package-type mixed-bed deionizer.

A mixed-bed deionizer contains a sulfonic-type cation exchanger and a quaternary-type strong-base anion exchanger. After an exhaustion cycle, the resins are hydraulically classified by means of an upflow backwash step. The density of the cation-exchange resin is greater than the density of the anion exchanger. The resins separate into two layers. A distributor is located at the interface of the two resin layers. The resins are regenerated simultaneously by passing a 4% NaOH solution downward through the anion-exchange layer and 2–5% H_2SO_4 upward through the cation-exchange layer. The two flows meet at this point and are discharged through the intermediate distributor. Regenerant quantities and concentrations are selected to yield a neutral regenerant effluent.

After regeneration and rinsing, the water level is lowered in the mixed-bed tank. The next step involves the introduction of air at the bottom distributor to intimately mix the cation- and anion-exchange resins. This, in effect, results in the formation of many separate deionized pairs in the unit. Therefore, leakage of cations and anions is virtually eliminated, resulting in a deionized water almost completely free of salts, acids, or bases.

Mixed-bed deionizers are commonly installed after two- or multiple-bed systems. Figure W-15 is a flow sheet of a high-purity water system used in an electronic components manufacturing plant. The system has provisions for the removal of suspended solids, dissolved organic matter, and ionized salts. Note the use of two mixed-bed deionizers and the recirculation line, which is installed to ensure high purity at all times.

Membrane Deionizing Process: The most recently developed process for deionization of water is *reverse osmosis* (RO). Raw water under pressure (200–600 psi) is forced through a semipermeable membrane. Newly developed membranes have controlled porosity which allows rejection of most dissolved salts, organic matter, and particulate matter while allowing the water to "permeate" through the membrane. A typical system will produce 75% product water and 25% waste stream. This means that for every gallon of raw-water input, the RO system will produce 3 qt of treated water. Approximately 2–15% of the ions present in the raw water will pass through the membrane and remain in the product water. Three types of membranes are currently used: the hollow fiber polyamide, the spiral-wound cellulose acetate sheet, and the tubular cellulose acetate module.

Depending on the quality of the raw water to be processed, various pretreatment methods may be required. If the raw water contains organics, a carbon filter may be employed to reduce the organic content before the water enters the RO unit. If the water contains turbidity, clarification-filtration may be used to reduce the load on the cartridge filter at the inlet to the system. A chemical feed system may be utilized to adjust the pH, kill bacteria, and prevent precipitation of such salts as $CaSO_4$. Best permeation rates are obtained at about 25°C. A flow sheet of a typical RO system is shown in Fig. W-16. RO is a very good method of reducing the salinity of brackish waters having salt contents in the range of 1,000–3,000 ppm. The permeate will contain about 100–300 ppm, which is within drinking water standards.

Another water treatment process for salt removal from brackish waters is *electrodialysis*. The raw water is pretreated for removal of suspended material, particularly iron and manganese. Electrodialysis cells utilize ion-exchange membranes, which were first developed in the early 1950s. These membranes are thin films and contain ion-exchange sites, as do ion-exchange resins. Cation-exchange membranes are negatively charged and permeable to cations. Anion membranes are permeable to anions and are negatively charged. These membranes are arranged in an alternate membrane stack, with adequate spacing between the membranes for the water flow. Direct current is applied to accomplish the electrodialysis. One stream of water flowing through the equipment is enriched in salt content, and the product stream is depleted of salts. When operating on brackish waters, the effluent salt content is usually in the range of 300 ppm TDS, $CaCO_3$ equivalents.

—A. Curtis Reents, *Techni-Chem, Inc., Cherry Valley, Ill.*

Waxes Waxes consist of high-molecular-weight hydrocarbons, carboxylic acids, alcohols, and esters. Waxes usually are harder, less greasy, and more brittle than fats and frequently are combinations of fatty acids with alcohols. The waxes of major commercial importance are listed with key properties in Table W-12.

Annual wax consumption in the United States alone is in the neighborhood of 2 billion lb, about 90% of which is made up of hydrocarbon waxes, the macro- and microcrystalline paraffin waxes. Other waxes, however, are important for specific uses and as additives (usually 1–5%) to paraffin-wax products for the enhancement of gloss and tensile strength and alteration of melting point and other physical properties. Well over half of the commercial and industrial wax products are blends of two or more waxes.

Waxes and wax-containing products serve a wide variety of uses, e.g., in floor coverings and polishes; adhesives; cosmetics; electrical insulations; leather finishes; matches; treated paper products, including waxed paper and milk cartons; molding and coating processes; shoe, automobile, and numerous other polishes; printing inks and varnishes; dental materials; explosives and pyrotechnics; crayons; candles; textile finishes; and rubber antioxidants.

Hydrocarbon waxes may be of natural origin or derived from petroleum and are of five main classes: (1) paraffin (macrocrystalline) waxes from petroleum, (2) microcrystalline waxes from petroleum, (3) high-heat distillates (lignitic tar–lignite paraffins), (4) synthetic hydrocarbons, and (5) natural earth waxes known as ozokerites. Petroleum-wax recovery is described under **Lubricating oils.**

Bayberry wax is a vegetable tallow (a triglyceride) often grouped with the waxes because of its physical waxy characteristics rather than chemical composition. Also known as *myrtle wax,* bayberry wax is used in Europe for soapmaking and in the United States as a blend in candle making. The material is derived from the berries of the *Myrica* plant. The product from Central America contains about 58% myristic acid, 36% palmitic acid, and 1.3% loeic acid plus minor ingredients.

Beeswax is formed and deposited by the honeybee, *Apis mellifera,* and consists mainly of a complex long-chain ester, *myricil palmitate,* $C_{15}H_{31}COOC_{30}H_{61}$, and *cerotic acid,* $C_{25}H_{51}COOH$. Commercially, it frequently is blended with paraffin, stearin, and vegetable waxes. Beeswax is easily dyed and is used in adhesives, candles, polishes, leather dressings, and cosmetics and as a protective coating for etching.

Candelilla wax, obtained from the shrubs *Pedilanthus pavonis* and *Euphorbia antisyphilitica* (Texas

TABLE W-12. *Major Waxes and Their Properties*

Wax	Sp gr Solid 15°C	25°C	Liquid 100°C	Mp, °C	Hardness, Shore at 25°C	Dielectric Constant, 10^3 Hz at 25°C	Volume Resistivity, 10^{13} Ω-cm	Saponification Value	Iodine Value
Bayberry	0.986	0.980	0.876	46.7–48.8	71–78	3.22–3.27	2–5	212	3–4
Beeswax (yellow)	0.967	0.958	0.846	63–64	43–50	2.87	5–12	91	10
Candelilla	0.984	0.980	0.855	67–69	100	2.50–2.63	120	51	16
Carnauba	0.998	0.997	0.842	86–93	100	2.66–2.83	60–120	83	10
Castor	1.00	0.989	0.896	86	93–99				
Ceresin	0.928	0.92	0.777	59–71	58–68	2.16–2.24	290	0	
Chlorinated paraffin (11% Cl₂)	0.860	...	*						
Japan	0.993	0.978	0.873	51–53	27–44	3	6–7	220	5–15
Jojoba	...	0.864							
Microcrystalline ASTM 150/155°F	0.913	0.901	0.768	66–69	41–48				
Montan	1.03	...	...	82–87	99	2.65	290	60–100	10–18
Ozokerite	0.936	0.929	0.885	61–78	58–70	2.4	30–40	0	
Paraffin, American, mp 121°F	0.896	0.880	0.752	49–50	...	2.2–2.4	300		
Polyethylene	0.922	...	...	100–104	100	2.2	140	6	
Spermaceti	0.95	0.92	0.812	44–47	...	3.6	3–4	127	5
Wool (lanolin)	0.95	0.93	...	36–43	...	...	...	98	15–30

*Decomposes.

and Mexico), consists of long-chain hydrocarbons and small amounts of esters. The wax principally is used as a substitute for carnauba wax or for blending with carnauba and beeswax. The majority of candelilla wax is used in furniture and shoe polishes.

Carnauba wax is derived from the leaves of the palm tree *Copernicia cerifera* (Brazil) and is a hard, high-melting, lustrous wax, consisting mainly of *ceryl palmitate*, $C_{25}H_{51}COOC_{30}H_{61}$. Over 70% of carnauba wax goes into floor polishes because of its self-polishing characteristics. It also is used in carbon-paper manufacture because of its non-smearing, nongreasy properties.

Castor wax is hydrogenated castor oil (essentially the glyceride of ricinoleic acid) and is a hard, non-greasy, white solid. Because it is odorless and tasteless, castor wax is valued as a coating material. It is also used in special lubricating greases and as a plasticizer in vinyl and cellulose plastics.

Ceresin is a refined ozokerite.

Cotton wax, which is contained in the cotton fiber (0.6%), is similar to carnauba wax and is a combination of C_{28}–C_{32} primary alcohols and C_{24}–C_{32} fatty acids.

Japan wax is derived from the outer skin of berries of plants of the genus *Rhus* (Japan and California) and is mainly palmitic acid in the form of glyceryl palmitate with stearic, oleic, and japonic acids. Also known as *sumac wax* and *Japan tallow*, it is used as an extender for beeswax and in polishes and candles.

Jojoba wax is a liquid wax derived from extraction of the seed beans of the evergreen shrub *Simmondsia californica* (southwest United States and Mexico). It consists of high-molecular-weight acids and alcohols and contains no glycerin. The wax is used in cosmetics, leather dressings, and lubricants.

Microcrystalline waxes are petroleum-derived. Unlike the paraffins, they are not straight-chain hydrocarbons but incorporate a branched-chain construction with a fine crystalline structure, resulting from treating the macrocrystalline paraffin waxes. They have higher melting points. A considerable amount of oil (1–15%) can be occluded within the wax without causing it to be oily or greasy.

Montan wax, also known as *lignite wax*, is produced mainly in Germany from lignite by solvent extraction. Oklahoma and Texas brown coals also contain the substance (up to 13%). The wax is used in polishes, leather finishes, phonograph records, insulation compounds, and candles. A syn-thetic substitute also is available, the octadecyl ether of vinyl alcohol, $C_{17}H_{34}CH_2 \cdot O \cdot CH:CHOH$.

Ozokerite, also known as *mineral wax* or *earth wax*, is a natural paraffin occurring in rocks found in central Europe and Utah. The substance is used in printing inks, crayons, sealing compounds, phonograph records, insulations, candles, polishes, and as an extender for beeswax. Ceresin is refined ozokerite.

Paraffin wax, a petroleum wax, is furnished in a variety of forms, including proprietarily formulated and treated waxes, such as chlorinated paraffin waxes. There are three main classifications: (1) soft paraffins, such as slack wax, (2) intermediate paraffins, such as scale wax, and (3) hard paraffins, such as block waxes. Most of the slack wax, which contains about 50% oil, is further processed into fully refined paraffin waxes. Scale wax (not to be confused with a variety of beeswax) is a partially refined paraffin wax that is softer and less expensive than the fully refined hard paraffin waxes and thus finds a wide variety of commercial and industrial uses. Hard paraffins are practically oil-free and offered in several grades, depending largely upon melting point.

Polyethylene of low molecular weight may be used for extending and modifying waxes.

Spermaceti wax is derived from sperm oil. It is a true wax and does not yield glycerin when saponified. The material is used mainly in ointments.

Wool wax (lanolin), obtained from wool grease, consists of esters of sterol alcohols combined with straight-chain fatty acids and with small quantities of free alcohols. About 95% of the content is fatty acid esters, but its emollient properties depend upon the free alcohols and acids. It is largely used in cosmetics, and several proprietary derivatives of lanolin also are used in this field.

REFERENCES

Brady, George S.: "Materials Handbook," 9th ed., McGraw-Hill, New York, 1956.

Mantell, Charles L. (ed.): "Engineering Materials Handbook," McGraw-Hill, New York, 1958.

Webbed Materials   (See **Fibers**; and **Papermaking and finishing**.)

Weighting Agent   (See **Waste treatment**.)

Welding   (See **Aluminum**; **Copper**; and **Oxygen**.)

Wet Classification   (See **Screening**.)

Wet-Process Phosphoric Acid   (See Fertilizers; and **Phosphoric acid**.)

Wetted-Wall Column   (See Absorption.)

Wetting Agents   [See Anionic surfactants (sulfur-bearing); and **Surfactants**.]

Wheat   (See Amino acids.)

Whey Processing   (See Membrane filters.)

White Metal   (See Tin.)

White Oil   (See Mineral oil.)

Whitening Agents   (See Detergents.)

Wohlwill Process   (See Gold.)

Wolfram   (See Tungsten.)

Wolframite   (See Indium; and Tungsten.)

Wood Chemicals   (See Resins; and Terpenes.)

Wood Preserving   (See Coal tar and derivatives.)

Wood Pulp   [See Acetate fibers; Cellulose ester plastics, organic; Pulp (wood) production and processing; and **Rayon**.]

Wool   (See Fibers; and Peptides and proteins.)

Wool Wax   (See Waxes.)

Wrought Iron   (See Iron and steel.)

Wüstite   (See Iron ores; and Iron- and steelmaking.)

Xanthates (See Flotation; and Rayon.)

Xenon (See Chemical elements.)

Xenotime (See Erbium; Gadolinium; Lutetium; Rare-earth elements and metals; Terbium; and Yttrium.)

X-Radiation (See Radioisotopes.)

Xylenes (See Hydrocarbons; and Petrochemical complex.)

Xylyl (See Radicals.)

Xylylene (See Parylene polymers.)

Yeasts The English *yeast* and the Dutch *gist* are derived from the Greek *zestos* (seethed, boiled). Taxonomically yeasts cannot be defined readily. They generally are included in the plant kingdom, belong to the division of *Thallophytae;* i.e., they have no roots, stems, or leaves. Yeasts do not contain chlorophyll and cannot generate energy from light and CO_2. Hence, they depend upon the presence of sugars and other nutrients in their environment and lead a saprophytic or parasitic life.

Yeasts generally are distinguished from fungi by the fact that they do not produce distinct conidia. Saccharomycetaceae include the ascoporogenous yeasts. These yeasts may produce spores in an ascus (or sac). The economic yeasts, which usually reproduce vegetatively by budding or by fission, belong to several genera within this family. The genus *Saccharomyces* comprises most of the culture yeasts, e.g., beer yeasts, distillers' yeasts, wine yeasts, bakers' yeasts, and dairy yeasts.

Most yeasts occur as individual, oval-shaped cells. The dimensions of a typical bakers' yeast cell are about 4–6×7–$10 \mu m$. The cells are enclosed in a semipermeable membrane, the cell wall, which consists largely of the polysaccharides glucan and mannan. A large nucleus and other internal structures, such as mitochondria, can be distinguished readily under the microscope. Yeasts commonly multiply by budding or fission, or they form pseudomycelia. Among the commercially used yeasts, the sherry yeasts are film formers.

Yeasts are facultative anaerobes; i.e., they can grow and ferment under both aerobic and anaerobic conditions. Under strictly anaerobic conditions, growth is limited to a few generations. In practice, small amounts of oxygen generally are available for yeast growth. Pasteur clearly recognized the different metabolic patterns of yeast under aerobic and anaerobic conditions. Specifically, he described fermentation as *vie sans air* (life without air). The well-known alcoholic fermentation, in which fermentable sugars are metabolized to ethanol and CO_2, had already been recognized by Gay-Lussac at the beginning of the nineteenth century

$$\underset{\text{Glucose}}{C_6H_{12}O_6} \rightarrow 2CO_2 + \underset{\text{Ethanol}}{2C_2H_5OH}$$

Pasteur proved the function of yeast in fermentation, an energy-yielding reaction in which 2 moles of adenosine triphosphate are formed for each mole of glucose fermented. In contrast, under aerobic conditions, yeasts switch to a respiratory metabolic pattern. This results in the complete oxidation of sugars to CO_2 and H_2O and yields greater free energy than the alcoholic fermentation.

Some yeasts do not ferment sugars or have only weak powers of fermentation. *Candida utilis* and some of the wine yeasts present during the initiation of most fermentations produce little or no ethanol. On the other hand, some yeasts have weak respiratory activity and grow best under anaerobic conditions, such as *S. uvarum* (formerly *S. carlsbergensis*), the lager beer yeast. All yeasts which can produce ethanol ferment glucose, but the ability to ferment other mono- or disaccharides differs appreciably. Yeasts occurring naturally in cereal mashes can generally ferment maltose but not lactose. The reverse is true of many dairy yeasts, such as *S. fragilis*.

Apart from a source of energy, e.g., sugar, yeasts require nitrogen, phosphate, sulfur compounds, and trace minerals for their growth. Ammonia, ammonium salts, urea, and amino acids are good sources of nitrogen. Biotin and sometimes other vitamins of the vitamin B complex are required for growth.

Yeasts produce a number of by-products during alcoholic fermentation. The most important of these are glycerol, higher aliphatic and aromatic alcohols, aldehydes, ketones, organic acids, and esters of alcohols and organic acids. These by-products give alcoholic beverages their pronounced flavor, and nearly all the flavor compounds identified can be found in all alcoholic beverages although in different concentrations. The formation of bread flavor is more complicated, resulting from the interaction of the by-products of alcoholic fermentation with other dough constituents during baking. For distilled alcoholic beverages, the flavor may be modified by fractional distillation and discarding some of the fractions.

The culture yeasts do not excrete enzymes into their surrounding medium and are therefore dependent on the presence of soluble low-molecular-weight energy sources. With some notable exceptions, yeasts do not ferment starches unless the starch is hydrolyzed enzymatically. The production of brewers' wort from cereal malt results in the formation of maltose and other saccharides from cereal starch which can be fermented by brewers' yeasts. Barley malt, wheat malt, and other cereals whose starches can be hydrolyzed to sugars are suitable substrates for the production of beer. In western countries, barley malt is the principal ingredient of brewer's mashes with the additional use of corn or rice as adjuncts. Beer (lager beer) is generally produced with a bottom-fermenting yeast (*S. uvarum*) and ale with a top-fermenting yeast (*S. cerevisiae*). In the Orient, rice is the starchy substrate for beer (or rice-wine production). The rice starch is generally hydrolyzed by growing amylase-excreting fungi, such as *Aspergillus oryzae*, or koji in the mash.

Brewers' yeast is generally produced in small, pure-culture yeast plants in the brewery for inoculation (pitching) into commercial worts. During fermentation the yeasts multiply (four- to eight-fold) and are harvested after the primary fermentation. A portion of this yeast is washed and used to inoculate 3–15 succeeding commercial fermentations. The cycle begins again with use of a fresh yeast inoculum from the pure culture plant. Excess yeast is recovered and dried for use in food or feed, or it is sold together with spent grains as a feed.

Distilled alcoholic beverages can be made from cereal beers, such as whiskey (from wheat), bourbon (corn), or rye. Brandy (the name is derived from *Brantwein*, burned wine) is made by distillation of grape wine. Vodka results from the fermentation and subsequent distillation of potato starch. It is low in its concentration of the above mentioned by-products of the alcoholic fermentation and consequently has a bland taste. Distillers' yeast remaining in the slops after distillation is not recovered as such but dried with the slops and sold as feed.

Wine is made by fermentation of grape juice or other fruit juices. Traditionally, fermentation is initiated by yeasts which occur naturally on the grapes and which can be found abundantly in and on winery equipment, particularly the grape crushers. During such natural fermentations, there is a succession of various species of yeast. Fermentation is usually initiated by yeasts with a low tolerance for ethanol, such as *Kloeckera apiculata, Hanseniaspora guillermondii* and *Candida*. These are followed by more tolerant yeasts, and *Saccharomyces* take over toward the end of the fermentation and may produce up to 16% ethanol by volume. In the United States and some other countries, the growth of natural yeasts is suppressed with sulfur dioxide, and pure culture yeasts (generally, *S. cerevisiae var. ellipsoideus*) are used to inoculate the musts. Such culture yeasts can be dried successfully and are available for use in wine making.

The fermentation of dairy products is of great importance in western countries. A fermented milk, kefir, is produced in the Balkan countries and Russia with the use of so-called kefir grains. They consist of a yeast, *Torulopsis holmii,* and lactic acid bacteria which excrete a polymeric material, kefiran, responsible for the particular consistency of the grains. The yeast and bacteria exist in a truly symbiotic relationship in these kefir grains.

S. cerevisiae is used all over the world for the production of yeast-raised baked goods, such as breads, rolls, and sweet goods. These yeasts ferment added sugars rapidly and usually have good maltose-fermenting ability. Maltose becomes available during dough fermentations through the amylolytic action of added malt on wheat starch. Since the yeast is killed on baking, the baking industry depends on yeast manufacturers to supply bakers' yeast in large quantities. In the United States about 350 million lb of bakers' yeast (at 30% solids) are produced annually. This yeast is grown in large fermenters under conditions of strict aerobicity. Large volumes of air are blown through the tanks (one volume of air per fermenter volume per minute) while the fermentable raw material (molasses) is added in small increments. Under these conditions little or no alcohol is formed, and 1 lb of yeast solids can be produced from 2 lb of fermentable sugars. The preferred raw materials are beet or cane molasses, but other materials such as enzymatically hydrolyzed starches may be used. The yeast is concentrated by centrifuging and pressing and sold as compressed yeast in the form of 1-lb extruded cakes (wax wrapped) or as crumbled yeast in plastic-lined paper bags. This compressed or crumbled yeast at about 30% solids requires refrigeration from the time of production to its final use in the bakery. Compressed yeast may be dried to a solids content of 92.5%; in this form, it is stable for 3 months. If it is protected by exclusion of air (nitrogen or vacuum packing), it is stable for more than 1 year.

Large amounts of yeast are dried after pasteurization to yield inactive dried yeast, which is used as a food and as a protein and vitamin supplement. The finely ground material of a moisture content of about 5% has approximately 45–50% protein (N $\times$ 6.25) and carries vitamins of the B complex, principally thiamine, riboflavin, and nicotinic acid. A great deal of this yeast is used in feed as a protein and vitamin supplement. Some of these yeasts are produced from "primary" grown bakers' yeasts by drum or spray drying.

Other sources are the spent yeast of breweries and *C. utilis* (torula) grown on spent sulfite liquor. The latter yeast is able to utilize pentoses, which may account for up to two-thirds of the fermentable sugars of waste sulfite liquors. Wood hydrolysates are used in other countries (mainly eastern Europe) for the production of fodder yeast.

Yeast protein is of good nutritional quality, being particularly high in L-lysine. The concentration of ribonucleic acids in yeast, 6–10%, limits its use in foods to about 20 g per person per day. Yeast hydrolysates and particularly yeast autolysates have an excellent flavor, resembling that of meat extract. They are used widely in soups and gravies as flavorants. See also **Peptides and proteins.**

REFERENCES
Cook, A. H. (ed.): "The Chemistry and Biology of Yeasts," Academic, New York, 1958.

Phaff, H. J., M. W. Miller, and E. M. Mrak: "The Life of Yeasts: Their Nature, Activity, Ecology, and Relation to Mankind," Harvard, Cambridge, Mass., 1966.

Reed, G., and H. J., Peppler: "Yeast Technology," Avi, Westport, Conn., 1973.

Rose, A. H., and J. S. Harrison (eds.): "The Yeasts," 3 vols., Academic, London, 1970.

Solomons, G. L.: "Materials and Methods in Fermentation," Academic, New York, 1969.

—Gerald Reed, *Universal Foods Corporation, Milwaukee, Wis.*

Yellow Brass (See **Copper.**)

YIG (See **Yttrium.**)

Ytterbium* (From *Ytterby,* a city and a near-by feldspar mine of the same name in the province of Stockholm.) **Yb** = 173.04 (at. wt.); 70 (at. no.). In 1878, de Marignac discovered ytterbium in Mosander's "pure" erbium oxide, but the ytterbium fraction he isolated was later found to contain scandium and lutetium.

Ytterbium is in group III, period 6, of the periodic table of the elements and is the second last (or the thirteenth) of the elements in the lanthanide series, which is commonly shown as the first extra row of elements below the main body of the periodic table. Because of its position in the periodic table, ytterbium exhibits two valence states, 3+ with thirteen $4f$ electrons and 2+ with fourteen $4f$ electrons, the maximum possible. As a result, ytterbium exhibits a variable chemical and

*Work performed in the Ames Laboratory of the U.S. Atomic Energy Commission. Contribution No. 3171.

alloying behavior, and the Yb^{++}/Yb^{3+} couple has an oxidation-reduction potential of 1.15 V in aqueous solutions.

Ytterbium melts at 815°C and boils at 1194°C. Its room-temperature density is 6.972 g/cm^3. Ytterbium is one of the least abundant elements; it ranks eleventh in the 17 elements of the rare-earth group and fifty-third of the naturally occurring elements. Metallic ytterbium's physical and chemical properties are similar to those of the alkaline-earth metals (Ca, Sr, Ba) and quite different from those of its rare-earth neighbors (Er, Tm, Lu), indicating that ytterbium is divalent in the metallic state.

The structure of ytterbium is face-centered cubic at room temperature (alpha Yb). It transforms to a body-centered cubic form (beta Yb) at high temperatures (792°C) and atmospheric pressure or at high pressure (39,000 atm) and room temperature. At pressures greater than 16,000 atm alpha Yb, which is a normal metallic conductor at standard pressure and temperature, becomes a semiconductor. Beta Yb, however, is a normal metallic conductor at all temperatures and pressures. When alpha Yb is cooled below 0°C, it transforms into a hexagonal form (alpha′ Yb), a phase only discovered in 1968.

Yb, which is silvery-gray in color, is stable in moist or dry air up to temperatures of 200°C. Above this temperature, the rate of oxidation increases with increasing temperature. Yb dissolves readily in dilute and concentrated mineral acids. Sodium and ammonium hydroxide solu-

indicates that YbO does not exist as a pure oxide phase. Yb_2O_3 melts near 2350°C and crystallizes in the cubic C-type oxide, which is formed by many of the R_2O_3 compounds. $YbCl_3$, $YbBr_3$, and YbI_3 are hygroscopic, usually exist as hydrated salts, and are soluble in water. YbF_3 is nonhygroscopic and insoluble in water. The dihalides are prepared by thermal decomposition of the corresponding ytterbium trihalide or by reduction with H_2 or Yb metal. Ytterbium dichloride dissolves in water, but the Yb^{++} ion is oxidized to the trivalent state by air with the liberation of hydrogen. Yb reacts with hydrogen to form an orthorhombic dihydride which is isostructural with the alkaline-earth dihydrides. A cubic dihydride, isostructural with the trivalent rare-earth dihydrides, can be prepared at high pressures. Yb forms three carbides, Yb_3C, Yb_2C_3, and YbC_2, which react with water to form hydrocarbons. The former yields methane gas and the latter two unsaturated hydrocarbons (mostly acetylene). B, Si, N, P, S, Se, and Te form at least one compound with ytterbium most of which are high-melting. Ytterbic hydroxide is insoluble and is precipitated by the addition of sodium, potassium, or ammonium hydroxide to an ytterbium-containing solution. Ytterbic sulfate is soluble in water, and its solubility decreases with increasing temperature. Ytterbous sulfate is prepared from crystalline ytterbic sulfate by methods similar to those used to prepare the dihalides (see above).

Ytterbium has 17 different isotopes, of which 7 are naturally occurring;

Isotope	168	170	171	172	173	174	176
Abundance, %	0.14	3.03	14.31	21.82	16.13	31.84	12.73

tions and sodium nitrite solution slowly corrode Yb at room temperature. Yb dissolves in liquid ammonia to give a dark blue color, which is similar to the behavior of the alkali and alkaline-earth metals. Yb forms an ytterbium trihalide by reaction with the corresponding halogen gases at room temperature; it exhibits alloying behaviors which are sometimes similar to those of the rare-earth metals and at other times like those of the alkaline-earth metals. Yb alloys with all the metals to the right of Cr and its cogeners, Mo and Re, in the periodic table to form intermetallic compounds. Extensive solid solutions are formed by Yb and Ca and by Yb and Eu.

Two ytterbium oxide compounds have been reported Yb_2O_3 and YbO, but recent evidence

Yb has a low toxicity rating.

Occurrence and Processing: The important minerals containing Yb are euxenite, gadolinite, monazite, and xenotime, the last two being the most important sources of Yb. Monazite is a phosphate containing thorium, light lanthanides, and small amounts of heavy lanthanides and Y. Although the heavy lanthanides are obtained as by-products from monazite, it is an important source of these elements. Major deposits are found in Australia, Brazil, India, South Africa, and the United States (Florida, Idaho, and Montana). Xenotime is also a rare-earth thorium phosphate but contains a higher proportion of heavy lanthanides than monazite. Xenotime is found along with monazite, but the fraction of xenotime in the ore is usually a few percent of the amount of monazite.

One of two methods is used to process monazite and xenotime. In the *acid process* the mineral is converted to a thorium–rare-earth sulfate mixture by heating in sulfuric acid. By fractional precipitation with oxalic acid, the light lanthanides are separated from thorium, yttrium, and the heavy lanthanides. The alternate *basic process* involves digestion of the monazite or xenotime in sodium hydroxide at high temperatures to form the insoluble rare-earth hydroxide in a trisodium phosphate solution. After filtering, the hydroxide is carefully neutralized to dissolve the light lanthanides, and the thorium, yttrium, and heavy lanthanides remain as an insoluble residue. In either process thorium is separated by anion exchange, extraction by nonaqueous solvents, or precipitation.

Ytterbium is easily separated from a mixture of yttrium and the heavy lanthanides by the sodium-amalgam-reduction technique. A buffered acid solution containing trivalent rare-earth ions is mixed with a liquid-sodium–mercury alloy, and Yb^{3+} is reduced to Yb^{++} and dissolved in the sodium-mercury alloy. The separated Yb-Na-Hg alloy is treated with hydrochloric acid, and Yb is extracted into the aqueous solution. The addition of oxalic acid precipitates ytterbium oxalate. Yb is regularly isolated from the other rare earths when they are separated from each other by ion exchange.

Yb metal is obtained by heating a mixture of lanthanum metal and ytterbium oxide in a high vacuum. Simultaneously, the Yb sublimes and is collected on a condenser plate while the lanthanum is oxidized to the sesquioxide.

Uses: The most important use of Yb and Yb compounds is in applied and fundamental research. Yb is used in magnetic-bubble-domain devices (ytterbium orthoferrite), in phosphors to convert infrared to visible light, as a laser with an output wavelength of 1.02 μm, and as a dopant in yttrium-aluminum garnets (YAG). Radioisotope [169]Yb is used in portable industrial and medical radiographic units. The largest amount of Yb consumed commercially, however, is in the mixed rare-earth compounds and metals. These uses are described under **Rare-earth elements and metals.**

—Karl A. Gschneidner, Jr., *Ames Laboratory, U.S. Atomic Energy Commission, Iowa State University, Ames, Iowa*

Yttrium (From *Ytterby,* a city and a near-by feldspar mine of the same name in the province of Stockholm.) $Y = 88.905$ (at. wt.); 39 (at. no.). A mineral, first noted near Ytterby in 1787 by C. A. Arrhenius, was called *yttrium earth* (or *oxide*) until 1794, when it was named *gadolinite* in honor of the Swedish chemist Johan Gadolin. In 1828, Fredrich Wöhler isolated a new element from gadolinite, naming it yttrium. In 1843, however, C. G. Mosander separated two other elements from the 1828 preparation and named them terbium (65) and erbium (68).

Yttrium is in group III, 5th period, of the periodic table of elements. However, except for total electrons, atomic weight, and density, most of the properties of this element are similar to the "heavy" rare-earth metals. In general, when the physical properties of Y are plotted against atomic number, Y falls between Gd (No. 64) and Er (No. 68) with an apparent atomic number of 64.5–67.5. For a summary of properties, see **Rare-earth elements and metals.**

The melting point of Y is 1520°C; boiling point, 3338°C, density 4.457 g/cm³. Yttrium, estimated at 33 ppm in the earth's crust, is potentially as plentiful as cobalt. Some estimates run considerably higher, placing its abundance in the range of Cu and Zn.

Yttrium metal retains a silvery-gray luster indefinitely in air and is stable up to about 400°C. The metal dissolves in most mineral acids but is relatively inert to a 1:1 mixture of concentrated HNO_3 and 48% HF. The period 1958–1962 saw 15,000–20,000 lb of metal produced and fabricated into rod, sheet, and tubing for atomic-reactor development programs. Yttrium metal is immiscible with liquid or solid uranium metal and alloys. The metal has a low thermal-neutron-absorption cross section and a low acute-toxicity rating. Because of its large atomic size among metal atoms, very few solid solutions are formed with other metals. Magnesium, thorium, and the other lanthanide elements are exceptions, however, in that they show nearly complete solid solubility with yttrium.

One natural isotope of yttrium is known, [89]Y. Fourteen artificial isotopes have been identified.

Yttrium has no absorption bands in the ultraviolet, visible, or near-infrared region of the spectrum. Emission spectrography is moderately sensitive for detecting very low levels of Y in minerals and chemicals. For macro amounts of Y (5–80%), the most useful analytical technique is x-ray fluorescence spectrography. Atomic absorption equipment also is a practical way of analyzing Y content accurately. There are no simple chemical procedures for analyzing Y because of its three outer valence electrons, which are characteristic of the lanthanide elements.

Yttrium has one valence of 3^+ and forms trivalent compounds with most inorganic anions and many organic radicals. Pure Y_2O_3 is white and has a melting point of $2410°C$ $(4370°F)$. It is the tenth most refractory oxide of the metallic elements and retains a body-centered cubic crystal structure from room temperature to its melting point. The oxide is stable in moist air. The heat of formation of Y_2O_3 at $25°C$, ΔH_c, is $455,500$ cal/mole, highest for any element in the periodic table.

Yttrium forms carbides which hydrolyze to form hydrocarbons. Yttrium nitride is unstable in water vapor. Yttrium silicide is a stable refractory material.

Yttrium has a strong affinity for hydrogen at elevated temperatures. At 1 atm in the range of $900–1200°C$, Y bonds with between one and two atoms of hydrogen forming one of the most stable hydrides known.

Minerals yielding Y also contain enrichments of the heavy rare-earth elements and varying amounts of the light lanthanides. Since 1966, a residue from uranium mining operations in the Blind River–Elliott Lake region of Ontario, Canada, has been the major source of Y. Xenotime, an yttrium phosphate, YPO_4, mineral from Malaysia, is another active source. Prior to 1965, most of the available Y originated from monazite. Certain apatite deposits have some of the calcium replaced with rare-earth elements unusually rich in Y content. Norway has been an active source of gadolinite, euxenite, and samarskite minerals, from which Y can be recovered.

Processing: The processing of minerals to extract Y is similar to that for other lanthanides. Mixed rare-earth elements are taken into solution by HNO_3 or H_2SO_4. Liquid-liquid organic ion-exchange solvent-extraction cells separate a pure Y fraction, which usually is precipitated as an oxalate and calcined to the oxide. Resin-packed ion-exchange columns can also be used. World capacity exists to produce pure Y chemical compounds at the rate of over 500,000 lb/yr.

Yttrium metal is produced by metallothermic reductions using Ca mixed with YCl_3 or YF_3 in a sealed container heated above $1550°C$. Although the reaction is complete within 1 hr, because of the high melting point and reactivity of Y, materials of containment and construction are a problem. Yttrium metal and alloys of Y-Co and Y-Mg have been electrowon from Y_2O_3 in a molten electrolyte of BaF_2-LiF-YF_3.

Uses: In the middle 1950s, one of the more important technical priorities was the development of metals and alloys compatible with the environment of nuclear reactors for the propulsion of submarines and aircraft. In 1958, when space exploration started, these efforts were greatly accelerated. One of the metals found to have outstanding properties for nuclear-reactor structural components was yttrium. Scant knowledge of mineral resources and necessary processes existed then. Pure metal and chemicals that were available had been produced as a by-product of refining monazite for thorium content and mixtures of the light rare-earth elements.

The U.S. Atomic Energy Commission Ames Laboratory at Iowa State University was called upon to refine yttrium ingots ready for fabricating into forms suitable for the reactor programs. The scientists in this laboratory met the challenge by producing about 10 tons of ductile Y by 1962. Most of this material was consumed, and knowledge of the workability, alloying behavior, and economic feasibility of using Y was accumulated and published.

Subsequently, other more available metals were selected, and almost no Y has been used in reactors since 1965. The cost effectiveness and mineral reserves of Y were key factors influencing this development.

In 1965, however, the discovery of a new red phosphor that made color television acceptable to the United States consumer established the first commercial use of Y. The phosphor base is yttrium oxide, Y_2O_3, of maximum purity, which is generally required of electronic materials functioning in a vacuum. The Y_2O_3 host matrix collects the energy from the cathode-ray-tube electron beam. Transfer of this energy to europium oxide, Eu_2O_3, in vacuum causes europium to emit a chromatically true visible red light. The europium-activated yttrium oxide red phosphor $(Eu:Y_2O_3)$ is used in nearly every color television set in the world.

The immediate acceptance of the red TV phosphor created a severe shortage of Y. Except for the small amount available from monazite, economical mineral sources were almost unknown and processing capacity limited. The television requirement for Y resulted in the discovery and development of Y mineral resources and the installation of solvent extraction plants. By 1969, the supply of Y was sufficient to meet all anticipated needs. Yttrium has proved to be abundant, and the processing of pure chemicals presents no technical or economic obstacles.

Another development in the electronics industry in the mid-1960s established the use of Y_2O_3 in YIGs and YAGs, synthetic *yttrium iron* and *yttrium aluminum garnets*. Pressed and sintered polycrystalline ceramics are made by a solid-state reaction of Y_2O_3 with iron oxide or aluminum oxide. Garnets operate in microwave bandpass (filters) circulators and isolators in telephone, radar, and space-communication networks.

The electronic use led to the development of single-crystal yttrium aluminum garnets which approach the brilliance and hardness of diamonds. These simulated diamonds are available in jewelry stores throughout the United States.

Yttrium oxide is the base for neodymium-doped laser crystals. A glass-ceramic composed of 90% Y_2O_3, 10% ThO_2, transparent in the visible and infrared spectra, has the highest softening point of any material discovered.

Yttrium oxide–stabilized zirconia, ZrO_2, refractory is available as cloth, fiber, or solid and is one of the most efficient insulating materials.

Yttrium oxide also is used in nickel-based superalloys produced by powder-metallurgy techniques in a process called *mechanical alloying*. Y_2O_3 is effective as a metallurgical dispersion hardening agent.

Yttrium metal is specified in several cobalt-base superalloys. The addition of Y improves the hot corrosion (sulfidation) resistance of the alloys at operating temperatures (2000–2200°F, 1093–1204°C). Yttrium in Fe-Cr-Al alloys improves workability and adds resistance to sag in heating-element alloys. Coatings containing Y metal applied to rotating turbine-engine parts show superior oxidation resistance.

One of the most promising applications for Y metal is in permanent magnets. When alloyed with five atoms of cobalt, YCo_5, the theoretical static-energy product $(BH)_{max}$ is 28.1 MGOe; dynamic energy $(BH)_{max}/4\pi d$ is 29.4×10^4 ergs/g. These values are second only to $PrCo_5$, the most powerful permanent-magnet material discovered.

Applications for Y which have emerged since 1965 make use of the unique refractory, optical, chemical stability, crystal structure, and atomic size of the element.

—Joseph G. Cannon, *Molybdenum Corporation of America, White Plains, N.Y.*

Z

Zeolites (See **Ceramics; Epoxy resins;** and **Ion-exchange resins.**)

Zeolitic Catalysts (See **Fluid catalytic cracking.**)

Ziegler Chemistry (See **Alcohol process, ALFOL;** and **Surfactants.**)

Zinc (G. *Zink.*) **Zn** = 65.37 (at. wt.); 30 (at. no.). Zinc is a blue-white metal that retains its luster in dry air but shows a slight tarnish when exposed to moist air or water. Zn is a member of group IIB (along with Cd and Hg) and is in period 4 of the periodic table. Melting point = 419.46°C; boiling point = 906°C; and density at 20°C = 7.133 g/cm^3.

There are five stable isotopes of Zn, 64, 66–68, and 70, and six unstable isotopes, 62, 63, 65, 69, 71, and 72. There are no known allotropic forms. See also **Chemical elements.**

Other major mechanical and physical properties of Zn are given in Table Z-1. The smelting and refining of Zn are described under **Zinc and zinc-lead smelting.** See also **Cadmium.**

Chemistry: Very pure Zn dissolves slowly in most acids or alkalis, but since solution is hastened by the presence of impurities, commercial Zn dissolves readily in a number of acids and alkalis; e.g., Zn and HCl provide a ready source of H_2 in the laboratory. In contact with Fe in the presence of water, Zn oxidizes readily. However, Zn is not decomposed by pure water. Concentrated HNO_3 first attacks Zn to form $Zn(NO_3)_2$, which is sparingly soluble, protecting the metal from further attack. In contact with acids and alkalis, Zn normally dissolves with the generation of H_2, but concentrated H_2SO_4 dissolves Zn with the evolution of SO_2.

The major oxide is ZnO, which results when Zn is ignited in air. Ignition of the carbonate, hydroxide, nitrate, oxalate, and sulfide also produce ZnO. Zinc hydroxide, $Zn(OH)_2$, which is amphoteric, is precipitated from Zn solutions with fixed alkalis. Zinc forms a number of salts, many of which are quite deliquescent and soluble in water. Insoluble Zn salts include the carbonate, arsenate, phosphate, oxalate, cyanide, ferrocyanide, and periodate. All Zn salts except ZnS are soluble in the fixed alkalis.

Organozinc: The diverse coordination behavior of Zn in inorganic compounds is also found in organic compounds of Zn.

The introduction of organometallic chemistry is generally associated with Frankland's prepara-

1178

TABLE Z-1. *Some Properties of Zinc*

Atomic volume, cm^3/g atom	14
Crystal structure	cph
Lattice constants at 20°C (68°F), Å:	
a	2.569
c	4.935
Closest approach of atoms, Å	2.659
Ionization potential, eV	9.391
Hardness, Mohs' scale	2.5
Tensile strength: 1,000 psi:	
Drawn	22–30
Cast	7–13
Latent heat of fusion, cal/g	24.4
Latent heat of vaporization, cal/g	419.5
Heat capacity, kcal/(g)(°C):	
At melting point	0.1199
800 °C	0.1076
1000 °C	0.1012
Viscosity, P:	
At 450°C	317×10^2
At 700°C	188×10^2
Thermal conductivity, kcal/(s)(cm)(°C):	
At 500°C	0.138
700°C	0.135
Volume change on fusion, % of solid	6.9
Resistivity at 20°C, $\mu\Omega$-cm	5.8
Temperature coefficient at 20°C	0.0037
Electrodeposited zinc:	
Density, g/cm^3	7.1
Hardness, Vickers	40–50

ASTM B-6 Specifications for Grades of Zinc

Type	Maximum Allowable Trace Elements, %		
	Pb	Fe	Cd
Special high grade	0.003	0.003	0.003
High grade	0.07	0.02	0.03
Intermediate	0.2	0.03	0.4
Brass special	0.6	0.03	0.5
Prime Western	1.6	0.08	

Properties of Zinc Casting Alloys

	ASTM-AG40A (XXIII) SAE 903	ASTM-AC41A (XXV) SAE 925
Tensile strength, psi:		
As cast	41,000	47,600
After 10 years indoors	35,000	39,300
Elongation in 2 in., %:		
As cast	10	7
After 10 years	16	13
Fatigue strength, psi	6,875	8,175
Compressive strength, psi	60,000	87,000
Hardness, Brinell (500 kg, 10-mm ball)	82	91
Coefficient of thermal expansion (32–212°F), in./(in.)(°F)	15.2×10^{-6}	15.2×10^{-6}

tion in 1848 of the first organozinc compound, dimethylzinc. Following this organozincs were employed extensively in the synthesis of other organometallic compounds and as useful intermediates in general organic synthesis. Also, the complex-forming ability of organozinc compounds was detected at a very early stage. In 1858 Wanklyn reported on the formation of the anionic complex sodium zinctriethyl, $NaZn(C_2H_5)_3$.

The strongly electrophilic character of zinc in its organic compounds appears to be a most significant aspect of organic zinc chemistry. Quite commonly, reactions of these compounds appear to involve complex formation as an initial step.

The catalytic effects of some organozincs on the trimerization of organic isocyanates suggests their industrial potential as catalytic agents.

Uses: Worldwide production of Zn (1970) was estimated at 5.3 million tons, of which the United States produced about 0.961 million tons. Other major zinc-producing nations include Canada, Australia, Mexico, Germany, Japan, Peru, and the Soviet Union.

The uses of zinc and its compounds fall into the following major classifications: (1) anticorrosion, (2) castings, (3) brass, (4) rolled zinc, and (5) zinc oxide.

The largest overall use of Zn is for anticorrosion, mainly for the protection of steel and iron. It has a relatively high electropotential (energy), as evidenced by its position in the galvanic and electromotive-force series. This characteristic makes possible zinc's wide use as a coating, anode, or pigment to provide electrolytic protection against corrosion. It also is the reason that Zn is the anode material for a number of types of batteries, fuel cells, and other power packs.

Zn is one of the most versatile materials available for coating other metals. It can be applied as a metal by hot-dip galvanizing, flake galvanizing, electrode position, sherardizing or cementation, scratching or rubbing, flame spraying or metalizing, plasma arc spraying, shot blasting, vacuum metalizing, and nonvacuum vapor condensation; and as the pigment of paints.

Zn coatings protect ferrous metals in two ways: (1) as a barrier because corrosion of their surface metal takes place only for a brief period, and (2) the insulation of the resultant relatively inert reaction-product film minimizes access of destructive elements to either the underlying Zn or basis metal. If the coating is broken or dissipated in small areas from long exposure, the Zn, because of its higher electropotential compared to ferrous

metals, sacrifices itself slowly to supply the needed protective current.

Hot-dip galvanizing is by far the most widely used method for Zn coating steel in the form of sheets, coils, wire, structurals, and hardware among other forms. Typical applications are roofing, siding, transmission towers, highway guardrails, light poles, culverts, and fencing. In 1970 over 452,000 tons of Zn were used for hot-dip galvanizing steel. The volume of Zn used for other coatings is of a lesser order.

Because of its inherent metallic characteristics and amenability to property improvement by alloying, Zn is extensively used in the casting field, primarily in the form of pressure die castings, used extensively in the automotive industry for grills, carburetors, instrument panels, hardware, and trim items. The appliance and heating fields call for a wide range of zinc die castings (see Fig. Z-1).

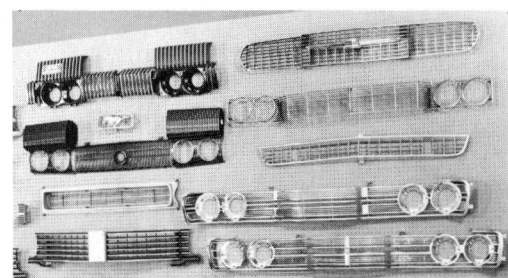

Fig. Z-1. Automotive grills demonstrate the size, complexity, and strength obtainable with zinc die castings.

Important constituents of consumer and industrial products are made of die-cast zinc. An increasing number of zinc die castings are used in business machines, cameras, optical instruments, and toys—indeed throughout the gamut of household items and devices used in modern living.

Zinc alloy is also used in slush, permanent-mold, sand, and centrifugal castings, as well as for stamping and forming dies. Approximately 462,000 tons of zinc alloy was used in the casting industry in 1970.

The ability of zinc to alloy readily with copper results in the production of a series of brasses varying from 3% Zn and 97% Cu to 45% Zn and 55% Cu. The average brass contains 30% Zn and 70% Cu. Varying amounts of other metals are also added for specific purposes. Red-brass pipe, 15% Zn and 85% Cu is specified for water and other piping in the construction field. Admiralty metal or aluminum brass is widely used for steam-condenser tubes in power plants and on ships. Brass is used for automobile radiator cores, tanks, and other parts. It is also used in electrical and hardware items. In 1970, 170,000 tons of zinc were used in brass products.

Wrought Zn in the form of sheet, strip, and rolled products is used in the construction industry for roofing leaders and gutters, terrazzo strip, nose bead, and similar building accessories. Because of its energy capacity wrought zinc is used in conventional dry cells, silver-zinc, and zinc-air batteries. It also finds application in photoengraving, lithography, and manufactured items that are spun or drawn. About 41,000 tons of wrought zinc were consumed in 1970.

The traditional markets for zinc oxide have been rubber, paints, and chemicals, but the use of zinc oxide for photocopy paper has created a major and expanding new market.

In paints, zinc oxide continues to make progress in the field of exterior latex systems, especially in alkyd-modified latex paints, which are increasingly favored. Because some mercury compounds are poisonous, the use of mercury is being reviewed, stimulating paint producers to reconsider zinc oxide for its role in mildew and fungus prevention as well as improved covering power and better retention. In 1970 about 206,000 tons of ZnO were produced.

REFERENCES

Boersma, J., and J. G. Noltes: "Organic Coordination Chemistry," International Lead Zinc Research Organization, Inc., New York, 1968.

Mathewson "Zinc: The Metal, Its Alloys and Compounds," ACS Monograph 142, Van Nostrand Reinhold, New York, 1959.

Zinc Institute, Inc.: A Mine to Market Outline: Zinc, New York, 1971.

—Ernest W. Horvick, *Zinc Institute, Inc., New York*

Zinc and Zinc-Lead Smelting Zinc and lead usually occur together in the form of their respective sulfides. Before the development of combined lead-zinc smelting processes, they were separated by the froth flotation process, i.e., the finely ground sulfides are treated with chemical reagents which cause one sulfide to be preferentially wetted. The nonwetted sulfide floats in an artificially prepared froth and can be taken off the top of the cell. The wetted sulfide sinks and remains on the bottom. In a typical lead-zinc flotation circuit, lead sulfide is floated and the zinc depressed in the first stage. The process is reversed in the second stage and the zinc sulfide floated.

Part of the gangue, or nonmetal-bearing components of the ore, remains at the bottom of the flotation cell at the end of the second stage. Thus, there is a concentration of metal values in the flotation process. The zinc and lead sulfide fractions so produced are dewatered to 6–8% moisture by thickening and filtering and are referred to as *zinc and lead concentrates.* See also **Flotation.**

Except for the combined lead-zinc smelting process, which is described here, the various lead-smelting processes, starting with lead concentrate, are described under **Lead.**

Zinc-Smelting Processes: Two major categories of zinc-smelting processes are described here, oxide reduction processes and combined lead-zinc smelting processes.

Oxide Reduction Processes: Zinc, like lead, can be produced by heating its oxide with carbon or carbon monoxide, but it is atypical in that this reduction takes place above the boiling point of zinc (906°C); hence the zinc has to be recovered as a vapor and condensed to the liquid metal.

The Horizontal Retort Process (Horizontal Distillation): This is the oldest zinc process but still produces about 12% of the world's zinc. Zinc oxide produced by roasting the sulfide in air is mixed with coal and charged to a battery of small retorts mounted in a producer-gas-fired furnace in which the retorts are heated to 1200°C over a 24- or 48-hr period.

The retorts are mounted in rows, usually four high. Each furnace side contains 300–400 retorts. The retorts are hand charged, and each has a refractory air-cooled condenser, from which the condensed zinc is tapped. A recent development is to use one condenser for a complete furnace side of retorts. In this mechanical condenser the zinc is condensed by a "rain" of molten zinc created by a mechanically driven rotor or stirrer.

The zinc sulfide is roasted on multihearth or fluid-bed roasters since a finely divided product is required; it can also be done on a Dwight Lloyd sinter machine, but in that case the product has to be crushed.

Vertical Retort Process: This was the first attempt to reduce the high labor requirement associated with the horizontal distillation process and improve on its arduous working conditions. In the vertical retort process, the roasted zinc sulfide (zinc oxide produced as for the horizontal distillation process) is mixed with coal and compressed into briquettes. These are heated to coke the coal, thus giving the briquette strength. The coked briquettes are charged continuously to a large vertical retort 35 ft high, 7 ft long, and 12 in. wide, constructed of silicon carbide brick. The retorts are heated by producer or natural gas fired into the combustion chamber around the retorts. Briquettes are continuously charged at the top and spent briquettes continuously removed by a screw extractor at the bottom. The zinc vapor is condensed in a mechanical condenser as in the horizontal distillation process.

The Electrothermic Process of the St. Joseph Lead Company: This process is based on an electrically heated shaft furnace. The zinc sulfide is roasted by a sintering process to give an agglomerated product, which is mixed with coke and charged to a cylindrical shaft furnace, 6 ft in diameter, equipped with eight carbon electrodes, four at the top and four at the bottom. There is a 24-ft vertical space between the electrode banks (total furnace height is 50 ft). The passage of the electric current raises the temperature of the charge so that zinc is reduced, distilled off, and condensed by pulling the vapor through a pool of cooled molten zinc in a refractory U-tube. The spent charge is withdrawn continuously from the bottom of the shaft and fresh charge continuously fed in at the top.

Combined Lead-Zinc Smelting Processes. Two processes in this category are described here, the Imperial Smelting Process* and the Kivcet Process.†

The Imperial Smelting Process: The ISP, shown in Figs. Z-2 and Z-3, uses a blast furnace capable of simultaneous production of lead and zinc. Recently, the process has been adapted for the recovery of copper. The limitation of smelting zinc in a blast furnace was the problem of condensing zinc vapor to metal in the presence of a CO_2-rich gas. The reaction

$$ZnO + CO \underset{850-950°C}{\overset{1000°}{\rightleftharpoons}} Zn \text{ vapor} + CO_2$$

is reversible. Hence in the presence of CO_2 zinc oxide would be produced rather than metallic zinc. This was overcome in the ISP by using a rain of molten lead to shock-chill the zinc vapor through the 850–950°C range and thus avoid reoxidation.

The charge to the ISP consists of sinter prepared on an updraft Dwight Lloyd sinter machine and metallurgical coke. The furnace has a rectangular cross-sectional shaft 20×10 ft and 30 ft high. The upper part is refractory-lined, and the lower

*A development of Imperial Smelting Corporation Limited, Avonmouth, England.

†A development of the U.S.S.R. Ministry of Nonferrous Metallurgy.

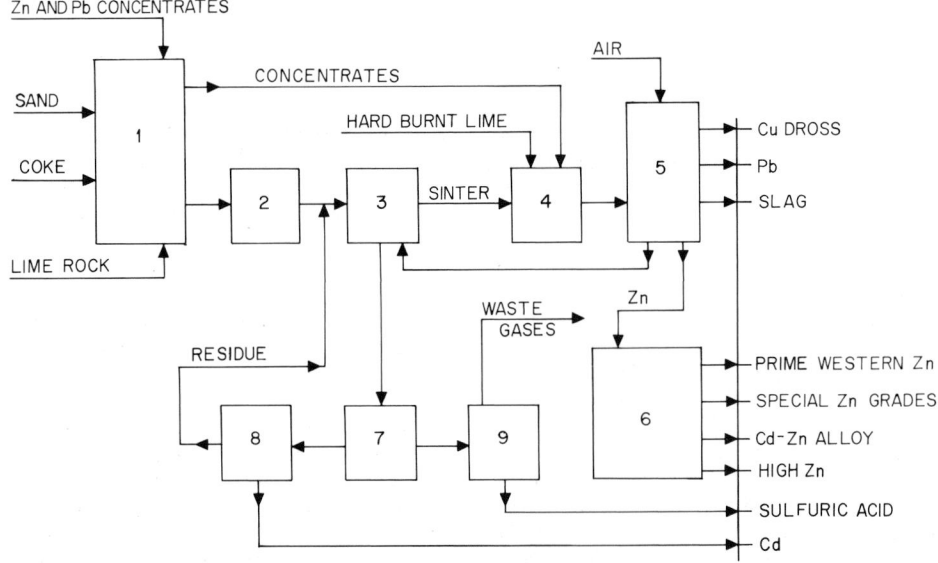

Fig. Z-2. Materials flow (schematic) in the Imperial Smelting Process for the simultaneous production of lead and zinc. (1) Materials handling, (2) proportioning plant, (3) updraft sinter plant, (4) charge preparation, (5) Imperial smelting furnace, (6) zinc refinery, (7) hot-gas processing, (8) cadmium plant, (9) acid plant.

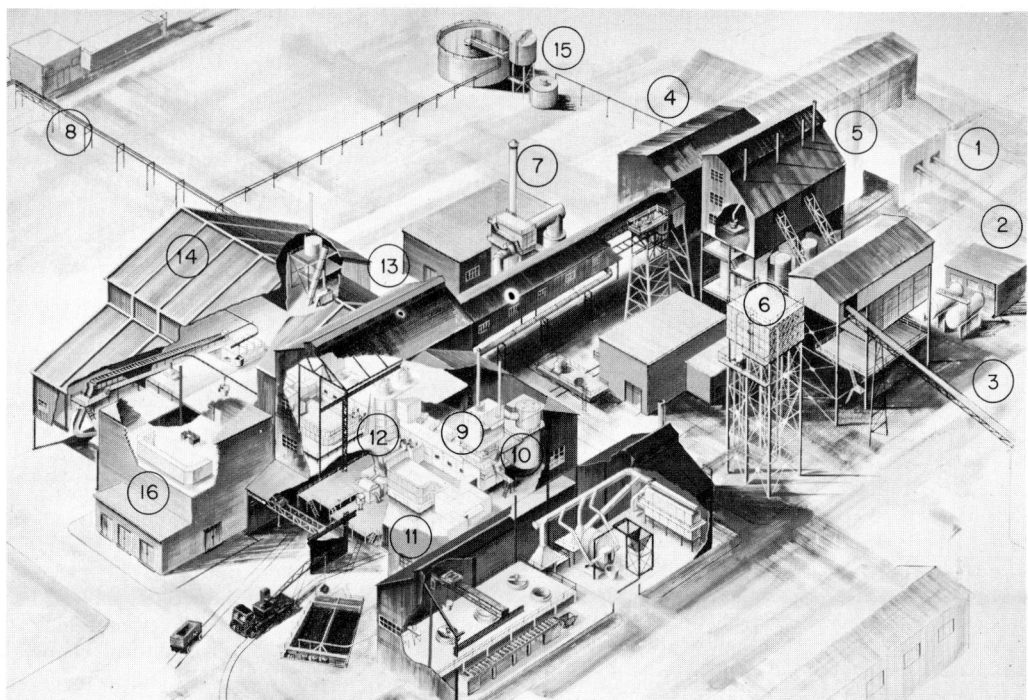

Fig. Z-3. A complete lead-zinc smelting plant, incorporating the Imperial Smelting Process, for Masinimport, Rumania. (Main contractors for this £5 million complex, Davy Powergas Ltd.) (1) Sinter input, (2) compressor house, (3) coke intake, (4) charge preparation, (5) coke preheating, (6) water tank, (7) air preheating, (8) oxygen storage, (9) condenser, (10) gas washer, (11) lead refinery, (12) furnace shaft, (13) dross treatment, (14) zinc casting, (15) effluent treatment, (16) storage.

part is formed by two rows of water jackets through which the water-cooled tuyeres enter. These admit preheated blast air (at 950°C) to the furnace. The coke in the charge provides heat to raise the temperature of the charge to over 1000°C to allow zinc reduction to take place and also provide a source of carbon for this reduction. The zinc vapor passes from the shaft to the condenser, which is a refractory-lined chamber full of molten lead. Rotors throw a rain of molten lead which condenses the zinc, giving a solution of zinc in lead. On cooling this solution, the zinc is liquated out and recovered. The zinc is collected in a holding bath and cast for sale. Since the reduction of lead oxide starts at a much lower temperature (600°C) than that of zinc, lead oxide is reduced almost immediately on entering the furnace by the carbon monoxide released in the reduction of zinc.

$$ZnO + C \rightarrow Zn \text{ vapor} + CO$$
$$PbO + CO \rightarrow Pb \text{ liquid} + CO_2$$

The molten lead passes down through the furnace and is tapped from the bottom together with the slag formed by the gangue constituents of the furnace charge. These are separated in a forehearth, and the lead is taken on for further refining.

It must be emphasized that the lead produced in the furnace is entirely separate from that used in the condenser. After the zinc has liquated out the condenser, lead is pumped back to the condenser for reuse.

Copper in the charge is concentrated in the lead tapped from the furnace and is recovered in the first stage of the lead-refining process.

The Kivcet Process: This is a combined lead-zinc-copper smelting process involving sulfide reduction. The finely ground sulfide concentrates are blown into a cyclone furnace by a stream of oxygen (see Fig. Z-4). The temperature of the furnace is raised by the combustion of the sulfur. Lead is immediately reduced and collects in a

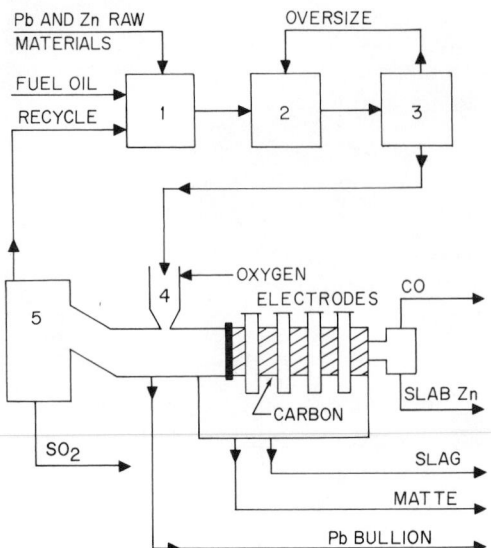

Fig. Z-4. Schematic representation of materials flow in the Kivcet process for combined lead-zinc-copper smelting. (1) Concentrate drier, (2) ball mill, (3) screen, (4) cyclone furnace, (5) gas cleaning. Electrodes are carbon. Sulfur dioxide goes to acid or sulfur recovery plant.

settling chamber. The reduced zinc is immediately reoxidized and enters the slag formed by the gangue constituents of the concentrates. This passes to another chamber, where zinc oxide is fumed off the slag by electric heating. Injection of the finely divided carbon causes immediate reduction to zinc vapor, which is condensed in a mechanical condenser similar to that used on the vertical retort process. Copper is recovered as a matte formed by reaction with the sulfur in the concentrate.

REFERENCES

Argall, George O., Jr.: Imperial Zinc-Lead Blast Furnace Now Smelts Variety of Metal Charges, *World Min.,* August 1966.

French Patent 1,514,811, March 17, 1967.

Mathewson C. H.: "Zinc: The Metal, Its Alloys and Compounds," ACS Monograph 142, Van Nostrand Reinhold, New York, 1959.

Morgan, S. W. K., and P. A. Temple: The Place of the Imperial Smelting Process in Non-ferrous Metallurgy, *J. Met.,* August 1967.

Report of United Nations Industrial Development Organisation Export Group Meeting on Lead and Zinc Production, April 28–May 2, 1969.

Sellwood, R. M.: Expansion of Zinc Production at Avonmouth, Robert Home Memorial Lecture, *Chem. Ind.,* Aug. 15, 1970.

Temple, P. A., and G. O. John: The Treatment of Complex Zinc/Lead/Copper Ores in the Imperial Smelting Process, *Can. Inst. Min. Metall. 8th Ann. Conf. Metall., Windsor, Ontario,* Aug. 25–27, 1969.

Woodward, O. H.: Review of the Broken Hill Lead-Silver-Zinc Industry, Broken Hill Mining Managers' Association in conjunction with The Australasian Institute of Mining and Metallurgy, 1965.

—H. E. Gibbs, Davy Powergas Ltd., London

Zinc Metal Pretreatment   (See **Conversion coatings.**)

Zircaloy   (See **Nuclear power plants.**)

Zircon   (See **Ceramics; Hafnium;** and **Plasma, chemical processing with.**)

Zirconium   (F. *zircone.*) **Zr** = 91.22 (at. wt.); 40 (at. no.). Pure elemental zirconium is a gray ductile metal. Zr is nearly identical to Hf in chemical behavior and similar to Ti, the other elements in group IVB of the periodic table. There are five natural isotopes, 90–92, 94, and 96. ^{95}Zr is a radioisotope. See also **Chemical elements.** The normal valence of Zr is 4$^+$, and its oxidation potential with respect to H$_2$ is −1.43 V. The metal usually has a thin oxide film that causes it to appear noble in most aqueous solutions.

Zr was discovered by M. H. Klaproth during his analysis of zircon (1789). The element proved very difficult to isolate. Berzelius produced a crude powder of Zr in 1824 by a bomb reduction of potassium fluorozirconate with potassium. Weiss and Neumann repeated and refined the Berzelius experiment in 1910, but their product assayed only 98% Zr. There was further experimental work by Lely and Hamburger, but the first ductile metal was produced by van Arkel, de Boer, and Fast, working at the University of Leyden and the Philips Lamp Works in Holland, about 1925. Practical production was made possible by W. J. Kroll of the U.S. Bureau of Mines, when he combined a vaporization-condensation technique to purify the tetrachloride and then made the reduction to metal, using molten Mg. In 1923, Coster and Hevesy found by x-ray techniques that most naturally occurring zircon contains about 2% hafnium; until the 1950s, the properties attributed to Zr were really those of an alloy of 98% Zr and 2% Hf. See also **Hafnium.**

Physical Properties: Important properties of Zr include density 6.5 g/cm^3; coefficient of linear expansion 5.8 × 10^{-6} per °C; thermal conductivity 0.05 cal/(s)(cm)(°C); specific heat 0.067 cal/(g)(°C); electrical resistivity 40 $\mu\Omega$-cm; modulus of elasticity 14×10^6 psi; ultimate strength at 20°C 64,000 psi; yield strength 44,000 psi; elongation in 2 in. at 20°C 21%; hardness, 90 on the Rockwell B scale and a Diamond Pyramid hardness of 175, using a 50 kg load. Zr may be bent around a radius three times the metal thickness. The foregoing data are based upon unalloyed Zr containing about 0.1% O_2.

In the course of their nuclear research, Kaufman and Utermeyer separated the Hf and Zr and found that Zr is nearly transparent to thermal neutrons. The neutron-absorption coefficient, previously believed to be greater than 2 barns, was found to be 0.18 barn once the Hf was reduced to a few hundred parts per million. A concentrated effort to exploit this low-cross-section property eventually resulted in the nearly universal use of Zr alloys as fuel cladding in military and civilian power reactors.

Crystallographically, metallic Zr is allotropic. The alpha (hexagonal close-packed) phase is stable up to about 863°C depending upon impurity content. Above this temperature, it transforms to a beta (body-centered cubic) phase. Oxygen, the most common impurity, tends to stabilize the alpha phase. Rapid cooling does not retain the high-temperature beta phase but does alter the morphology of the alpha transformed from it.

Zircon, $ZrSiO_4$: This is the principal ore for the production of Zr metal. It occurs in many regions as a beach sand usually mixed with silica, ilmenite, and rutile. The deposits are mined by the floating-dredge technique, with early stages of the beneficiation right on the dredge. Most of the silica is removed by spiral concentrators, leaving the ilmenite and rutile to be removed by magnetic and electrostatic separators. Only the purest concentrates are useful for metal production, the others being used for refractories. There are several techniques used at each stage of the metal-production sequence. The ore can be opened by caustic fusion, carbiding in an electric furnace, or direct chlorination, the last being currently favored. If the carbide is produced, it is chlorinated in a simple exothermic reaction to produce the water-soluble $ZrCl_4$. The Hf is removed from most of the Zr processed in a many-stage liquid-liquid extraction. The most common process uses a dilute HCl solution for the aqueous phase and methyl isobutyl ketone as the solvent phase. Ammonium thiocyanate is used to complex the Zr, and the Hf is extracted by the solvent. The proc-

ess is quite efficient, Zr containing less than 50 ppm Hf commonly being produced. Another system utilizing a nitric acid aqueous phase and tributyl phosphate as a solvent has also been used commercially. After the separation the metals are precipitated as sulfates or hydroxides and calcined to pure ZrO_2.

The oxide is blended with carbon and again chlorinated to obtain $ZrCl_4$, a white powder that is extremely hydroscopic. $ZrCl_4$ may be further purified by sublimation or evaporation from a salt solution. Gaseous $ZrCl_4$ is reacted with molten Mg to form the Zr metal and $MgCl_2$. Vacuum distillation removes the $MgCl_2$ and residual Mg, leaving a grey metal with porous appearance, hence the common name *sponge* for virgin Zr.

In an alternate method of reduction, developed commercially in Canada, a pure ZrF_4 is formed from the aqueous discharge of the Hf separation process. After drying, the anhydrous ZrF_4 is reduced by Ca in a CaF_2-lined reactor to form a solid Zr nugget suitable for direct melting.

Sodium has also been used commercially as a reductant for $ZrCl_4$. A small amount of metal is further refined by the van Arkel–de Boer crystal-bar process to yield metal with only a few hundred ppm total impurities; sponge metal is typically 99.8% Zr with O_2 and Fe as the major impurities.

Sponge metal is vacuum-arc-melted into ingots that are forged and worked into all common forms of metal products. The metal is easily formed, machined, and ground by conventional techniques. It can be welded if care is taken to shield the molten pool from air with either Ar or He. Zr metal is produced commercially in the United States, Canada, France, Japan, and the Soviet Union. United States production capacity (1971) was about 5 million lb/yr. Zirconium oxide is widely used in abrasives. See also **Abrasives.**

Chemical Properties: Zr and Hf are chemically nearly identical, forming compounds of the same type and with the same properties. Both elements form extremely stable compounds of high negative free energies. There also is a broad similarity between Zr and Ti compounds. Zr, Hf, and Ti form a continuous series of solid solutions with one another in the metallic state.

Zr is rapidly attacked by HF, most fluorides, aqua regia, and wet Cl_2. Oxidation becomes significant above 500°C and limits the use of Zr in air beyond that range. Finely divided Zr burns rapidly with a brilliant white light. This property is useful in flash powder, explosive primers, and pyrotechnics but does constitute a hazard in machining, sawing, and grinding where fine chips are generated; such debris should be gathered and burned in small quantities and not allowed to accumulate. Shredded Zr foil made possible the miniature flashbulb and the flash cube, since a tiny amount of Zr gives off the same amount of light as a much larger quantity of Mg. Zr also is used as a getter element in vacuum tubes and as an x-ray filter. A plasma process based on the dissociation of Zr and its subsequent treatment to produce zirconium dioxide and sodium silicate is described under **Plasma, chemical processing with.**

Alloys: The principal use of Zr is as the base metal in an alloy of 98% Zr, 1.5% Sn, 0.35% Fe-Cr-Ni, and 0.15% O_2. This alloy optimizes the low neutron cross section, reliable corrosion resistance in 650°F (343°C) water, and strength required of the cladding of the uranium fuel in a water-cooled nuclear power reactor. Because of this unique combination of qualities, Zr is the economic choice over stainless steel. About 90% of the Zr produced is used for this purpose.

Zirconium's corrosion resistance to a wide range of solutions has increased its use as a material of construction in the chemical process industries. Unlike many other metals, Zr resists corrosion by both strong acids and strong alkalis. Zr is totally resistant to caustic solutions at all temperatures and concentrations. The resistance to mineral acids is broad but not complete. Corrosion rates of less than 10 mils/yr are obtained in H_2SO_4 up to 55% boiling, HNO_3 in all concentrations to boiling, HCl up to 30% boiling, and H_3PO_4 up to 60% boiling.

REFERENCES

Blumenthal, W. B.: "The Chemical Behavior of Zirconium," Van Nostrand Reinhold, New York, 1958.
LaQue, F., and H. R. Copson: "Corrosion Resistance of Metals and Alloys," pp. 666–675, Van Nostrand Reinhold, New York, 1963.
Lustman, B., and F. Kerze: "The Metallurgy of Zirconium," McGraw-Hill, New York, 1955.

—John H. Schemel, *AMAX Specialty Metals Division, American Metal Climax, Inc., Akron, N.Y.*

Zirconium Dioxide   (See **Plasma, chemical processing with.**)

Zone Refining   (See **Germanium.**)

Zorgite   (See **Selenium.**)

Zwitterionic Surfactants   (See **Detergents;** and **Surfactants.**)

classified index

The structure of the classified index is shown by the accompanying table. Each entry of this encyclopedia is listed once and only once in the classified index. Cross references are used extensively throughout the book and these references, coupled with the detailed subject index at the end of the volume, provide a webbing to hold all topics together even though all topics are entered into the book alphabetically. Admittedly, some of the decisions involved in placing topics under the major categories of the accompanying table were arbitrary. This situation could have been avoided only by listing some entries under several categories. Such multiple listings would have resulted in a much longer and less convenient topical classification. Some of the major cross reference entries in the volume are included, but again never duplicated in the classified index. These are shown in italics. Seldom can a classified index render full justice to the content of a volume and, therefore, extensive use of the subject index is encouraged.

Main Topic	*Title of Category in Classified Index*
MATERIALS	**Chemical Elements**
	Organic Compounds—Structural Groups
	Organic Compounds (Not elsewhere classified)
	Major Raw Materials
ENERGY,	
ENGINEERING,	
AND EQUIPMENT	**Physical Operations and Energy Systems**
	Chemical Processes
	Chemical and Engineering Terms
	Atomic and Nuclear Chemical Technology
	Biochemical Technology
PRODUCTS	**Plastics, Resins, Fibers, Foams,**
	and Elastomers
	Functional and End-products

Smaltite
Smithsonite
Sphalerite
Spodumene
Stannite
Stibnite
Sylvinite
Wolframite
Wood
Wustite
Xenotime

Refined or manufactured

Activated aluminas
Activated carbon
Aluminas
Alums
Ammonia
Carbon black
Carbon dioxide
Carbon monoxide
Caustic potash
Caustic soda
Charcoal
Coke
Graphite
Hydrochloric acid
Hydrogen cyanide
Hydrogen fluoride
Hydrogen peroxide
Hydrogen sulfide
Lampblack
Nitric acid
Phosphoric acid
Pitch
Soda ash
Sodium chloride
Sulfuric acid
Tallow
Titanium dioxide
Turpentine
Water

**Physical operations
and energy systems** ⸺

Absorption
Adsorption
Aeration
Agglomeration
Agitation
Annealing
Attrition
Blending
Boilers
Bonding
Brazing
Briquetting
Calendering
Casting
Centrifuging
Chemisorption

Chilling
Clarification
Classification
Compacting
Compressor systems
Condensing
Cooling systems
Crushing
Crutching
Crystallization
Cyclones
Debarking
Decanting
Decolorizing
Defiberization
Degasification
Degreasing
Dehydration
Deionization
Dephlegmation
Desalting
Desorption
Dewatering
Dialysis
Diffusional operations
Distillation
Drying, gases
Drying, solids
Electric furnaces
Electrodialysis
Encapsulating
Energy systems, process
Evaporation
Expression
Extraction
Extruding
Filtration
Flaking
Flashing
Flotation
Forging
Fractionation
Frasch process
Freeze-drying
Furnaces
Graining
Granulation
Grinding
Hardening
Heat exchangers
Heat treating
Homogenization
Hydration
Hydrolysis
Hydrometallurgy
Jigging
Leaching
Liquid-solids separations
Melting
Membrane filters
Milling

Mixing, fluids
Mixing, pastes
Mixing, solids/solids
Molding
Molecular sieve
Mulling
Normalizing
Nozzles
Osmosis
Pasteurization
Pelletizing
Powder metallurgy
Prilling
Refrigeration
Safe handling of chemicals
Screening
Sedimentation
Separation operations
Settling
Sewage treatment
Shaking
Sintering
Size reduction
Sizing
Solvent extraction
Spinning
Spray-drying
Sterilization
Sublimation
Tabling
Tempering
Thickeners
Tumbling
Ultrasonic cleaning
Vacuum distillation
Vacuum filtration
Vapometallurgy
Vapor degreasing
Vaporization
Washing
Welding
Wet classification

Chemical processes ⸺

Acetylation
Alcohol process, ALFOL
Alcoholysis
Aldol condensation
Alkylation
Amination
Ammoxidation
Analyzers, process
Anodizing
Bayer process
Beneficiation, ore
Betterton-Kroll process
Biosynthesis
Bleaching
Calcining
Carbonization
Carburizing

Chemical and engineering terms

Atomic and nuclear chemical technology ———

Biochemical technology * ———

*Some related topics, such as amino acids, enzymes, nucleic acids, and peptides and proteins are listed elsewhere.

EDITOR'S NOTE: Page numbers in bold type indicate main entry on topic.